Applications

Business and Economics

Life Sciences

Social and Behavioral Science

Physical Sciences

CALCULUS

An Applied Approach

CALCULUS

An Applied Approach

Ninth Edition

Ron Larson
The Pennsylvania State University
The Behrend College

With the assistance of David C. Falvo
The Pennsylvania State University
The Behrend College

BROOKS/COLE
CENGAGE Learning™

Australia • Brazil • Japan • Korea • Mexico • Singapore • Spain • United Kingdom • United States

BROOKS/COLE
CENGAGE Learning™

Calculus: An Applied Approach
Ninth Edition

Ron Larson

Vice President, Editorial Director: P.J. Boardman
Publisher: Richard Stratton
Senior Development Editor: Laura Wheel
Senior Editorial Assistant: Haeree Chang
Associate Media Editor: Andrew Coppola
Senior Marketing Manager: Barb Bartoszek
Marketing Coordinator: Michael Ledesma
Marketing Communications Manager: Mary Anne Payumo
Content Project Manager: Jill Quinn
Senior Art Director: Jill Ort Haskell
Manufacturing Planner: Doug Bertke
Rights Acquisition Specialist: Shalice Shah-Caldwell
Text Designer: Larson Texts, Inc.
Cover Designer: Larson Texts, Inc.
Compositor: Larson Texts, Inc.
Cover Image: *Torus with Cross-Cap* by Helaman Ferguson.
Photograph by Ed Bernik photo/video.
Used with permission.

"The inscription on the base of my polished bronze *Torus with Cross-Cap* sculpture, $3x = x + h$ but $2x \neq h$, may seem a bit strange at first glance. The addition is of topological objects, h for a handle, x for a cross-cap. These are not numbers: the equation means that adding three cross-caps to a surface is equivalent to adding one cross-cap and a handle. In this topological "calculus" the usual cancellation law does not hold."

—Helaman Ferguson

Library of Congress Control Number: 2011935167

Student Edition:
ISBN-13: 978-1-133-10928-0
ISBN-10: 1-133-10928-4

Brooks/Cole
20 Channel Center Street
Boston, MA 02210
USA

Printed in China
2 3 4 5 6 7 15 14 13 12

Contents

7 Functions of Several Variables 419

8 Trigonometric Functions 505

9 Probability and Calculus 563

*Available at the text-specific website *www.cengagebrain.com*

Preface

Welcome to the Ninth Edition of *Calculus: An Applied Approach*! I am always excited about a new edition, but with this edition, I am even more excited. I had a single goal in mind with this revision—to provide you with a book that is both real and relevant. This book has a bright business-oriented design that complements the multitude of business and life sciences applications found throughout.

The theme for the revision is **"IT'S ALL ABOUT YOU."** The pedagogy of the book is rock solid and is based on years of teaching, years of writing, and years of feedback from instructors and students. Please pay special attention to the study aids with a red **U**. These study aids will help you learn calculus, use technology, refresh your algebra skills, and prepare for tests. For an overview of these aids, check out CALCULUS & YOU on page 0.

In each exercise set, quiz, and test, be sure to notice the reference to **CalcChat.com.** At this free site, you can download a step-by-step solution to any odd-numbered exercise. Also, you can talk to a tutor, free, during the hours posted at the site (20 hours a week in the summer, 40 hours a week during the school year).

New To This Edition

NEW Chapter Opener

Each *Chapter Opener* highlights a real-life problem from an example in the chapter, showing a graph related to the data and describing the math concept used to solve the problem.

NEW Section Opener

Each *Section Opener* highlights a real-life problem in the exercises, showing a graph for the situation with a description of how you will use the math of the section to solve the problem.

NEW SUMMARIZE

The *Summarize* feature at the end of each section helps you organize the lesson's key concepts into a concise summary, providing you with a valuable study tool.

NEW HOW DO YOU SEE IT? Exercise

The *How Do You See It?* exercise in each section presents a real-life problem that you will solve by visual inspection using the concepts learned in the lesson.

60. HOW DO YOU SEE IT? The graph shows the cost and revenue equations for a product.

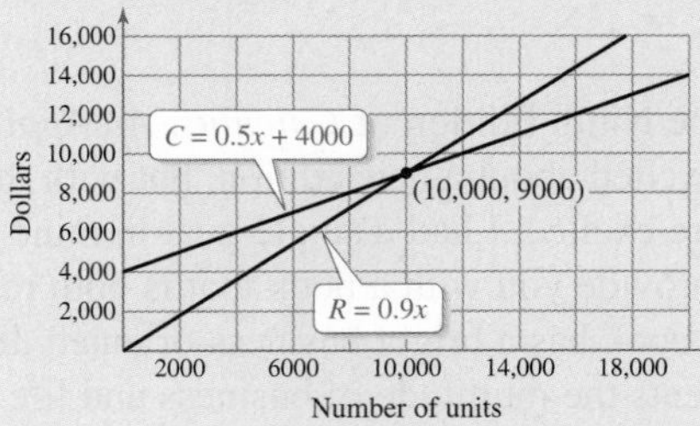

(a) For what numbers of units sold is there a loss for the company?

(b) For what number of units sold does the company break even?

(c) For what numbers of units sold is there a profit for the company?

REVISED Exercise Sets

The exercise sets have been carefully and extensively examined to ensure they are rigorous, relevant, and cover all topics suggested by our users. The exercises have been reorganized and titled so you can better see the connections between examples and exercises. Multi-step, real-life exercises reinforce problem-solving skills and mastery of concepts by giving you the opportunity to apply the concepts in real-life situations.

Calc Chat

For the past several years, an independent website—**CalcChat.com**—has been maintained to provide free solutions to all odd-numbered problems in the text. Thousands of students have visited the site for practice and help with their homework. For this edition, information from **CalcChat.com**, including which solutions students accessed most often, was used to help guide the revision of the exercises.

Table of Contents Changes

I have added Chapter 11 (Differential Equations), which was previously Appendix C. Chapter 0 (Precalculus Review) has been moved to Appendix A. Based on feedback from users, old Section 6.2 (Partial Fractions and Logistic Growth) has been removed.

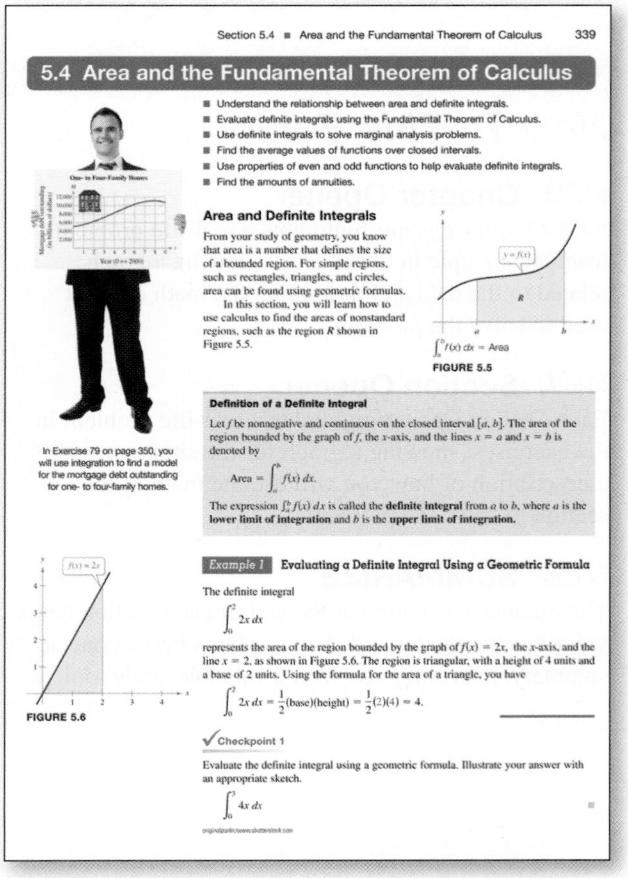
Section 5.4 ■ Area and the Fundamental Theorem of Calculus 339

5.4 Area and the Fundamental Theorem of Calculus

- Understand the relationship between area and definite integrals.
- Evaluate definite integrals using the Fundamental Theorem of Calculus.
- Use definite integrals to solve marginal analysis problems.
- Find the average values of functions over closed intervals.
- Use properties of even and odd functions to help evaluate definite integrals.
- Find the amounts of annuities.

Area and Definite Integrals

From your study of geometry, you know that area is a number that defines the size of a bounded region. For simple regions, such as rectangles, triangles, and circles, area can be found using geometric formulas.

In this section, you will learn how to use calculus to find the areas of nonstandard regions, such as the region R shown in Figure 5.5.

$\int_a^b f(x)\,dx = \text{Area}$

FIGURE 5.5

In Exercise 79 on page 350, you will use integration to find a model for the mortgage debt outstanding for one- to four-family homes.

Definition of a Definite Integral

Let f be nonnegative and continuous on the closed interval $[a, b]$. The area of the region bounded by the graph of f, the x-axis, and the lines $x = a$ and $x = b$ is denoted by

$$\text{Area} = \int_a^b f(x)\,dx.$$

The expression $\int_a^b f(x)\,dx$ is called the **definite integral** from a to b, where a is the **lower limit of integration** and b is the **upper limit of integration.**

Example 1 Evaluating a Definite Integral Using a Geometric Formula

The definite integral

$$\int_0^2 2x\,dx$$

represents the area of the region bounded by the graph of $f(x) = 2x$, the x-axis, and the line $x = 2$, as shown in Figure 5.6. The region is triangular, with a height of 4 units and a base of 2 units. Using the formula for the area of a triangle, you have

$$\int_0^2 2x\,dx = \frac{1}{2}(\text{base})(\text{height}) = \frac{1}{2}(2)(4) = 4.$$

FIGURE 5.6

Checkpoint 1

Evaluate the definite integral using a geometric formula. Illustrate your answer with an appropriate sketch.

$$\int_0^3 4x\,dx$$

Trusted Features

Section Objectives

A bulleted list of learning objectives provides you the opportunity to preview what will be presented in the upcoming section.

Definitions and Theorems

All definitions and theorems are highlighted for emphasis and easy recognition.

Checkpoint

Paired with every example, the *Checkpoint* problems encourage immediate practice and check your understanding of the concepts presented in the example. Answers to all *Checkpoint* problems appear at the back of the text to reinforce understanding of the skill sets learned.

Business Capsule

Business Capsules appear at the end of selected sections. These capsules and their accompanying research project highlight business situations related to the mathematical concepts covered in the chapter.

STUDY TIP

These hints and tips can be used to reinforce or expand upon concepts, help you learn how to study mathematics, caution you about common errors, address special cases, or show alternative or additional steps to a solution of an example.

TECH TUTOR

The *Tech Tutor* gives suggestions for effectively using tools such as calculators, graphing calculators, and spreadsheet programs to help deepen your understanding of concepts, ease lengthy calculations, and provide alternate solution methods for verifying answers obtained by hand.

ALGEBRA TUTOR

The *Algebra Tutor* appears throughout each chapter and offers algebraic support at point of use. This support is revisited in a two-page algebra review at the end of the chapter, where additional details of example solutions with explanations are provided.

Business Capsule

CitiKitty, Inc. was founded in 2005 by 26-year-old Rebecca Rescate after she moved into a small apartment in New York City with no place to hide her cat's litter box. Finding no easy-to-use cat toilet training kit, she created one, and CitiKitty was born with an initial investment of $20,000. Today the company flourishes with an expanded product line. Revenues in 2010 reached $350,000.

83. Research Project Use your school's library, the Internet, or some other reference source to find information about the start-up costs of beginning a business, such as the example above. Write a short paper about the company.

71. Project: Number of Stores For a project analyzing the numbers of Tiffany & Co. stores from 2000 through 2009, visit this text's website at *www.cengagebrain.com*. *(Source: Tiffany & Co.)*

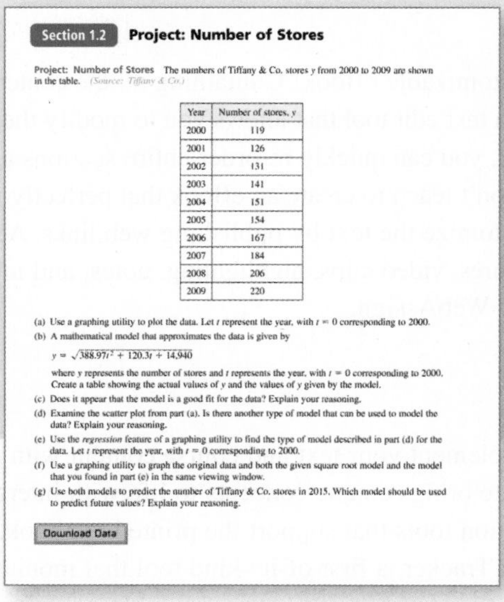

Section 1.2 Project: Number of Stores

Project: Number of Stores The numbers of Tiffany & Co. stores y from 2000 to 2009 are shown in the table. *(Source: Tiffany & Co.)*

Year	Number of stores, y
2000	119
2001	126
2002	131
2003	141
2004	151
2005	154
2006	167
2007	184
2008	206
2009	220

(a) Use a graphing utility to plot the data. Let t represent the year, with $t = 0$ corresponding to 2000.

(b) A mathematical model that approximates the data is given by

$$y = \sqrt{388.97t^2 + 120.3t + 14{,}940}$$

where y represents the number of stores and t represents the year, with $t = 0$ corresponding to 2000. Create a table showing the actual values of y and the values of y given by the model.

(c) Does it appear that the model is a good fit for the data? Explain your reasoning.

(d) Examine the scatter plot from part (a). Is there another type of model that can be used to model the data? Explain your reasoning.

(e) Use the *regression* feature of a graphing utility to find the type of model described in part (d) for the data. Let t represent the year, with $t = 0$ corresponding to 2000.

(f) Use a graphing utility to graph the original data and both the given square root model and the model that you found in part (e) in the same viewing window.

(g) Use both models to predict the number of Tiffany & Co. stores in 2015. Which model should be used to predict future values? Explain your reasoning.

Download Data

SKILLS WARM UP

The *Skills Warm Up* appears at the beginning of the exercise set for each section. These problems help you review previously learned skills that you will use in solving the section exercises.

Project

The projects at the end of selected sections involve in-depth applied exercises in which you will work with large, real-life data sets, often creating or analyzing models. These projects are offered online at *www.cengagebrain.com*.

Instructor Resources

Print

Complete Solutions Manual

ISBN-13: 978-1-133-36432-0

The *Complete Solutions Manual* provides worked-out solutions for all exercises in the text, including Checkpoints, Quiz Yourself, Test Yourself, and Tech Tutors.

Media

PowerLecture

ISBN-13: 978-1-133-36473-3

This comprehensive CD-ROM provides dynamic media tools designed to help you teach. PowerLecture includes Solution Builder, Diploma Computerized Testing, Microsoft® Powerpoint® lecture slides, and all art from the text.

Solution Builder

www.cengage.com/solutionbuilder

This online instructor database offers complete worked-out solutions of all exercises in the text. Solution Builder allows you to create customized, secure solutions printouts (in PDF format) matched exactly to the problems you assign in class.

Diploma Computerized Testing

Diploma is an easy-to-use assessment software containing hundreds of algorithmic questions derived from the text exercises. With Diploma, you can quickly create, customize, and deliver tests in both print and online formats. Diploma is available on the PowerLecture CD.

ENHANCED WebAssign

www.webassign.net

WebAssign's homework delivery system lets you deliver, collect, grade, and record assignments via the web. Enhanced WebAssign includes Cengage YouBook interactive eBook, Personal Study Plans, a Show My Work feature, Answer Evaluator, quizzes, videos, and more!

Cengage YouBook

YouBook is an interactive and customizable eBook! Containing all the content from the printed text, YouBook features a text edit tool that allows you to modify the textbook narrative as needed. With YouBook, you can quickly re-order entire sections and chapters or hide any content you don't teach to create an eBook that perfectly matches your syllabus. You can further customize the text by publishing web links. Additional media assets include: animated figures, video clips, highlighting, notes, and more! YouBook is available in Enhanced WebAssign.

CourseMate

www.cengagebrain.com

Interested in a simple way to complement your text and course content with study and practice materials? CourseMate brings course concepts to life with interactive learning, study, and exam preparation tools that support the printed textbook. CourseMate includes Engagement Tracker, a first-of-its-kind tool that monitors student engagement. Watch student comprehension soar as your class works with both the printed text and text-specific website.

Student Resources

Print

Student Solutions Manual

ISBN-13: 978-1-133-11279-2

The *Student Solutions Manual* provides complete worked-out solutions to all odd-numbered exercises in the text. In addition, the solutions of all Checkpoint, Quiz Yourself, Test Yourself, and Tech Tutor exercises are included.

Media

CourseMate

www.cengagebrain.com

Interested in a simple way to complement your text and course content with study and practice materials? CourseMate brings course concepts to life with interactive learning, study, and exam preparation tools that support the printed textbook. CourseMate includes: an interactive eBook, quizzes, flashcards, Excel guide, note-taking guide, videos, and more!

ENHANCED WebAssign

www.webassign.net

Enhanced WebAssign is an online homework system that lets instructors deliver, collect, grade, and record assignments via the web. Enhanced WebAssign includes Cengage YouBook interactive eBook, Personal Study Plans, a Show My Work feature, Answer Evaluator, quizzes, videos, and more! Be sure to check with your instructor to find out if Enhanced WebAssign is required for your course.

CengageBrain.com

To access additional course materials and companion resources, please visit *www.cengagebrain.com*. At the CengageBrain.com home page, search for the ISBN of your title (from the back cover of your book) using the search box at the top of the page. This will take you to the product page where free companion resources can be found.

Acknowledgements

I would like to thank my colleagues who have helped me develop this program. Their encouragement, criticisms, and suggestions have been invaluable to me.

Reviewers

Nasri Abdel-Aziz, *State University of New York College of Environmental Sciences and Forestry*
Alejandro Acuna, *Central New Mexico Community College*
Dona Boccio, *Queensborough Community College*
George Bradley, *Duquesne University*
Andrea Marchese, *Pace University*
Benselamonyuy Ntatin, *Austin Peay State University*
Maijian Qian, *California State University, Fullerton*
Judy Smalling, *St. Petersburg College*
Eddy Stringer, *Tallahassee Community College*

I would also like to thank the following reviewers, who have given me many useful insights to this and previous editions.

Carol Achs, *Mesa Community College;* Lateef Adelani, *Harris-Stowe State University, Saint Louis;* Frederick Adkins, *Indiana University of Pennsylvania;* Polly Amstutz, *University of Nebraska at Kearney;* George Anastassiou, *University of Memphis;* Judy Barclay, *Cuesta College;* Jean Michelle Benedict, *Augusta State University;* David Bregenzer, *Utah State University;* Ben Brink, *Wharton County Junior College;* Mary Chabot, *Mt. San Antonio College;* Jimmy Chang, *St. Petersburg College;* Joseph Chance, *University of Texas—Pan American;* John Chuchel, *University of California;* Derron Coles, *Oregon State University;* Miriam E. Connellan, *Marquette University;* William Conway, *University of Arizona;* Karabi Datta, *Northern Illinois University;* Keng Deng, *University of Louisiana at Lafayette;* Roger A. Engle, *Clarion University of Pennsylvania;* David French, *Tidewater Community College;* Randy Gallaher, *Lewis & Clark Community College;* Perry Gillespie, *Fayetteville State University;* Jose Gimenez, *Temple University;* Betty Givan, *Eastern Kentucky University;* Walter J. Gleason, *Bridgewater State College;* Shane Goodwin, *Brigham Young University of Idaho;* Mark Greenhalgh, *Fullerton College;* Harvey Greenwald, *California Polytechnic State University;* Karen Hay, *Mesa Community College;* Raymond Heitmann, *University of Texas at Austin;* Larry Hoehn, *Austin Peay State University;* William C. Huffman, *Loyola University of Chicago;* Arlene Jesky, *Rose State College;* Raja Khoury, *Collin County Community College;* Ronnie Khuri, *University of Florida;* Bernadette Kocyba, *J. Sergeant Reynolds Community College;* Duane Kouba, *University of California—Davis;* James A. Kurre, *The Pennsylvania State University;* Melvin Lax, *California State University—Long Beach;* Norbert Lerner, *State University of New York at Cortland;* Yuhlong Lio, *University of South Dakota;* Peter J. Livorsi, *Oakton Community College;* Ivan Loy, *Front Range Community College;* Peggy Luczak, *Camden County College;* Lewis D. Ludwig, *Denison University;* Samuel A. Lynch, *Southwest Missouri State University;* Augustine Maison, *Eastern Kentucky University;* Kevin McDonald, *Mt. San Antonio College;* Earl H. McKinney, *Ball State University;* Randall McNiece, *San Jacinto College;* Philip R. Montgomery, *University of Kansas;* John Nardo, *Oglethorpe University;* Mike Nasab, *Long Beach City College;* Karla Neal, *Louisiana State University;* James Osterburg, *University of Cincinnati;* Darla Ottman, *Elizabethtown Community & Technical College;* William Parzynski, *Montclair State University;* Scott Perkins, *Lake Sumter Community College;* Laurie Poe, *Santa Clara University;*

Adelaida Quesada, *Miami Dade College—Kendall;* Brooke P. Quinlan, *Hillsborough Community College;* David Ray, *University of Tennessee at Martin;* Rita Richards, *Scottsdale Community College;* Stephen B. Rodi, *Austin Community College;* Carol Rychly, *Augusta State University;* Yvonne Sandoval-Brown, *Pima Community College;* Richard Semmler, *Northern Virginia Community College—Annandale;* Bernard Shapiro, *University of Massachusetts—Lowell;* Mike Shirazi, *Germanna Community College;* Rick Simon, *University of La Verne;* Jane Y. Smith, *University of Florida;* Marvin Stick, *University of Massachusetts—Lowell;* DeWitt L. Sumners, *Florida State University;* Devki Talwar, *Indiana University of Pennsylvania;* Linda Taylor, *Northern Virginia Community College;* Stephen Tillman, *Wilkes University;* Jay Wiestling, *Palomar College;* Jonathan Wilkin, *Northern Virginia Community College;* Carol G. Williams, *Pepperdine University;* John Williams, *St. Petersburg College;* Ted Williamson, *Montclair State University;* Melvin R. Woodard, *Indiana University of Pennsylvania;* Carlton Woods, *Auburn University at Montgomery;* Jan E. Wynn, *Brigham Young University;* Robert A. Yawin, *Springfield Technical Community College;* Charles W. Zimmerman, *Robert Morris College*

My thanks to Robert Hostetler, The Pennsylvania State University, The Behrend College, Bruce Edwards, University of Florida, and David Heyd, The Pennsylvania State University, The Behrend College, for their significant contributions to previous editions of this text.

I would also like to thank the staff at Larson Texts, Inc. who assisted with proofreading the manuscript, preparing and proofreading the art package, and checking and typesetting the supplements.

On a personal level, I am grateful to my spouse, Deanna Gilbert Larson, for her love, patience, and support. Also, a special thanks goes to R. Scott O'Neil.

If you have suggestions for improving this text, please feel free to write to me. Over the past two decades I have received many useful comments from both instructors and students, and I value these comments very highly.

Ron Larson, Ph.D.
Professor of Mathematics
Penn State University
www.RonLarson.com

CALCULUS & YOU

Every feature in this text is designed to help you learn calculus. Whenever you see a red U, pay special attention to the study aid. These study aids represent years of experience in teaching students *just like you*. Ron Larson

STUDY TIP

The expressions for $f(g(x))$ and $g(f(x))$ are different in Example 5. In general, the composite of f with g is not the same as the composite of g with f.

The *Study Tips* occur at point of use throughout the text. They represent **common questions** that students ask me, **insights** into understanding concepts, and **alternative ways to look at concepts**. For instance, the *Study Tip* at the left provides insight into the importance of order when working with composite functions.

TECH TUTOR

If you have access to a symbolic differentiation utility, try using it to confirm the derivatives shown in this section.

The *Tech Tutors* give suggestions on how you can use various types of technology to help understand the material. This includes **graphing calculators**, **computer graphing programs**, and **spreadsheet programs** such as Excel. For instance, the *Tech Tutor* at the left points out that some calculators and some computer programs are capable of symbolic differentiation.

ALGEBRA TUTOR *xy*

For help in evaluating the expressions in Example 2, see the review of order of operations on page 105.

Throughout years of teaching, I have found that the greatest stumbling block to success in calculus is a weakness in algebra. Each time you see an *Algebra Tutor*, please read it carefully. Then, flip ahead to the referenced page and give yourself a chance to enjoy a brief **algebra refresher**. It will be time well spent.

HOW DO YOU SEE IT?

The *How Do You See It?* question in each exercise set helps you **visually summarize concepts** without messy computations.

SUMMARIZE

The *Summarize* outline at the end of each section asks you to write each learning objective in **your own words**.

SKILLS WARM UP

The *Skills Warm Up* exercises that precede each exercise set will help you **review previously learned skills**.

SUMMARY AND STUDY STRATEGIES

The *Summary and Study Strategies,* coupled with the Review Exercises are designed to help you organize your thoughts as you **prepare for a chapter test**.

QUIZ YOURSELF

The *Quiz Yourself* occurs midway in each chapter. Take each of these quizzes as you would **take a quiz in class**.

TEST YOURSELF

The *Test Yourself* occurs at the end of each chapter. All questions are answered so you can **check your progress**.

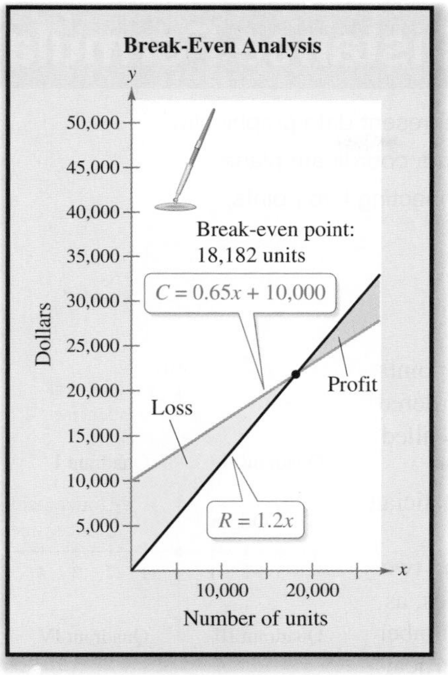

Example 5 on page 15 shows how the point of intersection of two graphs can be used to find the break-even point for a company manufacturing and selling a product.

1 Functions, Graphs, and Limits

1.1 The Cartesian Plane and the Distance Formula

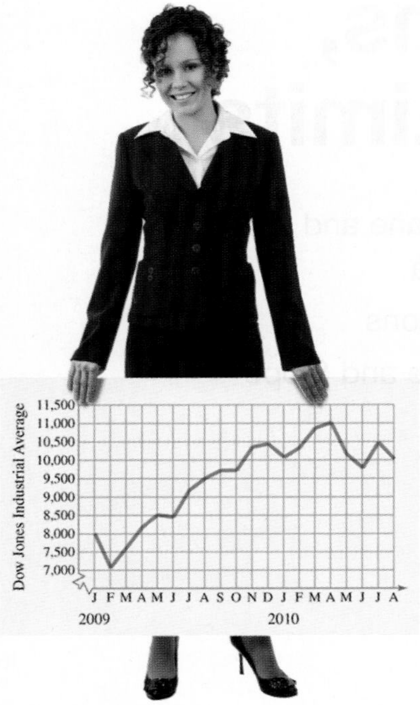

In Exercise 29 on page 9, you will use a line graph to estimate the Dow Jones Industrial Average.

- Plot points in a coordinate plane and represent data graphically.
- Find the distance between two points in a coordinate plane.
- Find the midpoint of a line segment connecting two points.
- Translate points in a coordinate plane.

The Cartesian Plane

Just as you can represent real numbers by points on a real number line, you can represent ordered pairs of real numbers by points in a plane called the **rectangular coordinate system,** or the **Cartesian plane,** after the French mathematician René Descartes (1596–1650).

The Cartesian plane is formed by using two real number lines intersecting at right angles, as shown in Figure 1.1. The horizontal real number line is usually called the ***x*-axis,** and the vertical real number line is usually called the ***y*-axis.** The point of intersection of these two axes is the **origin,** and the two axes divide the plane into four parts called **quadrants.**

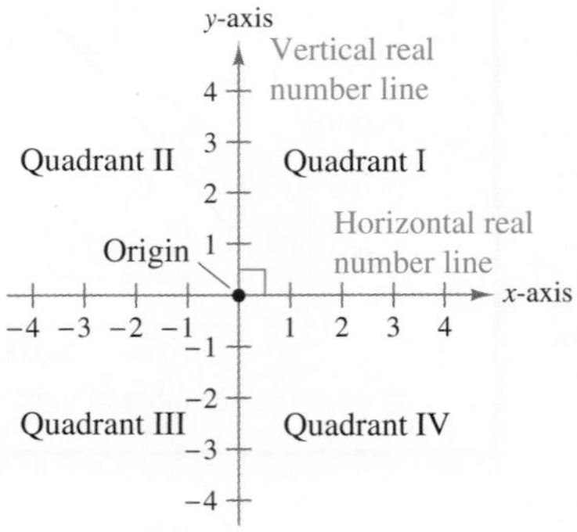

The Cartesian Plane

FIGURE 1.1

Each point in the plane corresponds to an **ordered pair** (x, y) of real numbers x and y, called **coordinates** of the point. The ***x*-coordinate** represents the directed distance from the y-axis to the point, and the ***y*-coordinate** represents the directed distance from the x-axis to the point, as shown in Figure 1.2.

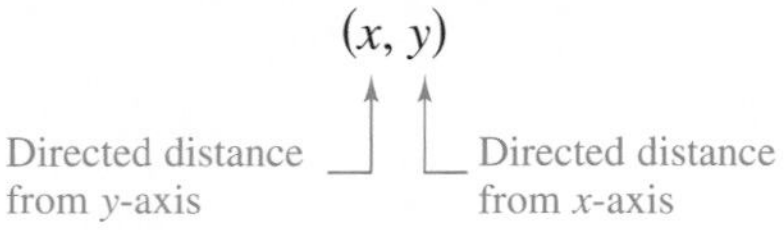

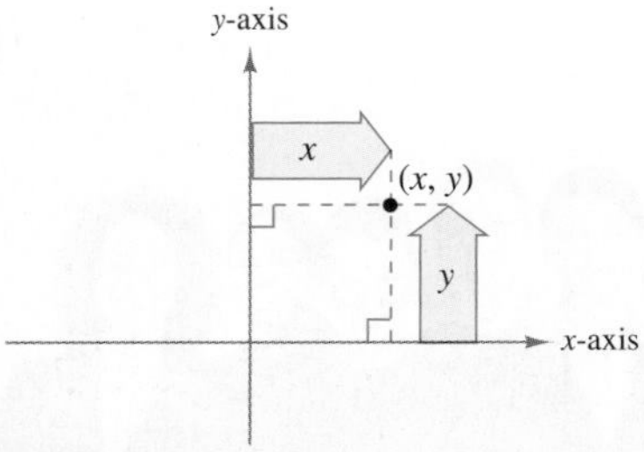

FIGURE 1.2

The notation (x, y) denotes both a point in the plane and an open interval on the real number line. The context will tell you which meaning is intended.

Example 1 Plotting Points in the Cartesian Plane

Plot the points

$$(-1, 2), \quad (3, 4), \quad (0, 0), \quad (3, 0), \quad \text{and} \quad (-2, -3).$$

SOLUTION To plot the point

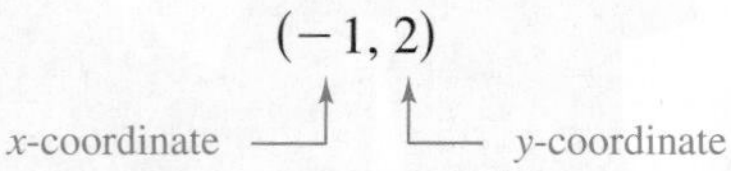

imagine a vertical line through -1 on the x-axis and a horizontal line through 2 on the y-axis. The intersection of these two lines is the point $(-1, 2)$. The other four points can be plotted in a similar way and are shown in Figure 1.3.

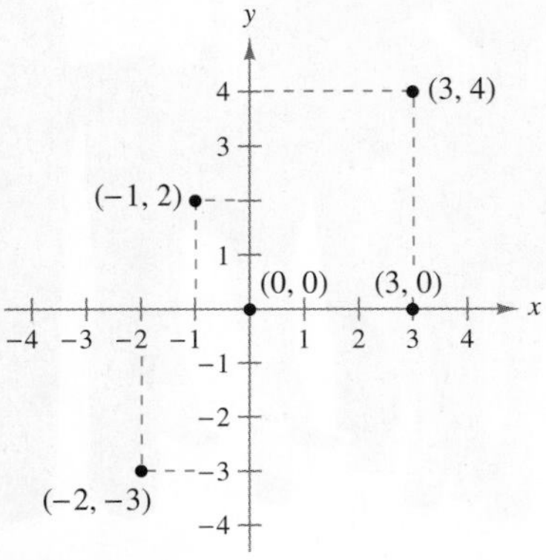

FIGURE 1.3

✓ Checkpoint 1

Plot the points

$$(-3, 2), \quad (4, -2), \quad (3, 1), \quad (0, -2), \quad \text{and} \quad (-1, -2).$$

Using a rectangular coordinate system allows you to visualize relationships between two variables. In Example 2, data is represented graphically by points plotted in a rectangular coordinate system. This type of graph is called a **scatter plot.**

Example 2 Sketching a Scatter Plot

The numbers E (in millions of people) of private-sector employees in the United States from 2000 through 2009 are shown in the table, where t represents the year. Sketch a scatter plot of the data. *(Source: U.S. Bureau of Labor Statistics)*

t	2000	2001	2002	2003	2004	2005	2006	2007	2008	2009
E	111	111	109	108	110	112	114	115	114	108

STUDY TIP

In Example 2, $t = 1$ could have been used to represent the year 2000. In that case, the horizontal axis would not have been broken, and the tick marks would have been labeled 1 through 10 (instead of 2000 through 2009).

SOLUTION To sketch a scatter plot of the data given in the table, you simply represent each pair of values by an ordered pair

(t, E)

and plot the resulting points, as shown in Figure 1.4. For instance, the first pair of values is represented by the ordered pair

(2000, 111).

Note that the break in the t-axis indicates that the numbers between 0 and 2000 have been omitted.

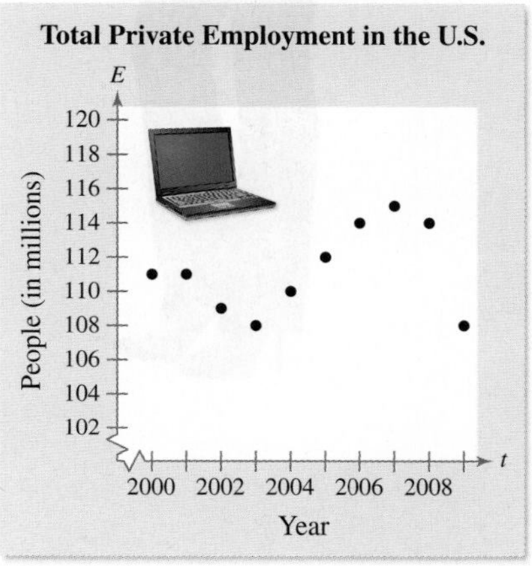

FIGURE 1.4

Checkpoint 2

The numbers E (in thousands of people) of federal employees in the United States from 2000 through 2009 are shown in the table, where t represents the year. Sketch a scatter plot of the data. *(Source: U.S. Bureau of Labor Statistics)*

t	2000	2001	2002	2003	2004	2005	2006	2007	2008	2009
E	2865	2764	2766	2761	2730	2732	2732	2734	2762	2828

The scatter plot in Example 2 is one way to represent the given data graphically. Another technique, a *bar graph*, is shown in Figure 1.5. Both graphical representations were created with a computer. If you have access to computer graphing software, try using it to represent graphically the data given in Example 2.

Another way to represent data is with a *line graph* (see Exercise 29).

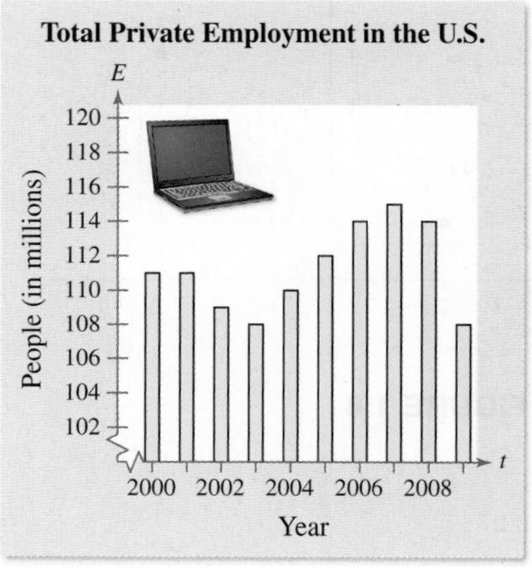

FIGURE 1.5

The Distance Formula

Recall from the Pythagorean Theorem that, for a right triangle with hypotenuse of length c and sides of lengths a and b, you have

$$a^2 + b^2 = c^2 \qquad \text{Pythagorean Theorem}$$

as shown in Figure 1.6. Note that the converse is also true. That is, if $a^2 + b^2 = c^2$, then the triangle is a right triangle.

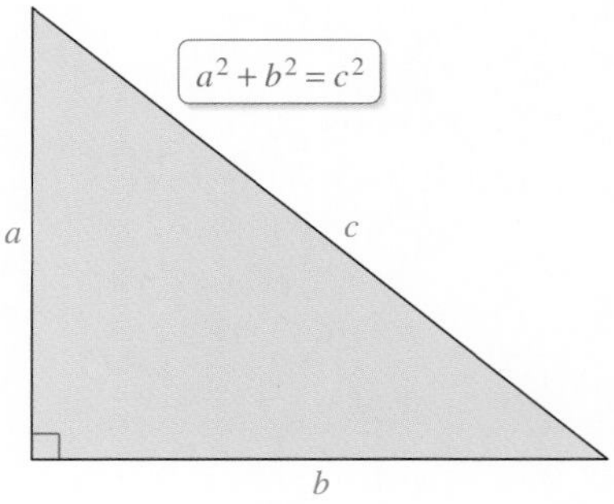

Pythagorean Theorem
FIGURE 1.6

Suppose you want to determine the distance d between two points

$$(x_1, y_1) \quad \text{and} \quad (x_2, y_2)$$

in the plane. With these two points, a right triangle can be formed, as shown in Figure 1.7. The length of the vertical side of the triangle is

$$|y_2 - y_1|$$

and the length of the horizontal side is

$$|x_2 - x_1|.$$

By the Pythagorean Theorem, you can write

$$d^2 = |x_2 - x_1|^2 + |y_2 - y_1|^2$$
$$d = \sqrt{|x_2 - x_1|^2 + |y_2 - y_1|^2}$$
$$d = \sqrt{(x_2 - x_1)^2 + (y_2 - y_1)^2}.$$

This result is the **Distance Formula.**

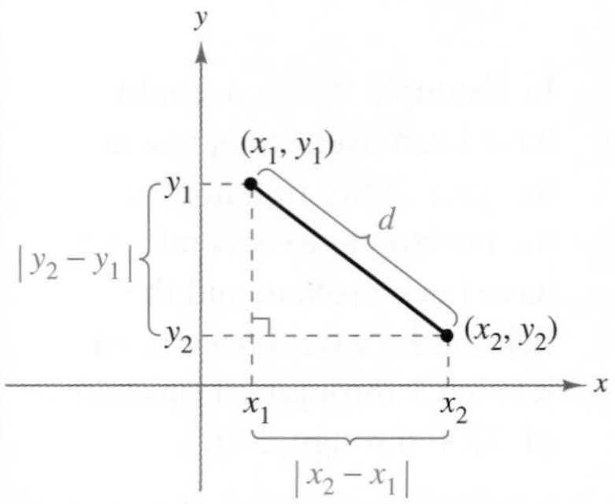

Distance Between Two Points
FIGURE 1.7

> **The Distance Formula**
>
> The distance d between the points (x_1, y_1) and (x_2, y_2) in the plane is
>
> $$d = \sqrt{(x_2 - x_1)^2 + (y_2 - y_1)^2}.$$

Example 3 Finding a Distance

Find the distance between the points $(-2, 1)$ and $(3, 4)$.

SOLUTION Let $(x_1, y_1) = (-2, 1)$ and $(x_2, y_2) = (3, 4)$. Then apply the Distance Formula as shown.

$$d = \sqrt{(x_2 - x_1)^2 + (y_2 - y_1)^2} \qquad \text{Distance Formula}$$
$$= \sqrt{[3 - (-2)]^2 + (4 - 1)^2} \qquad \text{Substitute for } x_1, y_1, x_2, \text{ and } y_2.$$
$$= \sqrt{(5)^2 + (3)^2} \qquad \text{Simplify.}$$
$$= \sqrt{34}$$
$$\approx 5.83 \qquad \text{Use a calculator.}$$

So, the distance between the points is about 5.83 units. Note in Figure 1.8 that a distance of 5.83 looks about right.

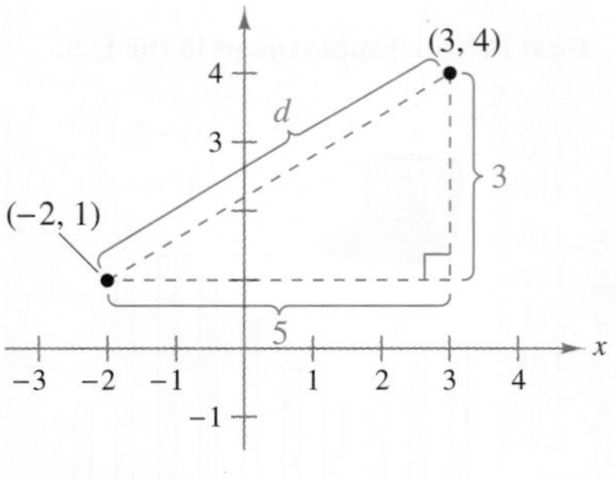

FIGURE 1.8

✓ Checkpoint 3

Find the distance between the points $(-2, 1)$ and $(2, 4)$. ■

Example 4 Verifying a Right Triangle

Use the Distance Formula to show that the points

$$(2, 1), \quad (4, 0), \quad \text{and} \quad (5, 7)$$

are vertices of a right triangle.

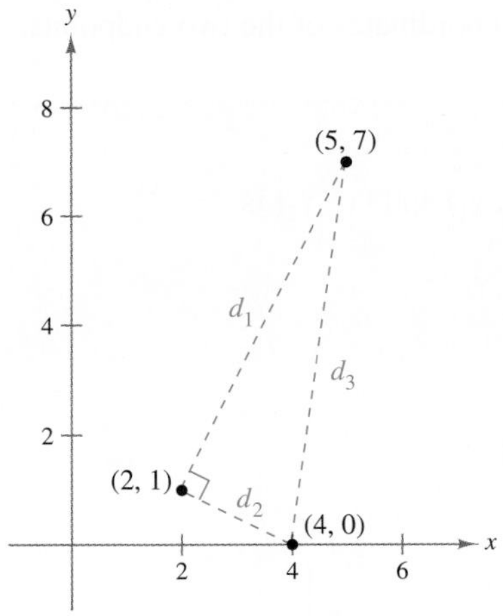

FIGURE 1.9

SOLUTION The three points are plotted in Figure 1.9. Using the Distance Formula, you can find the lengths of the three sides as shown below.

$$d_1 = \sqrt{(5-2)^2 + (7-1)^2} = \sqrt{9+36} = \sqrt{45}$$
$$d_2 = \sqrt{(4-2)^2 + (0-1)^2} = \sqrt{4+1} = \sqrt{5}$$
$$d_3 = \sqrt{(5-4)^2 + (7-0)^2} = \sqrt{1+49} = \sqrt{50}$$

Because

$$d_1^2 + d_2^2 = 45 + 5 = 50 = d_3^2$$

you can apply the converse of the Pythagorean Theorem to conclude that the triangle must be a right triangle.

✓Checkpoint 4

Use the Distance Formula to show that the points $(2, -1)$, $(5, 5)$, and $(6, -3)$ are vertices of a right triangle. ■

The figures provided with Examples 3 and 4 were not really essential to the solution. *Nevertheless*, it is strongly recommended that you develop the habit of including sketches with your solutions—even when they are not required.

Example 5 Finding the Length of a Pass

In a football game, a quarterback throws a pass from the 5-yard line, 20 yards from one sideline. The pass is caught by a wide receiver on the 45-yard line, 50 yards from the same sideline, as shown in Figure 1.10. How long was the pass?

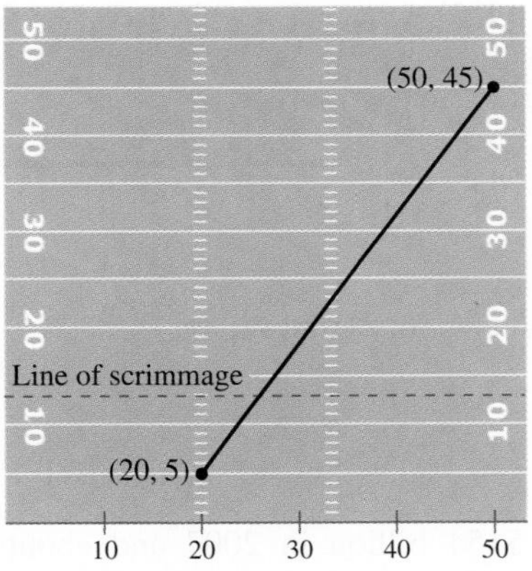

FIGURE 1.10

SOLUTION You can find the length of the pass by finding the distance between the points (20, 5) and (50, 45).

$$d = \sqrt{(50-20)^2 + (45-5)^2} \qquad \text{Distance Formula}$$
$$= \sqrt{900 + 1600}$$
$$= 50 \qquad \text{Simplify.}$$

So, the pass was 50 yards long.

✓Checkpoint 5

A quarterback throws a pass from the 10-yard line, 10 yards from one sideline. The pass is caught by a wide receiver on the 30-yard line, 25 yards from the same sideline. How long was the pass? ■

STUDY TIP

In Example 5, the scale along the goal line showing distance from the sideline does not normally appear on a football field. However, when you use coordinate geometry to solve real-life problems, you are free to place the coordinate system in any way that is convenient to the solution of the problem.

The Midpoint Formula

To find the **midpoint** of the line segment that joins two points in a coordinate plane, you can simply find the average values of the respective coordinates of the two endpoints.

The Midpoint Formula

The midpoint of the segment joining the points (x_1, y_1) and (x_2, y_2) is

$$\text{Midpoint} = \left(\frac{x_1 + x_2}{2}, \frac{y_1 + y_2}{2}\right).$$

Example 6 Finding a Segment's Midpoint

Find the midpoint of the line segment joining the points

$$(-5, -3) \quad \text{and} \quad (9, 3)$$

as shown in Figure 1.11.

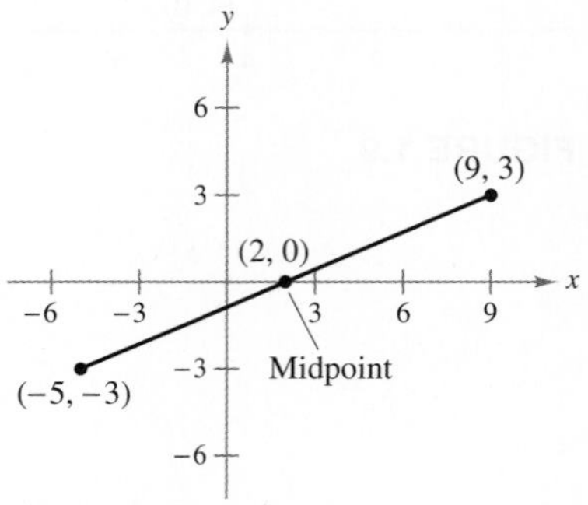

FIGURE 1.11

SOLUTION Let $(x_1, y_1) = (-5, -3)$ and $(x_2, y_2) = (9, 3)$.

$$\text{Midpoint} = \left(\frac{x_1 + x_2}{2}, \frac{y_1 + y_2}{2}\right) = \left(\frac{-5 + 9}{2}, \frac{-3 + 3}{2}\right) = (2, 0)$$

✓ Checkpoint 6

Find the midpoint of the line segment joining the points $(-6, 2)$ and $(2, 8)$. ■

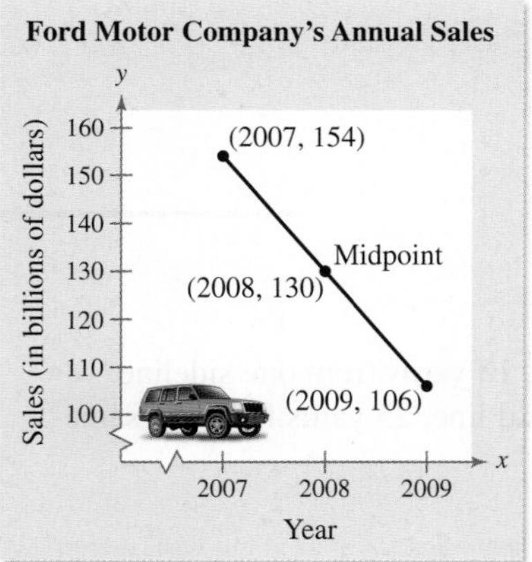

FIGURE 1.12

Example 7 Estimating Annual Sales

Ford Motor Company had annual sales of about \$154 billion in 2007 and about \$106 billion in 2009. Without knowing any additional information, what would you estimate the 2008 annual sales to have been? *(Source: Ford Motor Co.)*

SOLUTION One solution to the problem is to assume that sales followed a linear pattern. Then you can estimate the 2008 sales by finding the midpoint of the segment connecting the points (2007, 154) and (2009, 106).

$$\text{Midpoint} = \left(\frac{2007 + 2009}{2}, \frac{154 + 106}{2}\right) = (2008, 130)$$

So, you would estimate the 2008 sales to have been \$130 billion, as shown in Figure 1.12. (The actual 2008 sales were \$129 billion.)

✓ Checkpoint 7

CVS Caremark Corporation had annual sales of about \$76 billion in 2007 and about \$99 billion in 2009. What would you estimate the 2008 annual sales to have been? *(Source: CVS Caremark Corp.)* ■

Translating Points in the Plane

Much of computer graphics consists of transformations of points in a coordinate plane. One type of transformation, a translation, is illustrated in Example 8. Other types of transformations include reflections, rotations, and stretches.

Many movies now use extensive computer graphics, much of which consists of transformations of points in two- and three-dimensional space. The photo above shows a character from *Avatar*. The movie's animators used computer graphics to design the scenery, characters, motion, and even the lighting throughout much of the film.

Example 8 Translating Points in the Plane

Figure 1.13(a) shows the vertices of a parallelogram. Find the vertices of the parallelogram after it has been translated four units to the right and two units down.

SOLUTION To translate each vertex four units to the right, add 4 to each x-coordinate. To translate each vertex two units down, subtract 2 from each y-coordinate.

Original Point	*Translated Point*
$(1, 0)$	$(1 + 4, 0 - 2) = (5, -2)$
$(3, 2)$	$(3 + 4, 2 - 2) = (7, 0)$
$(3, 6)$	$(3 + 4, 6 - 2) = (7, 4)$
$(1, 4)$	$(1 + 4, 4 - 2) = (5, 2)$

The translated parallelogram is shown in Figure 1.13(b).

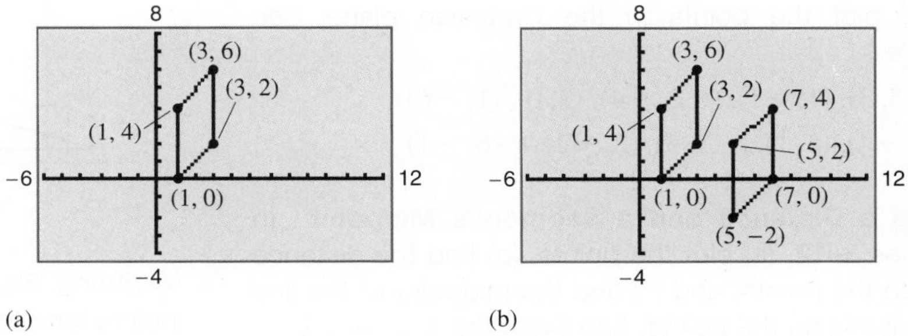

FIGURE 1.13

✓ Checkpoint 8

Find the vertices of the parallelogram in Example 8 after it has been translated two units to the left and four units down. ■

SUMMARIZE (Section 1.1)

1. Describe the Cartesian plane *(page 2)*. For examples of plotting points in the Cartesian plane, see Examples 1 and 2.
2. State the Distance Formula *(page 4)*. For examples of using the Distance Formula to find the distance between two points, see Examples 3, 4, and 5.
3. State the Midpoint Formula *(page 6)*. For examples of using the Midpoint Formula to find the midpoint of a line segment, see Examples 6 and 7.
4. Describe how to translate points in the Cartesian plane *(page 7)*. For an example of translating points in the Cartesian plane, see Example 8.

SKILLS WARM UP 1.1

The following warm-up exercises involve skills that were covered in a previous course. You will use these skills in the exercise set for this section. For additional help, review Appendix Section A.3.

In Exercises 1–6, simplify the expression.

1. $\sqrt{(3-6)^2+[1-(-5)]^2}$

2. $\sqrt{(-2-0)^2+[-7-(-3)]^2}$

3. $\dfrac{5+(-4)}{2}$

4. $\dfrac{-3+(-1)}{2}$

5. $\sqrt{27}+\sqrt{12}$

6. $\sqrt{8}-\sqrt{18}$

In Exercises 7–10, solve for x or y.

7. $\sqrt{(3-x)^2+(7-4)^2}=\sqrt{45}$

8. $\sqrt{(6-2)^2+(-2-y)^2}=\sqrt{52}$

9. $\dfrac{x+(-5)}{2}=7$

10. $\dfrac{-7+y}{2}=-3$

Exercises 1.1

See www.CalcChat.com for worked-out solutions to odd-numbered exercises.

Plotting Points in the Cartesian Plane In Exercises 1 and 2, plot the points in the Cartesian plane. *See Example 1.*

1. $(-5, 3), (1, -1), (-2, -4), (2, 0), (1, -6)$

2. $(0, -4), (5, 1), (-3, 5), (2, -2), (-6, -1)$

Finding a Distance and a Segment's Midpoint In Exercises 3–12, (a) plot the points, (b) find the distance between the points, and (c) find the midpoint of the line segment joining the points. *See Examples 1, 3, and 6.*

3. $(3, 1), (5, 5)$

4. $(-3, 2), (3, -2)$

5. $\left(\frac{1}{2}, 1\right), \left(-\frac{3}{2}, -5\right)$

6. $\left(\frac{2}{3}, -\frac{1}{3}\right), \left(\frac{5}{6}, 1\right)$

7. $(2, 2), (4, 14)$

8. $(-3, 7), (1, -1)$

9. $(-5, -2), (7, 3)$

10. $(2, -12), (8, -4)$

11. $(0, -4.8), (0.5, 6)$

12. $(5.2, 6.4), (-2.7, 1.8)$

Verifying a Right Triangle In Exercises 13–16, (a) find the length of each side of the right triangle and (b) show that these lengths satisfy the Pythagorean Theorem. *See Example 4.*

13.

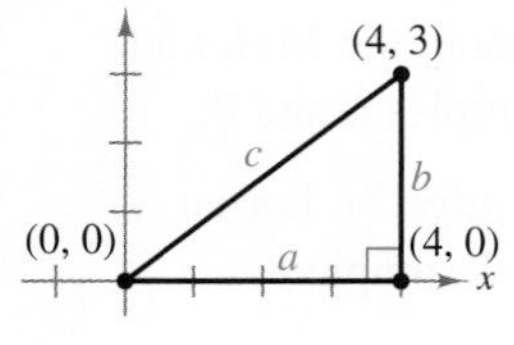

14.

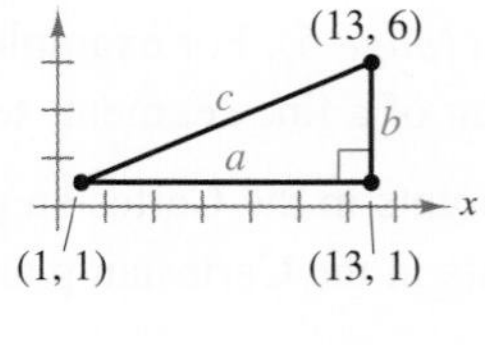

15.

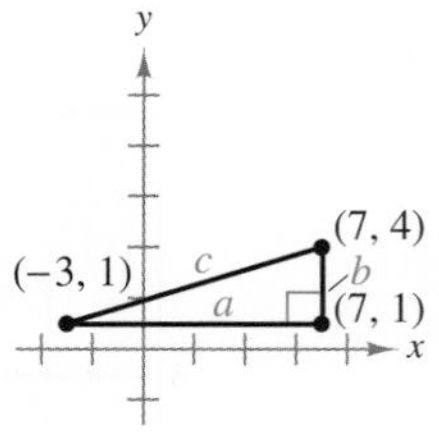

16.

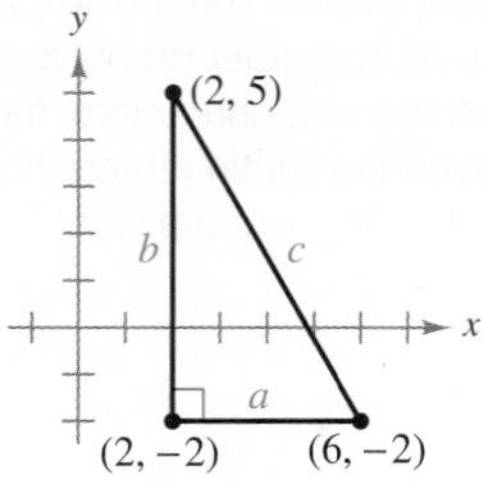

Verifying Figures In Exercises 17–20, show that the points form the vertices of the given figure. (A rhombus is a quadrilateral whose sides have the same length.)

	Vertices	*Figure*
17.	$(0, 1), (3, 7), (4, -1)$	Right triangle
18.	$(1, -3), (3, 2), (-2, 4)$	Isosceles triangle
19.	$(0, 0), (1, 2), (2, 1), (3, 3)$	Rhombus
20.	$(0, 1), (3, 7), (4, 4), (1, -2)$	Parallelogram

Finding Values In Exercises 21 and 22, find the value(s) of x such that the distance between the points is 5.

21. $(1, 0), (x, -4)$

22. $(2, -1), (x, 2)$

Finding Values In Exercises 23 and 24, find the value(s) of y such that the distance between the points is 8.

23. $(0, 0), (3, y)$

24. $(5, 1), (5, y)$

The answers to the odd-numbered and selected even-numbered exercises are given in the back of the text. Worked-out solutions to the odd-numbered exercises are given in the *Student Solutions Manual.*

25. Sports A soccer player passes the ball from a point that is 18 yards from an endline and 12 yards from a sideline. The pass is received by a teammate who is 42 yards from the same endline and 50 yards from the same sideline, as shown in the figure. How long is the pass?

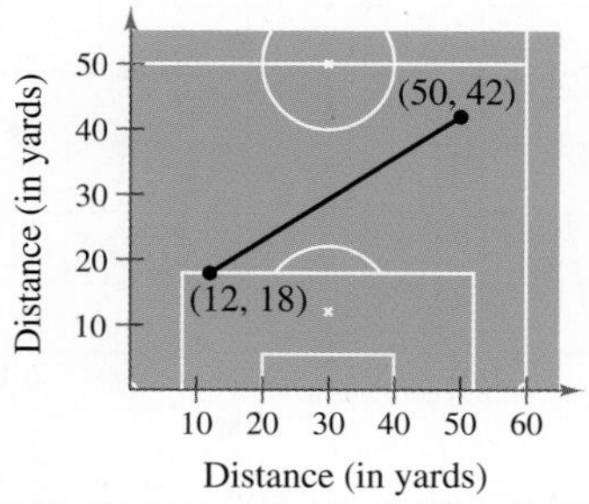

26. Sports The first soccer player in Exercise 25 passes the ball to another teammate who is 37 yards from the same endline and 33 yards from the same sideline. How long is the pass?

Graphing Data In Exercises 27 and 28, use a graphing utility to graph a scatter plot, a bar graph, and a line graph to represent the data. Describe any trends that appear.

27. Consumer Trends The numbers (in millions) of cable high-speed Internet customers in the United States for 2000 through 2009 are shown in the table. *(Source: National Cable & Telecommunications Association)*

Year	2000	2001	2002	2003	2004
Customers	4.0	7.3	11.6	16.5	21.0

Year	2005	2006	2007	2008	2009
Customers	25.4	28.9	35.7	39.3	41.8

28. Consumer Trends The numbers (in millions) of cellular telephone subscribers in the United States for 2000 through 2009 are shown in the table. *(Source: CTIA-The Wireless Association)*

Year	2000	2001	2002	2003	2004
Subscribers	109.5	128.4	140.8	158.7	182.1

Year	2005	2006	2007	2008	2009
Subscribers	207.9	233.0	255.4	270.3	285.6

29. Dow Jones Industrial Average The graph shows the Dow Jones Industrial Average for common stocks. *(Source: Dow Jones, Inc.)*

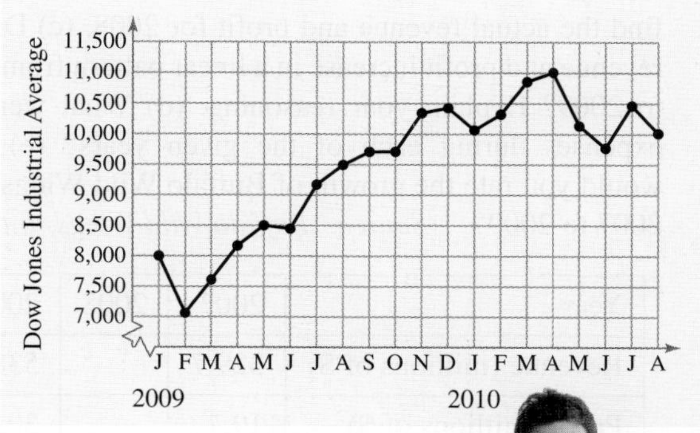

(a) Estimate the Dow Jones Industrial Average for March 2009, July 2009, and July 2010.

(b) Estimate the percent increase or decrease in the Dow Jones Industrial Average from April 2010 to May 2010.

30. Home Sales The graph shows the median sales prices (in thousands of dollars) of existing one-family homes sold in the United States from 1994 through 2009. *(Source: National Association of Realtors)*

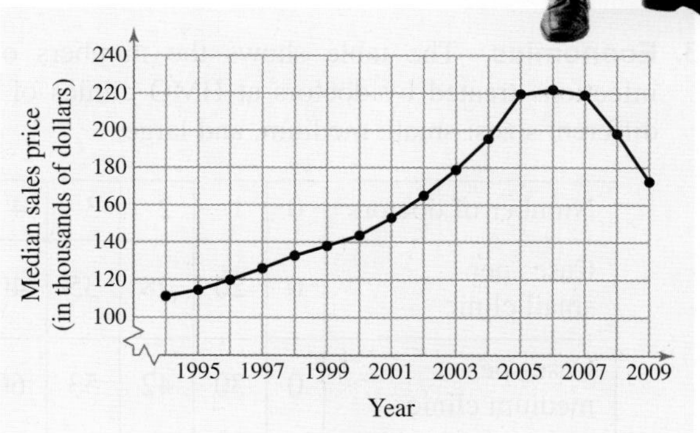

(a) Estimate the median sales prices of existing one-family homes for 1996, 2003, and 2008.

(b) Estimate the percent increase or decrease in the median value of existing one-family homes from 2001 to 2002.

The symbol [graphing calculator icon] indicates an exercise in which you are instructed to use graphing technology or a symbolic computer algebra system. The solutions of other exercises may also be facilitated by use of appropriate technology.

31. Revenue and Profit The revenues and profits of Buffalo Wild Wings for 2007 and 2009 are shown in the table. (a) Use the Midpoint Formula to estimate the revenue and profit in 2008. (b) Then use your school's library, the Internet, or some other reference source to find the actual revenue and profit for 2008. (c) Did the revenue and profit increase in a linear pattern from 2007 to 2009? Explain your reasoning. (d) What were the expenses during each of the given years? (e) How would you rate the growth of Buffalo Wild Wings from 2007 to 2009? *(Source: Buffalo Wild Wings, Inc.)*

Year	2007	2008	2009
Revenue (millions of \$)	329.7		538.9
Profit (millions of \$)	19.7		30.7

32. Revenue and Profit The revenues and profits of Cablevision Systems Corporation for 2007 and 2009 are shown in the table. (a) Use the Midpoint Formula to estimate the revenue and profit in 2008. (b) Then use your school's library, the Internet, or some other reference source to find the actual revenue and profit for 2008. (c) Did the revenue and profit increase in a linear pattern from 2007 to 2009? Explain your reasoning. (d) What were the expenses during each of the given years? (e) How would you rate the growth of Cablevision Systems Corporation from 2007 to 2009? *(Source: Cablevision Systems Corporation)*

Year	2007	2008	2009
Revenue (millions of \$)	6484.5		7773.3
Profit (millions of \$)	23.7		285.6

33. Economics The table shows the numbers of ear infections treated by doctors at HMO clinics of three different sizes: small, medium, and large.

Number of doctors	0	1	2	3	4
Cases per small clinic	0	20	28	35	40
Cases per medium clinic	0	30	42	53	60
Cases per large clinic	0	35	49	62	70

(a) On the same coordinate plane, show the relationship between doctors and treated ear infections using *three* line graphs, where the number of doctors is on the horizontal axis and the number of ear infections treated is on the vertical axis.

(b) Compare the three relationships.

(Source: Adapted from Taylor, Economics, *Fifth Edition)*

34. HOW DO YOU SEE IT? The scatter plot shows the numbers of pickup trucks sold in a city from 2006 to 2011.

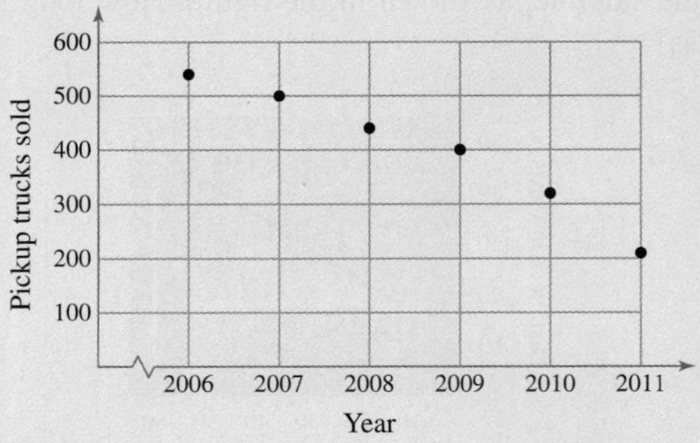

(a) In what year were 500 pickup trucks sold?

(b) About how many pickup trucks were sold in 2007?

(c) Describe the pattern shown by the data.

Translating Points in the Plane In Exercises 35 and 36, use the translation and the graph to find the vertices of the figure after it has been translated. *See Example 8.*

35. 3 units left and 5 units down

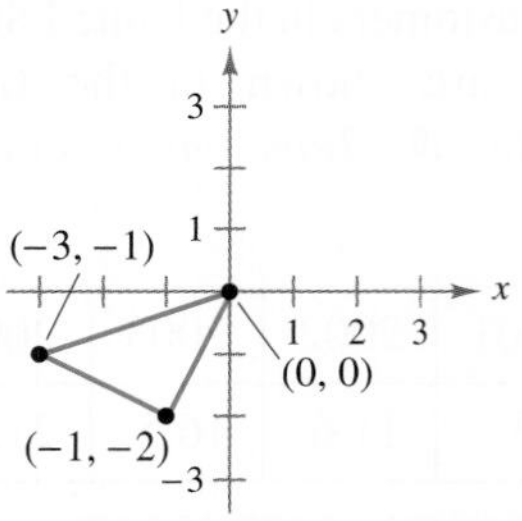

36. 2 units right and 4 units up

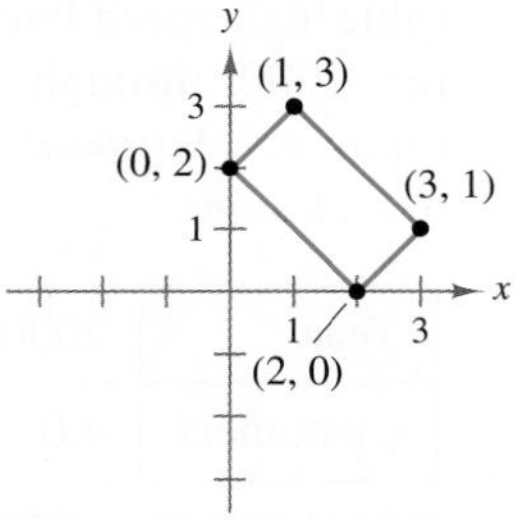

37. Using the Midpoint Formula Use the Midpoint Formula repeatedly to find the three points that divide the segment joining (x_1, y_1) and (x_2, y_2) into four equal parts.

38. Using the Midpoint Formula Use Exercise 37 to find the points that divide the line segment joining the given points into four equal parts.

(a) $(1, -2), (4, -1)$ (b) $(-2, -3), (0, 0)$

39. Using the Midpoint Formula Show that $\left(\frac{1}{3}[2x_1 + x_2], \frac{1}{3}[2y_1 + y_2]\right)$ is one of the points of trisection of the line segment joining (x_1, y_1) and (x_2, y_2). Then, find the second point of trisection by finding the midpoint of the segment joining

$$\left(\frac{1}{3}[2x_1 + x_2], \frac{1}{3}[2y_1 + y_2]\right) \quad \text{and} \quad (x_2, y_2).$$

40. Using the Midpoint Formula Use Exercise 39 to find the points of trisection of the line segment joining the given points.

(a) $(1, -2), (4, 1)$ (b) $(-2, -3), (0, 0)$

1.2 Graphs of Equations

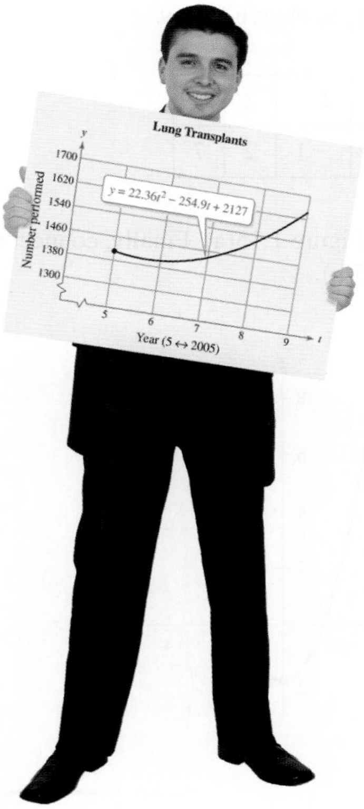

In Exercise 62 on page 21, you will use a mathematical model to analyze the number of lung transplants in the United States.

- Sketch graphs of equations by hand.
- Find the x- and y-intercepts of graphs of equations.
- Write the standard forms of equations of circles.
- Find the points of intersection of two graphs.
- Use mathematical models to model and solve real-life problems.

The Graph of an Equation

In Section 1.1, you used a coordinate system to represent graphically the relationship between two quantities. There, the graphical picture consisted of a collection of points in a coordinate plane (see Example 2 in Section 1.1).

Frequently, a relationship between two quantities is expressed as an equation. For instance, degrees on the Fahrenheit scale are related to degrees on the Celsius scale by the equation

$$F = \frac{9}{5}C + 32.$$

In this section, you will study some basic procedures for sketching the graphs of such equations. The **graph** of an equation is the set of all points that are solutions of the equation.

Example 1 Sketching the Graph of an Equation

Sketch the graph of $y = 7 - 3x$.

SOLUTION The simplest way to sketch the graph of an equation is the *point-plotting method*. With this method, you construct a table of values that consists of several solution points of the equation, as shown in the table below. For instance, when $x = 0$,

$$y = 7 - 3(0) = 7$$

which implies that $(0, 7)$ is a solution point of the equation.

x	0	1	2	3	4
$y = 7 - 3x$	7	4	1	-2	-5

From the table, it follows that $(0, 7)$, $(1, 4)$, $(2, 1)$, $(3, -2)$, and $(4, -5)$ are solution points of the equation. After plotting these points, you can see that they appear to lie on a line, as shown in Figure 1.14. The graph of the equation is the line that passes through the five plotted points.

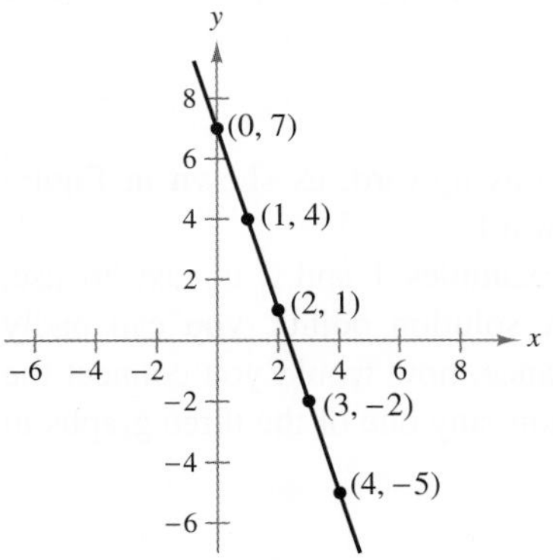

Solution Points for $y = 7 - 3x$
FIGURE 1.14

✓ Checkpoint 1

Sketch the graph of $y = 2x - 1$. ■

STUDY TIP

Even though the sketch shown in Figure 1.14 is referred to as the graph of $y = 7 - 3x$, it actually represents only a *portion* of the graph. The entire graph is a line that would extend off the page.

TECH TUTOR

On most graphing utilities, the display screen is two thirds as high as it is wide. On such screens, you can obtain a graph with a true geometric perspective by using a *square setting*—one in which

$$\frac{\text{Ymax} - \text{Ymin}}{\text{Xmax} - \text{Xmin}} = \frac{2}{3}.$$

One such setting is shown below. Notice that the x and y tick marks are equally spaced on a square setting, but not on a *standard setting*.

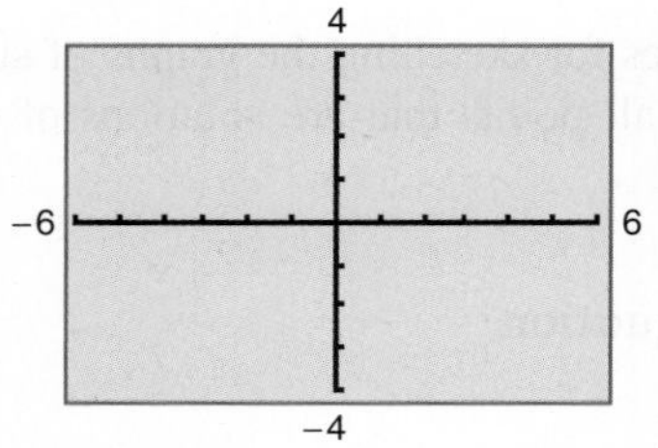

Example 2 Sketching the Graph of an Equation

Sketch the graph of $y = x^2 - 2$.

SOLUTION Begin by constructing a table of values, as shown below.

x	-2	-1	0	1	2	3
$y = x^2 - 2$	2	-1	-2	-1	2	7

Next, plot the points given in the table, as shown in Figure 1.15(a). Finally, connect the points with a smooth curve, as shown in Figure 1.15(b).

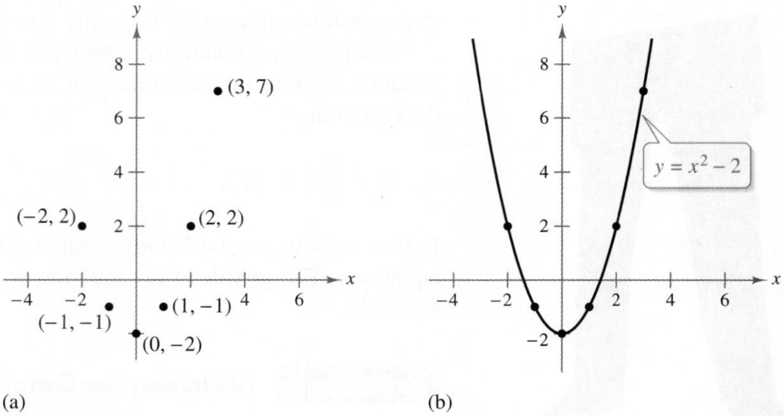

FIGURE 1.15

✓Checkpoint 2

Sketch the graph of $y = x^2 - 4$. ■

The graph shown in Example 2 is a **parabola.** The graph of any second-degree equation of the form

$$y = ax^2 + bx + c, \quad a \neq 0$$

has a similar shape. If $a > 0$, then the parabola opens upward, as shown in Figure 1.15(b), and if $a < 0$, then the parabola opens downward.

The point-plotting technique demonstrated in Examples 1 and 2 is easy to use, but it does have some shortcomings. With too few solution points, you can badly misrepresent the graph of a given equation. For instance, how would you connect the four points in Figure 1.16? Without further information, any one of the three graphs in Figure 1.17 would be reasonable.

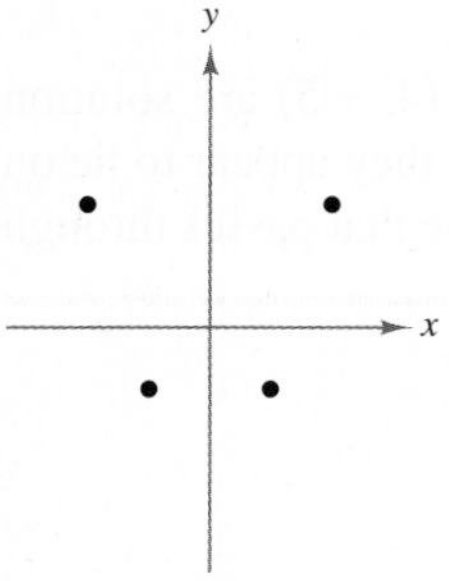

FIGURE 1.16

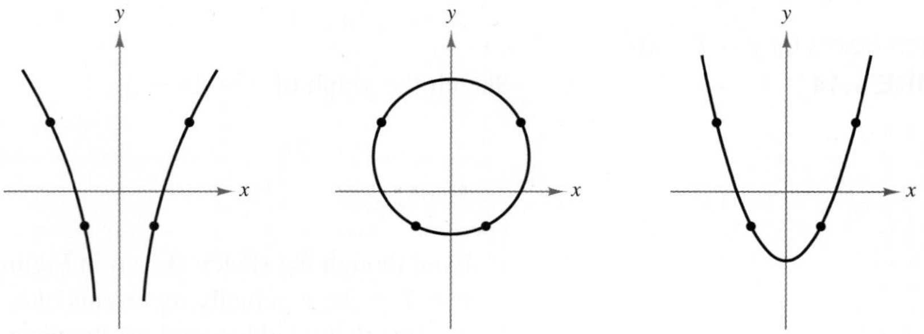

FIGURE 1.17

ALGEBRA TUTOR

Finding intercepts involves solving equations. For a review of some techniques for solving equations, see page 71.

Intercepts of a Graph

Some solution points have zero as either the x-coordinate or the y-coordinate. These points are called **intercepts** because they are the points at which the graph intersects the x- or y-axis.

Some texts denote the x-intercept as the x-coordinate of the point

$(a, 0)$

rather than the point itself. Unless it is necessary to make a distinction, when the term *intercept* is used in this text, it will mean either the point or the coordinate.

A graph may have no intercepts or several intercepts, as shown in Figure 1.18.

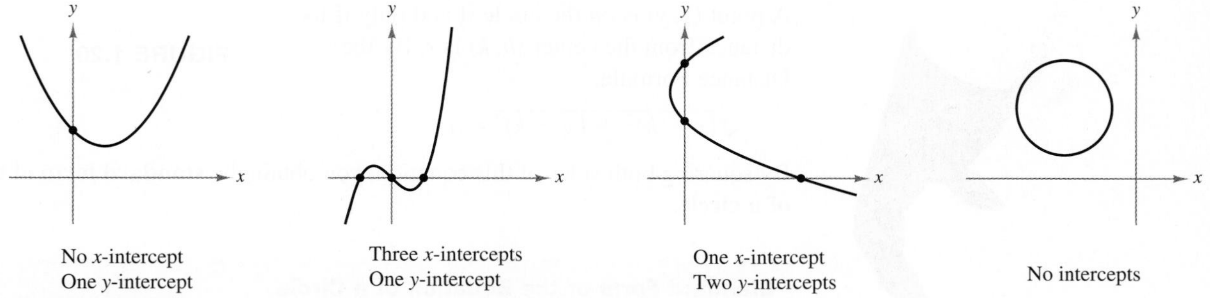

FIGURE 1.18

TECH TUTOR

Some graphing utilities have a built-in program that can find the x-intercepts of a graph. If your graphing utility has this feature, try using it to find the x-intercept of the graph of the equation in Example 3. (Your utility may call this the *root* or *zero* feature.)

Finding Intercepts

1. To find **x-intercepts,** let y be zero and solve the equation for x.
2. To find **y-intercepts,** let x be zero and solve the equation for y.

Example 3 Finding x- and y-Intercepts

Find the x- and y-intercepts of the graph of $y = x^3 - 4x$.

SOLUTION To find the x-intercepts, let y be zero and solve for x.

$$x^3 - 4x = 0 \quad \text{Let } y \text{ be zero.}$$
$$x(x^2 - 4) = 0 \quad \text{Factor out common monomial factor.}$$
$$x(x + 2)(x - 2) = 0 \quad \text{Factor.}$$
$$x = 0, -2, \text{ or } 2 \quad \text{Solve for } x.$$

Because this equation has three solutions, you can conclude that the graph has three x-intercepts:

$(0, 0)$, $(-2, 0)$, and $(2, 0)$. x-intercepts

To find the y-intercepts, let x be zero and solve for y. Doing this produces

$$y = x^3 - 4x = 0^3 - 4(0) = 0.$$

This equation has only one solution, so the graph has one y-intercept:

$(0, 0)$. y-intercept

(See Figure 1.19.)

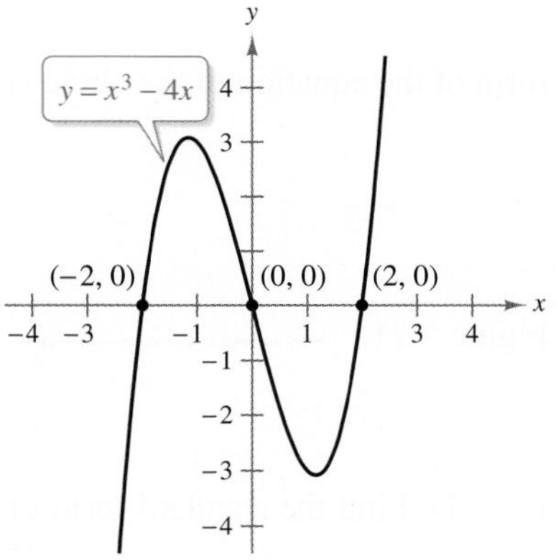

FIGURE 1.19

✓ Checkpoint 3

Find the x- and y-intercepts of the graph of $y = x^2 - 2x - 3$. ■

Circles

Throughout this course, you will learn to recognize several types of graphs from their equations. For instance, you should recognize that the graph of a second-degree equation of the form

$$y = ax^2 + bx + c, \quad a \neq 0$$

is a parabola (see Example 2). Another easily recognized graph is that of a **circle.**

Consider the circle shown in Figure 1.20. A point (x, y) is on the circle if and only if its distance from the center (h, k) is r. By the Distance Formula,

$$\sqrt{(x-h)^2 + (y-k)^2} = r.$$

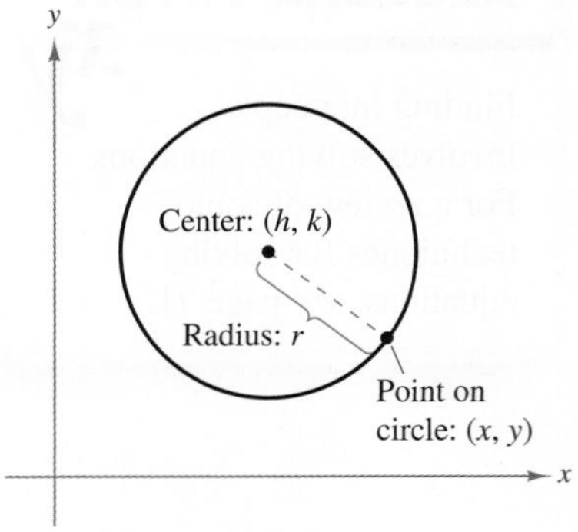

FIGURE 1.20

By squaring both sides of this equation, you obtain the **standard form of the equation of a circle.**

Standard Form of the Equation of a Circle

The **standard form of the equation of a circle** is

$$(x-h)^2 + (y-k)^2 = r^2. \qquad \text{Center at } (h, k)$$

The point (h, k) is the **center** of the circle, and the positive number r is the **radius** of the circle. The standard form of the equation of a circle whose center is the origin, $(h, k) = (0, 0)$, is

$$x^2 + y^2 = r^2. \qquad \text{Center at } (0, 0)$$

Example 4 Finding the Equation of a Circle

The point $(3, 4)$ lies on a circle whose center is at $(-1, 2)$. Find the standard form of the equation of this circle and sketch its graph.

SOLUTION The radius of the circle is the distance between $(-1, 2)$ and $(3, 4)$.

$$\begin{aligned} r &= \sqrt{[3-(-1)]^2 + (4-2)^2} && \text{Distance Formula} \\ &= \sqrt{(4)^2 + (2)^2} && \text{Simplify.} \\ &= \sqrt{16+4} && \text{Simplify.} \\ &= \sqrt{20} && \text{Radius} \end{aligned}$$

Using $(h, k) = (-1, 2)$ and $r = \sqrt{20}$, the standard form of the equation of the circle is

$$\begin{aligned} (x-h)^2 + (y-k)^2 &= r^2 && \text{Equation of a circle} \\ [x-(-1)]^2 + (y-2)^2 &= \left(\sqrt{20}\right)^2 && \text{Substitute for } h, k, \text{ and } r. \\ (x+1)^2 + (y-2)^2 &= 20. && \text{Write in standard form.} \end{aligned}$$

The graph of the equation of the circle is shown in Figure 1.21.

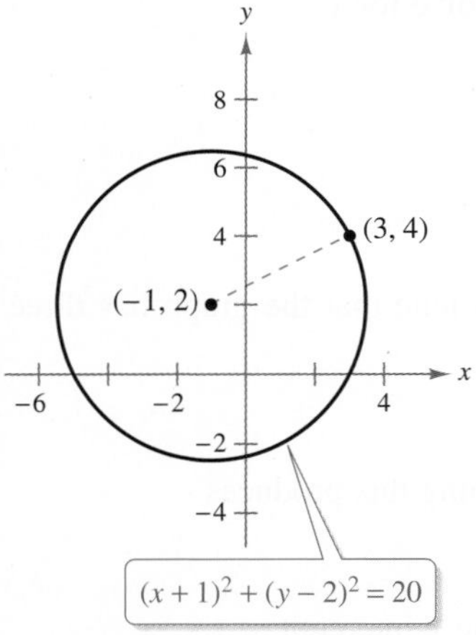

FIGURE 1.21

✓ Checkpoint 4

The point $(1, 5)$ lies on a circle whose center is at $(-2, 1)$. Find the standard form of the equation of this circle and sketch its graph. ■

Points of Intersection

An ordered pair that is a solution of two different equations is called a **point of intersection** of the graphs of the two equations. For instance, Figure 1.22 shows that the graphs of

$$y = x^2 - 3 \quad \text{and} \quad y = x - 1$$

have two points of intersection: $(2, 1)$ and $(-1, -2)$. To find the points analytically, set the two y-values equal to each other and solve the equation

$$x^2 - 3 = x - 1$$

for x.

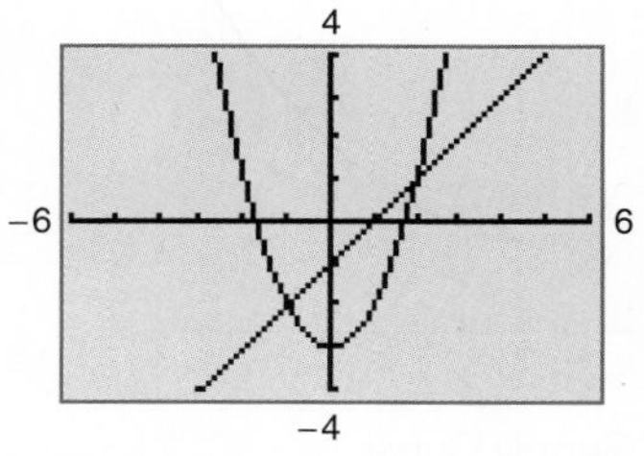

FIGURE 1.22

> **STUDY TIP**
>
> You can check the points of intersection in Figure 1.22 by verifying that the points are solutions of *both* of the original equations or by using the *intersect* feature of a graphing utility.

A common business application that involves points of intersection is **break-even analysis.** The marketing of a new product typically requires an initial investment. When sufficient units have been sold so that the total revenue has offset the total cost, the sale of the product has reached the **break-even point.** The **total cost** of producing x units of a product is denoted by C, and the **total revenue** from the sale of x units of the product is denoted by R. So, you can find the break-even point by setting the cost C equal to the revenue R and solving for x. In other words, the break-even point corresponds to the point of intersection of the cost and revenue graphs.

Example 5 Finding a Break-Even Point

A company manufactures a product at a cost of \$0.65 per unit and sells the product for \$1.20 per unit. The company's initial investment to produce the product was \$10,000. Will the company break even when it sells 18,000 units? How many units must the company sell to break even?

SOLUTION The total cost of producing x units of the product is given by

$$C = 0.65x + 10{,}000. \qquad \text{Cost equation}$$

The total revenue from the sale of x units is given by

$$R = 1.2x. \qquad \text{Revenue equation}$$

To find the break-even point, set the cost equal to the revenue and solve for x.

$$R = C \qquad \text{Set revenue equal to cost.}$$

$$1.2x = 0.65x + 10{,}000 \qquad \text{Substitute for } R \text{ and } C.$$

$$0.55x = 10{,}000 \qquad \text{Subtract } 0.65x \text{ from each side.}$$

$$x = \frac{10{,}000}{0.55} \qquad \text{Divide each side by 0.55.}$$

$$x \approx 18{,}182 \qquad \text{Use a calculator.}$$

So, the company will not break even when it sells 18,000 units. The company must sell 18,182 units before it breaks even. This result is shown graphically in Figure 1.23. Note in Figure 1.23 that sales less than 18,182 units correspond to a loss for the company ($R < C$), whereas sales greater than 18,182 units correspond to a profit for the company ($R > C$).

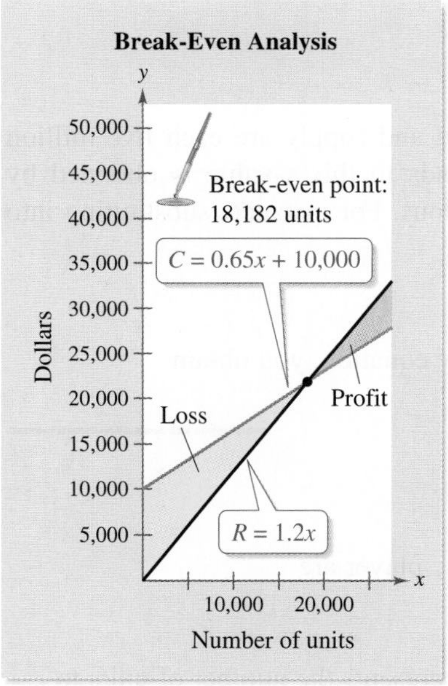

FIGURE 1.23

✓Checkpoint 5

How many units must the company in Example 5 sell to break even when the selling price is \$1.45 per unit?

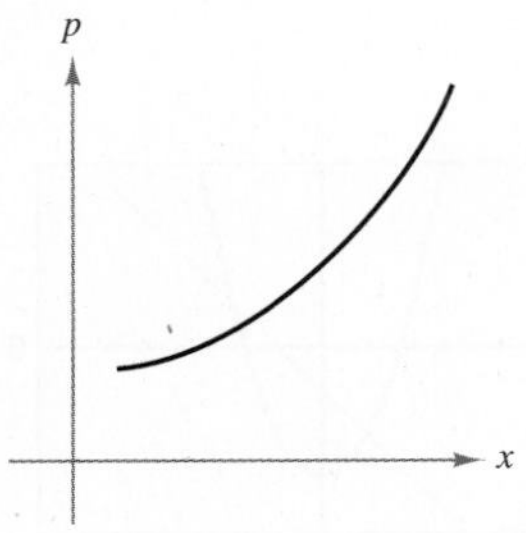

Supply Curve
FIGURE 1.24

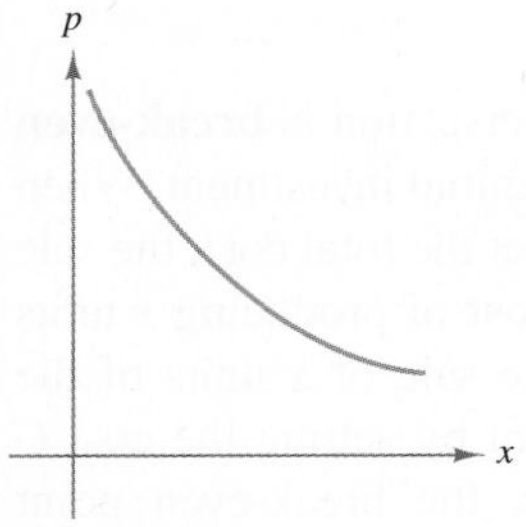

Demand Curve
FIGURE 1.25

Two types of equations that economists use to analyze a market are supply and demand equations. A **supply equation** shows the relationship between the unit price p of a product and the quantity supplied x. The graph of a supply equation is called a **supply curve.** (See Figure 1.24.) A typical supply curve rises because producers of a product want to sell more units when the unit price is higher.

A **demand equation** shows the relationship between the unit price p of a product and the quantity demanded x. The graph of a demand equation is called a **demand curve.** (See Figure 1.25.) A typical demand curve tends to show a decrease in the quantity demanded with each increase in price.

In an ideal situation, with no other factors present to influence the market, the production level should stabilize at the point of intersection of the graphs of the supply and demand equations. This point is called the **equilibrium point.** The x-coordinate of the equilibrium point is called the **equilibrium quantity** and the p-coordinate is called the **equilibrium price.** (See Figure 1.26.) You can find the equilibrium point by setting the demand equation equal to the supply equation and solving for x.

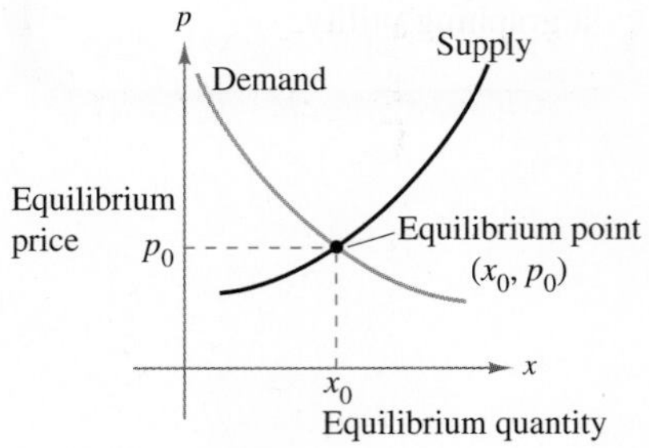

Equilibrium Point
FIGURE 1.26

Example 6 Finding the Equilibrium Point

The demand and supply equations for an e-book reader are given by

$p = 195 - 5.8x$ Demand equation

$p = 150 + 3.2x$ Supply equation

where p is the price in dollars and x represents the number of units in millions. Find the equilibrium point for this market.

SOLUTION Begin by setting the demand equation equal to the supply equation.

$195 - 5.8x = 150 + 3.2x$ Set equations equal to each other.

$45 - 5.8x = 3.2x$ Subtract 150 from each side.

$45 = 9x$ Add $5.8x$ to each side.

$5 = x$ Divide each side by 9.

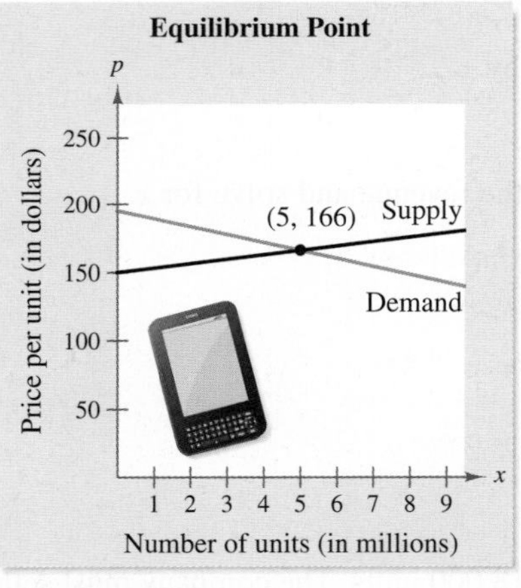

FIGURE 1.27

So, the equilibrium point occurs when the demand and supply are each five million units. (See Figure 1.27.) The price that corresponds to this x-value is obtained by substituting $x = 5$ into either of the original equations. For instance, substituting into the demand equation produces

$$p = 195 - 5.8(5) = 195 - 29 = \$166.$$

Note that when you substitute $x = 5$ into the supply equation, you obtain

$$p = 150 + 3.2(5) = 150 + 16 = \$166.$$

✓Checkpoint 6

The demand and supply equations for a Blu-ray disc player are

$$p = 136 - 3.5x \quad \text{and} \quad p = 112 + 2.5x$$

respectively, where p is the price in dollars and x represents the number of units in millions. Find the equilibrium point for this market. ■

Mathematical Models

ALGEBRA TUTOR *xy*

For help in evaluating the expressions in Example 7, see the review of order of operations on page 70.

In this text, you will see many examples of the use of equations as **mathematical models** of real-life phenomena. In developing a mathematical model to represent actual data, you should strive for two (often conflicting) goals—accuracy and simplicity.

Example 7 Using Mathematical Models

The table shows the annual sales (in millions of dollars) for Dollar Tree and 99 Cents Only Stores from 2005 through 2009. In 2009, the publication *Value Line* listed the projected 2010 sales for the companies as \$5770 million and \$1430 million, respectively. How do you think these projections were obtained? *(Source: Dollar Tree, Inc. and 99 Cents Only Stores)*

Year	2005	2006	2007	2008	2009
t	5	6	7	8	9
Dollar Tree	3393.9	3969.4	4242.6	4644.9	5231.2
99 Cents Only Stores	1023.6	1104.7	1199.4	1302.9	1355.2

SOLUTION The projections were obtained by using past sales to predict future sales. The past sales were modeled by equations that were found by a statistical procedure called least squares regression analysis. The models for the two companies are

$$S = 10.764t^2 + 284.31t + 1757.3, \quad 5 \le t \le 9 \qquad \text{Dollar Tree}$$

and

$$S = -3.486t^2 + 134.94t + 430.4, \quad 5 \le t \le 9. \qquad \text{99 Cents Only Stores}$$

Using $t = 10$ to represent 2010, you can predict the 2010 sales to be

$$S = 10.764(10)^2 + 284.31(10) + 1757.3 \approx 5676.8 \qquad \text{Dollar Tree}$$

and

$$S = -3.486(10)^2 + 134.94(10) + 430.4 \approx 1431.2. \qquad \text{99 Cents Only Stores}$$

These two projections are close to those projected by *Value Line*. The graphs of the two models are shown in Figure 1.28.

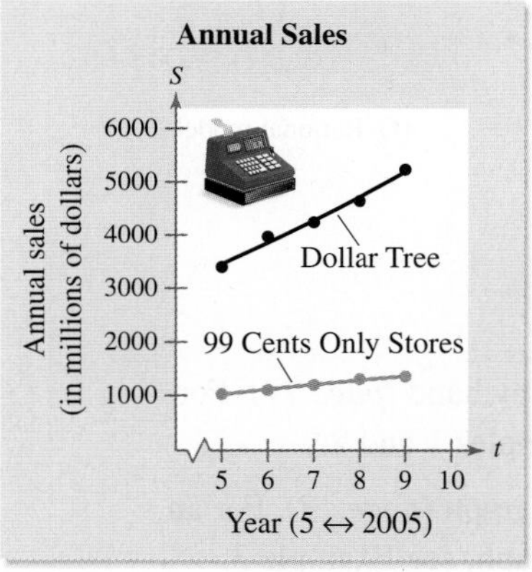

FIGURE 1.28

✓ Checkpoint 7

The table shows the annual revenues (in millions of dollars) for BJ's Wholesale Club from 2005 through 2009. In 2009, the publication *Value Line* listed the projected 2010 revenues for BJ's Wholesale Club as \$11,150 million. How does this projection compare with the projection obtained using the model below? *(Source: BJ's Wholesale Club, Inc.)*

$$S = -17.393t^2 + 845.59t + 4097.7, \quad 5 \le t \le 9$$

Year	2005	2006	2007	2008	2009
t	5	6	7	8	9
Revenues	7949.9	8480.3	9005.0	10,027.0	10,187.0

To test the accuracy of a model, you can compare the actual data with the values given by the model. Try doing this for each model in Example 7.

Much of your study of calculus will center around the behavior of the graphs of mathematical models. Figure 1.29 shows the graphs of six basic algebraic equations. Familiarity with these graphs will help you in the creation and use of mathematical models.

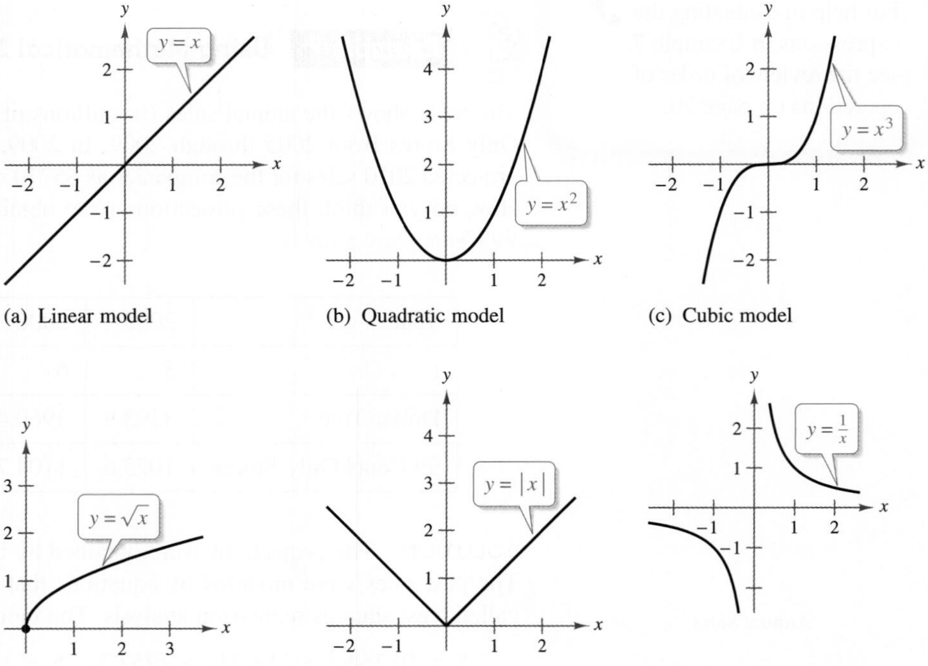

(a) Linear model (b) Quadratic model (c) Cubic model

(d) Square root model (e) Absolute value model (f) Rational model

FIGURE 1.29

SUMMARIZE (Section 1.2)

1. Describe how to sketch the graph of an equation by hand *(page 11)*. For examples of sketching a graph by hand, see Examples 1 and 2.
2. Describe how to find the *x*- and *y*-intercepts of a graph *(page 13)*. For an example of finding the *x*- and *y*-intercepts of a graph, see Example 3.
3. State the standard form of the equation of a circle *(page 14)*. For an example of finding the standard form of the equation of a circle, see Example 4.
4. Describe how to find a point of intersection of the graphs of two equations *(page 15)*. For examples of finding points of intersection, see Examples 5 and 6.
5. Describe break-even analysis *(page 15)*. For an example of break-even analysis, see Example 5.
6. Describe supply equations and demand equations *(page 16)*. For examples of a supply equation and a demand equation, see Example 6.
7. Describe a mathematical model *(page 17)*. For an example of a mathematical model, see Example 7.

SKILLS WARM UP 1.2 The following warm-up exercises involve skills that were covered in a previous course. You will use these skills in the exercise set for this section. For additional help, review Appendix Sections A.3 and A.4.

In Exercises 1–6, solve for y.

1. $5y - 12 = x$

2. $-y = 15 - x$

3. $x^3y + 2y = 1$

4. $x^2 + x - y^2 - 6 = 0$

5. $(x - 2)^2 + (y + 1)^2 = 9$

6. $(x + 6)^2 + (y - 5)^2 = 81$

In Exercises 7–10, evaluate the expression for the given value of x.

	Expression	*x-Value*
7.	$y = 5x$	$x = -2$
8.	$y = 3x - 4$	$x = 3$
9.	$y = 2x^2 + 1$	$x = 2$
10.	$y = x^2 + 2x - 7$	$x = -4$

In Exercises 11–14, factor the expression.

11. $x^2 - 3x + 2$

12. $x^2 + 5x + 6$

13. $y^2 - 3y + \frac{9}{4}$

14. $y^2 - 7y + \frac{49}{4}$

Exercises 1.2

See www.CalcChat.com for worked-out solutions to odd-numbered exercises.

Matching In Exercises 1–6, match the equation with its graph. [The graphs are labeled (a)–(f).]

1. $y = x - 2$

2. $y = -\frac{1}{2}x + 2$

3. $y = x^2 + 2x$

4. $y = \sqrt{9 - x^2}$

5. $y = |x| - 2$

6. $y = x^3 - x$

(a)

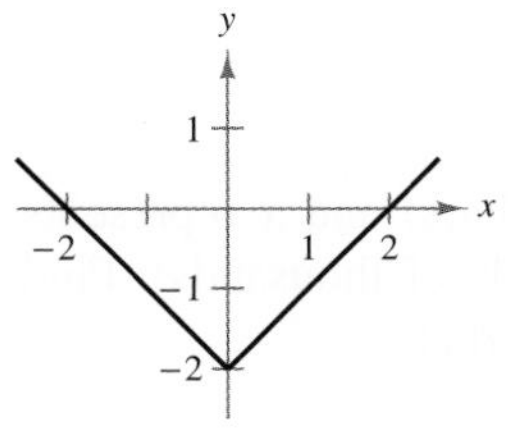

(b)

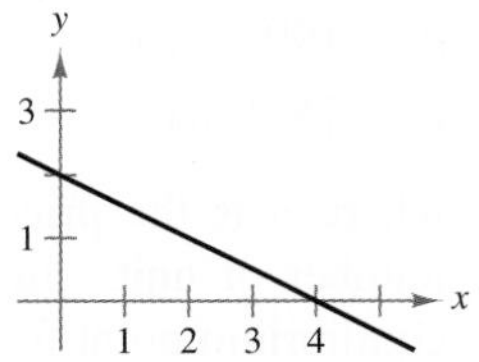

(c)

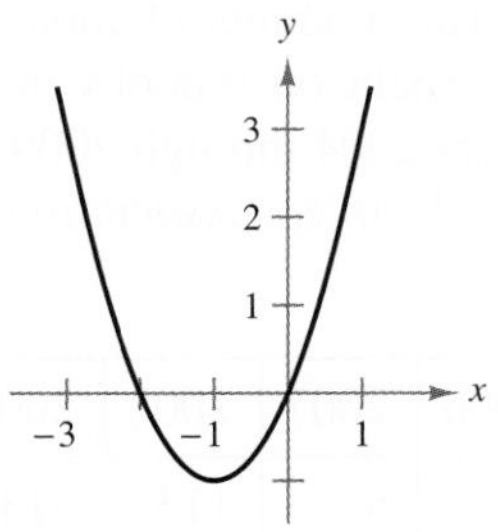

(d)

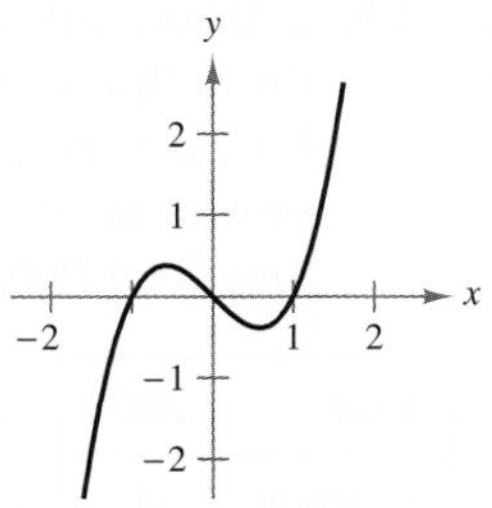

(e)

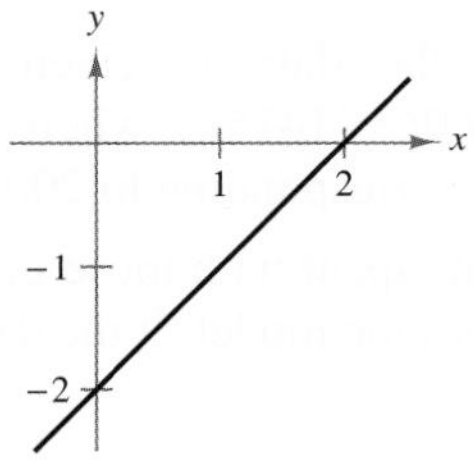

(f)

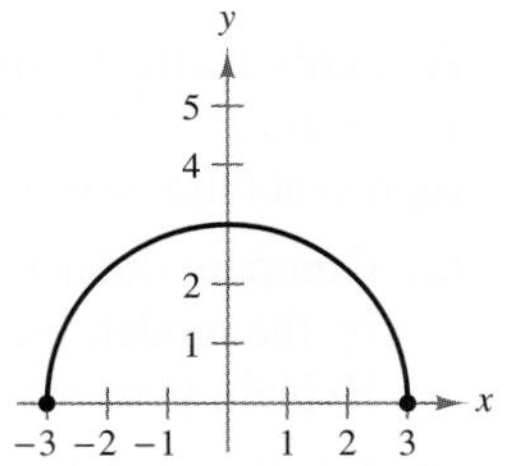

Sketching the Graph of an Equation In Exercises 7–22, sketch the graph of the equation. Use a graphing utility to verify your results. *See Examples 1, 2, and 3.*

7. $y = 2x + 3$

8. $y = -3x + 2$

9. $y = x^2 - 3$

10. $y = x^2 + 6$

11. $y = (x - 1)^2$

12. $y = (5 - x)^2$

13. $y = x^3 + 2$

14. $y = 1 - x^3$

15. $y = -\sqrt{x - 1}$

16. $y = \sqrt{x + 1}$

17. $y = |x + 1|$

18. $y = -|x - 2|$

19. $y = \dfrac{1}{x - 3}$

20. $y = \dfrac{1}{x + 1}$

21. $x = y^2 - 4$

22. $x = 4 - y^2$

Finding *x*- and *y*-Intercepts In Exercises 23–32, find the x- and y-intercepts of the graph of the equation. *See Example 3.*

23. $2x - y - 3 = 0$

24. $4x - 2y - 5 = 0$

25. $y = x^2 + x - 2$

26. $y = x^2 - 4x + 3$

27. $y = \sqrt{4 - x^2}$

28. $y = \sqrt{x + 9}$

29. $y = \dfrac{x^2 - 4}{x - 2}$

30. $y = \dfrac{x^2 + 3x}{2x}$

31. $x^2y - x^2 + 4y = 0$

32. $2x^2y + 8y - x^2 = 1$

Finding the Equation of a Circle In Exercises 33–40, find the standard form of the equation of the circle and sketch its graph. *See Example 4.*

33. Center: $(0, 0)$; radius: 4

34. Center: $(0, 0)$; radius: 5

35. Center: $(2, -1)$; radius: 3

36. Center: $(-4, 3)$; radius: 2

37. Center: $(-1, 1)$; solution point: $(-1, 5)$

38. Center: $(3, -2)$; solution point: $(-1, 1)$

39. Endpoints of a diameter: $(-6, -8)$, $(6, 8)$

40. Endpoints of a diameter: $(0, -4)$, $(6, 4)$

Finding Points of Intersection In Exercises 41–48, find the points of intersection (if any) of the graphs of the equations. Use a graphing utility to check your results.

41. $y = -x + 2,\ y = 2x - 1$

42. $y = -x + 7,\ y = \frac{3}{2}x - 8$

43. $y = -x^2 + 15,\ y = 3x + 11$

44. $y = x^2 + 2,\ y = x + 4$

45. $y = x^3,\ y = 2x$

46. $y = \sqrt{x},\ y = x$

47. $y = x^4 - 2x^2 + 1,\ y = 1 - x^2$

48. $y = x^3 - 2x^2 + x - 1,\ y = -x^2 + 3x - 1$

Break-Even Point In Exercises 49–54, C represents the total cost (in dollars) of producing x units of a product and R represents the total revenue (in dollars) from the sale of x units. How many units must the company sell to break even? *See Example 5.*

49. $C = 0.85x + 35{,}000,\ R = 1.55x$

50. $C = 6x + 500{,}000,\ R = 35x$

51. $C = 8650x + 250{,}000,\ R = 9950x$

52. $C = 2.5x + 10{,}000,\ R = 4.9x$

53. $C = 6x + 5000,\ R = 10x$

54. $C = 130x + 12{,}600,\ R = 200x$

55. Break-Even Analysis You are setting up a part-time business with an initial investment of \$15,000. The unit cost of the product is \$11.80, and the selling price is \$19.30.

(a) Find equations for the total cost C (in dollars) and total revenue R (in dollars) for x units.

(b) Find the break-even point by finding the point of intersection of the cost and revenue equations.

(c) How many units would yield a profit of \$1000?

56. Break-Even Analysis A 2010 Honda Accord costs \$28,695 with a gasoline engine. A 2010 Toyota Camry costs \$29,720 with a hybrid engine. The Accord gets 21 miles per gallon of gasoline and the Camry gets 34 miles per gallon of gasoline. Assume that the price of gasoline is \$2.719. *(Source: Consumer Reports)*

(a) Show that the cost C_g (in dollars) of driving the Honda Accord x miles is

$$C_g = 28{,}695 + \frac{2.719x}{21}$$

and the cost C_h (in dollars) of driving the Toyota Camry x miles is

$$C_h = 29{,}720 + \frac{2.719x}{34}.$$

(b) Find the break-even point. That is, find the mileage at which the hybrid-powered Toyota Camry becomes more economical than the gasoline-powered Honda Accord.

57. Supply and Demand The demand and supply equations for a handheld video game system are given by

$p = 240 - 4x$ Demand equation

$p = 135 + 3x$ Supply equation

where p is the price (in dollars) and x represents the number of units (in thousands). Find the equilibrium point for this market.

58. Supply and Demand The demand and supply equations for an MP3 player are given by

$p = 190 - 15x$ Demand equation

$p = 75 + 8x$ Supply equation

where p is the price (in dollars) and x represents the number of units (in hundreds of thousands). Find the equilibrium point for this market.

59. E-Book Spending The amounts of money y (in millions of dollars) spent on e-books in the United States in the years 2004 through 2009 are shown in the table. *(Source: Association of American Publishers)*

Year	2004	2005	2006	2007	2008	2009
Amount	30	44	54	67	113	313

A mathematical model for the data is given by $y = 8.148t^3 - 139.71t^2 + 789.0t - 1416$, where t represents the year, with $t = 4$ corresponding to 2004.

(a) Compare the actual amounts spent with those given by the model. How well does the model fit the data? Explain your reasoning.

(b) Use the model to predict the amount spent in 2014.

60. **HOW DO YOU SEE IT?** The graph shows the cost and revenue equations for a product.

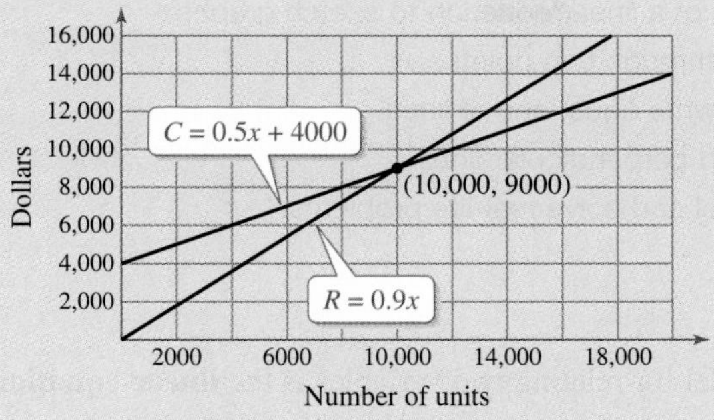

(a) For what numbers of units sold is there a loss for the company?

(b) For what number of units sold does the company break even?

(c) For what numbers of units sold is there a profit for the company?

61. Associate's Degrees A mathematical model for the numbers of associate's degrees conferred y (in thousands) from 2004 through 2008 is given by the equation $y = -1.50t^2 + 38.1t + 539$, where t represents the year, with $t = 4$ corresponding to 2004. *(Source: National Center for Education Statistics)*

(a) Use the model to complete the table.

Year	2004	2005	2006	2007	2008	2012
Degrees						

(b) This model was created using actual data from 2004 through 2008. How accurate do you think the model is in predicting the number of associate's degrees conferred in 2012? Explain your reasoning.

(c) Using this model, what is the prediction for the number of associate's degrees conferred in 2016? Do you think this prediction is valid?

62. Lung Transplants A mathematical model for the numbers of lung transplants performed in the United States in the years 2005 through 2009 is given by $y = 22.36t^2 - 254.9t + 2127$, where y is the number of transplants and t represents the year, with $t = 5$ corresponding to 2005. *(Source: Organ Procurement and Transplantation Network)*

(a) Use a graphing utility or a spreadsheet to complete the table.

Year	2005	2006	2007	2008	2009
Transplants					

(b) Use your school's library, the Internet, or some other reference source to find the actual numbers of lung transplants in the years 2005 through 2009. Compare the actual numbers with those given by the model. How well does the model fit the data? Explain your reasoning.

(c) Using this model, what is the prediction for the number of transplants in 2015? Do you think this prediction is valid? What factors could affect this model's accuracy?

63. Making a Conjecture Use a graphing utility to graph the equation $y = cx + 1$ for $c = 1, 2, 3, 4,$ and 5. Then make a conjecture about the x-coefficient and the graph of the equation.

64. Break-Even Point Define the break-even point for a business marketing a new product. Give examples of a linear cost equation and a linear revenue equation for which the break-even point is 10,000 units.

Finding Intercepts In Exercises 65–70, use a graphing utility to graph the equation and approximate the x- and y-intercepts of the graph.

65. $y = 0.24x^2 + 1.32x + 5.36$

66. $y = -0.56x^2 - 5.34x + 6.25$

67. $y = \sqrt{0.3x^2 - 4.3x + 5.7}$

68. $y = \sqrt{-1.21x^2 + 2.34x + 5.6}$

69. $y = \dfrac{0.2x^2 + 1}{0.1x + 2.4}$

70. $y = \dfrac{0.4x - 5.3}{0.4x^2 + 5.3}$

71. Project: Number of Stores For a project analyzing the numbers of Tiffany & Co. stores from 2000 through 2009, visit this text's website at *www.cengagebrain.com*. *(Source: Tiffany & Co.)*

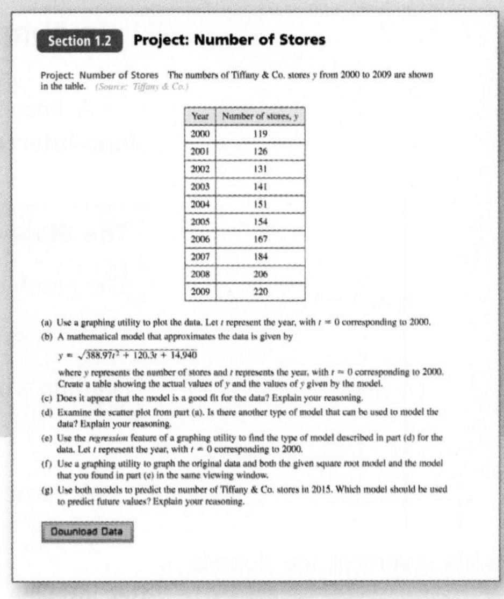

Section 1.2 Project: Number of Stores

Project: Number of Stores The numbers of Tiffany & Co. stores y from 2000 to 2009 are shown in the table. *(Source: Tiffany & Co.)*

Year	Number of stores, y
2000	119
2001	126
2002	131
2003	141
2004	151
2005	154
2006	167
2007	184
2008	206
2009	220

(a) Use a graphing utility to plot the data. Let t represent the year, with $t = 0$ corresponding to 2000.

(b) A mathematical model that approximates the data is given by

$$y = \sqrt{388.97t^2 + 120.3t + 14,940}$$

where y represents the number of stores and t represents the year, with $t = 0$ corresponding to 2000. Create a table showing the actual values of y and the values of y given by the model.

(c) Does it appear that the model is a good fit for the data? Explain your reasoning.

(d) Examine the scatter plot from part (a). Is there another type of model that can be used to model the data? Explain your reasoning.

(e) Use the *regression* feature of a graphing utility to find the type of model described in part (d) for the data. Let t represent the year, with $t = 0$ corresponding to 2000.

(f) Use a graphing utility to graph the original data and both the given square root model and the model that you found in part (e) in the same viewing window.

(g) Use both models to predict the number of Tiffany & Co. stores in 2015. Which model should be used to predict future values? Explain your reasoning.

Download Data

1.3 Lines in the Plane and Slope

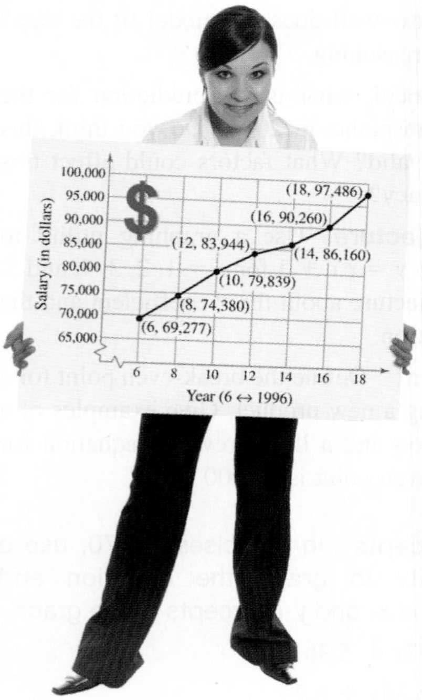

In Exercise 87 on page 32, you will use slope to analyze the average salaries of senior high school principals.

- Use the slope-intercept form of a linear equation to sketch graphs.
- Find slopes of lines passing through two points.
- Use the point-slope form to write equations of lines.
- Find equations of parallel and perpendicular lines.
- Use linear equations to model and solve real-life problems.

Using Slope

The simplest mathematical model for relating two variables is the **linear equation**

$$y = mx + b. \qquad \text{Linear equation}$$

This equation is called *linear* because its graph is a line. (In this text, the term *line* is used to mean *straight line*.) By letting $x = 0$, you can see that the line crosses the y-axis at

$$y = b$$

as shown in Figure 1.30. In other words, the y-intercept is $(0, b)$. The steepness or slope of the line is m.

$$y = mx + b$$

Slope (m) — y-intercept (b)

The **slope** of a nonvertical line is the number of units the line rises (or falls) vertically for each unit of horizontal change from left to right, as shown in Figure 1.30.

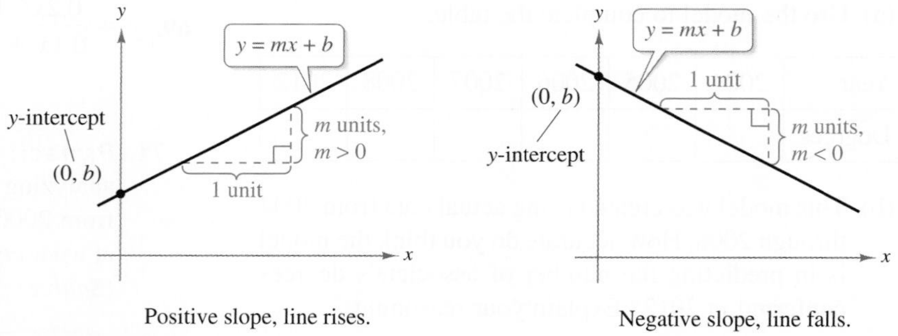

FIGURE 1.30

A linear equation that is written in the form $y = mx + b$ is said to be written in **slope-intercept form.**

> **The Slope-Intercept Form of the Equation of a Line**
>
> The graph of the equation
>
> $$y = mx + b$$
>
> is a line whose slope is m and whose y-intercept is $(0, b)$.

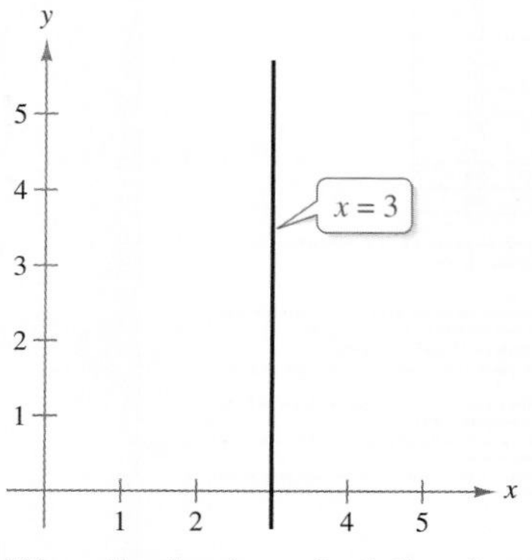

When the line is vertical, the slope is undefined.

FIGURE 1.31

A vertical line has an equation of the form

$$x = a. \qquad \text{Vertical line}$$

Because the equation of a vertical line cannot be written in the form $y = mx + b$, it follows that the slope of a vertical line is undefined, as indicated in Figure 1.31.

Once you have determined the slope and the y-intercept of a line, it is a relatively simple matter to sketch its graph.

Example 1 Graphing a Linear Equation

Sketch the graph of each linear equation.

a. $y = 2x + 1$

b. $y = 2$

c. $x + y = 2$

SOLUTION

a. This equation is written in slope-intercept form, $y = mx + b$. Because $b = 1$, the y-intercept is $(0, 1)$. Moreover, because the slope is $m = 2$, the line *rises* two units for each unit the line moves to the right, as shown in Figure 1.32(a).

b. By writing this equation in slope-intercept form

$$y = (0)x + 2$$

you can see that the y-intercept is $(0, 2)$ and the slope is zero. A zero slope implies that the line is horizontal—that is, it doesn't rise *or* fall, as shown in Figure 1.32(b).

c. By writing this equation in slope-intercept form

$$x + y = 2 \quad \text{Write original equation.}$$
$$y = -x + 2 \quad \text{Subtract } x \text{ from each side.}$$
$$y = (-1)x + 2 \quad \text{Write in slope-intercept form.}$$

you can see that the y-intercept is $(0, 2)$. Moreover, because the slope is $m = -1$, this line *falls* one unit for each unit the line moves to the right, as shown in Figure 1.32(c).

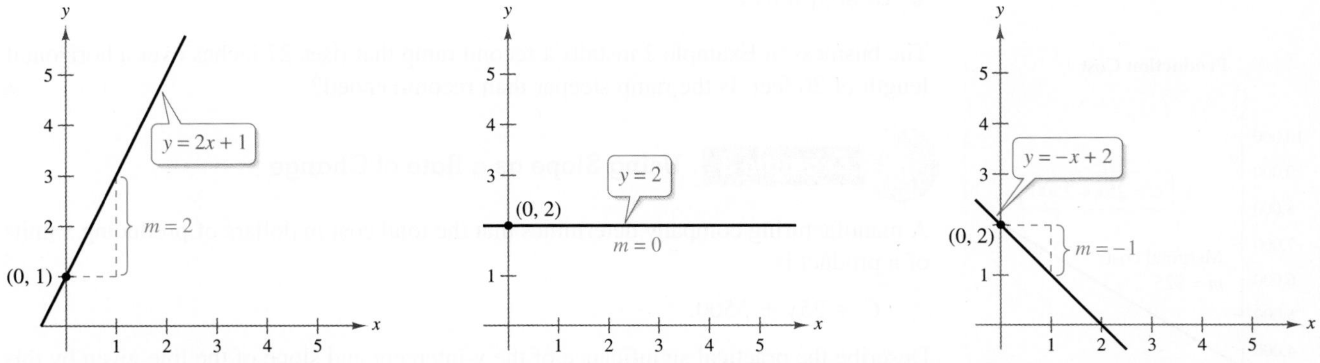

(a) When m is positive, the line rises from left to right.

(b) When m is zero, the line is horizontal.

(c) When m is negative, the line falls from left to right.

FIGURE 1.32

✓ Checkpoint 1

Sketch the graph of each linear equation.

a. $y = 4x - 2$

b. $x = 1$

c. $2x + y = 6$ ■

In real-life problems, the slope of a line can be interpreted as either a *ratio* or a *rate*. If the x-axis and y-axis have the same unit of measure, then the slope has no units and is a **ratio.** If the x-axis and y-axis have different units of measure, then the slope is a **rate** or **rate of change.**

Example 2 Using Slope as a Ratio

The maximum recommended slope of a wheelchair ramp is $\frac{1}{12} \approx 0.083$. A business installs a wheelchair ramp that rises 22 inches over a horizontal length of 24 feet, as shown in Figure 1.33. Is the ramp steeper than recommended? *(Source: ADA Standards for Accessible Design)*

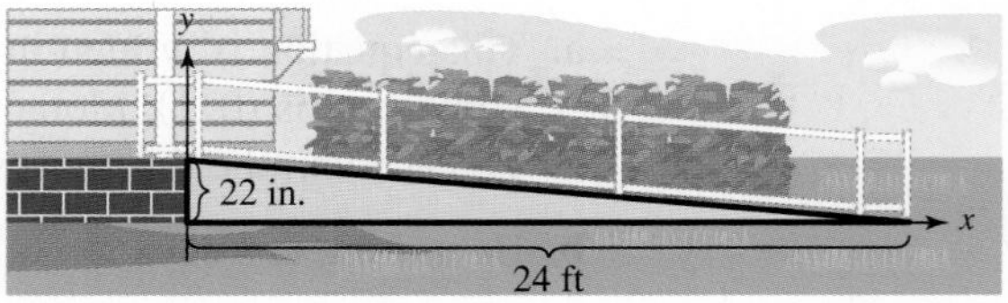

FIGURE 1.33

SOLUTION The horizontal length of the ramp is 24 feet or $12(24) = 288$ inches. So, the slope of the ramp is

$$\text{Slope} = \frac{\text{vertical change}}{\text{horizontal change}} = \frac{22 \text{ in.}}{288 \text{ in.}} \approx 0.076.$$

So, the slope is not steeper than recommended.

Checkpoint 2

The business in Example 2 installs a second ramp that rises 27 inches over a horizontal length of 26 feet. Is the ramp steeper than recommended?

Example 3 Using Slope as a Rate of Change

A manufacturing company determines that the total cost in dollars of producing x units of a product is

$$C = 25x + 3500.$$

Describe the practical significance of the y-intercept and slope of the line given by this equation.

SOLUTION The y-intercept $(0, 3500)$ tells you that the cost of producing zero units is \$3500. This is the **fixed cost** of production—it includes costs that must be paid regardless of the number of units produced. The slope, which is $m = 25$, tells you that the cost of producing each unit is \$25, as shown in Figure 1.34. Economists call the cost per unit the **marginal cost.** If the production increases by one unit, then the "margin" or extra amount of cost is \$25.

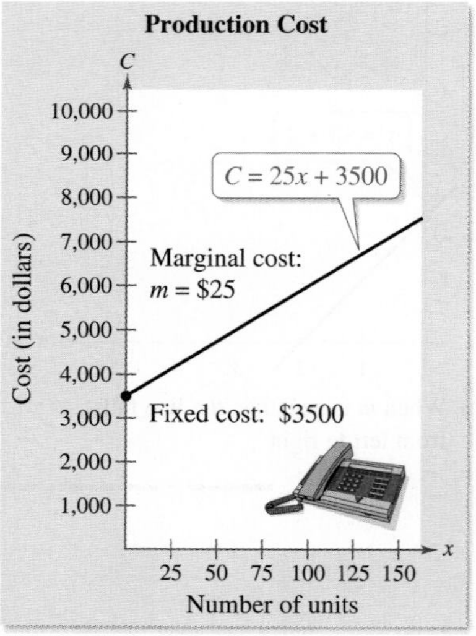

FIGURE 1.34

Checkpoint 3

A small business determines that the value of a copier t years after its purchase is $V = -300t + 1500$. Describe the practical significance of the y-intercept and slope of the line given by this equation.

Finding the Slope of a Line

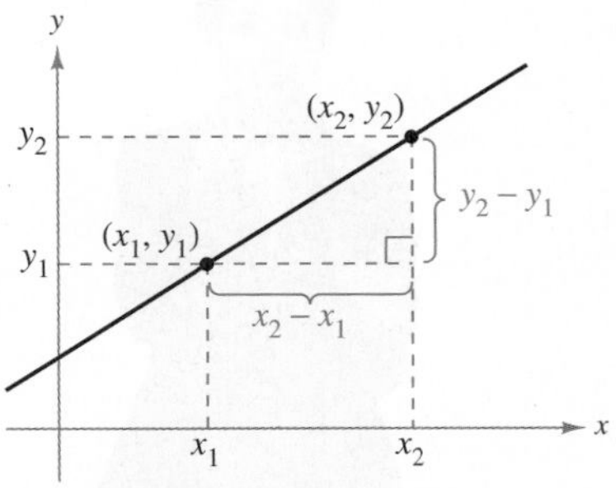

FIGURE 1.35

Given an equation of a nonvertical line, you can find its slope by writing the equation in slope-intercept form. When you are not given an equation, you can still find the slope of a line. For instance, suppose you want to find the slope of the line passing through the points (x_1, y_1) and (x_2, y_2), as shown in Figure 1.35. As you move from left to right along this line, a change of $(y_2 - y_1)$ units in the vertical direction corresponds to a change of $(x_2 - x_1)$ units in the horizontal direction. These two changes are denoted by the symbols

$$\Delta y = y_2 - y_1 = \text{the change in } y$$

and

$$\Delta x = x_2 - x_1 = \text{the change in } x.$$

(The symbol Δ is the Greek capital letter delta, and the symbols Δy and Δx are read as "delta y" and "delta x.") The ratio of Δy to Δx represents the slope of the line that passes through the points (x_1, y_1) and (x_2, y_2).

$$\text{Slope} = \frac{\Delta y}{\Delta x} = \frac{y_2 - y_1}{x_2 - x_1}$$

Be sure you see that Δx represents a single number, not the product of two numbers (Δ and x). The same is true for Δy.

The Slope of a Line Passing Through Two Points

The **slope** m of the nonvertical line passing through the points (x_1, y_1) and (x_2, y_2) is

$$m = \frac{\Delta y}{\Delta x} = \frac{y_2 - y_1}{x_2 - x_1}$$

where $x_1 \neq x_2$. Slope is not defined for vertical lines.

When this formula is used for slope, the *order of subtraction* is important. Given two points on a line, you may label either one of them as (x_1, y_1) and the other as (x_2, y_2). However, once you have done this, you must form the numerator and denominator using the same order of subtraction.

$$\underbrace{m = \frac{y_2 - y_1}{x_2 - x_1}}_{\text{Correct}} \qquad \underbrace{m = \frac{y_1 - y_2}{x_1 - x_2}}_{\text{Correct}} \qquad \underbrace{m = \frac{y_2 - y_1}{x_1 - x_2}}_{\text{Incorrect}}$$

For instance, the slope of the line passing through the points (3, 4) and (5, 7) can be calculated as

$$m = \frac{7 - 4}{5 - 3} = \frac{3}{2}$$

or

$$m = \frac{4 - 7}{3 - 5} = \frac{-3}{-2} = \frac{3}{2}.$$

Example 4 Finding the Slope of a Line

Find the slope of the line passing through each pair of points.

a. $(-2, 0)$ and $(3, 1)$ **b.** $(-1, 2)$ and $(2, 2)$

c. $(0, 4)$ and $(1, -1)$ **d.** $(3, 4)$ and $(3, 1)$

SOLUTION

a. Letting $(x_1, y_1) = (-2, 0)$ and $(x_2, y_2) = (3, 1)$, you obtain a slope of

$$m = \frac{y_2 - y_1}{x_2 - x_1} = \frac{1 - 0}{3 - (-2)} = \frac{1}{5}$$

← Difference in y-values
← Difference in x-values

as shown in Figure 1.36(a).

b. The slope of the line passing through $(-1, 2)$ and $(2, 2)$ is

$$m = \frac{2 - 2}{2 - (-1)} = \frac{0}{3} = 0.$$

See Figure 1.36(b).

c. The slope of the line passing through $(0, 4)$ and $(1, -1)$ is

$$m = \frac{-1 - 4}{1 - 0} = \frac{-5}{1} = -5.$$

See Figure 1.36(c).

d. The slope of the vertical line passing through $(3, 4)$ and $(3, 1)$ is not defined because division by zero is undefined. [See Figure 1.36(d).]

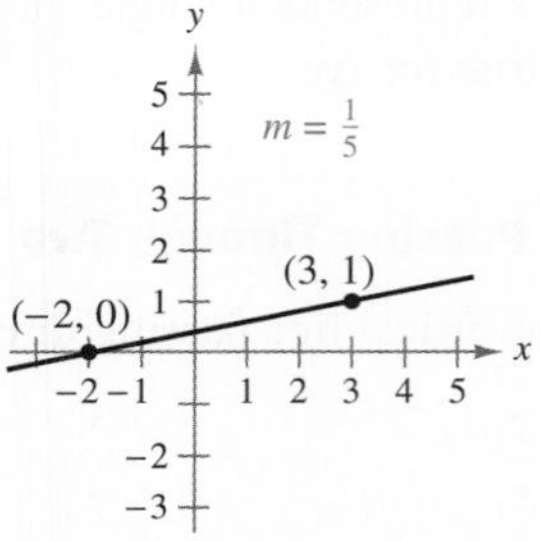

(a) Positive slope, line rises from left to right.

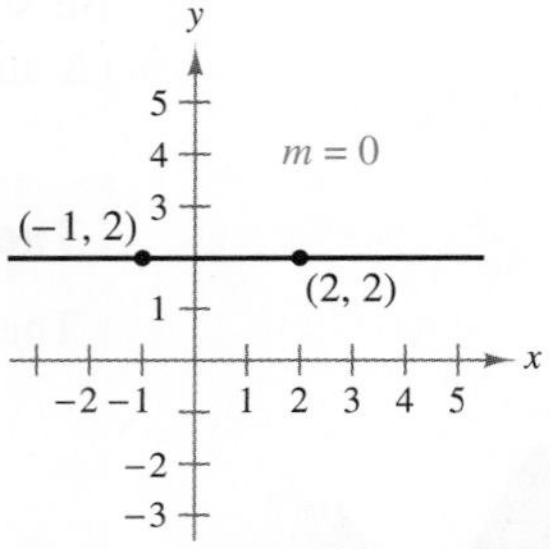

(b) Zero slope, line is horizontal.

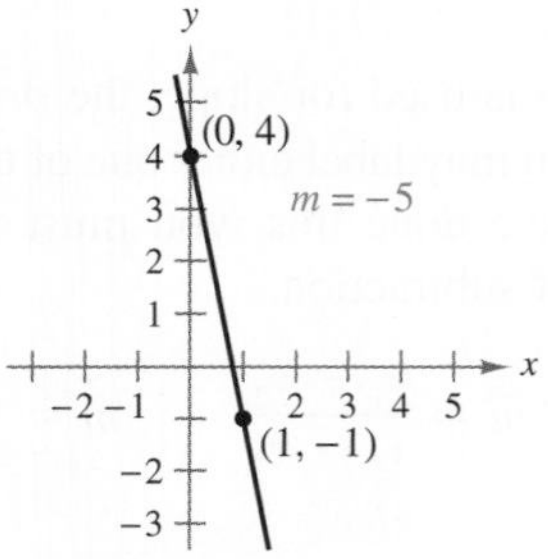

(c) Negative slope, line falls from left to right.

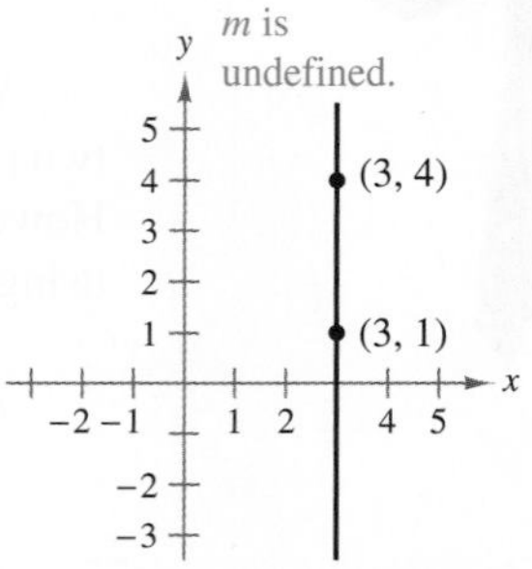

(d) Vertical line, undefined slope.

FIGURE 1.36

✓ Checkpoint 4

Find the slope of the line passing through each pair of points.

a. $(-3, 2)$ and $(5, 18)$

b. $(-2, 1)$ and $(-4, 2)$

c. $(2, -4)$ and $(-2, -4)$

Writing Linear Equations

When you know the slope of a line and the coordinates of one point on the line, you can find an equation for the line. For instance, in Figure 1.37, let (x_1, y_1) be a point on the line whose slope is m. If (x, y) is any other point on the line, then

$$\frac{y - y_1}{x - x_1} = m.$$

This equation, involving the variables x and y, can be rewritten in the form

$$y - y_1 = m(x - x_1)$$

which is the **point-slope form** of the equation of a line.

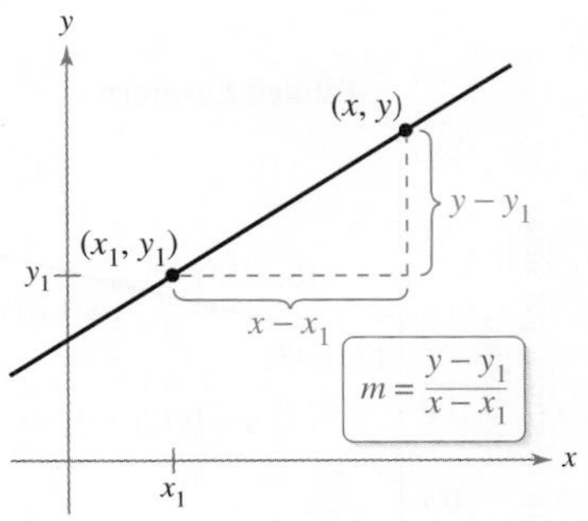

Any two points on a nonvertical line can be used to determine the slope of the line.

FIGURE 1.37

Point-Slope Form of the Equation of a Line

The equation of the line with slope m passing through the point (x_1, y_1) is

$$y - y_1 = m(x - x_1).$$

The point-slope form is most useful for *finding* the equation of a nonvertical line. You should remember this formula—it is used throughout the text.

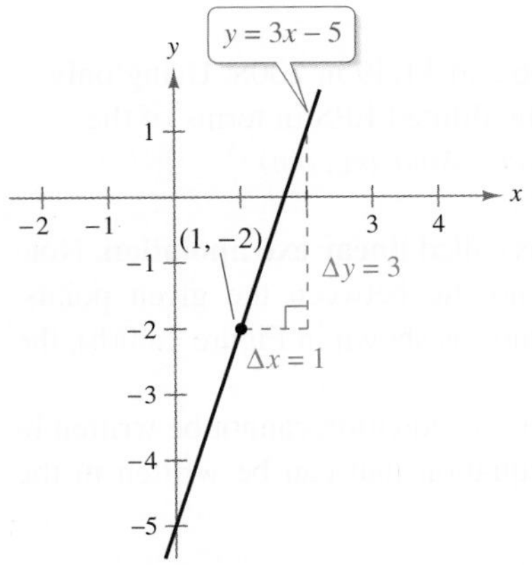

FIGURE 1.38

Example 5 Using the Point-Slope Form

Find the equation of the line that has a slope of 3 and passes through the point $(1, -2)$.

SOLUTION Use the point-slope form with $m = 3$ and $(x_1, y_1) = (1, -2)$.

$$y - y_1 = m(x - x_1) \qquad \text{Point-slope form}$$

$$y - (-2) = 3(x - 1) \qquad \text{Substitute for } m, x_1, \text{ and } y_1.$$

$$y + 2 = 3x - 3 \qquad \text{Simplify.}$$

$$y = 3x - 5 \qquad \text{Write in slope-intercept form.}$$

The slope-intercept form of the equation of the line is $y = 3x - 5$. The graph of this line is shown in Figure 1.38.

✓ Checkpoint 5

Find the equation of the line that has a slope of 2 and passes through the point $(-1, 2)$.

The point-slope form can be used to find an equation of the line passing through points (x_1, y_1) and (x_2, y_2). To do this, first find the slope of the line

$$m = \frac{y_2 - y_1}{x_2 - x_1}, \quad x_1 \neq x_2$$

and then use the point-slope form to obtain the equation

$$y - y_1 = \frac{y_2 - y_1}{x_2 - x_1}(x - x_1). \qquad \text{Two-point form}$$

This is sometimes called the **two-point form** of the equation of a line.

Example 6 Predicting Diluted Earnings Per Share

The diluted earnings per share (diluted EPS) for Tim Hortons, Inc. were \$1.43 in 2007 and \$1.55 in 2008. Using only this information, write a linear equation that gives the diluted EPS in terms of the year. Then predict the diluted EPS for 2009. *(Source: Tim Hortons, Inc.)*

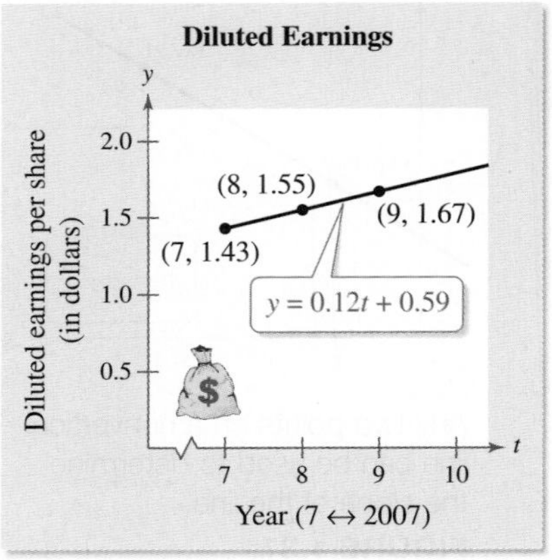

FIGURE 1.39

SOLUTION Let $t = 7$ represent 2007. Then the two given values are represented by the data points

$$(7, 1.43) \quad \text{and} \quad (8, 1.55).$$

The slope of the line through these points is

$$m = \frac{1.55 - 1.43}{8 - 7} = 0.12.$$

Using the point-slope form, you can find the equation that relates the diluted EPS y and the year t to be

$$y = 0.12t + 0.59.$$

Using $t = 9$ to represent 2009, you can predict the 2009 diluted EPS to be

$$y = 0.12(9) + 0.59 = 1.08 + 0.59 = 1.67.$$

According to this equation, the diluted EPS in 2009 was \$1.67, as shown in Figure 1.39. (In this case, the prediction is fairly good—the actual diluted EPS in 2009 was \$1.64.)

✓ Checkpoint 6

The diluted EPS for Amazon.com was \$0.45 in 2006 and \$1.49 in 2008. Using only this information, write a linear equation that gives the diluted EPS in terms of the year. Then predict the diluted EPS for 2009. *(Source: Amazon.com)* ■

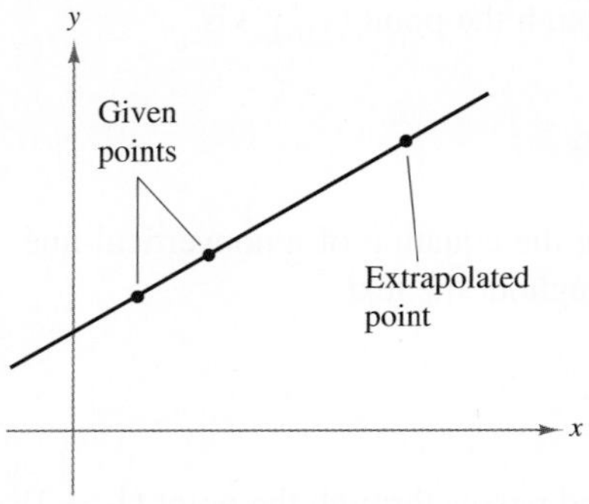

(a) Linear Extrapolation

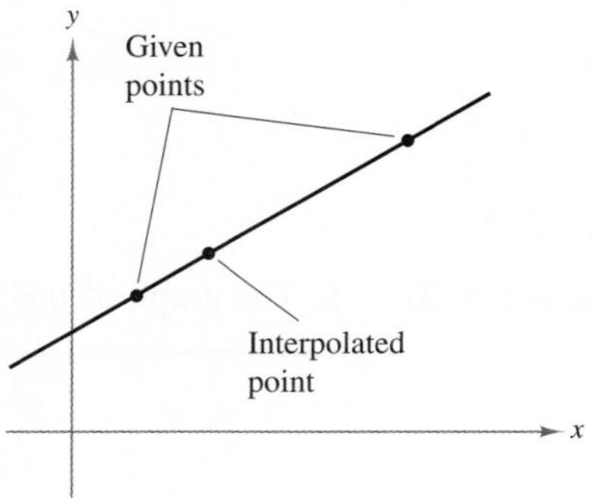

(b) Linear Interpolation

FIGURE 1.40

The prediction method illustrated in Example 6 is called **linear extrapolation.** Note in Figure 1.40(a) that an extrapolated point does not lie between the given points. When the estimated point lies between two given points, as shown in Figure 1.40(b), the procedure is called **linear interpolation.**

Because the slope of a vertical line is not defined, its equation cannot be written in slope-intercept form. However, every line has an equation that can be written in the **general form**

$$Ax + By + C = 0 \qquad \text{General form}$$

where A and B are not both zero. For instance, the vertical line given by $x = a$ can be represented by the general form

$$x - a = 0. \qquad \text{General form of vertical line}$$

The five most common forms of equations of lines are summarized below.

Equations of Lines

1. General form: $Ax + By + C = 0$
2. Vertical line: $x = a$
3. Horizontal line: $y = b$
4. Slope-intercept form: $y = mx + b$
5. Point-slope form: $y - y_1 = m(x - x_1)$

Parallel and Perpendicular Lines

Slope can be used to decide whether two nonvertical lines in a plane are parallel, perpendicular, or neither.

Parallel and Perpendicular Lines

1. Two distinct nonvertical lines are **parallel** if and only if their slopes are equal. That is, $m_1 = m_2$.
2. Two nonvertical lines are **perpendicular** if and only if their slopes are negative reciprocals of each other. That is, $m_1 = -1/m_2$.

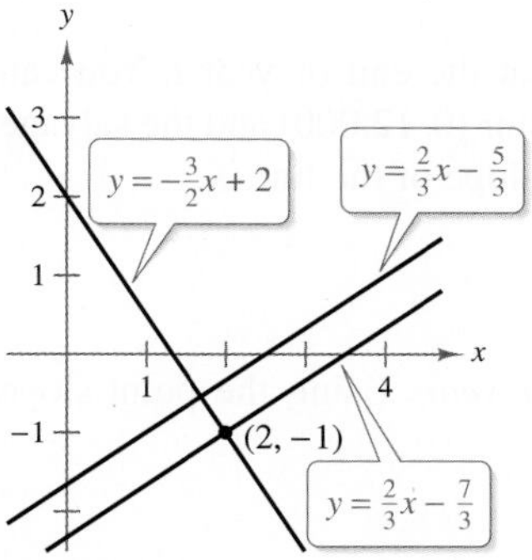

FIGURE 1.41

Example 7 Finding Parallel and Perpendicular Lines

Find equations of the lines that pass through the point $(2, -1)$ and are (a) parallel to the line $2x - 3y = 5$ and (b) perpendicular to the line $2x - 3y = 5$.

SOLUTION By writing the given equation in slope-intercept form

$$2x - 3y = 5 \quad \text{Write original equation.}$$

$$-3y = -2x + 5 \quad \text{Subtract } 2x \text{ from each side.}$$

$$y = \frac{2}{3}x - \frac{5}{3} \quad \text{Write in slope-intercept form.}$$

you can see that it has a slope of $m = \frac{2}{3}$, as shown in Figure 1.41.

a. Any line parallel to the given line must also have a slope of $\frac{2}{3}$. So, the line through $(2, -1)$ that is parallel to the given line has the following equation.

$$y - (-1) = \frac{2}{3}(x - 2) \quad \text{Write in point-slope form.}$$

$$y + 1 = \frac{2}{3}x - \frac{4}{3} \quad \text{Simplify.}$$

$$y = \frac{2}{3}x - \frac{4}{3} - 1 \quad \text{Solve for } y.$$

$$y = \frac{2}{3}x - \frac{7}{3} \quad \text{Write in slope-intercept form.}$$

b. Any line perpendicular to the given line must have a slope of $-\frac{3}{2}$ (because $-\frac{3}{2}$ is the negative reciprocal of $\frac{2}{3}$). So, the line through $(2, -1)$ that is perpendicular to the given line has the following equation.

$$y - (-1) = -\frac{3}{2}(x - 2) \quad \text{Write in point-slope form.}$$

$$y + 1 = -\frac{3}{2}x + 3 \quad \text{Simplify.}$$

$$y = -\frac{3}{2}x + 3 - 1 \quad \text{Solve for } y.$$

$$y = -\frac{3}{2}x + 2 \quad \text{Write in slope-intercept form.}$$

TECH TUTOR

On a graphing utility, lines will not appear to have the correct slopes unless you use a viewing window that has a *square setting*. For instance, try graphing the lines in Example 7 using the standard setting $-10 \le x \le 10$ and $-10 \le y \le 10$. Then reset the viewing window with the square setting $-9 \le x \le 9$ and $-6 \le y \le 6$. On which setting do the lines $y = \frac{2}{3}x - \frac{5}{3}$ and $y = -\frac{3}{2}x + 2$ appear to be perpendicular?

✓ Checkpoint 7

Find equations of the lines that pass through the point $(2, 1)$ and are (a) parallel to the line $2x - 4y = 5$ and (b) perpendicular to the line $2x - 4y = 5$. ■

Extended Application: Linear Depreciation

Most business expenses can be deducted the same year they occur. One exception to this is the cost of property that has a useful life of more than 1 year, such as buildings, cars, or equipment. Such costs must be **depreciated** over the useful life of the property. When the *same amount* is depreciated each year, the procedure is called **linear depreciation** or **straight-line depreciation.** The *book value* is the difference between the original value and the total amount of depreciation accumulated to date.

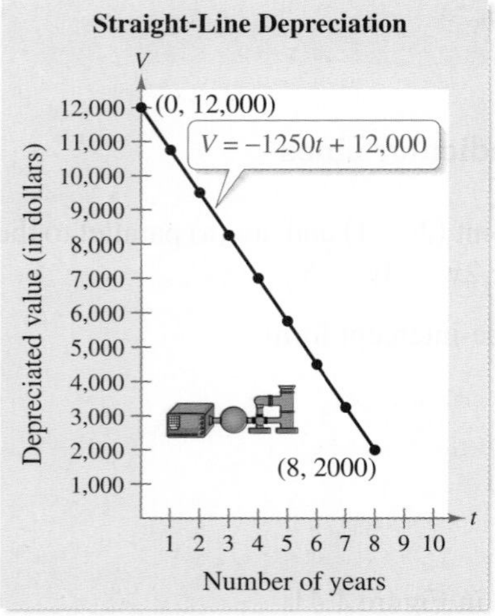

FIGURE 1.42

Example 8 Depreciating Equipment

Your company has purchased a $12,000 machine that has a useful life of 8 years. The salvage value at the end of 8 years is $2000. Write a linear equation that describes the book value of the machine each year.

SOLUTION Let V represent the value of the machine at the end of year t. You can represent the initial value of the machine by the ordered pair $(0, 12{,}000)$ and the salvage value of the machine by the ordered pair $(8, 2000)$. The slope of the line is

$$m = \frac{2000 - 12{,}000}{8 - 0} = \frac{-\$1250}{1 \text{ year}} \qquad m = \frac{V_2 - V_1}{t_2 - t_1}$$

which represents the annual depreciation in *dollars per year*. Using the point-slope form, you can write the equation of the line as shown.

$$V - 12{,}000 = -1250(t - 0) \qquad \text{Write in point-slope form.}$$

$$V = -1250t + 12{,}000 \qquad \text{Write in slope-intercept form.}$$

The graph of this equation is shown in Figure 1.42.

✓ Checkpoint 8

Write a linear equation for the machine in Example 8 when the salvage value at the end of 8 years is $1000.

SUMMARIZE (Section 1.3)

1. State the slope-intercept form of the equation of a line *(page 22)*. For an example of an equation in slope-intercept form, see Example 1.
2. Explain how to decide whether the slope of a line is a ratio or a rate of change *(page 24)*. For an example of a slope that is a ratio, see Example 2. For an example of a slope that is a rate of change, see Example 3.
3. Explain how to find the slope of a line through two points *(page 25)*. For an example of finding the slope of a line through two points, see Example 4.
4. State the point-slope form of the equation of a line *(page 27)*. For examples of equations in point-slope form, see Examples 5 and 6.
5. Explain how to decide whether two lines are parallel, perpendicular, or neither *(page 29)*. For an example, see Example 7.
6. Describe a real-life example of how a linear equation can be used to analyze the depreciation of property *(page 30, Example 8)*.

SKILLS WARM UP 1.3 The following warm-up exercises involve skills that were covered in a previous course. You will use these skills in the exercise set for this section. For additional help, review Appendix Section A.3.

In Exercises 1 and 2, simplify the expression.

1. $\dfrac{5-(-2)}{-3-4}$

2. $\dfrac{-7-(-10)}{4-1}$

3. Evaluate $-\dfrac{1}{m}$ when $m = -3$.

4. Evaluate $-\dfrac{1}{m}$ when $m = \dfrac{6}{7}$.

In Exercises 5–10, solve for y in terms of x.

5. $-4x + y = 7$

6. $3x - y = 7$

7. $y - 2 = 3(x - 4)$

8. $y - (-5) = -1[x - (-2)]$

9. $y - (-3) = \dfrac{4-(-3)}{2-1}(x-2)$

10. $y - 1 = \dfrac{-3-1}{-7-(-1)}[x-(-1)]$

Exercises 1.3

See www.CalcChat.com for worked-out solutions to odd-numbered exercises.

Estimating Slope In Exercises 1–4, estimate the slope of the line.

1.

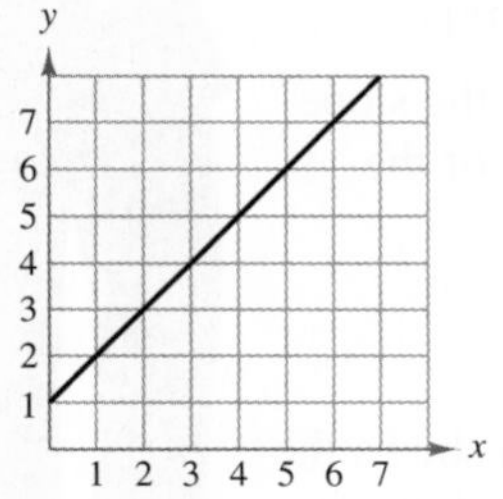

2.

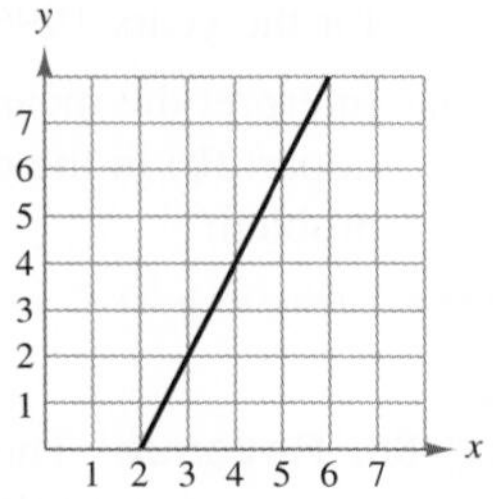

3.

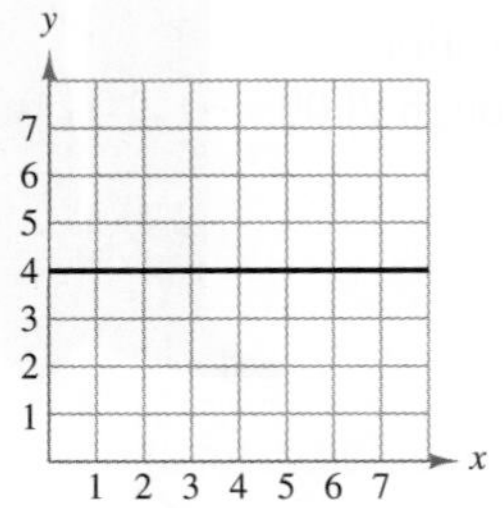

4.

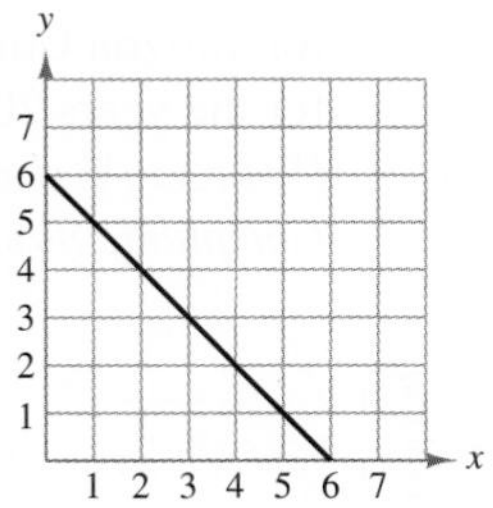

Finding the Slope and *y*-Intercept In Exercises 5–16, find the slope and y-intercept (if possible) of the equation of the line.

5. $y = x + 7$

6. $y = 4x + 3$

7. $5x + y = 20$

8. $2x + y = 40$

9. $7x + 6y = 30$

10. $2x + 3y = 9$

11. $3x - y = 15$

12. $2x - 3y = 24$

13. $x = 4$

14. $x + 5 = 0$

15. $y - 4 = 0$

16. $y + 1 = 0$

Graphing a Linear Equation In Exercises 17–26, sketch the graph of the equation. Use a graphing utility to verify your result. *See Example 1.*

17. $y = -2$

18. $y = -4$

19. $y = -2x + 1$

20. $y = 3x - 2$

21. $6x + 3y = 18$

22. $4x + 5y = 20$

23. $2x - y - 3 = 0$

24. $x + 2y + 6 = 0$

25. $3x + 5y + 15 = 0$

26. $-5x + 2y - 20 = 0$

Finding the Slope of a Line In Exercises 27–40, find the slope of the line passing through the pair of points. *See Example 4.*

27. $(0, -3), (9, 0)$

28. $(-2, 0), (1, 4)$

29. $(3, -4), (5, 2)$

30. $(1, 2), (-2, 2)$

31. $(4, -1), (2, 7)$

32. $\left(\frac{11}{3}, -2\right), \left(\frac{11}{3}, -10\right)$

33. $(-8, -3), (-8, -5)$

34. $(2, -1), (-2, -5)$

35. $(-2, 1), (4, -3)$

36. $(3, -5), (-2, -5)$

37. $\left(\frac{1}{4}, -2\right), \left(-\frac{3}{8}, 1\right)$

38. $\left(-\frac{3}{2}, -5\right), \left(\frac{5}{6}, 4\right)$

39. $\left(\frac{2}{3}, \frac{5}{2}\right), \left(\frac{1}{4}, -\frac{5}{6}\right)$

40. $\left(\frac{7}{8}, \frac{3}{4}\right), \left(\frac{5}{4}, -\frac{1}{4}\right)$

Finding Points on a Line In Exercises 41–48, use the point on the line and the slope of the line to find three additional points through which the line passes. (There are many correct answers.)

	Point	*Slope*		*Point*	*Slope*
41.	$(2, 1)$	$m = 0$	**42.**	$(-3, -1)$	$m = 0$
43.	$(1, 7)$	$m = -3$	**44.**	$(7, -2)$	$m = 2$
45.	$(6, -4)$	$m = \frac{2}{3}$	**46.**	$(-1, -6)$	$m = -\frac{1}{2}$
47.	$(-8, 1)$	m is undefined.			
48.	$(-3, 4)$	m is undefined.			

Using Slope In Exercises 49–52, use the concept of slope to determine whether the three points are collinear.

49. $(-2, 1), (-1, 0), (2, -2)$

50. $(0, 4), (7, -6), (-5, 11)$

51. $(-2, -1), (0, 3), (2, 7)$

52. $(4, 1), (-2, -2), (8, 3)$

Using the Point-Slope Form In Exercises 53–62, find the equation of the line that passes through the given point and has the given slope. Then use the equation to sketch the line. *See Example 5.*

	Point	*Slope*		*Point*	*Slope*
53.	$(0, 3)$	$m = \frac{3}{4}$	**54.**	$(0, 0)$	$m = \frac{2}{3}$
55.	$(-1, 2)$	m is undefined.			
56.	$(0, 4)$	m is undefined.			
57.	$(-2, 7)$	$m = 0$	**58.**	$(-2, 4)$	$m = 0$
59.	$(0, -2)$	$m = -4$	**60.**	$(-1, -4)$	$m = -2$
61.	$\left(0, \frac{2}{3}\right)$	$m = \frac{3}{4}$	**62.**	$\left(0, -\frac{2}{3}\right)$	$m = \frac{1}{6}$

Writing an Equation of a Line In Exercises 63–74, find the equation of the line that passes through the points. Then use the equation to sketch the line.

63. $(4, 3), (0, -5)$ **64.** $(-3, -4), (1, 4)$

65. $(0, 0), (-1, 3)$ **66.** $(-3, 6), (1, 2)$

67. $(2, 3), (2, -2)$ **68.** $(6, 1), (10, 1)$

69. $(3, -1), (-2, -1)$ **70.** $(2, 5), (2, -10)$

71. $\left(-\frac{1}{3}, 1\right), \left(-\frac{2}{3}, \frac{5}{6}\right)$ **72.** $\left(\frac{7}{8}, \frac{3}{4}\right), \left(\frac{5}{4}, -\frac{1}{4}\right)$

73. $\left(-\frac{1}{2}, 4\right), \left(\frac{1}{2}, 8\right)$ **74.** $(4, -1), \left(\frac{1}{4}, -5\right)$

75. Write an equation of the vertical line with x-intercept at 3.

76. Write an equation of the horizontal line through $(0, -5)$.

77. Write an equation of the line with y-intercept at -10 and parallel to all horizontal lines.

78. Write an equation of the line with x-intercept at -5 and parallel to all vertical lines.

Finding Parallel and Perpendicular Lines In Exercises 79–86, find equations of the lines through the given point (a) parallel to the given line and (b) perpendicular to the given line. Then use a graphing utility to graph all three equations in the same viewing window using a square setting. *See Example 7.*

	Point	*Line*		*Point*	*Line*
79.	$(-3, 2)$	$x + y = 7$	**80.**	$(2, 1)$	$4x - 2y = 3$
81.	$\left(-\frac{2}{3}, \frac{7}{8}\right)$	$3x + 4y = 7$	**82.**	$\left(\frac{7}{8}, \frac{3}{4}\right)$	$5x + 3y = 0$
83.	$(-1, 0)$	$y + 3 = 0$	**84.**	$(2, 5)$	$y + 4 = 0$
85.	$(1, 1)$	$x - 2 = 0$	**86.**	$(12, -3)$	$x + 4 = 0$

87. Average Salary The graph shows the average salaries (in dollars) of senior high school principals from 1996 through 2008. *(Source: Educational Research Service)*

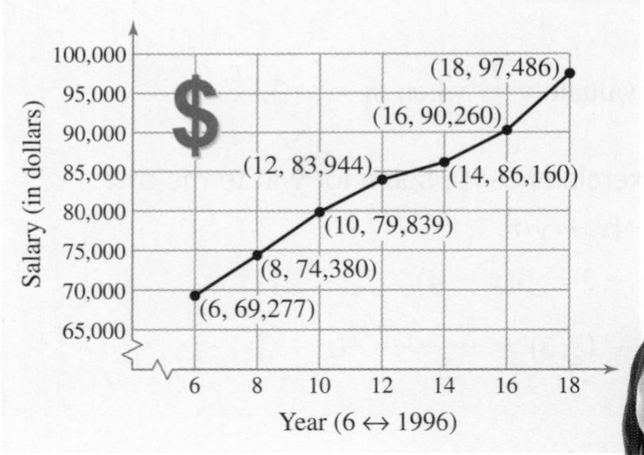

(a) Determine the time periods in which the average salary increased the greatest and the least.

(b) Find the slope of the line segment connecting the points for the years 1996 and 2008.

(c) Interpret the meaning of the slope in part (b) in the context of the problem.

88. Revenue The graph shows the revenue (in billions of dollars) for Verizon Communications for the years 2003 through 2009. *(Source: Verizon Communications)*

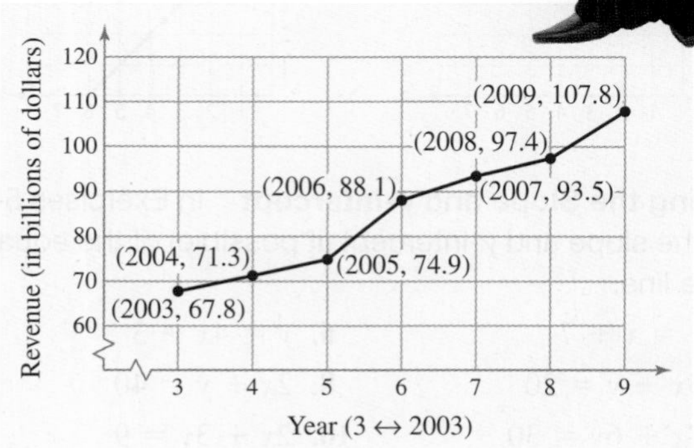

(a) Determine the years in which the revenue increased the greatest and the least.

(b) Find the slope of the line segment connecting the points for the years 2003 and 2009.

(c) Interpret the meaning of the slope in part (b) in the context of the problem.

89. Temperature Conversion Write a linear equation that expresses the relationship between the temperature in degrees Celsius C and the temperature in degrees Fahrenheit F. Use the fact that water freezes at 0°C (32°F) and boils at 100°C (212°F).

90. Chemistry Use the result of Exercise 89 to answer the following.

(a) A person has a temperature of 102.2°F. What is this temperature on the Celsius scale?

(b) The temperature in a room is 76°F. What is this temperature on the Celsius scale?

91. Population The resident population of Wisconsin (in thousands) was 5511 in 2004 and 5655 in 2009. Assume that the relationship between the population y and the year t is linear. Let $t = 0$ represent 2000. *(Source: U.S. Census Bureau)*

(a) Write a linear model for the data. What is the slope and what does it tell you about the population?

(b) Use the model to estimate the population in 2006.

(c) Use your school's library, the Internet, or some other reference source to find the actual population in 2006. How close was your estimate?

(d) Do you think your model could be used to predict the population in 2015? Explain.

92. Linear Depreciation A small business purchases a piece of equipment for $1025. After 5 years the equipment will be outdated, having no value.

(a) Write a linear equation giving the value y (in dollars) of the equipment in terms of the time t (in years), $0 \le t \le 5$.

(b) Use a graphing utility to graph the equation.

(c) Move the cursor along the graph and estimate (to two-decimal-place accuracy) the value of the equipment when $t = 3$.

(d) Move the cursor along the graph and estimate (to two-decimal-place accuracy) the time when the value of the equipment will be $600.

93. Personal Income Personal income (in billions of dollars) in the United States was 9937 in 2004 and 12,175 in 2009. Assume that the relationship between the personal income y and the time t (in years) is linear. Let $t = 0$ represent 2000. *(Source: U.S. Bureau of Economic Analysis)*

(a) Write a linear model for the data.

(b) Estimate the personal income in 2006.

(c) Estimate the personal income in 2011.

(d) Use your school's library, the Internet, or some other reference source to find the actual personal income in 2006 and 2011. How close were your estimates?

94. Choosing a Job As a salesperson, you receive a monthly salary of $2000, plus a commission of 7% of sales. You are offered a new job at $2300 per month, plus a commission of 5% of sales.

(a) Write linear equations for your monthly wage W (in dollars) in terms of your monthly sales S (in dollars) for your current job and for your job offer.

(b) Use a graphing utility to graph each equation and find the point of intersection. What does it signify?

(c) You think you can sell $20,000 worth of a product per month. Should you change jobs? Explain.

95. Profit Analysis A business manufactures a product at a cost of $50 per unit and sells the product for $120 per unit. The company's initial investment to produce the product is $350,000.

(a) Write linear equations giving the total cost C (in dollars) of producing x units and the revenue R (in dollars) from selling x units.

(b) Use the formula for profit, $P = R - C$, to write an equation for the profit derived from x units produced and sold.

(c) The company estimates it can sell 13,000 units. What will be the company's profit or loss?

96. HOW DO YOU SEE IT? Match the description of the situation with its graph. Then write the equation of the line. [The graphs are labeled (i), (ii), (iii), and (iv).]

(a) You are paying $10 per week to repay a $100 loan.

(b) An employee is paid $12.50 per hour plus $1.50 for each unit produced per hour.

(c) A sales representative receives $30 per day for food plus $0.51 for each mile traveled.

(d) A computer that was purchased for $600 depreciates $100 per year.

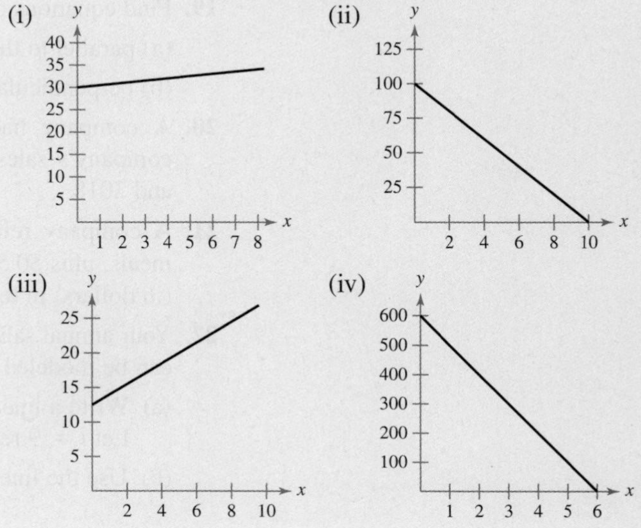

QUIZ YOURSELF

See www.CalcChat.com for worked-out solutions to odd-numbered exercises.

Take this quiz as you would take a quiz in class. When you are done, check your work against the answers given in the back of the book.

In Exercises 1–3, (a) plot the points, (b) find the distance between the points, and (c) find the midpoint of the line segment joining the points.

1. $(3, -2), (-3, 1)$ **2.** $\left(\frac{1}{4}, -\frac{3}{2}\right), \left(\frac{1}{2}, 2\right)$ **3.** $(-12, 4), (6, -2)$

4. Use the Distance Formula to show that the points $(4, 0)$, $(2, 1)$, and $(-1, -5)$ are vertices of a right triangle.

5. The resident population of Georgia (in thousands) was 9534 in 2007 and 9829 in 2009. Use the Midpoint Formula to estimate the population in 2008. *(Source: U.S. Census Bureau)*

In Exercises 6–8, sketch the graph of the equation and label the intercepts.

6. $y = 5x + 2$ **7.** $y = x^2 + x - 6$ **8.** $y = |x - 3|$

In Exercises 9–11, find the standard form of the equation of the circle and sketch its graph.

9. Center: $(0, 0)$; radius: 9

10. Center: $(-1, 0)$; radius: 6

11. Center: $(2, -2)$; solution point: $(-1, 2)$

12. A business manufactures a product at a cost of \$4.55 per unit and sells the product for \$7.19 per unit. The company's initial investment to produce the product was \$12,500. How many units must the company sell to break even?

In Exercises 13–15, find the equation of the line that passes through the given point and has the given slope. Then use the equation to sketch the line.

13. $(0, -3)$; $m = \frac{1}{2}$ **14.** $(1, 1)$; $m = 2$ **15.** $(6, 5)$; $m = -\frac{1}{3}$

In Exercises 16–18, find the equation of the line that passes through the points. Then use the equation to sketch the line.

16. $(1, -1), (-4, 5)$ **17.** $(-2, 3), (-2, 2)$ **18.** $\left(\frac{5}{2}, 2\right), (0, 2)$

19. Find equations of the lines that pass through the point $(3, -5)$ and are

(a) parallel to the line $x + 4y = -2$.

(b) perpendicular to the line $x + 4y = -2$.

20. A company had sales of \$1,330,000 in 2007 and \$1,800,000 in 2011. The company's sales can be modeled by a linear equation. Predict the sales in 2015 and 2018.

21. A company reimburses its sales representatives \$175 per day for lodging and meals, plus \$0.55 per mile driven. Write a linear equation giving the daily cost C (in dollars) in terms of x, the number of miles driven.

22. Your annual salary was \$34,600 in 2009 and \$37,800 in 2011. Assume your salary can be modeled by a linear equation.

(a) Write a linear equation giving your salary S (in dollars) in terms of the year t. Let $t = 9$ represent 2009.

(b) Use the linear model to predict your salary in 2015.

1.4 Functions

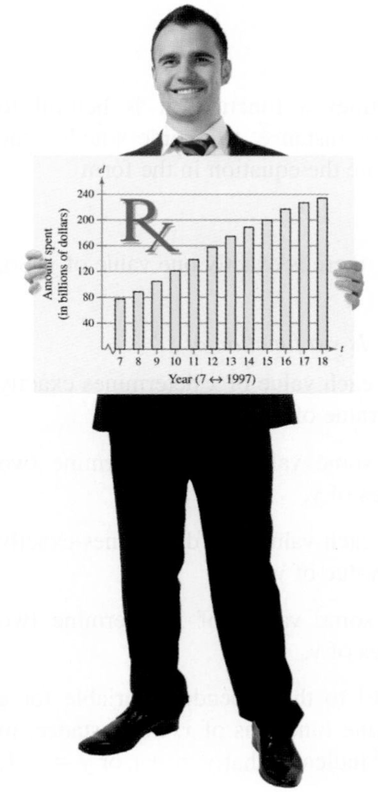

In Exercise 67 on page 46, you will use a function to estimate the amounts spent on prescription drugs in the United States.

- Decide whether the relationship between two variables is a function.
- Find the domains and ranges of functions.
- Use function notation and evaluate functions.
- Combine functions to create other functions.
- Find inverse functions algebraically.

Functions

In many common relationships between two variables, the value of one of the variables depends on the value of the other variable. Here are some examples.

1. The sales tax on an item depends on its selling price.
2. The distance an object moves in a given amount of time depends on its speed.
3. The area of a circle depends on its radius.

Consider the relationship between the total revenue R and the sale of x units of a product sold for \$1.25 per unit. This relationship can be expressed by the equation

$$R = 1.25x.$$

In this equation, the value of R depends on the choice of x. Because of this, R is the **dependent variable** and x is the **independent variable.**

Most of the relationships that you will study in this course have the property that for a given value of the independent variable, there corresponds exactly one value of the dependent variable. Such a relationship is a **function.**

Definition of Function

A **function** is a relationship between two variables such that to each value of the independent variable there corresponds exactly one value of the dependent variable.

The **domain** of the function is the set of all values of the independent variable for which the function is defined. The **range** of the function is the set of all values taken on by the dependent variable.

As illustrated in Figure 1.43, a function can be thought of as a machine that inputs values of the independent variable and outputs values of the dependent variable.

Although functions can be described by various means such as tables, graphs, and diagrams, they are most often specified by formulas or equations. For instance, the equation

$$y = 4x^2 + 3$$

describes y as a function of x. For this function, x is the independent variable and y is the dependent variable.

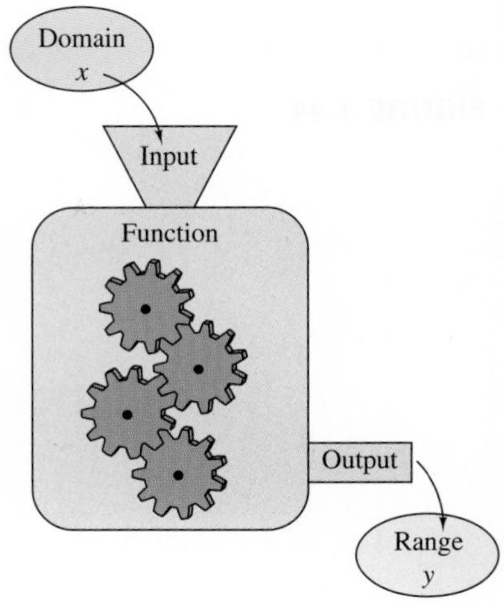

FIGURE 1.43

Example 1 Deciding Whether Equations Are Functions

Decide whether each equation defines y as a function of x.

a. $x + y = 1$ **b.** $x^2 + y^2 = 1$

c. $x^2 + y = 1$ **d.** $x + y^2 = 1$

SOLUTION To decide whether an equation defines a function, it is helpful to isolate the dependent variable on the left side. For instance, to decide whether the equation $x + y = 1$ defines y as a function of x, write the equation in the form

$$y = 1 - x.$$

From this form, you can see that for any value of x, there is exactly one value of y. So, y is a function of x.

Original Equation	*Rewritten Equation*	*Test: Is y a function of x?*
a. $x + y = 1$	$y = 1 - x$	Yes, each value of x determines exactly one value of y.
b. $x^2 + y^2 = 1$	$y = \pm\sqrt{1 - x^2}$	No, some values of x determine two values of y.
c. $x^2 + y = 1$	$y = 1 - x^2$	Yes, each value of x determines exactly one value of y.
d. $x + y^2 = 1$	$y = \pm\sqrt{1 - x}$	No, some values of x determine two values of y.

Note that the equations that assign two values ($\pm$) to the dependent variable for a given value of the independent variable do not define functions of x. For instance, in part (b), when $x = 0$, the equation $y = \pm\sqrt{1 - x^2}$ indicates that $y = +1$ or $y = -1$. Figure 1.44 shows the graphs of the four equations.

TECH TUTOR

Many graphing utilities have an *equation editor* feature that requires an equation to be written in "$y =$" form in order to be entered. So, the procedure used in Example 1, isolating the dependent variable on the left side, is also useful for graphing equations with a graphing utility. Note that to graph an equation in which y is not a function of x, such as a circle, you usually have to enter two or more equations into the graphing utility.

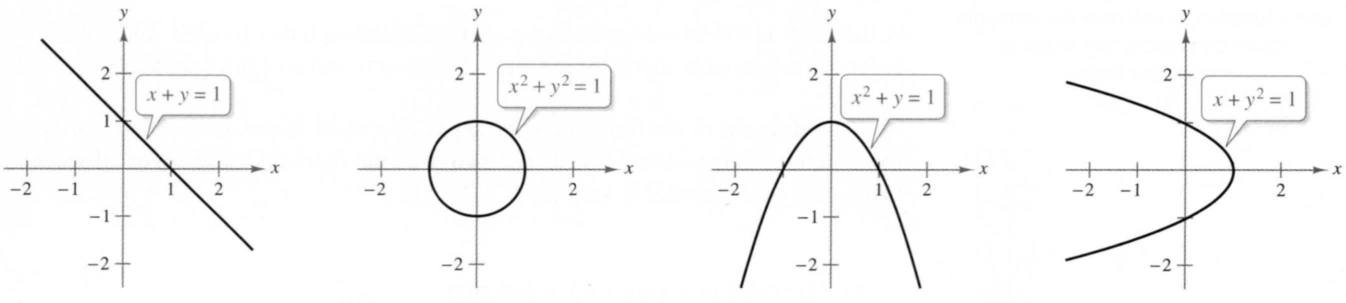

(a) (b) (c) (d)

FIGURE 1.44

✓ Checkpoint 1

Decide whether each equation defines y as a function of x.

a. $x - y = 1$ **b.** $x^2 + y^2 = 4$ **c.** $y^2 + x = 2$ **d.** $x^2 - y = 0$ ■

When the graph of a function is sketched, the standard convention is to let the horizontal axis represent the independent variable. When this convention is used, the test described in Example 1 has a nice graphical interpretation called the **Vertical Line Test.** This test states that if every vertical line intersects the graph of an equation at most once, then the equation defines y as a function of x. For instance, in Figure 1.44, the graphs in parts (a) and (c) pass the Vertical Line Test, but those in parts (b) and (d) do not.

The Domain and Range of a Function

The domain of a function may be described explicitly, or it may be *implied* by an equation used to define the function. For example, the function given by

$$y = \frac{1}{x^2 - 4}$$

has an implied domain that consists of all real numbers x except $x = \pm 2$. These two values are excluded from the domain because division by zero is undefined.

Another type of implied domain is that used to avoid even roots of negative numbers, as indicated in Example 2.

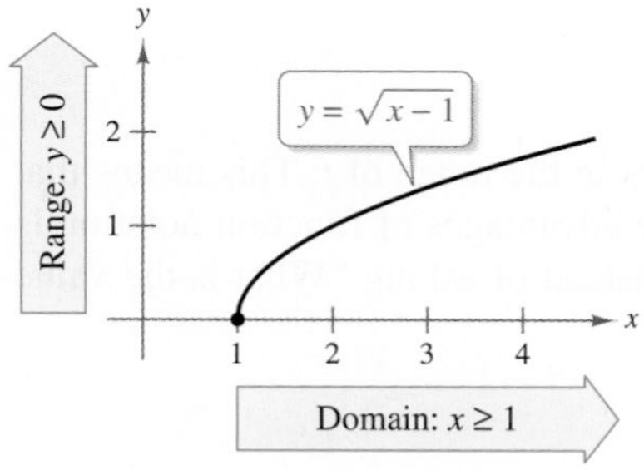

(a)

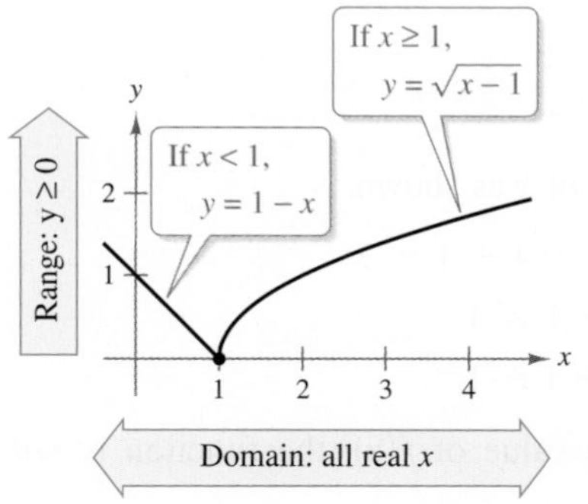

(b)

FIGURE 1.45

Example 2 Finding the Domain and Range of a Function

Find the domain and range of each function.

a. $y = \sqrt{x - 1}$

b. $y = \begin{cases} 1 - x, & x < 1 \\ \sqrt{x - 1}, & x \geq 1 \end{cases}$

SOLUTION

a. Because $\sqrt{x - 1}$ is not defined for $x - 1 < 0$ (that is, for $x < 1$), it follows that the domain of the function is the interval

$$x \geq 1 \text{ or } [1, \infty).$$

To find the range, observe that $\sqrt{x - 1}$ is never negative. Moreover, as x takes on the various values in the domain, y takes on all nonnegative values. So, the range is the interval

$$y \geq 0 \text{ or } [0, \infty)$$

as shown in Figure 1.45(a).

b. Because this function is defined for $x < 1$ *and* for $x \geq 1$, the domain is the entire set of real numbers. This function is called a **piecewise-defined function** because it is defined by two or more equations over a specified domain. When $x \geq 1$, the function behaves as in part (a). For $x < 1$, the values of $1 - x$ are positive. So, the range of the function is

$$y \geq 0 \text{ or } [0, \infty)$$

as shown in Figure 1.45(b).

✓ Checkpoint 2

Find the domain and range of each function.

a. $y = \sqrt{x + 1}$ **b.** $y = \begin{cases} x^2, & x \leq 0 \\ \sqrt{x}, & x > 0 \end{cases}$

A function is **one-to-one** when to each value of the dependent variable in the range there corresponds exactly one value of the independent variable in the domain. For instance, the function in Example 2(a) is one-to-one, whereas the function in Example 2(b) is not one-to-one.

Geometrically, a function is one-to-one when every horizontal line intersects the graph of the function at most once. This geometrical interpretation is the **Horizontal Line Test** for one-to-one functions. So, a graph that represents a one-to-one function must satisfy *both* the Vertical Line Test and the Horizontal Line Test.

Function Notation

When using an equation to define a function, you generally isolate the dependent variable on the left. For instance, writing the equation $x + 2y = 1$ as

$$y = \frac{1 - x}{2}$$

indicates that y is the dependent variable. In **function notation,** this equation has the form

$$f(x) = \frac{1 - x}{2}. \qquad \text{Function notation}$$

The independent variable is x, and the name of the function is "f." The symbol $f(x)$ is read as "f of x," and it denotes the value of the dependent variable. For instance, the value of f when $x = 3$ is

$$f(3) = \frac{1 - (3)}{2} = \frac{-2}{2} = -1.$$

The value $f(3)$ is called a **function value,** and it lies in the range of f. This means that the point $(3, f(3))$ lies on the graph of f. One of the advantages of function notation is that it allows you to be less wordy. For instance, instead of asking "What is the value of y when $x = 3$?" you can ask "What is $f(3)$?"

Example 3 Evaluating a Function

Find the values of the function

$$f(x) = 2x^2 - 4x + 1$$

when x is -1, 0, and 2. Is f one-to-one?

SOLUTION You can evaluate f at the given values of x as shown.

$$x = -1: \quad f(-1) = 2(-1)^2 - 4(-1) + 1 = 2 + 4 + 1 = 7$$
$$x = 0: \quad f(0) = 2(0)^2 - 4(0) + 1 = 0 - 0 + 1 = 1$$
$$x = 2: \quad f(2) = 2(2)^2 - 4(2) + 1 = 8 - 8 + 1 = 1$$

Because two different values of x yield the same value of $f(x)$, the function is *not* one-to-one, as shown in Figure 1.46.

STUDY TIP

You can use the Horizontal Line Test to determine whether the function in Example 3 is one-to-one. Because the line $y = 1$ intersects the graph of the function twice, the function is *not* one-to-one.

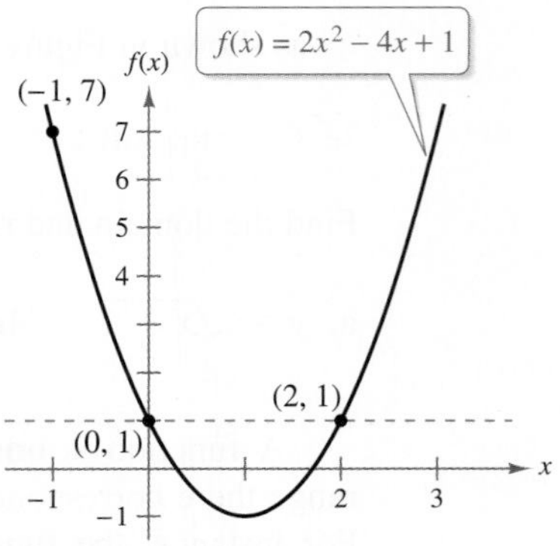

FIGURE 1.46

✓ Checkpoint 3

Find the values of $f(x) = x^2 - 5x + 1$ when x is 0, 1, and 4. Is f one-to-one? ■

Example 3 suggests that the role of the variable x in the equation

$$f(x) = 2x^2 - 4x + 1$$

is simply that of a placeholder. Informally, f could be defined by the equation

$$f(\quad) = 2(\quad)^2 - 4(\quad) + 1.$$

To evaluate $f(-2)$, simply place -2 in each set of parentheses.

$$f(-2) = 2(-2)^2 - 4(-2) + 1 = 8 + 8 + 1 = 17$$

Although f is often used as a convenient function name and x as the independent variable, you can use other symbols. For instance, the following equations all define the same function.

$f(x) = x^2 - 4x + 7$ Function name is f, independent variable is x.

$f(t) = t^2 - 4t + 7$ Function name is f, independent variable is t.

$g(s) = s^2 - 4s + 7$ Function name is g, independent variable is s.

Example 4 Evaluating a Function

Given $f(x) = x^2 + 7$, evaluate each expression.

a. $f(x + \Delta x)$ **b.** $\dfrac{f(x + \Delta x) - f(x)}{\Delta x}$

SOLUTION

a. To evaluate f at $x + \Delta x$, substitute $x + \Delta x$ for x in the original function.

$$\begin{aligned} f(x + \Delta x) &= (x + \Delta x)^2 + 7 \\ &= x^2 + 2x\,\Delta x + (\Delta x)^2 + 7 \end{aligned}$$

b. Using the result of part (a), you can write

$$\begin{aligned} \frac{f(x + \Delta x) - f(x)}{\Delta x} &= \frac{[(x + \Delta x)^2 + 7] - (x^2 + 7)}{\Delta x} \\ &= \frac{x^2 + 2x\,\Delta x + (\Delta x)^2 + 7 - x^2 - 7}{\Delta x} \\ &= \frac{2x\,\Delta x + (\Delta x)^2}{\Delta x} \\ &= \frac{\cancel{\Delta x}(2x + \Delta x)}{\cancel{\Delta x}} \\ &= 2x + \Delta x, \quad \Delta x \neq 0. \end{aligned}$$

✓ **Checkpoint 4**

Given $f(x) = x^2 + 3$, evaluate each expression.

a. $f(x + \Delta x)$ **b.** $\dfrac{f(x + \Delta x) - f(x)}{\Delta x}$

In Example 4(b), the expression

$\dfrac{f(x + \Delta x) - f(x)}{\Delta x}$ Difference quotient

is called a *difference quotient* and has a special significance in calculus. You will learn more about this in Chapter 2.

Combinations of Functions

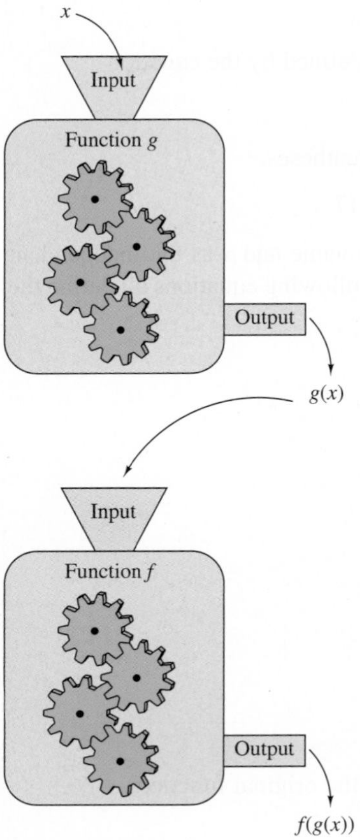

FIGURE 1.47

Two functions can be combined in various ways to create new functions. For instance, given $f(x) = 2x - 3$ and $g(x) = x^2 + 1$, you can form the following functions.

$$f(x) + g(x) = (2x - 3) + (x^2 + 1) = x^2 + 2x - 2 \qquad \text{Sum}$$

$$f(x) - g(x) = (2x - 3) - (x^2 + 1) = -x^2 + 2x - 4 \qquad \text{Difference}$$

$$f(x)g(x) = (2x - 3)(x^2 + 1) = 2x^3 - 3x^2 + 2x - 3 \qquad \text{Product}$$

$$\frac{f(x)}{g(x)} = \frac{2x - 3}{x^2 + 1} \qquad \text{Quotient}$$

You can combine two functions in yet another way called a **composition.** The resulting function is called a **composite function.** For instance, given $f(x) = x^2$ and $g(x) = x + 1$, the composite of f with g is

$$f(g(x)) = f(x + 1) = (x + 1)^2.$$

This composition is denoted by $f \circ g$ and is read as "f composed with g."

Definition of Composite Function

Let f and g be functions. The function given by

$$(f \circ g)(x) = f(g(x))$$

is the **composite** of f with g. The **domain** of $f \circ g$ is the set of all x in the domain of g such that $g(x)$ is in the domain of f, as indicated in Figure 1.47.

The composite of f with g may not be equal to the composite of g with f, as shown in the next example.

Example 5 Forming Composite Functions

Given $f(x) = 2x - 3$ and $g(x) = x^2 + 1$, find each composite function.

a. $f(g(x))$ **b.** $g(f(x))$

SOLUTION

a. The composite of f with g is given by

$$\begin{aligned} f(g(x)) &= 2(g(x)) - 3 && \text{Evaluate } f \text{ at } g(x). \\ &= 2(x^2 + 1) - 3 && \text{Substitute } x^2 + 1 \text{ for } g(x). \\ &= 2x^2 + 2 - 3 && \text{Distributive Property} \\ &= 2x^2 - 1. && \text{Simplify.} \end{aligned}$$

b. The composite of g with f is given by

$$\begin{aligned} g(f(x)) &= (f(x))^2 + 1 && \text{Evaluate } g \text{ at } f(x). \\ &= (2x - 3)^2 + 1 && \text{Substitute } 2x - 3 \text{ for } f(x). \\ &= 4x^2 - 12x + 9 + 1 && \text{Expand.} \\ &= 4x^2 - 12x + 10. && \text{Simplify.} \end{aligned}$$

STUDY TIP

The expressions for $f(g(x))$ and $g(f(x))$ are different in Example 5. In general, the composite of f with g is not the same as the composite of g with f.

✓ Checkpoint 5

Given $f(x) = 2x + 1$ and $g(x) = x^2 + 2$, find each composite function.

a. $f(g(x))$ **b.** $g(f(x))$ ■

Inverse Functions

Informally, the inverse function of f is another function g that "undoes" what f has done. For instance, subtraction can be used to undo addition, and division can be used to undo multiplication.

$$x \xrightarrow{\;f\;} f(x) \xrightarrow{\;g\;} g(f(x)) = x$$

Definition of Inverse Function

Let f and g be two functions such that

$$f(g(x)) = x \text{ for each } x \text{ in the domain of } g$$

and

$$g(f(x)) = x \text{ for each } x \text{ in the domain of } f.$$

Under these conditions, the function g is the **inverse function** of f. The function g is denoted by f^{-1}, which is read as "f-inverse." So,

$$f(f^{-1}(x)) = x \qquad \text{and} \qquad f^{-1}(f(x)) = x.$$

The domain of f must be equal to the range of f^{-1}, and the range of f must be equal to the domain of f^{-1}.

Do not be confused by the use of the superscript -1 to denote the inverse function f^{-1}. In this text, whenever f^{-1} is written, it *always* refers to the inverse function of f and *not* to the reciprocal of $f(x)$.

Example 6 **Finding Inverse Functions Informally**

Find the inverse function of each function informally.

a. $f(x) = 2x$ **b.** $f(x) = x + 4$

SOLUTION

a. The function f *multiplies* each input by 2. To "undo" this function, you need to *divide* each input by 2. So, the inverse function of $f(x) = 2x$ is

$$f^{-1}(x) = \frac{x}{2}.$$

b. The function f *adds* 4 to each input. To "undo" this function, you need to *subtract* 4 from each input. So, the inverse function of $f(x) = x + 4$ is

$$f^{-1}(x) = x - 4.$$

Check that f and f^{-1} are inverse functions by showing that $f(f^{-1}(x)) = x$ and $f^{-1}(f(x)) = x$.

✓Checkpoint 6

Find the inverse function of each function informally.

a. $f(x) = \frac{1}{5}x$ **b.** $f(x) = x - 6$ ■

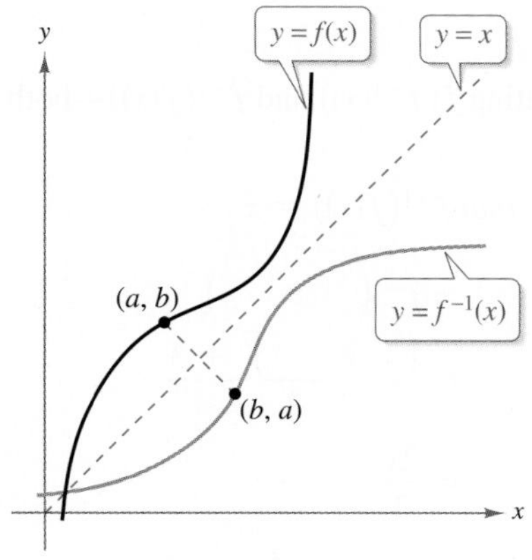

The graph of f^{-1} is a reflection of the graph of f in the line $y = x$.
FIGURE 1.48

The graphs of f and f^{-1} are mirror images of each other (with respect to the line $y = x$), as shown in Figure 1.48. Try using a graphing utility with a *square setting* to confirm this for each of the functions given in Example 6.

The functions in Example 6 are simple enough so that their inverse functions can be found by inspection. The next example demonstrates a strategy for finding the inverse functions of more complicated functions.

Example 7 Finding an Inverse Function

Find the inverse function of $f(x) = \sqrt{2x - 3}$.

SOLUTION Begin by replacing $f(x)$ with y. Then, interchange x and y and solve for y.

$$f(x) = \sqrt{2x-3} \qquad \text{Write original function.}$$
$$y = \sqrt{2x-3} \qquad \text{Replace } f(x) \text{ with } y.$$
$$x = \sqrt{2y-3} \qquad \text{Interchange } x \text{ and } y.$$
$$x^2 = 2y - 3 \qquad \text{Square each side.}$$
$$x^2 + 3 = 2y \qquad \text{Add 3 to each side.}$$
$$\frac{x^2+3}{2} = y \qquad \text{Divide each side by 2.}$$

So, the inverse function has the form

$$f^{-1}(\quad) = \frac{(\quad)^2 + 3}{2}.$$

Using x as the independent variable, you can write

$$f^{-1}(x) = \frac{x^2+3}{2}, \quad x \geq 0.$$

In Figure 1.49, note that the domain of f^{-1} coincides with the range of f.

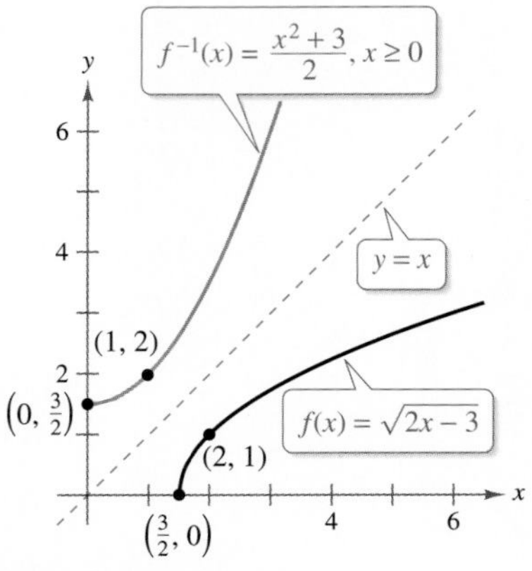

FIGURE 1.49

✓ Checkpoint 7

Find the inverse function of $f(x) = x^2 + 2$ for $x \geq 0$. ■

TECH TUTOR

A graphing utility can help you check that the graphs of f and f^{-1} are reflections of each other in the line $y = x$. To do this, graph $y = f(x)$, $y = f^{-1}(x)$, and $y = x$ in the same viewing window, using a square setting.

After you have found an inverse function, you should check your results. You can check your results *graphically* by observing that the graphs of f and f^{-1} are reflections of each other in the line

$$y = x.$$

You can check your results *algebraically* by evaluating $f(f^{-1}(x))$ and $f^{-1}(f(x))$—both should be equal to x.

Check that $f(f^{-1}(x)) = x$

$$f(f^{-1}(x)) = f\left(\frac{x^2+3}{2}\right) = \sqrt{2\left(\frac{x^2+3}{2}\right) - 3} = \sqrt{x^2} = x, \quad x \geq 0$$

Check that $f^{-1}(f(x)) = x$

$$f^{-1}(f(x)) = f^{-1}\left(\sqrt{2x-3}\right) = \frac{\left(\sqrt{2x-3}\right)^2 + 3}{2} = \frac{2x}{2} = x, \quad x \geq \frac{3}{2}$$

Not every function has an inverse function. In fact, for a function to have an inverse function, it must be one-to-one.

Example 8 A Function That Has No Inverse Function

Show that the function

$$f(x) = x^2 - 1$$

has no inverse function. (Assume that the domain of f is the set of all real numbers.)

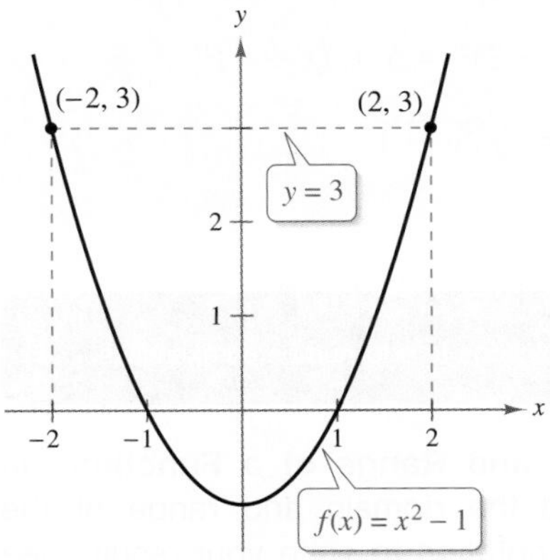

f is not one-to-one and has no inverse function.

FIGURE 1.50

SOLUTION Begin by sketching the graph of f, as shown in Figure 1.50. Note that

$$f(2) = (2)^2 - 1 = 3$$

and

$$f(-2) = (-2)^2 - 1 = 3.$$

So, f does not pass the Horizontal Line Test, which implies that it is not one-to-one and therefore has no inverse function. The same conclusion can be obtained by trying to find the inverse function of f algebraically.

$f(x) = x^2 - 1$	Write original function.
$y = x^2 - 1$	Replace $f(x)$ with y.
$x = y^2 - 1$	Interchange x and y.
$x + 1 = y^2$	Add 1 to each side.
$\pm\sqrt{x + 1} = y$	Take square root of each side.

The last equation does not define y as a function of x, and so f has no inverse function.

✓ Checkpoint 8

Show that the function

$$f(x) = x^2 + 4$$

has no inverse function.

SUMMARIZE (Section 1.4)

1. State the definition of a function *(page 35)*. For an example of a function, see Example 1.
2. Explain the meanings of domain and range *(page 37)*. For an example of a domain and a range, see Example 2.
3. Explain the meaning of function notation *(page 38)*. For examples of function notation, see Examples 3 and 4.
4. State the definition of a composite function *(page 40)*. For an example of a composite function, see Example 5.
5. State the definition of an inverse function *(page 41)*. For examples of inverse functions, see Examples 6 and 7.
6. State when a function does not have an inverse function *(page 43)*. For an example of a function that does not have an inverse function, see Example 8.

SKILLS WARM UP 1.4

The following warm-up exercises involve skills that were covered in a previous course. You will use these skills in the exercise set for this section. For additional help, review Appendix Sections A.3 and A.5.

In Exercises 1–6, simplify the expression.

1. $5(-1)^2 - 6(-1) + 9$

2. $(-2)^3 + 7(-2)^2 - 10$

3. $(x - 2)^2 + 5x - 10$

4. $(3 - x) + (x + 3)^3$

5. $\dfrac{1}{1 - (1 - x)}$

6. $1 + \dfrac{x - 1}{x}$

In Exercises 7–12, solve for y in terms of x.

7. $2x + y - 6 = 11$

8. $5y - 6x^2 - 1 = 0$

9. $(y - 3)^2 = 5 + (x + 1)^2$

10. $y^2 - 4x^2 = 2$

11. $x = \dfrac{2y - 1}{4}$

12. $x = \sqrt[3]{2y - 1}$

Exercises 1.4

See www.CalcChat.com for worked-out solutions to odd-numbered exercises.

Deciding Whether Equations Are Functions In Exercises 1–8, decide whether the equation defines y as a function of x. *See Example 1.*

1. $x^2 + y^2 = 16$

2. $x + y^2 = 4$

3. $\frac{1}{2}x - 6y = -3$

4. $3x - 2y + 5 = 0$

5. $x^2 + y = 4$

6. $x^2 + y^2 + 2x = 0$

7. $y = |x + 2|$

8. $x^2y - x^2 + 4y = 0$

Vertical Line Test In Exercises 9–12, use the Vertical Line Test to determine whether y is a function of x.

9. $x^2 + y^2 = 9$

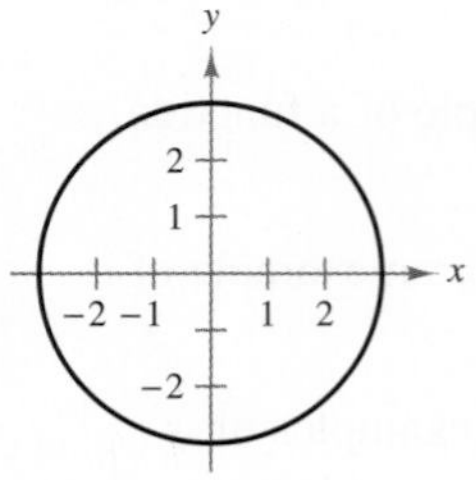

10. $x - xy + y + 1 = 0$

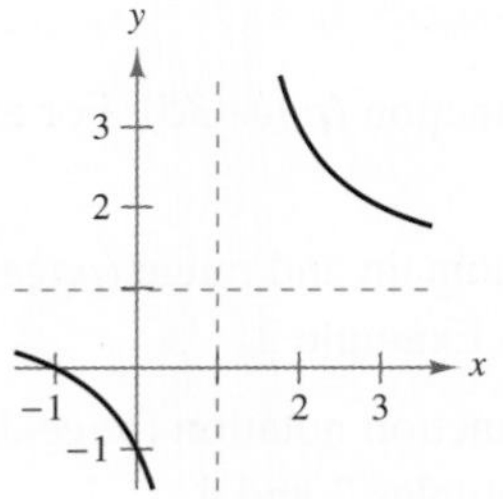

11. $x^2 = xy - 1$

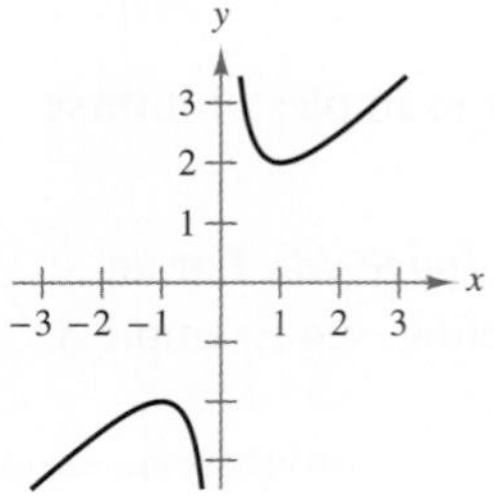

12. $x = |y|$

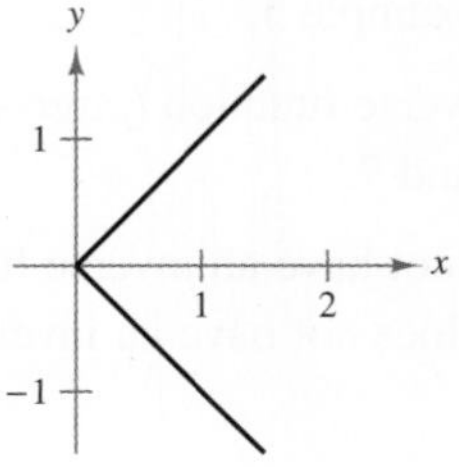

Finding the Domain and Range of a Function In Exercises 13–16, find the domain and range of the function. Use interval notation to write your result. *See Example 2.*

13. $f(x) = x^3$

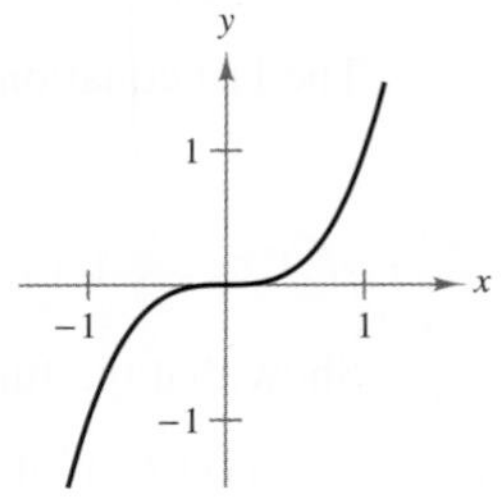

14. $f(x) = \sqrt{2x - 3}$

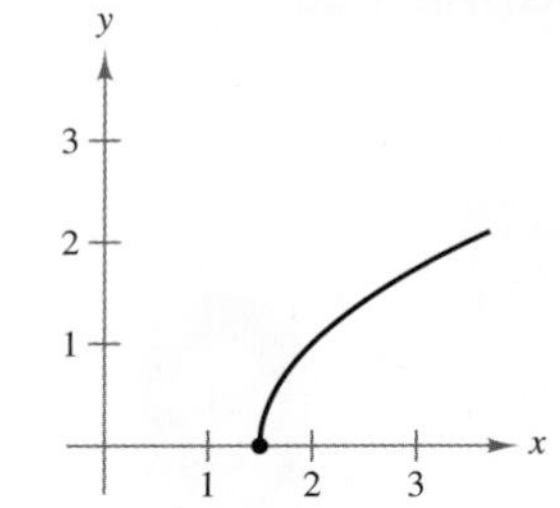

15. $f(x) = 4 - x^2$

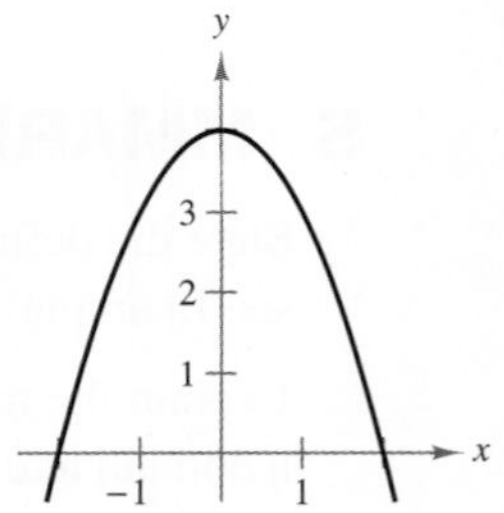

16. $f(x) = |x - 2|$

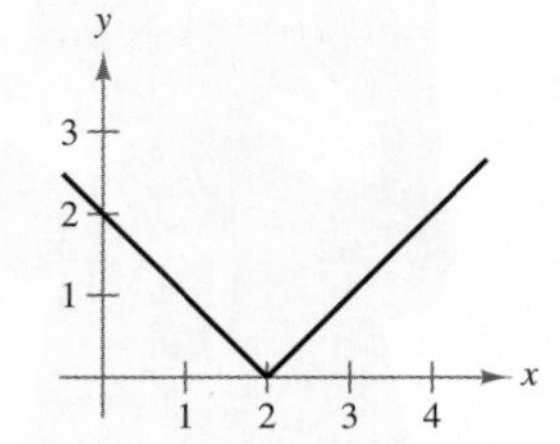

Finding the Domain and Range of a Function In Exercises 17–24, find the domain and range of the function. Use a graphing utility to verify your results. *See Example 2.*

17. $f(x) = 2x^2 - 5x + 1$

18. $f(x) = 5x^3 + 6x^2 - 1$

19. $f(x) = \dfrac{|x|}{x}$

20. $f(x) = \sqrt{9 - x^2}$

21. $f(x) = \dfrac{x}{\sqrt{x - 4}}$

22. $f(x) = \begin{cases} 3x + 2, & x < 0 \\ 2 - x, & x \geq 0 \end{cases}$

23. $f(x) = \dfrac{x - 2}{x + 4}$

24. $f(x) = \dfrac{x^2}{1 - x}$

Evaluating a Function In Exercises 25–28, evaluate the function at the specified values of the independent variable. Simplify the results. *See Example 3.*

25. $f(x) = 3x - 2$

(a) $f(0)$ (b) $f(5)$ (c) $f(x - 1)$

26. $f(x) = x^2 - 4x + 1$

(a) $f(-1)$ (b) $f\left(\frac{1}{2}\right)$ (c) $f(c + 2)$

27. $g(x) = 1/x$

(a) $g\left(\frac{1}{4}\right)$ (b) $g(-4)$ (c) $g(x + 4)$

28. $f(x) = |x| + 4$

(a) $f(-2)$ (b) $f(2)$ (c) $f(x + 2)$

Evaluating a Function In Exercises 29–34, evaluate the difference quotient and simplify the result. *See Example 4.*

29. $f(x) = x^2 - 5x + 2$

$\dfrac{f(x + \Delta x) - f(x)}{\Delta x}$

30. $h(x) = x^2 + x + 3$

$\dfrac{h(2 + \Delta x) - h(2)}{\Delta x}$

31. $g(x) = \sqrt{x + 1}$

$\dfrac{g(x + \Delta x) - g(x)}{\Delta x}$

32. $f(x) = \dfrac{1}{\sqrt{x}}$

$\dfrac{f(x) - f(2)}{x - 2}$

33. $f(x) = \dfrac{1}{x - 2}$

$\dfrac{f(x + \Delta x) - f(x)}{\Delta x}$

34. $f(x) = \dfrac{1}{x + 4}$

$\dfrac{f(x + \Delta x) - f(x)}{\Delta x}$

Combinations of Functions In Exercises 35–38, find (a) $f(x) + g(x)$, (b) $f(x) - g(x)$, (c) $f(x) \cdot g(x)$, (d) $f(x)/g(x)$, (e) $f(g(x))$, and (f) $g(f(x))$, if defined. *See Example 5.*

35. $f(x) = 2x - 5$

$g(x) = 5$

36. $f(x) = x^2 + 5$

$g(x) = \sqrt{1 - x}$

37. $f(x) = x^2 + 1$

$g(x) = x - 1$

38. $f(x) = \dfrac{x}{x + 1}$

$g(x) = x^3$

39. Composite Functions Given $f(x) = \sqrt{x}$ and $g(x) = x^2 - 1$, find the composite functions.

(a) $f(g(1))$ (b) $g(f(1))$

(c) $g(f(0))$ (d) $f(g(-4))$

(e) $f(g(x))$ (f) $g(f(x))$

40. Composite Functions Given $f(x) = 1/x$ and $g(x) = x^2 - 1$, find the composite functions.

(a) $f(g(2))$ (b) $g(f(2))$

(c) $f(g(-3))$ (d) $g\left(f\left(1/\sqrt{2}\right)\right)$

(e) $f(g(x))$ (f) $g(f(x))$

Finding Inverse Functions Informally In Exercises 41–44, find the inverse function of *f* informally. Verify your results by showing that $f(f^{-1}(x)) = x$ and $f^{-1}(f(x)) = x$. *See Example 6.*

41. $f(x) = 4x$

42. $f(x) = \frac{1}{3}x$

43. $f(x) = x + 12$

44. $f(x) = x - 3$

Finding an Inverse Function In Exercises 45–56, find the inverse function of *f*. *See Example 7.*

45. $f(x) = 2x - 3$

46. $f(x) = 7 - x$

47. $f(x) = \frac{3}{2}x + 1$

48. $f(x) = -6x - 4$

49. $f(x) = x^5$

50. $f(x) = x^3$

51. $f(x) = \dfrac{1}{x}$

52. $f(x) = -\dfrac{2}{x}$

53. $f(x) = \sqrt{9 - x^2}, \quad 0 \le x \le 3$

54. $f(x) = \sqrt{x^2 - 4}, \quad x \ge 2$

55. $f(x) = x^{2/3}, \quad x \ge 0$

56. $f(x) = x^{3/5}$

Determine Whether a Function Is One-to-One In Exercises 57–62, use a graphing utility to graph the function. Then use the Horizontal Line Test to determine whether the function is one-to-one. If it is, find its inverse function.

57. $f(x) = 3 - 7x$

58. $f(x) = \sqrt{x - 2}$

59. $f(x) = x^2$

60. $f(x) = x^4$

61. $f(x) = |x + 3|$

62. $f(x) = -5$

63. Graphing a Function Use the graph of $f(x) = \sqrt{x}$ below to sketch the graph of each function.

(a) $y = \sqrt{x} + 2$

(b) $y = -\sqrt{x}$

(c) $y = \sqrt{x - 2}$

(d) $y = \sqrt{x + 3}$

(e) $y = \sqrt{x - 4} - 1$

(f) $y = 2\sqrt{x}$

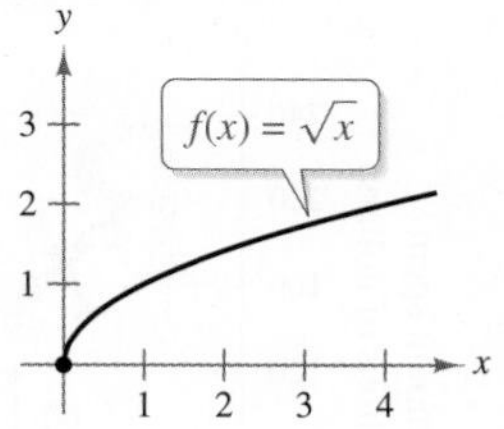

64. Graphing a Function Use the graph of $f(x) = |x|$ below to sketch the graph of each function.

(a) $y = |x| + 3$

(b) $y = -\frac{1}{2}|x|$

(c) $y = |x - 2|$

(d) $y = |x + 1| - 1$

(e) $y = 2|x|$

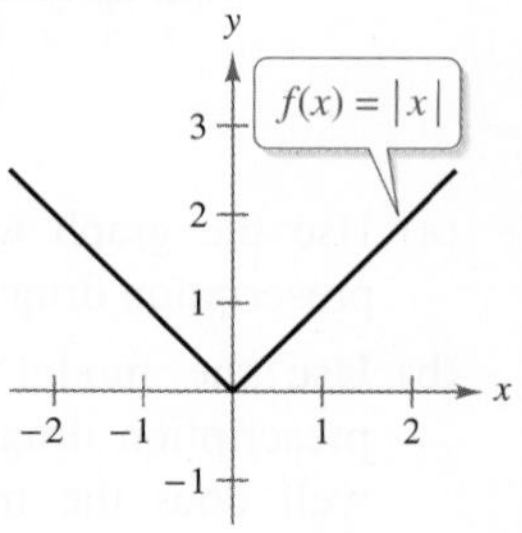

65. Writing a Function Use the graph of $f(x) = x^2$ to write an equation for each function whose graph is shown.

(a)

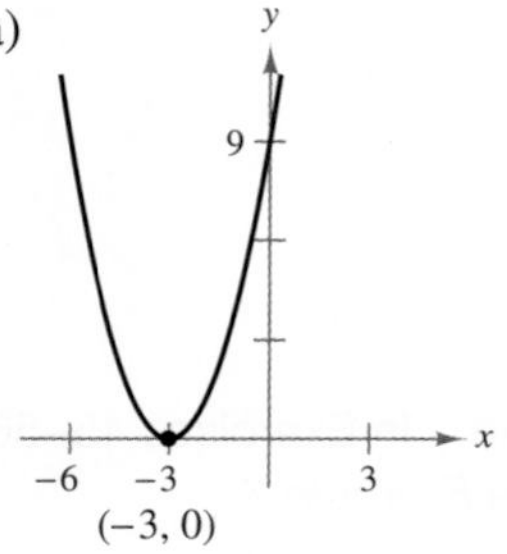

(b)

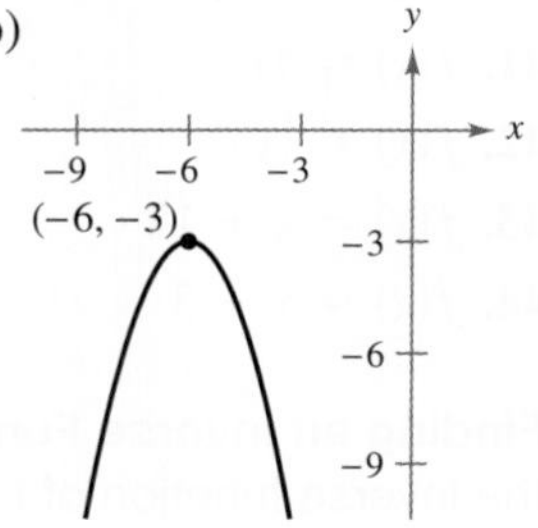

66. Writing a Function Use the graph of $f(x) = x^3$ to write an equation for each function whose graph is shown.

(a)

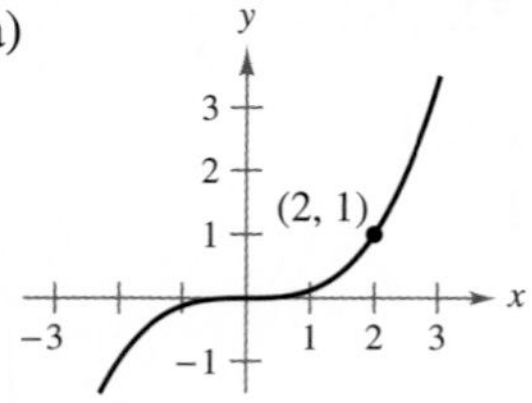

(b)

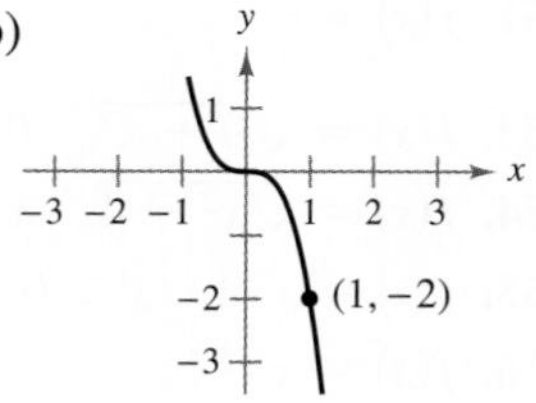

67. Prescription Drugs The amounts d (in billions of dollars) spent on prescription drugs in the United States from 1997 through 2008 (see figure) can be approximated by the model

$$d(t) = \begin{cases} 15.35t - 32.2, & 7 \le t \le 11 \\ -0.687t^2 + 33.57t - 146.5, & 12 \le t \le 18 \end{cases}$$

where t represents the year, with $t = 7$ corresponding to 1997. *(Source: U.S. Centers for Medicare & Medicaid Services)*

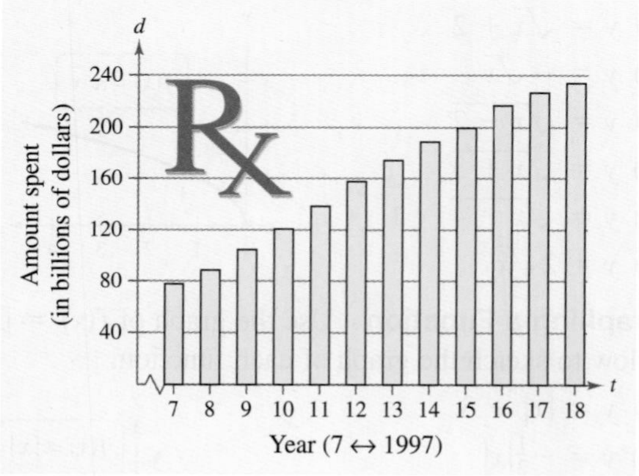

(a) Use the graph to estimate the amounts spent on prescription drugs in 2000, 2003, and 2007.

(b) Use the model to find the amounts spent on prescription drugs in 2000, 2003, and 2007. How well does the model fit the data? Explain your reasoning.

68. HOW DO YOU SEE IT? An electronically controlled thermostat in a home is programmed to lower the temperature automatically during the night. The temperature in the house T (in degrees Fahrenheit) is given in terms of t, the time in hours on a 24-hour clock (see figure).

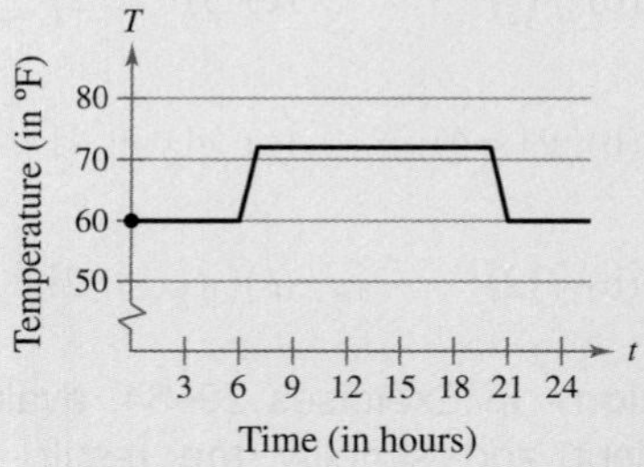

(a) Explain why T is a function of t.

(b) Approximate $T(4)$ and $T(15)$.

(c) The thermostat is reprogrammed to produce a temperature H for which

$$H(t) = T(t - 1).$$

How does this change the temperature?

(d) The thermostat is reprogrammed to produce a temperature H for which

$$H(t) = T(t) - 1.$$

How does this change the temperature?

69. Owning a Business You own two restaurants. From 2005 through 2011, the sales R_1 (in thousands of dollars) for one restaurant can be modeled by

$$R_1 = 690 - 8t - 0.8t^2, \quad t = 5, 6, 7, 8, 9, 10, 11$$

where $t = 1$ represents 2001. During the same seven-year period, the sales R_2 (in thousands of dollars) for the second restaurant can be modeled by

$$R_2 = 458 + 0.78t, \quad t = 5, 6, 7, 8, 9, 10, 11.$$

Write a function that represents the total sales for the two restaurants. Use a graphing utility to graph the total sales function.

70. Births and Deaths From 1990 to 2007, the total numbers of births B (in thousands) and deaths D (in thousands) in the United States can be approximated by the models

$$B(t) = -0.151t^3 + 8.00t^2 - 85.0t + 4176$$

and

$$D(t) = -1.25t^2 + 38.5t + 2136$$

where t represents the year, with $t = 0$ corresponding to 1990. Find $B(t) - D(t)$ and interpret this function. *(Source: U.S. National Center for Health Statistics)*

71. Cost The inventor of a new game believes that the variable cost for producing the game is \$1.95 per unit. The fixed cost is \$6000.

(a) Express the total cost C as a function of x, the number of games sold.

(b) Find a formula for the average cost per unit

$$\overline{C} = \frac{C}{x}.$$

(c) The selling price for each game is \$4.95. How many units must be sold before the average cost per unit falls below the selling price?

72. Demand The demand function for a commodity is

$$p = \frac{14.75}{1 + 0.01x}, \quad x \geq 0$$

where p is the price per unit and x is the number of units sold.

(a) Find x as a function of p.

(b) Find the number of units sold when the price is \$10.

73. Cost The weekly cost C of producing x units in a manufacturing process is given by $C(x) = 70x + 500$. The number of units x produced in t hours is given by $x(t) = 40t$.

(a) Find and interpret $C(x(t))$.

(b) Find the cost of 4 hours of production.

(c) After how much time does the cost of production reach \$18,000?

74. Revenue For groups of 80 or more people, a charter bus company determines the rate r (in dollars per person) according to the formula

$$r = 15 - 0.05(n - 80), \quad n \geq 80$$

where n is the number of people.

(a) Express the revenue R for the bus company as a function of n.

(b) Complete the table.

n	100	125	150	175	200	225	250
R							

(c) Would you use the formula for the rate? Explain your reasoning.

75. Cost, Revenue, and Profit A company invests \$98,000 for equipment to produce a new product. Each unit of the product costs \$12.30 and is sold for \$17.98. Let x be the number of units produced and sold.

(a) Write the total cost C as a function of x.

(b) Write the revenue R as a function of x.

(c) Write the profit P as a function of x.

76. Profit A manufacturer charges \$90 per unit for units that cost \$60 to produce. To encourage large orders from distributors, the manufacturer will reduce the price by \$0.01 per unit for each unit in excess of 100 units. (For example, an order of 101 units would have a price of \$89.99 per unit, and an order of 102 units would have a price of \$89.98 per unit.) This price reduction is discontinued when the price per unit drops to \$75.

(a) Express the price per unit p as a function of the order size x.

(b) Express the profit P as a function of the order size x.

Graphing Functions In Exercises 77–82, use a graphing utility to graph the function. Then use the *zoom* and *trace* features to find the zeros of the function. Is the function one-to-one?

77. $f(x) = 9x - 4x^2$

78. $h(x) = 6x^3 - 12x^2 + 4$

79. $g(t) = \dfrac{t + 3}{1 - t}$

80. $f(x) = 2\left(3x^2 - \dfrac{6}{x}\right)$

81. $g(x) = x^2\sqrt{x^2 - 4}$

82. $f(x) = (x + 3)^3$

Business Capsule

CitiKitty, Inc. was founded in 2005 by 26-year-old Rebecca Rescate after she moved into a small apartment in New York City with no place to hide her cat's litter box. Finding no easy-to-use cat toilet training kit, she created one, and CitiKitty was born with an initial investment of \$20,000. Today the company flourishes with an expanded product line. Revenues in 2010 reached \$350,000.

83. Research Project Use your school's library, the Internet, or some other reference source to find information about the start-up costs of beginning a business, such as the example above. Write a short paper about the company.

Courtesy of CitiKitty Inc.

1.5 Limits

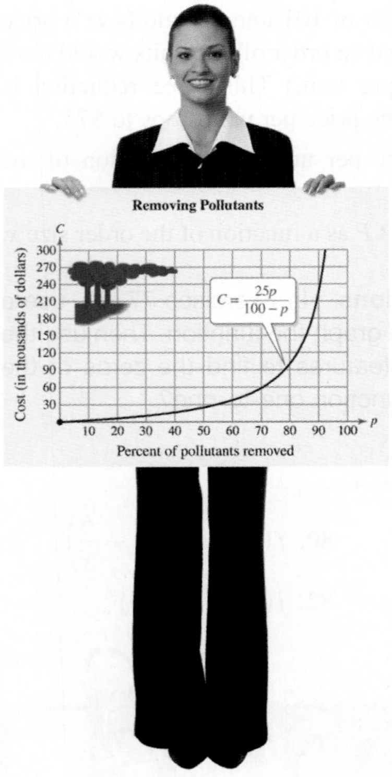

In Exercise 75 on page 59, you will use a limit to analyze the cost of removing pollutants from a small lake.

- Find limits of functions graphically and numerically.
- Understand the definition of the limit of a function and use the properties of limits to evaluate limits of functions.
- Use different analytic techniques to evaluate limits of functions.
- Evaluate one-sided limits.
- Recognize unbounded behavior of functions.

The Limit of a Function

In everyday language, people refer to a speed limit, a wrestler's weight limit, the limit of one's endurance, or stretching a spring to its limit. These phrases all suggest that a limit is a bound, which on some occasions may not be reached but on other occasions may be reached or exceeded.

Consider a spring that will break only when a weight of 10 pounds or more is attached. To determine how far the spring will stretch without breaking, you could attach increasingly heavier weights and measure the spring length s for each weight w, as shown in Figure 1.51. If the spring length approaches a value of L, then it is said that "the limit of s as w approaches 10 is L." A mathematical limit is much like the limit of a spring. The notation for a limit is

$$\lim_{x \to c} f(x) = L$$

which is read as "the limit of $f(x)$ as x approaches c is L."

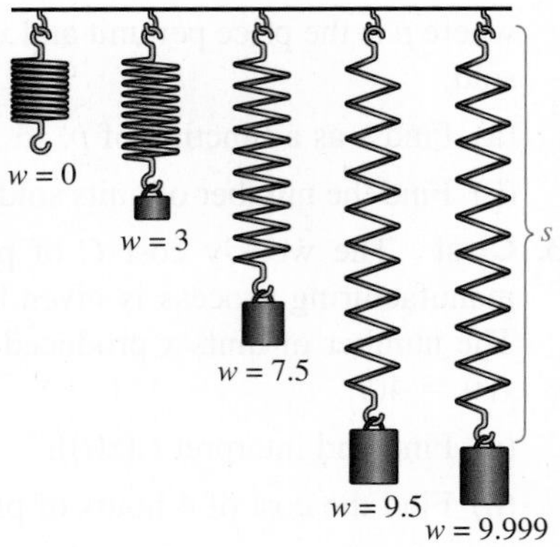

What is the limit of s as w approaches 10 pounds?

FIGURE 1.51

Example 1 Finding a Limit

Find the limit: $\lim_{x \to 1} (x^2 + 1)$.

SOLUTION Let $f(x) = x^2 + 1$. From the graph of f in Figure 1.52, it appears that $f(x)$ approaches 2 as x approaches 1 from either side, and you can write

$$\lim_{x \to 1} (x^2 + 1) = 2.$$

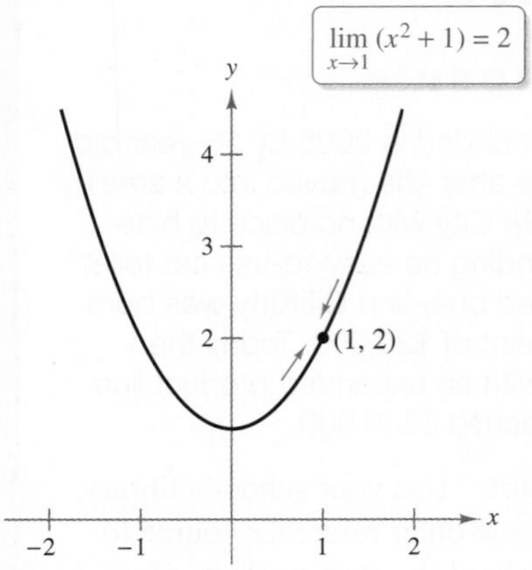

FIGURE 1.52

The table yields the same conclusion. Notice that as x gets closer and closer to 1, $f(x)$ gets closer and closer to 2.

x approaches 1. → ← *x* approaches 1.

x	0.900	0.990	0.999	1.000	1.001	1.010	1.100
$f(x)$	1.810	1.980	1.998	2.000	2.002	2.020	2.210

$f(x)$ approaches 2. → ← $f(x)$ approaches 2.

✓ Checkpoint 1

Find the limit: $\lim_{x \to 1} (2x + 4)$. ■

Example 2 Finding Limits Graphically and Numerically

Find the limit.

$$\lim_{x\to 1} f(x)$$

a. $f(x) = \dfrac{x^2 - 1}{x - 1}$ **b.** $f(x) = \dfrac{|x - 1|}{x - 1}$ **c.** $f(x) = \begin{cases} x, & x \neq 1 \\ 0, & x = 1 \end{cases}$

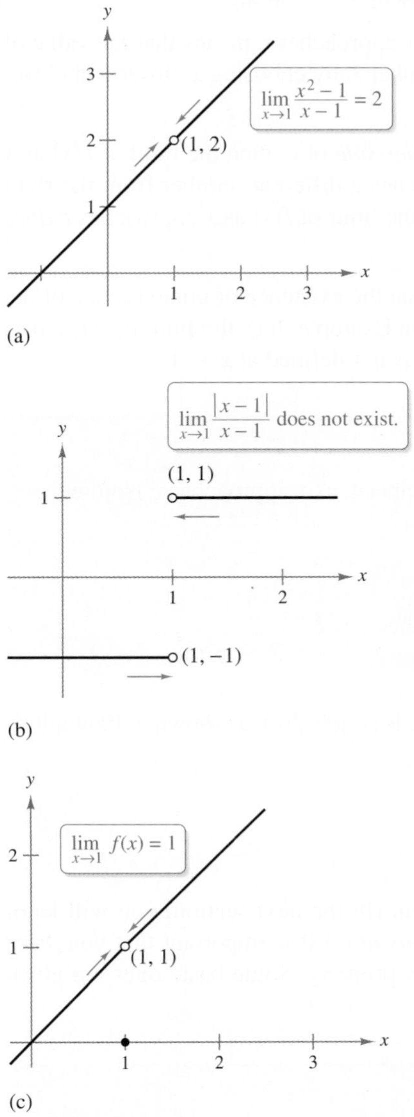

FIGURE 1.53

SOLUTION

a. From the graph of f in Figure 1.53(a), it appears that $f(x)$ approaches 2 as x approaches 1 from either side. A missing point is denoted by the open dot on the graph. This conclusion is reinforced by the table. Be sure you see that *it does not matter that $f(x)$ is undefined when $x = 1$.* The limit depends only on values of $f(x)$ near 1, not at 1.

x approaches 1. → ← x approaches 1.

x	0.900	0.990	0.999	1.000	1.001	1.010	1.100
$f(x)$	1.900	1.990	1.999	?	2.001	2.010	2.100

$f(x)$ approaches 2. → ← $f(x)$ approaches 2.

b. From the graph of f in Figure 1.53(b), you can see that $f(x) = -1$ for all values to the left of $x = 1$ and $f(x) = 1$ for all values to the right of $x = 1$. So, $f(x)$ is approaching a different value from the left of $x = 1$ than it is from the right of $x = 1$. In such situations, you can conclude that *the limit does not exist.* This conclusion is reinforced by the table.

x approaches 1. → ← x approaches 1.

x	0.900	0.990	0.999	1.000	1.001	1.010	1.100
$f(x)$	−1.000	−1.000	−1.000	?	1.000	1.000	1.000

$f(x)$ approaches −1. → ← $f(x)$ approaches 1.

c. From the graph of f in Figure 1.53(c), it appears that $f(x)$ approaches 1 as x approaches 1 from either side. This conclusion is reinforced by the table. It does not matter that $f(1) = 0$. The limit depends only on values of $f(x)$ near 1, not at 1.

x approaches 1. → ← x approaches 1.

x	0.900	0.990	0.999	1.000	1.001	1.010	1.100
$f(x)$	0.900	0.990	0.999	?	1.001	1.010	1.100

$f(x)$ approaches 1. → ← $f(x)$ approaches 1.

✓ Checkpoint 2

Find the limit.

$$\lim_{x\to 2} f(x)$$

a. $f(x) = \dfrac{x^2 - 4}{x - 2}$ **b.** $f(x) = \dfrac{|x - 2|}{x - 2}$ **c.** $f(x) = \begin{cases} x^2, & x \neq 2 \\ 0, & x = 2 \end{cases}$ ■

TECH TUTOR

You can use a graphing utility to estimate the limit of a function. One way is to estimate the limit numerically by creating a table of values using the *table* feature. Another way is to graph the function and use the *zoom* and *trace* features to estimate the limit graphically.

Definition of the Limit of a Function and Properties of Limits

There are three important ideas to learn from Examples 1 and 2.

1. Saying that the limit of $f(x)$ approaches L as x approaches c means that the value of $f(x)$ may be made *arbitrarily close* to the number L by choosing x closer and closer to c.

2. If $f(x)$ approaches the same number from *either side* of c, then the limit of $f(x)$ as x approaches c exists. However, if $f(x)$ approaches a different number from the right side of c than it does from the left side, then the limit of $f(x)$ as x approaches c *does not exist.* [See Example 2(b).]

3. The value of $f(x)$ when $x = c$ has no bearing on the existence or nonexistence of the limit of $f(x)$ as x approaches c. For instance, in Example 2(a), the limit of $f(x)$ exists as x approaches 1 even though the function f is not defined at $x = 1$.

Definition of the Limit of a Function

If $f(x)$ becomes arbitrarily close to a single number L as x approaches c from either side, then

$$\lim_{x \to c} f(x) = L$$

which is read as "the **limit** of $f(x)$ as x approaches c is L."

Many times the limit of $f(x)$ as x approaches c is simply $f(c)$, as shown in Example 1. Whenever the limit of $f(x)$ as x approaches c is

$$\lim_{x \to c} f(x) = f(c) \qquad \text{Substitute } c \text{ for } x.$$

the limit can be evaluated by **direct substitution.** (In the next section, you will learn that a function that has this property is *continuous at c*.) It is important that you learn to recognize the types of functions that have this property. Some basic ones are given in the following list.

Some Basic Limits

Let b and c be real numbers, and let n be a positive integer.

1. $\lim_{x \to c} b = b$ **2.** $\lim_{x \to c} x = c$ **3.** $\lim_{x \to c} x^n = c^n$ **4.** $\lim_{x \to c} \sqrt[n]{x} = \sqrt[n]{c}$

In Property 4, if n is even, then c must be positive.

Example 3 Evaluating Basic Limits

a. $\lim_{x \to 2} 3 = 3$ **b.** $\lim_{x \to -4} x = -4$

c. $\lim_{x \to 2} x^3 = 2^3 = 8$ **d.** $\lim_{x \to 9} \sqrt{x} = \sqrt{9} = 3$

✓ Checkpoint 3

Find the limit.

a. $\lim_{x \to 1} 5$ **b.** $\lim_{x \to 6} x$ **c.** $\lim_{x \to 5} x^2$ **d.** $\lim_{x \to -8} \sqrt[3]{x}$ ■

By combining the basic limits from the preceding page with the properties of limits shown below, you can find limits for a wide variety of algebraic functions.

Properties of Limits

Let b and c be real numbers, let n be a positive integer, and let f and g be functions with the following limits.

$\lim_{x \to c} f(x) = L$ and $\lim_{x \to c} g(x) = K$

1. Scalar multiple: $\lim_{x \to c} [bf(x)] = bL$
2. Sum or difference: $\lim_{x \to c} [f(x) \pm g(x)] = L \pm K$
3. Product: $\lim_{x \to c} [f(x) \cdot g(x)] = LK$
4. Quotient: $\lim_{x \to c} \dfrac{f(x)}{g(x)} = \dfrac{L}{K}$, provided $K \neq 0$
5. Power: $\lim_{x \to c} [f(x)]^n = L^n$
6. Radical: $\lim_{x \to c} \sqrt[n]{f(x)} = \sqrt[n]{L}$

In Property 6, if n is even, then L must be positive.

TECH TUTOR

Symbolic computer algebra systems are capable of evaluating limits. Try using a computer algebra system to evaluate the limit given in Example 4.

Example 4 Finding the Limit of a Polynomial Function

Find the limit: $\lim_{x \to 2} (x^2 + 2x - 3)$.

SOLUTION

$$\lim_{x \to 2} (x^2 + 2x - 3) = \lim_{x \to 2} x^2 + \lim_{x \to 2} 2x - \lim_{x \to 2} 3 \quad \text{Apply Property 2.}$$
$$= \lim_{x \to 2} x^2 + 2 \lim_{x \to 2} x - \lim_{x \to 2} 3 \quad \text{Apply Property 1.}$$
$$= 2^2 + 2(2) - 3 \quad \text{Use direct substitution.}$$
$$= 5$$

✓Checkpoint 4

Find the limit: $\lim_{x \to 1} (2x^2 - x + 4)$. ■

In Example 4, note that the limit (as $x \to 2$) of the *polynomial function*

$$p(x) = x^2 + 2x - 3$$

is simply the value of p at $x = 2$.

$$\lim_{x \to 2} p(x) = p(2) = 2^2 + 2(2) - 3 = 4 + 4 - 3 = 5$$

This is an illustration of the following important result, which states that the limit of a polynomial function can be evaluated by direct substitution.

The Limit of a Polynomial Function

If p is a polynomial function and c is any real number, then

$$\lim_{x \to c} p(x) = p(c).$$

David Davis/www.shutterstock.com

Techniques for Evaluating Limits

You have learned several techniques for evaluating limits. Another technique is shown in Example 5.

Example 5 Finding the Limit of a Function

Find the limit.

$$\lim_{x \to 1} \frac{x^3 - 1}{x - 1}$$

SOLUTION Note that the numerator and denominator are zero when $x = 1$. This implies that $x - 1$ is a factor of both, and you can divide out this common factor.

$$\frac{x^3 - 1}{x - 1} = \frac{(x - 1)(x^2 + x + 1)}{x - 1} \qquad \text{Factor numerator.}$$

$$= \frac{\cancel{(x - 1)}(x^2 + x + 1)}{\cancel{x - 1}} \qquad \text{Divide out common factor.}$$

$$= x^2 + x + 1, \quad x \neq 1 \qquad \text{Simplify.}$$

So, the rational function $(x^3 - 1)/(x - 1)$ and the polynomial function $x^2 + x + 1$ agree for all values of x other than $x = 1$, and you can apply the Replacement Theorem.

$$\lim_{x \to 1} \frac{x^3 - 1}{x - 1} = \lim_{x \to 1} (x^2 + x + 1) = 1^2 + 1 + 1 = 3$$

Figure 1.54 illustrates this result graphically. Note that the two graphs are identical except that the graph of g contains the point $(1, 3)$, whereas this point is missing on the graph of f. (In the graph of f in Figure 1.54, the missing point is denoted by an open dot.)

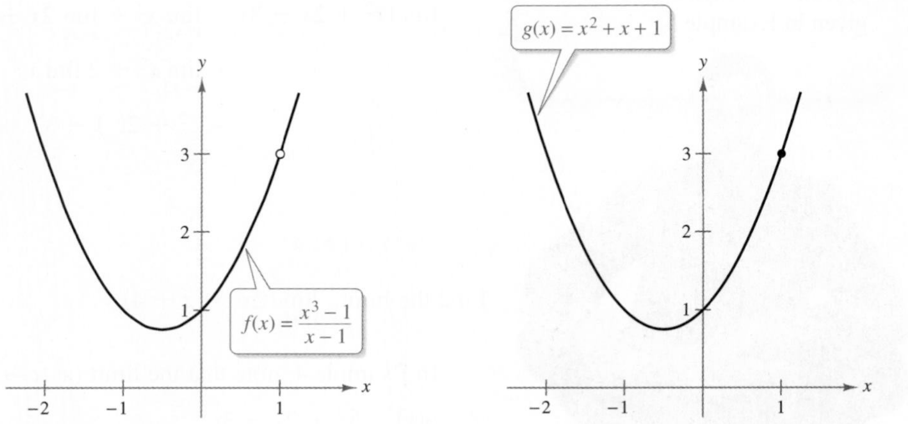

FIGURE 1.54

✓Checkpoint 5

Find the limit.

$$\lim_{x \to 2} \frac{x^3 - 8}{x - 2}$$

■

The technique used to evaluate the limit in Example 5 is called the **dividing out technique**. The validity of this technique stems from the fact that when two functions agree at all but a single number c, the functions must have identical limit behavior at $x = c$. This technique is further demonstrated in the next example.

Example 6 Using the Dividing Out Technique

For $\lim_{x\to-3} \dfrac{x^2 + x - 6}{x + 3}$, direct substitution fails because both the numerator and the denominator are zero when $x = -3$.

$$\lim_{x\to-3} \frac{x^2 + x - 6}{x + 3} \qquad \begin{array}{l} \leftarrow \lim_{x\to-3} (x^2 + x - 6) = 0 \\ \leftarrow \lim_{x\to-3} (x + 3) = 0 \end{array}$$

Because the limits of both the numerator and the denominator are zero when $x = -3$, they must have a *common factor* of $x + 3$. So, for all $x \neq -3$, you can divide out this factor and find the limit as shown.

$$\begin{aligned}
\lim_{x\to-3} \frac{x^2 + x - 6}{x + 3} &= \lim_{x\to-3} \frac{(x - 2)(x + 3)}{x + 3} && \text{Factor numerator.} \\
&= \lim_{x\to-3} \frac{(x - 2)\cancel{(x + 3)}}{\cancel{x + 3}} && \text{Divide out common factor.} \\
&= \lim_{x\to-3} (x - 2) && \text{Simplify.} \\
&= -3 - 2 && \text{Direct substitution} \\
&= -5 && \text{Simplify.}
\end{aligned}$$

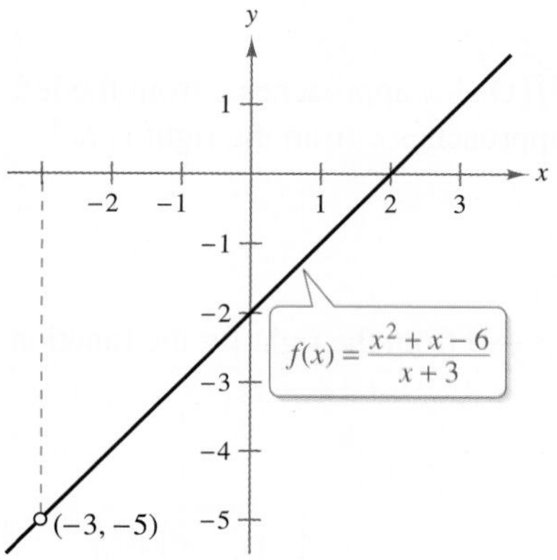

f is undefined when $x = -3$.
FIGURE 1.55

This result is shown graphically in Figure 1.55. Note that the graph of *f* coincides with the graph of $g(x) = x - 2$, except that the graph of *f* has a hole at $(-3, -5)$.

✓Checkpoint 6

Find the limit: $\lim_{x\to3} \dfrac{x^2 + x - 12}{x - 3}$. ■

> **STUDY TIP**
>
> When you try to evaluate a limit and both the numerator and denominator are zero, remember that you must rewrite the fraction so that the new denominator does not have 0 as its limit. One way to do this is to divide out common factors, as shown in Example 6. Another technique is to rationalize the numerator, as shown in Example 7.

Example 7 Finding a Limit of a Function

Find the limit: $\lim_{x\to0} \dfrac{\sqrt{x + 1} - 1}{x}$.

SOLUTION Direct substitution fails because both the numerator and the denominator are zero when $x = 0$. Rewrite the fraction by rationalizing the numerator.

$$\begin{aligned}
\frac{\sqrt{x + 1} - 1}{x} &= \left(\frac{\sqrt{x + 1} - 1}{x}\right)\left(\frac{\sqrt{x + 1} + 1}{\sqrt{x + 1} + 1}\right) \\
&= \frac{(x + 1) - 1}{x\left(\sqrt{x + 1} + 1\right)} \\
&= \frac{\cancel{x}}{\cancel{x}\left(\sqrt{x + 1} + 1\right)} \\
&= \frac{1}{\sqrt{x + 1} + 1}, \quad x \neq 0
\end{aligned}$$

Now, using the Replacement Theorem, you can evaluate the limit as shown.

$$\lim_{x\to0} \frac{\sqrt{x + 1} - 1}{x} = \lim_{x\to0} \frac{1}{\sqrt{x + 1} + 1} = \frac{1}{\sqrt{0 + 1} + 1} = \frac{1}{1 + 1} = \frac{1}{2}$$

✓Checkpoint 7

Find the limit: $\lim_{x\to0} \dfrac{\sqrt{x + 4} - 2}{x}$. ■

One-Sided Limits

In Example 2(b), you saw that one way in which a limit can fail to exist is when a function approaches a different value from the left of c than it approaches from the right of c. This type of behavior can be described more concisely with the concept of a **one-sided limit.**

$$\lim_{x\to c^-} f(x) = L \qquad \text{Limit from the left}$$

$$\lim_{x\to c^+} f(x) = L \qquad \text{Limit from the right}$$

The first of these two limits is read as "the limit of $f(x)$ as x approaches c from the left is L." The second is read as "the limit of $f(x)$ as x approaches c from the right is L."

Example 8 Finding One-Sided Limits

Find the limit as $x \to 0$ from the left and the limit as $x \to 0$ from the right for the function

$$f(x) = \frac{|2x|}{x}.$$

SOLUTION From the graph of f, shown in Figure 1.56, you can see that

$$f(x) = -2$$

for all $x < 0$. So, the limit from the left is

$$\lim_{x\to 0^-} \frac{|2x|}{x} = -2. \qquad \text{Limit from the left}$$

Because

$$f(x) = 2$$

for all $x > 0$, the limit from the right is

$$\lim_{x\to 0^+} \frac{|2x|}{x} = 2. \qquad \text{Limit from the right}$$

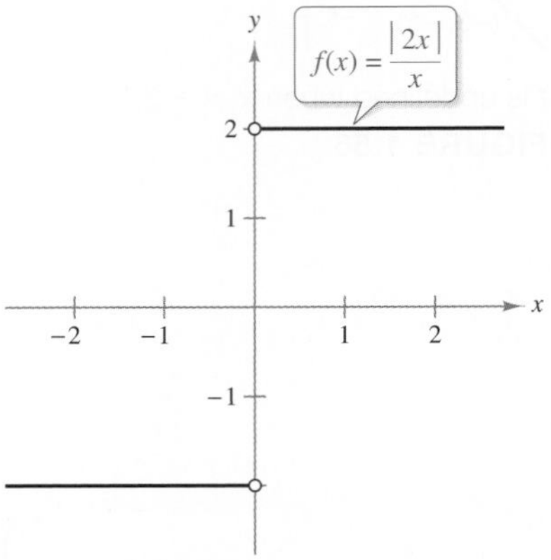

FIGURE 1.56

TECH TUTOR

On most graphing utilities, the absolute value function is denoted by *abs*. You can verify the result in Example 8 by graphing

$$y = \frac{\text{abs}(2x)}{x}$$

in the viewing window $-3 \le x \le 3$ and $-3 \le y \le 3$.

✓Checkpoint 8

Find each limit.

a. $\lim_{x\to 2^-} \frac{|x-2|}{x-2}$

b. $\lim_{x\to 2^+} \frac{|x-2|}{x-2}$

In Example 8, note that the function approaches different limits from the left and from the right. In such cases, the limit of $f(x)$ as $x \to c$ does not exist. For the limit of a function to exist as $x \to c$, *both* one-sided limits must exist and must be equal.

Existence of a Limit

If f is a function and c and L are real numbers, then

$$\lim_{x\to c} f(x) = L$$

if and only if both the left and right limits are equal to L.

Example 9 Finding One-Sided Limits

Find the limit of $f(x)$ as x approaches 1.

$$f(x) = \begin{cases} 4 - x, & x < 1 \\ 4x - x^2, & x > 1 \end{cases}$$

SOLUTION Remember that you are concerned about the value of f near $x = 1$ rather than at $x = 1$. So, for $x < 1$, $f(x)$ is given by

$$4 - x$$

and you can use direct substitution to obtain

$$\lim_{x\to 1^-} f(x) = \lim_{x\to 1^-} (4 - x) = 4 - 1 = 3.$$

For $x > 1$, $f(x)$ is given by

$$4x - x^2$$

and you can use direct substitution to obtain

$$\lim_{x\to 1^+} f(x) = \lim_{x\to 1^+} (4x - x^2) = 4(1) - 1^2 = 4 - 1 = 3.$$

Because both one-sided limits exist and are equal to 3, it follows that

$$\lim_{x\to 1} f(x) = 3.$$

The graph in Figure 1.57 confirms this conclusion.

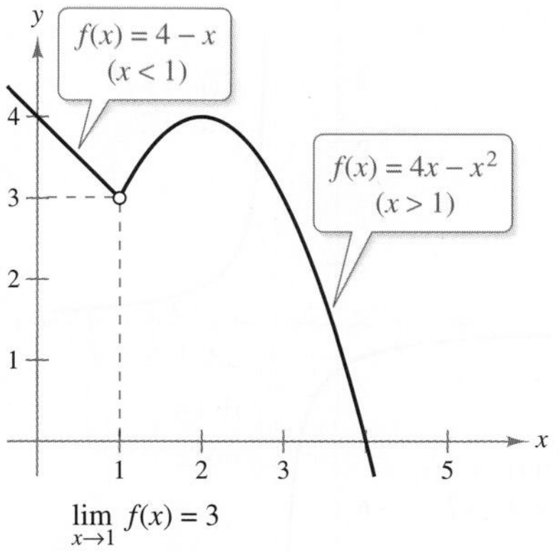

FIGURE 1.57

✓Checkpoint 9

Find the limit of $f(x)$ as x approaches 0.

$$f(x) = \begin{cases} x^2 + 1, & x < 0 \\ 2x + 1, & x > 0 \end{cases}$$

■

Example 10 Comparing One-Sided Limits

An overnight delivery service charges \$18 for the first pound and \$2 for each additional pound. Let x represent the weight of a parcel and let $f(x)$ represent the shipping cost.

$$f(x) = \begin{cases} 18, & 0 < x \le 1 \\ 20, & 1 < x \le 2 \\ 22, & 2 < x \le 3 \end{cases}$$

Show that the limit of $f(x)$ as $x \to 2$ does not exist.

SOLUTION The graph of f is shown in Figure 1.58. The limit of $f(x)$ as x approaches 2 from the left is

$$\lim_{x\to 2^-} f(x) = 20$$

whereas the limit of $f(x)$ as x approaches 2 from the right is

$$\lim_{x\to 2^+} f(x) = 22.$$

Because these one-sided limits are not equal, the limit of $f(x)$ as $x \to 2$ does not exist.

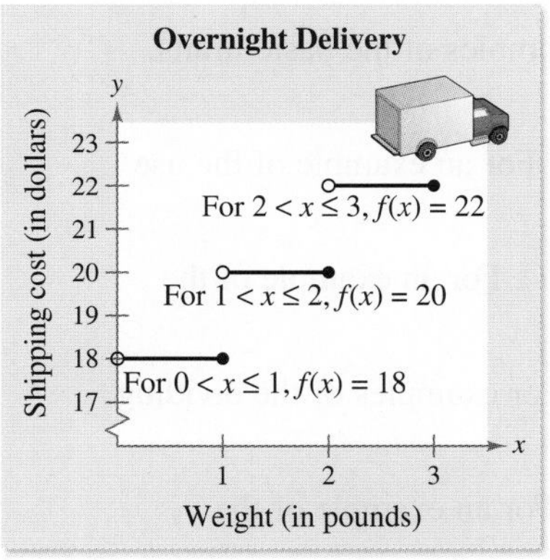

FIGURE 1.58

✓Checkpoint 10

Show that the limit of $f(x)$ as $x \to 1$ does not exist in Example 10.

■

Unbounded Behavior

Example 10 shows a limit that fails to exist because the limits from the left and right differ. Another important way in which a limit can fail to exist is when $f(x)$ increases or decreases without bound as x approaches c.

Example 11 An Unbounded Function

Find the limit (if possible): $\lim_{x\to 2} \dfrac{3}{x-2}$.

STUDY TIP

The equal sign in the statement $\lim_{x\to c^+} f(x) = \infty$ does not mean that the limit exists. On the contrary, it tells you how the limit *fails to exist* by denoting the unbounded behavior of $f(x)$ as x approaches c.

SOLUTION From Figure 1.59, you can see that $f(x)$ decreases without bound as x approaches 2 from the left and $f(x)$ increases without bound as x approaches 2 from the right. Symbolically, you can write this as

$$\lim_{x\to 2^-} \frac{3}{x-2} = -\infty$$

and

$$\lim_{x\to 2^+} \frac{3}{x-2} = \infty.$$

Because f is unbounded as x approaches 2, the limit does not exist.

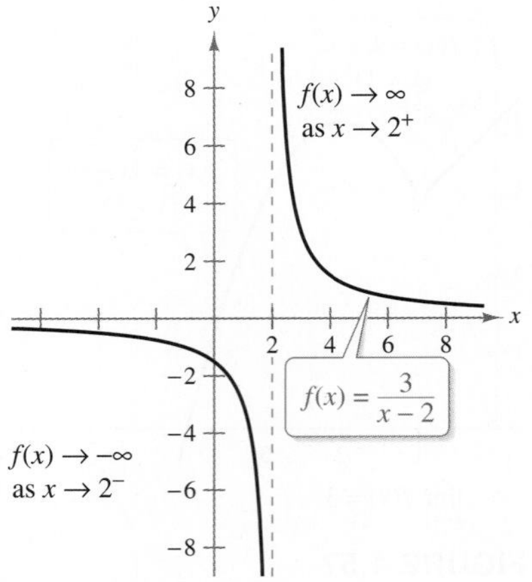

FIGURE 1.59

✓ Checkpoint 11

Find the limit (if possible): $\lim_{x\to -2} \dfrac{5}{x+2}$. ■

SUMMARIZE (Section 1.5)

1. State the definition of the limit of a function *(page 50)*. For examples of limits, see Examples 1 and 2.
2. Make a list of the basic limits *(page 50)*. For examples of the basic limits, see Example 3.
3. Make a list of the properties of limits *(page 51)*. For an example of the use of these properties, see Example 4.
4. State the limit of a polynomial function *(page 51)*. For an example of the limit of a polynomial function, see Example 4.
5. Describe the dividing out technique *(page 52)*. For examples of the dividing out technique, see Examples 5 and 6.
6. Describe the rationalizing technique *(page 53)*. For an example of the rationalizing technique, see Example 7.
7. Describe a one-sided limit *(page 54)*. For examples of one-sided limits, see Examples 8, 9, and 10.
8. Describe the limit $\lim_{x\to c} f(x)$ when $f(x)$ increases without bound as x approaches c *(page 56)*. For an example of an unbounded function, see Example 11.

holbox/www.shutterstock.com

SKILLS WARM UP 1.5 The following warm-up exercises involve skills that were covered in a previous course or in earlier sections. You will use these skills in the exercise set for this section. For additional help, review Appendix Section A.3, and Section 1.4.

In Exercises 1–4, simplify the expression by factoring.

1. $\dfrac{2x^3 + x^2}{6x}$ **2.** $\dfrac{x^5 + 9x^4}{x^2}$ **3.** $\dfrac{x^2 - 3x - 28}{x - 7}$ **4.** $\dfrac{x^2 + 11x + 30}{x + 5}$

In Exercises 5–8, evaluate the expression and simplify.

5. $f(x) = x^2 - 3x + 3$

(a) $f(-1)$ (b) $f(c)$ (c) $f(x + h)$

6. $f(x) = \begin{cases} 2x - 2, & x < 1 \\ 3x + 1, & x \geq 1 \end{cases}$

(a) $f(-1)$ (b) $f(3)$ (c) $f(t^2 + 1)$

7. $f(x) = x^2 - 2x + 2$ $\quad \dfrac{f(1 + h) - f(1)}{h}$

8. $f(x) = 4x$ $\quad \dfrac{f(2 + h) - f(2)}{h}$

In Exercises 9–12, find the domain and range of the function and sketch its graph.

9. $h(x) = -\dfrac{5}{x}$ **10.** $g(x) = \sqrt{25 - x^2}$ **11.** $f(x) = |x - 3|$ **12.** $f(x) = \dfrac{|x|}{x}$

In Exercises 13 and 14, determine whether y is a function of x.

13. $9x^2 + 4y^2 = 49$ **14.** $2x^2y + 8x = 7y$

Exercises 1.5

See www.CalcChat.com for worked-out solutions to odd-numbered exercises.

Finding Limits Graphically In Exercises 1–4, use the graph to find the limit. *See Examples 1 and 2.*

1.

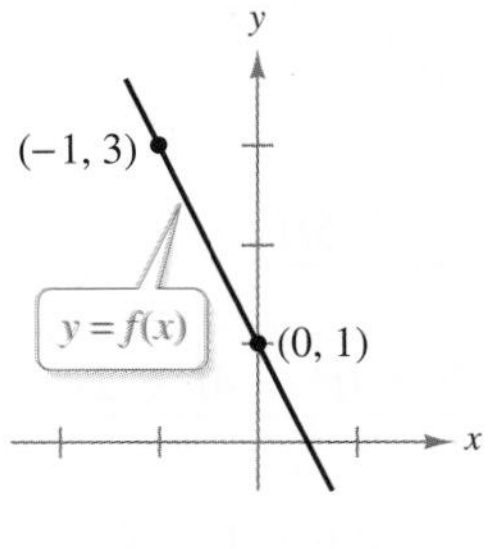

(a) $\lim_{x \to 0} f(x)$

(b) $\lim_{x \to -1} f(x)$

2.

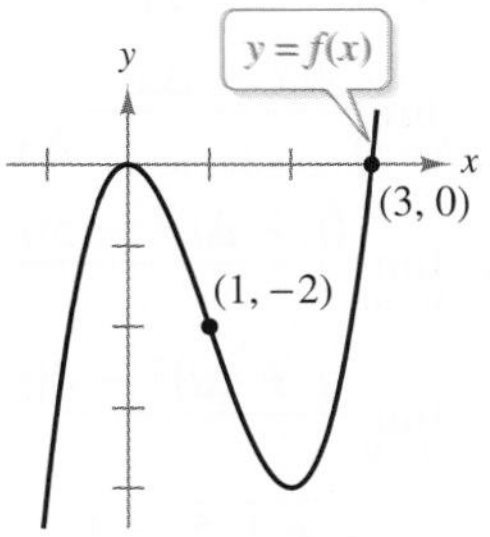

(a) $\lim_{x \to 1} f(x)$

(b) $\lim_{x \to 3} f(x)$

3.

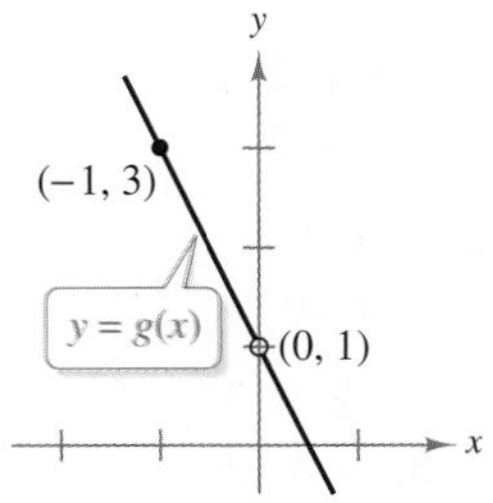

(a) $\lim_{x \to 0} g(x)$

(b) $\lim_{x \to -1} g(x)$

4.

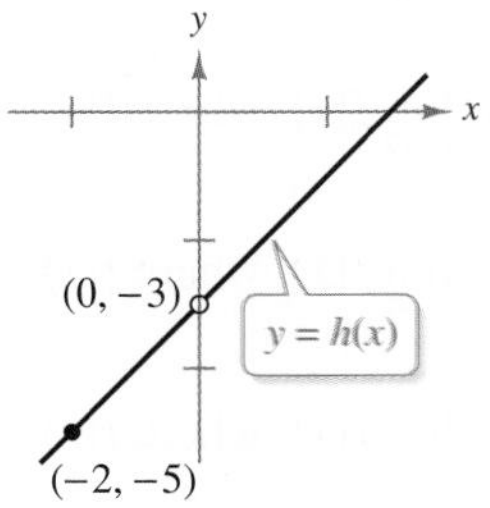

(a) $\lim_{x \to -2} h(x)$

(b) $\lim_{x \to 0} h(x)$

Finding Limits Numerically In Exercises 5–12, complete the table and use the result to estimate the limit. Use a graphing utility to graph the function to confirm your result. *See Examples 1 and 2.*

5. $\lim_{x \to 2} (2x + 5)$

x	1.9	1.99	1.999	2	2.001	2.01	2.1
$f(x)$				?			

6. $\lim_{x \to 2} (x^2 - 3x + 1)$

x	1.9	1.99	1.999	2	2.001	2.01	2.1
$f(x)$				?			

7. $\lim_{x \to 2} \dfrac{x - 2}{x^2 - 4}$

x	1.9	1.99	1.999	2	2.001	2.01	2.1
$f(x)$				?			

8. $\lim_{x\to 2} \dfrac{x-2}{x^2-3x+2}$

x	1.9	1.99	1.999	2	2.001	2.01	2.1
$f(x)$				?			

9. $\lim_{x\to 0} \dfrac{\sqrt{x+1}-1}{x}$

x	−0.1	−0.01	−0.001	0	0.001	0.01	0.1
$f(x)$				?			

10. $\lim_{x\to 0} \dfrac{\sqrt{x+2}-\sqrt{2}}{x}$

x	−0.1	−0.01	−0.001	0	0.001	0.01	0.1
$f(x)$				?			

11. $\lim_{x\to -4} \dfrac{\dfrac{1}{x+4}-\dfrac{1}{4}}{x}$

x	−4.1	−4.01	−4.001	−4	−3.999	−3.99	−3.9
$f(x)$				?			

12. $\lim_{x\to -2} \dfrac{\dfrac{1}{2}-\dfrac{1}{x+2}}{2x}$

x	−2.1	−2.01	−2.001	−2	−1.999	−1.99	−1.9
$f(x)$				?			

Evaluating Basic Limits In Exercises 13–20, find the limit. *See Example 3.*

13. $\lim_{x\to 3} 6$

14. $\lim_{x\to 5} 4$

15. $\lim_{x\to -2} x$

16. $\lim_{x\to 10} x$

17. $\lim_{x\to 7} x^2$

18. $\lim_{x\to 3} x^3$

19. $\lim_{x\to 16} \sqrt{x}$

20. $\lim_{x\to -1} \sqrt[3]{x}$

Operations with Limits In Exercises 21 and 22, find the limit of (a) $f(x)+g(x)$, (b) $f(x)g(x)$, and (c) $f(x)/g(x)$, as x approaches c.

21. $\lim_{x\to c} f(x) = 3$

$\lim_{x\to c} g(x) = 9$

22. $\lim_{x\to c} f(x) = \frac{3}{2}$

$\lim_{x\to c} g(x) = \frac{1}{2}$

Operations with Limits In Exercises 23 and 24, find the limit of (a) $\sqrt{f(x)}$, (b) $[3f(x)]$, and (c) $[f(x)]^2$, as x approaches c.

23. $\lim_{x\to c} f(x) = 16$

24. $\lim_{x\to c} f(x) = 9$

Using Properties of Limits In Exercises 25–36, find the limit using direct substitution. *See Examples 3 and 4.*

25. $\lim_{x\to -3} (2x+5)$

26. $\lim_{x\to 0} (3x-2)$

27. $\lim_{x\to 1} (1-x^2)$

28. $\lim_{x\to 2} (-x^2+x-2)$

29. $\lim_{x\to 3} \sqrt{x+6}$

30. $\lim_{x\to 4} \sqrt[3]{x+4}$

31. $\lim_{x\to -3} \dfrac{2}{x+2}$

32. $\lim_{x\to -2} \dfrac{3x+1}{2-x}$

33. $\lim_{x\to -2} \dfrac{x^2-1}{2x}$

34. $\lim_{x\to 7} \dfrac{5x}{x+2}$

35. $\lim_{x\to 5} \dfrac{\sqrt{x+11}+6}{x}$

36. $\lim_{x\to 12} \dfrac{\sqrt{x-3}-2}{x}$

Finding Limits In Exercises 37–58, find the limit (if it exists). *See Examples 5, 6, 7, 9, and 11.*

37. $\lim_{x\to -3} \dfrac{x^2-9}{x+3}$

38. $\lim_{x\to -1} \dfrac{2x^2-x-3}{x+1}$

39. $\lim_{x\to 2} \dfrac{2-x}{x^2-4}$

40. $\lim_{t\to 1} \dfrac{t^2+t-2}{t^2-1}$

41. $\lim_{x\to -2} \dfrac{x^3+8}{x+2}$

42. $\lim_{x\to -1} \dfrac{x^3+1}{x+1}$

43. $\lim_{\Delta x\to 0} \dfrac{2(x+\Delta x)-2x}{\Delta x}$

44. $\lim_{\Delta x\to 0} \dfrac{4(x+\Delta x)-5-(4x-5)}{\Delta x}$

45. $\lim_{\Delta t\to 0} \dfrac{(t+\Delta t)^2-5(t+\Delta t)-(t^2-5t)}{\Delta t}$

46. $\lim_{\Delta t\to 0} \dfrac{(t+\Delta t)^2-4(t+\Delta t)+2-(t^2-4t+2)}{\Delta t}$

47. $\lim_{x\to 4} \dfrac{\sqrt{x+5}-3}{x-4}$

48. $\lim_{x\to 3} \dfrac{\sqrt{x+1}-2}{x-3}$

49. $\lim_{x\to 0} \dfrac{\sqrt{x+5}-\sqrt{5}}{x}$

50. $\lim_{x\to 0} \dfrac{\sqrt{x+2}-\sqrt{2}}{x}$

51. $\lim_{x\to 2} f(x)$, where $f(x) = \begin{cases} 4-x, & x \neq 2 \\ 0, & x = 2 \end{cases}$

52. $\lim_{x\to 1} f(x)$, where $f(x) = \begin{cases} x^2+2, & x \neq 1 \\ 1, & x = 1 \end{cases}$

53. $\lim_{x\to 3} f(x)$, where $f(x) = \begin{cases} \frac{1}{3}x-2, & x \le 3 \\ -2x+5, & x > 3 \end{cases}$

54. $\lim_{s\to 1} f(s)$, where $f(s) = \begin{cases} s, & s \le 1 \\ 1-s, & s > 1 \end{cases}$

55. $\lim_{x \to -4} \frac{2}{x + 4}$

56. $\lim_{x \to 5} \frac{4}{x - 5}$

57. $\lim_{x \to 2} \frac{x - 2}{x^2 - 4x + 4}$

58. $\lim_{t \to 4} \frac{t + 4}{t^2 - 16}$

Finding One-Sided Limits In Exercises 59 and 60, use a graph to find the limit from the left and the limit from the right. *See Example 8.*

59. $\lim_{x \to -3^-} \frac{|x + 3|}{x + 3}$

$\lim_{x \to -3^+} \frac{|x + 3|}{x + 3}$

60. $\lim_{x \to 6^-} \frac{|x - 6|}{x - 6}$

$\lim_{x \to 6^+} \frac{|x - 6|}{x - 6}$

Graphical, Numerical, and Analytic Analysis In Exercises 61–64, use a graphing utility to graph the function and estimate the limit. Use a table to reinforce your conclusion. Then find the limit by analytic methods.

61. $\lim_{x \to 1^-} \frac{2}{x^2 - 1}$

62. $\lim_{x \to 1^+} \frac{5}{1 - x}$

63. $\lim_{x \to -2^-} \frac{1}{x + 2}$

64. $\lim_{x \to 0^-} \frac{x + 1}{x}$

Finding Limits Graphically In Exercises 65–70, use the graph to find the limit (if it exists).

(a) $\lim_{x \to c^+} f(x)$ (b) $\lim_{x \to c^-} f(x)$ (c) $\lim_{x \to c} f(x)$

65.

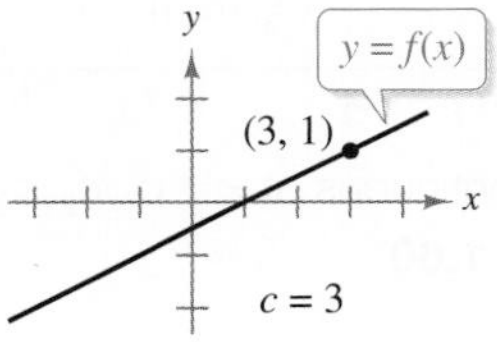

66.

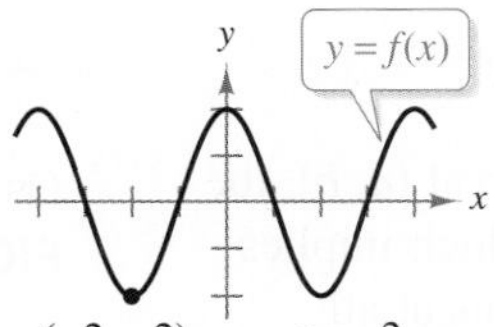

67.

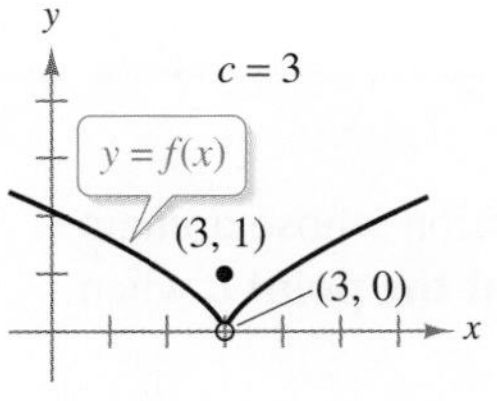

68.

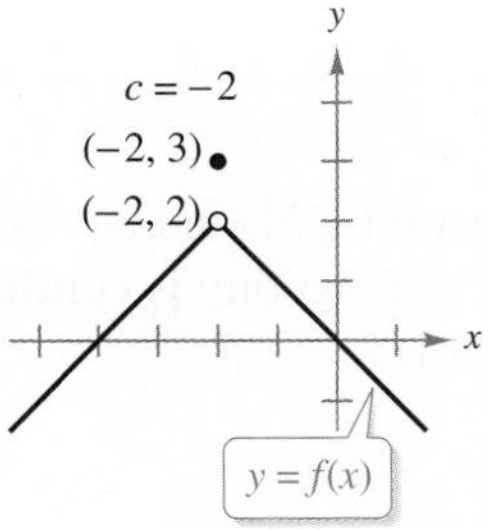

69.

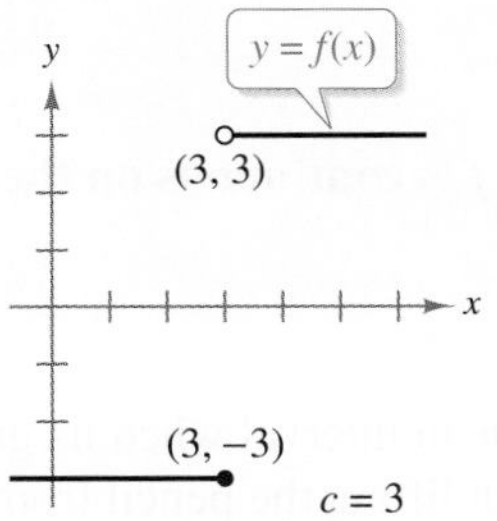

70.

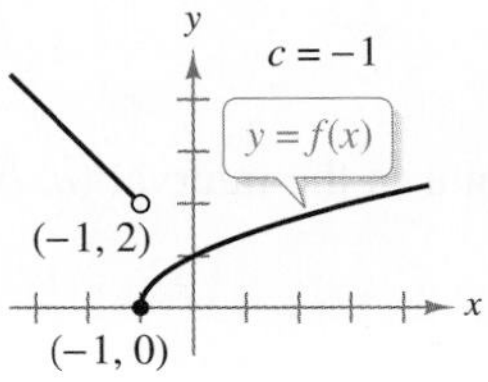

Estimating Limits In Exercises 71–74, use a graphing utility to estimate the limit (if it exists).

71. $\lim_{x \to 2} \frac{x^2 - 5x + 6}{x^2 - 4x + 4}$

72. $\lim_{x \to 1} \frac{x^2 + 6x - 7}{x^3 - x^2 + 2x - 2}$

73. $\lim_{x \to -4} \frac{x^3 + 4x^2 + x + 4}{2x^2 + 7x - 4}$

74. $\lim_{x \to -2} \frac{4x^3 + 7x^2 + x + 6}{3x^2 - x - 14}$

75. **Environment** The cost C (in thousands of dollars) of removing $p\%$ of the pollutants from the water in a small lake is given by

$$C = \frac{25p}{100 - p}, \quad 0 \le p < 100.$$

(a) Find the cost of removing 50% of the pollutants.

(b) What percent of the pollutants can be removed for $100 thousand?

(c) Evaluate $\lim_{p \to 100^-} C$. Explain your results.

76. **HOW DO YOU SEE IT?** The graph shows the cost C (in dollars) of making x photocopies at a copy shop.

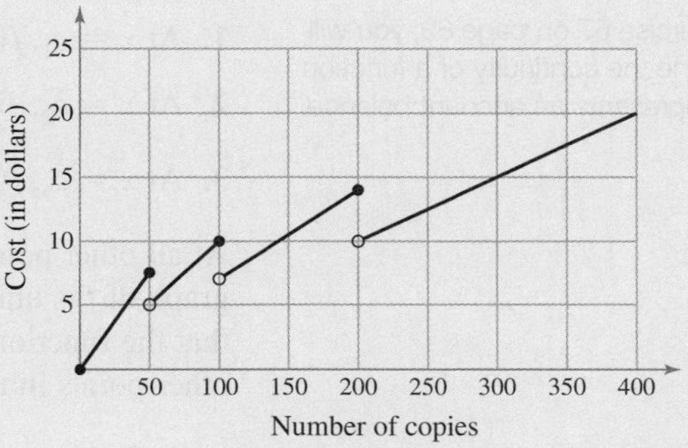

(a) Does $\lim_{x \to 50} C$ exist? Explain your reasoning.

(b) Does $\lim_{x \to 150} C$ exist? Explain your reasoning.

(c) You have to make 200 photocopies. Would it be better to make 200 or 201? Explain your reasoning.

77. **Compound Interest** Consider a certificate of deposit that pays 10% (annual percentage rate) on an initial deposit of $1000. The balance A after 10 years is $A = 1000(1 + 0.1x)^{10/x}$, where x is the length of the compounding period (in years).

(a) Use a graphing utility to graph A, where $0 \le x \le 1$.

(b) Use the *zoom* and *trace* features to estimate the balance for quarterly compounding and daily compounding.

(c) Use the *zoom* and *trace* features to estimate $\lim_{x \to 0^+} A$. What do you think this limit represents? Explain your reasoning.

1.6 Continuity

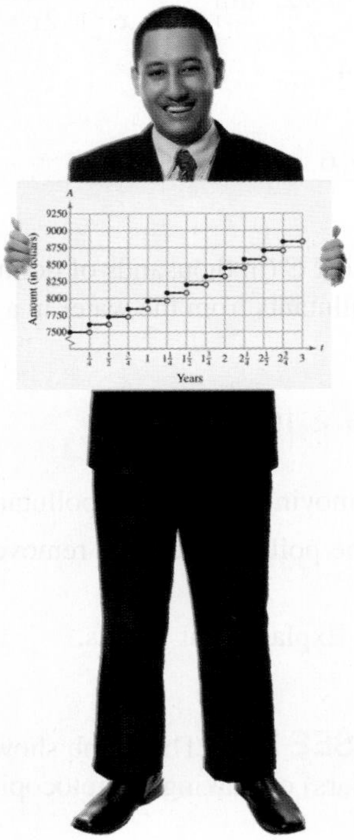

In Exercise 67 on page 69, you will examine the continuity of a function that represents an account balance.

- Determine the continuity of functions.
- Determine the continuity of functions on a closed interval.
- Use the greatest integer function to model and solve real-life problems.
- Use compound interest models to solve real-life problems.

Continuity

In mathematics, the term "continuous" has much the same meaning as it has in everyday use. To say that a function is continuous at

$$x = c$$

means that there is no interruption in the graph of f at c. That is, the graph of f

1. is unbroken at c.
2. has no holes, jumps, or gaps.

As simple as this concept may seem, its precise definition eluded mathematicians for many years. In fact, it was not until the early 1800's that a precise definition was finally developed.

Before looking at this definition, consider the function whose graph is shown in Figure 1.60. This figure identifies three values of x at which the function f is not continuous.

1. At $x = c_1$, $f(c_1)$ is not defined.
2. At $x = c_2$, $\lim_{x \to c_2} f(x)$ does not exist.
3. At $x = c_3$, $f(c_3) \neq \lim_{x \to c_3} f(x)$.

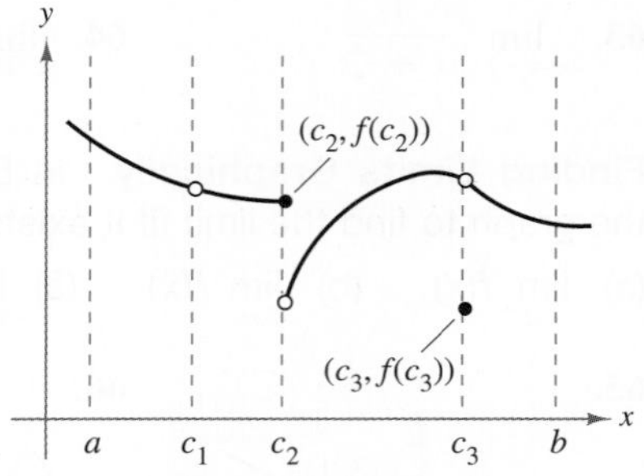

f is not continuous at $x = c_1, c_2, c_3$.

FIGURE 1.60

At all other points in the interval (a, b), the graph of f is uninterrupted, which implies that the function f is continuous at all other points in the interval (a, b).

Definition of Continuity

Let c be a number in the interval (a, b), and let f be a function whose domain contains the interval (a, b). The function f is **continuous at the point c** when the following conditions are true.

1. $f(c)$ is defined.
2. $\lim_{x \to c} f(x)$ exists.
3. $\lim_{x \to c} f(x) = f(c)$.

If f is continuous at every point in the interval (a, b), then f is **continuous on the open interval (a, b).**

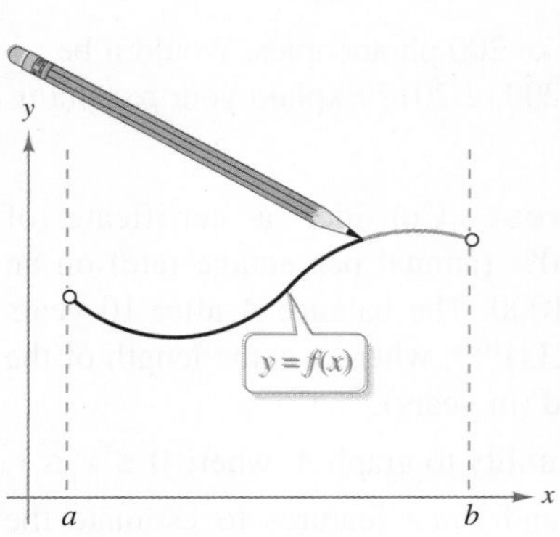

On the interval (a, b), the graph of f can be traced with a pencil.

FIGURE 1.61

Informally, you can say that a function is continuous on an interval when its graph on the interval can be traced using a pencil and paper without lifting the pencil from the paper, as shown in Figure 1.61.

TECH TUTOR

Most graphing utilities can draw graphs in two different modes: *connected mode* and *dot mode*. The *connected mode* works well as long as the function is continuous on the entire interval represented by the viewing window. If, however, the function is not continuous at one or more x-values in the viewing window, then the *connected mode* may try to "connect" parts of the graph that should not be connected. For instance, try graphing the function $y_1 = (x + 3)/(x - 2)$ in the viewing window $-8 \le x \le 8$ and $-6 \le y \le 6$ in *connected mode* and then in *dot mode*.

Continuity of Polynomial and Rational Functions

1. A polynomial function is continuous at every real number.
2. A rational function is continuous at every number in its domain.

Example 1 Determining Continuity of a Polynomial Function

Discuss the continuity of each function.

a. $f(x) = x^2 - 2x + 3$ **b.** $f(x) = x^3 - x$ **c.** $f(x) = x^4 - 2x^2 + 1$

SOLUTION Each of these functions is a *polynomial function*. So, each is continuous on the entire real number line, as indicated in Figure 1.62.

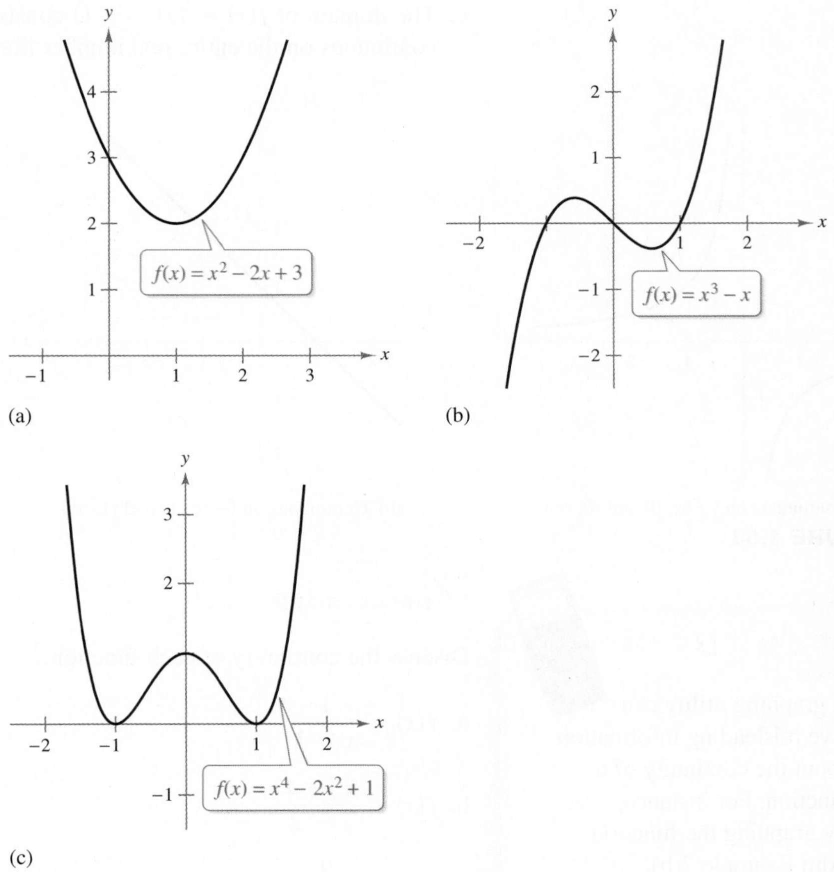

All three functions are continuous on $(-\infty, \infty)$.

FIGURE 1.62

✓ Checkpoint 1

Discuss the continuity of each function.

a. $f(x) = x^2 + x + 1$ **b.** $f(x) = x^3 + x$ **c.** $f(x) = x^4$ ■

Polynomial functions are one of the most important types of functions used in calculus. Be sure you see from Example 1 that the graph of a polynomial function is continuous on the entire real number line and therefore has no holes, jumps, or gaps. Rational functions, on the other hand, need not be continuous on the entire real number line, as shown in Example 2.

Example 2 Determining Continuity of a Rational Function

Discuss the continuity of each function.

a. $f(x) = \dfrac{1}{x}$ **b.** $f(x) = \dfrac{x^2 - 1}{x - 1}$ **c.** $f(x) = \dfrac{1}{x^2 + 1}$

SOLUTION Each of these functions is a rational function and is therefore continuous at every number in its domain.

a. The domain of $f(x) = 1/x$ consists of all real numbers except $x = 0$. So, this function is continuous on the intervals $(-\infty, 0)$ and $(0, \infty)$. [See Figure 1.63(a).]

b. The domain of $f(x) = (x^2 - 1)/(x - 1)$ consists of all real numbers except $x = 1$. So, this function is continuous on the intervals $(-\infty, 1)$ and $(1, \infty)$. [See Figure 1.63(b).]

c. The domain of $f(x) = 1/(x^2 + 1)$ consists of all real numbers. So, this function is continuous on the entire real number line. [See Figure 1.63(c).]

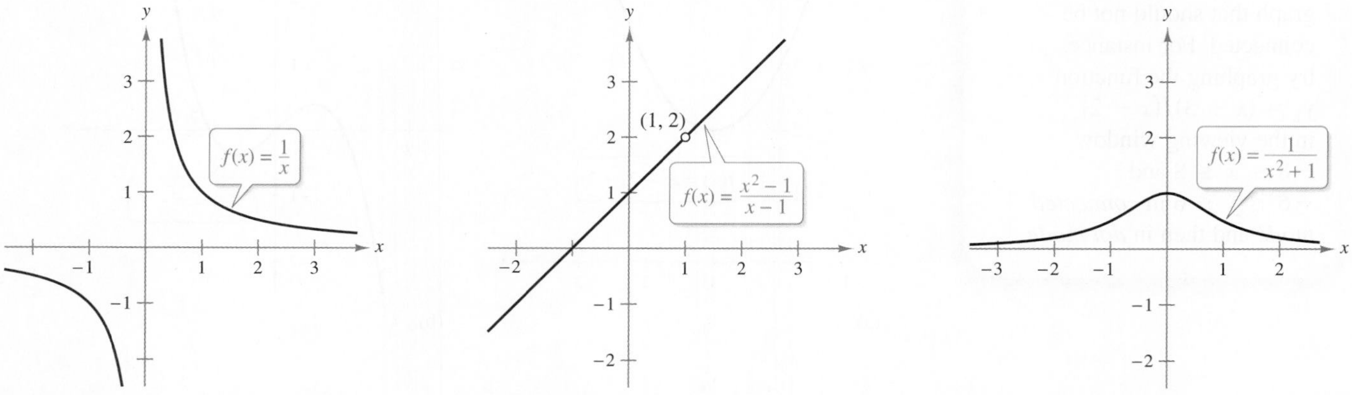

(a) Continuous on $(-\infty, 0)$ and $(0, \infty)$ (b) Continuous on $(-\infty, 1)$ and $(1, \infty)$ (c) Continuous on $(-\infty, \infty)$

FIGURE 1.63

✓Checkpoint 2

Discuss the continuity of each function.

a. $f(x) = \dfrac{1}{x - 1}$

b. $f(x) = \dfrac{x^2 - 4}{x - 2}$

c. $f(x) = \dfrac{1}{x^2 + 2}$

TECH TUTOR

A graphing utility can give misleading information about the continuity of a function. For instance, try graphing the function from Example 2(b), $f(x) = (x^2 - 1)/(x - 1)$, in a standard viewing window. On most graphing utilities, the graph appears to be continuous at every real number. However, because $x = 1$ is not in the domain of f, you know that f is not continuous at $x = 1$. You can verify this on a graphing utility using the *trace* or *table* feature.

Consider an open interval I that contains a real number c. If a function f is defined on I (except possibly at c), and f is not continuous at c, then f is said to have a **discontinuity** at c. Discontinuities fall into two categories: **removable** and **nonremovable.** A discontinuity at c is called removable when f can be made continuous by appropriately defining (or redefining) $f(c)$. For instance, the function in Example 2(b) has a removable discontinuity at $(1, 2)$. To remove the discontinuity, all you need to do is redefine the function so that $f(1) = 2$.

A discontinuity at $x = c$ is nonremovable when the function cannot be made continuous at $x = c$ by defining or redefining the function at $x = c$. For instance, the function in Example 2(a) has a nonremovable discontinuity at $x = 0$.

Continuity on a Closed Interval

The intervals discussed in Examples 1 and 2 are open. To discuss continuity on a closed interval, you can use the concept of one-sided limits, as defined in Section 1.5.

Definition of Continuity on a Closed Interval

Let f be defined on a closed interval $[a, b]$. If f is continuous on the open interval (a, b) and

$$\lim_{x\to a^+} f(x) = f(a) \quad \text{and} \quad \lim_{x\to b^-} f(x) = f(b)$$

then f is **continuous on the closed interval $[a, b]$.** Moreover, f is **continuous from the right** at a and **continuous from the left** at b.

Similar definitions can be made to cover continuity on intervals of the form $(a, b]$ and $[a, b)$, or on infinite intervals. For instance, the function

$$f(x) = \sqrt{x}$$

is continuous on the infinite interval $[0, \infty)$.

Example 3 Examining Continuity at an Endpoint

Discuss the continuity of

$$f(x) = \sqrt{3 - x}.$$

SOLUTION Notice that the domain of f is the set $(-\infty, 3]$. Moreover, f is continuous from the left at $x = 3$ because

$$\begin{aligned}\lim_{x\to 3^-} f(x) &= \lim_{x\to 3^-} \sqrt{3 - x} \\ &= \sqrt{3 - 3} \\ &= 0 \\ &= f(3).\end{aligned}$$

For all $x < 3$, the function f satisfies the three conditions for continuity. So, you can conclude that f is continuous on the interval $(-\infty, 3]$, as shown in Figure 1.64.

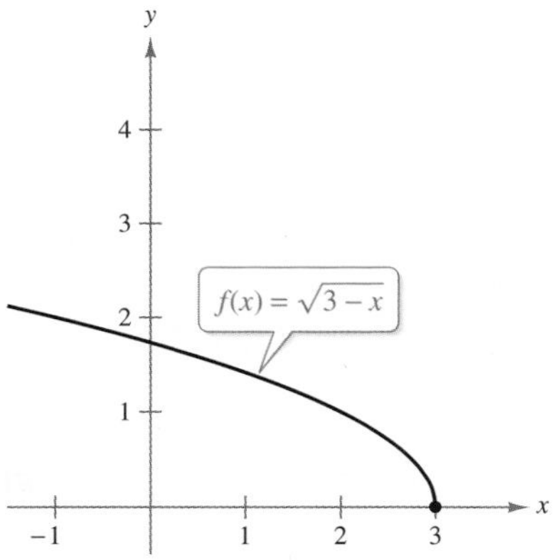

FIGURE 1.64

STUDY TIP

When working with radical functions of the form

$$f(x) = \sqrt{g(x)}$$

remember that the domain of f coincides with the solution of $g(x) \geq 0$.

Checkpoint 3

Discuss the continuity of

$$f(x) = \sqrt{x - 2}.$$

■

cristovao/Shutterstock.com

Example 4 Examining Continuity on a Closed Interval

Discuss the continuity of

$$g(x) = \begin{cases} 5 - x, & -1 \le x \le 2 \\ x^2 - 1, & 2 < x \le 3 \end{cases}.$$

SOLUTION The polynomial functions

$$5 - x$$

and

$$x^2 - 1$$

are continuous on the intervals $[-1, 2]$ and $(2, 3]$, respectively. So, to conclude that g is continuous on the entire interval

$$[-1, 3]$$

you need only to check the behavior of g when $x = 2$. You can do this by taking the one-sided limits when $x = 2$.

$$\lim_{x \to 2^-} g(x) = \lim_{x \to 2^-} (5 - x) = 5 - 2 = 3 \qquad \text{Limit from the left}$$

and

$$\lim_{x \to 2^+} g(x) = \lim_{x \to 2^+} (x^2 - 1) = 2^2 - 1 = 3 \qquad \text{Limit from the right}$$

Because these two limits are equal,

$$\lim_{x \to 2} g(x) = g(2) = 3.$$

So, g is continuous at $x = 2$ and, consequently, it is continuous on the entire interval

$$[-1, 3].$$

The graph of g is shown in Figure 1.65.

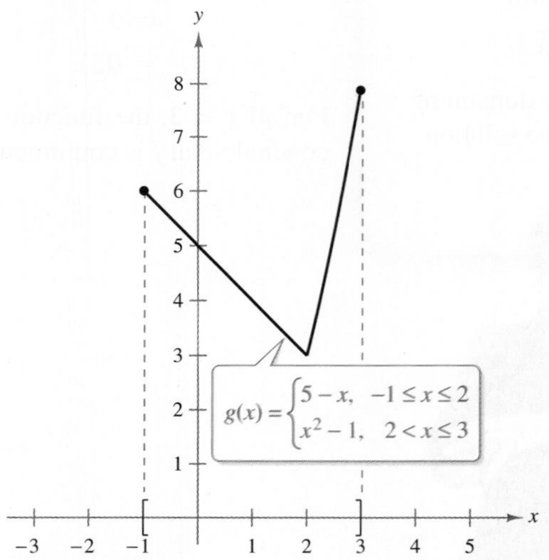

FIGURE 1.65

✓Checkpoint 4

Discuss the continuity of

$$f(x) = \begin{cases} x + 2, & -1 \le x < 3 \\ 14 - x^2, & 3 \le x \le 5 \end{cases}.$$

■

The Greatest Integer Function

Some functions that are used in business applications are **step functions.** For instance, the function in Example 10 in Section 1.5 is a step function. The **greatest integer function** is another example of a step function. This function is denoted by

$$[\![x]\!] = \text{greatest integer less than or equal to } x.$$

For example,

$$[\![-2.1]\!] = \text{greatest integer less than or equal to } -2.1 = -3$$
$$[\![-2]\!] = \text{greatest integer less than or equal to } -2 = -2$$
$$[\![1.5]\!] = \text{greatest integer less than or equal to } 1.5 = 1.$$

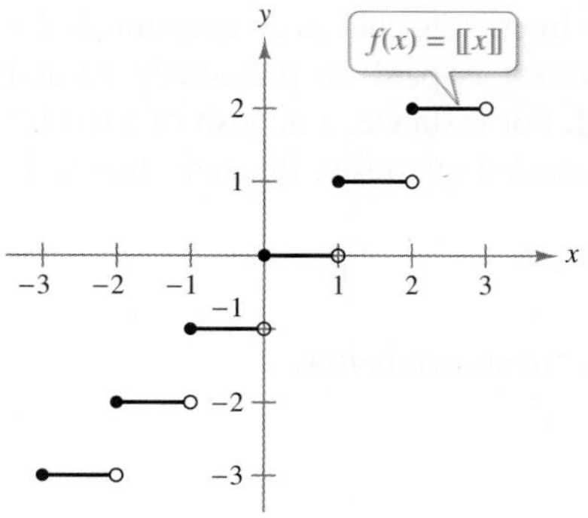

Greatest Integer Function
FIGURE 1.66

Note that the graph of the greatest integer function (Figure 1.66) jumps up one unit at each integer. This implies that the function is not continuous at each integer.

In real-life applications, the domain of the greatest integer function is often restricted to nonnegative values of x. In such cases, this function serves the purpose of **truncating** the decimal portion of x. For example, 1.345 is truncated to 1 and 3.57 is truncated to 3. That is,

$$[\![1.345]\!] = 1 \quad \text{and} \quad [\![3.57]\!] = 3.$$

Example 5 Modeling a Cost Function

A bookbinding company produces 10,000 books in an eight-hour shift. The fixed cost *per shift* amounts to \$5000, and the unit cost per book is \$3. Using the greatest integer function, you can write the cost of producing x books as

$$C = 5000\left(1 + \left[\!\!\left[\frac{x-1}{10{,}000}\right]\!\!\right]\right) + 3x.$$

Sketch the graph of this cost function.

SOLUTION Note that during the first eight-hour shift,

$$\left[\!\!\left[\frac{x-1}{10{,}000}\right]\!\!\right] = 0, \quad 1 \le x \le 10{,}000$$

which implies

$$C = 5000\left(1 + \left[\!\!\left[\frac{x-1}{10{,}000}\right]\!\!\right]\right) + 3x = 5000 + 3x.$$

During the second eight-hour shift,

$$\left[\!\!\left[\frac{x-1}{10{,}000}\right]\!\!\right] = 1, \quad 10{,}001 \le x \le 20{,}000$$

which implies

$$C = 5000\left(1 + \left[\!\!\left[\frac{x-1}{10{,}000}\right]\!\!\right]\right) + 3x$$
$$= 10{,}000 + 3x.$$

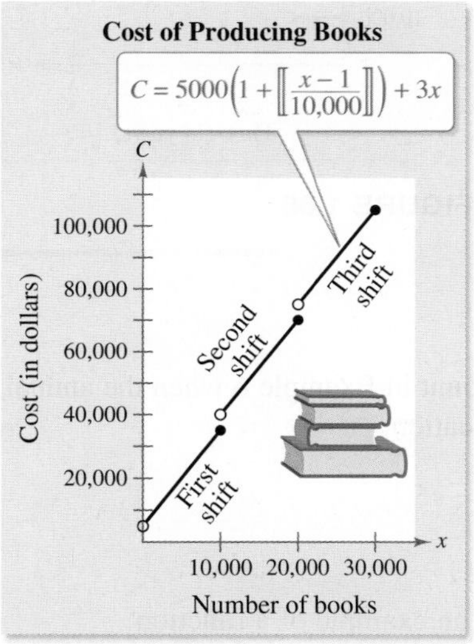

FIGURE 1.67

The graph of C is shown in Figure 1.67. Note the graph's discontinuities at $x = 10{,}000$, 20,000, and 30,000.

✓ Checkpoint 5

Use a graphing utility to graph the cost function in Example 5. ■

TECH TUTOR

You can use a spreadsheet or the *table* feature of a graphing utility to create a table. Try doing this for the data shown at the right. (Consult the user's manual of a spreadsheet software program for specific instructions on how to create a table.)

Extended Application: Compound Interest

Banks and other financial institutions differ on how interest is paid to an account. If the interest is added to the account so that future interest is paid on previously earned interest, then the interest is said to be **compounded.** For instance, a deposit of \$10,000 is made in an account that pays 6% interest, compounded quarterly. Because the 6% is the annual interest rate, the quarterly rate is

$$\tfrac{1}{4}(0.06) = 0.015$$

or 1.5%. The balances during the first five quarters are shown below.

Quarter	*Balance*
1st	\$10,000.00
2nd	$10{,}000.00 + (0.015)(10{,}000.00) = \$10{,}150.00$
3rd	$10{,}150.00 + (0.015)(10{,}150.00) = \$10{,}302.25$
4th	$10{,}302.25 + (0.015)(10{,}302.25) = \$10{,}456.78$
5th	$10{,}456.78 + (0.015)(10{,}456.78) = \$10{,}613.63$

Example 6 Graphing Compound Interest

Sketch the graph of the balance in the account described above.

SOLUTION Let A represent the balance in the account and let t represent the time, in years. You can use the greatest integer function to represent the balance, as shown.

$$A = 10{,}000(1 + 0.015)^{[\![4t]\!]}$$

From the graph shown in Figure 1.68, notice that the function has a discontinuity at each quarter. That is, A has discontinuities at

$$t = \frac{1}{4}, t = \frac{1}{2}, t = \frac{3}{4}, t = 1,$$

and

$$t = \frac{5}{4}.$$

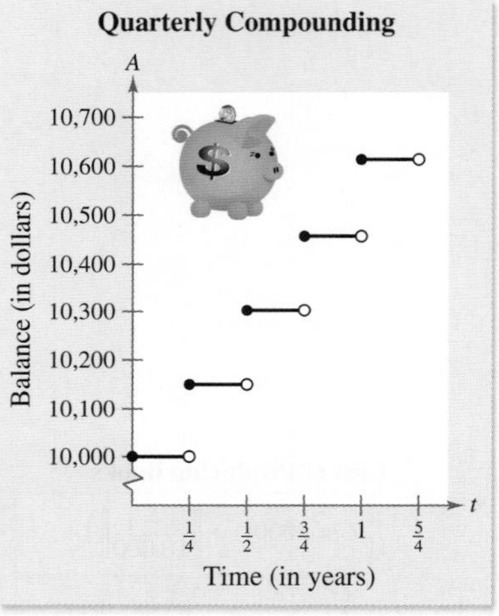

FIGURE 1.68

✓Checkpoint 6

Write an equation that gives the balance of the account in Example 6 when the annual interest rate is 3%. Then sketch the graph of the equation. ■

SUMMARIZE (Section 1.6)

1. State the definition of continuity *(page 60)*. For an example of a function that is continuous at every real number, see Example 1.
2. State the definition of continuity on a closed interval *(page 63)*. For an example of a function that is continuous on a closed interval, see Example 4.
3. State the definition of the greatest integer function *(page 65)*. For real-life examples of the greatest integer function, see Examples 5 and 6.

SKILLS WARM UP 1.6

The following warm-up exercises involve skills that were covered in a previous course or in earlier sections. You will use these skills in the exercise set for this section. For additional help, review Appendix Sections A.4 and A.5, and Section 1.5.

In Exercises 1–4, simplify the expression.

1. $\dfrac{x^2 + 6x + 8}{x^2 - 6x - 16}$

2. $\dfrac{x^2 - 5x - 6}{x^2 - 9x + 18}$

3. $\dfrac{2x^2 - 2x - 12}{4x^2 - 24x + 36}$

4. $\dfrac{x^3 - 16x}{x^3 + 2x^2 - 8x}$

In Exercises 5–8, solve for x.

5. $x^2 + 7x = 0$

6. $x^2 + 4x - 5 = 0$

7. $3x^2 + 8x + 4 = 0$

8. $x^3 + 5x^2 - 24x = 0$

In Exercises 9 and 10, find the limit.

9. $\lim_{x \to 3} (2x^2 - 3x + 4)$

10. $\lim_{x \to -2} (3x^3 - 8x + 7)$

Exercises 1.6

See www.CalcChat.com for worked-out solutions to odd-numbered exercises.

Determining Continuity In Exercises 1–10, determine whether the function is continuous on the entire real number line. Explain your reasoning. *See Examples 1 and 2.*

1. $f(x) = 5x^3 - x^2 + 2$

2. $f(x) = (x^2 - 1)^3$

3. $f(x) = \dfrac{3}{x^2 - 16}$

4. $f(x) = \dfrac{1}{9 - x^2}$

5. $f(x) = \dfrac{1}{4 + x^2}$

6. $f(x) = \dfrac{3x}{x^2 + 1}$

7. $f(x) = \dfrac{2x - 1}{x^2 - 8x + 15}$

8. $f(x) = \dfrac{x + 4}{x^2 - 6x + 5}$

9. $g(x) = \dfrac{x^2 - 4x + 4}{x^2 - 4}$

10. $g(x) = \dfrac{x^2 - 9x + 20}{x^2 - 16}$

Determining Continuity In Exercises 11–40, describe the interval(s) on which the function is continuous. Explain why the function is continuous on the interval(s). If the function has a discontinuity, identify the conditions of continuity that are not satisfied. *See Examples 1, 2, 3, 4, and 5.*

11. $f(x) = \dfrac{x^2 - 1}{x}$

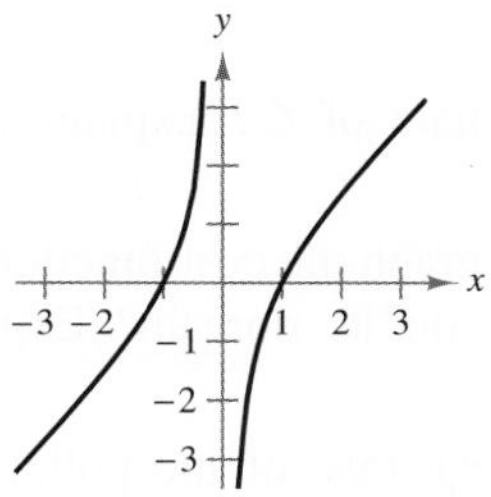

12. $f(x) = \dfrac{1}{x^2 - 4}$

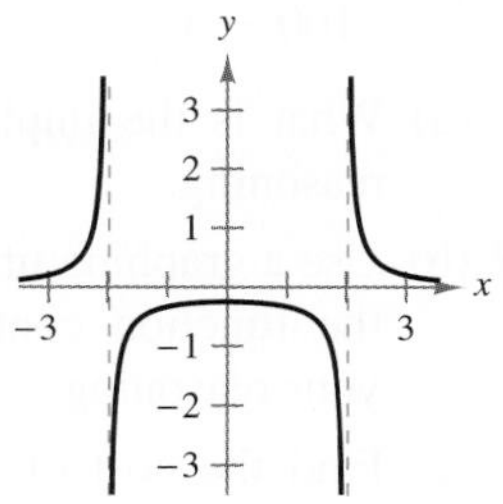

13. $f(x) = \dfrac{x^2 - 1}{x + 1}$

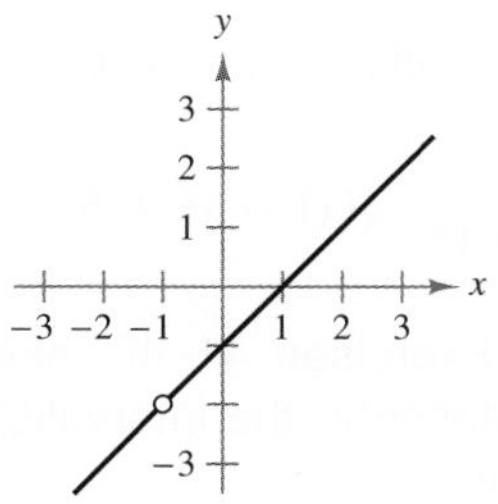

14. $f(x) = \dfrac{x^3 - 8}{x - 2}$

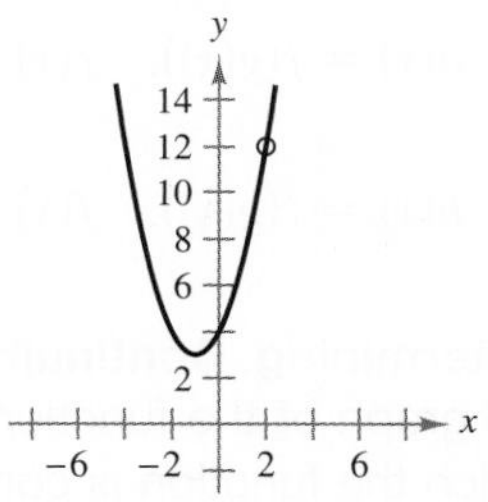

15. $f(x) = x^2 - 2x + 1$

16. $f(x) = 3 - 2x - x^2$

17. $f(x) = \dfrac{x}{x^2 - 1}$

18. $f(x) = \dfrac{x - 3}{x^2 - 9}$

19. $f(x) = \dfrac{x}{x^2 + 1}$

20. $f(x) = \dfrac{6}{x^2 + 3}$

21. $f(x) = \dfrac{x - 5}{x^2 - 9x + 20}$

22. $f(x) = \dfrac{x - 1}{x^2 + x - 2}$

23. $f(x) = \sqrt{4 - x}$

24. $f(x) = \sqrt{x - 1}$

25. $f(x) = \sqrt{x} + 2$

26. $f(x) = 3 - \sqrt{x}$

27. $f(x) = \begin{cases} -2x + 3, & -1 \le x \le 1 \\ x^2, & 1 < x \le 3 \end{cases}$

28. $f(x) = \begin{cases} \frac{1}{2}x + 1, & -3 \le x \le 2 \\ 3 - x, & 2 < x \le 4 \end{cases}$

29. $f(x) = \begin{cases} 3 + x, & x \le 2 \\ x^2 + 1, & x > 2 \end{cases}$

30. $f(x) = \begin{cases} x^2 - 4, & x \le 0 \\ 3x + 1, & x > 0 \end{cases}$

31. $f(x) = \dfrac{|x + 1|}{x + 1}$

32. $f(x) = \dfrac{|4 - x|}{4 - x}$

33. $f(x) = x\sqrt{x + 3}$

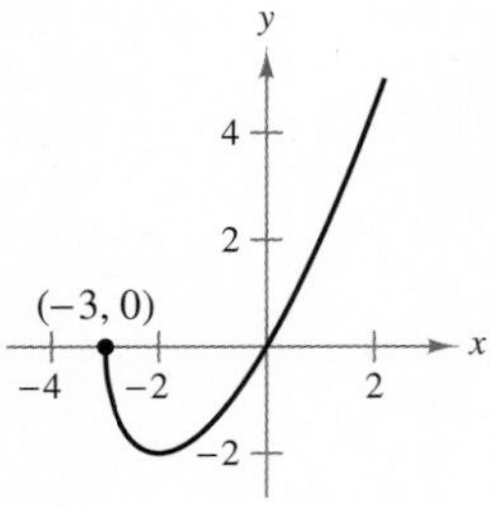

34. $f(x) = \dfrac{x + 1}{\sqrt{x}}$

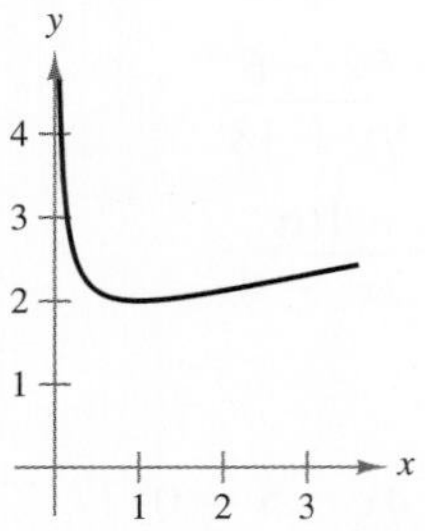

35. $f(x) = [\![2x]\!] + 1$

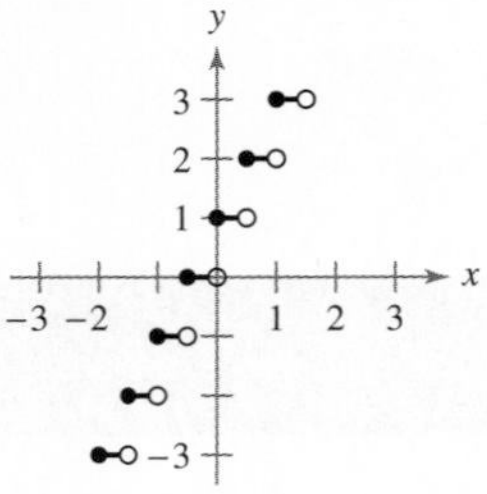

36. $f(x) = \dfrac{[\![x]\!]}{2} + x$

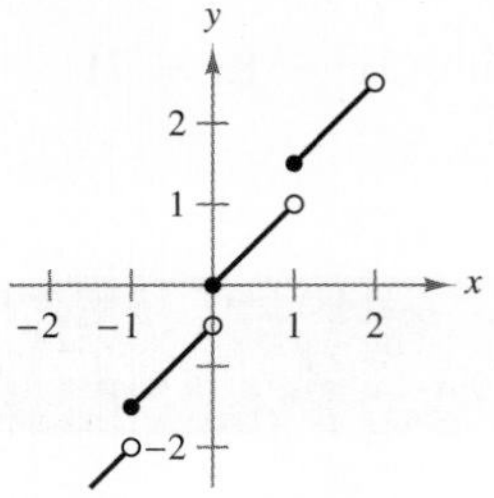

37. $f(x) = [\![x - 1]\!]$

38. $f(x) = x - [\![x]\!]$

39. $h(x) = f(g(x)),\quad f(x) = \dfrac{1}{\sqrt{x}},\quad g(x) = x - 1, x > 1$

40. $h(x) = f(g(x)),\quad f(x) = \dfrac{1}{x - 1},\quad g(x) = x^2 + 5$

Determining Continuity In Exercises 41–46, sketch the graph of the function and describe the interval(s) on which the function is continuous.

41. $f(x) = \dfrac{x^2 - 16}{x - 4}$

42. $f(x) = \dfrac{2x^2 + x}{x}$

43. $f(x) = \dfrac{x + 4}{3x^2 - 12}$

44. $f(x) = \dfrac{x^3 + x}{x}$

45. $f(x) = \begin{cases} x^2 + 1, & x < 0 \\ x - 1, & x \geq 0 \end{cases}$

46. $f(x) = \begin{cases} x^2 - 4, & x \leq 0 \\ 2x + 4, & x > 0 \end{cases}$

Determining Continuity on a Closed Interval In Exercises 47–50, discuss the continuity of the function on the closed interval. If there are any discontinuities, determine whether they are removable.

Function	*Interval*
47. $f(x) = x^2 - 4x - 5$	$[-1, 5]$
48. $f(x) = \dfrac{5}{x^2 + 1}$	$[-2, 2]$
49. $f(x) = \dfrac{1}{x - 2}$	$[1, 4]$
50. $f(x) = \dfrac{x}{x^2 - 4x + 3}$	$[0, 4]$

Finding Discontinuities In Exercises 51–56, use a graphing utility to graph the function. Use the graph to determine any x-value(s) at which the function is not continuous. Explain why the function is not continuous at the x-value(s).

51. $h(x) = \dfrac{1}{x^2 - x - 2}$

52. $k(x) = \dfrac{x - 4}{x^2 - 5x + 4}$

53. $f(x) = \begin{cases} 2x - 4, & x \leq 3 \\ x^2 - 2x, & x > 3 \end{cases}$

54. $f(x) = \begin{cases} 3x - 1, & x \leq 1 \\ x + 1, & x > 1 \end{cases}$

55. $f(x) = x - 2[\![x]\!]$

56. $f(x) = [\![2x - 1]\!]$

Making a Function Continuous In Exercises 57 and 58, find the constant a (Exercise 57) and the constants a and b (Exercise 58) such that the function is continuous on the entire real number line.

57. $f(x) = \begin{cases} x^3, & x \leq 2 \\ ax^2, & x > 2 \end{cases}$

58. $f(x) = \begin{cases} 2, & x \leq -1 \\ ax + b, & -1 < x < 3 \\ -2, & x \geq 3 \end{cases}$

Writing In Exercises 59 and 60, use a graphing utility to graph the function on the interval $[-4, 4]$. Does the graph of the function appear to be continuous on this interval? Is the function in fact continuous on $[-4, 4]$? Write a short paragraph about the importance of examining a function analytically as well as graphically.

59. $f(x) = \dfrac{x^2 + x}{x}$

60. $f(x) = \dfrac{x^3 - 8}{x - 2}$

61. Environmental Cost The cost C (in millions of dollars) of removing x percent of the pollutants emitted from the smokestack of a factory can be modeled by

$$C = \frac{2x}{100 - x}.$$

(a) What is the implied domain of C? Explain your reasoning.

(b) Use a graphing utility to graph the cost function. Is the function continuous on its domain? Explain your reasoning.

(c) Find the cost of removing 75% of the pollutants from the smokestack.

62. HOW DO YOU SEE IT? The graph shows the number of gallons G of gasoline in a person's car after t days.

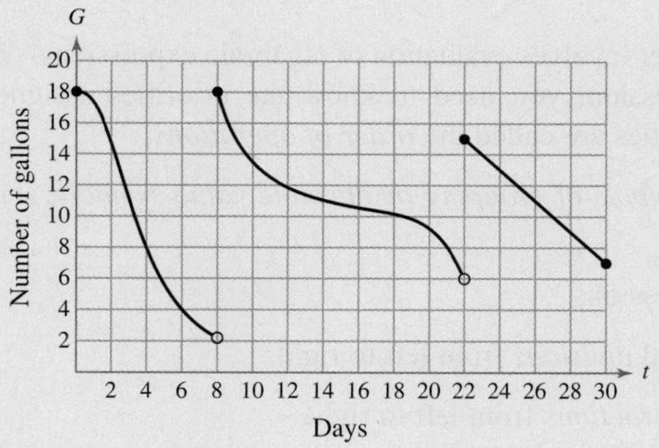

(a) On what days is the graph not continuous?

(b) What do you think happens on these days?

63. Biology The gestation period of rabbits is about 29 to 35 days. Therefore, the population of a form (rabbits' home) can increase dramatically in a short period of time. The table gives the population of a form, where t is the time in months and N is the rabbit population.

t	0	1	2	3	4	5	6
N	2	8	10	14	10	15	12

Graph the population as a function of time. Find any points of discontinuity in the function. Explain your reasoning.

64. Owning a Franchise You have purchased a franchise. You have determined a linear model for your revenue as a function of time. Is the model a continuous function? Would your actual revenue be a continuous function of time? Explain your reasoning.

65. Consumer Awareness The United States Postal Service first class mail rates for sending a letter are \$0.44 for the first ounce and \$0.20 for each additional ounce or fraction thereof up to 3.5 ounces. A model for the cost C (in dollars) of a first class mailing that weighs 3.5 ounces or less is given below. *(Source: United States Postal Service)*

$$C(x) = \begin{cases} 0.44, & 0 \le x \le 1 \\ 0.64, & 1 < x \le 2 \\ 0.84, & 2 < x \le 3 \\ 1.04, & 3 < x \le 3.5 \end{cases}$$

(a) Use a graphing utility to graph the function and then discuss its continuity. At what values is the function not continuous? Explain your reasoning.

(b) Find the cost of mailing a 2.5-ounce letter.

66. Health Food A co-op health food store charges \$3.50 for the first pound of organically grown peanuts and \$1.90 for each additional pound or fraction thereof.

(a) Use the greatest integer function to create a model for the cost C for x pounds of organically grown peanuts.

(b) Use a graphing utility to graph the function and then discuss its continuity.

67. Compound Interest A deposit of \$7500 is made in an account that pays 6% compounded quarterly. The amount A in the account after t years is

$$A = 7500(1.015)^{[\![4t]\!]}, \quad t \ge 0.$$

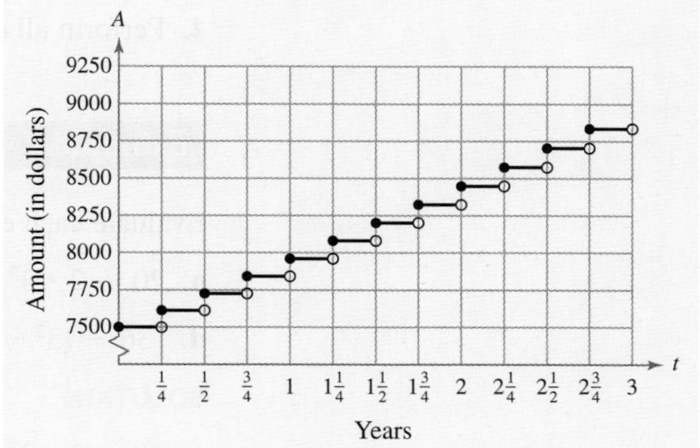

(a) Is the graph continuous? Explain your reasoning.

(b) What is the balance after 2 years?

(c) What is the balance after 7 years?

68. Salary Contract A union contract guarantees a 9% yearly increase for 5 years. For a current salary of \$28,500, the salaries for the next 5 years are given by

$$S = 28{,}500(1.09)^{[\![t]\!]}$$

where $t = 0$ represents the present year.

(a) Use the greatest integer function of a graphing utility to graph the salary function and then discuss its continuity.

(b) Find the salary during the fifth year (when $t = 5$).

69. Inventory Management The number of units in inventory in a small company is

$$N = 25\left(2\left[\!\!\left[\frac{t+2}{2}\right]\!\!\right] - t\right), \quad 0 \le t \le 12$$

where the real number t is the time in months.

(a) Use the greatest integer function of a graphing utility to graph this function and then discuss its continuity.

(b) How often must the company replenish its inventory?

ALGEBRA TUTOR

Order of Operations

Much of the algebra in this chapter involves evaluation of algebraic expressions. When you evaluate an algebraic expression, you need to know the priorities assigned to different operations. These priorities are called the *order of operations*.

1. Perform operations inside *symbols of grouping or absolute value symbols,* starting with the innermost symbol.
2. Evaluate all *exponential* expressions.
3. Perform all *multiplications* and *divisions* from left to right.
4. Perform all *additions* and *subtractions* from left to right.

Example 1 Using Order of Operations

Evaluate each expression.

a. $20 - 2 \cdot 3^2$ **b.** $3 + 8 \div 2 \cdot 2$ **c.** $7 - [(5 \cdot 3) + 2^3]$

d. $[36 \div (3^2 \cdot 2)] + 6$ **e.** $36 - [3^2 \cdot (2 \div 6)]$ **f.** $10 - 2(8 + |5 - 7|)$

SOLUTION

a.

$20 - 2 \cdot 3^2 = 20 - 2 \cdot 9$	Evaluate exponential expression.
$= 20 - 18$	Multiply.
$= 2$	Subtract.

b.

$3 + 8 \div 2 \cdot 2 = 3 + 4 \cdot 2$	Divide.
$= 3 + 8$	Multiply.
$= 11$	Add.

c.

$7 - [(5 \cdot 3) + 2^3] = 7 - [15 + 2^3]$	Multiply inside parentheses.
$= 7 - [15 + 8]$	Evaluate exponential expression.
$= 7 - 23$	Add inside brackets.
$= -16$	Subtract.

d.

$[36 \div (3^2 \cdot 2)] + 6 = [36 \div (9 \cdot 2)] + 6$	Evaluate exponential expression inside parentheses.
$= [36 \div 18] + 6$	Multiply inside parentheses.
$= 2 + 6$	Divide inside brackets.
$= 8$	Add.

e.

$36 - [3^2 \cdot (2 \div 6)] = 36 - \left[3^2 \cdot \frac{1}{3}\right]$	Divide inside parentheses.
$= 36 - \left[9 \cdot \frac{1}{3}\right]$	Evaluate exponential expression.
$= 36 - 3$	Multiply inside brackets.
$= 33$	Subtract.

f.

$10 - 2(8 + \lvert 5 - 7 \rvert) = 10 - 2(8 + \lvert -2 \rvert)$	Subtract inside absolute value symbols.
$= 10 - 2(8 + 2)$	Evaluate absolute value.
$= 10 - 2(10)$	Add inside parentheses.
$= 10 - 20$	Multiply.
$= -10$	Subtract.

TECH TUTOR

Most scientific and graphing calculators use the same order of operations listed above. Try entering the expressions in Example 1 into your calculator. Do you get the same results?

Solving Equations

A second algebraic skill used in this chapter is solving an equation in one variable.

1. To solve a *linear equation,* you can add or subtract the same quantity to or from each side of the equation. You can also multiply or divide each side of the equation by the same *nonzero* quantity.
2. To solve a *quadratic equation,* you can take the square root of each side, use factoring, or use the Quadratic Formula.
3. To solve a *radical equation,* isolate the radical on one side of the equation and square each side of the equation.
4. To solve an *absolute value equation,* use the definition of absolute value to rewrite the equation as two equations.

Example 2 Solving Equations

Solve each equation.

a. $3x - 3 = 5x - 7$ **b.** $2x^2 = 10$

c. $2x^2 + 5x - 6 = 6$ **d.** $\sqrt{2x - 7} = 5$

e. $|3x + 6| = 9$

SOLUTION

a. $3x - 3 = 5x - 7$ Write original (linear) equation.

$-3 = 2x - 7$ Subtract $3x$ from each side.

$4 = 2x$ Add 7 to each side.

$2 = x$ Divide each side by 2.

b. $2x^2 = 10$ Write original (quadratic) equation.

$x^2 = 5$ Divide each side by 2.

$x = \pm\sqrt{5}$ Take the square root of each side.

c. $2x^2 + 5x - 6 = 6$ Write original (quadratic) equation.

$2x^2 + 5x - 12 = 0$ Write in general form.

$(2x - 3)(x + 4) = 0$ Factor.

$2x - 3 = 0 \Rightarrow x = \frac{3}{2}$ Set first factor equal to zero.

$x + 4 = 0 \Rightarrow x = -4$ Set second factor equal to zero.

d. $\sqrt{2x - 7} = 5$ Write original (radical) equation.

$2x - 7 = 25$ Square each side.

$2x = 32$ Add 7 to each side.

$x = 16$ Divide each side by 2.

e. $|3x + 6| = 9$ Write original (absolute value) equation.

$3x + 6 = -9$ or $3x + 6 = 9$ Rewrite equivalent equations.

$3x = -15$ or $3x = 3$ Subtract 6 from each side.

$x = -5$ or $x = 1$ Divide each side by 3.

STUDY TIP

Solving radical equations can sometimes lead to *extraneous solutions* (those that do not satisfy the original equation). For instance, squaring both sides of the following equation yields two possible solutions, one of which is extraneous.

$$\sqrt{x} = x - 2$$
$$x = x^2 - 4x + 4$$
$$0 = x^2 - 5x + 4$$
$$= (x - 4)(x - 1)$$

$x - 4 = 0 \Rightarrow x = 4$ (solution)

$x - 1 = 0 \Rightarrow x = 1$ (extraneous)

SUMMARY AND STUDY STRATEGIES

After studying this chapter, you should have acquired the following skills.
The exercise numbers are keyed to the Review Exercises that begin on page 74.
Answers to odd-numbered Review Exercises are given in the back of the text.*

Section 1.1	Review Exercises
■ Plot points in a coordinate plane.	*1, 2*
■ Find the distance between two points in a coordinate plane. $d = \sqrt{(x_2 - x_1)^2 + (y_2 - y_1)^2}$	*3–8*
■ Find the midpoint of a line segment connecting two points. $\text{Midpoint} = \left(\frac{x_1 + x_2}{2}, \frac{y_1 + y_2}{2}\right)$	*9–14*
■ Interpret real-life data that is presented graphically.	*15, 16*
■ Translate points in a coordinate plane.	*17, 18*

Section 1.2	
■ Sketch graphs of equations by hand.	*19–28*
■ Find the x- and y-intercepts of graphs of equations.	*29–32*
■ Find the standard forms of equations of circles. $(x - h)^2 + (y - k)^2 = r^2$	*33–36*
■ Find the points of intersection of two graphs.	*37–40*
■ Find the break-even point for a business. The break-even point occurs when the revenue R is equal to the cost C.	*41, 42*
■ Find the equilibrium points of supply equations and demand equations. The equilibrium point is the point of intersection of the graphs of the supply and demand equations.	*43*
■ Use mathematical models to model and solve real-life problems.	*44*

Section 1.3	
■ Use the slope-intercept form of a linear equation to sketch graphs of lines. $y = mx + b$	*45–52*
■ Find slopes of lines passing through two points. $m = \frac{y_2 - y_1}{x_2 - x_1}$	*53–56*
■ Use the point-slope form to write equations of lines and sketch graphs of lines. $y - y_1 = m(x - x_1)$	*57–64*

* A wide range of valuable study aids are available to help you master the material in this chapter. The *Student Solutions Manual* includes step-by-step solutions to all odd-numbered exercises to help you review and prepare. The student website at *www.cengagebrain.com* offers algebra help and a *Graphing Technology Guide*, which contains step-by-step commands and instructions for a wide variety of graphing calculators.

Section 1.3 (continued)	Review Exercises
■ Find equations of parallel and perpendicular lines. Parallel lines: $m_1 = m_2$ Perpendicular lines: $m_1 = -\frac{1}{m_2}$	*65, 66*
■ Use linear equations to solve real-life problems such as predicting future sales or creating a linear depreciation schedule.	*67, 68*

Section 1.4

■ Use the Vertical Line Test to decide whether the relationship between two variables is a function. If every vertical line intersects the graph of an equation at most once, then the equation defines y as a function of x.	*69–72*
■ Find the domains and ranges of functions.	*73–78*
■ Use function notation and evaluate functions.	*79, 80*
■ Combine functions to create other functions.	*81, 82*
■ Use the Horizontal Line Test to determine whether functions have inverse functions. If they do, find the inverse functions. A function is one-to-one when every horizontal line intersects the graph of the function at most once. For a function to have an inverse function, it must be one-to-one.	*83–88*

Section 1.5

■ Use a table to estimate limits.	*89–92*
■ Determine whether limits exist. If they do, find the limits.	*93–110*

Section 1.6

■ Determine whether functions are continuous at a point, on an open interval, and on a closed interval.	*111–120*
■ Determine the constant such that f is continuous.	*121, 122*
■ Use analytic and graphical models of real-life data to solve real-life problems.	*123–126*

Study Strategies

- **Use a Graphing Utility** A graphing calculator or graphing software for a computer can help you in this course in two important ways. As an *exploratory device,* a graphing utility allows you to learn concepts by allowing you to compare graphs of equations. For instance, sketching the graphs of $y = x^2$, $y = x^2 + 1$, and $y = x^2 - 1$ helps confirm that adding (or subtracting) a constant to (or from) a function shifts the graph of the function vertically. As a *problem-solving tool,* a graphing utility frees you of some of the drudgery of sketching complicated graphs by hand. The time that you save can be spent using mathematics to solve real-life problems.
- **Use the Skills Warm-Up Exercises** Each exercise set in this text begins with a set of skills warm-up exercises. You should begin each homework session by quickly working all of these exercises. (All are answered in the back of the text.) The "old" skills covered in these exercises are needed to master the "new" skills in the section exercise set. The skills warm-up exercises remind you that mathematics is cumulative—to be successful in this course, you must retain "old" skills.
- **Use the Additional Study Aids** The additional study aids were prepared specifically to help you master the concepts discussed in the text. They are the *Student Solutions Manual,* the student website, and the *Graphing Technology Guide*.

Review Exercises

See www.CalcChat.com for worked-out solutions to odd-numbered exercises.

Plotting Points in the Cartesian Plane In Exercises 1 and 2, plot the points in the Cartesian plane.

1. $(2, 3), (0, 6), (-5, 1), (4, -3), (-3, -1)$
2. $(1, -4), (-1, -2), (6, -5), (-2, 0), (5, 5)$

Finding a Distance In Exercises 3–8, find the distance between the two points.

3. $(0, 0), (5, 2)$
4. $(1, 2), (4, 3)$
5. $(-1, 3), (-4, 6)$
6. $(6, 8), (-3, 7)$
7. $\left(\frac{1}{4}, -8\right), \left(\frac{3}{4}, -6\right)$
8. $(-0.6, 3), (4, -1.8)$

Finding a Segment's Midpoint In Exercises 9–14, find the midpoint of the line segment connecting the two points.

9. $(5, 6), (9, 2)$
10. $(0, 0), (-4, 8)$
11. $(-10, 4), (-6, 8)$
12. $(7, -9), (-3, 5)$
13. $\left(-1, \frac{1}{5}\right), \left(6, \frac{3}{5}\right)$
14. $(6, 1.2), (-3.2, 5)$

Revenues, Costs, and Profits In Exercises 15 and 16, use the graph below, which gives the revenues, costs, and profits for Google from 2005 through 2009. *(Source: Google, Inc.)*

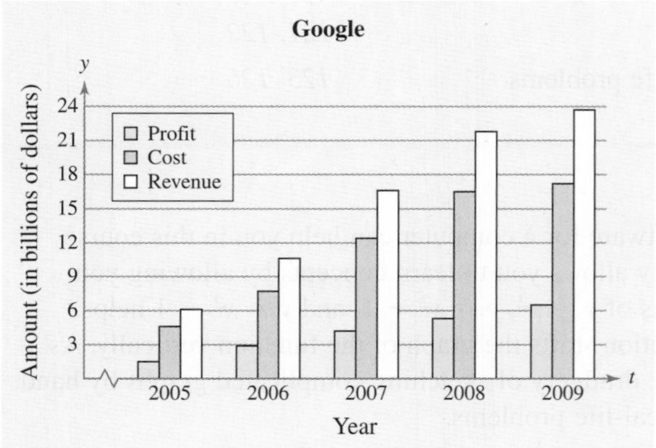

15. Write an equation that relates the revenue R, cost C, and profit P. Explain the relationship between the heights of the bars and the equation.
16. Estimate the revenue, cost, and profit for Google for each year.

Translating Points in the Plane In Exercises 17 and 18, use the translation and the graph to find the vertices of the figure after it has been translated.

17. 3 units left and 4 units up

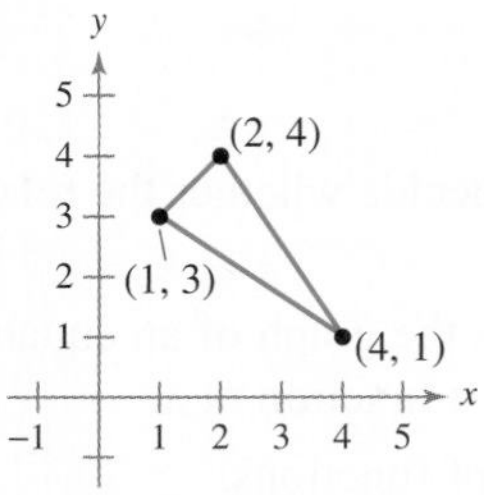

18. 4 units right and 1 unit down

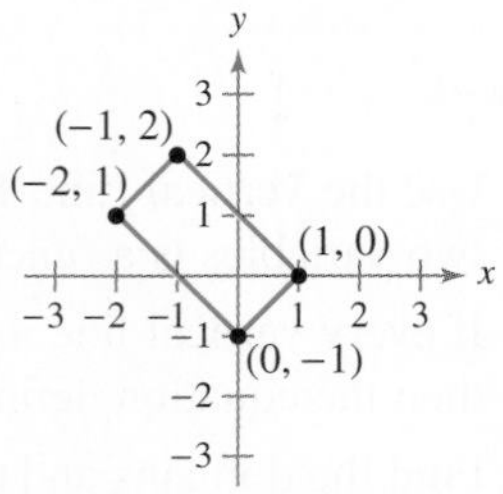

Sketching the Graph of an Equation In Exercises 19–28, sketch the graph of the equation.

19. $y = 4x - 12$
20. $y = 4 - 3x$
21. $y = x^2 + 5$
22. $y = 1 - x^2$
23. $y = |4 - x|$
24. $y = |2x - 3|$
25. $y = x^3 + 4$
26. $y = 2x^3 - 1$
27. $y = \sqrt{4x + 1}$
28. $y = \sqrt{2x}$

Finding *x*- and *y*-Intercepts In Exercises 29–32, find the x- and y-intercepts of the graph of the equation. Use a graphing utility to verify your results.

29. $4x + y + 3 = 0$
30. $3x - y + 6 = 0$
31. $y = x^2 + 2x - 8$
32. $y = (x - 1)^3 + 2(x - 1)^2$

Finding the Equation of a Circle In Exercises 33–36, find the standard form of the equation of the circle and sketch its graph.

33. Center: $(0, 0)$; radius: 8
34. Center: $(-5, -2)$; radius: 6
35. Center: $(0, 0)$; solution point: $\left(2, \sqrt{5}\right)$
36. Center: $(3, -4)$; solution point: $(-1, -1)$

Finding Points of Intersection In Exercises 37–40, find the point(s) of intersection (if any) of the graphs of the equations. Use a graphing utility to check your results.

37. $y = 2x + 13, \quad y = -5x - 1$
38. $y = x^2 - 5, \quad y = x + 1$
39. $y = x^3, \quad y = x$
40. $y = -x^2 + 4, \quad y = 2x - 1$

41. Break-Even Analysis A student organization wants to raise money by having a T-shirt sale. Each shirt costs \$8. The silk screening costs \$200 for the design, plus \$2 per shirt. Each shirt will sell for \$14.

(a) Find equations for the total cost C and the total revenue R for selling x shirts.

(b) Find the break-even point.

42. Break-Even Analysis You are starting a part-time business. You make an initial investment of \$6000. The unit cost of the product is \$6.50, and the selling price is \$13.90.

(a) Find equations for the total cost C and the total revenue R for selling x units of the product.

(b) Find the break-even point.

43. Supply and Demand The demand and supply equations for a cordless screwdriver are given by

$p = 91.4 - 0.009x$ Demand equation

$p = 6.4 + 0.008x$ Supply equation

where p is the price in dollars and x represents the number of units. Find the equilibrium point for this market.

44. Wind Energy The table shows the annual amounts of U.S. consumption W (in trillion Btu) of wind energy for the years 2003 through 2008. *(Source: U.S. Energy Information Administration)*

Year	2003	2004	2005
Consumption	115	142	178

Year	2006	2007	2008
Consumption	264	341	546

A mathematical model for the data is given by

$W = 3.870t^3 - 45.04t^2 + 205.7t - 204$

where t represents the year, with $t = 3$ corresponding to 2003.

(a) Compare the actual consumptions with those given by the model. How well does the model fit the data? Explain your reasoning.

(b) Use the model to predict the consumption in 2014.

Finding the Slope and y-Intercept In Exercises 45–52, find the slope and y-intercept (if possible) of the equation of the line. Then sketch the graph of the equation.

45. $y = -x + 12$

46. $y = 3x - 1$

47. $3x + y = -2$

48. $2x - 4y = -8$

49. $y = -\frac{5}{3}$

50. $x = -3$

51. $-2x - 5y - 5 = 0$

52. $3.2x - 0.8y + 5.6 = 0$

Finding the Slope of a Line In Exercises 53–56, find the slope of the line passing through the pair of points.

53. $(0, 0), (7, 6)$

54. $(-1, 5), (-5, 7)$

55. $(10, 17), (-11, -3)$

56. $(-11, -3), (-1, -3)$

Using the Point-Slope Form In Exercises 57–60, find the equation of the line that passes through the given point and has the given slope. Then use the equation to sketch the line.

Point	*Slope*
57. $(3, -1)$	$m = -2$
58. $(-3, -3)$	$m = \frac{1}{2}$
59. $(1.5, -4)$	$m = 0$
60. $(8, 2)$	m is undefined.

Writing an Equation of a Line In Exercises 61–64, find the equation of the line that passes through the points. Then use the equation to sketch the line.

61. $(1, -7), (7, 5)$

62. $(2, 4), (8, 12)$

63. $(5, 7), (5, 14)$

64. $(4, -3), (-2, -3)$

Finding Parallel and Perpendicular Lines In Exercises 65 and 66, find the equation of the line passing through the given point and satisfying the given condition.

65. Point: $(-3, 6)$

(a) Slope is $\frac{7}{8}$.

(b) Parallel to the line $4x + 2y = 7$

(c) Passes through the origin

(d) Perpendicular to the line $3x - 2y = 2$

66. Point: $(1, -3)$

(a) Parallel to the x-axis

(b) Perpendicular to the x-axis

(c) Parallel to the line $-4x + 5y = -3$

(d) Perpendicular to the line $5x - 2y = 3$

67. Demand When a wholesaler sold a product at \$32 per unit, sales were 750 units per week. After a price increase of \$5 per unit, however, the sales dropped to 700 units per week. Assume that the relationship between the price p and the units sold per week x is linear.

(a) Write a linear equation expressing x in terms of p.

(b) Predict the number of units sold per week at a price of \$34.50 per unit.

(c) Predict the number of units sold per week at a price of \$42 per unit.

68. Linear Depreciation A printing company purchases an advanced color copier/printer for $117,000. After 9 years, the equipment will be obsolete and have no value.

(a) Write a linear equation giving the value v (in dollars) of the equipment in terms of the time t (in years), $0 \le t \le 9$.

(b) Use a graphing utility to graph the equation.

(c) Move the cursor along the graph and estimate (to two-decimal-place accuracy) the value of the equipment after 4 years.

(d) Move the cursor along the graph and estimate (to two-decimal-place accuracy) the time when the value of the equipment will be $84,000.

Vertical Line Test In Exercises 69–72, use the Vertical Line Test to determine whether y is a function of x.

69. $y = -x^2 + 2$

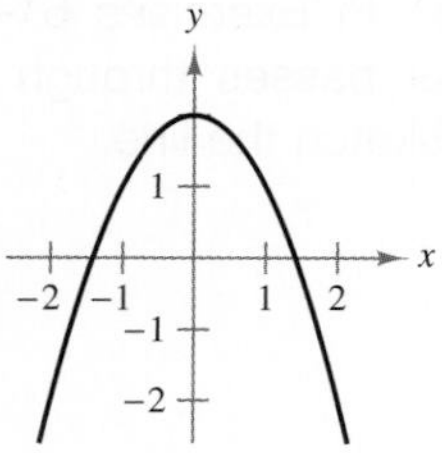

70. $x^2 + y^2 = 4$

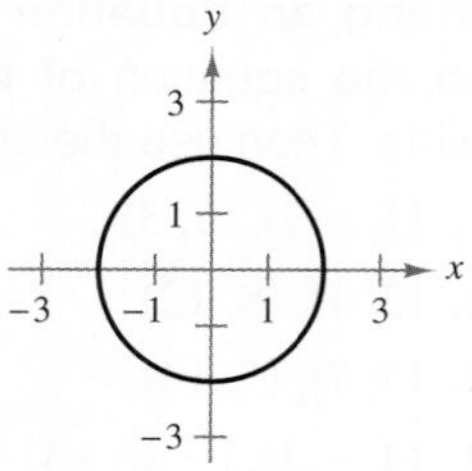

71. $y^2 - \frac{1}{4}x^2 = 4$

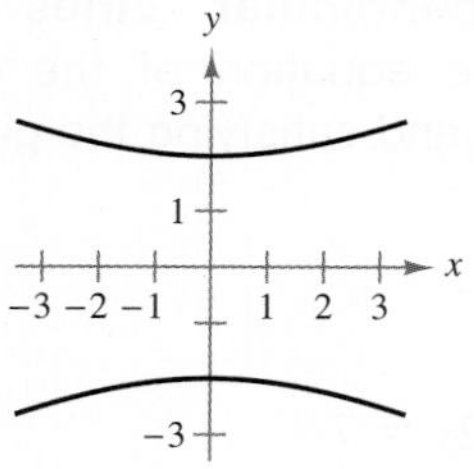

72. $y = |x + 4|$

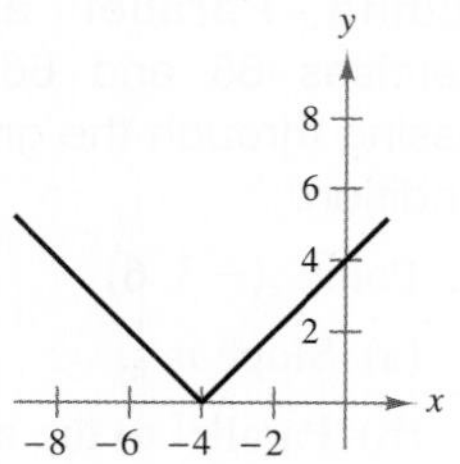

Finding the Domain and Range of a Function In Exercises 73–78, find the domain and range of the function. Use a graphing utility to verify your results.

73. $f(x) = x^3 + 2x^2 - x + 2$

74. $f(x) = 2$

75. $f(x) = \sqrt{x + 1}$

76. $f(x) = -|x| + 3$

77. $f(x) = \dfrac{x - 3}{x^2 + x - 12}$

78. $f(x) = \begin{cases} 6 - x, & x < 2 \\ 3x - 2, & x \ge 2 \end{cases}$

Evaluating a Function In Exercises 79 and 80, evaluate the function at the specified values of the independent variable. Simplify the result.

79. $f(x) = 3x + 4$

(a) $f(1)$ (b) $f(-5)$ (c) $f(x + 1)$

80. $f(x) = x^2 + 4x + 3$

(a) $f(0)$ (b) $f(3)$ (c) $f(x - 1)$

Combinations of Functions In Exercises 81 and 82, find (a) $f(x) + g(x)$, (b) $f(x) - g(x)$, (c) $f(x) \cdot g(x)$, (d) $f(x)/g(x)$, (e) $f(g(x))$, and (f) $g(f(x))$, if defined.

81. $f(x) = 1 + x^2$, $g(x) = 2x - 1$

82. $f(x) = 2x - 3$, $g(x) = \sqrt{x + 1}$

Determine Whether a Function Is One-to-One In Exercises 83–88, use a graphing utility to graph the function. Then use the Horizontal Line Test to determine whether the function is one-to-one. If it is, find its inverse function.

83. $f(x) = 4x - 3$

84. $f(x) = \frac{3}{2}x$

85. $f(x) = -x^2 + \frac{1}{2}$

86. $f(x) = x^3 - 1$

87. $f(x) = |x + 1|$

88. $f(x) = 6$

Finding Limits Numerically In Exercises 89–92, complete the table and use the result to estimate the limit. Use a graphing utility to graph the function to confirm your result.

89. $\lim_{x \to 1} (4x - 3)$

x	0.9	0.99	0.999	1	1.001	1.01	1.1
$f(x)$				?			

90. $\lim_{x \to 3} \dfrac{x - 3}{x^2 - 2x - 3}$

x	2.9	2.99	2.999	3	3.001	3.01	3.1
$f(x)$				?			

91. $\lim_{x \to 0} \dfrac{\sqrt{x + 6} - 6}{x}$

x	-0.1	-0.01	-0.001	0	0.001	0.01	0.1
$f(x)$				?			

92. $\lim_{x \to 7} \dfrac{\frac{1}{x - 7} - \frac{1}{7}}{x}$

x	6.9	6.99	6.999	7	7.001	7.01	7.1
$f(x)$				?			

Finding Limits In Exercises 93–110, find the limit (if it exists).

93. $\lim_{x\to 3} 8$

94. $\lim_{x\to 4} x^2$

95. $\lim_{x\to 2} (5x - 3)$

96. $\lim_{x\to 5} (2x + 4)$

97. $\lim_{x\to -1} \frac{x + 3}{6x + 1}$

98. $\lim_{t\to 3} \frac{t}{t + 5}$

99. $\lim_{t\to 0} \frac{t^2 + 1}{t}$

100. $\lim_{t\to 2} \frac{t + 1}{t - 2}$

101. $\lim_{x\to -2} \frac{x + 2}{x^2 - 4}$

102. $\lim_{x\to 3^-} \frac{x^2 - 9}{x - 3}$

103. $\lim_{x\to 0^+} \left(x - \frac{1}{x}\right)$

104. $\lim_{x\to 1/2} \frac{2x - 1}{6x - 3}$

105. $\lim_{x\to 0} \frac{[1/(x - 2)] - 1}{x}$

106. $\lim_{s\to 0} \frac{\left(1/\sqrt{1 + s}\right) - 1}{s}$

107. $\lim_{x\to 0} f(x)$, where $f(x) = \begin{cases} x + 5, & x \neq 0 \\ 3, & x = 0 \end{cases}$

108. $\lim_{x\to -2} f(x)$, where $f(x) = \begin{cases} \frac{1}{2}x + 5, & x < -2 \\ -x + 2, & x \geq -2 \end{cases}$

109. $\lim_{\Delta x\to 0} \frac{(x + \Delta x)^3 - (x + \Delta x) - (x^3 - x)}{\Delta x}$

110. $\lim_{\Delta x\to 0} \frac{1 - (x + \Delta x)^2 - (1 - x^2)}{\Delta x}$

Determining Continuity In Exercises 111–120, describe the interval(s) on which the function is continuous. Explain why the function is continuous on the interval(s). If the function has a discontinuity, identify the conditions of continuity that are not satisfied.

111. $f(x) = x + 6$

112. $f(x) = x^2 + 3x + 2$

113. $f(x) = \frac{1}{(x + 4)^2}$

114. $f(x) = \frac{x + 2}{x}$

115. $f(x) = \frac{3}{x + 1}$

116. $f(x) = \frac{x + 1}{2x + 2}$

117. $f(x) = [\![x + 3]\!]$

118. $f(x) = [\![x]\!] - 2$

119. $f(x) = \begin{cases} x, & x \leq 0 \\ x + 1, & x > 0 \end{cases}$

120. $f(x) = \begin{cases} x, & x \leq 0 \\ x^2, & x > 0 \end{cases}$

Making a Function Continuous In Exercises 121 and 122, find the constant a such that the function is continuous on the entire real number line.

121. $f(x) = \begin{cases} -x + 1, & x \leq 3 \\ ax - 8, & x > 3 \end{cases}$

122. $f(x) = \begin{cases} x + 1, & x < 1 \\ 2x + a, & x \geq 1 \end{cases}$

123. Consumer Awareness The cost C (in dollars) of purchasing x bottles of vitamins at a whole foods store is shown below.

$$C(x) = \begin{cases} 5.99x, & 0 < x \leq 5 \\ 4.99x, & 5 < x \leq 10 \\ 3.99x, & 10 < x \leq 15 \\ 2.99x, & x > 15 \end{cases}$$

(a) Use a graphing utility to graph the function and then discuss its continuity. At what values is the function not continuous? Explain your reasoning.

(b) Find the cost of purchasing 10 bottles.

124. Salary Contract A union contract guarantees a 10% salary increase yearly for 3 years. For a current salary of \$28,000, the salaries S (in thousands of dollars) for the next 3 years are given by

$$S(t) = \begin{cases} 28.00, & 0 < t \leq 1 \\ 30.80, & 1 < t \leq 2 \\ 33.88, & 2 < t \leq 3 \end{cases}$$

where $t = 0$ represents the present year. Does the limit of S exist as t approaches 2? Explain your reasoning.

125. Consumer Awareness A pay-as-you-go cellular phone charges \$1 for the first minute and \$0.10 for each additional minute or fraction thereof.

(a) Use the greatest integer function to create a model for the cost C of a phone call lasting t minutes.

(b) Use a graphing utility to graph the function and then discuss its continuity.

126. Recycling A recycling center pays \$0.50 for each pound of aluminum cans. Twenty-four aluminum cans weigh one pound. A mathematical model for the amount A paid by the recycling center is

$$A = \frac{1}{2}\left[\!\!\left[\frac{x}{24}\right]\!\!\right]$$

where x is the number of cans.

(a) Use a graphing utility to graph the function and then discuss its continuity.

(b) How much does the recycling center pay out for 1500 cans?

TEST YOURSELF

See www.CalcChat.com for worked-out solutions to odd-numbered exercises.

Take this test as you would take a test in class. When you are done, check your work against the answers given in the back of the book.

In Exercises 1–3, (a) find the distance between the points, (b) find the midpoint of the line segment joining the points, (c) find the slope of the line passing through the points, (d) find an equation of the line passing through the points, and (e) sketch the graph of the equation.

1. $(1, -1), (-4, 4)$ **2.** $\left(\frac{5}{2}, 2\right), (0, 2)$ **3.** $(2, 3), (-4, 1)$

4. The demand and supply equations for a product are $p = 65 - 2.1x$ and $p = 43 + 1.9x$, respectively, where p is the price (in dollars) and x represents the number of units, in thousands. Find the equilibrium point for this market.

In Exercises 5–7, find the slope and y-intercept (if possible) of the equation of the line. Then sketch the graph of the equation.

5. $y = \frac{1}{5}x - 2$ **6.** $x - \frac{7}{4} = 0$ **7.** $-x - 0.4y + 2.5 = 0$

8. Write an equation of the line that passes through $(-1, -7)$ and is perpendicular to $-4x + y = 8$.

9. Write an equation of the line that passes through $(2, 1)$ and is parallel to $5x - 2y = 8$.

In Exercises 10–12, (a) graph the function and label the intercepts, (b) determine the domain and range of the function, (c) find the value of the function when x is -3, -2, and 3, and (d) determine whether the function is one-to-one.

10. $f(x) = 2x + 5$ **11.** $f(x) = x^2 - x - 2$ **12.** $f(x) = \sqrt{x + 5}$

In Exercises 13 and 14, find the inverse function of f.

13. $f(x) = 4x + 6$ **14.** $f(x) = \sqrt[3]{8 - 3x}$

In Exercises 15–18, find the limit (if it exists).

15. $\lim_{x \to 0} \frac{x - 2}{x + 2}$ **16.** $\lim_{x \to 5} \frac{x + 5}{x - 5}$

17. $\lim_{x \to -3} \frac{x^2 + 2x - 3}{x^2 + 4x + 3}$ **18.** $\lim_{x \to 0} \frac{\sqrt{x + 9} - 3}{x}$

In Exercises 19–21, describe the interval(s) on which the function is continuous. Explain why the function is continuous on the interval(s). If the function has a discontinuity, identify the conditions of continuity that are not satisfied.

19. $f(x) = \frac{x^2 - 16}{x - 4}$ **20.** $f(x) = \sqrt{5 - x}$ **21.** $f(x) = \begin{cases} 1 - x, & x < 1 \\ x - x^2, & x \geq 1 \end{cases}$

22. The table lists the numbers of unemployed workers y (in thousands) in the United States for selected years. A mathematical model for the data is given by

$$y = 193.898t^3 - 3080.32t^2 + 15{,}478.5t - 16{,}925$$

where t represents the year, with $t = 4$ corresponding to 2004. *(Source: U.S. Bureau of Labor Statistics)*

(a) Compare the actual numbers of unemployed workers with those given by the model. How well does the model fit the data? Explain your reasoning.

(b) Use the model to predict the number of unemployed workers in 2014.

t	4	5	6
y	8149	7591	7001

t	7	8	9
y	7078	8924	14,265

Table for 22

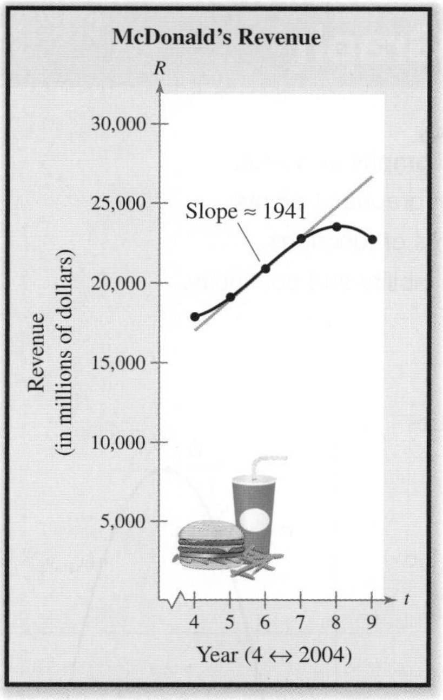

Example 11 on page 99 shows how differentiation can be used to find the rate of change in a company's revenue.

2 Differentiation

junjie/www.shutterstock.com
Kurhan/www.shutterstock.com

2.1 The Derivative and the Slope of a Graph

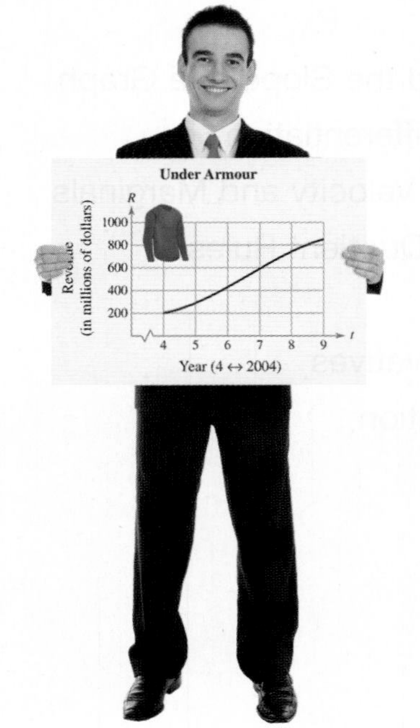

In Exercise 13 on page 89, you will estimate and interpret the slope of the graph of a revenue function.

- Identify tangent lines to a graph at a point.
- Approximate the slopes of tangent lines to graphs at points.
- Use the limit definition to find the slopes of graphs at points.
- Use the limit definition to find the derivatives of functions.
- Describe the relationship between differentiability and continuity.

Tangent Line to a Graph

Calculus is a branch of mathematics that studies rates of change of functions. In this course, you will learn that rates of change have many applications in real life. In Section 1.3, you learned how the slope of a line indicates the rate at which the line rises or falls. For a line, this rate (or slope) is the same at every point on the line. For graphs other than lines, the rate at which the graph rises or falls changes from point to point. For instance, in Figure 2.1, the parabola is rising more quickly at the point (x_1, y_1) than it is at the point (x_2, y_2). At the vertex (x_3, y_3), the graph levels off, and at the point (x_4, y_4), the graph is falling.

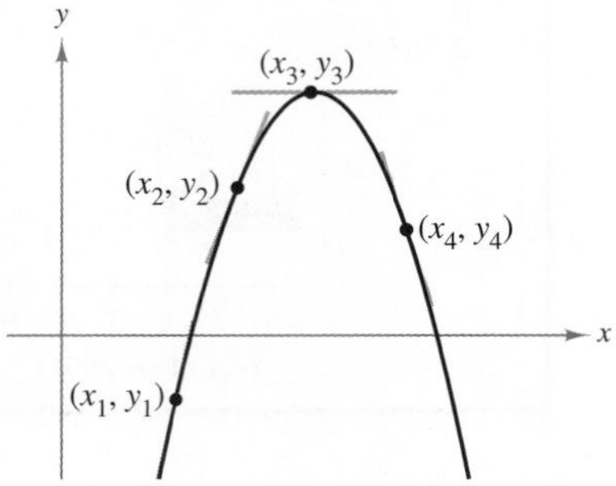

The slope of a nonlinear graph changes from one point to another.
FIGURE 2.1

To determine the rate at which a graph rises or falls at a *single point,* you can find the slope of the **tangent line** at the point. In simple terms, the tangent line to the graph of a function f at a point $P(x_1, y_1)$ is the line that best approximates the graph at that point, as shown in Figure 2.1. Figure 2.2 shows other examples of tangent lines.

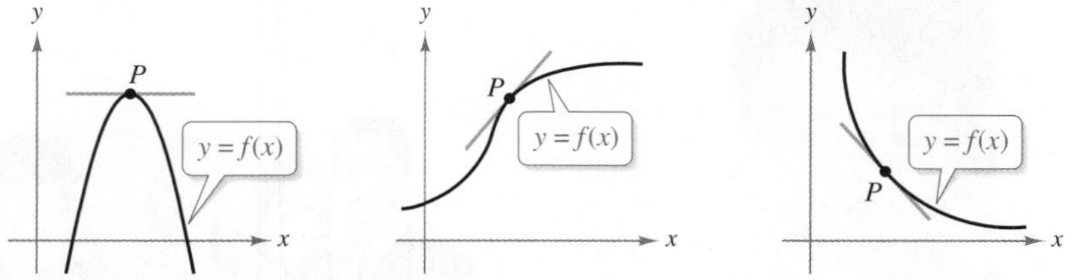

Tangent Line to a Graph at a Point
FIGURE 2.2

When Isaac Newton (1642–1727) was working on the "tangent line problem," he realized that it is difficult to define precisely what is meant by a tangent to a general curve. From geometry, you know that a line is tangent to a circle when the line intersects the circle at only one point, as shown in Figure 2.3. Tangent lines to a noncircular graph, however, can intersect the graph at more than one point. For instance, in the second graph in Figure 2.2, when the tangent line is extended, it intersects the graph at a point other than the point of tangency. In this section, you will see how the notion of a limit can be used to define a general tangent line.

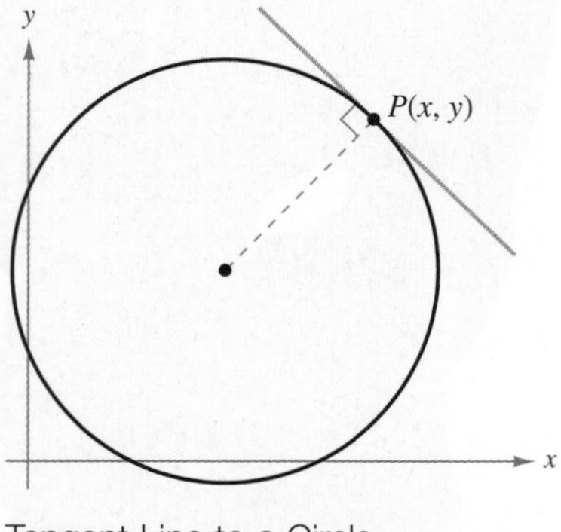

Tangent Line to a Circle
FIGURE 2.3

Slope of a Graph

Because a tangent line approximates the graph at a point, the problem of finding the slope of a graph at a point becomes one of finding the slope of the tangent line at the point.

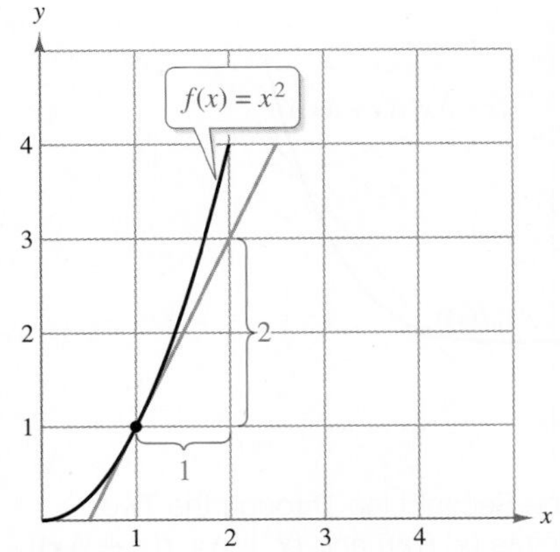

FIGURE 2.4

Example 1 Approximating the Slope of a Graph

Use the graph in Figure 2.4 to approximate the slope of the graph of

$$f(x) = x^2$$

at the point (1, 1).

SOLUTION From the graph of

$$f(x) = x^2$$

you can see that the tangent line at (1, 1) rises approximately two units for each unit change in x. So, the slope of the tangent line at (1, 1) is given by

$$\text{Slope} = \frac{\Delta y}{\Delta x} = \frac{\text{change in } y}{\text{change in } x} \approx \frac{2}{1} = 2.$$

Because the tangent line at the point (1, 1) has a slope of about 2, you can conclude that the graph has a slope of about 2 at the point (1, 1).

STUDY TIP

When visually approximating the slope of a graph, note that the scales on the horizontal and vertical axes may differ. When this happens (as it frequently does in applications), the slope of the tangent line is distorted, and you must be careful to account for the difference in scales.

✓ Checkpoint 1

Use the graph to approximate the slope of the graph of

$$f(x) = x^3$$

at the point (1, 1).

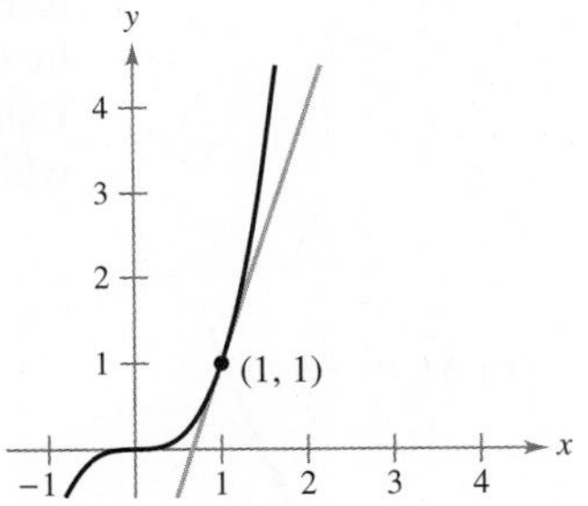

Example 2 Interpreting Slope

Figure 2.5 graphically depicts the average monthly temperature (in degrees Fahrenheit) in Duluth, Minnesota. Estimate the slope of this graph at the indicated point and give a physical interpretation of the result. *(Source: National Oceanic and Atmospheric Administration)*

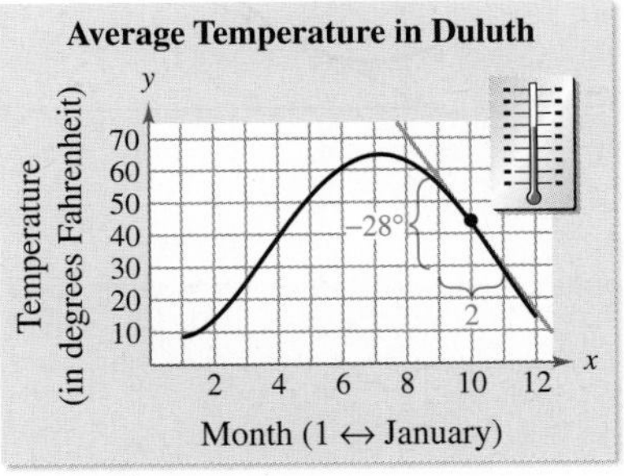

FIGURE 2.5

SOLUTION From the graph, you can see that the tangent line at the given point falls approximately 28 units for each two-unit change in x. So, you can estimate the slope at the given point to be

$$\text{Slope} = \frac{\Delta y}{\Delta x} = \frac{\text{change in } y}{\text{change in } x} \approx \frac{-28}{2} = -14 \text{ degrees per month.}$$

This means that you can expect the average daily temperatures in November to be about 14 degrees *lower* than the corresponding temperatures in October.

✓ Checkpoint 2

In Figure 2.5, for which months do the slopes of the tangent lines appear to be positive? Negative? Interpret these slopes in the context of the problem.

Slope and the Limit Process

In Examples 1 and 2, you approximated the slope of a graph at a point by making a careful graph and then "eyeballing" the tangent line at the point of tangency. A more precise method of approximating the slope of a tangent line makes use of a **secant line** through the point of tangency and a second point on the graph, as shown in Figure 2.6. If $(x, f(x))$ is the point of tangency and

$$(x + \Delta x, f(x + \Delta x))$$

is a second point on the graph of f, then the slope of the secant line through the two points is

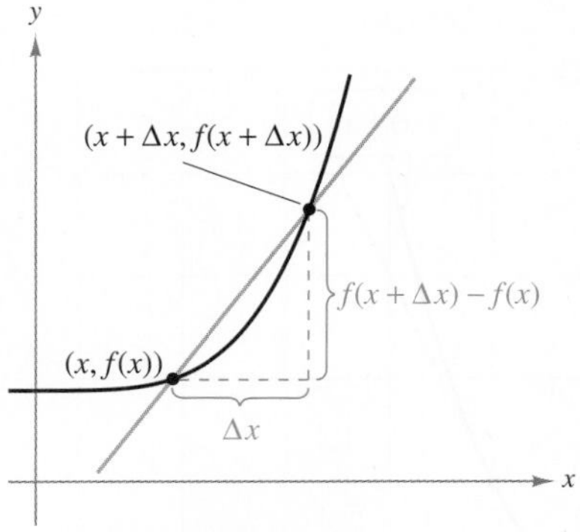

The Secant Line Through the Two Points $(x, f(x))$ and $(x + \Delta x, f(x + \Delta x))$

FIGURE 2.6

$$m = \frac{y_2 - y_1}{x_2 - x_1} \quad \text{Formula for slope}$$

$$m_{\text{sec}} = \frac{f(x + \Delta x) - f(x)}{(x + \Delta x) - x} \quad \frac{\text{Change in } y}{\text{Change in } x}$$

$$m_{\text{sec}} = \frac{f(x + \Delta x) - f(x)}{\Delta x}. \quad \text{Slope of secant line}$$

The right side of this equation is called the **difference quotient.** The denominator Δx is the **change in** x, and the numerator is the **change in** y. The beauty of this procedure is that you obtain more and more accurate approximations of the slope of the tangent line by choosing points closer and closer to the point of tangency, as shown in Figure 2.7. Using the limit process, you can find the *exact* slope of the tangent line at $(x, f(x))$, which is also the slope of the graph of f at $(x, f(x))$.

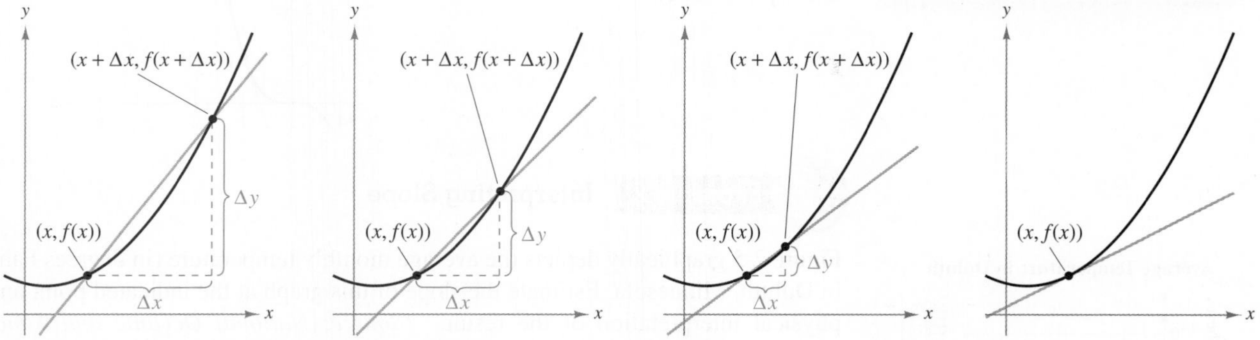

FIGURE 2.7 As Δx approaches 0, the secant lines approach the tangent line.

STUDY TIP

The variable Δx is used to represent the change in x in the definition of the slope of a graph. Other variables may also be used. Sometimes this definition is written as

$$m = \lim_{h \to 0} \frac{f(x + h) - f(x)}{h}.$$

Definition of the Slope of a Graph

The **slope** m of the graph of f at the point

$$(x, f(x))$$

is equal to the slope of the tangent line to the graph of f at $(x, f(x))$, and is given by

$$m = \lim_{\Delta x \to 0} m_{\text{sec}} = \lim_{\Delta x \to 0} \frac{f(x + \Delta x) - f(x)}{\Delta x}$$

provided this limit exists.

ALGEBRA TUTOR *xy*

For help in evaluating the expressions in Examples 3–6, see the review of simplifying fractional expressions on page 158.

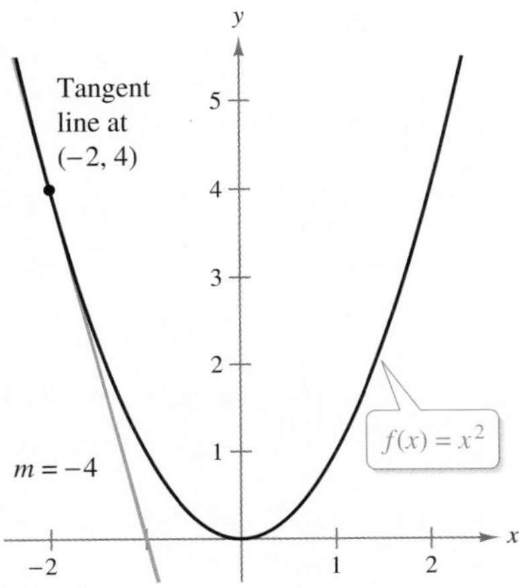

FIGURE 2.8

Example 3 Finding Slope by the Limit Process

Find the slope of the graph of

$$f(x) = x^2$$

at the point $(-2, 4)$.

SOLUTION Begin by finding an expression that represents the slope of a secant line at the point $(-2, 4)$.

$$\begin{aligned}
m_{\text{sec}} &= \frac{f(-2 + \Delta x) - f(-2)}{\Delta x} && \text{Set up difference quotient.}\\
&= \frac{(-2 + \Delta x)^2 - (-2)^2}{\Delta x} && \text{Use } f(x) = x^2.\\
&= \frac{4 - 4\,\Delta x + (\Delta x)^2 - 4}{\Delta x} && \text{Expand terms.}\\
&= \frac{-4\,\Delta x + (\Delta x)^2}{\Delta x} && \text{Simplify.}\\
&= \frac{\Delta x(-4 + \Delta x)}{\Delta x} && \text{Factor and divide out.}\\
&= -4 + \Delta x, \quad \Delta x \neq 0 && \text{Simplify.}
\end{aligned}$$

Next, take the limit of m_{sec} as $\Delta x \to 0$.

$$m = \lim_{\Delta x \to 0} m_{\text{sec}} = \lim_{\Delta x \to 0} (-4 + \Delta x) = -4 + 0 = -4$$

So, the graph of f has a slope of -4 at the point $(-2, 4)$, as shown in Figure 2.8.

✓ Checkpoint 3

Find the slope of the graph of

$$f(x) = x^2$$

at the point $(2, 4)$. ■

Example 4 Finding the Slope of a Graph

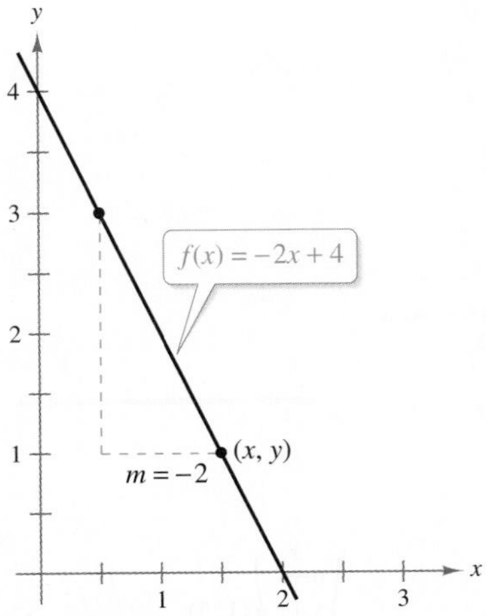

FIGURE 2.9

Find the slope of the graph of $f(x) = -2x + 4$.

SOLUTION You know from your study of linear functions that the line given by $f(x) = -2x + 4$ has a slope of -2, as shown in Figure 2.9. This conclusion is consistent with the limit definition of slope.

$$\begin{aligned}
m &= \lim_{\Delta x \to 0} \frac{f(x + \Delta x) - f(x)}{\Delta x}\\
&= \lim_{\Delta x \to 0} \frac{[-2(x + \Delta x) + 4] - (-2x + 4)}{\Delta x}\\
&= \lim_{\Delta x \to 0} \frac{-2x - 2\,\Delta x + 4 + 2x - 4}{\Delta x}\\
&= \lim_{\Delta x \to 0} \frac{-2\Delta x}{\Delta x}\\
&= -2
\end{aligned}$$

✓ Checkpoint 4

Find the slope of the graph of $f(x) = 2x + 5$. ■

Example 5 Finding a Formula for the Slope of a Graph

Find a formula for the slope of the graph of $f(x) = x^2 + 1$. What are the slopes at the points $(-1, 2)$ and $(2, 5)$?

SOLUTION

$$\begin{aligned}
m_{\text{sec}} &= \frac{f(x + \Delta x) - f(x)}{\Delta x} && \text{Set up difference quotient.}\\
&= \frac{[(x + \Delta x)^2 + 1] - (x^2 + 1)}{\Delta x} && \text{Use } f(x) = x^2 + 1.\\
&= \frac{x^2 + 2x\,\Delta x + (\Delta x)^2 + 1 - x^2 - 1}{\Delta x} && \text{Expand terms.}\\
&= \frac{2x\,\Delta x + (\Delta x)^2}{\Delta x} && \text{Simplify.}\\
&= \frac{\cancel{\Delta x}(2x + \Delta x)}{\cancel{\Delta x}} && \text{Factor and divide out.}\\
&= 2x + \Delta x, \quad \Delta x \neq 0 && \text{Simplify.}
\end{aligned}$$

Next, take the limit of m_{sec} as $\Delta x \to 0$.

$$\begin{aligned}
m &= \lim_{\Delta x \to 0} m_{\text{sec}}\\
&= \lim_{\Delta x \to 0} (2x + \Delta x)\\
&= 2x + 0\\
&= 2x
\end{aligned}$$

Using the formula $m = 2x$, you can find the slopes at the specified points. At $(-1, 2)$ the slope is $m = 2(-1) = -2$, and at $(2, 5)$ the slope is $m = 2(2) = 4$. The graph of f is shown in Figure 2.10.

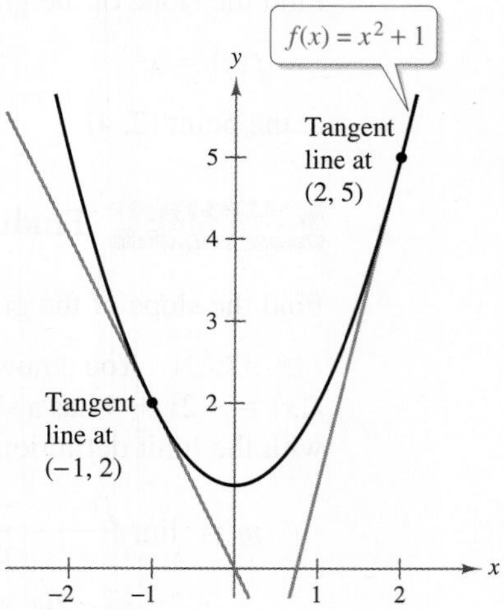

FIGURE 2.10

✓ Checkpoint 5

Find a formula for the slope of the graph of

$$f(x) = 4x^2 + 1.$$

What are the slopes at the points $(0, 1)$ and $(1, 5)$?

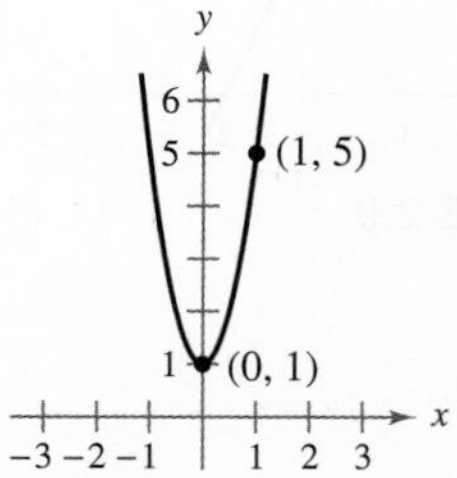

The Derivative of a Function

In Example 5, you started with the function

$$f(x) = x^2 + 1$$

and used the limit process to derive another function, $m = 2x$, that represents the slope of the graph of f at the point $(x, f(x))$. This derived function is called the **derivative** of f at x. It is denoted by $f'(x)$, which is read as "f prime of x."

STUDY TIP

The notation dy/dx is read as "the derivative of y with respect to x," and using limit notation, you can write

$$\frac{dy}{dx} = \lim_{\Delta x \to 0} \frac{\Delta y}{\Delta x}$$

$$= \lim_{\Delta x \to 0} \frac{f(x + \Delta x) - f(x)}{\Delta x}$$

$$= f'(x).$$

Definition of the Derivative

The **derivative of f at x** is given by

$$f'(x) = \lim_{\Delta x \to 0} \frac{f(x + \Delta x) - f(x)}{\Delta x}$$

provided this limit exists. A function is **differentiable** at x when its derivative exists at x. The process of finding derivatives is called **differentiation.**

In addition to $f'(x)$, other notations can be used to denote the derivative of $y = f(x)$. The most common are

$$\frac{dy}{dx}, \quad y', \quad \frac{d}{dx}[f(x)], \quad \text{and} \quad D_x[y].$$

Example 6 **Finding a Derivative**

Find the derivative of

$$f(x) = 3x^2 - 2x.$$

SOLUTION

$$\begin{aligned}
f'(x) &= \lim_{\Delta x \to 0} \frac{f(x + \Delta x) - f(x)}{\Delta x} \\
&= \lim_{\Delta x \to 0} \frac{[3(x + \Delta x)^2 - 2(x + \Delta x)] - (3x^2 - 2x)}{\Delta x} \\
&= \lim_{\Delta x \to 0} \frac{3x^2 + 6x\,\Delta x + 3(\Delta x)^2 - 2x - 2\,\Delta x - 3x^2 + 2x}{\Delta x} \\
&= \lim_{\Delta x \to 0} \frac{6x\,\Delta x + 3(\Delta x)^2 - 2\,\Delta x}{\Delta x} \\
&= \lim_{\Delta x \to 0} \frac{\cancel{\Delta x}(6x + 3\,\Delta x - 2)}{\cancel{\Delta x}} \\
&= \lim_{\Delta x \to 0} (6x + 3\,\Delta x - 2) \\
&= 6x + 3(0) - 2 \\
&= 6x - 2
\end{aligned}$$

So, the derivative of $f(x) = 3x^2 - 2x$ is

$$f'(x) = 6x - 2.$$

✓ Checkpoint 6

Find the derivative of

$$f(x) = x^2 - 5x.$$ ■

In many applications, it is convenient to use a variable other than x as the independent variable. Example 7 shows a function that uses t as the independent variable.

> **TECH TUTOR**
>
> You can use a graphing utility to confirm the result given in Example 7. One way to do this is to choose a point on the graph of $y = 2/t$, such as $(1, 2)$, and find the equation of the tangent line at that point. Using the derivative found in the example, you know that the slope of the tangent line when $t = 1$ is $m = -2$. This means that the tangent line at the point $(1, 2)$ is
>
> $$y - y_1 = m(t - t_1)$$
> $$y - 2 = -2(t - 1)$$
> $$y = -2t + 4.$$
>
> By graphing $y = 2/t$ and $y = -2t + 4$ in the same viewing window, as shown below, you can confirm that the line is tangent to the graph at the point $(1, 2)$.
>
> 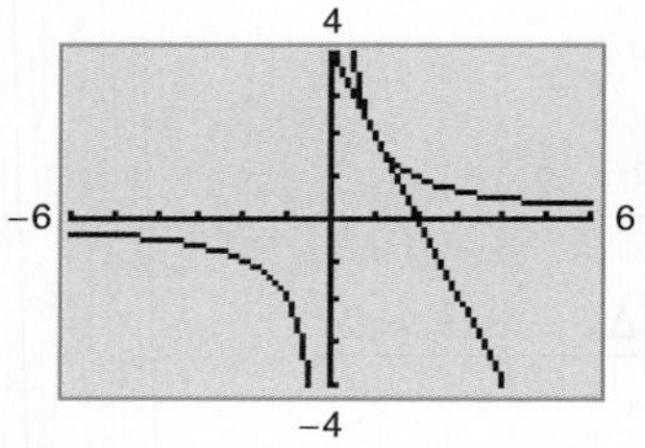
>

Example 7 Finding a Derivative

Find the derivative of y with respect to t for the function

$$y = \frac{2}{t}.$$

SOLUTION Consider $y = f(t)$, and use the limit process as shown.

$$\frac{dy}{dt} = \lim_{\Delta t \to 0} \frac{f(t + \Delta t) - f(t)}{\Delta t} \qquad \text{Set up difference quotient.}$$

$$= \lim_{\Delta t \to 0} \frac{\dfrac{2}{t + \Delta t} - \dfrac{2}{t}}{\Delta t} \qquad \text{Use } f(t) = 2/t.$$

$$= \lim_{\Delta t \to 0} \frac{\dfrac{2t - 2(t + \Delta t)}{t(t + \Delta t)}}{\Delta t} \qquad \text{Combine fractions in numerator.}$$

$$= \lim_{\Delta t \to 0} \frac{2t - 2t - 2\Delta t}{\Delta t(t)(t + \Delta t)} \qquad \text{Expand terms in numerator.}$$

$$= \lim_{\Delta t \to 0} \frac{-2\,\cancel{\Delta t}}{\cancel{\Delta t}(t)(t + \Delta t)} \qquad \text{Factor and divide out.}$$

$$= \lim_{\Delta t \to 0} \frac{-2}{t(t + \Delta t)} \qquad \text{Simplify.}$$

$$= \frac{-2}{t(t + 0)} \qquad \text{Direct substitution}$$

$$= -\frac{2}{t^2} \qquad \text{Simplify.}$$

So, the derivative of y with respect to t is

$$\frac{dy}{dt} = -\frac{2}{t^2}.$$

✓Checkpoint 7

Find the derivative of y with respect to t for the function $y = 4/t$. ■

Remember that the derivative of a function gives you a formula for finding the slope of the tangent line at any point on the graph of the function. For instance, in Example 7 the slope of the tangent line to the graph of f at the point $(1, 2)$ is given by

$$f'(1) = -\frac{2}{1^2} = -2.$$

To find the slopes of the graph at other points, substitute the t-coordinate of the point into the derivative, as shown below.

Point	*t-Coordinate*	*Slope*
$(2, 1)$	$t = 2$	$m = f'(2) = -\frac{2}{2^2} = -\frac{1}{2}$
$(-2, -1)$	$t = -2$	$m = f'(-2) = -\frac{2}{(-2)^2} = -\frac{1}{2}$

Differentiability and Continuity

Not every function is differentiable. Figure 2.11 shows some common situations in which a function will not be differentiable at a point—vertical tangent lines, discontinuities, and sharp turns in the graph. Each of the functions shown in Figure 2.11 is differentiable at every value of x *except* $x = 0$.

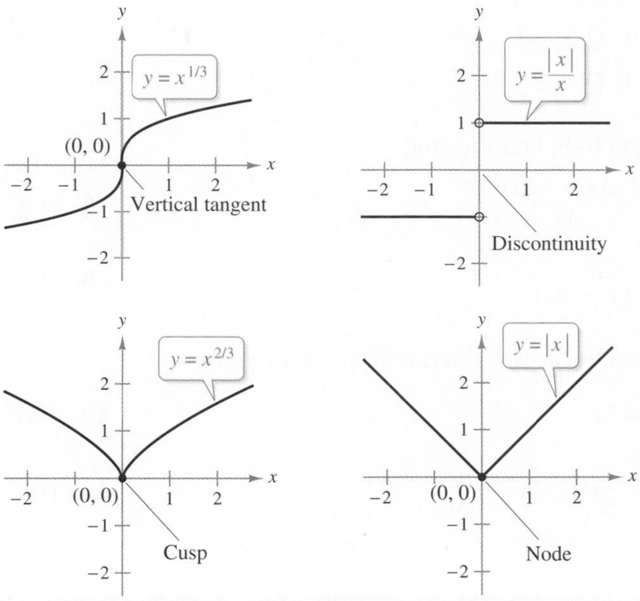

Functions That Are Not Differentiable at $x = 0$
FIGURE 2.11

In Figure 2.11, you can see that all but one of the functions are continuous at $x = 0$ but none are differentiable there. This shows that continuity is not a strong enough condition to guarantee differentiability. On the other hand, if a function is differentiable at a point, then it must be continuous at that point. This important result is stated in the following theorem.

Differentiability Implies Continuity

If a function f is differentiable at $x = c$, then f is continuous at $x = c$.

SUMMARIZE (Section 2.1)

1. Describe a tangent line and how it can be used to approximate the slope of a graph at a point *(page 80)*. For an example of a tangent line, see Example 1.
2. State the definition of the slope of a graph using the limit process *(page 82)*. For examples of finding the slope of a graph using the limit process, see Examples 3, 4, and 5.
3. State the definition of the derivative of a function *(page 85)*. For examples of the derivative of a function, see Examples 6 and 7.
4. Describe the relationship between differentiability and continuity *(page 87)*. For an example showing that continuity does not guarantee differentiability, see Figure 2.11.

SKILLS WARM UP 2.1 The following warm-up exercises involve skills that were covered in earlier sections. You will use these skills in the exercise set for this section. For additional help, review Sections 1.3, 1.4, and 1.5.

In Exercises 1–4, find an equation of the line containing P and Q.

1. $P(2, 1),\ Q(2, 4)$

2. $P(2, 2),\ Q(-5, 2)$

3. $P(2, 0),\ Q(3, -1)$

4. $P(3, 5),\ Q(-1, -7)$

In Exercises 5–8, find the limit.

5. $\lim_{\Delta x \to 0} \dfrac{2x\Delta x + (\Delta x)^2}{\Delta x}$

6. $\lim_{\Delta x \to 0} \dfrac{3x^2\Delta x + 3x(\Delta x)^2 + (\Delta x)^3}{\Delta x}$

7. $\lim_{\Delta x \to 0} \dfrac{1}{x(x + \Delta x)}$

8. $\lim_{\Delta x \to 0} \dfrac{(x + \Delta x)^2 - x^2}{\Delta x}$

In Exercises 9–12, find the domain of the function.

9. $f(x) = 3x$

10. $f(x) = \dfrac{1}{x - 1}$

11. $f(x) = \dfrac{1}{5}x^3 - 2x^2 + \dfrac{1}{3}x - 1$

12. $f(x) = \dfrac{6x}{x^3 + x}$

Exercises 2.1

See www.CalcChat.com for worked-out solutions to odd-numbered exercises.

Sketching Tangent Lines In Exercises 1–6, trace the graph and sketch the tangent lines at (x_1, y_1) and (x_2, y_2).

1.

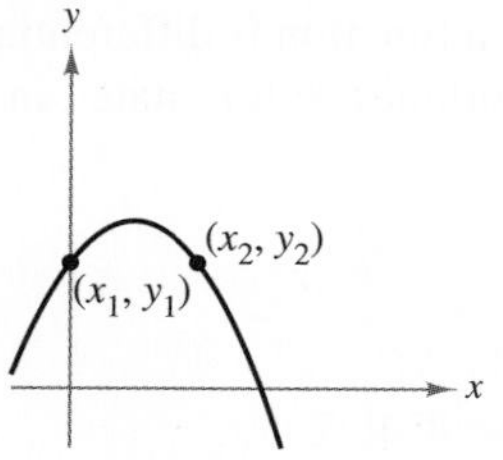

2.

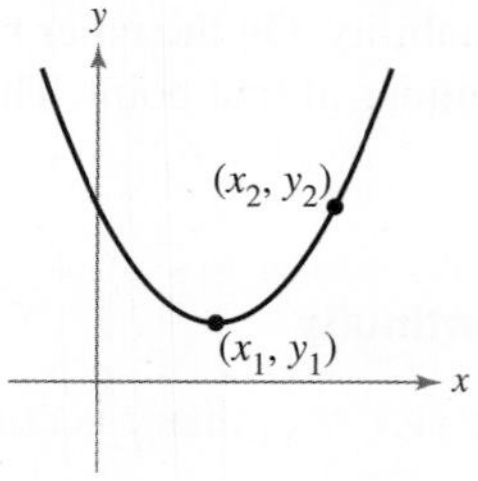

3.

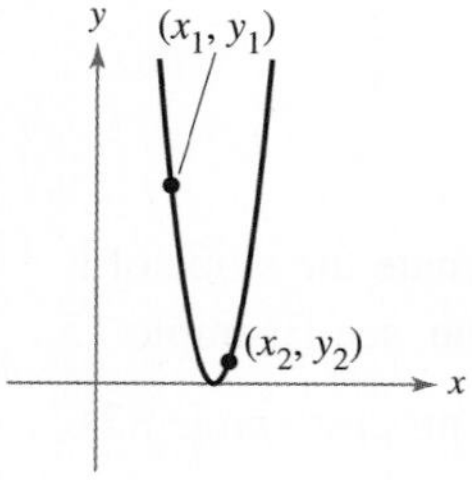

4.

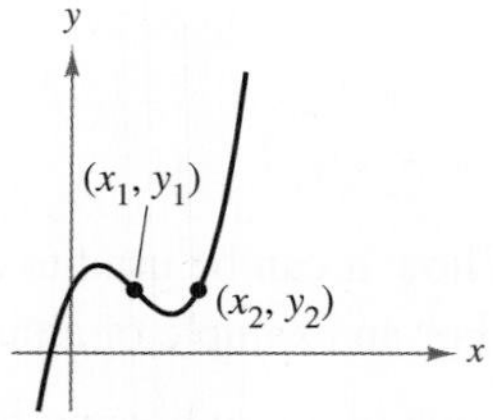

5.

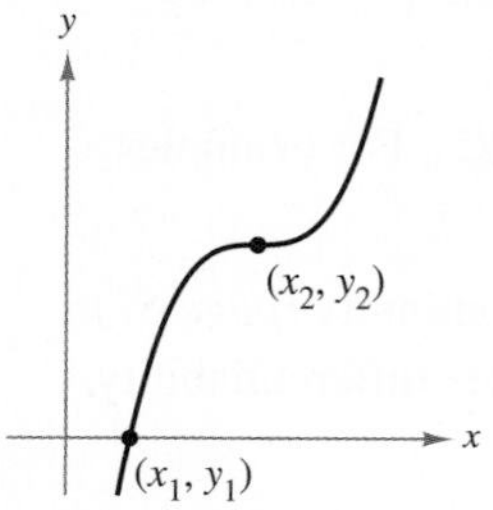

6.

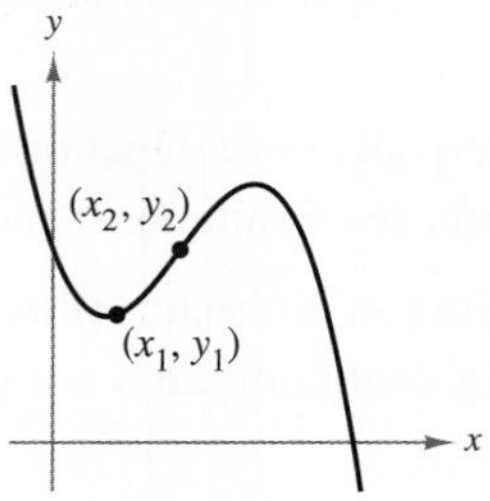

Approximating the Slope of a Graph In Exercises 7–12, estimate the slope of the graph at the point (x, y). (Each square on the grid is 1 unit by 1 unit.) *See Example 1.*

7.

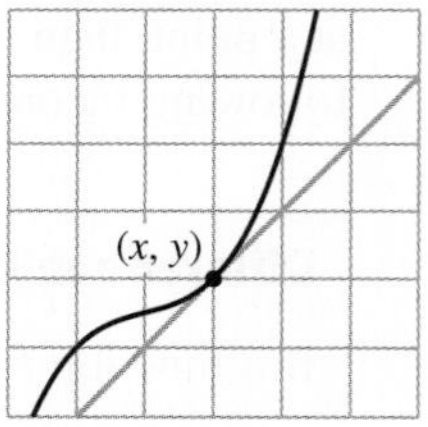

8.

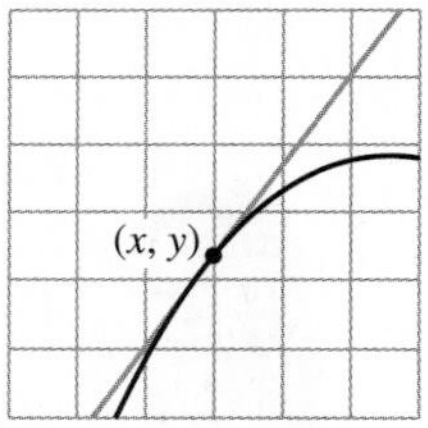

9.

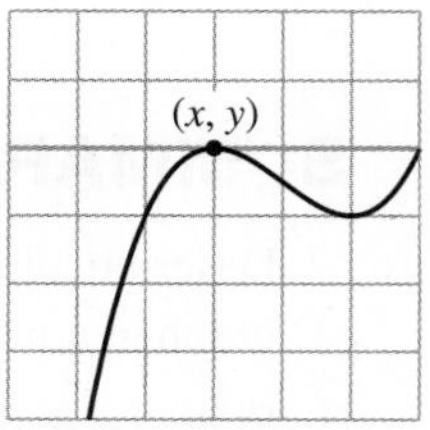

10.

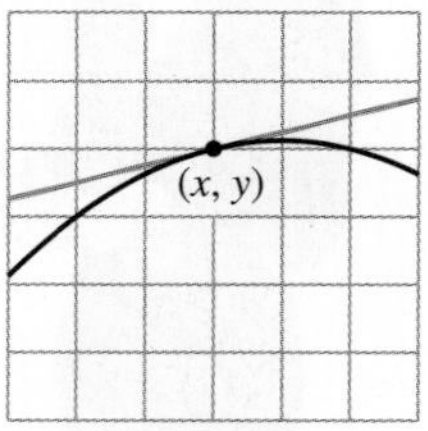

11.

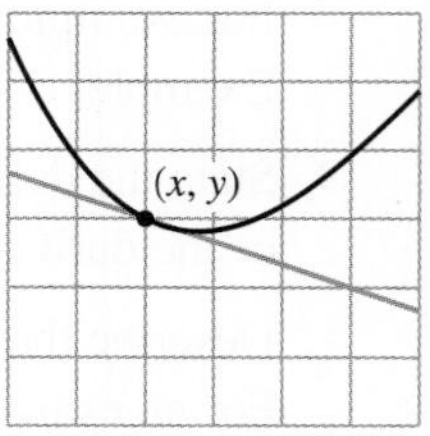

12.

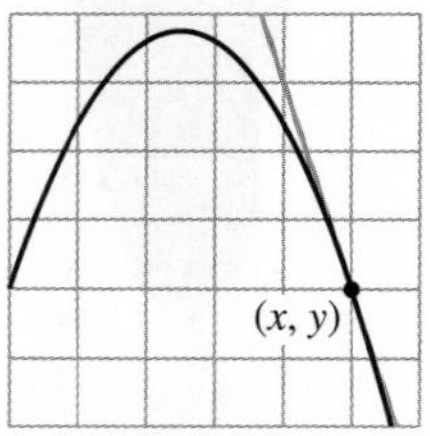

13. Revenue The graph represents the revenue R (in millions of dollars) for Under Armour from 2004 through 2009, where t represents the year, with $t = 4$ corresponding to 2004. Estimate and interpret the slopes of the graph for the years 2005 and 2007. *(Source: Under Armour, Inc.)*

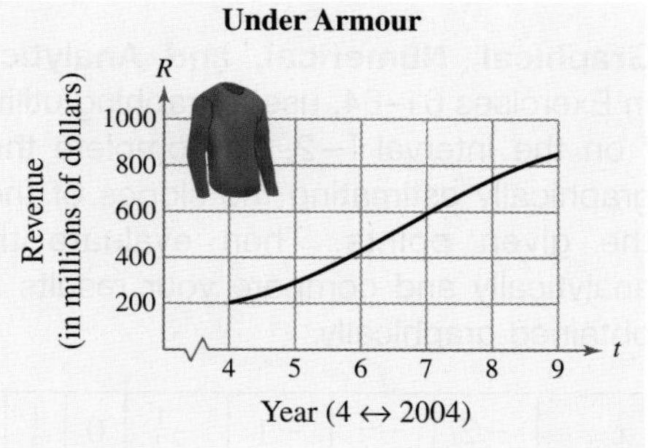

14. Sales The graph represents the sales S (in millions of dollars) for Scotts Miracle-Gro Company from 2003 through 2009, where t represents the year, with $t = 3$ corresponding to 2003. Estimate and interpret the slopes of the graph for the years 2006 and 2008. *(Source: Scotts Miracle-Gro Company)*

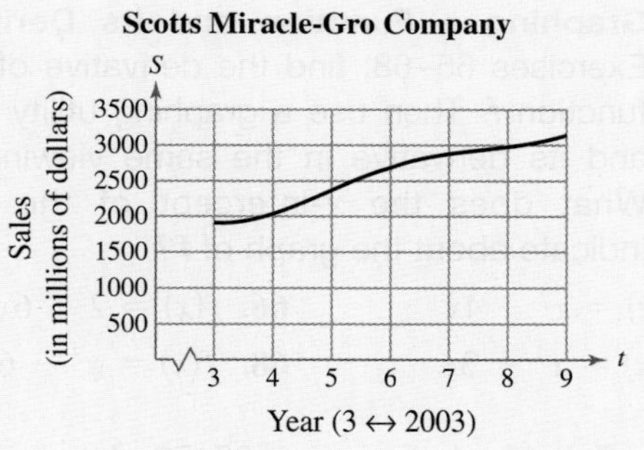

15. Temperature The graph represents the average monthly temperature F (in degrees Fahrenheit) in Blacksburg, Virginia for one year, where t represents the month, with $t = 1$ corresponding to January, $t = 2$ corresponding to February, and so on. Estimate and interpret the slopes of the graph at $t = 3$, 7, and 10. *(Source: National Oceanic and Atmospheric Administration)*

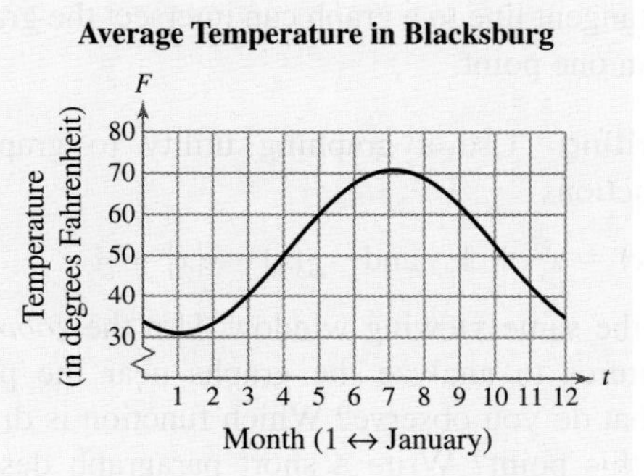

16. HOW DO YOU SEE IT? Two long distance runners starting out side by side begin a 10,000-meter run. Their distances are given by $s = f(t)$ and $s = g(t)$, where s is measured in thousands of meters and t is measured in minutes.

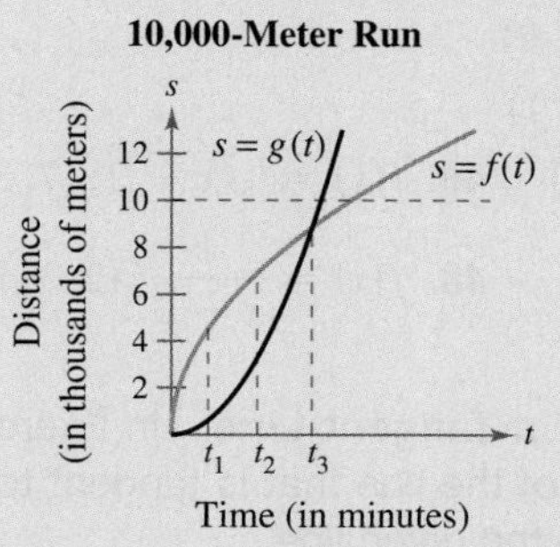

(a) Which runner is running faster at t_1?

(b) What conclusion can you make regarding their rates at t_2?

(c) What conclusion can you make regarding their rates at t_3?

(d) Which runner finishes the race first? Explain.

Finding the Slope of a Graph In Exercises 17–26, use the limit definition to find the slope of the tangent line to the graph of f at the given point. *See Examples 3, 4, and 5.*

17. $f(x) = -1$; $(0, -1)$

18. $f(x) = 6$; $(-2, 6)$

19. $f(x) = 8 - 3x$; $(2, 2)$

20. $f(x) = 6x + 3$; $(1, 9)$

21. $f(x) = 2x^2 - 3$; $(2, 5)$

22. $f(x) = 4 - x^2$; $(2, 0)$

23. $f(x) = x^3 - x$; $(2, 6)$

24. $f(x) = x^3 + 2x$; $(1, 3)$

25. $f(x) = 2\sqrt{x}$; $(4, 4)$

26. $f(x) = \sqrt{x + 1}$; $(8, 3)$

Finding a Derivative In Exercises 27–40, use the limit definition to find the derivative of the function. *See Examples 6 and 7.*

27. $f(x) = 3$

28. $f(x) = -2$

29. $f(x) = -5x$

30. $f(x) = 4x + 1$

31. $g(s) = \frac{1}{3}s + 2$

32. $h(t) = 6 - \frac{1}{2}t$

33. $f(x) = 4x^2 - 5x$

34. $f(x) = 2x^2 + 7x$

35. $h(t) = \sqrt{t - 1}$

36. $f(x) = \sqrt{x + 2}$

37. $f(t) = t^3 - 12t$

38. $f(t) = t^3 + t^2$

39. $f(x) = \dfrac{1}{x + 2}$

40. $g(s) = \dfrac{1}{s - 1}$

Finding an Equation of a Tangent Line In Exercises 41–48, use the limit definition to find an equation of the tangent line to the graph of f at the given point. Then verify your results by using a graphing utility to graph the function and its tangent line at the point.

41. $f(x) = \frac{1}{2}x^2;\ (2, 2)$ **42.** $f(x) = -x^2;\ (-1, -1)$

43. $f(x) = (x - 1)^2;\ (-2, 9)$

44. $f(x) = 2x^2 - 1;\ (0, -1)$

45. $f(x) = \sqrt{x} + 1;\ (4, 3)$ **46.** $f(x) = \sqrt{x + 2};\ (7, 3)$

47. $f(x) = \dfrac{1}{x};\ (1, 1)$ **48.** $f(x) = \dfrac{1}{x - 3};\ (2, -1)$

Finding an Equation of a Tangent Line In Exercises 49–52, find an equation of the line that is tangent to the graph of f and parallel to the given line.

Function	*Line*
49. $f(x) = -\frac{1}{4}x^2$	$x + y = 0$
50. $f(x) = x^2 - 7$	$2x + y = 0$
51. $f(x) = -\frac{1}{3}x^3$	$9x + y - 6 = 0$
52. $f(x) = x^2 - x$	$x + 2y - 6 = 0$

Determining Differentiability In Exercises 53–58, describe the x-values at which the function is differentiable. Explain your reasoning.

53. $y = |x + 3|$

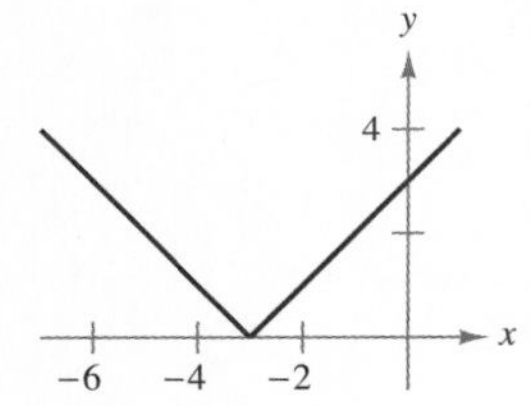

54. $y = |x^2 - 9|$

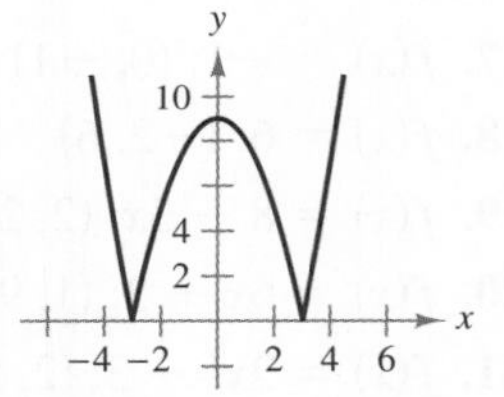

55. $y = (x - 3)^{2/3}$

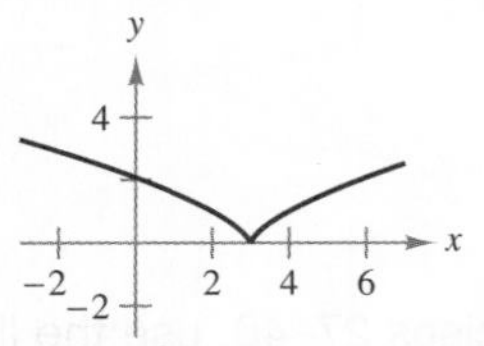

56. $y = \sqrt{x - 1}$

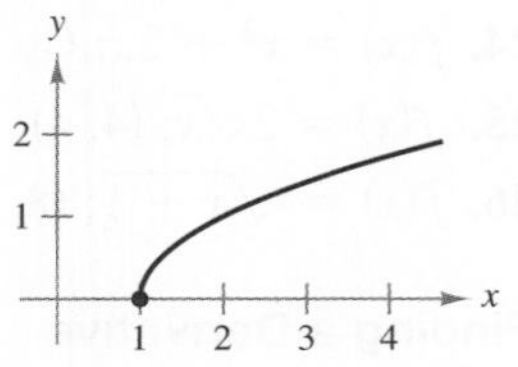

57. $y = \dfrac{x^2}{x^2 - 4}$

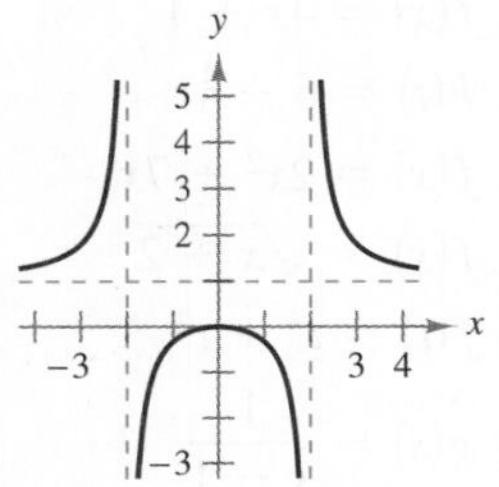

58. $y = \begin{cases} x^3 + 3, & x < 0 \\ x^3 - 3, & x \ge 0 \end{cases}$

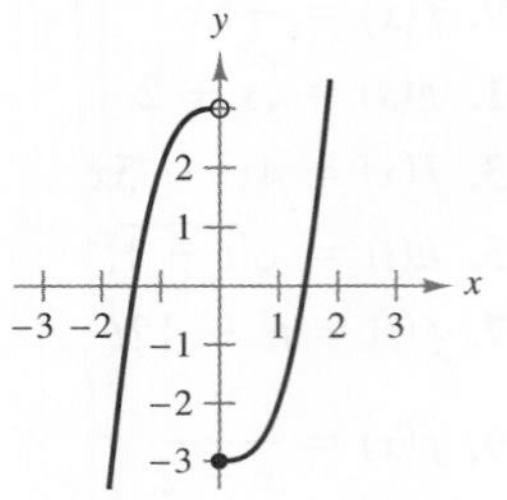

Writing a Function Using Derivatives In Exercises 59 and 60, identify a function f that has the given characteristics. Then sketch the function.

59. $f(0) = 2;\ f'(x) = -3$ for $-\infty < x < \infty$

60. $f(-2) = f(4) = 0;\ f'(1) = 0;$
$f'(x) < 0$ for $x < 1$; $f'(x) > 0$ for $x > 1$

Graphical, Numerical, and Analytic Analysis In Exercises 61–64, use a graphing utility to graph f on the interval $[-2, 2]$. Complete the table by graphically estimating the slopes of the graph at the given points. Then evaluate the slopes analytically and compare your results with those obtained graphically.

x	-2	$-\frac{3}{2}$	-1	$-\frac{1}{2}$	0	$\frac{1}{2}$	1	$\frac{3}{2}$	2
$f(x)$									
$f'(x)$									

61. $f(x) = \frac{1}{4}x^3$ **62.** $f(x) = \frac{3}{4}x^2$

63. $f(x) = -\frac{1}{2}x^3$ **64.** $f(x) = -\frac{3}{2}x^2$

Graphing a Function and Its Derivative In Exercises 65–68, find the derivative of the given function f. Then use a graphing utility to graph f and its derivative in the same viewing window. What does the x-intercept of the derivative indicate about the graph of f?

65. $f(x) = x^2 - 4x$ **66.** $f(x) = 2 + 6x - x^2$

67. $f(x) = x^3 - 3x$ **68.** $f(x) = x^3 - 6x^2$

True or False? In Exercises 69–72, determine whether the statement is true or false. If it is false, explain why or give an example that shows it is false.

69. The slope of the graph of $y = x^2$ is different at every point on the graph of f.

70. If a function is continuous at a point, then it is differentiable at that point.

71. If a function is differentiable at a point, then it is continuous at that point.

72. A tangent line to a graph can intersect the graph at more than one point.

73. Writing Use a graphing utility to graph the two functions

$$f(x) = x^2 + 1 \quad \text{and} \quad g(x) = |x| + 1$$

in the same viewing window. Use the *zoom* and *trace* features to analyze the graphs near the point $(0, 1)$. What do you observe? Which function is differentiable at this point? Write a short paragraph describing the geometric significance of differentiability at a point.

2.2 Some Rules for Differentiation

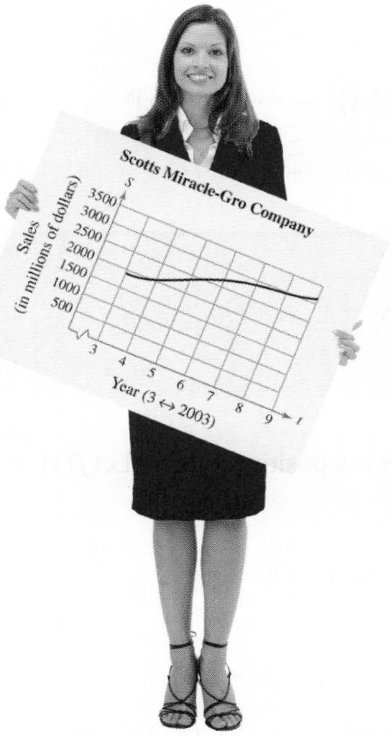

In Exercise 74 on page 102, you will use differentiation to find the rate of change in a company's sales.

- Find the derivatives of functions using the Constant Rule.
- Find the derivatives of functions using the Power Rule.
- Find the derivatives of functions using the Constant Multiple Rule.
- Find the derivatives of functions using the Sum and Difference Rules.
- Use derivatives to answer questions about real-life situations.

The Constant Rule

In Section 2.1, you found derivatives by the limit process. This process is tedious, even for simple functions, but fortunately there are rules that greatly simplify differentiation. These rules allow you to calculate derivatives without the *direct* use of limits.

The Constant Rule

The derivative of a constant function is zero. That is,

$$\frac{d}{dx}[c] = 0, \quad c \text{ is a constant.}$$

PROOF Let $f(x) = c$. Then, by the limit definition of the derivative, you can write

$$f'(x) = \lim_{\Delta x \to 0} \frac{f(x + \Delta x) - f(x)}{\Delta x} = \lim_{\Delta x \to 0} \frac{c - c}{\Delta x} = \lim_{\Delta x \to 0} 0 = 0.$$

So, $\frac{d}{dx}[c] = 0$.

Note in Figure 2.12 that the Constant Rule is equivalent to saying that the slope of a horizontal line is zero. An interpretation of the Constant Rule says that the tangent line to a constant function is the function itself. For instance, the equation of the tangent line to $f(x) = 4$ at $x = -1$ is

$$y = 4.$$

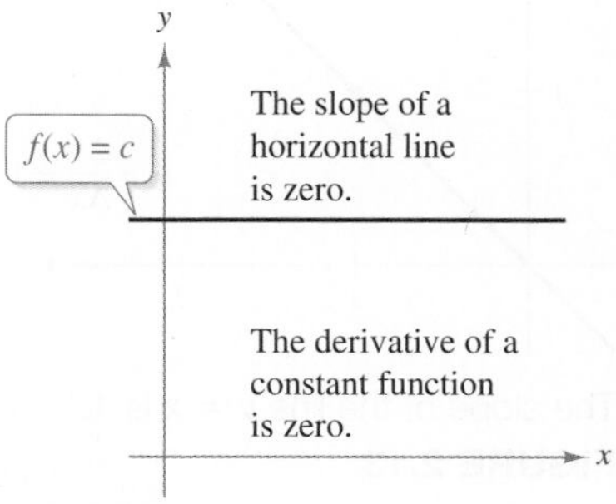

FIGURE 2.12

Example 1 Finding Derivatives of Constant Functions

a. $\frac{d}{dx}[7] = 0$

b. If $f(x) = 0$, then $f'(x) = 0$.

c. If $y = 2$, then $\frac{dy}{dx} = 0$.

d. If $g(t) = -\frac{3}{2}$, then $g'(t) = 0$.

✓Checkpoint 1

Find the derivative of each function.

a. $f(x) = -2$ **b.** $y = \pi$ **c.** $g(w) = \sqrt{5}$ **d.** $s(t) = 320.5$ ■

EDHAR/Shutterstock.com

The Power Rule

The binomial expansion process is used in proving a special case of the Power Rule.

$$(x + \Delta x)^2 = x^2 + 2x\,\Delta x + (\Delta x)^2$$

$$(x + \Delta x)^3 = x^3 + 3x^2\,\Delta x + 3x(\Delta x)^2 + (\Delta x)^3$$

$$(x + \Delta x)^n = x^n + nx^{n-1}\,\Delta x + \underbrace{\frac{n(n-1)x^{n-2}}{2}(\Delta x)^2 + \cdots + (\Delta x)^n}_{(\Delta x)^2 \text{ is a factor of these terms.}}$$

The (Simple) Power Rule

$$\frac{d}{dx}[x^n] = nx^{n-1}, \quad n \text{ is any real number.}$$

PROOF This proof is limited to the case in which n is a positive integer. Let $f(x) = x^n$. Using the binomial expansion, you can write

$$f'(x) = \lim_{\Delta x \to 0} \frac{f(x + \Delta x) - f(x)}{\Delta x} \qquad \text{Definition of derivative}$$

$$= \lim_{\Delta x \to 0} \frac{(x + \Delta x)^n - x^n}{\Delta x}$$

$$= \lim_{\Delta x \to 0} \frac{x^n + nx^{n-1}\,\Delta x + \dfrac{n(n-1)x^{n-2}}{2}(\Delta x)^2 + \cdots + (\Delta x)^n - x^n}{\Delta x}$$

$$= \lim_{\Delta x \to 0} \left[nx^{n-1} + \frac{n(n-1)x^{n-2}}{2}(\Delta x) + \cdots + (\Delta x)^{n-1}\right]$$

$$= nx^{n-1} + 0 + \cdots + 0$$

$$= nx^{n-1}.$$

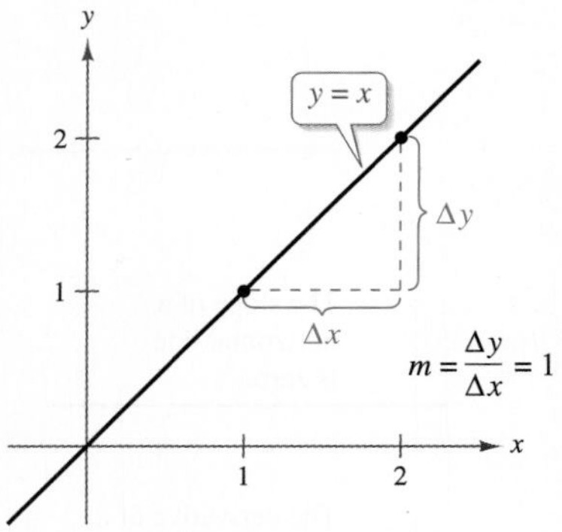

The slope of the line $y = x$ is 1.
FIGURE 2.13

For the Power Rule, the case in which $n = 1$ is worth remembering as a separate differentiation rule. That is,

$$\frac{d}{dx}[x] = 1. \qquad \text{The derivative of } x \text{ is 1.}$$

This rule is consistent with the fact that the slope of the line given by $y = x$ is 1. (See Figure 2.13.)

Example 2 Applying the Power Rule

Original Function	*Derivative*
a. $f(x) = x^3$	$f'(x) = 3x^2$
b. $y = \dfrac{1}{x^2} = x^{-2}$	$\dfrac{dy}{dx} = (-2)x^{-3} = -\dfrac{2}{x^3}$
c. $g(t) = t$	$g'(t) = 1$

✓ Checkpoint 2

Find the derivative of each function.

a. $f(x) = x^4$ **b.** $y = \dfrac{1}{x^3}$ **c.** $g(w) = w^2$ ■

In Example 2(b), note that *before* differentiating, you should rewrite $1/x^2$ as x^{-2}. Rewriting is the first step in *many* differentiation problems.

Original Function:		Rewrite:		Differentiate:		Simplify:
$y = \dfrac{1}{x^2}$	⟹	$y = x^{-2}$	⟹	$\dfrac{dy}{dx} = (-2)x^{-3}$	⟹	$\dfrac{dy}{dx} = -\dfrac{2}{x^3}$

Remember that the derivative of a function f is another function that gives the slope of the graph of f at any point at which f is differentiable. So, you can use the derivative to find slopes, as shown in Example 3.

Example 3 Finding the Slope of a Graph

Find the slopes of the graph of

$$f(x) = x^2$$

at $x = -2, -1, 0, 1,$ and 2.

SOLUTION Begin by using the Power Rule to find the derivative of f.

$$f'(x) = 2x \qquad \text{Derivative}$$

You can use the derivative to find the slopes of the graph of f, as shown.

x-Value	*Slope of Graph of f*
$x = -2$	$m = f'(-2) = 2(-2) = -4$
$x = -1$	$m = f'(-1) = 2(-1) = -2$
$x = 0$	$m = f'(0) = 2(0) = 0$
$x = 1$	$m = f'(1) = 2(1) = 2$
$x = 2$	$m = f'(2) = 2(2) = 4$

The graph of f is shown in Figure 2.14.

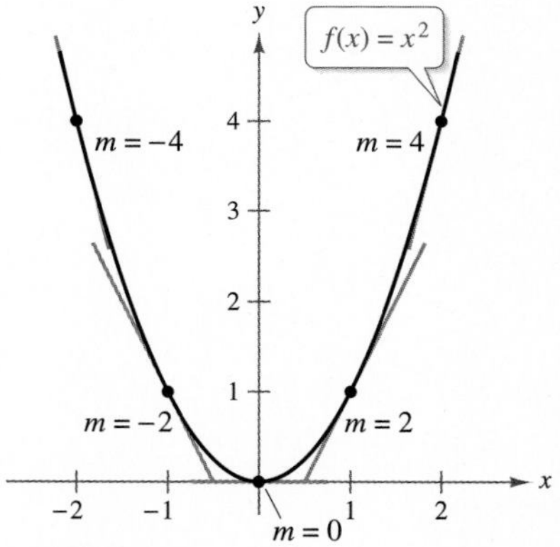

FIGURE 2.14

✓ Checkpoint 3

Find the slopes of the graph of

$$f(x) = x^3$$

at $x = -1, 0,$ and 1.

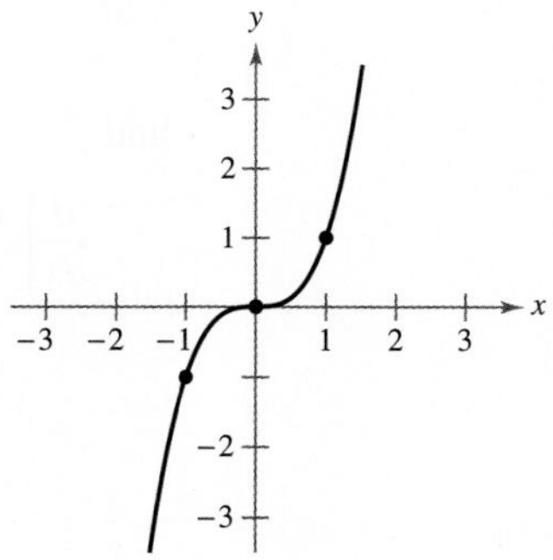

The Constant Multiple Rule

To prove the Constant Multiple Rule, the following property of limits is used.

$$\lim_{x\to a} cg(x) = c\left[\lim_{x\to a} g(x)\right]$$

The Constant Multiple Rule

If f is a differentiable function of x, and c is a real number, then

$$\frac{d}{dx}[cf(x)] = cf'(x), \quad c \text{ is a constant.}$$

PROOF Apply the definition of the derivative to produce

$$\begin{aligned}\frac{d}{dx}[cf(x)] &= \lim_{\Delta x\to 0}\frac{cf(x+\Delta x)-cf(x)}{\Delta x} && \text{Definition of derivative}\\ &= \lim_{\Delta x\to 0} c\left[\frac{f(x+\Delta x)-f(x)}{\Delta x}\right]\\ &= c\left[\lim_{\Delta x\to 0}\frac{f(x+\Delta x)-f(x)}{\Delta x}\right]\\ &= cf'(x).\end{aligned}$$

Informally, the Constant Multiple Rule states that constants can be factored out of the differentiation process.

$$\frac{d}{dx}[cf(x)] = c\frac{d}{dx}[f(x)] = cf'(x).$$

The usefulness of this rule is often overlooked, especially when the constant appears in the denominator, as shown below.

$$\begin{aligned}\frac{d}{dx}\left[\frac{f(x)}{c}\right] &= \frac{d}{dx}\left[\frac{1}{c}f(x)\right]\\ &= \frac{1}{c}\left(\frac{d}{dx}[f(x)]\right)\\ &= \frac{1}{c}f'(x).\end{aligned}$$

To use the Constant Multiple Rule efficiently, look for constants that can be factored out *before* differentiating. For example,

$$\begin{aligned}\frac{d}{dx}[5x^2] &= 5\frac{d}{dx}[x^2] && \text{Factor out 5.}\\ &= 5(2x) && \text{Differentiate.}\\ &= 10x && \text{Simplify.}\end{aligned}$$

and

$$\begin{aligned}\frac{d}{dx}\left[\frac{x^2}{5}\right] &= \frac{1}{5}\left(\frac{d}{dx}[x^2]\right) && \text{Factor out } \tfrac{1}{5}.\\ &= \frac{1}{5}(2x) && \text{Differentiate.}\\ &= \frac{2}{5}x. && \text{Simplify.}\end{aligned}$$

TECH TUTOR

If you have access to a symbolic differentiation utility, try using it to confirm the derivatives shown in this section.

Example 4 Using the Power and Constant Multiple Rules

Find the derivative of (a) $y = 2x^{1/2}$ and (b) $f(t) = \dfrac{4t^2}{5}$.

SOLUTION

a. Using the Constant Multiple Rule and the Power Rule, you can write

$$\frac{dy}{dx} = \frac{d}{dx}[2x^{1/2}] = \underbrace{2\frac{d}{dx}[x^{1/2}]}_{\text{Constant Multiple Rule}} = \underbrace{2\left(\frac{1}{2}x^{-1/2}\right)}_{\text{Power Rule}} = x^{-1/2} = \frac{1}{\sqrt{x}}.$$

b. Begin by rewriting $f(t)$ as

$$f(t) = \frac{4t^2}{5} = \frac{4}{5}t^2.$$

Then, use the Constant Multiple Rule and the Power Rule to obtain

$$f'(t) = \frac{d}{dt}\left[\frac{4}{5}t^2\right] = \frac{4}{5}\left(\frac{d}{dt}[t^2]\right) = \frac{4}{5}(2t) = \frac{8}{5}t.$$

✓ Checkpoint 4

Find the derivative of (a) $y = 4x^2$ and (b) $f(x) = 16x^{1/2}$. ■

You may find it helpful to combine the Constant Multiple Rule and the Power Rule into one combined rule.

$$\frac{d}{dx}[cx^n] = cnx^{n-1}, \quad n \text{ is a real number, } c \text{ is a constant.}$$

For instance, in Example 4(b), you can apply this combined rule to obtain

$$\frac{d}{dt}\left[\frac{4}{5}t^2\right] = \left(\frac{4}{5}\right)(2)(t) = \frac{8}{5}t.$$

The three functions in the next example are simple, yet errors are frequently made in differentiating functions involving constant multiples of the first power of x. Keep in mind that

$$\frac{d}{dx}[cx] = c, \quad c \text{ is a constant.}$$

Example 5 Applying the Constant Multiple Rule

Original Function	*Derivative*
a. $y = -\dfrac{3x}{2}$	$y' = -\dfrac{3}{2}$
b. $y = 3\pi x$	$y' = 3\pi$
c. $y = -\dfrac{x}{2}$	$y' = -\dfrac{1}{2}$

✓ Checkpoint 5

Find the derivative of (a) $y = \dfrac{t}{4}$ and (b) $y = -\dfrac{2x}{5}$. ■

Parentheses can play an important role in the use of the Constant Multiple Rule and the Power Rule. In Example 6, be sure you understand the mathematical conventions involving the use of parentheses.

Example 6 Using Parentheses When Differentiating

	Original Function	*Rewrite*	*Differentiate*	*Simplify*
a.	$y = \frac{5}{2x^3}$	$y = \frac{5}{2}(x^{-3})$	$y' = \frac{5}{2}(-3x^{-4})$	$y' = -\frac{15}{2x^4}$
b.	$y = \frac{5}{(2x)^3}$	$y = \frac{5}{8}(x^{-3})$	$y' = \frac{5}{8}(-3x^{-4})$	$y' = -\frac{15}{8x^4}$
c.	$y = \frac{7}{3x^{-2}}$	$y = \frac{7}{3}(x^2)$	$y' = \frac{7}{3}(2x)$	$y' = \frac{14x}{3}$
d.	$y = \frac{7}{(3x)^{-2}}$	$y = 63(x^2)$	$y' = 63(2x)$	$y' = 126x$

✓ Checkpoint 6

Find the derivative of each function.

a. $y = \frac{9}{4x^2}$

b. $y = \frac{9}{(4x)^2}$

When differentiating functions involving radicals, you should rewrite the function with rational exponents. For instance, you should rewrite

$$y = \sqrt[3]{x} \quad \text{as} \quad y = x^{1/3}$$

and you should rewrite

$$y = \frac{1}{\sqrt[3]{x^4}} \quad \text{as} \quad y = x^{-4/3}.$$

Example 7 Differentiating Radical Functions

	Original Function	*Rewrite*	*Differentiate*	*Simplify*
a.	$y = \sqrt{x}$	$y = x^{1/2}$	$y' = \left(\frac{1}{2}\right)x^{-1/2}$	$y' = \frac{1}{2\sqrt{x}}$
b.	$y = \frac{1}{2\sqrt[3]{x^2}}$	$y = \frac{1}{2}x^{-2/3}$	$y' = \frac{1}{2}\left(-\frac{2}{3}\right)x^{-5/3}$	$y' = -\frac{1}{3x^{5/3}}$
c.	$y = \sqrt{2x}$	$y = \sqrt{2}(x^{1/2})$	$y' = \sqrt{2}\left(\frac{1}{2}\right)x^{-1/2}$	$y' = \frac{1}{\sqrt{2x}}$

✓ Checkpoint 7

Find the derivative of each function.

a. $y = \sqrt{5x}$

b. $y = \sqrt[4]{x}$

The Sum and Difference Rules

To differentiate $y = 3x + 2x^3$, you would probably write

$$y' = 3 + 6x^2$$

without questioning your answer. The validity of differentiating a sum or difference of functions term by term is given by the Sum and Difference Rules.

The Sum and Difference Rules

The derivative of the sum or difference of two differentiable functions is the sum or difference of their derivatives.

$$\frac{d}{dx}[f(x) + g(x)] = f'(x) + g'(x) \qquad \text{Sum Rule}$$

$$\frac{d}{dx}[f(x) - g(x)] = f'(x) - g'(x) \qquad \text{Difference Rule}$$

PROOF Let $h(x) = f(x) + g(x)$. Then, you can prove the Sum Rule as shown.

$$\begin{aligned} h'(x) &= \lim_{\Delta x \to 0} \frac{h(x + \Delta x) - h(x)}{\Delta x} \qquad \text{Definition of derivative} \\ &= \lim_{\Delta x \to 0} \frac{f(x + \Delta x) + g(x + \Delta x) - f(x) - g(x)}{\Delta x} \\ &= \lim_{\Delta x \to 0} \frac{f(x + \Delta x) - f(x) + g(x + \Delta x) - g(x)}{\Delta x} \\ &= \lim_{\Delta x \to 0} \left[\frac{f(x + \Delta x) - f(x)}{\Delta x} + \frac{g(x + \Delta x) - g(x)}{\Delta x} \right] \\ &= \lim_{\Delta x \to 0} \frac{f(x + \Delta x) - f(x)}{\Delta x} + \lim_{\Delta x \to 0} \frac{g(x + \Delta x) - g(x)}{\Delta x} \\ &= f'(x) + g'(x) \end{aligned}$$

So,

$$\frac{d}{dx}[f(x) + g(x)] = f'(x) + g'(x).$$

The Difference Rule can be proved in a similar manner.

The Sum and Difference Rules can be extended to the sum or difference of any finite number of functions. For instance, if $y = f(x) + g(x) + h(x)$, then $y' = f'(x) + g'(x) + h'(x)$.

STUDY TIP

Look back at Example 6 on page 85. Notice that the example asks for the derivative of the difference of two functions. Compare the result with the one obtained in Example 8(b) at the right.

Example 8 Using the Sum and Difference Rules

Original Function	*Derivative*
a. $y = x^3 + 4x^2$	$y' = 3x^2 + 8x$
b. $f(x) = 3x^2 - 2x$	$f'(x) = 6x - 2$

✓ Checkpoint 8

Find the derivative of each function.

a. $f(x) = 2x^2 + 5x$ **b.** $y = x^4 - 2x$ ■

R. Gino Santa Maria/Shutterstock.com

With the differentiation rules listed in this section, you can differentiate any polynomial function.

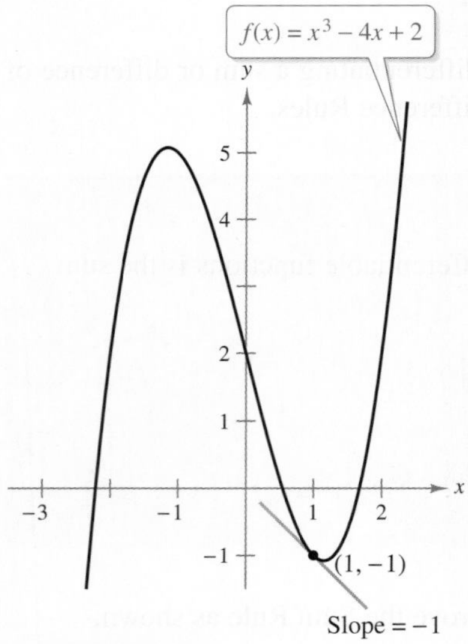

FIGURE 2.15

Example 9 Finding the Slope of a Graph

Find the slope of the graph of $f(x) = x^3 - 4x + 2$ at the point $(1, -1)$.

SOLUTION The derivative of $f(x)$ is

$$f'(x) = 3x^2 - 4.$$

So, the slope of the graph of f at $(1, -1)$ is

$$\text{Slope} = f'(1) = 3(1)^2 - 4 = 3 - 4 = -1$$

as shown in Figure 2.15.

✓ Checkpoint 9

Find the slope of the graph of $f(x) = x^2 - 5x + 1$ at the point $(2, -5)$. ■

Example 9 illustrates the use of the derivative for determining the shape of a graph. A rough sketch of the graph of $f(x) = x^3 - 4x + 2$ might lead you to think that the point $(1, -1)$ is a minimum point of the graph. After finding the slope at this point to be -1, however, you can conclude that the minimum point (where the slope is 0) is farther to the right. (You will study techniques for finding minimum and maximum points in Section 3.2.)

Example 10 Finding an Equation of a Tangent Line

Find an equation of the tangent line to the graph of

$$g(x) = -\frac{1}{2}x^4 + 3x^3 - 2x$$

at the point $\left(-1, -\frac{3}{2}\right)$.

SOLUTION The derivative of $g(x)$ is $g'(x) = -2x^3 + 9x^2 - 2$, which implies that the slope of the graph at the point $\left(-1, -\frac{3}{2}\right)$ is

$$\begin{aligned}\text{Slope} &= g'(-1)\\ &= -2(-1)^3 + 9(-1)^2 - 2\\ &= 2 + 9 - 2\\ &= 9\end{aligned}$$

as shown in Figure 2.16. Using the point-slope form, you can write the equation of the tangent line at $\left(-1, -\frac{3}{2}\right)$ as shown.

$$y - \left(-\frac{3}{2}\right) = 9[x - (-1)] \qquad \text{Point-slope form}$$

$$y + \frac{3}{2} = 9x + 9 \qquad \text{Simplify.}$$

$$y = 9x + \frac{15}{2} \qquad \text{Equation of tangent line}$$

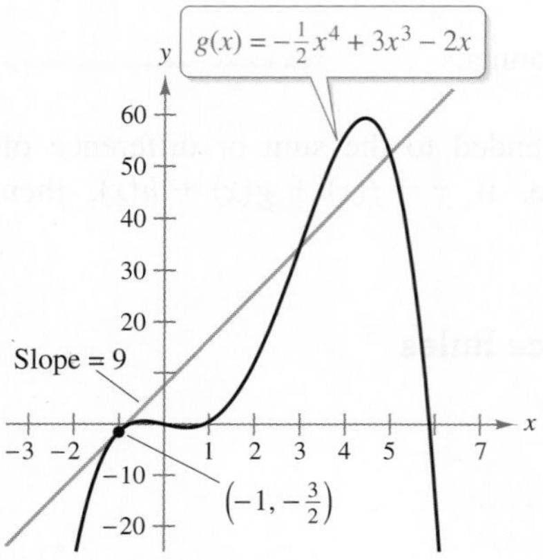

FIGURE 2.16

✓ Checkpoint 10

Find an equation of the tangent line to the graph of $f(x) = -x^2 + 3x - 2$ at the point $(2, 0)$. ■

Application

There are many applications of the derivative that you will study in this textbook. In Example 11, you will use a derivative to find the rate of change of a company's revenue with respect to time.

Example 11 Modeling Revenue

From 2004 through 2009, the revenue R (in millions of dollars) for McDonald's can be modeled by

$$R = -130.769t^3 + 2296.47t^2 - 11{,}493.5t + 35{,}493, \quad 4 \le t \le 9$$

where t represents the year, with $t = 4$ corresponding to 2004. At what rate was McDonald's revenue changing in 2006? *(Source: McDonald's Corporation)*

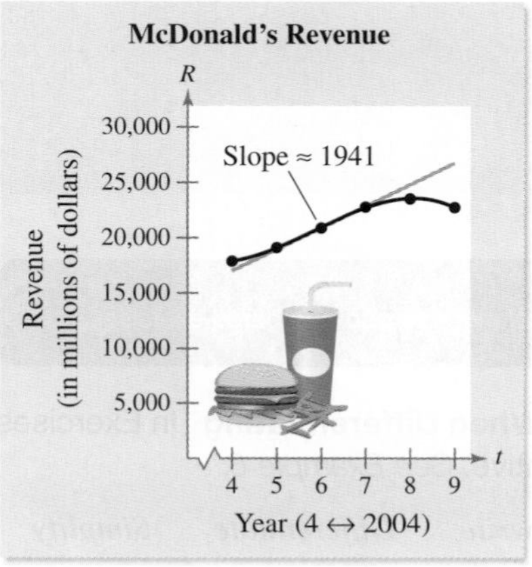

FIGURE 2.17

SOLUTION One way to answer this question is to find the derivative of the revenue model with respect to time.

$$\frac{dR}{dt} = -392.307t^2 + 4592.94t - 11{,}493.5, \quad 4 \le t \le 9$$

In 2006 (at $t = 6$), the rate of change of the revenue with respect to time is given by

$$\frac{dR}{dt} = -392.307(6)^2 + 4592.94(6) - 11{,}493.5 \approx 1941.$$

Because R is measured in millions of dollars and t is measured in years, it follows that the derivative dR/dt is measured in millions of dollars per year. So, at the end of 2006, McDonald's revenue was increasing at a rate of about \$1941 million per year, as shown in Figure 2.17.

✓Checkpoint 11

From 2000 through 2010, the sales per share S (in dollars) for Microsoft Corporation can be modeled by

$$S = 0.0330t^2 + 0.208t + 2.13, \quad 0 \le t \le 10$$

where t represents the year, with $t = 0$ corresponding to 2000. At what rate was Microsoft's sales per share changing in 2002? *(Source: Microsoft Corporation)* ■

SUMMARIZE (Section 2.2)

1. State the Constant Rule *(page 91)*. For an example of the Constant Rule, see Example 1.
2. State the Power Rule *(page 92)*. For examples of the Power Rule, see Examples 2 and 3.
3. State the Constant Multiple Rule *(page 94)*. For examples of the Constant Multiple Rule, see Examples 4, 5, 6, and 7.
4. State the Sum Rule *(page 97)*. For an example of the Sum Rule, see Example 8.
5. State the Difference Rule *(page 97)*. For an example of the Difference Rule, see Example 8.
6. Describe a real-life example of how differentiation can be used to analyze the rate of change of a company's revenue *(page 99, Example 11)*.

SKILLS WARM UP 2.2 The following warm-up exercises involve skills that were covered in a previous course. You will use these skills in the exercise set for this section. For additional help, review Appendix Sections A.3 and A.4.

In Exercises 1 and 2, evaluate each expression when $x = 2$.

1. (a) $2x^2$ (b) $(2x)^2$ (c) $2x^{-2}$

2. (a) $\frac{1}{(3x)^2}$ (b) $\frac{1}{4x^3}$ (c) $\frac{(2x)^{-3}}{4x^{-2}}$

In Exercises 3–6, simplify the expression.

3. $4(3)x^3 + 2(2)x$

4. $\frac{1}{2}(3)x^2 - \frac{3}{2}x^{1/2}$

5. $\left(\frac{1}{4}\right)x^{-3/4}$

6. $\frac{1}{3}(3)x^2 - 2\left(\frac{1}{2}\right)x^{-1/2} + \frac{1}{3}x^{-2/3}$

In Exercises 7–10, solve the equation.

7. $3x^2 + 2x = 0$

8. $x^3 - x = 0$

9. $x^2 + 8x - 20 = 0$

10. $x^2 - 10x - 24 = 0$

Exercises 2.2

See www.CalcChat.com for worked-out solutions to odd-numbered exercises.

Finding Derivatives In Exercises 1–24, find the derivative of the function. *See Examples 1, 2, 4, 5, and 8.*

1. $y = 3$

2. $f(x) = -8$

3. $y = x^5$

4. $f(x) = \frac{1}{x^5}$

5. $h(x) = 3x^3$

6. $h(x) = 2x^5$

7. $y = \frac{2x^3}{3}$

8. $g(t) = \frac{3t^2}{4}$

9. $f(x) = 4x$

10. $g(x) = 3x$

11. $y = 8 - x^3$

12. $y = t^2 - 6$

13. $f(x) = 4x^2 - 3x$

14. $g(x) = x^2 + 4x^3$

15. $f(t) = -3t^2 + 2t - 4$

16. $y = x^3 - 9x^2 + 2$

17. $s(t) = t^3 - 2t + 4$

18. $y = 2x^3 - x^2 + 3x - 1$

19. $g(x) = x^{2/3}$

20. $h(x) = x^{5/2}$

21. $y = 4t^{4/3}$

22. $f(x) = 10x^{1/2}$

23. $y = 4x^{-2} + 2x^2$

24. $s(t) = 4t^{-1} + 1$

Using Parentheses When Differentiating In Exercises 25–30, find the derivative. *See Example 6.*

	Function	*Rewrite*	*Differentiate*	*Simplify*
25.	$y = \frac{2}{7x^4}$			
26.	$y = \frac{2}{3x^2}$			
27.	$y = \frac{1}{(4x)^3}$			
28.	$y = \frac{\pi}{(3x)^2}$			
29.	$y = \frac{4}{(2x)^{-5}}$			
30.	$y = \frac{4x}{x^{-3}}$			

Differentiating Radical Functions In Exercises 31–36, find the derivative. *See Example 7.*

	Function	*Rewrite*	*Differentiate*	*Simplify*
31.	$y = 6\sqrt{x}$			
32.	$y = \frac{3\sqrt{x}}{4}$			
33.	$y = \frac{1}{5\sqrt[5]{x}}$			
34.	$y = \frac{3}{2\sqrt[4]{x^3}}$			
35.	$y = \sqrt{3x}$			
36.	$y = \sqrt[3]{6x^2}$			

Finding the Slope of a Graph In Exercises 37–44, find the slope of the graph of the function at the given point. *See Examples 3 and 9.*

37. $y = x^{3/2}$

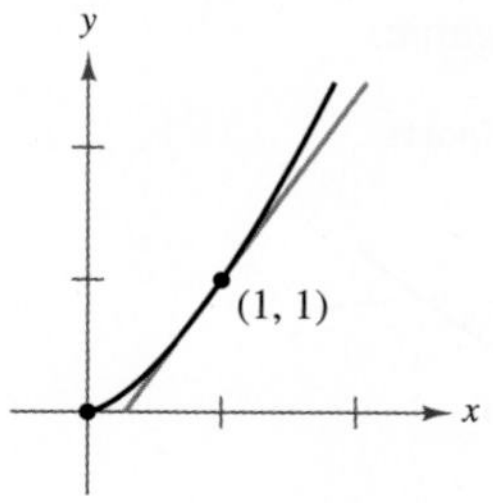

38. $y = x^{-1}$

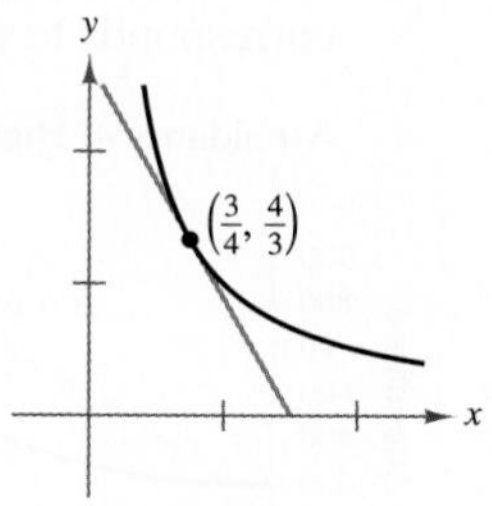

39. $f(t) = t^2$; $(4, 16)$

40. $f(x) = x^{-1/3}$; $\left(8, \frac{1}{2}\right)$

41. $f(x) = 2x^3 + 8x^2 - x - 4$; $(-1, 3)$

42. $f(x) = 3x^4 - 5x^3 + 6x^2 - 10x$; $(1, -6)$

43. $f(x) = -\frac{1}{2}x(1 + x^2)$; $(1, -1)$

44. $f(x) = 3(5 - x)^2$; $(5, 0)$

Finding an Equation of a Tangent Line In Exercises 45–50, (a) find an equation of the tangent line to the graph of the function at the given point, (b) use a graphing utility to graph the function and its tangent line at the point, and (c) use the derivative feature of a graphing utility to confirm your results. *See Example 10.*

45. $y = -2x^4 + 5x^2 - 3$; $(1, 0)$

46. $y = x^3 + x$; $(-1, -2)$

47. $f(x) = \sqrt[3]{x} + \sqrt[5]{x}$; $(1, 2)$

48. $f(x) = \dfrac{1}{\sqrt[3]{x^2}} - x$; $(-1, 2)$

49. $y = 3x\left(x^2 - \dfrac{2}{x}\right)$; $(2, 18)$

50. $y = (2x + 1)^2$; $(0, 1)$

Finding Derivatives In Exercises 51–62, find $f'(x)$.

51. $f(x) = x^2 - \dfrac{4}{x} - 3x^{-2}$

52. $f(x) = x^2 - 3x - 3x^{-2} + 5x^{-3}$

53. $f(x) = x^2 - 2x - \dfrac{2}{x^4}$

54. $f(x) = x^2 + 4x + \dfrac{1}{x}$

55. $f(x) = x^{4/5} + x$

56. $f(x) = x^{1/3} - 1$

57. $f(x) = x(x^2 + 1)$

58. $f(x) = (x^2 + 2x)(x + 1)$

59. $f(x) = \dfrac{2x^3 - 4x^2 + 3}{x^2}$

60. $f(x) = \dfrac{2x^2 - 3x + 1}{x}$

61. $f(x) = \dfrac{4x^3 - 3x^2 + 2x + 5}{x^2}$

62. $f(x) = \dfrac{-6x^3 + 3x^2 - 2x + 1}{x}$

Finding Horizontal Tangent Lines In Exercises 63–66, determine the point(s), if any, at which the graph of the function has a horizontal tangent line.

63. $y = -x^4 + 3x^2 - 1$

64. $y = x^3 + 3x^2$

65. $y = \frac{1}{2}x^2 + 5x$

66. $y = x^2 + 2x$

Exploring Relationships In Exercises 67 and 68, (a) sketch the graphs of f and g, (b) find $f'(1)$ and $g'(1)$, (c) sketch the tangent line to each graph at $x = 1$, and (d) explain the relationship between f' and g'.

67. $f(x) = x^3$
$g(x) = x^3 + 3$

68. $f(x) = x^2$
$g(x) = 3x^2$

Exploring Relationships In Exercises 69–72, the relationship between f and g is given. Explain the relationship between f' and g'.

69. $g(x) = f(x) + 6$

70. $g(x) = 2f(x)$

71. $g(x) = -5f(x)$

72. $g(x) = 3f(x) - 1$

73. Revenue The revenue R (in millions of dollars) for Under Armour from 2004 through 2009 can be modeled by

$$R = -5.1509t^3 + 103.166t^2 - 526.15t + 985.4$$

where t is the year, with $t = 4$ corresponding to 2004. *(Source: Under Armour, Inc.)*

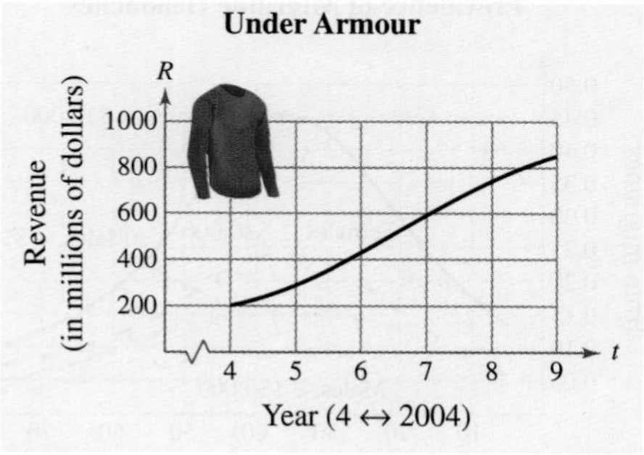

(a) Find the slopes of the graph for the years 2005 and 2007.

(b) Compare your results with those obtained in Exercise 13 in Section 2.1.

(c) Interpret the slope of the graph in the context of the problem.

74. Sales The sales S (in millions of dollars) for Scotts Miracle-Gro Company from 2003 through 2009 can be modeled by $S = 5.45682t^4 - 136.9359t^3 + 1219.018t^2 - 4294.73t + 7078.4$, where t is the year, with $t = 3$ corresponding to 2003. *(Source: Scotts Miracle-Gro Company)*

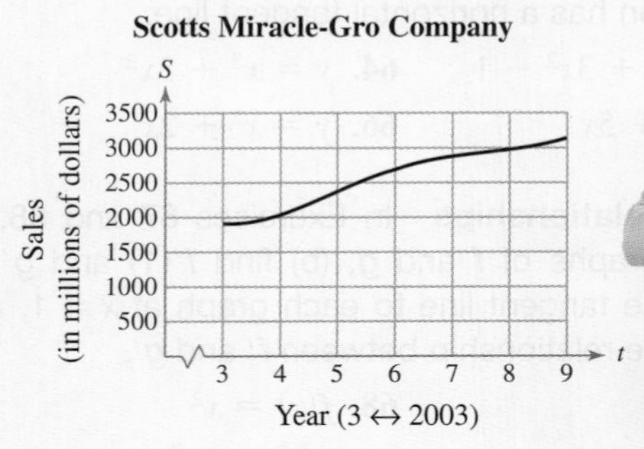

(a) Find the slopes of the graph for the years 2006 and 2008.

(b) Compare your results with those obtained in Exercise 14 in Section 2.1.

(c) Interpret the slope of the graph in the context of the problem.

75. Psychology: Migraine Prevalence The graph illustrates the prevalence of migraine headaches in males and females in selected income groups. *(Source: Adapted from Sue/Sue/Sue,* Understanding Abnormal Behavior, *Seventh Edition)*

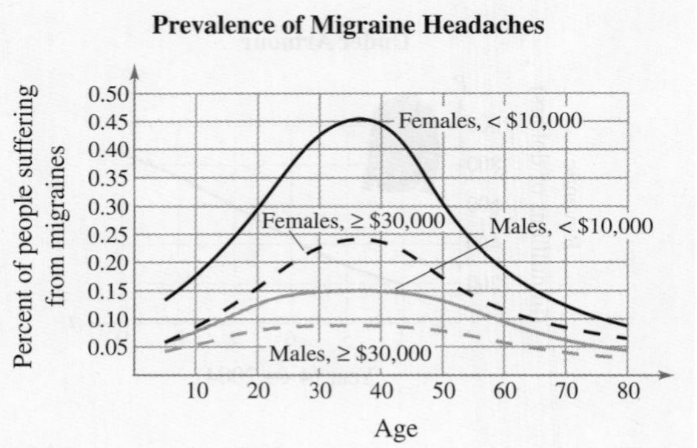

(a) Write a short paragraph describing your general observations about the prevalence of migraines in females and males with respect to age group and income bracket.

(b) Describe the graph of the derivative of each curve, and explain the significance of each derivative. Include an explanation of the units of the derivatives, and indicate the time intervals in which the derivatives would be positive and negative.

76. HOW DO YOU SEE IT? The attendance for four high school basketball games is given by $s = f(t)$, and the attendance for four high school football games is given by $s = g(t)$, where $t = 1$ corresponds to the first game.

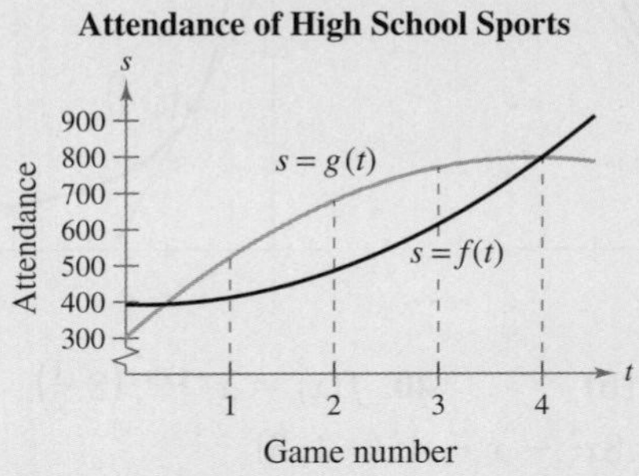

(a) Which attendance rate, f' or g', is greater at game 1?

(b) What conclusion can you make regarding the attendance rates, f' and g', at game 3?

(c) What conclusion can you make regarding the attendance rates, f' and g', at game 4?

(d) Which sport do you think would have a greater attendance for game 5? Explain your reasoning.

77. Cost The marginal cost for manufacturing an electrical component is \$7.75 per unit, and the fixed cost is \$500. Write the cost C as a function of x, the number of units produced. Show that the derivative of this cost function is a constant and is equal to the marginal cost.

78. Political Fundraiser A politician raises funds by selling tickets to a dinner for \$500. The politician pays \$150 for each dinner and has fixed costs of \$7000 to rent a dining hall and wait staff. Write the profit P as a function of x, the number of dinners sold. Show that the derivative of the profit function is a constant and is equal to the increase in profit from each dinner sold.

Finding Horizontal Tangent Lines In Exercises 79 and 80, use a graphing utility to graph f and f' over the given interval. Determine any points at which the graph of f has horizontal tangents.

	Function	*Interval*
79.	$f(x) = 4.1x^3 - 12x^2 + 2.5x$	$[0, 3]$
80.	$f(x) = x^3 - 1.4x^2 - 0.96x + 1.44$	$[-2, 2]$

True or False? In Exercises 81 and 82, determine whether the statement is true or false. If it is false, explain why or give an example that shows it is false.

81. If $f'(x) = g'(x)$, then $f(x) = g(x)$.

82. If $f(x) = g(x) + c$, then $f'(x) = g'(x)$.

2.3 Rates of Change: Velocity and Marginals

- Find the average rates of change of functions over intervals.
- Find the instantaneous rates of change of functions at points.
- Find the marginal revenues, marginal costs, and marginal profits for products.

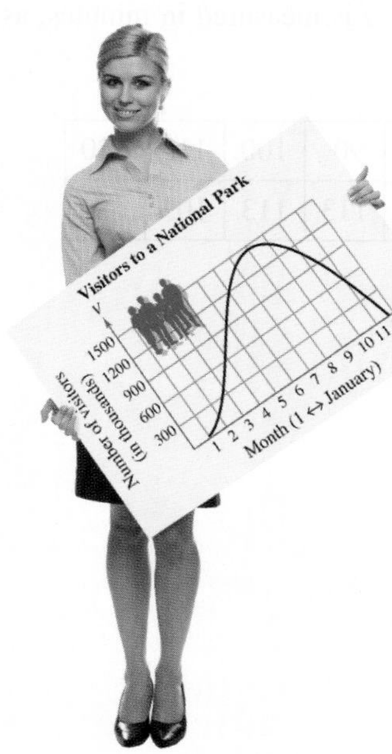

In Exercise 13 on page 114, you will use the graph of a function to estimate the rate of change of the number of visitors to a national park.

Average Rate of Change

In Sections 2.1 and 2.2, you studied the two primary applications of derivatives.

1. **Slope** The derivative of f is a function that gives the slope of the graph of f at a point $(x, f(x))$.
2. **Rate of Change** The derivative of f is a function that gives the rate of change of $f(x)$ with respect to x at the point $(x, f(x))$.

In this section, you will see that there are many real-life applications of rates of change. A few are velocity, acceleration, population growth rates, unemployment rates, production rates, and water flow rates. Although rates of change often involve change with respect to time, you can investigate the rate of change of one variable with respect to any other related variable.

When determining the rate of change of one variable with respect to another, you must be careful to distinguish between *average* and *instantaneous* rates of change. The distinction between these two rates of change is comparable to the distinction between the slope of the secant line through two points on a graph and the slope of the tangent line at one point on the graph.

Definition of Average Rate of Change

If $y = f(x)$, then the **average rate of change** of y with respect to x on the interval $[a, b]$ is

$$\text{Average rate of change} = \frac{f(b) - f(a)}{b - a} = \frac{\Delta y}{\Delta x}.$$

Note that $f(a)$ is the value of the function at the *left* endpoint of the interval, $f(b)$ is the value of the function at the *right* endpoint of the interval, and $b - a$ is the width of the interval, as shown in Figure 2.18.

STUDY TIP

In real-life problems, it is important to list the units of measure for a rate of change. The units for $\Delta y/\Delta x$ are "y-units" per "x-units." For example, if y is measured in miles and x is measured in hours, then $\Delta y/\Delta x$ is measured in *miles per hour*.

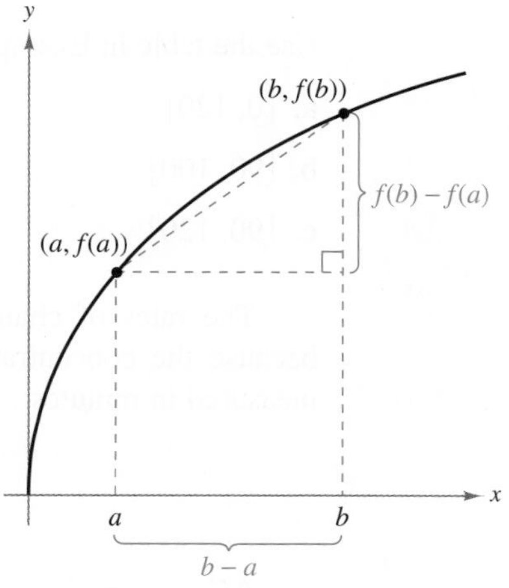

FIGURE 2.18

Example 1 Medicine

The concentration C (in milligrams per milliliter) of a drug in a patient's bloodstream is monitored over 10-minute intervals for 2 hours, where t is measured in minutes, as shown in the table.

t	0	10	20	30	40	50	60	70	80	90	100	110	120
C	0	2	17	37	55	73	89	103	111	113	113	103	68

Find the average rate of change of C over each interval.

a. [0, 10]

b. [0, 20]

c. [100, 110]

SOLUTION

a. For the interval [0, 10], the average rate of change is

Value of C at right endpoint
Value of C at left endpoint

$$\frac{\Delta C}{\Delta t} = \frac{2 - 0}{10 - 0} = \frac{2}{10} = 0.2 \text{ milligram per milliliter per minute.}$$

Width of interval

b. For the interval [0, 20], the average rate of change is

$$\frac{\Delta C}{\Delta t} = \frac{17 - 0}{20 - 0} = \frac{17}{20} = 0.85 \text{ milligram per milliliter per minute.}$$

c. For the interval [100, 110], the average rate of change is

$$\frac{\Delta C}{\Delta t} = \frac{103 - 113}{110 - 100} = \frac{-10}{10} = -1 \text{ milligram per milliliter per minute.}$$

Notice in Figure 2.19 that the average rate of change is positive when the concentration increases and negative when the concentration decreases.

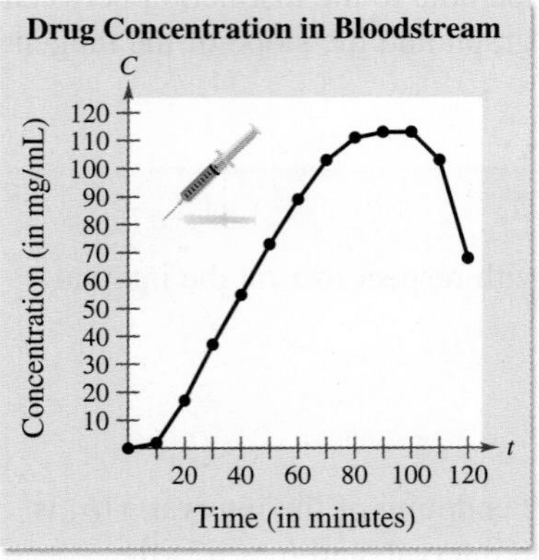

FIGURE 2.19

✓ Checkpoint 1

Use the table in Example 1 to find the average rate of change of C over each interval.

a. [0, 120]

b. [90, 100]

c. [90, 120] ■

The rates of change in Example 1 are in milligrams per milliliter per minute because the concentration is measured in milligrams per milliliter and the time is measured in minutes.

Concentration is measured in milligrams per milliliter.
Rate of change is measured in milligrams per milliliter per minute.

$$\frac{\Delta C}{\Delta t} = \frac{2 - 0}{10 - 0} = \frac{2}{10} = 0.2 \text{ milligram per milliliter per minute}$$

Time is measured in minutes.

A common application of an average rate of change is to find the **average velocity** of an object that is moving in a straight line. That is,

$$\text{Average velocity} = \frac{\text{change in distance}}{\text{change in time}}.$$

This formula is demonstrated in Example 2.

Example 2 Finding an Average Velocity

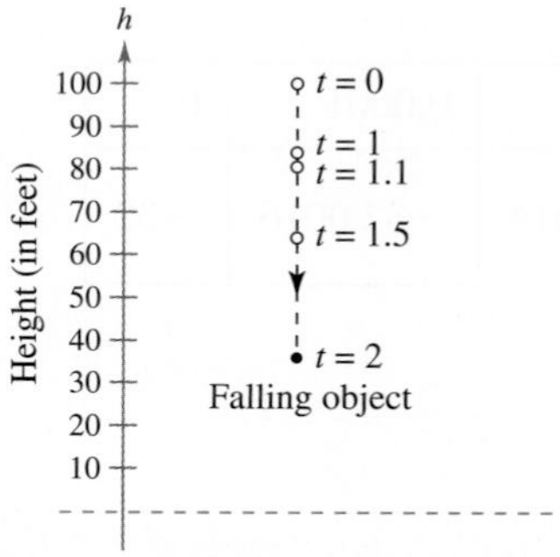

Some falling objects have considerable air resistance. Other falling objects have negligible air resistance. When modeling a falling-body problem, you must decide whether to account for air resistance or neglect it.

FIGURE 2.20

A free-falling object is dropped from a height of 100 feet. *Neglecting air resistance,* the height h (in feet) of the object at time t (in seconds) is given by

$$h = -16t^2 + 100. \qquad \text{(See Figure 2.20.)}$$

Find the average velocity of the object over each interval.

a. [1, 2] **b.** [1, 1.5] **c.** [1, 1.1]

SOLUTION You can use the position equation $h = -16t^2 + 100$ to determine the heights at

$$t = 1, 1.1, 1.5, \text{ and } 2$$

as shown in the table.

t (in seconds)	0	1	1.1	1.5	2
h (in feet)	100	84	80.64	64	36

a. For the interval [1, 2], the object falls from a height of 84 feet to a height of 36 feet. So, the average velocity is

$$\frac{\Delta h}{\Delta t} = \frac{36 - 84}{2 - 1} = \frac{-48}{1} = -48 \text{ feet per second.}$$

b. For the interval [1, 1.5], the average velocity is

$$\frac{\Delta h}{\Delta t} = \frac{64 - 84}{1.5 - 1} = \frac{-20}{0.5} = -40 \text{ feet per second.}$$

c. For the interval [1, 1.1], the average velocity is

$$\frac{\Delta h}{\Delta t} = \frac{80.64 - 84}{1.1 - 1} = \frac{-3.36}{0.1} = -33.6 \text{ feet per second.}$$

✓ Checkpoint 2

The height h (in feet) of a free-falling object at time t (in seconds) is given by

$$h = -16t^2 + 180.$$

Find the average velocity of the object over each interval.

a. [0, 1] **b.** [1, 2] **c.** [2, 3] ■

STUDY TIP

In Example 2, the average velocities are negative because the object is moving downward.

Instantaneous Rate of Change and Velocity

Suppose in Example 2 you wanted to find the rate of change of h at the instant $t = 1$ second. Such a rate is called an **instantaneous rate of change.** You can approximate the instantaneous rate of change at $t = 1$ by calculating the average rate of change over smaller and smaller intervals of the form $[1, 1 + \Delta t]$, as shown in the table. From the table, it seems reasonable to conclude that the instantaneous rate of change of the height at $t = 1$ is -32 feet per second.

Δt approaches 0.

Δt	1	0.5	0.1	0.01	0.001	0.0001	0
$\dfrac{\Delta h}{\Delta t}$	-48	-40	-33.6	-32.16	-32.016	-32.0016	-32

$\dfrac{\Delta h}{\Delta t}$ approaches -32.

STUDY TIP

The limit in the definition of instantaneous rate of change is the same as the limit in the definition of the derivative of f at x. This is the second major interpretation of the derivative—as an *instantaneous rate of change in one variable with respect to another*. Recall that the first interpretation of the derivative is as the slope of the graph of f at x.

Definition of Instantaneous Rate of Change

The **instantaneous rate of change** (or simply **rate of change**) of $y = f(x)$ at x is the limit of the average rate of change on the interval

$[x, x + \Delta x]$

as Δx approaches 0.

$$\lim_{\Delta x \to 0} \frac{\Delta y}{\Delta x} = \lim_{\Delta x \to 0} \frac{f(x + \Delta x) - f(x)}{\Delta x}$$

If y is a distance and x is time, then the rate of change is a **velocity.**

Example 3 Finding an Instantaneous Rate of Change

Find the velocity of the object in Example 2 at $t = 1$.

SOLUTION From Example 2, you know that the height of the falling object is given by

$h = -16t^2 + 100.$ Position function

By taking the derivative of this position function, you obtain the velocity function.

$h'(t) = -32t$ Velocity function

The velocity function gives the velocity at *any* time. So, at $t = 1$, the velocity is

$$\begin{aligned} h'(1) &= -32(1) \\ &= -32 \text{ feet per second.} \end{aligned}$$

✓Checkpoint 3

The height of the object in Checkpoint 2 is given by

$h = -16t^2 + 180.$

Find the velocities of the object at

a. $t = 1.75.$

b. $t = 2.$ ■

Andresr/Shutterstock.com

The general **position function** for a free-falling object, neglecting air resistance, is

$$h = -16t^2 + v_0 t + h_0 \qquad \text{Position function}$$

where h is the height (in feet), t is the time (in seconds), v_0 is the initial velocity (in feet per second), and h_0 is the initial height (in feet). Remember that the model assumes that positive velocities indicate upward motion and negative velocities indicate downward motion. The derivative

$$h' = -32t^2 + v_0 \qquad \text{Velocity function}$$

is the **velocity function.** The absolute value of the velocity is the **speed** of the object.

Example 4 Finding the Velocity of a Diver

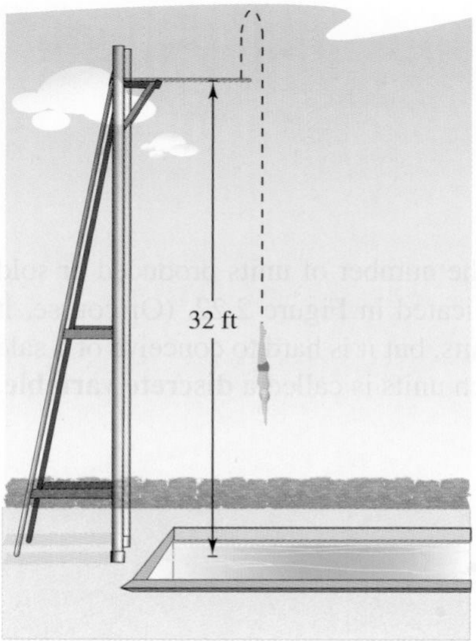

FIGURE 2.21

At time $t = 0$, a diver jumps from a diving board that is 32 feet high, as shown in Figure 2.21. Because the diver's initial velocity is 16 feet per second, the position of the diver is given by

$$h = -16t^2 + 16t + 32. \qquad \text{Position function}$$

a. When does the diver hit the water?

b. What is the diver's velocity at impact?

SOLUTION

a. To find the time at which the diver hits the water, let $h = 0$ and solve for t.

$$-16t^2 + 16t + 32 = 0 \qquad \text{Set } h \text{ equal to 0.}$$
$$-16(t^2 - t - 2) = 0 \qquad \text{Factor out common factor.}$$
$$-16(t + 1)(t - 2) = 0 \qquad \text{Factor.}$$
$$t = -1 \text{ or } t = 2 \qquad \text{Solve for } t.$$

The solution $t = -1$ does not make sense in the problem because it would mean that the diver hits the water 1 second before jumping. So, you can conclude that the diver hits the water at $t = 2$ seconds.

b. The velocity at time t is given by the derivative

$$h' = -32t + 16. \qquad \text{Velocity function}$$

The velocity at time $t = 2$ is

$$h' = -32(2) + 16 = -48 \text{ feet per second.}$$

✓ Checkpoint 4

At time $t = 0$, a diver jumps from a diving board that is 12 feet high with initial velocity 16 feet per second. The diver's position function is $h = -16t^2 + 16t + 12$.

a. When does the diver hit the water?

b. What is the diver's velocity at impact? ■

In Example 4, note that the diver's initial velocity is $v_0 = 16$ feet per second (upward) and the diver's initial height is $h_0 = 32$ feet.

Initial velocity is 16 feet per second.

Initial height is 32 feet.

$$h = -16t^2 + 16t + 32$$

Rates of Change in Economics: Marginals

Another important use of rates of change is in the field of economics. Economists refer to *marginal profit, marginal revenue,* and *marginal cost* as the rates of change of the profit, revenue, and cost with respect to x, the number of units produced or sold. An equation that relates these three quantities is

$$P = R - C$$

where P, R, and C represent the following quantities.

$P =$ total profit, $R =$ total revenue, and $C =$ total cost

The derivatives of these quantities are called the **marginal profit, marginal revenue,** and **marginal cost,** respectively.

$$\frac{dP}{dx} = \text{marginal profit}$$

$$\frac{dR}{dx} = \text{marginal revenue}$$

$$\frac{dC}{dx} = \text{marginal cost}$$

In many business and economics problems, the number of units produced or sold is restricted to nonnegative integer values, as indicated in Figure 2.22. (Of course, it could happen that a sale involves half or quarter units, but it is hard to conceive of a sale involving $\sqrt{2}$ units.) The variable that denotes such units is called a **discrete variable.**

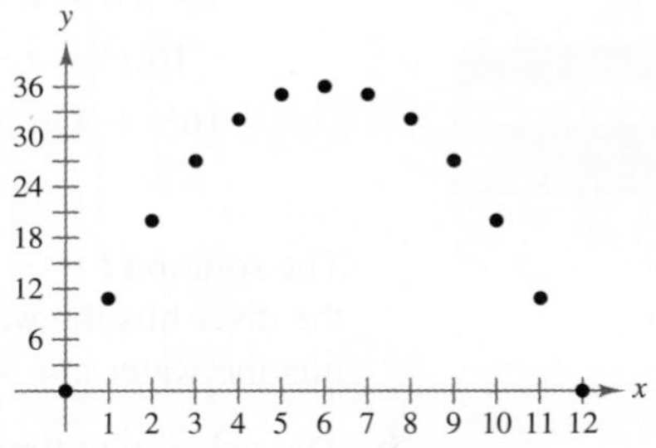

Function of a Discrete Variable

FIGURE 2.22

To analyze a function of a discrete variable x, you can temporarily assume that x is a **continuous variable** and is able to take on any real value in a given interval, as indicated in Figure 2.23. Then, you can use the methods of calculus to find the x-value that corresponds to the marginal revenue, maximum profit, minimum cost, or whatever is called for. Finally, you should round the solution to the nearest sensible x-value—cents, dollars, units, or days, depending on the context of the problem.

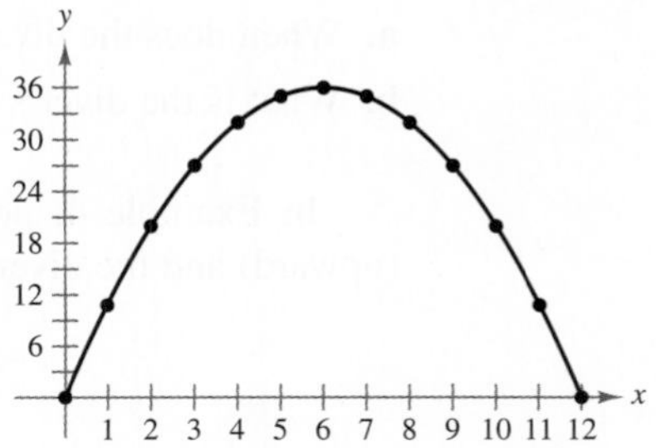

Function of a Continuous Variable

FIGURE 2.23

Example 5 Finding the Marginal Profit

The profit derived from selling x units of an alarm clock is given by

$$P = 0.0002x^3 + 10x.$$

a. Find the marginal profit for a production level of 50 units.

b. Compare the marginal profit with the actual gain in profit obtained by increasing the production level from 50 to 51 units.

SOLUTION

a. The profit is $P = 0.0002x^3 + 10x$. The marginal profit is given by the derivative

$$\frac{dP}{dx} = 0.0006x^2 + 10.$$

When $x = 50$, the marginal profit is

$$\frac{dP}{dx} = 0.0006(50)^2 + 10 \qquad \text{Substitute 50 for } x.$$
$$= 0.0006(2500) + 10$$
$$= 1.5 + 10$$
$$= \$11.50 \text{ per unit.} \qquad \text{Marginal profit for } x = 50$$

b. For $x = 50$, the actual profit is

$$P = 0.0002(50)^3 + 10(50) \qquad \text{Substitute 50 for } x.$$
$$= 0.0002(125{,}000) + 500$$
$$= 25 + 500$$
$$= \$525.00 \qquad \text{Actual profit for } x = 50$$

and for $x = 51$, the actual profit is

$$P = 0.0002(51)^3 + 10(51) \qquad \text{Substitute 51 for } x.$$
$$= 0.0002(132{,}651) + 510$$
$$\approx 26.53 + 510$$
$$= \$536.53. \qquad \text{Actual profit for } x = 51$$

So, the additional profit obtained by increasing production from 50 to 51 units is

$$536.53 - 525.00 = \$11.53. \qquad \text{Extra profit for one unit}$$

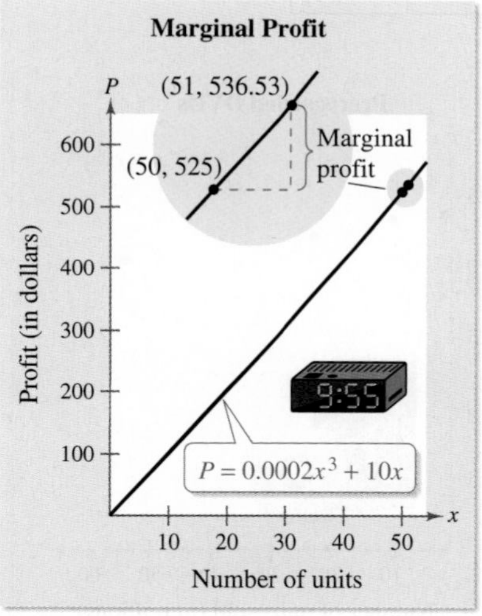

FIGURE 2.24

Note that the actual profit increase of \$11.53 (when x increases from 50 to 51 units) can be approximated by the marginal profit of \$11.50 per unit (when $x = 50$), as shown in Figure 2.24.

✓Checkpoint 5

Find the marginal profit in Example 5 for a production level of 100 units. Compare this with the actual gain in profit by increasing production from 100 to 101 units. ■

STUDY TIP

In Example 5, the marginal profit gives a good approximation of the actual change in profit because the graph of P is nearly straight over the interval $50 \le x \le 51$. You will study more about the use of marginals to approximate actual changes in Section 3.8.

The profit function in Example 5 is unusual in that the profit continues to increase as long as the number of units sold increases. In practice, it is more common to encounter situations in which sales can be increased only by lowering the price per item. Such reductions in price will ultimately cause the profit to decline.

The number of units x that consumers are willing to purchase at a given price per unit p is given by the **demand function**

$$p = f(x).$$ Demand function

The total revenue R is then related to the price per unit and the quantity demanded (or sold) by the equation

$$R = xp.$$ Revenue function

Example 6 Finding a Demand Function

The table shows the numbers x (in millions) of prerecorded high-definition DVDs sold in the United States and the average unit prices p (in dollars) from 2006 through 2009. Use this information to find the demand function and the total revenue function. *(Source: SNL Kagan)*

Year	2006	2007	2008	2009
x	1.2	9.8	22.7	54.0
p	23.75	23.38	22.21	20.43

SOLUTION Begin by making a scatter plot of the data using the ordered pairs (x, p), as shown in Figure 2.25. From the graph, it appears that a linear model would be a good fit for the data. To find a linear model for the demand function, use any two points, such as (1.2, 23.75) and (54.0, 20.43). The slope of the line through these points is

$$m = \frac{20.43 - 23.75}{54.0 - 1.2} = \frac{-3.32}{52.8} \approx -0.063.$$

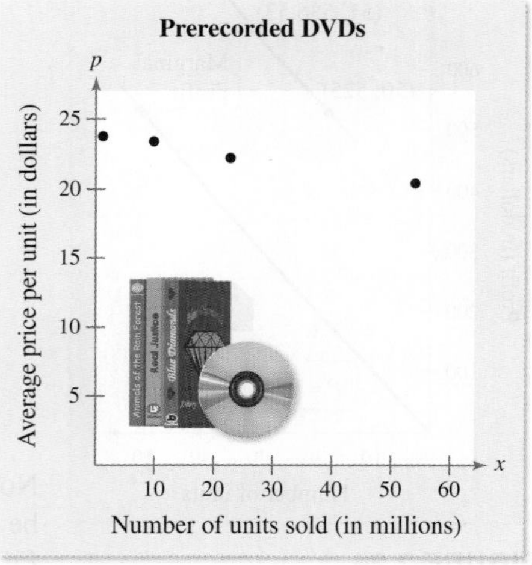

FIGURE 2.25

Using the point-slope form of a line, you can approximate the equation of the demand function to be $p = -0.063x + 23.83$. The total revenue function for prerecorded high-definition DVDs is

$$R = xp = x(-0.063x + 23.83) = -0.063x^2 + 23.83x.$$

✓ Checkpoint 6

Repeat Example 6 given that in 2010 an estimated 104.2 million prerecorded high-definition DVDs were sold at an average unit price of \$17.88. To find a linear model for the demand function, use the points (1.2, 23.75) and (104.2, 17.88). *(Source: SNL Kagan)* ■

TECH TUTOR

Another way to find a linear model for the demand function in Example 6 is to use the *linear regression* feature of a graphing utility or a spreadsheet software program. Use a graphing utility or a spreadsheet software program to find the demand function and compare your results with those in Example 6. You will learn more about linear regression in Section 7.7. (Consult the user's manual of a graphing utility or a spreadsheet software program for specific instructions.)

Example 7 Finding the Marginal Revenue

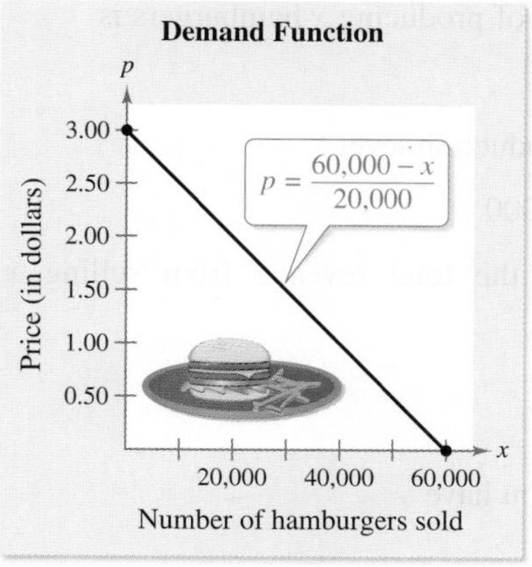

As the price decreases, more hamburgers are sold.

FIGURE 2.26

A fast-food restaurant has determined that the monthly demand for its hamburgers is given by

$$p = \frac{60{,}000 - x}{20{,}000}.$$

Figure 2.26 shows that as the price decreases, the quantity demanded increases. The table shows the demands for hamburgers at various prices.

x	60,000	50,000	40,000	30,000	20,000	10,000	0
p	\$0.00	\$0.50	\$1.00	\$1.50	\$2.00	\$2.50	\$3.00

Find the increase in revenue per hamburger for monthly sales of 20,000 hamburgers. In other words, find the marginal revenue when $x = 20{,}000$.

SOLUTION Because the demand is given by

$$p = \frac{60{,}000 - x}{20{,}000}$$

and the revenue is given by $R = xp$, you have

$$R = xp \qquad \text{Formula for revenue}$$

$$= x\left(\frac{60{,}000 - x}{20{,}000}\right) \qquad \text{Substitute for } p.$$

$$= \frac{1}{20{,}000}(60{,}000x - x^2). \qquad \text{Revenue function}$$

By differentiating, you can find the marginal revenue to be

$$\frac{dR}{dx} = \frac{1}{20{,}000}(60{,}000 - 2x).$$

So, at $x = 20{,}000$, the marginal revenue is

$$\frac{dR}{dx} = \frac{1}{20{,}000}(60{,}000 - 2x) \qquad \text{Marginal revenue}$$

$$= \frac{1}{20{,}000}[60{,}000 - 2(20{,}000)] \qquad \text{Substitute 20,000 for } x.$$

$$= \frac{1}{20{,}000}(60{,}000 - 40{,}000) \qquad \text{Multiply.}$$

$$= \frac{1}{20{,}000}(20{,}000) \qquad \text{Subtract.}$$

$$= \$1 \text{ per unit.} \qquad \text{Marginal revenue when } x = 20{,}000$$

So, for monthly sales of 20,000 hamburgers, you can conclude that the increase in revenue per hamburger is \$1.

STUDY TIP

Writing a demand function in the form $p = f(x)$ is a convention used in economics. From a consumer's point of view, it might seem more reasonable to think that the quantity demanded is a function of the price. Mathematically, however, the two points of view are equivalent because a typical demand function is one-to-one and so has an inverse function. For instance, in Example 7, you could write the demand function as $x = 60{,}000 - 20{,}000p$.

✓ Checkpoint 7

Find the revenue function and marginal revenue for a demand function of

$$p = 2000 - 4x.$$

Find the marginal revenue when $x = 250$. ■

Example 8 Finding the Marginal Profit

For the fast-food restaurant in Example 7, the cost of producing x hamburgers is

$$C = 5000 + 0.56x, \quad 0 \le x \le 50{,}000.$$

Find the profit and the marginal profit for each production level.

a. $x = 20{,}000$ **b.** $x = 24{,}400$ **c.** $x = 30{,}000$

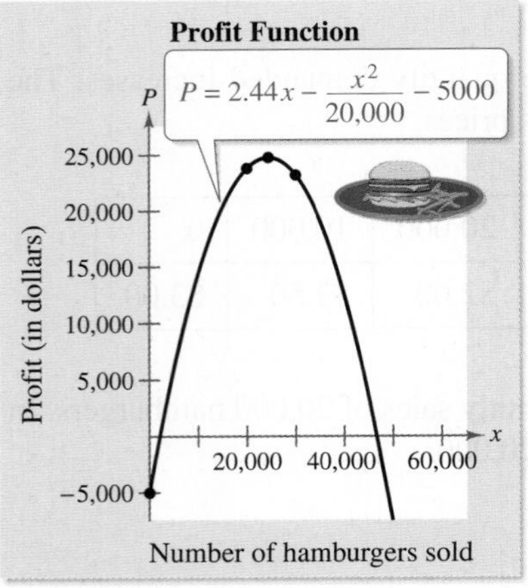

FIGURE 2.27

SOLUTION From Example 7, you know that the total revenue from selling x hamburgers is

$$R = \frac{1}{20{,}000}(60{,}000x - x^2).$$

Because the total profit is given by $P = R - C$, you have

$$\begin{aligned} P &= \frac{1}{20{,}000}(60{,}000x - x^2) - (5000 + 0.56x) \\ &= 3x - \frac{x^2}{20{,}000} - 5000 - 0.56x \\ &= 2.44x - \frac{x^2}{20{,}000} - 5000. \end{aligned}$$ See Figure 2.27.

So, the marginal profit is

$$\frac{dP}{dx} = 2.44 - \frac{x}{10{,}000}.$$

Using these formulas, you can compute the profit and marginal profit.

	Production	*Profit*	*Marginal Profit*
a.	$x = 20{,}000$	$P = \$23{,}800.00$	$2.44 - \frac{20{,}000}{10{,}000} = \$0.44$ per unit
b.	$x = 24{,}400$	$P = \$24{,}768.00$	$2.44 - \frac{24{,}400}{10{,}000} = \$0.00$ per unit
c.	$x = 30{,}000$	$P = \$23{,}200.00$	$2.44 - \frac{30{,}000}{10{,}000} = -\$0.56$ per unit

✓ Checkpoint 8

From Example 8, compare the marginal profit when 10,000 units are produced with the actual increase in profit from 10,000 units to 10,001 units. ■

SUMMARIZE (Section 2.3)

1. State the definition of average rate of change *(page 103)*. For examples of average rate of change, see Examples 1 and 2.
2. State the definition of instantaneous rate of change *(page 106)*. For examples of an instantaneous rate of change, see Examples 3 and 4.
3. Describe a real-life example of how rates of change can be used in the field of economics *(pages 109–112, Examples 5, 6, 7, and 8)*.

SKILLS WARM UP 2.3

The following warm-up exercises involve skills that were covered in earlier sections. You will use these skills in the exercise set for this section. For additional help, review Sections 2.1 and 2.2.

In Exercises 1–4, evaluate the expression.

1. $\dfrac{-63 - (-105)}{21 - 7}$

2. $\dfrac{-37 - 54}{16 - 3}$

3. $\dfrac{24 - 33}{9 - 6}$

4. $\dfrac{40 - 16}{18 - 8}$

In Exercises 5–12, find the derivative of the function.

5. $y = 4x^2 - 2x + 7$

6. $y = -3t^3 + 2t^2 - 8$

7. $s = -16t^2 + 24t + 30$

8. $y = -16x^2 + 54x + 70$

9. $A = \frac{1}{10}(-2r^3 + 3r^2 + 5r)$

10. $y = \frac{1}{9}(6x^3 - 18x^2 + 63x - 15)$

11. $y = 12x - \dfrac{x^2}{5000}$

12. $y = 138 + 74x - \dfrac{x^3}{10{,}000}$

Exercises 2.3

See www.CalcChat.com for worked-out solutions to odd-numbered exercises.

1. Research and Development The table shows the amounts A (in billions of dollars) spent on Research and Development in the United States from 1980 through 2008, where t is the year, with $t = 0$ corresponding to 1980. Approximate the average rate of change of A during each period. *(Source: U.S. National Science Foundation)*

(a) 1980–1985 (b) 1985–1990
(c) 1990–1995 (d) 1995–2000
(e) 2000–2005 (f) 1980–2008
(g) 1990–2008 (h) 2000–2008

t	0	1	2	3	4	5	6	7
A	63	72	81	90	102	115	120	126

t	8	9	10	11	12	13	14
A	134	142	152	161	165	166	169

t	15	16	17	18	19	20	21
A	184	197	212	226	245	267	277

t	22	23	24	25	26	27	28
A	276	288	299	322	347	373	398

2. Trade Deficit The graph shows the values I (in billions of dollars) of goods imported to the United States and the values E (in billions of dollars) of goods exported from the United States from 1980 through 2009. Approximate the average rates of change of I and E during each period. *(Source: U.S. International Trade Administration)*

(a) Imports: 1980–1990 (b) Exports: 1980–1990
(c) Imports: 1990–2000 (d) Exports: 1990–2000
(e) Imports: 2000–2009 (f) Exports: 2000–2009
(g) Imports: 1980–2009 (h) Exports: 1980–2009

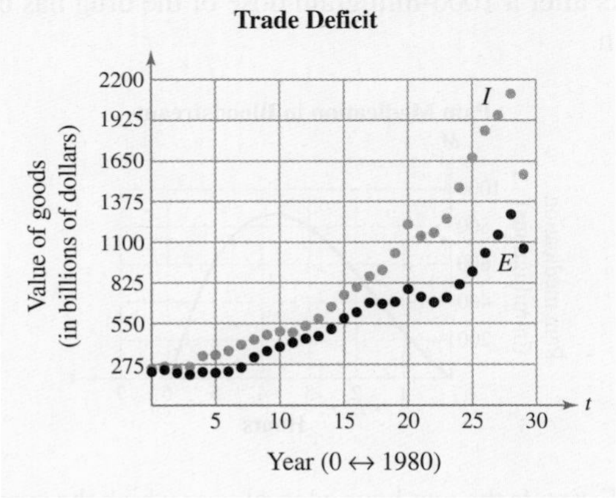

Finding Rates of Change In Exercises 3–12, use a graphing utility to graph the function and find its average rate of change over the interval. Compare this rate with the instantaneous rates of change at the endpoints of the interval.

3. $f(t) = 3t + 5; [1, 2]$ **4.** $h(x) = 2 - x; [0, 2]$

5. $h(x) = x^2 - 4x + 2; [-2, 2]$

6. $f(x) = x^2 - 6x - 1; [-1, 3]$

7. $f(x) = 3x^{4/3}; [1, 8]$ **8.** $f(x) = x^{3/2}; [1, 4]$

9. $f(x) = \frac{1}{x}; [1, 4]$ **10.** $f(x) = \frac{1}{\sqrt{x}}; [1, 4]$

11. $g(x) = x^4 - x^2 + 2; [1, 3]$

12. $g(x) = x^3 - 1; [-1, 1]$

13. Consumer Trends The graph shows the number of visitors V (in thousands) to a national park during a one-year period, where $t = 1$ represents January.

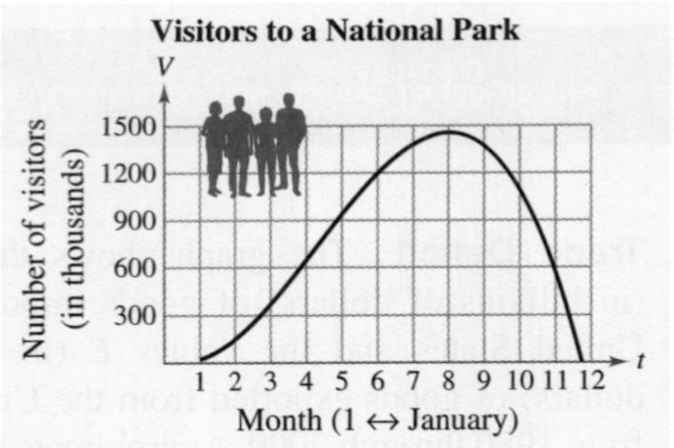

(a) Estimate the rate of change of V over the interval $[9, 12]$ and explain your results.

(b) Over what interval is the average rate of change approximately equal to the rate of change at $t = 8$? Explain your reasoning.

14. Medicine The graph shows the estimated number of milligrams of a pain medication M in the bloodstream t hours after a 1000-milligram dose of the drug has been given.

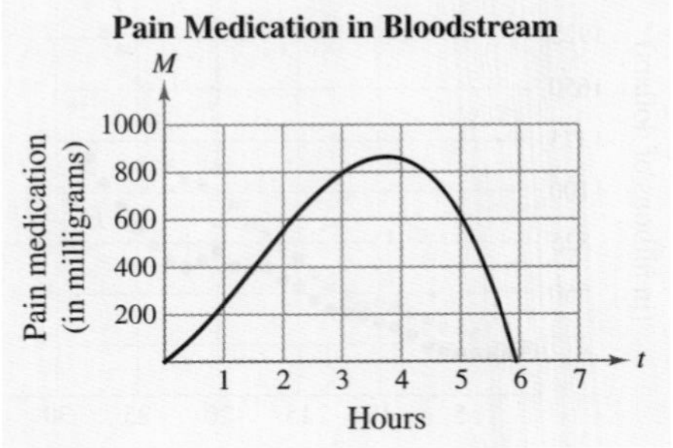

(a) Estimate the one-hour interval over which the average rate of change is the greatest.

(b) Over what interval is the average rate of change approximately equal to the rate of change at $t = 4$? Explain your reasoning.

15. Velocity The height s (in feet) at time t (in seconds) of a ball thrown upward from the top of a building is given by

$$s = -16t^2 + 30t + 250.$$

Find the average velocity over each indicated interval and compare this velocity with the instantaneous velocity at the endpoints of the interval.

(a) $[0, 1]$ (b) $[1, 2]$ (c) $[2, 3]$ (d) $[3, 4]$

16. Chemistry: Wind Chill At 0° Celsius, the heat loss H (in kilocalories per square meter per hour) from a person's body can be modeled by

$$H = 33(10\sqrt{v} - v + 10.45)$$

where v is the wind speed (in meters per second).

(a) Find $\frac{dH}{dv}$ and interpret its meaning in this situation.

(b) Find the rates of change of H when $v = 2$ and $v = 5$.

17. Velocity The height s (in feet) at time t (in seconds) of a silver dollar dropped from the top of a building is given by $s = -16t^2 + 555$.

(a) Find the average velocity over the interval $[2, 3]$.

(b) Find the instantaneous velocities when $t = 2$ and $t = 3$.

(c) How long will it take the coin to hit the ground?

(d) Find the velocity of the coin when it hits the ground.

18. Velocity A ball is thrown straight down from the top of a 210-foot building with an initial velocity of -18 feet per second.

(a) Find the position and velocity functions for the ball.

(b) Find the average velocity over the interval $[1, 2]$.

(c) Find the instantaneous velocities when $t = 1$ and $t = 2$.

(d) How long will it take the ball to hit the ground?

(e) Find the velocity of the ball when it hits the ground.

Marginal Cost In Exercises 19–22, find the marginal cost for producing x units. (The cost is measured in dollars.)

19. $C = 205{,}000 + 9800x$

20. $C = 150{,}000 + 7x^3$

21. $C = 55{,}000 + 470x - 0.25x^2, 0 \le x \le 940$

22. $C = 100(9 + 3\sqrt{x})$

Marginal Revenue In Exercises 23–26, find the marginal revenue for producing x units. (The revenue is measured in dollars.)

23. $R = 50x - 0.5x^2$ **24.** $R = 30x - x^2$

25. $R = -6x^3 + 8x^2 + 200x$

26. $R = 50(20x - x^{3/2})$

Marginal Profit In Exercises 27–30, find the marginal profit for producing x units. (The profit is measured in dollars.)

27. $P = -2x^2 + 72x - 145$

28. $P = -0.25x^2 + 2000x - 1{,}250{,}000$

29. $P = 0.0013x^3 + 12x$

30. $P = -0.5x^3 + 30x^2 - 164.25x - 1000$

31. Marginal Cost The cost C (in dollars) of producing x units of a product is given by

$$C = 3.6\sqrt{x} + 500.$$

(a) Find the additional cost when the production increases from 9 to 10 units.

(b) Find the marginal cost when $x = 9$.

(c) Compare the results of parts (a) and (b).

32. Marginal Revenue The revenue R (in dollars) from renting x apartments can be modeled by

$$R = 2x(900 + 32x - x^2).$$

(a) Find the additional revenue when the number of rentals is increased from 14 to 15.

(b) Find the marginal revenue when $x = 14$.

(c) Compare the results of parts (a) and (b).

33. Marginal Profit The profit P (in dollars) from selling x laptop computers is given by

$$P = -0.04x^2 + 25x - 1500.$$

(a) Find the additional profit when the sales increase from 150 to 151 units.

(b) Find the marginal profit when $x = 150$.

(c) Compare the results of parts (a) and (b).

34. Marginal Profit The profit P (in dollars) from selling x units of a product is given by

$$P = 36{,}000 + 2048\sqrt{x} - \frac{1}{8x^2}, \quad 150 \le x \le 275.$$

Find the marginal profit for each of the following sales.

(a) $x = 150$ (b) $x = 175$ (c) $x = 200$

(d) $x = 225$ (e) $x = 250$ (f) $x = 275$

35. Population Growth The population P (in thousands) of Japan from 1980 through 2010 can be modeled by

$$P = -15.56t^2 + 802.1t + 117{,}001$$

where t is the year, with $t = 0$ corresponding to 1980. *(Source: U.S. Census Bureau)*

(a) Evaluate P for $t = 0, 5, 10, 15, 20, 25,$ and 30. Explain these values.

(b) Determine the population growth rate, dP/dt.

(c) Evaluate dP/dt for the same values as in part (a). Explain your results.

36. Health The temperature T (in degrees Fahrenheit) of a person during an illness can be modeled by the equation

$$T = -0.0375t^2 + 0.3t + 100.4$$

where t is time in hours since the person started to show signs of a fever.

(a) Use a graphing utility to graph the function. Be sure to choose an appropriate window.

(b) Do the slopes of the tangent lines appear to be positive or negative? What does this tell you?

(c) Evaluate the function for $t = 0, 4, 8,$ and 12.

(d) Find dT/dt and explain its meaning in this situation.

(e) Evaluate dT/dt for $t = 0, 4, 8,$ and 12.

37. Economics Use the information in the table to find the models and answer the questions below.

Quantity produced and sold (Q)	Price (p)	Total revenue (TR)	Marginal revenue (MR)
0	160	0	—
2	140	280	130
4	120	480	90
6	100	600	50
8	80	640	10
10	60	600	−30

(a) Use the *regression* feature of a graphing utility to find a quadratic model that relates the total revenue (TR) to the quantity produced and sold (Q).

(b) Using derivatives, find a model for marginal revenue from the model you found in part (a).

(c) Calculate the marginal revenue for all values of Q using your model in part (b), and compare these values with the actual values given. How good is your model? *(Source: Adapted from Taylor,* Economics, *Fifth Edition)*

38. Profit The monthly demand function p and cost function C for x newspapers at a newsstand are given by

$$p = 5 - 0.001x \quad \text{and} \quad C = 35 + 1.5x.$$

(a) Find the monthly revenue R as a function of x.

(b) Find the monthly profit P as a function of x.

(c) Complete the table.

x	600	1200	1800	2400	3000
dR/dx					
dP/dx					
P					

39. Marginal Profit When the price of a glass of lemonade at a lemonade stand was \$1.75, 400 glasses were sold. When the price was lowered to \$1.50, 500 glasses were sold. Assume that the demand function is linear and that the marginal and fixed costs are \$0.10 and \$25, respectively.

(a) Find the profit P as a function of x, the number of glasses of lemonade sold.

(b) Use a graphing utility to graph P, and comment about the slopes of P when $x = 300$ and $x = 700$.

(c) Find the marginal profits when 300 glasses of lemonade are sold and when 700 glasses of lemonade are sold.

40. Marginal Profit When the admission price for a baseball game was \$6 per ticket, 36,000 tickets were sold. When the price was raised to \$7, only 33,000 tickets were sold. Assume that the demand function is linear and that the marginal and fixed costs for the ballpark owners are \$0.20 and \$85,000, respectively.

(a) Find the profit P as a function of x, the number of tickets sold.

(b) Use a graphing utility to graph P, and comment about the slopes of P when $x = 18{,}000$ and $x = 36{,}000$.

(c) Find the marginal profits when 18,000 tickets are sold and when 36,000 tickets are sold.

41. Fuel Cost A car is driven 15,000 miles a year and gets x miles per gallon. Assume that the average fuel cost is \$2.95 per gallon.

(a) Find the annual cost of fuel C as a function of x.

(b) Find dC/dx and explain its meaning in this situation.

(c) Use the functions to complete the table.

x	10	15	20	25	30	35	40
C							
dC/dx							

(d) Who would benefit more from a 1 mile per gallon increase in fuel efficiency—the driver who gets 15 miles per gallon or the driver who gets 35 miles per gallon? Explain.

42. Gasoline Sales The number N of gallons of regular unleaded gasoline sold by a gasoline station at a price of p dollars per gallon is given by $N = f(p)$.

(a) Describe the meaning of $f'(2.959)$.

(b) Is $f'(2.959)$ usually positive or negative? Explain.

43. Dow Jones Industrial Average The table shows the year-end closing prices p of the Dow Jones Industrial Average (DJIA) from 1995 through 2009, where t is the year, with $t = 5$ corresponding to 1995. *(Source: Dow Jones Industrial Average)*

t	5	6	7	8
p	5117.12	6448.26	7908.24	9181.43

t	9	10	11	12
p	11,497.12	10,786.85	10,021.50	8341.63

t	13	14	15	16
p	10,453.92	10,783.01	10,717.50	12,463.15

t	17	18	19
p	13,264.82	8776.39	10,428.05

(a) Determine the average rate of change in the value of the DJIA from 1995 through 2009.

(b) Estimate the instantaneous rate of change in 1998 by finding the average rate of change from 1996 to 2000.

(c) Estimate the instantaneous rate of change in 1998 by finding the average rate of change from 1997 to 1999.

(d) Compare your answers for parts (b) and (c). Which interval do you think produced the best estimate for the instantaneous rate of change in 1998?

44. HOW DO YOU SEE IT? Many populations in nature exhibit logistic growth, which consists of four phases, as shown in the figure. Describe the rate of growth of the population in each phase, and give possible reasons as to why the rates might be changing from phase to phase. *(Source: Adapted from Levine/Miller,* Biology: Discovering Life, *Second Edition)*

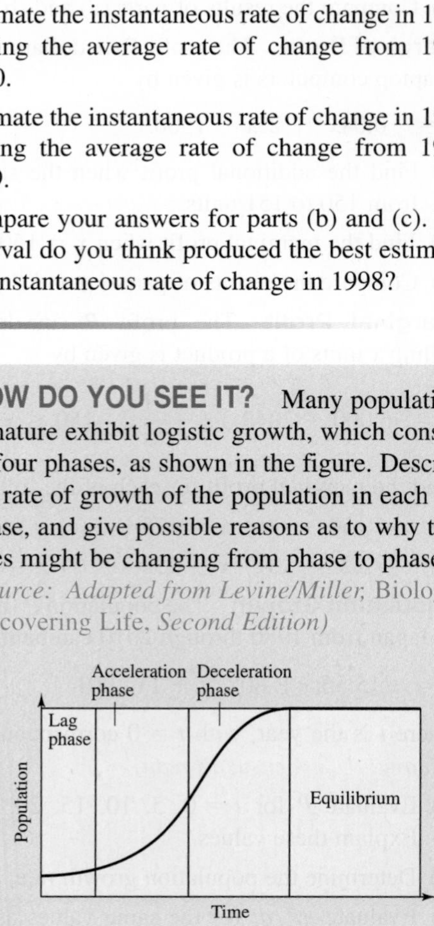

2.4 The Product and Quotient Rules

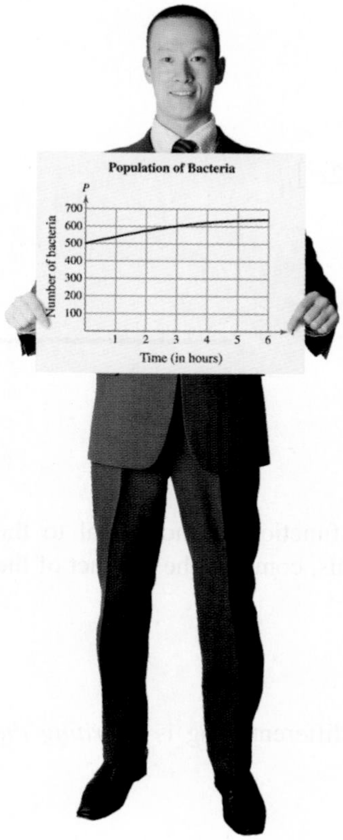

In Exercise 63 on page 125, you will use the Quotient Rule to find the rate of change of a population of bacteria.

- Find the derivatives of functions using the Product Rule.
- Find the derivatives of functions using the Quotient Rule.
- Use derivatives to answer questions about real-life situations.

The Product Rule

In Section 2.2, you saw that the derivative of a sum or difference of two functions is simply the sum or difference of their derivatives. The rules for the derivative of a product or quotient of two functions are not as simple.

The Product Rule

The derivative of the product of two differentiable functions is equal to the first function times the derivative of the second plus the second function times the derivative of the first.

$$\frac{d}{dx}[f(x)g(x)] = f(x)g'(x) + g(x)f'(x)$$

PROOF Some mathematical proofs, such as the proof of the Sum Rule, are straightforward. Others involve clever steps that may not appear to follow clearly from a prior step. The proof below involves such a step—adding and subtracting the same quantity. (This step is shown in color.) Let $F(x) = f(x)g(x)$.

$$\begin{aligned}
F'(x) &= \lim_{\Delta x \to 0} \frac{F(x + \Delta x) - F(x)}{\Delta x} \\
&= \lim_{\Delta x \to 0} \frac{f(x + \Delta x)g(x + \Delta x) - f(x)g(x)}{\Delta x} \\
&= \lim_{\Delta x \to 0} \frac{f(x + \Delta x)g(x + \Delta x) - f(x + \Delta x)g(x) + f(x + \Delta x)g(x) - f(x)g(x)}{\Delta x} \\
&= \lim_{\Delta x \to 0} \left[f(x + \Delta x) \frac{g(x + \Delta x) - g(x)}{\Delta x} + g(x) \frac{f(x + \Delta x) - f(x)}{\Delta x} \right] \\
&= \lim_{\Delta x \to 0} f(x + \Delta x) \frac{g(x + \Delta x) - g(x)}{\Delta x} + \lim_{\Delta x \to 0} g(x) \frac{f(x + \Delta x) - f(x)}{\Delta x} \\
&= \lim_{\Delta x \to 0} f(x + \Delta x) \cdot \lim_{\Delta x \to 0} \frac{g(x + \Delta x) - g(x)}{\Delta x} + \lim_{\Delta x \to 0} g(x) \cdot \lim_{\Delta x \to 0} \frac{f(x + \Delta x) - f(x)}{\Delta x} \\
&= f(x)g'(x) + g(x)f'(x)
\end{aligned}$$

STUDY TIP

Rather than trying to remember the formula for the Product Rule, it can be more helpful to remember its verbal statement:

> *the first function times the derivative of the second plus the second function times the derivative of the first.*

Example 1 Using the Product Rule

Find the derivative of $y = (3x - 2x^2)(5 + 4x)$.

SOLUTION Using the Product Rule, you can write

$$\frac{dy}{dx} = \overbrace{(3x - 2x^2)}^{\text{First}}\overbrace{\frac{d}{dx}[5 + 4x]}^{\text{Derivative of second}} + \overbrace{(5 + 4x)}^{\text{Second}}\overbrace{\frac{d}{dx}[3x - 2x^2]}^{\text{Derivative of first}}$$
$$= (3x - 2x^2)(4) + (5 + 4x)(3 - 4x)$$
$$= (12x - 8x^2) + (15 - 8x - 16x^2)$$
$$= 15 + 4x - 24x^2.$$

✓Checkpoint 1

Find the derivative of $y = (4x + 3x^2)(6 - 3x)$. ■

In general, the derivative of the product of two functions is not equal to the product of the derivatives of the two functions. To see this, compare the product of the derivatives of

$$f(x) = 3x - 2x^2 \quad \text{and} \quad g(x) = 5 + 4x$$

with the derivative found in Example 1.

In the next example, notice that the first step in differentiating is *rewriting the original function.*

TECH TUTOR

If you have access to a symbolic differentiation utility, try using it to confirm several of the derivatives in this section.

Example 2 Using the Product Rule

Find the derivative of $f(x) = \left(\frac{1}{x} + 1\right)(x - 1)$.

SOLUTION Rewrite the function. Then use the Product Rule to find the derivative.

$$f(x) = \left(\frac{1}{x} + 1\right)(x - 1) \qquad \text{Write original function.}$$
$$= (x^{-1} + 1)(x - 1) \qquad \text{Rewrite function.}$$
$$f'(x) = (x^{-1} + 1)\frac{d}{dx}[x - 1] + (x - 1)\frac{d}{dx}[x^{-1} + 1] \qquad \text{Product Rule}$$
$$= (x^{-1} + 1)(1) + (x - 1)(-x^{-2})$$
$$= \frac{1}{x} + 1 - \frac{x - 1}{x^2}$$
$$= \frac{x + x^2 - x + 1}{x^2} \qquad \text{Write with common denominator.}$$
$$= \frac{x^2 + 1}{x^2} \qquad \text{Simplify.}$$

✓Checkpoint 2

Find the derivative of

$$f(x) = \left(\frac{1}{x} + 1\right)(2x + 1).$$ ■

You now have two differentiation rules that deal with products—the Constant Multiple Rule and the Product Rule. The difference between these two rules is that the Constant Multiple Rule is used when one of the factors is a constant

$$F(x) = \overbrace{c}^{\text{Constant}}\;\overbrace{f(x)}^{\text{Variable quantity}} \qquad \text{Use Constant Multiple Rule.}$$

whereas the Product Rule is used when both of the factors are variable quantities

$$F(x) = \overbrace{f(x)}^{\text{Variable quantity}}\;\overbrace{g(x)}^{\text{Variable quantity}}. \qquad \text{Use Product Rule.}$$

The next example compares these two rules.

Example 3 Comparing Differentiation Rules

Find the derivative of each function.

a. $y = 2x(x^2 + 3x)$

b. $y = 2(x^2 + 3x)$

SOLUTION

a. Because both factors are variable quantities, use the Product Rule.

$$y = 2x(x^2 + 3x)$$

$$\frac{dy}{dx} = (2x)\frac{d}{dx}[x^2 + 3x] + (x^2 + 3x)\frac{d}{dx}[2x] \qquad \text{Product Rule}$$

$$= (2x)(2x + 3) + (x^2 + 3x)(2)$$

$$= 4x^2 + 6x + 2x^2 + 6x$$

$$= 6x^2 + 12x$$

STUDY TIP

You could calculate the derivative in Example 3(a) without the Product Rule. For instance,

$$y = 2x(x^2 + 3x) = 2x^3 + 6x^2$$

and

$$\frac{dy}{dx} = 6x^2 + 12x.$$

b. Because one of the factors is a constant, use the Constant Multiple Rule.

$$y = 2(x^2 + 3x)$$

$$\frac{dy}{dx} = 2\frac{d}{dx}[x^2 + 3x] \qquad \text{Constant Multiple Rule}$$

$$= 2(2x + 3)$$

$$= 4x + 6$$

✓ Checkpoint 3

Find the derivative of each function.

a. $y = 3x(2x^2 + 5x)$

b. $y = 3(2x^2 + 5x)$ ■

The Product Rule can be extended to products that have more than two factors. For example, if f, g, and h are differentiable functions of x, then

$$\frac{d}{dx}[f(x)g(x)h(x)] = f'(x)g(x)h(x) + f(x)g'(x)h(x) + f(x)g(x)h'(x).$$

The Quotient Rule

In Section 2.2, you saw that by using the Constant Rule, the Power Rule, the Constant Multiple Rule, and the Sum and Difference Rules, you were able to differentiate any polynomial function. By combining these rules with the Quotient Rule, you can now differentiate any *rational* function.

STUDY TIP

As suggested for the Product Rule, it can be more helpful to remember the verbal statement of the Quotient Rule rather than trying to remember the formula for the rule.

The Quotient Rule

The derivative of the quotient of two differentiable functions is equal to the denominator times the derivative of the numerator minus the numerator times the derivative of the denominator, all divided by the square of the denominator.

$$\frac{d}{dx}\left[\frac{f(x)}{g(x)}\right] = \frac{g(x)f'(x) - f(x)g'(x)}{[g(x)]^2}, \quad g(x) \neq 0$$

PROOF Begin by letting

$$F(x) = \frac{f(x)}{g(x)}.$$

As in the proof of the Product Rule, a key step in this proof is adding and subtracting the same quantity.

$$\begin{aligned}
F'(x) &= \lim_{\Delta x \to 0} \frac{F(x + \Delta x) - F(x)}{\Delta x} \\
&= \lim_{\Delta x \to 0} \frac{\dfrac{f(x + \Delta x)}{g(x + \Delta x)} - \dfrac{f(x)}{g(x)}}{\Delta x} \\
&= \lim_{\Delta x \to 0} \left[\left(\frac{g(x)f(x + \Delta x)}{g(x)g(x + \Delta x)} - \frac{f(x)g(x + \Delta x)}{g(x)g(x + \Delta x)}\right) \div \Delta x\right] \\
&= \lim_{\Delta x \to 0} \left[\frac{g(x)f(x + \Delta x) - f(x)g(x + \Delta x)}{g(x)g(x + \Delta x)} \cdot \frac{1}{\Delta x}\right] \\
&= \lim_{\Delta x \to 0} \frac{g(x)f(x + \Delta x) - f(x)g(x + \Delta x)}{\Delta x g(x)g(x + \Delta x)} \\
&= \lim_{\Delta x \to 0} \frac{g(x)f(x + \Delta x) - f(x)g(x) + f(x)g(x) - f(x)g(x + \Delta x)}{\Delta x g(x)g(x + \Delta x)} \\
&= \frac{\displaystyle\lim_{\Delta x \to 0} \frac{g(x)[f(x + \Delta x) - f(x)]}{\Delta x} - \lim_{\Delta x \to 0} \frac{f(x)[g(x + \Delta x) - g(x)]}{\Delta x}}{\displaystyle\lim_{\Delta x \to 0} [g(x)g(x + \Delta x)]} \\
&= \frac{g(x)\left[\displaystyle\lim_{\Delta x \to 0} \frac{f(x + \Delta x) - f(x)}{\Delta x}\right] - f(x)\left[\displaystyle\lim_{\Delta x \to 0} \frac{g(x + \Delta x) - g(x)}{\Delta x}\right]}{\displaystyle\lim_{\Delta x \to 0} [g(x)g(x + \Delta x)]} \\
&= \frac{g(x)f'(x) - f(x)g'(x)}{[g(x)]^2}
\end{aligned}$$

From the Quotient Rule, you can see that the derivative of a quotient is not, in general, the quotient of the derivatives. That is,

$$\frac{d}{dx}\left[\frac{f(x)}{g(x)}\right] \neq \frac{f'(x)}{g'(x)}.$$

Yuri Arcurs/Shutterstock.com

ALGEBRA TUTOR

When applying the Quotient Rule, it is suggested that you enclose all factors and derivatives in symbols of grouping, such as parentheses. Also, pay special attention to the subtraction required in the numerator. For help in simplifying expressions like the one in Example 4, see the *Chapter 2 Algebra Tutor* on page 159, Example 2(d).

Example 4 Using the Quotient Rule

Find the derivative of $y = \dfrac{x-1}{2x+3}$.

SOLUTION Apply the Quotient Rule, as shown.

$$\begin{aligned}\frac{dy}{dx} &= \frac{(2x+3)\frac{d}{dx}[x-1] - (x-1)\frac{d}{dx}[2x+3]}{(2x+3)^2}\\ &= \frac{(2x+3)(1)-(x-1)(2)}{(2x+3)^2}\\ &= \frac{2x+3-2x+2}{(2x+3)^2}\\ &= \frac{5}{(2x+3)^2}\end{aligned}$$

✓Checkpoint 4

Find the derivative of $y = \dfrac{x+4}{5x-2}$. ■

Example 5 Finding an Equation of a Tangent Line

Find an equation of the tangent line to the graph of

$$y = \frac{2x^2-4x+3}{2-3x}$$

at $x = 1$.

SOLUTION Apply the Quotient Rule, as shown.

$$\begin{aligned}\frac{dy}{dx} &= \frac{(2-3x)\frac{d}{dx}[2x^2-4x+3] - (2x^2-4x+3)\frac{d}{dx}[2-3x]}{(2-3x)^2}\\ &= \frac{(2-3x)(4x-4)-(2x^2-4x+3)(-3)}{(2-3x)^2}\\ &= \frac{-12x^2+20x-8-(-6x^2+12x-9)}{(2-3x)^2}\\ &= \frac{-12x^2+20x-8+6x^2-12x+9}{(2-3x)^2}\\ &= \frac{-6x^2+8x+1}{(2-3x)^2}\end{aligned}$$

When $x = 1$, the value of the function is $y = -1$ and the slope is $m = 3$. Using the point-slope form of a line, you can find the equation of the tangent line to be $y = 3x - 4$. The graph of the function and the tangent line is shown in Figure 2.28.

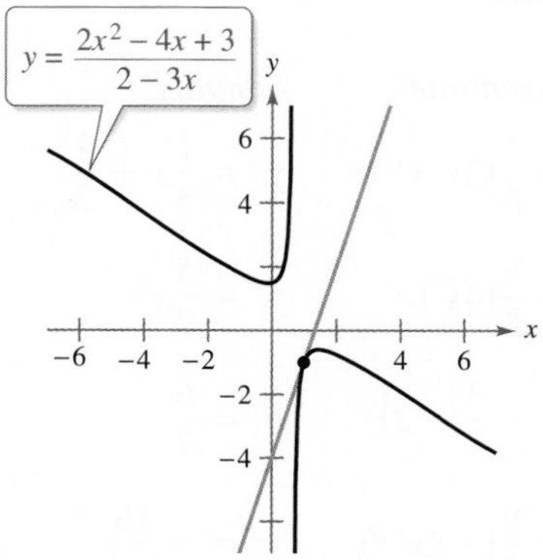

FIGURE 2.28

✓Checkpoint 5

Find an equation of the tangent line to the graph of

$$y = \frac{x^2-4}{2x+5}$$

at $x = 0$. Sketch the line tangent to the graph at $x = 0$. ■

STUDY TIP

Note in Example 6 that much of the work in obtaining the final form of the derivative occurs *after* applying the Quotient Rule. In general, direct application of differentiation rules often yields results that are not in simplified form. Note that two characteristics of simplified form are the absence of negative exponents and the combining of like terms.

Example 6 Rewriting Before Differentiating

Find the derivative of

$$y = \frac{3 - (1/x)}{x + 5}.$$

SOLUTION Begin by rewriting the function. Then apply the Quotient Rule and simplify the result.

$$y = \frac{3 - (1/x)}{x + 5}$$ Write original function.

$$= \frac{x[3 - (1/x)]}{x(x + 5)}$$ Multiply numerator and denominator by x.

$$= \frac{3x - 1}{x^2 + 5x}$$ Rewrite.

$$\frac{dy}{dx} = \frac{(x^2 + 5x)(3) - (3x - 1)(2x + 5)}{(x^2 + 5x)^2}$$ Apply Quotient Rule.

$$= \frac{(3x^2 + 15x) - (6x^2 + 13x - 5)}{(x^2 + 5x)^2}$$

$$= \frac{-3x^2 + 2x + 5}{(x^2 + 5x)^2}$$ Simplify.

✓ Checkpoint 6

Find the derivative of $y = \dfrac{3 - (2/x)}{x + 4}$. ■

Not every quotient needs to be differentiated by the Quotient Rule. For instance, each quotient in the next example can be considered as the product of a constant and a function of x. In such cases, the Constant Multiple Rule is more efficient than the Quotient Rule.

STUDY TIP

To see the benefit of using the Constant Multiple Rule for some quotients, try using the Quotient Rule to differentiate the functions in Example 7. You should obtain the same results, but with more work.

Example 7 Using the Constant Multiple Rule

	Original Function	*Rewrite*	*Differentiate*	*Simplify*
a.	$y = \dfrac{x^2 + 3x}{6}$	$y = \dfrac{1}{6}(x^2 + 3x)$	$y' = \dfrac{1}{6}(2x + 3)$	$y' = \dfrac{1}{3}x + \dfrac{1}{2}$
b.	$y = \dfrac{5x^4}{8}$	$y = \dfrac{5}{8}x^4$	$y' = \dfrac{5}{8}(4x^3)$	$y' = \dfrac{5}{2}x^3$
c.	$y = \dfrac{-3(3x - 2x^2)}{7x}$	$y = -\dfrac{3}{7}(3 - 2x)$	$y' = -\dfrac{3}{7}(-2)$	$y' = \dfrac{6}{7}$
d.	$y = \dfrac{9}{5x^2}$	$y = \dfrac{9}{5}(x^{-2})$	$y' = \dfrac{9}{5}(-2x^{-3})$	$y' = -\dfrac{18}{5x^3}$

✓ Checkpoint 7

Find the derivative of each function.

a. $y = \dfrac{x^2 + 4x}{5}$ **b.** $y = \dfrac{3x^4}{4}$ ■

Application

Example 8 Rate of Change of Systolic Blood Pressure

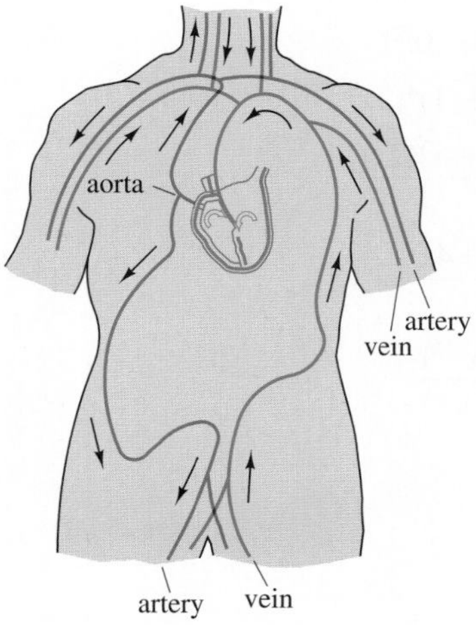

As blood moves from the heart through the major arteries out to the capillaries and back through the veins, the systolic blood pressure continuously drops. Consider a person whose systolic blood pressure P (in millimeters of mercury) is given by

$$P = \frac{25t^2 + 125}{t^2 + 1}, \quad 0 \le t \le 10$$

where t is measured in seconds. At what rate is the blood pressure changing 5 seconds after blood leaves the heart?

SOLUTION Begin by applying the Quotient Rule.

$$P = \frac{25t^2 + 125}{t^2 + 1} \qquad \text{Write original function.}$$

$$\frac{dP}{dt} = \frac{(t^2 + 1)(50t) - (25t^2 + 125)(2t)}{(t^2 + 1)^2} \qquad \text{Quotient Rule}$$

$$= \frac{50t^3 + 50t - 50t^3 - 250t}{(t^2 + 1)^2}$$

$$= -\frac{200t}{(t^2 + 1)^2} \qquad \text{Simplify.}$$

When $t = 5$, the rate of change is

$$\frac{dP}{dt} = -\frac{200(5)}{26^2} \approx -1.48 \text{ millimeters per second.}$$

So, the pressure is *dropping* at a rate of 1.48 millimeters per second at $t = 5$ seconds.

✓ Checkpoint 8

In Example 8, find the rate at which systolic blood pressure is changing at each time shown in the table below. Describe the changes in blood pressure as the blood moves away from the heart.

t	0	1	2	3	4	5	6	7
$\frac{dP}{dt}$								

SUMMARIZE (Section 2.4)

1. State the Product Rule *(page 117)*. For examples of the Product Rule, see Examples 1, 2, and 3.
2. State the Quotient Rule *(page 120)*. For examples of the Quotient Rule, see Examples 4, 5, and 6.
3. Describe a real-life example of how the Quotient Rule can be used to analyze the rate of change of systolic blood pressure *(page 123, Example 8)*.

Michal Kowalski/Shutterstock.com

SKILLS WARM UP 2.4

The following warm-up exercises involve skills that were covered in a previous course or earlier sections. You will use these skills in the exercise set for this section. For additional help, review Appendix Sections A.4 and A.5, and Section 2.2.

In Exercises 1–10, simplify the expression.

1. $(x^2 + 1)(2) + (2x + 7)(2x)$

2. $(2x - x^3)(8x) + (4x^2)(2 - 3x^2)$

3. $x(4)(x^2 + 2)^3(2x) + (x^2 + 4)(1)$

4. $x^2(2)(2x + 1)(2) + (2x + 1)^4(2x)$

5. $\dfrac{(2x + 7)(5) - (5x + 6)(2)}{(2x + 7)^2}$

6. $\dfrac{(x^2 - 4)(2x + 1) - (x^2 + x)(2x)}{(x^2 - 4)^2}$

7. $\dfrac{(x^2 + 1)(2) - (2x + 1)(2x)}{(x^2 + 1)^2}$

8. $\dfrac{(1 - x^4)(4) - (4x - 1)(-4x^3)}{(1 - x^4)^2}$

9. $(x^{-1} + x)(2) + (2x - 3)(-x^{-2} + 1)$

10. $\dfrac{(1 - x^{-1})(1) - (x - 4)(x^{-2})}{(1 - x^{-1})^2}$

In Exercises 11–14, find $f'(2)$.

11. $f(x) = 3x^2 - x + 4$

12. $f(x) = -x^3 + x^2 + 8x$

13. $f(x) = \dfrac{1}{x}$

14. $f(x) = x^2 - \dfrac{1}{x^2}$

Exercises 2.4

See www.CalcChat.com for worked-out solutions to odd-numbered exercises.

Using the Product Rule In Exercises 1–10, use the Product Rule to find the derivative of the function. *See Examples 1, 2, and 3.*

1. $f(x) = (2x - 3)(1 - 5x)$

2. $g(x) = (x - 4)(x + 2)$

3. $f(x) = (6x - x^2)(4 + 3x)$

4. $f(x) = (x^2 + 1)(2x + 5)$

5. $f(x) = x(x^2 + 3)$

6. $f(x) = x^2(3x^3 - 1)$

7. $h(x) = \left(\dfrac{2}{x} - 3\right)(x^2 + 7)$

8. $f(x) = (3 - x)\left(\dfrac{4}{x^2} - 5\right)$

9. $g(x) = (x^2 - 4x + 3)(x - 2)$

10. $g(x) = (x^2 - 2x + 1)(x^3 - 1)$

Using the Quotient Rule In Exercises 11–20, use the Quotient Rule to find the derivative of the function. *See Examples 4 and 6.*

11. $h(x) = \dfrac{x}{x - 5}$

12. $h(x) = \dfrac{x^2}{x + 3}$

13. $f(t) = \dfrac{2t^2 - 3}{3t + 1}$

14. $f(x) = \dfrac{x + 1}{x - 1}$

15. $f(t) = \dfrac{t^2 - 1}{t + 4}$

16. $g(x) = \dfrac{4x - 5}{x^2 - 1}$

17. $f(x) = \dfrac{x^2 + 6x + 5}{2x - 1}$

18. $f(x) = \dfrac{4x^2 - x + 1}{x + 2}$

19. $f(x) = \dfrac{6 + (2/x)}{3x - 1}$

20. $f(x) = \dfrac{5 - (1/x^2)}{x + 2}$

Using the Constant Multiple Rule In Exercises 21–30, find the derivative of the function. *See Example 7.*

Original Function	*Rewrite*	*Differentiate*	*Simplify*
21. $f(x) = \dfrac{x^3 + 6x}{3}$			
22. $f(x) = \dfrac{3x^2}{7}$			
23. $y = \dfrac{x^2 + 2x}{3}$			
24. $y = \dfrac{4x^{3/2}}{x}$			
25. $y = \dfrac{7}{3x^3}$			
26. $y = \dfrac{4}{5x^2}$			
27. $y = \dfrac{4x^2 - 3x}{8\sqrt{x}}$			
28. $y = \dfrac{5(3x^2 + 5x)}{8x}$			
29. $y = \dfrac{x^2 - 4x + 3}{2(x - 1)}$			
30. $y = \dfrac{x^2 - 4}{4(x + 2)}$			

Finding Derivatives In Exercises 31–44, find the derivative of the function. State which differentiation rule(s) you used to find the derivative.

31. $f(x) = (x^3 - 3x)(2x^2 + 3x + 5)$
32. $h(t) = (t^5 - 1)(4t^2 - 7t - 3)$
33. $f(x) = \dfrac{x^3 + 3x + 2}{x^2 - 1}$
34. $f(x) = (x^5 - 3x)\left(\dfrac{1}{x^2}\right)$
35. $f(x) = \dfrac{x^2 - x - 20}{x + 4}$
36. $h(t) = \dfrac{t + 2}{t^2 + 5t + 6}$
37. $g(t) = (2t^3 - 1)^2$
38. $f(x) = \sqrt[3]{x}(x + 1)$
39. $g(s) = \dfrac{s^2 - 2s + 5}{\sqrt{s}}$
40. $f(x) = \dfrac{x + 1}{\sqrt{x}}$
41. $f(x) = \dfrac{(x - 2)(3x + 1)}{4x + 2}$
42. $f(x) = \dfrac{(x + 1)(2x - 7)}{2x + 1}$
43. $g(x) = \left(\dfrac{x - 3}{x + 4}\right)(x^2 + 2x + 1)$
44. $f(x) = (3x^3 + 4x)(x - 5)(x + 1)$

Finding an Equation of a Tangent Line In Exercises 45–52, find an equation of the tangent line to the graph of the function at the given point. Then use a graphing utility to graph the function and the tangent line in the same viewing window. *See Example 5.*

Function	*Point*
45. $f(x) = (5x + 2)(x^2 + x)$	$(-1, 0)$
46. $h(x) = (x^2 - 1)^2$	$(-2, 9)$
47. $f(x) = x^3(x^2 - 4)$	$(1, -3)$
48. $f(x) = \sqrt{x}(x - 3)$	$(9, 18)$
49. $f(x) = \dfrac{x - 2}{x + 1}$	$\left(1, -\dfrac{1}{2}\right)$
50. $f(x) = \dfrac{2x + 1}{x - 1}$	$(2, 5)$
51. $f(x) = \dfrac{(3x - 2)(6x + 5)}{2x - 3}$	$(-1, -1)$
52. $g(x) = \dfrac{(x + 2)(x^2 + x)}{x - 4}$	$(1, -2)$

Finding Horizontal Tangent Lines In Exercises 53–56, find the point(s), if any, at which the graph of f has a horizontal tangent line.

53. $f(x) = \dfrac{x^2}{x - 1}$
54. $f(x) = \dfrac{x^2}{x^2 + 1}$
55. $f(x) = \dfrac{x^4}{x^3 + 1}$
56. $f(x) = \dfrac{x^4 + 3}{x^2 + 1}$

Graphing a Function and Its Derivative In Exercises 57–60, use a graphing utility to graph f and f' on the interval $[-2, 2]$.

57. $f(x) = x(x + 1)$
58. $f(x) = x^2(x + 1)$
59. $f(x) = x(x + 1)(x - 1)$
60. $f(x) = x^2(x + 1)(x - 1)$

Demand In Exercises 61 and 62, use the demand function to find the rate of change in the demand x for the given price p.

61. $x = 275\left(1 - \dfrac{3p}{5p + 1}\right)$, $p = \$4$
62. $x = 300 - p - \dfrac{2p}{p + 1}$, $p = \$3$

63. **Population Growth** A population of bacteria is introduced into a culture. The number of bacteria P can be modeled by

$$P = 500\left(1 + \frac{4t}{50 + t^2}\right)$$

where t is the time (in hours). Find the rate of change of the population at $t = 2$.

64. **Quality Control** The percent P of defective parts produced by a new employee t days after the employee starts work can be modeled by

$$P = \frac{t + 1750}{50(t + 2)}.$$

Find the rates of change of P at (a) $t = 1$ and (b) $t = 10$.

65. **Environment** The model

$$f(t) = \frac{t^2 - t + 1}{t^2 + 1}$$

measures the level of oxygen in a pond, where t is the time (in weeks) after organic waste is dumped into the pond. Find the rates of change of f with respect to t at (a) $t = 0.5$, (b) $t = 2$, and (c) $t = 8$. Interpret the meaning of these values.

66. Physical Science The temperature T (in degrees Fahrenheit) of food placed in a refrigerator is modeled by

$$T = 10\left(\frac{4t^2 + 16t + 75}{t^2 + 4t + 10}\right)$$

where t is the time (in hours). What is the initial temperature of the food? Find the rates of change of T with respect to t at (a) $t = 1$, (b) $t = 3$, (c) $t = 5$, and (d) $t = 10$. Interpret the meaning of these values.

67. Cost The cost C of producing x units of a product is given by $C = x^3 - 15x^2 + 87x - 73$ for $4 \le x \le 9$.

(a) Use a graphing utility to graph the marginal cost function and the average cost function, C/x, in the same viewing window.

(b) Find the point of intersection of the graphs of dC/dx and C/x. Does this point have any significance?

68. HOW DO YOU SEE IT? The advertising manager for a new product determines that P percent of the potential market is aware of the product t weeks after the advertising campaign begins.

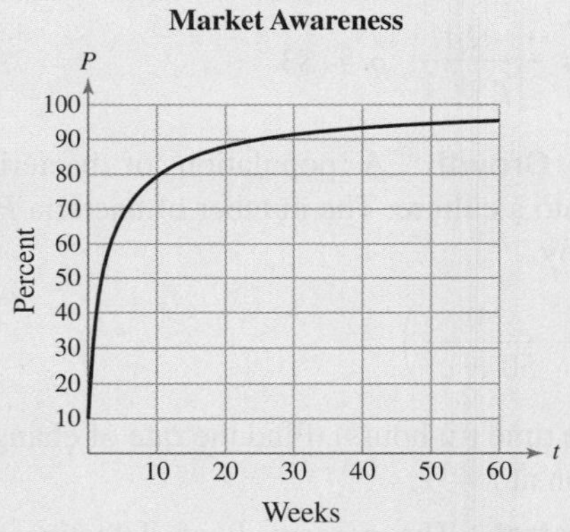

(a) What happens to the percent of people who are aware of the product in the long run?

(b) What happens to the rate of change of the percent of people who are aware of the product in the long run?

69. Inventory Replenishment The ordering and transportation cost C per unit (in thousands of dollars) of the components used in manufacturing a product is given by

$$C = 100\left(\frac{200}{x^2} + \frac{x}{x + 30}\right), \quad 1 \le x$$

where x is the order size (in hundreds). Find the rate of change of C with respect to x for each order size. What do these rates of change imply about increasing the size of an order? Of the given order sizes, which would you choose? Explain.

(a) $x = 10$ (b) $x = 15$ (c) $x = 20$

Gareth Davies/Getty Images

70. Managing a Store You are managing a store and have been adjusting the price of an item. You have found that you make a profit of \$50 when 10 units are sold, \$60 when 12 units are sold, and \$65 when 14 units are sold.

(a) Use the *regression* feature of a graphing utility to find a quadratic model that relates the profit P to the number of units sold x.

(b) Use a graphing utility to graph P.

(c) Find the point on the graph at which the marginal profit is zero. Interpret this point in the context of the problem.

Using Relationships In Exercises 71–74, use the given information to find $f'(2)$.

$g(2) = 3$ and $g'(2) = -2$

$h(2) = -1$ and $h'(2) = 4$

71. $f(x) = 2g(x) + h(x)$ **72.** $f(x) = 3 - g(x)$

73. $f(x) = g(x)h(x)$ **74.** $f(x) = \dfrac{g(x)}{h(x)}$

Business Capsule

In 1978 Ben Cohen and Jerry Greenfield used their combined life savings of \$8000 to convert an abandoned gas station in Burlington, Vermont into their first ice cream shop. Today, Ben & Jerry's Homemade Holdings, Inc. has almost 800 scoop shops in 25 countries. The company's three-part mission statement emphasizes product quality, economic reward, and a commitment to the community. Ben & Jerry's contributes a minimum of \$1.8 million annually through corporate philanthropy that is primarily employee led.

75. Research Project Use your school's library, the Internet, or some other reference source to find information on a company that is noted for its philanthropy and community commitment. (One such business is described above.) Write a short paper about the company.

QUIZ YOURSELF

See www.CalcChat.com for worked-out solutions to odd-numbered exercises.

Take this quiz as you would take a quiz in class. When you are done, check your work against the answers given in the back of the book.

In Exercises 1–3, use the limit definition to find the derivative of the function. Then find the slope of the tangent line to the graph of *f* at the given point.

1. $f(x) = 5x + 3;\ (-2, -7)$

2. $f(x) = \sqrt{x + 3};\ (1, 2)$

3. $f(x) = x^2 - 2x;\ (3, 3)$

In Exercises 4–13, find the derivative of the function.

4. $f(x) = 12$

5. $f(x) = 19x + 9$

6. $f(x) = 5 - 3x^2$

7. $f(x) = 12x^{1/4}$

8. $f(x) = 4x^{-2}$

9. $f(x) = 2\sqrt{x}$

10. $f(x) = \dfrac{2x + 3}{3x + 2}$

11. $f(x) = (x^2 + 1)(-2x + 4)$

12. $f(x) = (x^2 + 3x + 4)(5x - 2)$

13. $f(x) = \dfrac{4x}{x^2 + 3}$

In Exercises 14–17, use a graphing utility to graph the function and find its average rate of change on the given interval. Compare this rate with the instantaneous rates of change at the endpoints of the interval.

14. $f(x) = x^2 - 3x + 1;\ [0, 3]$

15. $f(x) = 2x^3 + x^2 - x + 4;\ [-1, 1]$

16. $f(x) = \dfrac{1}{2x};\ [2, 5]$

17. $f(x) = \sqrt[3]{x};\ [8, 27]$

18. The profit P (in dollars) from selling x units of a product is given by

$$P = -0.0125x^2 + 16x - 600.$$

(a) Find the additional profit when sales increase from 175 to 176 units.

(b) Find the marginal profit when $x = 175$.

(c) Compare the results of parts (a) and (b).

In Exercises 19 and 20, find an equation of the tangent line to the graph of *f* at the given point. Then use a graphing utility to graph the function and the equation of the tangent line in the same viewing window.

19. $f(x) = 5x^2 + 6x - 1;\ (-1, -2)$

20. $f(x) = (x^2 + 1)(4x - 3);\ (1, 2)$

21. From 2003 through 2009, the sales per share S (in dollars) for Columbia Sportswear can be modeled by

$$S = -0.13556t^3 + 1.8682t^2 - 4.351t + 23.52, \quad 3 \le t \le 9$$

where t represents the year, with $t = 3$ corresponding to 2003. *(Source: Columbia Sportswear Company)*

(a) Find the rate of change of the sales per share with respect to the year.

(b) At what rate were the sales per share changing in 2004? in 2007? in 2008?

2.5 The Chain Rule

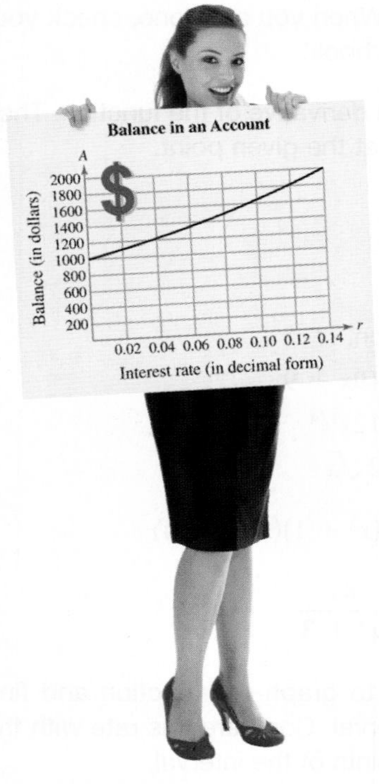

In Exercise 71 on page 136, you will use the General Power Rule to find the rate of change of the balance in an account.

- Find derivatives using the Chain Rule.
- Find derivatives using the General Power Rule.
- Write derivatives in simplified form.
- Use derivatives to answer questions about real-life situations.
- Review the basic differentiation rules for algebraic functions.

The Chain Rule

In this section, you will study one of the most powerful rules of differential calculus—the **Chain Rule.** This differentiation rule deals with composite functions and adds versatility to the rules presented in Sections 2.2 and 2.4. For example, compare the functions below. Those on the left can be differentiated without the Chain Rule, whereas those on the right are best done with the Chain Rule.

Without the Chain Rule	*With the Chain Rule*
$y = x^2 + 1$	$y = \sqrt{x^2 + 1}$
$y = x + 1$	$y = (x + 1)^{-1/2}$
$y = 3x + 2$	$y = (3x + 2)^5$
$y = \dfrac{x + 5}{x^2 + 2}$	$y = \left(\dfrac{x + 5}{x^2 + 2}\right)^2$
$y = \dfrac{x + 1}{x}$	$y = \sqrt{\dfrac{x + 1}{x}}$

The Chain Rule

If $y = f(u)$ is a differentiable function of u, and $u = g(x)$ is a differentiable function of x, then $y = f(g(x))$ is a differentiable function of x, and

$$\frac{dy}{dx} = \frac{dy}{du} \cdot \frac{du}{dx}$$

or, equivalently,

$$\frac{d}{dx}[f(g(x))] = f'(g(x))g'(x).$$

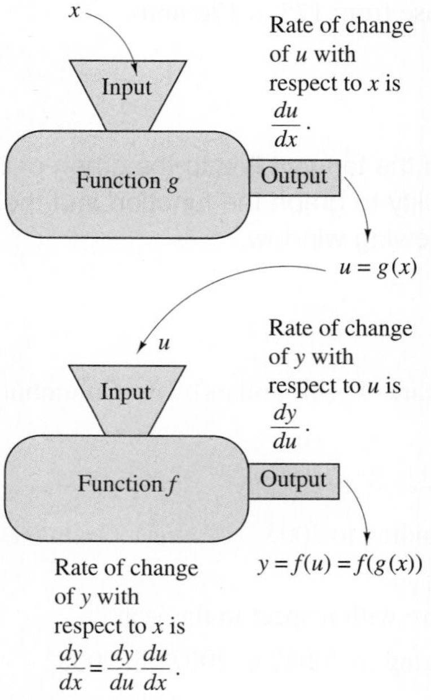

FIGURE 2.29

Basically, the Chain Rule states that if y changes dy/du times as fast as u, and u changes du/dx times as fast as x, then y changes

$$\frac{dy}{du} \cdot \frac{du}{dx}$$

times as fast as x, as illustrated in Figure 2.29. One advantage of the

$$\frac{dy}{dx}$$

notation for derivatives is that it helps you remember differentiation rules, such as the Chain Rule. For instance, in the formula

$$\frac{dy}{dx} = \frac{dy}{du} \cdot \frac{du}{dx}$$

you can imagine that the du's divide out.

When applying the Chain Rule, it is helpful to think of the composite function $y = f(g(x))$ or $y = f(u)$ as having two parts—an *inside* and an *outside*—as illustrated below.

Inside

$$y = f(g(x)) = f(u)$$

Outside

The Chain Rule tells you that the derivative of $y = f(u)$ is the derivative of the outer function (at the inner function u) *times* the derivative of the inner function. That is,

$$y' = f'(u) \cdot u'.$$

Example 1 Decomposing Composite Functions

Write each function as the composition of two functions.

a. $y = \dfrac{1}{x + 1}$ **b.** $y = \sqrt{3x^2 - x + 1}$

SOLUTION There is more than one correct way to decompose each function. One way for each is shown below.

$y = f(g(x))$	$u = g(x)$ *(inside)*	$y = f(u)$ *(outside)*
a. $y = \dfrac{1}{x + 1}$	$u = x + 1$	$y = \dfrac{1}{u}$
b. $y = \sqrt{3x^2 - x + 1}$	$u = 3x^2 - x + 1$	$y = \sqrt{u}$

✓ Checkpoint 1

Write each function as the composition of two functions, where $y = f(g(x))$.

a. $y = \dfrac{1}{\sqrt{x + 1}}$ **b.** $y = (x^2 + 2x + 5)^3$ ■

Example 2 Using the Chain Rule

Find the derivative of $y = (x^2 + 1)^3$.

SOLUTION To apply the Chain Rule, you need to identify the inside function u.

$$y = \overbrace{(x^2 + 1)}^{u}{}^3 = u^3$$

The inside function is $u = x^2 + 1$. By the Chain Rule, you can write the derivative as shown.

$$\frac{dy}{dx} = \overbrace{3(x^2 + 1)^2}^{\frac{dy}{du}}\overbrace{(2x)}^{\frac{du}{dx}} = 6x(x^2 + 1)^2$$

✓ Checkpoint 2

Find the derivative of $y = (x^3 + 1)^2$. ■

STUDY TIP

Try checking the result of Example 2 by expanding the function to obtain

$$y = x^6 + 3x^4 + 3x^2 + 1$$

and finding the derivative. Do you obtain the same answer?

The General Power Rule

The function in Example 2 illustrates one of the most common types of composite functions—a power function of the form

$$y = [u(x)]^n.$$

The rule for differentiating such functions is called the **General Power Rule,** and it is a special case of the Chain Rule.

> **The General Power Rule**
>
> If $y = [u(x)]^n$, where u is a differentiable function of x and n is a real number, then
>
> $$\frac{dy}{dx} = n[u(x)]^{n-1}\frac{du}{dx}$$
>
> or, equivalently,
>
> $$\frac{d}{dx}[u^n] = nu^{n-1}u'.$$

PROOF Apply the Chain Rule and the Simple Power Rule as shown.

$$\begin{aligned}\frac{dy}{dx} &= \frac{dy}{du}\cdot\frac{du}{dx}\\ &= \frac{d}{du}[u^n]\frac{du}{dx}\\ &= nu^{n-1}\frac{du}{dx}\end{aligned}$$

TECH TUTOR

If you have access to a symbolic differentiation utility, try using it to confirm the result of Example 3.

Yuri Arcurs/Shutterstock.com

Example 3 Using the General Power Rule

Find the derivative of

$$y = (3x - 2x^2)^3.$$

SOLUTION To apply the General Power Rule, you need to identify the inside function u.

$$y = \overbrace{(3x - 2x^2)}^{u}{}^3 = u^3$$

The inside function is

$$u = 3x - 2x^2.$$

So, by the General Power Rule,

$$\begin{aligned}\frac{dy}{dx} &= \overbrace{3(3x - 2x^2)^2}^{n\quad u^{n-1}}\overbrace{\frac{d}{dx}[3x - 2x^2]}^{u'}\\ &= 3(3x - 2x^2)^2(3 - 4x).\end{aligned}$$

✓ Checkpoint 3

Find the derivative of

$$y = (x^2 + 3x)^4.$$

■

Example 4 Finding an Equation of a Tangent Line

Find an equation of the tangent line to the graph of

$$y = \sqrt[3]{(x^2 + 4)^2}$$

at $x = 2$.

SOLUTION Begin by rewriting the function in rational exponent form.

$$y = (x^2 + 4)^{2/3} \qquad \text{Rewrite original function.}$$

Then, using the inside function, $u = x^2 + 4$, apply the General Power Rule.

$$\frac{dy}{dx} = \underbrace{\frac{2}{3}}_{n}\underbrace{(x^2 + 4)^{-1/3}}_{u^{n-1}}\underbrace{(2x)}_{u'} \qquad \text{Apply General Power Rule.}$$

$$= \frac{4x(x^2 + 4)^{-1/3}}{3}$$

$$= \frac{4x}{3\sqrt[3]{x^2 + 4}} \qquad \text{Write in radical form.}$$

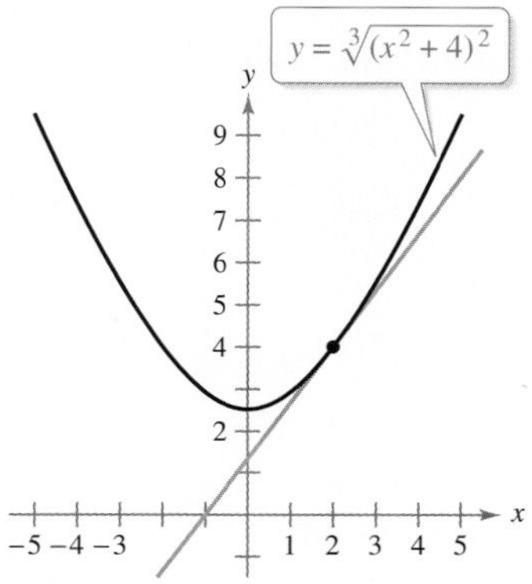

FIGURE 2.30

When $x = 2$, $y = 4$ and the slope of the line tangent to the graph at $(2, 4)$ is $\frac{4}{3}$. Using the point-slope form, you can find the equation of the tangent line to be $y = \frac{4}{3}x + \frac{4}{3}$. The graph of the function and the tangent line is shown in Figure 2.30.

✓Checkpoint 4

Find an equation of the tangent line to the graph of $y = \sqrt[3]{(x + 4)^2}$ at $x = 4$. ■

STUDY TIP

The derivative of a quotient can sometimes be found more easily with the General Power Rule than with the Quotient Rule. This is especially true when the numerator is a constant, as shown in Example 5.

Example 5 Differentiating a Quotient with a Constant Numerator

Find the derivative of

$$y = \frac{5}{(4x - 3)^2}.$$

SOLUTION Begin by rewriting the function in rational exponent form.

$$y = 5(4x - 3)^{-2} \qquad \text{Rewrite original function.}$$

Then, using the inside function, $u = 4x - 3$, apply the General Power Rule.

$$\frac{dy}{dx} = 5\underbrace{(-2)}_{n}\underbrace{(4x - 3)^{-3}}_{u^{n-1}}\underbrace{(4)}_{u'} \qquad \text{Apply General Power Rule.}$$

(5: Constant Multiple Rule)

$$= -40(4x - 3)^{-3} \qquad \text{Simplify.}$$

$$= \frac{-40}{(4x - 3)^3} \qquad \text{Write with positive exponent.}$$

✓Checkpoint 5

Find the derivative of $y = \dfrac{4}{2x + 1}$. ■

Simplification Techniques

Throughout this chapter, writing derivatives in simplified form has been emphasized. The reason for this is that most applications of derivatives require a simplified form. The next two examples illustrate some useful simplification techniques.

ALGEBRA TUTOR

In Example 6, note that you subtract exponents when factoring. That is, when $(1-x^2)^{-1/2}$ is factored out of $(1-x^2)^{1/2}$, the *remaining* factor has an exponent of $\frac{1}{2}-\left(-\frac{1}{2}\right)=1$. So, $(1-x^2)^{1/2}$ is equal to the product of $(1-x^2)^{-1/2}$ and $(1-x^2)^1$. For help in simplifying expressions like the one in Example 6, see the *Chapter 2 Algebra Tutor* on pages 158 and 159.

Example 6 Simplifying by Factoring Out Least Powers

Find the derivative of $y = x^2\sqrt{1-x^2}$.

SOLUTION

$$y = x^2\sqrt{1-x^2} \quad \text{Write original function.}$$
$$= x^2(1-x^2)^{1/2} \quad \text{Rewrite function.}$$
$$y' = x^2\frac{d}{dx}\left[(1-x^2)^{1/2}\right] + (1-x^2)^{1/2}\frac{d}{dx}\left[x^2\right] \quad \text{Product Rule}$$
$$= x^2\left[\frac{1}{2}(1-x^2)^{-1/2}(-2x)\right] + (1-x^2)^{1/2}(2x) \quad \text{General Power Rule}$$
$$= -x^3(1-x^2)^{-1/2} + 2x(1-x^2)^{1/2} \quad \text{Simplify.}$$
$$= x(1-x^2)^{-1/2}\left[-x^2(1) + 2(1-x^2)\right] \quad \text{Factor.}$$
$$= x(1-x^2)^{-1/2}(2-3x^2) \quad \text{Simplify.}$$
$$= \frac{x(2-3x^2)}{\sqrt{1-x^2}} \quad \text{Write in radical form.}$$

✓Checkpoint 6

Find and simplify the derivative of $y = x^2\sqrt{x^2+1}$. ■

STUDY TIP

In Example 7, try to find $f'(x)$ by applying the Quotient Rule to

$$f(x) = \frac{(3x-1)^2}{(x^2+3)^2}.$$

Which method do you prefer?

Example 7 Differentiating a Quotient Raised to a Power

Find the derivative of

$$f(x) = \left(\frac{3x-1}{x^2+3}\right)^2.$$

SOLUTION

$$f'(x) = \underset{n}{2}\underbrace{\left(\frac{3x-1}{x^2+3}\right)}_{u^{n-1}}\underbrace{\frac{d}{dx}\left[\frac{3x-1}{x^2+3}\right]}_{u'} \quad \text{General Power Rule}$$
$$= \left[\frac{2(3x-1)}{x^2+3}\right]\left[\frac{(x^2+3)(3)-(3x-1)(2x)}{(x^2+3)^2}\right] \quad \text{Quotient Rule}$$
$$= \frac{2(3x-1)(3x^2+9-6x^2+2x)}{(x^2+3)^3} \quad \text{Multiply.}$$
$$= \frac{2(3x-1)(-3x^2+2x+9)}{(x^2+3)^3} \quad \text{Simplify.}$$

✓Checkpoint 7

Find the derivative of

$$f(x) = \left(\frac{x+1}{x-5}\right)^2.$$

■

Application

Example 8 Finding Rates of Change

From 2000 through 2009, the revenue per share R (in dollars) for U.S. Cellular can be modeled by

$$R = (-0.009t^2 + 0.39t + 4.3)^2, \quad 0 \le t \le 9$$

where t is the year, with $t = 0$ corresponding to 2000. Use the model to approximate the rates of change in the revenue per share in 2001, 2002, and 2005. Would U.S. Cellular stockholders have been satisfied with the performance of this stock from 2000 through 2009? *(Source: U.S. Cellular)*

SOLUTION The rate of change in R is given by the derivative dR/dt. You can use the General Power Rule to find the derivative.

$$\frac{dR}{dt} = 2(-0.009t^2 + 0.39t + 4.3)(-0.018t + 0.39)$$

$$= (-0.036t + 0.78)(-0.009t^2 + 0.39t + 4.3)$$

In 2001, the revenue per share was changing at a rate of

$$[-0.036(1) + 0.78][-0.009(1)^2 + 0.39(1) + 4.3] \approx \$3.48 \text{ per year.}$$

In 2002, the revenue per share was changing at a rate of

$$[-0.036(2) + 0.78][-0.009(2)^2 + 0.39(2) + 4.3] \approx \$3.57 \text{ per year.}$$

In 2005, the revenue per share was changing at a rate of

$$[-0.036(5) + 0.78][-0.009(5)^2 + 0.39(5) + 4.3] \approx \$3.62 \text{ per year.}$$

The graph of the revenue per share function R is shown in Figure 2.31. So, most stockholders would have been satisfied with the performance of this stock.

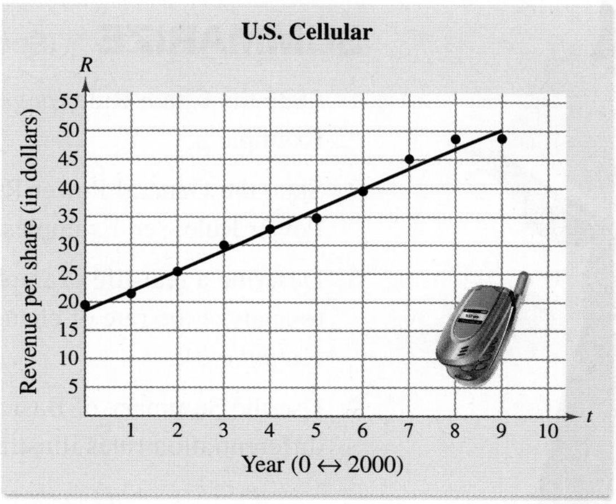

FIGURE 2.31

✓ Checkpoint 8

From 2000 through 2009, the sales per share S (in dollars) for Dollar Tree can be modeled by

$$S = (0.010t^2 + 0.27t + 3.1)^2, \quad 0 \le t \le 9$$

where t is the year, with $t = 0$ corresponding to 2000. Use the model to approximate the rate of change of sales per share in 2005. *(Source: Dollar Tree, Inc.)* ■

Review of Basic Differentiation Rules

You now have all the rules you need to differentiate *any* algebraic function. For your convenience, they are summarized below.

Summary of Basic Differentiation Rules

Let u and v be differentiable functions of x.

1. Constant Rule $\quad \frac{d}{dx}[c] = 0, \quad c$ is a constant.
2. Constant Multiple Rule $\quad \frac{d}{dx}[cu] = c\frac{du}{dx}, \quad c$ is a constant.
3. Sum and Difference Rules $\quad \frac{d}{dx}[u \pm v] = \frac{du}{dx} \pm \frac{dv}{dx}$
4. Product Rule $\quad \frac{d}{dx}[uv] = u\frac{dv}{dx} + v\frac{du}{dx}$
5. Quotient Rule $\quad \frac{d}{dx}\left[\frac{u}{v}\right] = \frac{v\frac{du}{dx} - u\frac{dv}{dx}}{v^2}$
6. Power Rules $\quad \frac{d}{dx}[x^n] = nx^{n-1}$

   $\frac{d}{dx}[u^n] = nu^{n-1}\frac{du}{dx}$
7. Chain Rule $\quad \frac{dy}{dx} = \frac{dy}{du} \cdot \frac{du}{dx}$

SUMMARIZE (Section 2.5)

1. State the Chain Rule *(page 128)*. For an example of the Chain Rule, see Example 2.
2. State the General Power Rule *(page 130)*. For examples of the General Power Rule, see Examples 3, 4, and 5.
3. Describe a real-life example of how the General Power Rule can be used to analyze the rate of change of a company's revenue per share *(page 133, Example 8)*.
4. Use the Summary of Basic Differentiation Rules to identify the differentiation rules illustrated by (a)–(f) below *(page 134)*.

(a) $\frac{d}{dx}[2x] = 2\frac{d}{dx}[x]$ (b) $\frac{d}{dx}[x^4] = 4x^3$

(c) $\frac{d}{dx}[8] = 0$ (d) $\frac{d}{dx}[x^2 + x] = \frac{d}{dx}[x^2] + \frac{d}{dx}[x]$

(e) $\frac{d}{dx}[x - x^3] = \frac{d}{dx}[x] - \frac{d}{dx}[x^3]$

(f) $\frac{d}{dx}[x(x + 1)] = (x)\frac{d}{dx}[x + 1] + (x + 1)\frac{d}{dx}[x]$

Kamenetskiy Konstantin/Shutterstock.com

SKILLS WARM UP 2.5 The following warm-up exercises involve skills that were covered in a previous course. You will use these skills in the exercise set for this section. For additional help, review Appendix Sections A.3 and A.4.

In Exercises 1–6, rewrite the expression with rational exponents.

1. $\sqrt[5]{(1-5x)^2}$ **2.** $\sqrt[4]{(2x-1)^3}$ **3.** $\dfrac{1}{\sqrt{4x^2+1}}$

4. $\dfrac{1}{\sqrt[3]{x-6}}$ **5.** $\dfrac{\sqrt{x}}{\sqrt[3]{1-2x}}$ **6.** $\dfrac{\sqrt{(3-7x)^3}}{2x}$

In Exercises 7–10, factor the expression.

7. $3x^3 - 6x^2 + 5x - 10$ **8.** $5x\sqrt{x} - x - 5\sqrt{x} + 1$

9. $4(x^2+1)^2 - x(x^2+1)^3$ **10.** $-x^5 + 3x^3 + x^2 - 3$

Exercises 2.5

See www.CalcChat.com for worked-out solutions to odd-numbered exercises.

Decomposing Composite Functions In Exercises 1–6, identify the inside function, $u = g(x)$, and the outside function, $y = f(u)$. *See Example 1.*

$y = f(g(x))$	$u = g(x)$	$y = f(u)$
1. $y = (6x-5)^4$		
2. $y = (x^2-2x+3)^3$		
3. $y = \sqrt{5x-2}$		
4. $y = \sqrt{1-x^2}$		
5. $y = \dfrac{1}{3x+1}$		
6. $y = \dfrac{1}{\sqrt{x^2-3}}$		

Using the Chain Rule In Exercises 7–12, find dy/du, du/dx, and dy/dx. *See Example 2.*

7. $y = u^2, u = 4x + 7$ **8.** $y = u^3, u = 3x^2 - 2$

9. $y = \sqrt{u}, u = 3 - x^2$ **10.** $y = 2\sqrt{u}, u = 5x + 9$

11. $y = u^{2/3}, u = 5x^4 - 2x$ **12.** $y = u^{-1}, u = x^3 + 2x^2$

Choosing a Differentiation Rule In Exercises 13–20, match the function with the rule that you would use to find the derivative *most efficiently*.

(a) Simple Power Rule (b) Constant Rule

(c) General Power Rule (d) Quotient Rule

13. $f(x) = \dfrac{2}{1-x^3}$ **14.** $f(x) = \dfrac{2x}{1-x^3}$

15. $f(x) = \sqrt[3]{8^2}$ **16.** $f(x) = \sqrt[3]{x^2}$

17. $f(x) = \dfrac{x^2+2}{x}$ **18.** $f(x) = \dfrac{\sqrt{x}}{x^3+2x-5}$

19. $f(x) = \dfrac{2}{x-2}$ **20.** $f(x) = \dfrac{5}{x^2+1}$

Using the General Power Rule In Exercises 21–36, use the General Power Rule to find the derivative of the function. *See Examples 3 and 5.*

21. $y = (2x-7)^3$ **22.** $g(x) = (4-2x)^3$

23. $f(x) = (5x - x^2)^{3/2}$ **24.** $y = (2x^3+1)^2$

25. $h(x) = (6x - x^3)^2$ **26.** $f(x) = (4x - x^2)^3$

27. $f(t) = \sqrt{t+1}$ **28.** $g(x) = \sqrt{5-3x}$

29. $s(t) = \sqrt{2t^2+5t+2}$ **30.** $y = \sqrt[3]{3x^3+4x}$

31. $y = \sqrt[3]{9x^2+4}$ **32.** $y = 2\sqrt{4-x^2}$

33. $f(x) = \dfrac{2}{(2-9x)^3}$ **34.** $g(x) = \dfrac{3}{\sqrt{(x^2+8x)^3}}$

35. $f(x) = \dfrac{1}{\sqrt{x^2+25}}$ **36.** $y = \dfrac{1}{\sqrt[3]{(4-x^3)^4}}$

Finding an Equation of a Tangent Line In Exercises 37–42, find an equation of the tangent line to the graph of f at the point $(2, f(2))$. Then use a graphing utility to graph the function and the tangent line in the same viewing window. *See Example 4.*

37. $f(x) = 2(x^2-1)^3$ **38.** $f(x) = 3(9x-4)^4$

39. $f(x) = \sqrt{4x^2-7}$ **40.** $f(x) = x\sqrt{x^2+5}$

41. $f(x) = \sqrt{x^2-2x+1}$ **42.** $f(x) = (4-3x^2)^{-2/3}$

Using Technology In Exercises 43–46, use a symbolic differentiation utility to find the derivative of the function. Graph the function and its derivative in the same viewing window. Describe the behavior of the function when the derivative is zero.

43. $f(x) = \dfrac{\sqrt{x}+1}{x^2+1}$ **44.** $f(x) = \sqrt{\dfrac{2x}{x+1}}$

45. $f(x) = \sqrt{\dfrac{x+1}{x}}$ **46.** $f(x) = \sqrt{x}(2-x^2)$

Finding Derivatives In Exercises 47–62, find the derivative of the function. State which differentiation rule(s) you used to find the derivative.

47. $y = \dfrac{1}{4 - x^2}$

48. $s(t) = \dfrac{1}{t^2 + 3t - 1}$

49. $y = -\dfrac{4}{(t + 2)^2}$

50. $f(x) = \dfrac{3x}{(x^3 - 4)^2}$

51. $f(x) = (2x - 1)(9 - 3x^2)$

52. $y = x^3(7x + 2)$

53. $y = \dfrac{1}{\sqrt{x + 2}}$

54. $g(x) = \dfrac{3}{\sqrt[3]{x^3 - 1}}$

55. $f(x) = x(3x - 9)^3$

56. $f(x) = x^3(x - 4)^2$

57. $y = x\sqrt{2x + 3}$

58. $y = t\sqrt{t + 1}$

59. $y = t^2\sqrt{t - 2}$

60. $y = \sqrt{x}(x - 2)^2$

61. $y = \left(\dfrac{6 - 5x}{x^2 - 1}\right)^2$

62. $y = \left(\dfrac{4x^2}{3 - x}\right)^3$

Finding an Equation of a Tangent Line In Exercises 63–70, find an equation of the tangent line to the graph of the function at the given point. Then use a graphing utility to graph the function and the tangent line in the same viewing window.

63. $f(t) = \dfrac{36}{(3 - t)^2}$; $(0, 4)$

64. $y = (4x^3 + 3)^2$; $(-1, 1)$

65. $f(x) = \sqrt[5]{3x^3 + 4x}$; $(2, 2)$

66. $s(x) = \dfrac{1}{\sqrt{x^2 - 3x + 4}}$; $\left(3, \frac{1}{2}\right)$

67. $f(t) = (t^2 - 9)\sqrt{t + 2}$; $(-1, -8)$

68. $y = \dfrac{2x}{\sqrt{x + 1}}$; $(3, 3)$

69. $f(x) = \dfrac{x + 1}{\sqrt{2x - 3}}$; $(2, 3)$

70. $y = \dfrac{x}{\sqrt{25 + x^2}}$; $(0, 0)$

71. Compound Interest You deposit \$1000 in an account with an annual interest rate of r (in decimal form) compounded monthly. At the end of 5 years, the balance A is

$$A = 1000\left(1 + \frac{r}{12}\right)^{60}.$$

Find the rates of change of A with respect to r when (a) $r = 0.08$, (b) $r = 0.10$, and (c) $r = 0.12$.

72. Biology The number N of bacteria in a culture after t days is modeled by

$$N = 400\left[1 - \frac{3}{(t^2 + 2)^2}\right].$$

Find the rate of change of N with respect to t when (a) $t = 0$, (b) $t = 1$, (c) $t = 2$, (d) $t = 3$, and (e) $t = 4$. (f) What can you conclude?

73. Depreciation The value V of a machine t years after it is purchased is inversely proportional to the square root of $t + 1$. The initial value of the machine is \$10,000.

(a) Write V as a function of t.

(b) Find the rate of depreciation when $t = 1$.

(c) Find the rate of depreciation when $t = 3$.

74. HOW DO YOU SEE IT? The cost C (in dollars) of producing x units of a product is

$C = 60x + 1350.$

For one week, management determined that the number of units produced x at the end of t hours was

$x = -1.6t^3 + 19t^2 - 0.5t - 1.$

The graph shows the cost C in terms of the time t.

Cost of Producing a Product

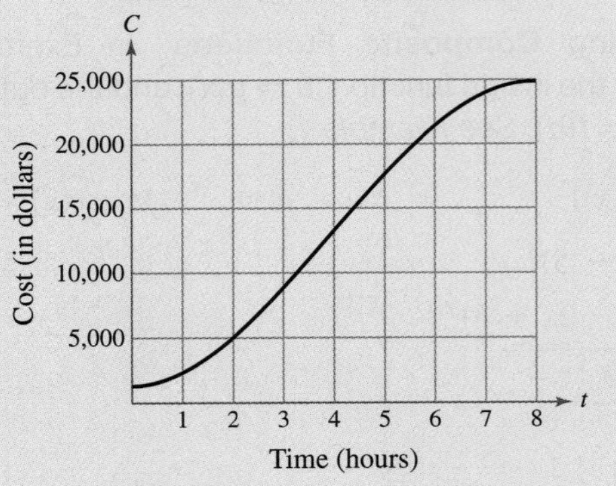

(a) Which is greater, the rate of change of the cost after 1 hour or the rate of change of the cost after 4 hours?

(b) Explain why the cost function is not increasing at a constant rate during the eight-hour shift.

75. Credit Card Rate The average annual rate r (in percent form) for commercial bank credit cards from 2003 through 2009 can be modeled by

$$r = \sqrt{2.8557t^4 - 72.792t^3 + 676.14t^2 - 2706t + 4096}$$

where t represents the year, with $t = 3$ corresponding to 2003. *(Source: Board of Governors of the Federal Reserve System)*

(a) Find the derivative of this model. Which differentiation rule(s) did you use?

(b) Use a graphing utility to graph the derivative on the interval $3 \le t \le 9$.

(c) Use the *trace* feature to find the year(s) during which the finance rate was changing the most.

(d) Use the *trace* feature to find the year(s) during which the finance rate was changing the least.

2.6 Higher-Order Derivatives

- Find higher-order derivatives.
- Find and use a position function to determine the velocity and acceleration of a moving object.

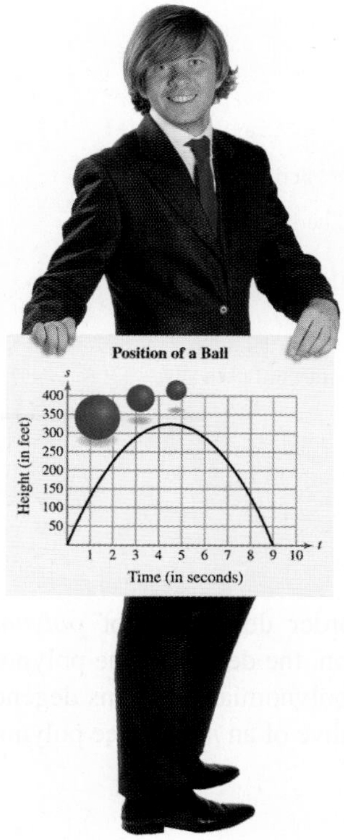

In Exercise 35 on page 142, you will use derivatives to find the velocity function and the acceleration function of a ball.

Second, Third, and Higher-Order Derivatives

The "standard" derivative f' is often called the **first derivative** of f. The derivative of f' is the **second derivative** of f and is denoted by f''.

$$\frac{d}{dx}[f'(x)] = f''(x) \qquad \text{Second derivative}$$

The derivative of f'' is the **third derivative** of f and is denoted by f'''.

$$\frac{d}{dx}[f''(x)] = f'''(x) \qquad \text{Third derivative}$$

By continuing this process, you obtain **higher-order derivatives** of f. Higher-order derivatives are denoted as follows.

Notation for Higher-Order Derivatives

1. 1st derivative:	y',	$f'(x)$,	$\frac{dy}{dx}$,	$\frac{d}{dx}[f(x)]$,	$D_x[y]$
2. 2nd derivative:	y'',	$f''(x)$,	$\frac{d^2y}{dx^2}$,	$\frac{d^2}{dx^2}[f(x)]$,	$D_x^2[y]$
3. 3rd derivative:	y''',	$f'''(x)$,	$\frac{d^3y}{dx^3}$,	$\frac{d^3}{dx^3}[f(x)]$,	$D_x^3[y]$
4. 4th derivative:	$y^{(4)}$,	$f^{(4)}(x)$,	$\frac{d^4y}{dx^4}$,	$\frac{d^4}{dx^4}[f(x)]$,	$D_x^4[y]$
5. nth derivative:	$y^{(n)}$,	$f^{(n)}(x)$,	$\frac{d^ny}{dx^n}$,	$\frac{d^n}{dx^n}[f(x)]$,	$D_x^n[y]$

Example 1 Finding Higher-Order Derivatives

Find the first five derivatives of

$$f(x) = 2x^4 - 3x^2.$$

SOLUTION

$$\begin{aligned} f(x) &= 2x^4 - 3x^2 && \text{Write original function.} \\ f'(x) &= 8x^3 - 6x && \text{First derivative} \\ f''(x) &= 24x^2 - 6 && \text{Second derivative} \\ f'''(x) &= 48x && \text{Third derivative} \\ f^{(4)}(x) &= 48 && \text{Fourth derivative} \\ f^{(5)}(x) &= 0 && \text{Fifth derivative} \end{aligned}$$

✓ Checkpoint 1

Find the first four derivatives of $f(x) = 6x^3 - 2x^2 + 1$. ■

Example 2 Finding Higher-Order Derivatives

Find the value of $g'''(2)$ for the function

$$g(t) = -t^4 + 2t^3 + t + 4.$$

SOLUTION Begin by differentiating three times.

$$g'(t) = -4t^3 + 6t^2 + 1 \quad \text{First derivative}$$
$$g''(t) = -12t^2 + 12t \quad \text{Second derivative}$$
$$g'''(t) = -24t + 12 \quad \text{Third derivative}$$

Then, evaluate the third derivative of g at $t = 2$.

$$g'''(2) = -24(2) + 12$$
$$= -36 \quad \text{Value of third derivative}$$

✓ Checkpoint 2

Find the value of $g'''(1)$ for $g(x) = x^4 - x^3 + 2x$. ■

TECH TUTOR

Higher-order derivatives of nonpolynomial functions can be difficult to find by hand. If you have access to a symbolic differentiation utility, try using it to find higher-order derivatives.

Examples 1 and 2 show how to find higher-order derivatives of *polynomial* functions. Note that with each successive differentiation, the degree of the polynomial drops by one. Eventually, higher-order derivatives of polynomial functions degenerate to a constant function. Specifically, the nth-order derivative of an nth-degree polynomial function

$$f(x) = a_n x^n + a_{n-1}x^{n-1} + \cdots + a_1 x + a_0$$

is the constant function

$$f^{(n)}(x) = n!a_n$$

where $n! = 1 \cdot 2 \cdot 3 \cdots n$. Each derivative of order higher than n is the zero function.

Example 3 Finding Higher-Order Derivatives

Find the first four derivatives of $y = x^{-1}$.

SOLUTION

$$y = x^{-1} = \frac{1}{x} \quad \text{Write original function.}$$
$$y' = (-1)x^{-2} = -\frac{1}{x^2} \quad \text{First derivative}$$
$$y'' = (-1)(-2)x^{-3} = \frac{2}{x^3} \quad \text{Second derivative}$$
$$y''' = (-1)(-2)(-3)x^{-4} = -\frac{6}{x^4} \quad \text{Third derivative}$$
$$y^{(4)} = (-1)(-2)(-3)(-4)x^{-5} = \frac{24}{x^5} \quad \text{Fourth derivative}$$

✓ Checkpoint 3

Find the fourth derivative of

$$y = \frac{1}{x^2}.$$ ■

David Davis/Shutterstock.com

Acceleration

In Section 2.3, you saw that the velocity of a free-falling object (neglecting air resistance) is given by the derivative of its position function. In other words, the rate of change of the position with respect to time is defined to be the velocity. In a similar way, the rate of change of the velocity with respect to time is defined to be the **acceleration** of the object.

STUDY TIP

Acceleration is measured in units of length per unit of time squared. For instance, if the velocity is measured in feet per second, then the acceleration is measured in "feet per second squared," or, more formally, in "feet per second per second."

$$s = f(t) \qquad \text{Position function}$$

$$\frac{ds}{dt} = f'(t) \qquad \text{Velocity function}$$

$$\frac{d^2s}{dt^2} = f''(t) \qquad \text{Acceleration function}$$

To find the position, velocity, or acceleration at a particular time t, substitute the given value of t into the appropriate function, as illustrated in Example 4.

Example 4 Finding Acceleration

A ball is thrown upward from the top of a 160-foot cliff, as shown in Figure 2.32. The initial velocity of the ball is 48 feet per second, which implies that the position function is

$$s = -16t^2 + 48t + 160$$

where the time t is measured in seconds. Find the height, velocity, and acceleration of the ball at $t = 3$.

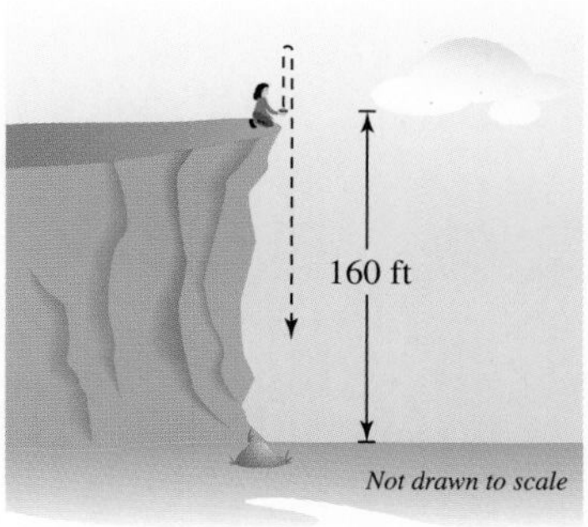

FIGURE 2.32

SOLUTION Begin by differentiating to find the velocity and acceleration functions.

$$s = -16t^2 + 48t + 160 \qquad \text{Position function}$$

$$\frac{ds}{dt} = -32t + 48 \qquad \text{Velocity function}$$

$$\frac{d^2s}{dt^2} = -32 \qquad \text{Acceleration function}$$

To find the height, velocity, and acceleration at $t = 3$, substitute $t = 3$ into each of the functions above.

$$\text{Height} = -16(3)^2 + 48(3) + 160 = 160 \text{ feet}$$

$$\text{Velocity} = -32(3) + 48 = -48 \text{ feet per second}$$

$$\text{Acceleration} = -32 \text{ feet per second squared}$$

✓ Checkpoint 4

A ball is thrown upward from the top of an 80-foot cliff with an initial velocity of 64 feet per second, which implies that the position function is

$$s = -16t^2 + 64t + 80$$

where the time t is measured in seconds. Find the height, velocity, and acceleration of the ball at $t = 2$. ■

The acceleration due to gravity on the surface of the moon is only about one-sixth that exerted on the surface of Earth.

In Example 4, notice that the acceleration of the ball is -32 feet per second squared at any time t. This constant acceleration is due to the gravitational force of Earth and is called the **acceleration due to gravity.** Note that the negative value indicates that the ball is being pulled *down*—toward Earth.

Although the acceleration exerted on a falling object is relatively constant near Earth's surface, it varies greatly throughout our solar system. Large planets exert a much greater gravitational pull than do small planets or moons. The next example describes the motion of a free-falling object on the moon.

Example 5 Finding Acceleration on the Moon

An astronaut standing on the surface of the moon throws a rock upward. The height s (in feet) of the rock is given by

$$s = -\frac{27}{10}t^2 + 27t + 6$$

where t is measured in seconds. How does the acceleration due to gravity on the moon compare with that on Earth?

SOLUTION

$$s = -\frac{27}{10}t^2 + 27t + 6 \qquad \text{Position function}$$

$$\frac{ds}{dt} = -\frac{27}{5}t + 27 \qquad \text{Velocity function}$$

$$\frac{d^2s}{dt^2} = -\frac{27}{5} \qquad \text{Acceleration function}$$

So, the acceleration at any time is

$$-\frac{27}{5} = -5.4 \text{ feet per second squared}$$

—about one-sixth of the acceleration due to gravity on Earth.

✓ Checkpoint 5

The position function on Earth, where s is measured in meters, t is measured in seconds, v_0 is the initial velocity in meters per second, and h_0 is the initial height in meters, is

$$s = -4.9t^2 + v_0t + h_0.$$

An object is thrown upward with an initial velocity of 2.2 meters per second from an initial height of 3.6 meters. What is the acceleration due to gravity on Earth in meters per second squared? ■

The position function described in Example 5 neglects air resistance, which is appropriate because the moon has no atmosphere—and *no air resistance*. This means that the position function for any free-falling object on the moon is given by

$$s = -\frac{27}{10}t^2 + v_0t + h_0$$

where s is the height (in feet), t is the time (in seconds), v_0 is the initial velocity (in feet per second), and h_0 is the initial height (in feet). For instance, the rock in Example 5 was thrown upward with an initial velocity of 27 feet per second and had an initial height of 6 feet. This position function is valid for all objects, whether heavy ones such as hammers or light ones such as feathers.

Example 6 Finding Velocity and Acceleration

The velocity v (in feet per second) of a certain automobile starting from rest is

$$v = \frac{80t}{t + 5} \qquad \text{Velocity function}$$

where t is the time (in seconds). The positions of the automobile at 10-second intervals are shown in Figure 2.33. Find the velocity and acceleration of the automobile at 10-second intervals from $t = 0$ to $t = 60$.

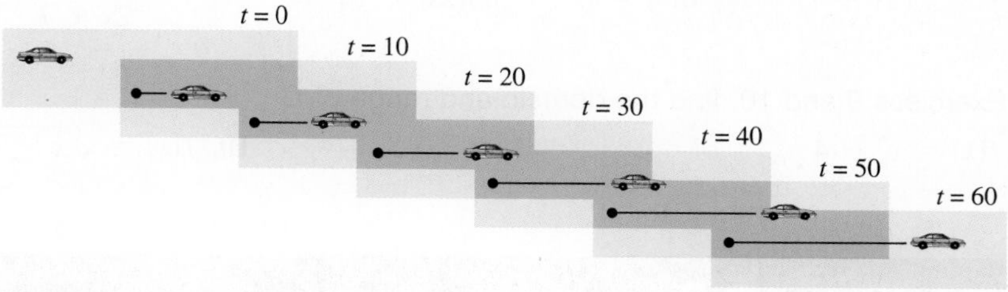

FIGURE 2.33

SOLUTION To find the acceleration function, differentiate the velocity function.

$$\frac{dv}{dt} = \frac{(t + 5)(80) - (80t)(1)}{(t + 5)^2} \qquad \text{Apply Quotient Rule.}$$

$$= \frac{400}{(t + 5)^2} \qquad \text{Acceleration function}$$

t (seconds)	0	10	20	30	40	50	60
v (ft/sec)	0	53.3	64.0	68.6	71.1	72.7	73.8
$\frac{dv}{dt}$ (ft/sec^2)	16	1.78	0.64	0.33	0.20	0.13	0.09

In the table, note that the acceleration approaches zero as the velocity levels off. This observation should agree with your experience—when riding in an accelerating automobile, you do not feel the velocity, but you do feel the acceleration. In other words, you feel changes in velocity.

✓ Checkpoint 6

Use a graphing utility to graph the velocity function and acceleration function in Example 6 in the same viewing window. Compare the graphs with the table in Example 6. As the velocity levels off, what does the acceleration approach? ■

SUMMARIZE (Section 2.6)

1. State the meaning of each derivative listed below *(page 137)*. For examples of higher-order derivatives, see Examples 1, 2, and 3.

   (a) y'' (b) $f^{(4)}(x)$ (c) $\frac{d^3y}{dx^3}$

2. Describe a real-life example of how higher-order derivatives can be used to analyze the velocity and acceleration of an object *(page 139, Examples 4, 5, and 6)*.

wavebreakmedia ltd/Shutterstock.com

SKILLS WARM UP 2.6

The following warm-up exercises involve skills that were covered in earlier sections. You will use these skills in the exercise set for this section. For additional help, review Sections 1.4 and 2.4.

In Exercises 1–4, solve the equation.

1. $-16t^2 + 24t = 0$
2. $-16t^2 + 80t + 224 = 0$
3. $-16t^2 + 128t + 320 = 0$
4. $-16t^2 + 9t + 1440 = 0$

In Exercises 5–8, find dy/dx.

5. $y = x^2(2x + 7)$
6. $y = (x^2 + 3x)(2x^2 - 5)$
7. $y = \dfrac{x^2}{2x + 7}$
8. $y = \dfrac{x^2 + 3x}{2x^2 - 5}$

In Exercises 9 and 10, find the domain and range of f.

9. $f(x) = x^2 - 4$
10. $f(x) = \sqrt{x - 7}$

Exercises 2.6

See www.CalcChat.com for worked-out solutions to odd-numbered exercises.

Finding Higher-Order Derivatives In Exercises 1–12, find the second derivative of the function. *See Examples 1 and 3.*

1. $f(x) = 9 - 2x$
2. $f(x) = 4x + 15$
3. $f(x) = x^2 + 7x - 4$
4. $f(x) = 3x^2 + 4x$
5. $g(t) = \frac{1}{3}t^3 - 4t^2 + 2t$
6. $f(x) = -2x^5 + 3x^4 + 8x$
7. $f(t) = \dfrac{3}{4t^2}$
8. $g(t) = \dfrac{8}{t}$
9. $f(x) = 3(2 - x^2)^3$
10. $y = 4(x^2 + 5x)^3$
11. $f(x) = \dfrac{x + 1}{x - 1}$
12. $g(t) = -\dfrac{4}{(t + 2)^2}$

Finding Higher-Order Derivatives In Exercises 13–18, find the third derivative of the function. *See Examples 1 and 3.*

13. $f(x) = x^5 - 3x^4$
14. $f(x) = x^4 - 2x^3$
15. $f(x) = 5x(x + 4)^3$
16. $f(x) = (x^3 - 6)^4$
17. $f(x) = \dfrac{3}{16x^2}$
18. $f(x) = -\dfrac{2}{x}$

Finding Higher-Order Derivatives In Exercises 19–24, find the given value. *See Example 2.*

	Function	*Value*
19.	$g(t) = 5t^4 + 10t^2 + 3$	$g''(2)$
20.	$f(x) = 9 - x^2$	$f''(-\sqrt{5})$
21.	$f(x) = \sqrt{4 - x}$	$f'''(-5)$
22.	$f(t) = \sqrt{2t + 3}$	$f'''(\frac{1}{2})$
23.	$f(x) = (x^3 - 2x)^3$	$f''(1)$
24.	$g(x) = (x^2 + 3x)^4$	$g''(-1)$

Finding Higher-Order Derivatives In Exercises 25–30, find the higher-order derivative. *See Examples 1 and 3.*

	Given	*Derivative*
25.	$f'(x) = 2x^2$	$f''(x)$
26.	$f''(x) = 20x^3 - 36x^2$	$f'''(x)$
27.	$f'''(x) = 2\sqrt{x - 1}$	$f^{(4)}(x)$
28.	$f'''(x) = 4x^{-4}$	$f^{(5)}(x)$
29.	$f^{(4)}(x) = (x^2 + 1)^2$	$f^{(6)}(x)$
30.	$f''(x) = 2x^2 + 7x - 12$	$f^{(5)}(x)$

Using Derivatives In Exercises 31–34, find the second derivative and solve the equation $f''(x) = 0$.

31. $f(x) = x^3 - 9x^2 + 27x - 27$
32. $f(x) = (x + 2)(x - 2)(x + 3)(x - 3)$
33. $f(x) = x\sqrt{x^2 - 1}$
34. $f(x) = \dfrac{x}{x^2 + 3}$

35. **Velocity and Acceleration** A ball is propelled straight upward from ground level with an initial velocity of 144 feet per second.
    (a) Write the position, velocity, and acceleration functions of the ball.
    (b) Find the height, velocity, and acceleration at $t = 3$.
    (c) When is the ball at its highest point? How high is this point?
    (d) How fast is the ball traveling when it hits the ground? How is this speed related to the initial velocity?

36. Velocity and Acceleration A brick becomes dislodged from the top of the Empire State Building (at a height of 1250 feet) and falls to the sidewalk below.

(a) Write the position, velocity, and acceleration functions of the brick.

(b) How long does it take the brick to hit the sidewalk?

(c) How fast is the brick traveling when it hits the sidewalk?

37. Velocity and Acceleration The velocity (in feet per second) of an automobile starting from rest is modeled by $ds/dt = 90t/(t + 10)$. Create a table showing the velocity and acceleration at 10-second intervals during the first minute of travel. What can you conclude?

38. Stopping Distance A car is traveling at a rate of 66 feet per second (45 miles per hour) when the brakes are applied. The position function for the car is given by $s = -8.25t^2 + 66t$, where s is measured in feet and t is measured in seconds. Use this function to complete the table showing the position, velocity, and acceleration for each given value of t. What can you conclude?

t	0	1	2	3	4
s					
$\frac{ds}{dt}$					
$\frac{d^2s}{dt^2}$					

39. Derivatives of Polynomial Functions Consider the function $f(x) = x^2 - 6x + 6$.

(a) Use a graphing utility to graph f, f', and f'' in the same viewing window.

(b) What is the relationship among the degree of f and the degrees of its successive derivatives?

(c) Repeat parts (a) and (b) for $f(x) = 3x^3 - 9x$.

(d) In general, what is the relationship among the degree of a polynomial function and the degrees of its successive derivatives?

40. HOW DO YOU SEE IT? The graph shows the position, velocity, and acceleration functions of a particle. Identify each function. Explain your reasoning.

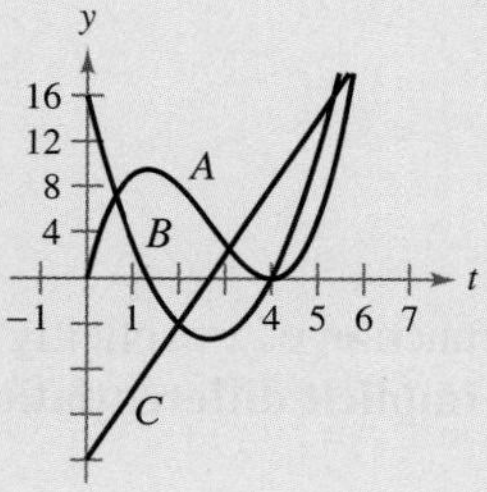

41. Modeling Data The table shows the revenues y (in millions of dollars) for eBay from 2004 to 2009, where t is the year, with $t = 4$ corresponding to 2004. *(Source: eBay Inc.)*

t	4	5	6	7	8	9
y	3271	4552	5970	7672	8541	8727

(a) Use a graphing utility to find a cubic model for the revenue $y(t)$ of eBay.

(b) Find the first and second derivatives of the function.

(c) Show that the revenue of eBay was increasing from 2005 to 2008.

(d) Find the year when the revenue was increasing at the greatest rate by solving $y''(t) = 0$.

42. Finding a Pattern Develop a general rule for

$$[xf(x)]^{(n)}$$

where f is a differentiable function of x.

True or False? In Exercises 43 and 44, determine whether the statement is true or false. If it is false, explain why or give an example that shows it is false.

43. If $y = f(x)g(x)$, then $y' = f'(x)g'(x)$.

44. If $f'(c)$ and $g'(c)$ are zero and $h(x) = f(x)g(x)$, then $h'(c) = 0$.

45. Project: Median Prices of U.S. Homes For a project analyzing the median prices of new privately owned homes in the United States from 2000 to 2009, visit this text's website at *www.cengagebrain.com.* *(Data Source: U.S. Census Bureau)*

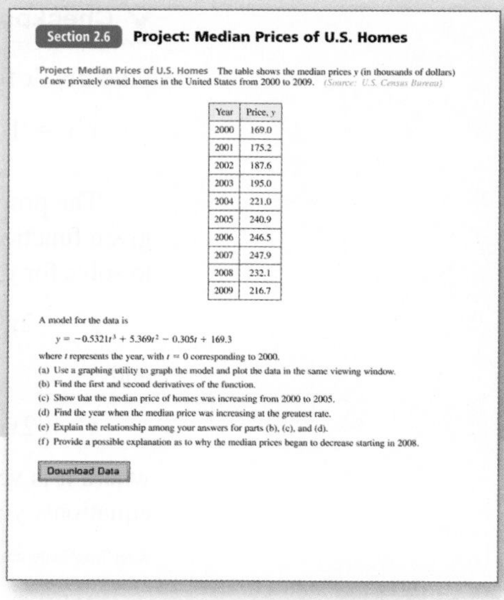

Section 2.6 **Project: Median Prices of U.S. Homes**

Project: Median Prices of U.S. Homes The table shows the median prices y (in thousands of dollars) of new privately owned homes in the United States from 2000 to 2009. *(Source: U.S. Census Bureau)*

Year	Price, y
2000	169.0
2001	175.2
2002	187.6
2003	195.0
2004	221.0
2005	240.9
2006	246.5
2007	247.9
2008	232.1
2009	216.7

A model for the data is

$$y = -0.5321t^3 + 5.369t^2 - 0.305t + 169.3$$

where t represents the year, with $t = 0$ corresponding to 2000.

(a) Use a graphing utility to graph the model and plot the data in the same viewing window.

(b) Find the first and second derivatives of the function.

(c) Show that the median price of homes was increasing from 2000 to 2005.

(d) Find the year when the median price was increasing at the greatest rate.

(e) Explain the relationship among your answers for parts (b), (c), and (d).

(f) Provide a possible explanation as to why the median prices began to decrease starting in 2008.

Download Data

2.7 Implicit Differentiation

- Find derivatives explicitly.
- Find derivatives implicitly.
- Use derivatives to answer questions about real-life situations.

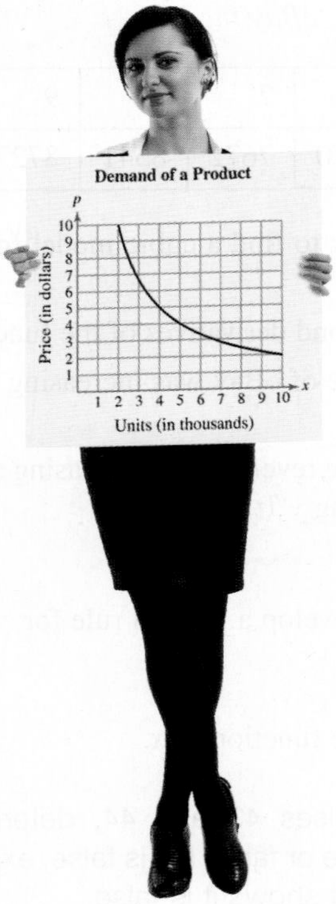

In Exercise 43 on page 150, you will use implicit differentiation to find the rate of change for a demand function.

Explicit and Implicit Functions

So far in this text, most functions involving two variables have been expressed in the **explicit form**

$$y = f(x).$$ Explicit form

That is, one of the two variables has been explicitly given in terms of the other. For example, in the equation

$$y = 3x - 5$$

the variable y is explicitly written as a function of x. Some functions, however, are not given explicitly and are only implied by a given equation, as shown in Example 1.

Example 1 Finding a Derivative Explicitly

Find dy/dx for the equation

$$xy = 1.$$

SOLUTION In this equation, y is **implicitly** defined as a function of x. One way to find dy/dx is first to solve the equation for y, then differentiate as usual.

$$xy = 1$$ Write original equation.

$$y = \frac{1}{x}$$ Solve for y.

$$= x^{-1}$$ Rewrite.

$$\frac{dy}{dx} = -x^{-2}$$ Differentiate with respect to x.

$$= -\frac{1}{x^2}$$ Simplify.

✓ Checkpoint 1

Find dy/dx for the equation

$$x^2y = 1.$$

■

The procedure shown in Example 1 works well whenever you can easily write the given function explicitly. You cannot, however, use this procedure when you are unable to solve for y as a function of x. For instance, how would you find dy/dx for the equations

$$x^2 - 2y^3 + 4y = 2$$

and

$$x^2 + 2xy - y^3 = 5$$

where it is very difficult to express y as a function of x explicitly? To differentiate such equations, you can use a procedure called **implicit differentiation.**

Implicit Differentiation

To understand how to find dy/dx implicitly, you must realize that the differentiation is taking place *with respect to x*. This means that when you differentiate terms involving x alone, you can differentiate as usual. *But* when you differentiate terms involving y, you must apply the Chain Rule because you are assuming that y is defined implicitly as a differentiable function of x. Study the next example carefully. Note in particular how the Chain Rule is used to introduce the dy/dx factors in Examples 2(b) and 2(d).

Example 2 Applying the Chain Rule

Differentiate each expression with respect to x.

a. $3x^2$ **b.** $2y^3$

c. $x + 3y$ **d.** xy^2

SOLUTION

a. The only variable in this expression is x. So, to differentiate with respect to x, you can use the Simple Power Rule and the Constant Multiple Rule to obtain

$$\frac{d}{dx}[3x^2] = 6x.$$

b. This case is different. The variable in the expression is y, and yet you are asked to differentiate with respect to x. To do this, assume that y is a differentiable function of x and use the Chain Rule.

$$\frac{d}{dx}\overbrace{[2y^3]}^{cu^n} = \overbrace{2}^{c}\overbrace{(3)}^{n}\overbrace{y^2}^{u^{n-1}}\overbrace{\frac{dy}{dx}}^{u'} \qquad \text{Chain Rule}$$

$$= 6y^2\frac{dy}{dx}$$

c. This expression involves both x and y. By the Sum Rule and the Constant Multiple Rule, you can write

$$\frac{d}{dx}[x + 3y] = 1 + 3\frac{dy}{dx}.$$

d. By the Product Rule and the Chain Rule, you can write

$$\frac{d}{dx}[xy^2] = x\frac{d}{dx}[y^2] + y^2\frac{d}{dx}[x] \qquad \text{Product Rule}$$

$$= x\left(2y\frac{dy}{dx}\right) + y^2(1) \qquad \text{Chain Rule}$$

$$= 2xy\frac{dy}{dx} + y^2.$$

✓Checkpoint 2

Differentiate each expression with respect to x.

a. $4x^3$

b. $3y^2$

c. $x + 5y$

d. xy^3

■

Guidelines for Implicit Differentiation

Consider an equation involving x and y in which y is a differentiable function of x. You can use the steps below to find dy/dx.

1. Differentiate both sides of the equation *with respect to x.*
2. Write the result so that all terms involving dy/dx are on the left side of the equation and all other terms are on the right side of the equation.
3. Factor dy/dx out of the terms on the left side of the equation.
4. Solve for dy/dx by dividing both sides of the equation by the left-hand factor that does not contain dy/dx.

In Example 3, note that implicit differentiation can produce an expression for dy/dx that contains both x and y.

Example 3 Using Implicit Differentiation

Find dy/dx for the equation $y^3 + y^2 - 5y - x^2 = -4$.

SOLUTION

1. Differentiate both sides of the equation with respect to x.

$$\frac{d}{dx}[y^3 + y^2 - 5y - x^2] = \frac{d}{dx}[-4]$$

$$\frac{d}{dx}[y^3] + \frac{d}{dx}[y^2] - \frac{d}{dx}[5y] - \frac{d}{dx}[x^2] = \frac{d}{dx}[-4]$$

$$3y^2\frac{dy}{dx} + 2y\frac{dy}{dx} - 5\frac{dy}{dx} - 2x = 0$$

2. Collect the dy/dx terms on the left side of the equation and move all other terms to the right side of the equation.

$$3y^2\frac{dy}{dx} + 2y\frac{dy}{dx} - 5\frac{dy}{dx} = 2x$$

3. Factor dy/dx out of the left side of the equation.

$$\frac{dy}{dx}(3y^2 + 2y - 5) = 2x$$

4. Solve for dy/dx by dividing by $(3y^2 + 2y - 5)$.

$$\frac{dy}{dx} = \frac{2x}{3y^2 + 2y - 5}$$

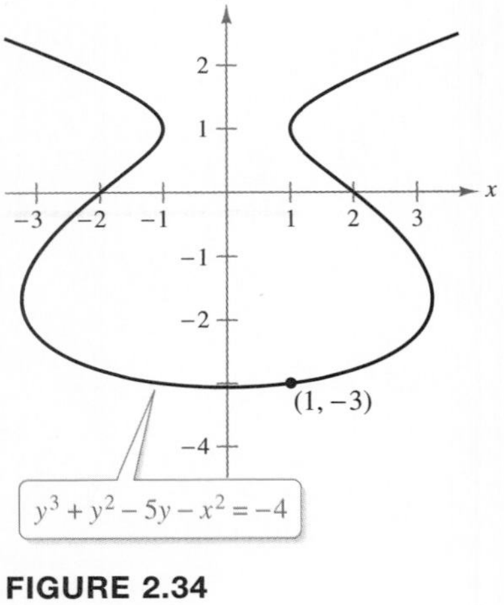

FIGURE 2.34

✓Checkpoint 3

Find dy/dx for the equation $y^2 + x^2 - 2y - 4x = 4$.

To see how you can use an implicit derivative, consider the graph shown in Figure 2.34. The derivative found in Example 3 gives a formula for the slope of the tangent line at a point on this graph. For instance, the slope at the point $(1, -3)$ is

$$\frac{dy}{dx} = \frac{2(1)}{3(-3)^2 + 2(-3) - 5} = \frac{1}{8}.$$

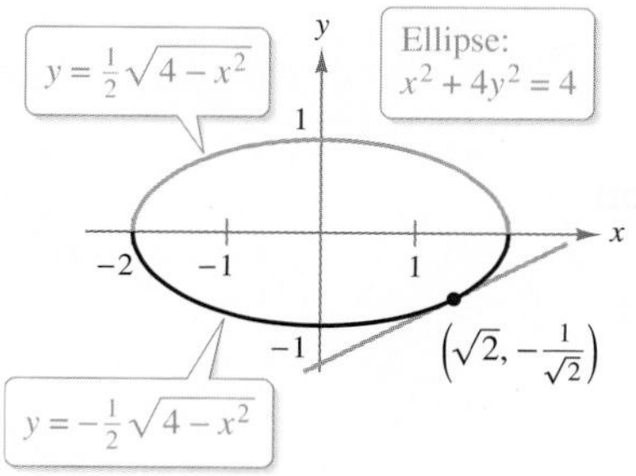

Slope of tangent line is $\frac{1}{2}$.

FIGURE 2.35

Example 4 Finding the Slope of a Graph Implicitly

Find the slope of the tangent line to the ellipse given by $x^2 + 4y^2 = 4$ at the point $(\sqrt{2}, -1/\sqrt{2})$, as shown in Figure 2.35.

SOLUTION

$$x^2 + 4y^2 = 4 \qquad \text{Write original equation.}$$

$$\frac{d}{dx}[x^2 + 4y^2] = \frac{d}{dx}[4] \qquad \text{Differentiate with respect to } x.$$

$$2x + 8y\frac{dy}{dx} = 0 \qquad \text{Implicit differentiation}$$

$$8y\frac{dy}{dx} = -2x \qquad \text{Subtract } 2x \text{ from each side.}$$

$$\frac{dy}{dx} = \frac{-2x}{8y} \qquad \text{Divide each side by } 8y.$$

$$\frac{dy}{dx} = -\frac{x}{4y} \qquad \text{Simplify.}$$

To find the slope at the given point, substitute $x = \sqrt{2}$ and $y = -1/\sqrt{2}$ into the derivative, as shown below.

$$\frac{dy}{dx} = -\frac{\sqrt{2}}{4(-1/\sqrt{2})}$$

$$= \frac{1}{2}$$

STUDY TIP

To see the benefit of implicit differentiation, try reworking Example 4 using the explicit function

$$y = -\frac{1}{2}\sqrt{4 - x^2}.$$

The graph of this function is the lower half of the ellipse.

✓Checkpoint 4

Find the slope of the tangent line to the circle $x^2 + y^2 = 25$ at the point $(3, -4)$. ■

Example 5 Finding the Slope of a Graph Implicitly

Find the slope of the graph of $2x^2 - y^2 = 1$ at the point $(1, 1)$.

SOLUTION Begin by finding dy/dx implicitly.

$$2x^2 - y^2 = 1 \qquad \text{Write original equation.}$$

$$4x - 2y\frac{dy}{dx} = 0 \qquad \text{Differentiate with respect to } x.$$

$$-2y\frac{dy}{dx} = -4x \qquad \text{Subtract } 4x \text{ from each side.}$$

$$\frac{dy}{dx} = \frac{2x}{y} \qquad \text{Divide each side by } -2y.$$

At the point $(1, 1)$, the slope of the graph is

$$\frac{dy}{dx} = \frac{2(1)}{1}$$

$$= 2$$

as shown in Figure 2.36. The graph is called a **hyperbola.**

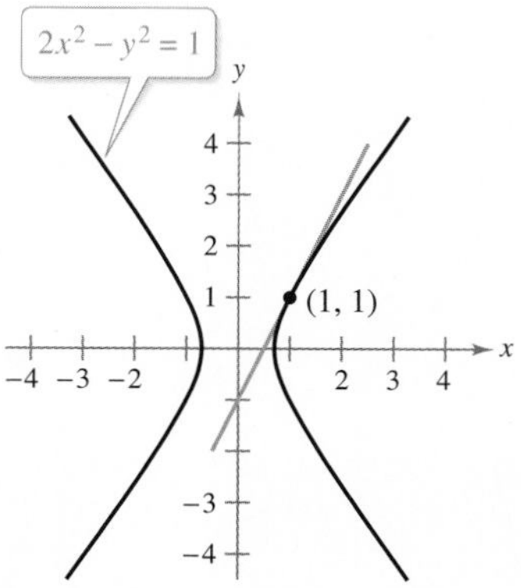

Hyperbola

FIGURE 2.36

✓Checkpoint 5

Find the slope of the graph of $x^2 - 9y^2 = 16$ at the point $(5, 1)$. ■

Application

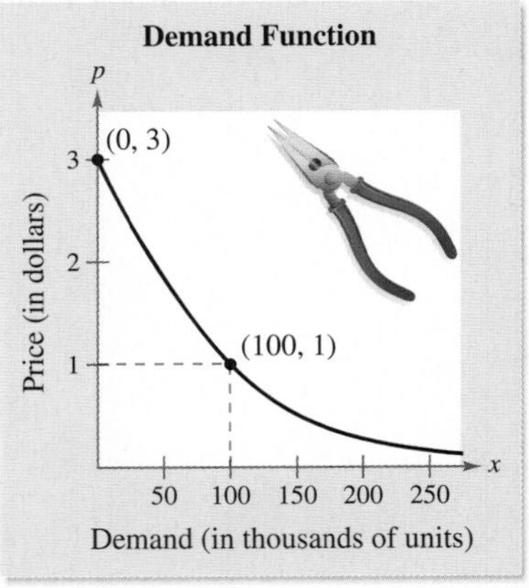

FIGURE 2.37

Example 6 Using a Demand Function

The demand function for a product is modeled by

$$p = \frac{3}{0.000001x^3 + 0.01x + 1}$$

where p is measured in dollars and x is measured in thousands of units, as shown in Figure 2.37. Find the rate of change of the demand x with respect to the price p when $x = 100$.

SOLUTION To simplify the differentiation, begin by rewriting the function. Then, differentiate *with respect to p*.

$$p = \frac{3}{0.000001x^3 + 0.01x + 1}$$

$$0.000001x^3 + 0.01x + 1 = \frac{3}{p}$$

$$0.000003x^2\frac{dx}{dp} + 0.01\frac{dx}{dp} = -\frac{3}{p^2}$$

$$(0.000003x^2 + 0.01)\frac{dx}{dp} = -\frac{3}{p^2}$$

$$\frac{dx}{dp} = -\frac{3}{p^2(0.000003x^2 + 0.01)}$$

When $x = 100$, the price is

$$p = \frac{3}{0.000001(100)^3 + 0.01(100) + 1} = \$1.$$

So, when $x = 100$ and $p = 1$, the rate of change of the demand with respect to the price is

$$\frac{dx}{dp} = -\frac{3}{(1)^2[0.000003(100)^2 + 0.01]} = -75.$$

This means that when $x = 100$, the demand is dropping at the rate of 75 thousand units for each dollar increase in price.

✓ Checkpoint 6

The demand function for a product is given by

$$p = \frac{2}{0.001x^2 + x + 1}.$$

Find dx/dp implicitly.

SUMMARIZE (Section 2.7)

1. State the guidelines for implicit differentiation *(page 146)*. For examples of implicit differentiation, see Examples 2, 3, 4, and 5.
2. Describe a real-life example of how implicit differentiation can be used to analyze the rate of change of a product's demand *(page 148, Example 6)*.

SKILLS WARM UP 2.7 The following warm-up exercises involve skills that were covered in a previous course. You will use these skills in the exercise set for this section. For additional help, review Appendix Section A.3.

In Exercises 1–6, solve the equation for *y*.

1. $x - \dfrac{y}{x} = 2$

2. $\dfrac{4}{x-3} = \dfrac{1}{y}$

3. $xy - x + 6y = 6$

4. $12 + 3y = 4x^2 + x^2y$

5. $x^2 + y^2 = 5$

6. $x = \pm\sqrt{6 - y^2}$

In Exercises 7–9, evaluate the expression at the given point.

7. $\dfrac{3x^2 - 4}{3y^2}$, $(2, 1)$

8. $\dfrac{x^2 - 2}{1 - y}$, $(0, -3)$

9. $\dfrac{5x}{3y^2 - 12y + 5}$, $(-1, 2)$

Exercises 2.7

See www.CalcChat.com for worked-out solutions to odd-numbered exercises.

Finding Derivatives In Exercises 1–12, find *dy/dx*. *See Examples 1 and 3.*

1. $xy = 4$

2. $3x^2 - y = 8x$

3. $y^2 = 1 - x^2,\ 0 \le x \le 1$

4. $y^3 = 4x^3 + 2x$

5. $x^2y^2 - 2x = 3$

6. $xy^2 + 4xy = 10$

7. $4y^2 - xy = 2$

8. $2xy^3 - x^2y = 2$

9. $\dfrac{xy - y^2}{y - x} = 1$

10. $\dfrac{2x + y}{x - 5y} = 1$

11. $\dfrac{2y - x}{y^2 - 3} = 5$

12. $\dfrac{4y^2}{y^2 - 9} = x^2$

Finding the Slope of a Graph Implicitly In Exercises 13–26, find the slope of the graph of the function at the given point. *See Examples 4 and 5.*

Equation	*Point*
13. $x^2 + y^2 = 16$	$(0, 4)$
14. $x^2 - y^2 = 25$	$(5, 0)$
15. $y + xy = 4$	$(-5, -1)$
16. $16x^2 + 25y^2 = 400$	$(5, 4)$
17. $x^3 - xy + y^2 = 4$	$(0, -2)$
18. $x^2y + y^2x = -2$	$(2, -1)$
19. $x^3y^3 - y = x$	$(0, 0)$
20. $x^3 + y^3 = 6xy$	$\left(\frac{4}{3}, \frac{8}{3}\right)$
21. $x^{1/2} + y^{1/2} = 9$	$(16, 25)$
22. $x^{2/3} + y^{2/3} = 5$	$(8, 1)$
23. $\sqrt{xy} = x - 2y$	$(4, 1)$
24. $(x + y)^3 = x^3 + y^3$	$(-1, 1)$
25. $y^2(x^2 + y^2) = 2x^2$	$(1, 1)$
26. $(x^2 + y^2)^2 = 8x^2y$	$(2, 2)$

Finding the Slope of a Graph Implicitly In Exercises 27–32, find the slope of the graph at the given point. *See Examples 4 and 5.*

27. $3x^2 - 2y + 5 = 0$

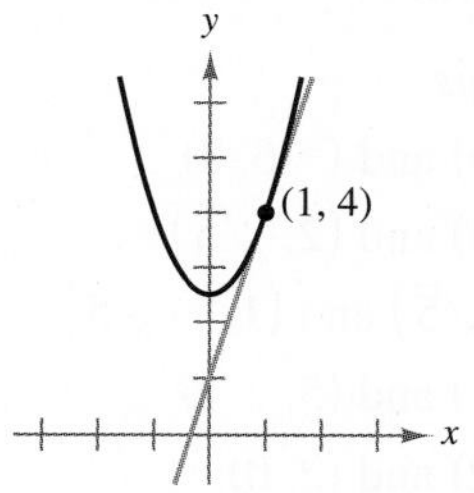

28. $4x^2 + 2y - 1 = 0$

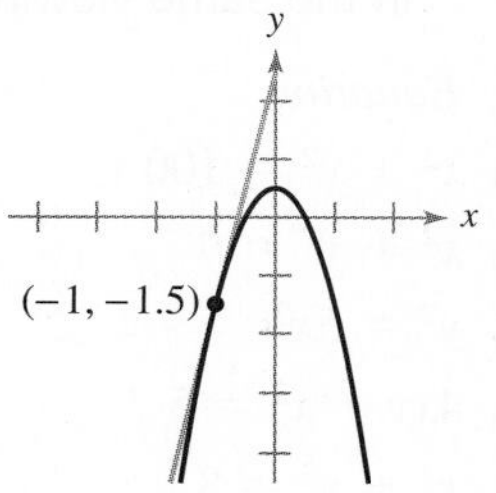

29. $x^2 + y^2 = 4$

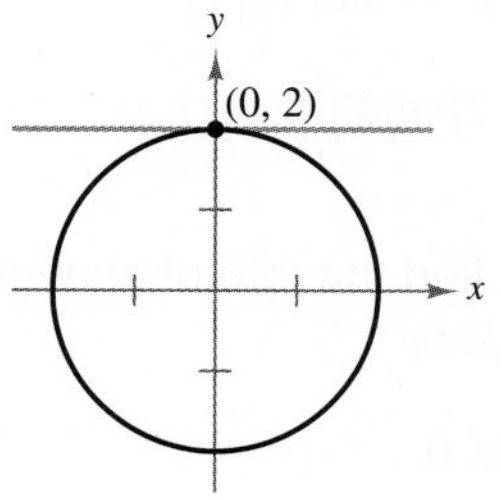

30. $4x^2 + 9y^2 = 36$

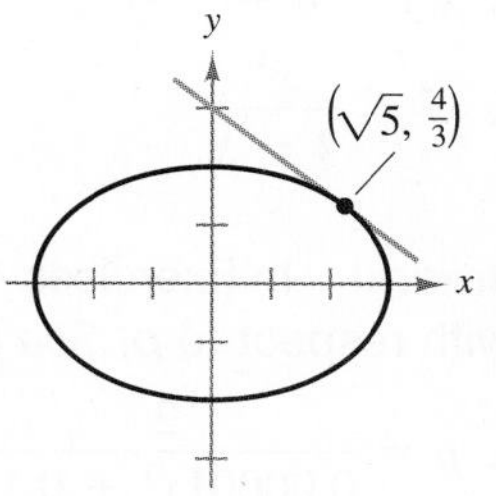

31. $(4 - x)y^2 = x^3$

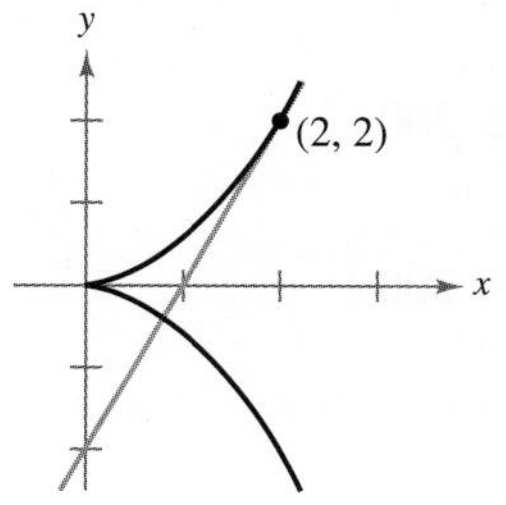

32. $x^2 - y^3 = 0$

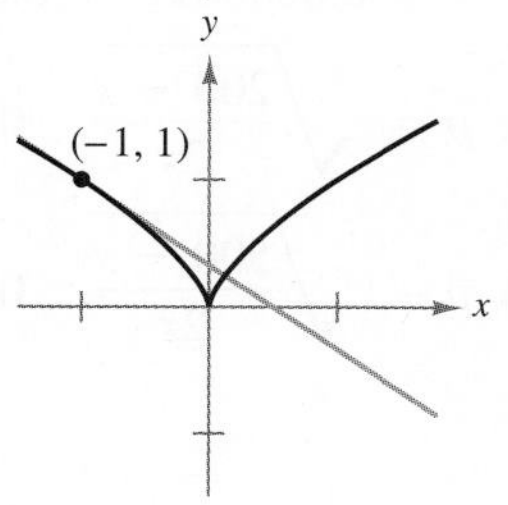

Finding Derivatives Implicitly and Explicitly In Exercises 33 and 34, find dy/dx implicitly and explicitly (the explicit functions are shown on the graph) and show that the results are equivalent. Use the graph to estimate the slope of the tangent line at the labeled point. Then verify your result analytically by evaluating dy/dx at the point.

33. $x - y^2 - 1 = 0$

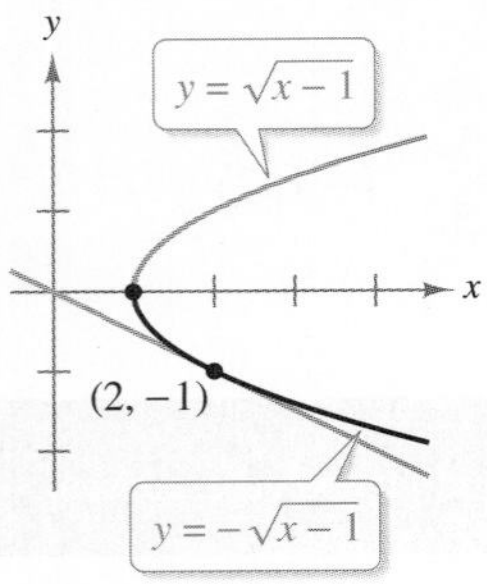

34. $4y^2 - x^2 = 7$

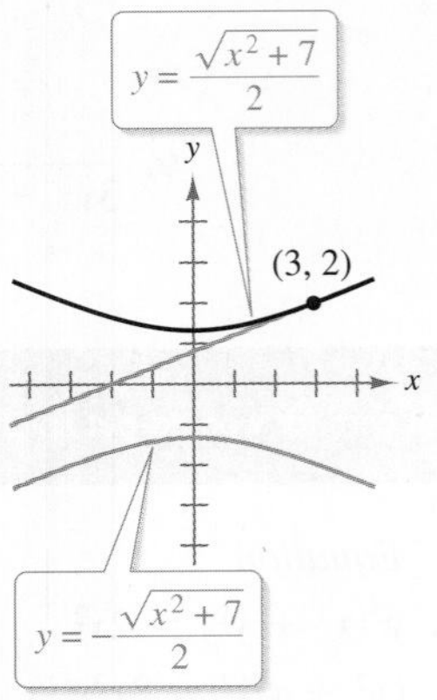

Finding an Equation of a Tangent Line In Exercises 35–42, find equations of the tangent lines to the graph at the given points. Use a graphing utility to graph the equation and the tangent lines in the same viewing window.

	Equation	*Points*
35.	$x^2 + y^2 = 100$	$(8, 6)$ and $(-6, 8)$
36.	$x^2 + y^2 = 9$	$(0, 3)$ and $(2, \sqrt{5})$
37.	$y^2 = 5x^3$	$(1, \sqrt{5})$ and $(1, -\sqrt{5})$
38.	$4xy + x^2 = 5$	$(1, 1)$ and $(5, -1)$
39.	$x^3 + y^3 = 8$	$(0, 2)$ and $(2, 0)$
40.	$x + y^3 = 6xy^3 - 1$	$(-1, 0)$ and $(0, -1)$
41.	$x^2y - 8 = -4y$	$(-2, 1)$ and $(6, \frac{1}{5})$
42.	$y^2 = \dfrac{x^3}{4 - x}$	$(2, 2)$ and $(2, -2)$

Demand In Exercises 43–46, find the rate of change of x with respect to p. *See Example 6.*

43. $p = \dfrac{2}{0.00001x^3 + 0.1x}, \quad x \geq 0$

44. $p = \dfrac{4}{0.000001x^2 + 0.05x + 1}, \quad x \geq 0$

45. $p = \sqrt{\dfrac{200 - x}{2x}}, \quad 0 < x \leq 200$

46. $p = \sqrt{\dfrac{500 - x}{2x}}, \quad 0 < x \leq 500$

47. Production Let x represent the units of labor and y the capital invested in a manufacturing process. When 135,540 units are produced, the relationship between labor and capital can be modeled by $100x^{0.75}y^{0.25} = 135{,}540$.

(a) Find the rate of change of y with respect to x when $x = 1500$ and $y = 1000$.

(b) The model used in this problem is called the *Cobb-Douglas production function.* Graph the model on a graphing utility and describe the relationship between labor and capital.

48. HOW DO YOU SEE IT? The graph shows the demand function for a product.

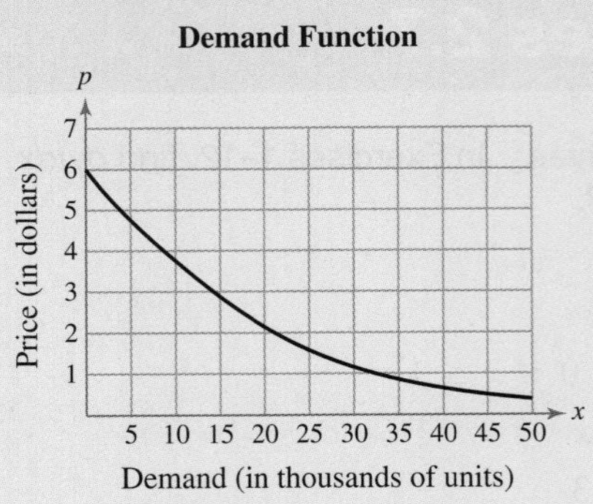

(a) What happens to the demand as the price increases?

(b) Over what interval is the rate of change of the demand with respect to the price decreasing?

49. Health: U.S. HIV/AIDS Epidemic The numbers (in thousands) of cases y of HIV/AIDS reported in the years 2004 through 2008 can be modeled by

$$y^2 - 1952.4 = 13.0345t^3 - 168.969t^2 + 465.66t$$

where t represents the year, with $t = 4$ corresponding to 2004. *(Source: U.S. Centers for Disease Control and Prevention)*

(a) Use a graphing utility to graph the model and describe the results.

(b) Use the graph to estimate the year during which the number of reported cases was decreasing at the greatest rate.

(c) Complete the table to estimate the year during which the number of reported cases was decreasing at the greatest rate. Compare this estimate with your answer in part (b).

t	4	5	6	7	8
y					
y'					

2.8 Related Rates

In Exercise 25 on page 157, you will use related rates to find the rate of change of the sales for a product.

- Examine related variables.
- Solve related-rate problems.

Related Variables

In this section, you will study problems involving variables that are changing with respect to time. If two or more such variables are related to each other, then their rates of change with respect to time are also related.

For instance, suppose that x and y are related by the equation

$$y = 2x.$$

If both variables are changing with respect to time, then their rates of change will also be related.

x and y are related. → $y = 2x$ ⇒ $\dfrac{dy}{dt} = 2\dfrac{dx}{dt}$ ← The rates of change of x and y are related.

In this simple example, you can see that because y always has twice the value of x, it follows that the rate of change of y with respect to time is always twice the rate of change of x with respect to time.

Example 1 Examining Two Rates That Are Related

The variables x and y are differentiable functions of t and are related by the equation

$$y = x^2 + 3.$$

When $x = 1$, $dx/dt = 2$. Find dy/dt when $x = 1$.

SOLUTION Use the Chain Rule to differentiate both sides of the equation with respect to t.

$$y = x^2 + 3 \qquad \text{Write original equation.}$$

$$\frac{d}{dt}[y] = \frac{d}{dt}[x^2 + 3] \qquad \text{Differentiate with respect to } t.$$

$$\frac{dy}{dt} = 2x\frac{dx}{dt} \qquad \text{Apply Chain Rule.}$$

When $x = 1$ and $dx/dt = 2$, you have

$$\frac{dy}{dt} = 2(1)(2)$$
$$= 4.$$

✓ Checkpoint 1

The variables x and y are differentiable functions of t and are related by the equation

$$y = x^3 + 2.$$

When $x = 1$, $dx/dt = 3$. Find dy/dt when $x = 1$. ■

Solving Related-Rate Problems

In Example 1, you were *given* the mathematical model.

Given equation: $y = x^2 + 3$

Given rate: $\dfrac{dx}{dt} = 2$ at $x = 1$

Find: $\dfrac{dy}{dt}$ at $x = 1$

In the next example, you must *create* a mathematical model from a verbal description.

Example 2 Changing Area

Total area increases as the outer radius increases.

A pebble is dropped into a calm pool of water, causing ripples in the form of concentric circles, as shown in the photo. The radius r of the outer ripple is increasing at a constant rate of 1 foot per second. When the radius is 4 feet, at what rate is the total area A of the disturbed water changing?

SOLUTION The variables r and A are related by the equation for the area of a circle, $A = \pi r^2$. To solve this problem, use the fact that the rate of change of the radius is given by dr/dt.

Equation: $A = \pi r^2$

Given rate: $\dfrac{dr}{dt} = 1$ at $r = 4$

Find: $\dfrac{dA}{dt}$ at $r = 4$

Using this model, you can proceed as in Example 1.

$$A = \pi r^2 \qquad \text{Write original equation.}$$

$$\frac{d}{dt}[A] = \frac{d}{dt}[\pi r^2] \qquad \text{Differentiate with respect to } t.$$

$$\frac{dA}{dt} = 2\pi r \frac{dr}{dt} \qquad \text{Apply Chain Rule.}$$

When $r = 4$ and $dr/dt = 1$, you have

$$\frac{dA}{dt} = 2\pi(4)(1) = 8\pi \qquad \text{Substitute 4 for } r \text{ and 1 for } dr/dt.$$

When the radius is 4 feet, the area is changing at a rate of 8π square feet per second.

✓ Checkpoint 2

As in Example 2, a pebble is dropped into the pool, but this time the radius r of the outer ripple is increasing at a rate of 2 feet per second. At what rate is the total area changing when the radius is 3 feet? ■

In Example 2, note that the radius changes at a *constant* rate ($dr/dt = 1$ for all t), but the area changes at a *nonconstant* rate.

When $r = 1$ ft	When $r = 2$ ft	When $r = 3$ ft	When $r = 4$ ft
$\dfrac{dA}{dt} = 2\pi$ ft²/sec	$\dfrac{dA}{dt} = 4\pi$ ft²/sec	$\dfrac{dA}{dt} = 6\pi$ ft²/sec	$\dfrac{dA}{dt} = 8\pi$ ft²/sec

The solution shown in Example 2 illustrates the steps for solving a related-rate problem.

Guidelines for Solving a Related-Rate Problem

1. Identify all *given* quantities and all quantities *to be determined.* If possible, make a sketch and label the quantities.
2. Write an equation that relates all variables whose rates of change are either given or to be determined.
3. Use the Chain Rule to differentiate both sides of the equation *with respect to time*.
4. *After* completing Step 3, substitute into the resulting equation all known values for the variables and their rates of change. Then solve for the required rate of change.

STUDY TIP

Be sure you notice the order of Steps 3 and 4 in the guidelines. Do not substitute the known values for the variables until after you have differentiated.

In Step 2 of the guidelines, note that you must write an equation that relates the given variables. To help you with this step, reference tables that summarize many common formulas are included in the appendices. For instance, the volume of a sphere of radius r is given by the formula

$$V = \frac{4}{3}\pi r^3$$

as listed in Appendix D.

The table below lists examples of the mathematical models for some common rates of change that can be used in the first step of the solution of a related-rate problem.

Verbal statement	Mathematical model
The velocity of a car after traveling for 1 hour is 50 miles per hour.	$x =$ distance traveled $\frac{dx}{dt} = 50$ when $t = 1$
Water is being pumped into a swimming pool at a rate of 10 cubic feet per minute.	$V =$ volume of water in pool $\frac{dV}{dt} = 10\ \text{ft}^3/\text{min}$
A population of bacteria is increasing at a rate of 2000 per hour.	$x =$ number in population $\frac{dx}{dt} = 2000$ bacteria per hour
Revenue is increasing at a rate of $4000 per month.	$R =$ revenue $\frac{dR}{dt} = 4000$ dollars per month
Profit is decreasing at a rate of $2500 per day.	$P =$ profit $\frac{dP}{dt} = -2500$ dollars per day

Example 3 Analyzing a Profit Function

A company's profit P (in dollars) from selling x units of a product can be modeled by

$$P = 500x - \frac{1}{4}x^2.$$ Model for profit

The sales are increasing at a rate of 10 units per day. Find the rate of change in the profit (in dollars per day) when 500 units have been sold.

SOLUTION Because the sales are increasing at a rate of 10 units per day, you know that at time t the rate of change is $dx/dt = 10$. So, the problem can be stated as shown.

Given rate: $\frac{dx}{dt} = 10$

Find: $\frac{dP}{dt}$ when $x = 500$

To find the rate of change of the profit, use the model for profit that relates the profit P and the units of the product sold x.

Equation: $P = 500x - \frac{1}{4}x^2$

By differentiating both sides of the equation with respect to t, you obtain

$$\frac{d}{dt}[P] = \frac{d}{dt}\left[500x - \frac{1}{4}x^2\right]$$ Differentiate with respect to t.

$$\frac{dP}{dt} = \left(500 - \frac{1}{2}x\right)\frac{dx}{dt}.$$ Apply Chain Rule.

When $x = 500$ units and $dx/dt = 10$, the rate of change in the profit is

$$\frac{dP}{dt} = \left[500 - \frac{1}{2}(500)\right](10) = (500 - 250)(10) = 250(10) = \$2500 \text{ per day.}$$

The graph of the profit function (in terms of x) is shown in Figure 2.38.

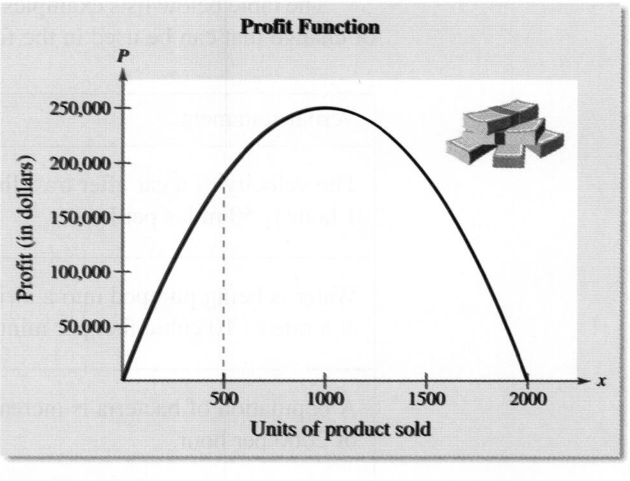

FIGURE 2.38

✓ Checkpoint 3

Find the rate of change in profit (in dollars per day) when 50 units have been sold, sales have increased at a rate of 10 units per day, and $P = 200x - \frac{1}{2}x^2$. ■

Example 4 Increasing Production

A company is increasing the production of a product at the rate of 200 units per week. The weekly demand function is modeled by

$$p = 100 - 0.001x$$

where p is the price per unit and x is the number of units produced in a week. Find the rate of change of the revenue with respect to time when the weekly production is 2000 units. Will the rate of change of the revenue be greater than \$20,000 per week?

SOLUTION Because production is increasing at a rate of 200 units per week, you know that at time t the rate of change is $dx/dt = 200$. So, the problem can be stated as shown.

$$\textit{Given rate: } \frac{dx}{dt} = 200$$

$$\textit{Find: } \frac{dR}{dt} \text{ when } x = 2000$$

To find the rate of change of the revenue, you must find an equation that relates the revenue R and the number of units produced x.

$$\textit{Equation: } R = xp = x(100 - 0.001x) = 100x - 0.001x^2$$

By differentiating both sides of the equation with respect to t, you obtain

$$R = 100x - 0.001x^2 \qquad \text{Write original equation.}$$

$$\frac{d}{dt}[R] = \frac{d}{dt}[100x - 0.001x^2] \qquad \text{Differentiate with respect to } t.$$

$$\frac{dR}{dt} = (100 - 0.002x)\frac{dx}{dt}. \qquad \text{Apply Chain Rule.}$$

Using $x = 2000$ and $dx/dt = 200$, you have

$$\begin{aligned}\frac{dR}{dt} &= [100 - 0.002(2000)](200)\\ &= \$19{,}200 \text{ per week.}\end{aligned}$$

No, the rate of change of the revenue will not be greater than \$20,000 per week.

✓ Checkpoint 4

Find the rate of change of the revenue with respect to time for the company in Example 4 when the weekly demand function is

$$p = 150 - 0.002x.$$

SUMMARIZE (Section 2.8)

1. Give a description of related variables *(page 151)*. For an example of two related variables, see Example 1.
2. State the guidelines for solving a related-rate problem *(page 153)*. For examples of solving related-rate problems, see Examples 2, 3, and 4.
3. Describe a real-life example of how related rates can be used to analyze the rate of change of a company's revenue *(page 155, Example 4)*.

SKILLS WARM UP 2.8 The following warm-up exercises involve skills that were covered in earlier sections. You will use these skills in the exercise set for this section. For additional help, review Section 2.7.

In Exercises 1–6, write a formula for the given quantity.

1. Area of a circle
2. Volume of a sphere
3. Surface area of a cube
4. Volume of a cube
5. Volume of a cone
6. Area of a triangle

In Exercises 7–10, find dy/dx by implicit differentiation.

7. $x^2 + y^2 = 9$ **8.** $3xy - x^2 = 6$ **9.** $x^2 + 2y + xy = 12$ **10.** $x + xy^2 - y^2 = xy$

Exercises 2.8

See www.CalcChat.com for worked-out solutions to odd-numbered exercises.

Examining Two Rates That Are Related In Exercises 1–4, assume that x and y are both differentiable functions of t. Use the given values to find (a) dy/dt and (b) dx/dt. See Example 1.

Equation	*Find*	*Given*
1. $y = \sqrt{x}$	(a) $\dfrac{dy}{dt}$ when $x = 4$,	$\dfrac{dx}{dt} = 3$
	(b) $\dfrac{dx}{dt}$ when $x = 25$,	$\dfrac{dy}{dt} = 2$
2. $y = 2(x^2 - 3x)$	(a) $\dfrac{dy}{dt}$ when $x = 3$,	$\dfrac{dx}{dt} = 2$
	(b) $\dfrac{dx}{dt}$ when $x = 1$,	$\dfrac{dy}{dt} = 5$
3. $xy = 4$	(a) $\dfrac{dy}{dt}$ when $x = 8$,	$\dfrac{dx}{dt} = 10$
	(b) $\dfrac{dx}{dt}$ when $x = 1$,	$\dfrac{dy}{dt} = -6$
4. $x^2 + y^2 = 25$	(a) $\dfrac{dy}{dt}$ when $x = 3, y = 4$,	$\dfrac{dx}{dt} = 8$
	(b) $\dfrac{dx}{dt}$ when $x = 4, y = 3$,	$\dfrac{dy}{dt} = -2$

5. Area The radius r of a circle is increasing at a rate of 3 inches per minute. Find the rates of change of the area when (a) $r = 6$ inches and (b) $r = 24$ inches.

6. Volume The radius r of a sphere is increasing at a rate of 3 inches per minute. Find the rates of change of the volume when (a) $r = 6$ inches and (b) $r = 24$ inches.

7. Area Let A be the area of a circle of radius r that is changing with respect to time. If dr/dt is constant, is dA/dt constant? Explain your reasoning.

8. Volume Let V be the volume of a sphere of radius r that is changing with respect to time. If dr/dt is constant, is dV/dt constant? Explain your reasoning.

9. Volume A spherical balloon is inflated with gas at a rate of 10 cubic feet per minute. How fast is the radius of the balloon changing at the instant the radius is (a) 1 foot and (b) 2 feet?

10. Volume The radius r of a right circular cone is increasing at a rate of 2 inches per minute. The height h of the cone is related to the radius by

$$h = 3r.$$

Find the rates of change of the volume when (a) $r = 6$ inches and (b) $r = 24$ inches.

11. Cost, Revenue, and Profit A company that manufactures sport supplements calculates that its costs and revenue can be modeled by the equations

$$C = 125{,}000 + 0.75x \quad \text{and} \quad R = 250x - \frac{1}{10}x^2$$

where x is the number of units of sport supplements produced in 1 week. Production during one particular week is 1000 units and is increasing at a rate of 150 units per week. Find the rates at which the (a) cost, (b) revenue, and (c) profit are changing.

12. Cost, Revenue, and Profit A company that manufactures pet toys calculates that its costs and revenue can be modeled by the equations

$$C = 75{,}000 + 1.05x \quad \text{and} \quad R = 500x - \frac{x^2}{25}$$

where x is the number of toys produced in 1 week. Production during one particular week is 5000 toys and is increasing at a rate of 250 toys per week. Find the rates at which the (a) cost, (b) revenue, and (c) profit are changing.

13. Volume All edges of a cube are expanding at a rate of 3 centimeters per second. How fast is the volume changing when each edge is (a) 1 centimeter and (b) 10 centimeters?

14. Surface Area All edges of a cube are expanding at a rate of 3 centimeters per second. How fast is the surface area changing when each edge is (a) 1 centimeter and (b) 10 centimeters?

15. Moving Point A point is moving along the graph of $y = x^2$ such that dx/dt is 3 inches per second. Find dy/dt for (a) $x = -3$, (b) $x = 0$, (c) $x = 1$, and (d) $x = 3$.

16. Moving Point A point is moving along the graph of $y = 1/(1 + x^2)$ such that dx/dt is 2 inches per second. Find dy/dt for (a) $x = -2$, (b) $x = 0$, (c) $x = 6$, and (d) $x = 10$.

17. Boating A boat is pulled by a winch on a dock, and the winch is 12 feet above the deck of the boat (see figure). The winch pulls the rope at a rate of 4 feet per second. Find the speed of the boat when 13 feet of rope is out. What happens to the speed of the boat as it gets closer and closer to the dock?

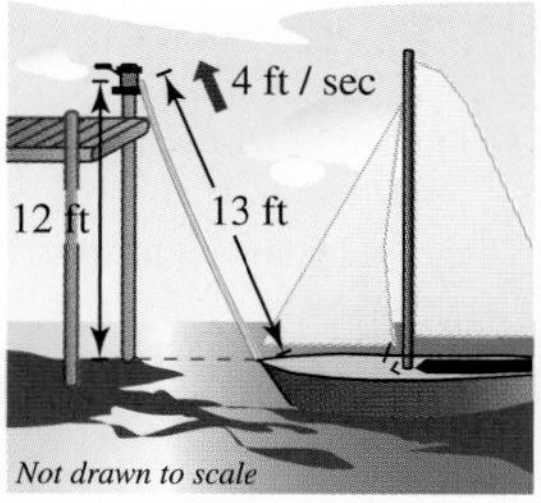

Figure for 17

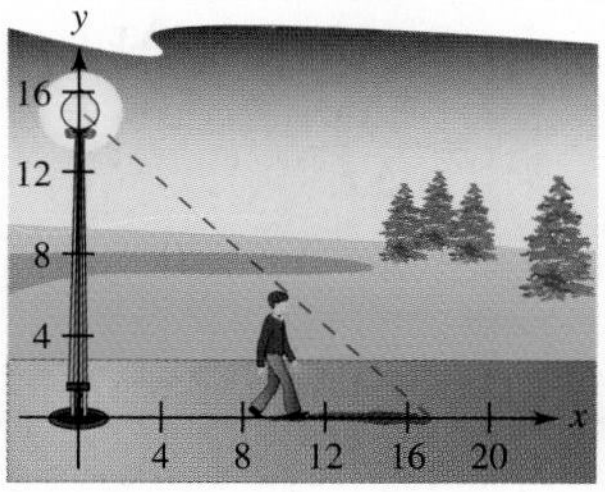

Figure for 18

18. Shadow Length A man 6 feet tall walks at a rate of 5 feet per second away from a light that is 15 feet above the ground (see figure).

(a) When he is 10 feet from the base of the light, at what rate is the tip of his shadow moving?

(b) When he is 10 feet from the base of the light, at what rate is the length of his shadow changing?

19. Air Traffic Control An airplane flying at an altitude of 6 miles passes directly over a radar antenna (see figure). When the airplane is 10 miles away ($s = 10$), the radar detects that the distance s is changing at a rate of 240 miles per hour. What is the speed of the airplane?

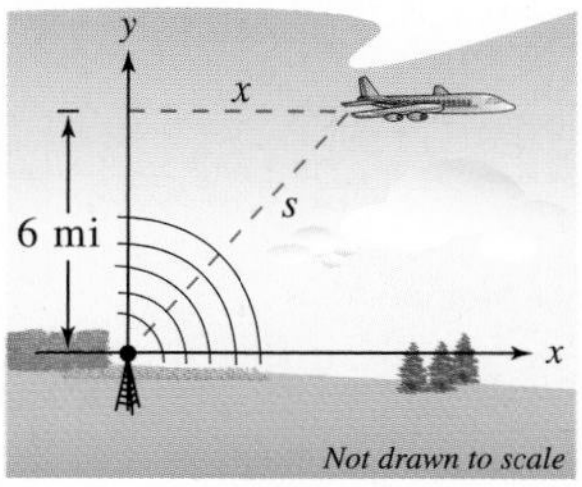

Figure for 19

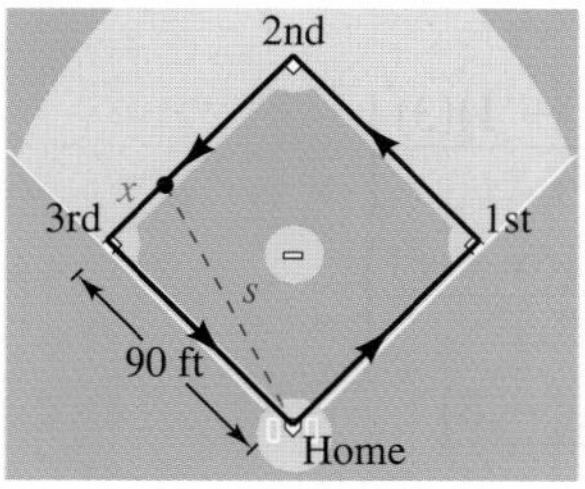

Figure for 20

20. Baseball A (square) baseball diamond has sides that are 90 feet long (see figure). A player 26 feet from third base is running at a speed of 30 feet per second. At what rate is the player's distance from home plate changing?

21. Air Traffic Control An air traffic controller spots two airplanes at the same altitude converging to a point as they fly at right angles to each other. One airplane is 150 miles from the point and has a speed of 450 miles per hour. The other is 200 miles from the point and has a speed of 600 miles per hour.

(a) At what rate is the distance between the planes changing?

(b) How much time does the controller have to get one of the airplanes on a different flight path?

22. Advertising Costs A retail sporting goods store estimates that weekly sales S and weekly advertising costs x are related by the equation $S = 2250 + 50x + 0.35x^2$. The current weekly advertising costs are \$1500, and these costs are increasing at a rate of \$125 per week. Find the current rate of change of weekly sales.

23. Environment An accident at an oil drilling platform is causing a circular oil slick. The slick is 0.08 foot thick, and when the radius of the slick is 150 feet, the radius is increasing at the rate of 0.5 foot per minute. At what rate (in cubic feet per minute) is oil flowing from the site of the accident?

24. Profit A company is increasing the production of a product at the rate of 25 units per week. The demand and cost functions for the product are given by $p = 50 - 0.01x$ and $C = 4000 + 40x - 0.02x^2$. Find the rate of change of the profit with respect to time when the weekly sales are $x = 800$ units. Use a graphing utility to graph the profit function, and use the *zoom* and *trace* features of the graphing utility to verify your result.

25. Sales The profit for a product is increasing at a rate of \$5600 per week. The demand and cost functions for the product are given by $p = 6000 - 25x$ and $C = 2400x + 5200$. Find the rate of change of sales with respect to time when the weekly sales are $x = 44$ units.

26. HOW DO YOU SEE IT? The graph shows the demand and supply equations for a product, where x represents the number of units (in thousands) and p is the price (in dollars). Using the graph, (a) determine whether dp/dt is positive or negative given that dx/dt is negative, and (b) determine whether dx/dt is positive or negative given that dp/dt is positive.

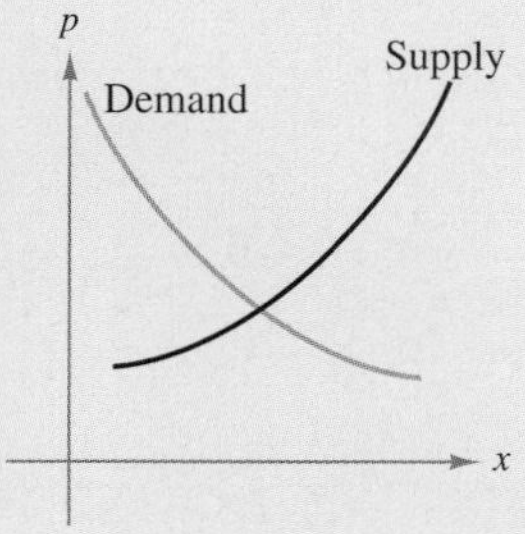

ALGEBRA TUTOR

TECH TUTOR

Symbolic algebra systems can simplify algebraic expressions. If you have access to such a system, try using it to simplify the expressions in this Algebra Tutor.

Simplifying Algebraic Expressions

To be successful in using derivatives, you must be good at simplifying algebraic expressions. Here are some helpful simplification techniques.

1. Combine *like terms*. This may involve expanding an expression by multiplying factors.
2. Divide out *common factors* in the numerator and denominator of an expression.
3. Factor an expression.
4. Rationalize a denominator.
5. Add, subtract, multiply, or divide fractions.

Example 1 Simplifying Fractional Expressions

a.

$$\frac{[3(x + \Delta x) + 5] - (3x + 5)}{\Delta x} = \frac{3x + 3\Delta x + 5 - 3x - 5}{\Delta x}$$ Multiply factors and remove parentheses.

$$= \frac{3\Delta x}{\Delta x}$$ Combine like terms.

$$= 3, \quad \Delta x \neq 0$$ Divide out common factors.

b.

$$\frac{(x + \Delta x)^2 - x^2}{\Delta x} = \frac{x^2 + 2x(\Delta x) + (\Delta x)^2 - x^2}{\Delta x}$$ Expand terms.

$$= \frac{2x(\Delta x) + (\Delta x)^2}{\Delta x}$$ Combine like terms.

$$= \frac{\Delta x(2x + \Delta x)}{\Delta x}$$ Factor.

$$= 2x + \Delta x, \quad \Delta x \neq 0$$ Divide out common factors.

c.

$$\frac{(x^2 - 1)(-2 - 2x) - (3 - 2x - x^2)(2)}{(x^2 - 1)^2}$$

$$= \frac{(-2x^2 - 2x^3 + 2 + 2x) - (6 - 4x - 2x^2)}{(x^2 - 1)^2}$$ Expand terms.

$$= \frac{-2x^2 - 2x^3 + 2 + 2x - 6 + 4x + 2x^2}{(x^2 - 1)^2}$$ Remove parentheses.

$$= \frac{-2x^3 + 6x - 4}{(x^2 - 1)^2}$$ Combine like terms.

d.

$$2\left(\frac{2x + 1}{3x}\right)\left[\frac{3x(2) - (2x + 1)(3)}{(3x)^2}\right]$$

$$= 2\left(\frac{2x + 1}{3x}\right)\left[\frac{6x - (6x + 3)}{(3x)^2}\right]$$ Multiply factors.

$$= \frac{2(2x + 1)(6x - 6x - 3)}{(3x)^3}$$ Multiply fractions and remove parentheses.

$$= \frac{2(2x + 1)(-3)}{3(9)x^3}$$ Combine like terms and factor.

$$= \frac{-2(2x + 1)}{9x^3}$$ Divide out common factors.

Example 2 **Simplifying Expressions with Powers**

Simplify each expression.

a. $(2x + 1)^2(6x + 1) + (3x^2 + x)(2)(2x + 1)(2)$

b. $(-1)(3x^2 - 2x)^{-2}(6x - 2)$

c. $(x)\left(\frac{1}{2}\right)(2x + 3)^{-1/2} + (2x + 3)^{1/2}(1)$

d. $\dfrac{x^2\left(\frac{1}{2}\right)(x^2 + 1)^{-1/2}(2x) - (x^2 + 1)^{1/2}(2x)}{x^4}$

SOLUTION

a. $(2x + 1)^2(6x + 1) + (3x^2 + x)(2)(2x + 1)(2)$

$= (2x + 1)[(2x + 1)(6x + 1) + (3x^2 + x)(2)(2)]$ Factor.

$= (2x + 1)[12x^2 + 8x + 1 + (12x^2 + 4x)]$ Multiply factors.

$= (2x + 1)(12x^2 + 8x + 1 + 12x^2 + 4x)$ Remove parentheses.

$= (2x + 1)(24x^2 + 12x + 1)$ Combine like terms.

b. $(-1)(3x^2 - 2x)^{-2}(6x - 2)$

$= \dfrac{(-1)(6x - 2)}{(3x^2 - 2x)^2}$ Rewrite as a fraction.

$= \dfrac{(-1)(2)(3x - 1)}{(3x^2 - 2x)^2}$ Factor.

$= \dfrac{-2(3x - 1)}{(3x^2 - 2x)^2}$ Multiply factors.

c. $(x)\left(\frac{1}{2}\right)(2x + 3)^{-1/2} + (2x + 3)^{1/2}(1)$

$= (2x + 3)^{-1/2}\left(\frac{1}{2}\right)[x + (2x + 3)(2)]$ Factor.

$= \dfrac{x + 4x + 6}{(2x + 3)^{1/2}(2)}$ Rewrite as a fraction.

$= \dfrac{5x + 6}{2(2x + 3)^{1/2}}$ Combine like terms.

d. $\dfrac{x^2\left(\frac{1}{2}\right)(x^2 + 1)^{-1/2}(2x) - (x^2 + 1)^{1/2}(2x)}{x^4}$

$= \dfrac{(x^3)(x^2 + 1)^{-1/2} - (x^2 + 1)^{1/2}(2x)}{x^4}$ Multiply factors.

$= \dfrac{(x^2 + 1)^{-1/2}(x)[x^2 - (x^2 + 1)(2)]}{x^4}$ Factor.

$= \dfrac{x[x^2 - (2x^2 + 2)]}{(x^2 + 1)^{1/2}x^4}$ Write with positive exponents.

$= \dfrac{x^2 - 2x^2 - 2}{(x^2 + 1)^{1/2}x^3}$ Divide out common factors and remove parentheses.

$= \dfrac{-x^2 - 2}{(x^2 + 1)^{1/2}x^3}$ Combine like terms.

STUDY TIP

All but one of the expressions in this Algebra Tutor are derivatives. Can you see what the original function is? Explain your reasoning.

SUMMARY AND STUDY STRATEGIES

After studying this chapter, you should have acquired the following skills.
The exercise numbers are keyed to the Review Exercises that begin on page 162.
Answers to odd-numbered Review Exercises are given in the back of the text.*

Section 2.1

	Review Exercises
■ Approximate the slope of the tangent line to a graph at a point.	*1–4*
■ Interpret the slope of a graph in a real-life setting.	*5–8*
■ Use the limit definition to find the slope of a graph at a point and the derivative of a function. $f'(x) = \lim_{\Delta x \to 0} \frac{f(x + \Delta x) - f(x)}{\Delta x}$	*9–24*
■ Use the graph of a function to recognize points at which the function is not differentiable.	*25–28*

Section 2.2

■ Use the Constant Rule for differentiation. $\frac{d}{dx}[c] = 0$	*29, 30*
■ Use the Power Rule for differentiation. $\frac{d}{dx}[x^n] = nx^{n-1}$	*31, 32*
■ Use the Constant Multiple Rule for differentiation. $\frac{d}{dx}[cf(x)] = cf'(x)$	*33–36*
■ Use the Sum and Difference Rules for differentiation. $\frac{d}{dx}[f(x) \pm g(x)] = f'(x) \pm g'(x)$	*37–40*
■ Use derivatives to find the slope of a graph.	*41–44*
■ Use derivatives to write equations of tangent lines.	*45–48*
■ Use derivatives to answer questions about real-life situations.	*49, 50*

Section 2.3

■ Find the average rate of change of a function over an interval and the instantaneous rate of change at a point. Average rate of change: $\frac{f(b) - f(a)}{b - a}$; Instantaneous rate of change: $\lim_{\Delta x \to 0} \frac{f(x + \Delta x) - f(x)}{\Delta x}$	*51–54*
■ Use derivatives to find the velocities of objects.	*55, 56*
■ Find the marginal revenues, marginal costs, and marginal profits for products.	*57–66*
■ Use derivatives to answer questions about real-life situations.	*67, 68*

* A wide range of valuable study aids are available to help you master the material in this chapter. The *Student Solutions Manual* includes step-by-step solutions to all odd-numbered exercises to help you review and prepare. The student website at *www.cengagebrain.com* offers algebra help and a *Graphing Technology Guide*, which contains step-by-step commands and instructions for a wide variety of graphing calculators.

Section 2.4	Review Exercises
■ Use the Product Rule for differentiation. $\frac{d}{dx}[f(x)g(x)] = f(x)g'(x) + g(x)f'(x)$	69–72
■ Use the Quotient Rule for differentiation. $\frac{d}{dx}\left[\frac{f(x)}{g(x)}\right] = \frac{g(x)f'(x) - f(x)g'(x)}{[g(x)]^2}$	73–76

Section 2.5

■ Use the General Power Rule for differentiation. $\frac{d}{dx}[u^n] = nu^{n-1}u'$	77–80
■ Use differentiation rules efficiently to find the derivative of any algebraic function, then simplify the result.	81–90
■ Use derivatives to answer questions about real-life situations. (Sections 2.1–2.5)	91, 92

Section 2.6

■ Find higher-order derivatives.	93–100
■ Find and use a position function to determine the velocity and acceleration of a moving object.	101, 102

Section 2.7

■ Find derivatives implicitly.	103–106
■ Use implicit differentiation to write equations of tangent lines.	107–110

Section 2.8

■ Solve related-rate problems.	111–114

Study Strategies

- **Simplify Your Derivatives** You may ask if you have to simplify your derivatives. The answer is "Yes, if you expect to use them." In the next chapter, you will see that almost all applications of derivatives require that the derivatives be written in simplified form. It is not difficult to see the advantage of a derivative in simplified form. Consider, for instance, the derivative of

$$f(x) = \frac{x}{\sqrt{x^2 + 1}}.$$

The "raw form" produced by the Quotient and Chain Rules

$$f'(x) = \frac{(x^2 + 1)^{1/2}(1) - (x)\left(\frac{1}{2}\right)(x^2 + 1)^{-1/2}(2x)}{\left(\sqrt{x^2 + 1}\right)^2}$$

is obviously much more difficult to use than the simplified form

$$f'(x) = \frac{1}{(x^2 + 1)^{3/2}}.$$

- **List Units of Measure in Applied Problems** When using derivatives in real-life applications, be sure to list the units of measure for each variable. For instance, if R is measured in dollars and t is measured in years, then the derivative dR/dt is measured in dollars per year.

Review Exercises

See www.CalcChat.com for worked-out solutions to odd-numbered exercises.

Approximating the Slope of a Graph In Exercises 1–4, estimate the slope of the graph at the point (x, y). (Each square on the grid is 1 unit by 1 unit.)

1.

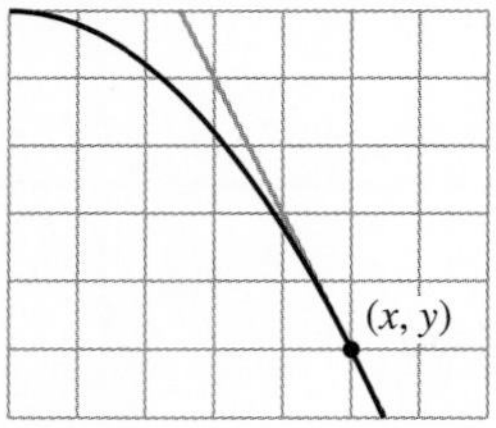

2.

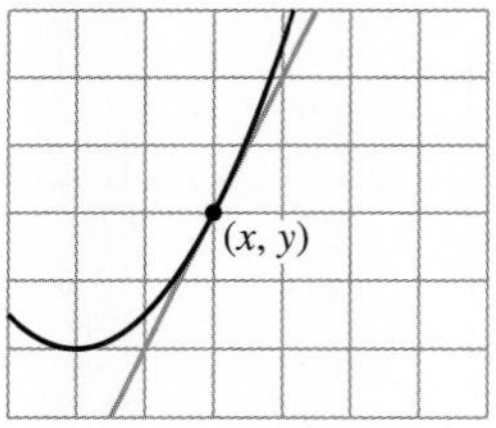

3.

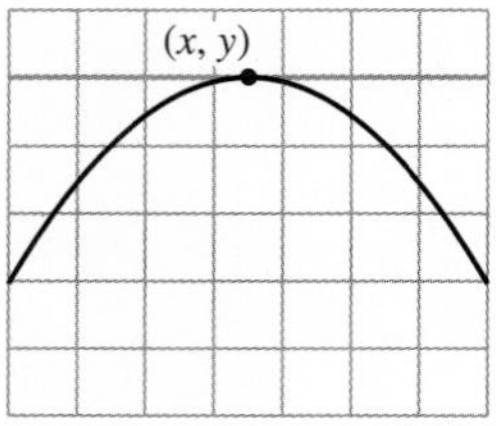

4.

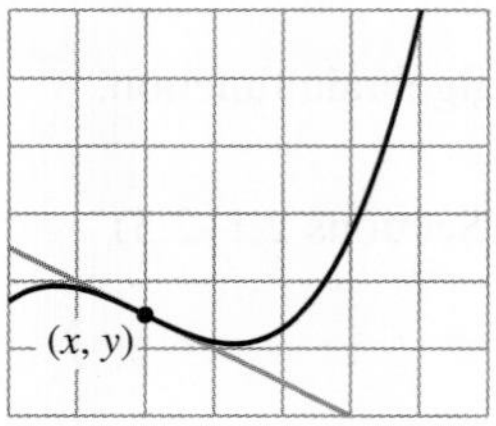

5. Sales The graph represents the sales S (in millions of dollars) for Tractor Supply Company for the years 2003 through 2009, where t represents the year, with $t = 3$ corresponding to 2003. Estimate and interpret the slopes of the graph for the years 2004 and 2007. *(Source: Tractor Supply Company)*

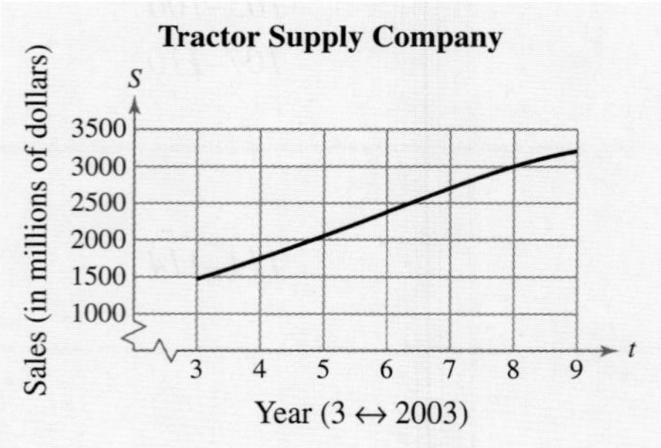

6. Farms The graph represents the amount of farm land L (in millions of acres) in the United States for the years 2004 through 2009, where t represents the year, with $t = 4$ corresponding to 2004. Estimate and interpret the slopes of the graph for the years 2005 and 2008. *(Source: U.S. Department of Agriculture)*

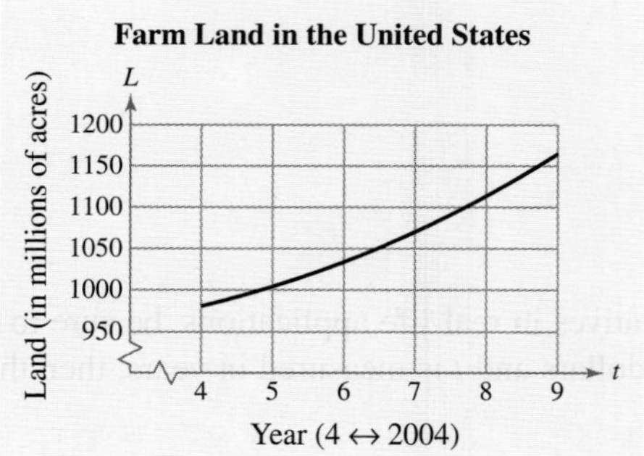

7. Consumer Trends The graph shows the number of visitors V (in thousands) to a national park during a one-year period, where $t = 1$ corresponds to January. Estimate and interpret the slopes of the graph at $t = 1$, 8, and 12.

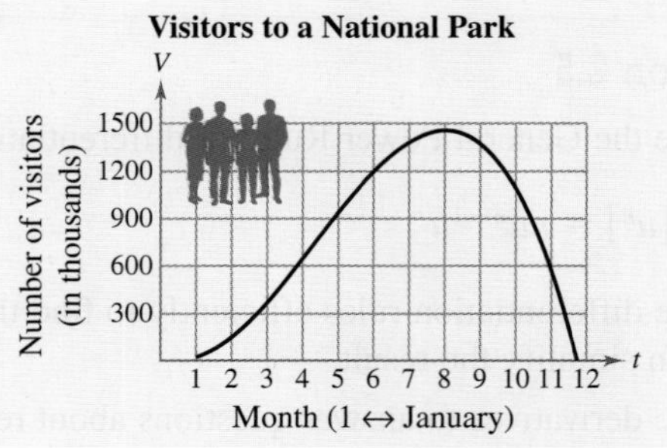

8. White-Water Rafting Two white-water rafters leave a campsite simultaneously and start downstream on a 9-mile trip. Their distances from the campsite are given by $s = f(t)$ and $s = g(t)$, where s is measured in miles and t is measured in hours.

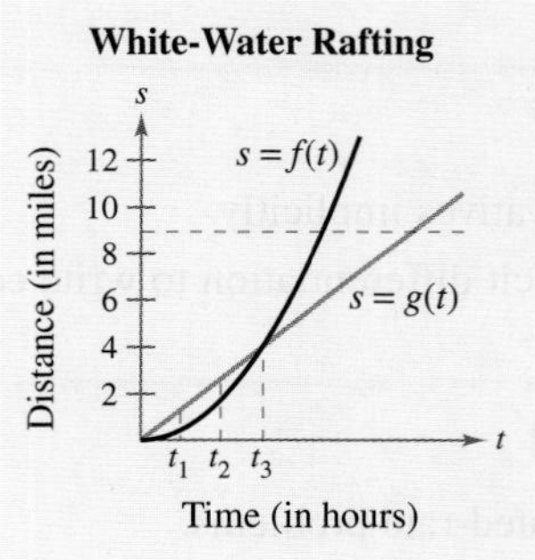

(a) Which rafter is traveling at a greater rate at t_1?

(b) What can you conclude about their rates at t_2?

(c) What can you conclude about their rates at t_3?

(d) Which rafter finishes the trip first? Explain your reasoning.

Finding the Slope of a Graph In Exercises 9–16, use the limit definition to find the slope of the tangent line to the graph of f at the given point.

9. $f(x) = -3x - 5;\ (-2, 1)$

10. $f(x) = 7x + 3;\ (-1, -4)$

11. $f(x) = x^2 - 4x;\ (1, -3)$

12. $f(x) = x^2 + 10;\ (2, 14)$

13. $f(x) = \sqrt{x + 9};\ (-5, 2)$

14. $f(x) = \sqrt{x - 1};\ (10, 3)$

15. $f(x) = \dfrac{1}{x - 5};\ (6, 1)$

16. $f(x) = \dfrac{1}{x + 4};\ (-3, 1)$

Finding a Derivative In Exercises 17–24, use the limit definition to find the derivative of the function.

17. $f(x) = 9x + 1$

18. $f(x) = 1 - 4x$

19. $f(x) = -\frac{1}{2}x^2 + 2x$

20. $f(x) = 4 - x^2$

21. $f(x) = \sqrt{x - 5}$

22. $f(x) = \sqrt{x} + 3$

23. $f(x) = \dfrac{5}{x}$

24. $f(x) = \dfrac{1}{x + 4}$

Determining Differentiability In Exercises 25–28, determine the *x*-values at which the function is differentiable. Explain your reasoning.

25. $y = \dfrac{x + 1}{x - 1}$

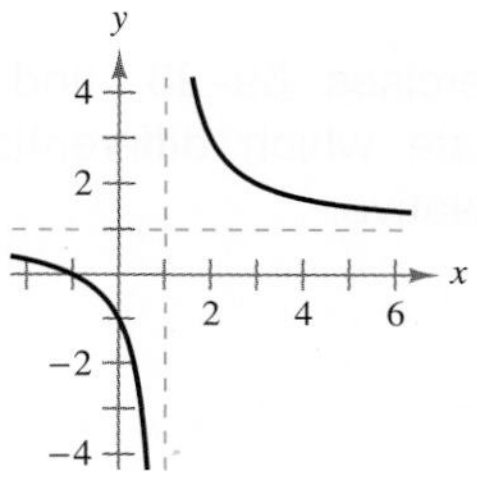

26. $y = -|x| + 3$

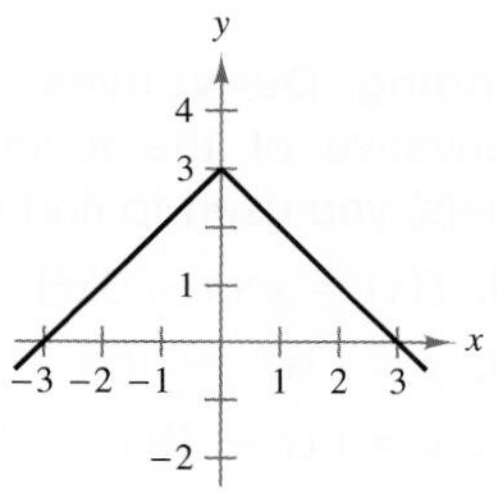

27. $y = \begin{cases} -x - 2, & x \le 0 \\ x^3 + 2, & x > 0 \end{cases}$

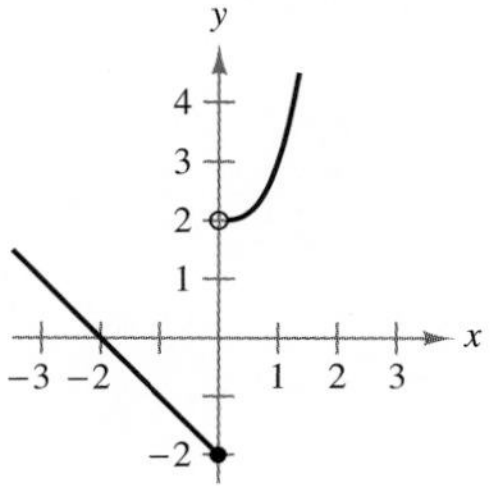

28. $y = (x + 1)^{2/3}$

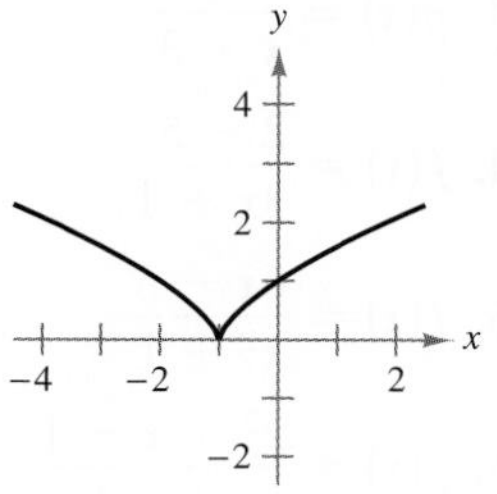

Finding Derivatives In Exercises 29–40, find the derivative of the function.

29. $y = -6$

30. $f(x) = 5$

31. $f(x) = x^3$

32. $h(x) = \dfrac{1}{x^4}$

33. $f(x) = 4x^2$

34. $g(t) = 6t^5$

35. $f(x) = \dfrac{2x^4}{5}$

36. $y = 3x^{2/3}$

37. $g(x) = 2x^4 + 3x^2$

38. $f(x) = 6x^2 - 4x$

39. $y = x^2 + 6x - 7$

40. $y = 2x^4 - 3x^3 + x$

Finding the Slope of a Graph In Exercises 41–44, find the slope of the graph of the function at the given point.

41. $f(x) = 2x^{-1/2}$; $(4, 1)$

42. $y = \dfrac{3}{2x} + 3$; $\left(\dfrac{1}{2}, 6\right)$

43. $g(x) = x^3 - 4x^2 - 6x + 8$; $(-1, 9)$

44. $y = 2x^4 - 5x^3 + 6x^2 - x$; $(1, 2)$

Finding an Equation of a Tangent Line In Exercises 45–48, (a) find an equation of the tangent line to the graph of the function at the given point, (b) use a graphing utility to graph the function and its tangent line at the point, and (c) use the *derivative* feature of a graphing utility to confirm your results.

45. $f(x) = 2x^2 - 3x + 1$; $(2, 3)$

46. $y = 11x^4 - 5x^2 + 1$; $(-1, 7)$

47. $f(x) = \sqrt{x} - \dfrac{1}{\sqrt{x}}$; $(1, 0)$

48. $f(x) = -x^2 - 4x - 4$; $(-4, -4)$

49. **Sales** The annual sales S (in millions of dollars) for Tractor Supply Company for the years 2003 through 2009 can be modeled by

$$S = -0.7500t^4 + 13.278t^3 - 74.50t^2 + 440.2t + 523$$

where t is the year, with $t = 3$ corresponding to 2003. *(Source: Tractor Supply Company)*

(a) Find the slopes of the graph for the years 2004 and 2007.

(b) Compare your results with those obtained in Exercise 5.

(c) Interpret the slope of the graph in the context of the problem.

50. **Farms** The amount of farm land L (in millions of acres) in the United States for the years 2004 through 2009 can be modeled by

$$L = 0.0991t^3 + 1.512t^2 + 4.01t + 933.9$$

where t is the year, with $t = 4$ corresponding to 2004. *(Source: U.S. Department of Agriculture)*

(a) Find the slopes of the graph for the years 2005 and 2008.

(b) Compare your results with those obtained in Exercise 6.

(c) Interpret the slope of the graph in the context of the problem.

Finding Rates of Change In Exercises 51–54, use a graphing utility to graph the function and find its average rate of change over the interval. Compare this rate with the instantaneous rates of change at the endpoints of the interval.

51. $f(t) = 4t + 3; [-3, 1]$

52. $f(x) = x^{2/3}; [1, 8]$

53. $f(x) = x^2 + 3x - 4; [0, 1]$

54. $f(x) = x^3 + x; [-2, 2]$

55. Velocity The height s (in feet) at time t (in seconds) of a ball thrown upward from the top of a 300-foot building with an initial velocity of 24 feet per second is given by

$$s = -16t^2 + 24t + 300.$$

(a) Find the average velocity on the interval $[1, 2]$.

(b) Find the instantaneous velocities when $t = 1$ and $t = 3$.

(c) How long will it take the ball to hit the ground?

(d) Find the velocity of the ball when it hits the ground.

56. Velocity A rock is dropped from a tower on the Brooklyn Bridge, 276 feet above the East River. Let t represent the time in seconds.

(a) Find the position and velocity functions for the rock.

(b) Find the average velocity over the interval $[0, 2]$.

(c) Find the instantaneous velocities at $t = 2$ and $t = 3$.

(d) How long will it take the rock to hit the water?

(e) Find the velocity of the rock when it hits the water.

Marginal Cost In Exercises 57–60, find the marginal cost for producing x units. (The cost is measured in dollars.)

57. $C = 2500 + 320x$

58. $C = 3x^3 + 24{,}000$

59. $C = 370 + 2.55\sqrt{x}$

60. $C = 475 + 5.25x^{2/3}$

Marginal Revenue In Exercises 61–64, find the marginal revenue for producing x units. (The revenue is measured in dollars.)

61. $R = 150x - 0.6x^2$

62. $R = 150x - \frac{3}{4}x^2$

63. $R = -4x^3 + 2x^2 + 100x$

64. $R = 4x + 10\sqrt{x}$

Marginal Profit In Exercises 65 and 66, find the marginal profit for producing x units. (The profit is measured in dollars.)

65. $P = -0.0002x^3 + 6x^2 - x - 2000$

66. $P = -\frac{1}{15}x^3 + 4000x^2 - 120x - 144{,}000$

67. Marginal Profit The profit P (in dollars) from selling x units of a product is given by

$$P = -0.05x^2 + 20x - 1000.$$

(a) Find the additional profit when the sales increase from 100 to 101 units.

(b) Find the marginal profit when $x = 100$ units.

(c) Compare the results of parts (a) and (b).

68. Population Growth The population P (in millions) of Brazil from 1980 through 2010 can be modeled by

$$P = -0.007t^2 + 2.78t + 123.6$$

where t represents the year, with $t = 0$ corresponding to 1980. *(Source: U.S. Census Bureau)*

(a) Evaluate P for $t = 0$, 5, 10, 15, 20, 25, and 30. Explain these values.

(b) Determine the population growth rate, dP/dt.

(c) Evaluate dP/dt for the same values as in part (a). Explain your results.

Finding Derivatives In Exercises 69–90, find the derivative of the function. State which differentiation rule(s) you used to find the derivative.

69. $f(x) = x^3(5 - 3x^2)$

70. $y = (3x^2 + 7)(x^2 - 2x)$

71. $y = (4x - 3)(x^3 - 2x^2)$

72. $s = \left(4 - \dfrac{1}{t^2}\right)(t^2 - 3t)$

73. $g(x) = \dfrac{x}{x + 3}$

74. $f(x) = \dfrac{2 - 5x}{3x + 1}$

75. $f(x) = \dfrac{6x - 5}{x^2 + 1}$

76. $f(x) = \dfrac{x^2 + x - 1}{x^2 - 1}$

77. $f(x) = (5x^2 + 2)^3$

78. $f(x) = \sqrt[3]{x^2 - 1}$

79. $h(x) = \dfrac{2}{\sqrt{x + 1}}$

80. $g(x) = \sqrt{x^6 - 12x^3 + 9}$

81. $g(x) = x\sqrt{x^2 + 1}$

82. $g(t) = \dfrac{t}{(1 - t)^3}$

83. $f(x) = x(1 - 4x^2)^2$

84. $f(x) = \left(x^2 + \dfrac{1}{x}\right)^5$

85. $h(x) = [x^2(2x + 3)]^3$

86. $f(x) = [(x - 2)(x + 4)]^2$

87. $f(x) = x^2(x - 1)^5$

88. $f(s) = s^3(s^2 - 1)^{5/2}$

89. $h(t) = \dfrac{\sqrt{3t + 1}}{(1 - 3t)^2}$

90. $g(x) = \dfrac{(3x + 1)^2}{(x^2 + 1)^2}$

91. Physical Science The temperature T (in degrees Fahrenheit) of food placed in a freezer can be modeled by

$$T = \frac{1300}{t^2 + 2t + 25}$$

where t is the time (in hours).

(a) Find the rates of change of T at $t = 1, 3, 5$, and 10.

(b) Graph the model on a graphing utility and describe the rate at which the temperature is changing.

92. Forestry According to the *Doyle Log Rule*, the volume V (in board-feet) of a log of length L (in feet) and diameter D (in inches) at the small end is

$$V = \left(\frac{D-4}{4}\right)^2 L.$$

Find the rates at which the volume is changing with respect to D for a 12-foot-long log whose smallest diameter is (a) 8 inches, (b) 16 inches, (c) 24 inches, and (d) 36 inches.

Finding Higher-Order Derivatives In Exercises 93–100, find the higher-order derivative.

	Given	*Derivative*
93.	$f(x) = 3x^2 + 7x + 1$	$f''(x)$
94.	$f'(x) = 5x^4 - 6x^2 + 2x$	$f'''(x)$
95.	$f'''(x) = -\dfrac{6}{x^4}$	$f^{(5)}(x)$
96.	$f(x) = \sqrt{x}$	$f^{(4)}(x)$
97.	$f'(x) = 7x^{5/2}$	$f''(x)$
98.	$f(x) = x^2 + \dfrac{3}{x}$	$f''(x)$
99.	$f''(x) = 6\sqrt[3]{x}$	$f'''(x)$
100.	$f'''(x) = 20x^4 - \dfrac{2}{x^3}$	$f^{(5)}(x)$

101. Athletics A person dives from a 30-foot platform with an initial velocity of 5 feet per second (upward).

(a) Find the position function of the diver.

(b) How long will it take the diver to hit the water?

(c) What is the diver's velocity at impact?

(d) What is the diver's acceleration at impact?

102. Velocity and Acceleration The position function of a particle is given by

$$s = \frac{1}{t^2 + 2t + 1}$$

where s is the height (in feet) and t is the time (in seconds). Find the velocity and acceleration functions.

Finding Derivatives In Exercises 103–106, use implicit differentiation to find dy/dx.

103. $x^2 + 3xy + y^3 = 10$

104. $x^2 + 9xy + y^2 = 0$

105. $y^2 - x^2 + 8x - 9y - 1 = 0$

106. $y^2 + x^2 - 6y - 2x - 5 = 0$

Finding an Equation of a Tangent Line In Exercises 107–110, use implicit differentiation to find an equation of the tangent line to the graph of the function at the given point.

	Equation	*Point*
107.	$y^2 = x - y$	$(2, 1)$
108.	$2\sqrt[3]{x} + 3\sqrt{y} = 10$	$(8, 4)$
109.	$y^2 - 2x = xy$	$(1, 2)$
110.	$y^3 - 2x^2y + 3xy^2 = -1$	$(0, -1)$

111. Area The radius r of a circle is increasing at a rate of 2 inches per minute. Find the rate of change of the area at (a) $r = 3$ inches and (b) $r = 10$ inches.

112. Moving Point A point is moving along the graph of $y = \sqrt{x}$ such that dx/dt is 3 centimeters per second. Find dy/dt for (a) $x = \frac{1}{4}$, (b) $x = 1$, and (c) $x = 4$.

113. Water Level A swimming pool is 40 feet long, 20 feet wide, 4 feet deep at the shallow end, and 9 feet deep at the deep end (see figure). Water is being pumped into the pool at the rate of 10 cubic feet per minute. How fast is the water level rising when there is 4 feet of water in the deep end?

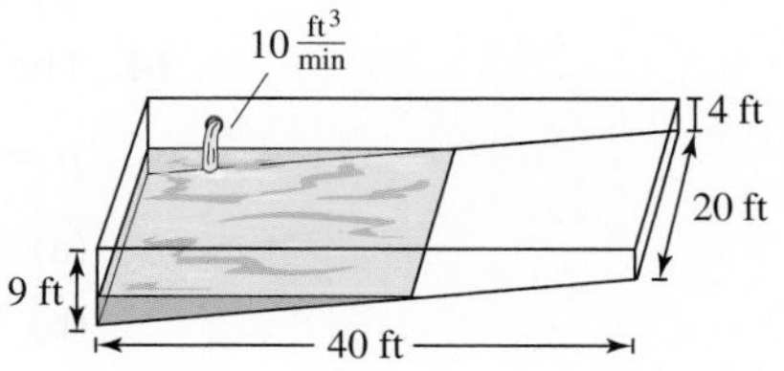

114. Profit The demand and cost functions for a product can be modeled by

$$p = 211 - 0.002x$$

and

$$C = 30x + 1{,}500{,}000$$

where x is the number of units produced.

(a) Write the profit function for this product.

(b) Find the marginal profit when 80,000 units are produced.

(c) Graph the profit function on a graphing utility and use the graph to determine the price you would charge for the product. Explain your reasoning.

TEST YOURSELF

See www.CalcChat.com for worked-out solutions to odd-numbered exercises.

Take this test as you would take a test in class. When you are done, check your work against the answers given in the back of the book.

In Exercises 1 and 2, use the limit definition to find the derivative of the function. Then find the slope of the tangent line to the graph of *f* at the given point.

1. $f(x) = x^2 + 1;\ (2, 5)$ **2.** $f(x) = \sqrt{x} - 2;\ (4, 0)$

In Exercises 3–11, find the derivative of the function.

3. $f(t) = t^3 + 2t$ **4.** $f(x) = 4x^2 - 8x + 1$ **5.** $f(x) = x^{3/2}$

6. $f(x) = (x + 3)(x^2 + 2x)$ **7.** $f(x) = -3x^{-3}$ **8.** $f(x) = \sqrt{x}(5 + x)$

9. $f(x) = (3x^2 + 4)^2$ **10.** $f(x) = \sqrt{1 - 2x}$ **11.** $f(x) = \dfrac{(5x - 1)^3}{x}$

12. Find an equation of the tangent line to the graph of

$$f(x) = x - \frac{1}{x}$$

at the point $(1, 0)$. Then use a graphing utility to graph the function and the tangent line in the same viewing window.

13. The annual sales S (in billions of dollars) of CVS Caremark for the years 2004 through 2009 can be modeled by

$$S = -1.3241t^3 + 26.562t^2 - 155.81t + 314.3$$

where t represents the year, with $t = 4$ corresponding to 2004. *(Source: CVS Caremark Corporation)*

(a) Approximate the average rate of change for the years from 2005 through 2008.

(b) Find the instantaneous rates of change of the model for the years 2005 and 2008.

(c) Interpret the results of parts (a) and (b) in the context of the problem.

14. The monthly demand and cost functions for a product are given by

$$p = 1700 - 0.016x \quad \text{and} \quad C = 715{,}000 + 240x.$$

(a) Write the profit function for this product.

(b) Find the rate of change of the profit when the monthly sales are $x = 700$ units.

In Exercises 15–17, find the third derivative of the function.

15. $f(x) = 2x^2 + 3x + 1$ **16.** $f(x) = \sqrt{3 - x}$ **17.** $f(x) = \dfrac{2x + 1}{2x - 1}$

18. A ball is thrown straight upward from a height of 75 feet above the ground with an initial velocity of 30 feet per second. Write the position, velocity, and acceleration functions of the ball. Find the height, velocity, and acceleration when $t = 2$.

In Exercises 19–21, use implicit differentiation to find dy/dx.

19. $x + xy = 6$ **20.** $y^2 + 2x - 2y + 1 = 0$ **21.** $x^2 - 2y^2 = 4$

22. The radius r of a right circular cylinder is increasing at a rate of 0.25 centimeter per minute. The height h of the cylinder is related to the radius by $h = 20r$. Find the rate of change of the volume when (a) $r = 0.5$ centimeter and (b) $r = 1$ centimeter.

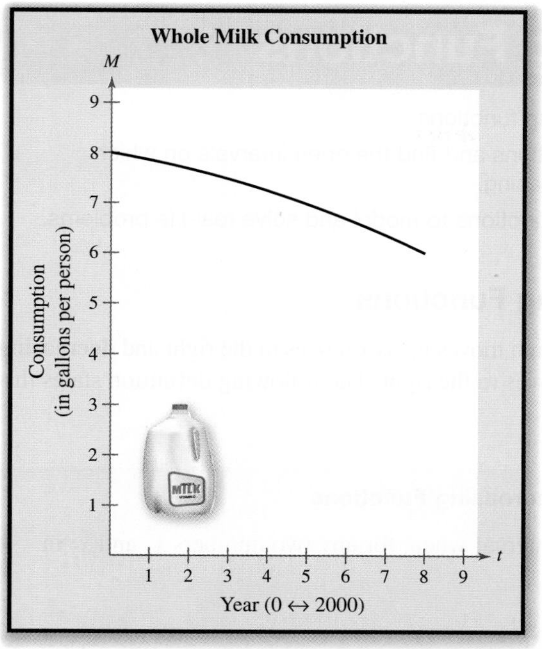

Example 2 on page 169 shows how the derivative can be used to show that milk consumption decreased in the United States from 2000 through 2008.

3 Applications of the Derivative

3.1 Increasing and Decreasing Functions

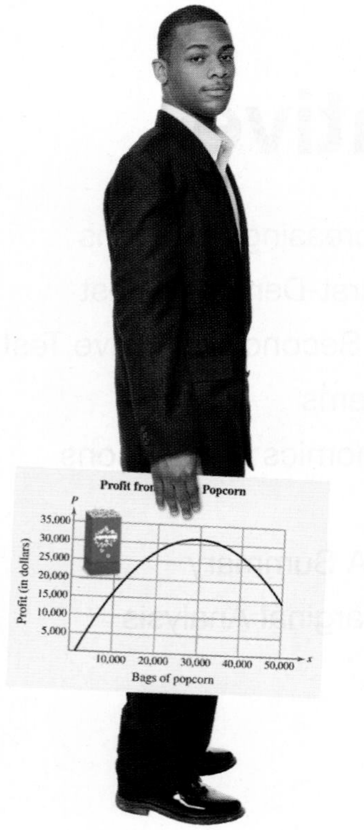

In Exercise 47 on page 176, you will use derivatives and critical numbers to find the intervals on which the profit from selling popcorn is increasing and decreasing.

- Test for increasing and decreasing functions.
- Find the critical numbers of functions and find the open intervals on which functions are increasing or decreasing.
- Use increasing and decreasing functions to model and solve real-life problems.

Increasing and Decreasing Functions

A function is **increasing** when its graph moves up as x moves to the right and **decreasing** when its graph moves down as x moves to the right. The following definition states this more formally.

Definition of Increasing and Decreasing Functions

A function f is **increasing** on an interval when, for any two numbers x_1 and x_2 in the interval,

$$x_2 > x_1 \quad \text{implies} \quad f(x_2) > f(x_1).$$

A function f is **decreasing** on an interval when, for any two numbers x_1 and x_2 in the interval,

$$x_2 > x_1 \quad \text{implies} \quad f(x_2) < f(x_1).$$

The function in Figure 3.1 is decreasing on the interval $(-\infty, a)$, constant on the interval (a, b), and increasing on the interval (b, ∞). Actually, from the definition of increasing and decreasing functions, the function shown in Figure 3.1 is decreasing on the interval $(-\infty, a]$ and increasing on the interval $[b, \infty)$. This text restricts the discussion to finding *open* intervals on which a function is increasing or decreasing.

The derivative of a function can be used to determine whether the function is increasing or decreasing on an interval.

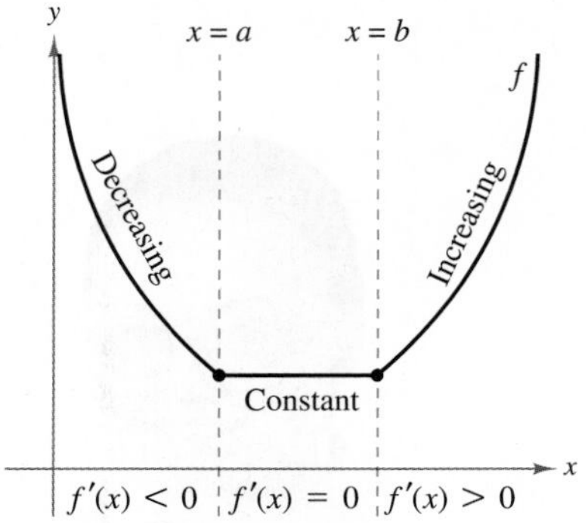

FIGURE 3.1

Test for Increasing and Decreasing Functions

Let f be differentiable on the interval (a, b).

1. If $f'(x) > 0$ for all x in (a, b), then f is increasing on (a, b).
2. If $f'(x) < 0$ for all x in (a, b), then f is decreasing on (a, b).
3. If $f'(x) = 0$ for all x in (a, b), then f is constant on (a, b).

STUDY TIP

The conclusions in the first two cases of testing for increasing and decreasing functions are valid even when $f'(x) = 0$ at a finite number of x-values in (a, b).

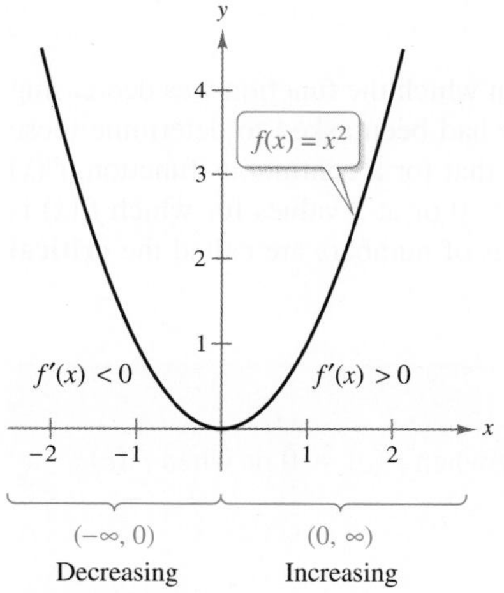

FIGURE 3.2

Example 1 Testing for Increasing and Decreasing Functions

Show that the function $f(x) = x^2$ is decreasing on the open interval $(-\infty, 0)$ and increasing on the open interval $(0, \infty)$.

SOLUTION The derivative of f is

$$f'(x) = 2x.$$

On the open interval $(-\infty, 0)$, the fact that x is negative implies that $f'(x) = 2x$ is also negative. So, by the test for a decreasing function, you can conclude that f is *decreasing* on this interval. Similarly, on the open interval $(0, \infty)$, the fact that x is positive implies that $f'(x) = 2x$ is also positive. So, it follows that f is *increasing* on this interval, as shown in Figure 3.2.

✓Checkpoint 1

Show that the function $f(x) = x^4$ is decreasing on the open interval $(-\infty, 0)$ and increasing on the open interval $(0, \infty)$. ■

Example 2 Modeling Consumption

From 2000 through 2008, the consumption M of whole milk in the United States (in gallons per person) can be modeled by

$$M = -0.015t^2 + 0.13t + 8.0, \quad 0 \le t \le 8$$

where $t = 0$ corresponds to 2000 (see Figure 3.3). Show that the consumption of whole milk was decreasing from 2000 to 2008. *(Source: U.S. Department of Agriculture)*

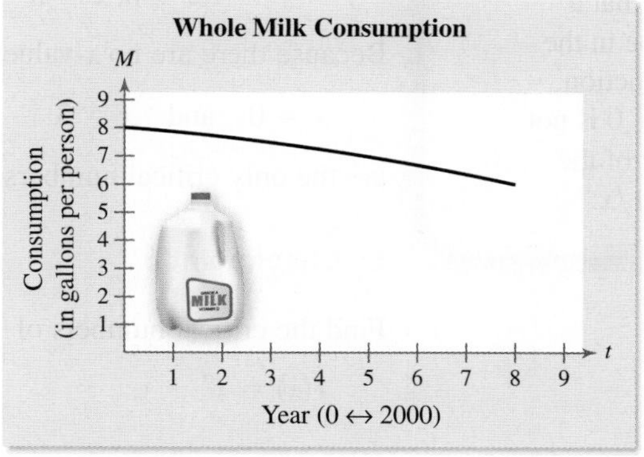

FIGURE 3.3

SOLUTION The derivative of this model is $dM/dt = -0.030t + 0.13$. For the open interval $(0, 8)$, the derivative is negative. So, the function is decreasing, which implies that the consumption of whole milk was decreasing during the given time period.

✓Checkpoint 2

From 2003 through 2008, the consumption F of fresh fruit in the United States (in pounds per person) can be modeled by

$$F = -0.7674t^2 + 2.872t + 277.87, \quad 3 \le t \le 8$$

where $t = 3$ corresponds to 2003. Show that the consumption of fresh fruit was decreasing from 2003 to 2008. *(Source: U.S. Department of Agriculture)* ■

Critical Numbers and Their Use

In Example 1, you were given two intervals: one on which the function was decreasing and one on which it was increasing. Suppose you had been asked to determine these intervals. To do this, you could have used the fact that for a continuous function, $f'(x)$ can change signs only at x-values for which $f'(x) = 0$ or at x-values for which $f'(x)$ is undefined, as shown in Figure 3.4. These two types of numbers are called the **critical numbers** of f.

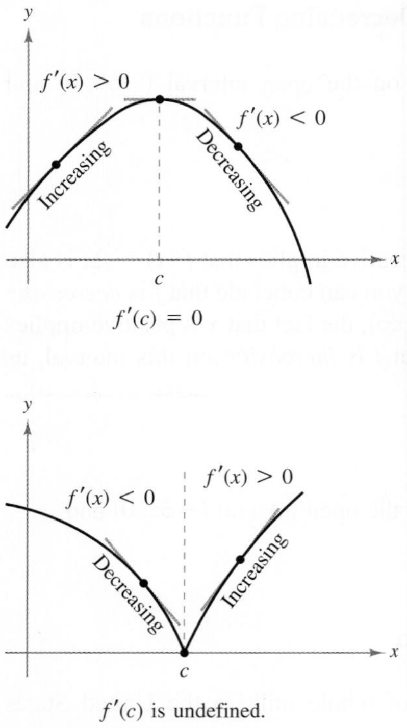

FIGURE 3.4

Definition of a Critical Number

If f is defined at c, then c is a critical number of f when $f'(c) = 0$ or when $f'(c)$ is undefined.

Example 3 Finding Critical Numbers

Find the critical numbers of

$$f(x) = 2x^3 - 9x^2.$$

SOLUTION Begin by differentiating the function.

$f(x) = 2x^3 - 9x^2$ Write original function.

$f'(x) = 6x^2 - 18x$ Differentiate.

To find the critical numbers of f, you must find all x-values for which $f'(x) = 0$ and all x-values for which $f'(x)$ is undefined.

$6x^2 - 18x = 0$ Set $f'(x)$ equal to 0.

$6x(x - 3) = 0$ Factor.

$x = 0, x = 3$ Critical numbers

Because there are no x-values for which f' is undefined, you can conclude that

$$x = 0 \quad \text{and} \quad x = 3$$

are the only critical numbers of f.

STUDY TIP

The definition of a critical number requires that a critical number be in the domain of the function. For instance, $x = 0$ is not a critical number of the function $f(x) = 1/x$.

✓ Checkpoint 3

Find the critical numbers of

$$f(x) = x^2 - x.$$

■

To determine the intervals on which a continuous function is increasing or decreasing, you can use the guidelines below.

Guidelines for Applying the Increasing/Decreasing Test

1. Find the derivative of f.
2. Locate the critical numbers of f and use these numbers to determine test intervals. That is, find all x for which $f'(x) = 0$ or $f'(x)$ is undefined.
3. Determine the sign of $f'(x)$ at one test value in each of the intervals.
4. Use the test for increasing and decreasing functions to decide whether f is increasing or decreasing on each interval.

Example 4 Intervals on Which *f* Is Increasing or Decreasing

Find the open intervals on which the function is increasing or decreasing.

$$f(x) = x^3 - \frac{3}{2}x^2$$

SOLUTION Begin by finding the derivative of f. Then set the derivative equal to zero and solve for the critical numbers.

$$f'(x) = 3x^2 - 3x \quad \text{Differentiate original function.}$$
$$3x^2 - 3x = 0 \quad \text{Set derivative equal to 0.}$$
$$3x(x - 1) = 0 \quad \text{Factor.}$$
$$x = 0, x = 1 \quad \text{Critical numbers}$$

Because there are no x-values for which f' is undefined, it follows that $x = 0$ and $x = 1$ are the only critical numbers. So, the intervals that need to be tested are

$$(-\infty, 0), (0, 1), \text{ and } (1, \infty). \quad \text{Test intervals}$$

The table summarizes the testing of these three intervals.

Interval	$-\infty < x < 0$	$0 < x < 1$	$1 < x < \infty$
Test value	$x = -1$	$x = \frac{1}{2}$	$x = 2$
Sign of $f'(x)$	$f'(-1) = 6 > 0$	$f'(\frac{1}{2}) = -\frac{3}{4} < 0$	$f'(2) = 6 > 0$
Conclusion	Increasing	Decreasing	Increasing

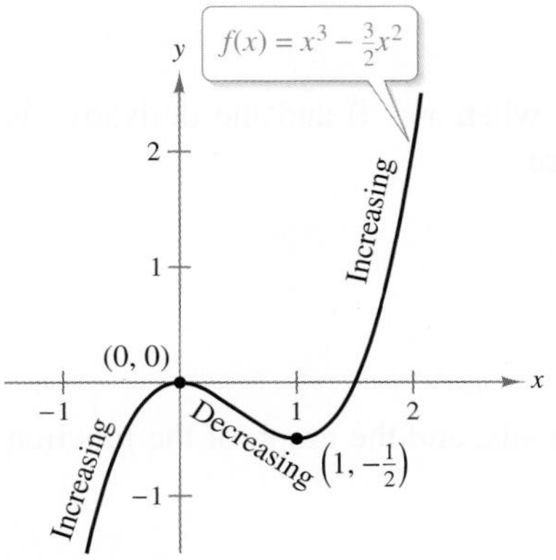

FIGURE 3.5

The graph of f is shown in Figure 3.5. Note that the test values in the intervals were chosen for convenience—other x-values could have been used.

✓ Checkpoint 4

Find the open intervals on which the function $f(x) = x^3 - 12x$ is increasing or decreasing.

TECH TUTOR

You can use the *trace* feature of a graphing utility to confirm the result of Example 4. Begin by graphing the function, as shown below. Then use the *trace* feature and move the cursor from left to right. In intervals on which the function is increasing, note that the y-values increase as the x-values increase, whereas in intervals on which the function is decreasing, the y-values decrease as the x-values increase.

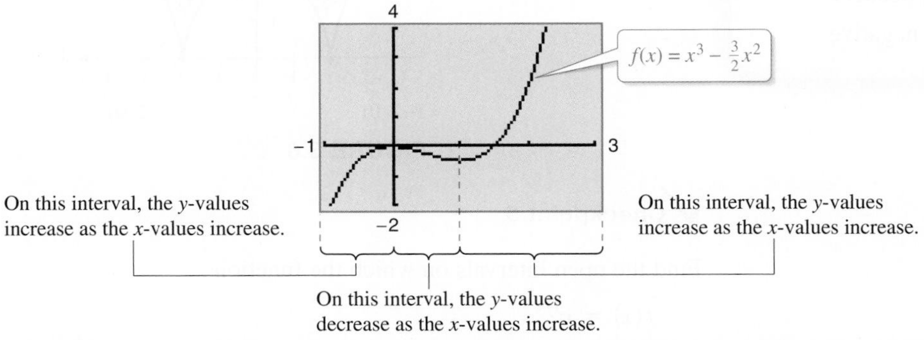

Not only is the function in Example 4 continuous on the entire real number line, it is also differentiable there. For such functions, the only critical numbers are those for which $f'(x) = 0$. The next example considers a continuous function that has *both* types of critical numbers—those for which $f'(x) = 0$ and those for which $f'(x)$ is undefined.

ALGEBRA TUTOR

For help on the algebra in Example 5, see Example 2(d) in the *Chapter 3 Algebra Tutor*, on page 243.

Example 5 Intervals on Which f Is Increasing or Decreasing

Find the open intervals on which the function

$$f(x) = (x^2 - 4)^{2/3}$$

is increasing or decreasing.

SOLUTION Begin by finding the derivative of the function.

$$f'(x) = \frac{2}{3}(x^2 - 4)^{-1/3}(2x) \quad \text{Differentiate.}$$

$$= \frac{4x}{3(x^2 - 4)^{1/3}} \quad \text{Simplify.}$$

From this, you can see that the derivative is zero when $x = 0$ and the derivative is undefined when $x = \pm 2$. So, the critical numbers are

$$x = -2, \quad x = 0, \quad \text{and} \quad x = 2. \quad \text{Critical numbers}$$

This implies that the test intervals are

$$(-\infty, -2), \quad (-2, 0), \quad (0, 2), \quad \text{and} \quad (2, \infty). \quad \text{Test intervals}$$

The table summarizes the testing of these four intervals, and the graph of the function is shown in Figure 3.6.

Interval	$-\infty < x < -2$	$-2 < x < 0$	$0 < x < 2$	$2 < x < \infty$
Test value	$x = -3$	$x = -1$	$x = 1$	$x = 3$
Sign of $f'(x)$	$f'(-3) < 0$	$f'(-1) > 0$	$f'(1) < 0$	$f'(3) > 0$
Conclusion	Decreasing	Increasing	Decreasing	Increasing

STUDY TIP

To test the intervals in the table in Example 5, it is not necessary to *evaluate* $f'(x)$ at each test value—you only need to determine its sign. For instance, you can determine the sign of $f'(-3)$ as shown.

$$f'(-3) = \frac{4(-3)}{3(9-4)^{1/3}} = \frac{\text{negative}}{\text{positive}} = \text{negative}$$

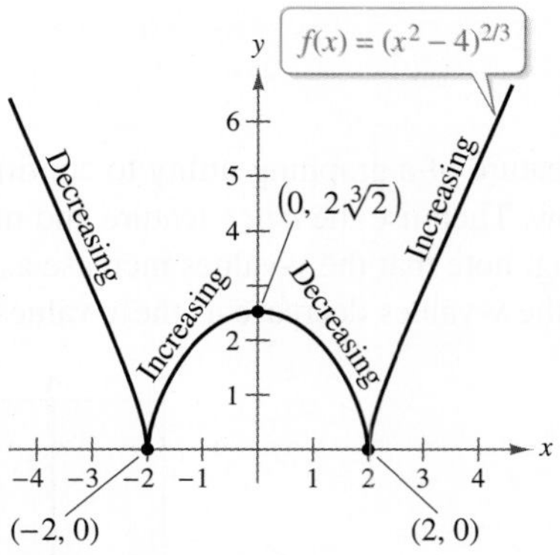

FIGURE 3.6

✓Checkpoint 5

Find the open intervals on which the function

$$f(x) = x^{2/3}$$

is increasing or decreasing.

The functions in Examples 1 through 5 are continuous on the entire real number line. If there are isolated x-values at which a function is not continuous, then these x-values should be used along with the critical numbers to determine the test intervals.

Example 6 **Testing a Function That Is Not Continuous**

The function

$$f(x) = \frac{x^4 + 1}{x^2}$$

is not continuous at $x = 0$. Because the derivative of f

$$f'(x) = \frac{2(x^4 - 1)}{x^3}$$

is zero at $x = \pm 1$, you should use the following numbers to determine the test intervals.

$x = -1, x = 1$ Critical numbers

$x = 0$ Discontinuity

After testing $f'(x)$, you can determine that f is decreasing on the intervals $(-\infty, -1)$ and $(0, 1)$, and increasing on the intervals $(-1, 0)$ and $(1, \infty)$, as shown in Figure 3.7.

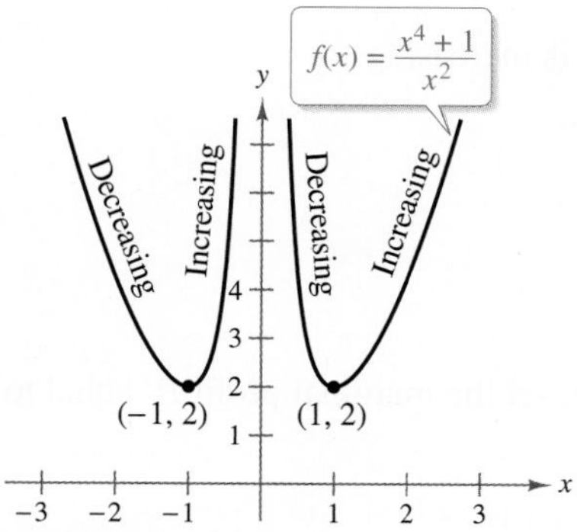

FIGURE 3.7

Checkpoint 6

Find the open invervals on which the function $f(x) = \dfrac{x^2 + 1}{x}$ is increasing or decreasing. ■

The converse of the test for increasing and decreasing functions is *not* true. For instance, it is possible for a function to be increasing on an interval even though its derivative is not positive at every point in the interval.

Example 7 **Testing an Increasing Function**

Show that $f(x) = x^3 - 3x^2 + 3x$ is increasing on the entire real number line.

SOLUTION From the derivative of f

$$f'(x) = 3x^2 - 6x + 3 = 3(x - 1)^2$$

you can see that the only critical number is $x = 1$. So, the test intervals are $(-\infty, 1)$ and $(1, \infty)$. The table summarizes the testing of these two intervals. From Figure 3.8, you can see that f is increasing on the entire real number line, even though $f'(1) = 0$. To convince yourself of this, look back at the definition of an increasing function.

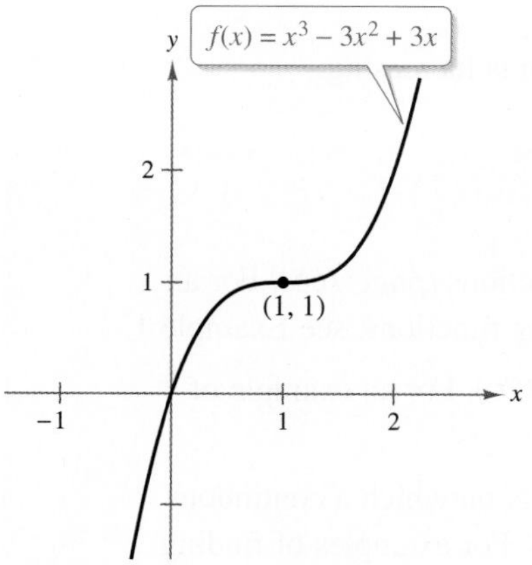

FIGURE 3.8

Interval	$-\infty < x < 1$	$1 < x < \infty$
Test value	$x = 0$	$x = 2$
Sign of $f'(x)$	$f'(0) = 3(-1)^2 > 0$	$f'(2) = 3(1)^2 > 0$
Conclusion	Increasing	Increasing

Checkpoint 7

Show that $f(x) = -x^3 + 2$ is decreasing on the entire real number line. ■

Application

Example 8 Profit Analysis

A national toy distributor determines the cost and revenue models for one of its games.

$$C = 2.4x - 0.0002x^2, \quad 0 \le x \le 6000$$
$$R = 7.2x - 0.001x^2, \quad 0 \le x \le 6000$$

Determine the interval on which the profit function is increasing.

SOLUTION The profit for producing x games is

$$\begin{aligned} P &= R - C \\ &= (7.2x - 0.001x^2) - (2.4x - 0.0002x^2) \\ &= 4.8x - 0.0008x^2. \end{aligned}$$

To find the interval on which the profit is increasing, set the marginal profit P' equal to zero and solve for x.

$$P' = 4.8 - 0.0016x \qquad \text{Differentiate profit function.}$$
$$4.8 - 0.0016x = 0 \qquad \text{Set } P' \text{ equal to 0.}$$
$$-0.0016x = -4.8 \qquad \text{Subtract 4.8 from each side.}$$
$$x = \frac{-4.8}{-0.0016} \qquad \text{Divide each side by } -0.0016.$$
$$x = 3000 \text{ games} \qquad \text{Simplify.}$$

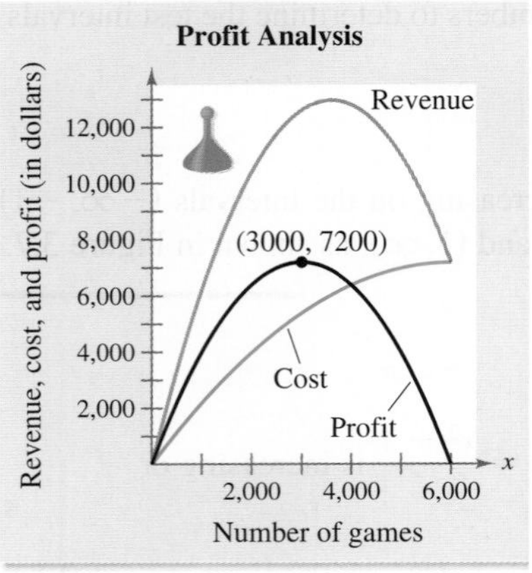

FIGURE 3.9

On the interval (0, 3000), P' is positive and the profit is *increasing*. On the interval (3000, 6000), P' is negative and the profit is *decreasing*. The graphs of the cost, revenue, and profit functions are shown in Figure 3.9.

✓ Checkpoint 8

A national distributor of pet toys determines the cost and revenue functions for one of its toys.

$$C = 1.2x - 0.0001x^2, \quad 0 \le x \le 6000$$
$$R = 3.6x - 0.0005x^2, \quad 0 \le x \le 6000$$

Determine the interval on which the profit function is increasing. ■

SUMMARIZE (Section 3.1)

1. State the test for increasing and decreasing functions *(page 168)*. For an example of testing for increasing and decreasing functions, see Example 1.
2. State the definition of a critical number *(page 170)*. For an example of finding a critical number, see Example 3.
3. State the guidelines for determining the intervals on which a continuous function is increasing or decreasing *(page 170)*. For examples of finding the intervals on which a function is increasing or decreasing, see Examples 4, 5, and 7.
4. Describe a real-life example of how testing for increasing and decreasing functions can be used to analyze the profit of a company *(page 174, Example 8)*.

SKILLS WARM UP 3.1

The following warm-up exercises involve skills that were covered in a previous course or earlier sections. You will use these skills in the exercise set for this section. For additional help, review Appendix Section A.3, and Section 1.4.

In Exercises 1–4, solve the equation.

1. $x^2 = 8x$ **2.** $15x = \frac{5}{8}x^2$ **3.** $\frac{x^2 - 25}{x^3} = 0$ **4.** $\frac{2x}{\sqrt{1 - x^2}} = 0$

In Exercises 5–8, find the domain of the function.

5. $y = \frac{x + 3}{x - 3}$ **6.** $y = \frac{2}{\sqrt{1 - x}}$ **7.** $y = \frac{2x + 1}{x^2 - 3x - 10}$ **8.** $y = \frac{3x}{\sqrt{9 - 3x^2}}$

In Exercises 9–12, evaluate the expression when $x = -2$, 0, and 2.

9. $-2(x + 1)(x - 1)$ **10.** $4(2x + 1)(2x - 1)$ **11.** $\frac{2x + 1}{(x - 1)^2}$ **12.** $\frac{-2(x + 1)}{(x - 4)^2}$

Exercises 3.1

See www.CalcChat.com for worked-out solutions to odd-numbered exercises.

Using Graphs In Exercises 1–4, use the graph to estimate the open intervals on which the function is increasing or decreasing.

1. $f(x) = -(x + 1)^2$

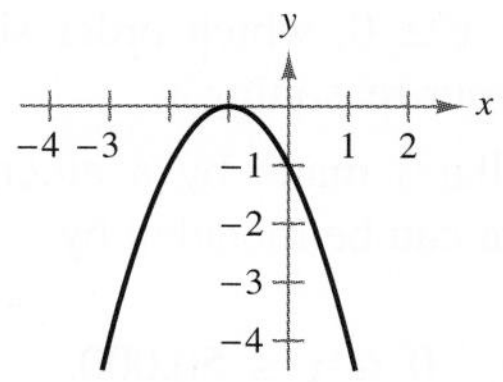

2. $f(x) = \frac{x^3}{4} - 3x$

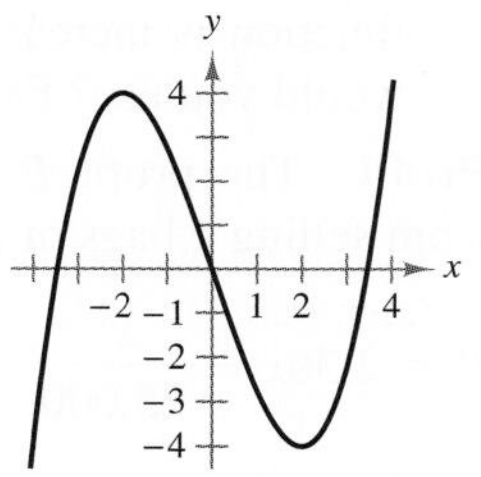

3. $f(x) = x^4 - 2x^2$

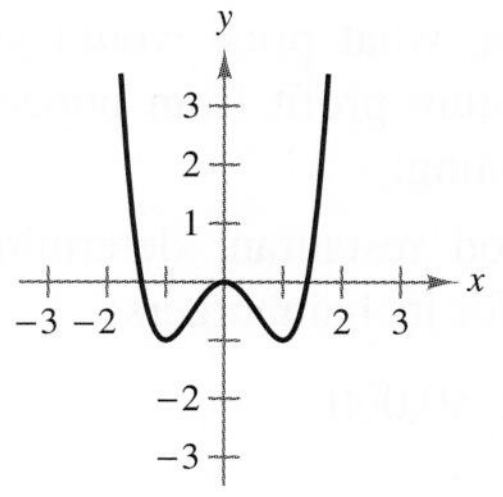

4. $f(x) = -(x^2 - 9)^{2/3}$

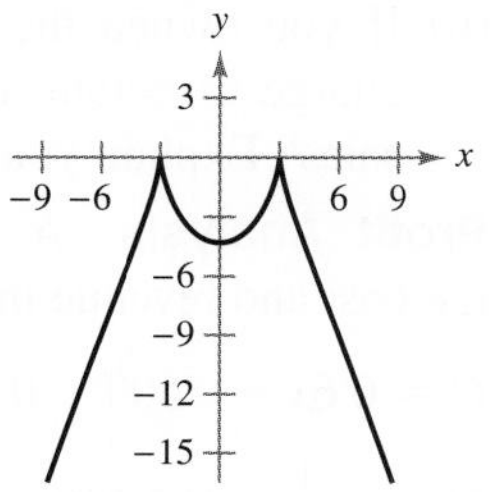

Finding Critical Numbers In Exercises 5–10, find the critical numbers of the function. *See Example 3.*

5. $f(x) = 4x^2 - 6x$ **6.** $f(x) = 3x^2 + 10$

7. $y = x^4 + 4x^3 + 8$ **8.** $g(x) = 2x^2 - 54x$

9. $f(x) = \sqrt{x^2 - 4}$

10. $y = \frac{x}{x^2 + 16}$

Intervals on Which *f* Is Increasing or Decreasing In Exercises 11–34, find the critical numbers and the open intervals on which the function is increasing or decreasing. Use a graphing utility to verify your results. *See Examples 4 and 5.*

11. $f(x) = 2x - 3$ **12.** $f(x) = 5 - 3x$

13. $y = x^2 - 6x$ **14.** $y = -x^2 + 2x$

15. $f(x) = -2x^2 + 4x + 3$ **16.** $f(x) = x^2 + 8x + 10$

17. $y = 3x^3 + 12x^2 + 15x$ **18.** $y = x^3 - 3x + 2$

19. $f(x) = x^4 - 2x^3$

20. $f(x) = \frac{1}{4}x^4 - 2x^2$

21. $g(x) = (x + 2)^2$

22. $y = (x - 2)^3$

23. $g(x) = -(x - 1)^2$

24. $y = x^3 - 6x^2$

25. $y = x^{1/3} + 1$

26. $y = x^{2/3} - 4$

27. $f(x) = \sqrt{x^2 - 1}$

28. $f(x) = \sqrt{9 - x^2}$

29. $g(x) = (x + 2)^{1/3}$

30. $g(x) = (x - 1)^{2/3}$

31. $f(x) = x\sqrt{x + 1}$

32. $h(x) = x\sqrt[3]{x - 1}$

33. $f(x) = \frac{x}{x^2 + 9}$

34. $f(x) = \frac{x^2}{x^2 + 4}$

Intervals on Which *f* Is Increasing or Decreasing In Exercises 35–42, find the critical numbers and the open intervals on which the function is increasing or decreasing. (*Hint:* Check for discontinuities.) Sketch the graph of the function. *See Example 6.*

35. $f(x) = \dfrac{x+4}{x-5}$

36. $f(x) = \dfrac{x}{x+1}$

37. $f(x) = \dfrac{2x}{16-x^2}$

38. $f(x) = \dfrac{x^2}{x^2-9}$

39. $y = \begin{cases} 4 - x^2, & x \le 0 \\ -2x, & x > 0 \end{cases}$

40. $y = \begin{cases} 3x + 1, & x \le 1 \\ 5 - x^2, & x > 1 \end{cases}$

41. $y = \begin{cases} 2x + 1, & x \le -1 \\ x^2 - 2, & x > -1 \end{cases}$

42. $y = \begin{cases} -x^3 + 1, & x \le 0 \\ -x^2 + 2x, & x > 0 \end{cases}$

43. Sales The sales S of Wal-Mart (in billions of dollars) from 2003 through 2009 can be modeled by

$$S = -1.598t^2 + 45.61t + 130.2, \quad 3 \le t \le 9$$

where t is the time in years, with $t = 3$ corresponding to 2003. Show that the sales were increasing from 2003 through 2009. *(Source: Wal-Mart Stores, Inc.)*

44. HOW DO YOU SEE IT? Plots of the relative numbers of N_2 (nitrogen) molecules that have a given velocity at each of three temperatures (in degrees Kelvin) are shown in the figure. Identify the differences in the average velocities (indicated by the peaks of the curves) for the three temperatures, and describe the intervals on which the velocity is increasing and decreasing for each of the three temperatures. *(Source: Adapted from Zumdahl, Chemistry, Seventh Edition)*

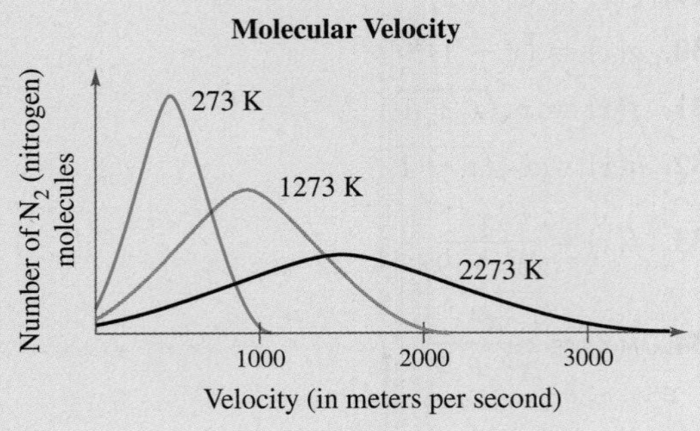

45. Medical Degrees The number y of medical degrees conferred in the United States from 1970 through 2008 can be modeled by

$$y = 0.692t^3 - 50.11t^2 + 1119.7t + 7894, \; 0 \le t \le 38$$

where t is the time in years, with $t = 0$ corresponding to 1970. *(Source: U.S. National Center for Education Statistics)*

(a) Use a graphing utility to graph the model. Then graphically estimate the years during which the model is increasing and the years during which it is decreasing.

(b) Use the test for increasing and decreasing functions to verify the result of part (a).

46. Cost The ordering and transportation cost C (in hundreds of dollars) for an automobile dealership is modeled by

$$C = 10\left(\frac{1}{x} + \frac{x}{x+3}\right), \quad x \ge 1$$

where x is the number of automobiles ordered.

(a) Find the intervals on which C is increasing or decreasing.

(b) Use a graphing utility to graph the cost function.

(c) Use the *trace* feature to determine the order sizes for which the cost is \$900. Assuming that the revenue function is increasing for $x \ge 0$, which order size would you use? Explain your reasoning.

47. Profit The profit P (in dollars) made by a cinema from selling x bags of popcorn can be modeled by

$$P = 2.36x - \frac{x^2}{25{,}000} - 3500, \quad 0 \le x \le 50{,}000.$$

(a) Find the intervals on which P is increasing and decreasing.

(b) If you owned the cinema, what price would you charge to obtain a maximum profit from popcorn sales? Explain your reasoning.

48. Profit Analysis A fast-food restaurant determines the cost and revenue models for its hamburgers.

$$C = 0.6x + 7500, \quad 0 \le x \le 50{,}000$$

$$R = \frac{1}{20{,}000}(65{,}000x - x^2), \quad 0 \le x \le 50{,}000$$

(a) Write the profit function for this situation.

(b) Determine the intervals on which the profit function is increasing and decreasing.

(c) Determine how many hamburgers the restaurant needs to sell to obtain a maximum profit. Explain your reasoning.

3.2 Extrema and the First-Derivative Test

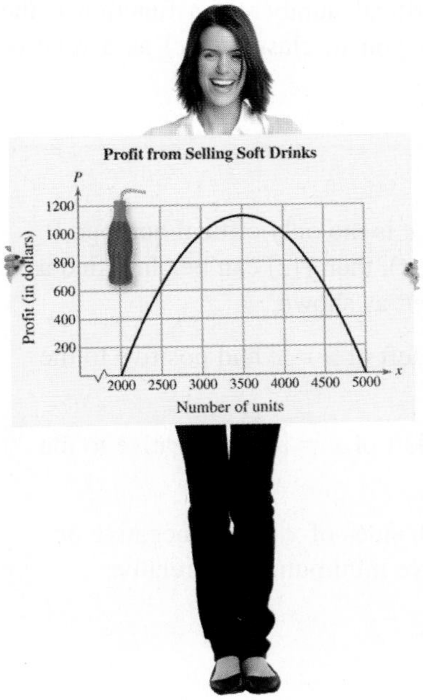

In Exercise 49 on page 185, you will use the First-Derivative Test to find the price of a soft drink that yields a maximum profit.

- Recognize the occurrence of relative extrema of functions.
- Use the First-Derivative Test to find the relative extrema of functions.
- Find absolute extrema of continuous functions on a closed interval.
- Find minimum and maximum values of real-life models and interpret the results in context.

Relative Extrema

You have used the derivative to determine the intervals on which a function is increasing or decreasing. In this section, you will examine the points at which a function changes from increasing to decreasing, or vice versa. At such a point, the function has a **relative extremum.** (The plural of extremum is *extrema.*) The **relative extrema** of a function include the **relative minima** and **relative maxima** of the function. For instance, the function shown in Figure 3.10 has a relative maximum at the left point and a relative minimum at the right point.

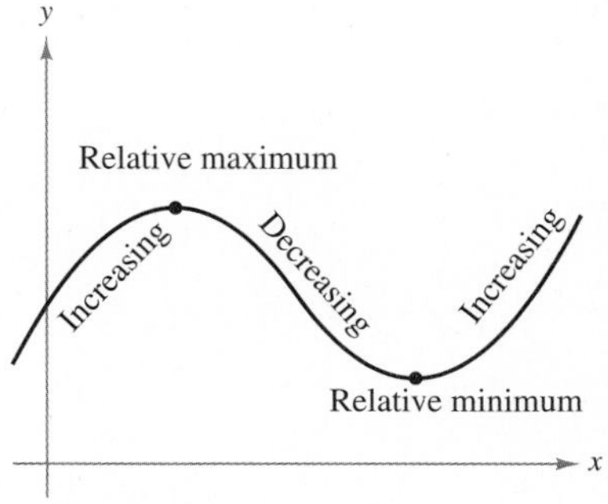

FIGURE 3.10

> **Definition of Relative Extrema**
>
> Let f be a function defined at c.
>
> **1.** $f(c)$ is a **relative maximum** of f when there exists an interval (a, b) containing c such that $f(x) \le f(c)$ for all x in (a, b).
>
> **2.** $f(c)$ is a **relative minimum** of f when there exists an interval (a, b) containing c such that $f(x) \ge f(c)$ for all x in (a, b).
>
> If $f(c)$ is a relative extremum of f, then the relative extremum is said to occur at $x = c$.

For a continuous function, the relative extrema must occur at critical numbers of the function, as shown in Figure 3.11.

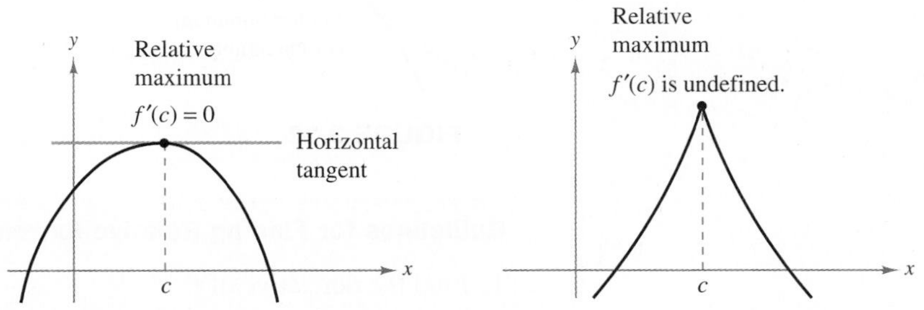

FIGURE 3.11

> **Occurrences of Relative Extrema**
>
> If f has a relative minimum or relative maximum at $x = c$, then c is a critical number of f. That is, either $f'(c) = 0$ or $f'(c)$ is undefined.

The First-Derivative Test

The discussion on the preceding page implies that in your search for relative extrema of a continuous function, you need to test only the critical numbers of the function. Once you have determined that c is a critical number of a function f, the **First-Derivative Test** for relative extrema enables you to classify $f(c)$ as a relative minimum, a relative maximum, or neither.

First-Derivative Test for Relative Extrema

Let f be continuous on the interval (a, b) in which c is the only critical number. If f is differentiable on the interval (except possibly at c), then $f(c)$ can be classified as a relative minimum, a relative maximum, or neither, as shown.

1. On the interval (a, b), if $f'(x)$ is negative to the left of $x = c$ and positive to the right of $x = c$, then $f(c)$ is a relative minimum.
2. On the interval (a, b), if $f'(x)$ is positive to the left of $x = c$ and negative to the right of $x = c$, then $f(c)$ is a relative maximum.
3. On the interval (a, b), if $f'(x)$ is positive on both sides of $x = c$ or negative on both sides of $x = c$, then $f(c)$ is neither a relative minimum nor a relative maximum.

A graphical interpretation of the First-Derivative Test is shown in Figure 3.12.

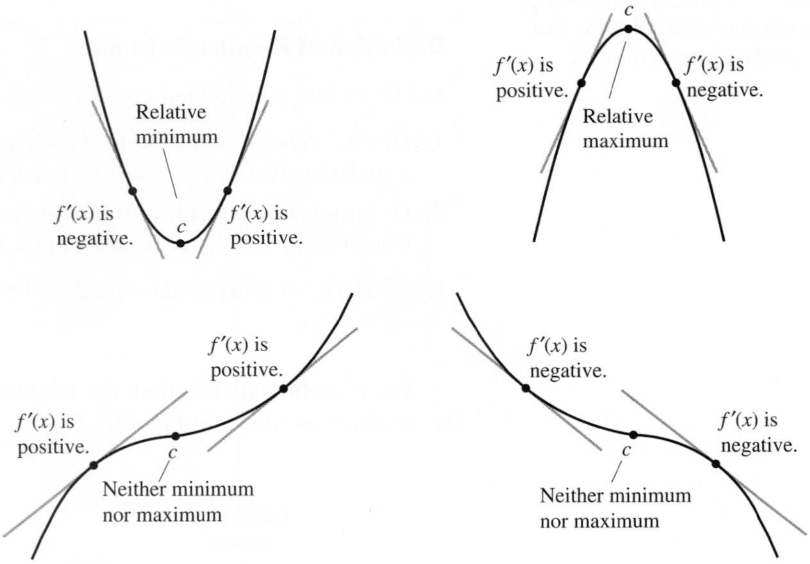

FIGURE 3.12

Guidelines for Finding Relative Extrema

1. Find the derivative of f.
2. Locate the critical numbers of f and use these numbers to determine the test intervals.
3. Test the sign of $f'(x)$ at an arbitrary number in each of the test intervals.
4. For each critical number c, use the First-Derivative Test to decide whether $f(c)$ is a relative minimum, a relative maximum, or neither.

Example 1 Finding Relative Extrema

TECH TUTOR

Some graphing calculators have a special feature that allows you to find the minimum or maximum of a function on an interval. Consult the user's manual for information on the *minimum* and *maximum* features of your graphing utility.

Find all relative extrema of the function

$$f(x) = 2x^3 - 3x^2 - 36x + 14.$$

SOLUTION Begin by finding the derivative of f.

$$f'(x) = 6x^2 - 6x - 36 \qquad \text{Differentiate.}$$

Next, find the critical numbers of f.

$$6x^2 - 6x - 36 = 0 \qquad \text{Set derivative equal to 0.}$$
$$6(x^2 - x - 6) = 0 \qquad \text{Factor out common factor.}$$
$$6(x - 3)(x + 2) = 0 \qquad \text{Factor.}$$
$$x = -2, x = 3 \qquad \text{Critical numbers}$$

Because $f'(x)$ is defined for all x, the only critical numbers of f are

$$x = -2 \quad \text{and} \quad x = 3. \qquad \text{Critical numbers}$$

Using these numbers, you can form the three test intervals

$$(-\infty, -2), (-2, 3), \quad \text{and} \quad (3, \infty). \qquad \text{Test intervals}$$

The testing of the three intervals is shown in the table.

Interval	$-\infty < x < -2$	$-2 < x < 3$	$3 < x < \infty$
Test value	$x = -3$	$x = 0$	$x = 4$
Sign of $f'(x)$	$f'(-3) = 36 > 0$	$f'(0) = -36 < 0$	$f'(4) = 36 > 0$
Conclusion	Increasing	Decreasing	Increasing

Using the First-Derivative Test, you can conclude that the critical number -2 yields a relative maximum $[f'(x)$ changes sign from positive to negative$]$, and the critical number 3 yields a relative minimum $[f'(x)$ changes sign from negative to positive$]$. The graph of f is shown in Figure 3.13. The relative maximum is

$$f(-2) = 58$$

and the relative minimum is

$$f(3) = -67.$$

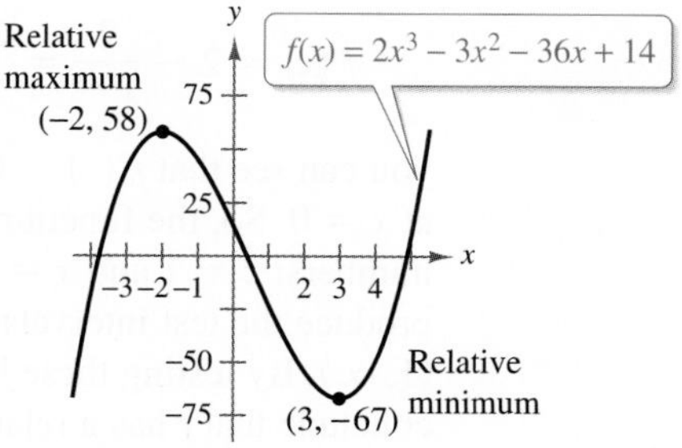

FIGURE 3.13

✓ Checkpoint 1

Find all relative extrema of

$$f(x) = 2x^3 - 6x + 1.$$

■

In Example 1, both critical numbers yielded relative extrema. In the next example, only one of the two critical numbers yields a relative extremum.

ALGEBRA TUTOR *xy*

For help on the algebra in Example 2, see Example 2(c) in the *Chapter 3 Algebra Tutor*, on page 243.

Example 2 Finding Relative Extrema

Find all relative extrema of the function $f(x) = x^4 - x^3$.

SOLUTION From the derivative of the function

$$f'(x) = 4x^3 - 3x^2 = x^2(4x - 3)$$

you can see that the function has only two critical numbers: $x = 0$ and $x = \frac{3}{4}$. These numbers produce the test intervals $(-\infty, 0)$, $\left(0, \frac{3}{4}\right)$, and $\left(\frac{3}{4}, \infty\right)$, which are tested in the table.

Interval	$-\infty < x < 0$	$0 < x < \frac{3}{4}$	$\frac{3}{4} < x < \infty$
Test value	$x = -1$	$x = \frac{1}{2}$	$x = 1$
Sign of $f'(x)$	$f'(-1) = -7 < 0$	$f'\left(\frac{1}{2}\right) = -\frac{1}{4} < 0$	$f'(1) = 1 > 0$
Conclusion	Decreasing	Decreasing	Increasing

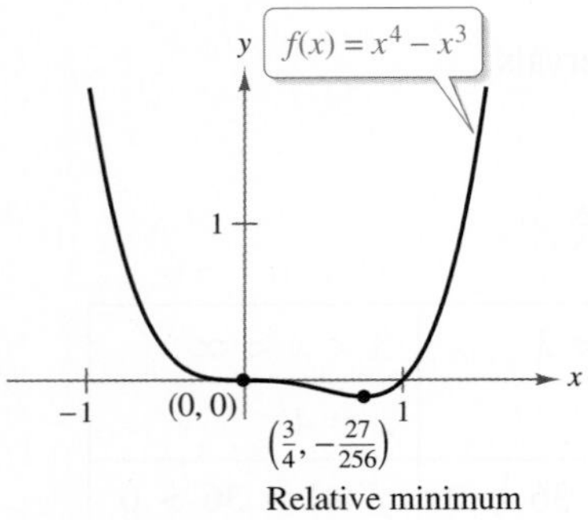

FIGURE 3.14

By the First-Derivative Test, it follows that f has a relative minimum at $x = \frac{3}{4}$, as shown in Figure 3.14. The relative minimum is

$$f\left(\frac{3}{4}\right) = -\frac{27}{256}.$$

Note that the critical number $x = 0$ does not yield a relative extremum because $f'(x)$ is negative on both sides of $x = 0$.

✓ Checkpoint 2

Find all relative extrema of $f(x) = x^4 - 4x^3$. ■

Example 3 Finding Relative Extrema

Find all relative extrema of the function

$$f(x) = 2x - 3x^{2/3}.$$

SOLUTION From the derivative of the function

$$f'(x) = 2 - \frac{2}{x^{1/3}} = \frac{2(x^{1/3} - 1)}{x^{1/3}}$$

you can see that $f'(1) = 0$ and f' is undefined at $x = 0$. So, the function has two critical numbers: $x = 1$ and $x = 0$. These numbers produce the test intervals $(-\infty, 0)$, $(0, 1)$, and $(1, \infty)$. By testing these intervals, you can conclude that f has a relative maximum at $(0, 0)$ and a relative minimum at $(1, -1)$, as shown in Figure 3.15.

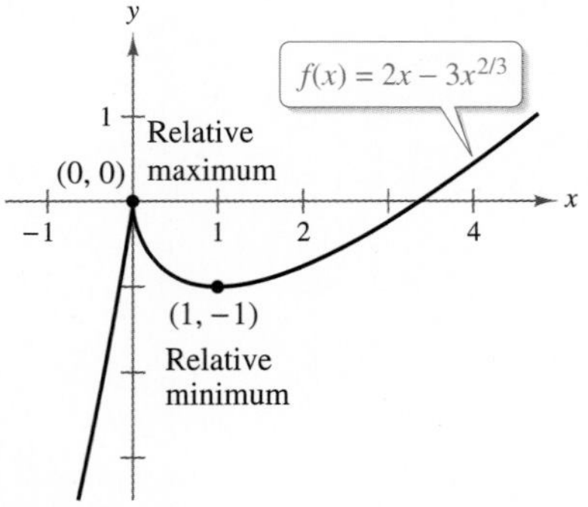

FIGURE 3.15

✓ Checkpoint 3

Find all relative extrema of $f(x) = 3x^{2/3} - 2x$. ■

Absolute Extrema

The terms *relative minimum* and *relative maximum* describe the *local* behavior of a function. To describe the *global* behavior of the function on an entire interval, you can use the terms **absolute maximum** and **absolute minimum.**

Definition of Absolute Extrema

Let f be defined on an interval I containing c.

1. $f(c)$ is an **absolute minimum of f** on I when $f(c) \leq f(x)$ for every x in I.
2. $f(c)$ is an **absolute maximum of f** on I when $f(c) \geq f(x)$ for every x in I.

The absolute minimum and absolute maximum values of a function on an interval are sometimes called simply the **minimum** and **maximum** of f on I.

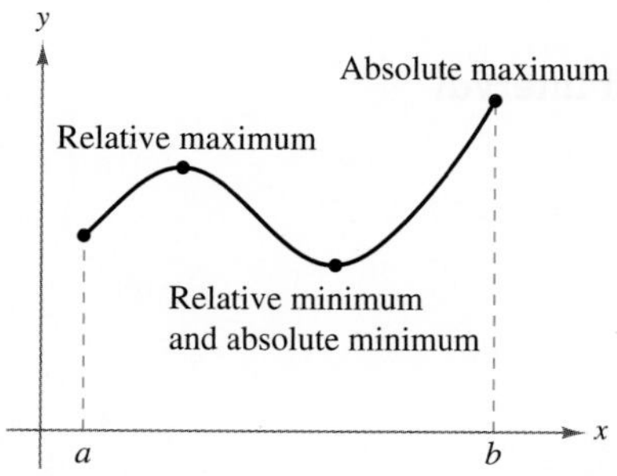

FIGURE 3.16

Be sure that you understand the distinction between relative extrema and absolute extrema. For instance, in Figure 3.16, the function has a relative minimum that also happens to be an absolute minimum on the interval $[a, b]$. The relative maximum of f, however, is not the absolute maximum on the interval $[a, b]$. The next theorem points out that if a continuous function has a closed interval as its domain, then it *must* have both an absolute minimum and an absolute maximum on the interval. From Figure 3.16, note that these extrema can occur at the endpoints of the interval.

Extreme Value Theorem

If f is continuous on a closed interval $[a, b]$, then f has both a minimum value and a maximum value on $[a, b]$.

Although a continuous function has just one minimum and one maximum value on a closed interval, either of these values can occur for more than one x-value. For instance, on the interval $[-3, 3]$, the function

$$f(x) = 9 - x^2$$

has a minimum value of zero at $x = -3$ *and* at $x = 3$, as shown in Figure 3.17.

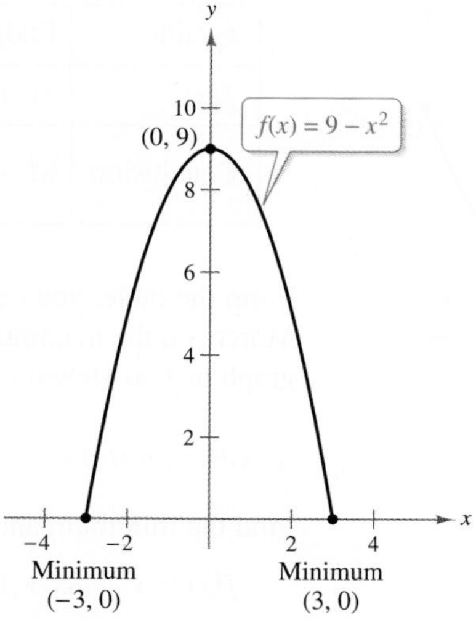

FIGURE 3.17

TECH TUTOR

A graphing utility can help you locate the extrema of a function on a closed interval. For instance, try using a graphing utility to confirm the results of Example 4. (Set the viewing window to $-1 \le x \le 6$ and $-8 \le y \le 4$.) Use the *trace* feature to check that the minimum y-value occurs at $x = 3$ and the maximum y-value occurs at $x = 0$.

When looking for extrema of a function on a *closed* interval, remember that you must consider the values of the function at the endpoints as well as at the critical numbers of the function. You can use the guidelines below to find extrema on a closed interval.

Guidelines for Finding Extrema on a Closed Interval

To find the extrema of a continuous function f on a closed interval $[a, b]$, use the following steps.

1. Find the critical numbers of f in the open interval (a, b).
2. Evaluate f at each of its critical numbers in (a, b).
3. Evaluate f at each endpoint, a and b.
4. The least of these values is the minimum, and the greatest is the maximum.

Example 4 Finding Extrema on a Closed Interval

Find the minimum and maximum values of

$$f(x) = x^2 - 6x + 2$$

on the interval $[0, 5]$.

SOLUTION Begin by differentiating the function.

$f(x) = x^2 - 6x + 2$ Write original function.

$f'(x) = 2x - 6$ Differentiate.

Next, find the critical numbers of f.

$2x - 6 = 0$ Set derivative equal to 0.

$2x = 6$ Add 6 to each side.

$x = 3$ Solve for x.

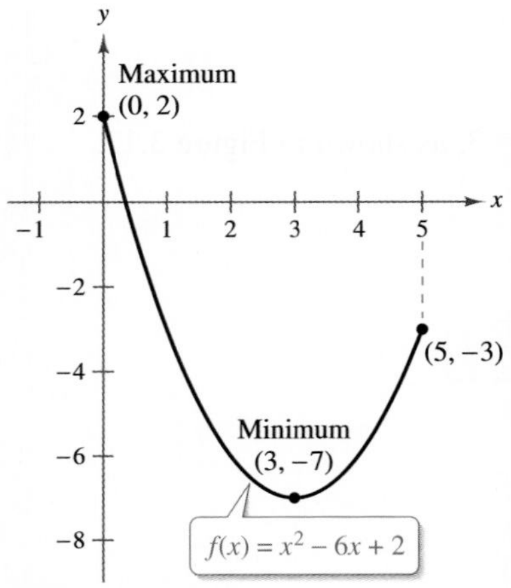

FIGURE 3.18

Because f' is defined for all x, you can conclude that the only critical number of f is $x = 3$. Because this number lies in the interval $[0, 5]$, you should evaluate f at this number *and* at the endpoints of the interval, as shown in the table.

x-value	Endpoint: $x = 0$	Critical number: $x = 3$	Endpoint: $x = 5$
$f(x)$	$f(0) = 2$	$f(3) = -7$	$f(5) = -3$
Conclusion	Maximum is 2.	Minimum is -7.	Neither maximum nor minimum

From the table, you can see that the minimum of f on the interval $[0, 5]$ is $f(3) = -7$. Moreover, the maximum of f on the interval $[0, 5]$ is $f(0) = 2$. This is confirmed by the graph of f, as shown in Figure 3.18.

✓Checkpoint 4

Find the minimum and maximum values of

$$f(x) = x^2 - 8x + 10$$

on the interval $[0, 7]$. Sketch the graph of f and label the minimum and maximum values. ■

Application

Finding the minimum and maximum values of a function is one of the most common applications of calculus.

Example 5 Finding the Maximum Profit

Recall the fast-food restaurant in Examples 7 and 8 in Section 2.3. The restaurant's profit function for hamburgers is given by

$$P = 2.44x - \frac{x^2}{20{,}000} - 5000, \quad 0 \le x \le 50{,}000.$$

Find the sales level that yields a maximum profit.

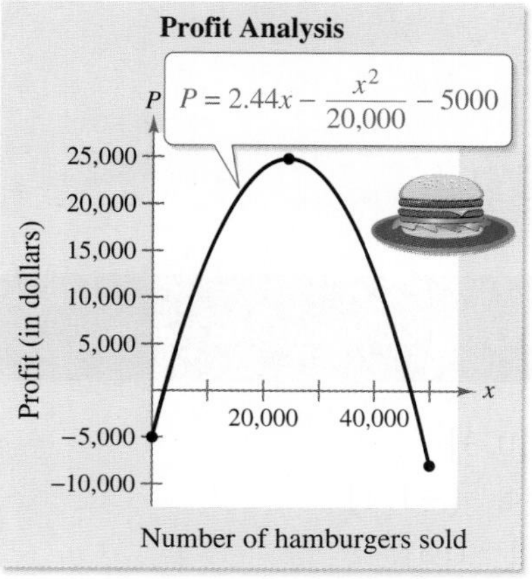

FIGURE 3.19

SOLUTION To begin, find an equation for marginal profit.

$$\frac{dP}{dx} = 2.44 - \frac{x}{10{,}000} \qquad \text{Find marginal profit.}$$

Next, set the marginal profit equal to zero and solve for x.

$$2.44 - \frac{x}{10{,}000} = 0 \qquad \text{Set marginal profit equal to 0.}$$

$$-\frac{x}{10{,}000} = -2.44 \qquad \text{Subtract 2.44 from each side.}$$

$$x = 24{,}400 \text{ hamburgers} \qquad \text{Critical number}$$

In Figure 3.19, you can see that the critical number $x = 24{,}400$ corresponds to the sales level that yields a maximum profit. To find the maximum profit, substitute $x = 24{,}400$ into the profit function.

$$P = 2.44x - \frac{x^2}{20{,}000} - 5000$$

$$= 2.44(24{,}400) - \frac{(24{,}400)^2}{20{,}000} - 5000$$

$$= \$24{,}768$$

✓Checkpoint 5

Verify the results of Example 5 by completing the table.

x (units)	24,000	24,200	24,300	24,400	24,500	24,600	24,800
P (profit)							

SUMMARIZE (Section 3.2)

1. State the First-Derivative Test *(page 178)*. For examples in which the First-Derivative Test is used, see Examples 1, 2, and 3.
2. State the guidelines for finding extrema on a closed interval *(page 182)*. For an example of finding the extrema of a function on a closed interval, see Example 4.
3. Describe a real-life example of how the First-Derivative Test can be used to find the sales level that yields a maximum profit for a company *(page 183, Example 5)*.

SKILLS WARM UP 3.2

The following warm-up exercises involve skills that were covered in earlier sections. You will use these skills in the exercise set for this section. For additional help, review Sections 2.2, 2.4, and 3.1.

In Exercises 1–6, solve the equation $f'(x) = 0$.

1. $f(x) = 4x^4 - 2x^2 + 1$
2. $f(x) = \frac{1}{3}x^3 - \frac{3}{2}x^2 - 10x$
3. $f(x) = 5x^{4/5} - 4x$
4. $f(x) = \frac{1}{2}x^2 - 3x^{5/3}$
5. $f(x) = \dfrac{x+4}{x^2+1}$
6. $f(x) = \dfrac{x-1}{x^2+4}$

In Exercises 7–10, use $g(x) = -x^5 - 2x^4 + 4x^3 + 2x - 1$ to determine the sign of the derivative.

7. $g'(-4)$
8. $g'(0)$
9. $g'(1)$
10. $g'(3)$

In Exercises 11 and 12, decide whether the function is increasing or decreasing on the given interval.

11. $f(x) = 2x^2 - 11x - 6, \quad (3, 6)$
12. $f(x) = x^3 + 2x^2 - 4x - 8, \quad (-2, 0)$

Exercises 3.2

See www.CalcChat.com for worked-out solutions to odd-numbered exercises.

Finding Relative Extrema In Exercises 1–12, find all relative extrema of the function. *See Examples 1, 2, and 3.*

1. $f(x) = -2x^2 + 4x + 3$
2. $f(x) = x^2 + 8x + 10$
3. $f(x) = x^2 - 6x$
4. $f(x) = -4x^2 + 4x + 1$
5. $f(x) = x^4 - 12x^3$
6. $g(x) = \frac{1}{5}x^5 - x$
7. $h(x) = -(x + 4)^3$
8. $h(x) = 2(x - 3)^3$
9. $f(x) = x^3 - 6x^2 + 15$
10. $f(x) = x^4 - 32x + 4$
11. $f(x) = 6x^{2/3} + 4x$
12. $f(x) = 3x - 36x^{1/3}$

Finding Relative Extrema In Exercises 13–18, use a graphing utility to find graphically all relative extrema of the function.

13. $f(x) = 2x - 6x^{2/3}$
14. $f(t) = (t - 1)^{1/3}$
15. $g(t) = t - \dfrac{1}{2t^2}$
16. $f(x) = x + \dfrac{1}{x}$
17. $f(x) = \dfrac{x}{x+1}$
18. $h(x) = \dfrac{6}{x^2+2}$

Finding Extrema on a Closed Interval In Exercises 19–30, find the absolute extrema of the function on the closed interval. Use a graphing utility to verify your results. *See Example 4.*

19. $f(x) = 2(3 - x), \quad [-1, 2]$
20. $f(x) = \frac{1}{3}(2x + 5), \quad [0, 5]$
21. $f(x) = 5 - 2x^2, \quad [0, 3]$
22. $f(x) = x^2 + 2x - 4, \quad [-1, 1]$
23. $f(x) = x^3 - 3x^2, \quad [-1, 3]$
24. $f(x) = x^3 - 12x, \quad [0, 4]$
25. $h(s) = \dfrac{1}{3-s}, \quad [0, 2]$
26. $h(t) = \dfrac{t}{t-2}, \quad [3, 5]$
27. $g(t) = \dfrac{t^2}{t^2+3}, \quad [-1, 1]$
28. $g(x) = 4\left(1 + \dfrac{1}{x} + \dfrac{1}{x^2}\right), \quad [-4, 5]$
29. $h(t) = (t - 1)^{2/3}, \quad [-7, 2]$
30. $g(x) = (x^2 - 4)^{2/3}, \quad [-6, 3]$

Determining Types of Extrema In Exercises 31–34, approximate the critical numbers of the function shown in the graph. Determine whether the function has a relative maximum, a relative minimum, an absolute maximum, an absolute minimum, or none of these at each critical number on the interval shown.

31.

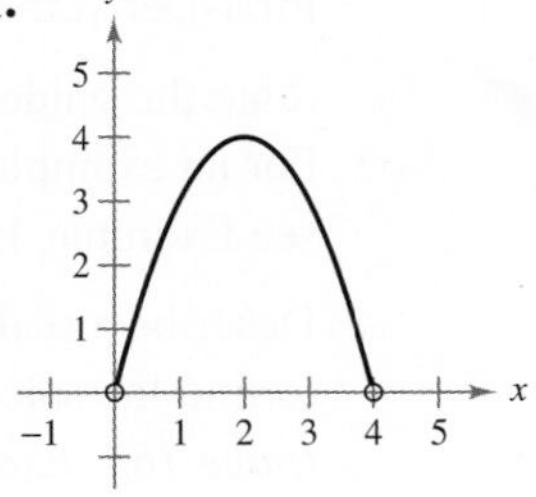

32.

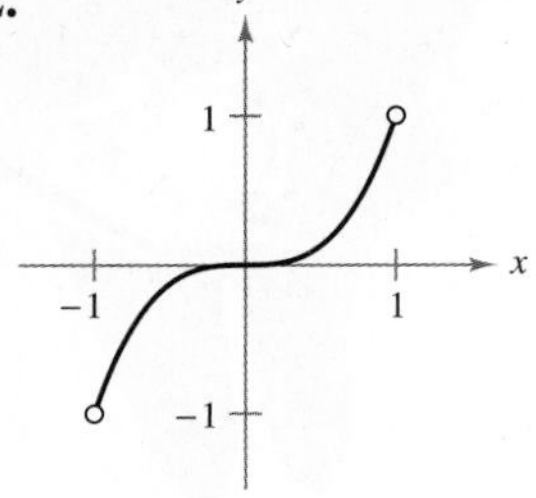

33.

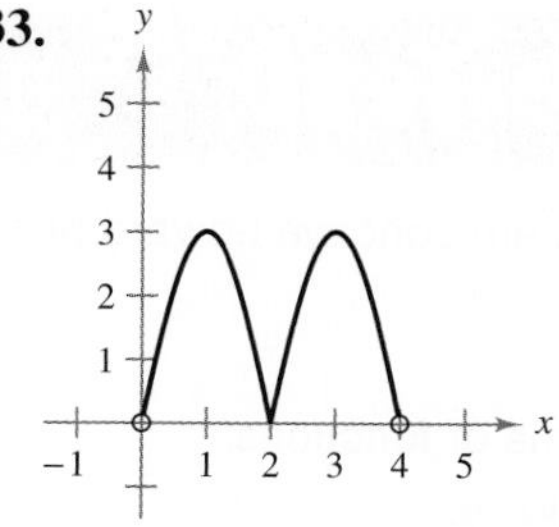

34.

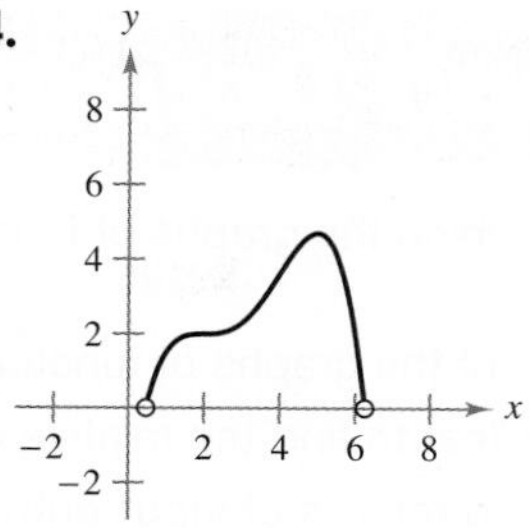

Finding Extrema on a Closed Interval In Exercises 35–38, use a graphing utility to find graphically the absolute extrema of the function on the closed interval.

35. $f(x) = 0.4x^3 - 1.8x^2 + x - 3, \quad [0, 5]$

36. $f(x) = 3.2x^5 + 5x^3 - 3.5x, \quad [0, 1]$

37. $f(x) = \frac{4}{3}x\sqrt{3 - x}, \quad [0, 3]$

38. $f(x) = 4\sqrt{x} - 2x + 1, \quad [0, 6]$

Finding Absolute Extrema In Exercises 39–42, find the absolute extrema of the function on the interval $[0, \infty)$.

39. $f(x) = x^2 + \dfrac{16}{x}$

40. $f(x) = \dfrac{8}{x + 1}$

41. $f(x) = \dfrac{2x}{x^2 + 4}$

42. $f(x) = 8 - \dfrac{4x}{x^2 + 1}$

Creating the Graph of a Function In Exercises 43 and 44, graph a function on the closed interval $[-2, 5]$ having the given characteristics. (There are many correct answers.)

43. Absolute maximum at $x = -2$

Absolute minimum at $x = 1$

Relative maximum at $x = 3$

44. Relative minimum at $x = -1$

Critical number at $x = 0$, but no extrema

Absolute maximum at $x = 2$

Absolute minimum at $x = 5$

45. Population The resident populations P (in millions) of the United States from 1790 through 2010 can be modeled by $P = 0.000006t^3 + 0.005t^2 + 0.14t + 4.6$, $-10 \le t \le 210$, where $t = 0$ corresponds to 1800. *(Source: U.S. Census Bureau)*

(a) Make a conjecture about the maximum and minimum populations of the United States from 1790 to 2010.

(b) Analytically find the maximum and minimum populations over the interval.

(c) Write a brief paragraph comparing your conjecture with your results in part (b).

46. HOW DO YOU SEE IT? The graph of the United States fertility rate shows the number of births per 1000 women in their lifetime according to the birth rate in that particular year. *(Source: U.S. National Center for Health Statistics)*

United States Fertility

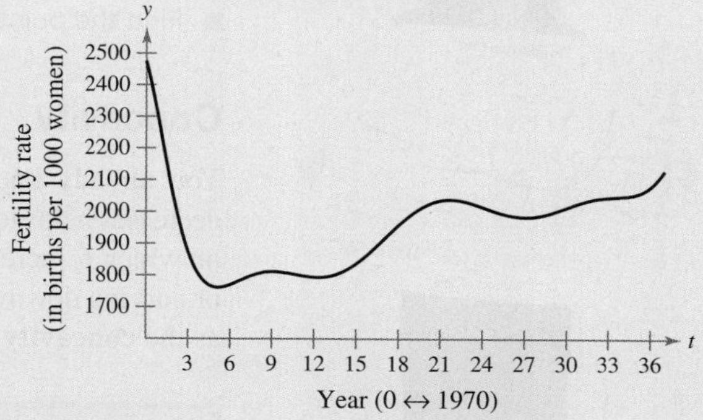

(a) Around what year was the fertility rate the highest, and to how many births per 1000 women did this rate correspond?

(b) During which time periods was the fertility rate increasing most rapidly? most slowly?

(c) During which time periods was the fertility rate decreasing most rapidly? most slowly?

(d) Give some possible real-life reasons for fluctuations in the fertility rate.

47. Cost A retailer determines that the cost C of ordering and storing x units of a product can be modeled by

$$C = 3x + \frac{20{,}000}{x}, \quad 0 < x \le 200.$$

The delivery truck can carry at most 200 units per order. Find the order size that will minimize the cost. Use a graphing utility to verify your result.

48. Medical Science Coughing forces the trachea (windpipe) to contract, which in turn affects the velocity of the air passing through the trachea. The velocity of the air during coughing can be modeled by

$$v = k(R - r)r^2, \quad 0 \le r < R$$

where k is a constant, R is the normal radius of the trachea, and r is the radius during coughing. What radius r will produce the maximum air velocity?

49. Profit When soft drinks are sold for \$1.00 per can at football games, approximately 6000 cans are sold. When the price is raised to \$1.20 per can, the quantity demanded drops to 5600. The initial cost is \$5000 and the cost per unit is \$0.50. Assuming that the demand function is linear, what number of units and what price will yield a maximum profit?

3.3 Concavity and the Second-Derivative Test

- Determine the intervals on which the graphs of functions are concave upward or concave downward.
- Find the points of inflection of the graphs of functions.
- Use the Second-Derivative Test to find the relative extrema of functions.
- Find the points of diminishing returns of input-output models.

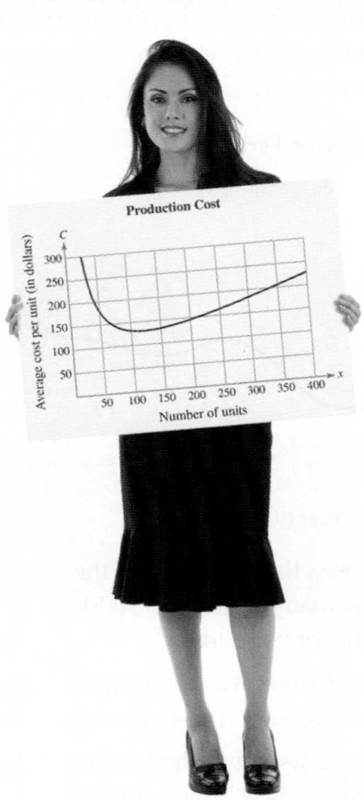

In Exercise 69 on page 194, you will use the Second-Derivative Test to find the production level that will minimize the average cost per unit.

Concavity

You already know that locating the intervals over which a function f increases or decreases helps to describe its graph. In this section, you will see that locating the intervals on which f' increases or decreases can determine where the graph of f is curving upward or curving downward. This property of curving upward or downward is defined formally as the **concavity** of the graph of the function.

Definition of Concavity

Let f be differentiable on an open interval I. The graph of f is

1. **concave upward** on I when f' is increasing on the interval.
2. **concave downward** on I when f' is decreasing on the interval.

In Figure 3.20, you can observe the following graphical interpretation of concavity.

1. A curve that is concave upward lies *above* its tangent line.
2. A curve that is concave downward lies *below* its tangent line.

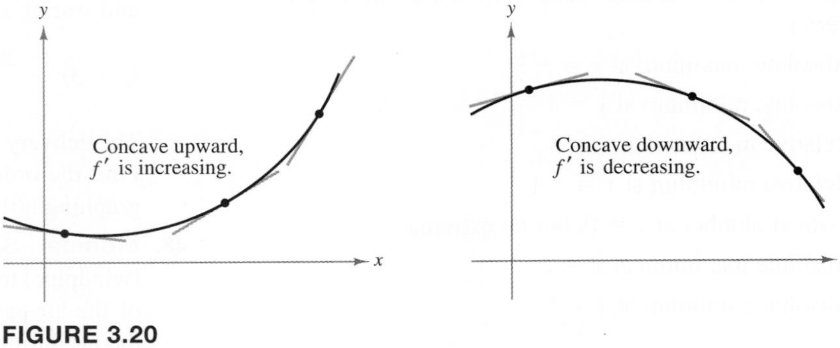

FIGURE 3.20

To find the open intervals on which the graph of a function is concave upward or concave downward, you can use the second derivative of the function as follows.

Test for Concavity

Let f be a function whose second derivative exists on an open interval I.

1. If $f''(x) > 0$ for all x in I, then the graph of f is concave upward on I.
2. If $f''(x) < 0$ for all x in I, then the graph of f is concave downward on I.

Example 1 **Determining Concavity**

a. The graph of the function

$$f(x) = x^2 \qquad \text{Original function}$$

is concave upward on the entire real number line because its second derivative

$$f''(x) = 2 \qquad \text{Second derivative}$$

is positive for all x. (See Figure 3.21.)

b. The graph of the function

$$f(x) = \sqrt{x} \qquad \text{Original function}$$

is concave downward for $x > 0$ because its second derivative

$$f''(x) = -\frac{1}{4}x^{-3/2} \qquad \text{Second derivative}$$

is negative for all $x > 0$. (See Figure 3.22.)

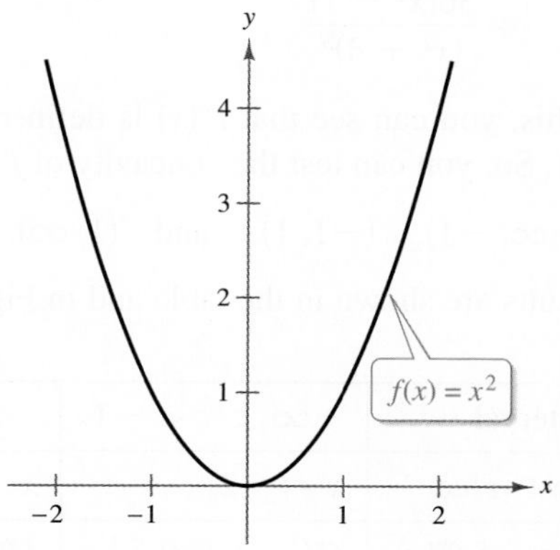

Concave Upward
FIGURE 3.21

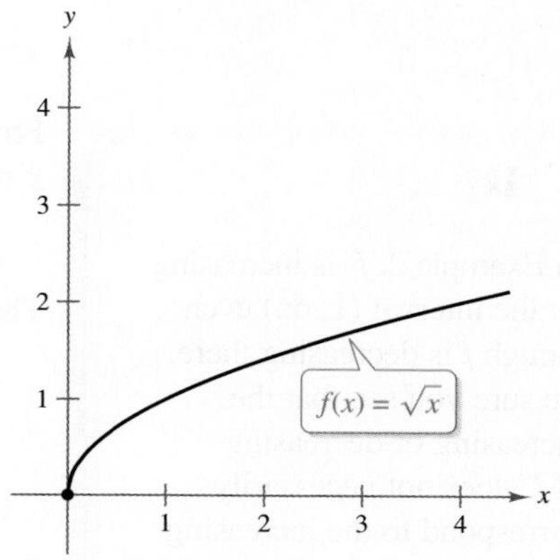

Concave Downward
FIGURE 3.22

✓ Checkpoint 1

Find the second derivative of f and discuss the concavity of its graph.

a. $f(x) = -2x^2$

b. $f(x) = -2\sqrt{x}$ ■

For a *continuous* function f, you can find the open intervals on which the graph of f is concave upward and concave downward as follows. [When there are x-values at which the function is not continuous, these values should be used, along with the points at which $f''(x) = 0$ or $f''(x)$ is undefined, to form the test intervals.]

Guidelines for Applying the Concavity Test

1. Locate the x-values at which $f''(x) = 0$ or $f''(x)$ is undefined.
2. Use these x-values to determine the test intervals.
3. Test the sign of $f''(x)$ in each test interval.

ALGEBRA TUTOR

For help on the algebra in Example 2, see Example 1(a) in the *Chapter 3 Algebra Tutor*, on page 242.

Example 2 Applying the Test for Concavity

Determine the open intervals on which the graph of

$$f(x) = \frac{6}{x^2 + 3}$$

is concave upward or concave downward.

SOLUTION Begin by finding the second derivative of f.

$$f(x) = 6(x^2 + 3)^{-1} \qquad \text{Rewrite original function.}$$

$$f'(x) = 6(-1)(x^2 + 3)^{-2}(2x) \qquad \text{Chain Rule}$$

$$= \frac{-12x}{(x^2 + 3)^2} \qquad \text{Simplify.}$$

$$f''(x) = \frac{(x^2 + 3)^2(-12) - (-12x)(2)(x^2 + 3)(2x)}{(x^2 + 3)^4} \qquad \text{Quotient Rule}$$

$$= \frac{-12(x^2 + 3) + 48x^2}{(x^2 + 3)^3} \qquad \text{Simplify.}$$

$$= \frac{36(x^2 - 1)}{(x^2 + 3)^3} \qquad \text{Simplify.}$$

From this, you can see that $f''(x)$ is defined for all real numbers and $f''(x) = 0$ when $x = \pm 1$. So, you can test the concavity of f by testing the intervals

$$(-\infty, -1), \quad (-1, 1), \quad \text{and} \quad (1, \infty). \qquad \text{Test intervals}$$

The results are shown in the table and in Figure 3.23.

Interval	$-\infty < x < -1$	$-1 < x < 1$	$1 < x < \infty$
Test value	$x = -2$	$x = 0$	$x = 2$
Sign of $f''(x)$	$f''(-2) > 0$	$f''(0) < 0$	$f''(2) > 0$
Conclusion	Concave upward	Concave downward	Concave upward

STUDY TIP

In Example 2, f' is increasing on the interval $(1, \infty)$ even though f is decreasing there. Be sure you see that the increasing or decreasing of f' does not necessarily correspond to the increasing or decreasing of f.

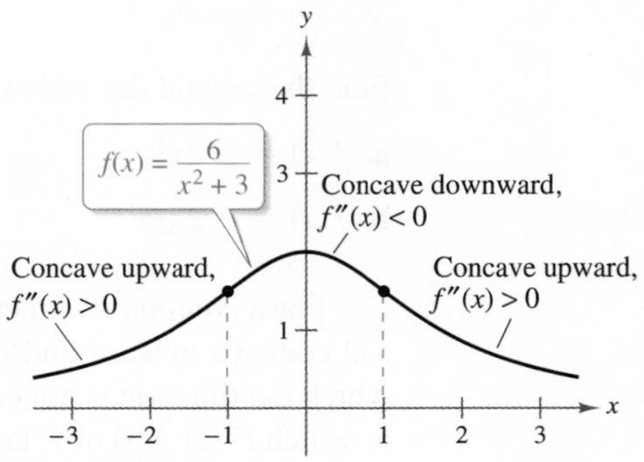

FIGURE 3.23

✓ Checkpoint 2

Determine the intervals on which the graph of

$$f(x) = \frac{12}{x^2 + 4}$$

is concave upward or concave downward.

Points of Inflection

If the tangent line to a graph exists at a point at which the concavity changes, then the point is a **point of inflection.** Three examples of inflection points are shown in Figure 3.24. (Note that the third graph has a vertical tangent line at its point of inflection.)

STUDY TIP

As shown in Figure 3.24, a graph crosses its tangent line at a point of inflection.

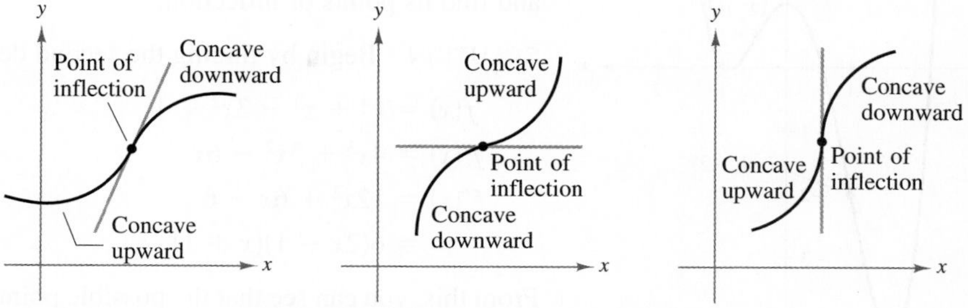

The graph *crosses* its tangent line at a point of inflection.

FIGURE 3.24

Definition of Point of Inflection

If the graph of a continuous function has a tangent line at a point where its concavity changes from upward to downward (or downward to upward), then the point is a **point of inflection.**

Because a point of inflection occurs where the concavity of a graph changes, it must be true that at such points the sign of f'' changes. So, to locate possible points of inflection, you need to determine only the values of x for which $f''(x) = 0$ or for which $f''(x)$ does not exist. This parallels the procedure for locating the relative extrema of f by determining the critical numbers of f.

Property of Points of Inflection

If $(c, f(c))$ is a point of inflection of the graph of f, then either $f''(c) = 0$ or $f''(c)$ is undefined.

Example 3 Finding a Point of Inflection

Discuss the concavity of the graph of $f(x) = 2x^3 + 1$ and find its point of inflection.

SOLUTION Differentiating twice produces the following.

$$f(x) = 2x^3 + 1 \qquad \text{Write original function.}$$
$$f'(x) = 6x^2 \qquad \text{Find first derivative.}$$
$$f''(x) = 12x \qquad \text{Find second derivative.}$$

Setting $f''(x) = 0$, you can determine that the only possible point of inflection occurs at $x = 0$. After testing the intervals $(-\infty, 0)$ and $(0, \infty)$, you can determine that the graph is concave downward on $(-\infty, 0)$ and concave upward on $(0, \infty)$. Because the concavity changes at $x = 0$, you can conclude that the graph of f has a point of inflection at $(0, 1)$, as shown in Figure 3.25.

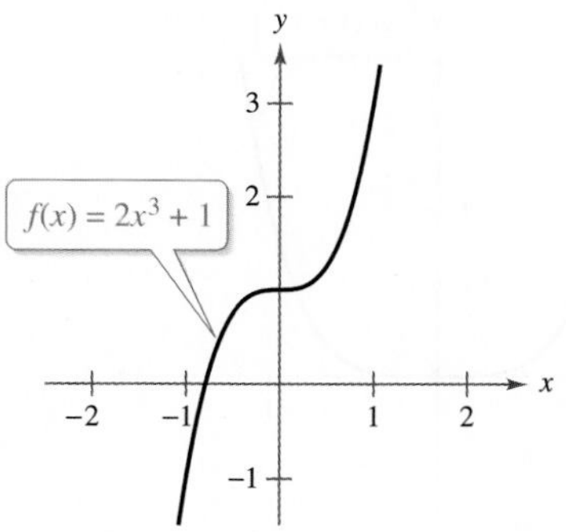

FIGURE 3.25

Checkpoint 3

Discuss the concavity of the graph of $f(x) = -x^3$ and find its point of inflection.

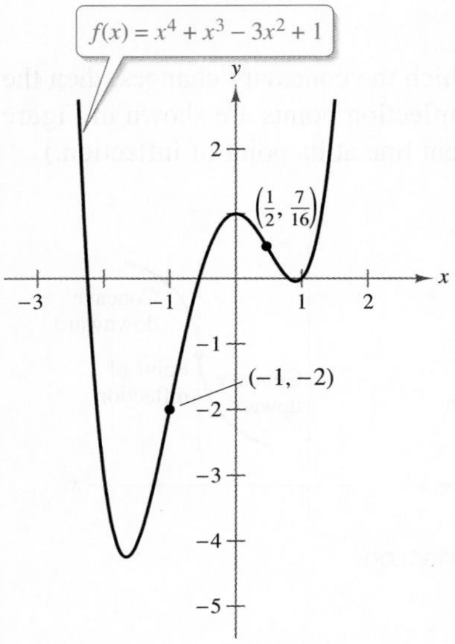

Two Points of Inflection
FIGURE 3.26

Example 4 Finding Points of Inflection

Discuss the concavity of the graph of

$$f(x) = x^4 + x^3 - 3x^2 + 1$$

and find its points of inflection.

SOLUTION Begin by finding the second derivative of f.

$f(x) = x^4 + x^3 - 3x^2 + 1$	Write original function.
$f'(x) = 4x^3 + 3x^2 - 6x$	Find first derivative.
$f''(x) = 12x^2 + 6x - 6$	Find second derivative.
$= 6(2x - 1)(x + 1)$	Factor.

From this, you can see that the possible points of inflection occur at $x = \frac{1}{2}$ and $x = -1$. After testing the intervals $(-\infty, -1)$, $\left(-1, \frac{1}{2}\right)$, and $\left(\frac{1}{2}, \infty\right)$, you can determine that the graph is concave upward on $(-\infty, -1)$, concave downward on $\left(-1, \frac{1}{2}\right)$, and concave upward on $\left(\frac{1}{2}, \infty\right)$. Because the concavity changes at $x = -1$ and $x = \frac{1}{2}$, you can conclude that the graph of f has points of inflection at these x-values, as shown in Figure 3.26. The points of inflection are

$$(-1, -2) \quad \text{and} \quad \left(\frac{1}{2}, \frac{7}{16}\right).$$

✓Checkpoint 4

Discuss the concavity of the graph of

$$f(x) = x^4 - 2x^3 + 1$$

and find its points of inflection. ■

It is possible for the second derivative to be zero at a point that is *not* a point of inflection. For example, compare the graphs of

$$f(x) = x^3 \quad \text{and} \quad g(x) = x^4$$

as shown in Figure 3.27. Both second derivatives are zero when $x = 0$, but only the graph of f has a point of inflection at $x = 0$. This shows that before concluding that a point of inflection exists at a value of x for which $f''(x) = 0$, you must test to be certain that the concavity actually changes at that point.

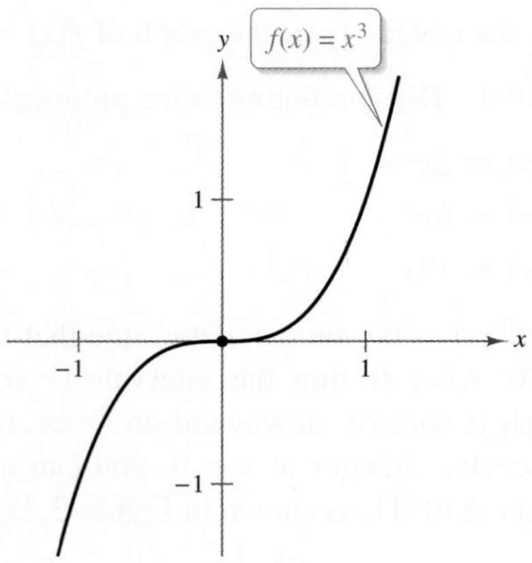

$f''(0) = 0$, and $(0, 0)$ is a point of inflection.

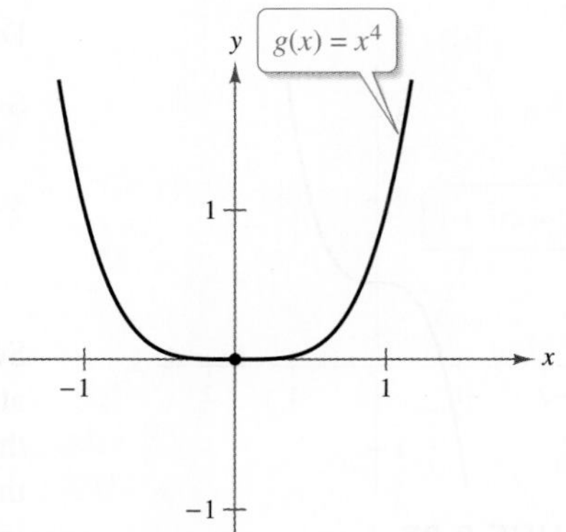

$g''(0) = 0$, but $(0, 0)$ is not a point of inflection.

FIGURE 3.27

The Second-Derivative Test

The second derivative can be used to perform a simple test for relative minima and relative maxima. If f is a function such that $f'(c) = 0$ and the graph of f is concave upward at $x = c$, then $f(c)$ is a relative minimum of f. Similarly, if f is a function such that $f'(c) = 0$ and the graph of f is concave downward at $x = c$, then $f(c)$ is a relative maximum of f, as shown in Figure 3.28.

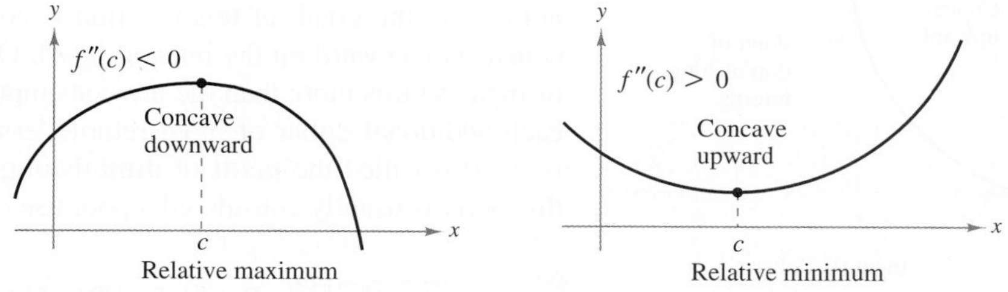

FIGURE 3.28

Second-Derivative Test

Let $f'(c) = 0$, and let f'' exist on an open interval containing c.

1. If $f''(c) > 0$, then $f(c)$ is a relative minimum.
2. If $f''(c) < 0$, then $f(c)$ is a relative maximum.
3. If $f''(c) = 0$, then the test fails. In such cases, you can use the First-Derivative Test to determine whether $f(c)$ is a relative minimum, a relative maximum, or neither.

Example 5 Using the Second-Derivative Test

Find the relative extrema of $f(x) = -3x^5 + 5x^3$.

SOLUTION Begin by finding the first derivative of f.

$$f'(x) = -15x^4 + 15x^2 = 15x^2(1 - x^2)$$

From this derivative, you can see that $x = 0$, $x = -1$, and $x = 1$ are the only critical numbers of f. Using the second derivative

$$f''(x) = -60x^3 + 30x = 30x(1 - 2x^2)$$

you can apply the Second-Derivative Test, as shown.

Point	$(-1, -2)$	$(0, 0)$	$(1, 2)$
Sign of $f''(x)$	$f''(-1) > 0$	$f''(0) = 0$	$f''(1) < 0$
Conclusion	Relative minimum	Test fails.	Relative maximum

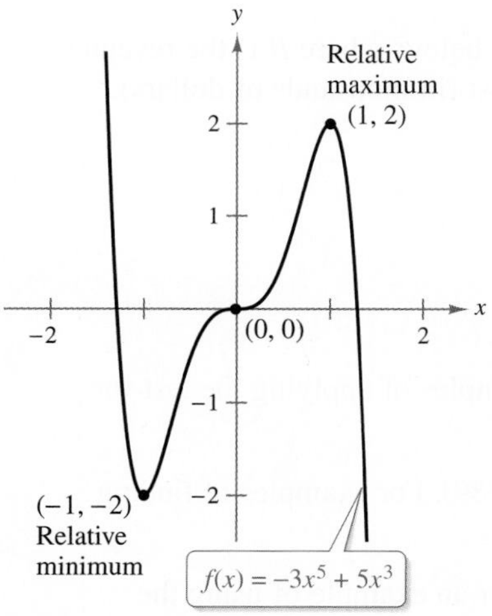

FIGURE 3.29

Because the Second-Derivative Test fails at $(0, 0)$, you can use the First-Derivative Test and observe that f is positive on both sides of $x = 0$. So, $(0, 0)$ is neither a relative minimum nor a relative maximum. A test for concavity would show that $(0, 0)$ is a point of inflection. The graph of f is shown in Figure 3.29.

✓Checkpoint 5

Find all relative extrema of $f(x) = x^4 - 4x^3 + 1$.

Extended Application: Diminishing Returns

In economics, the notion of concavity is related to the concept of **diminishing returns.** Consider a function

$$\underset{\text{Output}}{y} = f(\underset{\text{Input}}{x})$$

where x measures input (in dollars) and y measures output (in dollars). In Figure 3.30, notice that the graph of this function is concave upward on the interval (a, c) and is concave downward on the interval (c, b). On the interval (a, c), each additional dollar of input returns more than the previous input dollar. By contrast, on the interval (c, b), each additional dollar of input returns less than the previous input dollar. The point $(c, f(c))$ is called the **point of diminishing returns.** An increased investment beyond this point is usually considered a poor use of capital.

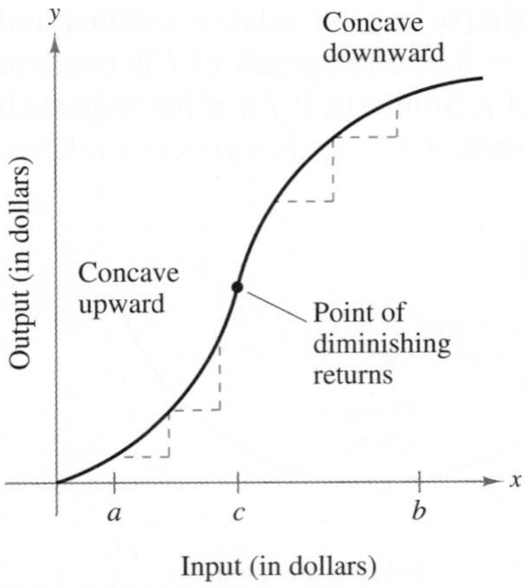

FIGURE 3.30

Example 6 Exploring Diminishing Returns

By increasing its advertising cost x (in thousands of dollars) for a product, a company discovers that it can increase the sales y (in thousands of dollars) according to the model

$$y = -\frac{1}{10}x^3 + 6x^2 + 400, \quad 0 \le x \le 40.$$

Find the point of diminishing returns for this product.

SOLUTION Begin by finding the first and second derivatives.

$$y' = 12x - \frac{3x^2}{10} \qquad \text{First derivative}$$

$$y'' = 12 - \frac{3x}{5} \qquad \text{Second derivative}$$

The second derivative is zero only when $x = 20$. By testing for concavity on the intervals $(0, 20)$ and $(20, 40)$, you can conclude that the graph has a point of diminishing returns when $x = 20$, as shown in Figure 3.31. So, the point of diminishing returns for this product occurs when $20,000 is spent on advertising.

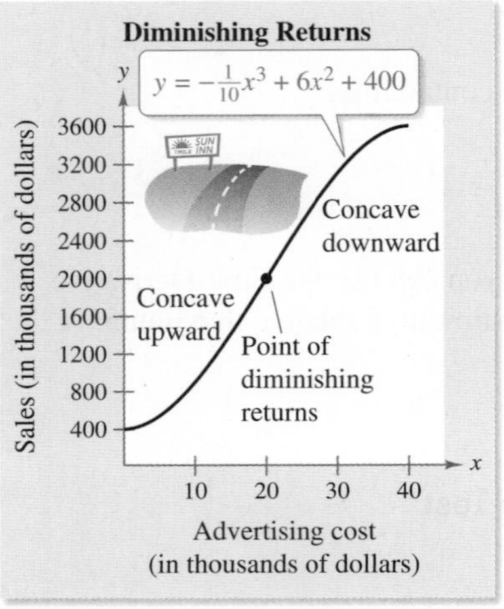

FIGURE 3.31

✓ Checkpoint 6

Find the point of diminishing returns for the model below, where R is the revenue (in thousands of dollars) and x is the advertising cost (in thousands of dollars).

$$R = \frac{1}{20{,}000}(450x^2 - x^3), \quad 0 \le x \le 300$$

SUMMARIZE (Section 3.3)

1. State the test for concavity *(page 186)*. For examples of applying the test for concavity, see Examples 1 and 2.
2. State the definition of point of inflection *(page 189)*. For examples of finding points of inflection, see Examples 3 and 4.
3. State the Second-Derivative Test *(page 191)*. For an example of using the Second-Derivative Test, see Example 5.
4. Describe a real-life example of how the second derivative can be used to find the point of diminishing returns for a product *(page 192, Example 6)*.

SKILLS WARM UP 3.3

The following warm-up exercises involve skills that were covered in earlier sections. You will use these skills in the exercise set for this section. For additional help, review Sections 2.2, 2.4, 2.5, 2.6, and 3.1.

In Exercises 1–6, find the second derivative of the function.

1. $f(x) = 4x^4 - 9x^3 + 5x - 1$

2. $g(s) = (s^2 - 1)(s^2 - 3s + 2)$

3. $g(x) = (x^2 + 1)^4$

4. $f(x) = (x - 3)^{4/3}$

5. $h(x) = \dfrac{4x + 3}{5x - 1}$

6. $f(x) = \dfrac{2x - 1}{3x + 2}$

In Exercises 7–10, find the critical numbers of the function.

7. $f(x) = 5x^3 - 5x + 11$

8. $f(x) = x^4 - 4x^3 - 10$

9. $g(t) = \dfrac{16 + t^2}{t}$

10. $h(x) = \dfrac{x^4 - 50x^2}{8}$

Exercises 3.3

See www.CalcChat.com for worked-out solutions to odd-numbered exercises.

Using Graphs In Exercises 1–4, state the signs of $f'(x)$ and $f''(x)$ on the interval (0, 2).

1.

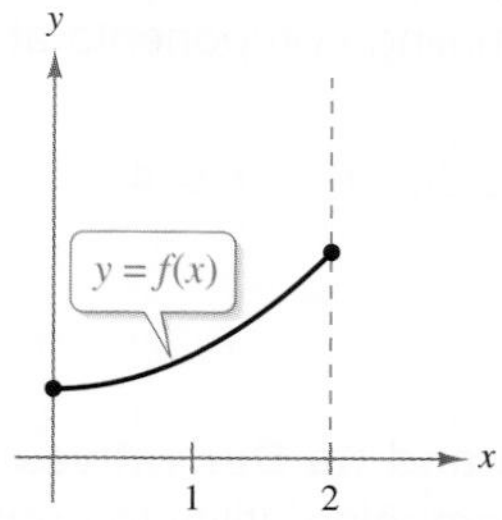

2.

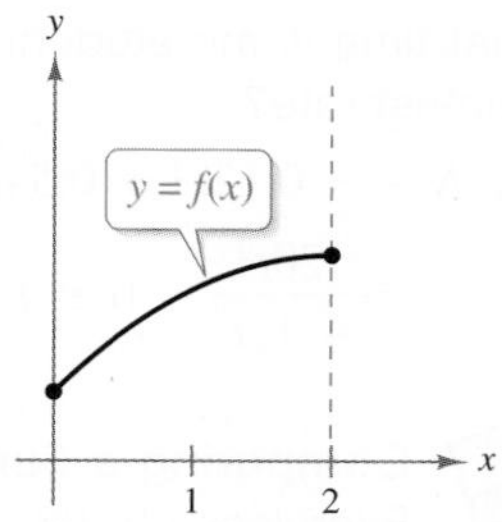

3.

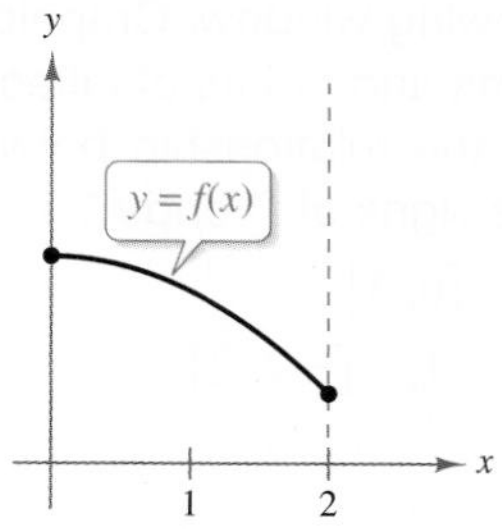

4.

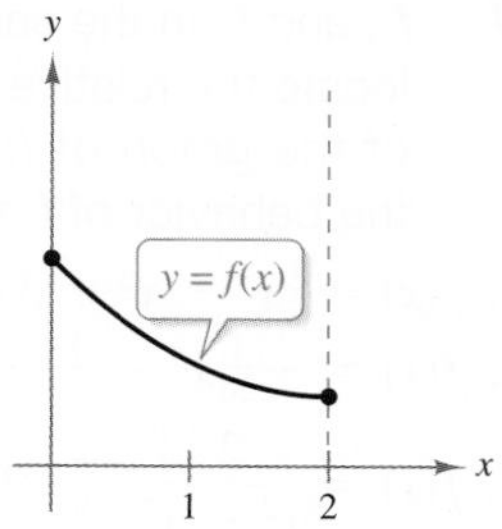

Applying the Test for Concavity In Exercises 5–12, determine the open intervals on which the graph of the function is concave upward or concave downward. *See Examples 1 and 2.*

5. $f(x) = -3x^2$

6. $f(x) = -5\sqrt{x}$

7. $y = -x^3 + 3x^2 - 2$

8. $y = -x^3 + 6x^2 - 9x - 1$

9. $f(x) = \dfrac{x^2 - 1}{2x + 1}$

10. $f(x) = \dfrac{x^2 + 4}{4 - x^2}$

11. $f(x) = \dfrac{24}{x^2 + 12}$

12. $f(x) = \dfrac{x^2}{x^2 + 1}$

Finding Points of Inflection In Exercises 13–20, discuss the concavity of the graph of the function and find the points of inflection. *See Examples 3 and 4.*

13. $f(x) = x^3 - 9x^2 + 24x - 18$

14. $f(x) = -4x^3 - 8x^2 + 32$

15. $f(x) = 2x^3 - 3x^2 - 12x + 5$

16. $f(x) = \frac{1}{2}x^4 + 2x^3$

17. $g(x) = 2x^4 - 8x^3 + 12x^2 + 12x$

18. $g(x) = x^5 + 5x^4 - 40x^2$

19. $f(x) = x(6 - x)^2$

20. $f(x) = (x - 1)^3(x - 5)$

Using the Second-Derivative Test In Exercises 21–34, find all relative extrema of the function. Use the Second-Derivative Test when applicable. *See Example 5.*

21. $f(x) = 6x - x^2$

22. $f(x) = 9x^2 - x^3$

23. $f(x) = x^3 - 5x^2 + 7x$

24. $f(x) = x^4 + 8x^3 - 6$

25. $f(x) = x^{2/3} - 3$

26. $f(x) = x + \dfrac{4}{x}$

27. $f(x) = \sqrt{x^2 + 1}$

28. $f(x) = \sqrt{2x^2 + 6}$

29. $f(x) = \sqrt{9 - x^2}$

30. $f(x) = \sqrt{4 - x^2}$

31. $f(x) = \dfrac{8}{x^2 + 2}$

32. $f(x) = \dfrac{x}{x^2 + 16}$

33. $f(x) = \dfrac{x}{x - 1}$

34. $f(x) = \dfrac{x}{x^2 - 1}$

Finding Relative Extrema In Exercises 35–38, use a graphing utility to estimate graphically all relative extrema of the function.

35. $f(x) = 5 + 3x^2 - x^3$ **36.** $f(x) = x^3 - 6x^2 + 7$

37. $f(x) = \frac{1}{2}x^4 - \frac{1}{3}x^3 - \frac{1}{2}x^2$

38. $f(x) = -\frac{1}{3}x^5 - \frac{1}{2}x^4 + x$

Using the Second-Derivative Test In Exercises 39–50, find all relative extrema and points of inflection. Then use a graphing utility to graph the function.

39. $f(x) = x^3 - 12x$ **40.** $f(x) = x^3 - 3x$

41. $g(x) = \sqrt{x} + \dfrac{4}{\sqrt{x}}$ **42.** $f(x) = x^3 - \frac{3}{2}x^2 - 6x$

43. $f(x) = \frac{1}{4}x^4 - 2x^2$ **44.** $f(x) = 2x^4 - 8x + 3$

45. $g(x) = (x - 2)(x + 1)^2$ **46.** $g(x) = (x - 6)(x + 2)^3$

47. $g(x) = x\sqrt{x + 3}$ **48.** $g(x) = x\sqrt{9 - x}$

49. $f(x) = \dfrac{4}{1 + x^2}$

50. $f(x) = \dfrac{2}{x^2 - 1}$

Creating a Function In Exercises 51–54, sketch a graph of a function f having the given characteristics. (There are many correct answers.)

51. $f(2) = f(4) = 0$
$f'(x) < 0$ if $x < 3$
$f'(3) = 0$
$f'(x) > 0$ if $x > 3$
$f''(x) > 0$

52. $f(2) = f(4) = 0$
$f'(x) > 0$ if $x < 3$
$f'(3)$ is undefined.
$f'(x) < 0$ if $x > 3$
$f''(x) > 0, x \neq 3$

53. $f(0) = f(2) = 0$
$f'(x) > 0$ if $x < 1$
$f'(1) = 0$
$f'(x) < 0$ if $x > 1$
$f''(x) < 0$

54. $f(0) = f(2) = 0$
$f'(x) < 0$ if $x < 1$
$f'(1) = 0$
$f'(x) > 0$ if $x > 1$
$f''(x) > 0$

Using Graphs In Exercises 55 and 56, use the graph to sketch the graph of f'. Find the intervals on which (a) $f'(x)$ is positive, (b) $f'(x)$ is negative, (c) f' is increasing, and (d) f' is decreasing. For each of these intervals, describe the corresponding behavior of f.

55.

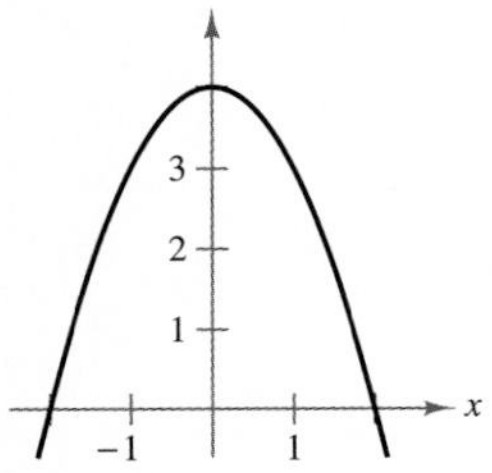

56.

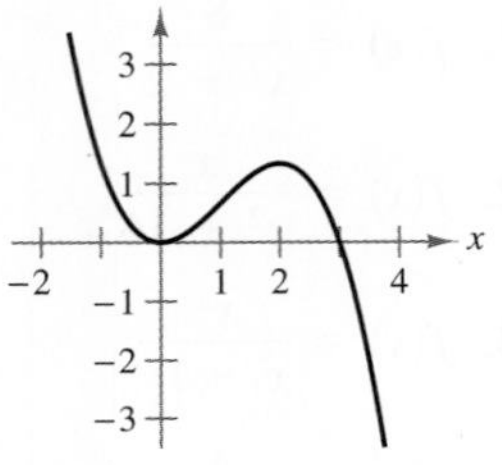

Using the First Derivative In Exercises 57–60, you are given f'. Find the intervals on which (a) $f'(x)$ is increasing or decreasing and (b) the graph of f is concave upward or concave downward. (c) Find the x-values of the relative extrema and inflection points of f.

57. $f'(x) = 2x + 5$ **58.** $f'(x) = 3x^2 - 2$

59. $f'(x) = -x^2 + 2x - 1$ **60.** $f'(x) = x^2 + x - 6$

Point of Diminishing Returns In Exercises 61 and 62, find the point of diminishing returns for the function. For each function, R is the revenue (in thousands of dollars) and x is the amount spent (in thousands of dollars) on advertising. Use a graphing utility to verify your results. *See Example 6.*

61. $R = \dfrac{1}{50{,}000}(600x^2 - x^3), \quad 0 \le x \le 400$

62. $R = -\dfrac{4}{9}x^3 + 4x^2 + 12, \quad 0 \le x \le 5$

Productivity In Exercises 63 and 64, consider a college student who works from 7 P.M. to 11 P.M. assembling mechanical components. The number N of components assembled after t hours is given by the function. At what time is the student assembling components at the greatest rate?

63. $N = -0.12t^3 + 0.54t^2 + 8.22t, \quad 0 \le t \le 4$

64. $N = \dfrac{20t^2}{4 + t^2}, \quad 0 \le t \le 4$

Comparing a Function and Its Derivatives In Exercises 65–68, use a graphing utility to graph f, f', and f'' in the same viewing window. Graphically locate the relative extrema and points of inflection of the graph of f. State the relationship between the behavior of f and the signs of f' and f''.

65. $f(x) = \frac{1}{2}x^3 - x^2 + 3x - 5, \quad [0, 3]$

66. $f(x) = -\frac{1}{20}x^5 - \frac{1}{12}x^2 - \frac{1}{3}x + 1, \quad [-2, 2]$

67. $f(x) = \dfrac{2}{x^2 + 1}, \quad [-3, 3]$

68. $f(x) = \dfrac{x^2}{x^2 + 1}, \quad [-3, 3]$

69. Average Cost A manufacturer has determined that the total cost C (in dollars) of operating a factory is $C = 0.5x^2 + 10x + 7200$, where x is the number of units produced. At what level of production will the average cost per unit be minimized? (The average cost per unit is C/x.)

70. Inventory Cost The cost C (in dollars) of ordering and storing x units is $C = 2x + 300{,}000/x$. What order size will produce a minimum cost?

71. Home Sales The median sales price p (in thousands of dollars) of new single-family houses sold in the United States from 1995 through 2009 can be modeled by $p = -0.02812t^4 + 1.177t^3 - 17.02t^2 + 108.7t - 115$, for $5 \le t \le 19$, where t is the year, with $t = 5$ corresponding to 1995. *(Source: U.S. Census Bureau)*

(a) Use a graphing utility to graph the model on the interval $[5, 19]$.

(b) Use the graph in part (a) to estimate the year corresponding to the absolute minimum sales price.

(c) Use the graph in part (a) to estimate the year corresponding to the absolute maximum sales price.

(d) During approximately which year was the rate of increase of the sales price the greatest? the least?

72. HOW DO YOU SEE IT? The graph shows the Dow Jones Industrial Average y on Black Monday, October 19, 1987, where $t = 0$ corresponds to 9:30 A.M., when the market opens, and $t = 6.5$ corresponds to 4 P.M., the closing time. *(Source: Wall Street Journal)*

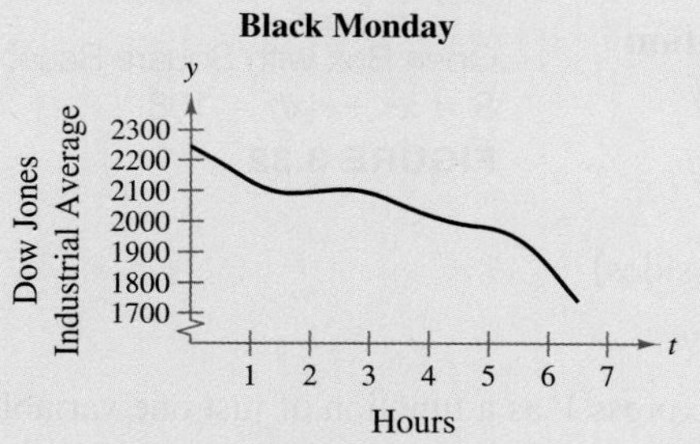

(a) Estimate the relative extrema and absolute extrema of the graph. Interpret your results in the context of the problem.

(b) Estimate the point of inflection of the graph on the interval $[1, 3]$. Interpret your result in the context of the problem.

73. Veteran Benefits From 1995 through 2008, the number v of veterans (in thousands) receiving compensation and pension benefits for service in the armed forces can be modeled by

$$v = -0.0687t^4 + 3.169t^3 - 45t^2 + 230.6t + 2950$$

for $5 \le t \le 18$, where t is the year, with $t = 5$ corresponding to 1995. *(Source: U.S. Department of Veterans Affairs)*

(a) Use a graphing utility to graph the model on the interval $[5, 18]$.

(b) Use the second derivative to determine the concavity of v.

(c) Find the point(s) of inflection of the graph of v.

(d) Interpret the meaning of the inflection point(s) of the graph of v.

74. Think About It Let S represent monthly sales of a new digital audio player. Write a statement describing S' and S'' for each of the following.

(a) The rate of change of sales is increasing.

(b) Sales are increasing, but at a greater rate.

(c) The rate of change of sales is steady.

(d) Sales are steady.

(e) Sales are declining, but at a lower rate.

(f) Sales have bottomed out and have begun to rise.

Business Capsule

While working in New York City in 2004, Matthew Corrin noticed an abundance of fresh food bars and decided that if someone could successfully brand one, that person could create the “Starbucks of the fresh food business.” With $275,000, he opened his first Freshii store in Toronto in 2005, and soon began developments for more. By the end of 2011, he will have 80–90 locations worldwide, with agreements signed for 400 additional stores in 25 cities and four countries. The mission of this eco-friendly chain is “to eliminate the excuse of people not eating fresh food because it isn’t convenient,” Corrin said.

75. Research Project Use your school’s library, the Internet, or some other reference source to research the financial history of a fast-growing company like the one discussed above. Gather data on the company’s costs and revenues over a period of time, and use a graphing utility to graph a scatter plot of the data. Fit models to the data. Do the models appear to be concave upward or downward? Do they appear to be increasing or decreasing? Discuss the implications of your answers.

3.4 Optimization Problems

■ Solve real-life optimization problems.

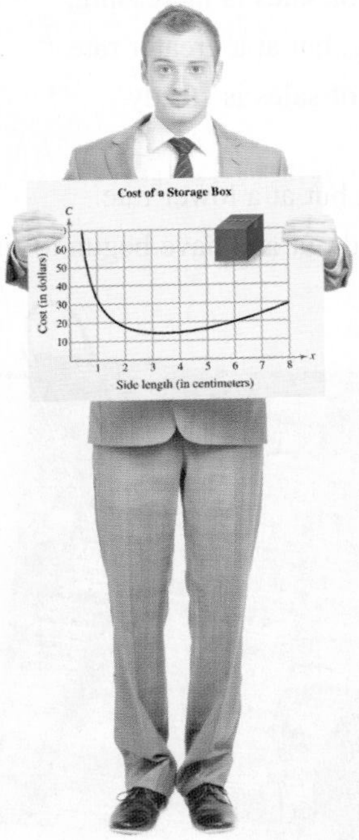

In Exercise 13 on page 202, you will use primary equations, secondary equations, and derivatives to find the dimensions of a box that will minimize the cost of making the box.

Solving Optimization Problems

One of the most common applications of calculus is the determination of optimum (minimum or maximum) values. Before learning a general method for solving optimization problems, consider the next example.

Example 1 Finding the Maximum Volume

A manufacturer wants to design an open box that has a square base and a surface area S of 108 square inches, as shown in Figure 3.32. What dimensions will produce a box with a maximum volume?

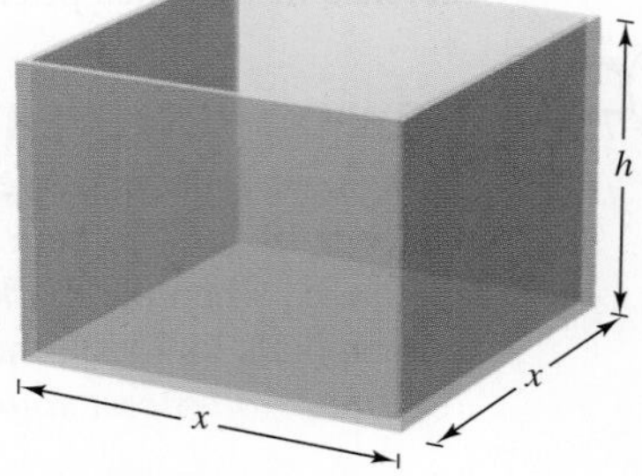

Open Box with Square Base: $S = x^2 + 4xh = 108$

FIGURE 3.32

SOLUTION Because the base of the box is square, the volume is

$$V = x^2h. \qquad \text{Primary equation}$$

This equation is called the **primary equation** because it gives a formula for the quantity to be optimized. The surface area of the box is

$$S = (\text{area of base}) + (\text{area of four sides})$$
$$108 = x^2 + 4xh. \qquad \text{Secondary equation}$$

Because V is to be optimized, it helps to express V as a function of just one variable. To do this, solve the secondary equation for h in terms of x to obtain

$$h = \frac{108 - x^2}{4x}$$

and substitute for h in the primary equation.

$$V = x^2h = x^2\left(\frac{108 - x^2}{4x}\right) = 27x - \frac{1}{4}x^3 \qquad \text{Function of one variable}$$

Before finding which x-value yields a maximum value of V, you need to determine the *feasible domain* of the function. That is, what values of x make sense in this problem? Because x must be nonnegative and the area of the base $(A = x^2)$ is at most 108, you can conclude that the feasible domain is

$$0 \le x \le \sqrt{108}. \qquad \text{Feasible domain}$$

Using the techniques described in the first three sections of this chapter, you can determine that $\left(\text{on the interval } 0 \le x \le \sqrt{108}\right)$ this function has an absolute maximum at $x = 6$ inches and $h = 3$ inches.

ALGEBRA TUTOR

For help on the algebra in Example 1, see Example 1(c) in the *Chapter 3 Algebra Tutor*, on page 242.

✓ Checkpoint 1

Use a graphing utility to graph the volume function

$$V = 27x - \tfrac{1}{4}x^3$$

from Example 1 on $0 \le x \le \sqrt{108}$. Verify that the function has an absolute maximum at $x = 6$. What is the maximum volume? ■

In studying Example 1, be sure that you understand the basic question that it asks. Remember that you are not ready to begin solving an optimization problem until you have clearly identified the problem. Once you are sure you understand what is being asked, you are ready to begin considering a method for solving the problem.

For instance, in Example 1, you should realize that there are infinitely many open boxes having 108 square inches of surface area. To begin solving this problem, you might ask yourself which basic shape would seem to yield a maximum volume. Should the box be tall, squat, or nearly cubical? You might even try calculating a few volumes, as shown in Figure 3.33, to see if you can get a better feeling for what the optimum dimensions should be.

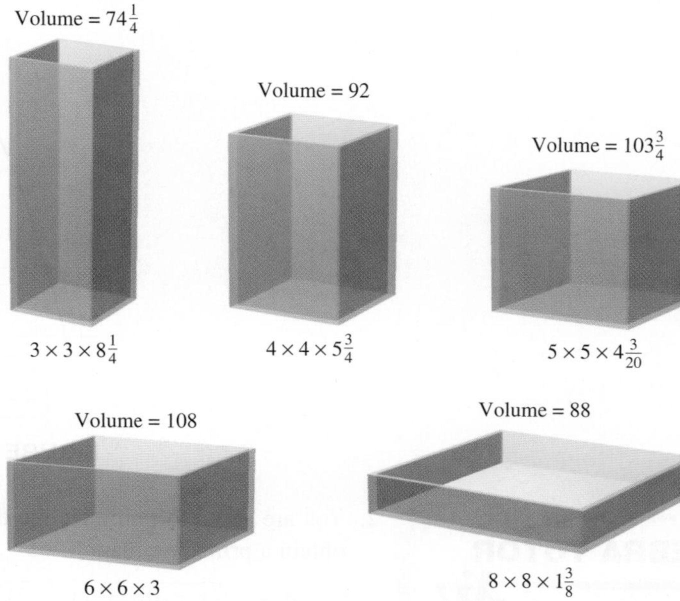

Which box has the greatest volume?
FIGURE 3.33

There are several steps in the solution of Example 1. The first step is to sketch a diagram and identify all *known* quantities and all quantities *to be determined*. The second step is to write a primary equation for the quantity to be optimized. Then, a secondary equation is used to rewrite the primary equation as a function of one variable. Finally, calculus is used to determine the optimum value. These steps are summarized below.

STUDY TIP

When performing Step 5, remember that to determine the maximum or minimum value of a continuous function f on a closed interval, you need to compare the values of f at its critical numbers with the values of f at the endpoints of the interval. The greatest of these values is the desired maximum, and the least is the desired minimum.

Guidelines for Solving Optimization Problems

1. Identify all given quantities and all quantities to be determined. If possible, make a sketch.
2. Write a **primary equation** for the quantity that is to be maximized or minimized. (A summary of several common formulas is given in Appendix D.)
3. Reduce the primary equation to one having a single independent variable. This may involve the use of a **secondary equation** that relates the independent variables of the primary equation.
4. Determine the feasible domain of the primary equation. That is, determine the values for which the stated problem makes sense.
5. Determine the desired maximum or minimum value by the calculus techniques discussed in Sections 3.1 through 3.3.

Example 2 Finding a Minimum Distance

Find the points on the graph of

$$y = 4 - x^2$$

that are closest to (0, 2).

SOLUTION

1. Figure 3.34 shows that there are two points at a minimum distance from (0, 2).

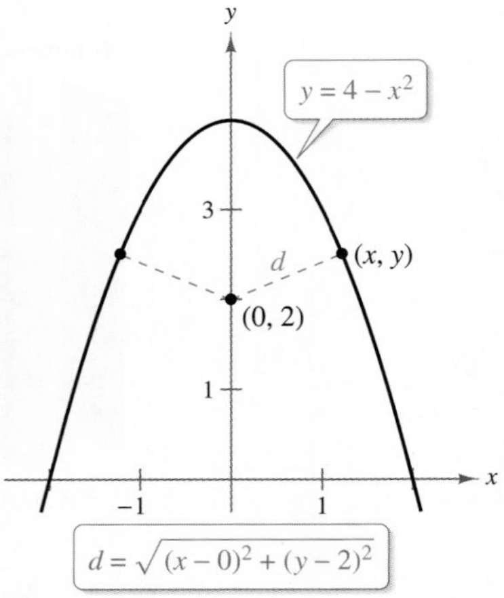

FIGURE 3.34

2. You are asked to minimize the distance d. So, you can use the Distance Formula to obtain a primary equation.

$$d = \sqrt{(x - 0)^2 + (y - 2)^2} \qquad \text{Primary equation}$$

3. Using the secondary equation $y = 4 - x^2$, you can rewrite the primary equation as a function of a single variable.

$$\begin{aligned} d &= \sqrt{x^2 + (4 - x^2 - 2)^2} && \text{Substitute } 4 - x^2 \text{ for } y. \\ &= \sqrt{x^2 + (2 - x^2)^2} && \text{Simplify.} \\ &= \sqrt{x^2 + 4 - 4x^2 + x^4} && \text{Expand binomial.} \\ &= \sqrt{x^4 - 3x^2 + 4} && \text{Combine like terms.} \end{aligned}$$

Because d is smallest when the expression under the radical is smallest, you simplify the problem by finding the minimum value of $f(x) = x^4 - 3x^2 + 4$.

4. The domain of f is the entire real number line.

5. To find the minimum value of $f(x)$, first find the critical numbers of f.

$$\begin{aligned} f'(x) &= 4x^3 - 6x && \text{Find derivative of } f. \\ 0 &= 4x^3 - 6x && \text{Set derivative equal to 0.} \\ 0 &= 2x(2x^2 - 3) && \text{Factor.} \\ x &= 0, x = \sqrt{\tfrac{3}{2}}, x = -\sqrt{\tfrac{3}{2}} && \text{Critical numbers} \end{aligned}$$

The First-Derivative Test verifies that $x = 0$ yields a relative maximum, whereas both $\sqrt{3/2}$ and $-\sqrt{3/2}$ yield a minimum. So, the points closest to (0, 2) are

$$\left(\sqrt{\tfrac{3}{2}}, \tfrac{5}{2}\right) \quad \text{and} \quad \left(-\sqrt{\tfrac{3}{2}}, \tfrac{5}{2}\right).$$

ALGEBRA TUTOR *xy*

For help on the algebra in Example 2, see Example 1(b) in the *Chapter 3 Algebra Tutor*, on page 242.

✓ Checkpoint 2

Find the points on the graph of $y = 4 - x^2$ that are closest to (0, 3). ■

Example 3 Finding a Minimum Area

A rectangular page will contain 24 square inches of print. The margins at the top and bottom of the page are $1\frac{1}{2}$ inches wide. The margins on each side are 1 inch wide. What should the dimensions of the page be to minimize the amount of paper used?

SOLUTION

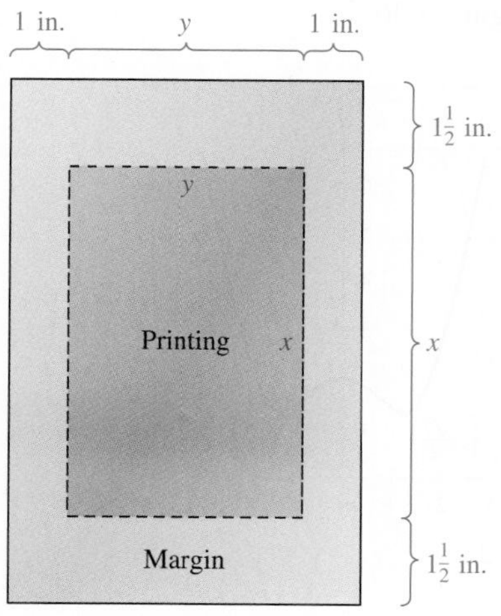

$A = (x + 3)(y + 2)$

FIGURE 3.35

1. A diagram of the page is shown in Figure 3.35.
2. Letting A be the area to be minimized, the primary equation is

$$A = (x + 3)(y + 2). \qquad \text{Primary equation}$$

3. The printed area inside the margins is given by

$$24 = xy. \qquad \text{Secondary equation}$$

Solving this equation for y produces

$$y = \frac{24}{x}.$$

By substituting this result into the primary equation, you obtain

$$A = (x + 3)\left(\frac{24}{x} + 2\right) \qquad \text{Write as a function of one variable.}$$

$$= (x + 3)\left(\frac{24 + 2x}{x}\right) \qquad \text{Rewrite second factor as a single fraction.}$$

$$= \frac{2x^2}{x} + \frac{30x}{x} + \frac{72}{x} \qquad \text{Multiply and separate into terms.}$$

$$= 2x + 30 + \frac{72}{x}. \qquad \text{Simplify.}$$

4. Because x must be positive, the feasible domain is $x > 0$.
5. To find the minimum area, begin by finding the critical numbers of A.

$$\frac{dA}{dx} = 2 - \frac{72}{x^2} \qquad \text{Find derivative of } A.$$

$$0 = 2 - \frac{72}{x^2} \qquad \text{Set derivative equal to 0.}$$

$$-2 = -\frac{72}{x^2} \qquad \text{Subtract 2 from each side.}$$

$$x^2 = 36 \qquad \text{Simplify.}$$

$$x = \pm 6 \qquad \text{Critical numbers}$$

Because $x = -6$ is not in the feasible domain, you need to consider only the critical number $x = 6$. Using the First-Derivative Test, it follows that A is a minimum when $x = 6$. So, the dimensions of the page should be

$$x + 3 = 6 + 3 = 9 \text{ inches} \quad \text{by} \quad y + 2 = \frac{24}{6} + 2 = 6 \text{ inches.}$$

✓ Checkpoint 3

A rectangular page will contain 54 square inches of print. The margins at the top and bottom of the page are $1\frac{1}{2}$ inches wide. The margins on each side are 1 inch wide. What should the dimensions of the page be to minimize the amount of paper used? ■

As applications go, the examples described in this section are fairly simple, and yet the resulting primary equations are quite complicated. Real-life applications often involve equations that are at least as complex as the ones in the examples. Remember that one of the main goals of this course is to enable you to use the power of calculus to analyze equations that at first glance seem formidable.

Also remember that once you have found the primary equation, you can use the graph of the equation to help solve the problem. For instance, the graphs of the primary equations in Examples 1 through 3 are shown in Figure 3.36.

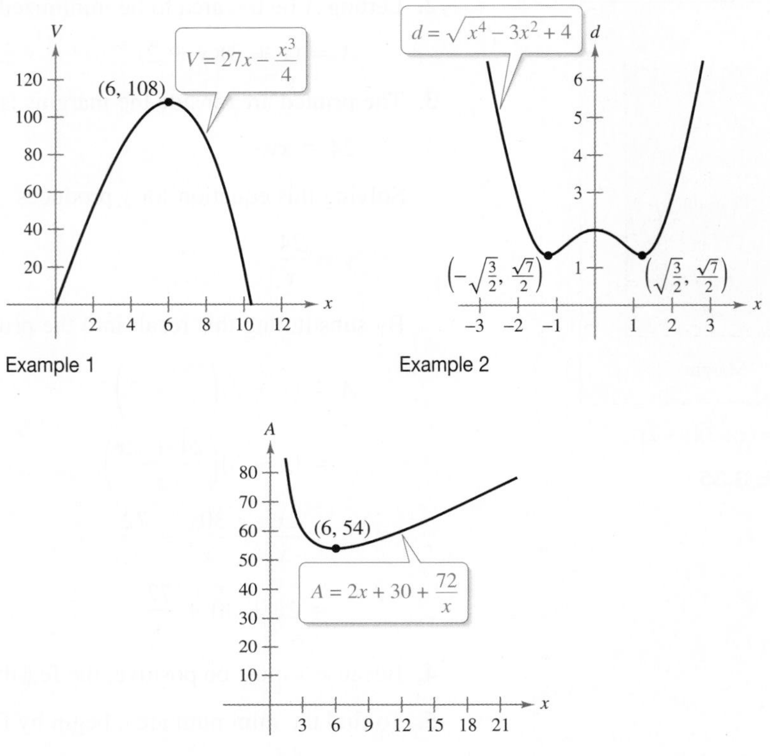

Example 3

FIGURE 3.36

SUMMARIZE (Section 3.4)

1. State what is meant by the primary equation of an optimization problem *(page 196)*. For examples of primary equations in optimization problems, see Examples 1, 2, and 3.
2. State what is meant by the feasible domain of a function *(page 196)*. For examples of feasible domains, see Examples 1, 2, and 3.
3. State what is meant by the secondary equation of an optimization problem *(page 197)*. For examples of secondary equations in optimization problems, see Examples 1, 2, and 3.
4. State the guidelines for solving optimization problems *(page 197)*. For examples of solving optimization problems, see Examples 2 and 3.
5. Describe a real-life example of how solving an optimization problem can be used to determine the dimensions of a page so that the amount of paper used is minimized *(page 199, Example 3)*.

SKILLS WARM UP 3.4 The following warm-up exercises involve skills that were covered in earlier sections. You will use these skills in the exercise set for this section. For additional help, review Section 3.1.

In Exercises 1–4, write a formula for the written statement.

1. The sum of one number and half a second number is 12.

2. The product of one number and twice another is 24.

3. The area of a rectangle is 24 square units.

4. The distance between two points is 10 units.

In Exercises 5–10, find the critical numbers of the function.

5. $y = x^2 + 6x - 9$

6. $y = 2x^3 - x^2 - 4x$

7. $y = 5x + \dfrac{125}{x}$

8. $y = 3x + \dfrac{96}{x^2}$

9. $y = \dfrac{x^2 + 1}{x}$

10. $y = \dfrac{x}{x^2 + 9}$

Exercises 3.4

See www.CalcChat.com for worked-out solutions to odd-numbered exercises.

Maximum Area In Exercises 1 and 2, find the length and width of a rectangle that has the given perimeter and a maximum area.

1. Perimeter: 100 meters **2.** Perimeter: P units

Minimum Perimeter In Exercises 3 and 4, find the length and width of a rectangle that has the given area and a minimum perimeter.

3. Area: 64 square feet **4.** Area: A square centimeters

5. Maximum Area A rancher has 200 feet of fencing to enclose two adjacent rectangular corrals (see figure). What dimensions should be used so that the enclosed area will be a maximum?

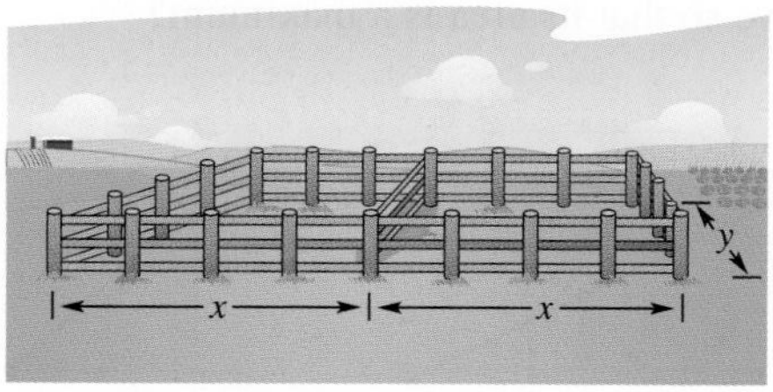

6. Minimum Dimensions A dairy farmer plans to enclose a rectangular pasture adjacent to a river. To provide enough grass for the herd, the pasture must contain 245,000 square meters. No fencing is required along the river. What dimensions will use the least amount of fencing?

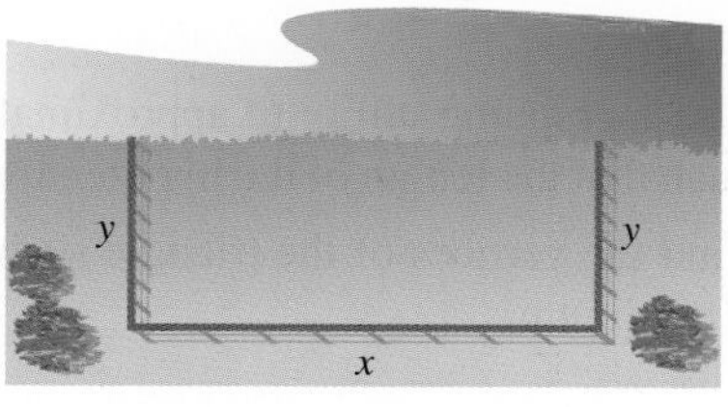

7. Maximum Volume

(a) Verify that each of the rectangular solids shown in the figure has a surface area of 150 square inches.

(b) Find the volume of each solid.

(c) Determine the dimensions of a rectangular solid (with a square base) of maximum volume if its surface area is 150 square inches.

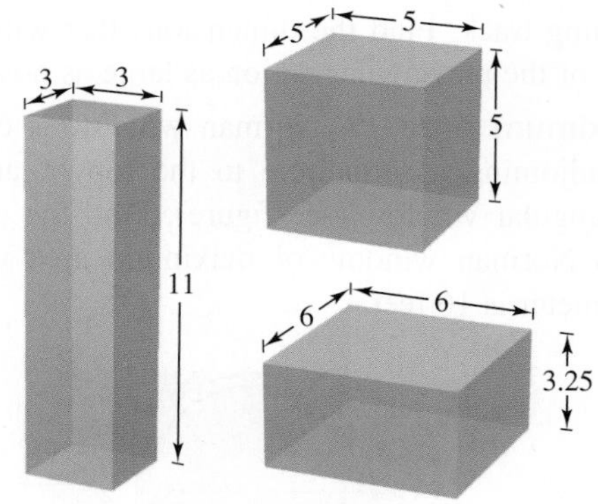

8. Maximum Volume A rectangular solid with a square base has a surface area of 337.5 square centimeters.

(a) Determine the dimensions that yield the maximum volume.

(b) Find the maximum volume.

9. Minimum Surface Area A rectangular solid with a square base has a volume of 8000 cubic inches.

(a) Determine the dimensions that yield the minimum surface area.

(b) Find the minimum surface area.

10. HOW DO YOU SEE IT? The graph shows the profit P (in thousands of dollars) of a company in terms of its advertising cost x (in thousands of dollars).

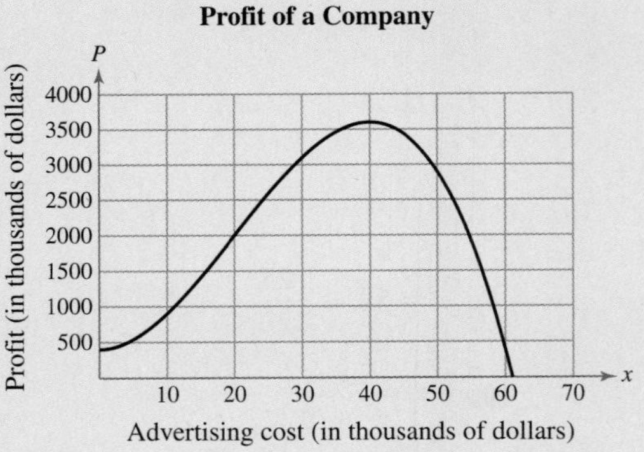

(a) Estimate the interval on which the profit is increasing.

(b) Estimate the interval on which the profit is decreasing.

(c) Estimate the amount of money the company should spend on advertising in order to yield a maximum profit.

(d) Estimate the point of diminishing returns.

11. Maximum Area An indoor physical fitness room consists of a rectangular region with a semicircle on each end. The perimeter of the room is to be a 200-meter running track. Find the dimensions that will make the area of the rectangular region as large as possible.

12. Maximum Area A Norman window is constructed by adjoining a semicircle to the top of an ordinary rectangular window (see figure). Find the dimensions of a Norman window of maximum area if the total perimeter is 16 feet.

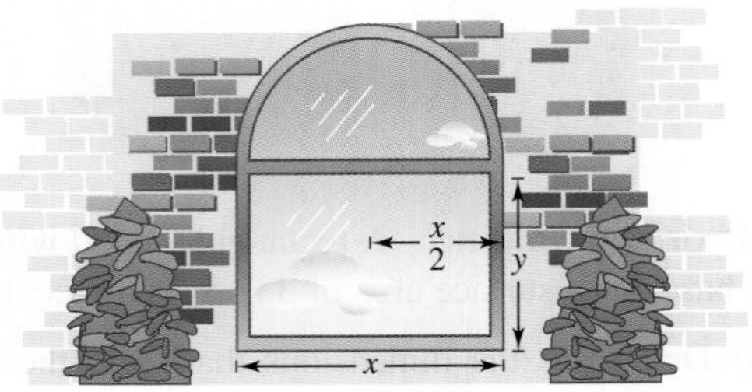

13. Minimum Cost A storage box with a square base must have a volume of 80 cubic centimeters. The top and bottom cost \$0.20 per square centimeter and the sides cost \$0.10 per square centimeter. Find the dimensions that will minimize cost.

14. Minimum Surface Area A net enclosure for golf practice is open at one end (see figure). The volume of the enclosure is $83\frac{1}{3}$ cubic meters. Find the dimensions that require the least amount of netting.

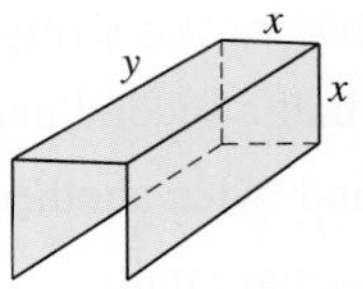

Figure for 14

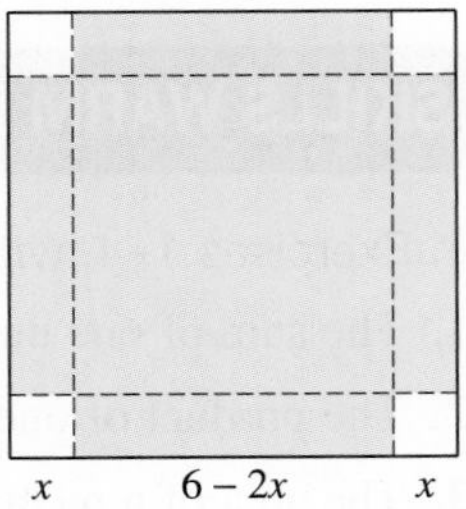

Figure for 15

15. Maximum Volume An open box is to be made from a six-inch by six-inch square piece of material by cutting equal squares from the corners and turning up the sides (see figure). Find the volume of the largest box that can be made.

16. Maximum Volume An open box is to be made from a three-foot by eight-foot rectangular piece of material by cutting equal squares from the corners and turning up the sides. Find the volume of the largest box that can be made in this manner.

17. Minimum Area A rectangular page is to contain 36 square inches of print. The margins at the top and bottom and on each side are to be $1\frac{1}{2}$ inches. Find the dimensions of the page that will minimize the amount of paper used.

18. Minimum Area A rectangular page is to contain 50 square inches of print. The margins at the top and bottom of the page are to be 2 inches wide. The margins on each side are to be 1 inch wide. Find the dimensions of the page that will minimize the amount of paper used.

19. Maximum Area A rectangle is bounded by the x- and y-axes and the graph of

$$y = \frac{1}{2}(6 - x)$$

(see figure). What length and width should the rectangle have so that its area is a maximum?

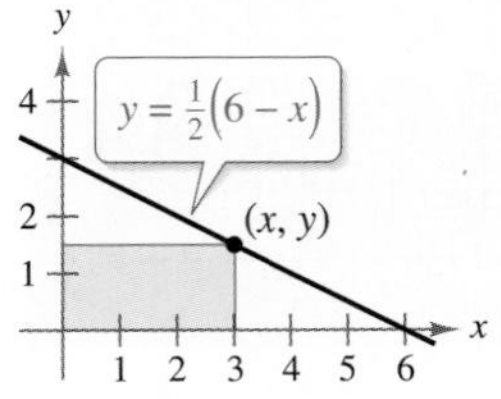

Figure for 19

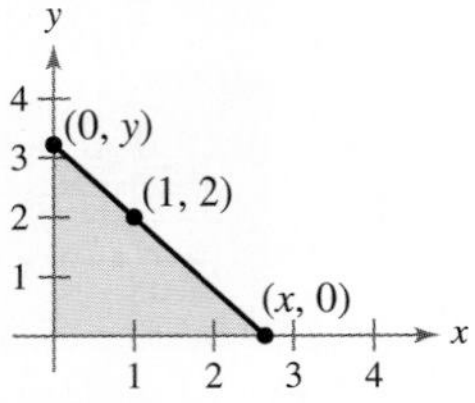

Figure for 20

20. Minimum Length and Minimum Area A right triangle is formed in the first quadrant by the x- and y-axes and a line through the point $(1, 2)$ (see figure).

(a) Write the length L of the hypotenuse as a function of x.

(b) Use a graphing utility to approximate x graphically such that the length of the hypotenuse is a minimum.

(c) Find the vertices of the triangle such that its area is a minimum.

21. Maximum Area A rectangle is bounded by the x-axis and the semicircle $y = \sqrt{25 - x^2}$ (see figure). What length and width should the rectangle have so that its area is a maximum?

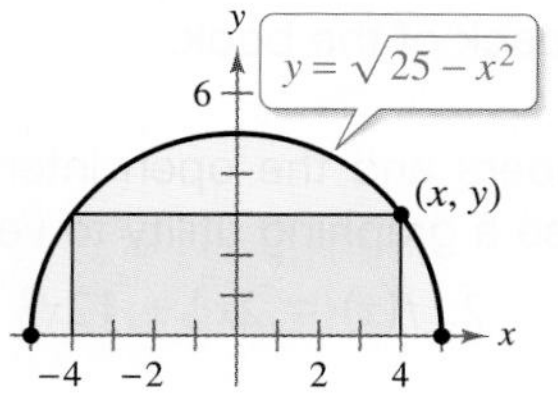

22. Maximum Area Find the dimensions of the largest rectangle that can be inscribed in a semicircle of radius r. (See Exercise 21.)

23. Minimum Surface Area You are designing a soft drink container that has the shape of a right circular cylinder. The container is supposed to hold 12 fluid ounces (1 fluid ounce is approximately 1.80469 cubic inches). Find the dimensions that will use a minimum amount of construction material.

24. Minimum Cost An energy drink container of the shape described in Exercise 23 must have a volume of 16 fluid ounces. The cost per square inch of constructing the top and bottom is twice the cost of constructing the lateral side. Find the dimensions that will minimize cost.

Finding a Minimum Distance In Exercises 25–28, find the points on the graph of the function that are closest to the given point. *See Example 2.*

25. $f(x) = x^2, \ \left(2, \frac{1}{2}\right)$ **26.** $f(x) = (x + 1)^2, \ (5, 3)$

27. $f(x) = \sqrt{x}, \ (4, 0)$ **28.** $f(x) = \sqrt{x - 8}, \ (12, 0)$

29. Maximum Volume A rectangular package to be sent by a postal service can have a maximum combined length and girth (perimeter of a cross section) of 108 inches. Find the dimensions of the package with maximum volume. Assume that the package's dimensions are x by x by y (see figure).

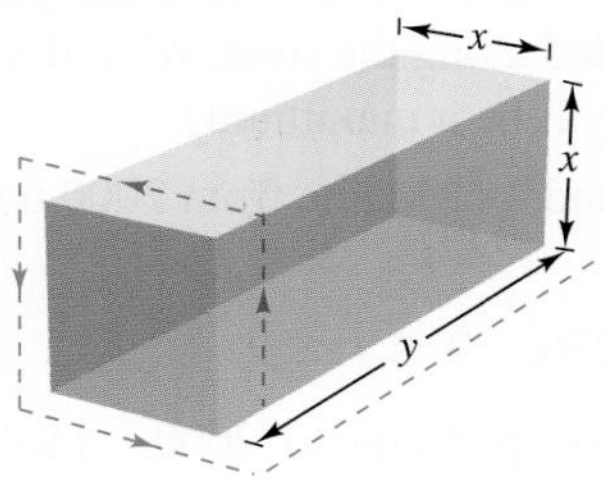

30. Minimum Surface Area A solid is formed by adjoining two hemispheres to the ends of a right circular cylinder. The total volume of the solid is 12 cubic inches. Find the radius of the cylinder that produces the minimum surface area.

31. Minimum Cost An industrial tank of the shape described in Exercise 30 must have a volume of 3000 cubic feet. The hemispherical ends cost twice as much per square foot of surface area as the sides. Find the dimensions that will minimize cost.

32. Minimum Time You are in a boat 2 miles from the nearest point on the coast. You are to go to point Q, located 3 miles down the coast and 1 mile inland (see figure). You can row at a rate of 2 miles per hour and you can walk at a rate of 4 miles per hour. Toward what point on the coast should you row in order to reach point Q in the least amount of time?

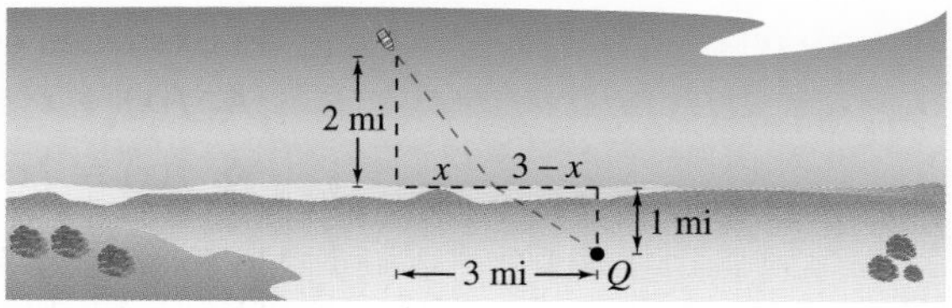

33. Minimum Area The sum of the circumference of a circle and the perimeter of a square is 16. Find the dimensions of the circle and square that produce a minimum total area.

34. Minimum Area The sum of the perimeters of an equilateral triangle and a square is 10. Find the dimensions of the triangle and square that produce a minimum total area.

35. Area Four feet of wire is to be used to form a square and a circle.

(a) Express the sum of the areas of the square and the circle as a function A of a side of the square x.

(b) What is the domain of A?

(c) Use a graphing utility to graph A on its domain.

(d) How much wire should be used for the square and how much for the circle in order to enclose the least total area? the greatest total area?

36. Maximum Yield A home gardener estimates that 16 apple trees will produce an average yield of 80 apples per tree. But because of the size of the garden, for each additional tree planted, the yield will decrease by four apples per tree. How many trees should be planted to maximize the total yield of apples? What is the maximum yield?

37. Farming A strawberry farmer will receive \$30 per bushel of strawberries during the first week of harvesting. Each week after that, the value will drop \$0.80 per bushel. The farmer estimates that there are approximately 120 bushels of strawberries in the fields, and that the crop is increasing at a rate of four bushels per week. When should the farmer harvest the strawberries to maximize their value? How many bushels of strawberries will yield the maximum value? What is the maximum value of the strawberries?

QUIZ YOURSELF

See www.CalcChat.com for worked-out solutions to odd-numbered exercises.

Take this quiz as you would take a quiz in class. When you are done, check your work against the answers given in the back of the book.

In Exercises 1–3, find the critical numbers and the open intervals on which the function is increasing or decreasing. Use a graphing utility to verify your results.

1. $f(x) = x^2 - 6x + 1$

2. $f(x) = 2x^3 + 12x^2$

3. $f(x) = \dfrac{x}{x^2 + 25}$

In Exercises 4–6, find all relative extrema of the function.

4. $f(x) = x^3 + 3x^2 - 5$

5. $f(x) = x^4 - 8x^2 + 3$

6. $f(x) = 2x^{2/3}$

In Exercises 7–9, find the absolute extrema of the function on the closed interval. Use a graphing utility to verify your results.

7. $f(x) = x^2 + 2x - 8, \quad [-2, 1]$

8. $f(x) = x^3 - 27x, \quad [-4, 4]$

9. $f(x) = \dfrac{x}{x^2 + 1}, \quad [0, 2]$

In Exercises 10 and 11, discuss the concavity of the graph of the function and find the points of inflection.

10. $f(x) = x^3 - 6x^2 + 7x$

11. $f(x) = x^4 - 24x^2$

In Exercises 12 and 13, use the Second-Derivative Test to find all relative extrema of the function.

12. $f(x) = 2x^3 + 3x^2 - 12x + 16$

13. $f(x) = 2x + \dfrac{18}{x}$

14. By increasing its advertising cost x for a product, a company discovers that it can increase the sales S according to the model

$$S = \frac{1}{3600}(360x^2 - x^3), \quad 0 \le x \le 240$$

where x and S are in thousands of dollars. Find the point of diminishing returns for this product.

15. A gardener has 200 feet of fencing to enclose a rectangular garden adjacent to a river (see figure). No fencing is needed along the river. What dimensions should be used so that the area of the garden will be a maximum?

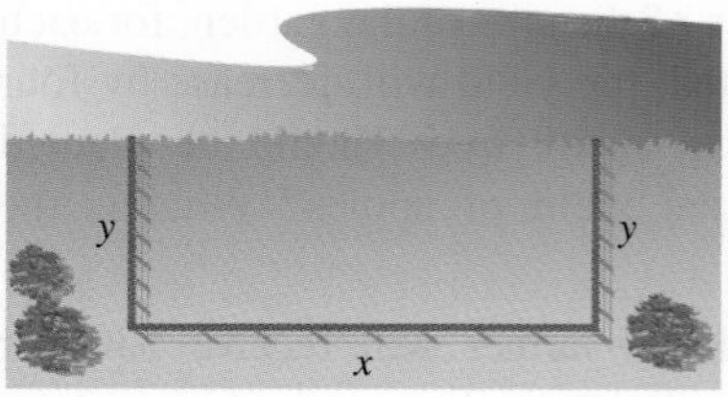

Figure for 15

16. The resident population P (in thousands) of Maine from 2000 through 2009 can be modeled by

$$P = 0.001t^3 - 0.64t^2 + 10.3t + 1276, \quad 0 \le t \le 9$$

where t is the year, with $t = 0$ corresponding to 2000. *(Source: U.S. Census Bureau)*

(a) During which year(s) was the population increasing? decreasing?

(b) During which year, from 2000 through 2009, was the population the greatest? the least?

3.5 Business and Economics Applications

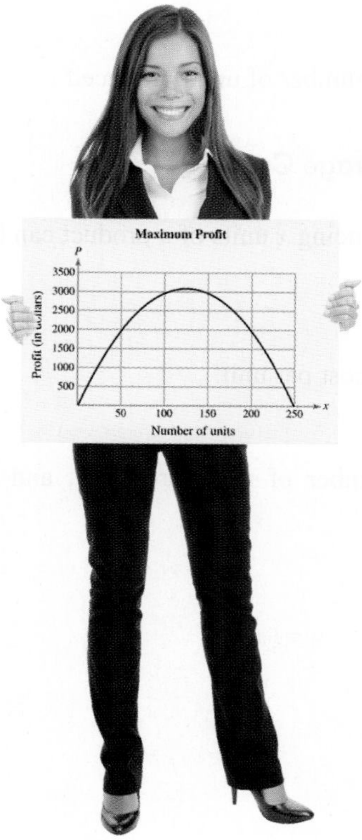

In Exercise 15 on page 212, you will use derivatives to find the price that yields a maximum profit.

- Solve business and economics optimization problems.
- Find the price elasticity of demand for demand functions.
- Recognize basic business terms and formulas.

Optimization in Business and Economics

The problems in this section are primarily optimization problems. So, the five-step procedure used in Section 3.4 is an appropriate strategy to follow.

Example 1 Finding the Maximum Revenue

A company has determined that its total revenue (in dollars) for a product can be modeled by

$$R = -x^3 + 450x^2 + 52{,}500x$$

where x is the number of units produced (and sold). What production level will yield a maximum revenue?

SOLUTION

1. A sketch of the revenue function is shown in Figure 3.37.

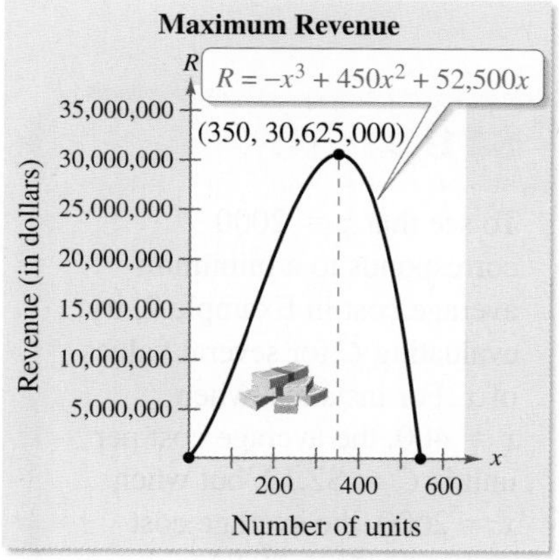

Maximum revenue occurs when $dR/dx = 0$.

FIGURE 3.37

2. The primary equation is the given revenue function.

$$R = -x^3 + 450x^2 + 52{,}500x$$

3. Because R is already given as a function of one variable, you do not need a secondary equation.

4. The feasible domain of the primary equation is

$$0 \le x \le 546. \qquad \text{Feasible domain}$$

This is determined by finding the x-intercepts of the revenue function, as shown in Figure 3.37.

5. To maximize the revenue, find the critical numbers.

$$\frac{dR}{dx} = -3x^2 + 900x + 52{,}500 = 0 \qquad \text{Set derivative equal to 0.}$$

$$-3(x - 350)(x + 50) = 0 \qquad \text{Factor.}$$

$$x = 350, x = -50 \qquad \text{Critical numbers}$$

The only critical number in the feasible domain is $x = 350$. From the graph of the function, you can see that the production level of 350 units corresponds to a maximum revenue.

✓ Checkpoint 1

Find the number of units that must be produced to maximize the revenue function $R = -x^3 + 150x^2 + 9375x$, where R is the total revenue (in dollars) and x is the number of units produced (and sold). What is the maximum revenue? ■

To study the effects of production levels on cost, one method economists use is the **average cost function** $\overline{C}$, which is defined as

$$\overline{C} = \frac{C}{x}$$ Average cost function

where $C = f(x)$ is the total cost function and x is the number of units produced.

Example 2 Finding the Minimum Average Cost

A company estimates that the cost (in dollars) of producing x units of a product can be modeled by

$$C = 800 + 0.04x + 0.0002x^2.$$

Find the production level that minimizes the average cost per unit.

SOLUTION

1. C represents the total cost, x represents the number of units produced, and $\overline{C}$ represents the average cost per unit.
2. The primary equation is

$$\overline{C} = \frac{C}{x}.$$ Primary equation

3. Substituting the given equation for C produces

$$\overline{C} = \frac{800 + 0.04x + 0.0002x^2}{x}$$ Substitute for C.

$$= \frac{800}{x} + 0.04 + 0.0002x.$$ Function of one variable

4. The feasible domain of this function is

$$x > 0$$ Feasible domain

because the company cannot produce a negative number of units.

5. You can find the critical numbers as shown.

$$\frac{d\overline{C}}{dx} = -\frac{800}{x^2} + 0.0002 = 0$$ Set derivative equal to 0.

$$0.0002 = \frac{800}{x^2}$$

$$x^2 = \frac{800}{0.0002}$$ Multiply each side by x^2 and divide each side by 0.0002.

$$x^2 = 4{,}000{,}000$$

$$x = \pm 2000$$ Critical numbers

By choosing the positive value of x and sketching the graph of $\overline{C}$, as shown in Figure 3.38, you can see that a production level of $x = 2000$ minimizes the average cost per unit.

STUDY TIP

To see that $x = 2000$ corresponds to a minimum average cost in Example 2, try evaluating $\overline{C}$ for several values of x. For instance, when $x = 400$, the average cost per unit is $\overline{C} = \$2.12$, but when $x = 2000$, the average cost per unit is $\overline{C} = \$0.84$.

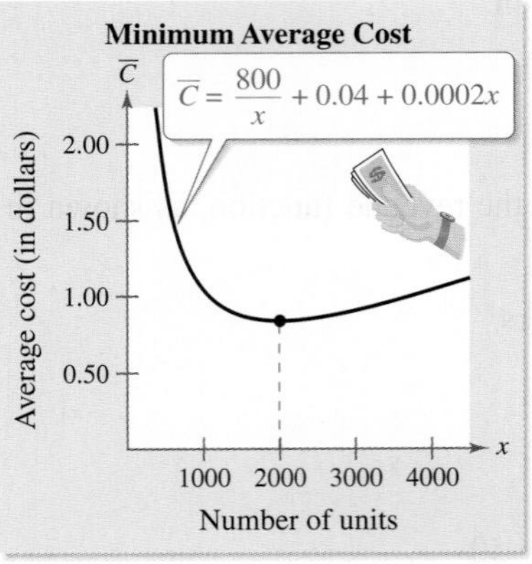

Minimum average cost occurs when $d\overline{C}/dx = 0$.

FIGURE 3.38

Checkpoint 2

Find the production level that minimizes the average cost per unit for the cost function

$$C = 400 + 0.05x + 0.0025x^2$$

where C is the cost (in dollars) of producing x units of a product.

Example 3 Finding the Maximum Revenue

A business sells 2000 units of a product per month at a price of \$10 each. It can sell 250 more items per month for each \$0.25 reduction in price. What price per unit will maximize the monthly revenue?

SOLUTION

1. Let x represent the number of units sold in a month, let p represent the price per unit, and let R represent the monthly revenue.

2. Because the revenue is to be maximized, the primary equation is

$$R = xp. \qquad \text{Primary equation}$$

3. A price of $p = \$10$ corresponds to $x = 2000$, and a price of $p = \$9.75$ corresponds to $x = 2250$. Using this information, you can use the two-point form to write the demand equation.

$$p - 10 = \frac{10 - 9.75}{2000 - 2250}(x - 2000) \qquad \text{Two-point form}$$

$$p - 10 = -0.001(x - 2000) \qquad \text{Simplify.}$$

$$p = -0.001x + 12 \qquad \text{Secondary equation}$$

Substituting this value into the revenue equation produces

$$R = x(-0.001x + 12) \qquad \text{Substitute for } p.$$

$$= -0.001x^2 + 12x. \qquad \text{Function of one variable}$$

4. The feasible domain of the revenue function is

$$0 \le x \le 12{,}000. \qquad \text{Feasible domain}$$

This is determined by finding the x-intercepts of the revenue function.

5. To maximize the revenue, find the critical numbers.

$$\frac{dR}{dx} = 12 - 0.002x = 0 \qquad \text{Set derivative equal to 0.}$$

$$-0.002x = -12$$

$$x = 6000 \qquad \text{Critical number}$$

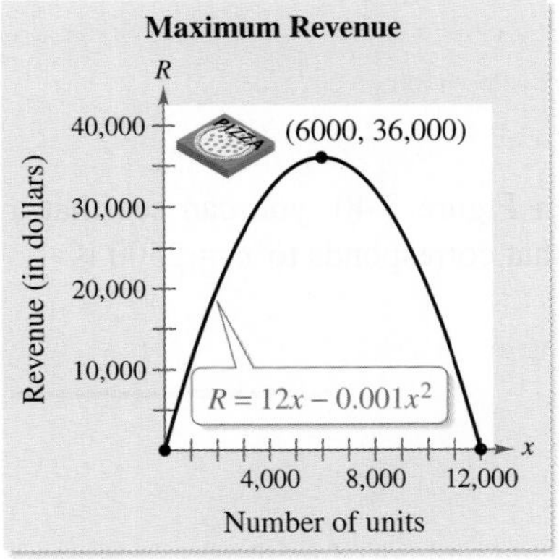

FIGURE 3.39

From the graph of R in Figure 3.39, you can see that this production level yields a maximum revenue. The price that corresponds to this production level is

$$p = 12 - 0.001x \qquad \text{Demand function}$$

$$= 12 - 0.001(6000) \qquad \text{Substitute 6000 for } x.$$

$$= \$6. \qquad \text{Price per unit}$$

✓Checkpoint 3

Find the price per unit that will maximize the monthly revenue for the business in Example 3 when it can sell only 200 more items per month for each \$0.25 reduction in price. ■

In Example 3, the revenue function was written as a function of x. It could also have been written as a function of p. That is,

$$R = 1000(12p - p^2).$$

By finding the critical numbers of this function, you can determine that the maximum revenue occurs at $p = 6$.

Example 4 Finding the Maximum Profit

ALGEBRA TUTOR *xy*

For help on the algebra in Example 4, see Example 2(b) in the *Chapter 3 Algebra Tutor*, on page 243.

The marketing department of a business has determined that the demand for a product can be modeled by $p = 50/\sqrt{x}$, where p is the price per unit (in dollars) and x is the number of units. The cost (in dollars) of producing x units is given by $C = 0.5x + 500$. What price will yield a maximum profit?

SOLUTION

1. Let R represent the revenue, P the profit, p the price per unit, x the number of units, and C the total cost of producing x units.
2. Because you are maximizing the profit, the primary equation is

$$P = R - C. \qquad \text{Primary equation}$$

3. Because the revenue is $R = xp$, you can write the profit function as

$$\begin{aligned} P &= R - C \\ &= xp - (0.5x + 500) && \text{Substitute for } R \text{ and } C. \\ &= x\left(\frac{50}{\sqrt{x}}\right) - 0.5x - 500 && \text{Substitute for } p. \\ &= 50\sqrt{x} - 0.5x - 500. && \text{Function of one variable} \end{aligned}$$

4. The feasible domain of the function is $127 < x \leq 7872$. (When x is less than 127 or greater than or equal to 7872, the profit is negative.)
5. To maximize the profit, find the critical numbers.

$$\begin{aligned} \frac{dP}{dx} = \frac{25}{\sqrt{x}} - 0.5 &= 0 && \text{Set derivative equal to 0.} \\ \frac{25}{\sqrt{x}} &= 0.5 && \text{Add 0.5 to each side.} \\ 50 &= \sqrt{x} && \text{Isolate } x\text{-term on one side.} \\ 2500 &= x && \text{Critical number} \end{aligned}$$

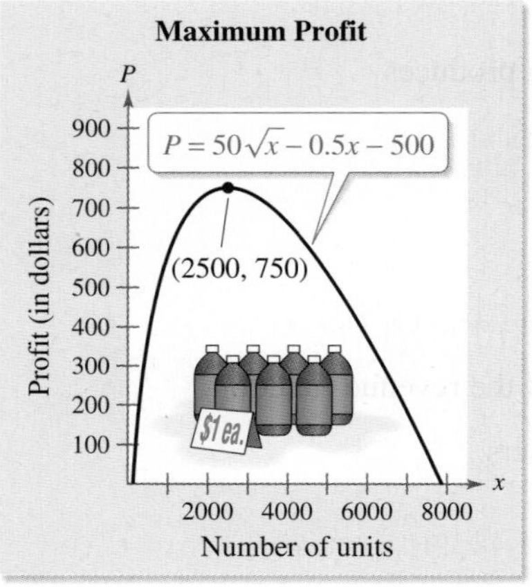

FIGURE 3.40

From the graph of the profit function shown in Figure 3.40, you can see that a maximum profit occurs at $x = 2500$. The price that corresponds to $x = 2500$ is

$$p = \frac{50}{\sqrt{x}} = \frac{50}{\sqrt{2500}} = \frac{50}{50} = \$1.00. \qquad \text{Price per unit}$$

✓ Checkpoint 4

Find the price that will maximize profit for the demand and cost functions

$$p = \frac{40}{\sqrt{x}} \quad \text{and} \quad C = 2x + 50$$

where p is the price per unit (in dollars), x is the number of units, and C is the cost (in dollars). ■

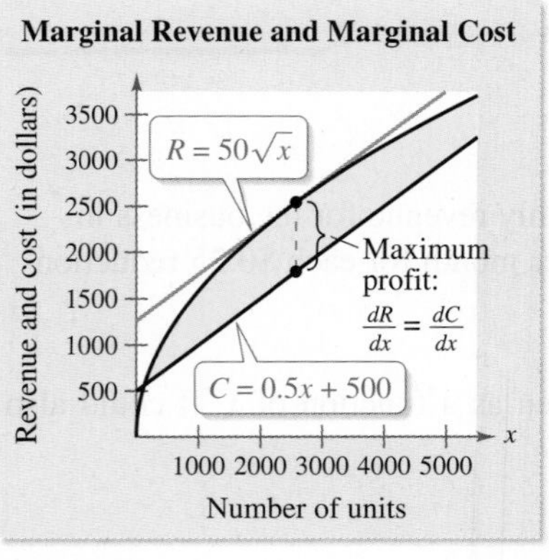

FIGURE 3.41

To find the maximum profit in Example 4, the equation $P = R - C$ was differentiated and set equal to zero. From the equation

$$\frac{dP}{dx} = \frac{dR}{dx} - \frac{dC}{dx} = 0$$

it follows that the maximum profit occurs when the marginal revenue is equal to the marginal cost, as shown in Figure 3.41.

Price Elasticity of Demand

STUDY TIP

In the discussion of price elasticity of demand, the price is assumed to decrease as the quantity demanded increases. So, the demand function $p = f(x)$ is decreasing and dp/dx is negative.

One way in which economists measure the responsiveness of consumers to a change in the price of a product is with **price elasticity of demand.** For example, a drop in the price of fresh tomatoes might result in a much greater demand for fresh tomatoes; such a demand is called **elastic.** On the other hand, the demand for items such as coffee and gasoline is relatively unresponsive to changes in price; the demand for such items is called **inelastic.**

More formally, the elasticity of demand is the percent change of a quantity demanded x, divided by the percent change in its price p. You can develop a formula for price elasticity of demand using the approximation

$$\frac{\Delta p}{\Delta x} \approx \frac{dp}{dx}$$

which is based on the definition of the derivative. Using this approximation, you can write

$$\begin{aligned} \text{Price elasticity of demand} &= \frac{\text{rate of change in demand}}{\text{rate of change in price}} \\ &= \frac{\Delta x/x}{\Delta p/p} \\ &= \frac{p/x}{\Delta p/\Delta x} \\ &\approx \frac{p/x}{dp/dx}. \end{aligned}$$

Definition of Price Elasticity of Demand

If $p = f(x)$ is a differentiable function, then the **price elasticity of demand** is given by

$$\eta = \frac{p/x}{dp/dx}$$

where η is the lowercase Greek letter eta. For a given price, the demand is **elastic** when $|\eta| > 1$, the demand is **inelastic** when $|\eta| < 1$, and the demand has **unit elasticity** when $|\eta| = 1$.

Price elasticity of demand is related to the total revenue function, as indicated in Figure 3.42 and the list below.

1. If the demand is *elastic*, then a decrease in price is accompanied by an increase in unit sales sufficient to increase the total revenue.
2. If the demand is *inelastic*, then a decrease in price is not accompanied by an increase in unit sales sufficient to increase the total revenue.

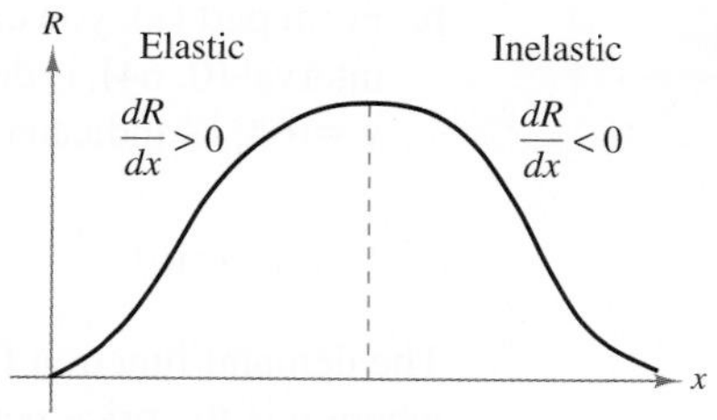

Revenue Curve
FIGURE 3.42

Example 5 Comparing Elasticity and Revenue

The demand function for a product is modeled by

$$p = 24 - 2\sqrt{x}, \quad 0 \le x \le 144$$

where p is the price per unit (in dollars) and x is the number of units. (See Figure 3.43.)

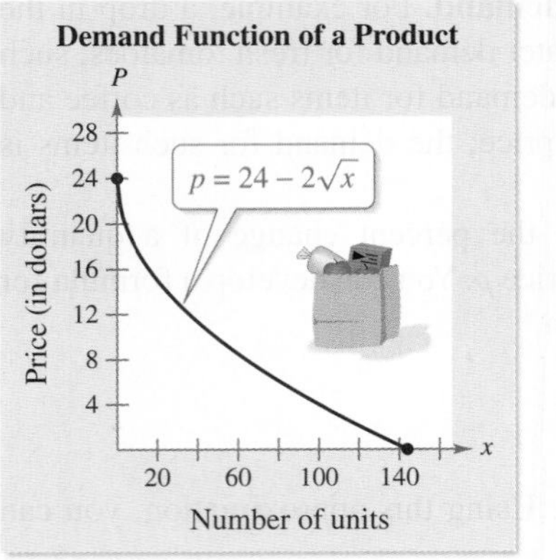

FIGURE 3.43

a. Determine when the demand is elastic, inelastic, and of unit elasticity.

b. Use the result of part (a) to describe the behavior of the revenue function.

SOLUTION

a. The price elasticity of demand is given by

$$\eta = \frac{p/x}{dp/dx} \qquad \text{Formula for price elasticity of demand}$$

$$= \frac{\frac{24 - 2\sqrt{x}}{x}}{\frac{-1}{\sqrt{x}}} \qquad \text{Substitute for } p/x \text{ and } dp/dx.$$

$$= \frac{\left(\frac{24 - 2\sqrt{x}}{x}\right)(-\sqrt{x})}{\left(\frac{-1}{\sqrt{x}}\right)(-\sqrt{x})} \qquad \text{Multiply numerator and denominator by } -\sqrt{x}.$$

$$= \frac{-24\sqrt{x} + 2x}{x} \qquad \text{Simplify.}$$

$$= -\frac{24\sqrt{x}}{x} + 2. \qquad \text{Rewrite as two fractions and simplify.}$$

The demand is of unit elasticity when $|\eta| = 1$. In the interval $[0, 144]$, the only solution of the equation

$$|\eta| = \left| -\frac{24\sqrt{x}}{x} + 2 \right| = 1 \qquad \text{Unit elasticity}$$

is $x = 64$. So, the demand is of unit elasticity when $x = 64$. For x-values in the interval $(0, 64)$,

$$|\eta| = \left| -\frac{24\sqrt{x}}{x} + 2 \right| > 1, \quad 0 < x < 64 \qquad \text{Elastic}$$

which implies that the demand is elastic when $0 < x < 64$. For x-values in the interval $(64, 144)$,

$$|\eta| = \left| -\frac{24\sqrt{x}}{x} + 2 \right| < 1, \quad 64 < x < 144 \qquad \text{Inelastic}$$

which implies that the demand is inelastic when $64 < x < 144$.

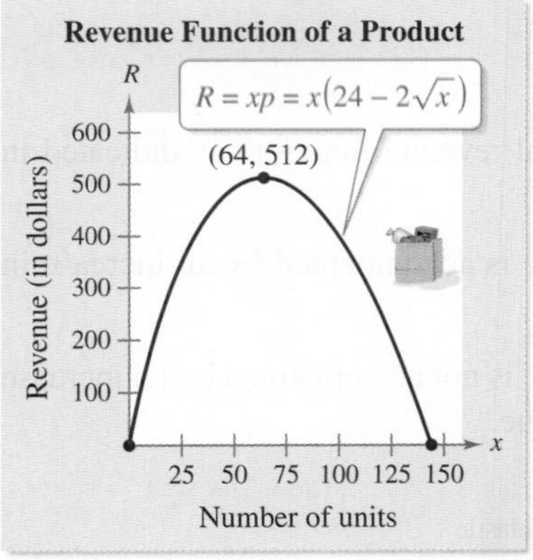

FIGURE 3.44

b. From part (a), you can conclude that the revenue function R is increasing on the open interval $(0, 64)$, is decreasing on the open interval $(64, 144)$, and is a maximum when $x = 64$, as indicated in Figure 3.44.

✓ Checkpoint 5

The demand function for a product is modeled by $p = 36 - 2\sqrt{x}$, $0 \le x \le 324$, where p is the price per unit (in dollars) and x is the number of units. Determine when the demand is elastic, inelastic, and of unit elasticity. ■

Business Terms and Formulas

This section concludes with a summary of the basic business terms and formulas used in this section. A summary of the graphs of the demand, revenue, cost, and profit functions is shown in Figure 3.45.

Summary of Business Terms and Formulas

x = number of units produced (or sold)	η = price elasticity of demand
p = price per unit	$= \dfrac{p/x}{dp/dx}$
R = total revenue from selling x units $= xp$	$\dfrac{dR}{dx}$ = marginal revenue
C = total cost of producing x units	$\dfrac{dC}{dx}$ = marginal cost
P = total profit from selling x units $= R - C$	$\dfrac{dP}{dx}$ = marginal profit
$\overline{C}$ = average cost per unit $= \dfrac{C}{x}$	

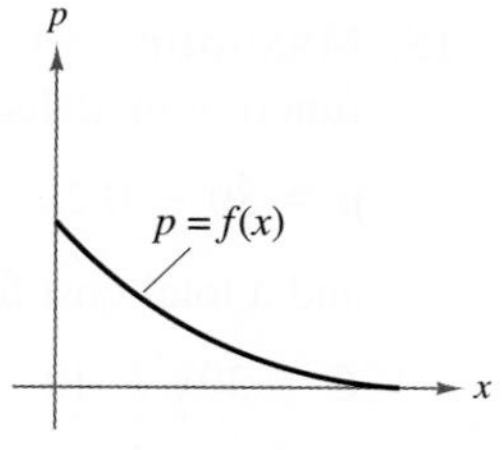

Demand Function
Quantity demanded increases as price decreases.

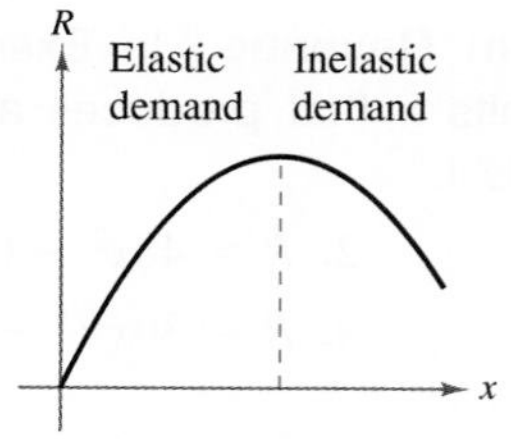

Revenue Function
The low prices required to sell more units eventually result in a decreasing revenue.

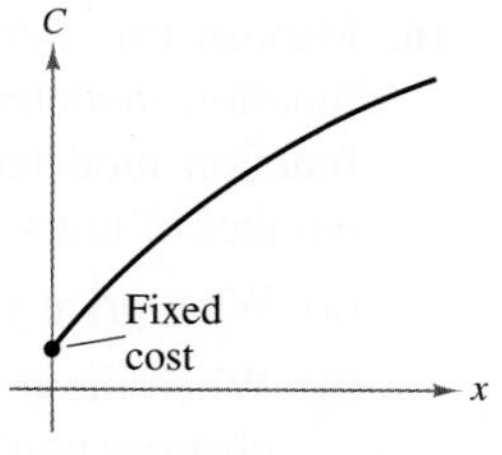

Cost Function
The total cost to produce x units includes the fixed cost.

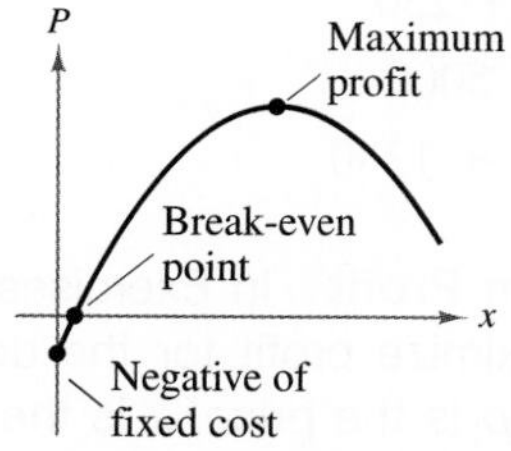

Profit Function
The break-even point occurs when $R = C$.

FIGURE 3.45

SUMMARIZE (Section 3.5)

1. Describe a real-life example of how optimization can be used to find the maximum revenue for a product *(page 205, Example 1)*.
2. State the definition of the average cost function *(page 206)*. For an example of an average cost function, see Example 2.
3. State the definition of price elasticity of demand *(page 209)*. For an example of price elasticity of demand, see Example 5.

SKILLS WARM UP 3.5

The following warm-up exercises involve skills that were covered in a previous course or earlier sections. You will use these skills in the exercise set for this section. For additional help, review Appendix Sections A.2 and A.3, and Section 2.3.

In Exercises 1–4, evaluate the expression for $x = 150$.

1. $\left|-\dfrac{300}{x} + 3\right|$ **2.** $\left|-\dfrac{600}{5x} + 2\right|$ **3.** $\left|\dfrac{(20x^{-1/2})/x}{-10x^{-3/2}}\right|$ **4.** $\left|\dfrac{(4000/x^2)/x}{-8000x^{-3}}\right|$

In Exercises 5–10, find the marginal revenue, marginal cost, or marginal profit.

5. $C = 650 + 1.2x + 0.003x^2$

6. $P = 0.01x^2 + 11x$

7. $P = -0.7x^2 + 7x - 50$

8. $C = 1700 + 4.2x + 0.001x^3$

9. $R = 14x - \dfrac{x^2}{2000}$

10. $R = 3.4x - \dfrac{x^2}{1500}$

Exercises 3.5

See www.CalcChat.com for worked-out solutions to odd-numbered exercises.

Finding the Maximum Revenue In Exercises 1–4, find the number of units x that produces a maximum revenue R. *See Example 1.*

1. $R = 800x - 0.2x^2$

2. $R = 48x^2 - 0.02x^3$

3. $R = 400x - x^2$

4. $R = 30x^{2/3} - 2x$

Finding the Minimum Average Cost In Exercises 5–8, find the number of units x that produces the minimum average cost per unit $\overline{C}$. *See Example 2.*

5. $C = 0.125x^2 + 20x + 5000$

6. $C = 0.001x^3 + 5x + 250$

7. $C = 2x^2 + 255x + 5000$

8. $C = 0.02x^3 + 55x^2 + 1380$

Finding the Maximum Profit In Exercises 9–12, find the price that will maximize profit for the demand and cost functions, where p is the price, x is the number of units, and C is the cost. *See Example 4.*

	Demand Function	*Cost Function*
9.	$p = 90 - x$	$C = 100 + 30x$
10.	$p = 70 - 0.01x$	$C = 8000 + 50x + 0.03x^2$
11.	$p = 50 - 0.1\sqrt{x}$	$C = 35x + 500$
12.	$p = \dfrac{24}{\sqrt{x}}$	$C = 0.4x + 600$

Average Cost In Exercises 13 and 14, use the cost function to find the production level at which the average cost is a minimum. For this production level, show that the marginal cost and average cost are equal. Use a graphing utility to graph the average cost function and verify your results.

13. $C = 2x^2 + 5x + 18$

14. $C = x^3 - 6x^2 + 13x$

15. Maximum Profit A commodity has a demand function modeled by

$$p = 80 - 0.2x$$

and a total cost function modeled by

$$C = 30x + 40$$

where x is the number of units.

(a) What price yields a maximum profit?

(b) When the profit is maximized, what is the average cost per unit?

16. Maximum Profit A commodity has a demand function modeled by $p = 100 - 0.5x$, and a total cost function modeled by $C = 50x + 37.5$, where x is the number of units.

(a) What price yields a maximum profit?

(b) When the profit is maximized, what is the average cost per unit?

Maximum Profit In Exercises 17 and 18, find the amount s spent on advertising (in thousands of dollars) that maximizes the profit P (in thousands of dollars). Find the point of diminishing returns.

17. $P = -2s^3 + 35s^2 - 100s + 200$

18. $P = -0.1s^3 + 6s^2 + 400$

19. Maximum Profit The cost per unit of producing an MP3 player is \$90. The manufacturer charges \$150 per unit for orders of 100 or less. To encourage large orders, however, the manufacturer reduces the charge by \$0.10 per player for each order in excess of 100 units. For instance, an order of 101 players would be \$149.90 per player, an order of 102 players would be \$149.80 per player, and so on. Find the largest order the manufacturer should allow to obtain a maximum profit.

20. Maximum Profit A real estate office handles a 50-unit apartment complex. When the rent is \$580 per month, all units are occupied. For each \$40 increase in rent, however, an average of one unit becomes vacant. Each occupied unit requires an average of \$45 per month for service and repairs. What rent should be charged to obtain a maximum profit?

21. Maximum Revenue When a wholesaler sold a product at \$40 per unit, sales were 300 units per week. After a price increase of \$5, however, the average number of units sold dropped to 275 per week. Assuming that the demand function is linear, what price per unit will yield a maximum total revenue?

22. Maximum Profit Assume that the amount of money deposited in a bank is proportional to the square of the interest rate the bank pays on the money. Furthermore, the bank can reinvest the money at 12% simple interest. Find the interest rate the bank should pay to maximize its profit.

23. Minimum Cost A power station is on one side of a river that is 0.5 mile wide, and a factory is 6 miles downstream on the other side of the river (see figure). It costs \$18 per foot to run overland power lines and \$25 per foot to run underwater power lines. Write a cost function for running the power lines from the power station to the factory. Use a graphing utility to graph your function. Estimate the value of x that minimizes the cost. Explain your results.

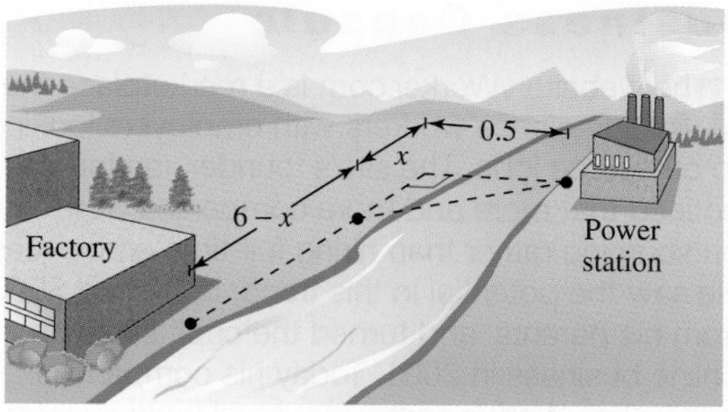

24. Minimum Cost An offshore oil well is 1 mile off the coast. The oil refinery is 2 miles down the coast. Laying pipe in the ocean is twice as expensive as laying it on land. Find the most economical path for the pipe from the well to the oil refinery.

Minimum Cost In Exercises 25 and 26, find the speed v, in miles per hour, that will minimize costs on a 110-mile delivery trip. The cost per hour for fuel is C dollars, and the driver is paid W dollars per hour. (Assume there are no costs other than wages and fuel.)

25. Fuel cost: $C = \dfrac{v^2}{300}$

Driver: $W = \$12$

26. Fuel cost: $C = \dfrac{v^2}{500}$

Driver: $W = \$9.50$

Elasticity In Exercises 27–32, find the price elasticity of demand for the demand function at the indicated x-value. Is the demand elastic, inelastic, or of unit elasticity at the indicated x-value? Use a graphing utility to graph the revenue function, and identify the intervals of elasticity and inelasticity.

	Demand Function	*Quantity Demanded*
27.	$p = 600 - 5x$	$x = 60$
28.	$p = 400 - 3x$	$x = 20$
29.	$p = 5 - 0.03x$	$x = 100$
30.	$p = 20 - 0.0002x$	$x = 30$
31.	$p = \dfrac{500}{x + 2}$	$x = 23$
32.	$p = \dfrac{500}{x^2} + 5$	$x = 5$

33. Elasticity The demand function for a product is modeled by

$$p = 20 - 0.02x, \quad 0 \le x \le 1000$$

where p is the price (in dollars) and x is the number of units.

(a) Determine when the demand is elastic, inelastic, and of unit elasticity.

(b) Use the result of part (a) to describe the behavior of the revenue function.

34. Elasticity The demand function for a product is given by $p = 800 - 4x$, $0 \le x \le 200$, where p is the price (in dollars) and x is the number of units.

(a) Determine when the demand is elastic, inelastic, and of unit elasticity.

(b) Use the result of part (a) to describe the behavior of the revenue function.

35. Minimum Cost The shipping and handling cost C of a manufactured product is modeled by

$$C = 4\left(\frac{25}{x^2} - \frac{x}{x - 10}\right), \quad 0 < x < 10$$

where C is measured in thousands of dollars and x is the number of units shipped (in hundreds). Use the *root* feature of a graphing utility to find the shipment size that minimizes the cost.

36. Minimum Cost The ordering and transportation cost C of the components used in manufacturing a product is modeled by

$$C = 8\left(\frac{2500}{x^2} - \frac{x}{x - 100}\right), \quad 0 < x < 100$$

where C is measured in thousands of dollars and x is the order size in hundreds. Use the *root* feature of a graphing utility to find the order size that minimizes the cost.

37. Revenue The demand for a car wash is $x = 900 - 45p$, where the current price is \$8. Can revenue be increased by lowering the price and thus attracting more customers? Use price elasticity of demand to determine your answer.

38. HOW DO YOU SEE IT? Match each graph with the function it best represents—a demand function, a revenue function, a cost function, or a profit function. Explain your reasoning. (The graphs are labeled *a–d*.)

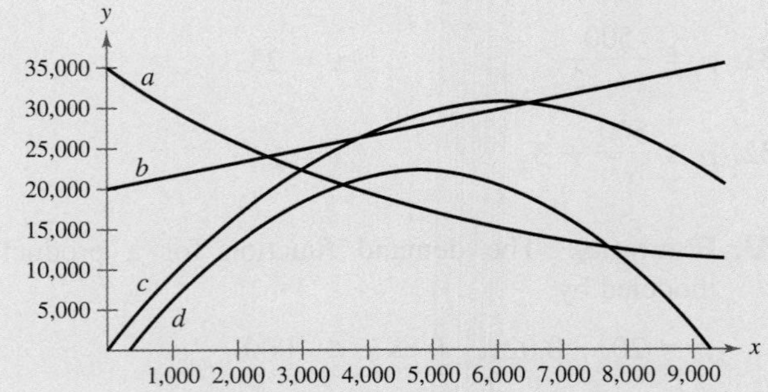

39. Sales The sales S (in millions of dollars per year) for The Clorox Company for the years 2001 through 2010 can be modeled by

$$S = -1.893t^3 + 41.03t^2 - 58.6t + 3972, \quad 1 \le t \le 10$$

where t represents the year, with $t = 1$ corresponding to 2001. *(Source: The Clorox Company)*

(a) During which year, from 2001 through 2010, were the company's sales increasing most rapidly?

(b) During which year were the sales increasing at the lowest rate?

(c) Find the rate of increase or decrease for each year in parts (a) and (b).

(d) Use a graphing utility to graph the sales function. Then use the *zoom* and *trace* features to confirm the results in parts (a), (b), and (c).

40. Sales The sales S (in billions of dollars) for Lockhead Martin Corporation from 2001 through 2010 can be modeled by

$$S = \frac{18.17 + 8.165t}{1 + 0.116t}, \quad 1 \le t \le 10$$

where t represents the year, with $t = 1$ corresponding to 2001. *(Source: Lockhead Martin Corporation)*

(a) During which year, from 2001 through 2010, were the company's sales the greatest? the least?

(b) During which year were the sales increasing at the greatest rate? decreasing at the greatest rate?

(c) Use a graphing utility to graph the revenue function. Then use the *zoom* and *trace* features to confirm the results in parts (a) and (b).

41. Demand A demand function is modeled by

$$x = \frac{a}{p^m}$$

where a is a constant and $m > 1$. Show that $\eta = -m$. In other words, show that a 1% increase in price results in an m% decrease in the quantity demanded.

42. Think About It Throughout this text, it is assumed that demand functions are decreasing. Can you think of a product that has an increasing demand function? That is, can you think of a product that becomes more in demand as its price increases? Explain your reasoning, and sketch a graph of the function.

Business Capsule

The website vWorker.com is a marketplace that links freelance workers with employers looking to outsource jobs. The site's founder, Ian Ippolito, noticed that more and more companies were outsourcing rather than hiring full-time employees. He saw the potential in this trend, borrowed \$5000 from his parents, and turned the capital into an online business in 2001. Today his company is known as vWorker.com—for virtual worker—and it connects more than 150,000 employers with over 300,000 workers worldwide. Posting and bidding are free, but vWorker takes a percentage of the workers' final earnings. Only eight years after the company's debut, annual revenues reached \$2.5 million in 2009.

43. Research Project Choose a company with an innovative product or service like the one described above. Use your school's library, the Internet, or some other reference source to research the history of the company. Collect data about the revenue that the product or service has generated, and find a mathematical model of the data. Summarize your findings.

3.6 Asymptotes

In Exercise 61 on page 225, you will use limits at infinity to find the limit of an average cost function as the number of units produced increases.

- Find the vertical asymptotes of functions and find infinite limits.
- Find the horizontal asymptotes of functions and find limits at infinity.
- Use asymptotes to answer questions about real-life situations.

Vertical Asymptotes and Infinite Limits

In the first three sections of this chapter, you studied ways in which you can use calculus to help analyze the graph of a function. In this section, you will study another valuable aid to curve sketching: the determination of vertical and horizontal asymptotes.

Recall from Section 1.5, Example 11, that the function

$$f(x) = \frac{3}{x-2}$$

is unbounded as x approaches 2 (see Figure 3.46).

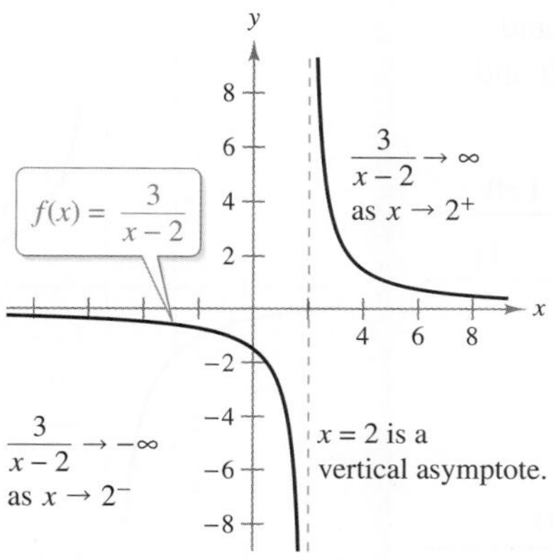

FIGURE 3.46

This type of behavior is described by saying that the line

$$x = 2 \qquad \text{Vertical asymptote}$$

is a **vertical asymptote** of the graph of f. The type of limit in which $f(x)$ approaches infinity (or negative infinity) as x approaches c from the left or from the right is an **infinite limit.** The infinite limits for the function $f(x) = 3/(x-2)$ can be written as

$$\lim_{x \to 2^-} \frac{3}{x-2} = -\infty$$

and

$$\lim_{x \to 2^+} \frac{3}{x-2} = \infty.$$

Definition of Vertical Asymptote

If $f(x)$ approaches infinity (or negative infinity) as x approaches c from the right or from the left, then the line

$$x = c$$

is a **vertical asymptote** of the graph of f.

One of the most common instances of a vertical asymptote is the graph of a *rational function*—that is, a function of the form $f(x) = p(x)/q(x)$, where $p(x)$ and $q(x)$ are polynomials. If c is a real number such that $q(c) = 0$ and $p(c) \neq 0$, then the graph of f has a vertical asymptote at $x = c$. Example 1 shows four cases.

TECH TUTOR

Use a spreadsheet or table to verify the results shown in Example 1. (Consult the user's manual of a spreadsheet software program for specific instructions on how to create a table.) For instance, in Example 1(a), notice that the values of $f(x) = 1/(x - 1)$ decrease and increase without bound as x gets closer and closer to 1 from the left and the right.

***x* Approaches 1 from the Left**

x	$f(x) = 1/(x - 1)$
0	-1
0.9	-10
0.99	-100
0.999	-1000
0.9999	$-10{,}000$

***x* Approaches 1 from the Right**

x	$f(x) = 1/(x - 1)$
2	1
1.1	10
1.01	100
1.001	1000
1.0001	10,000

Example 1 Finding Infinite Limits

	Limit from the left	*Limit from the right*	
a.	$\displaystyle\lim_{x\to1^-} \frac{1}{x-1} = -\infty$	$\displaystyle\lim_{x\to1^+} \frac{1}{x-1} = \infty$	See Figure 3.47(a).
b.	$\displaystyle\lim_{x\to1^-} \frac{-1}{x-1} = \infty$	$\displaystyle\lim_{x\to1^+} \frac{-1}{x-1} = -\infty$	See Figure 3.47(b).
c.	$\displaystyle\lim_{x\to1^-} \frac{-1}{(x-1)^2} = -\infty$	$\displaystyle\lim_{x\to1^+} \frac{-1}{(x-1)^2} = -\infty$	See Figure 3.47(c).
d.	$\displaystyle\lim_{x\to1^-} \frac{1}{(x-1)^2} = \infty$	$\displaystyle\lim_{x\to1^+} \frac{1}{(x-1)^2} = \infty$	See Figure 3.47(d).

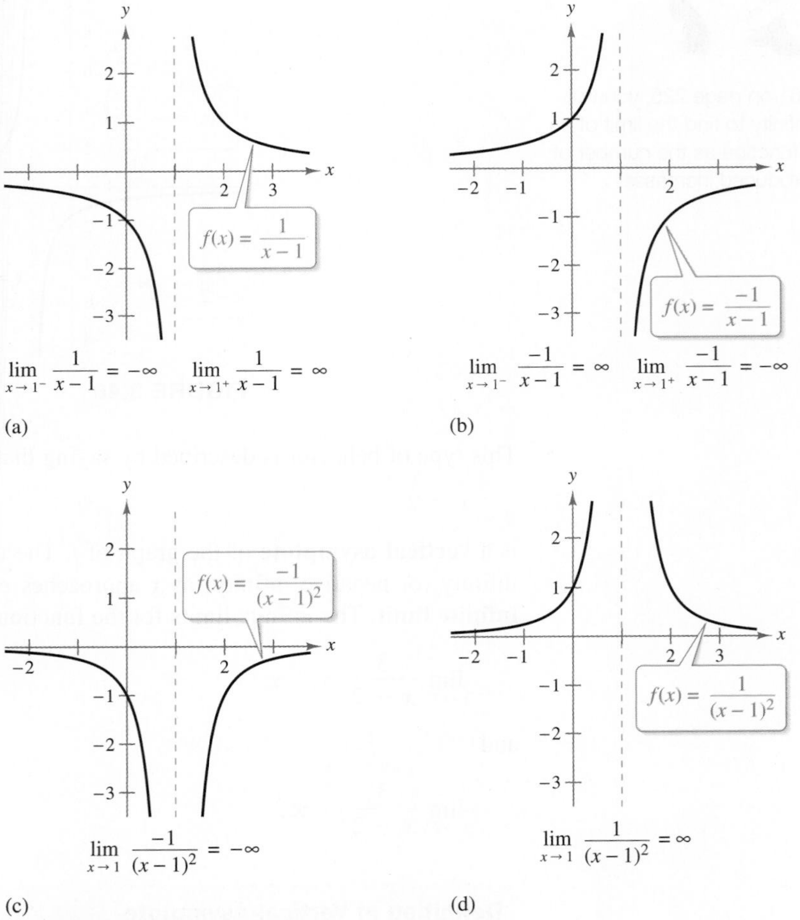

FIGURE 3.47

✓Checkpoint 1

Find each limit.

a. $\displaystyle\lim_{x\to2^-} \frac{1}{x-2}$ **b.** $\displaystyle\lim_{x\to2^+} \frac{1}{x-2}$ **c.** $\displaystyle\lim_{x\to-3^-} \frac{1}{x+3}$ **d.** $\displaystyle\lim_{x\to-3^+} \frac{1}{x+3}$ ■

Each of the graphs in Example 1 has only one vertical asymptote. As shown in the next example, the graph of a rational function can have more than one vertical asymptote.

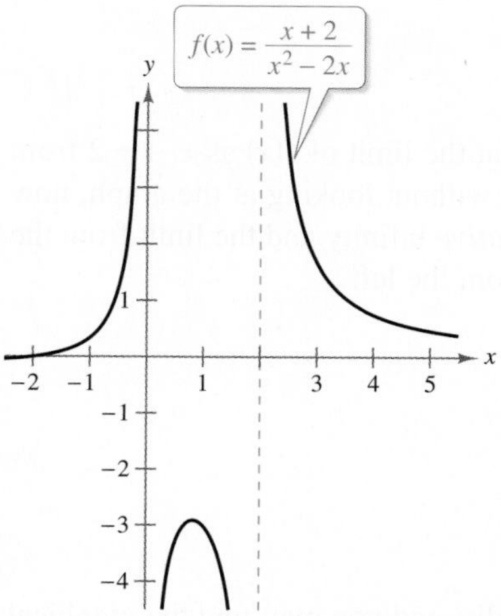

Vertical Asymptotes at $x = 0$ and $x = 2$

FIGURE 3.48

Example 2 Finding Vertical Asymptotes

Determine all vertical asymptotes of the graph of

$$f(x) = \frac{x + 2}{x^2 - 2x}.$$

SOLUTION The possible vertical asymptotes correspond to the x-values for which the denominator is zero.

$$x^2 - 2x = 0 \quad \text{Set denominator equal to 0.}$$

$$x(x - 2) = 0 \quad \text{Factor.}$$

$$x = 0, x = 2 \quad \text{Zeros of denominator}$$

Because the numerator of f is not zero at either of these x-values, you can conclude that the graph of f has two vertical asymptotes—one at $x = 0$ and one at $x = 2$, as shown in Figure 3.48.

✓ Checkpoint 2

Determine all vertical asymptotes of the graph of

$$f(x) = \frac{x + 4}{x^2 - 4x}.$$

■

Example 3 Finding Vertical Asymptotes

Determine all vertical asymptotes of the graph of

$$f(x) = \frac{x^2 + 2x - 8}{x^2 - 4}.$$

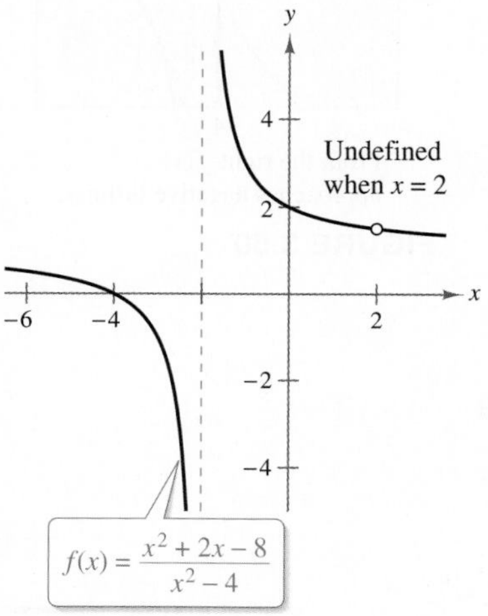

Vertical Asymptote at $x = -2$

FIGURE 3.49

SOLUTION First factor the numerator and denominator. Then divide out common factors.

$$f(x) = \frac{x^2 + 2x - 8}{x^2 - 4} \quad \text{Write original function.}$$

$$= \frac{(x + 4)(x - 2)}{(x + 2)(x - 2)} \quad \text{Factor numerator and denominator.}$$

$$= \frac{(x + 4)\cancel{(x - 2)}}{(x + 2)\cancel{(x - 2)}} \quad \text{Divide out common factors.}$$

$$= \frac{x + 4}{x + 2}, \quad x \neq 2 \quad \text{Simplify.}$$

For all values of x other than $x = 2$, the graph of this simplified function is the same as the graph of f. So, you can conclude that the graph of f has only one vertical asymptote. This occurs at $x = -2$, as shown in Figure 3.49.

✓ Checkpoint 3

Determine all vertical asymptotes of the graph of

$$f(x) = \frac{x^2 + 4x + 3}{x^2 - 9}.$$

■

TECH TUTOR

When you use a graphing utility to graph a function that has a vertical asymptote, the utility may try to connect separate branches of the graph. For instance, the figure below shows the graph of

$$f(x) = \frac{3}{x - 2}$$

on a graphing calculator.

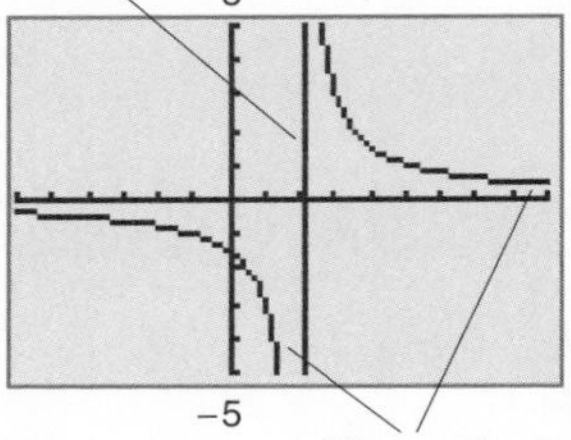

From Example 3, you know that the graph of

$$f(x) = \frac{x^2 + 2x - 8}{x^2 - 4}$$

has a vertical asymptote at $x = -2$. This implies that the limit of $f(x)$ as $x \to -2$ from the right (or from the left) is either ∞ or $-\infty$. But without looking at the graph, how can you determine that the limit from the left is *negative* infinity and the limit from the right is *positive* infinity? That is, why is the limit from the left

$$\lim_{x \to -2^-} \frac{x^2 + 2x - 8}{x^2 - 4} = -\infty \qquad \text{Limit from the left}$$

and why is the limit from the right

$$\lim_{x \to -2^+} \frac{x^2 + 2x - 8}{x^2 - 4} = \infty? \qquad \text{Limit from the right}$$

It is cumbersome to determine these limits analytically, and you may find the graphical method shown in Example 4 to be more efficient.

Example 4 Determining Infinite Limits

Find the limits.

$$\lim_{x \to 1^-} \frac{x^2 - 3x}{x - 1} \quad \text{and} \quad \lim_{x \to 1^+} \frac{x^2 - 3x}{x - 1}$$

SOLUTION Begin by considering the function

$$f(x) = \frac{x^2 - 3x}{x - 1}.$$

Because the denominator is zero when $x = 1$ and the numerator is not zero when $x = 1$, it follows that the graph of the function has a vertical asymptote at $x = 1$. This implies that each of the given limits is either ∞ or $-\infty$. To determine which, use a graphing utility to graph the function, as shown in Figure 3.50. From the graph, you can see that the limit from the left is positive infinity and the limit from the right is negative infinity. That is,

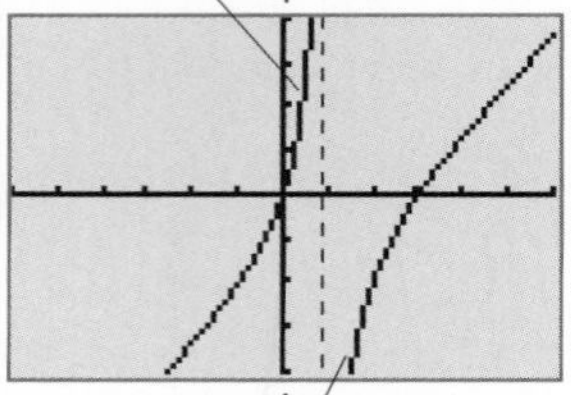

FIGURE 3.50

$$\lim_{x \to 1^-} \frac{x^2 - 3x}{x - 1} = \infty \qquad \text{Limit from the left}$$

and

$$\lim_{x \to 1^+} \frac{x^2 - 3x}{x - 1} = -\infty. \qquad \text{Limit from the right}$$

✓Checkpoint 4

Find the limits.

$$\lim_{x \to 2^-} \frac{x^2 - 4x}{x - 2} \quad \text{and} \quad \lim_{x \to 2^+} \frac{x^2 - 4x}{x - 2}$$

■

In Example 4, try evaluating $f(x)$ at x-values that are just barely to the left of 1. You will find that you can make the values of $f(x)$ arbitrarily large by choosing x sufficiently close to 1. For instance, $f(0.99999) \approx 199{,}999$.

Horizontal Asymptotes and Limits at Infinity

Another type of limit, called a **limit at infinity,** specifies a finite value approached by a function as x increases (or decreases) without bound.

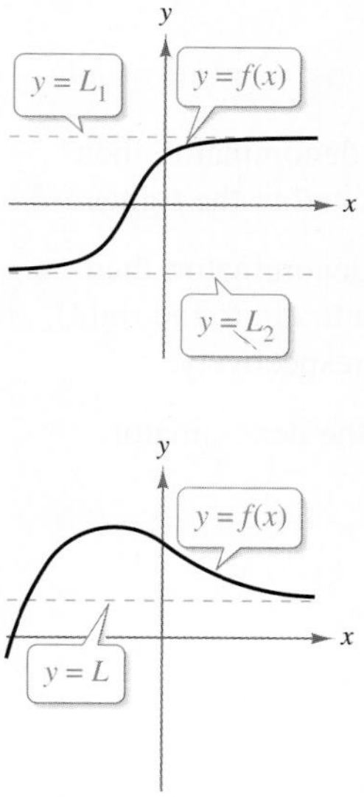

FIGURE 3.51

Definition of Horizontal Asymptote

If f is a function and L_1 and L_2 are real numbers, then the statements

$$\lim_{x\to\infty} f(x) = L_1 \quad \text{and} \quad \lim_{x\to-\infty} f(x) = L_2$$

denote **limits at infinity.** The lines $y = L_1$ and $y = L_2$ are **horizontal asymptotes** of the graph of f.

Figure 3.51 shows two ways in which the graph of a function can approach one or more horizontal asymptotes. Note that it is possible for the graph of a function to cross its horizontal asymptote.

Limits at infinity share many of the properties of limits discussed in Section 1.5. When finding horizontal asymptotes, you can use the property that

$$\lim_{x\to\infty} \frac{1}{x^r} = 0, \quad r > 0 \quad \text{and} \quad \lim_{x\to-\infty} \frac{1}{x^r} = 0, \quad r > 0.$$

(The second limit assumes that x^r is defined when $x < 0$.)

Example 5 Finding Limits at Infinity

Find the limit: $\lim_{x\to\infty} \left(5 - \frac{2}{x^2}\right)$.

SOLUTION

$$\begin{aligned}\lim_{x\to\infty} \left(5 - \frac{2}{x^2}\right) &= \lim_{x\to\infty} 5 - \lim_{x\to\infty} \frac{2}{x^2} \\ &= \lim_{x\to\infty} 5 - 2\left(\lim_{x\to\infty} \frac{1}{x^2}\right) \\ &= 5 - 2(0) \\ &= 5\end{aligned}$$

$\lim_{x\to\infty} [f(x) - g(x)] = \lim_{x\to\infty} f(x) - \lim_{x\to\infty} g(x)$

$\lim_{x\to\infty} cf(x) = c \lim_{x\to\infty} f(x)$

You can verify this limit by sketching the graph of

$$f(x) = 5 - \frac{2}{x^2}$$

as shown in Figure 3.52. Note that the graph has $y = 5$ as a horizontal asymptote to the right. By evaluating the limit

$$\lim_{x\to-\infty} \left(5 - \frac{2}{x^2}\right)$$

you can show that $y = 5$ is also a horizontal asymptote to the left.

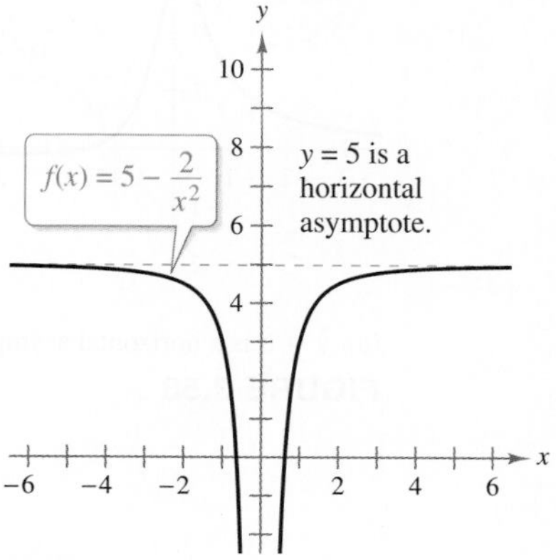

FIGURE 3.52

Checkpoint 5

Find the limit: $\lim_{x\to\infty} \left(2 + \frac{5}{x^2}\right)$.

There is an easy way to determine whether the graph of a *rational* function has a horizontal asymptote. This shortcut is based on a comparison of the degrees of the numerator and denominator of the rational function.

Horizontal Asymptotes of Rational Functions

Let $f(x) = p(x)/q(x)$ be a rational function.

1. If the degree of the numerator is less than the degree of the denominator, then $y = 0$ is a horizontal asymptote of the graph of f (to the left and to the right).
2. If the degree of the numerator is equal to the degree of the denominator, then $y = a/b$ is a horizontal asymptote of the graph of f (to the left and to the right), where a and b are the leading coefficients of $p(x)$ and $q(x)$, respectively.
3. If the degree of the numerator is greater than the degree of the denominator, then the graph of f has no horizontal asymptote.

Example 6 Finding Horizontal Asymptotes

Find the horizontal asymptote of the graph of each function.

a. $y = \dfrac{-2x + 3}{3x^2 + 1}$ **b.** $y = \dfrac{-2x^2 + 3}{3x^2 + 1}$ **c.** $y = \dfrac{-2x^3 + 3}{3x^2 + 1}$

SOLUTION

a. Because the degree of the numerator is less than the degree of the denominator, $y = 0$ is a horizontal asymptote. [See Figure 3.53(a).]

b. Because the degree of the numerator is equal to the degree of the denominator, the line $y = -\frac{2}{3}$ is a horizontal asymptote. [See Figure 3.53(b).]

c. Because the degree of the numerator is greater than the degree of the denominator, the graph has no horizontal asymptote. [See Figure 3.53(c).]

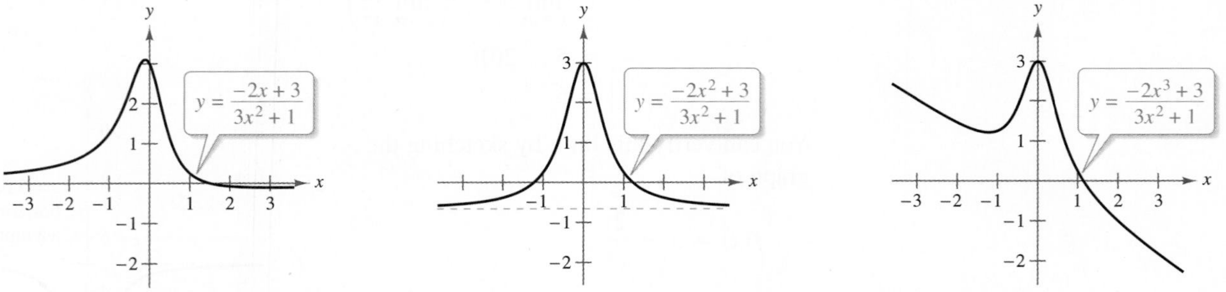

(a) $y = 0$ is a horizontal asymptote. (b) $y = -\frac{2}{3}$ is a horizontal asymptote. (c) No horizontal asymptote

FIGURE 3.53

✓ Checkpoint 6

Find the horizontal asymptote of the graph of each function.

a. $y = \dfrac{2x + 1}{4x^2 + 5}$ **b.** $y = \dfrac{2x^2 + 1}{4x^2 + 5}$ **c.** $y = \dfrac{2x^3 + 1}{4x^2 + 5}$

Some functions have two horizontal asymptotes: one to the right and one to the left (see Exercises 59 and 60).

Applications of Asymptotes

There are many examples of asymptotic behavior in real life. For instance, Example 7 describes the asymptotic behavior of an average cost function.

Example 7 Modeling Average Cost

A small business invests \$5000 in a new product. In addition to this initial investment, the product will cost \$0.50 per unit to produce.

a. Find the average cost per unit when 1000 units are produced.

b. Find the average cost per unit when 10,000 units are produced.

c. Find the average cost per unit when 100,000 units are produced.

d. What is the limit of the average cost as the number of units produced increases?

SOLUTION From the given information, you can model the total cost C (in dollars) by

$$C = 0.5x + 5000 \qquad \text{Total cost function}$$

where x is the number of units produced. This implies that the average cost function is

$$\overline{C} = \frac{C}{x} = 0.5 + \frac{5000}{x}. \qquad \text{Average cost function}$$

a. When only 1000 units are produced, the average cost per unit is

$$\overline{C} = 0.5 + \frac{5000}{1000} \qquad \text{Substitute 1000 for } x.$$

$$= \$5.50. \qquad \text{Average cost for 1000 units}$$

b. When 10,000 units are produced, the average cost per unit is

$$\overline{C} = 0.5 + \frac{5000}{10{,}000} \qquad \text{Substitute 10,000 for } x.$$

$$= \$1.00. \qquad \text{Average cost for 10,000 units}$$

c. When 100,000 units are produced, the average cost per unit is

$$\overline{C} = 0.5 + \frac{5000}{100{,}000} \qquad \text{Substitute 100,000 for } x.$$

$$= \$0.55. \qquad \text{Average cost for 100,000 units}$$

d. As x approaches infinity, the limiting average cost per unit is

$$\lim_{x\to\infty} \left(0.5 + \frac{5000}{x}\right) = \$0.50.$$

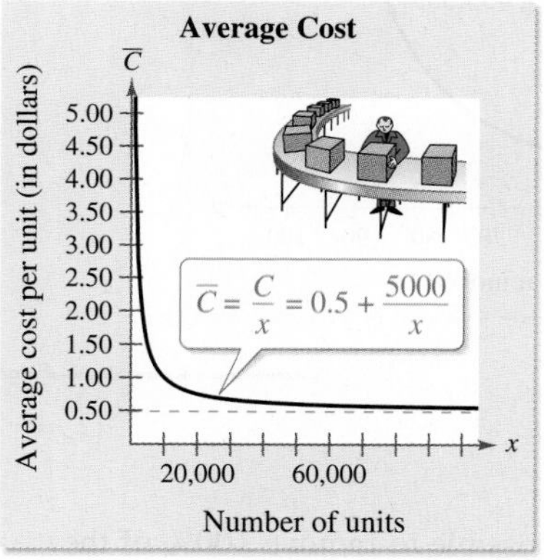

As $x \to \infty$, the average cost per unit approaches \$0.50.

FIGURE 3.54

As shown in Figure 3.54, this example points out one of the major problems faced by small businesses. That is, it is difficult to have competitively low prices when the production level is low.

✓ Checkpoint 7

A small business invests \$25,000 in a new product. In addition, the product will cost \$0.75 per unit to produce. Find the cost function and the average cost function. What is the limit of the average cost function as the number of units produced increases? ■

In Example 7, suppose that the small business had made an initial investment of \$50,000. How would this change the answers to the questions? Would it change the average cost of producing x units? Would it change the limiting average cost per unit?

Example 8 Modeling Smokestack Emission

A manufacturing plant has determined that the cost C (in dollars) of removing $p\%$ of the smokestack pollutants of its main smokestack is modeled by

$$C = \frac{80{,}000p}{100 - p}, \quad 0 \le p < 100.$$

What is the vertical asymptote of this function? What does the vertical asymptote mean to the plant owners?

SOLUTION The graph of the cost function is shown in Figure 3.55. From the graph, you can see that $p = 100$ is the vertical asymptote. This means that as the plant attempts to remove higher and higher percents of the pollutants, the cost increases dramatically. For instance, the cost of removing 85% of the pollutants is

$$C = \frac{80{,}000(85)}{100 - 85} \approx \$453{,}333 \qquad \text{Cost for 85\% removal}$$

but the cost of removing 90% is

$$C = \frac{80{,}000(90)}{100 - 90} = \$720{,}000. \qquad \text{Cost for 90\% removal}$$

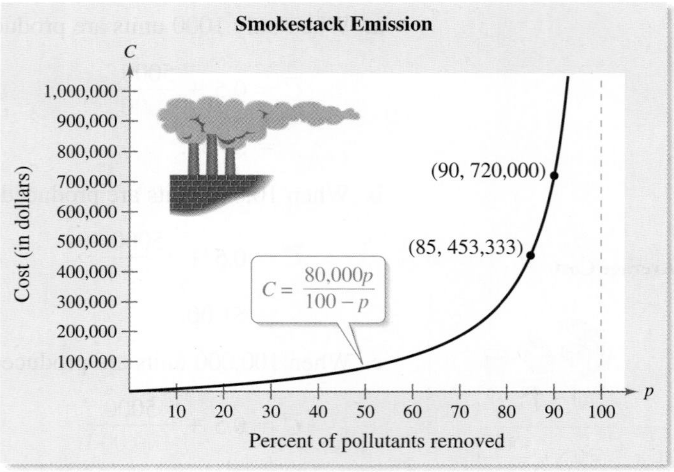

FIGURE 3.55

✓ Checkpoint 8

According to the cost function in Example 8, is it possible to remove 100% of the smokestack pollutants? Why or why not? ■

SUMMARIZE (Section 3.6)

1. State the definition of vertical asymptote *(page 215)*. For examples of vertical asymptotes, see Examples 1, 2, and 3.
2. State the definition of horizontal asymptote *(page 219)*. For examples of horizontal asymptotes, see Example 6.
3. Describe a real-life example of how asymptotic behavior can be used to analyze the average cost for a new product *(page 221, Example 7)*.

SKILLS WARM UP 3.6 The following warm-up exercises involve skills that were covered in earlier sections. You will use these skills in the exercise set for this section. For additional help, review Sections 1.5, 2.3, and 3.5.

In Exercises 1–8, find the limit.

1. $\lim_{x\to 2}(x+1)$

2. $\lim_{x\to -1}(3x+4)$

3. $\lim_{x\to -3}\dfrac{2x^2+x-15}{x+3}$

4. $\lim_{x\to 2}\dfrac{3x^2-8x+4}{x-2}$

5. $\lim_{x\to 2^+}\dfrac{x^2-5x+6}{x^2-4}$

6. $\lim_{x\to 1^-}\dfrac{x^2-6x+5}{x^2-1}$

7. $\lim_{x\to 0^+}\sqrt{x}$

8. $\lim_{x\to 1^+}\left(x+\sqrt{x-1}\right)$

In Exercises 9–12, find the average cost and the marginal cost.

9. $C = 150 + 3x$

10. $C = 1900 + 1.7x + 0.002x^2$

11. $C = 0.005x^2 + 0.5x + 1375$

12. $C = 760 + 0.05x$

Exercises 3.6

See www.CalcChat.com for worked-out solutions to odd-numbered exercises.

Vertical and Horizontal Asymptotes In Exercises 1–8, find the vertical and horizontal asymptotes.

1. $f(x) = \dfrac{x^2+1}{x^2}$

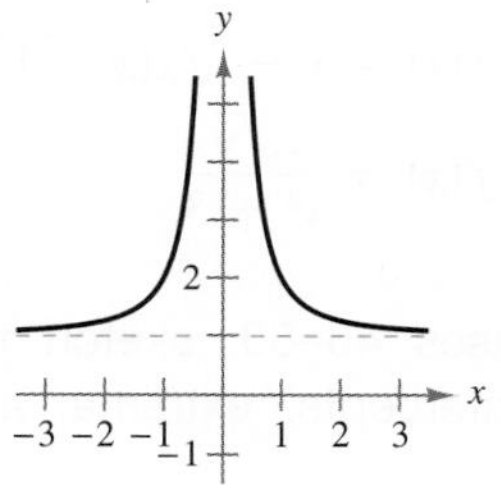

2. $f(x) = \dfrac{4}{(x-2)^3}$

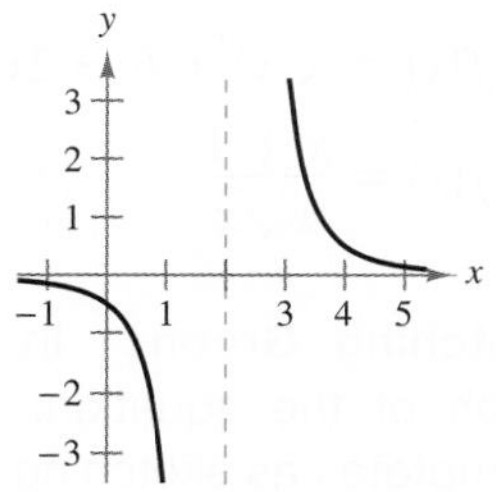

3. $f(x) = \dfrac{x^2-2}{x^2-x-2}$

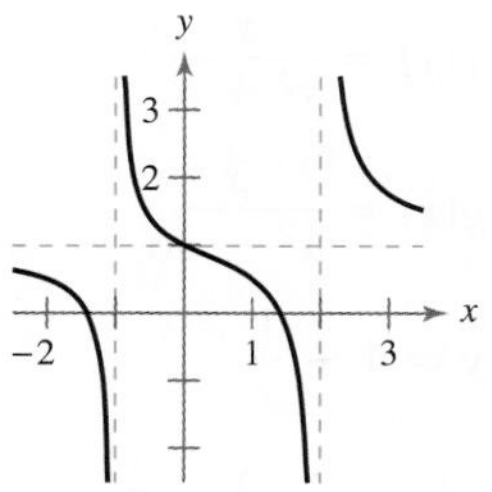

4. $y = \dfrac{x+1}{x+2}$

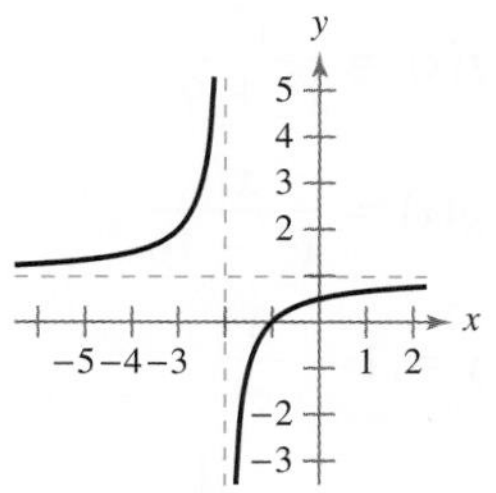

5. $f(x) = \dfrac{3x^2}{2(x^2+1)}$

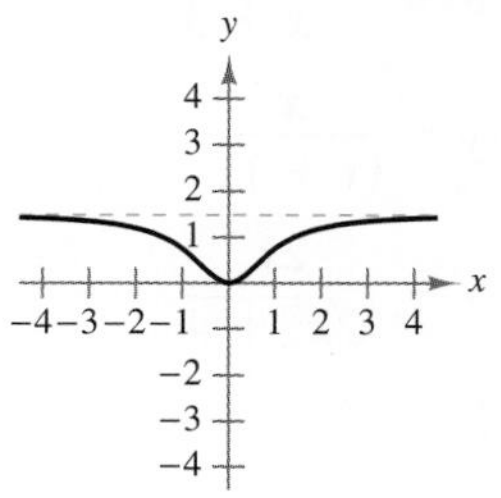

6. $f(x) = \dfrac{-4x}{x^2+4}$

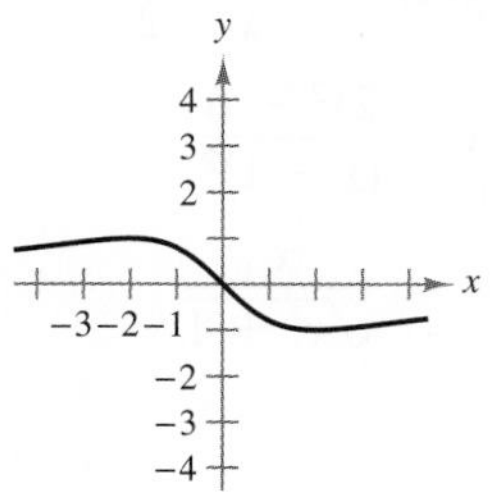

7. $f(x) = \dfrac{x^2-1}{2x^2-8}$

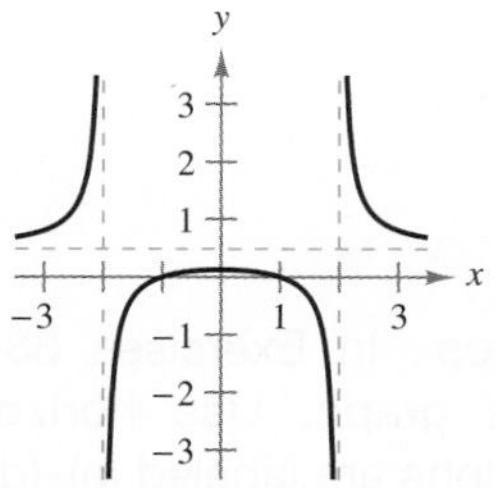

8. $f(x) = \dfrac{x^2+1}{x^3-8}$

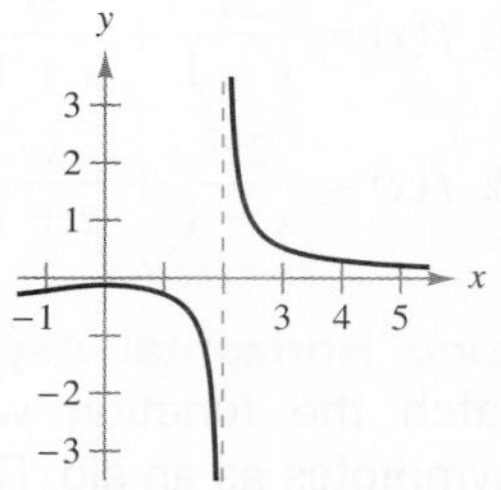

Finding Vertical Asymptotes In Exercises 9–14, determine all vertical asymptotes of the graph of the function. *See Examples 2 and 3.*

9. $f(x) = \dfrac{x-3}{x^2+3x}$

10. $f(x) = \dfrac{x}{x^2+6x}$

11. $f(x) = \dfrac{x^2-8x+15}{x^2-9}$

12. $f(x) = \dfrac{x^2+2x-35}{x^2-25}$

13. $f(x) = \dfrac{2x^2-x-3}{2x^2-11x+12}$

14. $f(x) = \dfrac{x^2+x-30}{4x^2-17x-15}$

Determining Infinite Limits In Exercises 15–20, use a graphing utility to find the limit. *See Example 4.*

15. $\lim_{x\to 6^+}\dfrac{1}{(x-6)^2}$

16. $\lim_{x\to -2^-}\dfrac{1}{x+2}$

17. $\lim_{x\to 3^+}\dfrac{x-4}{x-3}$

18. $\lim_{x\to 1^+}\dfrac{2+x}{1-x}$

19. $\lim_{x\to -1^-}\dfrac{x^2+1}{x^2-1}$

20. $\lim_{x\to 5^+}\dfrac{2x-3}{x^2-25}$

Finding Limits at Infinity In Exercises 21–24, find the limit. *See Example 5.*

21. $\lim_{x\to\infty}\left(1+\frac{1}{x}\right)$ **22.** $\lim_{x\to-\infty}\left(6-\frac{3}{x}\right)$

23. $\lim_{x\to-\infty}\left(7+\frac{4}{x^2}\right)$ **24.** $\lim_{x\to\infty}\left(10-\frac{8}{x^2}\right)$

Finding Horizontal Asymptotes In Exercises 25–32, find the horizontal asymptote of the graph of the function. *See Example 6.*

25. $f(x)=\frac{4x-3}{2x+1}$

26. $f(x)=\frac{5x^2+1}{10x^3-3x^2+7}$

27. $f(x)=\frac{3x}{4x^2-1}$

28. $f(x)=\frac{2x^2-5x-12}{1-6x-8x^2}$

29. $f(x)=\frac{5x^2}{x+3}$

30. $f(x)=\frac{x^3-2x^2+3x+1}{x^2-3x+2}$

31. $f(x)=\frac{2x}{x-1}+\frac{3x}{x+1}$

32. $f(x)=\frac{2x^2}{x-1}+\frac{3x}{x+1}$

Using Horizontal Asymptotes In Exercises 33–36, match the function with its graph. Use horizontal asymptotes as an aid. [The graphs are labeled (a)–(d).]

(a)

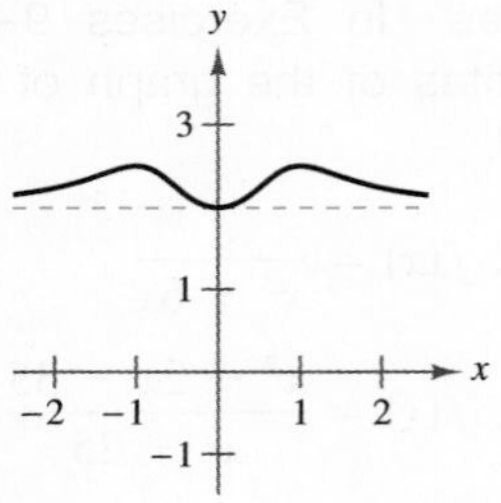

(b)

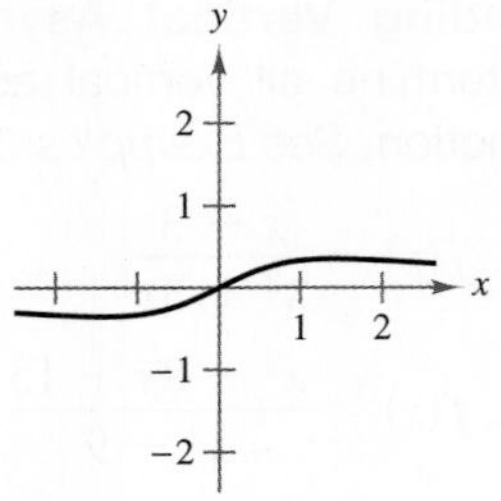

(c)

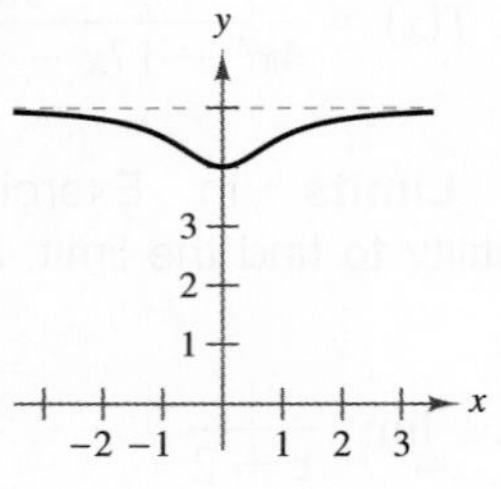

(d)

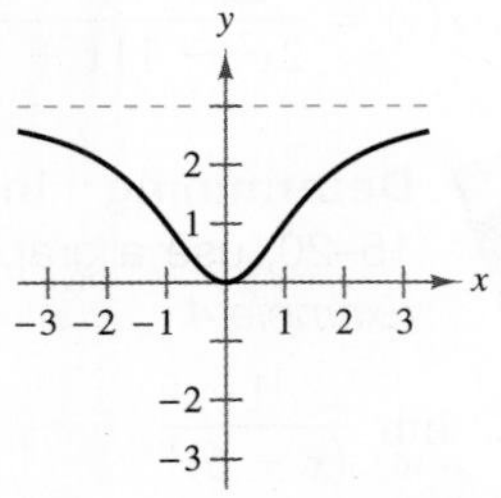

33. $f(x)=\frac{3x^2}{x^2+2}$ **34.** $f(x)=\frac{x}{x^2+2}$

35. $f(x)=2+\frac{x^2}{x^4+1}$ **36.** $f(x)=5-\frac{1}{x^2+1}$

Finding Limits at Infinity In Exercises 37 and 38, find $\lim_{x\to\infty} h(x)$, if possible.

37. $f(x)=5x^3-3$

(a) $h(x)=\frac{f(x)}{x^2}$ (b) $h(x)=\frac{f(x)}{x^3}$ (c) $h(x)=\frac{f(x)}{x^4}$

38. $f(x)=3x^2+7$

(a) $h(x)=\frac{f(x)}{x}$ (b) $h(x)=\frac{f(x)}{x^2}$ (c) $h(x)=\frac{f(x)}{x^3}$

Finding Limits at Infinity In Exercises 39 and 40, find each limit, if possible.

39. (a) $\lim_{x\to\infty}\frac{x^2+2}{x^3-1}$ (b) $\lim_{x\to\infty}\frac{x^2+2}{x^2-1}$ (c) $\lim_{x\to\infty}\frac{x^2+2}{x-1}$

40. (a) $\lim_{x\to\infty}\frac{4-5x}{2x^3+6}$ (b) $\lim_{x\to\infty}\frac{4-5x}{2x+6}$ (c) $\lim_{x\to\infty}\frac{4-5x^2}{2x+6}$

Estimating Limits at Infinity In Exercises 41–44, use a graphing utility or spreadsheet software program to complete the table. Then use the result to estimate the limit of $f(x)$ as x approaches infinity.

x	10^0	10^1	10^2	10^3	10^4	10^5	10^6
$f(x)$							

41. $f(x)=\sqrt{x^3+6}-2x$ **42.** $f(x)=x-\sqrt{x(x-1)}$

43. $f(x)=\frac{x+1}{x\sqrt{x}}$ **44.** $f(x)=\frac{\sqrt{x}}{x^2+3}$

Sketching Graphs In Exercises 45–60, sketch the graph of the equation. Use intercepts, extrema, and asymptotes as sketching aids.

45. $y=\frac{3x}{1-x}$ **46.** $y=\frac{x-3}{x-2}$

47. $f(x)=\frac{x^2}{x^2+9}$ **48.** $f(x)=\frac{x}{x^2+4}$

49. $g(x)=\frac{x^2}{x^2-16}$ **50.** $g(x)=\frac{x}{x^2-36}$

51. $y=1-\frac{3}{x^2}$ **52.** $y=1+\frac{1}{x}$

53. $f(x)=\frac{1}{x^2-x-2}$ **54.** $f(x)=\frac{x-2}{x^2-4x+3}$

55. $g(x)=\frac{x^2-x-2}{x-2}$ **56.** $g(x)=\frac{x^2-9}{x+3}$

57. $y=\frac{2x^2-6}{(x-1)^2}$ **58.** $y=\frac{x}{(x+1)^2}$

59. $y=\frac{x}{\sqrt{x^2+1}}$ **60.** $y=\frac{2x}{\sqrt{x^2+4}}$

61. Average Cost The cost C (in dollars) of producing x units of a product is $C = 1.15x + 6000$.

(a) Find the average cost function $\overline{C}$.

(b) Find $\overline{C}$ when $x = 600$ and when $x = 6000$.

(c) Determine the limit of the average cost function as x approaches infinity. Interpret the limit in the context of the problem.

62. Average Cost The cost C (in dollars) for a company to recycle x tons of material is $C = 1.25x + 10{,}500$.

(a) Find the average cost function $\overline{C}$.

(b) Find $\overline{C}$ when $x = 100$ and when $x = 1000$.

(c) Determine the limit of the average cost function as x approaches infinity. Interpret the limit in the context of the problem.

63. Average Profit The cost C and revenue R functions (in dollars) for producing and selling x units of a product are $C = 34.5x + 15{,}000$ and $R = 69.9x$.

(a) Find the average profit function

$$\overline{P} = \frac{R - C}{x}.$$

(b) Find the average profits when x is 1000, 10,000, and 100,000.

(c) What is the limit of the average profit function as x approaches infinity? Explain your reasoning.

64. HOW DO YOU SEE IT? The graph shows the temperature T (in degrees Fahrenheit) of an apple pie t seconds after it is removed from an oven.

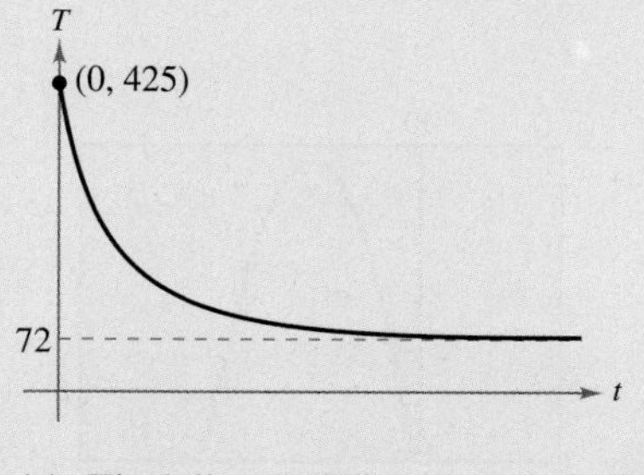

(a) Find $\lim_{t \to 0^+} T$. What does this limit represent?

(b) Find $\lim_{t \to \infty} T$. What does this limit represent?

65. Seizing Drugs The cost C (in millions of dollars) for the federal government to seize $p\%$ of an illegal drug as it enters the country is modeled by

$$C = \frac{528p}{100 - p}, \quad 0 \le p < 100.$$

(a) Find the costs of seizing 25%, 50%, and 75%.

(b) Find the limit of C as $p \to 100^-$. Interpret the limit in the context of the problem. Use a graphing utility to verify your result.

66. Removing Pollutants The cost C (in dollars) of removing $p\%$ of the air pollutants in the stack emission of a utility company that burns coal is modeled by

$$C = \frac{85{,}000p}{100 - p}, \quad 0 \le p < 100.$$

(a) Find the costs of removing 15%, 50%, and 95%.

(b) Find the limit of C as $p \to 100^-$. Interpret the limit in the context of the problem. Use a graphing utility to verify your result.

67. Learning Curve Psychologists have developed mathematical models to predict performance P (the percent of correct responses in decimal form) as a function of n, the number of times a task is performed. One such model is

$$P = \frac{0.5 + 0.9(n - 1)}{1 + 0.9(n - 1)}, \quad n > 0.$$

(a) Use a spreadsheet software program to complete the table for the model.

n	1	2	3	4	5	6	7	8	9	10
P										

(b) Find the limit as n approaches infinity.

(c) Use a graphing utility to graph this learning curve, and interpret the graph in the context of the problem.

68. Project: Alternative-Fueled Vehicles
For a project analyzing the number of alternative-fueled vehicles in use in the United States, visit this text's website at *www.cengagebrain.com*. *(Source: U.S. Energy Information Administration)*

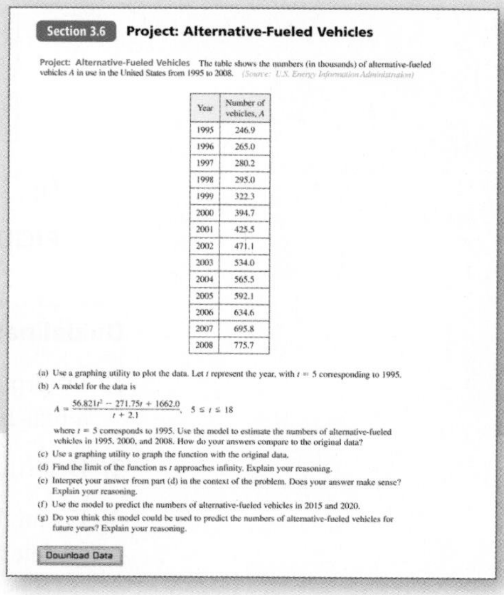

Section 3.6 **Project: Alternative-Fueled Vehicles**

Project: Alternative-Fueled Vehicles The table shows the numbers (in thousands) of alternative-fueled vehicles A in use in the United States from 1995 to 2008. *(Source: U.S. Energy Information Administration)*

Year	Number of vehicles, A
1995	246.9
1996	265.0
1997	280.2
1998	295.0
1999	322.3
2000	394.7
2001	425.5
2002	471.1
2003	534.0
2004	565.5
2005	592.1
2006	634.6
2007	695.8
2008	775.7

(a) Use a graphing utility to plot the data. Let t represent the year, with $t = 5$ corresponding to 1995.

(b) A model for the data is

$$A = \frac{56.821t^2 - 271.75t + 1662.0}{t + 2.1}, \quad 5 \le t \le 18$$

where $t = 5$ corresponds to 1995. Use the model to estimate the numbers of alternative-fueled vehicles in 1995, 2000, and 2008. How do your answers compare to the original data?

(c) Use a graphing utility to graph the function with the original data.

(d) Find the limit of the function as t approaches infinity. Explain your reasoning.

(e) Interpret your answer from part (d) in the context of the problem. Does your answer make sense? Explain your reasoning.

(f) Use the model to predict the numbers of alternative-fueled vehicles in 2015 and 2020.

(g) Do you think this model could be used to predict the numbers of alternative-fueled vehicles for future years? Explain your reasoning.

Download Data

3.7 Curve Sketching: A Summary

- Analyze the graphs of functions.
- Recognize the graphs of simple polynomial functions.

In Exercise 45 on page 234, you will analyze the graph of the Social Security average monthly benefits to determine whether the model is a good fit for the data.

Summary of Curve-Sketching Techniques

It would be difficult to overstate the importance of using graphs in mathematics. Descartes's introduction of analytic geometry contributed significantly to the rapid advances in calculus that began during the mid-seventeenth century.

So far, you have studied several concepts that are useful in analyzing the graph of a function.

- x-intercepts and y-intercepts (Section 1.2)
- Domain and range (Section 1.4)
- Continuity (Section 1.6)
- Differentiability (Section 2.1)
- Relative extrema (Section 3.2)
- Concavity (Section 3.3)
- Points of inflection (Section 3.3)
- Vertical asymptotes (Section 3.6)
- Horizontal asymptotes (Section 3.6)

When you are sketching the graph of a function, either by hand or with a graphing utility, remember that you cannot normally show the *entire* graph. The decision as to which part of the graph to show is crucial. For instance, which of the viewing windows in Figure 3.56 better represents the graph of

$$f(x) = x^3 - 25x^2 + 74x - 20?$$

Figure 3.56(a) gives a more complete view of the graph, but the context of the problem might indicate that Figure 3.56(b) is better.

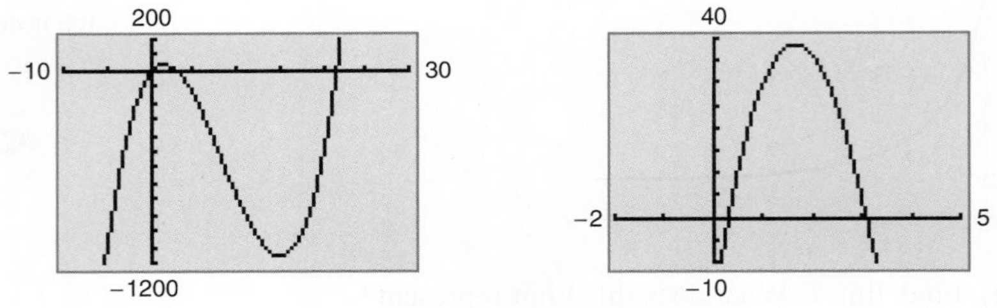

(a) (b)

FIGURE 3.56

Guidelines for Analyzing the Graph of a Function

1. Determine the domain and range of the function. When the function models a real-life situation, consider the context.
2. Determine the intercepts and asymptotes of the graph.
3. Locate the x-values at which $f'(x)$ and $f''(x)$ are zero or undefined. Use the results to determine where the relative extrema and the points of inflection occur.

TECH TUTOR

In Example 1, you are able to find the zeros of f, f', and f'' algebraically (by factoring). When this is not feasible, you can use a graphing utility to find the zeros. For instance, the function

$$g(x) = x^3 + 3x^2 - 9x + 6$$

is similar to the function in the example, but it does not factor with integer coefficients. Using a graphing utility, you can determine that the function has only one x-intercept, $x \approx -5.0275$.

Example 1 Analyzing a Graph

Analyze the graph of

$$f(x) = x^3 + 3x^2 - 9x + 5.$$

SOLUTION The y-intercept occurs at $(0, 5)$. Because this function factors as

$$f(x) = (x - 1)^2(x + 5) \qquad \text{Factored form}$$

the x-intercepts occur at $(-5, 0)$ and $(1, 0)$. The first derivative is

$$f'(x) = 3x^2 + 6x - 9 \qquad \text{First derivative}$$
$$= 3(x - 1)(x + 3). \qquad \text{Factored form}$$

So, the critical numbers of f are $x = 1$ and $x = -3$. The second derivative of f is

$$f''(x) = 6x + 6 \qquad \text{Second derivative}$$
$$= 6(x + 1) \qquad \text{Factored form}$$

which implies that the second derivative is zero when $x = -1$. By testing the values of $f'(x)$ and $f''(x)$, as shown in the table, you can see that f has one relative minimum, one relative maximum, and one point of inflection. The graph of f is shown in Figure 3.57.

	$f(x)$	$f'(x)$	$f''(x)$	Characteristics of graph
x in $(-\infty, -3)$		+	−	Increasing, concave downward
$x = -3$	32	0	−	Relative maximum
x in $(-3, -1)$		−	−	Decreasing, concave downward
$x = -1$	16	−	0	Point of inflection
x in $(-1, 1)$		−	+	Decreasing, concave upward
$x = 1$	0	0	+	Relative minimum
x in $(1, \infty)$		+	+	Increasing, concave upward

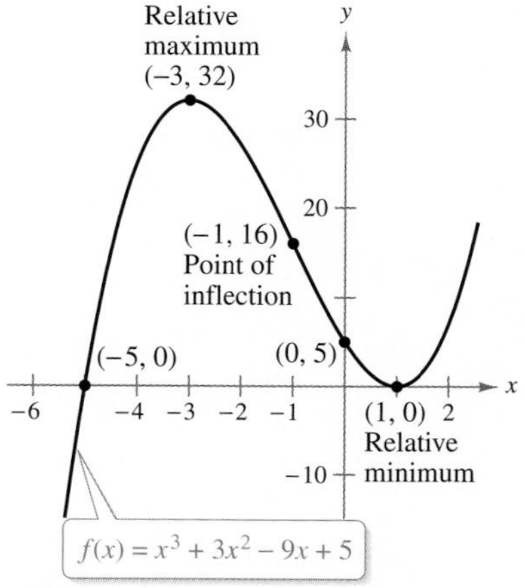

FIGURE 3.57

✓ Checkpoint 1

Analyze the graph of

$$f(x) = -x^3 + 3x^2 + 9x - 27.$$

Example 2 Analyzing a Graph

Analyze the graph of

$$f(x) = x^4 - 12x^3 + 48x^2 - 64x.$$

SOLUTION One of the intercepts occurs at (0, 0). Because this function factors as

$$f(x) = x(x^3 - 12x^2 + 48x - 64)$$
$$= x(x - 4)^3 \qquad \text{Factored form}$$

a second x-intercept occurs at (4, 0). The first derivative is

$$f'(x) = 4x^3 - 36x^2 + 96x - 64 \qquad \text{First derivative}$$
$$= 4(x - 1)(x - 4)^2. \qquad \text{Factored form}$$

So, the critical numbers of f are $x = 1$ and $x = 4$. The second derivative of f is

$$f''(x) = 12x^2 - 72x + 96 \qquad \text{Second derivative}$$
$$= 12(x - 4)(x - 2) \qquad \text{Factored form}$$

which implies that the second derivative is zero when $x = 2$ and $x = 4$. By testing the values of $f'(x)$ and $f''(x)$, as shown in the table, you can see that f has one relative minimum and two points of inflection. The graph is shown in Figure 3.58.

	$f(x)$	$f'(x)$	$f''(x)$	Characteristics of graph
x in $(-\infty, 1)$		$-$	$+$	Decreasing, concave upward
$x = 1$	-27	0	$+$	Relative minimum
x in $(1, 2)$		$+$	$+$	Increasing, concave upward
$x = 2$	-16	$+$	0	Point of inflection
x in $(2, 4)$		$+$	$-$	Increasing, concave downward
$x = 4$	0	0	0	Point of inflection
x in $(4, \infty)$		$+$	$+$	Increasing, concave upward

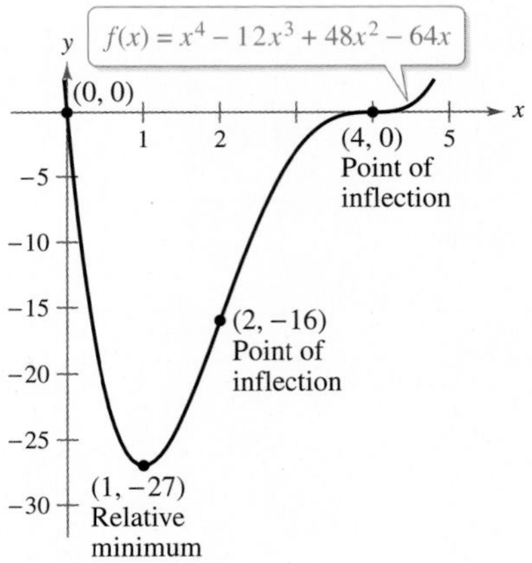

FIGURE 3.58

✓ Checkpoint 2

Analyze the graph of

$$f(x) = x^4 - 4x^3 + 5.$$

■

The fourth-degree polynomial function in Example 2 has one relative minimum and no relative maxima. In general, a polynomial function of degree n can have *at most* $n - 1$ relative extrema, and *at most* $n - 2$ points of inflection. Moreover, polynomial functions of even degree must have at least one relative extremum.

Example 3 Analyzing a Graph

Analyze the graph of

$$f(x) = \frac{x^2 - 2x + 4}{x - 2}.$$

SOLUTION The y-intercept occurs at $(0, -2)$. Using the Quadratic Formula on the numerator, you can see that there are no x-intercepts. Because the denominator is zero when $x = 2$ (and the numerator is not zero when $x = 2$), it follows that $x = 2$ is a vertical asymptote of the graph. There are no horizontal asymptotes because the degree of the numerator is greater than the degree of the denominator. The first derivative is

$$f'(x) = \frac{(x-2)(2x-2) - (x^2 - 2x + 4)}{(x-2)^2} \quad \text{First derivative}$$

$$= \frac{x(x-4)}{(x-2)^2}. \quad \text{Factored form}$$

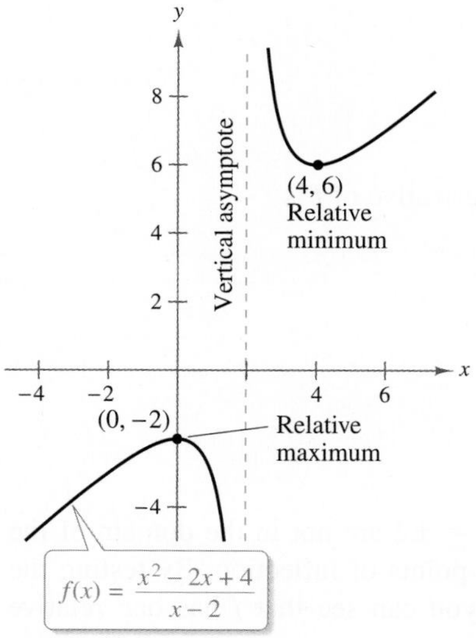

FIGURE 3.59

So, the critical numbers of f are $x = 0$ and $x = 4$. The second derivative is

$$f''(x) = \frac{(x-2)^2(2x-4) - (x^2 - 4x)(2)(x-2)}{(x-2)^4} \quad \text{Second derivative}$$

$$= \frac{(x-2)(2x^2 - 8x + 8 - 2x^2 + 8x)}{(x-2)^4}$$

$$= \frac{8}{(x-2)^3}. \quad \text{Factored form}$$

Because the second derivative has no zeros and because $x = 2$ is not in the domain of the function, you can conclude that the graph has no points of inflection. By testing the values of $f'(x)$ and $f''(x)$, as shown in the table, you can see that f has one relative minimum and one relative maximum. The graph of f is shown in Figure 3.59.

	$f(x)$	$f'(x)$	$f''(x)$	Characteristics of graph
x in $(-\infty, 0)$		+	−	Increasing, concave downward
$x = 0$	−2	0	−	Relative maximum
x in $(0, 2)$		−	−	Decreasing, concave downward
$x = 2$	Undef.	Undef.	Undef.	Vertical asymptote
x in $(2, 4)$		−	+	Decreasing, concave upward
$x = 4$	6	0	+	Relative minimum
x in $(4, \infty)$		+	+	Increasing, concave upward

✓ Checkpoint 3

Analyze the graph of

$$f(x) = \frac{x^2}{x - 1}.$$

Example 4 Analyzing a Graph

Analyze the graph of

$$f(x) = \frac{2(x^2 - 9)}{x^2 - 4}.$$

SOLUTION Begin by writing the function in factored form.

$$f(x) = \frac{2(x-3)(x+3)}{(x-2)(x+2)} \qquad \text{Factored form}$$

The y-intercept is $\left(0, \frac{9}{2}\right)$, and the x-intercepts are $(-3, 0)$ and $(3, 0)$. The graph of f has vertical asymptotes at $x = \pm 2$ and a horizontal asymptote at $y = 2$. The first derivative is

$$f'(x) = \frac{2[(x^2-4)(2x) - (x^2-9)(2x)]}{(x^2-4)^2} \qquad \text{First derivative}$$

$$= \frac{2(2x^3 - 8x - 2x^3 + 18x)}{(x^2-4)^2} \qquad \text{Multiply.}$$

$$= \frac{20x}{(x^2-4)^2}. \qquad \text{Factored form}$$

So, the critical number of f is $x = 0$. The second derivative of f is

$$f''(x) = \frac{(x^2-4)^2(20) - (20x)(2)(x^2-4)(2x)}{(x^2-4)^4} \qquad \text{Second derivative}$$

$$= \frac{20(x^2-4)(x^2-4-4x^2)}{(x^2-4)^4}$$

$$= -\frac{20(3x^2+4)}{(x^2-4)^3}. \qquad \text{Factored form}$$

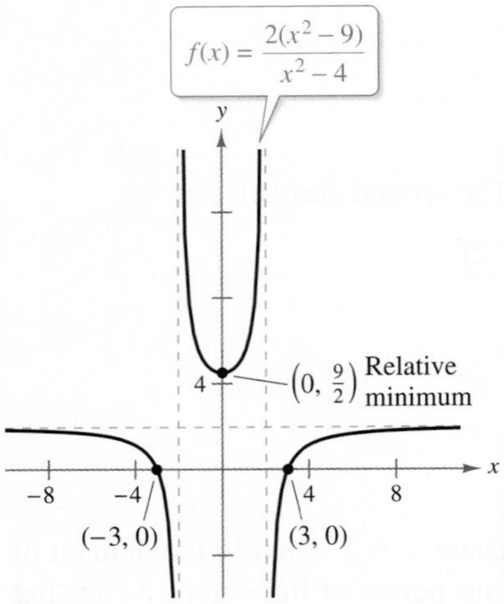

FIGURE 3.60

Because the second derivative has no zeros and $x = \pm 2$ are not in the domain of the function, you can conclude that the graph has no points of inflection. By testing the values of $f'(x)$ and $f''(x)$, as shown in the table, you can see that f has one relative minimum. The graph of f is shown in Figure 3.60.

	$f(x)$	$f'(x)$	$f''(x)$	Characteristics of graph
x in $(-\infty, -2)$		$-$	$-$	Decreasing, concave downward
$x = -2$	Undef.	Undef.	Undef.	Vertical asymptote
x in $(-2, 0)$		$-$	$+$	Decreasing, concave upward
$x = 0$	$\frac{9}{2}$	0	$+$	Relative minimum
x in $(0, 2)$		$+$	$+$	Increasing, concave upward
$x = 2$	Undef.	Undef.	Undef.	Vertical asymptote
x in $(2, \infty)$		$+$	$-$	Increasing, concave downward

✓ Checkpoint 4

Analyze the graph of

$$f(x) = \frac{x^2 + 1}{x^2 - 1}.$$

Example 5 **Analyzing a Graph**

Analyze the graph of

$$f(x) = 2x^{5/3} - 5x^{4/3}.$$

SOLUTION Begin by writing the function in factored form.

$$f(x) = x^{4/3}(2x^{1/3} - 5) \qquad \text{Factored form}$$

One of the intercepts is (0, 0). A second x-intercept occurs when $2x^{1/3} - 5 = 0$.

$$2x^{1/3} - 5 = 0$$
$$2x^{1/3} = 5$$
$$x^{1/3} = \frac{5}{2}$$
$$x = \left(\frac{5}{2}\right)^3$$
$$x = \frac{125}{8}$$

The first derivative is

$$f'(x) = \frac{10}{3}x^{2/3} - \frac{20}{3}x^{1/3} \qquad \text{First derivative}$$
$$= \frac{10}{3}x^{1/3}(x^{1/3} - 2). \qquad \text{Factored form}$$

So, the critical numbers of f are $x = 0$ and $x = 8$. The second derivative is

$$f''(x) = \frac{20}{9}x^{-1/3} - \frac{20}{9}x^{-2/3} \qquad \text{Second derivative}$$
$$= \frac{20}{9}x^{-2/3}(x^{1/3} - 1)$$
$$= \frac{20(x^{1/3} - 1)}{9x^{2/3}}. \qquad \text{Factored form}$$

So, possible points of inflection occur at $x = 1$ and when $x = 0$. By testing the values of $f'(x)$ and $f''(x)$, as shown in the table, you can see that f has one relative maximum, one relative minimum, and one point of inflection. The graph of f is shown in Figure 3.61.

	$f(x)$	$f'(x)$	$f''(x)$	Characteristics of graph
x in $(-\infty, 0)$		+	−	Increasing, concave downward
$x = 0$	0	0	Undef.	Relative maximum
x in $(0, 1)$		−	−	Decreasing, concave downward
$x = 1$	−3	−	0	Point of inflection
x in $(1, 8)$		−	+	Decreasing, concave upward
$x = 8$	−16	0	+	Relative minimum
x in $(8, \infty)$		+	+	Increasing, concave upward

TECH TUTOR

Some graphing utilities will not graph the function in Example 5 properly when the function is entered as

$f(x) = 2x\text{^}(5/3) - 5x\text{^}(4/3)$.

To correct for this, you can enter the function as

$f(x) = 2(\sqrt[3]{x})\text{^}5 - 5(\sqrt[3]{x})\text{^}4$.

Try entering both functions into a graphing utility to see whether both functions produce correct graphs.

ALGEBRA TUTOR

For help on the algebra in Example 5, see Example 2(a) in the *Chapter 3 Algebra Tutor*, on page 243.

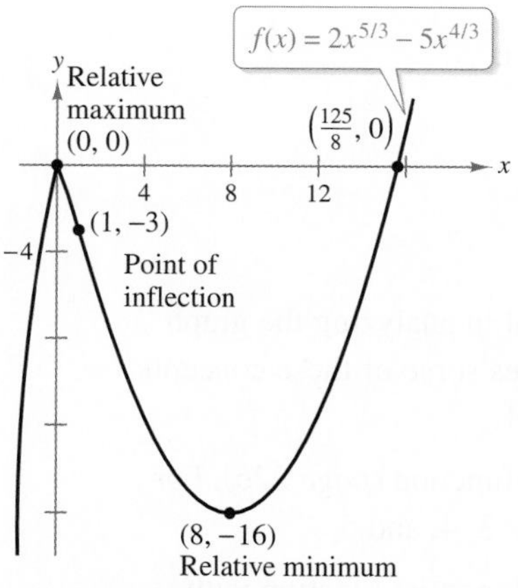

FIGURE 3.61

✓Checkpoint 5

Analyze the graph of $f(x) = 2x^{3/2} - 6x^{1/2}$. ■

Summary of Simple Polynomial Graphs

A summary of the graphs of polynomial functions of degrees 0, 1, 2, and 3 is shown in Figure 3.62. Because of their simplicity, lower-degree polynomial functions are commonly used as mathematical models.

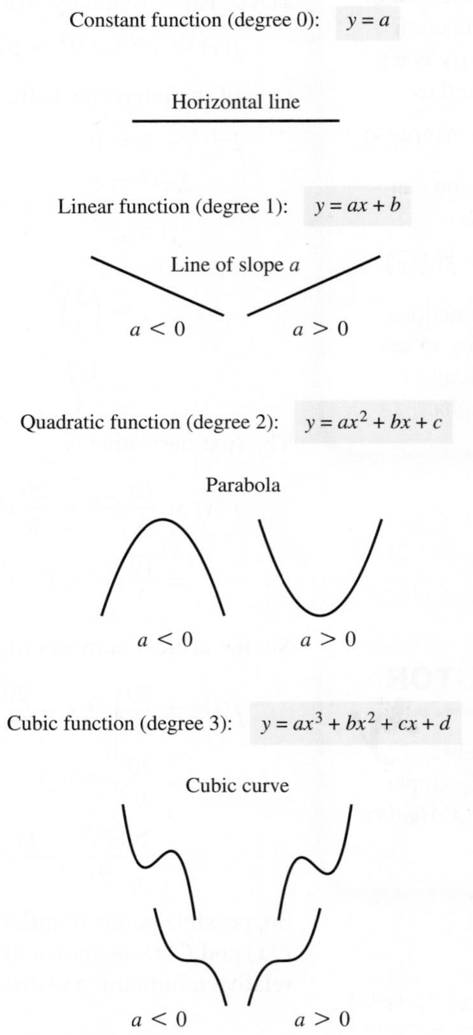

FIGURE 3.62

STUDY TIP

The graph of any cubic polynomial has one point of inflection. The slope of the graph at the point of inflection may be zero or nonzero.

SUMMARIZE (Section 3.7)

1. List the concepts you have learned that are useful in analyzing the graph of a function *(page 226)*. For an example that uses some of these concepts to analyze the graph of a function, see Example 1.
2. State the guidelines for analyzing the graph of a function *(page 226)*. For examples that use these guidelines, see Examples 3, 4, and 5.
3. State a general rule relating the degree n of a polynomial function with (a) the number of relative extrema and (b) the number of points of inflection *(page 229)*. For an example where this rule can be used to analyze the graph of a polynomial function, see Example 2.

SKILLS WARM UP 3.7 The following warm-up exercises involve skills that were covered in earlier sections. You will use these skills in the exercise set for this section. For additional help, review Sections 3.1 and 3.6.

In Exercises 1–4, find the vertical and horizontal asymptotes of the graph.

1. $f(x) = \dfrac{1}{x^2}$ **2.** $f(x) = \dfrac{8}{(x-2)^2}$ **3.** $f(x) = \dfrac{40x}{x+3}$ **4.** $f(x) = \dfrac{x^2-3}{x^2-4x+3}$

In Exercises 5–10, determine the open intervals on which the function is increasing or decreasing.

5. $f(x) = x^2 + 4x + 2$ **6.** $f(x) = -x^2 - 8x + 1$

7. $f(x) = x^3 - 3x + 1$ **8.** $f(x) = \dfrac{-x^3 + x^2 - 1}{x^2}$

9. $f(x) = \dfrac{x-2}{x-1}$ **10.** $f(x) = -x^3 - 4x^2 + 3x + 2$

Exercises 3.7

See www.CalcChat.com for worked-out solutions to odd-numbered exercises.

Analyzing a Graph In Exercises 1–22, analyze and sketch the graph of the function. Label any intercepts, relative extrema, points of inflection, and asymptotes. *See Examples 1, 2, 3, 4, and 5.*

1. $y = -x^2 - 2x + 3$ **2.** $y = 2x^2 - 4x + 1$

3. $y = x^3 - 4x^2 + 6$ **4.** $y = -x^3 + x - 2$

5. $y = 2 - x - x^3$ **6.** $y = x^3 + 3x^2 + 3x + 2$

7. $y = 3x^4 + 4x^3$ **8.** $y = x^4 - 2x^2$

9. $y = x^4 - 8x^3 + 18x^2 - 16x + 5$

10. $y = x^4 - 4x^3 + 16x - 16$

11. $y = \dfrac{x^2+1}{x}$ **12.** $y = \dfrac{x+2}{x}$

13. $y = \dfrac{x^2-6x+12}{x-4}$ **14.** $y = \dfrac{x^2+4x+7}{x+3}$

15. $y = \dfrac{x^2+1}{x^2-9}$ **16.** $y = \dfrac{2x}{x^2-1}$

17. $y = 3x^{2/3} - x^2$ **18.** $y = x^{5/3} - 5x^{2/3}$

19. $y = x\sqrt{9-x}$ **20.** $y = x\sqrt{4-x^2}$

21. $y = \begin{cases} x^2+1, & x \le 0 \\ 1-2x, & x > 0 \end{cases}$

22. $y = \begin{cases} x^2+4, & x < 0 \\ 4-x, & x \ge 0 \end{cases}$

Graphing a Function In Exercises 23–36, use a graphing utility to graph the function. Choose a window that allows all relative extrema and points of inflection to be identified on the graph.

23. $y = 3x^3 - 9x + 1$ **24.** $y = -4x^3 + 6x^2$

25. $y = x^5 - 5x$ **26.** $y = (x-1)^5$

27. $y = \dfrac{5-3x}{x-2}$ **28.** $y = \dfrac{x}{x^2+1}$

29. $y = 1 - x^{2/3}$ **30.** $y = (1-x)^{2/3}$

31. $y = x^{4/3}$ **32.** $y = x^{-1/3}$

33. $y = \dfrac{x}{\sqrt{x^2-4}}$

34. $y = \dfrac{x-3}{x}$

35. $y = \dfrac{x^3}{x^3-1}$

36. $y = \dfrac{x^4}{x^4-1}$

Interpreting a Graph In Exercises 37–40, use the graph of f' or f'' to sketch the graph of f. (There are many correct answers.)

37.

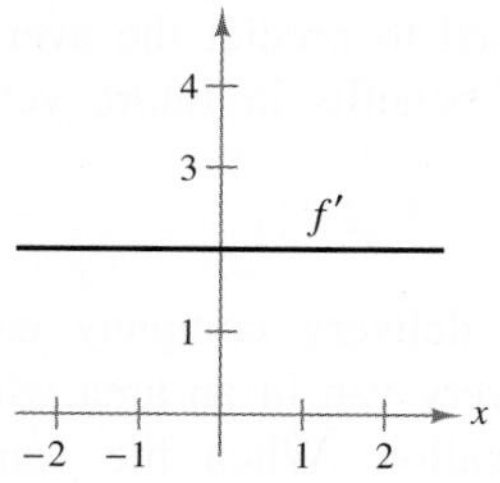

38.

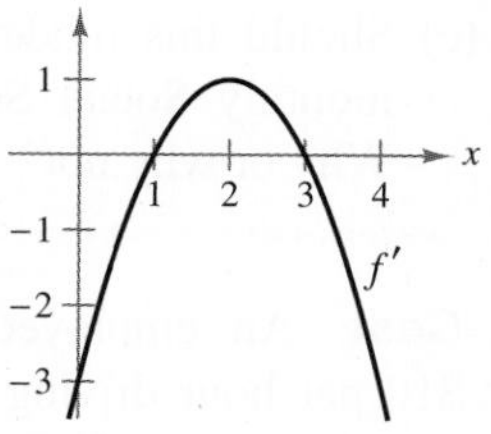

39.

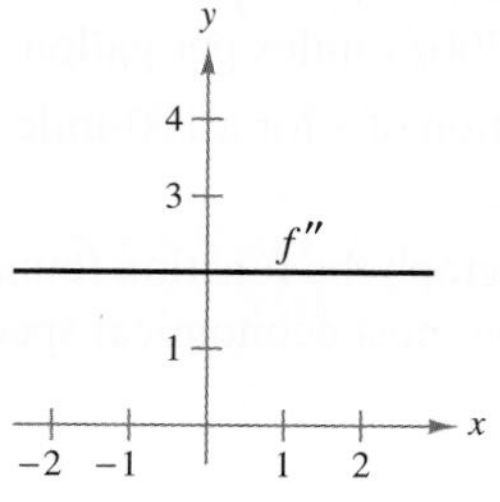

40.

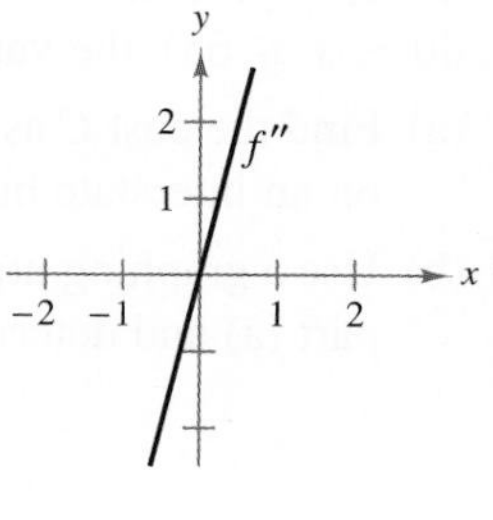

Sketching a Function In Exercises 41 and 42, sketch a graph of a function f having the given characteristics. (There are many correct answers.)

41. $f(-2) = f(0) = 0$
$f'(x) > 0$ if $x < -1$
$f'(x) < 0$ if $-1 < x < 0$
$f'(x) > 0$ if $x > 0$
$f'(-1) = f'(0) = 0$

42. $f(-1) = f(3) = 0$
$f'(1)$ is undefined.
$f'(x) < 0$ if $x < 1$
$f'(x) > 0$ if $x > 1$
$f''(x) < 0, x \neq 1$
$\lim_{x \to \infty} f(x) = 4$

Creating a Function In Exercises 43 and 44, create a function whose graph has the given characteristics. (There are many correct answers.)

43. Vertical asymptote: $x = 5$
Horizontal asymptote: $y = 0$

44. Vertical asymptote: $x = -3$
Horizontal asymptote: None

45. Social Security The table lists the average monthly Social Security benefits B (in dollars) for retired workers aged 62 and over from 2002 through 2009. A model for the data is

$$B = \frac{815.6 + 110.96t}{1 + 0.09t - 0.0033t^2}, \quad 2 \le t \le 9$$

where $t = 2$ corresponds to 2002. *(Source: U.S. Social Security Administration)*

t	2	3	4	5	6	7	8	9
B	895	922	955	1002	1044	1079	1153	1164

(a) Use a graphing utility to create a scatter plot of the data and graph the model in the same viewing window. How well does the model fit the data?

(b) Use the model to predict the average monthly benefit in 2014.

(c) Should this model be used to predict the average monthly Social Security benefits in future years? Why or why not?

46. Cost An employee of a delivery company earns \$10 per hour driving a delivery van in an area where gasoline costs \$2.80 per gallon. When the van is driven at a constant speed s (in miles per hour, with $40 \le s \le 65$), the van gets $700/s$ miles per gallon.

(a) Find the cost C as a function of s for a 100-mile trip on an interstate highway.

(b) Use a graphing utility to graph the function found in part (a) and determine the most economical speed.

47. Meteorology The monthly average high temperature T (in degrees Fahrenheit) for Boston, Massachusetts can be modeled by

$$T = \frac{30.83 - 2.861t + 0.181t^2}{1 - 0.206t + 0.0139t^2}, \quad 1 \le t \le 12$$

where t is the month, with $t = 1$ corresponding to January. Use a graphing utility to graph the model and find all absolute extrema. Interpret the meaning of these values in the context of the problem. *(Source: National Climatic Data Center)*

48. HOW DO YOU SEE IT? The graph shows a company's profits P for the years 1990 through 2010, where t is the year, with $t = 0$ corresponding to 2000.

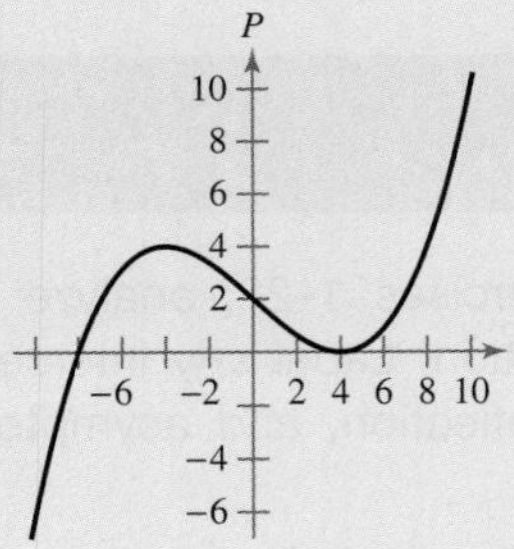

(a) For which values of t is P' zero? positive? negative? Interpret the meanings of these values in the context of the problem.

(b) For which values of t is P'' zero? positive? negative? Interpret the meanings of these values in the context of the problem.

Writing In Exercises 49 and 50, use a graphing utility to graph the function. Explain why there is no vertical asymptote when a superficial examination of the function may indicate that there should be one.

49. $h(x) = \dfrac{6 - 2x}{3 - x}$

50. $g(x) = \dfrac{x^2 + x - 2}{x - 1}$

51. Discovery Consider the function

$$f(x) = \frac{x^2 - 2x + 4}{x - 2}.$$

(a) Show that $f(x)$ can be rewritten as

$$f(x) = x + \frac{4}{x - 2}.$$

(b) Use a graphing utility to graph f and the line $y = x$. How do the two graphs compare as you zoom out?

(c) Use the results of part (b) to describe what is meant by a "slant asymptote."

3.8 Differentials and Marginal Analysis

- Find the differentials of functions.
- Use differentials in economics to approximate changes in revenue, cost, and profit.
- Find the differential of a function using differentiation formulas.

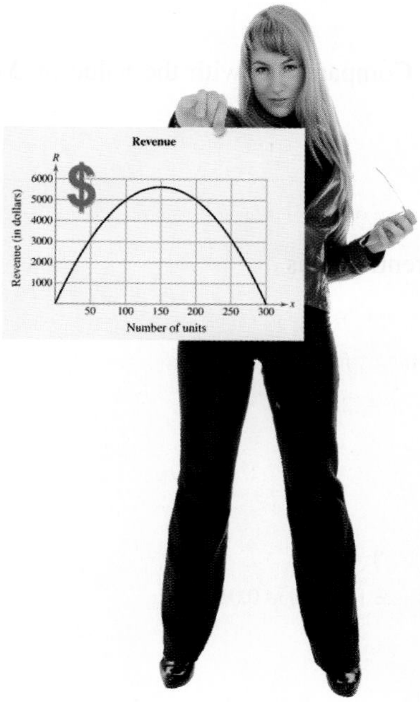

In Exercise 35 on page 241, you will use differentials to approximate the change in revenue for a one-unit increase in sales of a product.

Differentials

When the derivative was defined in Section 2.1 as the limit of the ratio $\Delta y/\Delta x$, it seemed natural to retain the quotient symbolism for the limit itself. So, the derivative of y with respect to x was denoted by

$$\frac{dy}{dx} = \lim_{\Delta x \to 0} \frac{\Delta y}{\Delta x}$$

even though dy/dx was not interpreted as the quotient of two separate quantities. In this section, you will see that the quantities dy and dx can be assigned meanings in such a way that their quotient, when $dx \neq 0$, is equal to the derivative of y with respect to x.

Definition of Differentials

Let $y = f(x)$ represent a differentiable function. The **differential of x** (denoted by dx) is any nonzero real number. The **differential of y** (denoted by dy) is $dy = f'(x)\,dx$.

In the definition of differentials, dx can have any nonzero value. In most applications, however, dx is chosen to be small, and this choice is denoted by $dx = \Delta x$.

One use of differentials is in approximating the change in $f(x)$ that corresponds to a change in x, as shown in Figure 3.63. This change is denoted by

$$\Delta y = f(x + \Delta x) - f(x). \qquad \text{Change in } y$$

In Figure 3.63, notice that as Δx gets smaller and smaller, the values of dy and Δy get closer and closer. That is, when Δx is small, $dy \approx \Delta y$. This **tangent line approximation** is the basis for most applications of differentials.

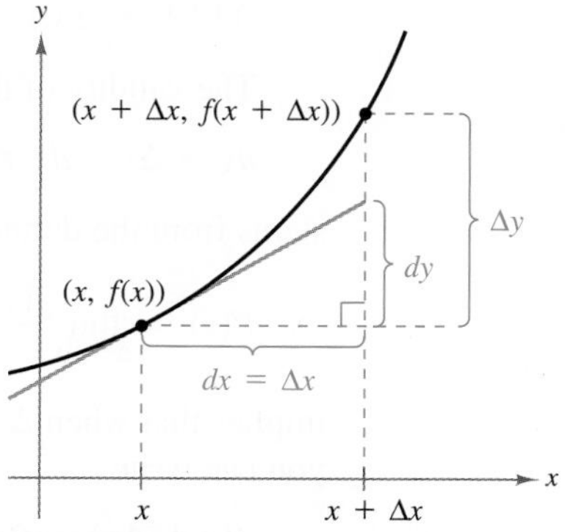

FIGURE 3.63

Note in Figure 3.63 that near the point of tangency, the graph of f is very close to the tangent line. This is the essence of the approximations used in this section. In other words, near the point of tangency, $dy \approx \Delta y$.

Example 1 Comparing Δy and dy

Consider the function given by

$$f(x) = x^2.$$

Find the value of dy when $x = 1$ and $dx = 0.01$. Compare this with the value of Δy when $x = 1$ and $\Delta x = 0.01$.

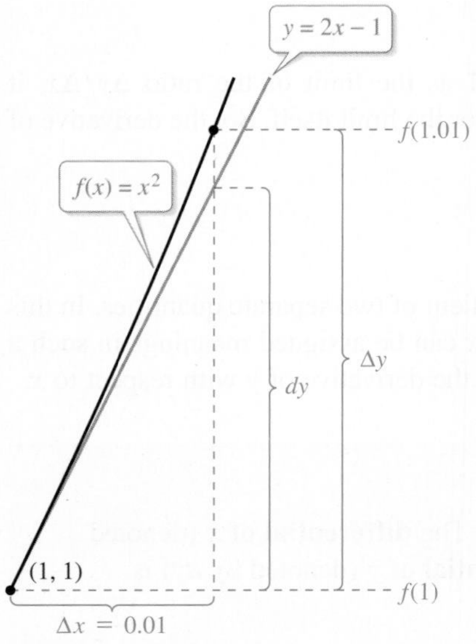

FIGURE 3.64

SOLUTION Begin by finding the derivative of f.

$$f'(x) = 2x \quad \text{Derivative of } f$$

When $x = 1$ and $dx = 0.01$, the value of the differential dy is

$$\begin{aligned} dy &= f'(x)\,dx && \text{Differential of } y \\ &= f'(1)(0.01) && \text{Substitute 1 for } x \text{ and 0.01 for } dx. \\ &= 2(1)(0.01) && \text{Use } f'(x) = 2x. \\ &= 0.02. && \text{Simplify.} \end{aligned}$$

When $x = 1$ and $\Delta x = 0.01$, the value of Δy is

$$\begin{aligned} \Delta y &= f(x + \Delta x) - f(x) && \text{Change in } y \\ &= f(1.01) - f(1) && \text{Substitute 1 for } x \text{ and 0.01 for } \Delta x. \\ &= (1.01)^2 - (1)^2 \\ &= 1.0201 - 1 \\ &= 0.0201. && \text{Simplify.} \end{aligned}$$

Note that $dy \approx \Delta y$, as shown in Figure 3.64.

✓Checkpoint 1

Find the value of dy when $x = 2$ and $dx = 0.01$ for $f(x) = x^4$. Compare this with the value of Δy when $x = 2$ and $\Delta x = 0.01$. ■

In Example 1, the tangent line to the graph of $f(x) = x^2$ at $x = 1$ is

$$y = 2x - 1 \quad \text{or} \quad g(x) = 2x - 1. \quad \text{Tangent line to the graph of } f \text{ at } x = 1$$

For x-values near 1, this line is close to the graph of f, as shown in Figure 3.64. For instance,

$$f(1.01) = 1.01^2 = 1.0201 \quad \text{and} \quad g(1.01) = 2(1.01) - 1 = 1.02.$$

The validity of the approximation

$$dy \approx \Delta y, \quad dx \neq 0$$

stems from the definition of the derivative. That is, the existence of the limit

$$f'(x) = \lim_{\Delta x \to 0} \frac{f(x + \Delta x) - f(x)}{\Delta x}$$

implies that when Δx is close to zero, then $f'(x)$ is close to the difference quotient. So, you can write

$$\begin{aligned} \frac{f(x + \Delta x) - f(x)}{\Delta x} &\approx f'(x) \\ f(x + \Delta x) - f(x) &\approx f'(x)\,\Delta x \\ \Delta y &\approx f'(x)\,\Delta x. \end{aligned}$$

Substituting dx for Δx and dy for $f'(x)\,dx$ produces $\Delta y \approx dy$.

Marginal Analysis

Differentials are used in economics to approximate changes in revenue, cost, and profit. Let $R = f(x)$ be the total revenue for selling x units of a product. When the number of units increases by 1, the change in x is $\Delta x = 1$, and the change in R is

$$\Delta R = f(x + \Delta x) - f(x) \approx dR = \frac{dR}{dx}\,dx.$$

In other words, you can use the differential dR to approximate the change in the revenue that accompanies the sale of one additional unit. Similarly, the differentials dC and dP can be used to approximate the changes in cost and profit that accompany the sale (or production) of one additional unit.

Example 2 Using Marginal Analysis

The demand function for a product is modeled by

$$p = 400 - x, \quad 0 \le x \le 400$$

where p is the price per unit (in dollars) and x is the number of units. Use differentials to approximate the change in revenue as sales increase from 149 units to 150 units. Compare this with the actual change in revenue.

SOLUTION Begin by finding the revenue function. Because the demand is given by $p = 400 - x$, the revenue is

$$\begin{aligned} R &= xp && \text{Formula for revenue} \\ &= x(400 - x) && \text{Use } p = 400 - x. \\ &= 400x - x^2. && \text{Multiply.} \end{aligned}$$

Next, find the marginal revenue, dR/dx.

$$\frac{dR}{dx} = 400 - 2x \qquad \text{Power Rule}$$

When $x = 149$ and $dx = \Delta x = 1$, the approximate change in the revenue is

$$\begin{aligned} \Delta R &\approx dR \\ &= \frac{dR}{dx}\,dx \\ &= (400 - 2x)\,dx \\ &= [400 - 2(149)](1) \\ &= \$102. \end{aligned}$$

When x increases from 149 to 150 and $R = f(x) = 400x - x^2$, the actual change in revenue is

$$\begin{aligned} \Delta R &= f(x + \Delta x) - f(x) \\ &= \left[400(150) - 150^2\right] - \left[400(149) - 149^2\right] \\ &= 37{,}500 - 37{,}399 \\ &= \$101. \end{aligned}$$

✓ Checkpoint 2

The demand function for a product is modeled by $p = 200 - x$, $0 \le x \le 200$, where p is the price per unit (in dollars) and x is the number of units. Use differentials to approximate the change in revenue as sales increase from 89 to 90 units. Compare this with the actual change in revenue. ■

Example 3 Using Marginal Analysis

The profit (in dollars) derived from selling x units of an item is modeled by

$$P = 0.0002x^3 + 10x.$$

Use the differential dP to approximate the change in profit when the production level changes from 50 to 51 units. Compare this with the actual gain in profit obtained by increasing the production level from 50 to 51 units.

STUDY TIP

Example 3 uses differentials to solve the same problem that was solved in Example 5 in Section 2.3. Look back at that solution. Which approach do you prefer?

SOLUTION The marginal profit is

$$\frac{dP}{dx} = 0.0006x^2 + 10.$$

When $x = 50$ and $dx = \Delta x = 1$, the approximate change in profit is

$$\begin{aligned}\Delta P &\approx dP\\ &= \frac{dP}{dx}\,dx\\ &= (0.0006x^2 + 10)\,dx\\ &= [0.0006(50)^2 + 10](1)\\ &= \$11.50.\end{aligned}$$

When x changes from 50 to 51 units and $P = f(x) = 0.0002x^3 + 10x$, the actual change in profit is

$$\begin{aligned}\Delta P &= f(x + \Delta x) - f(x)\\ &= [(0.0002)(51)^3 + 10(51)] - [(0.0002)(50)^3 + 10(50)]\\ &\approx 536.53 - 525.00\\ &= \$11.53.\end{aligned}$$

These values are shown graphically in Figure 3.65.

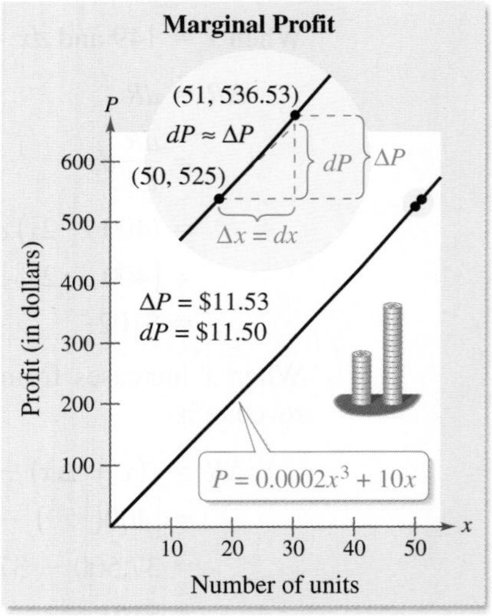

FIGURE 3.65

✓Checkpoint 3

Use the differential dP to approximate the change in profit for the profit function in Example 3 when the production level changes from 40 to 41 units. Compare this with the actual gain in profit obtained by increasing the production level from 40 to 41 units. ■

Formulas for Differentials

You can use the definition of differentials to rewrite each differentiation rule in **differential form.**

Differential Forms of Differentiation Rules

Constant Multiple Rule:	$d[cu] = c\,du$
Sum or Difference Rule:	$d[u \pm v] = du \pm dv$
Product Rule:	$d[uv] = u\,dv + v\,du$
Quotient Rule:	$d\left[\dfrac{u}{v}\right] = \dfrac{v\,du - u\,dv}{v^2}$
Constant Rule:	$d[c] = 0$
Power Rule:	$d[x^n] = nx^{n-1}\,dx$

The next example compares the derivatives and differentials of several simple functions.

Example 4 Finding Differentials

Function	*Derivative*	*Differential*
a. $y = x^2$	$\dfrac{dy}{dx} = 2x$	$dy = 2x\,dx$
b. $y = \dfrac{3x + 2}{5}$	$\dfrac{dy}{dx} = \dfrac{3}{5}$	$dy = \dfrac{3}{5}\,dx$
c. $y = 2x^2 - 3x$	$\dfrac{dy}{dx} = 4x - 3$	$dy = (4x - 3)\,dx$
d. $y = \dfrac{1}{x}$	$\dfrac{dy}{dx} = -\dfrac{1}{x^2}$	$dy = -\dfrac{1}{x^2}\,dx$

✓Checkpoint 4

Find the differential dy of each function.

a. $y = 4x^3$ **b.** $y = \dfrac{2x + 1}{3}$

c. $y = 3x^2 - 2x$ **d.** $y = \dfrac{1}{x^2}$ ■

SUMMARIZE (Section 3.8)

1. State the definition of differentials *(page 235)*. For an example of a differential, see Example 1.
2. Explain what is meant by marginal analysis *(page 237)*. For examples of marginal analysis, see Examples 2 and 3.
3. State the differential forms of the differential rules *(page 239)*. For an example that uses the differential forms of the differential rules, see Example 4.

SKILLS WARM UP 3.8 The following warm-up exercises involve skills that were covered in earlier sections. You will use these skills in the exercise set for this section. For additional help, review Sections 2.2 and 2.4.

In Exercises 1–12, find the derivative.

1. $C = 44 + 0.09x^2$

2. $C = 250 + 0.15x$

3. $R = x(1.25 + 0.02\sqrt{x})$

4. $R = x(15.5 - 1.55x)$

5. $P = -0.03x^{1/3} + 1.4x - 2250$

6. $P = -0.02x^2 + 25x - 1000$

7. $A = \frac{1}{4}\sqrt{3}x^2$

8. $A = 6x^2$

9. $C = 2\pi r$

10. $P = 4w$

11. $S = 4\pi r^2$

12. $P = 2x + \sqrt{2}x$

In Exercises 13–16, write a formula for the quantity.

13. Area A of a circle of radius r

14. Area A of a square of side x

15. Volume V of a cube of edge x

16. Volume V of a sphere of radius r

Exercises 3.8

See www.CalcChat.com for worked-out solutions to odd-numbered exercises.

Comparing Δy and dy In Exercises 1–6, compare the values of dy and Δy for the function. *See Example 1.*

	Function	*x-Value*	*Differential of x*
1.	$f(x) = 0.5x^3$	$x = 2$	$\Delta x = dx = 0.1$
2.	$f(x) = 1 - 2x^2$	$x = 0$	$\Delta x = dx = -0.1$
3.	$f(x) = x^4 + 1$	$x = -1$	$\Delta x = dx = 0.01$
4.	$f(x) = 2x + 1$	$x = 1$	$\Delta x = dx = 0.01$
5.	$f(x) = 3\sqrt{x}$	$x = 4$	$\Delta x = dx = 0.1$
6.	$f(x) = 6x^{4/3}$	$x = -1$	$\Delta x = dx = 0.01$

Finding Differentials In Exercises 7–12, let $x = 2$ and complete the table for the function.

$dx = \Delta x$	dy	Δy	$\Delta y - dy$	$\frac{dy}{\Delta y}$
1.000				
0.500				
0.100				
0.010				
0.001				

7. $y = x^2$

8. $y = x^5$

9. $y = \frac{1}{x^2}$

10. $y = \frac{1}{x}$

11. $y = \sqrt[4]{x}$

12. $y = \sqrt{x}$

Marginal Analysis In Exercises 13–18, use differentials to approximate the change in cost, revenue, or profit corresponding to an increase in sales of one unit. For instance, in Exercise 13, approximate the change in cost as x increases from 12 units to 13 units. *See Examples 2 and 3.*

	Function	*x-Value*
13.	$C = 0.05x^2 + 4x + 10$	$x = 12$
14.	$C = 0.025x^2 + 8x + 5$	$x = 10$
15.	$R = 30x - 0.15x^2$	$x = 75$
16.	$R = 50x - 1.5x^2$	$x = 15$
17.	$P = -0.5x^3 + 2500x - 6000$	$x = 50$
18.	$P = -x^2 + 60x - 100$	$x = 25$

Finding Differentials In Exercises 19–28, find the differential dy. *See Example 4.*

19. $y = 6x^4$

20. $y = \frac{8 - 4x}{3}$

21. $y = 3x^2 - 4$

22. $y = 3x^{2/3}$

23. $y = (4x - 1)^3$

24. $y = (x^2 + 3)(2x + 4)^2$

25. $y = \frac{x + 1}{2x - 1}$

26. $y = \frac{x}{x^2 + 1}$

27. $y = \sqrt{9 - x^2}$

28. $y = \sqrt[3]{6x^2}$

Finding an Equation of a Tangent Line In Exercises 29–32, find an equation of the tangent line to the function at the given point. Then find the function values and the tangent line values at $f(x + \Delta x)$ and $y(x + \Delta x)$ for $\Delta x = -0.01$ and 0.01.

Function	*Point*
29. $f(x) = 2x^3 - x^2 + 1$	$(-2, -19)$
30. $f(x) = 3x^2 - 1$	$(2, 11)$
31. $f(x) = \dfrac{x}{x^2 + 1}$	$(0, 0)$
32. $f(x) = \sqrt{25 - x^2}$	$(3, 4)$

33. Profit The profit P for a company producing x units is

$$P = (500x - x^2) - \left(\frac{1}{2}x^2 - 77x + 3000\right).$$

(a) Use differentials to approximate the change in profit when the production level changes from 115 to 120 units.

(b) Compare this with the actual change in profit.

34. Revenue The revenue R for a company selling x units is $R = 900x - 0.1x^2$.

(a) Use differentials to approximate the change in revenue as the sales increase from 3000 units to 3100 units.

(b) Compare this with the actual change in revenue.

35. Demand The demand function for a product is modeled by $p = 75 - 0.25x$, where p is the price per unit (in dollars) and x is the number of units.

(a) Use differentials to approximate the change in revenue as sales increase from 7 units to 8 units.

(b) Repeat part (a) as sales increase from 70 units to 71 units.

36. HOW DO YOU SEE IT? The graph shows the profit P (in dollars) from selling x units of an item. Use the graph to determine which is greater, the change in profit when the production level changes from 400 to 401 units or the change in profit when the production level changes from 900 to 901 units. Explain your reasoning.

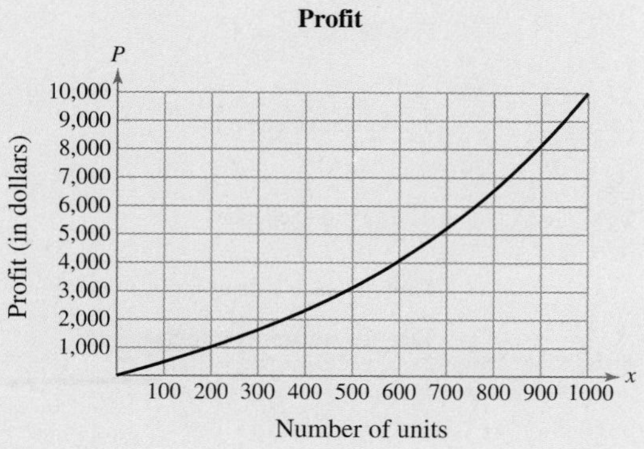

37. Biology: Wildlife Management A state game commission introduces 50 deer into newly acquired state game lands. The population N of the herd can be modeled by

$$N = \frac{10(5 + 3t)}{1 + 0.04t}$$

where t is the time in years. Use differentials to approximate the change in the herd size from $t = 5$ to $t = 6$.

38. Medical Science The concentration C (in milligrams per milliliter) of a drug in a patient's bloodstream t hours after injection into muscle tissue is modeled by

$$C = \frac{3t}{27 + t^3}.$$

Use differentials to approximate the change in the concentration when t changes from $t = 1$ to $t = 1.5$.

39. Marginal Analysis A retailer has determined that the monthly sales x of a watch are 150 units when the price is \$50, but decrease to 120 units when the price is \$60. Assume that the demand is a linear function of the price. Find the revenue R as a function of x and approximate the change in revenue for a one-unit increase in sales when $x = 141$. Make a sketch showing dR and ΔR.

40. Marginal Analysis The demand x for a web camera is 30,000 units per month when the price is \$25 and 40,000 units when the price is \$20. The initial investment is \$275,000 and the cost per unit is \$17. Assume that the demand is a linear function of the price. Find the profit P as a function of x and approximate the change in profit for a one-unit increase in sales when $x = 28{,}000$. Make a sketch showing dP and ΔP.

Error Propagation In Exercises 41 and 42, use the following information. Given the error in a measurement (Δx), the *propagated error* (Δy) can be approximated by the differential dy. The ratio dy/y is the *relative error*, which corresponds to a *percentage error* of $dy/y \times 100\%$.

41. Area The side of a square measures 6 inches, with a possible error of $\pm\frac{1}{16}$ inch. Estimate the propagated error and the percentage error in computing the area of the square.

42. Volume The radius of a sphere measures 6 inches, with a possible error of ± 0.02 inch. Estimate the propagated error and the percentage error in computing the volume of the sphere.

True or False? In Exercises 43 and 44, determine whether the statement is true or false. If it is false, explain why or give an example that shows it is false.

43. If $y = x + c$, then $dy = dx$.

44. If $y = ax + b$, then $\Delta y/\Delta x = dy/dx$.

ALGEBRA TUTOR

Solving Equations

Much of the algebra in Chapter 3 involves simplifying algebraic expressions (see pages 158 and 159) and solving algebraic equations (see page 71). The Algebra Tutor on page 71 illustrates some of the basic techniques for solving equations. On these two pages, you can review some of the more complicated techniques for solving equations.

When solving an equation, remember that your basic goal is to isolate the variable on one side of the equation. To do this, you use inverse operations. For instance, to isolate x in

$$x - 2 = 0$$

you add 2 to each side of the equation, because *addition* is the inverse operation of *subtraction*. To isolate x in

$$\sqrt{x} = 2$$

you square each side of the equation, because *squaring* is the inverse operation of *taking the square root.*

Example 1 Solving Equations

Solve each equation.

a. $\dfrac{36(x^2 - 1)}{(x^2 + 3)^3} = 0$ **b.** $0 = 2x(2x^2 - 3)$ **c.** $\dfrac{dV}{dx} = 0$, where $V = 27x - \dfrac{1}{4}x^3$

SOLUTION

a. $\dfrac{36(x^2 - 1)}{(x^2 + 3)^3} = 0$ — Example 2, page 188

$36(x^2 - 1) = 0$ — A fraction is equal to zero only when its numerator is zero.

$x^2 - 1 = 0$ — Divide each side by 36.

$x^2 = 1$ — Add 1 to each side.

$x = \pm 1$ — Take the square root of each side.

b. $0 = 2x(2x^2 - 3)$ — Example 2, page 198

$2x = 0 \Rightarrow x = 0$ — Set first factor equal to zero.

$2x^2 - 3 = 0 \Rightarrow x = \pm\sqrt{\frac{3}{2}}$ — Set second factor equal to zero.

c. $V = 27x - \dfrac{1}{4}x^3$ — Example 1, page 196

$\dfrac{dV}{dx} = 27 - \dfrac{3}{4}x^2$ — Find derivative of V.

$0 = 27 - \dfrac{3}{4}x^2$ — Set derivative equal to 0.

$\dfrac{3}{4}x^2 = 27$ — Add $\frac{3}{4}x^2$ to each side.

$x^2 = 36$ — Multiply each side by $\frac{4}{3}$.

$x = \pm 6$ — Take the square root of each side.

Example 2 Solving Equations

Solve each equation.

a. $\dfrac{20(x^{1/3}-1)}{9x^{2/3}}=0$ **b.** $\dfrac{25}{\sqrt{x}}-0.5=0$

c. $x^2(4x-3)=0$ **d.** $\dfrac{4x}{3(x^2-4)^{1/3}}=0$

e. $g'(x)=0$, where $g(x)=(x-2)(x+1)^2$

SOLUTION

a. $\dfrac{20(x^{1/3}-1)}{9x^{2/3}}=0$ Example 5, page 231

$20(x^{1/3}-1)=0$ A fraction is equal to zero only when its numerator is zero.

$x^{1/3}-1=0$ Divide each side by 20.

$x^{1/3}=1$ Add 1 to each side.

$x=1$ Cube each side.

b. $\dfrac{25}{\sqrt{x}}-0.5=0$ Example 4, page 208

$\dfrac{25}{\sqrt{x}}=0.5$ Add 0.5 to each side.

$25=0.5\sqrt{x}$ Multiply each side by $\sqrt{x}$.

$50=\sqrt{x}$ Divide each side by 0.5.

$2500=x$ Square both sides.

c. $x^2(4x-3)=0$ Example 2, page 180

$x^2=0 \Rightarrow x=0$ Set first factor equal to zero.

$4x-3=0 \Rightarrow x=\frac{3}{4}$ Set second factor equal to zero.

d. $\dfrac{4x}{3(x^2-4)^{1/3}}=0$ Example 5, page 172

$4x=0$ A fraction is equal to zero only when its numerator is zero.

$x=0$ Divide each side by 4.

e. $g(x)=(x-2)(x+1)^2$ Exercise 45, page 194

$g'(x)=(x-2)(2)(x+1)+(x+1)^2(1)$ Find derivative of g.

$(x-2)(2)(x+1)+(x+1)^2(1)=0$ Set derivative equal to zero.

$(x+1)[2(x-2)+(x+1)]=0$ Factor.

$(x+1)(2x-4+x+1)=0$ Multiply factors.

$(x+1)(3x-3)=0$ Combine like terms.

$x+1=0 \Rightarrow x=-1$ Set first factor equal to zero.

$3x-3=0 \Rightarrow x=1$ Set second factor equal to zero.

SUMMARY AND STUDY STRATEGIES

After studying this chapter, you should have acquired the following skills. The exercise numbers are keyed to the Review Exercises that begin on page 246. Answers to odd-numbered Review Exercises are given in the back of the text.*

Section 3.1	Review Exercises
■ Find the critical numbers of a function.	*1–6*
c is a critical number of f when $f'(c) = 0$ or $f'(c)$ is undefined.	
■ Find the open intervals on which a function is increasing or decreasing.	*7–12*
f is increasing when $f'(x) > 0$.	
f is decreasing when $f'(x) < 0$.	
■ Find intervals on which a real-life model is increasing or decreasing.	*13, 14*

Section 3.2

- Use the First-Derivative Test to find the relative extrema of a function. *15–24*
- Find the absolute extrema of a continuous function on a closed interval. *25–32*
- Find minimum and maximum values of a real-life model and interpret the results in context. *33, 34*

Section 3.3

- Find the open intervals on which the graph of a function is concave upward or concave downward. *35–38*
  f is concave upward when $f''(x) > 0$.
  f is concave downward when $f''(x) < 0$.
- Find the points of inflection of the graph of a function. *39–42*
- Use the Second-Derivative Test to find the relative extrema of a function. *43–48*
- Find the point of diminishing returns of an input-output model. *49, 50*

Section 3.4

- Solve real-life optimization problems. *51–54*

Section 3.5

- Solve business and economics optimization problems. *55–60*
- Find the price elasticity of demand for a demand function. *61, 62*

* A wide range of valuable study aids are available to help you master the material in this chapter. The *Student Solutions Manual* includes step-by-step solutions to all odd-numbered exercises to help you review and prepare. The student website at *www.cengagebrain.com* offers algebra help and a *Graphing Technology Guide*, which contains step-by-step commands and instructions for a wide variety of graphing calculators.

Section 3.6	Review Exercises
■ Find the vertical asymptotes of a function.	*63–66*
■ Find infinite limits.	*67–70*
■ Find the horizontal asymptotes of a function.	*71–74*
■ Use asymptotes to answer questions about real-life situations.	*75–78*

Section 3.7	
■ Analyze the graph of a function.	*79–92*

Section 3.8	
■ Use differentials to approximate changes in a function.	*93–96*
■ Use differentials in economics to approximate changes in cost, revenue, and profit.	*97–102*
■ Find the differential of a function using differentiation formulas.	*103–108*
■ Use differentials to approximate changes in real-life models.	*109–111*

Study Strategies

- **Solve Problems Graphically, Analytically, and Numerically** When analyzing the graph of a function, use a variety of problem-solving strategies. For instance, if you were asked to analyze the graph of

  $f(x) = x^3 - 4x^2 + 5x - 4$

  you could begin *graphically*. That is, you could use a graphing utility to find a viewing window that appears to show the important characteristics of the graph. From the graph shown below, the function appears to have one relative minimum, one relative maximum, and one point of inflection.

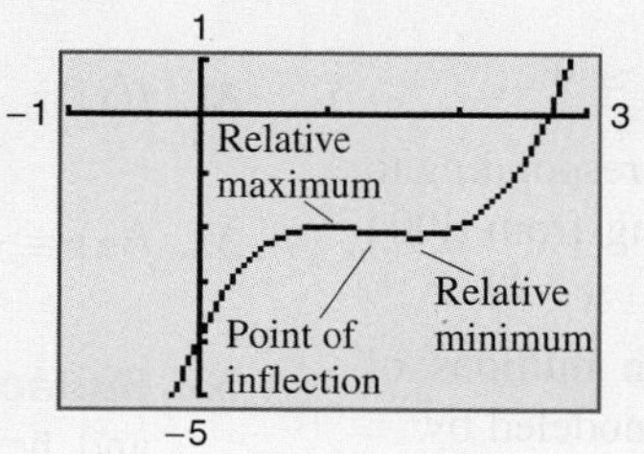

  Next, you could use calculus to *analyze* the graph. Because the derivative of f is

  $f'(x) = 3x^2 - 8x + 5 = (3x - 5)(x - 1)$

  the critical numbers of f are $x = \frac{5}{3}$ and $x = 1$. By the First-Derivative Test, you can conclude that $x = \frac{5}{3}$ yields a relative minimum and $x = 1$ yields a relative maximum. Because

  $f''(x) = 6x - 8$

  you can conclude that $x = \frac{4}{3}$ yields a point of inflection. Finally, you could analyze the graph *numerically*. For instance, you could construct a table of values and observe that f is increasing on the interval $(-\infty, 1)$, decreasing on the interval $\left(1, \frac{5}{3}\right)$, and increasing on the interval $\left(\frac{5}{3}, \infty\right)$.

- **Problem-Solving Strategies** When you get stuck while trying to solve an optimization problem, consider the strategies below.

  1. *Draw a Diagram.* If feasible, draw a diagram that represents the problem. Label all known values and unknown values on the diagram.
  2. *Solve a Simpler Problem.* Simplify the problem, or write several simple examples of the problem. For instance, if you are asked to find the dimensions that will produce a maximum area, try calculating the areas of several examples.
  3. *Rewrite the Problem in Your Own Words.* Rewriting a problem can help you understand it better.
  4. *Guess and Check.* Try guessing the answer, then check your guess in the statement of the original problem. By refining your guesses, you may be able to think of a general strategy for solving the problem.

Review Exercises

See www.CalcChat.com for worked-out solutions to odd-numbered exercises.

Finding Critical Numbers In Exercises 1–6, find the critical numbers of the function.

1. $f(x) = -x^2 + 2x + 4$
2. $y = 3x^2 + 18x$
3. $y = 4x^3 - 108x$
4. $f(x) = x^4 - 8x^2 + 13$
5. $g(x) = (x - 1)^2(x - 3)$
6. $h(x) = \sqrt{x}(x - 3)$

Intervals on Which *f* Is Increasing or Decreasing In Exercises 7–12, find the critical numbers and the open intervals on which the function is increasing or decreasing. Use a graphing utility to verify your results.

7. $f(x) = x^2 + x - 2$
8. $g(x) = (x + 2)^3$
9. $f(x) = -x^3 + 6x^2 - 2$
10. $y = x^3 - 12x^2$
11. $y = (x - 1)^{2/3}$
12. $y = 2x^{1/3} - 3$

13. **Revenue** The revenue R of Chipotle Mexican Grill (in millions of dollars) from 2004 through 2009 can be modeled by

$$R = 6.268t^2 + 136.07t - 191.3, \quad 4 \le t \le 9$$

where t is the time in years, with $t = 4$ corresponding to 2004. Show that the sales were increasing from 2004 through 2009. *(Source: Chipotle Mexican Grill, Inc.)*

14. **Revenue** The revenue R of Cintas (in millions of dollars) from 2000 through 2010 can be modeled by

$$R = -5.5778t^3 + 67.524t^2 + 45.22t + 1969.2$$

for $0 \le t \le 10$, where t is the time in years, with $t = 0$ corresponding to 2000. *(Source: Cintas Corporation)*

(a) Use a graphing utility to graph the model. Then graphically estimate the years during which the revenue was increasing and the years during which the revenue was decreasing.

(b) Use the test for increasing and decreasing functions to verify the result of part (a).

Finding Relative Extrema In Exercises 15–24, use the First-Derivative Test to find all relative extrema of the function. Use a graphing utility to verify your result.

15. $f(x) = 4x^3 - 6x^2 - 2$
16. $f(x) = \frac{1}{4}x^4 - 8x$
17. $g(x) = x^2 - 16x + 12$
18. $h(x) = 4 + 10x - x^2$
19. $h(x) = 2x^2 - x^4$
20. $s(x) = x^4 - 8x^2 + 3$
21. $f(x) = \dfrac{6}{x^2 + 1}$
22. $f(x) = \dfrac{2}{x^2 - 1}$
23. $h(x) = \dfrac{x^2}{x - 2}$
24. $g(x) = x - 6\sqrt{x}, \quad x > 0$

Finding Extrema on a Closed Interval In Exercises 25–32, find the absolute extrema of the function on the closed interval. Use a graphing utility to verify your result.

25. $f(x) = x^2 + 5x + 6; \quad [-3, 0]$
26. $f(x) = x^4 - 2x^3; \quad [0, 2]$
27. $f(x) = x^3 - 12x + 1; \quad [-4, 4]$
28. $f(x) = x^3 + 2x^2 - 3x + 4; \quad [-3, 2]$
29. $f(x) = 2\sqrt{x} - x; \quad [0, 9]$
30. $f(x) = \dfrac{x}{\sqrt{x^2 + 1}}; \quad [0, 2]$
31. $f(x) = \dfrac{2x}{x^2 + 1}; \quad [-1, 2]$
32. $f(x) = \dfrac{8}{x} + x; \quad [1, 4]$

33. **Surface Area** A right circular cylinder of radius r and height h has a volume of 25 cubic inches (see figure). The total surface area of the cylinder in terms of r is given by

$$S = 2\pi r\left(r + \frac{25}{\pi r^2}\right).$$

Find the radius that will minimize the surface area. Use a graphing utility to verify your result.

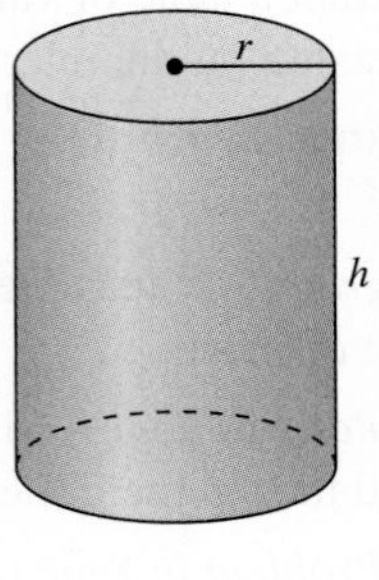

34. Profit The profit P (in dollars) made by a company from selling x tablet computers can be modeled by

$$P = 1.64x - \frac{x^2}{15{,}000} - 2500.$$

Find the number of units sold that will yield a maximum profit. What is the maximum profit?

Applying the Test for Concavity In Exercises 35–38, determine the open intervals on which the graph of the function is concave upward or concave downward.

35. $f(x) = (x - 2)^3$

36. $h(x) = x^5 - 10x^2$

37. $g(x) = \frac{1}{4}(-x^4 + 8x^2 - 12)$

38. $h(x) = x^3 - 6x$

Finding Points of Inflection In Exercises 39–42, discuss the concavity of the graph of the function and find the points of inflection.

39. $f(x) = \frac{1}{2}x^4 - 4x^3$

40. $f(x) = \frac{1}{4}x^4 - 2x^2 - x$

41. $f(x) = x^3(x - 3)^2$

42. $f(x) = (x - 1)^2(x - 3)$

Using the Second-Derivative Test In Exercises 43–48, use the Second-Derivative Test to find all relative extrema of the function.

43. $f(x) = x^3 - 6x^2 + 12x$

44. $f(x) = x^4 - 32x^2 + 12$

45. $f(x) = x^5 - 5x^3$

46. $f(x) = x(x^2 - 3x - 9)$

47. $f(x) = 2x^2(1 - x^2)$

48. $f(x) = x - 4\sqrt{x + 1}$

Point of Diminishing Returns In Exercises 49 and 50, find the point of diminishing returns for the function. For each function, R is the revenue (in thousands of dollars) and x is the amount spent (in thousands of dollars) on advertising. Use a graphing utility to verify your result.

49. $R = \frac{1}{1500}(150x^2 - x^3), \quad 0 \le x \le 100$

50. $R = -\frac{2}{3}(x^3 - 12x^2 - 6), \quad 0 \le x \le 8$

51. Minimum Perimeter Find the length and width of a rectangle that has an area of 225 square meters and a minimum perimeter.

52. Maximum Volume A rectangular solid with a square base has a surface area of 432 square centimeters.

(a) Determine the dimensions that yield the maximum volume.

(b) Find the maximum volume.

53. Maximum Volume An open box is to be made from a 10-inch by 16-inch rectangular piece of material by cutting equal squares from the corners and turning up the sides (see figure). Find the volume of the largest box that can be made.

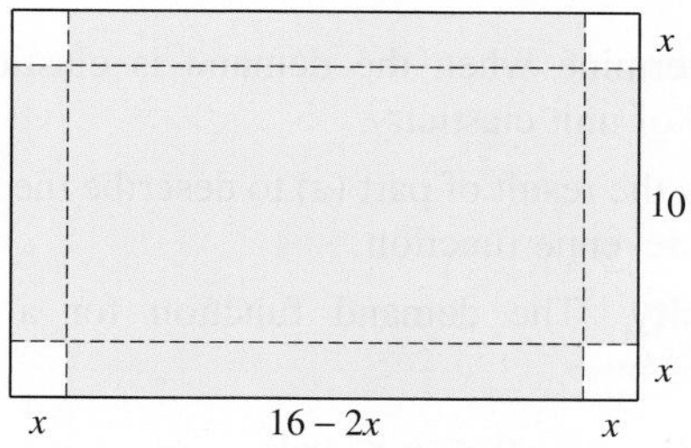

54. Minimum Area A rectangular page is to contain 108 square inches of print. The margins at the top and bottom of the page are to be $\frac{3}{4}$ inch wide. The margins on each side are to be 1 inch wide. Find the dimensions of the page that will minimize the amount of paper used.

Finding the Maximum Revenue In Exercises 55 and 56, find the number of units x that produces a maximum revenue R.

55. $R = 450x - 0.25x^2$

56. $R = 36x^2 - 0.05x^3$

Finding the Minimum Average Cost In Exercises 57 and 58, find the number of units x that produces the minimum average cost per unit $\overline{C}$.

57. $C = 0.2x^2 + 10x + 4500$

58. $C = 0.03x^3 + 30x + 3840$

59. Maximum Profit A commodity has a demand function modeled by

$$p = 36 - 4x$$

and a total cost function modeled by

$$C = 2x^2 + 6$$

where x is the number of units.

(a) What price yields a maximum profit?

(b) When the profit is maximized, what is the average cost per unit?

60. Maximum Profit The profit P (in thousands of dollars) for a company in terms of the amount s spent on advertising (in thousands of dollars) can be modeled by

$$P = -4s^3 + 72s^2 - 240s + 500.$$

Find the amount of advertising that maximizes the profit. Find the point of diminishing returns.

61. Elasticity The demand function for a product is modeled by

$p = 60 - 0.04x, \quad 0 \leq x \leq 1500$

where p is the price (in dollars) and x is the number of units.

(a) Determine when the demand is elastic, inelastic, and of unit elasticity.

(b) Use the result of part (a) to describe the behavior of the revenue function.

62. Elasticity The demand function for a product is modeled by

$p = 960 - x, \quad 0 \leq x \leq 960$

where p is the price (in dollars) and x is the number of units.

(a) Determine when the demand is elastic, inelastic, and of unit elasticity.

(b) Use the result of part (a) to describe the behavior of the revenue function.

Finding Vertical Asymptotes In Exercises 63–66, determine all vertical asymptotes of the graph of the function.

63. $f(x) = \dfrac{x + 4}{x^2 + 7x}$

64. $f(x) = \dfrac{x - 1}{x^2 - 4}$

65. $f(x) = \dfrac{x^2 - 16}{2x^2 + 9x + 4}$

66. $f(x) = \dfrac{x^2 + 6x + 9}{x^2 - 5x - 24}$

Determining Infinite Limits In Exercises 67–70, use a graphing utility to find the limit.

67. $\lim\limits_{x \to 0^+} \left(x - \dfrac{1}{x^3}\right)$

68. $\lim\limits_{x \to 0^-} \left(3 + \dfrac{1}{x}\right)$

69. $\lim\limits_{x \to -1^+} \dfrac{x^2 - 2x + 1}{x + 1}$

70. $\lim\limits_{x \to 3^-} \dfrac{3x^2 + 1}{x^2 - 9}$

Finding Horizontal Asymptotes In Exercises 71–74, find the horizontal asymptote of the graph of the function.

71. $f(x) = \dfrac{2x^2}{3x^2 + 5}$

72. $f(x) = \dfrac{3x^2 - 2x + 3}{x + 1}$

73. $f(x) = \dfrac{3x}{x^2 + 1}$

74. $f(x) = \dfrac{x}{x - 2} + \dfrac{2x}{x + 2}$

75. Average Cost The cost C (in dollars) of producing x units of a product is

$C = 0.75x + 4000.$

(a) Find the average cost function $\overline{C}$.

(b) Find $\overline{C}$ when $x = 100$ and when $x = 1000$.

(c) Determine the limit of the average cost function as x approaches infinity. Interpret the limit in the context of the problem.

76. Average Cost The cost C (in dollars) of producing x units of a product is

$C = 1.50x + 8000.$

(a) Find the average cost function $\overline{C}$.

(b) Find $\overline{C}$ when $x = 1000$ and when $x = 10{,}000$.

(c) Determine the limit of the average cost function as x approaches infinity. Interpret the limit in the context of the problem.

77. Seizing Drugs The cost C (in millions of dollars) for the federal government to seize $p\%$ of an illegal drug as it enters the country is modeled by

$C = \dfrac{250p}{100 - p}, \quad 0 \leq p < 100.$

(a) Find the costs of seizing 20%, 50%, and 90%.

(b) Find the limit of C as $p \to 100^-$. Interpret the limit in the context of the problem. Use a graphing utility to verify your result.

78. Removing Pollutants The cost C (in dollars) of removing $p\%$ of the air pollutants in the stack emission of a utility company that burns coal is modeled by

$C = \dfrac{160{,}000p}{100 - p}, \quad 0 \leq p < 100.$

(a) Find the costs of seizing 25%, 50%, and 75%.

(b) Find the limit of C as $p \to 100^-$. Interpret the limit in the context of the problem. Use a graphing utility to verify your result.

Analyzing a Graph In Exercises 79–90, analyze and sketch the graph of the function. Label any intercepts, relative extrema, points of inflection, and asymptotes.

79. $f(x) = 4x - x^2$

80. $f(x) = 4x^3 - x^4$

81. $f(x) = x^3 - 6x^2 + 3x + 10$

82. $f(x) = -x^3 + 3x^2 + 9x - 2$

83. $f(x) = x^4 - 4x^3 + 16x - 16$

84. $f(x) = x^5 + 1$

85. $f(x) = x\sqrt{16 - x^2}$

86. $f(x) = x^2\sqrt{9 - x^2}$

87. $f(x) = \dfrac{x + 1}{x - 1}$

88. $f(x) = \dfrac{x - 1}{3x^2 + 1}$

89. $f(x) = 3x^{2/3} - 2x$

90. $f(x) = x^{4/5}$

91. **Bacteria** The data in the table show the number N of bacteria in a culture at time t, where t is measured in days.

t	1	2	3	4	5	6	7	8
N	25	200	804	1756	2296	2434	2467	2473

A model for these data is

$$N = \frac{24{,}670 - 35{,}153t + 13{,}250t^2}{100 - 39t + 7t^2}, \quad 1 \le t \le 8.$$

(a) Use a graphing utility to create a scatter plot of the data and graph the model in the same viewing window. How well does the model fit the data?

(b) Use the model to predict the number of bacteria in the culture after 10 days.

(c) Should this model be used to predict the number of bacteria in the culture after a few months? Why or why not?

92. **Meteorology** The monthly average high temperatures T (in degrees Fahrenheit) in New York City can be modeled by

$$T = \frac{31.6 - 1.822t + 0.0984t^2}{1 - 0.194t + 0.0131t^2}, \quad 1 \le t \le 12$$

where t is the month, with $t = 1$ corresponding to January. Use a graphing utility to graph the model and find all absolute extrema. Interpret the meaning of these values in the context of the problem. *(Source: National Climatic Data Center)*

Comparing Δy and dy In Exercises 93–96, compare the values of dy and Δy for the function.

	Function	*x-Value*	*Differential of x*
93.	$f(x) = 2x^2$	$x = 2$	$\Delta x = dx = 0.01$
94.	$f(x) = x^4 + 3$	$x = 1$	$\Delta x = dx = 0.1$
95.	$f(x) = 6x - x^3$	$x = 3$	$\Delta x = dx = 0.1$
96.	$f(x) = 5x^{3/2}$	$x = 9$	$\Delta x = dx = 0.01$

Marginal Analysis In Exercises 97–102, use differentials to approximate the change in cost, revenue, or profit corresponding to an increase in sales of one unit. For instance, in Exercise 97, approximate the change in cost as x increases from 10 to 11.

	Function	*x-Value*
97.	$C = 40x^2 + 1225$	$x = 10$
98.	$C = 1.5\sqrt[3]{x} + 500$	$x = 125$
99.	$R = 6.25x + 0.4x^{3/2}$	$x = 225$
100.	$R = 80x - 0.35x^2$	$x = 80$
101.	$P = 0.003x^2 + 0.019x - 1200$	$x = 750$
102.	$P = -0.2x^3 + 3000x - 7500$	$x = 50$

Finding Differentials In Exercises 103–108, find the differential dy.

103. $y = 0.5x^3$

104. $y = 7x^4 + 2x^2$

105. $y = (3x^2 - 2)^3$

106. $y = \sqrt{36 - x^2}$

107. $y = \dfrac{2 - x}{x + 5}$

108. $y = \dfrac{3x^2}{x - 4}$

109. **Profit** The profit P (in dollars) for a company producing x units is

$$P = -0.8x^2 + 324x - 2000.$$

(a) Use differentials to approximate the change in profit when the production level changes from 100 to 101 units.

(b) Compare this with the actual change in profit.

110. **Demand** The demand function for a product is

$$p = 108 - 0.2x$$

where p is the price per unit (in dollars) and x is the number of units.

(a) Use differentials to approximate the change in revenue as sales increase from 20 units to 21 units.

(b) Repeat part (a) when sales increase from 40 units to 41 units.

111. **Physiology: Body Surface Area** The body surface area (BSA) of a 180-centimeter-tall (about six-foot-tall) person is modeled by

$$B = 0.1\sqrt{5w}$$

where B is the BSA (in square meters) and w is the weight (in kilograms). Use differentials to approximate the change in the person's BSA when the person's weight changes from 90 kilograms to 95 kilograms.

TEST YOURSELF

See www.CalcChat.com for worked-out solutions to odd-numbered exercises.

Take this test as you would take a test in class. When you are done, check your work against the answers given in the back of the book.

In Exercises 1–3, find the critical numbers and the open intervals on which the function is increasing or decreasing .

1. $f(x) = 3x^2 - 4$ **2.** $f(x) = x^3 - 12x$ **3.** $f(x) = (x - 5)^4$

In Exercises 4–6, use the First-Derivative Test to find all relative extrema of the function.

4. $f(x) = \frac{1}{3}x^3 - 9x + 4$ **5.** $f(x) = 2x^4 - 4x^2 - 5$ **6.** $f(x) = \frac{5}{x^2 + 2}$

In Exercises 7–9, find the absolute extrema of the function on the closed interval. Use a graphing utility to verify your result.

7. $f(x) = x^2 + 6x + 8,\ [-4, 0]$ **8.** $f(x) = 12\sqrt{x} - 4x,\ [0, 5]$

9. $f(x) = \frac{6}{x} + \frac{x}{2},\ [1, 6]$

In Exercises 10 and 11, determine the open intervals on which the graph of the function is concave upward or concave downward.

10. $f(x) = x^5 - 80x^2$ **11.** $f(x) = \frac{20}{3x^2 + 8}$

In Exercises 12 and 13, discuss the concavity of the graph of the function and find the points of inflection.

12. $f(x) = x^4 + 6$ **13.** $f(x) = x^4 - 54x^2 + 230$

In Exercises 14 and 15, use the Second-Derivative Test to find all relative extrema of the function.

14. $f(x) = x^3 - 6x^2 - 36x + 50$ **15.** $f(x) = \frac{3}{5}x^5 - 9x^3$

In Exercises 16–18, find the vertical and horizontal asymptotes of the graph of the function.

16. $f(x) = \frac{3x + 2}{x - 5}$ **17.** $f(x) = \frac{2x^2}{x^2 + 3}$ **18.** $f(x) = \frac{2x^2 - 5}{x - 1}$

In Exercises 19–21, analyze and sketch the graph of the function. Label any intercepts, relative extrema, points of inflection, and asymptotes.

19. $y = -x^3 + 3x^2 + 9x - 2$ **20.** $y = x^5 - 5x$ **21.** $y = \frac{x}{x^2 - 4}$

In Exercises 22–24, find the differential *dy*.

22. $y = 5x^2 - 3$ **23.** $y = \frac{1 - x}{x + 3}$ **24.** $y = (x + 4)^3$

25. The demand function for a product is modeled by

$$p = 280 - 0.4x, \quad 0 \le x \le 700$$

where p is the price per unit (in dollars) and x is the number of units. Determine when the demand is elastic, inelastic, and of unit elasticity.

4 Exponential and Logarithmic Functions

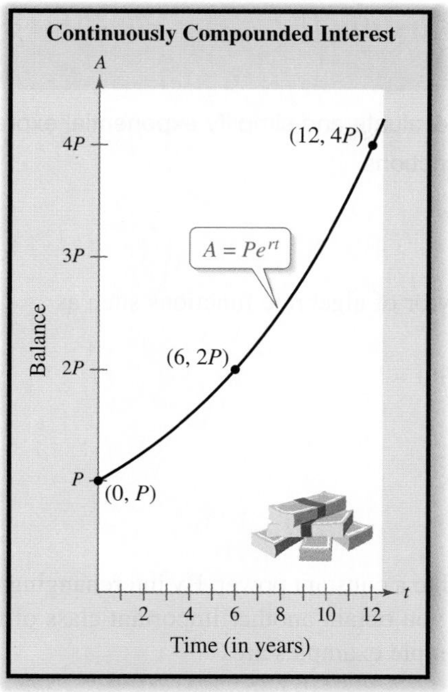

Example 3 on page 297 shows how an exponential growth model can be used to find the annual interest rate of an account.

4.1 Exponential Functions

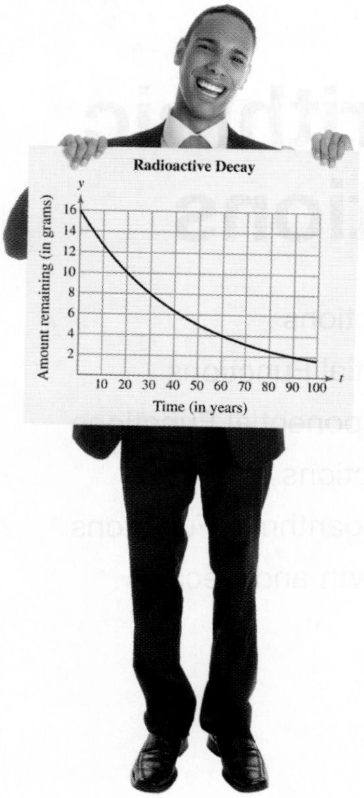

In Exercise 5 on page 256, you will evaluate an exponential function to find the remaining amount of a radioactive material.

- Use the properties of exponents to evaluate and simplify exponential expressions.
- Sketch the graphs of exponential functions.

Exponential Functions

You are already familiar with the behavior of algebraic functions such as

$$f(x) = x^2$$
$$g(x) = \sqrt{x} = x^{1/2}$$

and

$$h(x) = \frac{1}{x} = x^{-1}$$

each of which involves a variable raised to a constant power. By interchanging roles and raising a constant to a variable power, you obtain another important class of functions called **exponential functions.** Some simple examples are

$$f(x) = 2^x$$
$$g(x) = \left(\frac{1}{10}\right)^x = \frac{1}{10^x}$$

and

$$h(x) = 3^{2x} = 9^x.$$

In general, you can use any positive number $a \neq 1$ as the base of an exponential function.

Definition of Exponential Function

If $a > 0$ and $a \neq 1$, then the **exponential function** with base a is given by

$$f(x) = a^x.$$

In the definition of an exponential function, the base $a = 1$ is excluded because it yields

$$f(x) = 1^x = 1.$$

This is a constant function, not an exponential function.

When working with exponential functions, the properties of exponents, shown below, are useful.

Properties of Exponents

Let a and b be positive real numbers.

1. $a^0 = 1$ **2.** $a^x a^y = a^{x+y}$ **3.** $\frac{a^x}{a^y} = a^{x-y}$

4. $(a^x)^y = a^{xy}$ **5.** $(ab)^x = a^x b^x$ **6.** $\left(\frac{a}{b}\right)^x = \frac{a^x}{b^x}$

7. $a^{-x} = \frac{1}{a^x}$

Example 1 Applying Properties of Exponents

a. $(2^2)(2^3) = 2^{2+3} = 2^5 = 32$ — Apply Property 2.

b. $(2^2)(2^{-3}) = 2^{2-3} = 2^{-1} = \frac{1}{2}$ — Apply Properties 2 and 7.

c. $(3^2)^3 = 3^{2(3)} = 3^6 = 729$ — Apply Property 4.

d. $\left(\frac{1}{3}\right)^{-2} = \left(\frac{3}{1}\right)^2 = \frac{3^2}{1^2} = 9$ — Apply Properties 7 and 6.

e. $\frac{3^2}{3^3} = 3^{2-3} = 3^{-1} = \frac{1}{3}$ — Apply Properties 3 and 7.

f. $(2^{1/2})(3^{1/2}) = [(2)(3)]^{1/2} = 6^{1/2} = \sqrt{6}$ — Apply Property 5.

✓Checkpoint 1

Simplify each expression using the properties of exponents.

a. $(3^2)(3^3)$ **b.** $(3^2)(3^{-1})$ **c.** $(2^3)^2$

d. $(1/2)^{-3}$ **e.** $2^2/2^3$ **f.** $(2^{1/2})(5^{1/2})$ ■

Although Example 1 demonstrates the properties of exponents with integer and rational exponents, it is important to realize that the properties hold for *all* real exponents. With a calculator, you can obtain approximations of a^x for any positive number a and any real number x. Here are some examples.

$$2^{-0.6} \approx 0.660, \qquad \pi^{0.75} \approx 2.360, \qquad (1.56)^{\sqrt{2}} \approx 1.876$$

Example 2 Dating Organic Material

In living organic material, the ratio of radioactive carbon isotopes to the total number of carbon atoms is about 1 to 10^{12}. When organic material dies, its radioactive carbon isotopes begin to decay, with a half-life of about 5715 years. This means that after 5715 years, the ratio of isotopes to atoms will have decreased to one-half the original ratio; after a second 5715 years, the ratio will have decreased to one-fourth of the original; and so on. Figure 4.1 shows this decreasing ratio. The formula for the ratio R of carbon isotopes to carbon atoms is $R = (1/10^{12})(1/2)^{t/5715}$, where t is the time in years. Find the value of R for each period of time.

a. 10,000 years **b.** 20,000 years **c.** 25,000 years

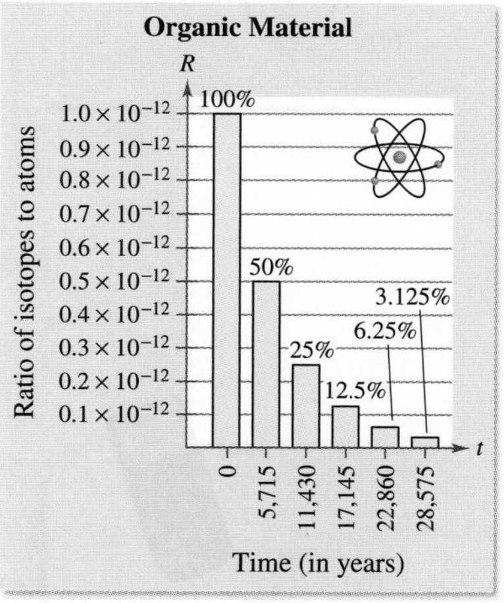

FIGURE 4.1

SOLUTION

a. $R = \left(\frac{1}{10^{12}}\right)\left(\frac{1}{2}\right)^{10{,}000/5715} \approx 2.973 \times 10^{-13}$ — Ratio for 10,000 years

b. $R = \left(\frac{1}{10^{12}}\right)\left(\frac{1}{2}\right)^{20{,}000/5715} \approx 8.842 \times 10^{-14}$ — Ratio for 20,000 years

c. $R = \left(\frac{1}{10^{12}}\right)\left(\frac{1}{2}\right)^{25{,}000/5715} \approx 4.821 \times 10^{-14}$ — Ratio for 25,000 years

✓Checkpoint 2

Use the formula for the ratio of carbon isotopes to carbon atoms in Example 2 to find the value of R for each period of time.

a. 5000 years **b.** 15,000 years **c.** 30,000 years ■

Graphs of Exponential Functions

The basic nature of the graph of an exponential function can be determined by the point-plotting method or by using a graphing utility.

Example 3 **Graphing Exponential Functions**

Sketch the graph of each exponential function.

a. $f(x) = 2^x$ **b.** $g(x) = \left(\frac{1}{2}\right)^x = 2^{-x}$ **c.** $h(x) = 3^x$

SOLUTION To sketch these functions by hand, you can begin by constructing a table of values, as shown below.

x	-3	-2	-1	0	1	2	3	4
$f(x) = 2^x$	$\frac{1}{8}$	$\frac{1}{4}$	$\frac{1}{2}$	1	2	4	8	16
$g(x) = 2^{-x}$	8	4	2	1	$\frac{1}{2}$	$\frac{1}{4}$	$\frac{1}{8}$	$\frac{1}{16}$
$h(x) = 3^x$	$\frac{1}{27}$	$\frac{1}{9}$	$\frac{1}{3}$	1	3	9	27	81

The graphs of the three functions are shown in Figure 4.2. Note that the graphs of $f(x) = 2^x$ and $h(x) = 3^x$ are increasing, whereas the graph of $g(x) = 2^{-x}$ is decreasing.

STUDY TIP

Note that a graph of the form $f(x) = a^x$, as shown in Example 3(a), is a reflection in the y-axis of the graph of the form $f(x) = a^{-x}$, as shown in Example 3(b).

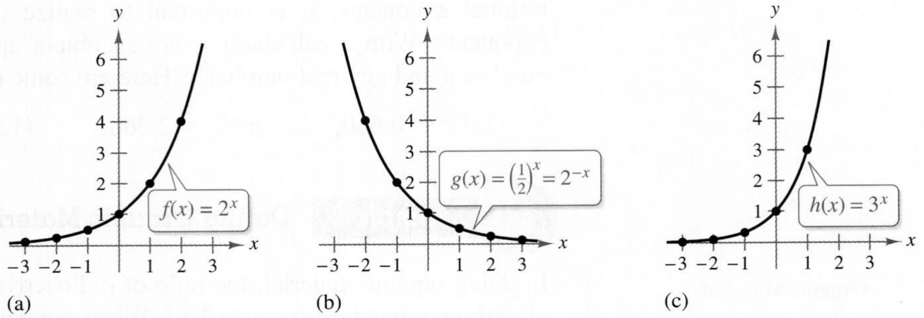

FIGURE 4.2

✓ **Checkpoint 3**

Sketch the graph of

$f(x) = 5^x$. ■

TECH TUTOR

Try graphing the functions

$f(x) = 2^x$ and $h(x) = 3^x$

in the same viewing window, as shown at the right. From the display, you can see that the graph of h is increasing more rapidly than the graph of f.

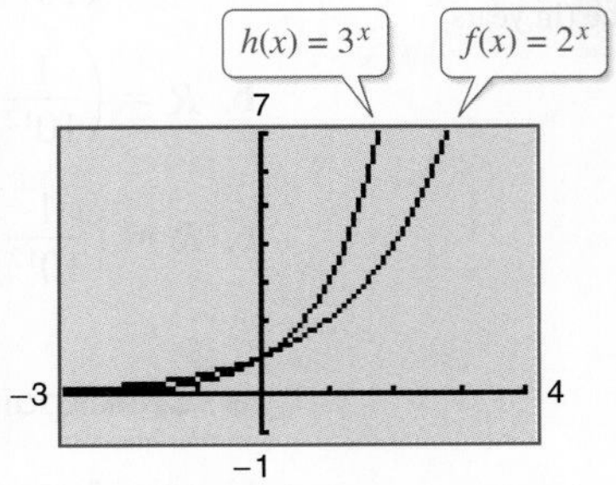

The forms of the graphs in Figure 4.2 are typical of the graphs of the exponential functions $y = a^{-x}$ and $y = a^{x}$, where $a > 1$. The basic characteristics of such graphs are summarized in Figure 4.3.

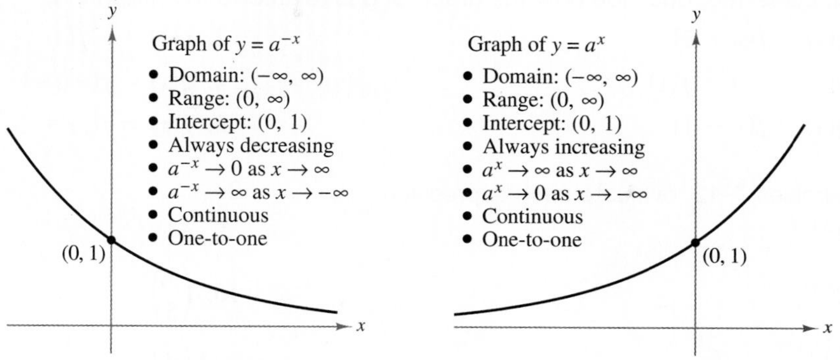

Characteristics of the Exponential Functions $y = a^{-x}$ and $y = a^{x}$ $(a > 1)$
FIGURE 4.3

Example 4 Graphing an Exponential Function

Sketch the graph of $f(x) = 3^{-x} - 1$.

SOLUTION Begin by creating a table of values, as shown below.

x	-2	-1	0	1	2
$f(x)$	$3^2 - 1 = 8$	$3^1 - 1 = 2$	$3^0 - 1 = 0$	$3^{-1} - 1 = -\frac{2}{3}$	$3^{-2} - 1 = -\frac{8}{9}$

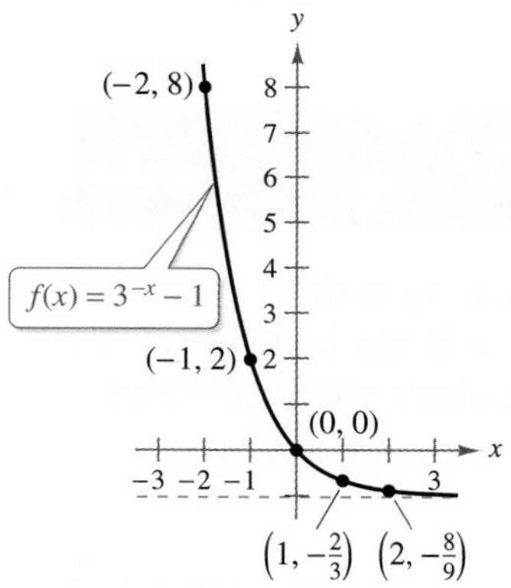

FIGURE 4.4

From the limit

$$\begin{aligned}\lim_{x\to\infty} (3^{-x} - 1) &= \lim_{x\to\infty} 3^{-x} - \lim_{x\to\infty} 1 \\ &= \lim_{x\to\infty} \frac{1}{3^x} - \lim_{x\to\infty} 1 \\ &= 0 - 1 \\ &= -1\end{aligned}$$

you can see that $y = -1$ is a horizontal asymptote of the graph. The graph is shown in Figure 4.4.

✓ Checkpoint 4

Sketch the graph of $f(x) = 2^{-x} + 1$.

SUMMARIZE (Section 4.1)

1. State the definition of an exponential function *(page 252)*. For examples of exponential functions, see Example 3.
2. State the properties of exponents *(page 252)*. For examples of the properties of exponents, see Examples 1 and 2.
3. State the basic characteristics of the graphs of the exponential functions $y = a^{-x}$ and $y = a^{x}$ *(page 255)*. For an example of the graph of an exponential function, see Example 4.

SKILLS WARM UP 4.1

The following warm-up exercises involve skills that were covered in earlier sections. You will use these skills in the exercise set for this section. For additional help, review Section 1.4.

In Exercises 1–6, describe how the graph of g is related to the graph of f.

1. $g(x) = f(x + 2)$
2. $g(x) = -f(x)$
3. $g(x) = -1 + f(x)$
4. $g(x) = f(-x)$
5. $g(x) = f(x - 1)$
6. $g(x) = f(x) + 2$

In Exercises 7–12, evaluate each expression.

7. $25^{3/2}$
8. $64^{3/4}$
9. $27^{2/3}$
10. $\left(\frac{1}{5}\right)^3$
11. $\left(\frac{1}{8}\right)^{1/3}$
12. $\left(\frac{5}{8}\right)^2$

In Exercises 13–18, solve for x.

13. $2x - 6 = 4$
14. $3x + 1 = 5$
15. $(x + 4)^2 = 25$
16. $(x - 2)^2 = 8$
17. $x^2 + 4x - 5 = 0$
18. $2x^2 - 3x + 1 = 0$

Exercises 4.1

See www.CalcChat.com for worked-out solutions to odd-numbered exercises.

Applying Properties of Exponents In Exercises 1–4, use the properties of exponents to simplify the expression. *See Example 1.*

1. (a) $(5^2)(5^3)$ (b) $(5^2)(5^{-3})$ (c) $(5^2)^2$ (d) 5^{-3}
2. (a) $\dfrac{5^3}{5^6}$ (b) $\left(\dfrac{1}{5}\right)^{-2}$ (c) $(8^{1/2})(2^{1/2})$ (d) $(32^{3/2})\left(\dfrac{1}{2}\right)^{3/2}$
3. (a) $\dfrac{5^3}{25^2}$ (b) $(9^{2/3})(3)(3^{2/3})$ (c) $[(25^{1/2})(5^2)]^{1/3}$ (d) $(8^2)(4^3)$
4. (a) $(4^3)(4^2)$ (b) $\left(\dfrac{1}{4}\right)^2(4^2)$ (c) $(4^6)^{1/2}$ (d) $[(8^{-1})(8^{2/3})]^3$

5. **Radioactive Decay** Beginning with 16 grams of a radioactive element whose half-life is 30 years, the mass y (in grams) remaining after t years is given by

$$y = 16\left(\frac{1}{2}\right)^{t/30}, \quad t \geq 0.$$

How much of the initial mass remains after 90 years?

6. **Radioactive Decay** Beginning with 23 grams of a radioactive element whose half-life is 45 years, the mass y (in grams) remaining after t years is given by

$$y = 23\left(\frac{1}{2}\right)^{t/45}, \quad t \geq 0.$$

How much of the initial mass remains after 150 years?

Graphing Exponential Functions In Exercises 7–18, sketch the graph of the function. *See Examples 3 and 4.*

7. $f(x) = 6^x$
8. $f(x) = 4^x$
9. $f(x) = \left(\frac{1}{5}\right)^x = 5^{-x}$
10. $f(x) = \left(\frac{1}{4}\right)^x = 4^{-x}$
11. $y = 2^{x-1}$
12. $y = 4^x + 3$
13. $y = -2^x$
14. $y = -5^x$
15. $y = 3^{-x^2}$
16. $y = 2^{-x^2}$
17. $s(t) = \frac{1}{4}(3^{-t})$
18. $s(t) = 2^{-t} + 3$

19. **Population Growth** The resident populations P (in millions) of the United States from 1995 through 2010 can be modeled by the exponential function

$$P(t) = 254.75(1.01)^t$$

where t is the time in years, with $t = 5$ corresponding to 1995. Use the model to estimate the populations in the years (a) 2013 and (b) 2020. *(Source: U.S. Census Bureau)*

20. Sales The sales S (in billions of dollars) for Walgreens from 2000 through 2010 can be modeled by the exponential function

$$S(t) = 22.52(1.125)^t$$

where t is the time in years, with $t = 0$ corresponding to 2000.

(a) Use the model to estimate the sales in 2014.

(b) Use the model to estimate the sales in 2018.

(Source: Walgreen Company)

21. Property Value A piece of property sells for \$64,000. The value of the property doubles every 15 years. A model for the value V of the property t years after the date of purchase is

$$V(t) = 64{,}000(2)^{t/15}.$$

Use the model to approximate the value of the property (a) 5 years and (b) 20 years after it is purchased.

22. HOW DO YOU SEE IT? Match the exponential function with its graph. Explain your reasoning. [The graphs are labeled (i), (ii), (iii), (iv), (v), and (vi).]

(i)

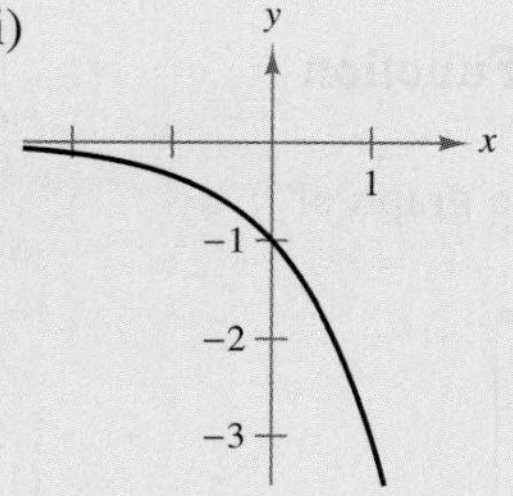

(ii)

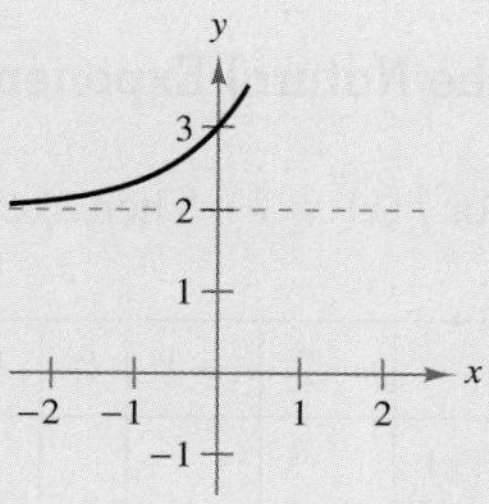

(iii)

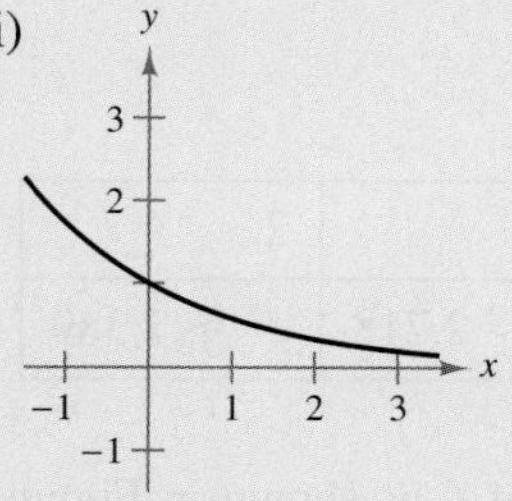

(iv)

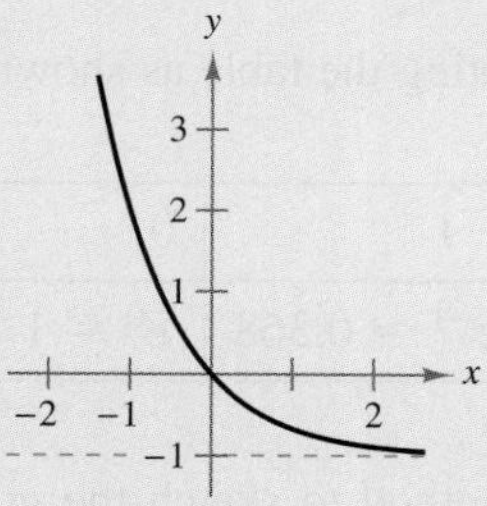

(v)

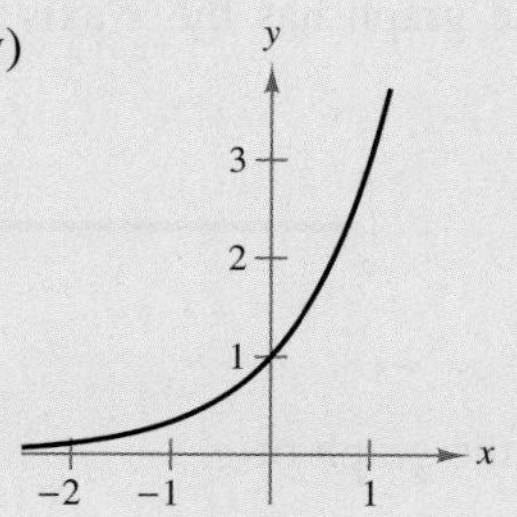

(vi)

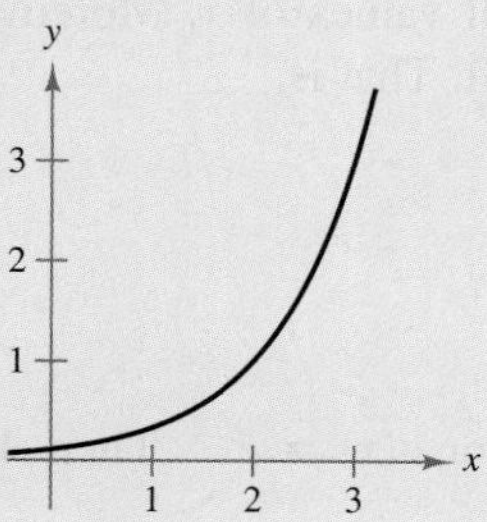

(a) $f(x) = 3^x$

(b) $f(x) = 3^{-x/2}$

(c) $f(x) = -3^x$

(d) $f(x) = 3^{x-2}$

(e) $f(x) = 3^{-x} - 1$

(f) $f(x) = 3^x + 2$

23. Inflation Rate With an annual rate of inflation of 4% over the next 10 years, the approximate cost C of goods or services during any year in the decade is given by

$$C(t) = P(1.04)^t,\ 0 \le t \le 10$$

where t is the time (in years) and P is the present cost. The price of an oil change for a car is presently \$24.95. Estimate the price 10 years from now.

24. Inflation Rate Repeat Exercise 23 using an annual rate of inflation of 10% over the next 10 years. The approximate cost C of goods or services is given by

$$C(t) = P(1.10)^t,\ 0 \le t \le 10.$$

25. Depreciation A car sells for \$28,000. The car depreciates such that each year it is worth $\frac{3}{4}$ of its value from the previous year. Find a model for the value V of the car after t years. Sketch a graph of the model and determine the value of the car 4 years after it is purchased.

26. Drug Concentration Immediately following an injection, the concentration of a drug in the bloodstream is 300 milligrams per milliliter. After t hours, the concentration is 75% of the level of the previous hour. Find a model for $C(t)$, the concentration of the drug after t hours. Sketch a graph of the model and determine the concentration of the drug after 8 hours.

27. School Nurses For the years 2001 through 2008, the average salaries y (in dollars) of school nurses in the public school system in the United States are shown in the table. *(Source: Educational Research Service)*

Year	2001	2002	2003	2004
Salary	37,188	38,221	39,165	40,201

Year	2005	2006	2007	2008
Salary	40,520	41,746	43,277	46,025

A model for this data is given by

$$y = 35{,}963(1.0279)^t$$

where t represents the year, with $t = 1$ corresponding to 2001.

(a) Compare the actual salaries with those given by the model. How well does the model fit the data? Explain your reasoning.

(b) Use a graphing utility to graph the model.

(c) Use the *zoom* and *trace* features of a graphing utility to predict the year during which the average salary of school nurses will reach \$54,000.

4.2 Natural Exponential Functions

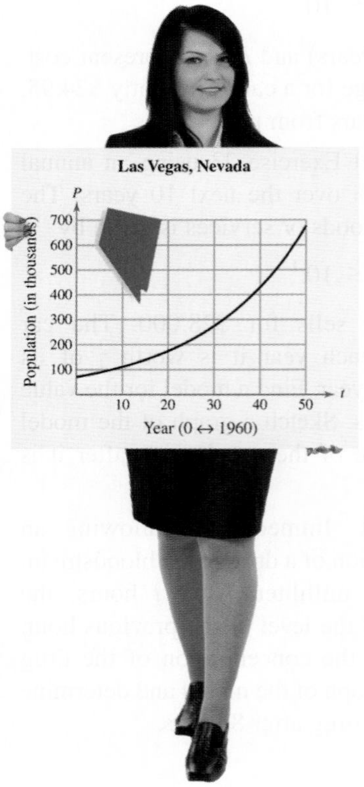

In Exercise 46 on page 266, you will evaluate a natural exponential function to find the population of Las Vegas, Nevada for several years.

- Evaluate and graph functions involving the natural exponential function.
- Solve compound interest problems.
- Solve present value problems.

Natural Exponential Functions

In Section 4.1, exponential functions were introduced using an unspecified base a. In calculus, the most convenient (or natural) choice for a base is the irrational number e, whose decimal approximation is

$$e \approx 2.71828182846.$$

Although this choice of base may seem unusual, its convenience will become apparent as the rules for differentiating exponential functions are developed in Section 4.3. In that development, you will encounter the limit used in the definition of e.

Limit Definition of e

The irrational number e is defined to be the limit of $(1 + x)^{1/x}$ as $x \to 0$. That is,

$$\lim_{x\to 0} (1 + x)^{1/x} = e.$$

Example 1 Graphing the Natural Exponential Function

Complete the table of values for $f(x) = e^x$. Then sketch the graph of f.

x	-2	-1	0	1	2
$f(x)$					

SOLUTION Begin by completing the table as shown.

x	-2	-1	0	1	2
$f(x)$	$e^{-2} \approx 0.135$	$e^{-1} \approx 0.368$	$e^0 = 1$	$e^1 \approx 2.718$	$e^2 \approx 7.389$

Then use the point-plotting method to sketch the graph of f, as shown in Figure 4.5. Note that e^x is positive for all values of x. Moreover, the graph has the x-axis as a horizontal asymptote to the left. That is,

$$\lim_{x\to -\infty} e^x = 0.$$

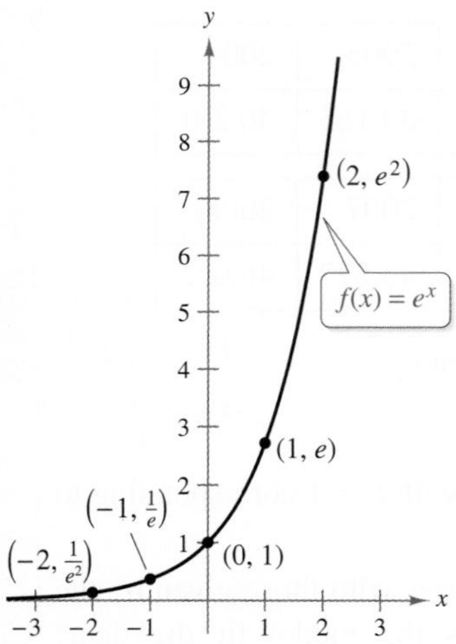

FIGURE 4.5

✓Checkpoint 1

Complete the table of values for $g(x) = e^{-x}$. Then sketch the graph of g.

x	-2	-1	0	1	2
$g(x)$					

Exponential functions are often used to model the growth of a quantity or a population. When the quantity's growth *is not* restricted, an exponential model is often used. When the quantity's growth *is* restricted, the best model is often a **logistic growth model** of the form

$$f(t) = \frac{a}{1 + be^{-kt}}.$$

Graphs of both types of population growth models are shown in Figure 4.6. The graph of a logistic growth model is called a *logistic curve*.

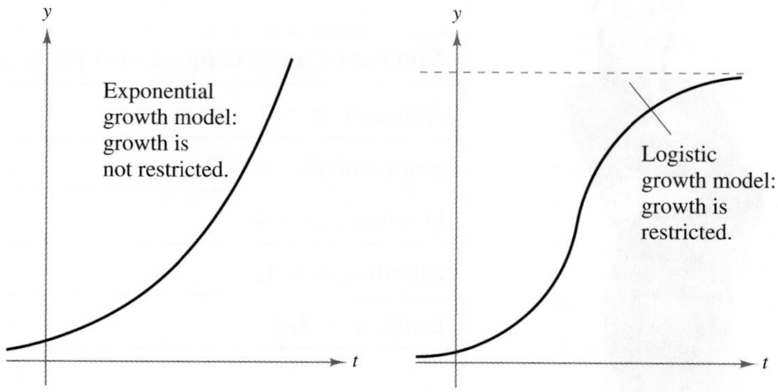

FIGURE 4.6

Example 2 Modeling a Population

A bacterial culture is growing according to the logistic growth model

$$y = \frac{1.25}{1 + 0.25e^{-0.4t}}, \quad t \geq 0$$

where y is the culture weight (in grams) and t is the time (in hours). Find the weight of the culture after 0 hours, 1 hour, and 10 hours. What is the limit of the model as t increases without bound?

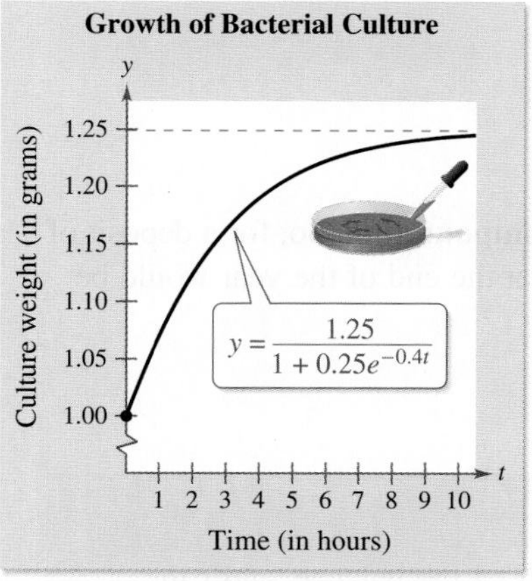

When a culture is grown in a dish, the size of the dish and the available food limit the culture's growth.

FIGURE 4.7

SOLUTION The graph of the model is shown in Figure 4.7.

$$y = \frac{1.25}{1 + 0.25e^{-0.4(0)}} = 1 \text{ gram} \qquad \text{Weight when } t = 0$$

$$y = \frac{1.25}{1 + 0.25e^{-0.4(1)}} \approx 1.071 \text{ grams} \qquad \text{Weight when } t = 1$$

$$y = \frac{1.25}{1 + 0.25e^{-0.4(10)}} \approx 1.244 \text{ grams} \qquad \text{Weight when } t = 10$$

As t approaches infinity, the limit of y is

$$\lim_{t \to \infty} \frac{1.25}{1 + 0.25e^{-0.4t}} = \lim_{t \to \infty} \frac{1.25}{1 + (0.25/e^{0.4t})} = \frac{1.25}{1 + 0} = 1.25.$$

So, as t increases without bound, the weight of the culture approaches 1.25 grams.

✓ Checkpoint 2

A bacterial culture is growing according to the model $y = 1.50/(1 + 0.2e^{-0.5t})$, $t \geq 0$, where y is the culture weight (in grams) and t is the time (in hours). Find the weight of the culture after 0 hours, 1 hour, and 10 hours. What is the limit of the model as t increases without bound? ■

Extended Application: Compound Interest

An amount of P dollars is deposited in an account at an annual interest rate of r (in decimal form). What is the balance after 1 year? The answer depends on the number of times the interest is compounded, according to the formula

$$A = P\left(1 + \frac{r}{n}\right)^n$$

where n is the number of compoundings per year. The balances for a deposit of \$1000 at 8%, for various compounding periods, are shown in the table.

Number of times compounded per year, n	Balance (in dollars), A
Annually, $n = 1$	$A = 1000\left(1 + \frac{0.08}{1}\right)^1 = \$1080.00$
Semiannually, $n = 2$	$A = 1000\left(1 + \frac{0.08}{2}\right)^2 = \$1081.60$
Quarterly, $n = 4$	$A = 1000\left(1 + \frac{0.08}{4}\right)^4 \approx \$1082.43$
Monthly, $n = 12$	$A = 1000\left(1 + \frac{0.08}{12}\right)^{12} \approx \$1083.00$
Daily, $n = 365$	$A = 1000\left(1 + \frac{0.08}{365}\right)^{365} \approx \$1083.28$

You may be surprised to discover that as n increases, the balance A approaches a limit, as indicated in the following development. In this development, let

$$x = \frac{r}{n}.$$

Then $x \to 0$ as $n \to \infty$, and you have

$$\begin{aligned} A &= \lim_{n\to\infty} P\left(1 + \frac{r}{n}\right)^n \\ &= P \lim_{n\to\infty} \left(1 + \frac{r}{n}\right)^n \\ &= P \lim_{n\to\infty} \left[\left(1 + \frac{r}{n}\right)^{n/r}\right]^r \\ &= P\left[\lim_{x\to 0} (1 + x)^{1/x}\right]^r \qquad \text{Substitute } x \text{ for } r/n. \\ &= Pe^r. \end{aligned}$$

This limit is the balance after 1 year of **continuous compounding.** So, for a deposit of \$1000 at 8%, compounded continuously, the balance at the end of the year would be

$$\begin{aligned} A &= 1000e^{0.08} \\ &\approx \$1083.29. \end{aligned}$$

Summary of Compound Interest Formulas

Let P be the amount deposited, t the number of years, A the balance, and r the annual interest rate (in decimal form).

1. Compounded n times per year: $A = P\left(1 + \dfrac{r}{n}\right)^{nt}$
2. Compounded continuously: $A = Pe^{rt}$

Williv/www.shutterstock.com

The average interest rates paid by banks on savings accounts have varied greatly during the past 30 years. At times savings accounts have earned as much as 12% annual interest, and at times they have earned less than 1%. The next example shows how the annual interest rate can affect the balance of an account.

Example 3 Finding Account Balances

You are creating a trust fund for your newborn nephew. You plan to deposit \$12,000 in an account, with instructions that the account be turned over to your nephew on his 25th birthday. Compare the balances in the account for each situation. Which account should you choose?

a. 7%, compounded continuously

b. 7%, compounded quarterly

c. 11%, compounded continuously

d. 11%, compounded quarterly

SOLUTION

a. $12{,}000e^{0.07(25)} \approx 69{,}055.23$ — 7%, compounded continuously

b. $12{,}000\left(1 + \frac{0.07}{4}\right)^{4(25)} \approx 68{,}017.87$ — 7%, compounded quarterly

c. $12{,}000e^{0.11(25)} \approx 187{,}711.58$ — 11%, compounded continuously

d. $12{,}000\left(1 + \frac{0.11}{4}\right)^{4(25)} \approx 180{,}869.07$ — 11%, compounded quarterly

The balances in the accounts for parts (a) and (c) are shown in Figure 4.8. Notice the dramatic difference between the balances at 7% and 11%. You should choose the account described in part (c) because it earns more money than the other accounts.

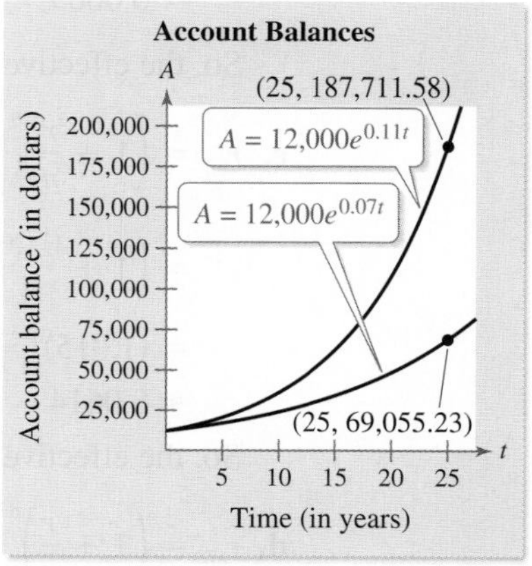

FIGURE 4.8

✓ Checkpoint 3

Find the balance in an account when \$2000 is deposited for 10 years at an interest rate of 9%, compounded as follows. Compare the results and make a general statement about compounding.

a. quarterly **b.** monthly

c. daily **d.** continuously ■

In Example 3, note that the interest earned depends on the frequency with which the interest is compounded. The annual percentage rate is called the **stated rate** or **nominal rate.** However, the nominal rate does not reflect the actual rate at which interest is earned, which means that the compounding produced an **effective rate** that is larger than the nominal rate. In general, the effective rate corresponding to a nominal rate of r that is compounded n times per year is

$$\text{Effective rate} = r_{eff} = \left(1 + \frac{r}{n}\right)^n - 1.$$

Example 4 Finding the Effective Rate of Interest

Find the effective rate of interest corresponding to a nominal rate of 6% per year compounded (a) annually, (b) semiannually, (c) quarterly, and (d) monthly.

SOLUTION

a. $r_{eff} = \left(1 + \frac{r}{n}\right)^n - 1$ Formula for effective rate of interest

$= \left(1 + \frac{0.06}{1}\right)^1 - 1$ Substitute for r and n.

$= 1.06 - 1$ Simplify.

$= 0.06$

So, the effective rate is 6% per year.

b. $r_{eff} = \left(1 + \frac{r}{n}\right)^n - 1$ Formula for effective rate of interest

$= \left(1 + \frac{0.06}{2}\right)^2 - 1$ Substitute for r and n.

$= (1.03)^2 - 1$ Simplify.

$= 0.0609$

So, the effective rate is 6.09% per year.

c. $r_{eff} = \left(1 + \frac{r}{n}\right)^n - 1$ Formula for effective rate of interest

$= \left(1 + \frac{0.06}{4}\right)^4 - 1$ Substitute for r and n.

$= (1.015)^4 - 1$ Simplify.

≈ 0.0614

So, the effective rate is about 6.14% per year.

d. $r_{eff} = \left(1 + \frac{r}{n}\right)^n - 1$ Formula for effective rate of interest

$= \left(1 + \frac{0.06}{12}\right)^{12} - 1$ Substitute for r and n.

$= (1.005)^{12} - 1$ Simplify.

≈ 0.0617

So, the effective rate is about 6.17% per year.

✓ Checkpoint 4

Repeat Example 4 using a nominal rate of 7%. ■

Present Value

In planning for the future, this problem often arises: "How much money P should be deposited now, at a fixed rate of interest r, in order to have a balance of A, t years from now?" The answer to this question is given by the **present value** of A.

To find the present value of a future investment, use the formula for compound interest as shown.

$$A = P\left(1 + \frac{r}{n}\right)^{nt} \qquad \text{Formula for compound interest}$$

Solving for P gives a present value of

$$P = \frac{A}{\left(1 + \frac{r}{n}\right)^{nt}} \quad \text{or} \quad P = \frac{A}{(1 + i)^N}$$

where $i = r/n$ is the interest rate per compounding period and $N = nt$ is the total number of compounding periods. You will learn another way to find the present value of a future investment in Section 6.1.

Example 5 Finding Present Value

An investor is purchasing a 10-year certificate of deposit that pays an annual percentage rate of 8%, compounded monthly. How much should the person invest in order to obtain a balance of $15,000 at maturity?

SOLUTION Here, $A = 15{,}000$, $r = 0.08$, $n = 12$, and $t = 10$. Using the formula for present value, you obtain

$$P = \frac{15{,}000}{\left(1 + \frac{0.08}{12}\right)^{12(10)}} \qquad \text{Substitute for } A, r, n, \text{ and } t.$$

$$\approx 6757.85. \qquad \text{Simplify.}$$

So, the person should invest $6757.85 in the certificate of deposit.

✓ Checkpoint 5

How much money should be deposited in an account paying 6% interest compounded monthly in order to have a balance of $20,000 after 3 years? ■

SUMMARIZE (Section 4.2)

1. State the limit definition of e *(page 258)*. For an example of a graph of a natural exponential function, see Example 1.
2. Describe a real-life example of how an exponential function can be used to model a population *(page 259, Example 2)*.
3. State the compound interest formulas for n compoundings per year and for continuous compounding *(page 260)*. For applications of these formulas, see Example 3.
4. State the formula for finding the effective rate of interest *(page 262)*. For an application of this formula, see Example 4.
5. State the formula for present value *(page 263)*. For an application of this formula, see Example 5.

SKILLS WARM UP 4.2

The following warm-up exercises involve skills that were covered in earlier sections. You will use these skills in the exercise set for this section. For additional help, review Sections 1.6 and 3.6.

In Exercises 1–4, discuss the continuity of the function.

1. $f(x) = \dfrac{3x^2 + 2x + 1}{x^2 + 1}$

2. $f(x) = \dfrac{x + 1}{x^2 - 4}$

3. $f(x) = \dfrac{x^2 - 6x + 5}{x^2 - 3}$

4. $g(x) = \dfrac{x^2 - 9x + 20}{x - 4}$

In Exercises 5–12, find the horizontal asymptote of the graph of the function.

5. $f(x) = \dfrac{25}{1 + 4x}$

6. $f(x) = \dfrac{16x}{3 + x^2}$

7. $f(x) = \dfrac{8x^3 + 2}{2x^3 + x}$

8. $f(x) = \dfrac{x}{2x}$

9. $f(x) = \dfrac{3}{2 + (1/x)}$

10. $f(x) = \dfrac{6}{1 + x^{-2}}$

11. $f(x) = 2^{-x}$

12. $f(x) = \dfrac{7}{1 + 5x}$

Exercises 4.2

See www.CalcChat.com for worked-out solutions to odd-numbered exercises.

Applying Properties of Exponents In Exercises 1–4, use the properties of exponents to simplify the expression.

1. (a) $(e^3)(e^4)$ (b) $(e^3)^4$
(c) $(e^3)^{-2}$ (d) e^0

2. (a) $\left(\dfrac{1}{e}\right)^{-2}$ (b) $\left(\dfrac{e^5}{e^2}\right)^{-1}$
(c) $\dfrac{e^5}{e^3}$ (d) $\dfrac{1}{e^{-3}}$

3. (a) $(e^2)^{5/2}$ (b) $(e^2)(e^{1/2})$
(c) $(e^{-2})^{-3}$ (d) $\dfrac{e^5}{e^{-2}}$

4. (a) $(e^{-3})^{2/3}$ (b) $\dfrac{e^4}{e^{-1/2}}$
(c) $(e^{-2})^{-4}$ (d) $(e^{-4})(e^{-3/2})$

Matching In Exercises 5–10, match the function with its graph. [The graphs are labeled (a)–(f).]

(a)

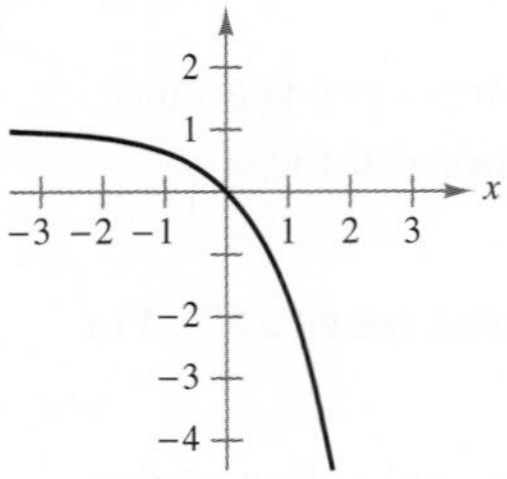

(b)

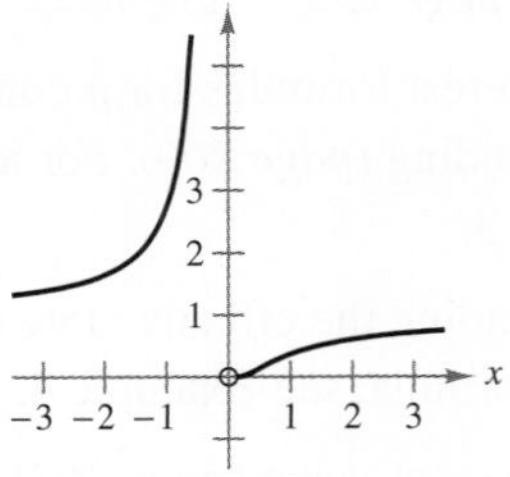

(c)

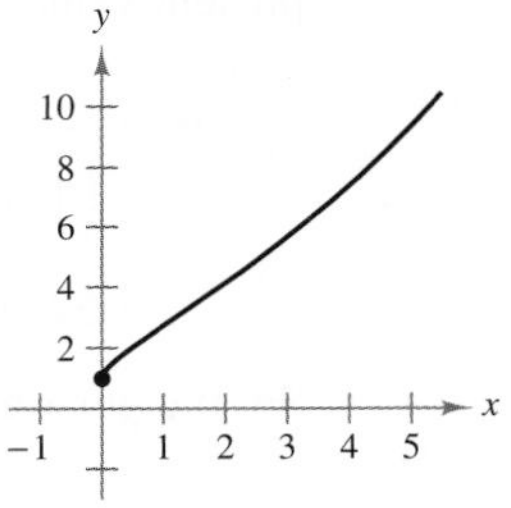

(d)

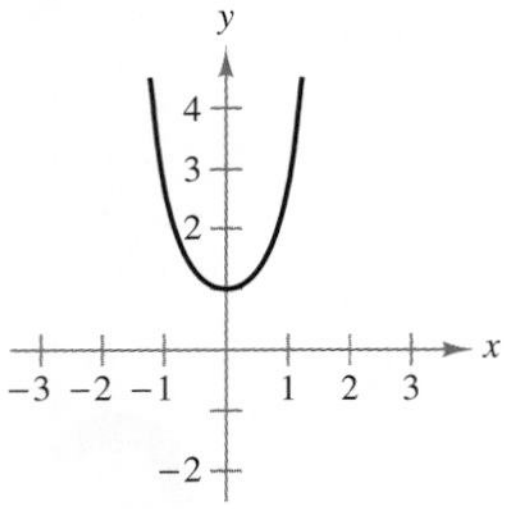

(e)

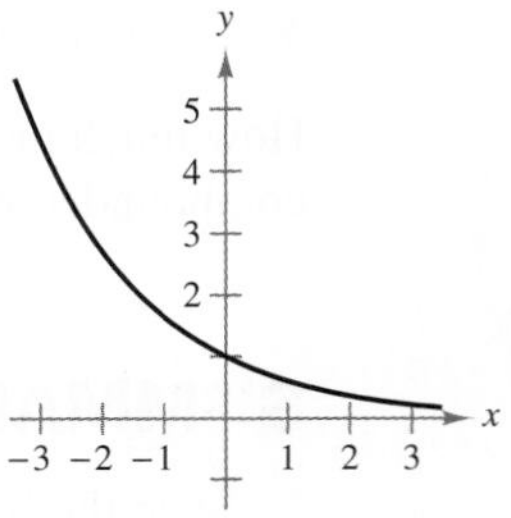

(f)

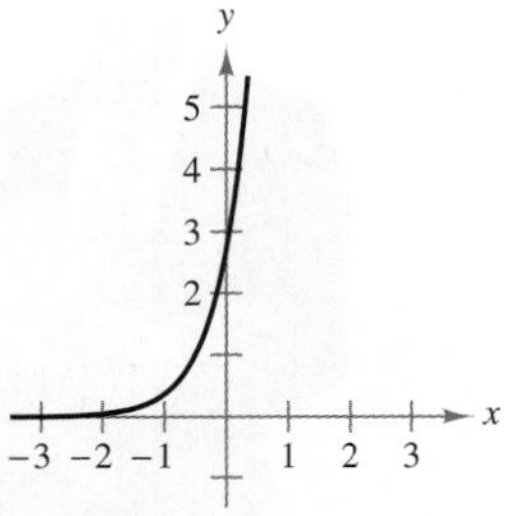

5. $f(x) = e^{2x+1}$

6. $f(x) = e^{-x/2}$

7. $f(x) = e^{x^2}$

8. $f(x) = e^{-1/x}$

9. $f(x) = e^{\sqrt{x}}$

10. $f(x) = -e^x + 1$

Graphing Natural Exponential Functions In Exercises 11–16, sketch the graph of the function. *See Example 1.*

11. $f(x) = e^{-x/3}$

12. $f(x) = e^{2x}$

13. $g(x) = e^x - 2$

14. $h(x) = e^{-x} + 5$

15. $g(x) = e^{1-x}$

16. $j(x) = e^{-x+2}$

Graphing Functions In Exercises 17–24, use a graphing utility to graph the function. Determine whether the function has any horizontal asymptotes and discuss the continuity of the function.

17. $N(t) = 500e^{-0.2t}$

18. $A(t) = 500e^{0.15t}$

19. $g(x) = \dfrac{2}{1 + e^{x^2}}$

20. $g(x) = \dfrac{10}{1 + e^{-x}}$

21. $f(x) = \dfrac{e^x + e^{-x}}{2}$

22. $f(x) = \dfrac{e^x - e^{-x}}{2}$

23. $f(x) = \dfrac{2}{1 + e^{1/x}}$

24. $f(x) = \dfrac{2}{1 + 2e^{-0.2x}}$

25. Graphing Exponential Functions Use a graphing utility to graph $f(x) = e^x$ and the given function in the same viewing window. How are the two graphs related?

(a) $g(x) = e^{x-2}$ (b) $h(x) = -\frac{1}{2}e^x$

(c) $q(x) = e^x + 3$

26. Graphing Logistic Growth Functions Use a graphing utility to graph the function. Describe the shape of the graph for very large and very small values of x.

(a) $f(x) = \dfrac{8}{1 + e^{-0.5x}}$ (b) $g(x) = \dfrac{8}{1 + e^{-0.5/x}}$

Finding Account Balances In Exercises 27–30, complete the table to determine the balance A for P dollars invested at rate r for t years, compounded n times per year. *See Example 3.*

n	1	2	4	12	365	Continuous compounding
A						

27. $P = \$1000$, $r = 3\%$, $t = 10$ years

28. $P = \$2500$, $r = 2.5\%$, $t = 20$ years

29. $P = \$1000$, $r = 4\%$, $t = 20$ years

30. $P = \$2500$, $r = 5\%$, $t = 40$ years

Finding Present Value In Exercises 31–34, complete the table to determine the amount of money P that should be invested at rate r to produce a final balance of \$100,000 in t years. *See Example 5.*

t	1	10	20	30	40	50
P						

31. $r = 4\%$, compounded continuously

32. $r = 3\%$, compounded continuously

33. $r = 5\%$, compounded monthly

34. $r = 6\%$, compounded daily

35. Trust Fund On the day of a child's birth, a deposit of \$20,000 is made in a trust fund that pays 8% interest, compounded continuously. Determine the balance in this account on the child's 21st birthday.

36. Trust Fund A deposit of \$10,000 is made in a trust fund that pays 7% interest, compounded continuously. It is specified that the balance will be given to the college from which the donor graduated after the money has earned interest for 50 years. How much will the college receive?

37. Effective Rate Find the effective rate of interest corresponding to a nominal rate of 9% per year compounded (a) annually, (b) semiannually, (c) quarterly, and (d) monthly.

38. Effective Rate Find the effective rate of interest corresponding to a nominal rate of 7.5% per year compounded (a) annually, (b) semiannually, (c) quarterly, and (d) monthly.

39. Present Value How much should be deposited in an account paying 7.2% interest compounded monthly in order to have a balance of \$8000 after 3 years?

40. Present Value How much should be deposited in an account paying 7.8% interest compounded monthly in order to have a balance of \$21,000 after 4 years?

41. Demand The demand function for a product is modeled by

$$p = 5000\left(1 - \frac{4}{4 + e^{-0.002x}}\right).$$

Find the price p (in dollars) of the product when the quantity demanded is (a) $x = 100$ units and (b) $x = 500$ units. (c) What is the limit of the price as x increases without bound?

42. Demand The demand function for a product is modeled by

$$p = 10{,}000\left(1 - \frac{3}{3 + e^{-0.001x}}\right).$$

Find the price p (in dollars) of the product when the quantity demanded is (a) $x = 1000$ units and (b) $x = 1500$ units. (c) What is the limit of the price as x increases without bound?

43. Probability The average time between incoming calls at a switchboard is 3 minutes. If a call has just come in, the probability that the next call will come within the next t minutes is $P(t) = 1 - e^{-t/3}$. Find the probability of each situation.

(a) A call comes in within $\frac{1}{2}$ minute.

(b) A call comes in within 2 minutes.

(c) A call comes in within 5 minutes.

44. Consumer Awareness An automobile gets 28 miles per gallon at speeds up to and including 50 miles per hour. At speeds greater than 50 miles per hour, the number of miles per gallon drops at the rate of 12% for each 10 miles per hour. If s is the speed (in miles per hour) and y is the number of miles per gallon, then $y = 28e^{0.6-0.012s}$, $s > 50$. Use this information and a spreadsheet to create a table showing the miles per gallon for $s = 50$, 55, 60, 65, and 70. What can you conclude?

45. Federal Debt The federal debts D (in billions of dollars) of the United States at the end of each year from 2000 through 2009 are shown in the table. *(Source: U.S. Office of Management and Budget)*

Year	2000	2001	2002	2003	2004
Debt	5629	5770	6198	6760	7355

Year	2005	2006	2007	2008	2009
Debt	7905	8451	8951	9986	11,876

A model for these data is given by $y = 5364.1e^{0.0796t}$, where t represents the year, with $t = 0$ corresponding to 2000.

(a) How well does the model fit the data?

(b) Find a linear model for the data. How well does the linear model fit the data? Which model, exponential or linear, is a better fit?

(c) Use both models to predict the year in which the federal debt will exceed 18,000 billion dollars.

46. Population The populations P (in thousands) of Las Vegas, Nevada from 1960 through 2009 can be modeled by $P = 70.751e^{0.0451t}$, where t is the time in years, with $t = 0$ corresponding to 1960. *(Source: U.S. Census Bureau)*

(a) Find the populations in 1960, 1970, 1980, 1990, 2000, and 2009.

(b) Explain why the change in population from 1960 to 1970 is not the same as the change in population from 1980 to 1990.

(c) Use the model to estimate the population in 2020.

47. Biology The population y of a bacterial culture is modeled by the logistic growth function $y = 925/(1 + e^{-0.3t})$, where t is the time in days.

(a) Use a graphing utility to graph the model.

(b) Does the population have a limit as t increases without bound? Explain your answer.

(c) How would the limit change if the model were $y = 1000/(1 + e^{-0.3t})$? Explain your answer. Draw some conclusions about this type of model.

48. HOW DO YOU SEE IT? The figure shows the graphs of $y = 2^x$, $y = e^x$, $y = 10^x$, $y = 2^{-x}$, $y = e^{-x}$, and $y = 10^{-x}$. Match each function with its graph. [The graphs are labeled (a)–(f).] Explain your reasoning.

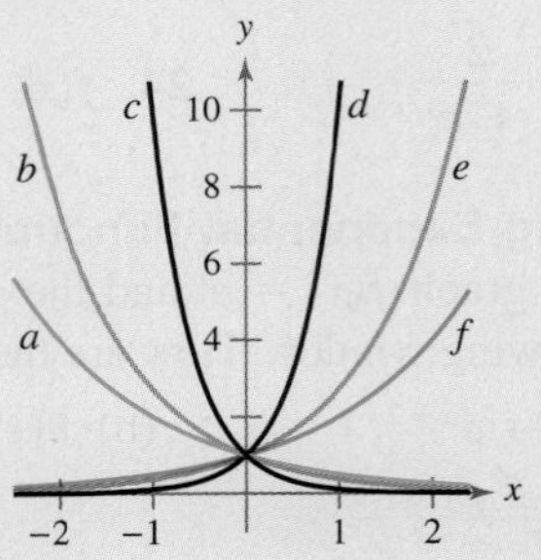

49. Learning Theory In a learning theory project, the proportion P of correct responses after n trials can be modeled by $P = 0.83/(1 + e^{-0.2n})$.

(a) Find the proportion of correct responses after 3 trials.

(b) Find the proportion of correct responses after 7 trials.

(c) Use a graphing utility to graph the model. Find the number of trials required for the proportion of correct responses to be 0.75.

(d) Does the proportion of correct responses have a limit as n increases without bound? Explain your reasoning.

50. Learning Theory In a typing class, the average number N of words per minute typed after t weeks of lessons can be modeled by $N = 95/(1 + 8.5e^{-0.12t})$.

(a) Find the average number of words per minute typed after 10 weeks.

(b) Find the average number of words per minute typed after 20 weeks.

(c) Use a graphing utility to graph the model. Find the number of weeks required to achieve an average of 70 words per minute.

(d) Does the number of words per minute have a limit as t increases without bound? Explain your reasoning.

51. Certificate of Deposit You want to invest $5000 in a certificate of deposit for 12 months. You are given the options below. Which would you choose? Explain.

(a) $r = 5.25\%$, quarterly compounding

(b) $r = 5\%$, monthly compounding

(c) $r = 4.75\%$, continuous compounding

4.3 Derivatives of Exponential Functions

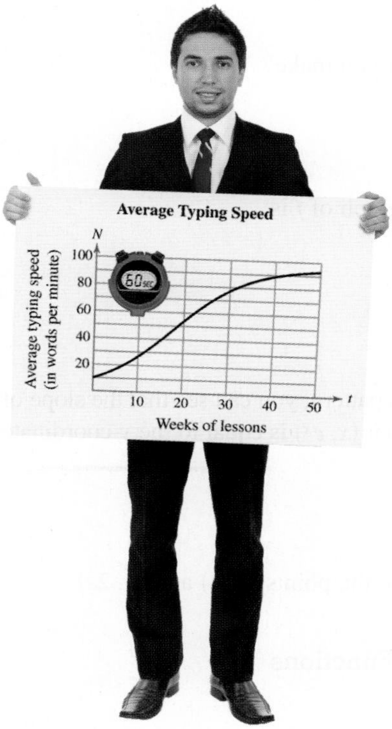

In Exercise 48 on page 274, you will use the derivative of an exponential function to find the rate of change of the average typing speed after 5, 10, and 30 weeks of lessons.

- Find the derivatives of natural exponential functions.
- Use calculus to analyze the graphs of real-life functions that involve the natural exponential function.
- Explore the normal probability density function.

Derivatives of Exponential Functions

In Section 4.2, it was stated that the most convenient base for exponential functions is the irrational number e. The convenience of this base stems primarily from the fact that the function

$$f(x) = e^x$$

is its own derivative. You will see that this is not true of other exponential functions of the form

$$y = a^x$$

where $a \neq e$. To verify that $f(x) = e^x$ is its own derivative, notice that the limit

$$\lim_{\Delta x \to 0} (1 + \Delta x)^{1/\Delta x} = e$$

implies that for small values of Δx,

$$e \approx (1 + \Delta x)^{1/\Delta x}$$

or

$$e^{\Delta x} \approx 1 + \Delta x.$$

This approximation is used in the following derivation.

$$\begin{aligned}
f'(x) &= \lim_{\Delta x \to 0} \frac{f(x + \Delta x) - f(x)}{\Delta x} && \text{Definition of derivative} \\
&= \lim_{\Delta x \to 0} \frac{e^{x+\Delta x} - e^x}{\Delta x} && \text{Use } f(x) = e^x. \\
&= \lim_{\Delta x \to 0} \frac{e^x(e^{\Delta x} - 1)}{\Delta x} && \text{Factor numerator.} \\
&= \lim_{\Delta x \to 0} \frac{e^x[(1 + \Delta x) - 1]}{\Delta x} && \text{Substitute } 1 + \Delta x \text{ for } e^{\Delta x}. \\
&= \lim_{\Delta x \to 0} \frac{e^x(\cancel{\Delta x})}{\cancel{\Delta x}} && \text{Divide out common factor.} \\
&= \lim_{\Delta x \to 0} e^x && \text{Simplify.} \\
&= e^x && \text{Evaluate limit.}
\end{aligned}$$

When u is a function of x, you can apply the Chain Rule to obtain the derivative of e^u with respect to x. Both formulas are summarized below.

Derivative of the Natural Exponential Function

Let u be a differentiable function of x.

1. $\dfrac{d}{dx}[e^x] = e^x$ **2.** $\dfrac{d}{dx}[e^u] = e^u \dfrac{du}{dx}$

Example 1 Finding Slopes of Tangent Lines

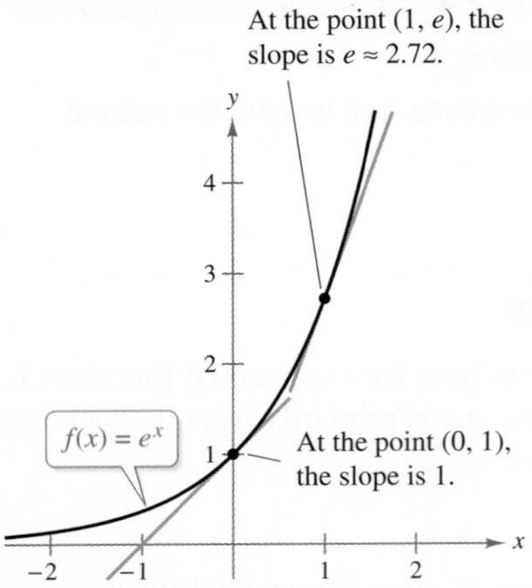

FIGURE 4.9

Find the slopes of the tangent lines to

$$f(x) = e^x$$

at the points (0, 1) and (1, e). What conclusion can you make?

SOLUTION Because the derivative of f is

$$f'(x) = e^x \qquad \text{Derivative}$$

it follows that the slope of the tangent line to the graph of f is

$$f'(0) = e^0 = 1 \qquad \text{Slope at point } (0, 1)$$

at the point (0, 1) and

$$f'(1) = e^1 = e \qquad \text{Slope at point } (1, e)$$

at the point (1, e), as shown in Figure 4.9. From this pattern, you can see that the slope of the tangent line to the graph of $f(x) = e^x$ at any point (x, e^x) is equal to the y-coordinate of the point.

✓ **Checkpoint 1**

Find the slopes of the tangent lines to $f(x) = 2e^x$ at the points (0, 2) and (1, $2e$). ■

STUDY TIP

In Example 2, notice that when you differentiate an exponential function, the exponent does not change. For instance, the derivative of $f(x) = e^{3x}$ is $f'(x) = 3e^{3x}$. In both f and f', the exponent is $3x$.

Example 2 Differentiating Exponential Functions

Differentiate each function.

a. $f(x) = e^{2x}$ **b.** $f(x) = e^{-3x^2}$

c. $f(x) = 6e^{x^3}$ **d.** $f(x) = e^{-x}$

SOLUTION

a. Let $u = 2x$. Then $du/dx = 2$, and you can apply the Chain Rule.

$$f'(x) = e^u \frac{du}{dx} = e^{2x}(2) = 2e^{2x}$$

b. Let $u = -3x^2$. Then $du/dx = -6x$, and you can apply the Chain Rule.

$$f'(x) = e^u \frac{du}{dx} = e^{-3x^2}(-6x) = -6xe^{-3x^2}$$

c. Let $u = x^3$. Then $du/dx = 3x^2$, and you can apply the Chain Rule.

$$f'(x) = 6e^u \frac{du}{dx} = 6e^{x^3}(3x^2) = 18x^2e^{x^3}$$

d. Let $u = -x$. Then $du/dx = -1$, and you can apply the Chain Rule.

$$f'(x) = e^u \frac{du}{dx} = e^{-x}(-1) = -e^{-x}$$

✓ **Checkpoint 2**

Differentiate each function.

a. $f(x) = e^{3x}$ **b.** $f(x) = e^{-2x^3}$

c. $f(x) = 4e^{x^2}$ **d.** $f(x) = e^{-2x}$ ■

The differentiation rules that you studied in Chapter 2 can be used with exponential functions, as shown in Example 3.

TECH TUTOR

If you have access to a symbolic differentiation utility, try using it to find the derivatives of the functions in Example 3.

Example 3 Differentiating Exponential Functions

Differentiate each function.

a. $f(x) = 4e$ **b.** $f(x) = e^{2x-1}$

c. $f(x) = xe^x$ **d.** $f(x) = \dfrac{e^x - e^{-x}}{2}$

e. $f(x) = \dfrac{e^x}{x}$ **f.** $f(x) = xe^x - e^x$

SOLUTION

a.
$$f(x) = 4e \qquad \text{Write original function.}$$
$$f'(x) = 0 \qquad \text{Constant Rule}$$

b.
$$f(x) = e^{2x-1} \qquad \text{Write original function.}$$
$$f'(x) = (e^{2x-1})(2) \qquad \text{Chain Rule}$$
$$= 2e^{2x-1} \qquad \text{Simplify.}$$

c.
$$f(x) = xe^x \qquad \text{Write original function.}$$
$$f'(x) = xe^x + e^x(1) \qquad \text{Product Rule}$$
$$= xe^x + e^x \qquad \text{Simplify.}$$

d.
$$f(x) = \frac{e^x - e^{-x}}{2} \qquad \text{Write original function.}$$
$$= \frac{1}{2}(e^x - e^{-x}) \qquad \text{Rewrite.}$$
$$f'(x) = \frac{1}{2}[e^x - e^{-x}(-1)] \qquad \text{Constant Multiple and Chain Rules}$$
$$= \frac{1}{2}(e^x + e^{-x}) \qquad \text{Simplify.}$$

e.
$$f(x) = \frac{e^x}{x} \qquad \text{Write original function.}$$
$$f'(x) = \frac{xe^x - e^x(1)}{x^2} \qquad \text{Quotient Rule}$$
$$= \frac{e^x(x-1)}{x^2} \qquad \text{Simplify.}$$

f.
$$f(x) = xe^x - e^x \qquad \text{Write original function.}$$
$$f'(x) = [xe^x + e^x(1)] - e^x \qquad \text{Product and Difference Rules}$$
$$= xe^x + e^x - e^x$$
$$= xe^x \qquad \text{Simplify.}$$

✓ Checkpoint 3

Differentiate each function.

a. $f(x) = 9e$ **b.** $f(x) = e^{3x+1}$ **c.** $f(x) = x^2e^x$

d. $f(x) = \dfrac{e^x + e^{-x}}{2}$ **e.** $f(x) = \dfrac{e^x}{x^2}$ **f.** $f(x) = x^2e^x - e^x$ ■

Applications

In Chapter 3, you learned how to use derivatives to analyze the graphs of functions. The next example applies those techniques to a function composed of exponential functions. In the example, notice that

$$e^a = e^b$$

implies that $a = b$.

Example 4 Analyzing a Catenary

When a telephone wire is hung between two poles, the wire forms a U-shaped curve called a **catenary.** For instance, the function

$$y = 30(e^{x/60} + e^{-x/60}), \quad -30 \le x \le 30$$

models the shape of a telephone wire strung between two poles that are 60 feet apart (x and y are measured in feet). Show that the lowest point on the wire is midway between the two poles. How much does the wire sag between the two poles?

SOLUTION First, find the derivative of the function.

$$y = 30(e^{x/60} + e^{-x/60}) \quad \text{Write original function.}$$
$$y' = 30\left[e^{x/60}\left(\tfrac{1}{60}\right) + e^{-x/60}\left(-\tfrac{1}{60}\right)\right] \quad \text{Derivative}$$
$$= 30\left(\tfrac{1}{60}\right)(e^{x/60} - e^{-x/60}) \quad \text{Factor out } \tfrac{1}{60}.$$
$$= \tfrac{1}{2}(e^{x/60} - e^{-x/60}) \quad \text{Simplify.}$$

To find the critical numbers, set the derivative equal to zero.

$$\tfrac{1}{2}(e^{x/60} - e^{-x/60}) = 0 \quad \text{Set derivative equal to 0.}$$
$$e^{x/60} - e^{-x/60} = 0 \quad \text{Multiply each side by 2.}$$
$$e^{x/60} = e^{-x/60} \quad \text{Add } e^{-x/60} \text{ to each side.}$$
$$\frac{x}{60} = -\frac{x}{60} \quad \text{If } e^a = e^b, \text{ then } a = b.$$
$$x = -x \quad \text{Multiply each side by 60.}$$
$$2x = 0 \quad \text{Add } x \text{ to each side.}$$
$$x = 0 \quad \text{Divide each side by 2.}$$

Using the First-Derivative Test, you can determine that the critical number $x = 0$ yields a relative minimum of the function. From the graph in Figure 4.10, you can see that this relative minimum is actually a minimum on the interval

$$[-30, 30].$$

So, you can conclude that the lowest point on the wire lies midway between the two poles. To find how much the wire sags between the two poles, you can compare its height at each pole with its height at the relative minimum.

$$y = 30(e^{-30/60} + e^{-(-30)/60}) \approx 67.7 \text{ feet} \quad \text{Height at left pole}$$
$$y = 30(e^{0/60} + e^{-(0)/60}) = 60 \text{ feet} \quad \text{Height at relative minimum}$$
$$y = 30(e^{30/60} + e^{-(30)/60}) \approx 67.7 \text{ feet} \quad \text{Height at right pole}$$

From this, you can see that the wire sags about 7.7 feet.

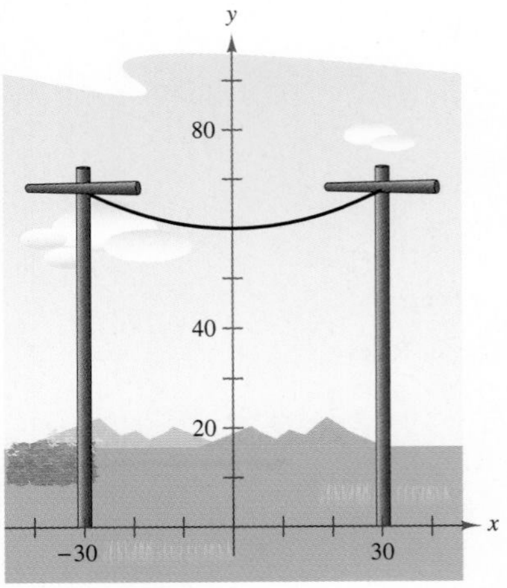

FIGURE 4.10

✓ Checkpoint 4

Use a graphing utility to graph the function in Example 4. Verify the minimum value. Use the information in the example to choose an appropriate viewing window.

Example 5 Finding a Maximum Revenue

The demand function for a product is modeled by

$$p = 56e^{-0.000012x} \qquad \text{Demand function}$$

where p is the price per unit (in dollars) and x is the number of units. What price will yield a maximum revenue? What is the maximum revenue at this price?

SOLUTION The revenue function is

$$R = xp = 56xe^{-0.000012x}. \qquad \text{Revenue function}$$

To find the maximum revenue *analytically*, you would first find the marginal revenue

$$\frac{dR}{dx} = 56xe^{-0.000012x}(-0.000012) + e^{-0.000012x}(56).$$

You would then set dR/dx equal to zero

$$56xe^{-0.000012x}(-0.000012) + e^{-0.000012x}(56) = 0$$

and solve for x. At this point, you can see that the analytical approach is rather cumbersome. In this problem, it is easier to use a *graphical* approach. After experimenting to find a reasonable viewing window, you can obtain a graph of R that is similar to that shown in Figure 4.11. Using the *maximum* feature, you can conclude that the maximum revenue occurs when x is about 83,333 units. To find the price that corresponds to this production level, substitute $x \approx 83{,}333$ into the demand function.

$$p \approx 56e^{-0.000012(83{,}333)} \approx \$20.60.$$

So, a price of about \$20.60 will yield a maximum revenue of

$$R \approx 56(83{,}333)e^{-0.000012(83{,}333)}$$
$$\approx \$1{,}716{,}771. \qquad \text{Maximum revenue}$$

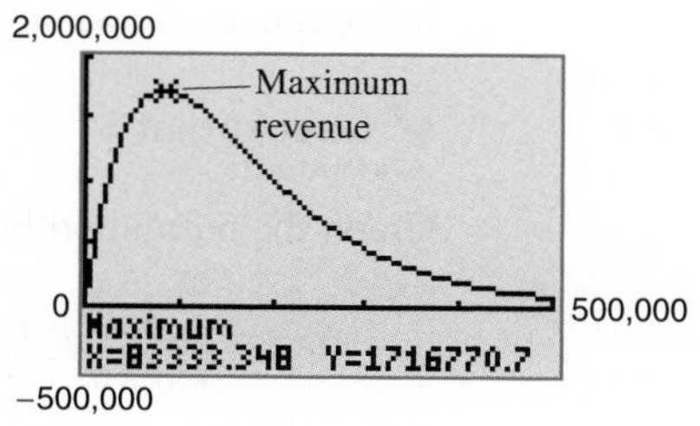

Use the *maximum* feature to approximate the x-value that corresponds to the maximum revenue.

FIGURE 4.11

✓ Checkpoint 5

The demand function for a product is modeled by

$$p = 50e^{-0.0000125x}$$

where p is the price per unit (in dollars) and x is the number of units. What price will yield a maximum revenue? What is the maximum revenue at this price? ■

Try solving Example 5 analytically. When you do this, you must solve the equation

$$56xe^{-0.000012x}(-0.000012) + e^{-0.000012x}(56) = 0.$$

Explain how you would solve this equation. What is the solution?

The Normal Probability Density Function

If you take a course in statistics or quantitative business analysis, you will spend quite a bit of time studying the characteristics and use of the **normal probability density function** given by

$$f(x) = \frac{1}{\sigma\sqrt{2\pi}}e^{-(x-\mu)^2/(2\sigma^2)}$$

where σ is the lowercase Greek letter sigma, and μ is the lowercase Greek letter mu. In this formula, σ represents the *standard deviation* of the probability distribution, and μ represents the *mean* of the probability distribution.

Example 6 Exploring a Probability Density Function

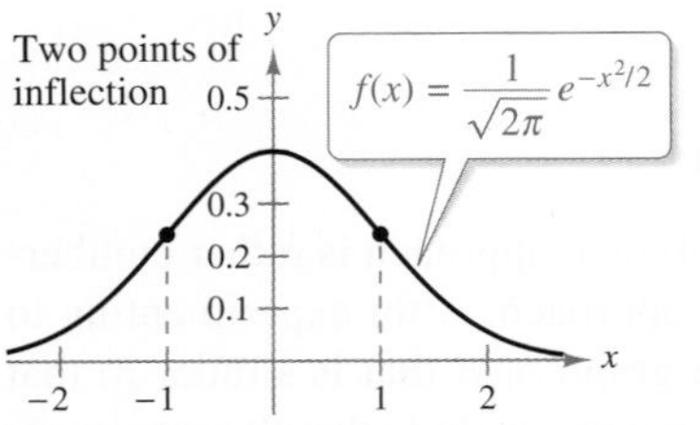

The graph of the normal probability density function is bell-shaped.

FIGURE 4.12

Show that the graph of the normal probability density function

$$f(x) = \frac{1}{\sqrt{2\pi}}e^{-x^2/2}$$

has points of inflection at $x = \pm 1$.

SOLUTION Begin by finding the second derivative of the function.

$$f'(x) = \frac{1}{\sqrt{2\pi}}(-x)e^{-x^2/2} \qquad \text{First derivative}$$

$$f''(x) = \frac{1}{\sqrt{2\pi}}[(-x)(-x)e^{-x^2/2} + (-1)e^{-x^2/2}] \qquad \text{Second derivative}$$

$$= \frac{1}{\sqrt{2\pi}}(e^{-x^2/2})(x^2 - 1) \qquad \text{Simplify.}$$

By setting the second derivative equal to 0, you can determine that $x = \pm 1$. By testing the concavity of the graph, you can then conclude that these x-values yield points of inflection, as shown in Figure 4.12.

✓ Checkpoint 6

Graph the normal probability density function

$$f(x) = \frac{1}{4\sqrt{2\pi}}e^{-x^2/32}$$

and approximate the points of inflection.

SUMMARIZE (Section 4.3)

1. State the derivative of the natural exponential function *(page 267)*. For examples of the derivative of the natural exponential function, see Examples 2 and 3.
2. Describe a real-life example of how a natural exponential function can be used to analyze the graph of a catenary *(page 270, Example 4)*.
3. Describe a real-life example of how a natural exponential function can be used to analyze a company's maximum revenue *(page 271, Example 5)*.
4. Describe a use of the natural exponential function in statistics *(page 272)*. For an example of the natural exponential function in statistics, see Example 6.

SKILLS WARM UP 4.3 The following warm-up exercises involve skills that were covered in a previous course or in earlier sections. You will use these skills in the exercise set for this section. For additional help, review Appendix Section A.4, and Sections 2.2, 2.4, and 3.2.

In Exercises 1–4, factor the expression.

1. $x^2e^x - \frac{1}{2}e^x$ **2.** $(xe^{-x})^{-1} + e^x$ **3.** $xe^x - e^{2x}$ **4.** $e^x - xe^{-x}$

In Exercises 5–8, find the derivative of the function.

5. $f(x) = \dfrac{3}{7x^2}$ **6.** $g(x) = 3x^2 - \dfrac{x}{6}$ **7.** $f(x) = (4x - 3)(x^2 + 9)$ **8.** $f(t) = \dfrac{t - 2}{\sqrt{t}}$

In Exercises 9 and 10, find the relative extrema of the function.

9. $f(x) = \frac{1}{8}x^3 - 2x$ **10.** $f(x) = x^4 - 2x^2 + 5$

Exercises 4.3

See www.CalcChat.com for worked-out solutions to odd-numbered exercises.

Differentiating Exponential Functions In Exercises 1–16, find the derivative of the function. *See Examples 2 and 3.*

1. $f(x) = 3e$ **2.** $f(x) = -5e$

3. $y = e^{5x}$ **4.** $y = e^{1-x}$

5. $y = e^{-x^2}$ **6.** $f(x) = e^{1/x}$

7. $f(x) = e^{-1/x^2}$ **8.** $g(x) = e^{\sqrt{x}}$

9. $f(x) = (x^2 + 1)e^{4x}$ **10.** $y = 4x^3e^{-x}$

11. $f(x) = \dfrac{2}{(e^x + e^{-x})^3}$ **12.** $f(x) = \dfrac{(e^x + e^{-x})^4}{2}$

13. $f(x) = \dfrac{e^x + 1}{e^x - 1}$ **14.** $f(x) = \dfrac{e^{2x}}{e^{2x} + 1}$

15. $y = xe^x - 4e^{-x}$ **16.** $y = x^2e^x - 2xe^x + 2e^x$

Finding the Slope of a Tangent Line In Exercises 17–20, find the slope of the tangent line to the exponential function at the point (0, 1).

17. $y = e^{4x}$

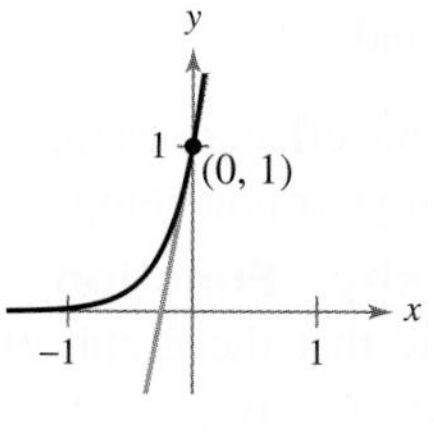

18. $y = e^{x/2}$

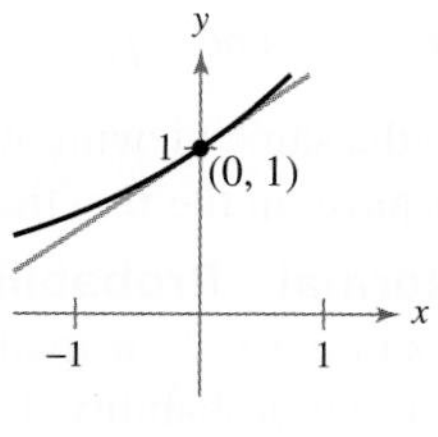

19. $y = e^{-3x}$

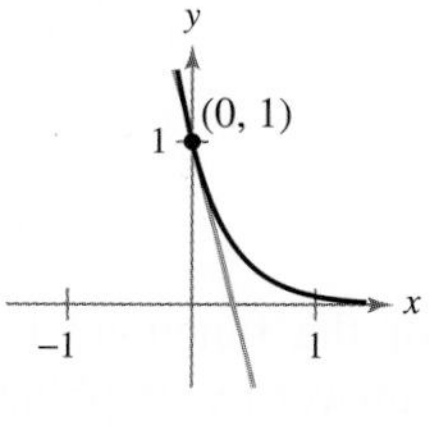

20. $y = e^{-x/2}$

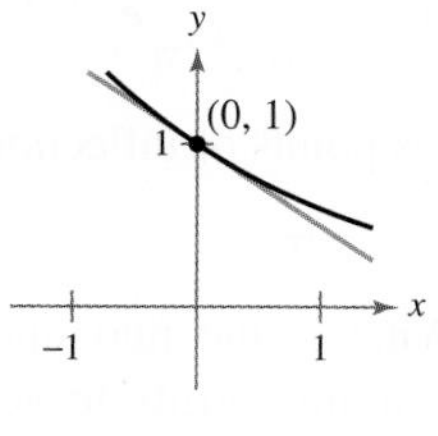

Finding an Equation of a Tangent Line In Exercises 21–26, find an equation of the tangent line to the graph of the function at the given point.

21. $y = e^{-2x+x^2}$, $(2, 1)$ **22.** $g(x) = e^{x^3}$, $\left(-1, \dfrac{1}{e}\right)$

23. $y = x^2e^{-x}$, $\left(2, \dfrac{4}{e^2}\right)$ **24.** $y = \dfrac{x}{e^{2x}}$, $\left(1, \dfrac{1}{e^2}\right)$

25. $y = (e^{2x} + 1)^3$, $(0, 8)$ **26.** $y = (e^{4x} - 2)^2$, $(0, 1)$

Finding Derivatives Implicitly In Exercises 27–30, find dy/dx implicitly.

27. $xe^y - 10x + 3y = 0$

28. $x^2y - e^y - 4 = 0$

29. $x^2e^{-x} + 2y^2 - xy = 0$

30. $e^{xy} + x^2 - y^2 = 10$

Finding Second Derivatives In Exercises 31–34, find the second derivative.

31. $f(x) = 2e^{3x} + 3e^{-2x}$ **32.** $f(x) = 5e^{-x} - 2e^{-5x}$

33. $f(x) = (1 + 2x)e^{4x}$ **34.** $f(x) = (3 + 2x)e^{-3x}$

Analyzing a Graph In Exercises 35–38, analyze and sketch the graph of the function. Label any relative extrema, points of inflection, and asymptotes.

35. $f(x) = \dfrac{1}{2 - e^{-x}}$ **36.** $f(x) = \dfrac{e^x - e^{-x}}{2}$

37. $f(x) = x^2e^{-x}$ **38.** $f(x) = xe^{-x}$

Solving Equations In Exercises 39–42, solve the equation for x.

39. $e^{-3x} = e$ **40.** $e^x = 1$

41. $e^{\sqrt{x}} = e^3$ **42.** $e^{-1/x} = e^{1/2}$

Depreciation In Exercises 43 and 44, the value V (in dollars) of an item is a function of the time t (in years).

(a) Sketch the function over the interval $[0, 10]$. Use a graphing utility to verify your graph.

(b) Find the rate of change of V when $t = 1$.

(c) Find the rate of change of V when $t = 5$.

(d) Use the values $(0, V(0))$ and $(10, V(10))$ to find the linear depreciation model for the item.

(e) Compare the exponential function and the model from part (d). What are the advantages of each?

43. $V = 15{,}000e^{-0.6286t}$ **44.** $V = 500{,}000e^{-0.2231t}$

45. Employment From 2000 through 2009, the numbers y (in millions) of employed people in the United States can be modeled by

$$y = 136.855 - 0.5841t + 0.31664t^2 - 0.002166e^t$$

where t represents the year, with $t = 0$ corresponding to 2000. *(Source: U.S. Bureau of Labor Statistics)*

(a) Use a graphing utility to graph the model.

(b) Use the graph to estimate the rates of change in the number of employed people in 2000, 2004, and 2009.

(c) Confirm the results from part (b) analytically.

46. HOW DO YOU SEE IT? The yield y (in pounds per acre) of an orchard at age t (in years) is modeled by

$$y = 7955.6e^{-0.0458/t}.$$

The graph is shown below.

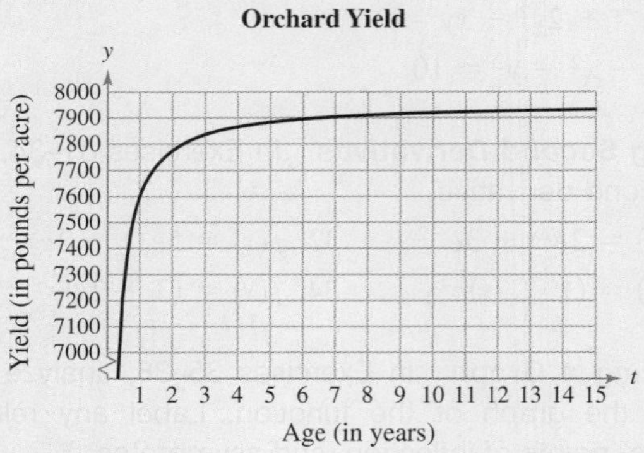

(a) What happens to the yield in the long run?

(b) What happens to the rate of change of the yield in the long run?

47. Compound Interest The balance A (in dollars) in a savings account is given by $A = 5000e^{0.08t}$, where t is measured in years. Find the rates at which the balance is changing when (a) $t = 1$ year, (b) $t = 10$ years, and (c) $t = 50$ years.

48. Learning Theory The average typing speed N (in words per minute) after t weeks of lessons is modeled by

$$N = \frac{95}{1 + 8.5e^{-0.12t}}.$$

Find the rates at which the typing speed is changing when (a) $t = 5$ weeks, (b) $t = 10$ weeks, and (c) $t = 30$ weeks.

49. Probability In a recent year, the mean SAT score for college-bound seniors on the mathematics portion was 516, with a standard deviation of 116. *(Source: The College Board)*

(a) Assuming the data can be modeled by a normal probability density function, find a model for these data.

(b) Use a graphing utility to graph the model. Be sure to choose an appropriate viewing window.

(c) Find the derivative of the model.

(d) Show that $f' > 0$ for $x < \mu$ and $f' < 0$ for $x > \mu$.

50. Probability A survey of a college freshman class has determined that the mean height of females in the class is 64 inches, with a standard deviation of 3.2 inches.

(a) Assuming the data can be modeled by a normal probability density function, find a model for these data.

(b) Use a graphing utility to graph the model. Be sure to choose an appropriate viewing window.

(c) Find the derivative of the model.

(d) Show that $f' > 0$ for $x < \mu$ and $f' < 0$ for $x > \mu$.

51. Normal Probability Density Function Use a graphing utility to graph the normal probability density function with $\mu = 0$ and $\sigma = 2, 3$, and 4 in the same viewing window. What effect does the standard deviation σ have on the function? Explain your reasoning.

52. Normal Probability Density Function Use a graphing utility to graph the normal probability density function with

$$\sigma = 1 \quad \text{and} \quad \mu = -2, \quad 1, \quad \text{and} \quad 3$$

in the same viewing window. What effect does the mean μ have on the function? Explain your reasoning.

53. Normal Probability Density Function Use Example 6 as a model to show that the graph of the normal probability density function with $\mu = 0$

$$f(x) = \frac{1}{\sigma\sqrt{2\pi}}e^{-x^2/(2\sigma^2)}$$

has points of inflection at

$$x = \pm\sigma.$$

What is the maximum value of the function? Use a graphing utility to verify your answer by graphing the function for several values of σ.

QUIZ YOURSELF

See www.CalcChat.com for worked-out solutions to odd-numbered exercises.

Take this quiz as you would take a quiz in class. When you are done, check your work against the answers given in the back of the book.

In Exercises 1–8, use properties of exponents to simplify the expression.

1. $4^3(4^2)$ **2.** $\left(\frac{1}{6}\right)^{-3}$

3. $\frac{3^8}{3^5}$ **4.** $(5^{1/2})(3^{1/2})$

5. $(e^2)(e^5)$ **6.** $(e^{2/3})(e^3)$

7. $\frac{e^2}{e^{-4}}$ **8.** $(e^{-1})^{-3}$

In Exercises 9–14, sketch the graph of the function.

9. $f(x) = 3^x - 2$

10. $f(x) = 5^{-x} + 2$

11. $f(x) = 6^{x-3}$

12. $f(x) = e^{x+2}$

13. $f(x) = e^x + 3$

14. $f(x) = e^{-2x} + 1$

15. After t years, the remaining mass y (in grams) of an initial mass of 35 grams of a radioactive element whose half-life is 80 years is given by

$$y = 35\left(\frac{1}{2}\right)^{t/80}, \quad t \ge 0.$$

How much of the initial mass remains after 50 years?

16. With an annual rate of inflation of 4.5% over the next 10 years, the approximate cost C of goods or services during any year in the decade is given by

$$C(t) = P(1.045)^t, \quad 0 \le t \le 10$$

where t is the time (in years) and P is the present cost. The price of a baseball game ticket is presently \$20. Estimate the price 10 years from now.

17. For $P = \$3000$, $r = 3.5\%$, and $t = 5$ years, find the balance in an account when interest is compounded (a) quarterly, (b) monthly, and (c) continuously.

18. How much should be deposited in an account paying 6% interest compounded monthly in order to have a balance of \$14,000 after 5 years?

In Exercises 19–22, find the derivative of the function.

19. $y = e^{5x}$ **20.** $y = e^{x-4}$

21. $y = 5e^{x+2}$ **22.** $y = 3e^x - xe^x$

23. Determine an equation of the tangent line to

$$y = e^{-2x}$$

at the point (0, 1).

24. Analyze and sketch the graph of

$$f(x) = 0.5x^2e^{-0.5x}.$$

Label any relative extrema, points of inflection, and asymptotes.

4.4 Logarithmic Functions

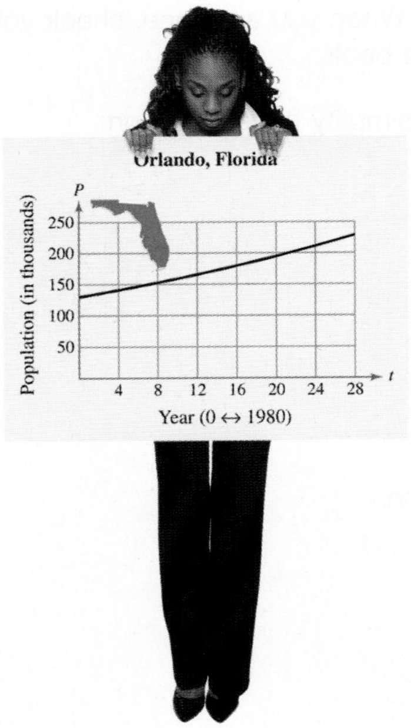

In Exercise 77 on page 283, you will solve a natural exponential equation to predict when the population of Orlando, Florida will reach 300,000.

- Sketch the graphs of natural logarithmic functions.
- Use properties of logarithms to simplify, expand, and condense logarithmic expressions.
- Use inverse properties of exponential and logarithmic functions to solve exponential and logarithmic equations.
- Use properties of natural logarithms to answer questions about real-life situations.

The Natural Logarithmic Function

From your previous algebra courses, you should be somewhat familiar with logarithms. For instance, the **common logarithm** $\log_{10} x$ is defined as

$$\log_{10} x = b \quad \text{if and only if} \quad 10^b = x.$$

The base of common logarithms is 10. In calculus, the most useful base for logarithms is the number *e*.

Definition of the Natural Logarithmic Function

The **natural logarithmic function,** denoted by $\ln x$, is defined as

$$\ln x = b \quad \text{if and only if} \quad e^b = x.$$

$\ln x$ is read as "el en of x" or as "the natural log of x."

This definition implies that the natural logarithmic function and the natural exponential function are inverse functions. So, every logarithmic equation can be written in an equivalent exponential form, and every exponential equation can be written in logarithmic form. Here are some examples.

Logarithmic form:	*Exponential form:*
$\ln 1 = 0$	$e^0 = 1$
$\ln e = 1$	$e^1 = e$
$\ln \frac{1}{e} = -1$	$e^{-1} = \frac{1}{e}$
$\ln 2 \approx 0.693$	$e^{0.693} \approx 2$
$\ln 0.1 \approx -2.303$	$e^{-2.303} \approx 0.1$

Because the functions $f(x) = e^x$ and $g(x) = \ln x$ are inverse functions, their graphs are reflections of each other in the line

$$y = x.$$

This reflective property is illustrated in Figure 4.13. The figure also contains a summary of several properties of the graph of the natural logarithmic function.

Notice that the domain of the natural logarithmic function is the set of *positive real numbers*—be sure you see that $\ln x$ is not defined for zero or for negative numbers. You can test this on your calculator. When you try evaluating

$$\ln(-1) \quad \text{or} \quad \ln 0$$

your calculator should indicate that the value is not a real number.

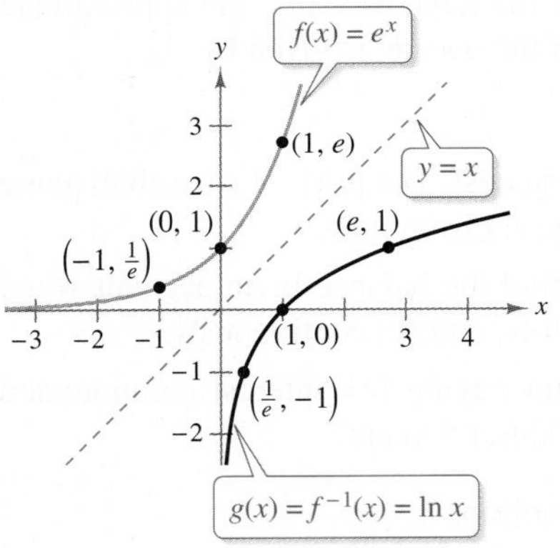

$g(x) = \ln x$

- Domain: $(0, \infty)$
- Range: $(-\infty, \infty)$
- Intercept: $(1, 0)$
- Always increasing
- $\ln x \to \infty$ as $x \to \infty$
- $\ln x \to -\infty$ as $x \to 0^+$
- Continuous
- One-to-one

FIGURE 4.13

David Gilder/www.shutterstock.com

Example 1 Graphing Logarithmic Functions

Sketch the graph of each function.

a. $f(x) = \ln(x + 1)$ **b.** $f(x) = 2\ln(x - 2)$

SOLUTION

a. Because the natural logarithmic function is defined only for positive values, the domain of the function is $x + 1 > 0$, or

$x > -1.$ Domain

To sketch the graph, begin by constructing a table of values, as shown below. Then plot the points in the table and connect them with a smooth curve, as shown in Figure 4.14(a).

x	-0.5	0	0.5	1	1.5	2
$\ln(x + 1)$	-0.693	0	0.405	0.693	0.916	1.099

b. The domain of this function is $x - 2 > 0$, or

$x > 2.$ Domain

A table of values for the function is shown below, and its graph is shown in Figure 4.14(b).

x	2.5	3	3.5	4	4.5	5
$2\ln(x - 2)$	-1.386	0	0.811	1.386	1.833	2.197

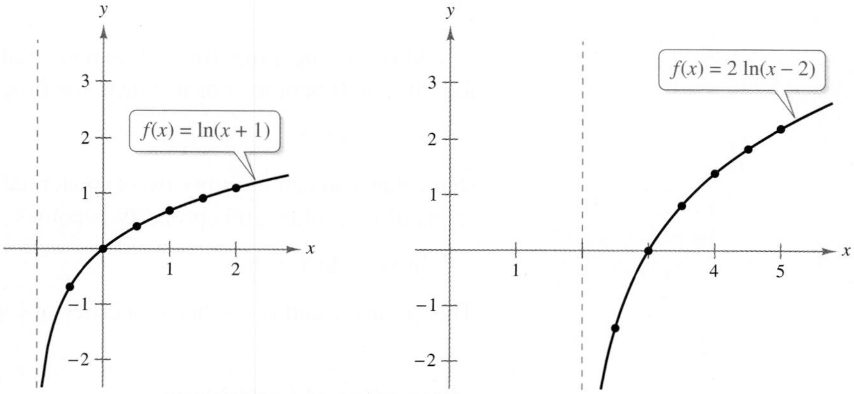

(a) (b)

FIGURE 4.14

✓ Checkpoint 1

Complete the table and sketch the graph of

$f(x) = \ln(x + 2).$

x	-1.5	-1	-0.5	0	0.5	1
$f(x)$						

■

TECH TUTOR

What happens when you take the logarithm of a negative number? Some graphing utilities do not give an error message for $\ln(-1)$. Instead, the graphing utility displays a complex number. For the purpose of this text, however, it is assumed that the domain of the logarithmic function is the set of positive real numbers.

STUDY TIP

How does the graph of $f(x) = \ln(x + 1)$ relate to the graph of $y = \ln x$? The graph of f is a translation of the graph of $y = \ln x$ one unit to the left.

Properties of Logarithmic Functions

Recall from Section 1.4 that inverse functions have the property that

$$f(f^{-1}(x)) = x \quad \text{and} \quad f^{-1}(f(x)) = x.$$

The properties listed below follow from the fact that the natural logarithmic function and the natural exponential function are inverse functions.

Inverse Properties of Logarithms and Exponents

1. $\ln e^x = x$ **2.** $e^{\ln x} = x$

Example 2 Applying Inverse Properties

Simplify each expression.

a. $\ln e^{\sqrt{2}}$ **b.** $e^{\ln 3x}$

SOLUTION

a. Because $\ln e^x = x$, it follows that

$$\ln e^{\sqrt{2}} = \sqrt{2}.$$

b. Because $e^{\ln x} = x$, it follows that

$$e^{\ln 3x} = 3x.$$

✓ Checkpoint 2

Simplify each expression.

a. $\ln e^3$ **b.** $e^{\ln(x+1)}$ ■

Most of the properties of exponential functions can be rewritten in terms of logarithmic functions. For instance, the property

$$e^x e^y = e^{x+y}$$

states that you can multiply two exponential expressions by adding their exponents. In terms of logarithms, this property becomes

$$\ln xy = \ln x + \ln y.$$

This property and two other properties of logarithms are summarized below.

Properties of Logarithms

1. $\ln xy = \ln x + \ln y$ **2.** $\ln \dfrac{x}{y} = \ln x - \ln y$ **3.** $\ln x^n = n \ln x$

STUDY TIP

There is no general property that can be used to rewrite $\ln(x + y)$. Specifically, $\ln(x + y)$ is not equal to $\ln x + \ln y$.

Rewriting a logarithm of a single quantity as the sum, difference, or multiple of logarithms is called *expanding* the logarithmic expression. The reverse procedure is called *condensing* a logarithmic expression.

TECH TUTOR

Try using a graphing utility to verify the results of Example 3(b). That is, try graphing the functions

$$y = \ln \sqrt{x^2 + 1}$$

and

$$y = \frac{1}{2}\ln(x^2 + 1).$$

Because these two functions are equivalent, their graphs should coincide.

Example 3 Expanding Logarithmic Expressions

Use the properties of logarithms to rewrite each expression as a sum, difference, or multiple of logarithms. (Assume $x > 0$ and $y > 0$.)

a. $\ln \frac{10}{9}$ **b.** $\ln \sqrt{x^2 + 1}$ **c.** $\ln \frac{xy}{5}$ **d.** $\ln[x^2(x + 1)]$

SOLUTION

a. $\ln \frac{10}{9} = \ln 10 - \ln 9$ Property 2

b. $\ln \sqrt{x^2 + 1} = \ln(x^2 + 1)^{1/2}$ Rewrite with rational exponent.

$= \frac{1}{2}\ln(x^2 + 1)$ Property 3

c. $\ln \frac{xy}{5} = \ln(xy) - \ln 5$ Property 2

$= \ln x + \ln y - \ln 5$ Property 1

d. $\ln[x^2(x + 1)] = \ln x^2 + \ln(x + 1)$ Property 1

$= 2 \ln x + \ln(x + 1)$ Property 3

✓Checkpoint 3

Use the properties of logarithms to rewrite each expression as a sum, difference, or multiple of logarithms. (Assume $x > 0$ and $y > 0$.)

a. $\ln \frac{2}{5}$ **b.** $\ln \sqrt[3]{x + 2}$ **c.** $\ln \frac{x}{5y}$ **d.** $\ln x(x + 1)^2$ ■

Example 4 Condensing Logarithmic Expressions

Use the properties of logarithms to rewrite each expression as the logarithm of a single quantity. (Assume $x > 0$ and $y > 0$.)

a. $\ln x + 2 \ln y$ **b.** $2 \ln(x + 2) - 3 \ln x$

SOLUTION

a. $\ln x + 2 \ln y = \ln x + \ln y^2$ Property 3

$= \ln xy^2$ Property 1

b. $2 \ln(x + 2) - 3 \ln x = \ln(x + 2)^2 - \ln x^3$ Property 3

$= \ln \frac{(x + 2)^2}{x^3}$ Property 2

✓Checkpoint 4

Use the properties of logarithms to rewrite each expression as the logarithm of a single quantity. (Assume $x > 0$ and $y > 0$.)

a. $4 \ln x + 3 \ln y$ **b.** $\ln(x + 1) - 2 \ln(x + 3)$ ■

Solving Exponential and Logarithmic Equations

To solve an exponential equation, first isolate the exponential expression. Then take the logarithm of each side of the equation and solve for the variable.

Example 5 Solving Exponential Equations

Solve each equation.

a. $e^x = 5$ **b.** $10 + e^{0.1t} = 14$

SOLUTION

a. $e^x = 5$	Write original equation.
$\ln e^x = \ln 5$	Take natural log of each side.
$x = \ln 5$	Inverse property: $\ln e^x = x$
b. $10 + e^{0.1t} = 14$	Write original equation.
$e^{0.1t} = 4$	Subtract 10 from each side.
$\ln e^{0.1t} = \ln 4$	Take natural log of each side.
$0.1t = \ln 4$	Inverse property: $\ln e^{0.1t} = 0.1t$
$t = 10 \ln 4$	Multiply each side by 10.

✓Checkpoint 5

Solve each equation.

a. $e^x = 6$ **b.** $5 + e^{0.2t} = 10$ ■

To solve a logarithmic equation, first isolate the logarithmic expression. Then exponentiate each side of the equation and solve for the variable.

Example 6 Solving Logarithmic Equations

Solve each equation.

a. $\ln x = 5$

b. $3 + 2 \ln x = 7$

SOLUTION

a. $\ln x = 5$	Write original equation.
$e^{\ln x} = e^5$	Exponentiate each side.
$x = e^5$	Inverse property: $e^{\ln x} = x$
b. $3 + 2 \ln x = 7$	Write original equation.
$2 \ln x = 4$	Subtract 3 from each side.
$\ln x = 2$	Divide each side by 2.
$e^{\ln x} = e^2$	Exponentiate each side.
$x = e^2$	Inverse property: $e^{\ln x} = x$

✓Checkpoint 6

Solve each equation.

a. $\ln x = 4$

b. $4 + 5 \ln x = 19$ ■

Application

Example 7 Finding Doubling Time

You deposit P dollars in an account whose annual interest rate is r, compounded continuously. How long will it take for your balance to double?

SOLUTION The balance in the account after t years is $A = Pe^{rt}$. So, the balance will have doubled when $Pe^{rt} = 2P$. To find the "doubling time," solve this equation for t.

$$Pe^{rt} = 2P \quad \text{Balance in account has doubled.}$$

$$e^{rt} = 2 \quad \text{Divide each side by } P.$$

$$\ln e^{rt} = \ln 2 \quad \text{Take natural log of each side.}$$

$$rt = \ln 2 \quad \text{Inverse property: } \ln e^{rt} = rt$$

$$t = \frac{1}{r}\ln 2 \quad \text{Divide each side by } r.$$

From this result, you can see that the time it takes for the balance to double is inversely proportional to the interest rate r. The table shows the doubling times for several interest rates. Notice that the doubling time decreases as the rate increases. The relationship between doubling time and the interest rate is shown graphically in Figure 4.15.

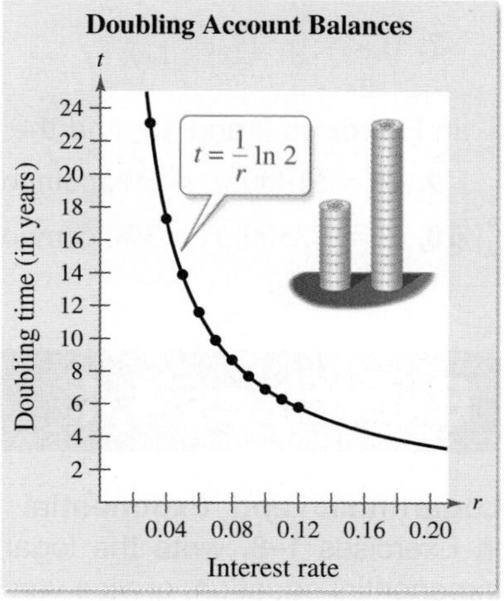

FIGURE 4.15

r	3%	4%	5%	6%	7%	8%	9%	10%	11%	12%
t	23.1	17.3	13.9	11.6	9.9	8.7	7.7	6.9	6.3	5.8

✓Checkpoint 7

Use the equation found in Example 7 to determine the amount of time it would take for your balance to double at an interest rate of 8.75%. ■

SUMMARIZE (Section 4.4)

1. State the definition of the natural logarithmic function *(page 276)*. For an example of graphing logarithmic functions, see Example 1.
2. State the inverse properties of logarithms and exponents *(page 278)*. For an example of applying these properties, see Example 2.
3. State the properties of logarithms *(page 278)*. For examples of using these properties to expand and condense logarithmic expressions, see Examples 3 and 4.
4. Identify the properties of logarithms and exponents used to solve the exponential and logarithmic equations in Examples 5 and 6 *(page 280)*.
5. Describe a real-life example of how a logarithm is used to determine how long it will take for an account balance to double *(page 281, Example 7)*.

SKILLS WARM UP 4.4

The following warm-up exercises involve skills that were covered in a previous course or in earlier sections. You will use these skills in the exercise set for this section. For additional help, review Appendix Section A.1, and Sections 1.4 and 4.2.

In Exercises 1–4, find the inverse function of f.

1. $f(x) = 5x$ **2.** $f(x) = x - 6$ **3.** $f(x) = 3x + 2$ **4.** $f(x) = \frac{3}{4}x - 9$

In Exercises 5–8, solve for x.

5. $0 < x + 4$ **6.** $0 < x^2 + 1$

7. $0 < \sqrt{x^2 - 1}$ **8.** $0 < x - 5$

In Exercises 9 and 10, find the balance in the account after 10 years.

9. $P = \$1900$, $r = 6\%$, compounded continuously

10. $P = \$2500$, $r = 3\%$, compounded continuously

Exercises 4.4

See www.CalcChat.com for worked-out solutions to odd-numbered exercises.

Logarithmic and Exponential Forms of Equations In Exercises 1–8, write the logarithmic equation as an exponential equation, or vice versa.

1. $\ln 2 = 0.6931 \ldots$ **2.** $\ln 9 = 2.1972 \ldots$

3. $\ln 0.2 = -1.6094 \ldots$ **4.** $\ln 0.05 = -2.9957 \ldots$

5. $e^0 = 1$ **6.** $e^2 = 7.3891 \ldots$

7. $e^{-3} = 0.0498 \ldots$ **8.** $e^{0.25} = 1.2840 \ldots$

Matching In Exercises 9–12, match the function with its graph. [The graphs are labeled (a)–(d).]

(a)

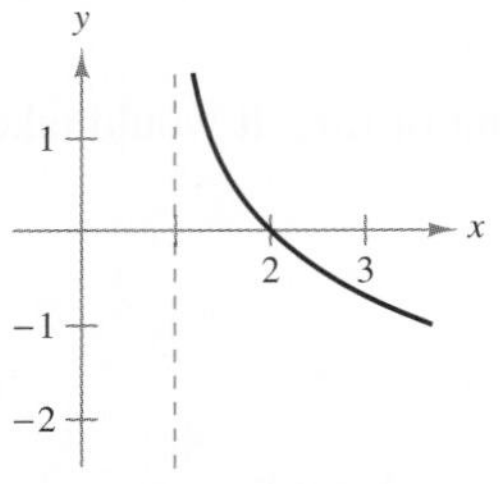

(b)

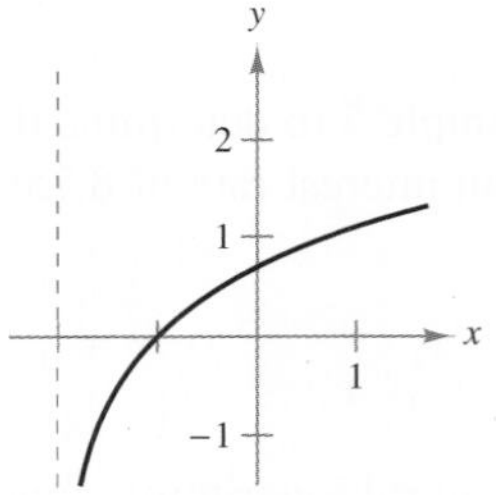

(c)

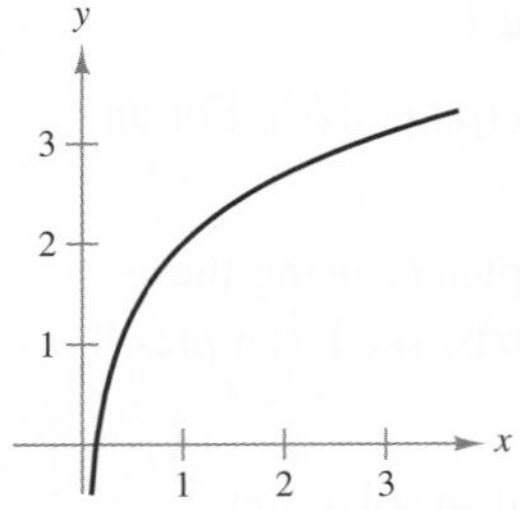

(d)

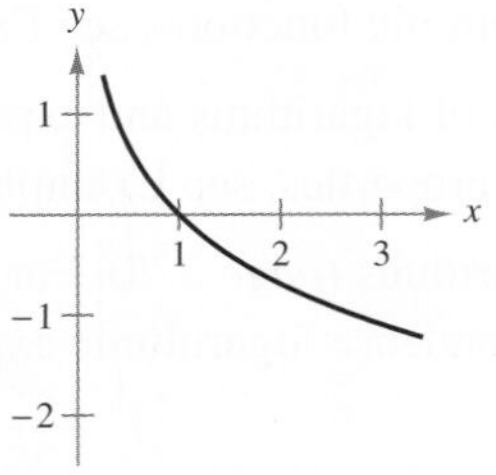

9. $f(x) = 2 + \ln x$

10. $f(x) = -\ln x$

11. $f(x) = \ln(x + 2)$

12. $f(x) = -\ln(x - 1)$

Graphing Logarithmic Functions In Exercises 13–18, sketch the graph of the function. *See Example 1.*

13. $y = \ln(x - 1)$ **14.** $y = \ln|x|$

15. $y = \ln 2x$ **16.** $y = 5 + \ln x$

17. $y = 3 \ln x$ **18.** $y = \frac{1}{4} \ln x$

Applying Inverse Properties In Exercises 19–24, apply the inverse properties of logarithmic and exponential functions to simplify the expression. *See Example 2.*

19. $\ln e^{x^2}$ **20.** $\ln e^{2x-1}$

21. $e^{\ln(5x+2)}$ **22.** $e^{\ln \sqrt{x}}$

23. $-1 + \ln e^{2x}$ **24.** $-8 + e^{\ln x^3}$

Expanding Logarithmic Expressions In Exercises 25–34, use the properties of logarithms to rewrite the expression as a sum, difference, or multiple of logarithms. *See Example 3.*

25. $\ln \frac{2}{3}$ **26.** $\ln \frac{1}{5}$

27. $\ln xyz$ **28.** $\ln \frac{xy}{z}$

29. $\ln \sqrt[3]{2x + 7}$

30. $\ln \sqrt{\frac{x^3}{x + 1}}$

31. $\ln[z(z - 1)^2]$

32. $\ln\left(x\sqrt[3]{x^2 + 1}\right)$

33. $\ln \frac{3x(x + 1)}{(2x + 1)^2}$

34. $\ln \frac{2x}{\sqrt{x^2 - 1}}$

Inverse Functions In Exercises 35–38, analytically show that the functions are inverse functions. Then use a graphing utility to show this graphically.

35. $f(x) = e^{2x}$, $g(x) = \ln \sqrt{x}$

36. $f(x) = e^{x} - 1$, $g(x) = \ln(x + 1)$

37. $f(x) = e^{2x-1}$, $g(x) = \frac{1}{2} + \ln \sqrt{x}$

38. $f(x) = e^{x/3}$, $g(x) = \ln x^3$

Using Properties of Logarithms In Exercises 39 and 40, use the properties of logarithms and the fact that $\ln 2 \approx 0.6931$ and $\ln 3 \approx 1.0986$ to approximate the logarithm. Then use a calculator to confirm your approximation.

39. (a) $\ln 6$ (b) $\ln \frac{3}{2}$ (c) $\ln 81$ (d) $\ln \sqrt{3}$

40. (a) $\ln 0.25$ (b) $\ln 24$ (c) $\ln \sqrt[3]{12}$ (d) $\ln \frac{1}{72}$

Condensing Logarithmic Expressions In Exercises 41–50, use the properties of logarithms to rewrite the expression as the logarithm of a single quantity. *See Example 4.*

41. $\ln(x - 2) - \ln(x + 2)$

42. $\ln(2x + 1) + \ln(2x - 1)$

43. $3 \ln x + 2 \ln y - 4 \ln z$

44. $4 \ln x + 6 \ln y - \ln z$

45. $4 \ln(x - 6) - \frac{1}{2} \ln(3x + 1)$

46. $2 \ln(5x + 3) + \frac{3}{2} \ln(x + 5)$

47. $3[\ln x + \ln(x + 3) - \ln(x + 4)]$

48. $\frac{1}{3}[2 \ln(x + 3) + \ln x - \ln(x^2 - 1)]$

49. $\frac{3}{2}[\ln x(x^2 + 1) - \ln(x + 1)]$

50. $2\left[\ln x + \frac{1}{4} \ln(x + 1)\right]$

Solving Exponential and Logarithmic Equations In Exercises 51–72, solve for x or t. *See Examples 5 and 6.*

51. $e^{\ln x} = 4$

52. $e^{\ln x^2} - 9 = 0$

53. $e^{x+1} = 4$

54. $e^{-0.5x} = 0.075$

55. $300e^{-0.2t} = 700$

56. $400e^{-0.0174t} = 1000$

57. $4e^{2x-1} - 1 = 5$

58. $2e^{-x+1} - 5 = 9$

59. $\ln x = 0$

60. $2 \ln x = 4$

61. $\ln 2x = 2.4$

62. $\ln 4x = 1$

63. $3 + 4 \ln x = 15$

64. $6 + 3 \ln x = 8$

65. $\ln x - \ln(x - 6) = 3$

66. $\ln x + \ln(x + 2) = 0$

67. $5^{2x} = 15$

68. $2^{1-x} = 6$

69. $500(1.07)^t = 1000$

70. $400(1.06)^t = 1300$

71. $\left(1 + \dfrac{0.07}{12}\right)^{12t} = 3$

72. $\left(1 + \dfrac{0.06}{12}\right)^{12t} = 5$

Compound Interest In Exercises 73 and 74, \$3000 is invested in an account at interest rate r, compounded continuously. Find the time required for the amount to (a) double and (b) triple.

73. $r = 0.085$

74. $r = 0.045$

75. Compound Interest A deposit of \$1000 is made in an account that earns interest at an annual rate of 5%. How long will it take for the balance to double when the interest is compounded (a) annually, (b) monthly, (c) daily, and (d) continuously?

76. Compound Interest Complete the table to determine the time t necessary for P dollars to triple when the interest is compounded continuously at rate r.

r	2%	4%	6%	8%	10%	12%	14%
t							

77. Population Growth The population P (in thousands) of Orlando, Florida from 1980 through 2009 can be modeled by $P = 130e^{0.0205t}$, where $t = 0$ corresponds to 1980. *(Source: U.S. Census Bureau)*

(a) What was the population of Orlando in 2009?

(b) In what year will Orlando have a population of 300,000?

78. Population Growth The population P (in thousands) of Phoenix, Arizona from 1980 through 2009 can be modeled by

$$P = 788e^{0.0248t}$$

where $t = 0$ corresponds to 1980. *(Source: U.S. Census Bureau)*

(a) What was the population of Phoenix in 2009?

(b) In what year will Phoenix have a population of 2,000,000?

Carbon Dating In Exercises 79–82, you are given the ratio of carbon atoms in a fossil. Use the information to estimate the age of the fossil. In living organic material, the ratio of radioactive carbon isotopes to the total number of carbon atoms is about 1 to 10^{12}. (See Example 2 in Section 4.1.) When organic material dies, its radioactive carbon isotopes begin to decay, with a half-life of about 5715 years. So, the ratio R of carbon isotopes to carbon-14 atoms is modeled by $R = 10^{-12}\left(\frac{1}{2}\right)^{t/5715}$, where t is the time (in years) and $t = 0$ represents the time when the organic material died.

79. $R = 0.32 \times 10^{-12}$

80. $R = 0.27 \times 10^{-12}$

81. $R = 0.22 \times 10^{-12}$

82. $R = 0.13 \times 10^{-12}$

83. Learning Theory Students in a mathematics class were given an exam and then retested monthly with equivalent exams. The average scores S (on a 100-point scale) for the class can be modeled by $S = 80 - 14 \ln(t + 1)$, $0 \le t \le 12$, where t is the time in months.

(a) What was the average score on the original exam?

(b) What was the average score after 4 months?

(c) After how many months was the average score 46?

84. HOW DO YOU SEE IT? The graph shows the percents of American males and females ages 20 and over who are no more than x inches tall. *(Source: National Center for Health Statistics)*

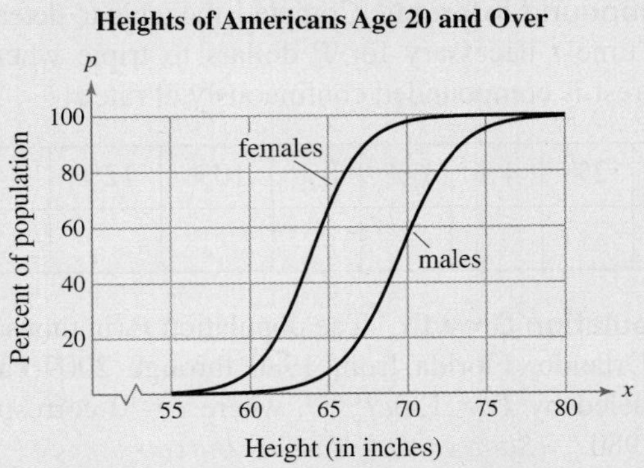

(a) Use the graph to determine the limit of each function as x approaches infinity. What do they mean?

(b) What is the median height of each sex?

85. Demand The demand function for a product is given by

$$p = 5000\left(1 - \frac{4}{4 + e^{-0.002x}}\right)$$

where p is the price per unit (in dollars) and x is the number of units sold. Find the numbers of units sold for prices of (a) $p = \$200$ and (b) $p = \$800$.

86. Demand The demand function for a product is given by

$$p = 10{,}000\left(1 - \frac{3}{3 + e^{-0.001x}}\right)$$

where p is the price per unit (in dollars) and x is the number of units sold. Find the numbers of units sold for prices of (a) $p = \$500$ and (b) $p = \$1500$.

87. Using a Property of Logarithms Demonstrate that

$$\frac{\ln x}{\ln y} \neq \ln\frac{x}{y} = \ln x - \ln y$$

by using a spreadsheet to complete the table.

x	y	$\frac{\ln x}{\ln y}$	$\ln\frac{x}{y}$	$\ln x - \ln y$
1	2			
3	4			
10	5			
4	0.5			

88. Finding Limits and Relative Extrema Use a spreadsheet to complete the table using

$$f(x) = \frac{\ln x}{x}.$$

x	1	5	10	10^2	10^4	10^6
$f(x)$						

(a) Use the table to estimate the limit: $\lim_{x\to\infty} f(x)$.

(b) Use a graphing utility to estimate the relative extrema of f.

Verifying Properties of Logarithms In Exercises 89 and 90, use a graphing utility to verify that the functions are equivalent for $x > 0$.

89. $f(x) = \ln\frac{x^2}{4}$

$g(x) = 2\ln x - \ln 4$

90. $f(x) = \ln\sqrt{x(x^2 + 1)}$

$g(x) = \frac{1}{2}[\ln x + \ln(x^2 + 1)]$

True or False? In Exercises 91–96, determine whether the statement is true or false given that $f(x) = \ln x$. If it is false, explain why or give an example that shows it is false.

91. $f(0) = 0$

92. $f(ax) = f(a) + f(x), \quad a > 0, x > 0$

93. $f(x - 2) = f(x) - f(2), \quad x > 2$

94. $\sqrt{f(x)} = \frac{1}{2}f(x)$

95. If $f(u) = 2f(v)$, then $v = u^2$.

96. If $f(x) < 0$, then $0 < x < 1$.

97. Finance You are investing P dollars at an annual interest rate of r, compounded continuously, for t years. Which of the following options would you choose to get the highest value of the investment? Explain your reasoning.

(a) Double the amount you invest.

(b) Double your interest rate.

(c) Double the number of years.

98. Think About It Are the times required for the investments in Exercises 75 and 76 to quadruple twice as long as the times for them to double? Give a reason for your answer and verify your answer algebraically.

99. Pursuit Curve Use a graphing utility to graph

$$y = 10\ln\left(\frac{10 + \sqrt{100 - x^2}}{10}\right) - \sqrt{100 - x^2}$$

over the interval (0, 10]. This graph is called a *tractrix* or *pursuit curve*. Use your school's library, the Internet, or some other reference source to find information about a tractrix. Explain how such a curve can arise in a real-life setting.

4.5 Derivatives of Logarithmic Functions

In Exercise 73 on page 292, you will use the derivative of a logarithmic function to find the rate of change of a demand function.

- Find the derivatives of natural logarithmic functions.
- Find the derivatives of exponential and logarithmic functions involving other bases.

Derivatives of Logarithmic Functions

Implicit differentiation can be used to develop the derivative of the natural logarithmic function.

$$y = \ln x \qquad \text{Natural logarithmic function}$$

$$e^y = x \qquad \text{Write in exponential form.}$$

$$\frac{d}{dx}[e^y] = \frac{d}{dx}[x] \qquad \text{Differentiate with respect to } x.$$

$$e^y \frac{dy}{dx} = 1 \qquad \text{Chain Rule}$$

$$\frac{dy}{dx} = \frac{1}{e^y} \qquad \text{Divide each side by } e^y.$$

$$\frac{dy}{dx} = \frac{1}{x} \qquad \text{Substitute } x \text{ for } e^y.$$

This result and its Chain Rule version are summarized below.

Derivative of the Natural Logarithmic Function

Let u be a differentiable function of x.

1. $\dfrac{d}{dx}[\ln x] = \dfrac{1}{x}$ **2.** $\dfrac{d}{dx}[\ln u] = \dfrac{1}{u}\dfrac{du}{dx}$

Example 1 Differentiating a Logarithmic Function

Find the derivative of

$$f(x) = \ln 2x.$$

SOLUTION Let $u = 2x$. Then $du/dx = 2$, and you can apply the Chain Rule as shown.

$$f'(x) = \frac{1}{u}\frac{du}{dx} \qquad \text{Chain Rule}$$

$$= \frac{1}{2x}(2)$$

$$= \frac{1}{x} \qquad \text{Simplify.}$$

✓ Checkpoint 1

Find the derivative of

$$f(x) = \ln 5x.$$

Example 2 Differentiating Logarithmic Functions

Find the derivative of each function.

a. $f(x) = \ln(2x^2 + 4)$ **b.** $f(x) = x \ln x$ **c.** $f(x) = \dfrac{\ln x}{x}$

SOLUTION

a.
$$f'(x) = \frac{1}{u}\frac{du}{dx} \qquad \text{Chain Rule}$$
$$= \frac{1}{2x^2 + 4}(4x) \qquad u = 2x^2 + 4,\ du/dx = 4x$$
$$= \frac{2x}{x^2 + 2} \qquad \text{Simplify.}$$

b.
$$f'(x) = x\frac{d}{dx}[\ln x] + (\ln x)\frac{d}{dx}[x] \qquad \text{Product Rule}$$
$$= x\left(\frac{1}{x}\right) + (\ln x)(1)$$
$$= 1 + \ln x \qquad \text{Simplify.}$$

c.
$$f'(x) = \frac{x\dfrac{d}{dx}[\ln x] - (\ln x)\dfrac{d}{dx}[x]}{x^2} \qquad \text{Quotient Rule}$$
$$= \frac{x\left(\dfrac{1}{x}\right) - \ln x}{x^2}$$
$$= \frac{1 - \ln x}{x^2} \qquad \text{Simplify.}$$

✓Checkpoint 2

Find the derivative of each function.

a. $f(x) = \ln(x^2 - 4)$ **b.** $f(x) = x^2 \ln x$ **c.** $f(x) = -\dfrac{\ln x}{x^2}$ ■

STUDY TIP

When you are differentiating logarithmic functions, it is often helpful to use the properties of logarithms to rewrite the function *before* differentiating. To see the advantage of rewriting before differentiating, try using the Chain Rule to differentiate $f(x) = \ln\sqrt{x + 1}$ and compare your work with that shown in Example 3.

Example 3 Rewriting Before Differentiating

$$f(x) = \ln\sqrt{x + 1} \qquad \text{Original function}$$
$$= \ln(x + 1)^{1/2} \qquad \text{Rewrite with rational exponent.}$$
$$= \frac{1}{2}\ln(x + 1) \qquad \text{Property of logarithms}$$
$$f'(x) = \frac{1}{2}\left(\frac{1}{x + 1}\right) \qquad \text{Differentiate.}$$
$$= \frac{1}{2(x + 1)} \qquad \text{Simplify.}$$

✓Checkpoint 3

Find the derivative of

$$f(x) = \ln \sqrt[3]{x + 1}.$$ ■

TECH TUTOR

A symbolic differentiation utility generally will not list the derivative of the logarithmic function in the form obtained in Example 4. Use a symbolic differentiation utility to find the derivative of the function in Example 4. Show that the two forms are equivalent by rewriting the answer obtained in Example 4.

Example 4 Rewriting Before Differentiating

Find the derivative of $f(x) = \ln[x(x^2 + 1)^2]$.

SOLUTION

$$\begin{aligned} f(x) &= \ln[x(x^2 + 1)^2] && \text{Write original function.} \\ &= \ln x + \ln(x^2 + 1)^2 && \text{Logarithmic properties} \\ &= \ln x + 2\ln(x^2 + 1) && \text{Logarithmic properties} \\ f'(x) &= \frac{1}{x} + 2\left(\frac{2x}{x^2 + 1}\right) && \text{Differentiate.} \\ &= \frac{1}{x} + \frac{4x}{x^2 + 1} && \text{Simplify.} \end{aligned}$$

✓ Checkpoint 4

Find the derivative of $f(x) = \ln[x^2\sqrt{x^2 + 1}]$. ■

Finding the derivative of the function in Example 4 without first rewriting would be a formidable task.

$$f'(x) = \frac{1}{x(x^2 + 1)^2} \frac{d}{dx}[x(x^2 + 1)^2]$$

You might try showing that this yields the same result obtained in Example 4, but be careful—the algebra is messy.

Example 5 Finding an Equation of a Tangent Line

Find an equation of the tangent line to the graph of $f(x) = 2 + 3x \ln x$ at the point $(1, 2)$.

SOLUTION Begin by finding the derivative of f.

$$\begin{aligned} f(x) &= 2 + 3x \ln x && \text{Write original function.} \\ f'(x) &= 3x\left(\frac{1}{x}\right) + (\ln x)(3) && \text{Differentiate.} \\ &= 3 + 3 \ln x && \text{Simplify.} \end{aligned}$$

The slope of the line tangent to the graph of f at $(1, 2)$ is

$$f'(1) = 3 + 3 \ln 1 = 3 + 3(0) = 3.$$

Using the point-slope form of a line, you can find the equation of the tangent line to be

$$y = 3x - 1.$$

The graph of the function and the tangent line are shown in Figure 4.16.

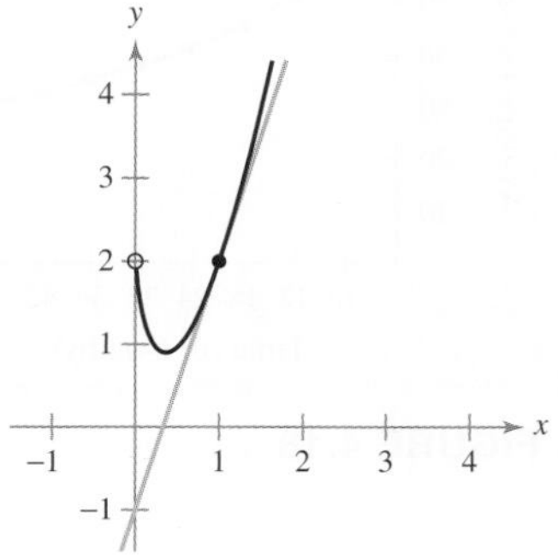

FIGURE 4.16

✓ Checkpoint 5

Find an equation of the tangent line to the graph of $f(x) = 4 \ln x$ at the point $(1, 0)$. ■

Example 6 Analyzing a Graph

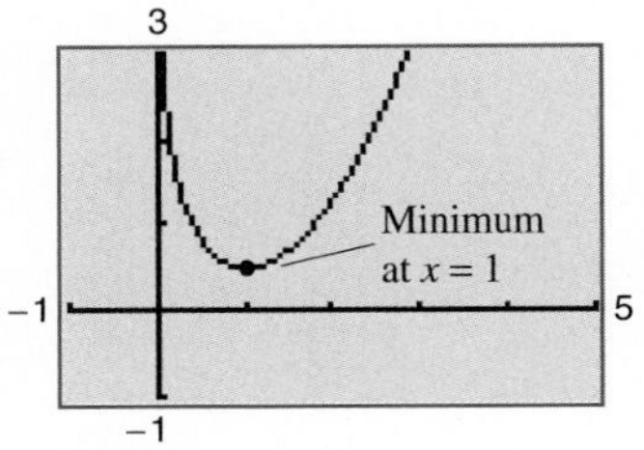

FIGURE 4.17

Analyze the graph of the function $f(x) = \frac{x^2}{2} - \ln x$.

SOLUTION From Figure 4.17, it appears that the function has a minimum at $x = 1$. To find the minimum analytically, find the critical numbers by setting the derivative of f equal to zero and solving for x.

$$f(x) = \frac{x^2}{2} - \ln x \qquad \text{Write original function.}$$

$$f'(x) = x - \frac{1}{x} \qquad \text{Differentiate.}$$

$$x - \frac{1}{x} = 0 \qquad \text{Set derivative equal to 0.}$$

$$x = \frac{1}{x} \qquad \text{Add } 1/x \text{ to each side.}$$

$$x^2 = 1 \qquad \text{Multiply each side by } x.$$

$$x = \pm 1 \qquad \text{Take square root of each side.}$$

Of these two possible critical numbers, only the positive one lies in the domain of f. By applying the First-Derivative Test, you can confirm that the function has a relative minimum at $x = 1$.

✓ Checkpoint 6

Determine the relative extrema of the function $f(x) = x - 2 \ln x$. ■

Example 7 Finding a Rate of Change

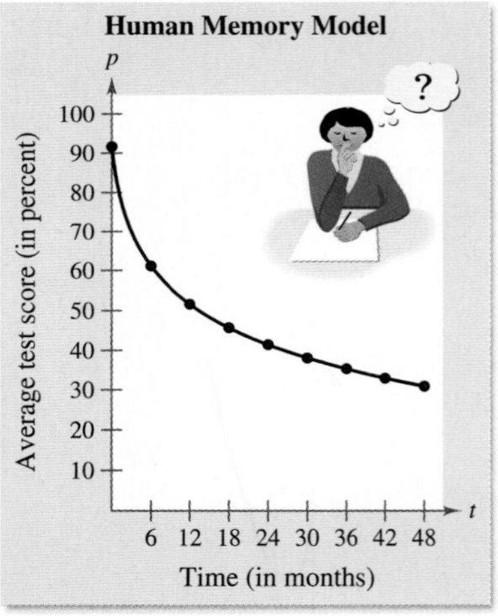

FIGURE 4.18

A group of 200 college students was tested every 6 months over a four-year period. The group was composed of students who took Spanish during the fall semester of their freshman year and did not take subsequent Spanish courses. The average test score p (in percent) is modeled by

$$p = 91.6 - 15.6 \ln(t + 1), \quad 0 \le t \le 48$$

where t is the time in months, as shown in Figure 4.18. At what rate was the average score changing after 1 year?

SOLUTION The rate of change is

$$\frac{dp}{dt} = -\frac{15.6}{t + 1}.$$

The rate of change when $t = 12$ is

$$\frac{dp}{dt} = -\frac{15.6}{12 + 1} = -\frac{15.6}{13} = -1.2.$$

This means that the average score was decreasing at the rate of 1.2% per month.

✓ Checkpoint 7

Suppose the average test score in Example 7 was modeled by

$$p = 92.3 - 16.9 \ln(t + 1), \quad 0 \le t \le 48$$

where t is the time in months. How would the rate at which the average test score was changing after 1 year compare with that of the model in Example 7? ■

Other Bases

This chapter began with a definition of a general exponential function

$$f(x) = a^x$$

where a is a positive number such that $a \neq 1$. The corresponding **logarithm to the base a** is defined by

$$\log_a x = b \quad \text{if and only if} \quad a^b = x.$$

As with the natural logarithmic function, the domain of the logarithmic function to the base a is the set of positive numbers.

Example 8 Evaluating Logarithms

a. $\log_2 8 = 3$ $\quad 2^3 = 8$

b. $\log_{10} 100 = 2$ $\quad 10^2 = 100$

c. $\log_{10} \frac{1}{10} = -1$ $\quad 10^{-1} = \frac{1}{10}$

d. $\log_3 81 = 4$ $\quad 3^4 = 81$

✓ Checkpoint 8

Evaluate each logarithm without using a calculator.

a. $\log_2 16$ **b.** $\log_{10} \frac{1}{100}$ **c.** $\log_2 \frac{1}{32}$ **d.** $\log_5 125$ ■

Most calculators have only two logarithm keys—a natural logarithm key denoted by (LN) and a common logarithm key denoted by (LOG). Logarithms to other bases can be evaluated with the following change-of-base formula.

$$\log_a x = \frac{\ln x}{\ln a}$$ Change-of-base formula

Example 9 Changing Bases to Evaluate Logarithms

Use the change-of-base formula and a calculator to evaluate each logarithm.

a. $\log_2 3$

b. $\log_3 6$

c. $\log_2(-1)$

SOLUTION In each case, use the change-of-base formula and a calculator.

a. $\log_2 3 = \dfrac{\ln 3}{\ln 2} \approx 1.585$ $\quad \log_a x = \dfrac{\ln x}{\ln a}$

b. $\log_3 6 = \dfrac{\ln 6}{\ln 3} \approx 1.631$ $\quad \log_a x = \dfrac{\ln x}{\ln a}$

c. $\log_2(-1)$ is not defined.

✓ Checkpoint 9

Use the change-of-base formula and a calculator to evaluate each logarithm.

a. $\log_2 5$ **b.** $\log_3 18$ **c.** $\log_4 80$ **d.** $\log_{16} 0.25$ ■

STUDY TIP

Remember that you can convert to base e using the formulas

$$a^x = e^{(\ln a)x}$$

and

$$\log_a x = \left(\frac{1}{\ln a}\right) \ln x.$$

To find derivatives of exponential or logarithmic functions to bases other than e, you can either convert to base e or use the differentiation rules shown below.

Other Bases and Differentiation

Let u be a differentiable function of x.

1. $\dfrac{d}{dx}[a^x] = (\ln a)a^x$
2. $\dfrac{d}{dx}[a^u] = (\ln a)a^u \dfrac{du}{dx}$
3. $\dfrac{d}{dx}[\log_a x] = \left(\dfrac{1}{\ln a}\right)\dfrac{1}{x}$
4. $\dfrac{d}{dx}[\log_a u] = \left(\dfrac{1}{\ln a}\right)\left(\dfrac{1}{u}\right)\dfrac{du}{dx}$

PROOF By definition, $a^x = e^{(\ln a)x}$. So, you can prove the first rule by letting $u = (\ln a)x$ and differentiating with base e to obtain

$$\frac{d}{dx}[a^x] = \frac{d}{dx}[e^{(\ln a)x}] = e^u \frac{du}{dx} = e^{(\ln a)x}(\ln a) = (\ln a)a^x.$$

Example 10 Finding a Rate of Change

Radioactive carbon isotopes have a half-life of 5715 years. An object contains 1 gram of the isotopes. The amount A (in grams) that will be present after t years is

$$A = \left(\frac{1}{2}\right)^{t/5715}.$$

At what rate is the amount changing when $t = 10{,}000$ years?

SOLUTION The derivative of A with respect to t is

$$\frac{dA}{dt} = \left(\ln \frac{1}{2}\right)\left(\frac{1}{2}\right)^{t/5715}\left(\frac{1}{5715}\right).$$

When $t = 10{,}000$, the rate at which the amount is changing is

$$\left(\ln \frac{1}{2}\right)\left(\frac{1}{2}\right)^{10{,}000/5715}\left(\frac{1}{5715}\right) \approx -0.000036$$

which implies that the amount of isotopes in the object is decreasing at the rate of 0.000036 gram per year.

✓ Checkpoint 10

Use a graphing utility to graph the model in Example 10. Describe the rate at which the amount is changing as time t increases. ■

SUMMARIZE (Section 4.5)

1. State the derivative of the natural logarithmic function *(page 285)*. For examples of the derivative of the natural logarithmic function, see Examples 1, 2, 3, and 4.
2. State the derivative of the logarithmic function to base *a (page 290)*. For an example of a derivative of a logarithmic function to base *a*, see Example 10.

visi.stock/www.shutterstock.com

SKILLS WARM UP 4.5 The following warm-up exercises involve skills that were covered in earlier sections. You will use these skills in the exercise set for this section. For additional help, review Sections 2.6, 2.7, and 4.4.

In Exercises 1–6, expand the logarithmic expression.

1. $\ln(x+1)^2$

2. $\ln x(x+1)$

3. $\ln \dfrac{x}{x+1}$

4. $\ln\left(\dfrac{x}{x-3}\right)^3$

5. $\ln \dfrac{4x(x-7)}{x^2}$

6. $\ln x^3(x+1)$

In Exercises 7 and 8, find dy/dx implicitly.

7. $y^2 + xy = 7$

8. $x^2y - xy^2 = 3x$

In Exercises 9 and 10, find the second derivative of f.

9. $f(x) = x^2(x+1) - 3x^3$

10. $f(x) = -\dfrac{1}{x^2}$

Exercises 4.5

See www.CalcChat.com for worked-out solutions to odd-numbered exercises.

Differentiating a Logarithmic Function In Exercises 1–22, find the derivative of the function. *See Examples 1, 2, 3, and 4.*

1. $y = \ln x^2$

2. $f(x) = \ln 7x$

3. $y = \ln(x^2 + 3)$

4. $f(x) = \ln(1 - x^2)$

5. $y = \ln\sqrt{x - 4}$

6. $y = \ln(1 - x)^{3/2}$

7. $y = (\ln x)^4$

8. $y = (\ln x^2)^2$

9. $f(x) = 2x \ln x$

10. $y = \dfrac{\ln x}{x^2}$

11. $y = \ln\left(x\sqrt{x^2 - 1}\right)$

12. $y = \ln[x(2x + 3)^2]$

13. $y = \ln \dfrac{x}{x+1}$

14. $y = \ln \dfrac{x}{x^2+1}$

15. $y = \ln \sqrt[3]{\dfrac{x-1}{x+1}}$

16. $y = \ln \sqrt{\dfrac{x+1}{x-1}}$

17. $y = \ln \dfrac{\sqrt{4+x^2}}{x}$

18. $y = \ln \dfrac{(6-x)^{3/2}}{x^{2/3}}$

19. $g(x) = e^{-x} \ln x$

20. $y = e^{x^2} \ln 4x^3$

21. $g(x) = \ln \dfrac{e^x + e^{-x}}{2}$

22. $f(x) = \ln \dfrac{1 + e^x}{1 - e^x}$

Evaluating Logarithms In Exercises 23–28, evaluate the logarithm without using a calculator. *See Example 8.*

23. $\log_5 25$

24. $\log_4 64$

25. $\log_3 \frac{1}{27}$

26. $\log_6 \frac{1}{36}$

27. $\log_7 49$

28. $\log_8 512$

Changing Bases to Evaluate Logarithms In Exercises 29–34, use the change-of-base formula and a calculator to evaluate the logarithm. *See Example 9.*

29. $\log_4 7$

30. $\log_6 10$

31. $\log_2 48$

32. $\log_5 12$

33. $\log_3 \frac{1}{2}$

34. $\log_7 \frac{2}{9}$

Differentiating Functions of Other Bases In Exercises 35–44, find the derivative of the function.

35. $y = 3^x$

36. $y = \left(\frac{1}{4}\right)^x$

37. $f(x) = \log_2 x$

38. $g(x) = \log_5 x$

39. $h(x) = 4^{2x-3}$

40. $f(x) = 10^{x^2}$

41. $y = \log_{10}(x^2 + 6x)$

42. $g(x) = \log_8(2x - 5)$

43. $y = x2^x$

44. $y = x3^{x+1}$

Finding an Equation of a Tangent Line In Exercises 45–52, find an equation of the tangent line to the graph of the function at the given point. *See Example 5.*

45. $y = \ln x^3$; $(1, 0)$

46. $y = \ln x^{5/2}$; $(1, 0)$

47. $y = x \ln x$; (e, e)

48. $y = \dfrac{\ln x}{x}$; $\left(e, \dfrac{1}{e}\right)$

49. $f(x) = \ln \dfrac{5(x+2)}{x}$; $\left(-\dfrac{5}{2}, 0\right)$

50. $f(x) = \ln\left(x\sqrt{x+3}\right)$; $\left(\frac{6}{5}, \frac{9}{10}\right)$

51. $y = \log_3 x$; $(27, 3)$

52. $g(x) = \log_{10} 2x$; $(5, 1)$

Finding Derivatives Implicitly In Exercises 53–56, find dy/dx implicitly.

53. $x^2 - 3 \ln y + y^2 = 10$

54. $\ln xy + 5x = 30$

55. $4x^3 + \ln y^2 + 2y = 2x$

56. $4xy + \ln(x^2y) = 7$

Finding an Equation of a Tangent Line In Exercises 57 and 58, use implicit differentiation to find an equation of the tangent line to the graph of the function at the given point.

57. $x + y - 1 = \ln(x^2 + y^2);\ (1, 0)$

58. $y^2 + \ln(xy) = 2;\ (e, 1)$

Finding Higher-Order Derivatives In Exercises 59–64, find the second derivative of the function.

59. $f(x) = x \ln \sqrt{x} + 2x$

60. $f(x) = 3 + 2 \ln x$

61. $f(x) = 2 + x^3 \ln x$

62. $f(x) = \dfrac{\ln x}{x^3} + x$

63. $f(x) = 5^x$

64. $f(x) = \log_{10} x$

65. Sound Intensity The relationship between the number of decibels β and the intensity of a sound I (in watts per square centimeter) is given by

$$\beta = 10 \log_{10}\left(\frac{I}{10^{-16}}\right).$$

Find the rate of change in the number of decibels when the intensity is 10^{-4} watt per square centimeter.

66. Chemistry The temperatures T (in °F) at which water boils at selected pressures p (in pounds per square inch) can be modeled by

$$T = 87.97 + 34.96 \ln p + 7.91\sqrt{p}.$$

Find the rate of change of the temperature when the pressure is 60 pounds per square inch.

Analyzing a Graph In Exercises 67–72, analyze and sketch the graph of the function. Label any relative extrema, points of inflection, and asymptotes. *See Example 6.*

67. $y = x - \ln x$

68. $y = \ln 2x - 2x^2$

69. $y = \dfrac{x}{\ln x}$

70. $y = \dfrac{\ln 5x}{x^2}$

71. $y = x^2 \ln \dfrac{x}{4}$

72. $y = (\ln x)^2$

Demand In Exercises 73 and 74, find dx/dp for the demand function. Interpret this rate of change for a price of \$10.

73. $x = \ln \dfrac{1000}{p}$

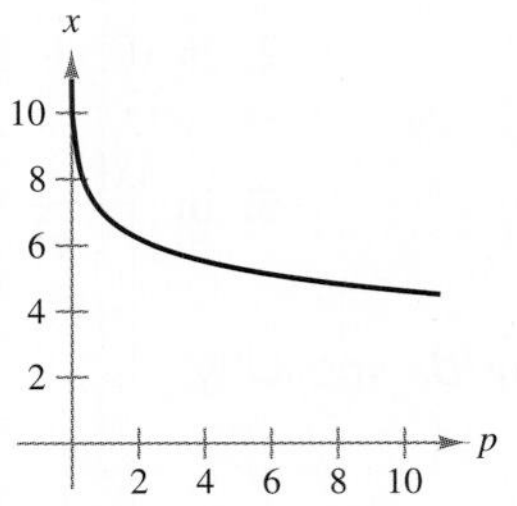

74. $x = \dfrac{500}{\ln(p^2 + 1)}$

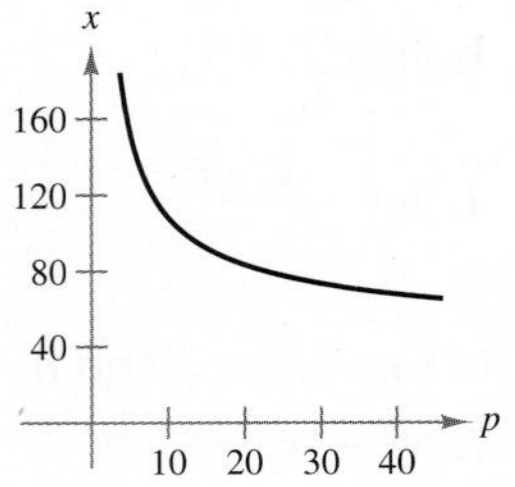

75. Demand Solve the demand function in Exercise 73 for p. Use the result to find dp/dx. Then find the rate of change when $p = \$10$. What is the relationship between this derivative and dx/dp?

76. Demand Solve the demand function in Exercise 74 for p. Use the result to find dp/dx. Then find the rate of change when $p = \$10$. What is the relationship between this derivative and dx/dp?

77. Minimum Average Cost The cost of producing x units of a product is modeled by

$$C = 500 + 300x - 300 \ln x, \quad x \geq 1.$$

(a) Find the average cost function $\overline{C}$.

(b) Find the minimum average cost analytically. Use a graphing utility to confirm your result.

78. Minimum Average Cost The cost of producing x units of a product is modeled by

$$C = 100 + 25x - 120 \ln x, \quad x \geq 1.$$

(a) Find the average cost function $\overline{C}$.

(b) Find the minimum average cost analytically. Use a graphing utility to confirm your result.

79. Consumer Trends The numbers of employees E (in thousands) at outpatient care centers from 2004 through 2009 are shown in the table.

Year	2004	2005	2006	2007	2008	2009
Employees	451	473	493	512	533	543

The data can be modeled by $E = 287 + 116.7 \ln t$, where $t = 4$ corresponds to 2004. *(Source: U.S. Bureau of Labor Statistics)*

(a) Use a graphing utility to plot the data and graph E over the interval $[4, 9]$.

(b) At what rate were the numbers of employees changing in 2006?

80. **HOW DO YOU SEE IT?** The graph shows the temperature T (in °C) of an object h hours after it is removed from a furnace.

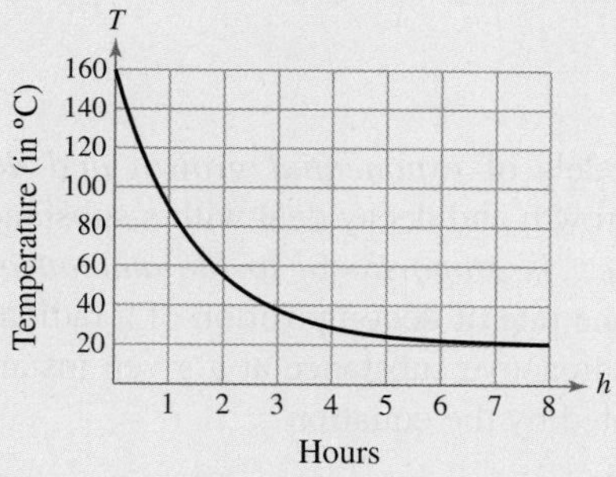

(a) Find $\lim_{h\to\infty} T$. What does this limit represent?

(b) When is the temperature changing most rapidly?

81. Home Mortgage The term t (in years) of a \$200,000 home mortgage at 7.5% interest can be approximated by

$$t = -13.375 \ln \frac{x - 1250}{x}, \quad x > 1250$$

where x is the monthly payment in dollars.

(a) Use a graphing utility to graph the model.

(b) Use the model to approximate the term of a home mortgage for which the monthly payment is \$1398.43. What is the total amount paid?

(c) Use the model to approximate the term of a home mortgage for which the monthly payment is \$1611.19. What is the total amount paid?

(d) Find the instantaneous rate of change of t with respect to x when $x = \$1398.43$ and $x = \$1611.19$.

(e) Write a short paragraph describing the benefit of the higher monthly payment.

82. Earthquake Intensity On the Richter scale, the magnitude R of an earthquake of intensity I is given by

$$R = \frac{\ln I - \ln I_0}{\ln 10}$$

where I_0 is the minimum intensity used for comparison. Assume $I_0 = 1$.

(a) Find the intensity of the March 11, 2011 earthquake in Japan for which $R = 9.0$.

(b) Find the intensity of the January 12, 2010 earthquake in Haiti for which $R = 7.0$.

(c) Find the factor by which the intensity is increased when the value of R is doubled.

(d) Find dR/dI.

83. Learning Theory Students in a learning theory study were given an exam and then retested monthly for 6 months with an equivalent exam. The data obtained in the study are shown in the table, where t is the time in months after the initial exam and s is the average score for the class.

t	1	2	3	4	5	6
s	84.2	78.4	72.1	68.5	67.1	65.3

(a) Use a graphing utility to find a logarithmic model for the average score s in terms of the time t.

(b) Use a graphing utility to plot the data and graph the model. How well does the model fit the data?

(c) Find the rate of change of s with respect to t when $t = 2$. Interpret the meaning of this rate of change in the context of the problem.

Business Capsule

While in college, Heikai Gani had a miserable experience trying to buy a new suit. With his friend Kyle Vucko, he created an online business model to deliver custom-tailored suits for men. With an initial investment of \$800,000, Indochino.com was born in 2007. Today the company has seven-digit revenues with over 17,000 customers. Indochino offers a perfect-fit promise; they will pay for alterations at a local tailor or issue a full refund.

84. Research Project Use your school's library, the Internet, or some other reference source to research information about an e-commerce company, such as the one discussed above. Collect data about the company (sales over a 10-year period, for example) and find a mathematical model that represents the data.

Courtesy of Indochino

4.6 Exponential Growth and Decay

■ Use exponential growth and decay to model real-life situations.

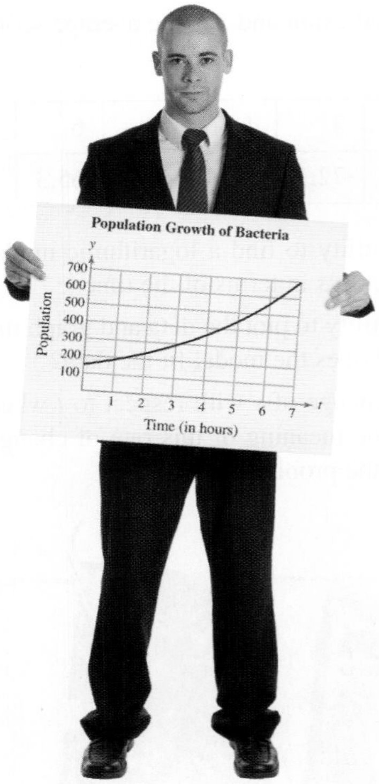

In Exercise 23 on page 300, you will use exponential growth to find the time it takes a population of bacteria to double.

Exponential Growth and Decay

In this section, you will learn to create models of *exponential growth and decay.* Real-life situations that involve exponential growth and decay deal with a substance or population whose *rate of change at any time t is proportional to the amount of the substance present at that time.* For example, the rate of decomposition of a radioactive substance is proportional to the amount of radioactive substance at a given instant. In its simplest form, this relationship is represented by the equation

Rate of change of y — is — proportional to y.

$$\frac{dy}{dt} = ky.$$

In this equation, k is a constant and y is a function of t. The solution of this equation is shown below.

> **Exponential Growth and Decay**
>
> If y is a positive quantity whose rate of change with respect to time is proportional to the quantity present at any time t, then y is of the form
>
> $$y = Ce^{kt}$$
>
> where C is the **initial value** and k is the **constant of proportionality. Exponential growth** is indicated by $k > 0$ and **exponential decay** by $k < 0$.

PROOF Because the rate of change of y is proportional to y, you can write

$$\frac{dy}{dt} = ky.$$

You can see that $y = Ce^{kt}$ is a solution of this equation by differentiating to obtain $dy/dt = kCe^{kt}$ and substituting.

$y = Ce^{kt}$	Original equation
$\frac{dy}{dt} = kCe^{kt}$	Differentiate.
$= k(Ce^{kt})$	Rewrite.
$= ky$	Substitute y for Ce^{kt}.

STUDY TIP

In the model $y = Ce^{kt}$, C is called the "initial value" because, when $t = 0$,

$$y = Ce^{k(0)}$$
$$= C(1)$$
$$= C.$$

Radioactive decay is measured in terms of **half-life,** the number of years required for half of the atoms in a sample of radioactive material to decay. The half-lives of some common radioactive isotopes are as shown.

Uranium (^{238}U)	4,470,000,000	years
Plutonium (^{239}Pu)	24,100	years
Carbon (^{14}C)	5,715	years
Radium (^{226}Ra)	1,599	years
Einsteinium (^{254}Es)	276	days
Nobelium (^{257}No)	25	seconds

Example 1 Modeling Radioactive Decay

A sample contains 1 gram of radium. Will more than 0.5 gram of radium remain after 1000 years?

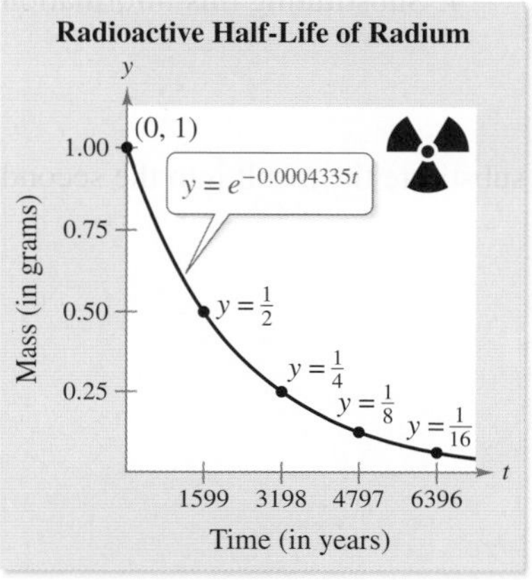

FIGURE 4.19

SOLUTION Let y represent the mass (in grams) of the radium in the sample. Because the rate of decay is proportional to y, you can conclude that y is of the form $y = Ce^{kt}$, where t is the time in years. From the given information, you know that $y = 1$ when $t = 0$. Substituting these values into the model produces

$$1 = Ce^{k(0)} \qquad \text{Substitute 1 for } y \text{ and 0 for } t.$$

which implies that $C = 1$. Because radium has a half-life of 1599 years, you know that $y = \frac{1}{2}$ when $t = 1599$. Substituting these values into the model allows you to solve for k.

$$y = e^{kt} \qquad \text{Exponential decay model}$$

$$\frac{1}{2} = e^{k(1599)} \qquad \text{Substitute } \tfrac{1}{2} \text{ for } y \text{ and 1599 for } t.$$

$$\ln \frac{1}{2} = 1599k \qquad \text{Take natural log of each side.}$$

$$\frac{1}{1599} \ln \frac{1}{2} = k \qquad \text{Divide each side by 1599.}$$

So, $k \approx -0.0004335$, and the exponential decay model is

$$y = e^{-0.0004335t}.$$

To find the amount of radium remaining in the sample after 1000 years, substitute $t = 1000$ into the model.

$$y = e^{-0.0004335(1000)} \approx 0.648 \text{ gram}$$

Yes, more than 0.5 gram of radium will remain after 1000 years. The graph of the model is shown in Figure 4.19.

✓Checkpoint 1

Use the model in Example 1 to determine the number of years required for a one-gram sample of radium to decay to 0.4 gram. ■

Instead of approximating the value of k in Example 1, you could leave the value exact and obtain

$$y = e^{[(1/1599)\ln(1/2)]t}$$

$$= e^{\ln[(1/2)^{(t/1599)}]}$$

$$= \left(\frac{1}{2}\right)^{t/1599}.$$

This version of the model clearly shows the "half-life." When $t = 1599$, the value of y is $\frac{1}{2}$; when $t = 2(1599)$, the value of y is $\frac{1}{4}$; and so on.

Guidelines for Modeling Exponential Growth and Decay

1. Use the given information to write *two* sets of conditions involving y and t.
2. Substitute the given conditions into the model $y = Ce^{kt}$ and use the results to solve for the constants C and k. (When one of the conditions involves $t = 0$, substitute that value first to solve for C.)
3. Use the model $y = Ce^{kt}$ to answer the question.

Example 2 Modeling Population Growth

ALGEBRA TUTOR

For help with the algebra in Example 2, see Example 1(c) in the *Chapter 4 Algebra Tutor* on page 302.

In a research experiment, a population of fruit flies is increasing in accordance with the exponential growth model. After 2 days, there are 100 flies, and after 4 days, there are 300 flies. How many flies will there be after 5 days?

SOLUTION Let y be the number of flies at time t. From the given information, you know that $y = 100$ when $t = 2$ and $y = 300$ when $t = 4$. Substituting this information into the model $y = Ce^{kt}$ produces

$$100 = Ce^{2k} \quad \text{and} \quad 300 = Ce^{4k}.$$

To solve for k, solve for C in the first equation and substitute the result into the second equation.

$$300 = Ce^{4k} \qquad \text{Second equation}$$

$$300 = \left(\frac{100}{e^{2k}}\right)e^{4k} \qquad \text{Substitute } 100/e^{2k} \text{ for } C.$$

$$\frac{300}{100} = e^{2k} \qquad \text{Divide each side by 100.}$$

$$\ln 3 = 2k \qquad \text{Take natural log of each side.}$$

$$\frac{1}{2}\ln 3 = k \qquad \text{Solve for } k.$$

Using $k = \frac{1}{2}\ln 3 \approx 0.5493$, you can determine that

$$C \approx \frac{100}{e^{2(0.5493)}} \approx 33.$$

So, the exponential growth model is

$$y = 33e^{0.5493t}$$

as shown in Figure 4.20. This implies that, after 5 days, the population is

$$y = 33e^{0.5493(5)} \approx 514 \text{ flies.}$$

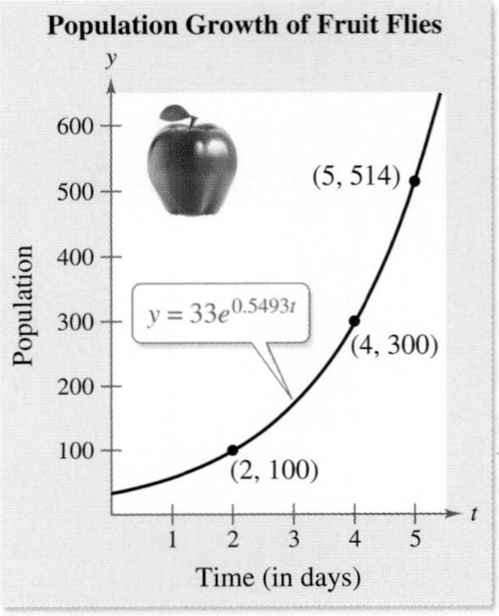

FIGURE 4.20

Checkpoint 2

Find the exponential growth model for a population of fruit flies for which there are 100 flies after 2 days and 400 flies after 4 days. ■

Example 3 Modeling Compound Interest

Money is deposited in an account for which the interest is compounded continuously. The balance in the account doubles in 6 years. What is the annual interest rate?

SOLUTION The balance A in an account with continuously compounded interest is given by the exponential growth model

$$A = Pe^{rt} \qquad \text{Exponential growth model}$$

where P is the original deposit, r is the annual interest rate (in decimal form), and t is the time (in years). From the given information, you know that

$$A = 2P$$

when $t = 6$, as shown in Figure 4.21. Use this information to solve for r.

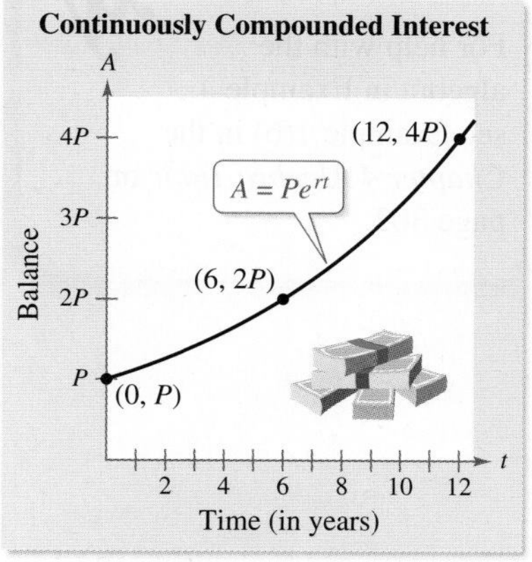

FIGURE 4.21

$A = Pe^{rt}$	Exponential growth model
$2P = Pe^{r(6)}$	Substitute $2P$ for A and 6 for t.
$2 = e^{6r}$	Divide each side by P.
$\ln 2 = 6r$	Take natural log of each side.
$\frac{1}{6}\ln 2 = r$	Divide each side by 6.

So, the annual interest rate is

$$r = \frac{1}{6}\ln 2 \approx 0.1155$$

or about 11.55%.

✓Checkpoint 3

Find the annual interest rate for an account whose balance doubles in 8 years and for which the interest is compounded continuously.

Each of the examples in this section uses the exponential growth model $y = Ce^{kt}$, in which the base is e. Exponential growth, however, can be modeled with any base. That is, the model

$$y = Ca^{bt}$$

also represents exponential growth. (To see this, note that the model can be written in the form $y = Ce^{(\ln a)bt}$.) In some real-life settings, bases other than e are more convenient. For instance, in Example 1, knowing that the half-life of radium is 1599 years, you can immediately write the exponential decay model as

$$y = \left(\frac{1}{2}\right)^{t/1599}.$$

Using this model, the amount of radium left in the sample after 1000 years is

$$y = \left(\frac{1}{2}\right)^{1000/1599} \approx 0.648 \text{ gram}$$

which is the same answer obtained in Example 1.

Example 4 Modeling Sales

ALGEBRA TUTOR *xy*

For help with the algebra in Example 4, see Example 1(b) in the *Chapter 4 Algebra Tutor* on page 302.

Four months after discontinuing advertising on national television, a manufacturer notices that sales have dropped from 100,000 MP3 players per month to 80,000. Using an exponential pattern of decline, what will the sales be after another 4 months?

SOLUTION Let y represent the number of MP3 players, let t represent the time (in months), and consider the exponential decay model

$$y = Ce^{kt}. \qquad \text{Exponential decay model}$$

From the given information, you know that $y = 100{,}000$ when $t = 0$. Using this information, you have

$$100{,}000 = Ce^0$$

which implies that $C = 100{,}000$. To solve for k, use the fact that $y = 80{,}000$ when $t = 4$.

$$\begin{aligned} y &= 100{,}000e^{kt} && \text{Exponential decay model} \\ 80{,}000 &= 100{,}000e^{k(4)} && \text{Substitute 80,000 for } y \text{ and 4 for } t. \\ 0.8 &= e^{4k} && \text{Divide each side by 100,000.} \\ \ln 0.8 &= 4k && \text{Take natural log of each side.} \\ \tfrac{1}{4}\ln 0.8 &= k && \text{Divide each side by 4.} \end{aligned}$$

So, $k = \frac{1}{4}\ln 0.8 \approx -0.0558$, which means that the model is

$$y = 100{,}000e^{-0.0558t}.$$

After four more months ($t = 8$), you can expect sales to drop to

$$\begin{aligned} y &= 100{,}000e^{-0.0558(8)} \\ &\approx 64{,}000 \text{ MP3 players} \end{aligned}$$

as shown in Figure 4.22.

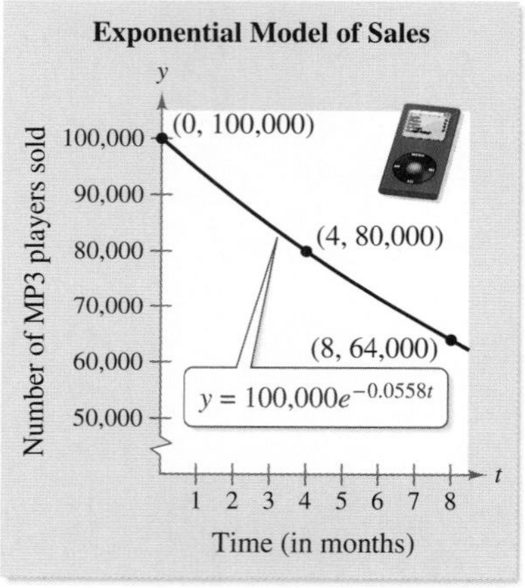

FIGURE 4.22

✓Checkpoint 4

Use the model in Example 4 to determine when sales will drop to 50,000 MP3 players. ■

SUMMARIZE (Section 4.6)

1. State the model used for exponential growth and decay *(page 294)*. For examples of the use of this model, see Examples 1, 2, 3, and 4.
2. State the guidelines for modeling exponential growth and decay *(page 296)*. For examples of the use of these guidelines, see Examples 2, 3, and 4.
3. Describe a real-life example of an exponential decay model *(pages 295 and 298, Examples 1 and 4)*.
4. Describe a real-life example of an exponential growth model *(pages 296 and 297, Examples 2 and 3)*.

SKILLS WARM UP 4.6

The following warm-up exercises involve skills that were covered in earlier sections. You will use these skills in the exercise set for this section. For additional help, review Sections 4.3 and 4.4.

In Exercises 1–4, solve the equation for k.

1. $12 = 24e^{4k}$

2. $10 = 3e^{5k}$

3. $25 = 16e^{-0.01k}$

4. $22 = 32e^{-0.02k}$

In Exercises 5–8, find the derivative of the function.

5. $y = 32e^{0.23t}$

6. $y = 18e^{0.072t}$

7. $y = 24e^{-1.4t}$

8. $y = 25e^{-0.001t}$

In Exercises 9–12, simplify the expression.

9. $e^{\ln 4}$

10. $4e^{\ln 3}$

11. $e^{\ln(2x+1)}$

12. $e^{\ln(x^2+1)}$

Exercises 4.6

See www.CalcChat.com for worked-out solutions to odd-numbered exercises.

Modeling Exponential Growth and Decay In Exercises 1–6, find the exponential function $y = Ce^{kt}$ that passes through the two given points.

1. $y = Ce^{kt}$

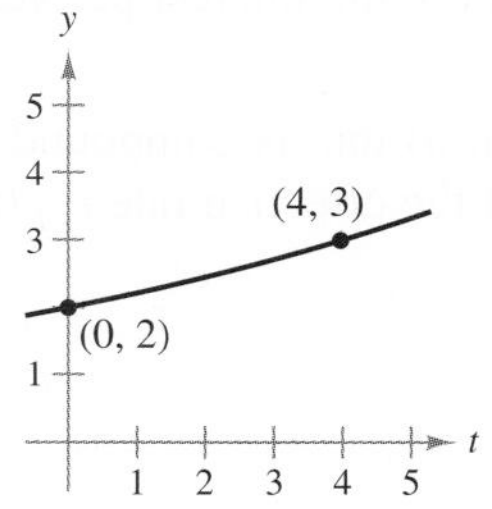

2. $y = Ce^{kt}$

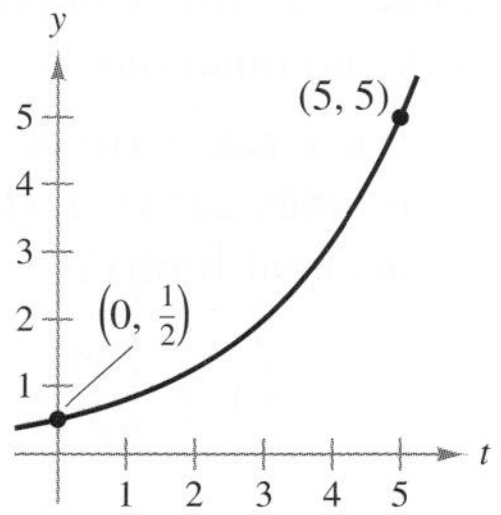

3. $y = Ce^{kt}$

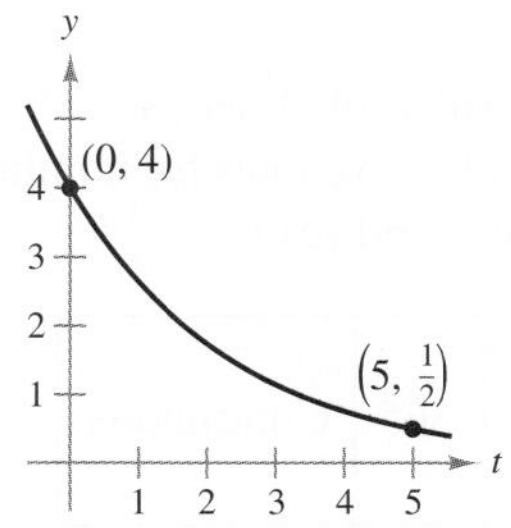

4. $y = Ce^{kt}$

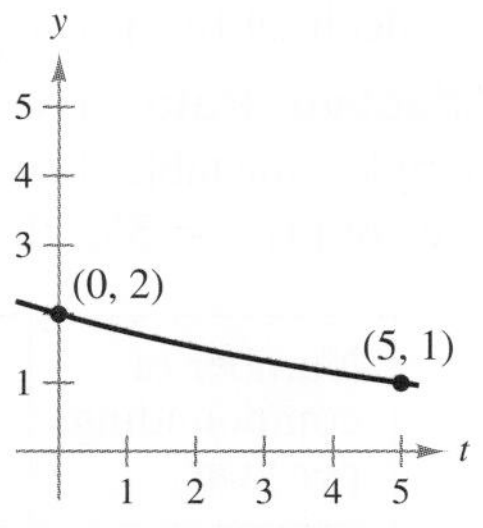

5. $y = Ce^{kt}$

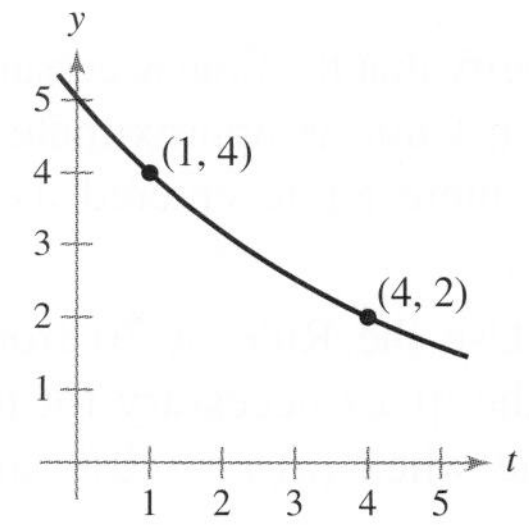

6. $y = Ce^{kt}$

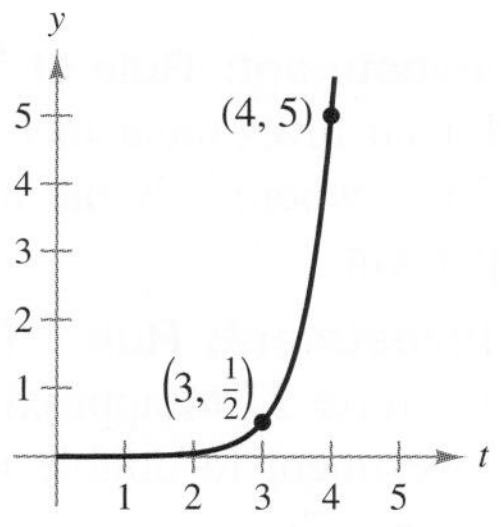

Determining Exponential Growth and Decay In Exercises 7–10, use the given information to write an exponential equation for y. Does the function represent exponential growth or exponential decay?

7. $\dfrac{dy}{dt} = 2y, \quad y = 10$ when $t = 0$

8. $\dfrac{dy}{dt} = -\dfrac{2}{3}y, \quad y = 20$ when $t = 0$

9. $\dfrac{dy}{dt} = -4y, \quad y = 30$ when $t = 0$

10. $\dfrac{dy}{dt} = 5.2y, \quad y = 18$ when $t = 0$

Modeling Radioactive Decay In Exercises 11–16, complete the table for each radioactive isotope. *See Example 1.*

	Isotope	*Half-life (in years)*	*Initial quantity*	*Amount after 1000 years*	*Amount after 10,000 years*
11.	^{226}Ra	1599	10 grams		
12.	^{226}Ra	1599		1.5 grams	
13.	^{14}C	5715			2 grams
14.	^{14}C	5715	3 grams		
15.	^{239}Pu	24,100		2.1 grams	
16.	^{239}Pu	24,100			0.4 gram

17. Radioactive Decay What percent of a present amount of radioactive radium (^{226}Ra) will remain after 900 years?

18. Radioactive Decay Find the half-life of a radioactive material for which 99.57% of the initial amount remains after 1 year.

19. Carbon Dating Carbon-14 (^{14}C) dating assumes that the carbon dioxide on the Earth today has the same radioactive content as it did centuries ago. If this is true, then the amount of ^{14}C absorbed by a tree that grew several centuries ago should be the same as the amount of ^{14}C absorbed by a similar tree today. A piece of ancient charcoal contains only 15% as much of the radioactive carbon as a piece of modern charcoal. How long ago was the tree burned to make the ancient charcoal? (The half-life of ^{14}C is 5715 years.)

20. Carbon Dating Repeat Exercise 19 for a piece of charcoal that contains 30% as much radioactive carbon as a modern piece.

Finding Exponential Models In Exercises 21 and 22, find exponential models $y_1 = Ce^{k_1 t}$ and $y_2 = C(2)^{k_2 t}$ that pass through the two given points. Compare the values of k_1 and k_2. Briefly explain your results.

21. $(0, 5), (12, 20)$ **22.** $(0, 8), \left(20, \frac{1}{2}\right)$

23. Population Growth The number of a certain type of bacteria increases continuously at a rate proportional to the number present. There are 150 bacteria at a given time and 450 bacteria 5 hours later.

(a) How many bacteria will there be 10 hours after the initial time?

(b) How long will it take for the population to double?

(c) Does the answer to part (b) depend on the starting time? Explain your reasoning.

24. HOW DO YOU SEE IT? The graph shows the populations (in thousands) of Cleveland, Ohio and Atlanta, Georgia from 2000 through 2008 using exponential models. *(Source: U.S. Census Bureau)*

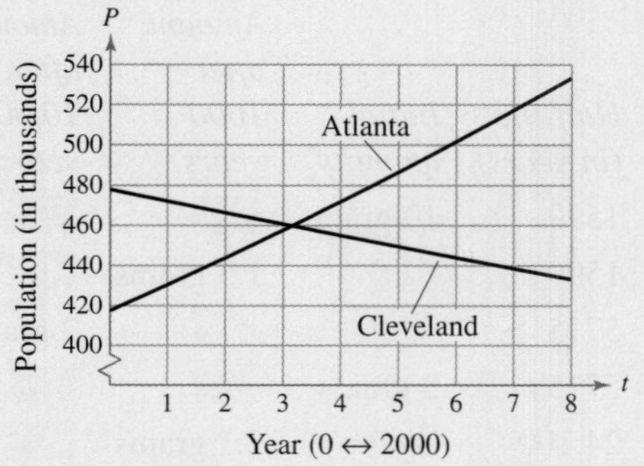

(a) Determine whether the population of each city is modeled by exponential growth or exponential decay. Explain your reasoning.

(b) Estimate the year when the two cities had the same population. What was this population?

Modeling Compound Interest In Exercises 25–32, complete the table for an account in which interest is compounded continuously. *See Example 3.*

	Initial investment	*Annual rate*	*Time to double*	*Amount after 10 years*	*Amount after 25 years*
25.	\$1000	12%			
26.	\$20,000	$10\frac{1}{2}$%			
27.	\$750		8 years		
28.	\$10,000		10 years		
29.	\$500			\$1292.85	
30.	\$2000				\$6008.33
31.		4.5%		\$10,000.00	
32.		2%		\$2000.00	

Finding Present Value In Exercises 33 and 34, determine the principal P that must be invested at interest rate r, compounded continuously, so that \$1,000,000 will be available for retirement in t years.

33. $r = 7.5\%, t = 40$

34. $r = 10\%, t = 25$

35. Effective Rate The effective rate of interest r_{eff} is the annual rate that will produce the same interest per year as the nominal rate r.

(a) For a rate r (in decimal form) that is compounded n times per year, show that the effective rate r_{eff} (in decimal form) is

$$r_{eff} = \left(1 + \frac{r}{n}\right)^n - 1.$$

(b) For a rate r (in decimal form) that is compounded continuously, show that the effective rate r_{eff} (in decimal form) is $r_{eff} = e^r - 1$.

36. Effective Rate Use the results of Exercise 35 to complete the table showing the effective rates for nominal rates of (a) $r = 5\%$, (b) $r = 6\%$, and (c) $r = 7\frac{1}{2}\%$.

Number of compoundings per year	4	12	365	Continuous
Effective yield				

37. Investment: Rule of 70 Verify that the time necessary for an investment to double in value is approximately $70/r$, where r is the annual interest rate entered as a percent.

38. Investment: Rule of 70 Use the Rule of 70 from Exercise 37 to approximate the times necessary for an investment to double in value when (a) $r = 10\%$ and (b) $r = 7\%$.

39. Depreciation A sports utility vehicle that costs \$21,500 new has a book value of \$13,600 after 2 years.

(a) Find a linear model for the value of the vehicle.

(b) Find an exponential model for the value of the vehicle.

(c) Find the book values of the vehicle after 1 year and after 4 years using each model.

(d) Use a graphing utility to graph the two models in the same viewing window. Which model depreciates faster in the first 2 years?

(e) Explain the advantages and disadvantages of using each model to a buyer and to a seller.

40. Population The table shows the populations P (in millions) of the United States from 1960 through 2010. *(Source: U.S. Census Bureau)*

Year	1960	1970	1980	1990	2000	2010
Population, P	181	205	228	250	282	309

(a) Use the 1960 and 1970 data to find an exponential model P_1. Let $t = 0$ represent 1960.

(b) Use a graphing utility to find an exponential model P_2 for all of the data. Let $t = 0$ represent 1960.

(c) Use a graphing utility to plot the data and graph both models in the same viewing window. Compare the actual data with the estimates from the models. Which model is more accurate?

41. Sales The cumulative sales S (in thousands of units) of a new product after it has been on the market for t years are modeled by $S = Ce^{k/t}$. During the first year, 5000 units were sold. The saturation point for the market is 30,000 units. That is, the limit of S as $t \to \infty$ is 30,000.

(a) Solve for C and k in the model.

(b) How many units will be sold after 5 years?

(c) Use a graphing utility to graph the sales function.

42. Sales The cumulative sales S (in thousands of units) of a new product after it has been on the market for t years are modeled by

$$S = 50(1 - e^{kt}).$$

During the first year, 8000 units were sold.

(a) Solve for k in the model.

(b) What is the saturation point for this product?

(c) How many units will be sold after 5 years?

(d) Use a graphing utility to graph the sales function.

43. Learning Curve The management of a factory finds that the maximum number of units a worker can produce in a day is 30. The learning curve for the number of units N produced per day after a new employee has worked for t days is modeled by

$$N = 30(1 - e^{kt}).$$

After 20 days on the job, a worker is producing 19 units in a day. How many days should pass before this worker is producing 25 units per day?

44. Learning Curve The management in Exercise 43 requires that a new employee be producing at least 20 units per day after 30 days on the job.

(a) Find a learning curve model that describes this minimum requirement.

(b) Find the number of days before a minimal achiever is producing 25 units per day.

45. Revenue A small business assumes that the demand function for one of its new products can be modeled by

$$p = Ce^{kx}.$$

When $p = \$45$, $x = 1000$ units, and when $p = \$40$, $x = 1200$ units.

(a) Solve for C and k in the model.

(b) Find the values of x and p that will maximize the revenue for this product.

46. Revenue Repeat Exercise 45 given that when $p = \$5$, $x = 300$ units, and when $p = \$4$, $x = 400$ units.

47. Project: Cell Phone Subscribers For a project analyzing the numbers of cell phone subscribers from 2000 to 2009, visit this text's website at *www.cengagebrain.com*. *(Source: CTIA-The Wireless Association)*

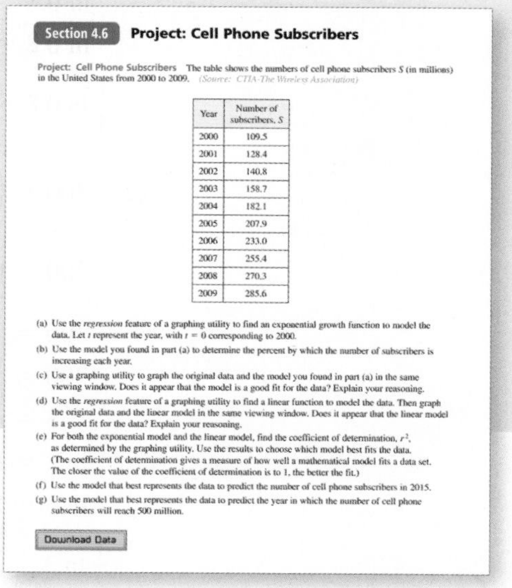

Section 4.6 **Project: Cell Phone Subscribers**

Project: Cell Phone Subscribers The table shows the numbers of cell phone subscribers S (in millions) in the United States from 2000 to 2009. *(Source: CTIA-The Wireless Association)*

Year	Number of subscribers, S
2000	109.5
2001	128.4
2002	140.8
2003	158.7
2004	182.1
2005	207.9
2006	233.0
2007	255.4
2008	270.3
2009	285.6

(a) Use the *regression* feature of a graphing utility to find an exponential growth function to model the data. Let t represent the year, with $t = 0$ corresponding to 2000.

(b) Use the model you found in part (a) to determine the percent by which the number of subscribers is increasing each year.

(c) Use a graphing utility to graph the original data and the model you found in part (a) in the same viewing window. Does it appear that the model is a good fit for the data? Explain your reasoning.

(d) Use the *regression* feature of a graphing utility to find a linear function to model the data. Then graph the original data and the linear model in the same viewing window. Does it appear that the linear model is a good fit for the data? Explain your reasoning.

(e) For both the exponential model and the linear model, find the coefficient of determination, r^2, as determined by the graphing utility. Use the results to choose which model best fits the data. (The coefficient of determination gives a measure of how well a mathematical model fits a data set. The closer the value of the coefficient of determination is to 1, the better the fit.)

(f) Use the model that best represents the data to predict the number of cell phone subscribers in 2015.

(g) Use the model that best represents the data to predict the year in which the number of cell phone subscribers will reach 500 million.

Download Data

ALGEBRA TUTOR

Solving Exponential and Logarithmic Equations

To find the extrema or points of inflection of an exponential or logarithmic function, you must know how to solve exponential and logarithmic equations. A few examples are given on page 280. Some additional examples are presented in this Algebra Tutor.

As with all equations, remember that your basic goal is to isolate the variable on one side of the equation. To do this, you use inverse operations. For instance, to isolate x in

$$e^x = 7$$

take the natural log of each side of the equation and use the property $\ln e^x = x$. Similarly, to isolate x in

$$\ln x = 5$$

exponentiate each side of the equation and use the property $e^{\ln x} = x$.

Example 1 **Solving Exponential Equations**

Solve each exponential equation.

a. $25 = 5e^{7t}$ **b.** $80{,}000 = 100{,}000e^{k(4)}$ **c.** $300 = \left(\dfrac{100}{e^{2k}}\right)e^{4k}$

SOLUTION

a.

$25 = 5e^{7t}$	Write original equation.
$5 = e^{7t}$	Divide each side by 5.
$\ln 5 = \ln e^{7t}$	Take natural log of each side.
$\ln 5 = 7t$	Apply the property $\ln e^a = a$.
$\frac{1}{7}\ln 5 = t$	Divide each side by 7.

b.

$80{,}000 = 100{,}000e^{k(4)}$	Example 4, page 298
$0.8 = e^{4k}$	Divide each side by 100,000.
$\ln 0.8 = \ln e^{4k}$	Take natural log of each side.
$\ln 0.8 = 4k$	Apply the property $\ln e^a = a$.
$\frac{1}{4}\ln 0.8 = k$	Divide each side by 4.

c.

$300 = \left(\dfrac{100}{e^{2k}}\right)e^{4k}$	Example 2, page 296
$300 = (100)\dfrac{e^{4k}}{e^{2k}}$	Rewrite product.
$300 = 100e^{4k-2k}$	To divide powers, subtract exponents.
$300 = 100e^{2k}$	Simplify.
$3 = e^{2k}$	Divide each side by 100.
$\ln 3 = \ln e^{2k}$	Take natural log of each side.
$\ln 3 = 2k$	Apply the property $\ln e^a = a$.
$\frac{1}{2}\ln 3 = k$	Divide each side by 2.

Example 2 **Solving Logarithmic Equations**

Solve each logarithmic equation.

a. $\ln x = 2$ **b.** $5 + 2 \ln x = 4$

c. $2 \ln 3x = 4$ **d.** $\ln x - \ln(x - 1) = 1$

SOLUTION

a.

$\ln x = 2$	Write original equation.
$e^{\ln x} = e^2$	Exponentiate each side.
$x = e^2$	Apply the property $e^{\ln a} = a$.

b.

$5 + 2 \ln x = 4$	Write original equation.
$2 \ln x = -1$	Subtract 5 from each side.
$\ln x = -\dfrac{1}{2}$	Divide each side by 2.
$e^{\ln x} = e^{-1/2}$	Exponentiate each side.
$x = e^{-1/2}$	Apply the property $e^{\ln a} = a$.

c.

$2 \ln 3x = 4$	Write original equation.
$\ln 3x = 2$	Divide each side by 2.
$e^{\ln 3x} = e^2$	Exponentiate each side.
$3x = e^2$	Apply the property $e^{\ln a} = a$.
$x = \dfrac{1}{3}e^2$	Divide each side by 3.

d.

$\ln x - \ln(x - 1) = 1$	Write original equation.
$\ln \dfrac{x}{x - 1} = 1$	$\ln m - \ln n = \ln(m/n)$
$e^{\ln[x/(x-1)]} = e^1$	Exponentiate each side.
$\dfrac{x}{x - 1} = e^1$	Apply the property $e^{\ln a} = a$.
$x = ex - e$	Multiply each side by $x - 1$.
$x - ex = -e$	Subtract ex from each side.
$x(1 - e) = -e$	Factor.
$x = \dfrac{-e}{1 - e}$	Divide each side by $1 - e$.
$x = \dfrac{e}{e - 1}$	Simplify.

STUDY TIP

Because the domain of a logarithmic function generally does not include all real numbers, be sure to check for extraneous solutions.

SUMMARY AND STUDY STRATEGIES

After studying this chapter, you should have acquired the following skills.
The exercise numbers are keyed to the Review Exercises that begin on page 306.
Answers to odd-numbered Review Exercises are given in the back of the text.*

Section 4.1	Review Exercises
■ Use the properties of exponents to evaluate and simplify exponential expressions. $a^0 = 1,\quad a^x a^y = a^{x+y},\quad \dfrac{a^x}{a^y} = a^{x-y},\quad (a^x)^y = a^{xy}$; $(ab)^x = a^x b^x,\quad \left(\dfrac{a}{b}\right)^x = \dfrac{a^x}{b^x},\quad a^{-x} = \dfrac{1}{a^x}$	*1, 2*
■ Sketch the graphs of exponential functions.	*3–8*
■ Use properties of exponents to answer questions about real-life situations.	*9–12*

Section 4.2	
■ Use the properties of exponents to evaluate and simplify natural exponential expressions.	*13, 14*
■ Sketch the graphs of natural exponential functions.	*15–18*
■ Solve compound interest problems. $A = P(1 + r/n)^{nt},\quad A = Pe^{rt}$	*19–24*
■ Solve effective rate of interest problems. $r_{eff} = (1 + r/n)^n - 1$	*25, 26*
■ Solve present value problems. $P = \dfrac{A}{(1 + r/n)^{nt}}$	*27, 28*
■ Answer questions involving the natural exponential function as a real-life model.	*29–34*

Section 4.3	
■ Find the derivatives of natural exponential functions. $\dfrac{d}{dx}[e^x] = e^x,\quad \dfrac{d}{dx}[e^u] = e^u\dfrac{du}{dx}$	*35–40*
■ Find equations of the tangent lines to the graphs of natural exponential functions.	*41–44*
■ Use calculus to analyze the graphs of functions that involve the natural exponential function.	*45–48*

Section 4.4	
■ Use the definition of the natural logarithmic function to write exponential equations in logarithmic form, and vice versa. $\ln x = b$ if and only if $e^b = x$.	*49–52*

* A wide range of valuable study aids are available to help you master the material in this chapter. The *Student Solutions Manual* includes step-by-step solutions to all odd-numbered exercises to help you review and prepare. The student website at *www.cengagebrain.com* offers algebra help and a *Graphing Technology Guide,* which contains step-by-step commands and instructions for a wide variety of graphing calculators.

Section 4.4 (continued)

	Review Exercises
■ Sketch the graphs of natural logarithmic functions.	53–56
■ Use properties of logarithms to simplify, expand, and condense logarithmic expressions. $\ln xy = \ln x + \ln y, \quad \ln \frac{x}{y} = \ln x - \ln y, \quad \ln x^n = n \ln x$	57–66
■ Use inverse properties of exponential and logarithmic functions to solve exponential and logarithmic equations. $\ln e^x = x, \quad e^{\ln x} = x$	67–80
■ Use properties of natural logarithms to answer questions about real-life situations.	81–84

Section 4.5

■ Find the derivatives of natural logarithmic functions. $\frac{d}{dx}[\ln x] = \frac{1}{x}, \quad \frac{d}{dx}[\ln u] = \frac{1}{u}\frac{du}{dx}$	85–98
■ Use the definition of logarithms to evaluate logarithmic expressions involving other bases. $\log_a x = b$ if and only if $a^b = x$	99–102
■ Use the change-of-base formula to evaluate logarithmic expressions involving other bases. $\log_a x = \frac{\ln x}{\ln a}$	103–106
■ Find the derivatives of exponential and logarithmic functions involving other bases. $\frac{d}{dx}[a^x] = (\ln a)a^x, \quad \frac{d}{dx}[a^u] = (\ln a)a^u \frac{du}{dx}$; $\frac{d}{dx}[\log_a x] = \left(\frac{1}{\ln a}\right)\frac{1}{x}, \quad \frac{d}{dx}[\log_a u] = \left(\frac{1}{\ln a}\right)\left(\frac{1}{u}\right)\frac{du}{dx}$	107–112
■ Use calculus to analyze the graphs of functions that involve the natural logarithmic function.	113–116
■ Use calculus to answer questions about real-life situations.	117, 118

Section 4.6

■ Use exponential growth and decay to model real-life situations.	119–132

Study Strategies

- **Classifying Differentiation Rules** Differentiation rules fall into two basic classes: (1) general rules that apply to all differentiable functions; and (2) specific rules that apply to special types of functions. At this point in the course, you have studied six general rules: the Constant Rule, the Constant Multiple Rule, the Sum Rule, the Difference Rule, the Product Rule, and the Quotient Rule. Although these rules were introduced in the context of algebraic functions, remember that they also can be used with exponential and logarithmic functions. You have also studied three specific rules: the Power Rule, the derivative of the natural exponential function, and the derivative of the natural logarithmic function. Each of these rules comes in two forms: the "simple" version, such as $D_x[e^x] = e^x$, and the Chain Rule version, such as $D_x[e^u] = e^u(du/dx)$.
- **To Memorize or Not to Memorize?** When studying mathematics, you need to memorize some formulas and rules. Much of this will come from practice—the formulas that you use most often will be committed to memory. Some formulas, however, are used only infrequently. With these, it is helpful to be able to *derive* the formula from a *known* formula. For instance, knowing the Log Rule for differentiation and the change-of-base formula, $\log_a x = (\ln x)/(\ln a)$, allows you to derive the formula for the derivative of a logarithmic function to base a.

Review Exercises

See www.CalcChat.com for worked-out solutions to odd-numbered exercises.

Applying Properties of Exponents In Exercises 1 and 2, use the properties of exponents to simplify the expression.

1. (a) $(4^5)(4^2)$ (b) $(7^2)^3$

(c) 2^{-4} (d) $\dfrac{3^8}{3^4}$

2. (a) $(5^4)(25^2)$ (b) $(9^{1/3})(3^{1/3})$

(c) $\left(\dfrac{1}{3}\right)^{-3}$ (d) $(6^4)(6^{-5})$

Graphing Exponential Functions In Exercises 3–8, sketch the graph of the function.

3. $f(x) = 9^{x/2}$ **4.** $g(x) = 16^{3x/2}$

5. $f(t) = \left(\frac{1}{6}\right)^t$ **6.** $g(t) = \left(\frac{1}{3}\right)^{-t}$

7. $f(x) = \left(\frac{1}{2}\right)^{2x} + 4$ **8.** $g(x) = \left(\frac{2}{3}\right)^{2x} + 1$

9. Population Growth The resident populations P (in thousands) of Wisconsin from 2000 through 2009 can be modeled by the exponential function

$$P(t) = 5382(1.0057)^t$$

where t is the time in years, with $t = 0$ corresponding to 2000. Use the model to estimate the populations in the years (a) 2016 and (b) 2025. *(Source: U.S. Census Bureau)*

10. Revenue The revenues R (in millions of dollars) for Panera Bread Company from 2000 through 2009 can be modeled by the exponential function

$$R(t) = 163.82(1.2924)^t$$

where t is the time in years, with $t = 0$ corresponding to 2000. Use the model to estimate the sales in the years (a) 2014 and (b) 2017. *(Source: Panera Bread Company)*

11. Property Value Suppose that the value of a piece of property doubles every 12 years. If you buy the property for \$55,000, its value t years after the date of purchase should be

$$V(t) = 55{,}000(2)^{t/12}.$$

Use the model to approximate the value of the property (a) 4 years and (b) 25 years after it is purchased.

12. Inflation Rate Suppose the annual rate of inflation averages 2% over the next 10 years. With this rate of inflation, the approximate cost C of goods or services during any year in the decade will be given by

$$C(t) = P(1.02)^t, \quad 0 \le t \le 10$$

where t is time in years and P is the present cost. If the cost of a graphing calculator is presently \$80, estimate the cost 10 years from now.

Applying Properties of Exponents In Exercises 13 and 14, use the properties of exponents to simplify the expression.

13. (a) $(e^5)^2$ (b) $\dfrac{e^3}{e^5}$

(c) $(e^4)(e^{3/2})$ (d) $(e^2)^{-4}$

14. (a) $(e^6)(e^{-3})$ (b) $(e^{-2})^{-5}$

(c) $\left(\dfrac{e^6}{e^2}\right)^{-1}$ (d) $(e^3)^{4/3}$

Graphing Natural Exponential Functions In Exercises 15–18, sketch the graph of the function.

15. $f(x) = e^{-x} + 1$

16. $g(x) = e^{2x} - 1$

17. $f(x) = 1 - e^x$

18. $g(x) = 2 + e^{x-1}$

Finding Account Balances In Exercises 19–22, complete the table to determine the balance A for P dollars invested at rate r for t years, compounded n times per year.

n	1	2	4	12	365	Continuous compounding
A						

19. $P = \$1000$, $r = 4\%$, $t = 5$ years

20. $P = \$7000$, $r = 6\%$, $t = 20$ years

21. $P = \$3000$, $r = 3.5\%$, $t = 10$ years

22. $P = \$4500$, $r = 2\%$, $t = 25$ years

Comparing Account Balances In Exercises 23 and 24, \$2000 is deposited in an account. Decide which account, (a) or (b), will have the greater balance after 10 years.

23. (a) 5%, compounded continuously

(b) 6%, compounded quarterly

24. (a) $6\frac{1}{2}\%$, compounded monthly

(b) $6\frac{1}{4}\%$, compounded continuously

25. Effective Rate Find the effective rate of interest corresponding to a nominal rate of 6% per year compounded (a) annually, (b) semiannually, (c) quarterly, and (d) monthly.

26. Effective Rate Find the effective rate of interest corresponding to a nominal rate of 8.25% per year compounded (a) annually, (b) semiannually, (c) quarterly, and (d) monthly.

27. Present Value How much should be deposited in an account paying 5% interest compounded quarterly in order to have a balance of $12,000 three years from now?

28. Present Value How much should be deposited in an account paying 8% interest compounded monthly in order to have a balance of $20,000 five years from now?

29. Demand The demand function for a product is modeled by

$$p = 12{,}500 - \frac{10{,}000}{2 + e^{-0.001x}}.$$

Find the price p (in dollars) of the product when the quantity demanded is (a) $x = 1000$ units and (b) $x = 2500$ units. (c) What is the limit of the price as x increases without bound?

30. Demand The demand function for a product is modeled by

$$p = 8000\left(1 - \frac{5}{5 + e^{-0.002x}}\right).$$

Find the price p (in dollars) of the product when the quantity demanded is (a) $x = 1000$ units and (b) $x = 2500$ units. (c) What is the limit of the price as x increases without bound?

31. Profit The net profits P (in millions of dollars) of Medco Health Solutions from 2000 through 2009 are shown in the table.

Year	2000	2001	2002	2003	2004
Profit	216.8	256.6	361.6	425.8	481.6

Year	2005	2006	2007	2008	2009
Profit	602.0	729.8	912.0	1102.9	1280.3

A model for this data is given by $P = 223.89e^{0.1979t}$, where t represents the year, with $t = 0$ corresponding to 2000. *(Source: Medco Health Solutions, Inc.)*

(a) How well does the model fit the data?

(b) Find a linear model for the data. How well does the linear model fit the data? Which model, exponential or linear, is a better fit?

(c) Use both models to predict the net profit in 2015.

32. Population The populations P (in thousands) of Albuquerque, New Mexico from 2000 through 2009 can be modeled by $P = 450e^{0.019t}$, where t is the time in years, with $t = 0$ corresponding to 2000. *(Source: U.S. Census Bureau)*

(a) Find the populations in 2000, 2005, and 2009.

(b) Use the model to estimate the population in 2020.

33. Biology A lake is stocked with 500 fish, and the fish population P begins to increase according to the logistic growth model

$$P = \frac{10{,}000}{1 + 19e^{-t/5}}, \quad t \geq 0$$

where t is measured in months.

(a) Find the number of fish in the lake after 4 months.

(b) Use a graphing utility to graph the model. Find the number of months it takes for the population of fish to reach 4000.

(c) Does the population have a limit as t increases without bound? Explain your reasoning.

34. Medicine On a college campus of 5000 students, the spread of a flu virus through the student body is modeled by

$$P = \frac{5000}{1 + 4999e^{-0.8t}}, \quad t \geq 0$$

where P is the total number of infected people and t is the time, measured in days.

(a) Find the number of students infected after 5 days.

(b) Use a graphing utility to graph the model. Find the number of days it takes for 2000 students to become infected with the flu.

(c) According to this model, will all the students on campus become infected with the flu? Explain your reasoning.

Differentiating Exponential Functions In Exercises 35–40, find the derivative of the function.

35. $y = 4e^{x^2}$

36. $y = 4e^{\sqrt{x}}$

37. $y = \dfrac{x}{e^{2x}}$

38. $y = x^2e^x$

39. $y = \dfrac{5}{1 + e^{2x}}$

40. $y = \dfrac{10}{1 - 2e^x}$

Finding an Equation of a Tangent Line In Exercises 41–44, find an equation of the tangent line to the graph of the function at the given point.

41. $y = e^{2-x}, \quad (2, 1)$

42. $y = e^{2x^2}, \quad (1, e^2)$

43. $y = x^2e^{-x}, \quad \left(1, \dfrac{1}{e}\right)$

44. $y = xe^x - e^x, \quad (1, 0)$

Analyzing a Graph In Exercises 45–48, analyze and sketch the graph of the function. Label any relative extrema, points of inflection, and asymptotes.

45. $f(x) = x^3e^x$

46. $f(x) = \dfrac{e^x}{x^2}$

47. $f(x) = \dfrac{1}{xe^x}$

48. $f(x) = \dfrac{x^2}{e^x}$

Logarithmic and Exponential Forms of Equations In Exercises 49–52, write the logarithmic equation as an exponential equation, or vice versa.

49. $\ln 12 = 2.4849\ldots$ **50.** $\ln 0.6 = -0.5108\ldots$

51. $e^{1.5} = 4.4816\ldots$ **52.** $e^{-4} = 0.0183\ldots$

Graphing Logarithmic Expressions In Exercises 53–56, sketch the graph of the function.

53. $y = \ln(4 - x)$ **54.** $y = \ln x - 3$

55. $y = \ln \frac{x}{3}$ **56.** $y = -2 \ln x$

Expanding Logarithmic Expressions In Exercises 57–62, use the properties of logarithms to rewrite the expression as a sum, difference, or multiple of logarithms.

57. $\ln \sqrt{x^2(x - 1)}$

58. $\ln \sqrt[3]{x^2 - 1}$

59. $\ln \frac{x^2}{(x + 1)^3}$

60. $\ln \frac{x^2}{x^2 + 1}$

61. $\ln\left(\frac{1 - x}{3x}\right)^3$

62. $\ln\left(\frac{x - 1}{x + 1}\right)^2$

Condensing Logarithmic Expressions In Exercises 63–66, use the properties of logarithms to rewrite the expression as the logarithm of a single quantity.

63. $\ln(2x + 5) + \ln(x - 3)$

64. $\frac{1}{3}\ln(x^2 - 6) - 2\ln(3x + 2)$

65. $4[\ln(x^3 - 1) + 2\ln x - \ln(x - 5)]$

66. $\frac{1}{2}[\ln x + 3\ln(x + 1) - \ln(x - 2)]$

Solving Exponential and Logarithmic Equations In Exercises 67–80, solve for x.

67. $e^{\ln x} = 3$ **68.** $e^{\ln(x+2)} = 5$

69. $\ln x = 3$ **70.** $\ln 5x = 2$

71. $\ln 2x - \ln(3x - 1) = 0$

72. $\ln x - \ln(x + 1) = 2$

73. $\ln x + \ln(x - 3) = 0$

74. $2\ln x + \ln(x - 2) = 0$

75. $e^{-1.386x} = 0.25$

76. $e^{-0.01x} - 5.25 = 0$

77. $e^{2x-1} - 6 = 0$

78. $4e^{2x-3} - 5 = 0$

79. $100(1.21)^x = 110$

80. $500(1.075)^{120x} = 100{,}000$

81. Compound Interest A deposit of \$400 is made in an account that earns interest at an annual rate of 2.5%. How long will it take for the balance to double when the interest is compounded (a) annually, (b) monthly, (c) daily, and (d) continuously?

82. Hourly Earnings The average hourly wages w (in dollars) for private industry employees in the United States from 1990 through 2009 can be modeled by

$$w = 10.2e^{0.0315t}$$

where $t = 0$ corresponds to 1990. *(Source: U.S. Bureau of Labor Statistics)*

(a) What was the average hourly wage in 2000?

(b) In what year will the average hourly wage be \$23?

83. Learning Theory Students in a psychology experiment were given an exam and then retested monthly with equivalent exams. The average scores S (on a 100-point scale) for the students can be modeled by

$$S = 75 - 6\ln(t + 1), \quad 0 \le t \le 12$$

where t is the time in months.

(a) What was the average score on the original exam?

(b) What was the average score after 4 months?

(c) After how many months was the average score 60?

84. Demand The demand function for a product is given by

$$p = 8000\left(1 - \frac{5}{5 + e^{-0.002x}}\right)$$

where p is the price per unit (in dollars) and x is the number of units sold. Find the numbers of units sold for prices of (a) $p = \$200$ and (b) $p = \$800$.

Differentiating a Logarithmic Function In Exercises 85–98, find the derivative of the function.

85. $f(x) = \ln 3x^2$ **86.** $y = \ln \sqrt{x}$

87. $y = \ln \frac{x(x - 1)}{x - 2}$ **88.** $y = \ln \frac{x^2}{x + 1}$

89. $f(x) = \ln e^{2x+1}$ **90.** $f(x) = \ln e^{x^2}$

91. $y = \frac{\ln x}{x^3}$ **92.** $y = \frac{x^2}{\ln x}$

93. $y = \ln(x^2 - 2)^{2/3}$

94. $y = \ln \sqrt[3]{x^3 + 1}$

95. $f(x) = \ln\left(x^2\sqrt{x + 1}\right)$

96. $f(x) = \ln \frac{x}{\sqrt{x + 1}}$

97. $y = \ln \frac{e^x}{1 + e^x}$

98. $y = \ln\left(e^{2x}\sqrt{e^{2x} - 1}\right)$

Evaluating Logarithms In Exercises 99–102, evaluate the logarithm without using a calculator.

99. $\log_6 36$

100. $\log_2 32$

101. $\log_{10} 1$

102. $\log_4 \frac{1}{64}$

Changing Bases to Evaluate Logarithms In Exercises 103–106, use the change-of-base formula and a calculator to evaluate the logarithm.

103. $\log_5 13$

104. $\log_4 18$

105. $\log_{16} 64$

106. $\log_4 125$

Differentiating Functions of Other Bases In Exercises 107–112, find the derivative of the function.

107. $y = 5^{2x+1}$

108. $y = 8^{x^3}$

109. $y = \log_3(2x - 1)$

110. $y = \log_{16}(x^2 - 3x)$

111. $y = \log_{10} \frac{3}{x}$

112. $y = \log_2 \frac{1}{x^2}$

Analyzing a Graph In Exercises 113–116, analyze and sketch the graph of the function. Label any relative extrema, points of inflection, and asymptotes.

113. $y = \ln(x + 3)$

114. $y = \dfrac{8 \ln x}{x^2}$

115. $y = \ln \dfrac{10}{x + 2}$

116. $y = \ln \dfrac{x^2}{9 - x^2}$

117. Music The numbers of download music singles D (in millions) from 2004 through 2009 can be modeled by

$$D = -1671.88 + 1282 \ln t$$

where $t = 4$ corresponds to 2004. Find the rates of change of the number of download music singles in 2005 and 2008. *(Source: Recording Industry Association of America)*

118. Minimum Average Cost The cost of producing x units of a product is modeled by

$$C = 200 + 75x - 300 \ln x, \quad x \geq 1.$$

(a) Find the average cost function $\overline{C}$.

(b) Find the minimum average cost analytically. Use a graphing utility to confirm your result.

Modeling Exponential Growth and Decay In Exercises 119 and 120, find the exponential function

$$y = Ce^{kt}$$

that passes through the two given points.

119. $(0, 3), (4, 1)$

120. $(1, 1), (5, 5)$

Modeling Radioactive Decay In Exercises 121–126, complete the table for each radioactive isotope.

	Isotope	*Half-life (in years)*	*Initial quantity*	*Amount after 1000 years*	*Amount after 10,000 years*
121.	^{226}Ra	1599	8 grams		
122.	^{226}Ra	1599		0.7 gram	
123.	^{14}C	5715			6 grams
124.	^{14}C	5715	5 grams		
125.	^{239}Pu	24,100		2.4 grams	
126.	^{239}Pu	24,100			7.1 grams

Modeling Compound Interest In Exercises 127–130, complete the table for an account in which interest is compounded continuously.

	Initial investment	*Annual rate*	*Time to double*	*Amount after 10 years*	*Amount after 25 years*
127.	\$600	8%			
128.	\$2000		7 years		
129.	\$15,000			\$18,321.04	
130.		4%		\$11,934.60	

131. Medical Science Soon after an injection, the concentration D (in milligrams per milliliter) of a drug in a patient's bloodstream is 500 milligrams per milliliter. After 6 hours, 50 milligrams per milliliter of the drug remains in the bloodstream.

(a) Find an exponential model for the concentration D after t hours.

(b) What is the concentration of the drug after 4 hours?

132. Population Growth The number of a certain type of bacteria increases continuously at a rate proportional to the number present. After 2 hours, there are 200 bacteria, and after 4 hours, there are 300 bacteria.

(a) Find an exponential model given the population P after t hours.

(b) How many bacteria will there be after 7 hours?

(c) How long will it take for the population to double?

TEST YOURSELF

See www.CalcChat.com for worked-out solutions to odd-numbered exercises.

Take this test as you would take a test in class. When you are done, check your work against the answers given in the back of the book.

In Exercises 1–4, use the properties of exponents to simplify the expression.

1. $3^2(3^{-2})$ **2.** $\left(\dfrac{2^3}{2^{-5}}\right)^{-1}$

3. $(e^{1/2})(e^4)$ **4.** $(e^3)^4$

In Exercises 5–10, sketch the graph of the function.

5. $f(x) = 5^{x-2}$ **6.** $f(x) = 4^{-x}$

7. $f(x) = e^{x-3}$ **8.** $f(x) = 8 + \ln x^2$

9. $f(x) = \ln(x - 5)$ **10.** $f(x) = 0.5 \ln x$

In Exercises 11–13, use the properties of logarithms to rewrite the expression as a sum, difference, or multiple of logarithms.

11. $\ln \dfrac{3}{2}$ **12.** $\ln \sqrt{x + y}$ **13.** $\ln \dfrac{x + 1}{y}$

In Exercises 14–16, use the properties of logarithms to rewrite the expression as the logarithm of a single quantity.

14. $\ln y + \ln(x + 1)$

15. $3 \ln x - 2 \ln(x - 1)$

16. $\ln x + 4 \ln y - \frac{1}{2} \ln(z + 4)$

In Exercises 17–19, solve the equation.

17. $e^{x-1} = 9$

18. $10e^{2x+1} = 900$

19. $50(1.06)^x = 1500$

20. A deposit of \$500 is made in an account that earns interest at an annual rate of 4%. How long will it take for the balance to double when the interest is compounded (a) annually, (b) monthly, (c) daily, and (d) continuously?

In Exercises 21–24, find the derivative of the function.

21. $y = e^{-3x} + 5$ **22.** $y = 7e^{x+2} + 2x$

23. $y = \ln(3 + x^2)$ **24.** $y = \ln \dfrac{5x}{x + 2}$

25. The revenues R (in millions of dollars) of skiing facilities in the United States from 2000 through 2008 can be modeled by

$$R = 1548e^{0.0617t}$$

where $t = 0$ corresponds to 2000. *(Source: U.S. Census Bureau)*

(a) Use this model to estimate the revenues in 2006.

(b) At what rate were the revenues changing in 2006?

26. What percent of a present amount of radioactive radium (^{226}Ra) will remain after 1200 years? (The half-life of ^{226}Ra is 1599 years.)

27. A population is growing continuously at the rate of 1.75% per year. Find the time necessary for the population to double in size.

5 Integration and Its Applications

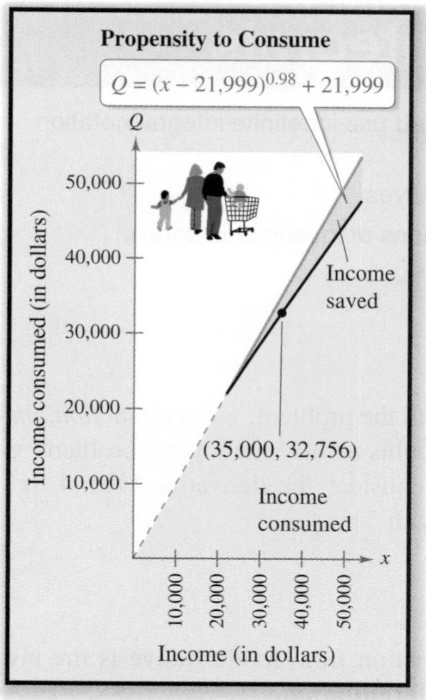

Example 8 on page 328 shows how integration can be used to analyze the marginal propensity to consume.

5.1 Antiderivatives and Indefinite Integrals

- Understand the definition of antiderivative and use indefinite integral notation for antiderivatives.
- Use basic integration rules to find antiderivatives.
- Use initial conditions to find particular solutions of indefinite integrals.
- Use antiderivatives to solve real-life problems.

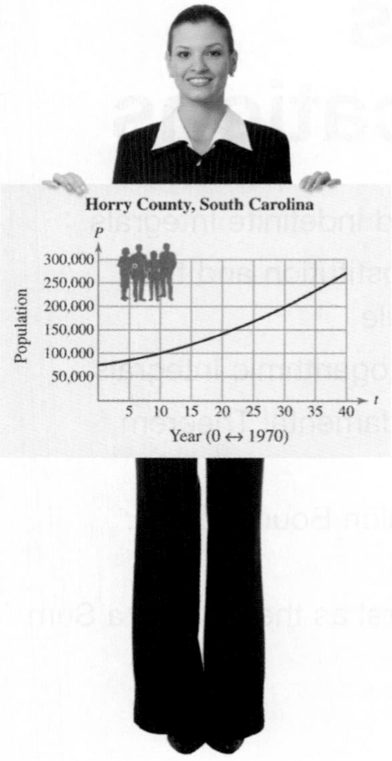

In Exercise 69 on page 321, you will use integration to find a model for the population of a county.

Antiderivatives

In Chapter 2, you were concerned primarily with the problem: *given a function, find its derivative*. Some important applications of calculus involve the inverse problem: *given a derivative, find the function*. For instance, consider the derivative $f'(x) = 3x^2$. To determine the function f, you might come up with

$$f(x) = x^3 \quad \text{because} \quad \frac{d}{dx}[x^3] = 3x^2.$$

This operation of determining the original function from its derivative is the inverse operation of differentiation. It is called **antidifferentiation.**

Definition of Antiderivative

A function F is an **antiderivative** of a function f when for every x in the domain of f, it follows that $F'(x) = f(x)$.

If $F(x)$ is an antiderivative of $f(x)$, then $F(x) + C$, where C is any constant, is also an antiderivative of $f(x)$. For example,

$$F(x) = x^3, \quad G(x) = x^3 - 5, \quad \text{and} \quad H(x) = x^3 + 0.3$$

are all antiderivatives of $3x^2$ because the derivative of each is $3x^2$. As it turns out, *all* antiderivatives of $3x^2$ are of the form $x^3 + C$. So, the process of antidifferentiation does not determine a single function, but rather a *family* of functions, each differing from the others by a constant.

STUDY TIP

In this text, the phrase "$F(x)$ is an antiderivative of $f(x)$" is used synonymously with "F is an antiderivative of f."

The antidifferentiation process is also called **integration** and is denoted by

$$\int \qquad \text{Integral sign}$$

which is called an **integral sign.** The symbol

$$\int f(x)\,dx \qquad \text{Indefinite integral}$$

is the **indefinite integral** of $f(x)$, and it denotes the family of antiderivatives of $f(x)$. That is, if $F'(x) = f(x)$ for all x, then you can write

Integral sign — Differential

$$\int f(x)\,dx = \underbrace{F(x) + C}$$

Integrand — Antiderivative

where $f(x)$ is the **integrand** and C is the **constant of integration.** The differential dx in the indefinite integral identifies the variable of integration. That is, the symbol $\int f(x)\,dx$ denotes the "antiderivative of f *with respect to* x" just as the symbol dy/dx denotes the "derivative of y *with respect to* x."

EDHAR/www.shutterstock.com

Finding Antiderivatives

The inverse relationship between the operations of integration and differentiation can be shown symbolically, as follows.

$$\frac{d}{dx}\left[\int f(x)\,dx\right] = f(x)$$ Differentiation is the inverse of integration.

$$\int f'(x)\,dx = f(x) + C$$ Integration is the inverse of differentiation.

This inverse relationship between integration and differentiation allows you to obtain integration formulas directly from differentiation formulas. The following summary lists the integration formulas that correspond to some of the differentiation formulas you have studied.

STUDY TIP

You will study the General Power Rule for integration in Section 5.2 and the Exponential and Log Rules in Section 5.3.

Basic Integration Rules

1. $\int k\,dx = kx + C$, k is a constant. — Constant Rule
2. $\int kf(x)\,dx = k\int f(x)\,dx$ — Constant Multiple Rule
3. $\int [f(x) + g(x)]\,dx = \int f(x)\,dx + \int g(x)\,dx$ — Sum Rule
4. $\int [f(x) - g(x)]\,dx = \int f(x)\,dx - \int g(x)\,dx$ — Difference Rule
5. $\int x^n\,dx = \frac{x^{n+1}}{n+1} + C, \quad n \neq -1$ — Simple Power Rule

Be sure you see that the Simple Power Rule has the restriction that n cannot be -1. So, you *cannot* use the Simple Power Rule to evaluate the integral

$$\int \frac{1}{x}\,dx.$$

To evaluate this integral, you need the Log Rule, which is described in Section 5.3.

Example 1 Finding Indefinite Integrals

STUDY TIP

Note in Example 1(b) that the integral $\int 1\,dx$ is usually shortened to the form $\int dx$.

Find each indefinite integral.

a. $\int \frac{1}{2}\,dx$ **b.** $\int 1\,dx$ **c.** $\int -5\,dt$

SOLUTION

a. $\int \frac{1}{2}\,dx = \frac{1}{2}x + C$ **b.** $\int 1\,dx = x + C$ **c.** $\int -5\,dt = -5t + C$

✓Checkpoint 1

Find each indefinite integral.

a. $\int 5\,dx$ **b.** $\int -1\,dr$ **c.** $\int 2\,dt$

TECH TUTOR

If you have access to a symbolic integration program, try using it to find antiderivatives.

Example 2 Finding an Indefinite Integral

$$\int 3x\,dx = 3\int x\,dx \qquad \text{Constant Multiple Rule}$$

$$= 3\int x^1\,dx \qquad \text{Rewrite } x \text{ as } x^1.$$

$$= 3\left(\frac{x^2}{2}\right) + C \qquad \text{Simple Power Rule with } n = 1$$

$$= \frac{3}{2}x^2 + C \qquad \text{Simplify.}$$

✓ Checkpoint 2

Find $\int 5x\,dx$. ■

In finding indefinite integrals, a strict application of the basic integration rules tends to produce cumbersome constants of integration. For instance, in Example 2, you could have written

$$\int 3x\,dx = 3\int x\,dx = 3\left(\frac{x^2}{2} + C\right) = \frac{3}{2}x^2 + 3C.$$

However, because C represents *any* constant, it is unnecessary to write $3C$ as the constant of integration. You can simply write $\frac{3}{2}x^2 + C$.

In Example 2, note that the general pattern of integration is similar to that of differentiation.

Original Integral: $\int 3x\,dx$ → Rewrite: $3\int x^1\,dx$ → Integrate: $3\left(\frac{x^2}{2}\right) + C$ → Simplify: $\frac{3}{2}x^2 + C$

Example 3 Rewriting Before Integrating

	Original Integral	*Rewrite*	*Integrate*	*Simplify*
a.	$\int \frac{1}{x^3}\,dx$	$\int x^{-3}\,dx$	$\frac{x^{-2}}{-2} + C$	$-\frac{1}{2x^2} + C$
b.	$\int \sqrt{x}\,dx$	$\int x^{1/2}\,dx$	$\frac{x^{3/2}}{3/2} + C$	$\frac{2}{3}x^{3/2} + C$

✓ Checkpoint 3

Find each indefinite integral.

a. $\int \frac{1}{x^2}\,dx$ **b.** $\int \sqrt[3]{x}\,dx$ ■

Remember that you can check your answer to an antidifferentiation problem by differentiating. For instance, in Example 3(b), you can confirm that $\frac{2}{3}x^{3/2} + C$ is the correct antiderivative by differentiating to obtain

$$\frac{d}{dx}\left[\frac{2}{3}x^{3/2} + C\right] = \left(\frac{2}{3}\right)\left(\frac{3}{2}\right)x^{1/2} = \sqrt{x}.$$

holbox/www.shutterstock.com

With the five basic integration rules, you can integrate *any* polynomial function, as demonstrated in the next example.

Example 4 Integrating Polynomial Functions

Find (a) $\int (x + 2)\, dx$ and (b) $\int (3x^4 - 5x^2 + x)\, dx$.

SOLUTION

a.
$$\int (x+2)\,dx = \int x\,dx + \int 2\,dx \qquad \text{Apply Sum Rule.}$$
$$= \frac{x^2}{2} + C_1 + 2x + C_2 \qquad \text{Apply Simple Power and Constant Rules.}$$
$$= \frac{x^2}{2} + 2x + C \qquad C = C_1 + C_2$$

The second line in this solution is usually omitted.

b.
$$\int (3x^4 - 5x^2 + x)\,dx = 3\left(\frac{x^5}{5}\right) - 5\left(\frac{x^3}{3}\right) + \frac{x^2}{2} + C$$
$$= \frac{3}{5}x^5 - \frac{5}{3}x^3 + \frac{1}{2}x^2 + C$$

✓Checkpoint 4

Find (a) $\int (x + 4)\, dx$ and (b) $\int (4x^3 - 5x + 2)\, dx$. ■

STUDY TIP

When integrating quotients, remember *not* to integrate the numerator and denominator separately. For instance, in Example 5, be sure you understand that

$$\int \frac{x-1}{\sqrt{x}}\,dx = \frac{2}{3}\sqrt{x}(x-3) + C$$

is not the same as

$$\frac{\int (x-1)\,dx}{\int \sqrt{x}\,dx} = \frac{\frac{1}{2}x^2 - x + C_1}{\frac{2}{3}x\sqrt{x} + C_2}.$$

Example 5 Rewriting Before Integrating

Find $\int \frac{x-1}{\sqrt{x}}\, dx$.

SOLUTION Begin by rewriting the quotient in the integrand as a difference. Then rewrite each term using rational exponents.

$$\int \frac{x-1}{\sqrt{x}}\,dx = \int \left(\frac{x}{\sqrt{x}} - \frac{1}{\sqrt{x}}\right) dx \qquad \text{Rewrite as a difference.}$$
$$= \int (x^{1/2} - x^{-1/2})\,dx \qquad \text{Rewrite using rational exponents.}$$
$$= \int x^{1/2}\,dx - \int x^{-1/2}\,dx \qquad \text{Apply Difference Rule.}$$
$$= \frac{x^{3/2}}{3/2} - \frac{x^{1/2}}{1/2} + C \qquad \text{Apply Simple Power Rule.}$$
$$= \frac{2}{3}x^{3/2} - 2x^{1/2} + C \qquad \text{Simplify.}$$
$$= \frac{2}{3}\sqrt{x}(x-3) + C \qquad \text{Factor.}$$

ALGEBRA TUTOR *xy*

For help on the algebra in Example 5, see Example 1(a) in the *Chapter 5 Algebra Tutor,* on page 366.

✓Checkpoint 5

Find $\int \frac{x+2}{\sqrt{x}}\, dx$. ■

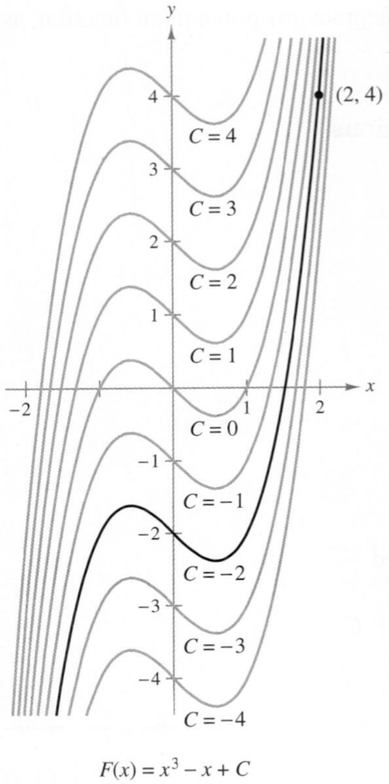

FIGURE 5.1

Particular Solutions

You have already seen that the equation $y = \int f(x)\,dx$ has many solutions, each differing from the others by a constant. This means that the graphs of any two antiderivatives of f are vertical translations of each other. For example, Figure 5.1 shows the graphs of several antiderivatives of the form

$$y = F(x) = \int (3x^2 - 1)\,dx = x^3 - x + C$$

for various integer values of C. Each of these antiderivatives is a solution of the *differential equation*

$$\frac{dy}{dx} = 3x^2 - 1.$$

A **differential equation** in x and y is an equation that involves x, y, and derivatives of y. The **general solution** of $dy/dx = 3x^2 - 1$ is $F(x) = x^3 - x + C$.

In many applications of integration, you are given enough information to determine a **particular solution.** To do this, you need to know the value of $F(x)$ for only one value of x. (This information is called an **initial condition.**) For example, in Figure 5.1, there is only one curve that passes through the point (2, 4). To find this curve, use the information below.

$F(x) = x^3 - x + C$ General solution

$F(2) = 4$ Initial condition

By using the initial condition in the general solution, you can determine that $F(2) = 2^3 - 2 + C = 4$, which implies that $C = -2$. So, the particular solution is

$F(x) = x^3 - x - 2.$ Particular solution

Example 6 Finding a Particular Solution

Find the general solution of

$$F'(x) = 2x - 2$$

and find the particular solution that satisfies the initial condition $F(1) = 2$.

SOLUTION Begin by integrating to find the general solution.

$F(x) = \int (2x - 2)\,dx$ Integrate $F'(x)$ to obtain $F(x)$.

$= x^2 - 2x + C$ General solution

Using the initial condition $F(1) = 2$, you can write

$$F(1) = 1^2 - 2(1) + C = 2$$

which implies that $C = 3$. So, the particular solution is

$F(x) = x^2 - 2x + 3.$ Particular solution

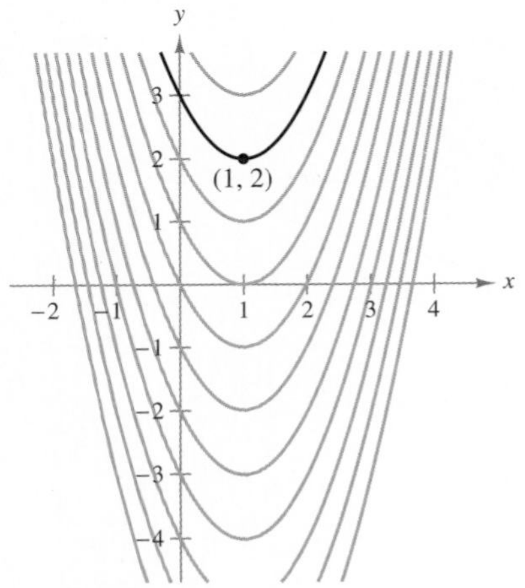

FIGURE 5.2

This solution is shown graphically in Figure 5.2. Note that each of the gray curves represents a solution of the equation $F'(x) = 2x - 2$. The black curve, however, is the only solution that passes through the point (1, 2), which means that $F(x) = x^2 - 2x + 3$ is the only solution that satisfies the initial condition.

✓Checkpoint 6

Find the general solution of $F'(x) = 4x + 2$, and find the particular solution that satisfies the initial condition $F(1) = 8$. ■

Applications

In Chapter 2, you used the general position function (neglecting air resistance) for a falling object

$$s(t) = -16t^2 + v_0t + s_0$$

where $s(t)$ is the height (in feet) and t is the time (in seconds). In the next example, integration is used to *derive* this function.

Example 7 Deriving a Position Function

A ball is thrown upward with an initial velocity of 64 feet per second from an initial height of 80 feet, as shown in Figure 5.3. Derive the position function giving the height s (in feet) as a function of the time t (in seconds). Will the ball be in the air for more than 5 seconds?

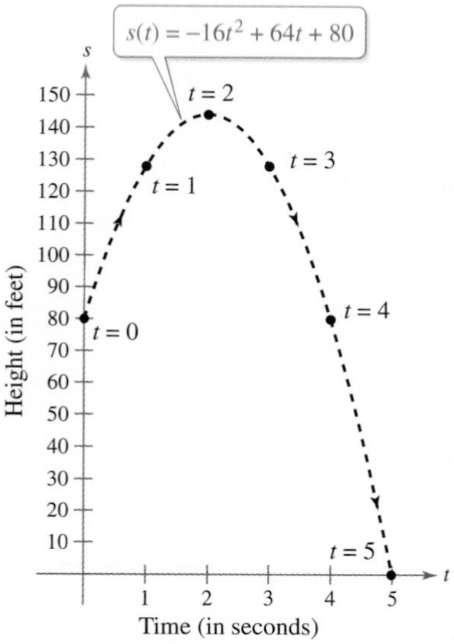

FIGURE 5.3

SOLUTION Let $t = 0$ represent the initial time. Then the two given conditions can be written as

$$s(0) = 80 \qquad \text{Initial height is 80 feet.}$$

$$s'(0) = 64. \qquad \text{Initial velocity is 64 feet per second.}$$

Because the acceleration due to gravity is -32 feet per second per second, you can integrate the acceleration function to find the velocity function, as shown.

$$s''(t) = -32 \qquad \text{Acceleration due to gravity}$$

$$s'(t) = \int -32\,dt \qquad \text{Integrate } s''(t) \text{ to obtain } s'(t).$$

$$= -32t + C_1 \qquad \text{Velocity function}$$

Using the initial velocity, you can conclude that $C_1 = 64$. Next, integrate the velocity function to find the position function.

$$s'(t) = -32t + 64 \qquad \text{Velocity function}$$

$$s(t) = \int (-32t + 64)\,dt \qquad \text{Integrate } s'(t) \text{ to obtain } s(t).$$

$$= -16t^2 + 64t + C_2 \qquad \text{Position function}$$

Using the initial height, it follows that $C_2 = 80$. So, the position function is given by

$$s(t) = -16t^2 + 64t + 80. \qquad \text{Position function}$$

To find the time when the ball hits the ground, set the position function equal to 0 and solve for t.

$$-16t^2 + 64t + 80 = 0 \qquad \text{Set } s(t) \text{ equal to zero.}$$

$$-16(t + 1)(t - 5) = 0 \qquad \text{Factor.}$$

$$t = -1, \quad t = 5 \qquad \text{Solve for } t.$$

Because the time must be positive, you can conclude that the ball hits the ground 5 seconds after it is thrown. So, the ball is not in the air for more than 5 seconds.

✓ Checkpoint 7

Derive the position function when a ball is thrown upward with an initial velocity of 32 feet per second from an initial height of 48 feet. When does the ball hit the ground? With what velocity does the ball hit the ground? ■

Example 8 Finding a Cost Function

The marginal cost of producing x units of a product is modeled by

$$\frac{dC}{dx} = 32 - 0.04x. \qquad \text{Marginal cost}$$

It costs \$50 to produce one unit. Find the total cost of producing 200 units.

SOLUTION To find the cost function, integrate the marginal cost function.

$$C = \int (32 - 0.04x)\,dx \qquad \text{Integrate } \frac{dC}{dx} \text{ to obtain } C.$$

$$= 32x - 0.04\left(\frac{x^2}{2}\right) + K$$

$$= 32x - 0.02x^2 + K \qquad \text{Cost function}$$

To solve for K, use the initial condition $C = 50$ when $x = 1$.

$$50 = 32(1) - 0.02(1)^2 + K \qquad \text{Substitute 50 for } C \text{ and 1 for } x.$$

$$18.02 = K \qquad \text{Solve for } K.$$

So, the total cost function is given by

$$C = 32x - 0.02x^2 + 18.02 \qquad \text{Cost function}$$

which implies that the cost of producing 200 units is

$$C = 32(200) - 0.02(200)^2 + 18.02$$

$$= \$5618.02.$$

STUDY TIP

In Example 8, note that K is used to represent the constant of integration rather than C. This is done to avoid confusion between the constant C and the cost function

$$C = 32x - 0.02x^2 + 18.02.$$

✓ Checkpoint 8

The marginal cost function for producing x units of a product is modeled by

$$\frac{dC}{dx} = 28 - 0.02x.$$

It costs \$40 to produce one unit. Find the total cost of producing 200 units. ■

SUMMARIZE (Section 5.1)

1. State the definition of antiderivative *(page 312)*. For examples of antiderivatives, see Examples 1, 2, 3, 4, and 5.
2. State the Constant Rule *(page 313)*. For an example of the Constant Rule, see Example 1.
3. State the Constant Multiple Rule *(page 313)*. For an example of the Constant Multiple Rule, see Example 2.
4. State the Sum Rule *(page 313)*. For an example of the Sum Rule, see Example 4.
5. State the Difference Rule *(page 313)*. For an example of the Difference Rule, see Example 5.
6. State the Simple Power Rule *(page 313)*. For examples of the Simple Power Rule, see Examples 2, 3, 4, and 5.
7. Describe a real-life example of how antidifferentiation can be used to find a cost function *(page 318, Example 8)*.

holbox/www.shutterstock.com

SKILLS WARM UP 5.1

The following warm-up exercises involve skills that were covered in a previous course or in earlier sections. You will use these skills in the exercise set for this section. For additional help, review Appendix Section A.3 and Section 1.2.

In Exercises 1–6, rewrite the expression using rational exponents.

1. $\dfrac{\sqrt{x}}{x}$

2. $\sqrt[3]{2x}(2x)$

3. $\sqrt{5x^3} + \sqrt{x^5}$

4. $\dfrac{1}{\sqrt{x}} + \dfrac{1}{\sqrt[3]{x^2}}$

5. $\dfrac{(x+1)^3}{\sqrt{x+1}}$

6. $\dfrac{\sqrt{x}}{\sqrt[3]{x}}$

In Exercises 7–10, let $(x, y) = (2, 2)$, and solve the equation for C.

7. $y = x^2 + 5x + C$

8. $y = 3x^3 - 6x + C$

9. $y = -16x^2 + 26x + C$

10. $y = -\frac{1}{4}x^4 - 2x^2 + C$

Exercises 5.1

See www.CalcChat.com for worked-out solutions to odd-numbered exercises.

Integration and Differentiation In Exercises 1–6, verify the statement by showing that the derivative of the right side is equal to the integrand on the left side.

1. $\displaystyle\int 4x\,dx = 2x^2 + C$

2. $\displaystyle\int 4x^3\,dx = x^4 + C$

3. $\displaystyle\int \left(-\frac{9}{x^4}\right) dx = \frac{3}{x^3} + C$

4. $\displaystyle\int \frac{4}{\sqrt{x}}\,dx = 8\sqrt{x} + C$

5. $\displaystyle\int \left(4x^3 - \frac{1}{x^2}\right) dx = x^4 + \frac{1}{x} + C$

6. $\displaystyle\int \left(1 - \frac{1}{\sqrt[3]{x^2}}\right) dx = x - 3\sqrt[3]{x} + C$

Finding Indefinite Integrals In Exercises 7–18, find the indefinite integral. Check your result by differentiating. *See Examples 1 and 2.*

7. $\displaystyle\int du$

8. $\displaystyle\int dr$

9. $\displaystyle\int 6\,dx$

10. $\displaystyle\int -4\,dx$

11. $\displaystyle\int 7x\,dx$

12. $\displaystyle\int 2x\,dx$

13. $\displaystyle\int 5t^2\,dt$

14. $\displaystyle\int 3t^4\,dt$

15. $\displaystyle\int 5x^{-3}\,dx$

16. $\displaystyle\int 4y^{-2}\,dy$

17. $\displaystyle\int y^{3/2}\,dy$

18. $\displaystyle\int v^{-1/2}\,dv$

Rewriting Before Integrating In Exercises 19–24, find the indefinite integral. *See Example 3.*

Original Integral	*Rewrite*	*Integrate*	*Simplify*
19. $\displaystyle\int \sqrt[3]{x^2}\,dx$			
20. $\displaystyle\int \frac{1}{x^4}\,dx$			
21. $\displaystyle\int \frac{1}{x\sqrt{x}}\,dx$			
22. $\displaystyle\int x(x^2+3)\,dx$			
23. $\displaystyle\int \frac{1}{2x^3}\,dx$			
24. $\displaystyle\int \frac{1}{(3x)^2}\,dx$			

Finding Indefinite Integrals In Exercises 25–36, find the indefinite integral. Check your result by differentiating. *See Examples 4 and 5.*

25. $\displaystyle\int (x+3)\,dx$

26. $\displaystyle\int (5-x)\,dx$

27. $\displaystyle\int (x^3+2)\,dx$

28. $\displaystyle\int (x^2-7)\,dx$

29. $\displaystyle\int (3x^3 - 6x^2 + 2)\,dx$

30. $\displaystyle\int (x^3 - 4x + 2)\,dx$

31. $\displaystyle\int (x^2 + 5x + 1)\,dx$

32. $\displaystyle\int (2x^4 - x^2 + 3)\,dx$

33. $\displaystyle\int \frac{2x^3 - 1}{x^3}\,dx$

34. $\displaystyle\int \frac{t^2 + 2}{t^2}\,dt$

35. $\displaystyle\int \frac{5x + 4}{\sqrt[3]{x}}\,dx$

36. $\displaystyle\int \frac{2x - 1}{\sqrt{x}}\,dx$

Interpreting a Graph In Exercises 37–40, the graph of the derivative of a function is given. Sketch the graphs of *two* functions that have the given derivative. (There is more than one correct answer.)

37.

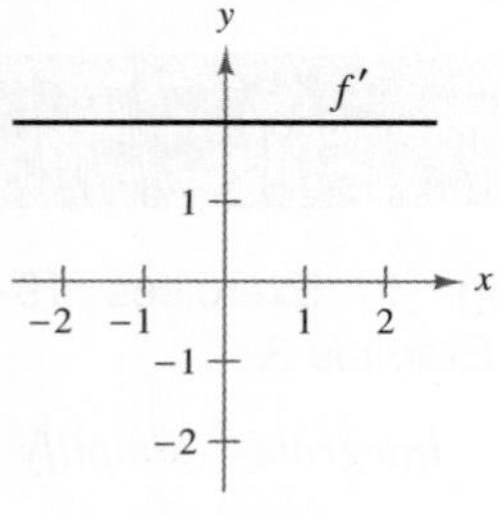

38.

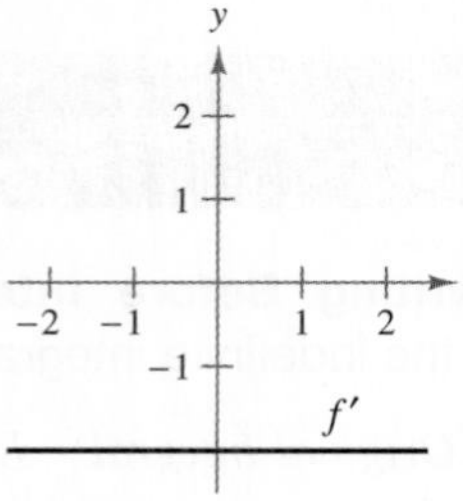

39.

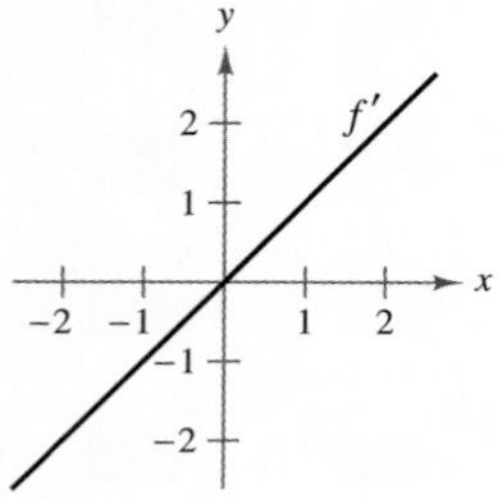

40.

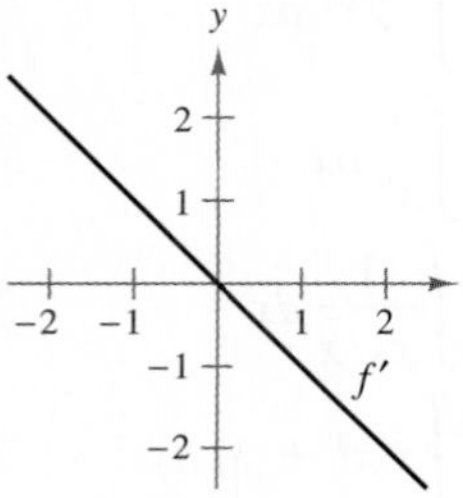

Finding Particular Solutions In Exercises 41–48, find the particular solution that satisfies the differential equation and the initial condition. *See Example 6.*

41. $f'(x) = 4x;\ f(0) = 6$

42. $f'(x) = 9x^2;\ f(0) = -1$

43. $f'(x) = 2x + 4;\ f(-2) = 3$

44. $f'(x) = \frac{1}{5}x - 2;\ f(10) = -10$

45. $f'(x) = 10x - 12x^3;\ f(3) = 2$

46. $f'(x) = 2\sqrt{x};\ f(4) = 12$

47. $f'(x) = \dfrac{2 - x}{x^3},\ x > 0;\ f(2) = \dfrac{3}{4}$

48. $f'(x) = \dfrac{x^2 - 5}{x^2},\ x > 0;\ f(1) = 2$

Finding Particular Solutions In Exercises 49–52, find a function f that satisfies the initial conditions.

49. $f''(x) = 2,\ f'(2) = 5,\ f(2) = 10$

50. $f''(x) = x^2,\ f'(0) = 6,\ f(0) = 3$

51. $f''(x) = x^{-2/3},\ f'(8) = 6,\ f(0) = 0$

52. $f''(x) = x^{-3/2},\ f'(1) = 2,\ f(9) = -4$

Finding a Cost Function In Exercises 53–56, find the cost function for the given marginal cost and fixed cost. *See Example 8.*

	Marginal Cost	*Fixed Cost* ($x = 0$)
53.	$\dfrac{dC}{dx} = 85$	\$5500
54.	$\dfrac{dC}{dx} = \dfrac{1}{50}x + 10$	\$1000
55.	$\dfrac{dC}{dx} = \dfrac{1}{20\sqrt{x}} + 4$	\$750
56.	$\dfrac{dC}{dx} = \dfrac{\sqrt[4]{x}}{10} + 10$	\$2300

Revenue and Demand In Exercises 57 and 58, find the revenue and demand functions for the given marginal revenue. (Use the fact that $R = 0$ when $x = 0$.)

57. $\dfrac{dR}{dx} = 225 - 3x$

58. $\dfrac{dR}{dx} = 310 - 4x$

Profit In Exercises 59–62, find the profit function for the given marginal profit and initial condition.

	Marginal Profit	*Initial Condition*
59.	$\dfrac{dP}{dx} = -18x + 1650$	$P(15) = \$22{,}725$
60.	$\dfrac{dP}{dx} = -40x + 250$	$P(5) = \$650$
61.	$\dfrac{dP}{dx} = -24x + 805$	$P(12) = \$8000$
62.	$\dfrac{dP}{dx} = -30x + 920$	$P(8) = \$6500$

Vertical Motion In Exercises 63–66, use $s''(t) = -32$ feet per second per second as the acceleration due to gravity. *See Example 7.*

63. The Grand Canyon is 6000 feet deep at the deepest part. A rock is dropped from this height. Express the height s (in feet) of the rock as a function of the time t (in seconds). How long will it take the rock to hit the canyon floor?

64. A ball is thrown upward with an initial velocity of 60 feet per second from an initial height of 16 feet. Express the height s (in feet) of the ball as a function of the time t (in seconds). How long will the ball be in the air?

65. With what initial velocity must an object be thrown upward from the ground to reach the height of the Washington Monument (550 feet)?

66. With what initial velocity must an object be thrown upward from a height of 5 feet to reach a maximum height of 230 feet?

67. Cost A company produces a product for which the marginal cost of producing x units is modeled by $dC/dx = 2x - 12$, and the fixed costs are \$125.

(a) Find the total cost function and the average cost function.

(b) Find the total cost of producing 50 units.

(c) In part (b), how much of the total cost is fixed? How much is variable? Give examples of fixed costs associated with the manufacture of a product. Give examples of variable costs.

68. Tree Growth An evergreen nursery usually sells a certain shrub after 6 years of growth and shaping. The growth rate during those 6 years is approximated by $dh/dt = 1.5t + 5$, where t is the time (in years) and h is the height (in centimeters). The seedlings are 12 centimeters tall when planted ($t = 0$).

(a) Find the height function.

(b) How tall are the shrubs when they are sold?

69. Population Growth The growth rate of the population of Horry County in South Carolina from 1970 through 2009 can be modeled by

$$\frac{dP}{dt} = 158.80t + 1758.6$$

where t is the time in years, with $t = 0$ corresponding to 1970. The county's population was 263,868 in 2009. *(Source: U.S. Census Bureau)*

(a) Find the model for Horry County's population.

(b) Use the model to predict the population in 2015. Does your answer seem reasonable? Explain your reasoning.

70. HOW DO YOU SEE IT? The graph shows the rate of change of the revenue of a company from 1990 through 2010.

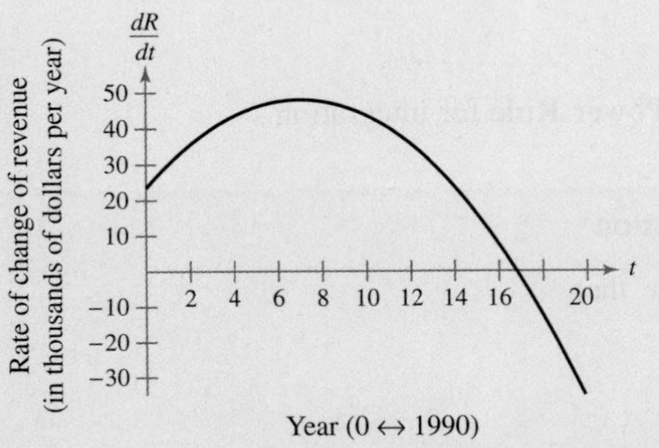

(a) Approximate the rate of change of the revenue in 1993. Explain your reasoning.

(b) Approximate the year when the revenue is maximum. Explain your reasoning.

71. Vital Statistics The rate of increase of the number of married couples M (in thousands) in the United States from 1980 through 2009 can be modeled by

$$\frac{dM}{dt} = -0.105t^2 + 14.02t + 217.8$$

where t is the time in years, with $t = 0$ corresponding to 1980. The number of married couples in 2009 was 60,844 thousand. *(Source: U.S. Census Bureau)*

(a) Find the model for the number of married couples in the United States.

(b) Use the model to predict the number of married couples in the United States in 2015. Does your answer seem reasonable? Explain your reasoning.

72. Internet Users The rate of growth of the number of Internet users I (in millions) in the world from 1991 through 2009 can be modeled by

$$\frac{dI}{dt} = 0.0556t^3 - 1.557t^2 + 25.70t - 59.2$$

where t is the time in years, with $t = 1$ corresponding to 1991. The number of Internet users in 2009 was 1833 million. *(Source: International Telecommunication Union)*

(a) Find the model for the number of Internet users in the world.

(b) Use the model to predict the number of Internet users in the world in 2015. Does your answer seem reasonable? Explain your reasoning.

73. Economics: Marginal Benefits and Costs The table gives the marginal benefit and marginal cost of producing x units of a product for a given company. (a) Plot the points in each column and use the *regression* feature of a graphing utility to find a linear model for marginal benefit and a quadratic model for marginal cost. (b) Use integration to find the benefit B and cost C equations. Assume $B(0) = 0$ and $C(0) = 425$. (c) Find the intervals in which the benefit exceeds the cost of producing x units. Make a recommendation for how many units the company should produce based on your findings. *(Source: Adapted from Taylor,* Economics, *Fifth Edition)*

Number of units	1	2	3	4	5
Marginal benefit	330	320	290	270	250
Marginal cost	150	120	100	110	120

Number of units	6	7	8	9	10
Marginal benefit	230	210	190	170	160
Marginal cost	140	160	190	250	320

5.2 Integration by Substitution and the General Power Rule

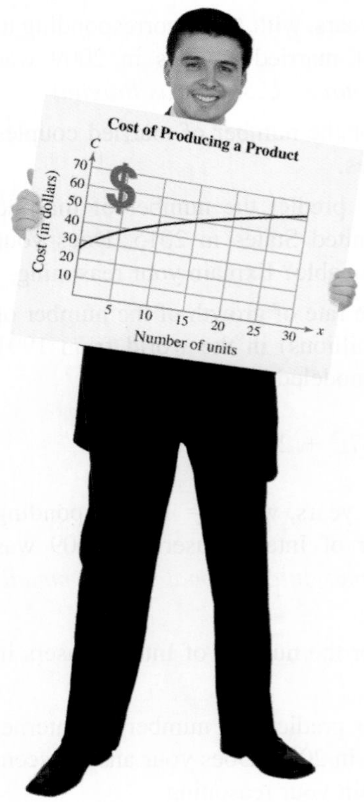

In Exercise 49 on page 330, you will use integration to find a model for the cost of producing a product.

- Use the General Power Rule to find indefinite integrals.
- Use substitution to find indefinite integrals.
- Use the General Power Rule to solve real-life problems.

The General Power Rule

In Section 5.1, you used the Simple Power Rule

$$\int x^n\,dx = \frac{x^{n+1}}{n+1} + C, \quad n \neq -1$$

to find antiderivatives of functions expressed as powers of x alone. In this section, you will study a technique for finding antiderivatives of more complicated functions.

To begin, consider how you might find the antiderivative of

$$2x(x^2 + 1)^3.$$

Because you are hunting for a function whose derivative is $2x(x^2 + 1)^3$, you might discover the antiderivative as shown.

$$\frac{d}{dx}[(x^2+1)^4] = 4(x^2+1)^3(2x) \qquad \text{Use Chain Rule.}$$

$$\frac{d}{dx}\left[\frac{(x^2+1)^4}{4}\right] = (x^2+1)^3(2x) \qquad \text{Divide both sides by 4.}$$

$$\frac{(x^2+1)^4}{4} + C = \int 2x(x^2+1)^3\,dx \qquad \text{Write in integral form.}$$

The key to this solution is the presence of the factor $2x$ in the integrand. In other words, this solution works because $2x$ is precisely the derivative of $(x^2 + 1)$. Letting $u = x^2 + 1$, you can write

$$\int \overbrace{(x^2+1)^3}^{u^3}\,\underbrace{2x\,dx}_{du} = \int u^3\,du$$

$$= \frac{u^4}{4} + C.$$

This is an example of the **General Power Rule** for integration.

General Power Rule for Integration

If u is a differentiable function of x, then

$$\int u^n \frac{du}{dx}\,dx = \int u^n\,du$$

$$= \frac{u^{n+1}}{n+1} + C, \quad n \neq -1.$$

When using the General Power Rule, you must first identify a factor u of the integrand that is raised to a power. Then, you must show that its derivative du/dx is also a factor of the integrand. This is demonstrated in Example 1.

Example 1 **Applying the General Power Rule**

Find each indefinite integral.

a. $\int 3(3x - 1)^4\, dx$ **b.** $\int (2x + 1)(x^2 + x)\, dx$

c. $\int 3x^2\sqrt{x^3 - 2}\, dx$ **d.** $\int \frac{-4x}{(1 - 2x^2)^2}\, dx$

SOLUTION

a. $\int 3(3x - 1)^4\, dx = \int \overbrace{(3x - 1)^4}^{u^n}\overbrace{(3)}^{\frac{du}{dx}}\, dx$ Let $u = 3x - 1$.

$= \frac{(3x - 1)^5}{5} + C$ General Power Rule

b. $\int (2x + 1)(x^2 + x)\, dx = \int \overbrace{(x^2 + x)}^{u^n}\overbrace{(2x + 1)}^{\frac{du}{dx}}\, dx$ Let $u = x^2 + x$.

$= \frac{(x^2 + x)^2}{2} + C$ General Power Rule

c. $\int 3x^2\sqrt{x^3 - 2}\, dx = \int \overbrace{(x^3 - 2)^{1/2}}^{u^n}\overbrace{(3x^2)}^{\frac{du}{dx}}\, dx$ Let $u = x^3 - 2$.

$= \frac{(x^3 - 2)^{3/2}}{3/2} + C$ General Power Rule

$= \frac{2}{3}(x^3 - 2)^{3/2} + C$ Simplify.

d. $\int \frac{-4x}{(1 - 2x^2)^2}\, dx = \int \overbrace{(1 - 2x^2)^{-2}}^{u^n}\overbrace{(-4x)}^{\frac{du}{dx}}\, dx$ Let $u = 1 - 2x^2$.

$= \frac{(1 - 2x^2)^{-1}}{-1} + C$ General Power Rule

$= -\frac{1}{1 - 2x^2} + C$ Simplify.

STUDY TIP

Example 1(b) illustrates a case of the General Power Rule that is sometimes overlooked—when the power is $n = 1$. In this case, the rule takes the form

$$\int u\frac{du}{dx}\, dx = \frac{u^2}{2} + C.$$

✓ Checkpoint 1

Find each indefinite integral.

a. $\int (3x^2 + 6)(x^3 + 6x)^2\, dx$ **b.** $\int 2x\sqrt{x^2 - 2}\, dx$ ■

Remember that you can verify the result of an indefinite integral by differentiating the function. For instance, you can check the answer to Example 1(a) as follows.

$\frac{d}{dx}\left[\frac{(3x - 1)^5}{5} + C\right] = \left(\frac{1}{5}\right)(5)(3x - 1)^4(3)$ Apply Chain Rule.

$= 3(3x - 1)^4$ Simplify.

Many times, part of the derivative du/dx is missing from the integrand, and in *some* cases you can make the necessary adjustments to apply the General Power Rule.

ALGEBRA TUTOR

For help on the algebra in Example 2, see Example 1(b) in the *Chapter 5 Algebra Tutor*, on page 366.

STUDY TIP

Try using the Chain Rule to check the result of Example 2. After differentiating

$$-\tfrac{1}{24}(3 - 4x^2)^3 + C$$

and simplifying, you should obtain the original integrand.

Example 2 Multiplying and Dividing by a Constant

Find $\displaystyle\int x(3 - 4x^2)^2\,dx$.

SOLUTION Let $u = 3 - 4x^2$. To apply the General Power Rule, you need to create $du/dx = -8x$ as a factor of the integrand. You can accomplish this by multiplying and dividing by the constant -8.

$$\begin{aligned}
\int x(3 - 4x^2)^2\,dx &= \int \left(-\frac{1}{8}\right)\overbrace{(3 - 4x^2)^2}^{u^n}\overbrace{(-8x)}^{\frac{du}{dx}}\,dx && \text{Multiply and divide by } -8.\\
&= -\frac{1}{8}\int (3 - 4x^2)^2(-8x)\,dx && \text{Factor } -\tfrac{1}{8} \text{ out of integrand.}\\
&= \left(-\frac{1}{8}\right)\left[\frac{(3 - 4x^2)^3}{3}\right] + C && \text{General Power Rule}\\
&= -\frac{(3 - 4x^2)^3}{24} + C && \text{Simplify.}
\end{aligned}$$

Checkpoint 2

Find $\displaystyle\int x^3(3x^4 + 1)^2\,dx$. ■

Example 3 Multiplying and Dividing by a Constant

Find $\displaystyle\int (x^2 + 2x)^3(x + 1)\,dx$.

SOLUTION Let $u = x^2 + 2x$. To apply the General Power Rule, you need to create $du/dx = 2x + 2$ as a factor of the integrand. You can accomplish this by multiplying and dividing by the constant 2.

$$\begin{aligned}
\int (x^2 + 2x)^3(x + 1)\,dx &= \int \left(\frac{1}{2}\right)\overbrace{(x^2 + 2x)^3}^{u^n}\overbrace{(2)(x + 1)}^{\frac{du}{dx}}\,dx && \text{Multiply and divide by 2.}\\
&= \frac{1}{2}\int (x^2 + 2x)^3(2x + 2)\,dx && \text{Rewrite integrand.}\\
&= \frac{1}{2}\left[\frac{(x^2 + 2x)^4}{4}\right] + C && \text{General Power Rule}\\
&= \frac{1}{8}(x^2 + 2x)^4 + C && \text{Simplify.}
\end{aligned}$$

Checkpoint 3

Find $\displaystyle\int (x^3 - 3x)^2(x^2 - 1)\,dx$. ■

STUDY TIP

In Example 4, be sure you see that you cannot factor variable quantities outside the integral sign. After all, if this were permissible, you could move the entire integrand outside the integral sign and eliminate the need for all integration rules except the rule

$$\int dx = x + C.$$

ALGEBRA TUTOR

For help on the algebra in Example 5, see Example 1(c) in the *Chapter 5 Algebra Tutor*, on page 366.

Example 4 A Failure of the General Power Rule

Find $\int -8(3 - 4x^2)^2\,dx$.

SOLUTION Let $u = 3 - 4x^2$. To apply the General Power Rule, you must create $du/dx = -8x$ as a factor of the integrand. In Examples 2 and 3, this was done by multiplying and dividing by a constant, and then factoring that constant out of the integrand. This strategy doesn't work with variables. That is,

$$\int -8(3 - 4x^2)^2\,dx \neq \frac{1}{x}\int (3 - 4x^2)^2(-8x)\,dx.$$

To find this indefinite integral, you can expand the integrand and use the Simple Power Rule.

$$\begin{aligned}\int -8(3 - 4x^2)^2\,dx &= \int (-72 + 192x^2 - 128x^4)\,dx\\ &= -72x + 64x^3 - \frac{128}{5}x^5 + C\end{aligned}$$

✓Checkpoint 4

Find $\int 2(3x^4 + 1)^2\,dx$. ■

When an integrand contains an extra constant factor that is not needed as part of du/dx, you can simply move the factor outside the integral sign, as shown in the next example.

Example 5 Applying the General Power Rule

Find $\int 7x^2\sqrt{x^3 + 1}\,dx$.

SOLUTION Let $u = x^3 + 1$. Then you need to create $du/dx = 3x^2$ by multiplying and dividing by 3. The constant factor $\frac{7}{3}$ is not needed as part of du/dx, and can be moved outside the integral sign.

$$\begin{aligned}\int 7x^2\sqrt{x^3 + 1}\,dx &= \int 7x^2(x^3 + 1)^{1/2}\,dx && \text{Rewrite with rational exponent.}\\ &= \int \frac{7}{3}(x^3 + 1)^{1/2}(3x^2)\,dx && \text{Multiply and divide by 3.}\\ &= \frac{7}{3}\int (x^3 + 1)^{1/2}(3x^2)\,dx && \text{Factor } \tfrac{7}{3} \text{ outside integral.}\\ &= \frac{7}{3}\left[\frac{(x^3 + 1)^{3/2}}{3/2}\right] + C && \text{General Power Rule}\\ &= \frac{14}{9}(x^3 + 1)^{3/2} + C && \text{Simplify.}\end{aligned}$$

✓Checkpoint 5

Find $\int 5x\sqrt{x^2 - 1}\,dx$. ■

Andresr/www.shutterstock.com

Substitution

The integration technique used in Examples 1, 2, 3, and 5 depends on your ability to recognize or create an integrand of the form

$$u^n \frac{du}{dx}.$$

With more complicated integrands, it is difficult to recognize the steps needed to fit the integrand to a basic integration formula. When this occurs, an alternative procedure called **substitution** or **change of variables** can be helpful. With this procedure, you completely rewrite the integral in terms of u and du. That is, if $u = f(x)$, then $du = f'(x)\,dx$, and the General Power Rule takes the form

$$\int u^n \frac{du}{dx}\,dx = \int u^n\,du.$$ General Power Rule

Example 6 **Integration by Substitution**

Find $\int \sqrt{1-3x}\,dx$.

SOLUTION Begin by letting $u = 1 - 3x$. Then, $du/dx = -3$ and $du = -3\,dx$. This implies that

$$dx = -\frac{1}{3}\,du$$

and you can find the indefinite integral as shown.

$$\int \sqrt{1-3x}\,dx = \int (1-3x)^{1/2}\,dx$$ Rewrite with rational exponent.

$$= \int u^{1/2}\left(-\frac{1}{3}\,du\right)$$ Substitute for x and dx.

$$= -\frac{1}{3}\int u^{1/2}\,du$$ Factor $-\frac{1}{3}$ out of integrand.

$$= \left(-\frac{1}{3}\right)\left(\frac{u^{3/2}}{3/2}\right) + C$$ Apply Power Rule.

$$= -\frac{2}{9}u^{3/2} + C$$ Simplify.

$$= -\frac{2}{9}(1-3x)^{3/2} + C$$ Substitute $1 - 3x$ for u.

You can check this result by differentiating.

$$\frac{d}{dx}\left[-\frac{2}{9}(1-3x)^{3/2} + C\right] = \left(-\frac{2}{9}\right)\left(\frac{3}{2}\right)(1-3x)^{1/2}(-3)$$

$$= \left(-\frac{1}{3}\right)(-3)(1-3x)^{1/2}$$

$$= \sqrt{1-3x}$$

✓ **Checkpoint 6**

Find $\int \sqrt{1-2x}\,dx$ by the method of substitution.

The basic steps for integration by substitution are outlined in the guidelines below.

Guidelines for Integration by Substitution

1. Let u be a function of x (usually part of the integrand).
2. Solve for x and dx in terms of u and du.
3. Convert the entire integral to u-variable form.
4. After integrating, rewrite the antiderivative as a function of x.
5. Check your answer by differentiating.

Example 7 Integration by Substitution

Find $\int x\sqrt{x^2 - 1}\, dx$.

SOLUTION Consider the substitution $u = x^2 - 1$, which produces

$$du = 2x\, dx.$$

To create $2x\, dx$ as part of the integral, multiply and divide by 2.

$$\int x\sqrt{x^2 - 1}\, dx = \frac{1}{2}\int \overbrace{(x^2 - 1)^{1/2}}^{u^{1/n}}\,\overbrace{2x\, dx}^{du} \qquad \text{Multiply and divide by 2.}$$

$$= \frac{1}{2}\int u^{1/2}\, du \qquad \text{Substitute for } x \text{ and } dx.$$

$$= \frac{1}{2}\left(\frac{u^{3/2}}{3/2}\right) + C \qquad \text{Apply Power Rule.}$$

$$= \frac{1}{3}u^{3/2} + C \qquad \text{Simplify.}$$

$$= \frac{1}{3}(x^2 - 1)^{3/2} + C \qquad \text{Substitute for } u.$$

You can check this result by differentiating.

$$\frac{d}{dx}\left[\frac{1}{3}(x^2 - 1)^{3/2} + C\right] = \frac{1}{3}\left(\frac{3}{2}\right)(x^2 - 1)^{1/2}(2x)$$

$$= \frac{1}{2}(2x)(x^2 - 1)^{1/2}$$

$$= x\sqrt{x^2 - 1}$$

✓ Checkpoint 7

Find $\int x\sqrt{x^2 + 4}\, dx$ by the method of substitution. ■

To become efficient at integration, you should learn to use *both* techniques discussed in this section. For simpler integrals, you should use pattern recognition and create du/dx by multiplying and dividing by an appropriate constant. For more complicated integrals, you should use a formal change of variables, as shown in Examples 6 and 7. For the integrals in this section's exercise set, try working several of the problems twice—once with pattern recognition and once using formal substitution.

EDHAR/www.shutterstock.com

Extended Application: Propensity to Consume

In 2009, the U.S. poverty level for a family of four was about \$22,000. Families at or below the poverty level tend to consume 100% of their income—that is, they use all their income to purchase necessities such as food, clothing, and shelter. As income level increases, the average consumption tends to drop below 100%. For instance, a family earning \$25,000 may be able to save \$500 and so consume only \$24,500 (98%) of their income. As the income increases, the ratio of consumption to savings tends to decrease. The rate of change of consumption with respect to income is called the **marginal propensity to consume.** *(Source: U.S. Census Bureau)*

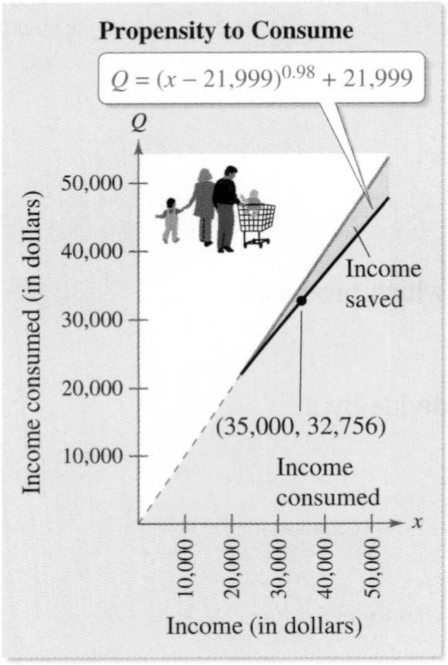

FIGURE 5.4

Example 8 Analyzing Consumption

For a family of four in 2009, the marginal propensity to consume income x (in dollars) can be modeled by

$$\frac{dQ}{dx} = \frac{0.98}{(x - 21{,}999)^{0.02}}, \quad x \geq 22{,}000$$

where Q represents the income consumed (in dollars). Use the model to estimate the amount consumed by a family of four whose 2009 income was \$35,000.

SOLUTION Begin by integrating dQ/dx to find a model for the consumption Q.

$$Q = \int \frac{0.98}{(x - 21{,}999)^{0.02}}\, dx \qquad \text{Integrate } \frac{dQ}{dx} \text{ to obtain } Q.$$

$$= \int 0.98(x - 21{,}999)^{-0.02}\, dx \qquad \text{Rewrite.}$$

$$= (x - 21{,}999)^{0.98} + C \qquad \text{General Power Rule}$$

To solve for C, use the initial condition that $Q = 22{,}000$ when $x = 22{,}000$.

$$22{,}000 = (22{,}000 - 21{,}999)^{0.98} + C$$

$$22{,}000 = 1 + C$$

$$21{,}999 = C$$

So, you can use the model $Q = (x - 21{,}999)^{0.98} + 21{,}999$ to estimate that a family of four with an income of $x = 35{,}000$ consumed about

$$Q = (35{,}000 - 21{,}999)^{0.98} + 21{,}999 \approx \$32{,}756.$$

The graph of Q is shown in Figure 5.4.

✓ Checkpoint 8

According to the model in Example 8, at what income level would a family of four consume \$32,000? ■

SUMMARIZE (Section 5.2)

1. State the General Power Rule for integration *(page 322)*. For examples of the General Power Rule, see Examples 1, 2, 3, and 5.
2. List the guidelines for integration by substitution *(page 327)*. For examples of integration by substitution, see Examples 6 and 7.
3. Describe a real-life example of how the General Power Rule can be used to analyze the marginal propensity to consume *(page 328, Example 8)*.

SKILLS WARM UP 5.2 The following warm-up exercises involve skills that were covered in earlier sections. You will use these skills in the exercise set for this section. For additional help, review Section 5.1.

In Exercises 1–9, find the indefinite integral.

1. $\int (2x^3 + 1)\, dx$ **2.** $\int (x^{1/2} + 3x - 4)\, dx$ **3.** $\int \frac{1}{x^2}\, dx$

4. $\int \frac{1}{3t^3}\, dt$ **5.** $\int (1 + 2t)t^{3/2}\, dt$ **6.** $\int \sqrt{x}(2x - 1)\, dx$

7. $\int \frac{5x^3 + 2}{x^2}\, dx$ **8.** $\int \frac{2x^2 - 5}{x^4}\, dx$ **9.** $\int \frac{8x^2 + 3}{\sqrt{x}}\, dx$

Exercises 5.2

See www.CalcChat.com for worked-out solutions to odd-numbered exercises.

Finding *u* and *du*/*dx* In Exercises 1–8, identify u and du/dx for the integral $\int u^n(du/dx)\, dx$.

1. $\int (5x^2 + 1)^2(10x)\, dx$ **2.** $\int (3 - 4x^2)^3(-8x)\, dx$

3. $\int \sqrt{1 - x^2}(-2x)\, dx$ **4.** $\int 3x^2\sqrt{x^3 + 1}\, dx$

5. $\int \left(4 + \frac{1}{x^2}\right)^5\left(\frac{-2}{x^3}\right) dx$ **6.** $\int \frac{1}{(1 + 2x)^2}(2)\, dx$

7. $\int (1 + \sqrt{x})^3\left(\frac{1}{2\sqrt{x}}\right) dx$ **8.** $\int (4 - \sqrt{x})^2\left(\frac{-1}{2\sqrt{x}}\right) dx$

Applying the General Power Rule In Exercises 9–34, find the indefinite integral. Check your result by differentiating. *See Examples 1, 2, 3, and 5.*

9. $\int (x - 1)^4\, dx$ **10.** $\int (x - 3)^{5/2}\, dx$

11. $\int (1 + 2x)^4(2)\, dx$ **12.** $\int (x^2 - 1)^3(2x)\, dx$

13. $\int (x^2 + 3x)(2x + 3)\, dx$

14. $\int (x^3 + 6x)^2(3x^2 + 6)\, dx$

15. $\int \sqrt{4x^2 - 5}\,(8x)\, dx$

16. $\int \sqrt[3]{1 - 2x^2}(-4x)\, dx$

17. $\int \frac{6x}{(3x^2 - 5)^4}\, dx$ **18.** $\int \frac{-12x^2}{(1 - 4x^3)^2}\, dx$

19. $\int x^2(2x^3 - 1)^4\, dx$ **20.** $\int x(1 - 2x^2)^3\, dx$

21. $\int t\sqrt{t^2 + 6}\, dt$ **22.** $\int t^4\sqrt[3]{t^5 - 9}\, dt$

23. $\int \frac{x^5}{(4 - x^6)^3}\, dx$ **24.** $\int \frac{x^2}{(x^3 - 1)^2}\, dx$

25. $\int (x^2 - 6x)^4(x - 3)\, dx$

26. $\int (4x^3 + 8x)^3(3x^2 + 2)\, dx$

27. $\int \frac{x + 1}{(x^2 + 2x - 3)^2}\, dx$ **28.** $\int \frac{x - 2}{\sqrt{x^2 - 4x + 3}}\, dx$

29. $\int 5x\sqrt[3]{1 - x^2}\, dx$ **30.** $\int 9x^3\sqrt{x^4 + 2}\, dx$

31. $\int \frac{6x}{(1 + x^2)^3}\, dx$ **32.** $\int \frac{4x + 6}{(x^2 + 3x + 7)^3}\, dx$

33. $\int \frac{-3}{\sqrt{2t + 3}}\, dt$ **34.** $\int \frac{3x^2}{\sqrt{1 - x^3}}\, dx$

Integration by Substitution In Exercises 35–42, use formal substitution to find the indefinite integral. Check your result by differentiating. *See Examples 6 and 7.*

35. $\int 12x(6x^2 - 1)^3\, dx$ **36.** $\int 3x^2(1 - x^3)^2\, dx$

37. $\int \sqrt[3]{4x + 3}\, dx$ **38.** $\int t\sqrt{t^2 + 1}\, dt$

39. $\int \frac{x}{\sqrt{x^2 + 25}}\, dx$ **40.** $\int \frac{3}{\sqrt{2x + 1}}\, dx$

41. $\int \frac{x^2 + 1}{\sqrt{x^3 + 3x + 4}}\, dx$ **42.** $\int \frac{x^2 + 3}{\sqrt[3]{x^3 + 9x}}\, dx$

Comparing Methods In Exercises 43–46, (a) perform the integration in two ways: once using the Simple Power Rule and once using the General Power Rule. (b) Explain the difference in the results. (c) Which method do you prefer? Explain your reasoning.

43. $\int (x - 1)^2\, dx$ **44.** $\int (3 - x)^2\, dx$

45. $\int x(x^2 - 1)^2\, dx$ **46.** $\int x(2x^2 + 1)^2\, dx$

47. Finding an Equation of a Function Find the equation of the function f whose graph passes through the point $(2, 10)$ and whose derivative is

$f'(x) = 2x(4x^2 - 10)^2.$

48. Finding an Equation of a Function Find the equation of the function f whose graph passes through the point $\left(0, \frac{7}{3}\right)$ and whose derivative is

$f'(x) = x\sqrt{1 - x^2}.$

49. Cost The marginal cost of a product is modeled by

$$\frac{dC}{dx} = \frac{4}{\sqrt{x + 1}}$$

where x is the number of units. When $x = 15$, $C = 50$.

(a) Find the cost function.

(b) Find the cost of producing 50 units.

50. Cost The marginal cost of a product is modeled by

$$\frac{dC}{dx} = \frac{12}{\sqrt[3]{12x + 1}}$$

where x is the number of units. When $x = 13$, $C = 100$.

(a) Find the cost function.

(b) Find the cost of producing 30 units.

Supply In Exercises 51 and 52, find the supply function $x = f(p)$ that satisfies the initial conditions.

51. $\dfrac{dx}{dp} = p\sqrt{p^2 - 25}$

$x = 600$ when $p = \$13$

52. $\dfrac{dx}{dp} = \dfrac{10}{\sqrt{p - 3}}$

$x = 100$ when $p = \$3$

Demand In Exercises 53 and 54, find the demand function $x = f(p)$ that satisfies the initial conditions.

53. $\dfrac{dx}{dp} = -\dfrac{6000p}{(p^2 - 16)^{3/2}}$

$x = 5000$ when $p = \$5$

54. $\dfrac{dx}{dp} = -\dfrac{400}{(0.02p - 1)^3}$

$x = 10{,}000$ when $p = \$100$

55. Gardening An evergreen nursery usually sells a type of shrub after 5 years of growth and shaping. The growth rate during those 5 years is approximated by

$$\frac{dh}{dt} = \frac{17.6t}{\sqrt{17.6t^2 + 1}}$$

where t is the time (in years) and h is the height (in inches). The seedlings are 6 inches tall when planted $(t = 0)$.

(a) Find the height function.

(b) How tall are the shrubs when they are sold?

56. HOW DO YOU SEE IT? The graph shows the rate of change of the revenue of a company from 1990 through 2010.

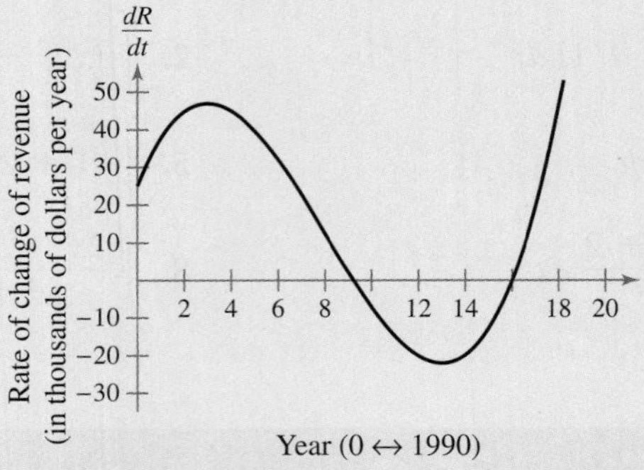

(a) Approximate the rate of change of the revenue in 2007. Explain your reasoning.

(b) Is $R(7) - R(6) > 0$? Explain your reasoning.

(c) Approximate the years in which the graph of the revenue is concave upward and the years in which it is concave downward. Approximate the years of any points of inflection.

Marginal Propensity to Consume In Exercises 57 and 58, (a) use the marginal propensity to consume, dQ/dx, to write Q as a function of x, where x is the income (in dollars) and Q is the income consumed (in dollars). Assume that families who have annual incomes of \$25,000 or less consume 100% of their income. (b) Use the result of part (a) and a spreadsheet to complete the table showing the income consumed and the income saved, $x - Q$, for various incomes. (c) Use a graphing utility to represent graphically the income consumed and saved. *See Example 8.*

x	25,000	50,000	100,000	150,000
Q				
$x - Q$				

57. $\dfrac{dQ}{dx} = \dfrac{0.95}{(x - 24{,}999)^{0.05}}, \quad x \geq 25{,}000$

58. $\dfrac{dQ}{dx} = \dfrac{0.93}{(x - 24{,}999)^{0.07}}, \quad x \geq 25{,}000$

Integration Using Technology In Exercises 59 and 60, use a symbolic integration utility to find the indefinite integral. Verify the result by differentiating.

59. $\displaystyle\int \frac{1}{\sqrt{x} + \sqrt{x + 1}}\, dx$

60. $\displaystyle\int \frac{x}{\sqrt{3x + 2}}\, dx$

5.3 Exponential and Logarithmic Integrals

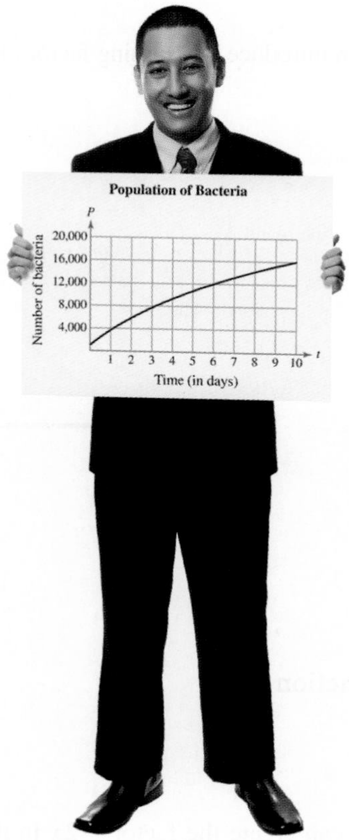

In Exercise 51 on page 337, you will use integration to find a model for a population of bacteria.

- Use the Exponential Rule to find indefinite integrals.
- Use the Log Rule to find indefinite integrals.

Using the Exponential Rule

Each of the differentiation rules for exponential functions has a corresponding integration rule.

> **Integrals of Exponential Functions**
>
> Let u be a differentiable function of x.
>
> $$\int e^x\,dx = e^x + C \qquad \text{Simple Exponential Rule}$$
>
> $$\int e^u \frac{du}{dx}\,dx = \int e^u\,du = e^u + C \qquad \text{General Exponential Rule}$$

Example 1 **Integrating Exponential Functions**

Find each indefinite integral.

a. $\int 2e^x\,dx$ **b.** $\int 2e^{2x}\,dx$ **c.** $\int (e^x + x)\,dx$

SOLUTION

a. $\int 2e^x\,dx = 2\int e^x\,dx$ Constant Multiple Rule

$= 2e^x + C$ Simple Exponential Rule

b. $\int 2e^{2x}\,dx = \int e^{2x}(2)\,dx$ Let $u = 2x$, then $\frac{du}{dx} = 2$.

$= \int e^u \frac{du}{dx}\,dx$ Substitute u and $\frac{du}{dx}$.

$= e^u + C$ General Exponential Rule

$= e^{2x} + C$ Substitute for u.

c. $\int (e^x + x)\,dx = \int e^x\,dx + \int x\,dx$ Sum Rule

$= e^x + \frac{x^2}{2} + C$ Simple Exponential and Power Rules

You can check each of these results by differentiating. For instance, in part (a),

$$\frac{d}{dx}[2e^x + C] = 2e^x.$$

✓ Checkpoint 1

Find each indefinite integral.

a. $\int 3e^x\,dx$ **b.** $\int 5e^{5x}\,dx$ **c.** $\int (e^x - x)\,dx$

leolintang/Shutterstock.com

TECH TUTOR

If you use a symbolic integration utility to find antiderivatives of exponential or logarithmic functions, you can easily obtain results that are beyond the scope of this course. For instance, the antiderivative of e^{x^2} involves the imaginary unit i and the probability function called "ERF." In this course, you are not expected to interpret or use such results.

Example 2 Integrating an Exponential Function

Find $\int e^{3x+1}\,dx$.

SOLUTION Let $u = 3x + 1$; then $du/dx = 3$. You can introduce the missing factor of 3 in the integrand by multiplying and dividing by 3.

$$\int e^{3x+1}\,dx = \frac{1}{3}\int e^{3x+1}(3)\,dx \qquad \text{Multiply and divide by 3.}$$

$$= \frac{1}{3}\int e^u \frac{du}{dx}\,dx \qquad \text{Substitute } u \text{ and } \frac{du}{dx}.$$

$$= \frac{1}{3}e^u + C \qquad \text{General Exponential Rule}$$

$$= \frac{1}{3}e^{3x+1} + C \qquad \text{Substitute for } u.$$

✓Checkpoint 2

Find $\int e^{2x+3}\,dx$. ■

Example 3 Integrating an Exponential Function

Find $\int 5xe^{-x^2}\,dx$.

SOLUTION Let $u = -x^2$; then $du/dx = -2x$. You can create the factor $-2x$ in the integrand by multiplying and dividing by -2.

$$\int 5xe^{-x^2}\,dx = \int \left(-\frac{5}{2}\right)e^{-x^2}(-2x)\,dx \qquad \text{Multiply and divide by } -2.$$

$$= -\frac{5}{2}\int e^{-x^2}(-2x)\,dx \qquad \text{Factor } -\tfrac{5}{2} \text{ out of the integrand.}$$

$$= -\frac{5}{2}\int e^u \frac{du}{dx}\,dx \qquad \text{Substitute } u \text{ and } \frac{du}{dx}.$$

$$= -\frac{5}{2}e^u + C \qquad \text{General Exponential Rule}$$

$$= -\frac{5}{2}e^{-x^2} + C \qquad \text{Substitute for } u.$$

ALGEBRA TUTOR

For help on the algebra in Example 3, see Example 1(d) in the *Chapter 5 Algebra Tutor*, on page 366.

✓Checkpoint 3

Find $\int 4xe^{x^2}\,dx$. ■

Remember that you cannot introduce a missing *variable* in the integrand. For instance, you cannot find $\int e^{x^2}\,dx$ by multiplying and dividing by $2x$ and then factoring $1/(2x)$ out of the integrand. That is,

$$\int e^{x^2}\,dx \neq \frac{1}{2x}\int e^{x^2}(2x)\,dx.$$

Using the Log Rule

When the Power Rules for integration were introduced in Sections 5.1 and 5.2, you saw that they work for powers other than $n = -1$.

$$\int x^n\,dx = \frac{x^{n+1}}{n+1} + C, \quad n \neq -1 \qquad \text{Simple Power Rule}$$

$$\int u^n \frac{du}{dx}\,dx = \int u^n\,du = \frac{u^{n+1}}{n+1} + C, \quad n \neq -1 \qquad \text{General Power Rule}$$

The Log Rule for integration allows you to integrate functions of the form $\int x^{-1}\,dx$ and $\int u^{-1}\,du$.

STUDY TIP

Notice the absolute values in the Log Rule. For those special cases in which u or x cannot be negative, you can omit the absolute value. For instance, in Example 4(b), it is not necessary to write the antiderivative as $\ln|x^2| + C$ because x^2 cannot be negative.

Log Rule for Integration

Let u be a differentiable function of x.

$$\int \frac{1}{x}\,dx = \ln|x| + C \qquad \text{Simple Log Rule}$$

$$\int \frac{du/dx}{u}\,dx = \int \frac{1}{u}\,du = \ln|u| + C \qquad \text{General Log Rule}$$

You can verify each of these rules by differentiating. For instance, to verify that $d/dx[\ln|x|] = 1/x$, notice that

$$\frac{d}{dx}[\ln x] = \frac{1}{x} \quad \text{and} \quad \frac{d}{dx}[\ln(-x)] = \frac{-1}{-x} = \frac{1}{x}.$$

Example 4 Using the Log Rule for Integration

Find each indefinite integral.

a. $\int \frac{4}{x}\,dx$ **b.** $\int \frac{2x}{x^2}\,dx$ **c.** $\int \frac{3}{3x+1}\,dx$

SOLUTION

a. $\int \frac{4}{x}\,dx = 4\int \frac{1}{x}\,dx$ — Constant Multiple Rule

$= 4\ln|x| + C$ — Simple Log Rule

b. $\int \frac{2x}{x^2}\,dx = \int \frac{du/dx}{u}\,dx$ — Let $u = x^2$; then $\frac{du}{dx} = 2x$.

$= \ln|u| + C$ — General Log Rule

$= \ln x^2 + C$ — Substitute for u.

c. $\int \frac{3}{3x+1}\,dx = \int \frac{du/dx}{u}\,dx$ — Let $u = 3x + 1$; then $\frac{du}{dx} = 3$.

$= \ln|u| + C$ — General Log Rule

$= \ln|3x + 1| + C$ — Substitute for u.

✓ Checkpoint 4

Find each indefinite integral.

a. $\int \frac{2}{x}\,dx$ **b.** $\int \frac{3x^2}{x^3}\,dx$ **c.** $\int \frac{2}{2x+1}\,dx$ ■

Yuri Arcurs/www.shutterstock.com

Example 5 **Using the Log Rule for Integration**

Find $\int \frac{1}{2x - 1}\,dx$.

SOLUTION Let $u = 2x - 1$; then $du/dx = 2$. You can create the necessary factor of 2 in the integrand by multiplying and dividing by 2.

$$\int \frac{1}{2x-1}\,dx = \frac{1}{2}\int \frac{2}{2x-1}\,dx \quad \text{Multiply and divide by 2.}$$

$$= \frac{1}{2}\int \frac{du/dx}{u}\,dx \quad \text{Substitute } u \text{ and } \frac{du}{dx}.$$

$$= \frac{1}{2}\ln|u| + C \quad \text{General Log Rule}$$

$$= \frac{1}{2}\ln|2x - 1| + C \quad \text{Substitute for } u.$$

✓Checkpoint 5

Find $\int \frac{1}{4x + 1}\,dx$. ■

Example 6 **Using the Log Rule for Integration**

Find $\int \frac{6x}{x^2 + 1}\,dx$.

SOLUTION Let $u = x^2 + 1$; then

$$\frac{du}{dx} = 2x.$$

You can create the necessary factor of $2x$ in the integrand by factoring a 3 out of the integrand.

$$\int \frac{6x}{x^2+1}\,dx = 3\int \frac{2x}{x^2+1}\,dx \quad \text{Factor 3 out of integrand.}$$

$$= 3\int \frac{du/dx}{u}\,dx \quad \text{Substitute } u \text{ and } \frac{du}{dx}.$$

$$= 3\ln|u| + C \quad \text{General Log Rule}$$

$$= 3\ln(x^2 + 1) + C \quad \text{Substitute for } u.$$

✓Checkpoint 6

Find $\int \frac{3x}{x^2 + 4}\,dx$. ■

ALGEBRA TUTOR *xy*

For help on the algebra at the right, see Example 2(d) in the *Chapter 5 Algebra Tutor*, on page 367.

Integrals to which the Log Rule can be applied are often given in disguised form. For instance, when a rational function has a numerator of degree greater than or equal to that of the denominator, you should use long division to rewrite the integrand. Here is an example.

$$\int \frac{x^2 + 6x + 1}{x^2 + 1}\,dx = \int \left(1 + \frac{6x}{x^2 + 1}\right)dx$$

$$= x + 3\ln(x^2 + 1) + C$$

The next example summarizes some additional situations in which it is helpful to rewrite the integrand in order to recognize the antiderivative.

ALGEBRA TUTOR

For help on the algebra in Example 7, see Example 2(a)–(c) in the *Chapter 5 Algebra Tutor*, on page 367.

Example 7 Rewriting Before Integrating

Find each indefinite integral.

a. $\displaystyle\int \frac{3x^2 + 2x - 1}{x^2}\,dx$ **b.** $\displaystyle\int \frac{1}{1 + e^{-x}}\,dx$ **c.** $\displaystyle\int \frac{x^2 + x + 1}{x - 1}\,dx$

SOLUTION

a. Begin by rewriting the integrand as the sum of three fractions.

$$\int \frac{3x^2 + 2x - 1}{x^2}\,dx = \int \left(\frac{3x^2}{x^2} + \frac{2x}{x^2} - \frac{1}{x^2}\right) dx$$
$$= \int \left(3 + \frac{2}{x} - \frac{1}{x^2}\right) dx$$
$$= 3x + 2\ln|x| + \frac{1}{x} + C$$

b. Begin by rewriting the integrand by multiplying and dividing by e^x.

$$\int \frac{1}{1 + e^{-x}}\,dx = \int \left(\frac{e^x}{e^x}\right)\frac{1}{1 + e^{-x}}\,dx$$
$$= \int \frac{e^x}{e^x + 1}\,dx$$
$$= \ln(e^x + 1) + C$$

c. Begin by dividing the numerator by the denominator.

$$\int \frac{x^2 + x + 1}{x - 1}\,dx = \int \left(x + 2 + \frac{3}{x - 1}\right) dx$$
$$= \frac{x^2}{2} + 2x + 3\ln|x - 1| + C$$

✓Checkpoint 7

Find each indefinite integral.

a. $\displaystyle\int \frac{4x^2 - 3x + 2}{x^2}\,dx$ **b.** $\displaystyle\int \frac{2}{e^{-x} + 1}\,dx$ **c.** $\displaystyle\int \frac{x^2 + 2x + 4}{x + 1}\,dx$

SUMMARIZE (Section 5.3)

1. State the Simple Exponential Rule *(page 331)*. For an example of the Simple Exponential Rule, see Example 1.
2. State the General Exponential Rule *(page 331)*. For examples of the General Exponential Rule, see Examples 2 and 3.
3. State the Simple Log Rule *(page 333)*. For an example of the Simple Log Rule, see Example 4.
4. State the General Log Rule *(page 333)*. For examples of the General Log Rule, see Examples 5 and 6.

SKILLS WARM UP 5.3

The following warm-up exercises involve skills that were covered in earlier sections. You will use these skills in the exercise set for this section. For additional help, review Section 5.1.

In Exercises 1–4, use long division to rewrite the quotient.

1. $\dfrac{x^2 + 4x + 2}{x + 2}$

2. $\dfrac{x^2 - 6x + 9}{x - 4}$

3. $\dfrac{x^3 + 4x^2 - 30x - 4}{x^2 - 4x}$

4. $\dfrac{x^4 - x^3 + x^2 + 15x + 2}{x^2 + 5}$

In Exercises 5–8, find the indefinite integral.

5. $\displaystyle\int \left(x^3 + \frac{1}{x^2}\right) dx$

6. $\displaystyle\int \frac{x^2 + 2x}{x}\, dx$

7. $\displaystyle\int \frac{x^3 + 4}{x^2}\, dx$

8. $\displaystyle\int \frac{x + 3}{x^3}\, dx$

Exercises 5.3

See www.CalcChat.com for worked-out solutions to odd-numbered exercises.

Integrating Exponential Functions In Exercises 1–12, use the Exponential Rule to find the indefinite integral. *See Examples 1, 2, and 3.*

1. $\displaystyle\int 2e^{2x}\, dx$

2. $\displaystyle\int -3e^{-3x}\, dx$

3. $\displaystyle\int e^{4x}\, dx$

4. $\displaystyle\int e^{-0.25x}\, dx$

5. $\displaystyle\int e^{5x-3}\, dx$

6. $\displaystyle\int e^{-x-1}\, dx$

7. $\displaystyle\int 9xe^{-x^2}\, dx$

8. $\displaystyle\int 3xe^{0.5x^2}\, dx$

9. $\displaystyle\int 5x^2e^{x^3}\, dx$

10. $\displaystyle\int -3x^3e^{-2x^4}\, dx$

11. $\displaystyle\int (2x + 1)e^{x^2+x}\, dx$

12. $\displaystyle\int (x - 4)e^{x^2-8x}\, dx$

Using the Log Rule for Integration In Exercises 13–30, use the Log Rule to find the indefinite integral. *See Examples 4, 5, and 6.*

13. $\displaystyle\int \frac{1}{x + 1}\, dx$

14. $\displaystyle\int \frac{1}{x - 5}\, dx$

15. $\displaystyle\int \frac{5}{5x + 2}\, dx$

16. $\displaystyle\int \frac{4}{4x - 7}\, dx$

17. $\displaystyle\int \frac{1}{3 - 2x}\, dx$

18. $\displaystyle\int \frac{1}{6x - 5}\, dx$

19. $\displaystyle\int \frac{2}{3x + 5}\, dx$

20. $\displaystyle\int \frac{5}{2x - 1}\, dx$

21. $\displaystyle\int \frac{x}{x^2 + 1}\, dx$

22. $\displaystyle\int \frac{x^2}{3 - x^3}\, dx$

23. $\displaystyle\int \frac{x^2}{x^3 + 1}\, dx$

24. $\displaystyle\int \frac{x}{x^2 + 4}\, dx$

25. $\displaystyle\int \frac{x + 3}{x^2 + 6x + 7}\, dx$

26. $\displaystyle\int \frac{x^2 + 2x + 3}{x^3 + 3x^2 + 9x + 1}\, dx$

27. $\displaystyle\int \frac{1}{x \ln x}\, dx$

28. $\displaystyle\int \frac{1}{x(\ln x)^2}\, dx$

29. $\displaystyle\int \frac{e^{-x}}{1 - e^{-x}}\, dx$

30. $\displaystyle\int \frac{e^x}{1 + e^x}\, dx$

Finding Indefinite Integrals In Exercises 31–46, use any basic integration formula or formulas to find the indefinite integral. State which integration formula(s) you used to find the integral.

31. $\displaystyle\int \frac{x^3 - 8x}{2x^2}\, dx$

32. $\displaystyle\int \frac{x - 1}{4x}\, dx$

33. $\displaystyle\int \frac{8x^3 + 3x^2 + 6}{x^3}\, dx$

34. $\displaystyle\int \frac{2x^3 - 6x^2 - 5x}{x^2}\, dx$

35. $\displaystyle\int \frac{e^{2x} + 2e^x + 1}{e^x}\, dx$

36. $\displaystyle\int \frac{e^{5x} - 3e^{3x} + e^x}{e^{3x}}\, dx$

37. $\displaystyle\int e^x\sqrt{1 - e^x}\, dx$

38. $\displaystyle\int (6x + e^x)\sqrt{3x^2 + e^x}\, dx$

39. $\displaystyle\int \frac{1 + e^{-x}}{1 + xe^{-x}}\, dx$

40. $\displaystyle\int \frac{2(e^x - e^{-x})}{(e^x + e^{-x})^2}\, dx$

41. $\displaystyle\int \frac{5}{e^{-5x} + 7}\, dx$

42. $\displaystyle\int \frac{3}{1 + e^{-3x}}\, dx$

43. $\displaystyle\int \frac{x^2 + 2x + 5}{x - 1}\,dx$

44. $\displaystyle\int \frac{x^3 - 36x + 3}{x + 6}\,dx$

45. $\displaystyle\int \frac{x - 3}{x + 3}\,dx$

46. $\displaystyle\int \frac{x^2 + x + 1}{x^2 + 1}\,dx$

Finding an Equation of a Function In Exercises 47–50, find the equation of the function f whose graph passes through the given point.

47. $f'(x) = \dfrac{1}{x^2}e^{2/x};\ (4, 6)$

48. $f'(x) = \dfrac{2}{1 + e^{-x}};\ (0, 3)$

49. $f'(x) = \dfrac{x^2 + 4x + 3}{x - 1};\ (2, 4)$

50. $f'(x) = \dfrac{x^3 - 4x^2 + 3}{x - 3};\ (4, -1)$

51. Biology A population P of bacteria is growing at the rate of

$$\frac{dP}{dt} = \frac{3000}{1 + 0.25t}$$

where t is the time (in days). When $t = 0$, the population is 1000.

(a) Find a model for the population.

(b) What is the population after 3 days?

(c) After how many days will the population be 12,000?

52. Biology Because of an insufficient oxygen supply, the trout population P in a lake is dying. The population's rate of change can be modeled by

$$\frac{dP}{dt} = -125e^{-t/20}$$

where t is the time (in days). When $t = 0$, the population is 2500.

(a) Find a model for the population.

(b) What is the population after 15 days?

(c) How long will it take for the entire trout population to die?

53. Demand The marginal price for the demand of a product can be modeled by

$$\frac{dp}{dx} = 0.1e^{-x/500}$$

where x is the quantity demanded. When the demand is 600 units, the price p is \$30.

(a) Find the demand function.

(b) Use a graphing utility to graph the demand function. Does price increase or decrease as demand increases?

(c) Use the *zoom* and *trace* features of the graphing utility to find the quantity demanded when the price is \$22.

54. HOW DO YOU SEE IT? The graph shows the rate of change of the revenue of a company from 1990 through 2010.

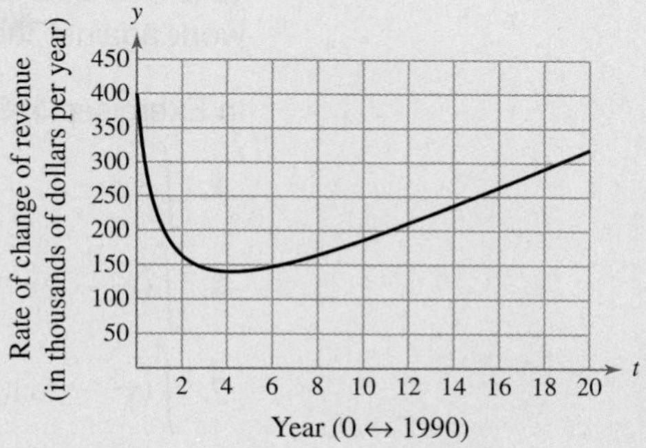

(a) Approximate the rate of change of the revenue in 2009. Explain your reasoning.

(b) Approximate the year when the rate of change of the revenue is the greatest. Explain your reasoning.

(c) Approximate the year when the revenue is maximum. Explain your reasoning.

55. Revenue The rate of change in revenue for Cablevision from 2002 through 2009 can be modeled by

$$\frac{dR}{dt} = 320.1e^{0.0993t}$$

where R is the revenue (in millions of dollars) and t is the time (in years), with $t = 2$ corresponding to 2002. In 2007, the revenue for Cablevision was \$6484.5 million. *(Source: Cablevision Systems Corporation)*

(a) Find a model for the revenue of Cablevision.

(b) Find Cablevision's revenue in 2009.

56. Revenue The rate of change in revenue for Under Armour from 2004 through 2009 can be modeled by

$$\frac{dR}{dt} = 13.897t + \frac{284.653}{t}$$

where R is the revenue (in millions of dollars) and t is the time (in years), with $t = 4$ corresponding to 2004. In 2008, the revenue for Under Armour was \$725.2 million. *(Source: Under Armour, Inc.)*

(a) Find a model for the revenue of Under Armour.

(b) Find Under Armour's revenue in 2006.

True or False In Exercises 57 and 58, determine whether the statement is true or false. If it is false, explain why or give an example that shows it is false.

57. $(\ln x)^{1/2} = \frac{1}{2}(\ln x)$

58. $\displaystyle\int \frac{1}{x}\,dx = \ln|ax| + C,\ a \neq 0$

QUIZ YOURSELF

See www.CalcChat.com for worked-out solutions to odd-numbered exercises.

Take this quiz as you would take a quiz in class. When you are done, check your work against the answers given in the back of the book.

In Exercises 1–9, find the indefinite integral. Check your result by differentiation.

1. $\int 3\,dx$ **2.** $\int 10x\,dx$ **3.** $\int \frac{1}{x^5}\,dx$

4. $\int (x^2 - 2x + 15)\,dx$ **5.** $\int (6x + 1)^3(6)\,dx$ **6.** $\int x(5x^2 - 2)^4\,dx$

7. $\int (x^2 - 5x)(2x - 5)\,dx$ **8.** $\int \frac{3x^2}{(x^3 + 3)^3}\,dx$ **9.** $\int \sqrt{5x + 2}\,dx$

In Exercises 10 and 11, find the particular solution that satisfies the differential equation and initial condition.

10. $f'(x) = 16x; f(0) = 1$ **11.** $f'(x) = 9x^2 + 4; f(1) = 5$

12. The marginal cost function for producing x units of a product is modeled by

$$\frac{dC}{dx} = 16 - 0.06x.$$

It costs \$25 to produce one unit. Find (a) the cost function C (in dollars), (b) the fixed cost (when $x = 0$), and (c) the total cost of producing 500 units.

13. Find the equation of the function f whose graph passes through the point $(0, 1)$ and whose derivative is

$$f'(x) = 2x^2 + 1.$$

14. The number of bolts B produced by a foundry changes according to the model

$$\frac{dB}{dt} = \frac{250t}{\sqrt{t^2 + 36}}, \quad 0 \le t \le 40$$

where t is the time (in hours). Find the number of bolts produced in (a) 8 hours and (b) 40 hours.

In Exercises 15–17, use the Exponential Rule to find the indefinite integral.

15. $\int 5e^{5x+4}\,dx$ **16.** $\int 3x^2e^{x^3}\,dx$ **17.** $\int (x - 3)e^{x^2-6x}\,dx$

In Exercises 18–20, use the Log Rule to find the indefinite integral.

18. $\int \frac{2}{2x - 1}\,dx$ **19.** $\int \frac{1}{3 - 8x}\,dx$ **20.** $\int \frac{x}{3x^2 + 4}\,dx$

21. The rate of change in sales for Advance Auto Parts from 2001 through 2009 can be modeled by

$$\frac{dS}{dt} = 26.32t + \frac{848.99}{t}$$

where S is the sales (in millions) and t is the time (in years), with $t = 1$ corresponding to 2001. In 2001, the sales for Advance Auto Parts were \$2517.6 million. *(Source: Advance Auto Parts, Inc.)*

(a) Find a model for the sales of Advance Auto Parts.

(b) Find the sales for Advance Auto Parts in 2008.

5.4 Area and the Fundamental Theorem of Calculus

- Understand the relationship between area and definite integrals.
- Evaluate definite integrals using the Fundamental Theorem of Calculus.
- Use definite integrals to solve marginal analysis problems.
- Find the average values of functions over closed intervals.
- Use properties of even and odd functions to help evaluate definite integrals.
- Find the amounts of annuities.

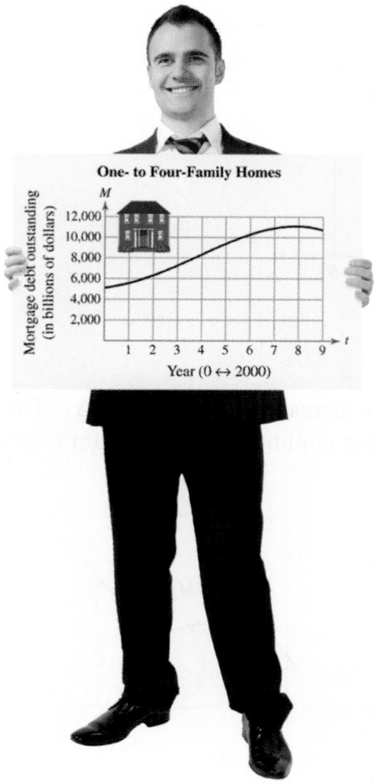

In Exercise 79 on page 350, you will use integration to find a model for the mortgage debt outstanding for one- to four-family homes.

Area and Definite Integrals

From your study of geometry, you know that area is a number that defines the size of a bounded region. For simple regions, such as rectangles, triangles, and circles, area can be found using geometric formulas.

In this section, you will learn how to use calculus to find the areas of nonstandard regions, such as the region R shown in Figure 5.5.

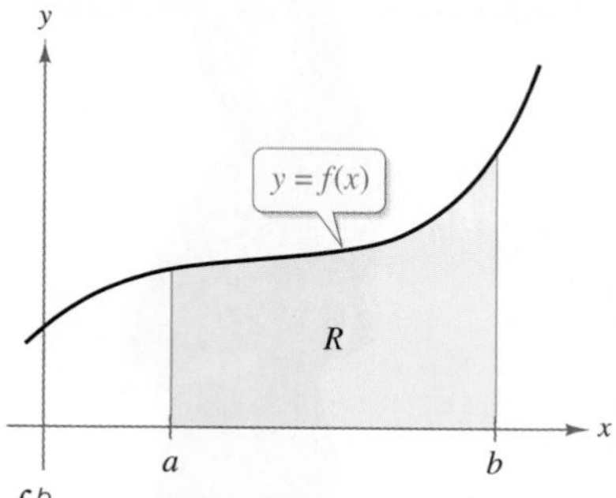

$\int_a^b f(x)\,dx = \text{Area}$

FIGURE 5.5

Definition of a Definite Integral

Let f be nonnegative and continuous on the closed interval $[a, b]$. The area of the region bounded by the graph of f, the x-axis, and the lines $x = a$ and $x = b$ is denoted by

$$\text{Area} = \int_a^b f(x)\,dx.$$

The expression $\int_a^b f(x)\,dx$ is called the **definite integral** from a to b, where a is the **lower limit of integration** and b is the **upper limit of integration.**

Example 1 Evaluating a Definite Integral Using a Geometric Formula

The definite integral

$$\int_0^2 2x\,dx$$

represents the area of the region bounded by the graph of $f(x) = 2x$, the x-axis, and the line $x = 2$, as shown in Figure 5.6. The region is triangular, with a height of 4 units and a base of 2 units. Using the formula for the area of a triangle, you have

$$\int_0^2 2x\,dx = \frac{1}{2}(\text{base})(\text{height}) = \frac{1}{2}(2)(4) = 4.$$

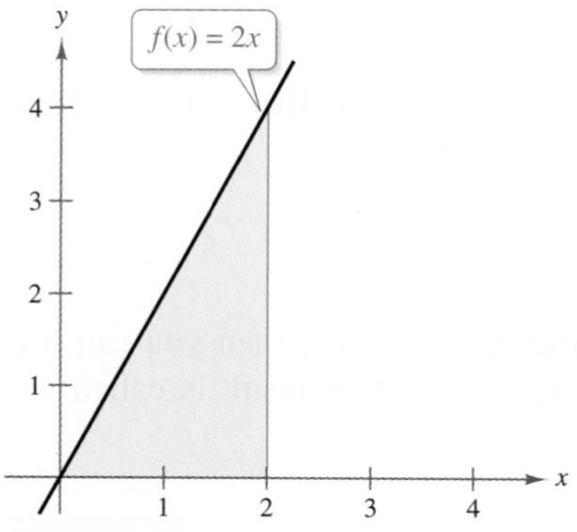

FIGURE 5.6

✓Checkpoint 1

Evaluate the definite integral using a geometric formula. Illustrate your answer with an appropriate sketch.

$$\int_0^3 4x\,dx$$

■

STUDY TIP

There are two basic ways to introduce the Fundamental Theorem of Calculus. One way uses an area function, as shown here. The other uses a summation process, as shown in Appendix B.

The Fundamental Theorem of Calculus

Consider the function A, which denotes the area of the region shown in Figure 5.7.

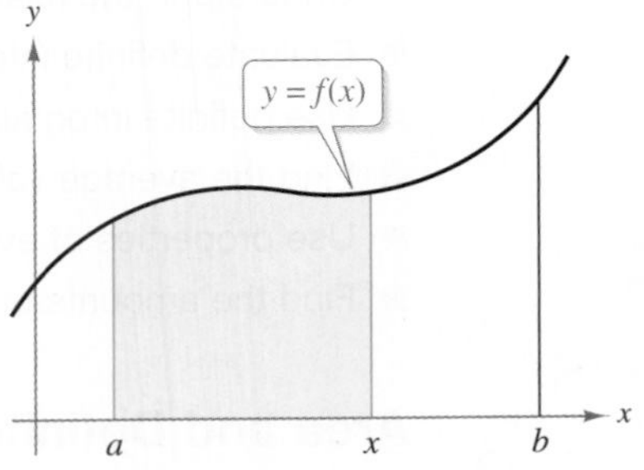

$A(x)$ = Area from a to x

FIGURE 5.7

To discover the relationship between A and f, let x increase by an amount Δx. This increases the area by ΔA. Let $f(m)$ and $f(M)$ denote the minimum and maximum values of f on the interval $[x, x + \Delta x]$.

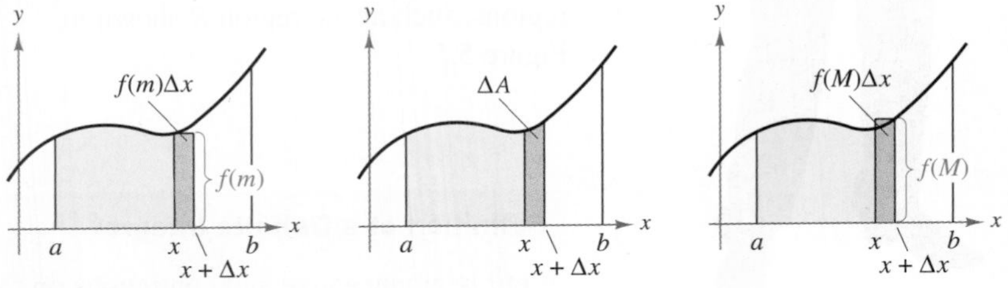

FIGURE 5.8

As indicated in Figure 5.8, you can write the inequality below.

$$f(m)\,\Delta x \le \Delta A \le f(M)\,\Delta x \qquad \text{See Figure 5.8.}$$

$$f(m) \le \frac{\Delta A}{\Delta x} \le f(M) \qquad \text{Divide each term by } \Delta x.$$

$$\lim_{\Delta x \to 0} f(m) \le \lim_{\Delta x \to 0} \frac{\Delta A}{\Delta x} \le \lim_{\Delta x \to 0} f(M) \qquad \text{Take limit of each term.}$$

$$f(x) \le A'(x) \le f(x) \qquad \text{Definition of derivative of } A(x)$$

So, $f(x) = A'(x)$, and $A(x) = F(x) + C$, where $F'(x) = f(x)$. Because $A(a) = 0$, it follows that $C = -F(a)$. So, $A(x) = F(x) - F(a)$, which implies that

$$A(b) = \int_a^b f(x)\,dx = F(b) - F(a).$$

This equation tells you that *if you can find an antiderivative for f,* then you can use the antiderivative to evaluate the definite integral $\int_a^b f(x)\,dx$. This result is called the **Fundamental Theorem of Calculus.**

The Fundamental Theorem of Calculus

If f is nonnegative and continuous on the closed interval $[a, b]$, then

$$\int_a^b f(x)\,dx = F(b) - F(a)$$

where F is any function such that $F'(x) = f(x)$ for all x in $[a, b]$.

Guidelines for Using the Fundamental Theorem of Calculus

1. The Fundamental Theorem of Calculus describes a way of *evaluating* a definite integral, not a procedure for finding antiderivatives.
2. In applying the Fundamental Theorem, it is helpful to use the notation

$$\int_a^b f(x)\,dx = F(x)\Big]_a^b = F(b) - F(a).$$

For instance, to evaluate $\int_1^3 x^3\,dx$, you can write

$$\begin{aligned}\int_1^3 x^3\,dx &= \frac{x^4}{4}\Big]_1^3\\ &= \frac{3^4}{4} - \frac{1^4}{4}\\ &= 20.\end{aligned}$$

3. The constant of integration C can be dropped because

$$\begin{aligned}\int_a^b f(x)\,dx &= \Big[F(x) + C\Big]_a^b\\ &= [F(b) + C] - [F(a) + C]\\ &= F(b) - F(a) + C - C\\ &= F(b) - F(a).\end{aligned}$$

In the development of the Fundamental Theorem of Calculus, f was assumed to be nonnegative on the closed interval $[a, b]$. As such, the definite integral was defined as an area. Now, with the Fundamental Theorem, the definition can be extended to include functions that are negative on all or part of the closed interval $[a, b]$. Specifically, if f is *any* function that is continuous on a closed interval $[a, b]$, then the **definite integral** of $f(x)$ from a to b is defined to be

$$\int_a^b f(x)\,dx = F(b) - F(a)$$

where F is an antiderivative of f. Remember that definite integrals do not necessarily represent areas and can be negative, zero, or positive.

STUDY TIP

Be sure you see the distinction between indefinite and definite integrals. The *indefinite integral*

$$\int f(x)\,dx$$

denotes a *family of functions*, each of which is an antiderivative of f, whereas the *definite integral*

$$\int_a^b f(x)\,dx$$

is a *number*.

Properties of Definite Integrals

Let f and g be continuous on the closed interval $[a, b]$.

1. $\int_a^b kf(x)\,dx = k\int_a^b f(x)\,dx$, k is a constant.
2. $\int_a^b [f(x) \pm g(x)]\,dx = \int_a^b f(x)\,dx \pm \int_a^b g(x)\,dx$
3. $\int_a^b f(x)\,dx = \int_a^c f(x)\,dx + \int_c^b f(x)\,dx, \quad a < c < b$
4. $\int_a^a f(x)\,dx = 0$
5. $\int_a^b f(x)\,dx = -\int_b^a f(x)\,dx$

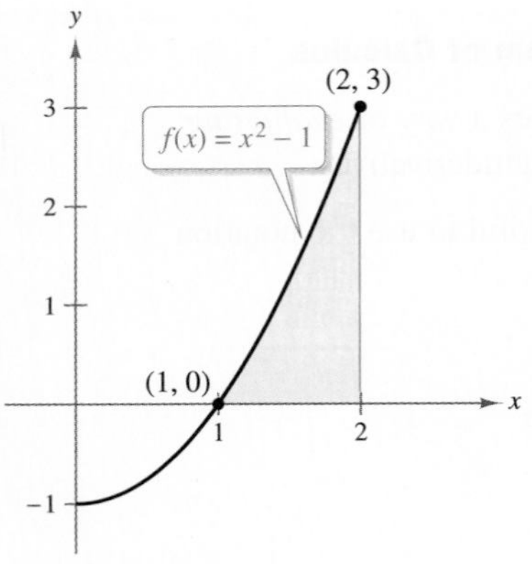

$\text{Area} = \int_1^2 (x^2 - 1)\,dx$

FIGURE 5.9

STUDY TIP

It is easy to make errors in signs when evaluating definite integrals. To avoid such errors, enclose the values of the antiderivative at the upper and lower limits of integration in separate sets of parentheses, as shown in Example 2.

Example 2 Finding Area by the Fundamental Theorem

Find the area of the region bounded by the x-axis and the graph of

$$f(x) = x^2 - 1, \quad 1 \le x \le 2.$$

SOLUTION Note that $f(x) \ge 0$ on the interval $1 \le x \le 2$, as shown in Figure 5.9. So, you can represent the area of the region by a definite integral. To find the area, use the Fundamental Theorem of Calculus.

$$\text{Area} = \int_1^2 (x^2 - 1)\,dx \qquad \text{Definition of definite integral}$$

$$= \left[\frac{x^3}{3} - x\right]_1^2 \qquad \text{Find antiderivative.}$$

$$= \left(\frac{2^3}{3} - 2\right) - \left(\frac{1^3}{3} - 1\right) \qquad \text{Apply Fundamental Theorem.}$$

$$= \frac{2}{3} - \left(-\frac{2}{3}\right)$$

$$= \frac{4}{3} \qquad \text{Simplify.}$$

So, the area of the region is $\frac{4}{3}$ square units.

✓Checkpoint 2

Find the area of the region bounded by the x-axis and the graph of

$$f(x) = x^2 + 1, \quad 2 \le x \le 3.$$

■

Example 3 Evaluating a Definite Integral

Evaluate the definite integral

$$\int_0^1 (4t + 1)^2\,dt$$

and sketch the region whose area is represented by the integral.

SOLUTION

$$\int_0^1 (4t + 1)^2\,dt = \frac{1}{4}\int_0^1 (4t + 1)^2(4)\,dt \qquad \text{Multiply and divide by 4.}$$

$$= \frac{1}{4}\left[\frac{(4t + 1)^3}{3}\right]_0^1 \qquad \text{Find antiderivative.}$$

$$= \frac{1}{4}\left[\left(\frac{5^3}{3}\right) - \left(\frac{1}{3}\right)\right] \qquad \text{Apply Fundamental Theorem.}$$

$$= \frac{1}{4}\left(\frac{124}{3}\right)$$

$$= \frac{31}{3} \qquad \text{Simplify.}$$

The region is shown in Figure 5.10.

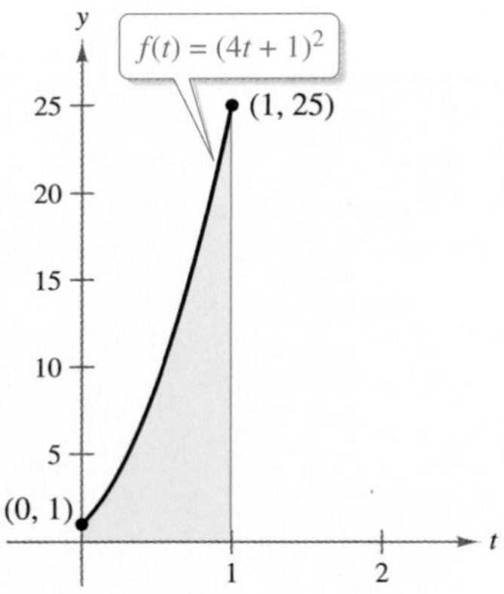

FIGURE 5.10

✓Checkpoint 3

Evaluate $\int_0^1 (2t + 3)^3\,dt$.

■

Example 4 Evaluating Definite Integrals

Evaluate each definite integral.

a. $\int_0^3 e^{2x}\,dx$ **b.** $\int_1^2 \frac{1}{x}\,dx$ **c.** $\int_1^4 -3\sqrt{x}\,dx$

SOLUTION

a. $\int_0^3 e^{2x}\,dx = \frac{1}{2}e^{2x}\Big]_0^3 = \frac{1}{2}(e^6 - e^0) \approx 201.21$

b. $\int_1^2 \frac{1}{x}\,dx = \ln x\Big]_1^2 = \ln 2 - \ln 1 = \ln 2 \approx 0.69$

c.

$$\begin{aligned}\int_1^4 -3\sqrt{x}\,dx &= -3\int_1^4 x^{1/2}\,dx && \text{Rewrite with rational exponent.}\\ &= -3\left[\frac{x^{3/2}}{3/2}\right]_1^4 && \text{Find antiderivative.}\\ &= -2x^{3/2}\Big]_1^4 \\ &= -2(4^{3/2} - 1^{3/2}) && \text{Apply Fundamental Theorem.}\\ &= -2(8-1)\\ &= -14 && \text{Simplify.}\end{aligned}$$

STUDY TIP

In Example 4(c), note that the value of a definite integral can be negative.

✓ Checkpoint 4

Evaluate each definite integral.

a. $\int_0^1 e^{4x}\,dx$ **b.** $\int_2^5 -\frac{1}{x}\,dx$

■

Example 5 Interpreting Absolute Value

Evaluate $\int_0^2 |2x - 1|\,dx$.

SOLUTION The region represented by the definite integral is shown in Figure 5.11. From the definition of absolute value, you can write

$$|2x - 1| = \begin{cases} -(2x-1), & x < \frac{1}{2} \\ 2x - 1, & x \ge \frac{1}{2}. \end{cases}$$

Using Property 3 of definite integrals, rewrite the integral as two definite integrals.

$$\begin{aligned}\int_0^2 |2x-1|\,dx &= \int_0^{1/2} -(2x-1)\,dx + \int_{1/2}^2 (2x-1)\,dx\\ &= \Big[-x^2 + x\Big]_0^{1/2} + \Big[x^2 - x\Big]_{1/2}^2\\ &= \left(-\frac{1}{4} + \frac{1}{2}\right) - (0+0) + (4-2) - \left(\frac{1}{4} - \frac{1}{2}\right)\\ &= \frac{5}{2}\end{aligned}$$

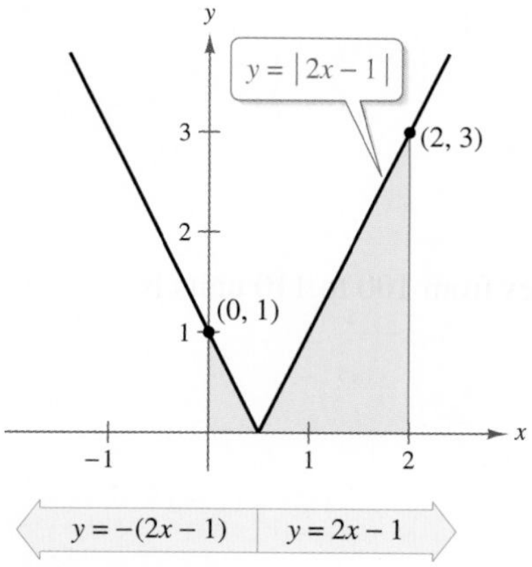

FIGURE 5.11

✓ Checkpoint 5

Evaluate $\int_0^5 |x - 2|\,dx$.

■

TECH TUTOR

Symbolic integration utilities can be used to evaluate definite integrals as well as indefinite integrals. If you have access to such a program, try using it to evaluate several of the definite integrals in this section.

Marginal Analysis

You have already studied *marginal analysis* in the context of derivatives and differentials (Sections 2.3 and 3.8). There, you were given a cost, revenue, or profit function, and you used the derivative to approximate the additional cost, revenue, or profit obtained by selling one additional unit. In this section, you will examine the reverse process. That is, you will be given the marginal cost, marginal revenue, or marginal profit and you will use a definite integral to find the exact increase or decrease in cost, revenue, or profit obtained by selling one or several additional units.

For instance, you are asked to find the additional revenue obtained by increasing sales from x_1 to x_2 units. When you know the revenue function R, you can find the additional revenue by subtracting $R(x_1)$ from $R(x_2)$. When you don't know R, you can use the marginal revenue function dR/dx to find the additional revenue by using a definite integral.

$$\int_{x_1}^{x_2} \frac{dR}{dx}\,dx = R(x_2) - R(x_1)$$

Example 6 Analyzing a Profit Function

The marginal profit for a product is modeled by

$$\frac{dP}{dx} = -0.0005x + 12.2.$$

a. Find the change in profit when sales increase from 100 to 101 units.

b. Find the change in profit when sales increase from 100 to 110 units.

SOLUTION

a. The change in profit obtained by increasing sales from 100 to 101 units is

$$\begin{aligned}\int_{100}^{101} \frac{dP}{dx}\,dx &= \int_{100}^{101} (-0.0005x + 12.2)\,dx\\ &= \Big[-0.00025x^2 + 12.2x\Big]_{100}^{101}\\ &\approx \$12.15.\end{aligned}$$

b. The change in profit obtained by increasing sales from 100 to 110 units is

$$\begin{aligned}\int_{100}^{110} \frac{dP}{dx}\,dx &= \int_{100}^{110} (-0.0005x + 12.2)\,dx\\ &= \Big[-0.00025x^2 + 12.2x\Big]_{100}^{110}\\ &\approx \$121.48.\end{aligned}$$

✓ Checkpoint 6

The marginal profit for a product is modeled by

$$\frac{dP}{dx} = -0.0002x + 14.2.$$

a. Find the change in profit when sales increase from 100 to 101 units.

b. Find the change in profit when sales increase from 100 to 110 units. ■

Average Value

The *average value* of a function on a closed interval is defined below.

> **Definition of the Average Value of a Function**
>
> If f is continuous on $[a, b]$, then the **average value** of f on $[a, b]$ is
>
> $$\text{Average value of } f \text{ on } [a, b] = \frac{1}{b-a}\int_a^b f(x)\,dx.$$

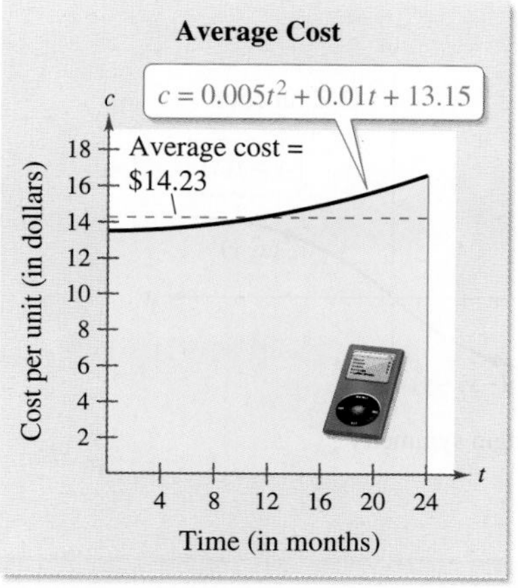

FIGURE 5.12

In Section 3.5, you studied the effects of production levels on cost using an average cost function. In the next example, you will study the effects of time on cost by using integration to find the average cost.

Example 7 Finding the Average Cost

The cost per unit c of producing MP3 players over a two-year period is modeled by

$$c = 0.005t^2 + 0.01t + 13.15, \quad 0 \le t \le 24$$

where t is the time (in months). Approximate the average cost per unit over the two-year period.

SOLUTION The average cost can be found by integrating c over the interval $[0, 24]$.

$$\begin{aligned}\text{Average cost per unit} &= \frac{1}{24}\int_0^{24} (0.005t^2 + 0.01t + 13.15)\,dt \\ &= \frac{1}{24}\left[\frac{0.005t^3}{3} + \frac{0.01t^2}{2} + 13.15t\right]_0^{24} \\ &= \frac{1}{24}(341.52) \\ &= \$14.23\end{aligned}$$

(See Figure 5.12.)

✓ Checkpoint 7

Find the average cost per unit over a two-year period when the cost per unit c of inline skates is given by

$$c = 0.005t^2 + 0.02t + 12.5, \quad 0 \le t \le 24$$

where t is the time (in months).

	A	B
1	t	$c = 0.005t^2 + 0.01t + 13.15$
2	0	13.15
3	1	13.165
4	2	13.19
5	3	13.225
6	4	13.27
7	5	13.325
8	6	13.39
9	7	13.465
10	8	13.55
11	9	13.645
12	10	13.75
13	11	13.865
14	12	13.99
15	13	14.125
16	14	14.27
17	15	14.425
18	16	14.59
19	17	14.765
20	18	14.95
21	19	15.145
22	20	15.35
23	21	15.565
24	22	15.79
25	23	16.025
26	24	16.27
27		
28	**Sum**	356.25
29	**Average**	14.25

You can use a spreadsheet, as shown at the left, to check the reasonableness of the average value found in Example 7. The spreadsheet assumes that one unit is produced each month, beginning with $t = 0$ and ending with $t = 24$. So, when $t = 0$, the cost is

$$\begin{aligned}c &= 0.005(0)^2 + 0.01(0) + 13.15 \\ &= \$13.15\end{aligned}$$

and when $t = 1$ the cost is

$$\begin{aligned}c &= 0.005(1)^2 + 0.01(1) + 13.15 \\ &= \$13.165\end{aligned}$$

and so on. Note in the spreadsheet that the cost increases each month, and the average of the 25 costs is \$14.25. So, you can conclude that the result of Example 7 is reasonable.

Even and Odd Functions

Several common functions have graphs that are symmetric with respect to the y-axis or the origin, as shown in Figure 5.13. If the graph of f is symmetric with respect to the y-axis, as in Figure 5.13(a), then

$$f(-x) = f(x) \qquad \text{Even function}$$

and f is called an **even** function. If the graph of f is symmetric with respect to the origin, as in Figure 5.13(b), then

$$f(-x) = -f(x) \qquad \text{Odd function}$$

and f is called an **odd** function.

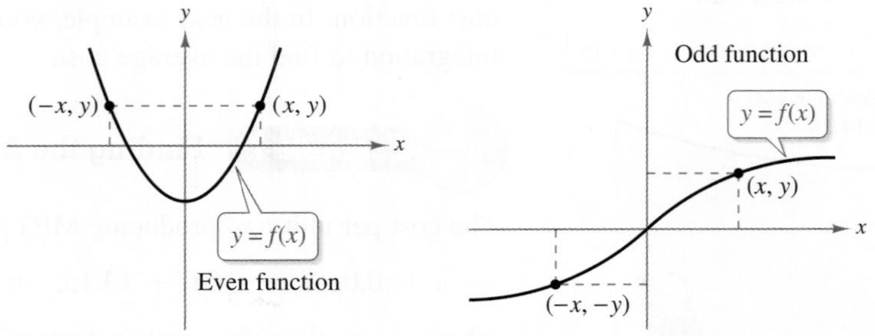

(a) y-axis symmetry (b) Origin symmetry

FIGURE 5.13

Integration of Even and Odd Functions

1. If f is an *even* function, then $\int_{-a}^{a} f(x)\,dx = 2\int_{0}^{a} f(x)\,dx$.

2. If f is an *odd* function, then $\int_{-a}^{a} f(x)\,dx = 0$.

Example 8 Integrating Even and Odd Functions

Evaluate each definite integral.

a. $\int_{-2}^{2} x^2\,dx$ **b.** $\int_{-2}^{2} x^3\,dx$

SOLUTION

a. Because $f(x) = x^2$ is an even function,

$$\int_{-2}^{2} x^2\,dx = 2\int_{0}^{2} x^2\,dx = 2\left[\frac{x^3}{3}\right]_0^2 = 2\left(\frac{8}{3} - 0\right) = \frac{16}{3}.$$

b. Because $f(x) = x^3$ is an odd function,

$$\int_{-2}^{2} x^3\,dx = 0.$$

✓ Checkpoint 8

Evaluate each definite integral.

a. $\int_{-1}^{1} x^4\,dx$ **b.** $\int_{-1}^{1} x^5\,dx$ ■

Annuity

A sequence of equal payments made at regular time intervals over a period of time is called an **annuity.** Some examples of annuities are payroll savings plans, monthly home mortgage payments, and individual retirement accounts. The **amount of an annuity** is the sum of the payments plus the interest earned.

Amount of an Annuity

If c represents a continuous income function in dollars per year (where t is the time in years), r represents the interest rate compounded continuously, and T represents the term of the annuity in years, then the **amount of an annuity** is

$$\text{Amount of an annuity} = e^{rT}\int_0^T c(t)e^{-rt}\,dt.$$

Example 9 Finding the Amount of an Annuity

You deposit \$2000 each year for 15 years in an individual retirement account (IRA) paying 5% interest. How much will you have in your IRA after 15 years?

SOLUTION The income function for your deposit is

$$c(t) = 2000.$$

So, the amount of the annuity after 15 years will be

$$\begin{aligned}\text{Amount of an annuity} &= e^{rT}\int_0^T c(t)e^{-rt}\,dt \\ &= e^{(0.05)(15)}\int_0^{15} 2000e^{-0.05t}\,dt \\ &= 2000e^{0.75}\left[-\frac{e^{-0.05t}}{0.05}\right]_0^{15} \\ &\approx \$44{,}680.00.\end{aligned}$$

✓Checkpoint 9

You deposit \$1000 each year in a savings account paying 4% interest. How much will be in the account after 10 years? ■

SUMMARIZE (Section 5.4)

1. State the definition of a definite integral *(page 339)*. For an example of a definite integral, see Example 1.
2. State the Fundamental Theorem of Calculus *(page 340)*. For examples of the Fundamental Theorem of Calculus, see Examples 2 and 3.
3. State the properties of definite integrals *(page 341)*. For examples of the properties, see Examples 4 and 5.
4. State the definition of the average value of a function *(page 345)*. For an example of finding the average value of a function, see Example 7.
5. State the rules for integrating even and odd functions *(page 346)*. For an example of integrating even and odd functions, see Example 8.

DUSAN ZIDAR/www.shutterstock.com

SKILLS WARM UP 5.4

The following warm-up exercises involve skills that were covered in earlier sections. You will use these skills in the exercise set for this section. For additional help, review Sections 5.1–5.3.

In Exercises 1–4, find the indefinite integral.

1. $\int (3x + 7)\, dx$ **2.** $\int \left(x^{3/2} + 2\sqrt{x}\right) dx$ **3.** $\int \frac{1}{5x}\, dx$ **4.** $\int e^{-6x}\, dx$

In Exercises 5–8, integrate the marginal function.

5. $\frac{dC}{dx} = 0.02x^{3/2} + 29{,}500$ **6.** $\frac{dR}{dx} = 9000 + 2x$

7. $\frac{dP}{dx} = 25{,}000 - 0.01x$ **8.** $\frac{dC}{dx} = 0.03x^2 + 4600$

Exercises 5.4

See www.CalcChat.com for worked-out solutions to odd-numbered exercises.

Evaluating a Definite Integral Using a Geometric Formula In Exercises 1–6, sketch the region whose area is represented by the definite integral. Then use a geometric formula to evaluate the integral. *See Example 1.*

1. $\int_0^2 3\, dx$ **2.** $\int_0^3 4\, dx$

3. $\int_0^4 x\, dx$ **4.** $\int_0^3 \frac{x}{3}\, dx$

5. $\int_{-3}^{3} \sqrt{9 - x^2}\, dx$ **6.** $\int_0^2 \sqrt{4 - x^2}\, dx$

Using Properties of Definite Integrals In Exercises 7 and 8, use the values $\int_0^5 f(x)\, dx = 6$ and $\int_0^5 g(x)\, dx = 2$ to evaluate the definite integral.

7. (a) $\int_0^5 [f(x) + g(x)]\, dx$ (b) $\int_0^5 [f(x) - g(x)]\, dx$

(c) $\int_0^5 -4f(x)\, dx$ (d) $\int_0^5 [f(x) - 3g(x)]\, dx$

8. (a) $\int_0^5 2g(x)\, dx$ (b) $\int_5^0 f(x)\, dx$

(c) $\int_5^5 f(x)\, dx$ (d) $\int_0^5 [f(x) - f(x)]\, dx$

Finding Area by the Fundamental Theorem In Exercises 9–16, find the area of the region. *See Example 2.*

9. $y = x - x^2$

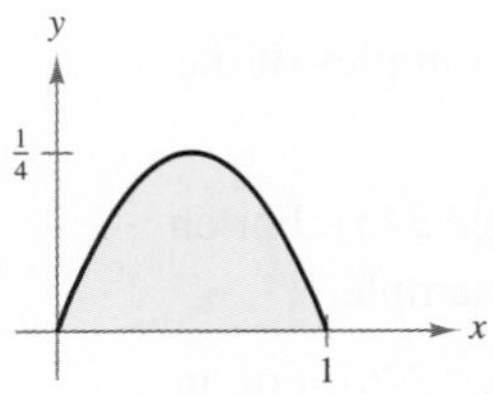

10. $y = 1 - x^4$

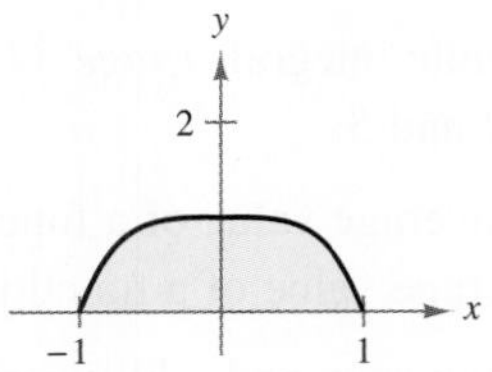

11. $y = \frac{1}{x^2}$

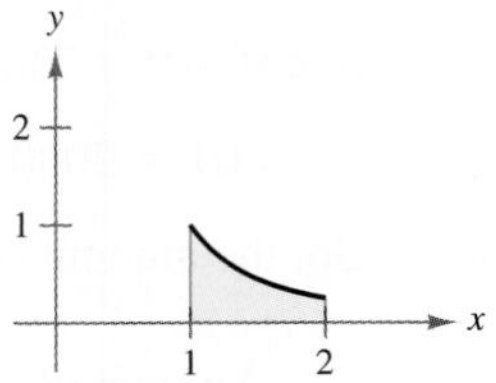

12. $y = \frac{2}{\sqrt{x}}$

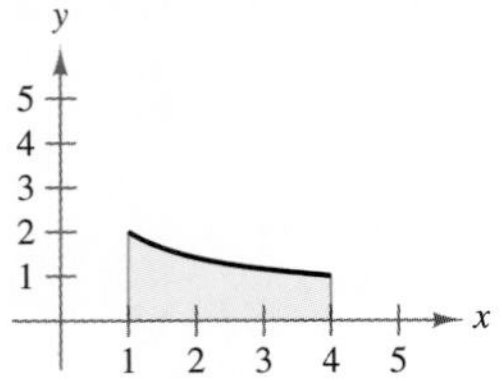

13. $y = 3e^{-x/2}$

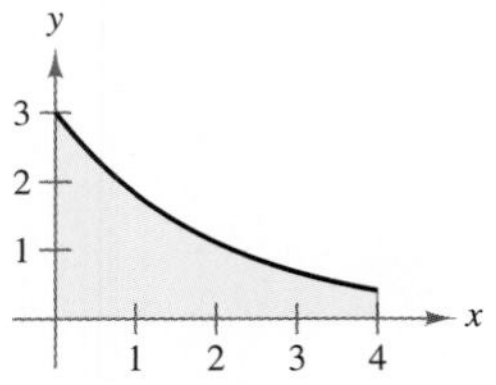

14. $y = 2e^{x/4}$

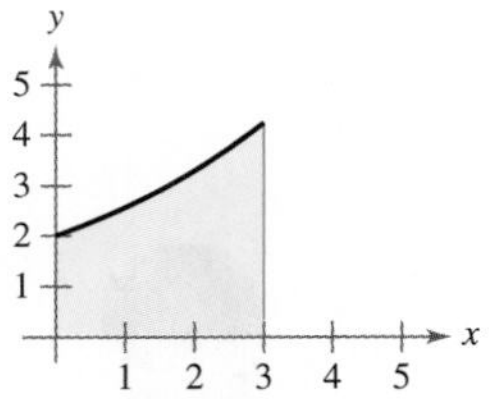

15. $y = \frac{x^2 + 4}{x}$

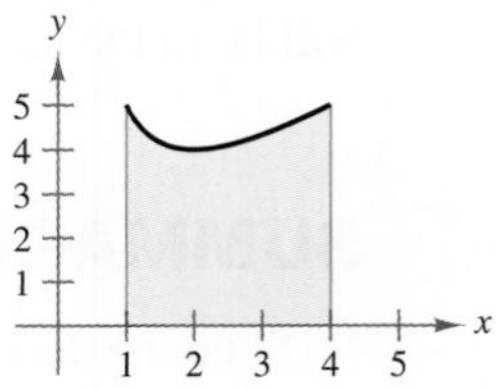

16. $y = \frac{x - 2}{x}$

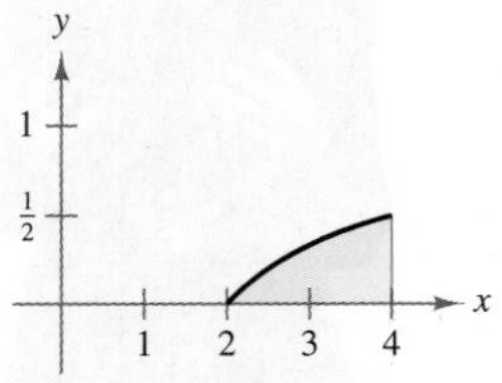

Evaluating a Definite Integral In Exercises 17–38, evaluate the definite integral. *See Examples 3 and 4.*

17. $\int_0^1 2x\, dx$ **18.** $\int_2^7 3v\, dv$

19. $\int_{-1}^{0} (x - 2)\, dx$ **20.** $\int_2^5 (-3x + 4)\, dx$

21. $\int_{-1}^{1} (3t + 4)^2\, dt$ **22.** $\int_0^1 (1 - 2x)^2\, dx$

23. $\int_0^3 (x-2)^3\,dx$

24. $\int_1^3 (x-3)^4\,dx$

25. $\int_{-1}^1 \left(\sqrt[3]{t}-2\right)dt$

26. $\int_1^4 \left(\sqrt{x}+x\right)dx$

27. $\int_{-1}^0 (t^{1/3}-t^{2/3})\,dt$

28. $\int_0^4 (x^{1/2}+x^{1/4})\,dx$

29. $\int_2^8 \frac{3}{x}\,dx$

30. $\int_1^4 -\frac{4}{x}\,dx$

31. $\int_0^4 \frac{1}{\sqrt{2x+1}}\,dx$

32. $\int_0^2 \frac{x}{\sqrt{1+2x^2}}\,dx$

33. $\int_1^2 e^{1-x}\,dx$

34. $\int_{-1}^1 (e^x-e^{-x})\,dx$

35. $\int_0^1 e^{2x}\sqrt{e^{2x}+1}\,dx$

36. $\int_0^1 \frac{e^{-x}}{\sqrt{e^{-x}+1}}\,dx$

37. $\int_0^2 \frac{x}{1+4x^2}\,dx$

38. $\int_0^1 \frac{e^{2x}}{e^{2x}+1}\,dx$

Interpreting Absolute Value In Exercises 39–42, evaluate the definite integral. *See Example 5.*

39. $\int_{-2}^1 |4x|\,dx$

40. $\int_{-1}^3 \left|\frac{x}{3}\right|dx$

41. $\int_2^8 |3x-9|\,dx$

42. $\int_0^3 |2x-3|\,dx$

Area of a Region In Exercises 43–46, find the area of the region bounded by the graphs of the equations. Use a graphing utility to verify your results.

43. $y=3x^2+1,\ y=0,\ x=0$, and $x=2$

44. $y=1+\sqrt{x},\ y=0,\ x=0$, and $x=4$

45. $y=4/x,\ y=0,\ x=1$, and $x=3$

46. $y=e^x,\ y=0,\ x=0$, and $x=2$

Marginal Analysis In Exercises 47–52, find the change in cost C, revenue R, or profit P, for the given marginal. In each case, assume that the number of units x increases by 3 from the specified value of x. *See Example 6.*

Marginal	*Number of Units, x*
47. $\frac{dC}{dx}=2.25$	$x=100$
48. $\frac{dC}{dx}=\frac{20{,}000}{x^2}$	$x=10$
49. $\frac{dR}{dx}=48-3x$	$x=12$
50. $\frac{dR}{dx}=75\left(20+\frac{900}{x}\right)$	$x=500$
51. $\frac{dP}{dx}=\frac{400-x}{150}$	$x=200$
52. $\frac{dP}{dx}=12.5\left(40-3\sqrt{x}\right)$	$x=125$

Average Value of a Function In Exercises 53–60, find the average value of the function on the interval. Then find all x-values in the interval for which the function is equal to its average value.

53. $f(x)=6x;\ [1,3]$

54. $f(x)=x^3;\ [0,2]$

55. $f(x)=4-x^2;\ [-2,2]$

56. $f(x)=x-2\sqrt{x};\ [0,4]$

57. $f(x)=2e^x;\ [-1,1]$

58. $f(x)=e^{x/4};\ [0,4]$

59. $f(x)=\frac{3}{x+2};\ [1,5]$

60. $f(x)=\frac{1}{(x-3)^2};\ [0,2]$

Integrating Even and Odd Functions In Exercises 61–64, evaluate the definite integral using the properties of even and odd functions. *See Example 8.*

61. $\int_{-1}^1 3x^4\,dx$

62. $\int_{-2}^2 (x^3-4x)\,dx$

63. $\int_{-1}^1 (2t^5-2t)\,dt$

64. $\int_{-2}^2 \left(\frac{1}{2}t^4+1\right)dt$

65. **Using Properties of Definite Integrals** Use the value $\int_0^1 x^2\,dx=\frac{1}{3}$ to evaluate each definite integral. Explain your reasoning.

(a) $\int_{-1}^0 x^2\,dx$

(b) $\int_{-1}^1 x^2\,dx$

(c) $\int_0^1 -x^2\,dx$

66. **Using Properties of Definite Integrals** Use the value $\int_0^4 x^3\,dx=64$ to evaluate each definite integral. Explain your reasoning.

(a) $\int_{-4}^0 x^3\,dx$

(b) $\int_{-4}^4 x^3\,dx$

(c) $\int_0^4 2x^3\,dx$

Finding the Amount of an Annuity In Exercises 67–70, find the amount of an annuity with income function $c(t)$, interest rate r, and term T. *See Example 9.*

67. $c(t)=\$250,\ r=8\%,\ T=6$ years

68. $c(t)=\$500,\ r=7\%,\ T=4$ years

69. $c(t)=\$1500,\ r=2\%,\ T=10$ years

70. $c(t)=\$2000,\ r=3\%,\ T=15$ years

Capital Accumulation In Exercises 71–74, you are given the rate of investment dI/dt. Find the capital accumulation over a five-year period by evaluating the definite integral

$$\text{Capital accumulation} = \int_0^5 \frac{dI}{dt}\,dt$$

where t is the time (in years).

71. $\frac{dI}{dt} = 500$ **72.** $\frac{dI}{dt} = 100t$

73. $\frac{dI}{dt} = 500\sqrt{t+1}$

74. $\frac{dI}{dt} = \frac{12{,}000t}{(t^2+2)^2}$

75. Cost The total cost of purchasing a piece of equipment and maintaining it for x years can be modeled by

$$C = 5000\left(25 + 3\int_0^x t^{1/4}\,dt\right).$$

Find the total cost after (a) 1 year, (b) 5 years, and (c) 10 years.

76. Depreciation A company purchases a new machine for which the rate of depreciation can be modeled by

$$\frac{dV}{dt} = 10{,}000(t-6), \quad 0 \le t \le 5$$

where V is the value of the machine after t years. Set up and evaluate the definite integral that yields the total loss of value of the machine over the first 3 years.

77. Compound Interest A deposit of \$2250 is made in a savings account at an annual interest rate of 6%, compounded continuously. Find the average balance in the account during the first 5 years.

78. HOW DO YOU SEE IT? A college graduate has two job offers. The starting salary for each is \$32,000, and after 8 years of service each will pay \$54,000. The salary increase for each offer is shown in the figure. From a strictly monetary viewpoint, which is the better offer? Explain your reasoning.

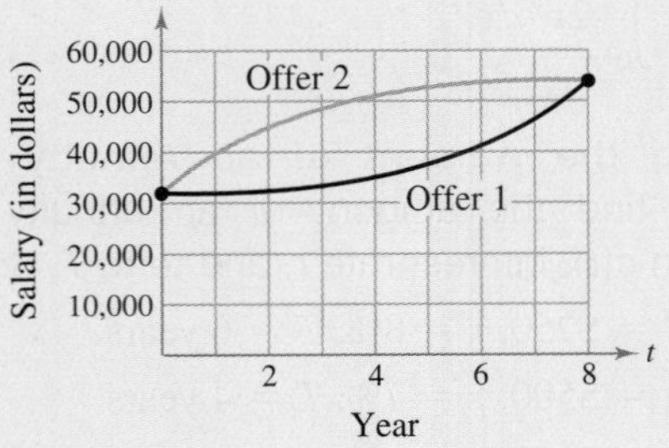

79. Mortgage Debt The rate of change of mortgage debt outstanding for one- to four-family homes in the United States from 2000 through 2009 can be modeled by

$$\frac{dM}{dt} = 547.56t - 69.459t^2 + 331.258e^{-t}$$

where M is the mortgage debt outstanding (in billions of dollars) and t is the year, with $t = 0$ corresponding to 2000. In 2000, the mortgage debt outstanding in the United States was \$5107 billion. *(Source: Board of Governors of the Federal Reserve System)*

(a) Write a model for the debt as a function of t.

(b) What was the average mortgage debt outstanding for 2000 through 2009?

Business Capsule

After losing her job as an account executive in 1985, Avis Yates Rivers used \$2500 to start a word processing business from the basement of her home. In 1996, as a spin-off, Ms. Yates Rivers established Technology Concepts Group. Today, this Somerset, New Jersey-based firm provides a wide range of information technology solutions. Revenues in 2010 were about \$3 million, and the company projects revenues of \$30 million in 2011 with the acquisition of equipment leasing expertise. Ms. Yates Rivers has become a nationally recognized leader, speaker, and advocate for minority- and women-owned small businesses.

80. Research Project Use your school's library, the Internet, or some other reference source to research a small company similar to the one described above. Describe the impact of different factors, such as start-up capital and market conditions, on a company's revenue.

5.5 The Area of a Region Bounded by Two Graphs

- Find the areas of regions bounded by two graphs.
- Find consumer and producer surpluses.
- Use the areas of regions bounded by two graphs to solve real-life problems.

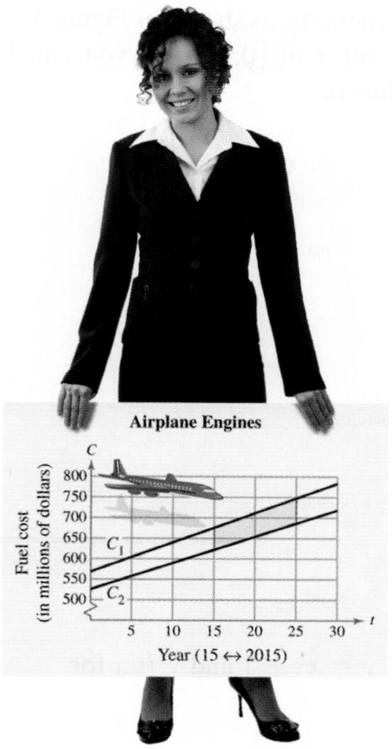

In Exercise 49 on page 358, you will use integration to find the amount saved on fuel costs by switching to more efficient airplane engines.

Area of a Region Bounded by Two Graphs

With a few modifications, you can extend the use of definite integrals from finding the area of a region *under a graph* to finding the area of a region *bounded by two graphs*. To see how this is done, consider the region bounded by the graphs of

$$f, \quad g, \quad x = a, \quad \text{and} \quad x = b$$

as shown in Figure 5.14. If the graphs of both f and g lie above the x-axis, then you can interpret the area of the region between the graphs as the area of the region under the graph of g subtracted from the area of the region under the graph of f, as shown in Figure 5.14.

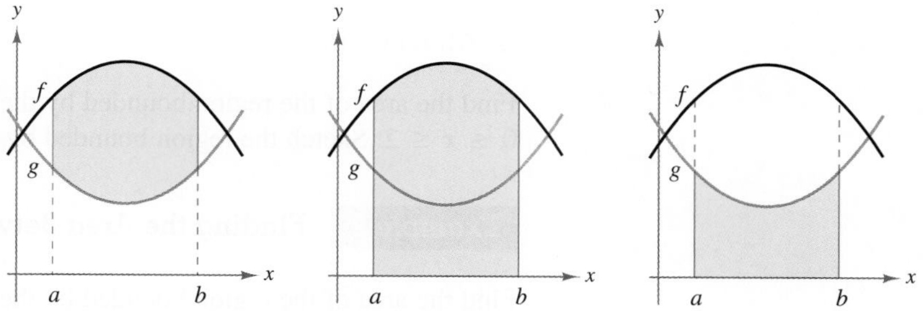

(Area between f and g) = (Area of region under f) − (Area of region under g)

$$\int_a^b [f(x) - g(x)]\,dx = \int_a^b f(x)\,dx - \int_a^b g(x)\,dx$$

FIGURE 5.14

Although Figure 5.14 depicts the graphs of f and g lying above the x-axis, this is not necessary, and the same integrand

$$[f(x) - g(x)]$$

can be used as long as both functions are continuous and $g(x) \le f(x)$ on the interval $[a, b]$.

Area of a Region Bounded by Two Graphs

If f and g are continuous on $[a, b]$ and $g(x) \le f(x)$ for all x in $[a, b]$, then the area of the region bounded by the graphs of

$$f, \quad g, \quad x = a, \quad \text{and} \quad x = b$$

(see Figure 5.15) is given by

$$A = \int_a^b [f(x) - g(x)]\,dx.$$

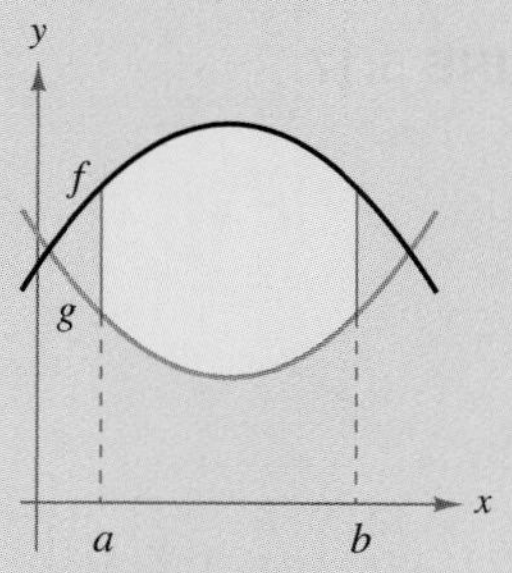

FIGURE 5.15

Example 1 Finding the Area Bounded by Two Graphs

Find the area of the region bounded by the graphs of $y = x^2 + 2$ and $y = x$ for $0 \le x \le 1$.

SOLUTION Begin by sketching the graphs of both functions, as shown in Figure 5.16. From the figure, you can see that $x \le x^2 + 2$ for all x in $[0, 1]$. So, you can let $f(x) = x^2 + 2$ and $g(x) = x$. Then find the area as shown.

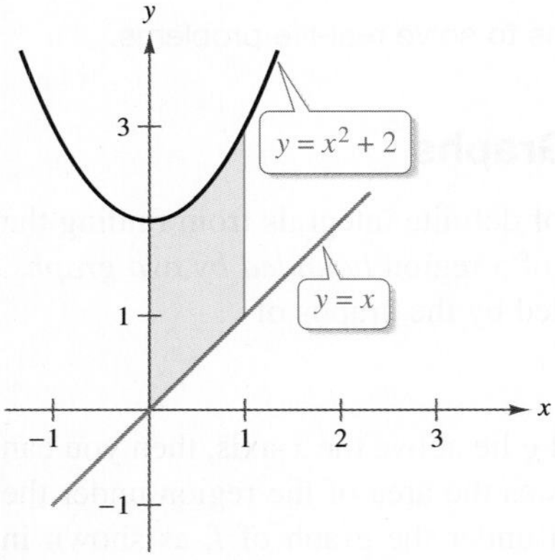

FIGURE 5.16

$$\text{Area} = \int_a^b [f(x) - g(x)]\,dx \quad \text{Area between } f \text{ and } g$$

$$= \int_0^1 [(x^2 + 2) - (x)]\,dx \quad \text{Substitute for } f \text{ and } g.$$

$$= \int_0^1 (x^2 - x + 2)\,dx$$

$$= \left[\frac{x^3}{3} - \frac{x^2}{2} + 2x\right]_0^1 \quad \text{Find antiderivative.}$$

$$= \frac{11}{6} \text{ square units} \quad \text{Apply Fundamental Theorem.}$$

✓ Checkpoint 1

Find the area of the region bounded by the graphs of $y = x^2 + 1$ and $y = x$ for $0 \le x \le 2$. Sketch the region bounded by the graphs. ■

Example 2 Finding the Area Between Intersecting Graphs

Find the area of the region bounded by the graphs of $y = 2 - x^2$ and $y = x$.

SOLUTION Because the values of a and b are not given, you must determine them by finding the x-coordinates of the points of intersection of the two graphs. To do this, equate the two functions and solve for x.

$$2 - x^2 = x \quad \text{Equate functions.}$$

$$-x^2 - x + 2 = 0 \quad \text{Write in general form.}$$

$$-(x + 2)(x - 1) = 0 \quad \text{Factor.}$$

$$x = -2, x = 1 \quad \text{Solve for } x.$$

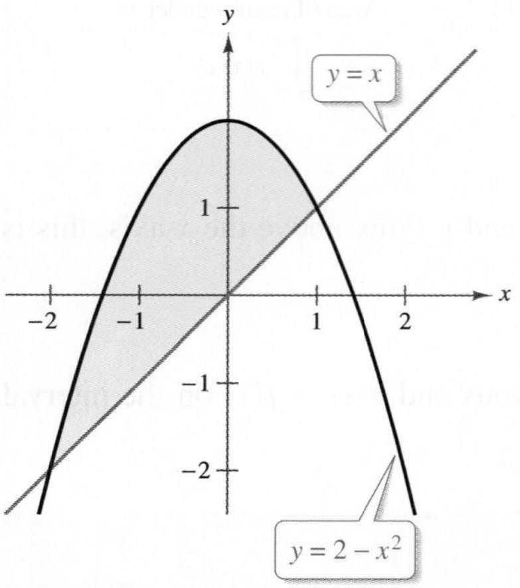

FIGURE 5.17

So, $a = -2$ and $b = 1$. In Figure 5.17, you can see that the graph of $f(x) = 2 - x^2$ lies above the graph of $g(x) = x$ for all x in the interval $[-2, 1]$.

$$\text{Area} = \int_a^b [f(x) - g(x)]\,dx \quad \text{Area between } f \text{ and } g$$

$$= \int_{-2}^1 [(2 - x^2) - (x)]\,dx \quad \text{Substitute for } f \text{ and } g.$$

$$= \int_{-2}^1 (-x^2 - x + 2)\,dx$$

$$= \left[-\frac{x^3}{3} - \frac{x^2}{2} + 2x\right]_{-2}^1 \quad \text{Find antiderivative.}$$

$$= \frac{9}{2} \text{ square units} \quad \text{Apply Fundamental Theorem.}$$

✓ Checkpoint 2

Find the area of the region bounded by the graphs of $y = 3 - x^2$ and $y = 2x$. ■

Example 3 Finding an Area Below the x-Axis

Find the area of the region bounded by the graph of

$$y = x^2 - 3x - 4$$

and the x-axis.

SOLUTION Begin by finding the x-intercepts of the graph. To do this, set the function equal to zero and solve for x.

$$x^2 - 3x - 4 = 0 \quad \text{Set function equal to 0.}$$

$$(x - 4)(x + 1) = 0 \quad \text{Factor.}$$

$$x = 4, x = -1 \quad \text{Solve for } x.$$

From Figure 5.18, you can see that $x^2 - 3x - 4 \le 0$ for all x in the interval $[-1, 4]$.

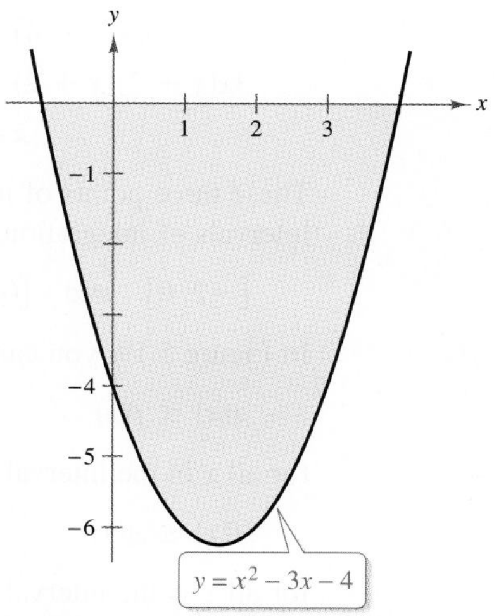

FIGURE 5.18

So, you can let

$$f(x) = 0 \quad \text{and} \quad g(x) = x^2 - 3x - 4$$

and find the area as shown.

$$\text{Area} = \int_a^b [f(x) - g(x)]\,dx \quad \text{Area between } f \text{ and } g$$

$$= \int_{-1}^{4} [(0) - (x^2 - 3x - 4)]\,dx \quad \text{Substitute for } f \text{ and } g.$$

$$= \int_{-1}^{4} (-x^2 + 3x + 4)\,dx$$

$$= \left[-\frac{x^3}{3} + \frac{3x^2}{2} + 4x\right]_{-1}^{4} \quad \text{Find antiderivative.}$$

$$= \frac{125}{6} \text{ square units} \quad \text{Apply Fundamental Theorem.}$$

✓Checkpoint 3

Find the area of the region bounded by the graph of

$$y = x^2 - x - 2$$

and the x-axis.

TECH TUTOR

Most graphing utilities can display regions that are bounded by two graphs. For instance, to graph the region in Example 3, set the viewing window to $-1 \le x \le 4$ and $-7 \le y \le 1$. Consult your user's manual for specific keystrokes on how to shade the graph. You should obtain the graph below.

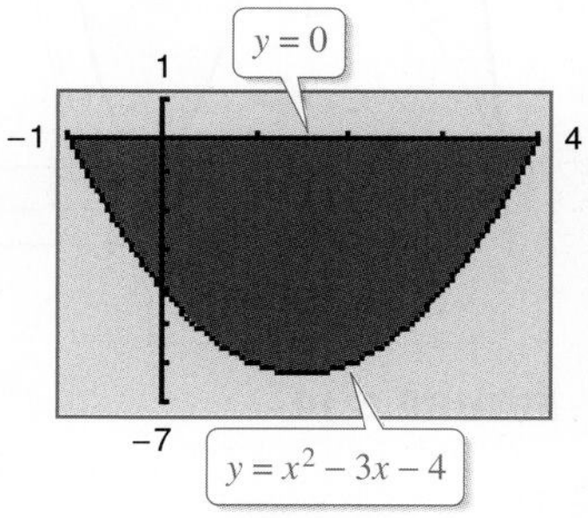

Sometimes two graphs intersect at more than two points. To determine the area of the region bounded by two such graphs, you must find *all* points of intersection and check to see which graph is above the other in each interval determined by the points.

Example 4 Using Multiple Points of Intersection

Find the area of the region bounded by the graphs of

$$f(x) = 3x^3 - x^2 - 10x \quad \text{and} \quad g(x) = -x^2 + 2x.$$

SOLUTION To find the points of intersection of the two graphs, set the functions equal to each other and solve for x.

$$f(x) = g(x) \qquad \text{Set } f(x) \text{ equal to } g(x).$$
$$3x^3 - x^2 - 10x = -x^2 + 2x \qquad \text{Substitute for } f(x) \text{ and } g(x).$$
$$3x^3 - 12x = 0 \qquad \text{Write in general form.}$$
$$3x(x^2 - 4) = 0$$
$$3x(x - 2)(x + 2) = 0 \qquad \text{Factor.}$$
$$x = 0, x = 2, x = -2 \qquad \text{Solve for } x.$$

These three points of intersection determine two intervals of integration:

$$[-2, 0] \quad \text{and} \quad [0, 2].$$

In Figure 5.19, you can see that

$$g(x) \le f(x)$$

for all x in the interval $[-2, 0]$, and that

$$f(x) \le g(x)$$

for all x in the interval $[0, 2]$. So, you must use two integrals to determine the area of the region bounded by the graphs of f and g: one for the interval $[-2, 0]$ and one for the interval $[0, 2]$.

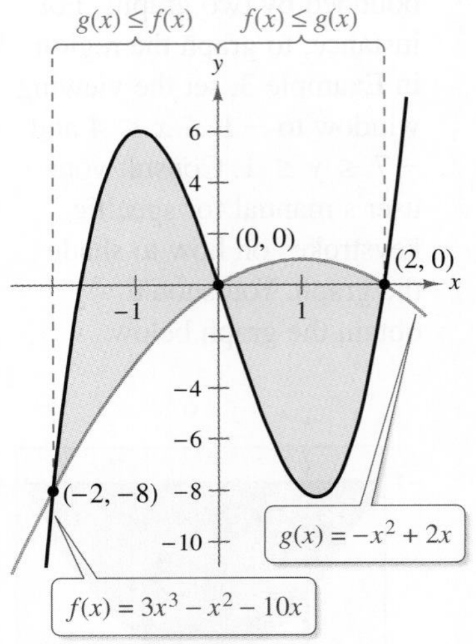

FIGURE 5.19

$$\text{Area} = \int_{-2}^{0} [f(x) - g(x)]\,dx + \int_{0}^{2} [g(x) - f(x)]\,dx$$
$$= \int_{-2}^{0} (3x^3 - 12x)\,dx + \int_{0}^{2} (-3x^3 + 12x)\,dx$$
$$= \left[\frac{3x^4}{4} - 6x^2\right]_{-2}^{0} + \left[-\frac{3x^4}{4} + 6x^2\right]_{0}^{2}$$
$$= (0 - 0) - (12 - 24) + (-12 + 24) - (0 + 0)$$
$$= 24$$

So, the region has an area of 24 square units.

STUDY TIP

It is easy to make an error when calculating areas such as the one in Example 4. To check your solution, make a sketch of the region on graph paper and then use the grid on the graph paper to approximate the area. Try doing this with the graph shown in Figure 5.19. Is your approximation close to 24 square units?

✓ Checkpoint 4

Find the area of the region bounded by the graphs of

$$f(x) = x^3 + 2x^2 - 3x \quad \text{and} \quad g(x) = x^2 + 3x.$$

Sketch a graph of the region. ■

Consumer Surplus and Producer Surplus

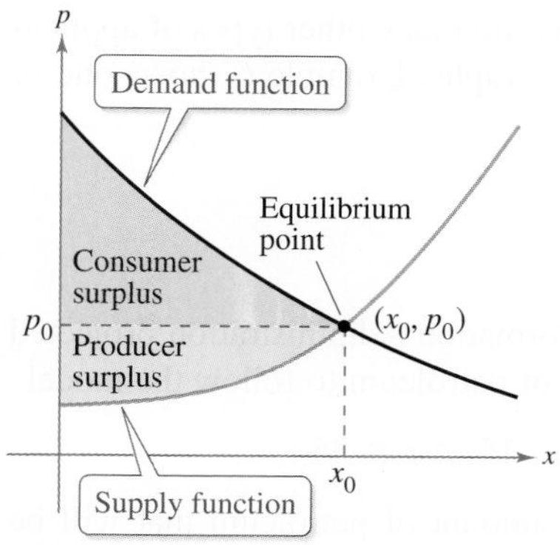

FIGURE 5.20

In Section 1.2, you learned that a demand function relates the price of a product to the consumer demand. You also learned that a supply function relates the price of a product to producers' willingness to supply the product. The point (x_0, p_0) at which a demand function $p = D(x)$ and a supply function $p = S(x)$ intersect is the equilibrium point.

Economists call the area of the region bounded by the graph of the demand function, the horizontal line $p = p_0$, and the vertical line $x = 0$ the **consumer surplus,** as shown in Figure 5.20. Consumer surplus is the difference between the amount consumers would be willing to pay and the actual amount paid for a product. The area of the region bounded by the graph of the supply function, the horizontal line $p = p_0$, and the vertical line $x = 0$ is called the **producer surplus,** as shown in Figure 5.20. Producer surplus is the difference between the amount a producer receives for selling a product and the minimum price needed to get the producer to supply the product.

Example 5 Finding Surpluses

The demand and supply functions for a product are modeled by

$$\textit{Demand: } p = -0.36x + 9 \quad \text{and} \quad \textit{Supply: } p = 0.14x + 2$$

where p is the price (in dollars) and x is the number of units (in millions). Find the consumer and producer surpluses for this product.

SOLUTION By equating the demand and supply functions, you can determine that the point of equilibrium occurs when $x = 14$ (million) and the price is \$3.96 per unit.

$$\begin{aligned}\text{Consumer surplus} &= \int_0^{14} (\text{demand function} - \text{price})\,dx \\ &= \int_0^{14} [(-0.36x + 9) - 3.96]\,dx \\ &= \Big[-0.18x^2 + 5.04x\Big]_0^{14} \\ &= 35.28\end{aligned}$$

The consumer surplus is \$35.28.

$$\begin{aligned}\text{Producer surplus} &= \int_0^{14} (\text{price} - \text{supply function})\,dx \\ &= \int_0^{14} [3.96 - (0.14x + 2)]\,dx \\ &= \Big[-0.07x^2 + 1.96x\Big]_0^{14} \\ &= 13.72\end{aligned}$$

The producer surplus is \$13.72. The consumer surplus and producer surplus are shown in Figure 5.21.

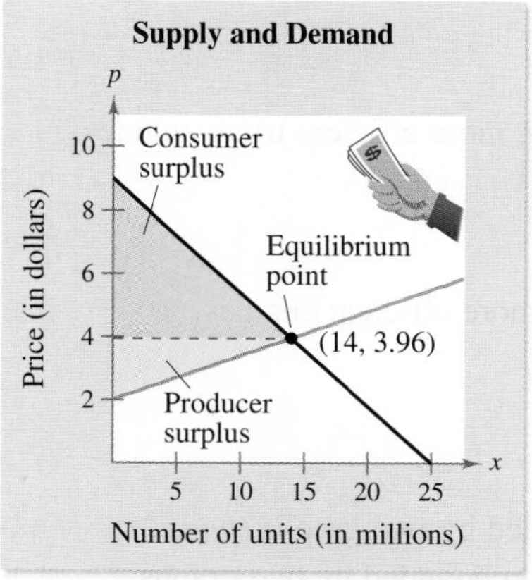

FIGURE 5.21

✓Checkpoint 5

The demand and supply functions for a product are modeled by

$$\textit{Demand: } p = -0.2x + 8 \quad \text{and} \quad \textit{Supply: } p = 0.1x + 2$$

where p is the price (in dollars) and x is the number of units (in millions). Find the consumer and producer surpluses for this product. ■

Application

In addition to consumer and producer surpluses, there are many other types of applications involving the area of a region bounded by two graphs. Example 6 shows one of these applications.

Example 6 Petroleum Consumption

In the *Annual Energy Outlook*, the U.S. Energy Information Administration projected the consumption C (in quadrillions of Btu per year) of petroleum to follow the model

$$C_1 = 0.00078t^3 - 0.0445t^2 + 0.917t + 35.49, \quad 15 \le t \le 35$$

where $t = 15$ corresponds to 2015. Determine the amount of petroleum that will be saved when the actual consumption follows the model

$$C_2 = 0.0067t^2 - 0.211t + 40.95, \quad 15 \le t \le 35.$$

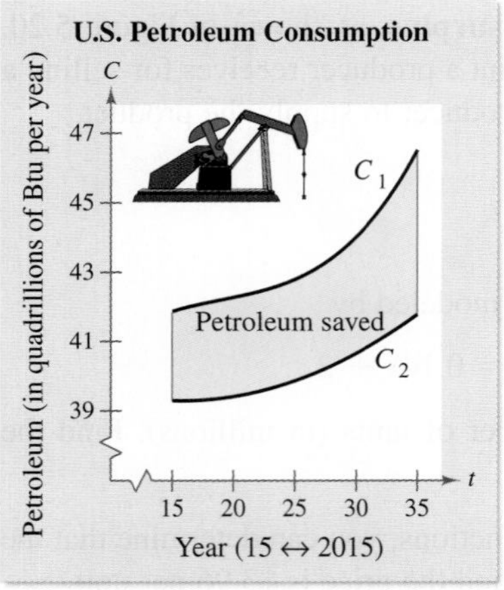

FIGURE 5.22

SOLUTION The petroleum saved can be represented as the area of the region between the graphs of C_1 and C_2, as shown in Figure 5.22.

$$\begin{aligned}\text{Petroleum saved} &= \int_{15}^{35} (C_1 - C_2)\, dt \\ &= \int_{15}^{35} (0.00078t^3 - 0.0512t^2 + 1.128t - 5.46)\, dt \\ &= \left[\frac{0.00078}{4}t^4 - \frac{0.0512}{3}t^3 + \frac{1.128}{2}t^2 - 5.46t\right]_{15}^{35} \\ &\approx 63.42\end{aligned}$$

So, about 63.42 quadrillion Btu of petroleum would be saved.

Checkpoint 6

The projected fuel cost C (in millions of dollars per year) for a trucking company from 2012 through 2024 is

$$C_1 = 2.21t + 5.6, \quad 12 \le t \le 24$$

where $t = 12$ corresponds to 2012. After purchasing more efficient truck engines, the company expects fuels costs to follow the model

$$C_2 = 2.04t + 4.7, \quad 12 \le t \le 24.$$

How much money will the company save with the more efficient engines?

SUMMARIZE (Section 5.5)

1. State the definition of the area of a region bounded by two graphs *(page 351)*. For examples of finding the area of a region bounded by two graphs, see Examples 1, 2, 3, and 4.
2. Describe a real-life example of how finding the area of a region bounded by two graphs can be used to find the consumer and producer surpluses for a product *(page 355, Example 5)*.
3. Describe a real-life example of how finding the area of a region bounded by two graphs can be used to analyze petroleum consumption *(page 356, Example 6)*.

SKILLS WARM UP 5.5 The following warm-up exercises involve skills that were covered in earlier sections. You will use these skills in the exercise set for this section. For additional help, review Section 1.2.

In Exercises 1–4, simplify the expression.

1. $(-x^2 + 4x + 3) - (x + 1)$

2. $(-2x^2 + 3x + 9) - (-x + 5)$

3. $(-x^3 + 3x^2 - 1) - (x^2 - 4x + 4)$

4. $(3x + 1) - (-x^3 + 9x + 2)$

In Exercises 5–8, find the points of intersection of the graphs.

5. $f(x) = x^2 - 4x + 4,\ g(x) = 4$

6. $f(x) = -3x^2,\ g(x) = 6 - 9x$

7. $f(x) = x^2,\ g(x) = -x + 6$

8. $f(x) = \frac{1}{2}x^3,\ g(x) = 2x$

Exercises 5.5

See www.CalcChat.com for worked-out solutions to odd-numbered exercises.

Finding the Area Bounded by Two Graphs In Exercises 1–8, find the area of the region. *See Examples 1, 2, 3, and 4.*

1. $f(x) = x^2 - 6x$
$g(x) = 0$

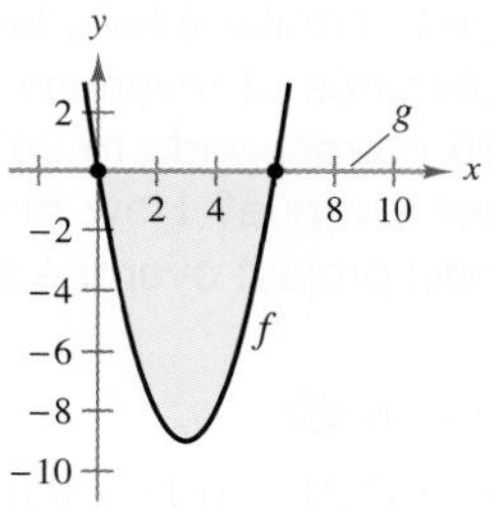

2. $f(x) = x^2 + 2x + 1$
$g(x) = 2x + 5$

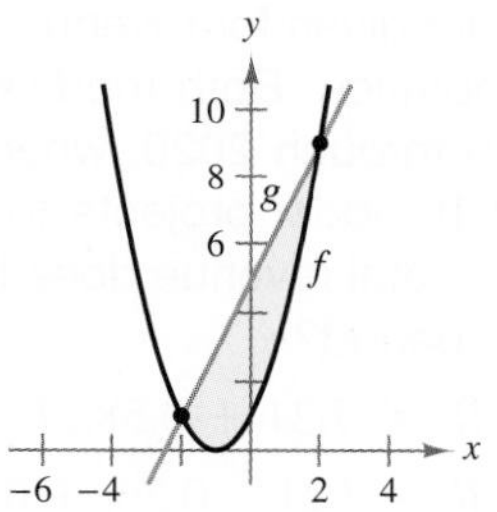

3. $f(x) = x^2 - 4x + 3$
$g(x) = -x^2 + 2x + 3$

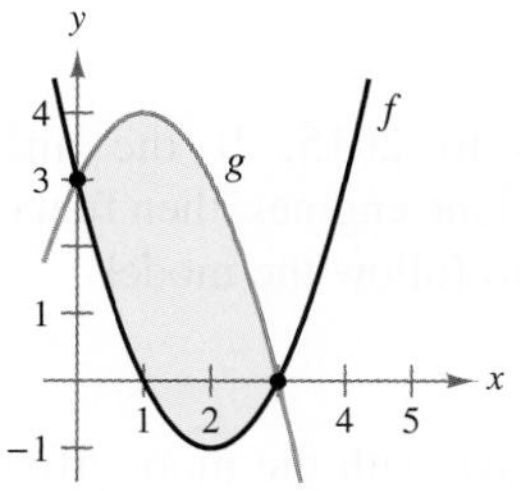

4. $f(x) = x^2$
$g(x) = x^3$

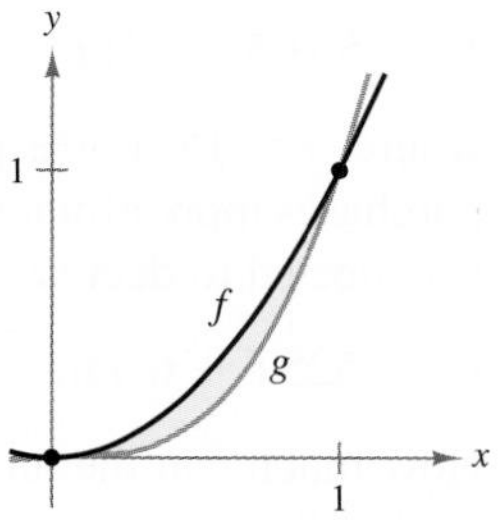

5. $f(x) = e^x - 1$
$g(x) = 0$

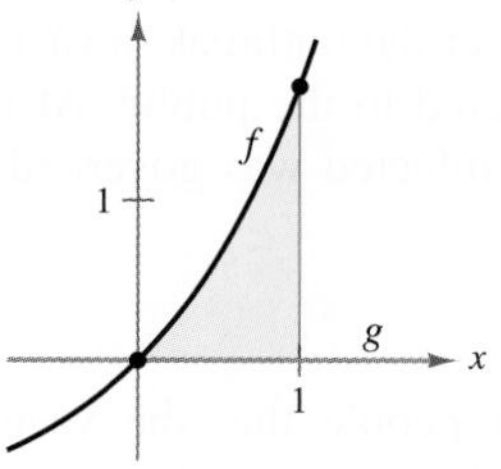

6. $f(x) = -x + 3$
$g(x) = \dfrac{2}{x}$

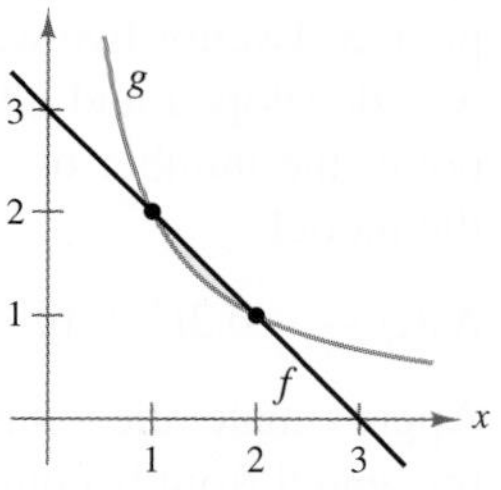

7. $f(x) = 3(x^3 - x)$
$g(x) = 0$

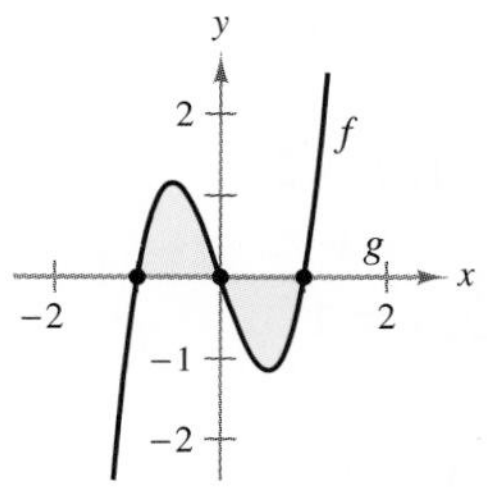

8. $f(x) = (x - 1)^3$
$g(x) = x - 1$

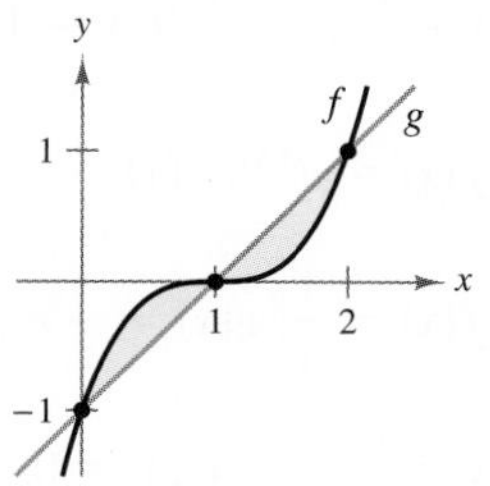

Finding the Region In Exercises 9–12, the integrand of the definite integral is a difference of two functions. Sketch the graph of each function and shade the region whose area is represented by the integral.

9. $\displaystyle\int_0^4 \left[(x + 1) - \tfrac{1}{2}x\right] dx$

10. $\displaystyle\int_{-1}^1 \left[(1 - x^2) - (x^2 - 1)\right] dx$

11. $\displaystyle\int_{-2}^2 \left[2x^2 - (x^4 - 2x^2)\right] dx$

12. $\displaystyle\int_{-4}^0 \left[(x - 6) - (x^2 + 5x - 6)\right] dx$

Think About It In Exercises 13 and 14, determine which value best approximates the area of the region bounded by the graphs of f and g. Make your selection based on a sketch of the region and not by performing any calculations.

13. $f(x) = x + 1,\ g(x) = (x - 1)^2$

(a) -2 (b) 2 (c) 10 (d) 4 (e) 8

14. $f(x) = 2 - \dfrac{1}{2}x,\ g(x) = 2 - \sqrt{x}$

(a) 1 (b) 6 (c) -3 (d) 3 (e) 4

Finding the Area Bounded by Two Graphs In Exercises 15–30, sketch the region bounded by the graphs of the functions and find the area of the region. *See Examples 1, 2, 3, and 4.*

15. $y = \dfrac{1}{x^2}, y = 0, x = 1, x = 5$

16. $y = 2x - 3, y = x^2 + 2x + 1, x = -2, x = 1$

17. $y = x^2 - 4x + 3, y = 3 + 4x - x^2$

18. $y = 4 - x^2, y = x^2$

19. $y = x^2 - 1, y = -x + 2, x = 0, x = 1$

20. $y = -x^3 + 3, y = x, x = -1, x = 1$

21. $f(x) = \sqrt[3]{x}, g(x) = x$

22. $f(x) = \sqrt{3x} + 1, g(x) = x + 1$

23. $f(x) = x^3 + 4x^2, g(x) = x + 4$

24. $f(x) = 1 - x, g(x) = x^4 - x$

25. $y = xe^{-x^2}, y = 0, x = 0, x = 1$

26. $y = \dfrac{e^{1/x}}{x^2}, y = 0, x = 1, x = 3$

27. $f(x) = e^{0.5x}, g(x) = -\dfrac{1}{x}, x = 1, x = 2$

28. $f(x) = \dfrac{1}{x}, g(x) = -e^x, x = \dfrac{1}{2}, x = 1$

29. $y = \dfrac{8}{x}, y = x^2, y = 0, x = 1, x = 4$

30. $y = x^2 - 2x + 1, y = x^2 - 10x + 25, y = 0$

Writing Integrals In Exercises 31–34, use a graphing utility to graph the region bounded by the graphs of the functions. Write the definite integral that represents the area of the region. (*Hint:* Multiple integrals may be necessary.)

31. $f(x) = 2x,\ g(x) = 4 - 2x,\ h(x) = 0$

32. $f(x) = x(x^2 - 3x + 3), g(x) = x^2$

33. $y = \dfrac{4}{x}, y = x, x = 1, x = 4$

34. $y = x^3 - 4x^2 + 1, y = x - 3$

Finding Area In Exercises 35–38, use a graphing utility to graph the region bounded by the graphs of the functions. Find the area of the region by hand.

35. $f(x) = x^2 - 4x, g(x) = 0$

36. $f(x) = 3 - 2x - x^2, g(x) = 0$

37. $f(x) = x^2 + 2x + 1, g(x) = x + 1$

38. $f(x) = -x^2 + 4x + 2, g(x) = x + 2$

Area of a Region In Exercises 39 and 40, use integration to find the area of the triangular region having the given vertices.

39. $(0, 0), (4, 0), (4, 4)$

40. $(0, 0), (4, 0), (6, 4)$

Consumer and Producer Surpluses In Exercises 41–46, find the consumer and producer surpluses by using the demand and supply functions, where p is the price (in dollars) and x is the number of units (in millions). *See Example 5.*

	Demand Function	*Supply Function*
41.	$p = 50 - 0.5x$	$p = 0.125x$
42.	$p = 300 - x$	$p = 100 + x$
43.	$p = 200 - 0.4x$	$p = 100 + 1.6x$
44.	$p = 975 - 23x$	$p = 42x$
45.	$p = 42 - 0.015x^2$	$p = 0.01x^2 + 2$
46.	$p = 62 - 0.3x$	$p = 0.002x^2 + 12$

Revenue In Exercises 47 and 48, two models, R_1 and R_2, are given for revenue (in billions of dollars) for a large corporation. Both models are estimates of revenues for 2015 through 2020, where $t = 15$ corresponds to 2015. Which model projects the greater revenue? How much more total revenue does that model project over the six-year period?

47. $R_1 = 7.21 + 0.58t, R_2 = 7.21 + 0.45t$

48. $R_1 = 7.21 + 0.26t + 0.02t^2, R_2 = 7.21 + 0.1t + 0.01t^2$

49. Fuel Cost The projected fuel cost C (in millions of dollars) for an airline from 2015 through 2025 is

$$C_1 = 568.5 + 7.15t$$

where $t = 15$ corresponds to 2015. If the airline purchases more efficient airplane engines, then fuel cost is expected to decrease and to follow the model

$$C_2 = 525.6 + 6.43t.$$

How much can the airline save with the more efficient engines? Explain your reasoning.

50. Health An epidemic was spreading such that t weeks after its outbreak it had infected

$$N_1(t) = 0.1t^2 + 0.5t + 150, \quad 0 \le t \le 50$$

people. Twenty-five weeks after the outbreak, a vaccine was developed and administered to the public. At that point, the number of people infected was governed by the model

$$N_2(t) = -0.2t^2 + 6t + 200.$$

Approximate the number of people that the vaccine prevented from becoming ill during the epidemic.

51. Consumer Trends For the years 1998 through 2008, the per capita consumption C of all fruit (in pounds) in the United States can be modeled by

$$C(t) = \begin{cases} -0.443t^2 + 5.02t + 277.7, & 8 \le t \le 12 \\ -0.775t^2 + 18.73t + 170.5, & 12 < t \le 18 \end{cases}$$

where t is the year, with $t = 8$ corresponding to 1998. *(Source: U.S. Department of Agriculture)*

(a) Use a graphing utility to graph this model.

(b) Suppose the fruit consumption from 2003 through 2008 had continued to follow the model for 1998 through 2002. How many more or fewer pounds of fruit would have been consumed from 2003 through 2008?

52. HOW DO YOU SEE IT? A state legislature is debating two proposals for eliminating the annual budget deficits by the year 2020. The rate of decrease of the deficits for each proposal is shown in the figure.

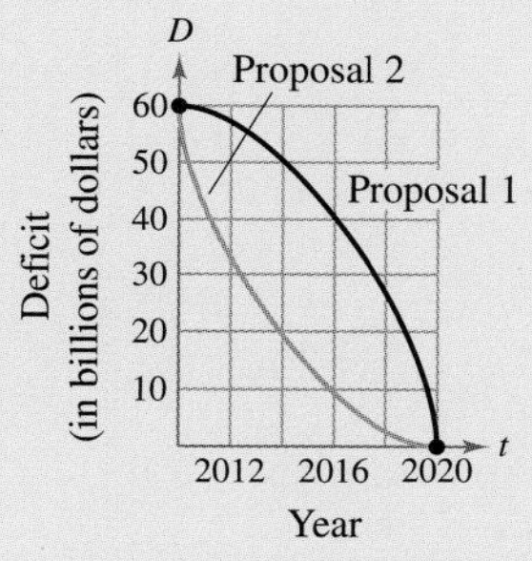

(a) What does the area between the two graphs represent?

(b) From the viewpoint of minimizing the cumulative state deficit, which is the better proposal? Explain your reasoning.

53. Cost, Revenue, and Profit The revenue from a manufacturing process (in millions of dollars) is projected to follow the model $R = 100$ for 10 years. Over the same period of time, the cost (in millions of dollars) is projected to follow the model $C = 60 + 0.2t^2$, where t is the time (in years). Approximate the profit over the 10-year period.

54. Cost, Revenue, and Profit Repeat Exercise 53 for revenue and cost models given by $R = 100 + 0.08t$ and $C = 60 + 0.2t^2$. Did the profit increase or decrease? Explain why.

55. Consumer and Producer Surpluses Factory orders for an air conditioner are about 6000 units per week when the price is \$331 and about 8000 units per week when the price is \$303. The supply function is given by $p = 0.0275x$. Find the consumer and producer surpluses. (Assume the demand function is linear.)

56. Consumer and Producer Surpluses Repeat Exercise 55 with a demand of about 6000 units per week when the price is \$325 and about 8000 units per week when the price is \$300. Find the consumer and producer surpluses. (Assume the demand function is linear.)

57. Lorenz Curve Economists use *Lorenz curves* to illustrate the distribution of income in a country. Letting x represent the percent of families in a country and y the percent of total income, the model $y = x$ would represent a country in which each family had the same income. The Lorenz curve, $y = f(x)$, represents the actual income distribution. The area between these two models, for $0 \le x \le 100$, indicates the "income inequality" of a country. In 2009, the Lorenz curve for the United States could be modeled by

$$y = (0.00061x^2 + 0.0224x + 1.666)^2, \quad 0 \le x \le 100$$

where x is measured from the poorest to the wealthiest families. Find the income inequality for the United States in 2009. *(Source: U.S. Census Bureau)*

58. Income Distribution Using the Lorenz curve in Exercise 57 and a spreadsheet, complete the table, which lists the percent of total income earned by each quintile in the United States in 2009.

Quintile	Lowest	2nd	3rd	4th	Highest
Percent					

59. Project: Social Security For a project analyzing the receipts and expenditures for the Old-Age and Survivors Insurance Trust Fund (Social Security Trust Fund) from 1990 through 2009, visit this text's website at *www.cengagebrain.com.* *(Source: Social Security Administration)*

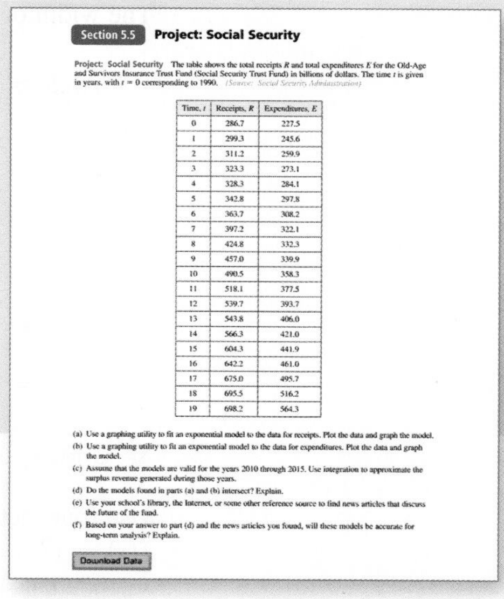

Section 5.5 **Project: Social Security**

Project: Social Security The table shows the total receipts R and total expenditures E for the Old-Age and Survivors Insurance Trust Fund (Social Security Trust Fund) in billions of dollars. The time t is given in years, with $t = 0$ corresponding to 1990. *(Source: Social Security Administration)*

Time, t	Receipts, R	Expenditures, E
0	286.7	227.5
1	299.3	245.6
2	311.2	259.9
3	323.3	273.1
4	328.3	284.1
5	342.8	297.8
6	363.7	308.2
7	397.2	322.1
8	424.8	332.3
9	457.0	339.9
10	490.5	358.3
11	518.1	377.5
12	539.7	393.7
13	543.8	406.0
14	566.3	421.0
15	604.3	441.9
16	642.2	461.0
17	675.0	495.7
18	695.5	516.2
19	698.2	564.3

(a) Use a graphing utility to fit an exponential model to the data for receipts. Plot the data and graph the model.

(b) Use a graphing utility to fit an exponential model to the data for expenditures. Plot the data and graph the model.

(c) Assume that the models are valid for the years 2010 through 2015. Use integration to approximate the surplus revenue generated during those years.

(d) Do the models found in parts (a) and (b) intersect? Explain.

(e) Use your school's library, the Internet, or some other reference source to find news articles that discuss the future of the fund.

(f) Based on your answer to part (d) and the news articles you found, will these models be accurate for long-term analysis? Explain.

Download Data

5.6 The Definite Integral as the Limit of a Sum

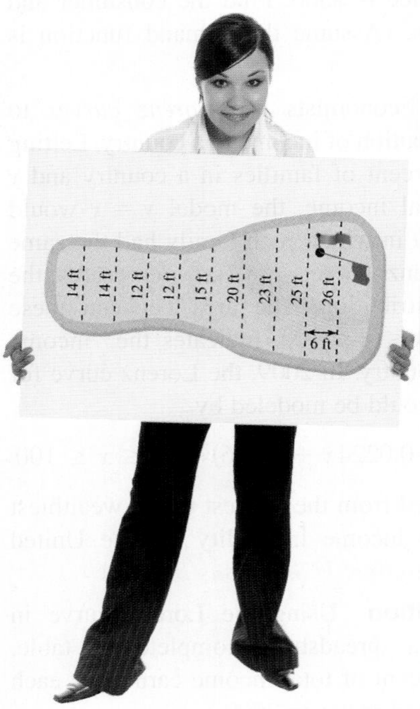

In Exercise 28 on page 365, you will use the Midpoint Rule to estimate the surface area of a golf green.

- Use the Midpoint Rule to approximate definite integrals.
- Understand the definite integral as the limit of a sum.

The Midpoint Rule

In Section 5.4, you learned that you cannot use the Fundamental Theorem of Calculus to evaluate a definite integral unless you can find an antiderivative of the integrand. When you cannot find an antiderivative of an integrand, you can use an approximation technique. One such technique, the **Midpoint Rule,** is demonstrated in Example 1.

Example 1 Approximating the Area of a Plane Region

Use the five rectangles in Figure 5.23 to approximate the area of the region bounded by the graph of $f(x) = -x^2 + 5$, the x-axis, and the lines $x = 0$ and $x = 2$.

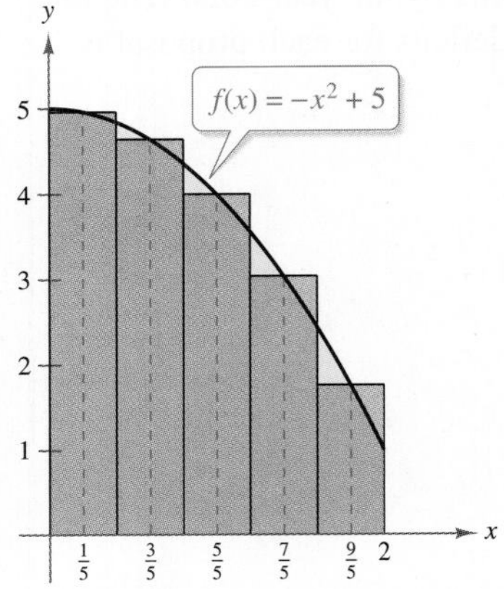

FIGURE 5.23

SOLUTION You can find the heights of the five rectangles by evaluating f at the midpoint of each of the following intervals.

$$\left[0, \frac{2}{5}\right], \quad \left[\frac{2}{5}, \frac{4}{5}\right], \quad \left[\frac{4}{5}, \frac{6}{5}\right], \quad \left[\frac{6}{5}, \frac{8}{5}\right], \quad \left[\frac{8}{5}, \frac{10}{5}\right]$$

The width of each rectangle is $\frac{2}{5}$. So, the sum of the five areas is

$$\begin{aligned}
\text{Area} &\approx \frac{2}{5}f\left(\frac{1}{5}\right) + \frac{2}{5}f\left(\frac{3}{5}\right) + \frac{2}{5}f\left(\frac{5}{5}\right) + \frac{2}{5}f\left(\frac{7}{5}\right) + \frac{2}{5}f\left(\frac{9}{5}\right) \\
&= \frac{2}{5}\left[f\left(\frac{1}{5}\right) + f\left(\frac{3}{5}\right) + f\left(\frac{5}{5}\right) + f\left(\frac{7}{5}\right) + f\left(\frac{9}{5}\right)\right] \\
&= \frac{2}{5}\left(\frac{124}{25} + \frac{116}{25} + \frac{100}{25} + \frac{76}{25} + \frac{44}{25}\right) \\
&= \frac{920}{125} \\
&= 7.36.
\end{aligned}$$

✓ Checkpoint 1

Use four rectangles to approximate the area of the region bounded by the graph of $f(x) = x^2 + 1$, the x-axis, $x = 0$, and $x = 2$. ■

TECH TUTOR

A program for several models of graphing utilities that uses the Midpoint Rule to approximate the definite integral

$$\int_a^b f(x)\,dx$$

can be found in Appendix E.

For the region in Example 1, you can find the exact area with a definite integral. That is,

$$\text{Area} = \int_0^2 (-x^2 + 5)\,dx = \frac{22}{3} \approx 7.33.$$

The approximation procedure used in Example 1 is the **Midpoint Rule.** You can use the Midpoint Rule to approximate *any* definite integral—not just those representing areas. The basic steps are summarized below.

Guidelines for Using the Midpoint Rule

To approximate the definite integral $\int_a^b f(x)\,dx$ with the Midpoint Rule, use the steps below.

1. Divide the interval $[a, b]$ into n subintervals, each of width

$$\Delta x = \frac{b-a}{n}.$$

2. Find the midpoint of each subinterval.

$$\text{Midpoints} = \{x_1, x_2, x_3, \ldots, x_n\}$$

3. Evaluate f at each midpoint and form the sum as shown.

$$\int_a^b f(x)\,dx \approx \frac{b-a}{n}[f(x_1) + f(x_2) + f(x_3) + \cdots + f(x_n)]$$

An important characteristic of the Midpoint Rule is that the approximation tends to improve as n increases. The table below shows the approximations for the area of the region described in Example 1 for various values of n. For example, when $n = 10$, the Midpoint Rule yields

$$\int_0^2 (-x^2 + 5)\,dx \approx \frac{2}{10}\left[f\left(\frac{1}{10}\right) + f\left(\frac{3}{10}\right) + \cdots + f\left(\frac{19}{10}\right)\right]$$
$$= 7.34.$$

n	5	10	15	20	25	30
Approximation	7.3600	7.3400	7.3363	7.3350	7.3344	7.3341

Note that as n increases, the approximation gets closer and closer to the exact value of the integral, which was found to be

$$\frac{22}{3} \approx 7.3333.$$

STUDY TIP

In Example 1, the Midpoint Rule is used to approximate an integral whose exact value can be found with the Fundamental Theorem of Calculus. This was done to illustrate the accuracy of the rule. In practice, of course, you would use the Midpoint Rule to approximate the values of definite integrals for which you cannot find an antiderivative. Examples 2 and 3 illustrate such integrals.

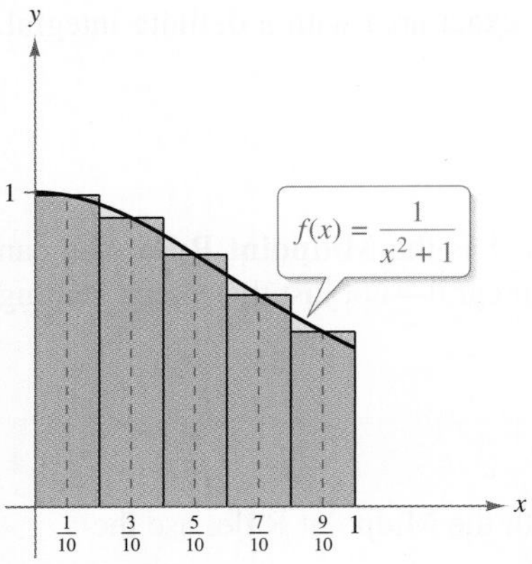

FIGURE 5.24

Example 2 Using the Midpoint Rule

Use the Midpoint Rule with $n = 5$ to approximate the area of the region bounded by the graph of

$$f(x) = \frac{1}{x^2 + 1}$$

the x-axis, and the lines $x = 0$ and $x = 1$.

SOLUTION The region is shown in Figure 5.24. With $n = 5$, the interval $[0, 1]$ is divided into five subintervals.

$$\left[0, \frac{1}{5}\right], \quad \left[\frac{1}{5}, \frac{2}{5}\right], \quad \left[\frac{2}{5}, \frac{3}{5}\right], \quad \left[\frac{3}{5}, \frac{4}{5}\right], \quad \left[\frac{4}{5}, 1\right]$$

The midpoints of these intervals are $\frac{1}{10}$, $\frac{3}{10}$, $\frac{5}{10}$, $\frac{7}{10}$, and $\frac{9}{10}$. Because each subinterval has a width of $\Delta x = (1 - 0)/5 = \frac{1}{5}$, you can approximate the value of the definite integral as shown.

$$\int_0^1 \frac{1}{x^2 + 1}\,dx \approx \frac{1}{5}\left(\frac{1}{1.01} + \frac{1}{1.09} + \frac{1}{1.25} + \frac{1}{1.49} + \frac{1}{1.81}\right)$$
$$\approx 0.786$$

The actual area of this region is $\pi/4 \approx 0.785$. So, the approximation is off by about 0.001.

Checkpoint 2

Use the Midpoint Rule with $n = 4$ to approximate the area of the region bounded by the graph of $f(x) = 1/(x^2 + 2)$, the x-axis, and the lines $x = 0$ and $x = 1$.

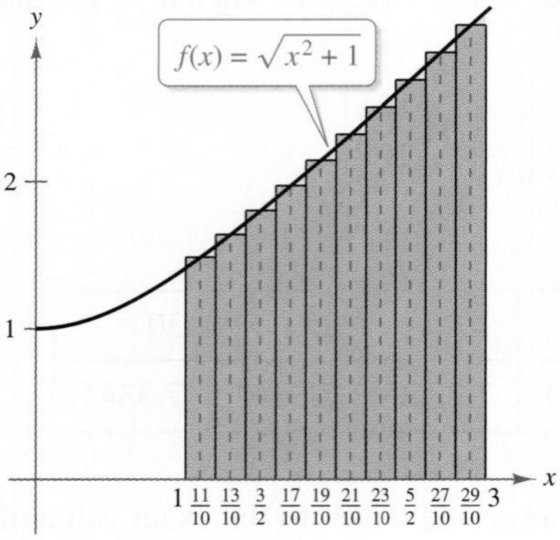

FIGURE 5.25

Example 3 Using the Midpoint Rule

Use the Midpoint Rule with $n = 10$ to approximate the area of the region bounded by the graph of $f(x) = \sqrt{x^2 + 1}$, the x-axis, and the lines $x = 1$ and $x = 3$.

SOLUTION The region is shown in Figure 5.25. After dividing the interval $[1, 3]$ into 10 subintervals, you can determine that the midpoints of these intervals are

$$\frac{11}{10}, \frac{13}{10}, \frac{3}{2}, \frac{17}{10}, \frac{19}{10}, \frac{21}{10}, \frac{23}{10}, \frac{5}{2}, \frac{27}{10}, \text{ and } \frac{29}{10}.$$

Because each subinterval has a width of $\Delta x = (3 - 1)/10 = \frac{1}{5}$, you can approximate the value of the definite integral as shown.

$$\int_1^3 \sqrt{x^2 + 1}\,dx \approx \frac{1}{5}\left[\sqrt{(1.1)^2 + 1} + \sqrt{(1.3)^2 + 1} + \cdots + \sqrt{(2.9)^2 + 1}\right]$$
$$\approx 4.504$$

It can be shown that the actual area is

$$\frac{1}{2}\left[3\sqrt{10} + \ln\left(3 + \sqrt{10}\right) - \sqrt{2} - \ln\left(1 + \sqrt{2}\right)\right] \approx 4.505.$$

So, the approximation is off by about 0.001.

STUDY TIP

The Midpoint Rule is necessary for solving certain real-life problems, such as measuring irregular areas like bodies of water (see Exercise 27).

Checkpoint 3

Use the Midpoint Rule with $n = 4$ to approximate the area of the region bounded by the graph of $f(x) = \sqrt{x^2 - 1}$, the x-axis, and the lines $x = 2$ and $x = 4$.

The Definite Integral as the Limit of a Sum

Consider the closed interval $[a, b]$, divided into n subintervals whose midpoints are x_i and whose widths are $\Delta x = (b - a)/n$. In this section, you have seen that the midpoint approximation

$$\int_a^b f(x)\,dx \approx f(x_1)\,\Delta x + f(x_2)\,\Delta x + f(x_3)\,\Delta x + \cdot \cdot \cdot + f(x_n)\,\Delta x$$
$$= [f(x_1) + f(x_2) + f(x_3) + \cdot \cdot \cdot + f(x_n)]\,\Delta x$$

becomes better and better as n increases. In fact, the limit of this sum as n approaches infinity is exactly equal to the definite integral. That is,

$$\int_a^b f(x)\,dx = \lim_{n\to\infty} [f(x_1) + f(x_2) + f(x_3) + \cdot \cdot \cdot + f(x_n)]\,\Delta x.$$

It can be shown that this limit is valid as long as x_i is *any* point in the ith interval.

Example 4 Approximating a Definite Integral

FIGURE 5.26

Use the Midpoint Rule program in Appendix E or a symbolic integration utility to approximate the definite integral

$$\int_0^1 e^{-x^2}\,dx.$$

SOLUTION Using the Midpoint Rule program (see Figure 5.26), you can complete the following table.

n	10	20	30	40	50
Approximation	0.7471	0.7469	0.7469	0.7468	0.7468

From the table, it appears that

$$\int_0^1 e^{-x^2}\,dx \approx 0.7468.$$

Using a symbolic integration utility, the value of the integral is approximately 0.7468241328.

✓Checkpoint 4

Use the Midpoint Rule program in Appendix E or a symbolic integration utility to approximate the definite integral

$$\int_0^1 e^{x^2}\,dx.$$

■

SUMMARIZE (Section 5.6)

1. Describe how to approximate the area of a region using rectangles *(page 360, Example 1).*
2. State the guidelines for using the Midpoint Rule *(page 361).* For examples of using these guidelines, see Examples 2 and 3.
3. State the definite integral as the limit of a sum *(page 363).*

elwynn/www.shutterstock.com

SKILLS WARM UP 5.6

The following warm-up exercises involve skills that were covered in a previous course or in earlier sections. You will use these skills in the exercise set for this section. For additional help, review Appendix Section A.2 and Section 3.6.

In Exercises 1–6, find the midpoint of the interval.

1. $\left[0, \frac{1}{3}\right]$

2. $\left[\frac{1}{10}, \frac{2}{10}\right]$

3. $\left[\frac{3}{20}, \frac{4}{20}\right]$

4. $\left[1, \frac{7}{6}\right]$

5. $\left[2, \frac{31}{15}\right]$

6. $\left[\frac{26}{9}, 3\right]$

In Exercises 7–10, find the limit.

7. $\displaystyle\lim_{x\to\infty} \frac{2x^2 + 4x - 1}{3x^2 - 2x}$

8. $\displaystyle\lim_{x\to\infty} \frac{4x + 5}{7x - 5}$

9. $\displaystyle\lim_{x\to\infty} \frac{x - 7}{x^2 + 1}$

10. $\displaystyle\lim_{x\to\infty} \frac{5x^3 + 1}{x^3 + x^2 + 4}$

Exercises 5.6

See www.CalcChat.com for worked-out solutions to odd-numbered exercises.

Approximating the Area of a Plane Region In Exercises 1–6, use the rectangles to approximate the area of the region. Compare your result with the exact area obtained using a definite integral. *See Example 1.*

1. $f(x) = -2x + 3, [0, 1]$

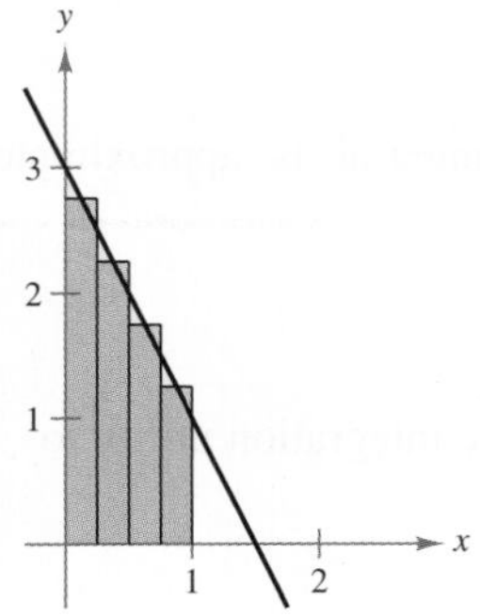

2. $f(x) = \dfrac{1}{x}, [1, 5]$

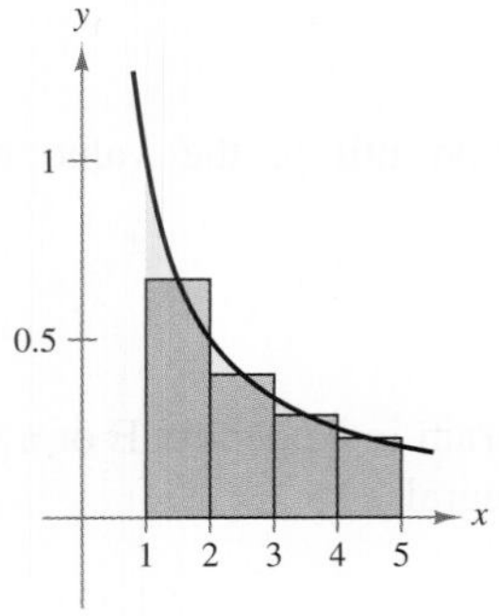

3. $f(x) = \sqrt{x}, [0, 1]$

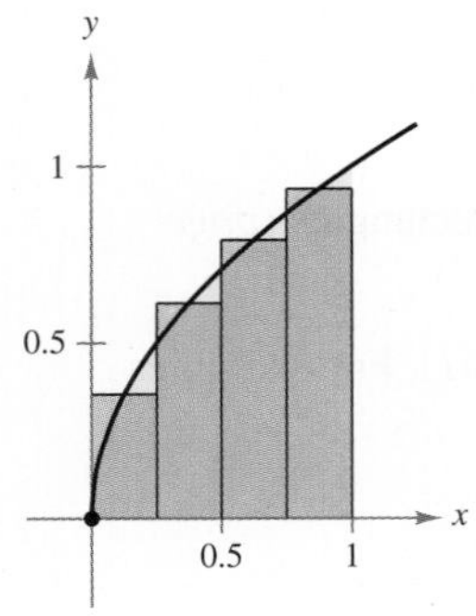

4. $f(x) = 1 - x^2, [-1, 1]$

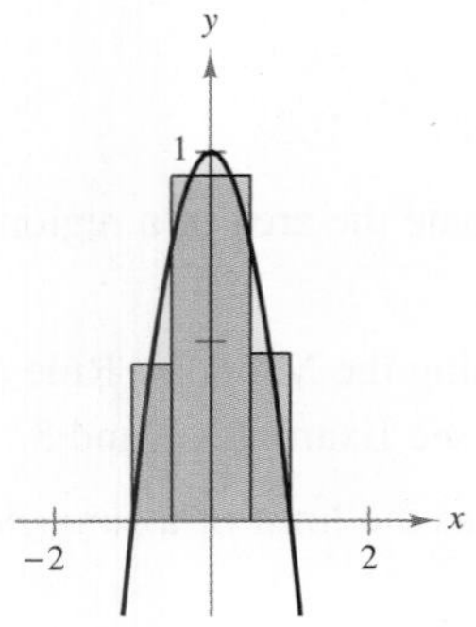

5. $f(x) = x^3 + 1, [0, 1]$

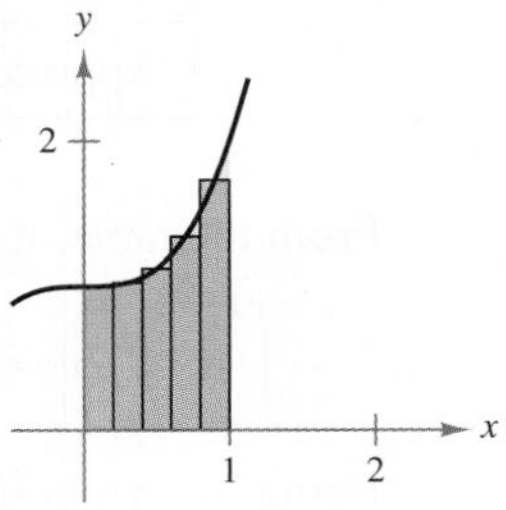

6. $f(x) = e^{-x/2}, [0, 3]$

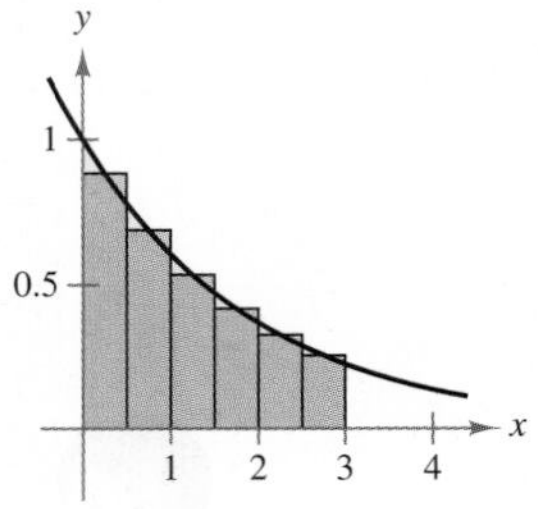

Using the Midpoint Rule In Exercises 7–12, use the Midpoint Rule with $n = 5$ to approximate the area of the region bounded by the graph of f and the x-axis over the interval. Sketch the region. *See Examples 2 and 3.*

Function	*Interval*
7. $f(x) = x^2$	$[1, 6]$
8. $f(x) = 4 - x^2$	$[-1, 0]$
9. $f(x) = x\sqrt{x + 4}$	$[0, 1]$
10. $f(x) = (x^2 + 1)^{2/3}$	$[0, 5]$
11. $f(x) = \dfrac{8}{x^2 + 1}$	$[-5, 5]$
12. $f(x) = \dfrac{5x}{x + 1}$	$[0, 2]$

Using the Midpoint Rule In Exercises 13–18, use the Midpoint Rule with $n = 4$ to approximate the area of the region bounded by the graph of f and the x-axis over the interval. Sketch the region. *See Examples 2 and 3.*

	Function	*Interval*
13.	$f(x) = 2x^2$	$[1, 3]$
14.	$f(x) = x^2 - x^3$	$[0, 1]$
15.	$f(x) = (x^2 - 4)^2$	$[-2, 2]$
16.	$f(x) = \sqrt{x^2 + 3}$	$[-1, 1]$
17.	$f(x) = \dfrac{1}{(x^2 + 1)^2}$	$[-1, 3]$
18.	$f(x) = \dfrac{2}{x^2 + 1}$	$[0, 2]$

Approximating a Definite Integral In Exercises 19–22, use the Midpoint Rule program in Appendix E or a symbolic integration utility to approximate the definite integral. If you use a Midpoint Rule program, complete the table. *See Example 4.*

n	10	20	30	40	50
Approximation					

19. $\displaystyle\int_0^4 \sqrt{2 + 3x^2}\,dx$ **20.** $\displaystyle\int_0^2 \frac{5}{x^3 + 1}\,dx$

21. $\displaystyle\int_1^3 x\sqrt[3]{x + 1}\,dx$ **22.** $\displaystyle\int_1^6 \frac{4}{\sqrt{1 + x^2}}\,dx$

Making a Closer Approximation In Exercises 23–26, use the Midpoint Rule program in Appendix E to approximate the definite integral. How large must n be to obtain an approximation that is correct to within 0.01?

23. $\displaystyle\int_0^4 (2x^2 + 3)\,dx$ **24.** $\displaystyle\int_1^2 (x^3 - 1)\,dx$

25. $\displaystyle\int_1^4 \frac{1}{x + 1}\,dx$ **26.** $\displaystyle\int_1^2 \sqrt{x + 2}\,dx$

27. Surface Area Use the Midpoint Rule to estimate the surface area of the pond shown in the figure.

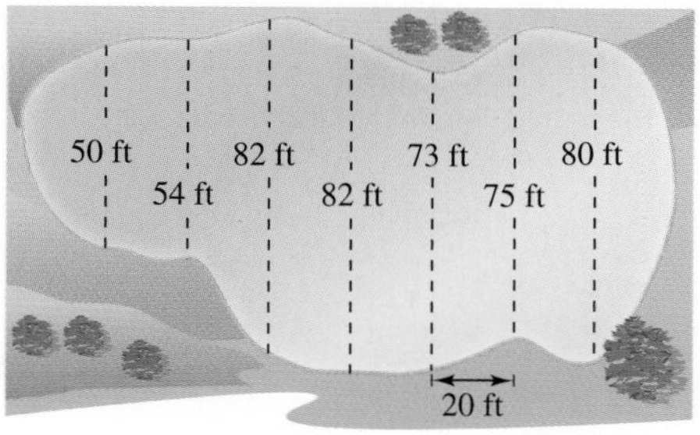

28. Surface Area Use the Midpoint Rule to estimate the surface area of the golf green shown in the figure.

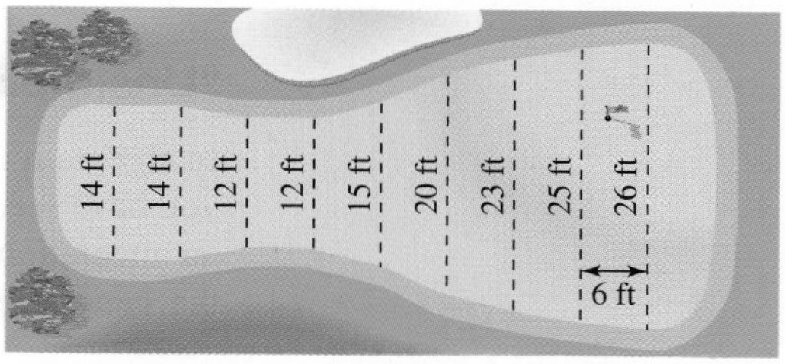

29. Surface Area Use the Midpoint Rule to estimate the surface area of the oil spill shown in the figure.

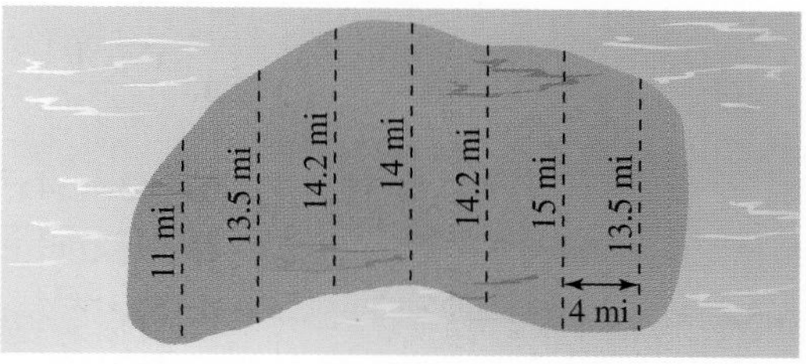

30. HOW DO YOU SEE IT? The graph shows three areas representing awake time, REM (rapid eye movement) sleep time, and non-REM sleep time, over a typical individual's lifetime. *(Source: Adapted from Bernstein/Clarke-Stewart/Roy/Wickens,* Psychology, *Seventh Edition)*

Sleep Patterns

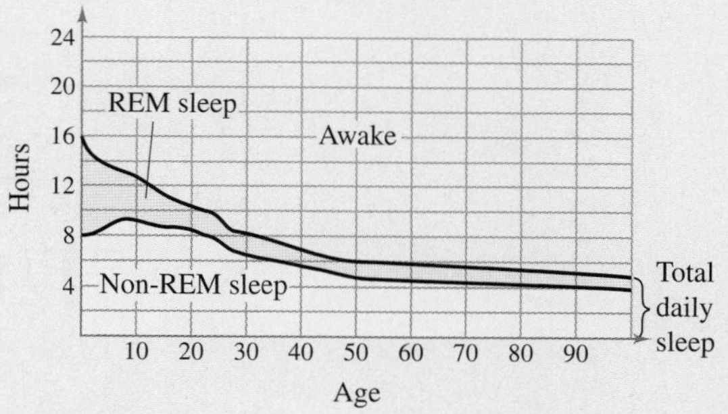

(a) Make generalizations about the amount of total sleep time (REM and non-REM) an individual gets as he or she gets older.

(b) How would you use the Midpoint Rule to estimate the amount of REM sleep time an individual gets between birth and age 10?

31. Numerical Approximation Use the Midpoint Rule with $n = 4$ to approximate π, where

$$\pi = \int_0^1 \frac{4}{1 + x^2}\,dx.$$

Then use a graphing utility to evaluate the definite integral. Compare your results.

ALGEBRA TUTOR

"Unsimplifying" an Algebraic Expression

In algebra it is often helpful to write an expression in its simplest form. In this chapter, you have seen that the reverse is often true in integration. That is, to fit an integrand to an integration formula, it often helps to "unsimplify" the expression. To do this, you use the same algebraic rules, but your goal is different. Here are some examples.

Example 1 Rewriting Algebraic Expressions

Rewrite each algebraic expression as indicated in the example.

a. $\dfrac{x-1}{\sqrt{x}}$

b. $x(3-4x^2)^2$

c. $7x^2\sqrt{x^3+1}$

d. $5xe^{-x^2}$

SOLUTION

a.
$$\begin{aligned}
\frac{x-1}{\sqrt{x}} &= \frac{x}{\sqrt{x}} - \frac{1}{\sqrt{x}} && \text{Example 5, page 315} \\
& && \text{Rewrite as two fractions.} \\
&= \frac{x^1}{x^{1/2}} - \frac{1}{x^{1/2}} && \text{Rewrite with rational exponents.} \\
&= x^{1-1/2} - x^{-1/2} && \text{Properties of exponents} \\
&= x^{1/2} - x^{-1/2} && \text{Simplify exponent.}
\end{aligned}$$

b.
$$\begin{aligned}
x(3-4x^2)^2 &= \frac{-8}{-8}x(3-4x^2)^2 && \text{Example 2, page 324} \\
& && \text{Multiply and divide by } -8. \\
&= \left(-\frac{1}{8}\right)(-8)x(3-4x^2)^2 && \text{Regroup.} \\
&= \left(-\frac{1}{8}\right)(3-4x^2)^2(-8x) && \text{Regroup.}
\end{aligned}$$

c.
$$\begin{aligned}
7x^2\sqrt{x^3+1} &= 7x^2(x^3+1)^{1/2} && \text{Example 5, page 325} \\
& && \text{Rewrite with rational exponent.} \\
&= \frac{3}{3}(7x^2)(x^3+1)^{1/2} && \text{Multiply and divide by 3.} \\
&= \frac{7}{3}(3x^2)(x^3+1)^{1/2} && \text{Regroup.} \\
&= \frac{7}{3}(x^3+1)^{1/2}(3x^2) && \text{Regroup.}
\end{aligned}$$

d.
$$\begin{aligned}
5xe^{-x^2} &= \frac{-2}{-2}(5x)e^{-x^2} && \text{Example 3, page 332} \\
& && \text{Multiply and divide by } -2. \\
&= \left(-\frac{5}{2}\right)(-2x)e^{-x^2} && \text{Regroup.} \\
&= \left(-\frac{5}{2}\right)e^{-x^2}(-2x) && \text{Regroup.}
\end{aligned}$$

Example 2 Rewriting Algebraic Expressions

Rewrite each algebraic expression.

a. $\dfrac{3x^2 + 2x - 1}{x^2}$

b. $\dfrac{1}{1 + e^{-x}}$

c. $\dfrac{x^2 + x + 1}{x - 1}$

d. $\dfrac{x^2 + 6x + 1}{x^2 + 1}$

SOLUTION

a. $\dfrac{3x^2 + 2x - 1}{x^2} = \dfrac{3x^2}{x^2} + \dfrac{2x}{x^2} - \dfrac{1}{x^2}$ — Example 7(a), page 335 Rewrite as separate fractions.

$= 3 + \dfrac{2}{x} - x^{-2}$ — Properties of exponents

$= 3 + 2\left(\dfrac{1}{x}\right) - x^{-2}$ — Regroup.

b. $\dfrac{1}{1 + e^{-x}} = \left(\dfrac{e^x}{e^x}\right)\dfrac{1}{1 + e^{-x}}$ — Example 7(b), page 335 Multiply and divide by e^x.

$= \dfrac{e^x}{e^x + e^x(e^{-x})}$ — Multiply.

$= \dfrac{e^x}{e^x + e^{x-x}}$ — Property of exponents

$= \dfrac{e^x}{e^x + e^0}$ — Simplify exponent.

$= \dfrac{e^x}{e^x + 1}$ — $e^0 = 1$

c. $\dfrac{x^2 + x + 1}{x - 1} = x + 2 + \dfrac{3}{x - 1}$ — Example 7(c), page 335 Use long division as shown below.

$$\begin{array}{r} x + 2 \\ x - 1 \enclose{longdiv}{x^2 + x + 1} \\ \underline{x^2 - x} \\ 2x + 1 \\ \underline{2x - 2} \\ 3 \end{array}$$

d. $\dfrac{x^2 + 6x + 1}{x^2 + 1} = 1 + \dfrac{6x}{x^2 + 1}$ — Bottom of page 334 Use long division as shown below.

$$\begin{array}{r} 1 \\ x^2 + 1 \enclose{longdiv}{x^2 + 6x + 1} \\ \underline{x^2 \qquad + 1} \\ 6x \end{array}$$

SUMMARY AND STUDY STRATEGIES

After studying this chapter, you should have acquired the following skills.
The exercise numbers are keyed to the Review Exercises that begin on page 370.
Answers to odd-numbered Review Exercises are given in the back of the text.*

Section 5.1	Review Exercises
■ Use basic integration rules to find indefinite integrals.	*1–14*

$$\int k\,dx = kx + C$$

$$\int kf(x)\,dx = k\int f(x)\,dx$$

$$\int [f(x) + g(x)]\,dx = \int f(x)\,dx + \int g(x)\,dx$$

$$\int [f(x) - g(x)]\,dx = \int f(x)\,dx - \int g(x)\,dx$$

$$\int x^n\,dx = \frac{x^{n+1}}{n+1} + C, \quad n \neq -1$$

■ Use initial conditions to find particular solutions of indefinite integrals.	*15–18*
■ Use antiderivatives to solve real-life problems.	*19, 20*

Section 5.2

■ Use the General Power Rule or integration by substitution to find indefinite integrals.	*21–32*

$$\int u^n \frac{du}{dx}\,dx = \int u^n\,du = \frac{u^{n+1}}{n+1} + C, \quad n \neq -1$$

■ Use the General Power Rule or integration by substitution to solve real-life problems.	*33, 34*

Section 5.3

■ Use the Exponential and Log Rules to find indefinite integrals.	*35–46*

$$\int e^x\,dx = e^x + C \qquad \int \frac{1}{x}\,dx = \ln|x| + C$$

$$\int e^u \frac{du}{dx}\,dx = \int e^u\,du = e^u + C \qquad \int \frac{du/dx}{u}\,dx = \int \frac{1}{u}\,du = \ln|u| + C$$

Section 5.4

■ Find the areas of regions using a geometric formula.	*47–50*
■ Use properties of definite integrals.	*51, 52*
■ Find the areas of regions bounded by the graph of a function and the x-axis.	*53–58*

* A wide range of valuable study aids are available to help you master the material in this chapter. The *Student Solutions Manual* includes step-by-step solutions to all odd-numbered exercises to help you review and prepare. The student website at *www.cengagebrain.com* offers algebra help and a *Graphing Technology Guide,* which contains step-by-step commands and instructions for a wide variety of graphing calculators.

Section 5.4 (continued)

	Review Exercises
■ Use the Fundamental Theorem of Calculus to evaluate definite integrals.	59–70
■ Find average values of functions over closed intervals.	71–76
■ Use properties of even and odd functions to help evaluate definite integrals.	77–80
■ Find amounts of annuities.	81, 82
■ Use definite integrals to solve marginal analysis problems.	83, 84
■ Use average values to solve real-life problems.	85, 86

$$\int_a^b f(x)\,dx = F(x)\Big]_a^b = F(b) - F(a), \quad \text{where} \quad F'(x) = f(x)$$

$$\text{Average value} = \frac{1}{b-a}\int_a^b f(x)\,dx$$

Even function: $f(-x) = f(x)$

If f is an *even* function, then $\int_{-a}^{a} f(x)\,dx = 2\int_0^a f(x)\,dx.$

Odd function: $f(-x) = -f(x)$

If f is an *odd* function, then $\int_{-a}^{a} f(x)\,dx = 0.$

Section 5.5

■ Find areas of regions bounded by two graphs.	87–94
■ Find consumer and producer surpluses.	95–98
■ Use the areas of regions bounded by two graphs to solve real-life problems.	99–102

$$A = \int_a^b [f(x) - g(x)]\,dx$$

Section 5.6

■ Use the Midpoint Rule to approximate values of definite integrals.	103–112
■ Use the Midpoint Rule to solve real-life problems.	113

$$\int_a^b f(x)\,dx \approx \frac{b-a}{n}[f(x_1) + f(x_2) + f(x_3) + \cdot \cdot \cdot + f(x_n)]$$

Study Strategies

■ **Indefinite and Definite Integrals** When evaluating integrals, remember that an indefinite integral is a *family of antiderivatives*, each differing by a constant C, whereas a definite integral is a *number*.

■ **Checking Antiderivatives by Differentiating** When finding an antiderivative, remember that you can check your result by differentiating. For example, you can confirm that the antiderivative

$$\int (3x^3 - 4x)\,dx = \frac{3}{4}x^4 - 2x^2 + C \quad \text{is correct by differentiating to obtain} \quad \frac{d}{dx}\left[\frac{3}{4}x^4 - 2x^2 + C\right] = 3x^3 - 4x.$$

Because the derivative is equal to the original integrand, you know that the antiderivative is correct.

■ **Grouping Symbols and the Fundamental Theorem** When using the Fundamental Theorem of Calculus to evaluate a definite integral, you can avoid sign errors by using grouping symbols. Here is an example.

$$\int_1^3 (x^3 - 9x)\,dx = \left[\frac{x^4}{4} - \frac{9x^2}{2}\right]_1^3 = \left[\frac{3^4}{4} - \frac{9(3^2)}{2}\right] - \left[\frac{1^4}{4} - \frac{9(1^2)}{2}\right] = \frac{81}{4} - \frac{81}{2} - \frac{1}{4} + \frac{9}{2} = -16$$

Review Exercises

See www.CalcChat.com for worked-out solutions to odd-numbered exercises.

Finding Indefinite Integrals In Exercises 1–14, find the indefinite integral. Check your result by differentiation.

1. $\int 16\,dx$
2. $\int -9\,dx$
3. $\int \frac{3}{5}x\,dx$
4. $\int 6x\,dx$
5. $\int 3x^2\,dx$
6. $\int 8x^3\,dx$
7. $\int (2x^2 + 5x)\,dx$
8. $\int (5 - 6x^2)\,dx$
9. $\int \frac{2}{3\sqrt[3]{x}}\,dx$
10. $\int 6x^{5/2}\,dx$
11. $\int \left(\sqrt[3]{x^4} + 3x\right)dx$
12. $\int \left(\frac{4}{\sqrt{x}} + \sqrt{x}\right)dx$
13. $\int \frac{2x^4 - 1}{\sqrt{x}}\,dx$
14. $\int \frac{1 - 3x}{x^2}\,dx$

Finding Particular Solutions In Exercises 15–18, find the particular solution that satisfies the differential equation and the initial condition.

15. $f'(x) = 12x;\ f(0) = -3$
16. $f'(x) = 3x + 1;\ f(2) = 6$
17. $f'(x) = 3x^2 - 8x;\ f(1) = 12$
18. $f'(x) = \sqrt{x};\ f(9) = 4$

19. **Vertical Motion** An object is projected upward from the ground with an initial velocity of 80 feet per second. Express the height s (in feet) of the object as a function of the time t (in seconds). How long will the object be in the air? (Use $s''(t) = -32$ feet per second per second as the acceleration due to gravity.)

20. **Revenue** A company produces a new product for which the rate of change of the revenue can be modeled by

$$\frac{dR}{dt} = 0.675t^{3/2}, \quad 0 \le t \le 225$$

where t is the time (in weeks). When $t = 0$, $R = 0$.

(a) Find a model for the revenue function.

(b) What is the revenue after 20 weeks?

(c) When will the weekly revenue be \$27,000?

Applying the General Power Rule In Exercises 21–32, find the indefinite integral. Check your result by differentiation.

21. $\int (x + 4)^3\,dx$
22. $\int (x - 6)^{4/3}\,dx$
23. $\int (5x + 1)^4(5)\,dx$
24. $\int (x^3 + 1)^2(3x^2)\,dx$
25. $\int (1 + 5x)^2\,dx$
26. $\int (6x - 2)^4\,dx$
27. $\int x^2(3x^3 + 1)^2\,dx$
28. $\int x(1 - 4x^2)^3\,dx$
29. $\int \frac{x^2}{(2x^3 - 5)^3}\,dx$
30. $\int \frac{x^2}{(x^3 - 4)^2}\,dx$
31. $\int \frac{1}{\sqrt{5x - 1}}\,dx$
32. $\int \frac{4x}{\sqrt{1 - 3x^2}}\,dx$

33. **Production** The rate of change of the output of a small sawmill is modeled by

$$\frac{dP}{dt} = 2t(0.001t^2 + 0.5)^{1/4}, \quad 0 \le t \le 40$$

where t is the time (in hours) and P is the output (in board-feet). Find the numbers of board-feet produced in (a) 6 hours and (b) 12 hours.

34. Cost The marginal cost for a catering service to cater to x people can be modeled by

$$\frac{dC}{dx} = \frac{5x}{\sqrt{x^2 + 1000}}.$$

When $x = 225$, the cost C (in dollars) is \$1136.06. Find the costs of catering to (a) 500 people and (b) 1000 people.

Using the Exponential and Log Rules In Exercises 35–46, use the Exponential Rule or the Log Rule to find the indefinite integral.

35. $\int 4e^{4x}\,dx$

36. $\int 3e^{-3x}\,dx$

37. $\int e^{-5x}\,dx$

38. $\int e^{6x}\,dx$

39. $\int 7xe^{3x^2}\,dx$

40. $\int (2t - 1)e^{t^2-t}\,dt$

41. $\int \frac{1}{x - 6}\,dx$

42. $\int \frac{1}{1 - 4x}\,dx$

43. $\int \frac{4}{6x - 1}\,dx$

44. $\int \frac{5}{2x + 3}\,dx$

45. $\int \frac{x^2}{1 - x^3}\,dx$

46. $\int \frac{x - 4}{x^2 - 8x}\,dx$

Evaluating a Definite Integral Using a Geometric Formula In Exercises 47–50, sketch the region whose area is given by the definite integral. Then use a geometric formula to evaluate the integral.

47. $\int_0^3 2\,dx$

48. $\int_0^6 \frac{x}{2}\,dx$

49. $\int_0^4 (4 - x)\,dx$

50. $\int_{-4}^{4} \sqrt{16 - x^2}\,dx$

51. Using Properties of Definite Integrals Given

$$\int_2^6 f(x)\,dx = 10 \quad \text{and} \quad \int_2^6 g(x)\,dx = 3$$

evaluate the definite integral.

(a) $\int_2^6 [f(x) + g(x)]\,dx$ (b) $\int_2^6 [f(x) - g(x)]\,dx$

(c) $\int_2^6 [2f(x) - 3g(x)]\,dx$ (d) $\int_2^6 5f(x)\,dx$

52. Using Properties of Definite Integrals Given

$$\int_0^3 f(x)\,dx = 4 \quad \text{and} \quad \int_3^6 f(x)\,dx = -1$$

evaluate the definite integral.

(a) $\int_0^6 f(x)\,dx$ (b) $\int_6^3 f(x)\,dx$

(c) $\int_4^4 f(x)\,dx$ (d) $\int_3^6 -10f(x)\,dx$

Finding Area by the Fundamental Theorem In Exercises 53–58, find the area of the region.

53. $f(x) = 4 - x^2$

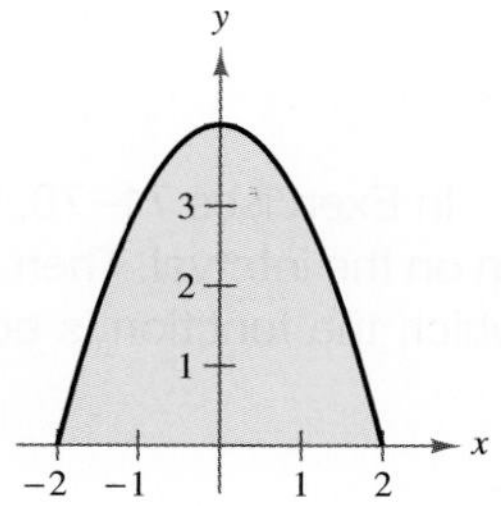

54. $f(x) = 9 - x^2$

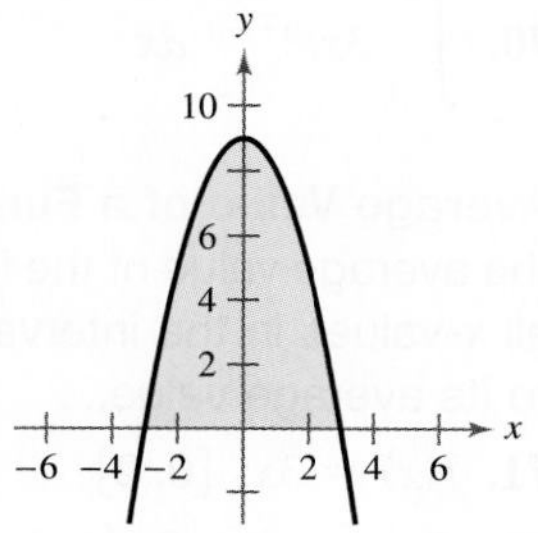

55. $f(x) = \frac{2}{x + 1}$

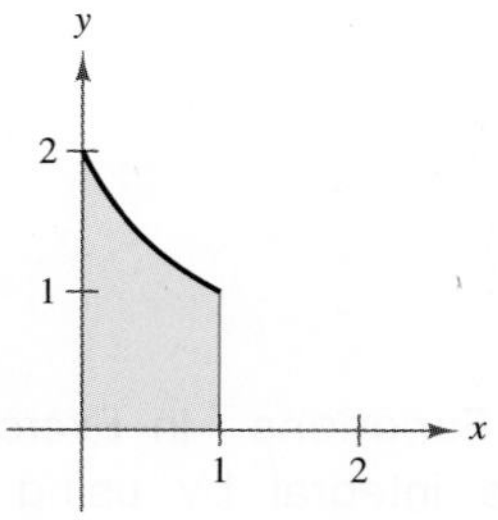

56. $f(x) = \frac{4}{\sqrt{x}}$

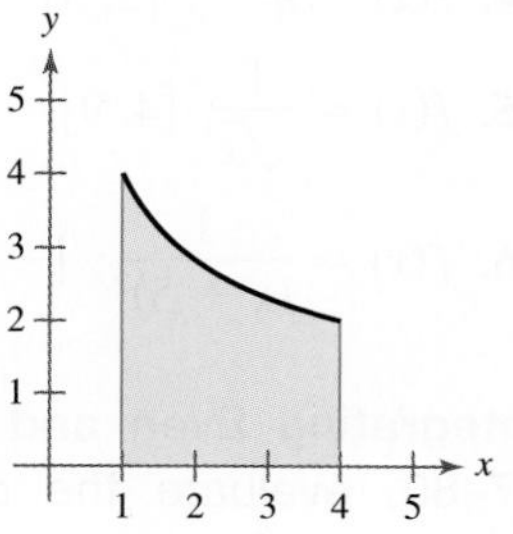

57. $f(x) = 2e^{x/2}$

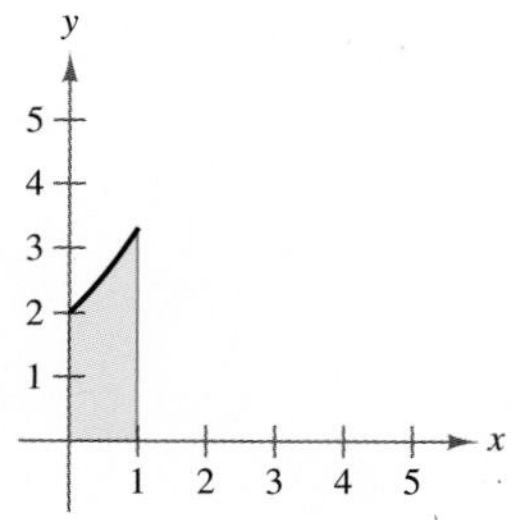

58. $f(x) = \frac{x - 1}{x}$

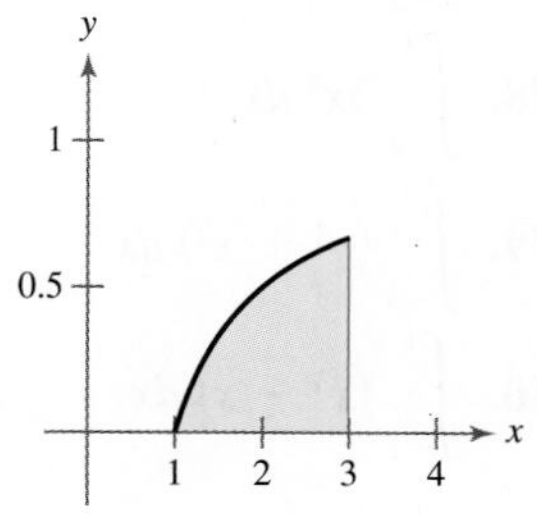

Evaluating a Definite Integral In Exercises 59–70, use the Fundamental Theorem of Calculus to evaluate the definite integral.

59. $\int_0^4 (2 + x)\,dx$

60. $\int_{-1}^1 (t^2 + 2)\,dt$

61. $\int_{-1}^1 (4t^3 - 2t)\,dt$

62. $\int_{-2}^2 (x^4 + 2x^2 - 5)\,dx$

63. $\int_{-2}^0 (x + 2)^3\,dx$

64. $\int_2^4 (2x - 3)^2\,dx$

65. $\int_0^3 \frac{1}{\sqrt{1 + x}}\,dx$

66. $\int_3^6 \frac{x}{3\sqrt{x^2 - 8}}\,dx$

67. $\int_3^9 \frac{5}{x}\,dx$

68. $\int_1^2 \left(\frac{1}{x^2} - \frac{1}{x^3}\right)dx$

69. $\int_0^{\ln 5} e^{x/5}\,dx$

70. $\int_{-1}^1 3xe^{x^2 - 1}\,dx$

Average Value of a Function In Exercises 71–76, find the average value of the function on the interval. Then find all x-values in the interval for which the function is equal to its average value.

71. $f(x) = 3x;\ [0, 2]$

72. $f(x) = x^2 + 2;\ [-3, 3]$

73. $f(x) = -2e^x;\ [0, 3]$

74. $f(x) = e^{5-x};\ [2, 5]$

75. $f(x) = \frac{1}{\sqrt{x}};\ [4, 9]$

76. $f(x) = \frac{1}{(x + 5)^2};\ [-1, 6]$

Integrating Even and Odd Functions In Exercises 77–80, evaluate the definite integral by using the properties of even and odd functions.

77. $\int_{-2}^2 6x^5\,dx$

78. $\int_{-4}^4 3x^4\,dx$

79. $\int_{-3}^3 (x^4 + x^2)\,dx$

80. $\int_{-1}^1 (x^3 - x)\,dx$

Finding the Amount of an Annuity In Exercises 81 and 82, find the amount of an annuity with income function $c(t)$, interest rate r, and term T.

81. $c(t) = \$3000,\ r = 6\%,\ T = 5$ years

82. $c(t) = \$1200,\ r = 7\%,\ T = 8$ years

83. Cost The marginal cost of serving an additional typical client at a law firm can be modeled by

$$\frac{dC}{dx} = 675 + 0.5x$$

where x is the number of clients. Find the change in cost C (in dollars) when x increases from 50 to 51 clients.

84. Profit The marginal profit obtained by selling x dollars of automobile insurance can be modeled by

$$\frac{dP}{dx} = 0.4\left(1 - \frac{5000}{x}\right), \quad x \geq 5000.$$

Find the change in the profit P (in dollars) when x increases from \$75,000 to \$100,000.

85. Compound Interest A deposit of \$500 is made in a savings account at an annual interest rate of 4%, compounded continuously. Find the average balance in the account during the first 2 years.

86. Revenue The rate of change in revenue for Texas Roadhouse from 2003 through 2009 can be modeled by

$$\frac{dR}{dt} = -11.5000t^2 + 142.140t - 294.91$$

where R is the revenue (in millions of dollars) and t is the time in years, with $t = 3$ corresponding to 2003. In 2006, the revenue for Texas Roadhouse was \$597.1 million. *(Source: Texas Roadhouse, Inc.)*

(a) Find the model for the revenue of Texas Roadhouse.

(b) What was the average revenue of Texas Roadhouse for 2003 through 2009?

Finding the Area Bounded by Two Graphs In Exercises 87–94, sketch the region bounded by the graphs of the functions and find the area of the region.

87. $y = \frac{1}{x^3},\ y = 0,\ x = 1,\ x = 3$

88. $y = x^2 + 4x - 5,\ y = 4x - 1$

89. $y = (x - 3)^2,\ y = 8 - (x - 3)^2$

90. $y = 4 - x,\ y = x^2 - 5x + 8,\ x = 0$

91. $y = \frac{4}{\sqrt{x + 1}},\ y = 0,\ x = 0,\ x = 8$

92. $y = \sqrt{x}(1 - x),\ y = 0$

93. $y = x,\ y = x^3$

94. $y = x^3 - 4x,\ y = -x^2 - 2x$

Consumer and Producer Surpluses In Exercises 95–98, find the consumer and producer surpluses by using the demand and supply functions, where p is the price (in dollars) and x is the number of units (in millions).

	Demand Function	*Supply Function*
95.	$p = 36 - 0.35x$	$p = 0.05x$
96.	$p = 200 - 0.2x$	$p = 50 + 1.3x$
97.	$p = 250 - x$	$p = 150 + x$
98.	$p = 500 - x$	$p = 1.25x + 162.5$

99. Revenue For the years 2015 through 2020, two models, R_1 and R_2, used to project the revenue (in millions of dollars) for a company are

$R_1 = 24.3 + 8.24t$

and

$R_2 = 21.6 + 9.36t$

where $t = 15$ corresponds to 2015. Which model projects the greater revenue? How much more total revenue does that model project over the six-year period?

100. Sales For the years 2000 through 2009, the sales (in millions of dollars) for Men's Wearhouse can be modeled by

$$R = \begin{cases} 23.596t^2 - 41.55t + 1310.5, & 0 \le t \le 6 \\ 38.7t^2 - 720.7t + 5261.2, & 6 < t \le 9 \end{cases}$$

where t is the year, with $t = 0$ corresponding to 2000. *(Source: Men's Wearhouse, Inc.)*

(a) Use a graphing utility to graph this model.

(b) Suppose the sales from 2007 through 2009 had continued to follow the model for 2000 through 2006. How much more or less would the sales have been for Men's Wearhouse?

101. Cost, Revenue, and Profit The revenue from a manufacturing process (in millions of dollars) is projected to follow the model

$R = 70$

for 10 years. Over the same period of time, the cost (in millions of dollars) is projected to follow the model

$C = 30 + 0.3t^2$

where t is the time (in years). Approximate the profit over the 10-year period.

102. Cost, Revenue, and Profit Repeat Exercise 101 for revenue and cost models given by

$R = 70 + 0.1t$

and

$C = 30 + 0.3t^2.$

Did the profit increase or decrease? Explain why.

Approximating the Area of a Plane Region In Exercises 103 and 104, use the rectangles to approximate the area of the region. Compare your result with the exact area obtained using a definite integral.

103. $f(x) = \dfrac{x}{3}$, $[0, 3]$

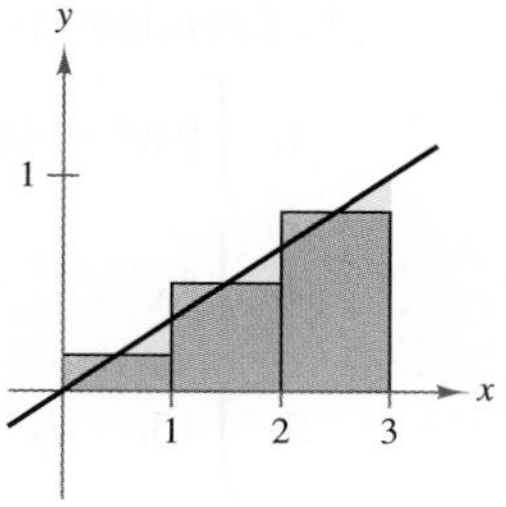

104. $f(x) = x^2 + 1$, $[0, 1]$

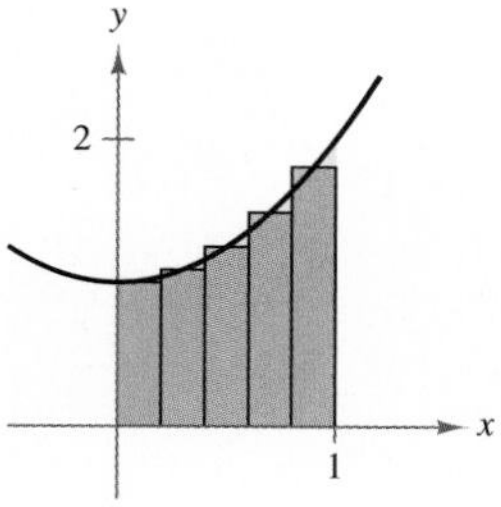

Using the Midpoint Rule In Exercises 105–108, use the Midpoint Rule with $n = 4$ to approximate the area of the region bounded by the graph of f and the x-axis over the interval. Sketch the region.

	Function	*Interval*
105.	$f(x) = x^2$	$[0, 2]$
106.	$f(x) = 2x - x^3$	$[0, 1]$
107.	$f(x) = (x^2 - 1)^2$	$[-1, 1]$
108.	$f(x) = \dfrac{3x}{x + 2}$	$[0, 4]$

Using the Midpoint Rule In Exercises 109–112, use the Midpoint Rule with $n = 6$ to approximate the area of the region bounded by the graph of f and the x-axis over the interval. Sketch the region.

	Function	*Interval*
109.	$f(x) = x + 3$	$[0, 3]$
110.	$f(x) = 9 - x^2$	$[-3, 3]$
111.	$f(x) = x\sqrt{x + 1}$	$[0, 2]$
112.	$f(x) = \dfrac{3}{x^2 + 1}$	$[-6, 6]$

113. Surface Area Use the Midpoint Rule to estimate the surface area of the swamp shown in the figure.

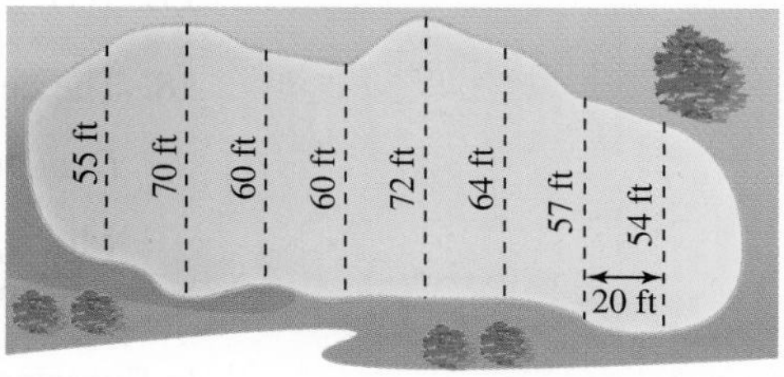

TEST YOURSELF

See www.CalcChat.com for worked-out solutions to odd-numbered exercises.

Take this test as you would take a test in class. When you are done, check your work against the answers given in the back of the book.

In Exercises 1–6, find the indefinite integral.

1. $\int (9x^2 - 4x + 13)\, dx$

2. $\int (x + 1)^2\, dx$

3. $\int 4x^3\sqrt{x^4 - 7}\, dx$

4. $\int \frac{5x - 6}{\sqrt{x}}\, dx$

5. $\int 15e^{3x}\, dx$

6. $\int \frac{3}{4x - 1}\, dx$

In Exercises 7 and 8, find the particular solution that satisfies the differential equation and initial condition.

7. $f'(x) = 6x - 5;\ f(-1) = 6$

8. $f'(x) = e^x + 1;\ f(0) = 1$

In Exercises 9–14, evaluate the definite integral.

9. $\int_0^1 16x\, dx$

10. $\int_{-3}^{3} (3 - 2x)\, dx$

11. $\int_{-1}^{1} (x^3 + x^2)\, dx$

12. $\int_{-1}^{2} \frac{2x}{\sqrt{x^2 + 1}}\, dx$

13. $\int_0^3 e^{4x}\, dx$

14. $\int_{-2}^{3} \frac{1}{x + 3}\, dx$

15. The rate of change in sales of PetSmart from 2000 through 2009 can be modeled by

$$\frac{dS}{dt} = 226.912e^{0.1013t}$$

where S is the sales (in millions of dollars) and t is the time (in years), with $t = 0$ corresponding to 2000. In 2004, the sales of PetSmart were \$3363.5 million. *(Source: PetSmart, Inc.)*

(a) Find the model for the sales of PetSmart.

(b) What were the average sales for 2000 through 2009?

In Exercises 16 and 17, sketch the region bounded by the graphs of the functions and find the area of the region.

16. $f(x) = 6,\ g(x) = x^2 - x - 6$

17. $f(x) = \sqrt[3]{x},\ g(x) = x^2$

18. The demand and supply functions for a product are modeled by

Demand: $p = -0.625x + 10$ and *Supply:* $p = 0.25x + 3$

where p is the price (in dollars) and x is the number of units (in millions). Find the consumer and producer surpluses for this product.

In Exercises 19 and 20, use the Midpoint Rule with $n = 4$ to approximate the area of the region bounded by the graph of f and the x-axis over the interval. Compare your result with the exact area. Sketch the region.

19. $f(x) = 3x^2,\quad [0, 1]$

20. $f(x) = x^2 + 1,\quad [-1, 1]$

6 Techniques of Integration

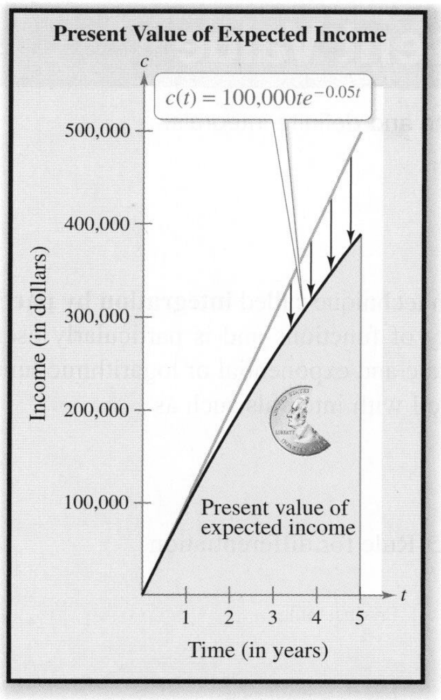

Example 7 on page 382 shows how integration by parts can be used to find the present value of a company's future income.

6.1 Integration by Parts and Present Value

- Use integration by parts to find indefinite and definite integrals.
- Find the present value of future income.

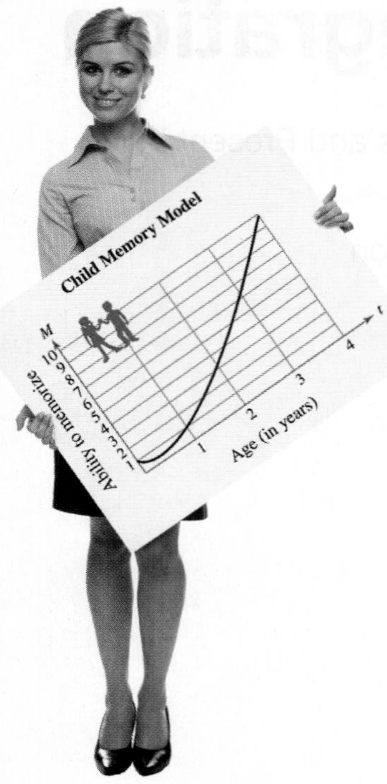

In Exercise 65 on page 384, you will use integration by parts to find the average value of a memory model for children.

Integration by Parts

In this section, you will study an integration technique called **integration by parts.** This technique can be applied to a wide variety of functions and is particularly useful for integrands involving the products of algebraic and exponential or logarithmic functions. For instance, integration by parts works well with integrals such as

$$\int x^2e^x\,dx \quad \text{and} \quad \int x \ln x\,dx.$$

Integration by parts is based on the Product Rule for differentiation.

$$\frac{d}{dx}[uv] = u\frac{dv}{dx} + v\frac{du}{dx} \qquad \text{Product Rule}$$

$$uv = \int u\frac{dv}{dx}\,dx + \int v\frac{du}{dx}\,dx \qquad \text{Integrate each side.}$$

$$uv = \int u\,dv + \int v\,du \qquad \text{Write in differential form.}$$

$$\int u\,dv = uv - \int v\,du \qquad \text{Rewrite.}$$

Integration by Parts

Let u and v be differentiable functions of x.

$$\int u\,dv = uv - \int v\,du$$

Note that the formula for integration by parts expresses the original integral in terms of another integral. Depending on the choices of u and dv, it may be easier to evaluate the second integral than the original one. Because the choices of u and dv are critical in the integration by parts process, the following guidelines are provided.

Guidelines for Integration by Parts

1. Try letting dv be the most complicated portion of the integrand that fits a basic integration rule. Then u will be the remaining factor(s) of the integrand.
2. Try letting u be the portion of the integrand whose derivative is a function simpler than u. Then dv will be the remaining factor(s) of the integrand.

Note that dv always includes the dx of the original integrand.

When using integration by parts, note that you can first choose dv or first choose u. After you choose, however, the choice of the other factor is determined—it must be the remaining portion of the integrand. Also note that dv must contain the differential dx of the original integral.

Example 1 Integration by Parts

Find $\int xe^x\,dx$.

SOLUTION To apply integration by parts, you must rewrite the original integral in the form $\int u\,dv$. That is, you must break $xe^x\,dx$ into two factors—one "part" representing u and the other "part" representing dv. There are several ways to do this.

$$\int \underbrace{(x)}_{u}\underbrace{(e^x\,dx)}_{dv} \qquad \int \underbrace{(e^x)}_{u}\underbrace{(x\,dx)}_{dv} \qquad \int \underbrace{(1)}_{u}\underbrace{(xe^x\,dx)}_{dv} \qquad \int \underbrace{(xe^x)}_{u}\underbrace{(dx)}_{dv}$$

The guidelines on the preceding page suggest the first option because $dv = e^x\,dx$ is the most complicated portion of the integrand that fits a basic integration formula *and* because the derivative of $u = x$ is simpler than x.

$$dv = e^x\,dx \quad \Rightarrow \quad v = \int dv = \int e^x\,dx = e^x$$

$$u = x \quad \Rightarrow \quad du = dx$$

Next, you can apply the integration by parts formula as shown.

$$\int u\,dv = uv - \int v\,du \qquad \text{Integration by parts formula}$$

$$\int xe^x\,dx = xe^x - \int e^x\,dx \qquad \text{Substitute.}$$

$$= xe^x - e^x + C \qquad \text{Integrate } \int e^x\,dx.$$

You can check this result by differentiating.

$$\frac{d}{dx}[xe^x - e^x + C] = xe^x + e^x(1) - e^x = xe^x$$

✓ **Checkpoint 1**

Find $\int xe^{2x}\,dx$. ■

In Example 1, notice that you do not need to include a constant of integration when solving $v = \int e^x\,dx = e^x$. To see why this is true, try replacing e^x by $e^x + C_1$ in the solution.

$$\int xe^x\,dx = x(e^x + C_1) - \int (e^x + C_1)\,dx$$

$$= xe^x + C_1x - e^x - C_1x + C$$

$$= xe^x - e^x + C$$

After integrating, you can see that the terms involving C_1 subtract out.

TECH TUTOR

If you have access to a symbolic integration utility, try using it to solve several of the exercises in this section. Note that the form of the integral may be slightly different from what you obtain when solving the exercise by hand.

STUDY TIP

To remember the integration by parts formula, you might like to use the "Z" pattern below. The top row represents the original integral, the diagonal row represents uv, and the bottom row represents the new integral.

Top row — Diagonal row — Bottom row

$$\int u\,dv = uv - \int v\,du$$

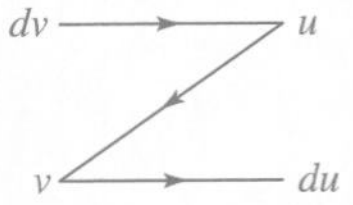

Example 2 Integration by Parts

Find $\int x^2 \ln x\,dx$.

SOLUTION In this case, x^2 is more easily integrated than $\ln x$. Furthermore, the derivative of $\ln x$ is simpler than $\ln x$. So, you should choose $dv = x^2\,dx$.

$$dv = x^2\,dx \quad \Rightarrow \quad v = \int dv = \int x^2\,dx = \frac{x^3}{3}$$

$$u = \ln x \quad \Rightarrow \quad du = \frac{1}{x}\,dx$$

Next, apply the integration by parts formula.

$$\int u\,dv = uv - \int v\,du \qquad \text{Integration by parts formula}$$

$$\int x^2 \ln x\,dx = \frac{x^3}{3}\ln x - \int \left(\frac{x^3}{3}\right)\left(\frac{1}{x}\right)dx \qquad \text{Substitute.}$$

$$= \frac{x^3}{3}\ln x - \frac{1}{3}\int x^2\,dx \qquad \text{Simplify.}$$

$$= \frac{x^3}{3}\ln x - \frac{x^3}{9} + C \qquad \text{Integrate.}$$

✓Checkpoint 2

Find $\int x \ln x\,dx$. ■

Example 3 Integrating by Parts with a Single Factor

Find $\int \ln x\,dx$.

SOLUTION This integrand is unusual because it has only one factor. In such cases, you should choose $dv = dx$ and choose u to be the single factor.

$$dv = dx \quad \Rightarrow \quad v = \int dv = \int dx = x$$

$$u = \ln x \quad \Rightarrow \quad du = \frac{1}{x}\,dx$$

Next, apply the integration by parts formula.

$$\int u\,dv = uv - \int v\,du \qquad \text{Integration by parts formula}$$

$$\int \ln x\,dx = x\ln x - \int (x)\left(\frac{1}{x}\right)dx \qquad \text{Substitute.}$$

$$= x\ln x - \int dx \qquad \text{Simplify.}$$

$$= x\ln x - x + C \qquad \text{Integrate.}$$

✓Checkpoint 3

Find $\int \ln 2x\,dx$. ■

Example 4 Using Integration by Parts Repeatedly

Find $\int x^2 e^x \, dx$.

SOLUTION The factors x^2 and e^x are both easy to integrate. Notice, however, that the derivative of x^2 becomes simpler, whereas the derivative of e^x does not. So, you should let $u = x^2$ and let $dv = e^x \, dx$.

$$dv = e^x \, dx \quad \Rightarrow \quad v = \int dv = \int e^x \, dx = e^x$$

$$u = x^2 \quad \Rightarrow \quad du = 2x \, dx$$

Next, apply the integration by parts formula.

$$\int x^2 e^x \, dx = x^2 e^x - \int 2xe^x \, dx$$

First application of integration by parts

This first use of integration by parts has succeeded in simplifying the original integral, but the integral on the right still doesn't fit a basic integration rule. To evaluate that integral, you can apply integration by parts again. This time, let $u = 2x$ and $dv = e^x \, dx$.

$$dv = e^x \, dx \quad \Rightarrow \quad v = \int dv = \int e^x \, dx = e^x$$

$$u = 2x \quad \Rightarrow \quad du = 2 \, dx$$

Next, apply the integration by parts formula.

$$\int x^2 e^x \, dx = x^2 e^x - \int 2xe^x \, dx$$

First application of integration by parts

$$= x^2 e^x - \left(2xe^x - \int 2e^x \, dx\right)$$

Second application of integration by parts

$$= x^2 e^x - 2xe^x + 2e^x + C$$

Integrate.

$$= e^x(x^2 - 2x + 2) + C$$

Simplify.

You can confirm this result by differentiating.

$$\frac{d}{dx}[e^x(x^2 - 2x + 2) + C] = e^x(2x - 2) + (x^2 - 2x + 2)(e^x)$$

$$= 2xe^x - 2e^x + x^2e^x - 2xe^x + 2e^x$$

$$= x^2 e^x$$

✓ Checkpoint 4

Find $\int x^3 e^x \, dx$. ■

When making repeated applications of integration by parts, you need to be careful not to interchange the substitutions in successive applications. For instance, in Example 4, the first substitution was $dv = e^x \, dx$ and $u = x^2$. If, in the second application, you had switched the substitution to $dv = 2x \, dx$ and $u = e^x$, you would have obtained

$$\int x^2 e^x \, dx = x^2 e^x - \int 2xe^x \, dx$$

$$= x^2 e^x - \left(x^2 e^x - \int x^2 e^x \, dx\right)$$

$$= \int x^2 e^x \, dx$$

thereby undoing the previous integration and returning to the *original* integral.

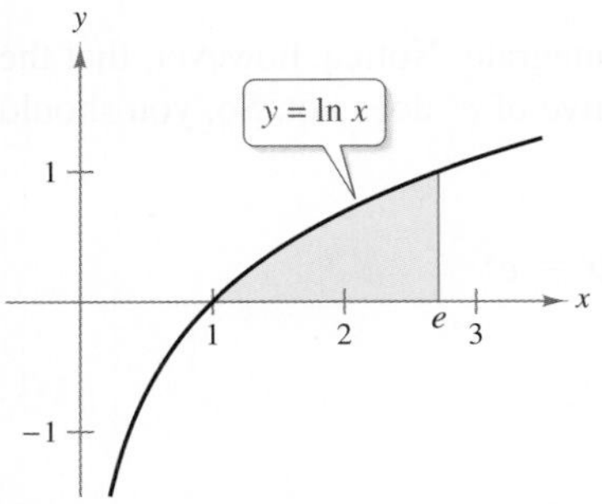

FIGURE 6.1

ALGEBRA TUTOR

For help on the algebra in Example 5, see Example 1 in the *Chapter 6 Algebra Tutor*, on page 412.

Example 5 Evaluating a Definite Integral

Evaluate $\int_1^e \ln x\, dx$.

SOLUTION Integration by parts was used to find the antiderivative of $\ln x$ in Example 3. Using this result, you can evaluate the definite integral as shown.

$$\int_1^e \ln x\, dx = \Big[x \ln x - x\Big]_1^e \qquad \text{Use result of Example 3.}$$

$$= (e \ln e - e) - (1 \ln 1 - 1) \qquad \text{Apply Fundamental Theorem.}$$

$$= (e - e) - (0 - 1)$$

$$= 1 \qquad \text{Simplify.}$$

The area represented by this definite integral is shown in Figure 6.1.

✓ Checkpoint 5

Evaluate $\int_0^1 x^2 e^x\, dx$. ■

Before starting the exercises in this section, remember that it is not enough to know *how* to use the various integration techniques. You also must know *when* to use them. Integration is first and foremost a problem of recognition—recognizing which formula or technique to apply to obtain an antiderivative. Often, a slight alteration of an integrand will necessitate the use of a different integration technique. Here are some examples.

Integral	*Technique*	*Antiderivative*
$\int x \ln x\, dx$	Integration by parts	$\frac{x^2}{2}\ln x - \frac{x^2}{4} + C$
$\int \frac{\ln x}{x}\, dx$	Power Rule: $\int u^n \frac{du}{dx}\, dx$	$\frac{(\ln x)^2}{2} + C$
$\int \frac{1}{x \ln x}\, dx$	Log Rule: $\int \frac{1}{u}\frac{du}{dx}\, dx$	$\ln\lvert\ln x\rvert + C$

As you gain experience in using integration by parts, your skill in determining u and dv will improve. The following summary lists several common integrals with suggestions for the choices of u and dv.

Summary of Common Integrals Using Integration by Parts

1. For integrals of the form

$$\int x^n e^{ax}\, dx$$

let $u = x^n$ and $dv = e^{ax}\, dx$. (See Examples 1 and 4.)

2. For integrals of the form

$$\int x^n \ln x\, dx$$

let $u = \ln x$ and $dv = x^n\, dx$. (See Examples 2 and 3.)

Present Value

Recall from Section 4.2 that the present value of a future payment is the amount that would have to be deposited today to produce the future payment. What is the present value of a future payment of $1000 one year from now? Because of inflation, $1000 today buys more than $1000 will buy a year from now. The definition below considers only the effect of inflation.

STUDY TIP

According to this definition, when the annual rate of inflation is 4%, the present value of $1000 one year from now is just $980.26.

Present Value

If c represents a continuous income function in dollars per year and the annual rate of inflation is r, then the actual total income over t_1 years is

$$\text{Actual income over } t_1 \text{ years} = \int_0^{t_1} c(t)\,dt$$

and its **present value** is

$$\text{Present value} = \int_0^{t_1} c(t)e^{-rt}\,dt.$$

Ignoring inflation, the equation for present value also applies to an interest-bearing account, where the annual interest rate r is compounded continuously and c is an income function in dollars per year.

Example 6 Finding Present Value

You have just won $1,000,000 in a state lottery. You will be paid an annuity of $50,000 a year for 20 years. When the annual rate of inflation is 6%, what is the present value of this income?

SOLUTION The income function for your winnings is given by $c(t) = 50{,}000$. So,

$$\begin{aligned}\text{Actual income} &= \int_0^{20} 50{,}000\,dt\\ &= \Big[50{,}000t\Big]_0^{20}\\ &= \$1{,}000{,}000.\end{aligned}$$

Because you do not receive this entire amount now, its present value is

$$\begin{aligned}\text{Present value} &= \int_0^{20} 50{,}000e^{-0.06t}\,dt\\ &= \left[\frac{50{,}000}{-0.06}e^{-0.06t}\right]_0^{20}\\ &\approx \$582{,}338.\end{aligned}$$

This present value represents the amount that the state must deposit now to cover your payments over the next 20 years. This shows why state lotteries are so profitable—for the states!

✓ Checkpoint 6

Find the present value of the income from the lottery ticket in Example 6 when the annual rate of inflation is 7%.

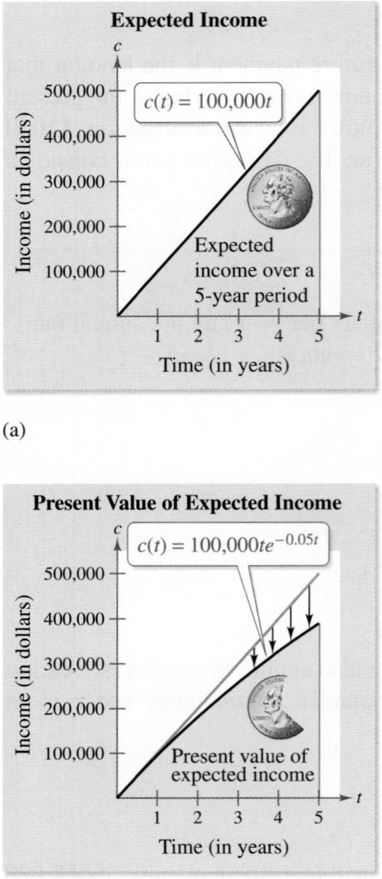

(a)

(b)

FIGURE 6.2

Example 7 Finding Present Value

A company expects its income during the next 5 years to be given by

$$c(t) = 100{,}000t, \quad 0 \le t \le 5. \qquad \text{See Figure 6.2(a).}$$

Assuming an annual inflation rate of 5%, can the company claim that the present value of this income is at least \$1 million?

SOLUTION The present value is

$$\text{Present value} = \int_0^5 100{,}000te^{-0.05t}\,dt = 100{,}000\int_0^5 te^{-0.05t}\,dt.$$

Using integration by parts, let $dv = e^{-0.05t}\,dt$.

$$dv = e^{-0.05t}\,dt \quad \Rightarrow \quad v = \int dv = \int e^{-0.05t}\,dt = -20e^{-0.05t}$$

$$u = t \quad \Rightarrow \quad du = dt$$

This implies that

$$\begin{aligned}\int te^{-0.05t}\,dt &= -20te^{-0.05t} + 20\int e^{-0.05t}\,dt\\ &= -20te^{-0.05t} - 400e^{-0.05t}\\ &= -20e^{-0.05t}(t + 20).\end{aligned}$$

So, the present value is

$$\begin{aligned}\text{Present value} &= 100{,}000\int_0^5 te^{-0.05t}\,dt \qquad \text{See Figure 6.2(b).}\\ &= 100{,}000\Big[-20e^{-0.05t}(t + 20)\Big]_0^5\\ &\approx \$1{,}059{,}961.\end{aligned}$$

Yes, the company can claim that the present value of its expected income during the next 5 years is at least \$1 million.

✓Checkpoint 7

A company expects its income during the next 10 years to be given by $c(t) = 20{,}000t$, for $0 \le t \le 10$. Assuming an annual inflation rate of 5%, what is the present value of this income? ■

SUMMARIZE (Section 6.1)

1. State the integration by parts formula *(page 376)*. For examples of using this formula, see Examples 1, 2, 3, 4, and 7.
2. State the guidelines for integration by parts *(page 376)*. For an example of using these guidelines, see Example 1.
3. Give a summary of the common integrals using integration by parts *(page 380)*. For examples of these common integrals, see Examples 1, 2, 3, and 4.
4. Describe a real-life example of how integration by parts can be used to find the present value of an annuity *(page 381, Example 6)*.

SKILLS WARM UP 6.1

The following warm-up exercises involve skills that were covered in earlier sections. You will use these skills in the exercise set for this section. For additional help, review Sections 4.3, 4.5, and 5.5.

In Exercises 1–6, find $f'(x)$.

1. $f(x) = \ln(x + 1)$

2. $f(x) = \ln(x^2 - 1)$

3. $f(x) = e^{x^3}$

4. $f(x) = e^{-x^2}$

5. $f(x) = x^2e^x$

6. $f(x) = xe^{-2x}$

In Exercises 7–10, find the area between the graphs of f and g.

7. $f(x) = -x^2 + 4,\ g(x) = x^2 - 4$

8. $f(x) = -x^2 + 2,\ g(x) = 1$

9. $f(x) = 4x,\ g(x) = x^2 - 5$

10. $f(x) = x^3 - 3x^2 + 2,\ g(x) = x - 1$

Exercises 6.1

See www.CalcChat.com for worked-out solutions to odd-numbered exercises.

Setting Up Integration by Parts In Exercises 1–4, identify u and dv for finding the integral using integration by parts. (Do not evaluate the integral.)

1. $\int xe^{3x}\,dx$

2. $\int x^2e^{3x}dx$

3. $\int x \ln 2x\,dx$

4. $\int \ln 4x\,dx$

Integration by Parts In Exercises 5–16, use integration by parts to find the indefinite integral. *See Examples 1, 2, 3, and 4.*

5. $\int xe^{3x}\,dx$

6. $\int xe^{-x}\,dx$

7. $\int x^3 \ln x\,dx$

8. $\int x^4 \ln x\,dx$

9. $\int \ln 2x\,dx$

10. $\int \ln x^2\,dx$

11. $\int x^2e^{-x}\,dx$

12. $\int x^2e^{2x}\,dx$

13. $\int \sqrt{x} \ln x\,dx$

14. $\int x^2\sqrt{x - 3}\,dx$

15. $\int 2x^2e^x\,dx$

16. $\int \frac{2x}{e^x}\,dx$

Finding Indefinite Integrals In Exercises 17–38, find the indefinite integral. (*Hint:* Integration by parts is not required for all the integrals.)

17. $\int e^{4x}\,dx$

18. $\int e^{-2x}\,dx$

19. $\int xe^{4x}\,dx$

20. $\int xe^{-2x}\,dx$

21. $\int \frac{x}{e^{x/4}}\,dx$

22. $\int \frac{1}{2}x^3e^x\,dx$

23. $\int t\ln(t + 1)\,dt$

24. $\int (x - 1)e^x\,dx$

25. $\int \frac{e^{1/t}}{t^2}\,dt$

26. $\int \frac{1}{x(\ln x)^3}\,dx$

27. $\int x(\ln x)^2\,dx$

28. $\int \ln 3x\,dx$

29. $\int \frac{(\ln x)^2}{x}\,dx$

30. $\int \frac{1}{x \ln 3x}\,dx$

31. $\int \frac{\ln x}{x^2}\,dx$

32. $\int \frac{\ln 2x}{x^2}\,dx$

33. $\int x\sqrt{x - 1}\,dx$

34. $\int \frac{x}{\sqrt{x - 1}}\,dx$

35. $\int x(x + 1)^2\,dx$

36. $\int \frac{x}{\sqrt{2 + 3x}}\,dx$

37. $\int \frac{xe^{2x}}{(2x + 1)^2}\,dx$

38. $\int \frac{x^3e^{x^2}}{(x^2 + 1)^2}\,dx$

Evaluating Definite Integrals In Exercises 39–46, use integration by parts to evaluate the definite integral. *See Example 5.*

39. $\int_1^e x^5 \ln x\, dx$

40. $\int_1^e 2x \ln x\, dx$

41. $\int_0^1 \ln(1 + 2x)\, dx$

42. $\int_0^4 \frac{x}{e^{x/2}}\, dx$

43. $\int_0^8 x\sqrt{x + 1}\, dx$

44. $\int_0^{12} \frac{x}{\sqrt{x + 4}}\, dx$

45. $\int_1^2 x^2 e^x\, dx$

46. $\int_0^2 \frac{x^2}{e^{3x}}\, dx$

Area of a Region In Exercises 47–52, find the area of the region bounded by the graphs of the equations. Use a graphing utility to verify your results.

47. $y = x^3e^x,\ y = 0,\ x = 0,\ x = 2$

48. $y = (x^2 - 1)e^x,\ y = 0,\ x = -1,\ x = 1$

49. $y = \frac{1}{9}xe^{-x/3},\ y = 0, x = 0, x = 3$

50. $y = x^{-3} \ln x, y = 0, x = e$

51. $y = x^2 \ln x,\ y = 0,\ x = 1,\ x = e$

52. $y = \frac{\ln x}{x^2},\ y = 0,\ x = 1,\ x = e$

Verifying Formulas In Exercises 53 and 54, use integration by parts to verify the formula.

53. $\int x^n \ln x\, dx = \frac{x^{n+1}}{(n + 1)^2}[-1 + (n + 1)\ln x] + C,$
$n \neq -1$

54. $\int x^n e^{ax}\, dx = \frac{x^n e^{ax}}{a} - \frac{n}{a}\int x^{n-1}e^{ax}\, dx,\ n > 0$

Using Formulas In Exercises 55–58, use the results of Exercises 53 and 54 to find the indefinite integral.

55. $\int x^2 e^{5x}\, dx$

56. $\int xe^{-3x}\, dx$

57. $\int x^{-4} \ln x\, dx$

58. $\int x^{3/2} \ln x\, dx$

Integration Using Technology In Exercises 59–62, use a symbolic integration utility to evaluate the integral.

59. $\int_0^2 t^3e^{-4t}\, dt$

60. $\int_1^4 (x^2 + 4) \ln x\, dx$

61. $\int_0^5 x^4(25 - x^2)^{3/2}\, dx$

62. $\int_1^e x^9 \ln x\, dx$

63. Demand A manufacturing company forecasts that the demand x (in units) for its product over the next 10 years can be modeled by

$x = 500(20 + te^{-0.1t}),\quad 0 \le t \le 10$

where t is the time in years.

(a) Use a graphing utility to decide whether the company is forecasting an increase or a decrease in demand over the decade.

(b) Find the total demand over the next 10 years.

(c) Find the average annual demand during the 10-year period.

64. Capital Campaign The board of trustees of a college is planning a five-year capital gifts campaign to raise money for the college. The goal is to have an annual gift income I that is modeled by

$I = 2000(375 + 68te^{-0.2t}),\quad 0 \le t \le 5$

where t is the time in years.

(a) Use a graphing utility to decide whether the board of trustees expects the gift income to increase or decrease over the five-year period.

(b) Find the expected total gift income over the five-year period.

(c) Determine the average annual gift income over the five-year period.

65. Memory Model A model for the ability M of a child to memorize, measured on a scale from 0 to 10, is

$M = 1 + 1.6t \ln t,\quad 0 < t \le 4$

where t is the child's age in years.

(a) Find the average value of this model between the child's first and second birthdays.

(b) Find the average value of this model between the child's third and fourth birthdays.

66. Revenue A company sells a seasonal product. The revenue R (in dollars) generated by sales of the product can be modeled by

$R = 410.5t^2e^{-t/30} + 25{,}000,\quad 0 \le t \le 365$

where t is the time in days.

(a) Find the average daily revenue during the first quarter, which is given by $0 \le t \le 90$.

(b) Find the average daily revenue during the fourth quarter, which is given by $274 \le t \le 365$.

(c) Find the total daily revenue during the year.

Finding Present Value In Exercises 67–72, find the present value of the income c (in dollars) over t_1 years at the given annual inflation rate r. *See Examples 6 and 7.*

67. $c = 5000,\ r = 4\%,\ t_1 = 4$ years

68. $c = 450,\ r = 4\%,\ t_1 = 10$ years

69. $c = 100{,}000 + 4000t,\ r = 5\%,\ t_1 = 10$ years

70. $c = 30{,}000 + 500t,\ r = 7\%,\ t_1 = 6$ years

71. $c = 1000 + 50e^{t/2},\ r = 6\%,\ t_1 = 4$ years

72. $c = 5000 + 25te^{t/10},\ r = 6\%,\ t_1 = 10$ years

73. Present Value You have just won \$1,500,000 in a state lottery. You will be paid an annuity of \$100,000 a year for 15 years. When the annual rate of inflation is 5%, what is the present value of this income?

74. Present Value You have just won \$65,000,000 in a lottery. You will be paid an annuity of \$2,500,000 a year for 26 years. When the annual rate of inflation is 3%, what is the present value of this income?

75. Present Value A company expects its income c during the next 4 years to be modeled by

$$c = 150{,}000 + 75{,}000t, \quad 0 \le t \le 4.$$

(a) Find the actual income for the business over the 4 years.

(b) Assuming an annual inflation rate of 4%, what is the present value of this income?

76. Present Value A professional athlete signs a three-year contract in which the earnings c can be modeled by $c = 500{,}000 + 125{,}000t$, where t represents the year

(a) Find the actual value of the athlete's contract.

(b) Assuming an annual inflation rate of 3%, what is the present value of the contract?

77. Present Value A professional athlete signs a four-year contract in which the earnings c can be modeled by $c = 3{,}000{,}000 + 750{,}000t$, where t represents the year.

(a) Find the actual value of the athlete's contract.

(b) Assuming an annual inflation rate of 5%, what is the present value of the contract?

78. HOW DO YOU SEE IT? The graphs of two equations show the expected income and the present value of the expected income for a company. Which graph represents the expected income and which graph represents the present value of the expected income? Explain your reasoning.

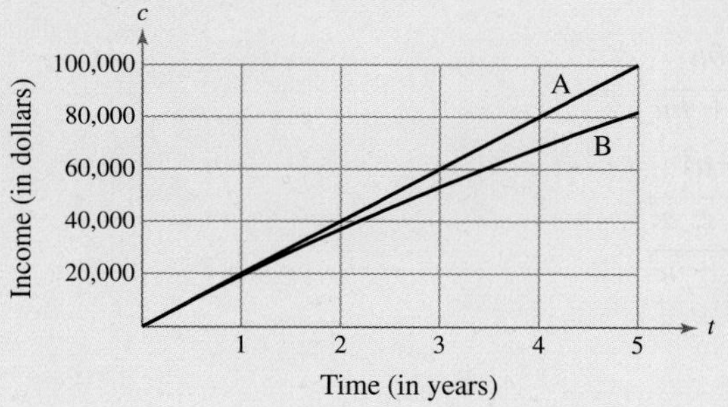

Future Value In Exercises 79 and 80, find the future value of the income (in dollars) given by $f(t)$ over t_1 years at annual interest rate r. If the function f represents a continuous investment over a period of t_1 years at an annual interest rate r (compounded continuously), then the future value of the investment is given by

$$\text{Future value} = e^{rt_1}\int_0^{t_1} f(t)e^{-rt}\,dt.$$

79. $f(t) = 3000,\ r = 8\%,\ t_1 = 10$ years

80. $f(t) = 3000e^{0.05t},\ r = 10\%,\ t_1 = 5$ years

81. Finance: Future Value Use the equation from Exercises 79 and 80 to calculate the following. *(Source: Adapted from Garman/Forgue,* Personal Finance, *Eighth Edition)*

(a) The future value of \$1200 saved each year for 10 years earning 7% interest

(b) A person who wishes to invest \$1200 each year finds one investment choice that is expected to pay 9% interest per year and another, riskier choice that may pay 10% interest per year. What is the difference in return (future value) if the investment is made for 15 years?

82. College Tuition Fund Assume your grandparents had continuously invested in a college fund according to the model

$$f(t) = 400t$$

for 18 years, at an annual interest rate of 7%.

(a) In 2010, the total cost of attending The Pennsylvania State University for 1 year was estimated to be \$26,276. Will the fund have grown enough to allow you to cover 4 years of expenses at The Pennsylvania State University? *(Source: The Pennsylvania State University)*

(b) In 2010, the total cost of attending The Ohio State University for 1 year was estimated to be \$23,604. Will the fund have grown enough to allow you to cover 4 years of expenses at The Ohio State University? *(Source: The Ohio State University)*

83. Midpoint Rule Use a program similar to the Midpoint Rule program in Appendix E with $n = 10$ to approximate

$$\int_1^4 \frac{4}{\sqrt{x} + \sqrt[3]{x}}\,dx.$$

84. Midpoint Rule Use a program similar to the Midpoint Rule program in Appendix E with $n = 12$ to approximate the area of the region bounded by the graphs of

$$y = \frac{10}{\sqrt{x}e^{x}},\ y = 0,\ x = 1, \text{ and } x = 4.$$

6.2 Integration Tables

- Use integration tables to find indefinite and definite integrals.
- Use reduction formulas to find indefinite integrals.
- Use integration tables to solve real-life problems.

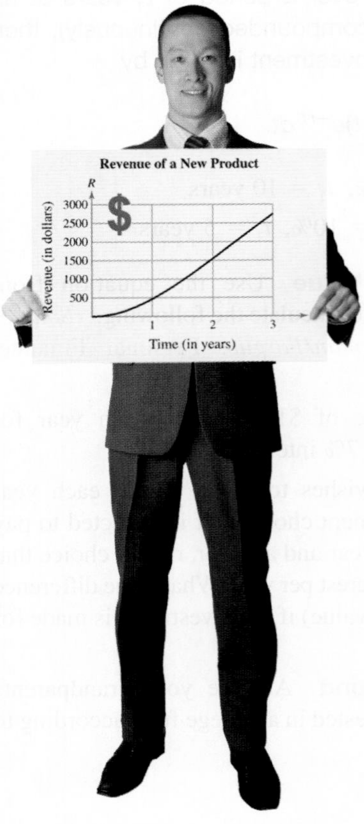

In Exercise 59 on page 392, you will use a formula from the integration table in Appendix C to find the total revenue of a new product during its first 2 years.

Integration Tables

You have studied several integration techniques that can be used with the basic integration formulas. Certainly these techniques and formulas do not cover every possible method for finding an antiderivative, but they do cover most of the important ones.

In this section, you will expand the list of integration formulas to form a table of integrals. As you add new integration formulas to the basic list, two effects occur. On one hand, it becomes increasingly difficult to memorize, or even become familiar with, the entire list of formulas. On the other hand, with a longer list you need fewer techniques for fitting an integral to one of the formulas on the list. The procedure of integrating by means of a long list of formulas is called **integration by tables.** (The table in Appendix C constitutes only a partial listing of integration formulas. Much longer lists exist, some of which contain several hundred formulas.)

Integration by tables should not be considered a trivial task. It requires considerable thought and insight, and it often requires substitution. Many people find a table of integrals to be a valuable supplement to the integration techniques discussed in this text. As you gain competence in the use of integration tables, you will improve in the use of the various integration techniques. In doing so, you should find that a combination of techniques and tables is the most versatile approach to integration.

Each integration formula in Appendix C can be developed using one or more of the techniques you have studied. You should try to verify several of the formulas. For instance, Formula 17

$$\int \frac{\sqrt{a+bu}}{u}\,du = 2\sqrt{a+bu} + a\int \frac{1}{u\sqrt{a+bu}}\,du \qquad \text{Formula 17}$$

can be verified using integration by parts, Formula 39

$$\int \frac{1}{1+e^u}\,du = u - \ln(1+e^u) + C \qquad \text{Formula 39}$$

can be verified using substitution, and Formula 44

$$\int (\ln u)^2\,du = u[2 - 2\ln u - (\ln u)^2] + C \qquad \text{Formula 44}$$

can be verified using integration by parts twice.

In the table of integrals in Appendix C, the formulas have been classified according to the form of the integrand. Several of the forms are listed below.

- Forms involving u^n
- Forms involving $a + bu$
- Forms involving $\sqrt{a + bu}$
- Forms involving $u^2 - a^2$
- Forms involving $\sqrt{u^2 \pm a^2}$
- Forms involving $\sqrt{a^2 - u^2}$
- Forms involving e^u
- Forms involving $\ln u$

leedsn/www.shutterstock.com

TECH TUTOR

Throughout this section, remember that a symbolic integration utility can be used instead of integration tables. If you have access to such a utility, try using it to find the indefinite integrals in Examples 1 and 2.

Example 1 Using Integration Tables

Find $\displaystyle\int \frac{x}{\sqrt{x-1}}\,dx.$

SOLUTION Because the expression inside the radical is linear, you should consider forms involving $\sqrt{a+bu}$, as in Formula 19.

$$\int \frac{u}{\sqrt{a+bu}}\,du = -\frac{2(2a-bu)}{3b^2}\sqrt{a+bu} + C \qquad \text{Formula 19}$$

Using this formula, let $a = -1$, $b = 1$, and $u = x$. Then $du = dx$, and you obtain

$$\int \frac{x}{\sqrt{x-1}}\,dx = -\frac{2(-2-x)}{3}\sqrt{x-1} + C \qquad \text{Substitute values of } a, b, \text{ and } u.$$

$$= \frac{2}{3}(2+x)\sqrt{x-1} + C. \qquad \text{Simplify.}$$

✓ Checkpoint 1

Use the integration table in Appendix C to find

$$\int \frac{x}{\sqrt{2+x}}\,dx.$$

■

Example 2 Using Integration Tables

Find $\displaystyle\int x\sqrt{x^4-9}\,dx.$

SOLUTION Because it is not clear which formula to use, you can begin by letting $u = x^2$ and $du = 2x\,dx$. With these substitutions, you can write the integral as shown.

$$\int x\sqrt{x^4-9}\,dx = \frac{1}{2}\int \sqrt{(x^2)^2-9}\,(2x)\,dx \qquad \text{Multiply and divide by 2.}$$

$$= \frac{1}{2}\int \sqrt{u^2-9}\,du \qquad \text{Substitute } u \text{ and } du.$$

Now, it appears that you can use Formula 23.

$$\int \sqrt{u^2-a^2}\,du = \frac{1}{2}\left(u\sqrt{u^2-a^2} - a^2\ln\left|u+\sqrt{u^2-a^2}\right|\right) + C$$

Letting $a = 3$, you obtain

$$\int x\sqrt{x^4-9}\,dx = \frac{1}{2}\int \sqrt{u^2-a^2}\,du$$

$$= \frac{1}{2}\left[\frac{1}{2}\left(u\sqrt{u^2-a^2} - a^2\ln\left|u+\sqrt{u^2-a^2}\right|\right)\right] + C$$

$$= \frac{1}{4}\left(x^2\sqrt{x^4-9} - 9\ln\left|x^2+\sqrt{x^4-9}\right|\right) + C.$$

✓ Checkpoint 2

Use the integration table in Appendix C to find

$$\int \frac{\sqrt{x^2+16}}{x}\,dx.$$

■

Example 3 Using Integration Tables

Find $\displaystyle\int \frac{1}{x\sqrt{x+1}}\,dx.$

SOLUTION Considering forms involving $\sqrt{a+bu}$, where $a = 1$, $b = 1$, and $u = x$, you can use Formula 15.

$$\int \frac{1}{u\sqrt{a+bu}}\,du = \frac{1}{\sqrt{a}}\ln\left|\frac{\sqrt{a+bu}-\sqrt{a}}{\sqrt{a+bu}+\sqrt{a}}\right| + C,\quad a > 0$$

So,

$$\begin{aligned}\int \frac{1}{x\sqrt{x+1}}\,dx &= \int \frac{1}{u\sqrt{a+bu}}\,du\\ &= \frac{1}{\sqrt{a}}\ln\left|\frac{\sqrt{a+bu}-\sqrt{a}}{\sqrt{a+bu}+\sqrt{a}}\right| + C\\ &= \ln\left|\frac{\sqrt{x+1}-1}{\sqrt{x+1}+1}\right| + C.\end{aligned}$$

✓ Checkpoint 3

Use the integration table in Appendix C to find $\displaystyle\int \frac{1}{x^2-4}\,dx.$ ■

Example 4 Using Integration Tables

Evaluate $\displaystyle\int_0^2 \frac{x}{1+e^{-x^2}}\,dx.$

SOLUTION Of the forms involving e^u, Formula 39

$$\int \frac{1}{1+e^u}\,du = u - \ln(1+e^u) + C$$

seems most appropriate. To use this formula, let $u = -x^2$ and $du = -2x\,dx$.

$$\begin{aligned}\int \frac{x}{1+e^{-x^2}}\,dx &= -\frac{1}{2}\int \frac{1}{1+e^{-x^2}}(-2x)\,dx\\ &= -\frac{1}{2}\int \frac{1}{1+e^u}\,du\\ &= -\frac{1}{2}[u - \ln(1+e^u)] + C\\ &= -\frac{1}{2}[-x^2 - \ln(1+e^{-x^2})] + C\\ &= \frac{1}{2}[x^2 + \ln(1+e^{-x^2})] + C\end{aligned}$$

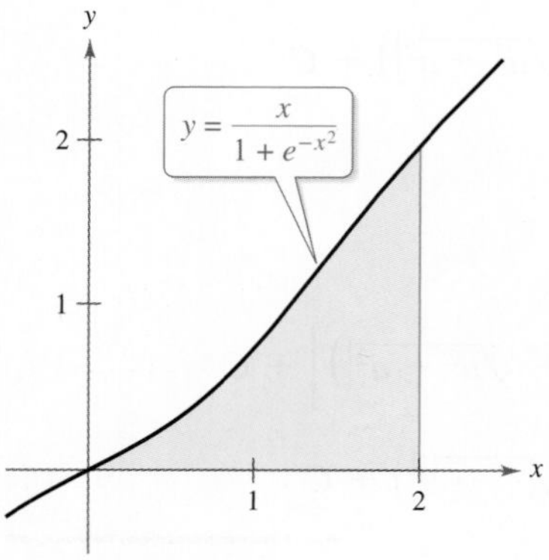

FIGURE 6.3

So, the value of the definite integral is

$$\int_0^2 \frac{x}{1+e^{-x^2}}\,dx = \frac{1}{2}\Big[x^2 + \ln(1+e^{-x^2})\Big]_0^2 \approx 1.66.$$

See Figure 6.3.

✓ Checkpoint 4

Use the integration table in Appendix C to evaluate $\displaystyle\int_0^1 \frac{x^2}{1+e^{x^3}}\,dx.$ ■

Reduction Formulas

Several of the formulas in the integration table have the form

$$\int f(x)\,dx = g(x) + \int h(x)\,dx$$

where the right side contains an integral. Such integration formulas are called **reduction formulas** because they reduce the original integral to the sum of a function and a simpler integral.

ALGEBRA TUTOR

For help on the algebra in Example 5, see Example 3 in the *Chapter 6 Algebra Tutor,* on page 413.

Example 5 Using a Reduction Formula

Find $\int x^2 e^x\,dx$.

SOLUTION Using Formula 38

$$\int u^n e^u\,du = u^n e^u - n\int u^{n-1} e^u\,du$$

you can let $u = x$ and $n = 2$. Then $du = dx$, and you can write

$$\int x^2 e^x\,dx = x^2 e^x - 2\int x e^x\,dx.$$

Then, using Formula 37

$$\int u e^u\,du = (u - 1)e^u + C$$

you can write

$$\begin{aligned}\int x^2 e^x\,dx &= x^2 e^x - 2\int x e^x\,dx \\ &= x^2 e^x - 2(x - 1)e^x + C \\ &= x^2 e^x - 2x e^x + 2e^x + C \\ &= e^x(x^2 - 2x + 2) + C.\end{aligned}$$

You can check this result by differentiating.

$$\begin{aligned}\frac{d}{dx}[e^x(x^2 - 2x + 2) + C] &= e^x(2x - 2) + (x^2 - 2x + 2)(e^x) \\ &= 2xe^x - 2e^x + x^2e^x - 2xe^x + 2e^x \\ &= x^2e^x\end{aligned}$$

✓Checkpoint 5

Use the integration table in Appendix C to find the indefinite integral $\int (\ln x)^2\,dx$. ■

TECH TUTOR

You have now studied two ways to find the indefinite integral in Example 5. Example 5 uses an integration table, and Example 4 in Section 6.1 uses integration by parts. A third way would be to use a symbolic integration utility.

Application

Integration can be used to find the probability that an event will occur. In such an application, the real-life situation is modeled by a *probability density function f*, and the probability that x will lie between a and b is represented by

$$P(a \le x \le b) = \int_a^b f(x)\,dx.$$

The probability $P(a \le x \le b)$ must be a number between 0 and 1.

Example 6 Finding a Probability

A psychologist finds that the probability that a participant in a memory experiment will recall between a and b percent (in decimal form) of the material is

$$P(a \le x \le b) = \int_a^b \frac{1}{e-2} x^2 e^x\,dx, \quad 0 \le a \le b \le 1.$$

Find the probability that a randomly chosen participant will recall between 0% and 87.5% of the material.

SOLUTION You can use the Constant Multiple Rule to rewrite the integral as

$$\frac{1}{e-2}\int_a^b x^2 e^x\,dx.$$

Note that the integrand is the same as the one in Example 5. Use the result of Example 5 to find the probability with $a = 0$ and $b = 0.875$.

$$\frac{1}{e-2}\int_0^{0.875} x^2 e^x\,dx = \frac{1}{e-2}\Big[e^x(x^2 - 2x + 2)\Big]_0^{0.875}$$
$$\approx 0.608$$

So, the probability is about 60.8%, as indicated in Figure 6.4.

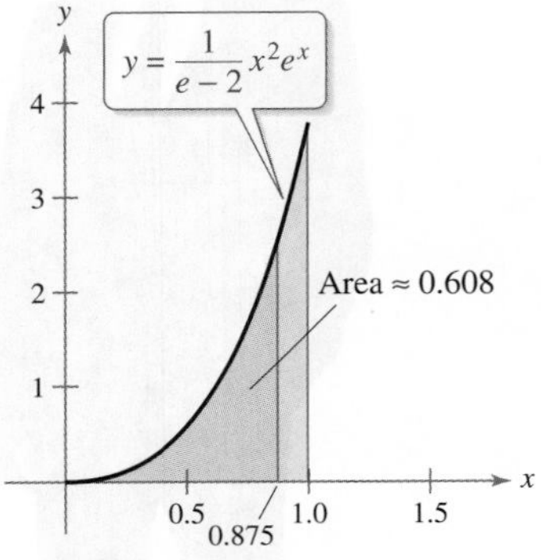

FIGURE 6.4

✓ Checkpoint 6

Use Example 6 to find the probability that a participant will recall between 0% and 62.5% of the material. ■

SUMMARIZE (Section 6.2)

1. Describe what is meant by integration by tables *(page 386)*. For examples of integration by tables, see Examples 1, 2, 3, and 4.
2. Describe what is meant by a reduction formula *(page 389)*. For an example of a reduction formula, see Example 5.
3. Describe a real-life example of how integration by tables can be used to analyze the results of a memory experiment *(page 390, Example 6)*.

Michal Kowalski/Shutterstock.com

SKILLS WARM UP 6.2

The following warm-up exercises involve skills that were covered in a previous course or in earlier sections. You will use these skills in the exercise set for this section. For additional help, review Appendix Section A.4 and Section 6.1.

In Exercises 1–4, expand the expression.

1. $(x + 4)^2$

2. $(x - 1)^2$

3. $\left(x + \frac{1}{2}\right)^2$

4. $\left(x - \frac{1}{3}\right)^2$

In Exercises 5 and 6, use integration by parts to find the indefinite integral.

5. $\int 2xe^x\,dx$

6. $\int 3x^2 \ln x\,dx$

Exercises 6.2

See www.CalcChat.com for worked-out solutions to odd-numbered exercises.

Using Integration Tables In Exercises 1–8, use the indicated formula from the integration table in Appendix C to find the indefinite integral. *See Examples 1, 2, and 3.*

1. $\int \frac{x}{(2 + 3x)^2}\,dx$, Formula 4

2. $\int \frac{1}{x(2 + 3x)^2}\,dx$, Formula 11

3. $\int \frac{x}{\sqrt{2 + 3x}}\,dx$, Formula 19

4. $\int \frac{4}{x^2 - 9}\,dx$, Formula 21

5. $\int \frac{2x}{\sqrt{x^4 - 9}}\,dx$, Formula 27

6. $\int x^2\sqrt{x^2 + 9}\,dx$, Formula 24

7. $\int x^3 e^{x^2}\,dx$, Formula 37

8. $\int \frac{x}{1 + e^{x^2}}\,dx$, Formula 39

Using Integration Tables In Exercises 9–36, use the integration table in Appendix C to find the indefinite integral. *See Examples 1, 2, 3, and 5.*

9. $\int \frac{1}{x(1 + x)}\,dx$

10. $\int \frac{1}{x(1 + x)^2}\,dx$

11. $\int \frac{1}{x\sqrt{x^2 + 9}}\,dx$

12. $\int \frac{1}{\sqrt{x^2 - 1}}\,dx$

13. $\int \frac{1}{x\sqrt{4 - x^2}}\,dx$

14. $\int \frac{\sqrt{x^2 - 9}}{x^2}\,dx$

15. $\int 3x \ln 3x\,dx$

16. $\int (\ln 5x)^2\,dx$

17. $\int \frac{6x}{1 + e^{3x^2}}\,dx$

18. $\int \frac{1}{1 + e^x}\,dx$

19. $\int x^2\sqrt{3 + x}\,dx$

20. $\int \frac{x}{x^4 - 9}\,dx$

21. $\int \frac{t^2}{(2 + 3t)^3}\,dt$

22. $\int \frac{\sqrt{3 + 4t}}{t}\,dt$

23. $\int \frac{1}{x\sqrt{3 + 4x}}\,dx$

24. $\int \sqrt{3 + x^2}\,dx$

25. $\int \frac{x^2}{1 + x}\,dx$

26. $\int \frac{1}{1 + e^{2x}}\,dx$

27. $\int \frac{x^2}{(3 + 2x)^5}\,dx$

28. $\int \frac{1}{x^2\sqrt{x^2 - 4}}\,dx$

29. $\int \frac{1}{x^2\sqrt{1 - x^2}}\,dx$

30. $\int \frac{2x}{(1 - 3x)^2}\,dx$

31. $\int 4x^2 \ln 2x\,dx$

32. $\int xe^{x^2}\,dx$

33. $\int \frac{x^2}{(3x - 5)^2}\,dx$

34. $\int \frac{1}{2x^2(2x - 1)^2}\,dx$

35. $\int \frac{\ln x}{x(4 + 3\ln x)}\,dx$

36. $\int (\ln x)^3\,dx$

Using Integration Tables In Exercises 37–44, use the integration table in Appendix C to evaluate the definite integral. *See Example 4.*

37. $\int_0^1 \frac{x}{\sqrt{1 + x}}\,dx$

38. $\int_0^5 \frac{x}{\sqrt{5 + 2x}}\,dx$

39. $\int_0^5 \frac{x}{(4 + x)^2}\,dx$

40. $\int_2^4 \frac{x^2}{(3x - 5)}\,dx$

41. $\int_0^4 \frac{6}{1 + e^{0.5x}}\,dx$

42. $\int_2^4 \sqrt{3 + x^2}\,dx$

43. $\int_1^2 x^3 \ln x^2\,dx$

44. $\int_0^3 \frac{x}{(1 + 3x)^4}\,dx$

Area of a Region In Exercises 45–50, use the integration table in Appendix C to find the exact area of the region bounded by the graphs of the equations. Use a graphing utility to verify your results.

45. $y = \dfrac{1}{(16 - x^2)^{3/2}},\ y = 0,\ x = -2,\ x = 2$

46. $y = \dfrac{2}{1 + e^{4x}},\ y = 0,\ x = 0,\ x = 1$

47. $y = \dfrac{1}{9x^2(2 + 3x)},\ y = 0,\ x = 1,\ x = 2$

48. $y = \dfrac{-e^x}{1 - e^{2x}},\ y = 0,\ x = 1,\ x = 2$

49. $y = x^2\sqrt{x^2 + 4},\ y = 0,\ x = \sqrt{5}$

50. $y = x \ln x^2,\ y = 0,\ x = 4$

Finding Indefinite Integrals Using Two Methods In Exercises 51–54, find the indefinite integral (a) using the integration table in Appendix C and (b) using integration by parts.

51. $\displaystyle\int \ln \frac{x}{3}\,dx$

52. $\displaystyle\int 4xe^{4x}\,dx$

53. $\displaystyle\int \frac{x}{\sqrt{7x - 3}}\,dx$

54. $\displaystyle\int 7x \ln 7x\,dx$

55. Probability The probability of recalling between a and b percent (in decimal form) of the material learned in a memory experiment is modeled by

$$P(a \le x \le b) = \int_a^b \frac{75}{14}\left(\frac{x}{\sqrt{4 + 5x}}\right) dx,$$

$0 \le a \le b \le 1.$

What are the probabilities of recalling (a) between 40% and 80% and (b) between 0% and 50% of the material?

56. Probability The probability of finding between a and b percent iron (in decimal form) in ore samples is modeled by

$$P(a \le x \le b) = \int_a^b 2x^3 e^{x^2}\,dx, \quad 0 \le a \le b \le 1.$$

What are the probabilities of finding (a) between 0% and 25% and (b) between 50% and 100% iron in a sample?

Population Growth In Exercises 57 and 58, use a graphing utility to graph the growth function. Use the integration table in Appendix C to find the average value of the growth function over the interval, where N is the size of a population and t is the time in days.

57. $N = \dfrac{5000}{1 + e^{4.8 - 1.9t}},\ [0, 2]$

58. $N = \dfrac{375}{1 + e^{4.20 - 0.25t}},\ [21, 28]$

59. Revenue The revenue (in dollars) for a new product is modeled by $R = 10{,}000[1 - 1/(1 + 0.1t^2)^{1/2}]$, where t is the time in years. Estimate the total revenue of the product over its first 2 years on the market.

60. HOW DO YOU SEE IT? The graph shows the rate of change of the sales of a new product.

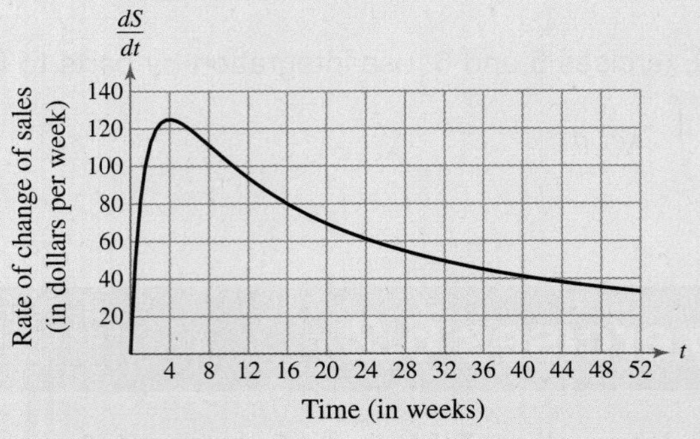

(a) Approximate the rate of change of the sales after 16 weeks. Explain your reasoning.

(b) Approximate the weeks for which the sales are increasing. Explain your reasoning.

61. Consumer and Producer Surpluses The demand and supply functions for a product are modeled by

Demand: $p = 60/\sqrt{x^2 + 81}$, *Supply:* $p = x/3$

where p is the price (in dollars) and x is the number of units (in millions). Find the consumer and producer surpluses for this product.

62. Project: Purchasing Power of the Dollar For a project analyzing the purchasing power of the dollar from 1983 through 2009, visit this text's website at *www.cengagebrain.com*. *(Source: U.S. Bureau of Labor Statistics)*

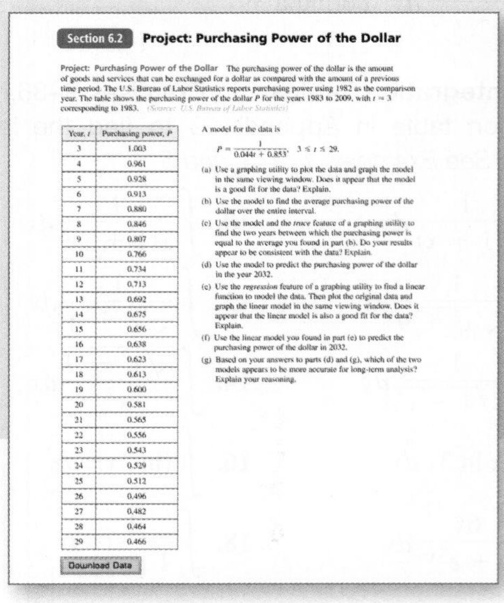

Section 6.2 Project: Purchasing Power of the Dollar

Project: Purchasing Power of the Dollar The purchasing power of the dollar is the amount of goods and services that can be exchanged for a dollar as compared with the amount of a previous time period. The U.S. Bureau of Labor Statistics reports purchasing power using 1982 as the comparison year. The table shows the purchasing power of the dollar P for the years 1983 to 2009, with $t = 3$ corresponding to 1983. *(Source: U.S. Bureau of Labor Statistics)*

Year, t	Purchasing power, P
3	1.003
4	0.961
5	0.928
6	0.913
7	0.880
8	0.846
9	0.807
10	0.766
11	0.734
12	0.713
13	0.692
14	0.675
15	0.656
16	0.638
17	0.623
18	0.613
19	0.600
20	0.581
21	0.565
22	0.556
23	0.543
24	0.529
25	0.512
26	0.496
27	0.482
28	0.464
29	0.466

Download Data

A model for the data is

$$P = \frac{1}{0.044t + 0.853}, \quad 3 \le t \le 29.$$

(a) Use a graphing utility to plot the data and graph the model in the same viewing window. Does it appear that the model is a good fit for the data? Explain.

(b) Use the model to find the average purchasing power of the dollar over the entire interval.

(c) Use the model and the *trace* feature of a graphing utility to find the two years between which the purchasing power is equal to the average you found in part (b). Do your results appear to be consistent with the data? Explain.

(d) Use the model to predict the purchasing power of the dollar in the year 2032.

(e) Use the *regression* feature of a graphing utility to find a linear function to model the data. Then plot the original data and graph the linear model in the same viewing window. Does it appear that the linear model is also a good fit for the data? Explain.

(f) Use the linear model you found in part (e) to predict the purchasing power of the dollar in 2032.

(g) Based on your answers to parts (d) and (g), which of the two models appears to be more accurate for long-term analysis? Explain your reasoning.

QUIZ YOURSELF

See www.CalcChat.com for worked-out solutions to odd-numbered exercises.

Take this quiz as you would take a quiz in class. When you are done, check your work against the answers given in the back of the book.

In Exercises 1–6, use integration by parts to find the indefinite integral.

1. $\int xe^{5x}\,dx$
2. $\int \ln x^3\,dx$
3. $\int (x+1)\ln x\,dx$
4. $\int x\sqrt{x+3}\,dx$
5. $\int x\ln\sqrt{x}\,dx$
6. $\int x^2 e^{-2x}\,dx$

7. A manufacturing company forecasts that the demand x (in units) for its product over the next 5 years can be modeled by

$x = 1000(45 + 20te^{-0.5t})$

where t is the time in years.

(a) Find the total demand over the next 5 years.

(b) Find the average annual demand during the 5-year period.

8. A small business expects its income c during the next 7 years to be given by

$c(t) = 32{,}000t, \quad 0 \le t \le 7.$

(a) Find the actual income for the business over the 7 years.

(b) Assuming an annual inflation rate of 3.3%, what is the present value of this income?

In Exercises 9–14, use the integration table in Appendix C to find the indefinite integral.

9. $\int \frac{x}{1+2x}\,dx$
10. $\int \frac{1}{x(0.1+0.2x)}\,dx$
11. $\int \frac{\sqrt{x^2-16}}{x^2}\,dx$
12. $\int \frac{1}{x\sqrt{4+9x}}\,dx$
13. $\int \frac{2x}{1+e^{4x^2}}\,dx$
14. $\int 2x(x^2+1)e^{x^2+1}\,dx$

15. The revenue (in millions of dollars) for a new product is modeled by

$R = \sqrt{144t^2 + 400}$

where t is the time in years.

(a) Estimate the total revenue of the product over its first 3 years on the market.

(b) Estimate the total revenue of the product over its first 6 years on the market.

In Exercises 16–21, evaluate the definite integral.

16. $\int_{-2}^{0} xe^{x/2}\,dx$
17. $\int_{1}^{2} 5x\ln x\,dx$
18. $\int_{0}^{8} \frac{x}{\sqrt{x+8}}\,dx$
19. $\int_{1}^{e} (\ln x)^2\,dx$
20. $\int_{2}^{3} \frac{1}{x^2\sqrt{9-x^2}}\,dx$
21. $\int_{4}^{6} \frac{2x}{x^4-4}\,dx$

6.3 Numerical Integration

- Use the Trapezoidal Rule to approximate definite integrals.
- Use Simpson's Rule to approximate definite integrals.
- Analyze the sizes of the errors when approximating definite integrals with the Trapezoidal Rule and Simpson's Rule.

In Exercise 43 on page 401, you will use Simpson's Rule to find the average median age of the U.S. resident population from 2001 through 2009.

The Trapezoidal Rule

In Section 5.6, you studied one technique for approximating the value of a *definite* integral—the Midpoint Rule. In this section, you will study two other approximation techniques: the **Trapezoidal Rule** and **Simpson's Rule.**

To develop the Trapezoidal Rule, consider a function f that is nonnegative and continuous on the closed interval $[a, b]$. To approximate the area represented by

$$\int_a^b f(x)dx$$

partition the interval into n subintervals, each of width

$$\Delta x = \frac{b-a}{n}. \qquad \text{Width of each subinterval}$$

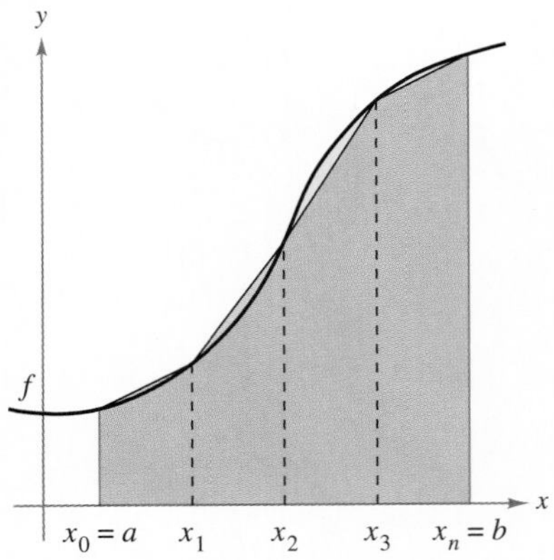

The area of the region can be approximated using four trapezoids.

FIGURE 6.5

Next, form n trapezoids, as shown in Figure 6.5. As you can see in Figure 6.6, the area of the first trapezoid is

$$\text{Area of first trapezoid} = \left(\frac{b-a}{n}\right)\left[\frac{f(x_0)+f(x_1)}{2}\right].$$

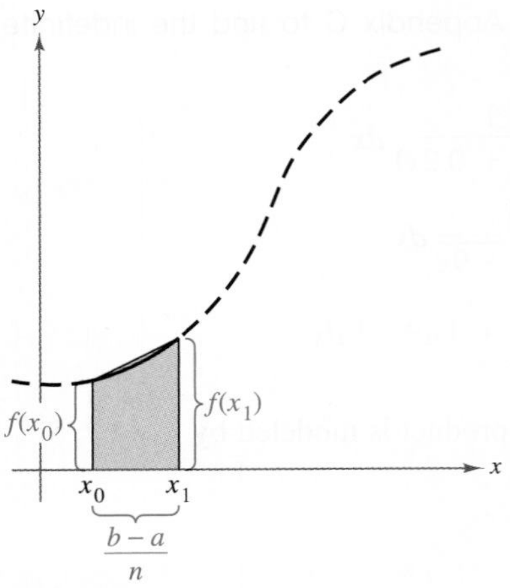

FIGURE 6.6

The areas of the other trapezoids follow a similar pattern, and the sum of the n areas is

$$\begin{aligned}\text{Area} &= \left(\frac{b-a}{n}\right)\left[\frac{f(x_0)+f(x_1)}{2} + \frac{f(x_1)+f(x_2)}{2} + \cdots + \frac{f(x_{n-1})+f(x_n)}{2}\right]\\ &= \left(\frac{b-a}{2n}\right)[f(x_0)+f(x_1)+f(x_1)+f(x_2)+\cdots+f(x_{n-1})+f(x_n)]\\ &= \left(\frac{b-a}{2n}\right)[f(x_0)+2f(x_1)+2f(x_2)+\cdots+2f(x_{n-1})+f(x_n)].\end{aligned}$$

Although this development assumes f to be continuous *and* nonnegative on $[a, b]$, the resulting formula is valid as long as f is continuous on $[a, b]$.

The Trapezoidal Rule

If f is continuous on $[a, b]$, then

$$\int_a^b f(x)dx \approx \left(\frac{b-a}{2n}\right)[f(x_0)+2f(x_1)+\cdots+2f(x_{n-1})+f(x_n)].$$

Note that the coefficients in the Trapezoidal Rule have the following pattern.

1 2 2 2 . . . 2 2 1

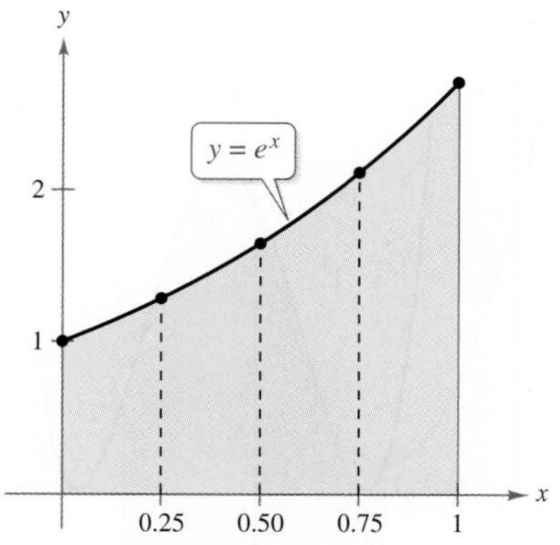

Four Subintervals

FIGURE 6.7

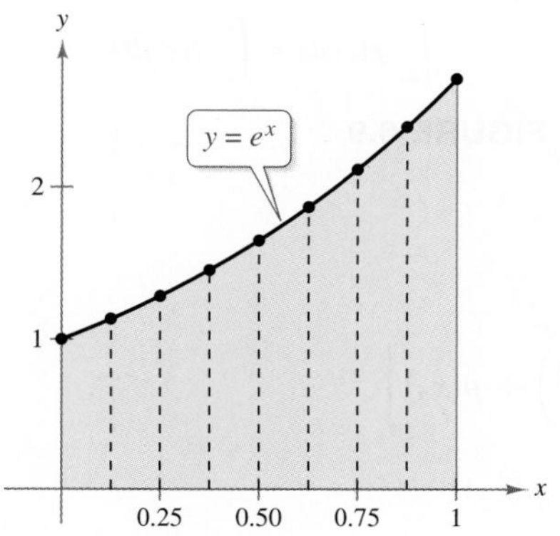

Eight Subintervals

FIGURE 6.8

Example 1 Using the Trapezoidal Rule

Use the Trapezoidal Rule to approximate $\int_0^1 e^x\,dx$. Compare the results for $n = 4$ and $n = 8$.

SOLUTION When $n = 4$, the width of each subinterval is

$$\frac{1-0}{4} = \frac{1}{4}$$

and the endpoints of the subintervals are

$$x_0 = 0, \quad x_1 = \frac{1}{4}, \quad x_2 = \frac{1}{2}, \quad x_3 = \frac{3}{4}, \quad \text{and} \quad x_4 = 1$$

as indicated in Figure 6.7. So, by the Trapezoidal Rule,

$$\int_0^1 e^x\,dx = \frac{1}{8}(e^0 + 2e^{0.25} + 2e^{0.5} + 2e^{0.75} + e^1)$$

$$\approx 1.7272. \qquad \text{Approximation using } n = 4$$

When $n = 8$, the width of each subinterval is

$$\frac{1-0}{8} = \frac{1}{8}$$

and the endpoints of the subintervals are

$$x_0 = 0, \quad x_1 = \frac{1}{8}, \quad x_2 = \frac{1}{4}, \quad x_3 = \frac{3}{8}, \quad x_4 = \frac{1}{2},$$

$$x_5 = \frac{5}{8}, \quad x_6 = \frac{3}{4}, \quad x_7 = \frac{7}{8}, \quad \text{and} \quad x_8 = 1$$

as indicated in Figure 6.8. So, by the Trapezoidal Rule,

$$\int_0^1 e^x\,dx = \frac{1}{16}(e^0 + 2e^{0.125} + 2e^{0.25} + \cdots + 2e^{0.875} + e^1)$$

$$\approx 1.7205. \qquad \text{Approximation using } n = 8$$

Of course, for *this particular* integral, you could have found an antiderivative and used the Fundamental Theorem of Calculus to find the exact value of the definite integral. The exact value is

$$\int_0^1 e^x\,dx = e - 1 \qquad \text{Exact value}$$

which is approximately 1.718282.

✓ Checkpoint 1

Use the Trapezoidal Rule with $n = 4$ to approximate

$$\int_0^1 e^{2x}\,dx.$$

■

TECH TUTOR

A symbolic integration utility can be used to evaluate a definite integral. Use a symbolic integration utility to approximate the integral $\int_0^1 e^{x^2}\,dx$.

There are two important points that should be made concerning the Trapezoidal Rule. First, the approximation tends to become more accurate as n increases. For instance, in Example 1, when $n = 16$, the Trapezoidal Rule yields an approximation of 1.7188. Second, although you could have used the Fundamental Theorem of Calculus to evaluate the integral in Example 1, this theorem cannot be used to evaluate an integral as simple as $\int_0^1 e^{x^2}\,dx$. Yet the Trapezoidal Rule can be easily applied to estimate this integral.

Simpson's Rule

One way to view the Trapezoidal Rule is to say that f is approximated by a first-degree polynomial on each subinterval. In Simpson's Rule, f is approximated by a second-degree polynomial on each subinterval.

To develop Simpson's Rule, partition the interval $[a, b]$ into an *even number n* of subintervals, each of width

$$\Delta x = \frac{b - a}{n}.$$

On the subinterval $[x_0, x_2]$, approximate the function f by the second-degree polynomial $p(x)$ that passes through the points

$$(x_0, f(x_0)), \quad (x_1, f(x_1)), \quad \text{and} \quad (x_2, f(x_2))$$

as shown in Figure 6.9. The Fundamental Theorem of Calculus can be used to show that

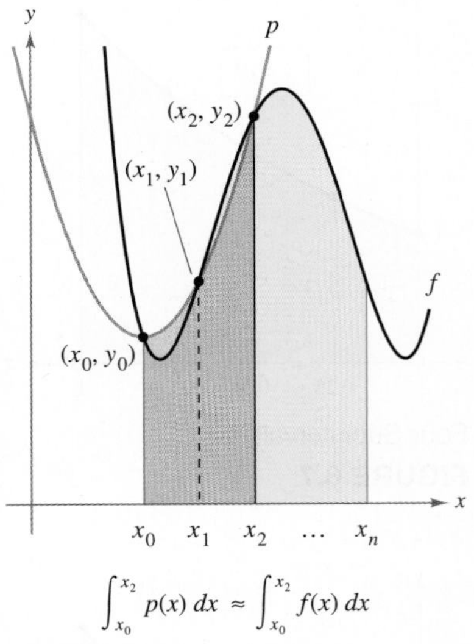

FIGURE 6.9

$$\begin{aligned}
\int_{x_0}^{x_2} f(x)\,dx &\approx \int_{x_0}^{x_2} p(x)\,dx \\
&= \left(\frac{x_2 - x_0}{6}\right)\left[p(x_0) + 4p\left(\frac{x_0 + x_2}{2}\right) + p(x_2)\right] \\
&= \frac{2[(b - a)/n]}{6}[p(x_0) + 4p(x_1) + p(x_2)] \\
&= \left(\frac{b - a}{3n}\right)[f(x_0) + 4f(x_1) + f(x_2)].
\end{aligned}$$

Repeating this process on the subintervals $[x_{i-2}, x_i]$ produces

$$\int_a^b f(x)\,dx \approx \left(\frac{b - a}{3n}\right)[f(x_0) + 4f(x_1) + f(x_2) + f(x_2) + 4f(x_3) + f(x_4) + \cdot \cdot \cdot + f(x_{n-2}) + 4f(x_{n-1}) + f(x_n)].$$

By grouping like terms, you can obtain the approximation shown below, which is known as Simpson's Rule. This rule is named after the English mathematician Thomas Simpson (1710–1761).

Simpson's Rule (*n* Is Even)

If f is continuous on $[a, b]$, then

$$\int_a^b f(x)\,dx \approx \left(\frac{b - a}{3n}\right)[f(x_0) + 4f(x_1) + 2f(x_2) + 4f(x_3) + \cdot \cdot \cdot + 4f(x_{n-1}) + f(x_n)].$$

Note that the coefficients in Simpson's Rule have the following pattern.

1 4 2 4 2 4 . . . 4 2 4 1

The Trapezoidal Rule and Simpson's Rule are necessary for solving certain real-life problems, such as approximating the present value of an income. You will see such problems in the exercise set for this section.

In Example 1, the Trapezoidal Rule was used to estimate the value of

$$\int_0^1 e^x\,dx.$$

The next example uses Simpson's Rule to approximate the same integral.

Example 2 Using Simpson's Rule

Use Simpson's Rule to approximate

$$\int_0^1 e^x\,dx.$$

Compare the results for $n = 4$ and $n = 8$.

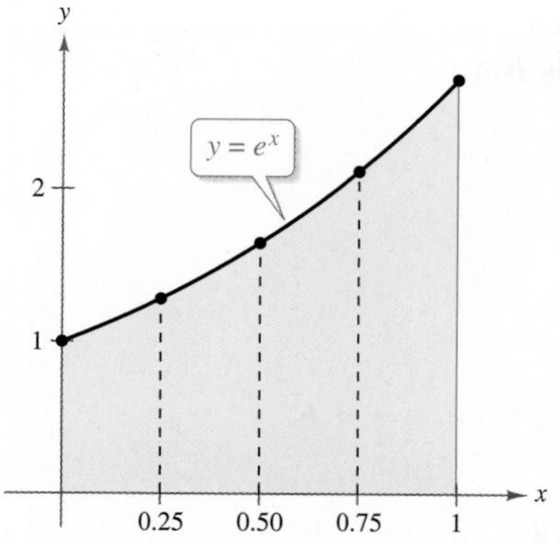

Four Subintervals
FIGURE 6.10

SOLUTION When $n = 4$, the width of each subinterval is

$$\frac{1-0}{4} = \frac{1}{4}$$

and the endpoints of the subintervals are

$$x_0 = 0,\quad x_1 = \frac{1}{4},\quad x_2 = \frac{1}{2},\quad x_3 = \frac{3}{4},\quad \text{and}\quad x_4 = 1$$

as indicated in Figure 6.10. So, by Simpson's Rule

$$\int_0^1 e^x dx = \frac{1}{12}(e^0 + 4e^{0.25} + 2e^{0.5} + 4e^{0.75} + e^1)$$

$$\approx 1.718319. \qquad \text{Approximation using } n = 4$$

When $n = 8$, the width of each subinterval is $(1 - 0)/8 = \frac{1}{8}$ and the endpoints of the subintervals are

$$x_0 = 0,\quad x_1 = \frac{1}{8},\quad x_2 = \frac{1}{4},\quad x_3 = \frac{3}{8},\quad x_4 = \frac{1}{2},$$

$$x_5 = \frac{5}{8},\quad x_6 = \frac{3}{4},\quad x_7 = \frac{7}{8},\quad \text{and}\quad x_8 = 1$$

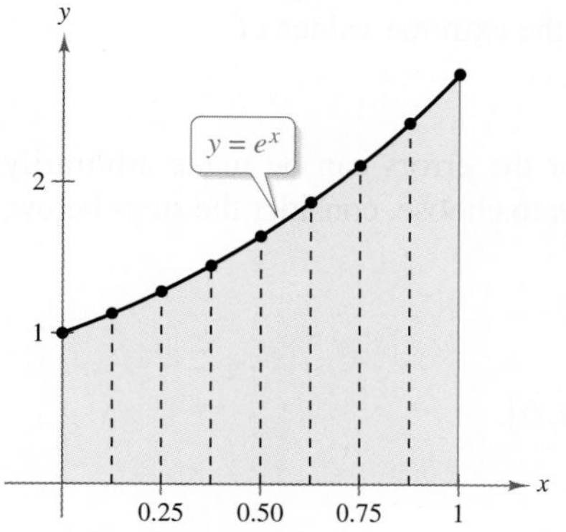

Eight Subintervals
FIGURE 6.11

as indicated in Figure 6.11. So, by Simpson's Rule

$$\int_0^1 e^x dx = \frac{1}{24}(e^0 + 4e^{0.125} + 2e^{0.25} + \cdots + 4e^{0.875} + e^1)$$

$$\approx 1.718284. \qquad \text{Approximation using } n = 8$$

Recall that the exact value of this integral is

$$\int_0^1 e^x dx = e - 1 \qquad \text{Exact value}$$

which is approximately

$$1.718282. \qquad \text{Approximate value}$$

So, with only eight subintervals, you obtained an approximation that is correct to the nearest 0.000002—an impressive result.

> **STUDY TIP**
>
> Comparing the results of Examples 1 and 2, you can see that for a given value of n, Simpson's Rule tends to be more accurate than the Trapezoidal Rule.

✓ Checkpoint 2

Use Simpson's Rule with $n = 4$ to approximate

$$\int_0^1 e^{2x}\,dx.$$

■

Error Analysis

TECH TUTOR

A program for several models of graphing utilities that uses Simpson's Rule to approximate the definite integral $\int_a^b f(x)\,dx$ can be found in Appendix E.

In Examples 1 and 2, you were able to calculate the exact value of the integral and compare that value with the approximations to see how good they were. In practice, you need to have a different way of telling how good an approximation is: such a way is provided in the next result.

Errors in the Trapezoidal Rule and Simpson's Rule

The errors E in approximating

$$\int_a^b f(x)\,dx$$

are as shown.

$$\textit{Trapezoidal Rule:}\quad |E| \le \frac{(b-a)^3}{12n^2}\left[\max|f''(x)|\right],\quad a \le x \le b$$

$$\textit{Simpson's Rule:}\quad |E| \le \frac{(b-a)^5}{180n^4}\left[\max|f^{(4)}(x)|\right],\quad a \le x \le b$$

This result indicates that the errors generated by the Trapezoidal Rule and Simpson's Rule have upper bounds dependent on the extreme values of

$$f''(x) \quad \text{and} \quad f^{(4)}(x)$$

in the interval $[a, b]$. Furthermore, the bounds for the errors can be made arbitrarily small by *increasing* n. To determine what value of n to choose, consider the steps below.

Trapezoidal Rule

1. Find $f''(x)$.
2. Find the maximum of $|f''(x)|$ on the interval $[a, b]$.
3. Set up the inequality
$$|E| \le \frac{(b-a)^3}{12n^2}\left[\max|f''(x)|\right].$$
4. For an error less than ϵ, solve for n in the inequality
$$\frac{(b-a)^3}{12n^2}\left[\max|f''(x)|\right] < \epsilon.$$
5. Partition $[a, b]$ into n subintervals and apply the Trapezoidal Rule.

Simpson's Rule

1. Find $f^{(4)}(x)$.
2. Find the maximum of $|f^{(4)}(x)|$ on the interval $[a, b]$.
3. Set up the inequality
$$|E| \le \frac{(b-a)^5}{180n^4}\left[\max|f^{(4)}(x)|\right].$$
4. For an error less than ϵ, solve for n in the inequality
$$\frac{(b-a)^5}{180n^4}\left[\max|f^{(4)}(x)|\right] < \epsilon.$$
5. Partition $[a, b]$ into n subintervals and apply Simpson's Rule.

ALGEBRA TUTOR

For help on the algebra in Example 3, see Example 4 in the Chapter 6 *Algebra Tutor*, on page 413.

Example 3 The Approximate Error in the Trapezoidal Rule

Use the Trapezoidal Rule to estimate the value of $\int_0^1 e^{-x^2}\,dx$ such that the error in the approximation of the integral is less than 0.01.

SOLUTION

1. Begin by finding the second derivative of $f(x) = e^{-x^2}$.

$$f(x) = e^{-x^2}$$
$$f'(x) = -2xe^{-x^2}$$
$$f''(x) = 4x^2e^{-x^2} - 2e^{-x^2}$$
$$= 2e^{-x^2}(2x^2 - 1)$$

2. f'' has only one critical number in the interval $[0, 1]$, and the maximum value of $|f''(x)|$ on this interval is $|f''(0)| = 2$.

3. The error E using the Trapezoidal Rule is bounded by

$$|E| \le \frac{(b-a)^3}{12n^2}(2) = \frac{1}{12n^2}(2) = \frac{1}{6n^2}.$$

4. To ensure that the approximation has an error of less than 0.01, you should choose n such that

$$\frac{1}{6n^2} < 0.01.$$

Solving for n, you can determine that n must be 5 or more.

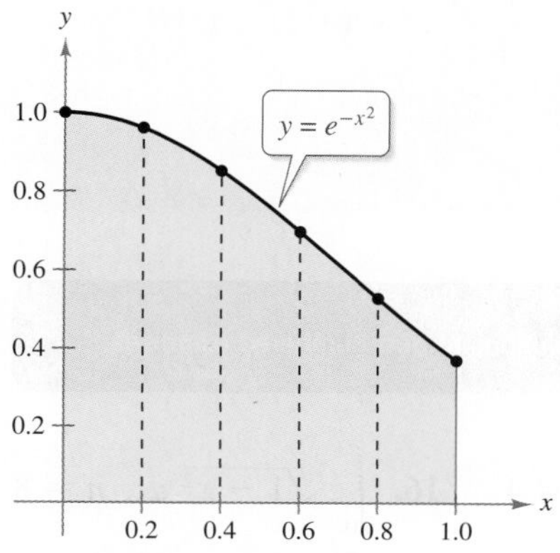

FIGURE 6.12

5. Partition $[0, 1]$ into five subintervals, as shown in Figure 6.12. Then apply the Trapezoidal Rule to obtain

$$\int_0^1 e^{-x^2}\,dx = \frac{1}{10}\left(\frac{1}{e^0} + \frac{2}{e^{0.04}} + \frac{2}{e^{0.16}} + \frac{2}{e^{0.36}} + \frac{2}{e^{0.64}} + \frac{1}{e^1}\right) \approx 0.744.$$

So, with an error less than 0.01, you know that

$$0.734 \le \int_0^1 e^{-x^2}\,dx \le 0.754.$$

✓ Checkpoint 3

Use the Trapezoidal Rule to estimate the value of

$$\int_0^1 \sqrt{1 + x^2}\,dx$$

such that the error in the approximation of the integral is less than 0.01. ■

SUMMARIZE (Section 6.3)

1. State the Trapezoidal Rule *(page 394)*. For an example of the Trapezoidal Rule, see Example 1.
2. State Simpson's Rule *(page 396)*. For an example of Simpson's Rule, see Example 2.
3. State the approximate errors in the Trapezoidal Rule and Simpson's Rule *(page 398)*. For an example of using the approximate error in the Trapezoidal Rule, see Example 3.

SKILLS WARM UP 6.3

The following warm-up exercises involve skills that were covered in a previous course or in earlier sections. You will use these skills in the exercise set for this section. For additional help, review Appendix Section A.1, and Sections 2.2, 2.6, 3.2, 4.3, and 4.5.

In Exercises 1–6, find the indicated derivative.

1. $f(x) = \dfrac{1}{x}$, $f''(x)$

2. $f(x) = \ln(2x + 1)$, $f^{(4)}(x)$

3. $f(x) = 2 \ln x$, $f^{(4)}(x)$

4. $f(x) = x^3 - 2x^2 + 7x - 12$, $f''(x)$

5. $f(x) = e^{2x}$, $f^{(4)}(x)$

6. $f(x) = e^{x^2}$, $f''(x)$

In Exercises 7 and 8, find the absolute maximum of f on the interval.

7. $f(x) = -x^2 + 6x + 9, [0, 4]$

8. $f(x) = \dfrac{8}{x^3}, [1, 2]$

In Exercises 9 and 10, solve for n.

9. $\dfrac{1}{4n^2} < 0.001$

10. $\dfrac{1}{16n^4} < 0.0001$

Exercises 6.3

See www.CalcChat.com for worked-out solutions to odd-numbered exercises.

Using the Trapezoidal Rule and Simpson's Rule In Exercises 1–10, use the Trapezoidal Rule and Simpson's Rule to approximate the value of the definite integral for the indicated value of n. Compare these results with the exact value of the definite integral. Round your answers to four decimal places. *See Examples 1 and 2.*

1. $\displaystyle\int_0^2 x^2\,dx,\ n = 4$

2. $\displaystyle\int_0^1 \left(\frac{x^2}{2} + 1\right) dx,\ n = 4$

3. $\displaystyle\int_0^2 e^{-4x}\,dx,\ n = 8$

4. $\displaystyle\int_1^3 (4 - x^2)\,dx,\ n = 4$

5. $\displaystyle\int_1^2 \frac{1}{x}\,dx,\ n = 8$

6. $\displaystyle\int_1^2 \frac{1}{x^2}\,dx,\ n = 4$

7. $\displaystyle\int_0^4 \sqrt{x}\,dx,\ n = 8$

8. $\displaystyle\int_0^8 \sqrt[3]{x}\,dx,\ n = 8$

9. $\displaystyle\int_0^1 xe^{3x^2}\,dx,\ n = 4$

10. $\displaystyle\int_0^2 x\sqrt{x^2 + 1}\,dx,\ n = 4$

Using the Trapezoidal Rule and Simpson's Rule In Exercises 11–20, approximate the value of the definite integral using (a) the Trapezoidal Rule and (b) Simpson's Rule for the indicated value of n. Round your answers to three decimal places. *See Examples 1 and 2.*

11. $\displaystyle\int_0^1 \frac{1}{1 + x^2}\,dx,\ n = 4$

12. $\displaystyle\int_0^4 \frac{8}{x^2 + 3}\,dx,\ n = 4$

13. $\displaystyle\int_0^4 \frac{1}{\sqrt[3]{x^2 + 1}}\,dx,\ n = 8$

14. $\displaystyle\int_0^2 \frac{1}{\sqrt{1 + x^3}}\,dx,\ n = 4$

15. $\displaystyle\int_0^2 \sqrt{1 + x^3}\,dx,\ n = 4$

16. $\displaystyle\int_0^1 \sqrt{1 - x^2}\,dx,\ n = 8$

17. $\displaystyle\int_0^1 e^{x^2}\,dx,\ n = 8$

18. $\displaystyle\int_0^2 e^{-x^2}\,dx,\ n = 4$

19. $\displaystyle\int_0^3 \frac{1}{2 - 2x + x^2}\,dx,\ n = 6$

20. $\displaystyle\int_0^3 \frac{x}{2 + x + x^2}\,dx,\ n = 6$

Present Value In Exercises 21 and 22, use the Simpson's Rule program in Appendix E with $n = 8$ to approximate the present value of the income $c(t)$ over t_1 years at the given annual interest rate r. Then use the integration capabilities of a graphing utility to approximate the present value. Compare the results. (Present value is defined in Section 6.1.)

21. $c(t) = 6000 + 200\sqrt{t},\ r = 7\%,\ t_1 = 4$

22. $c(t) = 200{,}000 + 15{,}000\sqrt[3]{t},\ r = 10\%,\ t_1 = 8$

Marginal Analysis In Exercises 23 and 24, use the Simpson's Rule program in Appendix E with $n = 4$ to approximate the change in revenue from the marginal revenue function dR/dx. In each case, assume that the number of units sold x increases from 14 to 16.

23. $\dfrac{dR}{dx} = 5\sqrt{8000 - x^3}$

24. $\dfrac{dR}{dx} = 50\sqrt{x}\sqrt{20 - x}$

Probability In Exercises 25–28, use the Simpson's Rule program in Appendix E with $n = 6$ to approximate the indicated normal probability. The standard normal probability density function is

$$f(x) = \frac{1}{\sqrt{2\pi}} e^{-x^2/2}.$$

If x is chosen at random from a population with this density, then the probability that x lies in the interval $[a, b]$ is

$$P(a \le x \le b) = \int_a^b f(x)\,dx.$$

25. $P(0 \le x \le 1)$ **26.** $P(0 \le x \le 2)$

27. $P(0 \le x \le 4)$ **28.** $P(0 \le x \le 1.5)$

Surveying In Exercises 29 and 30, use the Simpson's Rule program in Appendix E to estimate the number of square feet of land in the lot, where x and y are measured in feet, as shown in the figures. In each case, the land is bounded by a stream and two straight roads.

29.

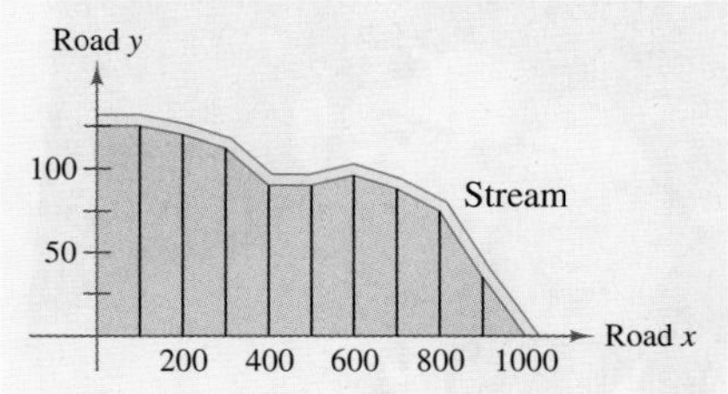

x	0	100	200	300	400	500
y	125	125	120	112	90	90

x	600	700	800	900	1000
y	95	88	75	35	0

30.

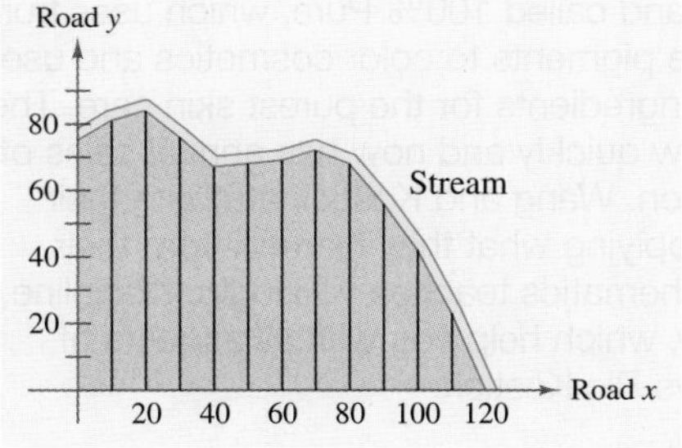

x	0	10	20	30	40	50	60
y	75	81	84	76	67	68	69

x	70	80	90	100	110	120
y	72	68	56	42	23	0

Error Analysis In Exercises 31–34, use the error formulas to find bounds for the error in approximating the definite integral using (a) the Trapezoidal Rule and (b) Simpson's Rule. Let $n = 4$.

31. $\displaystyle\int_0^2 (x^2 + 2x)\,dx$ **32.** $\displaystyle\int_0^1 \frac{1}{x+1}\,dx$

33. $\displaystyle\int_0^1 e^{x^3}\,dx$ **34.** $\displaystyle\int_0^1 e^{2x^2}\,dx$

Error Analysis In Exercises 35–38, use the error formulas to find n such that the error in the approximation of the definite integral is less than 0.0001 using (a) the Trapezoidal Rule and (b) Simpson's Rule. *See Example 3.*

35. $\displaystyle\int_0^2 x^4\,dx$ **36.** $\displaystyle\int_1^3 \frac{1}{x}\,dx$

37. $\displaystyle\int_1^3 e^{2x}\,dx$ **38.** $\displaystyle\int_3^5 \ln x\,dx$

Using Simpson's Rule In Exercises 39–42, use the Simpson's Rule program in Appendix E with $n = 100$ to approximate the definite integral.

39. $\displaystyle\int_1^4 x\sqrt{x+4}\,dx$

40. $\displaystyle\int_1^4 x^2\sqrt{x+4}\,dx$

41. $\displaystyle\int_2^5 10xe^{-x}\,dx$

42. $\displaystyle\int_2^5 10x^2e^{-x}\,dx$

43. Median Age The table shows the median ages of the U.S. resident population for the years 2001 through 2009. *(Source: U.S. Census Bureau)*

Year	2001	2002	2003	2004	2005
Median age	35.5	35.7	35.9	36.0	36.2

Year	2006	2007	2008	2009
Median age	36.3	36.5	36.7	36.8

(a) Use Simpson's Rule to estimate the average median age over the time period.

(b) A model for the data is

$$A = 35.4 + 0.16t - 0.000004e^t, \quad 1 \le t \le 9$$

where A is the median age and t is the year, with $t = 1$ corresponding to 2001. Use integration to find the average median age over the time period.

(c) Compare the results of parts (a) and (b).

44. Electricity The table shows the residential prices of electricity (in cents per kilowatt-hour) for the years 2001 through 2009. *(Source: U.S. Energy Information Administration)*

Year	2001	2002	2003	2004	2005
Price	8.58	8.44	8.72	8.95	9.45

Year	2006	2007	2008	2009
Price	10.40	10.65	11.26	11.55

(a) Use Simpson's Rule to estimate the average residential price of electricity over the time period.

(b) A model for the data is

$$E = 8.4 - 1.39t + 0.291t^2 - 0.0160t^3 + 1.27097\sqrt{t}$$

for $1 \le t \le 9$, where E is the residential price of electricity (in cents per kilowatt-hour) and t is the year, with $t = 1$ corresponding to 2001. Use integration to find the average residential price of electricity over the time period.

(c) Compare the results of parts (a) and (b).

45. Medicine A body assimilates a 12-hour cold tablet at a rate modeled by $dC/dt = 8 - \ln(t^2 - 2t + 4)$, $0 \le t \le 12$, where dC/dt is measured in milligrams per hour and t is the time in hours. Use Simpson's Rule with $n = 8$ to estimate the total amount of the drug absorbed into the body during the 12 hours.

46. HOW DO YOU SEE IT? The graph shows the weekly revenue (in thousands of dollars) for a company.

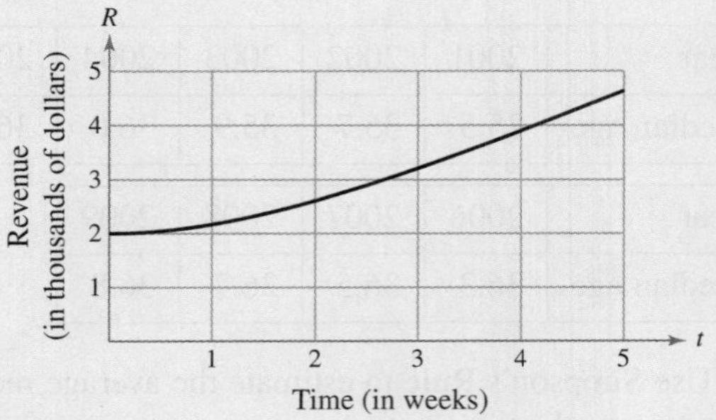

(a) Which gives a more accurate approximation of the total weekly revenue for the first 4 weeks using the Trapezoidal Rule, $n = 8$ or $n = 16$?

(b) Which gives a more accurate approximation of the total weekly revenue for the first 4 weeks, the Trapezoidal Rule with $n = 8$ or Simpson's Rule with $n = 8$?

47. Consumer Trends The rate of change in the number of subscribers S to a newly introduced magazine is modeled by

$$\frac{dS}{dt} = 1000t^2e^{-t}, \quad 0 \le t \le 6$$

where t is the time in years. Use Simpson's Rule with $n = 12$ to estimate the total increase in the number of subscribers during the first 6 years.

48. Using Simpson's Rule Prove that Simpson's Rule is exact when used to approximate the integral of a cubic polynomial function, and demonstrate the result for

$$\int_0^1 x^3\,dx$$

with $n = 2$.

Business Capsule

Susie Wang and Ric Kostick graduated in 2002 from the University of California at Berkeley with degrees in mathematics. Together they launched a cosmetics brand called 100% Pure, which uses fruit and vegetable pigments to color cosmetics and uses only organic ingredients for the purest skin care. The company grew quickly and now has annual sales of over \$15 million. Wang and Kostick attribute their success to applying what they learned from their studies. "Mathematics teaches you logic, discipline, and accuracy, which help you with all aspects of daily life," says Ric Kostick.

49. Research Project Use your school's library, the Internet, or some other reference source to research the opportunity cost of attending graduate school for 2 years to receive a Masters of Business Administration (MBA) degree rather than working for 2 years with a bachelor's degree. Write a short paper describing these costs.

6.4 Improper Integrals

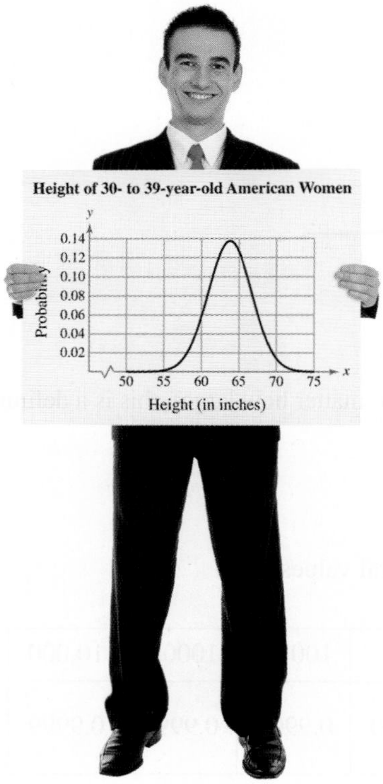

In Exercise 27 on page 411, you will evaluate an improper integral to determine the probability that a 30- to 39-year-old woman is 6 feet or taller.

- Recognize improper integrals.
- Evaluate improper integrals with infinite limits of integration.
- Use improper integrals to solve real-life problems.
- Find the present value of a perpetuity.

Improper Integrals

The definition of a definite integral

$$\int_a^b f(x)\,dx$$

requires that the interval $[a, b]$ be finite. Furthermore, the Fundamental Theorem of Calculus, by which you have been evaluating definite integrals, requires that f be continuous on $[a, b]$. Some integrals do not satisfy these requirements because of one of the conditions below.

1. One or both of the limits of integration are infinite.
2. The function f has an infinite discontinuity in the interval $[a, b]$.

Integrals having either of these characteristics are called **improper integrals.** In this section, you will study integrals where one or both limits of integration are infinite. For instance, the integral

$$\int_0^{\infty} e^{-x}\,dx$$

is improper because one limit of integration is infinite, as indicated in Figure 6.13.

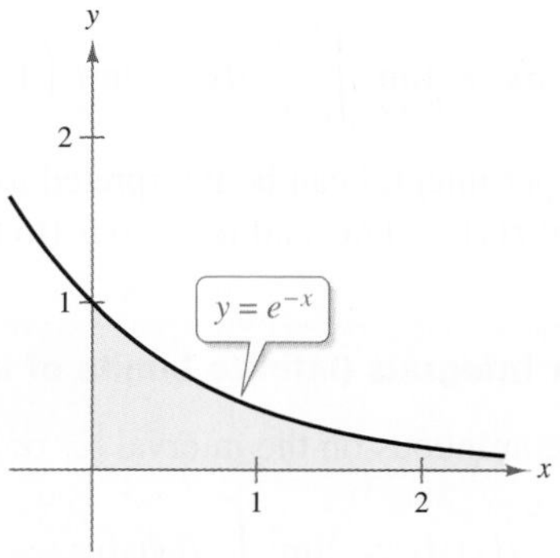

FIGURE 6.13

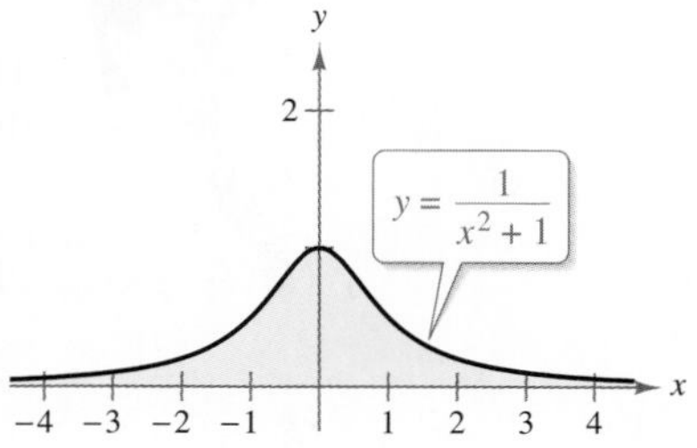

FIGURE 6.14

Similarly, the integral

$$\int_{-\infty}^{\infty} \frac{1}{x^2 + 1}\,dx$$

is improper because both limits of integration are infinite, as indicated in Figure 6.14.

The integrals

$$\int_1^5 \frac{1}{\sqrt{x - 1}}\,dx \quad \text{and} \quad \int_{-2}^{2} \frac{1}{(x + 1)^2}\,dx$$

are improper because their integrands have an **infinite discontinuity**—that is, they approach infinity somewhere in the interval of integration. Evaluating an integral whose integrand has an infinite discontinuity is beyond the scope of this text.

Integrals with Infinite Limits of Integration

To see how to evaluate an improper integral, consider the integral shown in Figure 6.15.

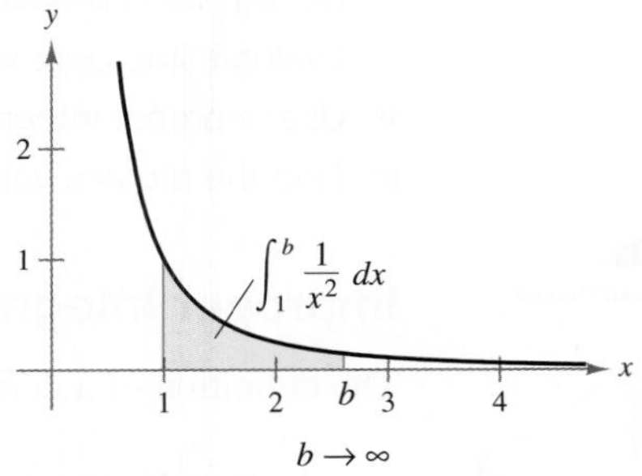

FIGURE 6.15

As long as b is a real number that is greater than 1 (no matter how large), this is a definite integral whose value is

$$\int_1^b \frac{1}{x^2}\,dx = \left[-\frac{1}{x}\right]_1^b = -\frac{1}{b} + 1 = 1 - \frac{1}{b}.$$

The table shows the values of this integral for several values of b.

b	2	5	10	100	1000	10,000
$\int_1^b \frac{1}{x^2}\,dx = 1 - \frac{1}{b}$	0.5000	0.8000	0.9000	0.9900	0.9990	0.9999

From this table, it appears that the value of the integral is approaching a limit as b increases without bound. This limit is denoted by the *improper integral* shown below.

$$\int_1^\infty \frac{1}{x^2}\,dx = \lim_{b\to\infty} \int_1^b \frac{1}{x^2}\,dx = \lim_{b\to\infty}\left(1 - \frac{1}{b}\right) = 1$$

This improper integral can be interpreted as the area of the *unbounded* region between the graph of $f(x) = 1/x^2$ and the x-axis (to the right of $x = 1$).

Improper Integrals (Infinite Limits of Integration)

1. If f is continuous on the interval $[a, \infty)$, then

$$\int_a^\infty f(x)\,dx = \lim_{b\to\infty} \int_a^b f(x)\,dx.$$

2. If f is continuous on the interval $(-\infty, b]$, then

$$\int_{-\infty}^b f(x)\,dx = \lim_{a\to-\infty} \int_a^b f(x)\,dx.$$

3. If f is continuous on the interval $(-\infty, \infty)$, then

$$\int_{-\infty}^\infty f(x)\,dx = \int_{-\infty}^c f(x)\,dx + \int_c^\infty f(x)\,dx$$

where c is any real number.

In the first two cases, if the limit exists, then the improper integral **converges;** otherwise, the improper integral **diverges.** In the third case, the integral on the left diverges when either one of the integrals on the right diverges.

Andresr/www.shutterstock.com

TECH TUTOR

Symbolic integration utilities evaluate improper integrals in much the same way that they evaluate definite integrals. Use a symbolic integration utility to evaluate the integral in Example 1.

Example 1 Evaluating an Improper Integral

Determine the convergence or divergence of $\displaystyle\int_1^{\infty} \frac{1}{x}\,dx$.

SOLUTION Begin by applying the definition of an improper integral.

$$\int_1^{\infty} \frac{1}{x}\,dx = \lim_{b\to\infty} \int_1^{b} \frac{1}{x}\,dx \qquad \text{Definition of improper integral}$$

$$= \lim_{b\to\infty} \Big[\ln x\Big]_1^b \qquad \text{Find antiderivative.}$$

$$= \lim_{b\to\infty} (\ln b - 0) \qquad \text{Apply Fundamental Theorem.}$$

$$= \infty \qquad \text{Evaluate limit.}$$

Because the limit is infinite, the improper integral diverges.

✓ Checkpoint 1

Determine the convergence or divergence of each improper integral.

a. $\displaystyle\int_1^{\infty} \frac{1}{x^3}\,dx$ **b.** $\displaystyle\int_1^{\infty} \frac{1}{\sqrt{x}}\,dx$

As you begin to work with improper integrals, you will find that integrals that appear to be similar can have very different values. For instance, consider the two improper integrals

$$\int_1^{\infty} \frac{1}{x}\,dx = \infty \qquad \text{Divergent integral}$$

and

$$\int_1^{\infty} \frac{1}{x^2}\,dx = 1. \qquad \text{Convergent integral}$$

The first integral diverges and the second converges to 1. Graphically, this means that the areas shown in Figure 6.16 are very different. The region lying between the graph of

$$y = \frac{1}{x}$$

and the x-axis (for $x \geq 1$) has an *infinite* area, and the region lying between the graph of

$$y = \frac{1}{x^2}$$

and the x-axis (for $x \geq 1$) has a *finite* area.

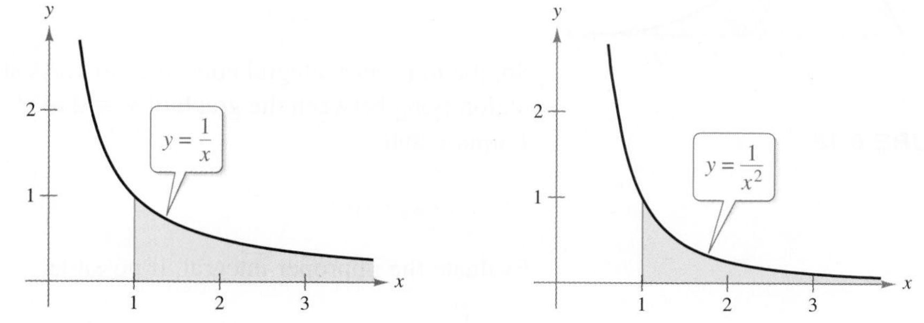

Diverges (infinite area) Converges (finite area)

FIGURE 6.16

Ljupco Smokovski/www.shutterstock.com

ALGEBRA TUTOR

For help on the algebra in Example 2, see Example 2(a) in the *Chapter 6 Algebra Tutor*, on page 412.

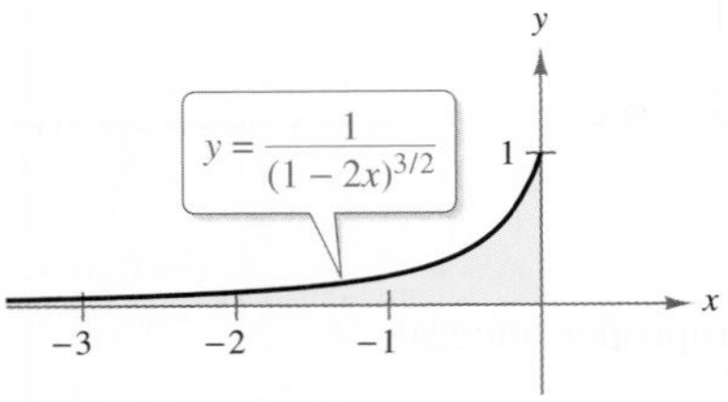

FIGURE 6.17

Example 2 Evaluating an Improper Integral

Evaluate the improper integral.

$$\int_{-\infty}^{0} \frac{1}{(1-2x)^{3/2}}\,dx$$

SOLUTION Begin by applying the definition of an improper integral.

$$\begin{aligned}
\int_{-\infty}^{0} \frac{1}{(1-2x)^{3/2}}\,dx &= \lim_{a\to-\infty} \int_{a}^{0} \frac{1}{(1-2x)^{3/2}}\,dx && \text{Definition of improper integral}\\
&= \lim_{a\to-\infty} \left[\frac{1}{\sqrt{1-2x}}\right]_a^0 && \text{Find antiderivative.}\\
&= \lim_{a\to-\infty} \left(1 - \frac{1}{\sqrt{1-2a}}\right) && \text{Apply Fundamental Theorem.}\\
&= 1 - 0 && \text{Evaluate limit.}\\
&= 1 && \text{Simplify.}
\end{aligned}$$

So, the improper integral converges to 1. As shown in Figure 6.17, this implies that the region lying between the graph of $y = 1/(1-2x)^{3/2}$ and the x-axis (for $x \le 0$) has an area of 1 square unit.

Checkpoint 2

Evaluate the improper integral, if possible.

$$\int_{-\infty}^{0} \frac{1}{(x-1)^2}\,dx$$

■

ALGEBRA TUTOR

For help on the algebra in Example 3, see Example 2(b) in the *Chapter 6 Algebra Tutor*, on page 412.

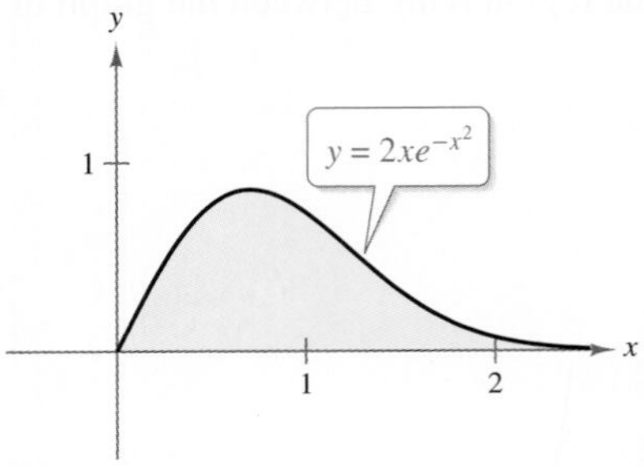

FIGURE 6.18

Example 3 Evaluating an Improper Integral

Evaluate the improper integral.

$$\int_{0}^{\infty} 2xe^{-x^2}\,dx$$

SOLUTION Begin by applying the definition of an improper integral.

$$\begin{aligned}
\int_{0}^{\infty} 2xe^{-x^2}\,dx &= \lim_{b\to\infty} \int_{0}^{b} 2xe^{-x^2}\,dx && \text{Definition of improper integral}\\
&= \lim_{b\to\infty} \left[-e^{-x^2}\right]_0^b && \text{Find antiderivative.}\\
&= \lim_{b\to\infty} (-e^{-b^2} + 1) && \text{Apply Fundamental Theorem.}\\
&= 0 + 1 && \text{Evaluate limit.}\\
&= 1 && \text{Simplify.}
\end{aligned}$$

So, the improper integral converges to 1. As shown in Figure 6.18, this implies that the region lying between the graph of $y = 2xe^{-x^2}$ and the x-axis (for $x \ge 0$) has an area of 1 square unit.

Checkpoint 3

Evaluate the improper integral, if possible.

$$\int_{-\infty}^{0} e^{2x}\,dx$$

■

Application

In Section 4.3, you studied the graph of the *normal probability density function*

$$f(x) = \frac{1}{\sigma\sqrt{2\pi}}e^{-(x-\mu)^2/(2\sigma^2)}.$$

This function is used in statistics to represent a population that is normally distributed with a mean of μ and a standard deviation of σ. Specifically, when an outcome x is chosen at random from the population, the probability that x will have a value between a and b is

$$P(a \le x \le b) = \int_a^b \frac{1}{\sigma\sqrt{2\pi}}e^{-(x-\mu)^2/(2\sigma^2)}dx.$$

As shown in Figure 6.19, the probability $P(-\infty < x < \infty)$ is

$$P(-\infty < x < \infty) = \int_{-\infty}^{\infty} \frac{1}{\sigma\sqrt{2\pi}}e^{-(x-\mu)^2/(2\sigma^2)}dx = 1.$$

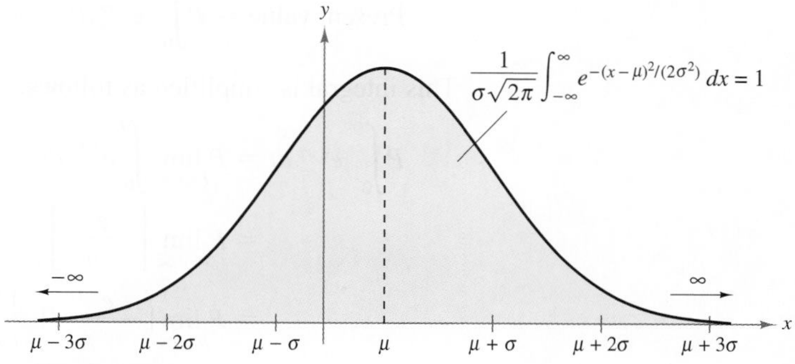

FIGURE 6.19

Example 4 Finding a Probability

The mean height of American men (from 20 to 29 years old) is 69 inches, and the standard deviation is 3 inches. A 20- to 29-year-old man is chosen at random from the population. What is the probability that he is 6 feet tall or taller? *(Source: U.S. National Center for Health Statistics)*

SOLUTION Note that the mean and standard deviation are given in inches and the height of the man chosen at random is given in feet. To calculate the probability, you need to use the same units for these quantities. Because it is easier to convert feet to inches, use 72 inches (1 foot = 12 inches) for the man's height. So, the probability can be written as $P(72 \le x < \infty)$. Using a mean of $\mu = 69$ and a standard deviation of $\sigma = 3$, the probability $P(72 \le x < \infty)$ is given by the improper integral

$$P(72 \le x < \infty) = \int_{72}^{\infty} \frac{1}{3\sqrt{2\pi}}e^{-(x-69)^2/18}dx.$$

Using a symbolic integration utility, you can approximate the value of this integral to be 0.158. So, the probability that the man is 6 feet tall or taller is about 15.8%.

Many professional basketball players are over $6\frac{1}{2}$ feet tall. When a man is chosen at random from the population, the probability that he is $6\frac{1}{2}$ feet tall or taller is less than half of one percent.

✓Checkpoint 4

Use Example 4 to find the probability that a 20- to 29-year-old man chosen at random from the population is 6 feet 6 inches tall or taller.

Present Value of a Perpetuity

Recall from Section 6.1 that for an interest-bearing account, the present value over t_1 years is

$$\text{Present value} = \int_0^{t_1} c(t)e^{-rt}\,dt$$

where c represents a continuous income function (in dollars per year) and the annual interest rate r is compounded continuously. If the size of an annuity's payment is a constant number of dollars P, then $c(t)$ is equal to P and the present value is

$$\text{Present value} = \int_0^{t_1} Pe^{-rt}\,dt = P\int_0^{t_1} e^{-rt}\,dt.$$ Present value of an annuity with payment P

Consider an annuity, such as a scholarship fund, that pays the same amount each year *forever*. Because the annuity continues indefinitely, the number of years t_1 approaches infinity. Such an annuity is called a **perpetual annuity** or a **perpetuity.** This situation can be represented by the following improper integral.

$$\text{Present value} = P\int_0^{\infty} e^{-rt}\,dt$$ Present value of a perpetuity with payment P

This integral is simplified as follows.

$$P\int_0^{\infty} e^{-rt}\,dt = P\lim_{b\to\infty}\int_0^{b} e^{-rt}\,dt$$ Definition of improper integral

$$= P\lim_{b\to\infty}\left[-\frac{e^{-rt}}{r}\right]_0^b$$ Find antiderivative.

$$= P\lim_{b\to\infty}\left(-\frac{e^{-rb}}{r} + \frac{1}{r}\right)$$ Apply Fundamental Theorem.

$$= P\left(0 + \frac{1}{r}\right)$$ Evaluate limit.

$$= \frac{P}{r}$$ Simplify.

So, the improper integral converges to P/r. As shown in Figure 6.20, this implies that the region lying between the graph of

$$y = Pe^{-rt}$$

and the t-axis for $t \geq 0$ has an area equal to the annual payment P divided by the annual interest rate r.

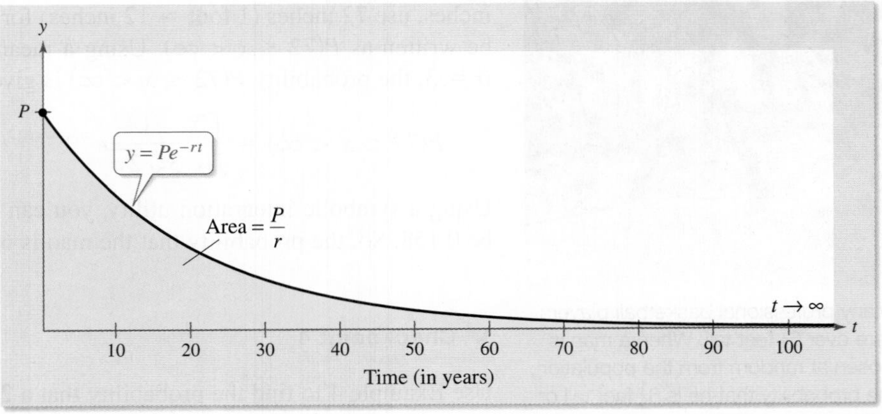

FIGURE 6.20

The present value of a perpetuity is defined as follows.

Present Value of a Perpetuity

If P represents the size of each annual payment in dollars and the annual interest rate is r (compounded continuously), then the present value of a perpetuity is

$$\text{Present value} = P\int_0^{\infty} e^{-rt}\,dt = \frac{P}{r}.$$

This definition is useful in determining the amount of money needed to start an endowment, such as a scholarship fund, as shown in Example 5.

Example 5 Finding Present Value

You want to start a scholarship fund at your alma mater. You plan to give one $9000 scholarship annually beginning one year from now, and you have at most $120,000 to start the fund. You also want the scholarship to be given out indefinitely. Assuming an annual interest rate of 8% (compounded continuously), do you have enough money for the scholarship fund?

SOLUTION To answer this question, you must find the present value of the scholarship fund. Because the scholarship is to be given out each year indefinitely, the time period is infinite. The fund is a perpetuity with $P = 9000$ and $r = 0.08$. The present value is

$$\begin{aligned}\text{Present value} &= \frac{P}{r}\\ &= \frac{9000}{0.08}\\ &= 112{,}500.\end{aligned}$$

The amount you need to start the scholarship fund is $112,500. Yes, you have enough money to start the scholarship fund.

✓Checkpoint 5

In Example 5, do you have enough money to start a scholarship fund that pays $10,000 annually? Explain why or why not.

SUMMARIZE (Section 6.4)

1. Describe the different types of improper integrals *(page 403)*. For examples of evaluating improper integrals, see Examples 1, 2, and 3.
2. Define the term *converges* as it applies to improper integrals *(page 404)*. For examples of improper integrals that converge, see Examples 2 and 3.
3. Define the term *diverges* as it applies to improper integrals *(page 404)*. For an example of an improper integral that diverges, see Example 1.
4. Describe a real-life example of how an improper integral can be used to find a probability *(page 407, Example 4)*.
5. Describe a real-life example of how an improper integral can be used to find the present value of a perpetuity *(page 409, Example 5)*.

SKILLS WARM UP 6.4 The following warm-up exercises involve skills that were covered in earlier sections. You will use these skills in the exercise set for this section. For additional help, review Sections 1.5, 4.1, and 4.4.

In Exercises 1–6, find the limit.

1. $\lim_{x\to 2}(2x+5)$

2. $\lim_{x\to 1}\left(\frac{1}{x}+2x^2\right)$

3. $\lim_{x\to -4}\frac{x+4}{x^2-16}$

4. $\lim_{x\to 0}\frac{x^2-2x}{x^3+3x^2}$

5. $\lim_{x\to 1}\frac{1}{\sqrt{x-1}}$

6. $\lim_{x\to -3}\frac{x^2+2x-3}{x+3}$

In Exercises 7–10, evaluate the expression (a) when $x = b$ and (b) when $x = 0$.

7. $\frac{4}{3}(2x-1)^3$

8. $\frac{1}{x-5}+\frac{3}{(x-2)^2}$

9. $\ln(5-3x^2)-\ln(x+1)$

10. $e^{3x^2}+e^{-3x^2}$

Exercises 6.4

See www.CalcChat.com for worked-out solutions to odd-numbered exercises.

Determining Whether an Integral Is Improper In Exercises 1–6, decide whether the integral is improper. Explain your reasoning.

1. $\int_0^1 \frac{dx}{3x-2}$

2. $\int_1^3 \frac{dx}{x^2}$

3. $\int_0^1 \frac{2x-5}{x^2-5x+6}\,dx$

4. $\int_1^\infty x^2\,dx$

5. $\int_0^5 e^{-x}\,dx$

6. $\int_{-\infty}^\infty \frac{1}{x^2+3}\,dx$

Evaluating an Improper Integral In Exercises 7–20, determine whether the improper integral diverges or converges. Evaluate the integral if it converges. *See Examples 1, 2, and 3.*

7. $\int_1^\infty \frac{1}{x^2}\,dx$

8. $\int_1^\infty \frac{1}{\sqrt[3]{x}}\,dx$

9. $\int_0^\infty e^{x/3}\,dx$

10. $\int_0^\infty \frac{5}{e^{2x}}\,dx$

11. $\int_5^\infty \frac{x}{\sqrt{x^2-16}}\,dx$

12. $\int_5^\infty \frac{1}{\sqrt{2x-1}}\,dx$

13. $\int_{-\infty}^0 e^{-x}\,dx$

14. $\int_{-\infty}^{-1} \frac{1}{x^2}\,dx$

15. $\int_1^\infty \frac{e^{\sqrt{x}}}{\sqrt{x}}\,dx$

16. $\int_{-\infty}^0 \frac{x}{x^2+1}\,dx$

17. $\int_{-\infty}^\infty 2xe^{-3x^2}\,dx$

18. $\int_{-\infty}^\infty x^2e^{-x^3}\,dx$

19. $\int_4^\infty \frac{1}{x(\ln x)^3}\,dx$

20. $\int_{-\infty}^0 \frac{1}{\sqrt[3]{1-x}}\,dx$

Area of a Region In Exercises 21–26, find the area of the unbounded shaded region.

21. $y = e^{-x}$

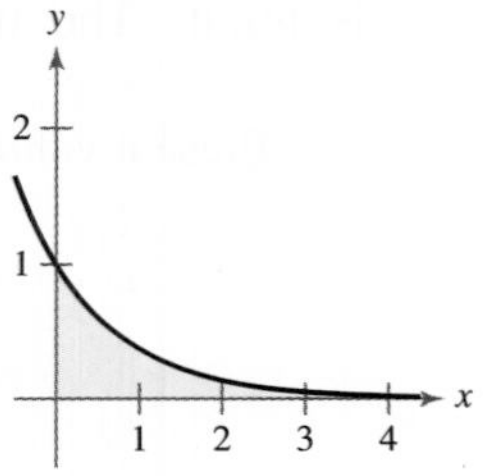

22. $y = e^{x/4}$

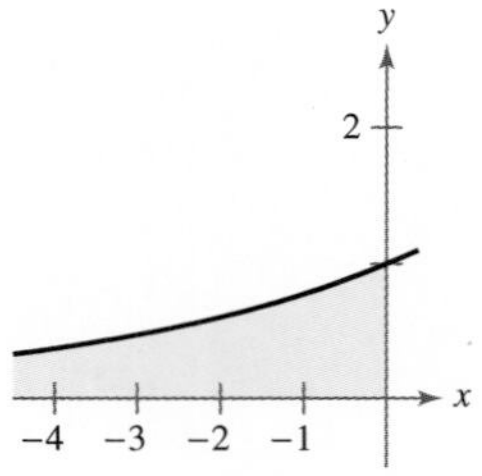

23. $y = -\frac{1}{x^3}$

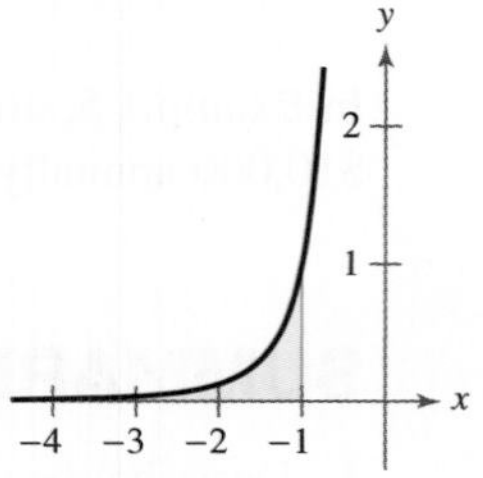

24. $y = \frac{5}{\sqrt{4-x}}$

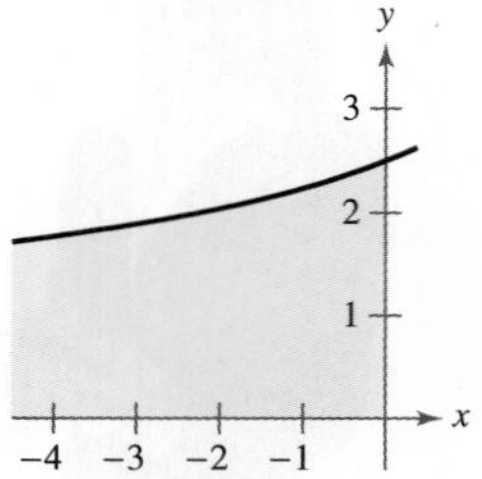

25. $y = \frac{6x}{x^2+1}$

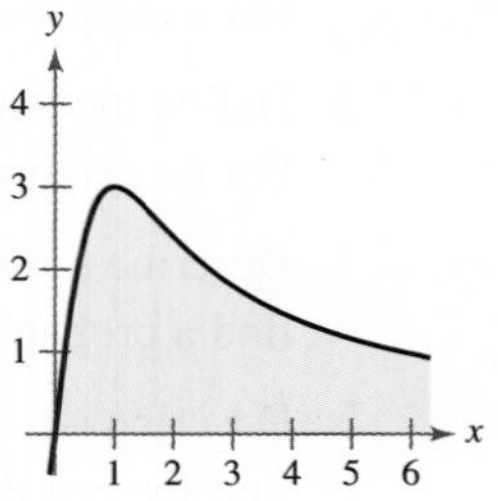

26. $y = \frac{16x}{x^2+4}$

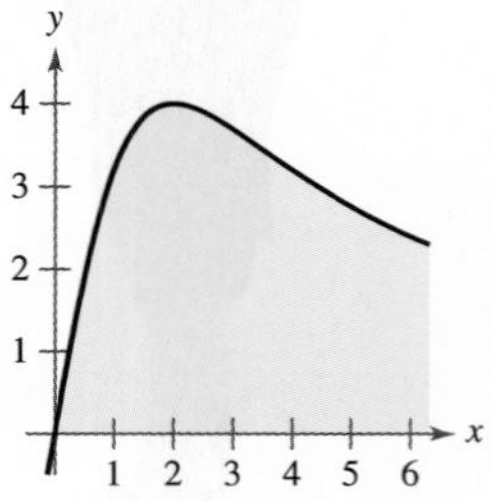

27. Women's Height The mean height of American women (from 30 to 39 years old) is 63.8 inches, and the standard deviation is 2.9 inches. Use a symbolic integration utility or a graphing utility to find the probability that a 30- to 39-year-old woman chosen at random is

(a) between 5 and 6 feet tall.

(b) 5 feet 8 inches or taller.

(c) 6 feet or taller.

(Source: U.S. National Center for Health Statistics)

28. HOW DO YOU SEE IT? The graph shows the probability density function for a car brand that has a mean fuel efficiency of 26 miles per gallon and a standard deviation of 2.4 miles per gallon.

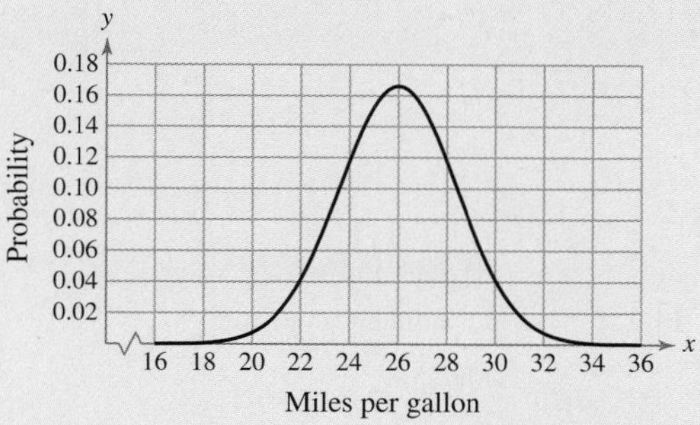

(a) Which is greater, the probability of choosing a car at random that gets between 26 and 28 miles per gallon or the probability of choosing a car at random that gets between 22 and 24 miles per gallon?

(b) Which is greater, the probability of choosing a car at random that gets between 20 and 22 miles per gallon or the probability of choosing a car at random that gets at least 30 miles per gallon?

29. Quality Control A company manufactures wooden yardsticks. The lengths of the yardsticks are normally distributed with a mean of 36 inches and a standard deviation of 0.2 inch. Use a symbolic integration utility or a graphing utility to find the probability that a yardstick chosen at random is

(a) longer than 35.5 inches.

(b) longer than 35.9 inches.

30. Quality Control A company manufactures compact fluorescent light bulbs. The life spans of the light bulbs are normally distributed with a mean of 9000 hours and a standard deviation of 500 hours. Use a symbolic integration utility or a graphing utility to find the probability that a light bulb chosen at random has a life span that is

(a) between 8000 and 10,000 hours.

(b) 11,000 hours or longer.

Endowment In Exercises 31 and 32, determine the amount of money required to set up a charitable endowment that pays the amount P each year indefinitely for the annual interest rate r compounded continuously. *See Example 5.*

31. $P = \$5000$, $r = 7.5\%$

32. $P = \$12{,}000$, $r = 6\%$

33. Scholarship Fund You want to start a scholarship fund at your alma mater. You plan to give one \$18,000 scholarship annually beginning one year from now, and you have at most \$400,000 to start the fund. You also want the scholarship to be given out indefinitely. Assuming an annual interest rate of 5% compounded continuously, do you have enough money for the scholarship fund?

34. Charitable Foundation A charitable foundation wants to help schools buy computers. The foundation plans to donate \$35,000 each year to one school beginning one year from now, and the foundation has at most \$500,000 to start the fund. The foundation wants the donation to be given out indefinitely. Assuming an annual interest rate of 8% compounded continuously, does the foundation have enough money to fund the donation?

35. Present Value A business is expected to yield a continuous flow of profit at the rate of \$500,000 per year. Assuming an annual interest rate of 9% compounded continuously, what is the present value of the business

(a) for 20 years?

(b) forever?

36. Present Value A farm is expected to yield a continuous flow of profit at the rate of \$75,000 per year. Assuming an annual interest rate of 8% compounded continuously, what is the present value of the farm

(a) for 20 years?

(b) forever?

Capitalized Cost In Exercises 37–40, the capitalized cost C of an asset is given by

$$C = C_0 + \int_0^n c(t)e^{-rt}\,dt$$

where C_0 is the original investment, t is the time in years, r is the annual interest rate compounded continuously, and $c(t)$ is the annual cost of maintenance (in dollars). Find the capitalized cost of an asset (a) for 5 years, (b) for 10 years, and (c) forever.

37. $C_0 = \$650{,}000$, $c(t) = 25{,}000$, $r = 10\%$

38. $C_0 = \$800{,}000$, $c(t) = 30{,}000$, $r = 4\%$

39. $C_0 = \$300{,}000$, $c(t) = 15{,}000t$, $r = 6\%$

40. $C_0 = \$650{,}000$, $c(t) = 25{,}000(1 + 0.08t)$, $r = 12\%$

ALGEBRA TUTOR

Algebra and Integration Techniques

Integration techniques involve many different algebraic skills. For a definite integral, you need a variety of algebraic skills to apply the Fundamental Theorem of Calculus and evaluate the resulting expression. Study the examples in this Algebra Tutor. Be sure that you understand the algebra used in each step.

Example 1 Evaluating an Expression

Evaluate the expression

$$(e \ln e - e) - (1 \ln 1 - 1).$$

SOLUTION Recall that

$$\ln e = 1 \quad \text{because} \quad e^1 = e$$

and

$$\ln 1 = 0 \quad \text{because} \quad e^0 = 1.$$

$(e \ln e - e) - (1 \ln 1 - 1)$	Example 5, page 380
$= [e(1) - e] - [1(0) - 1]$	Logarithmic properties
$= (e - e) - (0 - 1)$	Multiply.
$= 0 - (-1)$	Simplify.
$= 1$	Simplify.

Example 2 Evaluating Expressions

Find the limit.

a. $\lim_{a \to -\infty} \left(1 - \dfrac{1}{\sqrt{1 - 2a}}\right)$

b. $\lim_{b \to \infty} (-e^{-b^2} + 1)$

SOLUTION

a. $\lim_{a \to -\infty} \left(1 - \dfrac{1}{\sqrt{1 - 2a}}\right)$	Example 2, page 406
$= \lim_{a \to -\infty} 1 - \lim_{a \to -\infty} \left(\dfrac{1}{\sqrt{1 - 2a}}\right)$	$\lim_{x \to -\infty} [f(x) - g(x)] = \lim_{x \to -\infty} f(x) - \lim_{x \to -\infty} g(x)$
$= 1 - 0$	Evaluate limits.
$= 1$	Simplify.
b. $\lim_{b \to \infty} (-e^{-b^2} + 1)$	Example 3, page 406
$= \lim_{b \to \infty} (-e^{-b^2}) + \lim_{b \to \infty} 1$	$\lim_{x \to \infty} [f(x) + g(x)] = \lim_{x \to \infty} f(x) + \lim_{x \to \infty} g(x)$
$= \lim_{b \to \infty} \left(\dfrac{1}{-e^{b^2}}\right) + \lim_{b \to \infty} 1$	Rewrite with positive exponent.
$= 0 + 1$	Evaluate limits.
$= 1$	Simplify.

Example 3 **Algebra and Integration Techniques**

Simplify the expression

$$x^2e^x - 2(x - 1)e^x.$$

SOLUTION

$$\begin{aligned} &x^2e^x - 2(x - 1)e^x && \text{Example 5, page 389} \\ &\quad = x^2e^x - 2(xe^x - e^x) && \text{Multiply factors.} \\ &\quad = x^2e^x - 2xe^x + 2e^x && \text{Multiply factors.} \\ &\quad = e^x(x^2 - 2x + 2) && \text{Factor.} \end{aligned}$$

Example 4 **Solving a Rational Inequality**

Solve the rational inequality

$$\frac{1}{6n^2} < 0.01$$

for n, where n is a positive integer.

SOLUTION

$$\begin{aligned} &\frac{1}{6n^2} < 0.01 && \text{Example 3, page 399} \\ &\frac{1}{6n^2} < \frac{1}{100} && \text{Rewrite decimal as a fraction.} \\ &\frac{100}{6n^2} < 1 && \text{Multiply each side by 100.} \\ &100 < 6n^2 && \text{Multiply each side by } 6n^2 \ (n > 0). \\ &\frac{100}{6} < n^2 && \text{Divide each side by 6.} \\ &\frac{50}{3} < n^2 && \text{Simplify.} \\ &\sqrt{\frac{50}{3}} < n && \text{Take positive square root of each side } (n > 0). \end{aligned}$$

Because n is a positive integer and

$$\sqrt{\frac{50}{3}} \approx 4.08$$

n must be 5 or more. You can check this result using a graphing utility. Let $y_1(x) = 1/6x^2$ and $y_2(x) = 0.01$. Then use the *intersect* feature (see figure) to determine that $x \approx 4.08$. So, the solution found algebraically is correct.

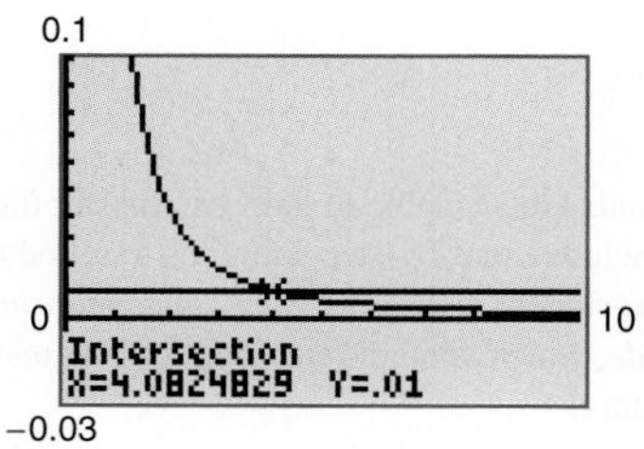

SUMMARY AND STUDY STRATEGIES

After studying this chapter, you should have acquired the following skills.
The exercise numbers are keyed to the Review Exercises that begin on page 416.
Answers to odd-numbered Review Exercises are given in the back of the text.*

Section 6.1	Review Exercises
■ Use integration by parts to find indefinite and definite integrals.	*1–12*
■ Find the present value of future income.	*13–18*

Section 6.1

Review Exercises

- Use integration by parts to find indefinite and definite integrals. *1–12*

$$\int u\,dv = uv - \int v\,du$$

For integrals of the form

$$\int x^n e^{ax}\,dx$$

let $u = x^n$ and $dv = e^{ax}\,dx$.

For integrals of the form

$$\int x^n \ln x\,dx$$

let $u = \ln x$ and $dv = x^n\,dx$.

- Find the present value of future income. *13–18*

$$\text{Actual income over } t_1 \text{ years} = \int_0^{t_1} c(t)\,dt$$

$$\text{Present value} = \int_0^{t_1} c(t)e^{-rt}\,dt$$

Section 6.2

- Use integration tables to find indefinite integrals. *19–32*
- Use integration tables to solve real-life problems. *33, 34*

Section 6.3

- Use the Trapezoidal Rule and Simpson's Rule to approximate definite integrals. *35–46*

Trapezoidal Rule:

$$\int_a^b f(x)\,dx \approx \left(\frac{b-a}{2n}\right)[f(x_0) + 2f(x_1) + \cdots + 2f(x_{n-1}) + f(x_n)]$$

Simpson's Rule:

$$\int_a^b f(x)\,dx \approx \left(\frac{b-a}{3n}\right)[f(x_0) + 4f(x_1) + 2f(x_2) + 4f(x_3) + \cdots + 4f(x_{n-1}) + f(x_n)]$$

* A wide range of valuable study aids are available to help you master the material in this chapter. The *Student Solutions Manual* includes step-by-step solutions to all odd-numbered exercises to help you review and prepare. The student website at *www.cengagebrain.com* offers algebra help and a *Graphing Technology Guide*, which contains step-by-step commands and instructions for a wide variety of graphing calculators.

Section 6.3 (continued)	Review Exercises
■ Analyze the sizes of the errors when approximating definite integrals with the Trapezoidal Rule and Simpson's Rule.	47–50

Errors in the Trapezoidal Rule:

$$|E| \le \frac{(b-a)^3}{12n^2}[\max|f''(x)|], \quad a \le x \le b$$

Errors in Simpson's Rule:

$$|E| \le \frac{(b-a)^5}{180n^4}[\max|f^{(4)}(x)|], \quad a \le x \le b$$

Section 6.4

■ Evaluate improper integrals with infinite limits of integration.	51–56

$$\int_a^{\infty} f(x)\,dx = \lim_{b\to\infty}\int_a^b f(x)\,dx$$

$$\int_{-\infty}^{b} f(x)\,dx = \lim_{a\to-\infty}\int_a^b f(x)\,dx$$

$$\int_{-\infty}^{\infty} f(x)\,dx = \int_{-\infty}^{c} f(x)\,dx + \int_c^{\infty} f(x)\,dx$$

■ Find the area of an unbounded region.	57–60
■ Find the present value of a perpetuity.	61–64

$$\text{Present value} = P\int_0^{\infty} e^{-rt}\,dt = \frac{P}{r}$$

Study Strategies

- **Use a Variety of Approaches** To be efficient at finding antiderivatives, you need to use a variety of approaches.
  1. Check to see whether the integral fits one of the basic integration formulas—you should have these formulas memorized.
  2. Try an integration technique such as substitution or integration by parts to rewrite the integral in a form that fits one of the basic integration formulas.
  3. Use a table of integrals.
  4. Use a symbolic integration utility.
- **Use Numerical Integration** When solving a definite integral, remember that you cannot apply the Fundamental Theorem of Calculus unless you can find an antiderivative of the integrand. This is not always possible—even with a symbolic integration utility. In such cases, you can use a numerical technique such as the Midpoint Rule, the Trapezoidal Rule, or Simpson's Rule to approximate the value of the integral.

Review Exercises

See www.CalcChat.com for worked-out solutions to odd-numbered exercises.

Integration by Parts In Exercises 1–8, use integration by parts to find the indefinite integral.

1. $\int \frac{\ln x}{\sqrt{x}}\,dx$
2. $\int x \ln 4x\,dx$
3. $\int (x+1)e^x\,dx$
4. $\int xe^{-3x}\,dx$
5. $\int x\sqrt{x-5}\,dx$
6. $\int \frac{x}{\sqrt{x+8}}\,dx$
7. $\int 2x^2e^{2x}\,dx$
8. $\int (\ln x)^3\,dx$

Evaluating Definite Integrals In Exercises 9–12, use integration by parts to evaluate the definite integral.

9. $\int_1^e 6x \ln x\,dx$
10. $\int_0^4 \ln(1+3x)\,dx$
11. $\int_0^1 \frac{x}{e^{x/4}}\,dx$
12. $\int_0^2 x^2e^{3x}\,dx$

Finding Present Value In Exercises 13–16, find the present value of the income given by c (in dollars) over t_1 years at the given annual inflation rate r.

13. $c = 20{,}000$, $r = 4\%$, $t_1 = 5$ years
14. $c = 10{,}000 + 1500t$, $r = 6\%$, $t_1 = 10$ years
15. $c = 24{,}000t$, $r = 5\%$, $t_1 = 10$ years
16. $c = 20{,}000 + 100e^{t/2}$, $r = 5\%$, $t_1 = 5$ years

17. **Present Value** A company expects its income c during the next 4 years to be modeled by

$$c = 200{,}000 + 50{,}000t, \quad 0 \le t \le 4.$$

(a) Find the actual income for the business over the 4 years.

(b) Assuming an annual inflation rate of 6%, what is the present value of this income?

18. **Present Value** A company expects its income c during the next 7 years to be modeled by

$$c = 400{,}000 + 175{,}000t, \quad 0 \le t \le 7.$$

(a) Find the actual income for the business over the 7 years.

(b) Assuming an annual inflation rate of 4%, what is the present value of this income?

Using Integration Tables In Exercises 19–22, use the indicated formula from the integration table in Appendix C to find the indefinite integral.

19. $\int \frac{x^2}{2+3x}\,dx$, Formula 6
20. $\int \frac{1}{1+e^{6x}}\,dx$, Formula 40
21. $\int \sqrt{x^2-16}\,dx$, Formula 23
22. $\int x^5 \ln x\,dx$, Formula 43

Using Integration Tables In Exercises 23–32, use the integration table in Appendix C to find the indefinite integral.

23. $\int \frac{x}{(2+3x)^2}\,dx$
24. $\int \frac{x}{\sqrt{2+3x}}\,dx$
25. $\int \frac{\sqrt{x^2+25}}{x}\,dx$
26. $\int \frac{1}{x(4+3x)}\,dx$
27. $\int \frac{1}{x^2-4}\,dx$
28. $\int (\ln 3x)^2\,dx$
29. $\int \frac{x}{\sqrt{1+x}}\,dx$
30. $\int \frac{1}{x^2\sqrt{16-x^2}}\,dx$
31. $\int \frac{\sqrt{1+x}}{x}\,dx$
32. $\int \frac{1}{(x^2-9)^2}\,dx$

33. **Probability** The probability of recalling between a and b percent (in decimal form) of the material learned in a memory experiment is modeled by

$$P(a \le x \le b) = \int_a^b \frac{96}{11}\left(\frac{x}{\sqrt{9+16x}}\right)dx,$$

$0 \le a \le b \le 1.$

What are the probabilities of recalling (a) between 0% and 80% and (b) between 0% and 50% of the material?

34. **Probability** The probability of locating between a and b percent of the oil and gas deposits (in decimal form) in a region is modeled by

$$P(a \le x \le b) = \int_a^b 1.5x^2e^{x^{1.5}}\,dx, \quad 0 \le a \le b \le 1.$$

What are the probabilities of locating (a) between 40% and 60% and (b) between 0% and 50% of the deposits?

Using the Trapezoidal Rule and Simpson's Rule In Exercises 35–40, use the Trapezoidal Rule and Simpson's Rule to approximate the value of the definite integral for the indicated value of n. Compare these results with the exact value of the definite integral. Round your answers to four decimal places.

35. $\int_1^3 \frac{1}{x^2}\,dx,\ n = 4$

36. $\int_0^2 (x^2 + 1)\,dx,\ n = 8$

37. $\int_1^2 \frac{1}{x^3}\,dx,\ n = 8$

38. $\int_1^2 x^3\,dx,\ n = 4$

39. $\int_0^4 e^{-x/2}\,dx,\ n = 4$

40. $\int_0^8 \sqrt{x + 3}\,dx,\ n = 8$

Using the Trapezoidal Rule and Simpson's Rule In Exercises 41–46, approximate the value of the definite integral using (a) the Trapezoidal Rule and (b) Simpson's Rule for the indicated value of n. Round your answers to three decimal places.

41. $\int_1^2 \frac{1}{1 + \ln x}\,dx,\ n = 4$

42. $\int_0^2 \frac{1}{\sqrt{1 + x^3}}\,dx,\ n = 8$

43. $\int_0^1 \frac{x^{3/2}}{2 - x^2}\,dx,\ n = 4$

44. $\int_0^1 e^{x^2}\,dx,\ n = 6$

45. $\int_0^8 \frac{3}{x^2 + 2}\,dx,\ n = 8$

46. $\int_0^1 \sqrt{1 - x}\,dx,\ n = 4$

Error Analysis In Exercises 47 and 48, use the error formulas to find bounds for the error in approximating the definite integral using (a) the Trapezoidal Rule and (b) Simpson's Rule.

47. $\int_0^2 e^{2x}\,dx,\ n = 4$

48. $\int_2^4 \frac{1}{x - 1}\,dx,\ n = 8$

Error Analysis In Exercises 49 and 50, use the error formulas to find n such that the error in the approximation of the definite integral is less than 0.0001 using (a) the Trapezoidal Rule and (b) Simpson's Rule.

49. $\int_0^3 x^2\,dx$

50. $\int_0^5 e^{x/5}\,dx$

Evaluating an Improper Integral In Exercises 51–56, determine whether the improper integral diverges or converges. Evaluate the integral if it converges.

51. $\int_{-\infty}^{-1} \frac{1}{x^5}\,dx$

52. $\int_1^{\infty} \frac{1}{\sqrt{x}}\,dx$

53. $\int_{-\infty}^{0} \frac{1}{\sqrt[3]{8 - x}}\,dx$

54. $\int_0^{\infty} e^{-2x}\,dx$

55. $\int_1^{\infty} \frac{\ln x}{x}\,dx$

56. $\int_0^{\infty} \frac{e^x}{1 + e^x}\,dx$

Area of a Region In Exercises 57–60, find the area of the unbounded shaded region.

57. $y = e^{-x/4}$

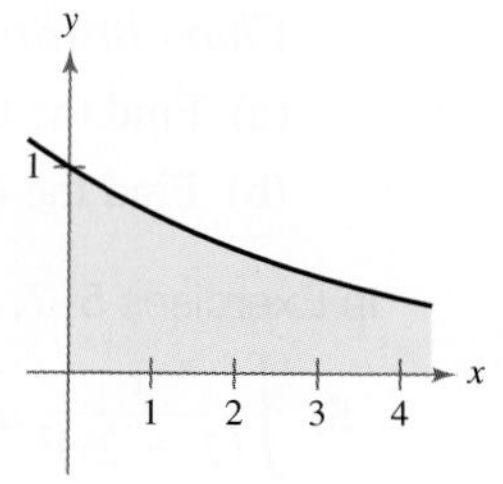

58. $y = \frac{2x}{x^2 + 2}$

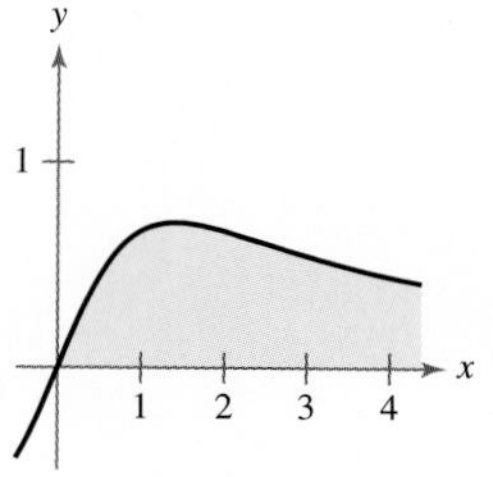

59. $y = 4xe^{-2x^2}$

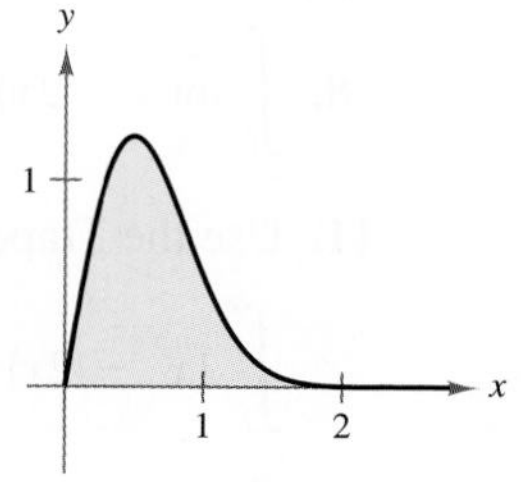

60. $y = \frac{3}{(1 - 3x)^{2/3}}$

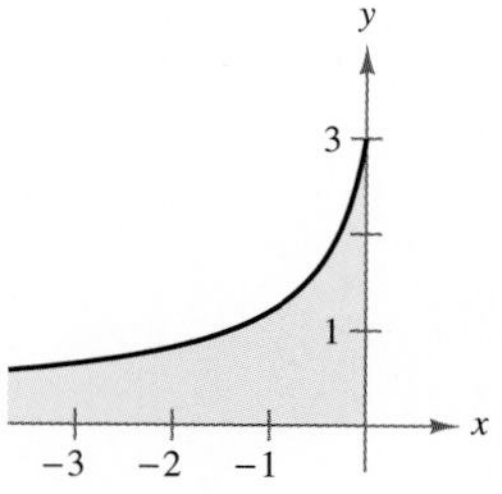

Endowment In Exercises 61 and 62, determine the amount of money required to set up a charitable endowment that pays the amount P each year indefinitely for the annual interest rate r compounded continuously.

61. $P = \$8000,\ r = 3\%$

62. $P = \$15{,}000,\ r = 5\%$

63. **Scholarship Fund** You want to start a scholarship fund at your alma mater. You plan to give one \$21,000 scholarship annually beginning one year from now and you have at most \$325,000 to start the fund. You also want the scholarship to be given out indefinitely. Assuming an annual interest rate of 7% compounded continuously, do you have enough money for the scholarship fund?

64. **Present Value** You are considering buying a franchise that yields a continuous income stream of \$100,000 per year. Assuming an annual interest rate of 6% compounded continuously, what is the present value of the franchise (a) for 15 years and (b) forever?

TEST YOURSELF

See www.CalcChat.com for worked-out solutions to odd-numbered exercises.

Take this test as you would take a test in class. When you are done, check your work against the answers given in the back of the book.

In Exercises 1–3, use integration by parts to find the indefinite integral.

1. $\int xe^{x+1}\,dx$ **2.** $\int 9x^2 \ln x\,dx$ **3.** $\int x^2 e^{-x/3}\,dx$

4. The revenue R (in millions of dollars) for P.F. Chang's China Bistro from 2001 through 2009 can be modeled by

$$R = 295.1 + 147.66\sqrt{t}\ln t, \quad 1 \le t \le 9$$

where t is the year, with $t = 1$ corresponding to 2001. *(Source: P.F. Chang's China Bistro)*

(a) Find the total revenue for the years 2001 through 2009.

(b) Find the average revenue for the years 2001 through 2009.

In Exercises 5–7, use the integration table in Appendix C to find the indefinite integral.

5. $\int \frac{x}{(7+2x)^2}\,dx$ **6.** $\int \frac{3x^2}{1+e^{x^3}}\,dx$ **7.** $\int \frac{2x^3}{\sqrt{1+5x^2}}\,dx$

In Exercises 8–10, use integration by parts or the integration table in Appendix C to evaluate the definite integral.

8. $\int_0^1 \ln(3-2x)\,dx$ **9.** $\int_3^6 \frac{x}{\sqrt{x-2}}\,dx$ **10.** $\int_{-3}^{-1} \frac{\sqrt{x^2+16}}{x}\,dx$

11. Use the Trapezoidal Rule with $n = 4$ to approximate

$$\int_2^5 (x^2 - 2x)\,dx.$$

Compare your result with the exact value of the definite integral.

12. Use Simpson's Rule with $n = 4$ to approximate

$$\int_0^1 9xe^{3x}\,dx.$$

Compare your result with the exact value of the definite integral.

In Exercises 13–15, determine whether the improper integral diverges or converges. Evaluate the integral if it converges.

13. $\int_0^{\infty} e^{-3x}\,dx$ **14.** $\int_0^9 \frac{2}{\sqrt{x}}\,dx$ **15.** $\int_{-\infty}^0 \frac{1}{(4x-1)^{2/3}}\,dx$

16. A magazine publisher offers two subscription plans. Plan A is a one-year subscription for \$19.95. Plan B is a lifetime subscription (lasting indefinitely) for \$149.

(a) A subscriber considers using plan A indefinitely. Assuming an annual inflation rate of 4%, find the present value of the money the subscriber will spend using plan A.

(b) Based on your answer to part (a), which plan should the subscriber use? Explain.

7 Functions of Several Variables

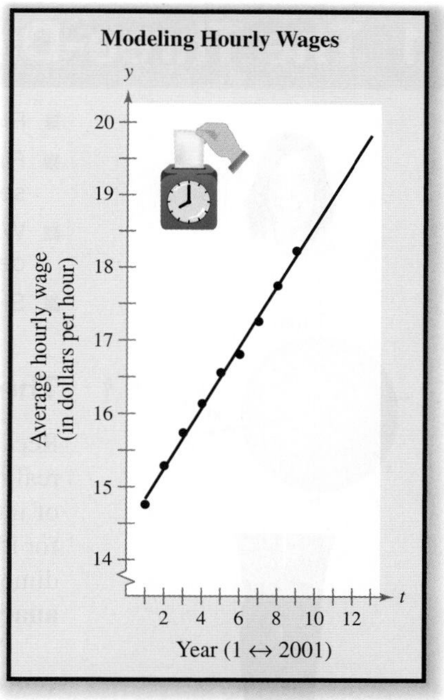

Example 3 on page 477 shows how least squares regression analysis can be used to find the best-fitting line that models hourly wages for production workers in manufacturing industries.

7.1 The Three-Dimensional Coordinate System

- Plot points in space.
- Find distances between points in space and find midpoints of line segments in space.
- Write the standard forms of the equations of spheres and find the centers and radii of spheres.
- Sketch the coordinate plane traces of surfaces.

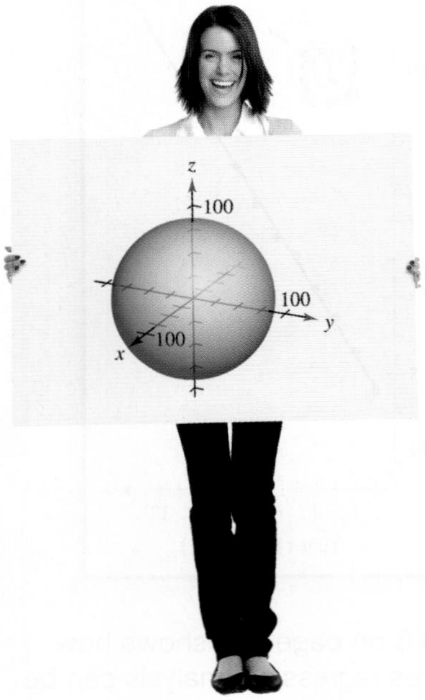

In Exercise 57 on page 426, you will model the shape of a spherical building using the standard equation of a sphere.

The Three-Dimensional Coordinate System

Recall from Section 1.1 that the Cartesian plane is determined by two perpendicular real number lines called the x-axis and the y-axis. These axes together with their point of intersection (the origin) allow you to develop a two-dimensional coordinate system for identifying points in a plane. To identify a point in space, you must introduce a third dimension to the model. The geometry of this three-dimensional model is called **solid analytic geometry.**

You can construct a **three-dimensional coordinate system** by passing a z-axis perpendicular to both the x- and y-axes at the origin. Figure 7.1 shows the positive portion of each coordinate axis. Taken as pairs, the axes determine three **coordinate planes:** the ***xy*-plane,** the ***xz*-plane,** and the ***yz*-plane.** These three coordinate planes separate the three-dimensional coordinate system into eight **octants.** The first octant is the one for which all three coordinates are positive. In this three-dimensional system, a point P in space is determined by an ordered triple (x, y, z), where x, y, and z are as follows.

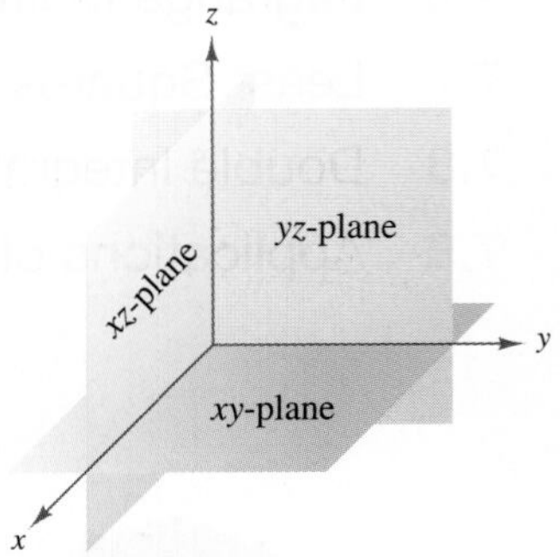

FIGURE 7.1

$x =$ directed distance from yz-plane to P

$y =$ directed distance from xz-plane to P

$z =$ directed distance from xy-plane to P

Example 1 Plotting Points in Space

Plot the points in the same three-dimensional coordinate system.

a. $(2, -3, 3)$ **b.** $(-2, 6, 2)$

c. $(1, 4, 0)$ **d.** $(2, 2, -3)$

SOLUTION To plot the point $(2, -3, 3)$, notice that

$$x = 2, \quad y = -3, \quad \text{and} \quad z = 3.$$

To help visualize the point, locate the point $(2, -3)$ in the xy-plane (denoted by a cross in Figure 7.2). The point lies three units above the cross. You can plot the other points in a similar manner, as shown in Figure 7.2.

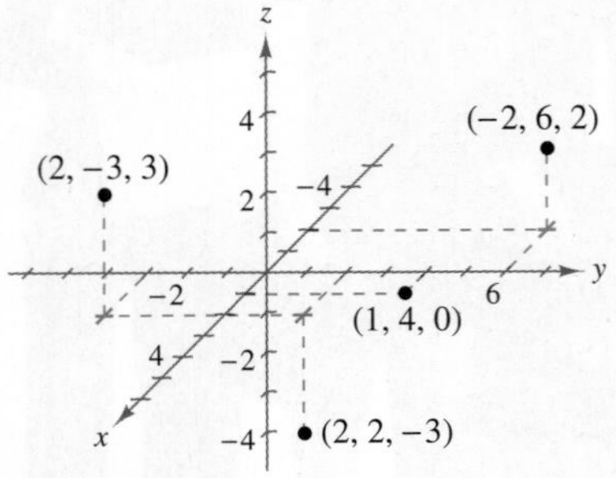

FIGURE 7.2

Checkpoint 1

Plot the points in the same three-dimensional coordinate system.

a. $(2, 5, 1)$ **b.** $(-2, -4, 3)$ **c.** $(4, 0, -5)$

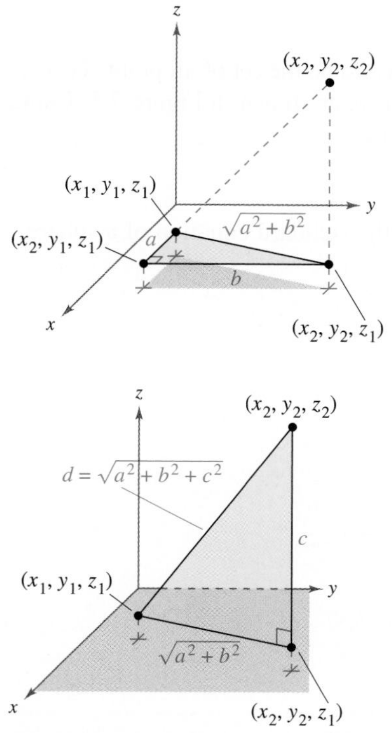

FIGURE 7.3

The Distance and Midpoint Formulas

Many of the formulas established for the two-dimensional coordinate system can be extended to three dimensions. For example, to find the distance between two points in space, you can use the Pythagorean Theorem twice, as shown in Figure 7.3. By doing this, you will obtain the formula for the distance between two points in space.

> **Distance Formula in Space**
>
> The distance between the points (x_1, y_1, z_1) and (x_2, y_2, z_2) is
>
> $$d = \sqrt{(x_2 - x_1)^2 + (y_2 - y_1)^2 + (z_2 - z_1)^2}.$$

Example 2 Finding the Distance Between Two Points

Find the distance between $(1, 0, 2)$ and $(2, 4, -3)$.

SOLUTION

$$d = \sqrt{(x_2 - x_1)^2 + (y_2 - y_1)^2 + (z_2 - z_1)^2} \qquad \text{Write Distance Formula.}$$
$$= \sqrt{(2 - 1)^2 + (4 - 0)^2 + (-3 - 2)^2} \qquad \text{Substitute.}$$
$$= \sqrt{1 + 16 + 25} \qquad \text{Simplify.}$$
$$= \sqrt{42} \qquad \text{Simplify.}$$

✓ Checkpoint 2

Find the distance between $(2, 3, -1)$ and $(0, 5, 3)$. ■

Notice the similarity between the Distance Formula in the plane and the Distance Formula in space. The Midpoint Formulas in the plane and in space are also similar.

> **Midpoint Formula in Space**
>
> The midpoint of the line segment joining the points (x_1, y_1, z_1) and (x_2, y_2, z_2) is
>
> $$\text{Midpoint} = \left(\frac{x_1 + x_2}{2}, \frac{y_1 + y_2}{2}, \frac{z_1 + z_2}{2}\right).$$

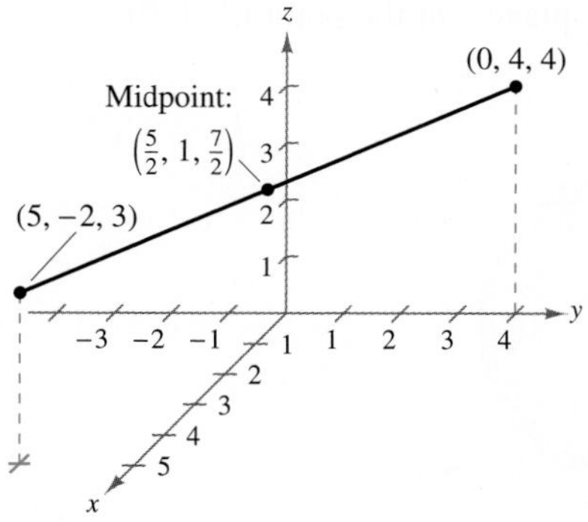

FIGURE 7.4

Example 3 Using the Midpoint Formula

Find the midpoint of the line segment joining

$(5, -2, 3)$ and $(0, 4, 4)$.

SOLUTION Using the Midpoint Formula, the midpoint is

$$\left(\frac{5 + 0}{2}, \frac{-2 + 4}{2}, \frac{3 + 4}{2}\right) = \left(\frac{5}{2}, 1, \frac{7}{2}\right)$$

as shown in Figure 7.4.

✓ Checkpoint 3

Find the midpoint of the line segment joining

$(3, -2, 0)$ and $(-8, 6, -4)$. ■

The Equation of a Sphere

A **sphere** with center at (h, k, l) and radius r is defined to be the set of all points (x, y, z) such that the distance between (x, y, z) and (h, k, l) is r, as shown in Figure 7.5. Using the Distance Formula, this condition can be written as

$$\sqrt{(x-h)^2+(y-k)^2+(z-l)^2}=r.$$

By squaring both sides of this equation, you obtain the standard equation of a sphere.

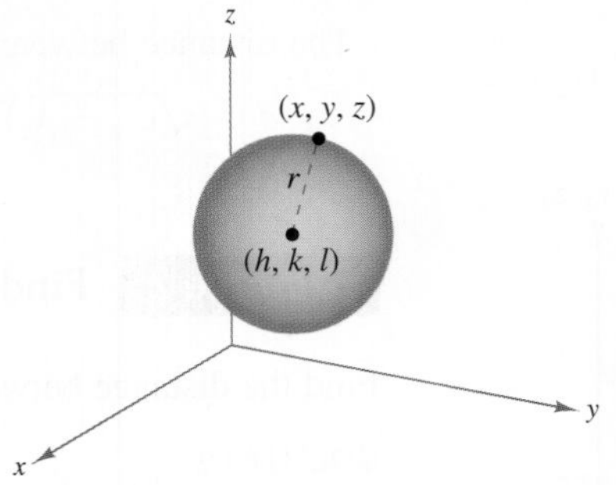

Sphere: Radius r, Center (h, k, l)

FIGURE 7.5

Standard Equation of a Sphere

The **standard equation of a sphere** with center at (h, k, l) and radius r is

$$(x-h)^2+(y-k)^2+(z-l)^2=r^2.$$

Example 4 Finding the Equation of a Sphere

Find the standard equation of the sphere with center at $(2, 4, 3)$ and radius 3. Does this sphere intersect the xy-plane?

SOLUTION

$$(x-h)^2+(y-k)^2+(z-l)^2=r^2 \quad \text{Write standard equation.}$$
$$(x-2)^2+(y-4)^2+(z-3)^2=3^2 \quad \text{Substitute.}$$
$$(x-2)^2+(y-4)^2+(z-3)^2=9 \quad \text{Simplify.}$$

In Figure 7.6, note that the center of the sphere lies three units above the xy-plane. The sphere has a radius of 3, so it must intersect the xy-plane—at the point $(2, 4, 0)$.

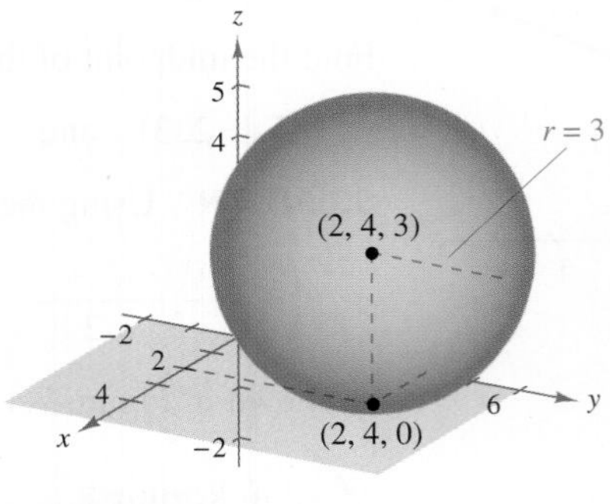

FIGURE 7.6

✓Checkpoint 4

Find the standard equation of the sphere with center at $(4, 3, 2)$ and radius 5. ■

Example 5 Finding the Equation of a Sphere

Find the equation of the sphere that has the points $(3, -2, 6)$ and $(-1, 4, 2)$ as endpoints of a diameter.

SOLUTION By the Midpoint Formula, the center of the sphere is

$$(h, k, l) = \left(\frac{3+(-1)}{2}, \frac{-2+4}{2}, \frac{6+2}{2}\right) \quad \text{Apply Midpoint Formula.}$$

$$= (1, 1, 4). \quad \text{Simplify.}$$

By the Distance Formula, the radius is

$$r = \sqrt{(3-1)^2 + (-2-1)^2 + (6-4)^2} \quad \text{Apply Distance Formula.}$$

$$= \sqrt{4+9+4} \quad \text{Simplify.}$$

$$= \sqrt{17}. \quad \text{Simplify.}$$

So, the standard equation of the sphere is

$$(x-h)^2 + (y-k)^2 + (z-l)^2 = r^2 \quad \text{Write formula for a sphere.}$$

$$(x-1)^2 + (y-1)^2 + (z-4)^2 = 17. \quad \text{Substitute.}$$

✓Checkpoint 5

Find the equation of the sphere that has the points $(-2, 5, 7)$ and $(4, 1, -3)$ as endpoints of a diameter. ■

Example 6 Finding the Center and Radius of a Sphere

Find the center and radius of the sphere whose equation is

$$x^2 + y^2 + z^2 - 2x + 4y - 6z + 8 = 0.$$

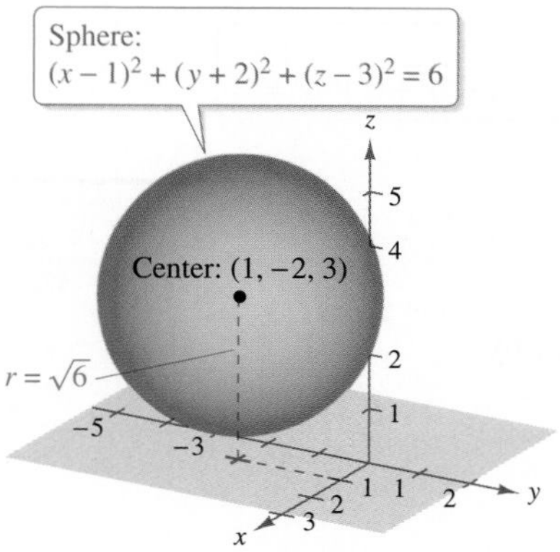

FIGURE 7.7

SOLUTION You can obtain the standard equation of the sphere by completing the square. To do this, begin by grouping terms with the same variable. Then add "the square of half the coefficient of each linear term" to each side of the equation. So, to complete the square of $(x^2 - 2x)$, add $\left[\frac{1}{2}(-2)\right]^2 = 1$ to each side. To complete the square of $(y^2 + 4y)$, add $\left[\frac{1}{2}(4)\right]^2 = 4$ to each side. To complete the square of $(z^2 - 6z)$, add $\left[\frac{1}{2}(-6)\right]^2 = 9$ to each side.

$$x^2 + y^2 + z^2 - 2x + 4y - 6z + 8 = 0$$

$$(x^2 - 2x + \quad) + (y^2 + 4y + \quad) + (z^2 - 6z + \quad) = -8$$

$$(x^2 - 2x + 1) + (y^2 + 4y + 4) + (z^2 - 6z + 9) = -8 + 1 + 4 + 9$$

$$(x-1)^2 + (y+2)^2 + (z-3)^2 = 6$$

So, the center of the sphere is $(1, -2, 3)$, and its radius is $\sqrt{6}$, as shown in Figure 7.7.

✓Checkpoint 6

Find the center and radius of the sphere whose equation is

$$x^2 + y^2 + z^2 + 6x - 8y + 2z - 10 = 0.$$

■

Note in Example 6 that the points satisfying the equation of the sphere are "surface points," not "interior points." In general, the collection of points satisfying an equation involving x, y, and z is called a **surface in space.**

Traces of Surfaces

Finding the intersection of a surface with one of the three coordinate planes (or with a plane parallel to one of the three coordinate planes) helps visualize the surface. Such an intersection is called a **trace** of the surface. For example, the xy-trace of a surface consists of all points that are common to both the surface *and* the xy-plane. Similarly, the xz-trace of a surface consists of all points that are common to both the surface and the xz-plane.

Example 7 Finding a Trace of a Surface

Sketch the xy-trace of the sphere given by $(x - 3)^2 + (y - 2)^2 + (z + 4)^2 = 5^2$.

SOLUTION To find the xy-trace of this surface, use the fact that every point in the xy-plane has a z-coordinate of zero. By substituting $z = 0$ into the original equation, the resulting equation will represent the intersection of the surface with the xy-plane.

$$(x - 3)^2 + (y - 2)^2 + (z + 4)^2 = 5^2 \quad \text{Write original equation.}$$
$$(x - 3)^2 + (y - 2)^2 + (0 + 4)^2 = 25 \quad \text{Let } z = 0 \text{ to find } xy\text{-trace.}$$
$$(x - 3)^2 + (y - 2)^2 + 16 = 25 \quad \text{Simplify.}$$
$$(x - 3)^2 + (y - 2)^2 = 9 \quad \text{Subtract 16 from each side.}$$
$$(x - 3)^2 + (y - 2)^2 = 3^2 \quad \text{Equation of circle}$$

From this equation, you can see that the xy-trace is a circle of radius 3, as shown in Figure 7.8.

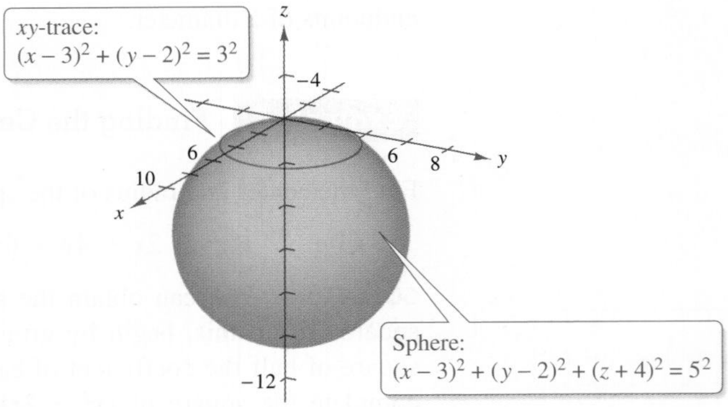

FIGURE 7.8

✓ Checkpoint 7

Find the equation of the xy-trace of the sphere given by

$$(x + 1)^2 + (y - 2)^2 + (z + 3)^2 = 5^2.$$

SUMMARIZE (Section 7.1)

1. State the Distance Formula in space *(page 421)*. For an example of the Distance Formula in space, see Example 2.
2. State the Midpoint Formula in space *(page 421)*. For an example of the Midpoint Formula in space, see Example 3.
3. State the standard equation of a sphere *(page 422)*. For examples of finding equations of spheres, see Examples 4 and 5.
4. Explain what is meant by the trace of a surface *(page 424)*. For an example of finding the trace of a surface, see Example 7.

SKILLS WARM UP 7.1 The following warm-up exercises involve skills that were covered in earlier sections. You will use these skills in the exercise set for this section. For additional help, review Sections 1.1 and 1.2.

In Exercises 1–4, find the distance between the points.

1. $(5, 1), (3, 5)$ **2.** $(2, 3), (-1, -1)$ **3.** $(-5, 4), (-5, -4)$ **4.** $(-3, 6), (-3, -2)$

In Exercises 5–8, find the midpoint of the line segment connecting the points.

5. $(2, 5), (6, 9)$ **6.** $(-1, -2), (3, 2)$ **7.** $(-6, 0), (6, 6)$ **8.** $(-4, 3), (2, -1)$

In Exercises 9 and 10, write the standard form of the equation of the circle.

9. Center: $(2, 3)$; radius: 2 **10.** Endpoints of a diameter: $(4, 0), (-2, 8)$

Exercises 7.1

See www.CalcChat.com for worked-out solutions to odd-numbered exercises.

Plotting Points in Space In Exercises 1–4, plot the points in the same three-dimensional coordinate system. *See Example 1.*

1. $(2, 1, 3), (-1, 2, 1), (3, -2, 5), \left(\frac{3}{2}, 4, -2\right)$

2. $(-5, -2, 2), (5, -2, -2), (1, 3, 1), (-2, 4, -3)$

3. $(0, 4, -5), (4, 0, 5), \left(-2, \frac{1}{2}, 0\right), \left(-\frac{1}{2}, 3, 1\right)$

4. $(-3, 0, -1), (2, -1, 1), (-1, -3, -2), (1, 3, 4)$

Finding Points in Space In Exercises 5–8, find the coordinates of the point.

5. The point is located three units behind the yz-plane, four units to the right of the xz-plane, and five units above the xy-plane.

6. The point is located seven units in front of the yz-plane, two units to the left of the xz-plane, and one unit below the xy-plane.

7. The point is located on the x-axis, 10 units in front of the yz-plane.

8. The point is located in the yz-plane, three units to the right of the xz-plane, and two units above the xy-plane.

9. Think About It What is the z-coordinate of any point in the xy-plane?

10. Think About It What is the y-coordinate of any point in the xz-plane?

Finding the Distance Between Two Points In Exercises 11–14, find the distance between the two points. *See Example 2.*

11. $(4, 1, 5), (8, 2, 6)$

12. $(-4, -1, 1), (2, -1, 5)$

13. $(-1, -5, 7), (-3, 4, -4)$

14. $(8, -2, 2), (8, -2, 4)$

Using the Midpoint Formula In Exercises 15–18, find the midpoint of the line segment joining the two points. *See Example 3.*

15. $(6, -4, 2), (-2, 1, 3)$ **16.** $(4, 0, -6), (8, 8, 20)$

17. $(-5, -2, 5), (6, 3, -7)$ **18.** $(0, -2, 5), (4, 2, 7)$

Using the Midpoint Formula In Exercises 19–22, find (x, y, z).

19.

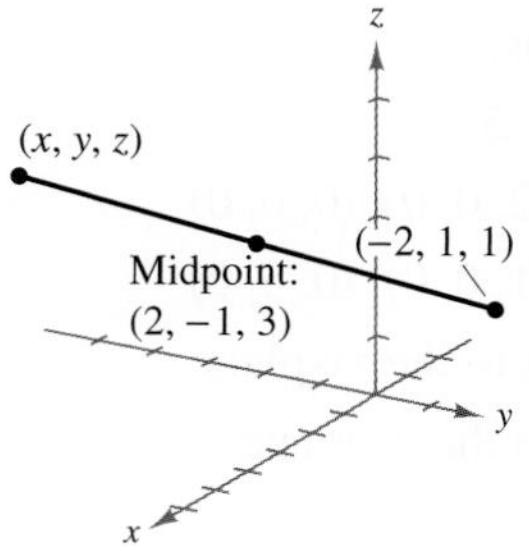

20.

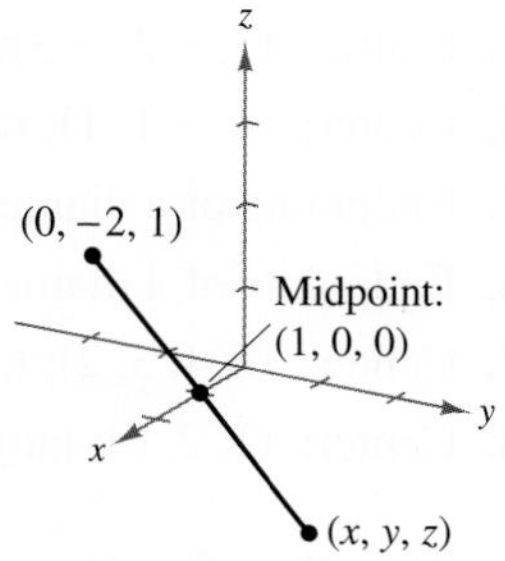

21.

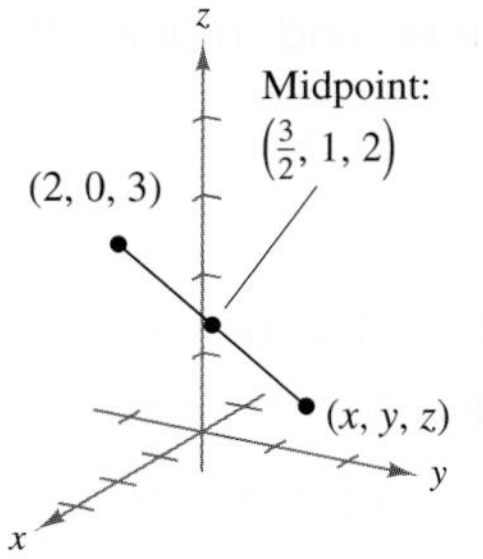

22.

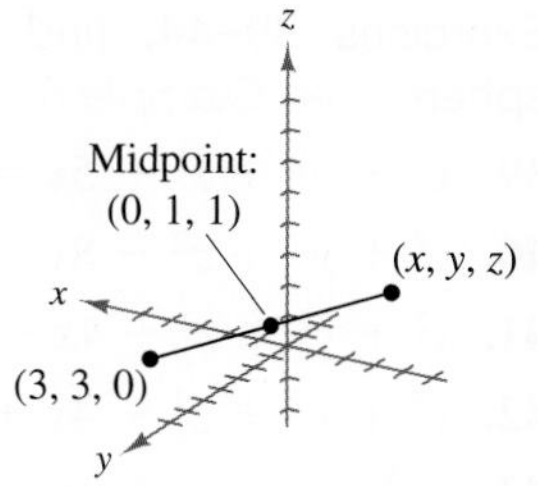

Identifying Triangles In Exercises 23–26, find the lengths of the sides of the triangle with the given vertices, and determine whether the triangle is a right triangle, an isosceles triangle, or neither.

23. $(0, 0, 0), (2, 2, 1), (2, -4, 4)$

24. $(5, 3, 4), (7, 1, 3), (3, 5, 3)$

25. $(-1, 0, -2), (-1, 5, 2), (-3, -1, 1)$

26. $(5, 0, 0), (0, 2, 0), (0, 0, -3)$

27. Think About It The triangle in Exercise 23 is translated five units downward along the z-axis. Determine the coordinates of the translated triangle.

28. Think About It The triangle in Exercise 24 is translated three units to the right along the y-axis. Determine the coordinates of the translated triangle.

Finding the Equation of a Sphere In Exercises 29–38, find the standard equation of the sphere. *See Examples 4 and 5.*

29.

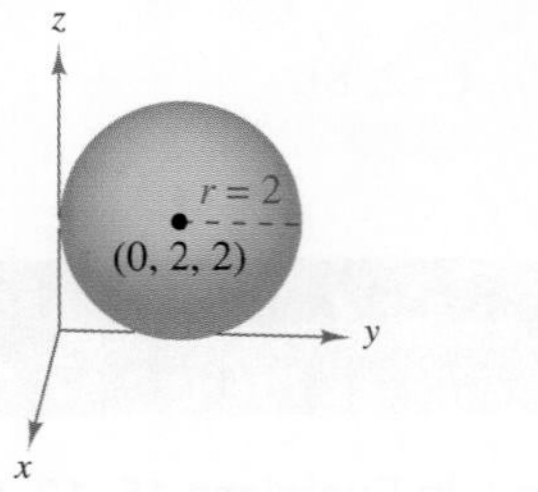

30.

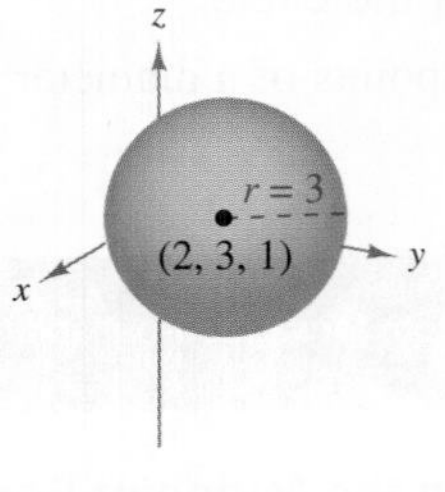

31.

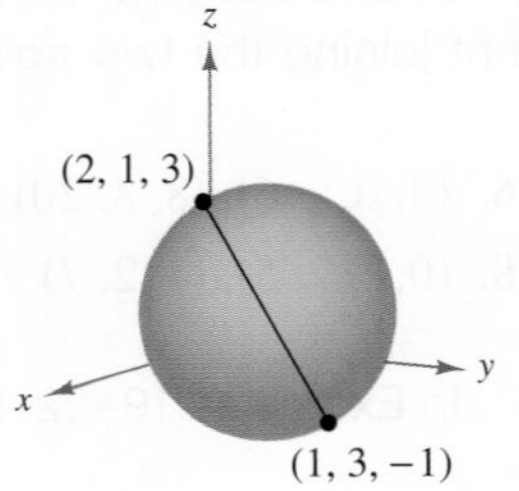

32.

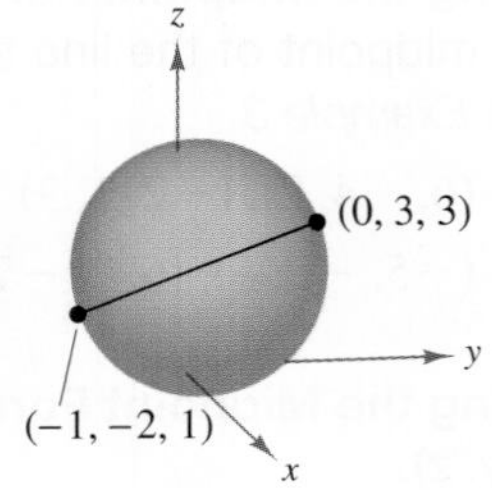

33. Center: $(3, -2, -3)$; radius: 4

34. Center: $(4, -1, 1)$; radius: 5

35. Endpoints of a diameter: $(2, 0, 0)$, $(0, 6, 0)$

36. Endpoints of a diameter: $(1, 0, 0)$, $(0, 5, 0)$

37. Center: $(-4, 3, 2)$; tangent to the xy-plane

38. Center: $(1, 2, 0)$; tangent to the yz-plane

Finding the Center and Radius of a Sphere In Exercises 39–44, find the center and radius of the sphere. *See Example 6.*

39. $x^2 + y^2 + z^2 - 5x = 0$

40. $x^2 + y^2 + z^2 - 8y = 0$

41. $x^2 + y^2 + z^2 + 4x - 2y + 8z - 4 = 0$

42. $x^2 + y^2 + z^2 - 4y + 6z + 4 = 0$

43. $2x^2 + 2y^2 + 2z^2 - 4x - 12y - 8z + 3 = 0$

44. $4x^2 + 4y^2 + 4z^2 - 8x + 16y + 11 = 0$

Finding the Trace of a Surface In Exercises 45–48, sketch the xy-trace of the sphere. *See Example 7.*

45. $(x - 1)^2 + (y - 3)^2 + (z - 2)^2 = 25$

46. $(x + 1)^2 + (y + 2)^2 + (z - 2)^2 = 16$

47. $x^2 + y^2 + z^2 - 6x - 10y + 6z + 30 = 0$

48. $x^2 + y^2 + z^2 - 4y + 2z - 60 = 0$

Finding the Trace of a Surface In Exercises 49–52, sketch the yz-trace of the sphere. *See Example 7.*

49. $x^2 + (y + 3)^2 + z^2 = 25$

50. $(x + 2)^2 + (y - 3)^2 + z^2 = 9$

51. $x^2 + y^2 + z^2 - 4x - 4y - 6z - 12 = 0$

52. $x^2 + y^2 + z^2 - 6x - 10y + 6z + 30 = 0$

Finding the Trace of a Surface In Exercises 53–56, sketch the trace of the intersection of each plane with the given sphere.

53. $x^2 + y^2 + z^2 = 25$

(a) $z = 3$ (b) $x = 4$

54. $x^2 + y^2 + z^2 = 169$

(a) $x = 5$ (b) $y = 12$

55. $x^2 + y^2 + z^2 - 4x - 6y + 9 = 0$

(a) $x = 2$ (b) $y = 3$

56. $x^2 + y^2 + z^2 - 8x - 6z + 16 = 0$

(a) $x = 4$ (b) $z = 3$

57. Architecture A spherical building has a diameter of 165 feet. The center of the building is placed at the origin of a three-dimensional coordinate system. What is the equation of the sphere that models the shape of the building?

58. HOW DO YOU SEE IT? Crystals are classified according to their symmetry.

(a) Crystals shaped like cubes are classified as isometric. The vertices of an isometric crystal mapped onto a three-dimensional coordinate system are shown in the figure. Determine (x, y, z).

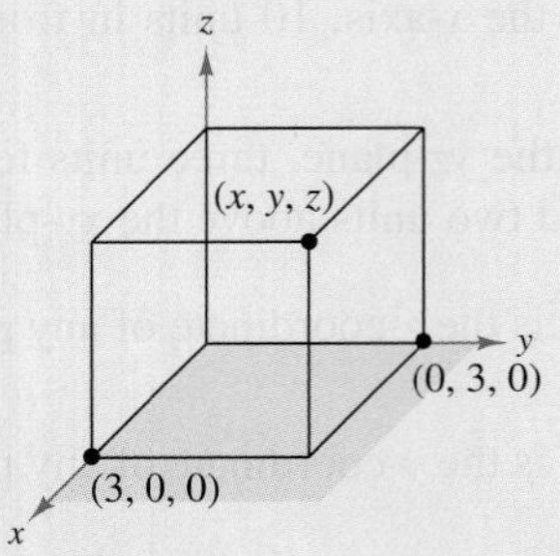

Figure for (a)

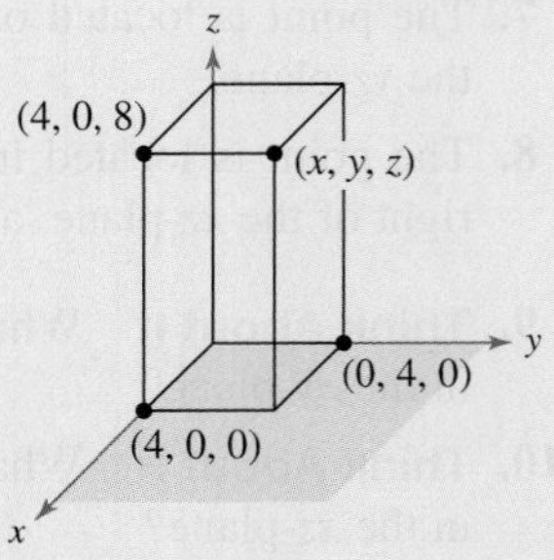

Figure for (b)

(b) Crystals shaped like rectangular prisms are classified as tetragonal. The vertices of a tetragonal crystal mapped onto a three-dimensional coordinate system are shown in the figure. Determine (x, y, z).

7.2 Surfaces in Space

In Exercise 49 on page 435, you will use the dimensions of Earth to write an equation of an ellipsoid that models its shape.

- Sketch planes in space.
- Draw planes in space with different numbers of intercepts.
- Classify quadric surfaces in space.

Equations of Planes in Space

In Section 7.1, you studied one type of surface in space—a sphere. In this section, you will study a second type—a plane in space. The **general equation of a plane** in space is

$$ax + by + cz = d.$$ General equation of a plane

Note the similarity of this equation to the general equation of a line in the plane. In fact, when you intersect the plane represented by this equation with each of the three coordinate planes, you will obtain traces that are lines, as shown in Figure 7.9.

In Figure 7.9, the points where the plane intersects the three coordinate axes are the x-, y-, and z-intercepts of the plane. By connecting these three points, you can form a triangular region, which helps you visualize the plane in space.

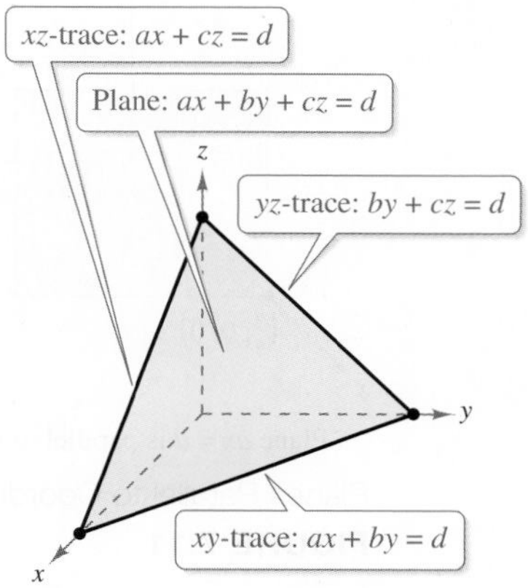

FIGURE 7.9

Example 1 Sketching a Plane in Space

Find the x-, y-, and z-intercepts of the plane given by

$$3x + 2y + 4z = 12.$$

Then sketch the plane.

SOLUTION To find the x-intercept, let both y and z be zero.

$$3x + 2(0) + 4(0) = 12 \quad \text{Substitute 0 for } y \text{ and } z.$$
$$3x = 12 \quad \text{Simplify.}$$
$$x = 4 \quad \text{Solve for } x.$$

So, the x-intercept is $(4, 0, 0)$. To find the y-intercept, let x and z be zero and conclude that $y = 6$. So, the y-intercept is $(0, 6, 0)$. Similarly, by letting x and y be zero, you can determine that $z = 3$ and that the z-intercept is $(0, 0, 3)$. Figure 7.10 shows the triangular portion of the plane formed by connecting the three intercepts

$$(4, 0, 0), \quad (0, 6, 0), \quad \text{and} \quad (0, 0, 3).$$

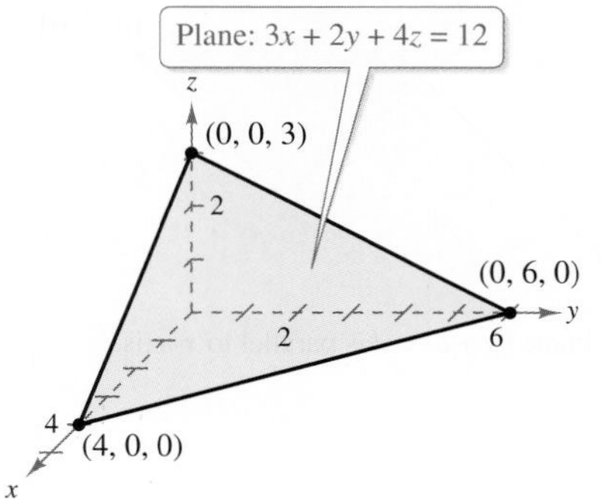

Sketch Made by Connecting Intercepts: (4, 0, 0), (0, 6, 0), (0, 0, 3)
FIGURE 7.10

✓ Checkpoint 1

Find the x-, y-, and z-intercepts of the plane given by

$$2x + 4y + z = 8.$$

Then sketch the plane. ■

Drawing Planes in Space

The planes shown in Figures 7.9 and 7.10 have three intercepts. When this occurs, you can draw the plane by sketching the triangular region formed by connecting the three intercepts.

It is possible for a plane in space to have fewer than three intercepts. This occurs when one or more of the coefficients in the equation $ax + by + cz = d$ is zero. Figure 7.11 shows some planes in space that have only one intercept, and Figure 7.12 shows some that have only two intercepts. In each figure, note the use of dashed lines and shading to give the illusion of three dimensions.

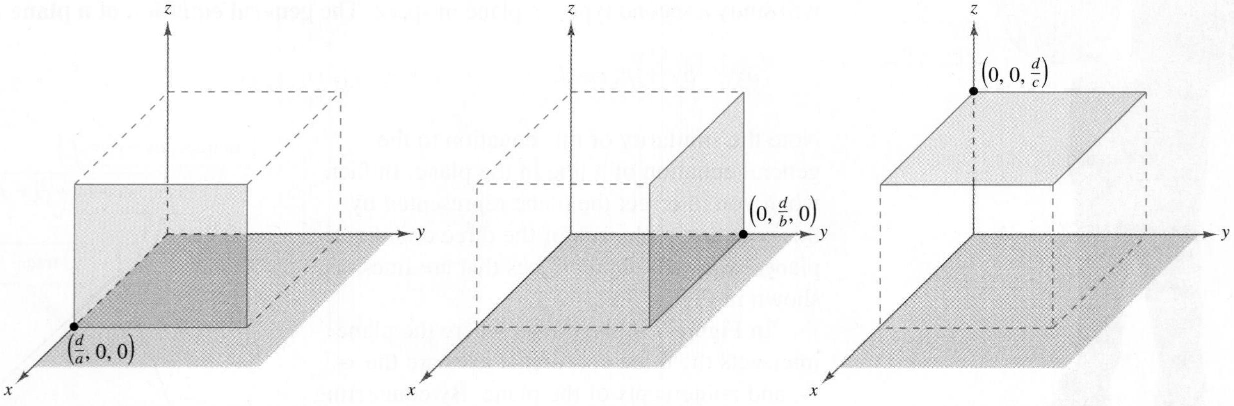

Plane $ax = d$ is parallel to yz-plane. Plane $by = d$ is parallel to xz-plane. Plane $cz = d$ is parallel to xy-plane.

Planes Parallel to Coordinate Planes
FIGURE 7.11

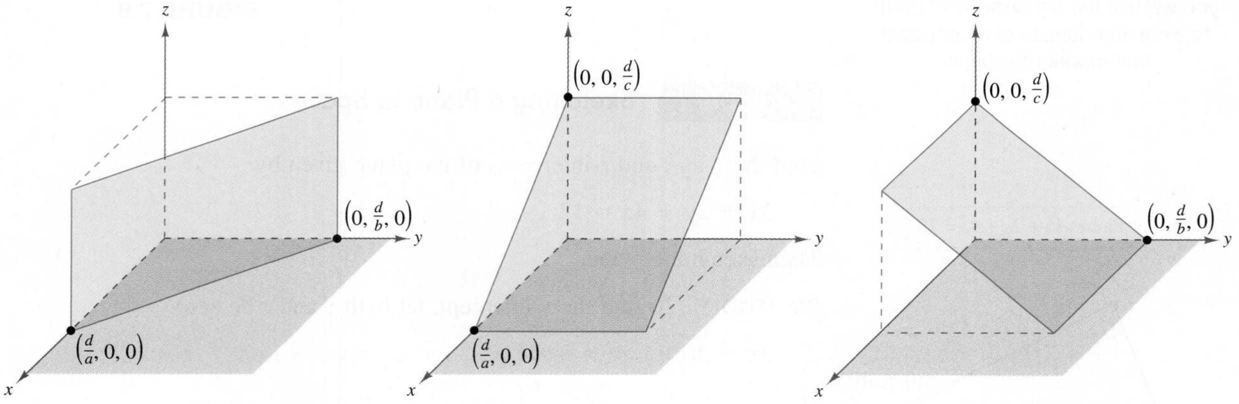

Plane $ax + by = d$ is parallel to z-axis. Plane $ax + cz = d$ is parallel to y-axis. Plane $by + cz = d$ is parallel to x-axis.

Planes Parallel to Coordinate Axes
FIGURE 7.12

When an equation of a plane has a missing variable, such as

$$2x + z = 1 \qquad \text{See Figure 7.13.}$$

the plane must be *parallel to the axis* represented by the missing variable, as shown in Figure 7.12. When two variables are missing from an equation of a plane, the plane is *parallel to the coordinate plane* represented by the missing variables, as shown in Figure 7.11.

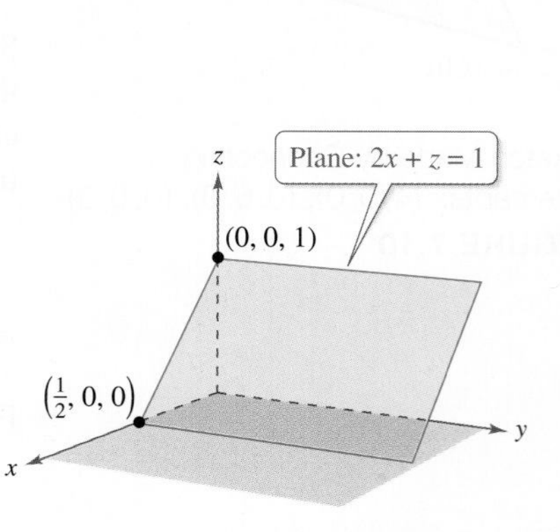

Plane $2x + z = 1$ is parallel to the y-axis.
FIGURE 7.13

Quadric Surfaces

A third common type of surface in space is a **quadric surface.** Quadric surfaces are the three-dimensional analogs of conic sections. The equation of a quadric surface in space is a second-degree equation in three variables, such as

$$Ax^2 + By^2 + Cz^2 + Dx + Ey + Fz + G = 0.$$ Second-degree equation

There are six basic types of quadric surfaces.

1. Elliptic cone
2. Elliptic paraboloid
3. Hyperbolic paraboloid
4. Ellipsoid
5. Hyperboloid of one sheet
6. Hyperboloid of two sheets

The six types are summarized on the next two pages. Notice that each surface is pictured with two types of three-dimensional sketches. The computer-generated sketches use traces with hidden lines to give the illusion of three dimensions. The artist-rendered sketches use shading to create the same illusion.

All of the quadric surfaces on the next two pages are centered at the origin and have axes along the coordinate axes. Moreover, only one of several possible orientations of each surface is shown. When the surface has a different center or is oriented along a different axis, its standard equation will change accordingly. For instance, the ellipsoid

$$\frac{x^2}{1^2} + \frac{y^2}{3^2} + \frac{z^2}{2^2} = 1$$

has (0, 0, 0) as its center, but the ellipsoid

$$\frac{(x-2)^2}{1^2} + \frac{(y+1)^2}{3^2} + \frac{(z-4)^2}{2^2} = 1$$

has (2, −1, 4) as its center. A computer-generated graph of the first ellipsoid is shown in Figure 7.14.

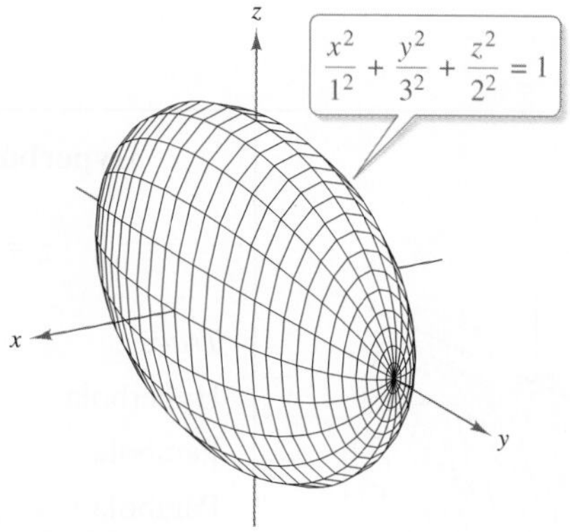

FIGURE 7.14

TECH TUTOR

If you have access to a three-dimensional graphing utility, try using it to graph the surface in Figure 7.14. When you do this, you will discover that sketching surfaces in space is not a simple task—even with a graphing utility.

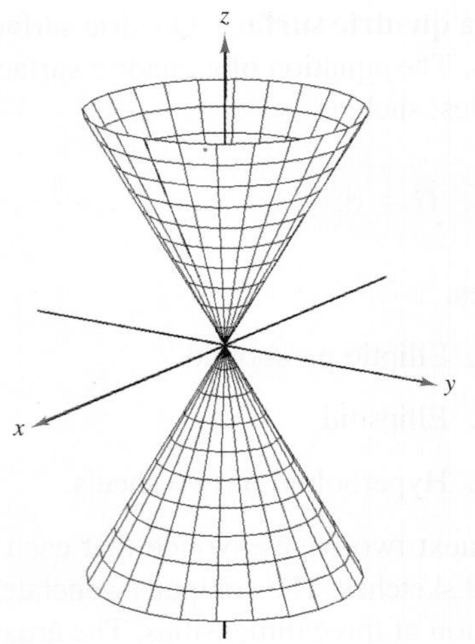

Elliptic Cone

$$\frac{x^2}{a^2} + \frac{y^2}{b^2} - \frac{z^2}{c^2} = 0$$

Trace	*Plane*
Ellipse	Parallel to *xy*-plane
Hyperbola	Parallel to *xz*-plane
Hyperbola	Parallel to *yz*-plane

The axis of the cone corresponds to the variable whose coefficient is negative. The traces in the coordinate planes parallel to this axis are intersecting lines.

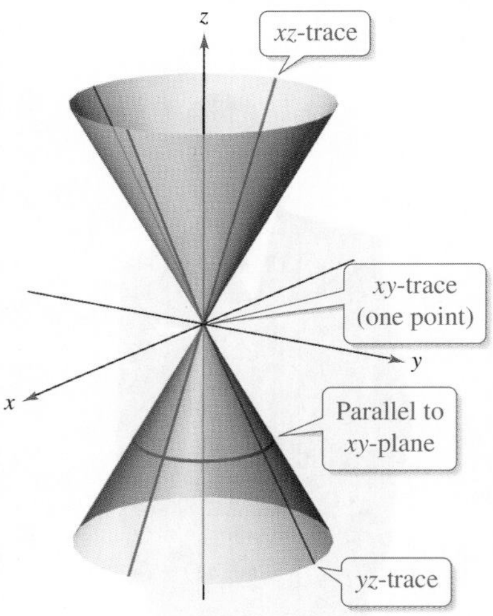

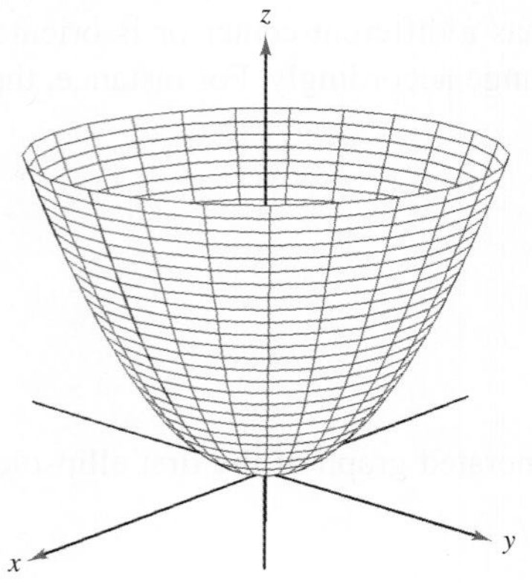

Elliptic Paraboloid

$$z = \frac{x^2}{a^2} + \frac{y^2}{b^2}$$

Trace	*Plane*
Ellipse	Parallel to *xy*-plane
Parabola	Parallel to *xz*-plane
Parabola	Parallel to *yz*-plane

The axis of the paraboloid corresponds to the variable raised to the first power.

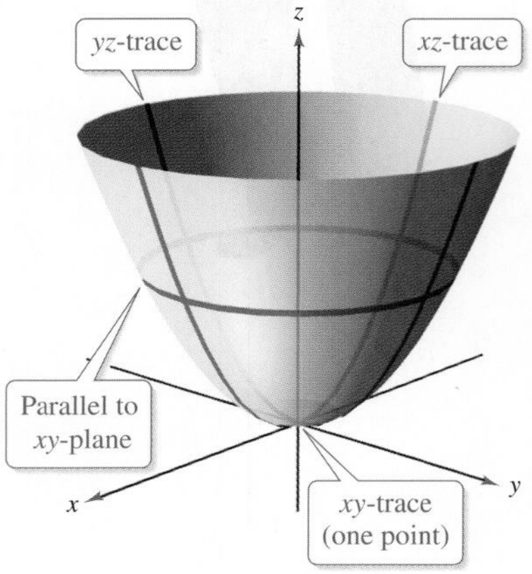

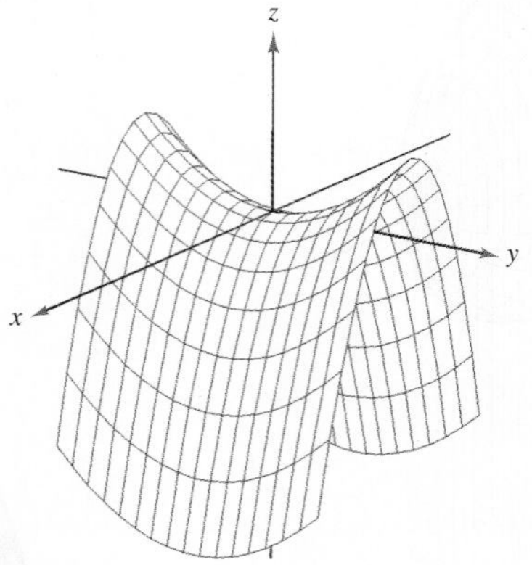

Hyperbolic Paraboloid

$$z = \frac{y^2}{b^2} - \frac{x^2}{a^2}$$

Trace	*Plane*
Hyperbola	Parallel to *xy*-plane
Parabola	Parallel to *xz*-plane
Parabola	Parallel to *yz*-plane

The axis of the paraboloid corresponds to the variable raised to the first power.

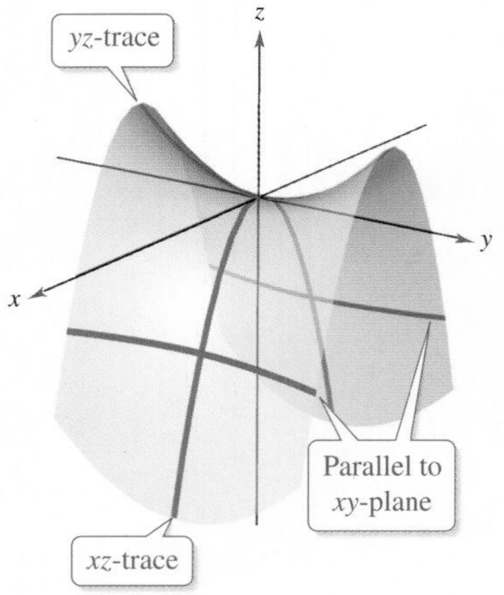

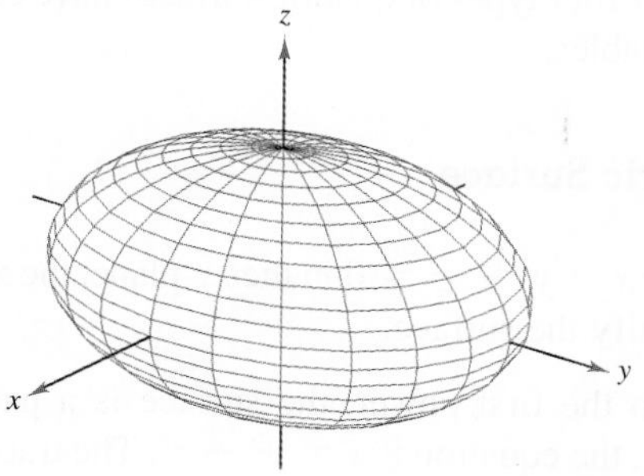

Ellipsoid

$$\frac{x^2}{a^2} + \frac{y^2}{b^2} + \frac{z^2}{c^2} = 1$$

Trace	*Plane*
Ellipse	Parallel to *xy*-plane
Ellipse	Parallel to *xz*-plane
Ellipse	Parallel to *yz*-plane

The surface is a sphere when the coefficients a, b, and c are equal and nonzero.

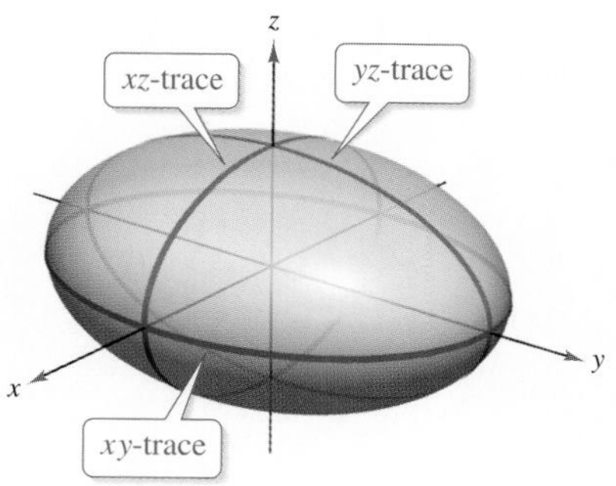

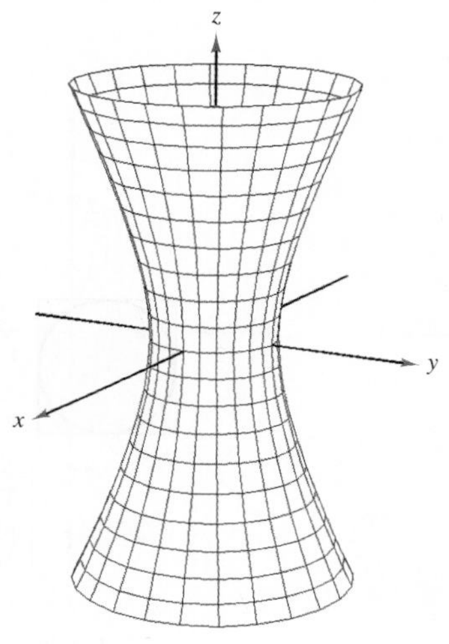

Hyperboloid of One Sheet

$$\frac{x^2}{a^2} + \frac{y^2}{b^2} - \frac{z^2}{c^2} = 1$$

Trace	*Plane*
Ellipse	Parallel to *xy*-plane
Hyperbola	Parallel to *xz*-plane
Hyperbola	Parallel to *yz*-plane

The axis of the hyperboloid corresponds to the variable whose coefficient is negative.

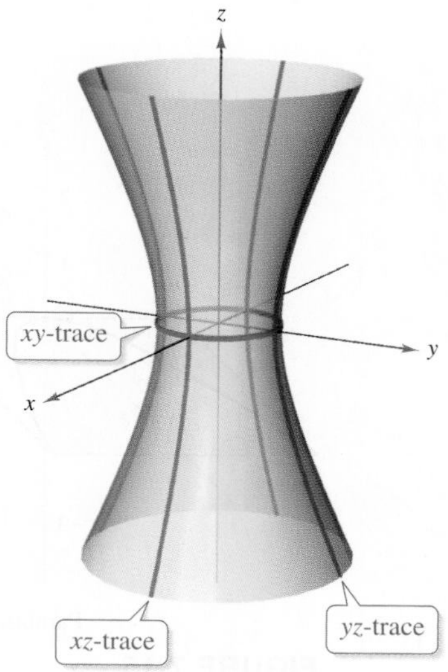

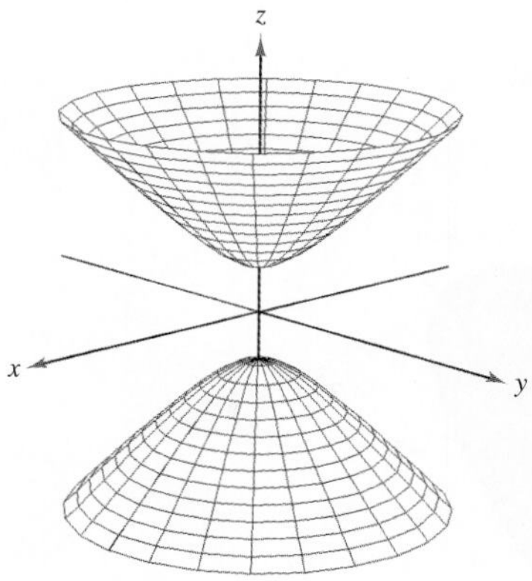

Hyperboloid of Two Sheets

$$\frac{z^2}{c^2} - \frac{x^2}{a^2} - \frac{y^2}{b^2} = 1$$

Trace	*Plane*
Ellipse	Parallel to *xy*-plane
Hyperbola	Parallel to *xz*-plane
Hyperbola	Parallel to *yz*-plane

The axis of the hyperboloid corresponds to the variable whose coefficient is positive. There is no trace in the coordinate plane perpendicular to this axis.

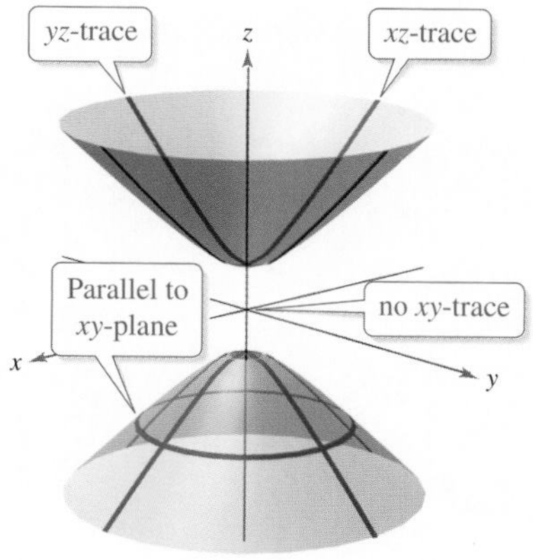

When classifying quadric surfaces, note that the two types of paraboloids have one variable raised to the first power. The other four types of quadric surfaces have equations that are of second degree in *all* three variables.

Example 2 Classifying a Quadric Surface

Describe the traces of the surface given by $x - y^2 - z^2 = 0$ in the xy-plane, the xz-plane, and the plane given by $x = 1$. Then classify the surface.

SOLUTION Because x is raised only to the first power, the surface is a paraboloid whose axis is the x-axis. In standard form, the equation is $x = y^2 + z^2$. The traces in the xy-plane, the xz-plane, and the plane given by $x = 1$ are as shown.

Trace in xy-plane $(z = 0)$:	$x = y^2$	Parabola
Trace in xz-plane $(y = 0)$:	$x = z^2$	Parabola
Trace in plane $x = 1$:	$y^2 + z^2 = 1$	Circle

These three traces are shown in Figure 7.15.

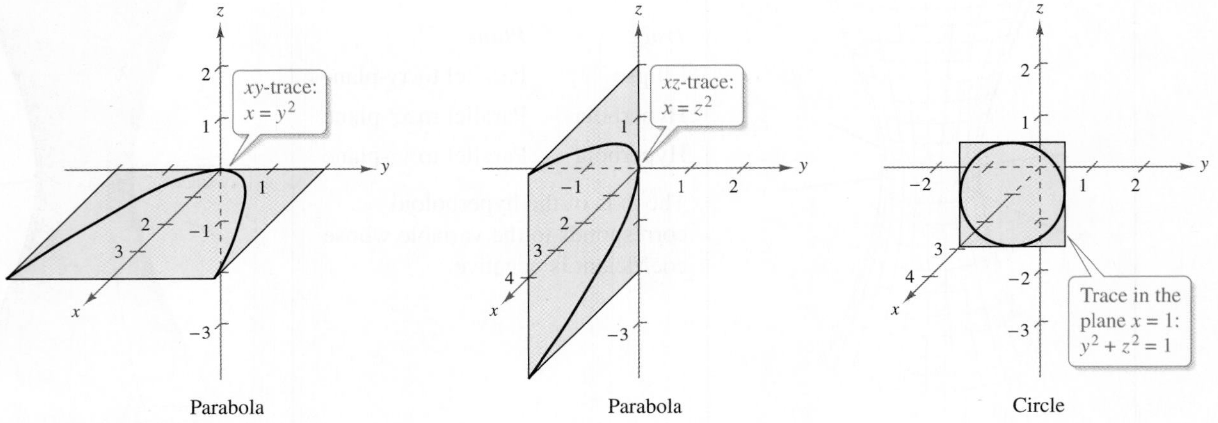

FIGURE 7.15

From the traces, you can see that the surface is an elliptic (or circular) paraboloid, as shown in Figure 7.16.

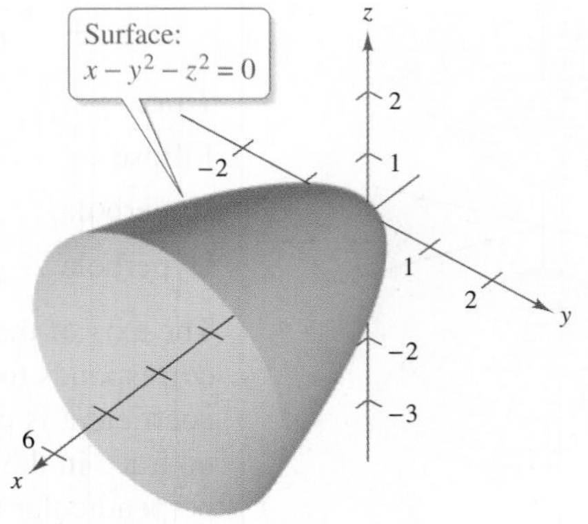

Elliptic Paraboloid
FIGURE 7.16

✓Checkpoint 2

Describe the traces of the surface given by $x^2 + y^2 - z^2 = 1$ in the xy-plane, the yz-plane, the xz-plane, and the plane given by $z = 3$. Then classify the surface. ■

Example 3 Classifying Quadric Surfaces

Classify the surface given by each equation.

a. $x^2 - 4y^2 - 4z^2 - 4 = 0$

b. $x^2 + 4y^2 + z^2 - 4 = 0$

SOLUTION

a. The equation $x^2 - 4y^2 - 4z^2 - 4 = 0$ can be written in standard form as

$$\frac{x^2}{4} - y^2 - z^2 = 1. \qquad \text{Standard form}$$

From the standard form, you can see that the graph is a hyperboloid of two sheets, with the x-axis as its axis, as shown in Figure 7.17(a).

b. The equation $x^2 + 4y^2 + z^2 - 4 = 0$ can be written in standard form as

$$\frac{x^2}{4} + y^2 + \frac{z^2}{4} = 1. \qquad \text{Standard form}$$

From the standard form, you can see that the graph is an ellipsoid, as shown in Figure 7.17(b).

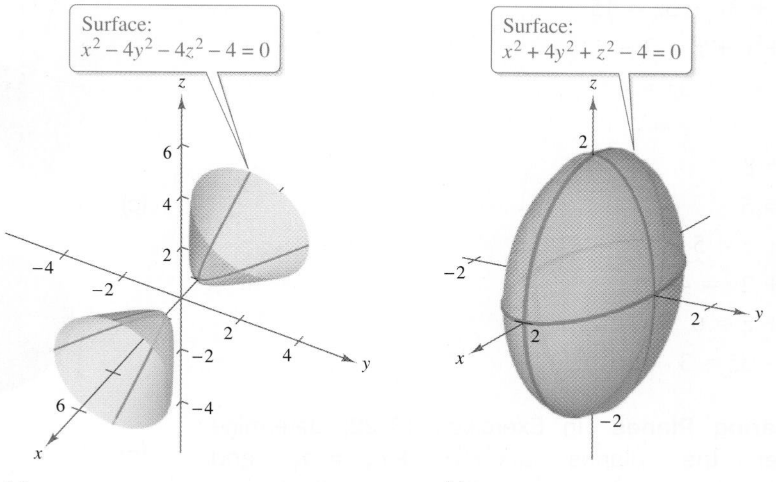

FIGURE 7.17

✓ Checkpoint 3

Classify the surface given by each equation.

a. $4x^2 + 9y^2 - 36z = 0$

b. $36x^2 + 16y^2 - 144z^2 = 0$ ■

SUMMARIZE (Section 7.2)

1. State the general equation of a plane in space *(page 427)*. For an example of sketching a plane in space, see Example 1.
2. List the six basic types of quadric surfaces *(page 429)*. For examples of classifying quadric surfaces, see Examples 2 and 3.

SKILLS WARM UP 7.2

The following warm-up exercises involve skills that were covered in earlier sections. You will use these skills in the exercise set for this section. For additional help, review Sections 1.2 and 7.1.

In Exercises 1–4, find the *x*- and *y*-intercepts of the function.

1. $3x + 4y = 12$

2. $6x + y = -8$

3. $-2x + y = -2$

4. $-x - y = 5$

In Exercises 5 and 6, write the equation of the sphere in standard form.

5. $16x^2 + 16y^2 + 16z^2 = 4$

6. $9x^2 + 9y^2 + 9z^2 = 36$

Exercises 7.2

See www.CalcChat.com for worked-out solutions to odd-numbered exercises.

Sketching a Plane in Space In Exercises 1–12, find the intercepts and sketch the graph of the plane. *See Example 1.*

1. $4x + 2y + 6z = 12$

2. $3x + 6y + 2z = 6$

3. $3x + 3y + 5z = 15$

4. $x + y + z = 3$

5. $2x - y + 3z = 4$

6. $2x - y + z = 4$

7. $z = 8$

8. $x = 5$

9. $y + z = 5$

10. $x + 2y = 4$

11. $x + z = 6$

12. $x - 3z = 3$

Comparing Planes In Exercises 13–22, determine whether the planes $a_1x + b_1y + c_1z = d_1$ and $a_2x + b_2y + c_2z = d_2$ are parallel, perpendicular, or neither. The planes are parallel when there exists a nonzero constant k such that $a_1 = ka_2$, $b_1 = kb_2$, and $c_1 = kc_2$, and are perpendicular when $a_1a_2 + b_1b_2 + c_1c_2 = 0$.

13. $5x - 3y + z = 4,\ x + 4y + 7z = 1$

14. $3x + y - 4z = 3,\ -9x - 3y + 12z = 4$

15. $x - 5y - z = 1,\ 5x - 25y - 5z = -3$

16. $x + 3y + 2z = 6,\ 4x - 12y + 8z = 24$

17. $x + 2y = 3,\ 4x + 8y = 5$

18. $x + 3y + z = 7,\ x - 5z = 0$

19. $2x + y = 3,\ 3x - 5z = 0$

20. $2x - z = 1,\ 4x + y + 8z = 10$

21. $x = 3,\ z = -1$

22. $x = -2,\ y = 4$

Matching In Exercises 23–28, match the equation with its graph. Then classify the quadric surface. [The graphs are labeled (a)–(f).]

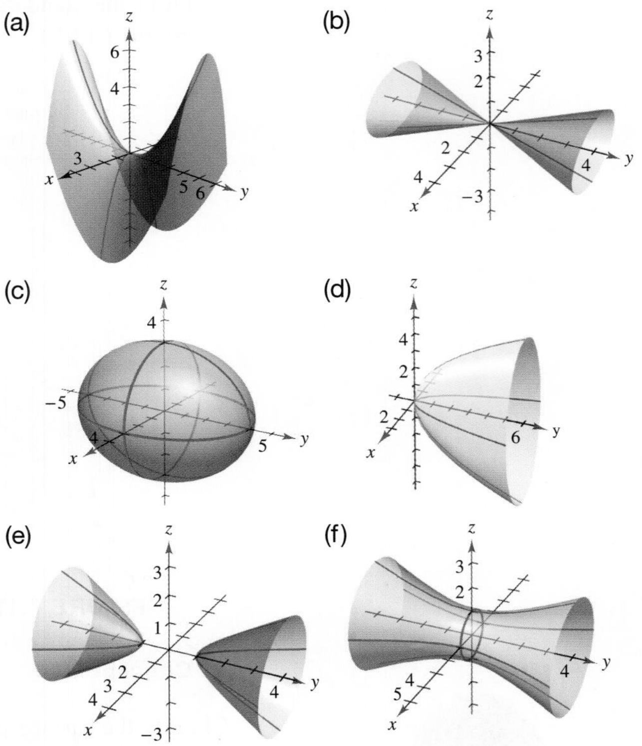

23. $\dfrac{x^2}{9} + \dfrac{y^2}{16} + \dfrac{z^2}{9} = 1$

24. $15x^2 - 4y^2 + 15z^2 = -4$

25. $4x^2 - y^2 + 4z^2 = 4$

26. $y^2 = 4x^2 + 9z^2$

27. $4x^2 - 4y + z^2 = 0$

28. $4x^2 - y^2 + 4z = 0$

Classifying a Quadric Surface In Exercises 29–34, describe the traces of the surface in the given planes. Then classify the surface. *See Example 2.*

29. $z = x^2 - y^2$

(a) xy-plane (b) $x = 3$ (c) xz-plane

30. $y = x^2 + z^2$

(a) xy-plane (b) $y = 1$ (c) yz-plane

31. $\frac{x^2}{4} + y^2 + z^2 = 1$

(a) xy-plane (b) xz-plane (c) yz-plane

32. $y^2 + z^2 - x^2 = 1$

(a) xy-plane (b) xz-plane (c) yz-plane

33. $z^2 - \frac{x^2}{9} - \frac{y^2}{16} = 1$

(a) xz-plane (b) $x = 2$ (c) $z = 4$

34. $y^2 + \frac{z^2}{4} - x^2 = 0$

(a) $y = -1$ (b) $z = 4$ (c) yz-plane

Classifying a Quadric Surface In Exercises 35–48, classify the quadric surface. *See Example 3.*

35. $x^2 + \frac{y^2}{4} + z^2 = 1$

36. $z^2 = x^2 + \frac{y^2}{4}$

37. $25x^2 + 25y^2 - z^2 = 5$

38. $z = 4x^2 + y^2$

39. $x^2 - y^2 + z = 0$

40. $z^2 - x^2 - \frac{y^2}{4} = 1$

41. $x^2 - y + z^2 = 0$

42. $9x^2 + 4y^2 - 8z^2 = 72$

43. $z^2 = 9x^2 + y^2$

44. $\frac{x^2}{9} + \frac{y^2}{16} + \frac{z^2}{16} = 1$

45. $2x^2 - y^2 + 2z^2 = -4$

46. $4y = x^2 + z^2$

47. $3z = -y^2 + x^2$

48. $z^2 = 2x^2 + 2y^2$

49. Physical Science Because of the forces caused by its rotation, Earth is actually an oblate ellipsoid rather than a sphere. The equatorial radius is 3963 miles and the polar radius is 3950 miles. Find an equation of the ellipsoid. Assume that the center of Earth is at the origin and the xy-trace ($z = 0$) corresponds to the equator.

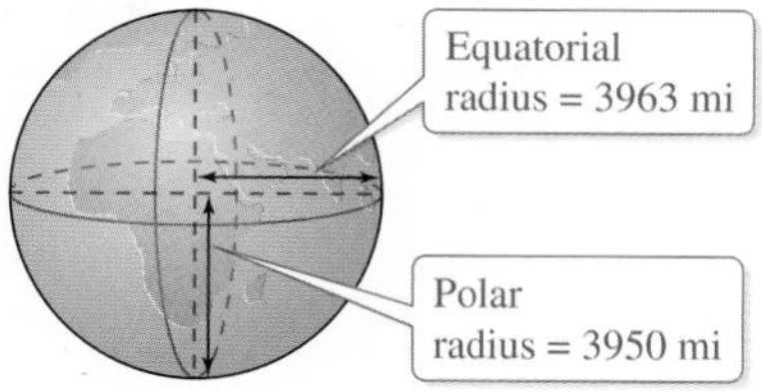

50. HOW DO YOU SEE IT? Each elliptic paraboloid below represents a statue that is modeled by the quadric surface $z = x^2 + y^2$. Match each of the four graphs with the point in space from which the statue is viewed. The four points are (0, 0, 20), (0, 20, 0), (20, 0, 0), and (10, 10, 20).

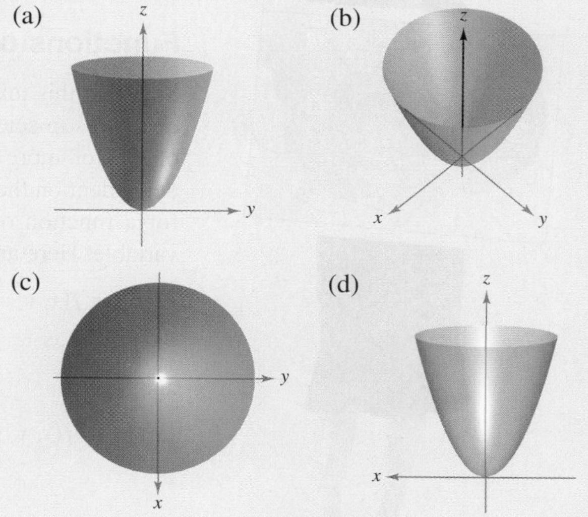

51. Modeling Data Personal consumption expenditures (in billions of dollars) for several types of recreation from 2004 through 2009 are shown in the table, where x is the expenditures on amusement parks and campgrounds, y is the expenditures on live entertainment (excluding sports), and z is the expenditures on spectator sports.

Year	2004	2005	2006	2007	2008	2009
x	33.1	34.9	37.4	40.6	43.0	41.8
y	13.2	13.8	14.9	15.0	15.4	14.5
z	15.5	16.3	17.8	19.5	20.5	20.7

A model for the data in the table is given by $-0.62x + 0.41y + z = 0.38$. *(Source: U.S. Bureau of Economic Analysis)*

(a) Complete a fourth row of the table using the model to approximate z for the given values of x and y. Compare the approximations with the actual values of z.

(b) According to this model, increases in expenditures of recreation types y and z would correspond to what kind of change in expenditures of recreation type x?

7.3 Functions of Several Variables

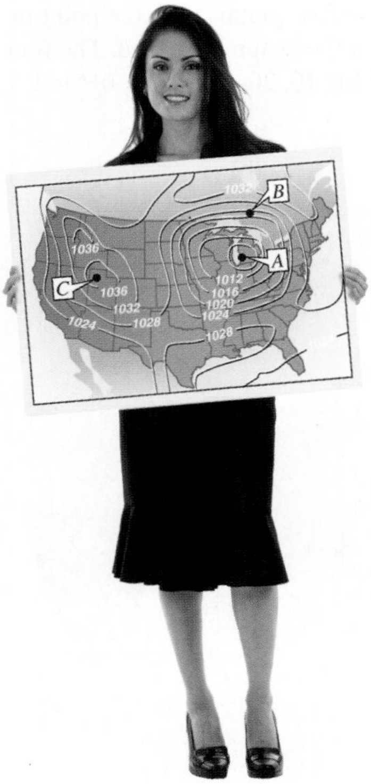

In Exercise 49 on page 443, you will read a weather map and identify areas of high and low pressure.

- Evaluate functions of several variables.
- Find the domains and ranges of functions of two variables.
- Read contour maps and sketch level curves of functions of two variables.
- Use functions of several variables to answer questions about real-life situations.

Functions of Several Variables

So far in this text, you have studied functions of a single independent variable. Many quantities in science, business, and technology, however, are functions not of one, but of two or more variables. For instance, the demand function for a product is often dependent on the price *and* the advertising, rather than on the price alone. The notation for a function of two or more variables is similar to that for a function of a single variable. Here are two examples.

$$z = f(\underbrace{x, y}_{\text{2 variables}}) = x^2 + xy \qquad \text{Function of two variables}$$

and

$$w = f(\underbrace{x, y, z}_{\text{3 variables}}) = x + 2y - 3z \qquad \text{Function of three variables}$$

Definition of a Function of Two Variables

Let D be a set of ordered pairs of real numbers. If to each ordered pair (x, y) in D there corresponds a unique real number $f(x, y)$, then f is called a **function of x and y.** The set D is the **domain** of f, and the corresponding set of values for $f(x, y)$ is the **range** of f. Functions of three, four, or more variables are defined similarly.

For the function given by

$$z = f(x, y)$$

x and y are called the **independent variables** and z is called the **dependent variable.**

Example 1 Evaluating Functions of Several Variables

a. For $f(x, y) = 2x^2 - y^2$, you can evaluate $f(2, 3)$ as shown.

$$f(2, 3) = 2(2)^2 - (3)^2 = 8 - 9 = -1$$

b. For $f(x, y, z) = e^x(y + z)$, you can evaluate $f(0, -1, 4)$ as shown.

$$f(0, -1, 4) = e^0(-1 + 4) = (1)(3) = 3$$

✓ Checkpoint 1

Find the indicated function values.

a. For $f(x, y) = x^2 + 2xy$, find $f(2, -1)$.

b. For $f(x, y, z) = \dfrac{2x^2z}{y^3}$, find $f(-3, 2, 1)$. ■

The Domain and Range of a Function of Two Variables

A function of two variables can be represented graphically as a surface in space by letting

$z = f(x, y)$. Function of two variables

When sketching the graph of a function of x and y, remember that even though the graph is three-dimensional, the domain of the function is two-dimensional—it consists of the points in the xy-plane for which the function is defined. As with functions of a single variable, unless specifically restricted, the domain of a function of two variables is assumed to be the set of all points (x, y) for which the defining equation has meaning. In other words, to each point (x, y) in the domain of f there corresponds a point (x, y, z) on the surface, and conversely, to each point (x, y, z) on the surface there corresponds a point (x, y) in the domain of f.

Example 2 Finding the Domain and Range of a Function

Find the domain and range of the function

$$f(x, y) = \sqrt{64 - x^2 - y^2}.$$

SOLUTION Because no restrictions are given, the domain is assumed to be the set of all points for which the defining equation makes sense.

$64 - x^2 - y^2 \geq 0$ Quantity inside radical must be nonnegative.

$-x^2 - y^2 \geq -64$ Subtract 64 from each side.

$x^2 + y^2 \leq 64$ Multiply each side by -1 and reverse the inequality symbol.

So, the domain of f is the set of all points that lie on or inside the circle given by

$x^2 + y^2 \leq 8^2$ Domain of the function

as shown in Figure 7.18. The range of f is the set

$0 \leq z \leq 8.$ Range of the function

As shown in Figure 7.19, the graph of the function is a hemisphere.

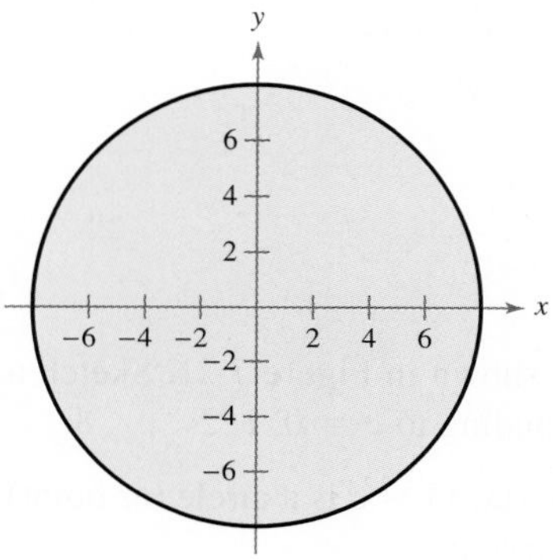

Domain of $f(x, y) = \sqrt{64 - x^2 - y^2}$

FIGURE 7.18

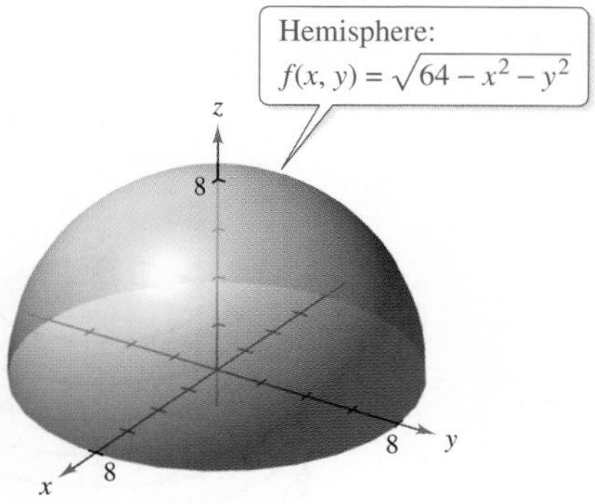

Domain: $x^2 + y^2 \leq 64$
Range: $0 \leq z \leq 8$

FIGURE 7.19

✓ Checkpoint 2

Consider the function

$$f(x, y) = \sqrt{9 - x^2 - y^2}.$$

a. Find the domain of f.

b. Find the range of f. ■

Contour Maps and Level Curves

A **contour map** of a surface is created by *projecting* traces, taken in evenly spaced planes that are parallel to the xy-plane, onto the xy-plane. Each projection is a **level curve** of the surface.

Contour maps are used to create weather, topographical, and population density maps. For instance, Figure 7.20(a) shows a graph of a "mountain and valley" surface given by $z = f(x, y)$. Each of the level curves in Figure 7.20(b) represents the intersection of the surface $z = f(x, y)$ with a plane $z = c$, where $c = 828, 830, \ldots, 854$.

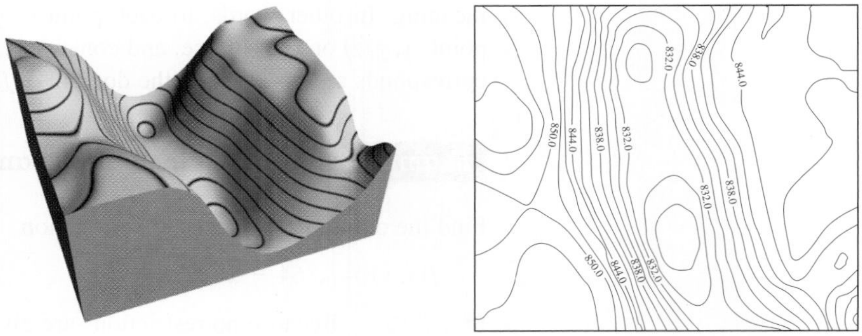

(a) Surface (b) Contour map

FIGURE 7.20

Example 3 Sketching a Contour Map

The hemisphere given by $f(x, y) = \sqrt{64 - x^2 - y^2}$ is shown in Figure 7.21. Sketch a contour map of this surface using level curves corresponding to $c = 0, 1, 2, \ldots, 8$.

SOLUTION For each value of c, the equation given by $f(x, y) = c$ is a circle (or point) in the xy-plane. For instance, when $c_1 = 0$, the level curve is

$$x^2 + y^2 = 8^2 \qquad \text{Circle of radius 8}$$

which is a circle of radius 8. Figure 7.22 shows the nine level curves for the hemisphere.

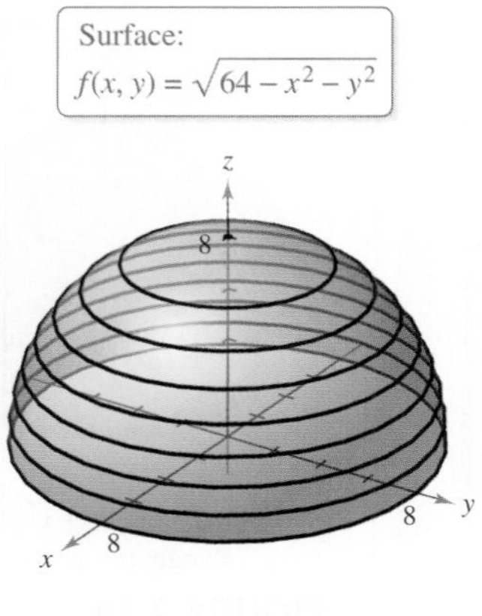

Hemisphere

FIGURE 7.21

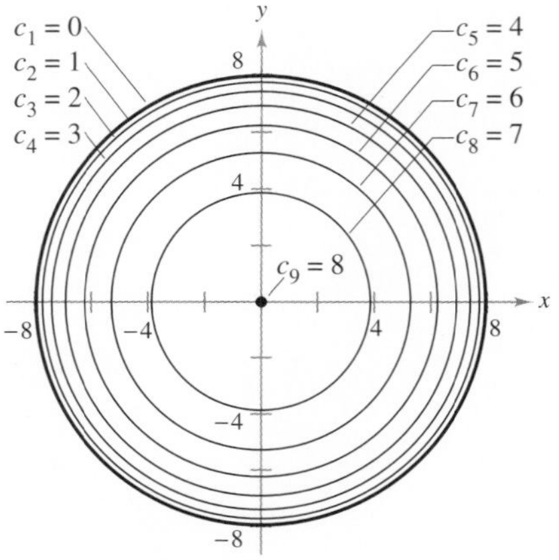

Contour map

FIGURE 7.22

✓ Checkpoint 3

Describe the level curves of $f(x, y) = \sqrt{9 - x^2 - y^2}$. Sketch the level curves for $c = 0, 1, 2$, and 3. ■

Applications

The **Cobb-Douglas production function** is used in economics to represent the numbers of units produced by varying amounts of labor and capital. Let x represent the number of units of labor and let y represent the number of units of capital. Then, the number of units produced is modeled by

$$f(x, y) = Cx^a y^{1-a}$$

where C and a are constants, with $0 < a < 1$.

Example 4 Using a Production Function

A manufacturer estimates that its production (measured in units of a product) can be modeled by $f(x, y) = 100x^{0.6}y^{0.4}$, where the labor x is measured in person-hours and the capital y is measured in thousands of dollars.

a. What is the production level when $x = 1000$ and $y = 500$?

b. What is the production level when $x = 2000$ and $y = 1000$?

c. How does doubling the amounts of labor and capital from part (a) to part (b) affect the production?

SOLUTION

a. When $x = 1000$ and $y = 500$, the production level is

$$f(1000, 500) = 100(1000)^{0.6}(500)^{0.4} \approx 75{,}786 \text{ units.}$$

b. When $x = 2000$ and $y = 1000$, the production level is

$$f(2000, 1000) = 100(2000)^{0.6}(1000)^{0.4} \approx 151{,}572 \text{ units.}$$

c. When the amounts of labor and capital are doubled, the production level also doubles. In Exercise 44, you are asked to show that this is characteristic of the Cobb-Douglas production function.

A contour map of this function is shown in Figure 7.23. Note that the level curves occur at increments of 10,000.

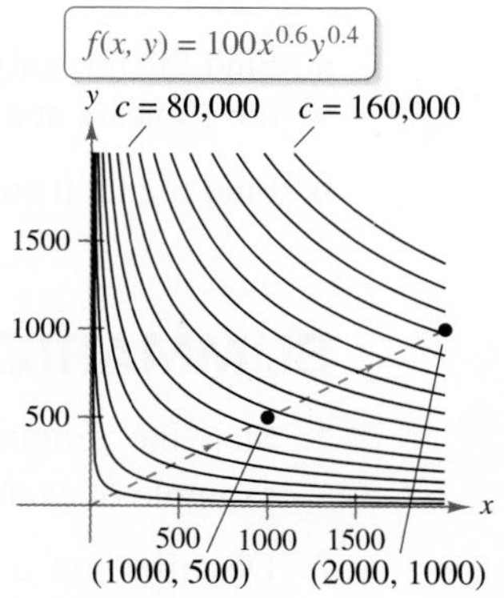

Level Curves (at Increments of 10,000)

FIGURE 7.23

✓ Checkpoint 4

Use the Cobb-Douglas production function in Example 4 to find the production levels when $x = 1500$ and $y = 1000$ and when $x = 1000$ and $y = 1500$. Use your results to determine which variable has a greater influence on production. ■

Example 5 Finding Monthly Payments

The monthly payment M for an installment loan of P dollars taken out over t years at an annual interest rate of r is given by

$$M = f(P, r, t) = \frac{\dfrac{Pr}{12}}{1 - \left[\dfrac{1}{1 + (r/12)}\right]^{12t}}.$$

a. Find the monthly payment for a home mortgage of \$100,000 taken out for 30 years at an annual interest rate of 7%.

b. Find the monthly payment for a car loan of \$22,000 taken out for 5 years at an annual interest rate of 8%.

SOLUTION

a. When $P = \$100{,}000$, $r = 0.07$, and $t = 30$, the monthly payment is

$$\begin{aligned} M &= f(100{,}000, 0.07, 30) \\ &= \frac{\dfrac{(100{,}000)(0.07)}{12}}{1 - \left[\dfrac{1}{1 + (0.07/12)}\right]^{12(30)}} \\ &\approx \$665.30. \end{aligned}$$

b. When $P = \$22{,}000$, $r = 0.08$, and $t = 5$, the monthly payment is

$$\begin{aligned} M &= f(22{,}000, 0.08, 5) \\ &= \frac{\dfrac{(22{,}000)(0.08)}{12}}{1 - \left[\dfrac{1}{1 + (0.08/12)}\right]^{12(5)}} \\ &\approx \$446.08. \end{aligned}$$

✓ Checkpoint 5

a. Find the monthly payment M for a home mortgage of \$100,000 taken out for 30 years at an annual interest rate of 3%.

b. Find the total amount of money you will pay for the mortgage. ■

SUMMARIZE (Section 7.3)

1. State the definition of a function of two variables *(page 436)*. For an example of evaluating a function of two variables, see Example 1.
2. Describe how a contour map of a surface is created *(page 438)*. For an example of sketching a contour map, see Example 3.
3. State the Cobb-Douglas production function *(page 439)*. For an example of using the Cobb-Douglas production function, see Example 4.
4. Describe a real-life example of how a function of several variables can be used to find the monthly payment for a loan *(page 440, Example 5)*.

SKILLS WARM UP 7.3 The following warm-up exercises involve skills that were covered in a previous course or in earlier sections. You will use these skills in the exercise set for this section. For additional help, review Appendix Section A.3 and Section 1.4.

In Exercises 1–4, evaluate the function when $x = -3$.

1. $f(x) = 5 - 2x$ **2.** $f(x) = -x^2 + 4x + 5$ **3.** $y = \sqrt{4x^2 - 3x + 4}$ **4.** $y = \sqrt[3]{34 - 4x + 2x^2}$

In Exercises 5–8, find the domain of the function.

5. $f(x) = 5x^2 + 3x - 2$ **6.** $g(x) = \dfrac{1}{2x} - \dfrac{2}{x + 3}$ **7.** $h(y) = \sqrt{y - 5}$ **8.** $f(y) = \sqrt{y^2 - 5}$

In Exercises 9 and 10, evaluate the expression.

9. $(476)^{0.65}$ **10.** $(251)^{0.35}$

Exercises 7.3

See www.CalcChat.com for worked-out solutions to odd-numbered exercises.

Evaluating Functions of Several Variables In Exercises 1–14, find the function values. *See Example 1.*

1. $f(x, y) = \dfrac{x}{y}$

(a) $f(3, 2)$ (b) $f(-1, 4)$ (c) $f(30, 5)$
(d) $f(5, y)$ (e) $f(x, 2)$ (f) $f(5, t)$

2. $f(x, y) = 4 - x^2 - 4y^2$

(a) $f(0, 0)$ (b) $f(0, 1)$ (c) $f(2, 3)$
(d) $f(1, y)$ (e) $f(x, 0)$ (f) $f(t, 1)$

3. $f(x, y) = xe^y$

(a) $f(5, 0)$ (b) $f(3, 2)$ (c) $f(2, -1)$
(d) $f(5, y)$ (e) $f(x, 2)$ (f) $f(t, t)$

4. $g(x, y) = \ln|x + y|$

(a) $g(2, 3)$ (b) $g(5, 6)$ (c) $g(e, 0)$
(d) $g(0, 1)$ (e) $g(2, -3)$ (f) $g(e, e)$

5. $h(x, y, z) = \dfrac{xy}{z}$

(a) $h(2, 3, 9)$ (b) $h(1, 0, 1)$

6. $f(x, y, z) = \sqrt{x + y + z}$

(a) $f(0, 5, 4)$ (b) $f(6, 8, -3)$

7. $V(r, h) = \pi r^2 h$

(a) $V(3, 10)$ (b) $V(5, 2)$

8. $F(r, N) = 500\left(1 + \dfrac{r}{12}\right)^N$

(a) $F(0.09, 60)$ (b) $F(0.14, 240)$

9. $A(P, r, t) = P\left[\left(1 + \dfrac{r}{12}\right)^{12t} - 1\right]\left(1 + \dfrac{12}{r}\right)$

(a) $A(100, 0.10, 10)$ (b) $A(275, 0.0925, 40)$

10. $A(P, r, t) = Pe^{rt}$

(a) $A(500, 0.10, 5)$ (b) $A(1500, 0.12, 20)$

11. $f(x, y) = \displaystyle\int_x^y (2t - 3)\, dt$

(a) $f(1, 2)$ (b) $f(1, 4)$

12. $g(x, y) = \displaystyle\int_x^y \frac{1}{t}\, dt$

(a) $g(4, 1)$ (b) $g(6, 3)$

13. $f(x, y) = x^2 - 2y$

(a) $f(x + \Delta x, y)$ (b) $\dfrac{f(x, y + \Delta y) - f(x, y)}{\Delta y}$

14. $f(x, y) = 3xy + y^2$

(a) $f(x + \Delta x, y)$ (b) $\dfrac{f(x, y + \Delta y) - f(x, y)}{\Delta y}$

Finding the Domain and Range of a Function In Exercises 15–30, find the domain and range of the function. *See Example 2.*

15. $f(x, y) = \sqrt{16 - x^2 - y^2}$

16. $z = \sqrt{4 - x^2 - y^2}$

17. $f(x, y) = x^2 + y^2$

18. $f(x, y) = x^2 + y^2 - 1$

19. $f(x, y) = e^{x/y}$

20. $f(x, y) = ye^{1/x}$

21. $g(x, y) = \ln(4 - x - y)$

22. $f(x, y) = \ln(x + y)$

23. $z = \sqrt{9 - 3x^2 - y^2}$

24. $z = \sqrt{4 - x^2 - 4y^2}$

25. $z = \dfrac{y}{x}$

26. $f(x, y) = \dfrac{x}{y}$

27. $f(x, y) = \dfrac{1}{xy}$

28. $g(x, y) = \dfrac{1}{x - y}$

29. $h(x, y) = x\sqrt{y}$

30. $f(x, y) = \sqrt{xy}$

Matching In Exercises 31–34, match the graph of the surface with one of the contour maps. [The contour maps are labeled (a)–(d).]

(a)

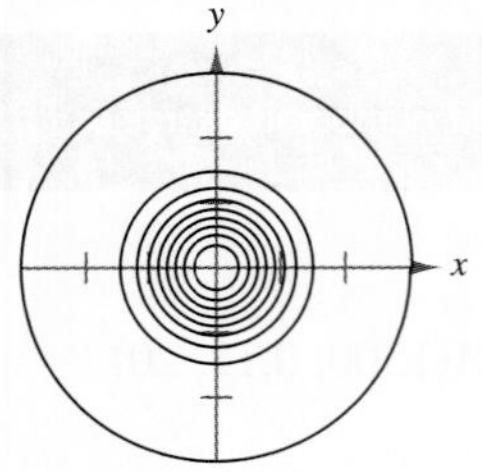

(b)

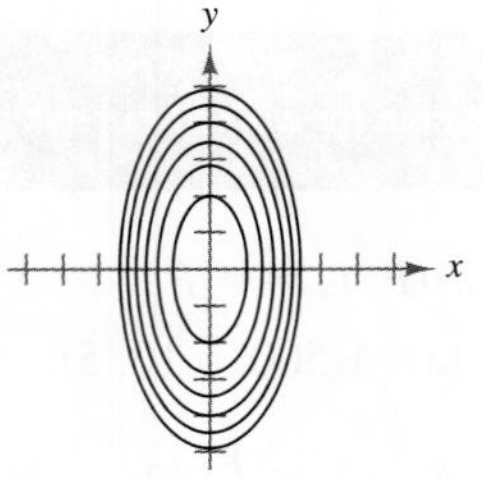

(c)

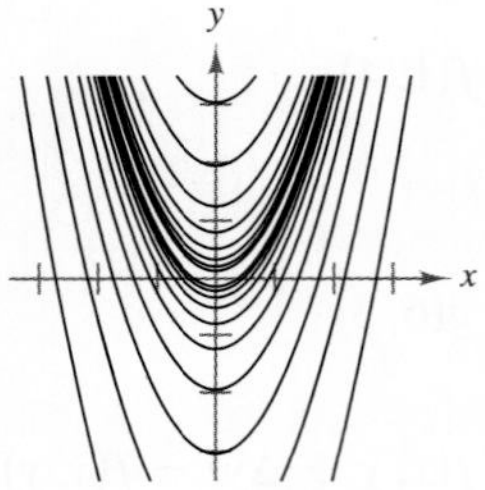

(d)

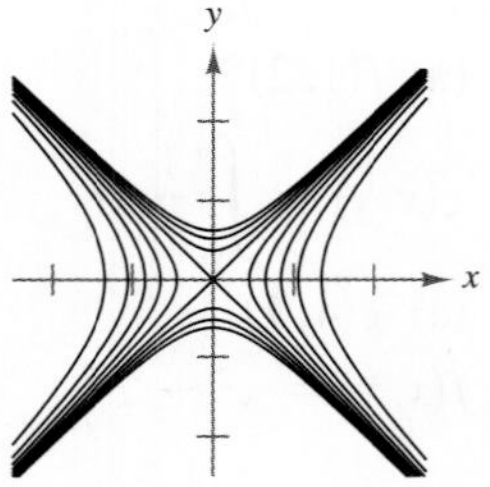

31. $f(x, y) = x^2 + \dfrac{y^2}{4}$

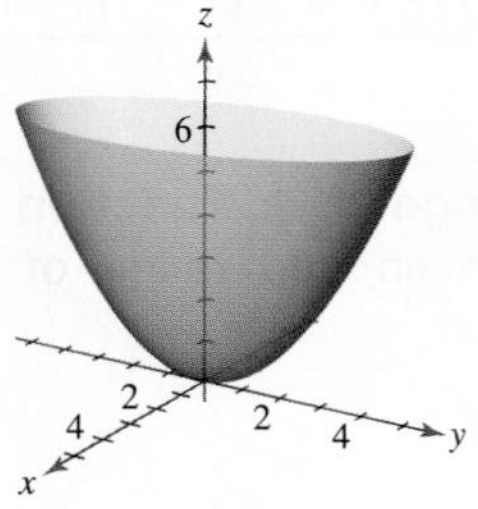

32. $f(x, y) = e^{1 - x^2 + y^2}$

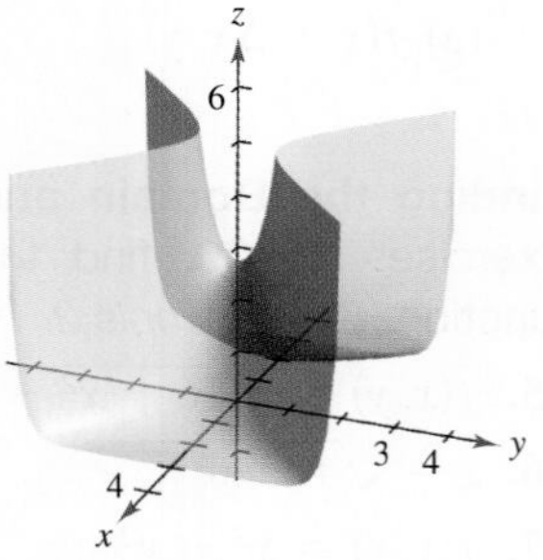

33. $f(x, y) = e^{1 - x^2 - y^2}$

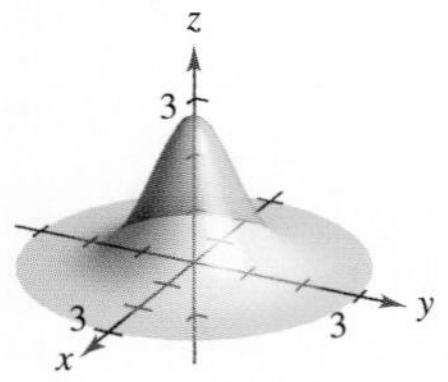

34. $f(x, y) = \ln|y - x^2|$

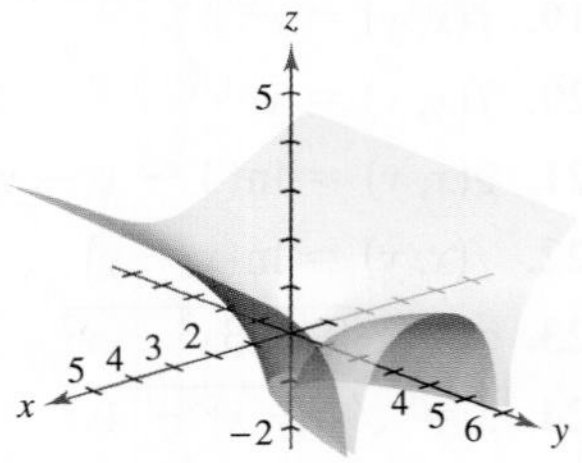

Sketching a Contour Map In Exercises 35–42, describe the level curves of the function. Sketch a contour map of the surface using level curves for the given *c*-values. *See Example 3.*

Function	*c-Values*
35. $z = x + y$	$c = -1, 0, 2, 4$
36. $z = 6 - 2x - 3y$	$c = 0, 2, 4, 6, 8, 10$
37. $z = \sqrt{25 - x^2 - y^2}$	$c = 0, 1, 2, 3, 4, 5$
38. $f(x, y) = x^2 + y^2$	$c = 0, 2, 4, 6, 8$
39. $f(x, y) = xy$	$c = \pm 1, \pm 2, . . . , \pm 6$
40. $z = e^{xy}$	$c = 1, 2, 3, 4, \frac{1}{2}, \frac{1}{3}, \frac{1}{4}$
41. $f(x, y) = \dfrac{x}{x^2 + y^2}$	$c = \pm\dfrac{1}{2}, \pm 1, \pm\dfrac{3}{2}, \pm 2$
42. $f(x, y) = \ln(x - y)$	$c = 0, \pm\frac{1}{2}, \pm 1, \pm\frac{3}{2}, \pm 2$

43. Cobb-Douglas Production Function A manufacturer estimates the Cobb-Douglas production function to be given by

$$f(x, y) = 100x^{0.75}y^{0.25}.$$

Estimate the production level when $x = 1500$ and $y = 1000$.

44. Cobb-Douglas Production Function Use the Cobb-Douglas production function (Example 4) to show that when both the number of units of labor and the number of units of capital are doubled, the production level is also doubled.

45. Profit A sporting goods manufacturer produces regulation soccer balls at two plants. The costs of producing x_1 units at location 1 and x_2 units at location 2 are given by

$$C_1(x_1) = 0.02x_1^2 + 4x_1 + 500$$

and

$$C_2(x_2) = 0.05x_2^2 + 4x_2 + 275$$

respectively. If the product sells for \$50 per unit, then the profit function for the product is given by

$$P(x_1, x_2) = 50(x_1 + x_2) - C_1(x_1) - C_2(x_2).$$

Find (a) $P(250, 150)$, (b) $P(300, 200)$, and (c) $P(600, 400)$.

46. Queuing Model The average amount of time that a customer waits in line for service is given by

$$W(x, y) = \frac{1}{x - y}, \quad y < x$$

where y is the average arrival rate and x is the average service rate (x and y are measured in the number of customers per hour). Evaluate W at each point.

(a) $(15, 10)$ (b) $(12, 9)$ (c) $(12, 6)$ (d) $(4, 2)$

47. Investment In 2011, an investment of \$2000 was made in a bond earning 10% compounded annually. The investor pays tax at rate R, and the annual rate of inflation is I. In the year 2021, the value V of the bond in constant 2011 dollars is given by

$$V(I, R) = 2000\left[\frac{1 + 0.10(1 - R)}{1 + I}\right]^{10}.$$

Use this function of two variables and a spreadsheet to complete the table.

Tax Rate	Inflation Rate		
	0	0.03	0.05
0			
0.28			
0.35			

48. Investment A principal of \$5000 is deposited in a savings account that earns an interest rate of r (in decimal form), compounded continuously. The amount $A(r, t)$ after t years is

$$A(r, t) = 5000e^{rt}.$$

Use this function of two variables and a spreadsheet to complete the table.

	Number of Years			
Rate	5	10	15	20
0.02				
0.04				
0.06				
0.08				

49. Meteorology Meteorologists measure the atmospheric pressure in millibars. From these observations they create weather maps on which the curves of equal atmospheric pressure (isobars) are drawn (see figure). On the map, the closer the isobars, the higher the wind speed. Match points A, B, and C with (a) highest pressure, (b) lowest pressure, and (c) highest wind velocity.

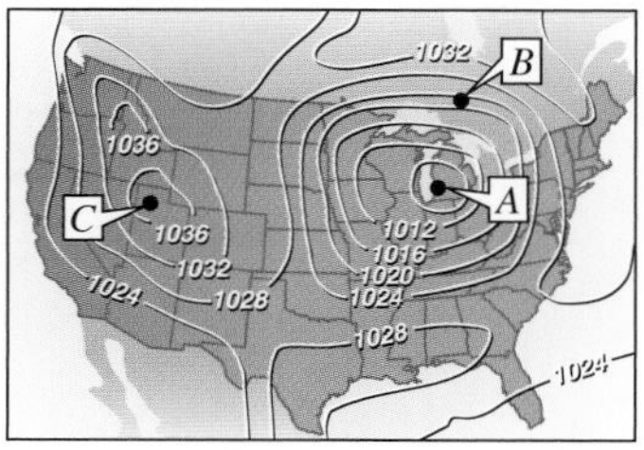

50. HOW DO YOU SEE IT? The contour map of the Southern Hemisphere shown in the figure was computer generated using data collected by satellite instrumentation. Color is used to show the "ozone hole" in Earth's atmosphere. The purple and blue areas represent the lowest levels of ozone, and the green areas represent the highest levels. *(Source: National Aeronautics and Space Administration)*

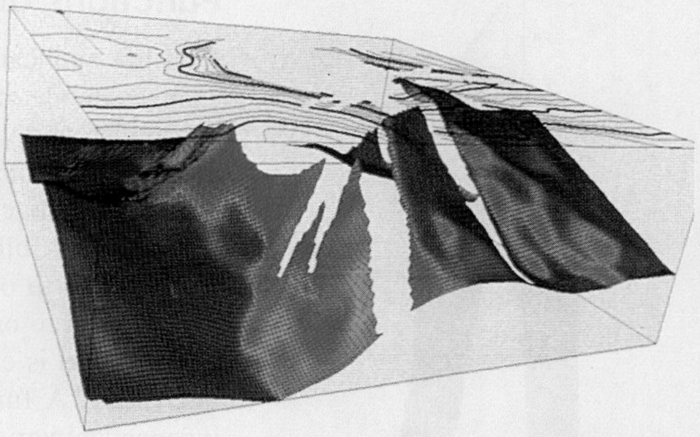

(a) Describe the areas that have the lowest levels of ozone.

(b) Do the level curves correspond to equally spaced ozone levels? Explain.

51. Earnings per Share The earnings per share z (in dollars) for Apple from 2005 through 2010 can be modeled by $z = 0.379x - 0.135y - 3.45$, where x is the sales (in billions of dollars) and y is the shareholder's equity (in billions of dollars). *(Source: Apple Inc.)*

(a) Find the earnings per share when $x = 20$ and $y = 10$.

(b) Which of the two variables in this model has the greater influence on the earnings per share? Explain.

52. Shareholder's Equity The shareholder's equity z (in millions of dollars) for Sketchers from 2001 through 2009 can be modeled by $z = 0.175x + 0.772y - 275$, where x is the sales (in millions of dollars) and y is the total assets (in millions of dollars). *(Source: Sketchers U.S.A. Inc.)*

(a) Find the shareholder's equity when $x = 1000$ and $y = 500$.

(b) Which of the two variables in this model has the greater influence on shareholder's equity? Explain.

53. Monthly Payments You are taking out a home mortgage for \$120,000, and you are given the options below. Find the monthly payment and the total amount of money you will pay for each mortgage. Which option would you choose? Explain your reasoning.

(a) A fixed annual rate of 8%, over a term of 20 years.

(b) A fixed annual rate of 7%, over a term of 30 years.

(c) A fixed annual rate of 7%, over a term of 15 years.

NASA

7.4 Partial Derivatives

- Find the first partial derivatives of functions of two variables.
- Find the slopes of surfaces in the x- and y-directions and use partial derivatives to answer questions about real-life situations.
- Find the partial derivatives of functions of several variables.
- Find higher-order partial derivatives.

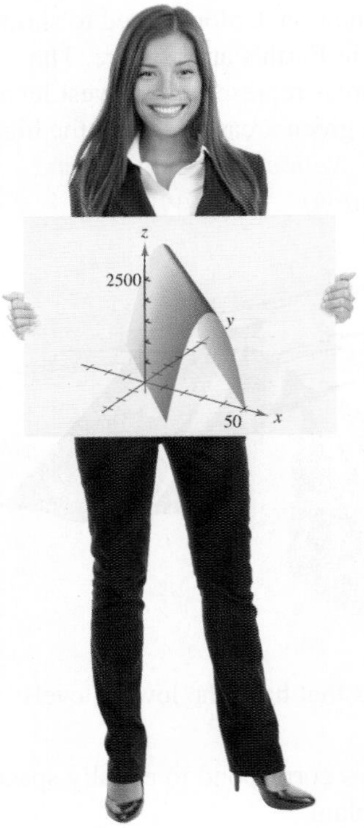

In Exercise 60 on page 453, you will use partial derivatives to find the marginal revenues of a pharmaceutical corporation at two locations that produce the same medicine.

Functions of Two Variables

Real-life applications of functions of several variables are often concerned with how changes in one of the variables will affect the values of the functions. For instance, an economist who wants to determine the effect of a tax increase on the economy might make calculations using different tax rates while holding all other variables, such as unemployment, constant.

You can follow a similar procedure to find the rate of change of a function f with respect to one of its independent variables. That is, you find the derivative of f with respect to one independent variable while holding the other variable(s) constant. This process is called **partial differentiation,** and each derivative is called a **partial derivative.** A function of several variables has as many partial derivatives as it has independent variables.

Partial Derivatives of a Function of Two Variables

If $z = f(x, y)$, then the **first partial derivatives of f with respect to x and y** are the functions $\partial z/\partial x$ and $\partial z/\partial y$, defined as shown.

$$\frac{\partial z}{\partial x} = \lim_{\Delta x \to 0} \frac{f(x + \Delta x, y) - f(x, y)}{\Delta x}$$ y is held constant.

$$\frac{\partial z}{\partial y} = \lim_{\Delta y \to 0} \frac{f(x, y + \Delta y) - f(x, y)}{\Delta y}$$ x is held constant.

STUDY TIP

The notation $\partial z/\partial x$ is read as "the partial derivative of z with respect to x," and $\partial z/\partial y$ is read as "the partial derivative of z with respect to y."

This definition indicates that if $z = f(x, y)$, then to find $\partial z/\partial x$, you *consider y to be constant* and differentiate with respect to x. Similarly, to find $\partial z/\partial y$, you *consider x to be constant* and differentiate with respect to y.

Example 1 Finding Partial Derivatives

Find $\partial z/\partial x$ and $\partial z/\partial y$ for the function $z = 3x - x^2y^2 + 2x^3y$.

SOLUTION

$$\frac{\partial z}{\partial x} = 3 - 2xy^2 + 6x^2y$$ Hold y constant and differentiate with respect to x.

$$\frac{\partial z}{\partial y} = -2x^2y + 2x^3$$ Hold x constant and differentiate with respect to y.

✓Checkpoint 1

Find $\dfrac{\partial z}{\partial x}$ and $\dfrac{\partial z}{\partial y}$ for $z = 2x^2 - 4x^2y^3 + y^4$. ■

Notation for First Partial Derivatives

The first partial derivatives of $z = f(x, y)$ are denoted by

$$\frac{\partial z}{\partial x} = f_x(x, y) = z_x = \frac{\partial}{\partial x}[f(x, y)]$$

and

$$\frac{\partial z}{\partial y} = f_y(x, y) = z_y = \frac{\partial}{\partial y}[f(x, y)].$$

The values of the first partial derivatives at the point (a, b) are denoted by

$$\left.\frac{\partial z}{\partial x}\right|_{(a,\,b)} = f_x(a, b)$$

and

$$\left.\frac{\partial z}{\partial y}\right|_{(a,\,b)} = f_y(a, b).$$

TECH TUTOR

Symbolic differentiation utilities can be used to find partial derivatives of a function of two variables. Try using a symbolic differentiation utility to find the first partial derivatives of the function in Example 2.

Example 2 Finding and Evaluating Partial Derivatives

Find the first partial derivatives of

$$f(x, y) = xe^{x^2y}$$

and evaluate each at the point $(1, \ln 2)$.

SOLUTION To find the first partial derivative with respect to x, hold y constant and differentiate using the Product Rule.

$$\begin{aligned} f_x(x, y) &= x\frac{\partial}{\partial x}[e^{x^2y}] + e^{x^2y}\frac{\partial}{\partial x}[x] && \text{Apply Product Rule.} \\ &= xe^{x^2y}(2xy) + e^{x^2y} && y \text{ is held constant.} \\ &= e^{x^2y}(2x^2y + 1) && \text{Simplify.} \end{aligned}$$

At the point $(1, \ln 2)$, the value of this derivative is

$$\begin{aligned} f_x(1, \ln 2) &= e^{(1)^2(\ln 2)}[2(1)^2(\ln 2) + 1] && \text{Substitute for } x \text{ and } y. \\ &= 2(2 \ln 2 + 1) && \text{Simplify.} \\ &\approx 4.773. && \text{Use a calculator.} \end{aligned}$$

To find the first partial derivative with respect to y, hold x constant and differentiate to obtain

$$\begin{aligned} f_y(x, y) &= xe^{x^2y}(x^2) && \text{Apply Constant Multiple Rule.} \\ &= x^3e^{x^2y}. && \text{Simplify.} \end{aligned}$$

At the point $(1, \ln 2)$, the value of this derivative is

$$\begin{aligned} f_y(1, \ln 2) &= (1)^3e^{(1)^2(\ln 2)} && \text{Substitute for } x \text{ and } y. \\ &= 2. && \text{Simplify.} \end{aligned}$$

✓ Checkpoint 2

Find the first partial derivatives of

$$f(x, y) = x^2y^3$$

and evaluate each at the point $(1, 2)$. ■

Graphical Interpretation of Partial Derivatives

Earlier in the text, you studied graphical interpretations of the derivative of a function of a single variable. There, you found that $f'(x_0)$ represents the slope of the tangent line to the graph of $y = f(x)$ at the point (x_0, y_0). The partial derivatives of a function of two variables also have useful graphical interpretations. Consider the function

$$z = f(x, y).$$ Function of two variables

As shown in Figure 7.24(a), the graph of this function is a surface in space. If $y = y_0$, then

$$z = f(x, y_0)$$ Function of one variable

is a function of one variable. The graph of this function is the curve that is the intersection of the plane $y = y_0$ and the surface $z = f(x, y)$. On this curve, the partial derivative

$$f_x(x, y_0)$$ Slope in x-direction

represents the slope in the plane $y = y_0$, as shown in Figure 7.24(a). Similarly, if $x = x_0$, then

$$z = f(x_0, y)$$ Function of one variable

is a function of one variable. Its graph is the intersection of the plane $x = x_0$ and the surface $z = f(x, y)$. On this curve, the partial derivative

$$f_y(x_0, y)$$ Slope in y-direction

represents the slope in the plane $x = x_0$, as shown in Figure 7.24(b).

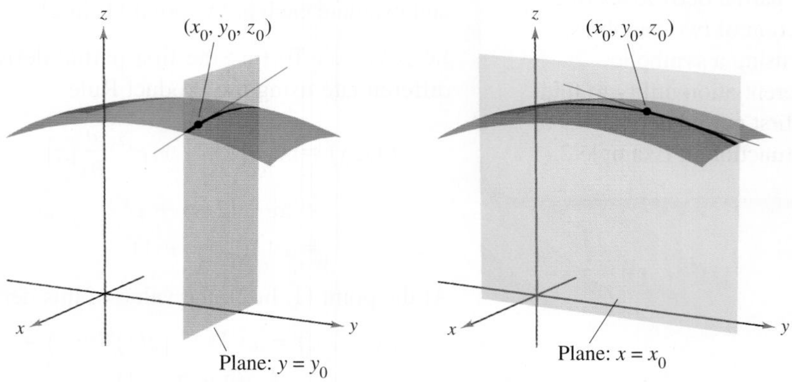

(a) $f_x(x, y_0)$ = slope in x-direction

(b) $f_y(x_0, y)$ = slope in y-direction

FIGURE 7.24

Informally, $f_x(x_0, y_0)$ and $f_y(x_0, y_0)$ at the point (x_0, y_0, z_0) denote the **slopes of the surface in the x- and y-directions,** respectively.

Guidelines for Finding the Slopes of a Surface at a Point

Let (x_0, y_0, z_0) be a point on the surface of

$$z = f(x, y).$$

1. Find the partial derivatives of f with respect to x and y.
2. The slope of the x-direction at (x_0, y_0, z_0) is $f_x(x_0, y_0)$.
3. The slope of the y-direction at (x_0, y_0, z_0) is $f_y(x_0, y_0)$.

Example 3 Finding Slopes in the x- and y-Directions

Find the slopes of the surface given by

$$f(x, y) = -\frac{x^2}{2} - y^2 + \frac{25}{8}$$

at the point $\left(\frac{1}{2}, 1, 2\right)$ in

a. the x-direction.

b. the y-direction.

SOLUTION

a. To find the slope in the x-direction, hold y constant and differentiate with respect to x to obtain

$$f_x(x, y) = -x. \qquad \text{Partial derivative with respect to } x$$

At the point $\left(\frac{1}{2}, 1, 2\right)$, the slope in the x-direction is

$$f_x\left(\tfrac{1}{2}, 1\right) = -\tfrac{1}{2} \qquad \text{Slope in } x\text{-direction}$$

as shown in Figure 7.25(a).

b. To find the slope in the y-direction, hold x constant and differentiate with respect to y to obtain

$$f_y(x, y) = -2y. \qquad \text{Partial derivative with respect to } y$$

At the point $\left(\frac{1}{2}, 1, 2\right)$, the slope in the y-direction is

$$f_y\left(\tfrac{1}{2}, 1\right) = -2 \qquad \text{Slope in } y\text{-direction}$$

as shown in Figure 7.25(b).

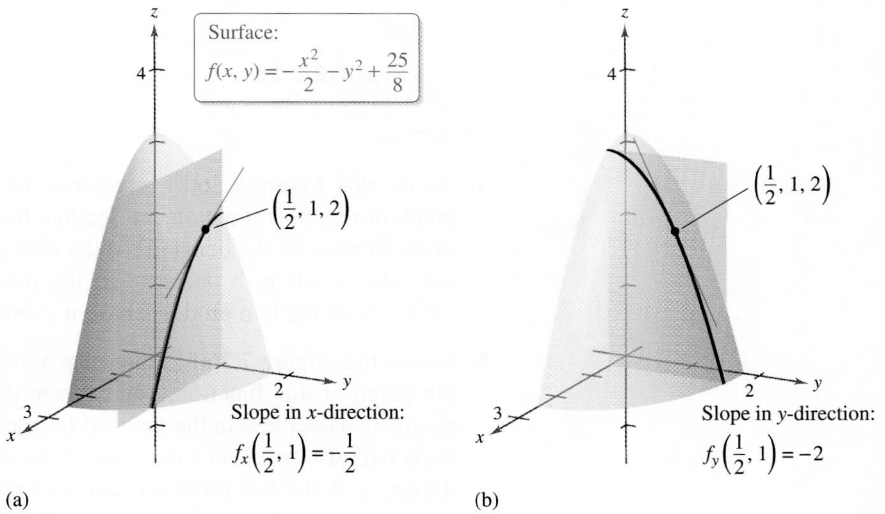

FIGURE 7.25

✓ Checkpoint 3

Find the slopes of the surface given by

$$f(x, y) = 4x^2 + 9y^2 + 36$$

at the point $(1, -1, 49)$ in

a. the x-direction.

b. the y-direction.

In 2010, Subway was chosen as the number one franchise by *Entrepreneur* magazine. In early 2011, Subway had more than 34,000 franchises worldwide. What type of product would be complementary to a Subway sandwich? What type of product would be a substitute?

Consumer products in the same market or in related markets can be classified as **complementary** or **substitute products.** When two products have a complementary relationship, an increase in the sale of one product will be accompanied by an increase in the sale of the other product. For instance, Blu-ray™ players and Blu-ray™ discs have a complementary relationship.

When two products have a substitute relationship, an increase in the sale of one product will be accompanied by a decrease in the sale of the other product. For instance, Blu-ray Disc™ players and DVD players both compete in the same home entertainment market, and you would expect a drop in the price of one to be a deterrent to the sale of the other.

Example 4 Examining Demand Functions

The demand functions for two products are represented by

$$x_1 = f(p_1, p_2) \quad \text{and} \quad x_2 = g(p_1, p_2)$$

where p_1 and p_2 are the prices per unit for the two products, and x_1 and x_2 are the numbers of units sold. The graphs of two different demand functions for x_1 are shown in Figure 7.26. Use them to classify the products as complementary or substitute products.

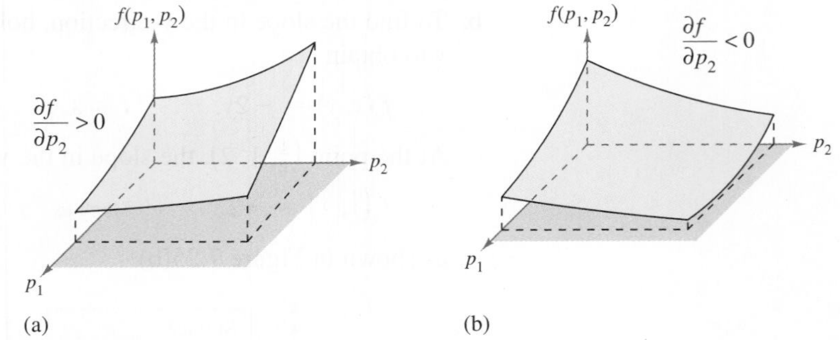

FIGURE 7.26

SOLUTION

a. Notice that Figure 7.26(a) represents the demand for the *first product*. From the graph of this function, you can see that for a fixed price p_1, an increase in p_2 results in an increase in the demand for the first product. Remember that an increase in p_2 will also result in a decrease in the demand for the second product. So, when $\partial f/\partial p_2 > 0$, the two products have a *substitute* relationship.

b. Notice that Figure 7.26(b) represents a different demand for the *first product*. From the graph of this function, you can see that for a fixed price p_1, an increase in p_2 results in a decrease in the demand for the first product. Remember that an increase in p_2 will also result in a decrease in the demand for the second product. So, when $\partial f/\partial p_2 < 0$, the two products have a *complementary* relationship.

✓ Checkpoint 4

Determine if the demand functions below describe a complementary or a substitute product relationship.

$$x_1 = 100 - 2p_1 + 1.5p_2$$
$$x_2 = 145 + \tfrac{1}{2}p_1 - \tfrac{3}{4}p_2$$

British Retail Photography/Alamy

Functions of Three Variables

The concept of a partial derivative can be extended naturally to functions of three or more variables. For instance, the function

$$w = f(x, y, z) \qquad \text{Function of three variables}$$

has three partial derivatives, each of which is formed by considering two of the variables to be constant. That is, to define the partial derivative of w with respect to x, consider y *and* z to be constant and write

$$\frac{\partial w}{\partial x} = f_x(x, y, z) = \lim_{\Delta x \to 0} \frac{f(x + \Delta x, y, z) - f(x, y, z)}{\Delta x}.$$

To define the partial derivative of w with respect to y, consider x *and* z to be constant and write

$$\frac{\partial w}{\partial y} = f_y(x, y, z) = \lim_{\Delta y \to 0} \frac{f(x, y + \Delta y, z) - f(x, y, z)}{\Delta y}.$$

To define the partial derivative of w with respect to z, consider x *and* y to be constant and write

$$\frac{\partial w}{\partial z} = f_z(x, y, z) = \lim_{\Delta z \to 0} \frac{f(x, y, z + \Delta z) - f(x, y, z)}{\Delta z}.$$

TECH TUTOR

A symbolic differentiation utility can be used to find the partial derivatives of a function of three or more variables. Try using a symbolic differentiation utility to find the partial derivative $f_y(x, y, z)$ for the function in Example 5.

Example 5 Finding Partial Derivatives of a Function

Find the three partial derivatives of the function

$$w = xe^{xy+2z}.$$

SOLUTION Holding y and z constant, you obtain

$$\begin{aligned}
\frac{\partial w}{\partial x} &= x\frac{\partial}{\partial x}[e^{xy+2z}] + e^{xy+2z}\frac{\partial}{\partial x}[x] && \text{Apply Product Rule.}\\
&= x(e^{xy+2z})(y) + e^{xy+2z}(1) && \text{Hold } y \text{ and } z \text{ constant.}\\
&= (xy + 1)e^{xy+2z}. && \text{Simplify.}
\end{aligned}$$

Holding x and z constant, you obtain

$$\begin{aligned}
\frac{\partial w}{\partial y} &= x\frac{\partial}{\partial y}[e^{xy+2z}] && \text{Apply Constant Multiple Rule.}\\
&= x(e^{xy+2z})(x) && \text{Hold } x \text{ and } z \text{ constant.}\\
&= x^2e^{xy+2z}. && \text{Simplify.}
\end{aligned}$$

Holding x and y constant, you obtain

$$\begin{aligned}
\frac{\partial w}{\partial z} &= x\frac{\partial}{\partial z}[e^{xy+2z}] && \text{Apply Constant Multiple Rule.}\\
&= x(e^{xy+2z})(2) && \text{Hold } x \text{ and } y \text{ constant.}\\
&= 2xe^{xy+2z}. && \text{Simplify.}
\end{aligned}$$

✓ Checkpoint 5

Find the three partial derivatives of the function

$$w = x^2y\ln(xz).$$

■

In Example 5, the Product Rule is used only when finding the partial derivative with respect to x. For $\partial w/\partial y$ and $\partial w/\partial z$, x is considered to be constant, so the Constant Multiple Rule is used.

Higher-Order Partial Derivatives

As with ordinary derivatives, it is possible to take second-, third-, and higher-order partial derivatives of a function of several variables, provided such derivatives exist. Higher-order derivatives are denoted by the order in which the differentiation occurs. For instance, there are four different ways to find a second partial derivative of $z = f(x, y)$.

1. $\dfrac{\partial}{\partial x}\left(\dfrac{\partial f}{\partial x}\right) = \dfrac{\partial^2 f}{\partial x^2} = f_{xx}$ Differentiate twice with respect to x.

2. $\dfrac{\partial}{\partial y}\left(\dfrac{\partial f}{\partial y}\right) = \dfrac{\partial^2 f}{\partial y^2} = f_{yy}$ Differentiate twice with respect to y.

3. $\dfrac{\partial}{\partial y}\left(\dfrac{\partial f}{\partial x}\right) = \dfrac{\partial^2 f}{\partial y \partial x} = f_{xy}$ Differentiate first with respect to x and then with respect to y.

4. $\dfrac{\partial}{\partial x}\left(\dfrac{\partial f}{\partial y}\right) = \dfrac{\partial^2 f}{\partial x \partial y} = f_{yx}$ Differentiate first with respect to y and then with respect to x.

The third and fourth cases are **mixed partial derivatives.** Notice that with the two types of notation for mixed partials, different conventions are used for indicating the order of differentiation. For instance, the partial derivative

$$\frac{\partial}{\partial y}\left(\frac{\partial f}{\partial x}\right) = \frac{\partial^2 f}{\partial y \partial x}$$ Right-to-left order

indicates differentiation with respect to x first, but the partial derivative

$$(f_y)_x = f_{yx}$$ Left-to-right order

indicates differentiation with respect to y first. To remember this, note that in each case you differentiate first with respect to the variable "nearest" f.

Example 6 Finding Second Partial Derivatives

Find the second partial derivatives of

$$f(x, y) = 3xy^2 - 2y + 5x^2y^2$$

and determine the value of $f_{xy}(-1, 2)$.

SOLUTION Begin by finding the first partial derivatives.

$$f_x(x, y) = 3y^2 + 10xy^2 \qquad f_y(x, y) = 6xy - 2 + 10x^2y$$

Then, differentiating with respect to x and y produces

$$f_{xx}(x, y) = 10y^2, \qquad f_{yy}(x, y) = 6x + 10x^2,$$

$$f_{xy}(x, y) = 6y + 20xy, \qquad f_{yx}(x, y) = 6y + 20xy.$$

Finally, the value of $f_{xy}(x, y)$ at the point $(-1, 2)$ is

$$f_{xy}(-1, 2) = 6(2) + 20(-1)(2) = 12 - 40 = -28.$$

✓ Checkpoint 6

Find the second partial derivatives of

$$f(x, y) = 4x^2y^2 + 2x + 4y^2.$$

■

Notice in Example 6 that the two mixed partials are equal. It can be shown that when a function has continuous second partial derivatives, then the order in which the partial derivatives are taken is irrelevant.

A function of two variables has two first partial derivatives and four second partial derivatives. For a function of three variables, there are three first partials

$$f_x, \quad f_y, \quad \text{and} \quad f_z$$

and nine second partials

$$f_{xx}, \quad f_{xy}, \quad f_{xz}, \quad f_{yx}, \quad f_{yy}, \quad f_{yz}, \quad f_{zx}, \quad f_{zy}, \quad \text{and} \quad f_{zz}$$

of which six are mixed partials. To find partial derivatives of order three and higher, follow the same pattern used to find second partial derivatives. For instance, if $z = f(x, y)$, then

$$z_{xxx} = \frac{\partial}{\partial x}\left(\frac{\partial^2 f}{\partial x^2}\right) = \frac{\partial^3 f}{\partial x^3} \quad \text{and} \quad z_{xxy} = \frac{\partial}{\partial y}\left(\frac{\partial^2 f}{\partial x^2}\right) = \frac{\partial^3 f}{\partial y \partial x^2}.$$

Example 7 Finding Second Partial Derivatives

Find the second partial derivatives of

$$f(x, y, z) = ye^x + x \ln z.$$

SOLUTION Begin by finding the first partial derivatives.

$$f_x(x, y, z) = ye^x + \ln z, \quad f_y(x, y, z) = e^x, \quad f_z(x, y, z) = \frac{x}{z}$$

Then, differentiate with respect to x, y, and z to find the nine second partial derivatives.

$$f_{xx}(x, y, z) = ye^x, \quad f_{xy}(x, y, z) = e^x, \quad f_{xz}(x, y, z) = \frac{1}{z}$$

$$f_{yx}(x, y, z) = e^x, \quad f_{yy}(x, y, z) = 0, \quad f_{yz}(x, y, z) = 0$$

$$f_{zx}(x, y, z) = \frac{1}{z}, \quad f_{zy}(x, y, z) = 0, \quad f_{zz}(x, y, z) = -\frac{x}{z^2}$$

✓Checkpoint 7

Find the second partial derivatives of $f(x, y, z) = xe^y + 2xz + y^2$. ■

SUMMARIZE (Section 7.4)

1. State the definition of partial derivatives of a function of two variables *(page 444)*. For an example of finding the partial derivatives of a function of two variables, see Example 1.
2. Describe the notation used for first partial derivatives *(page 445)*. For examples of this notation, see Examples 1 and 2.
3. State the guidelines for finding the slopes of a surface at a point *(page 446)*. For an example of finding slopes, see Example 3.
4. Describe a real-life example of how partial derivatives can be used to examine the demand functions of two products *(page 448, Example 4)*.
5. Explain how to find the partial derivatives of a function of three variables *(page 449)*. For an example of finding the partial derivatives of a function of three variables, see Example 5.
6. List the different ways to find the second partial derivatives of a function of two variables *(page 450)*. For an example of finding the second partial derivatives of a function of two variables, see Example 6.

SKILLS WARM UP 7.4

The following warm-up exercises involve skills that were covered in earlier sections. You will use these skills in the exercise set for this section. For additional help, review Sections 2.2, 2.4, 2.5, 4.3, and 4.5.

In Exercises 1–8, find the derivative of the function.

1. $f(x) = \sqrt{x^2 + 3}$
2. $g(x) = (3 - x^2)^3$
3. $g(t) = te^{2t+1}$
4. $f(x) = e^{2x}\sqrt{1 - e^{2x}}$
5. $f(x) = \ln(3 - 2x)$
6. $u(t) = \ln\sqrt{t^3 - 6t}$
7. $g(x) = \dfrac{5x^2}{(4x - 1)^2}$
8. $f(x) = \dfrac{(x + 2)^3}{(x^2 - 9)^2}$

In Exercises 9 and 10, evaluate the derivative at the point (2, 4).

9. $f(x) = x^2e^{x-2}$
10. $g(x) = x\sqrt{x^2 - x + 2}$

Exercises 7.4

See www.CalcChat.com for worked-out solutions to odd-numbered exercises.

Finding Partial Derivatives In Exercises 1–14, find the first partial derivatives. *See Example 1.*

1. $z = 3x + 5y - 1$
2. $z = x^2 - 2y$
3. $f(x, y) = 3x - 6y^2$
4. $f(x, y) = x + 4y^{3/2}$
5. $f(x, y) = \dfrac{x}{y}$
6. $f(x, y) = \dfrac{xy}{x^2 + y^2}$
7. $f(x, y) = \sqrt{x^2 + y^2}$
8. $z = x\sqrt{y}$
9. $z = x^2e^{2y}$
10. $z = xe^{x+y}$
11. $h(x, y) = e^{-(x^2+y^2)}$
12. $g(x, y) = e^{x/y}$
13. $z = \ln\dfrac{x + y}{x - y}$
14. $g(x, y) = \ln(x^2 + y^2)$

Finding and Evaluating Partial Derivatives In Exercises 15–22, find the first partial derivatives and evaluate each at the given point. *See Example 2.*

Function	*Point*
15. $f(x, y) = 3x^2 + xy - y^2$	(2, 1)
16. $f(x, y) = x^2 - 3xy + y^2$	(1, −1)
17. $f(x, y) = e^{3xy}$	(0, 4)
18. $f(x, y) = e^xy^2$	(0, 2)
19. $f(x, y) = \dfrac{xy}{x - y}$	(2, −2)
20. $f(x, y) = \dfrac{4xy}{\sqrt{x^2 + y^2}}$	(1, 0)
21. $f(x, y) = \ln(3x + 5y)$	(1, 0)
22. $f(x, y) = \ln\sqrt{xy}$	(−1, −1)

Finding Slopes in the *x*- and *y*-Directions In Exercises 23–26, find the slopes of the surface at the given point in (a) the *x*-direction and (b) the *y*-direction. *See Example 3.*

23. $z = xy$
(1, 2, 2)

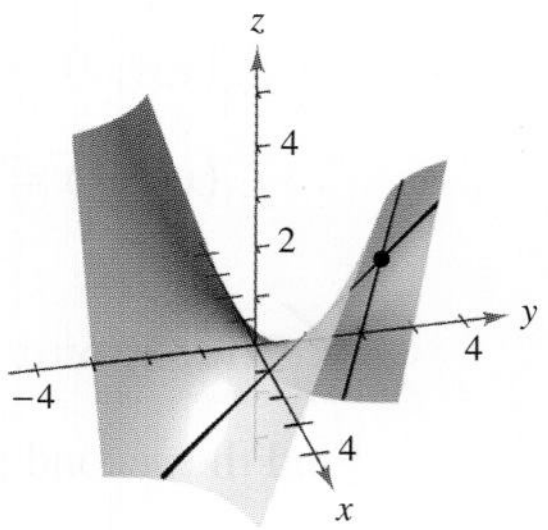

24. $z = \sqrt{25 - x^2 - y^2}$
(3, 0, 4)

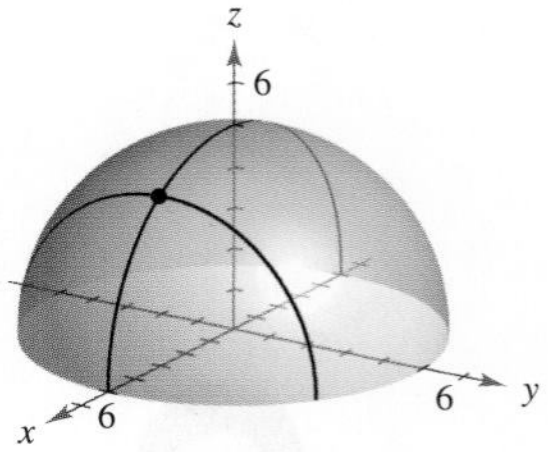

25. $z = 4 - x^2 - y^2$
(1, 1, 2)

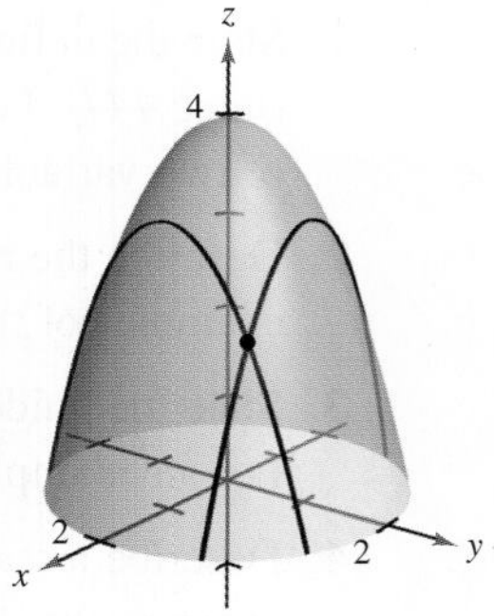

26. $z = x^2 - y^2$
(−2, 1, 3)

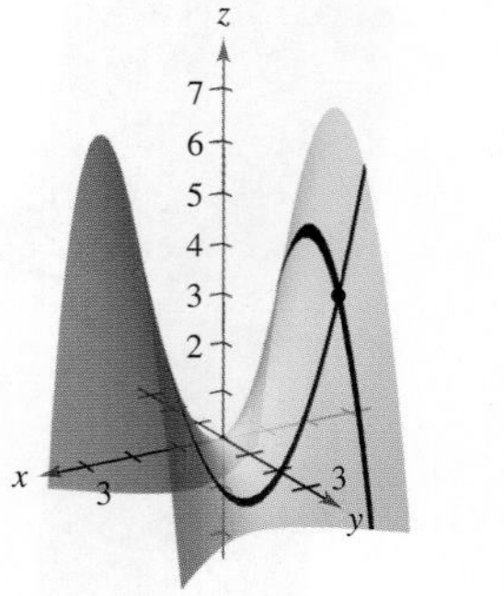

Finding Partial Derivatives In Exercises 27–30, find the first partial derivatives. *See Example 5.*

27. $w = xy^2z^4$
28. $w = x^3yz^2$
29. $w = \dfrac{2z}{x + y}$
30. $w = \dfrac{xy}{x + y + z}$

Finding and Evaluating Partial Derivatives In Exercises 31–38, find the first partial derivatives and evaluate each at the given point.

Function	*Point*
31. $w = 2xz^2 + 3xyz - 6y^2z$	$(1, -1, 2)$
32. $w = 3x^2y - 5xyz + 10yz^2$	$(3, 4, -2)$
33. $w = \sqrt{x^2 + y^2 + z^2}$	$(2, -1, 2)$
34. $w = \sqrt{3x^2 + y^2 - 2z^2}$	$(1, -2, 1)$
35. $w = y^3z^2e^{2x^2}$	$(\frac{1}{2}, -1, 2)$
36. $w = xye^{z^2}$	$(2, 1, 0)$
37. $w = \ln(5x + 2y^3 - 3z)$	$(4, 1, -1)$
38. $w = \ln\sqrt{x^2 + y^2 + z^2}$	$(3, 0, 4)$

Using First Partial Derivatives In Exercises 39–42, find values of x and y such that $f_x(x, y) = 0$ and $f_y(x, y) = 0$ simultaneously.

39. $f(x, y) = x^2 + 4xy + y^2 - 4x + 16y + 3$

40. $f(x, y) = 3x^3 - 12xy + y^3$

41. $f(x, y) = \dfrac{1}{x} + \dfrac{1}{y} + xy$

42. $f(x, y) = \ln(x^2 + y^2 + 1)$

Finding Second Partial Derivatives In Exercises 43–50, find the four second partial derivatives. *See Example 6.*

43. $z = x^3 - 4y^2$

44. $z = 2x^2 + y^5$

45. $z = x^2 - 2xy + 3y^2$

46. $z = y^3 - 4xy^2 - 1$

47. $z = (3x^4 - 2y^3)^3$

48. $z = \sqrt{9 - x^2 - y^2}$

49. $z = \dfrac{x^2 - y^2}{2xy}$

50. $z = \dfrac{x}{x + y}$

Finding and Evaluating Second Partial Derivatives In Exercises 51–54, find the four second partial derivatives and evaluate each at the given point.

Function	*Point*
51. $f(x, y) = x^4 - 3x^2y^2 + y^2$	$(1, 0)$
52. $f(x, y) = x^3 + 2xy^3 - 3y$	$(3, 2)$
53. $f(x, y) = y^3e^{x^2}$	$(1, -1)$
54. $f(x, y) = x^2e^y$	$(-1, 0)$

Finding Second Partial Derivatives In Exercises 55–58, find the nine second partial derivatives. *See Example 7.*

55. $w = x^2 - 3xy + 4yz + z^3$

56. $w = x^2y^3 + 2xyz - 3yz$

57. $w = \dfrac{4xz}{x + y}$

58. $w = \dfrac{xy}{x + y + z}$

59. Marginal Cost A company manufactures two models of bicycles: a mountain bike and a racing bike. The cost function for producing x mountain bikes and y racing bikes is given by

$$C = 10\sqrt{xy} + 149x + 189y + 675.$$

(a) Find the marginal costs ($\partial C/\partial x$ and $\partial C/\partial y$) when $x = 120$ and $y = 160$.

(b) When additional production is required, which model of bicycle results in the cost increasing at a higher rate? How can this be determined from the cost model?

60. Marginal Revenue A pharmaceutical corporation has two locations that produce the same over-the-counter medicine. If x_1 and x_2 are the numbers of units produced at location 1 and location 2, respectively, then the total revenue for the product is given by

$$R = 200x_1 + 200x_2 - 4x_1^2 - 8x_1x_2 - 4x_2^2.$$

When $x_1 = 4$ and $x_2 = 12$, find

(a) the marginal revenue for location 1, $\partial R/\partial x_1$.

(b) the marginal revenue for location 2, $\partial R/\partial x_2$.

61. Marginal Productivity Consider the Cobb-Douglas production function

$$f(x, y) = 200x^{0.7}y^{0.3}.$$

When $x = 1000$ and $y = 500$, find

(a) the marginal productivity of labor, $\partial f/\partial x$.

(b) the marginal productivity of capital, $\partial f/\partial y$.

62. Marginal Productivity Repeat Exercise 61 for the production function given by $f(x, y) = 100x^{0.75}y^{0.25}$.

Complementary and Substitute Products In Exercises 63 and 64, determine whether the demand functions describe complementary or substitute product relationships. Using the notation of Example 4, let x_1 and x_2 be the demands for two products whose prices are p_1 and p_2, respectively. *See Example 4.*

63. $x_1 = 150 - 2p_1 - \frac{5}{2}p_2, \quad x_2 = 350 - \frac{3}{2}p_1 - 3p_2$

64. $x_1 = 150 - 2p_1 + 1.8p_2, \quad x_2 = 350 + \frac{3}{4}p_1 - 1.9p_2$

65. Expenditures The expenditures z (in billions of dollars) for spectator sports from 2004 through 2009 can be modeled by

$$z = 0.62x - 0.41y + 0.38$$

where x is the expenditures on amusement parks and campgrounds, and y is the expenditures on live entertainment (excluding sports), both in billions of dollars. *(Source: U.S. Bureau of Economic Analysis)*

(a) Find $\partial z/\partial x$ and $\partial z/\partial y$.

(b) Interpret the partial derivatives in the context of the problem.

66. Shareholder's Equity The shareholder's equity z (in millions of dollars) for Skechers from 2001 through 2009 can be modeled by

$$z = 0.175x - 0.772y - 275$$

where x is the sales (in millions of dollars) and y is the total assets (in millions of dollars). *(Source: Skechers U.S.A. Inc.)*

(a) Find $\partial z/\partial x$ and $\partial z/\partial y$.

(b) Interpret the partial derivatives in the context of the problem.

67. Psychology Early in the twentieth century, an intelligence test called the *Stanford-Binet Test* (more commonly known as the *IQ test*) was developed. In this test, an individual's mental age M is divided by the individual's chronological age C and the quotient is multiplied by 100. The result is the individual's IQ.

$$IQ(M, C) = \frac{M}{C} \times 100$$

Find the partial derivatives of IQ with respect to M and with respect to C. Evaluate the partial derivatives at the point $(12, 10)$ and interpret the result. *(Source: Adapted from Bernstein/Clark-Stewart/Roy/Wickens,* Psychology, *Fourth Edition)*

68. HOW DO YOU SEE IT? Use the graph of the surface to determine the sign of each partial derivative. Explain your reasoning.

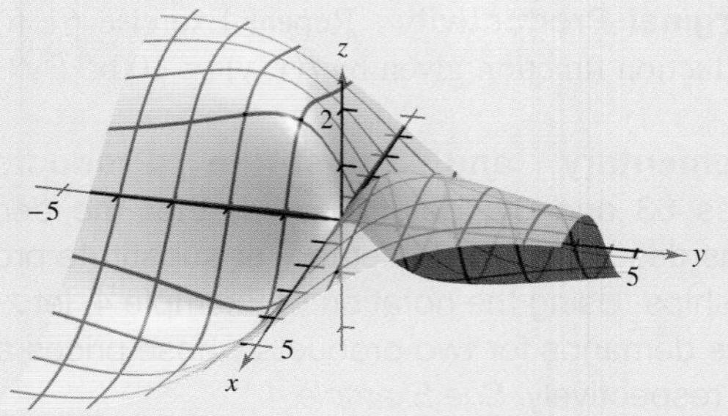

(a) $f_x(4, 1)$ (b) $f_y(4, 1)$

(c) $f_x(-1, -2)$ (d) $f_y(-1, -2)$

69. Investment The value of an investment of \$1000 earning 10% interest compounded annually is

$$V(I, R) = 1000\left[\frac{1 + 0.10(1 - R)}{1 + I}\right]^{10}$$

where I is the annual rate of inflation and R is the tax rate for the person making the investment. Calculate $V_I(0.03, 0.28)$ and $V_R(0.03, 0.28)$. Determine whether the tax rate or the rate of inflation is the greater "negative" influence on the growth of the investment.

70. Think About It Let N be the number of applicants to a university, p the charge for food and housing at the university, and t the tuition. Suppose that N is a function of p and t such that $\partial N/\partial p < 0$ and $\partial N/\partial t < 0$. How would you interpret the fact that both partials are negative?

71. Marginal Utility The utility function $U = f(x, y)$ is a measure of the utility (or satisfaction) derived by a person from the consumption of two products x and y. Suppose the utility function is given by $U = -5x^2 + xy - 3y^2$.

(a) Determine the marginal utility of product x.

(b) Determine the marginal utility of product y.

(c) When $x = 2$ and $y = 3$, should a person consume one more unit of product x or one more unit of product y? Explain your reasoning.

(d) Use a three-dimensional graphing utility to graph the function. Interpret the marginal utilities of products x and y graphically.

Business Capsule

In 1996, twin sisters Izzy and Coco Tihanyi started Surf Diva, a surf school and apparel company for women and girls, in La Jolla, California. To advertise their business, they would donate surf lessons and give the surf report on local radio stations in exchange for air time. Today, they have schools and surf camps in Los Angeles, San Diego, and Costa Rica. Their clothing line can be found in the Surf Diva Boutique as well as other surf and specialty shops, sporting goods stores, and airport gift shops.

72. Research Project Use your school's library, the Internet, or some other reference source to research a company that increased the demand for its product by creative advertising. Write a paper about the company. Use graphs to show how a change in demand is related to a change in the marginal utility of a product or service.

Courtesy of Surf Diva

7.5 Extrema of Functions of Two Variables

- Understand the relative extrema of functions of two variables.
- Use the First-Partials Test to find the relative extrema of functions of two variables.
- Use the Second-Partials Test to find the relative extrema of functions of two variables.
- Use relative extrema to answer questions about real-life situations.

In Exercise 43 on page 462, you will find the dimensions of a rectangular package of maximum volume that can be sent by a shipping company.

Relative Extrema

Earlier in the text, you learned how to use derivatives to find the relative minimum and relative maximum values of a function of a single variable. In this section, you will learn how to use partial derivatives to find the relative minimum and relative maximum values of a function of two variables.

Relative Extrema of a Function of Two Variables

Let f be a function defined on a region containing (x_0, y_0). The function f has a **relative maximum** at (x_0, y_0) when there is a circular region R centered at (x_0, y_0) such that

$$f(x, y) \le f(x_0, y_0) \qquad f \text{ has a relative maximum at } (x_0, y_0).$$

for all (x, y) in R. The function f has a **relative minimum** at (x_0, y_0) when there is a circular region R centered at (x_0, y_0) such that

$$f(x, y) \ge f(x_0, y_0) \qquad f \text{ has a relative minimum at } (x_0, y_0).$$

for all (x, y) in R.

To say that f has a relative maximum at (x_0, y_0) means that the point (x_0, y_0, z_0) is at least as high as all nearby points on the graph of $z = f(x, y)$. Similarly, f has a relative minimum at (x_0, y_0) when (x_0, y_0, z_0) is at least as low as all nearby points on the graph. (See Figure 7.27.)

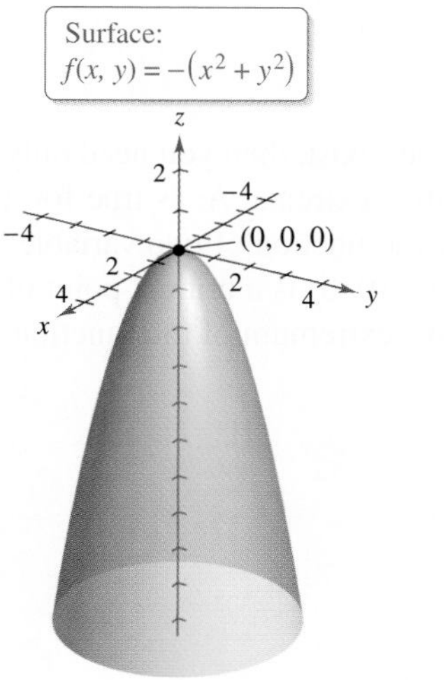

f has an absolute maximum at (0, 0, 0).

FIGURE 7.28

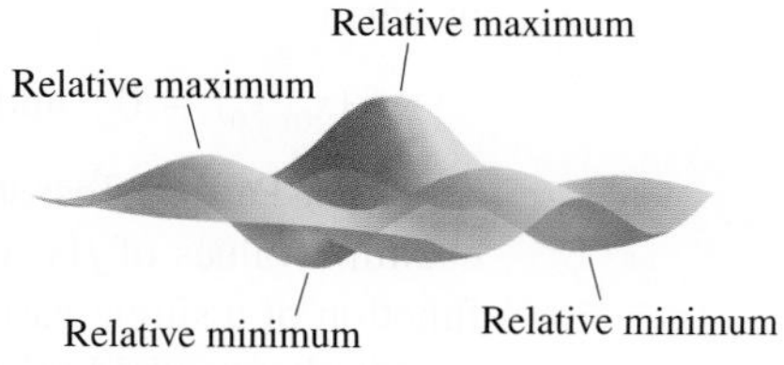

Relative Extrema

FIGURE 7.27

As in single-variable calculus, you need to distinguish between relative extrema and absolute extrema of a function of two variables. The number $f(x_0, y_0)$ is an absolute maximum of f in the region R when it is greater than or equal to all other function values in the region. (An absolute minimum of f in a region is defined similarly.) For instance, the function

$$f(x, y) = -(x^2 + y^2)$$

is a paraboloid, opening downward, with vertex at (0, 0, 0). (See Figure 7.28.) The number

$$f(0, 0) = 0$$

is an absolute maximum of the function over the entire xy-plane.

The First-Partials Test for Relative Extrema

To locate the relative extrema of a function of two variables, you can use a procedure that is similar to the First-Derivative Test used for functions of a single variable.

First-Partials Test for Relative Extrema

If f has a relative extremum at (x_0, y_0) on an open region R in the xy-plane, and the first partial derivatives of f exist in R, then

$$f_x(x_0, y_0) = 0$$

and

$$f_y(x_0, y_0) = 0$$

as shown in Figure 7.29.

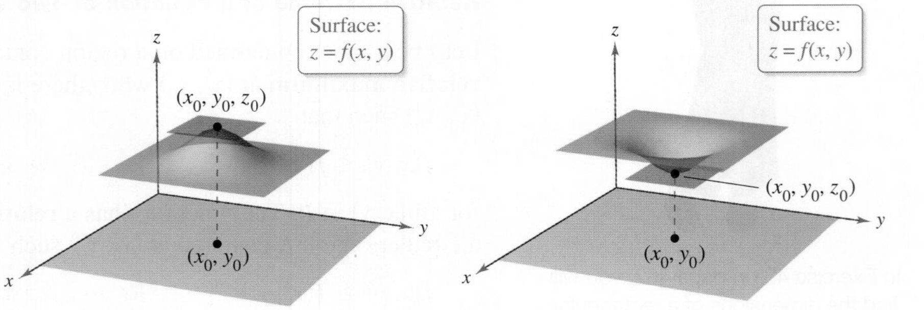

Relative maximum Relative minimum

FIGURE 7.29

An *open* region in the xy-plane is similar to an open interval on the real number line. For instance, the region R consisting of the interior of the circle $x^2 + y^2 = 1$ is an open region. If the region R consists of the interior of the circle *and* the points on the circle, then it is a *closed* region.

A point (x_0, y_0) is a **critical point** of f when $f_x(x_0, y_0)$ or $f_y(x_0, y_0)$ is undefined or when

$$f_x(x_0, y_0) = 0 \quad \text{and} \quad f_y(x_0, y_0) = 0. \qquad \text{Critical point}$$

The First-Partials Test states that if the first partial derivatives exist, then you need only examine values of $f(x, y)$ at critical points to find the relative extrema. As is true for a function of a single variable, however, the critical points of a function of two variables do not always yield relative extrema. For instance, the point $(0, 0)$ is a critical point of the surface shown in Figure 7.30, but $f(0, 0)$ is not a relative extremum of the function. Such points are called **saddle points** of the function.

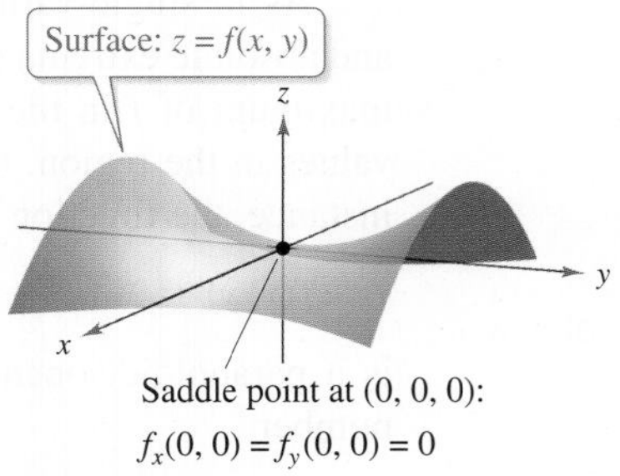

FIGURE 7.30

Example 1 Finding Relative Extrema

Find the relative extrema of

$$f(x, y) = 2x^2 + y^2 + 8x - 6y + 20.$$

SOLUTION Begin by finding the first partial derivatives of f.

$$f_x(x, y) = 4x + 8 \quad \text{and} \quad f_y(x, y) = 2y - 6$$

Because these partial derivatives are defined for all points in the xy-plane, the only critical points are those for which both first partial derivatives are zero. To locate these points, set $f_x(x, y)$ and $f_y(x, y)$ equal to 0, and solve the resulting system of equations.

$4x + 8 = 0$ Set $f_x(x, y)$ equal to 0.

$2y - 6 = 0$ Set $f_y(x, y)$ equal to 0.

The solution of this system is $x = -2$ and $y = 3$. So, the point $(-2, 3)$ is the only critical number of f. From the graph of the function, shown in Figure 7.31, you can see that this critical point yields a relative minimum of the function. So, the function has only one relative extremum, which is

$f(-2, 3) = 3.$ Relative minimum

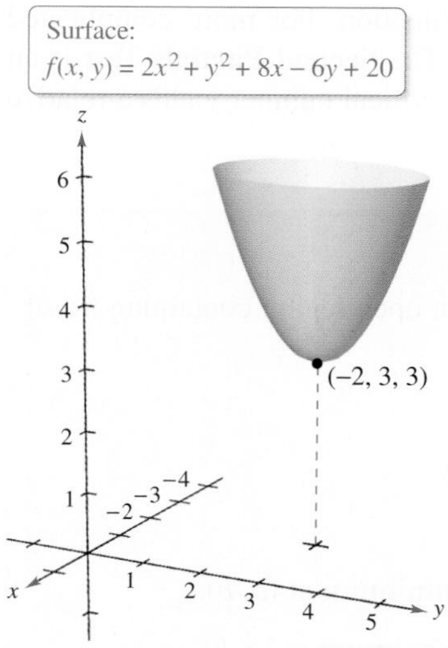

FIGURE 7.31

✓ Checkpoint 1

Find the relative extrema of

$$f(x, y) = x^2 + 2y^2 + 16x - 8y + 8.$$ ■

Example 1 shows a relative minimum occurring at one type of critical point—the type for which both $f_x(x, y)$ and $f_y(x, y)$ are zero. The next example shows a relative maximum that occurs at the other type of critical point—the type for which either $f_x(x, y)$ or $f_y(x, y)$ is undefined.

Example 2 Finding Relative Extrema

Find the relative extrema of

$$f(x, y) = 1 - (x^2 + y^2)^{1/3}.$$

SOLUTION Begin by finding the first partial derivatives of f.

$$f_x(x, y) = -\frac{2x}{3(x^2 + y^2)^{2/3}} \quad \text{and} \quad f_y(x, y) = -\frac{2y}{3(x^2 + y^2)^{2/3}}$$

These partial derivatives are defined for all points in the xy-plane *except* the point $(0, 0)$. So, $(0, 0)$ is a critical point of f. Moreover, this is the only critical point, because there are no other values of x and y for which either partial derivative is undefined or for which both partial derivatives are zero. From the graph of the function, shown in Figure 7.32, you can see that this critical point yields a relative maximum of the function. So, the function has only one relative extremum, which is

$f(0, 0) = 1.$ Relative maximum

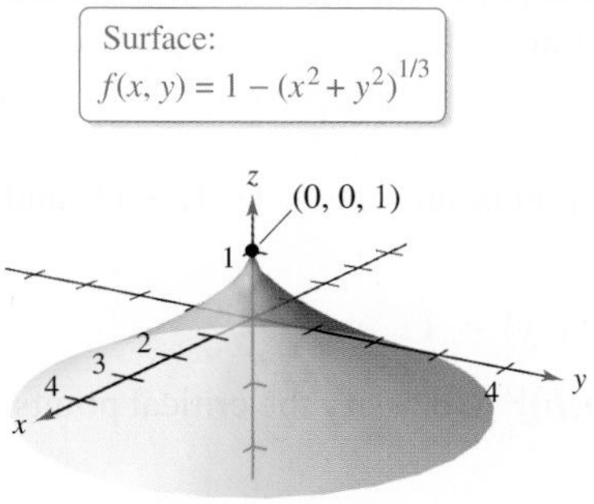

$f_x(x, y)$ and $f_y(x, y)$ are undefined at $(0, 0)$.
FIGURE 7.32

✓ Checkpoint 2

Find the relative extrema of

$$f(x, y) = \sqrt{1 - \frac{x^2}{16} - \frac{y^2}{4}}.$$ ■

The Second-Partials Test for Relative Extrema

For functions such as those in Examples 1 and 2, you can determine the *types* of extrema at the critical points by sketching the graph of the function. For more complicated functions, a graphical approach is not so easy to use. The **Second-Partials Test** is an analytical test that can be used to determine whether a critical number yields a relative minimum, a relative maximum, or neither.

Second-Partials Test for Relative Extrema

Let f have continuous second partial derivatives on an open region containing (a, b) for which

$$f_x(a, b) = 0 \quad \text{and} \quad f_y(a, b) = 0.$$

To test for relative extrema of f, consider the quantity

$$d = f_{xx}(a, b)\, f_{yy}(a, b) - [f_{xy}(a, b)]^2.$$

1. If $d > 0$ and $f_{xx}(a, b) > 0$, then f has a **relative minimum** at (a, b).

2. If $d > 0$ and $f_{xx}(a, b) < 0$, then f has a **relative maximum** at (a, b).

3. If $d < 0$, then $(a, b, f(a, b))$ is a **saddle point.**

4. The test gives no information when $d = 0$.

Note in the Second-Partials Test that if $d > 0$, then $f_{xx}(a, b)$ and $f_{yy}(a, b)$ must have the same sign. So, you can replace $f_{xx}(a, b)$ with $f_{yy}(a, b)$ in the first two parts of the test.

ALGEBRA TUTOR

For help in solving the system of equations

$$y - x^3 = 0$$
$$x - y^3 = 0$$

in Example 3, see Example 1(a) in the *Chapter 7 Algebra Tutor*, on page 496.

Example 3 Applying the Second-Partials Test

Find the relative extrema and saddle points of $f(x, y) = xy - \frac{1}{4}x^4 - \frac{1}{4}y^4$.

SOLUTION Begin by finding the critical points of f. Because

$$f_x(x, y) = y - x^3 \quad \text{and} \quad f_y(x, y) = x - y^3$$

are defined for all points in the xy-plane, the only critical points are those for which both first partial derivatives are zero. By solving the equations

$$y - x^3 = 0 \quad \text{and} \quad x - y^3 = 0$$

simultaneously, you can determine that the critical points are $(1, 1)$, $(-1, -1)$, and $(0, 0)$. Furthermore, because

$$f_{xx}(x, y) = -3x^2, \quad f_{yy}(x, y) = -3y^2, \quad \text{and} \quad f_{xy}(x, y) = 1$$

you can use the quantity $d = f_{xx}(a, b) f_{yy}(a, b) - [f_{xy}(a, b)]^2$ to classify the critical points, as shown.

Critical Point	d	$f_{xx}(x, y)$	*Conclusion*
$(1, 1)$	$(-3)(-3) - 1 = 8$	-3	Relative maximum
$(-1, -1)$	$(-3)(-3) - 1 = 8$	-3	Relative maximum
$(0, 0)$	$(0)(0) - 1 = -1$	0	Saddle point

The graph of f is shown in Figure 7.33.

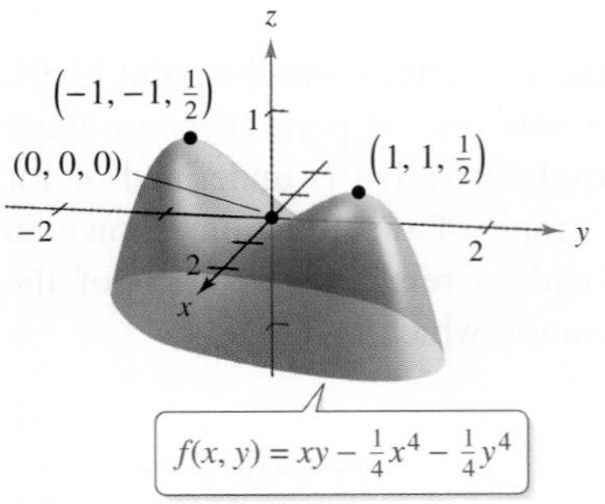

FIGURE 7.33

Checkpoint 3

Find the relative extrema and saddle points of $f(x, y) = \dfrac{y^2}{16} - \dfrac{x^2}{4}$. ■

Applications

Example 4 Finding a Maximum Profit

A company makes two substitute products whose demand functions are given by

$$x_1 = 200(p_2 - p_1) \qquad \text{Demand for product 1}$$
$$x_2 = 500 + 100p_1 - 180p_2 \qquad \text{Demand for product 2}$$

where p_1 and p_2 are the prices per unit (in dollars) and x_1 and x_2 are the numbers of units sold. The costs of producing the two products are \$0.50 and \$0.75 per unit, respectively. Find the prices that will yield a maximum profit.

SOLUTION The cost function is

$$\begin{aligned} C &= 0.5x_1 + 0.75x_2 && \text{Write cost function.} \\ &= 0.5(200)(p_2 - p_1) + 0.75(500 + 100p_1 - 180p_2) && \text{Substitute.} \\ &= 375 - 25p_1 - 35p_2. && \text{Simplify.} \end{aligned}$$

The revenue function is

$$\begin{aligned} R &= p_1x_1 + p_2x_2 && \text{Write revenue function.} \\ &= p_1(200)(p_2 - p_1) + p_2(500 + 100p_1 - 180p_2) && \text{Substitute.} \\ &= -200p_1^2 - 180p_2^2 + 300p_1p_2 + 500p_2. && \text{Simplify.} \end{aligned}$$

This implies that the profit function is

$$\begin{aligned} P &= R - C && \text{Write profit function.} \\ &= -200p_1^2 - 180p_2^2 + 300p_1p_2 + 500p_2 - (375 - 25p_1 - 35p_2) \\ &= -200p_1^2 - 180p_2^2 + 300p_1p_2 + 25p_1 + 535p_2 - 375. \end{aligned}$$

Next, find the first partial derivatives of P.

$$\frac{\partial P}{\partial p_1} = -400p_1 + 300p_2 + 25 \qquad \frac{\partial P}{\partial p_2} = 300p_1 - 360p_2 + 535$$

By setting the first partial derivatives equal to zero and solving the equations

$$\begin{aligned} -400p_1 + 300p_2 + 25 &= 0 \\ 300p_1 - 360p_2 + 535 &= 0 \end{aligned}$$

simultaneously, you can conclude that the solution is $p_1 \approx \$3.14$ and $p_2 \approx \$4.10$. From the graph of P shown in Figure 7.34, you can see that this critical number yields a maximum. So, the maximum profit is

$$P(p_1, p_2) \approx P(3.14, 4.10) = \$761.48.$$

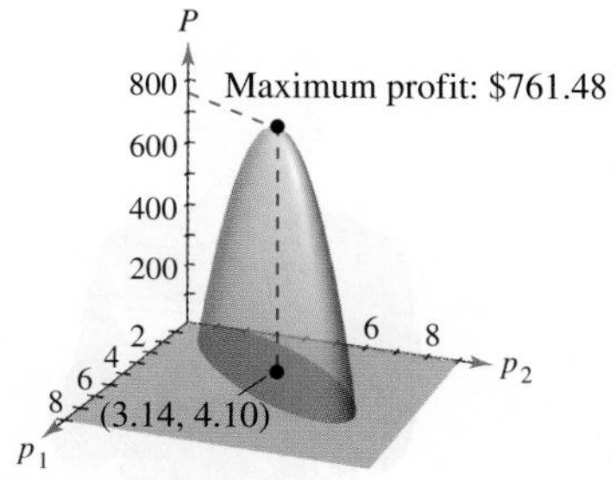

FIGURE 7.34

STUDY TIP

In Example 4, you can check that the two products are substitutes by observing that x_1 increases as p_2 increases and x_2 increases as p_1 increases.

ALGEBRA TUTOR

For help in solving the system of equations in Example 4, see Example 1(b) in the *Chapter 7 Algebra Tutor*, on page 496.

✓ Checkpoint 4

Find the prices that will yield a maximum profit for the products in Example 4 when the costs of producing the two products are \$0.75 and \$0.50 per unit, respectively. ■

In Example 4, to convince yourself that the maximum profit is \$761.48, try substituting other prices, such as $p_1 = \$2$ and $p_2 = \$3$, into the profit function. For each pair of prices, you will obtain a profit that is less than \$761.48.

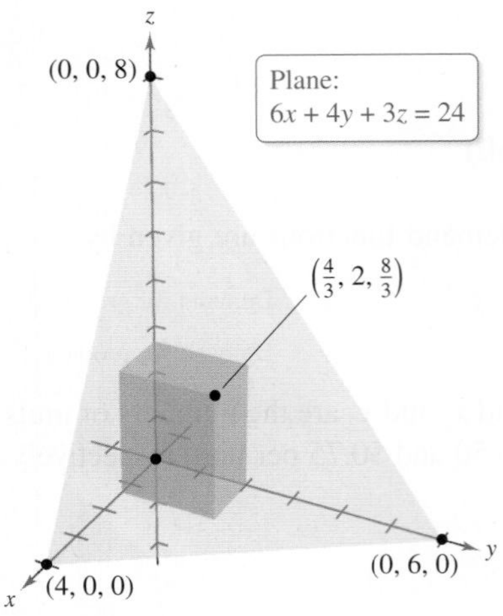

FIGURE 7.35

ALGEBRA TUTOR

For help in solving the system of equations

$$y(24 - 12x - 4y) = 0$$
$$x(24 - 6x - 8y) = 0$$

in Example 5, see Example 2(a) in the *Chapter 7 Algebra Tutor*, on page 497.

Example 5 Finding a Maximum Volume

A rectangular box is resting on the xy-plane with one vertex at the origin. The opposite vertex lies in the plane

$$6x + 4y + 3z = 24$$

as shown in Figure 7.35. Find the maximum volume of such a box.

SOLUTION Let x, y, and z represent the length, width, and height of the box. Because one vertex of the box lies in the plane given by $6x + 4y + 3z = 24$ or

$$z = \tfrac{1}{3}(24 - 6x - 4y) \qquad \text{Solve for } z.$$

you can write the volume of the box as a function of two variables.

$$\begin{aligned} V &= xyz && \text{Volume = (width)(length)(height)} \\ &= xy\left(\tfrac{1}{3}\right)(24 - 6x - 4y) && \text{Substitute for } z. \\ &= \tfrac{1}{3}(24xy - 6x^2y - 4xy^2) && \text{Simplify.} \end{aligned}$$

Next, find the first partial derivatives of V.

$$\begin{aligned} V_x &= \tfrac{1}{3}(24y - 12xy - 4y^2) && \text{Partial with respect to } x \\ &= \tfrac{1}{3}y(24 - 12x - 4y) && \text{Factor.} \\ V_y &= \tfrac{1}{3}(24x - 6x^2 - 8xy) && \text{Partial with respect to } y \\ &= \tfrac{1}{3}x(24 - 6x - 8y) && \text{Factor.} \end{aligned}$$

By solving the equations

$$\begin{aligned} \tfrac{1}{3}y(24 - 12x - 4y) &= 0 && \text{Set } V_x \text{ equal to 0.} \\ \tfrac{1}{3}x(24 - 6x - 8y) &= 0 && \text{Set } V_y \text{ equal to 0.} \end{aligned}$$

simultaneously, you can conclude that the solutions are $(0, 0)$, $(0, 6)$, $(4, 0)$, and $\left(\tfrac{4}{3}, 2\right)$. Using the Second-Partials Test, you can determine that the maximum volume occurs when the width is $x = \tfrac{4}{3}$ and the length is $y = 2$. For these values, the height of the box is

$$z = \tfrac{1}{3}\left[24 - 6\left(\tfrac{4}{3}\right) - 4(2)\right] = \tfrac{8}{3}.$$

So, the maximum volume is

$$V = xyz = \left(\tfrac{4}{3}\right)(2)\left(\tfrac{8}{3}\right) = \tfrac{64}{9} \text{ cubic units.}$$

✓ Checkpoint 5

Find the maximum volume of a box that is resting on the xy-plane with one vertex at the origin and the opposite vertex in the plane $2x + 4y + z = 8$. ■

SUMMARIZE (Section 7.5)

1. State the definition of relative extrema of a function of two variables *(page 455)*. For examples of relative extrema, see Examples 1 and 2.
2. State the First-Partials Test for relative extrema *(page 456)*. For examples of using the First-Partials Test, see Examples 1 and 2.
3. State the Second-Partials Test for relative extrema *(page 458)*. For examples of using the Second-Partials Test, see Examples 3 and 5.
4. Describe a real-life example of how relative extrema can be used to find a company's maximum profit *(page 459, Example 4)*.

SKILLS WARM UP 7.5 The following warm-up exercises involve skills that were covered in earlier sections. You will use these skills in the exercise set for this section. For additional help, review Section 7.4.

In Exercises 1–8, solve the system of equations.

1. $\begin{cases} 5x = 15 \\ 3x - 2y = 5 \end{cases}$

2. $\begin{cases} \frac{1}{2}y = 3 \\ -x + 5y = 19 \end{cases}$

3. $\begin{cases} x + y = 5 \\ x - y = -3 \end{cases}$

4. $\begin{cases} x + y = 8 \\ 2x - y = 4 \end{cases}$

5. $\begin{cases} 2x - y = 8 \\ 3x - 4y = 7 \end{cases}$

6. $\begin{cases} 2x - 4y = 14 \\ 3x + y = 7 \end{cases}$

7. $\begin{cases} x^2 + x = 0 \\ 2yx + y = 0 \end{cases}$

8. $\begin{cases} 3y^2 + 6y = 0 \\ xy + x + 2 = 0 \end{cases}$

In Exercises 9–14, find all first and second partial derivatives of the function.

9. $z = 4x^3 - 3y^2$

10. $z = 2x^5 - y^3$

11. $z = x^4 - \sqrt{xy} + 2y$

12. $z = 2x^2 - 3xy + y^2$

13. $z = ye^{xy^2}$

14. $z = xe^{xy}$

Exercises 7.5

See www.CalcChat.com for worked-out solutions to odd-numbered exercises.

Applying the Second-Partials Test In Exercises 1–18, find the critical points, relative extrema, and saddle points of the function. *See Examples 1, 2, and 3.*

1. $f(x, y) = x^2 - y^2 + 4x - 8y - 11$

2. $f(x, y) = x^2 + y^2 + 2x - 6y + 6$

3. $f(x, y) = \sqrt{x^2 + y^2 + 1}$

4. $f(x, y) = \sqrt{25 - (x - 2)^2 - y^2}$

5. $f(x, y) = (x - 1)^2 + (y - 3)^2$

6. $f(x, y) = 9 - (x - 3)^2 - (y + 2)^2$

7. $f(x, y) = 2x^2 + 2xy + y^2 + 2x - 3$

8. $f(x, y) = -x^2 - 5y^2 + 8x - 10y - 13$

9. $f(x, y) = -5x^2 + 4xy - y^2 + 16x + 10$

10. $f(x, y) = x^2 + 6xy + 10y^2 - 4y + 4$

11. $f(x, y) = 3x^2 + 2y^2 - 6x - 4y + 16$

12. $f(x, y) = -3x^2 - 2y^2 + 3x - 4y + 5$

13. $f(x, y) = -x^3 + 4xy - 2y^2 + 1$

14. $f(x, y) = x^2 - 3xy - y^2$

15. $f(x, y) = \dfrac{1}{2}xy$

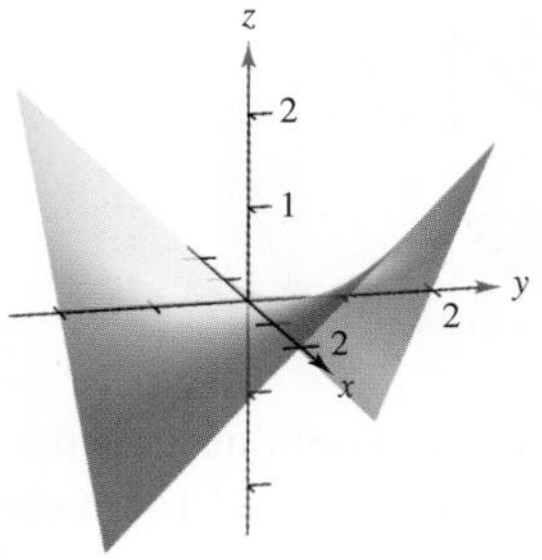

16. $f(x, y) = x + y + 2xy - x^2 - y^2$

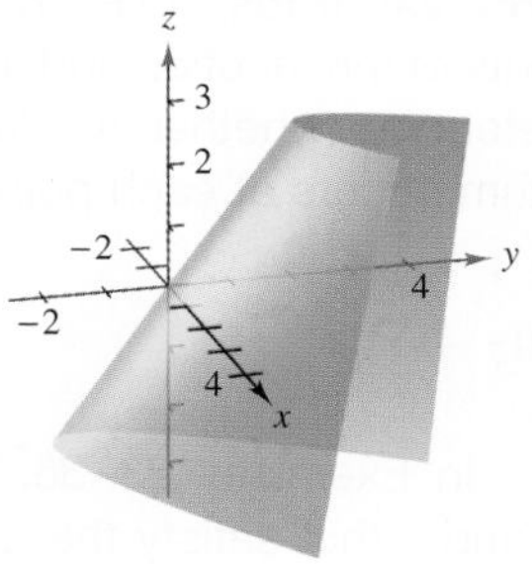

17. $f(x, y) = (x + y)e^{1 - x^2 - y^2}$

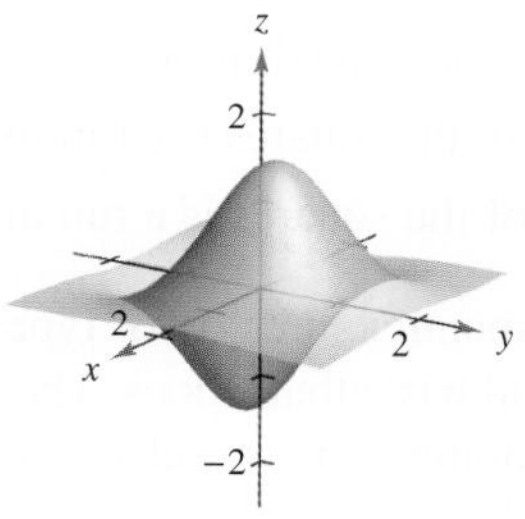

18. $f(x, y) = 3e^{-(x^2 + y^2)}$

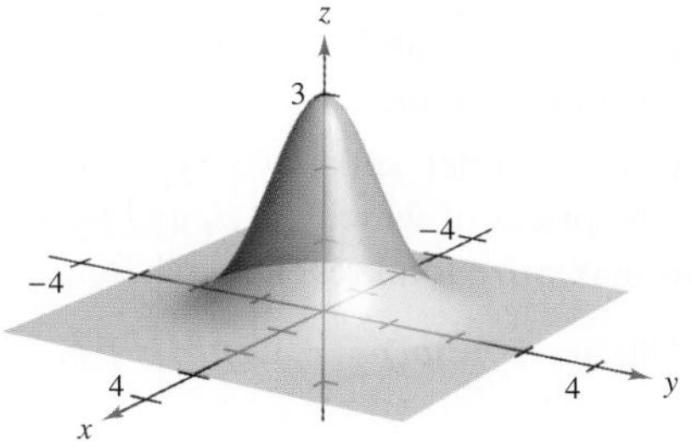

Think About It In Exercises 19–24, determine whether there is a relative maximum, a relative minimum, a saddle point, or insufficient information to determine the nature of the function $f(x, y)$ at the critical point (x_0, y_0).

19. $f_{xx}(x_0, y_0) = 9,\ f_{yy}(x_0, y_0) = 4,\ f_{xy}(x_0, y_0) = 6$

20. $f_{xx}(x_0, y_0) = -3,\ f_{yy}(x_0, y_0) = -8,\ f_{xy}(x_0, y_0) = 2$

21. $f_{xx}(x_0, y_0) = -9,\ f_{yy}(x_0, y_0) = 6,\ f_{xy}(x_0, y_0) = 10$

22. $f_{xx}(x_0, y_0) = 25,\ f_{yy}(x_0, y_0) = 8,\ f_{xy}(x_0, y_0) = 10$

23. $f_{xx}(x_0, y_0) = 5,\ f_{yy}(x_0, y_0) = 5,\ f_{xy}(x_0, y_0) = 3$

24. $f_{xx}(x_0, y_0) = 8,\ f_{yy}(x_0, y_0) = 7,\ f_{xy}(x_0, y_0) = 9$

Analyzing a Function In Exercises 25–30, find the critical points, relative extrema, and saddle points of the function. List the critical points for which the Second-Partials Test fails.

25. $f(x, y) = (xy)^2$

26. $f(x, y) = \sqrt{x^2 + y^2}$

27. $f(x, y) = x^3 + y^3$

28. $f(x, y) = x^3 + y^3 - 3x^2 + 6y^2 + 3x + 12y + 7$

29. $f(x, y) = x^{2/3} + y^{2/3}$

30. $f(x, y) = (x^2 + y^2)^{2/3}$

Analyzing a Function of Three Variables In Exercises 31 and 32, find the critical points of the function and, from the form of the function, determine whether a relative maximum or a relative minimum occurs at each point.

31. $f(x, y, z) = (x - 1)^2 + (y + 3)^2 + z^2$

32. $f(x, y, z) = 6 - [x(y + 2)(z - 1)]^2$

Finding Positive Numbers In Exercises 33–36, find three positive numbers x, y, and z that satisfy the given conditions.

33. The sum is 45 and the product is a maximum.

34. The sum is 32 and $P = xy^2z$ is a maximum.

35. The sum is 60 and the sum of the squares is a minimum.

36. The sum is 2 and the sum of the squares is a minimum.

37. Revenue A company manufactures two types of sneakers: running shoes and basketball shoes. The total revenue from x_1 units of running shoes and x_2 units of basketball shoes is

$$R = -5x_1^2 - 8x_2^2 - 2x_1x_2 + 42x_1 + 102x_2$$

where x_1 and x_2 are in thousands of units. Find x_1 and x_2 so as to maximize the revenue.

38. Revenue A retail outlet sells two types of riding lawn mowers, the prices of which are p_1 and p_2. Find p_1 and p_2 so as to maximize total revenue, where

$$R = 515p_1 + 805p_2 + 1.5p_1p_2 - 1.5p_1^2 - p_2^2.$$

Revenue In Exercises 39 and 40, find p_1 and p_2, the prices per unit (in dollars), so as to maximize the total revenue

$$R = x_1p_1 + x_2p_2$$

where x_1 and x_2 are the numbers of units sold, for a retail outlet that sells two competitive products with the given demand functions. See Example 4.

39. $x_1 = 1000 - 2p_1 + p_2,\ x_2 = 1500 + 2p_1 - 1.5p_2$

40. $x_1 = 1000 - 4p_1 + 2p_2,\ x_2 = 900 + 4p_1 - 3p_2$

41. Profit A corporation manufactures a product for a high-performance automobile engine at two locations. The cost of producing x_1 units at location 1 is

$$C_1 = 0.05x_1^2 + 15x_1 + 5400$$

and the cost of producing x_2 units at location 2 is

$$C_2 = 0.03x_2^2 + 15x_2 + 6100.$$

The demand function for the product is

$$p = 225 - 0.4(x_1 + x_2)$$

and the total revenue function is

$$R = [225 - 0.4(x_1 + x_2)](x_1 + x_2).$$

Find the production levels at the two locations that will maximize the profit

$$P = R - C_1 - C_2.$$

42. Profit A corporation manufactures candles at two locations. The cost of producing x_1 units at location 1 is

$$C_1 = 0.02x_1^2 + 4x_1 + 500$$

and the cost of producing x_2 units at location 2 is

$$C_2 = 0.05x_2^2 + 4x_2 + 275.$$

The candles sell for $15 per unit. Find the quantity that should be produced at each location to maximize the profit

$$P = 15(x_1 + x_2) - C_1 - C_2.$$

43. Volume Find the dimensions of a rectangular package of maximum volume that may be sent by a shipping company, assuming that the sum of the length and the girth (perimeter of a cross section) cannot exceed 96 inches.

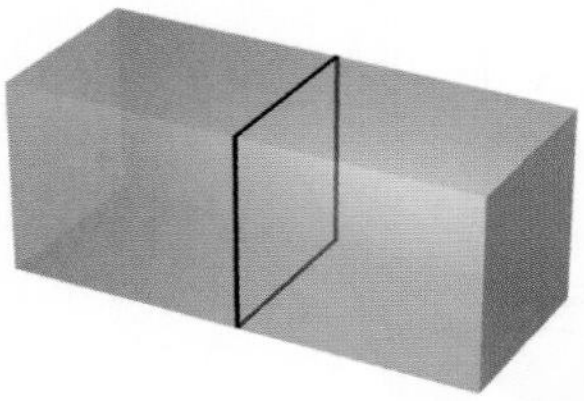

44. Volume Repeat Exercise 43, assuming that the sum of the length and the girth cannot exceed 144 inches.

45. Cost A manufacturer makes an open-top wooden crate having a volume of 18 cubic feet. Material costs are \$0.20 per square foot for the base and \$0.15 per square foot for the sides. Find the dimensions that minimize the cost of each crate. What is the minimum cost?

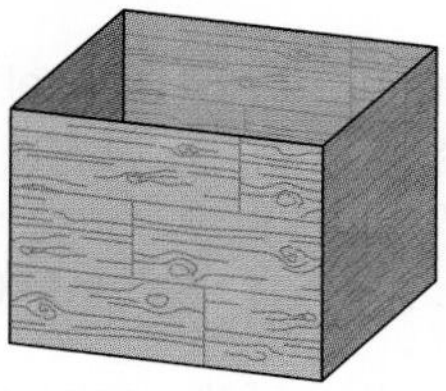

46. Cost A home improvement contractor is painting the walls and ceiling of a rectangular room. The volume of the room is 1584 cubic feet. The cost of wall paint is \$0.06 per square foot and the cost of ceiling paint is \$0.11 per square foot. Find the room dimensions that will minimize the cost of the paint. What is the minimum cost of the paint?

47. Cost An automobile manufacturer has determined that its annual labor and equipment cost (in millions of dollars) can be modeled by

$$C(x, y) = 2x^2 + 3y^2 - 15x - 20y + 4xy + 39$$

where x is the amount spent per year on labor and y is the amount spent per year on equipment (both in millions of dollars). Find the values of x and y that minimize the annual labor and equipment cost. What is this cost?

48. Medicine In order to treat a certain bacterial infection, a combination of two drugs is being tested. Studies have shown that the duration of the infection in laboratory tests can be modeled by

$$D(x, y) = x^2 + 2y^2 - 18x - 24y + 2xy + 120$$

where x is the dosage of the first drug and y is the dosage of the second drug (both in hundreds of milligrams). Find the amount of each drug necessary to minimize the duration of the infection.

49. Biology A lake is to be stocked with smallmouth and largemouth bass. Let x represent the number of smallmouth bass and let y represent the number of largemouth bass. The weight of each fish is dependent on the population densities. After a six-month period, the weight of a single smallmouth bass is given by

$$W_1 = 3 - 0.002x - 0.001y$$

and the weight of a single largemouth bass is given by

$$W_2 = 4.5 - 0.004x - 0.005y.$$

Assuming that no fish die during the six-month period, how many smallmouth and largemouth bass should be stocked in the lake so that the *total* weight T of bass in the lake is a maximum?

50. HOW DO YOU SEE IT? The figure shows the level curves for a function $f(x, y)$. What, if anything, can be said about f at the points A, B, C, and D? Explain your reasoning.

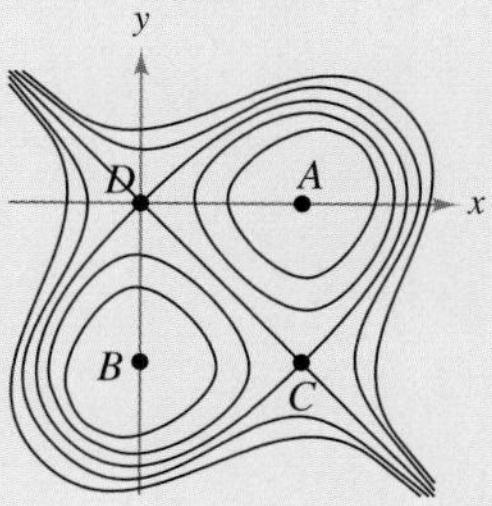

51. Hardy-Weinberg Law Common blood types are determined genetically by the three alleles A, B, and O. (An allele is any of a group of possible mutational forms of a gene.) A person whose blood type is AA, BB, or OO is homozygous. A person whose blood type is AB, AO, or BO is heterozygous. The Hardy-Weinberg Law states that the proportion P of heterozygous individuals in any given population is modeled by

$$P(p, q, r) = 2pq + 2pr + 2qr$$

where p represents the percent of allele A in the population, q represents the percent of allele B in the population, and r represents the percent of allele O in the population. Use the fact that $p + q + r = 1$ (the sum of the three must equal 100%) to show that the maximum proportion of heterozygous individuals in any population is $\frac{2}{3}$.

52. Shannon Diversity Index One way to measure species diversity is to use the Shannon diversity index H. A habitat consists of three species A, B, and C, and its Shannon diversity index is

$$H = -x \ln x - y \ln y - z \ln z$$

where x is the percent of species A in the habitat, y is the percent of species B in the habitat, and z is the percent of species C in the habitat. Use the fact that $x + y + z = 1$ (the sum of the three must equal 100%) to show that the maximum value of H occurs when

$$x = y = z = \tfrac{1}{3}.$$

What is the maximum value of H?

True or False? In Exercises 53 and 54, determine whether the statement is true or false. If it is false, explain why or give an example that shows it is false.

53. A saddle point always occurs at a critical point.

54. If $f(x, y)$ has a relative maximum at (x_0, y_0, z_0), then $f_x(x_0, y_0) = f_y(x_0, y_0) = 0$.

QUIZ YOURSELF

See www.CalcChat.com for worked-out solutions to odd-numbered exercises.

Take this quiz as you would take a quiz in class. When you are done, check your work against the answers given in the back of the book.

In Exercises 1–3, (a) plot the points in a three-dimensional coordinate system, (b) find the distance between the points, and (c) find the coordinates of the midpoint of the line segment joining the points.

1. $(1, 3, 2), (-1, 2, 0)$ **2.** $(-1, 3, 4), (5, 1, -6)$ **3.** $(0, -3, 3), (3, 0, -3)$

In Exercises 4 and 5, find the standard equation of the sphere.

4. Center: $(2, -1, 3)$; radius: 4

5. Endpoints of a diameter: $(0, 3, 1), (2, 5, -5)$

6. Find the center and radius of the sphere whose equation is

$$x^2 + y^2 + z^2 - 8x - 2y - 6z - 23 = 0.$$

In Exercises 7–9, find the intercepts and sketch the graph of the plane.

7. $2x + 3y + z = 6$ **8.** $x - 2z = 4$ **9.** $y = 3$

In Exercises 10–12, classify the quadric surface.

10. $\dfrac{x^2}{4} + \dfrac{y^2}{9} + \dfrac{z^2}{16} = 1$ **11.** $z^2 - x^2 - y^2 = 25$ **12.** $81z - 9x^2 - y^2 = 0$

In Exercises 13–15, find $f(1, 0)$ and $f(4, -1)$.

13. $f(x, y) = x - 9y^2$ **14.** $f(x, y) = \sqrt{4x^2 + y}$ **15.** $f(x, y) = \ln(x - 2y)$

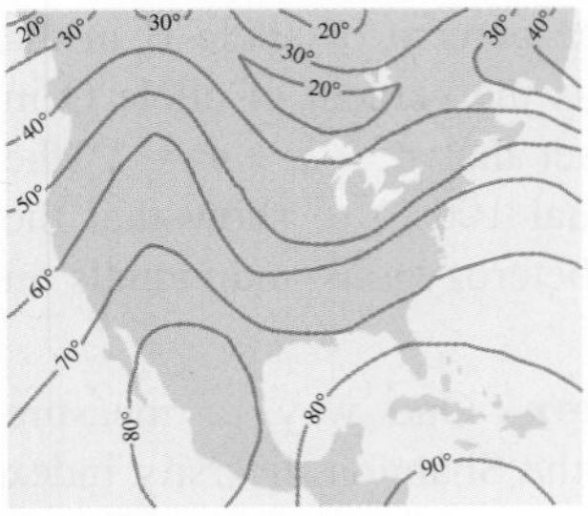

Figure for 16

16. The contour map shows level curves of equal temperature (isotherms), measured in degrees Fahrenheit, across North America on a spring day. Use the map to find the approximate range of temperatures in

(a) the Great Lakes region.

(b) the United States.

(c) Mexico.

In Exercises 17–20, find the first partial derivatives and evaluate each at the point $(-2, 3)$.

17. $f(x, y) = x^2 + 2y^2 - 3x - y + 1$ **18.** $f(x, y) = \dfrac{3x - y^2}{x + y}$

19. $f(x, y) = x^3e^{2y}$ **20.** $f(x, y) = \ln(2x + 7y)$

In Exercises 21 and 22, find the critical points, relative extrema, and saddle points of the function.

21. $f(x, y) = 3x^2 + y^2 - 2xy - 6x + 2y$

22. $f(x, y) = -x^3 + 4xy - 2y^2 + 1$

23. A company manufactures two types of wood burning stoves: a freestanding model and a fireplace-insert model. The total cost (in thousands of dollars) for producing x freestanding stoves and y fireplace-insert stoves can be modeled by

$$C(x, y) = \tfrac{1}{16}x^2 + y^2 - 10x - 40y + 820.$$

Find the values of x and y that minimize the total cost. What is this cost?

7.6 Lagrange Multipliers

- Understand the Method of Lagrange Multipliers.
- Use Lagrange multipliers to solve constrained optimization problems.

In Exercise 37 on page 472, you will use Lagrange multipliers to find the dimensions that will minimize the cost of fencing in two corrals.

Lagrange Multipliers with One Constraint

In Example 5 in Section 7.5, you were asked to find the dimensions of the rectangular box of maximum volume that would fit in the first octant beneath the plane

$$6x + 4y + 3z = 24$$

as shown again in Figure 7.36. Another way of stating this problem is to say that you are asked to find the maximum of

$$V = xyz \qquad \text{Objective function}$$

subject to the constraint

$$6x + 4y + 3z - 24 = 0. \qquad \text{Constraint}$$

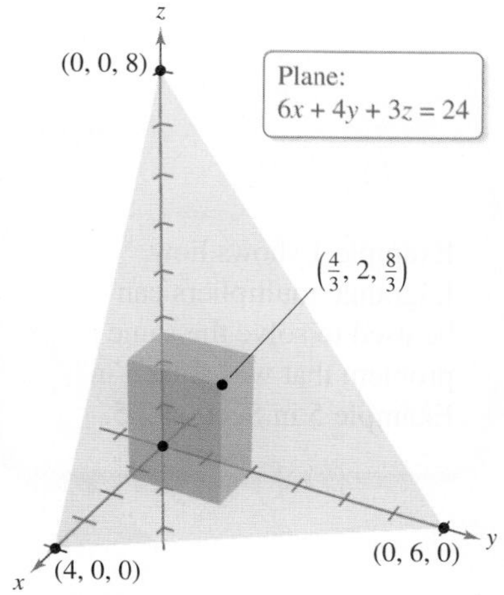

FIGURE 7.36

This type of problem is called a **constrained optimization** problem. In Section 7.5, you answered this question by solving for z in the constraint equation and then rewriting V as a function of two variables.

In this section, you will study a different (and often better) way to solve constrained optimization problems. This method involves the use of variables called **Lagrange multipliers,** named after the French mathematician Joseph Louis Lagrange (1736–1813).

Method of Lagrange Multipliers

If $f(x, y)$ has a maximum or minimum subject to the constraint $g(x, y) = 0$, then it will occur at one of the critical numbers of the function F defined by

$$F(x, y, \lambda) = f(x, y) - \lambda g(x, y).$$

The variable λ (the lowercase Greek letter lambda) is called a **Lagrange multiplier.** To find the minimum or maximum of f, use the following steps.

1. Solve the following system of equations.

$$F_x(x, y, \lambda) = 0 \qquad F_y(x, y, \lambda) = 0 \qquad F_\lambda(x, y, \lambda) = 0$$

2. Evaluate f at each solution point obtained in the first step. The greatest value yields the maximum of f subject to the constraint $g(x, y) = 0$, and the least value yields the minimum of f subject to the constraint $g(x, y) = 0$.

When using the Method of Lagrange Multipliers for functions of three variables, F has the form

$$F(x, y, z, \lambda) = f(x, y, z) - \lambda g(x, y, z).$$

The system of equations used in Step 1 is

$$F_x(x, y, z, \lambda) = 0 \quad F_y(x, y, z, \lambda) = 0 \quad F_z(x, y, z, \lambda) = 0 \quad F_\lambda(x, y, z, \lambda) = 0.$$

The Method of Lagrange Multipliers gives you a way of finding critical points but does not tell you whether these points yield minima, maxima, or neither. To make this distinction, you must rely on the context of the problem.

Constrained Optimization Problems

Example 1 Using Lagrange Multipliers

Find the maximum of

$$V = xyz \qquad \text{Objective function}$$

subject to the constraint

$$6x + 4y + 3z - 24 = 0. \qquad \text{Constraint}$$

SOLUTION First, let $f(x, y, z) = xyz$ and $g(x, y, z) = 6x + 4y + 3z - 24$. Then, define a new function F as

$$F(x, y, z, \lambda) = f(x, y, z) - \lambda g(x, y, z)$$
$$= xyz - \lambda(6x + 4y + 3z - 24).$$

To find the critical numbers of F, begin by finding the partial derivatives of F with respect to x, y, z, and λ. Then, set the partial derivatives equal to zero.

$$F_x(x, y, z, \lambda) = yz - 6\lambda \quad \Rightarrow \quad yz - 6\lambda = 0$$
$$F_y(x, y, z, \lambda) = xz - 4\lambda \quad \Rightarrow \quad xz - 4\lambda = 0$$
$$F_z(x, y, z, \lambda) = xy - 3\lambda \quad \Rightarrow \quad xy - 3\lambda = 0$$
$$F_\lambda(x, y, z, \lambda) = -6x - 4y - 3z + 24 \quad \Rightarrow \quad -6x - 4y - 3z + 24 = 0$$

Solving for λ in the first equation produces

$$yz - 6\lambda = 0 \quad \Rightarrow \quad \lambda = \frac{yz}{6}.$$

Substituting for λ in the second and third equations produces the following.

$$xz - 4\left(\frac{yz}{6}\right) = 0 \quad \Rightarrow \quad y = \frac{3}{2}x$$
$$xy - 3\left(\frac{yz}{6}\right) = 0 \quad \Rightarrow \quad z = 2x$$

Next, substitute for y and z in the equation $F_\lambda(x, y, z, \lambda) = 0$ and solve for x.

$$F_\lambda(x, y, z, \lambda) = 0$$
$$-6x - 4y - 3z + 24 = 0$$
$$-6x - 4\left(\tfrac{3}{2}x\right) - 3(2x) + 24 = 0$$
$$-18x = -24$$
$$x = \tfrac{4}{3}$$

Using this x-value, you can conclude that the critical values are $x = \frac{4}{3}$, $y = 2$, and $z = \frac{8}{3}$, which implies that the maximum is

$$V = xyz \qquad \text{Write objective function.}$$
$$= \left(\frac{4}{3}\right)(2)\left(\frac{8}{3}\right) \qquad \text{Substitute values of } x, y, \text{ and } z.$$
$$= \frac{64}{9} \text{ cubic units.} \qquad \text{Maximum volume}$$

✓ Checkpoint 1

Find the maximum volume of $V = xyz$ subject to the constraint

$$2x + 4y + z - 8 = 0.$$

■

STUDY TIP

Example 1 shows how Lagrange multipliers can be used to solve the same problem that was solved in Example 5 in Section 7.5.

ALGEBRA TUTOR *xy*

The most difficult aspect of many Lagrange multiplier problems is the complicated algebra needed to solve the system of equations arising from

$F(x, y, \lambda) = f(x, y) - \lambda g(x, y).$

There is no general way to proceed in every case, so you should study the examples carefully, and refer to the *Chapter 7 Algebra Tutor* on pages 496 and 497.

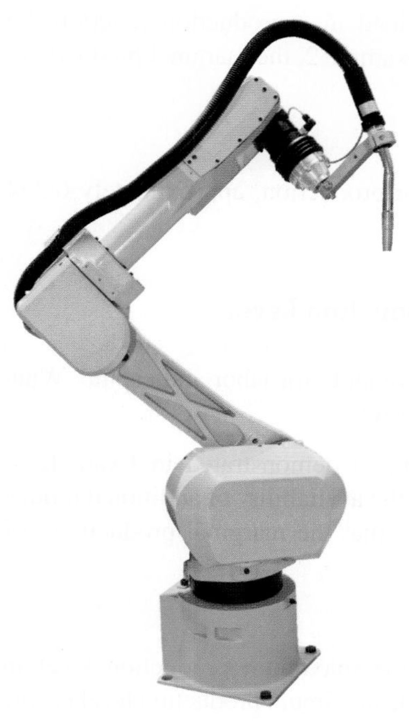

For some industrial applications, a simple robot can cost more than a year's wages and benefits for one employee. So, manufacturers must carefully balance the amount of money spent on labor and capital.

TECH TUTOR

You can use a spreadsheet to solve constrained optimization problems. Try using a spreadsheet to solve the problem in Example 2. (Consult the user's manual of a spreadsheet software program for specific instructions on how to solve a constrained optimization problem.)

Example 2 Finding a Maximum Production Level

A manufacturer's production is modeled by the Cobb-Douglas function

$$f(x, y) = 100x^{3/4}y^{1/4} \qquad \text{Objective function}$$

where x represents the units of labor (at \$150 per unit) and y represents the units of capital (at \$250 per unit). The total costs for labor and capital cannot exceed \$50,000. Will the manufacturer's maximum production level exceed 16,000 units?

SOLUTION Because total labor and capital expenses cannot exceed \$50,000, the constraint is

$$150x + 250y = 50{,}000 \qquad \text{Constraint}$$
$$150x + 250y - 50{,}000 = 0. \qquad \text{Write in general form.}$$

To find the maximum production level, begin by writing the function

$$F(x, y, \lambda) = 100x^{3/4}y^{1/4} - \lambda(150x + 250y - 50{,}000).$$

Then find the partial derivatives of F with respect to x, y, and λ.

$$F_x(x, y, \lambda) = 75x^{-1/4}y^{1/4} - 150\lambda$$
$$F_y(x, y, \lambda) = 25x^{3/4}y^{-3/4} - 250\lambda$$
$$F_\lambda(x, y, \lambda) = -150x - 250y + 50{,}000$$

Next, set the partial derivatives equal to zero to obtain the following system of equations.

$$75x^{-1/4}y^{1/4} - 150\lambda = 0 \qquad \text{Equation 1}$$
$$25x^{3/4}y^{-3/4} - 250\lambda = 0 \qquad \text{Equation 2}$$
$$-150x - 250y + 50{,}000 = 0 \qquad \text{Equation 3}$$

By solving for λ in the first equation

$$75x^{-1/4}y^{1/4} - 150\lambda = 0 \qquad \text{Equation 1}$$
$$\lambda = \tfrac{1}{2}x^{-1/4}y^{1/4} \qquad \text{Solve for } \lambda.$$

and substituting for λ in Equation 2, you obtain

$$25x^{3/4}y^{-3/4} - 250\left(\tfrac{1}{2}\right)x^{-1/4}y^{1/4} = 0 \qquad \text{Substitute in Equation 2.}$$
$$25x - 125y = 0 \qquad \text{Multiply by } x^{1/4}y^{3/4}.$$
$$x = 5y. \qquad \text{Solve for } x.$$

So, $x = 5y$. By substituting for x in Equation 3, you obtain

$$-150(5y) - 250y + 50{,}000 = 0 \qquad \text{Substitute in Equation 3.}$$
$$-1000y = -50{,}000 \qquad \text{Simplify.}$$
$$y = 50 \qquad \text{Solve for } y.$$

When $y = 50$ units of capital, it follows that $x = 5(50) = 250$ units of labor. So, the maximum production level is

$$f(250, 50) = 100(250)^{3/4}(50)^{1/4} \qquad \text{Substitute for } x \text{ and } y.$$
$$\approx 16{,}719 \text{ units.} \qquad \text{Maximum production level}$$

You can conclude that the maximum production level will exceed 16,000 units.

✓ Checkpoint 2

In Example 2, suppose that each labor unit costs \$200 and each capital unit costs \$250. Find the maximum production level when labor and capital cannot exceed \$50,000.

Economists call the Lagrange multiplier obtained in a production function the **marginal productivity of money.** For instance, in Example 2, the marginal productivity of money when $x = 250$ and $y = 50$ is

$$\lambda = \frac{1}{2}x^{-1/4}y^{1/4} = \frac{1}{2}(250)^{-1/4}(50)^{1/4} \approx 0.334.$$

This means that for each additional dollar spent on production, approximately 0.334 additional unit of the product can be produced.

Example 3 Finding a Maximum Production Level

The manufacturer in Example 2 now has $70,000 available for labor and capital. What is the maximum number of units that can be produced?

SOLUTION You could rework the entire problem, as demonstrated in Example 2. However, because the only change in the problem is the availability of additional money to spend on labor and capital, you can use the fact that the marginal productivity of money is

$$\lambda \approx 0.334.$$

Because an additional $20,000 is available and the maximum production level in Example 2 was 16,719 units, you can conclude that the maximum production level is now

$$16{,}719 + (0.334)(20{,}000) \approx 23{,}400 \text{ units.}$$

Try using the procedure demonstrated in Example 2 to confirm this result.

Checkpoint 3

The manufacturer in Example 2 now has $80,000 available for labor and capital. What is the maximum number of units that can be produced?

TECH TUTOR

You can use a three-dimensional graphing utility to confirm graphically the results of Examples 2 and 3. Begin by graphing the surface $f(x, y) = 100x^{3/4}y^{1/4}$. Then graph the vertical plane given by $150x + 250y = 50{,}000$. As shown below, the maximum production level corresponds to the highest point on the intersection of the surface and the plane.

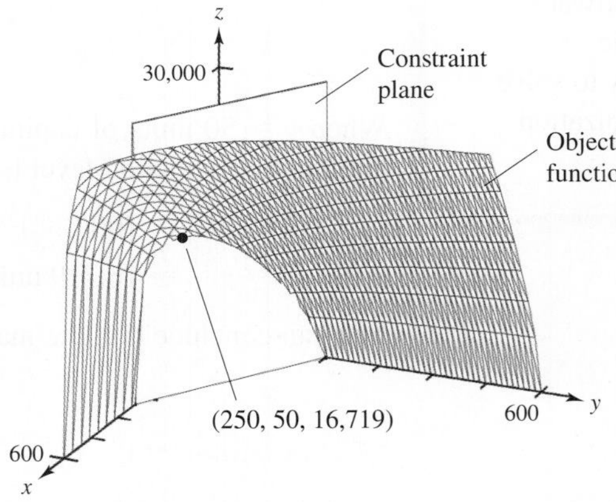

> **STUDY TIP**
>
> The constrained optimization problem in Example 4 on this page is represented graphically in Figure 7.37. The graph of the objective function is a paraboloid and the graph of the constraint is a vertical plane. In the "unconstrained" optimization problem on page 459, the maximum profit occurred at the vertex of the paraboloid. In this "constrained" problem, however, the maximum profit corresponds to the highest point on the curve that is the intersection of the paraboloid and the vertical "constraint" plane.

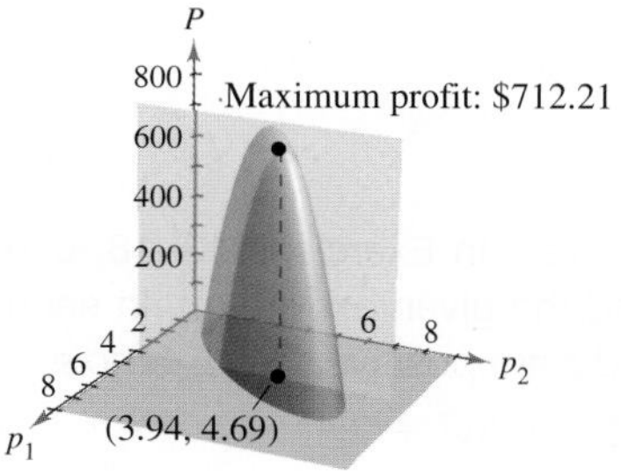

FIGURE 7.37

In Example 4 in Section 7.5, you found the maximum profit for two substitute products whose demand functions are given by

$$x_1 = 200(p_2 - p_1) \qquad \text{Demand for product 1}$$
$$x_2 = 500 + 100p_1 - 180p_2. \qquad \text{Demand for product 2}$$

With this model, the total demand, $x_1 + x_2$, is completely determined by the prices p_1 and p_2. In many real-life situations, this assumption is too simplistic; regardless of the prices of the substitute brands, the annual total demands for some products, such as toothpaste, are relatively constant. In such situations, the total demand is **limited,** and variations in price do not affect the total demand as much as they affect the market share of the substitute brands.

Example 4 Finding a Maximum Profit

A company makes two substitute products whose demand functions are given by

$$x_1 = 200(p_2 - p_1) \qquad \text{Demand for product 1}$$
$$x_2 = 500 + 100p_1 - 180p_2 \qquad \text{Demand for product 2}$$

where p_1 and p_2 are the prices per unit (in dollars) and x_1 and x_2 are the numbers of units sold. The costs of producing the two products are \$0.50 and \$0.75 per unit, respectively. The total demand is limited to 200 units per year. Find the prices that will yield a maximum profit.

SOLUTION From Example 4 in Section 7.5, the profit function is modeled by

$$P = -200p_1^2 - 180p_2^2 + 300p_1p_2 + 25p_1 + 535p_2 - 375.$$

The total demand for the two products is

$$\begin{aligned} x_1 + x_2 &= 200(p_2 - p_1) + 500 + 100p_1 - 180p_2 \\ &= 200p_2 - 200p_1 + 500 + 100p_1 - 180p_2 \\ &= -100p_1 + 20p_2 + 500. \end{aligned}$$

Because the total demand is limited to 200 units, the constraint is

$$-100p_1 + 20p_2 + 500 = 200. \qquad \text{Constraint}$$

Using Lagrange multipliers, you can determine that the maximum profit occurs when $p_1 \approx \$3.94$ and $p_2 \approx \$4.69$. This corresponds to an annual profit of about \$712.21.

✓Checkpoint 4

In Example 4, find the prices that will yield a maximum profit when the total demand is limited to 250 units per year.

SUMMARIZE (Section 7.6)

1. Explain the Method of Lagrange Multipliers *(page 465)*. For an example of how to use Lagrange multipliers, see Example 1.
2. Describe a real-life example of using Lagrange multipliers to find a manufacturer's maximum production level *(page 467, Example 2)*.
3. Describe a real-life example of using Lagrange multipliers to find a company's maximum profit *(page 469, Example 4)*.

SKILLS WARM UP 7.6 The following warm-up exercises involve skills that were covered in earlier sections. You will use these skills in the exercise set for this section. For additional help, review Section 7.4.

In Exercises 1–6, solve the system of linear equations.

1. $\begin{cases} 4x - 6y = 3 \\ 2x + 3y = 2 \end{cases}$

2. $\begin{cases} 6x - 6y = 5 \\ -3x - y = 1 \end{cases}$

3. $\begin{cases} 5x - y = 25 \\ x - 5y = 15 \end{cases}$

4. $\begin{cases} 4x - 9y = 5 \\ -x + 8y = -2 \end{cases}$

5. $\begin{cases} 2x - y + z = 3 \\ 2x + 2y + z = 4 \\ -x + 2y + 3z = -1 \end{cases}$

6. $\begin{cases} -x - 4y + 6z = -2 \\ x - 3y - 3z = 4 \\ 3x + y + 3z = 0 \end{cases}$

In Exercises 7–10, find all first partial derivatives.

7. $f(x, y) = x^2y + xy^2$
8. $f(x, y) = 25(xy + y^2)^2$
9. $f(x, y, z) = x(x^2 - 2xy + yz)$
10. $f(x, y, z) = z(xy + xz + yz)$

Exercises 7.6

See www.CalcChat.com for worked-out solutions to odd-numbered exercises.

Using Lagrange Multipliers In Exercises 1–12, use Lagrange multipliers to find the given extremum. In each case, assume that x and y are positive. *See Example 1.*

1. Maximize $f(x, y) = xy$
   Constraint: $x + y = 10$
2. Maximize $f(x, y) = xy$
   Constraint: $x + 3y = 6$
3. Minimize $f(x, y) = x^2 + y^2$
   Constraint: $x + y - 8 = 0$
4. Minimize $f(x, y) = x^2 + y^2$
   Constraint: $-2x - 4y + 5 = 0$
5. Maximize $f(x, y) = x^2 - y^2$
   Constraint: $2y - x^2 = 0$
6. Minimize $f(x, y) = x^2 - y^2$
   Constraint: $x - 2y + 6 = 0$
7. Maximize $f(x, y) = 2x + 2xy + y$
   Constraint: $2x + y = 100$
8. Minimize $f(x, y) = 3x + y + 10$
   Constraint: $x^2y = 6$
9. Maximize $f(x, y) = \sqrt{6 - x^2 - y^2}$
   Constraint: $x + y - 2 = 0$
10. Minimize $f(x, y) = \sqrt{x^2 + y^2}$
    Constraint: $2x + 4y - 15 = 0$
11. Maximize $f(x, y) = e^{xy}$
    Constraint: $x^2 + y^2 - 8 = 0$
12. Minimize $f(x, y) = 2x + y$
    Constraint: $xy = 32$

Using Lagrange Multipliers In Exercises 13–18, use Lagrange multipliers to find the given extremum. In each case, assume that x, y, and z are positive. *See Example 1.*

13. Minimize $f(x, y, z) = 2x^2 + 3y^2 + 2z^2$
    Constraint: $x + y + z - 24 = 0$
14. Maximize $f(x, y, z) = xyz$
    Constraint: $x + y + z - 6 = 0$
15. Minimize $f(x, y, z) = x^2 + y^2 + z^2$
    Constraint: $x + y + z = 1$
16. Minimize $f(x, y) = x^2 - 8x + y^2 - 12y + 48$
    Constraint: $x + y = 8$
17. Maximize $f(x, y, z) = x + y + z$
    Constraint: $x^2 + y^2 + z^2 = 1$
18. Maximize $f(x, y, z) = x^2y^2z^2$
    Constraint: $x^2 + y^2 + z^2 = 1$

Finding Positive Numbers In Exercises 19–22, find three positive numbers x, y, and z that satisfy the given conditions.

19. The sum is 60 and the product is a maximum.
20. The sum is 80 and $P = x^2yz$ is a maximum.
21. The sum is 120 and the sum of the squares is a minimum.
22. The sum is 36 and the sum of the cubes is a minimum.

Finding Distance In Exercises 23–26, find the minimum distance from the curve or surface to the given point. (*Hint:* Start by minimizing the square of the distance.)

23. Line: $x + y = 6$, $(0, 0)$

Minimize $d^2 = x^2 + y^2$

24. Circle: $(x - 4)^2 + y^2 = 4$, $(0, 10)$

Minimize $d^2 = x^2 + (y - 10)^2$

25. Plane: $x + y + z = 1$, $(2, 1, 1)$

Minimize $d^2 = (x - 2)^2 + (y - 1)^2 + (z - 1)^2$

26. Cone: $z = \sqrt{x^2 + y^2}$, $(4, 0, 0)$

Minimize $d^2 = (x - 4)^2 + y^2 + z^2$

27. Volume A rectangular box is resting on the xy-plane with one vertex at the origin. The opposite vertex lies in the plane $2x + 3y + 5z = 90$. Find the dimensions that maximize the volume. (*Hint:* Maximize $V = xyz$ subject to the constraint $2x + 3y + 5z - 90 = 0$.)

28. Volume Find the dimensions of the rectangular package of largest volume subject to the constraint that the sum of the length and the girth cannot exceed 108 inches (see figure). (*Hint:* Maximize $V = xyz$ subject to the constraint $x + 2y + 2z = 108$.)

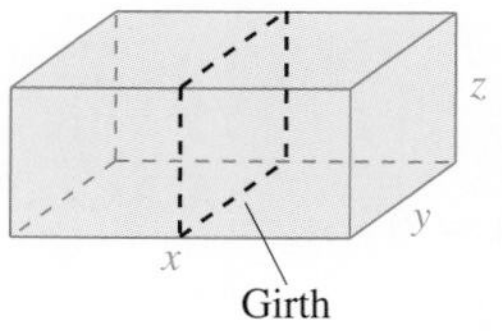

Figure for 28

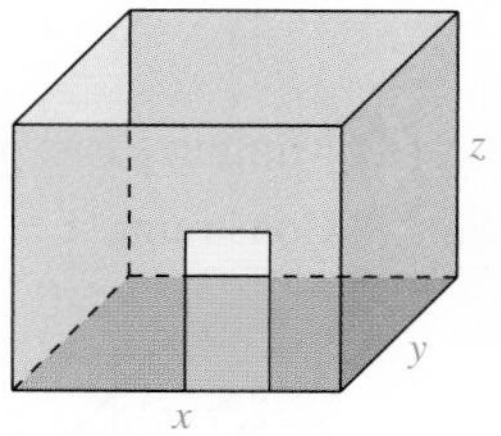

Figure for 29

29. Cost In redecorating an office, the cost for new carpeting is \$3 per square foot and the cost of wallpapering a wall is \$1 per square foot. Find the dimensions of the largest office that can be redecorated for \$1296 (see figure). (*Hint:* Maximize $V = xyz$ subject to $3xy + 2xz + 2yz = 1296$.)

30. Cost A cargo container (in the shape of a rectangular solid) must have a volume of 480 cubic feet. Find the dimensions of the container that has a minimum cost, if the bottom will cost \$5 per square foot to construct and the sides and top will cost \$3 per square foot to construct.

31. Cost A manufacturer has an order for 1000 units of fine paper that can be produced at two locations. Let x_1 and x_2 be the numbers of units produced at the two locations. The cost function is modeled by

$$C = 0.25x_1^2 + 25x_1 + 0.05x_2^2 + 12x_2.$$

Find the number of units that should be produced at each location to minimize the cost.

32. Cost A manufacturer has an order for 2000 units of all-terrain vehicle tires that can be produced at two locations. Let x_1 and x_2 be the numbers of units produced at the two plants. The cost function is modeled by

$$C = 0.25x_1^2 + 10x_1 + 0.15x_2^2 + 12x_2.$$

Find the number of units that should be produced at each location to minimize the cost.

33. Production The production function for a company is given by

$$f(x, y) = 100x^{0.25}y^{0.75}$$

where x is the number of units of labor (at \$48 per unit) and y is the number of units of capital (at \$36 per unit). The total cost for labor and capital cannot exceed \$100,000.

(a) Find the maximum production level for this manufacturer.

(b) Find the marginal productivity of money.

(c) Use the marginal productivity of money to find the maximum number of units that can be produced when \$125,000 is available for labor and capital.

(d) Use the marginal productivity of money to find the maximum number of units that can be produced when \$350,000 is available for labor and capital.

34. Production Repeat Exercise 33 for the production function given by

$$f(x, y) = 100x^{0.6}y^{0.4}.$$

35. Least-Cost Rule The production function for a company is given by

$$f(x, y) = 100x^{0.7}y^{0.3}$$

where x is the number of units of labor (at \$50 per unit) and y is the number of units of capital (at \$100 per unit). Management sets a production goal of 20,000 units.

(a) Find the numbers of units of labor and capital needed to meet the production goal while minimizing the cost.

(b) Show that the conditions of part (a) are met when

$$\frac{\text{Marginal productivity of labor}}{\text{Marginal productivity of capital}} = \frac{\text{Unit price of labor}}{\text{Unit price of capital}}.$$

This proportion is called the *Least-Cost Rule* (or *Equimarginal Rule*).

36. Least-Cost Rule Repeat Exercise 35 for the production function given by

$$f(x, y) = 100x^{0.4}y^{0.6}.$$

37. Construction A rancher plans to use an existing stone wall and the side of a barn as a boundary for two adjacent rectangular corrals (see figure). Fencing for the perimeter costs \$10 per foot. To separate the corrals, a fence that costs \$4 per foot will divide the region. The total area of the two corrals is to be 6000 square feet.

(a) Find the dimensions that will minimize the cost of the fencing.

(b) What is the minimum cost?

38. Office Space Partitions will be used in an office to form four equal work areas with a total area of 360 square feet (see figure). The partitions that are x feet long cost \$100 per foot and the partitions that are y feet long cost \$120 per foot.

(a) Find the dimensions x and y that will minimize the cost of the partitions.

(b) What is the minimum cost?

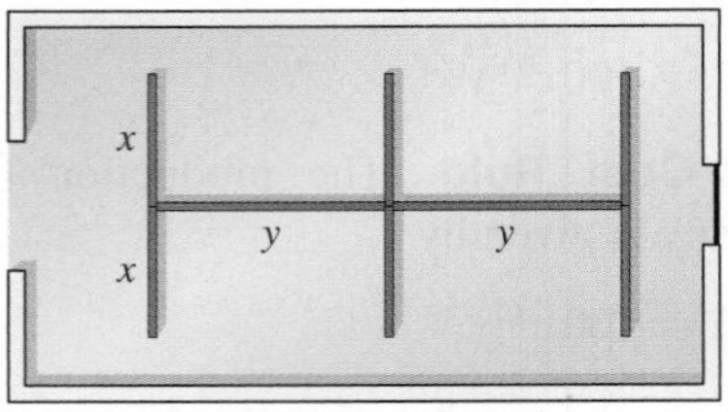

39. Biology A microbiologist must prepare a culture medium in which to grow a certain type of bacteria. The proportion of salt in this medium is given by

$$S = 12xyz$$

where x, y, and z are the amounts (in liters) of the three nutrient solutions to be mixed in the medium. For the bacteria to grow, the medium must be 13% salt. Nutrient solutions x, y, and z cost \$1, \$2, and \$3 per liter, respectively. How much of each nutrient solution should be used to minimize the cost of the culture medium?

40. Biology Repeat Exercise 39 for a salt-content model given by

$$S = 0.01x^2y^2z^2.$$

41. Nutrition The number of grams of your favorite ice cream can be modeled by

$$G(x, y, z) = 0.05x^2 + 0.16xy + 0.25z^2$$

where x is the number of fat grams, y is the number of carbohydrate grams, and z is the number of protein grams. Find the maximum number of grams of ice cream you can eat without consuming more than 400 calories. Assume that there are 9 calories per fat gram, 4 calories per carbohydrate gram, and 4 calories per protein gram.

42. HOW DO YOU SEE IT? The graphs show the constraint and several level curves of the objective function. Use the graph to approximate the indicated extrema.

(a) Maximize $z = xy$

Constraint: $2x + y = 4$

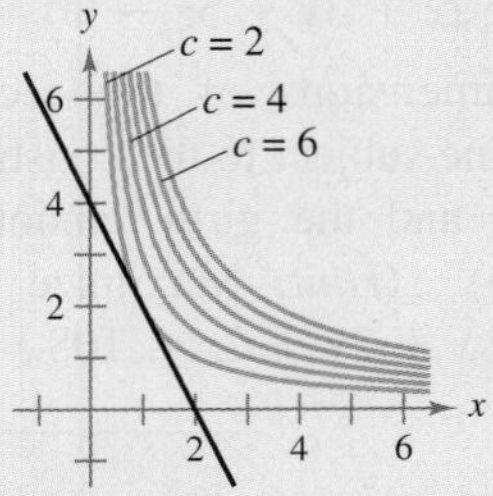

(b) Minimize $z = x^2 + y^2$

Constraint: $x + y - 4 = 0$

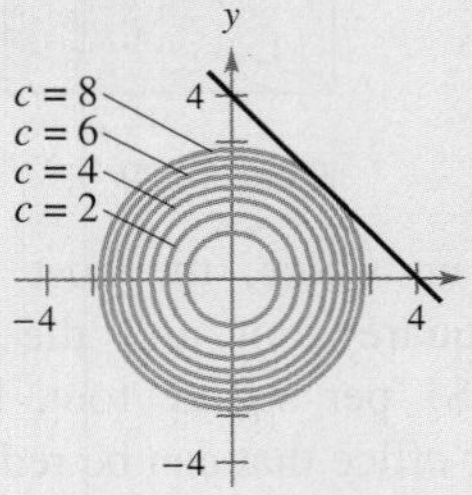

43. Advertising A private golf club is determining how to spend its \$2700 advertising budget. The club knows from prior experience that the number of responses A is given by

$$A = 0.0001t^2pr^{1.5}$$

where t is the number of cable television ads, p is the number of newspaper ads, and r is the number of radio ads. A cable television ad costs \$30, a newspaper ad costs \$12, and a radio ad costs \$15.

(a) How much should be spent on each type of advertising to obtain the maximum number of responses? (Assume the golf club uses each type of advertising.)

(b) What is the maximum number of responses expected?

7.7 Least Squares Regression Analysis

- Find the sum of the squared errors for mathematical models.
- Find the least squares regression lines for data.

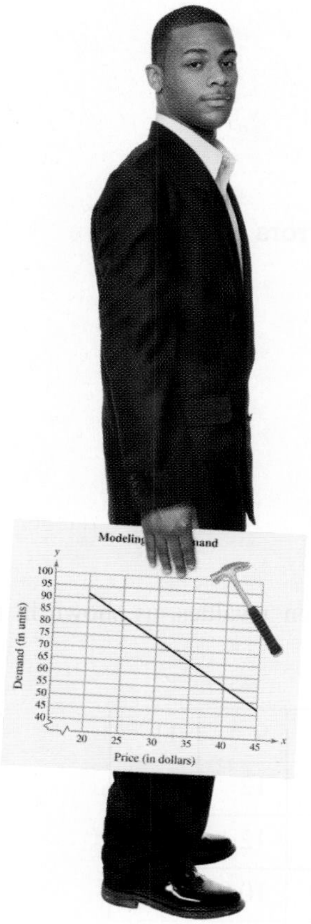

In Exercise 14 on page 478, you will find the least squares regression line that models the demand of a tool at a hardware store in terms of the price.

Measuring the Accuracy of a Mathematical Model

When seeking a mathematical model to fit data, the goals are simplicity and accuracy. For instance, a simple linear model for the points shown in Figure 7.38(a) is

$$f(x) = 1.9x - 5. \qquad \text{Linear model}$$

Figure 7.38(b), however, shows that by choosing the slightly more complicated quadratic model

$$g(x) = 0.20x^2 - 0.7x + 1 \qquad \text{Quadratic model}$$

you can obtain significantly greater accuracy.*

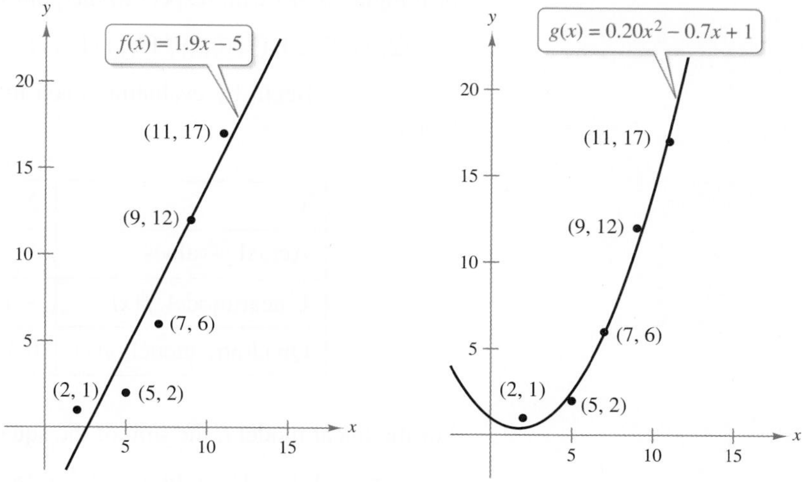

(a) (b)

FIGURE 7.38

To measure how well the model

$$y = f(x)$$

fits a collection of points, sum the squares of the differences between the actual y-values and the model's y-values. This sum is called the **sum of the squared errors** and is denoted by S. Graphically, S can be interpreted as the sum of the squares of the vertical distances between the graph of f and the given points in the plane, as shown in Figure 7.39. If the model is a perfect fit, then

$$S = 0.$$

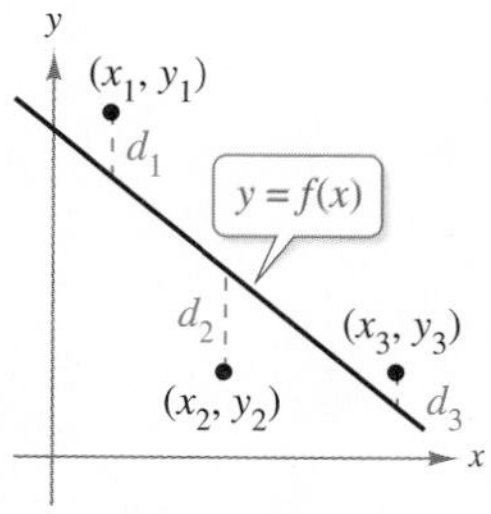

Sum of the squared errors: $S = d_1^2 + d_2^2 + d_3^2$

FIGURE 7.39

However, when a perfect fit is not feasible, you should use a model that minimizes S.

---

* An analytic method for finding a quadratic model for a collection of data is not given in this text. You can perform this task using a graphing utility or a spreadsheet software program that has a built-in program for finding the least squares regression quadratic.

Definition of the Sum of the Squared Errors

The **sum of the squared errors** for the model $y = f(x)$ with respect to the points

$$(x_1, y_1), (x_2, y_2), \ldots, (x_n, y_n)$$

is given by

$$S = [f(x_1) - y_1]^2 + [f(x_2) - y_2]^2 + \cdots + [f(x_n) - y_n]^2.$$

Example 1 Finding the Sum of the Squared Errors

Find the sum of the squared errors for the linear model

$$f(x) = 1.9x - 5 \qquad \text{Linear model}$$

and the quadratic model

$$g(x) = 0.20x^2 - 0.7x + 1 \qquad \text{Quadratic model}$$

(see Figure 7.38) with respect to the points

$$(2, 1), (5, 2), (7, 6), (9, 12), (11, 17).$$

SOLUTION Begin by evaluating each model at the given x-values, as shown in the table.

x	2	5	7	9	11
Actual y-values	1	2	6	12	17
Linear model, $f(x)$	−1.2	4.5	8.3	12.1	15.9
Quadratic model, $g(x)$	0.4	2.5	5.9	10.9	17.5

For the linear model f, the sum of the squared errors is

$$S = (-1.2 - 1)^2 + (4.5 - 2)^2 + (8.3 - 6)^2 + (12.1 - 12)^2 + (15.9 - 17)^2$$
$$= 17.6.$$

Similarly, the sum of the squared errors for the quadratic model g is

$$S = (0.4 - 1)^2 + (2.5 - 2)^2 + (5.9 - 6)^2 + (10.9 - 12)^2 + (17.5 - 17)^2$$
$$= 2.08.$$

✓Checkpoint 1

Find the sum of the squared errors for the linear model

$$f(x) = 2.9x - 6 \qquad \text{Linear model}$$

and the quadratic model

$$g(x) = 0.20x^2 + 0.5x - 1 \qquad \text{Quadratic model}$$

with respect to the points

$$(2, 1), (4, 5), (6, 9), (8, 16), (10, 24).$$

Then decide which model is a better fit. ■

In Example 1, note that the sum of the squared errors for the quadratic model is less than the sum of the squared errors for the linear model, which confirms that the quadratic model is a better fit.

Least Squares Regression Line

The sum of the squared errors can be used to determine which of several models is the best fit for a collection of data. In general, if the sum of the squared errors of f is less than the sum of the squared errors of g, then f is said to be a better fit for the data than g. In regression analysis, you consider all possible models of a certain type. The one that is defined to be the best-fitting model is the one with the least sum of the squared errors. Example 2 shows how to use the optimization techniques described in Section 7.5 to find the best-fitting linear model for a collection of data.

ALGEBRA TUTOR *xy*

For help in solving the system of equations in Example 2, see Example 2(b) in the *Chapter 7 Algebra Tutor*, on page 497.

Example 2 Finding the Best Linear Model

Find the values of a and b such that the linear model

$$f(x) = ax + b$$

has a minimum sum of the squared errors for the points

$$(-3, 0), (-1, 1), (0, 2), (2, 3).$$

SOLUTION The sum of the squared errors is

$$\begin{aligned} S &= [f(x_1) - y_1]^2 + [f(x_2) - y_2]^2 + [f(x_3) - y_3]^2 + [f(x_4) - y_4]^2 \\ &= (-3a + b - 0)^2 + (-a + b - 1)^2 + (b - 2)^2 + (2a + b - 3)^2 \\ &= 14a^2 - 4ab + 4b^2 - 10a - 12b + 14. \end{aligned}$$

To find the values of a and b for which S is a minimum, you can use the techniques described in Section 7.5. That is, find the partial derivatives of S.

$$\frac{\partial S}{\partial a} = 28a - 4b - 10 \qquad \text{Differentiate with respect to } a.$$

$$\frac{\partial S}{\partial b} = -4a + 8b - 12 \qquad \text{Differentiate with respect to } b.$$

Next, set each partial derivative equal to zero.

$$28a - 4b - 10 = 0 \qquad \text{Set } \partial S/\partial a \text{ equal to 0.}$$

$$-4a + 8b - 12 = 0 \qquad \text{Set } \partial S/\partial b \text{ equal to 0.}$$

The solution of this system of linear equations is

$$a = \frac{8}{13} \quad \text{and} \quad b = \frac{47}{26}.$$

So, the best-fitting linear model for the given points is

$$f(x) = \frac{8}{13}x + \frac{47}{26}.$$

The graph of this model is shown in Figure 7.40.

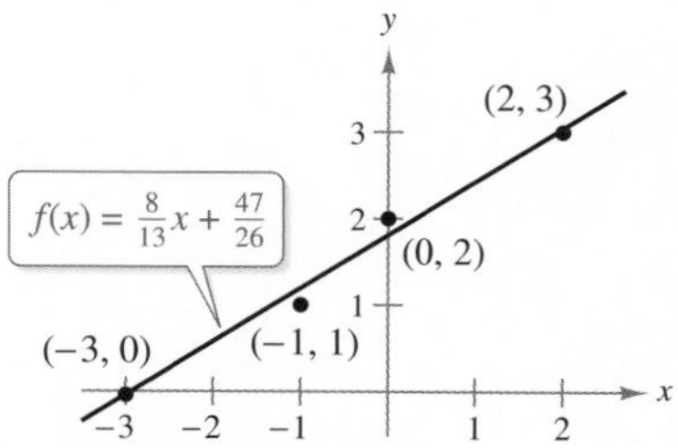

FIGURE 7.40

✓ Checkpoint 2

Find the values of a and b such that the linear model

$$f(x) = ax + b$$

has a minimum sum of the squared errors for the points

$$(-2, 0), (0, 2), (2, 5), (4, 7).$$

■

TECH TUTOR

Most graphing utilities and spreadsheet software programs have a built-in linear regression program. When you run such a program, the "r-value" gives a measure of how well the model fits the data. The closer the value of $|r|$ is to 1, the better the fit. You should use a graphing utility or a spreadsheet software program to verify your solutions to the exercises.

The line in Example 2 is called the **least squares regression line** for the given data. The solution shown in Example 2 can be generalized to find a formula for the least squares regression line. Consider the linear model

$$f(x) = ax + b$$

and the points

$$(x_1, y_1), (x_2, y_2), \ldots, (x_n, y_n).$$

The sum of the squared errors is

$$\begin{aligned} S &= [f(x_1) - y_1]^2 + [f(x_2) - y_2]^2 + \cdots + [f(x_n) - y_n]^2 \\ &= (ax_1 + b - y_1)^2 + (ax_2 + b - y_2)^2 + \cdots + (ax_n + b - y_n)^2. \end{aligned}$$

To minimize S, set the partial derivatives $\partial S/\partial a$ and $\partial S/\partial b$ equal to zero and solve for a and b. The results are summarized below.

The Least Squares Regression Line

The **least squares regression line** for the points

$$(x_1, y_1), (x_2, y_2), \ldots, (x_n, y_n)$$

is $f(x) = ax + b$, where

$$a = \frac{n\sum_{i=1}^{n} x_i y_i - \sum_{i=1}^{n} x_i \sum_{i=1}^{n} y_i}{n\sum_{i=1}^{n} x_i^2 - \left(\sum_{i=1}^{n} x_i\right)^2} \quad \text{and} \quad b = \frac{1}{n}\left(\sum_{i=1}^{n} y_i - a\sum_{i=1}^{n} x_i\right).$$

The summation notation

$$\sum_{i=1}^{n} x_i$$

where Σ is the Greek letter sigma, is used to indicate the sum of the numbers

$$x_1 + x_2 + \cdots + x_n.$$

Similarly,

$$\sum_{i=1}^{n} x_i y_i = x_1 y_1 + x_2 y_2 + \ldots + x_n y_n, \quad \sum_{i=1}^{n} x_i^2 = x_1^2 + x_2^2 + \ldots + x_n^2,$$

and so on.

In the formula for the least squares regression line, note that if the x-values are symmetrically spaced about zero, then

$$\sum_{i=1}^{n} x_i = 0$$

and the formulas for a and b simplify to

$$a = \frac{n\sum_{i=1}^{n} x_i y_i}{n\sum_{i=1}^{n} x_i^2} \quad \text{and} \quad b = \frac{1}{n}\sum_{i=1}^{n} y_i.$$

Note also that only the *development* of the least squares regression line involves partial derivatives. The *application* of this formula is a matter of computing the values of a and b. This task is performed much more simply on a calculator or a computer than by hand.

Example 3 Modeling Hourly Wages

The average hourly wages y (in dollars per hour) for production workers in manufacturing industries from 2001 through 2009 are shown in the table. Find the least squares regression line for the data and use the result to estimate the average hourly wage in 2013. *(Source: U.S. Bureau of Labor Statistics)*

Year	2001	2002	2003	2004	2005	2006	2007	2008	2009
y	14.76	15.29	15.74	16.14	16.56	16.81	17.26	17.75	18.23

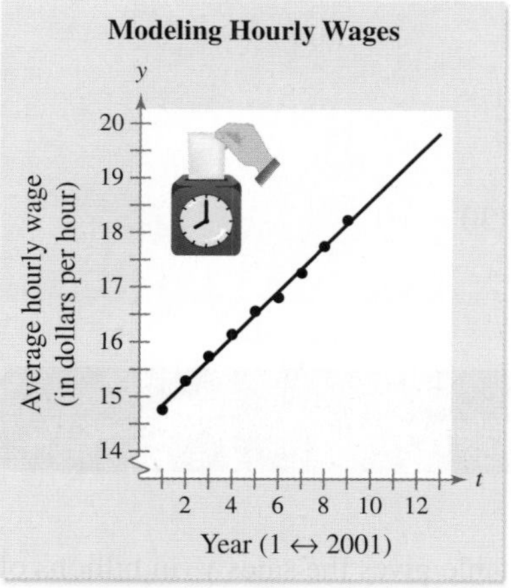

FIGURE 7.41

SOLUTION Let t represent the year, with $t = 1$ corresponding to 2001. Then, you need to find the linear model that best fits the points

$$(1, 14.76), (2, 15.29), (3, 15.74), (4, 16.14), (5, 16.56), (6, 16.81), (7, 17.26), (8, 17.75), (9, 18.23).$$

Using a calculator with a built-in least squares regression program, you can determine that the best-fitting line is

$$y = 0.416t + 14.42. \qquad \text{Best-fitting line}$$

With this model, you can estimate the 2013 average hourly wage, using $t = 13$, to be

$$\begin{aligned} y &= 0.416(13) + 14.42 \\ &= 19.828 \\ &\approx \$19.83 \text{ per hour.} \end{aligned}$$

This result is shown graphically in Figure 7.41.

✓ Checkpoint 3

The numbers of cellular phone subscribers y (in thousands) for the years 2000 through 2009 are shown in the table. Find the least squares regression line for the data and use the result to estimate the number of subscribers in 2013. Let t represent the year, with $t = 0$ corresponding to 2000. *(Source: CTIA-The Wireless Association)*

Year	2000	2001	2002	2003	2004
y	190,478	128,375	140,767	158,722	182,140

Year	2005	2006	2007	2008	2009
y	207,896	233,041	255,396	270,334	285,646

SUMMARIZE (Section 7.7)

1. State the definition of the sum of the squared errors *(page 474)*. For an example of finding the sum of the squared errors, See Example 1.
2. State the definition of the least squares regression line *(page 476)*. For an example of finding the least squares regression line, see Example 2.
3. Describe a real-life example of modeling hourly wages using the least squares regression line *(page 477, Example 3)*.

David Gilder/Shutterstock.com

SKILLS WARM UP 7.7 The following warm-up exercises involve skills that were covered in a previous course or in earlier sections. You will use these skills in the exercise set for this section. For additional help, review Appendix Section A.3 and Section 7.4.

In Exercises 1 and 2, evaluate the expression.

1. $(2.5 - 1)^2 + (3.25 - 2)^2 + (4.1 - 3)^2$

2. $(1.1 - 1)^2 + (2.08 - 2)^2 + (2.95 - 3)^2$

In Exercises 3 and 4, find the partial derivatives of S.

3. $S = a^2 + 6b^2 - 4a - 8b - 4ab + 6$

4. $S = 4a^2 + 9b^2 - 6a - 4b - 2ab + 8$

In Exercises 5–10, evaluate the sum.

5. $\sum_{i=1}^{5} i$

6. $\sum_{i=1}^{6} 2i$

7. $\sum_{i=1}^{4} \frac{1}{i}$

8. $\sum_{i=1}^{3} i^2$

9. $\sum_{i=1}^{6} (2 - i)^2$

10. $\sum_{i=1}^{5} (30 - i^2)$

Exercises 7.7

See www.CalcChat.com for worked-out solutions to odd-numbered exercises.

Finding the Sum of the Squared Errors In Exercises 1–4, find the sum of the squared errors for the linear model $f(x)$ and the quadratic model $g(x)$ using the given points. *See Example 1.*

1. $f(x) = 1.6x + 6$, $g(x) = 0.29x^2 + 2.2x + 6$
$(-3, 2), (-2, 2), (-1, 4), (0, 6), (1, 8)$

2. $f(x) = -0.7x + 2$, $g(x) = 0.06x^2 - 0.7x + 1$
$(-3, 4), (-1, 2), (1, 1), (3, 0)$

3. $f(x) = -3.3x + 11$, $g(x) = -1.25x^2 + 0.5x + 10$
$(0, 10), (1, 9), (2, 6), (3, 0)$

4. $f(x) = 2.0x - 3$, $g(x) = 0.14x^2 + 1.3x - 3$
$(-1, -4), (1, -3), (2, 0), (4, 5), (6, 9)$

Finding the Least Squares Regression Line In Exercises 5–8, find the least squares regression line for the given points. Then plot the points and sketch the regression line. *See Example 2.*

5. $(-2, -1), (0, 0), (2, 3)$

6. $(-3, 0), (-1, 1), (1, 1), (3, 2)$

7. $(-2, 4), (-1, 1), (0, -1), (1, -3)$

8. $(-5, -3), (-4, -2), (-2, -1), (-1, 1)$

Finding the Least Squares Regression Line In Exercises 9–12, use the regression capabilities of a graphing utility or a spreadsheet to find the least squares regression line for the given points.

9. $(-4, -1), (-2, 0), (2, 4), (4, 5)$

10. $(-5, 1), (1, 3), (2, 3), (2, 5)$

11. $(0, 6), (4, 3), (5, 0), (8, -4), (10, -5)$

12. $(-10, 10), (-5, 8), (3, 6), (7, 4), (5, 0)$

13. Sales The table gives the sales y (in billions of dollars) for Best Buy from 2002 through 2009. *(Source: Best Buy Company, Inc.)*

Year	2002	2003	2004	2005
Sales, y	20.9	24.5	27.4	30.8

Year	2006	2007	2008	2009
Sales, y	35.9	40.0	45.0	49.7

(a) Use the regression capabilities of a graphing utility or a spreadsheet to find the least squares regression line for the data. Let $t = 2$ represent 2002.

(b) Estimate the sales in 2014.

(c) In what year will the sales be \$85 billion?

14. Demand A hardware retailer wants to know the demand y for a tool as a function of price x. The monthly sales for four different prices of the tool are listed in the table.

Price, x	\$25	\$30	\$35	\$40
Demand, y	82	75	67	55

(a) Use the regression capabilities of a graphing utility or a spreadsheet to find the least squares regression line for the data.

(b) Estimate the demand when the price is \$32.95.

(c) What price will create a demand of 83 tools?

15. Agriculture An agronomist used four test plots to determine the relationship between the wheat yield y (in bushels per acre) and the amount of fertilizer x (in pounds per acre). The results are shown in the table.

Fertilizer, x	100	150	200	250
Yield, y	35	44	50	56

(a) Use the regression capabilities of a graphing utility or a spreadsheet to find the least squares regression line for the data.

(b) Estimate the yield for a fertilizer application of 160 pounds per acre.

16. HOW DO YOU SEE IT? Match the regression equation with the appropriate graph. Explain your reasoning. (Note that the x- and y-axes are broken.)

(a) $y = 0.22x - 7.5$ (b) $y = -0.35x + 11.5$

(c) $y = 0.09x + 19.8$ (d) $y = -1.29x + 89.8$

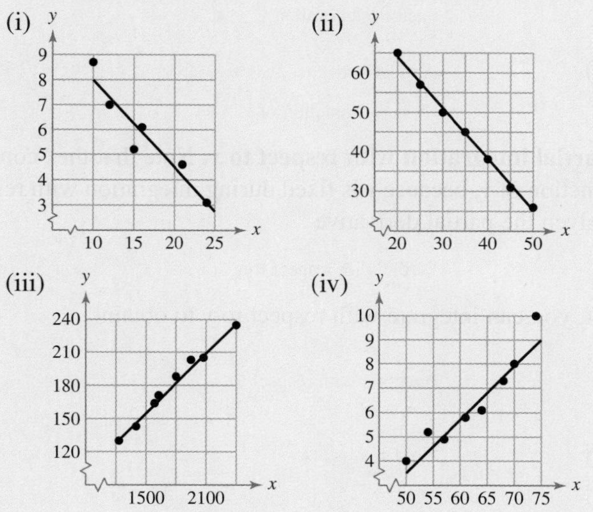

Determining Correlation In Exercises 17–22, plot the points and determine whether the data have positive, negative, or no linear correlation (see figures below). Then use a graphing utility to find the value of r and confirm your result. The number r is called the *correlation coefficient*. It is a measure of how well the model fits the data. Correlation coefficients vary between -1 and 1, and the closer $|r|$ is to 1, the better the model.

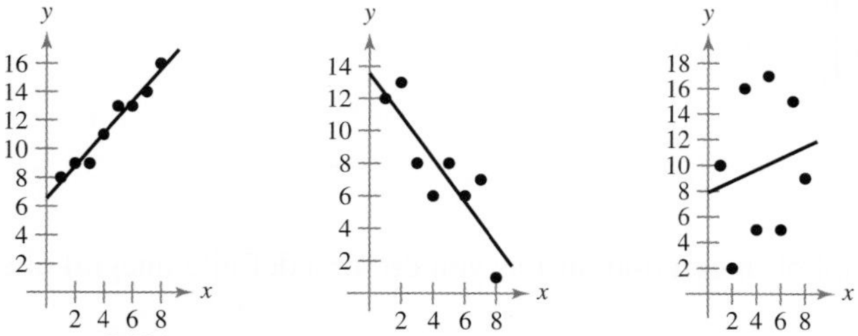

$r = 0.981$ Positive correlation

$r = -0.866$ Negative correlation

$r = 0.190$ No correlation

17. (1, 4), (2, 6), (3, 8), (4, 11), (5, 13), (6, 15)

18. (1, 7.5), (2, 7), (3, 7), (4, 6), (5, 5), (6, 4.9)

19. (1, 3), (2, 6), (3, 2), (4, 3), (5, 9), (6, 1)

20. (0.5, 2), (0.75, 1.75), (1, 3), (1.5, 3.2), (2, 3.7), (2.6, 4)

21. (1, 36), (2, 10), (3, 0), (4, 4), (5, 16), (6, 36)

22. (0.5, 9), (1, 8.5), (1.5, 7), (2, 5.5), (2.5, 5), (3, 3.5)

True or False? In Exercises 23–28, determine whether the statement is true or false. If it is false, explain why or give an example that shows it is false.

23. Data that are modeled by

$$y = 3.29x - 4.17$$

have a negative correlation.

24. Data that are modeled by

$$y = -0.238x + 25$$

have a negative correlation.

25. When the correlation coefficient is $r \approx -0.98781$, the model is a good fit.

26. A correlation coefficient of $r \approx 0.201$ implies that the data have no correlation.

27. A linear regression model with a positive correlation coefficient will have a slope that is greater than 0.

28. When the correlation coefficient for a linear regression model is close to -1, the regression line cannot be used to describe the data.

29. Project: Financial Data For a project analyzing the revenues per share, revenues, and shareholder's equity of McDonald's from 2000 through 2009, visit the text's website at *www.cengagebrain.com*. *(Source: McDonald's Corporation)*

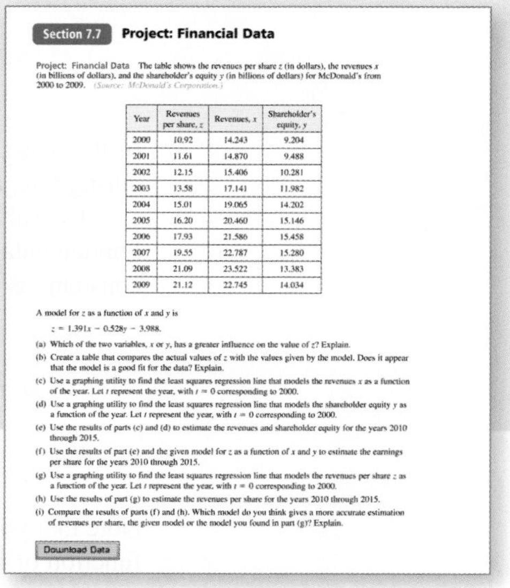

Section 7.7 **Project: Financial Data**

Project: Financial Data The table shows the revenues per share z (in dollars), the revenues x (in billions of dollars), and the shareholder's equity y (in billions of dollars) for McDonald's from 2000 to 2009. *(Source: McDonald's Corporation)*

Year	Revenues per share, z	Revenues, x	Shareholder's equity, y
2000	10.92	14.243	9.204
2001	11.61	14.870	9.488
2002	12.15	15.406	10.281
2003	13.58	17.141	11.982
2004	15.01	19.065	14.202
2005	16.20	20.460	15.146
2006	17.93	21.586	15.458
2007	19.55	22.787	15.280
2008	21.09	23.522	13.383
2009	21.12	22.745	14.034

A model for z as a function of x and y is

$$z = 1.391x - 0.528y - 3.988.$$

(a) Which of the two variables, x or y, has a greater influence on the value of z? Explain.

(b) Create a table that compares the actual values of z with the values given by the model. Does it appear that the model is a good fit for the data? Explain.

(c) Use a graphing utility to find the least squares regression line that models the revenues x as a function of the year. Let t represent the year, with $t = 0$ corresponding to 2000.

(d) Use a graphing utility to find the least squares regression line that models the shareholder equity y as a function of the year. Let t represent the year, with $t = 0$ corresponding to 2000.

(e) Use the results of parts (c) and (d) to estimate the revenues and shareholder equity for the years 2010 through 2015.

(f) Use the results of part (e) and the given model for z as a function of x and y to estimate the earnings per share for the years 2010 through 2015.

(g) Use a graphing utility to find the least squares regression line that models the revenues per share z as a function of the year. Let t represent the year, with $t = 0$ corresponding to 2000.

(h) Use the results of part (g) to estimate the revenues per share for the years 2010 through 2015.

(i) Compare the results of parts (f) and (h). Which model do you think gives a more accurate estimation of revenues per share, the given model or the model you found in part (g)? Explain.

Download Data

7.8 Double Integrals and Area in the Plane

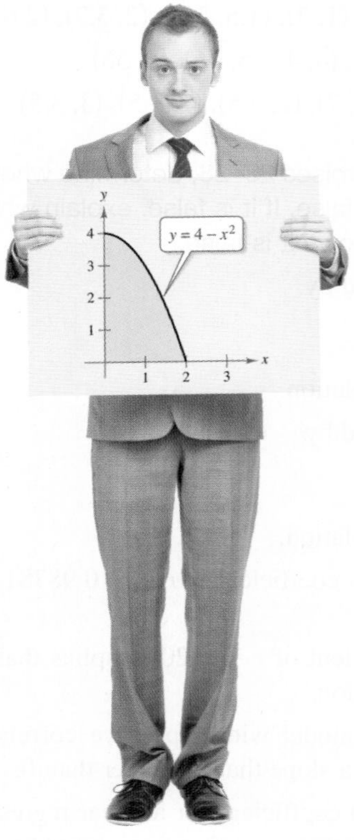

In Exercise 29 on page 487, you will use a double integral to find the area of a region.

- Evaluate double integrals.
- Use double integrals to find the areas of regions.

Double Integrals

In Section 7.4, you learned that it is meaningful to differentiate functions of several variables with respect to one variable while holding the other variable(s) constant. You can *integrate* functions of several variables by a similar procedure. For instance, if you are given the partial derivative

$$f_x(x, y) = 2xy \qquad \text{Partial with respect to } x$$

then, by holding y constant, you can integrate with respect to x to obtain

$$\begin{aligned} f(x, y) &= \int f_x(x, y)\,dx && \text{Integrate with respect to } x. \\ &= \int 2xy\,dx && \text{Hold } y \text{ constant.} \\ &= y\int 2x\,dx && \text{Factor out constant } y. \\ &= y(x^2) + C(y) && \text{Antiderivative of } 2x \text{ is } x^2. \\ &= x^2y + C(y). && C(y) \text{ is a function of } y. \end{aligned}$$

This procedure is called **partial integration with respect to** $\boldsymbol{x}$. Note that the "constant of integration" $C(y)$ is a function of y, because y is fixed during integration with respect to x. Similarly, if you are given the partial derivative

$$f_y(x, y) = x^2 + 2 \qquad \text{Partial with respect to } y$$

then, by holding x constant, you can integrate with respect to y to obtain

$$\begin{aligned} f(x, y) &= \int f_y(x, y)\,dy && \text{Integrate with respect to } y. \\ &= \int (x^2 + 2)\,dy && \text{Hold } x \text{ constant.} \\ &= (x^2 + 2)\int dy && \text{Factor out constant } x^2 + 2. \\ &= (x^2 + 2)(y) + C(x) && \text{Antiderivative of 1 is } y. \\ &= x^2y + 2y + C(x). && C(x) \text{ is a function of } x. \end{aligned}$$

In this case, the "constant of integration" $C(x)$ is a function of x, because x is fixed during integration with respect to y.

To evaluate a definite integral of a function of several variables, you can apply the Fundamental Theorem of Calculus to one variable while holding the other variable(s) constant, as shown.

$$\int_1^{2y} 2xy\,dx = x^2y\Big]_1^{2y} = (2y)^2y - (1)^2y = 4y^3 - y$$

x is the variable of integration and y is fixed. (points to dx)

Replace x by the limits of integration. (points to $2y$ and 1)

The result is a function of y. (points to $4y^3 - y$)

Note that you omit the constant of integration, just as you do for a definite integral of a function of one variable.

Lasse Kristensen/Shutterstock.com

Example 1 **Finding Partial Integrals**

a. $$\int_1^x (2x^2y^{-2} + 2y)\,dy = \left[\frac{-2x^2}{y} + y^2\right]_1^x$$ Hold x constant.

$$= \left(\frac{-2x^2}{x} + x^2\right) - \left(\frac{-2x^2}{1} + 1\right)$$

$$= 3x^2 - 2x - 1$$

b. $$\int_y^{5y} \sqrt{x - y}\,dx = \left[\frac{2}{3}(x - y)^{3/2}\right]_y^{5y}$$ Hold y constant.

$$= \frac{2}{3}[(5y - y)^{3/2} - (y - y)^{3/2}]$$

$$= \frac{16}{3}y^{3/2}$$

✓ Checkpoint 1

Find each partial integral.

a. $\displaystyle\int_1^x (4xy + y^3)\,dy$ **b.** $\displaystyle\int_y^{y2} \frac{1}{x + y}\,dx$ ■

STUDY TIP

Notice that the difference between the two types of double integrals is the order in which the integration is performed, *dy dx* or *dx dy*.

In Example 1(a), note that the definite integral defines a function of x and can *itself* be integrated. An "integral of an integral" is called a **double integral.** With a function of two variables, there are two types of double integrals.

$$\int_a^b \int_{g_1(x)}^{g_2(x)} f(x, y)\,dy\,dx = \int_a^b \left[\int_{g_1(x)}^{g_2(x)} f(x, y)\,dy\right] dx$$

$$\int_a^b \int_{g_1(y)}^{g_2(y)} f(x, y)\,dx\,dy = \int_a^b \left[\int_{g_1(y)}^{g_2(y)} f(x, y)\,dx\right] dy$$

TECH TUTOR

A symbolic integration utility can be used to evaluate double integrals. To do this, you need to enter the integrand, then integrate twice—once with respect to one of the variables and then with respect to the other variable. Use a symbolic integration utility to evaluate the double integral in Example 2.

Example 2 **Evaluating a Double Integral**

$$\int_1^2 \int_0^x (2xy + 3)\,dy\,dx = \int_1^2 \left[\int_0^x (2xy + 3)\,dy\right] dx$$

$$= \int_1^2 \left[xy^2 + 3y\right]_0^x dx$$

$$= \int_1^2 (x^3 + 3x)\,dx$$

$$= \left[\frac{x^4}{4} + \frac{3x^2}{2}\right]_1^2$$

$$= \left(\frac{2^4}{4} + \frac{3(2^2)}{2}\right) - \left(\frac{1^4}{4} + \frac{3(1^2)}{2}\right)$$

$$= \frac{33}{4}$$

✓ Checkpoint 2

Evaluate $\displaystyle\int_1^2 \int_0^x (5x^2y - 2)\,dy\,dx.$ ■

Finding Area with a Double Integral

One of the simplest applications of a double integral is finding the area of a plane region. For instance, consider the region R that is bounded by

$$a \le x \le b \quad \text{and} \quad g_1(x) \le y \le g_2(x)$$

as shown in Figure 7.42. Using the techniques described in Section 5.5, you know that the area of R is

$$\int_a^b [g_2(x) - g_1(x)]\, dx. \qquad \text{Area of } R$$

This same area is also given by the double integral

$$\int_a^b \int_{g_1(x)}^{g_2(x)} dy\, dx \qquad \text{Area of } R$$

because

$$\int_a^b \int_{g_1(x)}^{g_2(x)} dy\, dx = \int_a^b \Big[\, y \,\Big]_{g_1(x)}^{g_2(x)} dx = \int_a^b [g_2(x) - g_1(x)]\, dx.$$

Region is bounded by
$a \le x \le b$
$g_1(x) \le y \le g_2(x)$

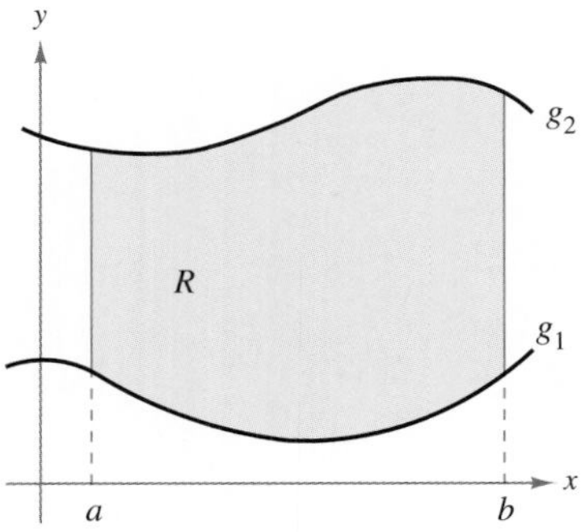

FIGURE 7.42

Figure 7.43 shows the two basic types of plane regions whose areas can be determined by a double integral.

STUDY TIP

To designate a double integral or an area of a region without specifying a particular order of integration, you can use the symbol

$$\iint_R dA$$

where $dA = dx\, dy$ or $dA = dy\, dx$.

Determining Area in the Plane by Double Integrals

Region is bounded by
$a \le x \le b$
$g_1(x) \le y \le g_2(x)$

Region is bounded by
$c \le y \le d$
$h_1(y) \le x \le h_2(y)$

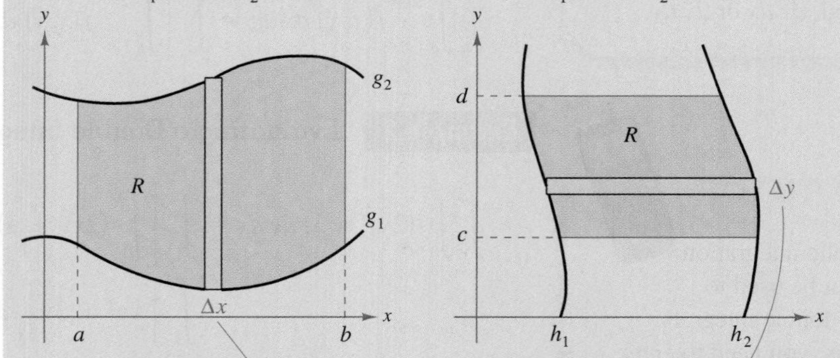

$$\text{Area} = \int_a^b \int_{g_1(x)}^{g_2(x)} dy\, dx \qquad \text{Area} = \int_c^d \int_{h_1(y)}^{h_2(y)} dx\, dy$$

FIGURE 7.43

STUDY TIP

In Figure 7.43, note that the horizontal or vertical orientation of the narrow rectangle indicates the order of integration. The "outer" variable of integration always corresponds to the width of the rectangle. Notice also that the outer limits of integration for a double integral are constant, whereas the inner limits may be functions of the outer variable.

Example 3 Finding Area with a Double Integral

Use a double integral to find the area of the rectangular region shown in Figure 7.44.

SOLUTION The bounds for x are $1 \le x \le 5$ and the bounds for y are $2 \le y \le 4$. So, the area of the region is

$$\int_1^5 \int_2^4 dy\,dx = \int_1^5 \Big[y\Big]_2^4 dx \qquad \text{Integrate with respect to } y.$$

$$= \int_1^5 (4 - 2)\,dx \qquad \text{Apply Fundamental Theorem of Calculus.}$$

$$= \int_1^5 2\,dx \qquad \text{Simplify.}$$

$$= \Big[2x\Big]_1^5 \qquad \text{Integrate with respect to } x.$$

$$= 10 - 2 \qquad \text{Apply Fundamental Theorem of Calculus.}$$

$$= 8 \text{ square units.} \qquad \text{Simplify.}$$

You can confirm this by noting that the rectangle measures two units by four units.

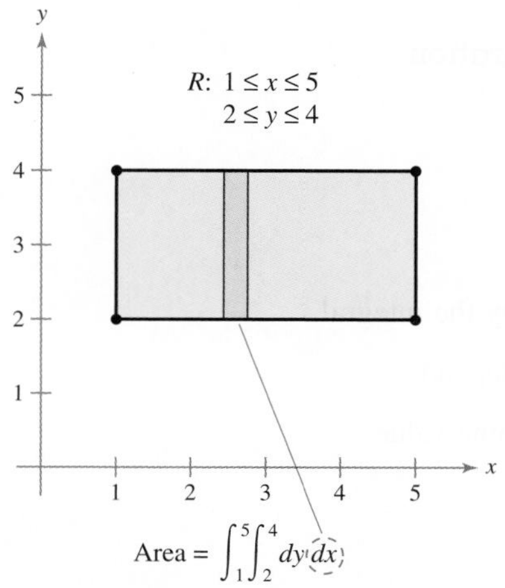

FIGURE 7.44

✓ Checkpoint 3

Use a double integral to find the area of the rectangular region shown in Example 3 by integrating with respect to x and then with respect to y. ■

Example 4 Finding Area with a Double Integral

Use a double integral to find the area of the region bounded by the graphs of

$y = x^2$ and $y = x^3$.

SOLUTION As shown in Figure 7.45, the two graphs intersect when $x = 0$ and $x = 1$. Choosing x to be the outer variable, the bounds for x are $0 \le x \le 1$. On the interval $0 \le x \le 1$, the region is bounded above by $y = x^2$ and below by $y = x^3$. So, the bounds for y are

$$x^3 \le y \le x^2.$$

This implies that the area of the region is

$$\int_0^1 \int_{x^3}^{x^2} dy\,dx = \int_0^1 \Big[y\Big]_{x^3}^{x^2} dx \qquad \text{Integrate with respect to } y.$$

$$= \int_0^1 (x^2 - x^3)\,dx \qquad \text{Apply Fundamental Theorem of Calculus.}$$

$$= \left[\frac{x^3}{3} - \frac{x^4}{4}\right]_0^1 \qquad \text{Integrate with respect to } x.$$

$$= \frac{1}{3} - \frac{1}{4} \qquad \text{Apply Fundamental Theorem of Calculus.}$$

$$= \frac{1}{12} \text{ square unit.} \qquad \text{Simplify.}$$

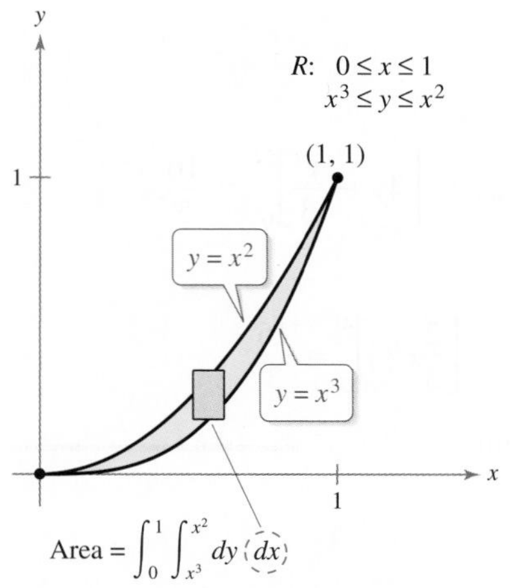

FIGURE 7.45

✓ Checkpoint 4

Use a double integral to find the area of the region bounded by the graphs of

$y = 2x$ and $y = x^2$. ■

In setting up double integrals, the most difficult task is likely to be determining the correct limits of integration. This can be simplified by making a sketch of the region R and identifying the appropriate bounds for x and y.

Example 5 Changing the Order of Integration

For the double integral

$$\int_0^2 \int_{y^2}^4 dx\, dy$$

a. sketch the region R whose area is represented by the integral,

b. rewrite the integral so that x is the outer variable, and

c. show that both orders of integration yield the same value.

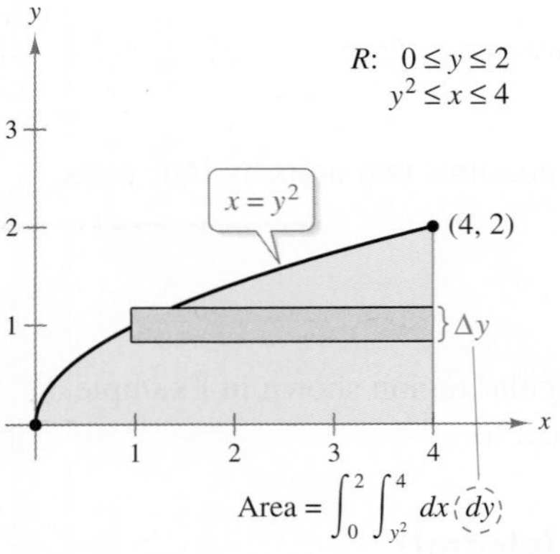

FIGURE 7.46

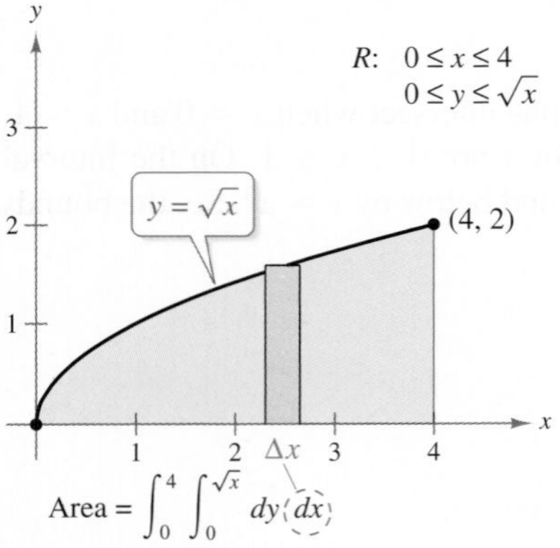

FIGURE 7.47

SOLUTION

a. From the limits of integration, you know that

$y^2 \le x \le 4$ Inner limits of integration

which means that the region R is bounded on the left by the parabola $x = y^2$ and on the right by the line $x = 4$. Furthermore, because

$0 \le y \le 2$ Outer limits of integration

you know that the region lies above the x-axis, as shown in Figure 7.46.

b. If you interchange the order of integration so that x is the outer variable, then x will have constant bounds of integration given by

$0 \le x \le 4.$ Outer limits of integration

By solving for y in the equation $x = y^2$, you can conclude that the bounds for y are

$0 \le y \le \sqrt{x}$ Inner limits of integration

as shown in Figure 7.47. So, with x as the outer variable, the integral can be written as

$$\int_0^4 \int_0^{\sqrt{x}} dy\, dx.$$

c. Integrating with respect to x, you have

$$\int_0^2 \int_{y^2}^4 dx\, dy = \int_0^2 \Big[x\Big]_{y^2}^4 dy = \int_0^2 (4 - y^2)\, dy = \left[4y - \frac{y^3}{3}\right]_0^2 = \frac{16}{3}.$$

Integrating with respect to y, you have

$$\int_0^4 \int_0^{\sqrt{x}} dy\, dx = \int_0^4 \Big[y\Big]_0^{\sqrt{x}} dx = \int_0^4 \sqrt{x}\, dx = \left[\frac{2}{3}x^{3/2}\right]_0^4 = \frac{16}{3}.$$

So, both orders of integration yield the same value.

✓ Checkpoint 5

For the double integral $\int_0^2 \int_{2y}^4 dx\, dy$,

a. sketch the region R whose area is represented by the integral,

b. rewrite the integral so that x is the outer variable, and

c. show that both orders of integration yield the same result.

Example 6 Finding Area with a Double Integral

Use a double integral to calculate the area denoted by

$$\int_R\int dA$$

where R is the region bounded by $y = x$ and $y = x^2 - x$.

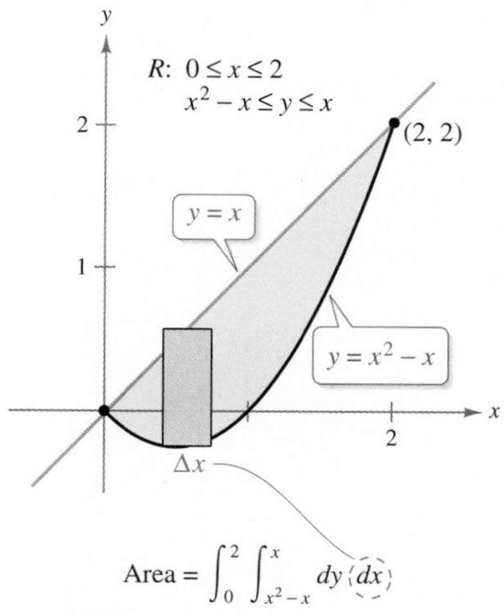

FIGURE 7.48

SOLUTION Begin by sketching the region R, as shown in Figure 7.48. From the sketch, you can see that vertical rectangles of width dx are more convenient than horizontal ones. So, x is the outer variable of integration and its constant bounds are $0 \le x \le 2$. This implies that the bounds for y are $x^2 - x \le y \le x$, and the area is given by

$$\int_R\int dA = \int_0^2 \int_{x^2-x}^{x} dy\, dx \qquad \text{Substitute bounds for region.}$$

$$= \int_0^2 \Big[y \Big]_{x^2-x}^{x} dx \qquad \text{Integrate with respect to } y.$$

$$= \int_0^2 [x - (x^2 - x)]\, dx \qquad \text{Apply Fundamental Theorem of Calculus.}$$

$$= \int_0^2 (2x - x^2)\, dx \qquad \text{Simplify.}$$

$$= \left[x^2 - \frac{x^3}{3} \right]_0^2 \qquad \text{Integrate with respect to } x.$$

$$= 4 - \frac{8}{3} \qquad \text{Apply Fundamental Theorem of Calculus.}$$

$$= \frac{4}{3} \text{ square units.} \qquad \text{Simplify.}$$

✓Checkpoint 6

Use a double integral to calculate the area denoted by

$$\int_R\int dA$$

where R is the region bounded by $y = 2x + 3$ and $y = x^2$. ■

As you are working the exercises for this section, you should be aware that the primary uses of double integrals will be discussed in Section 7.9. Double integrals by way of areas in the plane have been introduced so that you can gain practice in finding the limits of integration. When setting up a double integral, remember that your first step should be to sketch the region R. After doing this, you have two choices of integration orders: $dx\,dy$ or $dy\,dx$.

SUMMARIZE (Section 7.8)

1. Describe a procedure for finding a partial integral with respect to one variable *(page 480)*. For an example of finding a partial integral with respect to x or to y, see Example 1.
2. Explain how to determine the area of a region in the plane using a double integral *(page 482)*. For examples of finding area using a double integral, see Examples 3, 4, and 5.

SKILLS WARM UP 7.8 The following warm-up exercises involve skills that were covered in earlier sections. You will use these skills in the exercise set for this section. For additional help, review Sections 5.2–5.5.

In Exercises 1–12, evaluate the definite integral.

1. $\int_0^1 dx$

2. $\int_0^2 3\,dy$

3. $\int_1^4 2x^2\,dx$

4. $\int_0^1 2x^3\,dx$

5. $\int_1^2 (x^3 - 2x + 4)\,dx$

6. $\int_0^2 (4 - y^2)\,dy$

7. $\int_1^2 \frac{2}{7x^2}\,dx$

8. $\int_1^4 \frac{2}{\sqrt{x}}\,dx$

9. $\int_0^2 \frac{2x}{x^2 + 1}\,dx$

10. $\int_2^e \frac{1}{y - 1}\,dy$

11. $\int_0^2 xe^{x^2+1}\,dx$

12. $\int_0^1 e^{-2y}\,dy$

In Exercises 13–16, sketch the region bounded by the graphs of the equations.

13. $y = x,\ y = 0,\ x = 3$

14. $y = x,\ y = 3,\ x = 0$

15. $y = 4 - x^2,\ y = 0,\ x = 0$

16. $y = x^2,\ y = 4x$

Exercises 7.8

See www.CalcChat.com for worked-out solutions to odd-numbered exercises.

Finding Partial Integrals In Exercises 1–10, find the partial integral. *See Example 1.*

1. $\int_0^x (2x - y)\,dy$

2. $\int_0^y (5x + 8y)\,dx$

3. $\int_x^{x^2} \frac{y}{x}\,dy$

4. $\int_1^{2y} \frac{y}{x}\,dx$

5. $\int_2^y (6x^2y + y^2)\,dx$

6. $\int_4^x (xy^3 + 4y)\,dy$

7. $\int_{x^3}^{\sqrt{x}} (x^2 + 3y^2)\,dy$

8. $\int_{-\sqrt{1-y^2}}^{\sqrt{1-y^2}} (x^2 + y^2)\,dx$

9. $\int_1^{e^y} \frac{y \ln x}{x}\,dx$

10. $\int_y^3 \frac{xy}{\sqrt{x^2 + 1}}\,dx$

Evaluating a Double Integral In Exercises 11–24, evaluate the double integral. *See Example 2.*

11. $\int_0^1 \int_0^2 (x + y)\,dy\,dx$

12. $\int_0^2 \int_0^2 (6 - x^2)\,dy\,dx$

13. $\int_0^3 \int_0^4 xy\,dx\,dy$

14. $\int_{-1}^1 \int_{-2}^2 (x^2 - y^2)\,dy\,dx$

15. $\int_0^2 \int_0^{6x^2} x^3\,dy\,dx$

16. $\int_0^2 \int_{3y^2-6y}^{2y-y^2} 3y\,dx\,dy$

17. $\int_0^1 \int_0^y (x + y)\,dx\,dy$

18. $\int_0^2 \int_0^{\sqrt{1-y^2}} -5xy\,dx\,dy$

19. $\int_0^1 \int_0^{3x} (3x^2 + 3y^2 + 1)\,dy\,dx$

20. $\int_0^1 \int_y^{2y} (1 + 2x^2 + 2y^2)\,dx\,dy$

21. $\int_0^1 \int_0^x \sqrt{1 - x^2}\,dy\,dx$

22. $\int_0^4 \int_0^x \frac{2}{x^2 + 1}\,dy\,dx$

23. $\int_0^\infty \int_0^\infty e^{-(x+y)/2}\,dy\,dx$

24. $\int_0^\infty \int_0^\infty xye^{-(x^2+y^2)}\,dx\,dy$

Finding Area with a Double Integral In Exercises 25–30, use a double integral to find the area of the specified region. *See Example 3.*

25.

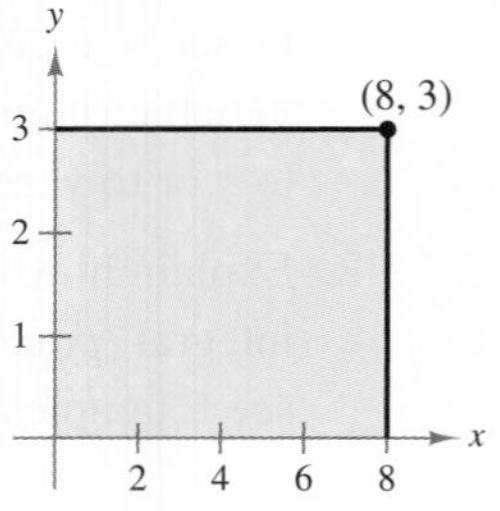

26.

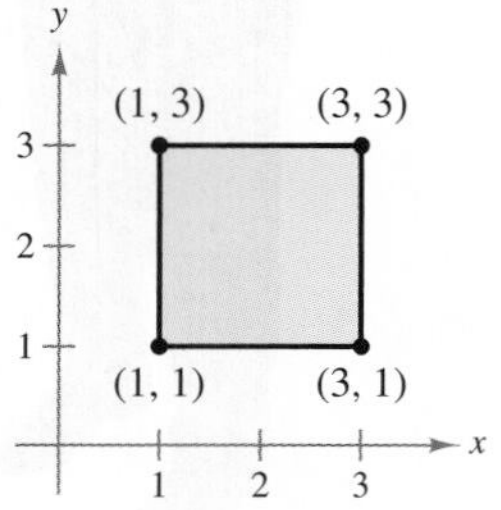

27.

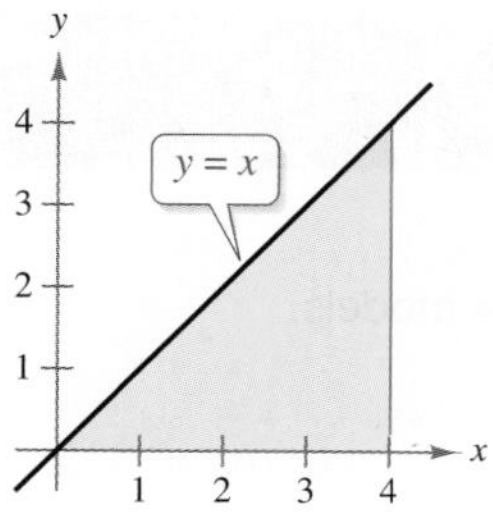

28.

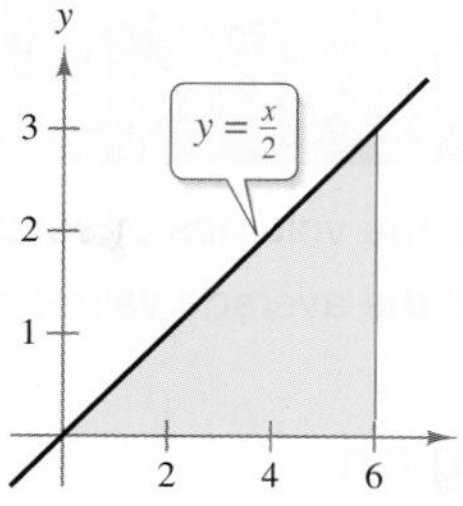

29.

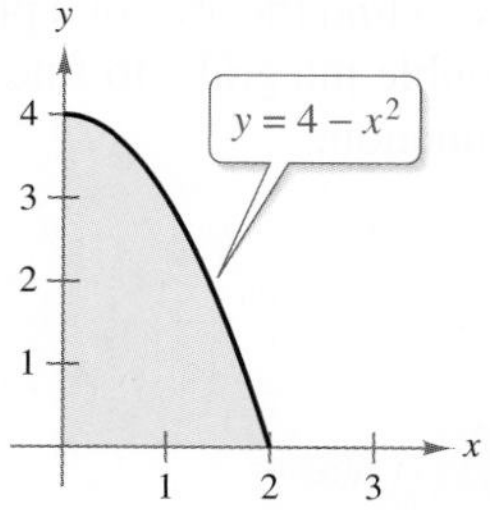

30.

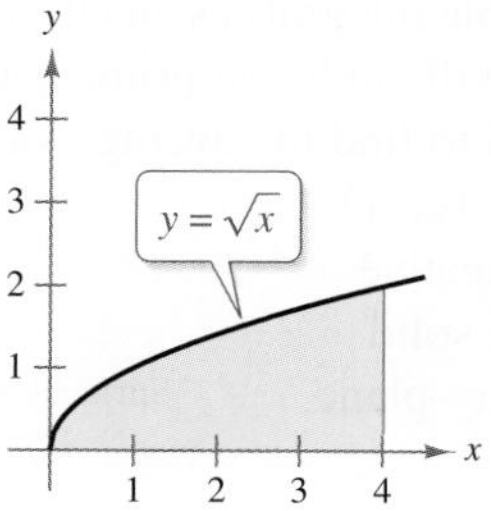

Finding Area with a Double Integral In Exercises 31–36, use a double integral to find the area of the region bounded by the graphs of the equations. *See Example 4.*

31. $y = 9 - x^2,\ y = 0$

32. $y = x^{3/2},\ y = x$

33. $2x - 3y = 0,\ x + y = 5,\ y = 0$

34. $xy = 9,\ y = x,\ y = 0,\ x = 9$

35. $y = x,\ y = 2x,\ x = 2$

36. $y = 4 - x^2,\ y = x + 2$

Changing the Order of Integration In Exercises 37–44, sketch the region R whose area is given by the double integral. Then change the order of integration and show that both orders yield the same value. *See Example 5.*

37. $\displaystyle\int_0^1\int_0^2 dy\,dx$

38. $\displaystyle\int_1^2\int_2^4 dx\,dy$

39. $\displaystyle\int_0^1\int_{2y}^2 dx\,dy$

40. $\displaystyle\int_0^4\int_0^{\sqrt{x}} dy\,dx$

41. $\displaystyle\int_0^2\int_{x/2}^1 dy\,dx$

42. $\displaystyle\int_0^4\int_{\sqrt{x}}^2 dy\,dx$

43. $\displaystyle\int_0^1\int_{y^2}^{\sqrt[3]{y}} dx\,dy$

44. $\displaystyle\int_{-2}^2\int_0^{4-y^2} dx\,dy$

45. Think About It Explain why you need to change the order of integration to evaluate the double integral. Then evaluate the double integral.

(a) $\displaystyle\int_0^3\int_y^3 e^{x^2}\,dx\,dy$ (b) $\displaystyle\int_0^2\int_x^2 e^{-y^2}\,dy\,dx$

46. HOW DO YOU SEE IT? Complete the double integrals so that each one represents the area of the region R (see figure).

(a) Area $= \displaystyle\int\int dx\,dy$ (b) Area $= \displaystyle\int\int dy\,dx$

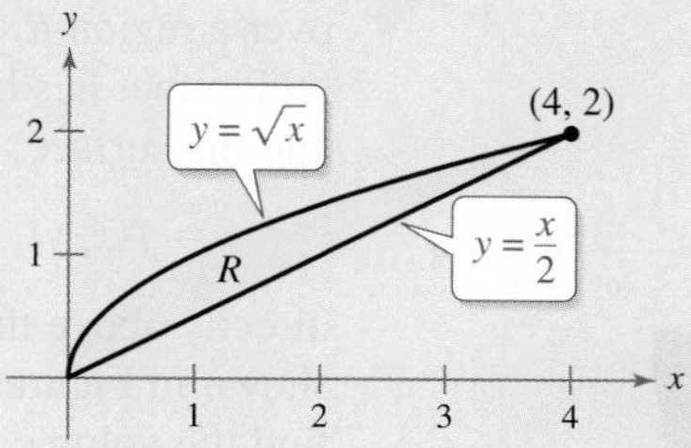

Evaluating a Double Integral In Exercises 47–54, use a symbolic integration utility to evaluate the double integral.

47. $\displaystyle\int_0^1\int_0^2 e^{-x^2-y^2}\,dx\,dy$

48. $\displaystyle\int_0^2\int_{x^2}^{2x} (x^3 + 3y^2)\,dy\,dx$

49. $\displaystyle\int_1^2\int_0^x e^{xy}\,dy\,dx$

50. $\displaystyle\int_1^2\int_y^{2y} \ln(x + y)\,dx\,dy$

51. $\displaystyle\int_0^1\int_x^1 \sqrt{1 - x^2}\,dy\,dx$

52. $\displaystyle\int_0^3\int_0^{x^2} \sqrt{x}\sqrt{1 + x}\,dy\,dx$

53. $\displaystyle\int_0^2\int_{\sqrt{4-x^2}}^{4-x^2/4} \frac{xy}{x^2 + y^2 + 1}\,dy\,dx$

54. $\displaystyle\int_0^4\int_0^y \frac{2}{(x + 1)(y + 1)}\,dx\,dy$

True or False? In Exercises 55 and 56, determine whether the statement is true or false. If it is false, explain why or give an example that shows it is false.

55. $\displaystyle\int_{-1}^1\int_{-2}^2 y\,dy\,dx = \int_{-1}^1\int_{-2}^2 y\,dx\,dy$

56. $\displaystyle\int_2^5\int_1^6 x\,dy\,dx = \int_1^6\int_2^5 x\,dx\,dy$

7.9 Applications of Double Integrals

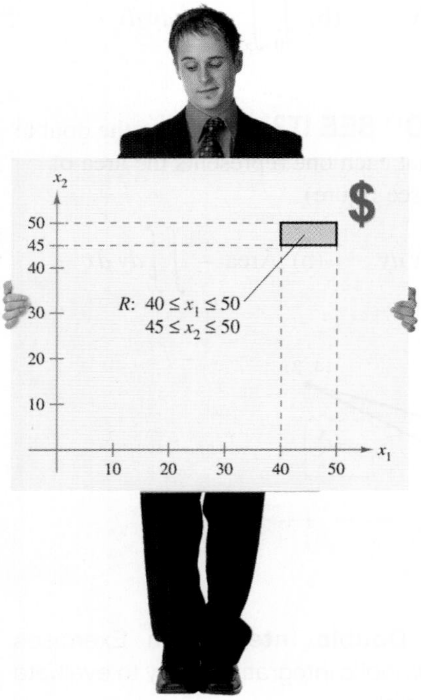

In Exercise 31 on page 495, you will use a double integral to find the average weekly profit of a company.

- Use double integrals to find the volumes of solids.
- Use double integrals to find the average values of real-life models.

Volume of a Solid Region

In Section 7.8, you used double integrals as an alternative way to find the area of a plane region. In this section, you will study the primary uses of double integrals: to find the volume of a solid region and to find the average value of a function.

Consider a function $z = f(x, y)$ that is continuous and nonnegative over a region R. Let S be the solid region that lies between the xy-plane and the surface

$$z = f(x, y)$$

directly above the region R, as shown in Figure 7.49. You can find the volume of S by integrating

$$f(x, y)$$

over the region R.

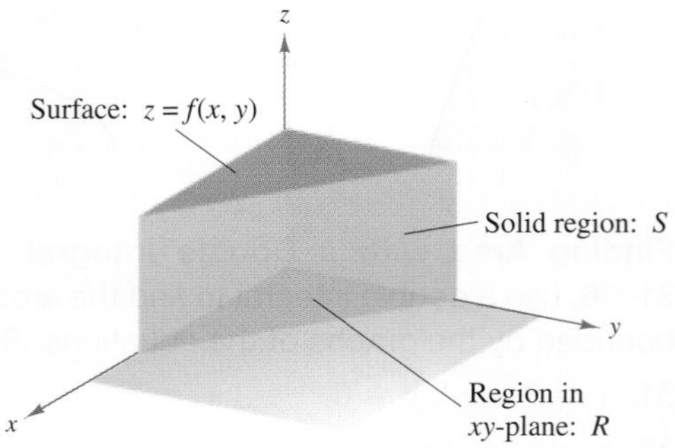

FIGURE 7.49

Determining Volume with Double Integrals

If R is a bounded region in the xy-plane and f is continuous and nonnegative over R, then the **volume of the solid** region between the surface

$$z = f(x, y)$$

and R is given by the double integral

$$\int_R \int f(x, y)\, dA$$

where $dA = dx\, dy$ or $dA = dy\, dx$.

You can use the following guidelines when finding the volume of a solid.

Guidelines for Finding the Volume of a Solid

1. Write the equation of the surface in the form

$$z = f(x, y)$$

and sketch the solid region.

2. Sketch the region R in the xy-plane and determine the order and limits of integration.
3. Evaluate the double integral

$$\int_R \int f(x, y)\, dA$$

using the order and limits determined in the second step.

Example 1 Finding the Volume of a Solid

Find the volume of the solid region bounded in the first octant by the plane

$$z = 2 - x - 2y.$$

SOLUTION

1. The equation of the surface is already in the form $z = f(x, y)$. A graph of the solid region is shown in Figure 7.50.

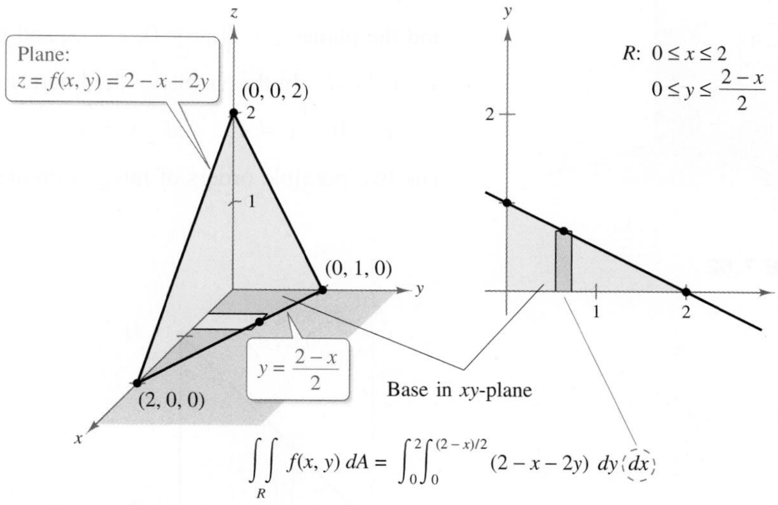

FIGURE 7.50

2. Sketch the region R in the xy-plane. In Figure 7.50, you can see that the region R is bounded by the lines $x = 0$, $y = 0$, and $y = \frac{1}{2}(2 - x)$. One way to set up the double integral is to choose x as the outer variable. With that choice, the bounds for x are $0 \le x \le 2$ and the bounds for y are $0 \le y \le \frac{1}{2}(2 - x)$.
3. The volume of the solid region is

$$\begin{aligned}
V &= \int_0^2 \int_0^{(2-x)/2} (2 - x - 2y)\, dy\, dx \\
&= \int_0^2 \Big[(2 - x)y - y^2\Big]_0^{(2-x)/2} dx \\
&= \int_0^2 \left\{(2 - x)\left(\frac{1}{2}\right)(2 - x) - \left[\frac{1}{2}(2 - x)\right]^2\right\} dx \\
&= \frac{1}{4}\int_0^2 (2 - x)^2\, dx \\
&= \frac{1}{4}\left[-\frac{1}{3}(2 - x)^3\right]_0^2 \\
&= \frac{2}{3} \text{ cubic unit.}
\end{aligned}$$

✓Checkpoint 1

Find the volume of the solid region bounded in the first octant by the plane $z = 4 - 2x - y$.

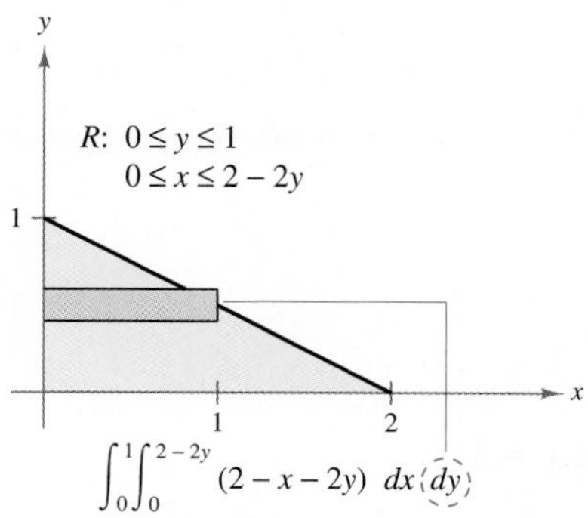

FIGURE 7.51

Example 1 uses $dy\,dx$ as the order of integration. The other order, $dx\,dy$, as indicated in Figure 7.51, produces the same result. Try verifying this.

In Example 1, the problem could be solved with either order of integration. Moreover, had you used the order $dx\,dy$, you would have obtained a double integral of comparable difficulty. There are, however, some occasions in which one order of integration is much more convenient than the other. Example 2 shows such a case.

Example 2 Comparing Different Orders of Integration

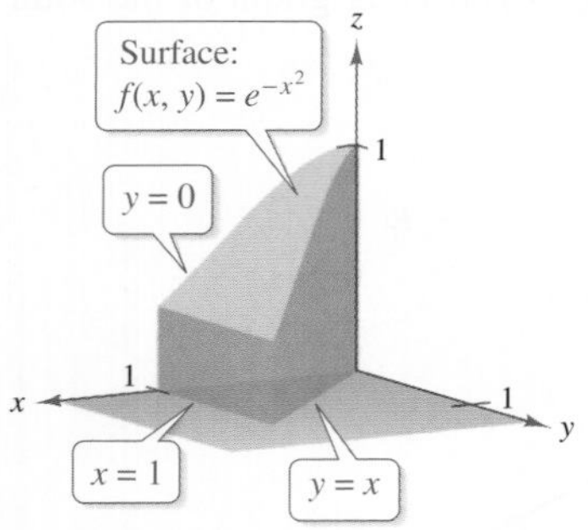

FIGURE 7.52

Find the volume of the solid region bounded by the surface

$$f(x, y) = e^{-x^2} \quad \text{Surface}$$

and the planes $z = 0$, $y = 0$, $y = x$, and $x = 1$, as shown in Figure 7.52.

SOLUTION In the xy-plane, the bounds of region R are the lines

$$y = 0, \quad x = 1, \quad \text{and} \quad y = x.$$

The two possible orders of integration are indicated in Figure 7.53.

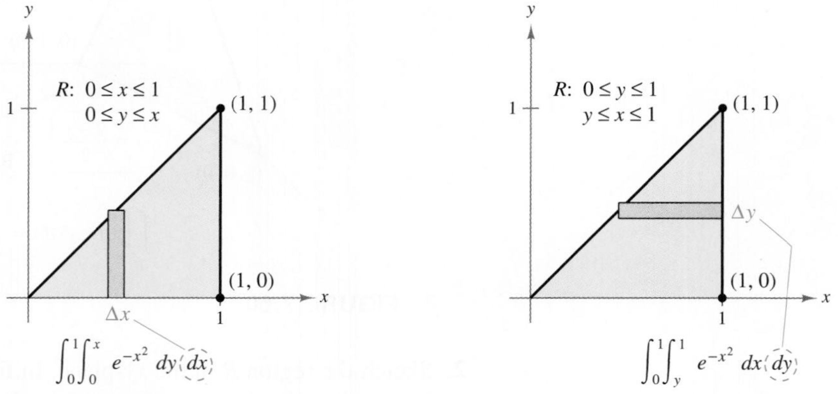

FIGURE 7.53

By setting up the corresponding integrals, you can see that the order $dy\,dx$ produces an integral that is easier to evaluate than the order $dx\,dy$.

$$\begin{aligned}
V &= \int_0^1 \int_0^x e^{-x^2}\, dy\, dx \\
&= \int_0^1 \Big[e^{-x^2} y \Big]_0^x dx \\
&= \int_0^1 xe^{-x^2}\, dx \\
&= \left[-\frac{1}{2} e^{-x^2} \right]_0^1 \\
&= -\frac{1}{2}\left(\frac{1}{e} - 1\right) \\
&\approx 0.316 \text{ cubic unit}
\end{aligned}$$

TECH TUTOR

Use a symbolic integration utility to evaluate the double integral in Example 2.

✓ Checkpoint 2

Find the volume under the surface

$$f(x, y) = e^{x^2}$$

bounded by the xz-plane and the planes $y = 2x$ and $x = 1$. ■

In the guidelines for finding the volume of a solid given at the beginning of this section, the first step suggests that you sketch the three-dimensional solid region. This is a good suggestion, but it is not always feasible and is not as important as making a sketch of the two-dimensional region R.

Example 3 Finding the Volume of a Solid

Find the volume of the solid bounded above by the surface

$$f(x, y) = 6x^2 - 2xy$$

and below by the plane region R shown in Figure 7.54.

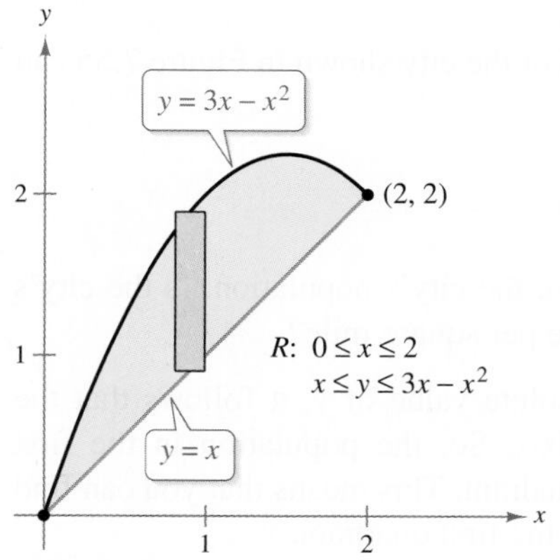

FIGURE 7.54

SOLUTION Because the region R is bounded by the parabola

$$y = 3x - x^2$$

and the line

$$y = x$$

the limits for y are $x \leq y \leq 3x - x^2$. The limits for x are $0 \leq x \leq 2$, and the volume of the solid is

$$\begin{aligned} V &= \int_0^2 \int_x^{3x-x^2} (6x^2 - 2xy)\, dy\, dx \\ &= \int_0^2 \Big[6x^2y - xy^2 \Big]_x^{3x-x^2} dx \\ &= \int_0^2 [(18x^3 - 6x^4 - 9x^3 + 6x^4 - x^5) - (6x^3 - x^3)]\, dx \\ &= \int_0^2 (4x^3 - x^5)\, dx \\ &= \Big[x^4 - \frac{x^6}{6} \Big]_0^2 \\ &= \frac{16}{3} \text{ cubic units.} \end{aligned}$$

✓ Checkpoint 3

Find the volume of the solid bounded above by the surface

$$f(x, y) = 4x^2 + 2xy$$

and below by the plane region bounded by $y = x^2$ and $y = 2x$.

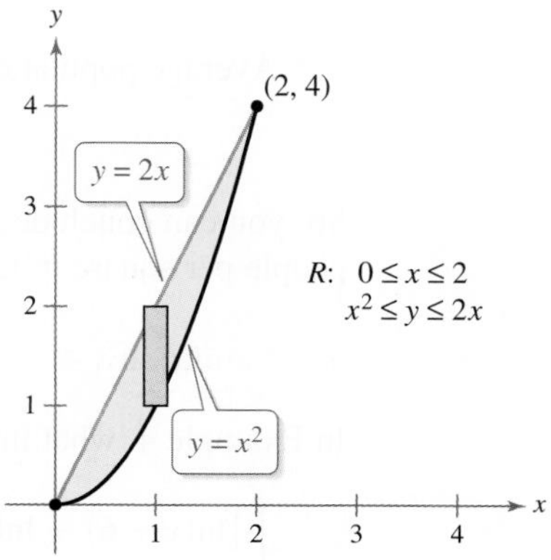

■

A *population density function*

$$p = f(x, y)$$

is a model that describes the density (in people per square unit) of a region. To find the population of a region R, evaluate the double integral

$$\int_R \int f(x, y)\, dA.$$

Example 4 Finding the Population of a City

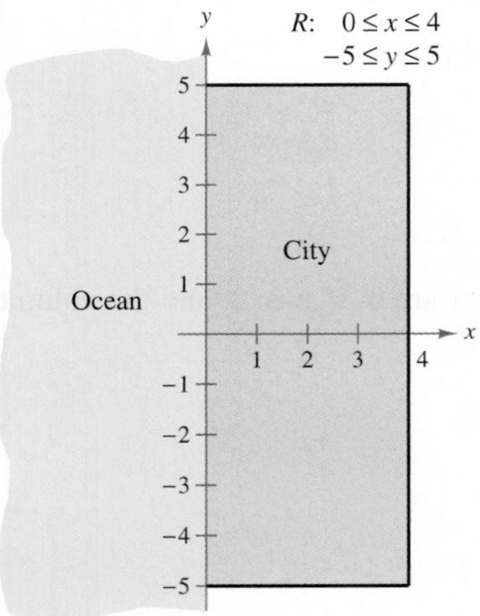

FIGURE 7.55

The population density (in people per square mile) of the city shown in Figure 7.55 can be modeled by

$$f(x, y) = \frac{50{,}000}{x + |y| + 1}$$

where x and y are measured in miles. Approximate the city's population. Is the city's average population density less than 10,000 people per square mile?

SOLUTION Because the model involves the absolute value of y, it follows that the population density is symmetrical about the x-axis. So, the population in the first quadrant is equal to the population in the fourth quadrant. This means that you can find the total population by doubling the population in the first quadrant.

$$\begin{aligned}
\text{Population} &= 2\int_0^4 \int_0^5 \frac{50{,}000}{x + y + 1}\, dy\, dx \\
&= 100{,}000 \int_0^4 \int_0^5 \frac{1}{x + y + 1}\, dy\, dx \\
&= 100{,}000 \int_0^4 \Big[\ln(x + y + 1)\Big]_0^5 dx \\
&= 100{,}000 \int_0^4 [\ln(x + 6) - \ln(x + 1)]\, dx \\
&= 100{,}000\Big[(x + 6)\ln(x + 6) - (x + 6) - (x + 1)\ln(x + 1) + (x + 1)\Big]_0^4 \\
&= 100{,}000\Big[(x + 6)\ln(x + 6) - (x + 1)\ln(x + 1) - 5\Big]_0^4 \\
&= 100{,}000[10\ln(10) - 5\ln(5) - 5 - 6\ln(6) + 5] \\
&\approx 422{,}810 \text{ people}
\end{aligned}$$

So, the city's population is about 422,810. Because the city covers a region 4 miles wide and 10 miles long, its area is 40 square miles. So, the average population density is

$$\begin{aligned}
\text{Average population density} &= \frac{422{,}810}{40} \\
&\approx 10{,}570 \text{ people per square mile.}
\end{aligned}$$

So, you can conclude that the city's average population density is not less than 10,000 people per square mile.

✓ Checkpoint 4

In Example 4, what integration technique was used to integrate

$$\int [\ln(x + 6) - \ln(x + 1)]\, dx?$$

■

Average Value of a Function over a Region

Average Value of a Function over a Region

If f is integrable over the plane region R with area A, then its **average value** over R is

$$\text{Average value} = \frac{1}{A}\int_R\int f(x, y)\,dA.$$

Example 5 Finding Average Profit

A manufacturer determines that the profit for selling x units of one product and y units of a second product is modeled by

$$P = -(x - 200)^2 - (y - 100)^2 + 5000.$$

The weekly sales for product 1 vary between 150 and 200 units, and the weekly sales for product 2 vary between 80 and 100 units. Estimate the average weekly profit for the two products.

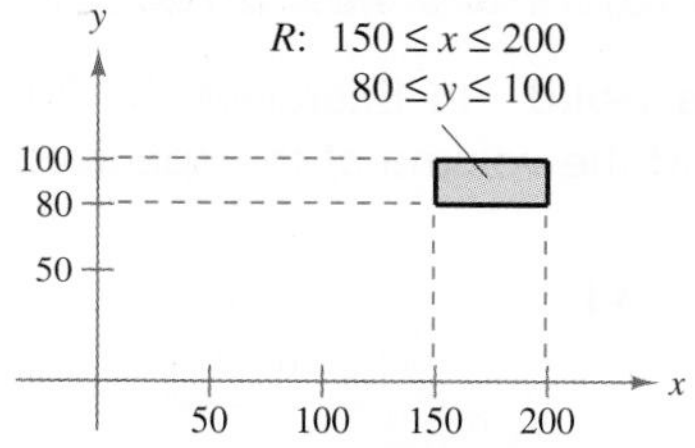

FIGURE 7.56

SOLUTION Because $150 \le x \le 200$ and $80 \le y \le 100$, you can estimate the weekly profit to be the average of the profit function over the rectangular region shown in Figure 7.56. Because the area of this rectangular region is $(50)(20) = 1000$, it follows that the average profit V is

$$\begin{aligned}
V &= \frac{1}{1000}\int_{150}^{200}\int_{80}^{100}\left[-(x - 200)^2 - (y - 100)^2 + 5000\right] dy\,dx \\
&= \frac{1}{1000}\int_{150}^{200}\left[-(x - 200)^2 y - \frac{(y - 100)^3}{3} + 5000y\right]_{80}^{100} dx \\
&= \frac{1}{1000}\int_{150}^{200}\left[-20(x - 200)^2 - \frac{292{,}000}{3}\right] dx \\
&= \frac{1}{3000}\left[-20(x - 200)^3 + 292{,}000x\right]_{150}^{200} \\
&\approx \$4033.
\end{aligned}$$

✓Checkpoint 5

Find the average value of $f(x, y) = 4 - \frac{1}{2}x - \frac{1}{2}y$ over the region $0 \le x \le 2$ and $0 \le y \le 2$. ■

SUMMARIZE (Section 7.9)

1. State the volume of a solid region using double integrals *(page 488)*. For examples of finding the volume of a solid, see Examples 1, 2, and 3.
2. Give the guidelines for finding the volume of a solid *(page 488)*. For examples of using these guidelines, see Examples 1, 2, and 3.
3. Describe a real-life example of how a double integral can be used to find a city's population *(page 492, Example 4)*.
4. State the average value of a function over a region *(page 493)*. For an example of finding the average value of a function, see Example 5.

SKILLS WARM UP 7.9

The following warm-up exercises involve skills that were covered in earlier sections. You will use these skills in the exercise set for this section. For additional help, review Sections 5.4 and 7.8.

In Exercises 1–4, sketch the region that is described.

1. $0 \le x \le 2,\ 0 \le y \le 1$

2. $1 \le x \le 3,\ 2 \le y \le 3$

3. $0 \le x \le 4,\ 0 \le y \le 2x - 1$

4. $0 \le x \le 2,\ 0 \le y \le x^2$

In Exercises 5–10, evaluate the double integral.

5. $\displaystyle\int_0^1 \int_1^2 dy\, dx$

6. $\displaystyle\int_0^3 \int_1^3 dx\, dy$

7. $\displaystyle\int_0^1 \int_0^x x\, dy\, dx$

8. $\displaystyle\int_0^4 \int_1^y y\, dx\, dy$

9. $\displaystyle\int_1^3 \int_x^{x^2} 2\, dy\, dx$

10. $\displaystyle\int_0^1 \int_x^{-x^2+2} dy\, dx$

Exercises 7.9

See www.CalcChat.com for worked-out solutions to odd-numbered exercises.

Finding the Volume of a Solid In Exercises 1–8, sketch the region of integration in the *xy*-plane and evaluate the double integral. *See Example 1.*

1. $\displaystyle\int_0^2 \int_0^1 (3x + 4y)\, dy\, dx$

2. $\displaystyle\int_0^3 \int_0^1 (2x + 6y)\, dy\, dx$

3. $\displaystyle\int_{-1}^1 \int_0^{\sqrt{1-x^2}} x^2y\, dy\, dx$

4. $\displaystyle\int_0^2 \int_0^{4-x^2} xy^2\, dy\, dx$

5. $\displaystyle\int_0^1 \int_{y^2}^{y} (x^2 + y^2)\, dx\, dy$

6. $\displaystyle\int_0^6 \int_{y/2}^{3} (x + y)\, dx\, dy$

7. $\displaystyle\int_{-a}^a \int_{-\sqrt{a^2-x^2}}^{\sqrt{a^2-x^2}} dy\, dx$

8. $\displaystyle\int_0^a \int_0^{\sqrt{a^2-x^2}} dy\, dx$

Comparing Different Orders of Integration In Exercises 9–12, set up the integral for both orders of integration and use the more convenient order to evaluate the integral over the region *R*. *See Example 2.*

9. $\displaystyle\int_R\int xy\, dA$

R: rectangle with vertices at (0, 0), (0, 5), (3, 5), (3, 0)

10. $\displaystyle\int_R\int x\, dA$

R: semicircle bounded by $y = \sqrt{25 - x^2}$ and $y = 0$

11. $\displaystyle\int_R\int \frac{y}{x^2 + y^2}\, dA$

R: triangle bounded by $y = x$, $y = 2x$, $x = 2$

12. $\displaystyle\int_R\int \frac{y}{1 + x^2}\, dA$

R: region bounded by $y = 0$, $y = \sqrt{x}$, $x = 4$

Finding the Volume of a Solid In Exercises 13–20, use a double integral to find the volume of the specified solid. *See Example 3.*

13.

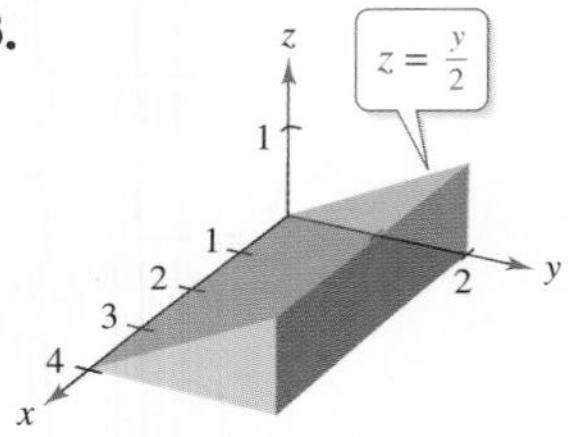

$0 \le x \le 4$
$0 \le y \le 2$

14.

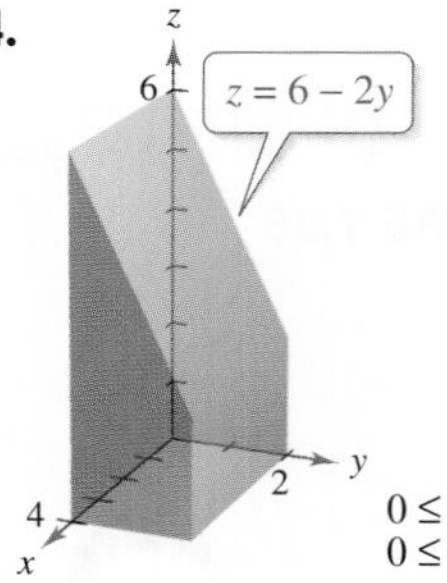

$0 \le x \le 4$
$0 \le y \le 2$

15.

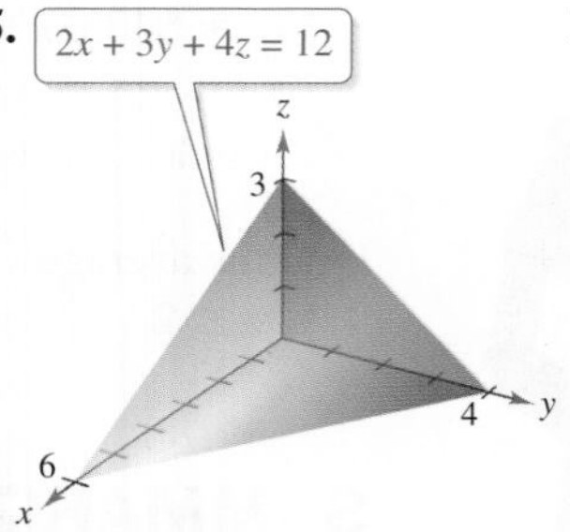

16.

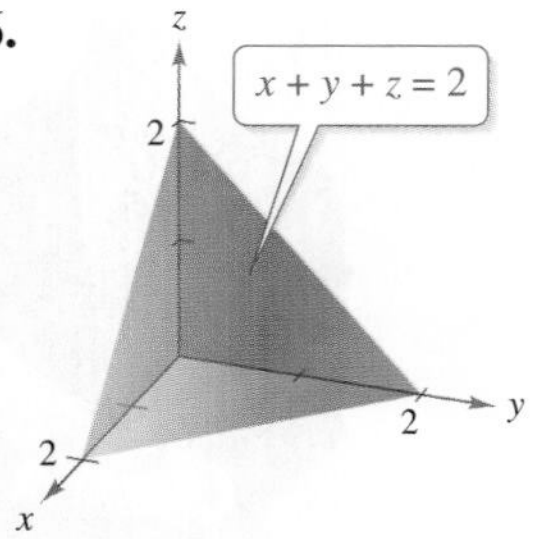

17.

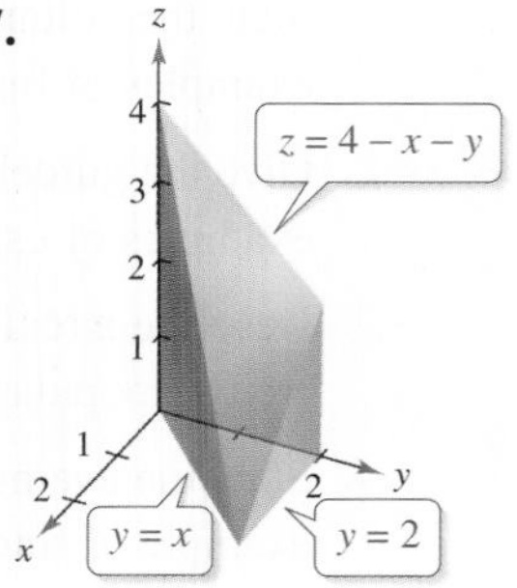

18.

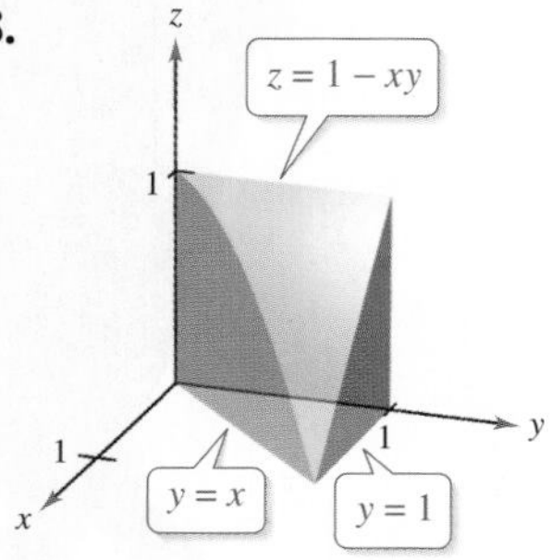

19.

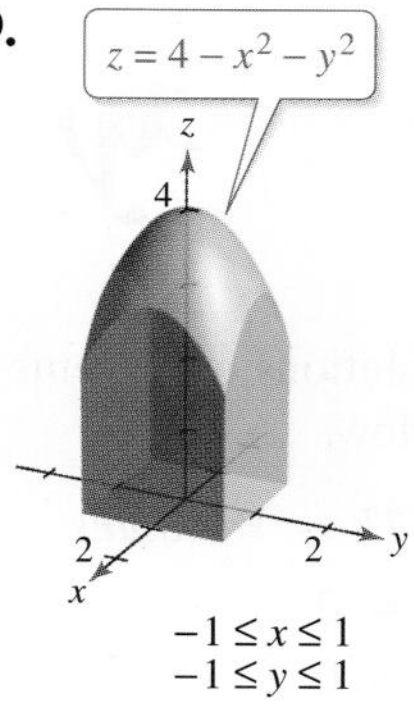

20.

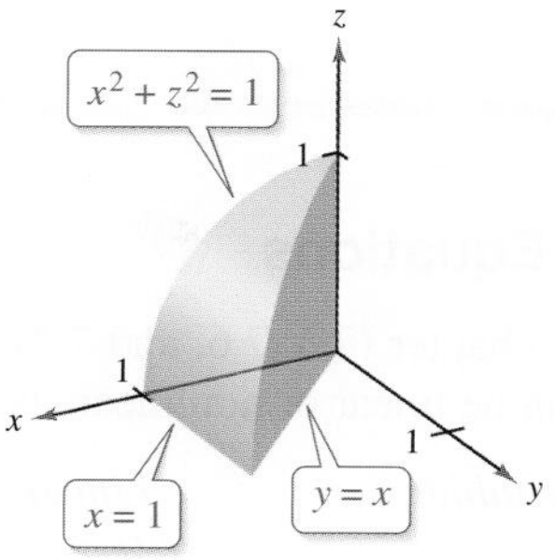

Finding the Volume of a Solid In Exercises 21–24, use a double integral to find the volume of the solid bounded by the graphs of the equations.

21. $z = xy,\ z = 0,\ y = 2x,\ y = 0,\ x = 0,\ x = 3$

22. $z = x,\ z = 0,\ y = x,\ y = 0,\ x = 0,\ x = 4$

23. $z = 9 - x^2,\ z = 0,\ y = x + 2,\ y = 0,\ x = 0, x = 2$

24. $z = x + y,\ x^2 + y^2 = 4$ (first octant)

25. Population Density The population density (in people per square mile) for a coastal town can be modeled by

$$f(x, y) = \frac{120{,}000}{(2 + x + y)^3}$$

where x and y are measured in miles. What is the population inside the rectangular area defined by the vertices

$(0, 0),\quad (2, 0),\quad (0, 2),\quad$ and $\quad (2, 2)$?

26. Population Density The population density (in people per square mile) for a coastal town on an island can be modeled by

$$f(x, y) = \frac{5000xe^y}{1 + 2x^2}$$

where x and y are measured in miles. What is the population inside the rectangular area defined by the vertices

$(0, 0),\quad (4, 0),\quad (0, -2),\quad$ and $\quad (4, -2)$?

Average Value of a Function over a Region In Exercises 27–30, find the average value of $f(x, y)$ over the region R. *See Example 5.*

27. $f(x, y) = y$

R: rectangle with vertices $(0, 0), (5, 0), (5, 3), (0, 3)$

28. $f(x, y) = xy$

R: rectangle with vertices $(0, 0), (4, 0), (4, 2), (0, 2)$

29. $f(x, y) = x^2 + y^2$

R: square with vertices $(0, 0), (2, 0), (2, 2), (0, 2)$

30. $f(x, y) = e^{x+y}$

R: triangle with vertices $(0, 0), (0, 1), (1, 1)$

31. Average Weekly Profit A firm's weekly profit (in dollars) in marketing two products is given by

$$P = 192x_1 + 576x_2 - x_1^2 - 5x_2^2 - 2x_1x_2 - 5000$$

where x_1 and x_2 represent the numbers of units of each product sold weekly. Estimate the average weekly profit when x_1 varies between 40 and 50 units and x_2 varies between 45 and 50 units.

32. Average Weekly Profit After a change in marketing, the weekly profit of the firm in Exercise 31 is given by

$$P = 200x_1 + 580x_2 - x_1^2 - 5x_2^2 - 2x_1x_2 - 7500.$$

Estimate the average weekly profit when x_1 varies between 55 and 65 units and x_2 varies between 50 and 60 units.

33. Average Revenue A company sells two products whose demand functions are given by

$$x_1 = 500 - 3p_1 \quad \text{and} \quad x_2 = 750 - 2.4p_2.$$

So, the total revenue is given by

$$R = x_1p_1 + x_2p_2.$$

Estimate the average revenue when price p_1 varies between \$50 and \$75 and price p_2 varies between \$100 and \$150.

34. HOW DO YOU SEE IT? The figure below shows Erie County, New York. Let $f(x, y)$ represent the total annual snowfall at the point (x, y) in the county, where R is the county. Interpret each of the following.

(a) $\displaystyle\iint_R f(x, y)\, dA$

(b) $\dfrac{\displaystyle\iint_R f(x, y)\, dA}{\displaystyle\iint_R dA}$

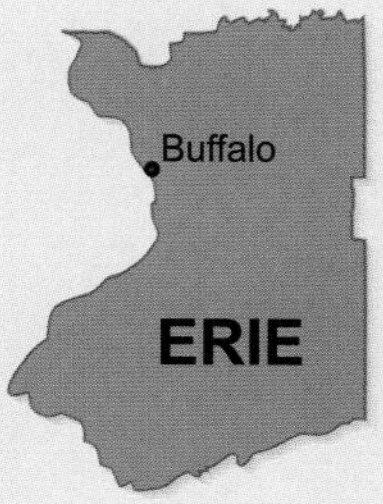

35. Average Production The Cobb-Douglas production function for an automobile manufacturer is

$$f(x, y) = 100x^{0.6}y^{0.4}$$

where x is the number of units of labor and y is the number of units of capital. Estimate the average production level when the number of units of labor x varies between 200 and 250 and the number of units of capital y varies between 300 and 325.

36. Average Production Repeat Exercise 35 for the production function given by $f(x, y) = x^{0.25}y^{0.75}$.

ALGEBRA TUTOR

Solving Systems of Equations

Three of the sections in this chapter (7.5, 7.6, and 7.7) involve solutions of systems of equations. These systems can be linear or nonlinear, as shown below.

Nonlinear System in Two Variables

$$\begin{cases} 4x + 3y = 6 \\ x^2 - y = 4 \end{cases}$$

Linear System in Three Variables

$$\begin{cases} -x + 2y + 4z = 2 \\ 2x - y + z = 0 \\ 6x + 2z = 3 \end{cases}$$

There are many techniques for solving a system of linear equations. Two of the more common ones are listed here.

1. *Substitution:* Solve for one of the variables in one of the equations and substitute the value into another equation.
2. *Elimination:* Add multiples of one equation to a second equation to eliminate a variable in the second equation.

Example 1 Solving Systems of Equations

Solve each system of equations.

a. $\begin{cases} y - x^3 = 0 \\ x - y^3 = 0 \end{cases}$ **b.** $\begin{cases} -400p_1 + 300p_2 = -25 \\ 300p_1 - 360p_2 = -535 \end{cases}$

SOLUTION

a. Example 3, page 458

$\begin{cases} y - x^3 = 0 \\ x - y^3 = 0 \end{cases}$ Equation 1, Equation 2

$y = x^3$ Solve for y in Equation 1.

$x - (x^3)^3 = 0$ Substitute x^3 for y in Equation 2.

$x - x^9 = 0$ $(x^m)^n = x^{mn}$

$x(x - 1)(x + 1)(x^2 + 1)(x^4 + 1) = 0$ Factor.

$x = 0$ Set factors equal to zero.

$x = 1$ Set factors equal to zero.

$x = -1$ Set factors equal to zero.

b. Example 4, page 459

$\begin{cases} -400p_1 + 300p_2 = -25 \\ 300p_1 - 360p_2 = -535 \end{cases}$ Equation 1, Equation 2

$p_2 = \frac{1}{12}(16p_1 - 1)$ Solve for p_2 in Equation 1.

$300p_1 - 360\left(\frac{1}{12}\right)(16p_1 - 1) = -535$ Substitute for p_2 in Equation 2.

$300p_1 - 30(16p_1 - 1) = -535$ Multiply factors.

$-180p_1 = -565$ Combine like terms.

$p_1 = \frac{113}{36} \approx 3.14$ Divide each side by -180.

$p_2 = \frac{1}{12}\left[16\left(\frac{113}{36}\right) - 1\right]$ Find p_2 by substituting p_1.

$p_2 \approx 4.10$ Solve for p_2.

Example 2 **Solving Systems of Equations**

Solve each system of equations.

a. $\begin{cases} y(24 - 12x - 4y) = 0 \\ x(24 - 6x - 8y) = 0 \end{cases}$

b. $\begin{cases} 28a - 4b = 10 \\ -4a + 8b = 12 \end{cases}$

SOLUTION

a. Example 5, page 460

Before solving this system of equations, factor 4 out of the first equation and factor 2 out of the second equation.

$$\begin{cases} y(24 - 12x - 4y) = 0 \\ x(24 - 6x - 8y) = 0 \end{cases}$$ Original Equation 1 / Original Equation 2

$$\begin{cases} y(4)(6 - 3x - y) = 0 \\ x(2)(12 - 3x - 4y) = 0 \end{cases}$$ Factor 4 out of Equation 1. / Factor 2 out of Equation 2.

$$\begin{cases} y(6 - 3x - y) = 0 \\ x(12 - 3x - 4y) = 0 \end{cases}$$ Equation 1 / Equation 2

In each equation, either factor can be 0, so you obtain four different linear systems. For the first system, substitute $y = 0$ into the second equation to obtain $x = 4$.

$$\begin{cases} y = 0 \\ 12 - 3x - 4y = 0 \end{cases}$$ (4, 0) is a solution.

You can solve the second system by the method of elimination.

$$\begin{cases} 6 - 3x - y = 0 \\ 12 - 3x - 4y = 0 \end{cases}$$ $\left(\frac{4}{3}, 2\right)$ is a solution.

The third system is already solved.

$$\begin{cases} y = 0 \\ x = 0 \end{cases}$$ (0, 0) is a solution.

You can solve the last system by substituting $x = 0$ into the first equation to obtain $y = 6$.

$$\begin{cases} 6 - 3x - y = 0 \\ x = 0 \end{cases}$$ (0, 6) is a solution.

b. Example 2, page 475

$\begin{cases} 28a - 4b = 10 \\ -4a + 8b = 12 \end{cases}$	Equation 1 / Equation 2
$-2a + 4b = 6$	Divide Equation 2 by 2.
$26a = 16$	Add new equation to Equation 1.
$a = \frac{8}{13}$	Divide each side by 26.
$28\left(\frac{8}{13}\right) - 4b = 10$	Substitute for a in Equation 1.
$b = \frac{47}{26}$	Solve for b.

SUMMARY AND STUDY STRATEGIES

After studying this chapter, you should have acquired the following skills.
The exercise numbers are keyed to the Review Exercises that begin on page 500.
Answers to odd-numbered Review Exercises are given in the back of the text.*

Section 7.1	Review Exercises
■ Plot points in space.	*1, 2*
■ Find the distance between two points in space. $d = \sqrt{(x_2 - x_1)^2 + (y_2 - y_1)^2 + (z_2 - z_1)^2}$	*3, 4*
■ Find the midpoint of a line segment in space. $\text{Midpoint} = \left(\frac{x_1 + x_2}{2}, \frac{y_1 + y_2}{2}, \frac{z_1 + z_2}{2}\right)$	*5, 6*
■ Write the standard forms of the equations of spheres. $(x - h)^2 + (y - k)^2 + (z - l)^2 = r^2$	*7–10*
■ Find the centers and radii of spheres.	*11, 12*
■ Sketch the coordinate plane traces of spheres.	*13, 14*

Section 7.2	
■ Sketch planes in space.	*15–18*
■ Classify quadric surfaces in space.	*19–26*

Section 7.3	
■ Evaluate functions of several variables.	*27, 28*
■ Find the domains and ranges of functions of two variables.	*29–32*
■ Sketch level curves of functions of two variables.	*33–36*
■ Use functions of several variables to answer questions about real-life situations.	*37–40*

Section 7.4	
■ Find the first partial derivatives of functions of several variables.	*41–50*
■ Find the slopes of surfaces in the x- and y-directions.	*51–54*
■ Find the second partial derivatives of functions of several variables.	*55–60*
■ Use partial derivatives to answer questions about real-life situations.	*61, 62*

Section 7.5	
■ Find the relative extrema of functions of two variables.	*63–70*
■ Use relative extrema to answer questions about real-life situations.	*71, 72*

* A wide range of valuable study aids are available to help you master the material in this chapter. The *Student Solutions Manual* includes step-by-step solutions to all odd-numbered exercises to help you review and prepare. The student website at *www.cengagebrain.com* offers algebra help and a *Graphing Technology Guide*, which contains step-by-step commands and instructions for a wide variety of graphing calculators.

Section 7.6

	Review Exercises
■ Use Lagrange multipliers to find extrema of functions of several variables.	73–78
■ Use Lagrange multipliers to answer questions about real-life situations.	79, 80

Section 7.7

■ Find the least squares regression line, $y = ax + b$, for data.	81, 82

$$a = \left[n\sum_{i=1}^{n} x_i y_i - \sum_{i=1}^{n} x_i \sum_{i=1}^{n} y_i\right] \Bigg/ \left[n\sum_{i=1}^{n} x_1^2 - \left(\sum_{i=1}^{n} x_i\right)^2\right], \quad b = \frac{1}{n}\left(\sum_{i=1}^{n} y_i - a\sum_{i=1}^{n} x_i\right)$$

■ Use least squares regression lines to model real-life data.	83, 84

Section 7.8

■ Evaluate double integrals.	85–88
■ Use double integrals to find the areas of regions.	89–92

Section 7.9

■ Use double integrals to find the volumes of solids.	93–98

Volume $= \int_R\int f(x, y)\, dA$

■ Use double integrals to find the average values of functions.	99–103

Average value $= \dfrac{1}{A}\int_R\int f(x, y)\, dA$

Study Strategies

■ **Comparing Two Dimensions with Three Dimensions** Many of the formulas and techniques in this chapter are generalizations of formulas and techniques used in earlier chapters of the text. Here are several examples.

Two-Dimensional Coordinate System	Three-Dimensional Coordinate System
Distance Formula $d = \sqrt{(x_2 - x_1)^2 + (y_2 - y_1)^2}$	*Distance Formula* $d = \sqrt{(x_2 - x_1)^2 + (y_2 - y_1)^2 + (z_2 - z_1)^2}$
Midpoint Formula Midpoint $= \left(\dfrac{x_1 + x_2}{2}, \dfrac{y_1 + y_2}{2}\right)$	*Midpoint Formula* Midpoint $= \left(\dfrac{x_1 + x_2}{2}, \dfrac{y_1 + y_2}{2}, \dfrac{z_1 + z_2}{2}\right)$
Equation of Circle $(x - h)^2 + (y - k)^2 = r^2$	*Equation of Sphere* $(x - h)^2 + (y - k)^2 + (z - l)^2 = r^2$
Equation of Line $ax + by = c$	*Equation of Plane* $ax + by + cz = d$
Derivative of $y = f(x)$ $\dfrac{dy}{dx} = \lim_{\Delta x \to 0} \dfrac{f(x + \Delta x) - f(x)}{\Delta x}$	*Partial Derivative of $z = f(x, y)$* $\dfrac{\partial z}{\partial x} = \lim_{\Delta x \to 0} \dfrac{f(x + \Delta x, y) - f(x, y)}{\Delta x}$
Area of Region $A = \int_a^b f(x)\, dx$	*Volume of Region* $V = \int_R\int f(x, y)\, dA$

Review Exercises

See www.CalcChat.com for worked-out solutions to odd-numbered exercises.

Plotting Points in Space In Exercises 1 and 2, plot the points in the same three-dimensional coordinate system.

1. $(2, -1, 4), (-1, 3, -3), (-2, -2, 1), (3, 1, 2)$

2. $(1, -2, -3), (-4, -3, 5), \left(4, \frac{5}{2}, 1\right), (-2, 2, 2)$

Finding the Distance Between Two Points In Exercises 3 and 4, find the distance between the two points.

3. $(1, 0, 2), (3, 5, 8)$

4. $(-4, 1, 5), (1, 3, 7)$

Using the Midpoint Formula In Exercises 5 and 6, find the midpoint of the line segment joining the two points.

5. $(2, 6, 4), (-4, 2, 8)$

6. $(5, 0, 7), (-1, -2, 9)$

Finding the Equation of a Sphere In Exercises 7–10, find the standard equation of the sphere.

7. Center: $(0, 1, 0)$; radius: 5

8. Center: $(4, -5, 3)$; radius: 10

9. Diameter endpoints: $(3, -4, -1), (1, 0, -5)$

10. Diameter endpoints: $(3, 4, 0), (5, 8, 2)$

Finding the Center and Radius of a Sphere In Exercises 11 and 12, find the center and radius of the sphere.

11. $x^2 + y^2 + z^2 - 8x + 4y - 6z - 20 = 0$

12. $x^2 + y^2 + z^2 + 4y - 10z - 7 = 0$

Finding the Trace of a Surface In Exercises 13 and 14, sketch the *xy*-trace of the sphere.

13. $(x + 2)^2 + (y - 1)^2 + (z - 3)^2 = 25$

14. $(x - 1)^2 + (y + 3)^2 + (z - 6)^2 = 72$

Sketching a Plane in Space In Exercises 15–18, find the intercepts and sketch the graph of the plane.

15. $x + 2y + 3z = 6$ **16.** $2y + z = 4$

17. $3x - 6z = 12$ **18.** $4x - y + 2z = 8$

Classifying a Quadric Surface In Exercises 19–26, classify the quadric surface.

19. $x^2 + y^2 + z^2 - 2x + 4y - 6z + 5 = 0$

20. $16x^2 + 16y^2 - 9z^2 = 0$

21. $x^2 + \dfrac{y^2}{16} + \dfrac{z^2}{9} = 1$

22. $x^2 - \dfrac{y^2}{16} - \dfrac{z^2}{9} = 1$

23. $z = \dfrac{x^2}{9} + y^2$

24. $-4x^2 + y^2 + z^2 = 4$

25. $z = \sqrt{x^2 + y^2}$

26. $z = x^2 - \dfrac{y^2}{4}$

Evaluating Functions of Several Variables In Exercises 27 and 28, find the function values.

27. $f(x, y) = xy^2$

(a) $f(2, 3)$ (b) $f(0, 1)$

(c) $f(-5, 7)$ (d) $f(-2, -4)$

28. $f(x, y) = \dfrac{x^2}{y}$

(a) $f(6, 9)$ (b) $f(8, 4)$ (c) $f(t, 2)$ (d) $f(r, r)$

Finding the Domain and Range of a Function In Exercises 29–32, find the domain and range of the function.

29. $f(x, y) = \sqrt{1 - x^2 - y^2}$

30. $f(x, y) = x^2 + y^2 - 3$

31. $f(x, y) = e^{xy}$

32. $f(x, y) = \dfrac{1}{x + y}$

Sketching a Contour Map In Exercises 33–36, describe the level curves of the function. Sketch a contour map of the surface using level curves for the given *c*-values.

	Function	*c-Values*
33.	$z = 10 - 2x - 5y$	$c = 0, 2, 4, 5, 10$
34.	$z = \sqrt{9 - x^2 - y^2}$	$c = 0, 1, 2, 3$
35.	$z = (xy)^2$	$c = 1, 4, 9, 12, 16$
36.	$z = y - x^2$	$c = 0, \pm 1, \pm 2$

37. Meteorology The contour map shown below represents the average yearly precipitation for Oklahoma. *(Source: National Climatic Data Center)*

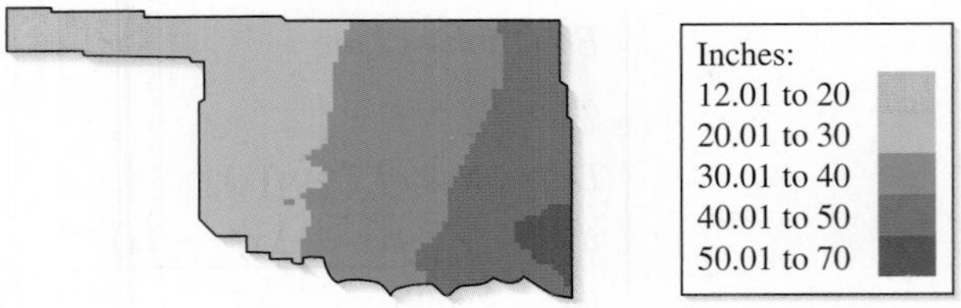

(a) Do the level curves correspond to equally spaced levels of precipitation? Explain.

(b) Describe how to obtain a more detailed contour map.

38. **Chemistry** The acidity of rainwater is measured in units called pH, and smaller pH values are increasingly acidic. The map shows the curves of equal pH and gives evidence that downwind of heavily industrialized areas, the acidity has been increasing. Using the level curves on the map, determine the direction of the prevailing winds in the northeastern United States.

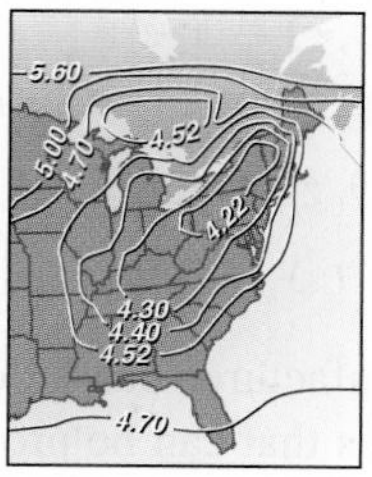

39. **Earnings per Share** The earnings per share z (in dollars) for Hewlett-Packard from 2003 through 2010 can be modeled by

$$z = -4.51 + 0.046x + 0.060y$$

where x is the sales (in billions of dollars) and y is the shareholder's equity (in billions of dollars). *(Source: Hewlett-Packard Company)*

(a) Find the earnings per share when $x = 100$ and $y = 40$.

(b) Which of the two variables in this model has the greater influence on the earnings per share? Explain.

40. **Shareholder's Equity** The shareholder's equity z (in billions of dollars) for Wal-Mart from 2000 through 2010 can be modeled by

$$z = 1.54 + 0.116x + 0.122y$$

where x is the net sales (in billions of dollars) and y is the total assets (in billions of dollars). *(Source: Wal-Mart Stores, Inc.)*

(a) Find the shareholder's equity when $x = 300$ and $y = 130$.

(b) Which of the two variables in this model has the greater influence on shareholder's equity? Explain.

Finding Partial Derivatives In Exercises 41–50, find the first partial derivatives.

41. $f(x, y) = x^2y + 3xy + 2x - 5y$

42. $f(x, y) = 4xy + xy^2 - 3x^2y$

43. $z = \dfrac{x^2}{y^2}$

44. $z = (xy + 2x + 4y)^2$

45. $f(x, y) = \ln(5x + 4y)$

46. $f(x, y) = \ln\sqrt{2x + 3y}$

47. $f(x, y) = xe^y + ye^x$

48. $f(x, y) = x^2e^{-2y}$

49. $w = xyz^2$

50. $w = 3xy - 5xz + 2yz$

Finding Slopes in the x- and y-Directions In Exercises 51–54, find the slopes of the surface at the given point in (a) the x-direction and (b) the y-direction.

51. $z = 3xy$
$(-2, -3, 18)$

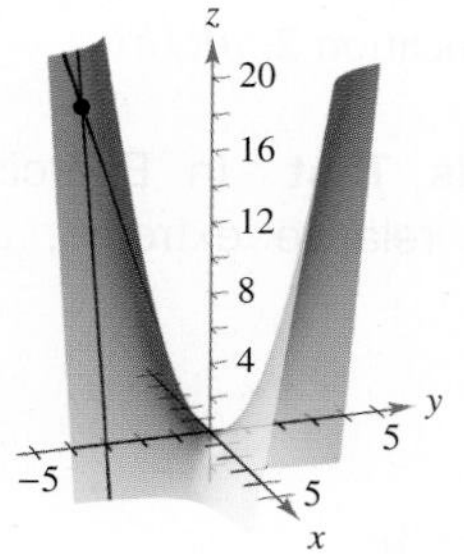

52. $z = y^2 - x^2$
$(1, 2, 3)$

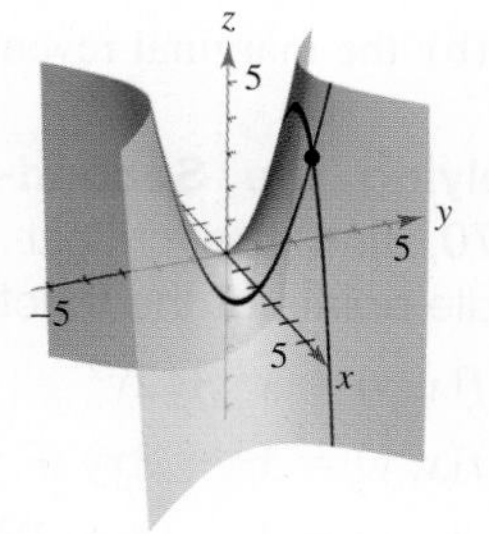

53. $z = 8 - x^2 - y^2$
$(1, 1, 6)$

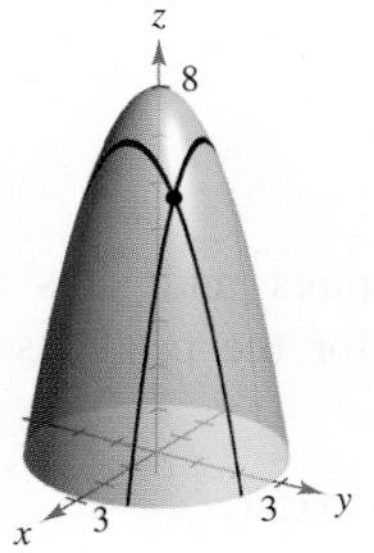

54. $z = \sqrt{100 - x^2 - y^2}$
$(0, 6, 8)$

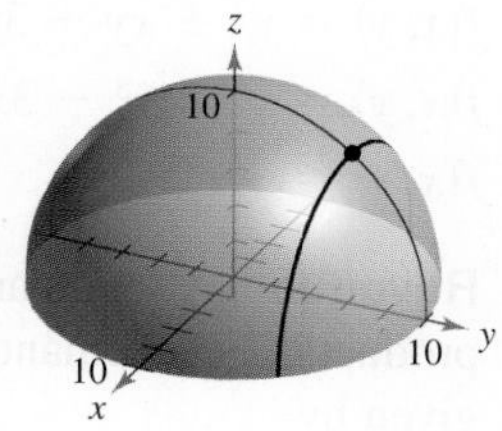

Finding Second Partial Derivatives In Exercises 55–60, find all second partial derivatives.

55. $f(x, y) = 3x^2 - xy + 2y^3$

56. $f(x, y) = \dfrac{y}{x + y}$

57. $f(x, y) = \sqrt{1 + x + y}$

58. $f(x, y) = x^2e^{-y^2}$

59. $f(x, y, z) = xy + 5x^2yz^3 - 3y^3z$

60. $f(x, y, z) = \dfrac{3yz}{x + z}$

61. **Marginal Cost** A company manufactures two models of skis: cross-country skis and downhill skis. The cost function for producing x pairs of cross-country skis and y pairs of downhill skis is given by

$$C = 15(xy)^{1/3} + 99x + 139y + 2293.$$

(a) Find the marginal costs ($\partial C/\partial x$ and $\partial C/\partial y$) when $x = 500$ and $y = 250$.

(b) When additional production is required, which model of skis results in the cost increasing at a higher rate? How can this be determined from the cost model?

62. Marginal Revenue At a baseball stadium, souvenir hats are sold at two locations. If x_1 and x_2 are the numbers of baseball hats sold at location 1 and location 2, respectively, then the total revenue for the hats is modeled by

$$R = 15x_1 + 16x_2 - \frac{1}{10}x_1^2 - \frac{1}{10}x_2^2 - \frac{1}{100}x_1x_2.$$

When $x_1 = 50$ and $x_2 = 40$, find

(a) the marginal revenue for location 1, $\partial R/\partial x_1$.

(b) the marginal revenue for location 2, $\partial R/\partial x_2$.

Applying the Second-Partials Test In Exercises 63–70, find the critical points, relative extrema, and saddle points of the function.

63. $f(x, y) = x^2 + 2y^2$

64. $f(x, y) = x^3 - 3xy + y^2$

65. $f(x, y) = 1 - (x + 2)^2 + (y - 3)^2$

66. $f(x, y) = e^x - x + y^2$

67. $f(x, y) = x^3 + y^2 - xy$

68. $f(x, y) = y^2 + xy + 3y - 2x + 5$

69. $f(x, y) = x^3 + y^3 - 3x - 3y + 2$

70. $f(x, y) = -x^2 - y^2$

71. Revenue A company manufactures and sells two products. The demand functions for the products are given by

$$p_1 = 100 - x_1 \quad \text{and} \quad p_2 = 200 - 0.5x_2$$

where p_1 and p_2 are the prices per unit (in dollars) and x_1 and x_2 are the numbers of units sold. The total revenue function is given by

$$R = x_1p_1 + x_2p_2.$$

Find x_1 and x_2 so as to maximize revenue.

72. Profit A company manufactures a product at two locations. The cost of producing x_1 units at location 1 is

$$C_1 = 0.03x_1^2 + 4x_1 + 300$$

and the cost of producing x_2 units at location 2 is

$$C_2 = 0.05x_2^2 + 7x_2 + 175.$$

The product sells for \$10 per unit. Find the quantity that should be produced at each location to maximize the profit

$$P = 10(x_1 + x_2) - C_1 - C_2.$$

Using Lagrange Multipliers In Exercises 73–78, use Lagrange multipliers to find the given extremum. In each case, assume that the variables are positive.

73. Maximize $f(x, y) = 2xy$.
Constraint: $2x + y = 12$

74. Maximize $f(x, y) = 2x + 3xy + y$.
Constraint: $x + 2y = 29$

75. Minimize $f(x, y) = x^2 + y^2$.
Constraint: $x + y = 4$

76. Minimize $f(x, y) = 3x^2 - y^2$.
Constraint: $2x - 2y + 5 = 0$

77. Maximize $f(x, y, z) = xyz$.
Constraint: $x + 2y + z - 4 = 0$

78. Maximize $f(x, y, z) = x^2z + yz$.
Constraint: $2x + y + z = 5$

79. Cost A manufacturer has an order for 1000 units of wooden benches that can be produced at two locations. Let x_1 and x_2 be the numbers of units produced at the two locations. The cost function is modeled by

$$C = 0.25x_1^2 + 10x_1 + 0.15x_2^2 + 12x_2.$$

Use Lagrange multipliers to find the number of units that should be produced at each location to minimize the cost.

80. Production The production function for a manufacturer is given by

$$f(x, y) = 4x + xy + 2y$$

where x is the number of units of labor (at \$20 per unit) and y is the number of units of capital (at \$4 per unit). The total cost for labor and capital cannot exceed \$2000. Use Lagrange multipliers to find the maximum production level for this manufacturer.

Finding the Least Squares Regression Line In Exercises 81 and 82, find the least squares regression line for the given points. Then plot the points and sketch the regression line.

81. $(-2, -3), (-1, -1), (1, 2), (3, 2)$

82. $(-3, -1), (-2, -1), (0, 0), (1, 1), (2, 1)$

83. Demand A store manager wants to know the demand y for a digital camera as a function of price x. The monthly sales for four different prices of the digital camera are listed in the table.

Price, x	\$80	\$90	\$100	\$110
Demand, y	140	117	91	63

(a) Use the regression capabilities of a graphing utility or a spreadsheet to find the least squares regression line for the data.

(b) Estimate the demand when the price is \$85.

(c) What price will create a demand of 200 cameras?

84. Work Force The number of men x (in millions) and the number of women y (in millions) in the labor force from 2001 through 2010 are shown in the table. *(Source: U.S. Bureau of Labor Statistics)*

Year	2001	2002	2003	2004	2005
Men, x	76.9	77.5	78.2	79.0	80.0
Women, y	66.8	67.4	68.3	68.4	69.3

Year	2006	2007	2008	2009	2010
Men, x	81.3	82.1	82.5	82.1	82.0
Women, y	70.2	71.0	71.8	72.0	71.9

(a) Use the regression capabilities of a graphing utility or a spreadsheet to find the least squares regression line for the data.

(b) Estimate the number of women in the labor force when there are 80 million men in the labor force.

Evaluating a Double Integral In Exercises 85–88, evaluate the double integral.

85. $\displaystyle\int_0^1\int_0^{1+x} (4x - 2y)\,dy\,dx$

86. $\displaystyle\int_{-3}^3\int_0^4 (x - y^2)\,dx\,dy$

87. $\displaystyle\int_1^2\int_1^{2y} \frac{x}{y^2}\,dx\,dy$

88. $\displaystyle\int_0^4\int_0^{\sqrt{16-x^2}} 2x\,dy\,dx$

Finding Area with a Double Integral In Exercises 89–92, use a double integral to find the area of the region bounded by the graphs of the equations.

89. $y = 9 - x^2,\ y = 5$

90. $y = \dfrac{4}{x},\ y = 0,\ x = 1,\ x = 4$

91. $y = \sqrt{x + 3},\ y = \dfrac{1}{3}x + 1$

92. $y = x^2 - 2x - 2,\ y = -x$

Finding the Volume of a Solid In Exercises 93–96, use a double integral to find the volume of the specified solid.

93.

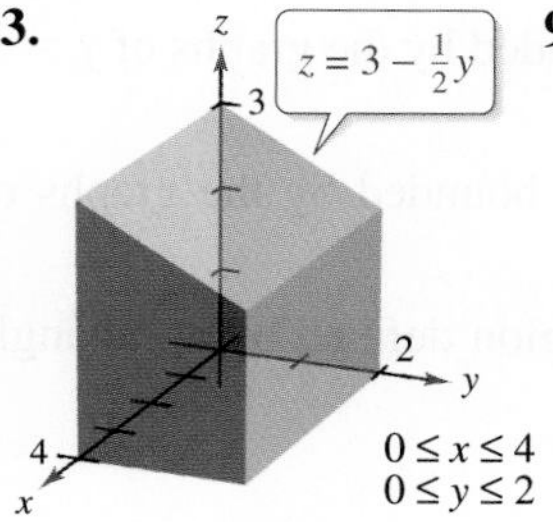

94.

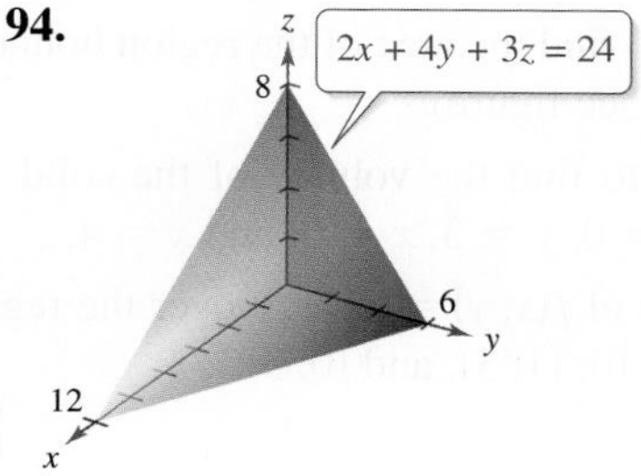

95.

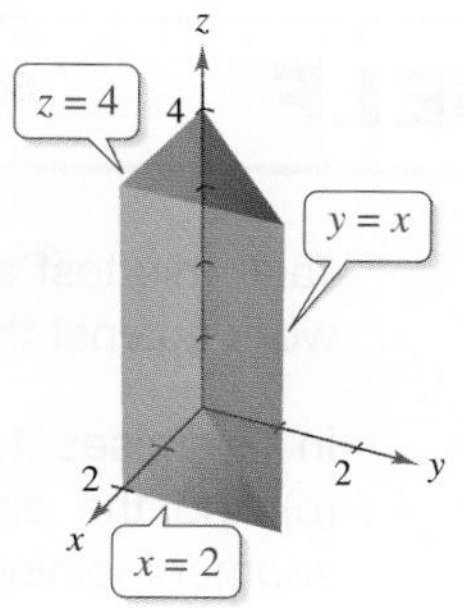

96.

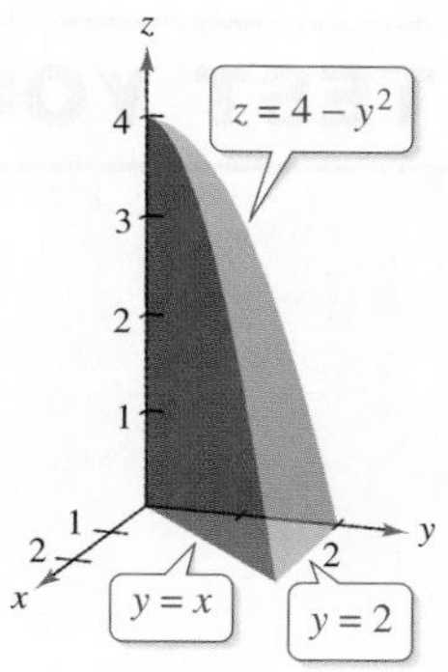

Finding the Volume of a Solid In Exercises 97 and 98, use a double integral to find the volume of the solid bounded by the graphs of the equations.

97. $z = (xy)^2,\ z = 0,\ y = 0,\ y = 4,\ x = 0,\ x = 4$

98. $z = x + y,\ z = 0,\ x = 0,\ x = 3,\ y = x,\ y = 0$

Average Value of a Function over a Region In Exercises 99 and 100, find the average value of $f(x, y)$ over the region R.

99. $f(x, y) = xy$

R: rectangle with vertices $(0, 0), (4, 0), (4, 3), (0, 3)$

100. $f(x, y) = x^2 + 2xy + y^2$

R: rectangle with vertices $(0, 0), (2, 0), (2, 5), (0, 5)$

101. Average Weekly Profit A firm's weekly profit (in dollars) in marketing two products is given by

$$P = 150x_1 + 400x_2 - x_1^2 - 5x_2^2 - 2x_1x_2 - 3000$$

where x_1 and x_2 represent the numbers of units of each product sold weekly. Estimate the average weekly profit when x_1 varies between 30 and 40 units and x_2 varies between 40 and 50 units.

102. Average Revenue A company sells two products whose demand functions are given by

$$x_1 = 500 - 2.5p_1 \quad\text{and}\quad x_2 = 750 - 3p_2.$$

So, the total revenue is given by

$$R = x_1p_1 + x_2p_2.$$

Estimate the average revenue when price p_1 varies between \$25 and \$50 and price p_2 varies between \$75 and \$125.

103. Real Estate The value of real estate (in dollars per square foot) for a city is given by

$$f(x, y) = 0.003x^{2/3}y^{3/4}$$

where x and y are measured in feet. What is the average value of real estate inside the rectangular area defined by the vertices $(0, 0), (5280, 0), (5280, 3960)$, and $(0, 3960)$?

TEST YOURSELF

See www.CalcChat.com for worked-out solutions to odd-numbered exercises.

Take this test as you would take a test in class. When you are done, check your work against the answers given in the back of the book.

In Exercises 1–3, (a) plot the points in a three-dimensional coordinate system, (b) find the distance between the points, and (c) find the midpoint of the line segment joining the points.

1. $(1, -3, 0), (3, -1, 0)$ **2.** $(-2, 2, 3), (-4, 0, 2)$ **3.** $(3, -7, 2), (5, 11, -6)$

4. Find the center and radius of the sphere whose equation is

$$x^2 + y^2 + z^2 - 20x + 10y - 10z + 125 = 0.$$

In Exercise 5–7, classify the quadric surface.

5. $4x^2 + 2y^2 - z^2 = 16$ **6.** $36x^2 + 9y^2 - 4z^2 = 0$

7. $4x^2 - y^2 - 16z = 0$

In Exercises 8–10, find $f(3, 3)$ and $f(1, 4)$.

8. $f(x, y) = x^2 + xy + 1$ **9.** $f(x, y) = \dfrac{x + 2y}{3x - y}$

10. $f(x, y) = xy \ln \dfrac{x}{y}$

In Exercises 11 and 12, find the first partial derivatives and evaluate each at the point $(10, -1)$.

11. $f(x, y) = 3x^2 + 9xy^2 - 2$ **12.** $f(x, y) = x\sqrt{x + y}$

In Exercises 13 and 14, find the critical points, relative extrema, and saddle points of the function.

13. $f(x, y) = 3x^2 + 4y^2 - 6x + 16y - 4$

14. $f(x, y) = 4xy - x^4 - y^4$

15. The production function for a company is given by

$$f(x, y) = 60x^{0.7}y^{0.3}$$

where x is the number of units of labor (at \$42 per unit) and y is the number of units of capital (at \$144 per unit). The total cost for labor and capital cannot exceed \$240,000. Use Lagrange multipliers to find the maximum production level for this manufacturer.

16. Find the least squares regression line for the points $(1, 2)$, $(3, 3)$, $(6, 4)$, $(8, 6)$, and $(11, 7)$.

In Exercises 17 and 18, evaluate the double integral.

17. $\displaystyle\int_0^1 \int_x^1 (30x^2y - 1)\, dy\, dx$ **18.** $\displaystyle\int_0^{\sqrt{e-1}} \int_0^{2y} \frac{1}{y^2 + 1}\, dx\, dy$

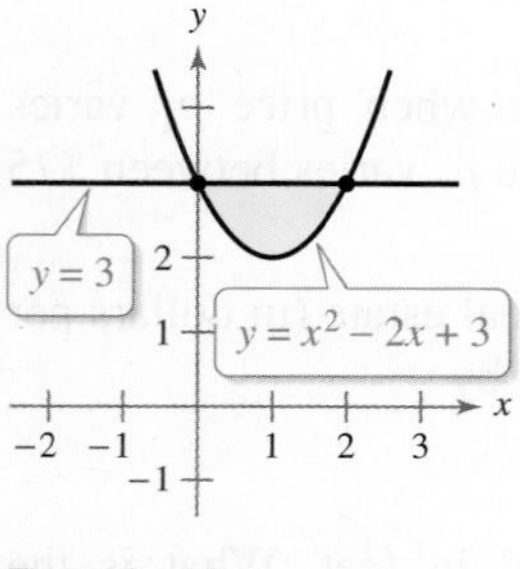

Figure for 19

19. Use a double integral to find the area of the region bounded by the graphs of $y = 3$ and $y = x^2 - 2x + 3$ (see figure).

20. Use a double integral to find the volume of the solid bounded by the graphs of $z = 8 - 2x$, $z = 0$, $y = 0$, $y = 3$, $x = 0$, and $x = 4$.

21. Find the average value of $f(x, y) = x^2 + y$ over the region defined by a rectangle with vertices $(0, 0)$, $(1, 0)$, $(1, 3)$, and $(0, 3)$.

8 Trigonometric Functions

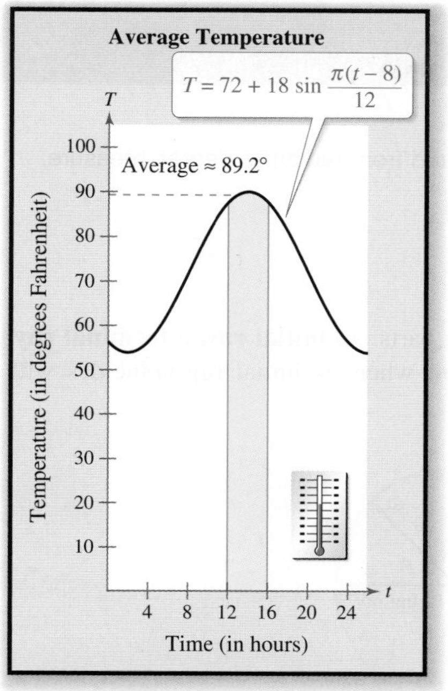

Example 10 on page 550 shows how integration and a trigonometric model can be used to find the average temperature during a four-hour period.

Yuri Arcurs/www.shutterstock.com
Kurhan/www.shutterstock.com

8.1 Radian Measure of Angles

- Find coterminal angles.
- Convert from degree to radian measure and from radian to degree measure.
- Use formulas relating to triangles.

In Exercise 45 on page 512, you will use properties of similar triangles to find the height of a streetlight.

Angles and Degree Measure

As shown in Figure 8.1, an **angle** has three parts: an **initial ray,** a **terminal ray,** and a **vertex.** An angle is in **standard position** when its initial ray coincides with the positive x-axis and its vertex is at the origin.

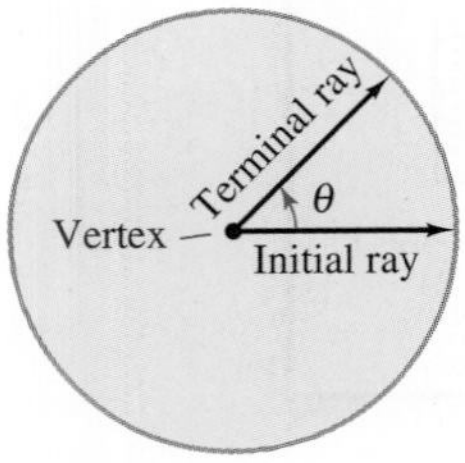

Standard Position of an Angle
FIGURE 8.1

Figure 8.2 shows the degree measures of several common angles. Note that θ (the lowercase Greek letter theta) is used to represent an angle and its measure. Angles whose measures are between 0° and 90° are **acute,** and angles whose measures are between 90° and 180° are **obtuse.** An angle whose measure is 90° is a **right** angle, and an angle whose measure is 180° is a **straight** angle.

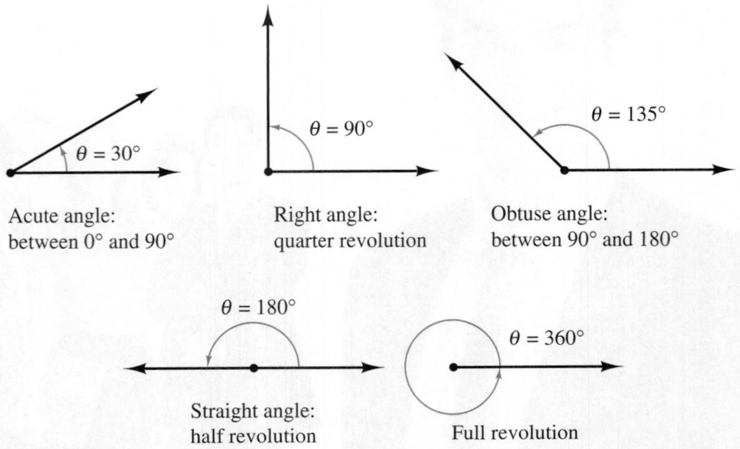

FIGURE 8.2

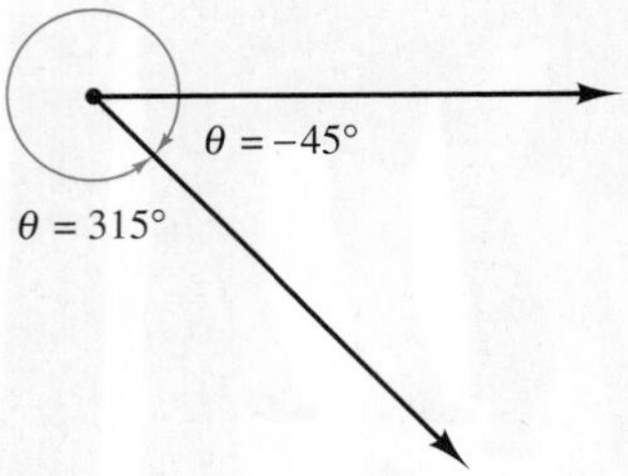

Coterminal Angles
FIGURE 8.3

Positive angles are measured *counterclockwise* beginning with the initial ray. Negative angles are measured *clockwise*. For instance, Figure 8.3 shows an angle whose measure is $-45°$.

Merely knowing where an angle's initial and terminal rays are located does not allow you to assign a measure to the angle. To measure an angle, you must know how the terminal ray was revolved. For example, Figure 8.3 shows that the angle measuring $-45°$ has the same terminal ray as the angle measuring 315°. Such angles are called **coterminal.**

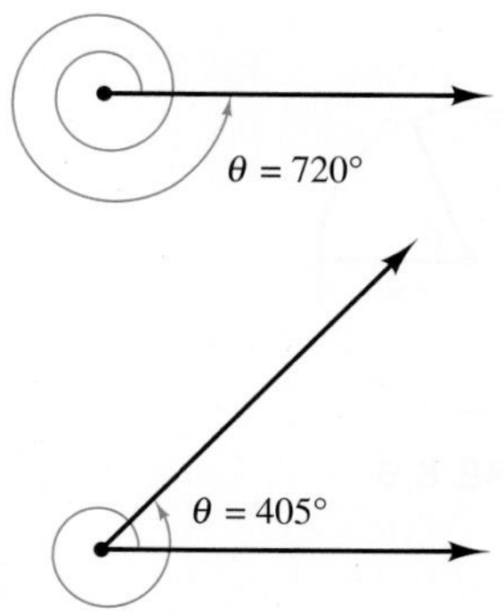

FIGURE 8.4

Although it may seem strange to consider angle measures that are larger than 360°, such angles have very useful applications in trigonometry. An angle that is larger than 360° is one whose terminal ray has revolved more than one full revolution counterclockwise. Figure 8.4 shows two angles measuring more than 360°. In a similar way, you can generate an angle whose measure is less than −360° by revolving a terminal ray more than one full revolution clockwise.

Example 1 Finding Coterminal Angles

For each angle, find a coterminal angle θ such that $0° \le \theta < 360°$.

a. 450° **b.** 750°

c. −160° **d.** −390°

SOLUTION

a. To find an angle coterminal to 450°, subtract 360°, as shown in Figure 8.5(a).

$$\theta = 450° - 360° = 90°$$

b. To find an angle that is coterminal to 750°, subtract 2(360°), as shown in Figure 8.5(b).

$$\theta = 750° - 2(360°) = 750° - 720° = 30°$$

c. To find an angle coterminal to −160°, add 360°, as shown in Figure 8.5(c).

$$\theta = -160° + 360° = 200°$$

d. To find an angle that is coterminal to −390°, add 2(360°), as shown in Figure 8.5(d).

$$\theta = -390° + 2(360°) = -390° + 720° = 330°$$

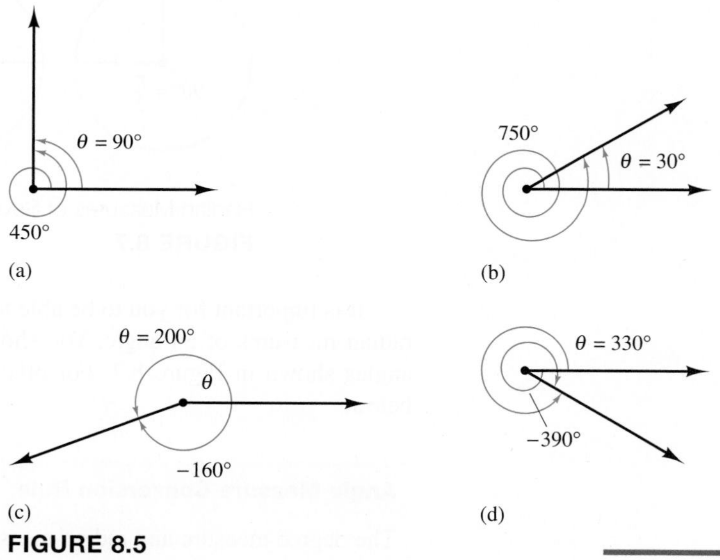

FIGURE 8.5

✓ Checkpoint 1

For each angle, find a coterminal angle θ such that $0° \le \theta < 360°$.

a. −210°

b. −330°

c. 495°

d. 390°

Radian Measure

A second way to measure angles is in terms of radians. To assign a radian measure to an angle θ, consider θ to be the central angle of a circular sector of radius 1, as shown in Figure 8.6. The radian measure of θ is then defined to be the length of the arc of the sector. Recall that the circumference of a circle is given by

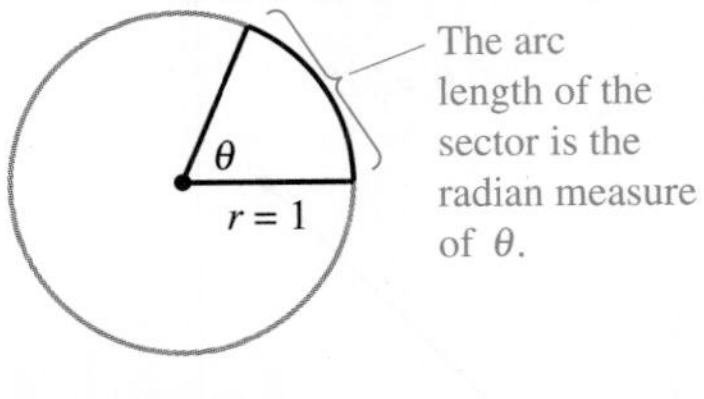

FIGURE 8.6

$$\text{Circumference} = (2\pi)(\text{radius}).$$

So, the circumference of a circle of radius 1 is simply 2π, and you can conclude that the radian measure of an angle measuring 360° is 2π. In other words,

$$360° = 2\pi \text{ radians}$$

or

$$180° = \pi \text{ radians}.$$

Figure 8.7 gives the radian measures of several common angles.

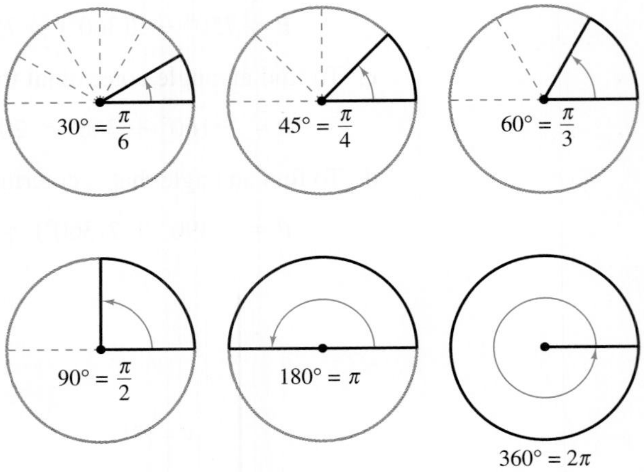

Radian Measures of Several Common Angles

FIGURE 8.7

It is important for you to be able to convert back and forth between the degree and radian measures of an angle. You should remember the conversions for the common angles shown in Figure 8.7. For other conversions, you can use the conversion rule below.

Angle Measure Conversion Rule

The degree measure and radian measure of an angle are related by the equation

$$180° = \pi \text{ radians}.$$

Conversions between degrees and radians can be done as follows.

1. To convert degrees to radians, multiply degrees by $\dfrac{\pi \text{ radians}}{180°}$.

2. To convert radians to degrees, multiply radians by $\dfrac{180°}{\pi \text{ radians}}$.

TECH TUTOR

Most calculators and graphing utilities have both degree and radian modes. You should learn how to use your calculator to convert from degrees to radians, and vice versa. Use a calculator or graphing utility to verify the results of Examples 2 and 3.

Example 2 Converting from Degrees to Radians

Convert each degree measure to radian measure.

a. $135°$ **b.** $40°$ **c.** $540°$ **d.** $-270°$

SOLUTION To convert from degree measure to radian measure, multiply the degree measure by $(\pi \text{ radians})/180°$.

a. $135° = (135 \text{ degrees})\left(\dfrac{\pi \text{ radians}}{180 \text{ degrees}}\right) = \dfrac{3\pi}{4} \text{ radians}$

b. $40° = (40 \text{ degrees})\left(\dfrac{\pi \text{ radians}}{180 \text{ degrees}}\right) = \dfrac{2\pi}{9} \text{ radian}$

c. $540° = (540 \text{ degrees})\left(\dfrac{\pi \text{ radians}}{180 \text{ degrees}}\right) = 3\pi \text{ radians}$

d. $-270° = (-270 \text{ degrees})\left(\dfrac{\pi \text{ radians}}{180 \text{ degrees}}\right) = -\dfrac{3\pi}{2} \text{ radians}$

✓ Checkpoint 2

Convert each degree measure to radian measure.

a. $225°$ **b.** $-45°$ **c.** $240°$ **d.** $150°$ ■

Although it is common to list radian measure in multiples of π, this is not necessary. For instance, when the degree measure of an angle is 79.3°, the radian measure is

$$79.3° = (79.3 \text{ degrees})\left(\frac{\pi \text{ radians}}{180 \text{ degrees}}\right) \approx 1.384 \text{ radians}.$$

Example 3 Converting from Radians to Degrees

Convert each radian measure to degree measure.

a. $-\dfrac{\pi}{2}$ **b.** $\dfrac{7\pi}{4}$ **c.** $\dfrac{11\pi}{6}$ **d.** $\dfrac{9\pi}{2}$

SOLUTION To convert from radian measure to degree measure, multiply the radian measure by $180°/(\pi \text{ radians})$.

a. $-\dfrac{\pi}{2} \text{ radians} = \left(-\dfrac{\pi}{2} \text{ radians}\right)\left(\dfrac{180 \text{ degrees}}{\pi \text{ radians}}\right) = -90°$

b. $\dfrac{7\pi}{4} \text{ radians} = \left(\dfrac{7\pi}{4} \text{ radians}\right)\left(\dfrac{180 \text{ degrees}}{\pi \text{ radians}}\right) = 315°$

c. $\dfrac{11\pi}{6} \text{ radians} = \left(\dfrac{11\pi}{6} \text{ radians}\right)\left(\dfrac{180 \text{ degrees}}{\pi \text{ radians}}\right) = 330°$

d. $\dfrac{9\pi}{2} \text{ radians} = \left(\dfrac{9\pi}{2} \text{ radians}\right)\left(\dfrac{180 \text{ degrees}}{\pi \text{ radians}}\right) = 810°$

✓ Checkpoint 3

Convert each radian measure to degree measure.

a. $\dfrac{5\pi}{3}$ **b.** $\dfrac{7\pi}{6}$ **c.** $\dfrac{3\pi}{2}$ **d.** $-\dfrac{3\pi}{4}$ ■

Triangles

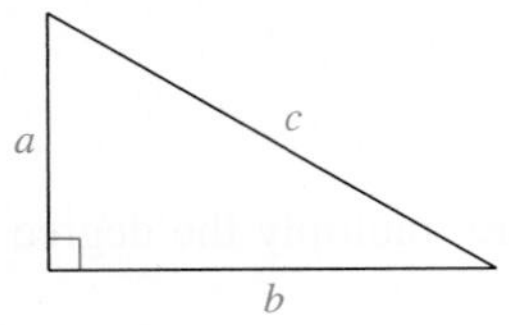

$a^2 + b^2 = c^2$

FIGURE 8.8

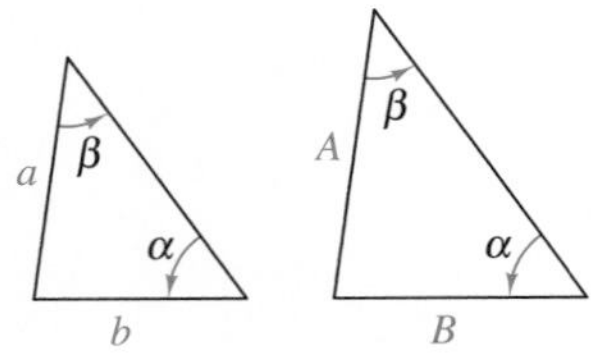

$\frac{a}{b} = \frac{A}{B}$

FIGURE 8.9

A Summary of Rules About Triangles

1. The sum of the angles of a triangle is 180°.
2. The sum of the two acute angles of a right triangle is 90°.
3. **Pythagorean Theorem** The sum of the squares of the legs of a right triangle is equal to the square of the hypotenuse, as shown in Figure 8.8.
4. **Similar Triangles** If two triangles are similar (have the same angle measures), then the ratios of the corresponding sides are equal, as shown in Figure 8.9.
5. The area of a triangle is equal to one-half the base times the height. That is, $A = \frac{1}{2}bh$.
6. Each angle of an equilateral triangle measures 60°.
7. Each acute angle of an isosceles right triangle measures 45°.
8. The altitude of an equilateral triangle bisects its base.

Example 4 Finding the Area of a Triangle

Find the area of an equilateral triangle with one-foot sides.

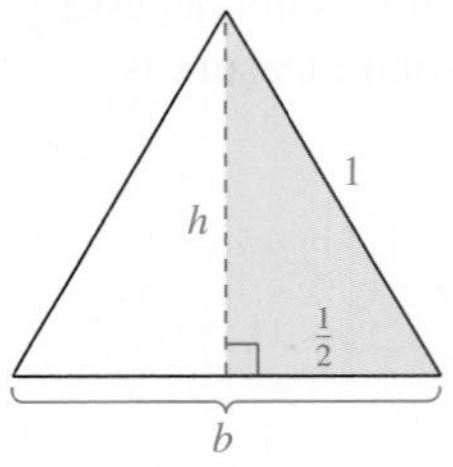

FIGURE 8.10

SOLUTION To use the formula $A = \frac{1}{2}bh$, you must first find the height of the triangle, as shown in Figure 8.10. To do this, apply the Pythagorean Theorem to the shaded portion of the triangle.

$$h^2 + \left(\frac{1}{2}\right)^2 = 1^2 \quad \text{Pythagorean Theorem}$$

$$h^2 = \frac{3}{4} \quad \text{Simplify.}$$

$$h = \frac{\sqrt{3}}{2} \quad \text{Solve for } h.$$

So, the area of the triangle is

$$A = \frac{1}{2}bh = \frac{1}{2}(1)\left(\frac{\sqrt{3}}{2}\right) = \frac{\sqrt{3}}{4} \text{ square foot.}$$

✓ Checkpoint 4

Find the area of an isosceles right triangle with a hypotenuse of $\sqrt{2}$ feet. ■

SUMMARIZE (Section 8.1)

1. Explain how to convert from degree measure to radian measure *(page 508)*. For an example of converting from degrees to radians, see Example 2.
2. Explain how to convert from radian measure to degree measure *(page 508)*. For an example of converting from radians to degrees, see Example 3.
3. State the formula for the area of a triangle *(page 510)*. For an example of finding the area of a triangle, see Example 4.

David Gilder/Shutterstock.com

SKILLS WARM UP 8.1 The following warm-up exercises involve skills that were covered in earlier sections. You will use these skills in the exercise set for this section. For additional help, review Section 1.1.

In Exercises 1 and 2, find the area of the triangle.

1. Base: 10 cm; height: 7 cm

2. Base: 4 in.; height: 6 in.

In Exercises 3–6, let *a* and *b* represent the lengths of the legs, and let *c* represent the length of the hypotenuse, of a right triangle. Solve for the missing side length.

3. $a = 5, b = 12$

4. $a = 3, c = 5$

5. $a = 8, c = 17$

6. $b = 8, c = 10$

In Exercises 7–10, let *a*, *b*, and *c* represent the side lengths of a triangle. Use the given information to determine whether the figure is a right triangle, an isosceles triangle, or an equilateral triangle.

7. $a = 4, b = 4, c = 4$

8. $a = 3, b = 3, c = 4$

9. $a = 12, b = 16, c = 20$

10. $a = 1, b = 1, c = \sqrt{2}$

Exercises 8.1

See www.CalcChat.com for worked-out solutions to odd-numbered exercises.

Finding Coterminal Angles In Exercises 1–6, determine two coterminal angles (one positive and one negative) for each angle. Give the answers in degrees. *See Example 1.*

1.

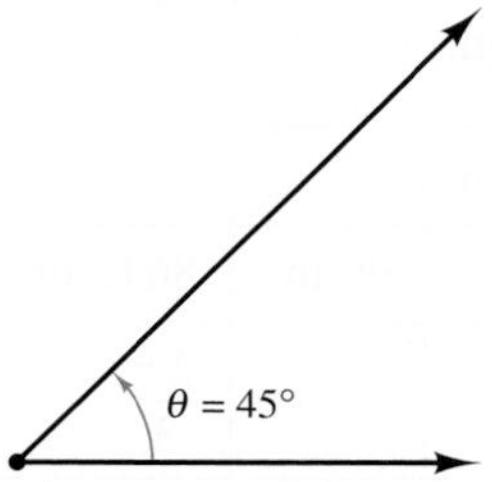

2.

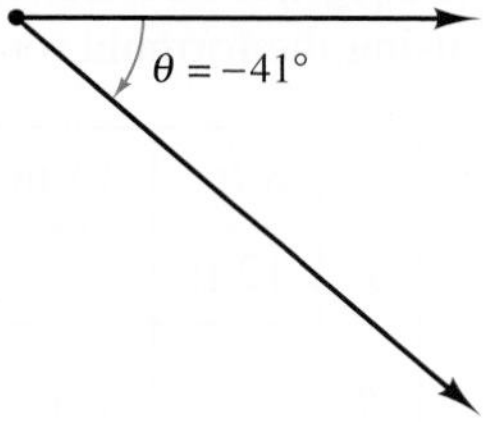

3.

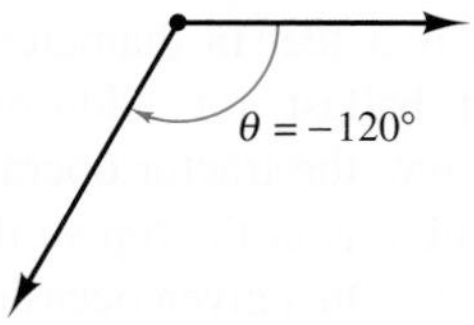

4.

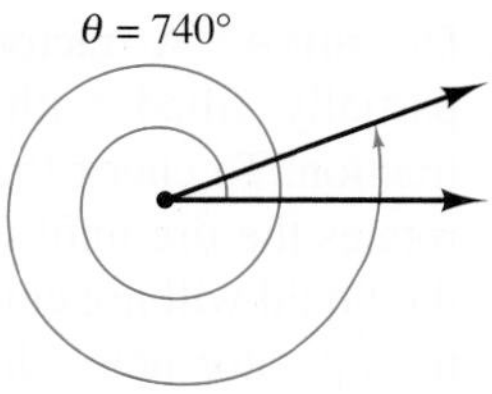

5.

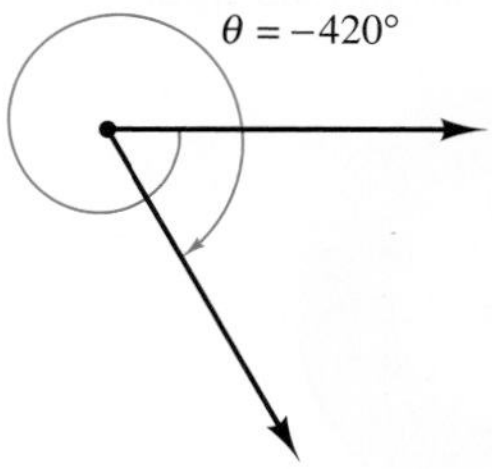

6.

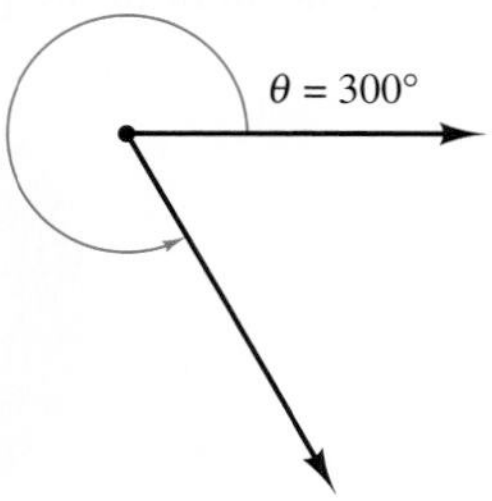

Finding Coterminal Angles In Exercises 7–10, determine two coterminal angles (one positive and one negative) for each angle. Give the answers in radians.

7.

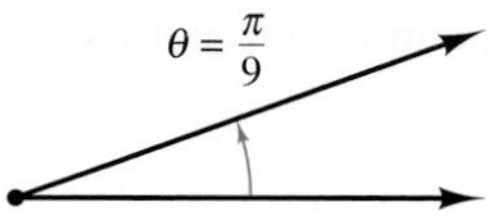

8.

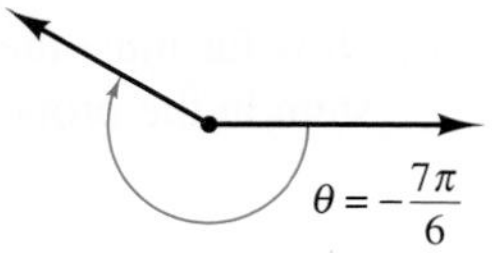

9.

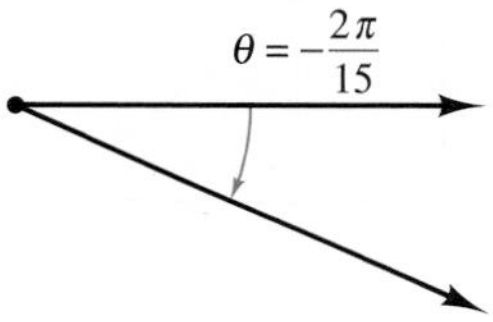

10.

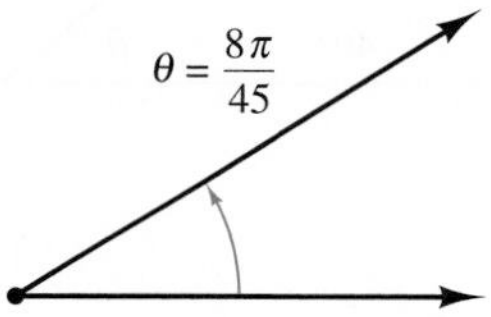

Converting from Degrees to Radians In Exercises 11–22, express the angle in radian measure as a multiple of π. Use a calculator to verify your result. *See Example 2.*

11. $30°$

12. $60°$

13. $270°$

14. $210°$

15. $675°$

16. $120°$

17. $-24°$

18. $-585°$

19. $-144°$

20. $-315°$

21. $330°$

22. $405°$

Converting from Radians to Degrees In Exercises 23–32, express the angle in degree measure. Use a calculator to verify your result. *See Example 3.*

23. $\frac{5\pi}{2}$

24. $\frac{5\pi}{4}$

25. $\frac{7\pi}{3}$

26. $\frac{\pi}{9}$

27. $-\frac{\pi}{12}$

28. $-\frac{7\pi}{12}$

29. $\frac{4\pi}{15}$

30. $-\frac{8\pi}{9}$

31. $\frac{19\pi}{6}$

32. $\frac{8\pi}{3}$

Analyzing Triangles In Exercises 33–40, solve the triangle for the indicated side and/or angle.

33.

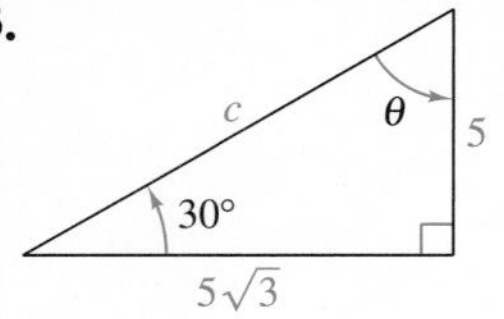

34.

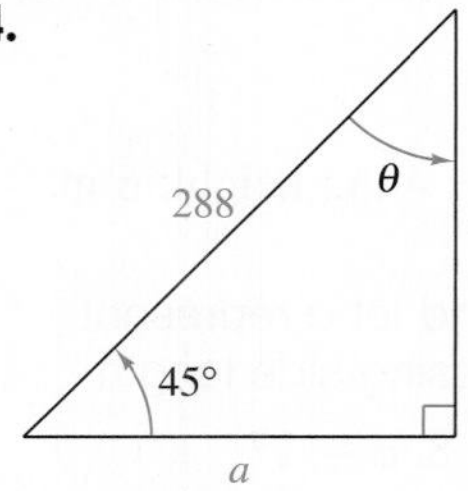

35.

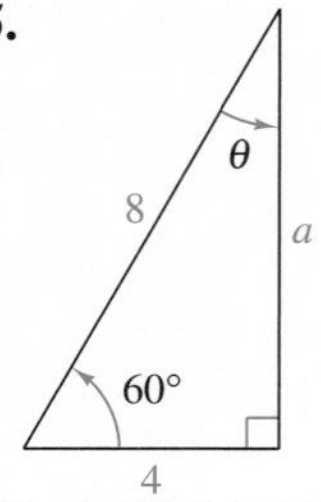

36.

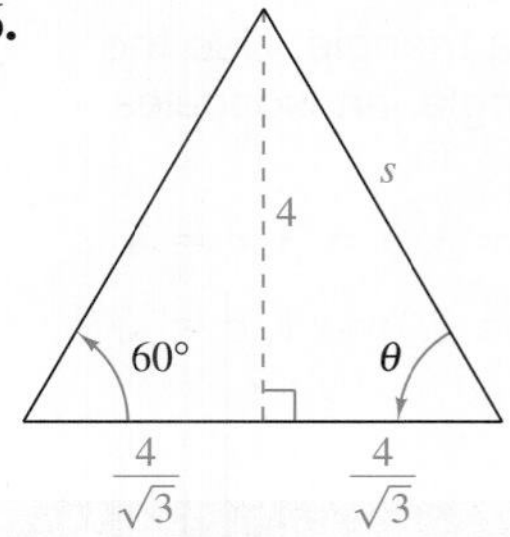

37.

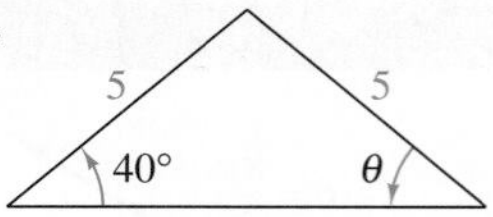

38.

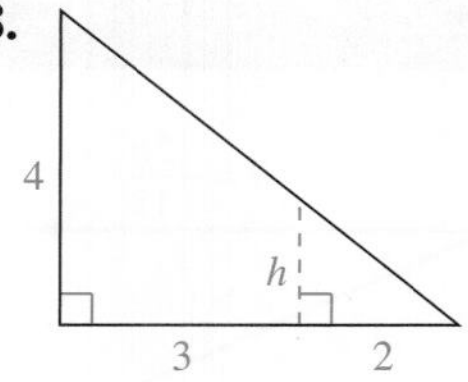

39.

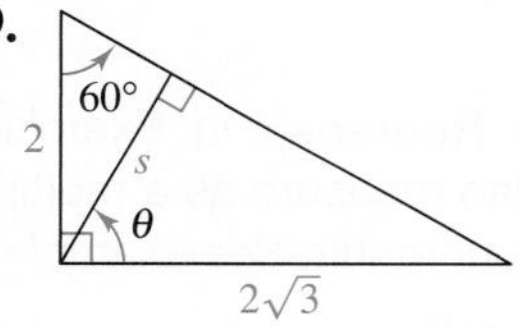

40.

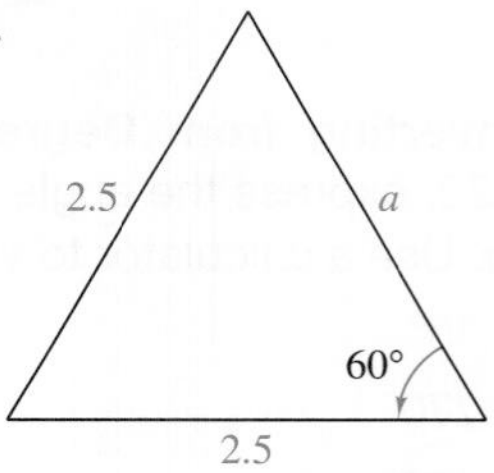

Finding the Area of an Equilateral Triangle In Exercises 41–44, find the area of the equilateral triangle with sides of length s. *See Example 4.*

41. $s = 4$ in.

42. $s = 8$ m

43. $s = 5$ ft

44. $s = 12$ cm

45. Height A person 6 feet tall standing 16 feet from a streetlight casts a shadow 8 feet long (see figure). What is the height of the streetlight?

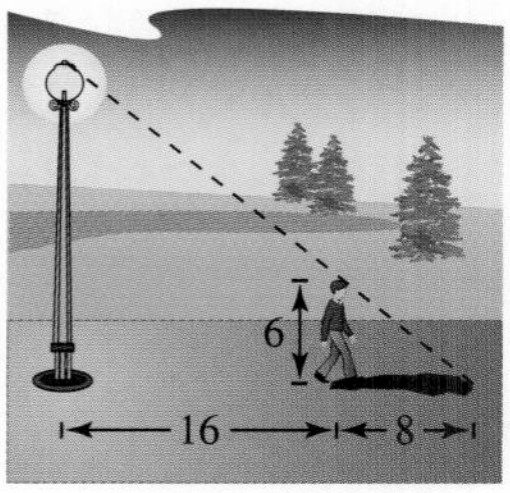

46. Length A guy wire is stretched from a broadcasting tower at a point 200 feet above the ground to an anchor 125 feet from the base (see figure). How long is the wire?

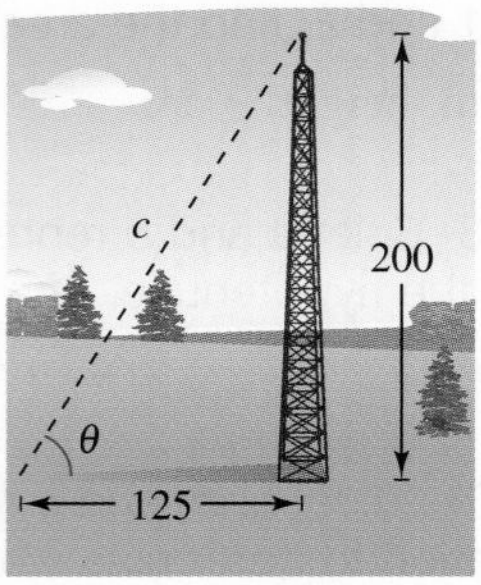

Arc Length In Exercises 47–50, use the following information, as shown in the figure. For a circle of radius r, a central angle θ (in radians) intercepts an arc of length s given by $s = r\theta$.

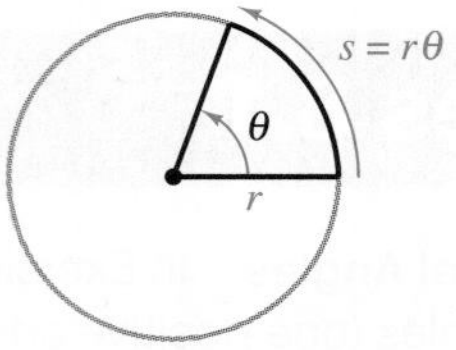

47. Using the Arc Length Formula Complete the table using the formula for arc length.

r	8 ft	15 in.	85 cm		
s	12 ft			96 in.	8642 mi
θ		1.6	$\frac{3\pi}{4}$	4	$\frac{2\pi}{3}$

48. Distance A tractor tire that is 5 feet in diameter is partially filled with a liquid ballast for additional traction. To check the air pressure, the tractor operator rotates the tire until the valve stem is at the top so that the liquid will not enter the gauge. On a given occasion, the operator notes that the tire must be rotated 80° to have the stem in the proper position (see figure).

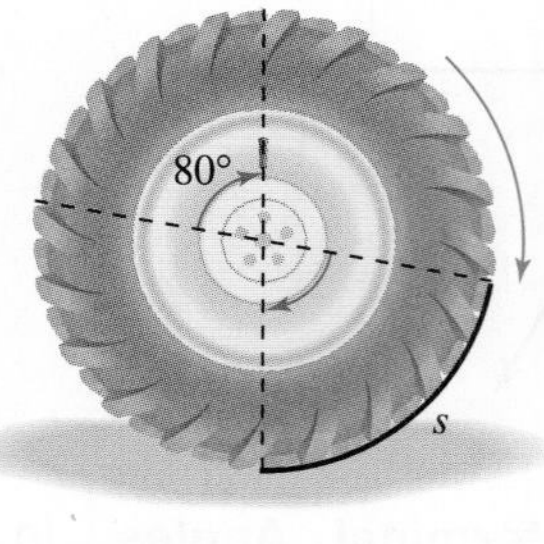

(a) What is the radius of the tractor tire?

(b) Find the radian measure of this rotation.

(c) How far must the tractor be moved to get the valve stem in the proper position?

49. Clock The minute hand on a clock is $3\frac{1}{2}$ inches long (see figure). Through what distance does the tip of the minute hand move in 25 minutes?

Figure for 49

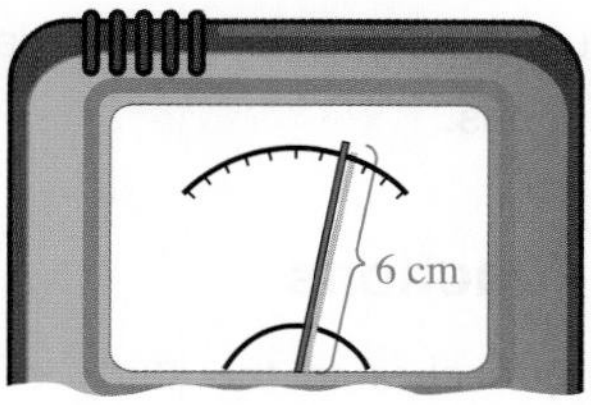

Figure for 50

50. Instrumentation The pointer of a voltmeter is 6 centimeters in length (see figure). Find the angle (in radians and degrees) through which the pointer rotates when it moves 2.5 centimeters on the scale.

51. Speed of Revolution A compact disc can have an angular speed of up to 3142 radians per minute.

(a) At this angular speed, how many revolutions per minute would the CD make?

(b) How long would it take the CD to make 10,000 revolutions?

52. HOW DO YOU SEE IT? Determine which angles in the figure are coterminal angles with Angle A. Explain your reasoning.

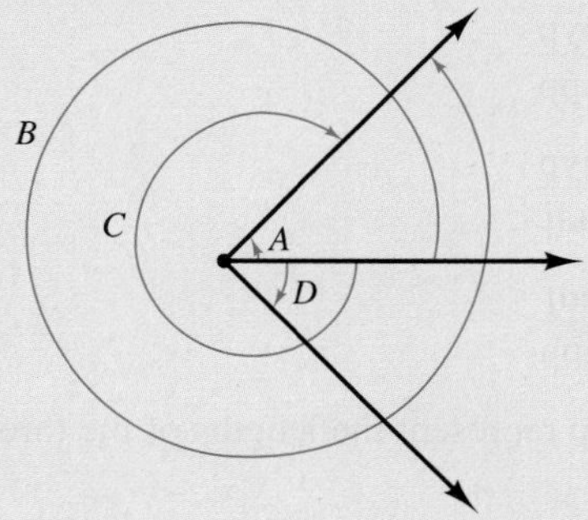

Area of a Sector of a Circle In Exercises 53 and 54, use the following information. A sector of a circle is the region bounded by two radii of the circle and their intercepted arc (see figure).

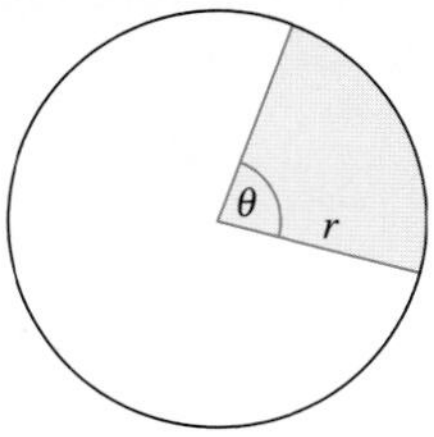

For a circle of radius r, the area A of a sector of the circle with central angle θ (in radians) is given by $A = \frac{1}{2}r^2\theta$.

53. Sprinkler System A sprinkler system on a farm is set to spray water over a distance of 70 feet and rotates through an angle of 120°. Find the area of the region.

54. Windshield Wiper A car's rear windshield wiper rotates 125°. The wiper mechanism has a total length of 25 inches and wipes the windshield over a distance of 14 inches. Find the area covered by the wiper.

True or False? In Exercises 55–58, determine whether the statement is true or false. If it is false, explain why or give an example that shows it is false.

55. An angle whose measure is 75° is obtuse.

56. $\theta = -35°$ is coterminal to 325°.

57. A right triangle can have one angle whose measure is 89°.

58. An angle whose measure is π radians is a straight angle.

Business Capsule

After a successful career as a critical care nurse, Grandee Ann Ray started Grand Ideas, a corporate specialty products firm in Charleston, South Carolina. The company offers innovative and unique custom-branded promotional products, apparel, and gifts for businesses and events. Ray started Grand Ideas from her home in 2001 with little more than a cell phone, a fax machine, and minimal inventory. Annual sales quickly surpassed $1.2 million. After weathering a national recession, today the company has rebounded to increased sales growth by adapting, streamlining, and aggressively pursuing a changing market.

59. Research Project Use your school's library, the Internet, or some other reference source to gather information on a company that offers unique products or services to its customers. Collect data about the revenue that the company has generated, and find a mathematical model of the data. Write a short paper that summarizes your findings.

8.2 The Trigonometric Functions

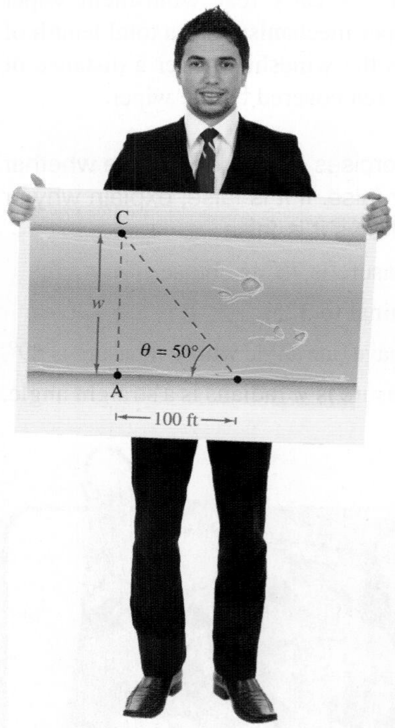

In Exercise 72 on page 523, you will use trigonometric functions to find the width of a river.

- Understand the definitions of the trigonometric functions.
- Understand the trigonometric identities.
- Evaluate trigonometric functions and solve right triangles.
- Solve trigonometric equations.

The Trigonometric Functions

There are two common approaches to the study of trigonometry. In one case the trigonometric functions are defined as ratios of two sides of a right triangle. In the other case these functions are defined in terms of a point on the terminal side of an arbitrary angle. The first approach is the one generally used in surveying, navigation, and astronomy, where a typical problem involves a triangle, three of whose six parts (sides and angles) are known and three of which are to be determined. The second approach is the one normally used in science and economics, where the periodic nature of the trigonometric functions is emphasized. In the definitions below, the six trigonometric functions

sine, **cosecant,** **cosine,** **secant,** **tangent,** and **cotangent**

are defined from both viewpoints. These six functions are normally abbreviated sin, csc, cos, sec, tan, and cot, respectively.

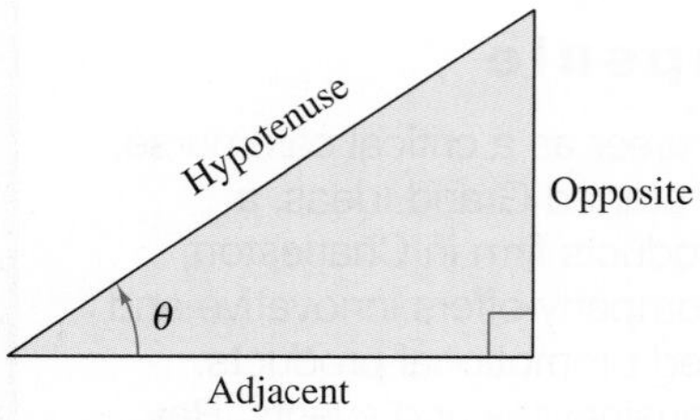

FIGURE 8.11

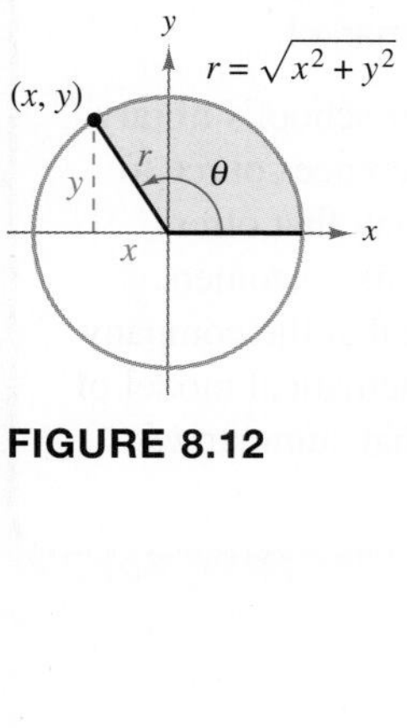

FIGURE 8.12

Definitions of the Trigonometric Functions

Right Triangle Definition: $0 < \theta < \frac{\pi}{2}$ (See Figure 8.11.)

$$\sin\theta = \frac{\text{opp}}{\text{hyp}} \qquad \csc\theta = \frac{\text{hyp}}{\text{opp}}$$

$$\cos\theta = \frac{\text{adj}}{\text{hyp}} \qquad \sec\theta = \frac{\text{hyp}}{\text{adj}}$$

$$\tan\theta = \frac{\text{opp}}{\text{adj}} \qquad \cot\theta = \frac{\text{adj}}{\text{opp}}$$

The abbreviations opp, adj, and hyp represent the lengths of the three sides of a right triangle.

opp = the length of the side *opposite* θ
adj = the length of the side *adjacent* to θ
hyp = the length of the *hypotenuse*

Circular Function Definition: Let θ be an angle in standard position with (x, y) a point on the terminal ray of θ and $r = \sqrt{x^2 + y^2} \neq 0$. (See Figure 8.12.)

$$\sin\theta = \frac{y}{r} \qquad \csc\theta = \frac{r}{y}$$

$$\cos\theta = \frac{x}{r} \qquad \sec\theta = \frac{r}{x}$$

$$\tan\theta = \frac{y}{x} \qquad \cot\theta = \frac{x}{y}$$

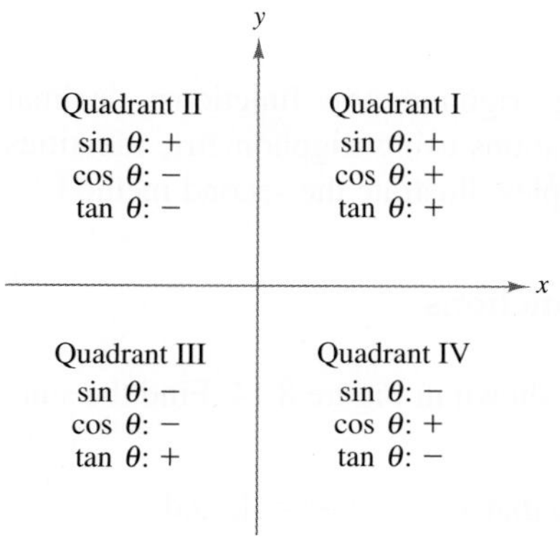

FIGURE 8.13

Trigonometric Identities

In the circular function definition of the six trigonometric functions, the value of r is always positive. From this, it follows that the signs of the trigonometric functions are determined from the signs of x and y, as shown in Figure 8.13.

The trigonometric *reciprocal identities* below are also direct consequences of the definitions.

$$\sin\theta = \frac{1}{\csc\theta} \qquad \cos\theta = \frac{1}{\sec\theta} \qquad \tan\theta = \frac{\sin\theta}{\cos\theta} = \frac{1}{\cot\theta}$$

$$\csc\theta = \frac{1}{\sin\theta} \qquad \sec\theta = \frac{1}{\cos\theta} \qquad \cot\theta = \frac{\cos\theta}{\sin\theta} = \frac{1}{\tan\theta}$$

Furthermore, because

$$\begin{aligned}\sin^2\theta + \cos^2\theta &= \left(\frac{y}{r}\right)^2 + \left(\frac{x}{r}\right)^2 \\ &= \frac{x^2 + y^2}{r^2} \\ &= \frac{r^2}{r^2} \\ &= 1\end{aligned}$$

you can obtain the Pythagorean Identity $\sin^2\theta + \cos^2\theta = 1$. Other trigonometric identities are listed below. In the list, ϕ is the lowercase Greek letter phi.

STUDY TIP

The symbol $\sin^2\theta$ is used to represent $(\sin\theta)^2$.

Trigonometric Identities

Pythagorean Identities

$\sin^2\theta + \cos^2\theta = 1 \qquad \tan^2\theta + 1 = \sec^2\theta \qquad \cot^2\theta + 1 = \csc^2\theta$

Reduction Formulas

$\sin(-\theta) = -\sin\theta \qquad \cos(-\theta) = \cos\theta \qquad \tan(-\theta) = -\tan\theta$

$\sin\theta = -\sin(\theta - \pi) \qquad \cos\theta = -\cos(\theta - \pi) \qquad \tan\theta = \tan(\theta - \pi)$

Sum or Difference of Two Angles

$\sin(\theta \pm \phi) = \sin\theta\cos\phi \pm \cos\theta\sin\phi$

$\cos(\theta \pm \phi) = \cos\theta\cos\phi \mp \sin\theta\sin\phi$

$$\tan(\theta \pm \phi) = \frac{\tan\theta \pm \tan\phi}{1 \mp \tan\theta\tan\phi}$$

Double Angle

$\sin 2\theta = 2\sin\theta\cos\theta$

$\cos 2\theta = 2\cos^2\theta - 1 = 1 - 2\sin^2\theta$

Half Angle

$\sin^2\theta = \frac{1}{2}(1 - \cos 2\theta)$

$\cos^2\theta = \frac{1}{2}(1 + \cos 2\theta)$

Although an angle can be measured in either degrees or radians, radian measure is preferred in calculus. So, all angles in the remainder of this chapter are assumed to be measured in radians unless otherwise indicated. In other words, sin 3 means the sine of 3 radians, and sin 3° means the sine of 3 degrees.

Evaluating Trigonometric Functions

There are two common methods of evaluating trigonometric functions: decimal approximations using a calculator and exact evaluations using trigonometric identities and formulas from geometry. The next three examples illustrate the second method.

Example 1 Evaluating Trigonometric Functions

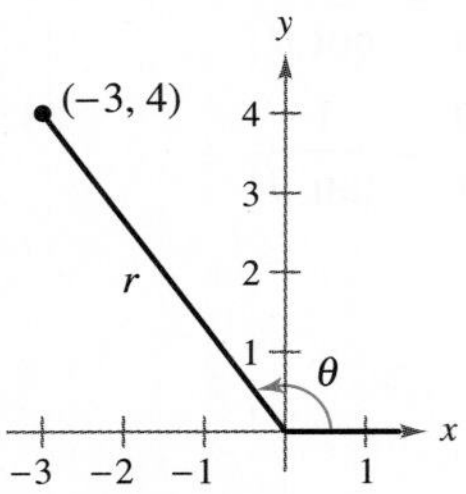

FIGURE 8.14

Let $(-3, 4)$ be a point on the terminal side of θ, as shown in Figure 8.14. Find the sine, cosine, and tangent of θ.

SOLUTION Referring to Figure 8.14, you can see that $x = -3$, $y = 4$, and

$$\begin{aligned} r &= \sqrt{x^2 + y^2} \\ &= \sqrt{(-3)^2 + 4^2} \\ &= \sqrt{25} \\ &= 5. \end{aligned}$$

So, the values of the sine, cosine, and tangent of θ are as shown.

$$\sin\theta = \frac{y}{r} = \frac{4}{5}$$

$$\cos\theta = \frac{x}{r} = -\frac{3}{5}$$

$$\tan\theta = \frac{y}{x} = -\frac{4}{3}$$

✓Checkpoint 1

Let $(\sqrt{3}, 1)$ be a point on the terminal side of θ, as shown in the figure. Find the sine, cosine, and tangent of θ.

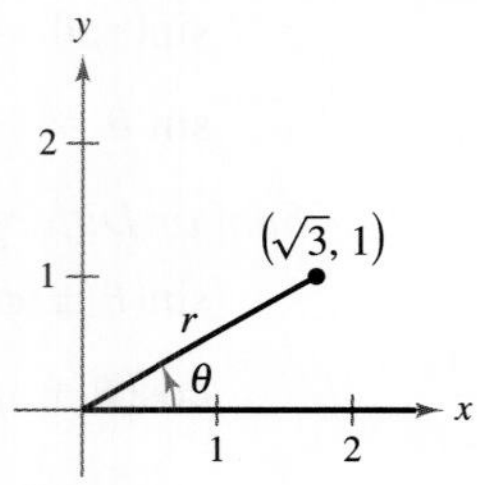

■

STUDY TIP

Learning the table of values at the right is worth the effort because doing so will increase both your efficiency and your confidence. Here is a pattern for the sine function that may help you remember the values.

θ	0	30°	45°
$\sin\theta$	$\frac{\sqrt{0}}{2}$	$\frac{\sqrt{1}}{2}$	$\frac{\sqrt{2}}{2}$

θ	60°	90°
$\sin\theta$	$\frac{\sqrt{3}}{2}$	$\frac{\sqrt{4}}{2}$

Reverse the order to get cosine values of the same angles.

The sines, cosines, and tangents of several common angles are listed in the table below. You should remember, or be able to derive, these values.

Trigonometric Values of Common Angles

θ (degrees)	0	30°	45°	60°	90°	180°	270°
θ (radians)	0	$\frac{\pi}{6}$	$\frac{\pi}{4}$	$\frac{\pi}{3}$	$\frac{\pi}{2}$	π	$\frac{3\pi}{2}$
$\sin\theta$	0	$\frac{1}{2}$	$\frac{\sqrt{2}}{2}$	$\frac{\sqrt{3}}{2}$	1	0	-1
$\cos\theta$	1	$\frac{\sqrt{3}}{2}$	$\frac{\sqrt{2}}{2}$	$\frac{1}{2}$	0	-1	0
$\tan\theta$	0	$\frac{\sqrt{3}}{3}$	1	$\sqrt{3}$	Undefined	0	Undefined

To extend the use of the values in the table on the preceding page, you can use the concept of a reference angle, as shown in Figure 8.15, together with the appropriate quadrant sign. The **reference angle θ'** for an angle θ is the smallest positive angle between the terminal side of θ and the x-axis. For instance, the reference angle for 135° is 45° and the reference angle for 210° is 30°.

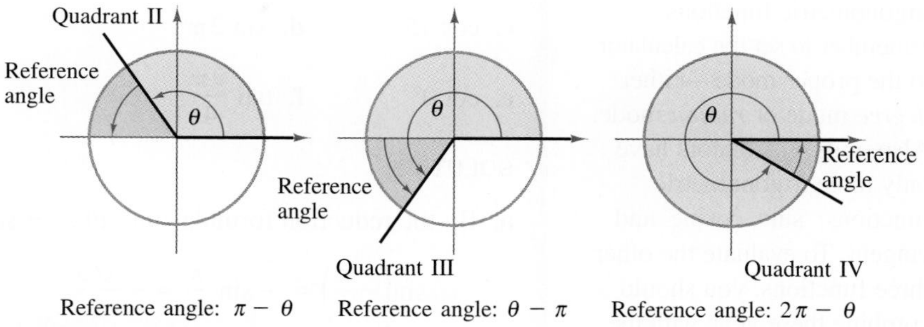

FIGURE 8.15

To find the value of a trigonometric function of any angle θ, first determine the function value for the associated reference angle θ'. Then, depending on the quadrant in which θ lies, prefix the appropriate sign to the function value.

Example 2 Evaluating Trigonometric Functions

Evaluate each trigonometric function.

a. $\sin \frac{3\pi}{4}$ **b.** $\tan 330°$ **c.** $\cos \frac{7\pi}{6}$

SOLUTION

a. Because the reference angle for $3\pi/4$ is $\pi/4$ and the sine is positive in the second quadrant, you can write

$$\sin \frac{3\pi}{4} = \sin \frac{\pi}{4}$$ Reference angle

$$= \frac{\sqrt{2}}{2}.$$ See Figure 8.16(a).

b. Because the reference angle for 330° is 30° and the tangent is negative in the fourth quadrant, you can write

$$\tan 330° = -\tan 30°$$ Reference angle

$$= -\frac{\sqrt{3}}{3}.$$ See Figure 8.16(b).

c. Because the reference angle for $7\pi/6$ is $\pi/6$ and the cosine is negative in the third quadrant, you can write

$$\cos \frac{7\pi}{6} = -\cos \frac{\pi}{6}$$ Reference angle

$$= -\frac{\sqrt{3}}{2}.$$ See Figure 8.16(c).

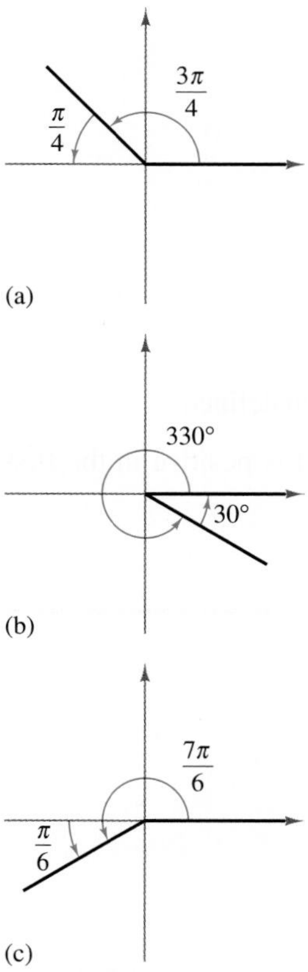

FIGURE 8.16

✓ Checkpoint 2

Evaluate each trigonometric function.

a. $\sin \frac{5\pi}{6}$ **b.** $\cos 135°$ **c.** $\tan \frac{5\pi}{3}$ ■

TECH TUTOR

When using a calculator to evaluate trigonometric functions, remember to set the calculator to the proper mode—either *degree* mode or *radian* mode. Also, most calculators have only three trigonometric functions: sine, cosine, and tangent. To evaluate the other three functions, you should combine these keys with the reciprocal key $\boxed{x^{-1}}$. For instance, you can evaluate $\sec(\pi/7)$ on most calculators using the following keystroke sequence in *radian* mode.

COS (π ÷ 7) x^{-1} ENTER

Example 3 Evaluating Trigonometric Functions

Evaluate each trigonometric function.

a. $\sin\left(-\dfrac{\pi}{3}\right)$ **b.** $\sec 60^\circ$

c. $\cos 15^\circ$ **d.** $\sin 2\pi$

e. $\cot 0^\circ$ **f.** $\tan \dfrac{9\pi}{4}$

SOLUTION

a. By the reduction formula $\sin(-\theta) = -\sin\theta$,

$$\sin\left(-\frac{\pi}{3}\right) = -\sin\frac{\pi}{3} = -\frac{\sqrt{3}}{2}.$$

b. By the reciprocal identity $\sec\theta = 1/\cos\theta$,

$$\sec 60^\circ = \frac{1}{\cos 60^\circ} = \frac{1}{1/2} = 2.$$

c. By the difference formula $\cos(\theta - \phi) = \cos\theta\cos\phi + \sin\theta\sin\phi$,

$$\begin{aligned}\cos 15^\circ &= \cos(45^\circ - 30^\circ)\\ &= (\cos 45^\circ)(\cos 30^\circ) + (\sin 45^\circ)(\sin 30^\circ)\\ &= \left(\frac{\sqrt{2}}{2}\right)\left(\frac{\sqrt{3}}{2}\right) + \left(\frac{\sqrt{2}}{2}\right)\left(\frac{1}{2}\right)\\ &= \frac{\sqrt{6}}{4} + \frac{\sqrt{2}}{4}\\ &= \frac{\sqrt{6}+\sqrt{2}}{4}.\end{aligned}$$

d. Because the reference angle for 2π is 0,

$$\sin 2\pi = \sin 0 = 0.$$

e. Using the reciprocal identity

$$\cot\theta = \frac{1}{\tan\theta}$$

and the fact that $\tan 0^\circ = 0$, you can conclude that $\cot 0^\circ$ is undefined.

f. Because the reference angle for $9\pi/4$ is $\pi/4$ and the tangent is positive in the first quadrant,

$$\tan\frac{9\pi}{4} = \tan\frac{\pi}{4} = 1.$$

✓ Checkpoint 3

Evaluate each trigonometric function.

a. $\sin\left(-\dfrac{\pi}{6}\right)$ **b.** $\csc 45^\circ$

c. $\cos 75^\circ$ **d.** $\cos 2\pi$

e. $\sec 0^\circ$ **f.** $\cot\dfrac{13\pi}{4}$ ■

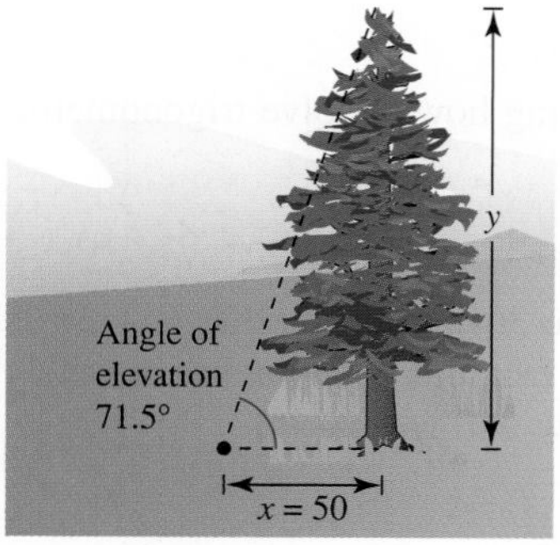

FIGURE 8.17

Example 4 Solving a Right Triangle

A surveyor is standing 50 feet from the base of a large tree, as shown in Figure 8.17. The surveyor measures the angle of elevation to the top of the tree as 71.5°. How tall is the tree?

SOLUTION From Figure 8.17, you can see that

$$\tan 71.5° = \frac{y}{x}$$

where $x = 50$ and y is the height of the tree. So, the height of the tree is

$$y = (x)(\tan 71.5°) \approx (50)(2.98868) \approx 149.4 \text{ feet.}$$

✓Checkpoint 4

Find the height of a building that casts a 75-foot shadow when the angle of elevation of the sun is 35°. ■

Some occupations, such as that of a military pilot, require excellent vision, including good depth perception and good peripheral vision.

Example 5 Calculating Peripheral Vision

To measure the extent of your peripheral vision, stand 1 foot from the corner of a room, facing the corner. Have a friend move an object along the wall until you can just barely see it. When the object is 2 feet from the corner, as shown in Figure 8.18, what is the total angle of your peripheral vision?

SOLUTION Let α represent the total angle of your peripheral vision. As shown in Figure 8.19, you can model the physical situation with an isosceles right triangle whose legs are $\sqrt{2}$ feet and whose hypotenuse is 2 feet. In the triangle, the angle θ is given by

$$\tan \theta = \frac{y}{x} = \frac{\sqrt{2}}{\sqrt{2} - 1} \approx 3.414.$$

Using the inverse tangent function of a calculator in *degree* mode and the following keystrokes

(TAN⁻¹) (( ) 3.414 ( )) (ENTER)

you can determine that $\theta \approx 73.7°$. So, $\alpha/2 \approx 180° - 73.7° = 106.3°$, which implies that $\alpha \approx 212.6°$. In other words, the total angle of your peripheral vision is about 212.6°.

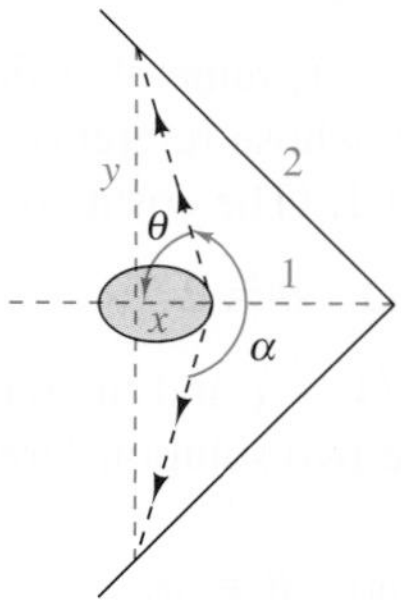

FIGURE 8.18

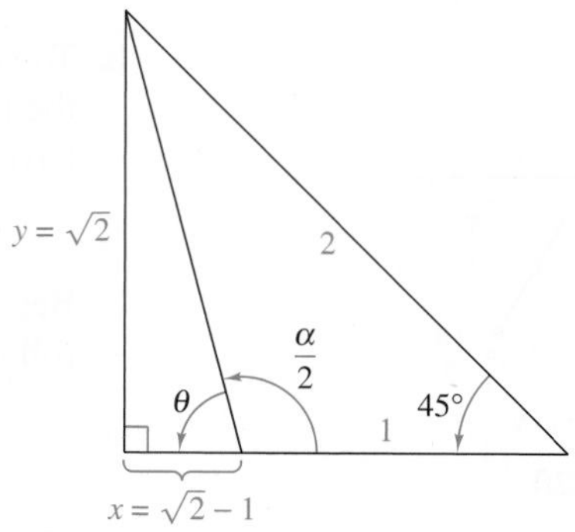

FIGURE 8.19

✓Checkpoint 5

When the object in Example 5 is 4 feet from the corner, find the total angle of your peripheral vision. ■

Ivonne Wierink/www.shutterstock.com

ALGEBRA TUTOR *xy*

For more examples of the algebra involved in solving trigonometric equations, see the *Chapter 8 Algebra Tutor*, on pages 554 and 555.

Solving Trigonometric Equations

An important part of the study of trigonometry is learning how to solve trigonometric equations. For example, consider the equation

$$\sin \theta = 0.$$

You know that $\theta = 0$ is one solution. Also, in Example 3(d), you saw that $\theta = 2\pi$ is another solution. But these are not the only solutions. In fact, this equation has infinitely many solutions. Any one of the values of θ shown below will work.

$$\ldots, -3\pi, -2\pi, -\pi, 0, \pi, 2\pi, 3\pi, \ldots$$

To simplify the situation, the search for solutions can be restricted to the interval

$$0 \le \theta \le 2\pi$$

as shown in Example 6.

Example 6 Solving Trigonometric Equations

Solve for θ in each equation. Assume $0 \le \theta \le 2\pi$.

a. $\sin \theta = -\dfrac{\sqrt{3}}{2}$ **b.** $\cos \theta = 1$ **c.** $\tan \theta = 1$

SOLUTION

a. To solve the equation $\sin \theta = -\sqrt{3}/2$, first remember that

$$\sin \frac{\pi}{3} = \frac{\sqrt{3}}{2}.$$

Because the sine is negative in the third and fourth quadrants, it follows that you are seeking values of θ in these quadrants that have a reference angle of $\pi/3$. The two angles fitting these criteria are

$$\theta = \pi + \frac{\pi}{3} = \frac{4\pi}{3}$$

and

$$\theta = 2\pi - \frac{\pi}{3} = \frac{5\pi}{3}$$

as indicated in Figure 8.20.

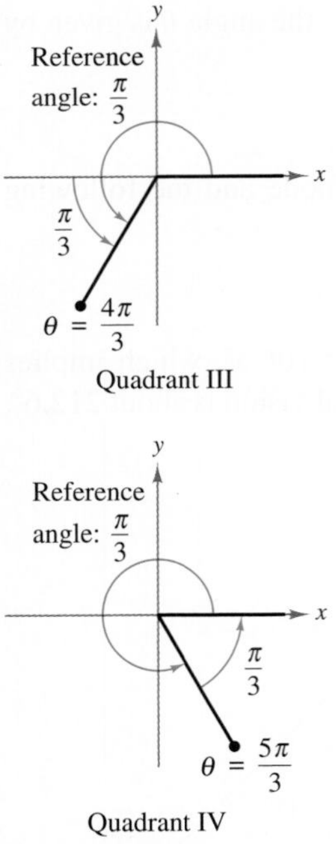

FIGURE 8.20

b. To solve $\cos \theta = 1$, remember that $\cos 0 = 1$ and note that in the interval $[0, 2\pi]$, the only angles whose reference angles are 0 are 0, π, and 2π. Of these, 0 and 2π have cosines of 1. (The cosine of π is -1.) So, the equation has two solutions:

$$\theta = 0 \quad \text{and} \quad \theta = 2\pi.$$

c. Because $\tan \pi/4 = 1$ and the tangent is positive in the first and third quadrants, it follows that the two solutions are

$$\theta = \frac{\pi}{4} \quad \text{and} \quad \theta = \pi + \frac{\pi}{4} = \frac{5\pi}{4}.$$

✓ Checkpoint 6

Solve for θ in each equation. Assume $0 \le \theta \le 2\pi$.

a. $\cos \theta = \dfrac{\sqrt{2}}{2}$ **b.** $\tan \theta = -\sqrt{3}$ **c.** $\sin \theta = -\dfrac{1}{2}$ ■

Example 7 **Solving a Trigonometric Equation**

Solve the equation for θ.

$$\cos 2\theta = 2 - 3\sin\theta, \quad 0 \le \theta \le 2\pi$$

SOLUTION You can use the double-angle identity $\cos 2\theta = 1 - 2\sin^2\theta$ to rewrite the original equation, as shown.

$$\begin{aligned} \cos 2\theta &= 2 - 3\sin\theta \\ 1 - 2\sin^2\theta &= 2 - 3\sin\theta \\ 0 &= 2\sin^2\theta - 3\sin\theta + 1 \\ 0 &= (2\sin\theta - 1)(\sin\theta - 1) \end{aligned}$$

Next, set each factor equal to zero. For $2\sin\theta - 1 = 0$, you have $\sin\theta = \frac{1}{2}$, which has solutions of

$$\theta = \frac{\pi}{6} \quad \text{and} \quad \theta = \frac{5\pi}{6}.$$

For $\sin\theta - 1 = 0$, you have $\sin\theta = 1$, which has a solution of

$$\theta = \frac{\pi}{2}.$$

So, for $0 \le \theta \le 2\pi$, the three solutions are

$$\theta = \frac{\pi}{6}, \ \frac{\pi}{2}, \ \text{and} \ \frac{5\pi}{6}.$$

✓ Checkpoint 7

Solve the equation for θ.

$$\sin 2\theta + \sin\theta = 0, \quad 0 \le \theta \le 2\pi$$

■

STUDY TIP

In Example 7, note that the expression $2\sin^2\theta - 3\sin\theta + 1$ is a quadratic in $\sin\theta$, and as such can be factored. For instance, when you let $x = \sin\theta$, the quadratic factors as $2x^2 - 3x + 1 = (2x - 1)(x - 1)$.

SUMMARIZE (Section 8.2)

1. State the circular function definition of the six trigonometric functions *(page 514)*. For an example of evaluating trigonometric functions, see Example 1.
2. Explain what is meant by a reference angle *(page 517)*. For an example of evaluating trigonometric functions using reference angles, see Example 2.
3. State the reduction formulas *(page 515)*. For an example of evaluating trigonometric functions using a reduction formula, see Example 3(a).
4. Describe a real-life example of how a trigonometric function can be used to find the height of a tree *(page 519, Example 4)*.
5. State the double-angle identities *(page 515)*. For an example of solving a trigonometric equation using a double-angle identity, see Example 7.

Andresr/www.shutterstock.com

SKILLS WARM UP 8.2

The following warm-up exercises involve skills that were covered in a previous course or in earlier sections. You will use these skills in the exercise set for this section. For additional help, review Appendix Section A.4 and Section 8.1.

In Exercises 1–4, convert the angle to radian measure.

1. $315°$ **2.** $-300°$ **3.** $-225°$ **4.** $390°$

In Exercises 5–8, solve for *x*.

5. $x^2 - x = 0$ **6.** $2x^2 + x = 0$ **7.** $x^2 - 2x = 3$ **8.** $x^2 - 5x = -6$

In Exercises 9–12, solve for *t*.

9. $\frac{2\pi}{24}(t - 4) = \frac{\pi}{2}$ **10.** $\frac{2\pi}{12}(t - 2) = \frac{\pi}{4}$ **11.** $\frac{2\pi}{365}(t - 10) = \frac{\pi}{4}$ **12.** $\frac{2\pi}{12}(t - 4) = \frac{\pi}{2}$

Exercises 8.2

See www.CalcChat.com for worked-out solutions to odd-numbered exercises.

Evaluating Trigonometric Functions In Exercises 1–6, determine all six trigonometric functions of the angle θ. *See Example 1.*

1.

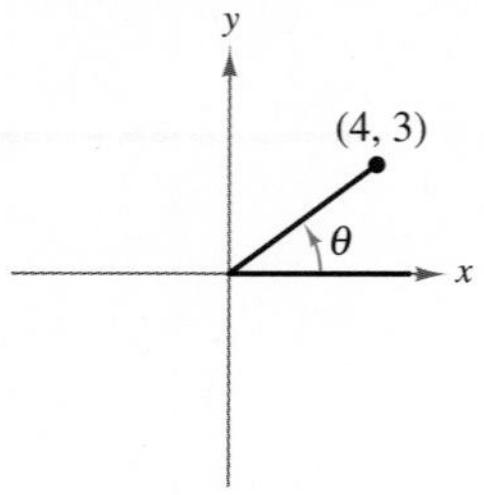

2.

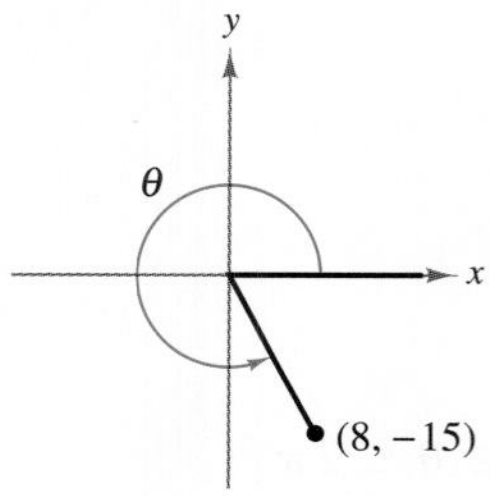

3.

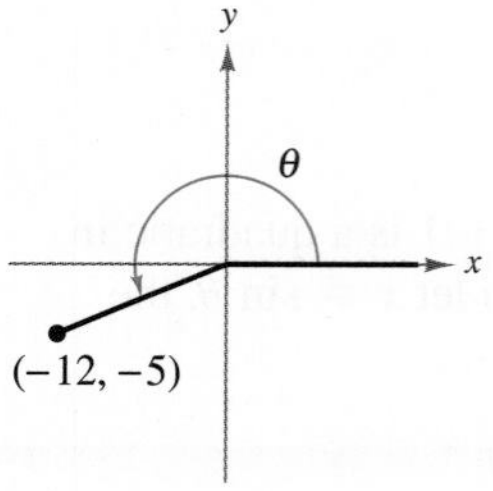

4.

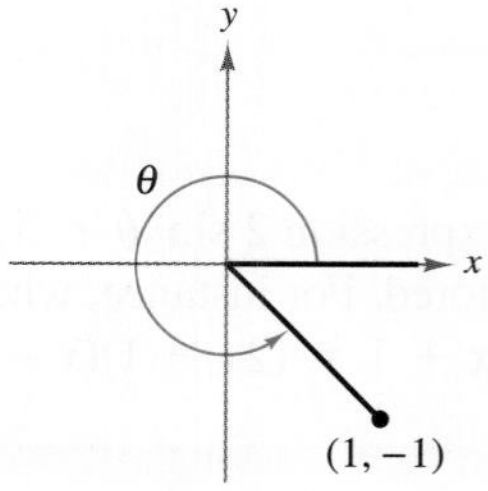

5.

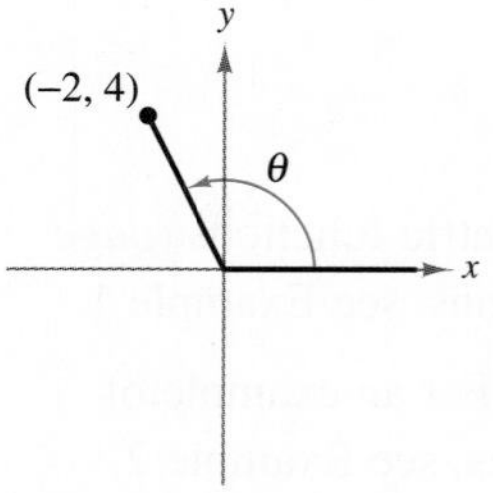

6.

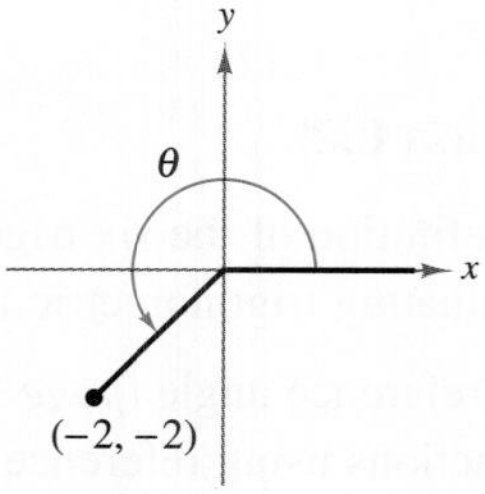

Finding Trigonometric Functions In Exercises 7–12, sketch a right triangle corresponding to the trigonometric function of the angle θ and find the other five trigonometric functions of θ.

7. $\sin \theta = \frac{1}{3}$ **8.** $\cot \theta = 5$

9. $\sec \theta = 2$ **10.** $\cos \theta = \frac{5}{7}$

11. $\tan \theta = 3$ **12.** $\csc \theta = 4.25$

Determining a Quadrant In Exercises 13–18, determine the quadrant in which θ lies.

13. $\sin \theta < 0, \cos \theta > 0$ **14.** $\sin \theta > 0, \cos \theta < 0$

15. $\sin \theta > 0, \sec \theta > 0$ **16.** $\cot \theta < 0, \cos \theta > 0$

17. $\csc \theta > 0, \tan \theta < 0$ **18.** $\cos \theta > 0, \tan \theta < 0$

Evaluating Trigonometric Functions In Exercises 19–32, evaluate the six trigonometric functions of the angle without using a calculator. *See Examples 2 and 3.*

19. $60°$ **20.** $-\frac{2\pi}{3}$

21. $\frac{\pi}{4}$ **22.** $\frac{5\pi}{4}$

23. $-\frac{\pi}{2}$ **24.** $150°$

25. $225°$ **26.** $\frac{4\pi}{3}$

27. $300°$ **28.** $210°$

29. $750°$ **30.** $510°$

31. $\frac{10\pi}{3}$ **32.** $\frac{17\pi}{3}$

Evaluating Trigonometric Functions In Exercises 33–42, use a calculator to evaluate the trigonometric function to four decimal places.

33. $\sin 10°$ **34.** $\csc 10°$

35. $\tan \frac{\pi}{9}$ **36.** $\tan \frac{10\pi}{9}$

37. $\cos(-110°)$ **38.** $\cos 250°$

39. $\tan 240°$ **40.** $\cot 210°$

41. $\sin(-0.65)$ **42.** $\tan 4.5$

Solving a Right Triangle In Exercises 43–48, solve for x, y, or r as indicated. *See Example 4.*

43. Solve for y.

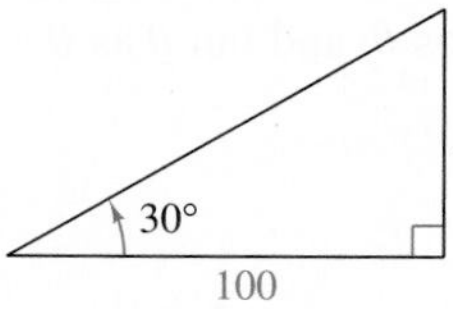

44. Solve for x.

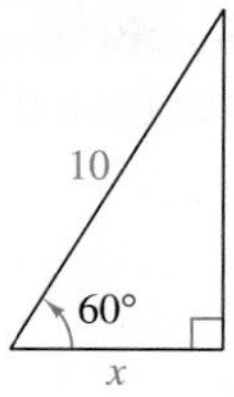

45. Solve for x.

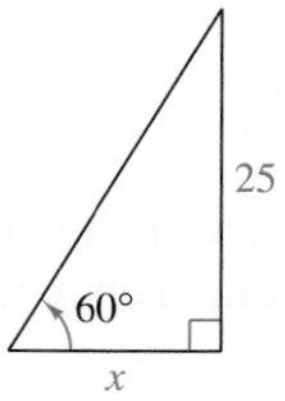

46. Solve for r.

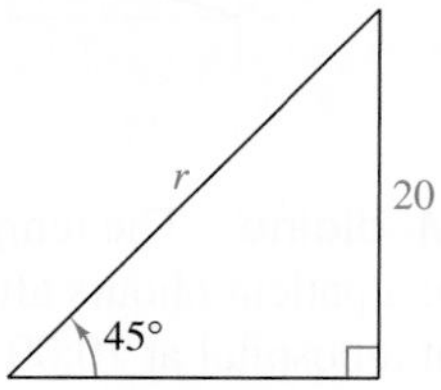

47. Solve for r.

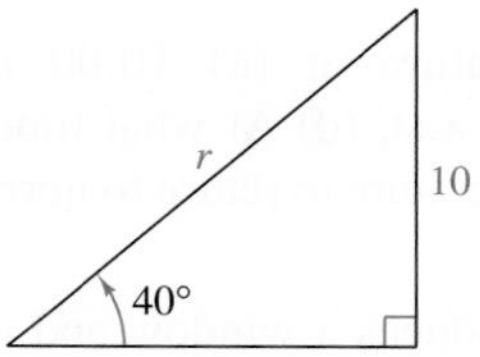

48. Solve for x.

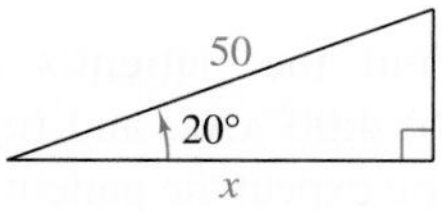

Solving Trigonometric Equations In Exercises 49–60, solve the equation for θ. Assume $0 \le \theta \le 2\pi$. *See Example 6.*

49. $\sin\theta = \frac{1}{2}$ **50.** $\cos\theta = -\frac{1}{2}$

51. $\tan\theta = \sqrt{3}$ **52.** $\cos\theta = -\frac{\sqrt{2}}{2}$

53. $\csc\theta = \frac{2\sqrt{3}}{3}$ **54.** $\cot\theta = -1$

55. $\sec\theta = 2$ **56.** $\sec\theta = -2$

57. $\sin\theta = -\frac{\sqrt{2}}{2}$ **58.** $\cot\theta = -\sqrt{3}$

59. $\sin\theta = \frac{\sqrt{3}}{2}$ **60.** $\tan\theta = \frac{\sqrt{3}}{3}$

Solving Trigonometric Equations In Exercises 61–70, solve the equation for θ. Assume $0 \le \theta \le 2\pi$. For some of the equations, you should use the trigonometric identities listed in this section. Use the *trace* feature of a graphing utility to verify your results. *See Example 7.*

61. $2\sin^2\theta = 1$ **62.** $\tan^2\theta = 3$

63. $\tan^2\theta - \tan\theta = 0$ **64.** $2\cos^2\theta - \cos\theta = 1$

65. $\sin 2\theta - \cos\theta = 0$

66. $\cos 2\theta + 3\cos\theta + 2 = 0$

67. $\sin\theta = \cos\theta$ **68.** $\sec\theta\csc\theta = 2\csc\theta$

69. $\cos^2\theta + \sin\theta = 1$ **70.** $\cos\frac{\theta}{2} - \cos\theta = 1$

71. Length A 20-foot ladder leaning against the side of a house makes a 75° angle with the ground (see figure). How far up the side of the house does the ladder reach?

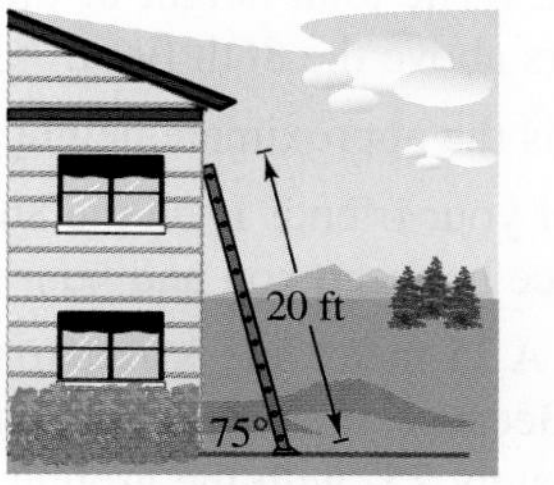

72. Width of a River A biologist wants to know the width w of a river in order to set instruments to study the pollutants in the water. From point A, the biologist walks downstream 100 feet and sights to point C. From this sighting it is determined that $\theta = 50°$ (see figure). How wide is the river?

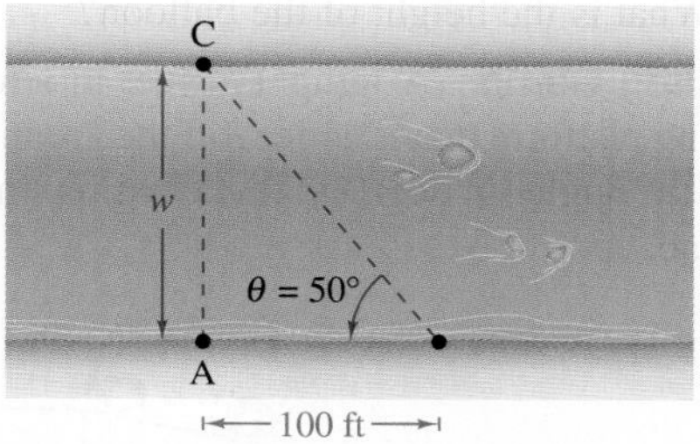

73. Distance An airplane flying at an altitude of 6 miles is on a flight path that passes directly over an observer (see figure). Let θ be the angle of elevation from the observer to the plane. Find the distance d from the observer to the plane when (a) $\theta = 30°$, (b) $\theta = 60°$, and (c) $\theta = 90°$.

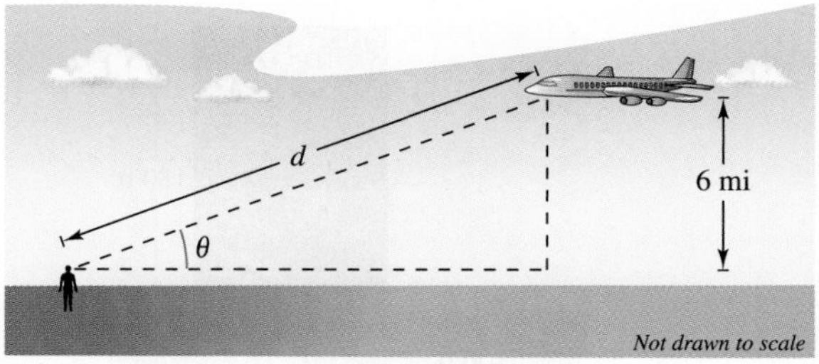

74. Skateboard Ramp A skateboard ramp with a height of 4 feet has an angle of elevation of 18° (see figure). How long is the skateboard ramp?

75. Empire State Building You are standing 45 meters from the base of the Empire State Building. You estimate that the angle of elevation to the top of the 86th floor is 82°. The total height of the building is another 123 meters above the 86th floor.

(a) What is the approximate height of the building?

(b) One of your friends is on the 86th floor. What is the distance between you and your friend?

76. Height A 25-meter line is used to tether a helium-filled balloon. Because of a breeze, the line makes an angle of approximately 75° with the ground.

(a) Draw the right triangle that gives a visual representation of the problem. Show the known side lengths and angles of the triangle and use a variable to indicate the height of the balloon.

(b) Use a trigonometric function to write an equation involving the unknown quantity.

(c) What is the height of the balloon?

77. Loading Ramp A ramp $17\frac{1}{2}$ feet in length rises to a loading platform that is $3\frac{1}{2}$ feet off the ground (see figure). Find the angle (in degrees) that the ramp makes with the ground.

78. Height The height of a building is 180 feet. Find the angle of elevation (in degrees) to the top of the building from a point 100 feet from the base of the building (see figure).

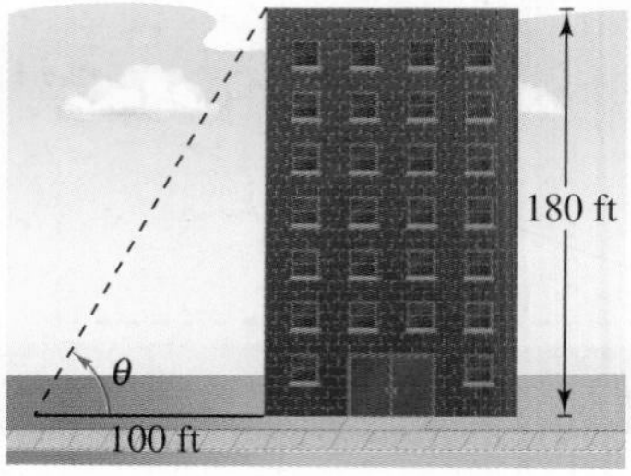

79. Height of a Mountain In traveling across flat land, you notice a mountain directly in front of you. Its angle of elevation (to the peak) is 3.5°. After you drive 13 miles closer to the mountain, the angle of elevation is 9°. Approximate the height of the mountain.

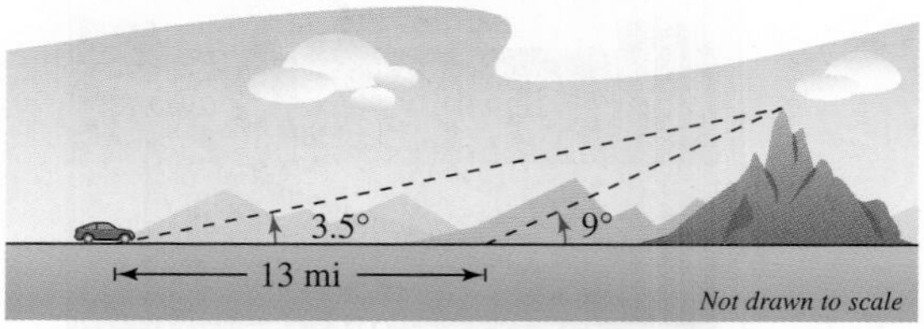

80. HOW DO YOU SEE IT? Consider an angle in standard position with $r = 12$ centimeters, as shown in the figure. Describe the changes in the values of x, y, $\sin\theta$, $\cos\theta$, and $\tan\theta$ as θ increases from 0° to 90°.

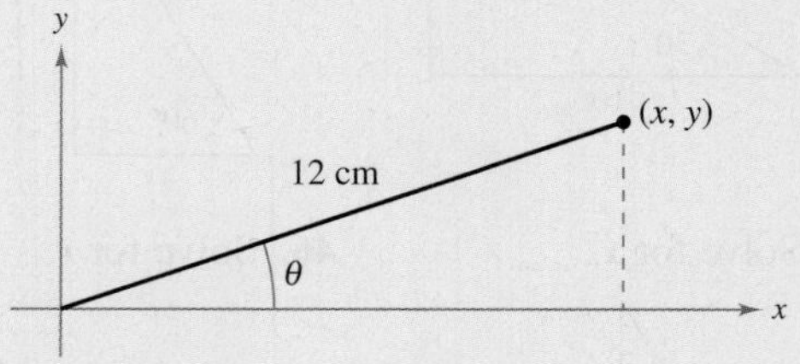

81. Medicine The temperature T (in degrees Fahrenheit) of a patient t hours after arriving at the emergency room of a hospital at 10:00 P.M. is given by

$$T(t) = 98.6 + 4\cos\frac{\pi t}{36}, \quad 0 \le t \le 18.$$

Find the patient's temperature at (a) 10:00 P.M., (b) 4:00 A.M., and (c) 10:00 A.M. (d) At what time do you expect the patient's temperature to return to normal? Explain your reasoning.

82. Sales A company that produces a window and door insulating kit forecasts monthly sales over the next 2 years to be

$$S = 23.1 + 0.442t + 4.3\sin\frac{\pi t}{6}$$

where S is measured in thousands of units and t is the time in months, with $t = 1$ corresponding to January 2011. Find the monthly sales for (a) February 2011, (b) February 2012, (c) September 2011, and (d) September 2012.

Graphing Functions In Exercises 83 and 84, use a graphing utility or a spreadsheet to complete the table. Then graph the function.

x	0	2	4	6	8	10
$f(x)$						

83. $f(x) = \frac{2}{5}x + 2\sin\frac{\pi x}{5}$ **84.** $f(x) = \frac{1}{2}(5 - x) + 3\cos\frac{\pi x}{5}$

True or False? In Exercises 85–88, determine whether the statement is true or false. If it is false, explain why or give an example that shows it is false.

85. $\sin 10° \csc 10° = 1$ **86.** $\frac{\sin 60°}{\sin 30°} = \sin 2°$

87. $\sin^2 45° - \cos^2 45° = 1$

88. Because $\sin(-t) = -\sin t$, it can be said that the sine of a negative angle is a negative number.

8.3 Graphs of Trigonometric Functions

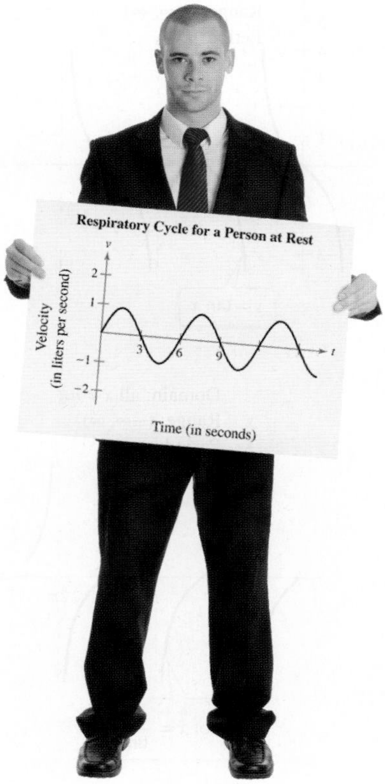

In Exercise 73 on page 533, you will use a trigonometric function to model the air flow of a person's respiratory cycle.

- Sketch graphs of trigonometric functions.
- Evaluate limits of trigonometric functions.
- Use trigonometric functions to model real-life situations.

Graphs of Trigonometric Functions

When you are sketching the graph of a trigonometric function, it is common to use x (rather than θ) as the independent variable. For instance, you can sketch the graph of

$$f(x) = \sin x$$

by constructing a table of values, plotting the resulting points, and connecting them with a smooth curve, as shown in Figure 8.21. Some examples of values are shown in the table below.

x	0	$\frac{\pi}{6}$	$\frac{\pi}{4}$	$\frac{\pi}{3}$	$\frac{\pi}{2}$	$\frac{2\pi}{3}$	$\frac{3\pi}{4}$	$\frac{5\pi}{6}$	π
$\sin x$	0.00	0.50	0.71	0.87	1.00	0.87	0.71	0.50	0.00

In Figure 8.21, note that the maximum value of $\sin x$ is 1 and the minimum value is -1. The **amplitude** of the sine function (or the cosine function) is defined to be half of the difference between its maximum and minimum values. So, the amplitude of $f(x) = \sin x$ is 1.

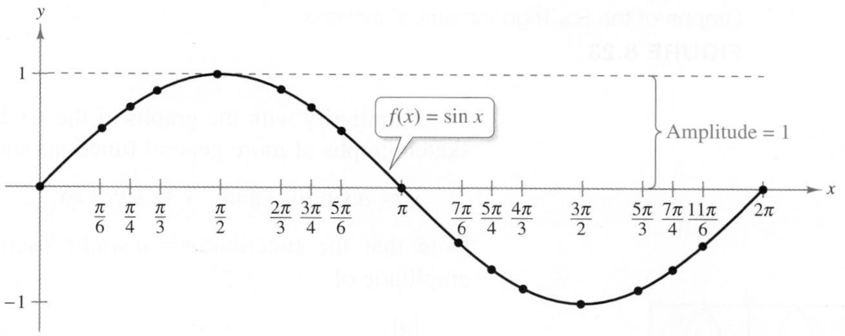

FIGURE 8.21

The periodic nature of the sine function becomes evident when you observe that as x increases beyond 2π, the graph repeats itself over and over, continuously oscillating about the x-axis. The **period** of the function is the distance (on the x-axis) between successive cycles, as shown in Figure 8.22. So, the period of $f(x) = \sin x$ is 2π.

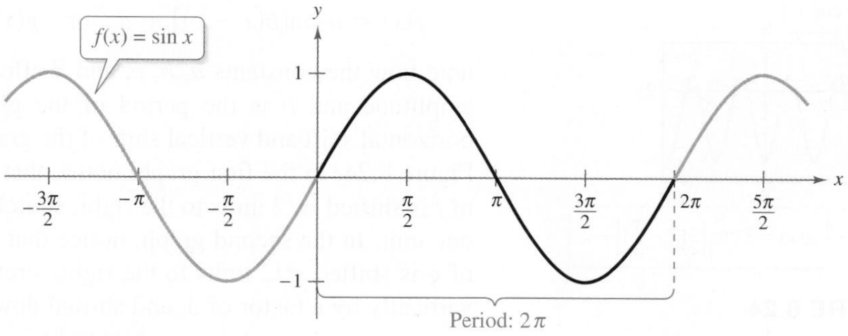

FIGURE 8.22

Benjamin Thorn/www.shutterstock.com

Figure 8.23 shows the graphs of at least one cycle of all six trigonometric functions.

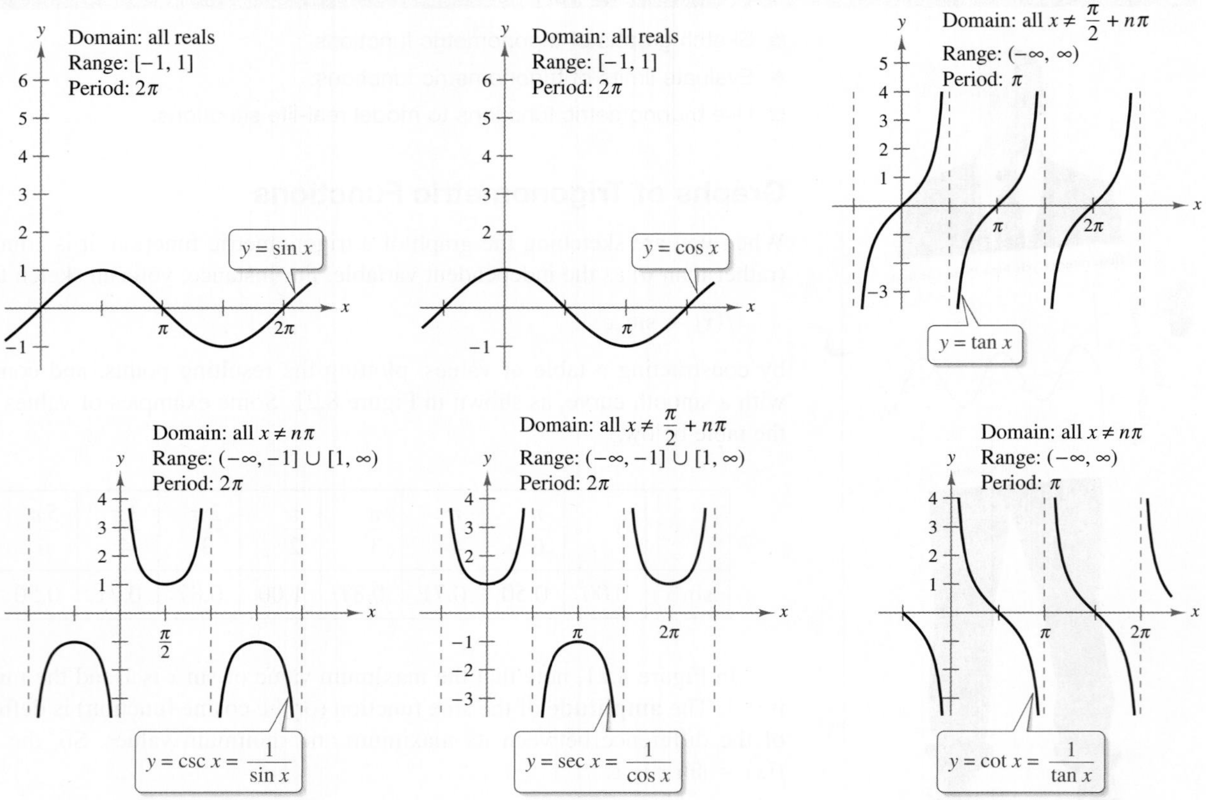

Graphs of the Six Trigonometric Functions
FIGURE 8.23

Familiarity with the graphs of the six basic trigonometric functions allows you to sketch graphs of more general functions such as

$$y = a \sin bx \quad \text{and} \quad y = a \cos bx.$$

Note that the function $y = a \sin bx$ oscillates between $-a$ and a and so has an amplitude of

$$|a|. \qquad \text{Amplitude of } y = a \sin bx$$

Furthermore, because $bx = 0$ when $x = 0$ and $bx = 2\pi$ when $x = 2\pi/b$, it follows that the function $y = a \sin bx$ has a period of

$$\frac{2\pi}{|b|}. \qquad \text{Period of } y = a \sin bx$$

When graphing general functions such as

$$f(x) = a \sin[b(x - c)] + d \quad \text{or} \quad g(x) = a \cos[b(x - c)] + d$$

note how the constants a, b, c, and d affect the graph. You already know that a is the amplitude and b is the period of the graph. The constants c and d determine the horizontal shift and vertical shift of the graph, respectively. Two examples are shown in Figure 8.24. In the first graph, notice that relative to the graph of $y = \sin x$, the graph of f is shifted $\pi/2$ units to the right, stretched vertically by a factor of 2, and shifted up one unit. In the second graph, notice that relative to the graph of $y = \cos x$, the graph of g is shifted $\pi/2$ units to the right, stretched horizontally by a factor of $\frac{1}{2}$, stretched vertically by a factor of 3, and shifted down two units.

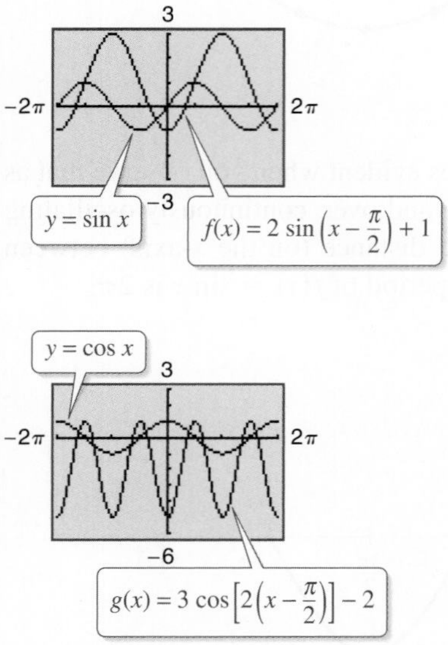

FIGURE 8.24

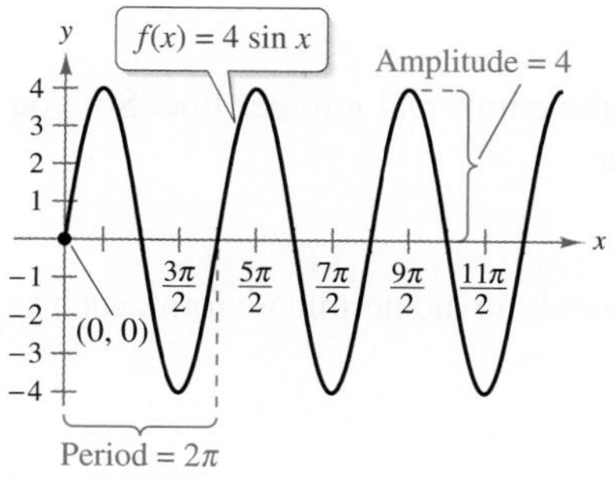

FIGURE 8.25

Example 1 Graphing a Trigonometric Function

Sketch the graph of $f(x) = 4 \sin x$.

SOLUTION The graph of $f(x) = 4 \sin x$ has the characteristics below.

Amplitude: 4

Period: 2π

Three cycles of the graph are shown in Figure 8.25, starting with the point (0, 0).

✓ Checkpoint 1

Sketch the graph of $g(x) = 2 \cos x$. ■

Example 2 Graphing a Trigonometric Function

Sketch the graph of $f(x) = 3 \cos 2x$.

SOLUTION The graph of $f(x) = 3 \cos 2x$ has the characteristics below.

Amplitude: 3

Period: $\dfrac{2\pi}{2} = \pi$

Almost three cycles of the graph are shown in Figure 8.26, starting with the maximum point (0, 3).

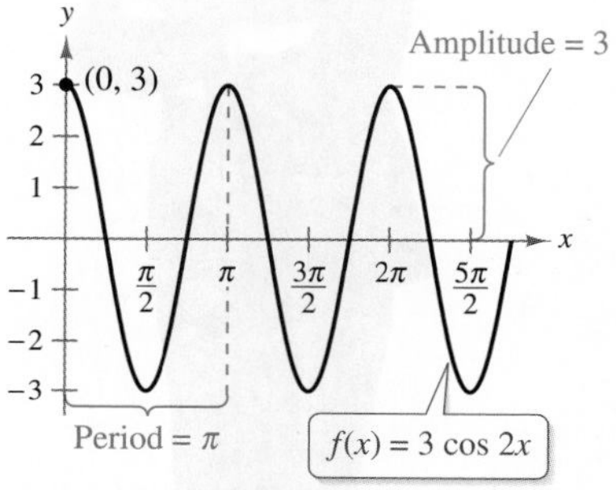

FIGURE 8.26

✓ Checkpoint 2

Sketch the graph of $g(x) = 2 \sin 4x$. ■

Example 3 Graphing a Trigonometric Function

Sketch the graph of $f(x) = -2 \tan 3x$.

SOLUTION The graph of this function has a period of $\pi/3$. The vertical asymptotes of this tangent function occur at

$$x = \ldots, \underbrace{-\frac{\pi}{6}, \frac{\pi}{6}}_{\text{Period} = \frac{\pi}{3}}, \frac{\pi}{2}, \frac{5\pi}{6}, \ldots$$

Several cycles of the graph are shown in Figure 8.27, starting with the vertical asymptote $x = -\pi/6$.

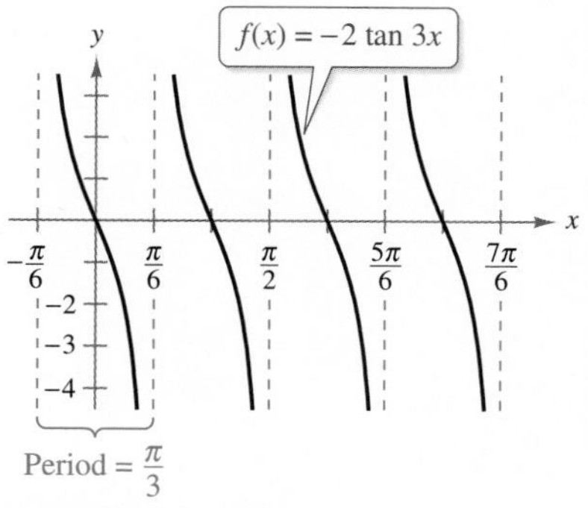

FIGURE 8.27

✓ Checkpoint 3

Sketch the graph of $g(x) = \tan 4x$. ■

Limits of Trigonometric Functions

The sine and cosine functions are continuous over the entire real number line. So, you can use direct substitution to evaluate a limit such as

$$\lim_{x \to 0} \sin x = \sin 0 = 0.$$

When direct substitution with a trigonometric limit yields an indeterminate form, such as

$$\frac{0}{0} \qquad \text{Indeterminate form}$$

you can rely on technology to help evaluate the limit. The next example examines the limit of a function that you will encounter again in Section 8.4.

Example 4 Evaluating a Trigonometric Limit

Use a calculator to evaluate the function

$$f(x) = \frac{\sin x}{x}$$

at several x-values near $x = 0$. Then use the result to estimate

$$\lim_{x \to 0} \frac{\sin x}{x}.$$

Use a graphing utility (set in *radian* mode) to confirm your result.

SOLUTION The table shows several values of the function at x-values near zero. (Note that the function is undefined when $x = 0$.)

x	-0.20	-0.15	-0.10	-0.05	0.05	0.10	0.15	0.20
$\frac{\sin x}{x}$	0.9933	0.9963	0.9983	0.9996	0.9996	0.9983	0.9963	0.9933

From the table, it appears that the limit is 1. That is,

$$\lim_{x \to 0} \frac{\sin x}{x} = 1.$$

Figure 8.28 shows the graph of

$$f(x) = \frac{\sin x}{x}.$$

From this graph, it appears that $f(x)$ gets closer and closer to 1 as x approaches zero (from either side).

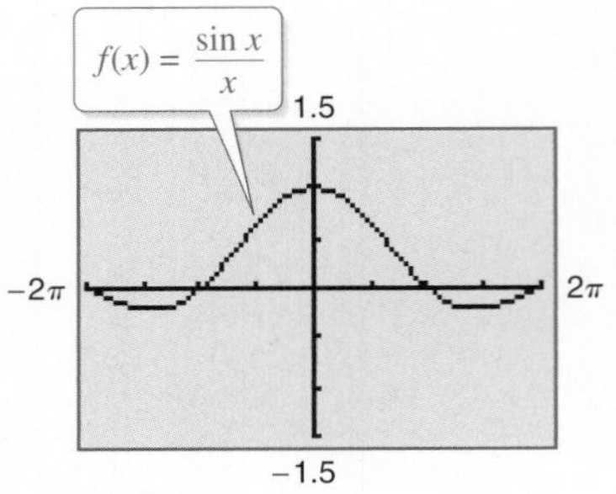

FIGURE 8.28

✓ Checkpoint 4

Use a calculator to evaluate the function

$$f(x) = \frac{1 - \cos x}{x}$$

at several x-values near $x = 0$. Then use the result to estimate

$$\lim_{x \to 0} \frac{1 - \cos x}{x}.$$

■

Andresr/www.shutterstock.com

Applications

There are many examples of periodic phenomena in both business and biology. Many businesses have cyclical sales patterns, and plant growth is affected by the day-night cycle. The next example describes the cyclical pattern followed by many types of predator-prey populations, such as coyotes and rabbits.

Example 5 Modeling Predator-Prey Cycles

The population P of a predator at time t (in months) is modeled by

$$P = 10{,}000 + 3000 \sin \frac{2\pi t}{24}, \quad t \geq 0$$

and the population p of its primary food source (its prey) is modeled by

$$p = 15{,}000 + 5000 \cos \frac{2\pi t}{24}, \quad t \geq 0.$$

Graph both models on the same set of axes and explain the oscillations in the size of each population.

SOLUTION Each function has a period of 24 months. The predator population has an amplitude of 3000 and oscillates about the line $y = 10{,}000$. The prey population has an amplitude of 5000 and oscillates about the line $y = 15{,}000$. The graphs of the two models are shown in Figure 8.29. The cycles of this predator-prey population are explained by the diagram below.

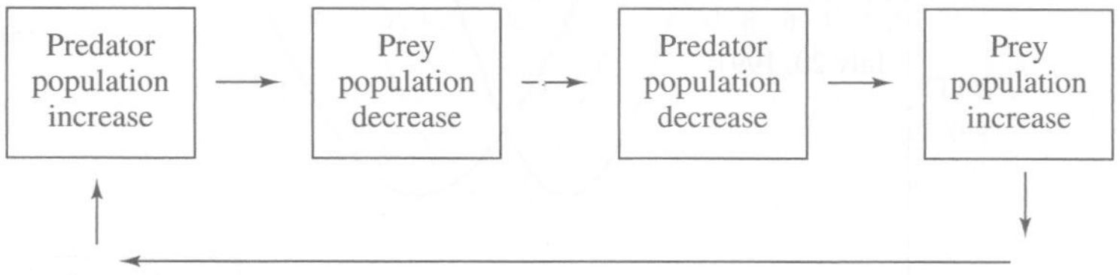

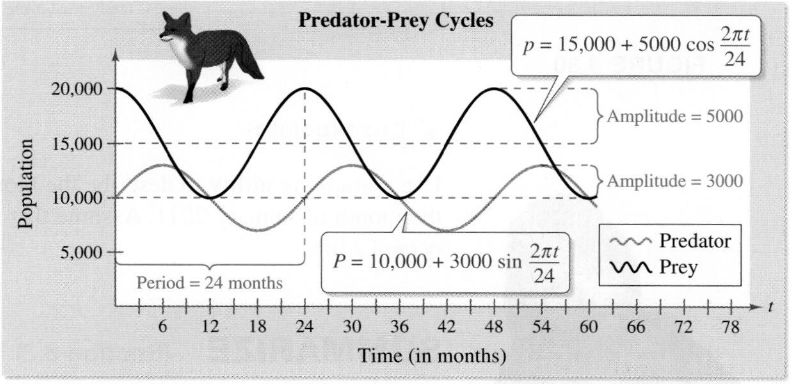

FIGURE 8.29

Checkpoint 5

Repeat Example 5 for the following models.

$$P = 12{,}000 + 2500 \sin \frac{2\pi t}{12}, \quad t \geq 0 \qquad \text{Predator population}$$

$$p = 18{,}000 + 6000 \sin \frac{2\pi t}{12}, \quad t \geq 0 \qquad \text{Prey population}$$

■

Example 6 Modeling Biorhythms

A theory that attempts to explain the ups and downs of everyday life states that each person has three cycles, which begin at birth. These three cycles can be modeled by sine waves

Physical (23 days): $P = \sin\dfrac{2\pi t}{23}, \quad t \geq 0$

Emotional (28 days): $E = \sin\dfrac{2\pi t}{28}, \quad t \geq 0$

Intellectual (33 days): $I = \sin\dfrac{2\pi t}{33}, \quad t \geq 0$

where t is the number of days since birth. Describe the biorhythms during the month of September 2011, for a person who was born on July 20, 1991.

SOLUTION Figure 8.30 shows the person's biorhythms during the month of September 2011. Note that September 1, 2011 was the 7348th day of the person's life.

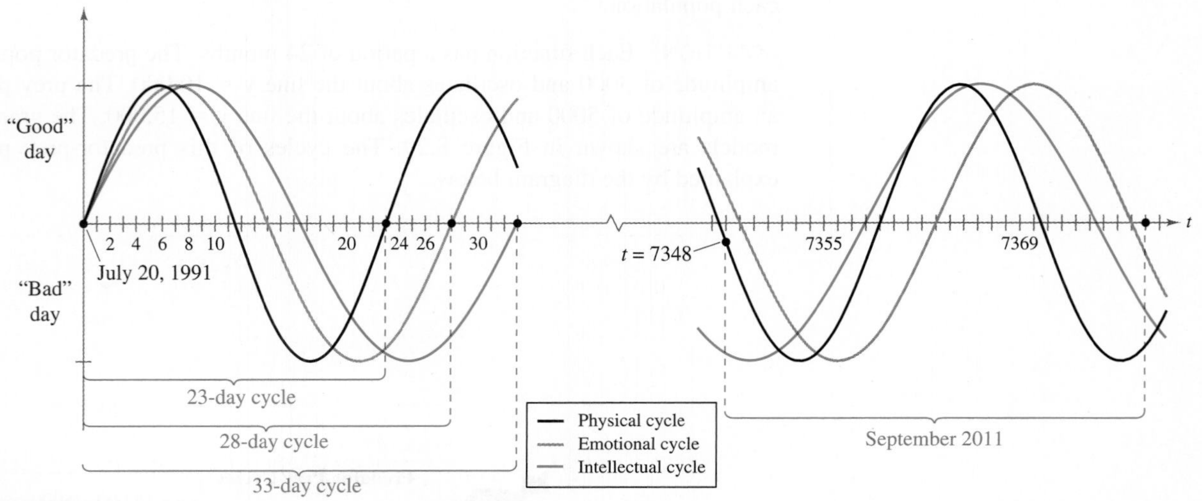

FIGURE 8.30

✓ Checkpoint 6

Use a graphing utility to describe the biorhythms of the person in Example 6 during the month of January 2011. Assume that January 1, 2011 is the 7105th day of the person's life. ■

SUMMARIZE (Section 8.3)

1. Describe the graph of $y = \sin x$ *(page 526)*. For an example of graphing a sine function, see Example 1.
2. Describe the graph of $y = \cos x$ *(page 526)*. For an example of graphing a cosine function, see Example 2.
3. Describe the graph of $y = \tan x$ *(page 526)*. For an example of graphing a tangent function, see Example 3.
4. Describe a real-life example of how the graphs of trigonometric functions can be used to analyze biorhythms *(page 530, Example 6)*.

SKILLS WARM UP 8.3

The following warm-up exercises involve skills that were covered in earlier sections. You will use these skills in the exercise set for this section. For additional help, review Sections 1.5 and 8.2.

In Exercises 1 and 2, find the limit.

1. $\lim_{x \to 2} (x^2 + 4x + 2)$

2. $\lim_{x \to 3} (x^3 - 2x^2 + 1)$

In Exercises 3–10, evaluate the trigonometric function without using a calculator.

3. $\cos \frac{\pi}{2}$

4. $\sin \pi$

5. $\tan \frac{5\pi}{4}$

6. $\cot \frac{2\pi}{3}$

7. $\sin \frac{11\pi}{6}$

8. $\cos \frac{5\pi}{6}$

9. $\cos \frac{5\pi}{3}$

10. $\sin \frac{4\pi}{3}$

In Exercises 11–18, use a calculator to evaluate the trigonometric function to four decimal places.

11. $\cos 15°$

12. $\sin 220°$

13. $\sin 275°$

14. $\cos 310°$

15. $\sin 103°$

16. $\cos 72°$

17. $\tan 327°$

18. $\tan 140°$

Exercises 8.3

See www.CalcChat.com for worked-out solutions to odd-numbered exercises.

Finding the Period and Amplitude In Exercises 1–14, find the period and amplitude of the trigonometric function.

1. $y = 2 \sin 2x$

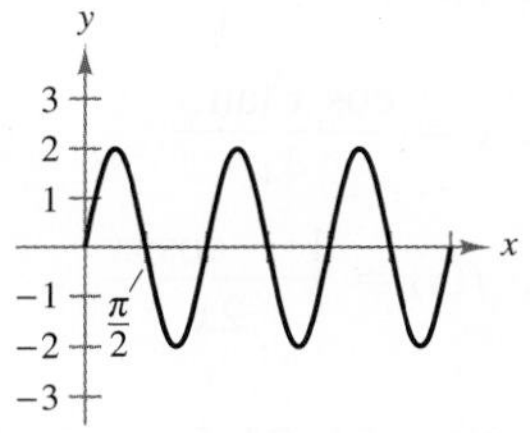

2. $y = 3 \cos 3x$

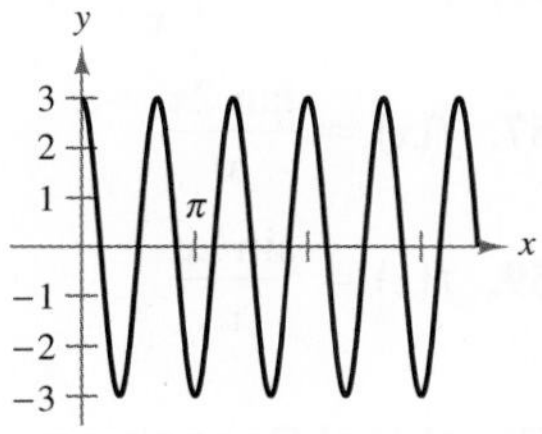

3. $y = \frac{3}{2} \cos \frac{x}{2}$

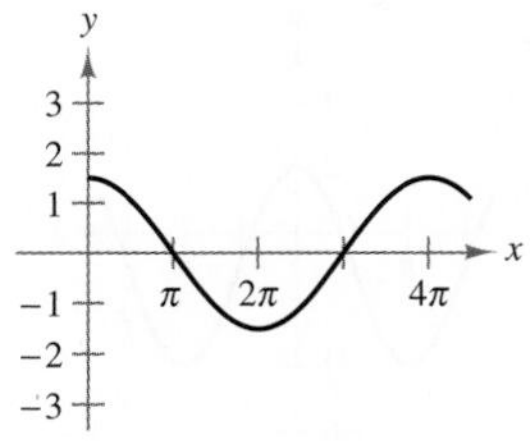

4. $y = -2 \sin \frac{x}{3}$

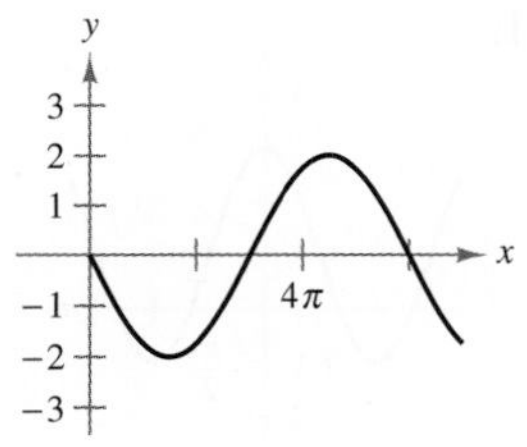

5. $y = \frac{1}{2} \cos \pi x$

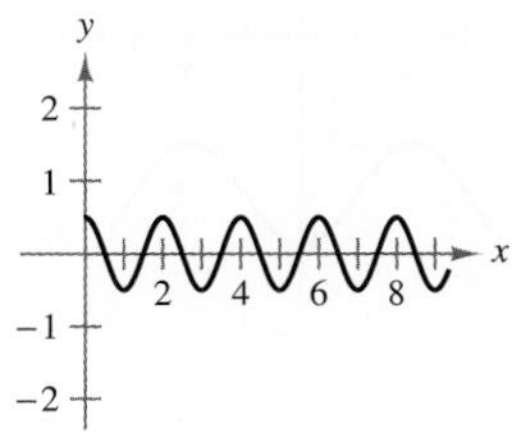

6. $y = \frac{5}{2} \cos \frac{\pi x}{2}$

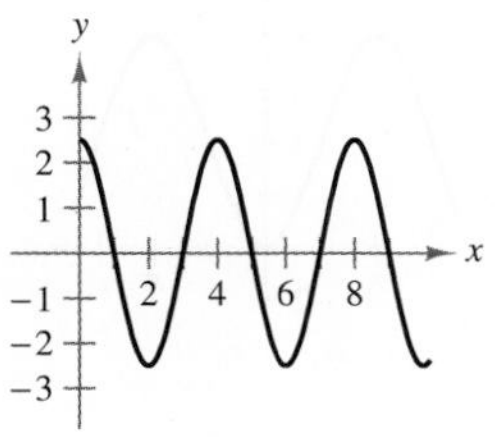

7. $y = -\sin 3x$

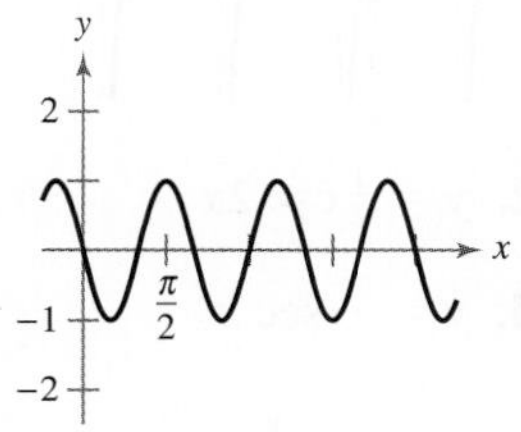

8. $y = -\cos \frac{2x}{3}$

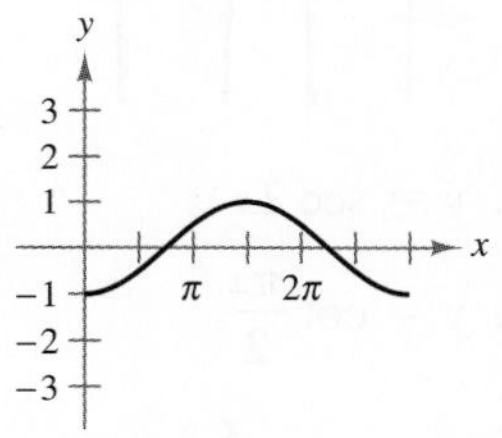

9. $y = -\frac{3}{2} \sin 6x$

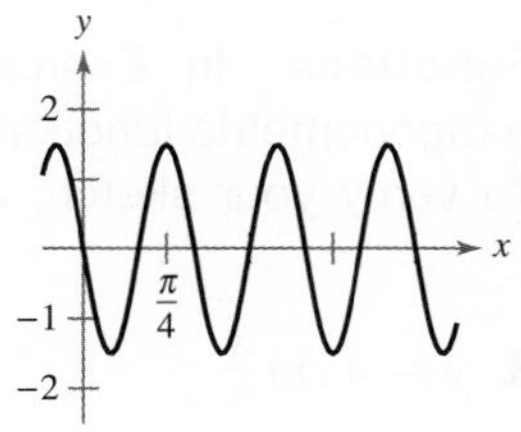

10. $y = \frac{1}{3} \sin 8x$

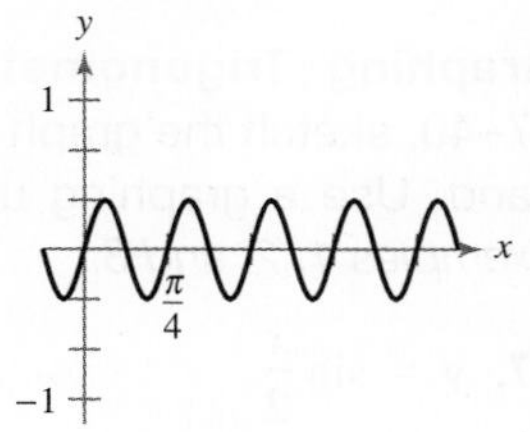

11. $y = \frac{1}{2} \sin \frac{2x}{3}$

12. $y = 5 \cos \frac{x}{4}$

13. $y = 3 \sin 4\pi x$

14. $y = \frac{2}{3} \cos \frac{\pi x}{10}$

Finding the Period In Exercises 15–20, find the period of the trigonometric function.

15. $y = 3 \tan x$

16. $y = 7 \tan 2\pi x$

17. $y = 3 \sec 5x$

18. $y = \csc 4x$

19. $y = \cot \frac{\pi x}{6}$

20. $y = 5 \tan \frac{2\pi x}{3}$

Matching In Exercises 21–26, match the trigonometric function with the correct graph and give the period of the function. [The graphs are labeled (a)–(f).]

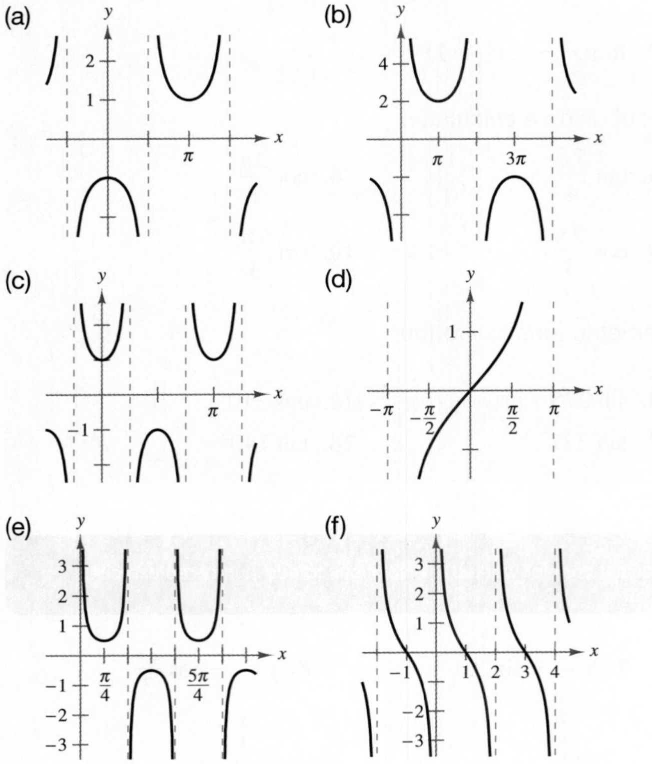

21. $y = \sec 2x$

22. $y = \frac{1}{2} \csc 2x$

23. $y = \cot \dfrac{\pi x}{2}$

24. $y = -\sec x$

25. $y = 2 \csc \dfrac{x}{2}$

26. $y = \tan \dfrac{x}{2}$

Graphing Trigonometric Functions In Exercises 27–40, sketch the graph of the trigonometric function by hand. Use a graphing utility to verify your sketch. *See Examples 1, 2, and 3.*

27. $y = \sin \dfrac{x}{2}$

28. $y = 4 \sin \dfrac{x}{3}$

29. $y = 2 \cos \dfrac{\pi x}{3}$

30. $y = \dfrac{3}{2} \cos \dfrac{2x}{3}$

31. $y = -2 \sin 6x$

32. $y = -3 \cos 4x$

33. $y = \cos 2\pi x$

34. $y = \dfrac{3}{2} \sin \dfrac{\pi x}{4}$

35. $y = 2 \tan x$

36. $y = 2 \cot x$

37. $y = \dfrac{1}{2} \tan \dfrac{\pi x}{2}$

38. $y = -\csc \dfrac{x}{3}$

39. $y = 2 \sec 4x$

40. $y = -\tan 3x$

Graphing Trigonometric Functions In Exercises 41–50, use a graphing utility to graph the trigonometric function.

41. $y = -\sin \dfrac{2\pi x}{3}$

42. $y = 10 \cos \dfrac{\pi x}{6}$

43. $y = \cot 2x$

44. $y = 3 \tan \pi x$

45. $y = \csc \dfrac{2x}{3}$

46. $y = \cot \dfrac{2x}{3}$

47. $y = 2 \sec 2x$

48. $y = \sec \pi x$

49. $y = \csc 2\pi x$

50. $y = -\tan x$

Evaluating Trigonometric Limits In Exercises 51–60, complete the table using a spreadsheet or a graphing utility (set in *radian* mode) to estimate $\lim_{x \to 0} f(x)$. Use a graphing utility to graph the function and confirm your result. *See Example 4.*

x	-0.1	-0.01	-0.001	0.001	0.01	0.1
$f(x)$						

51. $f(x) = \dfrac{\sin 4x}{2x}$

52. $f(x) = \dfrac{\sin 2x}{\sin 3x}$

53. $y = \dfrac{\tan 3x}{\tan 4x}$

54. $f(x) = \dfrac{1 - \cos 2x}{x}$

55. $f(x) = \dfrac{3(1 - \cos x)}{x}$

56. $f(x) = \dfrac{2 \sin(x/4)}{x}$

57. $f(x) = \dfrac{\tan 2x}{x}$

58. $y = \dfrac{\cos x \tan x}{4x}$

59. $f(x) = \dfrac{\sin^2 x}{x}$

60. $f(x) = \dfrac{1 - \cos^2 x}{2x}$

Graphical Reasoning In Exercises 61–64, find a and d for $f(x) = a \cos x + d$ such that the graph of f matches the figure.

61.

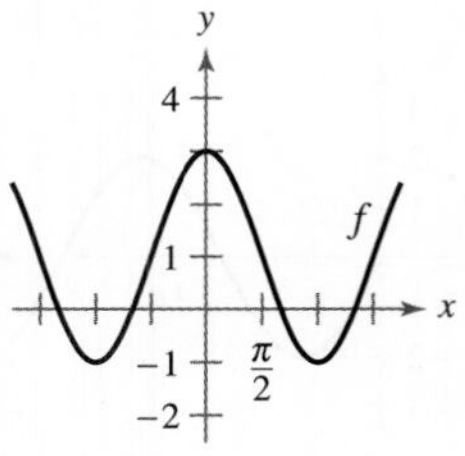

62.

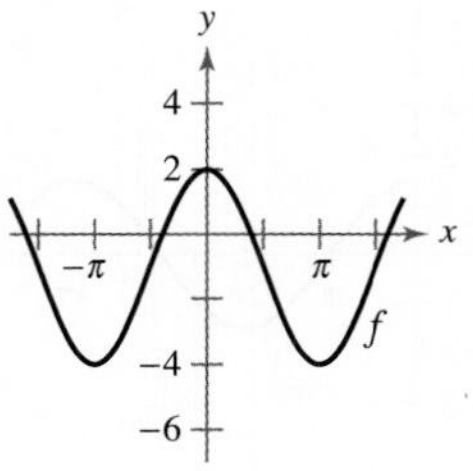

63.

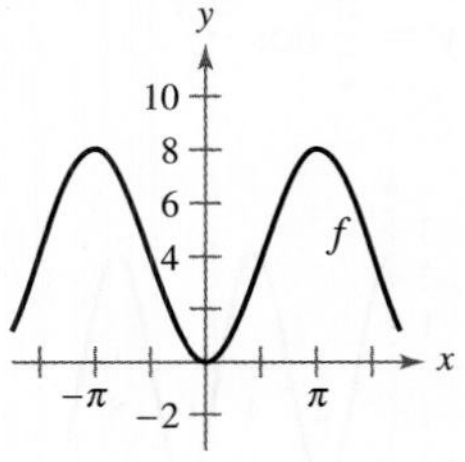

64.

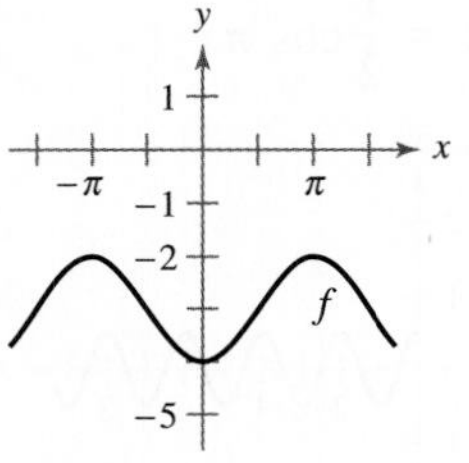

Phase Shift In Exercises 65–68, match the function with the correct graph. [The graphs are labeled (a)–(d).]

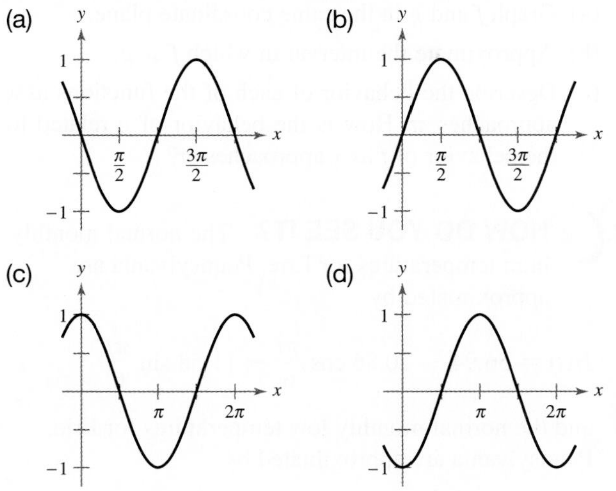

65. $y = \sin x$

66. $y = \sin\left(x - \dfrac{\pi}{2}\right)$

67. $y = \sin(x - \pi)$

68. $y = \sin\left(x - \dfrac{3\pi}{2}\right)$

69. Biology: Predator-Prey Cycle The population P of a predator at time t (in months) is modeled by

$$P = 8000 + 2500 \sin \frac{2\pi t}{24}$$

and the population p of its prey is modeled by

$$p = 12{,}000 + 4000 \cos \frac{2\pi t}{24}.$$

(a) Use a graphing utility to graph both models in the same viewing window.

(b) Explain the oscillations in the size of each population.

70. Biology: Predator-Prey Cycle The population P of a predator at time t (in months) is modeled by

$$P = 5700 + 1200 \sin \frac{2\pi t}{24}$$

and the population p of its prey is modeled by

$$p = 9800 + 2750 \cos \frac{2\pi t}{24}.$$

(a) Use a graphing utility to graph both models in the same viewing window.

(b) Explain the oscillations in the size of each population.

71. Biorhythms For a person born on July 20, 1991, use the biorhythm cycles given in Example 6 to calculate this person's three energy levels on December 31, 2015. Assume this is the 8930th day of the person's life.

72. Biorhythms Use your birthday and the biorhythm cycles given in Example 6 to calculate your three energy levels on December 31, 2015.

73. Health For a person at rest, the velocity v (in liters per second) of air flow into and out of the lungs during a respiratory cycle is given by

$$v = 0.9 \sin \frac{\pi t}{3}$$

where t is the time (in seconds). Inhalation occurs when $v > 0$, and exhalation occurs when $v < 0$.

(a) Find the time for one full respiratory cycle.

(b) Find the number of cycles per minute.

(c) Use a graphing utility to graph the velocity function.

74. Health After a person exercises for a few minutes, the velocity v (in liters per second) of air flow into and out of the lungs during a respiratory cycle is given by

$$v = 1.75 \sin \frac{\pi t}{2}$$

where t is the time (in seconds). Inhalation occurs when $v > 0$, and exhalation occurs when $v < 0$.

(a) Find the time for one full respiratory cycle.

(b) Find the number of cycles per minute.

(c) Use a graphing utility to graph the velocity function.

75. Music When tuning a piano, a technician strikes a tuning fork for the A above middle C and sets up wave motion that can be approximated by

$$y = 0.001 \sin 880\pi t$$

where t is the time (in seconds).

(a) What is the period p of this function?

(b) What is the frequency f of this note $(f = 1/p)$?

(c) Use a graphing utility to graph this function.

76. Health The function

$$P = 100 - 20 \cos(5\pi t/3)$$

approximates the blood pressure P (in millimeters of mercury) at time t (in seconds) for a person at rest.

(a) Find the period of the function.

(b) Find the number of heartbeats per minute.

(c) Use a graphing utility to graph the pressure function.

77. Construction Workers The number W (in thousands) of construction workers employed in the United States during 2010 can be modeled by

$$W = 5488 + 347.6\sin(0.45t + 4.153)$$

where t is the time (in months), with $t = 1$ corresponding to January. *(Source: U.S. Bureau of Labor Statistics)*

(a) Use a graphing utility to graph W.

(b) Did the number of construction workers exceed 5.5 million in 2010? If so, during which month(s)?

78. Sales The snowmobile sales S (in units) at a dealership are modeled by

$$S = 58.3 + 32.5\cos\frac{\pi t}{6}$$

where t is the time (in months), with $t = 1$ corresponding to January.

(a) Use a graphing utility to graph S.

(b) Will the sales exceed 75 units during any month? If so, during which month(s)?

79. Physics Use the graphs below to answer each question.

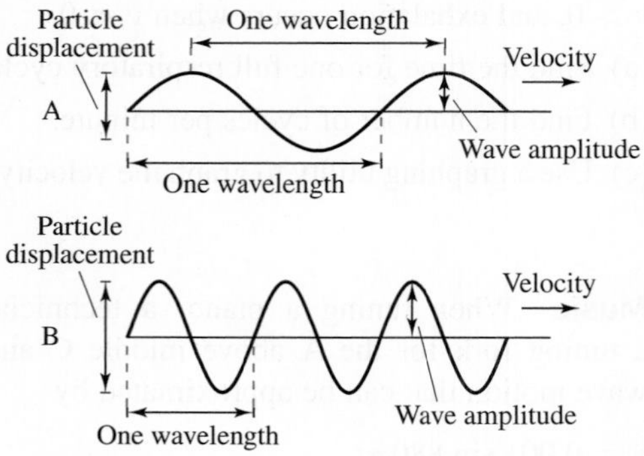

(a) Which graph (A or B) has a longer wavelength, or period?

(b) Which graph (A or B) has a greater amplitude?

(c) The *frequency* of a graph is the number of oscillations or cycles that occur during a given period of time. Which graph (A or B) has a greater frequency?

(d) Based on the definition of frequency in part (c), how are frequency and period related?

(Source: Adapted from Shipman/Wilson/Todd, An Introduction to Physical Science, *Eleventh Edition)*

80. Think About It Consider the function given by $y = \cos bx$ on the interval $(0, 2\pi)$.

(a) Sketch the graph of

$$y = \cos bx$$

for $b = \frac{1}{2}$, 2, and 3.

(b) How does the value of b affect the graph?

(c) How many complete cycles occur between 0 and 2π for each value of b?

81. Think About It Consider the functions given by $f(x) = 2\sin x$ and $g(x) = 0.5\csc x$ on the interval $(0, \pi)$.

(a) Graph f and g in the same coordinate plane.

(b) Approximate the interval in which $f > g$.

(c) Describe the behavior of each of the functions as x approaches π. How is the behavior of g related to the behavior of f as x approaches π?

82. HOW DO YOU SEE IT? The normal monthly high temperatures for Erie, Pennsylvania are approximated by

$$H(t) = 56.94 - 20.86\cos\frac{\pi t}{6} - 11.58\sin\frac{\pi t}{6}$$

and the normal monthly low temperatures for Erie, Pennsylvania are approximated by

$$L(t) = 41.80 - 17.13\cos\frac{\pi t}{6} - 13.39\sin\frac{\pi t}{6}$$

where t is the time (in months), with $t = 1$ corresponding to January. *(Source: National Climatic Data Center)*

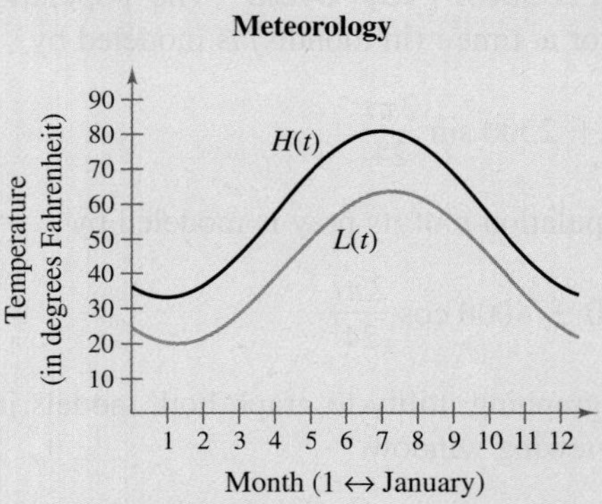

(a) During what part of the year is the difference between the normal high and low temperatures greatest? When is it smallest?

(b) The sun is the farthest north in the sky around June 21, but the graph shows the highest temperatures at a later date. Approximate the lag time of the temperatures relative to the position of the sun.

True or False? In Exercises 83–86, determine whether the statement is true or false. If it is false, explain why or give an example that shows it is false.

83. The amplitude of $f(x) = -3\cos 2x$ is -3.

84. The period of $f(x) = 5\cot\left(-\frac{4x}{3}\right)$ is $\frac{3\pi}{2}$.

85. $\lim_{x \to 0} \frac{\sin 5x}{3x} = \frac{5}{3}$

86. One solution of $\tan x = 1$ is $\frac{5\pi}{4}$.

QUIZ YOURSELF

See www.CalcChat.com for worked-out solutions to odd-numbered exercises.

Take this quiz as you would take a quiz in class. When you are done, check your work against the answers given in the back of the book.

In Exercises 1–4, express the angle in radian measure as a multiple of π. Use a calculator to verify your result.

1. $15°$ **2.** $105°$ **3.** $-80°$ **4.** $35°$

In Exercises 5–8, express the angle in degree measure. Use a calculator to verify your result.

5. $\dfrac{2\pi}{3}$ **6.** $\dfrac{4\pi}{15}$ **7.** $-\dfrac{4\pi}{3}$ **8.** $\dfrac{11\pi}{12}$

In Exercises 9–14, evaluate the trigonometric function *without* using a calculator.

9. $\sin\left(-\dfrac{\pi}{4}\right)$ **10.** $\cos 240°$

11. $\tan \dfrac{5\pi}{6}$ **12.** $\cot 45°$

13. $\sec(-60°)$ **14.** $\csc \dfrac{3\pi}{2}$

In Exercises 15–17, solve the equation for θ. Assume $0 \le \theta \le 2\pi$.

15. $\tan \theta - 1 = 0$

16. $\cos^2 \theta - 2 \cos \theta + 1 = 0$

17. $\sin^2 \theta = 3 \cos^2 \theta$

In Exercises 18–20, find the indicated side and/or angle.

18.

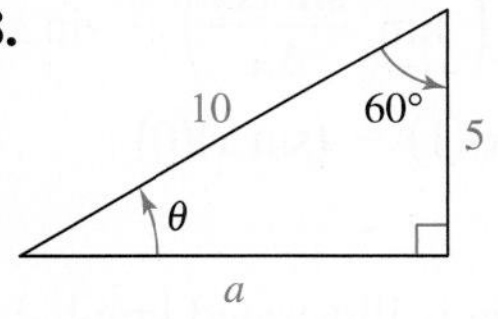

19.

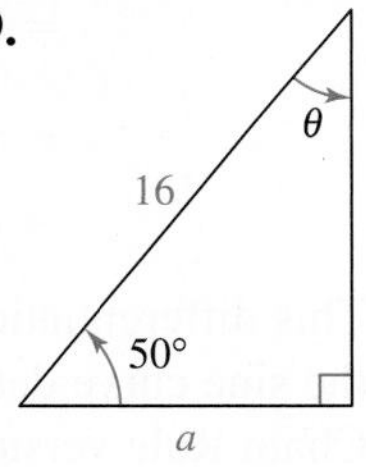

20.

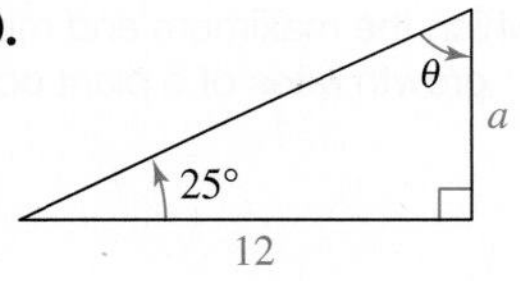

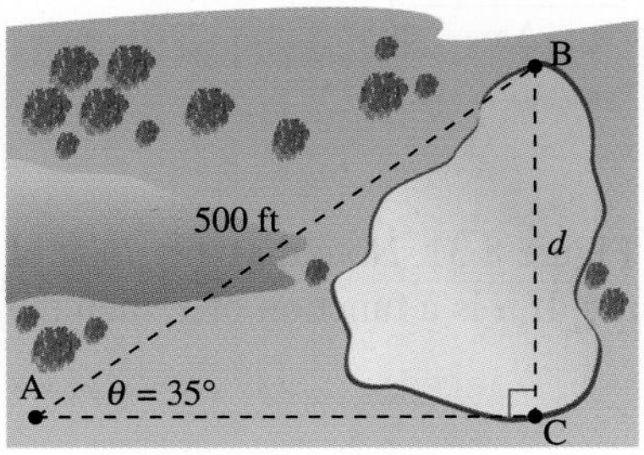

Figure for 21

21. A map maker needs to determine the distance d across a small lake. The distance from point A to point B is 500 feet and the angle θ is 35° (see figure). What is d?

In Exercises 22–24, (a) sketch the graph and (b) determine the period of the function.

22. $y = -3 \sin \dfrac{3x}{4}$ **23.** $y = -2 \cos 4x$ **24.** $y = \tan \dfrac{\pi x}{3}$

25. A company that produces snowboards forecasts monthly sales for 1 year to be

$$S = 53.5 + 40.5 \cos \frac{\pi t}{6}$$

where S is the sales (in thousands of dollars) and t is the time (in months), with $t = 1$ corresponding to January.

(a) Use a graphing utility to graph S.

(b) Use the graph to determine the months of maximum and minimum sales.

8.4 Derivatives of Trigonometric Functions

- Find derivatives of trigonometric functions.
- Find the relative extrema of trigonometric functions.
- Use derivatives of trigonometric functions to answer questions about real-life situations.

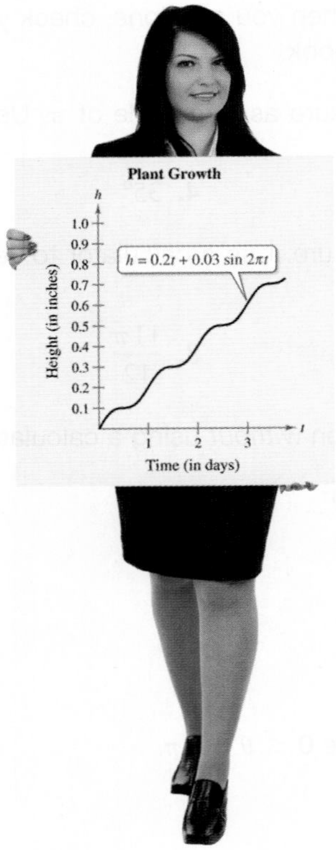

In Exercise 71 on page 543, you will use derivatives of trigonometric functions to find the times of day at which the maximum and minimum growth rates of a plant occur.

Derivatives of Trigonometric Functions

In Example 4 and Checkpoint 4 in the preceding section, you looked at two important trigonometric limits:

$$\lim_{\Delta x \to 0} \frac{\sin \Delta x}{\Delta x} = 1$$

and

$$\lim_{\Delta x \to 0} \frac{1 - \cos \Delta x}{\Delta x} = 0.$$

These two limits are used in the development of the derivative of the sine function.

$$\begin{aligned}
\frac{d}{dx}[\sin x] &= \lim_{\Delta x \to 0} \frac{\sin(x + \Delta x) - \sin x}{\Delta x} \\
&= \lim_{\Delta x \to 0} \frac{\sin x \cos \Delta x + \cos x \sin \Delta x - \sin x}{\Delta x} \\
&= \lim_{\Delta x \to 0} \left[\frac{\cos x \sin \Delta x - (\sin x)(1 - \cos \Delta x)}{\Delta x} \right] \\
&= \lim_{\Delta x \to 0} \left[(\cos x)\left(\frac{\sin \Delta x}{\Delta x}\right) - (\sin x)\left(\frac{1 - \cos \Delta x}{\Delta x}\right) \right] \\
&= \cos x \left(\lim_{\Delta x \to 0} \frac{\sin \Delta x}{\Delta x} \right) - \sin x \left(\lim_{\Delta x \to 0} \frac{1 - \cos \Delta x}{\Delta x} \right) \\
&= (\cos x)(1) - (\sin x)(0) \\
&= \cos x
\end{aligned}$$

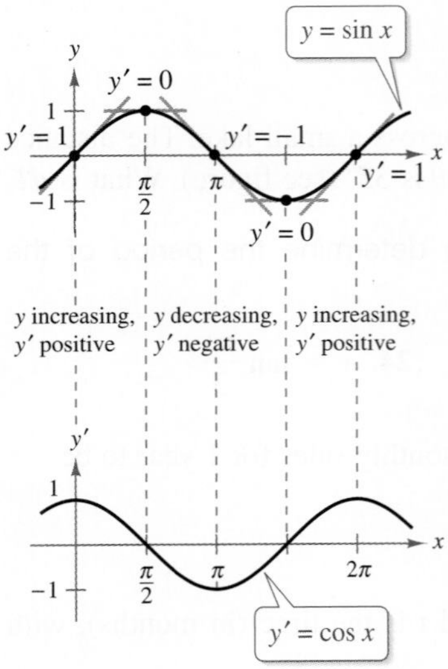

$\frac{d}{dx}[\sin x] = \cos x$

FIGURE 8.31

This differentiation rule is illustrated graphically in Figure 8.31. Note that the *slope* of the sine curve determines the *value* of the cosine curve. If u is a function of x, then the Chain Rule version of this differentiation rule is

$$\frac{d}{dx}[\sin u] = \cos u \frac{du}{dx}.$$

The Chain Rule versions of the differentiation rules for all six trigonometric functions are listed below. To help you remember these differentiation rules, note that each trigonometric function that begins with a "c" has a negative sign in its derivative.

Derivatives of the Six Basic Trigonometric Functions

$$\frac{d}{dx}[\sin u] = \cos u \frac{du}{dx} \qquad \frac{d}{dx}[\cos u] = -\sin u \frac{du}{dx}$$

$$\frac{d}{dx}[\tan u] = \sec^2 u \frac{du}{dx} \qquad \frac{d}{dx}[\cot u] = -\csc^2 u \frac{du}{dx}$$

$$\frac{d}{dx}[\sec u] = \sec u \tan u \frac{du}{dx} \qquad \frac{d}{dx}[\csc u] = -\csc u \cot u \frac{du}{dx}$$

Example 1 Differentiating Trigonometric Functions

Differentiate each function.

a. $y = \sin 2x$ **b.** $y = \cos(x - 1)$ **c.** $y = \tan 3x$

SOLUTION

a. Letting $u = 2x$, you obtain $u' = 2$, and the derivative is

$$\frac{dy}{dx} = \cos u \frac{du}{dx} = \cos 2x \frac{d}{dx}[2x] = (\cos 2x)(2) = 2 \cos 2x.$$

b. Letting $u = x - 1$, you can see that $u' = 1$. So, the derivative is

$$\frac{dy}{dx} = -\sin u \frac{du}{dx} = -\sin(x - 1) \frac{d}{dx}[x - 1] = -\sin(x - 1)(1) = -\sin(x - 1).$$

c. Letting $u = 3x$, you have $u' = 3$, which implies that

$$\frac{dy}{dx} = \sec^2 u \frac{du}{dx} = \sec^2 3x \frac{d}{dx}[3x] = (\sec^2 3x)(3) = 3 \sec^2 3x.$$

Example 2 Differentiating a Trigonometric Function

Differentiate the function

$$f(x) = \cos 3x^2.$$

SOLUTION Letting $u = 3x^2$, you obtain

$$f'(x) = -\sin u \frac{du}{dx} \qquad \text{Apply Cosine Differentiation Rule.}$$

$$= -\sin 3x^2 \frac{d}{dx}[3x^2] \qquad \text{Substitute } 3x^2 \text{ for } u.$$

$$= -(\sin 3x^2)(6x)$$

$$= -6x \sin 3x^2. \qquad \text{Simplify.}$$

Example 3 Differentiating a Trigonometric Function

Differentiate the function

$$f(x) = \tan^4 x.$$

SOLUTION Begin by rewriting the function.

$$f(x) = \tan^4 x \qquad \text{Write original function.}$$

$$= (\tan x)^4 \qquad \text{Rewrite.}$$

$$f'(x) = 4(\tan x)^3 \frac{d}{dx}[\tan x] \qquad \text{Apply Power Rule.}$$

$$= 4 \tan^3 x \sec^2 x \qquad \text{Apply Tangent Differentiation Rule.}$$

TECH TUTOR

When you use a symbolic differentiation utility to differentiate trigonometric functions, you can easily obtain results that appear to be different from those you would obtain by hand. Try using a symbolic differentiation utility to differentiate the function in Example 3. How does your result compare with the given solution?

✓ Checkpoints 1, 2, and 3

Differentiate each function.

a. $y = \cos 4x$ **b.** $y = \sin(x^2 - 1)$ **c.** $y = \tan \dfrac{x}{2}$

d. $y = 2 \cos x^3$ **e.** $y = \sin^3 x$

Example 4 Differentiating a Trigonometric Function

Differentiate $y = \csc \frac{x}{2}$.

SOLUTION

$$y = \csc \frac{x}{2} \qquad \text{Write original function.}$$

$$\frac{dy}{dx} = -\csc \frac{x}{2} \cot \frac{x}{2} \frac{d}{dx}\left[\frac{x}{2}\right] \qquad \text{Apply Cosecant Differentiation Rule.}$$

$$= -\frac{1}{2} \csc \frac{x}{2} \cot \frac{x}{2} \qquad \text{Simplify.}$$

✓Checkpoint 4

Differentiate each function.

a. $y = \sec 4x$ **b.** $y = \cot x^2$ ■

STUDY TIP

Notice that all of the differentiation rules that you learned in earlier chapters in the text can be applied to trigonometric functions. For instance, Example 5 uses the General Power Rule and Example 6 uses the Product Rule.

Example 5 Differentiating a Trigonometric Function

Differentiate $f(t) = \sqrt{\sin 4t}$.

SOLUTION Begin by rewriting the function in rational exponent form. Then apply the General Power Rule to find the derivative.

$$f(t) = (\sin 4t)^{1/2} \qquad \text{Rewrite with rational exponent.}$$

$$f'(t) = \left(\frac{1}{2}\right)(\sin 4t)^{-1/2} \frac{d}{dt}[\sin 4t] \qquad \text{Apply General Power Rule.}$$

$$= \left(\frac{1}{2}\right)(\sin 4t)^{-1/2}(4 \cos 4t) \qquad \text{Apply Sine Differentiation Rule.}$$

$$= \frac{2 \cos 4t}{\sqrt{\sin 4t}} \qquad \text{Simplify.}$$

✓Checkpoint 5

Differentiate each function.

a. $f(x) = \sqrt{\cos 2x}$ **b.** $f(x) = \sqrt[3]{\tan 3x}$ ■

Example 6 Differentiating a Trigonometric Function

Differentiate $y = x \sin x$.

SOLUTION Using the Product Rule, you can write

$$y = x \sin x \qquad \text{Write original function.}$$

$$\frac{dy}{dx} = x\frac{d}{dx}[\sin x] + \sin x \frac{d}{dx}[x] \qquad \text{Apply Product Rule.}$$

$$= x \cos x + \sin x. \qquad \text{Simplify.}$$

✓Checkpoint 6

Differentiate each function.

a. $y = x^2 \cos x$ **b.** $y = t \sin 2t$ ■

Relative Extrema of Trigonometric Functions

Recall that the critical numbers of a function $y = f(x)$ are the x-values for which $f'(x) = 0$ or $f'(x)$ is undefined.

Example 7 Finding Relative Extrema

Find the relative extrema of

$$y = \frac{x}{2} - \sin x$$

in the interval $(0, 2\pi)$.

SOLUTION To find the relative extrema of the function, begin by finding its critical numbers. The derivative of y is

$$\frac{dy}{dx} = \frac{1}{2} - \cos x.$$

By setting the derivative equal to zero, you obtain $\cos x = \frac{1}{2}$. So, in the interval $(0, 2\pi)$, the critical numbers are $x = \pi/3$ and $x = 5\pi/3$. Using the First-Derivative Test, you can conclude that $\pi/3$ yields a relative minimum and $5\pi/3$ yields a relative maximum, as shown in Figure 8.32.

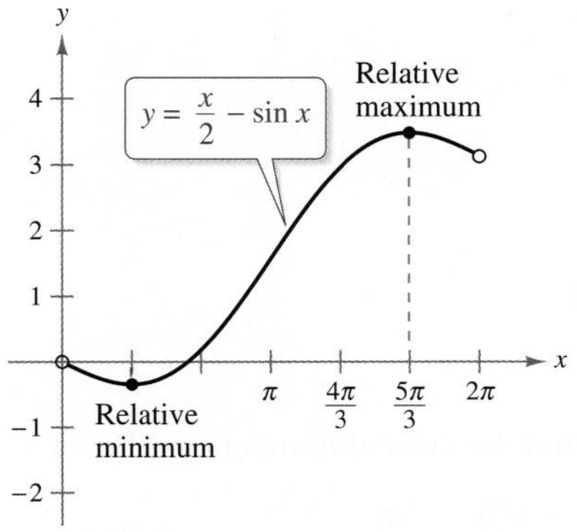

FIGURE 8.32

✓ **Checkpoint 7**

Find the relative extrema of

$$y = \frac{x}{2} - \cos x$$

in the interval $(0, 2\pi)$. ■

Example 8 Finding Relative Extrema

Find the relative extrema of $f(x) = 2 \sin x - \cos 2x$ in the interval $(0, 2\pi)$.

SOLUTION

$f(x) = 2 \sin x - \cos 2x$	Write original function.
$f'(x) = 2 \cos x + 2 \sin 2x$	Differentiate.
$0 = 2 \cos x + 2 \sin 2x$	Set derivative equal to 0.
$0 = 2 \cos x + 4 \cos x \sin x$	Identity: $\sin 2x = 2 \cos x \sin x$
$0 = 2(\cos x)(1 + 2 \sin x)$	Factor.

From this, you can see that the critical numbers occur when $\cos x = 0$ and when $\sin x = -\frac{1}{2}$. So, in the interval $(0, 2\pi)$, the critical numbers are

$$x = \frac{\pi}{2}, \frac{3\pi}{2}, \frac{7\pi}{6}, \frac{11\pi}{6}.$$

Using the First-Derivative Test, you can determine that $(\pi/2, 3)$ and $(3\pi/2, -1)$ are relative maxima, and $\left(7\pi/6, -\frac{3}{2}\right)$ and $\left(11\pi/6, -\frac{3}{2}\right)$ are relative minima, as shown in Figure 8.33.

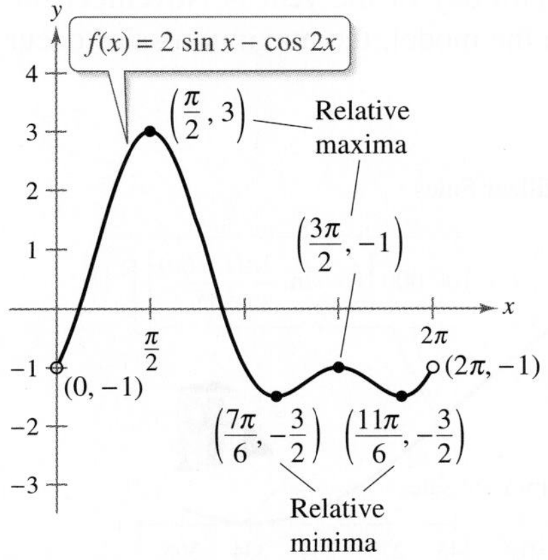

FIGURE 8.33

✓ **Checkpoint 8**

Find the relative extrema of $y = \frac{1}{2} \sin 2x + \cos x$ on the interval $(0, 2\pi)$. ■

Applications

Example 9 Modeling Seasonal Sales

A fertilizer manufacturer finds that the sales of one of its fertilizer brands follow a seasonal pattern that can be modeled by

$$F = 100{,}000\left[1 + \sin\frac{2\pi(t - 60)}{365}\right], \quad t \geq 0$$

where F is the amount sold (in pounds) and t is the time (in days), with $t = 1$ corresponding to January 1. On which day of the year is the maximum amount of fertilizer sold?

SOLUTION The derivative of the model is

$$\frac{dF}{dt} = 100{,}000\left(\frac{2\pi}{365}\right)\cos\frac{2\pi(t - 60)}{365}.$$

Setting this derivative equal to zero produces

$$\cos\frac{2\pi(t - 60)}{365} = 0.$$

Because cosine is zero at $\pi/2$ and $3\pi/2$, you can find the critical numbers as shown.

$$\frac{2\pi(t - 60)}{365} = \frac{\pi}{2} \qquad\qquad \frac{2\pi(t - 60)}{365} = \frac{3\pi}{2}$$

$$2\pi(t - 60) = \frac{365\pi}{2} \qquad\qquad 2\pi(t - 60) = \frac{(365)(3\pi)}{2}$$

$$t - 60 = \frac{365}{4} \qquad\qquad t - 60 = \frac{3(365)}{4}$$

$$t = \frac{365}{4} + 60 \qquad\qquad t = \frac{3(365)}{4} + 60$$

$$t \approx 151 \qquad\qquad t \approx 334$$

The 151st day of the year is May 31 and the 334th day of the year is November 30. From Figure 8.34, you can see that, according to the model, the maximum sales occur on May 31.

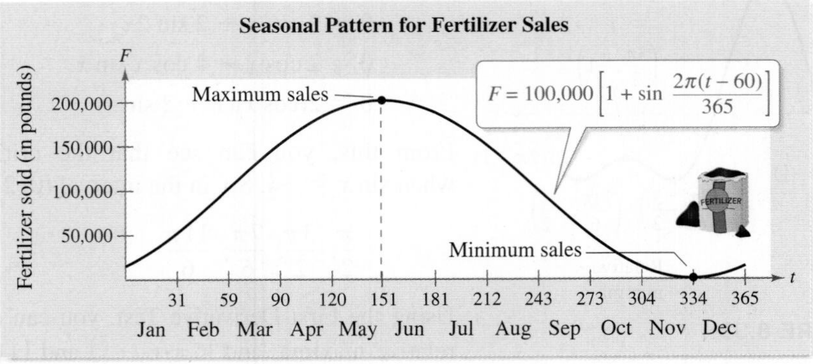

FIGURE 8.34

Checkpoint 9

Using the model from Example 9, find the rate at which sales are changing when $t = 59$.

TECH TUTOR

Because of the difficulty of solving some trigonometric equations, it can be difficult to find the critical numbers of a trigonometric function. For instance, consider the function

$$f(x) = 2\sin x - \cos 3x.$$

Setting the derivative of this function equal to zero produces

$$2\cos x + 3\sin 3x = 0.$$

This equation is difficult to solve analytically. So, it is difficult to find the relative extrema of f analytically. With a graphing utility, however, you can estimate the relative extrema graphically using the *zoom* and *trace* features. You can also try other approximation techniques, such as Newton's Method, which is discussed in Section 10.6.

Example 10 Modeling Temperature Change

The temperature T (in degrees Fahrenheit) during a given 24-hour period can be modeled by

$$T = 70 + 15\sin\frac{\pi(t-8)}{12}, \quad t \ge 0$$

where t is the time (in hours), with $t = 0$ corresponding to midnight, as shown in Figure 8.35. Find the rate at which the temperature is changing at 6 A.M.

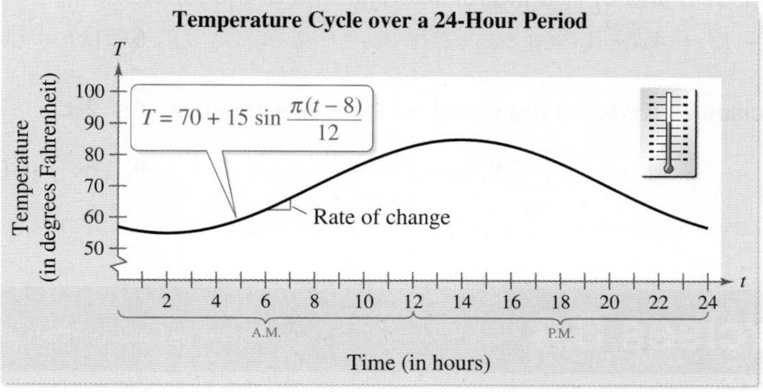

FIGURE 8.35

SOLUTION The rate of change of the temperature is given by the derivative

$$\frac{dT}{dt} = \frac{15\pi}{12}\cos\frac{\pi(t-8)}{12}.$$

Because 6 A.M. corresponds to $t = 6$, the rate of change at 6 A.M. is

$$\begin{aligned}\frac{15\pi}{12}\cos\left(\frac{-2\pi}{12}\right) &= \frac{5\pi}{4}\cos\left(-\frac{\pi}{6}\right)\\ &= \frac{5\pi}{4}\left(\frac{\sqrt{3}}{2}\right)\\ &\approx 3.4° \text{ per hour.}\end{aligned}$$

✓ Checkpoint 10

In Example 10, find the rate at which the temperature is changing at 8 P.M. ■

SUMMARIZE (Section 8.4)

1. State the Sine, Tangent, and Secant Differentiation Rules *(page 536)*. For examples of using these rules, see Examples 1, 3, 5, and 6.
2. State the Cosine, Cotangent, and Cosecant Differentiation Rules *(page 536)*. For examples of using these rules, see Examples 1, 2, and 4.
3. Explain how to find the relative extrema of a trigonometric function *(page 539)*. For examples of finding the relative extrema of trigonometric functions, see Examples 7 and 8.
4. Describe a real-life example of how the derivative of a trigonometric function can be used to analyze the rate of change of temperature *(page 541, Example 10)*.

SKILLS WARM UP 8.4

The following warm-up exercises involve skills that were covered in earlier sections. You will use these skills in the exercise set for this section. For additional help, review Sections 2.2, 2.4, 2.5, 3.2, and 8.2.

In Exercises 1–4, find the derivative of the function.

1. $f(x) = 3x^3 - 2x^2 + 4x - 7$

2. $g(x) = (x^3 + 4)^4$

3. $f(x) = (x - 1)(x^2 + 2x + 3)$

4. $g(x) = \dfrac{2x}{x^2 + 5}$

In Exercises 5 and 6, find the relative extrema of the function.

5. $f(x) = x^2 + 4x + 1$

6. $f(x) = \frac{1}{3}x^3 - 4x + 2$

In Exercises 7–10, solve the equation for x. Assume $0 \le x \le 2\pi$.

7. $\sin x = \dfrac{\sqrt{3}}{2}$

8. $\cos x = -\dfrac{1}{2}$

9. $\cos \dfrac{x}{2} = 0$

10. $\sin \dfrac{x}{2} = -\dfrac{\sqrt{2}}{2}$

Exercises 8.4

See www.CalcChat.com for worked-out solutions to odd-numbered exercises.

Differentiating Trigonometric Functions In Exercises 1–28, find the derivative of the trigonometric function. *See Examples 1, 2, 3, 4, 5, and 6.*

1. $y = \sin \dfrac{x}{3}$

2. $f(x) = \cos 2x$

3. $y = \tan 4x^3$

4. $y = \sin(3x + 1)$

5. $f(t) = \tan 5t$

6. $g(x) = 3 \cos x^4$

7. $y = \sin^2 x$

8. $y = \cos^4 x$

9. $y = \sec \pi x$

10. $y = \frac{1}{2} \csc 2x$

11. $y = \cot(2x + 1)$

12. $f(t) = \cos \dfrac{2}{t}$

13. $y = \sqrt{\tan 2x}$

14. $y = \sqrt[3]{\sin 6x}$

15. $f(t) = t^2 \cos t$

16. $y = (x + 3) \csc x$

17. $g(t) = \dfrac{\cos t}{t}$

18. $f(x) = \dfrac{\sin x}{x}$

19. $y = e^{x^2} \sec x$

20. $y = e^{-2x^2} \cot x$

21. $y = \cos 3x + \sin^2 x$

22. $y = \csc^2 x - \sec 3x$

23. $y = x \sin \dfrac{1}{x}$

24. $y = x^2 \cos \dfrac{1}{x}$

25. $y = 2 \tan^2 4x$

26. $y = -\sin^4 2x$

27. $y = e^{2x} \sin 2x$

28. $y = e^{-x} \cos \dfrac{x}{2}$

Differentiating Trigonometric Functions In Exercises 29–40, find the derivative of the function and simplify your answer by using the trigonometric identities listed in Section 8.2.

29. $y = \cos^2 x$

30. $y = \frac{1}{4} \sin^2 2x$

31. $y = \cos^2 x - \sin^2 x$

32. $y = \dfrac{x}{2} + \dfrac{\sin 2x}{4}$

33. $y = \sin^2 x - \cos 2x$

34. $y = 3 \sin x - 2 \sin^3 x$

35. $y = \tan x - x$

36. $y = \cot x + x$

37. $y = \dfrac{\sin^3 x}{3} - \dfrac{\sin^5 x}{5}$

38. $y = \dfrac{\sec^7 x}{7} - \dfrac{\sec^5 x}{5}$

39. $y = \ln(\sin^2 x)$

40. $y = \frac{1}{2}\ln(\cos^2 x)$

Finding an Equation of a Tangent Line In Exercises 41–48, find an equation of the tangent line to the graph of the function at the given point.

	Function	*Point*
41.	$y = \tan x$	$\left(-\dfrac{\pi}{4}, -1\right)$
42.	$y = \sec x$	$\left(\dfrac{\pi}{3}, 2\right)$
43.	$y = \sin 4x$	$(\pi, 0)$
44.	$y = \csc^2 x$	$\left(\dfrac{\pi}{2}, 1\right)$
45.	$y = \cot x$	$\left(\dfrac{3\pi}{4}, -1\right)$
46.	$y = \sin x \cos x$	$\left(\dfrac{3\pi}{2}, 0\right)$
47.	$y = \ln(\sin x + 2)$	$\left(\dfrac{3\pi}{2}, 0\right)$
48.	$y = \sqrt{\sin x}$	$\left(\dfrac{\pi}{6}, \dfrac{\sqrt{2}}{2}\right)$

Implicit Differentiation In Exercises 49 and 50, use implicit differentiation to find dy/dx and evaluate the derivative at the given point.

Function	*Point*
49. $\sin x + \cos 2y = 1$	$\left(\frac{\pi}{2}, \frac{\pi}{4}\right)$
50. $\tan(x + y) = x$	$(0, 0)$

Slope of a Tangent Line In Exercises 51–56, find the slope of the tangent line to the given sine function at the origin. Compare this value with the number of complete cycles in the interval $[0, 2\pi]$.

51. $y = \sin \frac{5x}{4}$

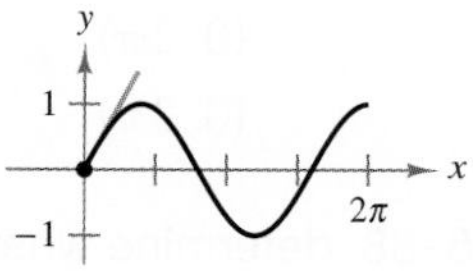

52. $y = \sin \frac{5x}{2}$

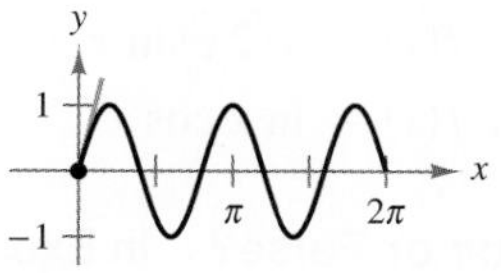

53. $y = \sin 2x$

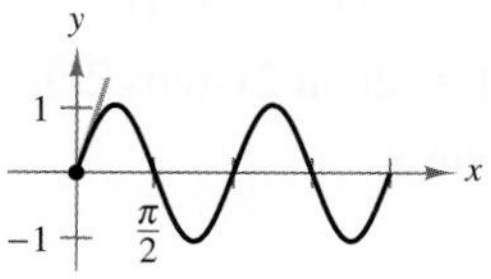

54. $y = \sin \frac{3x}{2}$

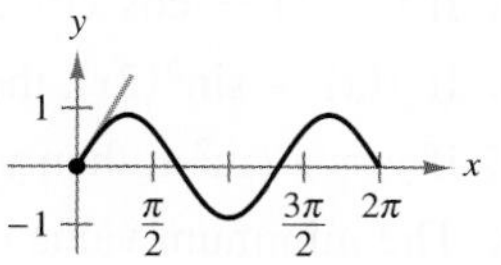

55. $y = \sin x$

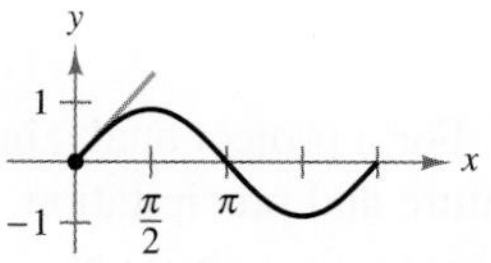

56. $y = \sin \frac{x}{2}$

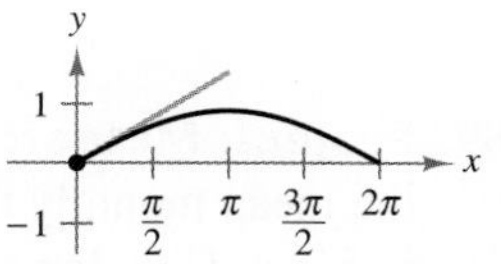

Finding Relative Extrema In Exercises 57–66, find the relative extrema of the trigonometric function in the interval $(0, 2\pi)$. Use a graphing utility to confirm your results. *See Examples 7 and 8.*

57. $y = \cos \frac{\pi x}{2}$ **58.** $y = -3 \sin \frac{x}{3}$

59. $y = \cos^2 x$ **60.** $y = \sec \frac{x}{2}$

61. $y = 2 \sin x + \sin 2x$ **62.** $y = 2 \cos x + \cos 2x$

63. $y = e^x \cos x$ **64.** $y = e^{-x} \sin x$

65. $y = x - 2 \sin x$ **66.** $y = \sin^2 x + \cos x$

Finding Second Derivatives In Exercises 67–70, find the second derivative of the trigonometric function.

67. $y = x \sin x$ **68.** $y = \sec x$

69. $y = \cos x^2$ **70.** $y = \csc^2 \pi x$

71. Biology Plants do not grow at constant rates during a normal 24-hour period because their growth is affected by sunlight. Suppose that the growth of a certain plant species in a controlled environment is given by the model

$$h = 0.2t + 0.03 \sin 2\pi t$$

where h is the height of the plant (in inches) and t is the time (in days), with $t = 0$ corresponding to midnight of day 1. During what time of day is the rate of growth of this plant

(a) a maximum? (b) a minimum?

72. Meteorology The normal average daily temperature in degrees Fahrenheit for a city is given by

$$T = 55 - 21 \cos \frac{2\pi(t - 32)}{365}$$

where t is the time (in days), with $t = 1$ corresponding to January 1. Find the expected date of

(a) the warmest day. (b) the coldest day.

73. Construction Workers The numbers W (in thousands) of construction workers employed in the United States during 2010 can be modeled by

$$W = 5488 + 347.6 \sin(0.45t + 4.153)$$

where t is the time (in months), with $t = 1$ corresponding to January. Approximate the month t in which the number of construction workers employed was a maximum. What was the maximum number of construction workers employed? *(Source: U.S. Bureau of Labor Statistics)*

74. Transportation Workers The numbers W (in thousands) of scenic and sightseeing transportation workers employed in the United States during 2010 can be modeled by

$$W = 27.8 + 8.25 \sin(0.561t - 2.5913)$$

where t is the time (in months), with $t = 1$ corresponding to January. Approximate the month t in which the number of scenic and sightseeing transportation workers employed was a maximum. What was the maximum number of scenic and sightseeing transportation workers employed? *(Source: U.S. Bureau of Labor Statistics)*

75. Meteorology The number of hours of daylight D in New Orleans can be modeled by

$$D = 12.12 + 1.87 \cos \frac{\pi(t + 5.83)}{6}$$

where t is the time (in months), with $t = 1$ corresponding to January. Approximate the month t in which New Orleans has the maximum number of daylight hours. What is this maximum number of daylight hours? *(Source: U.S. Naval Observatory)*

76. Tides Throughout the day, the depth of water D (in meters) at the end of a dock varies with the tides. The depth for one particular day can be modeled by

$$D = 3.5 + 1.5 \cos \frac{\pi t}{6}, \quad 0 \le t \le 24$$

where t is the time (in hours), with $t = 0$ corresponding to midnight.

(a) Determine dD/dt.

(b) Evaluate dD/dt for $t = 4$ and $t = 20$, and interpret your results.

(c) Find the time(s) when the water depth is the greatest and the time(s) when the water depth is the least.

(d) What is the greatest depth? What is the least depth? Did you have to use calculus to determine these depths? Explain your reasoning.

77. Think About It Rewrite the trigonometric function in terms of sine and/or cosine and then differentiate to prove the following differentiation rules.

(a) $\dfrac{d}{dx}[\tan x] = \sec^2 x$

(b) $\dfrac{d}{dx}[\sec x] = \sec x \tan x$

(c) $\dfrac{d}{dx}[\cot x] = -\csc^2 x$

(d) $\dfrac{d}{dx}[\csc x] = -\csc x \cot x$

78. HOW DO YOU SEE IT? The graph shows the height h (in feet) above ground of a seat on a Ferris wheel at time t (in seconds).

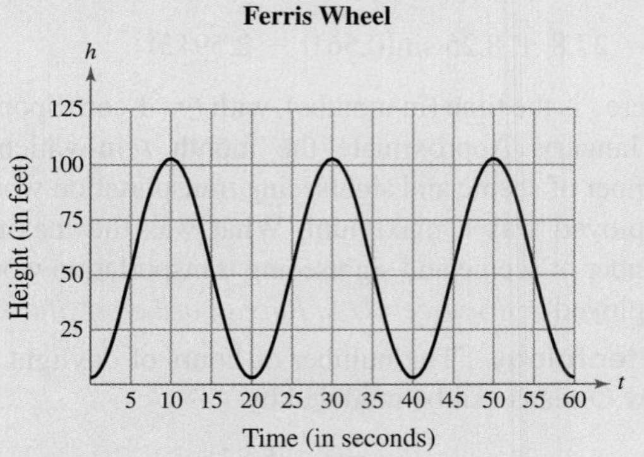

(a) What is the period of the model? What does the period tell you about the ride?

(b) Find the intervals on which the height is increasing and decreasing.

(c) Estimate the relative extrema in the interval [0, 60].

Analyzing a Function In Exercises 79–84, use a graphing utility to (a) graph f and f' in the same viewing window over the specified interval, (b) find the critical numbers of f, (c) find the interval(s) on which f' is positive and the interval(s) on which f' is negative, and (d) find the relative extrema in the interval. Note the behavior of f in relation to the sign of f'.

Function	*Interval*
79. $f(t) = t^2 \sin t$	$(0, 2\pi)$
80. $f(x) = \dfrac{x^2 - 2}{\sin x} - 5x$	$(0, 2\pi)$
81. $f(x) = \sin x - \frac{1}{3}\sin 3x + \frac{1}{5}\sin 5x$	$(0, \pi)$
82. $f(x) = x \sin x$	$(0, \pi)$
83. $f(x) = \sqrt{2x}\sin x$	$(0, 2\pi)$
84. $f(x) = \ln x \cos x$	$(0, 2\pi)$

True or False? In Exercises 85–88, determine whether the statement is true or false. If it is false, explain why or give an example that shows it is false.

85. If $y = (1 - \cos x)^{1/2}$, then $y' = \frac{1}{2}(1 - \cos x)^{-1/2}$.

86. If $f(x) = \sin^2(2x)$, then $f'(x) = 2(\sin 2x)(\cos 2x)$.

87. If $y = x \sin^2 x$, then $y' = 2x \sin x$.

88. The minimum value of

$$y = 3 \sin x + 2$$

is -1.

89. Project: Meteorology For a project analyzing the mean monthly temperature and precipitation in Sioux City, Iowa, visit this text's website at *www.cengagebrain.com*. *(Source: National Oceanic and Atmospheric Administration)*

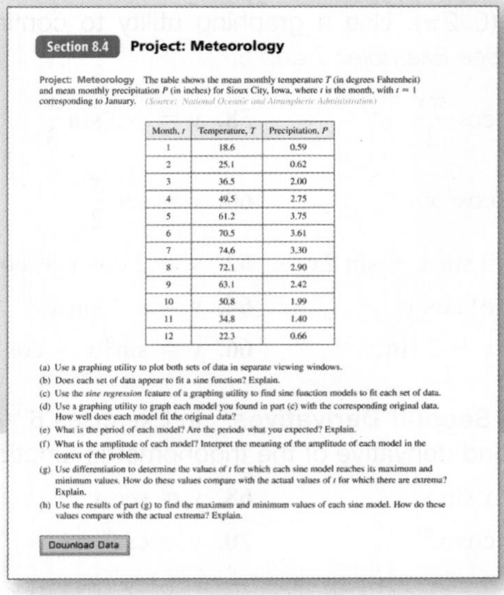

Section 8.4 **Project: Meteorology**

Project: Meteorology The table shows the mean monthly temperature T (in degrees Fahrenheit) and mean monthly precipitation P (in inches) for Sioux City, Iowa, where t is the month, with $t = 1$ corresponding to January. *(Source: National Oceanic and Atmospheric Administration)*

Month, t	Temperature, T	Precipitation, P
1	18.6	0.59
2	25.1	0.62
3	36.5	2.00
4	49.5	2.75
5	61.2	3.75
6	70.5	3.61
7	74.6	3.30
8	72.1	2.90
9	63.1	2.42
10	50.8	1.99
11	34.8	1.40
12	22.3	0.66

(a) Use a graphing utility to plot both sets of data in separate viewing windows.

(b) Does each set of data appear to fit a sine function? Explain.

(c) Use the *sine regression* feature of a graphing utility to find sine function models to fit each set of data.

(d) Use a graphing utility to graph each model you found in part (c) with the corresponding original data. How well does each model fit the original data?

(e) What is the period of each model? Are the periods what you expected? Explain.

(f) What is the amplitude of each model? Interpret the meaning of the amplitude of each model in the context of the problem.

(g) Use differentiation to determine the values of t for which each sine model reaches its maximum and minimum values. How do these values compare with the actual values of t for which there are extrema? Explain.

(h) Use the results of part (g) to find the maximum and minimum values of each sine model. How do these values compare with the actual extrema? Explain.

Download Data

8.5 Integrals of Trigonometric Functions

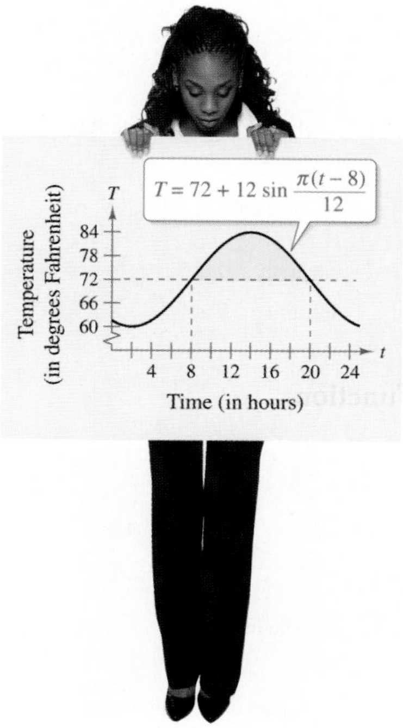

In Exercise 61 on page 553, you will use integrals of trigonometric functions to find the cost of cooling a house.

- Learn the trigonometric integration rules that correspond directly to differentiation rules.
- Integrate the six basic trigonometric functions.
- Use trigonometric integrals to solve real-life problems.

Trigonometric Integrals

For each trigonometric differentiation rule, there is a corresponding integration rule. For instance, corresponding to the differentiation rule

$$\frac{d}{dx}[\sin u] = \cos u \frac{du}{dx}$$

is the integration rule

$$\int \cos u \, du = \sin u + C$$

and corresponding to the differentiation rule

$$\frac{d}{dx}[\cos u] = -\sin u \frac{du}{dx}$$

is the integration rule

$$\int \sin u \, du = -\cos u + C.$$

Integrals Involving Trigonometric Functions

Differentiation Rule	*Integration Rule*
$\frac{d}{dx}[\sin u] = \cos u \frac{du}{dx}$	$\int \cos u \, du = \sin u + C$
$\frac{d}{dx}[\cos u] = -\sin u \frac{du}{dx}$	$\int \sin u \, du = -\cos u + C$
$\frac{d}{dx}[\tan u] = \sec^2 u \frac{du}{dx}$	$\int \sec^2 u \, du = \tan u + C$
$\frac{d}{dx}[\sec u] = \sec u \tan u \frac{du}{dx}$	$\int \sec u \tan u \, du = \sec u + C$
$\frac{d}{dx}[\cot u] = -\csc^2 u \frac{du}{dx}$	$\int \csc^2 u \, du = -\cot u + C$
$\frac{d}{dx}[\csc u] = -\csc u \cot u \frac{du}{dx}$	$\int \csc u \cot u \, du = -\csc u + C$

STUDY TIP

Note that the list above gives you formulas for integrating only two of the six trigonometric functions: the sine function and the cosine function. The list does not show you how to integrate the other four trigonometric functions. Rules for integrating those functions are discussed later in this section.

TECH TUTOR

If you have access to a symbolic integration utility, try using it to integrate the functions in Examples 1, 2, and 3. Does your utility give the same results that are given in the examples?

Example 1 Integrating a Trigonometric Function

Find $\int 2 \cos x \, dx$.

SOLUTION Let $u = x$. Then $du = dx$.

$$\begin{aligned}
\int 2 \cos x \, dx &= 2\int \cos x \, dx && \text{Apply Constant Multiple Rule.}\\
&= 2\int \cos u \, du && \text{Substitute for } x \text{ and } dx.\\
&= 2 \sin u + C && \text{Integrate.}\\
&= 2 \sin x + C && \text{Substitute for } u.
\end{aligned}$$

Example 2 Integrating a Trigonometric Function

Find $\int 3x^2 \sin x^3 \, dx$.

SOLUTION Let $u = x^3$. Then $du = 3x^2 \, dx$.

$$\begin{aligned}
\int 3x^2 \sin x^3 \, dx &= \int (\sin x^3) 3x^2 \, dx && \text{Rewrite integrand.}\\
&= \int \sin u \, du && \text{Substitute for } x^3 \text{ and } 3x^2\,dx.\\
&= -\cos u + C && \text{Integrate.}\\
&= -\cos x^3 + C && \text{Substitute for } u.
\end{aligned}$$

✓ Checkpoints 1 and 2

Find (a) $\int 5 \sin x \, dx$ and (b) $\int 4x^3 \cos x^4 \, dx$. ■

Example 3 Integrating a Trigonometric Function

Find $\int \sec 3x \tan 3x \, dx$.

SOLUTION Let $u = 3x$. Then $du = 3 \, dx$.

$$\begin{aligned}
\int \sec 3x \tan 3x \, dx &= \frac{1}{3}\int (\sec 3x \tan 3x) 3 \, dx && \text{Multiply and divide by 3.}\\
&= \frac{1}{3}\int \sec u \tan u \, du && \text{Substitute for } 3x \text{ and } 3\,dx.\\
&= \frac{1}{3} \sec u + C && \text{Integrate.}\\
&= \frac{1}{3} \sec 3x + C && \text{Substitute for } u.
\end{aligned}$$

✓ Checkpoint 3

Find $\int \sec^2 5x \, dx$. ■

Example 4 Integrating a Trigonometric Function

Find $\int e^x \sec^2 e^x \, dx$.

SOLUTION Let $u = e^x$. Then $du = e^x \, dx$.

$$\int e^x \sec^2 e^x \, dx = \int (\sec^2 e^x)e^x \, dx \quad \text{Rewrite integrand.}$$

$$= \int \sec^2 u \, du \quad \text{Substitute for } e^x \text{ and } e^x\,dx.$$

$$= \tan u + C \quad \text{Integrate.}$$

$$= \tan e^x + C \quad \text{Substitute for } u.$$

✓Checkpoint 4

Find $\int 2 \csc 2x \cot 2x \, dx$. ■

The next two examples use the General Power Rule for integration and the General Log Rule for integration. Recall from Chapter 5 that these rules are

$$\int u^n \frac{du}{dx} \, dx = \frac{u^{n+1}}{n+1} + C, \quad n \neq -1 \quad \text{General Power Rule}$$

and

$$\int \frac{du/dx}{u} \, dx = \ln|u| + C. \quad \text{General Log Rule}$$

The key to using these two rules is identifying the proper substitution for u.

STUDY TIP

Remember to check your answers to integration problems by differentiating. In Example 5, for instance, try differentiating the answer

$$y = \frac{1}{12} \sin^3 4x + C.$$

You should obtain the original integrand, as shown.

$$y' = \frac{1}{12} 3(\sin 4x)^2(\cos 4x)4$$
$$= \sin^2 4x \cos 4x$$

Example 5 Using the General Power Rule

Find $\int \sin^2 4x \cos 4x \, dx$.

SOLUTION Let $u = \sin 4x$. Then $du/dx = 4 \cos 4x$.

$$\int \sin^2 4x \cos 4x \, dx = \frac{1}{4} \int \overbrace{(\sin 4x)^2}^{u^2} \overbrace{(4 \cos 4x)}^{du/dx} \, dx \quad \text{Rewrite integrand.}$$

$$= \frac{1}{4} \int u^2 \, du \quad \text{Substitute for } \sin 4x \text{ and } 4 \cos 4x\,dx.$$

$$= \frac{1}{4} \frac{u^3}{3} + C \quad \text{Integrate.}$$

$$= \frac{1}{4} \frac{(\sin 4x)^3}{3} + C \quad \text{Substitute for } u.$$

$$= \frac{1}{12} \sin^3 4x + C \quad \text{Simplify.}$$

✓Checkpoint 5

Find $\int \cos^3 2x \sin 2x \, dx$. ■

Example 6 Using the Log Rule

Find $\int \frac{\sin x}{\cos x}\, dx.$

SOLUTION Let $u = \cos x$. Then $du/dx = -\sin x$.

$$\int \frac{\sin x}{\cos x}\, dx = -\int \frac{-\sin x}{\cos x}\, dx \quad \text{Rewrite integrand.}$$

$$= -\int \frac{du/dx}{u}\, dx \quad \text{Substitute for } \cos x \text{ and } -\sin x.$$

$$= -\ln|u| + C \quad \text{Apply Log Rule.}$$

$$= -\ln|\cos x| + C \quad \text{Substitute for } u.$$

✓Checkpoint 6

Find $\int \frac{\cos x}{\sin x}\, dx.$ ■

Example 7 Evaluating a Definite Integral

Evaluate $\int_0^{\pi/4} \cos 2x\, dx.$

SOLUTION

$$\int_0^{\pi/4} \cos 2x\, dx = \left[\frac{1}{2}\sin 2x\right]_0^{\pi/4} = \frac{1}{2} - 0 = \frac{1}{2}$$

✓Checkpoint 7

Find $\int_0^{\pi/2} \sin 2x\, dx.$ ■

Example 8 Finding Area by Integration

Find the area of the region bounded by the x-axis and the graph of

$$y = \sin x$$

for $0 \le x \le \pi$.

SOLUTION As indicated in Figure 8.36, this area is given by

$$\text{Area} = \int_0^{\pi} \sin x\, dx = \Big[-\cos x\Big]_0^{\pi} = -(-1) - (-1) = 2.$$

So, the region has an area of 2 square units.

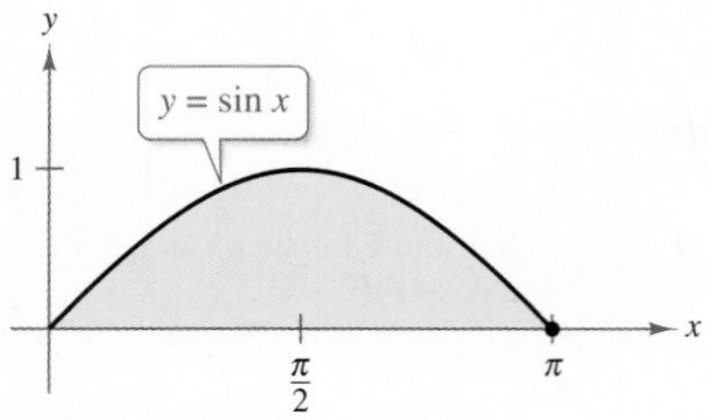

FIGURE 8.36

✓Checkpoint 8

Find the area of the region bounded by the graphs of

$$y = \cos x \quad \text{and} \quad y = 0$$

for $0 \le x \le \frac{\pi}{2}$. ■

Integrals of the Six Basic Trigonometric Functions

At the beginning of this section, the integration rules for the sine and cosine functions were listed. Now, using the result of Example 6, you have an integration rule for the tangent function. That rule is

$$\int \tan x\,dx = \int \frac{\sin x}{\cos x}\,dx = -\ln|\cos x| + C.$$

Integration formulas for the other three trigonometric functions can be developed in a similar way. For instance, to obtain the integration formula for the secant function, you can integrate as shown.

$$\begin{aligned}
\int \sec x\,dx &= \int \frac{\sec x(\sec x + \tan x)}{\sec x + \tan x}\,dx \\
&= \int \frac{\sec^2 x + \sec x \tan x}{\sec x + \tan x}\,dx \\
&= \int \frac{du/dx}{u}\,dx && \text{Substitute: } u = \sec x + \tan x. \\
&= \ln|u| + C && \text{Apply Log Rule.} \\
&= \ln|\sec x + \tan x| + C && \text{Substitute for } u.
\end{aligned}$$

The integrals of the six basic trigonometric functions are summarized below.

Integrals of the Six Basic Trigonometric Functions

$$\int \sin u\,du = -\cos u + C \qquad \int \cos u\,du = \sin u + C$$

$$\int \tan u\,du = -\ln|\cos u| + C \qquad \int \sec u\,du = \ln|\sec u + \tan u| + C$$

$$\int \cot u\,du = \ln|\sin u| + C \qquad \int \csc u\,du = \ln|\csc u - \cot u| + C$$

Example 9 Integrating a Trigonometric Function

Find $\int \tan 4x\,dx$.

SOLUTION Let $u = 4x$. Then $du = 4\,dx$.

$$\begin{aligned}
\int \tan 4x\,dx &= \frac{1}{4}\int (\tan 4x)4\,dx && \text{Rewrite integrand.} \\
&= \frac{1}{4}\int \tan u\,du && \text{Substitute for } 4x \text{ and } 4\,dx. \\
&= -\frac{1}{4}\ln|\cos u| + C && \text{Integrate.} \\
&= -\frac{1}{4}\ln|\cos 4x| + C && \text{Substitute for } u.
\end{aligned}$$

✓ Checkpoint 9

Find $\int \sec 2x\,dx$.

■

Application

In the next example, recall from Section 5.4 that the average value of a function f over an interval $[a, b]$ is given by

$$\frac{1}{b-a}\int_a^b f(x)\,dx.$$

Example 10 Finding an Average Temperature

The temperature T (in degrees Fahrenheit) during a 24-hour period can be modeled by

$$T = 72 + 18 \sin \frac{\pi(t-8)}{12}$$

where t is the time (in hours), with $t = 0$ corresponding to midnight. Will the average temperature during the four-hour period from noon to 4 P.M. be greater than 85°?

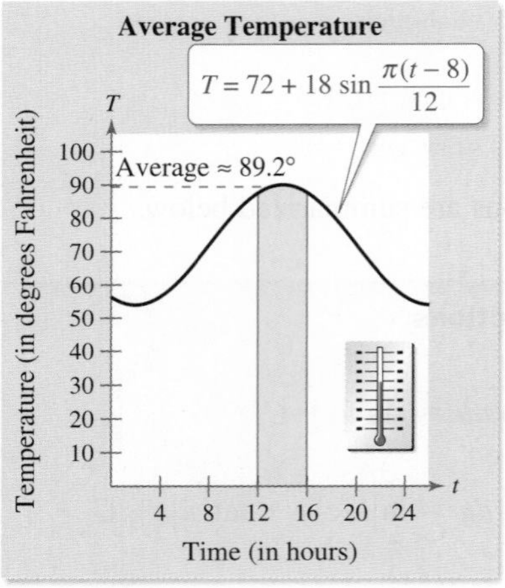

FIGURE 8.37

SOLUTION To find the average temperature A, use the formula for the average value of a function on an interval.

$$\begin{aligned}
A &= \frac{1}{4}\int_{12}^{16}\left[72 + 18\sin\frac{\pi(t-8)}{12}\right]dt \\
&= \frac{1}{4}\left[72t + 18\left(\frac{12}{\pi}\right)\left(-\cos\frac{\pi(t-8)}{12}\right)\right]_{12}^{16} \\
&= \frac{1}{4}\left[72(16) + 18\left(\frac{12}{\pi}\right)\left(\frac{1}{2}\right) - 72(12) + 18\left(\frac{12}{\pi}\right)\left(\frac{1}{2}\right)\right] \\
&= \frac{1}{4}\left(288 + \frac{216}{\pi}\right) \\
&= 72 + \frac{54}{\pi} \\
&\approx 89.2^\circ
\end{aligned}$$

So, the average temperature is about 89.2°, as indicated in Figure 8.37. You can conclude that the average temperature from noon to 4 P.M. will be greater than 85°.

✓Checkpoint 10

Use the function in Example 10 to find the average temperature from 9 A.M. to noon. ■

SUMMARIZE (Section 8.5)

1. For each trigonometric integration rule on page 545, state the corresponding differentiation rule *(page 545)*. For examples of using these integration rules, see Examples 1–8.
2. State the integration rules for the six basic trigonometric functions *(page 549)*. For examples of using these integration rules, see Examples 1, 2, 5, 7, 8, and 9.
3. Describe a real-life example of how the integral of a trigonometric function can be used to find an average temperature *(page 550, Example 10)*.

SKILLS WARM UP 8.5

The following warm-up exercises involve skills that were covered in earlier sections. You will use these skills in the exercise set for this section. For additional help, review Sections 5.4 and 8.2.

In Exercises 1–8, evaluate the trigonometric function.

1. $\cos \frac{5\pi}{4}$

2. $\sin \frac{7\pi}{6}$

3. $\sin\left(-\frac{\pi}{3}\right)$

4. $\cos\left(-\frac{\pi}{6}\right)$

5. $\tan \frac{5\pi}{6}$

6. $\cot \frac{5\pi}{3}$

7. $\sec \pi$

8. $\cos \frac{\pi}{2}$

In Exercises 9–16, simplify the expression using the trigonometric identities.

9. $\sin x \sec x$

10. $\csc x \cos x$

11. $\cos^2 x(\sec^2 x - 1)$

12. $\sin^2 x(\csc^2 x - 1)$

13. $\sec x \sin\left(\frac{\pi}{2} - x\right)$

14. $\cot x \cos\left(\frac{\pi}{2} - x\right)$

15. $\cot x \sec x$

16. $\cot x(\sin^2 x)$

In Exercises 17–20, evaluate the definite integral.

17. $\int_0^4 (x^2 + 3x - 4)\, dx$

18. $\int_{-1}^1 (1 - x^2)\, dx$

19. $\int_0^2 x(4 - x^2)\, dx$

20. $\int_0^1 x(9 - x^2)\, dx$

Exercises 8.5

See www.CalcChat.com for worked-out solutions to odd-numbered exercises.

Integrating Trigonometric Functions In Exercises 1–32, find the indefinite integral. *See Examples 1, 2, 3, 4, 5, 6, and 9.*

1. $\int 4 \sin x\, dx$

2. $\int 8 \cos x\, dx$

3. $\int \sin 2x\, dx$

4. $\int \cos 6x\, dx$

5. $\int 4x^3 \cos x^4\, dx$

6. $\int 2x \sin x^2\, dx$

7. $\int \sec^2 \frac{x}{5}\, dx$

8. $\int \csc^2 4x\, dx$

9. $\int \sec 2x \tan 2x\, dx$

10. $\int \csc \frac{x}{3} \cot \frac{x}{3}\, dx$

11. $\int \tan^3 x \sec^2 x\, dx$

12. $\int \sqrt{\cot x} \csc^2 x\, dx$

13. $\int \frac{\sec^2 x}{\tan x}\, dx$

14. $\int \frac{\csc^2 x}{\cot x}\, dx$

15. $\int \frac{\sec x \tan x}{\sec x - 1}\, dx$

16. $\int \frac{\cos t}{1 + \sin t}\, dt$

17. $\int \frac{\sin x}{1 + \cos x}\, dx$

18. $\int \frac{1 - \cos \theta}{\theta - \sin \theta}\, d\theta$

19. $\int \cot \pi x\, dx$

20. $\int \tan 5x\, dx$

21. $\int \csc 2x\, dx$

22. $\int \sec \frac{x}{2}\, dx$

23. $\int \sec^6 \frac{x}{4} \tan \frac{x}{4}\, dx$

24. $\int \csc^3 5x \cot 5x\, dx$

25. $\int \frac{\csc^2 x}{\cot^3 x}\, dx$

26. $\int \frac{\sin x}{\cos^2 x}\, dx$

27. $\int e^x \sin e^x\, dx$

28. $\int e^{-x} \tan e^{-x}\, dx$

29. $\int e^{\sin \pi x} \cos \pi x\, dx$

30. $\int e^{\sec 4x} \sec 4x \tan 4x\, dx$

31. $\int (\sin x + \cos x)^2\, dx$

32. $\int (1 + \tan \theta)^2\, d\theta$

Integration by Parts In Exercises 33–38, use integration by parts to find the indefinite integral.

33. $\int x \cos 2x \, dx$

34. $\int x \sin 5x \, dx$

35. $\int 6x \sec^2 x \, dx$

36. $\int \frac{\theta}{3} \sec \theta \tan \theta \, d\theta$

37. $\int t \csc 3t \cot 3t \, dt$

38. $\int 4x \csc^2 x \, dx$

Evaluating Definite Integrals In Exercises 39–46, evaluate the definite integral. *See Example 7.*

39. $\int_0^{\pi/4} \cos \frac{4x}{3} \, dx$

40. $\int_0^{\pi/6} \sin 6x \, dx$

41. $\int_{\pi/2}^{2\pi/3} \sec^2 \frac{x}{2} \, dx$

42. $\int_0^{\pi/2} (x + \cos x) \, dx$

43. $\int_{\pi/12}^{\pi/4} \csc 2x \cot 2x \, dx$

44. $\int_0^{\pi/8} \sin 2x \cos 2x \, dx$

45. $\int_0^1 \tan(1 - x) \, dx$

46. $\int_0^{\pi/4} \sec x \tan x \, dx$

Finding Area by Integration In Exercises 47–52, find the area of the region. *See Example 8.*

47. $y = \cos \frac{x}{4}$

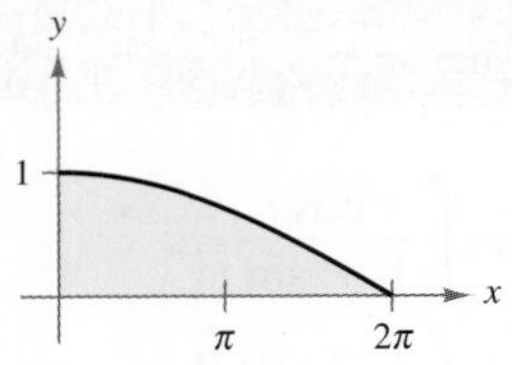

48. $y = \tan x$

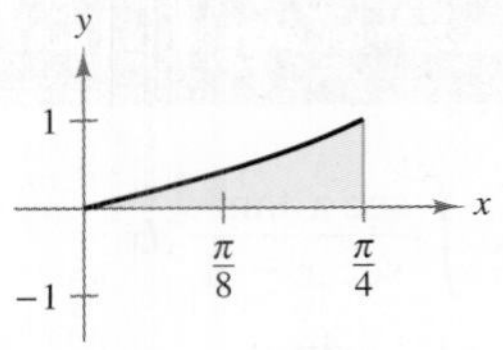

49. $y = x + \sin x$

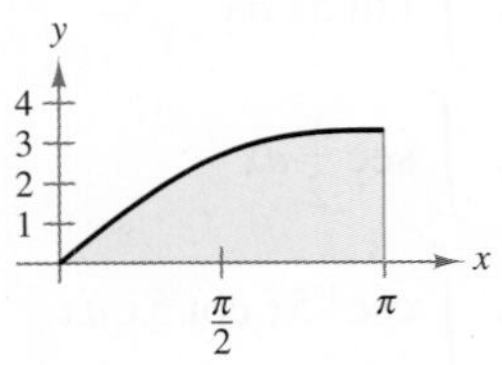

50. $y = \frac{x}{2} + \cos x$

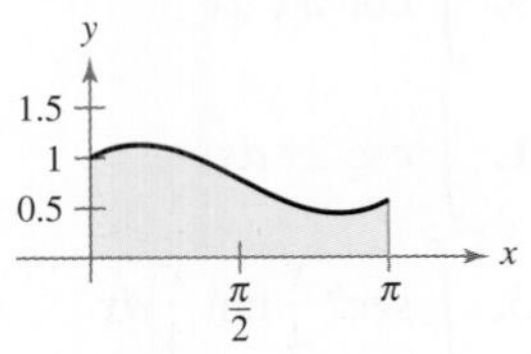

51. $y = \sin x + \cos 2x$

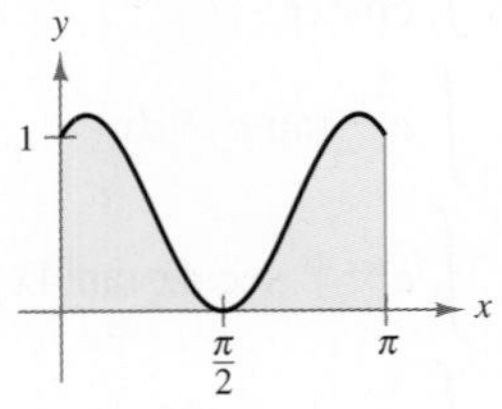

52. $y = 2 \sin x + \sin 2x$

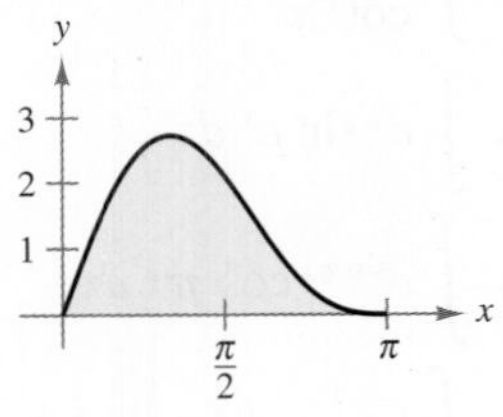

Finding the Area Bounded by Two Graphs In Exercises 53–56, sketch the region bounded by the graphs of the functions and find the area of the region.

53. $y = \cos x, y = 2 - \cos x, x = 0, x = 2\pi$

54. $y = \sin x, y = \cos 2x, x = -\frac{\pi}{2}, x = \frac{\pi}{6}$

55. $y = 2 \sin x, y = \tan x, x = 0, x = \frac{\pi}{3}$

56. $y = \sec^2 \frac{x}{4}, y = 4 - x^2, x = -1, x = 1$

57. Meteorology The average monthly precipitation P (in inches), including rain, snow, and ice, for Sacramento, California can be modeled by

$$P = 2.47 \sin(0.40t + 1.80) + 2.08, \quad 0 \le t \le 12$$

where t is the time (in months), with $t = 1$ corresponding to January. Find the total annual precipitation for Sacramento. *(Source: National Climatic Data Center)*

58. Meteorology The average monthly precipitation P (in inches), including rain, snow, and ice, for Bismarck, North Dakota can be modeled by

$$P = 1.07 \sin(0.59t + 3.94) + 1.52, \quad 0 \le t \le 12$$

where t is the time (in months), with $t = 1$ corresponding to January. Find the total annual precipitation for Bismarck. *(Source: National Climatic Data Center)*

59. Consumer Trends Energy consumption in the United States is seasonal. The primary residential energy consumption Q (in trillion Btu) in the United States during 2009 can be modeled by

$$Q = 936 + 737.3 \cos(0.31t + 0.928), \quad 0 \le t \le 12$$

where t is the time (in months), with $t = 1$ corresponding to January. Find the average primary residential energy consumption during

(a) the first quarter ($0 \le t \le 3$).

(b) the fourth quarter ($9 \le t \le 12$).

(c) the entire year ($0 \le t \le 12$).

(Source: U. S. Energy Information Administration)

60. Construction Workers The number W (in thousands) of construction workers employed in the United States during 2010 can be modeled by

$$W = 5488 + 347.6 \sin(0.45t - 4.153)$$

where t is the time (in months), with $t = 1$ corresponding to January. Find the average number of construction workers during

(a) the first quarter ($0 \le t \le 3$).

(b) the second quarter ($3 \le t \le 6$).

(c) the entire year ($0 \le t \le 12$).

(Source: U.S. Bureau of Labor Statistics)

61. Cost The temperature T (in degrees Fahrenheit) in a house is given by

$$T = 72 + 12 \sin \frac{\pi(t - 8)}{12}$$

where t is the time (in hours), with $t = 0$ corresponding to midnight. The hourly cost of cooling a house is \$0.30 per degree.

(a) Find the cost C of cooling this house between 8 A.M. and 8 P.M., when the thermostat is set at 72°F (see figure) by evaluating the integral

$$C = 0.3 \int_8^{20} \left[72 + 12 \sin \frac{\pi(t - 8)}{12} - 72\right] dt.$$

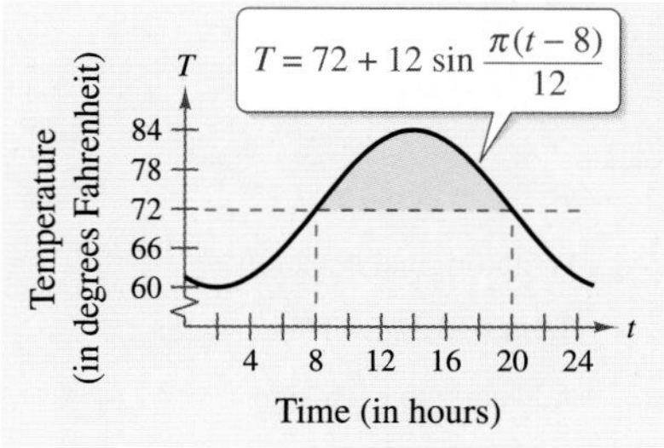

(b) Find the savings realized by resetting the thermostat to 78°F (see figure) by evaluating the integral

$$C = 0.3 \int_{10}^{18} \left[72 + 12 \sin \frac{\pi(t - 8)}{12} - 78\right] dt.$$

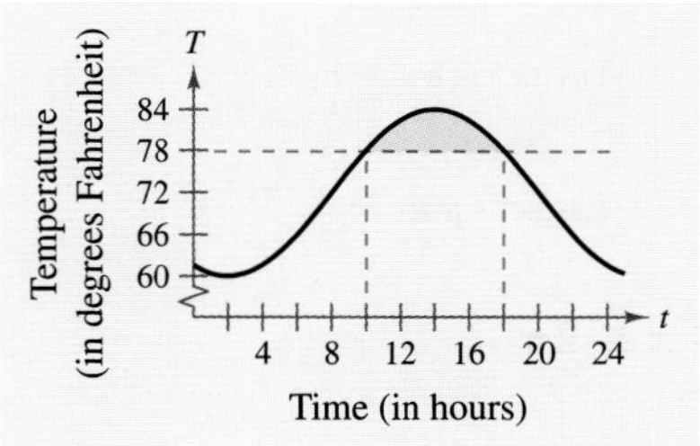

62. Water Supply The flow rate R (in thousands of gallons per hour) of water at a pumping station during a day can be modeled by

$$R = 53 + 7 \sin\left(\frac{\pi t}{6} + 3.6\right) + 9 \cos\left(\frac{\pi t}{12} + 8.9\right),$$

$$0 \le t \le 24$$

where t is the time in hours, with $t = 0$ corresponding to midnight.

(a) Find the average hourly flow rate from midnight to noon $(0 \le t \le 12)$.

(b) Find the average hourly flow rate from noon to midnight $(12 \le t \le 24)$.

(c) Find the total volume of water pumped in one day.

63. Sales In Example 9 in Section 8.4, the sales of a seasonal product can be modeled by

$$F = 100{,}000\left[1 + \sin \frac{2\pi(t - 60)}{365}\right], \quad t \ge 0$$

where F is the amount sold (in pounds) and t is the time (in days), with $t = 1$ corresponding to January 1. The manufacturer of this product wants to set up a manufacturing schedule to produce a uniform amount each day. What should this amount be? (Assume that there are 200 production days during the year.)

64. HOW DO YOU SEE IT? The graph shows the sales S (in thousands of units) of a seasonal product, where t is the time (in months), with $t = 1$ corresponding to January.

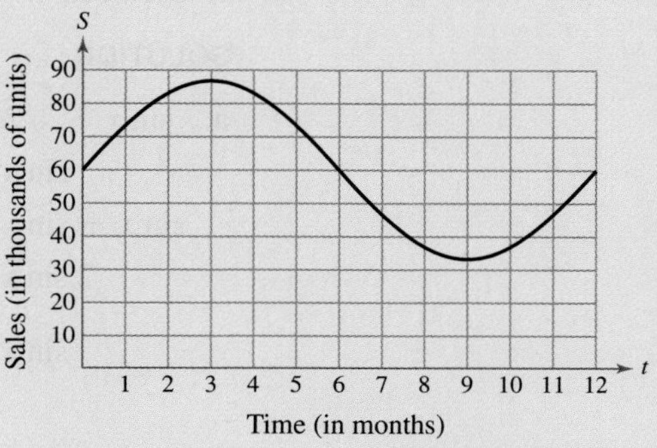

(a) Which is greater, the average monthly sales from January through March, or the average monthly sales from October through December? Explain your reasoning.

(b) Estimate the average monthly sales for the entire year. Explain your reasoning.

Using Simpson's Rule In Exercises 65 and 66, use the Simpson's Rule program in Appendix E to approximate the integral.

	Integral	n
65.	$\int_0^{\pi/2} \sqrt{x} \sin x \, dx$	8
66.	$\int_0^{\pi} \sqrt{1 + \cos^2 x} \, dx$	20

True or False? In Exercises 67 and 68, determine whether the statement is true or false. If it is false, explain why or give an example that shows it is false.

67. $\int_a^b \sin x \, dx = \int_a^{b+2\pi} \sin x \, dx$

68. $4 \int \sin x \cos x \, dx = 0$

ALGEBRA TUTOR

Solving Trigonometric Equations

Solving a trigonometric equation requires the use of trigonometry, but it also requires the use of algebra. Some examples of solving trigonometric equations were presented on pages 520 and 521. Here are several others.

Example 1 Solving Trigonometric Equations

Solve each trigonometric equation. Assume $0 \le x \le 2\pi$.

a. $\sin x + \sqrt{2} = -\sin x$

b. $3\tan^2 x = 1$

c. $\cot x \cos^2 x = 2\cot x$

SOLUTION

a. $\sin x + \sqrt{2} = -\sin x$ Write original equation.

$\sin x = -\sin x - \sqrt{2}$ Subtract $\sqrt{2}$ from each side.

$\sin x + \sin x = -\sqrt{2}$ Add $\sin x$ to each side.

$2\sin x = -\sqrt{2}$ Combine like terms.

$\sin x = -\dfrac{\sqrt{2}}{2}$ Divide each side by 2.

$$x = \frac{5\pi}{4}, \frac{7\pi}{4}, \quad 0 \le x \le 2\pi$$

b. $3\tan^2 x = 1$ Write original equation.

$\tan^2 x = \dfrac{1}{3}$ Divide each side by 3.

$\tan x = \pm\dfrac{\sqrt{3}}{3}$ Extract square roots.

$$x = \frac{\pi}{6}, \frac{5\pi}{6}, \frac{7\pi}{6}, \frac{11\pi}{6}, \quad 0 \le x \le 2\pi$$

c. $\cot x \cos^2 x = 2\cot x$ Write original equation.

$\cot x \cos^2 x - 2\cot x = 0$ Subtract $2\cot x$ from each side.

$\cot x(\cos^2 x - 2) = 0$ Factor.

Setting each factor equal to zero, you obtain the solutions in the interval $0 \le x \le 2\pi$ as shown.

$\cot x = 0$ and $\cos^2 x - 2 = 0$

$x = \dfrac{\pi}{2}, \dfrac{3\pi}{2}$ $\qquad \cos^2 x = 2$

$\cos x = \pm\sqrt{2}$

No solution is obtained from $\cos x = \pm\sqrt{2}$ because $\pm\sqrt{2}$ are outside the range of the cosine function. So, the equation has two solutions

$$x = \frac{\pi}{2} \quad \text{and} \quad x = \frac{3\pi}{2}$$

in the interval $0 \le x \le 2\pi$.

Example 2 **Solving Trigonometric Equations**

Solve each trigonometric equation in the interval $[0, 2\pi]$.

a. $2\sin^2 x - \sin x - 1 = 0$

b. $2\sin^2 x + 3\cos x - 3 = 0$

c. $\sin t - \cos 2t = 0$

SOLUTION

a.

$$2\sin^2 x - \sin x - 1 = 0 \quad \text{Write original equation.}$$

$$(2\sin x + 1)(\sin x - 1) = 0 \quad \text{Factor.}$$

Set each factor equal to zero. The solutions in the interval $[0, 2\pi]$ are

$2\sin x + 1 = 0$ and $\sin x - 1 = 0$

$\sin x = -\frac{1}{2}$ $\sin x = 1$

$x = \frac{7\pi}{6}, \frac{11\pi}{6}$ $x = \frac{\pi}{2}$.

b.

$$2\sin^2 x + 3\cos x - 3 = 0 \quad \text{Write original equation.}$$

$$2(1 - \cos^2 x) + 3\cos x - 3 = 0 \quad \text{Pythagorean Identity}$$

$$2 - 2\cos^2 x + 3\cos x - 3 = 0 \quad \text{Multiply.}$$

$$-2\cos^2 x + 3\cos x - 1 = 0 \quad \text{Combine like terms.}$$

$$2\cos^2 x - 3\cos x + 1 = 0 \quad \text{Multiply each side by } -1.$$

$$(2\cos x - 1)(\cos x - 1) = 0 \quad \text{Factor.}$$

Set each factor equal to zero. The solutions in the interval $[0, 2\pi]$ are

$2\cos x - 1 = 0$ and $\cos x - 1 = 0$

$2\cos x = 1$ $\cos x = 1$

$\cos x = \frac{1}{2}$ $x = 0, 2\pi$.

$x = \frac{\pi}{3}, \frac{5\pi}{3}$

c.

$$\sin t - \cos 2t = 0 \quad \text{Write original equation.}$$

$$\sin t - (1 - 2\sin^2 t) = 0 \quad \text{Double-Angle Identity}$$

$$\sin t - 1 + 2\sin^2 t = 0 \quad \text{Remove parentheses.}$$

$$2\sin^2 t + \sin t - 1 = 0 \quad \text{Rewrite.}$$

$$(2\sin t - 1)(\sin t + 1) = 0 \quad \text{Factor.}$$

Set each factor equal to zero. The solutions in the interval $[0, 2\pi]$ are

$2\sin t - 1 = 0$ and $\sin t + 1 = 0$

$2\sin t = 1$ $\sin t = -1$

$\sin t = \frac{1}{2}$ $t = \frac{3\pi}{2}$.

$t = \frac{\pi}{6}, \frac{5\pi}{6}$

SUMMARY AND STUDY STRATEGIES

After studying this chapter, you should have acquired the following skills.
The exercise numbers are keyed to the Review Exercises that begin on page 558.
Answers to odd-numbered Review Exercises are given in the back of the text.*

Section 8.1	Review Exercises
■ Find coterminal angles.	*1–4*
■ Convert from degree to radian measure and from radian to degree measure. π radians $= 180°$	*5–16*
■ Use formulas relating to triangles.	*17–20*
■ Use formulas relating to triangles to solve real-life problems.	*21, 22*

Section 8.2

■ Evaluate trigonometric functions. *23–32*

Right Triangle Definition: $0 < \theta < \dfrac{\pi}{2}$

$$\sin\theta = \frac{\text{opp}}{\text{hyp}} \qquad \csc\theta = \frac{\text{hyp}}{\text{opp}}$$

$$\cos\theta = \frac{\text{adj}}{\text{hyp}} \qquad \sec\theta = \frac{\text{hyp}}{\text{adj}}$$

$$\tan\theta = \frac{\text{opp}}{\text{adj}} \qquad \cot\theta = \frac{\text{adj}}{\text{opp}}$$

Circular Function Definition: Let θ be an angle in standard position with (x, y) a point on the terminal ray of θ and $r = \sqrt{x^2 + y^2} \neq 0$.

$$\sin\theta = \frac{y}{r} \qquad \csc\theta = \frac{r}{y}$$

$$\cos\theta = \frac{x}{r} \qquad \sec\theta = \frac{r}{x}$$

$$\tan\theta = \frac{y}{x} \qquad \cot\theta = \frac{x}{y}$$

■ Use a calculator to approximate values of trigonometric functions.	*33–40*
■ Solve right triangles.	*41–44*
■ Solve trigonometric equations.	*45–50*
■ Use right triangles to solve real-life problems.	*51, 52*

Section 8.3

■ Find the period and amplitude of trigonometric functions.	*53–56*
■ Sketch graphs of trigonometric functions.	*57–64*
■ Use trigonometric functions to model real-life situations.	*65, 66*

---

* A wide range of valuable study aids are available to help you master the material in this chapter. The *Student Solutions Manual* includes step-by-step solutions to all odd-numbered exercises to help you review and prepare. The student website at *www.cengagebrain.com* offers algebra help and a *Graphing Technology Guide*, which contains step-by-step commands and instructions for a wide variety of graphing calculators.

Section 8.4

Review Exercises

- Find derivatives of trigonometric functions. *67–84*

$$\frac{d}{dx}[\sin u] = \cos u \frac{du}{dx} \qquad \frac{d}{dx}[\cos u] = -\sin u \frac{du}{dx}$$

$$\frac{d}{dx}[\tan u] = \sec^2 u \frac{du}{dx} \qquad \frac{d}{dx}[\cot u] = -\csc^2 u \frac{du}{dx}$$

$$\frac{d}{dx}[\sec u] = \sec u \tan u \frac{du}{dx} \qquad \frac{d}{dx}[\csc u] = -\csc u \cot u \frac{du}{dx}$$

- Find the equations of tangent lines to graphs of trigonometric functions. *85–90*
- Find the relative extrema of trigonometric functions. *91–96*
- Use derivatives of trigonometric functions to answer questions about real-life situations. *97, 98*

Section 8.5

- Find indefinite integrals of trigonometric functions. *99–110*

$$\int \cos u\, du = \sin u + C \qquad \int \sin u\, du = -\cos u + C$$

$$\int \sec^2 u\, du = \tan u + C \qquad \int \sec u \tan u\, du = \sec u + C$$

$$\int \csc^2 u\, du = -\cot u + C \qquad \int \csc u \cot u\, du = -\csc u + C$$

$$\int \tan u\, du = -\ln|\cos u| + C \qquad \int \sec u\, du = \ln|\sec u + \tan u| + C$$

$$\int \cot u\, du = \ln|\sin u| + C \qquad \int \csc u\, du = \ln|\csc u - \cot u| + C$$

- Evaluate definite integrals of trigonometric functions. *111–118*
- Find the areas of regions in the plane. *119–122*
- Use trigonometric integrals to solve real-life problems. *123–126*

Study Strategies

- **Degree and Radian Modes** When using a computer or calculator to evaluate or graph a trigonometric function, be sure that you use the proper mode—*radian* mode or *degree* mode.
- **Checking the Form of an Answer** Because of the abundance of trigonometric identities, solutions of problems in this chapter can take a variety of forms. For instance, the expressions $-\ln|\cot x| + C$ and $\ln|\tan x| + C$ are equivalent. So, when you are checking your solutions with those given in the back of the text, remember that your solution might be correct, even if its form doesn't agree precisely with that given in the text.
- **Using Technology** Throughout this chapter, remember that technology can help you *graph* trigonometric functions, *evaluate* trigonometric functions, *differentiate* trigonometric functions, and *integrate* trigonometric functions. Consider, for instance, the difficulty of sketching the graph of the function below *without* using a graphing utility.

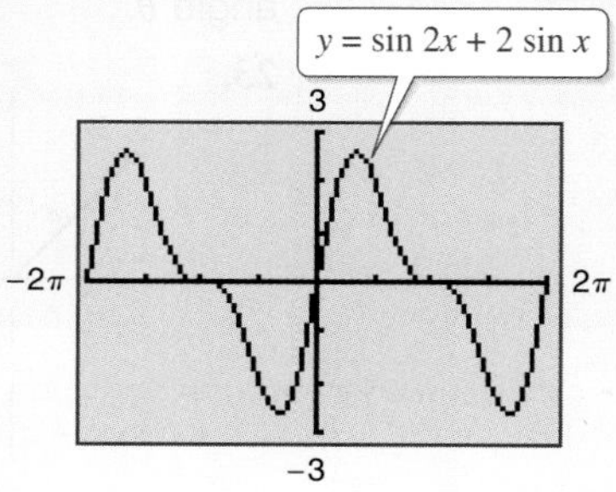

Review Exercises

See www.CalcChat.com for worked-out solutions to odd-numbered exercises.

Finding Coterminal Angles In Exercises 1–4, determine two coterminal angles (one positive and one negative) for each angle. Give the answers in degrees.

1.

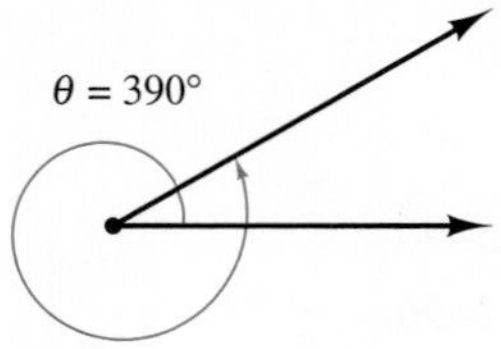

2.

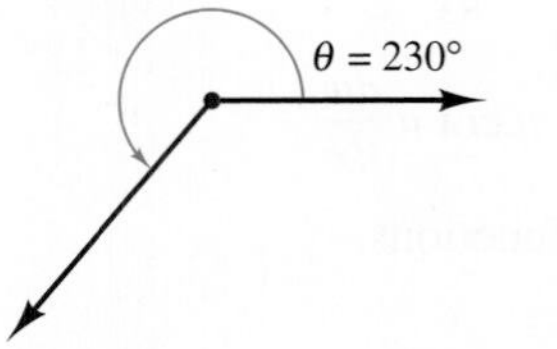

3. $\theta = -405°$

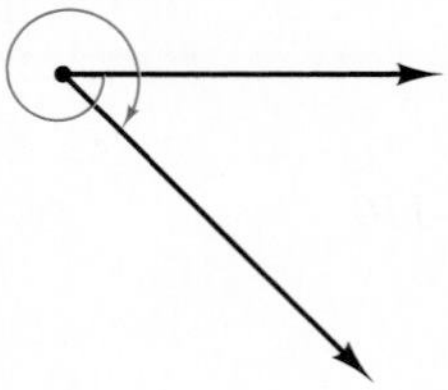

4.

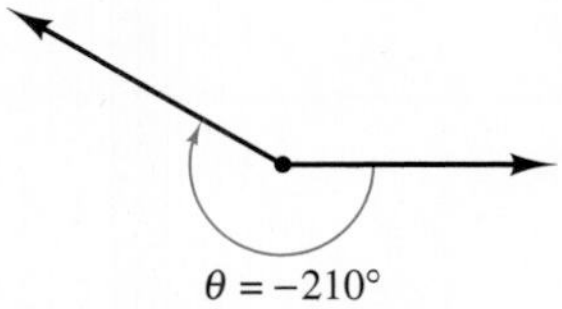

Converting from Degrees to Radians In Exercises 5–12, express the angle in radian measure as a multiple of π. Use a calculator to verify your result.

5. 340° **6.** 300°

7. $-60°$ **8.** $-30°$

9. $-480°$ **10.** $-540°$

11. 110° **12.** 320°

Converting from Radians to Degrees In Exercises 13–16, express the angle in degree measure. Use a calculator to verify your result.

13. $\dfrac{4\pi}{3}$

14. $\dfrac{5\pi}{6}$

15. $-\dfrac{2\pi}{3}$

16. $-\dfrac{11\pi}{6}$

Using Triangles In Exercises 17–20, solve the triangle for the indicated side and/or angle.

17.

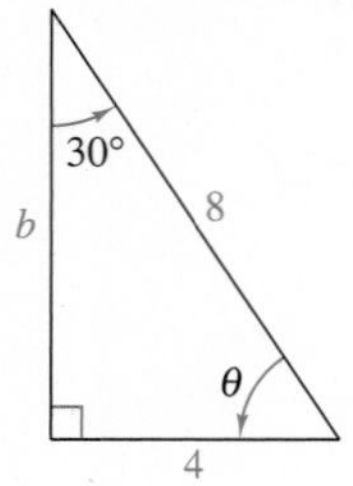

18.

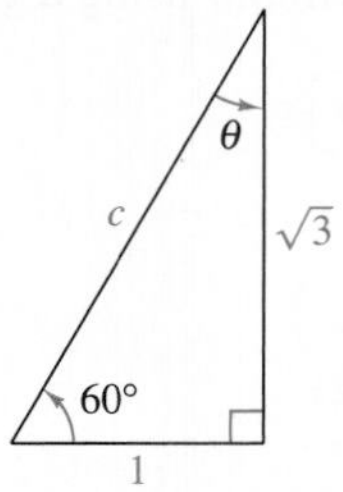

19.

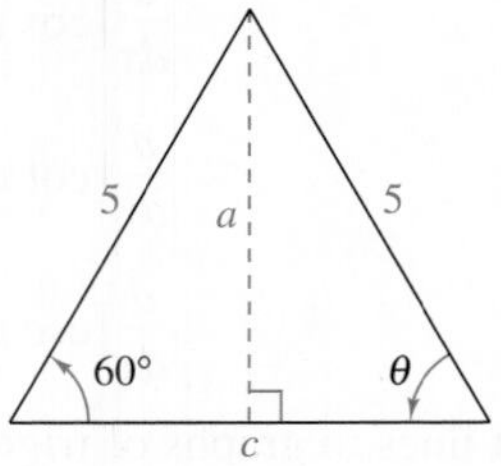

20.

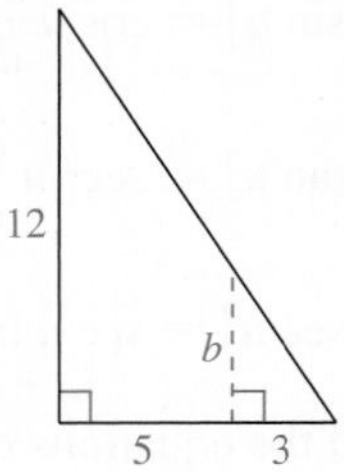

21. Height A ladder of length 16 feet leans against the side of a house. The bottom of the ladder is 4.4 feet from the house (see figure). Find the height h of the top of the ladder.

22. Length To stabilize a 75-foot tower for a radio antenna, a guy wire must be attached from the top of the tower to an anchor 50 feet from the base (see figure). How long is the wire?

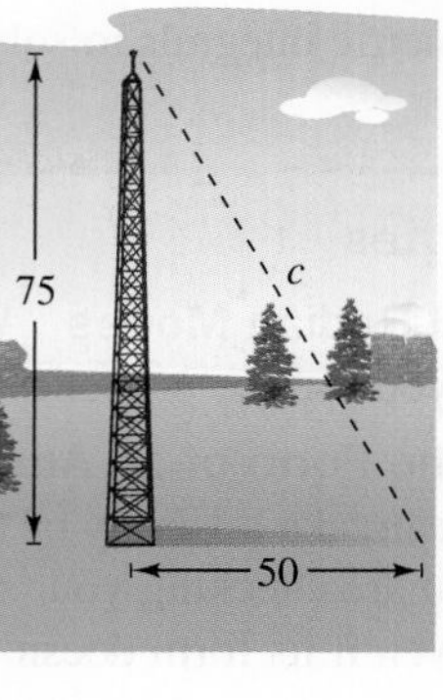

Evaluating Trigonometric Functions In Exercises 23 and 24, determine all six trigonometric functions of the angle θ.

23.

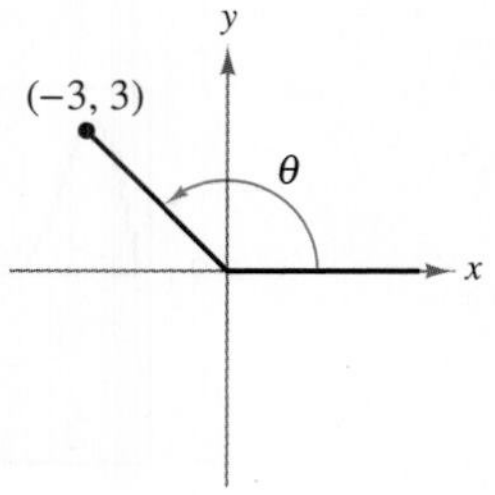

24.

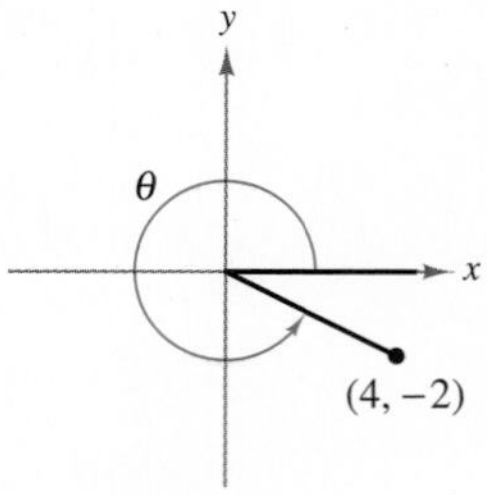

Evaluating Trigonometric Functions In Exercises 25–32, evaluate the six trigonometric functions of the angle without using a calculator.

25. $-45°$
26. $240°$
27. $\dfrac{5\pi}{3}$
28. $-\dfrac{4\pi}{3}$
29. $-225°$
30. $180°$
31. $-\dfrac{11\pi}{6}$
32. $\dfrac{5\pi}{2}$

Evaluating Trigonometric Functions In Exercises 33–40, use a calculator to evaluate the trigonometric function to four decimal places.

33. $\tan 33°$
34. $\cot 216°$
35. $\sec \dfrac{12\pi}{5}$
36. $\csc \dfrac{2\pi}{9}$
37. $\sin\left(-\dfrac{\pi}{9}\right)$
38. $\cos\left(-\dfrac{3\pi}{7}\right)$
39. $\cos 105°$
40. $\sin 224°$

Solving a Right Triangle In Exercises 41–44, solve for x, y, or r as indicated.

41. Solve for r.

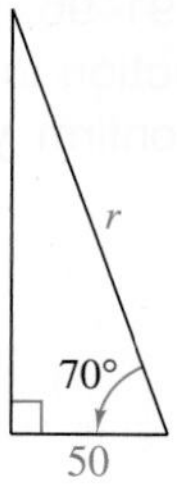

42. Solve for y.

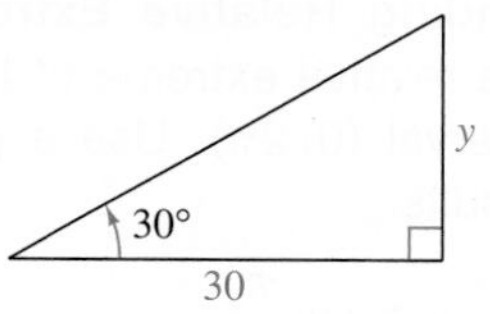

43. Solve for x.

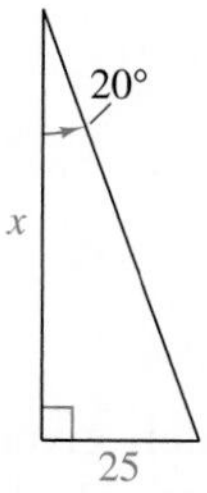

44. Solve for r.

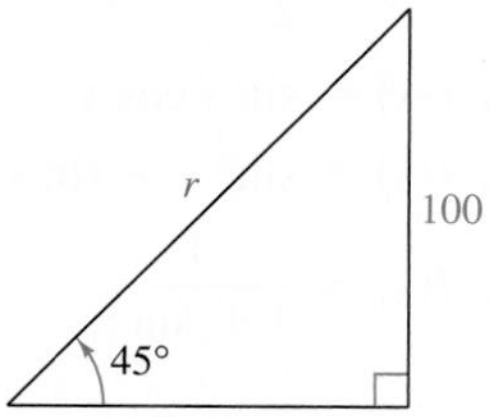

Solving Trigonometric Equations In Exercises 45–50, solve the equation for θ. Assume $0 \le \theta \le 2\pi$. For some of the equations, you should use the trigonometric identities listed in Section 8.2. Use the *trace* feature of a graphing utility to verify your results.

45. $2\cos\theta + 1 = 0$
46. $2\cos^2\theta = 1$
47. $2\sin^2\theta + 3\sin\theta + 1 = 0$
48. $\cos^3\theta = \cos\theta$
49. $\sec^2\theta - \sec\theta - 2 = 0$
50. $2\sec^2\theta + \tan^2\theta - 3 = 0$

51. **Height** The length of the shadow of a tree is 125 feet when the angle of elevation of the sun is 33° (see figure). Approximate the height h of the tree.

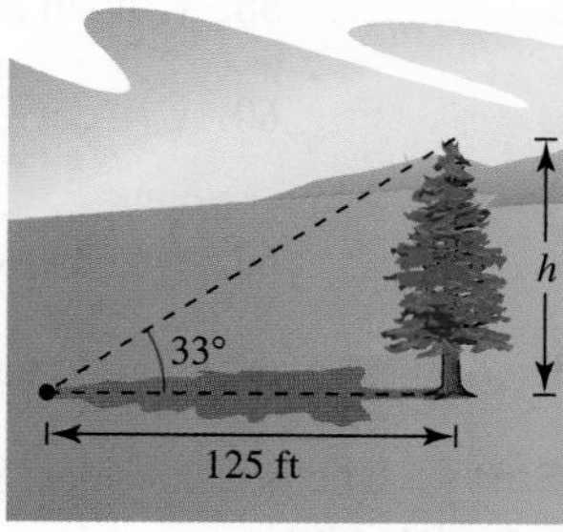

52. **Length** A guy wire runs from the ground to a cell tower. The wire is attached to the cell tower 150 feet above the ground. The angle formed between the wire and the ground is 43° (see figure).

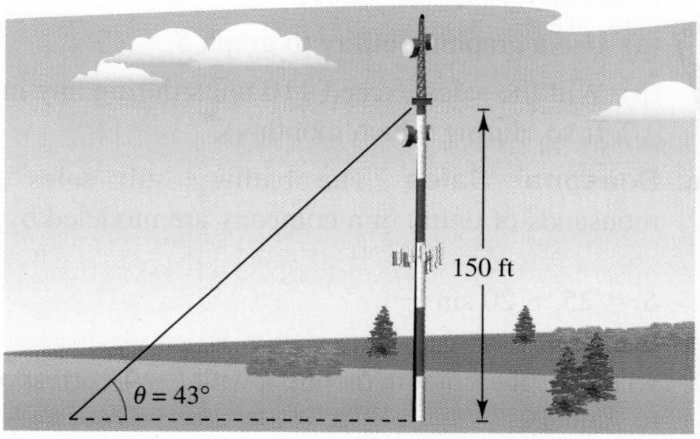

(a) How long is the guy wire?

(b) How far from the base of the tower is the guy wire anchored to the ground?

Finding the Period and Amplitude In Exercises 53–56, find the period and amplitude of the trigonometric function.

53. $y = -2\sin 4x$

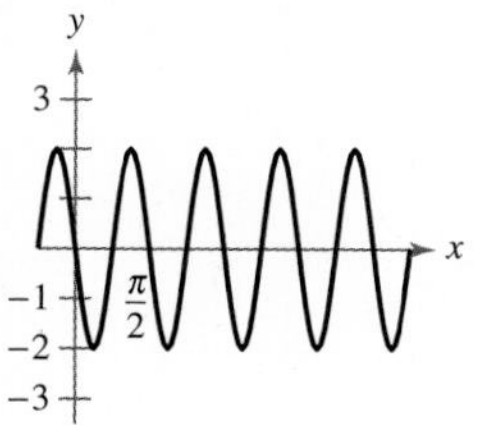

54. $y = \cos 2x$

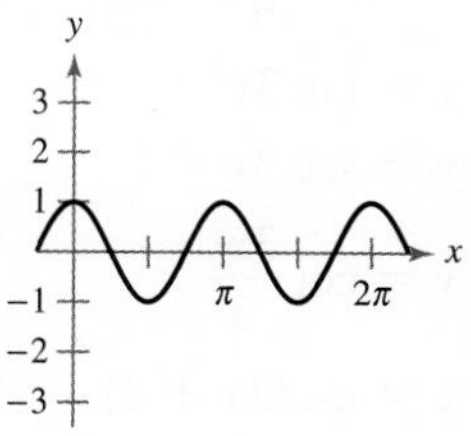

55. $y = -2\cos\dfrac{3x}{2}$

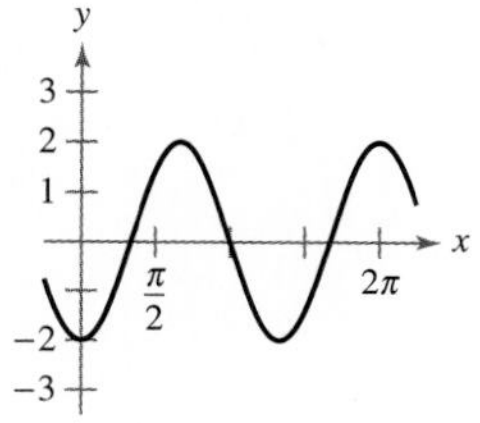

56. $y = \sin\dfrac{x}{2}$

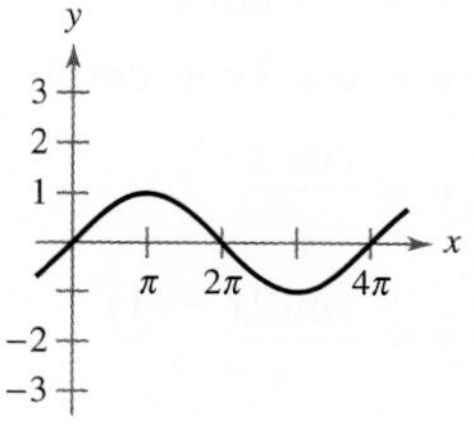

Graphing Trigonometric Functions In Exercises 57–64, sketch the graph of the trigonometric function by hand.

57. $y = 2\cos 6x$

58. $y = \sin 2\pi x$

59. $y = \frac{1}{3}\tan x$

60. $y = \cot \frac{x}{2}$

61. $y = 3\sin \frac{2x}{5}$

62. $y = 8\cos \frac{x}{4}$

63. $y = \sec 2\pi x$

64. $y = 3\csc 2x$

65. Seasonal Sales The jet ski sales S (in units) of a company are modeled by

$$S = 74 - 40\cos \frac{\pi t}{6}$$

where t is the time (in months), with $t = 1$ corresponding to January.

(a) Use a graphing utility to graph S.

(b) Will the sales exceed 110 units during any month? If so, during which month(s)?

66. Seasonal Sales The bathing suit sales S (in thousands of units) of a company are modeled by

$$S = 25 + 20\sin \frac{\pi t}{6}$$

where t is the time (in months), with $t = 1$ corresponding to January.

(a) Use a graphing utility to graph S.

(b) Will the sales exceed 42,000 units during any month? If so, during which month(s)?

Differentiating Trigonometric Functions In Exercises 67–84, find the derivative of the trigonometric function.

67. $y = \sin 5\pi x$

68. $y = \cos \frac{x}{4}$

69. $y = \tan 3x^3$

70. $y = \sec 3x$

71. $y = \cot \frac{3x^2}{5}$

72. $y = \csc(3x + 4)$

73. $y = \sqrt{\cos 2\pi x}$

74. $y = \sqrt[3]{\sec 5x}$

75. $y = -x\tan x$

76. $y = \csc 3x + \cot 3x$

77. $y = \frac{\cos x}{x^2}$

78. $y = \frac{\sin(2x - 1)}{x + 3}$

79. $y = \sin^2 x + x$

80. $y = x\cos x - \sin x$

81. $y = \csc^4 x$

82. $y = \sec^2 2x$

83. $y = e^x \cot x$

84. $y = e^{\sin x}$

Finding an Equation of a Tangent Line In Exercises 85–90, find an equation of the tangent line to the graph of the function at the given point.

	Function	*Point*
85.	$y = \cos 2x$	$\left(\frac{\pi}{4}, 0\right)$
86.	$y = \tan 2x$	$\left(\frac{5\pi}{8}, 1\right)$
87.	$y = \csc x$	$\left(\frac{3\pi}{2}, -1\right)$
88.	$y = \sin 2x$	$\left(\frac{\pi}{2}, 0\right)$
89.	$y = \frac{1}{2}\sin^2 x$	$\left(\frac{\pi}{2}, \frac{1}{2}\right)$
90.	$y = -x\cos x$	(π, π)

Finding Relative Extrema In Exercises 91–96, find the relative extrema of the trigonometric function in the interval $(0, 2\pi)$. Use a graphing utility to confirm your results.

91. $y = \sin \frac{\pi x}{4}$

92. $y = \cos \frac{3x}{2}$

93. $f(x) = \frac{x}{2} + \cos x$

94. $f(x) = \sin x \cos x$

95. $f(x) = \sin^2 x + \sin x$

96. $f(x) = \frac{1}{2 + \sin x}$

97. Seasonal Sales Refer to the model given in Exercise 65. Approximate the month t in which the sales of jet skis were a maximum. What was the maximum number of jet skis sold?

98. Seasonal Sales Refer to the model given in Exercise 66.

(a) Approximate the month t in which the sales of bathing suits were a maximum. What was the maximum number of bathing suits sold?

(b) Approximate the month t in which the sales of bathings suits were a minimum. What was the minimum number of bathing suits sold?

Integrating Trigonometric Functions In Exercises 99–110, find the indefinite integral.

99. $\int \sin 3x \, dx$

100. $\int \cos \frac{x}{4} \, dx$

101. $\int \csc 5x \cot 5x \, dx$

102. $\int 2x \sec^2 x^2 \, dx$

103. $\int x \csc^2 x^2 \, dx$

104. $\int \sec 8x \tan 8x \, dx$

105. $\int \frac{\sec^2 2x}{\tan 2x} \, dx$

106. $\int \frac{\cos x}{\sin^3 x} \, dx$

107. $\int \tan 3x \, dx$

108. $\int \csc \frac{3x}{4} \, dx$

109. $\int \sin^3 x \cos x \, dx$

110. $\int e^{\cos 3x} \sin 3x \, dx$

Evaluating Definite Integrals In Exercises 111–118, evaluate the definite integral.

111. $\int_0^{\pi} (1 + \sin x) \, dx$

112. $\int_{-\pi}^{\pi} \frac{1}{2}(1 + \cos 2x) \, dx$

113. $\int_{-\pi/6}^{\pi/6} \sec^2 x \, dx$

114. $\int_{\pi/6}^{\pi/2} \csc^2 x \, dx$

115. $\int_{-\pi/3}^{\pi/3} 4 \sec x \tan x \, dx$

116. $\int_{\pi/6}^{\pi/3} \csc x \cot x \, dx$

117. $\int_{-\pi/2}^{\pi/2} (2x + \cos x) \, dx$

118. $\int_0^{\pi} 2x \sin x^2 \, dx$

Finding Area by Integration In Exercises 119–122, find the area of the region.

119. $y = \sin 3x$

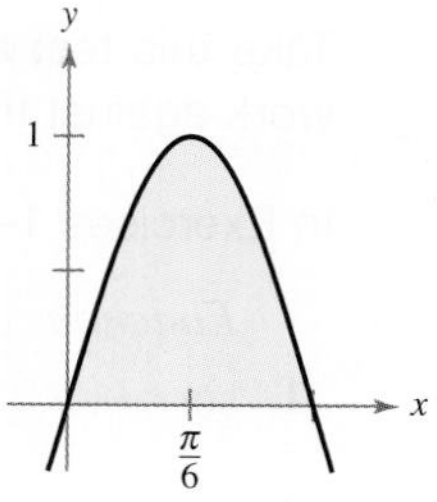

120. $y = \cot x$

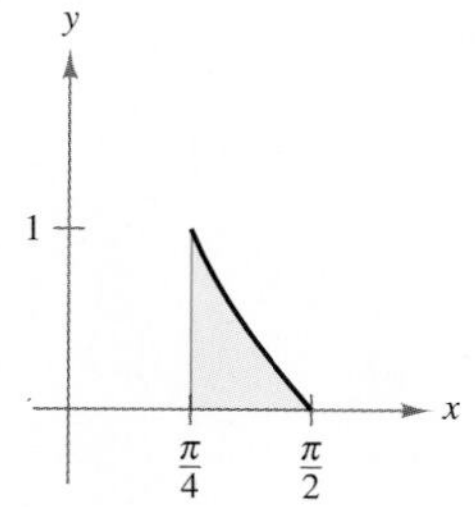

121. $y = 2 \sin x + \cos 3x$

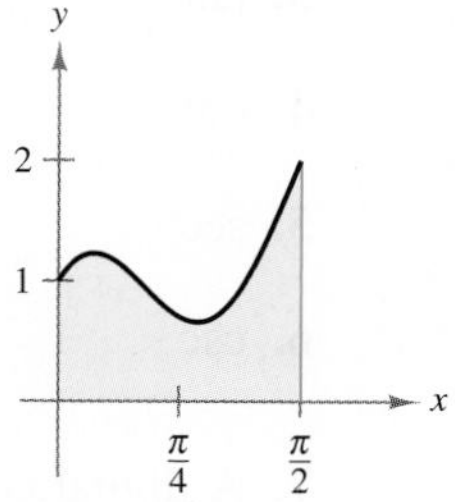

122. $y = 2 \cos x + \cos 2x$

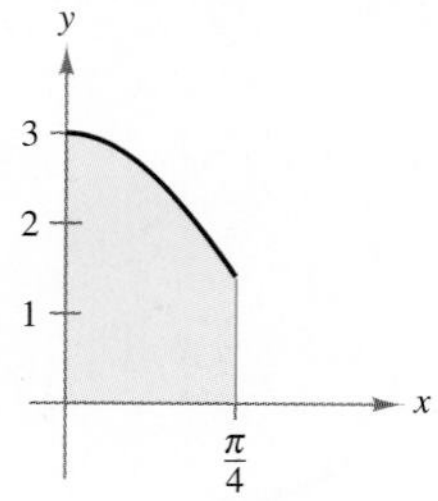

123. Meteorology The average monthly precipitation P (in inches), including rain, snow, and ice, for Dodge City, Kansas can be modeled by

$$P = 1.27 \sin(0.58t - 2.05) + 1.98, \quad 0 \le t \le 12$$

where t is the time (in months), with $t = 1$ corresponding to January. Find the total annual precipitation for Dodge City. *(Source: National Climatic Data Center)*

124. Health For a person at rest, the velocity v (in liters per second) of air flow into and out of the lungs during a respiratory cycle is given by

$$v = 0.9 \sin \frac{\pi t}{3}$$

where t is the time (in seconds). Find the volume in liters of air inhaled during one cycle by integrating this function over the interval $[0, 3]$.

125. Sales The sales S (in billions of dollars) for Lowe's for the years 2000 through 2009 can be modeled by

$$S = 15.31 \sin(0.37t - 1.27) + 33.66, \quad 0 \le t \le 9$$

where t is the year, with $t = 0$ corresponding to 2000. Find the average sales for Lowe's from 2000 through 2009. *(Source: Lowe's Companies, Inc.)*

126. Electricity The oscillating current in an electrical circuit can be modeled by

$$I = 2 \sin(60\pi t) + \cos(120\pi t)$$

where I is measured in amperes and t is measured in seconds. Find the average current for the time intervals (a) $0 \le t \le \frac{1}{240}$, (b) $0 \le t \le \frac{1}{60}$, and (c) $0 \le t \le \frac{1}{30}$.

TEST YOURSELF

See www.CalcChat.com for worked-out solutions to odd-numbered exercises.

Take this test as you would take a test in class. When you are done, check your work against the answers given in the back of the book.

In Exercises 1–6, copy and complete the table. Use a calculator if necessary.

	Function	θ *(deg)*	θ *(rad)*	*Function Value*
1.	sin	67.5°	▭	▭
2.	cos	▭	$\frac{\pi}{5}$	▭
3.	tan	15°	▭	▭
4.	cot	▭	$-\frac{\pi}{6}$	▭
5.	sec	−40°	▭	▭
6.	csc	▭	$-\frac{5\pi}{4}$	▭

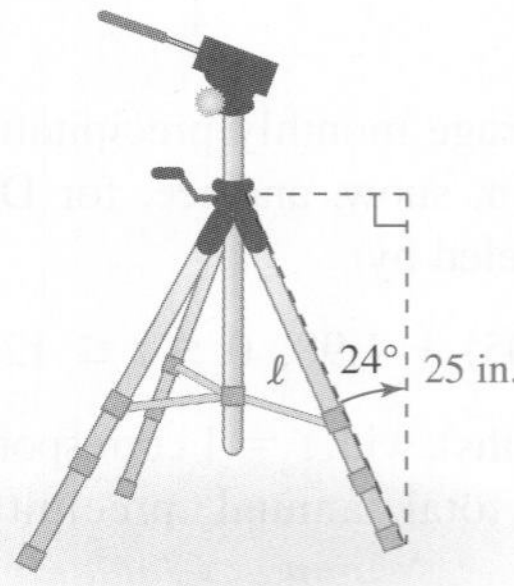

Figure for 7

7. A digital camera tripod has a height of 25 inches, and an angle of 24° is formed between the vertical and the leg of length ℓ (see figure). What is ℓ?

In Exercises 8–10, solve the equation for θ. Assume $0 \le \theta \le 2\pi$.

8. $2 \sin \theta - \sqrt{2} = 0$ **9.** $\cos^2 \theta - \sin^2 \theta = 0$ **10.** $\csc \theta = \sqrt{3} \sec \theta$

In Exercises 11–13, sketch the graph of the trigonometric function by hand.

11. $y = 3 \sin 2x$ **12.** $y = 4 \cos 3\pi x$ **13.** $y = \cot \frac{\pi x}{5}$

In Exercises 14–16, (a) find the derivative of the trigonometric function and (b) find the relative extrema of the trigonometric function in the interval $(0, 2\pi)$.

14. $y = \cos x - \cos^2 x$ **15.** $y = \sec\left(x - \frac{\pi}{4}\right)$ **16.** $y = \frac{1}{3 - \sin(x + \pi)}$

In Exercises 17–19, find the indefinite integral.

17. $\int \sin 5x \, dx$ **18.** $\int \sec^2 \frac{x}{4} \, dx$ **19.** $\int x \csc x^2 \, dx$

In Exercises 20–23, evaluate the definite integral.

20. $\int_{1/4}^{1/2} \cos \pi x \, dx$ **21.** $\int_0^{\pi} \sec^2 \frac{x}{3} \tan \frac{x}{3} \, dx$

22. $\int_{3\pi/4}^{5\pi/6} \csc 2x \cot 2x \, dx$ **23.** $\int_{\pi/4}^{3\pi/8} \sin 4x \cos 4x \, dx$

24. The monthly sales S (in thousands of dollars) of a company that produces insect repellent can be modeled by

$$S = 20.3 - 17.2 \cos \frac{\pi t}{6}$$

where t is the time (in months), with $t = 1$ corresponding to January.

(a) Find the total sales during the year $(0 \le t \le 12)$.

(b) Find the average monthly sales from April through October $(3 \le t \le 10)$.

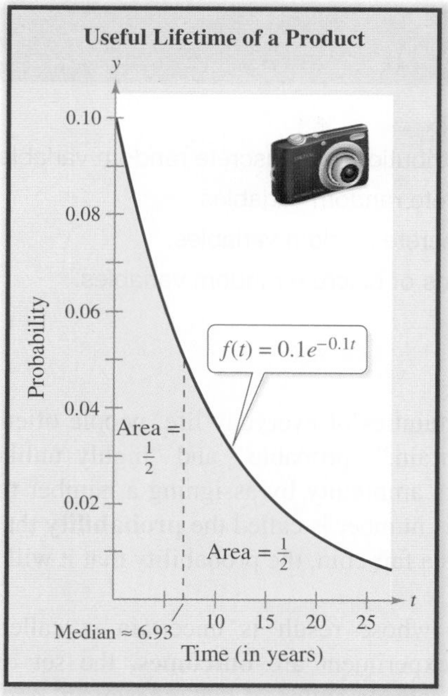

Example 4 on page 583 shows how calculus can be used to find the mean and median useful lifetimes of a product.

9 Probability and Calculus

9.1 Discrete Probability

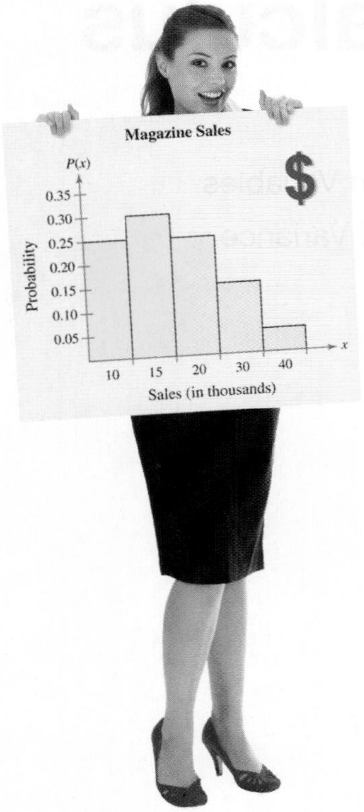

In Exercise 29 on page 572, you will find the expected value and standard deviation of the sales of a weekly magazine.

- Describe sample spaces for experiments.
- Assign values to, and form frequency distributions for, discrete random variables.
- Find the probability distributions for discrete random variables.
- Find the expected values or means of discrete random variables.
- Find the variances and standard deviations of discrete random variables.

Sample Spaces

When assigning measurements to the uncertainties of everyday life, people often use ambiguous terminology such as "fairly certain," "probable," and "highly unlikely." Probability theory allows you to remove this ambiguity by assigning a number to the likelihood of the occurrence of an event. This number is called the **probability** that the event will occur. For example, when you toss a fair coin, the probability that it will land heads up is one-half or 0.5.

In probability theory, any happening whose result is uncertain is called an **experiment.** The possible results of the experiment are **outcomes,** the set of all possible outcomes of the experiment is the **sample space** of the experiment, and any subcollection of a sample space is an **event.**

For instance, consider an experiment in which a coin is tossed. The sample space of this experiment consists of two outcomes: either the coin will land heads up (denoted by H) or it will land tails up (denoted by T). So, the sample space S is

$$S = \{H, T\}. \qquad \text{Sample space}$$

In this text, all outcomes of a sample space are assumed to be equally likely. For instance, when a coin is tossed, H and T are assumed to be equally likely.

Example 1 Finding a Sample Space

An experiment consists of tossing a six-sided die.

a. What is the sample space?

b. Describe the event corresponding to a number greater than 2 turning up.

SOLUTION

a. The sample space S consists of six outcomes, which can be represented by the numbers 1 through 6. That is,

$$S = \{1, 2, 3, 4, 5, 6\}. \qquad \text{Sample space}$$

Note that each of the outcomes in the sample space is equally likely.

b. The event E corresponding to a number greater than 2 turning up is a subset of S. That is,

$$E = \{3, 4, 5, 6\}. \qquad \text{Event}$$

✓Checkpoint 1

An experiment consists of tossing two six-sided dice.

a. What is the sample space?

b. Describe the event corresponding to a sum greater than or equal to seven points when the dice are tossed. ■

Yuri Arcurs/Shutterstock.com

ALGEBRA TUTOR

For examples of how to count the number of ways an event can happen, see the *Chapter 9 Algebra Tutor* on pages 590 and 591.

Discrete Random Variables

A function that assigns a numerical value to each of the outcomes in a sample space is called a **random variable.** For instance, when two coins are tossed, the sample space is $S = \{HH, HT, TH, TT\}$. These possible outcomes can be assigned the numbers 2, 1, and 0, depending on the number of heads in the outcome.

Definition of Discrete Random Variable

Let S be a sample space. A **random variable** is a function x that assigns a numerical value to each outcome in S. When the set of values taken on by the random variable is finite, the random variable is **discrete.** The number of times a specific value of x occurs is the **frequency** of x and is denoted by $n(x)$.

Example 2 Finding Frequencies

Three coins are tossed. A random variable assigns the number 0, 1, 2, or 3 to each possible outcome, depending on the number of heads in the outcome.

$$S = \{HHH, HHT, HTH, HTT, THH, THT, TTH, TTT\}$$
$$\downarrow \quad \downarrow \quad \downarrow \quad \downarrow \quad \downarrow \quad \downarrow \quad \downarrow \quad \downarrow$$
$$3 \quad 2 \quad 2 \quad 1 \quad 2 \quad 1 \quad 1 \quad 0$$

Find the frequencies of 0, 1, 2, and 3. Then use a bar graph to represent the result.

SOLUTION To find the frequencies, simply count the number of occurrences of each value of the random variable, as shown in the table.

Random variable, x	0	1	2	3
Frequency of x, $n(x)$	1	3	3	1

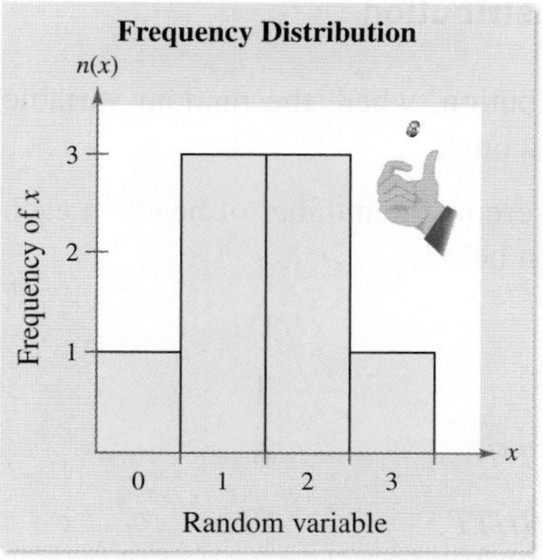

FIGURE 9.1

This table is called a **frequency distribution** of the random variable. The result is shown graphically by the bar graph in Figure 9.1.

✓ Checkpoint 2

Two coins are tossed. A random variable assigns the number 0, 1, or 2 to each possible outcome, depending on the number of heads in the outcome.

$$S = \{HH, HT, TH, TT\}$$
$$\downarrow \quad \downarrow \quad \downarrow \quad \downarrow$$
$$2 \quad 1 \quad 1 \quad 0$$

Find the frequencies of 0, 1, and 2. Then use a bar graph to represent the result. ■

STUDY TIP

In Example 2, note that the sample space consists of *eight* outcomes, each of which is *equally likely*. The sample space does *not* consist of the outcomes "zero heads," "one head," "two heads," and "three heads." You cannot consider these events to be outcomes because they are not equally likely.

Discrete Probability

The probability of a random variable x is

$$P(x) = \frac{\text{Frequency of } x}{\text{Number of outcomes in } S} = \frac{n(x)}{n(S)}$$

where $n(S)$ is the number of equally likely outcomes in the sample space. By this definition, it follows that the probability of an event must be a number between 0 and 1. That is, $0 \le P(x) \le 1$.

The collection of probabilities corresponding to the values of the random variable is called the **probability distribution** of the random variable. If the range of a discrete random variable consists of m different values

$$\{x_1, x_2, x_3, \ldots, x_m\}$$

then the sum of the probabilities of x_i is 1. This can be written as

$$P(x_1) + P(x_2) + P(x_3) + \cdots + P(x_m) = 1.$$

Example 3 Finding a Probability Distribution

Five coins are tossed. Graph the probability distribution, where the random variable represents the number of heads in each possible outcome.

SOLUTION Let x be the random variable that represents the number of heads in each possible outcome. The possible outcomes are shown below.

x	*Outcome*	$n(x)$
0	*TTTTT*	1
1	*HTTTT, THTTT, TTHTT, TTTHT, TTTTH*	5
2	*HHTTT, HTHTT, HTTHT, HTTTH, THHTT THTHT, THTTH, TTHHT, TTHTH, TTTHH*	10
3	*HHHTT, HHTHT, HHTTH, HTHHT, HTHTH HTTHH, THHHT, THHTH, THTHH, TTHHH*	10
4	*HHHHT, HHHTH, HHTHH, HTHHH, THHHH*	5
5	*HHHHH*	1

The number of outcomes in the sample space is $n(S) = 32$. The probability of each value of the random variable is shown in the table.

Random variable, x	0	1	2	3	4	5
Probability, $P(x)$	$\frac{1}{32}$	$\frac{5}{32}$	$\frac{10}{32}$	$\frac{10}{32}$	$\frac{5}{32}$	$\frac{1}{32}$

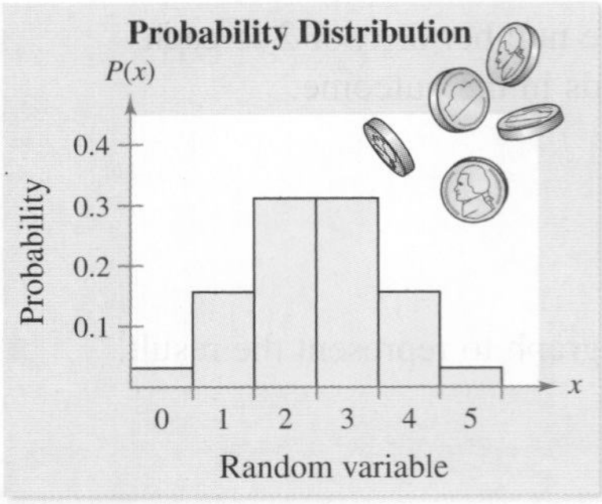

FIGURE 9.2

A graph of this probability distribution is shown in Figure 9.2. Note that values of the random variable are represented by intervals on the x-axis. Observe that the sum of the probabilities is 1.

✓Checkpoint 3

Two six-sided dice are tossed. Graph the probability distribution, where the random variable represents the sum of the points in each possible outcome. ■

Expected Value

Suppose you repeated the coin-tossing experiment in Example 3 several times. On the average, how many heads would you expect to turn up? From Figure 9.2, it seems reasonable that the average number of heads would be $2\frac{1}{2}$. This "average" is the **expected value** of the random variable.

> **Definition of Expected Value**
>
> If the range of a discrete random variable consists of m different values $\{x_1, x_2, x_3, \ldots, x_m\}$, then the **expected value** of the random variable is
>
> $$E(x) = x_1P(x_1) + x_2P(x_2) + x_3P(x_3) + \cdots + x_mP(x_m).$$
>
> The expected value is also called the **mean** of the random variable and is usually denoted by μ (the lowercase Greek letter mu). Because the mean often occurs near the center of the values in the range of the random variable, it is called a **measure of central tendency.**

Example 4 Finding an Expected Value

Five coins are tossed. Find the expected value of the number of heads that will turn up.

SOLUTION Using the results of Example 3, you obtain the expected value as shown.

$$E(x) = \overbrace{(0)\left(\tfrac{1}{32}\right)}^{\text{0 Heads}} + \overbrace{(1)\left(\tfrac{5}{32}\right)}^{\text{1 Head}} + \overbrace{(2)\left(\tfrac{10}{32}\right)}^{\text{2 Heads}} + \overbrace{(3)\left(\tfrac{10}{32}\right)}^{\text{3 Heads}} + \overbrace{(4)\left(\tfrac{5}{32}\right)}^{\text{4 Heads}} + \overbrace{(5)\left(\tfrac{1}{32}\right)}^{\text{5 Heads}} = \tfrac{80}{32} = 2.5$$

✓ Checkpoint 4

Two six-sided dice are tossed. Find the expected value of the sum of the points. ■

Example 5 Finding an Expected Value

Over a period of 1 year (225 selling days), a sales representative sold from zero to six units per day, as shown in Figure 9.3. From these data, what is the average number of units per day the sales representative should expect to sell?

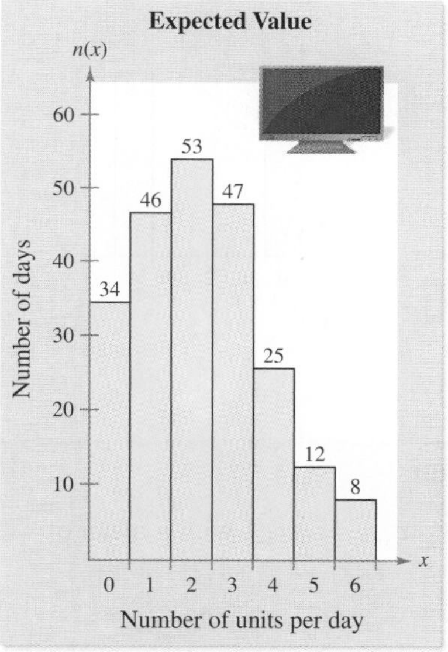

FIGURE 9.3

SOLUTION One way to answer this question is to calculate the expected value of the number of units.

$$\begin{aligned} E(x) &= (0)\left(\tfrac{34}{225}\right) + (1)\left(\tfrac{46}{225}\right) + (2)\left(\tfrac{53}{225}\right) + (3)\left(\tfrac{47}{225}\right) + (4)\left(\tfrac{25}{225}\right) + (5)\left(\tfrac{12}{225}\right) + (6)\left(\tfrac{8}{225}\right) \\ &= \frac{501}{225} \\ &\approx 2.23 \text{ units per day} \end{aligned}$$

✓ Checkpoint 5

Over a period of 1 year, a salesperson worked 6 days a week (312 selling days) and sold from zero to six units per day. Using the data in the table shown below, what is the average number of units per day the sales representative should expect to sell?

Number of units per day	0	1	2	3	4	5	6
Number of days	39	60	75	62	48	18	10

■

Variance and Standard Deviation

The expected value or mean gives a measure of the average value assigned by a random variable. But the mean does not tell the whole story. For instance, all three of the distributions shown below have a mean of 2.

Distribution 1

Random variable, x	0	1	2	3	4
Frequency of x, $n(x)$	2	2	2	2	2

Distribution 2

Random variable, x	0	1	2	3	4
Frequency of x, $n(x)$	0	3	4	3	0

Distribution 3

Random variable, x	0	1	2	3	4
Frequency of x, $n(x)$	5	0	0	0	5

Even though each distribution has the same mean, the patterns of the distributions are quite different. In the first distribution, each value has the same frequency [see Figure 9.4(a)]. In the second, the values are clustered about the mean [see Figure 9.4(b)]. In the third distribution, the values are far from the mean [see Figure 9.4(c)].

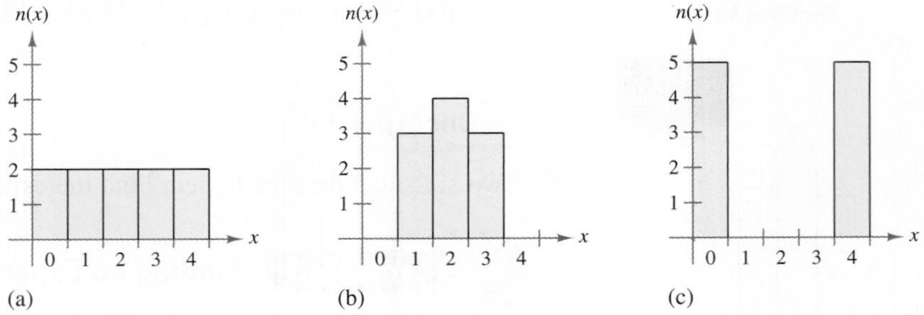

FIGURE 9.4

> **Definitions of Variance and Standard Deviation**
>
> Consider a random variable whose range is $\{x_1, x_2, x_3, \ldots, x_m\}$ with a mean of μ. The **variance** of the random variable is
>
> $$V(x) = (x_1 - \mu)^2 P(x_1) + (x_2 - \mu)^2 P(x_2) + \cdots + (x_m - \mu)^2 P(x_m).$$
>
> The **standard deviation** of the random variable is
>
> $$\sigma = \sqrt{V(x)}$$
>
> (σ is the lowercase Greek letter sigma).

When the standard deviation is small, most of the values of the random variable are clustered near the mean. As the standard deviation becomes larger, the distribution becomes more and more spread out. For instance, in the three distributions shown in Figure 9.4, you would expect the second to have the smallest standard deviation and the third to have the largest. This is confirmed in Example 6.

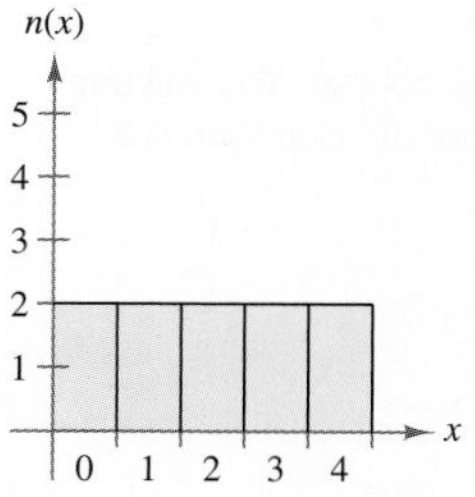

(a) Mean = 2; standard deviation ≈ 1.41

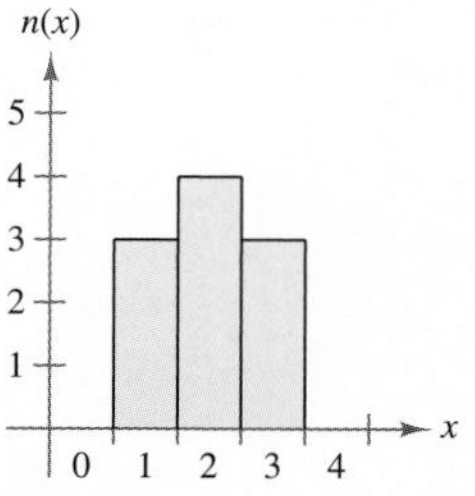

(b) Mean = 2; standard deviation ≈ 0.77

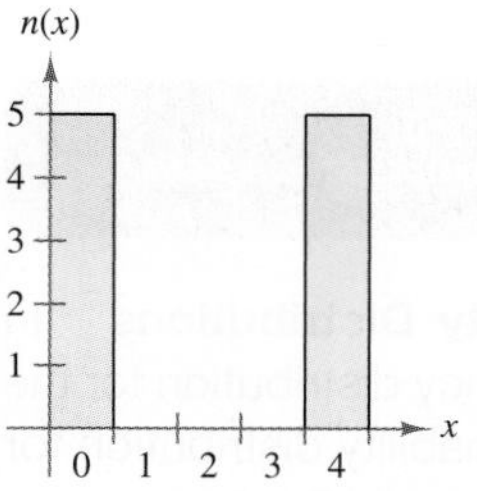

(c) Mean = 2; standard deviation = 2

FIGURE 9.5

Example 6 Finding Variance and Standard Deviation

Find the variance and standard deviation of each of the three distributions shown on the preceding page.

SOLUTION

a. For Distribution 1, the mean is $\mu = 2$, the variance is

$$V(x) = (0-2)^2\left(\tfrac{2}{10}\right) + (1-2)^2\left(\tfrac{2}{10}\right) + (2-2)^2\left(\tfrac{2}{10}\right) + (3-2)^2\left(\tfrac{2}{10}\right) + (4-2)^2\left(\tfrac{2}{10}\right)$$
$$= 2 \qquad \text{Variance}$$

and the standard deviation is $\sigma = \sqrt{2} \approx 1.41$.

b. For Distribution 2, the mean is $\mu = 2$, the variance is

$$V(x) = (0-2)^2\left(\tfrac{0}{10}\right) + (1-2)^2\left(\tfrac{3}{10}\right) + (2-2)^2\left(\tfrac{4}{10}\right) + (3-2)^2\left(\tfrac{3}{10}\right) + (4-2)^2\left(\tfrac{0}{10}\right)$$
$$= 0.6 \qquad \text{Variance}$$

and the standard deviation is $\sigma = \sqrt{0.6} \approx 0.77$.

c. For Distribution 3, the mean is $\mu = 2$, the variance is

$$V(x) = (0-2)^2\left(\tfrac{5}{10}\right) + (1-2)^2\left(\tfrac{0}{10}\right) + (2-2)^2\left(\tfrac{0}{10}\right) + (3-2)^2\left(\tfrac{0}{10}\right) + (4-2)^2\left(\tfrac{5}{10}\right)$$
$$= 4 \qquad \text{Variance}$$

and the standard deviation is $\sigma = \sqrt{4} = 2$.

As you can see in Figure 9.5, the second distribution has the smallest standard deviation and the third distribution has the largest.

✓ Checkpoint 6

Find the variance and standard deviation of the distribution shown in the table. Then graph the distribution.

Random variable, x	0	1	2	3	4
Frequency of x, $n(x)$	1	2	4	2	1

■

SUMMARIZE (Section 9.1)

1. State the meanings of probability, experiment, sample space, and event *(page 564)*. For an example of finding a sample space, see Example 1.
2. State the definition of a discrete random variable *(page 565)*. For an example of finding the frequencies of a discrete random variable, see Example 2.
3. State the definition of a probability distribution *(page 566)*. For an example of finding a probability distribution, see Example 3.
4. State the definition of expected value *(page 567)*. For examples of finding an expected value, see Examples 4 and 5.
5. State the definitions of variance and standard deviation *(page 568)*. For an example of finding variance and standard deviation, see Example 6.

SKILLS WARM UP 9.1 The following warm-up exercises involve skills that were covered in a previous course. You will use these skills in the exercise set for this section. For additional help, review Appendix Sections A.3 and A.5.

In Exercises 1 and 2, solve for x.

1. $\frac{1}{9} + \frac{2}{3} + \frac{2}{9} = x$

2. $\frac{1}{3} + \frac{5}{12} + \frac{1}{8} + \frac{1}{12} + \frac{x}{24} = 1$

In Exercises 3–6, evaluate the expression.

3. $0\left(\frac{1}{16}\right) + 1\left(\frac{3}{16}\right) + 2\left(\frac{8}{16}\right) + 3\left(\frac{3}{16}\right) + 4\left(\frac{1}{16}\right)$

4. $0\left(\frac{1}{12}\right) + 1\left(\frac{2}{12}\right) + 2\left(\frac{6}{12}\right) + 3\left(\frac{2}{12}\right) + 4\left(\frac{1}{12}\right)$

5. $(0 - 1)^2\left(\frac{1}{4}\right) + (1 - 1)^2\left(\frac{1}{2}\right) + (2 - 1)^2\left(\frac{1}{4}\right)$

6. $(0 - 2)^2\left(\frac{1}{12}\right) + (1 - 2)^2\left(\frac{2}{12}\right) + (2 - 2)^2\left(\frac{6}{12}\right) + (3 - 2)^2\left(\frac{2}{12}\right) + (4 - 2)^2\left(\frac{1}{12}\right)$

In Exercises 7–10, write the fraction as a percent. Round your answers to 2 decimal places, if necessary.

7. $\frac{3}{8}$ **8.** $\frac{9}{11}$ **9.** $\frac{13}{24}$ **10.** $\frac{112}{256}$

Exercises 9.1

See www.CalcChat.com for worked-out solutions to odd-numbered exercises.

Finding Sample Spaces and Events In Exercises 1–4, list or describe the elements in the specified set. *See Example 1.*

1. Coin Toss A coin is tossed three times.

(a) The sample space S

(b) The event A that at least two heads occur

(c) The event B that no more than one head occurs

2. Coin Toss A coin is tossed. When a head occurs, the coin is tossed again; otherwise, a die is tossed.

(a) The sample space S

(b) The event A that 4, 5, or 6 occurs on the die

(c) The event B that two heads occur

3. Poll Three people are asked their opinions on a political issue. They can answer "In favor" (I), "Opposed" (O), or "Undecided" (U).

(a) The sample space S

(b) The event A that at least two people are in favor

(c) The event B that no more than one person is opposed

4. Credit Card Fraud Four cases of credit card fraud are examined. The method of fraud is "stolen card" (S), "counterfeit card" (C), "mail order" (M), or "other" (O).

(a) The sample space S

(b) The event A that at least three cases are mail order fraud

(c) The event B that no more than one case is counterfeit card fraud

Finding Frequency and Probability Distributions In Exercises 5 and 6, (a) find the frequency distribution for the random variable and (b) find the probability distribution for the random variable. *See Examples 2 and 3.*

5. Coin Toss Four coins are tossed. A random variable assigns the number 0, 1, 2, 3, or 4 to each possible outcome, depending on the number of heads in the outcome.

6. Exam Three students answer a true-false question on an examination. A random variable assigns the number 0, 1, 2, or 3 to each possible outcome, depending on the number of answers of *true* among the three students.

7. Random Selection In a class of 72 students, 44 are girls and, of these, 12 are going to college. Of the 28 boys in the class, 9 are going to college. A student is selected at random from the class. What is the probability that the person chosen is

(a) going to college?

(b) not going to college?

(c) a girl who is not going to college?

8. Random Selection A card is chosen at random from a standard 52-card deck of playing cards. What is the probability that the card is

(a) red?

(b) a 5?

(c) black and a face card?

Determining a Missing Probability In Exercises 9 and 10, find the missing value of the probability distribution.

9.

x	0	1	2	3	4
$P(x)$	0.20	0.35	0.15	?	0.05

10.

x	0	1	2	3	4	5
$P(x)$	0.05	?	0.25	0.30	0.15	0.10

Identifying Probability Distributions In Exercises 11–14, determine whether the table represents a probability distribution. If it is a probability distribution, sketch its graph. If it is not a probability distribution, state any properties that are not satisfied.

11.

x	0	1	2	3
$P(x)$	0.10	0.45	0.30	0.15

12.

x	0	1	2	3	4	5
$P(x)$	0.05	0.30	0.10	0.40	0.15	0.20

13.

x	0	1	2	3	4
$P(x)$	$\frac{12}{50}$	$\frac{20}{50}$	$\frac{8}{50}$	$\frac{10}{50}$	$-\frac{5}{50}$

14.

x	0	1	2	3	4	5
$P(x)$	$\frac{8}{30}$	$\frac{2}{30}$	$\frac{6}{30}$	$\frac{3}{30}$	$\frac{4}{30}$	$\frac{7}{30}$

Using Probability Distributions In Exercises 15–18, sketch a graph of the probability distribution and find the required probabilities.

15.

x	0	1	2	3	4
$P(x)$	$\frac{1}{20}$	$\frac{3}{20}$	$\frac{6}{20}$	$\frac{6}{20}$	$\frac{4}{20}$

(a) $P(1 \le x \le 3)$
(b) $P(x \ge 2)$

16.

x	0	1	2	3	4
$P(x)$	$\frac{8}{20}$	$\frac{6}{20}$	$\frac{3}{20}$	$\frac{2}{20}$	$\frac{1}{20}$

(a) $P(x \le 2)$
(b) $P(x > 2)$

17.

x	0	1	2	3	4	5
$P(x)$	0.041	0.189	0.247	0.326	0.159	0.038

(a) $P(x \le 3)$
(b) $P(x > 3)$

18.

x	0	1	2	3
$P(x)$	0.027	0.189	0.441	0.343

(a) $P(1 \le x \le 2)$
(b) $P(x < 2)$

19. Health The table shows the probability distribution of the numbers of AIDS cases diagnosed in the United States in 2009 by age group. *(Source: Centers for Disease Control and Prevention)*

Age, a	14 and under	15–24	25–34	35–44
$P(a)$	0.004	0.198	0.262	0.255

Age, a	45–54	55–64	65 and over
$P(a)$	0.195	0.069	0.017

(a) Sketch the probability distribution.
(b) Find the probability that an individual diagnosed with AIDS was from 15 to 44 years of age.
(c) Find the probability that an individual diagnosed with AIDS was at least 35 years of age.
(d) Find the probability that an individual diagnosed with AIDS was at most 24 years of age.

20. Children The table shows the probability distribution of the numbers of children per family in the United States in 2009. *(Source: U.S. Census Bureau)*

Children, x	0	1	2	3 or more
$P(x)$	0.548	0.193	0.167	0.092

(a) Sketch the probability distribution.
(b) Find the probability that a family has at least 2 children.
(c) Find the probability that a family has at most 2 children.
(d) Find the probability that a family has at least 1 child.

Rob Byron/www.shutterstock.com

21. Biology Consider a couple who have four children. Assume that it is equally likely that each child is a girl or a boy.

(a) Complete the set to form the sample space consisting of 16 elements.

$S = \{gggg, gggb, ggbg, . . .\}$

(b) Complete the table, in which the random variable x is the number of girls in the family.

x	0	1	2	3	4
$P(x)$					

(c) Use the table in part (b) to sketch the graph of the probability distribution.

(d) Use the table in part (b) to find the probability that at least one of the children is a boy.

22. Die Toss Consider the experiment of tossing a 6-sided die twice.

(a) Complete the set to form the sample space of 36 elements. Note that each element is an ordered pair in which the entries are the numbers of points on the first and second tosses, respectively.

$S = \{(1, 1), (1, 2), . . . , (2, 1), (2, 2), . . .\}$

(b) Complete the table, in which the random variable x is the sum of the two rolls.

x	2	3	4	5	6	7	8	9	10	11	12
$P(x)$											

(c) Use the table in part (b) to sketch the graph of the probability distribution.

(d) Use the table in part (b) to find $P(10 \le x \le 12)$.

Finding Expected Value, Variance, and Standard Deviation In Exercises 23–26, find the expected value, variance, and standard deviation for the given probability distribution. *See Examples 4, 5, and 6.*

23.

x	1	2	3	4	5
$P(x)$	$\frac{1}{16}$	$\frac{3}{16}$	$\frac{8}{16}$	$\frac{3}{16}$	$\frac{1}{16}$

24.

x	1	2	3	4	5
$P(x)$	$\frac{4}{10}$	$\frac{2}{10}$	$\frac{2}{10}$	$\frac{1}{10}$	$\frac{1}{10}$

25.

x	-3	-1	0	3	5
$P(x)$	$\frac{1}{5}$	$\frac{1}{5}$	$\frac{1}{5}$	$\frac{1}{5}$	$\frac{1}{5}$

26.

x	-5000	-2500	300
$P(x)$	0.008	0.052	0.940

Finding Mean, Variance, and Standard Deviation In Exercises 27 and 28, find the mean, variance, and standard deviation of the discrete random variable x. *See Examples 4, 5, and 6.*

27. Die Toss x is (a) the number of points when a four-sided die is tossed once and (b) the sum of the points when the four-sided die is tossed twice.

28. Coin Toss x is the number of heads when a coin is tossed four times.

29. Revenue A publishing company introduces a new weekly magazine that sells for \$4.95 on the newsstand. The marketing group of the company estimates that sales x (in thousands) will be approximated by the following probability function.

x	10	15	20	30	40
$P(x)$	0.25	0.30	0.25	0.15	0.05

(a) Find $E(x)$ and σ.

(b) Find the expected revenue.

30. Personal Income The probability distribution of the random variable x, the annual income of a family (in thousands of dollars) in a certain section of a large city, is shown in the table. Find $E(x)$ and σ.

x	30	40	50	60	80
$P(x)$	0.10	0.20	0.50	0.15	0.05

31. Insurance An insurance company needs to determine the annual premium required to break even on fire protection policies with a face value of \$90,000. The random variable x is the claim size on these policies, and the analysis is restricted to the losses \$30,000, \$60,000, and \$90,000. The probability distribution of x is as shown in the table. What premium should customers be charged for the company to break even?

x	0	30,000	60,000	90,000
$P(x)$	0.995	0.0036	0.0011	0.0003

32. Insurance An insurance company needs to determine the annual premium required to break even for collision protection for cars with a value of \$10,000. The random variable x is the claim size on these policies, and the analysis is restricted to the losses \$1000, \$5000, and \$10,000. The probability distribution of x is as shown in the table. What premium should customers be charged for the company to break even?

x	0	1000	5000	10,000
$P(x)$	0.936	0.040	0.020	0.004

33. Baseball A baseball fan examined the record of a favorite baseball player's performance during his last season. The numbers of games in which the player had zero, one, two, three, and four hits are recorded in the table shown below.

Number of hits	0	1	2	3	4
Frequency	40	61	40	17	2

(a) Complete the table below, where x is the number of hits.

x	0	1	2	3	4
$P(x)$					

(b) Use the table in part (a) to sketch the graph of the probability distribution.

(c) Use the table in part (a) to find $P(1 \leq x \leq 3)$.

(d) Determine the mean, variance, and standard deviation. Explain your results.

34. HOW DO YOU SEE IT? News Station A says that there is a 40% chance of rain for the next three days. News Station B says that there is a 30% chance of rain for the next three days. Determine which graph represents the probability distribution for each news station. Explain your reasoning.

(i)

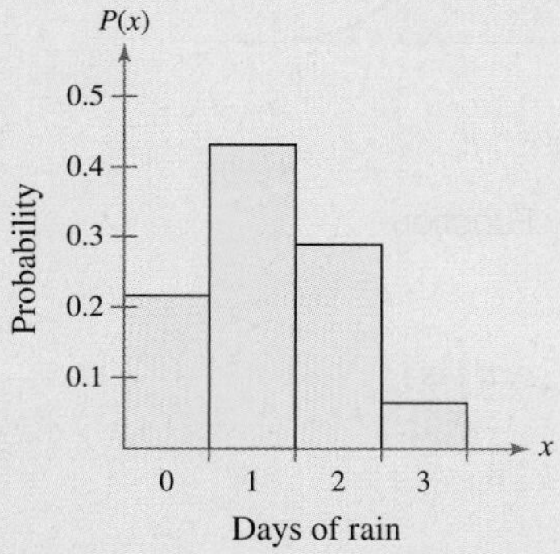

(ii)

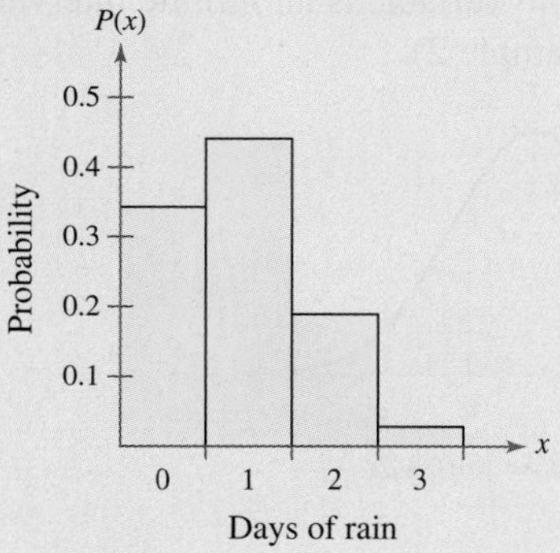

Games of Chance If x is a player's net gain in a game of chance, then $E(x)$ is usually negative. This value gives the average amount per game the player can expect to lose over the long run. In Exercises 35 and 36, find the player's expected net gain for one play of the specified game.

35. Roulette In roulette, the wheel has the 38 numbers 00, 0, 1, 2, . . . , 34, 35, and 36, marked on equally spaced slots. If a player bets \$1 on a number and wins, then the player keeps the dollar and receives an additional \$35. Otherwise, the dollar is lost.

36. Raffle A service organization is selling \$2 raffle tickets as part of a fundraising program. The first prize is a boat valued at \$2950, and the second prize is a camping tent valued at \$400. In addition to the first and second prizes, there are 25 \$20 gift certificates to be awarded. The number of tickets sold is 3000.

37. Market Analysis A sporting goods company has decided on two possible cities in which to open a new store. Management estimates that city 1 will yield \$20 million in revenues if successful and will lose \$4 million if not, whereas city 2 will yield \$50 million in revenues if successful and will lose \$9 million if not. City 1 has a 0.3 probability of being successful and city 2 has a 0.2 probability of being successful. In which city should the sporting goods company open the new store with respect to the expected return from each store?

38. Market Analysis Repeat Exercise 37 for the case in which the probabilities of city 1 and city 2 being successful are 0.4 and 0.25, respectively.

39. Project: Education For a project analyzing the educational attainment of people in the United States by age, visit this text's website at *www.cengagebrain.com*. *(Data Source: U.S. Census Bureau)*

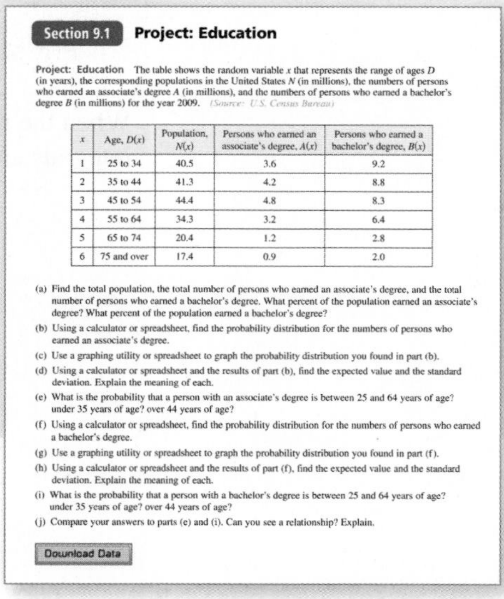

Section 9.1 Project: Education

Project: Education The table shows the random variable x that represents the range of ages D (in years), the corresponding populations in the United States N (in millions), the numbers of persons who earned an associate's degree A (in millions), and the numbers of persons who earned a bachelor's degree B (in millions) for the year 2009. *(Source: U.S. Census Bureau)*

x	Age, $D(x)$	Population, $N(x)$	Persons who earned an associate's degree, $A(x)$	Persons who earned a bachelor's degree, $B(x)$
1	25 to 34	40.5	3.6	9.2
2	35 to 44	41.3	4.2	8.8
3	45 to 54	44.4	4.8	8.3
4	55 to 64	34.3	3.2	6.4
5	65 to 74	20.4	1.2	2.8
6	75 and over	17.4	0.9	2.0

(a) Find the total population, the total number of persons who earned an associate's degree, and the total number of persons who earned a bachelor's degree. What percent of the population earned an associate's degree? What percent of the population earned a bachelor's degree?

(b) Using a calculator or spreadsheet, find the probability distribution for the numbers of persons who earned an associate's degree.

(c) Use a graphing utility or spreadsheet to graph the probability distribution you found in part (b).

(d) Using a calculator or spreadsheet and the results of part (b), find the expected value and the standard deviation. Explain the meaning of each.

(e) What is the probability that a person with an associate's degree is between 25 and 64 years of age? under 35 years of age? over 44 years of age?

(f) Using a calculator or spreadsheet, find the probability distribution for the numbers of persons who earned a bachelor's degree.

(g) Use a graphing utility or spreadsheet to graph the probability distribution you found in part (f).

(h) Using a calculator or spreadsheet and the results of part (f), find the expected value and the standard deviation. Explain the meaning of each.

(i) What is the probability that a person with a bachelor's degree is between 25 and 64 years of age? under 35 years of age? over 44 years of age?

(j) Compare your answers to parts (e) and (i). Can you see a relationship? Explain.

Download Data

9.2 Continuous Random Variables

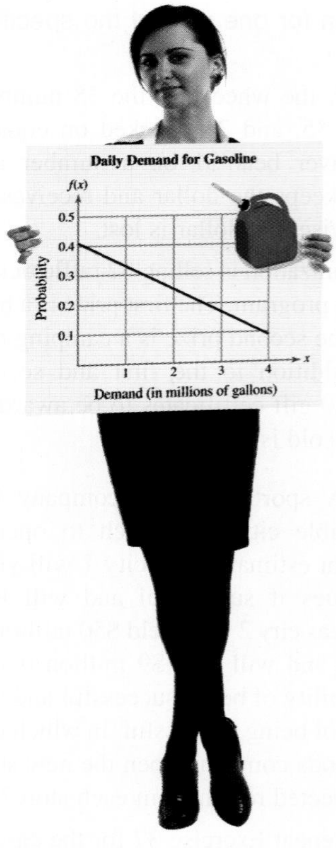

In Exercise 26 on page 579, you will use a probability density function to find the probability of the daily demand for gasoline.

- Verify continuous probability density functions and use continuous probability density functions to find probabilities.
- Use continuous probability density functions to answer questions about real-life situations.

Continuous Random Variables

In many applications of probability, it is useful to consider a random variable whose range is an interval on the real number line. Such a random variable is **continuous.** For instance, the random variable that measures the height of a person in a population is continuous.

To define the probability of an event involving a continuous random variable, you cannot simply count the number of ways the event can occur (as you can with a discrete random variable). Rather, you need to define a function f called a **probability density function.**

Definition of Probability Density Function

Consider a function f of a continuous random variable x whose range is the interval $[a, b]$. The function is a **probability density function** when it is nonnegative and continuous on the interval $[a, b]$ and when

$$\int_a^b f(x)\,dx = 1.$$

See Figure 9.6.

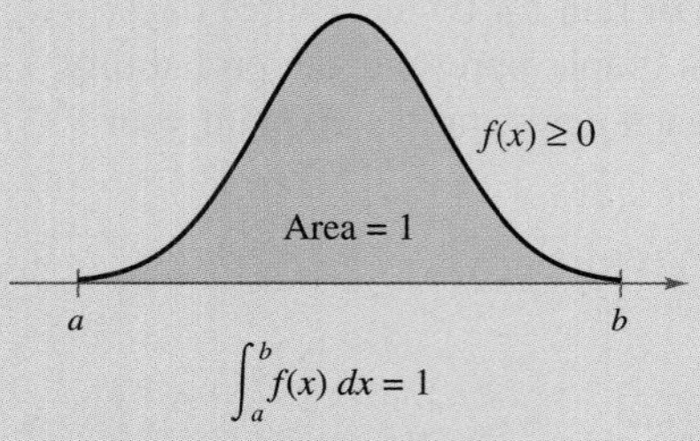

Probability Density Function
FIGURE 9.6

The probability that x lies in the interval $[c, d]$ is

$$P(c \le x \le d) = \int_c^d f(x)\,dx.$$

See Figure 9.7.

When the range of the continuous random variable is an infinite interval, the integrals are improper integrals (see Example 2).

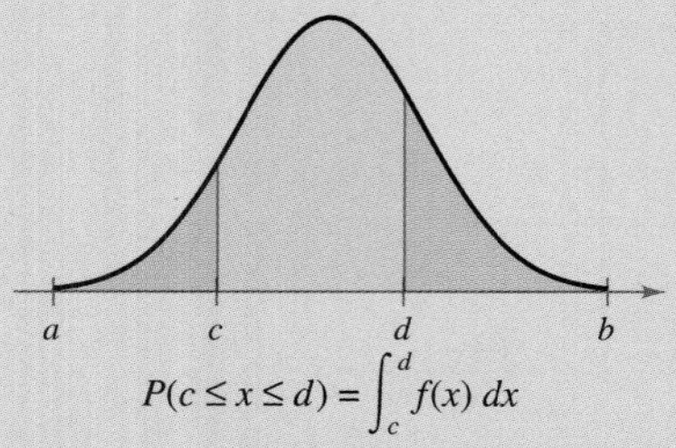

Probability that x lies in the interval $[c, d]$
FIGURE 9.7

Example 1 Verifying a Probability Density Function

Show that

$$f(x) = 12x(1 - x)^2$$

is a probability density function over the interval [0, 1].

SOLUTION Begin by observing that f is continuous and nonnegative on the interval [0, 1].

$$f(x) = 12x(1 - x)^2 \geq 0, \quad 0 \leq x \leq 1 \qquad f(x) \text{ is nonnegative on } [0, 1].$$

Next, evaluate the integral below.

$$\int_0^1 12x(1 - x)^2\,dx = 12\int_0^1 (x^3 - 2x^2 + x)\,dx \qquad \text{Expand polynomial.}$$

$$= 12\left[\frac{x^4}{4} - \frac{2x^3}{3} + \frac{x^2}{2}\right]_0^1 \qquad \text{Integrate.}$$

$$= 12\left(\frac{1}{4} - \frac{2}{3} + \frac{1}{2}\right) \qquad \text{Apply Fundamental Theorem of Calculus.}$$

$$= 1 \qquad \text{Simplify.}$$

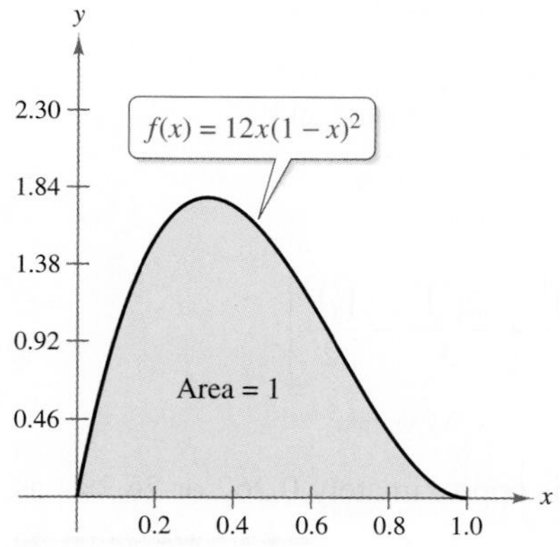

FIGURE 9.8

Because this value is 1, you can conclude that f is a probability density function over the interval [0, 1]. The graph of f is shown in Figure 9.8.

✓ Checkpoint 1

Show that $f(x) = \frac{1}{2}x$ is a probability density function over the interval [0, 2]. ■

The next example deals with an infinite interval and its corresponding improper integral.

Example 2 Verifying a Probability Density Function

Show that

$$f(t) = 0.1e^{-0.1t}$$

is a probability density function over the infinite interval $[0, \infty)$.

SOLUTION Begin by observing that f is continuous and nonnegative on the interval $[0, \infty)$.

$$f(t) = 0.1e^{-0.1t} \geq 0, \quad t \geq 0 \qquad f(t) \text{ is nonnegative on } [0, \infty).$$

Next, evaluate the integral below.

$$\int_0^{\infty} 0.1e^{-0.1t}\,dt = \lim_{b\to\infty}\left[-e^{-0.1t}\right]_0^b \qquad \text{Improper integral}$$

$$= \lim_{b\to\infty}(-e^{-0.1b} + 1) \qquad \text{Evaluate limit.}$$

$$= 1$$

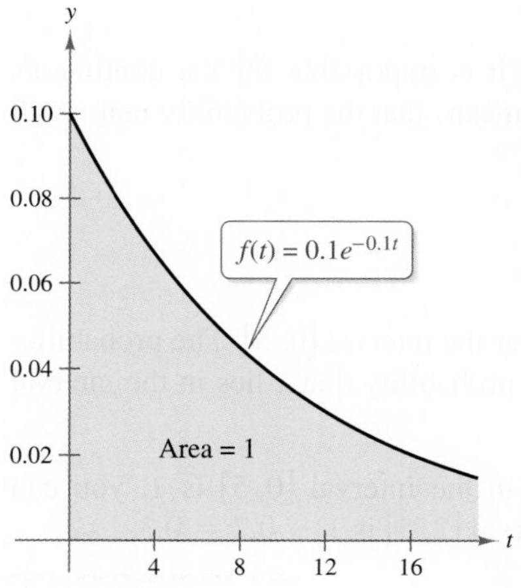

FIGURE 9.9

Because this value is 1, you can conclude that f is a probability density function over the interval $[0, \infty)$. The graph of f is shown in Figure 9.9.

✓ Checkpoint 2

Show that $f(x) = 2e^{-2x}$ is a probability density function over the interval $[0, \infty)$. ■

Example 3 Finding a Probability

For the probability density function in Example 1

$$f(x) = 12x(1 - x)^2$$

find the probability that x lies in the interval $\frac{1}{2} \le x \le \frac{3}{4}$.

SOLUTION

$$\begin{aligned} P\left(\tfrac{1}{2} \le x \le \tfrac{3}{4}\right) &= 12\int_{1/2}^{3/4} x(1-x)^2\,dx && \text{Integrate } f(x) \text{ over } \left[\tfrac{1}{2}, \tfrac{3}{4}\right]. \\ &= 12\int_{1/2}^{3/4} (x^3 - 2x^2 + x)\,dx && \text{Expand polynomial.} \\ &= 12\left[\frac{x^4}{4} - \frac{2x^3}{3} + \frac{x^2}{2}\right]_{1/2}^{3/4} && \text{Integrate.} \\ &= 12\left[\frac{\left(\frac{3}{4}\right)^4}{4} - \frac{2\left(\frac{3}{4}\right)^3}{3} + \frac{\left(\frac{3}{4}\right)^2}{2} - \frac{\left(\frac{1}{2}\right)^4}{4} + \frac{2\left(\frac{1}{2}\right)^3}{3} - \frac{\left(\frac{1}{2}\right)^2}{2}\right] \\ &\approx 0.262 && \text{Approximate.} \end{aligned}$$

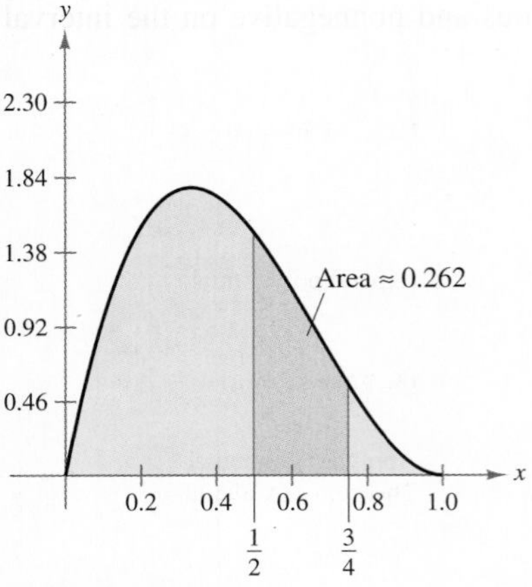

FIGURE 9.10

So, the probability that x lies in the interval $\left[\frac{1}{2}, \frac{3}{4}\right]$ is approximately 0.262 or 26.2%, as indicated in Figure 9.10.

✓Checkpoint 3

Find the probability that x lies in the interval $\frac{1}{2} \le x \le 1$ for the probability density function in Checkpoint 1. ■

In Example 3, the probability that x lies in any of the intervals $\frac{1}{2} < x < \frac{3}{4}$, $\frac{1}{2} \le x < \frac{3}{4}$, or $\frac{1}{2} < x \le \frac{3}{4}$ is the same. In other words, the inclusion of either endpoint adds nothing to the probability. This demonstrates an important difference between discrete and continuous random variables. For a continuous random variable, the probability that x will be precisely one value (such as 0.5) is considered to be zero, because

$$P(0.5 \le x \le 0.5) = \int_{0.5}^{0.5} f(x)\,dx = 0.$$

You should not interpret this result to mean that it is impossible for the continuous random variable x to have the value 0.5. It simply means that the probability that x will have this *exact* value is insignificant.

Example 4 Finding a Probability

Consider a probability density function defined over the interval [0, 5]. The probability that x lies in the interval [0, 2] is 0.7. What is the probability that x lies in the interval [2, 5]?

SOLUTION Because the probability that x lies in the interval [0, 5] is 1, you can conclude that the probability that x lies in the interval [2, 5] is $1 - 0.7 = 0.3$.

✓Checkpoint 4

A probability density function is defined over the interval [0, 4]. The probability that x lies in [0, 1] is 0.6. What is the probability that x lies in [1, 4]? ■

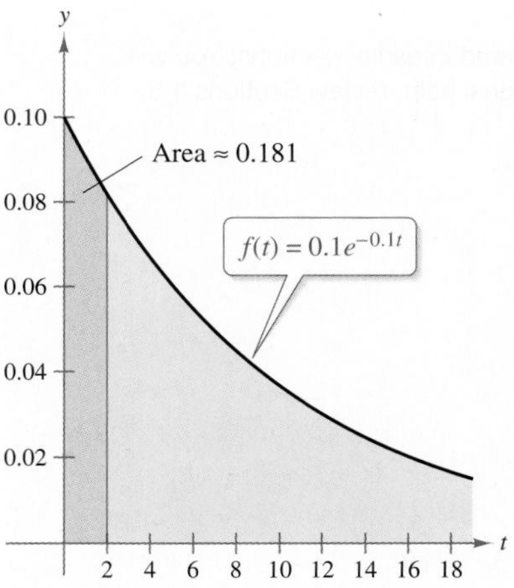

(a) $P(0 \le t \le 2) \approx 0.181$

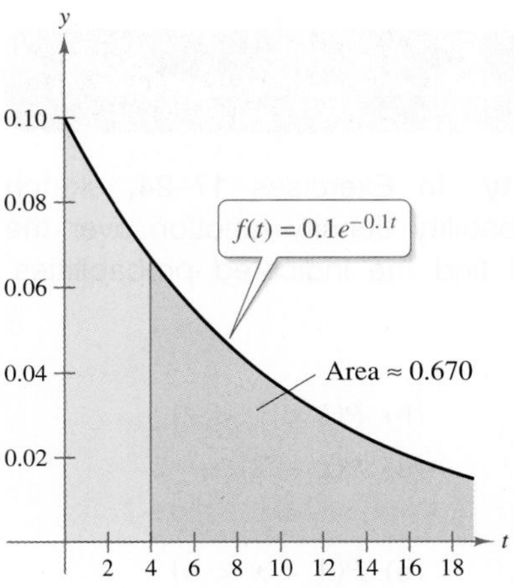

(b) $P(4 < t < \infty) \approx 0.670$

FIGURE 9.11

Application

The probability density function in Example 5 has the form

$$f(x) = ae^{-ax}, \quad 0 \le x < \infty.$$

This probability density function is called the *exponential probability density function*. In Section 9.3, you will learn more about this probability density function and others.

Example 5 Modeling the Lifetime of a Product

The useful lifetime (in years) of a product is modeled by the probability density function

$$f(t) = 0.1e^{-0.1t}$$

for $0 \le t < \infty$. Find the probability that a randomly selected unit will have a lifetime falling in each interval.

a. No more than 2 years **b.** More than 4 years

SOLUTION

a. The probability that the unit will last no more than 2 years is

$$P(0 \le t \le 2) = 0.1\int_0^2 e^{-0.1t}\,dt \qquad \text{Integrate } f(t) \text{ over } [0, 2].$$

$$= \Big[-e^{-0.1t}\Big]_0^2 \qquad \text{Find antiderivative.}$$

$$= -e^{-0.2} + 1 \qquad \text{Apply Fundamental Theorem of Calculus.}$$

$$\approx 0.181. \qquad \text{Approximate.}$$

b. The probability that the unit will last more than 4 years is

$$P(4 < t < \infty) = 0.1\int_4^\infty e^{-0.1t}\,dt \qquad \text{Integrate } f(t) \text{ over } [4, \infty).$$

$$= \lim_{b\to\infty}\Big[-e^{-0.1t}\Big]_4^b \qquad \text{Improper integral}$$

$$= \lim_{b\to\infty}(-e^{-0.1b} + e^{-0.4}) \qquad \text{Evaluate limit.}$$

$$= e^{-0.4}$$

$$\approx 0.670. \qquad \text{Approximate.}$$

These two probabilities are illustrated graphically in Figure 9.11.

✓Checkpoint 5

For the function in Example 5, find the probability that a randomly selected unit will have a lifetime of more than 2 years, but no more than 4 years. ■

SUMMARIZE (Section 9.2)

1. State the definition of a probability density function *(page 574)*. For examples of finding a probability, see Examples 1, 2, 3, and 4.
2. Describe a real-life example of how a probability density function can be used to analyze the useful lifetime of a product *(page 577, Example 5)*.

SKILLS WARM UP 9.2

The following warm-up exercises involve skills that were covered in earlier sections. You will use these skills in the exercise set for this section. For additional help, review Sections 1.6, 5.4, and 6.4.

In Exercises 1–4, determine whether f is continuous and nonnegative on the given interval.

1. $f(x) = \dfrac{1}{x}, \quad [1, 4]$

2. $f(x) = x^2 - 1, \quad [0, 1]$

3. $f(x) = 3 - x, \quad [1, 5]$

4. $f(x) = e^{-x}, \quad [0, 1]$

In Exercises 5–7, evaluate the definite integral.

5. $\displaystyle\int_0^4 \frac{1}{4}\, dx$

6. $\displaystyle\int_1^2 \frac{2 - x}{2}\, dx$

7. $\displaystyle\int_0^\infty 3e^{-3t}\, dt$

Exercises 9.2

See www.CalcChat.com for worked-out solutions to odd-numbered exercises.

Identifying Probability Density Functions In Exercises 1–10, use a graphing utility to graph the function. Then determine whether the function f represents a probability density function over the given interval. If f is not a probability density function, identify the condition(s) that is (are) not satisfied. *See Examples 1 and 2.*

1. $f(x) = \frac{1}{8}, \quad [0, 8]$

2. $f(x) = \frac{1}{5}, \quad [0, 4]$

3. $f(x) = \dfrac{4 - x}{8}, \quad [0, 4]$

4. $f(x) = \dfrac{x}{18}, \quad [0, 6]$

5. $f(x) = \frac{1}{7}e^{-x/7}, \quad [0, 3]$

6. $f(x) = \frac{1}{6}e^{-x/6}, \quad [0, \infty)$

7. $f(x) = 2\sqrt{4 - x}, \quad [0, 2]$

8. $f(x) = 12x^2(1 - x), \quad [0, 2]$

9. $f(x) = \frac{2}{9}x(3 - x), \quad [0, 3]$

10. $f(x) = 0.4e^{-0.4x}, \quad [0, \infty)$

Making a Probability Density Function In Exercises 11–16, find the constant k such that the function f is a probability density function over the given interval.

11. $f(x) = kx, \quad [1, 4]$

12. $f(x) = kx^3, \quad [0, 4]$

13. $f(x) = k(4 - x^2), \quad [-2, 2]$

14. $f(x) = k\sqrt{x}(1 - x), \quad [0, 1]$

15. $f(x) = ke^{-x/2}, \quad [0, \infty)$

16. $f(x) = \dfrac{k}{b - a}, \quad [a, b]$

Finding a Probability In Exercises 17–24, sketch the graph of the probability density function over the indicated interval and find the indicated probabilities. *See Examples 3 and 4.*

17. $f(x) = \frac{1}{3}, \quad [0, 3]$

(a) $P(0 < x < 2)$ (b) $P(1 < x < 2)$

(c) $P(1 < x < 3)$ (d) $P(x = 2)$

18. $f(x) = \frac{1}{10}, \quad [0, 10]$

(a) $P(0 < x < 6)$ (b) $P(4 < x < 6)$

(c) $P(8 < x < 10)$ (d) $P(x \geq 2)$

19. $f(x) = \dfrac{x}{50}, \quad [0, 10]$

(a) $P(0 < x < 6)$ (b) $P(4 < x < 6)$

(c) $P(8 < x < 10)$ (d) $P(x \geq 2)$

20. $f(x) = \dfrac{2x}{25}, \quad [0, 5]$

(a) $P(0 < x < 3)$ (b) $P(1 < x < 3)$

(c) $P(3 < x < 5)$ (d) $P(x \geq 1)$

21. $f(x) = \frac{3}{16}\sqrt{x}, \quad [0, 4]$

(a) $P(0 < x < 2)$ (b) $P(2 < x < 4)$

(c) $P(1 < x < 3)$ (d) $P(x \leq 3)$

22. $f(x) = \dfrac{5}{4(x + 1)^2}, \quad [0, 4]$

(a) $P(0 < x < 2)$ (b) $P(2 < x < 4)$

(c) $P(1 < x < 3)$ (d) $P(x \leq 3)$

23. $f(t) = \frac{1}{3}e^{-t/3}, \quad [0, \infty)$

(a) $P(t < 2)$ (b) $P(t \geq 2)$

(c) $P(1 < t < 4)$ (d) $P(t = 3)$

24. $f(t) = \frac{3}{256}(16 - t^2), \quad [-4, 4]$

(a) $P(t < -2)$

(b) $P(t > 2)$

(c) $P(-1 < t < 1)$

(d) $P(t > -2)$

25. Waiting Time Buses arrive and depart from a college every 30 minutes. The probability density function for the waiting time t (in minutes) for a person arriving at the bus stop is

$f(t) = \frac{1}{30}, \quad [0, 30].$

Find the probabilities that the person will wait (a) no more than 5 minutes and (b) at least 18 minutes.

26. Demand The daily demand for gasoline x (in millions of gallons) in a city is described by the probability density function

$f(x) = 0.41 - 0.08x, \quad [0, 4].$

Find the probabilities that the daily demand for gasoline will be (a) no more than 3 million gallons and (b) at least 2 million gallons.

27. Learning Theory The time t (in hours) required for a new employee to learn to successfully operate a machine in a manufacturing process is described by the probability density function

$f(t) = \frac{5}{324}t\sqrt{9 - t}, \quad [0, 9].$

Find the probabilities that a new employee will learn to operate the machine (a) in less than 3 hours and (b) in more than 4 hours but less than 8 hours.

28. HOW DO YOU SEE IT? The graph shows two models for predicting the weekly demand for a new product.

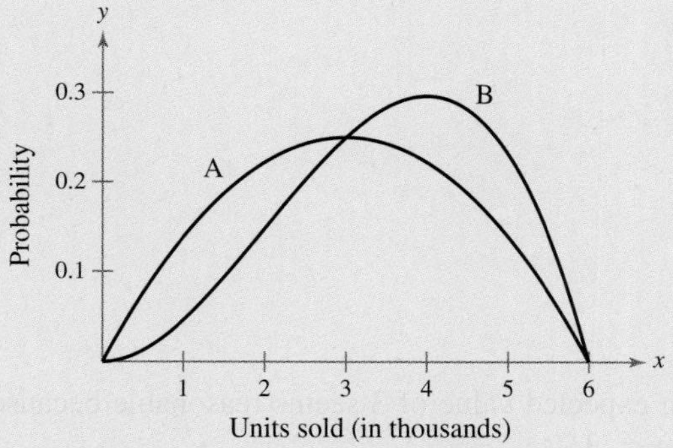

(a) Which model predicts a higher probability of selling between 0 and 2000 units?

(b) Which model predicts a higher probability of selling between 4000 and 6000 units?

Using the Exponential Density Function In Exercises 29–32, find the required probabilities using the *exponential density function*

$$f(t) = \frac{1}{\lambda}e^{-t/\lambda}, \quad [0, \infty).$$

29. Waiting Time The waiting time t (in minutes) for service at the checkout at a grocery store is exponentially distributed with $\lambda = 3$. Find the probabilities of waiting (a) less than 2 minutes, (b) more than 2 minutes but less than 4 minutes, and (c) at least 2 minutes.

30. Waiting Time The length of time t (in hours) required to unload trucks at a depot is exponentially distributed with $\lambda = \frac{3}{4}$. Find the probabilities that the trucks can be unloaded (a) in less than 1 hour, (b) in more than 1 hour but less than 2 hours, and (c) in at most 3 hours.

31. Useful Life The lifetime t (in years) of a battery is exponentially distributed with $\lambda = 5$. Find the probabilities that the lifetime of a given battery will be (a) less than 6 years, (b) more than 2 years but less than 6 years, and (c) more than 8 years.

32. Useful Life The time t (in years) until failure of a component in a machine is exponentially distributed with $\lambda = 3.5$. Find the probabilities that the lifetime of a given component will be (a) less than 1 year, (b) more than 2 years but less than 4 years, and (c) at least 5 years.

33. Meteorology A meteorologist predicts that the amount of rainfall x (in inches) expected for a certain coastal community during a hurricane has the probability density function

$$f(x) = \frac{\pi}{30}\sin\frac{\pi x}{15}, \quad 0 \le x \le 15.$$

Find and interpret the probabilities.

(a) $P(0 \le x \le 10)$

(b) $P(10 \le x \le 15)$

(c) $P(0 \le x < 5)$

(d) $P(12 \le x \le 15)$

34. Demand The weekly demand x (in tons) for a certain product is described by the probability density function

$f(x) = \frac{1}{36}xe^{-x/6}, \quad [0, \infty).$

Find and interpret the probabilities.

(a) $P(x < 6)$

(b) $P(6 < x < 12)$

(c) $P(x > 12) = 1 - P(x \le 12)$

35. Demand Given the conditions in Exercise 34, determine the number of tons that should be ordered each week so that the demand can be met for 90% of the weeks.

9.3 Expected Value and Variance

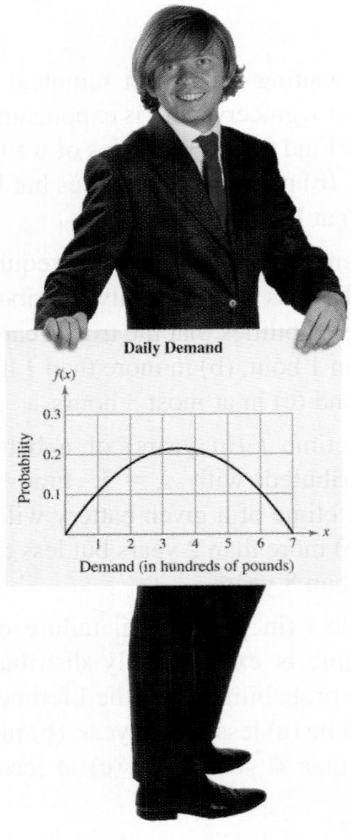

In Exercise 39 on page 588, you will use integration to find the mean, standard deviation, and median of the daily demand for a product.

- Find the expected values or means of continuous probability density functions.
- Find the variances and standard deviations of continuous probability density functions.
- Find the medians of continuous probability density functions.
- Use special probability density functions to answer questions about real-life situations.

Expected Value

In Section 9.1, you studied the concepts of expected value (or mean), variance, and standard deviation of *discrete* random variables. In this section, you will extend these concepts to *continuous* random variables.

Definition of Expected Value

If f is a probability density function of a continuous random variable x over the interval $[a, b]$, then the **expected value** or **mean** of x is

$$\mu = E(x) = \int_a^b xf(x)\,dx.$$

Example 1 Finding Average Weekly Demand

The weekly demand for a product is modeled by the probability density function

$$f(x) = \frac{1}{36}(-x^2 + 6x), \quad 0 \le x \le 6$$

where x is the number of units sold (in thousands). Find the expected weekly demand for this product.

SOLUTION

$$\begin{aligned}\mu &= E(x)\\ &= \frac{1}{36}\int_0^6 x(-x^2 + 6x)\,dx\\ &= \frac{1}{36}\int_0^6 (-x^3 + 6x^2)\,dx\\ &= \frac{1}{36}\left[\frac{-x^4}{4} + 2x^3\right]_0^6\\ &= 3\end{aligned}$$

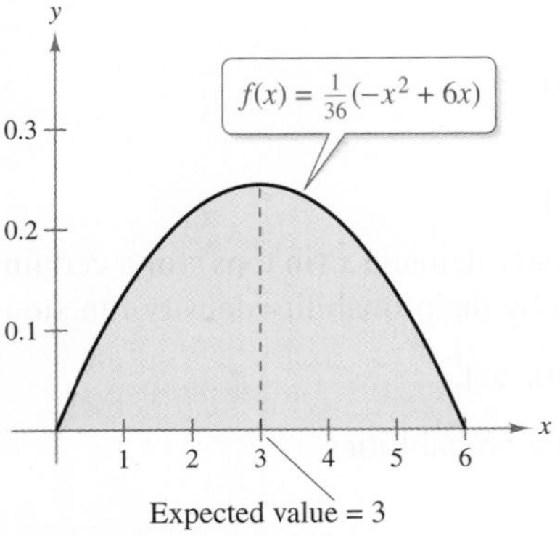

FIGURE 9.12

In Figure 9.12, you can see that an expected value of 3 seems reasonable because the region is symmetric about the line $x = 3$.

Checkpoint 1

Find the expected value of the probability density function

$$f(x) = \tfrac{1}{32}(3x)(4 - x)$$

on the interval $[0, 4]$. ■

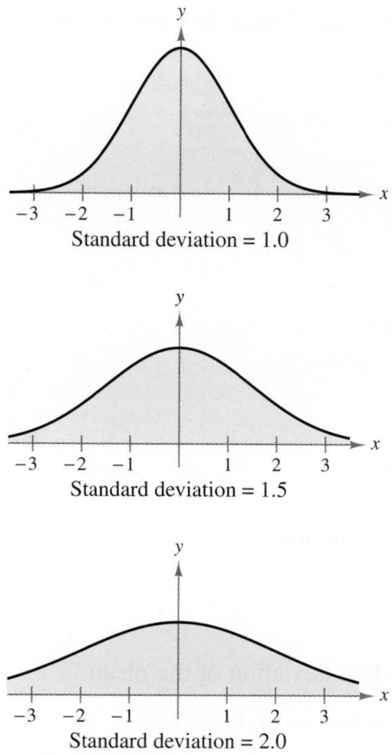

FIGURE 9.13

Variance and Standard Deviation

Definitions of Variance and Standard Deviation

If f is a probability density function of a continuous random variable x over the interval $[a, b]$, then the **variance** of x is

$$V(x) = \int_a^b (x - \mu)^2 f(x)\, dx$$

where μ is the mean of x. The **standard deviation** of x is

$$\sigma = \sqrt{V(x)}.$$

Recall from Section 9.1 that distributions that are clustered about the mean tend to have smaller standard deviations than distributions that are more dispersed. For instance, the probability density distributions shown in Figure 9.13 have a mean of $\mu = 0$, but they have different standard deviations. Because the first distribution is clustered more toward the mean, its standard deviation is the smallest of the three.

Example 2 Finding Variance and Standard Deviation

Find the variance and standard deviation of the probability density function

$$f(x) = 2 - 2x, \quad 0 \le x \le 1.$$

SOLUTION Begin by finding the mean.

$$\begin{aligned}\mu &= \int_0^1 x(2 - 2x)\, dx \\ &= \int_0^1 (2x - 2x^2)\, dx \\ &= \left[x^2 - \frac{2x^3}{3}\right]_0^1 \\ &= \frac{1}{3} \qquad \text{Mean}\end{aligned}$$

Next, apply the formula for variance.

$$\begin{aligned}V(x) &= \int_0^1 \left(x - \frac{1}{3}\right)^2 (2 - 2x)\, dx \\ &= \int_0^1 \left(-2x^3 + \frac{10x^2}{3} - \frac{14x}{9} + \frac{2}{9}\right) dx \\ &= \left[-\frac{x^4}{2} + \frac{10x^3}{9} - \frac{7x^2}{9} + \frac{2x}{9}\right]_0^1 \\ &= \frac{1}{18} \qquad \text{Variance}\end{aligned}$$

Finally, you can conclude that the standard deviation is

$$\sigma = \sqrt{\frac{1}{18}} \approx 0.236. \qquad \text{Standard deviation}$$

✓Checkpoint 2

Find the variance and standard deviation of the probability density function in Checkpoint 1.

The integral for variance can be difficult to evaluate. The following alternative formula is sometimes simpler.

Alternative Formula for Variance

If f is a probability density function of a continuous random variable x over the interval $[a, b]$, then the **variance** of x is

$$V(x) = \int_a^b x^2 f(x)\, dx - \mu^2$$

where μ is the mean of x.

Example 3 Using the Alternative Formula

Find the standard deviation of the probability density function

$$f(x) = \frac{2}{\pi(x^2 - 2x + 2)}, \quad 0 \le x \le 2.$$

What percent of the distribution lies within one standard deviation of the mean?

SOLUTION Begin by using a symbolic integration utility to find the mean.

$$\mu = \int_0^2 \left[\frac{2}{\pi(x^2 - 2x + 2)}\right](x)\, dx$$
$$= 1 \qquad \text{Mean}$$

Next, use a symbolic integration utility to find the variance.

$$V(x) = \int_0^2 \left[\frac{2}{\pi(x^2 - 2x + 2)}\right](x^2)\, dx - 1^2$$
$$\approx 0.273 \qquad \text{Variance}$$

This implies that the standard deviation is

$$\sigma \approx \sqrt{0.273}$$
$$\approx 0.522. \qquad \text{Standard deviation}$$

To find the percent of the distribution that lies within one standard deviation of the mean, integrate the probability density function between

$$\mu - \sigma = 0.478$$

and

$$\mu + \sigma = 1.522.$$

Using a symbolic integration utility, you obtain

$$\int_{0.478}^{1.522} \frac{2}{\pi(x^2 - 2x + 2)}\, dx \approx 0.613.$$

So, about 61.3% of the distribution lies within one standard deviation of the mean. This result is illustrated in Figure 9.14.

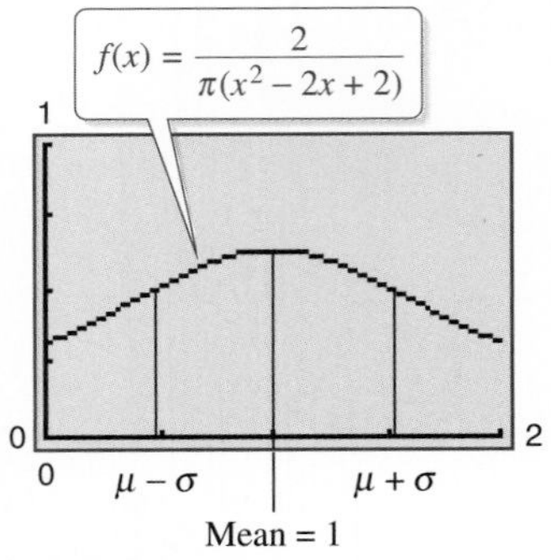

FIGURE 9.14

✓ Checkpoint 3

Use a symbolic integration utility to find the percent of the distribution in Example 3 that lies within 1.5 standard deviations of the mean.

Median

The mean of a probability density function is an example of a *measure of central tendency*. Another useful measure of central tendency is the **median.**

Definition of Median

If f is a probability density function of a continuous random variable x over the interval $[a, b]$, then the **median** of x is the number m such that

$$\int_a^m f(x)\,dx = 0.5.$$

Example 4 Comparing Mean and Median

In Example 5 in Section 9.2, the probability density function

$$f(t) = 0.1e^{-0.1t}, \quad 0 \le t < \infty$$

was used to model the useful lifetime of a product. Find the mean and median useful lifetimes.

SOLUTION Using integration by parts or a symbolic integration utility, you can find the mean to be

$$\begin{aligned}\mu &= \int_0^\infty 0.1te^{-0.1t}\,dt \\ &= 10 \text{ years.} \qquad \text{Mean}\end{aligned}$$

The median is given by

$$\begin{aligned}\int_0^m 0.1e^{-0.1t}\,dt &= 0.5 \\ \Big[-e^{-0.1t}\Big]_0^m &= 0.5 \\ -e^{-0.1m} + 1 &= 0.5 \\ -e^{-0.1m} &= -0.5 \\ e^{-0.1m} &= 0.5 \\ -0.1m &= \ln 0.5 \\ m &= -10 \ln 0.5 \\ m &\approx 6.93 \text{ years.} \qquad \text{Median}\end{aligned}$$

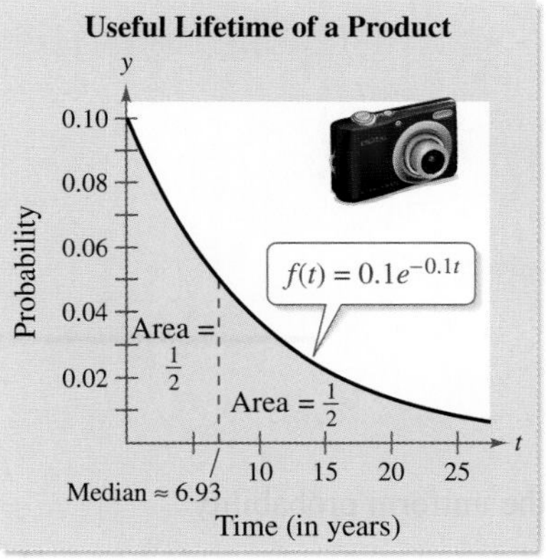

FIGURE 9.15

From this, you can see that the mean and median of a probability distribution can be quite different. Using the mean, the "average" lifetime of a product is 10 years, but using the median, the "average" lifetime is approximately 6.93 years. In Figure 9.15, note that half of the products have useful lifetimes of approximately 6.93 years or less. You can confirm this by evaluating the integral

$$\int_0^{-10 \ln 0.5} 0.1e^{-0.1t}\,dt$$

using a symbolic integration utility.

✓ Checkpoint 4

Find the mean and median of the probability density function

$$f(x) = 2e^{-2x}, \quad 0 \le x < \infty.$$

Special Probability Density Functions

The remainder of this section describes three common types of probability density functions: uniform, exponential, and normal. The **uniform probability density function** is defined as

$$f(x) = \frac{1}{b-a}, \quad a \le x \le b. \qquad \text{Uniform probability density function}$$

This probability density function represents a continuous random variable for which each outcome is equally likely.

Example 5 Analyzing a Probability Density Function

Find the expected value and standard deviation of the uniform probability density function

$$f(x) = \frac{1}{8}, \quad 0 \le x \le 8.$$

SOLUTION The expected value (or mean) is

$$\mu = \int_0^8 \frac{1}{8} x \, dx$$
$$= \left[\frac{x^2}{16}\right]_0^8$$
$$= 4. \qquad \text{Expected value}$$

The variance is

$$V(x) = \int_0^8 \frac{1}{8} x^2 \, dx - 4^2$$
$$= \left[\frac{x^3}{24}\right]_0^8 - 16$$
$$\approx 5.333. \qquad \text{Variance}$$

The standard deviation is

$$\sigma \approx \sqrt{5.333}$$
$$\approx 2.309. \qquad \text{Standard deviation}$$

The graph of f is shown in Figure 9.16.

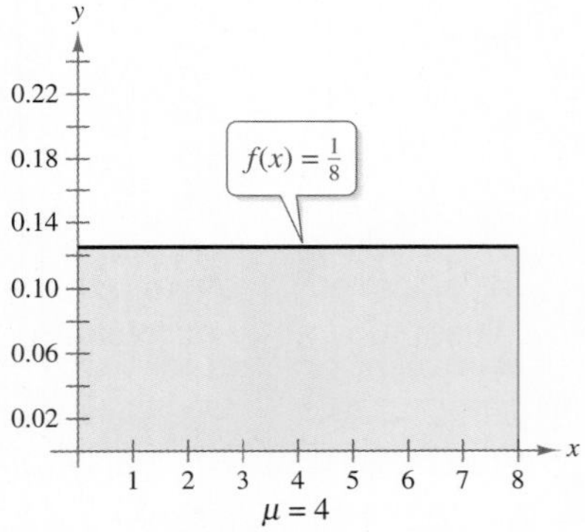

Uniform Probability Density Function
FIGURE 9.16

✓Checkpoint 5

Find the expected value and standard deviation of the uniform probability density function

$$f(x) = \tfrac{1}{2}, \quad 0 \le x \le 2.$$

It can be shown that the uniform probability density function has an expected value of

$$\mu = \tfrac{1}{2}(a + b) \qquad \text{Expected value}$$

and a variance of

$$V(x) = \tfrac{1}{12}(b - a)^2. \qquad \text{Variance}$$

Try using these formulas to confirm the results of Example 5.

The second special type of probability density function is the **exponential probability density function,** which has the form

$$f(x) = ae^{-ax}, \quad 0 \le x < \infty. \qquad \text{Exponential probability density function, } a > 0$$

This function has an expected value of

$$\mu = \frac{1}{a} \qquad \text{Expected value}$$

and a variance of

$$V(x) = \frac{1}{a^2}. \qquad \text{Variance}$$

The function in Example 4 is an exponential probability density function. Use the formula $\mu = 1/a$ to confirm the mean found in Example 4.

The third special type of probability density function (and the most widely used) is the **normal probability density function** given by

$$f(x) = \frac{1}{\sigma\sqrt{2\pi}}e^{-(x-\mu)^2/(2\sigma^2)}, \quad -\infty < x < \infty. \qquad \text{Normal probability density function}$$

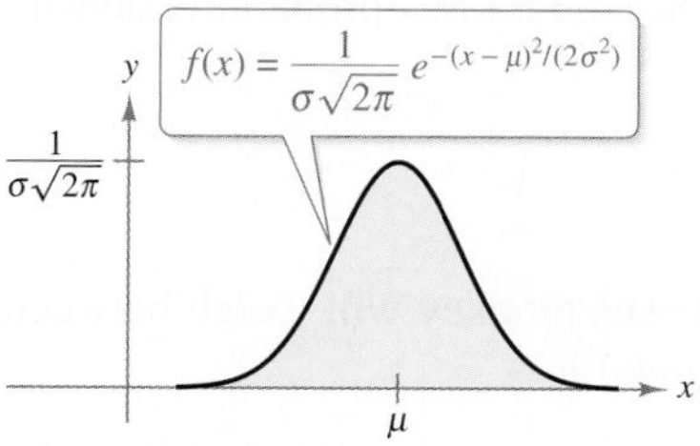

Normal Probability Density Function
FIGURE 9.17

The expected value of this function is μ, and the standard deviation is σ. Figure 9.17 shows the graph of a typical normal probability density function.

A normal probability density function for which

$$\mu = 0 \quad \text{and} \quad \sigma = 1$$

is called a **standard normal probability density function.**

TECH TUTOR

Finding the antiderivative of the normal probability density function analytically is beyond the scope of this text. You can use a graphing utility, a symbolic integration utility, or a spreadsheet to find this antiderivative. Use one of these tools to evaluate the definite integrals in Example 6.

Example 6 Finding a Probability

In 2010, the scores on the Graduate Management Admission Test (GMAT) could be modeled by a normal probability density function with a mean of $\mu = 540.4$ and a standard deviation of $\sigma = 120.72$. A person who took the GMAT is selected at random. What is the probability that the person scored between 600 and 700? What is the probability that the person scored between 700 and 800? *(Source: Graduate Management Admission Council)*

SOLUTION Using a symbolic integration utility, the first probability is

$$P(600 \le x \le 700) = \int_{600}^{700} \frac{1}{120.72\sqrt{2\pi}}e^{-(x-540.4)^2/(2\cdot 120.72^2)}\,dx$$
$$\approx 0.218.$$

So, the probability of choosing a person who scored between 600 and 700 is about 21.8%. In a similar way, you can find the probability of choosing a person who scored between 700 and 800. By evaluating the integral

$$P(700 \le x \le 800) = \int_{700}^{800} \frac{1}{120.72\sqrt{2\pi}}e^{-(x-540.4)^2/(2\cdot 120.72^2)}\,dx$$
$$\approx 0.077.$$

you can see that the probability is about 7.7%.

STUDY TIP

In Example 6, note that probabilities connected with normal distributions can be evaluated with a table of values, with a symbolic integration utility, or with a spreadsheet.

✓ Checkpoint 6

Find the probability that a person selected at random scored between 400 and 600 on the test described in Example 6.

In addition to being popular zoo animals, rhesus monkeys are commonly used in medical and behavioral research. Research on rhesus monkeys led to the discovery of the Rh factor in human red blood cells.

Example 7 Modeling Body Weight

Assume the weights of adult male rhesus monkeys are normally distributed with a mean of 17 pounds and a standard deviation of 4 pounds. In a typical population of adult male rhesus monkeys, what percent of the monkeys would have weights within one standard deviation of the mean?

SOLUTION For this population, $\mu = 17$ and $\sigma = 4$. So, the normal probability density function is

$$f(x) = \frac{1}{4\sqrt{2\pi}} e^{-(x-17)^2/32}.$$

The probability that a randomly chosen adult male rhesus monkey will weigh between 13 and 21 pounds (that is, within 4 pounds of 17 pounds) is

$$P(13 \le x \le 21) = \int_{13}^{21} \frac{1}{4\sqrt{2\pi}} e^{-(x-17)^2/32}\, dx$$
$$\approx 0.683.$$

So, about 68% of the adult male rhesus monkeys have weights that lie within one standard deviation of the mean, as shown in Figure 9.18.

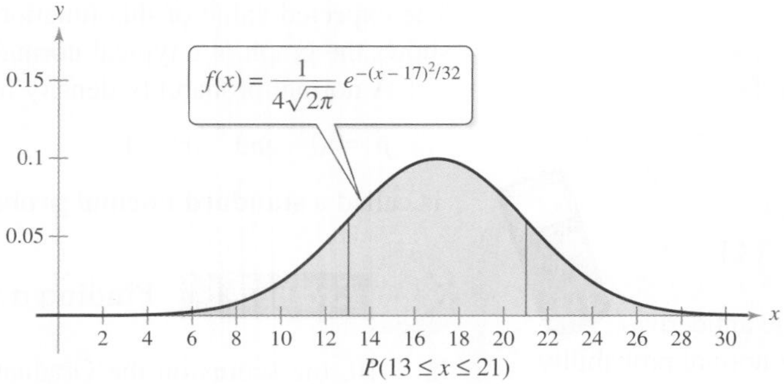

FIGURE 9.18

✓Checkpoint 7

Use the results of Example 7 to find the probability that an adult male rhesus monkey chosen at random will weigh more than 21 pounds. ■

The result described in Example 7 can be generalized to all normal distributions. That is, in any normal distribution, the probability that x lies within one standard deviation of the mean is about 68%. For normal distributions, 95.4% of the x-values lie within two standard deviations of the mean, and almost all (99.7%) of the x-values lie within three standard deviations of the mean.

SUMMARIZE (Section 9.3)

1. State the definition of expected value for a continuous random variable *(page 580)*. For an example of finding an expected value, see Example 1.
2. State the definitions of variance and standard deviation for a continuous random variable *(page 581)*. For an example of finding variance and standard deviation, see Example 2.
3. List the three special probability density functions discussed in this section *(pages 584 and 585)*. For examples of these functions, see Examples 4, 5, 6, and 7.

SKILLS WARM UP 9.3

The following warm-up exercises involve skills that were covered in earlier sections. You will use these skills in the exercise set for this section. For additional help, review Sections 5.4 and 9.2.

In Exercises 1 and 2, solve for *m*.

1. $\int_0^m \frac{1}{16}\,dx = 0.5$

2. $\int_0^m \frac{1}{3}e^{-t/3}\,dt = 0.5$

In Exercises 3 and 4, evaluate the definite integral.

3. $\int_0^2 \frac{x^2}{2}\,dx$

4. $\int_1^2 x(4-2x)\,dx$

In Exercises 5 and 6, find the indicated probability using the given probability density function.

5. $f(x) = \frac{1}{8}$, $[0, 8]$

(a) $P(x \le 2)$

(b) $P(3 < x < 7)$

6. $f(x) = 6x - 6x^2$, $[0, 1]$

(a) $P\left(x \le \frac{1}{2}\right)$

(b) $P\left(\frac{1}{4} \le x \le \frac{3}{4}\right)$

Exercises 9.3

See www.CalcChat.com for worked-out solutions to odd-numbered exercises.

Finding Mean, Variance, and Standard Deviation In Exercises 1–12, use the given probability density function over the indicated interval to find the (a) mean, (b) variance, and (c) standard deviation of the random variable. (d) Then sketch the graph of the density function and locate the mean on the graph. *See Examples 1, 2, and 3.*

1. $f(x) = \frac{1}{3}$, $[0, 3]$

2. $f(x) = \frac{1}{4}$, $[0, 4]$

3. $f(t) = \frac{t}{18}$, $[0, 6]$

4. $f(x) = \frac{x}{8}$, $[0, 4]$

5. $f(x) = \frac{4}{3x^2}$, $[1, 4]$

6. $f(x) = \frac{5}{2}x^{3/2}$, $[0, 1]$

7. $f(x) = 1 - \frac{1}{2}x$, $[0, 2]$

8. $f(x) = \frac{1}{2} - \frac{1}{8}x$, $[0, 4]$

9. $f(x) = 6x(1 - x)$, $[0, 1]$

10. $f(x) = \frac{3}{32}x(4 - x)$, $[0, 4]$

11. $f(x) = \frac{3}{16}\sqrt{4 - x}$, $[0, 4]$

12. $f(x) = \frac{1}{18}\sqrt{9 - x}$, $[0, 9]$

Finding the Mean and Median In Exercises 13–18, find the mean and median of the probability density function. *See Example 4.*

13. $f(x) = \frac{1}{11}$, $[0, 11]$

14. $f(x) = 0.05$, $[0, 20]$

15. $f(x) = 4(1 - 2x)$, $\left[0, \frac{1}{2}\right]$

16. $f(x) = \frac{4}{3} - \frac{2}{3}x$, $[0, 1]$

17. $f(t) = \frac{1}{9}e^{-t/9}$, $[0, \infty)$

18. $f(t) = \frac{2}{5}e^{-2t/5}$, $[0, \infty)$

Special Probability Density Functions In Exercises 19–24, identify the probability density function. Then find the mean, variance, and standard deviation without integrating. *See Example 5.*

19. $f(x) = \frac{1}{10}$, $[0, 10]$

20. $f(x) = \frac{1}{6}$, $[0, 6]$

21. $f(x) = \frac{1}{8}e^{-x/8}$, $[0, \infty)$

22. $f(x) = \frac{5}{3}e^{-5x/3}$, $[0, \infty)$

23. $f(x) = \frac{1}{11\sqrt{2\pi}}e^{-(x-100)^2/242}$, $(-\infty, \infty)$

24. $f(x) = \frac{1}{6\sqrt{2\pi}}e^{-(x-30)^2/72}$, $(-\infty, \infty)$

Finding Probabilities Using Technology In Exercises 25–30, use a symbolic integration utility to find the mean, standard deviation, and given probability.

Function	*Probability*
25. $f(x) = \dfrac{1}{\sqrt{2\pi}} e^{-x^2/2}$	$P(0 \le x \le 0.85)$
26. $f(x) = \dfrac{1}{\sqrt{2\pi}} e^{-x^2/2}$	$P(-1.21 \le x \le 1.21)$
27. $f(x) = \frac{1}{6} e^{-x/6}$	$P(x \ge 2.23)$
28. $f(x) = \frac{3}{4} e^{-3x/4}$	$P(x \ge 0.27)$
29. $f(x) = \dfrac{1}{2\sqrt{2\pi}} e^{-(x-8)^2/8}$	$P(3 \le x \le 13)$
30. $f(x) = \dfrac{1}{1.5\sqrt{2\pi}} e^{-(x-2)^2/4.5}$	$P(-2.5 \le x \le 2.5)$

Finding Probabilities In Exercises 31 and 32, let x be a random variable that is normally distributed with the given mean and standard deviation. Find the indicated probabilities using a symbolic integration utility. *See Example 6.*

31. $\mu = 50, \sigma = 10$

(a) $P(x > 55)$

(b) $P(x > 60)$

(c) $P(x < 60)$

(d) $P(30 < x < 55)$

32. $\mu = 70, \sigma = 14$

(a) $P(x > 65)$

(b) $P(x < 98)$

(c) $P(x < 49)$

(d) $P(56 < x < 75)$

33. Transportation The arrival time t (in minutes) of a bus at a bus stop is uniformly distributed between 10:00 A.M. and 10:10 A.M.

(a) Find the probability density function for the random variable t.

(b) Find the mean and standard deviation of the arrival times.

(c) What is the probability that you will miss the bus if you arrive at the bus stop at 10:03 A.M.?

34. Transportation Repeat Exercise 33 for a bus that arrives between 10:00 A.M. and 10:05 A.M.

35. Useful Life The time t (in years) until failure of an appliance is exponentially distributed with a mean of 2 years.

(a) Find the probability density function for the random variable t.

(b) Find the probability that the appliance will fail in less than 1 year.

36. Useful Life The time t (in years) until failure of a printer is exponentially distributed with a mean of 4 years.

(a) Find the probability density function for the random variable t.

(b) Find the probability that the printer will fail in more than 1 year but less than 3 years.

37. Waiting Time The waiting time t (in minutes) for service in a store is exponentially distributed with a mean of 5 minutes.

(a) Find the probability density function for the random variable t.

(b) Find the probability that the waiting time is within one standard deviation of the mean.

38. License Renewal The time t (in minutes) spent at a driver's license renewal center is exponentially distributed with a mean of 15 minutes.

(a) Find the probability density function for the random variable t.

(b) Find the probability that the time spent is within one standard deviation of the mean.

39. Demand The daily demand x for a certain product (in hundreds of pounds) is a random variable with the probability density function

$$f(x) = \frac{6}{343}x(7 - x), \quad [0, 7].$$

(a) Find the mean and standard deviation of the demand.

(b) Find the median of the demand.

(c) Find the probability that the demand is within one standard deviation of the mean.

40. Demand Repeat Exercise 39 for the probability density function

$$f(x) = \frac{3}{256}(x - 2)(10 - x), \quad [2, 10].$$

41. Consumer Trends The number of coupons x used by a customer in a grocery store is a random variable with the probability density function

$$f(x) = \frac{2x + 1}{12}, \quad [0, 3].$$

Find the expected number of coupons a customer will use.

42. Cost The daily cost x (in dollars) of electricity in a city is a random variable with the probability density function

$$f(x) = 0.28e^{-0.28x}, \quad [0, \infty).$$

Find the median daily cost of electricity.

43. Demand The daily demand x for water (in millions of gallons) in a town is a random variable with the probability density function

$$f(x) = \frac{1}{9}xe^{-x/3}, \quad [0, \infty).$$

(a) Find the mean and standard deviation of the demand.

(b) Find the probability that the demand is greater than 4 million gallons on a given day.

44. HOW DO YOU SEE IT? In chemistry, the probability of finding an electron at a particular position is greatest close to the nucleus and drops off rapidly as the distance from the nucleus increases. The graph displays the probability of finding the electron at points along a line drawn from the nucleus outward in any direction for the hydrogen 1s orbital. Make a sketch of this graph, and add to your sketch an indication of where you think the median might be. *(Source: Adapted from Zumdahl,* Chemistry, *Seventh Edition)*

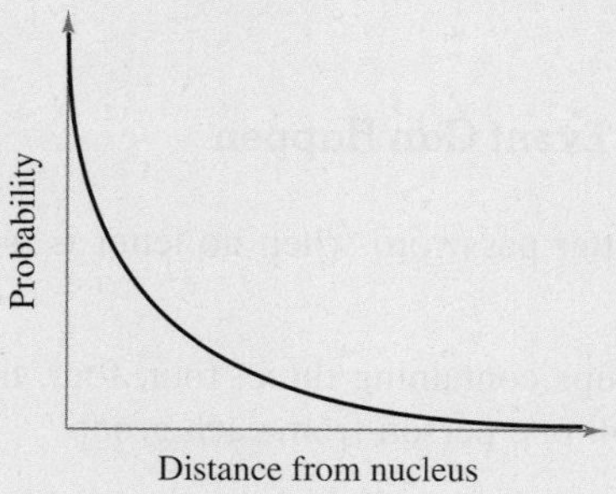

45. Useful Life The lifetime of a certain type of battery is normally distributed with a mean of 400 hours and a standard deviation of 24 hours. You purchased one of the batteries, and its useful life was 340 hours.

(a) How far, in standard deviations, did the useful life of your battery fall short of the expected life?

(b) Use a symbolic integration utility to approximate the percent of all other batteries of this type with useful lives that exceed that of your battery.

46. Education The scores on a national exam are normally distributed with a mean of 150 and a standard deviation of 16. You scored 174 on the exam.

(a) By how much, in standard deviations, did your score exceed the national mean?

(b) Use a symbolic integration utility to approximate the percent of those who took the exam who had scores lower than yours.

47. Useful Life A storage battery has an expected lifetime of 4.5 years with a standard deviation of 0.5 year. Assume that the useful lives of these batteries are normally distributed.

(a) Use a computer or graphing utility and Simpson's Rule (with $n = 12$) to approximate the probability that a given battery will last for 4 to 5 years.

(b) Will 10% of the batteries last less than 3 years?

48. Wages The employees of a large corporation are paid an average wage of \$14.50 per hour with a standard deviation of \$1.50. Assume that these wages are normally distributed.

(a) Use a computer or graphing utility and Simpson's Rule (with $n = 10$) to approximate the percent of employees who earn hourly wages of \$11 to \$14.

(b) Are 20% of the employees paid more than \$16 per hour?

49. Medical Science A medical research team has determined that for a group of 500 females, the length of pregnancy from conception to birth varies according to an approximately normal distribution with a mean of 266 days and a standard deviation of 16 days.

(a) Use a graphing utility to graph the distribution.

(b) Use a symbolic integration utility to approximate the probability that a pregnancy will last from 240 days to 280 days.

(c) Use a symbolic integration utility to approximate the probability that a pregnancy will last more than 280 days.

50. Education For high school graduates from 2008 through 2010, the scores on the ACT Test can be modeled by a normal probability density function with a mean of 21.1 and a standard deviation of 5.1. *(Source: ACT, Inc.)*

(a) Use a graphing utility to graph the distribution.

(b) Use a symbolic integration utility to approximate the probability that a person who took the ACT scored between 24 and 36.

(c) Use a symbolic integration utility to approximate the probability that a person who took the ACT scored more than 26.

51. Fuel Mileage Assume the fuel mileages of all 2011 model vehicles are normally distributed with a mean of 21.0 miles per gallon and a standard deviation of 5.6 miles per gallon. *(Source: U.S. Environmental Protection Agency)*

(a) Use a graphing utility to graph the distribution.

(b) Use a symbolic integration utility to approximate the probability that a vehicle's fuel mileage is between 25 and 30 miles per gallon.

(c) Use a symbolic integration utility to approximate the probability that a vehicle's fuel mileage is less than 18 miles per gallon.

ALGEBRA TUTOR

Using Counting Principles

In discrete probability, one of the basic skills is being able to count the number of ways an event can happen. To do this, the strategies below can be helpful.

1. *The Fundamental Counting Principle:* The number of ways that two or more events can occur is the product of the numbers of ways each event can occur by itself. These ways can be listed graphically using a tree diagram.
2. *Permutations:* The number of permutations of n elements is $n!$.
3. *Combinations:* The number of combinations of n elements taken r at a time is

$$_{n}C_{r} = \frac{n!}{(n-r)!r!}.$$

TECH TUTOR

Most graphing utilities have a *factorial* key. Consult your user's manual for specific keystrokes for your graphing utility.

In the strategies above, note that $n!$ (read as "n factorial") is defined as

$$n! = 1 \cdot 2 \cdot 3 \cdot 4 \cdots (n-1) \cdot n$$

where n is a positive integer. As a special case, zero factorial is defined as

$$0! = 1.$$

Example 1 Counting the Ways an Event Can Happen

a. How many ways can you form a five-letter password when no letter is used more than once?

b. Your class is divided into five work groups containing three, four, four, three, and five people. How many ways can you poll one person from each group?

c. In how many orders can seven runners finish a race when there are no ties?

d. You have 12 phone calls to return. In how many orders can you return them?

SOLUTION

a. For the first letter of the password, you have 26 choices. For the second letter, you have 25 choices. For the third letter, you have 24 choices, and so on.

$$\text{Number of ways} = 26 \cdot 25 \cdot 24 \cdot 23 \cdot 22 \qquad \text{Counting Principle}$$
$$= 7{,}893{,}600 \qquad \text{Multiply.}$$

b. When choosing one person from the first group, you have 3 choices. From the second group, you have 4 choices, and so on.

$$\text{Number of ways} = 3 \cdot 4 \cdot 4 \cdot 3 \cdot 5 \qquad \text{Counting Principle}$$
$$= 720 \qquad \text{Multiply.}$$

c. The solution is given by the number of permutations of the seven runners.

$$\text{Number of ways} = 7! \qquad \text{Permutation}$$
$$= 5040 \qquad \text{Use a calculator.}$$

d. The solution is given by the number of permutations of the 12 phone calls.

$$\text{Number of ways} = 12! \qquad \text{Permutation}$$
$$= 479{,}001{,}600 \qquad \text{Use a calculator.}$$

Example 2 Counting the Ways an Event Can Happen

How many different ways can you choose a three-person group

a. from a class of 20 people?

b. from a class of 40 people?

SOLUTION

a. The number of ways to choose a three-person group from a class of 20 is given by the number of combinations of 20 elements taken three at a time.

$$\begin{aligned}
\text{Number of ways} &= {}_{20}C_3 && \text{Combination} \\
&= \frac{20!}{17!\,3!} && \text{Formula for combination} \\
&= \frac{20 \cdot 19 \cdot 18 \cdot 17!}{17! \cdot 3!} \\
&= \frac{20 \cdot 19 \cdot 18}{3 \cdot 2 \cdot 1} && \text{Divide out common factors.} \\
&= 20 \cdot 19 \cdot 3 && \text{Divide out common factors.} \\
&= 1140 && \text{Multiply.}
\end{aligned}$$

b. The number of ways to choose a three-person group from a class of 40 is given by ${}_{40}C_3$, which is 9880. You can verify this answer using a graphing utility, as shown in the figure.

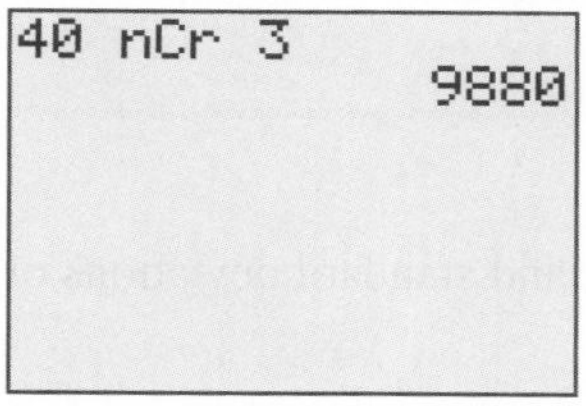

TECH TUTOR

Most graphing utilities have a *combination* key. Consult your user's manual for specific keystrokes for your graphing utility.

Example 3 Counting the Ways an Event Can Happen

To test for defective units, you are choosing a sample of 10 from a manufacturing production of 2000 units. How many different samples of 10 are possible?

SOLUTION The solution is given by the number of combinations of 2000 elements taken 10 at a time.

$$\begin{aligned}
\text{Number of ways} &= {}_{2000}C_{10} && \text{Combination} \\
&= \frac{2000!}{1990!\,10!} && \text{Formula for combination} \\
&\approx 2.76 \times 10^{26} && \text{Use a calculator.} \\
&= 276{,}000{,}000{,}000{,}000{,}000{,}000{,}000{,}000
\end{aligned}$$

From these examples, you can see that combinations and permutations can be very large numbers.

SUMMARY AND STUDY STRATEGIES

After studying this chapter, you should have acquired the following skills.
The exercise numbers are keyed to the Review Exercises that begin on page 593.
Answers to odd-numbered Review Exercises are given in the back of the text.*

Section 9.1	Review Exercises
■ Describe sample spaces and events for experiments.	*1–4*
■ Form frequency distributions for discrete random variables.	*5, 6*
■ Find probability distributions for discrete random variables.	*7, 8*
■ Determine whether tables represent probability distributions.	*9, 10*
■ Use probability distributions to find probabilities.	*11, 12*
■ Find the expected values (or means), variances, and standard deviations of discrete random variables.	*13–16*

$$\mu = E(x) = x_1P(x_1) + x_2P(x_2) + x_3P(x_3) + \cdot \cdot \cdot + x_mP(x_m)$$
$$V(x) = (x_1 - \mu)^2P(x_1) + \cdot \cdot \cdot + (x_m - \mu)^2P(x_m), \quad \sigma = \sqrt{V(x)}$$

■ Find expected values, variances, and standard deviations in real-life situations.	*17–20*

Section 9.2

■ Verify probability density functions.	*21–28*
■ Use probability density functions to find probabilities.	*29–34*

$$P(c \le x \le d) = \int_c^d f(x)\, dx$$

■ Use probability density functions to answer questions about real-life situations.	*35, 36*

Section 9.3

■ Find the means, variances, and standard deviations of continuous probability density functions.	*37–42*

$$\mu = E(x) = \int_a^b xf(x)\, dx, \quad V(x) = \int_a^b (x - \mu)^2 f(x)\, dx, \quad \sigma = \sqrt{V(x)}$$

■ Find the medians of probability density functions.	*43–46*

$$\int_a^m f(x)\, dx = 0.5$$

■ Find the means, variances, and standard deviations of special probability density functions.	*47–52*
■ Use continuous probability density functions to answer questions about real-life situations.	*53–60*

Study Strategies

■ **Using Technology** Integrals that arise with continuous probability density functions tend to be difficult to evaluate by hand. When evaluating such integrals, use a symbolic integration utility or a numerical integration technique such as Simpson's Rule with a programmable calculator.

* A wide range of valuable study aids are available to help you master the material in this chapter. The *Student Solutions Manual* includes step-by-step solutions to all odd-numbered exercises to help you review and prepare. The student website at *www.cengagebrain.com* offers algebra help and a *Graphing Technology Guide*, which contains step-by-step commands and instructions for a wide variety of graphing calculators.

Review Exercises

See www.CalcChat.com for worked-out solutions to odd-numbered exercises.

Finding Sample Spaces and Events In Exercises 1–4, list or describe the elements in the specified set.

1. **Die Toss** A 12-sided die is tossed once.
   (a) The sample space S
   (b) The event A that a number greater than 8 is tossed
   (c) The event B that an odd number is tossed
2. **Letters** A letter from the word *calculus* is selected.
   (a) The sample space S
   (b) The event A that the letter is a vowel
   (c) The event B that the letter is a c
3. **Lottery** A three-digit number is drawn in a lottery. Each digit is a number from 0 to 9.
   (a) The sample space S
   (b) The event A that the three-digit number starts with a 1
   (c) The event B that the three-digit number is divisible by 50
4. **Cards** One card is drawn from a standard 52-card deck of playing cards.
   (a) The sample space S
   (b) The event A that the card is red
   (c) The event B that the card is a black 5

Frequency Distributions In Exercises 5 and 6, complete the table to form the frequency distribution of the random variable x. Then construct a bar graph to represent the result.

5. **Bar Code** A computer randomly selects a three-digit bar code. Each digit can be 0 or 1, and x is the number of 1's in the bar code.

x	0	1	2	3
$n(x)$				

6. **Kittens** A cat has a litter of four kittens. Let x represent the number of male kittens.

x	0	1	2	3	4
$n(x)$					

7. **Bar Code** Use the frequency distribution in Exercise 5 to find the probability distribution for the random variable.
8. **Kittens** Use the frequency distribution in Exercise 6 to find the probability distribution for the random variable.

Identifying Probability Distributions In Exercises 9 and 10, determine whether the table represents a probability distribution. If it is a probability distribution, sketch its graph. If it is not a probability distribution, state any properties that are not satisfied.

9.

x	0	1	2	3
$P(x)$	0.25	0.45	0.20	0.15

10.

x	0	1	2	3	4	5
$P(x)$	$\frac{1}{25}$	$\frac{3}{25}$	$\frac{5}{25}$	$\frac{7}{25}$	$\frac{8}{25}$	$\frac{1}{25}$

Using Probability Distributions In Exercises 11 and 12, sketch a graph of the probability distribution and find the required probabilities.

11.

x	1	2	3	4	5
$P(x)$	$\frac{1}{18}$	$\frac{7}{18}$	$\frac{5}{18}$	$\frac{3}{18}$	$\frac{2}{18}$

(a) $P(2 \le x \le 4)$
(b) $P(x \ge 3)$

12.

x	−2	−1	1	3	5
$P(x)$	$\frac{1}{11}$	$\frac{2}{11}$	$\frac{4}{11}$	$\frac{3}{11}$	$\frac{1}{11}$

(a) $P(x < 0)$
(b) $P(x > 1)$

Finding Expected Value, Variance, and Standard Deviation In Exercises 13–16, find the expected value, variance, and standard deviation for the given probability distribution.

13.

x	0	1	2	3	4
$P(x)$	$\frac{1}{10}$	$\frac{3}{10}$	$\frac{2}{10}$	$\frac{3}{10}$	$\frac{1}{10}$

14.

x	1	2	3	4	5
$P(x)$	$\frac{1}{8}$	$\frac{1}{8}$	$\frac{2}{8}$	$\frac{3}{8}$	$\frac{1}{8}$

15.

x	0	1	2	3
$P(x)$	0.006	0.240	0.614	0.140

16.

x	0	1	2
$P(x)$	0.310	0.685	0.005

17. Revenue A publishing company introduces a new weekly magazine that sells for \$3.95. The marketing group of the company estimates that sales x (in thousands) will be approximated by the following probability function.

x	10	15	20	30	40
$P(x)$	0.10	0.20	0.50	0.15	0.05

(a) Find $E(x)$ and σ.

(b) Find the expected revenue.

18. Raffle A service organization is selling \$5 raffle tickets as part of a fundraising program. The first and second prizes are \$3000 and \$1000, respectively. In addition to the first and second prizes, there are 50 \$20 gift certificates to be awarded. The number of tickets sold is 2000. Find the player's expected net gain when one ticket is purchased.

19. Consumer Trends The probability distribution of the random variable x, the number of cars per household in a city, is shown in the table. Find the expected value, variance, and standard deviation of x.

x	0	1	2	3	4	5
$P(x)$	0.10	0.28	0.39	0.17	0.04	0.02

20. Vital Statistics The probability distribution of the random variable x, the number of children per family in a city, is shown in the table. Find the expected value, variance, and standard deviation of x.

x	0	1	2	3	4
$P(x)$	0.12	0.31	0.43	0.12	0.02

Identifying Probability Density Functions In Exercises 21–28, use a graphing utility to graph the function. Then determine whether the function f represents a probability density function over the given interval. If f is not a probability density function, identify the condition(s) that is (are) not satisfied.

21. $f(x) = \frac{1}{12}$, $[0, 12]$

22. $f(x) = \frac{1}{8}$, $[1, 8]$

23. $f(x) = \frac{1}{4}(3 - x)$, $[0, 4]$

24. $f(x) = \frac{3}{4}x^2(2 - x)$, $[0, 2]$

25. $f(x) = \dfrac{1}{4\sqrt{x}}$, $[1, 9]$

26. $f(x) = 8.75x^{3/2}(1 - x)$, $[0, 2]$

27. $f(x) = \frac{1}{8}e^{-x/8}$, $[0, 8]$

28. $f(x) = \frac{1}{5}e^{-x/5}$, $[0, \infty)$

Finding a Probability In Exercises 29–34, sketch the graph of the probability density function over the indicated interval and find the indicated probabilities.

29. $f(x) = \frac{1}{16}$, $[0, 16]$

(a) $P(0 < x < 5)$

(b) $P(12 < x < 13)$

(c) $P(x \geq 5)$

(d) $P(8 < x < 12)$

30. $f(x) = \dfrac{x}{32}$, $[0, 8]$

(a) $P(0 < x < 3)$

(b) $P(7 < x < 8)$

(c) $P(4 < x < 6)$

(d) $P(x \geq 6)$

31. $f(x) = \frac{1}{50}(10 - x)$, $[0, 10]$

(a) $P(0 < x < 2)$

(b) $P(x \geq 7)$

(c) $P(x \leq 5)$

(d) $P(8 < x < 9)$

32. $f(x) = \frac{1}{36}(9 - x^2)$, $[-3, 3]$

(a) $P(x < 2)$

(b) $P(x > -2)$

(c) $P(-1 < x < 2)$

(d) $P(x > 2)$

33. $f(x) = \dfrac{2}{(x + 1)^2}$, $[0, 1]$

(a) $P\left(0 < x < \frac{1}{2}\right)$

(b) $P\left(\frac{1}{4} < x < \frac{3}{4}\right)$

(c) $P\left(x \geq \frac{1}{2}\right)$

(d) $P\left(\frac{1}{10} < x < \frac{3}{10}\right)$

34. $f(x) = \dfrac{3}{128}\sqrt{x}$, $[0, 16]$

(a) $P(4 < x < 9)$

(b) $P(4 < x < 16)$

(c) $P(x < 9)$

(d) $P(0 < x < 12)$

35. Waiting Time The waiting time t (in minutes) for patients arriving at a health clinic is described by the probability density function

$$f(t) = \tfrac{1}{12}e^{-t/12}, \quad [0, \infty).$$

Find the probabilities that a patient will wait

(a) no more than 5 minutes.

(b) between 9 and 12 minutes.

36. Medicine The time t (in days) until recovery after a certain medical procedure is described by the probability density function

$$f(t) = \frac{1}{4\sqrt{t-4}}, \quad [5, 13].$$

Find the probability that a patient will take (a) no more than 6 days to recover and (b) at least 8 days to recover.

Finding Mean, Variance, and Standard Deviation In Exercises 37–42, use the given probability density function over the indicated interval to find the (a) mean, (b) variance, and (c) standard deviation of the random variable. (d) Then sketch the graph of the density function and locate the mean on the graph.

37. $f(x) = \frac{1}{7}, \ [0, 7]$

38. $f(x) = \frac{x}{72}, \ [0, 12]$

39. $f(x) = \frac{3}{2x^2}, \ [1, 3]$

40. $f(x) = \frac{8-x}{32}, \ [0, 8]$

41. $f(x) = \frac{2}{9}x(3-x), \ [0, 3]$

42. $f(x) = \frac{3}{16}\sqrt{x}, \ [0, 4]$

Finding the Median In Exercises 43–46, find the median of the probability density function.

43. $f(x) = \frac{1}{14}, \ [0, 14]$

44. $f(x) = \frac{3}{2} - x, \ [0, 1]$

45. $f(x) = 0.25e^{-x/4}, \ [0, \infty)$

46. $f(x) = \frac{5}{6}e^{-5x/6}, \ [0, \infty)$

Special Probability Density Functions In Exercises 47–52, identify the probability density function. Then find the mean, variance, and standard deviation without integrating.

47. $f(x) = \frac{1}{2}, \ [0, 2]$

48. $f(x) = \frac{1}{9}, \ [0, 9]$

49. $f(x) = \frac{1}{6}e^{-x/6}, \ [0, \infty)$

50. $f(x) = \frac{4}{5}e^{-4x/5}, \ [0, \infty)$

51. $f(x) = \frac{1}{3\sqrt{2\pi}}e^{-(x-16)^2/18}, \ (-\infty, \infty)$

52. $f(x) = \frac{1}{5\sqrt{2\pi}}e^{-(x-40)^2/50}, \ (-\infty, \infty)$

53. Transportation The arrival time t (in minutes) of a train at a train station is uniformly distributed between 7:00 A.M. and 7:20 A.M.

(a) Find the probability density function for the random variable t.

(b) Find the mean and standard deviation of the arrival time.

(c) What is the probability that you will miss the train if you arrive at the train station at 7:04 A.M.?

54. Transportation Repeat Exercise 53 for a train that arrives between 7:00 A.M. and 7:15 A.M.

55. Waiting Time The waiting time t (in minutes) for service at the checkout at a grocery store is exponentially distributed with a mean of 15 minutes.

(a) Find the probability density function for the random variable t.

(b) Find the probability that the waiting time is less than 10 minutes.

56. Useful Life The time t (in hours) until failure of a mechanical unit is exponentially distributed with a mean of 350 hours.

(a) Find the probability density function for the random variable t.

(b) Find the probability that the mechanical unit will fail in more than 400 hours but less than 500 hours.

57. Demand The daily demand x for a certain product (in hundreds of pounds) is a random variable with the probability density function

$$f(x) = \frac{3}{256}x(8-x), \quad [0, 8].$$

(a) Find the mean and standard deviation of the demand.

(b) Find the median of the demand.

(c) Find the probability that the demand is within one standard deviation of the mean.

58. Demand Repeat Exercise 57 for the probability density function $f(x) = \frac{6}{125}x(5-x), [0, 5]$.

59. Heart Transplants Assume the waiting times for heart transplants are normally distributed with a mean of 168 days and a standard deviation of 25 days. *(Source: Organ Procurement and Transplant Network)*

(a) Use a graphing utility to graph the distribution.

(b) Use a symbolic integration utility to approximate the probability that a waiting time is between 70 and 105 days.

(c) Use a symbolic integration utility to approximate the probability that a waiting time is more than 120 days.

60. Botany In a botany experiment, plants are grown in a nutrient solution. The heights of the plants are found to be normally distributed with a mean of 42 centimeters and a standard deviation of 3 centimeters.

(a) Use a graphing utility to graph the distribution.

(b) Use a symbolic integration utility to approximate the probability that the height of a plant is between 40 and 45 centimeters.

(c) Use a symbolic integration utility to approximate the probability that the height of a plant is less than 50 centimeters.

TEST YOURSELF

See www.CalcChat.com for worked-out solutions to odd-numbered exercises.

Take this test as you would take a test in class. When you are done, check your work against the answers given in the back of the book.

1. Four students answer a true-false question on an exam. The random variable x is the number of answers of *true* among the four students.

(a) Write the sample space for the possible outcomes.

(b) Find the frequency distribution for the random variable x.

(c) Find the probability distribution for the random variable x.

2. A card is chosen at random from a standard 52-card deck of playing cards. What is the probability that the card will be red and not a face card?

In Exercises 3 and 4, sketch a graph of the probability distribution and find the indicated probabilities.

3.

x	1	2	3	4
$P(x)$	$\frac{3}{16}$	$\frac{7}{16}$	$\frac{1}{16}$	$\frac{5}{16}$

(a) $P(x < 3)$ (b) $P(x \geq 3)$

4.

x	7	8	9	10	11
$P(x)$	0.21	0.13	0.19	0.42	0.05

(a) $P(7 \leq x \leq 10)$ (b) $P(x > 8)$

In Exercises 5 and 6, find the expected value, variance, and standard deviation for the given probability distribution.

5.

x	0	1	2	3
$P(x)$	$\frac{2}{10}$	$\frac{1}{10}$	$\frac{4}{10}$	$\frac{3}{10}$

6.

x	-2	-1	0	1	2
$P(x)$	0.141	0.305	0.257	0.063	0.234

In Exercises 7–9, use a graphing utility to graph the function. Then determine whether the function f represents a probability density function over the given interval. If f is not a probability density function, identify the condition(s) that is (are) not satisfied.

7. $f(x) = \dfrac{1}{16}$, $[0, 8]$ **8.** $f(x) = \dfrac{3 - x}{6}$, $[-1, 1]$ **9.** $f(x) = \dfrac{3}{4}e^{-3x/4}$, $[0, \infty)$

In Exercises 10–12, find the indicated probabilities for the probability density function.

10. $f(x) = \dfrac{2x}{9}$, $[0, 3]$ (a) $P(0 \leq x \leq 1)$ (b) $P(2 \leq x \leq 3)$

11. $f(x) = 4(x - x^3)$, $[0, 1]$ (a) $P(0 < x < 0.5)$ (b) $P(0.25 \leq x < 1)$

12. $f(x) = 2xe^{-x^2}$, $[0, \infty)$ (a) $P(x < 1)$ (b) $P(x \geq 1)$

In Exercises 13–15, find the mean, variance, and standard deviation of the probability density function.

13. $f(x) = \frac{1}{14}$, $[0, 14]$ **14.** $f(x) = 3x - \frac{3}{2}x^2$, $[0, 1]$

15. $f(x) = e^{-x}$, $[0, \infty)$

16. An *intelligence quotient* or IQ is a number that is meant to measure intelligence. The IQs of students in a school are normally distributed with a mean of 110 and a standard deviation of 10. Use a symbolic integration utility to find the probability that a student selected at random will have an IQ within one standard deviation of the mean.

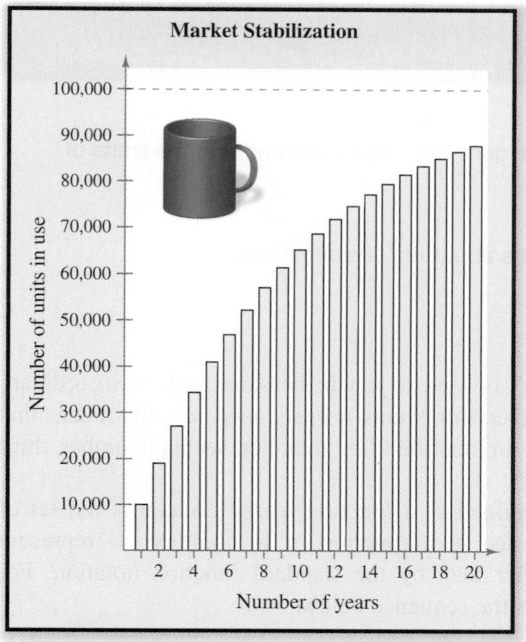

Example 8 on page 614 shows how a convergent series can be used to find the stabilization point of a product.

10 Series and Taylor Polynomials

10.1 Sequences

- Find the terms of sequences.
- Determine the convergence or divergence of sequences and find the limits of convergent sequences.
- Find patterns for sequences.
- Use sequences to answer questions about real-life situations.

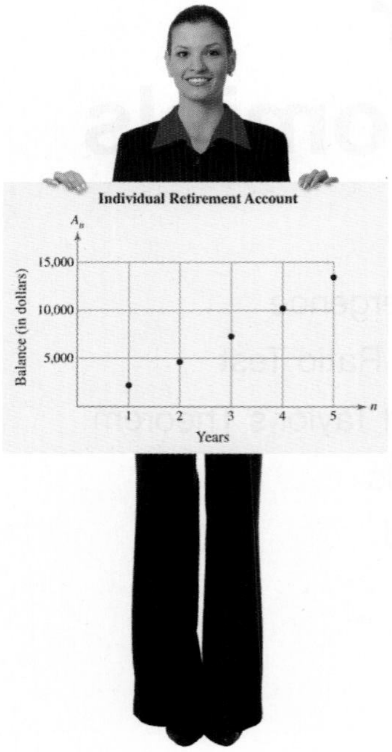

In Exercise 63 on page 605, you will use a sequence to find the balance in an account.

Sequences

In mathematics, the word "sequence" is used in much the same way as in ordinary English. To say that a collection of objects or events is *in sequence* usually means that the collection is ordered so that it has an identified first member, second member, third member, and so on.

Mathematically, a sequence is defined as a function whose domain is the set of positive integers. Although a sequence is a function, it is common to represent sequences by subscript notation rather than by the standard function notation. For instance, the equation $a_n = 2^n$ defines the sequence below.

$$\begin{array}{cccccc} a_1, & a_2, & a_3, & a_4, & \ldots, & a_n, \ldots \\ \downarrow & \downarrow & \downarrow & \downarrow & & \downarrow \\ 2, & 4, & 8, & 16, & \ldots, & 2^n, \ldots \end{array}$$

Definition of Sequence

A **sequence** is a function whose domain is the set of positive integers. The function values $a_1, a_2, a_3, \ldots, a_n, \ldots$ are the **terms** of the sequence. The number a_n is the ***n*th term** of the sequence, and the entire sequence is denoted by $\{a_n\}$.

STUDY TIP

Occasionally it is convenient to begin subscripting a sequence with zero. In such cases, you can write

$$a_0, a_1, a_2, \ldots, a_n, \ldots.$$

Example 1 Finding Terms of a Sequence

Write the first four terms of (a) $a_n = 2n + 1$ and (b) $b_n = 3/(n + 1)$.

SOLUTION

a. The first four terms of the sequence whose nth term is $a_n = 2n + 1$ are

$$\begin{array}{cccc} a_1, & a_2, & a_3, & a_4 \\ \downarrow & \downarrow & \downarrow & \downarrow \\ 2(1) + 1 = 3, & 2(2) + 1 = 5, & 2(3) + 1 = 7, & 2(4) + 1 = 9. \end{array}$$

b. The first four terms of the sequence whose nth term is $b_n = \dfrac{3}{n + 1}$ are

$$\begin{array}{cccc} b_1, & b_2, & b_3, & b_4 \\ \downarrow & \downarrow & \downarrow & \downarrow \\ \dfrac{3}{(1) + 1} = \dfrac{3}{2}, & \dfrac{3}{(2) + 1} = \dfrac{3}{3}, & \dfrac{3}{(3) + 1} = \dfrac{3}{4}, & \dfrac{3}{(4) + 1} = \dfrac{3}{5}. \end{array}$$

✓ Checkpoint 1

Write the first four terms, starting with $n = 1$, of (a) $a_n = 3n - 1$ and (b) $b_n = n/(n^2 + 1)$. ■

The Limit of a Sequence

The primary focus of this chapter is sequences whose terms approach limiting values. Such sequences are said to **converge.** If the limit of a sequence does not exist, then the sequence **diverges.** For instance, the terms of the sequence

$$\frac{1}{2}, \frac{1}{4}, \frac{1}{8}, \frac{1}{16}, \ldots, \frac{1}{2^n}, \ldots$$

approach 0 as n increases. You can write this limit as $\lim\limits_{n\to\infty} a_n = \lim\limits_{n\to\infty} \frac{1}{2^n} = 0.$

Although there are technical differences, you can for the most part operate with limits of sequences just as you did with limits of continuous functions in Section 3.6. For instance, to evaluate the limit of the sequence whose nth term is

$$a_n = \frac{2n}{n+1}$$

you can write

$$\lim_{n\to\infty} \frac{2n}{n+1} = \lim_{n\to\infty} \frac{2}{1+(1/n)} \qquad \text{Divide numerator and denominator by } n.$$

$$= \frac{2}{1+0} \qquad \text{Take limit as } n \to \infty.$$

$$= 2. \qquad \text{Limit of sequence}$$

TECH TUTOR

Symbolic algebra utilities are capable of evaluating the limit of a sequence. Use a symbolic algebra utility to evaluate the limits in Example 2.

Example 2 Finding the Limit of a Sequence

Find the limit of each sequence (if it exists) as n approaches infinity.

a. $a_n = 3 + (-1)^n$ **b.** $a_n = \dfrac{n}{1-2n}$ **c.** $a_n = \dfrac{2^n}{2^n - 1}$

SOLUTION

a. The terms of the sequence whose nth term is $a_n = 3 + (-1)^n$ alternate between 2 and 4.

$$a_1 = 2, \quad a_2 = 4, \quad a_3 = 2, \quad a_4 = 4, \ldots$$

So, the limit as $n \to \infty$ does not exist, and the sequence diverges.

b. The limit of the sequence whose nth term is $a_n = n/(1 - 2n)$ is

$$\lim_{n\to\infty} \frac{n}{1-2n} = \lim_{n\to\infty} \frac{1}{(1/n) - 2} \qquad \text{Divide numerator and denominator by } n.$$

$$= -\frac{1}{2}. \qquad \text{Take limit as } n \to \infty.$$

So, the sequence converges to $-\frac{1}{2}$.

c. The limit of the sequence whose nth term is $a_n = 2^n/(2^n - 1)$ is

$$\lim_{n\to\infty} \frac{2^n}{2^n - 1} = \lim_{n\to\infty} \frac{1}{1 - (1/2^n)} \qquad \text{Divide numerator and denominator by } 2^n.$$

$$= 1. \qquad \text{Take limit as } n \to \infty.$$

So, the sequence converges to 1.

✓ Checkpoint 2

Find the limit (if it exists) of (a) $a_n = n/(1 - n^2)$ and (b) $b_n = 2^{n+1}/(2^n + 1)$ as n approaches infinity. ■

David Davis/www.shutterstock.com

In this chapter you will learn that many important sequences in calculus involve factorials. If n is a positive integer, then **n factorial** is defined as

$$n! = 1 \cdot 2 \cdot 3 \cdot 4 \cdot \cdot \cdot (n-1) \cdot n. \qquad n \text{ factorial}$$

As a special case, $0!$ is defined to be 1.

$0! = 1$
$1! = 1$
$2! = 1 \cdot 2 = 2$
$3! = 1 \cdot 2 \cdot 3 = 6$
$4! = 1 \cdot 2 \cdot 3 \cdot 4 = 24$
$5! = 1 \cdot 2 \cdot 3 \cdot 4 \cdot 5 = 120$
$6! = 1 \cdot 2 \cdot 3 \cdot 4 \cdot 5 \cdot 6 = 720$

ALGEBRA TUTOR

For help in simplifying factorial expressions, see Example 1 in the *Chapter 10 Algebra Tutor,* on page 652.

Factorials follow the same conventions for order of operations as exponents. That is, just as $2x^3$ and $(2x)^3$ imply different orders of operations, $2n!$ and $(2n)!$ imply different orders, as shown.

$$2n! = 2(n!) = 2(1 \cdot 2 \cdot 3 \cdot 4 \cdot \cdot \cdot n)$$
$$(2n)! = 1 \cdot 2 \cdot 3 \cdot 4 \cdot \cdot \cdot n \cdot (n+1) \cdot \cdot \cdot (2n)$$

TECH TUTOR

Most graphing utilities have a *factorial* key. Consult your user's manual for specific keystrokes for your graphing utility.

Try evaluating $n!$ for several values of n. You will find that n does not have to be very large before $n!$ becomes huge. For instance,

$$10! = 3{,}628{,}800.$$

Example 3 Finding the Limit of a Sequence

Find the limit of the sequence whose nth term is

$$a_n = \frac{(-1)^n}{n!}.$$

SOLUTION One way to determine the limit is to write several terms of the sequence and look for a pattern.

$$\begin{array}{cccccc} a_1, & a_2, & a_3, & a_4, & a_5, & a_6 \\ \downarrow & \downarrow & \downarrow & \downarrow & \downarrow & \downarrow \\ -\frac{1}{1}, & \frac{1}{2}, & -\frac{1}{6}, & \frac{1}{24}, & -\frac{1}{120}, & \frac{1}{720} \end{array}$$

From these terms, it is clear that the denominator is increasing without bound while the numerator is bounded. So, you can write

$$\lim_{n\to\infty} \frac{(-1)^n}{n!} = 0.$$

This result is shown graphically in Figure 10.1. Note that the terms of the sequence alternate between positive and negative values.

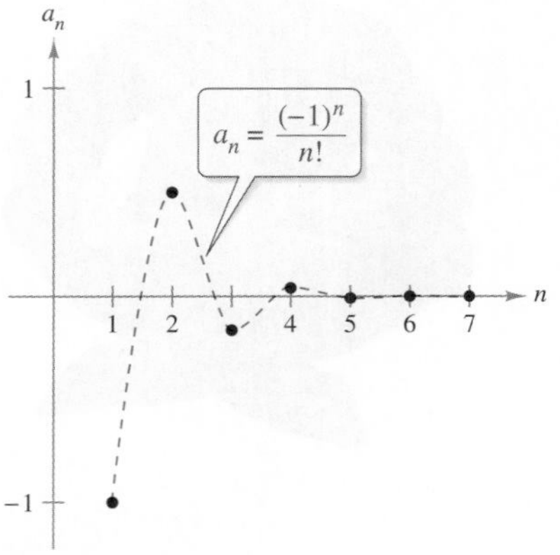

FIGURE 10.1

TECH TUTOR

Try using a symbolic algebra utility to evaluate the limit in Example 3.

✓ Checkpoint 3

Find the limit of the sequence whose nth term is

$$a_n = \frac{(-1)^{n+1}}{(n+1)!}.$$

■

Pattern Recognition for Sequences

Sometimes the first several terms of a sequence are listed without the *n*th term. In such cases, you need to discover a *pattern* in the sequence and find a formula for the *n*th term. Once the *n*th term has been specified, you can investigate the convergence or divergence of the sequence.

Example 4 **Finding a Pattern for a Sequence**

Determine an *n*th term for the sequence and then decide whether the sequence converges or diverges.

$$\frac{1}{3}, \frac{1}{9}, \frac{1}{27}, \frac{1}{81}, \ldots$$

SOLUTION Begin by observing that the numerators are the constant 1 and the denominators are successive powers of 3. So, you can write the *n*th term as

$$a_n = \frac{1}{3^n}.$$

In the terms of the sequence, it is clear that the denominator is increasing without bound while the numerator is bounded, as shown in the table.

n	1	5	10	20
$\frac{1}{3^n}$	$\frac{1}{3}$	$\frac{1}{3^5} = \frac{1}{243}$	$\frac{1}{3^{10}} = \frac{1}{59{,}049}$	$\frac{1}{3^{20}} = \frac{1}{3{,}486{,}784{,}401}$

So, you can write

$$\lim_{n\to\infty} \frac{1}{3^n} = 0$$

which means that the sequence converges to zero.

✓Checkpoint 4

Determine an *n*th term for the sequence and then decide whether the sequence converges or diverges.

$$1, \frac{1}{2}, \frac{1}{3}, \frac{1}{4}, \frac{1}{5}, \ldots$$

■

Searching for a pattern for the *n*th term of a sequence can be difficult. It helps to consider the patterns below.

nth Term	*Terms*	*Type of Sequence*
$(-1)^n$	$-1, 1, -1, 1, -1, 1, -1, 1, \ldots$	Changes in sign
$(-1)^{n+1}$	$1, -1, 1, -1, 1, -1, 1, -1, \ldots$	Changes in sign
$an + b$	$a + b, 2a + b, 3a + b, 4a + b, \ldots$	Arithmetic
ar^{n-1}	$a, ar, ar^2, ar^3, ar^4, \ldots$	Geometric
$n!$	$1, 2, 6, 24, 120, 720, \ldots$	Factorial
n^p	$1, 2^p, 3^p, 4^p, 5^p, 6^p, \ldots$	Power

For instance, because the sequence

$$4, 7, 10, 13, 16, \ldots$$

is arithmetic (the difference between consecutive terms, 3, is the same), an *n*th term for the sequence is $3n + 1$.

Without a specific rule for the nth term of a sequence, it is not possible to determine the convergence or divergence of the sequence; knowing the first several terms is not enough. For instance, the first three terms of the four sequences below are identical. Yet, from their nth terms, you can determine that two of the sequences converge to zero, one converges to $\frac{1}{9}$, and one diverges.

$$\{a_n\} = \left\{\frac{1}{2}, \frac{1}{4}, \frac{1}{8}, \frac{1}{16}, \ldots, \frac{1}{2^n}, \ldots\right\}$$

$$\{b_n\} = \left\{\frac{1}{2}, \frac{1}{4}, \frac{1}{8}, \frac{1}{15}, \ldots, \frac{6}{(n+1)(n^2-n+6)}, \ldots\right\}$$

$$\{c_n\} = \left\{\frac{1}{2}, \frac{1}{4}, \frac{1}{8}, \frac{7}{62}, \ldots, \frac{n^2-3n+3}{9n^2-25n+18}, \ldots\right\}$$

$$\{d_n\} = \left\{\frac{1}{2}, \frac{1}{4}, \frac{1}{8}, 0, \ldots, \frac{-n(n+1)(n-4)}{6(n^2+3n-2)}, \ldots\right\}$$

So, when only the first several terms of a sequence are given, there are many possible patterns that can be used to write a formula for the nth term. In such a situation, remember that your decision as to whether the sequence converges or diverges depends on your description of the nth term.

Example 5 Finding a Pattern for a Sequence

Determine an nth term for the sequence

$$-\frac{1}{1}, \frac{3}{2}, -\frac{7}{6}, \frac{15}{24}, -\frac{31}{120}, \ldots.$$

SOLUTION Begin by observing that the numerators are 1 less than 2^n.

$$2^1 - 1 = 1$$
$$2^2 - 1 = 3$$
$$2^3 - 1 = 7$$
$$2^4 - 1 = 15$$
$$2^5 - 1 = 31$$

So, you can generate the numerators by the rule

$$2^n - 1, \quad n = 1, 2, 3, 4, 5, \ldots.$$

Factoring the denominators produces

$$1 = 1!$$
$$2 = 1 \cdot 2 = 2!$$
$$6 = 1 \cdot 2 \cdot 3 = 3!$$
$$24 = 1 \cdot 2 \cdot 3 \cdot 4 = 4!$$
$$120 = 1 \cdot 2 \cdot 3 \cdot 4 \cdot 5 = 5!$$

So, the denominators can be represented by $n!$. Finally, because the signs alternate, you can write

$$a_n = (-1)^n\left(\frac{2^n - 1}{n!}\right)$$

as one possible formula for the nth term of this sequence.

✓ Checkpoint 5

Determine an nth term for the sequence $\frac{1}{2}, -\frac{4}{6}, \frac{9}{24}, -\frac{16}{120}, \ldots.$ ■

Application

There are many applications of sequences in business and economics. The next example involves the balance in an account for which the interest is compounded monthly. The terms of the sequence are the balances at the end of the first month, the end of the second month, and so on.

Example 6 Finding Balances

A deposit of \$1000 is made in an account that earns 3% interest, compounded monthly. The balance in the account after n months is given by

$$A_n = 1000\left(1 + \frac{0.03}{12}\right)^n.$$

a. Write the first three terms of the sequence.

b. Find the balance in the account after four years by computing the 48th term of the sequence.

SOLUTION

a. The first three terms of the sequence are shown below.

$$A_1 = 1000\left(1 + \frac{0.03}{12}\right)^1 = \$1002.50 \qquad \text{One-month balance}$$

$$A_2 = 1000\left(1 + \frac{0.03}{12}\right)^2 = \$1005.01 \qquad \text{Two-month balance}$$

$$A_3 = 1000\left(1 + \frac{0.03}{12}\right)^3 = \$1007.52 \qquad \text{Three-month balance}$$

b. The 48th term of the sequence is

$$A_{48} = 1000\left(1 + \frac{0.03}{12}\right)^{48} = \$1127.33. \qquad \text{Four-year balance}$$

✓Checkpoint 6

A deposit of \$1000 is made in an account that earns 3% interest, compounded quarterly. The balance in the account after n quarters is given by

$$A_n = 1000\left(1 + \frac{0.03}{4}\right)^n.$$

a. Write the first three terms of the sequence.

b. Find the balance in the account after four years by computing the 16th term of the sequence.

SUMMARIZE (Section 10.1)

1. State the definition of a sequence *(page 598)*. For an example of finding the terms of a sequence, see Example 1.
2. Explain how to determine the convergence or divergence of a sequence *(page 599)*. For examples of finding the limit of a sequence, see Examples 2 and 3.
3. Describe a real-life example of how a sequence can be used to represent the monthly balances in an account. *(page 603, Example 6).*

holbox/www.shutterstock.com

SKILLS WARM UP 10.1

The following warm-up exercises involve skills that were covered in a previous course or in earlier sections. You will use these skills in the exercise set for this section. For additional help, review Appendix Sections A.4 and A.5, and Sections 1.5 and 3.6.

In Exercises 1–4, find the limit.

1. $\lim_{x\to\infty} \frac{1}{x^3}$ **2.** $\lim_{x\to\infty} \frac{2x^2}{x^2+1}$ **3.** $\lim_{x\to\infty} \frac{x^3-1}{x^2+2}$ **4.** $\lim_{x\to\infty} \frac{1}{2^{x-1}}$

In Exercises 5–8, simplify the expression.

5. $\frac{n^2-4}{n^2+2n}$ **6.** $\frac{n^2+n-12}{n^2-16}$ **7.** $\frac{3}{n}+\frac{1}{n^3}$ **8.** $\frac{1}{n-1}+\frac{1}{n+2}, \quad n \ge 2$

Exercises 10.1

See www.CalcChat.com for worked-out solutions to odd-numbered exercises.

Finding the Terms of a Sequence In Exercises 1–10, write the first five terms of the sequence. *See Example 1.*

1. $a_n = 2n - 1$ **2.** $a_n = 5n + 2$

3. $a_n = 3^n$ **4.** $a_n = \left(-\frac{1}{2}\right)^n$

5. $a_n = \frac{n}{n+1}$ **6.** $a_n = \frac{n-1}{n^2+2}$

7. $a_n = \frac{3^n}{n!}$ **8.** $a_n = \frac{3n!}{(n-1)!}$

9. $a_n = \frac{(-1)^n}{n^2}$ **10.** $a_n = 5 - \frac{1}{n} + \frac{1}{n^2}$

Finding the Limit of a Sequence In Exercises 11–30, find the limit of the sequence (if it exists) as n approaches infinity. Then state whether the sequence converges or diverges. *See Examples 2 and 3.*

11. $a_n = \frac{5}{n}$ **12.** $a_n = \frac{n}{2}$

13. $a_n = 3 - \frac{1}{2^n}$ **14.** $a_n = 5 - \frac{1}{4^n}$

15. $a_n = (0.5)^n$ **16.** $a_n = \frac{1}{n^{3/2}}$

17. $a_n = \frac{n+1}{n}$ **18.** $a_n = \frac{n+1}{n^2-3}$

19. $a_n = \frac{n^2+3n-4}{2n^2+n-3}$ **20.** $a_n = \frac{\sqrt{n}}{\sqrt{n+1}}$

21. $a_n = \frac{n^2-25}{n+5}$ **22.** $a_n = \frac{n+2}{n^2+1}$

23. $a_n = \frac{1+(-1)^n}{n}$ **24.** $a_n = 1 + (-1)^n$

25. $a_n = \frac{n!}{n}$ **26.** $a_n = \frac{n!}{(n+1)!}$

27. $a_n = \frac{(n+1)!}{n!}$ **28.** $a_n = \frac{(n-2)!}{n!}$

29. $a_n = (-1)^n\left(\frac{n}{n+1}\right)$ **30.** $a_n = (-1)^n\frac{n}{n^2+1}$

Using Graphs to Determine Convergence In Exercises 31 and 32, use the graph of the sequence to decide whether the sequence converges or diverges. Then verify your result analytically.

31.

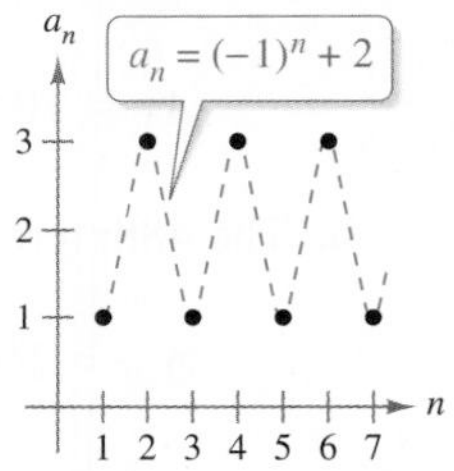

32.

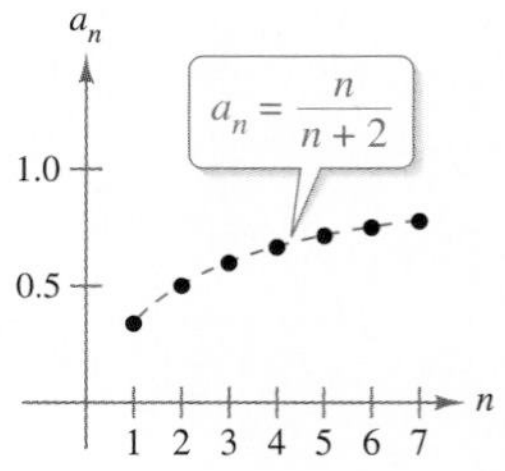

Finding a Pattern for a Sequence In Exercises 33–46, write an expression for the nth term of the sequence. (There is more than one correct answer.) *See Examples 4 and 5.*

33. $1, 4, 7, 10, \ldots$ **34.** $3, 7, 11, 15, \ldots$

35. $-1, 4, 9, 14, \ldots$ **36.** $1, \frac{1}{4}, \frac{1}{9}, \frac{1}{16}, \ldots$

37. $\frac{2}{3}, \frac{3}{4}, \frac{4}{5}, \frac{5}{6}, \ldots$ **38.** $2, \frac{3}{3}, \frac{4}{5}, \frac{5}{7}, \frac{6}{9}, \ldots$

39. $2, -1, \frac{1}{2}, -\frac{1}{4}, \frac{1}{8}, \ldots$ **40.** $\frac{1}{3}, \frac{2}{9}, \frac{4}{27}, \frac{8}{81}, \ldots$

41. $2, 1+\frac{1}{2}, 1+\frac{1}{3}, 1+\frac{1}{4}, \ldots$

42. $1+\frac{1}{2}, 1+\frac{1}{4}, 1+\frac{1}{8}, 1+\frac{1}{16}, \ldots$

43. $-2, 2, -2, 2, \ldots$ **44.** $2, -4, 6, -8, 10, \ldots$

45. $4, \frac{8}{2}, \frac{12}{6}, \frac{16}{24}, \frac{20}{120}, \ldots$ **46.** $-\frac{3}{2}, \frac{9}{6}, -\frac{27}{24}, \frac{81}{120}, \ldots$

Using Arithmetic Sequences In Exercises 47–50, write the next two terms of the arithmetic sequence. Describe the pattern you used to find these terms.

47. $2, 5, 8, 11, \ldots$ **48.** $\frac{7}{2}, 4, \frac{9}{2}, 5, \ldots$

49. $1, \frac{5}{3}, \frac{7}{3}, 3, \ldots$ **50.** $\frac{1}{2}, \frac{5}{4}, 2, \frac{11}{4}, \ldots$

Using Geometric Sequences In Exercises 51–54, write the next two terms of the geometric sequence. Describe the pattern you used to find these terms.

51. $3, -\frac{3}{2}, \frac{3}{4}, -\frac{3}{8}, . . .$

52. $5, 10, 20, 40, . . .$

53. $2, 6, 18, 54, . . .$

54. $9, 6, 4, \frac{8}{3}, . . .$

Identifying Sequences In Exercises 55–58, determine whether the sequence is arithmetic or geometric. Then write the nth term of the sequence.

55. $20, 10, 5, \frac{5}{2}, . . .$

56. $100, 92, 84, 76, . . .$

57. $\frac{8}{3}, \frac{10}{3}, 4, \frac{14}{3}, . . .$

58. $378, -126, 42, -14, . . .$

Finding a Sequence In Exercises 59 and 60, give an example of a sequence satisfying the given condition. (There is more than one correct answer.)

59. Converges to $\frac{3}{4}$

60. Converges to 100

61. Compound Interest Consider the sequence $\{A_n\}$, whose nth term is given by

$$A_n = P\left(1 + \frac{r}{12}\right)^n$$

where P is the principal, A_n is the balance in the account after n months, and r is the annual percentage rate (in decimal form). Write the first 10 terms of the sequence for $P = \$9000$ and $r = 0.06$.

62. Compound Interest Consider the sequence $\{A_n\}$, whose nth term is given by $A_n = P(1 + r)^n$, where P is the principal, A_n is the balance in the account after n years, and r is the annual percentage rate (in decimal form). Write the first 10 terms of the sequence for $P = \$5000$ and $r = 0.08$.

63. Individual Retirement Account A deposit of \$2000 is made each year in an account that earns 11% interest compounded annually. The balance in the account after n years is given by

$$A_n = 2000(11)[(1.1)^n - 1].$$

(a) Write the first six terms of the sequence.

(b) Find the balance in the account after 20 years by computing the 20th term of the sequence.

(c) Find the balance in the account after 40 years by computing the 40th term of the sequence.

64. Investment A deposit of \$100 is made each month in an account that earns 6% interest, compounded monthly. The balance in the account after n months is given by

$$A_n = 100(201)[(1.005)^n - 1].$$

(a) Write the first six terms of this sequence.

(b) Find the balance in the account after 5 years by computing the 60th term of the sequence.

(c) Find the balance in the account after 20 years by computing the 240th term of the sequence.

65. Carbon Dioxide The average concentration levels a_n (in parts per million) of carbon dioxide (CO_2) in Earth's atmosphere for selected years since 1980 are shown in the table, where n is the year, with $n = 0$ corresponding to 1980. *(Source: National Oceanic & Atmospheric Administration)*

n	0	5	10	15	20	25	30
a_n	339	345	354	360	369	379	389

(a) Use the *regression* feature of a graphing utility to find a model of the form $a_n = kn + b$ for the data. Use a graphing utility to plot the points and graph the model.

(b) Use the model to predict the average concentration level of CO_2 in the year 2020.

66. HOW DO YOU SEE IT? The graphs of two sequences are shown below. Which graph represents a sequence with alternating signs? Explain.

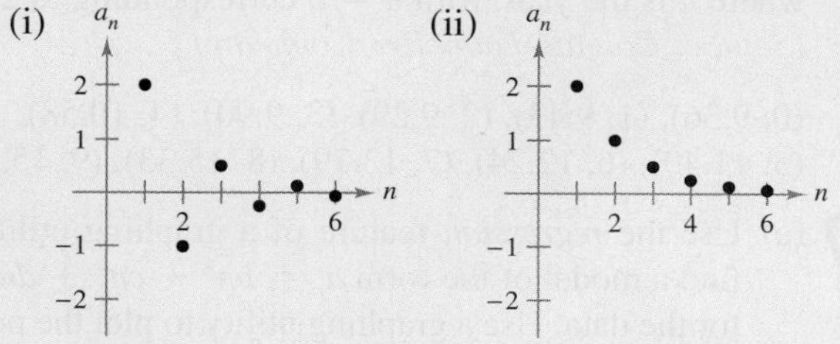

67. Number of Stores The numbers of Best Buy stores in the United States from 2000 through 2010 are shown in the table, where a_n is the number of stores and n is the year, with $n = 0$ corresponding to 2000. *(Source: Best Buy Companies, Inc.)*

n	0	1	2	3	4	5
a_n	357	419	481	548	608	668

n	6	7	8	9	10
a_n	742	822	923	1023	1069

(a) Use the *regression* feature of a graphing utility to find a model of the form $a_n = kn + b$ for the data. Use a graphing utility to plot the points and graph the model.

(b) Use the model to predict the number of stores in 2016.

68. Cost The average costs per patient per day at community hospitals from 2001 through 2008 are shown in the table, where a_n is the average cost (in dollars) and n is the year, with $n = 1$ corresponding to 2001. *(Source: Health Forum)*

n	1	2	3	4	5	6	7	8
a_n	1217	1290	1379	1450	1522	1612	1690	1782

(a) Use the *regression* feature of a graphing utility to find a model of the form $a_n = kn + b$ for the data. Use a graphing utility to plot the points and graph the model.

(b) Use the model to predict the cost in 2014.

69. Sales The sales a_n (in billions of dollars) of Colgate-Palmolive Company from 2000 through 2009 are shown below as ordered pairs of the form (n, a_n), where n is the year, with $n = 0$ corresponding to 2000. *(Source: Colgate-Palmolive Company)*

(0, 9.36), (1, 9.43), (2, 9.29), (3, 9.90), (4, 10.58), (5, 11.40), (6, 12.24), (7, 13.79), (8, 15.33), (9, 15.33)

(a) Use the *regression* feature of a graphing utility to find a model of the form $a_n = bn^3 + cn^2 + dn + f$ for the data. Use a graphing utility to plot the points and graph the model.

(b) Use the model to predict the sales in 2015.

70. Federal Debt The federal debt for the years 2000 through 2010 is approximated by the model

$$a_n = 0.0099n^3 - 0.079n^2 + 0.60n + 5.4, \quad n = 0, 1, 2, \ldots, 10$$

where a_n is the federal debt (in trillions of dollars) and n is the year, with $n = 0$ corresponding to 2000. *(Source: U.S. Office of Management and Budget)*

(a) Write the terms of this finite sequence.

(b) Construct a bar graph that represents the sequence.

71. Physical Science A ball is dropped from a height of 12 feet, and on each rebound it rises to $\frac{2}{3}$ its preceding height.

(a) Write an expression for the height of the nth rebound.

(b) Compute the heights for the first 6 rebounds.

(c) Determine the convergence or divergence of the sequence of rebounds. If it converges, find the limit.

72. Budget Analysis A government program that currently costs taxpayers \$1.3 billion per year is to be cut back by 15% per year.

(a) Write an expression for the amount budgeted for this program after n years.

(b) Compute the budget amounts for the first 4 years.

(c) Determine the convergence or divergence of the sequence of reduced budgets. If it converges, find the limit.

73. Salary A person accepts a position with a company at a salary of \$32,800 for the first year. The person is guaranteed a raise of 5% per year for the next 3 years.

(a) Determine the person's salary during the fourth year of employment.

(b) Assuming the raises continue, determine the convergence or divergence of the sequence of salaries. Explain why it converges or diverges.

74. Think About It Consider the sequence whose nth term a_n is given by

$$a_n = \left(1 + \frac{1}{n}\right)^n.$$

Demonstrate that the terms of this sequence approach e by finding a_1, a_{10}, a_{100}, a_{1000}, and $a_{10,000}$.

75. Project: Revenues For a project analyzing the revenues for Amazon.com from 2000 through 2010, visit this text's website at *www.cengagebrain.com.* *(Data Source: Amazon.com)*

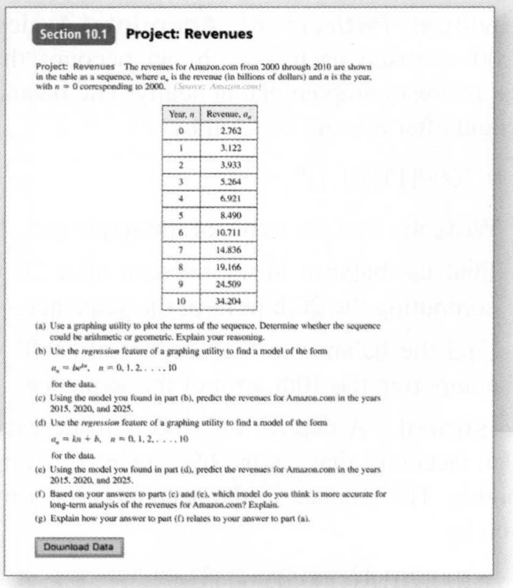

Section 10.1 **Project: Revenues**

Project: Revenues The revenues for Amazon.com from 2000 through 2010 are shown in the table as a sequence, where a_n is the revenue (in billions of dollars) and n is the year, with $n = 0$ corresponding to 2000. *(Source: Amazon.com)*

Year, n	Revenue, a_n
0	2.762
1	3.122
2	3.933
3	5.264
4	6.921
5	8.490
6	10.711
7	14.836
8	19.166
9	24.509
10	34.204

(a) Use a graphing utility to plot the terms of the sequence. Determine whether the sequence could be arithmetic or geometric. Explain your reasoning.

(b) Use the *regression* feature of a graphing utility to find a model of the form

$a_n = be^{kn}, \quad n = 0, 1, 2, \ldots, 10$

for the data.

(c) Using the model you found in part (b), predict the revenues for Amazon.com in the years 2015, 2020, and 2025.

(d) Use the *regression* feature of a graphing utility to find a model of the form

$a_n = kn + b, \quad n = 0, 1, 2, \ldots, 10$

for the data.

(e) Using the model you found in part (d), predict the revenues for Amazon.com in the years 2015, 2020, and 2025.

(f) Based on your answers to parts (c) and (e), which model do you think is more accurate for long-term analysis of the revenues for Amazon.com? Explain.

(g) Explain how your answer to part (f) relates to your answer to part (a).

Download Data

10.2 Series and Convergence

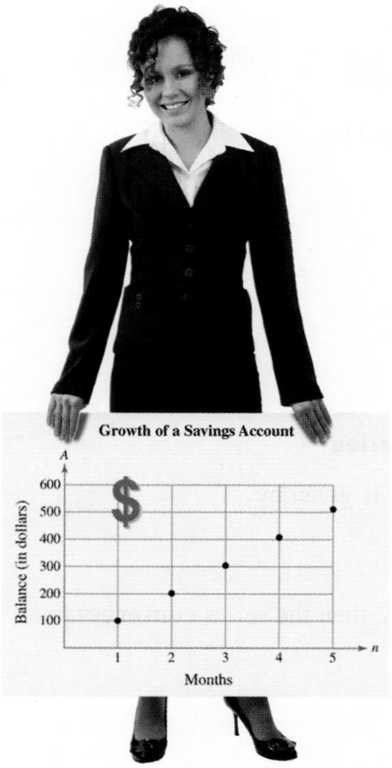

In Exercise 51 on page 617, you will use a series to model and find the balance in an account after 5 years.

- Write finite sums using sigma notation.
- Find the partial sums of series and determine the convergence or divergence of infinite series.
- Use the nth-Term Test for Divergence to show that series diverge.
- Find the nth partial sums of geometric series and determine the convergence or divergence of geometric series.
- Use geometric series to model and solve real-life problems.

Sigma Notation

In this section, you will study infinite summations. The decimal representation of $\frac{1}{3}$ is a simple example of an infinite summation.

$$\begin{aligned}\frac{1}{3} &= 0.33333\ldots \\ &= 0.3 + 0.03 + 0.003 + 0.0003 + 0.00003 + \cdots \\ &= \frac{3}{10} + \frac{3}{10^2} + \frac{3}{10^3} + \frac{3}{10^4} + \frac{3}{10^5} + \cdots \\ &= \sum_{n=1}^{\infty} \frac{3}{10^n}\end{aligned}$$

The last notation is called **sigma notation** or **summation notation.**

> **Sigma Notation**
>
> The finite sum $a_1 + a_2 + a_3 + \cdots + a_n$ can be written as
>
> $$\sum_{i=1}^{n} a_i.$$
>
> The letter i is the **index of summation,** and 1 and n are the lower and upper limits of summation, respectively.

Although i, j, k, and n are commonly used as indices of summation, any letter can be used. Moreover, the upper and lower limits of summation can be any two integers.

> **ALGEBRA TUTOR**
>
> For help in rewriting expressions that use sigma notation, see Example 2 in the *Chapter 10 Algebra Tutor,* on page 653.

Example 1 Using Sigma Notation

Sum	*Σ Notation*
a. $1 + 2 + 3 + 4 + 5 + 6$	$\sum_{i=1}^{6} i$
b. $3(1) + 3\left(\frac{1}{2}\right) + 3\left(\frac{1}{2}\right)^2 + \cdots + 3\left(\frac{1}{2}\right)^n$	$\sum_{k=0}^{n} 3\left(\frac{1}{2}\right)^k$

✓ Checkpoint 1

Use sigma notation to write the sum

$$4\left(-\frac{1}{2}\right) + 4\left(\frac{1}{4}\right) + 4\left(-\frac{1}{8}\right) + 4\left(\frac{1}{16}\right).$$

Karlova Irina/www.shutterstock.com

Infinite Series

The sum of all the terms of the infinite sequence $\{a_n\}$ is called an **infinite series** (or simply a **series**) and is denoted by

$$\sum_{n=1}^{\infty} a_n = a_1 + a_2 + a_3 + \cdot \cdot \cdot + a_n + \cdot \cdot \cdot . \qquad \text{Infinite series}$$

The **sequence of partial sums** of the series is denoted by

$$S_1 = a_1$$
$$S_2 = a_1 + a_2$$
$$S_3 = a_1 + a_2 + a_3$$
$$\vdots$$
$$S_n = a_1 + a_2 + a_3 + \cdot \cdot \cdot + a_n.$$

STUDY TIP

As you study this chapter, it is important to distinguish between an infinite series and a sequence. A sequence is an ordered collection of numbers

$$a_1, a_2, a_3, . . ., a_n, . . .$$

whereas a series is an infinite sum of terms from a sequence

$$a_1 + a_2 + \cdot \cdot \cdot + a_n + \cdot \cdot \cdot .$$

Convergence and Divergence of an Infinite Series

For the infinite series $\sum_{n=1}^{\infty} a_n$, the ***n*th partial sum** is given by

$$S_n = a_1 + a_2 + a_3 + \cdot \cdot \cdot + a_n. \qquad n\text{th partial sum}$$

If the sequence of partial sums $\{S_n\}$ converges to S, then the series **converges** to S. This limit is denoted by

$$\lim_{n\to\infty} S_n = \sum_{n=1}^{\infty} a_n = S$$

and S is called the **sum of the series.** If $\{S_n\}$ diverges, then the series **diverges.**

TECH TUTOR

Symbolic algebra utilities can be used to evaluate infinite sums. Use a symbolic algebra utility to evaluate the sum in Example 2(a).

Example 2 Determining Convergence or Divergence

a. The infinite series

$$\sum_{n=1}^{\infty} \frac{1}{2^n} = \frac{1}{2} + \frac{1}{4} + \frac{1}{8} + \frac{1}{16} + \cdot \cdot \cdot$$

has the partial sums listed below.

$$S_1 = \frac{1}{2}, \quad S_2 = \frac{3}{4}, \quad S_3 = \frac{7}{8}, \quad S_4 = \frac{15}{16}, \quad . . . , \quad S_n = \frac{2^n - 1}{2^n}$$

Because the limit $\lim_{n\to\infty} S_n$ is 1, it follows that the infinite series converges and its sum is 1. So, you can write

$$\sum_{n=1}^{\infty} \frac{1}{2^n} = \frac{1}{2} + \frac{1}{4} + \frac{1}{8} + \frac{1}{16} + \cdot \cdot \cdot = 1.$$

b. The infinite series

$$\sum_{n=1}^{\infty} 1 = 1 + 1 + 1 + 1 + \cdot \cdot \cdot$$

diverges because $S_n = n$ and the sequence of partial sums $\{S_n\}$ diverges.

✓ Checkpoint 2

Determine the convergence or divergence of the series $\sum_{n=1}^{\infty} \frac{1}{4^n}$. ■

Tom Wang/www.shutterstock.com

The properties below are useful in determining the sums of infinite series.

Properties of Infinite Series

For the convergent infinite series

$$\sum_{n=1}^{\infty} a_n = A$$

and

$$\sum_{n=1}^{\infty} b_n = B$$

and the real number c, the following properties are true.

1. $\displaystyle\sum_{n=1}^{\infty} ca_n = c\sum_{n=1}^{\infty} a_n = cA$

2. $\displaystyle\sum_{n=1}^{\infty} (a_n + b_n) = \sum_{n=1}^{\infty} a_n + \sum_{n=1}^{\infty} b_n = A + B$

3. $\displaystyle\sum_{n=1}^{\infty} (a_n - b_n) = \sum_{n=1}^{\infty} a_n - \sum_{n=1}^{\infty} b_n = A - B$

Example 3 Using Properties of Infinite Series

Find the sum of each infinite series.

a. $\displaystyle\sum_{n=1}^{\infty} \frac{5}{2^n}$ **b.** $\displaystyle\sum_{n=1}^{\infty} \left(\frac{1}{2^n} + \frac{1}{2^{n+1}}\right)$ **c.** $\displaystyle\sum_{n=1}^{\infty} \left(\frac{1}{2^n} - \frac{1}{2^{n+1}}\right)$

SOLUTION Begin by noting from Example 2(a) that

$$\sum_{n=1}^{\infty} \frac{1}{2^n} = 1.$$

So, you can conclude that

$$\sum_{n=1}^{\infty} \frac{1}{2^{n+1}} = \sum_{n=1}^{\infty} \frac{1}{2}\left(\frac{1}{2^n}\right) = \frac{1}{2}\sum_{n=1}^{\infty} \frac{1}{2^n} = \frac{1}{2}(1) = \frac{1}{2}.$$

a. Using Property 1, you can write

$$\sum_{n=1}^{\infty} \frac{5}{2^n} = 5\sum_{n=1}^{\infty} \frac{1}{2^n} = 5(1) = 5.$$

b. Using Property 2, you can write

$$\sum_{n=1}^{\infty} \left(\frac{1}{2^n} + \frac{1}{2^{n+1}}\right) = \sum_{n=1}^{\infty} \frac{1}{2^n} + \sum_{n=1}^{\infty} \frac{1}{2^{n+1}} = 1 + \frac{1}{2} = \frac{3}{2}.$$

c. Using Property 3, you can write

$$\sum_{n=1}^{\infty} \left(\frac{1}{2^n} - \frac{1}{2^{n+1}}\right) = \sum_{n=1}^{\infty} \frac{1}{2^n} - \sum_{n=1}^{\infty} \frac{1}{2^{n+1}} = 1 - \frac{1}{2} = \frac{1}{2}.$$

✓Checkpoint 3

Use the properties of infinite series to find the sum of each infinite series.

a. $\displaystyle\sum_{n=1}^{\infty} \frac{3}{4^n}$ **b.** $\displaystyle\sum_{n=1}^{\infty} \left(\frac{1}{4^n} + \frac{1}{2^n}\right)$ **c.** $\displaystyle\sum_{n=1}^{\infty} \left(\frac{1}{4^n} - \frac{1}{2^n}\right)$

The *n*th-Term Test for Divergence

With an infinite series, there are two primary questions.

1. Does the series converge or does it diverge?
2. When the series converges, to what value does it converge?

A simple test for divergence gives a partial answer to the first question.

STUDY TIP

Be sure you see that the *n*th-Term Test is a test for *divergence*, not for convergence. That is, when the *n*th term does not converge to zero, you know that the series diverges. When the *n*th term does converge to zero, the series may or may not converge.

nth-Term Test for Divergence

Consider the infinite series $\sum_{n=1}^{\infty} a_n$. If

$$\lim_{n\to\infty} a_n \neq 0$$

then the series diverges.

Example 4 Using the nth-Term Test for Divergence

Use the *n*th-Term Test to determine whether each series diverges.

a. $\sum_{n=1}^{\infty} 2^n$ **b.** $\sum_{n=1}^{\infty} \frac{1}{2^n}$ **c.** $\sum_{n=1}^{\infty} \frac{n!}{2n! + 1}$

SOLUTION

a. By the *n*th-Term Test, the infinite series

$$\sum_{n=1}^{\infty} 2^n = 2 + 4 + 8 + 16 + \cdot \cdot \cdot$$

diverges because

$$\lim_{n\to\infty} 2^n = \infty.$$

b. The *n*th-Term Test tells you nothing about the infinite series

$$\sum_{n=1}^{\infty} \frac{1}{2^n} = \frac{1}{2} + \frac{1}{4} + \frac{1}{8} + \frac{1}{16} + \cdot \cdot \cdot$$

because

$$\lim_{n\to\infty} \frac{1}{2^n} = 0.$$

From Example 2(a), you know that this series converges. The point here is that you cannot deduce this from the *n*th-Term Test.

c. The infinite series

$$\sum_{n=1}^{\infty} \frac{n!}{2n! + 1} = \frac{1}{3} + \frac{2}{5} + \frac{6}{13} + \frac{24}{49} + \frac{120}{241} + \cdot \cdot \cdot$$

diverges because

$$\lim_{n\to\infty} \frac{n!}{2n! + 1} = \lim_{n\to\infty} \frac{1}{2 + (1/n!)} = \frac{1}{2 + 0} = \frac{1}{2}.$$

✓ Checkpoint 4

Use the *n*th-Term Test to determine whether each series diverges.

a. $\sum_{n=1}^{\infty} \frac{2^n}{2^{n+1} + 1}$ **b.** $\sum_{n=1}^{\infty} \frac{n^2}{n^2 + 1}$ ■

Geometric Series

If a is a nonzero real number, then the infinite series

$$\sum_{n=0}^{\infty} ar^n = a + ar + ar^2 + \cdots + ar^n + \cdots$$ Geometric series

is called a **geometric series** with ratio r, $r \neq 0$. Note that the first term of this series is

$ar^0 = a.$

When the index begins with $n = 1$, the first term is

$ar^1 = ar.$

*n*th Partial Sum of a Geometric Series

The nth partial sum of the geometric series $\sum_{n=0}^{\infty} ar^n$ is

$$S_n = \frac{a(1 - r^{n+1})}{1 - r}, \quad r \neq 1.$$

Example 5 Finding an *n*th Partial Sum

Find the third, fifth, and tenth partial sums of the geometric series

$$\sum_{n=0}^{\infty} 3\left(\frac{1}{4}\right)^n = 3 + \frac{3}{4} + \frac{3}{4^2} + \frac{3}{4^3} + \cdots.$$

SOLUTION For this geometric series,

$a = 3$ and $r = \frac{1}{4}$.

Because the index begins with $n = 0$, the nth partial sum is

$$\begin{aligned} S_n &= \frac{a(1 - r^{n+1})}{1 - r} \\ &= \frac{3[1 - (1/4)^{n+1}]}{1 - (1/4)} \\ &= \frac{3[1 - (1/4)^{n+1}]}{3/4} \\ &= 4\left[1 - \left(\frac{1}{4}\right)^{n+1}\right] \\ &= 4 - \left(\frac{1}{4}\right)^n. \end{aligned}$$

Using this formula, you can find the third, fifth, and tenth partial sums as shown.

$S_3 = 4 - \left(\frac{1}{4}\right)^3 \approx 3.984$ Third partial sum

$S_5 = 4 - \left(\frac{1}{4}\right)^5 \approx 3.999$ Fifth partial sum

$S_{10} = 4 - \left(\frac{1}{4}\right)^{10} \approx 4.000$ Tenth partial sum

✓ Checkpoint 5

Find the 5th, 50th, and 500th partial sums of the geometric series

$$\sum_{n=0}^{\infty} 5\left(\frac{1}{10}\right)^n.$$

■

When applying the formula for the nth partial sum of a geometric series, be sure to check that the index begins with $n = 0$. When it begins with some other number, you will have to adjust S_n accordingly. Here is an example.

$$\sum_{n=1}^{10} ar^n = -a + \sum_{n=0}^{10} ar^n = -a + \frac{a(1 - r^{11})}{1 - r}$$

The same type of adjustment is used in the next example.

Example 6 Finding an Annuity Balance

A deposit of \$50 is made every month for 2 years in a savings account that pays 6%, compounded monthly. What is the balance in the account at the end of 2 years?

SOLUTION Recall from Section 4.2 that the compound interest formula is

$$A = P\left(1 + \frac{r}{n}\right)^{nt}. \qquad \text{Formula for compound interest}$$

To find the balance in the account after 24 months, consider each of the 24 deposits separately. The first deposit will gain interest for 24 months, and its balance will be

$$A_{24} = 50\left(1 + \frac{0.06}{12}\right)^{24} = 50(1.005)^{24}.$$

The second deposit will gain interest for 23 months, and its balance will be

$$A_{23} = 50\left(1 + \frac{0.06}{12}\right)^{23} = 50(1.005)^{23}.$$

Continuing this process, you will find that the last deposit will gain interest for only 1 month, and its balance will be

$$A_1 = 50\left(1 + \frac{0.06}{12}\right)^{1} = 50(1.005).$$

The total balance resulting from the 24 deposits will be

$$\begin{aligned} A &= A_1 + A_2 + \cdot \cdot \cdot + A_{24} \\ &= \sum_{n=1}^{24} A_n \\ &= \sum_{n=1}^{24} 50(1.005)^n. \end{aligned}$$

Noting that the index begins with $n = 1$, you can use the formula for the nth partial sum to find the balance.

$$\begin{aligned} A &= \sum_{n=1}^{24} 50(1.005)^n \\ &= -50 + \sum_{n=0}^{24} 50(1.005)^n \\ &= -50 + \frac{50(1 - 1.005^{25})}{1 - 1.005} \\ &\approx \$1277.96 \end{aligned}$$

✓ Checkpoint 6

A deposit of \$20 is made every month for 4 years in an account that pays 3% compounded monthly. What is the balance in the account at the end of 4 years?

STUDY TIP

You can apply the test for convergence or divergence of a geometric series regardless of the beginning value of the index of summation. When the series converges, however, the formula for its sum is valid only for a geometric series whose index begins at $n = 0$. When the index begins with some other number, you will have to adjust the formula, as shown in Example 7(c).

Convergence of a Geometric Series

Consider the geometric series given by

$$\sum_{n=0}^{\infty} ar^n.$$

If $|r| \geq 1$, then the series diverges. If $|r| < 1$, then the series converges to the sum

$$\sum_{n=0}^{\infty} ar^n = \frac{a}{1-r}, \quad |r| < 1.$$

Example 7 Determining Convergence or Divergence

Decide whether each series converges or diverges.

a. $\displaystyle\sum_{n=0}^{\infty} \left(-\frac{1}{2}\right)^n$ **b.** $\displaystyle\sum_{n=0}^{\infty} \left(\frac{3}{2}\right)^n$ **c.** $\displaystyle\sum_{n=1}^{\infty} \frac{4}{3^n}$

SOLUTION

a. For this geometric series, $a = 1$ and $r = -\frac{1}{2}$. Because $|r| < 1$, it follows that the series converges. Moreover, because the index begins with $n = 0$, you can apply the formula for the sum of a geometric series to conclude that

$$\sum_{n=0}^{\infty} \left(-\frac{1}{2}\right)^n = \frac{a}{1-r} = \frac{1}{1-(-1/2)} = \frac{1}{3/2} = \frac{2}{3}.$$

b. For this geometric series,

$$a = 1 \quad \text{and} \quad r = \frac{3}{2}.$$

Because $|r| > 1$, it follows that the series diverges.

c. By rewriting this geometric series as

$$\sum_{n=1}^{\infty} 4\left(\frac{1}{3}\right)^n$$

you can see that $a = 4$ and $r = \frac{1}{3}$. Because $|r| < 1$, the series converges. To find the sum of the series, note that the index begins with $n = 1$, and then adjust the formula for the sum as shown.

$$\begin{aligned}\sum_{n=1}^{\infty} 4\left(\frac{1}{3}\right)^n &= -4 + \sum_{n=0}^{\infty} 4\left(\frac{1}{3}\right)^n \\ &= -4 + \frac{4}{1-(1/3)} \\ &= -4 + \frac{4}{2/3} \\ &= -4 + 6 \\ &= 2\end{aligned}$$

So, the series converges to 2.

✓ Checkpoint 7

Decide whether each series converges or diverges.

a. $\displaystyle\sum_{n=0}^{\infty} \left(\frac{2}{5}\right)^n$ **b.** $\displaystyle\sum_{n=0}^{\infty} 4\left(\frac{3}{2}\right)^n$ **c.** $\displaystyle\sum_{n=1}^{\infty} \frac{5}{4^n}$ ■

Applications

Example 8 Modeling Market Stabilization

A manufacturer sells 10,000 units of a product each year. In any given year, each unit has a 10% chance of breaking. That is, after 1 year you expect that only 9000 of the previous year's 10,000 units will still be in use. During the next year, this number will drop by an additional 10% to 8100, and so on. How many units will be in use after 20 years? Is the number of units in use stabilizing? If so, what is the stabilization point?

SOLUTION You can model this situation with a geometric series, as shown.

End of Year	*Number of Units in Use*
0	$10{,}000$
1	$10{,}000 + 10{,}000(0.9)$
2	$10{,}000 + 10{,}000(0.9) + 10{,}000(0.9)^2$
3	$10{,}000 + 10{,}000(0.9) + 10{,}000(0.9)^2 + 10{,}000(0.9)^3$

After 20 years, the number of units in use will be

$$\sum_{n=0}^{20} 10{,}000(0.9)^n = \frac{10{,}000[1 - (0.9)^{21}]}{1 - 0.9} \approx 89{,}058.$$

As indicated in Figure 10.2, the number of units in use is approaching a stabilization point of

$$\sum_{n=0}^{\infty} 10{,}000(0.9)^n = \frac{10{,}000}{1 - 0.9} = 100{,}000 \text{ units.}$$

Note in Figure 10.2 that the number of years represents how many years have passed since the beginning of year zero. For instance, at the end of year zero, 1 year has passed; at the end of year one, 2 years have passed; and so on.

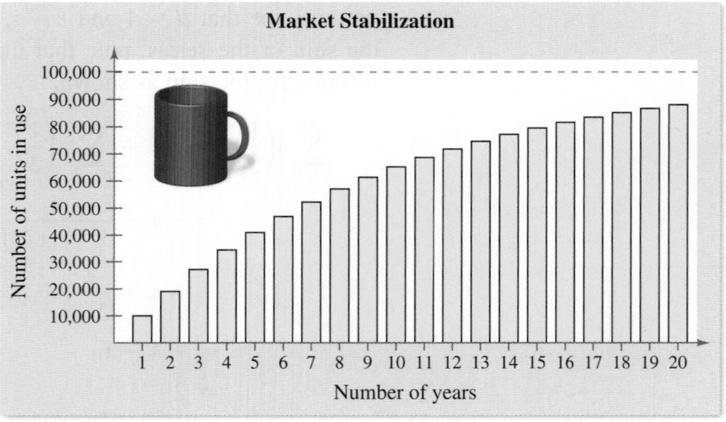

FIGURE 10.2

✓ Checkpoint 8

Repeat Example 8 when the manufacturer sells 10,000 units and any given unit has a 25% chance of breaking. Find the number of units that will still be in use after 40 years. ■

Example 9 Modeling a Bouncing Ball

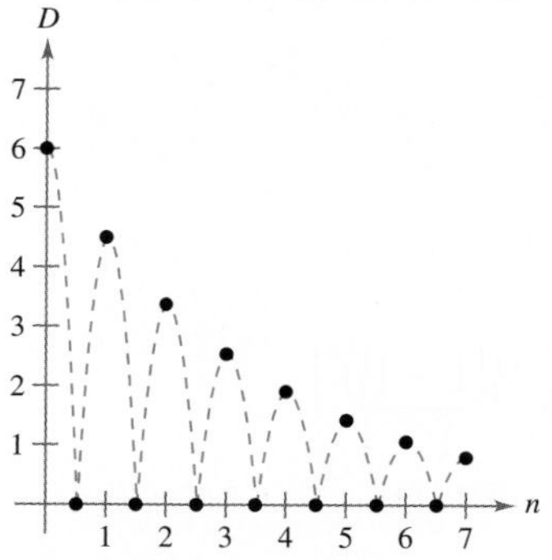

FIGURE 10.3

A ball is dropped from a height of 6 feet and begins to bounce. The height of each bounce is $\frac{3}{4}$ that of the preceding bounce, as shown in Figure 10.3. Find the total vertical distance traveled by the ball.

SOLUTION When the ball hits the ground the first time, it has traveled a distance of $D_1 = 6$. Between the first and second times it hits the ground, it travels an additional distance of

$$D_2 = \underbrace{6\left(\frac{3}{4}\right)}_{\text{Up}} + \underbrace{6\left(\frac{3}{4}\right)}_{\text{Down}} = 12\left(\frac{3}{4}\right).$$

Between the second and third times the ball hits the ground, it travels an additional distance of

$$D_3 = \underbrace{6\left(\frac{3}{4}\right)\left(\frac{3}{4}\right)}_{\text{Up}} + \underbrace{6\left(\frac{3}{4}\right)\left(\frac{3}{4}\right)}_{\text{Down}} = 12\left(\frac{3}{4}\right)^2.$$

By continuing this process, you can determine that the total vertical distance is

$$\begin{aligned} D &= 6 + 12\left(\frac{3}{4}\right) + 12\left(\frac{3}{4}\right)^2 + \cdots \\ &= 6 - 12 + 12 + 12\left(\frac{3}{4}\right) + 12\left(\frac{3}{4}\right)^2 + \cdots \\ &= -6 + \sum_{n=0}^{\infty} 12\left(\frac{3}{4}\right)^n \\ &= -6 + \frac{12}{1 - (3/4)} \\ &= -6 + 48 \\ &= 42 \text{ feet.} \end{aligned}$$

✓ Checkpoint 9

Find the total vertical distance the ball travels in Example 9 when it is dropped from 20 feet and bounces $\frac{3}{4}$ of the height of the preceding bounce. ■

SUMMARIZE (Section 10.2)

1. Explain how to write a finite sum using sigma notation *(page 607)*. For an example of using sigma notation, see Example 1.
2. Explain how to determine the convergence or divergence of an infinite series *(page 608)*. For an example of determining the convergence or divergence of an infinite series, see Example 2.
3. State the nth-Term Test for divergence *(page 610)*. For an example of using the nth-Term Test, see Example 4.
4. State the nth partial sum of a geometric series *(page 611)*. For an example of finding an nth partial sum, see Example 5.
5. Explain how to determine the convergence or divergence of a geometric series *(page 613)*. For an example of determining the convergence or divergence of a geometric series, see Example 7.

SKILLS WARM UP 10.2

The following warm-up exercises involve skills that were covered in a previous course or in earlier sections. You will use these skills in the exercise set for this section. For additional help, review Appendix Section A.5 and Sections 3.6 and 10.1.

In Exercises 1 and 2, add the fractions.

1. $\frac{1}{2} + \frac{1}{3} + \frac{1}{4} + \frac{1}{5}$

2. $1 + \frac{3}{4} + \frac{4}{6} + \frac{5}{8}$

In Exercises 3–6, evaluate the expression.

3. $\dfrac{1 - \left(\frac{1}{2}\right)^5}{1 - \frac{1}{2}}$

4. $\dfrac{3\left[1 - \left(\frac{1}{3}\right)^4\right]}{1 - \frac{1}{3}}$

5. $\dfrac{2\left[1 - \left(\frac{1}{4}\right)^3\right]}{1 - \frac{1}{4}}$

6. $\dfrac{\frac{1}{2}\left[1 - \left(\frac{1}{2}\right)^5\right]}{1 - \frac{1}{2}}$

In Exercises 7–10, find the limit.

7. $\lim_{n\to\infty} \dfrac{3n}{4n + 1}$

8. $\lim_{n\to\infty} \dfrac{3n}{n^2 + 1}$

9. $\lim_{n\to\infty} \dfrac{n!}{n! - 3}$

10. $\lim_{n\to\infty} \dfrac{2n! + 1}{4n! - 1}$

Exercises 10.2

See www.CalcChat.com for worked-out solutions to odd-numbered exercises.

Using Sigma Notation In Exercises 1–4, use sigma notation to write the sum. *See Example 1.*

1. $2 + 4 + 6 + 8 + 10$

2. $\frac{3}{2} + \frac{3}{4} + \frac{3}{8} + \frac{3}{16} + \frac{3}{32} + \frac{3}{64}$

3. $\frac{1}{1} + \frac{1}{3} + \frac{1}{5} + \frac{1}{7} + \frac{1}{9} + \frac{1}{11} + \frac{1}{13}$

4. $5\left(-\frac{1}{4}\right) + 5\left(\frac{1}{16}\right) + 5\left(-\frac{1}{64}\right)$

Finding Partial Sums In Exercises 5–8, write the first five terms of the sequence of partial sums.

5. $\sum_{n=1}^{\infty} \dfrac{1}{n^2} = 1 + \dfrac{1}{4} + \dfrac{1}{9} + \dfrac{1}{16} + \dfrac{1}{25} + \cdots$

6. $\sum_{n=1}^{\infty} \dfrac{3}{2^{n-1}} = 3 + \dfrac{3}{2} + \dfrac{3}{4} + \dfrac{3}{8} + \dfrac{3}{16} + \cdots$

7. $\sum_{n=1}^{\infty} (-1)^{n+1}\dfrac{3^n}{2^{n-1}} = 3 - \dfrac{9}{2} + \dfrac{27}{4} - \dfrac{81}{8} + \dfrac{243}{16} - \cdots$

8. $\sum_{n=1}^{\infty} \dfrac{(-1)^{n+1}}{n!} = 1 - \dfrac{1}{2} + \dfrac{1}{6} - \dfrac{1}{24} + \dfrac{1}{120} - \cdots$

Verifying Divergence In Exercises 9–16, verify that the infinite series diverges. *See Examples 2, 4, and 7.*

9. $\sum_{n=0}^{\infty} 5\left(\dfrac{5}{2}\right)^n = 5 + \dfrac{25}{2} + \dfrac{125}{4} + \dfrac{625}{8} + \cdots$

10. $\sum_{n=0}^{\infty} \left(\dfrac{4}{3}\right)^n = 1 + \dfrac{4}{3} + \dfrac{16}{9} + \dfrac{64}{27} + \cdots$

11. $\sum_{n=0}^{\infty} 1000(1.055)^n = 1000 + 1055 + 1113.025 + \cdots$

12. $\sum_{n=0}^{\infty} 2(-1.03)^n = 2 - 2.06 + 2.1218 - \cdots$

13. $\sum_{n=1}^{\infty} \dfrac{n}{n+1} = \dfrac{1}{2} + \dfrac{2}{3} + \dfrac{3}{4} + \dfrac{4}{5} + \cdots$

14. $\sum_{n=1}^{\infty} \dfrac{n}{2n+3} = \dfrac{1}{5} + \dfrac{2}{7} + \dfrac{3}{9} + \dfrac{4}{11} + \cdots$

15. $\sum_{n=1}^{\infty} \dfrac{n^2}{n^2+1} = \dfrac{1}{2} + \dfrac{4}{5} + \dfrac{9}{10} + \dfrac{16}{17} + \cdots$

16. $\sum_{n=1}^{\infty} \dfrac{n}{\sqrt{n^2+1}} = \dfrac{1}{\sqrt{2}} + \dfrac{2}{\sqrt{5}} + \dfrac{3}{\sqrt{10}} + \dfrac{4}{\sqrt{17}} + \cdots$

Verifying Convergence In Exercises 17–20, verify that the geometric series converges. *See Examples 2, 4, and 7.*

17. $\sum_{n=0}^{\infty} 2\left(\dfrac{3}{4}\right)^n = 2 + \dfrac{3}{2} + \dfrac{9}{8} + \dfrac{27}{32} + \dfrac{81}{128} + \cdots$

18. $\sum_{n=0}^{\infty} 4\left(-\dfrac{1}{3}\right)^n = 4 - \dfrac{4}{3} + \dfrac{4}{9} - \dfrac{4}{27} + \dfrac{4}{81} - \cdots$

19. $\sum_{n=0}^{\infty} (0.9)^n = 1 + 0.9 + 0.81 + 0.729 + \cdots$

20. $\sum_{n=0}^{\infty} (-0.6)^n = 1 - 0.6 + 0.36 - 0.216 + \cdots$

Using Properties of Infinite Series In Exercises 21–28, find the sum of the infinite series. *See Examples 3 and 7.*

21. $\sum_{n=0}^{\infty} \dfrac{7}{3^n}$

22. $\sum_{n=0}^{\infty} \dfrac{5}{4^n}$

23. $\sum_{n=0}^{\infty} 5\left(\dfrac{3}{8}\right)^n$

24. $\sum_{n=0}^{\infty} 6\left(\dfrac{4}{5}\right)^n$

25. $\sum_{n=0}^{\infty} \left(\dfrac{1}{2^n} - \dfrac{1}{3^n}\right)$

26. $\sum_{n=0}^{\infty} \left(\dfrac{1}{3^n} + \dfrac{1}{4^n}\right)$

27. $\sum_{n=0}^{\infty} [(0.7)^n + (0.9)^n]$

28. $\sum_{n=0}^{\infty} [(0.4)^n - (0.8)^n]$

Finding an *n*th Partial Sum In Exercises 29–32, find the indicated partial sums for the geometric series. *See Example 5.*

Geometric Series	*Partial Sums*
29. $\sum_{n=0}^{\infty} 3\left(\frac{1}{8}\right)^n$	S_4, S_6, S_{10}
30. $\sum_{n=0}^{\infty} 6\left(\frac{3}{4}\right)^n$	S_3, S_8, S_{30}
31. $\sum_{n=0}^{\infty} 5\left(-\frac{1}{2}\right)^n$	S_5, S_7, S_{10}
32. $\sum_{n=0}^{\infty} 2\left(-\frac{2}{3}\right)^n$	S_9, S_{12}, S_{18}

Determining Convergence or Divergence In Exercises 33–42, determine the convergence or divergence of the series. Use a symbolic algebra utility to verify your result. *See Examples 2, 4, and 7.*

33. $\sum_{n=1}^{\infty} \frac{n+10}{10n+1}$

34. $\sum_{n=0}^{\infty} \frac{4}{2^n}$

35. $\sum_{n=1}^{\infty} \frac{n!+1}{n!}$

36. $\sum_{n=1}^{\infty} \frac{n+1}{2n-1}$

37. $\sum_{n=1}^{\infty} \frac{3n-1}{2n+1}$

38. $\sum_{n=0}^{\infty} \frac{1}{6^n}$

39. $\sum_{n=0}^{\infty} (1.075)^n$

40. $\sum_{n=1}^{\infty} \frac{2^n}{100}$

41. $\sum_{n=0}^{\infty} \frac{8}{5^n}$

42. $\sum_{n=0}^{\infty} n!$

Using Geometric Series In Exercises 43–46, the repeating decimal is expressed as a geometric series. Find the sum of the geometric series and write the decimal as the ratio of two integers.

43. $0.\overline{4} = 0.4 + 0.04 + 0.004 + 0.0004 + \cdots$

44. $0.\overline{9} = 0.9 + 0.09 + 0.009 + 0.0009 + \cdots$

45. $0.\overline{81} = 0.81 + 0.0081 + 0.000081 + \cdots$

46. $0.\overline{21} = 0.21 + 0.0021 + 0.000021 + \cdots$

47. Sales A company produces a new product for which it estimates the annual sales to be 8000 units. Suppose that in any given year 10% of the units (regardless of age) will become inoperative.

(a) How many units will be in use after n years?

(b) Find the market stabilization level of the product.

48. Sales Repeat Exercise 47 with the assumption that 25% of the units will become inoperative each year.

49. Physical Science A ball is dropped from a height of 16 feet. Each time it drops h feet, it rebounds $0.81h$ feet. Find the total vertical distance traveled by the ball.

50. Physical Science The ball in Exercise 49 takes the times listed below for each fall, where t is measured in seconds.

$$\begin{aligned} s_1 &= -16t^2 + 16 & s_1 &= 0 \text{ when } t = 1 \\ s_2 &= -16t^2 + 16(0.81) & s_2 &= 0 \text{ when } t = 0.9 \\ s_3 &= -16t^2 + 16(0.81)^2 & s_3 &= 0 \text{ when } t = (0.9)^2 \\ s_4 &= -16t^2 + 16(0.81)^3 & s_4 &= 0 \text{ when } t = (0.9)^3 \\ &\vdots & &\vdots \\ s_n &= -16t^2 + 16(0.81)^{n-1} & s_n &= 0 \text{ when } t = (0.9)^{n-1} \end{aligned}$$

Beginning with s_2, the ball takes the same amount of time to bounce up as it does to fall, and so the total time elapsed before it comes to rest is given by

$$t = 1 + 2\sum_{n=1}^{\infty} (0.9)^n.$$

Find this total time.

51. Annuity A deposit of \$100 is made at the beginning of each month for 5 years in an account that pays 9% interest, compounded monthly. The balance A in the account at the end of 5 years is

$$A = 100\left(1 + \frac{0.09}{12}\right)^1 + 100\left(1 + \frac{0.09}{12}\right)^2 + \cdots + 100\left(1 + \frac{0.09}{12}\right)^{60}.$$

(a) Use sigma notation to write the balance in the account at the end of 5 years.

(b) Find the balance in the account at the end of 5 years.

52. Annuity A deposit of \$150 is made at the beginning of each month for 4 years in an account that pays 3% interest, compounded monthly. The balance A in the account at the end of 4 years is

$$A = 150\left(1 + \frac{0.03}{12}\right)^1 + 150\left(1 + \frac{0.03}{12}\right)^2 + \cdots + 150\left(1 + \frac{0.03}{12}\right)^{48}.$$

(a) Use sigma notation to write the balance in the account at the end of 4 years.

(b) Find the balance in the account at the end of 4 years.

53. Annuity A deposit of P dollars is made every month for t years in an account that pays an annual interest rate of r%, compounded monthly. Let

$$N = 12t$$

be the total number of deposits. Show that the balance in the account after t years is

$$A = P\left[\left(1 + \frac{r}{12}\right)^N - 1\right]\left(1 + \frac{12}{r}\right), \quad t > 0.$$

54. HOW DO YOU SEE IT? The graph shows the revenues (in black) and costs (in red) of a company for the past five years. Is the total profit over the five-year period positive or negative? Explain.

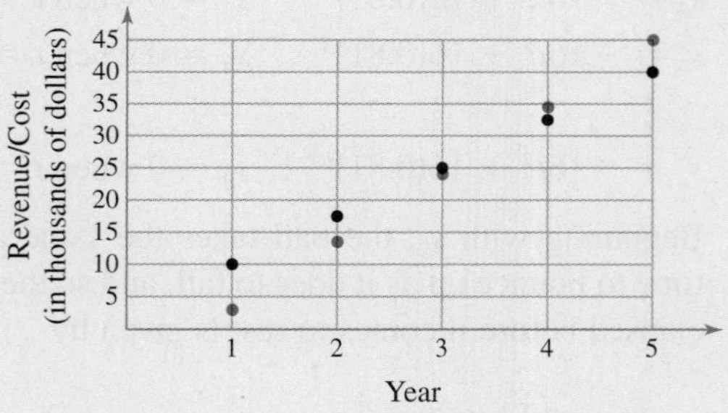

55. Consumer Trends: Multiplier Effect The annual spending by tourists in a resort city is \$500 million. Approximately 75% of that revenue is again spent in the resort city, and of that amount approximately 75% is again spent in the resort city. If this pattern continues, write the geometric series that gives the total amount of spending generated by the \$500 million and find the sum of the series.

56. Consumer Trends: Multiplier Effect Repeat Exercise 55 assuming the percent of the revenue that is spent in the city each time is 60%.

57. Depreciation A company buys a machine for \$225,000 that depreciates at a rate of 30% per year. Find a formula for the value of the machine after n years. What is its value after 5 years?

58. Depreciation Repeat Exercise 57 assuming the machine depreciates at a rate of 25% per year.

59. Salary You accept a job that pays a salary of \$40,000 the first year. During the next 39 years, you will receive a 4% raise each year. What would be your total compensation over the 40-year period?

60. Salary You go to work at a company that pays \$0.01 for the first day, \$0.02 for the second day, \$0.04 for the third day, and so on. If the daily wage keeps doubling, what would your total income be for working

(a) 29 days?

(b) 30 days?

(c) 31 days?

61. Probability: Coin Toss A fair coin is tossed until a head appears. The probability that the first head appears on the nth toss is given by

$$P = \left(\frac{1}{2}\right)^n, \quad n \geq 1.$$

Show that $\sum_{n=1}^{\infty} \left(\frac{1}{2}\right)^n = 1.$

62. Probability: Coin Toss Use a symbolic algebra utility to estimate the expected number of tosses required until the first head occurs in the experiment in Exercise 61.

Probability In Exercises 63 and 64, the random variable n represents the number of units of a product sold per day in a store. The probability distribution of n is given by $P(n)$. Find the probability that two units are sold in a given day $[P(2)]$ and show that

$P(0) + P(1) + P(2) + P(3) + \cdot \cdot \cdot = 1.$

63. $P(n) = \frac{1}{2}\left(\frac{1}{2}\right)^n$

64. $P(n) = \frac{1}{3}\left(\frac{2}{3}\right)^n$

65. Environment A factory is polluting a river such that at every mile down river from the factory an environmental expert finds 15% less pollutant than at the preceding mile. If the pollutant's concentration is 500 parts per million at the factory, what is its concentration 12 miles down river?

66. Finance: Annuity The simplest kind of annuity is a straight-line annuity, which pays a fixed amount per month until the annuitant dies. Suppose that, when he turns 65, Bob wants to purchase a straight-line annuity that has a premium of \$100,000 and pays \$880 per month. Use sigma notation to represent each scenario below, and give the numerical amount that the summation represents. *(Source: Adapted from Garman/Forgue,* Personal Finance, *Eighth Edition)*

(a) Suppose Bob dies 10 months after he takes out the annuity. How much will he have collected up to that point?

(b) Suppose Bob lives the average number of months beyond age 65 for a man (168 months). How much more or less than the \$100,000 will he have collected?

Finding Sums In Exercises 67–70, use a symbolic algebra utility to evaluate the summation.

67. $\sum_{n=1}^{\infty} n^2\left(\frac{1}{2}\right)^n$

68. $\sum_{n=1}^{\infty} 2n^3\left(\frac{1}{5}\right)^n$

69. $\sum_{n=1}^{\infty} \frac{1}{(2n)!}$

70. $\sum_{n=1}^{\infty} n\left(\frac{4}{11}\right)^n$

True or False? In Exercises 71 and 72, determine whether the statement is true or false. If it is false, explain why or give an example that shows it is false.

71. If $\lim_{n\to\infty} a_n = 0$, then $\sum_{n=1}^{\infty} a_n$ converges.

72. If $|r| < 1$, then

$$\sum_{n=1}^{\infty} ar^n = \frac{a}{1-r}.$$

10.3 p-Series and the Ratio Test

- Determine the convergence or divergence of p-series.
- Use the Ratio Test to determine the convergence or divergence of series.

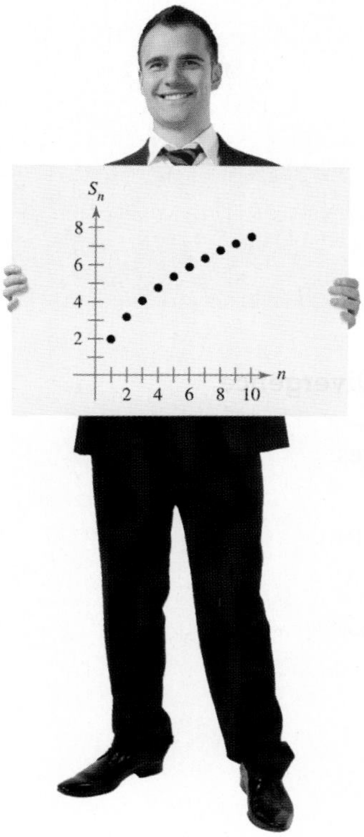

In Exercises 41–46 on page 625, you will use the graph of a sequence of partial sums to help determine whether a series converges or diverges.

p-Series

In Section 10.2, you studied geometric series. In this section you will study another common type of series called a ***p*-series.**

Definition of *p*-Series

Let p be a positive constant. An infinite series of the form

$$\sum_{n=1}^{\infty} \frac{1}{n^p} = \frac{1}{1^p} + \frac{1}{2^p} + \frac{1}{3^p} + \cdots$$

is a ***p*-series.** For $p = 1$, the series

$$\sum_{n=1}^{\infty} \frac{1}{n} = 1 + \frac{1}{2} + \frac{1}{3} + \cdots$$

is the **harmonic series.**

Example 1 **Classifying Infinite Series**

Classify each infinite series.

a. $\sum_{n=1}^{\infty} \frac{1}{n^3}$ **b.** $\sum_{n=1}^{\infty} \frac{1}{\sqrt{n}}$ **c.** $\sum_{n=1}^{\infty} \frac{1}{3^n}$

SOLUTION

a. The infinite series

$$\sum_{n=1}^{\infty} \frac{1}{n^3} = \frac{1}{1^3} + \frac{1}{2^3} + \frac{1}{3^3} + \cdots$$

is a p-series with $p = 3$.

b. The infinite series

$$\sum_{n=1}^{\infty} \frac{1}{\sqrt{n}} = \frac{1}{1^{1/2}} + \frac{1}{2^{1/2}} + \frac{1}{3^{1/2}} + \cdots$$

is a p-series with $p = \frac{1}{2}$.

c. The infinite series

$$\sum_{n=1}^{\infty} \frac{1}{3^n} = \frac{1}{3^1} + \frac{1}{3^2} + \frac{1}{3^3} + \cdots$$

is *not* a p-series. It is a geometric series.

✓Checkpoint 1

Classify each infinite series.

a. $\sum_{n=1}^{\infty} \frac{1}{n^{\pi}}$ **b.** $\sum_{n=1}^{\infty} \frac{1}{n\sqrt{n}}$ **c.** $\sum_{n=1}^{\infty} \frac{1}{2^n}$ ■

Some infinite p-series converge and others diverge. With the test below, you can determine the convergence or divergence of a p-series.

Test for Convergence of a p-Series

Consider the p-series

$$\sum_{n=1}^{\infty} \frac{1}{n^p} = \frac{1}{1^p} + \frac{1}{2^p} + \frac{1}{3^p} + \cdots.$$

1. The series diverges when $0 < p \leq 1$.

2. The series converges when $p > 1$.

Example 2 Determining Convergence or Divergence

Determine whether each p-series converges or diverges.

a. $\displaystyle\sum_{n=1}^{\infty} \frac{1}{n^{0.9}}$

b. $\displaystyle\sum_{n=1}^{\infty} \frac{1}{n}$

c. $\displaystyle\sum_{n=1}^{\infty} \frac{1}{n^{1.1}}$

SOLUTION

a. For the p-series

$$\sum_{n=1}^{\infty} \frac{1}{n^{0.9}} = \frac{1}{1^{0.9}} + \frac{1}{2^{0.9}} + \frac{1}{3^{0.9}} + \cdots$$

$p = 0.9$. Because $p \leq 1$, you can conclude that the series diverges.

b. For the p-series

$$\sum_{n=1}^{\infty} \frac{1}{n} = \frac{1}{1} + \frac{1}{2} + \frac{1}{3} + \cdots$$

$p = 1$, which means that the series is the *harmonic* series. Because $p \leq 1$, you can conclude that the series diverges.

c. For the p-series

$$\sum_{n=1}^{\infty} \frac{1}{n^{1.1}} = \frac{1}{1^{1.1}} + \frac{1}{2^{1.1}} + \frac{1}{3^{1.1}} + \cdots$$

$p = 1.1$. Because $p > 1$, you can conclude that the series converges.

✓ Checkpoint 2

Determine whether each p-series converges or diverges.

a. $\displaystyle\sum_{n=1}^{\infty} \frac{1}{n\sqrt{n}}$ **b.** $\displaystyle\sum_{n=1}^{\infty} \frac{1}{n^{2.5}}$ **c.** $\displaystyle\sum_{n=1}^{\infty} \frac{1}{n^{1/10}}$ ■

In Example 2, notice that the p-Series Test tells you only whether the series diverges or converges. It does not give a formula for the sum of a convergent p-series. To approximate such a sum, you can use a computer to evaluate several partial sums.

The Ratio Test

At this point, you have studied two convergence tests: one for a geometric series and one for a p-series. The next test is more general: it can be applied to infinite series that do not happen to be geometric series or p-series.

STUDY TIP

Although the Ratio Test is listed with an index that begins with $n = 1$, it can be applied to an infinite series with a beginning index of $n = 0$ (or any other integer).

The Ratio Test

Let $\sum_{n=1}^{\infty} a_n$ be an infinite series with nonzero terms.

1. The series converges when $\lim_{n\to\infty} \left|\frac{a_{n+1}}{a_n}\right| < 1$.

2. The series diverges when $\lim_{n\to\infty} \left|\frac{a_{n+1}}{a_n}\right| > 1$ or $\lim_{n\to\infty} \left|\frac{a_{n+1}}{a_n}\right| = \infty$.

3. The test is inconclusive when $\lim_{n\to\infty} \left|\frac{a_{n+1}}{a_n}\right| = 1$.

The Ratio Test is particularly useful for series that converge rapidly. Series involving factorial or exponential functions are frequently of this type.

Example 3 Using the Ratio Test

Determine the convergence or divergence of the infinite series

$$\sum_{n=0}^{\infty} \frac{2^n}{n!} = \frac{1}{1} + \frac{2}{1} + \frac{4}{2} + \frac{8}{6} + \frac{16}{24} + \frac{32}{120} + \cdot \cdot \cdot .$$

SOLUTION Using the Ratio Test with

$$a_n = \frac{2^n}{n!}$$

you obtain

$$\begin{aligned}
\lim_{n\to\infty} \left|\frac{a_{n+1}}{a_n}\right| &= \lim_{n\to\infty} \left[\frac{2^{n+1}}{(n+1)!} \div \frac{2^n}{n!}\right] \\
&= \lim_{n\to\infty} \left[\frac{2^{n+1}}{(n+1)!} \cdot \frac{n!}{2^n}\right] \\
&= \lim_{n\to\infty} \left[\frac{2(2^n)}{(n+1)n!} \cdot \frac{n!}{2^n}\right] \\
&= \lim_{n\to\infty} \frac{2}{n+1} \\
&= 0.
\end{aligned}$$

STUDY TIP

A step frequently used in applications of the Ratio Test involves simplifying quotients of factorials. In Example 3, for instance, notice that

$$\frac{n!}{(n+1)!} = \frac{n!}{(n+1)n!} = \frac{1}{n+1}.$$

Because this limit is less than 1, you can conclude that the series converges. Using a computer, you can approximate the sum of the series to be

$$S \approx S_{10} \approx 7.39. \qquad \text{Approximate using 10th partial sum.}$$

✓ Checkpoint 3

Determine the convergence or divergence of the infinite series

$$\sum_{n=0}^{\infty} \frac{3^n}{n!}.$$

Example 3 indicates that as n approaches infinity, the sequence $\{n!\}$ increases faster than $\{2^n\}$. In the table below, you can see that although the factorial sequence $\{n!\}$ has a slow start, it quickly overpowers the exponential sequence $\{2^n\}$.

n	0	1	2	3	4	5	6	7	8	9
2^n	1	2	4	8	16	32	64	128	256	512
$n!$	1	1	2	6	24	120	720	5040	40,320	362,880

From this table, you can also see that the sequence $\{n\}$ approaches infinity more slowly than the sequence $\{2^n\}$. This is further demonstrated in Example 4.

Example 4 Using the Ratio Test

Determine the convergence or divergence of the infinite series

$$\sum_{n=1}^{\infty} \frac{n}{2^n} = \frac{1}{2} + \frac{2}{4} + \frac{3}{8} + \frac{4}{16} + \frac{5}{32} + \frac{6}{64} + \cdots.$$

SOLUTION Using the Ratio Test with $a_n = n/2^n$, you obtain

$$\begin{aligned}
\lim_{n\to\infty}\left|\frac{a_{n+1}}{a_n}\right| &= \lim_{n\to\infty}\left(\frac{n+1}{2^{n+1}} \div \frac{n}{2^n}\right)\\
&= \lim_{n\to\infty}\left(\frac{n+1}{2^{n+1}} \cdot \frac{2^n}{n}\right)\\
&= \lim_{n\to\infty}\frac{n+1}{2n}\\
&= \frac{1}{2}.
\end{aligned}$$

Because this limit is less than 1, you can conclude that the series converges. Using a computer, you can determine that the sum of the series is $S = 2$.

✓Checkpoint 4

Determine the convergence or divergence of the infinite series

$$\sum_{n=1}^{\infty} \frac{5^n}{n^2}.$$

■

When applying the Ratio Test, remember that when the limit of $|a_{n+1}/a_n|$ as $n \to \infty$ is 1, the test does not tell you whether the series converges or diverges. This type of result often occurs with series that converge or diverge slowly. For instance, when you apply the Ratio Test to the harmonic series in which

$$a_n = \frac{1}{n}$$

you obtain

$$\lim_{n\to\infty}\left|\frac{a_{n+1}}{a_n}\right| = \lim_{n\to\infty}\frac{1/(n+1)}{1/n} = \lim_{n\to\infty}\frac{n}{n+1} = 1.$$

So, from the Ratio Test, you cannot conclude that the harmonic series diverges. (The Ratio Test is also inconclusive for any p-series.) From the p-Series Test, you know that it diverges.

Example 5 Using the Ratio Test

Determine the convergence or divergence of the infinite series

$$\sum_{n=1}^{\infty} \frac{2^n}{n^2} = \frac{2}{1} + \frac{4}{4} + \frac{8}{9} + \frac{16}{16} + \frac{32}{25} + \frac{64}{36} + \cdots.$$

SOLUTION Using the Ratio Test with $a_n = 2^n/n^2$, you obtain

$$\begin{aligned}
\lim_{n\to\infty} \left|\frac{a_{n+1}}{a_n}\right| &= \lim_{n\to\infty} \left[\frac{2^{n+1}}{(n+1)^2} \div \frac{2^n}{n^2}\right] \\
&= \lim_{n\to\infty} \left[\frac{2^{n+1}}{(n+1)^2} \cdot \frac{n^2}{2^n}\right] \\
&= \lim_{n\to\infty} 2\left(\frac{n}{n+1}\right)^2 \\
&= 2.
\end{aligned}$$

Because this limit is greater than 1, you can conclude that the series diverges.

✓ Checkpoint 5

Determine the convergence or divergence of the infinite series

$$\sum_{n=1}^{\infty} \frac{n!}{10^n}.$$

■

Summary of Tests for Series

Test	*Series*	*Converges*	*Diverges*
nth-Term	$\sum_{n=1}^{\infty} a_n$	No test	$\lim_{n\to\infty} a_n \neq 0$
Geometric Series	$\sum_{n=0}^{\infty} ar^n$	$\lvert r\rvert < 1$	$\lvert r\rvert \geq 1$
p-Series	$\sum_{n=1}^{\infty} \frac{1}{n^p}$	$p > 1$	$0 < p \leq 1$
Ratio	$\sum_{n=1}^{\infty} a_n$	$\lim_{n\to\infty} \left\lvert\frac{a_{n+1}}{a_n}\right\rvert < 1$	$\lim_{n\to\infty} \left\lvert\frac{a_{n+1}}{a_n}\right\rvert > 1$ or $\lim_{n\to\infty} \left\lvert\frac{a_{n+1}}{a_n}\right\rvert = \infty$

SUMMARIZE (Section 10.3)

1. State the definition of p-series *(page 619)*. For an example of classifying infinite series, see Example 1.
2. State the test for convergence of a p-series *(page 620)*. For an example of determining the convergence or divergence of a p-series, see Example 2.
3. State the Ratio Test *(page 621)*. For examples of using the Ratio Test, see Examples 3, 4, and 5.

SKILLS WARM UP 10.3

The following warm-up exercises involve skills that were covered in earlier sections. You will use these skills in the exercise set for this section. For additional help, review Sections 3.6, 10.1, and 10.2.

In Exercises 1–4, simplify the expression.

1. $\dfrac{n!}{(n+1)!}$

2. $\dfrac{(n+1)!}{n!}$

3. $\dfrac{3^{n+1}}{n+1} \cdot \dfrac{n}{3^n}$

4. $\dfrac{(n+1)^2}{(n+1)!} \cdot \dfrac{n!}{n^2}$

In Exercises 5–8, find the limit.

5. $\lim\limits_{n\to\infty} \dfrac{(n+1)^2}{n^2}$

6. $\lim\limits_{n\to\infty} \dfrac{5^{n+1}}{5^n}$

7. $\lim\limits_{n\to\infty} \left(\dfrac{5}{n+1} \div \dfrac{5}{n}\right)$

8. $\lim\limits_{n\to\infty} \left(\dfrac{(n+1)^3}{3^{n+1}} \div \dfrac{n^3}{3^n}\right)$

In Exercises 9 and 10, decide whether the series is geometric.

9. $\displaystyle\sum_{n=1}^{\infty} \frac{1}{4^n}$

10. $\displaystyle\sum_{n=1}^{\infty} \frac{1}{n^4}$

Exercises 10.3

See www.CalcChat.com for worked-out solutions to odd-numbered exercises.

Classifying Infinite Series In Exercises 1–8, determine whether the series is a p-series. *See Example 1.*

1. $\displaystyle\sum_{n=1}^{\infty} \frac{1}{n^2}$

2. $\displaystyle\sum_{n=1}^{\infty} \frac{1}{n^6}$

3. $\displaystyle\sum_{n=1}^{\infty} \frac{1}{5^n}$

4. $\displaystyle\sum_{n=1}^{\infty} n^{-3/4}$

5. $\displaystyle\sum_{n=1}^{\infty} \frac{1}{n^n}$

6. $\displaystyle\sum_{n=1}^{\infty} \frac{1}{n+1}$

7. $1 + \dfrac{1}{\sqrt[3]{2}} + \dfrac{1}{\sqrt[3]{3}} + \dfrac{1}{\sqrt[3]{4}} + \cdots$

8. $1 + \frac{1}{4} + \frac{1}{16} + \frac{1}{64} + \cdots$

Determining Convergence or Divergence In Exercises 9–18, determine the convergence or divergence of the p-series. *See Example 2.*

9. $\displaystyle\sum_{n=1}^{\infty} \frac{1}{n^3}$

10. $\displaystyle\sum_{n=1}^{\infty} \frac{1}{n^{1/3}}$

11. $\displaystyle\sum_{n=1}^{\infty} \frac{1}{\sqrt[5]{n}}$

12. $\displaystyle\sum_{n=1}^{\infty} \frac{1}{n^{4/3}}$

13. $\displaystyle\sum_{n=1}^{\infty} \frac{1}{n^{1.03}}$

14. $\displaystyle\sum_{n=1}^{\infty} \frac{1}{n^{\pi/2}}$

15. $1 + \dfrac{1}{\sqrt{2}} + \dfrac{1}{\sqrt{3}} + \dfrac{1}{\sqrt{4}} + \cdots$

16. $1 + \frac{1}{4} + \frac{1}{9} + \frac{1}{16} + \frac{1}{25} + \cdots$

17. $1 + \dfrac{1}{\sqrt[3]{4}} + \dfrac{1}{\sqrt[3]{9}} + \dfrac{1}{\sqrt[3]{16}} + \cdots$

18. $1 + \frac{1}{16} + \frac{1}{81} + \frac{1}{256} + \frac{1}{625} + \cdots$

Using the Ratio Test In Exercises 19–32, use the Ratio Test to determine the convergence or divergence of the series. *See Examples 3, 4, and 5.*

19. $\displaystyle\sum_{n=0}^{\infty} \frac{6^n}{n!}$

20. $\displaystyle\sum_{n=0}^{\infty} \frac{n!}{3^n}$

21. $\displaystyle\sum_{n=1}^{\infty} n\left(\frac{3}{4}\right)^n$

22. $\displaystyle\sum_{n=1}^{\infty} n\left(\frac{3}{2}\right)^n$

23. $\displaystyle\sum_{n=1}^{\infty} \frac{n}{4^n}$

24. $\displaystyle\sum_{n=1}^{\infty} \frac{n^2}{2^n}$

25. $\displaystyle\sum_{n=1}^{\infty} \frac{2^n}{n^5}$

26. $\displaystyle\sum_{n=0}^{\infty} (-1)^n e^{-n}$

27. $\displaystyle\sum_{n=0}^{\infty} \frac{(-1)^n 2^n}{n!}$

28. $\displaystyle\sum_{n=0}^{\infty} \frac{4n}{n!}$

29. $\displaystyle\sum_{n=0}^{\infty} \frac{4^n}{3^n+1}$

30. $\displaystyle\sum_{n=0}^{\infty} \frac{3^n}{n+1}$

31. $\displaystyle\sum_{n=0}^{\infty} \frac{n5^n}{n!}$

32. $\displaystyle\sum_{n=1}^{\infty} \frac{2n!}{n^5}$

Using the Ratio Test with a p-Series In Exercises 33–36, verify that the Ratio Test is inconclusive for the p-series.

33. $\displaystyle\sum_{n=1}^{\infty} \frac{1}{n^{3/2}}$

34. $\displaystyle\sum_{n=1}^{\infty} \frac{1}{n^{1/2}}$

35. $\displaystyle\sum_{n=1}^{\infty} \frac{1}{n^3}$

36. $\displaystyle\sum_{n=1}^{\infty} \frac{1}{n^4}$

Maximum Error of a p-Series In Exercises 37–40, for $p \geq 3$, the sum S of a convergent p-series differs from its nth partial sum S_n by no more than

$$\frac{1}{(p-1)n^{p-1}}.$$

Approximate the sum of the p-series by finding its nth partial sum. Then find the maximum error of your approximation.

37. $\sum_{n=1}^{\infty} \frac{1}{n^3}$, S_4

38. $\sum_{n=1}^{\infty} \frac{1}{n^4}$, S_4

39. $\sum_{n=1}^{\infty} \frac{1}{n^{7/2}}$, S_{10}

40. $\sum_{n=1}^{\infty} \frac{1}{n^7}$, S_5

Matching In Exercises 41–46, match the series with the graph of its sequence of partial sums. [The graphs are labeled (a)–(f).] Then determine the convergence or divergence of the series.

(a)
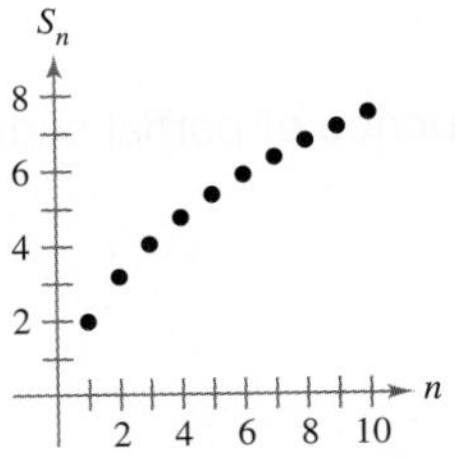

(b)
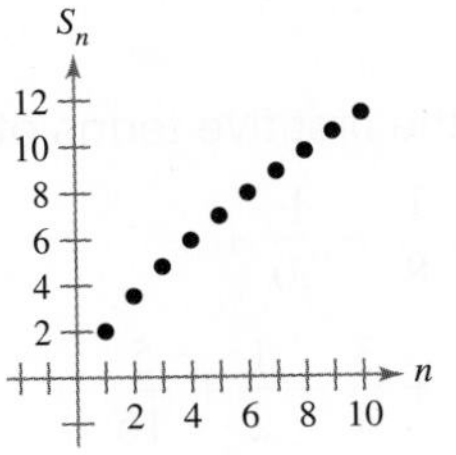

(c)
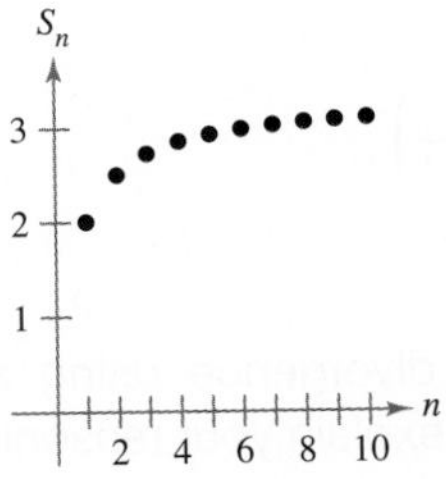

(d)
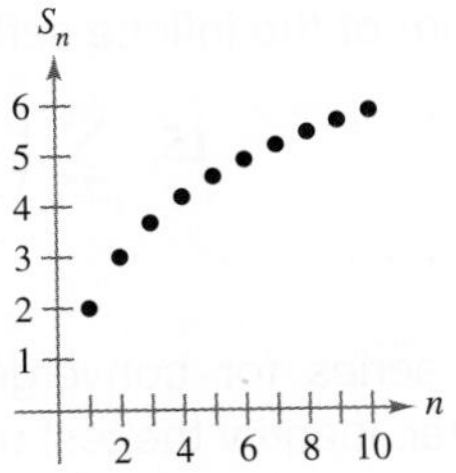

(e)
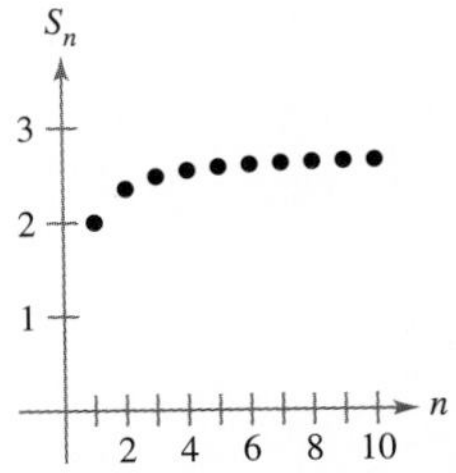

(f)
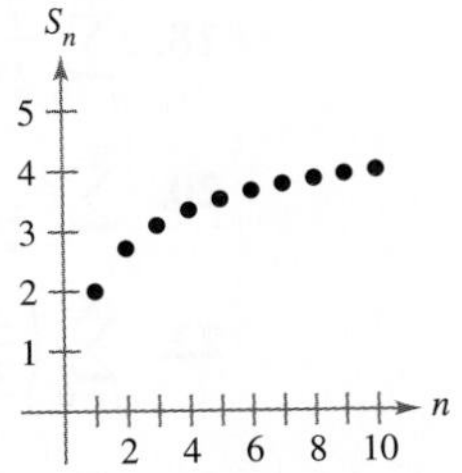

41. $\sum_{n=1}^{\infty} \frac{2}{\sqrt[4]{n^3}}$

42. $\sum_{n=1}^{\infty} \frac{2}{n}$

43. $\sum_{n=1}^{\infty} \frac{2}{\sqrt{n^5}}$

44. $\sum_{n=1}^{\infty} \frac{2}{\sqrt[5]{n^2}}$

45. $\sum_{n=1}^{\infty} \frac{2}{n\sqrt{n}}$

46. $\sum_{n=1}^{\infty} \frac{2}{n^2}$

Determining Convergence or Divergence In Exercises 47–64, test the series for convergence or divergence using any appropriate test from this chapter. Identify the test used and explain your reasoning. If the series converges, find the sum whenever possible.

47. $\sum_{n=1}^{\infty} \frac{2n}{n+1}$

48. $\sum_{n=1}^{\infty} \frac{5}{n}$

49. $\sum_{n=1}^{\infty} \frac{4}{n\sqrt[4]{n}}$

50. $\sum_{n=1}^{\infty} \frac{1}{n\sqrt[3]{n}}$

51. $\sum_{n=1}^{\infty} \left(-\frac{3}{4}\right)^n$

52. $\sum_{n=0}^{\infty} \left(\frac{5}{6}\right)^n$

53. $\sum_{n=0}^{\infty} \frac{(-1)^n 2^n}{3^n}$

54. $\sum_{n=2}^{\infty} \ln n$

55. $\sum_{n=1}^{\infty} \left(\frac{1}{n^2} - \frac{1}{n^3}\right)$

56. $\sum_{n=1}^{\infty} \frac{n3^n}{n!}$

57. $\sum_{n=0}^{\infty} \left(\frac{5}{4}\right)^n$

58. $\sum_{n=1}^{\infty} n(0.4)^n$

59. $\sum_{n=1}^{\infty} \frac{n!}{3^{n-1}}$

60. $\sum_{n=1}^{\infty} \frac{n}{3n-2}$

61. $\sum_{n=1}^{\infty} \frac{n}{\sqrt{n^2+1}}$

62. $\sum_{n=1}^{\infty} 2e^{-n}$

$\sum_{n=1}^{\infty} \frac{2^n}{5^{n-1}}$

64. $\sum_{n=1}^{\infty} \frac{n^2}{n^2+1}$

65. ...nk About It A friend in your calculus class tells te...hat the following series converges because the frien...e very small and approach 0 rapidly. Is your ...rect? Explain.

$$\frac{1}{10{,}000} + \frac{1}{\ldots 01} + \frac{1}{10{,}002} + \cdots$$

66. **HOW DO ... SEE IT?** The graphs show the sequences of ...l sums of the p-series

$\sum_{n=1}^{\infty} \frac{1}{n^{0.4}}$ and $\sum_{n=1}^{\infty} \frac{\ldots}{n}$

Using the p-Series Test, ... the second series converge... series diverges and ...in how the graphs show this.

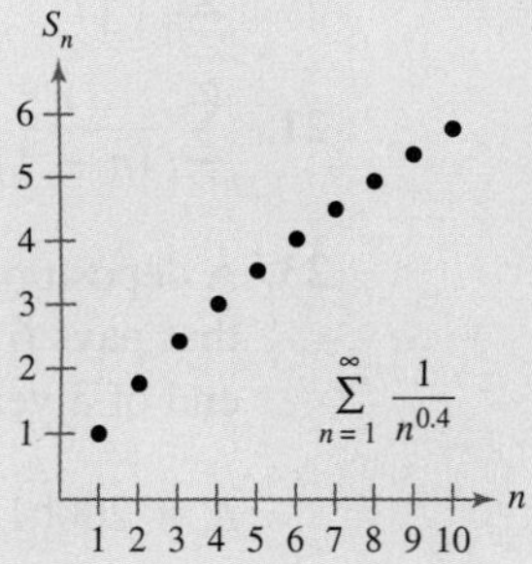

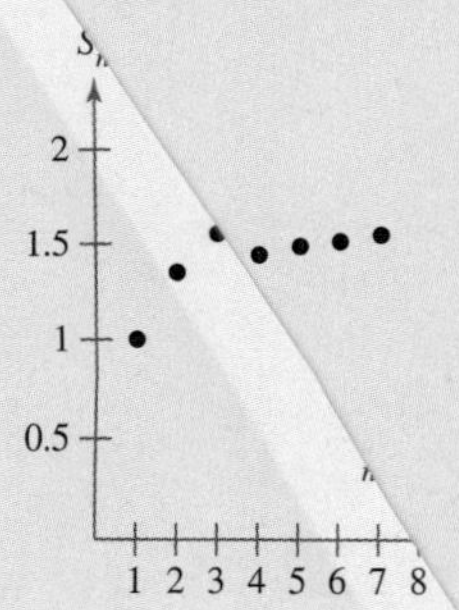

QUIZ YOURSELF

See www.CalcChat.com for worked-out solutions to odd-numbered exercises.

Take this quiz as you would take a quiz in class. When you are done, check your work against the answers given in the back of the book.

In Exercises 1–4, write the first five terms of the sequence.

1. $a_n = \left(-\frac{1}{4}\right)^n$

2. $a_n = \frac{n + }{n \; 3}$

3. $a_n = 5(-1)^n$

4. $a_n = \frac{- 2}{n!}$

In Exercises 5–8, find the limit of the seque e (if it exists) as n approaches infinity. Then state whether the sequence conve s or diverges.

5. $a_n = \frac{3}{\sqrt{n}}$

6. $a_n = \frac{n}{2n + 3}$

7. $a_n = \frac{2}{(n + 1)!}$

8. $a_n = \frac{(-1)^n}{2}$

In Exercises 9–11, write an pression for the nth term of the sequence. (There is more than one correct ar er.)

9. $0, \frac{1}{4}, \frac{2}{9}, \frac{3}{16}, \ldots$

10. $-3, \sqrt{3}, -\sqrt[3]{3}, \sqrt[4]{3}, \ldots$

11. $\frac{1}{2}, 2, \frac{1}{2}, 2, \ldots$

In Exercises 12 13, write the first five terms of the sequence of partial sums.

12. $\sum_{n=1}^{\infty} \frac{n - 1}{n!} \; 0 + \frac{1}{2} + \frac{1}{3} + \frac{1}{8} + \frac{1}{30} + \cdots$

13. $\sum_{n=}^{\infty} \quad ^{+1}\frac{n}{2^{n-1}} = 1 - 1 + \frac{3}{4} - \frac{1}{2} + \frac{5}{16} - \cdots$

In E ses 14–16, find the sum of the infinite series.

$\sum_{n=0} 4\left(\frac{2}{3}\right)^n$

15. $\sum_{n=0}^{\infty} \left(\frac{1}{3^n} - \frac{1}{3^{n+1}}\right)$

. $5 + 0.5 + 0.05 + 0.005 + \cdots$

In Exercises 17–22, test the series for convergence or divergence using any appropriate test from this chapter. Identify the test used and explain your reasoning.

17. $\sum_{n=1}^{\infty} \frac{2n^2 - 1}{n^2 + 1}$

18. $\sum_{n=0}^{\infty} \frac{7}{2^n}$

19. $\sum_{n=0}^{\infty} \left(\frac{5}{3}\right)^n$

20. $\sum_{n=1}^{\infty} \frac{1}{\sqrt[3]{n^5}}$

21. $\sum_{n=1}^{\infty} \frac{n}{(n - 1)!}$

22. $\sum_{n=0}^{\infty} \left(\frac{2}{3}\right)^n n!$

23. A deposit of \$200 is made at the beginning of each month for 3 years in an account that pays 6% interest, compounded monthly. The balance A in the account at the end of 3 years is

$$A = 200\left(1 + \frac{0.06}{12}\right)^1 + 200\left(1 + \frac{0.06}{12}\right)^2 + \cdots + 200\left(1 + \frac{0.06}{12}\right)^{36}.$$

(a) Use sigma notation to write the balance in the account at the end of 3 years.

(b) Find the balance in the account at the end of 3 years.

10.4 Power Series and Taylor's Theorem

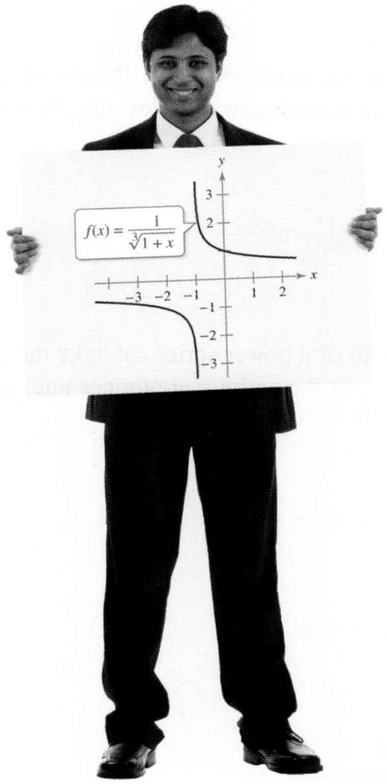

In Exercise 34 on page 635, you will use a power series to write a binomial series to represent a radical function.

- ■ Recognize power series.
- ■ Find the radii of convergence of power series.
- ■ Use Taylor's Theorem to find power series for functions.
- ■ Use the basic list of power series to find power series for functions.

Power Series

In the preceding two sections, you studied infinite series whose terms are constants. In this section, you will study infinite series that have variable terms. Specifically, you will study a type of infinite series that is called a **power series.** Informally, you can think of a power series as a "very long" polynomial.

Definition of Power Series

If x is a variable, then an infinite series of the form

$$\sum_{n=0}^{\infty} a_n x^n = a_0 + a_1 x + a_2 x^2 + a_3 x^3 + \cdot\cdot\cdot + a_n x^n + \cdot\cdot\cdot$$

is called a **power series.** More generally, an infinite series of the form

$$\sum_{n=0}^{\infty} a_n (x-c)^n = a_0 + a_1(x-c) + a_2(x-c)^2 + \cdot\cdot\cdot + a_n(x-c)^n + \cdot\cdot\cdot$$

is called a **power series centered at c,** where c is a constant.

The index of a power series usually begins with $n = 0$. In such cases, assume that

$$(x-c)^0 = 1$$

even for $x = c$.

Example 1 **Power Series**

a. The following power series is centered at 0.

$$\sum_{n=0}^{\infty} \frac{x^n}{n!} = 1 + x + \frac{x^2}{2!} + \frac{x^3}{3!} + \cdot\cdot\cdot$$

b. The following power series is centered at 1.

$$\sum_{n=1}^{\infty} \frac{(x-1)^n}{n} = (x-1) + \frac{(x-1)^2}{2} + \frac{(x-1)^3}{3} + \cdot\cdot\cdot$$

c. The following power series is centered at -1.

$$\sum_{n=1}^{\infty} \frac{(x+1)^n}{n} = (x+1) + \frac{(x+1)^2}{2} + \frac{(x+1)^3}{3} + \cdot\cdot\cdot$$

✓ Checkpoint 1

Identify the center of each power series.

a. $\sum_{n=1}^{\infty} \frac{(x-2)^n}{n^2}$ **b.** $\sum_{n=1}^{\infty} \frac{(x+3)^n}{3^n}$ **c.** $\sum_{n=1}^{\infty} \frac{(-1)^n x^{2n}}{(2n)!}$ ■

Radius of Convergence of a Power Series

A power series in x can be viewed as a function of x,

$$f(x) = \sum_{n=0}^{\infty} a_n(x - c)^n$$

where the *domain* of f is the set of all x for which the power series converges. Determining this domain is one of the primary problems associated with power series. Of course, every power series converges at its center c because

$$\begin{aligned} f(c) &= \sum_{n=0}^{\infty} a_n(c - c)^n \\ &= a_0(1) + 0 + 0 + \cdot \cdot \cdot \\ &= a_0. \end{aligned}$$

So, c is always in the domain of f. In fact, the domain of a power series can take three basic forms: a single point, an interval centered at c, or the entire real number line, as shown in Figure 10.4.

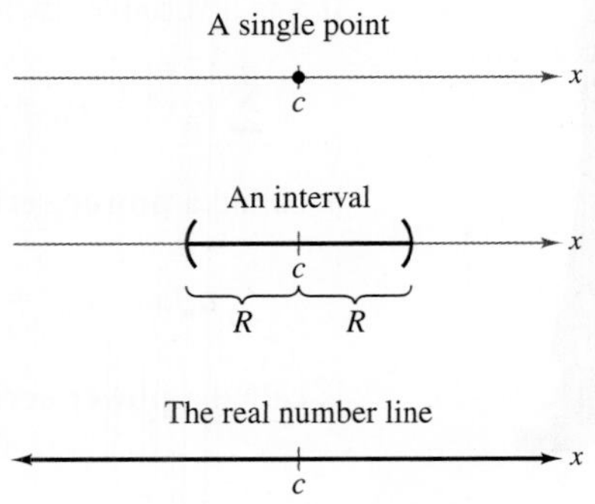

The domain of a power series can take three basic forms: a single point, an interval centered at c, or the entire real number line.

FIGURE 10.4

Convergence of a Power Series

For a power series centered at c, precisely one of the following is true.

1. The series converges only at c.
2. There exists a positive real number R such that the series converges for $|x - c| < R$ and diverges for $|x - c| > R$.
3. The series converges for all x.

The number R is the **radius of convergence** of the power series. If the series converges only at c, then $R = 0$, and if the series converges for all x, then $R = \infty$.

In the second case, the series converges in the interval $(c - R, c + R)$ and diverges in the intervals $(-\infty, c - R)$ and $(c + R, \infty)$. Determining the convergence or divergence at the endpoints

$$c - R \quad \text{and} \quad c + R$$

can be difficult, and, except for simple cases, the endpoint question is left open. To find the radius of convergence of a power series, use the Ratio Test, as demonstrated in Examples 2 and 3.

Example 2 Finding the Radius of Convergence

Find the radius of convergence of the power series

$$\sum_{n=0}^{\infty} \frac{x^n}{n!}.$$

SOLUTION For this power series, $a_n = 1/n!$. So, you have

$$\begin{aligned}\lim_{n\to\infty}\left|\frac{a_{n+1}x^{n+1}}{a_n x^n}\right| &= \lim_{n\to\infty}\left|\frac{x^{n+1}/(n+1)!}{x^n/n!}\right| \\ &= \lim_{n\to\infty}\left|\frac{x}{n+1}\right| \\ &= 0.\end{aligned}$$

So, by the Ratio Test, this series converges for all x, and the radius of convergence is $R = \infty$.

✓ Checkpoint 2

Find the radius of convergence of the power series $\sum_{n=0}^{\infty} \frac{x^{2n}}{(2n)!}$. ■

Example 3 Finding the Radius of Convergence

Find the radius of convergence of the power series

$$\sum_{n=0}^{\infty} \frac{(-1)^n(x+1)^n}{2^n}.$$

SOLUTION For this power series, $a_n = (-1)^n/2^n$. So, you have

$$\begin{aligned}\lim_{n\to\infty}\left|\frac{a_{n+1}(x+1)^{n+1}}{a_n(x+1)^n}\right| &= \lim_{n\to\infty}\left|\frac{(-1)^{n+1}(x+1)^{n+1}/2^{n+1}}{(-1)^n(x+1)^n/2^n}\right| \\ &= \lim_{n\to\infty}\left|\frac{(-1)(x+1)}{2}\right| \\ &= \lim_{n\to\infty}\left|\frac{x+1}{2}\right| \\ &= \left|\frac{x+1}{2}\right|.\end{aligned}$$

By the Ratio Test, this series will converge as long as $|(x+1)/2| < 1$ or $|x+1| < 2$. So, the radius of convergence is $R = 2$. Because the series is centered at $x = -1$, it will converge in the interval $(-3, 1)$, as shown in Figure 10.5.

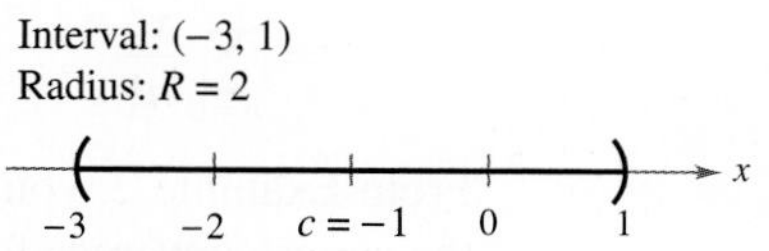

FIGURE 10.5

✓ Checkpoint 3

Find the radius of convergence of the power series

$$\sum_{n=0}^{\infty} \frac{(-1)^n(x-2)^n}{3^n}.$$

■

Taylor and Maclaurin Series

The problem of finding a power series for a given function is answered by **Taylor's Theorem,** named after the English mathematician Brook Taylor (1685–1731). This theorem shows how to use derivatives of a function f to write the power series for f.

Taylor's Theorem

If f is represented by a power series centered at c, then the power series has the form

$$f(x) = \sum_{n=0}^{\infty} \frac{f^{(n)}(c)(x-c)^n}{n!}$$

$$= f(c) + \frac{f'(c)(x-c)}{1!} + \frac{f''(c)(x-c)^2}{2!} + \frac{f'''(c)(x-c)^3}{3!} + \cdot \cdot \cdot .$$

This power series is called the **Taylor series** for $f(x)$ centered at c. When the series is centered at 0, it is called a **Maclaurin series,** after the Scottish mathematician Colin Maclaurin (1698–1746).

Example 4 Finding a Maclaurin Series

Find the power series for

$$f(x) = e^x$$

centered at 0. What is the radius of convergence of the series?

SOLUTION Begin by finding several derivatives of f and evaluating each at $c = 0$.

$f(x) = e^x$ $\quad f(0) = 1$ $\quad$ Write original function.

$f'(x) = e^x$ $\quad f'(0) = 1$ $\quad$ Find first derivative.

$f''(x) = e^x$ $\quad f''(0) = 1$ $\quad$ Find second derivative.

$f'''(x) = e^x$ $\quad f'''(0) = 1$ $\quad$ Find third derivative.

$f^{(4)}(x) = e^x$ $\quad f^{(4)}(0) = 1$ $\quad$ Find fourth derivative.

$f^{(5)}(x) = e^x$ $\quad f^{(5)}(0) = 1$ $\quad$ Find fifth derivative.

From this pattern, you can see that $f^{(n)}(0) = 1$. So, by Taylor's Theorem,

$$e^x = f(0) + f'(0)x + \frac{f''(0)x^2}{2!} + \frac{f'''(0)x^3}{3!} + \cdot \cdot \cdot$$

$$= 1 + x + \frac{x^2}{2!} + \frac{x^3}{3!} + \cdot \cdot \cdot$$

$$= \sum_{n=0}^{\infty} \frac{x^n}{n!}.$$

From Example 2, you know that the radius of convergence is $R = \infty$. In other words, the series converges for all values of x.

✓ Checkpoint 4

Find the power series for

$$f(x) = e^{-x}$$

centered at 0. What is the radius of convergence of the series? ■

Example 5 Finding a Taylor Series

Find the power series for

$$f(x) = \frac{1}{x}$$

centered at 1. Then use the result to evaluate $f\left(\frac{1}{2}\right)$.

SOLUTION Successive differentiation of $f(x)$ produces the pattern below.

$f(x) = x^{-1}$	$f(1) = 1 = 0!$	Write original function.
$f'(x) = -x^{-2}$	$f'(1) = -1 = -(1!)$	Find first derivative.
$f''(x) = 2x^{-3}$	$f''(1) = 2 = 2!$	Find second derivative.
$f'''(x) = -6x^{-4}$	$f'''(1) = -6 = -(3!)$	Find third derivative.
$f^{(4)}(x) = 24x^{-5}$	$f^{(4)}(1) = 24 = 4!$	Find fourth derivative.
$f^{(5)}(x) = -120x^{-6}$	$f^{(5)}(1) = -120 = -(5!)$	Find fifth derivative.

From this pattern, you can see that $f^{(n)}(1) = (-1)^n n!$. So, by Taylor's Theorem,

$$\begin{aligned}\frac{1}{x} &= f(1) + f'(1)(x-1) + \frac{f''(1)(x-1)^2}{2!} + \frac{f'''(1)(x-1)^3}{3!} + \cdots \\ &= 1 - (x-1) + \frac{2!(x-1)^2}{2!} - \frac{3!(x-1)^3}{3!} + \frac{4!(x-1)^4}{4!} - \cdots \\ &= 1 - (x-1) + (x-1)^2 - (x-1)^3 + (x-1)^4 - \cdots \\ &= \sum_{n=0}^{\infty} (-1)^n (x-1)^n.\end{aligned}$$

To evaluate the series when $x = \frac{1}{2}$, use the formula for the sum of a geometric series.

$$f\left(\frac{1}{2}\right) = \sum_{n=0}^{\infty} (-1)^n \left(\frac{1}{2} - 1\right)^n = \sum_{n=0}^{\infty} \left(\frac{1}{2}\right)^n = \frac{1}{1 - (1/2)} = 2$$

✓Checkpoint 5

Find the power series for $f(x) = \ln x$, centered at 1.

In Example 5, the radius of convergence of the series is $R = 1$, and its interval of convergence is $(0, 2)$. (It is possible to show that the series diverges when $x = 0$ and when $x = 2$.) Figure 10.6 compares the graph of $f(x) = 1/x$ and the graph of the Taylor series for f. In the figure, note that the domains are different. In other words, the power series in Example 5 represents f only in the interval $(0, 2)$.

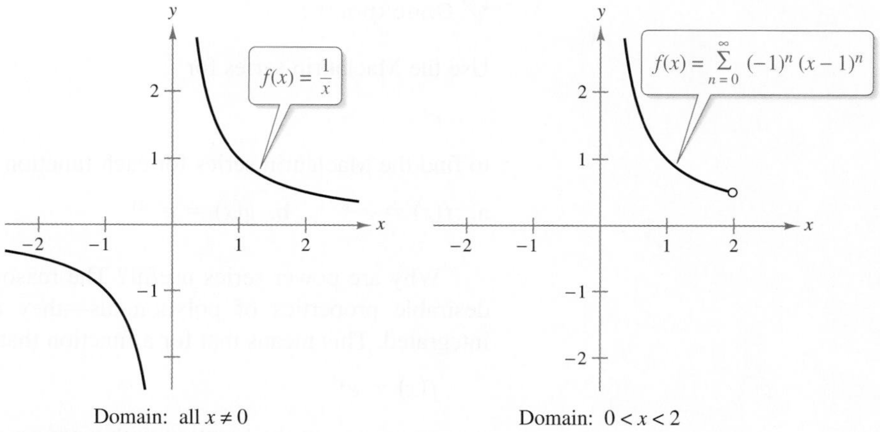

FIGURE 10.6

Example 6 Finding a Maclaurin Series

Find the Maclaurin series for each function.

a. $f(x) = e^{x^2}$

b. $f(x) = e^{-x^2}$

SOLUTION

a. To use Taylor's Theorem directly, you would have to calculate successive derivatives of $f(x) = e^{x^2}$. By calculating the first two derivatives

$$f'(x) = 2xe^{x^2}$$

and

$$f''(x) = (4x^2 + 2)e^{x^2}$$

you can see that this task would be very tedious. Fortunately, there is a simpler way to find the power series. From Example 4, you already know that the Maclaurin series for e^x is

$$e^x = 1 + x + \frac{x^2}{2!} + \frac{x^3}{3!} + \frac{x^4}{4!} + \cdots.$$

So, to find the Maclaurin series for e^{x^2}, you can substitute x^2 for x in the series for e^x. Doing this produces

$$\begin{aligned} e^{x^2} &= 1 + x^2 + \frac{(x^2)^2}{2!} + \frac{(x^2)^3}{3!} + \frac{(x^2)^4}{4!} + \cdots \\ &= 1 + x^2 + \frac{x^4}{2!} + \frac{x^6}{3!} + \frac{x^8}{4!} + \cdots \\ &= \sum_{n=0}^{\infty} \frac{x^{2n}}{n!}. \end{aligned}$$

b. Using an approach similar to the one used in part (a), you can find the Maclaurin series for e^{-x^2} by substituting $-x^2$ for x in the series for e^x. Doing this produces

$$\begin{aligned} e^{-x^2} &= 1 + (-x^2) + \frac{(-x^2)^2}{2!} + \frac{(-x^2)^3}{3!} + \frac{(-x^2)^4}{4!} + \cdots \\ &= 1 - x^2 + \frac{x^4}{2!} - \frac{x^6}{3!} + \frac{x^8}{4!} - \cdots \\ &= \sum_{n=0}^{\infty} \frac{(-1)^n x^{2n}}{n!}. \end{aligned}$$

✓ Checkpoint 6

Use the Maclaurin series for

$$e^x$$

to find the Maclaurin series for each function.

a. $f(x) = e^{2x}$ **b.** $g(x) = e^{-2x}$ ■

Why are power series useful? The reason is that power series share many of the desirable properties of polynomials—they can be easily differentiated and easily integrated. This means that for a function that cannot be easily integrated, such as

$$f(x) = e^{x^2}$$

you can represent the function with a power series and then integrate the power series.

A Basic List of Power Series

Example 6 illustrates an important point in determining power series representations of functions. Although Taylor's Theorem is applicable to a wide variety of functions, it is often tedious to use because of the complexity of finding derivatives. The most practical use of Taylor's Theorem is in developing power series for a *basic list* of common functions. Then, from the basic list, you can determine power series for other functions by the operations of addition, subtraction, multiplication, division, differentiation, integration, and composition with known power series.

Power Series for Common Functions

Function	*Interval of Convergence*
$\frac{1}{x} = 1 - (x-1) + (x-1)^2 - (x-1)^3 + (x-1)^4 - \cdots + (-1)^n(x-1)^n + \cdots$	$0 < x < 2$
$\frac{1}{x+1} = 1 - x + x^2 - x^3 + x^4 - x^5 + \cdots + (-1)^n x^n + \cdots$	$-1 < x < 1$
$\ln x = (x-1) - \frac{(x-1)^2}{2} + \frac{(x-1)^3}{3} - \frac{(x-1)^4}{4} + \cdots + \frac{(-1)^{n-1}(x-1)^n}{n} + \cdots$	$0 < x \le 2$
$e^x = 1 + x + \frac{x^2}{2!} + \frac{x^3}{3!} + \frac{x^4}{4!} + \cdots + \frac{x^n}{n!} + \cdots$	$-\infty < x < \infty$
$(1+x)^k = 1 + kx + \frac{k(k-1)x^2}{2!} + \frac{k(k-1)(k-2)x^3}{3!} + \frac{k(k-1)(k-2)(k-3)x^4}{4!} + \cdots$	$-1 < x < 1$

The last series in the list above is called a **binomial series.** Example 7 illustrates the use of such a series.

Example 7 Using the Basic List of Power Series

Find the power series for

$$g(x) = \sqrt[3]{1+x}$$

centered at zero.

SOLUTION Begin by rewriting g as

$$g(x) = (1+x)^{1/3}.$$

Using the binomial series

$$(1+x)^k = 1 + kx + \frac{k(k-1)x^2}{2!} + \frac{k(k-1)(k-2)x^3}{3!} + \cdots$$

with $k = \frac{1}{3}$, you can write

$$(1+x)^{1/3} = 1 + \frac{x}{3} - \frac{2x^2}{3^2 2!} + \frac{2 \cdot 5x^3}{3^3 3!} - \frac{2 \cdot 5 \cdot 8x^4}{3^4 4!} + \cdots$$

which converges for $-1 < x < 1$.

✓ Checkpoint 7

Use the basic list of power series to find the power series for

$$g(x) = \sqrt{1+x}$$

centered at zero. ■

Example 8 Using the Basic List of Power Series

Find the power series for each function.

a. $f(x) = 1 + 2e^x$, centered at 0

b. $f(x) = e^{2x+1}$, centered at 0

c. $f(x) = \ln 2x$, centered at 1

SOLUTION

a. To find the power series for this function, multiply the series for e^x by 2 and add 1.

$$\begin{aligned} 1 + 2e^x &= 1 + 2\left(1 + x + \frac{x^2}{2!} + \frac{x^3}{3!} + \frac{x^4}{4!} + \cdots\right) \\ &= 1 + 2 + 2x + \frac{2x^2}{2!} + \frac{2x^3}{3!} + \frac{2x^4}{4!} + \cdots \\ &= 1 + \sum_{n=0}^{\infty} \frac{2x^n}{n!} \end{aligned}$$

b. To find the power series for $e^{2x+1} = e^{2x}e$, substitute $2x$ for x in the series for e^x and multiply the result by e.

$$\begin{aligned} e^{2x+1} &= e^{2x}e \\ &= e\left[1 + (2x) + \frac{(2x)^2}{2!} + \frac{(2x)^3}{3!} + \frac{(2x)^4}{4!} + \cdots\right] \\ &= e\sum_{n=0}^{\infty} \frac{2^n x^n}{n!} \end{aligned}$$

c. To find the power series for $\ln 2x$ centered at 1, use the properties of logarithms.

$$\begin{aligned} \ln 2x &= \ln 2 + \ln x \\ &= \ln 2 + (x-1) - \frac{(x-1)^2}{2} + \frac{(x-1)^3}{3} - \frac{(x-1)^4}{4} + \cdots \\ &= \ln 2 + \sum_{n=1}^{\infty} \frac{(-1)^{n-1}(x-1)^n}{n} \end{aligned}$$

✓ Checkpoint 8

Use the basic list of power series to find the power series for each function.

a. $f(x) = 1 - e^{-x}$, centered at 0

b. $f(x) = e^{-x+2}$, centered at 0

c. $f(x) = \ln x$, centered at 1

SUMMARIZE (Section 10.4)

1. State the definition of power series *(page 627)*. For an example of a power series, see Example 1.
2. State the definition of convergence of a power series *(page 628)*. For examples of finding the radius of convergence of a power series, see Examples 2 and 3.
3. State Taylor's Theorem *(page 630)*. For examples of using Taylor's Theorem to find a power series for a function, see Examples 4, 5, and 6.

Francesco Ridolfi/Shutterstock.com

SKILLS WARM UP 10.4

The following warm-up exercises involve skills that were covered in earlier sections. You will use these skills in the exercise set for this section. For additional help, review Sections 1.4, 2.6, 4.3, 4.5, and 10.1.

In Exercises 1–4, find $f(g(x))$ and $g(f(x))$.

1. $f(x) = x^2, \quad g(x) = x - 1$

2. $f(x) = 3x, \quad g(x) = 2x + 1$

3. $f(x) = \sqrt{x + 4}, \quad g(x) = x^2$

4. $f(x) = e^x, \quad g(x) = x^2$

In Exercises 5–8, find $f'(x)$, $f''(x)$, $f'''(x)$, and $f^{(4)}(x)$.

5. $f(x) = 5e^x$

6. $f(x) = \ln x$

7. $f(x) = 3e^{2x}$

8. $f(x) = \ln 2x$

In Exercises 9 and 10, simplify the expression.

9. $\dfrac{3^n}{n!} \div \dfrac{3^{n+1}}{(n+1)!}$

10. $\dfrac{n!}{(n+2)!} \div \dfrac{(n+1)!}{(n+3)!}$

Exercises 10.4

See www.CalcChat.com for worked-out solutions to odd-numbered exercises.

Using Power Series In Exercises 1–4, identify the center of each power series. Then write the first five terms of the power series. *See Example 1.*

1. $\displaystyle\sum_{n=0}^{\infty} \left(\frac{x}{4}\right)^n$

2. $\displaystyle\sum_{n=1}^{\infty} \frac{(-1)^n (x-2)^n}{3^n}$

3. $\displaystyle\sum_{n=0}^{\infty} \frac{(-1)^{n+1}(x+1)^n}{n!}$

4. $\displaystyle\sum_{n=1}^{\infty} \frac{(-1)^n x^n}{(n-1)!}$

Finding the Radius of Convergence In Exercises 5–24, find the radius of convergence of the power series. *See Examples 2 and 3.*

5. $\displaystyle\sum_{n=0}^{\infty} \left(\frac{x}{2}\right)^n$

6. $\displaystyle\sum_{n=0}^{\infty} \left(\frac{x}{5}\right)^n$

7. $\displaystyle\sum_{n=1}^{\infty} \frac{(-1)^n x^n}{n}$

8. $\displaystyle\sum_{n=1}^{\infty} (-1)^{n+1} n x^n$

9. $\displaystyle\sum_{n=0}^{\infty} \frac{x^{5n}}{n!}$

10. $\displaystyle\sum_{n=0}^{\infty} \frac{(3x)^n}{n!}$

11. $\displaystyle\sum_{n=0}^{\infty} n!\left(\frac{x}{2}\right)^n$

12. $\displaystyle\sum_{n=0}^{\infty} \frac{(-1)^n x^n}{(n+1)(n+2)}$

13. $\displaystyle\sum_{n=1}^{\infty} \frac{(-1)^{n+1} x^n}{4^n}$

14. $\displaystyle\sum_{n=0}^{\infty} \frac{(-1)^n n!(x-4)^n}{3^n}$

15. $\displaystyle\sum_{n=1}^{\infty} \frac{(-1)^{n+1}(x-5)^n}{n5^n}$

16. $\displaystyle\sum_{n=0}^{\infty} \frac{(x-2)^{n+1}}{(n+1)3^{n+1}}$

17. $\displaystyle\sum_{n=0}^{\infty} \frac{(-1)^{n+1}(x-1)^{n+1}}{n+1}$

18. $\displaystyle\sum_{n=1}^{\infty} \frac{(-1)^{n+1}(x-2)^n}{n2^n}$

19. $\displaystyle\sum_{n=1}^{\infty} \frac{(x-3)^{n-1}}{3^{n-1}}$

20. $\displaystyle\sum_{n=1}^{\infty} \frac{(-1)^n x^{2n+1}}{2n+1}$

21. $\displaystyle\sum_{n=1}^{\infty} \frac{n}{n+1}(-2x)^{n-1}$

22. $\displaystyle\sum_{n=0}^{\infty} \frac{(-1)^n x^{2n}}{n!}$

23. $\displaystyle\sum_{n=0}^{\infty} \frac{x^{2n+1}}{(2n+1)!}$

24. $\displaystyle\sum_{n=0}^{\infty} \frac{n!x^n}{(n+1)!}$

Finding Taylor and Maclaurin Series In Exercises 25–32, apply Taylor's Theorem to find the power series (centered at c) for the function. Then find the radius of convergence. *See Examples 4 and 5.*

	Function	*Center*
25.	$f(x) = e^x$	$c = 1$
26.	$f(x) = e^{-x}$	$c = 1$
27.	$f(x) = e^{3x}$	$c = 0$
28.	$f(x) = e^{-4x}$	$c = 0$
29.	$f(x) = \dfrac{1}{x+1}$	$c = 0$
30.	$f(x) = \dfrac{1}{2-x}$	$c = 0$
31.	$f(x) = \sqrt{x}$	$c = 1$
32.	$f(x) = \sqrt{x}$	$c = 4$

Using the Basic List of Power Series In Exercises 33–36, use the basic list of power series to find the binomial series (centered at $c = 0$) for the function. Then find the radius of convergence. *See Example 7.*

33. $f(x) = \dfrac{1}{(1+x)^2}$

34. $f(x) = \dfrac{1}{\sqrt[3]{1+x}}$

35. $f(x) = \dfrac{1}{(1+x)^3}$

36. $f(x) = \sqrt[4]{1+x}$

Finding the Radius of Convergence In Exercises 37–40, find the radius of convergence of (a) $f(x)$, (b) $f'(x)$, (c) $f''(x)$, and (d) $\int f(x)\,dx$.

37. $f(x) = \sum_{n=0}^{\infty} \left(\frac{x}{2}\right)^n$

38. $f(x) = \sum_{n=1}^{\infty} \frac{x^n}{n5^n}$

39. $f(x) = \sum_{n=0}^{\infty} \frac{(x+1)^{n+1}}{n+1}$

40. $f(x) = \sum_{n=1}^{\infty} \frac{(-1)^{n+1}(x-1)^n}{n}$

Finding Power Series In Exercises 41–52, find the power series for the function using the suggested method. Use the basic list of power series for common functions on page 633. *See Example 8.*

41. Use the power series for e^x.

$f(x) = e^{x^3}$

42. Use the series found in Exercise 41.

$f(x) = e^{-x^3}$

43. Differentiate the series found in Exercise 41.

$f(x) = 3x^2 e^{x^3}$

44. Use the power series for e^x and e^{-x}.

$f(x) = \dfrac{e^x + e^{-x}}{2}$

45. Use the power series for $\dfrac{1}{x+1}$.

$f(x) = \dfrac{1}{x^4 + 1}$

46. Use the power series for $\dfrac{1}{x+1}$.

$f(x) = \dfrac{2x}{x+1}$

47. Use the power series for $\dfrac{1}{x+1}$.

$f(x) = \dfrac{3x}{x^2+1}$

48. Integrate the series for $\dfrac{1}{x+1}$.

$f(x) = \ln(x+1)$

49. Integrate the series for $\dfrac{1}{x}$.

$f(x) = \ln x$

50. Differentiate the series found in Exercise 44.

$f(x) = \dfrac{e^x - e^{-x}}{2}$

51. Differentiate the series for $-\dfrac{1}{x}$.

$f(x) = \dfrac{1}{x^2}$

52. Differentiate the power series for e^x term by term and use the resulting series to show that

$$\frac{d}{dx}[e^x] = e^x.$$

53. Think About It Consider the geometric series

$$\sum_{n=0}^{\infty} \left(\frac{x}{4}\right)^n.$$

(a) Find the sum of the series when $x = \frac{5}{2}$. Use a graphing utility to graph the first six terms of the sequence of partial sums and the horizontal line representing the sum of the series.

(b) Repeat part (a) for $x = -\frac{5}{2}$.

(c) Compare the rates of convergence of the partial sums with the sums of the series in parts (a) and (b). How do the plots of the partial sums differ as they converge toward the sum of the series?

54. HOW DO YOU SEE IT? Match the graph of the first 10 terms of the sequence of partial sums of the series

$$g(x) = \sum_{n=0}^{\infty} \left(\frac{x}{3}\right)^n$$

with the indicated value of the function. [The graphs are labeled (i), (ii), (iii), and (iv).] Explain how you made your choice.

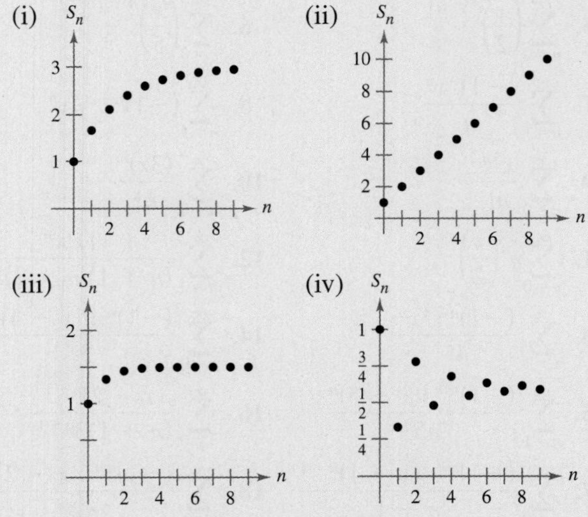

(a) $g(1)$ (b) $g(2)$

(c) $g(3)$ (d) $g(-2)$

10.5 Taylor Polynomials

- Find Taylor polynomials for functions.
- Use Taylor polynomials to determine the maximum errors of approximations and to approximate definite integrals.

In Exercise 4 on page 642, you will use Taylor polynomials to approximate an exponential function.

Taylor Polynomials and Approximation

In Section 10.4, you saw that it is sometimes possible to obtain an *exact* power series representation of a function. For example, the function

$$f(x) = e^{-x}$$

can be represented exactly by the power series

$$e^{-x} = \sum_{n=0}^{\infty} \frac{(-1)^n}{n!}x^n$$
$$= 1 - x + \frac{x^2}{2!} - \frac{x^3}{3!} + \frac{x^4}{4!} - \cdots + \frac{(-1)^n x^n}{n!} + \cdots.$$

The problem with using this power series is that the exactness of its representation depends on the summation of an infinite number of terms. In practice, this is not feasible, and you must be content with a finite summation that approximates the function rather than representing it exactly. For instance, consider the sequence of partial sums below.

$S_0(x) = 1$ Zeroth-degree Taylor polynomial

$S_1(x) = 1 - x$ First-degree Taylor polynomial

$S_2(x) = 1 - x + \frac{x^2}{2!}$ Second-degree Taylor polynomial

$S_3(x) = 1 - x + \frac{x^2}{2!} - \frac{x^3}{3!}$ Third-degree Taylor polynomial

$S_4(x) = 1 - x + \frac{x^2}{2!} - \frac{x^3}{3!} + \frac{x^4}{4!}$ Fourth-degree Taylor polynomial

$S_5(x) = 1 - x + \frac{x^2}{2!} - \frac{x^3}{3!} + \frac{x^4}{4!} - \frac{x^5}{5!}$ Fifth-degree Taylor polynomial

$S_6(x) = 1 - x + \frac{x^2}{2!} - \frac{x^3}{3!} + \frac{x^4}{4!} - \frac{x^5}{5!} + \frac{x^6}{6!}$ Sixth-degree Taylor polynomial

$\vdots$

$S_n(x) = 1 - x + \frac{x^2}{2!} - \cdots + \frac{(-1)^n x^n}{n!}$ nth-degree Taylor polynomial

Each of these polynomial approximations of e^{-x} is a **Taylor polynomial** for e^{-x}. As n approaches infinity, the graphs of these Taylor polynomials become closer and closer approximations of the graph of $f(x) = e^{-x}$.

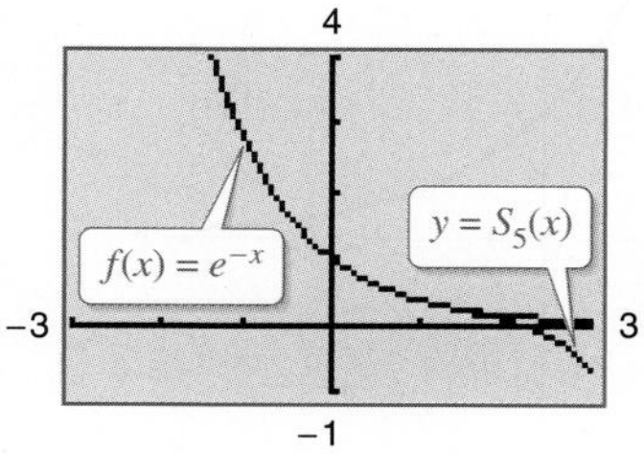

FIGURE 10.7

TECH TUTOR

Symbolic differentiation utilities are programmed to find Taylor polynomials. Try using a symbolic differentiation utility to compute the fifth-degree Taylor polynomial for $f(x) = e^{-x}$. After finding the Taylor polynomial for $f(x) = e^{-x}$, you can use a graphing utility to compare graphs of this Taylor polynomial with the graph of $f(x) = e^{-x}$, as shown in Figure 10.7.

For example, Figure 10.8 shows the graphs of

$$S_1, \quad S_2, \quad S_3, \quad \text{and} \quad S_4.$$

Notice in the figure that the closer x is to the center of convergence ($x = 0$ in this case), the better the polynomial approximates e^{-x}. This conclusion is reinforced by the table below. From Section 10.4, you know that the power series for e^{-x},

$$e^{-x} = 1 - x + \frac{x^2}{2!} - \frac{x^3}{3!} + \frac{x^4}{4!} - \frac{x^5}{5!} + \cdot \cdot \cdot + \frac{(-1)^n x^n}{n!} + \cdot \cdot \cdot \quad \text{Power series for } e^{-x}$$

converges for *all* x. From the figure and table, however, you can see that the farther x is from zero, the more terms you need to obtain a good approximation.

x	0	0.5	1.0	1.5	2.0
$S_1 = 1 - x$	1	0.5000	0	−0.5000	−1.0
$S_2 = 1 - x + \frac{x^2}{2!}$	1	0.6250	0.5000	0.6250	1.0
$S_3 = 1 - x + \frac{x^2}{2!} - \frac{x^3}{3!}$	1	0.6042	0.3333	0.0625	−0.3333
$S_4 = 1 - x + \frac{x^2}{2!} - \frac{x^3}{3!} + \frac{x^4}{4!}$	1	0.6068	0.3750	0.2734	0.3333
e^{-x}	1	0.6065	0.3679	0.2231	0.1353

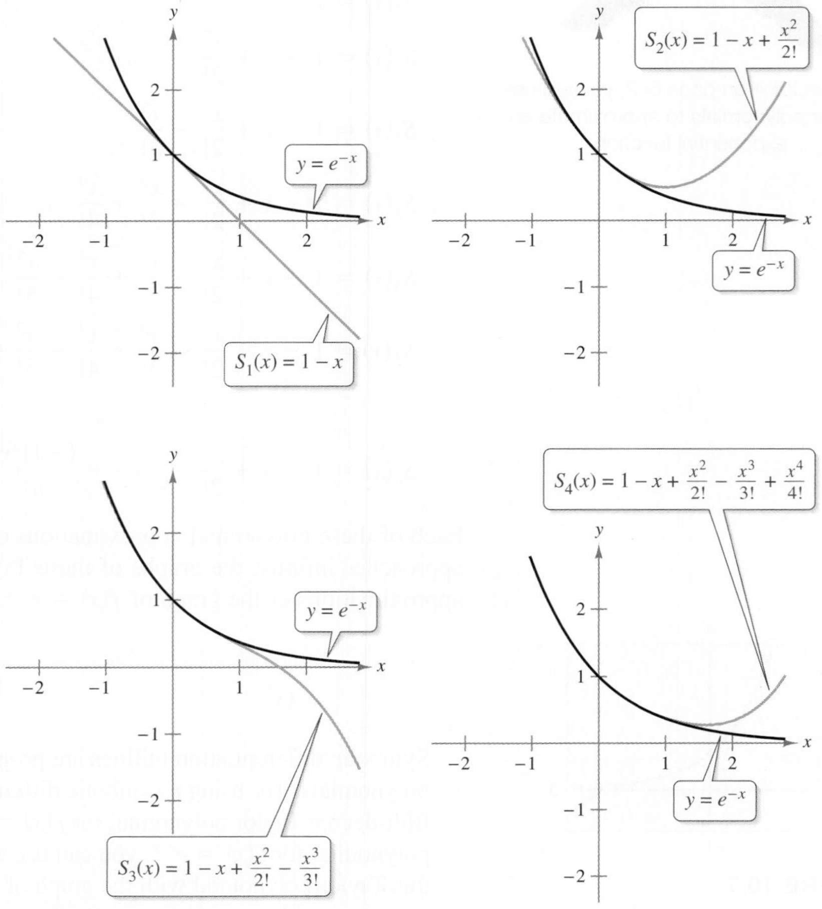

FIGURE 10.8

Example 1 Finding Taylor Polynomials

Find a Taylor polynomial that is a reasonable approximation of

$$f(x) = e^{-x^2}$$

in the interval $[-1, 1]$. What degree of Taylor polynomial should you use?

SOLUTION To begin, you need to define what constitutes a "reasonable" approximation. Suppose you decide that you want the values of $S_n(x)$ and e^{-x^2} to differ by no more than 0.01 in the interval $[-1, 1]$. To answer the question, you can compute Taylor polynomials of higher and higher degree and then graphically compare them with

$$f(x) = e^{-x^2}.$$

For instance, the fourth-, sixth-, and eighth-degree Taylor polynomials for f are as shown.

$$S_4(x) = 1 - x^2 + \frac{x^4}{2}$$ Fourth-degree Taylor polynomial

$$S_6(x) = 1 - x^2 + \frac{x^4}{2} - \frac{x^6}{6}$$ Sixth-degree Taylor polynomial

$$S_8(x) = 1 - x^2 + \frac{x^4}{2} - \frac{x^6}{6} + \frac{x^8}{24}$$ Eighth-degree Taylor polynomial

To compare $S_4(x)$ and e^{-x^2} graphically, use a graphing utility to graph both equations in the same viewing window, as shown in Figure 10.9. Then, use the *trace* feature of the graphing utility to compare the y-values at -1 and 1. When you do this, you will find that the y-values differ by more than 0.01.

Next, perform the same comparison with $S_6(x)$ and e^{-x^2}. With these two graphs, the y-values still differ by more than 0.01 in the interval $[-1, 1]$. Finally, by graphically comparing $S_8(x)$ and e^{-x^2}, you can determine that their graphs differ by less than 0.01 in the interval $[-1, 1]$. To convince yourself of this, try evaluating e^{-x^2} and $S_8(x)$ for several x-values. From the table, you can see that the values of e^{-x^2} and $S_8(x)$ differ by less than 0.01.

x	-1.0	-0.5	0	0.5	1.0
$S_8(x)$	0.375	0.779	1	0.779	0.375
e^{-x^2}	0.368	0.779	1	0.779	0.368

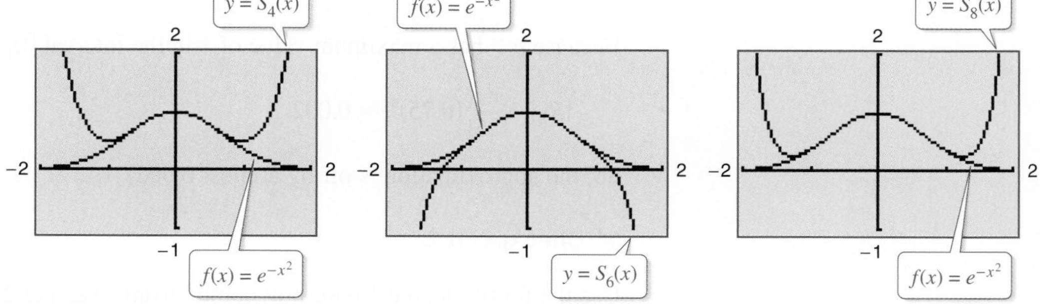

FIGURE 10.9

✓Checkpoint 1

Find the 12th-degree Taylor polynomial for $f(x) = e^{2x^3}$ centered at zero. Then use a graphing utility to graph the function and the polynomial in the same viewing window.

Taylor's Theorem with Remainder

In Example 1, you studied a graphical procedure for determining how well a Taylor polynomial approximates a function f. The theorem below gives you an analytic procedure for determining this.

Taylor's Theorem with Remainder

Let f have derivatives up through order $n + 1$ for every x in an interval I containing c. Then, for all x in I,

$$f(x) = S_n(x) + R_n$$

where $S_n(x)$ is the nth-degree Taylor polynomial for f and where R_n is given by

$$R_n = \frac{f^{(n+1)}(z)}{(n+1)!}(x - c)^{n+1}$$

for some number z between c and x. The value R_n is called the **nth remainder** of $f(x)$.

Although this theorem appears to give a formula for the exact remainder, note that it does not specify which value of z should be used to find R_n. In other words, the practical application of this theorem lies not in calculating R_n, but in finding bounds for R_n.

Example 2 Using a Taylor Polynomial Approximation

Approximate $e^{-0.75}$ using a fourth-degree Taylor polynomial, and determine the maximum error of the approximation.

SOLUTION Using the fourth-degree Taylor polynomial for e^{-x}, you obtain the approximation as shown.

$$e^{-x} \approx 1 - x + \frac{x^2}{2!} - \frac{x^3}{3!} + \frac{x^4}{4!}$$

$$e^{-0.75} \approx 1 - 0.75 + \frac{(0.75)^2}{2!} - \frac{(0.75)^3}{3!} + \frac{(0.75)^4}{4!} \approx 0.474$$

By Taylor's Theorem with Remainder, the error in this approximation is

$$R_4 = -\frac{e^{-z}}{5!}(0.75)^5, \quad 0 \le z \le 0.75.$$

Because e^{-z} has a maximum value of 1 in the interval $[0, 0.75]$, it follows that

$$|R_4| \le \frac{1}{5!}(0.75)^5 \approx 0.002.$$

So, the approximation is off by at most 0.002.

✓ Checkpoint 2

Use the fourth-degree Taylor polynomial from Example 2 to approximate $e^{-0.5}$ and determine the maximum error of the approximation.

Earlier in the text, you looked at different ways to approximate a definite integral: the Midpoint Rule, the Trapezoidal Rule, and Simpson's Rule. The next example shows how you can use a power series to approximate a definite integral.

R. Gino Santa Maria/www.shutterstock.com

Example 3 Approximating a Definite Integral

Use the eighth-degree Taylor polynomial for e^{-x^2} to approximate the definite integral.

$$\int_0^1 e^{-x^2}dx$$

SOLUTION The eighth-degree Taylor polynomial for e^{-x^2} is

$$S_8(x) = 1 - x^2 + \frac{x^4}{2} - \frac{x^6}{6} + \frac{x^8}{24}.$$

Using this polynomial, you can write the approximation as shown.

$$\begin{aligned}\int_0^1 e^{-x^2}\,dx &\approx \int_0^1 S_8(x)\,dx \\ &= \int_0^1 \left(1 - x^2 + \frac{x^4}{2} - \frac{x^6}{6} + \frac{x^8}{24}\right)dx \\ &= \left[x - \frac{x^3}{3} + \frac{x^5}{5(2)} - \frac{x^7}{7(6)} + \frac{x^9}{9(24)}\right]_0^1 \\ &= 1 - \frac{1}{3} + \frac{1}{10} - \frac{1}{42} + \frac{1}{216} \\ &\approx 0.747\end{aligned}$$

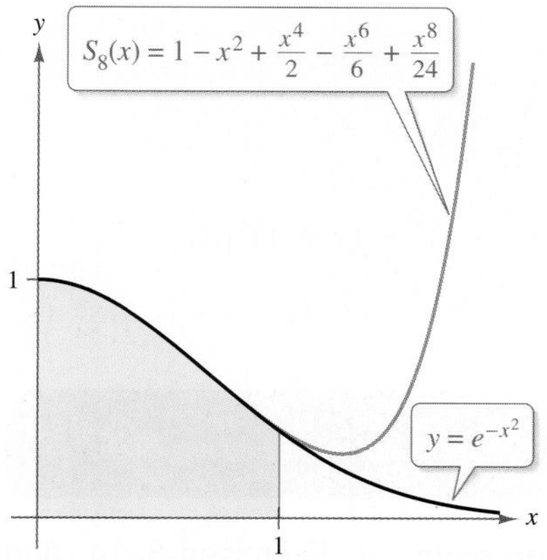

FIGURE 10.10

This result is shown graphically in Figure 10.10.

✓ Checkpoint 3

Use the 12th-degree Taylor polynomial from Checkpoint 1 to approximate

$$\int_0^1 e^{2x^3}\,dx.$$

When using a Taylor polynomial to approximate a function on an interval, remember that as you move farther away from the center of the Taylor polynomial, you must use a polynomial of higher and higher degree to obtain an indicated accuracy. For definite integrals, it helps to locate the center of the Taylor polynomial at the midpoint of the interval of integration. For instance, to approximate

$$\int_0^2 e^{-x^2}\,dx$$

you could use a Taylor polynomial centered at 1. When you try this with an eighth-degree Taylor polynomial, you will obtain an approximation of 0.882.

SUMMARIZE (Section 10.5)

1. Explain how a Taylor polynomial can be used to approximate a function *(page 637)*. For an example of finding a Taylor polynomial, see Example 1.
2. Explain how to determine the maximum error of a Taylor polynomial approximation using Taylor's Theorem with Remainder *(page 640)*. For an example of using Taylor's Theorem with Remainder, see Example 2.
3. Explain how a Taylor polynomial can be used to approximate a definite integral *(page 641)*. For an example of using a Taylor polynomial to approximate a definite integral, see Example 3.

SKILLS WARM UP 10.5

The following warm-up exercises involve skills that were covered in earlier sections. You will use these skills in the exercise set for this section. For additional help, review Sections 5.4 and 10.4.

In Exercises 1–6, find a power series representation for the function.

1. $f(x) = e^{3x}$

2. $f(x) = e^{-3x}$

3. $f(x) = \dfrac{4}{x}$

4. $f(x) = \ln 5x$

5. $f(x) = (1 + x)^{1/4}$

6. $f(x) = \sqrt{1 + x}$

In Exercises 7–10, evaluate the definite integral.

7. $\displaystyle\int_0^1 (1 - x + x^2 - x^3 + x^4)\,dx$

8. $\displaystyle\int_0^{1/2} \left(1 + \frac{x}{3} - \frac{x^2}{9} + \frac{5x^3}{27}\right) dx$

9. $\displaystyle\int_1^2 \left[(x - 1) - \frac{(x-1)^2}{2} + \frac{(x-1)^3}{3}\right] dx$

10. $\displaystyle\int_1^{3/2} [1 - (x - 1) + (x - 1)^2 - (x - 1)^3]\,dx$

Exercises 10.5

See www.CalcChat.com for worked-out solutions to odd-numbered exercises.

Using Taylor Polynomials In Exercises 1 and 2, use a spreadsheet to complete the table using the indicated Taylor polynomials as approximations of the function f.

1.

x	0	$\frac{1}{4}$	$\frac{1}{2}$	$\frac{3}{4}$	1
$f(x) = e^{x/2}$					
$1 + \frac{x}{2}$					
$1 + \frac{x}{2} + \frac{x^2}{8}$					
$1 + \frac{x}{2} + \frac{x^2}{8} + \frac{x^3}{48}$					
$1 + \frac{x}{2} + \frac{x^2}{8} + \frac{x^3}{48} + \frac{x^4}{384}$					

2.

x	0	$\frac{1}{4}$	$\frac{1}{2}$	$\frac{3}{4}$	1
$f(x) = \ln(x^2 + 1)$					
x^2					
$x^2 - \frac{x^4}{2}$					
$x^2 - \frac{x^4}{2} + \frac{x^6}{3}$					
$x^2 - \frac{x^4}{2} + \frac{x^6}{3} - \frac{x^8}{4}$					

Finding Taylor Polynomials In Exercises 3–14, find the Taylor polynomials (centered at zero) of degrees (a) 1, (b) 2, (c) 3, and (d) 4. *See Example 1.*

3. $f(x) = e^x$

4. $f(x) = e^{-x/2}$

5. $f(x) = e^{2x}$

6. $f(x) = e^{3x}$

7. $f(x) = \ln(x + 1)$

8. $f(x) = \ln(2x + 1)$

9. $f(x) = \sqrt{x + 1}$

10. $f(x) = \sqrt[3]{x + 1}$

11. $f(x) = \dfrac{1}{(x + 1)^2}$

12. $f(x) = \dfrac{1}{(x + 1)^3}$

13. $f(x) = \dfrac{x}{x + 1}$

14. $f(x) = \dfrac{4}{x + 1}$

Finding Taylor Polynomials In Exercises 15 and 16, use a symbolic differentiation utility to find the Taylor polynomials (centered at zero) of degrees (a) 2, (b) 4, (c) 6, and (d) 8.

15. $f(x) = \dfrac{1}{1 + x^2}$

16. $f(x) = e^{-x^2}$

Finding Taylor Polynomials In Exercises 17 and 18, use a symbolic differentiation utility to find the fourth-degree Taylor polynomial (centered at zero).

17. $f(x) = \dfrac{1}{\sqrt[3]{x + 1}}$

18. $f(x) = xe^x$

Matching In Exercises 19–22, match the Taylor polynomial approximation of the function

$f(x) = e^{-x^2/2}$

with its graph. [The graphs are labeled (a)–(d).] Use a graphing utility to verify your results.

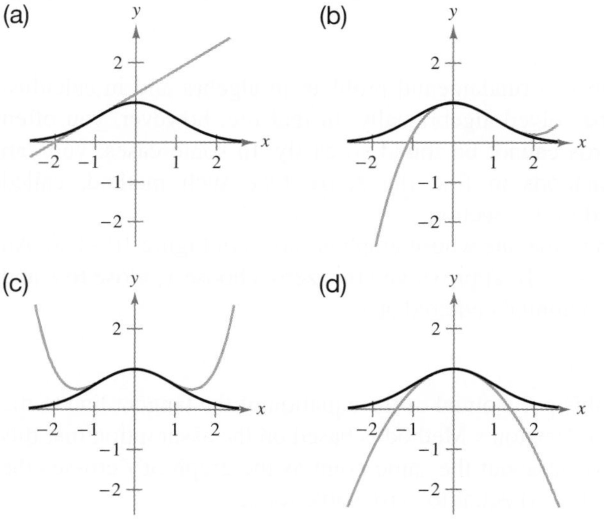

19. $y = -\frac{1}{2}x^2 + 1$

20. $y = \frac{1}{8}x^4 - \frac{1}{2}x^2 + 1$

21. $y = e^{-1/2}[(x + 1) + 1]$

22. $y = e^{-1/2}\left[\frac{1}{3}(x - 1)^3 - (x - 1) + 1\right]$

Using a Taylor Polynomial Approximation In Exercises 23–26, use a sixth-degree Taylor polynomial centered at c for the function f to obtain the required approximation. Then determine the maximum error of the approximation. *See Example 2.*

	Function	*Center*	*Approximation*
23.	$f(x) = e^{-x}$	$c = 0$	$f\left(\frac{1}{2}\right)$
24.	$f(x) = e^{4x}$	$c = 0$	$f\left(\frac{1}{4}\right)$
25.	$f(x) = \ln x$	$c = 2$	$f\left(\frac{3}{2}\right)$
26.	$f(x) = \sqrt{x}$	$c = 4$	$f(5)$

Finding the Maximum Error In Exercises 27 and 28, determine the maximum error guaranteed by Taylor's Theorem with Remainder when the fifth-degree Taylor polynomial is used to approximate f in the given interval.

	Function	*Center*	*Interval*
27.	$f(x) = e^{-x}$	$c = 0$	$[0, 1]$
28.	$f(x) = \dfrac{1}{x}$	$c = 1$	$\left[1, \dfrac{3}{2}\right]$

Using the Maximum Error In Exercises 29 and 30, determine the degree of the Taylor polynomial centered at c required to approximate f in the given interval to an accuracy of ± 0.001.

	Function	*Center*	*Interval*
29.	$f(x) = e^x$	$c = 1$	$[0, 2]$
30.	$f(x) = \dfrac{1}{x}$	$c = 1$	$\left[1, \dfrac{3}{2}\right]$

Approximating a Definite Integral In Exercises 31–34, use a sixth-degree Taylor polynomial centered at zero to approximate the definite integral. *See Example 3.*

	Function	*Approximation*
31.	$f(x) = e^{-x^3}$	$\int_0^1 e^{-x^3}\,dx$
32.	$f(x) = \ln(x^2 + 1)$	$\int_{-1/4}^{1/4} \ln(x^2 + 1)\,dx$
33.	$f(x) = \dfrac{1}{\sqrt{1 + x^2}}$	$\int_0^{1/2} \dfrac{1}{\sqrt{1 + x^2}}\,dx$
34.	$f(x) = \dfrac{1}{\sqrt[3]{1 + x^2}}$	$\int_0^{1/2} \dfrac{1}{\sqrt[3]{1 + x^2}}\,dx$

35. Comparing Taylor Polynomials Consider the functions $f(x) = e^x$ and $g(x) = xe^x$.

(a) Find the Taylor polynomial of degree 4 for f and the Taylor polynomial of degree 5 for g. What is the relationship between them?

(b) Use the result from part (a) to find the Taylor polynomial for $h(x) = x^2e^x$. What is the degree of this polynomial?

(c) Use the result from part (a) to find the Taylor polynomial for $n(x) = e^x/x$. What is the degree of this polynomial?

36. HOW DO YOU SEE IT? The figure shows first-, second-, and third-degree polynomial approximations P_1, P_2, and P_3 of a function f. Label the graphs of P_1, P_2, and P_3. Explain your reasoning.

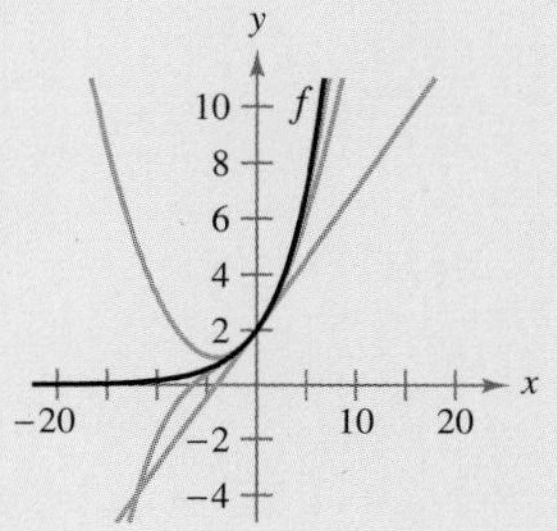

10.6 Newton's Method

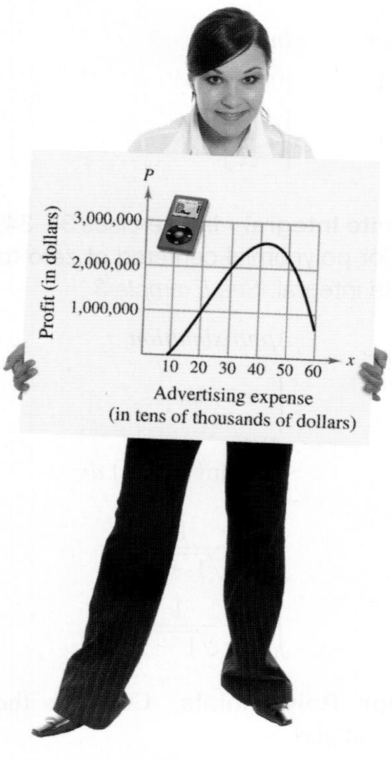

In Exercise 46 on page 651, you will use Newton's Method to find the smaller advertising amount that will yield a profit of \$2,500,000.

- Use Newton's Method to approximate the zeros of functions.
- Understand the situations in which Newton's Method may not converge.

Newton's Method

Finding the zeros of a function is a fundamental problem in algebra and in calculus. Sometimes this problem can be solved algebraically. In real life, however, you often encounter functions whose zeros cannot be found so easily. In some cases, you can use various approximation methods to find the zeros. One such method, called **Newton's Method,** is described in this section.

Consider a function *f*, such as the one whose graph is shown in Figure 10.11(a). An actual zero of the function is $x = c$. To approximate this zero, choose x_1 close to c and form the first-degree Taylor polynomial centered at x_1.

$$S_1(x) = f(x_1) + f'(x_1)(x - x_1) \qquad \text{First-degree Taylor polynomial}$$

Graphically, you can interpret this polynomial as the equation of the tangent line to the graph of f at the point $(x_1, f(x_1))$. Newton's Method is based on the assumption that this tangent line will cross the x-axis at about the same point as the graph of f crosses the x-axis. With this assumption, set $S_1(x)$ equal to zero, solve for x,

$$f(x_1) + f'(x_1)(x - x_1) = 0$$
$$f'(x_1)(x - x_1) = -f(x_1)$$
$$x - x_1 = -\frac{f(x_1)}{f'(x_1)}$$
$$x = x_1 - \frac{f(x_1)}{f'(x_1)}$$

and use the resulting value as a new, and hopefully better, approximation of the actual zero c. So, from the approximation x_1, you form a second approximation,

$$x_2 = x_1 - \frac{f(x_1)}{f'(x_1)}. \qquad \text{Second estimate}$$

To obtain an even better approximation, you can use x_2 to calculate x_3,

$$x_3 = x_2 - \frac{f(x_2)}{f'(x_2)}. \qquad \text{Third estimate [see Figure 10.11(b)].}$$

Repeated application of this process is called Newton's Method.

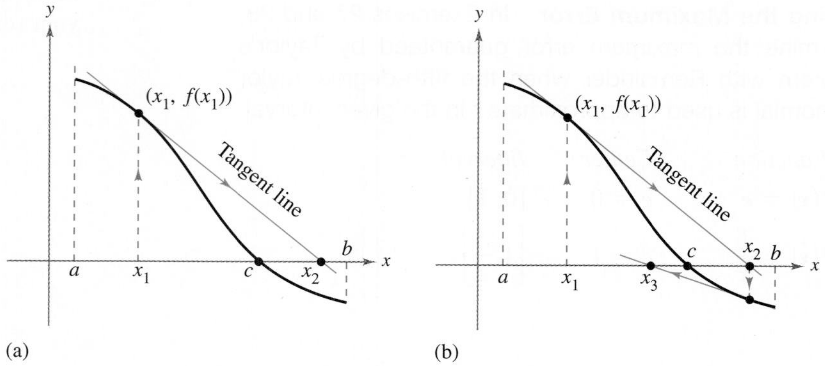

The x-intercept of the tangent line approximates the zero of f.

FIGURE 10.11

For many functions, just a few iterations of Newton's Method will produce approximations having very small errors. This method is summarized below.

Newton's Method

Let c be a zero of f and let f be differentiable on an open interval containing c. To approximate c, use the following steps.

1. Make an initial approximation x_1 that is close to c. (A graph is helpful.)
2. Determine a new approximation using the formula

$$x_{n+1} = x_n - \frac{f(x_n)}{f'(x_n)}.$$

3. When $|x_n - x_{n+1}|$ is within the desired accuracy, let x_{n+1} serve as the final approximation. Otherwise, return to Step 2 and calculate a new approximation.

Each successive application of this procedure is called an **iteration.**

Example 1 Using Newton's Method

Use three iterations of Newton's Method to approximate a zero of

$$f(x) = x^2 - 2.$$

Use $x_1 = 1$ as the initial guess.

SOLUTION The first derivative of f is $f'(x) = 2x$. So, the iterative formula for Newton's Method is

$$x_{n+1} = x_n - \frac{f(x_n)}{f'(x_n)}$$
$$= x_n - \frac{x_n^2 - 2}{2x_n}.$$

The calculations for three iterations are shown in the table. The first iteration is depicted graphically in Figure 10.12.

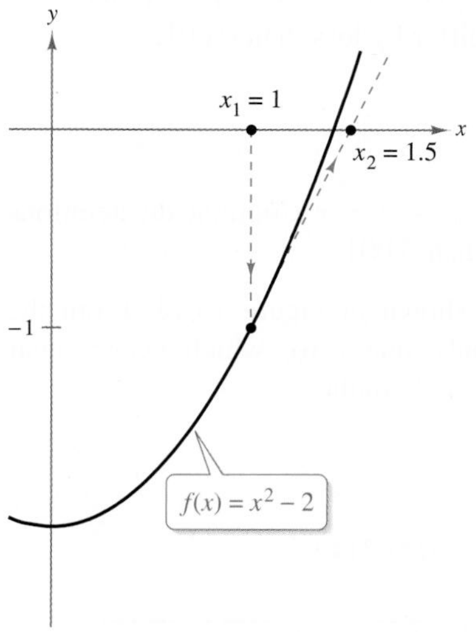

FIGURE 10.12

n	x_n	$f(x_n)$	$f'(x_n)$	$\frac{f(x_n)}{f'(x_n)}$	$x_n - \frac{f(x_n)}{f'(x_n)}$
1	1.000000	−1.000000	2.000000	−0.500000	1.500000
2	1.500000	0.250000	3.000000	0.083333	1.416667
3	1.416667	0.006945	2.833334	1.002451	1.414216
4	1.414216				

So, the approximation is 1.414216. For this particular function, you can easily determine the exact zero to be $\sqrt{2} \approx 1.414214$. So, after only three iterations of Newton's Method, you can obtain an approximation that is within 0.000002 of an actual root.

Checkpoint 1

Use three iterations of Newton's Method to approximate a zero of $f(x) = x^2 - 3$. Use $x_1 = 1$ as the initial guess.

Example 2 Using Newton's Method

Use Newton's Method to approximate the zeros of $f(x) = 2x^3 + x^2 - x + 1$. Continue the iterations until two successive approximations differ by less than 0.001.

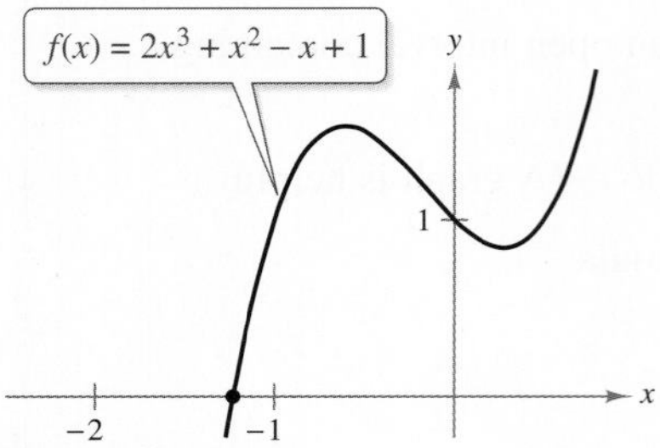

FIGURE 10.13

SOLUTION Begin by sketching a graph of f, as shown in Figure 10.13. From the graph, you can observe that the function has only one zero, which occurs near $x_1 = -1.2$. Next, differentiate f and form the iterative formula

$$x_{n+1} = x_n - \frac{f(x_n)}{f'(x_n)} = x_n - \frac{2x_n^3 + x_n^2 - x_n + 1}{6x_n^2 + 2x_n - 1}.$$

From the table, you can estimate the zero of f to be -1.23375.

n	x_n	$f(x_n)$	$f'(x_n)$	$\frac{f(x_n)}{f'(x_n)}$	$x_n - \frac{f(x_n)}{f'(x_n)}$
1	−1.20000	0.18400	5.24000	0.03511	−1.23511
2	−1.23511	−0.00771	5.68276	−0.00136	−1.23375
3	−1.23375	−0.00001	5.66533	−0.00000	−1.23375
4	−1.23375				

✓ Checkpoint 2

Use Newton's Method to approximate the zeros of $f(x) = x^3 + 2x + 1$. Continue the iterations until two successive approximations differ by less than 0.001. ■

Example 3 Using Newton's Method

Use Newton's Method to approximate the zeros of $f(x) = e^x + x$. Continue the iterations until two successive approximations differ by less than 0.001.

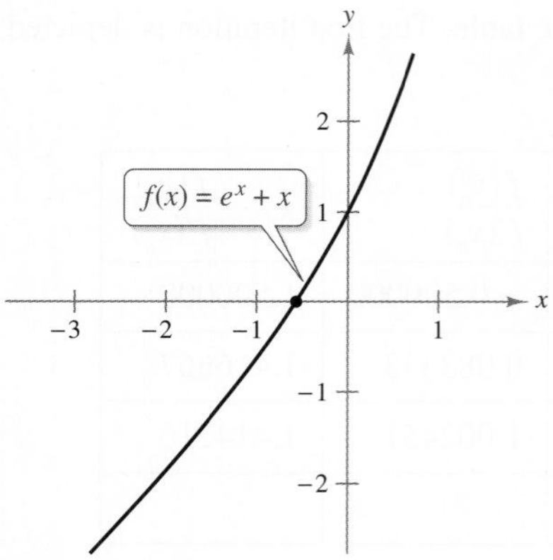

FIGURE 10.14

SOLUTION Begin by sketching a graph of f, as shown in Figure 10.14. From the graph, you can observe that the function has only one zero, which occurs near $x_1 = -0.5$. Next, differentiate f and form the iterative formula

$$x_{n+1} = x_n - \frac{f(x_n)}{f'(x_n)} = x_n - \frac{e^{x_n} + x_n}{e^{x_n} + 1}.$$

From the table, you can estimate the zero of f to be -0.567143.

n	x_n	$f(x_n)$	$f'(x_n)$	$\frac{f(x_n)}{f'(x_n)}$	$x_n - \frac{f(x_n)}{f'(x_n)}$
1	−0.500000	0.106531	1.606531	0.066311	−0.566311
2	−0.566311	0.001305	1.567616	0.000832	−0.567143
3	−0.567143				

✓ Checkpoint 3

Use Newton's Method to approximate the zeros of $f(x) = x^2 - e^x + 2$. Continue the iterations until two successive approximations differ by less than 0.001. ■

STUDY TIP

Newton's Method is necessary for working with models of real-life data, such as models of average cost. You will see such models in the exercise set for this section.

Example 4 Finding a Point of Intersection

Estimate the point of intersection of the graphs of

$$y = e^{-x^2} \quad \text{and} \quad y = x.$$

Use Newton's Method and continue the iterations until two successive approximations differ by less than 0.0001.

SOLUTION The point of intersection of the two graphs occurs when $e^{-x^2} = x$, which implies that

$$0 = x - e^{-x^2}.$$

To use Newton's Method, let

$$f(x) = x - e^{-x^2}$$

and employ the iterative formula

$$x_{n+1} = x_n - \frac{f(x_n)}{f'(x_n)} = x_n - \frac{x_n - e^{-x_n^2}}{1 + 2x_n e^{-x_n^2}}.$$

The table shows three iterations of Newton's Method beginning with an initial approximation of $x_1 = 0.5$.

n	x_n	$f(x_n)$	$f'(x_n)$	$\frac{f(x_n)}{f'(x_n)}$	$x_n - \frac{f(x_n)}{f'(x_n)}$
1	0.500000	−0.27880	1.77880	−0.15673	0.65673
2	0.65673	0.00706	1.85331	0.00381	0.65292
3	0.65292	0.00000	1.85261	0.00000	0.65292
4	0.65292				

So, you can estimate that the point of intersection occurs when $x \approx 0.65292$, as shown in Figure 10.15.

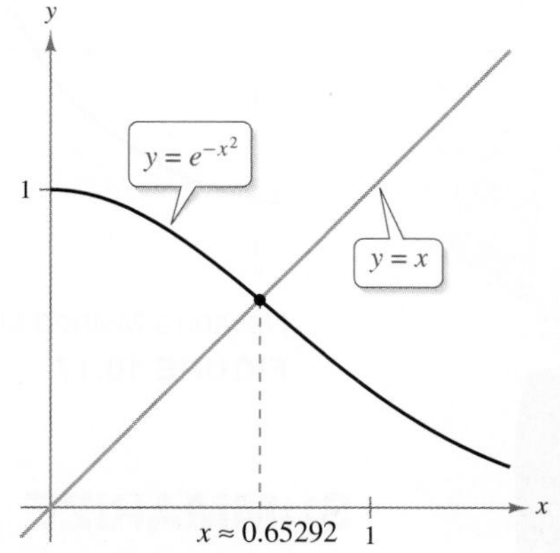

FIGURE 10.15

✓Checkpoint 4

Estimate the point of intersection of the graphs of $y = e^{-x}$ and $y = x$. Use Newton's Method and continue the iterations until two successive approximations differ by less than 0.001.

Convergence of Newton's Method

When, as in Examples 1, 2, and 3, the approximations approach a zero of a function, Newton's Method is said to **converge.** You should know, however, that Newton's Method does not always converge. Two situations in which it may not converge are

1. when $f'(x_n) = 0$ for some n. (See Figure 10.16.)

2. when $\lim_{n\to\infty} x_n$ does not exist. (See Figure 10.17.)

The type of problem illustrated in Figure 10.16 can usually be overcome with a better choice of x_1. The problem illustrated in Figure 10.17, however, is usually more serious. For instance, Newton's Method does not converge for any choice of x_1 (other than the actual zero) for the function

$$f(x) = x^{1/3}.$$

Using a graphing utility, try running the program discussed at the left to approximate the zero of this function. When you do, you will find that the program fails.

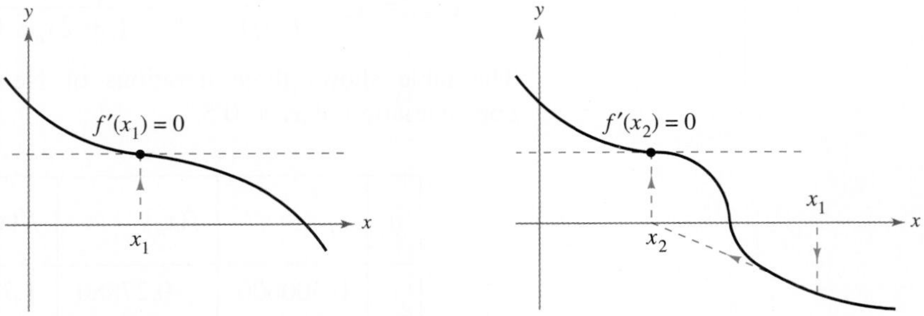

Newton's Method fails to converge when $f'(x_n) = 0$.
FIGURE 10.16

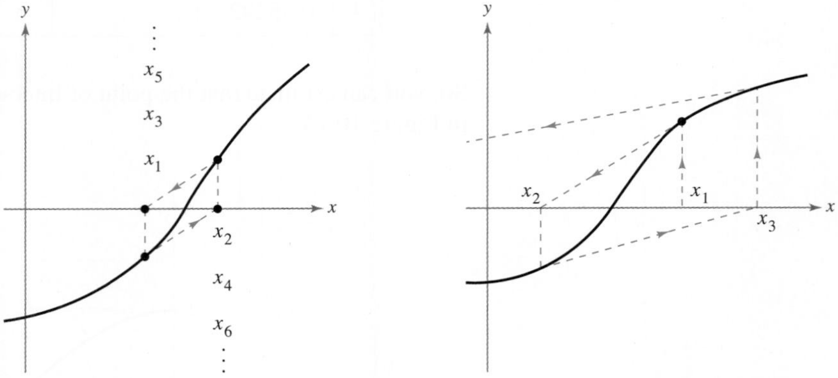

Newton's Method fails to converge for every x-value other than the actual zero of f.
FIGURE 10.17

TECH TUTOR

Technology can be used in a variety of ways to approximate zeros of functions. Earlier in the text, you saw how the *zoom* feature of a graphing utility could be used to approximate a zero graphically.

Another way to use technology to approximate the zeros of a function is to program Newton's Method into a graphing utility. Appendix E lists the programs for several models of graphing utilities.

Before running the program, enter the function into the utility. Then graph the function and estimate one of its zeros. When you run the program, you will be prompted to enter the estimate as x. After two successive approximations differ by less than 0.0000000001, the program will display the refined approximation.

SUMMARIZE (Section 10.6)

1. Describe Newton's Method for approximating the zeros of a function *(page 645)*. For examples of using Newton's Method to approximate the zeros of a function, see Examples 1, 2, and 3.
2. Discuss the convergence of Newton's Method *(page 648)*. For examples of when Newton's Method fails to converge, see Figures 10.16 and 10.17.

SKILLS WARM UP 10.6 The following warm-up exercises involve skills that were covered in a previous course or in earlier sections. You will use these skills in the exercise set for this section. For additional help, review Appendix Section A.2 and Sections 1.2, 2.2, and 4.3.

In Exercises 1–4, evaluate f and f' at the given x-value.

1. $f(x) = x^2 - 2x - 1, \quad x = 2.4$

2. $f(x) = x^3 - 2x^2 + 1, \quad x = -0.6$

3. $f(x) = e^{2x} - 2, \quad x = 0.35$

4. $f(x) = e^{x^2} - 7x + 3, \quad x = 1.4$

In Exercises 5–8, solve for x.

5. $|x - 5| \le 0.1$

6. $|4 - 5x| \le 0.01$

7. $\left|2 - \frac{x}{3}\right| \le 0.01$

8. $|2x + 7| \le 0.01$

In Exercises 9 and 10, find the point(s) of intersection of the graphs of the two equations.

9. $y = x^2 - x - 2, \quad y = 2x - 1$

10. $y = x^2, \quad y = x + 1$

Exercises 10.6

See www.CalcChat.com for worked-out solutions to odd-numbered exercises.

Using Newton's Method In Exercises 1 and 2, complete two iterations of Newton's Method for the function using the given initial estimate. *See Example 1.*

1. $f(x) = x^2 - 5, x_1 = 2$

2. $f(x) = 2x^2 - 3, x_1 = 1$

Using Newton's Method In Exercises 3–12, use Newton's Method to approximate the indicated zero(s) of the function. Continue the iterations until two successive approximations differ by less than 0.001. Then find the zero(s) using a graphing utility and compare the results. *See Examples 2 and 3.*

3. $f(x) = x^3 + x - 1$

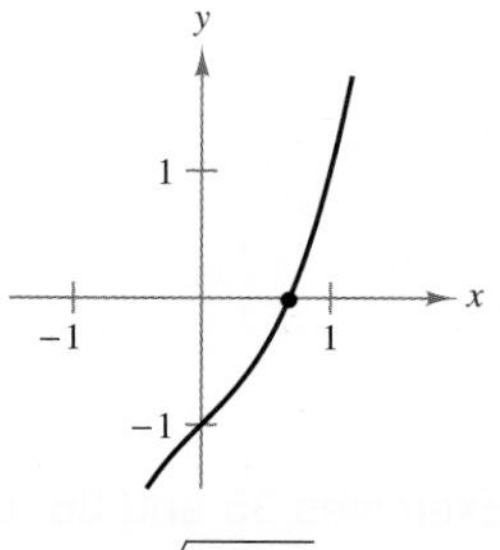

4. $f(x) = x^5 + x - 1$

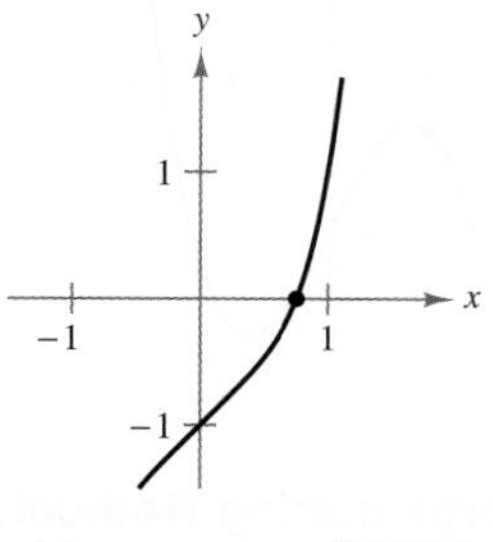

5. $y = 5\sqrt{x - 1} - 2x$

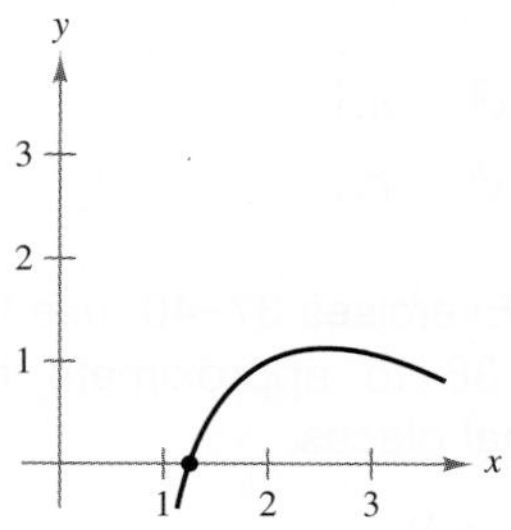

6. $f(x) = x - 2\sqrt{x + 1}$

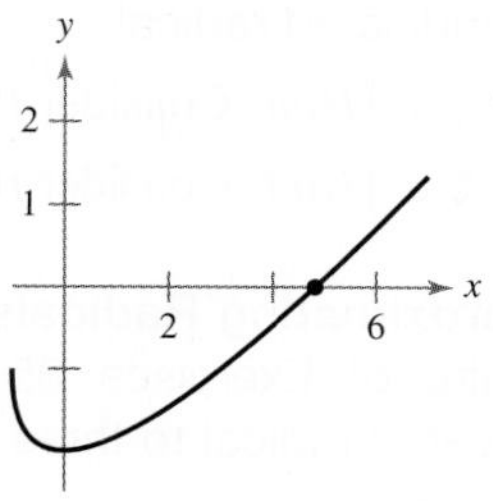

7. $f(x) = \ln x + x$

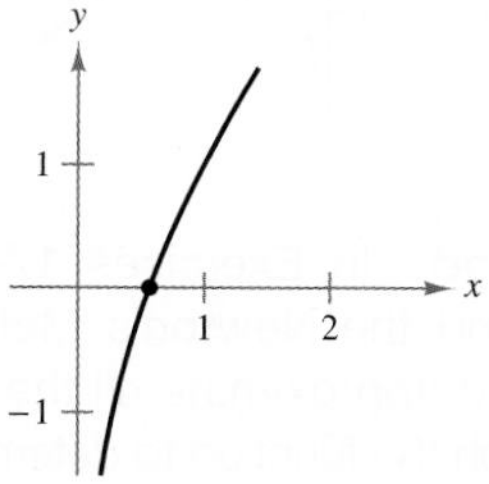

8. $f(x) = \ln x - \frac{1}{x}$

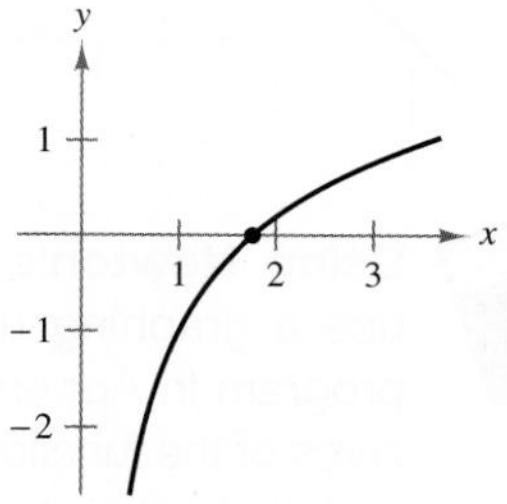

9. $f(x) = e^{x/2} + x^2 - 3$

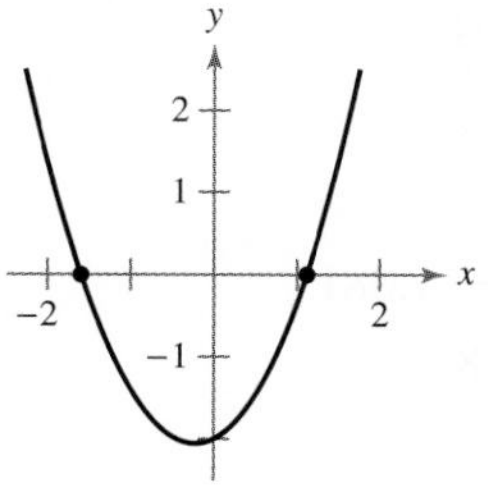

10. $f(x) = e^{-x^2} - x^2$

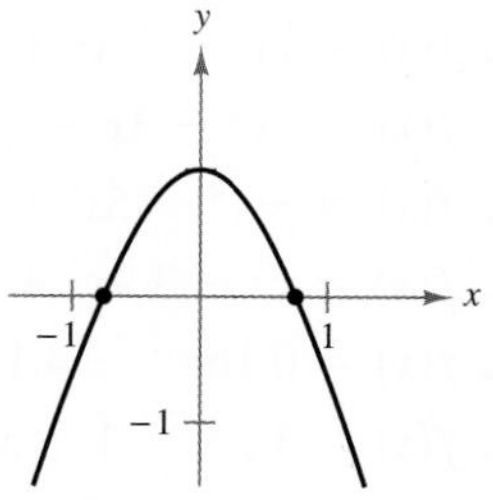

11. $y = x^4 + x^3 - 1$

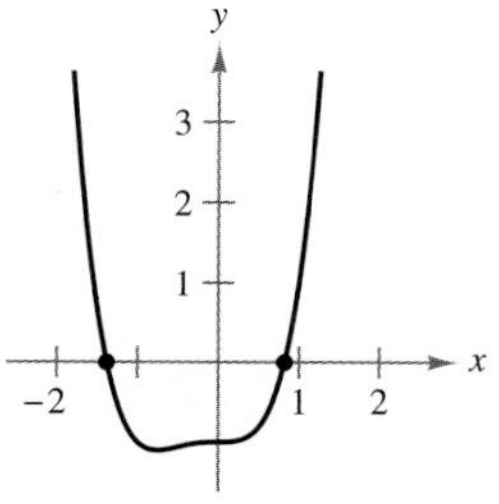

12. $f(x) = x^3 - 3x + 1$

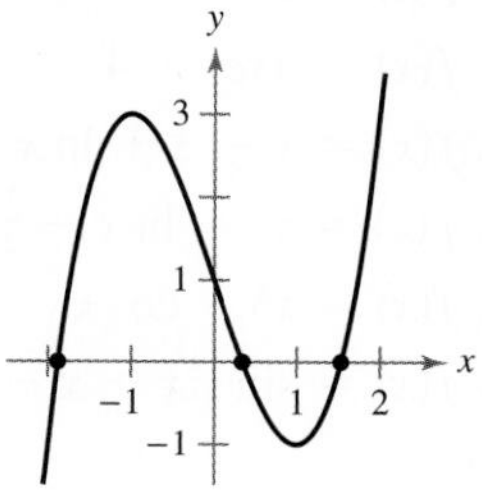

Finding a Point of Intersection In Exercises 13–16, use Newton's Method to estimate the point of intersection of the graphs to three decimal places. Continue the iterations until two successive approximations differ by less than 0.001. *See Example 4.*

13. $f(x) = 3 - x$

$g(x) = \dfrac{1}{x^2 + 1}$

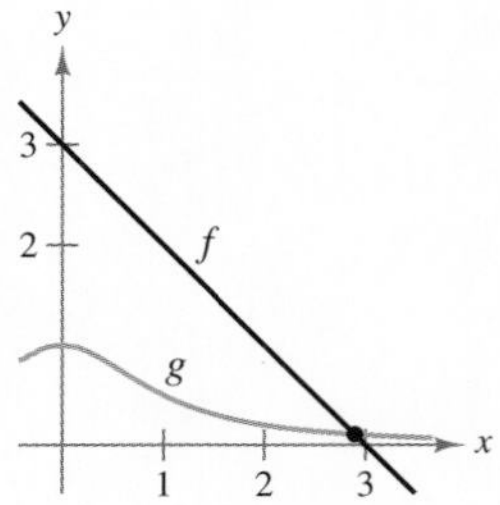

14. $f(x) = 2x$

$g(x) = \sqrt{x + 4}$

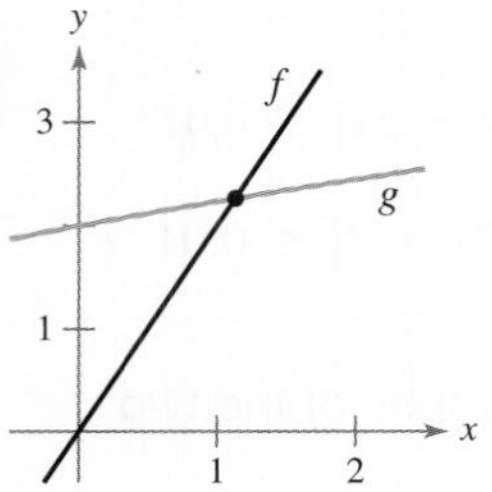

15. $f(x) = x$

$g(x) = e^{-x}$

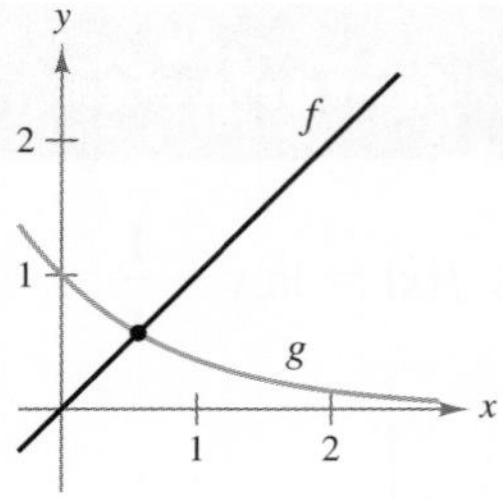

16. $f(x) = -x + 2$

$g(x) = \ln x$

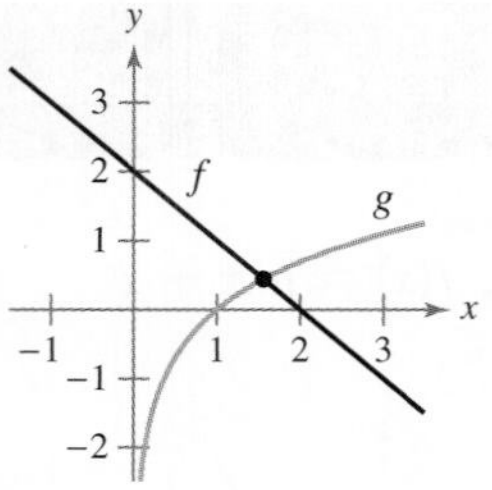

Using Newton's Method In Exercises 17–30, use a graphing utility and the Newton's Method program in Appendix E to approximate all the real zeros of the function. Graph the function to determine an initial estimate of a zero.

17. $f(x) = \frac{1}{4}x^3 - 3x^2 + \frac{3}{4}x - 2$

18. $f(x) = \frac{1}{2}x^3 - \frac{11}{3}x^2 + \frac{10}{3}x + \frac{32}{3}$

19. $f(x) = \frac{1}{2}x^4 - 3x - 3$

20. $f(x) = -x^4 + 5x^2 - 5$

21. $f(x) = x^3 - 3.9x^2 + 4.79x - 1.881$

22. $f(x) = 0.16e^{x^2} - 4.1x^3 + 6.8$

23. $f(x) = 3\sqrt{x - 1} - x$

24. $f(x) = x^4 - 10x^2 - 11$

25. $f(x) = 2x + 4 - e^{3x}$

26. $f(x) = 5xe^x - 4$

27. $f(x) = x - 3 + \ln x$

28. $f(x) = x^2 - \ln x - \frac{3}{2}$

29. $f(x) = x^3 - \cos x$

30. $f(x) = \sin \pi x + x - 1$

Convergence of Newton's Method In Exercises 31–34, apply Newton's Method using the indicated initial estimate. Then explain why the method fails.

31. $y = 2x^3 - 6x^2 + 6x - 1, \quad x_1 = 1$

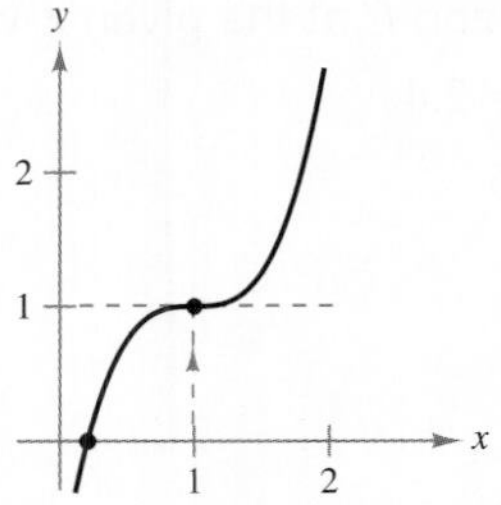

32. $y = 4x^3 - 12x^2 + 12x - 3, \quad x_1 = \frac{3}{2}$

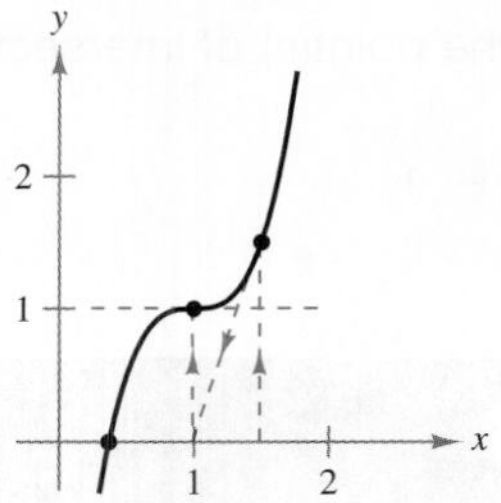

33. $y = -x^3 + 3x^2 - x + 1, \quad x_1 = 1$

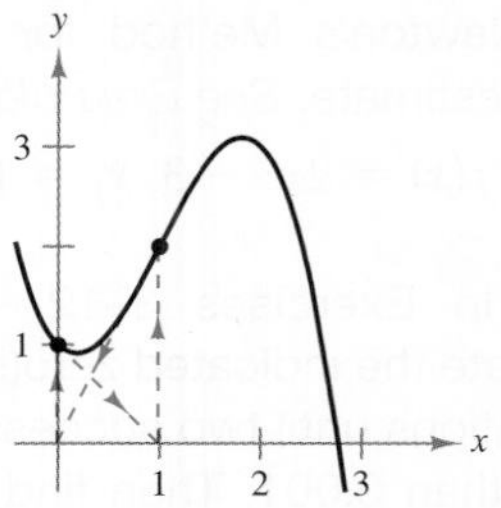

34. $y = x^3 - 2x - 2, \quad x_1 = 0$

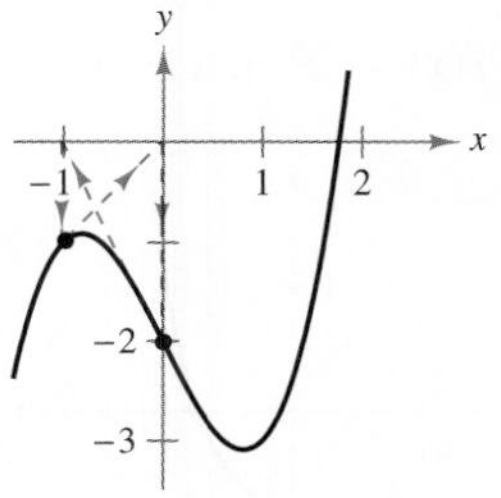

Approximating Radicals In Exercises 35 and 36, use Newton's Method to obtain a general rule for approximating the indicated radical.

35. $\sqrt{a}$ [*Hint:* Consider $f(x) = x^2 - a$.]

36. $\sqrt[n]{a}$ [*Hint:* Consider $f(x) = x^n - a$.]

Approximating Radicals In Exercises 37–40, use the results of Exercises 35 and 36 to approximate the indicated radical to three decimal places.

37. $\sqrt{7}$

38. $\sqrt{6}$

39. $\sqrt[4]{6}$

40. $\sqrt[3]{15}$

41. Approximating Reciprocals Use Newton's Method to show that the equation $x_{n+1} = x_n(2 - ax_n)$ can be used to approximate $1/a$ if x_1 is an initial guess of the reciprocal of a. Note that this method of approximating reciprocals uses only the operations of multiplication and subtraction. [*Hint:* Consider $f(x) = (1/x) - a$.]

42. Approximating Reciprocals Use the result of Exercise 41 to approximate (a) $\frac{1}{3}$ and (b) $\frac{1}{11}$ to three decimal places.

Using Newton's Method In Exercises 43–49, some typical problems from previous chapters are given. In each case, use Newton's Method to approximate the solution.

43. Minimum Distance Find the point on the graph of $f(x) = 4 - x^2$ that is closest to the point $(1, 0)$.

44. Minimum Distance Find the point on the graph of $f(x) = x^2$ that is closest to the point $(4, -3)$.

45. Minimum Time You are in a boat 2 miles from the nearest point on the coast (see figure). You are to go to a point Q, which is 3 miles down the coast and 1 mile inland. You can row at 3 miles per hour and walk at 4 miles per hour. Toward what point on the coast should you row in order to reach point Q in the least amount of time?

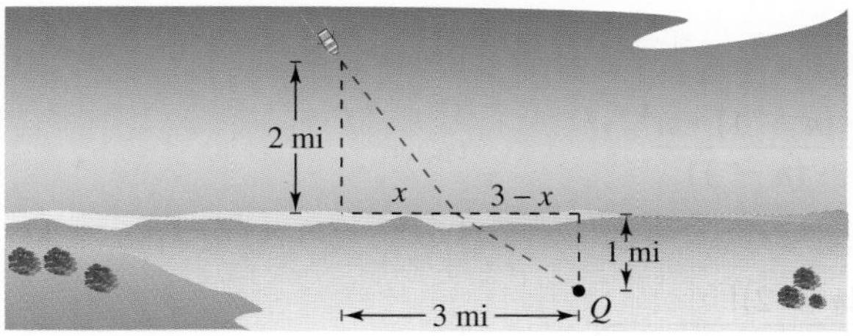

46. Advertising Costs A company that produces portable digital audio players estimates that the profit P (in dollars) for selling a particular model is

$$P = -76x^3 + 4830x^2 - 320{,}000, \quad 0 \le x \le 60$$

where x is the advertising expense (in tens of thousands of dollars) (see figure). According to this model, find the smaller of the two advertising amounts that yield a profit P of \$2,500,000.

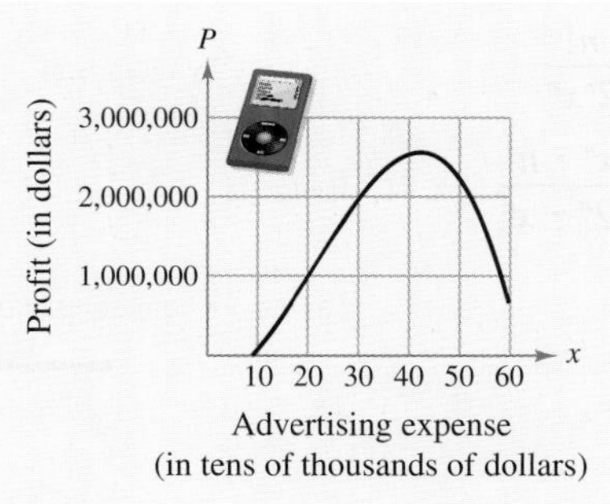

47. Average Cost A company estimates that the cost C (in dollars) of producing x units of a product is given by

$$C = 0.0001x^3 + 0.02x^2 + 0.4x + 800.$$

Find the production level that minimizes the average cost per unit.

48. Medicine The concentration C (in milligrams per milliliter) of an antibiotic in the bloodstream t hours after injection into muscle tissue is given by

$$C = \frac{3t^2 + t}{50 + t^3}.$$

When is the concentration the greatest?

49. Cost The ordering and transportation cost C (in thousands of dollars) of the components used in manufacturing a product is given by

$$C = 100\left(\frac{200}{x^2} + \frac{x}{x + 30}\right), \quad x \ge 1$$

where x is the order size (in hundreds). Find the order size that minimizes the cost.

50. HOW DO YOU SEE IT? For what value(s) will Newton's Method fail to converge for the function shown in the graph? Explain your reasoning.

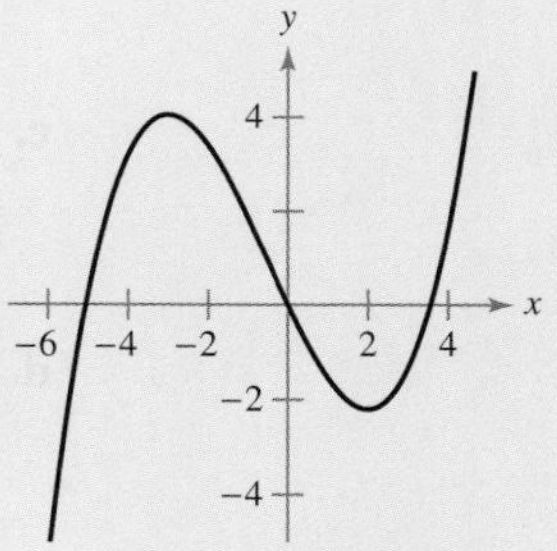

True or False? In Exercises 51 and 52, determine whether the statement is true or false. If it is false, explain why or give an example that shows it is false.

51. The zeros of

$$f(x) = \frac{p(x)}{q(x)}$$

coincide with the zeros of $p(x)$.

52. The roots of

$$\sqrt{f(x)} = 0$$

coincide with the roots of $f(x) = 0$.

ALGEBRA TUTOR

Example 1 **Simplifying Factorial Expressions**

Simplify each expression.

a. $\dfrac{8!}{2! \cdot 6!}$ **b.** $\dfrac{2! \cdot 6!}{3! \cdot 5!}$ **c.** $\dfrac{n!}{(n-1)!}$ **d.** $\dfrac{(2n+2)!}{(2n+4)!}$

e. $\dfrac{x^{n+1}}{(n+1)!} \div \dfrac{x^n}{n!}$ **f.** $\dfrac{2^{n+1}x^{n+1}}{(n+1)!} \div \dfrac{2^n x^n}{n!}$

SOLUTION

a. $\dfrac{8!}{2! \cdot 6!} = \dfrac{1 \cdot 2 \cdot 3 \cdot 4 \cdot 5 \cdot 6 \cdot 7 \cdot 8}{1 \cdot 2 \cdot 1 \cdot 2 \cdot 3 \cdot 4 \cdot 5 \cdot 6}$ Factor.

$= \dfrac{7 \cdot 8}{1 \cdot 2}$ Divide out common factors.

$= \dfrac{56}{2}$ Multiply.

$= 28$ Divide.

b. $\dfrac{2! \cdot 6!}{3! \cdot 5!} = \dfrac{1 \cdot 2 \cdot 1 \cdot 2 \cdot 3 \cdot 4 \cdot 5 \cdot 6}{1 \cdot 2 \cdot 3 \cdot 1 \cdot 2 \cdot 3 \cdot 4 \cdot 5}$ Factor.

$= \dfrac{6}{3}$ Divide out common factors.

$= 2$ Divide.

c. $\dfrac{n!}{(n-1)!} = \dfrac{1 \cdot 2 \cdot 3 \cdots (n-1) \cdot n}{1 \cdot 2 \cdot 3 \cdots (n-1)}$ Factor.

$= n$ Divide out common factors.

d. $\dfrac{(2n+2)!}{(2n+4)!} = \dfrac{(2n+2)!}{(2n+2)!(2n+3)(2n+4)}$ Factor.

$= \dfrac{1}{(2n+3)(2n+4)}$ Divide out common factors.

e. $\dfrac{x^{n+1}}{(n+1)!} \div \dfrac{x^n}{n!} = \dfrac{x^{n+1}}{(n+1)!} \cdot \dfrac{n!}{x^n}$ Multiply by reciprocal.

$= \dfrac{x \cdot x^n}{n!(n+1)} \cdot \dfrac{n!}{x^n}$ Factor.

$= \dfrac{x}{n+1}$ Divide out common factors.

f. $\dfrac{2^{n+1}x^{n+1}}{(n+1)!} \div \dfrac{2^n x^n}{n!} = \dfrac{2^{n+1}x^{n+1}}{(n+1)!} \cdot \dfrac{n!}{2^n x^n}$ Multiply by reciprocal.

$= \dfrac{2 \cdot 2^n \cdot x \cdot x^n \cdot n!}{n!(n+1) \cdot 2^n \cdot x^n}$ Factor.

$= \dfrac{2x}{n+1}$ Divide out common factors.

Example 2 Rewriting Expressions with Sigma Notation

Rewrite each expression.

a. $\sum_{i=1}^{5} 3i$ **b.** $\sum_{k=3}^{6} (1 + k^2)$ **c.** $\sum_{i=0}^{8} \frac{1}{i!}$

d. $\sum_{n=1}^{\infty} 3\left(\frac{1}{2}\right)^n$ **e.** $\sum_{n=1}^{20} 6(1.01)^n$ **f.** $\sum_{n=1}^{10} \left(n + \frac{1}{3^n}\right)$

SOLUTION

a.
$$\begin{aligned}\sum_{i=1}^{5} 3i &= 3(1) + 3(2) + 3(3) + 3(4) + 3(5)\\ &= 3(1 + 2 + 3 + 4 + 5)\\ &= 3(15)\\ &= 45\end{aligned}$$

b.
$$\begin{aligned}\sum_{k=3}^{6} (1 + k^2) &= (1 + 3^2) + (1 + 4^2) + (1 + 5^2) + (1 + 6^2)\\ &= 10 + 17 + 26 + 37\\ &= 90\end{aligned}$$

c.
$$\begin{aligned}\sum_{i=0}^{8} \frac{1}{i!} &= \frac{1}{0!} + \frac{1}{1!} + \frac{1}{2!} + \frac{1}{3!} + \frac{1}{4!} + \frac{1}{5!} + \frac{1}{6!} + \frac{1}{7!} + \frac{1}{8!}\\ &= 1 + 1 + \frac{1}{2} + \frac{1}{6} + \frac{1}{24} + \frac{1}{120} + \frac{1}{720} + \frac{1}{5040} + \frac{1}{40{,}320}\\ &\approx 2.71828\end{aligned}$$

d.
$$\begin{aligned}\sum_{n=1}^{\infty} 3\left(\frac{1}{2}\right)^n &= \sum_{n=0}^{\infty} 3\left(\frac{1}{2}\right)^{n+1}\\ &= \sum_{n=0}^{\infty} 3\left(\frac{1}{2}\right)\left(\frac{1}{2}\right)^n\\ &= \frac{3}{2}\sum_{n=0}^{\infty}\left(\frac{1}{2}\right)^n\end{aligned}$$

e.
$$\begin{aligned}\sum_{n=1}^{20} 6(1.01)^n &= 6(1.01)^1 + 6(1.01)^2 + \cdots + 6(1.01)^{20}\\ &= -6 + 6 + 6(1.01)^1 + 6(1.01)^2 + \cdots + 6(1.01)^{20}\\ &= -6 + 6(1.01)^0 + 6(1.01)^1 + 6(1.01)^2 + \cdots + 6(1.01)^{20}\\ &= -6 + \sum_{n=0}^{20} 6(1.01)^n\end{aligned}$$

f.
$$\begin{aligned}\sum_{n=1}^{10} \left(n + \frac{1}{3^n}\right) &= \sum_{n=1}^{10} n + \sum_{n=1}^{10} \frac{1}{3^n}\\ &= \sum_{n=1}^{10} n + \sum_{n=1}^{10} \left(\frac{1}{3}\right)^n\\ &= \sum_{n=1}^{10} n - 1 + \sum_{n=0}^{10} \left(\frac{1}{3}\right)^n\end{aligned}$$

SUMMARY AND STUDY STRATEGIES

After studying this chapter, you should have acquired the following skills. The exercise numbers are keyed to the Review Exercises that begin on page 656. Answers to odd-numbered Review Exercises are given in the back of the text.*

Section 10.1	Review Exercises
■ Find the terms of sequences.	*1–6*
■ Determine the convergence or divergence of sequences and find the limits of convergent sequences.	*7–16*
■ Find the *n*th terms of sequences.	*17–22*
■ Use sequences to answer questions about real-life situations.	*23–26*

Section 10.2

■ Find the partial sums of series. $S_1 = a_1,\quad S_2 = a_1 + a_2,\quad S_3 = a_1 + a_2 + a_3, \ldots,\quad S_n = a_1 + a_2 + a_3 + \cdots + a_n$	*27–30*
■ Determine the convergence or divergence of infinite series.	*31–34*
■ Use the *n*th-Term Test to show that series diverge. The series $\sum_{n=1}^{\infty} a_n$ diverges when $\lim_{n\to\infty} a_n \neq 0$.	*35–38*
■ Find the *n*th partial sums of geometric series. $S_n = \frac{a(1 - r^{n+1})}{1 - r},\quad r \neq 1$	*39, 40*
■ Determine the convergence or divergence of geometric series. If $\lvert r\rvert \geq 1$, then the series *diverges*. If $\lvert r\rvert < 1$, then the series *converges* and its sum is $\frac{a}{1 - r}$.	*41–46*
■ Use geometric series to model and solve real-life situations.	*47–54*

Section 10.3

■ Determine the convergence or divergence of *p*-series. The series diverges when $0 < p \leq 1$. The series converges when $p > 1$.	*55–58*
■ Use the Ratio Test to determine the convergence or divergence of series. The series converges when $\lim_{n\to\infty} \left\lvert \frac{a_{n+1}}{a_n} \right\rvert < 1$. The series diverges when $\lim_{n\to\infty} \left\lvert \frac{a_{n+1}}{a_n} \right\rvert > 1$.	*59–64*
■ Match series with the graphs of their sequences of partial sums.	*65–68*

* A wide range of valuable study aids are available to help you master the material in this chapter. The *Student Solutions Manual* includes step-by-step solutions to all odd-numbered exercises to help you review and prepare. The student website at *www.cengagebrain.com* offers algebra help and a *Graphing Technology Guide*, which contains step-by-step commands and instructions for a wide variety of graphing calculators.

Section 10.4 — Review Exercises

- Find the centers of power series and write the first five terms. *69–72*
- Find the radii of convergence of power series. *73–80*
- Use Taylor's Theorem to find power series for functions. *81–86*

$$f(x) = \sum_{n=0}^{\infty} \frac{f^{(n)}(c)(x-c)^n}{n!} = f(c) + \frac{f'(c)(x-c)}{1!} + \frac{f''(c)(x-c)^2}{2!} + \cdots$$

- Use the basic list of power series to find power series for functions. *87–94*

Section 10.5

- Find Taylor polynomials for functions. *95–100*
- Use Taylor polynomials to approximate the values of functions at points and determine the maximum errors of approximation. *101–104*
- Approximate definite integrals using Taylor polynomials. *105–108*

Section 10.6

- Use Newton's Method to approximate the zeros of functions. *109–114*

$$x_{n+1} = x_n - \frac{f(x_n)}{f'(x_n)}$$

- Use Newton's Method to approximate points of intersection of graphs. *115–118*
- Use a graphing utility and the Newton's Method program to approximate the zeros of functions. *119, 120*

Study Strategies

- **Using the List of Basic Power Series** To be efficient at finding Taylor series or Taylor polynomials, learn how to use the basic list below.

Function	*Interval of Convergence*
$\frac{1}{x} = 1 - (x-1) + (x-1)^2 - (x-1)^3 + (x-1)^4 - \cdots + (-1)^n(x-1)^n + \cdots$	$0 < x < 2$
$\frac{1}{x+1} = 1 - x + x^2 - x^3 + x^4 - x^5 + \cdots + (-1)^n x^n + \cdots$	$-1 < x < 1$
$\ln x = (x-1) - \frac{(x-1)^2}{2} + \frac{(x-1)^3}{3} - \frac{(x-1)^4}{4} + \cdots + \frac{(-1)^{n-1}(x-1)^n}{n} + \cdots$	$0 < x \le 2$
$e^x = 1 + x + \frac{x^2}{2!} + \frac{x^3}{3!} + \frac{x^4}{4!} + \cdots + \frac{x^n}{n!} + \cdots$	$-\infty < x < \infty$
$(1+x)^k = 1 + kx + \frac{k(k-1)x^2}{2!} + \frac{k(k-1)(k-2)x^3}{3!} + \frac{k(k-1)(k-2)(k-3)x^4}{4!} + \cdots$	$-1 < x < 1$

- **Using Technology to Approximate Zeros** Newton's Method is only one way that technology can be used to approximate the zeros of a function. Another way is to use the *zoom* and *trace* features of a graphing utility, as shown below. (Compare this with the procedure described in the Tech Tutor on page 648.)

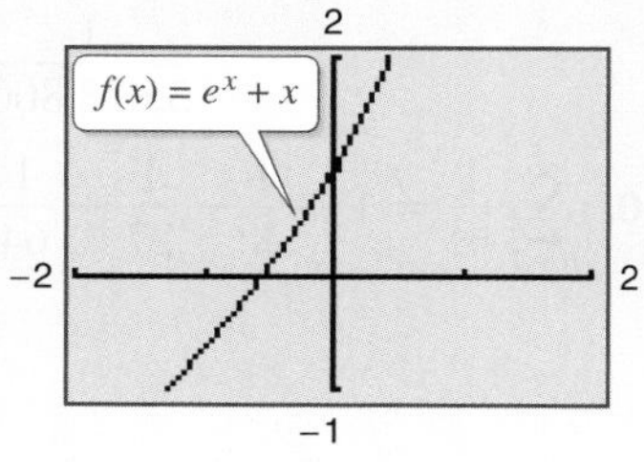

Original screen

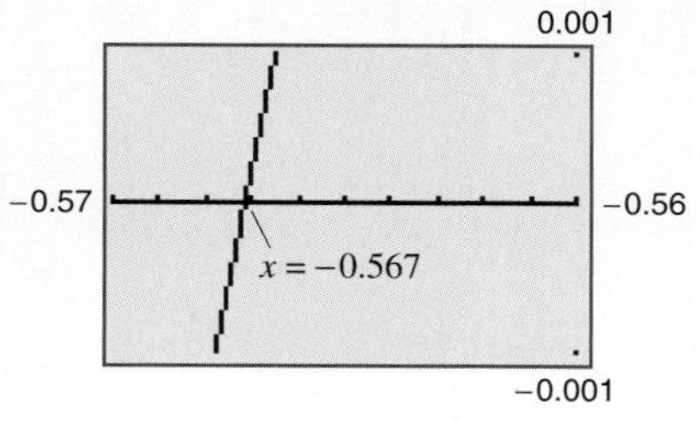

Screen after zooming five times

Review Exercises

See www.CalcChat.com for worked-out solutions to odd-numbered exercises.

Finding Terms of a Sequence In Exercises 1–6, write the first five terms of the sequence.

1. $a_n = 3n + 4$

2. $a_n = 5^n$

3. $a_n = \left(\frac{3}{2}\right)^n$

4. $a_n = \frac{n-2}{n^2+2}$

5. $a_n = \frac{4^n}{n!}$

6. $a_n = \frac{(-1)^n}{n^3}$

Finding the Limit of a Sequence In Exercises 7–16, find the limit of the sequence (if it exists) as n approaches infinity. Then state whether the sequence converges or diverges.

7. $a_n = \frac{n+1}{n^2}$

8. $a_n = \frac{5n+2}{n}$

9. $a_n = \frac{n^3}{n^2+1}$

10. $a_n = 10e^{-n}$

11. $a_n = 5 + \frac{1}{3^n}$

12. $a_n = \frac{n}{\sqrt{n^2+1}}$

13. $a_n = \frac{1}{n^{4/3}}$

14. $a_n = 3 + (-1)^n$

15. $a_n = \frac{n!}{(n-3)!}$

16. $a_n = \frac{(n-1)!}{(n+1)!}$

Finding a Pattern for a Sequence In Exercises 17–22, write an expression for the nth term of the sequence. (There is more than one correct answer.)

17. 7, 9, 11, 13, 15, . . .

18. 1, 5, 9, 13, 17, . . .

19. $1, \frac{1}{2}, \frac{1}{6}, \frac{1}{24}, \frac{1}{120}, \ldots$

20. $\frac{1}{2}, \frac{2}{5}, \frac{3}{10}, \frac{4}{17}, \ldots$

21. $\frac{1}{3}, -\frac{2}{9}, \frac{4}{27}, -\frac{8}{81}, \ldots$

22. $1, \frac{2}{5}, \frac{4}{25}, \frac{8}{125}, \frac{16}{625}, \ldots$

23. Compound Interest Consider the sequence $\{A_n\}$, whose nth term is given by

$$A_n = P\left[1 + \frac{r}{12}\right]^n$$

where P is the principal, A_n is the balance in the account after n months, and r is the annual percentage rate (in decimal form). Write the first 10 terms of the sequence for $P = \$1000$ and $r = 0.03$.

24. Compound Interest Consider the sequence $\{A_n\}$, whose nth term is given by

$$A_n = P(1 + r)^n$$

where P is the principal, A_n is the balance in the account after n years, and r is the annual percentage rate (in decimal form). Write the first 10 terms of the sequence for $P = \$2500$ and $r = 0.04$.

25. Physical Science A ball is dropped from a height of 16 feet, and on each rebound it rises to $\frac{3}{4}$ of its preceding height.

(a) Write an expression for the height of the nth rebound.

(b) Compute the heights for the first 5 rebounds.

(c) Determine the convergence or divergence of the sequence of rebounds. If it converges, find the limit.

26. Sales A mail-order company sells \$15,000 worth of products during its first year. The company's goal is to increase sales by \$10,000 each year.

(a) Write an expression for the amount of sales during the nth year.

(b) Compute the sales for the first 6 years.

(c) Determine the convergence or divergence of the sequence of sales. If it converges, find the limit.

Finding Partial Sums In Exercises 27–30, write the first five terms of the sequence of partial sums.

27. $\sum_{n=1}^{\infty} \left(\frac{3}{2}\right)^n = \frac{3}{2} + \frac{9}{4} + \frac{27}{8} + \frac{81}{16} + \frac{243}{32} + \cdots$

28. $\sum_{n=1}^{\infty} \frac{(-1)^{n+1}}{2n} = \frac{1}{2} - \frac{1}{4} + \frac{1}{6} - \frac{1}{8} + \frac{1}{10} - \cdots$

29. $\sum_{n=1}^{\infty} \frac{(-1)^{n+1}}{(2n)!} = \frac{1}{2} - \frac{1}{24} + \frac{1}{720} - \frac{1}{40{,}320} + \frac{1}{3{,}628{,}800} - \cdots$

30. $\sum_{n=1}^{\infty} \frac{1}{n^3} = 1 + \frac{1}{8} + \frac{1}{27} + \frac{1}{64} + \frac{1}{125} + \cdots$

Determining Convergence or Divergence In Exercises 31–34, determine the convergence or divergence of the infinite series.

31. $\sum_{n=1}^{\infty} \frac{n^2+1}{n(n+1)}$

32. $\sum_{n=0}^{\infty} \left(\frac{1}{3}\right)^n$

33. $\sum_{n=0}^{\infty} 2(0.25)^{n+1}$

34. $\sum_{n=1}^{\infty} \frac{\sqrt{n^3}}{n}$

Using the *n*th-Term Test In Exercises 35–38, use the *n*th-Term Test to verify that the series diverges.

35. $\sum_{n=1}^{\infty} \frac{2n}{n+5}$

36. $\sum_{n=2}^{\infty} \frac{n^3}{1-n^3}$

37. $\sum_{n=1}^{\infty} \frac{n^2}{n^2+1}$

38. $\sum_{n=1}^{\infty} \frac{n}{\sqrt{4n^2+1}}$

Finding an *n*th Partial Sum In Exercises 39 and 40, find the indicated partial sums for the geometric series.

	Geometric Series	*Partial Sums*
39.	$\sum_{n=0}^{\infty} 2\left(\frac{2}{3}\right)^n$	S_3, S_5, S_{10}
40.	$\sum_{n=0}^{\infty} 3\left(-\frac{1}{5}\right)^n$	S_4, S_9, S_{20}

Determining Convergence or Divergence In Exercises 41–46, determine whether the geometric series converges or diverges. If it converges, find its sum.

41. $\sum_{n=0}^{\infty} \left(\frac{5}{4}\right)^n$

42. $\sum_{n=0}^{\infty} 3\left(-\frac{4}{3}\right)^n$

43. $\sum_{n=0}^{\infty} \frac{1}{4}(4)^n$

44. $\sum_{n=0}^{\infty} 4\left(-\frac{1}{4}\right)^n$

45. $\sum_{n=0}^{\infty} [(0.5)^n + (0.2)^n]$

46. $\sum_{n=0}^{\infty} [(1.5)^n + (0.2)^n]$

47. Sales A company produces a new product for which it estimates the annual sales to be 9000 units. Suppose that in any given year 15% of the units (regardless of age) will become inoperative.

(a) How many units will be in use after n years?

(b) Find the market stabilization level of the product.

48. Annuity A deposit of \$200 is made at the beginning of each month for 3 years in an account that pays 6% interest, compounded monthly. The balance A in the account at the end of 3 years is

$$A = 200\left(1+\frac{0.06}{12}\right)^1 + 200\left(1+\frac{0.06}{12}\right)^2 + \cdots + 200\left(1+\frac{0.06}{12}\right)^{36}.$$

(a) Use sigma notation to write the balance in the account at the end of 3 years.

(b) Find the balance in the account at the end of 3 years.

Multiplier Effect In Exercises 49–52, use the following information. A tax rebate has been given to property owners by the state government with the anticipation that each property owner spends approximately $p\%$ of the rebate, and in turn each recipient of this amount spends $p\%$ of what they receive, and so on. For the given tax rebate, find the total amount put back into the state's economy, if this effect continues without end.

	Tax Rebate	$p\%$
49.	\$500	75%
50.	\$250	80%
51.	\$600	72.5%
52.	\$450	77.5%

53. Depreciation A company buys a machine for \$120,000 that depreciates at a rate of 30% per year.

(a) Find a formula for the value of the machine after n years.

(b) Find the value of the machine after 5 years.

54. Salary You accept a job that pays a salary of \$50,000 the first year. During the next 39 years, you will receive a 5.5% raise each year. What would be your total compensation over the 40-year period?

Determining Convergence or Divergence In Exercises 55–58, determine the convergence or divergence of the *p*-series.

55. $\sum_{n=1}^{\infty} \frac{1}{n^4}$

56. $\sum_{n=1}^{\infty} 2n^{-2/3}$

57. $\sum_{n=1}^{\infty} \frac{1}{n\sqrt[4]{n}}$

58. $\sum_{n=1}^{\infty} \frac{1}{n^e}$

Using the Ratio Test In Exercises 59–64, use the Ratio Test to determine the convergence or divergence of the series.

59. $\sum_{n=1}^{\infty} \frac{n4^n}{n!}$ **60.** $\sum_{n=0}^{\infty} \frac{n!}{4^n}$

61. $\sum_{n=1}^{\infty} \frac{(-1)^n 3^n}{n}$ **62.** $\sum_{n=1}^{\infty} \frac{2^n}{n}$

63. $\sum_{n=1}^{\infty} \frac{n2^n}{n!}$

64. $\sum_{n=1}^{\infty} \frac{2n}{1-4^n}$

Matching In Exercises 65–68, match the series with the graph of its sequence of partial sums. [The graphs are labeled (a)–(d).] Then determine the convergence or divergence of the series.

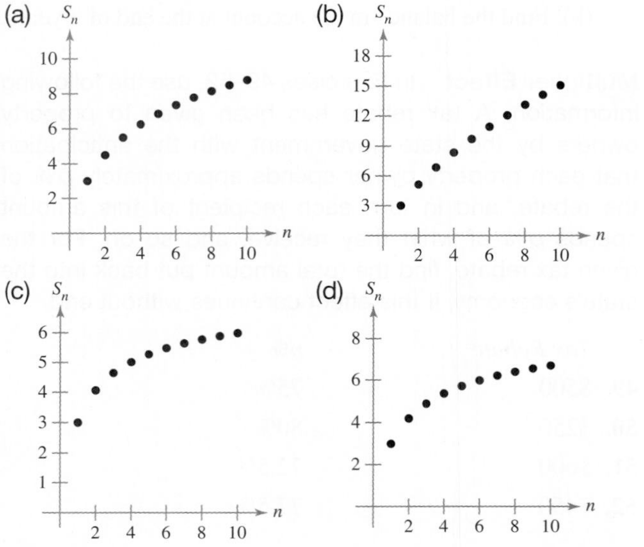

65. $\sum_{n=1}^{\infty} \frac{3}{\sqrt{n}}$ **66.** $\sum_{n=1}^{\infty} \frac{3}{\sqrt[3]{n^4}}$

67. $\sum_{n=1}^{\infty} \frac{3}{n}$ **68.** $\sum_{n=1}^{\infty} \frac{3}{\sqrt{n^3}}$

Using Power Series In Exercises 69–72, identify the center of each power series. Then write the first five terms of the power series.

69. $\sum_{n=1}^{\infty} \frac{(x-2)^n}{2^n}$

70. $\sum_{n=0}^{\infty} \frac{(x+1)^n}{n!}$

71. $\sum_{n=1}^{\infty} \frac{(-1)^n x^n}{3^n}$

72. $\sum_{n=0}^{\infty} \frac{(-1)^{n+1} x^n}{(n+1)!}$

Finding the Radius of Convergence In Exercises 73–80, find the radius of convergence of the power series.

73. $\sum_{n=0}^{\infty} \left(\frac{x}{10}\right)^n$ **74.** $\sum_{n=0}^{\infty} (2x)^n$

75. $\sum_{n=0}^{\infty} \frac{(4x)^n}{n!}$ **76.** $\sum_{n=0}^{\infty} \frac{(-1)^n (x-2)^n}{(n+1)^2}$

77. $\sum_{n=1}^{\infty} \frac{3^n (x-2)^n}{n}$

78. $\sum_{n=0}^{\infty} n!(x-3)^n$

79. $\sum_{n=0}^{\infty} \frac{(x-2)^n}{2^n}$

80. $\sum_{n=0}^{\infty} \frac{(-1)^n x^{2n}}{(2n)!}$

Finding Taylor and Maclaurin Series In Exercises 81–86, apply Taylor's Theorem to find the power series (centered at c) for the function. Then find the radius of convergence.

Function	*Center*
81. $f(x) = e^{-0.5x}$	$c = 0$
82. $f(x) = e^{-x/3}$	$c = 0$
83. $f(x) = \frac{1}{\sqrt{x}}$	$c = 1$
84. $f(x) = \frac{1}{x}$	$c = -1$
85. $f(x) = \sqrt[4]{1+x}$	$c = 0$
86. $f(x) = \frac{1}{(1+x)^3}$	$c = 0$

Using the Basic List of Power Series In Exercises 87–94, use the basic list of power series for common functions on page 633 to find the series (centered at $c = 0$) for the function.

87. $f(x) = \ln(x+2)$

88. $f(x) = e^{2x+1}$

89. $f(x) = (1+x^2)^2$

90. $f(x) = \frac{1}{x^3+1}$

91. $f(x) = x^2 e^x$

92. $f(x) = \frac{e^x}{x^2}$

93. $f(x) = \frac{x^2}{x+1}$

94. $f(x) = \frac{\sqrt{x}}{x+1}$

Finding Taylor Polynomials In Exercises 95–100, find the Taylor polynomials (centered at zero) of degrees (a) 1, (b) 2, (c) 3, and (d) 4.

95. $f(x) = e^{-4x}$

96. $f(x) = e^{x+1}$

97. $f(x) = \ln(x + 2)$

98. $f(x) = \sqrt{x + 2}$

99. $f(x) = \dfrac{1}{(x + 3)^2}$

100. $f(x) = \dfrac{1}{x + 5}$

Using a Taylor Polynomial Approximation In Exercises 101–104, use a sixth-degree Taylor polynomial centered at c for the function f to obtain the required approximation. Then determine the maximum error of the approximation.

	Function	*Center*	*Approximation*
101.	$f(x) = e^{x/3}$	$c = 0$	$f(1.25)$
102.	$f(x) = \dfrac{1}{\sqrt{x}}$	$c = 1$	$f(1.15)$
103.	$f(x) = \ln(1 + x)$	$c = 1$	$f(1.5)$
104.	$f(x) = e^{x-1}$	$c = 0$	$f(1.75)$

Approximating a Definite Integral In Exercises 105–108, use a sixth-degree Taylor polynomial centered at zero to approximate the definite integral.

	Function	*Approximation*
105.	$f(x) = \sqrt{1 + x^3}$	$\displaystyle\int_0^{0.3} \sqrt{1 + x^3}\,dx$
106.	$f(x) = e^{-x^2}$	$\displaystyle\int_0^{0.5} e^{-x^2}\,dx$
107.	$f(x) = \ln(x^2 + 1)$	$\displaystyle\int_0^{0.75} \ln(x^2 + 1)\,dx$
108.	$f(x) = \sqrt{1 + x}$	$\displaystyle\int_0^{0.5} \sqrt{1 + x}\,dx$

Using Newton's Method In Exercises 109–114, use Newton's Method to approximate the indicated zero(s) of the function. Continue the iterations until two successive approximations differ by less than 0.001. Then find the zero(s) using a graphing utility, and compare the results.

109. $f(x) = x^3 + 3x - 1$

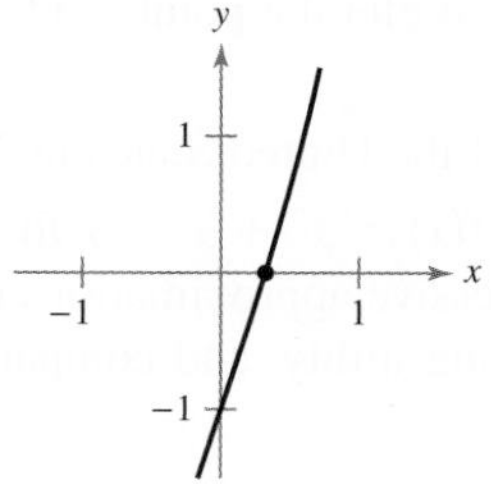

110. $f(x) = x^3 + 2x + 1$

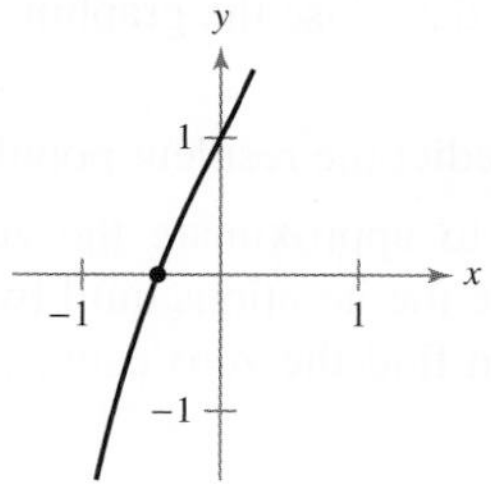

111. $f(x) = \sqrt{x + 4} + x$

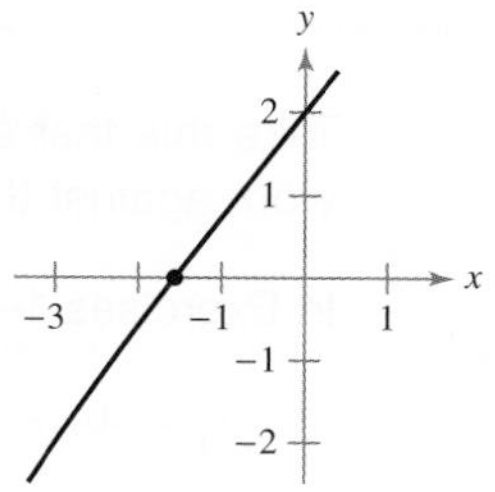

112. $f(x) = \ln\dfrac{x}{3} + \dfrac{x}{3}$

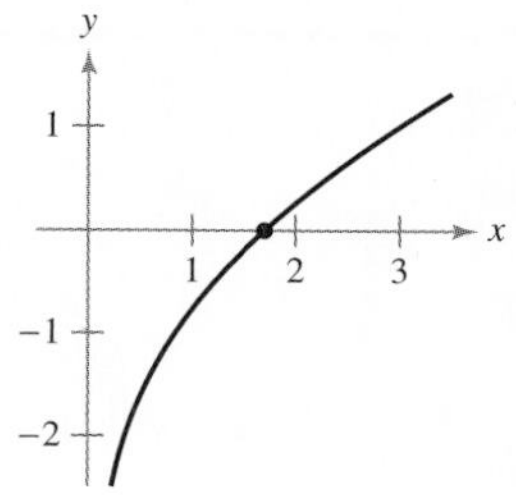

113. $f(x) = x^4 + x - 3$

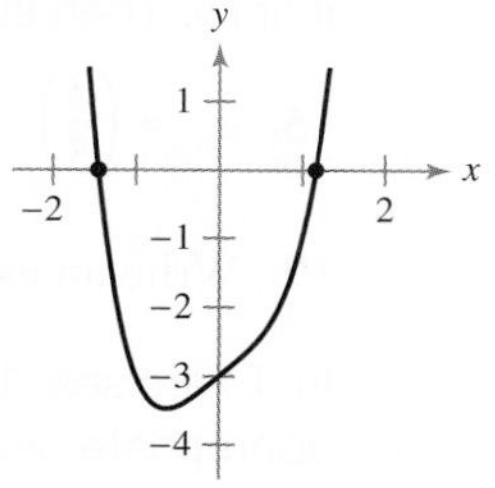

114. $f(x) = x^4 + x^2 - 1$

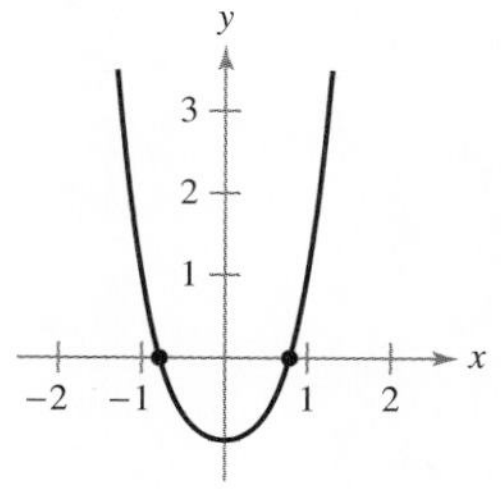

Finding a Point of Intersection In Exercises 115–118, use Newton's Method to estimate the point of intersection of the graphs to three decimal places. Continue the iterations until two successive approximations differ by less than 0.001.

115. $f(x) = x^5$
$g(x) = x + 3$

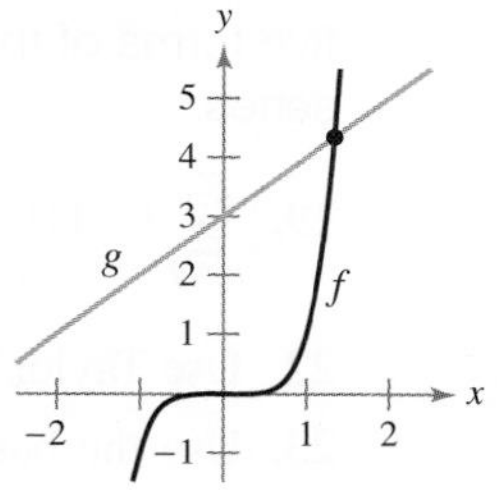

116. $f(x) = 1 - x$
$g(x) = x^5 + 2$

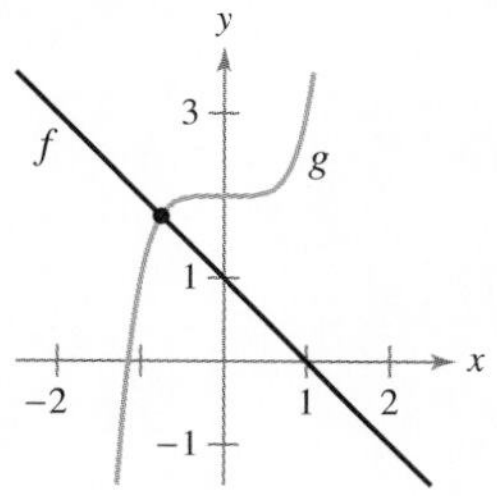

117. $f(x) = x^3$
$g(x) = e^{-x}$

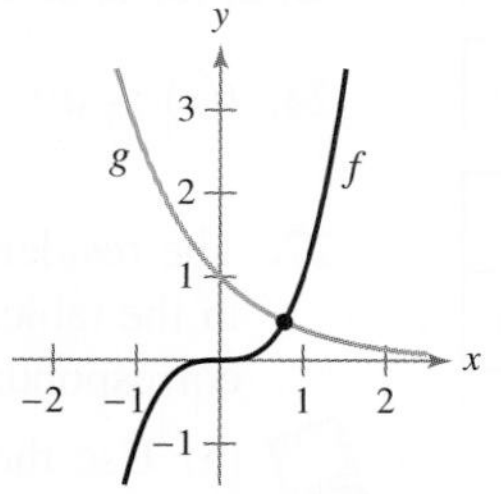

118. $f(x) = 2x^2$
$g(x) = 5e^{-x}$

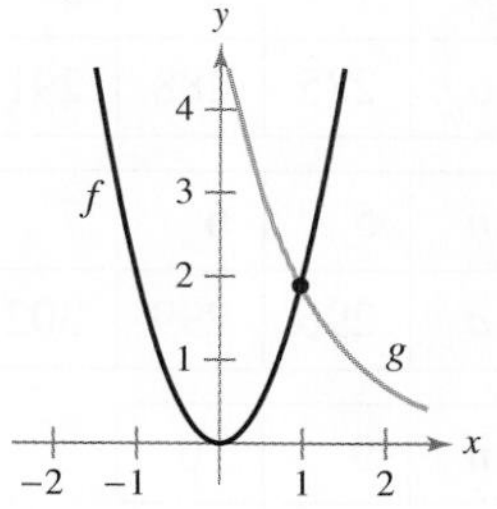

Using Newton's Method In Exercises 119 and 120, use a graphing utility and the Newton's Method program in Appendix E to approximate all the real zeros of the function. Graph the function to determine an initial estimate of a zero.

119. $f(x) = x^3 + 2x^2 - x + 5$

120. $f(x) = x^4 + x^3 - 3x^2 + 2$

TEST YOURSELF

See www.CalcChat.com for worked-out solutions to odd-numbered exercises.

Take this test as you would take a test in class. When you are done, check your work against the answers given in the back of the book.

In Exercises 1–4, write the first five terms of the sequence.

1. $a_n = 4n - 2$ **2.** $a_n = 2^n$ **3.** $a_n = \dfrac{n+3}{n}$ **4.** $a_n = \dfrac{(-1)^n}{n!}$

In Exercises 5–8, find the limit of the sequence (if it exists) as n approaches infinity. Then state whether the sequence converges or diverges.

5. $a_n = \left(\dfrac{3}{5}\right)^n$ **6.** $a_n = \dfrac{n^2}{3n^2 + 4}$ **7.** $a_n = \dfrac{(-1)^{n+1}}{6}$ **8.** $a_n = \dfrac{4n}{(n-1)!}$

9. Write an expression for the nth term of the sequence $\frac{1}{2}, -\frac{2}{5}, \frac{3}{10}, -\frac{4}{17}, \frac{5}{26}, \ldots$.

In Exercises 10–17, test the series for convergence or divergence using any appropriate test from this chapter. Identify the test used and explain your reasoning.

10. $\displaystyle\sum_{n=1}^{\infty} \frac{4^n}{n!}$ **11.** $\displaystyle\sum_{n=1}^{\infty} \frac{n+1}{n-3}$ **12.** $\displaystyle\sum_{n=0}^{\infty} \frac{2}{5^n}$ **13.** $\displaystyle\sum_{n=1}^{\infty} \frac{\sqrt[3]{n}}{\sqrt{n}}$

14. $\displaystyle\sum_{n=0}^{\infty} 2\left(\frac{5}{3}\right)^n$ **15.** $\displaystyle\sum_{n=1}^{\infty} \frac{4}{\sqrt[4]{n^5}}$ **16.** $\displaystyle\sum_{n=0}^{\infty} 5\left(-\frac{1}{6}\right)^n$ **17.** $\displaystyle\sum_{n=1}^{\infty} \frac{(n+1)!}{5^n}$

18. Find the sum of the infinite series $\displaystyle\sum_{n=0}^{\infty} \left(\frac{2}{5^n} - \frac{1}{7^{n+1}}\right)$.

In Exercises 19–21, (a) identify the center of each power series, (b) write the first five terms of the power series, and (c) find the radius of convergence of the power series.

19. $\displaystyle\sum_{n=0}^{\infty} (-1)^{n+1}\left(\frac{x}{3}\right)^n$ **20.** $\displaystyle\sum_{n=0}^{\infty} \frac{x^n}{(n+1)!}$ **21.** $\displaystyle\sum_{n=0}^{\infty} \frac{(-1)^n(x-3)^n}{(n+4)^2}$

22. Use Taylor's Theorem to find the power series (centered at zero) for $f(x) = e^{3x+1}$.

23. Use the basic list of power series for common functions on page 633 to find the power series for $f(x) = (1 + x)^{2/3}$.

In Exercises 24–26, find the Taylor polynomials (centered at zero) of degrees (a) 1, (b) 2, (c) 3, and (d) 4.

24. $f(x) = e^{x/2}$ **25.** $f(x) = \dfrac{2}{x+2}$ **26.** $f(x) = \dfrac{1}{(x+4)^2}$

n	1	2	3	4	5	6	7	8	9	10
a_n	285	288	291	293	296	299	302	305	307	309

Table for 27

27. The resident populations of the United States from 2001 through 2010 are shown in the table, where a_n is the population (in millions) and n is the year, with $n = 1$ corresponding to 2001. *(Source: U.S. Census Bureau)*

(a) Use the *regression* feature of a graphing utility to find a model of the form $a_n = kn + b$ for the data. Use the graphing utility to plot the points and graph the model.

(b) Use the model to predict the resident population of the United States in 2020.

28. Use Newton's Method to approximate the zero of $f(x) = x^3 + x - 3$ to three decimal places. Continue the iterations until two successive approximations differ by less than 0.001. Then find the zero using a graphing utility, and compare the results.

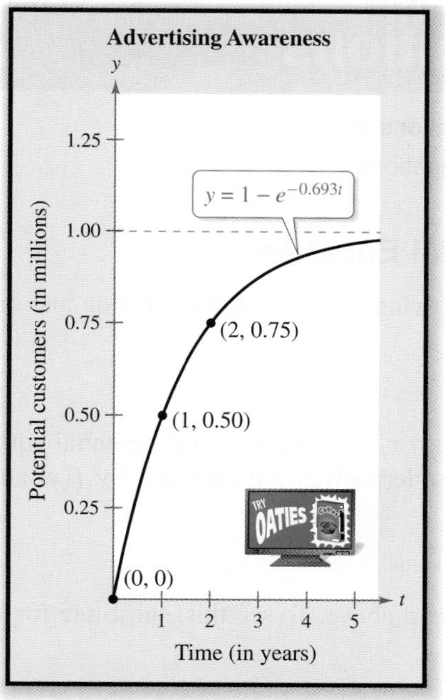

Example 2 on page 681 shows how a differential equation can be used to find a model for the advertising awareness of a product.

11 Differential Equations

11.1 Solutions of Differential Equations

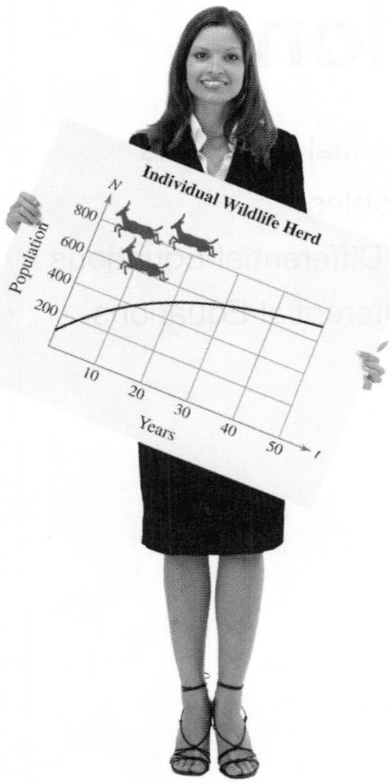

In Exercise 39 on page 667, you will use a differential equation and initial conditions to find an equation for the population of a wildlife herd.

- Find general solutions of differential equations.
- Find particular solutions of differential equations.

General Solution of a Differential Equation

A **differential equation** is an equation involving a differentiable function and one or more of its derivatives. For instance,

$$y' + 2y = 0 \qquad \text{Differential equation}$$

is a differential equation. A function $y = f(x)$ is a **solution** of a differential equation if the equation is satisfied when y and its derivatives are replaced by $f(x)$ and its derivatives. For example,

$$y = e^{-2x} \qquad \text{Solution of differential equation}$$

is a solution of the differential equation shown above. To see this, substitute for y and $y' = -2e^{-2x}$ in the original equation.

$$\begin{aligned} y' + 2y &= -2e^{-2x} + 2(e^{-2x}) \qquad \text{Substitute for } y \text{ and } y'. \\ &= 0 \end{aligned}$$

In the same way, you can show that $y = 2e^{-2x}$, $y = -3e^{-2x}$, and $y = \frac{1}{2}e^{-2x}$ are also solutions of the differential equation. In fact, each function given by

$$y = Ce^{-2x} \qquad \text{General solution}$$

where C is a real number, is a solution of the equation. This family of solutions is called the **general solution** of the differential equation.

Example 1 Verifying Solutions

Determine whether each function is a solution of the differential equation

$$y'' - y = 0.$$

a. $y = Ce^{x}$ **b.** $y = Ce^{-x}$

SOLUTION

a. Because $y' = Ce^{x}$ and $y'' = Ce^{x}$, it follows that

$$\begin{aligned} y'' - y &= Ce^{x} - Ce^{x} \\ &= 0. \end{aligned}$$

So, $y = Ce^{x}$ is a solution.

b. Because $y' = -Ce^{-x}$ and $y'' = Ce^{-x}$, it follows that

$$\begin{aligned} y'' - y &= Ce^{-x} - Ce^{-x} \\ &= 0. \end{aligned}$$

So, $y = Ce^{-x}$ is also a solution.

✓ Checkpoint 1

Determine whether

$$y = Ce^{4x}$$

is a solution of the differential equation $y' = y$. ■

EDHAR/Shutterstock.com

Particular Solutions and Initial Conditions

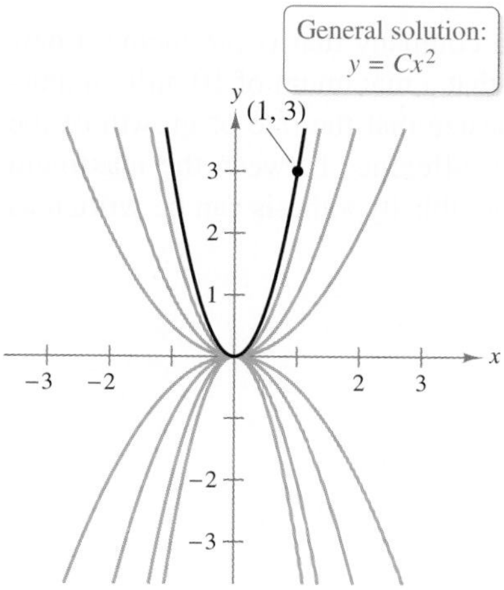

Solution Curves for $xy' - 2y = 0$
FIGURE 11.1

A **particular solution** of a differential equation is any solution that is obtained by assigning specific values to the arbitrary constant(s) in the general solution.*

Geometrically, the general solution of a differential equation represents a family of curves known as **solution curves.** For instance, the general solution of the differential equation $xy' - 2y = 0$ is

$$y = Cx^2. \qquad \text{General solution}$$

Figure 11.1 shows several solution curves corresponding to different values of C.

Particular solutions of a differential equation are obtained from **initial conditions** placed on the unknown function and its derivatives. For instance, in Figure 11.1, suppose you want to find the particular solution whose graph passes through the point $(1, 3)$. This initial condition can be written as

$$y = 3 \quad \text{when} \quad x = 1. \qquad \text{Initial condition}$$

Substituting these values into the general solution produces $3 = C(1)^2$, which implies that $C = 3$. So, the particular solution is

$$y = 3x^2. \qquad \text{Particular solution}$$

Example 2 Finding a Particular Solution

For the differential equation

$$xy' - 3y = 0$$

verify that $y = Cx^3$ is a solution. Then find the particular solution determined by the initial condition $y = 2$ when $x = -3$.

SOLUTION You know that $y = Cx^3$ is a solution because $y' = 3Cx^2$ and

$$\begin{aligned} xy' - 3y &= x(3Cx^2) - 3(Cx^3) \\ &= 3Cx^3 - 3Cx^3 \\ &= 0. \end{aligned}$$

Furthermore, the initial condition $y = 2$ when $x = -3$ yields

$$\begin{aligned} y &= Cx^3 && \text{General solution} \\ 2 &= C(-3)^3 && \text{Substitute initial condition.} \\ -\frac{2}{27} &= C && \text{Solve for } C. \end{aligned}$$

and you can conclude that the particular solution is

$$y = -\frac{2x^3}{27}. \qquad \text{Particular solution}$$

Try checking this solution by substituting for y and y' in the original differential equation.

STUDY TIP

To determine a particular solution, the number of initial conditions must match the number of constants in the general solution.

✓ Checkpoint 2

For the differential equation $xy' - 2y = 0$, verify that $y = Cx^2$ is a solution. Then find the particular solution determined by the initial condition $y = 1$ when $x = 4$.

*Some differential equations have solutions other than those given by their general solutions. These are called **singular solutions.** In this brief discussion of differential equations, singular solutions will not be discussed.

Example 3 Finding a Particular Solution

You are working in the marketing department of a company that is producing a new cereal product to be sold nationally. You determine that a maximum of 10 million units of the product could be sold in a year. You hypothesize that the rate of growth of the sales x (in millions of units) is proportional to the difference between the maximum sales and the current sales. As a differential equation, this hypothesis can be written as

$$\frac{dx}{dt} = k(10 - x), \quad 0 \le x \le 10.$$

Rate of change of x — is proportional to — the difference between 10 and x.

The general solution of this differential equation is

$$x = 10 - Ce^{-kt} \qquad \text{General solution}$$

where t is the time in years. After 1 year, 250,000 units have been sold. Sketch the graph of the sales function over a 10-year period.

SOLUTION Because the product is new, you can assume that $x = 0$ when $t = 0$. So, you have *two* initial conditions.

$$x = 0 \quad \text{when} \quad t = 0 \qquad \text{First initial condition}$$

$$x = 0.25 \quad \text{when} \quad t = 1 \qquad \text{Second initial condition}$$

Substituting the first initial condition into the general solution produces

$$0 = 10 - Ce^{-k(0)}$$

which implies that

$$C = 10.$$

Substituting the second initial condition into the general solution produces

$$0.25 = 10 - 10e^{-k(1)}$$

$$-9.75 = -10e^{-k}$$

$$0.975 = e^{-k}$$

$$-\ln 0.975 = k$$

which implies that

$$k = -\ln 0.975 \approx 0.0253.$$

So, the particular solution is

$$x = 10 - 10e^{-0.0253t}. \qquad \text{Particular solution}$$

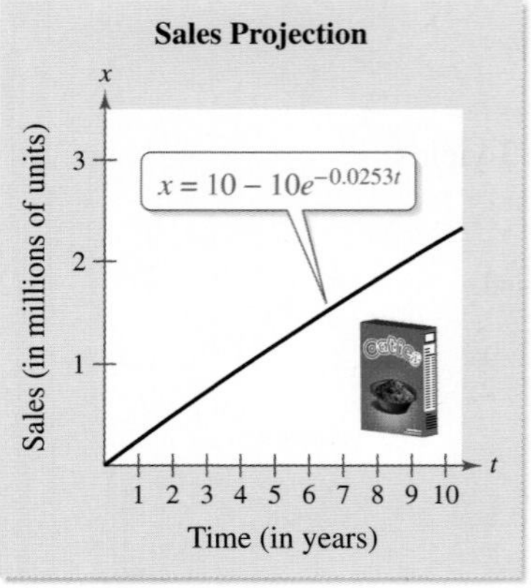

FIGURE 11.2

The table shows the annual sales during the first 10 years, and the graph of the solution is shown in Figure 11.2.

t	1	2	3	4	5	6	7	8	9	10
x	0.25	0.49	0.73	0.96	1.19	1.41	1.62	1.83	2.04	2.24

✓ Checkpoint 3

Repeat Example 3 using the initial conditions $x = 0$ when $t = 0$ and $x = 0.3$ when $t = 1$.

In the first three examples in this section, each solution was given in explicit form, such as $y = f(x)$. Sometimes you will encounter solutions for which it is more convenient to write the solution in implicit form, as shown in Example 4.

Example 4 Sketching Graphs of Solutions

Given that

$$2y^2 - x^2 = C \qquad \text{General solution}$$

is the general solution of the differential equation

$$2yy' - x = 0$$

sketch the particular solutions represented by $C = 0, \pm 1$, and ± 4.

SOLUTION The particular solutions represented by $C = 0, \pm 1$, and ± 4 are shown in Figure 11.3.

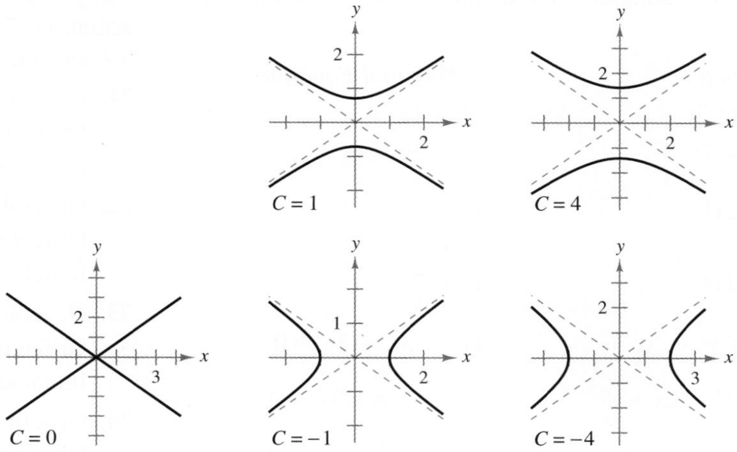

Graphs of Five Particular Solutions
FIGURE 11.3

✓ Checkpoint 4

Given that

$$y = Cx^2$$

is the general solution of

$$xy' - 2y = 0$$

sketch the particular solutions represented by $C = 1$, $C = 2$, and $C = 4$. ■

SUMMARIZE (Section 11.1)

1. Explain how to verify a solution of a differential equation *(page 662)*. For an example of verifying a solution, see Example 1.
2. Describe the difference between a general solution of a differential equation and a particular solution *(pages 662 and 663)*. For an example of a general solution of a differential equation and a particular solution, see Example 2.
3. Describe a real-life example of how a differential equation can be used to model the sales of a company's product *(page 664, Example 3)*.

SKILLS WARM UP 11.1 The following warm-up exercises involve skills that were covered in earlier sections. You will use these skills in the exercise set for this section. For additional help, review Sections 2.2, 2.6, 4.3, and 4.4.

In Exercises 1–4, find the first and second derivatives of the function.

1. $y = 3x^2 + 2x + 1$ **2.** $y = -2x^3 - 8x + 4$ **3.** $y = -3e^{2x}$ **4.** $y = -3e^{x^2}$

In Exercises 5 and 6, solve for k.

5. $0.5 = 9 - 9e^{-k}$ **6.** $14.75 = 25 - 25e^{-2k}$

Exercises 11.1

See www.CalcChat.com for worked-out solutions to odd-numbered exercises.

Verifying Solutions In Exercises 1–12, verify that the function is a solution of the differential equation. *See Example 1.*

	Solution	*Differential Equation*
1.	$y = Ce^{4x}$	$y' = 4y$
2.	$y = e^{-2x}$	$y' + 2y = 0$
3.	$y = 2x^3$	$y' - \frac{3}{x}y = 0$
4.	$y = 4x^2$	$y' - \frac{2}{x}y = 0$
5.	$y = Cx^2 - 3x$	$xy' - 3x - 2y = 0$
6.	$y = x^2 + 2x + \frac{C}{x}$	$xy' + y = x(3x + 4)$
7.	$y = x \ln x + Cx + 4$	$x(y' - 1) - (y - 4) = 0$
8.	$y = Ce^{x-x^2}$	$y' + (2x - 1)y = 0$
9.	$y = x^2$	$x^2y'' - 2y = 0$
10.	$y = e^{x^3}$	$y'' - 3x^2y' - 6xy = 0$
11.	$y = C_1 \sin x - C_2 \cos x$	$y'' + y = 0$
12.	$y = C_1e^{4x} + C_2e^{-x}$	$y'' - 3y' - 4y = 0$

Determining Solutions In Exercises 13–16, determine whether the function is a solution of the differential equation $y^{(4)} - 16y = 0$.

13. $y = e^{-2x}$ **14.** $y = 5 \ln x$

15. $y = \frac{4}{x}$ **16.** $y = 3 \sin 2x$

Determining Solutions In Exercises 17–20, determine whether the function is a solution of the differential equation $y''' - 3y' + 2y = 0$.

17. $y = \frac{2}{9}xe^{-2x}$

18. $y = \cos x$

19. $y = xe^x$

20. $y = x \ln x$

Finding a Particular Solution In Exercises 21–24, verify that the general solution satisfies the differential equation. Then find the particular solution that satisfies the initial condition. *See Example 2.*

21. General solution: $y = Ce^{-2x}$
Differential equation: $y' + 2y = 0$
Initial condition: $y = 3$ when $x = 0$

22. General solution: $2x^2 + 3y^2 = C$
Differential equation: $2x + 3yy' = 0$
Initial condition: $y = 2$ when $x = 1$

23. General solution: $y = C_1 + C_2 \ln x$
Differential equation: $xy'' + y' = 0$
Initial condition: $y = 5$ and $y' = 0.5$ when $x = 1$

24. General solution: $y = C_1e^{4x} + C_2e^{-3x}$
Differential equation: $y'' - y' - 12y = 0$
Initial condition: $y = 5$ and $y' = 6$ when $x = 0$

Sketching Graphs of Solutions In Exercises 25 and 26, the general solution of the differential equation is given. Sketch the particular solutions that correspond to the indicated values of C. *See Example 4.*

	General Solution	*Differential Equation*	*C-Values*
25.	$y = C(x + 2)^2$	$(x + 2)y' - 2y = 0$	$0, \pm1, \pm2$
26.	$y = Ce^{-x}$	$y' + y = 0$	$0, \pm1, \pm4$

Finding General Solutions In Exercises 27–34, use integration to find the general solution of the differential equation.

27. $\frac{dy}{dx} = 3x^2$ **28.** $\frac{dy}{dx} = 2x^3 - 3x$

29. $\frac{dy}{dx} = \frac{1}{1 + x}$ **30.** $\frac{dy}{dx} = \frac{x - 2}{x}$

31. $\frac{dy}{dx} = x\sqrt{x^2 + 6}$ **32.** $\frac{dy}{dx} = \frac{x}{1 + x^2}$

33. $\frac{dy}{dx} = \cos 4x$ **34.** $\frac{dy}{dx} = 4 \sin x$

Finding a Particular Solution In Exercises 35–38, some of the curves corresponding to different values of C in the general solution of the differential equation are shown in the figure. Find the particular solution that passes through the point plotted on the graph.

35. $y^2 = Cx^3$

$2xy' - 3y = 0$

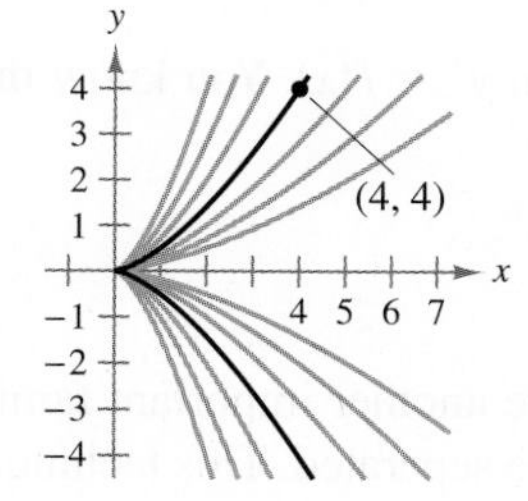

36. $2x^2 - y^2 = C$

$yy' - 2x = 0$

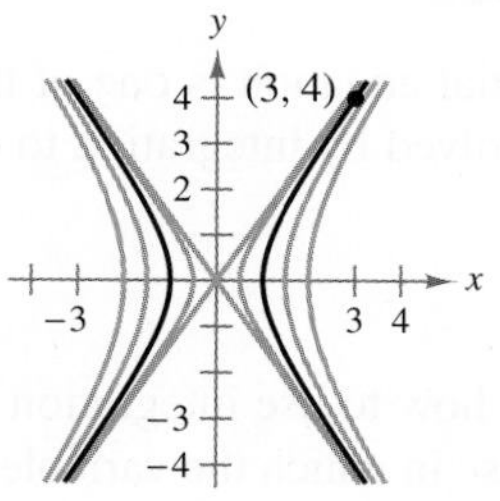

37. $y = Ce^x$

$y' - y = 0$

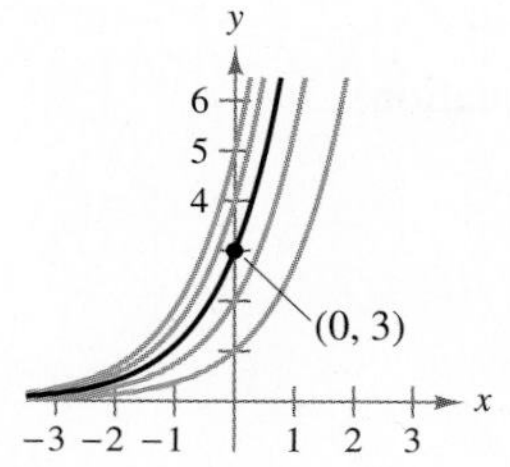

38. $y^2 = 2Cx$

$2xy' - y = 0$

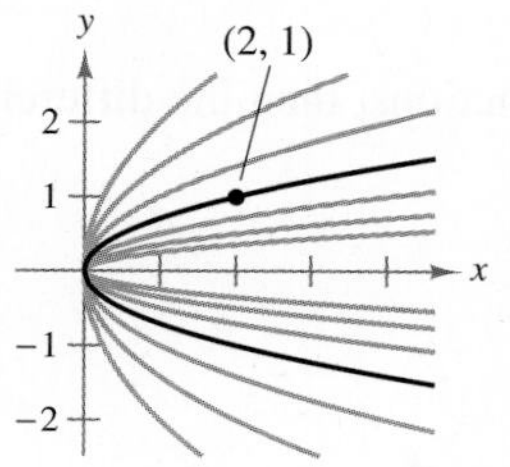

39. Biology The limiting capacity of the habitat of a wildlife herd is 750. The growth rate dN/dt of the herd is proportional to the unutilized opportunity for growth, as described by the differential equation

$$\frac{dN}{dt} = k(750 - N).$$

The general solution of this differential equation is

$$N = 750 - Ce^{-kt}.$$

When $t = 0$, the population of the herd is 100. After 2 years, the population has grown to 160.

(a) Write the population N as a function of t.

(b) What is the population of the herd after 4 years?

40. Investment The rate of growth of an investment is proportional to the amount in the investment at any time t. That is,

$$\frac{dA}{dt} = kA.$$

The general solution of this differential equation is

$$A = Ce^{kt}.$$

The initial investment is \$1000, and after 10 years the balance is \$3320.12. What is the particular solution?

41. Safety Assume that the rate of change per hour in the number of miles s of road cleared by a snowplow is inversely proportional to the depth h of the snow. This rate of change is described by the differential equation

$$\frac{ds}{dh} = \frac{k}{h}.$$

Show that

$$s = 25 - \frac{13}{\ln 3} \ln \frac{h}{2}$$

is a solution of this differential equation.

42. HOW DO YOU SEE IT? The graph shows a solution of one of the following differential equations. Determine the correct equation. Explain your reasoning.

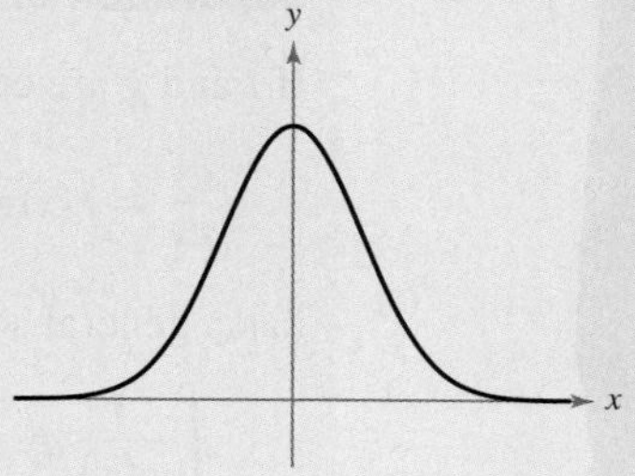

(i) $y' = xy$ (ii) $y' = \dfrac{4x}{y}$

(iii) $y' = -4xy$ (iv) $y' = 4 - xy$

43. Verifying a Solution Show that $y = a + Ce^{k(1-b)t}$ is a solution of the differential equation

$$y = a + b(y - a) + \left(\frac{1}{k}\right)\left(\frac{dy}{dt}\right)$$

where k is a constant.

44. Using a Solution The function $y = Ce^{kx}$ is a solution of the differential equation

$$\frac{dy}{dx} = 0.07y.$$

Is it possible to determine C or k from the information given? If so, find its value.

True or False? In Exercises 45 and 46, determine whether the statement is true or false. If it is false, explain why or give an example that shows it is false.

45. A differential equation can have more than one solution.

46. If $y = f(x)$ is a solution of a differential equation, then $y = f(x) + C$ is also a solution.

11.2 Separation of Variables

- Use separation of variables to solve differential equations.
- Use differential equations to model and solve real-life problems.

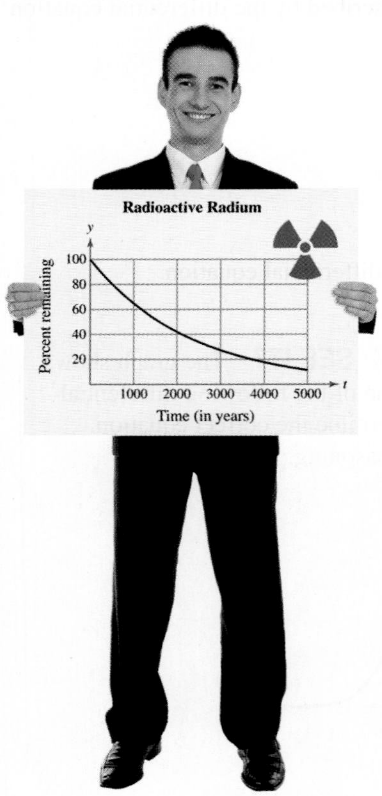

In Exercise 45 on page 673, you will use separation of variables to find the percent of a present amount of radioactive radium remaining after 25 years.

Separation of Variables

The simplest type of differential equation is one of the form $y' = f(x)$. You know that this type of equation can be solved by integration to obtain

$$y = \int f(x)\,dx.$$

In this section, you will learn how to use integration to solve another important family of differential equations—those in which the variables can be separated. This technique is called **separation of variables.**

Separation of Variables

If f and g are continuous functions, then the differential equation

$$\frac{dy}{dx} = f(x)g(y)$$

has a general solution of

$$\int \frac{1}{g(y)}\,dy = \int f(x)\,dx + C.$$

Essentially, the technique of separation of variables is just what its name implies. For a differential equation involving x and y, you separate the variables by grouping the x variables on one side and the y variables on the other. After separating variables, integrate each side to obtain the general solution.

TECH TUTOR

You can use a symbolic integration utility to solve a differential equation that has separable variables. Use a symbolic integration utility to solve the differential equation in Example 1.

Example 1 Solving a Differential Equation

Find the general solution of

$$\frac{dy}{dx} = \frac{x}{y^2 + 1}.$$

SOLUTION Begin by separating variables, then integrate each side.

$$\frac{dy}{dx} = \frac{x}{y^2 + 1} \qquad \text{Differential equation}$$

$$(y^2 + 1)\,dy = x\,dx \qquad \text{Separate variables.}$$

$$\int (y^2 + 1)\,dy = \int x\,dx \qquad \text{Integrate each side.}$$

$$\frac{y^3}{3} + y = \frac{x^2}{2} + C \qquad \text{General solution}$$

✓ Checkpoint 1

Find the general solution of $\dfrac{dy}{dx} = \dfrac{x^2}{y}$. ■

Edyta Pawlowska/Shutterstock.com

Example 2 Solving a Differential Equation

Find the general solution of

$$\frac{dy}{dx} = \frac{x}{y}.$$

SOLUTION Begin by separating variables, then integrate each side.

$$\frac{dy}{dx} = \frac{x}{y} \qquad \text{Differential equation}$$

$$y\,dy = x\,dx \qquad \text{Separate variables.}$$

$$\int y\,dy = \int x\,dx \qquad \text{Integrate each side.}$$

$$\frac{y^2}{2} = \frac{x^2}{2} + C_1 \qquad \text{Find antiderivative of each side.}$$

$$y^2 = x^2 + C \qquad \text{Multiply each side by 2.}$$

So, the general solution is $y^2 = x^2 + C$. Note that C_1 is used as a temporary constant of integration in anticipation of multiplying each side of the equation by 2 to produce the constant C.

✓Checkpoint 2

Find the general solution of

$$\frac{dy}{dx} = \frac{x+1}{y}.$$

Example 3 Solving a Differential Equation

Find the general solution of $e^y \dfrac{dy}{dx} = 2x$. Use a graphing utility to graph several solutions.

SOLUTION Begin by separating variables, then integrate each side.

$$e^y \frac{dy}{dx} = 2x \qquad \text{Differential equation}$$

$$e^y\,dy = 2x\,dx \qquad \text{Separate variables.}$$

$$\int e^y\,dy = \int 2x\,dx \qquad \text{Integrate each side.}$$

$$e^y = x^2 + C \qquad \text{Find antiderivative of each side.}$$

By taking the natural logarithm of each side, you can write the general solution as

$$y = \ln(x^2 + C). \qquad \text{General solution}$$

The graphs of the particular solutions given by $C = 0$, 5, 10, and 15 are shown in Figure 11.4.

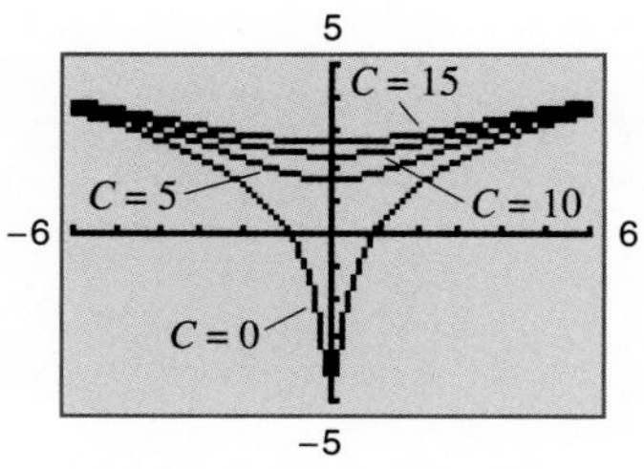

FIGURE 11.4

✓Checkpoint 3

Find the general solution of

$$2y\frac{dy}{dx} = -2x.$$

Use a graphing utility to graph the particular solutions given by $C = 1$, 2, and 4.

Example 4 Finding a Particular Solution

Solve the differential equation $xe^{x^2} + yy' = 0$ subject to the initial condition $y = 1$ when $x = 0$.

SOLUTION

$$xe^{x^2} + yy' = 0 \quad \text{Differential equation}$$

$$y\frac{dy}{dx} = -xe^{x^2} \quad \text{Subtract } xe^{x^2} \text{ from each side.}$$

$$y\,dy = -xe^{x^2}\,dx \quad \text{Separate variables.}$$

$$\int y\,dy = \int -xe^{x^2}\,dx \quad \text{Integrate each side.}$$

$$\frac{y^2}{2} = -\frac{1}{2}e^{x^2} + C_1 \quad \text{Find antiderivative of each side.}$$

$$y^2 = -e^{x^2} + C \quad \text{Multiply each side by 2.}$$

To find the particular solution, substitute the initial condition values to obtain

$$(1)^2 = -e^{(0)^2} + C.$$

This implies that $1 = -1 + C$, or $C = 2$. So, the particular solution that satisfies the initial condition is

$$y^2 = -e^{x^2} + 2. \quad \text{Particular solution}$$

Checkpoint 4

Solve the differential equation $e^x + yy' = 0$ subject to the initial condition $y = 2$ when $x = 0$. ■

Example 5 Solving a Differential Equation

Example 3 in Section 11.1 uses the differential equation

$$\frac{dx}{dt} = k(10 - x), \quad 0 \le x \le 10$$

to model the sales of a new product. Solve this differential equation.

SOLUTION

$$\frac{dx}{dt} = k(10 - x) \quad \text{Differential equation}$$

$$\frac{1}{10 - x}\,dx = k\,dt \quad \text{Separate variables.}$$

$$\int \frac{1}{10 - x}\,dx = \int k\,dt \quad \text{Integrate each side.}$$

$$-\ln(10 - x) = kt + C_1 \quad \text{Find antiderivatives.}$$

$$\ln(10 - x) = -kt - C_1 \quad \text{Multiply each side by } -1.$$

$$10 - x = e^{-kt - C_1} \quad \text{Exponentiate each side.}$$

$$x = 10 - Ce^{-kt} \quad \text{Solve for } x.$$

STUDY TIP

In Example 5, the context of the original model indicates that $(10 - x)$ is positive. So, when you integrate $1/(10 - x)$, you can write $-\ln(10 - x)$, rather than $-\ln|10 - x|$.

Also note in Example 5 that the solution agrees with the one that was given in Example 3 in Section 11.1.

Checkpoint 5

Solve the differential equation $\dfrac{dy}{dx} = k(65 - y)$ for $0 \le y \le 65$. ■

Application

Example 6 Corporate Investing

A corporation invests part of its receipts at a rate of P dollars per year in a fund for future corporate expansion. The fund earns r percent interest per year compounded continuously. The rate of growth of the amount A in the fund is

$$\frac{dA}{dt} = rA + P$$

where t is the time (in years). Solve the differential equation for A as a function of t, where $A = 0$ when $t = 0$.

SOLUTION You can solve the differential equation using separation of variables.

$$\frac{dA}{dt} = rA + P \qquad \text{Differential equation}$$

$$dA = (rA + P)\,dt \qquad \text{Differential form}$$

$$\frac{dA}{rA + P} = dt \qquad \text{Separate variables.}$$

$$\frac{1}{r}\ln|rA + P| = t + C_1 \qquad \text{Integrate.}$$

$$\ln(rA + P) = rt + C_2 \qquad \text{Assume } rA + P > 0 \text{ and multiply each side by } r.$$

$$rA + P = e^{rt + C_2} \qquad \text{Exponentiate each side.}$$

$$A = \frac{C_3e^{rt} - P}{r} \qquad \text{Solve for } A.$$

$$A = Ce^{rt} - \frac{P}{r} \qquad \text{General solution}$$

Using $A = 0$ when $t = 0$, you find the value of C.

$$0 = Ce^{r(0)} - \frac{P}{r} \quad \Longrightarrow \quad C = \frac{P}{r}$$

So, the differential equation for A as a function of t can be written as

$$A = \frac{P}{r}(e^{rt} - 1).$$

✓Checkpoint 6

Use the result of Example 6 to find A when $P = \$550{,}000$, $r = 5.9\%$, and $t = 25$ years. ■

SUMMARIZE (Section 11.2)

1. Explain how to use separation of variables to solve a differential equation *(page 668)*. For examples of solving a differential equation using separation of variables, see Examples 1, 2, 3, 4, and 5.
2. Describe a real-life example of how separation of variables can be used to solve a differential equation that models corporate investing *(page 671, Example 6)*.

Helder Almeida/Shutterstock.com

SKILLS WARM UP 11.2

The following warm-up exercises involve skills that were covered in earlier sections. You will use these skills in the exercise set for this section. For additional help, review Sections 4.4, 5.2, and 5.3.

In Exercises 1–6, find the indefinite integral and check your result by differentiating.

1. $\int x^{3/2}\,dx$ **2.** $\int (t^3 - t^{1/3})\,dt$ **3.** $\int \frac{2}{x-5}\,dx$

4. $\int \frac{y}{2y^2+1}\,dy$ **5.** $\int e^{2y}\,dy$ **6.** $\int xe^{1-x^2}\,dx$

In Exercises 7–10, solve the equation for *C* or *k*.

7. $(3)^2 - 6(3) = 1 + C$ **8.** $(-1)^2 + (-2)^2 = C$

9. $10 = 2e^{2k}$ **10.** $(6)^2 - 3(6) = e^{-k}$

Exercises 11.2

See www.CalcChat.com for worked-out solutions to odd-numbered exercises.

Separation of Variables In Exercises 1–6, decide whether the variables in the differential equation can be separated.

1. $\frac{dy}{dx} = \frac{x}{y+3}$ **2.** $\frac{dy}{dx} = \frac{x+1}{x}$

3. $\frac{dy}{dx} = \frac{1}{x} + 1$ **4.** $\frac{dy}{dx} = \frac{x}{x+y}$

5. $\frac{dy}{dx} = x - y$ **6.** $x\frac{dy}{dx} = \frac{1}{y}$

Solving a Differential Equation In Exercises 7–26, use separation of variables to find the general solution of the differential equation. *See Examples 1 and 2.*

7. $\frac{dy}{dx} = 2x$ **8.** $\frac{dy}{dx} = \frac{1}{x}$

9. $\frac{dr}{ds} = 0.05r$ **10.** $\frac{dr}{ds} = 0.05s$

11. $\frac{dy}{dx} = \frac{x-1}{y^3}$ **12.** $\frac{dy}{dx} = \frac{x^2-3}{y^2}$

13. $3y^2\frac{dy}{dx} = 1$ **14.** $\frac{dy}{dx} = x^2y$

15. $x^2 + 4y\frac{dy}{dx} = 0$ **16.** $(1+y)\frac{dy}{dx} - 4x = 0$

17. $y' - xy = 0$ **18.** $y' - y = 5$

19. $e^y\frac{dy}{dt} = 3t^2 + 1$ **20.** $e^x(y' + 1) = 1$

21. $\frac{dy}{dx} = \sqrt{1-y}$ **22.** $y' = \frac{x}{y} - \frac{x}{1+y}$

23. $(2+x)y' = 2y$ **24.** $y' - y(x+1) = 0$

25. $y\frac{dy}{dx} = \sin x$ **26.** $y\frac{dy}{dx} = 6\cos(\pi x)$

Solving a Differential Equation In Exercises 27–30, (a) find the general solution of the differential equation and (b) use a graphing utility to graph the particular solutions given by $C = 1$, 2, and 4. *See Example 3.*

27. $\frac{dy}{dx} = x$ **28.** $\frac{dy}{dx} = -\frac{2x}{y}$

29. $\frac{dy}{dx} = y + 3$ **30.** $\frac{dy}{dx} = 0.25x(4 - y)$

Finding a Particular Solution In Exercises 31–38, use the initial condition to find the particular solution of the differential equation. *See Example 4.*

	Differential Equation	*Initial Condition*
31.	$yy' - e^x = 0$	$y = 4$ when $x = 0$
32.	$\sqrt{x} + \sqrt{y}\,y' = 0$	$y = 4$ when $x = 1$
33.	$x(y+4) + y' = 0$	$y = -5$ when $x = 0$
34.	$\frac{dy}{dx} = x^2(1+y)$	$y = 3$ when $x = 0$
35.	$\sqrt{x^2 - 16}\,y' = 5x$	$y = -2$ when $x = 5$
36.	$y' = e^{x-2y}$	$y = 0$ when $x = 0$
37.	$\frac{dy}{dx} = y\cos x$	$y = 1$ when $x = 0$
38.	$\frac{dy}{dx} = 2xy\sin x^2$	$y = 1$ when $x = 0$

Finding an Equation In Exercises 39 and 40, find an equation of the graph that passes through the point and has the specified slope. Then graph the equation.

39. Point: $(-1, 1)$

Slope: $y' = -\frac{9x}{16y}$

40. Point: $(8, 2)$

Slope: $y' = \frac{2y}{3x}$

Velocity In Exercises 41 and 42, solve the differential equation to find velocity v as a function of time t if $v = 0$ when $t = 0$. The differential equation models the motion of two people on a toboggan after consideration of the forces of gravity, friction, and air resistance.

41. $12.5\dfrac{dv}{dt} = 43.2 - 1.25v$

42. $12.5\dfrac{dv}{dt} = 43.2 - 1.75v$

43. Biology: Cell Growth The growth rate of a spherical cell with volume V is proportional to its surface area S. For a sphere, the surface area and volume are related by $S = kV^{2/3}$. So, a model for the cell's growth is

$$\frac{dV}{dt} = kV^{2/3}.$$

Solve this differential equation.

44. HOW DO YOU SEE IT? The differential equation

$$\frac{dN}{dt} = k(30 - N)$$

represents the rate of change of the number of units produced per day by a new employee, where N is the number of units and t is the time (in days). The general solution of this differential equation is

$$N = 30 - 30e^{-kt}.$$

The graphs below show the particular solutions for $k = 0.3$, 0.6, and 1. Match the value of k with each graph. Explain your reasoning.

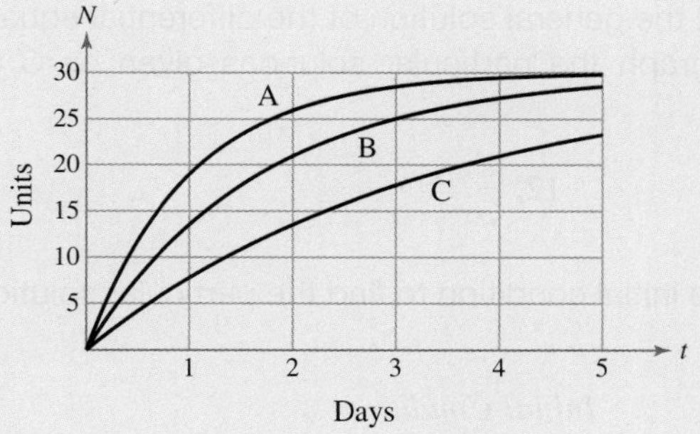

45. Radioactive Decay The rate of decomposition of radioactive radium is proportional to the amount present at any time. The half-life of radioactive radium is 1599 years. What percent of a present amount will remain after 25 years?

46. Radioactive Decay The rate of decomposition of radioactive einsteinium is proportional to the amount present at any time. The half-life of radioactive einsteinium is 276 days. After 100 days, 0.5 gram remains. What was the initial amount?

47. Weight Gain A calf that weighed 60 pounds at birth gains weight at the rate

$$\frac{dw}{dt} = k(1200 - w)$$

where w is the weight (in pounds) and t is the time (in years).

(a) Solve this differential equation.

(b) Use a graphing utility to graph the particular solutions for

$k = 0.8$, 0.9, and 1.

(c) The animal is sold when its weight reaches 800 pounds. Find the time of sale for each of the models in part (b).

(d) What is the maximum weight of the animal for each of the models in part (b)?

Business Capsule

After finding that the camera he wanted was sold out at a local store, Jack Abraham was inspired to start Milo.com. Named after his dog, the site shows buyers which nearby stores currently have a product in stock. This benefits not only shoppers but also retailers, as Milo drives foot traffic into their stores. In just one year, Milo.com grew to cover more than 140 retailers in 50,000 locations across the United States. In 2010, the company was bought by eBay, where Abraham now leads the local division.

48. Research Project Use your school's library, the Internet, or some other reference source to gather information about a company that offers innovative products or services. Collect data about the revenue that the company has generated and find a mathematical model of the data. Write a short paper that summarizes your findings.

Courtesy of Milo

QUIZ YOURSELF

See www.CalcChat.com for worked-out solutions to odd-numbered exercises.

Take this quiz as you would take a quiz in class. When you are done, check your work against the answers given in the back of the book.

In Exercises 1–4, verify that the function is a solution of the differential equation.

	Solution	*Differential Equation*
1.	$y = Ce^{-x/2}$	$2y' + y = 0$
2.	$y = C_1 \cos x + C_2 \sin x$	$y'' + y = 0$
3.	$y = \dfrac{1}{x}$	$xy'' + 2y' = 0$
4.	$y = \dfrac{x^3}{5} - x + C\sqrt{x}$	$2xy' - y = x^3 - x$

In Exercises 5 and 6, verify that the general solution satisfies the differential equation. Then find the particular solution that satisfies the initial condition.

5. General solution: $y = C_1 \sin 3x + C_2 \cos 3x$
Differential equation: $y'' + 9y = 0$
Initial condition: $y = 2$ and $y' = 1$ when $x = \pi/6$

6. General solution: $y = C_1 x + C_2 x^3$
Differential equation: $x^2y'' - 3xy' + 3y = 0$
Initial condition: $y = 0$ and $y' = 4$ when $x = 2$

In Exercises 7–10, use separation of variables to find the general solution of the differential equation.

7. $\dfrac{dy}{dx} = -4x + 4$

8. $y' = (x + 2)(y - 1)$

9. $y\dfrac{dy}{dx} = \dfrac{1}{2x + 1}$

10. $\dfrac{dy}{dx} = \dfrac{x}{3y^2 + 1}$

In Exercises 11 and 12, (a) find the general solution of the differential equation and (b) use a graphing utility to graph the particular solutions given by $C = 0$ and $C = \pm 1$.

11. $\dfrac{dy}{dx} = \dfrac{x^2 + 1}{2y}$

12. $\dfrac{dy}{dx} = \dfrac{y}{x - 3}$

In Exercises 13 and 14, use the initial condition to find the particular solution of the differential equation.

	Differential Equation	*Initial Condition*
13.	$y' + 2y - 1 = 0$	$y = 1$ when $x = 0$
14.	$\dfrac{dy}{dx} = y \sin \pi x$	$y = -3$ when $x = \dfrac{1}{2}$

15. Find an equation of the graph that passes through the point $(0, 2)$ and has a slope of $y' = 3x^2y$. Then graph the equation.

16. Ignoring resistance, a sailboat starting from rest accelerates at a rate proportional to the difference between the velocities of the wind and the boat. With a 20-knot wind, this acceleration is described by the differential equation $dv/dt = k(20 - v)$, where v is the velocity of the boat (in knots) and t is the time (in hours). After half an hour, the boat is moving at 10 knots. Write the velocity as a function of time.

11.3 First-Order Linear Differential Equations

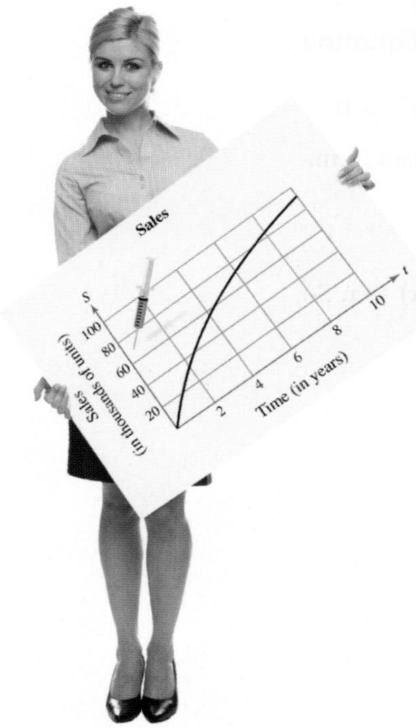

In Exercise 35 on page 678, you will use a first-order linear differential equation to find the sales of a biomedical syringe for 10 years.

- Solve first-order linear differential equations.
- Use first-order linear differential equations to model and solve real-life problems.

First-Order Linear Differential Equations

In this section, you will see how to solve a very important class of differential equations—first-order linear differential equations. The term "first-order" refers to the fact that the highest-order derivative of y in the equation is the first derivative.

Definition of a First-Order Linear Differential Equation

A **first-order linear differential equation** is an equation of the form

$$y' + P(x)y = Q(x)$$

where P and Q are functions of x. An equation that is written in this form is said to be in **standard form.**

To solve a linear differential equation, write it in standard form to identify the functions $P(x)$ and $Q(x)$. Then integrate $P(x)$ and form the expression

$$u(x) = e^{\int P(x)\,dx} \qquad \text{Integrating factor}$$

which is called an **integrating factor.** The general solution of the equation is

$$y = \frac{1}{u(x)}\int Q(x)u(x)\,dx. \qquad \text{General solution}$$

Example 1 Solving a Linear Differential Equation

Find the general solution of $y' + y = e^x$.

SOLUTION For this equation, $P(x) = 1$ and $Q(x) = e^x$. So, the integrating factor is

$$\begin{aligned} u(x) &= e^{\int P(x)dx} \\ &= e^{\int dx} \\ &= e^x. \qquad \text{Integrating factor} \end{aligned}$$

This implies that the general solution is

$$\begin{aligned} y &= \frac{1}{u(x)}\int Q(x)u(x)\,dx \\ &= \frac{1}{e^x}\int e^x(e^x)\,dx \\ &= e^{-x}\left(\frac{1}{2}e^{2x} + C\right) \\ &= \frac{1}{2}e^x + Ce^{-x}. \qquad \text{General solution} \end{aligned}$$

✓ Checkpoint 1

Find the general solution of

$$y' - y = 10.$$

■

In Example 1, the differential equation was given in standard form. For equations that are not written in standard form, you should first convert to standard form so that you can identify the functions $P(x)$ and $Q(x)$.

Example 2 Solving a Linear Differential Equation

Find the general solution of $xy' - 2y = x^2$. Assume $x > 0$.

SOLUTION Begin by writing the equation in standard form.

$$y' + \left(-\frac{2}{x}\right)y = x \qquad \text{Standard form, } y' + P(x)y = Q(x)$$

In this form, you can see that $P(x) = -2/x$ and $Q(x) = x$. So,

$$\begin{aligned}\int P(x)\,dx &= -\int \frac{2}{x}\,dx\\ &= -2\ln x\\ &= -\ln x^2\end{aligned}$$

which implies that the integrating factor is

$$\begin{aligned}u(x) &= e^{\int P(x)\,dx}\\ &= e^{-\ln x^2}\\ &= \frac{1}{e^{\ln x^2}}\\ &= \frac{1}{x^2}. \qquad \text{Integrating factor}\end{aligned}$$

This implies that the general solution is

$$\begin{aligned}y &= \frac{1}{u(x)}\int Q(x)u(x)\,dx && \text{Form of general solution}\\ &= \frac{1}{1/x^2}\int x\left(\frac{1}{x^2}\right)dx && \text{Substitute.}\\ &= x^2\int \frac{1}{x}\,dx && \text{Simplify.}\\ &= x^2(\ln x + C). && \text{General solution}\end{aligned}$$

TECH TUTOR

From Example 2, you can see that it can be difficult to solve a linear differential equation. Fortunately, the task is greatly simplified by symbolic integration utilities. Use a symbolic integration utility to find the particular solution of the differential equation in Example 2, given the initial condition $y = 1$ when $x = 1$.

✓ Checkpoint 2

Find the general solution of $xy' - y = x$. Assume $x > 0$. ■

Guidelines for Solving a Linear Differential Equation

1. Write the equation in standard form.

   $$y' + P(x)y = Q(x)$$

2. Find the integrating factor.

   $$u(x) = e^{\int P(x)\,dx}$$

3. Evaluate the integral below to find the general solution.

   $$y = \frac{1}{u(x)}\int Q(x)u(x)\,dx$$

Application

Example 3 Intravenous Feeding

Glucose is added intravenously to the bloodstream at the rate of q units per minute, and the body removes glucose from the bloodstream at a rate proportional to the amount present. Assume that A is the amount of glucose in the bloodstream at time t and that the rate of change of the amount of glucose is

$$\frac{dA}{dt} = q - kA$$

where k is a constant. Find the general solution of the differential equation.

SOLUTION In standard form, this linear differential equation is

$$\frac{dA}{dt} + kA = q \qquad \text{Standard form}$$

which implies that $P(t) = k$ and $Q(t) = q$. So, the integrating factor is

$$\begin{aligned} u(t) &= e^{\int P(t)\,dt} \\ &= e^{\int k\,dt} \\ &= e^{kt} \qquad \text{Integrating factor} \end{aligned}$$

and the general solution is

$$\begin{aligned} A &= \frac{1}{u(t)} \int Q(t)u(t)\,dt \\ &= e^{-kt} \int qe^{kt}dt \\ &= e^{-kt}\left(\frac{q}{k}e^{kt} + C\right) \\ &= \frac{q}{k} + Ce^{-kt}. \qquad \text{General solution} \end{aligned}$$

✓ Checkpoint 3

Use the general solution

$$A = \frac{q}{k} + Ce^{-kt}$$

from Example 3 to find the particular solution determined by the initial condition $A = 0$ when $t = 0$. (Assume $k = 0.05$ and $q = 0.05$.) ■

SUMMARIZE (Section 11.3)

1. State the definition of a first-order linear differential equation *(page 675)*. For examples of solving a first-order linear differential equation, see Examples 1 and 2.
2. State the guidelines for solving a first-order linear differential equation *(page 676)*. For examples of solving a first-order linear differential equation, see Examples 1 and 2.
3. Describe a real-life example of how a first-order linear differential equation can be used to analyze intravenous feeding *(page 677, Example 3)*.

Sean Nel/Shutterstock.com

SKILLS WARM UP 11.3

The following warm-up exercises involve skills that were covered in earlier sections. You will use these skills in the exercise set for this section. For additional help, review Sections 4.2, 4.4, and 5.1–5.3.

In Exercises 1–4, simplify the expression.

1. $e^{-x}(e^{2x} + e^x)$ **2.** $\frac{1}{e^{-x}}(e^{-x} + e^{2x})$ **3.** $e^{-\ln x^3}$ **4.** $e^{2\ln x + x}$

In Exercises 5–10, find the indefinite integral.

5. $\int 4e^{2x}\,dx$ **6.** $\int xe^{3x^2}\,dx$ **7.** $\int \frac{1}{2x+5}\,dx$

8. $\int \frac{x+1}{x^2+2x+3}\,dx$ **9.** $\int (4x-3)^2\,dx$ **10.** $\int x(1-x^2)^2\,dx$

Exercises 11.3

See www.CalcChat.com for worked-out solutions to odd-numbered exercises.

Writing in Standard Form In Exercises 1–6, write the first-order linear differential equation in standard form.

1. $x^3 - 2x^2y' + 3y = 0$ **2.** $y' - 5(2x - y) = 0$

3. $xy' + y = xe^x$ **4.** $xy' + y = x^3y$

5. $y + 1 = (x - 1)y'$ **6.** $x = x^2(y' + y)$

Solving a Linear Differential Equation In Exercises 7–18, find the general solution of the first-order linear differential equation. *See Examples 1 and 2.*

7. $\frac{dy}{dx} + 3y = 6$ **8.** $\frac{dy}{dx} - 5y = 15$

9. $\frac{dy}{dx} - y = e^{4x}$ **10.** $\frac{dy}{dx} + 3y = e^{-3x}$

11. $\frac{dy}{dx} = \frac{x^2 + 3}{x}$

12. $\frac{dy}{dx} = \frac{e^{-2x}}{1 + e^{-2x}}$

13. $y' + 2xy = 10x$

14. $y' + 5y = e^{5x}$

15. $(x - 1)y' + y = x^2 - 1$

16. $xy' + y = x^2 + 1$

17. $x^3y' + 2y = e^{1/x^2}$

18. $xy' + y = x^2 \ln x$

Using Two Methods In Exercises 19–22, solve for *y* in two ways.

19. $y' + y = 4$

20. $y' - 3y = -2$

21. $y' - 2xy = 2x$

22. $y' + 4xy = x$

Matching In Exercises 23–26, match the differential equation with its solution without solving the differential equation. Explain your reasoning.

Differential Equation	*Solution*
23. $y' - 2x = 0$	(a) $y = Ce^{x^2}$
24. $y' - 2y = 0$	(b) $y = -\frac{1}{2} + Ce^{x^2}$
25. $y' - 2xy = 0$	(c) $y = x^2 + C$
26. $y' - 2xy = x$	(d) $y = Ce^{2x}$

Finding a Particular Solution In Exercises 27–34, find the particular solution that satisfies the initial condition.

Differential Equation	*Initial Condition*
27. $y' + y = 6e^x$	$y = 3$ when $x = 0$
28. $y' + 2y = e^{-2x}$	$y = 4$ when $x = 1$
29. $xy' + y = 0$	$y = 2$ when $x = 2$
30. $y' + y = x$	$y = 4$ when $x = 0$
31. $y' + 3x^2y = 3x^2$	$y = 6$ when $x = 0$
32. $y' + (2x - 1)y = 0$	$y = 2$ when $x = 1$
33. $xy' + 2y = 3x^2 - 5x$	$y = 3$ when $x = -1$
34. $2xy' - y = x^3 - x$	$y = 2$ when $x = 4$

35. Sales The rate of change (in thousands of units) in sales *S* of a biomedical syringe is modeled by

$$\frac{dS}{dt} = 0.2(100 - S) + 0.2t$$

where *t* is the time (in years). Solve this differential equation and use the result to complete the table.

t	0	1	2	3	4	5	6	7	8	9	10
S	0										

36. HOW DO YOU SEE IT? The rate of change in the spread of a rumor in a school is modeled by

$$\frac{dP}{dt} = k(1 - P)$$

where P is the percent (in decimal form) of the students who have heard the rumor and t is the time (in hours), with $t = 0$ corresponding to 8:00 A.M. The graph shows the particular solution for this differential equation.

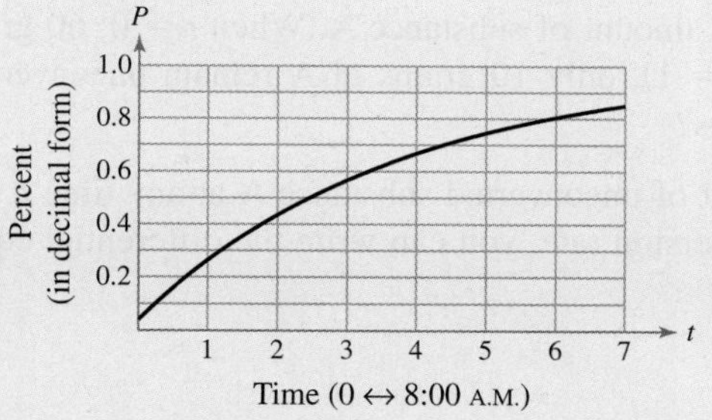

(a) What percent of the students have heard the rumor by 8:00 A.M.?

(b) At what time have 50% of the students heard the rumor?

(c) What percent of the students have heard the rumor by 3:00 P.M.?

37. Vertical Motion A falling object encounters air resistance that is proportional to its velocity v. The acceleration due to gravity is -9.8 meters per second per second. The rate of change in velocity is

$$\frac{dv}{dt} = kv - 9.8.$$

Solve this differential equation to find v as a function of time t.

38. Velocity A booster rocket carrying an observation satellite is launched into space. The rocket and satellite have mass m and are subject to air resistance proportional to the velocity v at any time t. A differential equation that models the velocity of the rocket and satellite is

$$m\frac{dv}{dt} = -mg - kv$$

where g is the acceleration due to gravity. Solve the differential equation for v as a function of t.

39. Learning Curve The management at a medical supply factory has found that the maximum number of units an employee can produce in a day is 40. The rate of increase in the number of units N produced with respect to time t (in days) by a new employee is proportional to $40 - N$. This rate of change of performance with respect to time can be modeled by

$$\frac{dN}{dt} = k(40 - N).$$

(a) Solve this differential equation.

(b) Find the particular solution for a new employee who produced 10 units on the first day at the factory and 19 units on the twentieth day.

40. Investment Let A be the amount in a fund earning interest for t years at the annual rate of r (in decimal form), compounded continuously. If a continuous cash flow of P dollars per year is withdrawn from the fund, then the rate of decrease of A is given by the differential equation

$$\frac{dA}{dt} = rA - P$$

where $A = A_0$ when $t = 0$.

(a) Solve this equation for A as a function of t.

(b) Use the result from part (a) to find A when $A_0 = \$2{,}000{,}000$, $r = 0.07$, $P = \$250{,}000$, and $t = 5$ years.

41. Project: Weight Loss For a project analyzing a person's weight loss, visit this text's website at *www.cengagebrain.com.* *(Data Source: The College Mathematics Journal)*

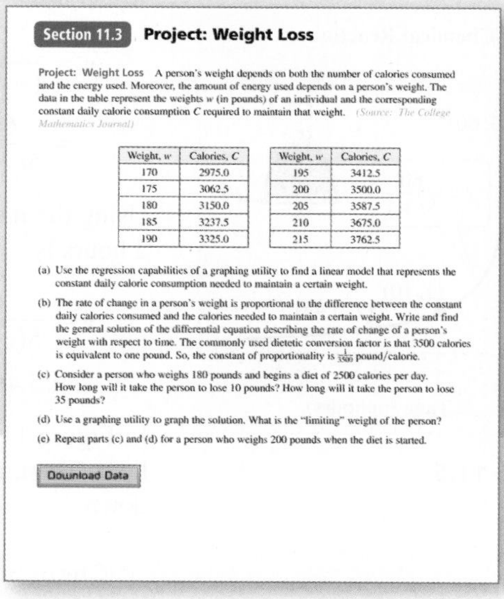

Section 11.3 Project: Weight Loss

Project: Weight Loss A person's weight depends on both the number of calories consumed and the energy used. Moreover, the amount of energy used depends on a person's weight. The data in the table represent the weights w (in pounds) of an individual and the corresponding constant daily calorie consumption C required to maintain that weight. *(Source: The College Mathematics Journal)*

Weight, w	Calories, C
170	2975.0
175	3062.5
180	3150.0
185	3237.5
190	3325.0

Weight, w	Calories, C
195	3412.5
200	3500.0
205	3587.5
210	3675.0
215	3762.5

(a) Use the regression capabilities of a graphing utility to find a linear model that represents the constant daily calorie consumption needed to maintain a certain weight.

(b) The rate of change in a person's weight is proportional to the difference between the constant daily calories consumed and the calories needed to maintain a certain weight. Write and find the general solution of the differential equation describing the rate of change of a person's weight with respect to time. The commonly used dietetic conversion factor is that 3500 calories is equivalent to one pound. So, the constant of proportionality is $\frac{1}{3500}$ pound/calorie.

(c) Consider a person who weighs 180 pounds and begins a diet of 2500 calories per day. How long will it take the person to lose 10 pounds? How long will it take the person to lose 35 pounds?

(d) Use a graphing utility to graph the solution. What is the "limiting" weight of the person?

(e) Repeat parts (c) and (d) for a person who weighs 200 pounds when the diet is started.

Download Data

11.4 Applications of Differential Equations

■ Use differential equations to model and solve real-life problems.

In Exercise 23 on page 686, you will use a differential equation and initial conditions to determine how many beavers there will be in a wetlands area after 10 years.

Applications of Differential Equations

Example 1 Modeling a Chemical Reaction

During a chemical reaction, substance A is converted into substance B at a rate that is proportional to the square of the amount of substance A. When $t = 0$, 60 grams of A are present, and after 1 hour ($t = 1$), only 10 grams of A remain unconverted. How much of A is present after 2 hours?

SOLUTION Let y be the amount of unconverted substance A at any time t. From the given assumption about the conversion rate, you can write the differential equation as shown.

$$\frac{dy}{dt} = ky^2$$

Rate of change of y — is proportional to — the square of y.

Using separation of variables *or* a symbolic integration utility, you can find the general solution to be

$$y = \frac{-1}{kt + C}.$$ General solution

To solve for the constants C and k, use the initial conditions. That is, because $y = 60$ when $t = 0$, you can determine that $C = -\frac{1}{60}$. Similarly, because $y = 10$ when $t = 1$, it follows that

$$10 = \frac{-1}{k - (1/60)}$$

which implies that $k = -\frac{1}{12}$. So, the particular solution is

$$y = \frac{-1}{(-1/12)t - (1/60)}$$ Substitute for k and C.

$$= \frac{60}{5t + 1}.$$ Particular solution

Using the model, you can determine that the amount of unconverted substance A after 2 hours is

$$y = \frac{60}{5(2) + 1}$$

$$\approx 5.45 \text{ grams.}$$

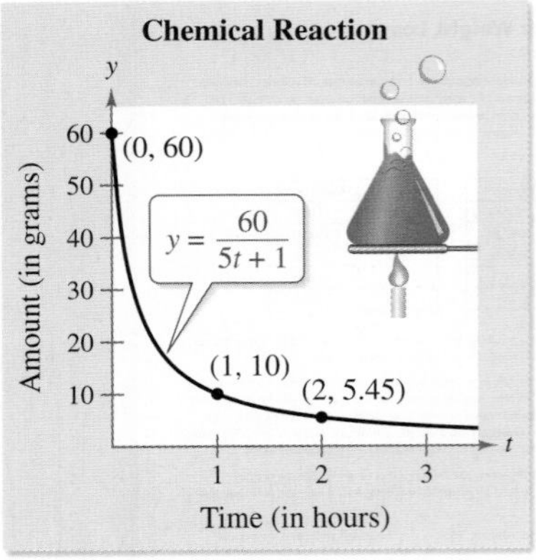

FIGURE 11.5

In Figure 11.5, note that the chemical conversion is occurring rapidly during the first hour. Then, as more and more of substance A is converted, the conversion rate slows down.

✓ Checkpoint 1

Use the chemical reaction model in Example 1 to find the amount y of substance A (in grams) as a function of t (in hours) given that $y = 40$ grams when $t = 0$ and $y = 5$ grams when $t = 2$.

Example 2 Modeling Advertising Awareness

The new cereal product from Example 3 in Section 11.1 is introduced through an advertising campaign to a population of 1 million potential customers. The rate at which the population hears about the product is assumed to be proportional to the number of people who are not yet aware of the product. By the end of 1 year, half of the population has heard of the product. How many will have heard of it by the end of 2 years?

SOLUTION Let y be the number (in millions) of people at time t who have heard of the product. This means that $(1 - y)$ is the number (in millions) of people who have not heard of it, and dy/dt is the rate at which the population hears about the product. From the given assumption, you can write the differential equation as shown.

$$\frac{dy}{dt} = k(1 - y)$$

Rate of change of y — is proportional to — the difference between 1 and y.

Using separation of variables *or* a symbolic integration utility, you can find the general solution to be

$$y = 1 - Ce^{-kt}. \qquad \text{General solution}$$

To solve for the constants C and k, use the initial conditions. That is, because $y = 0$ when $t = 0$, you can determine that $C = 1$. Similarly, because $y = 0.5$ when $t = 1$, it follows that $0.5 = 1 - e^{-k}$, which implies that

$$k = -\ln 0.5 \approx 0.693.$$

So, the particular solution is

$$y = 1 - e^{-0.693t}. \qquad \text{Particular solution}$$

This model is shown graphically in Figure 11.6. Using the model, you can determine that the number of people who have heard of the product after 2 years is

$$\begin{aligned} y &= 1 - e^{-0.693(2)} \\ &\approx 0.75 \text{ or } 750{,}000 \text{ people.} \end{aligned}$$

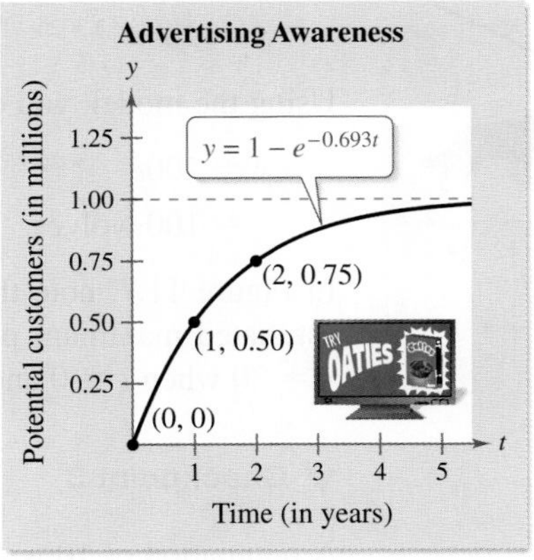

FIGURE 11.6

Checkpoint 2

Repeat Example 2 given that by the end of 1 year, only one-fourth of the population have heard of the product. ■

Earlier in the text, you studied two models for population growth: *exponential growth*, which assumes that the rate of change of y is proportional to y, and *logistic growth,* which assumes that the rate of change of y is proportional to y and $1 - y/L$, where L is the population limit.

The next example describes a third type of growth model called a **Gompertz growth model.** This model assumes that the rate of change of y is proportional to y and the natural log of L/y, where L is the population limit.

ALGEBRA TUTOR

For help with the algebra in solving for C in Example 3, see Example 1 in the *Chapter 11 Algebra Tutor,* on page 688. For help with the algebra in solving for k in Example 3, see Example 3 in the *Chapter 11 Algebra Tutor,* on page 689.

Example 3 Modeling Population Growth

A population of 20 wolves has been introduced into a national park. The forest service estimates that the maximum population the park can sustain is 200 wolves. After 3 years, the population is estimated to be 40 wolves. If the population follows a Gompertz growth model, how many wolves will there be 10 years after their introduction?

SOLUTION Let y be the number of wolves at any time t. From the given assumption about the rate of growth of the population, you can write the differential equation as shown.

$$\frac{dy}{dt} = ky \ln \frac{200}{y}$$

Rate of change of y | is proportional to | the product of y and | the log of the ratio of 200 and y.

Using separation of variables *or* a symbolic integration utility, you can find the general solution to be

$$y = 200e^{-Ce^{-kt}}. \qquad \text{General solution}$$

To solve for the constants C and k, use the initial conditions. That is, because $y = 20$ when $t = 0$, you can determine that

$$C = \ln 10$$
$$\approx 2.3026.$$

Similarly, because $y = 40$ when $t = 3$, it follows that

$$40 = 200e^{-2.3026e^{-k(3)}}$$

which implies that $k \approx 0.1194$. So, the particular solution is

$$y = 200e^{-2.3026e^{-0.1194t}}. \qquad \text{Particular solution}$$

Using the model, you can estimate the wolf population after 10 years to be

$$y = 200e^{-2.3026e^{-0.1194(10)}}$$
$$\approx 100 \text{ wolves.}$$

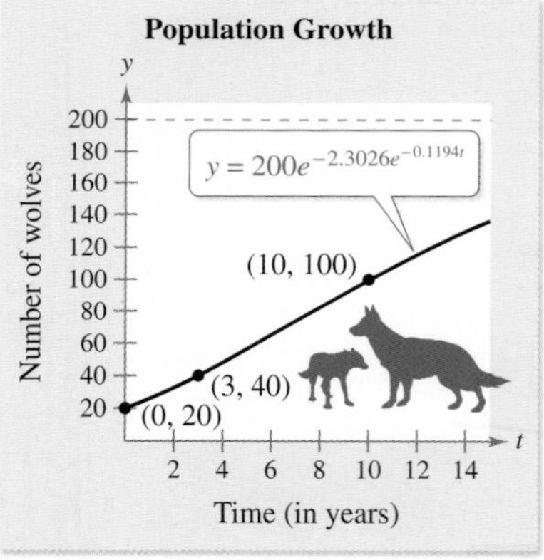

FIGURE 11.7

In Figure 11.7, note that after 10 years the population has reached about half of the estimated maximum population. Try checking the growth model to see that it yields $y = 20$ when $t = 0$ and $y = 40$ when $t = 3$.

✓Checkpoint 3

A population of 10 wolves has been introduced into a national park. The forest service estimates that the maximum population the park can sustain is 150 wolves. After 3 years, the population is estimated to be 25 wolves. If the population follows a Gompertz growth model, how many wolves will there be 10 years after their introduction? ■

In genetics, a commonly used hybrid selection model is based on the differential equation

$$\frac{dy}{dt} = ky(1 - y)(a - by).$$

In this model, y represents the portion of the population that has a certain characteristic and t represents the time (measured in generations). The numbers a, b, and k are constants that depend on the genetic characteristic that is being studied.

ALGEBRA TUTOR

For help with the algebra in solving for C in Example 4, see Example 2 in the *Chapter 11 Algebra Tutor,* on page 688. For help with the algebra in solving for k in Example 4, see Example 4 in the *Chapter 11 Algebra Tutor,* on page 689.

Example 4 Modeling Hybrid Selection

You are studying a population of beetles to determine how quickly characteristic D will pass from one generation to the next. At the beginning of your study ($t = 0$), you find that half the population has characteristic D. After four generations ($t = 4$), you find that 80% of the population has characteristic D. Use the hybrid selection model above with $a = 2$ and $b = 1$ to find the percent of the population that will have characteristic D after 10 generations.

SOLUTION Using $a = 2$ and $b = 1$, the differential equation for the hybrid selection model is

$$\frac{dy}{dt} = ky(1 - y)(2 - y).$$

Using separation of variables *or* a symbolic integration utility, you can find the general solution to be

$$\frac{y(2 - y)}{(1 - y)^2} = Ce^{2kt}. \qquad \text{General solution}$$

To solve for the constants C and k, use the initial conditions. That is, because $y = 0.5$ when $t = 0$, you can determine that $C = 3$. Similarly, because $y = 0.8$ when $t = 4$, it follows that

$$\frac{0.8(1.2)}{(0.2)^2} = 3e^{8k}$$

which implies that

$$k = \frac{1}{8}\ln 8 \approx 0.2599.$$

So, the particular solution is

$$\frac{y(2 - y)}{(1 - y)^2} = 3e^{0.5199t}. \qquad \text{Particular solution}$$

Using the model, you can estimate the percent of the population that will have characteristic D after 10 generations to be given by

$$\frac{y(2 - y)}{(1 - y)^2} = 3e^{0.5199(10)}.$$

Using a symbolic algebra utility, you can solve this equation for y to obtain $y \approx 0.96$. The graph of the model is shown in Figure 11.8.

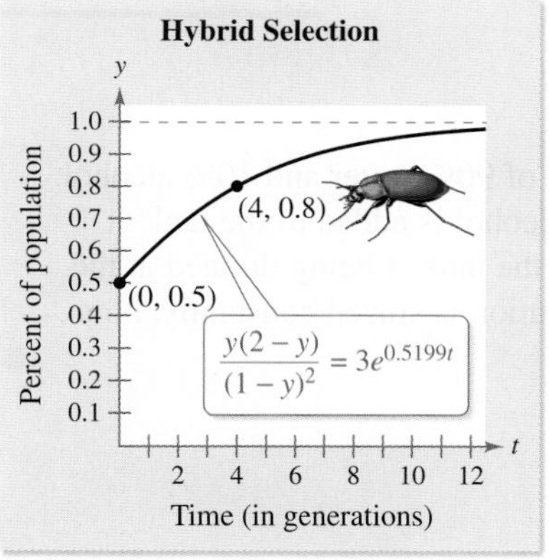

FIGURE 11.8

✓Checkpoint 4

Repeat Example 4 given that only 25% of the population has characteristic D when $t = 0$ and 50% of the population has characteristic D when $t = 4$.

Example 5 Modeling a Chemical Mixture

4 gal/min

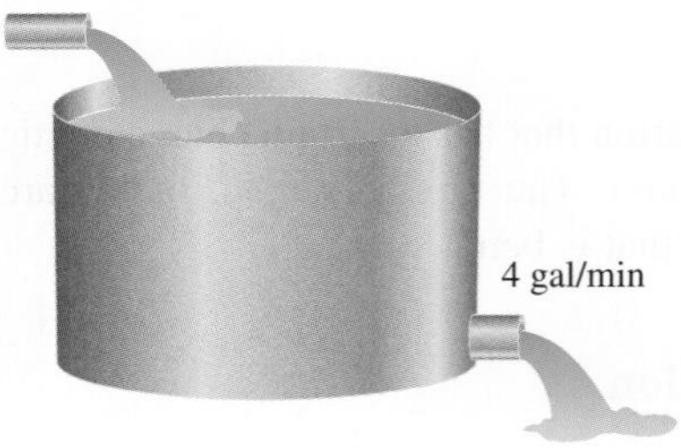

FIGURE 11.9

A tank contains 40 gallons of a solution composed of 90% water and 10% alcohol. A second solution containing half water and half alcohol is added to the tank at the rate of 4 gallons per minute. At the same time, the tank is being drained at the rate of 4 gallons per minute, as shown in Figure 11.9. Assuming that the solution is stirred constantly, will there be at least 14 gallons of alcohol in the tank after 10 minutes?

SOLUTION Let y be the number of gallons of alcohol in the tank at any time t. The *percent* of alcohol in the 40-gallon tank at any time is $y/40$. Moreover, because 4 gallons of solution are being drained each minute, the rate of change of y is

$$\frac{dy}{dt} = -4\left(\frac{y}{40}\right) + 2$$

Rate of change of y — is equal to the amount of alcohol draining out — plus the amount of alcohol entering.

where 2 represents the number of gallons of alcohol entering each minute in the 50% solution. In standard form, this linear differential equation is

$$y' + \frac{1}{10}y = 2. \qquad \text{Standard form}$$

Using an integrating factor *or* a symbolic integration utility, you can find the general solution to be

$$y = 20 + Ce^{-t/10}. \qquad \text{General solution}$$

Because $y = 4$ when $t = 0$, you can conclude that $C = -16$. So, the particular solution is

$$y = 20 - 16e^{-t/10}. \qquad \text{Particular solution}$$

Using this model, you can determine that the amount of alcohol in the tank when $t = 10$ is

$$y = 20 - 16e^{-(10)/10} \approx 14.1 \text{ gallons.}$$

Yes, there will be at least 14 gallons of alcohol in the tank after 10 minutes.

✓Checkpoint 5

A tank contains 50 gallons of a solution composed of 90% water and 10% alcohol. A second solution containing half water and half alcohol is added to the tank at the rate of 5 gallons per minute. At the same time, the tank is being drained at the rate of 5 gallons per minute. Assuming that the solution is stirred constantly, how much alcohol will be in the tank after 10 minutes? ■

SUMMARIZE (Section 11.4)

1. Describe a real-life example of how a differential equation can be used to model a chemical reaction *(page 680, Example 1)*.
2. Describe a real-life example of how a differential equation can be used to model population growth *(page 682, Example 3)*.
3. Describe a real-life example of how a differential equation can be used to model hybrid selection *(page 683, Example 4)*.

Michal Kowalski/Shutterstock.com

SKILLS WARM UP 11.4

The following warm-up exercises involve skills that were covered in earlier sections. You will use these skills in the exercise set for this section. For additional help, review Sections 4.6, 11.2, and 11.3.

In Exercises 1–4, use separation of variables to find the general solution of the differential equation.

1. $\dfrac{dy}{dx} = 3x$

2. $2y\dfrac{dy}{dx} = 3$

3. $\dfrac{dy}{dx} = 2xy$

4. $\dfrac{dy}{dx} = \dfrac{x-4}{4y^3}$

In Exercises 5–8, use an integrating factor to solve the first-order linear differential equation.

5. $y' + 2y = 4$

6. $y' + 2y = e^{-2x}$

7. $y' + xy = x$

8. $xy' + 2y = x^2$

In Exercises 9 and 10, write the equation that models the statement.

9. The rate of change of y with respect to x is proportional to the square of x.

10. The rate of change of x with respect to t is proportional to the difference of x and t.

Exercises 11.4

See www.CalcChat.com for worked-out solutions to odd-numbered exercises.

Chemical Reaction In Exercises 1 and 2, use the chemical reaction model described in Example 1 to find the amount y (in grams) as a function of time t (in hours). Then use a graphing utility to graph the function.

1. $y = 45$ grams when $t = 0$; $y = 4$ grams when $t = 2$

2. $y = 75$ grams when $t = 0$; $y = 12$ grams when $t = 1$

Advertising Awareness In Exercises 3 and 4, use the advertising awareness model described in Example 2 to find the number of people y (in millions) aware of the product as a function of time t (in years).

3. $y = 0$ when $t = 0$; $y = 0.75$ when $t = 1$

4. $y = 0$ when $t = 0$; $y = 0.9$ when $t = 2$

Population Growth In Exercises 5 and 6, use the Gompertz growth model described in Example 3 to find the population y as a function of time t (in years).

5. $y = 100$ when $t = 0$; $y = 150$ when $t = 2$

6. $y = 30$ when $t = 0$; $y = 60$ when $t = 4$

Hybrid Selection In Exercises 7 and 8, use the hybrid selection model described in Example 4 to find the percent y (in decimal form) of the population that has the indicated characteristics as a function of time t (in generations).

7. $y = 0.1$ when $t = 0$; $y = 0.4$ when $t = 4$

8. $y = 0.6$ when $t = 0$; $y = 0.75$ when $t = 2$

Finding a Particular Solution In Exercises 9–14, assume that the rate of change in y is proportional to y. Solve the resulting differential equation $dy/dx = ky$ and find the particular solution that passes through the points.

9. $(0, 1), (3, 2)$

10. $(0, 4), (1, 6)$

11. $(0, 4), (4, 1)$

12. $(0, 60), (5, 30)$

13. $(2, 2), (3, 4)$

14. $(1, 4), (2, 1)$

15. Chemical Reaction During a chemical reaction, a compound changes into another compound at a rate proportional to the unchanged amount y. Write the differential equation for the chemical reaction model. Find the particular solution when the initial amount of the original compound is 20 grams and the amount remaining after 1 hour is 16 grams.

16. Chemical Reaction Using the result of Exercise 15, when will 75% of the compound have been changed? When will 95% of the compound have been changed?

17. Population Growth The rate of change of the population of a city is proportional to the population P at any time t (in years). In 2000, the population was 200,000, and the constant of proportionality was 0.015. Estimate the population of the city in the year 2020.

18. Fruit Flies The rate of change of an experimental population of fruit flies is proportional to the population P at any time t (in days). There were 100 flies after the second day of the experiment and 300 flies after the fourth day. Approximately how many flies were in the original population?

19. Chemistry A wet towel hung from a clothesline to dry loses moisture through evaporation at a rate proportional to its moisture content. After 1 hour, the towel has lost 40% of its original moisture content. How long will it take the towel to lose 80% of its original moisture content?

20. Meteorology The barometric pressure y (in inches of mercury) at an altitude of x miles above sea level decreases at a rate proportional to the current pressure according to the model

$$\frac{dy}{dx} = -0.2y$$

where $y = 29.92$ inches when $x = 0$. Find the barometric pressure (a) at the top of Mt. St. Helens (8364 feet) and (b) at the top of Mt. McKinley (20,320 feet).

21. Sales Growth The rate of change in sales S (in thousands of units) of a new product is proportional to the difference between L and S at any time t (in years), where L is the maximum number of units of the new product available. When $t = 0$, $S = 0$. Write and solve the differential equation for this sales model.

22. Sales Growth Use the result of Exercise 21 to find the particular solutions when (a) $L = 100$ and $S = 25$ when $t = 2$, and (b) $L = 500$ and $S = 50$ when $t = 1$.

23. Biology A population of eight beavers has been introduced into a new wetlands area. Biologists estimate that the maximum population the wetlands can sustain is 60 beavers. After 3 years, the population is 15 beavers. The population follows a Gompertz growth model. How many beavers will there be in the wetlands after 10 years?

24. Biology A population of 30 rabbits has been introduced into a new region. It is estimated that the maximum population the region can sustain is 400 rabbits. After 1 year, the population is estimated to be 90 rabbits. The population follows a Gompertz growth model. How many rabbits will there be after 3 years?

25. Biology At any time t (in years), the rate of growth of the population N of deer in a state park is proportional to the product of N and $L - N$, where $L = 500$ is the maximum number of deer the park can maintain.

(a) Use a symbolic integration utility to find the general solution.

(b) Find the particular solution given the conditions $N = 100$ when $t = 0$ and $N = 200$ when $t = 4$.

(c) Find N when $t = 1$.

(d) Find t when $N = 350$.

26. Biology At any time t (in years), the rate of growth of the population N of fish in a pond is proportional to the product of N and $L - N$, where $L = 1000$ is the maximum number of fish the pond can maintain.

(a) Use a symbolic integration utility to find the general solution.

(b) Find the particular solution given the conditions $N = 200$ when $t = 0$ and $N = 500$ when $t = 2$.

(c) Find N when $t = 1$.

(d) Find t when $N = 700$.

27. Chemical Mixture A 100-gallon tank is full of a solution containing 25 pounds of a concentrate. Starting at time $t = 0$, distilled water is admitted to the tank at the rate of 5 gallons per minute, and the well-stirred solution is withdrawn at the same rate.

(a) Find the amount Q of the concentrate in the solution as a function of t by solving the differential equation

$$Q' = -5\left(\frac{Q}{100}\right).$$

(b) Find the time required for the amount of concentrate in the tank to reach 15 pounds.

28. Chemical Mixture A 200-gallon tank is half full of distilled water. At time $t = 0$, a solution containing 0.5 pound of concentrate per gallon enters the tank at the rate of 5 gallons per minute, and the well-stirred mixture is withdrawn at the same rate.

(a) Find the amount Q of the concentrate in the solution as a function of t by solving the differential equation

$$Q' = -5\left(\frac{Q}{100}\right) + \frac{5}{2}.$$

(b) Find the amount of concentrate in the tank after 30 minutes.

29. Population Growth When predicting population growth, demographers must consider birth and death rates as well as the net change caused by the difference between the rates of immigration and emigration. Let P be the population at time t and let N be the net increase per unit time due to the difference between immigration and emigration. So, the rate of growth of the population is given by

$$\frac{dP}{dt} = kP + N, \quad N \text{ is constant.}$$

Solve the differential equation to find P as a function of t.

30. HOW DO YOU SEE IT? In a learning theory project, the rate of change in the percent P (in decimal form) of correct responses after n trials can be modeled by

$$\frac{dP}{dn} = kP(1 - P).$$

The graph shows the particular solutions for two different groups.

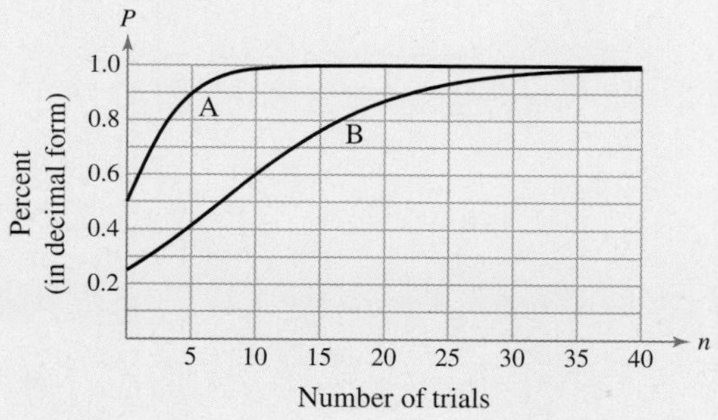

(a) What was the percent of correct responses before any trials for each group?

(b) What is the limit of P as t approaches infinity for each group?

(c) After how many trials are 75% of the responses correct for each group?

31. Investment A large corporation starts at time $t = 0$ to invest part of its profit at a rate of P dollars per year in a fund for future expansion. Assume that the fund earns r percent interest per year compounded continuously. The rate of growth of the amount A in the fund is given by

$$\frac{dA}{dt} = rA + P$$

where $A = 0$ when $t = 0$, and r is in decimal form. Solve this differential equation for A as a function of t.

Investment In Exercises 32–34, use the result of Exercise 31.

32. Find A for each situation.

(a) $P = \$100{,}000$, $r = 0.12$, and $t = 5$ years

(b) $P = \$250{,}000$, $r = 0.15$, and $t = 10$ years

33. Find P if the corporation needs \$120,000,000 in 8 years and the fund earns 8% interest compounded continuously.

34. Find t if the corporation needs \$800,000 and it can invest \$75,000 per year in a fund earning 13% interest compounded continuously.

Medical Science In Exercises 35–38, a medical researcher wants to determine the concentration C (in moles per liter) of a tracer drug injected into a moving fluid with flow R (in liters per minute). Solve this problem by considering a single-compartment dilution model (see figure). Assume that the fluid is continuously mixed and that the volume V (in liters) of fluid in the compartment is constant.

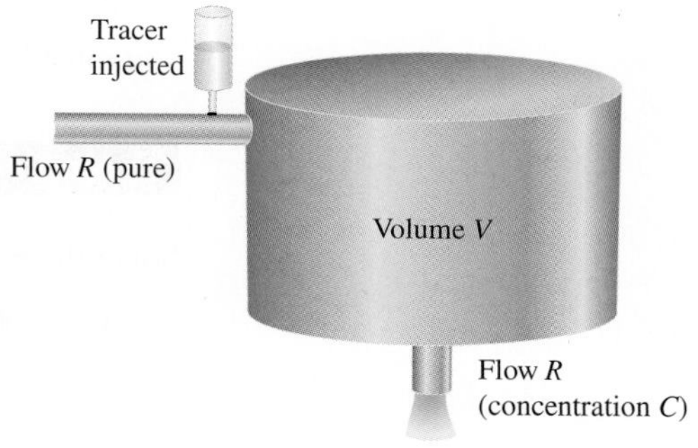

35. Mixture If the tracer is injected instantaneously at time $t = 0$, then the concentration of the fluid in the compartment begins diluting according to the differential equation

$$\frac{dC}{dt} = \left(-\frac{R}{V}\right)C, \quad C = C_0 \text{ when } t = 0.$$

(a) Solve this differential equation to find the concentration as a function of time.

(b) Find the limit of C as $t \to \infty$.

36. Mixture Use the solution of the differential equation in Exercise 35 to find the concentration as a function of time. Then use a graphing utility to graph the function.

(a) $V = 2$ liters, $R = 0.5$ L/min, and $C_0 = 0.6$ mol/L

(b) $V = 2$ liters, $R = 1.5$ L/min, and $C_0 = 0.6$ mol/L

37. Mixture In Exercises 35 and 36, it was assumed that there was a single initial injection of the tracer drug into the compartment. Now consider the case in which the tracer is continuously injected (beginning at $t = 0$) at a constant rate of Q mol/min. The concentration of the fluid in the compartment begins diluting according to the differential equation

$$\frac{dC}{dt} = \frac{Q}{V} - \left(\frac{R}{V}\right)C, \quad C = 0 \text{ when } t = 0.$$

(a) Solve this differential equation to find the concentration as a function of time.

(b) Find the limit of C as $t \to \infty$.

38. Mixture Use the solution of the differential equation in Exercise 37 to find the concentration as a function of time. Then use a graphing utility to graph the function.

(a) $Q = 2$ mol/min, $V = 2$ liters, and $R = 0.5$ L/min

(b) $Q = 1$ mol/min, $V = 2$ liters, and $R = 1.0$ L/min

ALGEBRA TUTOR

Solving Equations

To solve for the constants in the general solution of a differential equation, you will need to use algebraic skills. For instance, to solve for C in the differential equation

$$y = 200e^{-Ce^{-kt}}$$

you must know how to solve an exponential equation, as shown in Example 1.

Example 1 Solving for C

Solve the exponential equation

$$y = 200e^{-Ce^{-kt}}$$

for C when $y = 20$ and $t = 0$.

SOLUTION

$y = 200e^{-Ce^{-kt}}$	Example 3, page 682
$20 = 200e^{-Ce^{-k(0)}}$	Substitute 20 for y and 0 for t.
$20 = 200e^{-C}$	$e^{-k(0)} = e^0 = 1$
$\frac{1}{10} = e^{-C}$	Divide each side by 200.
$\frac{1}{10} = \frac{1}{e^C}$	Definition of negative exponent
$e^C = 10$	Cross-multiply.
$\ln e^C = \ln 10$	Take natural log of each side.
$C = \ln 10$	Apply the property $\ln e^a = a$.
$C \approx 2.3026$	Approximate.

Example 2 Solving for C

Solve the exponential equation

$$\frac{y(2 - y)}{(1 - y)^2} = Ce^{2kt}$$

for C when $y = 0.5$ and $t = 0$.

SOLUTION

$\frac{y(2 - y)}{(1 - y)^2} = Ce^{2kt}$	Example 4, page 683
$\frac{0.5(2 - 0.5)}{(1 - 0.5)^2} = Ce^{2k(0)}$	Substitute 0.5 for y and 0 for t.
$\frac{0.75}{(0.5)^2} = C$	Simplify.
$\frac{0.75}{0.25} = C$	Evaluate power.
$3 = C$	Divide.

Example 3 **Solving for *k***

Solve the exponential equation

$$y = 200e^{-2.3026e^{-kt}}$$

for k when $y = 40$ and $t = 3$.

SOLUTION

$y = 200e^{-2.3026e^{-kt}}$ — Example 3, page 682

$40 = 200e^{-2.3026e^{-k(3)}}$ — Substitute 40 for y and 3 for t.

$\frac{1}{5} = e^{-2.3026e^{-3k}}$ — Divide each side by 200.

$\ln \frac{1}{5} = -2.3026e^{-3k}$ — Take natural log of each side.

$\frac{\ln(1/5)}{-2.3026} = e^{-3k}$ — Divide each side by -2.3026.

$\ln\left[\frac{\ln(1/5)}{-2.3026}\right] = \ln e^{-3k}$ — Take natural log of each side.

$\ln\left[\frac{\ln(1/5)}{-2.3026}\right] = -3k$ — Apply the property $\ln e^a = a$.

$-\frac{1}{3}\ln\left[\frac{\ln(1/5)}{-2.3026}\right] = k$ — Multiply each side by $-\frac{1}{3}$.

$0.1194 \approx k$ — Approximate.

Example 4 **Solving for *k***

Solve the exponential equation

$$\frac{y(2 - y)}{(1 - y)^2} = 3e^{2kt}$$

for k when $y = 0.8$ and $t = 4$.

SOLUTION

$\frac{y(2 - y)}{(1 - y)^2} = 3e^{2kt}$ — Example 4, page 683

$\frac{0.8(2 - 0.8)}{(1 - 0.8)^2} = 3e^{2k(4)}$ — Substitute 0.8 for y and 4 for t.

$\frac{0.8(1.2)}{(0.2)^2} = 3e^{8k}$ — Simplify.

$24 = 3e^{8k}$ — Evaluate fraction.

$8 = e^{8k}$ — Divide each side by 3.

$\ln 8 = \ln e^{8k}$ — Take natural log of each side.

$\ln 8 = 8k$ — Apply the property $\ln e^a = a$.

$\frac{1}{8}\ln 8 = k$ — Multiply each side by $\frac{1}{8}$.

$0.2599 \approx k$ — Approximate.

SUMMARY AND STUDY STRATEGIES

After studying this chapter, you should have acquired the following skills.
The exercise numbers are keyed to the Review Exercises that begin on page 691.
Answers to odd-numbered Review Exercises are given in the back of the text.*

Section 11.1	Review Exercises
■ Verify solutions of differential equations.	*1–14*
■ Find particular solutions of differential equations.	*15–20*
■ Use integration to find general solutions of differential equations.	*21–26*

Section 11.2

- Decide whether the variables in differential equations can be separated. *27–30*
- Use separation of variables to solve differential equations. *31–50*

  If f and g are continuous functions, then the differential equation

  $$\frac{dy}{dx} = f(x)g(y)$$

  has a general solution of

  $$\int \frac{1}{g(y)}\,dy = \int f(x)\,dx + C.$$

- Use differential equations to model real-life problems and use separation of variables to solve them. *51, 52*

Section 11.3

- Write first-order linear differential equations in standard form. *53–56*
- Solve first-order linear differential equations. *57–66*

  A first-order linear differential equation is an equation of the form

  $y' + P(x)y = Q(x)$

  where P and Q are functions of x. An equation that is written in this form is said to be in standard form.

Section 11.4

- Use differential equations to model and solve real-life problems. *67–77*

Study Strategies

- **Using Technology** Throughout this chapter, remember that technology can help you *solve* a differential equation and *graph* a particular solution.

* A wide range of valuable study aids are available to help you master the material in this chapter. The *Student Solutions Manual* includes step-by-step solutions to all odd-numbered exercises to help you review and prepare. The student website at *www.cengagebrain.com* offers algebra help and a *Graphing Technology Guide*, which contains step-by-step commands and instructions for a wide variety of graphing calculators.

Review Exercises

See www.CalcChat.com for worked-out solutions to odd-numbered exercises.

Verifying Solutions In Exercises 1–10, verify that the function is a solution of the differential equation.

	Solution	*Differential Equation*
1.	$y = Ce^{x/2}$	$2y' = y$
2.	$y = e^{-x}$	$3y' + 4y = e^{-x}$
3.	$y = 3e^{x^2}$	$y' - 2xy = 0$
4.	$y = 2e^{2x}$	$y'' - y' - 2y = 0$
5.	$y = x^2$	$\frac{1}{2}y' - \frac{1}{x}y = 0$
6.	$y = \frac{1}{x}$	$xy'' + 2y' = 0$
7.	$y = \frac{x^3}{5} - x + C\sqrt{x}$	$2xy' - y = x^3 - x$
8.	$y = \cos x + 3\sin x$	$y'' + y = 0$
9.	$y = C_1e^x + C_2e^{-x}$	$y'' - y = 0$
10.	$y = C_1e^{x/2} + C_2e^{-2x}$	$2y'' + 3y' - 2y = 0$

Determining Solutions In Exercises 11–14, determine whether the function is a solution of the differential equation

$xy' - 2y = x^3e^x$.

11. $y = x^2$

12. $y = \ln x$

13. $y = x^2e^x$

14. $y = x^2e^x - 5x^2$

Finding a Particular Solution In Exercises 15 and 16, verify that the general solution satisfies the differential equation. Then find the particular solution that satisfies the initial condition.

15. General solution: $y = Ce^{-5x}$
Differential equation: $y' + 5y = 0$
Initial condition: $y = 1$ when $x = 0$

16. General solution: $y = Ce^{-x^2}$
Differential equation: $y' + 2xy = 0$
Initial condition: $y = -2$ when $x = 0$

Sketching Graphs of Solutions In Exercises 17 and 18, the general solution of the differential equation is given. Sketch the particular solutions that correspond to the indicated values of C.

	General Solution	*Differential Equation*	*C-Values*
17.	$y = Ce^{3/x}$	$x^2y' + 3y = 0$	1, 2, 4
18.	$y = \frac{C}{x+1}$	$(x+1)y' + y = 0$	$0, \pm 1, \pm 2$

Finding a Particular Solution In Exercises 19 and 20, some of the curves corresponding to different values of C in the general solution of the differential equation are shown in the figure. Find the particular solution that passes through the point plotted on the graph.

19. $y = C(x^2 + 1)$
$(x^2 + 1)y' = 2xy$

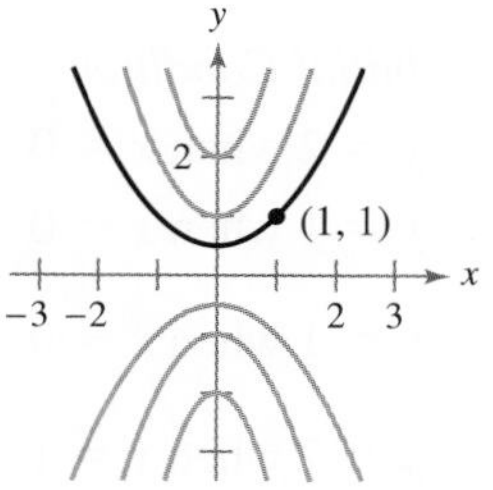

20. $y = Ce^x - 3$
$y' - y = 3$

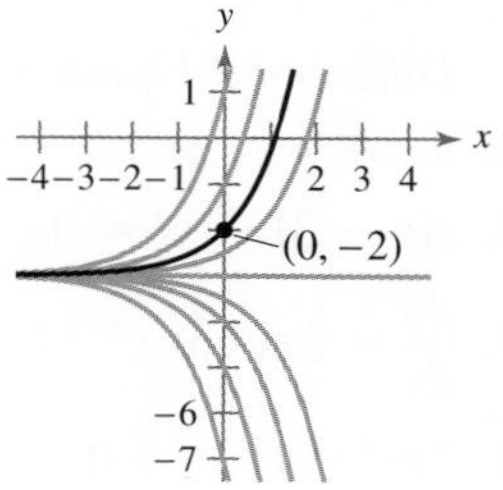

Finding General Solutions In Exercises 21–26, use integration to find the general solution of the differential equation.

21. $\frac{dy}{dx} = 2x^2 + 5$

22. $\frac{dy}{dx} = x^3 - 2x$

23. $\frac{dy}{dx} = 3\cos x$

24. $\frac{dy}{dx} = \sin 4x$

25. $\frac{dy}{dx} = \frac{x+3}{x}$

26. $\frac{dy}{dx} = 3e^{-x/3}$

Separation of Variables In Exercises 27–30, decide whether the variables in the differential equation can be separated.

27. $\frac{dy}{dx} = \frac{y}{x+3}$

28. $\frac{dy}{dx} = \frac{y}{x+y}$

29. $\frac{dy}{dx} = \frac{x^2+y}{y}$

30. $\frac{dy}{dx} = \frac{1}{y} - 1$

Solving a Differential Equation In Exercises 31–42, use separation of variables to find the general solution of the differential equation.

31. $\frac{dy}{dx} = 4x$

32. $\frac{dy}{dx} = \frac{2}{x}$

33. $4y^3\frac{dy}{dx} = 5$

34. $\frac{dy}{dx} = 6xy$

35. $y' + 2xy^2 = 0$

36. $y' - y = 12$

37. $y' = (x+1)(y+1)$

38. $(3 + 2y)\frac{dy}{dx} = 2x$

39. $\frac{dy}{dx} = -\frac{y+2}{2x^3}$

40. $\frac{dy}{dx} = \frac{y}{x} - \frac{y}{x+1}$

41. $\frac{dy}{dx} = \frac{\cos x}{y}$

42. $y' = \frac{\sin x}{y^2}$

Solving a Differential Equation In Exercises 43 and 44, (a) find the general solution of the differential equation and (b) use a graphing utility to graph the particular solutions given by $C = 0$, ± 1, and ± 2.

43. $\dfrac{dy}{dx} = 3x^2$ **44.** $\dfrac{dy}{dx} = 0.5x(2 - y)$

Finding a Particular Solution In Exercises 45–48, use the initial condition to find the particular solution of the differential equation.

	Differential Equation	*Initial Condition*
45.	$yy' + e^x = 0$	$y = 2$ when $x = 0$
46.	$x(y - 3) + y' = 0$	$y = 1$ when $x = 0$
47.	$\dfrac{dy}{dx} = 2xy \cos x^2$	$y = 1$ when $x = 0$
48.	$2xy' - \ln x^2 = 0$	$y = 2$ when $x = 1$

Finding an Equation In Exercises 49 and 50, find an equation of the graph that passes through the point and has the specified slope. Then graph the equation.

49. Point: $(1, 1)$

Slope: $y' = \dfrac{6x}{5y}$

50. Point: $(-2, -1)$

Slope: $y' = y^2x$

51. Radioactive Decay The rate of decomposition of radioactive carbon is proportional to the amount present at any time. The half-life of radioactive carbon is 5715 years. What percent of a present amount will remain after 1000 years?

52. Advertising Awareness A newly opened chiropractor's office is introduced through an advertising campaign to a population of 120,000 potential customers. The rate at which the population hears about the new office is assumed to be proportional to the number of people who are not yet aware of it. This is described by the differential equation

$$\frac{dy}{dt} = k(120 - y)$$

where y is the number of people (in thousands) who have heard of the new office, and t is the time (in months). After 6 months, half of the population have heard of the new office. How many will have heard of it after 12 months?

Writing in Standard Form In Exercises 53–56, write the first-order linear differential equation in standard form.

53. $x^4 + 4x^2y' - 4y = 0$

54. $y' + 10(2x + y) = 0$

55. $x = 2x^3(y' - y)$ **56.** $y - 2 = (x + 2)y'$

Solving a Linear Differential Equation In Exercises 57–64, find the general solution of the first-order linear differential equation.

57. $\dfrac{dy}{dx} + 4y = 8$

58. $\dfrac{dy}{dx} - 10y = 20$

59. $\dfrac{dy}{dx} - \dfrac{y}{x} = 2x - 3$

60. $\dfrac{dy}{dx} - \dfrac{4y}{x} = 3x + 2$

61. $y' + 6y = e^{2x}$

62. $y' + 2xy = 4x$

63. $\dfrac{dy}{dx} + \dfrac{y}{x} = 3x + 4$

64. $\dfrac{dy}{dx} + \dfrac{2y}{x} = 3x + 1$

Finding a Particular Solution In Exercises 65 and 66, find the particular solution that satisfies the initial condition.

	Differential Equation	*Initial Condition*
65.	$y' + y = 6$	$y = 3$ when $x = 0$
66.	$y' + 3xy = 3x$	$y = 1$ when $x = 0$

67. Safety Assume that the rate of change per hour in the number of miles s of road cleared by a snowplow is inversely proportional to the depth h of the snow. That is,

$$\frac{ds}{dh} = \frac{k}{h}.$$

Find s as a function of h for $s = 25$ miles when $h = 2$ inches and $s = 12$ miles when $h = 6$ inches $(2 < h < 15)$.

68. Chemistry A wet shirt hung from a clothesline to dry loses moisture through evaporation at a rate proportional to its moisture content. After 1 hour, the shirt has lost 60% of its original moisture content. How long will it take the shirt to lose 90% of its original moisture content?

69. Economics: Pareto's Law According to the economist Vilfredo Pareto (1848–1923), the rate of decrease in the number of people y in a stable economy having an income of at least x dollars is directly proportional to the number of such people and inversely proportional to their income x. This is modeled by the differential equation

$$\frac{dy}{dx} = -k\frac{y}{x}.$$

Solve the differential equation to find y as a function of x.

70. Economics: Pareto's Law In 2009, 24.1 million people in the United States earned at least \$75,000 and 109.8 million people earned at least \$25,000 (see figure). Assume that Pareto's Law holds and use the result of Exercise 69 to determine the number of people (in millions) who earned (a) at least \$20,000 and (b) at least \$100,000. *(Source: U.S. Census Bureau)*

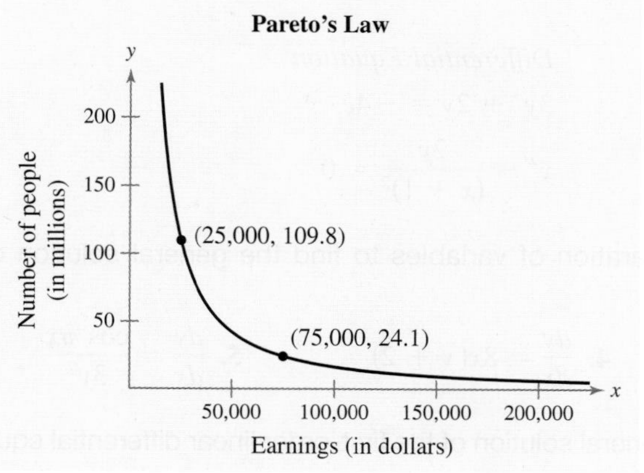

71. Chemical Reaction During a chemical reaction, a compound changes into another compound at a rate proportional to the cube root of the unchanged amount y. Write the differential equation for the chemical reaction model. Find the particular solution when the initial amount of the original compound is 27 grams and the amount remaining after 1 hour is 8 grams.

72. Chemical Reaction Using the result of Exercise 71, when will 80% of the compound have been changed? When will 99% of the compound have been changed?

Chemistry: Newton's Law of Cooling In Exercises 73 and 74, use Newton's Law of Cooling, which states that the rate of change in the temperature of an object is proportional to the difference between the temperature T of the object and the temperature T_0 of the surrounding environment. This is described by the differential equation

$$\frac{dT}{dt} = k(T - T_0).$$

73. A steel ingot whose temperature is 1500°F is placed in a room whose temperature is a constant 90°F. One hour later, the temperature of the ingot is 1120°F. What is the temperature of the ingot 5 hours after it is placed in the room?

74. Food at a temperature of 70°F is placed in a freezer that is set at 0°F. After 1 hour, the temperature of the food is 48°F.

(a) Find the temperature of the food after it has been in the freezer 6 hours.

(b) How long will it take the food to cool to a temperature of 10°F?

75. Biology A population of 12 pelicans has been introduced into a new wetlands area. Biologists estimate that the maximum population the wetlands can sustain is 64 pelicans. After 3 years, the population is 28 pelicans. The population follows a Gompertz growth model. How many pelicans will there be in the wetlands after 8 years?

76. Chemical Mixture A tank contains 30 gallons of a solution composed of 80% water and 20% alcohol. A second solution containing half water and half alcohol is added to the tank at the rate of 6 gallons per minute. At the same time, the well-stirred solution is withdrawn at the same rate, as shown in the figure.

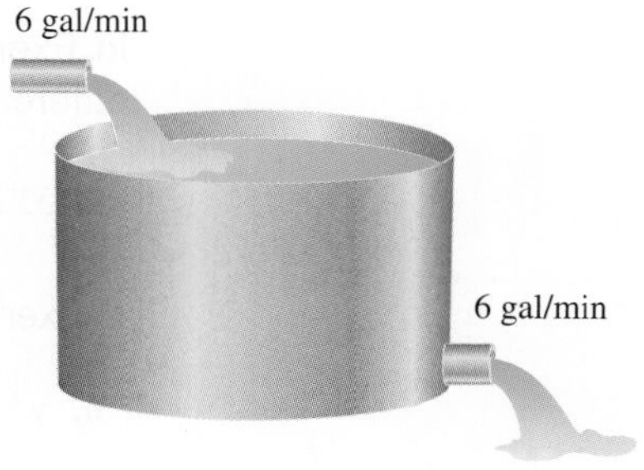

(a) Find the amount y of alcohol in the solution as a function of t by solving the differential equation

$$\frac{dy}{dt} = -6\left(\frac{y}{30}\right) + 3.$$

(b) Find the amount of alcohol in the tank after 10 minutes.

77. Chemical Mixture A tank contains 20 gallons of a solution composed of 90% water and 10% alcohol. A second solution containing half water and half alcohol is added to the tank at the rate of 2 gallons per minute. At the same time, the well-stirred solution is withdrawn at the same rate.

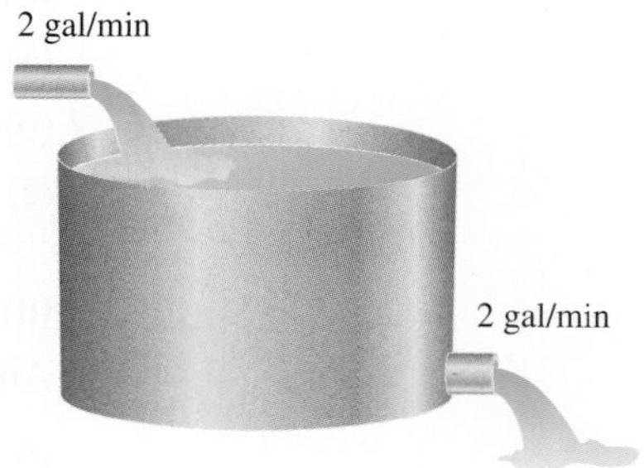

(a) Find the amount y of alcohol in the solution as a function of t by solving the differential equation

$$\frac{dy}{dt} = -2\left(\frac{y}{20}\right) + 1.$$

(b) Find the amount of alcohol in the tank after 10 minutes.

TEST YOURSELF

See www.CalcChat.com for worked-out solutions to odd-numbered exercises.

Take this test as you would take a test in class. When you are done, check your work against the answers given in the back of the book.

In Exercises 1 and 2, verify that the function is a solution of the differential equation.

Solution	*Differential Equation*
1. $y = e^{-2x}$	$3y' + 2y = -4e^{-2x}$
2. $y = \dfrac{1}{x+1}$	$y'' - \dfrac{2y}{(x+1)^2} = 0$

In Exercises 3–5, use separation of variables to find the general solution of the differential equation.

3. $yy' = x$ **4.** $\dfrac{dy}{dx} = 8x(y + 2)$ **5.** $\dfrac{dy}{dx} = \dfrac{\cos \pi x}{3y^2}$

In Exercises 6–8, find the general solution of the first-order linear differential equation.

6. $y' + 5y = 2$ **7.** $y' - 2y = e^{2x}$ **8.** $y' = \dfrac{x^2 - y}{x}$

In Exercises 9–11, use the initial condition to find the particular solution of the differential equation.

Differential Equation	*Initial Condition*
9. $y' + x^2y - x^2 = 0$	$y = 0$ when $x = 0$
10. $y'e^{-x^2} = 2xy$	$y = e$ when $x = 0$
11. $x\dfrac{dy}{dx} = \dfrac{\ln x}{7}$	$y = -2$ when $x = 1$

12. A lamb that weighed 10 pounds at birth gains weight at the rate of

$$\frac{dw}{dt} = k(200 - w)$$

where w is the weight (in pounds) and t is the time (in years).

(a) Solve the differential equation.

(b) Use a graphing utility to graph the particular solutions for $k = 0.8, 0.9$, and 1.

(c) The animal is sold when its weight reaches 150 pounds. Find the time of sale for each of the models in part (b).

(d) What is the maximum weight of the animal for each of the models in part (b)?

13. An infectious disease spreads through a large population according to the model

$$\frac{dy}{dt} = \frac{1 - y}{4}$$

where y is the percent (in decimal form) of the population exposed to the disease, and t is the time (in years).

(a) Solve this differential equation, assuming $y(0) = 0$.

(b) Find the number of years it will take for half of the population to be exposed to the disease.

(c) Find the percent of the population that will have been exposed to the disease after 4 years.

Appendices

Appendices D and E are located on the website that accompanies this text at *www.cengagebrain.com*.

A Precalculus Review

A.1 The Real Number Line and Order

- Represent, classify, and order real numbers.
- Use inequalities to represent sets of real numbers.
- Solve inequalities.
- Use inequalities to model and solve real-life problems.

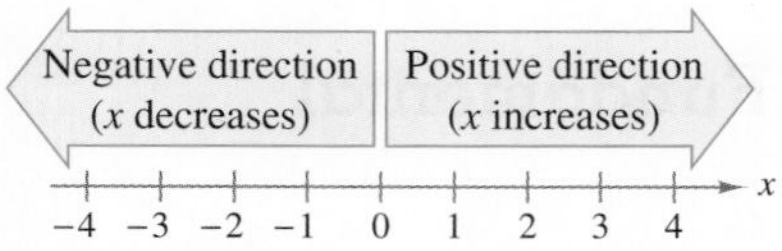

The Real Number Line

FIGURE A.1

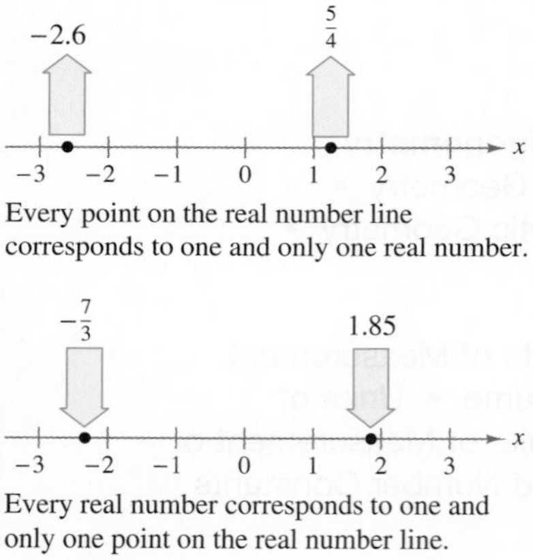

FIGURE A.2

The Real Number Line

Real numbers can be represented with a coordinate system called the **real number line** (or x-axis), as shown in Figure A.1. The **positive direction** (to the right) is denoted by an arrowhead and indicates the direction of increasing values of x. The real number corresponding to a particular point on the real number line is called the **coordinate** of the point. As shown in Figure A.1, it is customary to label those points whose coordinates are integers.

The point on the real number line corresponding to zero is called the **origin.** Numbers to the right of the origin are **positive,** and numbers to the left of the origin are **negative.** The term **nonnegative** describes a number that is either positive or zero.

The importance of the real number line is that it provides you with a conceptually perfect picture of the real numbers. That is, each point on the real number line corresponds to one and only one real number, and each real number corresponds to one and only one point on the real number line. This type of relationship is called a **one-to-one correspondence** and is illustrated in Figure A.2.

Each of the four points in Figure A.2 corresponds to a real number that can be expressed as the ratio of two integers.

$$-2.6 = -\tfrac{13}{5} \qquad \tfrac{5}{4} \qquad -\tfrac{7}{3} \qquad 1.85 = \tfrac{37}{20}$$

Such numbers are called **rational.** Rational numbers have either terminating or infinitely repeating decimal representations.

Terminating Decimals	*Infinitely Repeating Decimals*
$\frac{2}{5} = 0.4$	$\frac{1}{3} = 0.333\ldots = 0.\overline{3}$*
$\frac{7}{8} = 0.875$	$\frac{12}{7} = 1.714285714285\ldots = 1.\overline{714285}$

Real numbers that are not rational are called **irrational,** and they cannot be represented as the ratio of two integers (or as terminating or infinitely repeating decimals). So, a decimal approximation is used to represent an irrational number. Some irrational numbers occur so frequently in applications that mathematicians have invented special symbols to represent them. For example, the symbols $\sqrt{2}$, π, and e represent irrational numbers whose decimal approximations are as shown. (See Figure A.3.)

$$\sqrt{2} \approx 1.4142135623$$
$$\pi \approx 3.1415926535$$
$$e \approx 2.7182818284$$

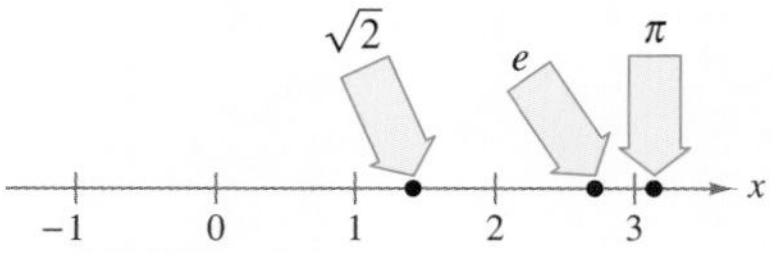

FIGURE A.3

*The bar indicates which digit or digits repeat infinitely.

Order and Intervals on the Real Number Line

One important property of the real numbers is that they are **ordered:** 0 is less than 1, -3 is less than -2.5, π is less than $\frac{22}{7}$, and so on. You can visualize this property on the real number line by observing that a is less than b if and only if a lies to the left of b on the real number line. Symbolically, "a is less than b" is denoted by the inequality $a < b$. For example, the inequality

$$\frac{3}{4} < 1$$

follows from the fact that $\frac{3}{4}$ lies to the left of 1 on the real number line, as shown in Figure A.4.

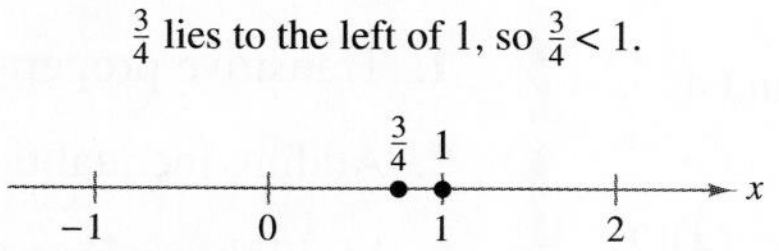

FIGURE A.4

When three real numbers a, x, and b are ordered such that $a < x$ and $x < b$, we say that x is **between** a and b and write

$a < x < b$. x is between a and b.

The set of *all* real numbers between a and b is called the **open interval** between a and b and is denoted by (a, b). An interval of the form (a, b) does not contain the "endpoints" a and b. Intervals that include their endpoints are called **closed** and are denoted by $[a, b]$. Intervals of the form $[a, b)$ and $(a, b]$ are neither open nor closed. Figure A.5 shows the nine types of intervals on the real number line.

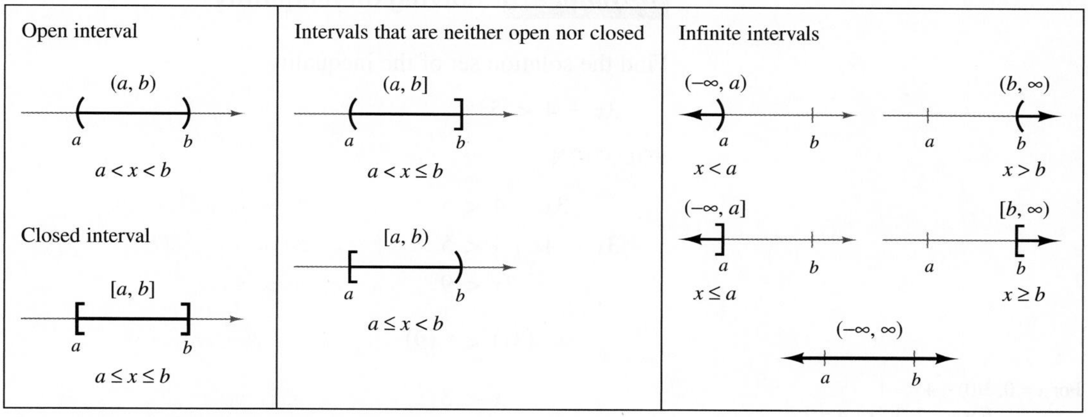

Intervals on the Real Number Line

FIGURE A.5

Note that a square bracket is used to denote "less than or equal to" ($\le$) or "greater than or equal to" ($\ge$). Furthermore, the symbols

∞ Positive infinity

and

$-\infty$ Negative infinity

denote **positive** and **negative infinity,** respectively. These symbols do not denote real numbers; they merely let you describe unbounded conditions more concisely. For instance, the interval $[b, \infty)$ is unbounded to the right because it includes *all* real numbers that are greater than or equal to b.

Solving Inequalities

In calculus, you are frequently required to "solve inequalities" involving variable expressions such as $3x - 4 < 5$. The number a is a **solution** of an inequality if the inequality is true when a is substituted for x. The set of all values of x that satisfy an inequality is called the **solution set** of the inequality. The following properties are useful for solving inequalities. (Similar properties are obtained when $<$ is replaced by $\leq$ and $>$ is replaced by $\geq$.)

STUDY TIP

Notice the differences between Properties 3 and 4. For example,

$-3 < 4 \Rightarrow (-3)(2) < (4)(2)$

and

$-3 < 4$
$\Rightarrow (-3)(-2) > (4)(-2).$

Properties of Inequalities

Let a, b, c, and d be real numbers.

1. Transitive property: $a < b$ and $b < c$ $\Rightarrow$ $a < c$
2. Adding inequalities: $a < b$ and $c < d$ $\Rightarrow$ $a + c < b + d$
3. Multiplying by a (positive) constant: $a < b$ $\Rightarrow$ $ac < bc, \quad c > 0$
4. Multiplying by a (negative) constant: $a < b$ $\Rightarrow$ $ac > bc, \quad c < 0$
5. Adding a constant: $a < b$ $\Rightarrow$ $a + c < b + c$
6. Subtracting a constant: $a < b$ $\Rightarrow$ $a - c < b - c$

Note that you *reverse the inequality* when you multiply by a negative number. For example, if $x < 3$, then $-4x > -12$. This principle also applies to division by a negative number. So, if $-2x > 4$, then $x < -2$.

Example 1 Solving an Inequality

Find the solution set of the inequality

$$3x - 4 < 5.$$

SOLUTION

$3x - 4 < 5$	Write original inequality.
$3x - 4 + 4 < 5 + 4$	Add 4 to each side.
$3x < 9$	Simplify.
$\frac{1}{3}(3x) < \frac{1}{3}(9)$	Multiply each side by $\frac{1}{3}$.
$x < 3$	Simplify.

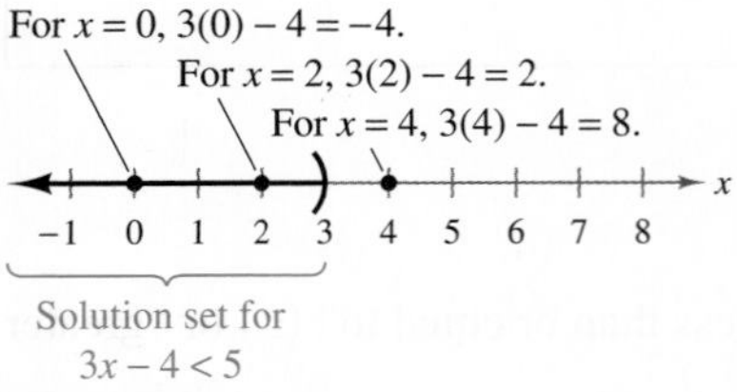

FIGURE A.6

So, the solution set is the interval $(-\infty, 3)$, as shown in Figure A.6. Once you have solved an inequality, it is a good idea to check some x-values in your solution set to see whether they satisfy the original inequality. You should also check some values outside your solution set to verify that they do *not* satisfy the inequality. For instance, Figure A.6 shows that when $x = 0$ or $x = 2$, the inequality is satisfied, but when $x = 4$, the inequality is not satisfied.

Checkpoint 1

Find the solution set of the inequality

$$2x - 3 < 7.$$

■

In Example 1, all five inequalities listed as steps in the solution have the same solution set, and they are called **equivalent inequalities.**

The inequality in Example 1 involves a first-degree polynomial. To solve inequalities involving polynomials of higher degree, you can use the fact that a polynomial can change signs *only* at its real zeros (the real numbers that make the polynomial zero). Between two consecutive real zeros, a polynomial must be entirely positive or entirely negative. This means that when the real zeros of a polynomial are put in order, they divide the real number line into **test intervals** in which the polynomial has no sign changes. That is, if a polynomial has the factored form

$$(x - r_1)(x - r_2), \ldots, (x - r_n), \qquad r_1 < r_2 < r_3 < \cdots < r_{n-1} < r_n$$

then the test intervals are

$$(-\infty, r_1), \quad (r_1, r_2), \quad \ldots, \quad (r_{n-1}, r_n), \quad \text{and} \quad (r_n, \infty).$$

For example, the polynomial

$$x^2 - x - 6 = (x - 3)(x + 2)$$

can change signs only at $x = -2$ and $x = 3$. To determine the sign of the polynomial in the intervals $(-\infty, -2)$, $(-2, 3)$, and $(3, \infty)$, you need to test only *one value* in each interval.

Example 2 Solving a Polynomial Inequality

$$x^2 < x + 6 \qquad \text{Original inequality}$$

$$x^2 - x - 6 < 0 \qquad \text{Polynomial form}$$

$$(x - 3)(x + 2) < 0 \qquad \text{Factor.}$$

So, the polynomial $x^2 - x - 6$ has $x = -2$ and $x = 3$ as its zeros. You can solve the inequality by testing the sign of the polynomial in each of the intervals $(-\infty, -2)$, $(-2, 3)$, and $(3, \infty)$. In each interval, choose a representative x-value and evaluate the polynomial.

Interval	*x-Value*	*Polynomial Value*	*Conclusion*
$(-\infty, -2)$	$x = -3$	$(-3)^2 - (-3) - 6 = 6$	Positive
$(-2, 3)$	$x = 0$	$(0)^2 - (0) - 6 = -6$	Negative
$(3, \infty)$	$x = 4$	$(4)^2 - (4) - 6 = 6$	Positive

From this you can conclude that the inequality is satisfied for all x-values in $(-2, 3)$. This implies that the solution of the inequality $x^2 < x + 6$ is the interval $(-2, 3)$, as shown in Figure A.7. Note that the original inequality contains a "less than" symbol. This means that the solution set does not contain the endpoints of the test interval $(-2, 3)$.

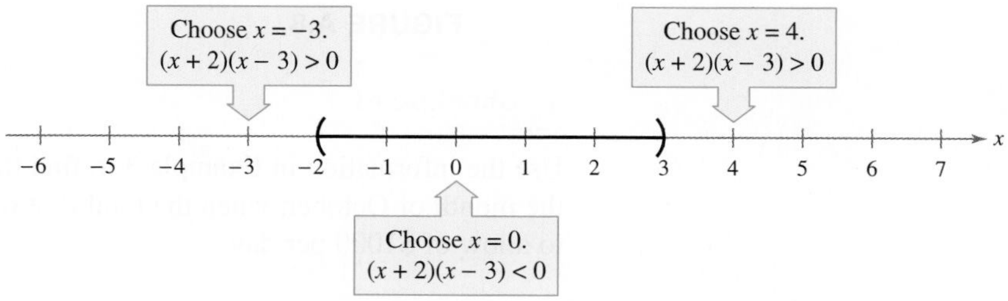

FIGURE A.7

✓ Checkpoint 2

Find the solution set of the inequality

$$x^2 > 3x + 10.$$

■

Application

Inequalities are frequently used to describe conditions that occur in business and science. For instance, the inequality

$$8.8 \le W \le 26.4$$

describes the typical weights W (in pounds) of adult rhesus monkeys. Example 3 shows how an inequality can be used to describe the production levels in a manufacturing plant.

Example 3 Production Levels

In addition to fixed overhead costs of \$500 per day, the cost of producing x units of an item is \$2.50 per unit. During the month of August, the total cost of production varied from a high of \$1325 to a low of \$1200 per day. Find the high and low *production levels* during the month.

SOLUTION Because it costs \$2.50 to produce one unit, it costs $2.5x$ to produce x units. Furthermore, because the fixed cost per day is \$500, the total daily cost C (in dollars) of producing x units is

$$C = 2.5x + 500.$$

Now, because the cost ranged from \$1200 to \$1325, you can write the following.

$$1200 \le 2.5x + 500 \le 1325 \qquad \text{Write original inequality.}$$

$$1200 - 500 \le 2.5x + 500 - 500 \le 1325 - 500 \qquad \text{Subtract 500 from each part.}$$

$$700 \le 2.5x \le 825 \qquad \text{Simplify.}$$

$$\frac{700}{2.5} \le \frac{2.5x}{2.5} \le \frac{825}{2.5} \qquad \text{Divide each part by 2.5.}$$

$$280 \le x \le 330 \qquad \text{Simplify.}$$

So, the daily production levels during the month of August varied from a low of 280 units to a high of 330 units, as shown in Figure A.8.

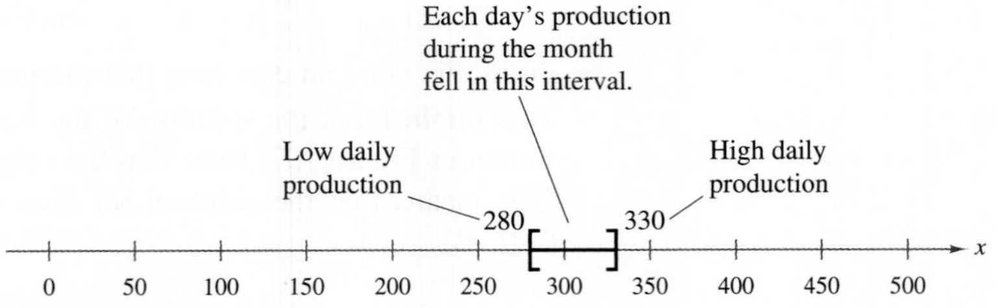

FIGURE A.8

✓Checkpoint 3

Use the information in Example 3 to find the high and low production levels during the month of October, when the total cost of production varied from a high of \$1500 to a low of \$1000 per day.

Exercises A.1

See www.CalcChat.com for worked-out solutions to odd-numbered exercises.

Classifying Real Numbers In Exercises 1–10, determine whether the real number is rational or irrational.

1. 0.25
2. -3678
3. $\dfrac{3\pi}{2}$
4. $3\sqrt{2} - 1$
5. $4.3\overline{451}$
6. $\dfrac{22}{7}$
7. $\sqrt[3]{64}$
8. $0.\overline{8177}$
9. $\sqrt[3]{60}$
10. $2e$

Checking Solutions In Exercises 11–14, determine whether each given value of x satisfies the inequality.

11. $5x - 12 > 0$
 (a) $x = 3$ (b) $x = -3$ (c) $x = \frac{5}{2}$
12. $x + 1 < \dfrac{x}{3}$
 (a) $x = 0$ (b) $x = 4$ (c) $x = -4$
13. $0 < \dfrac{x-2}{4} < 2$
 (a) $x = 4$ (b) $x = 10$ (c) $x = 0$
14. $-1 < \dfrac{3-x}{2} \le 1$
 (a) $x = 0$ (b) $x = 1$ (c) $x = 5$

Solving an Inequality In Exercises 15–28, solve the inequality. Then graph the solution set on the real number line. *See Examples 1 and 2.*

15. $x - 5 \ge 7$
16. $2x > 3$
17. $4x + 1 < 2x$
18. $2x + 7 < 3$
19. $4 - 2x < 3x - 1$
20. $x - 4 \le 2x + 1$
21. $-4 < 2x - 3 < 4$
22. $0 \le x + 3 < 5$
23. $\dfrac{3}{4} > x + 1 > \dfrac{1}{4}$
24. $-1 < -\dfrac{x}{3} < 1$
25. $\dfrac{x}{2} + \dfrac{x}{3} > 5$
26. $\dfrac{x}{2} - \dfrac{x}{3} > 5$
27. $2x^2 - x < 6$
28. $2x^2 + 1 < 9x - 3$

Writing Inequalities In Exercises 29–32, use inequality notation to describe the subset of real numbers.

29. **Earnings Per Share** A company expects its earnings per share E for the next quarter to be no less than \$4.10 and no more than \$4.25.

30. **Production** The estimated daily oil production p at a refinery is greater than 2 million barrels but less than 2.4 million barrels.

31. **Survey** According to a survey, the percent p of Americans who now conduct most of their banking transactions online is no more than 40%.

32. **Income** The net income I of a company is expected to be no less than \$239 million.

33. **Physiology** The maximum heart rate of a person in normal health is related to the person's age by the equation

$$r = 220 - A$$

where r is the maximum heart rate (in beats per minute) and A is the person's age (in years). Some physiologists recommend that during physical activity, a sedentary person should strive to increase his or her heart rate to at least 60% of the maximum heart rate, and a highly fit person should strive to increase his or her heart rate to at most 90% of the maximum heart rate. Use inequality notation to express the range of the target heart rate for physical activity for a 20-year-old.

34. **Annual Operating Costs** A utility company has a fleet of vans. The annual operating cost C (in dollars) of each van is estimated to be

$$C = 0.35m + 2500$$

where m is the number of miles driven. What number of miles will yield an annual operating cost that is less than \$13,000?

35. **Profit** The revenue for selling x units of a product is

$$R = 115.95x$$

and the cost of producing x units is

$$C = 95x + 750.$$

To obtain a profit, the revenue must be *greater than* the cost. For what values of x will this product return a profit?

36. **Sales** A doughnut shop sells a dozen doughnuts for \$4.50. Beyond the fixed cost of \$220 per day, it costs \$2.75 for enough materials and labor to produce each dozen doughnuts. During the month of January, the daily profit varies between \$60 and \$270. Between what levels (in dozens) do the daily sales vary?

True or False? In Exercises 37 and 38, determine whether each statement is true or false, given $a < b$.

37. (a) $-2a < -2b$
 (b) $a + 2 < b + 2$
 (c) $6a < 6b$
 (d) $\dfrac{1}{a} < \dfrac{1}{b}$

38. (a) $a - 4 < b - 4$
 (b) $4 - a < 4 - b$
 (c) $-3b < -3a$
 (d) $\dfrac{a}{4} < \dfrac{b}{4}$

A.2 Absolute Value and Distance on the Real Number Line

- Find the absolute values of real numbers and understand the properties of absolute value.
- Find the distance between two numbers on the real number line.
- Define intervals on the real number line.
- Use intervals to model and solve real-life problems and find the midpoint of an interval.

Absolute Value of a Real Number

TECH TUTOR

Absolute value expressions can be evaluated on a graphing utility. When an expression such as $|3 - 8|$ is evaluated, parentheses should surround the expression, as in abs(3 − 8).

Definition of Absolute Value

The **absolute value** of a real number a is

$$|a| = \begin{cases} a, & \text{if } a \geq 0 \\ -a, & \text{if } a < 0. \end{cases}$$

At first glance, it may appear from this definition that the absolute value of a real number can be negative, but this is not possible. For example, let $a = -3$. Then, because $-3 < 0$, you have

$$|a| = |-3| = -(-3) = 3.$$

The following properties are useful for working with absolute values.

Properties of Absolute Value

1. Multiplication: $|ab| = |a||b|$
2. Division: $\left|\frac{a}{b}\right| = \frac{|a|}{|b|}, \quad b \neq 0$
3. Power: $|a^n| = |a|^n$
4. Square root: $\sqrt{a^2} = |a|$

Be sure you understand the fourth property in this list. A common error in algebra is to imagine that by squaring a number and then taking the square root, you come back to the original number. But this is true only if the original number is nonnegative. For instance, if $a = 2$, then

$$\sqrt{2^2} = \sqrt{4} = 2$$

but if $a = -2$, then

$$\sqrt{(-2)^2} = \sqrt{4} = 2.$$

The reason for this is that (by definition) the square root symbol

$$\sqrt{}$$

denotes only the nonnegative root.

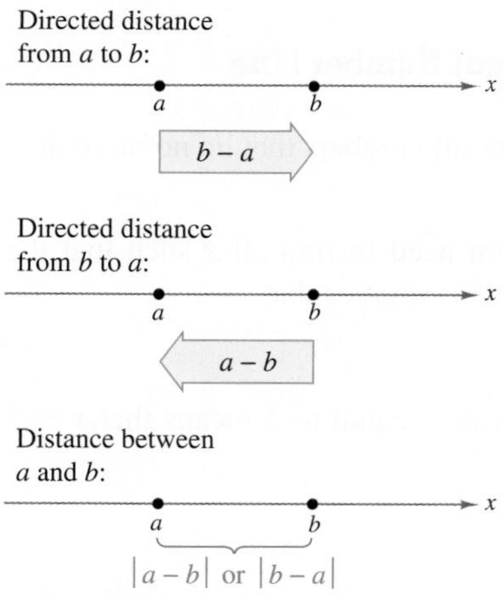

FIGURE A.9

Distance on the Real Number Line

Consider two distinct points on the real number line, as shown in Figure A.9.

1. The **directed distance from *a* to *b*** is

   $b - a.$

2. The **directed distance from *b* to *a*** is

   $a - b.$

3. The **distance between *a* and *b*** is

   $|a - b|$ or $|b - a|.$

In Figure A.9, note that because b is to the right of a, the directed distance from a to b (moving to the right) is positive. Moreover, because a is to the left of b, the directed distance from b to a (moving to the left) is negative. The distance *between* two points on the real number line can never be negative.

Distance Between Two Points on the Real Number Line

The distance d between points x_1 and x_2 on the real number line is given by

$$d = |x_2 - x_1| = \sqrt{(x_2 - x_1)^2}.$$

Note that the order of subtraction with x_1 and x_2 does not matter because

$$|x_2 - x_1| = |x_1 - x_2| \quad \text{and} \quad (x_2 - x_1)^2 = (x_1 - x_2)^2.$$

Example 1 Finding Distance on the Real Number Line

Determine the distance between -3 and 4 on the real number line. What is the directed distance from -3 to 4? What is the directed distance from 4 to -3?

SOLUTION The distance between -3 and 4 is given by

$$|-3 - 4| = |-7| = 7 \qquad |a - b|$$

or

$$|4 - (-3)| = |7| = 7 \qquad |b - a|$$

as shown in Figure A.10.

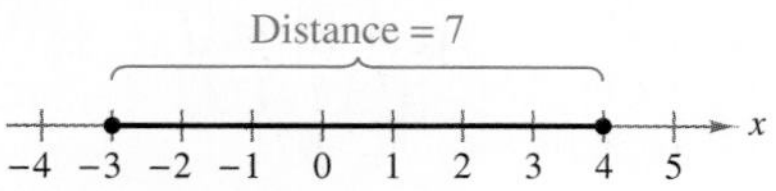

FIGURE A.10

The directed distance from -3 to 4 is

$$4 - (-3) = 7. \qquad b - a$$

The directed distance from 4 to -3 is

$$-3 - 4 = -7. \qquad a - b$$

✓ Checkpoint 1

Determine the distance between -2 and 6 on the real number line. What is the directed distance from -2 to 6? What is the directed distance from 6 to -2?

Intervals Defined by Absolute Value

Example 2 Defining an Interval on the Real Number Line

Find the interval on the real number line that contains all numbers that lie no more than two units from 3.

SOLUTION Let x be any point in this interval. You need to find all x such that the distance between x and 3 is less than or equal to 2. This implies that

$$|x - 3| \le 2.$$

Requiring the absolute value of $x - 3$ to be less than or equal to 2 means that $x - 3$ must lie between -2 and 2. So, you can write

$$-2 \le x - 3 \le 2.$$

Solving this pair of inequalities, you have

$$-2 + 3 \le x - 3 + 3 \le 2 + 3$$

$$1 \le x \le 5. \quad \text{Solution set}$$

So, the interval is $[1, 5]$, as shown in Figure A.11.

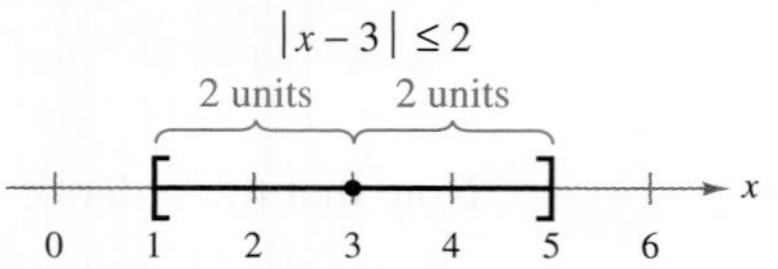

FIGURE A.11

✓Checkpoint 2

Find the interval on the real number line that contains all numbers that lie no more than four units from 6. ■

Two Basic Types of Inequalities Involving Absolute Value

Let a and d be real numbers, where $d > 0$.

$|x - a| \le d$ if and only if $a - d \le x \le a + d$.

$|x - a| \ge d$ if and only if $x \le a - d$ or $a + d \le x$.

Inequality	*Interpretation*	*Graph*
$\lvert x - a\rvert \le d$	All numbers x whose distance from a is less than or equal to d.	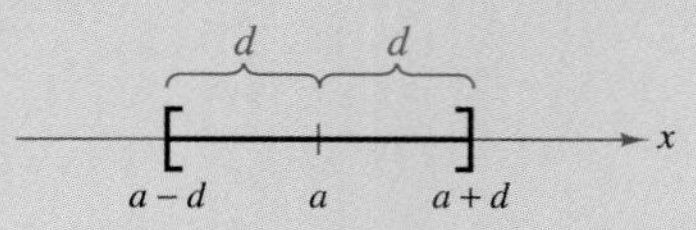
$\lvert x - a\rvert \ge d$	All numbers x whose distance from a is greater than or equal to d.	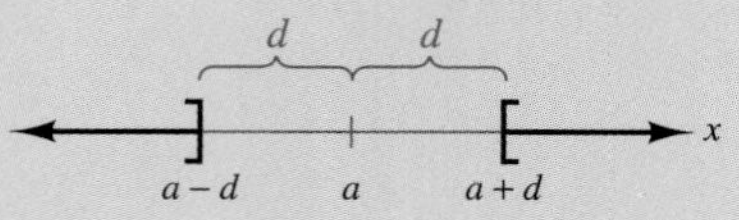

Be sure you see that inequalities of the form $|x - a| \ge d$ have solution sets consisting of two intervals. To describe the two intervals without using absolute values, you must use *two* separate inequalities, connected by an "or" to indicate union.

Application

Example 3 Quality Control

A large manufacturer hired a quality control firm to determine the reliability of a product. Using statistical methods, the firm determined that the manufacturer could expect $0.35\% \pm 0.17\%$ of the units to be defective. The manufacturer offers a money-back guarantee on this product. How much should be budgeted to cover the refunds on 100,000 units? (Assume that the retail price is \$8.95.) Will the manufacturer have to establish a refund budget greater than \$5000?

SOLUTION Let r represent the percent of defective units (in decimal form). You know that r will differ from 0.0035 by at most 0.0017.

$$0.0035 - 0.0017 \le r \le 0.0035 + 0.0017$$
$$0.0018 \le r \le 0.0052 \quad \text{Figure A.12(a)}$$

Now, letting x be the number of defective units out of 100,000, it follows that $x = 100{,}000r$ and you have

$$0.0018(100{,}000) \le 100{,}000r \le 0.0052(100{,}000)$$
$$180 \le x \le 520. \quad \text{Figure A.12(b)}$$

Finally, letting C be the cost of refunds, you have $C = 8.95x$. So, the total cost of refunds for 100,000 units should fall within the interval given by

$$180(8.95) \le 8.95x \le 520(8.95)$$
$$\$1611 \le C \le \$4654. \quad \text{Figure A.12(c)}$$

The manufacturer will *not* have to establish a refund budget greater than \$5000.

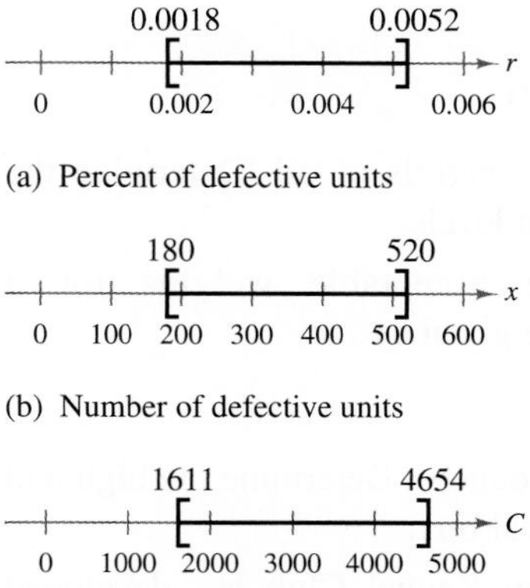

(a) Percent of defective units

(b) Number of defective units

(c) Cost of refunds

FIGURE A.12

✓ Checkpoint 3

Use the information in Example 3 to determine how much should be budgeted to cover refunds on 250,000 units.

In Example 3, the manufacturer should expect to spend between \$1611 and \$4654 for refunds. Of course, the safer budget figure for refunds would be the higher of these estimates. From a statistical point of view, however, the most representative estimate would be the average of these two extremes. Graphically, the average of two numbers is the **midpoint** of the interval with the two numbers as endpoints, as shown in Figure A.13.

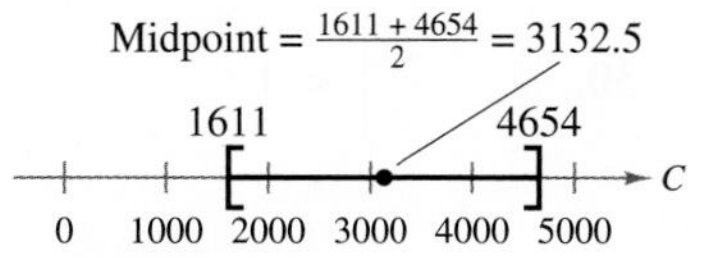

FIGURE A.13

Midpoint of an Interval

The **midpoint** of the interval with endpoints a and b is found by taking the average of the endpoints.

$$\text{Midpoint} = \frac{a + b}{2}$$

Exercises A.2

See www.CalcChat.com for worked-out solutions to odd-numbered exercises.

Finding Distance on the Real Number Line In Exercises 1–6, determine (a) the distance between a and b, (b) the directed distance from a to b, and (c) the directed distance from b to a. *See Example 1.*

1. $a = 126, b = 75$ **2.** $a = -126, b = -75$

3. $a = 9.34, b = -5.65$ **4.** $a = -2.05, b = 4.25$

5. $a = \frac{16}{5}, b = \frac{112}{75}$ **6.** $a = -\frac{18}{5}, b = \frac{61}{15}$

Describing Intervals Using Absolute Value In Exercises 7–18, use absolute values to describe the given interval (or pair of intervals) on the real number line.

7. $[-2, 2]$ **8.** $(-3, 3)$

9. $(-\infty, -2) \cup (2, \infty)$

10. $(-\infty, -3] \cup [3, \infty)$

11. $[2, 8]$

12. $(-7, -1)$

13. $(-\infty, 0) \cup (4, \infty)$

14. $(-\infty, 20) \cup (24, \infty)$

15. All numbers *less than* three units from 5

16. All numbers *more than* five units from 2

17. y is at most *two units* from a.

18. y is *less than* h units from c.

Solving an Inequality In Exercises 19–34, solve the inequality. Then graph the solution set on the real number line. *See Example 2.*

19. $|x| < 4$ **20.** $|2x| < 6$

21. $\left|\frac{x}{2}\right| > 3$ **22.** $|3x| > 12$

23. $|x - 5| < 2$ **24.** $|3x + 1| \geq 4$

25. $\left|\frac{x - 3}{2}\right| \geq 5$ **26.** $|2x + 1| < 5$

27. $|10 - x| > 4$ **28.** $|25 - x| \geq 20$

29. $|9 - 2x| < 1$ **30.** $\left|1 - \frac{2x}{3}\right| < 1$

31. $|x - a| \leq b,\ b > 0$

32. $|2x - a| \geq b,\ b > 0$

33. $\left|\frac{3x - a}{4}\right| < 2b,\ b > 0$

34. $\left|a - \frac{5x}{2}\right| > b,\ b > 0$

Finding a Midpoint In Exercises 35–40, find the midpoint of the given interval.

35. $[8, 24]$ **36.** $[7.3, 12.7]$

37. $[-6.85, 9.35]$ **38.** $[-4.6, -1.3]$

39. $\left[-\frac{1}{2}, \frac{3}{4}\right]$ **40.** $\left[\frac{5}{6}, \frac{5}{2}\right]$

41. Stock Price A stock market analyst predicts that over the next year, the price p of a stock will not change from its current price of \$33.15 by more than \$2. Use absolute values to write this prediction as an inequality.

42. Production The estimated daily production x at a refinery is given by

$$|x - 200{,}000| \leq 25{,}000$$

where x is measured in barrels of oil. Determine the high and low production levels.

43. Manufacturing The acceptable weights for a 20-ounce cereal box are given by

$$|x - 20| \leq 0.75$$

where x is measured in ounces. Determine the high and low weights for the cereal box.

44. Weight The American Kennel Club has developed guidelines for judging the features of various breeds of dogs. To not receive a penalty, the guidelines specify that the weights for male collies must satisfy the inequality

$$\left|\frac{w - 67.5}{7.5}\right| \leq 1$$

where w is the weight (in pounds). Determine the interval on the real number line in which these weights lie. *(Source: The American Kennel Club, Inc.)*

Budget Variance In Exercises 45–48, (a) use absolute value notation to represent the two intervals in which expenses must lie if they are to be within \$500 and within 5% of the specified budget amount and (b) using the more stringent constraint, determine whether the given expense is at variance with the budget restriction.

	Item	*Budget*	*Expense*
45.	Utilities	\$4750.00	\$5116.37
46.	Insurance	\$15,000.00	\$14,695.00
47.	Maintenance	\$20,000.00	\$22,718.35
48.	Taxes	\$7500.00	\$8691.00

49. Quality Control In determining the reliability of a product, a manufacturer determines that it should expect 0.05% ± 0.01% of the units to be defective. The manufacturer offers a money-back guarantee on this product. How much should be budgeted to cover the refunds on 150,000 units? (Assume that the retail price is \$195.99.)

A.3 Exponents and Radicals

- Evaluate expressions involving exponents or radicals.
- Simplify expressions with exponents.
- Find the domains of algebraic expressions.

Expressions Involving Exponents or Radicals

Properties of Exponents

1. Whole-number exponents: $x^n = \underbrace{x \cdot x \cdot x \cdot \cdot \cdot x}_{n \text{ factors}}$
2. Zero exponent: $x^0 = 1, \quad x \neq 0$
3. Negative exponents: $x^{-n} = \dfrac{1}{x^n}, \quad x \neq 0$
4. Radicals (principal nth root): $\sqrt[n]{x} = a \Rightarrow x = a^n$
5. Rational exponents $(1/n)$: $x^{1/n} = \sqrt[n]{x}$
6. Rational exponents (m/n): $x^{m/n} = (x^{1/n})^m = \left(\sqrt[n]{x}\right)^m$
   $x^{m/n} = (x^m)^{1/n} = \sqrt[n]{x^m}$
7. Special convention (square root): $\sqrt[2]{x} = \sqrt{x}$

STUDY TIP

If n is even, then the principal nth root is positive. For example, $\sqrt{4} = +2$ and $\sqrt[4]{81} = +3$.

Example 1 Evaluating Expressions

Expression	x-Value	Substitution
a. $y = -2x^2$	$x = 4$	$y = -2(4^2) = -2(16) = -32$
b. $y = 3x^{-3}$	$x = -1$	$y = 3(-1)^{-3} = \dfrac{3}{(-1)^3} = \dfrac{3}{-1} = -3$
c. $y = (-x)^2$	$x = \dfrac{1}{2}$	$y = \left(-\dfrac{1}{2}\right)^2 = \dfrac{1}{4}$
d. $y = \dfrac{2}{x^{-2}}$	$x = 3$	$y = \dfrac{2}{3^{-2}} = 2(3^2) = 18$

Checkpoint 1

Evaluate $y = 4x^{-2}$ for $x = 3$. ■

Example 2 Evaluating Expressions

Expression	x-Value	Substitution
a. $y = 2x^{1/2}$	$x = 4$	$y = 2\sqrt{4} = 2(2) = 4$
b. $y = \sqrt[3]{x^2}$	$x = 8$	$y = 8^{2/3} = (8^{1/3})^2 = 2^2 = 4$

Checkpoint 2

Evaluate $y = 4x^{1/3}$ for $x = 8$. ■

Operations with Exponents

TECH TUTOR

Graphing utilities perform the established order of operations when evaluating an expression. To see this, try entering the expressions

$$1200\left(1 + \frac{0.09}{12}\right)^{12\cdot 6}$$

and

$$1200 \times 1 + \left(\frac{0.09}{12}\right)^{12\cdot 6}$$

into your graphing utility to see that the expressions result in different values.

Operations with Exponents

1. Multiplying like bases: $x^n x^m = x^{n+m}$ Add exponents.
2. Dividing like bases: $\dfrac{x^n}{x^m} = x^{n-m}$ Subtract exponents.
3. Removing parentheses: $(xy)^n = x^n y^n$

   $\left(\dfrac{x}{y}\right)^n = \dfrac{x^n}{y^n}$

   $(x^n)^m = x^{nm}$
4. Special conventions: $-x^n = -(x^n), \quad -x^n \neq (-x)^n$

   $cx^n = c(x^n), \quad cx^n \neq (cx)^n$

   $x^{n^m} = x^{(n^m)}, \quad x^{n^m} \neq (x^n)^m$

Example 3 **Simplifying Expressions with Exponents**

Simplify each expression.

a. $2x^2(x^3)$ **b.** $(3x)^2\sqrt[3]{x}$ **c.** $\dfrac{3x^2}{(x^{1/2})^3}$

d. $\dfrac{5x^4}{(x^2)^3}$ **e.** $x^{-1}(2x^2)$ **f.** $\dfrac{-\sqrt{x}}{5x^{-1}}$

SOLUTION

a. $2x^2(x^3) = 2x^{2+3} = 2x^5$ $\quad x^n x^m = x^{n+m}$

b. $(3x)^2\sqrt[3]{x} = 9x^2x^{1/3} = 9x^{2+(1/3)} = 9x^{7/3}$ $\quad x^n x^m = x^{n+m}$

c. $\dfrac{3x^2}{(x^{1/2})^3} = 3\left(\dfrac{x^2}{x^{3/2}}\right) = 3x^{2-(3/2)} = 3x^{1/2}$ $\quad (x^n)^m = x^{nm}, \ \dfrac{x^n}{x^m} = x^{n-m}$

d. $\dfrac{5x^4}{(x^2)^3} = \dfrac{5x^4}{x^6} = 5x^{4-6} = 5x^{-2} = \dfrac{5}{x^2}$ $\quad (x^n)^m = x^{nm}, \ \dfrac{x^n}{x^m} = x^{n-m}$

e. $x^{-1}(2x^2) = 2x^{-1}x^2 = 2x^{-1+2} = 2x$ $\quad x^n x^m = x^{n+m}$

f. $\dfrac{-\sqrt{x}}{5x^{-1}} = -\dfrac{1}{5}\left(\dfrac{x^{1/2}}{x^{-1}}\right) = -\dfrac{1}{5}x^{(1/2)+1} = -\dfrac{1}{5}x^{3/2}$ $\quad \dfrac{x^n}{x^m} = x^{n-m}$

✓ **Checkpoint 3**

Simplify each expression.

a. $3x^2(x^4)$

b. $(2x)^3\sqrt{x}$

c. $\dfrac{4x^2}{(x^{1/3})^2}$

■

Note in Example 3 that one characteristic of simplified expressions is the absence of negative exponents. Another characteristic of simplified expressions is that sums and differences are written in *factored form*. To do this, you can use the **Distributive Property.**

$$abx^n + acx^{n+m} = ax^n(b + cx^m)$$

Study the next example carefully to be sure that you understand the concepts involved in the factoring process.

Example 4 Simplifying by Factoring

Simplify each expression by factoring.

a. $2x^2 - x^3$ **b.** $2x^3 + x^2$ **c.** $2x^{1/2} + 4x^{5/2}$ **d.** $2x^{-1/2} + 3x^{5/2}$

SOLUTION

a. $2x^2 - x^3 = x^2(2 - x)$

b. $2x^3 + x^2 = x^2(2x + 1)$

c. $2x^{1/2} + 4x^{5/2} = 2x^{1/2}(1 + 2x^2)$

d. $2x^{-1/2} + 3x^{5/2} = x^{-1/2}(2 + 3x^3) = \dfrac{2 + 3x^3}{\sqrt{x}}$

✓Checkpoint 4

Simplify each expression by factoring.

a. $x^3 - 2x$

b. $2x^{1/2} + 8x^{3/2}$ ■

STUDY TIP

To check that a simplified expression is equivalent to the original expression, try substituting values for x into each expression.

Many algebraic expressions obtained in calculus occur in unsimplified form. For instance, the two expressions shown in the following example are the result of an operation in calculus called *differentiation*. [The first is the derivative of $2(x + 1)^{3/2}(2x - 3)^{5/2}$, and the second is the derivative of $2(x + 1)^{1/2}(2x - 3)^{5/2}$.]

Example 5 Simplifying by Factoring

a.
$$\begin{aligned}
&3(x + 1)^{1/2}(2x - 3)^{5/2} + 10(x + 1)^{3/2}(2x - 3)^{3/2}\\
&\quad = (x + 1)^{1/2}(2x - 3)^{3/2}[3(2x - 3) + 10(x + 1)]\\
&\quad = (x + 1)^{1/2}(2x - 3)^{3/2}(6x - 9 + 10x + 10)\\
&\quad = (x + 1)^{1/2}(2x - 3)^{3/2}(16x + 1)
\end{aligned}$$

b.
$$\begin{aligned}
&(x + 1)^{-1/2}(2x - 3)^{5/2} + 10(x + 1)^{1/2}(2x - 3)^{3/2}\\
&\quad = (x + 1)^{-1/2}(2x - 3)^{3/2}[(2x - 3) + 10(x + 1)]\\
&\quad = (x + 1)^{-1/2}(2x - 3)^{3/2}(2x - 3 + 10x + 10)\\
&\quad = (x + 1)^{-1/2}(2x - 3)^{3/2}(12x + 7)\\
&\quad = \frac{(2x - 3)^{3/2}(12x + 7)}{(x + 1)^{1/2}}
\end{aligned}$$

✓Checkpoint 5

Simplify the expression by factoring.

$$(x + 2)^{1/2}(3x - 1)^{3/2} + 4(x + 2)^{-1/2}(3x - 1)^{5/2}$$ ■

Example 6 shows some additional types of expressions that can occur in calculus. [The expression in Example 6(d) is an antiderivative of $(x+1)^{2/3}(2x+3)$, and the expression in Example 6(e) is the derivative of $(x+2)^3/(x-1)^3$.]

TECH TUTOR

A graphing utility offers several ways to calculate rational exponents and radicals. You should be familiar with the x-squared key $\boxed{x^2}$. This key squares the value of an expression.

For rational exponents or exponents other than 2, use the $\boxed{\wedge}$ key.

For radical expressions, you can use the square root key $\boxed{\sqrt{\ }}$, the cube root key $\boxed{\sqrt[3]{\ }}$, or the xth root key $\boxed{\sqrt[x]{\ }}$. Consult your graphing utility user's guide for specific keystrokes you can use to evaluate rational exponents and radical expressions.

Example 6 Factors Involving Quotients

Simplify each expression by factoring.

a. $\dfrac{3x^2+x^4}{2x}$

b. $\dfrac{\sqrt{x}+x^{3/2}}{x}$

c. $(9x+2)^{-1/3}+18(9x+2)$

d. $\dfrac{3}{5}(x+1)^{5/3}+\dfrac{3}{4}(x+1)^{8/3}$

e. $\dfrac{3(x+2)^2(x-1)^3-3(x+2)^3(x-1)^2}{[(x-1)^3]^2}$

SOLUTION

a. $\dfrac{3x^2+x^4}{2x}=\dfrac{x^2(3+x^2)}{2x}=\dfrac{x^{2-1}(3+x^2)}{2}=\dfrac{x(3+x^2)}{2}$

b. $\dfrac{\sqrt{x}+x^{3/2}}{x}=\dfrac{x^{1/2}(1+x)}{x}=\dfrac{1+x}{x^{1-(1/2)}}=\dfrac{1+x}{\sqrt{x}}$

c. $(9x+2)^{-1/3}+18(9x+2)=(9x+2)^{-1/3}[1+18(9x+2)^{4/3}]$

$$=\frac{1+18(9x+2)^{4/3}}{\sqrt[3]{9x+2}}$$

d. $\dfrac{3}{5}(x+1)^{5/3}+\dfrac{3}{4}(x+1)^{8/3}=\dfrac{12}{20}(x+1)^{5/3}+\dfrac{15}{20}(x+1)^{8/3}$

$$=\frac{3}{20}(x+1)^{5/3}[4+5(x+1)]$$

$$=\frac{3}{20}(x+1)^{5/3}(4+5x+5)$$

$$=\frac{3}{20}(x+1)^{5/3}(5x+9)$$

e. $\dfrac{3(x+2)^2(x-1)^3-3(x+2)^3(x-1)^2}{[(x-1)^3]^2}$

$$=\frac{3(x+2)^2(x-1)^2[(x-1)-(x+2)]}{(x-1)^6}$$

$$=\frac{3(x+2)^2(x-1-x-2)}{(x-1)^{6-2}}$$

$$=\frac{-9(x+2)^2}{(x-1)^4}$$

✓ Checkpoint 6

Simplify the expression by factoring.

$$\frac{5x^3+x^6}{3x}$$

Domain of an Algebraic Expression

When working with algebraic expressions involving x, you face the potential difficulty of substituting a value of x for which the expression is not defined (does not produce a real number). For example, the expression $\sqrt{2x+3}$ is *not defined* when $x = -2$ because

$$\sqrt{2(-2)+3} = \sqrt{-1}$$

is not a real number.

The set of all values for which an expression is defined is called its **domain.** So, the domain of $\sqrt{2x+3}$ is the set of all values of x such that $\sqrt{2x+3}$ is a real number. In order for $\sqrt{2x+3}$ to represent a real number, it is necessary that

$$2x + 3 \geq 0. \qquad \text{Expression must be nonnegative.}$$

In other words, $\sqrt{2x+3}$ is defined only for those values of x that lie in the interval $\left[-\frac{3}{2}, \infty\right)$, as shown in Figure A.14.

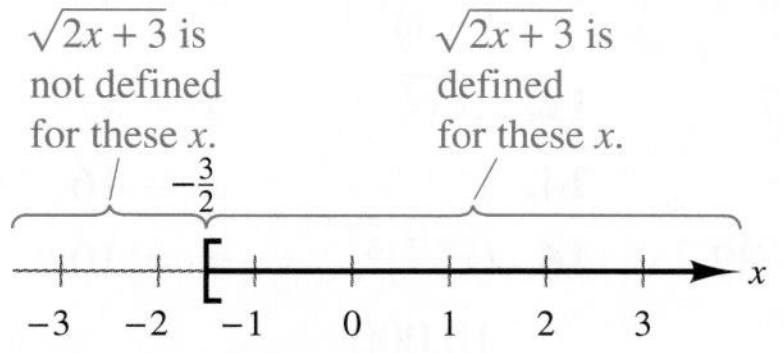

FIGURE A.14

Example 7 **Finding the Domain of an Expression**

Find the domain of each expression.

a. $\sqrt{3x-2}$

b. $\dfrac{1}{\sqrt{3x-2}}$

c. $\sqrt[3]{9x+1}$

SOLUTION

a. The domain of $\sqrt{3x-2}$ consists of all x such that

$$3x - 2 \geq 0 \qquad \text{Expression must be nonnegative.}$$

which implies that $x \geq \frac{2}{3}$. So, the domain is $\left[\frac{2}{3}, \infty\right)$.

b. The domain of $1/\sqrt{3x-2}$ is the same as the domain of $\sqrt{3x-2}$, except that $1/\sqrt{3x-2}$ is not defined when $3x - 2 = 0$. Because this occurs when $x = \frac{2}{3}$, the domain is $\left(\frac{2}{3}, \infty\right)$.

c. Because $\sqrt[3]{9x+1}$ is defined for all real numbers, its domain is $(-\infty, \infty)$.

Checkpoint 7

Find the domain of each expression.

a. $\sqrt{x-2}$

b. $\dfrac{1}{\sqrt{x-2}}$

c. $\sqrt[3]{x-2}$

Exercises A.3

See www.CalcChat.com for worked-out solutions to odd-numbered exercises.

Evaluating Expressions In Exercises 1–20, evaluate the expression for the given value of x. *See Examples 1 and 2.*

	Expression	*x-Value*		*Expression*	*x-Value*
1.	$-2x^3$	$x = 3$	**2.**	$\dfrac{x^2}{3}$	$x = 6$
3.	$4x^{-3}$	$x = 2$	**4.**	$7x^{-2}$	$x = 5$
5.	$\dfrac{1 + x^{-1}}{x^{-1}}$	$x = 3$	**6.**	$x - 4x^{-2}$	$x = 3$
7.	$3x^2 - 4x^3$	$x = -2$	**8.**	$5(-x)^3$	$x = 3$
9.	$6x^0 - (6x)^0$	$x = 10$	**10.**	$\dfrac{1}{(-x)^{-3}}$	$x = 4$
11.	$\sqrt[3]{x^2}$	$x = 27$	**12.**	$\sqrt{x^3}$	$x = \frac{1}{9}$
13.	$x^{-1/2}$	$x = 4$	**14.**	$x^{-3/4}$	$x = 16$
15.	$x^{-2/5}$	$x = -32$	**16.**	$(x^{2/3})^3$	$x = 10$
17.	$500x^{60}$	$x = 1.01$	**18.**	$\dfrac{10{,}000}{x^{120}}$	$x = 1.1$
19.	$\sqrt[3]{x}$	$x = -54$	**20.**	$\sqrt[6]{x}$	$x = 325$

Simplifying Expressions with Exponents In Exercises 21–30, simplify the expression. *See Example 3.*

21. $6y^{-2}(2y^4)^{-3}$ **22.** $z^{-3}(3z^4)$

23. $10(x^2)^2$ **24.** $(4x^3)^2$

25. $\dfrac{7x^2}{x^{-3}}$ **26.** $\dfrac{x^{-3}}{\sqrt{x}}$

27. $\dfrac{10(x + y)^3}{4(x + y)^{-2}}$ **28.** $\left(\dfrac{12s^2}{9s}\right)^3$

29. $\dfrac{3x\sqrt{x}}{x^{1/2}}$ **30.** $\left(\sqrt[3]{x^2}\right)^3$

Simplifying Radicals In Exercises 31–36, simplify by removing all possible factors from the radical.

31. $\sqrt{8}$ **32.** $\sqrt[3]{\frac{16}{27}}$

33. $\sqrt[3]{54x^5}$ **34.** $\sqrt[4]{(3x^2y^3)^4}$

35. $\sqrt[3]{144x^9y^{-4}z^5}$ **36.** $\sqrt[4]{32xy^5z^{-8}}$

Simplifying by Factoring In Exercises 37–44, simplify each expression by factoring. *See Examples 4, 5, and 6.*

37. $3x^3 - 12x$

38. $8x^4 - 6x^2$

39. $2x^{5/2} + x^{-1/2}$

40. $5x^{3/2} - x^{-3/2}$

41. $3x(x + 1)^{3/2} - 6(x + 1)^{1/2}$

42. $2x(x - 1)^{5/2} - 4(x - 1)^{3/2}$

43. $\dfrac{5x^6 + x^3}{3x^2}$

44. $\dfrac{(x + 1)(x - 1)^2 - (x - 1)^3}{(x + 1)^2}$

Finding the Domain of an Expression In Exercises 45–52, find the domain of the expression. *See Example 7.*

45. $\sqrt{x - 4}$ **46.** $\sqrt{5 - 2x}$

47. $\sqrt{x^2 + 3}$ **48.** $\sqrt{4x^2 + 1}$

49. $\dfrac{1}{\sqrt[3]{x - 4}}$

50. $\dfrac{1}{\sqrt[3]{x + 4}}$

51. $\dfrac{\sqrt{x + 2}}{1 - x}$

52. $\dfrac{1}{\sqrt{2x + 3}} + \sqrt{6 - 4x}$

Compound Interest In Exercises 53–56, a certificate of deposit has a principal of P dollars and an annual percentage rate of r (expressed as a decimal) compounded n times per year. The balance A in the account is given by

$$A = P\left(1 + \frac{r}{n}\right)^N$$

where N is the number of compoundings. Use a graphing utility to find the balance in the account.

53. $P = \$10{,}000, \quad r = 6.5\%, \quad n = 12, \quad N = 120$

54. $P = \$7000, \quad r = 5\%, \quad n = 365, \quad N = 1000$

55. $P = \$5000, \quad r = 5.5\%, \quad n = 4, \quad N = 60$

56. $P = \$8000, \quad r = 7\%, \quad n = 6, \quad N = 90$

57. Period of a Pendulum The period of a pendulum is

$$T = 2\pi\sqrt{\frac{L}{32}}$$

where T is the period (in seconds) and L is the length (in feet) of the pendulum. Find the period of a pendulum whose length is 4 feet.

58. Annuity After n annual payments of P dollars have been made into an annuity earning an annual percentage rate of r compounded annually, the balance A is given by

$$A = P(1 + r) + P(1 + r)^2 + \cdot \cdot \cdot + P(1 + r)^n.$$

Rewrite this formula by completing the following factorization.

$$A = P(1 + r)(\qquad)$$

A.4 Factoring Polynomials

- Use special products and factorization techniques to factor polynomials.
- Use synthetic division to factor polynomials of degree three or more.
- Use the Rational Zero Theorem to find the real zeros of polynomials.

Factorization Techniques

The **Fundamental Theorem of Algebra** states that every nth-degree polynomial

$$a_n x^n + a_{n-1}x^{n-1} + \cdot \cdot \cdot + a_1 x + a_0, \quad a_n \neq 0$$

has precisely n **zeros.** (The zeros may be repeated or imaginary.) The zeros of a polynomial in x are the values of x that make the polynomial zero. The problem of finding the zeros of a polynomial is equivalent to the problem of factoring the polynomial into linear factors.

Special Products and Factorization Techniques

Quadratic Formula	*Example*
$ax^2 + bx + c = 0 \Rightarrow x = \dfrac{-b \pm \sqrt{b^2 - 4ac}}{2a}$	$x^2 + 3x - 1 = 0 \Rightarrow x = \dfrac{-3 \pm \sqrt{13}}{2}$
Special Products	*Examples*
$x^2 - a^2 = (x - a)(x + a)$	$x^2 - 9 = (x - 3)(x + 3)$
$x^3 - a^3 = (x - a)(x^2 + ax + a^2)$	$x^3 - 8 = (x - 2)(x^2 + 2x + 4)$
$x^3 + a^3 = (x + a)(x^2 - ax + a^2)$	$x^3 + 64 = (x + 4)(x^2 - 4x + 16)$
$x^4 - a^4 = (x - a)(x + a)(x^2 + a^2)$	$x^4 - 16 = (x - 2)(x + 2)(x^2 + 4)$
Binomial Theorem	*Examples*
$(x + a)^2 = x^2 + 2ax + a^2$	$(x + 3)^2 = x^2 + 6x + 9$
$(x - a)^2 = x^2 - 2ax + a^2$	$(x^2 - 5)^2 = x^4 - 10x^2 + 25$
$(x + a)^3 = x^3 + 3ax^2 + 3a^2x + a^3$	$(x + 2)^3 = x^3 + 6x^2 + 12x + 8$
$(x - a)^3 = x^3 - 3ax^2 + 3a^2x - a^3$	$(x - 1)^3 = x^3 - 3x^2 + 3x - 1$
$(x + a)^4 = x^4 + 4ax^3 + 6a^2x^2 + 4a^3x + a^4$	$(x + 2)^4 = x^4 + 8x^3 + 24x^2 + 32x + 16$
$(x - a)^4 = x^4 - 4ax^3 + 6a^2x^2 - 4a^3x + a^4$	$(x - 4)^4 = x^4 - 16x^3 + 96x^2 - 256x + 256$

$$(x + a)^n = x^n + nax^{n-1} + \frac{n(n-1)}{2!}a^2x^{n-2} + \frac{n(n-1)(n-2)}{3!}a^3x^{n-3} + \cdot \cdot \cdot + na^{n-1}x + a^n{}^*$$

$$(x - a)^n = x^n - nax^{n-1} + \frac{n(n-1)}{2!}a^2x^{n-2} - \frac{n(n-1)(n-2)}{3!}a^3x^{n-3} + \cdot \cdot \cdot \pm na^{n-1}x \mp a^n$$

Factoring by Grouping	*Example*
$acx^3 + adx^2 + bcx + bd = ax^2(cx + d) + b(cx + d)$	$3x^3 - 2x^2 - 6x + 4 = x^2(3x - 2) - 2(3x - 2)$
$= (ax^2 + b)(cx + d)$	$= (x^2 - 2)(3x - 2)$

* The factorial symbol ! is defined as follows: $0! = 1,\ 1! = 1,\ 2! = 2 \cdot 1 = 2,\ 3! = 3 \cdot 2 \cdot 1 = 6,\ 4! = 4 \cdot 3 \cdot 2 \cdot 1 = 24$, and so on.

Example 1 Applying the Quadratic Formula

Use the Quadratic Formula to find all real zeros of each polynomial.

a. $4x^2 + 6x + 1$

b. $x^2 + 6x + 9$

c. $2x^2 - 6x + 5$

SOLUTION

a. Using $a = 4$, $b = 6$, and $c = 1$, you can write

$$\begin{aligned} x = \frac{-b \pm \sqrt{b^2 - 4ac}}{2a} &= \frac{-6 \pm \sqrt{6^2 - 4(4)(1)}}{2(4)} \\ &= \frac{-6 \pm \sqrt{36 - 16}}{8} \\ &= \frac{-6 \pm \sqrt{20}}{8} \\ &= \frac{-6 \pm 2\sqrt{5}}{8} \\ &= \frac{2\left(-3 \pm \sqrt{5}\right)}{2(4)} \\ &= \frac{-3 \pm \sqrt{5}}{4}. \end{aligned}$$

So, there are two real zeros:

$$x = \frac{-3 - \sqrt{5}}{4} \approx -1.309 \quad \text{and} \quad x = \frac{-3 + \sqrt{5}}{4} \approx -0.191.$$

b. In this case, $a = 1$, $b = 6$, and $c = 9$, and the Quadratic Formula yields

$$x = \frac{-b \pm \sqrt{b^2 - 4ac}}{2a} = \frac{-6 \pm \sqrt{36 - 36}}{2} = -\frac{6}{2} = -3.$$

So, there is one (repeated) real zero: $x = -3$.

c. For this quadratic equation, $a = 2$, $b = -6$, and $c = 5$. So,

$$x = \frac{-b \pm \sqrt{b^2 - 4ac}}{2a} = \frac{6 \pm \sqrt{36 - 40}}{4} = \frac{6 \pm \sqrt{-4}}{4}.$$

Because $\sqrt{-4}$ is imaginary, there are no real zeros.

✓ Checkpoint 1

Use the Quadratic Formula to find all real zeros of each polynomial.

a. $2x^2 + 4x + 1$

b. $x^2 - 8x + 16$

c. $2x^2 - x + 5$ ■

The zeros in Example 1(a) are irrational, and the zeros in Example 1(c) are imaginary. In both of these cases the quadratic is said to be **irreducible** because it cannot be factored into linear factors with rational coefficients. The next example shows how to find the zeros associated with *reducible* quadratics. In this example, factoring is used to find the zeros of each quadratic. Try using the Quadratic Formula to obtain the same zeros.

Recall that the zeros of a polynomial in x are the values of x that make the polynomial zero. To find the zeros, factor the polynomial into linear factors and set each factor equal to zero. For instance, the zeros of $(x - 2)(x - 3)$ occur when $x - 2 = 0$ and $x - 3 = 0$.

Example 2 Finding Real Zeros by Factoring

Find all the real zeros of each quadratic polynomial.

a. $x^2 - 5x + 6$ **b.** $x^2 - 6x + 9$ **c.** $2x^2 + 5x - 3$

SOLUTION

a. Because

$$x^2 - 5x + 6 = (x - 2)(x - 3)$$

the zeros are $x = 2$ and $x = 3$.

b. Because

$$x^2 - 6x + 9 = (x - 3)^2$$

the only zero is $x = 3$.

c. Because

$$2x^2 + 5x - 3 = (2x - 1)(x + 3)$$

the zeros are $x = \frac{1}{2}$ and $x = -3$.

✓ Checkpoint 2

Find all the real zeros of each quadratic polynomial.

a. $x^2 - 2x - 15$ **b.** $x^2 + 2x + 1$ **c.** $2x^2 - 7x + 6$ ■

Example 3 Finding the Domain of a Radical Expression

Find the domain of $\sqrt{x^2 - 3x + 2}$.

Values of $\sqrt{x^2 - 3x + 2}$

x	$\sqrt{x^2 - 3x + 2}$
0	$\sqrt{2}$
1	0
1.5	Undefined
2	0
3	$\sqrt{2}$

SOLUTION Because

$$x^2 - 3x + 2 = (x - 1)(x - 2)$$

you know that the zeros of the quadratic are $x = 1$ and $x = 2$. So, you need to test the sign of the quadratic in the three intervals $(-\infty, 1)$, $(1, 2)$, and $(2, \infty)$, as shown in Figure A.15. After testing each of these intervals, you can see that the quadratic is negative in the center interval and positive in the outer two intervals. Moreover, because the quadratic is zero when $x = 1$ and $x = 2$, you can conclude that the domain of $\sqrt{x^2 - 3x + 2}$ is

$$(-\infty, 1] \cup [2, \infty). \qquad \text{Domain}$$

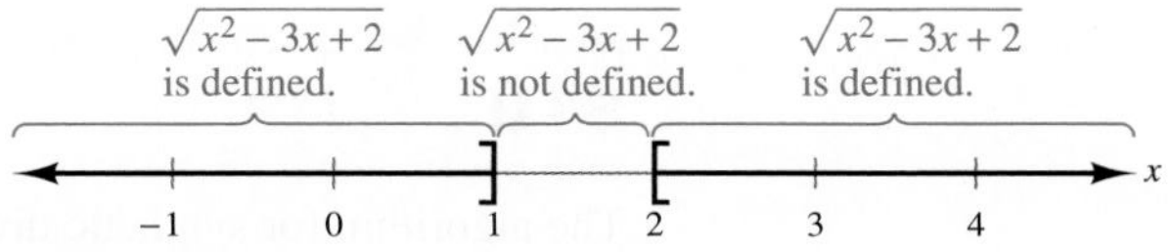

FIGURE A.15

✓ Checkpoint 3

Find the domain of

$$\sqrt{x^2 + x - 2}.$$

■

Factoring Polynomials of Degree Three or More

It can be difficult to find the zeros of polynomials of degree three or more. However, if one of the zeros of a polynomial is known, then you can use that zero to reduce the degree of the polynomial. For example, if you know that $x = 2$ is a zero of

$$x^3 - 4x^2 + 5x - 2$$

then you know that $(x - 2)$ is a factor, and you can use long division to factor the polynomial as shown.

$$\begin{aligned} x^3 - 4x^2 + 5x - 2 &= (x - 2)(x^2 - 2x + 1) \\ &= (x - 2)(x - 1)(x - 1) \end{aligned}$$

As an alternative to long division, many people prefer to use **synthetic division** to reduce the degree of a polynomial.

Synthetic Division for a Cubic Polynomial

Given: $x = x_1$ is a zero of $ax^3 + bx^2 + cx + d$.

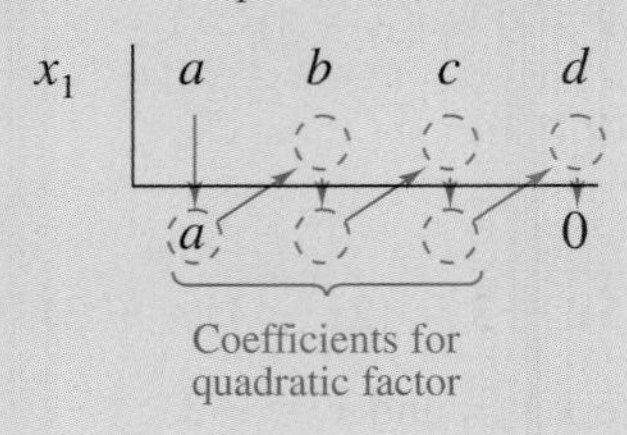

Vertical pattern: *Add terms.*

Diagonal pattern: *Multiply by x_1.*

Performing synthetic division on the polynomial

$$x^3 - 4x^2 + 5x - 2$$

using the given zero, $x = 2$, produces the following.

$$\begin{array}{r|rrrr} 2 & 1 & -4 & 5 & -2 \\ & & 2 & -4 & 2 \\ \hline & 1 & -2 & 1 & 0 \end{array}$$

$(x - 2)(x^2 - 2x + 1) = x^3 - 4x^2 + 5x - 2$

When you use synthetic division, remember to take *all* coefficients into account—*even when some of them are zero.* For instance, when you know that $x = -2$ is a zero of $x^3 + 3x + 14$, you can apply synthetic division as shown.

$$\begin{array}{r|rrrr} -2 & 1 & 0 & 3 & 14 \\ & & -2 & 4 & -14 \\ \hline & 1 & -2 & 7 & 0 \end{array}$$

$(x + 2)(x^2 - 2x + 7) = x^3 + 3x + 14$

STUDY TIP

The algorithm for synthetic division given above works *only* for divisors of the form $x - x_1$. Remember that $x + x_1 = x - (-x_1)$.

The Rational Zero Theorem

There is a systematic way to find the *rational* zeros of a polynomial. You can use the **Rational Zero Theorem** (also called the Rational Root Theorem).

> **Rational Zero Theorem**
>
> If a polynomial
>
> $$a_nx^n + a_{n-1}x^{n-1} + \cdot \cdot \cdot + a_1x + a_0$$
>
> has integer coefficients, then every *rational* zero is of the form
>
> $$x = \frac{p}{q}$$
>
> where p is a factor of a_0, and q is a factor of a_n.

Example 4 Using the Rational Zero Theorem

Find all real zeros of the polynomial.

$$2x^3 + 3x^2 - 8x + 3$$

SOLUTION

$$(2)x^3 + 3x^2 - 8x + (3)$$

Factors of constant term: $\pm1, \pm3$

Factors of leading coefficient: $\pm1, \pm2$

The possible rational zeros are the factors of the constant term divided by the factors of the leading coefficient.

$$1, -1, 3, -3, \frac{1}{2}, -\frac{1}{2}, \frac{3}{2}, -\frac{3}{2}$$

By testing these possible zeros, you can see that $x = 1$ works.

$$2(1)^3 + 3(1)^2 - 8(1) + 3 = 2 + 3 - 8 + 3 = 0$$

Now, by synthetic division you have the following.

1	2	3	−8	3
		2	5	−3
	2	5	−3	0

$(x - 1)(2x^2 + 5x - 3) = 2x^3 + 3x^2 - 8x + 3$

Finally, by factoring the quadratic

$$2x^2 + 5x - 3 = (2x - 1)(x + 3)$$

you have

$$2x^3 + 3x^2 - 8x + 3 = (x - 1)(2x - 1)(x + 3)$$

and you can conclude that the zeros are $x = 1$, $x = \frac{1}{2}$, and $x = -3$.

STUDY TIP

In Example 4, you can check that the zeros are correct by substituting into the original polynomial.

Check that $x = 1$ is a zero.

$$2(1)^3 + 3(1)^2 - 8(1) + 3$$
$$= 2 + 3 - 8 + 3$$
$$= 0$$

Check that $x = \frac{1}{2}$ is a zero.

$$2\left(\frac{1}{2}\right)^3 + 3\left(\frac{1}{2}\right)^2 - 8\left(\frac{1}{2}\right) + 3$$
$$= \frac{1}{4} + \frac{3}{4} - 4 + 3$$
$$= 0$$

Check that $x = -3$ is a zero.

$$2(-3)^3 + 3(-3)^2 - 8(-3) + 3$$
$$= -54 + 27 + 24 + 3$$
$$= 0$$

✓ Checkpoint 4

Find all real zeros of the polynomial.

$$2x^3 - 3x^2 - 3x + 2$$

■

Exercises A.4

See www.CalcChat.com for worked-out solutions to odd-numbered exercises.

Applying the Quadratic Formula In Exercises 1–8, use the Quadratic Formula to find all real zeros of the second-degree polynomial. *See Example 1.*

1. $6x^2 - 7x + 1$
2. $8x^2 - 2x - 1$
3. $4x^2 - 12x + 9$
4. $9x^2 + 12x + 4$
5. $y^2 + 4y + 1$
6. $y^2 + 5y - 2$
7. $2x^2 + 3x - 4$
8. $3x^2 - 8x - 4$

Factoring Polynomials In Exercises 9–18, write the second-degree polynomial as the product of two linear factors.

9. $x^2 - 4x + 4$
10. $x^2 + 10x + 25$
11. $4x^2 + 4x + 1$
12. $9x^2 - 12x + 4$
13. $3x^2 - 4x + 1$
14. $2x^2 - x - 1$
15. $3x^2 - 5x + 2$
16. $4x^2 + 19x + 12$
17. $x^2 - 4xy + 4y^2$
18. $x^2 - xy - 2y^2$

Factoring Polynomials In Exercises 19–34, completely factor the polynomial.

19. $81 - y^4$
20. $x^4 - 16$
21. $x^3 - 8$
22. $y^3 - 64$
23. $y^3 + 64$
24. $z^3 + 125$
25. $x^3 - y^3$
26. $(x - a)^3 + b^3$
27. $x^3 - 4x^2 - x + 4$
28. $x^3 - x^2 - x + 1$
29. $2x^3 - 3x^2 + 4x - 6$
30. $x^3 - 5x^2 - 5x + 25$
31. $2x^3 - 4x^2 - x + 2$
32. $x^3 - 7x^2 - 4x + 28$
33. $x^4 - 15x^2 - 16$
34. $2x^4 - 49x^2 - 25$

Finding Real Zeros by Factoring In Exercises 35–54, find all real zeros of the polynomial. *See Example 2.*

35. $x^2 - 5x$
36. $2x^2 - 3x$
37. $x^2 - 9$
38. $x^2 - 25$
39. $x^2 - 3$
40. $x^2 - 8$
41. $(x - 3)^2 - 9$
42. $(x + 1)^2 - 36$
43. $x^2 + x - 2$
44. $x^2 + 5x + 6$
45. $x^2 - 5x - 6$
46. $x^2 + x - 20$
47. $3x^2 + 5x + 2$
48. $2x^2 - x - 1$
49. $x^3 + 64$
50. $x^3 - 216$
51. $x^4 - 16$
52. $x^4 - 625$
53. $x^3 - x^2 - 4x + 4$
54. $2x^3 + x^2 + 6x + 3$

Finding the Domain of a Radical Expression In Exercises 55–60, find the domain of the expression. *See Example 3.*

55. $\sqrt{x^2 - 4}$
56. $\sqrt{4 - x^2}$
57. $\sqrt{x^2 - 7x + 12}$
58. $\sqrt{x^2 - 8x + 15}$
59. $\sqrt{5x^2 + 6x + 1}$
60. $\sqrt{3x^2 - 10x + 3}$

Using Synthetic Division In Exercises 61–64, use synthetic division to complete the indicated factorization.

61. $x^3 - 3x^2 - 6x - 2 = (x + 1)(\quad)$
62. $x^3 - 2x^2 - x + 2 = (x - 2)(\quad)$
63. $2x^3 - x^2 - 2x + 1 = (x + 1)(\quad)$
64. $x^4 - 16x^3 + 96x^2 - 256x + 256 = (x - 4)(\quad)$

Using the Rational Zero Theorem In Exercises 65–74, use the Rational Zero Theorem to find all real zeros of the polynomial. *See Example 4.*

65. $x^3 - x^2 - 10x - 8$
66. $x^3 - 7x - 6$
67. $x^3 - 6x^2 + 11x - 6$
68. $x^3 + 2x^2 - 5x - 6$
69. $6x^3 - 11x^2 - 19x - 6$
70. $18x^3 - 9x^2 - 8x + 4$
71. $x^3 - 3x^2 - 3x - 4$
72. $2x^3 - x^2 - 13x - 6$
73. $4x^3 + 11x^2 + 5x - 2$
74. $3x^3 + 4x^2 - 13x + 6$

75. Production Level The minimum average cost of producing x units of a product occurs when the production level is set at the (positive) solution of

$$0.0003x^2 - 1200 = 0.$$

How many solutions does this equation have? Find and interpret the solution(s) in the context of the problem. What production level will minimize the average cost?

76. Profit The profit P (in dollars) from sales is given by

$$P = -200x^2 + 2000x - 3800$$

where x is the number of units sold per day (in hundreds). Determine the interval for x such that the profit will be greater than \$1000.

A.5 Fractions and Rationalization

- Simplify rational expressions.
- Add and subtract rational expressions.
- Simplify rational expressions involving radicals.
- Rationalize numerators and denominators of rational expressions.

Simplifying Rational Expressions

In this section, you will review operations involving fractional expressions such as

$$\frac{2}{x}, \quad \frac{x^2 + 2x - 4}{x + 6}, \quad \text{and} \quad \frac{1}{\sqrt{x^2 + 1}}.$$

The first two expressions have polynomials as both numerator and denominator and are called **rational expressions.** A rational expression is **proper** when the degree of the numerator is less than the degree of the denominator. For example,

$$\frac{x}{x^2 + 1}$$

is proper. If the degree of the numerator is greater than or equal to the degree of the denominator, then the rational expression is **improper.** For example,

$$\frac{x^2}{x^2 + 1} \quad \text{and} \quad \frac{x^3 + 2x + 1}{x + 1}$$

are both improper.

A fraction is in simplest form when its numerator and denominator have no factors in common aside from ± 1. To write a fraction in simplest form, divide out common factors.

$$\frac{a \cdot c}{b \cdot c} = \frac{a}{b}, \quad c \neq 0$$

The key to success in simplifying rational expressions lies in your ability to factor polynomials. When simplifying rational expressions, be sure to factor each polynomial completely before concluding that the numerator and denominator have no common factors.

Example 1 Simplifying a Rational Expression

Write $\dfrac{12 + x - x^2}{2x^2 - 9x + 4}$ in simplest form.

SOLUTION

$$\frac{12 + x - x^2}{2x^2 - 9x + 4} = \frac{(4 - x)(3 + x)}{(2x - 1)(x - 4)} \qquad \text{Factor completely.}$$

$$= \frac{-\cancel{(x - 4)}(3 + x)}{(2x - 1)\cancel{(x - 4)}} \qquad (4 - x) = -(x - 4)$$

$$= -\frac{3 + x}{2x - 1}, \quad x \neq 4 \qquad \text{Divide out common factors.}$$

STUDY TIP

To simplify a rational expression, it may be necessary to change the sign of a factor by factoring out (-1), as shown in Example 1.

✓ Checkpoint 1

Write $\dfrac{x^2 + 8x - 20}{x^2 + 11x + 10}$ in simplest form.

Operations with Fractions

Operations with Fractions

1. Add fractions (find a common denominator):

$$\frac{a}{b}+\frac{c}{d}=\frac{a}{b}\left(\frac{d}{d}\right)+\frac{c}{d}\left(\frac{b}{b}\right)=\frac{ad}{bd}+\frac{bc}{bd}=\frac{ad+bc}{bd},\quad b\neq 0, d\neq 0$$

2. Subtract fractions (find a common denominator):

$$\frac{a}{b}-\frac{c}{d}=\frac{a}{b}\left(\frac{d}{d}\right)-\frac{c}{d}\left(\frac{b}{b}\right)=\frac{ad}{bd}-\frac{bc}{bd}=\frac{ad-bc}{bd},\quad b\neq 0, d\neq 0$$

3. Multiply fractions:

$$\left(\frac{a}{b}\right)\left(\frac{c}{d}\right)=\frac{ac}{bd},\quad b\neq 0, d\neq 0$$

4. Divide fractions (invert and multiply):

$$\frac{a/b}{c/d}=\left(\frac{a}{b}\right)\left(\frac{d}{c}\right)=\frac{ad}{bc},\quad b\neq 0, c\neq 0, d\neq 0$$

$$\frac{a/b}{c}=\frac{a/b}{c/1}=\left(\frac{a}{b}\right)\left(\frac{1}{c}\right)=\frac{a}{bc},\quad b\neq 0, c\neq 0$$

5. Divide out common factors:

$$\frac{\not{a}b}{\not{a}c}=\frac{b}{c},\quad a\neq 0, c\neq 0$$

$$\frac{ab+ac}{ad}=\frac{\not{a}(b+c)}{\not{a}d}=\frac{b+c}{d},\quad a\neq 0, d\neq 0$$

Example 2 **Adding and Subtracting Rational Expressions**

Perform each indicated operation and simplify.

a. $x+\frac{1}{x}$ **b.** $\frac{1}{x+1}-\frac{2}{2x-1}$

SOLUTION

a. $x+\frac{1}{x}=\frac{x^2}{x}+\frac{1}{x}$ Write with common denominator.

$=\frac{x^2+1}{x}$ Add fractions.

b.
$$\frac{1}{x+1}-\frac{2}{2x-1}=\frac{(2x-1)}{(x+1)(2x-1)}-\frac{2(x+1)}{(x+1)(2x-1)}$$
$$=\frac{2x-1-2x-2}{2x^2+x-1}$$
$$=\frac{-3}{2x^2+x-1}$$

✓ Checkpoint 2

Perform each indicated operation and simplify.

a. $x+\frac{2}{x}$ **b.** $\frac{2}{x+1}-\frac{1}{2x+1}$

■

In adding (or subtracting) fractions whose denominators have no common factors, it is convenient to use the following pattern.

$$\frac{a}{b} + \frac{c}{d} = \frac{a}{b} \times \frac{c}{d} = \frac{ad + bc}{bd}$$

For instance, in Example 2(b), you could have used this pattern as shown.

$$\frac{1}{x+1} - \frac{2}{2x-1} = \frac{(2x-1) - 2(x+1)}{(x+1)(2x-1)}$$

$$= \frac{2x - 1 - 2x - 2}{(x+1)(2x-1)}$$

$$= \frac{-3}{2x^2 + x - 1}$$

In Example 2, the denominators of the rational expressions have no common factors. When the denominators do have common factors, it is best to find the least common denominator before adding or subtracting. For instance, when adding

$$\frac{1}{x} \quad \text{and} \quad \frac{2}{x^2}$$

you can recognize that the least common denominator is x^2 and write

$$\frac{1}{x} + \frac{2}{x^2} = \frac{x}{x^2} + \frac{2}{x^2}$$ Write with common denominator.

$$= \frac{x+2}{x^2}.$$ Add fractions.

This is further demonstrated in Example 3.

Example 3 **Adding Rational Expressions**

Add the rational expressions.

$$\frac{x}{x^2 - 1} + \frac{3}{x+1}$$

SOLUTION Because $x^2 - 1 = (x+1)(x-1)$, the least common denominator is $x^2 - 1$.

$$\frac{x}{x^2 - 1} + \frac{3}{x+1} = \frac{x}{(x-1)(x+1)} + \frac{3}{x+1}$$ Factor.

$$= \frac{x}{(x-1)(x+1)} + \frac{3(x-1)}{(x-1)(x+1)}$$ Write with common denominator.

$$= \frac{x + 3(x-1)}{(x-1)(x+1)}$$ Add fractions.

$$= \frac{x + 3x - 3}{(x-1)(x+1)}$$ Multiply.

$$= \frac{4x - 3}{x^2 - 1}$$ Simplify.

✓Checkpoint 3

Add the rational expressions.

$$\frac{x}{x^2 - 4} + \frac{2}{x - 2}$$

■

Example 4 Subtracting Rational Expressions

Subtract the rational expressions.

$$\frac{1}{2(x^2 + 2x)} - \frac{1}{4x}$$

SOLUTION In this case, the least common denominator is $4x(x + 2)$.

$$\frac{1}{2(x^2 + 2x)} - \frac{1}{4x} = \frac{1}{2x(x + 2)} - \frac{1}{2(2x)} \qquad \text{Factor.}$$

$$= \frac{2}{2(2x)(x + 2)} - \frac{x + 2}{2(2x)(x + 2)} \qquad \text{Write with common denominator.}$$

$$= \frac{2 - (x + 2)}{4x(x + 2)} \qquad \text{Subtract fractions.}$$

$$= \frac{2 - x - 2}{4x(x + 2)} \qquad \text{Remove parentheses.}$$

$$= \frac{-\cancel{x}}{4\cancel{x}(x + 2)} \qquad \text{Divide out common factor.}$$

$$= \frac{-1}{4(x + 2)}, \quad x \neq 0 \qquad \text{Simplify.}$$

✓ Checkpoint 4

Subtract the rational expressions.

$$\frac{1}{3(x^2 + 2x)} - \frac{1}{3x}$$

Example 5 Combining Three Rational Expressions

Perform the operations and simplify.

$$\frac{3}{x - 1} - \frac{2}{x} + \frac{x + 3}{x^2 - 1}$$

SOLUTION Using the factored denominators $(x - 1)$, x, and $(x + 1)(x - 1)$, you can see that the least common denominator is $x(x + 1)(x - 1)$.

$$\frac{3}{x - 1} - \frac{2}{x} + \frac{x + 3}{x^2 - 1} = \frac{3(x)(x + 1)}{x(x + 1)(x - 1)} - \frac{2(x + 1)(x - 1)}{x(x + 1)(x - 1)} + \frac{(x + 3)(x)}{x(x + 1)(x - 1)}$$

$$= \frac{3(x)(x + 1) - 2(x + 1)(x - 1) + (x + 3)(x)}{x(x + 1)(x - 1)}$$

$$= \frac{3x^2 + 3x - 2x^2 + 2 + x^2 + 3x}{x(x + 1)(x - 1)}$$

$$= \frac{2x^2 + 6x + 2}{x(x + 1)(x - 1)}$$

$$= \frac{2(x^2 + 3x + 1)}{x(x + 1)(x - 1)}$$

✓ Checkpoint 5

Perform the operations and simplify.

$$\frac{4}{x} - \frac{2}{x^2} + \frac{4}{x + 3}$$

Expressions Involving Radicals

In calculus, the operation of differentiation tends to produce "messy" expressions when applied to fractional expressions. This is especially true when the fractional expressions involve radicals. When differentiation is used, it is important to be able to simplify these expressions in order to obtain more manageable forms. The expressions in Example 6 are the results of differentiation. In each case, note how much *simpler* the simplified form is than the original form.

Example 6 Simplifying an Expression with Radicals

Simplify each expression.

a. $$\frac{\sqrt{x+1} - \dfrac{x}{2\sqrt{x+1}}}{x+1}$$

b. $$\left(\frac{1}{x+\sqrt{x^2+1}}\right)\left(1+\frac{2x}{2\sqrt{x^2+1}}\right)$$

SOLUTION

a. $$\frac{\sqrt{x+1} - \dfrac{x}{2\sqrt{x+1}}}{x+1} = \frac{\dfrac{2(x+1)}{2\sqrt{x+1}} - \dfrac{x}{2\sqrt{x+1}}}{x+1}$$ Write with common denominator.

$$= \frac{\dfrac{2x+2-x}{2\sqrt{x+1}}}{\dfrac{x+1}{1}}$$ Subtract fractions.

$$= \frac{x+2}{2\sqrt{x+1}}\left(\frac{1}{x+1}\right)$$ To divide, invert and multiply.

$$= \frac{x+2}{2(x+1)^{3/2}}$$ Multiply.

b. $$\left(\frac{1}{x+\sqrt{x^2+1}}\right)\left(1+\frac{2x}{2\sqrt{x^2+1}}\right) = \left(\frac{1}{x+\sqrt{x^2+1}}\right)\left(1+\frac{x}{\sqrt{x^2+1}}\right)$$

$$= \left(\frac{1}{x+\sqrt{x^2+1}}\right)\left(\frac{\sqrt{x^2+1}}{\sqrt{x^2+1}}+\frac{x}{\sqrt{x^2+1}}\right)$$

$$= \left(\frac{1}{x+\sqrt{x^2+1}}\right)\left(\frac{x+\sqrt{x^2+1}}{\sqrt{x^2+1}}\right)$$

$$= \frac{1}{\sqrt{x^2+1}}$$

✓ Checkpoint 6

Simplify each expression.

a. $$\frac{\sqrt{x+2} - \dfrac{x}{4\sqrt{x+2}}}{x+2}$$

b. $$\left(\frac{1}{x+\sqrt{x^2+4}}\right)\left(1+\frac{x}{\sqrt{x^2+4}}\right)$$

Rationalization Techniques

In working with quotients involving radicals, it is often convenient to move the radical expression from the denominator to the numerator, or vice versa. For example, you can move $\sqrt{2}$ from the denominator to the numerator in the following quotient by multiplying by $\sqrt{2}/\sqrt{2}$.

Radical in Denominator		*Rationalize*		*Radical in Numerator*
$\dfrac{1}{\sqrt{2}}$	⇨	$\dfrac{1}{\sqrt{2}}\left(\dfrac{\sqrt{2}}{\sqrt{2}}\right)$	⇨	$\dfrac{\sqrt{2}}{2}$

This process is called **rationalizing the denominator.** A similar process is used to **rationalize the numerator.**

> **STUDY TIP**
>
> The success of the second and third rationalizing techniques stems from the following.
>
> $$(\sqrt{a}-\sqrt{b})(\sqrt{a}+\sqrt{b}) = a - b$$

Rationalizing Techniques

1. When the denominator is $\sqrt{a}$, multiply by $\dfrac{\sqrt{a}}{\sqrt{a}}$.
2. When the denominator is $\sqrt{a}-\sqrt{b}$, multiply by $\dfrac{\sqrt{a}+\sqrt{b}}{\sqrt{a}+\sqrt{b}}$.
3. When the denominator is $\sqrt{a}+\sqrt{b}$, multiply by $\dfrac{\sqrt{a}-\sqrt{b}}{\sqrt{a}-\sqrt{b}}$.

The same guidelines apply to rationalizing numerators.

Example 7 Rationalizing Denominators and Numerators

Rationalize the denominator or numerator.

a. $\dfrac{3}{\sqrt{12}}$ **b.** $\dfrac{\sqrt{x+1}}{2}$ **c.** $\dfrac{1}{\sqrt{5}+\sqrt{2}}$ **d.** $\dfrac{1}{\sqrt{x}-\sqrt{x+1}}$

SOLUTION

a. $$\frac{3}{\sqrt{12}} = \frac{3}{2\sqrt{3}} = \frac{3}{2\sqrt{3}}\left(\frac{\sqrt{3}}{\sqrt{3}}\right) = \frac{3\sqrt{3}}{2(3)} = \frac{\sqrt{3}}{2}$$

b. $$\frac{\sqrt{x+1}}{2} = \frac{\sqrt{x+1}}{2}\left(\frac{\sqrt{x+1}}{\sqrt{x+1}}\right) = \frac{x+1}{2\sqrt{x+1}}$$

c. $$\frac{1}{\sqrt{5}+\sqrt{2}} = \frac{1}{\sqrt{5}+\sqrt{2}}\left(\frac{\sqrt{5}-\sqrt{2}}{\sqrt{5}-\sqrt{2}}\right) = \frac{\sqrt{5}-\sqrt{2}}{5-2} = \frac{\sqrt{5}-\sqrt{2}}{3}$$

d. $$\frac{1}{\sqrt{x}-\sqrt{x+1}} = \frac{1}{\sqrt{x}-\sqrt{x+1}}\left(\frac{\sqrt{x}+\sqrt{x+1}}{\sqrt{x}+\sqrt{x+1}}\right)$$
$$= \frac{\sqrt{x}+\sqrt{x+1}}{x-(x+1)}$$
$$= -\sqrt{x}-\sqrt{x+1}$$

✓Checkpoint 7

Rationalize the denominator or numerator.

a. $\dfrac{5}{\sqrt{8}}$ **b.** $\dfrac{\sqrt{x+2}}{4}$ **c.** $\dfrac{1}{\sqrt{6}-\sqrt{3}}$ **d.** $\dfrac{1}{\sqrt{x}+\sqrt{x+2}}$

Exercises A.5

See www.CalcChat.com for worked-out solutions to odd-numbered exercises.

Simplifying a Rational Expression In Exercises 1–4, write the rational expression in simplest form. *See Example 1.*

1. $\dfrac{x^2 - 7x + 12}{x^2 + 3x - 18}$ **2.** $\dfrac{x^2 - 5x - 6}{x^2 + 11x + 10}$

3. $\dfrac{x^2 + 3x - 10}{2x^2 - x - 6}$ **4.** $\dfrac{3x^2 + 13x + 12}{x^2 - 4x - 21}$

Adding and Subtracting Rational Expressions In Exercises 5–16, perform the indicated operations and simplify. *See Examples 2, 3, 4, and 5.*

5. $\dfrac{x}{x - 2} + \dfrac{3}{x - 2}$ **6.** $\dfrac{5x + 10}{2x - 1} - \dfrac{2x + 10}{2x - 1}$

7. $x - \dfrac{3}{x}$ **8.** $3x + \dfrac{2}{x^2}$

9. $\dfrac{2}{x - 3} + \dfrac{5x}{3x + 4}$ **10.** $\dfrac{3}{3x - 1} - \dfrac{1}{x + 2}$

11. $\dfrac{2}{x^2 - 4} - \dfrac{1}{x - 2}$ **12.** $\dfrac{5}{x^2 - 9} + \dfrac{x}{x + 3}$

13. $\dfrac{x}{x^2 + x - 2} - \dfrac{1}{x + 2}$ **14.** $\dfrac{2}{x + 1} + \dfrac{3x - 2}{x^2 - 2x - 3}$

15. $\dfrac{2}{x^2 + 1} - \dfrac{1}{x} + \dfrac{1}{x^3 + x}$ **16.** $\dfrac{3}{x + 2} + \dfrac{3}{x - 2} + \dfrac{1}{x^2 - 4}$

Simplifying an Expression with Radicals In Exercises 17–28, simplify the expression. *See Example 6.*

17. $\dfrac{-x}{(x + 1)^{3/2}} + \dfrac{2}{(x + 1)^{1/2}}$

18. $2\sqrt{x}(x - 2) + \dfrac{(x - 2)^2}{2\sqrt{x}}$

19. $\dfrac{2 - t}{2\sqrt{1 + t}} - \sqrt{1 + t}$

20. $-\dfrac{\sqrt{x^2 + 1}}{x^2} + \dfrac{1}{\sqrt{x^2 + 1}}$

21. $\left(2x\sqrt{x^2 + 1} - \dfrac{x^3}{\sqrt{x^2 + 1}}\right) \div (x^2 + 1)$

22. $\left(\sqrt{x^3 + 1} - \dfrac{3x^3}{2\sqrt{x^3 + 1}}\right) \div (x^3 + 1)$

23. $\dfrac{(x^2 + 2)^{1/2} - x^2(x^2 + 2)^{-1/2}}{x^2}$

24. $\dfrac{x(x + 1)^{-1/2} - (x + 1)^{1/2}}{x^2}$

25. $\dfrac{\dfrac{\sqrt{x + 1}}{\sqrt{x}} - \dfrac{\sqrt{x}}{\sqrt{x + 1}}}{2(x + 1)}$

26. $\dfrac{\dfrac{2x^2}{3(x^2 - 1)^{2/3}} - (x^2 - 1)^{1/3}}{x^2}$

27. $\dfrac{-x^2}{(2x + 3)^{3/2}} + \dfrac{2x}{(2x + 3)^{1/2}}$

28. $\dfrac{-x}{2(3 + x^2)^{3/2}} + \dfrac{3}{(3 + x^2)^{1/2}}$

Rationalizing Denominators and Numerators In Exercises 29–42, rationalize the denominator or numerator and simplify. *See Example 7.*

29. $\dfrac{2}{\sqrt{10}}$ **30.** $\dfrac{3}{\sqrt{21}}$

31. $\dfrac{4x}{\sqrt{x - 1}}$ **32.** $\dfrac{5y}{\sqrt{y + 7}}$

33. $\dfrac{49(x - 3)}{\sqrt{x^2 - 9}}$ **34.** $\dfrac{10(x + 2)}{\sqrt{x^2 - x - 6}}$

35. $\dfrac{5}{\sqrt{14} - 2}$ **36.** $\dfrac{13}{6 + \sqrt{10}}$

37. $\dfrac{1}{\sqrt{6} + \sqrt{5}}$ **38.** $\dfrac{x}{\sqrt{2} + \sqrt{3}}$

39. $\dfrac{2}{\sqrt{x} + \sqrt{x - 2}}$ **40.** $\dfrac{10}{\sqrt{x} + \sqrt{x + 5}}$

41. $\dfrac{\sqrt{x + 2} - \sqrt{2}}{x}$ **42.** $\dfrac{\sqrt{x + 1} - 1}{x}$

43. Installment Loan The monthly payment M (in dollars) for an installment loan is given by the formula

$$M = P\left[\frac{r/12}{1 - \left(\dfrac{1}{(r/12) + 1}\right)^N}\right]$$

where P is the amount of the loan (in dollars), r is the annual percentage rate (in decimal form), and N is the number of monthly payments. Enter the formula into a graphing utility, and use it to find the monthly payment for a loan of \$10,000 at an annual percentage rate of 7.5% ($r = 0.075$) for 5 years ($N = 60$ monthly payments).

44. Inventory A retailer has determined that the cost C (in dollars) of ordering and storing x units of a product is

$$C = 6x + \frac{900{,}000}{x}.$$

(a) Write the expression for cost as a single fraction.

(b) Which order size should the retailer place: 240 units, 387 units, or 480 units? Explain your reasoning.

B Alternative Introduction to the Fundamental Theorem of Calculus

In this appendix, a summation process is used to provide an alternative development of the definite integral. It is intended that this supplement follow Section 5.3 in the text. If used, this appendix should replace the material preceding Example 2 in Section 5.4. Example 1 below shows how the area of a region in the plane can be approximated by the use of rectangles.

Example 1 Using Rectangles to Approximate the Area of a Region

Use the four rectangles shown in Figure B.1 to approximate the area of the region lying between the graph of

$$f(x) = \frac{x^2}{2}$$

and the x-axis, between $x = 0$ and $x = 4$.

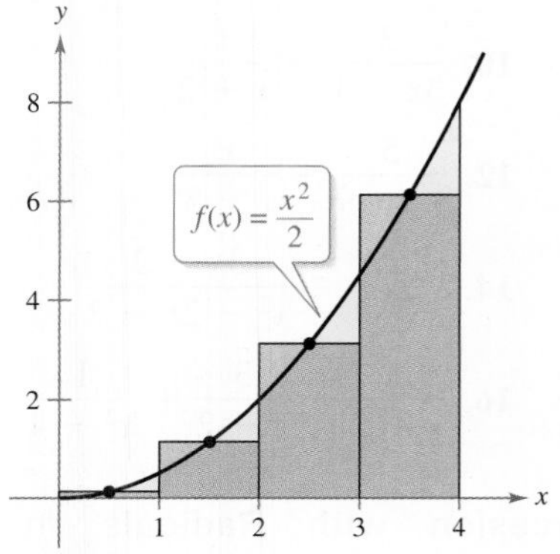

FIGURE B.1

SOLUTION You can find the heights of the rectangles by evaluating the function f at each of the midpoints of the subintervals

$$[0, 1], \quad [1, 2], \quad [2, 3], \quad [3, 4].$$

Because the width of each rectangle is 1, the sum of the areas of the four rectangles is

$$\begin{aligned} S &= \overbrace{(1)}^{\text{width}}\overbrace{f\left(\frac{1}{2}\right)}^{\text{height}} + \overbrace{(1)}^{\text{width}}\overbrace{f\left(\frac{3}{2}\right)}^{\text{height}} + \overbrace{(1)}^{\text{width}}\overbrace{f\left(\frac{5}{2}\right)}^{\text{height}} + \overbrace{(1)}^{\text{width}}\overbrace{f\left(\frac{7}{2}\right)}^{\text{height}} \\ &= \frac{1}{8} + \frac{9}{8} + \frac{25}{8} + \frac{49}{8} \\ &= \frac{84}{8} \\ &= 10.5. \end{aligned}$$

So, you can approximate the area of the region to be 10.5 square units.

STUDY TIP

The approximation technique used in Example 1 is called the *Midpoint Rule*. The Midpoint Rule is discussed further in Section 5.6.

The procedure shown in Example 1 can be generalized. Let f be a continuous function defined on the closed interval $[a, b]$. To begin, partition the interval into n subintervals, each of width

$$\Delta x = \frac{b-a}{n}$$

as shown.

$$a = x_0 < x_1 < x_2 < \cdot \cdot \cdot < x_{n-1} < x_n = b$$

In each subinterval $[x_{i-1}, x_i]$, choose an arbitrary point c_i and form the sum

$$S = f(c_1)\,\Delta x + f(c_2)\,\Delta x + \cdot \cdot \cdot + f(c_{n-1})\,\Delta x + f(c_n)\,\Delta x.$$

This type of summation is called a **Riemann sum** and is often written using summation notation, as shown below.

$$S = \sum_{i=1}^{n} f(c_i)\,\Delta x, \quad x_{i-1} \le c_i \le x_i$$

For the Riemann sum in Example 1, the interval is $[a, b] = [0, 4]$, the number of subintervals is $n = 4$, the width of each subinterval is $\Delta x = 1$, and the point c_i in each subinterval is its midpoint. So, you can write the approximation in Example 1 as

$$\begin{aligned} S &= \sum_{i=1}^{n} f(c_i)\,\Delta x \\ &= \sum_{i=1}^{4} f(c_i)(1) \\ &= \frac{1}{8} + \frac{9}{8} + \frac{25}{8} + \frac{49}{8} \\ &= \frac{84}{8}. \end{aligned}$$

Example 2 Using a Riemann Sum to Approximate Area

Use a Riemann sum to approximate the area of the region bounded by the graph of

$$f(x) = -x^2 + 2x$$

and the x-axis, for $0 \le x \le 2$. In the Riemann sum, let $n = 6$ and choose c_i to be the left endpoint of each subinterval.

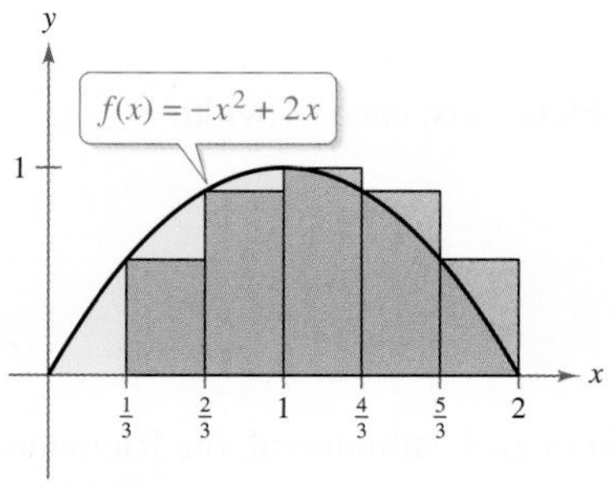

FIGURE B.2

SOLUTION Subdivide the interval $[0, 2]$ into six subintervals, each of width

$$\begin{aligned} \Delta x &= \frac{2-0}{6} \\ &= \frac{1}{3} \end{aligned}$$

as shown in Figure B.2. Because c_i is the left endpoint of each subinterval, the Riemann sum is given by

$$\begin{aligned} S &= \sum_{i=1}^{n} f(c_i)\,\Delta x \\ &= \left[f(0) + f\left(\frac{1}{3}\right) + f\left(\frac{2}{3}\right) + f(1) + f\left(\frac{4}{3}\right) + f\left(\frac{5}{3}\right)\right]\left(\frac{1}{3}\right) \\ &= \left[0 + \frac{5}{9} + \frac{8}{9} + 1 + \frac{8}{9} + \frac{5}{9}\right]\left(\frac{1}{3}\right) \\ &= \frac{35}{27} \text{ square units.} \end{aligned}$$

Example 2 illustrates an important point. If a function f is continuous and nonnegative over the interval $[a, b]$, then the Riemann sum

$$S = \sum_{i=1}^{n} f(c_i)\,\Delta x$$

can be used to approximate the area of the region bounded by the graph of f and the x-axis, between $x = a$ and $x = b$. Moreover, for a given interval, as the number of subintervals increases, the approximation to the actual area will improve. This is illustrated in the next two examples by using Riemann sums to approximate the area of a triangle.

Example 3 Approximating the Area of a Triangle

Use a Riemann sum to approximate the area of the triangular region bounded by the graph of

$$f(x) = 2x$$

and the x-axis, $0 \le x \le 3$. Use a partition of six subintervals and choose c_i to be the left endpoint of each subinterval.

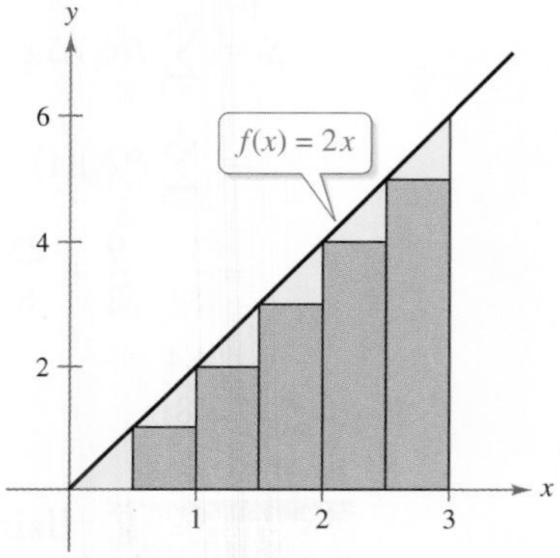

FIGURE B.3

SOLUTION Subdivide the interval $[0, 3]$ into six subintervals, each of width

$$\Delta x = \frac{3-0}{6} = \frac{1}{2}$$

as shown in Figure B.3. Because c_i is the left endpoint of each subinterval, the Riemann sum is given by

$$\begin{aligned} S &= \sum_{i=1}^{n} f(c_i)\,\Delta x \\ &= \left[f(0) + f\left(\frac{1}{2}\right) + f(1) + f\left(\frac{3}{2}\right) + f(2) + f\left(\frac{5}{2}\right)\right]\left(\frac{1}{2}\right) \\ &= [0 + 1 + 2 + 3 + 4 + 5]\left(\frac{1}{2}\right) \\ &= \frac{15}{2} \text{ square units.} \end{aligned}$$

TECH TUTOR

Most graphing utilities are able to sum the first n terms of a sequence. Try using a graphing utility to verify the right Riemann sum in Example 3.

The approximations in Examples 2 and 3 are called **left Riemann sums,** because c_i was chosen to be the left endpoint of each subinterval. Using the right endpoints in Example 3, the **right Riemann sum** is $\frac{21}{2}$. Note that the exact area of the triangular region in Example 3 is

$$\text{Area} = \frac{1}{2}(\text{base})(\text{height})$$
$$= \frac{1}{2}(3)(6)$$
$$= 9 \text{ square units.}$$

So, the left Riemann sum gives an approximation that is less than the actual area, and the right Riemann sum gives an approximation that is greater than the actual area.

In Example 4, you will see that the approximation improves as the number of subintervals increases.

Example 4 Increasing the Number of Subintervals

Let $f(x) = 2x$, $0 \leq x \leq 3$. Use a graphing utility to determine the left and right Riemann sums for $n = 10$, $n = 100$, and $n = 1000$ subintervals.

SOLUTION A graphing utility program for this problem is shown in Figure B.4. [Note that the function $f(x) = 2x$ is entered as Y1.]

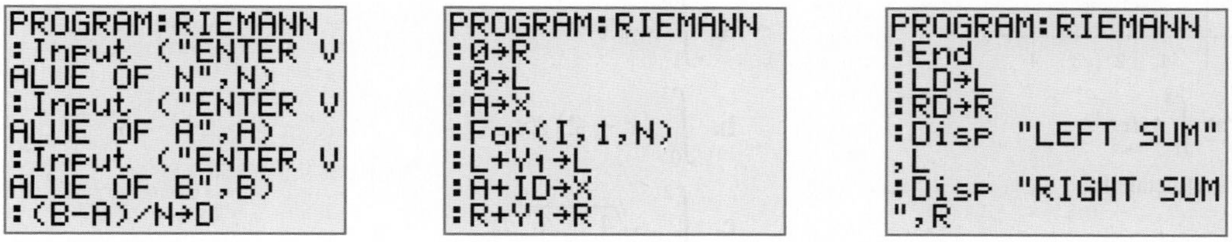

FIGURE B.4

Running this program for $n = 10$, $n = 100$, and $n = 1000$ gives the results shown in the table.

n	Left Reimann sum	Right Reimann sum
10	8.100	9.900
100	8.910	9.090
1000	8.991	9.009

From the results of Example 4, it appears that the Riemann sums are approaching the limit 9 as n approaches infinity. It is this observation that motivates the definition of a **definite integral.** In this definition, consider the partition of $[a, b]$ into n subintervals of equal width $\Delta x = (b - a)/n$, as shown.

$$a = x_0 < x_1 < x_2 < \cdot \cdot \cdot < x_{n-1} < x_n = b$$

Moreover, consider c_i to be an arbitrary point in the ith subinterval $[x_{i-1}, x_i]$. To say that the number of subintervals n approaches infinity is equivalent to saying that the width, Δx, of the subintervals approaches zero.

Definition of Definite Integral

If f is a continuous function defined on the closed interval $[a, b]$, then the **definite integral of f on $[a, b]$** is

$$\int_a^b f(x)\,dx = \lim_{\Delta x \to 0} \sum_{i=1}^{n} f(c_i)\,\Delta x$$

$$= \lim_{n \to \infty} \sum_{i=1}^{n} f(c_i)\,\Delta x.$$

If f is continuous and nonnegative on the interval $[a, b]$, then the definite integral of f on $[a, b]$ gives the area of the region bounded by the graph of f, the x-axis, and the vertical lines $x = a$ and $x = b$.

Evaluation of a definite integral by its limit definition can be difficult. However, there are times when a definite integral can be solved by recognizing that it represents the area of a common type of geometric figure.

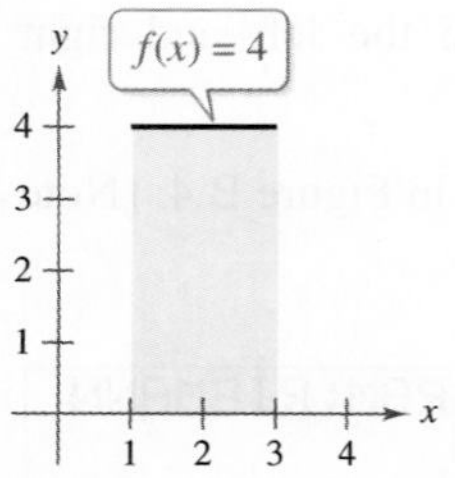

(a) $\int_1^3 4\,dx$

Rectangle

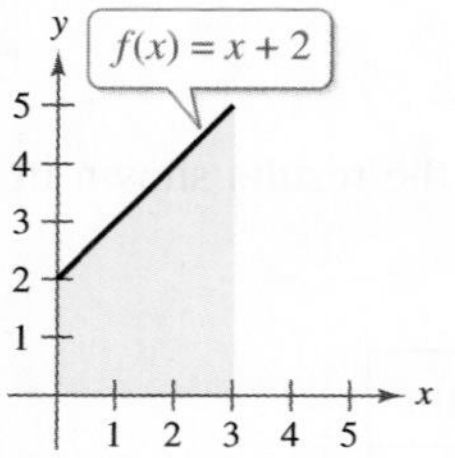

(b) $\int_0^3 (x + 2)\,dx$

Trapezoid

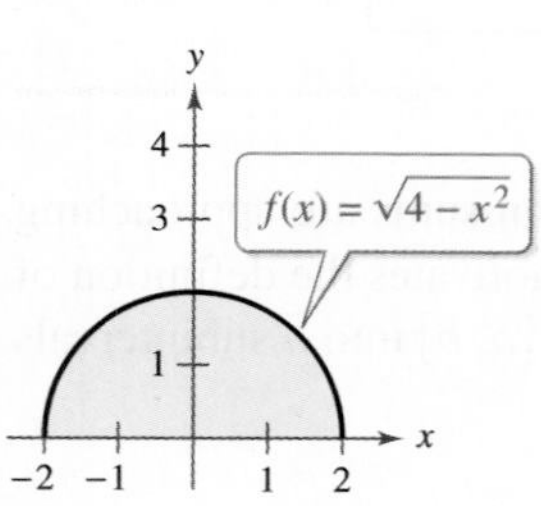

(c) $\int_{-2}^{2} \sqrt{4 - x^2}\,dx$

Semicircle

FIGURE B.5

Example 5 The Areas of Common Geometric Figures

Sketch the region corresponding to each of the definite integrals. Then evaluate each definite integral using a geometric formula.

a. $\int_1^3 4\,dx$

b. $\int_0^3 (x + 2)\,dx$

c. $\int_{-2}^{2} \sqrt{4 - x^2}\,dx$

SOLUTION A sketch of each region is shown in Figure B.5.

a. The region associated with this definite integral is a rectangle of height 4 and width 2. Moreover, because the function $f(x) = 4$ is continuous and nonnegative on the interval $[1, 3]$, you can conclude that the area of the rectangle is given by the definite integral. So, the value of the definite integral is

$$\int_1^3 4\,dx = 4(2) = 8 \text{ square units.}$$

b. The region associated with this definite integral is a trapezoid with an altitude of 3 and parallel bases of lengths 2 and 5. The formula for the area of a trapezoid is $\frac{1}{2}h(b_1 + b_2)$, and so you have

$$\int_0^3 (x + 2)\,dx = \frac{1}{2}(3)(2 + 5)$$

$$= \frac{21}{2} \text{ square units.}$$

c. The region associated with this definite integral is a semicircle of radius 2. The formula for the area of a semicircle is $\frac{1}{2}\pi r^2$, and so you have

$$\int_{-2}^{2} \sqrt{4 - x^2}\,dx = \frac{1}{2}\pi(2^2)$$

$$= 2\pi \text{ square units.}$$

For some simple functions, it is possible to evaluate definite integrals by the Riemann sum definition. In the next example, you will use the fact that the sum of the first n integers is given by the formula

$$1 + 2 + \cdot \cdot \cdot + n = \sum_{i=1}^{n} i = \frac{n(n+1)}{2}$$

See Exercise 29.

to compute the area of the triangular region in Examples 3 and 4.

Example 6 Evaluating a Definite Integral by Its Definition

Evaluate $\int_0^3 2x\, dx$.

SOLUTION Let

$$\Delta x = \frac{b-a}{n} = \frac{3}{n}$$

and choose c_i to be the right endpoint of each subinterval,

$$c_i = \frac{3i}{n}.$$

Then you have

$$\begin{aligned}\int_0^3 2x\, dx &= \lim_{\Delta x \to 0} \sum_{i=1}^{n} f(c_i)\Delta x \\ &= \lim_{n\to\infty} \sum_{i=1}^{n} 2\left(i\frac{3}{n}\right)\left(\frac{3}{n}\right) \\ &= \lim_{n\to\infty} \frac{18}{n^2} \sum_{i=1}^{n} i \\ &= \lim_{n\to\infty} \left(\frac{18}{n^2}\right)\left(\frac{n(n+1)}{2}\right) \\ &= \lim_{n\to\infty} \left(9 + \frac{9}{n}\right).\end{aligned}$$

This limit can be evaluated in the same way that you calculated horizontal asymptotes in Section 3.6. In particular, as n approaches infinity, you see that $9/n$ approaches 0, and the limit above is 9. So, you can conclude that

$$\int_0^3 2x\, dx = 9.$$

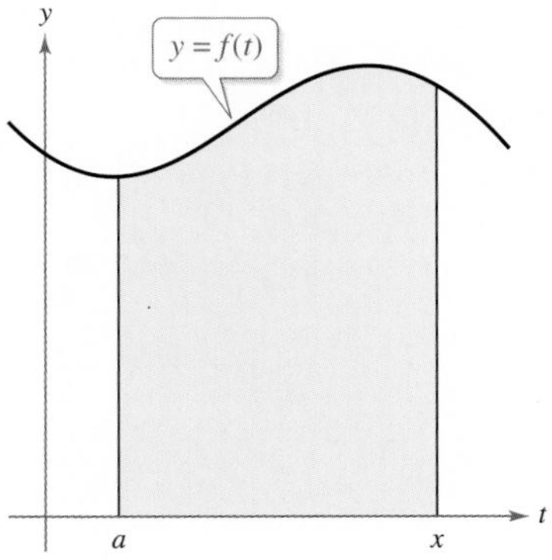

FIGURE B.6

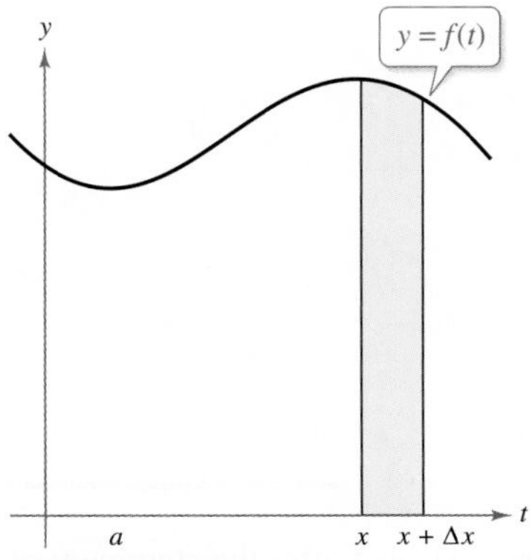

FIGURE B.7

From Example 6, you can see that it can be difficult to evaluate the definite integral of even a simple function by using Riemann sums. A computer can help in calculating these sums for large values of n, but this procedure would give only an approximation of the definite integral. Fortunately, the **Fundamental Theorem of Calculus** provides a technique for evaluating definite integrals using antiderivatives, and for this reason it is often thought to be the most important theorem in calculus. In the remainder of this appendix, you will see how derivatives and integrals are related via the Fundamental Theorem of Calculus.

To simplify the discussion, assume that f is a continuous nonnegative function defined on the interval $[a, b]$. Let $A(x)$ be the area of the region under the graph of f from a to x, as indicated in Figure B.6. The area under the shaded region in Figure B.7 is

$$A(x + \Delta x) - A(x).$$

If Δx is small, then this area is approximated by the area of the rectangle of height $f(x)$ and width Δx. So, you have

$$A(x + \Delta x) - A(x) \approx f(x)\,\Delta x.$$

Dividing by Δx produces

$$f(x) \approx \frac{A(x + \Delta x) - A(x)}{\Delta x}.$$

By taking the limit as Δx approaches 0, you can see that

$$\begin{aligned} f(x) &= \lim_{\Delta x \to 0} \frac{A(x + \Delta x) - A(x)}{\Delta x} \\ &= A'(x) \end{aligned}$$

and you can establish the fact that the area function $A(x)$ is an antiderivative of f. Although it was assumed that f is continuous and nonnegative, this development is valid when the function f is simply continuous on the closed interval $[a, b]$. This result is used in the proof of the Fundamental Theorem of Calculus.

Fundamental Theorem of Calculus

If f is a continuous function on the closed interval $[a, b]$, then

$$\int_a^b f(x)dx = F(b) - F(a)$$

where F is any function such that $F'(x) = f(x)$.

PROOF From the discussion above, you know that

$$\int_a^x f(x)\,dx = A(x)$$

and in particular,

$$A(a) = \int_a^a f(x)\,dx = 0$$

and

$$A(b) = \int_a^b f(x)\,dx.$$

If F is *any* antiderivative of f, then you know that F differs from A by a constant. That is,

$$A(x) = F(x) + C.$$

So,

$$\begin{aligned} \int_a^b f(x)\,dx &= A(b) - A(a) \\ &= [F(b) + C] - [F(a) + C] \\ &= F(b) + C - F(a) - C \\ &= F(b) - F(a). \end{aligned}$$

You are now ready to continue Section 5.4, on page 341, just after the statement of the Fundamental Theorem of Calculus.

Exercises B

See www.CalcChat.com for worked-out solutions to odd-numbered exercises.

Using Rectangles to Approximate the Area of a Region In Exercises 1 and 2, use the rectangles to approximate the area of the region. *See Example 1.*

1. $y = x + 1$

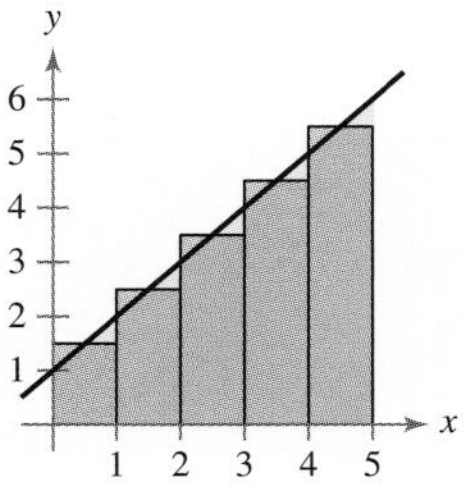

2. $y = 4 - x^2$

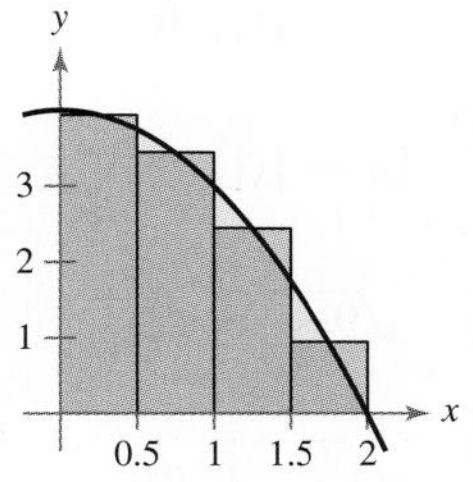

Using a Riemann Sum to Approximate Area In Exercises 3–8, use the left Riemann sum and the right Riemann sum to approximate the area of the region using the indicated number of subintervals. *See Examples 2 and 3.*

3. $y = \sqrt{x}$

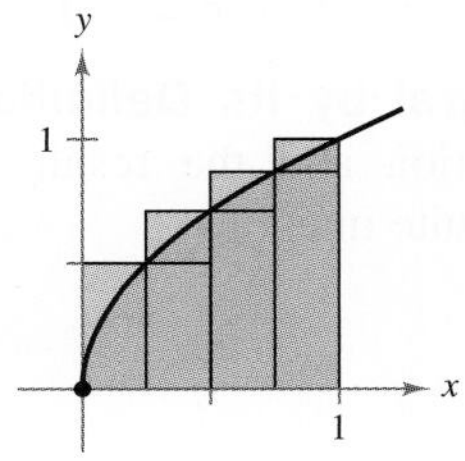

4. $y = \sqrt{x} + 1$

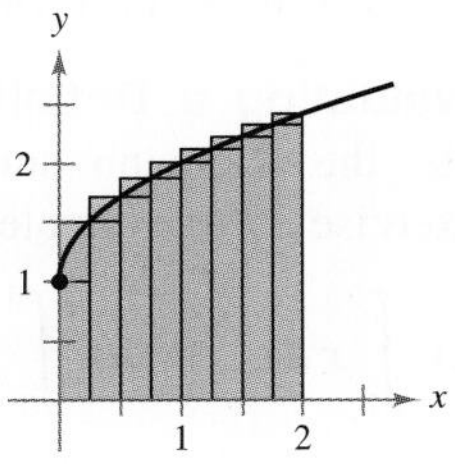

5. $y = \dfrac{1}{x}$

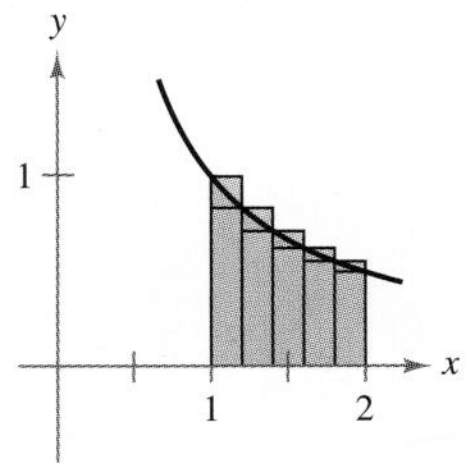

6. $y = \dfrac{1}{x - 2}$

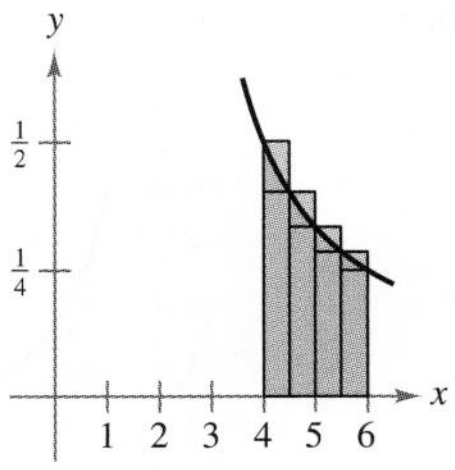

7. $y = \sqrt{1 - x^2}$

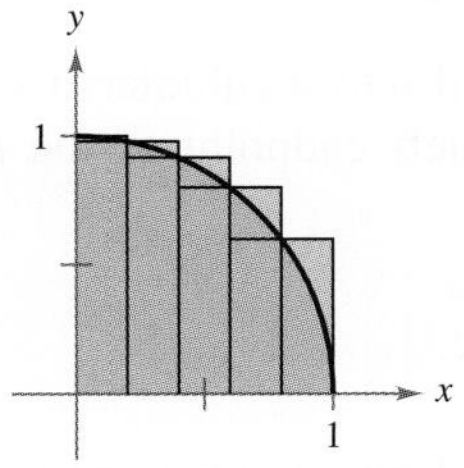

8. $y = \sqrt{x + 1}$

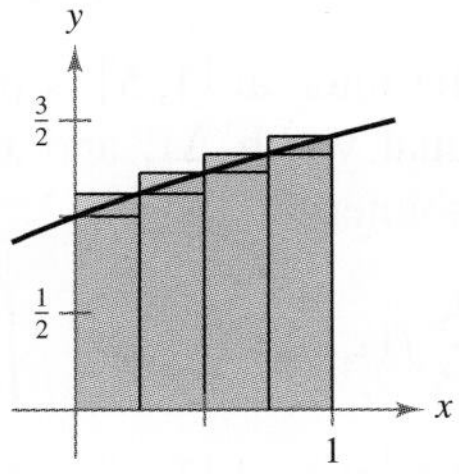

9. Comparing Riemann Sums Consider a triangle of area 2 bounded by the graphs of $y = x$, $y = 0$, and $x = 2$.

(a) Sketch the graph of the region.

(b) Divide the interval $[0, 2]$ into n equal subintervals and show that the endpoints of the subintervals are

$$0 < 1\left(\frac{2}{n}\right) < \cdot \cdot \cdot < (n - 1)\left(\frac{2}{n}\right) < n\left(\frac{2}{n}\right).$$

(c) Show that the left Riemann sum is

$$S_L = \sum_{i=1}^{n} \left[(i - 1)\left(\frac{2}{n}\right)\right]\left(\frac{2}{n}\right).$$

(d) Show that the right Riemann sum is

$$S_R = \sum_{i=1}^{n} \left[i\left(\frac{2}{n}\right)\right]\left(\frac{2}{n}\right).$$

(e) Complete the table below.

n	5	10	50	100
Left sum, S_L				
Right sum, S_R				

(f) Show that $\lim_{n\to\infty} S_L = \lim_{n\to\infty} S_R = 2$.

10. Comparing Riemann Sums Consider a trapezoid of area 4 bounded by the graphs of $y = x$, $y = 0$, $x = 1$, and $x = 3$.

(a) Sketch the graph of the region.

(b) Divide the interval $[1, 3]$ into n equal subintervals and show that the endpoints of the subintervals are

$$1 < 1 + 1\left(\frac{2}{n}\right) < \cdot \cdot \cdot < 1 + (n - 1)\left(\frac{2}{n}\right) < 1 + n\left(\frac{2}{n}\right).$$

(c) Show that the left Riemann sum is

$$S_L = \sum_{i=1}^{n} \left[1 + (i - 1)\left(\frac{2}{n}\right)\right]\left(\frac{2}{n}\right).$$

(d) Show that the right Riemann sum is

$$S_R = \sum_{i=1}^{n} \left[1 + i\left(\frac{2}{n}\right)\right]\left(\frac{2}{n}\right).$$

(e) Complete the table below.

n	5	10	50	100
Left sum, S_L				
Right sum, S_R				

(f) Show that $\lim_{n\to\infty} S_L = \lim_{n\to\infty} S_R = 4$.

Writing a Definite Integral In Exercises 11–18, set up a definite integral that yields the area of the region. (Do not evaluate the integral.)

11. $f(x) = 3$

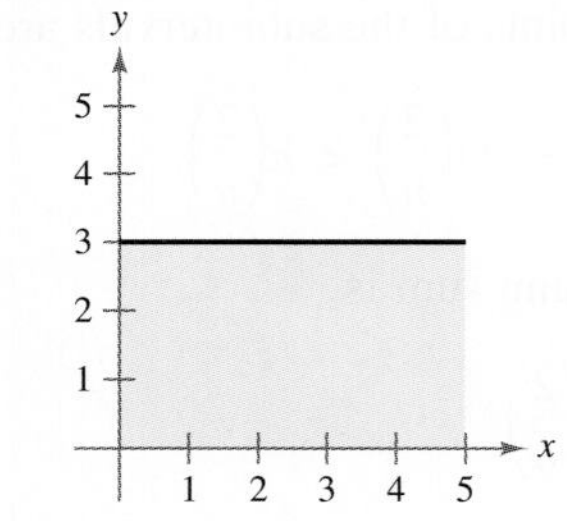

12. $f(x) = 4 - 2x$

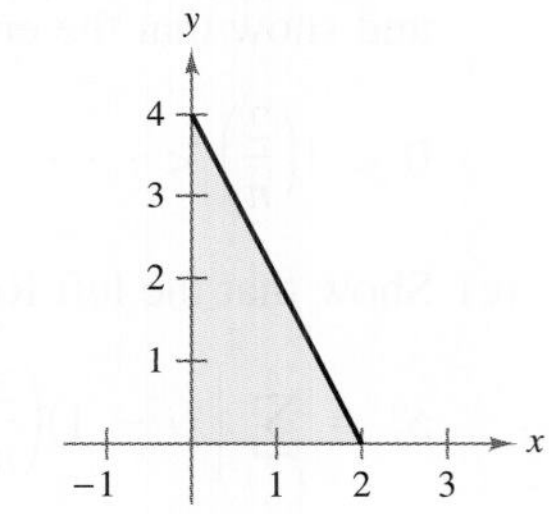

13. $f(x) = 4 - |x|$

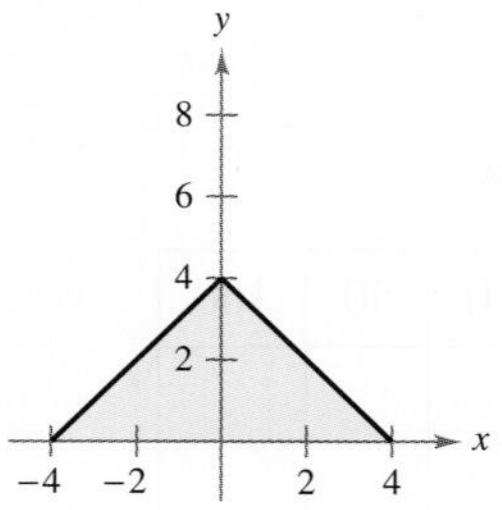

14. $f(x) = x^2$

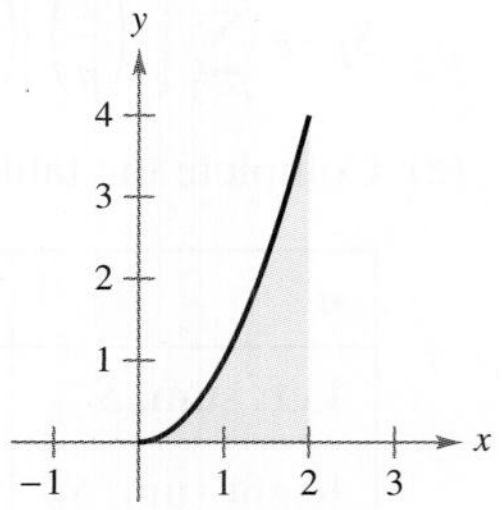

15. $f(x) = 4 - x^2$

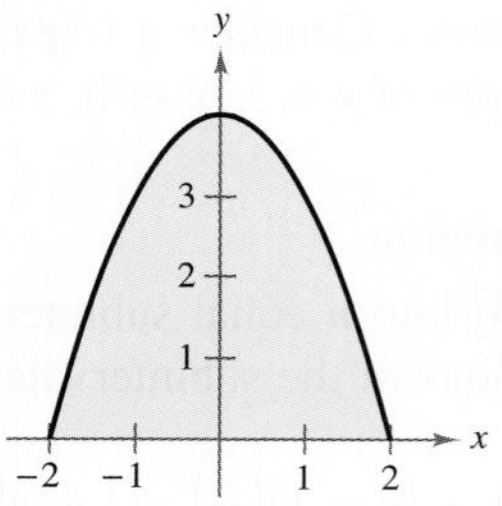

16. $f(x) = \dfrac{1}{x^2 + 1}$

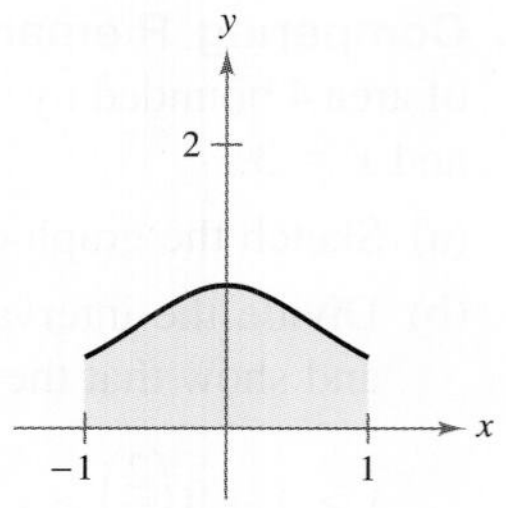

17. $f(x) = \sqrt{x + 1}$

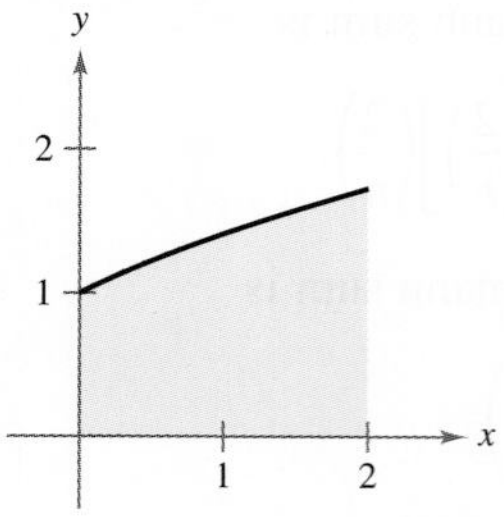

18. $f(x) = (x^2 + 1)^2$

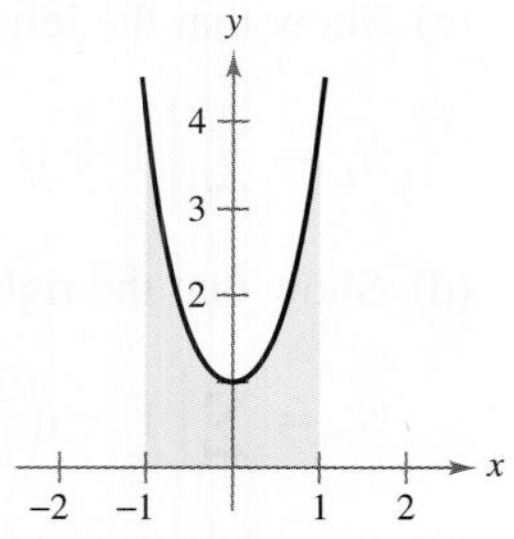

Finding Areas of Common Geometric Figures In Exercises 19–28, sketch the region whose area is given by the definite integral. Then use a geometric formula to evaluate the integral ($a > 0, r > 0$). *See Example 5.*

19. $\displaystyle\int_0^3 4\,dx$

20. $\displaystyle\int_{-a}^{a} 4\,dx$

21. $\displaystyle\int_0^4 x\,dx$

22. $\displaystyle\int_0^4 \frac{x}{2}\,dx$

23. $\displaystyle\int_0^2 (2x + 5)\,dx$

24. $\displaystyle\int_0^5 (5 - x)\,dx$

25. $\displaystyle\int_{-1}^{1} (1 - |x|)\,dx$

26. $\displaystyle\int_{-a}^{a} (a - |x|)\,dx$

27. $\displaystyle\int_{-3}^{3} \sqrt{9 - x^2}\,dx$

28. $\displaystyle\int_{-r}^{r} \sqrt{r^2 - x^2}\,dx$

29. Proving a Sum Show that $\displaystyle\sum_{i=1}^{n} i = \frac{n(n+1)}{2}$.

(*Hint:* Add the two sums below.)

$$S = 1 + 2 + 3 + \cdot\cdot\cdot + (n - 2) + (n - 1) + n$$

$$S = n + (n - 1) + (n - 2) + \cdot\cdot\cdot + 3 + 2 + 1$$

30. Evaluating a Definite Integral by Its Definition Use the Riemann sum definition and the result of Exercise 29 to evaluate the definite integrals.

(a) $\displaystyle\int_1^2 x\,dx$ (b) $\displaystyle\int_0^4 3x\,dx$

Comparing a Sum with an Integral In Exercises 31 and 32, use the figure to fill in the blank with the symbol <, >, or =.

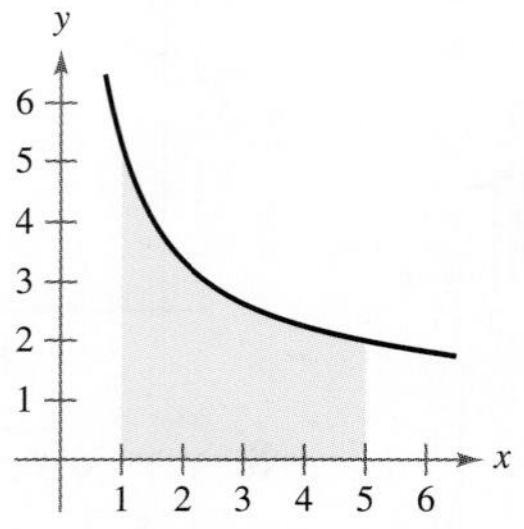

31. The interval $[1, 5]$ is partitioned into n subintervals of equal width Δx, and x_i is the left endpoint of the ith subinterval.

$$\sum_{i=1}^{n} f(x_i)\,\Delta x \quad \boxed{} \quad \int_1^5 f(x)\,dx$$

32. The interval $[1, 5]$ is partitioned into n subintervals of equal width Δx, and x_i is the right endpoint of the ith subinterval.

$$\sum_{i=1}^{n} f(x_i)\,\Delta x \quad \boxed{} \quad \int_1^5 f(x)\,dx$$

C Formulas

C.1 Differentiation and Integration Formulas

■ Use differentiation and integration tables to supplement differentiation and integration techniques.

Differentiation Formulas

1. $\frac{d}{dx}[cu] = cu'$

2. $\frac{d}{dx}[u \pm v] = u' \pm v'$

3. $\frac{d}{dx}[uv] = uv' + vu'$

4. $\frac{d}{dx}\left[\frac{u}{v}\right] = \frac{vu' - uv'}{v^2}$

5. $\frac{d}{dx}[c] = 0$

6. $\frac{d}{dx}[u^n] = nu^{n-1}u'$

7. $\frac{d}{dx}[x] = 1$

8. $\frac{d}{dx}[\ln u] = \frac{u'}{u}$

9. $\frac{d}{dx}[e^u] = e^u u'$

10. $\frac{d}{dx}[\sin u] = (\cos u)u'$

11. $\frac{d}{dx}[\cos u] = -(\sin u)u'$

12. $\frac{d}{dx}[\tan u] = (\sec^2 u)u'$

13. $\frac{d}{dx}[\cot u] = -(\csc^2 u)u'$

14. $\frac{d}{dx}[\sec u] = (\sec u \tan u)u'$

15. $\frac{d}{dx}[\csc u] = -(\csc u \cot u)u'$

Integration Formulas

Forms Involving u^n

1. $\int u^n\,du = \frac{u^{n+1}}{n+1} + C, \quad n \neq -1$

2. $\int \frac{1}{u}\,du = \ln|u| + C$

Forms Involving $a + bu$

3. $\int \frac{u}{a+bu}\,du = \frac{1}{b^2}(bu - a\ln|a+bu|) + C$

4. $\int \frac{u}{(a+bu)^2}\,du = \frac{1}{b^2}\left(\frac{a}{a+bu} + \ln|a+bu|\right) + C$

5. $\int \frac{u}{(a+bu)^n}\,du = \frac{1}{b^2}\left[\frac{-1}{(n-2)(a+bu)^{n-2}} + \frac{a}{(n-1)(a+bu)^{n-1}}\right] + C, \quad n \neq 1, 2$

6. $\int \frac{u^2}{a+bu}\,du = \frac{1}{b^3}\left[-\frac{bu}{2}(2a - bu) + a^2\ln|a+bu|\right] + C$

7. $\int \frac{u^2}{(a+bu)^2}\,du = \frac{1}{b^3}\left(bu - \frac{a^2}{a+bu} - 2a\ln|a+bu|\right) + C$

8. $\int \frac{u^2}{(a+bu)^3}\,du = \frac{1}{b^3}\left[\frac{2a}{a+bu} - \frac{a^2}{2(a+bu)^2} + \ln|a+bu|\right] + C$

9. $\int \frac{u^2}{(a+bu)^n}\,du = \frac{1}{b^3}\left[\frac{-1}{(n-3)(a+bu)^{n-3}} + \frac{2a}{(n-2)(a+bu)^{n-2}} - \frac{a^2}{(n-1)(a+bu)^{n-1}}\right] + C, \quad n \neq 1, 2, 3$

10. $\int \frac{1}{u(a+bu)}\,du = \frac{1}{a}\ln\left|\frac{u}{a+bu}\right| + C$

Integration Formulas (continued)

11. $\displaystyle\int \frac{1}{u(a+bu)^2}\,du = \frac{1}{a}\left(\frac{1}{a+bu} + \frac{1}{a}\ln\left|\frac{u}{a+bu}\right|\right) + C$

12. $\displaystyle\int \frac{1}{u^2(a+bu)}\,du = -\frac{1}{a}\left(\frac{1}{u} + \frac{b}{a}\ln\left|\frac{u}{a+bu}\right|\right) + C$

13. $\displaystyle\int \frac{1}{u^2(a+bu)^2}\,du = -\frac{1}{a^2}\left[\frac{a+2bu}{u(a+bu)} + \frac{2b}{a}\ln\left|\frac{u}{a+bu}\right|\right] + C$

Forms Involving $\sqrt{a+bu}$

14. $\displaystyle\int u^n\sqrt{a+bu}\,du = \frac{2}{b(2n+3)}\left[u^n(a+bu)^{3/2} - na\int u^{n-1}\sqrt{a+bu}\,du\right]$

15. $\displaystyle\int \frac{1}{u\sqrt{a+bu}}\,du = \frac{1}{\sqrt{a}}\ln\left|\frac{\sqrt{a+bu}-\sqrt{a}}{\sqrt{a+bu}+\sqrt{a}}\right| + C, \quad a > 0$

16. $\displaystyle\int \frac{1}{u^n\sqrt{a+bu}}\,du = \frac{-1}{a(n-1)}\left[\frac{\sqrt{a+bu}}{u^{n-1}} + \frac{(2n-3)b}{2}\int \frac{1}{u^{n-1}\sqrt{a+bu}}\,du\right], \quad n \neq 1$

17. $\displaystyle\int \frac{\sqrt{a+bu}}{u}\,du = 2\sqrt{a+bu} + a\int \frac{1}{u\sqrt{a+bu}}\,du$

18. $\displaystyle\int \frac{\sqrt{a+bu}}{u^n}\,du = \frac{-1}{a(n-1)}\left[\frac{(a+bu)^{3/2}}{u^{n-1}} + \frac{(2n-5)b}{2}\int \frac{\sqrt{a+bu}}{u^{n-1}}\,du\right], \quad n \neq 1$

19. $\displaystyle\int \frac{u}{\sqrt{a+bu}}\,du = -\frac{2(2a-bu)}{3b^2}\sqrt{a+bu} + C$

20. $\displaystyle\int \frac{u^n}{\sqrt{a+bu}}\,du = \frac{2}{(2n+1)b}\left(u^n\sqrt{a+bu} - na\int \frac{u^{n-1}}{\sqrt{a+bu}}\,du\right)$

Forms Involving $u^2 - a^2,\ a > 0$

21. $\displaystyle\int \frac{1}{u^2-a^2}\,du = -\int \frac{1}{a^2-u^2}\,du = \frac{1}{2a}\ln\left|\frac{u-a}{u+a}\right| + C$

22. $\displaystyle\int \frac{1}{(u^2-a^2)^n}\,du = \frac{-1}{2a^2(n-1)}\left[\frac{u}{(u^2-a^2)^{n-1}} + (2n-3)\int \frac{1}{(u^2-a^2)^{n-1}}\,du\right], \quad n \neq 1$

Forms Involving $\sqrt{u^2 \pm a^2},\ a > 0$

23. $\displaystyle\int \sqrt{u^2 \pm a^2}\,du = \frac{1}{2}\left(u\sqrt{u^2 \pm a^2} \pm a^2\ln\left|u + \sqrt{u^2 \pm a^2}\right|\right) + C$

24. $\displaystyle\int u^2\sqrt{u^2 \pm a^2}\,du = \frac{1}{8}\left[u(2u^2 \pm a^2)\sqrt{u^2 \pm a^2} - a^4\ln\left|u + \sqrt{u^2 \pm a^2}\right|\right] + C$

25. $\displaystyle\int \frac{\sqrt{u^2+a^2}}{u}\,du = \sqrt{u^2+a^2} - a\ln\left|\frac{a+\sqrt{u^2+a^2}}{u}\right| + C$

26. $\displaystyle\int \frac{\sqrt{u^2 \pm a^2}}{u^2}\,du = \frac{-\sqrt{u^2 \pm a^2}}{u} + \ln\left|u + \sqrt{u^2 \pm a^2}\right| + C$

27. $\displaystyle\int \frac{1}{\sqrt{u^2 \pm a^2}}\,du = \ln\left|u + \sqrt{u^2 \pm a^2}\right| + C$

28. $\displaystyle\int \frac{1}{u\sqrt{u^2+a^2}}\,du = \frac{-1}{a}\ln\left|\frac{a+\sqrt{u^2+a^2}}{u}\right| + C$

29. $\displaystyle\int \frac{u^2}{\sqrt{u^2 \pm a^2}}\,du = \frac{1}{2}\left(u\sqrt{u^2 \pm a^2} \mp a^2 \ln\left|u + \sqrt{u^2 \pm a^2}\right|\right) + C$

30. $\displaystyle\int \frac{1}{u^2\sqrt{u^2 \pm a^2}}\,du = \mp\frac{\sqrt{u^2 \pm a^2}}{a^2 u} + C$

31. $\displaystyle\int \frac{1}{(u^2 \pm a^2)^{3/2}}\,du = \frac{\pm u}{a^2\sqrt{u^2 \pm a^2}} + C$

Forms Involving $\sqrt{a^2 - u^2}$, $a > 0$

32. $\displaystyle\int \frac{\sqrt{a^2 - u^2}}{u}\,du = \sqrt{a^2 - u^2} - a\ln\left|\frac{a + \sqrt{a^2 - u^2}}{u}\right| + C$

33. $\displaystyle\int \frac{1}{u\sqrt{a^2 - u^2}}\,du = \frac{-1}{a}\ln\left|\frac{a + \sqrt{a^2 - u^2}}{u}\right| + C$

34. $\displaystyle\int \frac{1}{u^2\sqrt{a^2 - u^2}}\,du = \frac{-\sqrt{a^2 - u^2}}{a^2 u} + C$

35. $\displaystyle\int \frac{1}{(a^2 - u^2)^{3/2}}\,du = \frac{u}{a^2\sqrt{a^2 - u^2}} + C$

Forms Involving e^u

36. $\displaystyle\int e^u\,du = e^u + C$

37. $\displaystyle\int ue^u\,du = (u - 1)e^u + C$

38. $\displaystyle\int u^n e^u\,du = u^n e^u - n\int u^{n-1}e^u\,du$

39. $\displaystyle\int \frac{1}{1 + e^u}\,du = u - \ln(1 + e^u) + C$

40. $\displaystyle\int \frac{1}{1 + e^{nu}}\,du = u - \frac{1}{n}\ln(1 + e^{nu}) + C$

Forms Involving $\ln u$

41. $\displaystyle\int \ln u\,du = u(-1 + \ln u) + C$

42. $\displaystyle\int u\ln u\,du = \frac{u^2}{4}(-1 + 2\ln u) + C$

43. $\displaystyle\int u^n \ln u\,du = \frac{u^{n+1}}{(n + 1)^2}[-1 + (n + 1)\ln u] + C, \quad n \neq -1$

44. $\displaystyle\int (\ln u)^2\,du = u[2 - 2\ln u + (\ln u)^2] + C$

45. $\displaystyle\int (\ln u)^n\,du = u(\ln u)^n - n\int (\ln u)^{n-1}\,du$

Forms Involving $\sin u$ or $\cos u$

46. $\displaystyle\int \sin u\,du = -\cos u + C$

47. $\displaystyle\int \cos u\,du = \sin u + C$

48. $\displaystyle\int \sin^2 u\,du = \frac{1}{2}(u - \sin u\cos u) + C$

49. $\displaystyle\int \cos^2 u\,du = \frac{1}{2}(u + \sin u\cos u) + C$

50. $\displaystyle\int \sin^n u\,du = -\frac{\sin^{n-1}u\cos u}{n} + \frac{n - 1}{n}\int \sin^{n-2}u\,du$

51. $\displaystyle\int \cos^n u\,du = \frac{\cos^{n-1}u\sin u}{n} + \frac{n - 1}{n}\int \cos^{n-2}u\,du$

52. $\displaystyle\int u\sin u\,du = \sin u - u\cos u + C$

53. $\displaystyle\int u\cos u\,du = \cos u + u\sin u + C$

54. $\displaystyle\int u^n \sin u\,du = -u^n\cos u + n\int u^{n-1}\cos u\,du$

Integration Formulas (continued)

55. $\int u^n \cos u\, du = u^n \sin u - n\int u^{n-1} \sin u\, du$

56. $\int \frac{1}{1 \pm \sin u}\, du = \tan u \mp \sec u + C$

57. $\int \frac{1}{1 \pm \cos u}\, du = -\cot u \pm \csc u + C$

58. $\int \frac{1}{\sin u \cos u}\, du = \ln|\tan u| + C$

Forms Involving tan *u*, cot *u*, sec *u*, or csc *u*

59. $\int \tan u\, du = -\ln|\cos u| + C$

60. $\int \cot u\, du = \ln|\sin u| + C$

61. $\int \sec u\, du = \ln|\sec u + \tan u| + C$

62. $\int \csc u\, du = \ln|\csc u - \cot u| + C$

63. $\int \tan^2 u\, du = -u + \tan u + C$

64. $\int \cot^2 u\, du = -u - \cot u + C$

65. $\int \sec^2 u\, du = \tan u + C$

66. $\int \csc^2 u\, du = -\cot u + C$

67. $\int \tan^n u\, du = \frac{\tan^{n-1} u}{n-1} - \int \tan^{n-2} u\, du, \quad n \neq 1$

68. $\int \cot^n u\, du = -\frac{\cot^{n-1} u}{n-1} - \int \cot^{n-2} u\, du, \quad n \neq 1$

69. $\int \sec^n u\, du = \frac{\sec^{n-2} u \tan u}{n-1} + \frac{n-2}{n-1}\int \sec^{n-2} u\, du, \quad n \neq 1$

70. $\int \csc^n u\, du = -\frac{\csc^{n-2} u \cot u}{n-1} + \frac{n-2}{n-1}\int \csc^{n-2} u\, du, \quad n \neq 1$

71. $\int \frac{1}{1 \pm \tan u}\, du = \frac{1}{2}(u \pm \ln|\cos u \pm \sin u|) + C$

72. $\int \frac{1}{1 \pm \cot u}\, du = \frac{1}{2}(u \mp \ln|\sin u \pm \cos u|) + C$

73. $\int \frac{1}{1 \pm \sec u}\, du = u + \cot u \mp \csc u + C$

74. $\int \frac{1}{1 \pm \csc u}\, du = u - \tan u \pm \sec u + C$

C.2 Formulas from Business and Finance

■ Summary of business and finance formulas

Formulas from Business

Basic Terms

x = number of units produced (or sold)

p = price per unit

R = total revenue from selling x units

C = total cost of producing x units

$\overline{C}$ = average cost per unit

P = total profit from selling x units

Basic Equations

$$R = xp \qquad \overline{C} = \frac{C}{x} \qquad P = R - C$$

Typical Graphs of Supply and Demand Curves

Supply curves increase as price increases and demand curves decrease as price increases. The equilibrium point occurs when the supply and demand curves intersect.

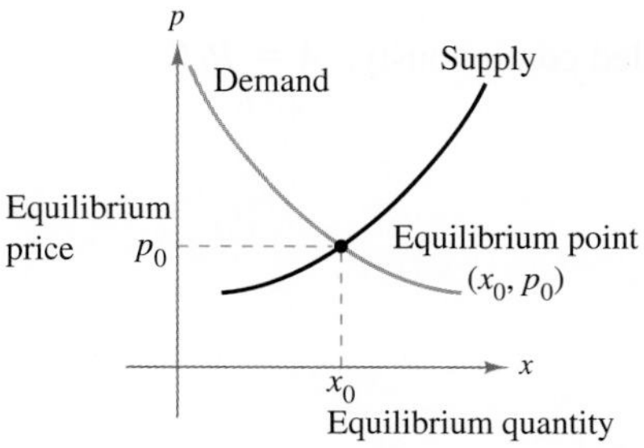

Demand Function: $p = f(x)$ = price required to sell x units

$$\eta = \frac{p/x}{dp/dx} = \text{price elasticity of demand}$$

(When $|\eta| < 1$, the demand is inelastic. When $|\eta| > 1$, the demand is elastic.)

Typical Graphs of Revenue, Cost, and Profit Functions

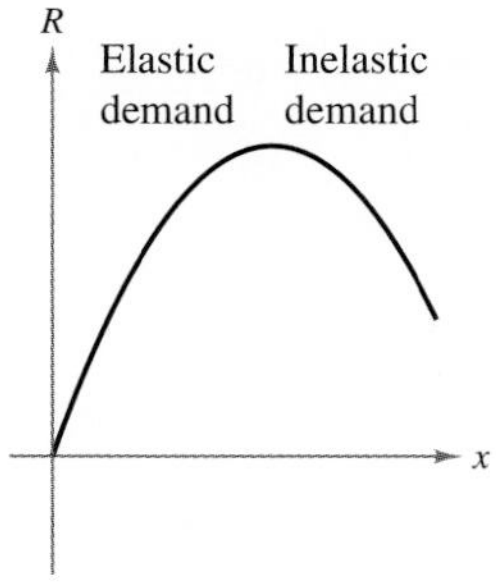

Revenue Function

The low prices required to sell more units eventually result in a decreasing revenue.

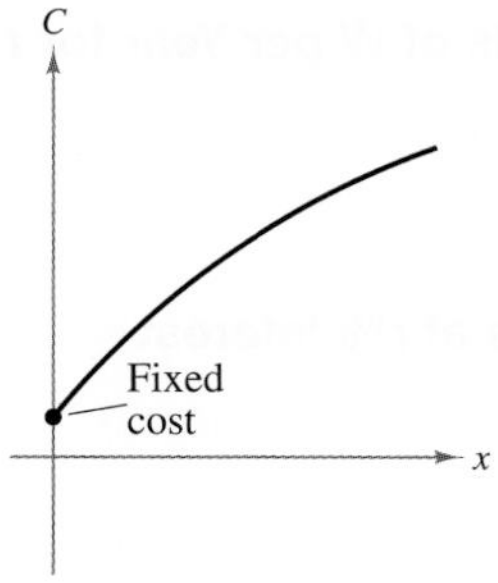

Cost Function

The total cost to produce x units includes the fixed cost.

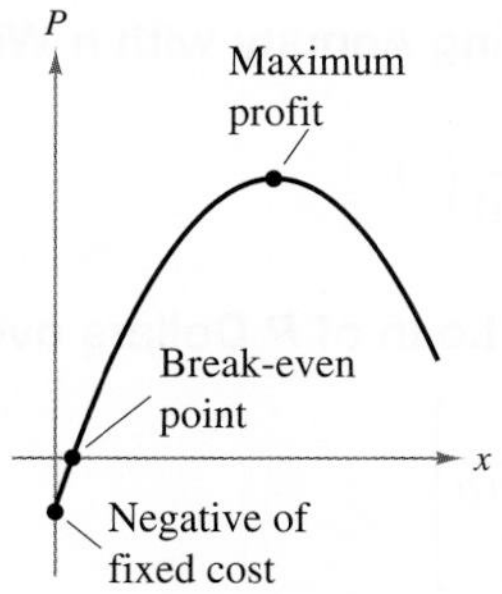

Profit Function

The break-even point occurs when $R = C$.

Formulas from Business (continued)

Marginals

$$\frac{dR}{dx} = \text{marginal revenue} \approx \text{the } \textit{extra} \text{ revenue from selling one additional unit}$$

$$\frac{dC}{dx} = \text{marginal cost} \approx \text{the } \textit{extra} \text{ cost of producing one additional unit}$$

$$\frac{dP}{dx} = \text{marginal profit} \approx \text{the } \textit{extra} \text{ profit from selling one additional unit}$$

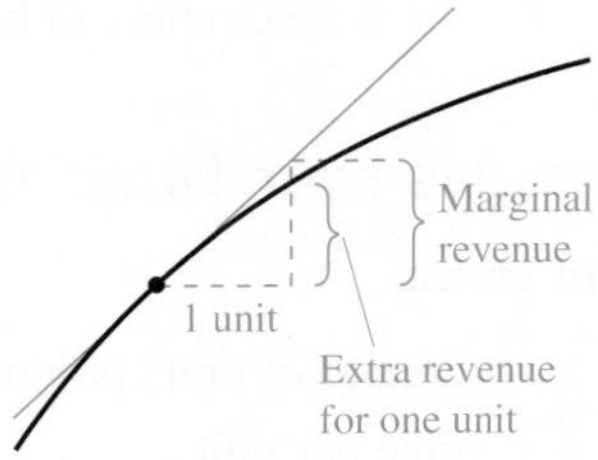

Revenue Function

Formulas from Finance

Basic Terms

P = amount of deposit

r = interest rate

n = number of times interest is compounded per year

t = number of years

A = balance after t years

Compound Interest Formulas

1. Balance when interest is compounded n times per year: $A = P\left(1 + \frac{r}{n}\right)^{nt}$
2. Balance when interest is compounded continuously: $A = Pe^{rt}$

Effective Rate of Interest

$$r_{eff} = \left(1 + \frac{r}{n}\right)^n - 1$$

Present Value of a Future Investment

$$P = \frac{A}{\left(1 + \frac{r}{n}\right)^{nt}}$$

Balance of an Increasing Annuity After *n* Deposits of *P* per Year for *t* Years

$$A = P\left[\left(1 + \frac{r}{n}\right)^{nt} - 1\right]\left(1 + \frac{n}{r}\right)$$

Initial Deposit for a Decreasing Annuity with *n* Withdrawals of *W* per Year for *t* Years

$$P = W\left(\frac{n}{r}\right)\left\{1 - \left[\frac{1}{1 + (r/n)}\right]^{nt}\right\}$$

Monthly Installment *M* for a Loan of *P* Dollars over *t* Years at *r*% Interest

$$M = P\left\{\frac{r/12}{1 - \left[\frac{1}{1 + (r/12)}\right]^{12t}}\right\}$$

Amount of an Annuity

$$e^{rT}\int_0^T c(t)e^{-rt}\,dt$$

$c(t)$ is the continuous income function in dollars per year and T is the term of the annuity in years.

Answers to Selected Exercises

Chapter 1

Section 1.1 *(page 8)*

Skills Warm Up *(page 8)*

1. $3\sqrt{5}$ **2.** $2\sqrt{5}$ **3.** $\frac{1}{2}$ **4.** -2 **5.** $5\sqrt{3}$
6. $-\sqrt{2}$ **7.** $x = -3, x = 9$
8. $y = -8, y = 4$ **9.** $x = 19$ **10.** $y = 1$

1.

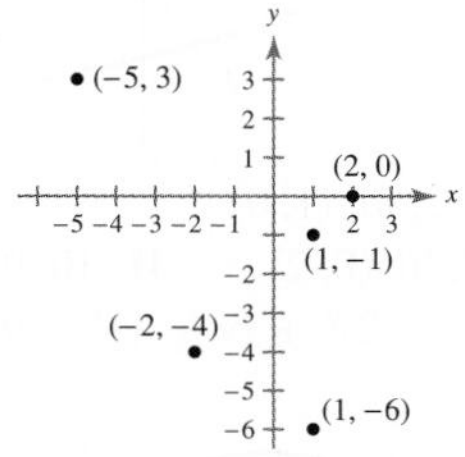

3. (a)

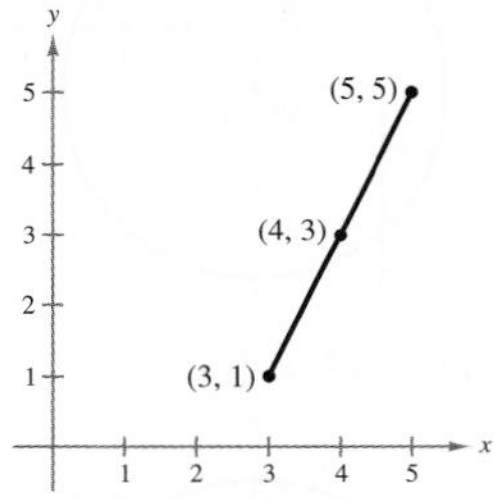

(b) $d = 2\sqrt{5}$
(c) Midpoint: $(4, 3)$

5. (a)

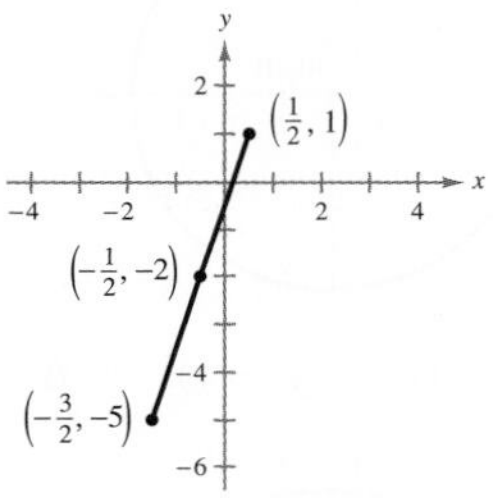

(b) $d = 2\sqrt{10}$
(c) Midpoint: $\left(-\frac{1}{2}, -2\right)$

7. (a)

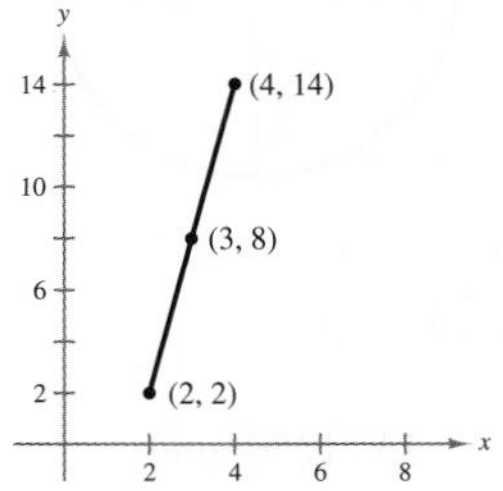

(b) $d = 2\sqrt{37}$
(c) Midpoint: $(3, 8)$

9. (a)

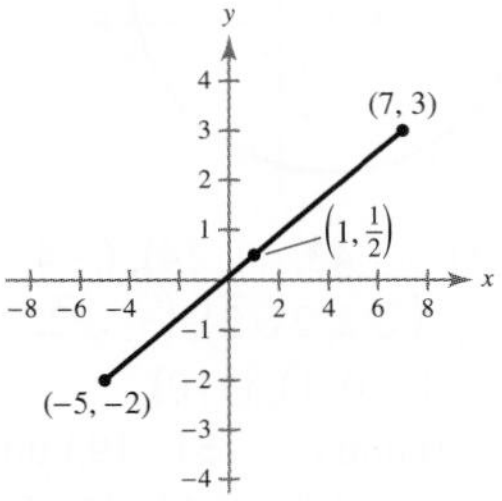

(b) $d = 13$
(c) Midpoint: $\left(1, \frac{1}{2}\right)$

11. (a)

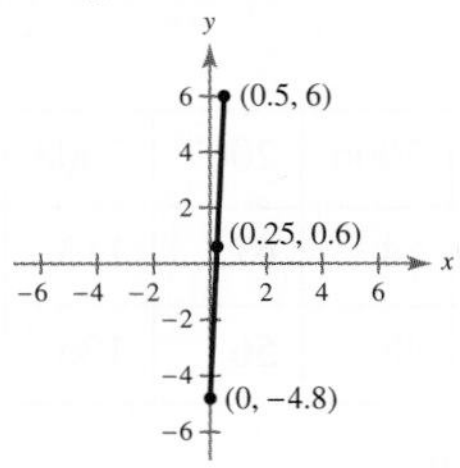

(b) $d = \sqrt{116.89}$ (c) Midpoint: $(0.25, 0.6)$

13. (a) $a = 4, b = 3, c = 5$
(b) $4^2 + 3^2 = 5^2$

15. (a) $a = 10, b = 3, c = \sqrt{109}$
(b) $10^2 + 3^2 = \left(\sqrt{109}\right)^2$

17. $d_1 = \sqrt{45}, d_2 = \sqrt{20},$
$d_3 = \sqrt{65}$
$d_1^2 + d_2^2 = d_3^2$

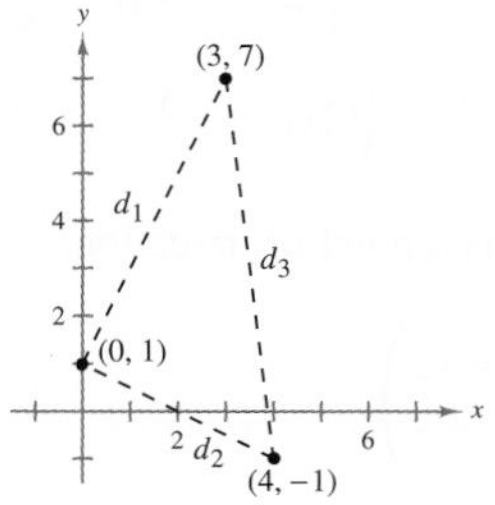

19. $d_1 = d_2 = d_3 = d_4$
$= \sqrt{5}$

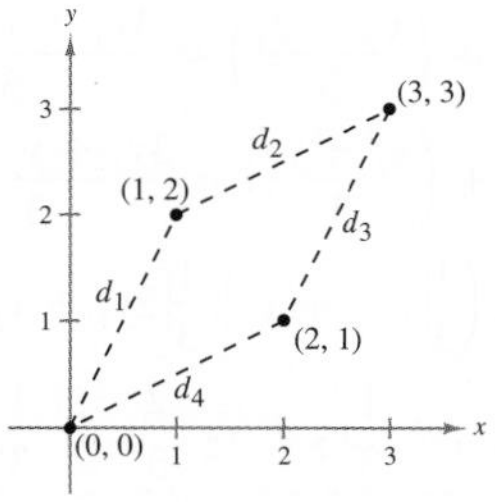

21. $x = 4, -2$ **23.** $y = \pm\sqrt{55}$ **25.** About 45 yards

27.

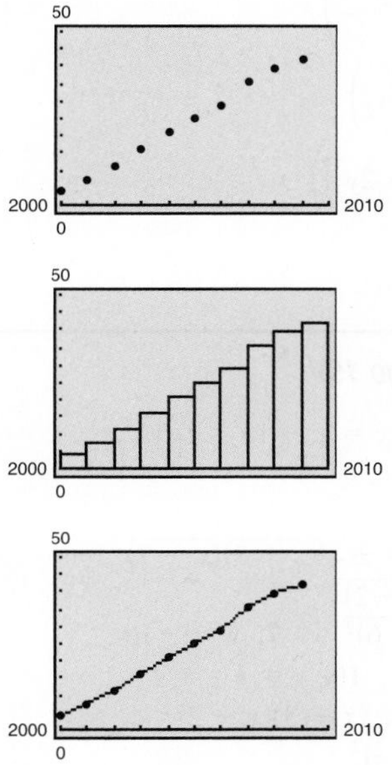

The number of cable high-speed Internet customers increases each year.

29. (a) 7600; 9200; 10,500 (b) 8.2% decrease

31. (a) Revenue: \$434.3 million
Profit: \$25.2 million
(b) Actual 2008 revenue: \$422.4 million
Actual 2008 profit: \$24.4 million
(c) Yes, the actual amounts were very close to those predicted by the Midpoint Formula.
(d) Expenses for 2007: \$310 million
Expenses for 2008: \$398 million
Expenses for 2009: \$508.2 million
(e) Answers will vary.

33. (a)

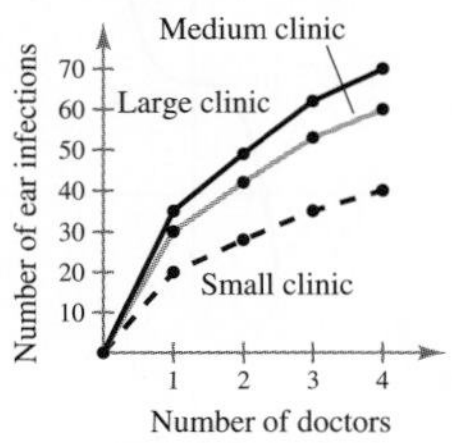

(b) The larger the clinic, the more patients a doctor can treat.

35. $(-6, -6), (-4, -7), (-3, -5)$

37. $\left(\frac{3x_1 + x_2}{4}, \frac{3y_1 + y_2}{4}\right), \left(\frac{x_1 + x_2}{2}, \frac{y_1 + y_2}{2}\right),$
$\left(\frac{x_1 + 3x_2}{4}, \frac{y_1 + 3y_2}{4}\right)$

39. $x_1 + \left(\frac{x_2 - x_1}{3}\right) = \frac{3x_1 + x_2 - x_1}{3} = \frac{1}{3}(2x_1 + x_2)$

$y_1 + \left(\frac{y_2 - y_1}{3}\right) = \frac{3y_1 + y_2 - y_1}{3} = \frac{1}{3}(2y_1 + y_2)$

So, $\left(\frac{1}{3}[2x_1 + x_2], \frac{1}{3}[2y_1 + y_2]\right)$ is a point of trisection.

$$\left(\frac{\frac{2}{3}x_1 + \frac{1}{3}x_2 + x_2}{2}, \frac{\frac{2}{3}y_1 + \frac{1}{3}y_2 + y_2}{2}\right)$$
$$= \left(\frac{\frac{2}{3}x_1 + \frac{4}{3}x_2}{2}, \frac{\frac{2}{3}y_1 + \frac{4}{3}y_2}{2}\right)$$
$$= \left(\frac{1}{3}x_1 + \frac{2}{3}x_2, \frac{1}{3}y_1 + \frac{2}{3}y_2\right)$$
$$= \left(\frac{1}{3}[x_1 + 2x_2], \frac{1}{3}[y_1 + 2y_2]\right)$$

Section 1.2 *(page 19)*

Skills Warm Up *(page 19)*

1. $y = \frac{1}{5}(x + 12)$ **2.** $y = x - 15$

3. $y = \frac{1}{x^3 + 2}$

4. $y = \pm\sqrt{x^2 + x - 6} = \pm\sqrt{(x + 3)(x - 2)}$

5. $y = -1 \pm \sqrt{9 - (x - 2)^2}$

6. $y = 5 \pm \sqrt{81 - (x + 6)^2}$ **7.** $y = -10$

8. $y = 5$ **9.** $y = 9$ **10.** $y = 1$

11. $(x - 2)(x - 1)$ **12.** $(x + 3)(x + 2)$

13. $\left(y - \frac{3}{2}\right)^2$ **14.** $\left(y - \frac{7}{2}\right)^2$

1. e **2.** b **3.** c **4.** f **5.** a **6.** d

7.

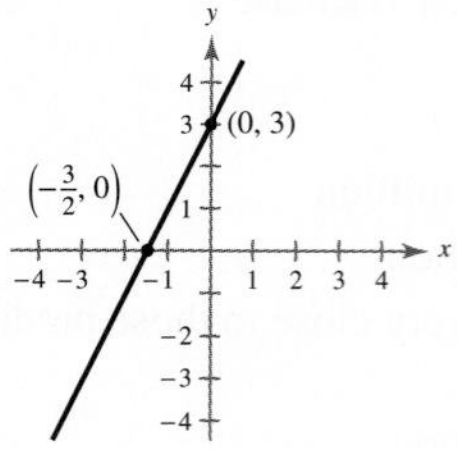

9.

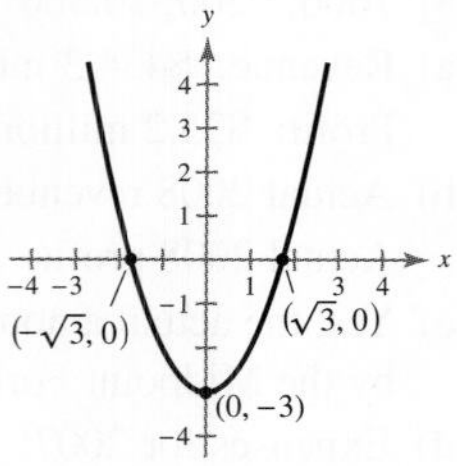

11.

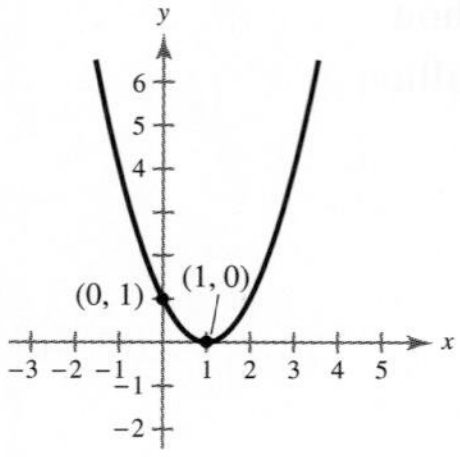

13.

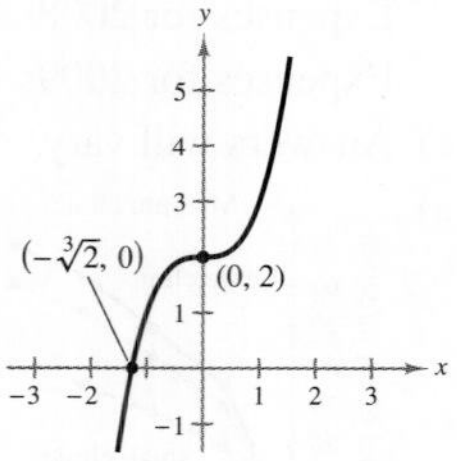

15.

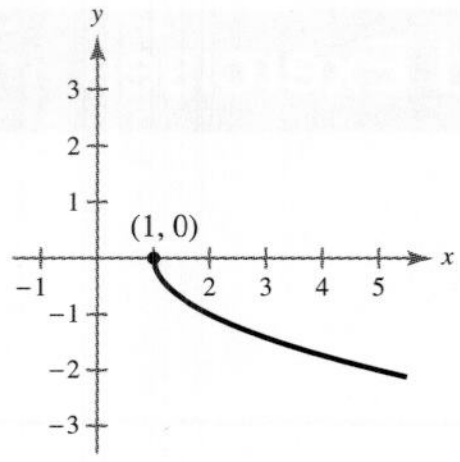

17.

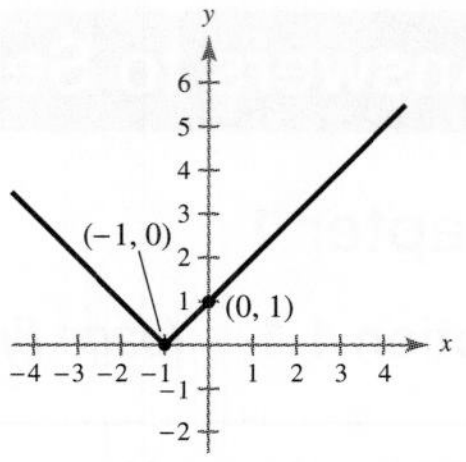

19.

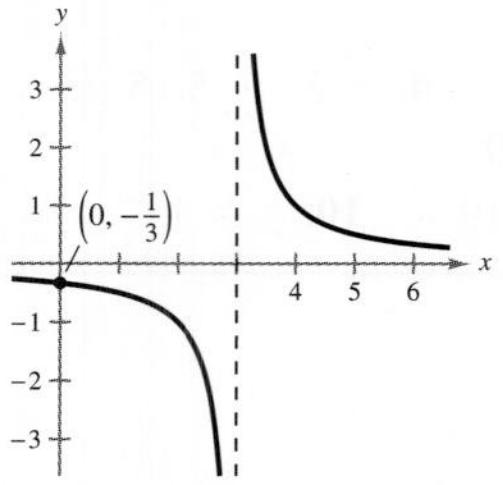

21.

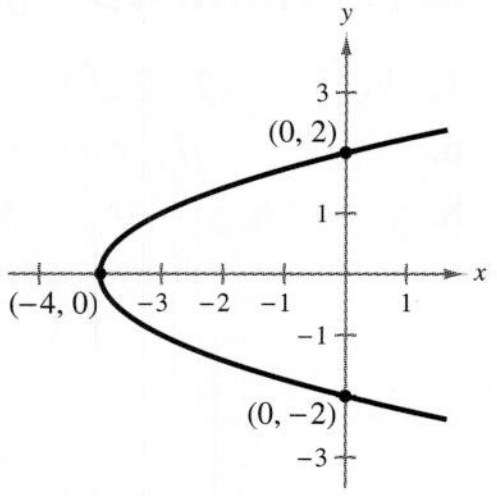

23. $(0, -3), \left(\frac{3}{2}, 0\right)$ **25.** $(0, -2), (-2, 0), (1, 0)$

27. $(-2, 0), (0, 2), (2, 0)$ **29.** $(-2, 0), (0, 2)$ **31.** $(0, 0)$

33. $x^2 + y^2 = 16$ **35.** $(x - 2)^2 + (y + 1)^2 = 9$

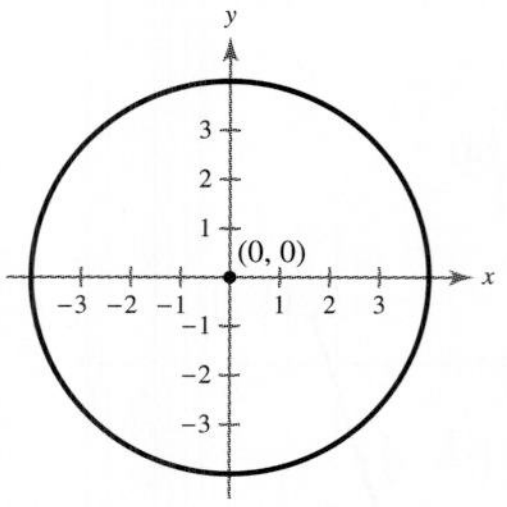

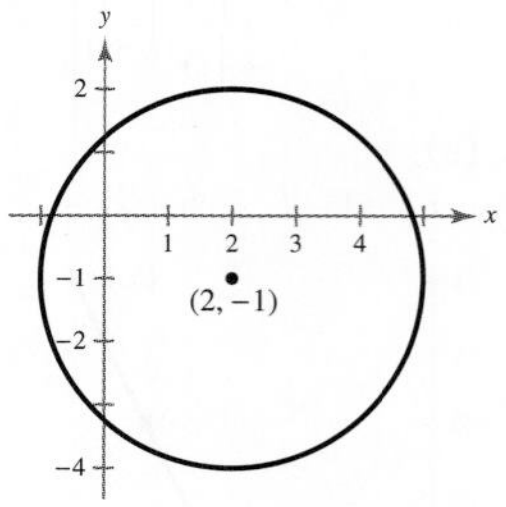

37. $(x + 1)^2 + (y - 1)^2 = 16$ **39.** $x^2 + y^2 = 100$

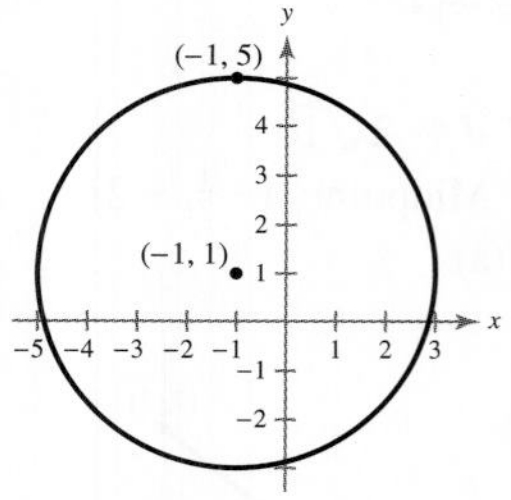

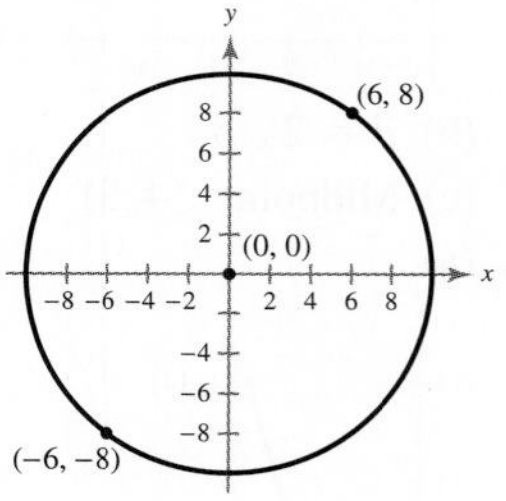

41. $(1, 1)$ **43.** $(1, 14), (-4, -1)$

45. $(0, 0), (\sqrt{2}, 2\sqrt{2}), (-\sqrt{2}, -2\sqrt{2})$

47. $(-1, 0), (0, 1), (1, 0)$

49. 50,000 units **51.** 193 units **53.** 1250 units

55. (a) $C = 11.8x + 15{,}000$; $R = 19.3x$
(b) 2000 units
(c) 2134 units

57. $(15, 180)$

59. (a)

Year	2004	2005	2006	2007	2008	2009
Amount	30	44	54	67	113	313
Model	26	55	48	56	126	308

The model fits the data well.
Answers will vary.

(b) \$4605 million

61. (a)

Year	2004	2005	2006
Degrees	667.4	692	713.6

Year	2007	2008	2012
Degrees	732.2	747.8	780.2

(b) Answers will vary.

(c) 764.6; No; the number of associate's degrees should keep increasing over time.

63. 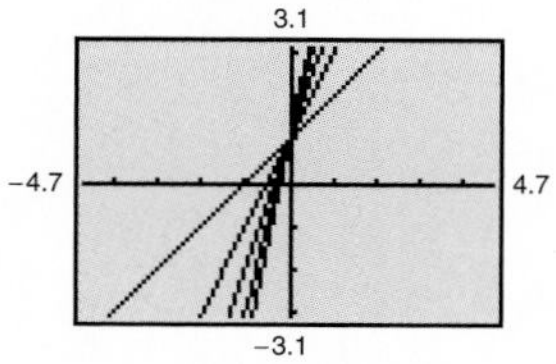

The greater the value of c, the steeper the line.

65. 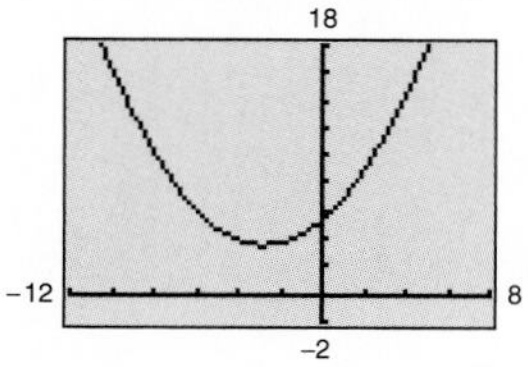

(0, 5.36)

67. 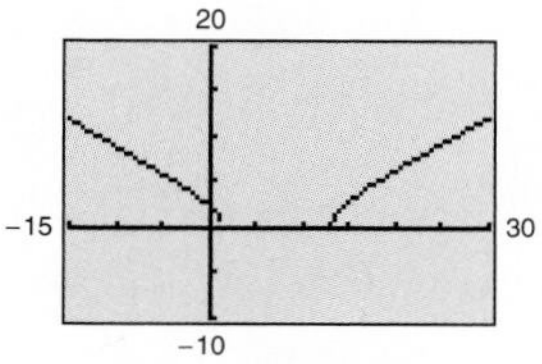

(1.4780, 0), (12.8553, 0), (0, 2.3875)

69. 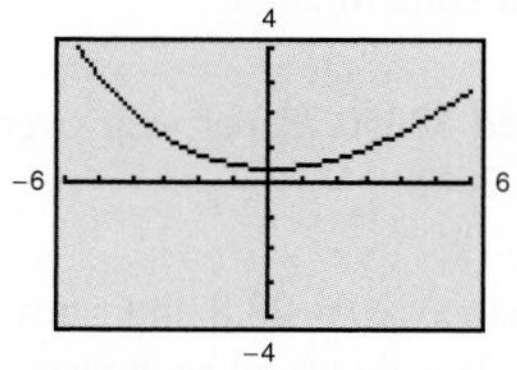

(0, 0.4167)

71. Answers will vary.

Section 1.3 *(page 31)*

> **Skills Warm Up** *(page 31)*
>
> **1.** -1 **2.** 1 **3.** $\frac{1}{3}$ **4.** $-\frac{7}{6}$
> **5.** $y = 4x + 7$ **6.** $y = 3x - 7$
> **7.** $y = 3x - 10$ **8.** $y = -x - 7$
> **9.** $y = 7x - 17$ **10.** $y = \frac{2}{3}x + \frac{5}{3}$

1. 1 **3.** 0 **5.** $m = 1$, (0, 7) **7.** $m = -5$, (0, 20)
9. $m = -\frac{7}{6}$, (0, 5) **11.** $m = 3$, (0, −15)
13. m is undefined; no y-intercept. **15.** $m = 0$, (0, 4)

17. 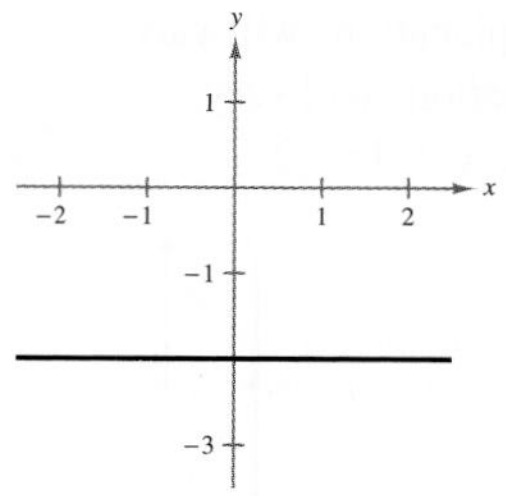

19. 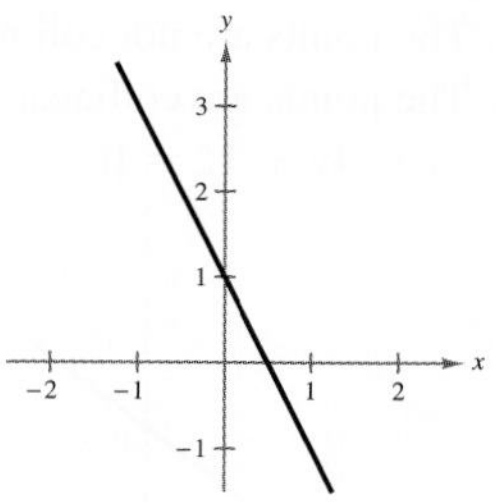

21. 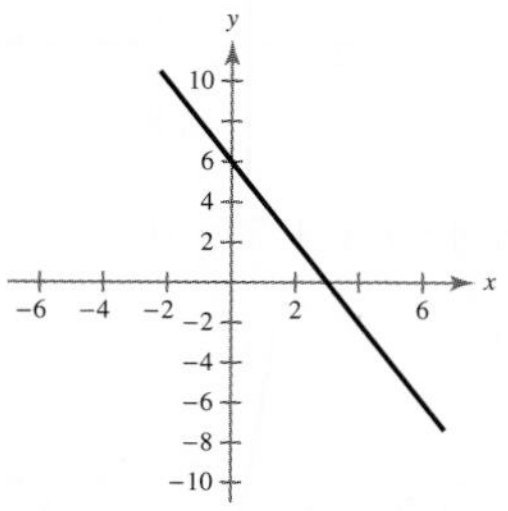

23. 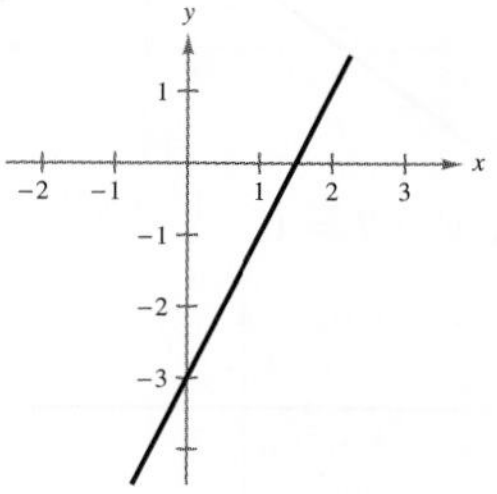

25. 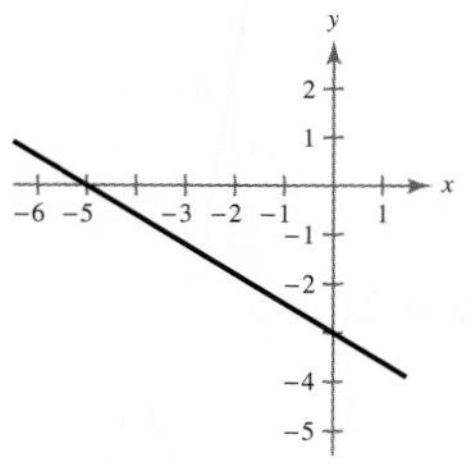

27. 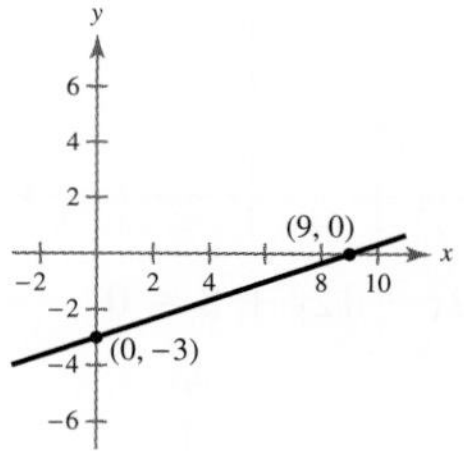

$m = \frac{1}{3}$

29. 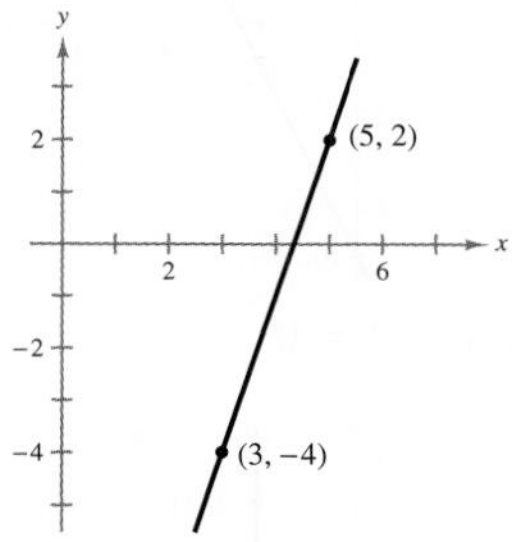

$m = 3$

31. 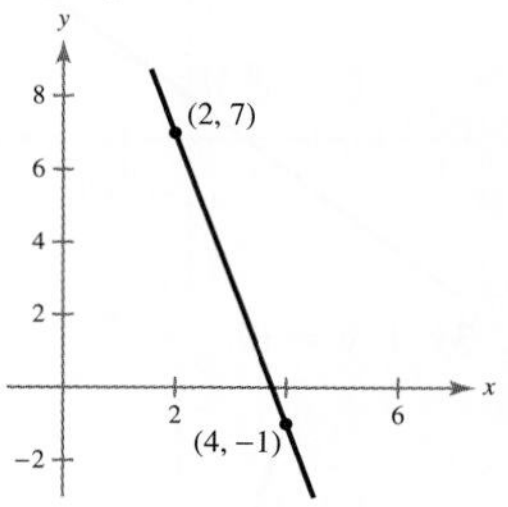

$m = -4$

33. 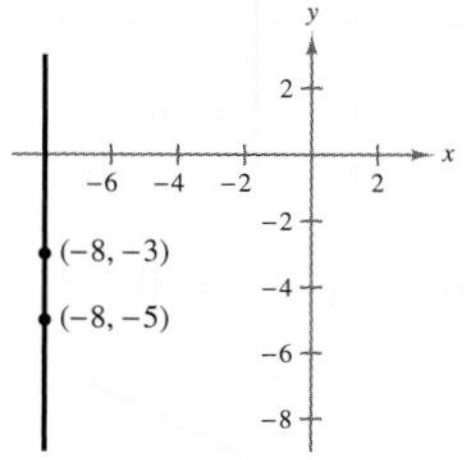

m is undefined.

35. 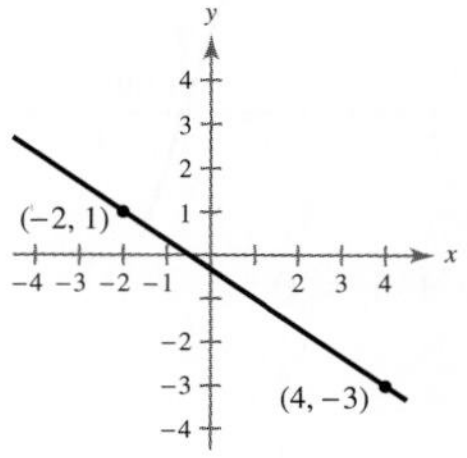

$m = -\frac{2}{3}$

37. 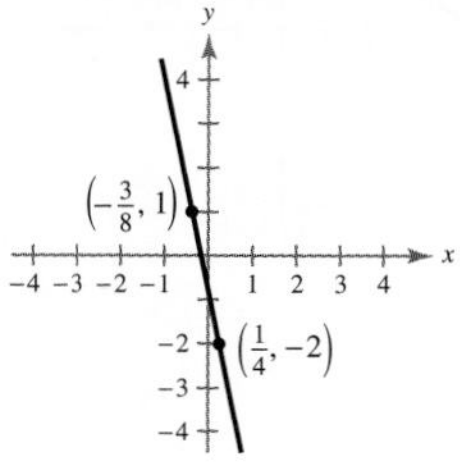

$m = -\frac{24}{5}$

39. 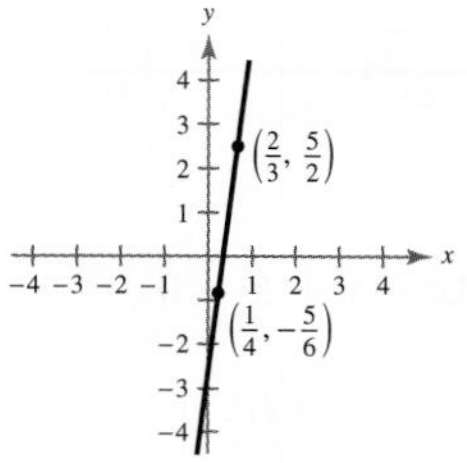

$m = 8$

41. (0, 1), (1, 1), (3, 1) **43.** (0, 10), (2, 4), (3, 1)
45. (3, −6), (9, −2), (12, 0) **47.** (−8, 0), (−8, 2), (−8, 3)

49. The points are not collinear. Explanations will vary.

51. The points are collinear. Explanations will vary.

53. $3x - 4y + 12 = 0$

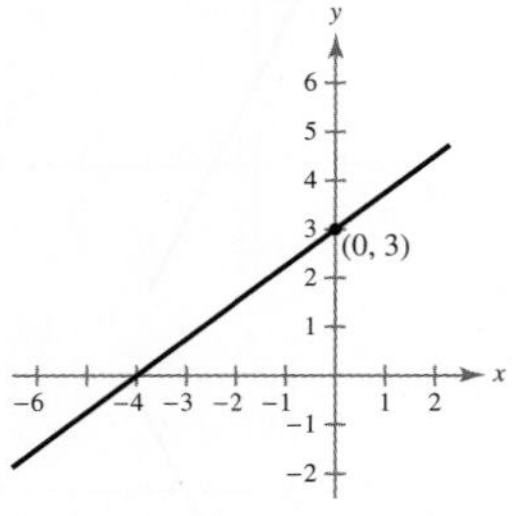

55. $x + 1 = 0$

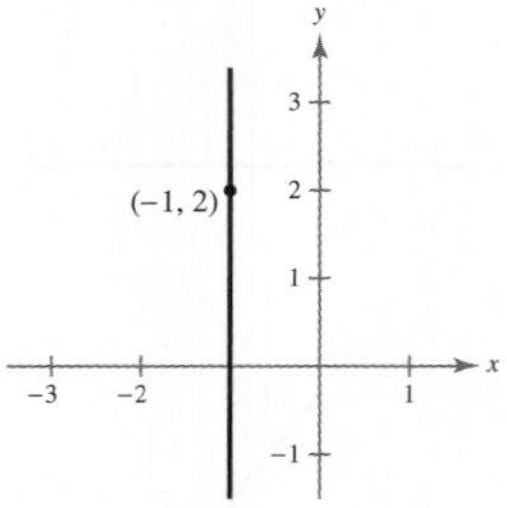

57. $y - 7 = 0$

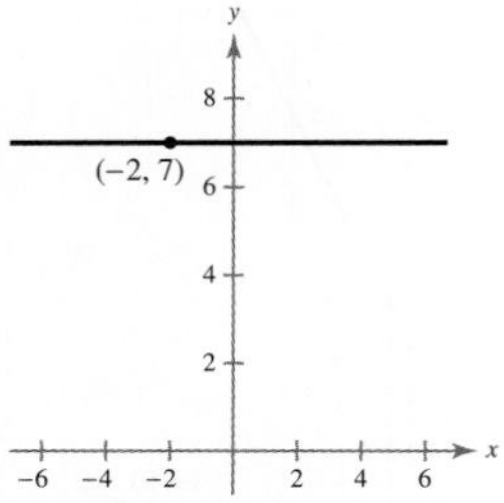

59. $4x + y + 2 = 0$

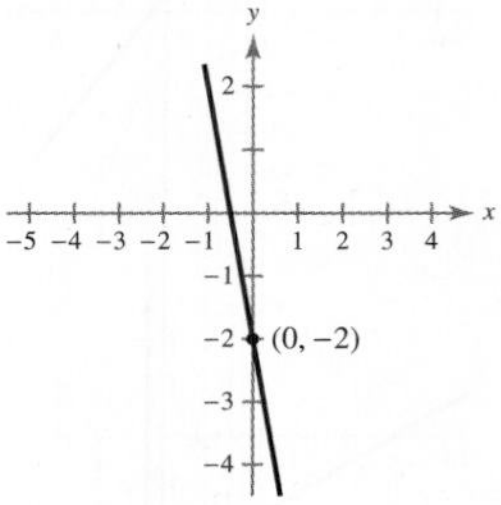

61. $9x - 12y + 8 = 0$

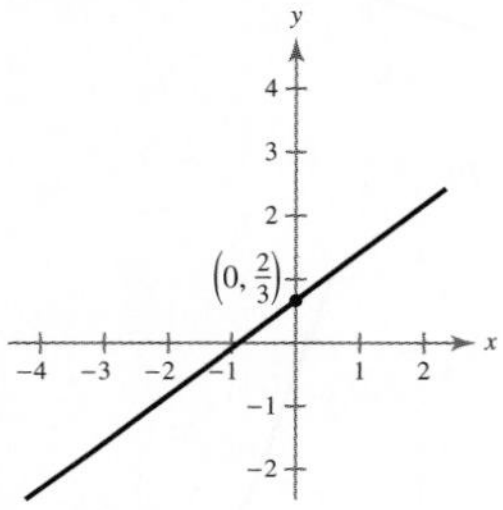

63. $y = 2x - 5$

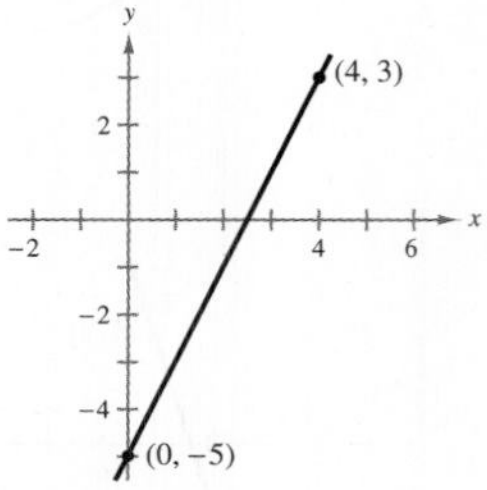

65. $3x + y = 0$

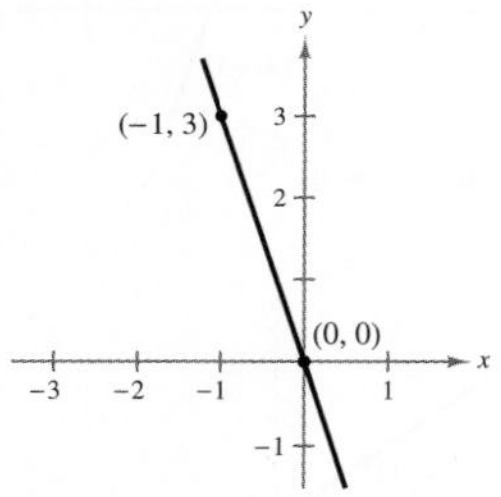

67. $x - 2 = 0$

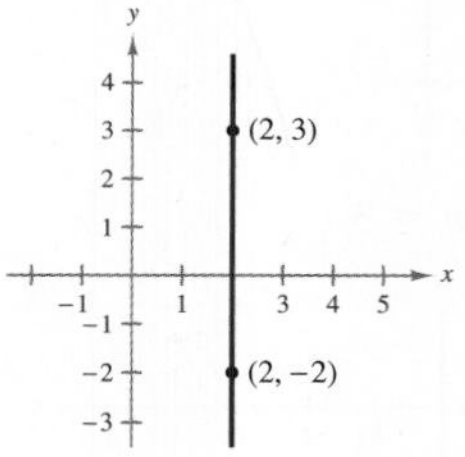

69. $y + 1 = 0$

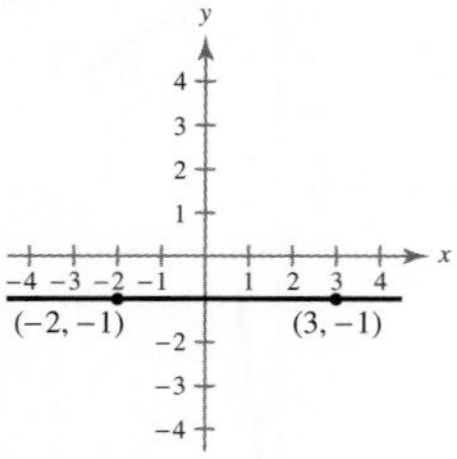

71. $3x - 6y + 7 = 0$

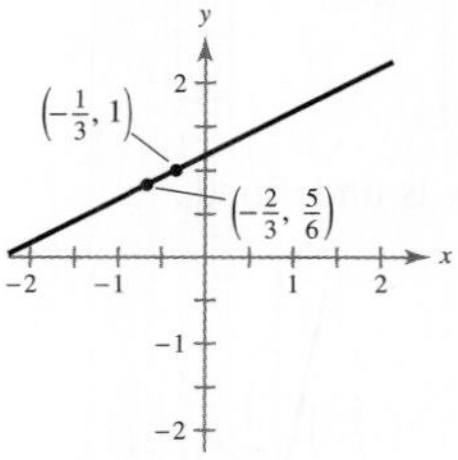

73. $4x - y + 6 = 0$

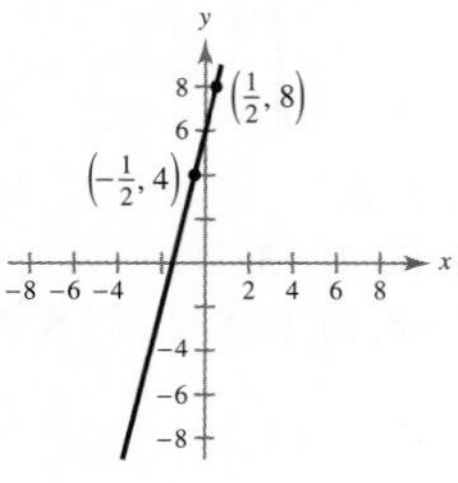

75. $x - 3 = 0$ **77.** $y + 10 = 0$

79. (a) $x + y + 1 = 0$ (b) $x - y + 5 = 0$

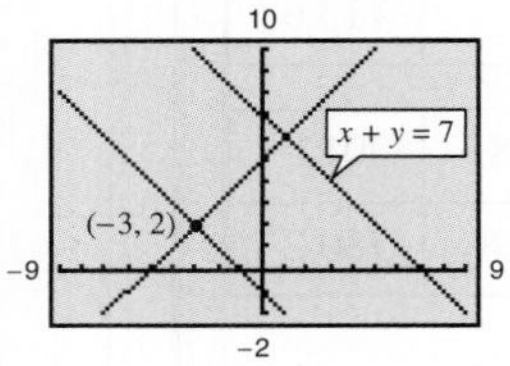

81. (a) $6x + 8y - 3 = 0$ (b) $96x - 72y + 127 = 0$

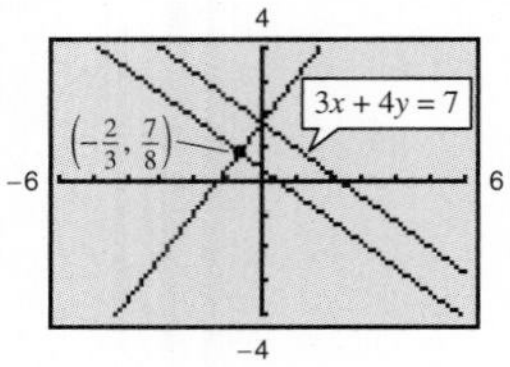

83. (a) $y = 0$ (b) $x + 1 = 0$

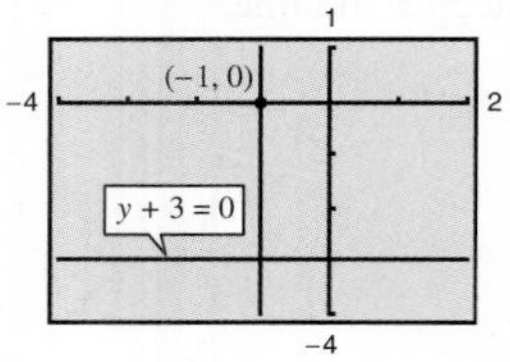

85. (a) $x - 1 = 0$ (b) $y - 1 = 0$

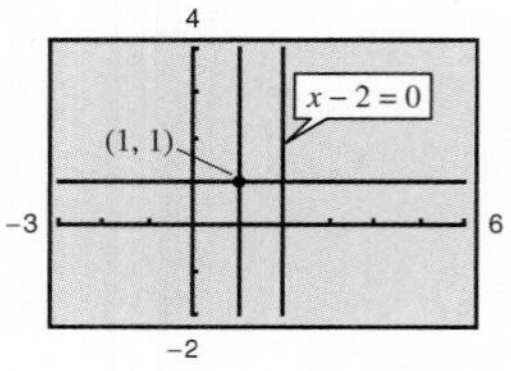

87. (a) The average salary increased the most from 2006 to 2008 and increased the least from 2002 to 2004.
(b) $m = 2350.75$
(c) The average salary increased $2350.75 per year over the 12 years between 1996 and 2008.

89. $F = \frac{9}{5}C + 32$ or $C = \frac{5}{9}F - \frac{160}{9}$

91. (a) $y = 28.8t + 5395.8$; The slope $m = 28.8$ indicates that the population increases by 28.8 thousand each year.
(b) 5568.6 thousand (5,568,600)
(c) 5572 thousand (5,572,000); The estimate was very close to the actual population.
(d) The model could possibly be used to predict the population in 2015 if the population continues to grow at the same linear rate.

93. (a) $y = 447.6t + 8146.6$ (b) $10,832.2 billion
(c) $13,070.2 billion (d) Answers will vary.

95. (a) $C = 50x + 350{,}000$; $R = 120x$
(b) $P = 70x - 350{,}000$ (c) $560,000 profit

Quiz Yourself *(page 34)*

1. (a)

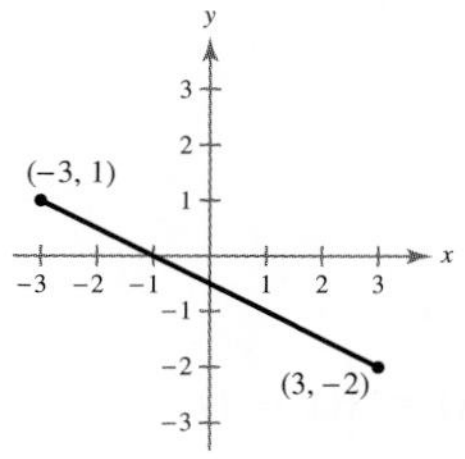

(b) $d = 3\sqrt{5}$
(c) Midpoint: $(0, -0.5)$

2. (a)

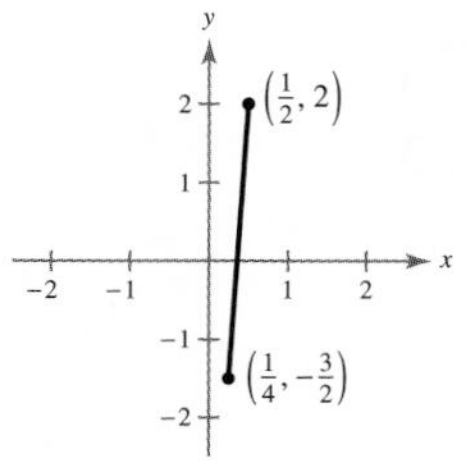

(b) $d = \sqrt{12.3125}$
(c) Midpoint: $\left(\frac{3}{8}, \frac{1}{4}\right)$

3. (a)

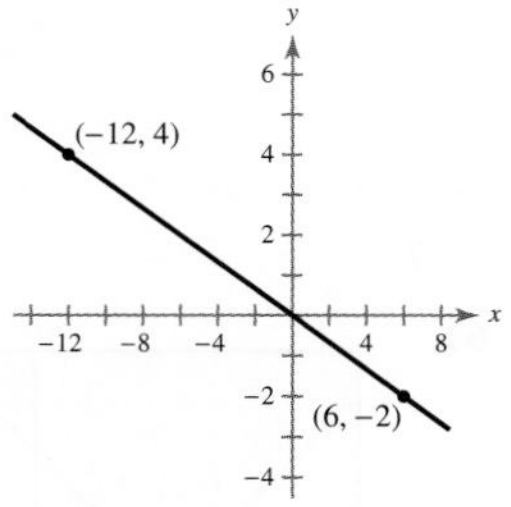

(b) $d = 6\sqrt{10}$
(c) Midpoint: $(-3, 1)$

4. $d_1 = \sqrt{5}$
$d_2 = \sqrt{45}$
$d_3 = \sqrt{50}$
$d_1^2 + d_2^2 = d_3^2$

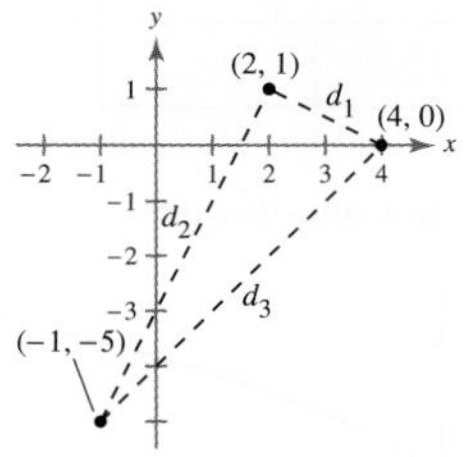

5. 9681.5 thousand

6.

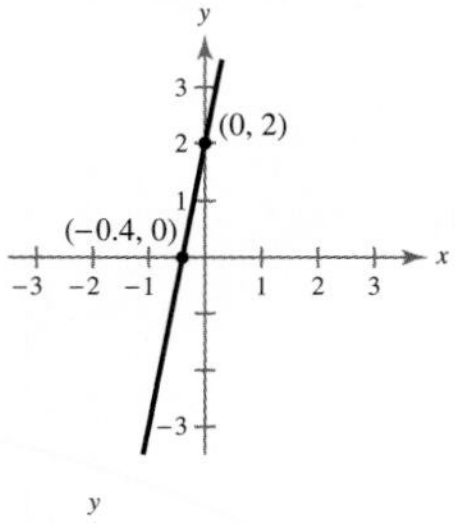

7.

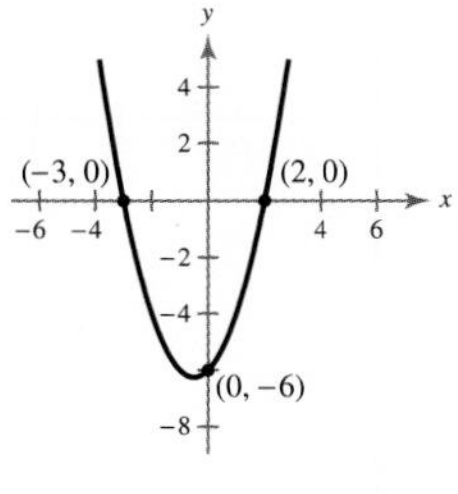

8.

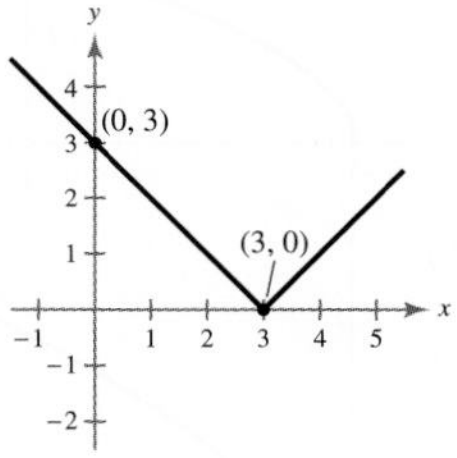

9. $x^2 + y^2 = 81$

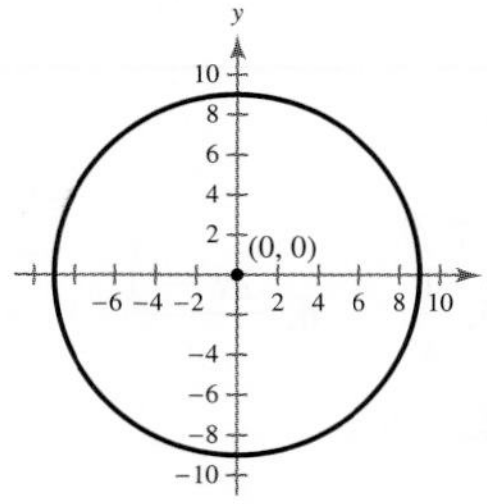

10. $(x + 1)^2 + y^2 = 36$

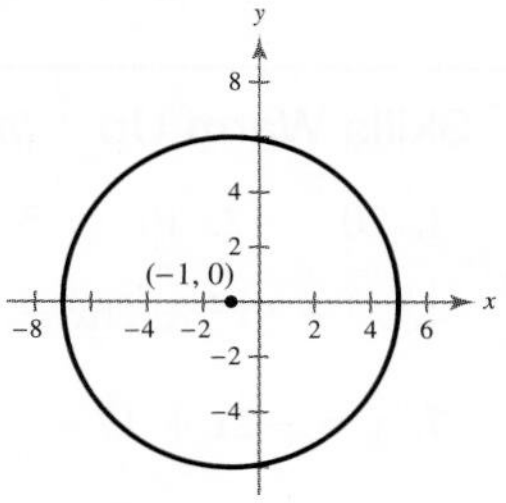

11. $(x - 2)^2 + (y + 2)^2 = 25$

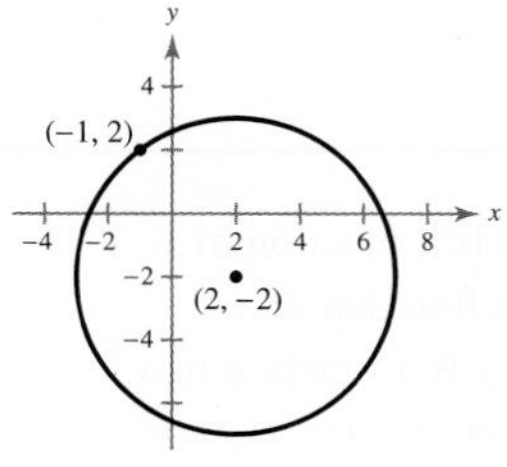

12. 4735 units

13. $y = \frac{1}{2}x - 3$

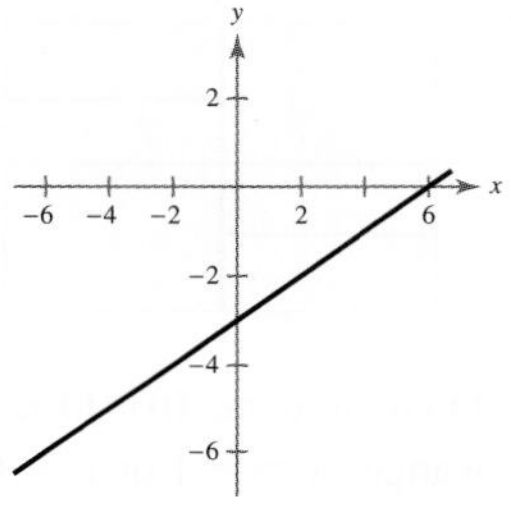

14. $y = 2x - 1$

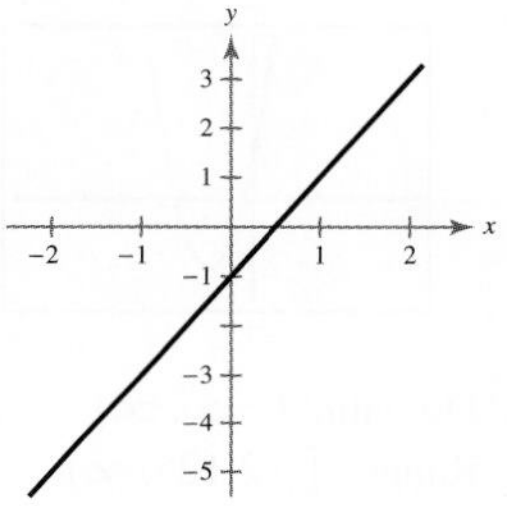

15. $y = -\frac{1}{3}x + 7$

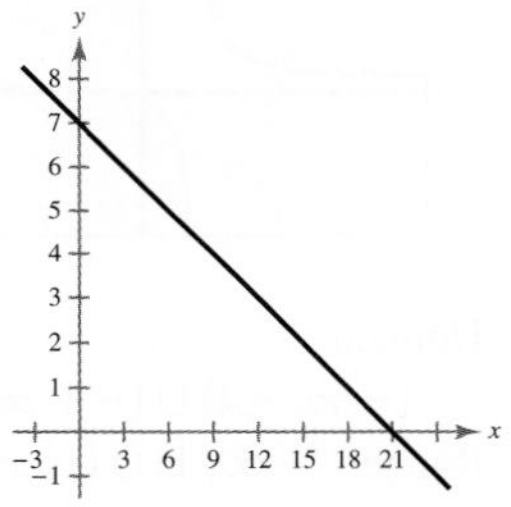

16. $y = -1.2x + 0.2$

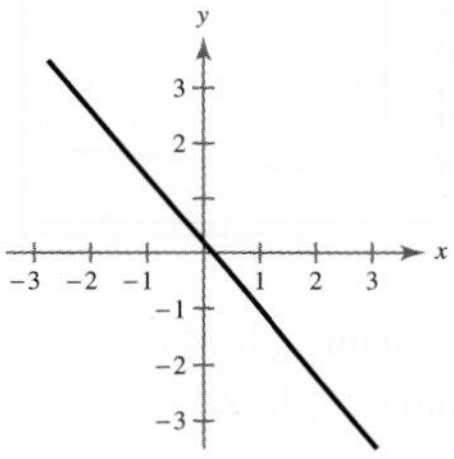

17. $x = -2$

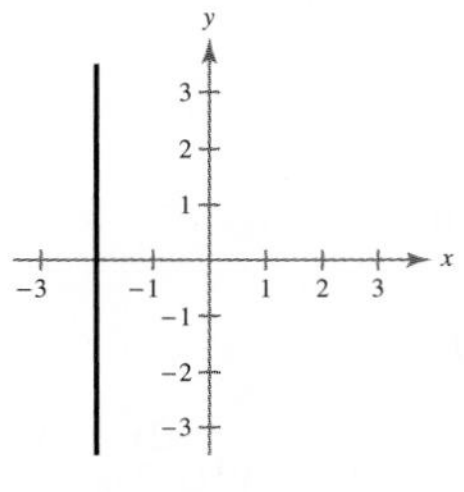

18. $y = 2$

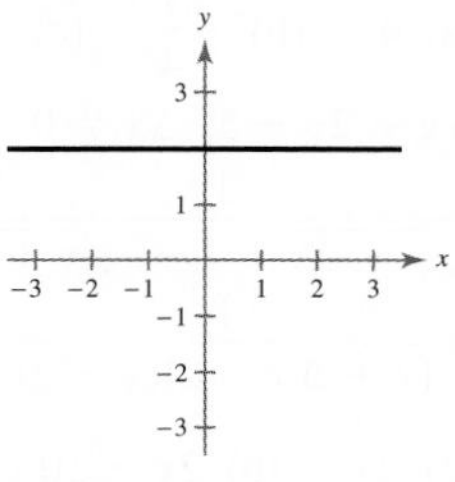

19. (a) $y = -0.25x - 4.25$ (b) $y = 4x - 17$
20. 2015: \$2,270,000; 2018: \$2,622,500
21. $C = 0.55x + 175$
22. (a) $S = 1600t + 20{,}200$ (b) \$44,200

Section 1.4 *(page 44)*

Skills Warm Up *(page 44)*

1. 20 **2.** 10 **3.** $x^2 + x - 6$

4. $x^3 + 9x^2 + 26x + 30$ **5.** $\frac{1}{x}$ **6.** $\frac{2x - 1}{x}$

7. $y = -2x + 17$ **8.** $y = \frac{6}{5}x^2 + \frac{1}{5}$

9. $y = 3 \pm \sqrt{5 + (x + 1)^2}$ **10.** $y = \pm\sqrt{4x^2 + 2}$

11. $y = 2x + \frac{1}{2}$ **12.** $y = \frac{x^3}{2} + \frac{1}{2}$

1. y is not a function of x. **3.** y is a function of x.
5. y is a function of x. **7.** y is a function of x.
9. y is not a function of x. **11.** y is a function of x.
13. Domain: $(-\infty, \infty)$
Range: $(-\infty, \infty)$
15. Domain: $(-\infty, \infty)$
Range: $(-\infty, 4]$

17.

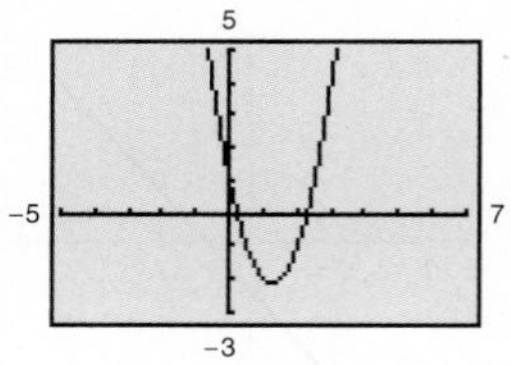

Domain: $(-\infty, \infty)$
Range: $[-2.125, \infty)$

19.

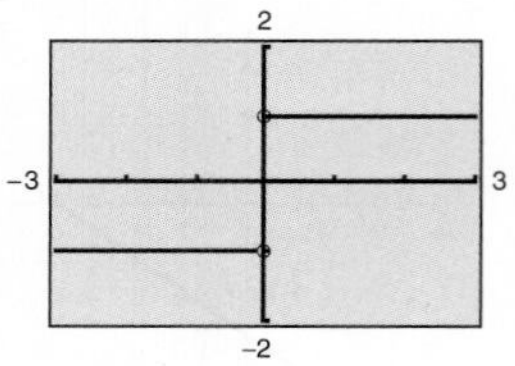

Domain: $(-\infty, 0) \cup (0, \infty)$
Range: $y = -1$ or $y = 1$

21.

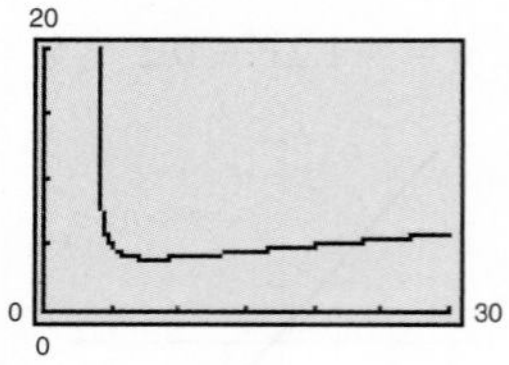

Domain: $(4, \infty)$
Range: $[4, \infty)$

23.

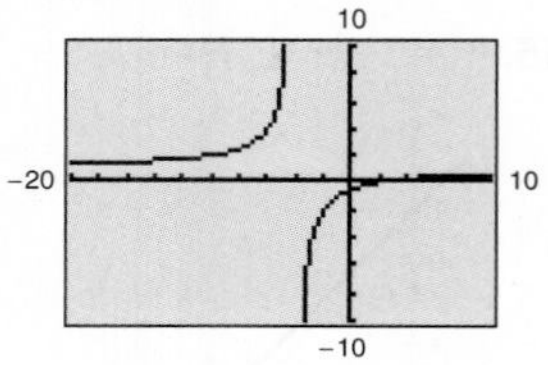

Domain:
$(-\infty, -4) \cup (-4, \infty)$
Range: $(-\infty, 1) \cup (1, \infty)$

25. (a) -2 (b) 13 (c) $3x - 5$

27. (a) 4 (b) $-\frac{1}{4}$ (c) $\frac{1}{x + 4}$

29. $\Delta x + 2x - 5,\ \Delta x \neq 0$

31. $\dfrac{1}{\sqrt{x + \Delta x + 1} + \sqrt{x + 1}},\ \Delta x \neq 0$

33. $-\dfrac{1}{(x + \Delta x - 2)(x - 2)},\ \Delta x \neq 0$

35. (a) $2x$ (b) $2x - 10$ (c) $10x - 25$ (d) $\frac{2x - 5}{5}$
(e) 5 (f) 5

37. (a) $x^2 + x$ (b) $x^2 - x + 2$
(c) $(x^2 + 1)(x - 1) = x^3 - x^2 + x - 1$
(d) $\frac{x^2 + 1}{x - 1}$ (e) $x^2 - 2x + 2$ (f) x^2

39. (a) 0 (b) 0 (c) -1 (d) $\sqrt{15}$
(e) $\sqrt{x^2 - 1}$ (f) $x - 1, x \geq 0$

41. $f^{-1}(x) = \frac{1}{4}x$
$f(f^{-1}(x)) = 4\left(\frac{1}{4}x\right) = x$
$f^{-1}(f(x)) = \frac{1}{4}(4x) = x$

43. $f^{-1}(x) = x - 12$
$f(f^{-1}(x)) = (x - 12) + 12 = x$
$f^{-1}(f(x)) = (x + 12) - 12 = x$

45. $f^{-1}(x) = \frac{x + 3}{2}$ **47.** $f^{-1}(x) = \frac{2}{3}(x - 1)$

49. $f^{-1}(x) = \sqrt[5]{x}$ **51.** $f^{-1}(x) = \frac{1}{x}$

53. $f^{-1}(x) = \sqrt{9 - x^2},\quad 0 \leq x \leq 3$

55. $f^{-1}(x) = x^{3/2},\quad x \geq 0$

57.

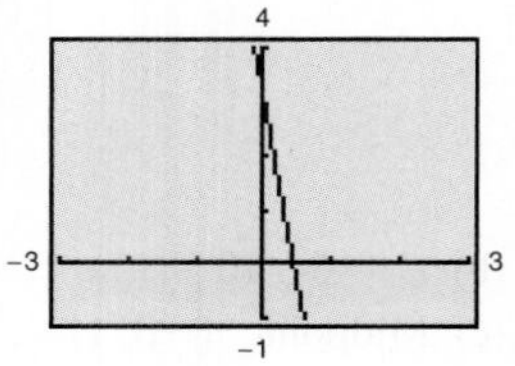

$f(x)$ is one-to-one. $f^{-1}(x) = \frac{3 - x}{7}$

59.

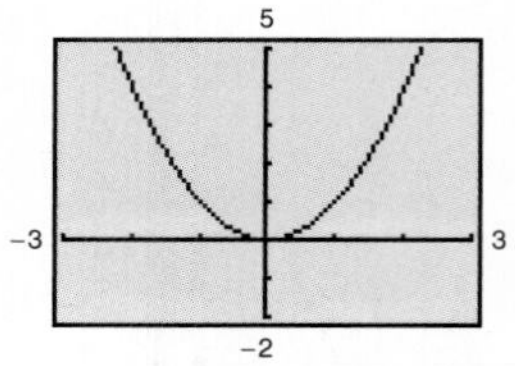

$f(x)$ is not one-to-one.

61.

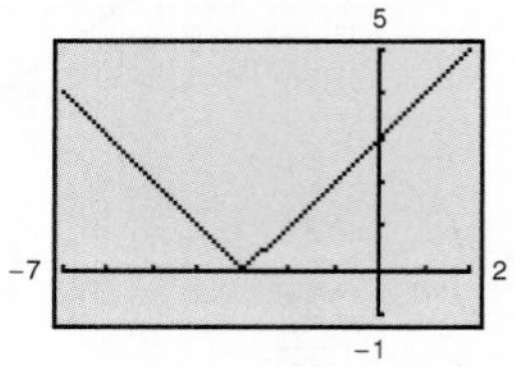

$f(x)$ is not one-to-one.

63. (a)

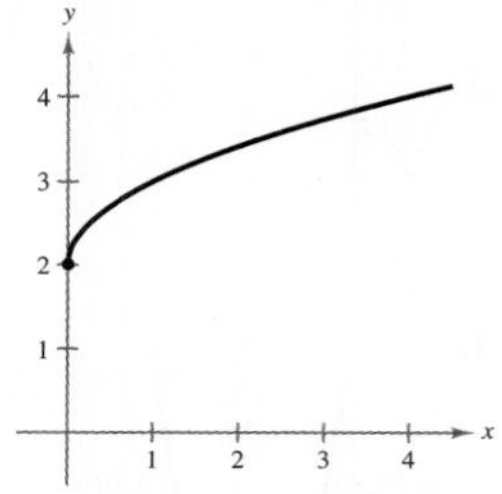

(b)

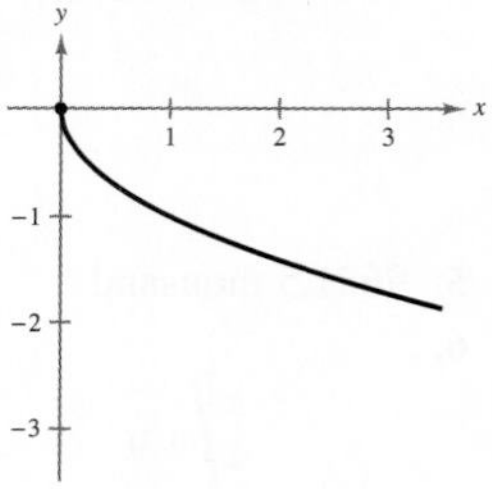

(c)

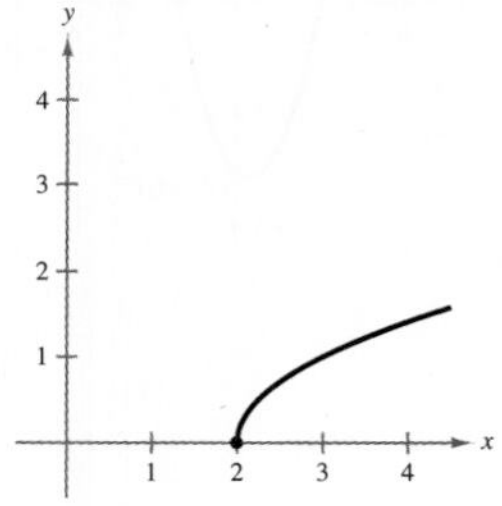

(d)

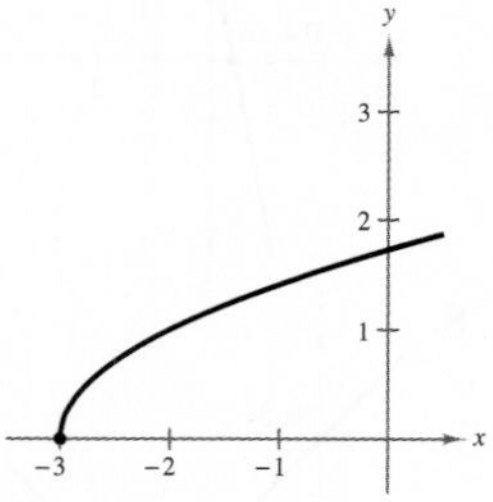

(e)

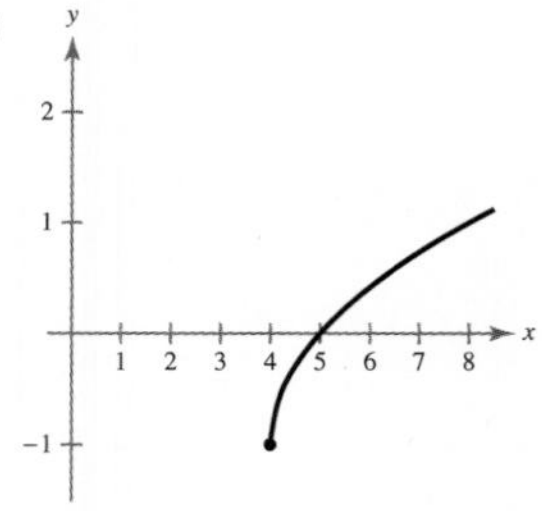

(f)

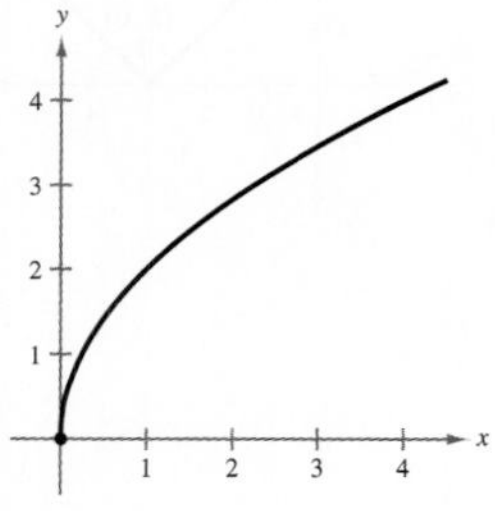

65. (a) $y = (x + 3)^2$ (b) $y = -(x + 6)^2 - 3$

67. (a) 2000: \$120 billion
2003: \$170 billion
2007: \$225 billion
(b) 2000: \$121.3 billion
2003: \$173.8 billion
2007: \$225.6 billion
The model fits the data well.

69. $R_T = R_1 + R_2 = -0.8t^2 - 7.22t + 1148$
$t = 5, 6, \ldots, 11$

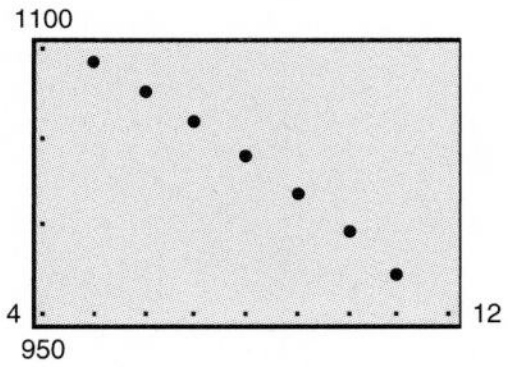

71. (a) $C(x) = 1.95x + 6000$
(b) $C = 1.95 + \dfrac{6000}{x}$
(c) More than 2000 units

73. (a) $C(x(t)) = 2800t + 500$
This function gives the production cost for t hours.
(b) \$11,700
(c) 6.25 hours

75. (a) $C(x) = 12.30x + 98{,}000$
(b) $R(x) = 17.98x$
(c) $P(x) = 5.68x - 98{,}000$

77.

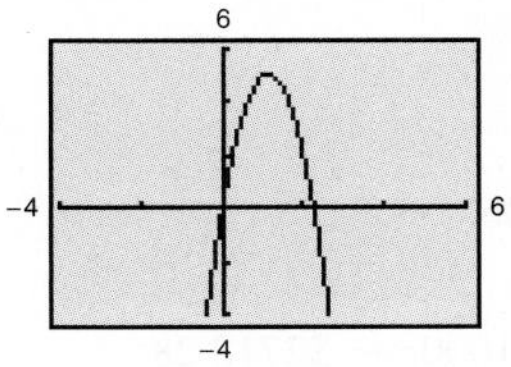

Zeros: $x = 0, \frac{9}{4}$
$f(x)$ is not one-to-one.

79.

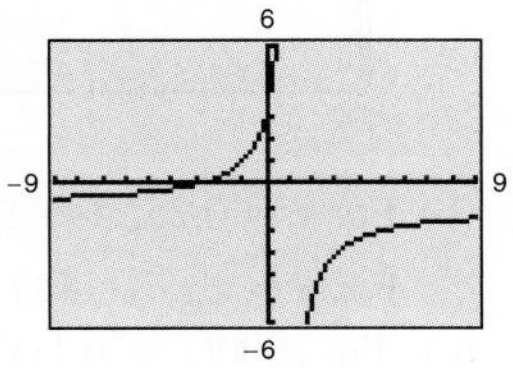

Zero: $t = -3$
$g(t)$ is one-to-one.

81.

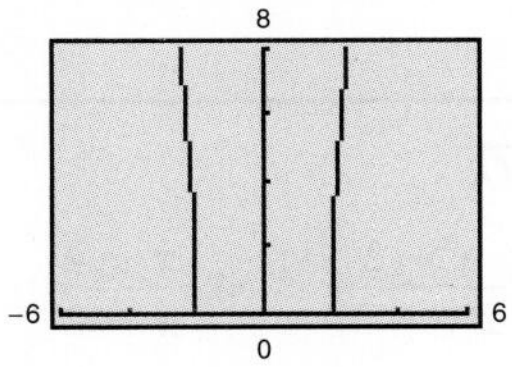

Zeros: ± 2
$g(x)$ is not one-to-one.

83. Answers will vary.

Section 1.5 *(page 57)*

Skills Warm Up *(page 57)*

1. $\frac{1}{3}x^2 + \frac{1}{6}x$ **2.** $x^2(x + 9)$ **3.** $x + 4$ **4.** $x + 6$

5. (a) 7 (b) $c^2 - 3c + 3$
(c) $x^2 + 2xh + h^2 - 3x - 3h + 3$

6. (a) -4 (b) 10 (c) $3t^2 + 4$ **7.** h **8.** 4

9. Domain: $(-\infty, 0) \cup (0, \infty)$
Range: $(-\infty, 0) \cup (0, \infty)$

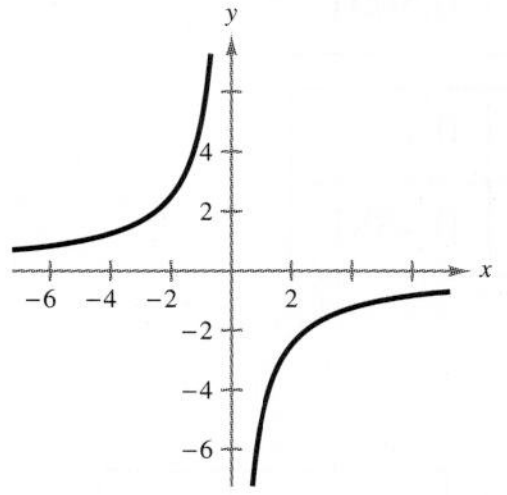

10. Domain: $[-5, 5]$
Range: $[0, 5]$

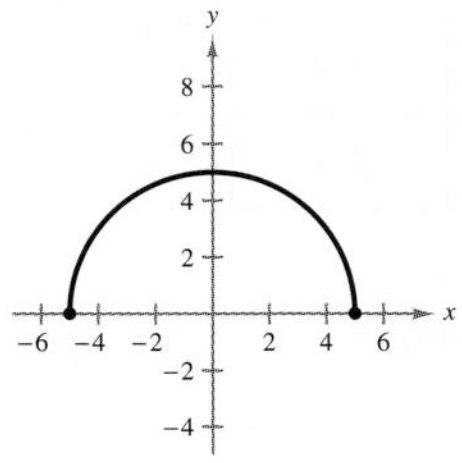

11. Domain: $(-\infty, \infty)$
Range: $[0, \infty)$

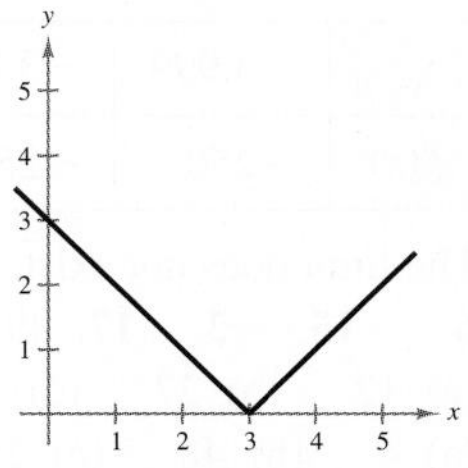

12. Domain:
$(-\infty, 0) \cup (0, \infty)$
Range: $-1, 1$

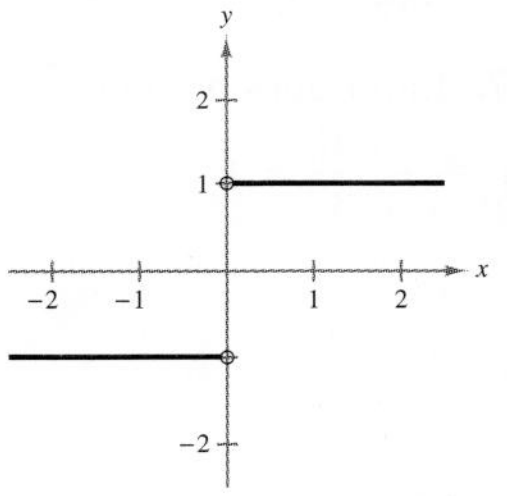

13. y is not a function of x.
14. y is a function of x.

1. (a) 1 (b) 3 **3.** (a) 1 (b) 3

5.

x	1.9	1.99	1.999	2
$f(x)$	8.8	8.98	8.998	?

x	2.001	2.01	2.1
$f(x)$	9.002	9.02	9.2

$\lim_{x \to 2}(2x + 5) = 9$

7.

x	1.9	1.99	1.999	2
$f(x)$	0.2564	0.2506	0.2501	?

x	2.001	2.01	2.1
$f(x)$	0.2499	0.2494	0.2439

$\lim_{x \to 2} \dfrac{x - 2}{x^2 - 4} = \dfrac{1}{4}$

9.

x	-0.1	-0.01	-0.001	0
$f(x)$	0.5132	0.5013	0.5001	?

x	0.001	0.01	0.1
$f(x)$	0.4999	0.4988	0.4881

$\lim_{x\to 0} \frac{\sqrt{x+1}-1}{x} = 0.5$

11.

x	-4.1	-4.01	-4.001	-4
$f(x)$	2.5	25	250	?

x	-3.999	-3.99	-3.9
$f(x)$	-250	-25	-2.5

The limit does not exist.

13. 6 **15.** -2 **17.** 49 **19.** 4

21. (a) 12 (b) 27 (c) $\frac{1}{3}$

23. (a) 4 (b) 48 (c) 256

25. -1 **27.** 0 **29.** 3 **31.** -2 **33.** $-\frac{3}{4}$ **35.** 2

37. -6 **39.** $-\frac{1}{4}$ **41.** 12 **43.** 2 **45.** $2t-5$

47. $\frac{1}{6}$ **49.** $\frac{1}{2\sqrt{5}}$ **51.** 2 **53.** -1

55. Limit does not exist. **57.** Limit does not exist.

59. $\lim_{x\to -3^-} \frac{|x+3|}{x+3} = -1$, $\lim_{x\to -3^+} \frac{|x+3|}{x+3} = 1$

61.

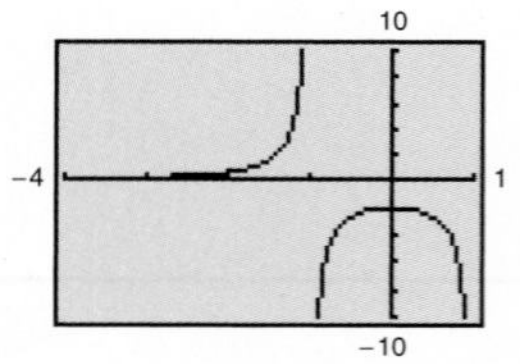

x	0	0.5	0.9	0.99
$f(x)$	-2	-2.67	-10.53	-100.5

x	0.999	0.9999	1
$f(x)$	-1000.5	$-10{,}000.5$	Undefined

$-\infty$

63.

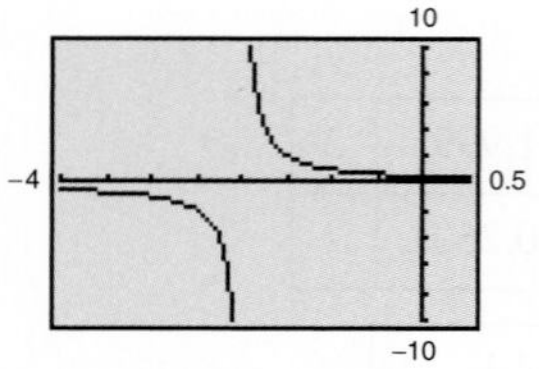

x	-3	-2.5	-2.1	-2.01
$f(x)$	-1	-2	-10	-100

x	-2.001	-2.0001	-2
$f(x)$	-1000	$-10{,}000$	Undefined

$-\infty$

65. (a) 1 (b) 1 (c) 1

67. (a) 0 (b) 0 (c) 0

69. (a) 3 (b) -3 (c) Limit does not exist.

71.

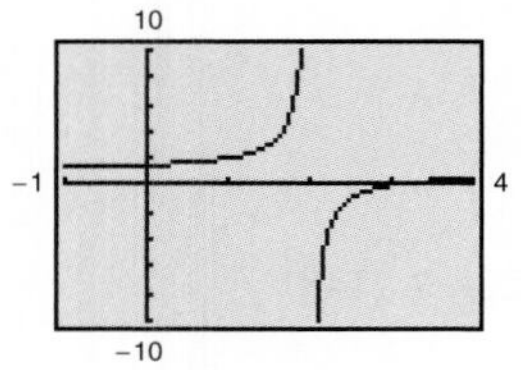

Limit does not exist.

73.

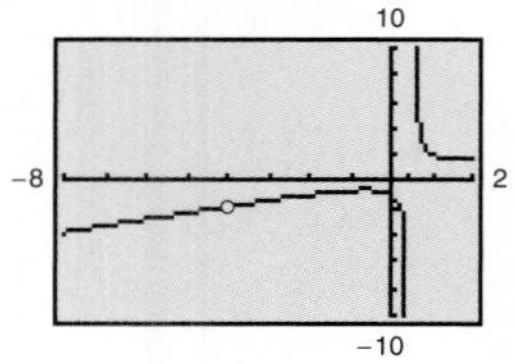

$-\frac{17}{9} \approx -1.8889$

75. (a) \$25 thousand (b) 80%

(c) ∞; The cost function increases without bound as x approaches 100 from the left. Therefore, according to the model, it is not possible to remove 100% of the pollutants.

77. (a)

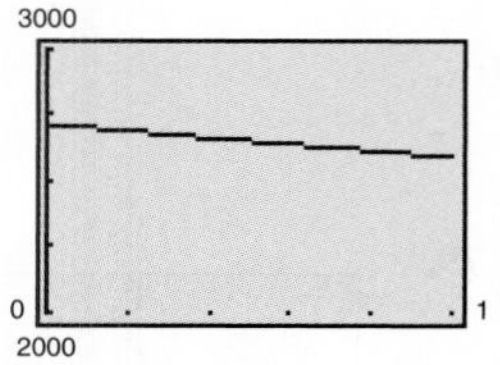

(b) For $x = 0.25$, $A \approx \$2685.06$.

For $x = \frac{1}{365}$, $A \approx \$2717.91$.

(c) $\lim_{x\to 0^+} 1000(1+0.1x)^{10/x} = 1000e \approx \$2718.28$; continuous compounding

Section 1.6 *(page 67)*

Skills Warm Up *(page 67)*

1. $\frac{x+4}{x-8}$ **2.** $\frac{x+1}{x-3}$ **3.** $\frac{x+2}{2(x-3)}$ **4.** $\frac{x-4}{x-2}$

5. $x = 0, -7$ **6.** $x = -5, 1$ **7.** $x = -\frac{2}{3}, -2$

8. $x = 0, 3, -8$ **9.** 13 **10.** -1

1. Continuous; The function is a polynomial.

3. Not continuous ($x \neq \pm 4$)

5. Continuous; The rational function's domain is the set of real numbers.

7. Not continuous ($x \neq 3$ and $x \neq 5$)

9. Not continuous ($x \neq \pm 2$)

11. $(-\infty, 0)$ and $(0, \infty)$; Explanations will vary. There is a discontinuity at $x = 0$, because $f(0)$ is not defined.

13. $(-\infty, -1)$ and $(-1, \infty)$; Explanations will vary. There is a discontinuity at $x = -1$, because $f(-1)$ is not defined.

15. $(-\infty, \infty)$; Explanations will vary.

17. $(-\infty, -1)$, $(-1, 1)$, and $(1, \infty)$; Explanations will vary. There are discontinuities at $x = \pm 1$, because $f(\pm 1)$ is not defined.

19. $(-\infty, \infty)$; Explanations will vary.

21. $(-\infty, 4)$, $(4, 5)$, and $(5, \infty)$; Explanations will vary. There are discontinuities at $x = 4$ and $x = 5$, because $f(4)$ and $f(5)$ are not defined.

23. $(-\infty, 4]$; Explanations will vary.

25. $[0, \infty)$; Explanations will vary.

27. $[-1, 3]$; Explanations will vary.

29. $(-\infty, \infty)$; Explanations will vary.

31. $(-\infty, -1)$ and $(-1, \infty)$; Explanations will vary. There is a discontinuity at $x = -1$, because $f(-1)$ is not defined.

33. $[-3, \infty)$; Explanations will vary.

35. Continuous on all intervals $\left(\frac{c}{2}, \frac{c}{2} + \frac{1}{2}\right)$, where c is an integer. Explanations will vary. There are discontinuities at $x = \frac{c}{2}$, where c is an integer, because $\lim_{x \to c} f\left(\frac{c}{2}\right)$ does not exist.

37. Continuous on all intervals $(c, c + 1)$, where c is an integer. Explanations will vary. There are discontinuities at $x = c$, where c is an integer, because $\lim_{x \to c} f(c)$ does not exist.

39. $(1, \infty)$; Explanations will vary.

41.

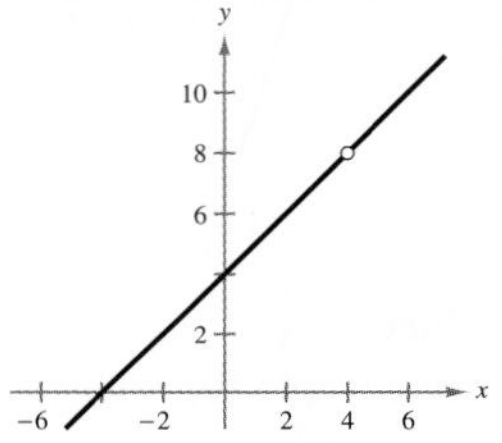

Continuous on $(-\infty, 4)$ and $(4, \infty)$

43.

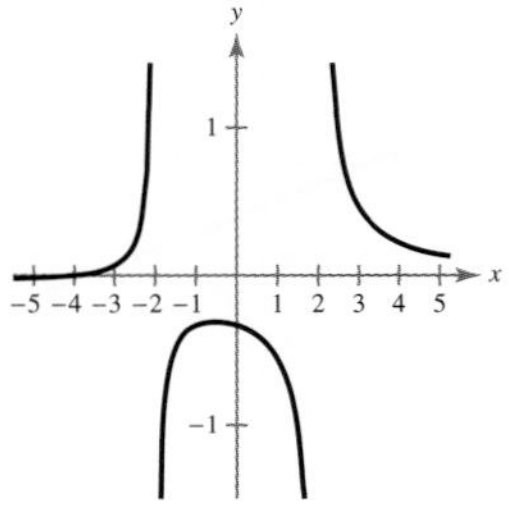

Continuous on $(-\infty, -2)$, $(-2, 2)$, and $(2, \infty)$

45.

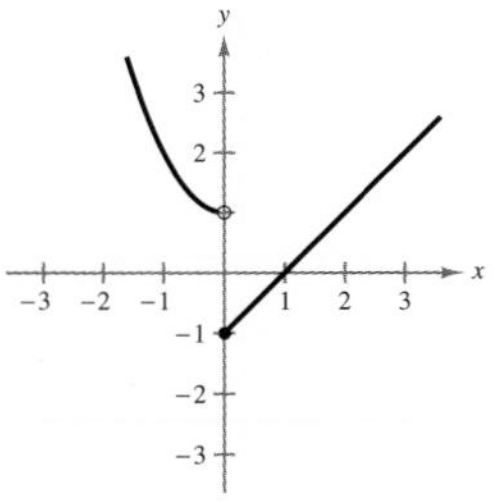

Continuous on $(-\infty, 0)$ and $(0, \infty)$

47. Continuous **49.** Nonremovable discontinuity at $x = 2$

51.

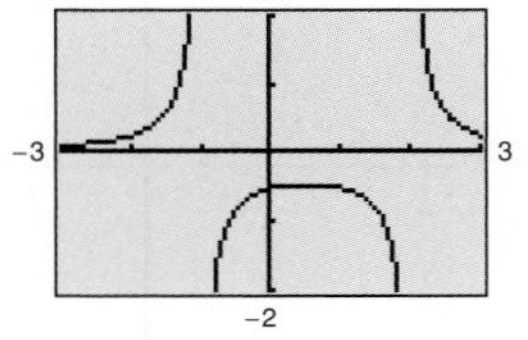

Not continuous at $x = 2$ and $x = -1$, because $f(-1)$ and $f(2)$ are not defined.

53.

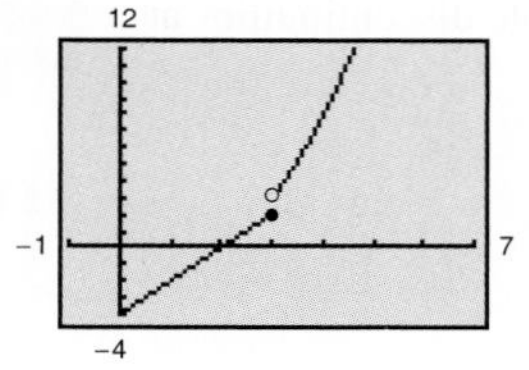

Not continuous at $x = 3$, because $\lim_{x \to 3} f(3)$ does not exist.

55.

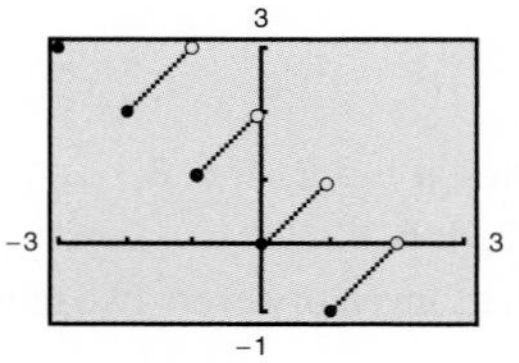

Not continuous at all integers c, because $\lim_{x \to c} f(c)$ does not exist.

57. $a = 2$

59.

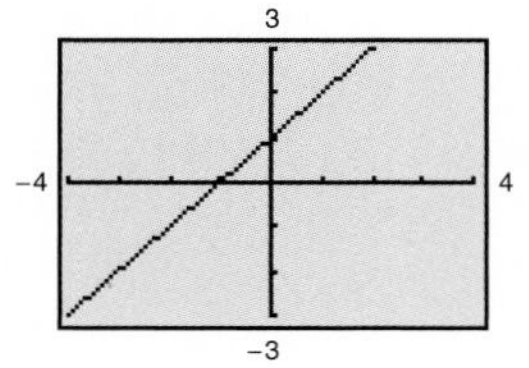

The graph of $f(x) = \frac{x^2 + x}{x}$ appears to be continuous on $[-4, 4]$, but f is not continuous at $x = 0$. Explanations will vary.

61. (a) $[0, 100)$; Explanations will vary.

(b)

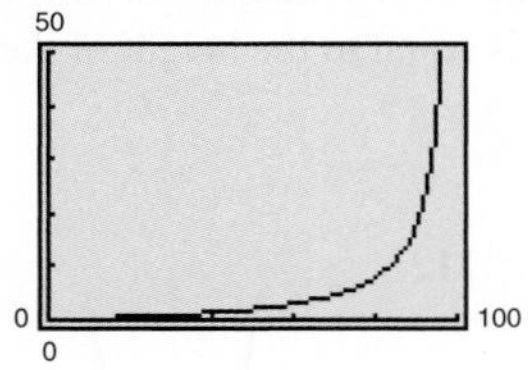

Continuous; Explanations will vary.

(c) \$6 million

63.

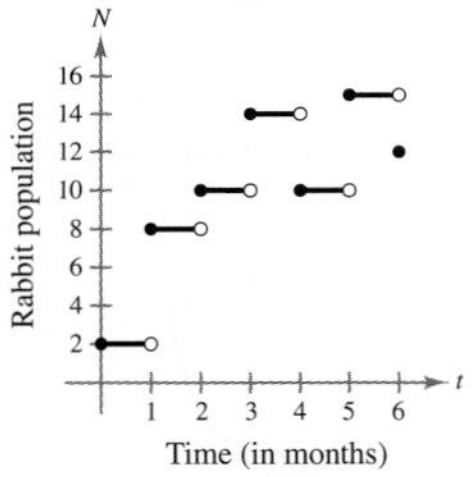

There are nonremovable discontinuities at $t = 1, 2, 3, 4, 5,$ and 6.

65. (a)

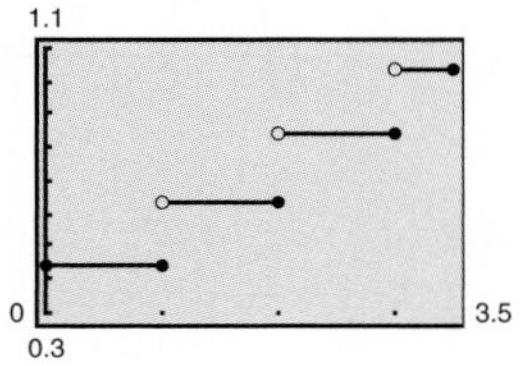

Discontinuities at $x = 1$, $x = 2$, $x = 3$

Explanations will vary.

(b) \$0.84

67. (a) The graph has nonremovable discontinuities at $t = \frac{1}{4}, \frac{1}{2}, \frac{3}{4}, 1, \frac{5}{4}, \ldots$.
(b) \$8448.69
(c) \$11,379.17

69. (a)

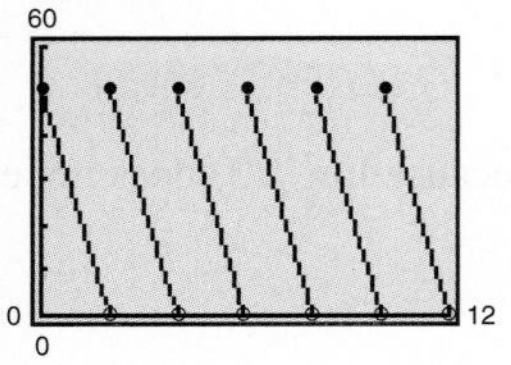

Nonremovable discontinuities at $t = 2, 4, 6, 8, \ldots$; N is not continuous at $t = 2, 4, 6, 8, \ldots$.
(b) The company must replenish its inventory every two months.

Review Exercises for Chapter 1 *(page 74)*

1.

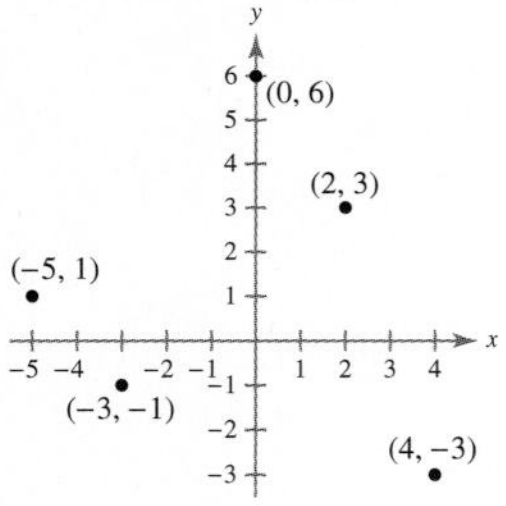

3. $\sqrt{29}$ **5.** $3\sqrt{2}$ **7.** $\dfrac{\sqrt{17}}{2}$ **9.** $(7, 4)$

11. $(-8, 6)$ **13.** $\left(\dfrac{5}{2}, \dfrac{2}{5}\right)$

15. $P = R - C$; The difference in the heights of the bars that represent revenue and cost is equal to the height of the bar that represents profit.

17. $(-2, 7), (-1, 8), (1, 5)$

19.

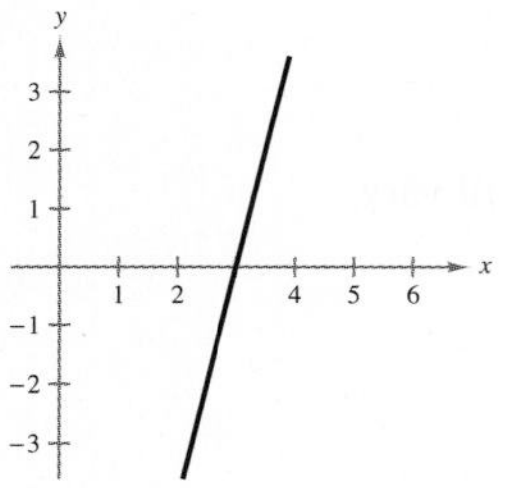

21.

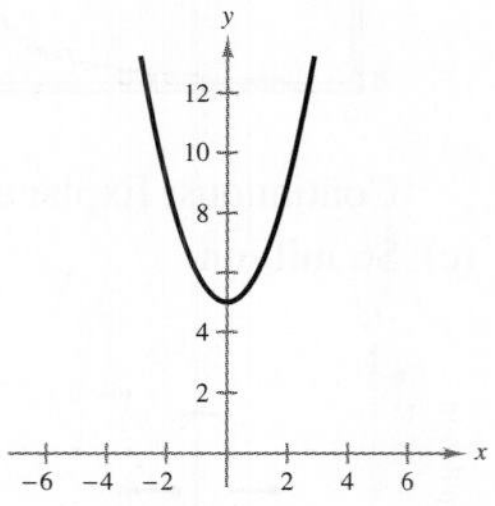

23.

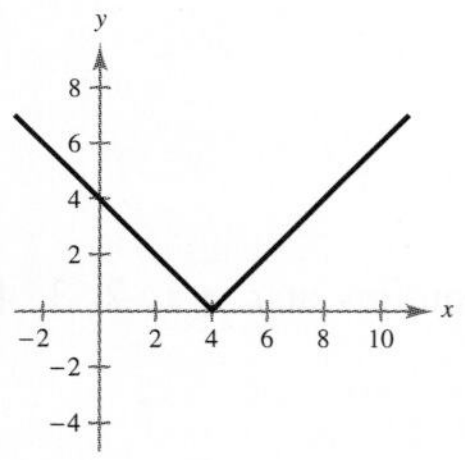

25.

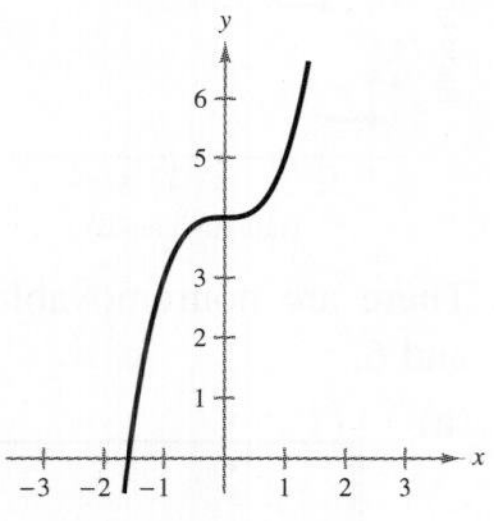

27.

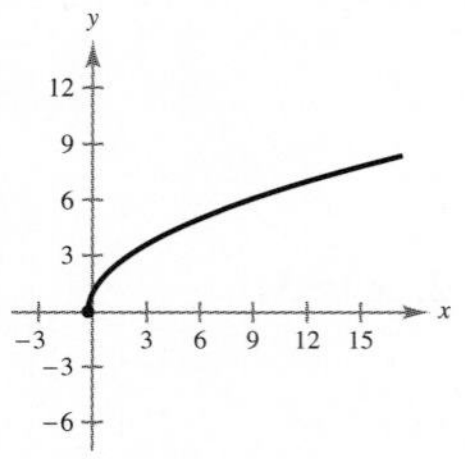

29. $\left(-\frac{3}{4}, 0\right), (0, -3)$ **31.** $(-4, 0), (2, 0), (0, -8)$

33. $x^2 + y^2 = 64$

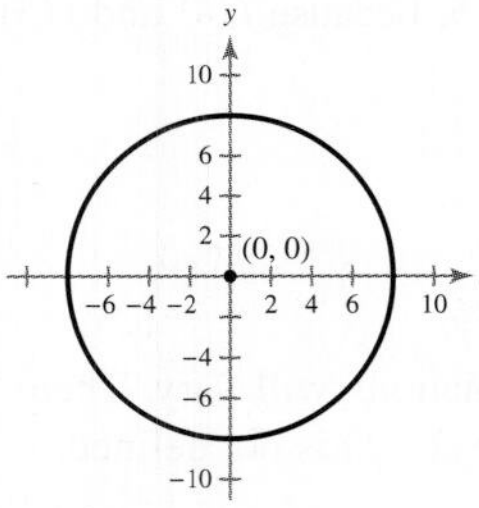

35. $x^2 + y^2 = 9$

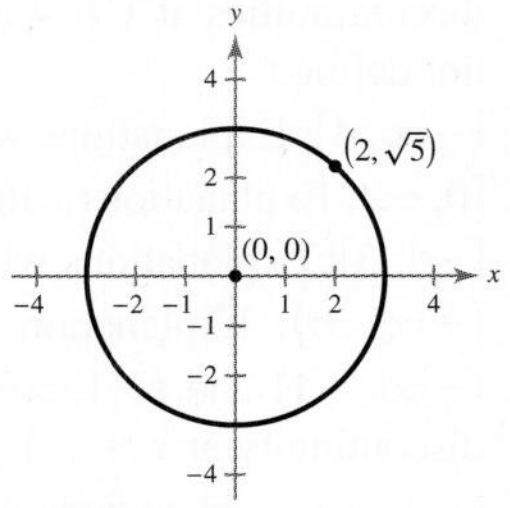

37. $(-2, 9)$ **39.** $(-1, -1), (0, 0), (1, 1)$

41. (a) $C = 10x + 200$
$R = 14x$
(b) 50 shirts

43. $p = \$46.40$
$x = 5000$ units

45. Slope: -1
y-intercept: $(0, 12)$

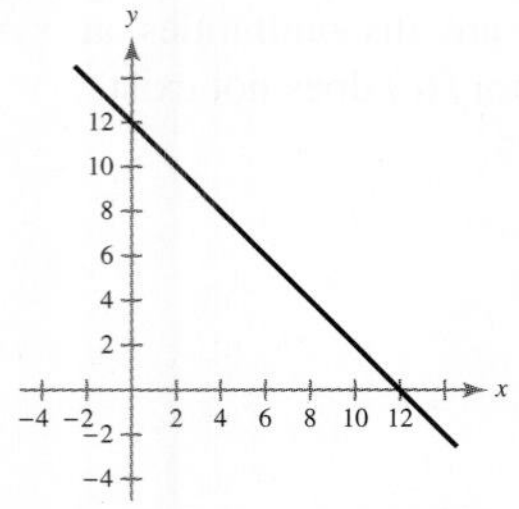

47. Slope: -3
y-intercept: $(0, -2)$

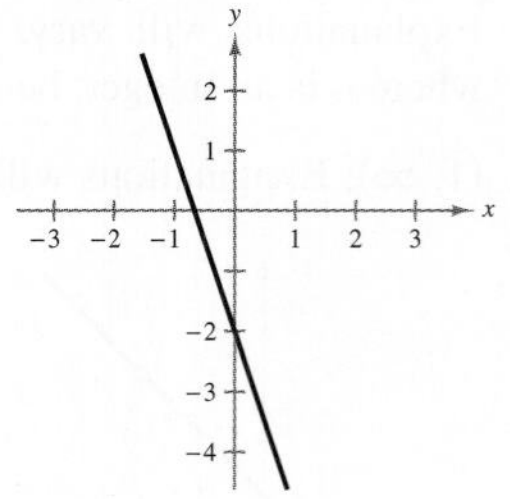

49. Slope: 0 (horizontal line)
y-intercept: $\left(0, -\frac{5}{3}\right)$

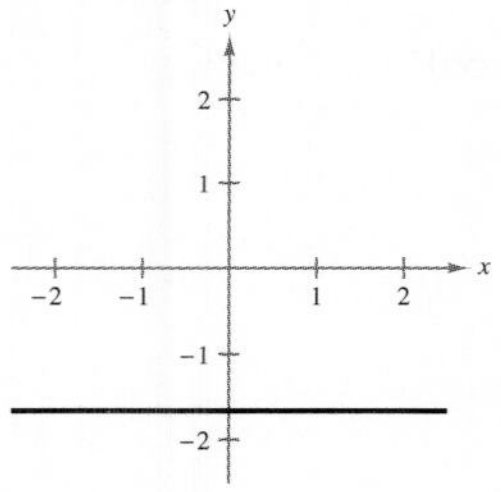

51. Slope: $-\frac{2}{5}$
y-intercept: $(0, -1)$

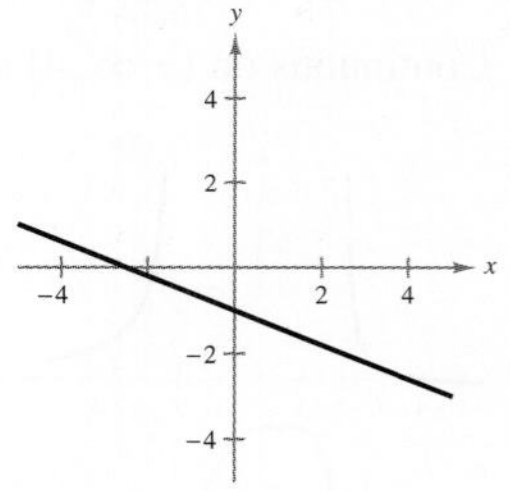

53. $\frac{6}{7}$ **55.** $\frac{20}{21}$

57. $y = -2x + 5$

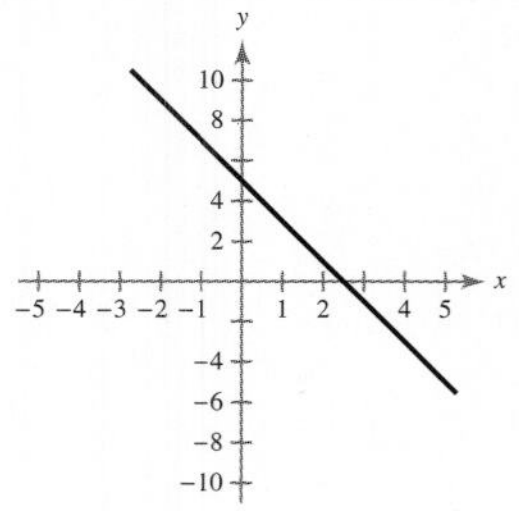

59. $y = -4$

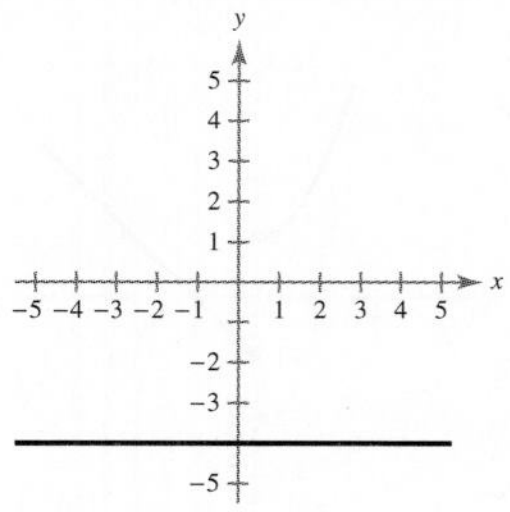

61. $y = 2x - 9$

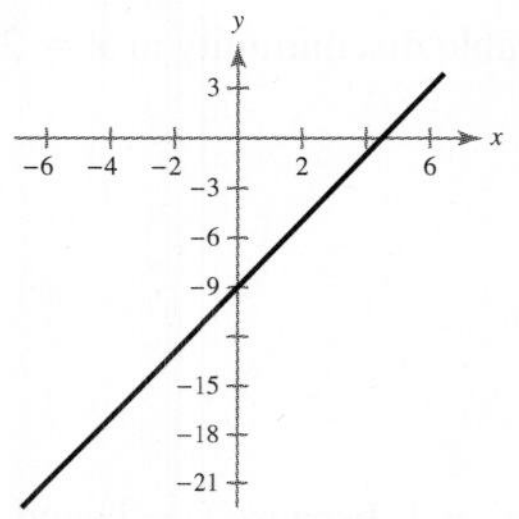

63. $x = 5$

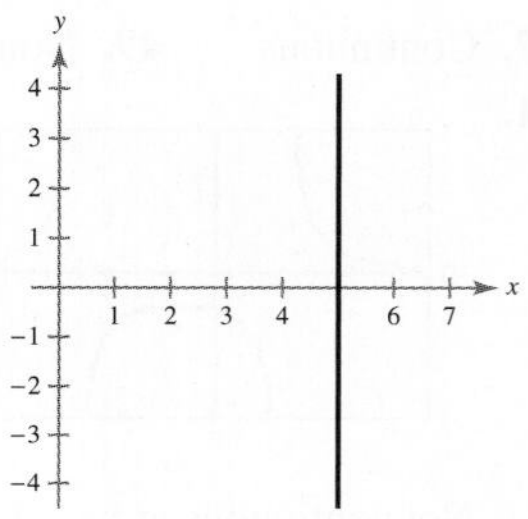

65. (a) $7x - 8y + 69 = 0$ (b) $2x + y = 0$
(c) $2x + y = 0$ (d) $2x + 3y - 12 = 0$

67. (a) $x = -10p + 1070$ (b) 725 units (c) 650 units

69. y is a function of x. **71.** y is not a function of x.

73. Domain: $(-\infty, \infty)$
Range: $(-\infty, \infty)$

75. Domain: $[-1, \infty)$
Range: $[0, \infty)$

77. Domain: $(-\infty, -4) \cup (-4, 3) \cup (3, \infty)$
Range: $(-\infty, 0) \cup \left(0, \frac{1}{7}\right) \cup \left(\frac{1}{7}, \infty\right)$

79. (a) 7 (b) -11 (c) $3x + 7$

81. (a) $x^2 + 2x$ (b) $x^2 - 2x + 2$ (c) $2x^3 - x^2 + 2x - 1$
(d) $\dfrac{1 + x^2}{2x - 1}$ (e) $4x^2 - 4x + 2$ (f) $2x^2 + 1$

83.

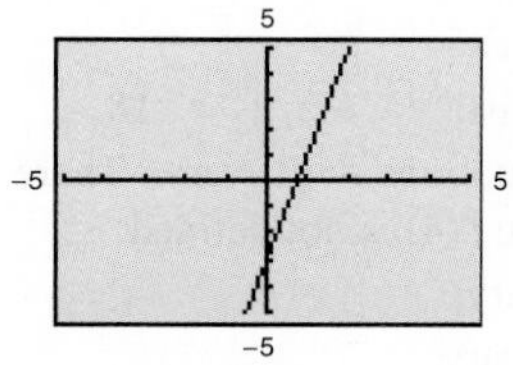

$f(x)$ is one-to-one.
$f^{-1}(x) = \frac{1}{4}(x + 3)$

85.

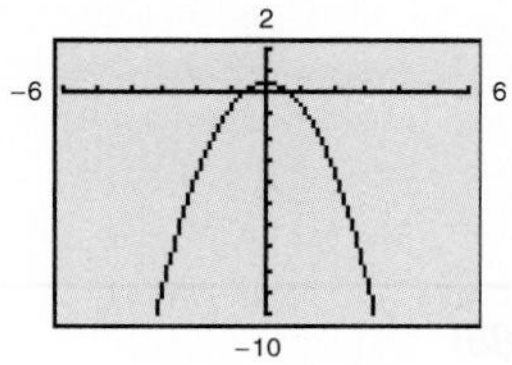

$f(x)$ does not have an inverse function.

87.

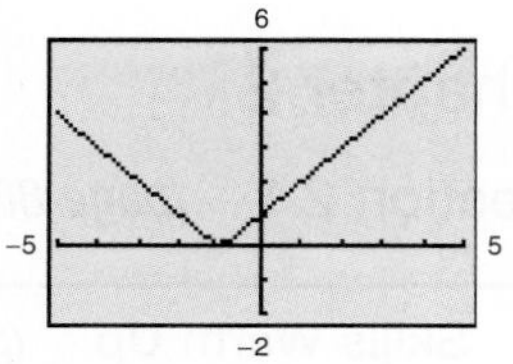

$f(x)$ does not have an inverse function.

89.

x	0.9	0.99	0.999	1	1.001	1.01	1.1
$f(x)$	0.6	0.96	0.996	?	1.004	1.04	1.4

$\lim_{x \to 1} (4x - 3) = 1$

91.

x	-0.1	-0.01	-0.001	0
$f(x)$	35.71	355.26	3550.71	?

x	0.001	0.01	0.1
$f(x)$	-3550.31	-354.85	-35.30

$\lim_{x \to 0} \dfrac{\sqrt{x + 6} - 6}{x}$ does not exist.

93. 8 **95.** 7 **97.** $-\frac{2}{5}$ **99.** Limit does not exist.

101. $-\frac{1}{4}$ **103.** $-\infty$ **105.** Limit does not exist.

107. 5 **109.** $3x^2 - 1$

111. $(-\infty, \infty)$; For any c on the real number line, $F(c)$ is defined, $\lim_{x \to c} f(x)$ exists, and $\lim_{x \to c} f(x) = f(c)$.

113. $(-\infty, -4)$ and $(-4, \infty)$; $f(-4)$ is undefined.

115. $(-\infty, -1)$ and $(-1, \infty)$; $f(-1)$ is undefined.

117. Continuous on all intervals $(c, c + 1)$, where c is an integer; $\lim_{x \to c} f(c)$ does not exist.

119. $(-\infty, 0)$ and $(0, \infty)$; $\lim_{x \to 0} f(x)$ does not exist.

121. $a = 2$

123. (a)

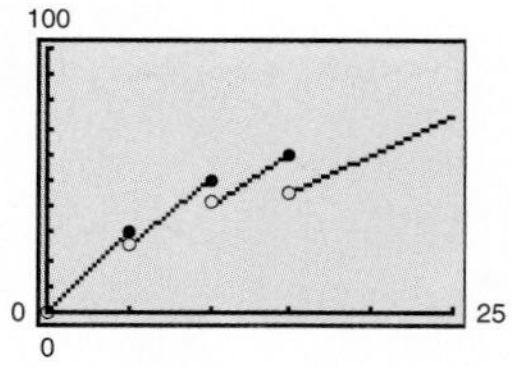

Explanations will vary. The function is defined for all values of x greater than zero. The function is discontinuous when $x = 5$, $x = 10$, and $x = 15$.
(b) \$49.90

125. (a) $C(t) = \begin{cases} 1 + 0.1[\![t]\!], & t > 0,\ t \text{ not an integer} \\ 1 + 0.1[\![t - 1]\!], & t > 0,\ t \text{ an integer} \end{cases}$
(b)

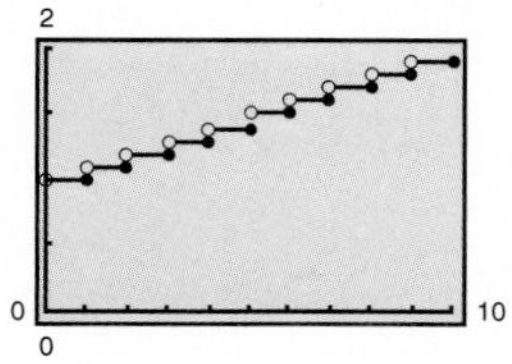

C is not continuous at $t = 1, 2, 3, \ldots$.

Chapter Test *(page 78)*

1. (a) $d = 5\sqrt{2}$ (b) Midpoint: $(-1.5, 1.5)$
(c) $m = -1$ (d) $y = -x$
(e)

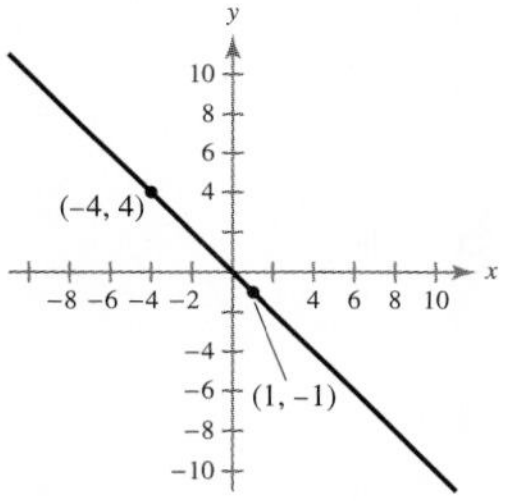

2. (a) $d = 2.5$ (b) Midpoint: $(1.25, 2)$
(c) $m = 0$ (d) $y = 2$
(e)

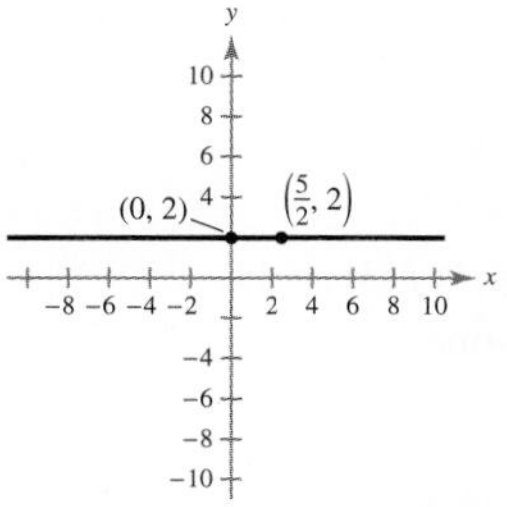

3. (a) $d = 2\sqrt{10}$ (b) Midpoint: $(-1, 2)$
(c) $m = \frac{1}{3}$ (d) $y = \frac{1}{3}x + \frac{7}{3}$
(e)

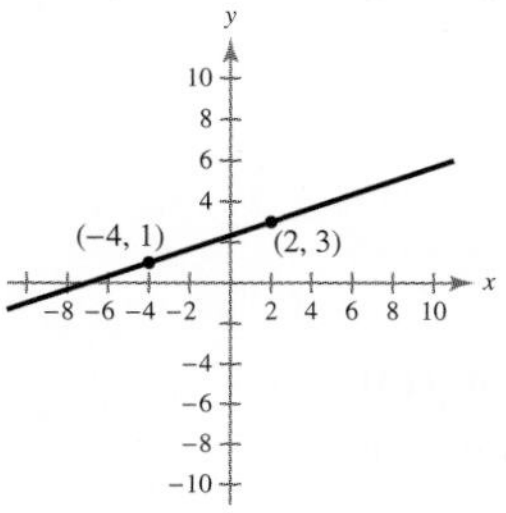

4. $(5.5, 53.45)$

5. $m = \frac{1}{5}$; $(0, -2)$

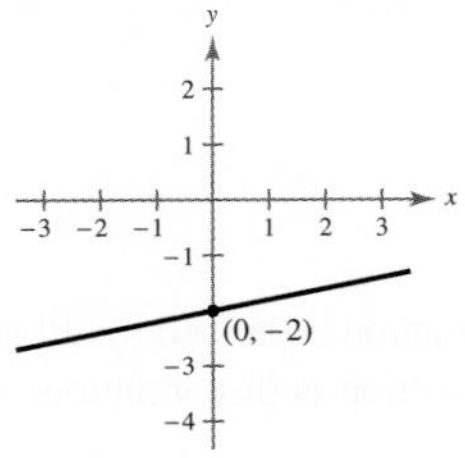

6. m is undefined; no y-intercept

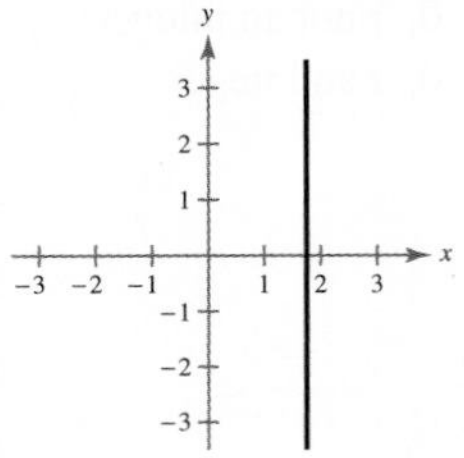

7. $m = -2.5$; $(0, 6.25)$

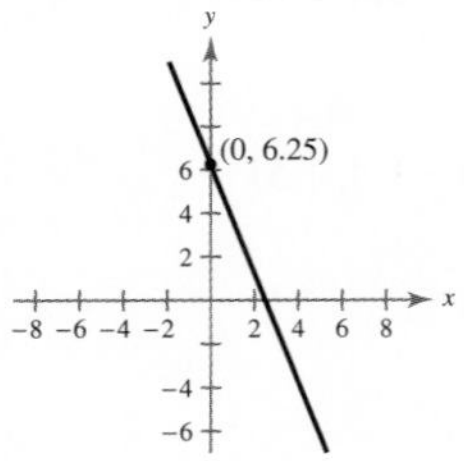

8. $y = -\frac{1}{4}x - \frac{29}{4}$ **9.** $y = \frac{5}{2}x - 4$

10. (a)

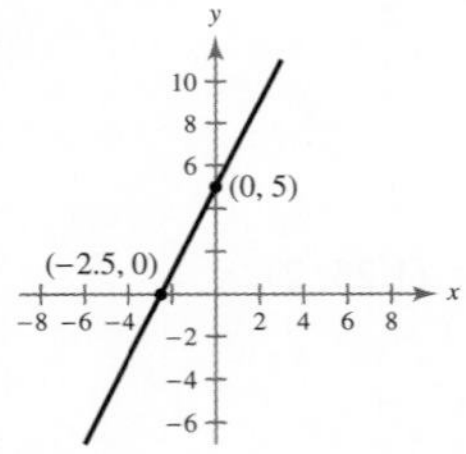

(b) Domain: $(-\infty, \infty)$
Range: $(-\infty, \infty)$

(c) $f(-3) = -1; f(-2) = 1; f(3) = 11$

(d) The function is one-to-one.

11. (a)

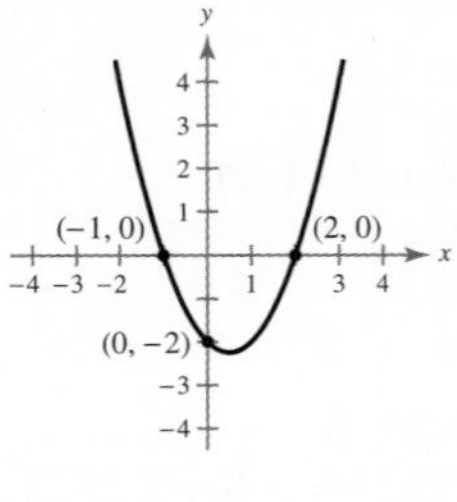

(b) Domain: $(-\infty, \infty)$
Range: $(-2.25, \infty)$

(c) $f(-3) = 10; f(-2) = 4; f(3) = 4$

(d) The function is not one-to-one.

12. (a)

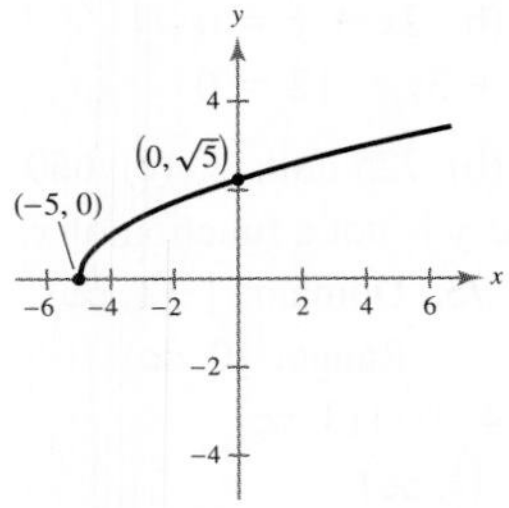

(b) Domain: $[-5, \infty)$
Range: $[0, \infty)$

(c) $f(-3) = \sqrt{2}; f(-2) = \sqrt{3}; f(3) = 2\sqrt{2}$

(d) The function is one-to-one.

13. $f^{-1}(x) = \frac{1}{4}x - \frac{3}{2}$ **14.** $f^{-1}(x) = -\frac{1}{3}x^3 + \frac{8}{3}$

15. -1 **16.** Limit does not exist. **17.** 2 **18.** $\frac{1}{6}$

19. $(-\infty, 4)$ and $(4, \infty)$; Explanations will vary. There is a discontinuity at $x = 4$, because $f(4)$ is not defined.

20. $(-\infty, 5]$; Explanations will vary.

21. $(-\infty, \infty)$; Explanations will vary.

22. (a) The model fits the data well. Explanations will vary.
(b) 128,087.392 thousand (128,087,392)

Chapter 2

Section 2.1 *(page 88)*

Skills Warm Up *(page 88)*

1. $x = 2$ **2.** $y = 2$ **3.** $y = -x + 2$

4. $y = 3x - 4$ **5.** $2x$ **6.** $3x^2$ **7.** $\frac{1}{x^2}$

8. $2x$ **9.** $(-\infty, \infty)$ **10.** $(-\infty, 1) \cup (1, \infty)$

11. $(-\infty, \infty)$ **12.** $(-\infty, 0) \cup (0, \infty)$

1.

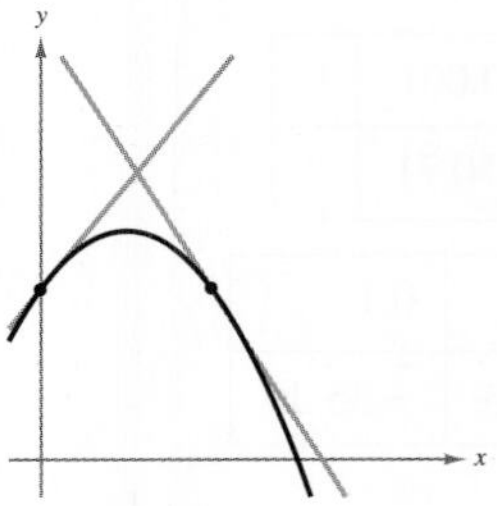

3.

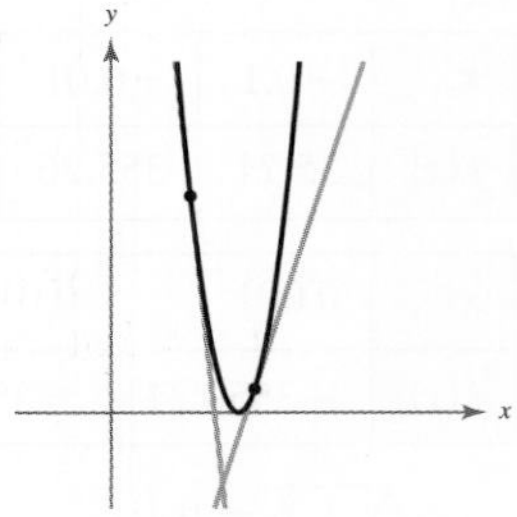

5.

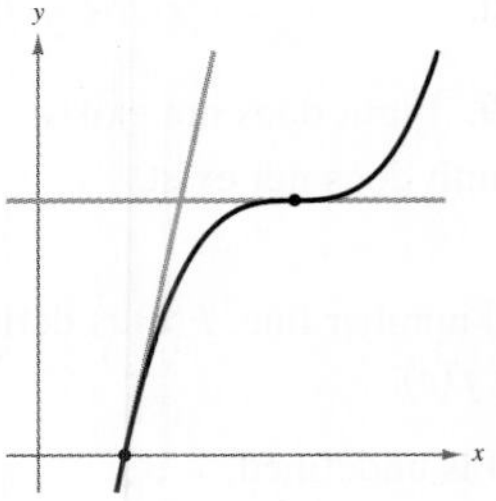

7. $m = 1$ **9.** $m = 0$ **11.** $m = -\frac{1}{3}$

13. 2005: $m \approx 119$
2007: $m \approx 161$
The slope is the rate of change of revenue at the given point in time.

15. $t = 3$: $m \approx 9$
$t = 7$: $m \approx 0$
$t = 10$: $m \approx -10$
The slope is the rate of change of the average temperature at the given point in time.

17. $f'(x) = 0$
$f'(0) = 0$

19. $f'(x) = -3$
$f'(2) = -3$

21. $f'(x) = 4x$
$f'(2) = 8$

23. $f'(x) = 3x^2 - 1$
$f'(2) = 11$

25. $f'(x) = \dfrac{1}{\sqrt{x}}$
$f'(4) = \dfrac{1}{2}$

27. $f(x) = 3$
$f(x + \Delta x) = 3$
$f(x + \Delta x) - f(x) = 0$
$\dfrac{f(x + \Delta x) - f(x)}{\Delta x} = 0$
$\lim\limits_{\Delta x \to 0} \dfrac{f(x + \Delta x) - f(x)}{\Delta x} = 0$

29. $f(x) = -5x$
$f(x + \Delta x) = -5x - 5\Delta x$
$f(x + \Delta x) - f(x) = -5\Delta x$
$\dfrac{f(x + \Delta x) - f(x)}{\Delta x} = -5$
$\lim\limits_{\Delta x \to 0} \dfrac{f(x + \Delta x) - f(x)}{\Delta x} = -5$

31. $g(s) = \dfrac{1}{3}s + 2$
$g(s + \Delta s) = \dfrac{1}{3}s + \dfrac{1}{3}\Delta s + 2$
$g(s + \Delta s) - g(s) = \dfrac{1}{3}\Delta s$
$\dfrac{g(s + \Delta s) - g(s)}{\Delta s} = \dfrac{1}{3}$
$\lim\limits_{\Delta s \to 0} \dfrac{g(s + \Delta s) - g(s)}{\Delta s} = \dfrac{1}{3}$

33. $f(x) = 4x^2 - 5x$
$f(x + \Delta x) = 4x^2 + 8x\Delta x + 4(\Delta x)^2 - 5x - 5\Delta x$
$f(x + \Delta x) - f(x) = 8x\Delta x + 4(\Delta x)^2 - 5\Delta x$
$\dfrac{f(x + \Delta x) - f(x)}{\Delta x} = 8x + 4\Delta x - 5$
$\lim\limits_{\Delta x \to 0} \dfrac{f(x + \Delta x) - f(x)}{\Delta x} = 8x - 5$

35. $h(t) = \sqrt{t - 1}$
$h(t + \Delta t) = \sqrt{t + \Delta t - 1}$
$h(t + \Delta t) - h(t) = \sqrt{t + \Delta t - 1} - \sqrt{t - 1}$
$\dfrac{h(t + \Delta t) - h(t)}{\Delta t} = \dfrac{1}{\sqrt{t + \Delta t - 1} + \sqrt{t - 1}}$
$\lim\limits_{\Delta t \to 0} \dfrac{h(t + \Delta t) - h(t)}{\Delta t} = \dfrac{1}{2\sqrt{t - 1}}$

37. $f(t) = t^3 - 12t$
$f(t + \Delta t) = t^3 + 3t^2\Delta t + 3t(\Delta t)^2 + (\Delta t)^3 - 12t - 12\Delta t$
$f(t + \Delta t) - f(t) = 3t^2\Delta t + 3t(\Delta t)^2 + (\Delta t)^3 - 12\Delta t$
$\dfrac{f(t + \Delta t) - f(t)}{\Delta t} = 3t^2 + 3t\Delta t + (\Delta t)^2 - 12$
$\lim\limits_{\Delta t \to 0} \dfrac{f(t + \Delta t) - f(t)}{\Delta t} = 3t^2 - 12$

39. $f(x) = \dfrac{1}{x + 2}$
$f(x + \Delta x) = \dfrac{1}{x + \Delta x + 2}$
$f(x + \Delta x) - f(x) = \dfrac{-\Delta x}{(x + \Delta x + 2)(x + 2)}$
$\dfrac{f(x + \Delta x) - f(x)}{\Delta x} = \dfrac{-1}{(x + \Delta x + 2)(x + 2)}$
$\lim\limits_{\Delta x \to 0} \dfrac{f(x + \Delta x) - f(x)}{\Delta x} = -\dfrac{1}{(x + 2)^2}$

41. $y = 2x - 2$

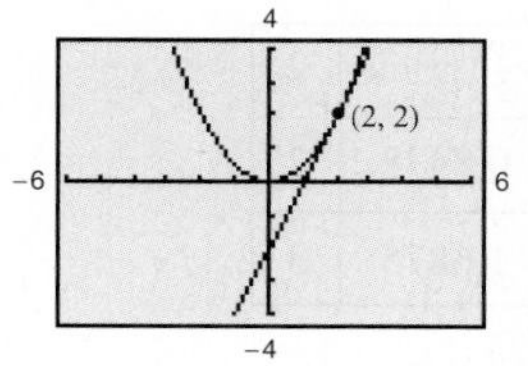

43. $y = -6x - 3$

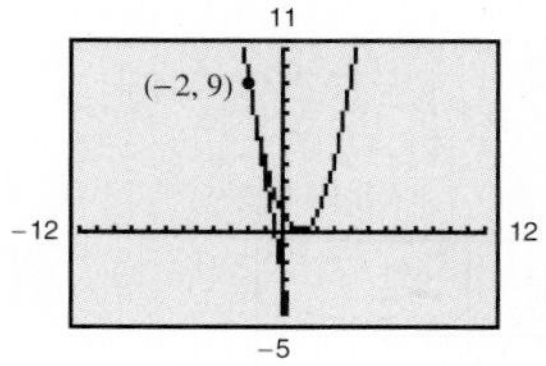

45. $y = \dfrac{x}{4} + 2$

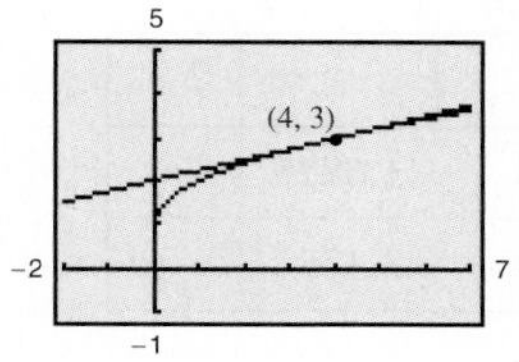

47. $y = -x + 2$

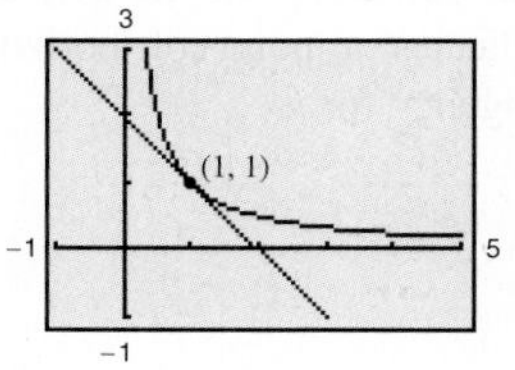

49. $y = -x + 1$

51. $y = -9x + 18$, $y = -9x - 18$

53. y is differentiable for all $x \neq -3$.
At $(-3, 0)$ the graph has a node.

55. y is differentiable for all $x \neq 3$.
At $(3, 0)$ the graph has a cusp.

57. y is differentiable for all $x = \pm 2$.
The function is not defined at $x = \pm 2$.

59. $f(x) = -3x + 2$

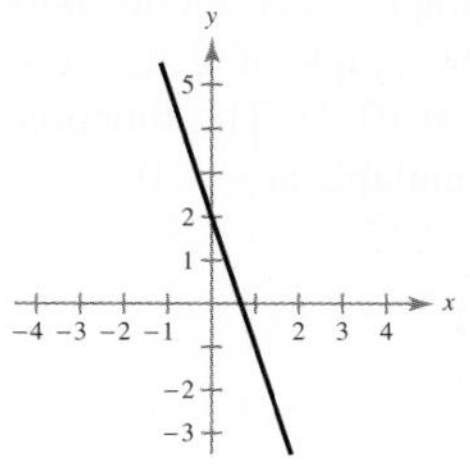

61. 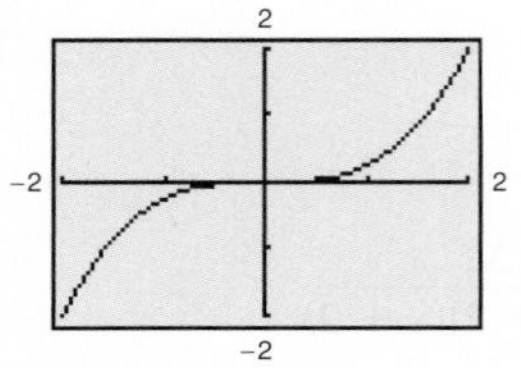

$f'(x) = \frac{3}{4}x^2$

x	-2	$-\frac{3}{2}$	-1	$-\frac{1}{2}$
$f(x)$	-2	-0.8438	-0.25	-0.0313
$f'(x)$	3	1.6875	0.75	0.1875

x	0	$\frac{1}{2}$	1	$\frac{3}{2}$	2
$f(x)$	0	0.0313	0.25	0.8438	2
$f'(x)$	0	0.1875	0.75	1.6875	3

63. 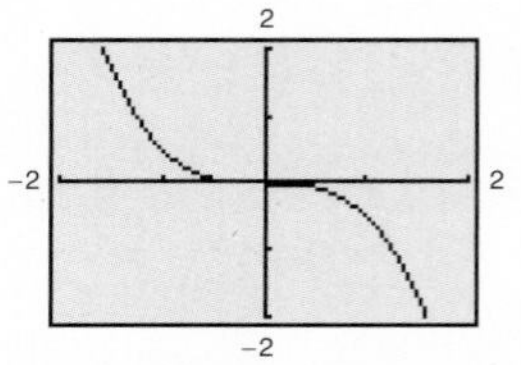

$f'(x) = -\frac{3}{2}x^2$

x	-2	$-\frac{3}{2}$	-1	$-\frac{1}{2}$
$f(x)$	4	1.6875	0.5	0.0625
$f'(x)$	-6	-3.375	-1.5	-0.375

x	0	$\frac{1}{2}$	1	$\frac{3}{2}$	2
$f(x)$	0	-0.0625	-0.5	-1.6875	-4
$f'(x)$	0	-0.375	-1.5	-3.375	-6

65. $f'(x) = 2x - 4$

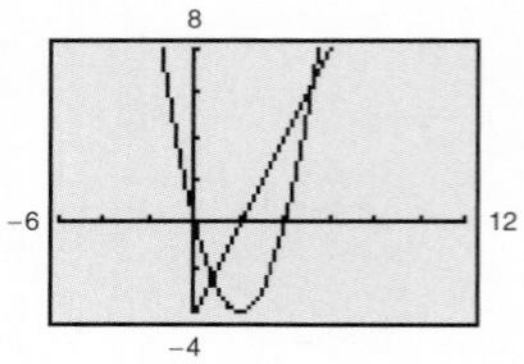

The x-intercept of the derivative indicates a point of horizontal tangency for f.

67. $f'(x) = 3x^2 - 3$

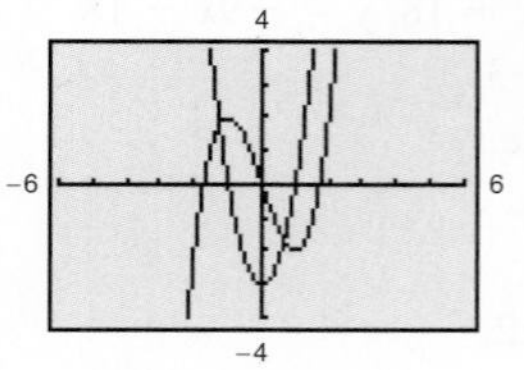

The x-intercepts of the derivative indicate points of horizontal tangency for f.

69. True **71.** True

73. 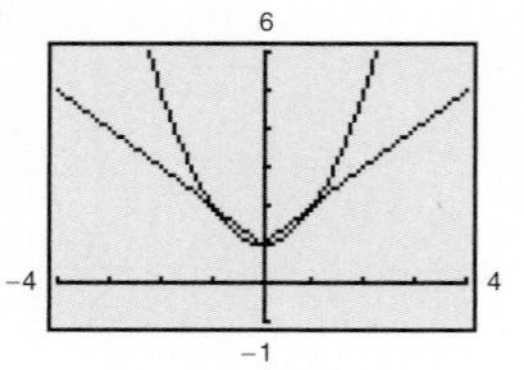

The graph of f is smooth at (0, 1), but the graph of g has a sharp point at (0, 1). The function f is differentiable at $x = 0$.

Section 2.2 *(page 100)*

Skills Warm Up *(page 100)*

1. (a) 8 (b) 16 (c) $\frac{1}{2}$

2. (a) $\frac{1}{36}$ (b) $\frac{1}{32}$ (c) $\frac{1}{64}$

3. $4x(3x^2 + 1)$ **4.** $\frac{3}{2}x^{1/2}(x^{3/2} - 1)$ **5.** $\frac{1}{4x^{3/4}}$

6. $x^2 - \frac{1}{x^{1/2}} + \frac{1}{3x^{2/3}}$ **7.** $0, -\frac{2}{3}$

8. $0, \pm 1$ **9.** $-10, 2$ **10.** $-2, 12$

1. 0 **3.** $5x^4$ **5.** $9x^2$ **7.** $2x^2$ **9.** 4 **11.** $-3x^2$

13. $8x - 3$ **15.** $-6t + 2$ **17.** $3t^2 - 2$ **19.** $\frac{2}{3\sqrt[3]{x}}$

21. $\frac{16}{3}t^{1/3}$ **23.** $-\frac{8}{x^3} + 4x$

25. Function: $y = \frac{2}{7x^4}$

Rewrite: $y = \frac{2}{7}x^{-4}$

Differentiate: $y' = -\frac{8}{7}x^{-5}$

Simplify: $y' = -\frac{8}{7x^5}$

27. Function: $y = \frac{1}{(4x)^3}$

Rewrite: $y = \frac{1}{64}x^{-3}$

Differentiate: $y' = -\frac{3}{64}x^{-4}$

Simplify: $y' = -\frac{3}{64x^4}$

29. Function: $y = \frac{4}{(2x)^{-5}}$

Rewrite: $y = 128x^5$

Differentiate: $y' = 128(5)x^4$

Simplify: $y' = 640x^4$

31. Function: $y = 6\sqrt{x}$

Rewrite: $y = 6x^{1/2}$

Differentiate: $y' = 6\left(\frac{1}{2}\right)x^{-1/2}$

Simplify: $y' = \frac{3}{\sqrt{x}}$

33. Function: $y = \frac{1}{5\sqrt[5]{x}}$

Rewrite: $y = \frac{1}{5}x^{-1/5}$

Differentiate: $y' = \frac{1}{5}\left(-\frac{1}{5}\right)x^{-6/5}$

Simplify: $y' = -\frac{1}{25\sqrt[5]{x^6}}$

35. Function: $y = \sqrt{3x}$

Rewrite: $y = (3x)^{1/2}$

Differentiate: $y' = \left(\frac{1}{2}\right)(3x)^{-1/2}(3)$

Simplify: $y' = \frac{3}{2\sqrt{3x}}$

37. $\frac{3}{2}$ **39.** 8 **41.** -11 **43.** -2

45. (a) $y = 2x - 2$
(b) and (c)

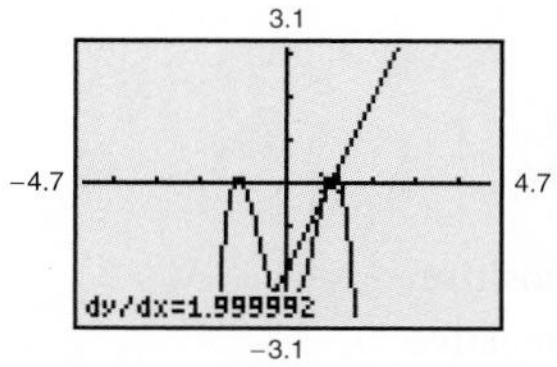

47. (a) $y = \frac{8}{15}x + \frac{22}{15}$
(b) and (c)

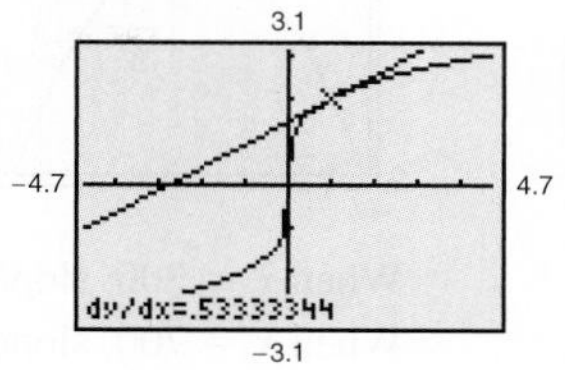

49. (a) $y = 36x - 54$
(b) and (c)

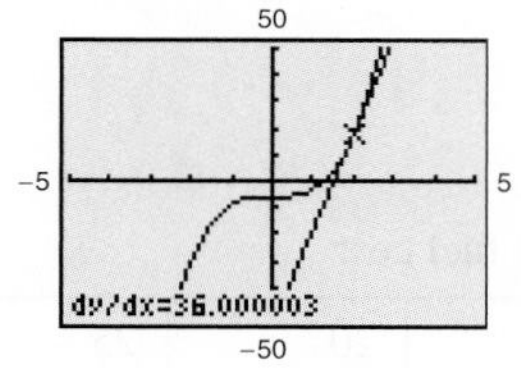

51. $2x + \dfrac{4}{x^2} + \dfrac{6}{x^3}$ **53.** $2x - 2 + \dfrac{8}{x^5}$ **55.** $\dfrac{4}{5x^{1/5}} + 1$

57. $3x^2 + 1$ **59.** $\dfrac{2x^3 - 6}{x^3}$ **61.** $\dfrac{4x^3 - 2x - 10}{x^3}$

63. $(0, -1), \left(-\dfrac{\sqrt{6}}{2}, \dfrac{5}{4}\right), \left(\dfrac{\sqrt{6}}{2}, \dfrac{5}{4}\right)$ **65.** $(-5, -12.5)$

67. (a)

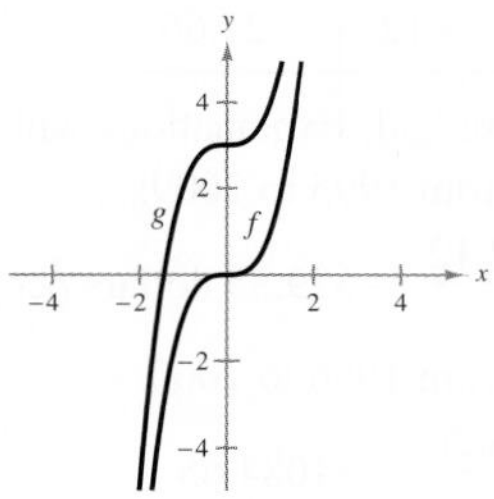

(b) $f'(1) = g'(1) = 3$

(c)

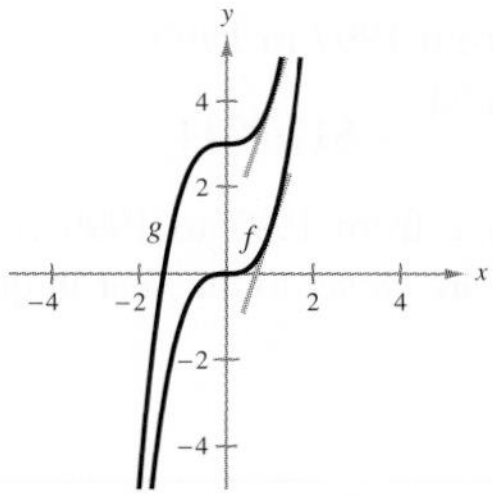

(d) $f' = g' = 3x^2$ for every value of x.

69. $f'(x) = g'(x)$ **71.** $-5f'(x) = g'(x)$

73. (a) 2005: $m \approx 119.2$; 2007: $m \approx 161$
(b) These results are close to the estimates in Exercise 13 in Section 2.1.
(c) The slope of the graph at time t is the rate at which sales are increasing in millions of dollars per year.

75. (a) The men and women who seem to suffer most from migraines are those between 30 and 40 years old. More females than males suffer from migraines. Fewer people whose incomes are greater than or equal to \$30,000 suffer from migraines than people whose incomes are less than \$10,000.
(b) The derivatives are positive up to approximately 37 years old and negative after about 37 years of age. The percent of adults suffering from migraines increases up to about 37 years old, then decreases. The units of the derivative are percents of adults suffering from migraines per year.

77. $C = 7.75x + 500$
$C' = 7.75$, which equals the marginal cost.

79.

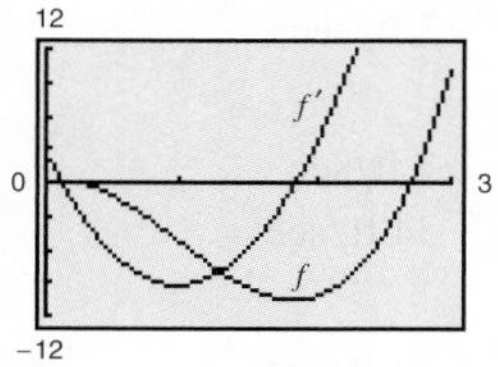

(0.11, 0.14), (1.84, −10.49)

81. False. Let $f(x) = x$ and $g(x) = x + 1$.

Section 2.3 *(page 113)*

Skills Warm Up *(page 113)*

1. 3 **2.** -7 **3.** -3 **4.** 2.4
5. $y' = 8x - 2$ **6.** $y' = -9t^2 + 4t$
7. $s' = -32t + 24$ **8.** $y' = -32x + 54$
9. $A' = -\frac{3}{5}r^2 + \frac{3}{5}r + \frac{1}{2}$ **10.** $y' = 2x^2 - 4x + 7$
11. $y' = 12 - \dfrac{x}{2500}$ **12.** $y' = 74 - \dfrac{3x^2}{10{,}000}$

1. (a) \$10.4 billion/yr (b) \$7.4 billion/yr
(c) \$6.4 billion/yr (d) \$16.6 billion/yr
(e) \$11 billion/yr (f) \$11.96 billion/yr
(g) \$13.67 billion/yr (h) \$16.38 billion/yr

3.

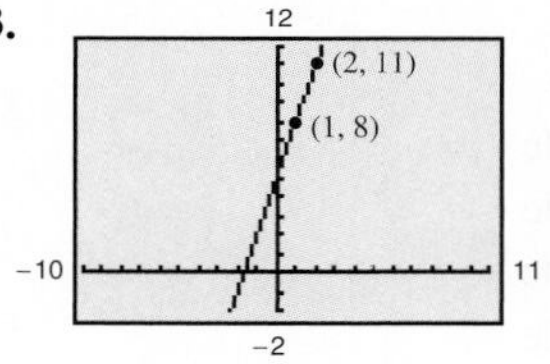

Average rate: 3
Instantaneous rates:
$f'(1) = f'(2) = 3$

5.

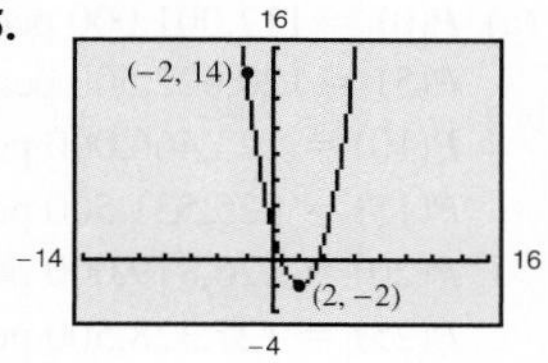

Average rate: -4
Instantaneous rates:
$h'(-2) = -8, h'(2) = 0$

7.

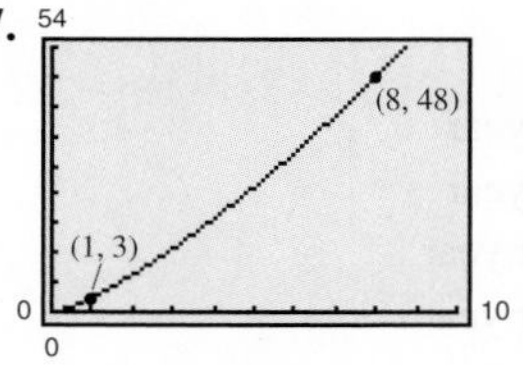

Average rate: $\frac{45}{7}$
Instantaneous rates:
$f'(1) = 4, f'(8) = 8$

9.

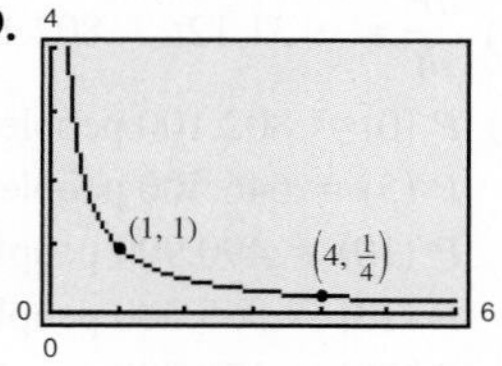

Average rate: $-\frac{1}{4}$
Instantaneous rates:
$f'(1) = -1, f'(4) = -\frac{1}{16}$

11.

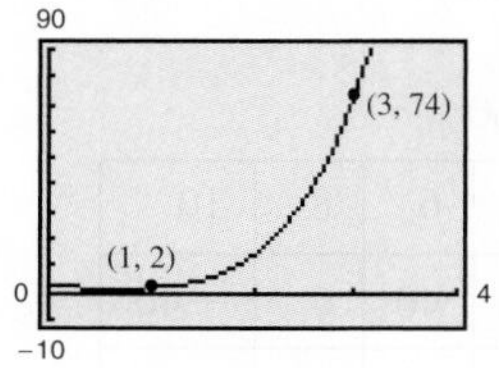

Average rate: 36
Instantaneous rates:
$g'(1) = 2, g'(3) = 102$

13. (a) -450
The number of visitors to the park is decreasing at an average rate of 450 thousand people per month from September to December.
(b) Answers will vary. The instantaneous rate of change at $t = 8$ is approximately 0.

15. (a) Average rate: 14 ft/sec
Instantaneous rates: $s'(0) = 30$ ft/sec
$s'(1) = -2$ ft/sec
(b) Average rate: -18 ft/sec
Instantaneous rates: $s'(1) = -2$ ft/sec
$s'(2) = -34$ ft/sec
(c) Average rate: -50 ft/sec
Instantaneous rates: $s'(2) = -34$ ft/sec
$s'(3) = -66$ ft/sec
(d) Average rate: -82 ft/sec
Instantaneous rates: $s'(3) = -66$ ft/sec
$s'(4) = -98$ ft/sec

17. (a) -80 ft/sec
(b) $s'(2) = -64$ ft/sec,
$s'(3) = -96$ ft/sec
(c) $\frac{\sqrt{555}}{4} \approx 5.89$ sec
(d) $-8\sqrt{555} \approx -188.5$ ft/sec

19. 9800 dollars **21.** $470 - 0.5x$ dollars, $0 \le x \le 940$

23. $50 - x$ dollars **25.** $-18x^2 + 16x + 200$ dollars

27. $-4x + 72$ dollars **29.** $0.0039x^2 + 12$ dollars

31. (a) \$0.58 (b) \$0.60
(c) The results are nearly the same.

33. (a) \$12.96 (b) \$13.00
(c) The results are nearly the same.

35. (a) $P(0) = 117{,}001{,}000$ people
$P(5) = 120{,}622{,}500$ people
$P(10) = 123{,}466{,}000$ people
$P(15) = 125{,}531{,}500$ people
$P(20) = 126{,}819{,}000$ people
$P(25) = 127{,}328{,}500$ people
$P(30) = 127{,}060{,}000$ people
The population is growing from 1980 to 2005. It then begins to decline.
(b) $\frac{dP}{dt} = -31.12t + 802.1$
(c) $P'(0) = 802{,}100$ people per year
$P'(5) = 646{,}500$ people per year
$P'(10) = 490{,}900$ people per year
$P'(15) = 335{,}300$ people per year
$P'(20) = 179{,}700$ people per year
$P'(25) = 24{,}100$ people per year
$P'(30) = -131{,}500$ people per year
The rate of growth is decreasing.

37. (a) $TR = -10Q^2 + 160Q$
(b) $(TR)' = MR = -20Q + 160$
(c)

Q	0	2	4	6	8	10
Model	160	120	80	40	0	-40
Table	–	130	90	50	10	-30

Answers will vary.

39. (a) $P = -0.0025x^2 + 2.65x - 25$
(b)
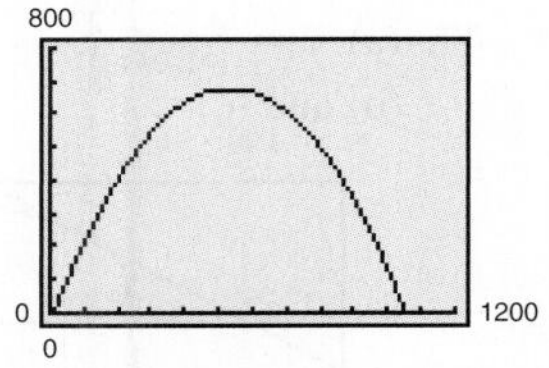

When $x = 300$, slope is positive.
When $x = 700$, slope is negative.
(c) $P'(300) = 1.15$
$P'(700) = -0.85$

41. (a) $C = \frac{44{,}250}{x}$
(b) $\frac{dC}{dx} = \frac{-44{,}250}{x^2}$
This is the rate of change of fuel cost.
(c)

x	10	15	20	25
C	4425.00	2950.00	2212.50	1770.00
dC/dx	-442.5	-196.67	-110.63	-70.80

x	30	35	40
C	1475.00	1264.29	1106.25
dC/dx	-49.17	-36.12	-27.66

(d) The driver who gets 15 mi/gal; Explanations will vary.

43. (a) Average rate of change from 1995 to 2009:
$\frac{\Delta p}{\Delta t} = \frac{10{,}428.05 - 5117.12}{19 - 5} \approx 379.35$ dollars per year
(b) Average rate of change from 1996 to 2000:
$\frac{\Delta p}{\Delta t} = \frac{10{,}786.85 - 6448.26}{10 - 6} \approx \$1084.65$
(c) Average rate of change from 1997 to 1999:
$\frac{\Delta p}{\Delta t} = \frac{11{,}497.12 - 7908.24}{9 - 7} \approx \$1794.44$
(d) The average rate of change from 1997 to 1999 is a better estimate because the data are closer to the year in question.

Section 2.4 *(page 124)*

Skills Warm Up *(page 124)*

1. $2(3x^2 + 7x + 1)$ **2.** $4x^2(6 - 5x^2)$
3. $8x^2(x^2 + 2)^3 + (x^2 + 4)$
4. $(2x)(2x + 1)[2x + (2x + 1)^3]$
5. $\frac{23}{(2x + 7)^2}$ **6.** $-\frac{x^2 + 8x + 4}{(x^2 - 4)^2}$
7. $-\frac{2(x^2 + x - 1)}{(x^2 + 1)^2}$ **8.** $\frac{4(3x^4 - x^3 + 1)}{(1 - x^4)^2}$
9. $\frac{4x^3 - 3x^2 + 3}{x^2}$ **10.** $\frac{x^2 - 2x + 4}{(x - 1)^2}$
11. 11 **12.** 0 **13.** $-\frac{1}{4}$ **14.** $\frac{17}{4}$

1. $f'(x) = (2x - 3)(-5) + (2)(1 - 5x) = -20x + 17$
3. $f'(x) = (6x - x^2)(3) + (6 - 2x)(4 + 3x)$
$= -9x^2 + 28x + 24$

5. $f'(x) = x(2x) + 1(x^2 + 3) = 3x^2 + 3$

7. $h'(x) = \left(\frac{2}{x} - 3\right)(2x) + \left(-\frac{2}{x^2}\right)(x^2 + 7) = -\frac{14}{x^2} - 6x + 2$

9. $g'(x) = (x^2 - 4x + 3)(1) + (2x - 4)(x - 2)$
$= 3x^2 - 12x + 11$

11. $h'(x) = \frac{(x - 5)(1) - (x)(1)}{(x - 5)^2} = -\frac{5}{(x - 5)^2}$

13. $f'(t) = \frac{(3t + 1)(4t) - (2t^2 - 3)3}{(3t + 1)^2} = \frac{6t^2 + 4t + 9}{(3t + 1)^2}$

15. $f'(t) = \frac{(t + 4)(2t) - (t^2 - 1)(1)}{(t + 4)^2} = \frac{t^2 + 8t + 1}{(t + 4)^2}$

17. $f'(x) = \frac{(2x - 1)(2x + 6) - (x^2 + 6x + 5)(2)}{(2x - 1)^2}$
$= \frac{2x^2 - 2x - 16}{(2x - 1)^2}$

19. $f'(x) = \frac{(3x - 1)\left(-\frac{2}{x^2}\right) - \left(6 + \frac{2}{x}\right)(3)}{(3x - 1)^2}$
$= \frac{-18x^2 - 12x + 2}{x^2(3x - 1)^2}$

21. Function: $f(x) = \frac{x^3 + 6x}{3}$

Rewrite: $f(x) = \frac{x^3}{3} + 2x$

Differentiate: $f'(x) = x^2 + 2$

Simplify: $f'(x) = x^2 + 2$

23. Function: $y = \frac{x^2 + 2x}{3}$

Rewrite: $y = \frac{1}{3}(x^2 + 2x)$

Differentiate: $y' = \frac{1}{3}(2x + 2)$

Simplify: $y' = \frac{2}{3}(x + 1)$

25. Function: $y = \frac{7}{3x^3}$

Rewrite: $y = \frac{7}{3}x^{-3}$

Differentiate: $y' = -7x^{-4}$

Simplify: $y' = -\frac{7}{x^4}$

27. Function: $y = \frac{4x^2 - 3x}{8\sqrt{x}}$

Rewrite: $y = \frac{1}{2}x^{3/2} - \frac{3}{8}x^{1/2},\ x \neq 0$

Differentiate: $y' = \frac{3}{4}x^{1/2} - \frac{3}{16}x^{-1/2}$

Simplify: $y' = \frac{3}{4}\sqrt{x} - \frac{3}{16\sqrt{x}}$

29. Function: $y = \frac{x^2 - 4x + 3}{2(x - 1)}$

Rewrite: $y = \frac{1}{2}(x - 3),\ x \neq 1$

Differentiate: $y' = \frac{1}{2}(1),\ x \neq 1$

Simplify: $y' = \frac{1}{2},\ x \neq 1$

31. $10x^4 + 12x^3 - 3x^2 - 18x - 15$; Product Rule

33. $\frac{x^4 - 6x^2 - 4x - 3}{(x^2 - 1)^2}$; Quotient Rule **35.** 1; Power Rule

37. $12t^2(2t^3 - 1)$; Product Rule

39. $\frac{3s^2 - 2s - 5}{2s^{3/2}}$; Quotient Rule

41. $\frac{12x^2 + 12x - 2}{(4x + 2)^2}$; Quotient Rule

43. $\frac{2x^3 + 11x^2 - 8x - 17}{(x + 4)^2}$; Quotient Rule

45. $y = 3x + 3$

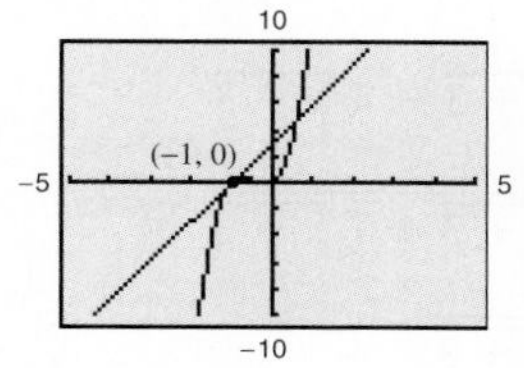

47. $y = -7x + 4$

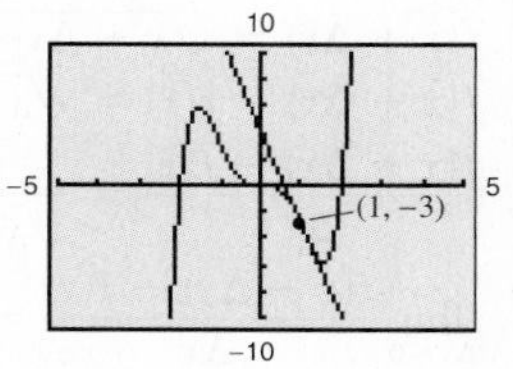

49. $y = \frac{3}{4}x - \frac{5}{4}$

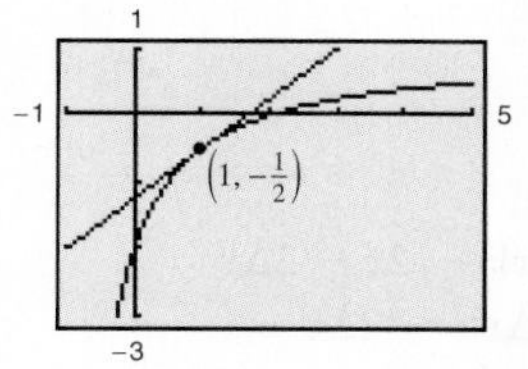

51. $y = \frac{31}{5}x + \frac{26}{5}$

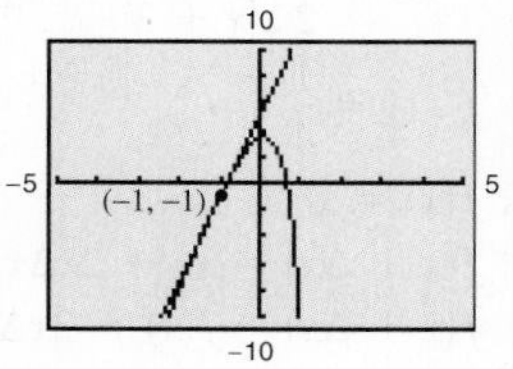

53. (0, 0), (2, 4) **55.** $(0, 0), \left(\sqrt[3]{-4}, -2.117\right)$

57.

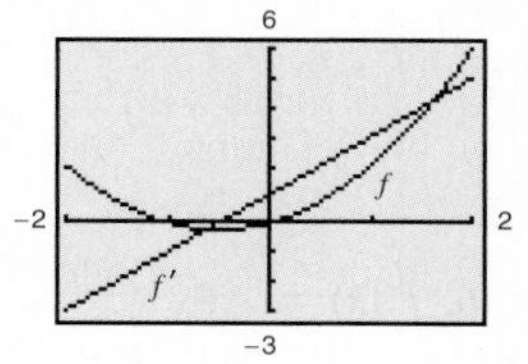

59.

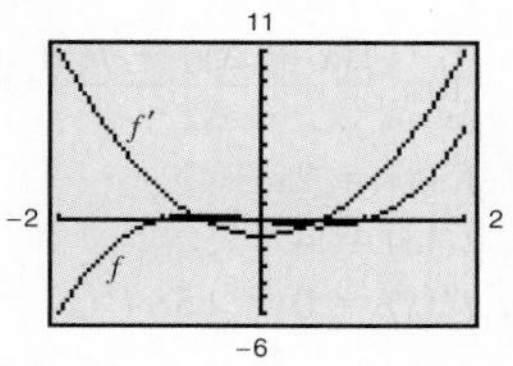

61. −\$1.87/unit **63.** 31.55 bacteria/hr

65. (a) −0.480/wk (b) 0.120/wk (c) 0.015/wk

Each rate in parts (a), (b), and (c) is the rate at which the level of oxygen in the pond is changing at that particular time.

67. (a)

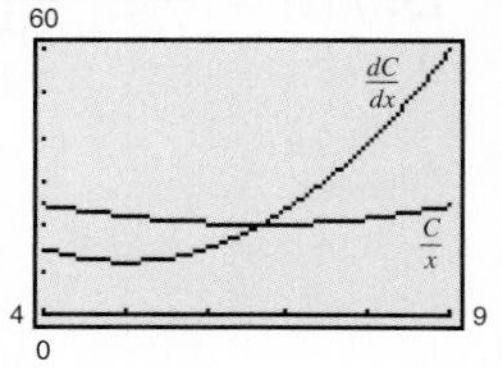

(b) At $x = 6.683$, $\frac{C}{x} = \frac{dC}{dx} \approx 20.50$.

So, the point of intersection is (6.683, 20.50). At this point the average cost is at a minimum.

69. (a) −38.125 (b) −10.37 (c) −3.80

Increasing the order size reduces the cost per item; Choices and explanations will vary.

71. $f'(2) = 0$ **73.** $f'(2) = 14$ **75.** Answers will vary.

Quiz Yourself *(page 127)*

1. $f(x) = 5x + 3$

$f(x + \Delta x) = 5x + 5\Delta x + 3$

$f(x + \Delta x) - f(x) = 5\Delta x$

$\dfrac{f(x + \Delta x) - f(x)}{\Delta x} = 5$

$\lim\limits_{\Delta x \to 0} \dfrac{f(x + \Delta x) - f(x)}{\Delta x} = 5$

$f'(x) = 5$

$f'(-2) = 5$

2. $f(x) = \sqrt{x + 3}$

$f(x + \Delta x) = \sqrt{x + \Delta x + 3}$

$f(x + \Delta x) - f(x) = \sqrt{x + \Delta x + 3} - \sqrt{x + 3}$

$\dfrac{f(x + \Delta x) - f(x)}{\Delta x} = \dfrac{1}{\sqrt{x + \Delta x + 3} + \sqrt{x + 3}}$

$\lim\limits_{\Delta x \to 0} \dfrac{f(x + \Delta x) - f(x)}{\Delta x} = \dfrac{1}{2\sqrt{x + 3}}$

$f'(x) = \dfrac{1}{2\sqrt{x + 3}}$

$f'(1) = \dfrac{1}{4}$

3. $f(x) = x^2 - 2x$

$f(x + \Delta x) = x^2 + 2x\Delta x + (\Delta x)^2 - 2x - 2\Delta x$

$f(x + \Delta x) - f(x) = 2x\Delta x + (\Delta x)^2 - 2\Delta x$

$\dfrac{f(x + \Delta x) - f(x)}{\Delta x} = 2x + \Delta x - 2$

$\lim\limits_{\Delta x \to 0} \dfrac{f(x + \Delta x) - f(x)}{\Delta x} = 2x - 2$

$f'(x) = 2x - 2$

$f'(3) = 4$

4. $f'(x) = 0$ **5.** $f'(x) = 19$ **6.** $f'(x) = -6x$

7. $f'(x) = \dfrac{3}{x^{3/4}}$ **8.** $f'(x) = -\dfrac{8}{x^3}$ **9.** $f'(x) = \dfrac{1}{\sqrt{x}}$

10. $f'(x) = -\dfrac{5}{(3x + 2)^2}$ **11.** $f'(x) = -6x^2 + 8x - 2$

12. $f'(x) = 15x^2 + 26x + 14$ **13.** $f'(x) = \dfrac{-4(x^2 - 3)}{(x^2 + 3)^2}$

14.

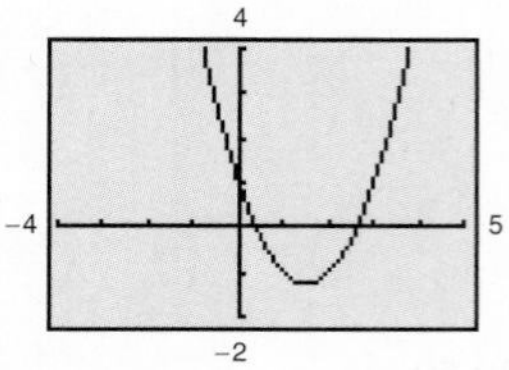

Average rate: 0

Instantaneous rates: $f'(0) = -3, f'(3) = 3$

15.

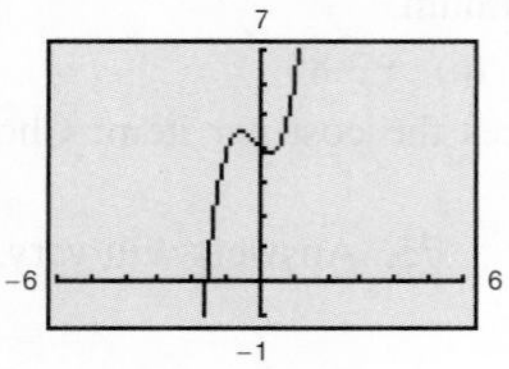

Average rate: 1

Instantaneous rates: $f'(-1) = 3, f'(1) = 7$

16.

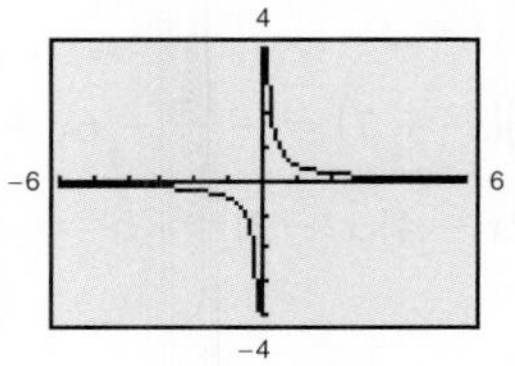

Average rate: $-\frac{1}{20}$

Instantaneous rates: $f'(2) = -\frac{1}{8}, f'(5) = -\frac{1}{50}$

17.

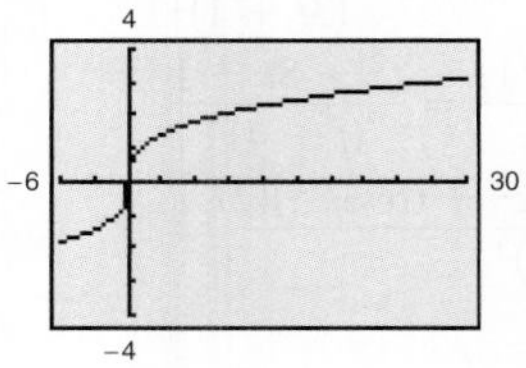

Average rate: $\frac{1}{19}$

Instantaneous rates: $f'(8) = \frac{1}{12}, f'(27) = \frac{1}{27}$

18. (a) \$11.61 (b) \$11.63

(c) The results are approximately equal.

19. $y = -4x - 6$ **20.** $y = 10x - 8$

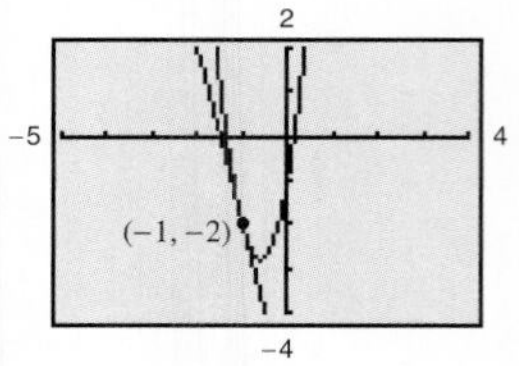

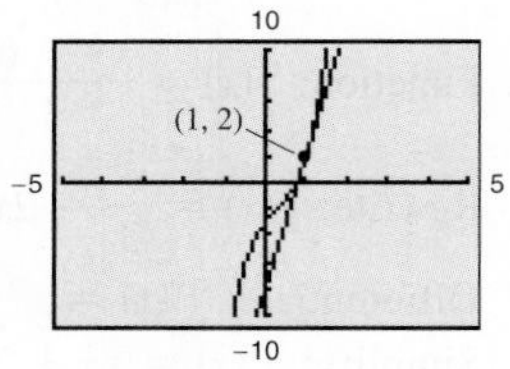

21. (a) $\dfrac{dS}{dt} = -0.40668t^2 + 3.7364t - 4.351$

(b) 2004: \$4.08772/yr

2007: \$1.87648/yr

2008: −\$0.48732/yr

Section 2.5 *(page 135)*

Skills Warm Up *(page 135)*

1. $(1 - 5x)^{2/5}$ **2.** $(2x - 1)^{3/4}$

3. $(4x^2 + 1)^{-1/2}$ **4.** $(x - 6)^{-1/3}$

5. $x^{1/2}(1 - 2x)^{-1/3}$ **6.** $(2x)^{-1}(3 - 7x)^{3/2}$

7. $(x - 2)(3x^2 + 5)$ **8.** $(x - 1)(5\sqrt{x} - 1)$

9. $(x^2 + 1)^2(4 - x - x^3)$

10. $(3 - x^2)(x - 1)(x^2 + x + 1)$

$y = f(g(x))$	$u = g(x)$	$y = f(u)$
1. $y = (6x - 5)^4$	$u = 6x - 5$	$y = u^4$
3. $y = \sqrt{5x - 2}$	$u = 5x - 2$	$y = \sqrt{u}$
5. $y = (3x + 1)^{-1}$	$u = 3x + 1$	$y = u^{-1}$

7. $\dfrac{dy}{du} = 2u$

$\dfrac{du}{dx} = 4$

$\dfrac{dy}{dx} = 32x + 56$

9. $\dfrac{dy}{du} = \dfrac{1}{2\sqrt{u}}$

$\dfrac{du}{dx} = -2x$

$\dfrac{dy}{dx} = -\dfrac{x}{\sqrt{3 - x^2}}$

11. $\dfrac{dy}{du} = \dfrac{2}{3u^{1/3}}$

$\dfrac{du}{dx} = 20x^3 - 2$

$\dfrac{dy}{dx} = \dfrac{40x^3 - 4}{3\sqrt[3]{5x^4 - 2x}}$

13. c **15.** b **17.** a **19.** c **21.** $6(2x - 7)^2$

23. $\dfrac{3\sqrt{5x - x^2}\,(5 - 2x)}{2}$ **25.** $6x(6 - x^2)(2 - x^2)$

27. $\dfrac{1}{2\sqrt{t + 1}}$ **29.** $\dfrac{4t + 5}{2\sqrt{2t^2 + 5t + 2}}$ **31.** $\dfrac{6x}{(9x^2 + 4)^{2/3}}$

33. $\dfrac{54}{(2 - 9x)^4}$ **35.** $\dfrac{-x}{(25 + x^2)^{3/2}}$

37. $y = 216x - 378$ **39.** $y = \frac{8}{3}x - \frac{7}{3}$

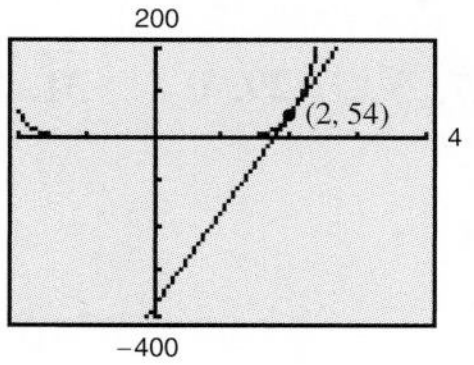

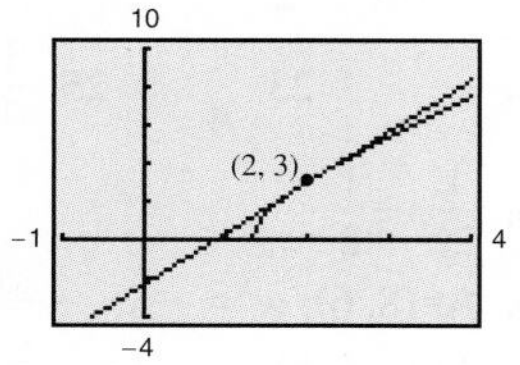

41. $y = x - 1$

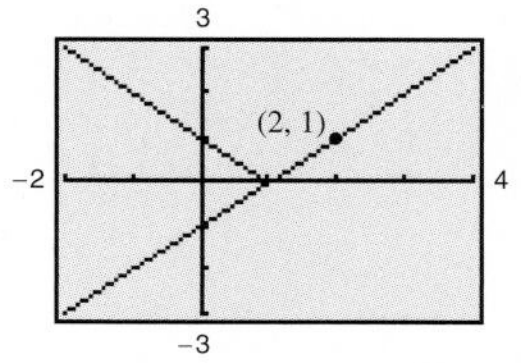

43. $f'(x) = \dfrac{1 - 3x^2 - 4x^{3/2}}{2\sqrt{x}(x^2 + 1)^2}$

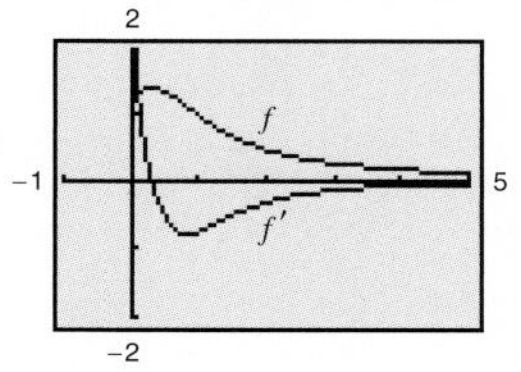

The zero of $f'(x)$ corresponds to the point on the graph of $f(x)$ where the tangent line is horizontal.

45. $f'(x) = -\dfrac{\sqrt{(x + 1)/x}}{2x(x + 1)}$

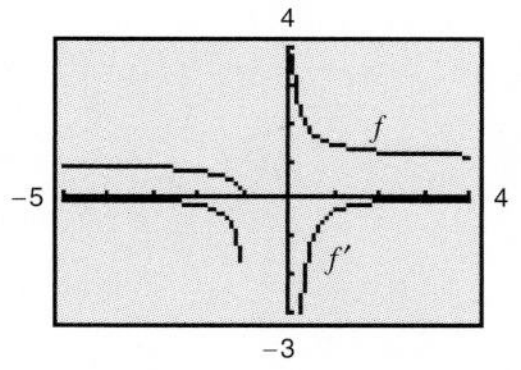

$f'(x)$ has no zeros.

In Exercises 47–61, the differentiation rule(s) used may vary. A sample answer is provided.

47. $\dfrac{2x}{(4 - x^2)^2}$; Chain Rule **49.** $\dfrac{8}{(t + 2)^3}$; Chain Rule

51. $-6(3x^2 - x - 3)$; Product Rule

53. $-\dfrac{1}{2(x + 2)^{3/2}}$; Power Rule

55. $27(x - 3)^2(4x - 3)$; Product Rule and Chain Rule

57. $\dfrac{3(x + 1)}{\sqrt{2x + 3}}$; Product Rule and Chain Rule

59. $\dfrac{t(5t - 8)}{2\sqrt{t - 2}}$; Product Rule and Chain Rule

61. $\dfrac{2(6 - 5x)(5x^2 - 12x + 5)}{(x^2 - 1)^3}$; Chain Rule and Quotient Rule

63. $y = \frac{8}{3}t + 4$ **65.** $y = \frac{1}{2}x + 1$

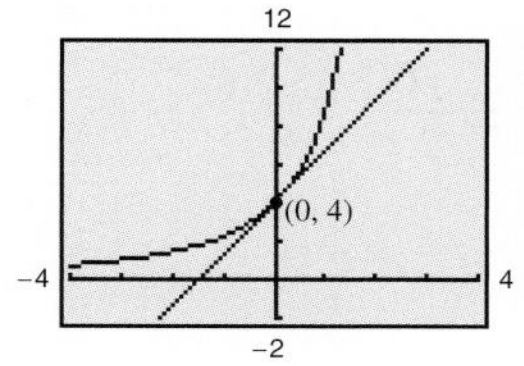

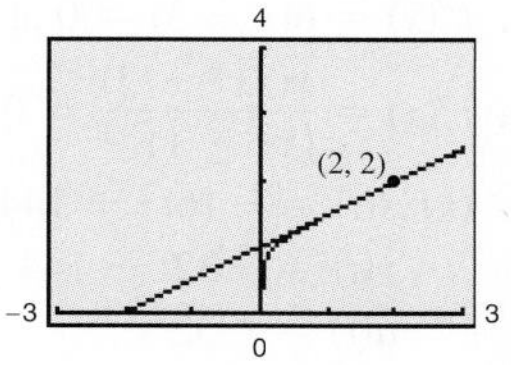

67. $y = -6t - 14$ **69.** $y = -2x + 7$

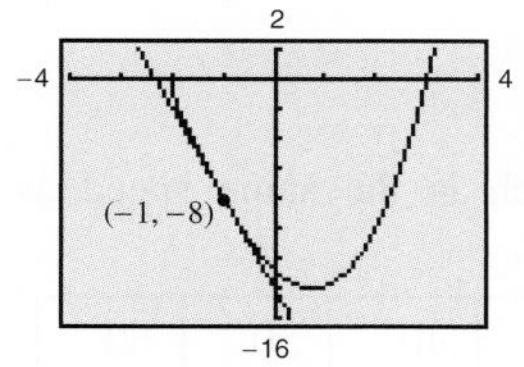

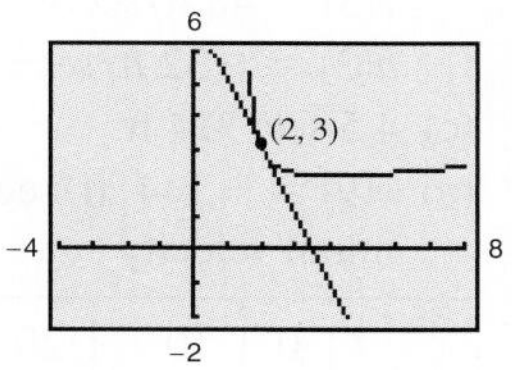

71. (a) \$74.00 per 1%
(b) \$81.59 per 1%
(c) \$89.94 per 1%

73. (a) $V = \dfrac{10{,}000}{\sqrt{t + 1}}$
(b) About −\$1767.77 per year
(c) −\$625.00 per year

75. (a) $r'(t) = \dfrac{11.4228t^3 - 218.376t^2 + 1352.28t - 2706}{2\sqrt{2.8557t^4 - 72.792t^3 + 676.14t^2 - 2706t + 4096}}$; Chain Rule

(b)

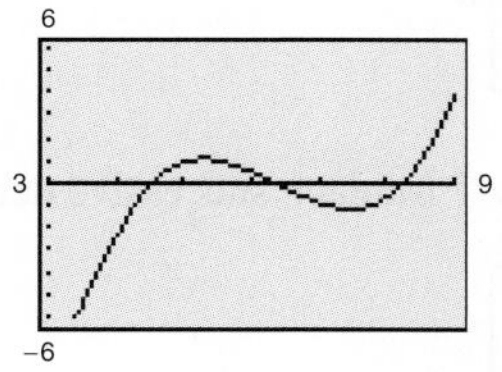

(c) $t = 3$ (d) $t \approx 4.52,\ t \approx 6.36,\ t \approx 8.24$

Section 2.6 *(page 142)*

Skills Warm Up *(page 142)*

1. $t = 0, \frac{3}{2}$ **2.** $t = -2, 7$ **3.** $t = -2, 10$

4. $t = \dfrac{9 \pm 3\sqrt{10{,}249}}{32}$ **5.** $\dfrac{dy}{dx} = 6x^2 + 14x$

6. $\dfrac{dy}{dx} = 8x^3 + 18x^2 - 10x - 15$

7. $\dfrac{dy}{dx} = \dfrac{2x(x + 7)}{(2x + 7)^2}$ **8.** $\dfrac{dy}{dx} = -\dfrac{6x^2 + 10x + 15}{(2x^2 - 5)^2}$

9. Domain: $(-\infty, \infty)$
Range: $[-4, \infty)$

10. Domain: $[7, \infty)$
Range: $[0, \infty)$

1. 0 **3.** 2 **5.** $2t - 8$ **7.** $\frac{9}{2t^4}$

9. $18(2 - x^2)(5x^2 - 2)$ **11.** $\frac{4}{(x-1)^3}$ **13.** $60x^2 - 72x$

15. $120x + 360$ **17.** $-\frac{9}{2x^5}$ **19.** 260 **21.** $-\frac{1}{648}$

23. 12 **25.** $4x$ **27.** $\frac{1}{\sqrt{x-1}}$ **29.** $12x^2 + 4$

31. $f''(x) = 6(x - 3) = 0$ at $x = 3$.

33. $f''(x) = \frac{x(2x^2 - 3)}{(x^2 - 1)^{3/2}} = 0$ at $x = \pm\frac{\sqrt{6}}{2}$.

35. (a) $s(t) = -16t^2 + 144t$
$v(t) = -32t + 144$
$a(t) = -32$
(b) $s(3) = 288$ ft
$v(3) = 48$ ft/sec
$a(3) = -32$ ft/sec^2
(c) 4.5 sec; 324 ft
(d) $v(9) = -144$ ft/sec, which is the same speed as the initial velocity

37.

t	0	10	20	30	40	50	60
$\frac{ds}{dt}$	0	45	60	67.5	72	75	77.1
$\frac{d^2s}{dt^2}$	9	2.25	1	0.56	0.36	0.25	0.18

As time increases, velocity increases and acceleration decreases.

39. (a)

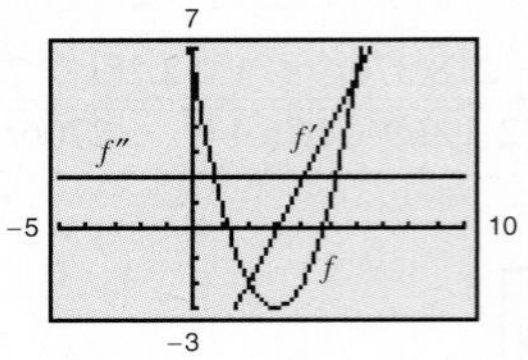

(b) The degree decreases by 1 for each successive derivative.

(c)

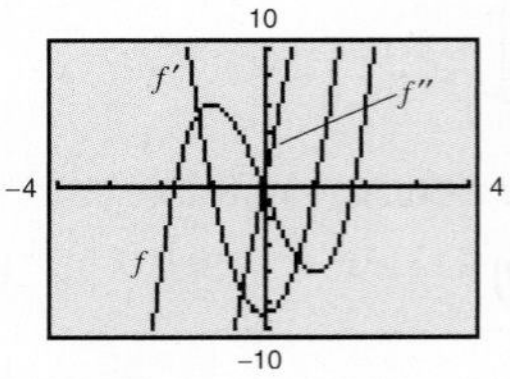

(d) The degree decreases by 1 for each successive derivative.

41. (a) $y = -68.991t^3 + 1208.34t^2 - 5445.4t + 10{,}145$
(b) $y'(t) = -206.973t^2 + 2416.68t - 5445.4$
$y''(t) = -413.946t + 2416.68$
(c) $y'(t) > 0$ on $[5, 8]$
(d) 2005 ($t \approx 5.84$)

43. False. The Product Rule is
$[f(x)g(x)]' = f(x)g'(x) + g(x)f'(x)$.

45. Answers will vary.

Section 2.7 *(page 149)*

Skills Warm Up *(page 149)*

1. $y = x^2 - 2x$ **2.** $y = \frac{x-3}{4}$
3. $y = 1, x \neq -6$ **4.** $y = -4, x \neq \pm\sqrt{3}$
5. $y = \pm\sqrt{5 - x^2}$ **6.** $y = \pm\sqrt{6 - x^2}$ **7.** $\frac{8}{3}$
8. $-\frac{1}{2}$ **9.** $\frac{5}{7}$

1. $-\frac{y}{x}$ **3.** $-\frac{x}{y}$ **5.** $\frac{1 - xy^2}{x^2y}$ **7.** $\frac{y}{8y - x}$ **9.** 0

11. $-\frac{1}{10y - 2}$ **13.** 0 **15.** $-\frac{1}{4}$ **17.** $\frac{1}{2}$ **19.** -1

21. $-\frac{5}{4}$ **23.** $\frac{1}{4}$ **25.** $\frac{1}{3}$ **27.** 3 **29.** 0 **31.** 2

33. $\frac{1}{2y}, -\frac{1}{2}$

35. At $(8, 6)$: $y = -\frac{4}{3}x + \frac{50}{3}$
At $(-6, 8)$: $y = \frac{3}{4}x + \frac{25}{2}$

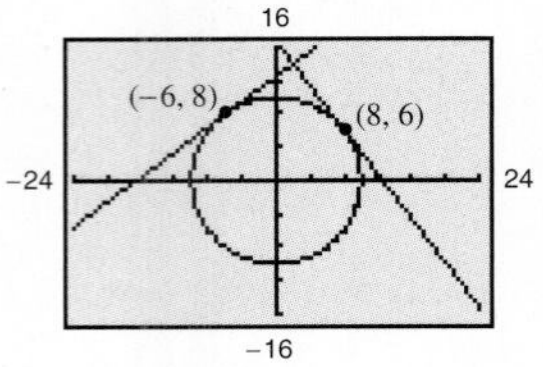

37. At $(1, \sqrt{5})$: $15x - 2\sqrt{5}y - 5 = 0$
At $(1, -\sqrt{5})$: $15x + 2\sqrt{5}y - 5 = 0$

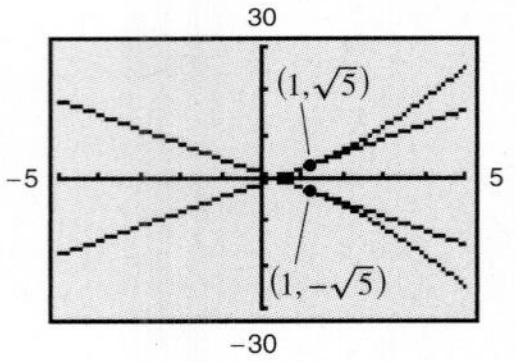

39. At $(0, 2)$: $y = 2$
At $(2, 0)$: $x = 2$

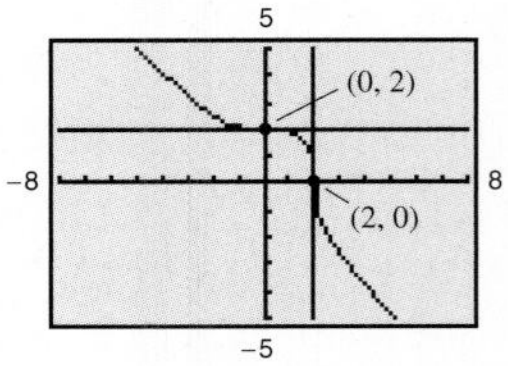

41. At $(-2, 1)$: $y = \frac{1}{2}x + 2$
At $(6, \frac{1}{5})$: $y = -0.06x + 0.56$

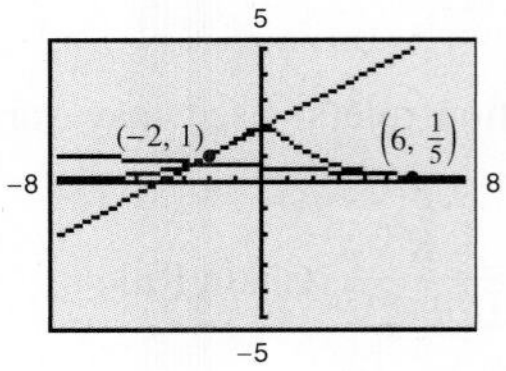

43. $-\frac{2}{p^2(0.00003x^2 + 0.1)}$ **45.** $-\frac{4xp}{2p^2 + 1}$

47. (a) -2

(b) 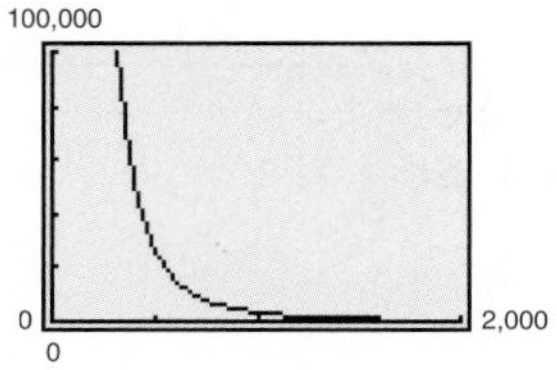

As more labor is used, less capital is available.

As more capital is used, less labor is available.

49. (a) 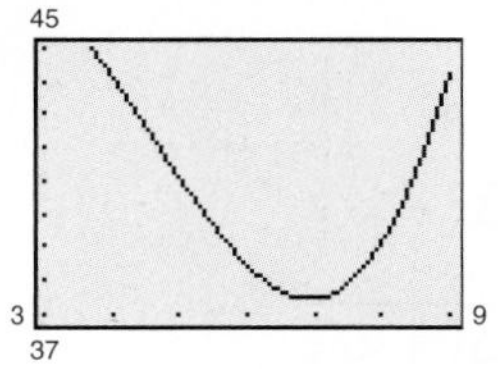

The number of cases of HIV/AIDS decreases from 2004 through 2007, then begins to increase.

(b) 2005

(c)

t	4	5	6	7	8
y	44.11	41.06	38.46	37.46	39.21
y'	-2.95	-3.00	-2.01	0.22	3.38

2005

Section 2.8 *(page 156)*

Skills Warm Up *(page 156)*

1. $A = \pi r^2$ **2.** $V = \frac{4}{3}\pi r^3$ **3.** $S = 6s^2$

4. $V = s^3$ **5.** $V = \frac{1}{3}\pi r^2 h$ **6.** $A = \frac{1}{2}bh$

7. $-\dfrac{x}{y}$ **8.** $\dfrac{2x - 3y}{3x}$ **9.** $-\dfrac{2x + y}{x + 2}$

10. $-\dfrac{y^2 - y + 1}{2xy - 2y - x}$

1. (a) $\frac{3}{4}$ (b) 20 **3.** (a) $-\frac{5}{8}$ (b) $\frac{3}{2}$

5. (a) 36π in.^{2}/min (b) 144π in.^{2}/min

7. If $\dfrac{dr}{dt}$ is constant, $\dfrac{dA}{dt} = 2\pi r \dfrac{dr}{dt}$ and so is proportional to r.

9. (a) $\dfrac{5}{2\pi}$ ft/min (b) $\dfrac{5}{8\pi}$ ft/min

11. (a) 112.5 dollars/wk
(b) 7500 dollars/wk
(c) 7387.5 dollars/wk

13. (a) 9 cm^3/sec (b) 900 cm^3/sec

15. (a) -18 in./sec (b) 0 in./sec
(c) 6 in./sec (d) 18 in./sec

17. -10.4 ft/sec; As $x \to 0$, $\dfrac{dx}{dt}$ increases. **19.** 300 mi/h

21. (a) -750 mi/h (b) 20 min

23. About 37.7 ft^3/min **25.** 4 units/wk

Review Exercises for Chapter 2 *(page 162)*

1. -2 **3.** 0

5. Answers will vary. Sample answer:
$t = 4$: slope $\approx$ \$290 million/yr; Sales were increasing by about \$290 million/yr in 2004.
$t = 7$: slope $\approx$ \$320 million/yr; Sales were increasing by about \$320 million/yr in 2007.

7. Answers will vary. Sample answer:
$t = 1$: slope $\approx$ 65 thousand visitors/month; The number of visitors to the national park is increasing at about 65,000 visitors per month in January.
$t = 8$: slope $\approx$ 0 visitors/month; The number of visitors to the national park is neither increasing nor decreasing in August.
$t = 12$: slope ≈ -1000 thousand visitors/month; The number of visitors to the national park is decreasing at about 1,000,000 visitors per month in December.

9. -3 **11.** -2 **13.** $\frac{1}{4}$ **15.** -1 **17.** 9

19. $-x + 2$ **21.** $\dfrac{1}{2\sqrt{x - 5}}$ **23.** $-\dfrac{5}{x^2}$

25. All values except $x = 1$; The function is not defined at $x = 1$.

27. All values except $x = 0$; A function is not differentiable at a discontinuity.

29. 0 **31.** $3x^2$ **33.** $8x$ **35.** $\dfrac{8x^3}{5}$ **37.** $8x^3 + 6x$

39. $2x + 6$ **41.** -0.125 **43.** 5

45. (a) $y = 5x - 7$
(b) and (c) 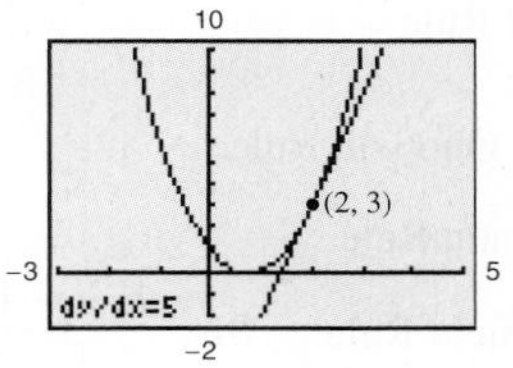

47. (a) $y = x - 1$
(b) and (c) 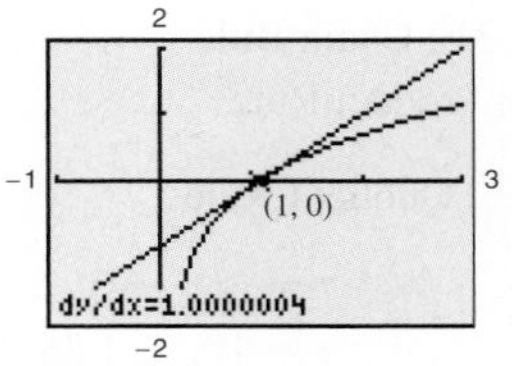

49. (a) 2004: $m \approx 290$
2007: $m \approx 320$
(b) Results should be similar.
(c) The slope shows the rate at which sales were increasing or decreasing in a particular year.

51. 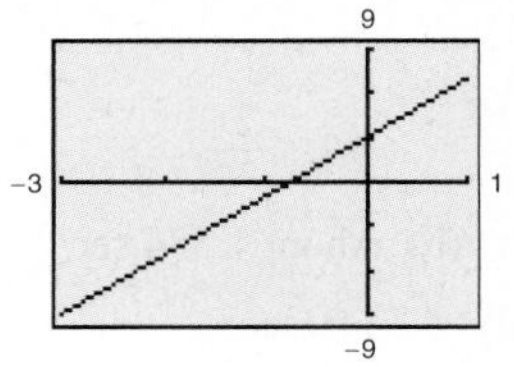

Average rate of change: 4
Instantaneous rate of change at $t = -3$: 4
Instantaneous rate of change at $t = 1$: 4

53. 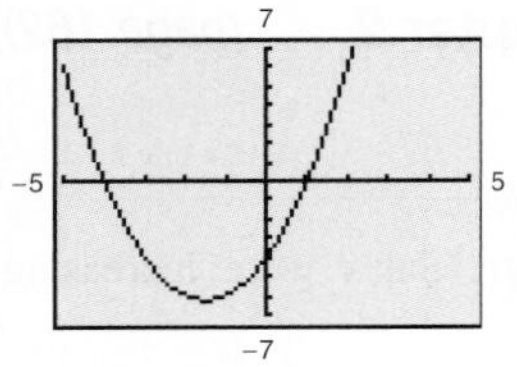

Average rate of change: 4
Instantaneous rate of change at $x = 0$: 3
Instantaneous rate of change at $x = 1$: 5

55. (a) -24 ft/sec (b) $t = 1$: -8 ft/sec; $t = 3$: -72 ft/sec
(c) 5.14 sec
(d) -140.5 ft/sec

57. $\frac{dC}{dx} = 320$ **59.** $\frac{dC}{dx} = \frac{1.275}{\sqrt{x}}$

61. $\frac{dR}{dx} = -1.2x + 150$ **63.** $\frac{dR}{dx} = -12x^2 + 4x + 100$

65. $\frac{dP}{dx} = -0.0006x^2 + 12x - 1$

67. (a) \$9.95 (b) \$10
(c) Parts (a) and (b) differ by only \$0.05.

In Exercises 69–89, the differentiation rule(s) used may vary. A sample answer is provided.

69. $15x^2(1 - x^2)$; Power Rule

71. $16x^3 - 33x^2 + 12x$; Product Rule

73. $\frac{3}{(x + 3)^2}$; Quotient Rule

75. $\frac{2(3 + 5x - 3x^2)}{(x^2 + 1)^2}$; Quotient Rule

77. $30x(5x^2 + 2)^2$; Chain Rule

79. $-\frac{1}{(x + 1)^{3/2}}$; Quotient Rule **81.** $\frac{2x^2 + 1}{\sqrt{x^2 + 1}}$; Product Rule

83. $80x^4 - 24x^2 + 1$; Product Rule

85. $18x^5(x + 1)(2x + 3)^2$; Chain Rule

87. $x(x - 1)^4(7x - 2)$; Product Rule

89. $\frac{3(9t + 5)}{2\sqrt{3t + 1}(1 - 3t)^3}$; Quotient Rule

91. (a) $t = 1$: -6.63 $t = 3$: -6.5
$t = 5$: -4.33 $t = 10$: -1.36
(b) 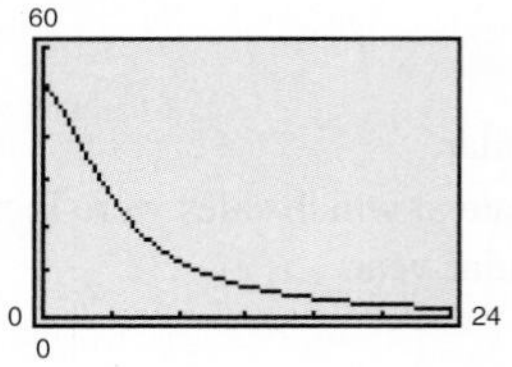

The rate of decrease is approaching zero.

93. 6 **95.** $-\frac{120}{x^6}$ **97.** $\frac{35x^{3/2}}{2}$ **99.** $\frac{2}{x^{2/3}}$

101. (a) $s(t) = -16t^2 + 5t + 30$ (b) About 1.534 sec
(c) About -44.09 ft/sec (d) -32 ft/sec^2

103. $-\frac{2x + 3y}{3(x + y^2)}$ **105.** $\frac{2x - 8}{2y - 9}$ **107.** $y = \frac{1}{3}x + \frac{1}{3}$

109. $y = \frac{4}{3}x + \frac{2}{3}$

111. (a) 12π square inches per minute
(b) 40π square inches per minute

113. $\frac{1}{64}$ ft/min

Test Yourself *(page 166)*

1. $f(x) = x^2 + 1$
$f(x + \Delta x) = x^2 + 2x\Delta x + \Delta x^2 + 1$
$f(x + \Delta x) - f(x) = 2x\Delta x + \Delta x^2$
$\frac{f(x + \Delta x) - f(x)}{\Delta x} = 2x + \Delta x$
$\lim_{\Delta x \to 0} \frac{f(x + \Delta x) - f(x)}{\Delta x} = 2x$
$f'(x) = 2x$
$f'(2) = 4$

2. $f(x) = \sqrt{x} - 2$
$f(x + \Delta x) = \sqrt{x + \Delta x} - 2$
$f(x + \Delta x) - f(x) = \sqrt{x + \Delta x} - \sqrt{x}$
$\frac{f(x + \Delta x) - f(x)}{\Delta x} = \frac{1}{\sqrt{x + \Delta x} + \sqrt{x}}$
$\lim_{\Delta x \to 0} \frac{f(x + \Delta x) - f(x)}{\Delta x} = \frac{1}{2\sqrt{x}}$
$f'(x) = \frac{1}{2\sqrt{x}}$
$f'(4) = \frac{1}{4}$

3. $f'(t) = 3t^2 + 2$ **4.** $f'(x) = 8x - 8$ **5.** $f'(x) = \frac{3\sqrt{x}}{2}$

6. $f'(x) = 3x^2 + 10x + 6$ **7.** $f'(x) = \frac{9}{x^4}$

8. $f'(x) = \frac{5 + x}{2\sqrt{x}} + \sqrt{x}$ **9.** $f'(x) = 36x^3 + 48x$

10. $f'(x) = -\frac{1}{\sqrt{1 - 2x}}$

11. $f'(x) = \frac{(10x + 1)(5x - 1)^2}{x^2} = 250x - 75 + \frac{1}{x^2}$

12. $y = 2x - 2$
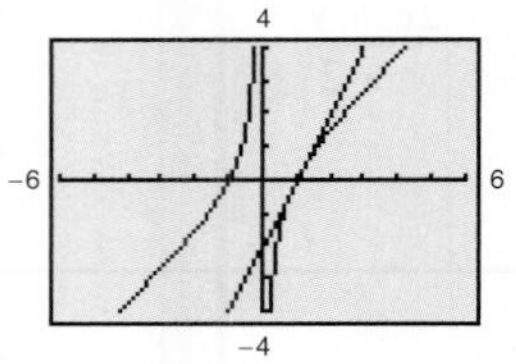

13. (a) \$18.69 billion/yr
(b) 2005: \$10.50 billion/yr
2008: \$14.95 billion/yr
(c) The annual sales of CVS Caremark from 2005 to 2008 increased on average by about \$18.69 billion/yr, and the instantaneous rates of change for 2005 and 2008 are \$10.50 billion/yr and \$14.95 billion/yr, respectively.

14. (a) $P = -0.016x^2 + 1460x - 715{,}000$ (b) \$1437.60

15. 0 **16.** $-\frac{3}{8(3 - x)^{5/2}}$ **17.** $-\frac{96}{(2x - 1)^4}$

18. $s(t) = -16t^2 + 30t + 75$ $s(2) = 71$ ft
$v(t) = -32t + 30$ $v(2) = -34$ ft/sec
$a(t) = -32$ $a(2) = -32$ ft/sec^2

19. $\frac{dy}{dx} = -\frac{1 + y}{x}$ **20.** $\frac{dy}{dx} = -\frac{1}{y - 1}$ **21.** $\frac{dy}{dx} = \frac{x}{2y}$

22. (a) 3.75π cm^3/min (b) 15π cm^3/min

Chapter 3

Section 3.1 *(page 175)*

Skills Warm Up *(page 175)*

1. $x = 0, x = 8$ **2.** $x = 0, x = 24$ **3.** $x = \pm 5$
4. $x = 0$ **5.** $(-\infty, 3) \cup (3, \infty)$ **6.** $(-\infty, 1)$
7. $(-\infty, -2) \cup (-2, 5) \cup (5, \infty)$ **8.** $(-\sqrt{3}, \sqrt{3})$
9. $x = -2$: -6
$x = 0$: 2
$x = 2$: -6
10. $x = -2$: 60
$x = 0$: -4
$x = 2$: 60
11. $x = -2$: $-\frac{1}{3}$
$x = 0$: 1
$x = 2$: 5
12. $x = -2$: $\frac{1}{18}$
$x = 0$: $-\frac{1}{8}$
$x = 2$: $-\frac{3}{2}$

1. Increasing on $(-\infty, -1)$
Decreasing on $(-1, \infty)$
3. Increasing on $(-1, 0)$ and $(1, \infty)$
Decreasing on $(-\infty, -1)$ and $(0, 1)$
5. $x = \frac{3}{4}$ **7.** $x = 0, x = -3$ **9.** $x = \pm 2$
11. No critical numbers
Increasing on $(-\infty, \infty)$

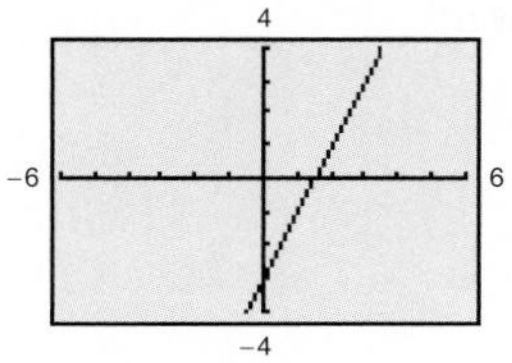

13. Critical number: $x = 3$
Decreasing on $(-\infty, 3)$
Increasing on $(3, \infty)$

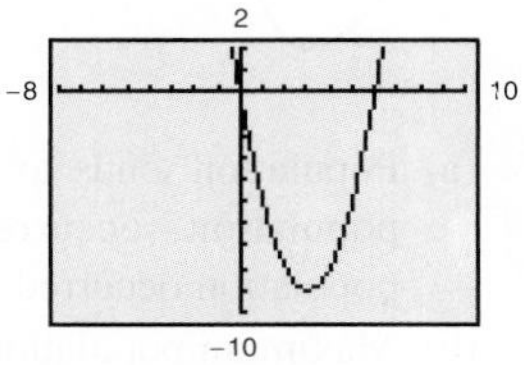

15. Critical number: $x = 1$
Increasing on $(-\infty, 1)$
Decreasing on $(1, \infty)$

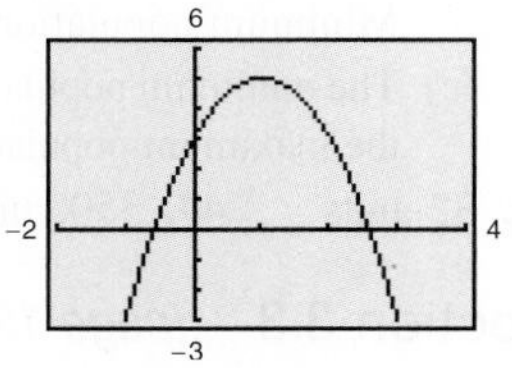

17. Critical numbers:
$x = -1, x = -\frac{5}{3}$
Increasing on $\left(-\infty, -\frac{5}{3}\right)$
and $(-1, \infty)$
Decreasing on $\left(-\frac{5}{3}, -1\right)$

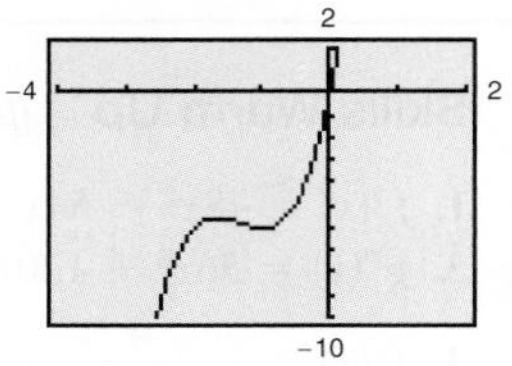

19. Critical numbers:
$x = 0, x = \frac{3}{2}$
Decreasing on $\left(-\infty, \frac{3}{2}\right)$
Increasing on $\left(\frac{3}{2}, \infty\right)$

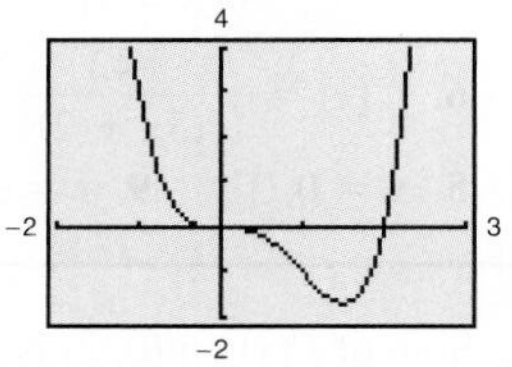

21. Critical number: $x = -2$
Decreasing on $(-\infty, -2)$
Increasing on $(-2, \infty)$

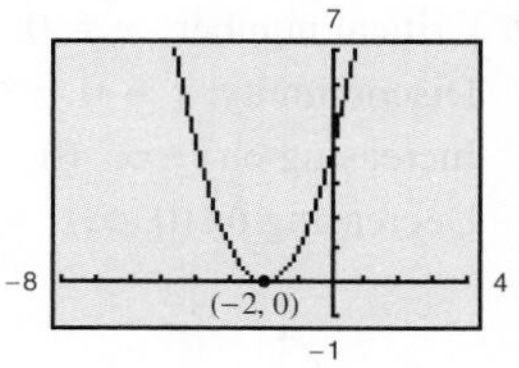

23. Critical number: $x = 1$
Increasing on $(-\infty, 1)$
Decreasing on $(1, \infty)$

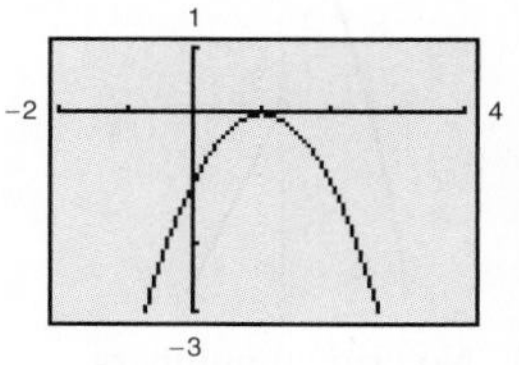

25. Critical number: $x = 0$
Increasing on $(-\infty, 0)$
and $(0, \infty)$

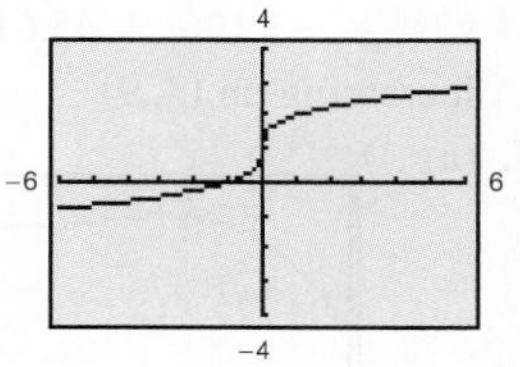

27. Critical numbers:
$x = -1, x = 1$
Decreasing on $(-\infty, -1)$
Increasing on $(1, \infty)$

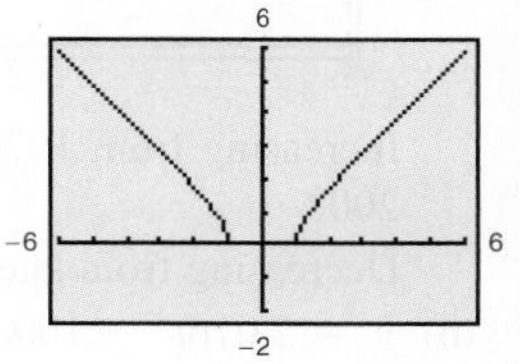

29. Critical number: $x = -2$
Increasing on $(-\infty, -2)$
and $(-2, \infty)$

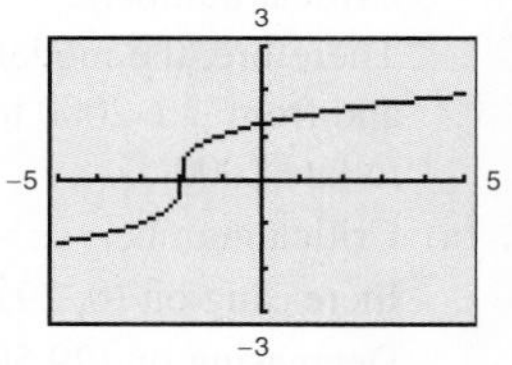

31. Critical numbers:
$x = -1, x = -\frac{2}{3}$
Decreasing on $\left(-1, -\frac{2}{3}\right)$
Increasing on $\left(-\frac{2}{3}, \infty\right)$

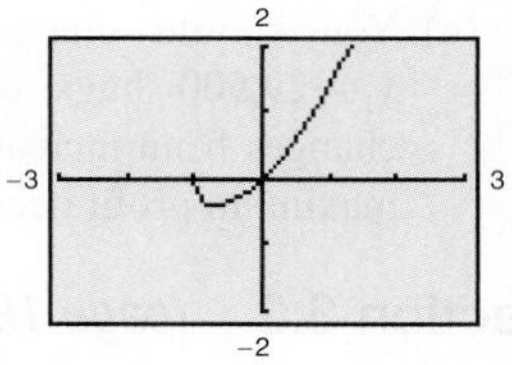

33. Critical numbers:
$x = -3, x = 3$
Decreasing on $(-\infty, -3)$
and $(3, \infty)$
Increasing on $(-3, 3)$

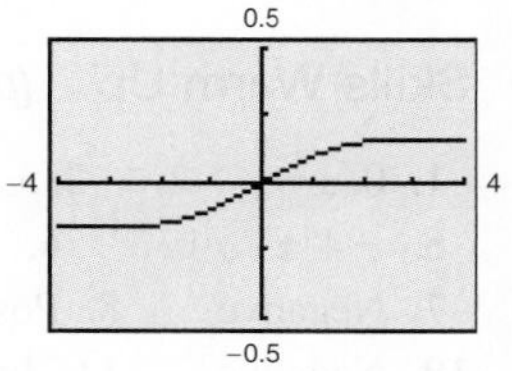

35. No critical numbers
Discontinuity: $x = 5$
Decreasing on $(-\infty, 5)$
and $(5, \infty)$

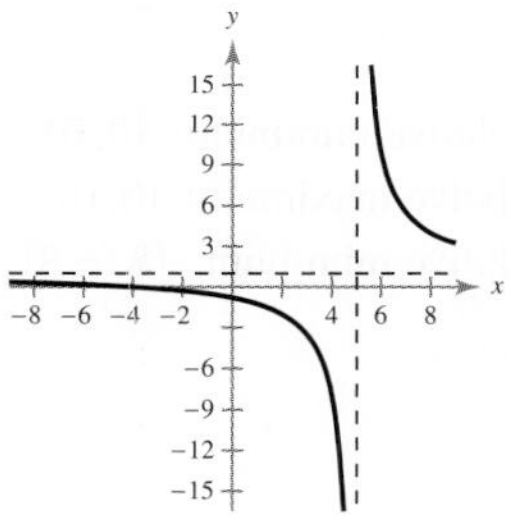

37. No critical numbers
Discontinuities: $x = \pm 4$
Increasing on $(-\infty, -4)$,
$(-4, 4)$, and $(4, \infty)$

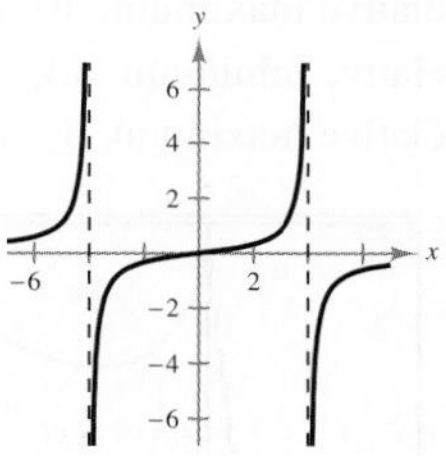

39. Critical number: $x = 0$
Discontinuity: $x = 0$
Increasing on $(-\infty, 0)$
Decreasing on $(0, \infty)$

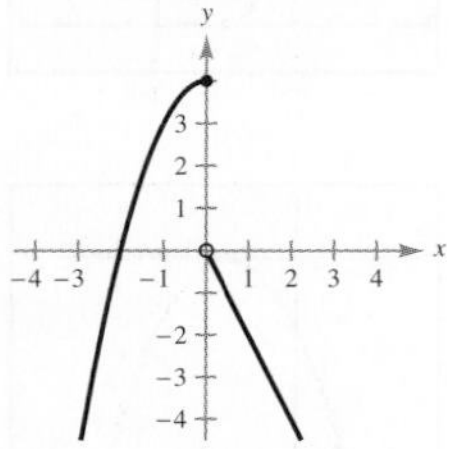

41. Critical numbers: $x = -1, 0$
Increasing on $(-\infty, -1)$ and $(0, \infty)$
Decreasing on $(-1, 0)$

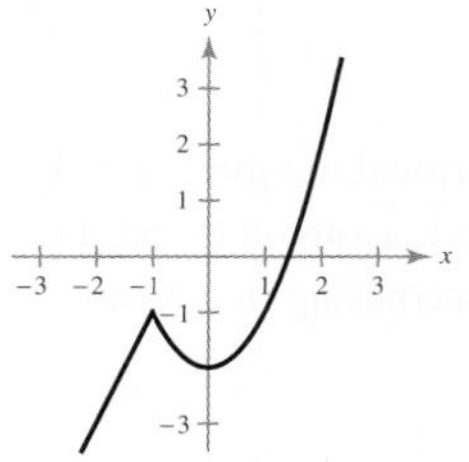

43. No critical numbers
$s'(t) = -3.196t + 45.61$
Increasing on $(3, 9)$

45. (a)

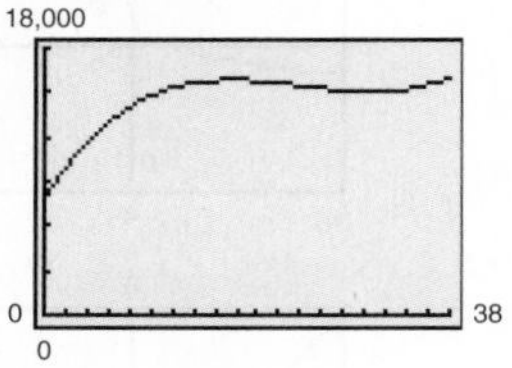

Increasing from 1970 to late 1987 and from late 2000 to 2008
Decreasing from late 1987 to late 2000

(b) $y' = 2.076t^2 - 100.22t + 1119.7$
Critical numbers: $t = 17.6, t = 30.7$
Therefore, the model is increasing from 1970 to late 1987 and from late 2000 to 2008 and decreasing from late 1987 to late 2000.

47. (a) Critical number: $x = 29{,}500$
Increasing on $(0, 29{,}500)$
Decreasing on $(29{,}500, 50{,}000)$

(b) You should charge the price that yields sales of $x = 29{,}500$ bags of popcorn. Because the function changes from increasing to decreasing at $x = 29{,}500$, the maximum profit occurs at this value.

Section 3.2 *(page 184)*

Skills Warm Up *(page 184)*

1. $0, \pm\frac{1}{2}$ **2.** $-2, 5$ **3.** 1 **4.** $0, 125$
5. $-4 \pm \sqrt{17}$ **6.** $1 \pm \sqrt{5}$
7. Negative **8.** Positive **9.** Positive
10. Negative **11.** Increasing **12.** Decreasing

1. Relative maximum: $(1, 5)$ **3.** Relative minimum: $(3, -9)$
5. Relative minimum: $(9, -2187)$ **7.** No relative extrema
9. Relative maximum: $(0, 15)$
Relative minimum: $(4, -17)$
11. Relative maximum: $(-1, 2)$; Relative minimum: $(0, 0)$
13.

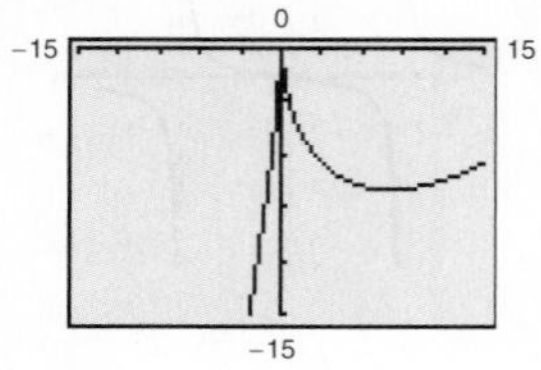

Relative maximum: $(0, 0)$
Relative minimum: $(8, -8)$

15.

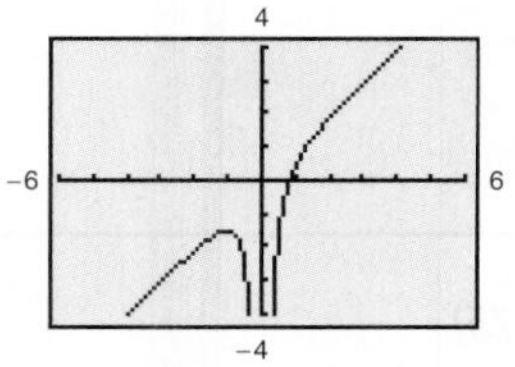

Relative maximum: $\left(-1, -\frac{3}{2}\right)$

17.

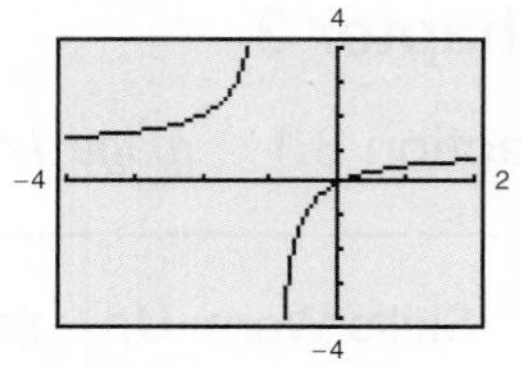

No relative extrema

19. Minimum: $(2, 2)$
Maximum: $(-1, 8)$
21. Maximum: $(0, 5)$
Minimum: $(3, -13)$
23. Minima: $(-1, -4), (2, -4)$
Maxima: $(0, 0), (3, 0)$
25. Maximum: $(2, 1)$
Minimum: $\left(0, \frac{1}{3}\right)$
27. Maxima: $\left(-1, \frac{1}{4}\right), \left(1, \frac{1}{4}\right)$
Minimum: $(0, 0)$
29. Maximum: $(-7, 4)$
Minimum: $(1, 0)$
31. 2, absolute maximum (and relative maximum)
33. 1, absolute maximum (and relative maximum);
2, absolute minimum (and relative minimum);
3, absolute maximum (and relative maximum)
35. Maximum: $(5, 7)$
Minimum: $(2.69, -5.55)$
37. Maximum: $(2, 2.\overline{6})$
Minima: $(0, 0), (3, 0)$
39. No relative maximum
Minimum: $(2, 12)$
41. Maximum: $\left(2, \frac{1}{2}\right)$
Minimum: $(0, 0)$
43. Answers will vary. Sample answer:

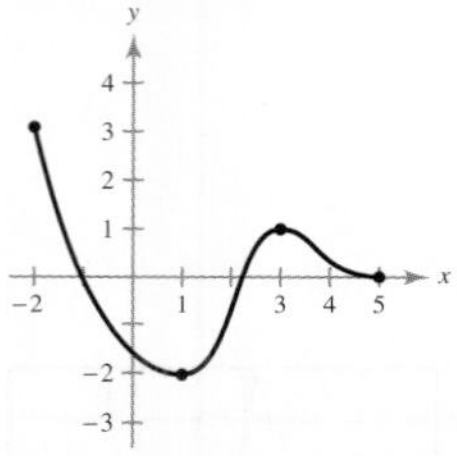

45. (a) Population tends to increase each year, so the minimum population occurred in 1790 and the maximum population occurred in 2010.
(b) Maximum population: 310.07 million
Minimum population: 3.69 million
(c) The minimum population was about 3.69 million in 1790 and the maximum population was about 310.07 million in 2010.

47. 82 units **49.** 3500 units, \$2.25

Section 3.3 *(page 193)*

Skills Warm Up *(page 193)*

1. $f''(x) = 48x^2 - 54x$ **2.** $g''(s) = 12s^2 - 18s + 2$
3. $g''(x) = 56x^6 + 120x^4 + 72x^2 + 8$
4. $f''(x) = \dfrac{4}{9(x-3)^{2/3}}$ **5.** $h''(x) = \dfrac{190}{(5x-1)^3}$
6. $f''(x) = -\dfrac{42}{(3x+2)^3}$ **7.** $x = \pm\dfrac{\sqrt{3}}{3}$
8. $x = 0, 3$ **9.** $t = \pm 4$ **10.** $x = 0, \pm 5$

1. Sign of $f'(x)$ on $(0, 2)$ is positive.
Sign of $f''(x)$ on $(0, 2)$ is positive.

3. Sign of $f'(x)$ on $(0, 2)$ is negative.
Sign of $f''(x)$ on $(0, 2)$ is negative.
5. Concave downward on $(-\infty, \infty)$
7. Concave upward on $(-\infty, 1)$
Concave downward on $(1, \infty)$
9. Concave upward on $\left(-\infty, -\frac{1}{2}\right)$
Concave downward on $\left(-\frac{1}{2}, \infty\right)$
11. Concave upward on $(-\infty, -2)$ and $(2, \infty)$
Concave downward on $(-2, 2)$
13. Concave downward on $(-\infty, 3)$
Concave upward on $(3, \infty)$
Point of inflection: $(3, 0)$
15. Concave downward on $\left(-\infty, \frac{1}{2}\right)$
Concave upward on $\left(\frac{1}{2}, \infty\right)$
Point of inflection: $\left(\frac{1}{2}, -\frac{3}{2}\right)$
17. Concave upward on $(-\infty, \infty)$
No inflection points
19. Concave downward on $(-\infty, 4)$
Concave upward on $(4, \infty)$
Inflection point: $(4, 16)$
21. Relative maximum: $(3, 9)$
23. Relative maximum: $(1, 3)$
Relative minimum: $\left(\frac{7}{3}, \frac{49}{27}\right)$
25. Relative minimum: $(0, -3)$
27. Relative minimum: $(0, 1)$
29. Relative maximum: $(0, 3)$
31. Relative maximum: $(0, 4)$
33. No relative extrema
35. Relative maximum: $(2, 9)$
Relative minimum: $(0, 5)$
37. Relative maximum: $(0, 0)$
Relative minima: $(-0.5, -0.052)$, $(1, -0.\overline{3})$
39. Relative maximum: $(-2, 16)$
Relative minimum: $(2, -16)$
Point of inflection: $(0, 0)$

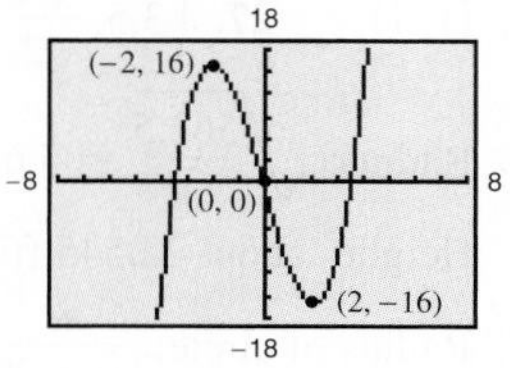

41. No relative maximum
Relative minimum: $(4, 4)$
Point of inflection:
$\left(12, \frac{8\sqrt{3}}{3}\right)$

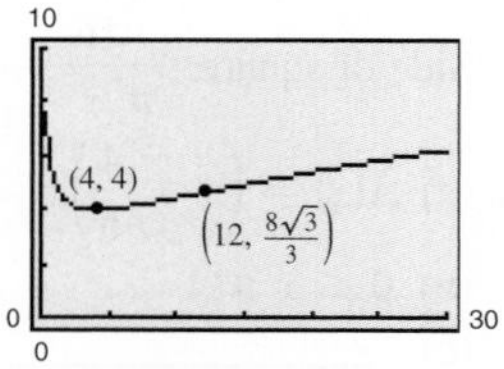

43. Relative maximum: $(0, 0)$
Relative minima: $(\pm 2, -4)$
Points of inflection:
$\left(\pm\frac{2\sqrt{3}}{3}, -\frac{20}{9}\right)$

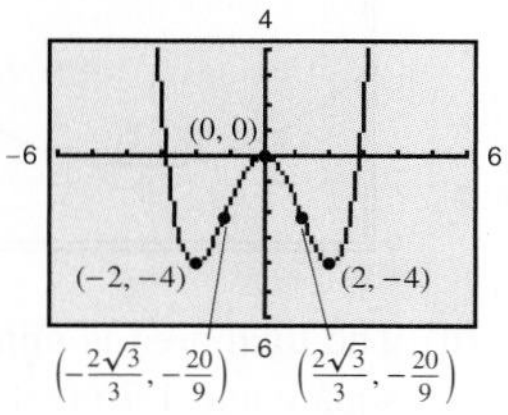

45. Relative maximum: $(-1, 0)$
Relative minimum: $(1, -4)$
Point of inflection: $(0, -2)$

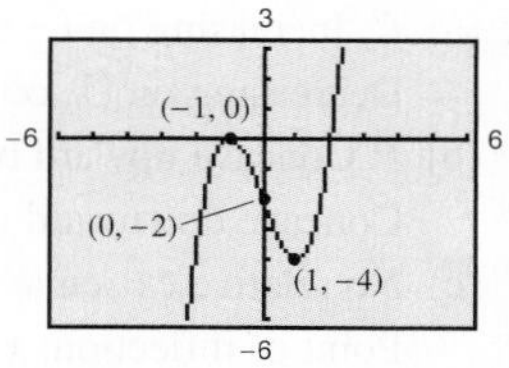

47. Relative minimum: $(-2, -2)$
No inflection points

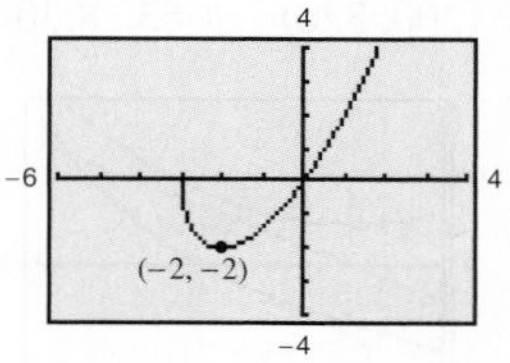

49. Relative maximum: $(0, 4)$
Points of inflection:
$\left(\pm\frac{\sqrt{3}}{3}, 3\right)$

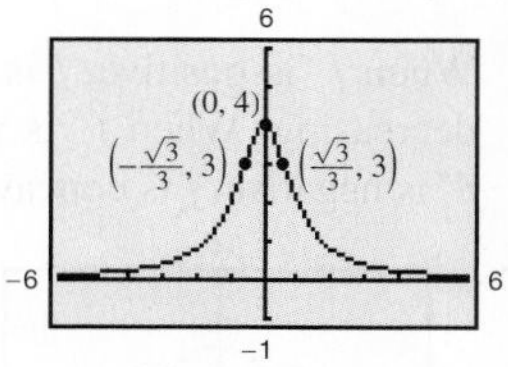

51.

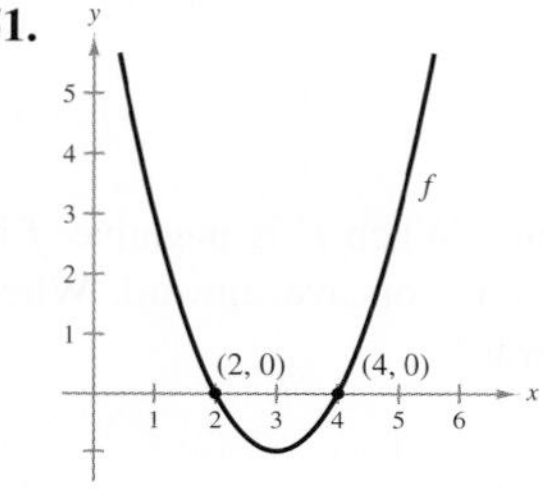

53.

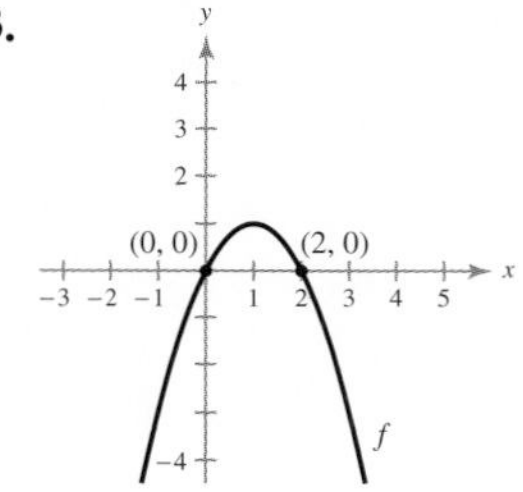

55.

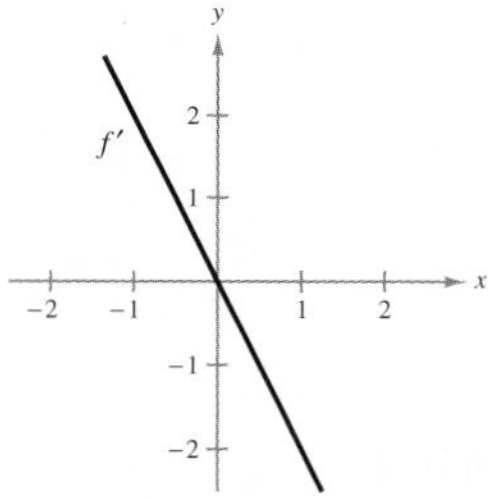

(a) f': Positive on $(-\infty, 0)$
f: Increasing on $(-\infty, 0)$
(b) f': Negative on $(0, \infty)$
f: Decreasing on $(0, \infty)$
(c) f': Not increasing
f: Not concave upward
(d) f': Decreasing on $(-\infty, \infty)$
f: Concave downward on $(-\infty, \infty)$
57. (a) f': Increasing on $(-\infty, \infty)$
(b) f: Concave upward on $(-\infty, \infty)$
(c) Relative minimum: $x = -2.5$
No inflection points

59. (a) f': Increasing on $(-\infty, 1)$
Decreasing on $(1, \infty)$
(b) f: Concave upward on $(-\infty, 1)$
Concave downward on $(1, \infty)$
(c) No relative extrema
Point of inflection: $x = 1$

61. $(200, 320)$ **63.** 8:30 P.M.

65.
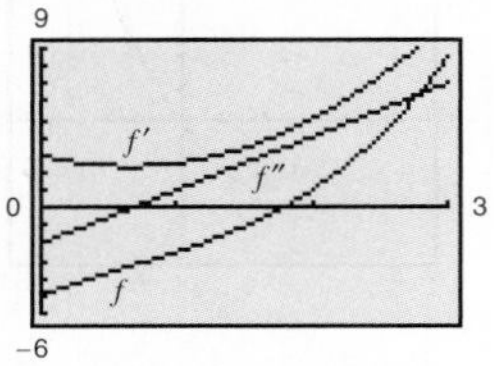

Relative minimum: $(0, -5)$
Relative maximum: $(3, 8.5)$
Point of inflection: $\left(\frac{2}{3}, -3.2963\right)$

When f' is positive, f is increasing. When f' is negative, f is decreasing. When f'' is positive, f is concave upward. When f'' is negative, f is concave downward.

67.
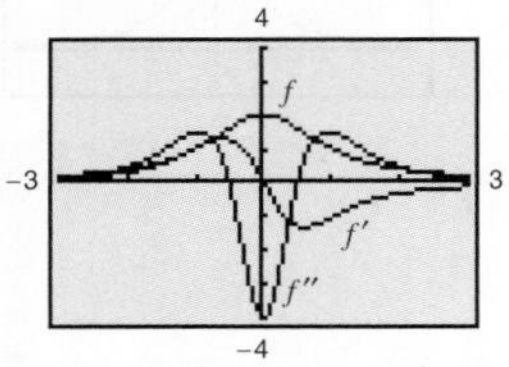

Relative maximum: $(0, 2)$
Points of inflection: $(0.58, 1.5), (-0.58, 1.5)$

When f' is positive, f is increasing. When f' is negative, f is decreasing. When f'' is positive, f is concave upward. When f'' is negative, f is concave downward.

69. 120 units

71. (a)
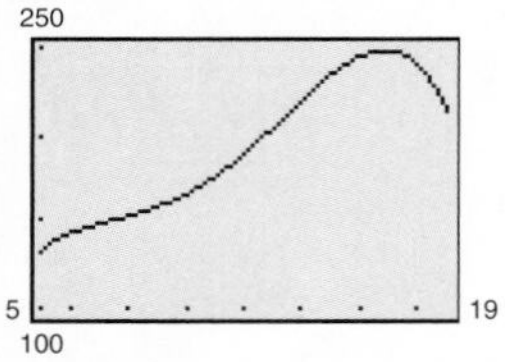

(b) 1995
(c) 2007
(d) Greatest: 2003
Least: 2009

73. (a)
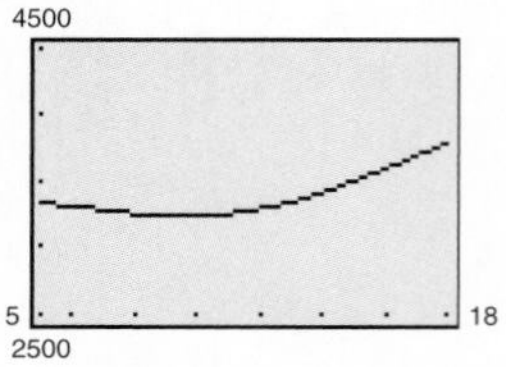

(b) Concave downward on $(5, 6.6517)$
Concave upward on $(6.6517, 16.4123)$
Concave downward on $(16.4123, 18)$
(c) Points of inflection: $(6.6517, 3291.0160)$ and $(16.4123, 3638.4227)$
(d) The first inflection point is where the change in the number of veterans receiving benefits starts to increase after it has been decreasing. The second inflection point is where the change in the number of veterans receiving benefits starts to decrease again.

75. Answers will vary.

Section 3.4 *(page 201)*

Skills Warm Up *(page 201)*

1. $x + \frac{1}{2}y = 12$ **2.** $2xy = 24$ **3.** $xy = 24$
4. $\sqrt{(x_2 - x_1)^2 + (y_2 - y_1)^2} = 10$
5. $x = -3$ **6.** $x = -\frac{2}{3}, 1$ **7.** $x = \pm 5$
8. $x = 4$ **9.** $x = \pm 1$ **10.** $x = \pm 3$

1. $l = w = 25$ m **3.** $l = w = 8$ ft
5. $x = 25$ ft, $y = \frac{100}{3}$ ft
7. (a) Proof
(b) $V_1 = 99$ in.^{3}
$V_2 = 125$ in.^{3}
$V_3 = 117$ in.^{3}
(c) 5 in. × 5 in. × 5 in.
9. (a) $l = w = h = 20$ in.
(b) 2400 in.^{2}
11. Width of rectangle: $\dfrac{100}{\pi} \approx 31.8$ m
Length of rectangle: 50 m
13. $l = w = 2\sqrt[3]{5} \approx 3.42$
$h = 4\sqrt[3]{5} \approx 6.84$
15. $V = 16$ in.^{3} **17.** 9 in. by 9 in.
19. Length: 3 units
Width: 1.5 units
21. Length: $5\sqrt{2}$ units
Width: $\dfrac{5\sqrt{2}}{2}$ units
23. Radius: about 1.51 in.
Height: about 3.02 in.
25. $(1, 1)$ **27.** $\left(3.5, \dfrac{\sqrt{14}}{2}\right)$ **29.** 18 in. × 18 in. × 36 in.
31. Radius: $\sqrt[3]{\dfrac{562.5}{\pi}} \approx 5.636$ ft
Height: about 22.545 ft
33. Radius of circle: $\dfrac{8}{\pi + 4}$
Side of square: $\dfrac{16}{\pi + 4}$
35. (a) $A(x) = \left(1 + \dfrac{4}{\pi}\right)x^2 - \dfrac{8}{\pi}x + \dfrac{4}{\pi}$
(b) $0 \le x \le 1$
(c)
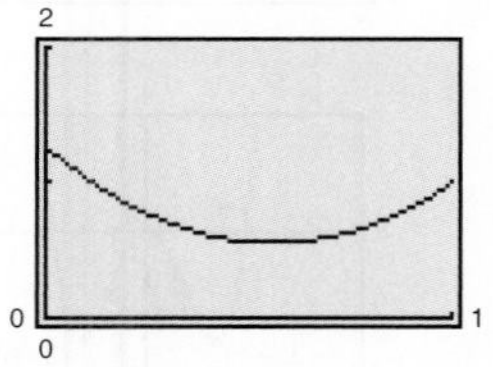

(d) The total area is minimum when 2.24 feet is used for the square and 1.76 feet is used for the circle.
The total area is maximum when all 4 feet is used for the circle.
37. 4.75 weeks; 135 bushels; \$3645

Quiz Yourself *(page 204)*

1. Critical number: $x = 3$
Increasing on $(3, \infty)$
Decreasing on $(-\infty, 3)$

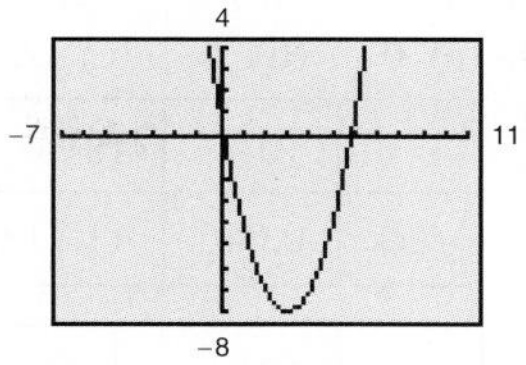

2. Critical numbers:
$x = -4, x = 0$
Increasing on
$(-\infty, -4)$ and $(0, \infty)$
Decreasing on $(-4, 0)$

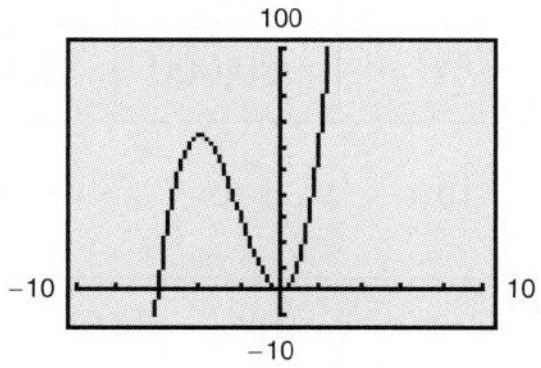

3. Critical numbers: $x = 5$, $x = -5$
Increasing on $(-5, 5)$
Decreasing on $(-\infty, -5)$ and $(5, \infty)$

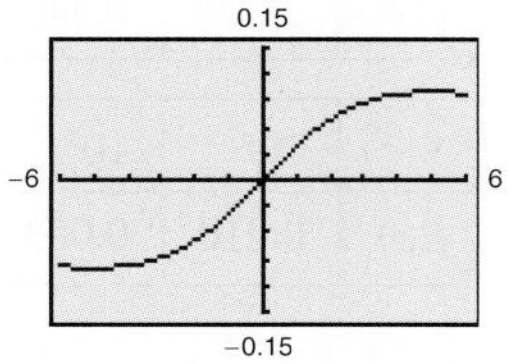

4. Relative minimum: $(0, -5)$
Relative maximum: $(-2, -1)$

5. Relative minima: $(2, -13), (-2, -13)$
Relative maximum: $(0, 3)$

6. Relative minimum: $(0, 0)$

7. Minimum: $(-1, -9)$
Maximum: $(1, -5)$

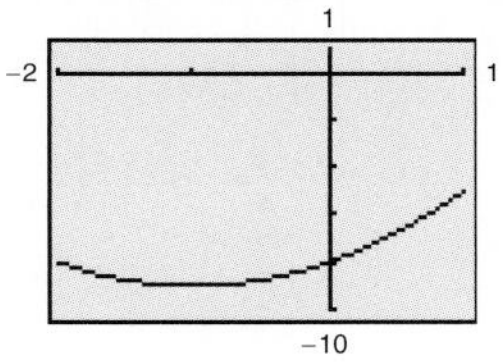

8. Minimum: $(3, -54)$
Maximum: $(-3, 54)$

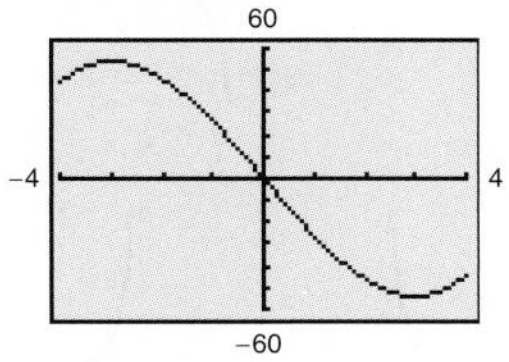

9. Minimum: $(0, 0)$
Maximum: $(1, 0.5)$

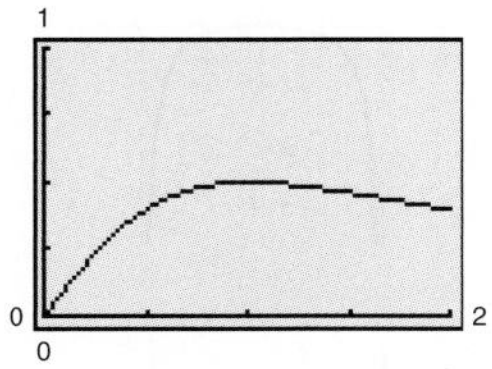

10. Point of inflection: $(2, -2)$
Concave downward on $(-\infty, 2)$
Concave upward on $(2, \infty)$

11. Points of inflection: $(-2, -80)$ and $(2, -80)$
Concave downward on $(-2, 2)$
Concave upward on $(-\infty, -2)$ and $(2, \infty)$

12. Relative minimum: $(1, 9)$
Relative maximum: $(-2, 36)$

13. Relative minimum: $(3, 12)$
Relative maximum: $(-3, -12)$

14. \$120,000 $(x = 120)$ **15.** 50 ft by 100 ft

16. (a) Increasing from 2000 to early 2008
Decreasing from early 2008 to 2009
(b) Early 2008; 2000

Section 3.5 *(page 212)*

Skills Warm Up *(page 212)*

1. 1 **2.** $\frac{6}{5}$ **3.** 2 **4.** $\frac{1}{2}$

5. $\dfrac{dC}{dx} = 1.2 + 0.006x$ **6.** $\dfrac{dP}{dx} = 0.02x + 11$

7. $\dfrac{dP}{dx} = -1.4x + 7$ **8.** $\dfrac{dC}{dx} = 4.2 + 0.003x^2$

9. $\dfrac{dR}{dx} = 14 - \dfrac{x}{1000}$ **10.** $\dfrac{dR}{dx} = 3.4 - \dfrac{x}{750}$

1. 2000 units **3.** 200 units **5.** 200 units

7. 50 units **9.** \$60 **11.** \$40

13. 3 units

$\overline{C}(3) = 17$; $\dfrac{dC}{dx} = 4x + 5$; when $x = 3$, $\dfrac{dC}{dx} = 17$

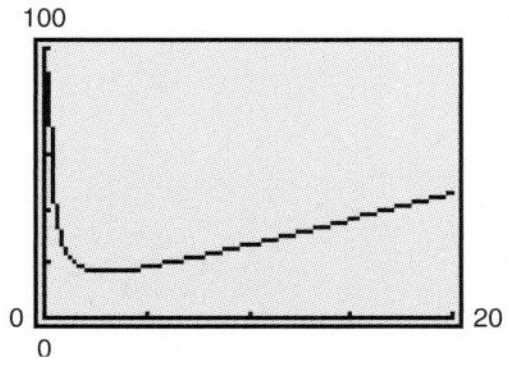

15. (a) \$55 (b) \$30.32

17. The maximum profit occurs when $s = 10$ (or \$10,000). The point of diminishing returns occurs at $s = \frac{35}{6}$ (or \$5833.33).

19. 350 players **21.** \$50

23. $C =$ cost under water + cost on land
$= 25(5280)\sqrt{x^2 + 0.25} + 18(5280)(6 - x)$
$= 132{,}000\sqrt{x^2 + 0.25} + 570{,}240 - 95{,}040x$

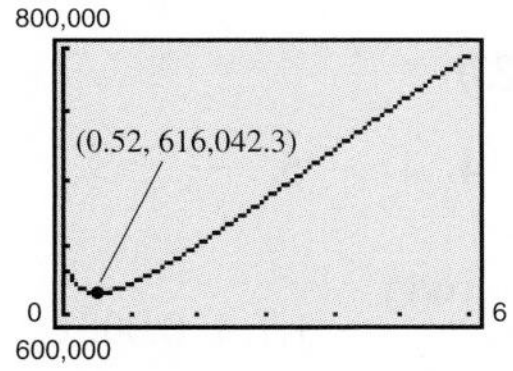

The line should run from the power station to a point across the river approximately 0.52 mile downstream.

$\left(\text{Exact: } x = \dfrac{9\sqrt{301}}{301} \text{ mi}\right)$

25. $v = 60$ mi/h

27. -1, unit elastic

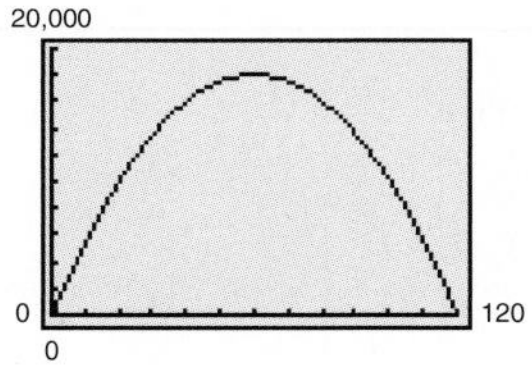

Elastic: $(0, 60)$
Inelastic: $(60, 120)$

29. $-\frac{2}{3}$, inelastic

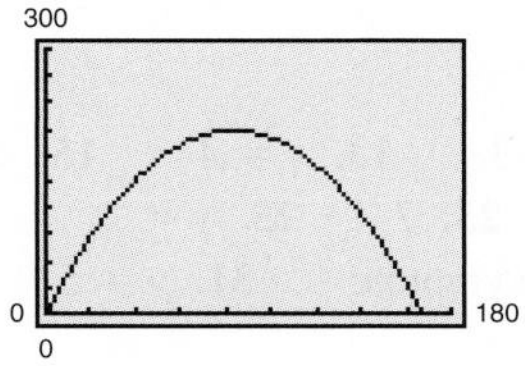

Elastic: $\left(0, 83\frac{1}{3}\right)$
Inelastic: $\left(83\frac{1}{3}, 166\frac{2}{3}\right)$

31. $-\frac{25}{23}$, elastic

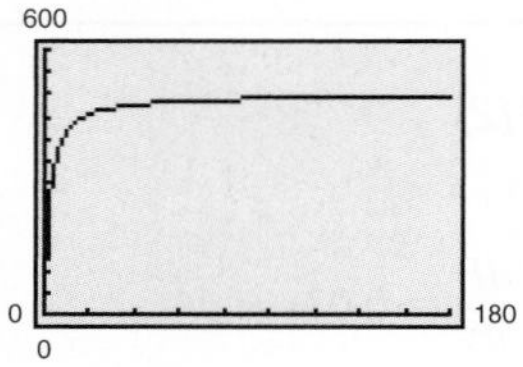

Elastic: $(0, \infty)$

33. (a) Elastic: $[0, 500)$
Unit elastic: $x = 500$
Inelastic: $(500, 1000]$
(b) The revenue function increases on the interval $[0, 500)$, then is flat at 500, and decreases on the interval $(500, 1000]$.

35. 500 units ($x = 5$)

37. No; when $p = 8$, $x = 540$ and $\eta = -\frac{2}{3}$.
Because $|\eta| = \frac{2}{3} < 1$, demand is inelastic.

39. (a) 2007
(b) 2001
(c) 2007: \$237.55 million/yr
2001: \$17.78 million/yr
(d)

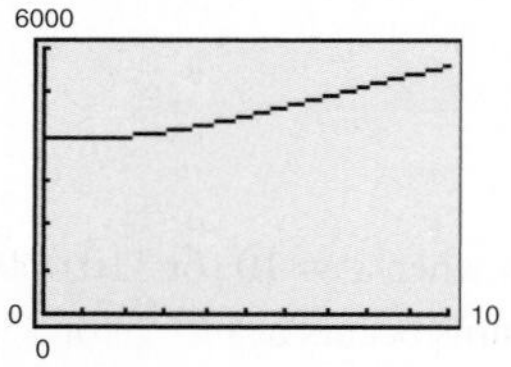

41. Proof **43.** Answers will vary.

Section 3.6 *(page 223)*

Skills Warm Up *(page 223)*

1. 3 **2.** 1 **3.** -11 **4.** 4 **5.** $-\frac{1}{4}$
6. -2 **7.** 0 **8.** 1

9. $\overline{C} = \dfrac{150}{x} + 3$, $\dfrac{dC}{dx} = 3$

10. $\overline{C} = \dfrac{1900}{x} + 1.7 + 0.002x$, $\dfrac{dC}{dx} = 1.7 + 0.004x$

11. $\overline{C} = 0.005x + 0.5 + \dfrac{1375}{x}$, $\dfrac{dC}{dx} = 0.01x + 0.5$

12. $\overline{C} = \dfrac{760}{x} + 0.05$, $\dfrac{dC}{dx} = 0.05$

1. Vertical asymptote: $x = 0$
Horizontal asymptote: $y = 1$

3. Vertical asymptotes: $x = -1, x = 2$
Horizontal asymptote: $y = 1$

5. Vertical asymptote: none
Horizontal asymptote: $y = \frac{3}{2}$

7. Vertical asymptotes: $x = \pm 2$
Horizontal asymptote: $y = \frac{1}{2}$

9. $x = 0, x = -3$ **11.** $x = -3$ **13.** $x = 4$ **15.** ∞
17. $-\infty$ **19.** ∞ **21.** 1 **23.** 7 **25.** $y = 2$
27. $y = 0$ **29.** No horizontal asymptote **31.** $y = 5$

33. d **34.** b **35.** a **36.** c

37. (a) ∞ (b) 5 (c) 0

39. (a) 0 (b) 1 (c) ∞

41.

x	10^0	10^1	10^2	10^3
$f(x)$	0.646	11.718	800.003	29,622.777

x	10^4	10^5	10^6
$f(x)$	980,000	31,422,776.6	998,000,000

$\lim_{x\to\infty} \sqrt{x^3 + 6} - 2x = \infty$

43.

x	10^0	10^1	10^2	10^3
$f(x)$	2.000	0.348	0.101	0.032

x	10^4	10^5	10^6
$f(x)$	0.010	0.003	0.001

$\lim_{x\to\infty} \dfrac{x + 1}{x\sqrt{x}} = 0$

45.

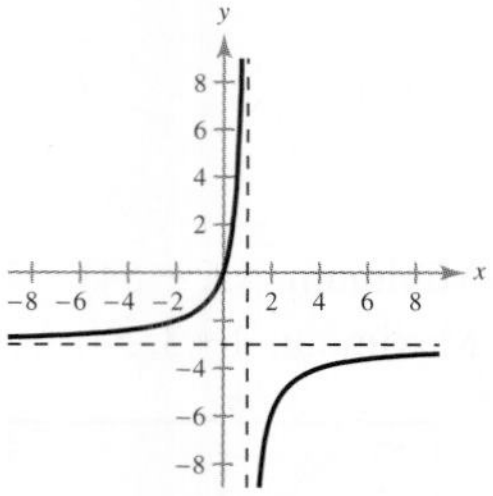

47.

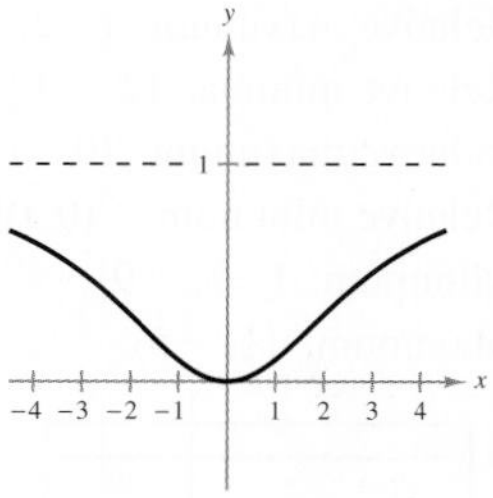

49.

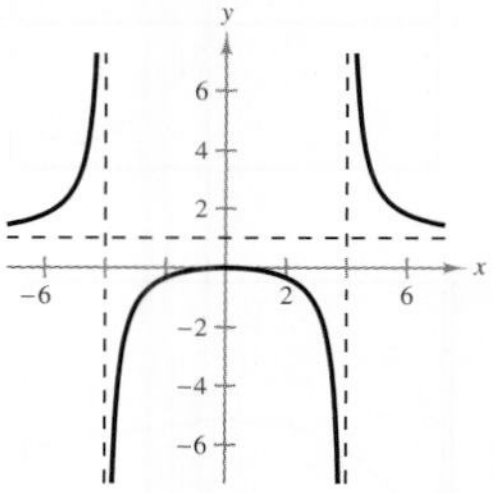

51.

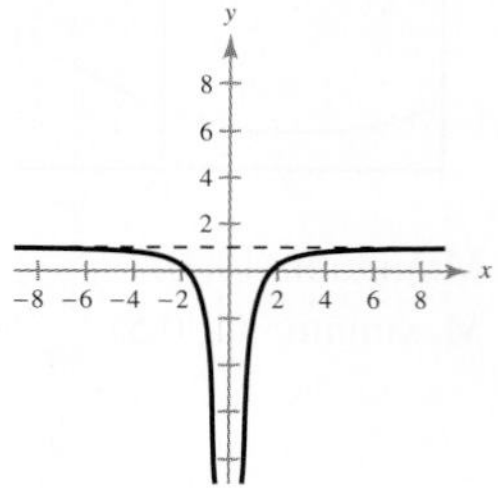

53.

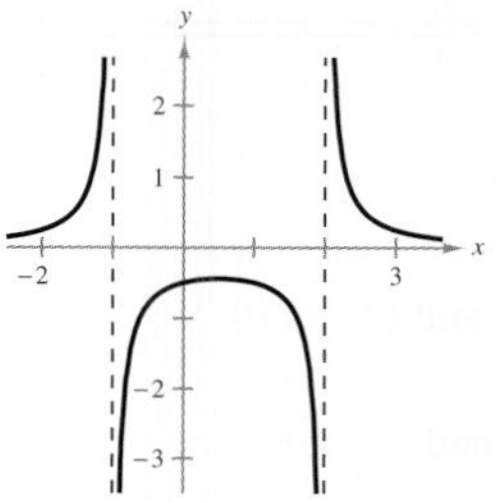

55.

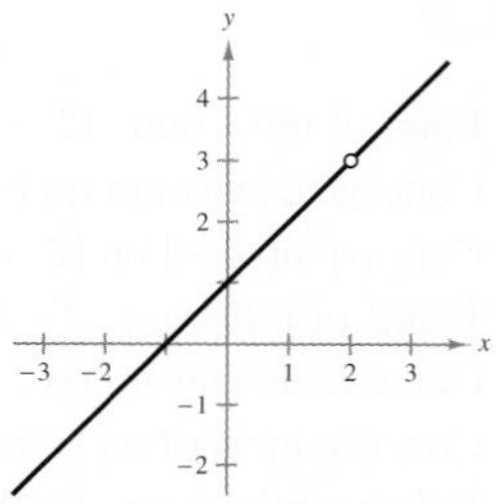

57.

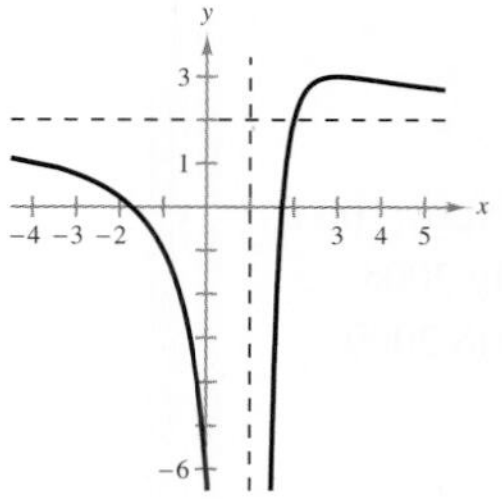

59.

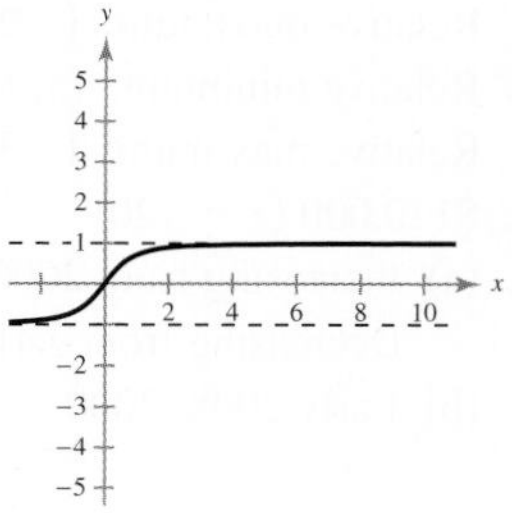

61. (a) $\overline{C} = 1.15 + \dfrac{6000}{x}$

(b) $\overline{C}(600) = 11.15$; $\overline{C}(6000) = 2.15$

(c) \$1.15; The cost approaches \$1.15 as the number of units produced increases.

63. (a) $\overline{P} = 35.4 - \dfrac{15{,}000}{x}$

(b) $\overline{P}(1000) = \$20.40$; $\overline{P}(10{,}000) = \$33.90$; $\overline{P}(100{,}000) = \$35.25$

(c) \$35.40; Explanations will vary.

65. (a) 25%: \$176 million; 50%: \$528 million; 75%: \$1584 million

(b) ∞; The limit does not exist, which means the cost increases without bound as the government approaches 100% seizure of the illegal drug entering the country.

67. (a)

n	1	2	3	4	5
P	0.5	0.74	0.82	0.86	0.89

n	6	7	8	9	10
P	0.91	0.92	0.93	0.94	0.95

(b) 1

(c)

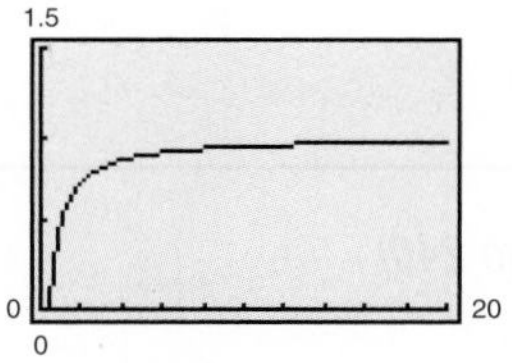

The percent of correct responses approaches 100% as the number of times the task is performed increases.

Section 3.7 *(page 233)*

Skills Warm Up *(page 233)*

1. Vertical asymptote: $x = 0$
Horizontal asymptote: $y = 0$

2. Vertical asymptote: $x = 2$
Horizontal asymptote: $y = 0$

3. Vertical asymptote: $x = -3$
Horizontal asymptote: $y = 40$

4. Vertical asymptotes: $x = 1, x = 3$
Horizontal asymptote: $y = 1$

5. Decreasing on $(-\infty, -2)$
Increasing on $(-2, \infty)$

6. Increasing on $(-\infty, -4)$
Decreasing on $(-4, \infty)$

7. Increasing on $(-\infty, -1)$ and $(1, \infty)$
Decreasing on $(-1, 1)$

8. Decreasing on $(-\infty, 0)$ and $(\sqrt[3]{2}, \infty)$
Increasing on $(0, \sqrt[3]{2})$

9. Increasing on $(-\infty, 1)$ and $(1, \infty)$

10. Decreasing on $(-\infty, -3)$ and $(\frac{1}{3}, \infty)$
Increasing on $(-3, \frac{1}{3})$

1.

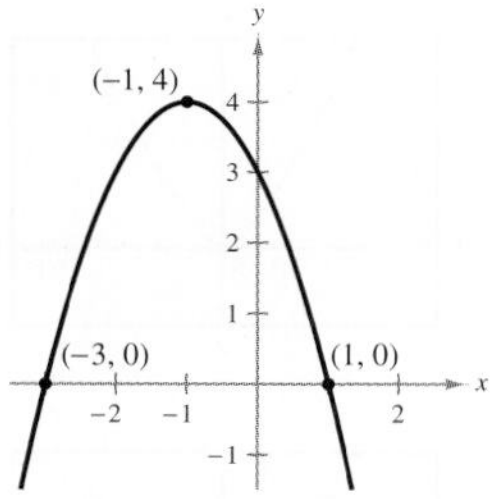

3.

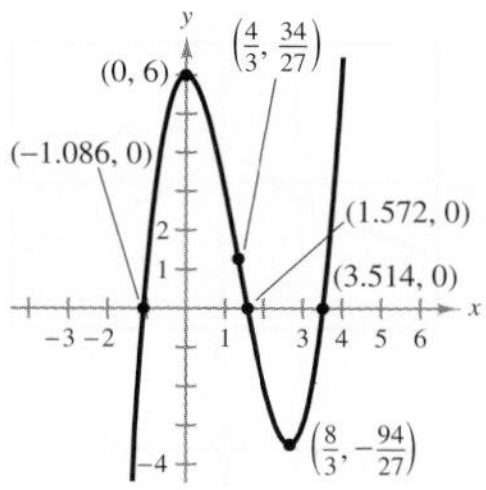

5.

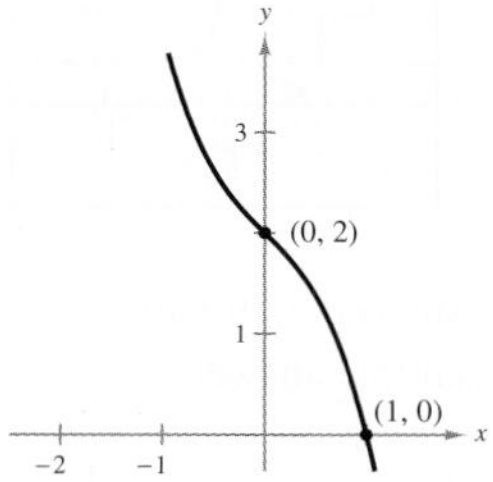

7.

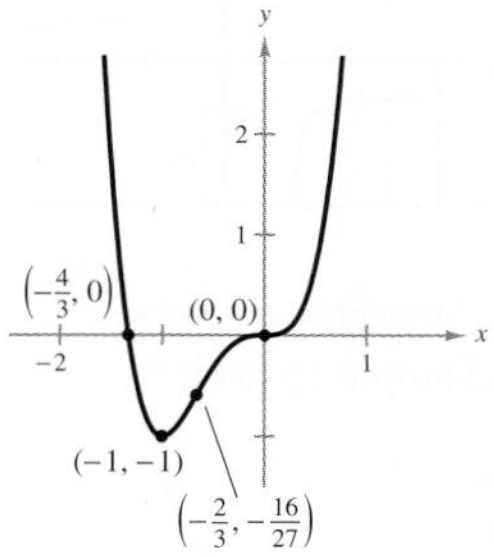

9.

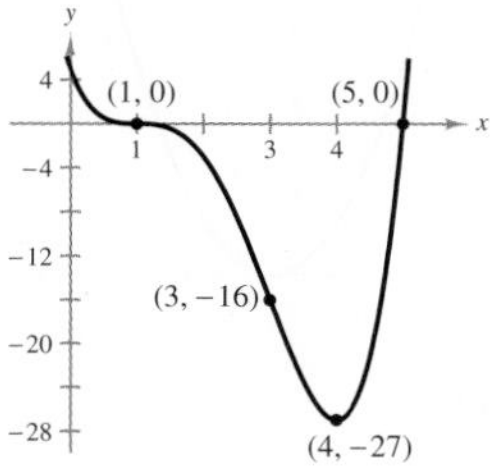

11.

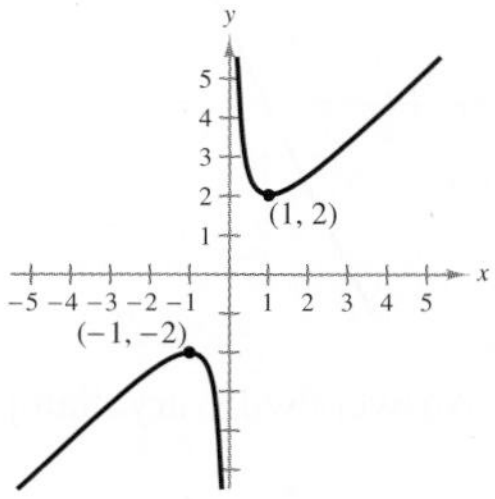

13.

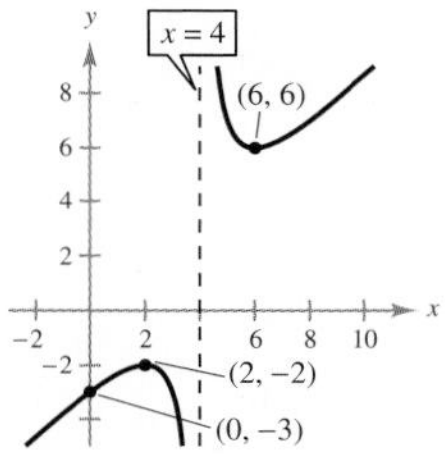

15.

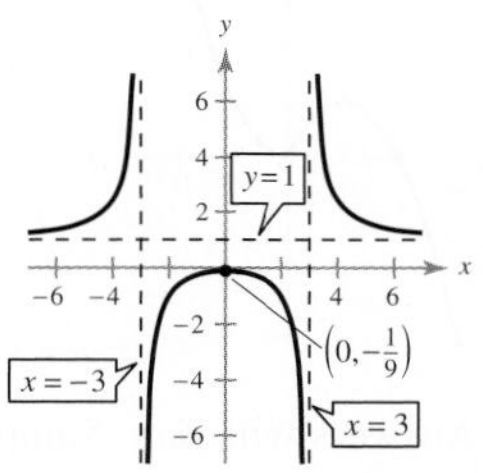

17.

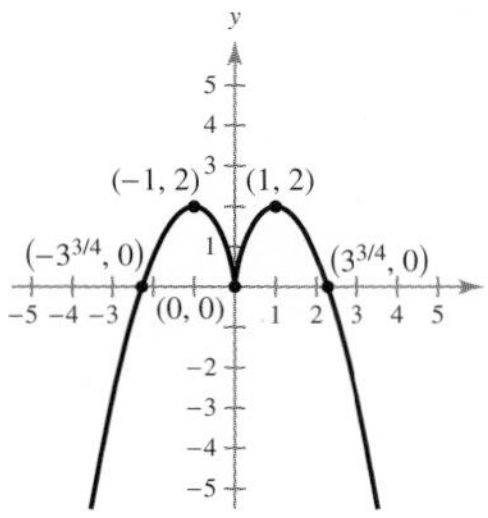

19.

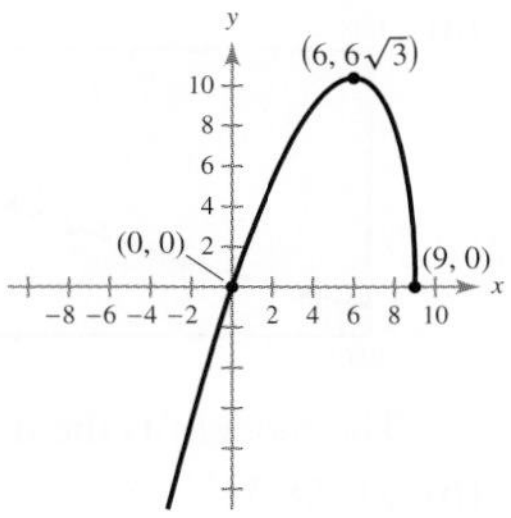

21.

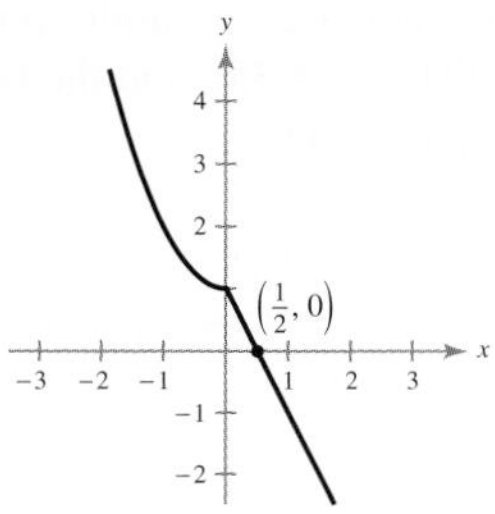

23.

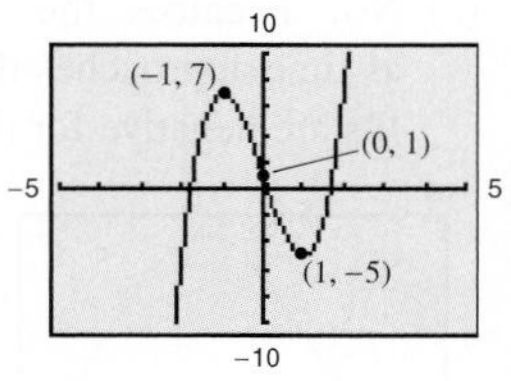

25.

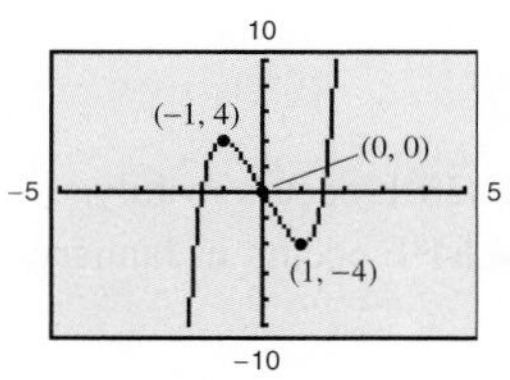

27.

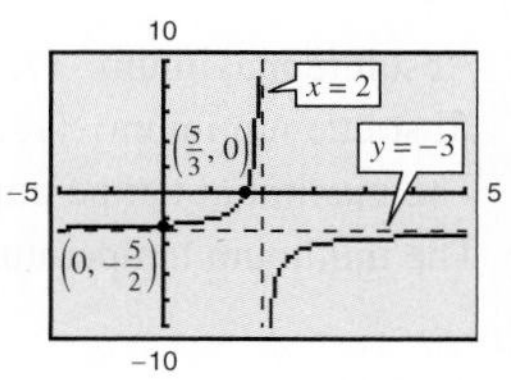

29.

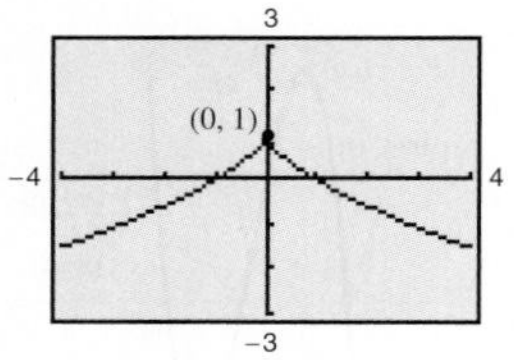

31.

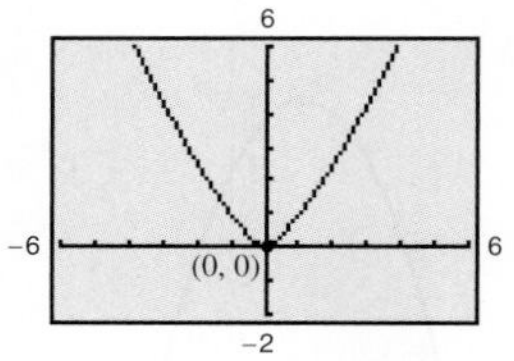

33.

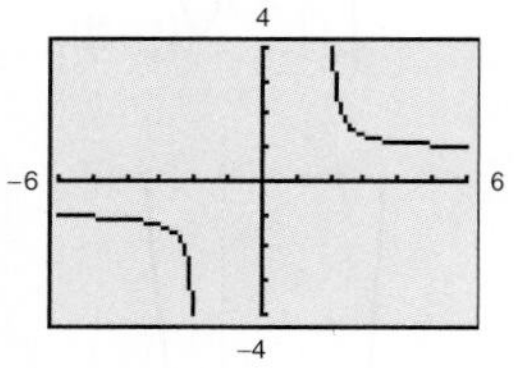

35.

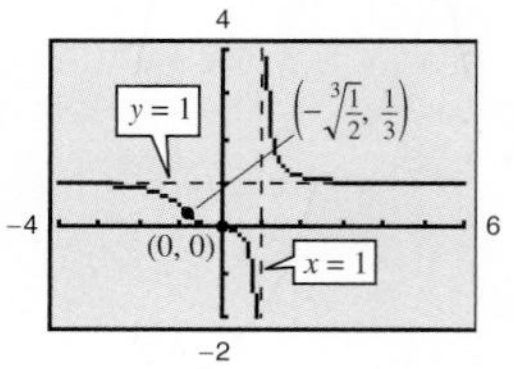

37. Answers will vary.
Sample answer:

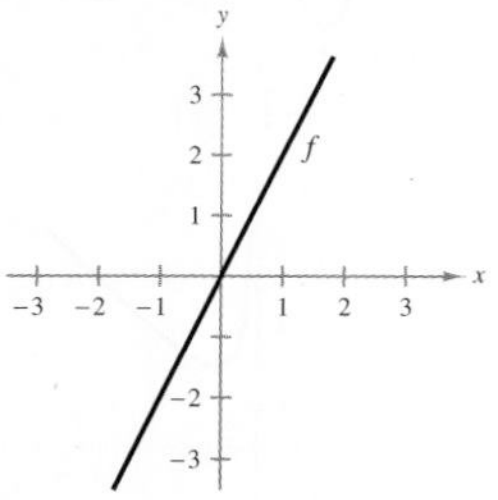

39. Answers will vary.
Sample answer:

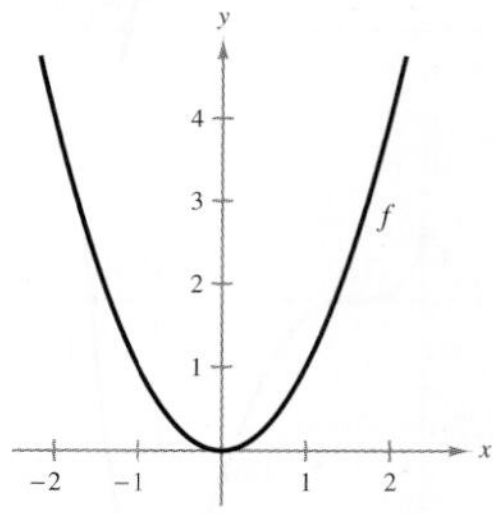

41. Answers will vary. Sample answer:

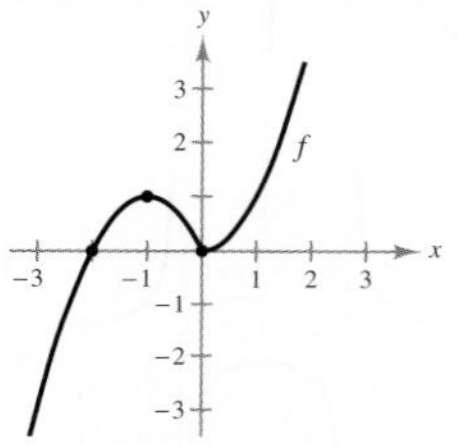

43. Answers will vary. Sample answer: $y = \dfrac{1}{x-5}$

45. (a)

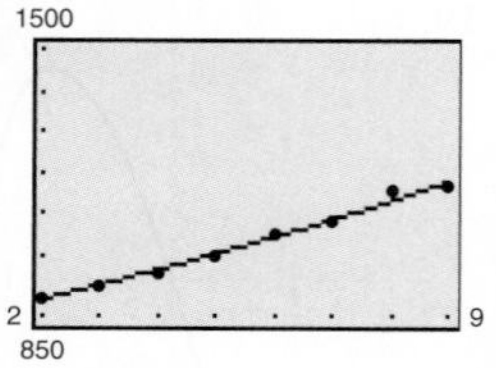

The model fits the data well.

(b) $1468.54

(c) No, because the benefits increase without bound as time approaches the year 2035 ($x = 35$), and the benefits are negative for the years past 2035.

47.

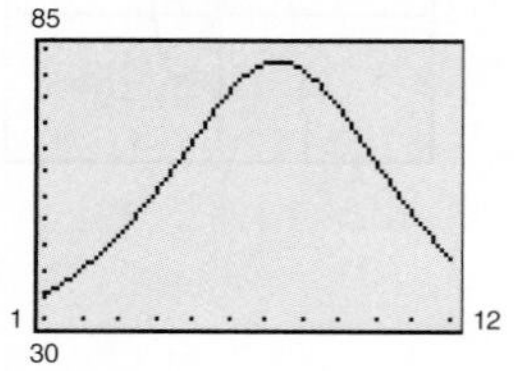

Absolute maximum: $(7, 82.28)$
Absolute minimum: $(1, 34.84)$
The maximum temperature of 82.28°F occurs in July.
The minimum temperature of 34.84°F occurs in January.

49.

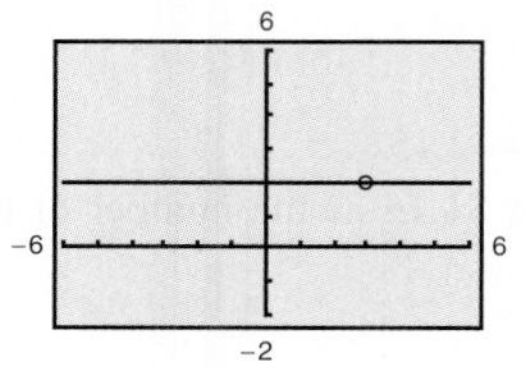

The rational function has the common factor $3 - x$ in the numerator and denominator. At $x = 3$, there is a hole in the graph, not a vertical asymptote.

51. (a) $f(x) = \dfrac{x^2 - 2x + 4}{x - 2} = \dfrac{x^2 - 2x}{x - 2} + \dfrac{4}{x - 2}$

$= \dfrac{x(x-2)}{x-2} + \dfrac{4}{x-2} = x + \dfrac{4}{x-2}$

(b)

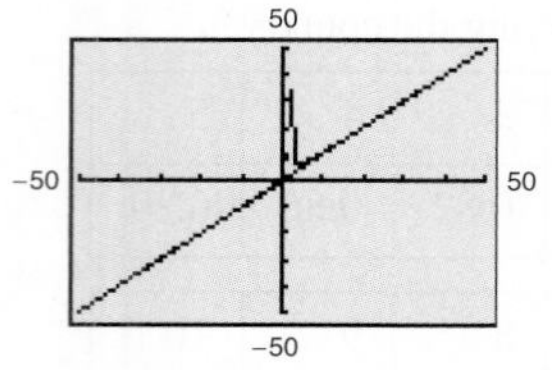

The graphs become almost identical as you zoom out.

(c) A slant asymptote is neither horizontal nor vertical. It is diagonal, following $y = x$.

Section 3.8 *(page 240)*

Skills Warm Up *(page 240)*

1. $\dfrac{dC}{dx} = 0.18x$ **2.** $\dfrac{dC}{dx} = 0.15$

3. $\dfrac{dR}{dx} = 1.25 + 0.03\sqrt{x}$ **4.** $\dfrac{dR}{dx} = 15.5 - 3.1x$

5. $\dfrac{dP}{dx} = -\dfrac{0.01}{\sqrt[3]{x^2}} + 1.4$ **6.** $\dfrac{dP}{dx} = -0.04x + 25$

7. $\dfrac{dA}{dx} = \dfrac{\sqrt{3}}{2}x$ **8.** $\dfrac{dA}{dx} = 12x$ **9.** $\dfrac{dC}{dr} = 2\pi$

10. $\dfrac{dP}{dw} = 4$ **11.** $\dfrac{dS}{dr} = 8\pi r$ **12.** $\dfrac{dP}{dx} = 2 + \sqrt{2}$

13. $A = \pi r^2$ **14.** $A = x^2$

15. $V = x^3$ **16.** $V = \frac{4}{3}\pi r^3$

1. $dy = 0.6$, $\Delta y = 0.6305$ **3.** $dy = -0.04$, $\Delta y \approx -0.0394$ **5.** $dy = 0.075$, $\Delta y \approx 0.0745$

7.

$dx = \Delta x$	dy	Δy	$\Delta y - dy$	$\dfrac{dy}{\Delta y}$
1.000	4.000	5.000	1.0000	0.8000
0.500	2.000	2.2500	0.2500	0.8889
0.100	0.400	0.4100	0.0100	0.9756
0.010	0.040	0.0401	0.0001	0.9975
0.001	0.004	0.0040	0.0000	1.0000

9.

$dx = \Delta x$	dy	Δy	$\Delta y - dy$	$\frac{dy}{\Delta y}$
1.000	-0.25000	-0.13889	0.11111	1.79999
0.500	-0.12500	-0.09000	0.03500	1.38889
0.100	-0.02500	-0.02324	0.00176	1.07573
0.010	-0.00250	-0.00248	0.00002	1.00806
0.001	-0.00025	-0.00025	0.00000	1.00000

11.

$dx = \Delta x$	dy	Δy	$\Delta y - dy$	$\frac{dy}{\Delta y}$
1.000	0.14865	0.12687	-0.02178	1.17167
0.500	0.07433	0.06823	-0.00610	1.08940
0.100	0.01487	0.01459	-0.00028	1.01919
0.010	0.00149	0.00148	-0.00001	1.00676
0.001	0.00015	0.00015	0.00000	1.00000

13. \$5.20 **15.** \$7.50 **17.** $-\$1250$ **19.** $dy = 24x^3\,dx$

21. $dy = 6x\,dx$ **23.** $dy = 12(4x - 1)^2\,dx$

25. $dy = -\frac{3}{(2x - 1)^2}\,dx$ **27.** $dy = \frac{-x}{\sqrt{9 - x^2}}\,dx$

29. $y = 28x + 37$
For $\Delta x = -0.01$, $f(x + \Delta x) = -19.281302$ and $y(x + \Delta x) = -19.28$
For $\Delta x = 0.01$, $f(x + \Delta x) = -18.721298$ and $y(x + \Delta x) = -18.72$

31. $y = x$
For $\Delta x = -0.01$, $f(x + \Delta x) = -0.009999$ and $y(x + \Delta x) = -0.01$
For $\Delta x = 0.01$, $f(x + \Delta x) = 0.009999$ and $y(x + \Delta x) = 0.01$

33. (a) $dP = \$1160$ (b) Actual: \$1122.50

35. (a) \$71.50 (b) \$40.00

37. Approximately 19 deer

39. $R = -\frac{1}{3}x^2 + 100x$; \$6

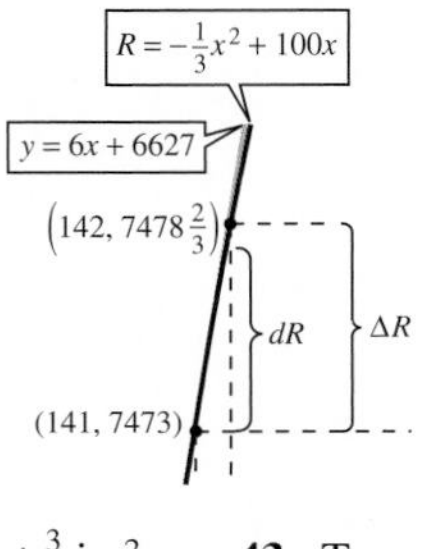

41. $\pm\frac{3}{4}$ in.^{2} **43.** True
2.08%

Review Exercises for Chapter 3 *(page 246)*

1. $x = 1$ **3.** $x = -3, x = 3$ **5.** $x = 1, x = \frac{7}{3}$

7. Critical number: $x = -\frac{1}{2}$
Increasing on $\left(-\frac{1}{2}, \infty\right)$
Decreasing on $\left(-\infty, -\frac{1}{2}\right)$

9. Critical numbers: $x = 0, x = 4$
Increasing on $(0, 4)$
Decreasing on $(-\infty, 0)$ and $(4, \infty)$

11. Critical number: $x = 1$
Increasing on $(1, \infty)$
Decreasing on $(-\infty, 1)$

13. The only critical number is $t \approx -10.85$. Any $t > -10.85$ produces a positive dR/dt, so the sales were increasing from 2004 to 2009.

15. Relative maximum: $(0, -2)$
Relative minimum: $(1, -4)$

17. Relative minimum: $(8, -52)$

19. Relative maxima: $(-1, 1), (1, 1)$
Relative minimum: $(0, 0)$

21. Relative maximum: $(0, 6)$

23. Relative maximum: $(0, 0)$
Relative minimum: $(4, 8)$

25. Maximum: $(0, 6)$
Minimum: $\left(-\frac{5}{2}, -\frac{1}{4}\right)$

27. Maxima: $(-2, 17), (4, 17)$
Minima: $(-4, -15), (2, -15)$

29. Maximum: $(1, 1)$
Minimum: $(9, -3)$

31. Maximum: $(1, 1)$
Minimum: $(-1, -1)$

33. $r \approx 1.58$ in.

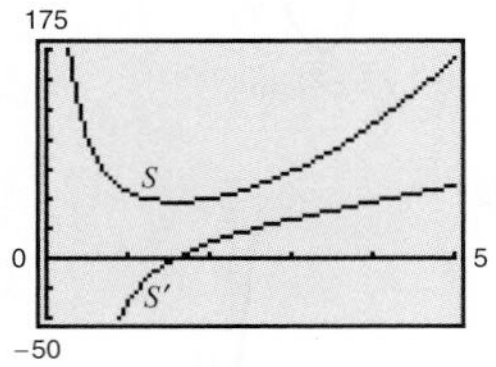

35. Concave upward on $(2, \infty)$
Concave downward on $(-\infty, 2)$

37. Concave upward on $\left(-\frac{2\sqrt{3}}{3}, \frac{2\sqrt{3}}{3}\right)$
Concave downward on $\left(-\infty, -\frac{2\sqrt{3}}{3}\right)$ and $\left(\frac{2\sqrt{3}}{3}, \infty\right)$

39. $(0, 0), (4, -128)$

41. $(0, 0), (1.0652, 4.5244), (2.5348, 3.5246)$

43. No relative extrema

45. Relative maximum: $\left(-\sqrt{3}, 6\sqrt{3}\right)$
Relative minimum: $\left(\sqrt{3}, -6\sqrt{3}\right)$

47. Relative maxima: $\left(-\frac{\sqrt{2}}{2}, \frac{1}{2}\right), \left(\frac{\sqrt{2}}{2}, \frac{1}{2}\right)$
Relative minimum: $(0, 0)$

49. $\left(50, 166\frac{2}{3}\right)$ **51.** $l = w = 15$ m **53.** 144 in.^{3}

55. $x = 900$ **57.** $x = 150$ **59.** (a) \$24 (b) \$8

61. (a) For $0 < x < 750$, $|\eta| > 1$ and the demand is elastic.
For $750 < x < 1500$, $|\eta| < 1$ and the demand is inelastic.
For $x = 750$, the demand has unit elasticity.
(b) From 0 to 750 units, revenue is increasing.
From 750 to 1500 units, revenue does not increase.

63. $x = -7, x = 0$ **65.** $x = -\frac{1}{2}$ **67.** $-\infty$

69. ∞ **71.** $y = \frac{2}{3}$ **73.** $y = 0$

CHAPTER 3

75. (a) $\overline{C} = 0.75 + \dfrac{4000}{x}$

(b) $\overline{C}(100) = 40.75$

$\overline{C}(1000) = 4.75$

(c) The limit is 0.75. As more and more units are produced, the average cost per unit will approach \$0.75.

77. (a) 20%: \$62.5 million

50%: \$250 million

90%: \$2250 million

(b) The limit is ∞, meaning that as the percent gets very close to 100, the cost grows without bound.

79.

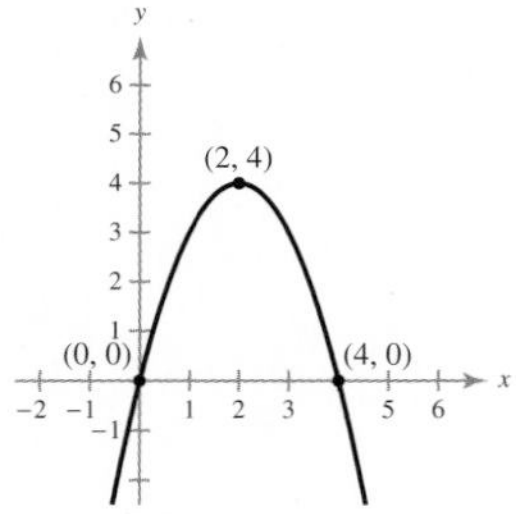

81.

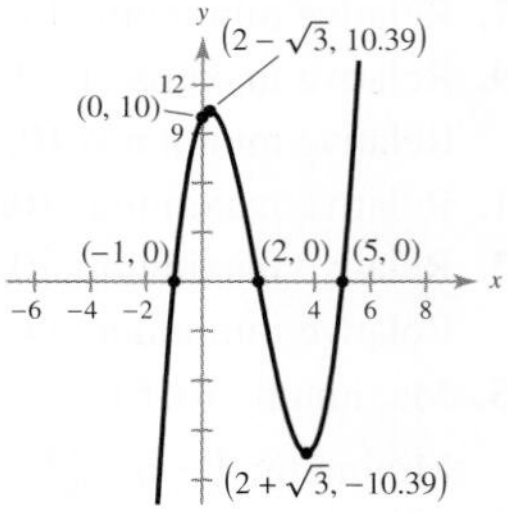

83.

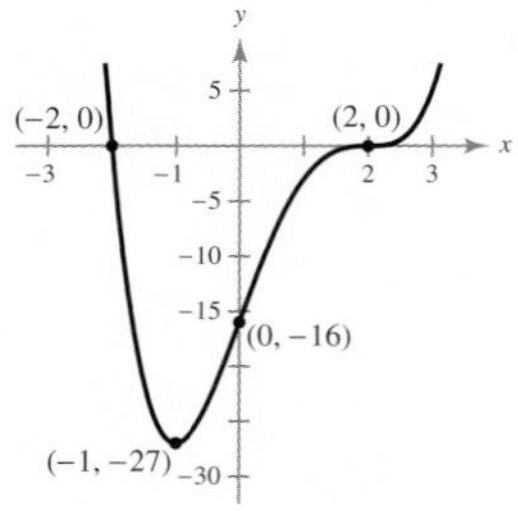

85.

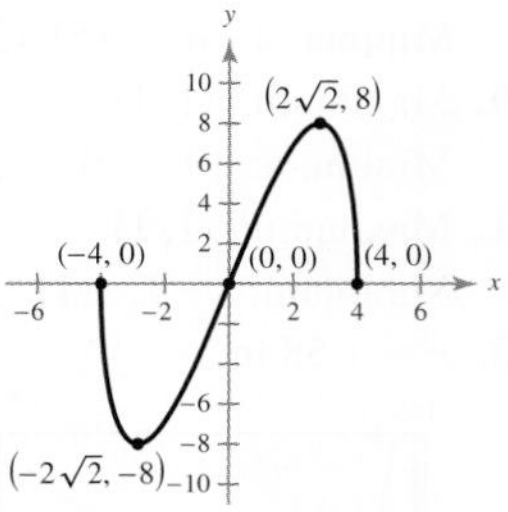

87.

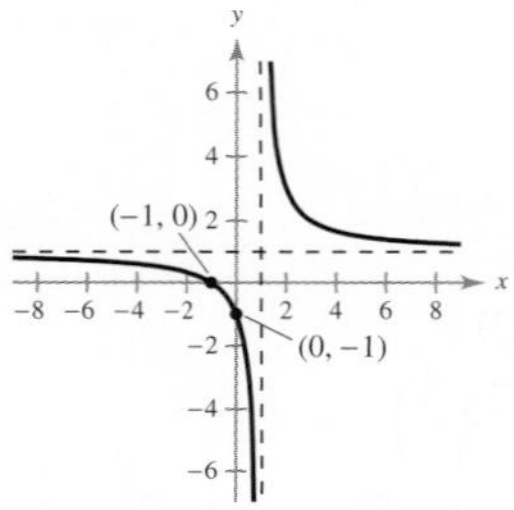

89.

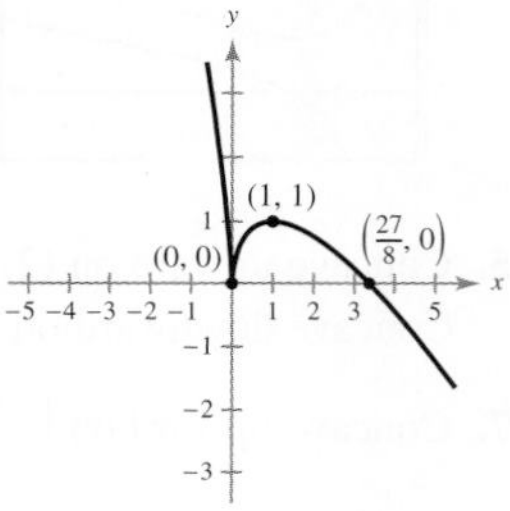

91. (a)

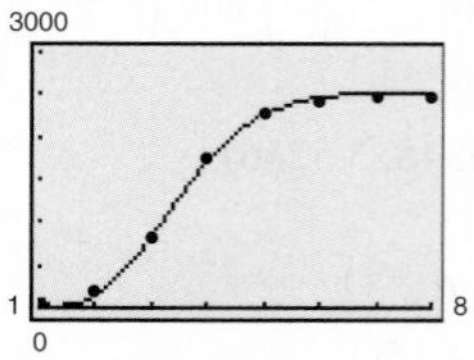

The model fits the data well.

(b) ≈ 2434 bacteria

(c) Answers will vary.

93. $dy = 0.08, \Delta y = 0.0802$ **95.** $dy = -2.1, \Delta y = -2.191$

97. \$800 **99.** \$15.25 **101.** $\approx$\$4.52

103. $dy = 1.5x^2\,dx$ **105.** $dy = 18x(3x^2 - 2)^2\,dx$

107. $dy = -\dfrac{7}{(x+5)^2}\,dx$

109. (a) \$164

(b) \$163.2, a difference of \$0.80.

111. $B = 0.1\sqrt{5w}$

$$\frac{dB}{dw} = \frac{0.05\sqrt{5}}{\sqrt{w}}$$

$$\Delta B \approx dB = \frac{0.05\sqrt{5}}{\sqrt{w}}\,dw = \frac{0.05\sqrt{5}}{\sqrt{90}}(5) \approx 0.059 \text{ m}^2$$

Test Yourself *(page 250)*

1. Critical number: $x = 0$
Increasing on $(0, \infty)$
Decreasing on $(-\infty, 0)$

2. Critical numbers: $x = -2, x = 2$
Increasing on $(-\infty, -2)$ and $(2, \infty)$
Decreasing on $(-2, 2)$

3. Critical number: $x = 5$
Increasing on $(5, \infty)$
Decreasing on $(-\infty, 5)$

4. Relative minimum: $(3, -14)$
Relative maximum: $(-3, 22)$

5. Relative minima: $(-1, -7)$ and $(1, -7)$
Relative maximum: $(0, -5)$

6. Relative maximum: $(0, 2.5)$

7. Minimum: $(-3, -1)$
Maximum: $(0, 8)$

8. Minimum: $(0, 0)$
Maximum: $(2.25, 9)$

9. Minimum: $\left(2\sqrt{3}, 2\sqrt{3}\right)$
Maximum: $(1, 6.5)$

10. Concave upward: $(2, \infty)$
Concave downward: $(-\infty, 2)$

11. Concave upward: $\left(-\infty, -\dfrac{2\sqrt{2}}{3}\right)$ and $\left(\dfrac{2\sqrt{2}}{3}, \infty\right)$
Concave downward: $\left(-\dfrac{2\sqrt{2}}{3}, \dfrac{2\sqrt{2}}{3}\right)$

12. No point of inflection
The graph is concave upward on its entire domain.

13. Concave upward: $(-\infty, -3)$ and $(3, \infty)$
Concave downward: $(-3, 3)$
Points of inflection: $(-3, -175)$ and $(3, -175)$

14. Relative minimum: $(6, -166)$
Relative maximum: $(-2, 90)$

15. Relative minimum: $(3, -97.2)$
Relative maximum: $(-3, 97.2)$

16. Vertical asymptote: $x = 5$
Horizontal asymptote: $y = 3$

17. Horizontal asymptote: $y = 2$

18. Vertical asymptote: $x = 1$

19.

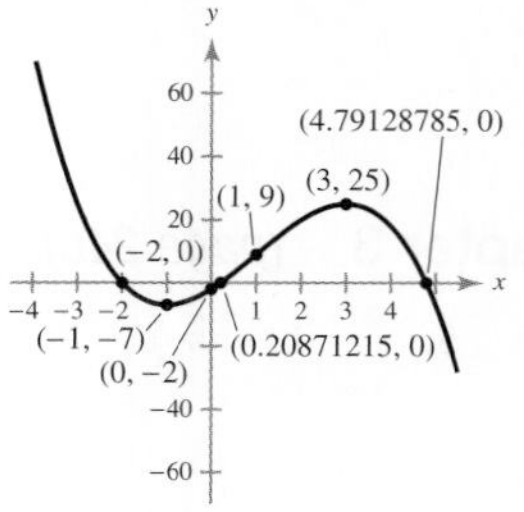

20.

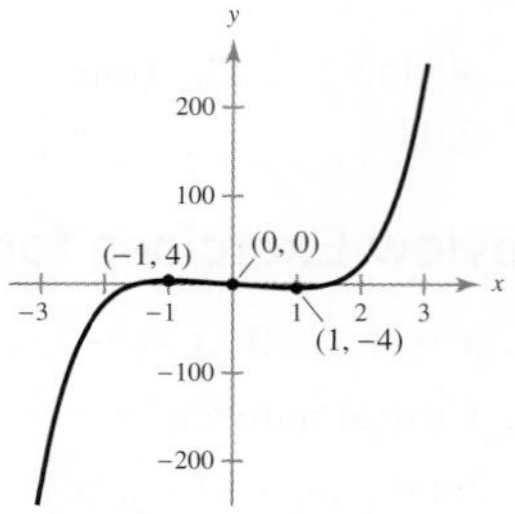

21.

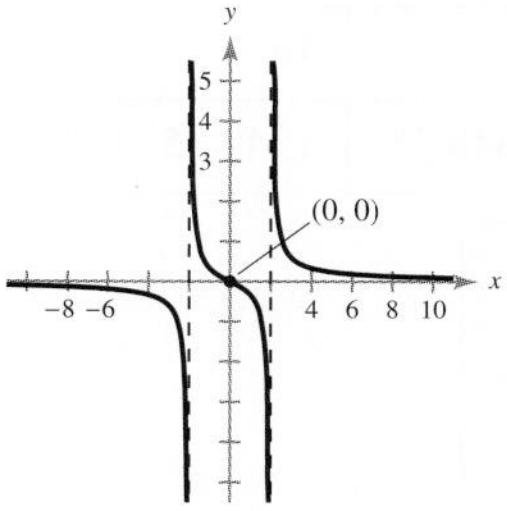

22. $dy = 10x\,dx$

23. $dy = \dfrac{-4}{(x+3)^2}\,dx$ **24.** $dy = 3(x+4)^2\,dx$

25. For $0 \le x < 350$, $|\eta| > 1$ and the demand is elastic.
For $350 < x \le 700$, $|\eta| < 1$ and the demand is inelastic.
For $x = 350$, the demand has unit elasticity.

Chapter 4

Section 4.1 *(page 256)*

Skills Warm Up *(page 256)*

1. Horizontal shift to the left two units
2. Reflection about the x-axis
3. Vertical shift down one unit
4. Reflection about the y-axis
5. Horizontal shift to the right one unit
6. Vertical shift up two units

7. 125 **8.** 22.63 **9.** 9 **10.** $\frac{1}{125}$
11. $\frac{1}{2}$ **12.** $\frac{25}{64}$ **13.** 5 **14.** $\frac{4}{3}$ **15.** $-9, 1$
16. $2 \pm 2\sqrt{2}$ **17.** $1, -5$ **18.** $\frac{1}{2}, 1$

1. (a) 3125 (b) $\frac{1}{5}$ (c) 625 (d) $\frac{1}{125}$
3. (a) $\frac{1}{5}$ (b) 27 (c) 5 (d) 4096
5. 2 g

7.

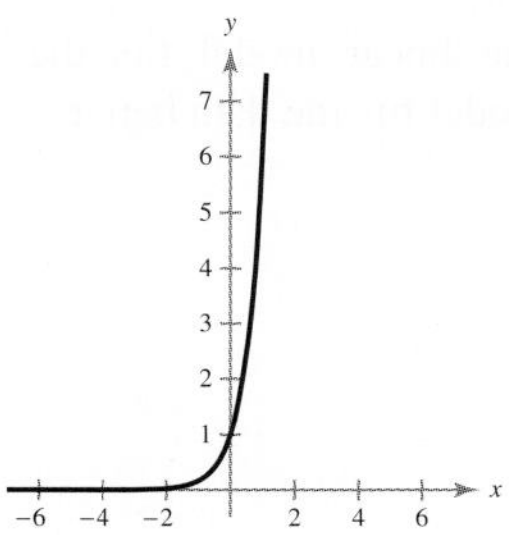

9.

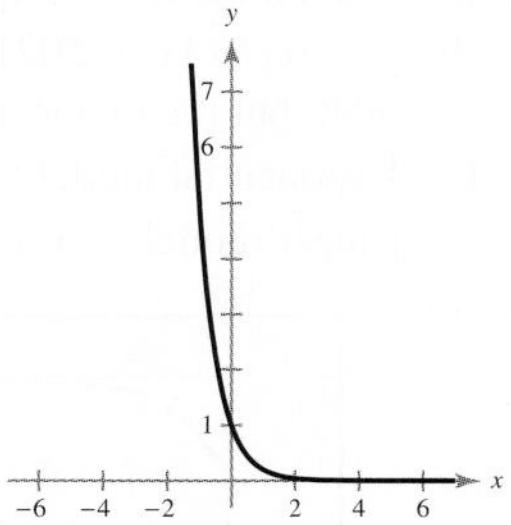

11.

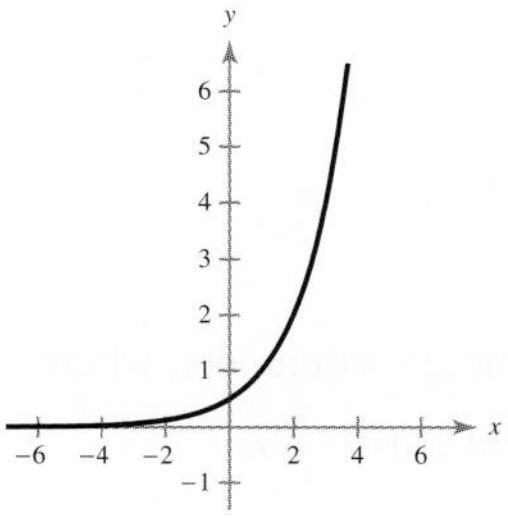

13.

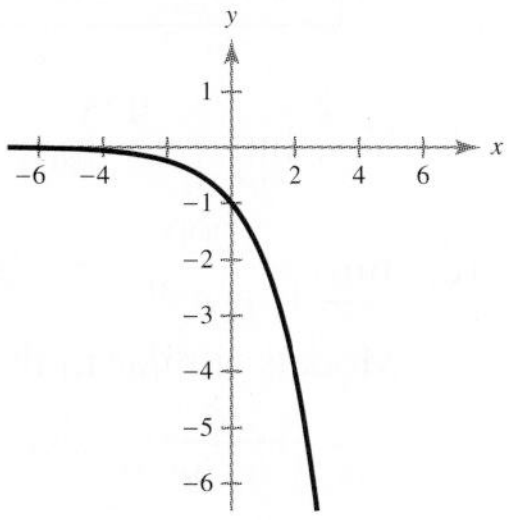

15.

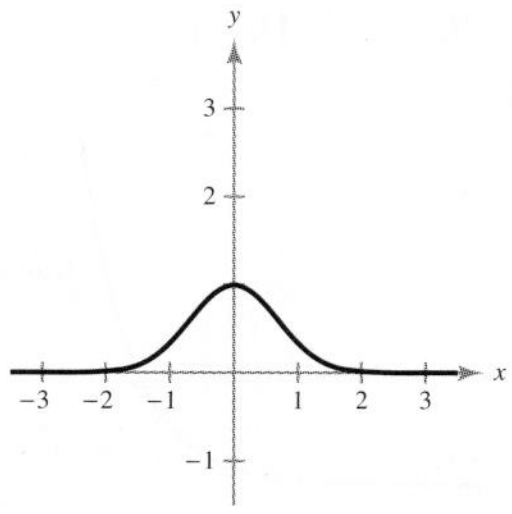

17.

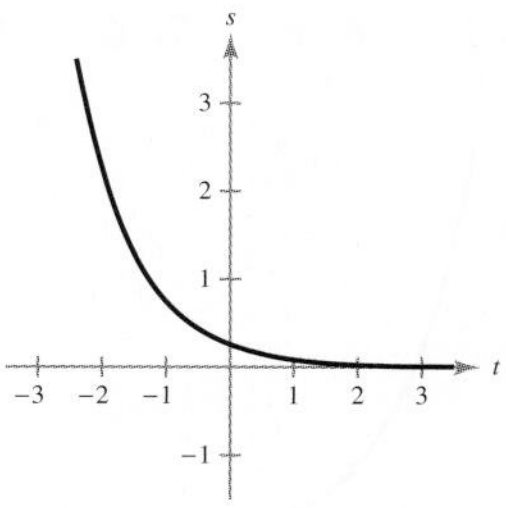

19. (a) $P(23) \approx 320.26$ million
(b) $P(30) \approx 343.36$ million

21. (a) $V(5) \approx \$80{,}634.95$ (b) $V(20) \approx \$161{,}269.89$

23. \$36.93

25. $V(t) = 28{,}000\left(\frac{3}{4}\right)^t$

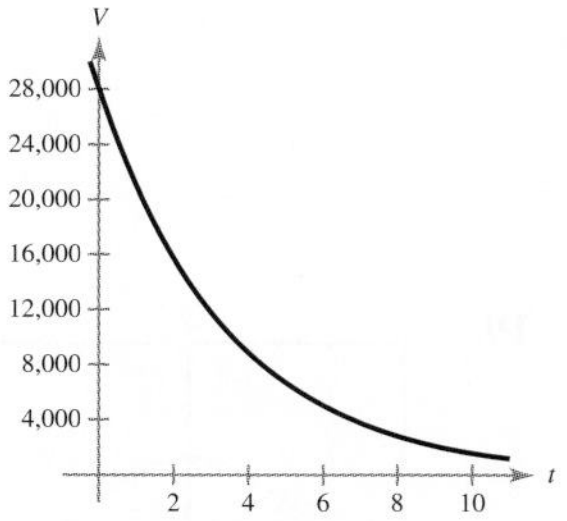

$V(4) = 28{,}000\left(\frac{3}{4}\right)^4 \approx 8859.38$

27. (a)

Year	2001	2002	2003	2004
Actual	37,188	38,221	39,165	40,201
Model	36,966	37,998	39,058	40,148

Year	2005	2006	2007	2008
Actual	40,520	41,746	43,277	46,025
Model	41,268	42,419	43,603	44,819

The model fits the data well. Explanations will vary.

(b)

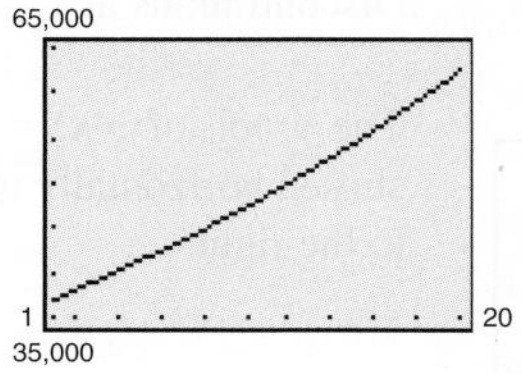

(c) 2014

Section 4.2 *(page 264)*

Skills Warm Up *(page 264)*

1. Continuous on $(-\infty, \infty)$
2. Discontinuous at $x = \pm 2$
3. Discontinuous at $x = \pm\sqrt{3}$
4. Removable discontinuity at $x = 4$

5. $y = 0$ **6.** $y = 0$ **7.** $y = 4$ **8.** $y = \frac{1}{2}$
9. $y = \frac{3}{2}$ **10.** $y = 6$ **11.** $y = 0$ **12.** $y = 0$

1. (a) e^7 (b) e^{12} (c) $\dfrac{1}{e^6}$ (d) 1

3. (a) e^5 (b) $e^{5/2}$ (c) e^6 (d) e^7

5. f **6.** e **7.** d **8.** b **9.** c **10.** a

11.

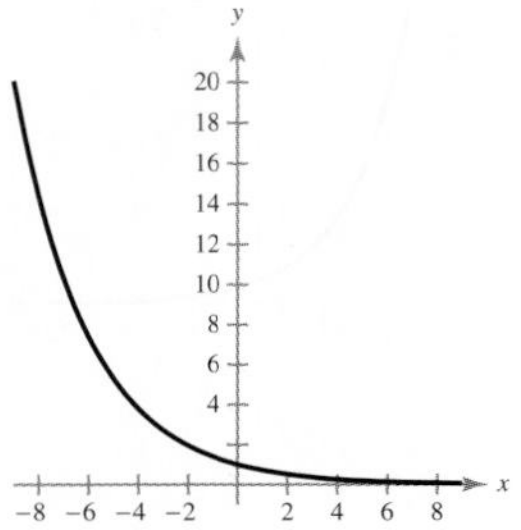

13.

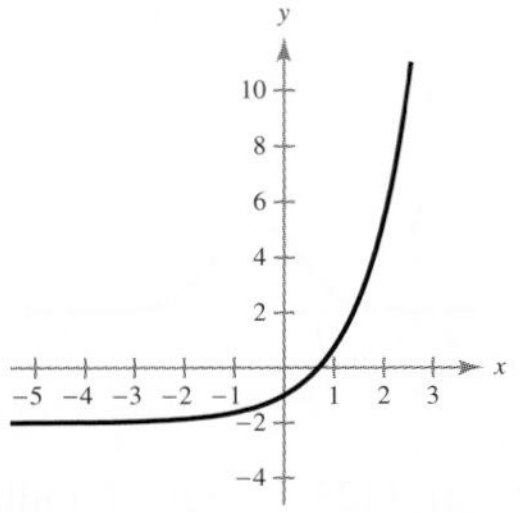

15.

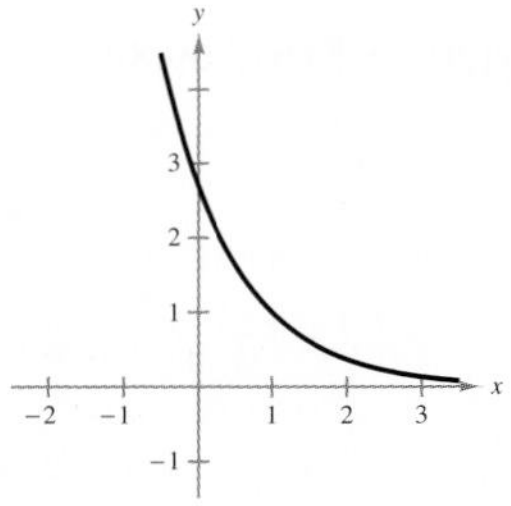

17.

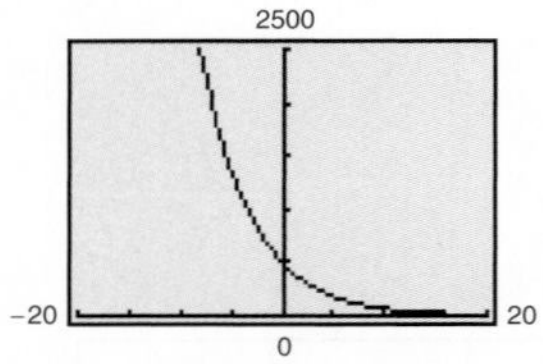

Horizontal asymptote: $N = 0$
Continuous on the entire real number line

19.

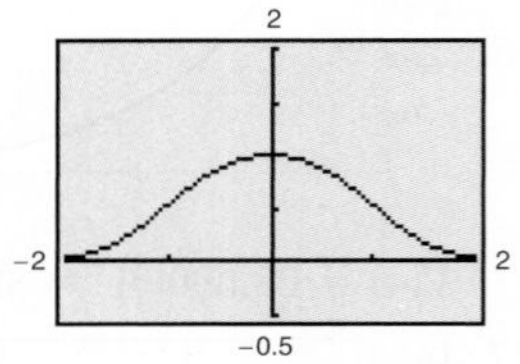

Horizontal asymptote: $g = 0$
Continuous on the entire real number line

21.

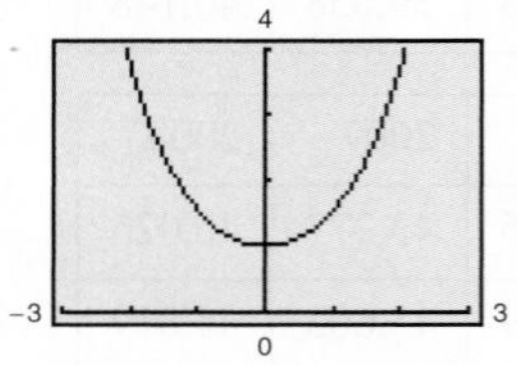

No horizontal asymptotes
Continuous on the entire real number line

23.

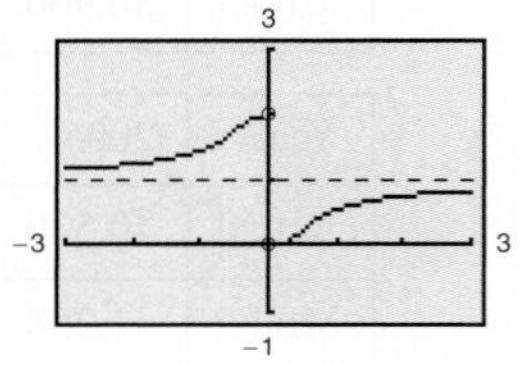

Horizontal asymptote: $y = 1$
Discontinuous at $x = 0$

25. (a)

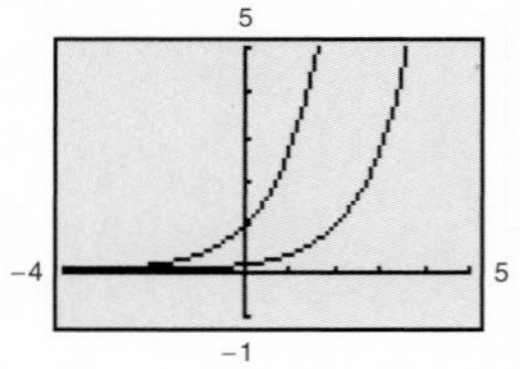

The graph of $g(x) = e^{x-2}$ is shifted horizontally two units to the right.

(b)

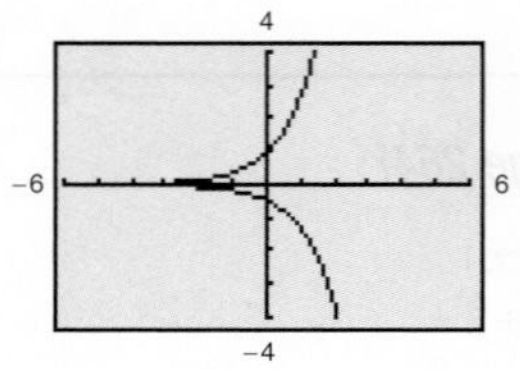

The graph of $h(x) = -\frac{1}{2}e^x$ decreases at a rate slower than the rate at which the graph of $f(x) = e^x$ increases.

(c)

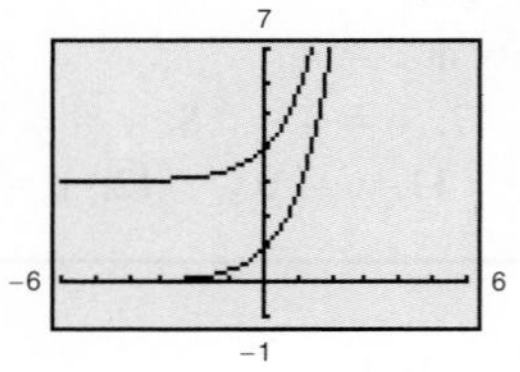

The graph of $q(x) = e^x + 3$ is shifted vertically three units upward.

27.

n	1	2	4	12
A	1343.92	1346.86	1348.35	1349.35

n	365	Continuous compounding
A	1349.84	1349.86

29.

n	1	2	4	12
A	2191.12	2208.04	2216.72	2222.58

n	365	Continuous compounding
A	2225.44	2225.54

31.

t	1	10	20
P	96,078.94	67,032.00	44,932.90

t	30	40	50
P	30,119.42	20,189.65	13,533.53

33.

t	1	10	20
P	95,132.82	60,716.10	36,864.45

t	30	40	50
P	22,382.66	13,589.88	8251.24

35. \$107,311.12

37. (a) 9% (b) 9.20% (c) 9.31% (d) 9.38%

39. \$6450.04

41. (a) \$849.53 (b) \$421.12 (c) $\lim_{x \to \infty} p = 0$

43. (a) 0.1535 (b) 0.4866 (c) 0.8111

45. (a) The model fits the data well.
(b) $y = 637.11x + 5021.1$; The linear model fits the data well, but the exponential model fits the data better.
(c) Exponential model: 2016
Linear model: 2021

47. (a)

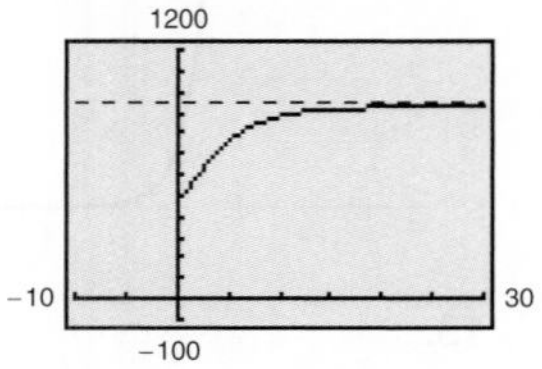

(b) Yes, $\lim_{t \to \infty} \dfrac{925}{1 + e^{-0.3t}} = 925$

(c) $\lim_{t \to \infty} \dfrac{1000}{1 + e^{-0.3t}} = 1000$

Models similar to this logistic growth model, where $y = \dfrac{a}{1 + be^{-ct}}$ have a limit of a as $t \to \infty$.

49. (a) 0.536 (b) 0.666
(c)
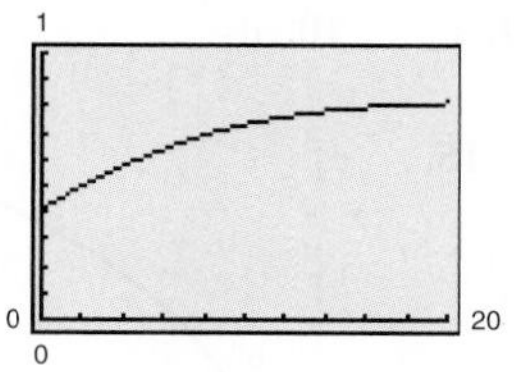

(d) Yes, $\lim_{n\to\infty} \frac{0.83}{1 + e^{-0.2n}} = 0.83$

51. Amounts earned:
(a) \$5267.71 (b) \$5255.81 (c) \$5243.23
You should choose the certificate of deposit in part (a) because it earns more money than the others.

Section 4.3 *(page 273)*

Skills Warm Up *(page 273)*

1. $\frac{1}{2}e^x(2x^2 - 1)$ **2.** $\frac{e^x(x + 1)}{x}$ **3.** $e^x(x - e^x)$

4. $e^{-x}(e^{2x} - x)$ **5.** $-\frac{6}{7x^3}$ **6.** $6x - \frac{1}{6}$

7. $6(2x^2 - x + 6)$ **8.** $\frac{t + 2}{2t^{3/2}}$

9. Relative maximum: $\left(-\frac{4\sqrt{3}}{3}, \frac{16\sqrt{3}}{9}\right)$

Relative minimum: $\left(\frac{4\sqrt{3}}{3}, -\frac{16\sqrt{3}}{9}\right)$

10. Relative maximum: $(0, 5)$
Relative minima: $(-1, 4), (1, 4)$

1. 0 **3.** $5e^{5x}$ **5.** $-2xe^{-x^2}$ **7.** $\frac{2}{x^3}e^{-1/x^2}$

9. $e^{4x}(4x^2 + 2x + 4)$ **11.** $-\frac{6(e^x - e^{-x})}{(e^x + e^{-x})^4}$

13. $-\frac{2e^x}{(e^x - 1)^2}$ **15.** $xe^x + e^x + 4e^{-x}$ **17.** 4 **19.** -3

21. $y = 2x - 3$ **23.** $y = \frac{4}{e^2}$ **25.** $y = 24x + 8$

27. $\frac{dy}{dx} = \frac{10 - e^y}{xe^y + 3}$ **29.** $\frac{dy}{dx} = \frac{e^{-x}(x^2 - 2x) + y}{4y - x}$

31. $6(3e^{3x} + 2e^{-2x})$ **33.** $32e^{4x}(x + 1)$

35.
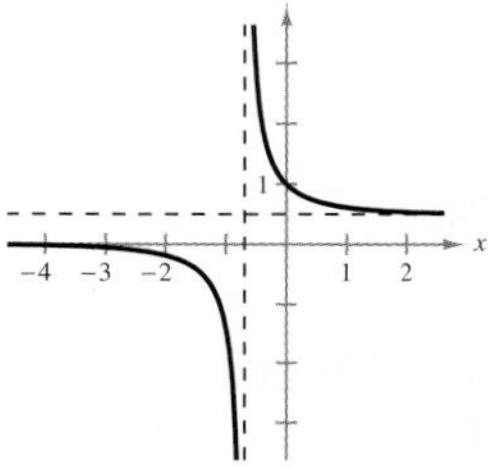

No relative extrema
No points of inflection
Horizontal asymptote to the right: $y = \frac{1}{2}$
Horizontal asymptote to the left: $y = 0$
Vertical asymptote: $x \approx -0.693$

37.
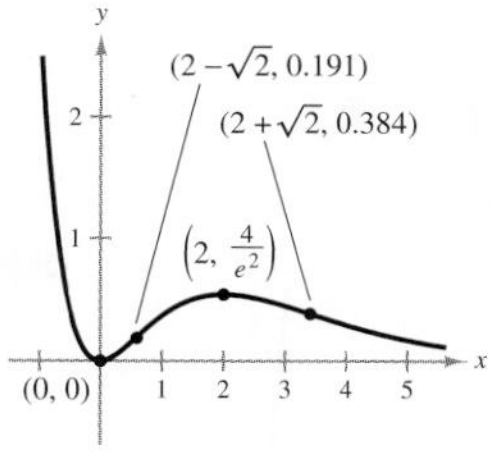

Relative minimum: $(0, 0)$
Relative maximum: $\left(2, \frac{4}{e^2}\right)$
Points of inflection:
$(2 - \sqrt{2}, 0.191)$,
$(2 + \sqrt{2}, 0.384)$

Horizontal asymptote to the right: $y = 0$

39. $x = -\frac{1}{3}$ **41.** $x = 9$

43. (a)
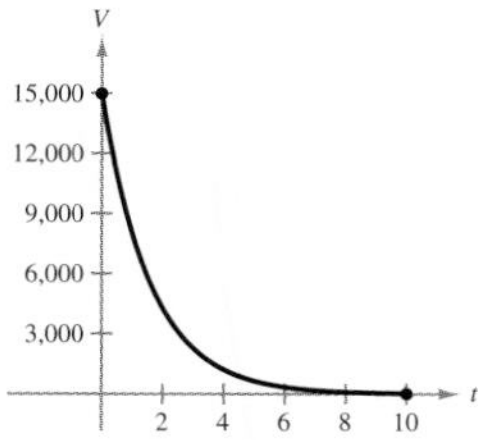

(b) $-\$5028.84$/yr
(c) $-\$406.89$/yr
(d) $V = -1497.2t + 15{,}000$
(e) In the exponential function, the initial rate of depreciation is greater than in the linear model. The linear model has a constant rate of depreciation.

45. (a)
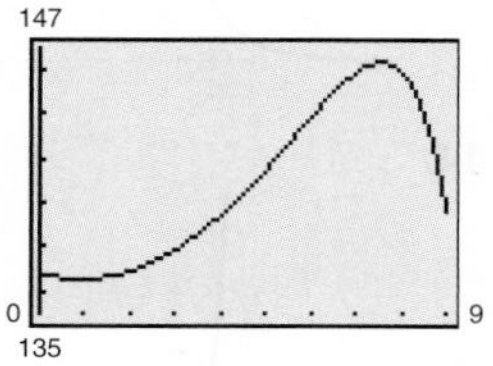

(b and c) 2000: -0.59 million people/yr
2004: 1.83 million people/yr
2009: -12.44 million people/yr

47. (a) \$433.31 per year
(b) \$890.22 per year
(c) \$21,839.26 per year

49. (a) $f(x) = \frac{1}{116\sqrt{2\pi}}e^{-(x-516)^2/26{,}912}$
(b)
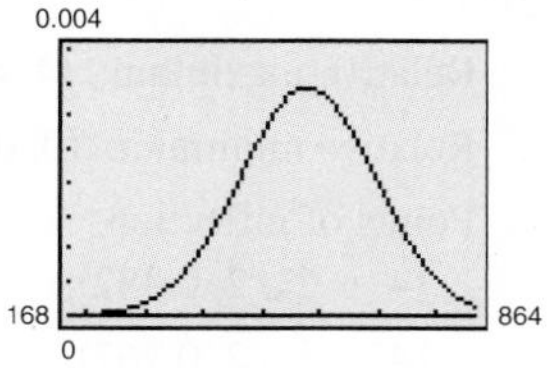

(c) $f'(x) = \frac{-1}{1{,}560{,}896\sqrt{2\pi}}(x - 516)e^{-(x-516)^2/26{,}912}$
(d) Answers will vary.

51.
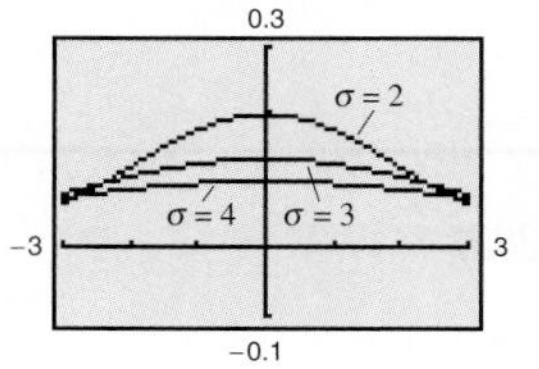

As σ increases, the graph becomes flatter.

53. Proof; maximum: $\left(0, \frac{1}{\sigma\sqrt{2\pi}}\right)$; answers will vary.

Sample answer:
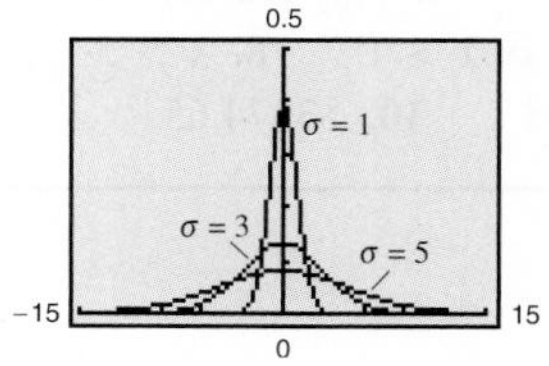

Quiz Yourself (page 275)

1. 1024 **2.** 216 **3.** 27 **4.** $\sqrt{15}$
5. e^7 **6.** $e^{11/3}$ **7.** e^6 **8.** e^3

9.

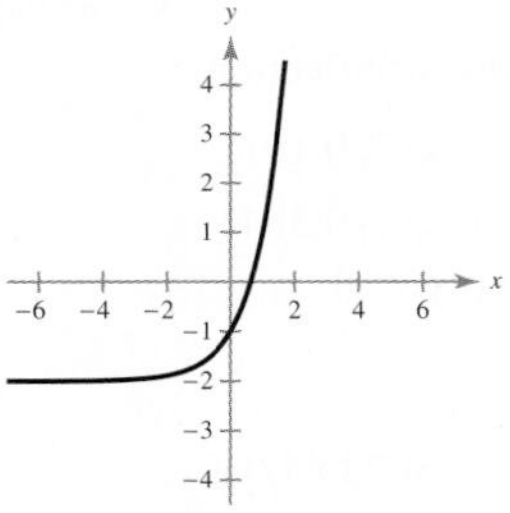

10.

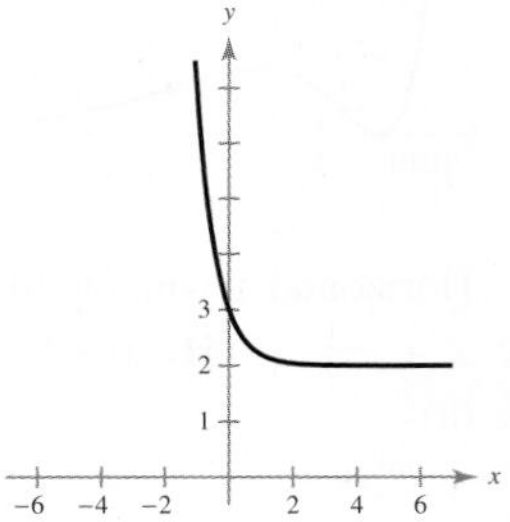

11.

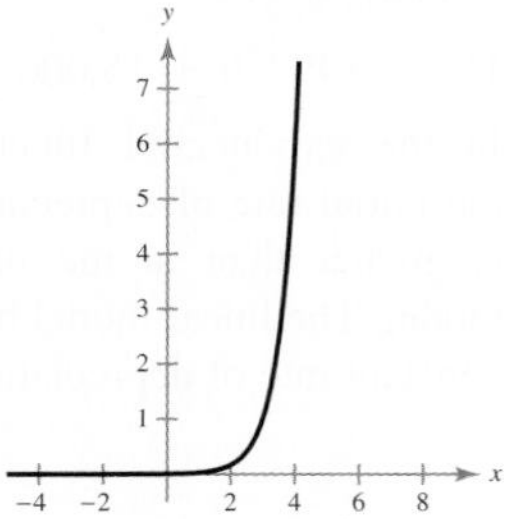

12.

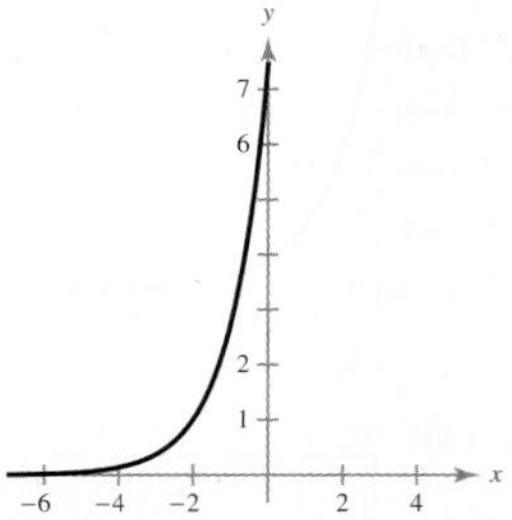

13.

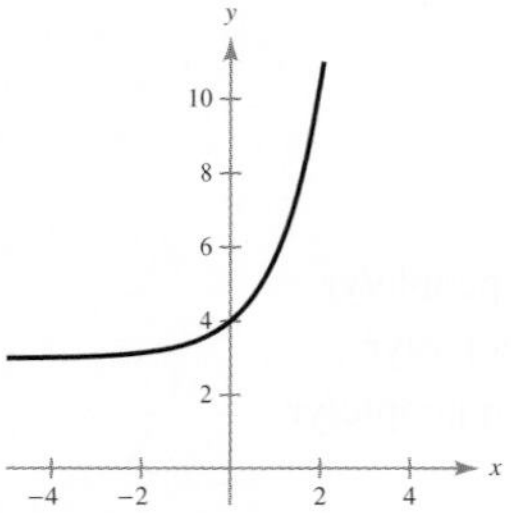

14.

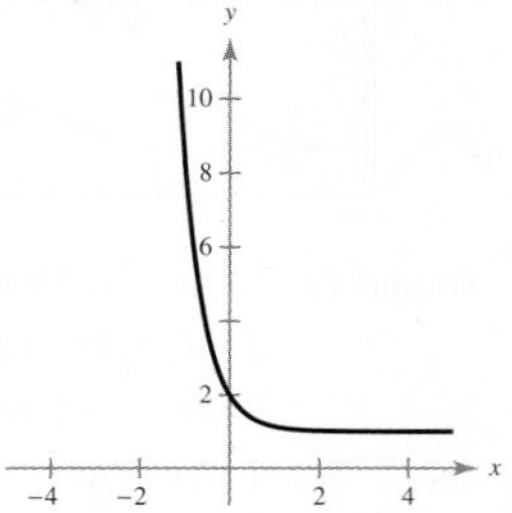

15. 22.69 grams **16.** \$31.06
17. (a) \$3571.02 (b) \$3572.83 (c) \$3573.74
18. \$10,379.21 **19.** $5e^{5x}$ **20.** e^{x-4} **21.** $5e^{x+2}$
22. $e^x(2 - x)$ **23.** $y = -2x + 1$

24.

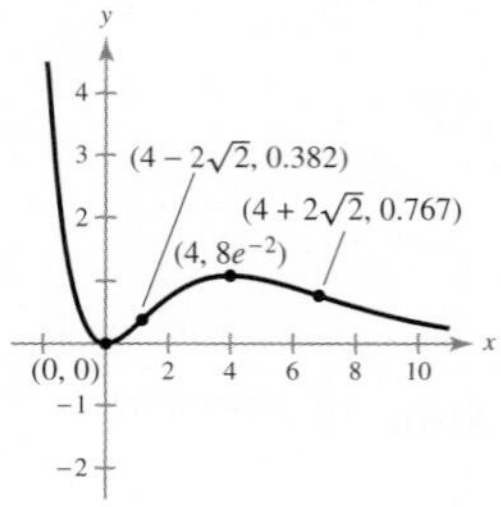

Relative maximum: $(4, 8e^{-2})$
Relative minimum: $(0, 0)$
Points of inflection:
$(4 - 2\sqrt{2}, 0.382)$,
$(4 + 2\sqrt{2}, 0.767)$
Horizontal asymptote to the right: $y = 0$

Section 4.4 (page 282)

Skills Warm Up (page 282)

1. $f^{-1}(x) = \frac{1}{5}x$ **2.** $f^{-1}(x) = x + 6$
3. $f^{-1}(x) = \dfrac{x - 2}{3}$ **4.** $f^{-1}(x) = \frac{4}{3}(x + 9)$
5. $x > -4$ **6.** Any real number x
7. $x < -1$ or $x > 1$ **8.** $x > 5$
9. \$3462.03 **10.** \$3374.65

1. $e^{0.6931. . .} = 2$ **3.** $e^{-1.6094. . .} = 0.2$ **5.** $\ln 1 = 0$
7. $\ln(0.0498. . .) = -3$ **9.** c **10.** d
11. b **12.** a

13.

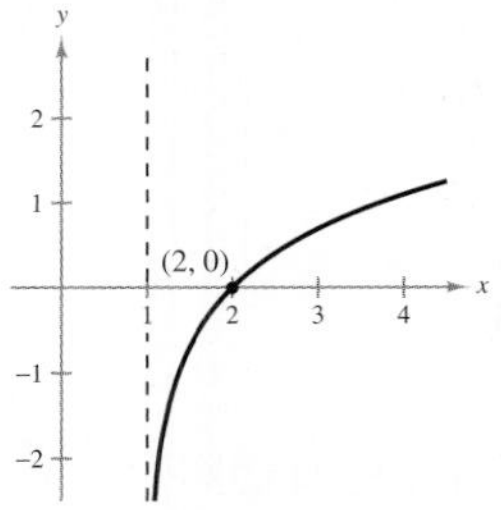

15.

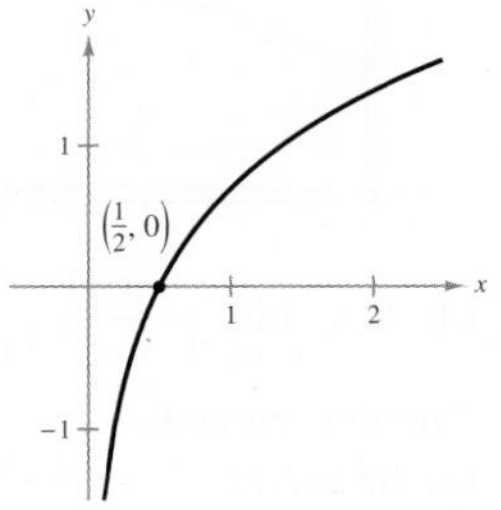

17.

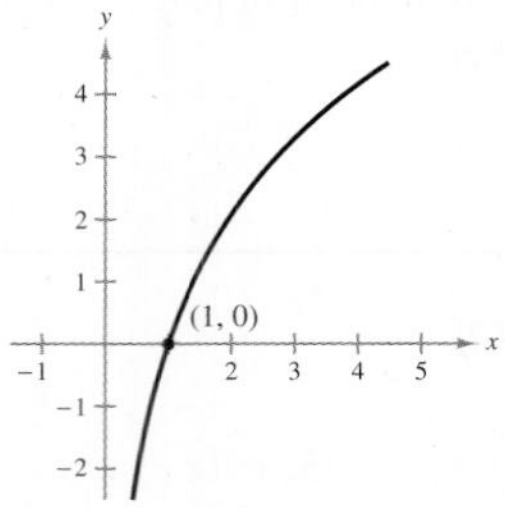

19. x^2 **21.** $5x + 2$ **23.** $2x - 1$
25. $\ln 2 - \ln 3$ **27.** $\ln x + \ln y + \ln z$
29. $\frac{1}{3}\ln(2x + 7)$ **31.** $\ln z + 2\ln(z - 1)$
33. $\ln 3 + \ln x + \ln(x + 1) - 2\ln(2x + 1)$
35. Answers will vary. **37.** Answers will vary.

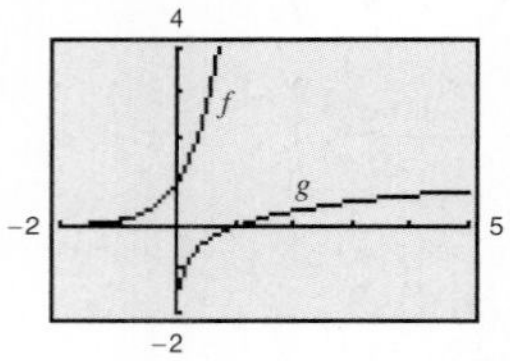

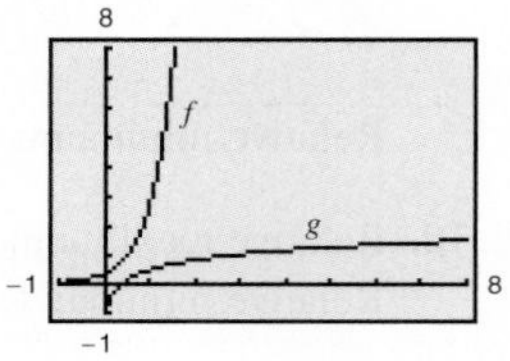

39. (a) 1.7917 (b) 0.4055 (c) 4.3944 (d) 0.5493
41. $\ln\dfrac{x - 2}{x + 2}$ **43.** $\ln\dfrac{x^3y^2}{z^4}$ **45.** $\dfrac{\ln(x - 6)^4}{\sqrt{3x + 1}}$
47. $\ln\left[\dfrac{x(x + 3)}{x + 4}\right]^3$ **49.** $\ln\left[\dfrac{x(x^2 + 1)}{x + 1}\right]^{3/2}$
51. $x = 4$ **53.** $x = \ln 4 - 1 \approx 0.3863$
55. $t = \dfrac{\ln 7 - \ln 3}{-0.2} \approx -4.2365$
57. $x = \frac{1}{2}\left(1 + \ln\frac{3}{2}\right) \approx 0.7027$ **59.** $x = 1$
61. $x = \dfrac{e^{2.4}}{2} \approx 5.5116$ **63.** $x = e^3 \approx 20.0855$
65. $x = \dfrac{6e^3}{e^3 - 1} \approx 6.314$ **67.** $x = \dfrac{\ln 15}{2\ln 5} \approx 0.8413$
69. $t = \dfrac{\ln 2}{\ln 1.07} \approx 10.2448$
71. $t = \dfrac{\ln 3}{12\ln[1 + (0.07/12)]} \approx 15.7402$
73. (a) 8.15 yr (b) 12.92 yr
75. (a) 14.21 yr (b) 13.89 yr (c) 13.86 yr (d) 13.86 yr
77. (a) $P(29) \approx 235{,}576$ (b) 2020
79. 9395 yr **81.** 12,484 yr
83. (a) 80 (b) 57.5 (c) 10 mo
85. (a) ≈896 units (b) ≈136 units

87.

x	y	$\dfrac{\ln x}{\ln y}$	$\ln \dfrac{x}{y}$	$\ln x - \ln y$
1	2	0	-0.6931	-0.6931
3	4	0.7925	-0.2877	-0.2877
10	5	1.4307	0.6931	0.6931
4	0.5	-2	2.0794	2.0794

89.

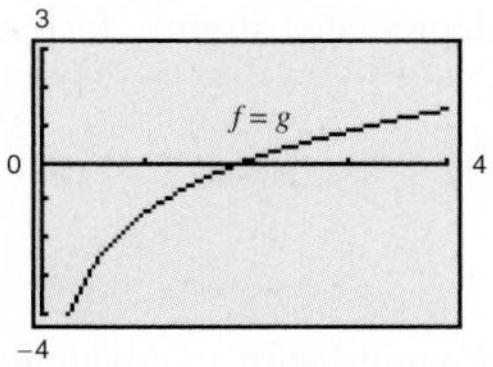

91. False. $f(x) = \ln x$ is undefined for $x \le 0$.

93. False. $f\left(\dfrac{x}{2}\right) = f(x) - f(2)$ **95.** False. $u = v^2$

97. Options (b) and (c) will give you the same amount, but it makes more sense to double the rate, not the time. So option (b) is better than option (c). If you are looking for a long-term investment, choose option (a).

99.

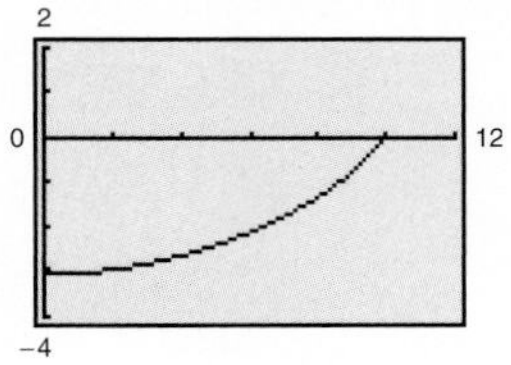

Answers will vary.

Section 4.5 *(page 291)*

Skills Warm Up *(page 291)*

1. $2 \ln(x + 1)$ **2.** $\ln x + \ln(x + 1)$
3. $\ln x - \ln(x + 1)$ **4.** $3[\ln x - \ln(x - 3)]$
5. $\ln 4 + \ln x + \ln(x - 7) - 2 \ln x$
6. $3 \ln x + \ln(x + 1)$
7. $-\dfrac{y}{x + 2y}$ **8.** $\dfrac{3 - 2xy + y^2}{x(x - 2y)}$
9. $-12x + 2$ **10.** $-\dfrac{6}{x^4}$

1. $\dfrac{2}{x}$ **3.** $\dfrac{2x}{x^2 + 3}$ **5.** $\dfrac{1}{2(x - 4)}$ **7.** $\dfrac{4}{x}(\ln x)^3$

9. $2 \ln x + 2$ **11.** $\dfrac{2x^2 - 1}{x(x^2 - 1)}$ **13.** $\dfrac{1}{x(x + 1)}$

15. $\dfrac{2}{3(x^2 - 1)}$ **17.** $-\dfrac{4}{x(4 + x^2)}$ **19.** $e^{-x}\left(\dfrac{1}{x} - \ln x\right)$

21. $\dfrac{e^x - e^{-x}}{e^x + e^{-x}}$ **23.** 2 **25.** -3 **27.** 2 **29.** 1.404

31. 5.585 **33.** -0.631 **35.** $(\ln 3)3^x$ **37.** $\dfrac{1}{x \ln 2}$

39. $(2 \ln 4)4^{2x-3}$ **41.** $\dfrac{2x + 6}{(x^2 + 6x) \ln 10}$ **43.** $2^x(1 + x \ln 2)$

45. $y = 3x - 3 = 3(x - 1)$ **47.** $y = 2x - e$

49. $y = -\frac{8}{5}x - 4$ **51.** $y = \dfrac{1}{27 \ln 3}x - \dfrac{1}{\ln 3} + 3$

53. $\dfrac{2xy}{3 - 2y^2}$ **55.** $\dfrac{y(1 - 6x^2)}{1 + y}$ **57.** $y = x - 1$

59. $\dfrac{1}{2x}$ **61.** $x(6 \ln x + 5)$ **63.** $(\ln 5)^2 5^x$

65. $\dfrac{d\beta}{dI} = \dfrac{10}{(\ln 10)I}$, so for $I = 10^{-4}$, the rate of change is about 43,429.4 db/W/cm².

67.

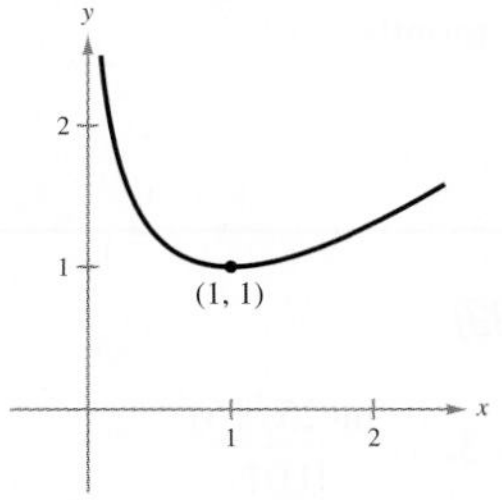

Relative minimum: $(1, 1)$

69.

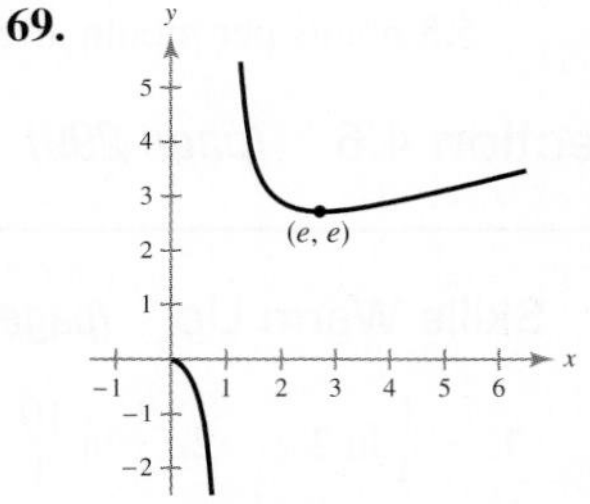

Discontinuity: $x = 1$
Relative minimum: (e, e)
Point of inflection: $\left(e^2, \dfrac{e^2}{2}\right)$

71.

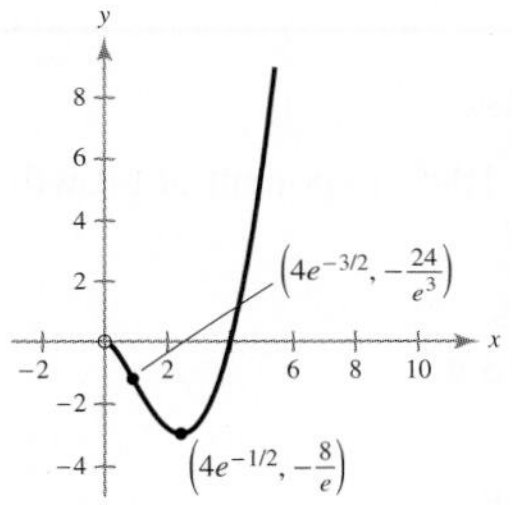

Relative minimum: $\left(4e^{-1/2}, \dfrac{-8}{e}\right)$
Point of inflection: $\left(4e^{-3/2}, \dfrac{-24}{e^3}\right)$

73. $-\dfrac{1}{p}, -\dfrac{1}{10}$

75. $p = 1000e^{-x}$

$\dfrac{dp}{dx} = -1000e^{-x}$

At $p = 10$, rate of change $= -10$.

$\dfrac{dp}{dx}$ and $\dfrac{dx}{dp}$ are reciprocals of each other.

77. (a) $\overline{C} = \dfrac{500 + 300x - 300 \ln x}{x}$
(b) Minimum of 279.15 at $e^{8/3}$

79. (a)

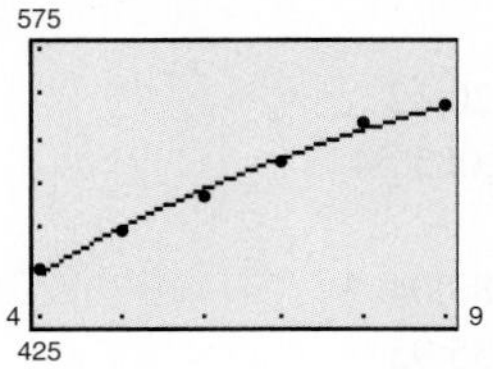

(b) 19.45 (thousand) per year

81. (a)

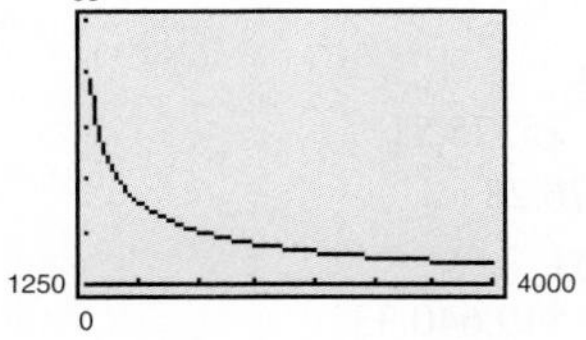

(b) $t \approx 30$; \$503,434.80 (c) $t \approx 20$; \$386,685.60
(d) ≈ -0.081; ≈ -0.029
(e) For a higher monthly payment, the term is shorter, and the total amount paid is smaller.

83. (a) $s(t) = 84.66 - 11.00 \ln t$

(b) 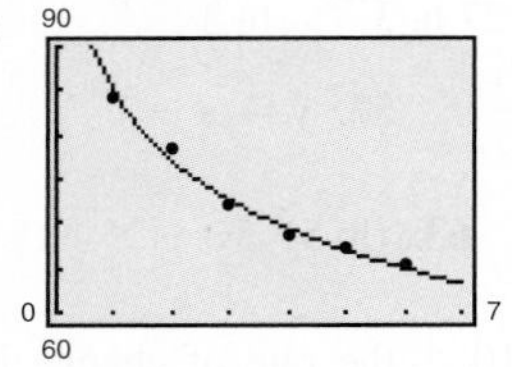

The model fits the data well.

(c) -5.5; The average score is decreasing at a rate of 5.5 points per month after 2 months.

Section 4.6 *(page 299)*

Skills Warm Up *(page 299)*

1. $-\frac{1}{4}\ln 2$ **2.** $\frac{1}{5}\ln\frac{10}{3}$ **3.** $-\frac{\ln(25/16)}{0.01}$

4. $-\frac{\ln(11/16)}{0.02}$ **5.** $7.36e^{0.23t}$ **6.** $1.296e^{0.072t}$

7. $-33.6e^{-1.4t}$ **8.** $-0.025e^{-0.001t}$ **9.** 4

10. 12 **11.** $2x + 1$ **12.** $x^2 + 1$

1. $y = 2e^{0.1014t}$ **3.** $y = 4e^{-0.4159t}$

5. $y = 4\sqrt[3]{2}e^{((\ln 0.5)/3)t}$ **7.** $y = 10e^{2t}$, exponential growth

9. $y = 30e^{-4t}$, exponential decay

11. *Amount after 1000 years:* 6.48 g
Amount after 10,000 years: 0.13 g

13. *Initial quantity:* 6.73 g
Amount after 1000 years: 5.96 g

15. *Initial quantity:* 2.16 g
Amount after 10,000 years: 1.62 g

17. 68% **19.** 15,642 yr

21. $k_1 = \frac{\ln 4}{12} \approx 0.1155$, so $y_1 = 5e^{0.1155t}$.

$k_2 = \frac{1}{6}$, so $y_2 = 5(2)^{t/6}$.

Explanations will vary.

23. (a) 1350 (b) $\frac{5\ln 2}{\ln 3} \approx 3.15$ hr

(c) No. Answers will vary.

25. *Time to double:* 5.78 yr
Amount after 10 years: \$3320.12
Amount after 25 years: \$20,085.54

27. *Annual rate:* 8.66%
Amount after 10 years: \$1783.04
Amount after 25 years: \$6535.95

29. *Annual rate:* 9.50%
Time to double: 7.30 yr
Amount after 25 years: \$5375.51

31. *Initial investment:* \$6376.28
Time to double: 15.40 yr
Amount after 25 years: \$19,640.33

33. \$49,787.07

35. (a) Answers will vary. (b) Answers will vary.

37. Answers will vary.

39. (a) $y = 21{,}500 - 3950x$

(b) $y = 21{,}500e^{-0.229x}$

(c) Linear Model:
after 1 year: \$17,550
after 4 years: \$5700
Exponential Model:
after 1 year: \$17,099.56
after 4 years: \$8602.50

(d) 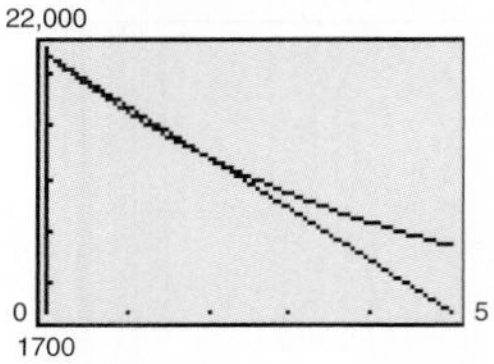

The exponential model depreciates slightly faster.

(e) After the second year, a buyer would gain an advantage by using the linear model, because it yields a lower value for the vehicle. A seller would want to use the exponential model, because it yields a higher value for the vehicle.

41. (a) $C = 30$
$k = \ln(\frac{1}{6}) \approx -1.7918$

(b) $30e^{-0.35836} = 20.9646$ thousand, or 20,965 units

(c) 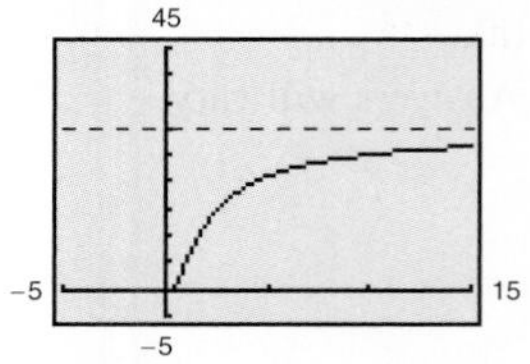

43. About 36 days

45. (a) $C \approx 81.090$, $k = \frac{\ln(45/40)}{-200} \approx -0.0005889$

(b) $x = 1/0.0005889 \approx 1698$ units, $p \approx \$29.83$

47. Answers will vary.

Review Exercises for Chapter 4 *(page 306)*

1. (a) 16,384 (b) 117,649 (c) 0.0625 (d) 81

3. 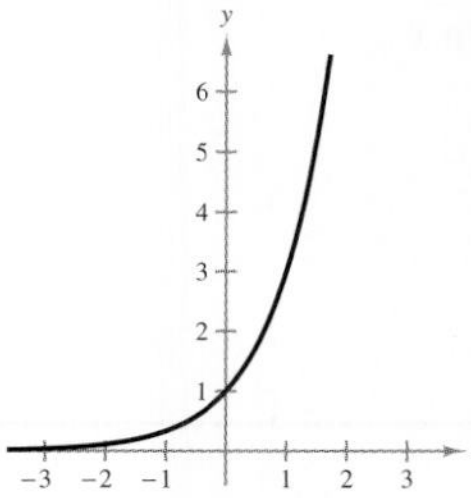

5. 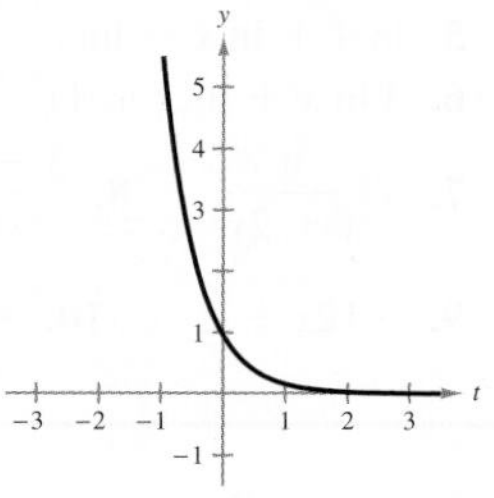

7. 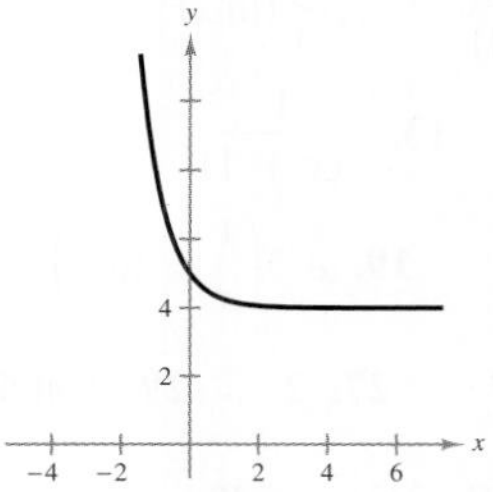

9. (a) 5894.39 (thousand) (b) 6203.76 (thousand)

11. (a) \$69,295.66 (b) \$233,081.88

13. (a) e^{10} (b) $\frac{1}{e^2}$ (c) $e^{11/2}$ (d) $\frac{1}{e^8}$

15.

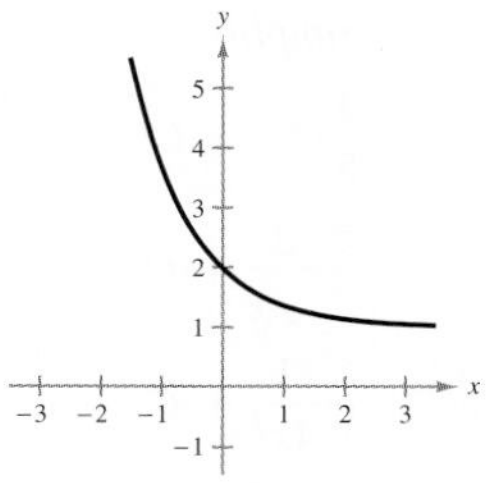

17.

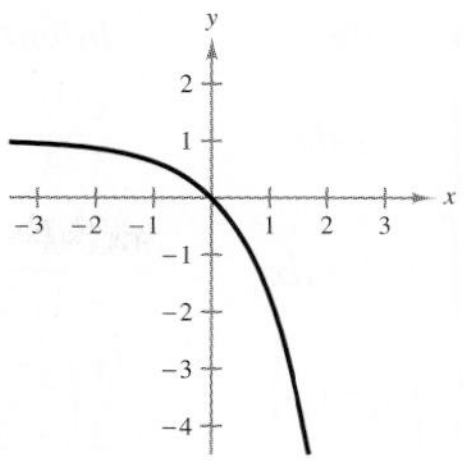

19.

n	1	2	4	12
A	\$1216.65	\$1218.99	\$1220.19	\$1221.00

n	365	Continuous compounding
A	\$1221.39	\$1221.40

21.

n	1	2	4	12
A	4231.80	4244.33	4250.73	4255.03

n	365	Continuous compounding
A	4257.13	4257.20

23. b **25.** (a) 6% (b) 6.09% (c) 6.14% (d) 6.17%

27. \$10,338.10

29. (a) \$8276.81 (b) \$7697.12 (c) \$7500

31. (a) The model fits the data very well.
(b) $y = 116.85x + 111.1$
The linear model fits the data moderately well.
The exponential model is a better fit.
(c) Exponential: \$4357.50 (million)
Linear: \$1863.85 (million)

33. (a) $P \approx 1049$ fish
(b)

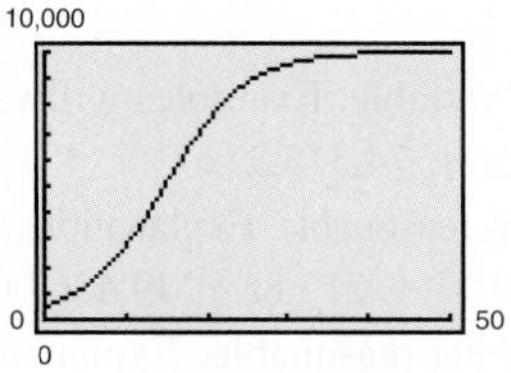

13 months

(c) Yes, P approaches 10,000 fish as t approaches ∞.

35. $8xe^{x^2}$ **37.** $\dfrac{1-2x}{e^{2x}}$ **39.** $-\dfrac{10e^{2x}}{(1+e^{2x})^2}$

41. $y = 3 - x$ **43.** $y = \dfrac{x}{e}$

45.

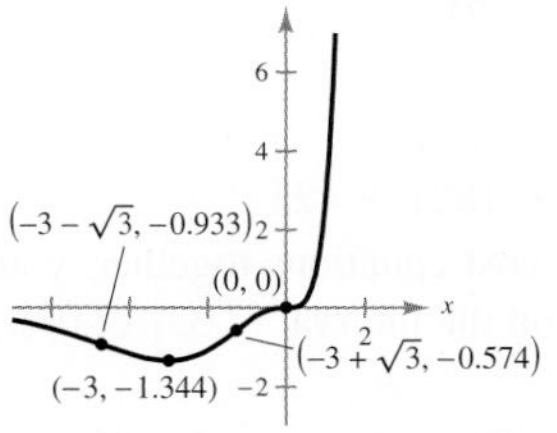

Relative minimum: $(-3, -1.344)$
Inflection points: $(0, 0)$, $(-3+\sqrt{3}, -0.574)$, and $(-3-\sqrt{3}, -0.933)$
Horizontal asymptote: $y = 0$

47.

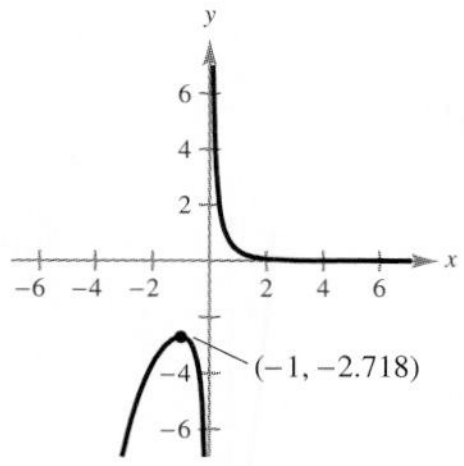

Relative maximum: $(-1, -2.718)$
Horizontal asymptote: $y = 0$
Vertical asymptote: $x = 0$

49. $e^{2.4849} \approx 12$ **51.** $\ln 4.4816 \approx 1.5$

53.

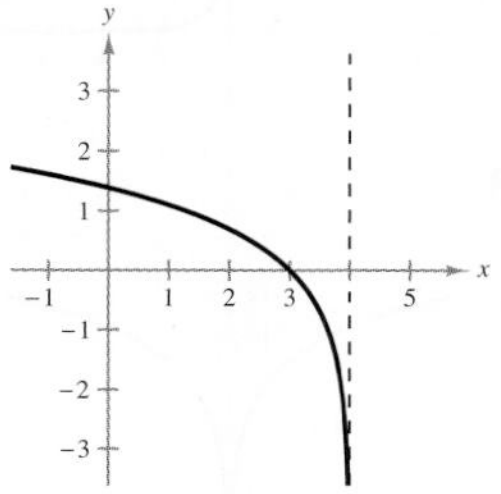

55.

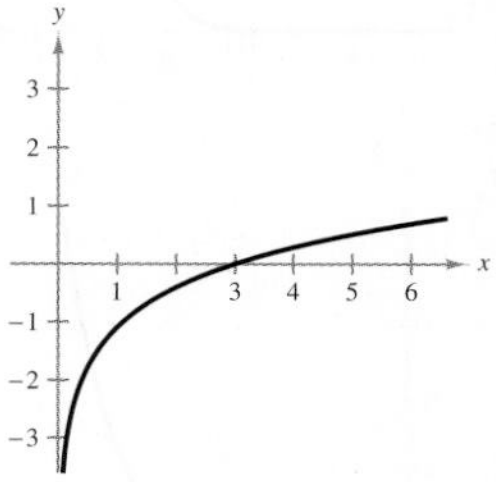

57. $\ln x + \frac{1}{2}\ln(x-1)$ **59.** $2\ln x - 3\ln(x+1)$

61. $3[\ln(1-x) - \ln 3 - \ln x]$

63. $\ln(2x^2 - x - 15)$ **65.** $4\ln\left(\dfrac{x^5 - x^2}{x-5}\right)$

67. 3 **69.** $e^3 \approx 20.09$ **71.** 1

73. $\dfrac{3+\sqrt{13}}{2} \approx 3.3028$ **75.** $-\dfrac{\ln(0.25)}{1.386} \approx 1.0002$

77. $\frac{1}{2}(\ln 6 + 1) \approx 1.3959$ **79.** $\dfrac{\ln 1.1}{\ln 1.21} = 0.5$

81. (a) ≈ 28.07 years (b) ≈ 27.75 years
(c) ≈ 27.73 years (d) ≈ 27.73 years

83. (a) 75 (b) 65.34 (c) ≈ 11 months

85. $\dfrac{2}{x}$ **87.** $\dfrac{1}{x} + \dfrac{1}{x-1} - \dfrac{1}{x-2} = \dfrac{x^2-4x+2}{x(x-2)(x-1)}$

89. 2 **91.** $\dfrac{1-3\ln x}{x^4}$ **93.** $\dfrac{4x}{3(x^2-2)}$

95. $\dfrac{2}{x} + \dfrac{1}{2(x+1)}$ **97.** $\dfrac{1}{1+e^x}$ **99.** 2 **101.** 0

103. 1.594 **105.** 1.500 **107.** $(2\ln 5)5^{2x+1}$

109. $\dfrac{2}{(2x-1)\ln 3}$ **111.** $\dfrac{-1}{\ln 10}\cdot\dfrac{1}{x} = -\dfrac{1}{x\ln 10}$

113.

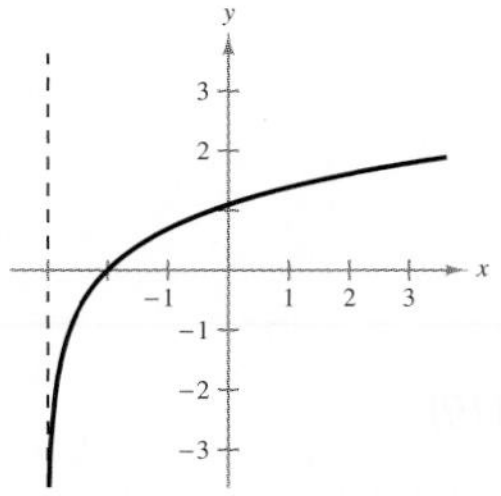

No relative extrema
No points of inflection

115.

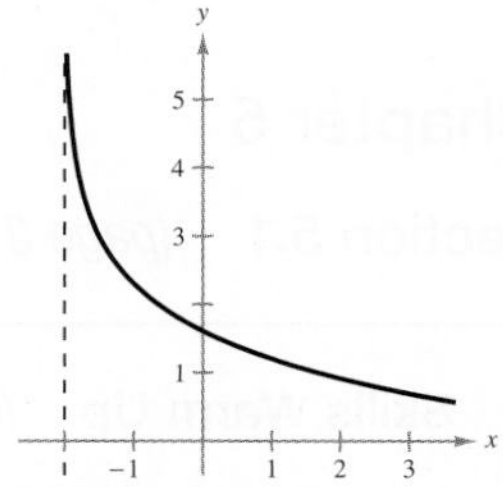

No relative extrema
No points of inflection

117. 2005: 256.4 (million)
2008: 160.25 (million)

119. $y = 3e^{-0.27465t}$ **121.** 5.19 g; 0.10 g

123. 20.18 g; 17.88 g **125.** 2.47 g; 1.85 g

127. 8.66 yr, \$1335.32, \$4433.43

129. 2%, 34.66 yr, \$24,730.82

131. (a) $D = 500e^{-0.38376t}$
(b) 107.72 milligrams per milliliter

Test Yourself *(page 310)*

1. 1 **2.** $\frac{1}{256}$ **3.** $e^{9/2}$ **4.** e^{12}

5.

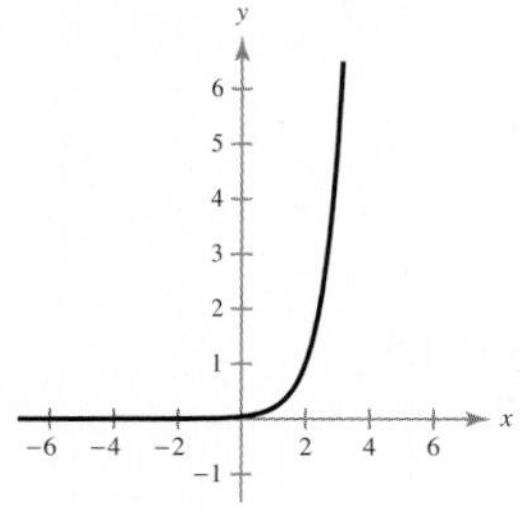

6.

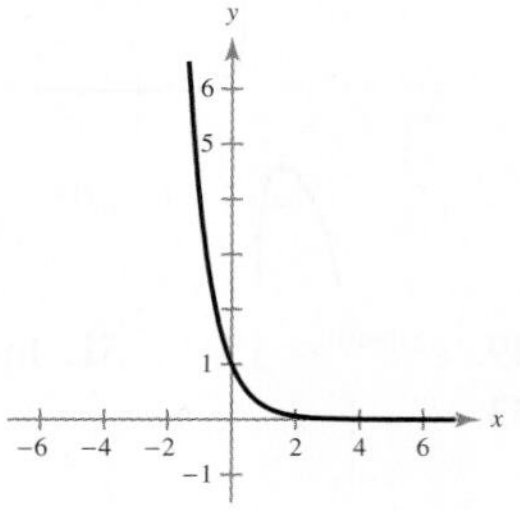

7.

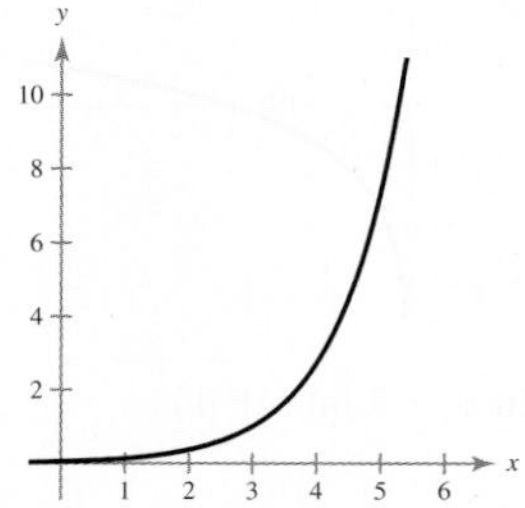

8.

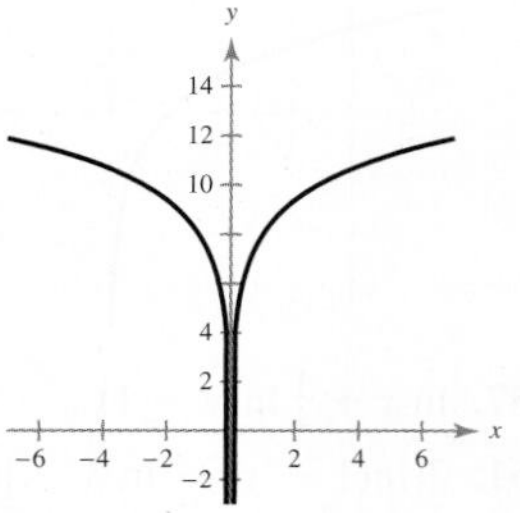

9.

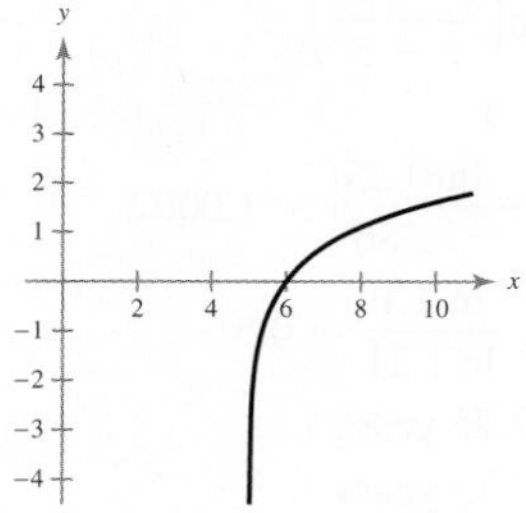

10.

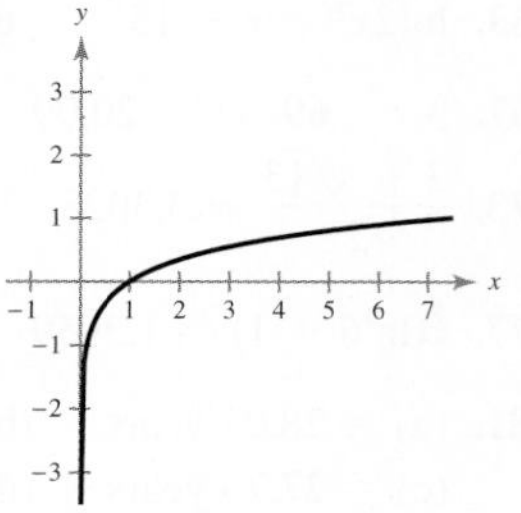

11. $\ln 3 - \ln 2$ **12.** $\frac{1}{2}\ln(x + y)$ **13.** $\ln(x + 1) - \ln y$

14. $\ln[y(x + 1)]$ **15.** $\ln\dfrac{x^3}{(x - 1)^2}$ **16.** $\ln\dfrac{xy^4}{\sqrt{(z + 4)}}$

17. $x \approx 3.197$ **18.** $x \approx 1.750$ **19.** $x \approx 58.371$

20. (a) 17.67 yr (b) 17.36 yr (c) 17.33 yr (d) 17.33 yr

21. $-3e^{-3x}$ **22.** $7e^{x+2} + 2$

23. $\dfrac{2x}{3 + x^2}$ **24.** $\dfrac{2}{x(x + 2)}$

25. (a) \$2241.54 million (b) \$138.30 million/yr

26. 59.4% **27.** 39.61 yr

Chapter 5

Section 5.1 *(page 319)*

Skills Warm Up *(page 319)*

1. $x^{-1/2}$ **2.** $(2x)^{4/3}$ **3.** $5^{1/2}x^{3/2} + x^{5/2}$

4. $x^{-1/2} + x^{-2/3}$ **5.** $(x + 1)^{5/2}$ **6.** $x^{1/6}$

7. -12 **8.** -10 **9.** 14 **10.** 14

1–5. Answers will vary. **7.** $u + C$ **9.** $6x + C$

11. $\frac{7}{2}x^2 + C$ **13.** $\frac{5}{3}t^3 + C$

15. $-\dfrac{5}{2x^2} + C$ **17.** $\dfrac{2}{5}y^{5/2} + C$

	Rewrite	*Integrate*	*Simplify*
19. $\int \sqrt[3]{x^2}\,dx$	$\int x^{2/3}\,dx$	$\dfrac{x^{5/3}}{5/3} + C$	$\dfrac{3}{5}x^{5/3} + C$
21.	$\int x^{-3/2}\,dx$	$\dfrac{x^{-1/2}}{-1/2} + C$	$-\dfrac{2}{\sqrt{x}} + C$
23.	$\dfrac{1}{2}\int x^{-3}\,dx$	$\dfrac{1}{2}\left(\dfrac{x^{-2}}{-2}\right) + C$	$-\dfrac{1}{4x^2} + C$

25. $\dfrac{x^2}{2} + 3x + C$ **27.** $\dfrac{1}{4}x^4 + 2x + C$

29. $\dfrac{3}{4}x^4 - 2x^3 + 2x + C$ **31.** $\dfrac{1}{3}x^3 + \dfrac{5}{2}x^2 + x + C$

33. $2x + \dfrac{1}{2x^2} + C$ **35.** $3x^{2/3}(x + 2) + C$

37.

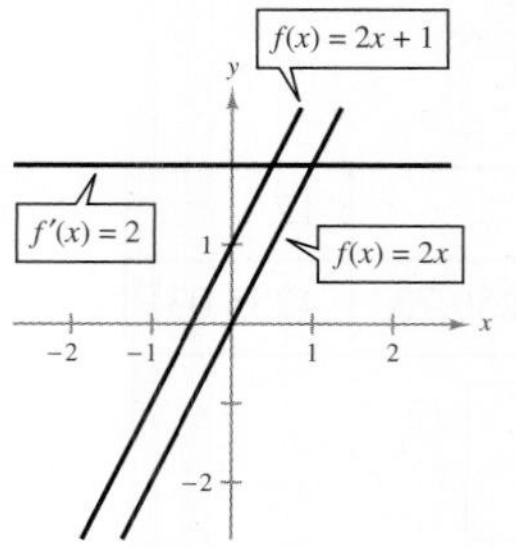

39.

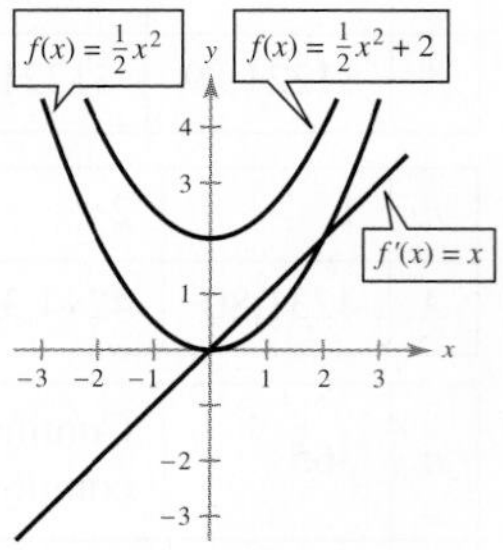

41. $f(x) = 2x^2 + 6$ **43.** $f(x) = x^2 + 4x + 7$

45. $f(x) = 5x^2 - 3x^4 + 200$ **47.** $f(x) = -\dfrac{1}{x^2} + \dfrac{1}{x} + \dfrac{1}{2}$

49. $f(x) = x^2 + x + 4$ **51.** $f(x) = \frac{9}{4}x^{4/3}$

53. $C = 85x + 5500$ **55.** $C = \frac{1}{10}\sqrt{x} + 4x + 750$

57. $R = 225x - \frac{3}{2}x^2$, $p = 225 - \frac{3}{2}x$

59. $P = -9x^2 + 1650x$ **61.** $P = -12x^2 + 805x + 68$

63. $s(t) = -16t^2 + 6000$; about 19.36 sec

65. $v_0 = 40\sqrt{22} \approx 187.62$ ft/sec

67. (a) $C = x^2 - 12x + 125$ (b) \$2025

$$\overline{C} = x - 12 + \frac{125}{x}$$

(c) \$125 is fixed. \$1900 is variable. Examples will vary.

69. (a) $P(t) = 79.4t^2 + 1758.6t + 74{,}515.2$

(b) 314,437.2; Yes, this seems reasonable. Explanations will vary.

71. (a) $M(t) = -0.035t^3 + 7.01t^2 + 217.8t + 49{,}486.005$

(b) 64,195.63; Yes, this seems reasonable. Explanations will vary.

73. (a)

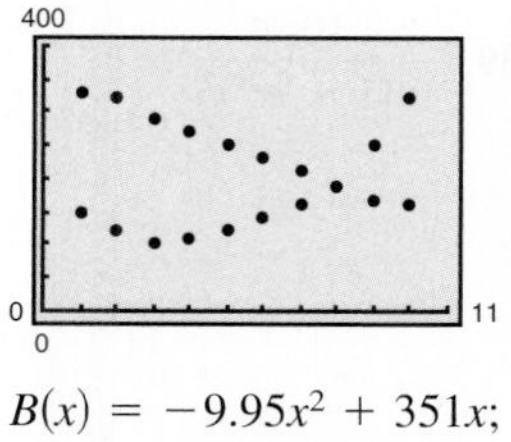

$$\frac{dB}{dx} = -19.9x + 351$$

$$\frac{dC}{dx} = 5.38x^2 - 40.6x + 182$$

(b) $B(x) = -9.95x^2 + 351x$;
$C(x) = 1.79x^3 - 20.3x^2 + 182x + 425$

(c) Graphing the benefit and cost equations together, you see that benefit exceeds cost on the interval (2.32, 12.00).

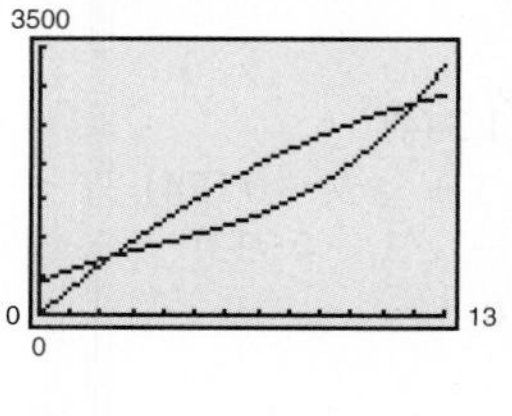

The company should produce from 3 to 11 units.

Section 5.2 *(page 329)*

Skills Warm Up *(page 329)*

1. $\frac{1}{2}x^4 + x + C$ **2.** $\frac{3}{2}x^2 + \frac{2}{3}x^{3/2} - 4x + C$

3. $-\frac{1}{x} + C$ **4.** $-\frac{1}{6t^2} + C$ **5.** $\frac{4}{7}t^{7/2} + \frac{2}{5}t^{5/2} + C$

6. $\frac{4}{5}x^{5/2} - \frac{2}{3}x^{3/2} + C$ **7.** $\frac{5x^3 - 4}{2x} + C$

8. $\frac{-6x^2 + 5}{3x^3} + C$ **9.** $\frac{2}{5}\sqrt{x}(8x^2 + 15) + C$

$\int u^n \frac{du}{dx}\,dx$	u	$\frac{du}{dx}$
1. $\int (5x^2 + 1)^2(10x)\,dx$	$5x^2 + 1$	$10x$
3. $\int \sqrt{1 - x^2}\,(-2x)\,dx$	$1 - x^2$	$-2x$
5. $\int \left(4 + \frac{1}{x^2}\right)^5\left(\frac{-2}{x^3}\right)dx$	$4 + \frac{1}{x^2}$	$-\frac{2}{x^3}$
7. $\int \left(1 + \sqrt{x}\right)^3\left(\frac{1}{2\sqrt{x}}\right)dx$	$1 + \sqrt{x}$	$\frac{1}{2\sqrt{x}}$

9. $\frac{1}{5}(x - 1)^5 + C$ **11.** $\frac{1}{5}(1 + 2x)^5 + C$

13. $\frac{1}{2}(x^2 + 3x)^2 + C$ **15.** $\frac{2}{3}(4x^2 - 5)^{3/2} + C$

17. $\frac{1}{3(5 - 3x^2)^3} + C$ **19.** $\frac{(2x^3 - 1)^5}{30} + C$

21. $\frac{1}{3}(t^2 + 6)^{3/2} + C$ **23.** $\frac{1}{12(x^6 - 4)^2} + C$

25. $\frac{1}{10}(x^2 - 6x)^5 + C$ **27.** $-\frac{1}{2(x^2 + 2x - 3)} + C$

29. $-\frac{15}{8}(1 - x^2)^{4/3} + C$ **31.** $-\frac{3}{2(1 + x^2)} + C$

33. $-3\sqrt{2t + 3} + C$ **35.** $\frac{1}{4}(6x^2 - 1)^4 + C$

37. $\frac{3}{16}(4x + 3)^{4/3} + C$ **39.** $\sqrt{x^2 + 25} + C$

41. $\frac{2}{3}\sqrt{x^3 + 3x + 4} + C$

43. (a) $\frac{1}{3}x^3 - x^2 + x + C_1 = \frac{1}{3}(x - 1)^3 + C_2$
(b) Answers differ by a constant: $C_1 = C_2 - \frac{1}{3}$
(c) Answers will vary.

45. (a) $\frac{1}{6}x^6 - \frac{1}{2}x^4 + \frac{1}{2}x^2 + C_1 = \frac{(x^2 - 1)^3}{6} + C_2$
(b) Answers differ by a constant: $C_1 = C_2 - \frac{1}{6}$
(c) Answers will vary.

47. $f(x) = \frac{1}{12}(4x^2 - 10)^3 - 8$

49. (a) $C = 8\sqrt{x + 1} + 18$ (b) \$75.13

51. $x = \frac{1}{3}(p^2 - 25)^{3/2} + 24$ **53.** $x = \frac{6000}{\sqrt{p^2 - 16}} + 3000$

55. (a) $h = \sqrt{17.6t^2 + 1} + 5$ (b) 26 in.

57. (a) $Q = (x - 24{,}999)^{0.95} + 24{,}999$

(b)

x	25,000	50,000	100,000	150,000
Q	25,000	40,067.14	67,786.18	94,512.29
$x - Q$	0	9932.86	32,213.82	55,487.71

(c)

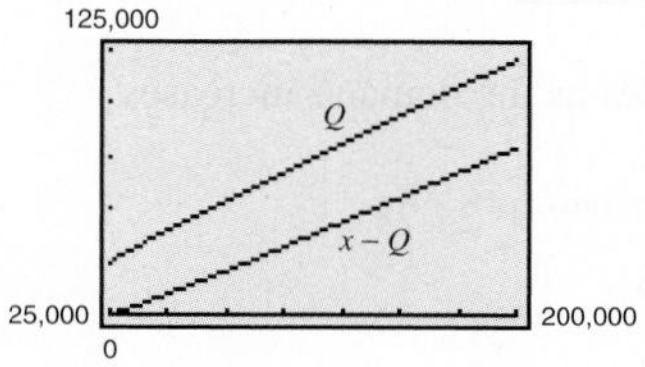

59. $-\frac{2}{3}x^{3/2} + \frac{2}{3}(x + 1)^{3/2} + C$

Section 5.3 *(page 336)*

Skills Warm Up *(page 336)*

1. $x + 2 - \frac{2}{x + 2}$ **2.** $x - 2 + \frac{1}{x - 4}$

3. $x + 8 + \frac{2x - 4}{x^2 - 4x}$ **4.** $x^2 - x - 4 + \frac{20x + 22}{x^2 + 5}$

5. $\frac{1}{4}x^4 - \frac{1}{x} + C$ **6.** $\frac{1}{2}x^2 + 2x + C$

7. $\frac{1}{2}x^2 - \frac{4}{x} + C$ **8.** $-\frac{1}{x} - \frac{3}{2x^2} + C$

1. $e^{2x} + C$ **3.** $\frac{1}{4}e^{4x} + C$ **5.** $\frac{1}{5}e^{5x-3} + C$

7. $-\frac{9}{2}e^{-x^2} + C$ **9.** $\frac{5}{3}e^{x^3} + C$ **11.** $e^{x^2+x} + C$

13. $\ln|x + 1| + C$ **15.** $\ln|5x + 2| + C$

17. $-\frac{1}{2}\ln|3 - 2x| + C$ **19.** $\frac{2}{3}\ln|3x + 5| + C$

21. $\ln\sqrt{x^2 + 1} + C$ **23.** $\frac{1}{3}\ln|x^3 + 1| + C$

25. $\frac{1}{2}\ln|x^2 + 6x + 7| + C$ **27.** $\ln|\ln x| + C$

29. $\ln|1 - e^{-x}| + C$

31. $\frac{1}{4}x^2 - 4\ln|x| + C$; General Power Rule and Log Rule

33. $8x + 3\ln|x| - \frac{3}{x^2} + C$; General Power Rule and Log Rule

35. $e^x + 2x - e^{-x} + C$;
Exponential Rule and General Power Rule

37. $-\frac{2}{3}(1 - e^x)^{3/2} + C$; Exponential Rule

39. $\ln|e^x + x| + C$; Log Rule

41. $\frac{1}{7}\ln(7e^{5x} + 1) + C$; Log Rule

43. $\frac{1}{2}x^2 + 3x + 8\ln|x - 1| + C$;
General Power Rule and Log Rule

45. $x - 6\ln|x + 3| + C$;
General Power Rule, General Log Rule

47. $f(x) = \frac{-e^{2/x}}{2} + \frac{e^{1/2}}{2} + 6$

49. $f(x) = \frac{1}{2}x^2 + 5x + 8\ln|x - 1| - 8$

51. (a) $P(t) = 1000[1 + \ln(1 + 0.25t)^{12}]$
(b) $P(3) \approx 7715$ bacteria (c) $t \approx 6$ days

53. (a) $p = -50e^{-x/500} + 45.06$
(b)

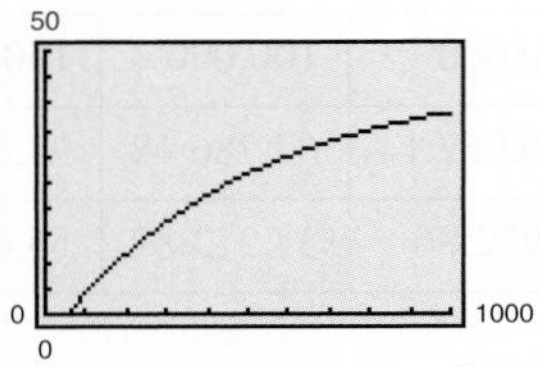

The price increases as the demand increases.
(c) 387

55. (a) $R(t) = 3223.56e^{0.0993t} + 24.78$
(b) $R(9) \approx \$7903.66$

57. False. $\ln x^{1/2} = \frac{1}{2}\ln x$

Quiz Yourself *(page 338)*

1. $3x + C$ **2.** $5x^2 + C$ **3.** $-\dfrac{1}{4x^4} + C$
4. $\dfrac{x^3}{3} - x^2 + 15x + C$ **5.** $\dfrac{(6x+1)^4}{4} + C$
6. $\dfrac{1}{50}(5x^2 - 2)^5 + C$ **7.** $\dfrac{(x^2 - 5x)^2}{2} + C$
8. $-\dfrac{1}{2(x^3 + 3)^2} + C$ **9.** $\dfrac{2}{15}(5x + 2)^{3/2} + C$
10. $f(x) = 8x^2 + 1$ **11.** $f(x) = 3x^3 + 4x - 2$
12. (a) $C = -0.03x^2 + 16x + 9.03$ (b) \$9.03 (c) \$509.03
13. $f(x) = \dfrac{2}{3}x^3 + x + 1$
14. (a) 1000 bolts (b) About 8612 bolts
15. $e^{5x+4} + C$ **16.** $e^{x^3} + C$
17. $\frac{1}{2}e^{(x^2 - 6x)} + C$ **18.** $\ln|2x - 1| + C$
19. $-\frac{1}{8}\ln|3 - 8x| + C$ **20.** $\frac{1}{6}\ln(3x^2 + 4) + C$
21. (a) $S(t) = 13.16t^2 + 848.99\ln(t) + 2504.44$
(b) \$5112.11 million

Section 5.4 *(page 348)*

Skills Warm Up *(page 348)*

1. $\frac{3}{2}x^2 + 7x + C$ **2.** $\frac{2}{5}x^{5/2} + \frac{4}{3}x^{3/2} + C$
3. $\dfrac{1}{5}\ln|x| + C$ **4.** $-\dfrac{1}{6e^{6x}} + C$
5. $C = 0.008x^{5/2} + 29{,}500x + C$
6. $R = x^2 + 9000x + C$
7. $P = 25{,}000x - 0.005x^2 + C$
8. $C = 0.01x^3 + 4600x + C$

1.

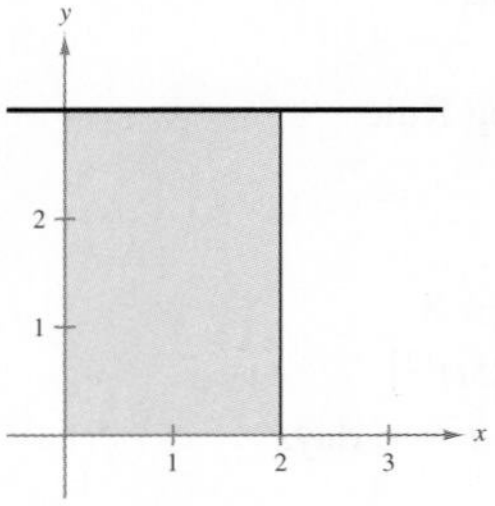

Area = 6

3.

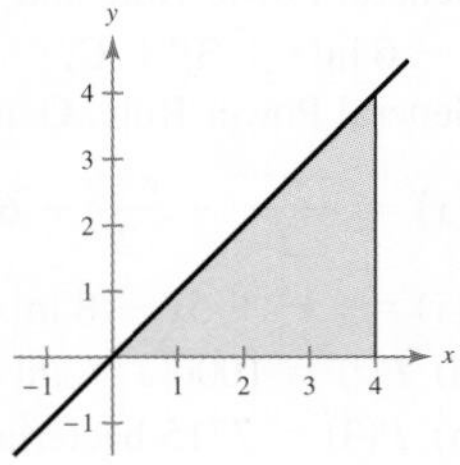

Area = 8

5.

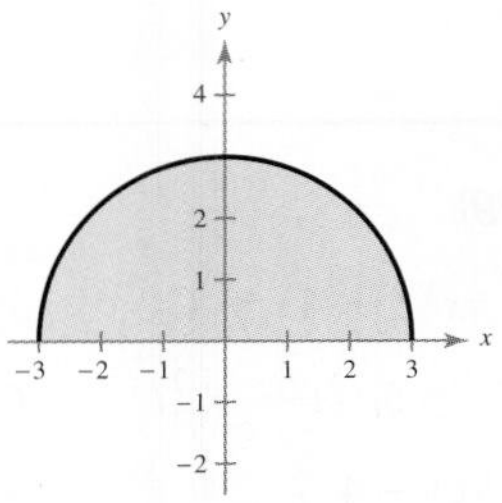

Area $= \dfrac{9\pi}{2}$

7. (a) 8 (b) 4 (c) -24 (d) 0
9. $\dfrac{1}{6}$ **11.** $\dfrac{1}{2}$ **13.** $6\left(1 - \dfrac{1}{e^2}\right)$ **15.** $8\ln 2 + \dfrac{15}{2}$
17. 1 **19.** $-\frac{5}{2}$ **21.** 38 **23.** $-\frac{15}{4}$ **25.** -4
27. $-\frac{27}{20}$ **29.** $6\ln 2 \approx 4.16$ **31.** 2
33. $-e^{-1} + 1 \approx 0.63$ **35.** $\frac{1}{3}\left[(e^2 + 1)^{3/2} - 2\sqrt{2}\right] \approx 7.157$
37. $\frac{1}{8}\ln 17 \approx 0.354$ **39.** 10 **41.** 39 **43.** 10
45. $4\ln 3 \approx 4.394$ **47.** \$6.75 **49.** \$22.50 **51.** \$3.97
53. Average = 12
$x = 2$
55. Average $= \dfrac{8}{3}$
$x = \pm\dfrac{2\sqrt{3}}{3} \approx \pm 1.155$
57. Average $= e - e^{-1} \approx 2.3504$
$x = \ln\left(\dfrac{e - e^{-1}}{2}\right) \approx 0.1614$
59. Average $= \frac{3}{4}\ln\frac{7}{3} \approx 0.6355$ **61.** $\frac{6}{5}$ **63.** 0
$x = 4/\ln(7/3) - 2 \approx 2.721$
65. (a) $\frac{1}{3}$ (b) $\frac{2}{3}$ (c) $-\frac{1}{3}$
Explanations will vary.
67. \$1925.23 **69.** \$16,605.21
71. \$2500 **73.** \$4565.65
75. (a) \$137,000 (b) \$214,720.93 (c) \$338,393.53
77. \$2623.94
79. (a) $M(t) = 273.78t^2 - 23.153t^3 - 331.258e^{-t} + 5438.258$
(b) \$8573.88 billion

Section 5.5 *(page 357)*

Skills Warm Up *(page 357)*

1. $-x^2 + 3x + 2$ **2.** $-2x^2 + 4x + 4$
3. $-x^3 + 2x^2 + 4x - 5$ **4.** $x^3 - 6x - 1$
5. (0, 4), (4, 4) **6.** (1, −3), (2, −12)
7. (−3, 9), (2, 4) **8.** (−2, −4), (0, 0), (2, 4)

1. 36 **3.** 9 **5.** $e - 2$ **7.** $\frac{3}{2}$

9.

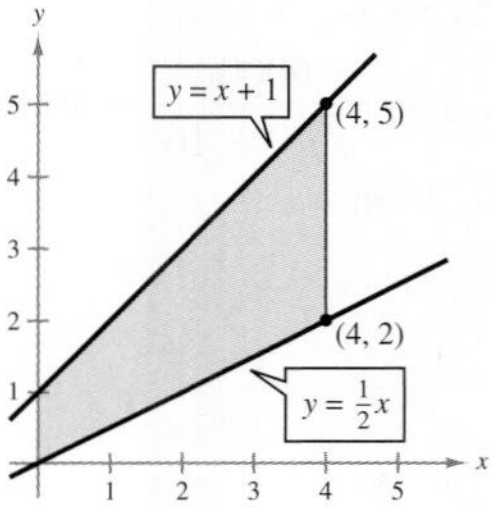

11.

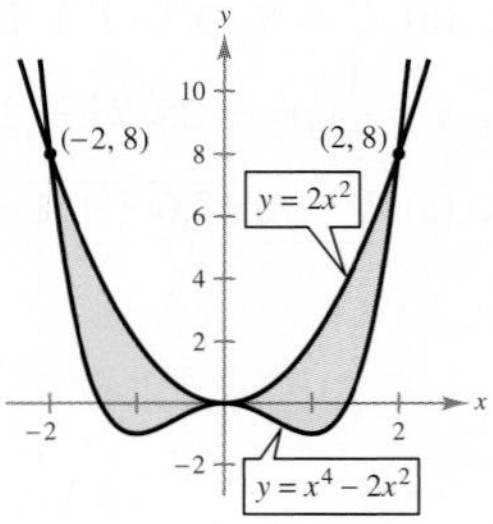

13. d

15.

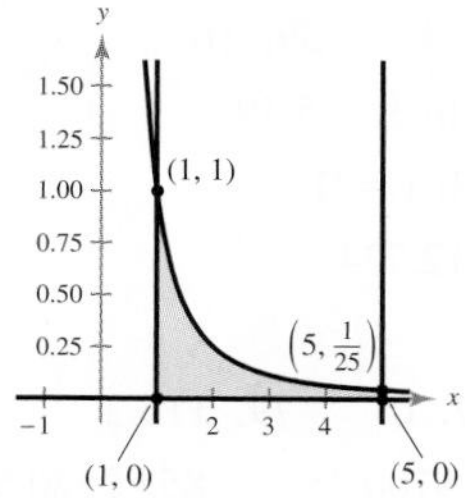

Area $= \frac{4}{5}$

17.

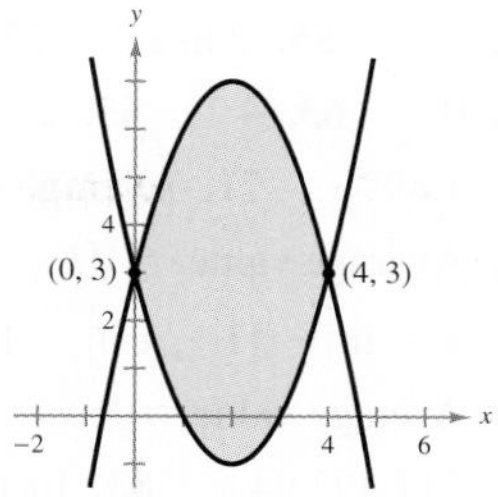

Area $= \frac{64}{3}$

19.

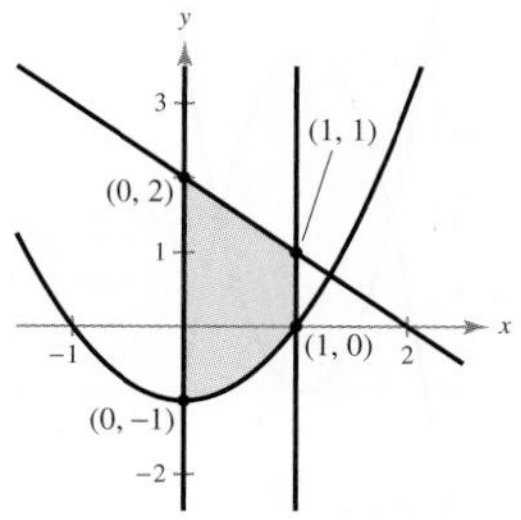

Area $= 2\frac{1}{6}$

21.

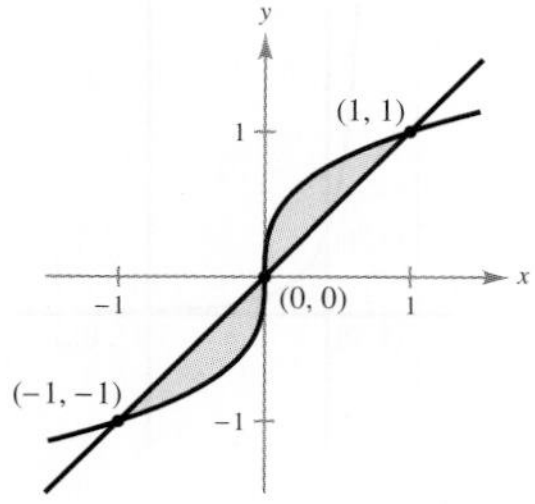

Area $= \frac{1}{2}$

23.

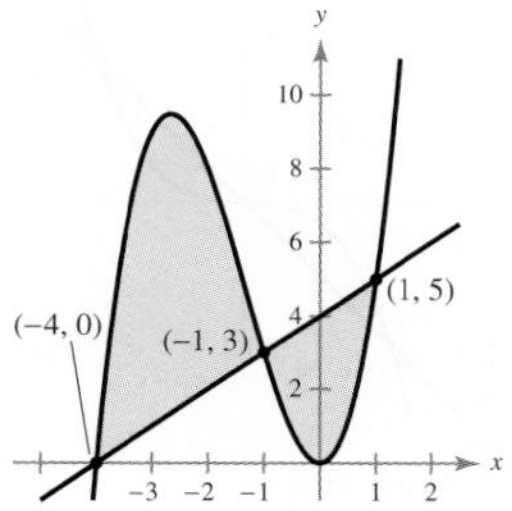

Area $= 21\frac{1}{12}$

25.

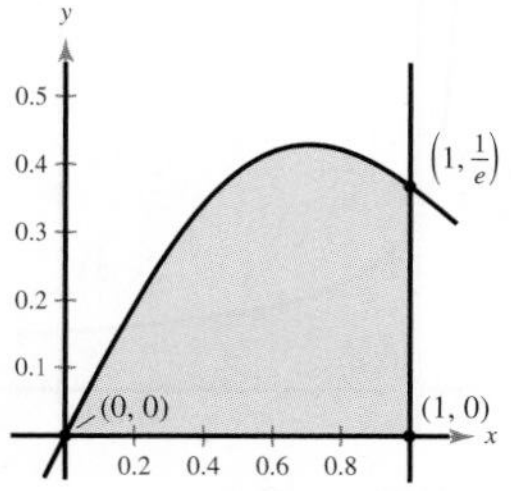

Area $= -\frac{1}{2}e^{-1} + \frac{1}{2}$

27.

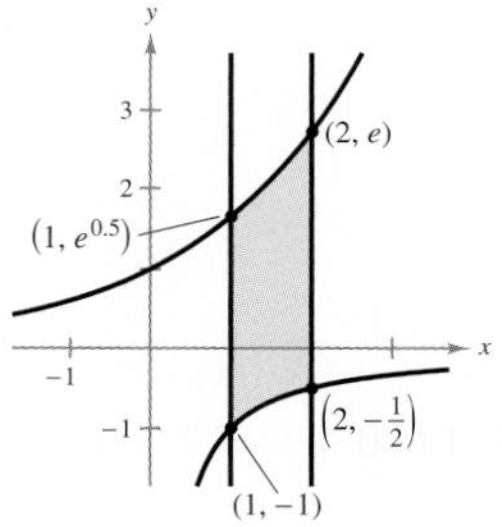

Area $= (2e + \ln 2) - 2e^{1/2}$

29.

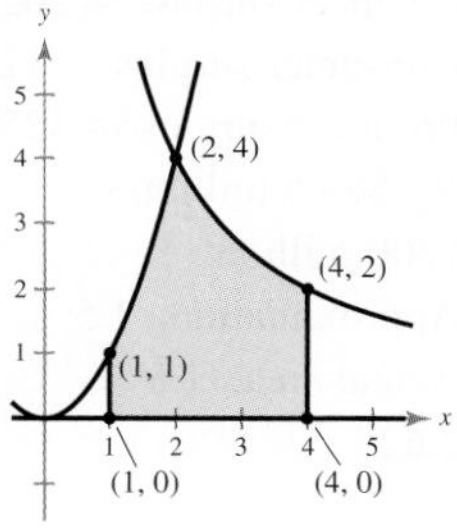

Area $= \frac{7}{3} + 8 \ln 2$

31.

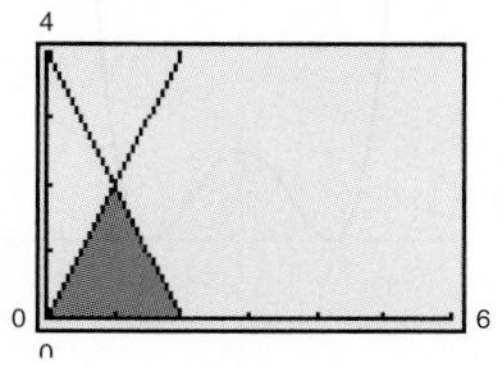

Area $= \int_0^1 2x\,dx + \int_1^2 (4 - 2x)\,dx$

33.

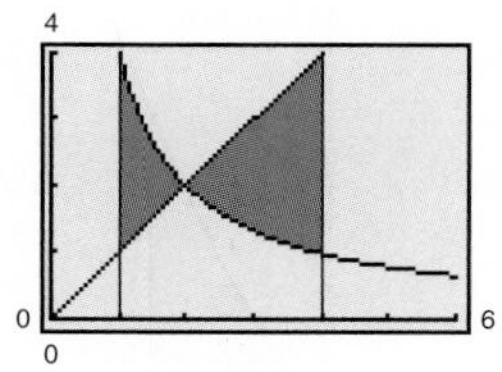

Area $= \int_1^2 \left(\frac{4}{x} - x\right) dx + \int_2^4 \left(x - \frac{4}{x}\right) dx$

35.

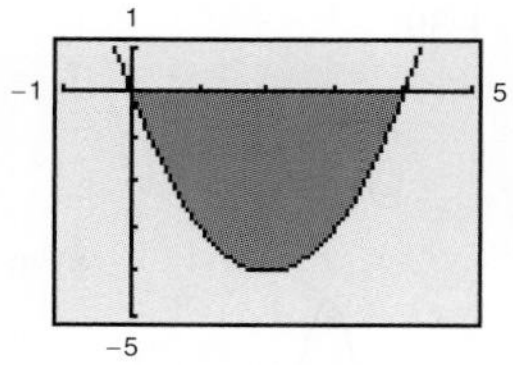

Area $= \frac{32}{3}$

37.

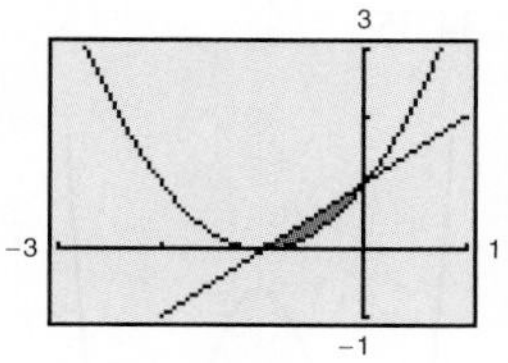

Area $= \frac{1}{6}$

39. 8

41. Consumer surplus = 1600
Producer surplus = 400

43. Consumer surplus = 500
Producer surplus = 2000

45. Consumer surplus = 640.00
Producer surplus ≈ 426.67

47. R_1; \$11.375 billion

49. \$573 million; Explanations will vary.

51. (a)

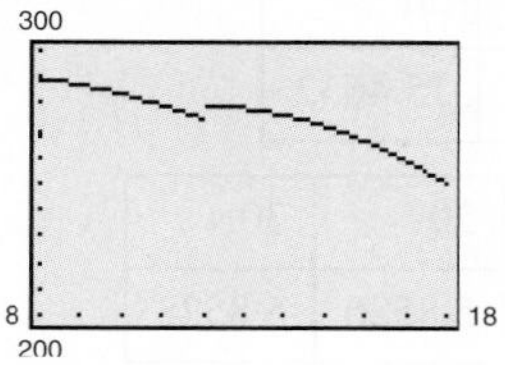

(b) 124.25 fewer pounds

53. \$333.33 million

55. CS = \$700,000
PS = \$1,375,000

57. 2077.10 **59.** Answers will vary.

Section 5.6 *(page 364)*

Skills Warm Up *(page 364)*

1. $\frac{1}{6}$ **2.** $\frac{3}{20}$ **3.** $\frac{7}{40}$ **4.** $\frac{13}{12}$ **5.** $\frac{61}{30}$ **6.** $\frac{53}{18}$
7. $\frac{2}{3}$ **8.** $\frac{4}{7}$ **9.** 0 **10.** 5

1. Approximation: 2
Exact area: 2

3. Approximation: 0.6730
Exact area: $\frac{2}{3} \approx 0.6667$

5. Approximation: 1.245
Exact area: $\frac{5}{4} = 1.25$

7. 71.25

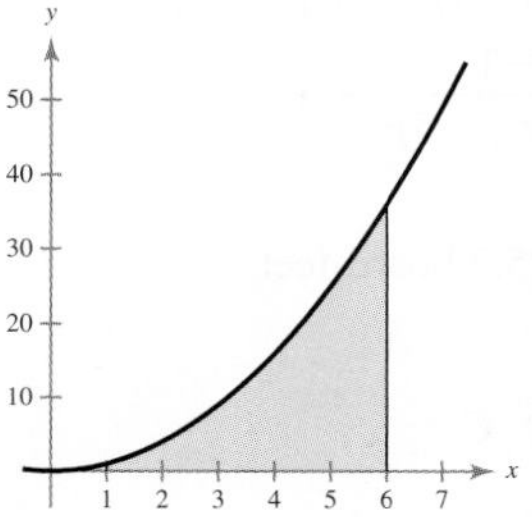

9. 1.079

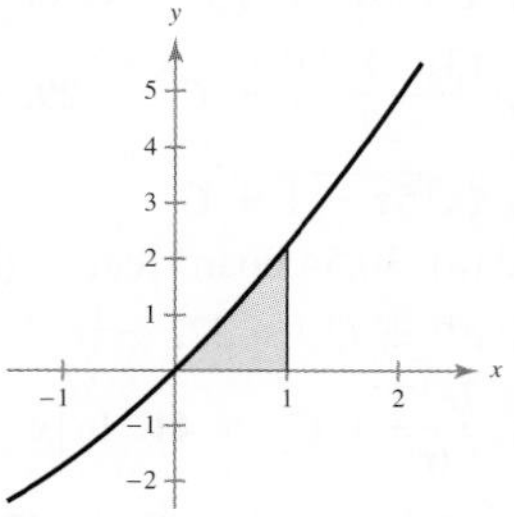

11. 24.28

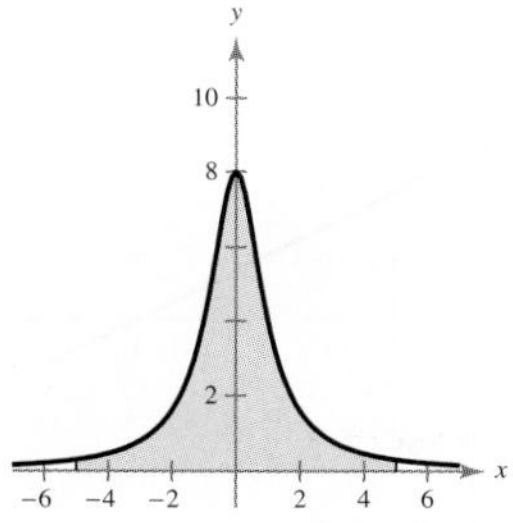

13. 17.25

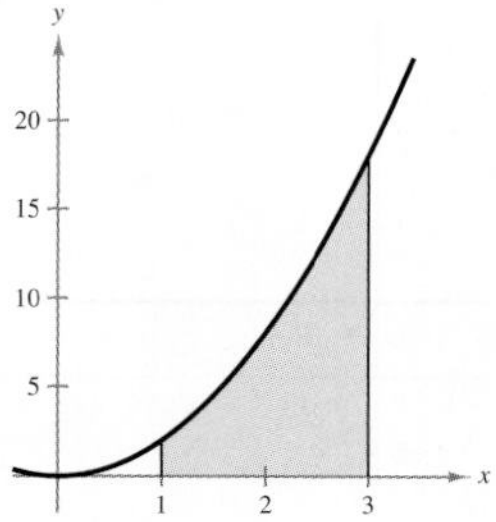

CHAPTER 5

15. 34.25

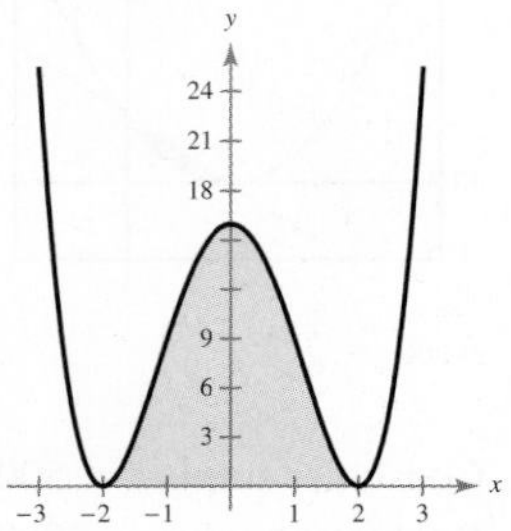

17. 1.39

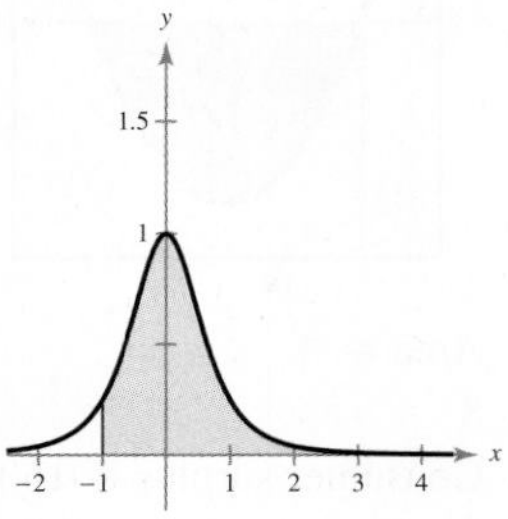

19.

n	10	20	30
Approximation	15.4543	15.4628	15.4644

n	40	50
Approximation	15.4650	15.4652

21.

n	10	20	30
Approximation	5.8520	5.8526	5.8528

n	40	50
Approximation	5.8528	5.8528

23. Area ≈ 54.6667, $n = 33$ **25.** Area ≈ 0.9163, $n = 3$

27. 9920 ft^2 **29.** 381.6 mi^2

31. Midpoint Rule: 3.1468
Graphing utility: 3.141593

Review Exercises for Chapter 5 *(page 370)*

1. $16x + C$ **3.** $\frac{3}{10}x^2 + C$ **5.** $x^3 + C$

7. $\frac{2}{3}x^3 + \frac{5}{2}x^2 + C$ **9.** $x^{2/3} + C$ **11.** $\frac{3}{7}x^{7/3} + \frac{3}{2}x^2 + C$

13. $\frac{4}{9}x^{9/2} - 2\sqrt{x} + C$ **15.** $6x^2 - 3$ **17.** $x^3 - 4x^2 + 15$

19. $s(t) = -16t^2 + 80t$
5 seconds

21. $\frac{1}{4}(x + 4)^4 + C$ **23.** $\frac{1}{5}(5x + 1)^5 + C$

25. $x + 5x^2 + \frac{25}{3}x^3 + C$ or $\frac{1}{15}(1 + 5x)^3 + C_1$

27. $\dfrac{(3x^3 + 1)^3}{27} + C$ **29.** $\dfrac{-1}{12(2x^3 - 5)^2} + C$

31. $\frac{2}{5}\sqrt{5x - 1} + C$

33. (a) 30.54 board-feet (b) 125.2 board-feet

35. $e^{4x} + C$ **37.** $-\frac{1}{5}e^{-5x} + C$

39. $\dfrac{7e^{3x^2}}{6} + C$ **41.** $\ln|x - 6| + C$

43. $\frac{2}{3}\ln|6x - 1| + C$ **45.** $-\frac{1}{3}\ln|1 - x^3| + C$

47.

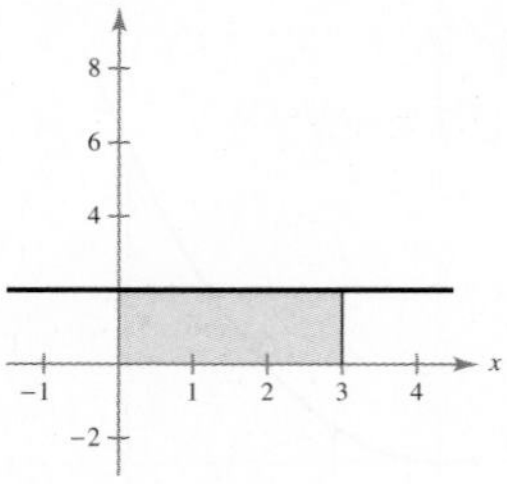

Area = 6

49.

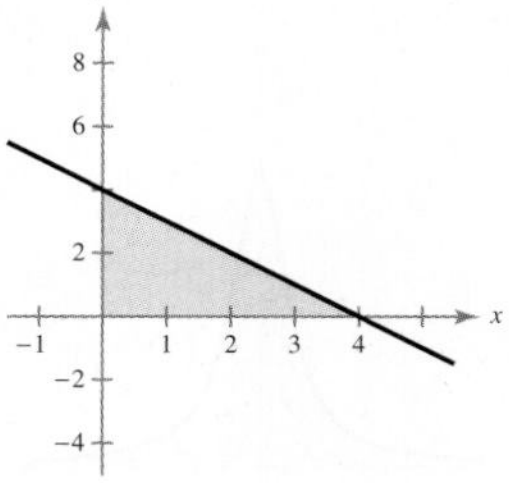

Area = 8

51. (a) 13 (b) 7 (c) 11 (d) 50

53. $\frac{32}{3}$ **55.** $2 \ln 2$ **57.** $4e^{1/2} - 4$ **59.** 16

61. 0 **63.** 4 **65.** 2 **67.** $5 \ln 3 \approx 5.49$

69. 1.899 **71.** Average value $= 3$; $x = 1$

73. Average value $= \frac{2}{3}(1 - e^3) \approx -12.724$;
$x = \ln\left[-\frac{1}{3}(1 - e^3)\right] \approx 1.850$

75. Average value $= \frac{2}{5}$; $x = \frac{25}{4}$ **77.** 0 **79.** 115.2

81. \$17,492.94 **83.** Increases by \$700.25 **85.** \$520.54

87.

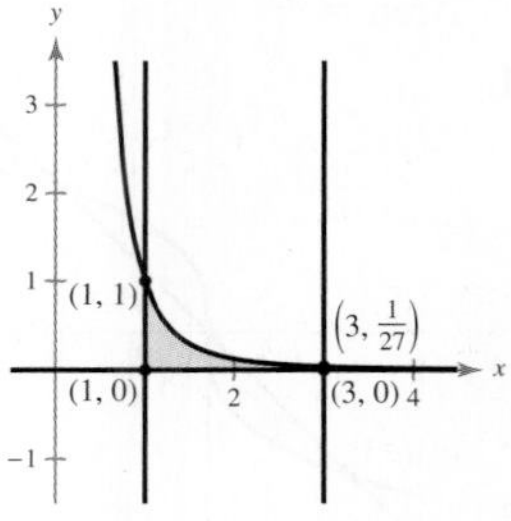

Area = $\frac{4}{9}$

89.

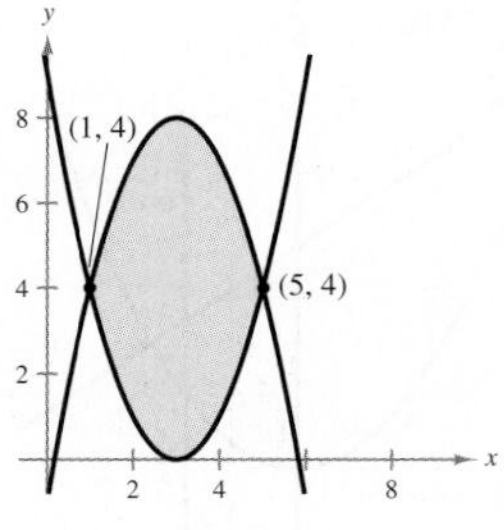

Area = $\frac{64}{3}$

91.

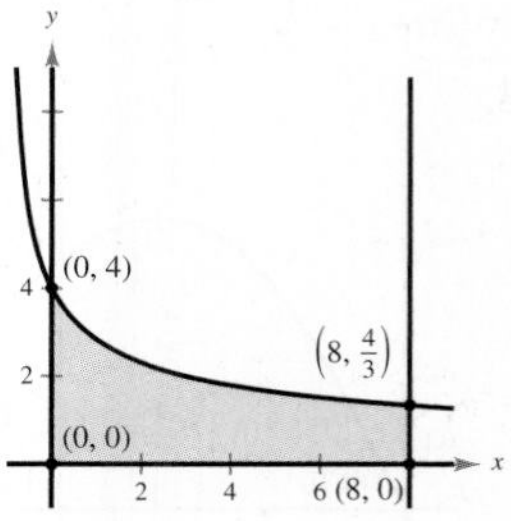

Area = 16

93.

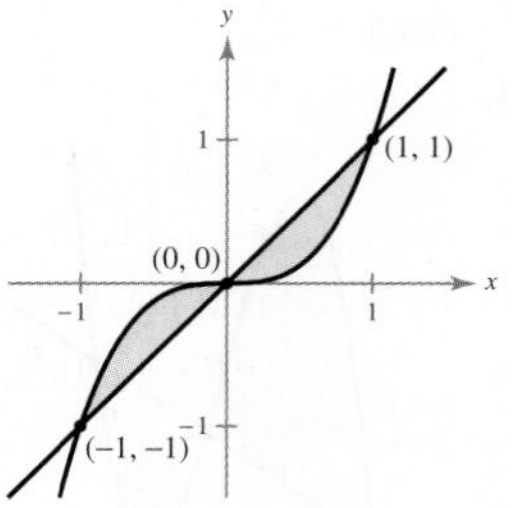

Area = $\frac{1}{2}$

95. Consumer surplus = 1417.5
Producer surplus = 202.5

97. Consumer surplus = 1250
Producer surplus = 1250

99. R_2; \$84.5 million

101. \$300 million

103. Approximation: 1.5
Actual area: 1.5

105. 2.625

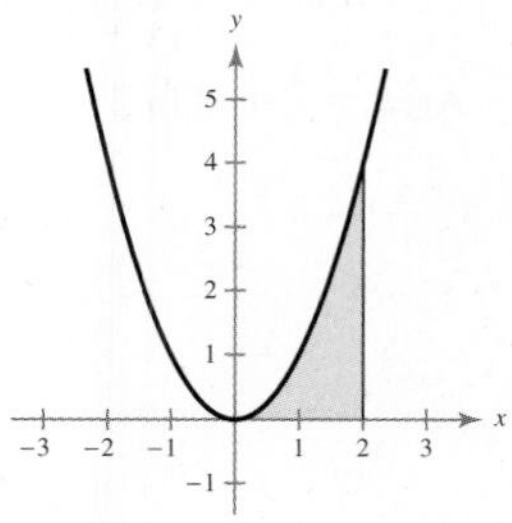

107. 1.070

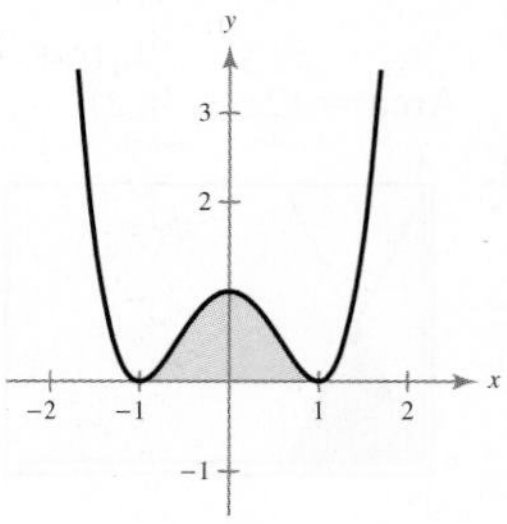

109. 13.5

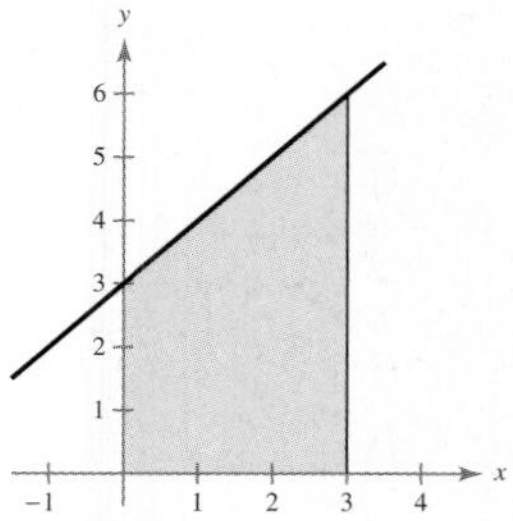

111. 3.032

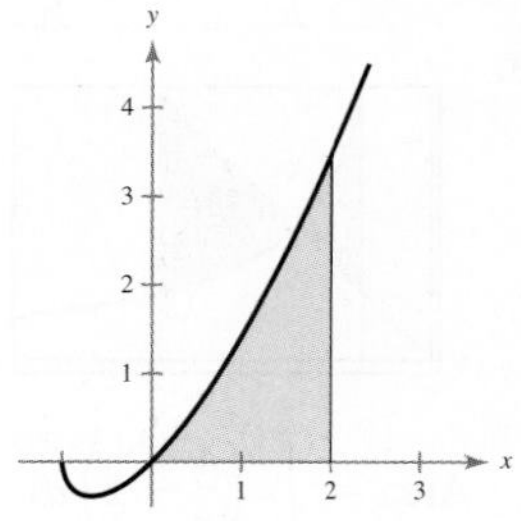

113. 9840 ft^2

Test Yourself *(page 374)*

1. $3x^3 - 2x^2 + 13x + C$ **2.** $\frac{(x+1)^3}{3} + C$

3. $\frac{2(x^4-7)^{3/2}}{3} + C$ **4.** $\frac{10x^{3/2}}{3} - 12x^{1/2} + C$

5. $5e^{3x} + C$ **6.** $\frac{3}{4}\ln|4x-1| + C$

7. $f(x) = 3x^2 - 5x - 2$ **8.** $f(x) = e^x + x$

9. 8 **10.** 18 **11.** $\frac{2}{3}$

12. $2\sqrt{5} - 2\sqrt{2} \approx 1.644$ **13.** $\frac{1}{4}(e^{12} - 1) \approx 40{,}688.4$

14. $\ln 6 \approx 1.792$

15. (a) $S(t) = 2240e^{0.1013t} + 4.3906,\ 0 \le t \le 9$
(b) \$3661.68 million

16.

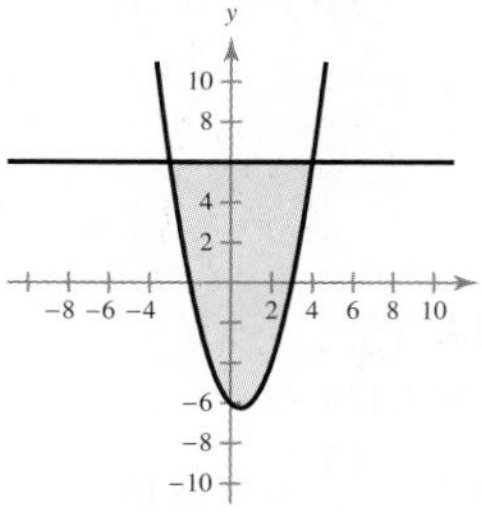

Area $= \frac{343}{6} \approx 57.167$

17.

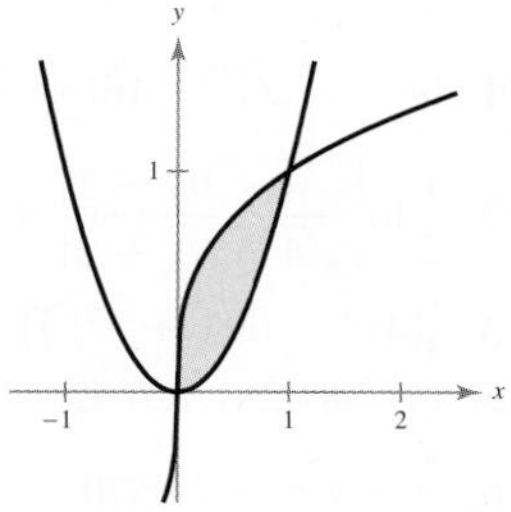

Area $= \frac{5}{12}$

18. Consumer surplus = 20 million
Producer surplus = 8 million

19. Midpoint Rule: $\frac{63}{64} \approx 0.9844$
Exact area: 1

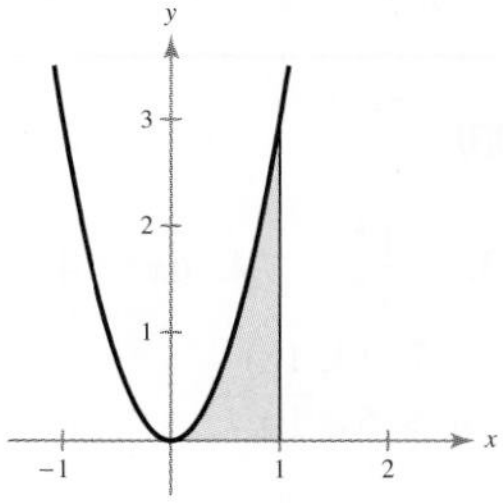

20. Midpoint Rule: $\frac{21}{8} = 2.625$
Exact area: $\frac{8}{3} = 2.\overline{6}$

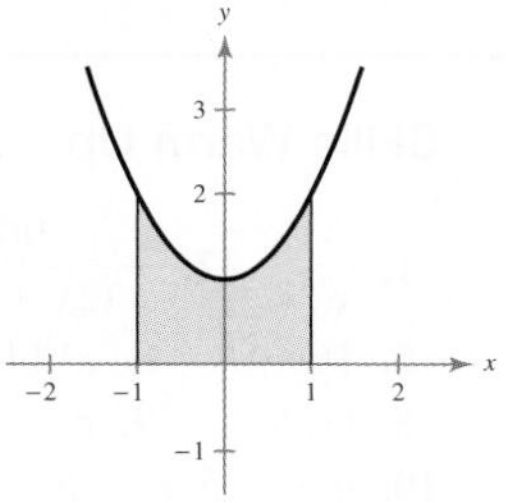

Chapter 6

Section 6.1 *(page 383)*

Skills Warm Up *(page 383)*

1. $\frac{1}{x+1}$ **2.** $\frac{2x}{x^2-1}$ **3.** $3x^2e^{x^3}$

4. $-2xe^{-x^2}$ **5.** $e^x(x^2+2x)$ **6.** $e^{-2x}(1-2x)$

7. $\frac{64}{3}$ **8.** $\frac{4}{3}$ **9.** 36 **10.** 8

1. $u = x;\ dv = e^{3x}\,dx$ **3.** $u = \ln 2x;\ dv = x\,dx$

5. $\frac{1}{3}xe^{3x} - \frac{1}{9}e^{3x} + C$ **7.** $\frac{x^4}{16}(4\ln x - 1) + C$

9. $x\ln 2x - x + C$ **11.** $-x^2e^{-x} - 2xe^{-x} - 2e^{-x} + C$

13. $\frac{2}{3}x^{3/2}\left(\ln x - \frac{2}{3}\right) + C$ **15.** $2x^2e^x - 4e^xx + 4e^x + C$

17. $\frac{1}{4}e^{4x} + C$ **19.** $\frac{1}{4}xe^{4x} - \frac{1}{16}e^{4x} + C$

21. $-4e^{-x/4}(x+4) + C$

23. $\frac{1}{2}t^2\ln(t+1) - \frac{1}{2}\ln(t+1) - \frac{1}{4}t^2 + \frac{1}{2}t + C$

25. $-e^{1/t} + C$ **27.** $\frac{1}{2}x^2(\ln x)^2 - \frac{1}{2}x^2\ln x + \frac{1}{4}x^2 + C$

29. $\frac{1}{3}(\ln x)^3 + C$ **31.** $-\frac{1}{x}(\ln x + 1) + C$

33. $\frac{2}{3}x(x-1)^{3/2} - \frac{4}{15}(x-1)^{5/2} + C$

35. $\frac{1}{4}x^4 + \frac{2}{3}x^3 + \frac{1}{2}x^2 + C$ **37.** $\frac{e^{2x}}{4(2x+1)} + C$

39. $\frac{5}{36}e^6 + \frac{1}{36} \approx 56.060$ **41.** $\frac{3}{2}\ln 3 - 1 \approx 0.648$

43. $\frac{1192}{15} \approx 79.467$ **45.** $e(2e - 1) \approx 12.060$

47. Area $= 2e^2 + 6 \approx 20.778$

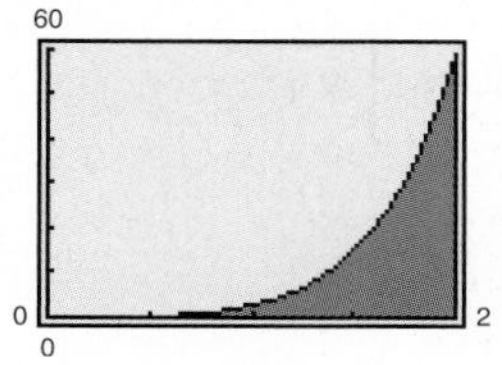

49. Area $= \frac{e-2}{e}$
≈ 0.2642

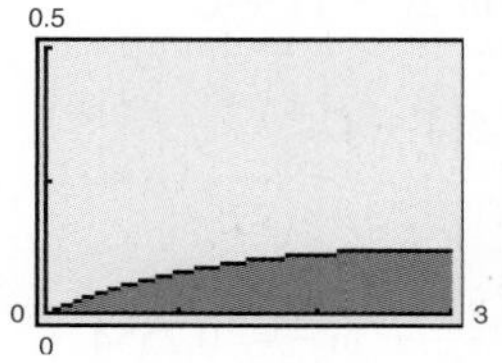

51. Area $= \frac{1}{9}(2e^3 + 1)$
≈ 4.575

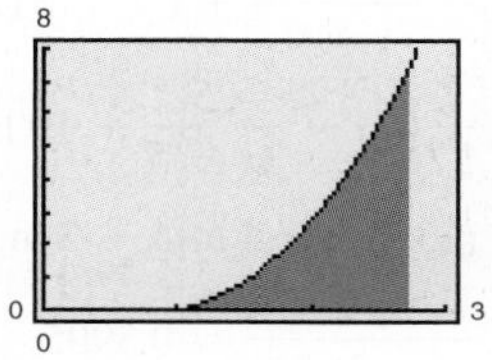

53. Proof **55.** $\frac{e^{5x}}{125}(25x^2 - 10x + 2) + C$

57. $-\frac{1}{9x^3}(1 + 3\ln x) + C$ **59.** $\frac{3}{128} - \frac{379}{128}e^{-8} \approx 0.022$

61. $\frac{1{,}171{,}875}{256}\pi \approx 14{,}381.070$

63.

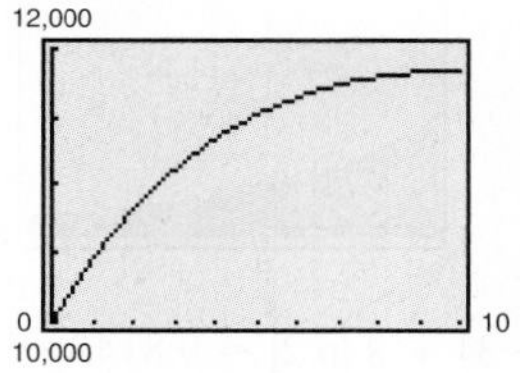

(a) Increase
(b) 113,212 units
(c) 11,321 units/yr

65. (a) $3.2\ln 2 - 0.2 \approx 2.018$
(b) $12.8\ln 4 - 7.2\ln 3 - 1.8 \approx 8.035$

67. \$18,482.03 **69.** \$931,265.10 **71.** \$4103.07

73. \$1,055,267 **75.** (a) \$1,200,000 (b) \$1,094,142.27

77. (a) \$18,000,000 (b) \$16,133,084 **79.** \$45,957.78

81. (a) \$17,378.62 (b) \$3681.26 **83.** ≈ 4.254

Section 6.2 *(page 391)*

Skills Warm Up *(page 391)*

1. $x^2 + 8x + 16$ **2.** $x^2 - 2x + 1$

3. $x^2 + x + \frac{1}{4}$ **4.** $x^2 - \frac{2}{3}x + \frac{1}{9}$

5. $2e^x(x-1) + C$ **6.** $x^3\ln x - \frac{x^3}{3} + C$

1. $\frac{1}{9}\left(\frac{2}{2+3x} + \ln|2+3x|\right) + C$

3. $\frac{2(3x-4)}{27}\sqrt{2+3x} + C$ **5.** $\ln\left(x^2 + \sqrt{x^4 - 9}\right) + C$

7. $\frac{1}{2}(x^2 - 1)e^{x^2} + C$ **9.** $\ln\left|\frac{x}{1+x}\right| + C$

11. $-\frac{1}{3}\ln\left|\frac{3 + \sqrt{x^2+9}}{x}\right| + C$

13. $-\frac{1}{2}\ln\left|\frac{2 + \sqrt{4 - x^2}}{x}\right| + C$ **15.** $\frac{3}{4}x^2[-1 + 2\ln(3x)] + C$

17. $3x^2 - \ln(1 + e^{3x^2}) + C$

19. $\frac{2}{35}(x+3)^{3/2}(5x^2 - 12x + 24) + C$

21. $\frac{1}{27}\left[\frac{4}{2+3t} - \frac{2}{(2+3t)^2} + \ln|2+3t|\right] + C$

23. $\frac{1}{\sqrt{3}}\ln\left|\frac{\sqrt{3+4x} - \sqrt{3}}{\sqrt{3+4x} + \sqrt{3}}\right| + C$

25. $-\frac{1}{2}x(2 - x) + \ln|x+1| + C$

27. $\frac{1}{8}\left[\frac{-1}{2(3+2x)^2} + \frac{2}{(3+2x)^3} - \frac{9}{4(3+2x)^4}\right] + C$

29. $-\frac{\sqrt{1-x^2}}{x} + C$ **31.** $\frac{4}{9}x^3(3\ln 2x - 1) + C$

33. $\frac{1}{27}\left(3x - \frac{25}{3x-5} + 10\ln|3x-5|\right) + C$

35. $\frac{1}{9}(3\ln x - 4\ln|4 + 3\ln x|) + C$

37. $\frac{-2\sqrt{2} + 4}{3} \approx 0.3905$ **39.** $-\frac{5}{9} + \ln\frac{9}{4} \approx 0.2554$

41. $12\left(2 + \ln\left|\frac{2}{1+e^2}\right|\right) \approx 6.7946$ **43.** $8\ln 2 - \frac{15}{8} \approx 3.6702$

45. Area $= \frac{1}{8\sqrt{3}} \approx 0.0722$ **47.** Area $= \frac{3\ln\frac{4}{5} + 1}{36} \approx 0.0092$

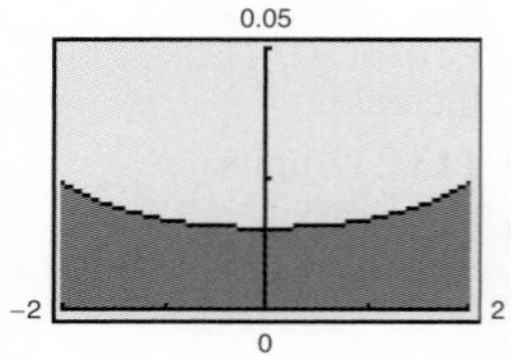

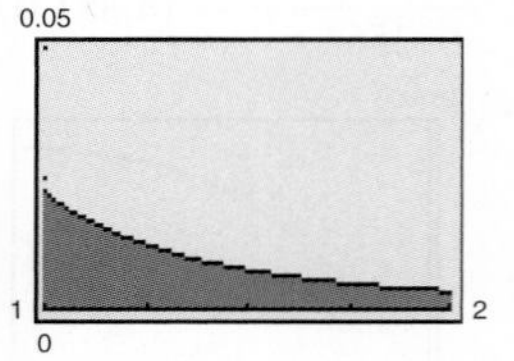

49. Area $= \frac{1}{4}\left[21\sqrt{5} - 8\ln\left(\sqrt{5} + 3\right) + 8\ln 2\right] \approx 9.8145$

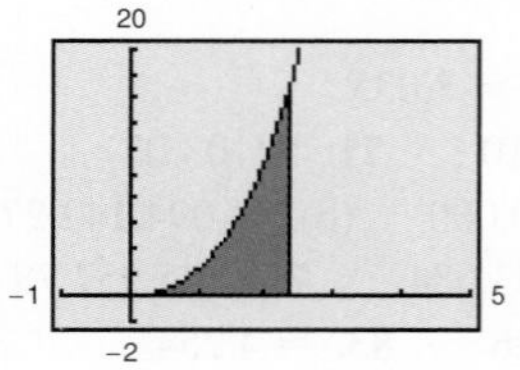

51. $x\left(\ln\frac{x}{3} - 1\right) + C$ **53.** $\frac{2}{147}(7x + 6)\sqrt{7x - 3} + C$

55. (a) 0.483 (b) 0.283

57.

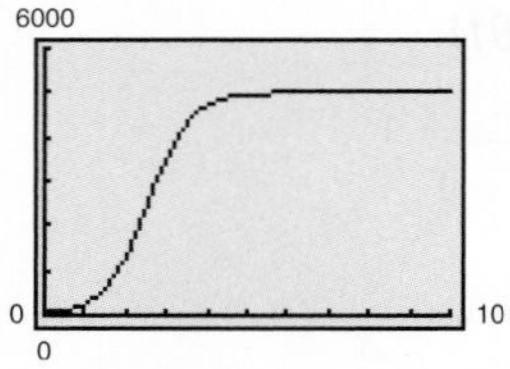

Average value: 401.40

59. \$1138.43

61. Consumer surplus: ≈ 17.92
Producer surplus: 24

Quiz Yourself *(page 393)*

1. $\frac{1}{5}xe^{5x} - \frac{1}{25}e^{5x} + C$ **2.** $3x\ln x - 3x + C$

3. $\frac{1}{2}x^2\ln x + x\ln x - \frac{1}{4}x^2 - x + C$

4. $\frac{2}{3}x(x+3)^{3/2} - \frac{4}{15}(x+3)^{5/2} + C$

5. $\frac{x^2}{4}\ln x - \frac{x^2}{8} + C$ **6.** $-\frac{1}{2}e^{-2x}\left(x^2 + x + \frac{1}{2}\right) + C$

7. (a) 282,016 units (b) 56,403 units

8. (a) \$784,000 (b) \$673,108.31

9. $\frac{1}{4}(2x - \ln|1 + 2x|) + C$ **10.** $10\ln\left|\frac{x}{0.1 + 0.2x}\right| + C$

11. $\ln\left|x + \sqrt{x^2 - 16}\right| - \frac{\sqrt{x^2 - 16}}{x} + C$

12. $\frac{1}{2}\ln\left|\frac{\sqrt{4+9x} - 2}{\sqrt{4+9x} + 2}\right| + C$

13. $\frac{1}{4}[4x^2 - \ln(1 + e^{4x^2})] + C$ **14.** $x^2e^{x^2+1} + C$

15. (a) \$84,281,126.52 (b) \$257,392,429.72

16. $\frac{8}{e} - 4 \approx -1.0570$ **17.** $10\ln 2 - \frac{15}{4} \approx 3.1815$

18. $\frac{64}{3}\left(\sqrt{2} - 1\right) \approx 8.8366$ **19.** $e - 2 \approx 0.7183$

20. $\frac{\sqrt{5}}{18} \approx 0.1242$ **21.** $\frac{1}{4}\left(\ln\frac{17}{19} - \ln\frac{7}{9}\right) \approx 0.0350$

Section 6.3 *(page 400)*

Skills Warm Up *(page 400)*

1. $\frac{2}{x^3}$ **2.** $-\frac{96}{(2x+1)^4}$ **3.** $-\frac{12}{x^4}$ **4.** $6x - 4$

5. $16e^{2x}$ **6.** $e^{x^2}(4x^2 + 2)$ **7.** (3, 18)

8. (1, 8) **9.** $n < -5\sqrt{10},\ n > 5\sqrt{10}$

10. $n < -5, n > 5$

	Trapezoidal Rule	*Simpson's Rule*	*Exact Value*
1.	2.7500	2.6667	2.6667
3.	0.2704	0.2512	0.2499
5.	0.6941	0.6932	0.6931
7.	5.2650	5.3046	5.3333
9.	3.8643	3.3022	3.1809

11. (a) 0.783 (b) 0.785 **13.** (a) 2.540 (b) 2.541

15. (a) 3.283 (b) 3.240 **17.** (a) 1.470 (b) 1.463

19. (a) 1.879 (b) 1.888 **21.** \$21,831.20; \$21,836.98

23. \$678.36 **25.** 0.3413 = 34.13%

27. 0.5000 = 50.00% **29.** 89,500 ft^2

31. (a) $|E| \le \frac{1}{12} \approx 0.0833$ (b) $|E| \le 0$

33. (a) $|E| \le \frac{5e}{64} \approx 0.212$ (b) $|E| \le \frac{13e}{1024} \approx 0.035$

35. (a) $n = 566$ (b) $n = 16$

37. (a) $n = 3280$ (b) $n = 60$ **39.** 19.5215 **41.** 3.6558

43. (a) 36.2 years (b) 36.2 years
(c) The results are the same.

45. 58.912 mg **47.** 1878 subscribers

49. Answers will vary.

Section 6.4 *(page 410)*

Skills Warm Up *(page 410)*

1. 9 **2.** 3 **3.** $-\frac{1}{8}$ **4.** Limit does not exist.
5. Limit does not exist. **6.** -4
7. (a) $\frac{32}{3}b^3 - 16b^2 + 8b - \frac{4}{3}$ (b) $-\frac{4}{3}$
8. (a) $\dfrac{b^2 - b - 11}{(b-2)^2(b-5)}$ (b) $\dfrac{11}{20}$
9. (a) $\ln\left(\dfrac{5 - 3b^2}{b+1}\right)$ (b) $\ln 5 \approx 1.609$
10. (a) $e^{-3b^2}(e^{6b^2} + 1)$ (b) 2

1. Improper; The integrand has an infinite discontinuity when $x = \frac{2}{3}$, and $0 \le \frac{2}{3} \le 1$.
3. Not improper; continuous on $[0, 1]$
5. Not improper; continuous on $[0, 5]$
7. Converges; 1 **9.** Diverges **11.** Diverges
13. Diverges **15.** Diverges **17.** Converges; 0
19. Converges; $\dfrac{1}{2(\ln 4)^2}$ **21.** 1 **23.** $\dfrac{1}{2}$ **25.** ∞
27. (a) 0.9026 (b) 0.0738 (c) 0.00235
29. (a) $\displaystyle\int_{35.5}^{\infty} f(x)\,dx \approx 0.9938$ (b) $\displaystyle\int_{35.9}^{\infty} f(x)\,dx \approx 0.6915$
31. \$66,666.67 **33.** Yes, $\$360{,}000 < \$400{,}000$.
35. (a) \$4,637,228 (b) \$5,555,556
37. (a) \$748,367.34 (b) \$808,030.14 (c) \$900,000.00
39. (a) \$453,901.30 (b) \$807,922.43 (c) \$4,466,666.67

Review Exercises for Chapter 6 *(page 416)*

1. $2\sqrt{x}\ln x - 4\sqrt{x} + C$ **3.** $xe^x + C$
5. $\frac{2}{15}(x-5)^{3/2}(3x+10) + C$ **7.** $x^2e^{2x} - xe^{2x} + \frac{1}{2}e^{2x} + C$
9. $3e^2 - \dfrac{3(e^2-1)}{2} \approx 12.584$ **11.** $16 - 20e^{-1/4} \approx 0.4240$
13. \$90,634.62 **15.** \$865,958.50
17. (a) \$1,200,000 (b) \$1,052,649.52
19. $\frac{1}{54}(9x^2 - 12x + 8\ln|3x+2|) + C$
21. $\dfrac{x}{2}\sqrt{x^2-16} - 8\ln\left(\sqrt{x^2-16} + x\right) + C$
23. $\dfrac{1}{9}\left(\dfrac{2}{2+3x} + \ln|2+3x|\right) + C$
25. $\sqrt{x^2+25} - 5\ln\left|\dfrac{5 + \sqrt{x^2+25}}{x}\right| + C$
27. $\dfrac{1}{4}\ln\left|\dfrac{x-2}{x+2}\right| + C$ **29.** $\dfrac{2}{3}(x-2)\sqrt{1+x} + C$
31. $2\sqrt{1+x} + \ln\left|\dfrac{\sqrt{1+x}-1}{\sqrt{1+x}+1}\right| + C$
33. (a) 0.675 (b) 0.290
35. Exact: $\frac{2}{3} \approx 0.6667$
Trapezoidal: 0.7050
Simpson's: 0.6715
37. Exact: $\frac{3}{8} = 0.375$
Trapezoidal: 0.3786
Simpson's: 0.3751
39. Exact: $2 - 2e^{-2} \approx 1.7293$
Trapezoidal: 1.7652
Simpson's: 1.7299
41. (a) 0.741 (b) 0.737
43. (a) 0.305 (b) 0.289 **45.** (a) 2.961 (b) 2.936
47. (a) $|E| \le \dfrac{e^4}{6} \approx 9.0997$ (b) $|E| \le \dfrac{e^4}{90} \approx 0.6066$
49. (a) $n = 214$ (b) $n = 2$
51. Converges, $-\frac{1}{4}$ **53.** Diverges **55.** Diverges
57. $A \approx 4$ **59.** $A = 1$ **61.** \$266,666.67 **63.** No

Test Yourself *(page 418)*

1. $xe^{x+1} - e^{x+1} + C$ **2.** $3x^3\ln x - x^3 + C$
3. $-3x^2e^{-x/3} - 18xe^{-x/3} - 54e^{-x/3} + C$
4. (a) $\approx$\$6494.47 million (b) $\approx$\$811.81 million
5. $\dfrac{1}{4}\left(\dfrac{7}{7+2x} + \ln|7+2x|\right) + C$ **6.** $x^3 - \ln(1 + e^{x^3}) + C$
7. $-\frac{2}{75}(2 - 5x^2)\sqrt{1+5x^2} + C$ **8.** $-1 + \frac{3}{2}\ln 3 \approx 0.6479$
9. $8\frac{2}{3}$ **10.** $4\ln\left[3\left(\sqrt{17} - 4\right)\right] + \sqrt{17} - 5 \approx -4.8613$
11. Exact: 18.0
Trapezoidal: 18.28
12. Simpson's Rule: 41.3606; Exact: 41.1711
13. Converges; $\frac{1}{3}$ **14.** Converges; 12 **15.** Diverges
16. (a) \$498.75 (b) Plan B, because $\$149 < \$498.75$.

Chapter 7

Section 7.1 *(page 425)*

Skills Warm Up *(page 425)*

1. $2\sqrt{5}$ **2.** 5 **3.** 8 **4.** 8 **5.** $(4, 7)$
6. $(1, 0)$ **7.** $(0, 3)$ **8.** $(-1, 1)$
9. $(x-2)^2 + (y-3)^2 = 4$
10. $(x-1)^2 + (y-4)^2 = 25$

1.

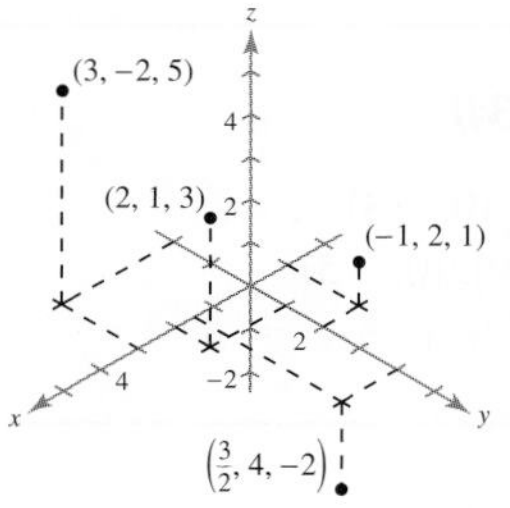

3.

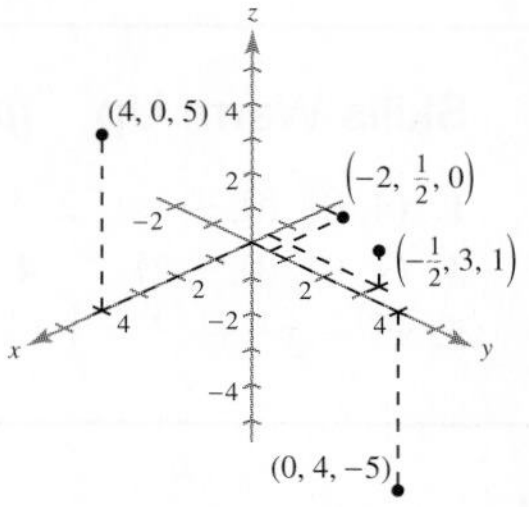

5. $(-3, 4, 5)$ **7.** $(10, 0, 0)$ **9.** 0 **11.** $3\sqrt{2}$
13. $\sqrt{206}$ **15.** $\left(2, -\frac{3}{2}, \frac{5}{2}\right)$ **17.** $\left(\frac{1}{2}, \frac{1}{2}, -1\right)$
19. $(6, -3, 5)$ **21.** $(1, 2, 1)$ **23.** $3, 3\sqrt{5}, 6$; right triangle
25. $\sqrt{41}, \sqrt{14}, \sqrt{41}$; isosceles triangle
27. $(0, 0, -5), (2, 2, -4), (2, -4, -1)$
29. $x^2 + (y-2)^2 + (z-2)^2 = 4$
31. $\left(x - \frac{3}{2}\right)^2 + (y-2)^2 + (z-1)^2 = \frac{21}{4}$
33. $(x-3)^2 + (y+2)^2 + (z+3)^2 = 16$
35. $(x-1)^2 + (y-3)^2 + z^2 = 10$
37. $(x+4)^2 + (y-3)^2 + (z-2)^2 = 4$
39. Center: $\left(\frac{5}{2}, 0, 0\right)$
Radius: $\frac{5}{2}$
41. Center: $(-2, 1, -4)$
Radius: 5
43. Center: $(1, 3, 2)$
Radius: $\dfrac{5\sqrt{2}}{2}$

45.

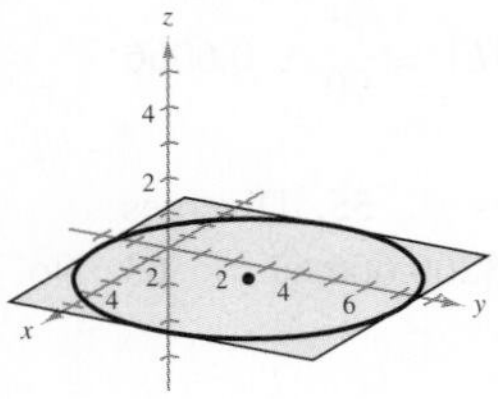

47.

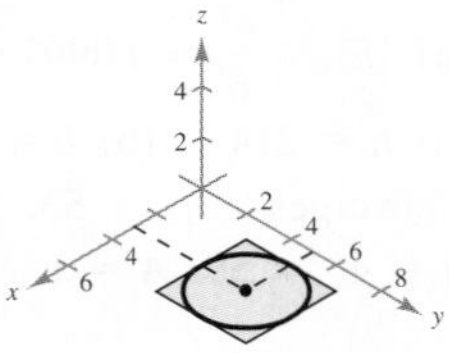

49.

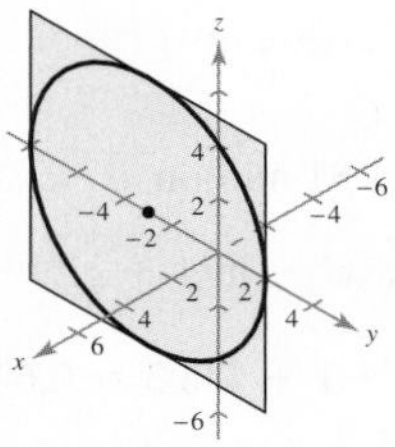

51.

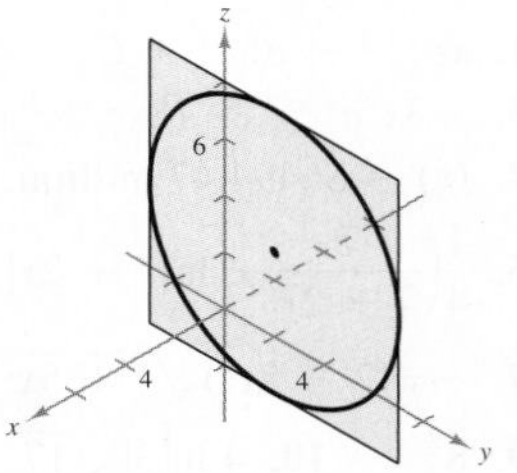

53. (a)

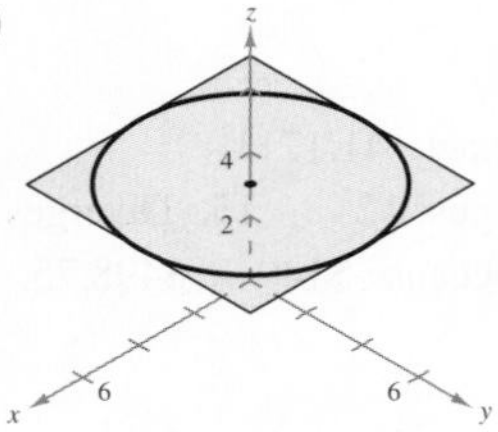

(b)

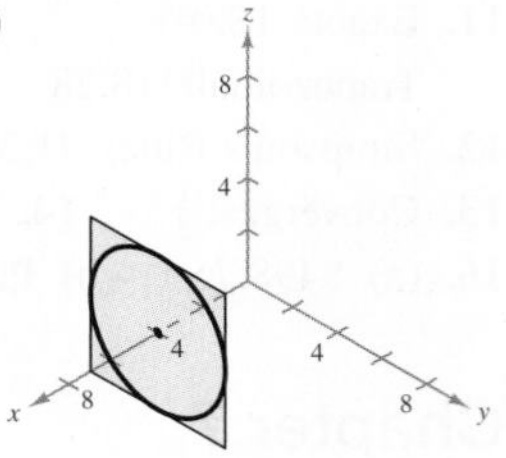

55. (a)

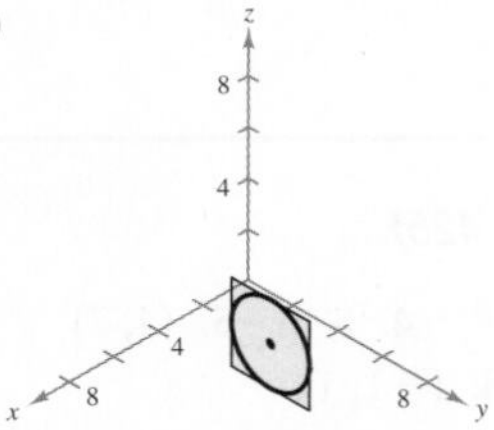

(b)

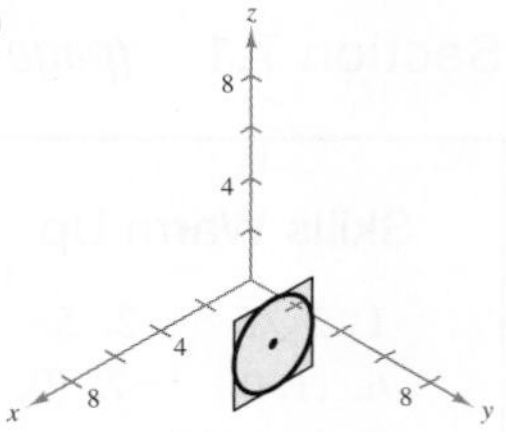

57. $x^2 + y^2 + z^2 = 6806.25$

Section 7.2 *(page 434)*

Skills Warm Up *(page 434)*

1. $(4, 0), (0, 3)$ **2.** $\left(-\frac{4}{3}, 0\right), (0, -8)$
3. $(1, 0), (0, -2)$ **4.** $(-5, 0), (0, -5)$
5. $x^2 + y^2 + z^2 = \frac{1}{4}$ **6.** $x^2 + y^2 + z^2 = 4$

1.

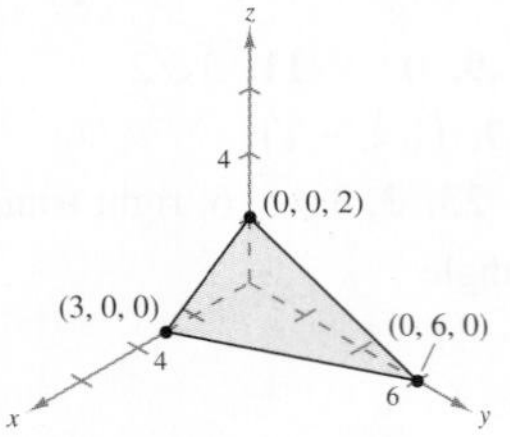

3.

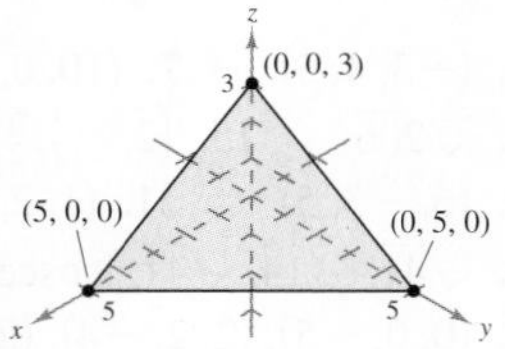

5.

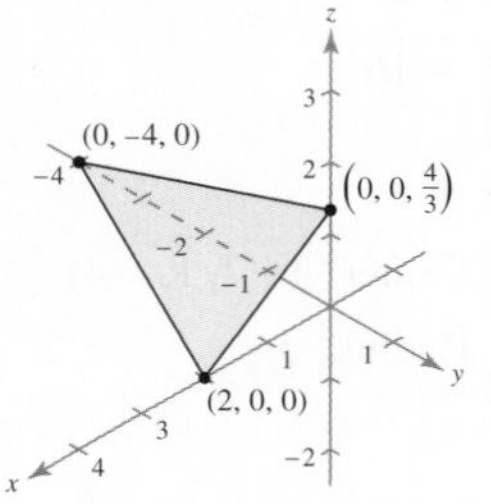

7.

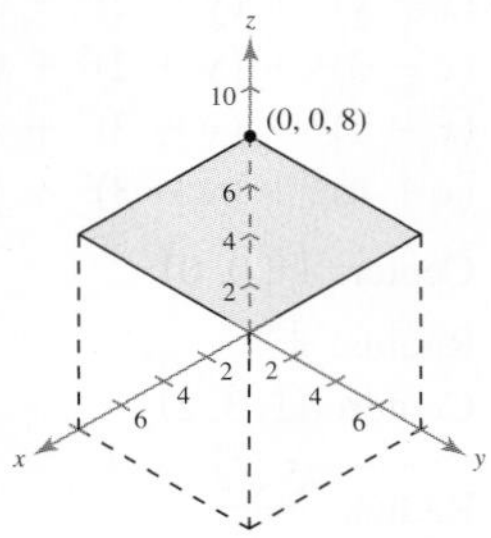

9.

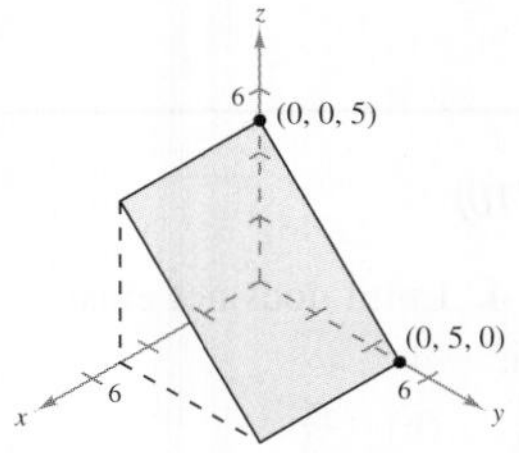

11.

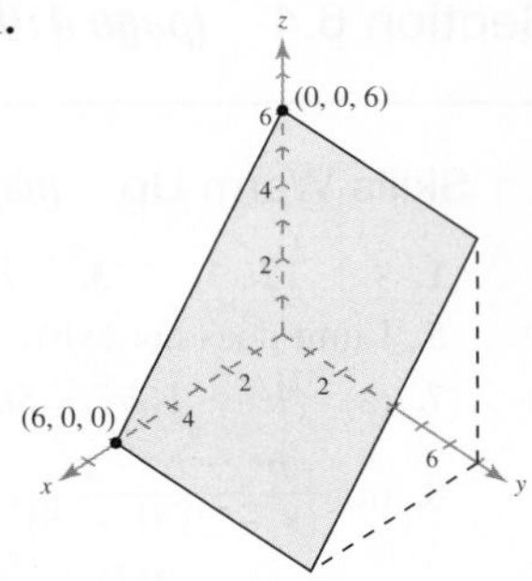

13. Perpendicular **15.** Parallel **17.** Parallel
19. Neither parallel nor perpendicular **21.** Perpendicular
23. c; Ellipsoid **24.** e; Hyperboloid of two sheets
25. f; Hyperboloid of one sheet **26.** b; Elliptic cone
27. d; Elliptic paraboloid **28.** a; Hyperbolic paraboloid
29. (a) $x = \pm y$; Lines
(b) $z = 9 - y^2$; Parabola
(c) $z = x^2$; Parabola
Hyperbolic parabola
31. (a) $\frac{x^2}{4} + y^2 = 1$; Ellipse (b) $\frac{x^2}{4} + z^2 = 1$; Ellipse
(c) $y^2 + z^2 = 1$; Circle
Ellipsoid
33. (a) $z^2 - \frac{x^2}{9} = 1$; Hyperbola
(b) $\frac{9}{13}z^2 - \frac{9}{208}y^2 = 1$; Hyperbola
(c) $\frac{x^2}{135} + \frac{y^2}{240} = 1$; Ellipse
Hyperboloid of two sheets
35. Ellipsoid **37.** Hyperboloid of one sheet
39. Hyperbolic paraboloid **41.** Elliptic paraboloid
43. Elliptic cone **45.** Hyperboloid of two sheets
47. Hyperbolic paraboloid
49. $\frac{x^2}{3963^2} + \frac{y^2}{3963^2} + \frac{z^2}{3950^2} = 1$
51. (a)

Year	2004	2005	2006
x	33.1	34.9	37.4
y	13.2	13.8	14.9
z (actual)	15.5	16.3	17.8
z (approximated)	15.5	16.4	17.5

Year	2007	2008	2009
x	40.6	43.0	41.8
y	15.0	15.4	14.5
z (actual)	19.5	20.5	20.7
z (approximated)	19.4	20.7	20.4

The approximated values of z are very close to the actual values.

(b) According to the model, increases in expenditures of recreation types y and z will correspond to an increase in expenditures of recreation type x.

Section 7.3 (page 441)

Skills Warm Up (page 441)

1. 11 **2.** -16 **3.** 7 **4.** 4 **5.** $(-\infty, \infty)$
6. $(-\infty, -3) \cup (-3, 0) \cup (0, \infty)$
7. $[5, \infty)$ **8.** $\left(-\infty, -\sqrt{5}\right] \cup \left[\sqrt{5}, \infty\right)$
9. 55.0104 **10.** 6.9165

1. (a) $\frac{3}{2}$ (b) $-\frac{1}{4}$ (c) 6 (d) $\frac{5}{y}$ (e) $\frac{x}{2}$ (f) $\frac{5}{t}$
3. (a) 5 (b) $3e^2$ (c) $2e^{-1}$ (d) $5e^y$ (e) xe^2 (f) te^t
5. (a) $\frac{2}{3}$ (b) 0 **7.** (a) 90π (b) 50π
9. (a) \$20,655.20 (b) \$1,397,672.67 **11.** (a) 0 (b) 6
13. (a) $x^2 + 2x\,\Delta x + (\Delta x)^2 - 2y$ (b) -2, $\Delta y \neq 0$
15. Domain: all points (x, y) inside and on the circle $x^2 + y^2 = 16$
Range: $[0, 4]$
17. Domain: all points (x, y)
Range: $[0, \infty)$
19. Domain: all points (x, y) such that $y \neq 0$
Range: $(0, \infty)$
21. Domain: the half-plane below the line $y = -x + 4$
Range: $(-\infty, \infty)$
23. Domain: all points (x, y) inside and on the ellipse $3x^2 + y^2 = 9$
Range: $[0, 3]$
25. Domain: all points (x, y) such that $x \neq 0$
Range: $(-\infty, \infty)$
27. Domain: all points (x, y) such that $x \neq 0$ and $y \neq 0$
Range: $(-\infty, 0)$ and $(0, \infty)$
29. Domain: all points (x, y) such that $y \geq 0$
Range: $(-\infty, \infty)$
31. b **32.** d **33.** a **34.** c
35. The level curves are parallel lines.

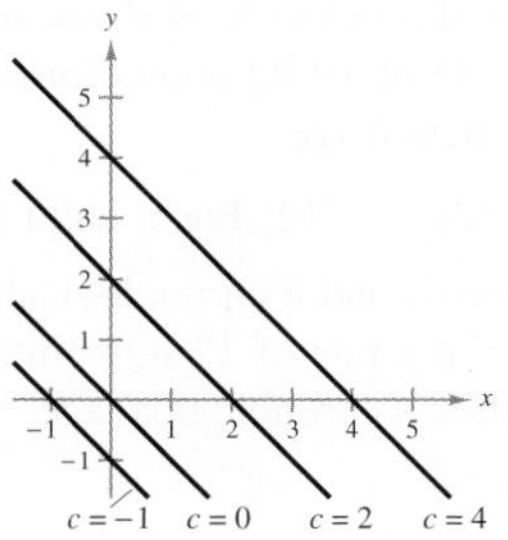

37. The level curves are circles.

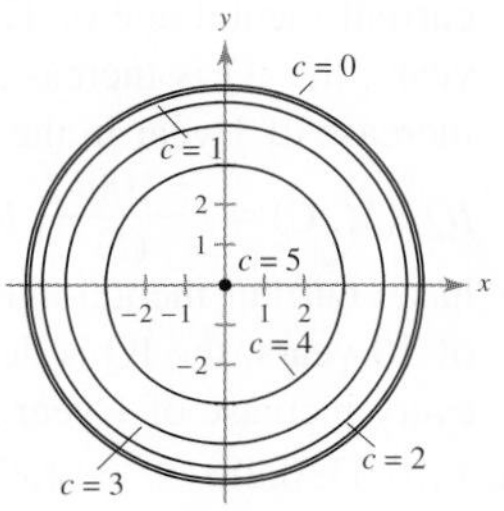

39. The level curves are hyperbolas.

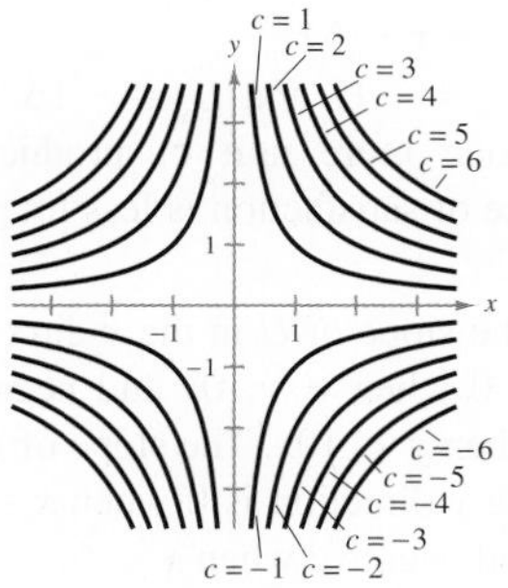

41. The level curves are circles.

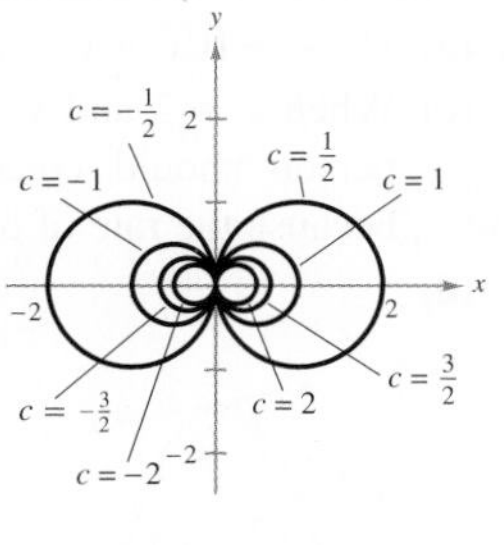

43. 135,540 units
45. (a) \$15,250 (b) \$18,425 (c) \$30,025

47.

R \ I	0	0.03	0.05
0	\$5187.48	\$3859.98	\$3184.67
0.28	\$4008.46	\$2982.67	\$2460.85
0.35	\$3754.27	\$2793.53	\$2304.80

49. (a) C (b) A (c) B
51. (a) \$2.78 earnings per share
(b) x; Explanations will vary. Sample answer: The x-variable has a greater influence on the earnings per share because the absolute value of its coefficient is larger than the absolute value of the coefficient of the y-term.
53. Option (a): \$1003.73, \$240,895.20;
Option (b): \$798.36, \$287,409.60;
Option (c): \$1078.59, \$194,146.20;
Answers will vary.

Section 7.4 (page 452)

Skills Warm Up (page 452)

1. $\frac{x}{\sqrt{x^2+3}}$ **2.** $-6x(3 - x^2)^2$ **3.** $e^{2t+1}(2t + 1)$
4. $\frac{e^{2x}(2 - 3e^{2x})}{\sqrt{1 - e^{2x}}}$ **5.** $-\frac{2}{3 - 2x}$ **6.** $\frac{3(t^2 - 2)}{2t(t^2 - 6)}$
7. $-\frac{10x}{(4x - 1)^3}$ **8.** $-\frac{(x + 2)^2(x^2 + 8x + 27)}{(x^2 - 9)^3}$
9. $f'(2) = 8$ **10.** $g'(2) = \frac{7}{2}$

1. $\frac{\partial z}{\partial x} = 3$; $\frac{\partial z}{\partial y} = 5$ **3.** $f_x(x, y) = 3$; $f_y(x, y) = -12y$
5. $f_x(x, y) = \frac{1}{y}$; $f_y(x, y) = -\frac{x}{y^2}$
7. $f_x(x, y) = \frac{x}{\sqrt{x^2 + y^2}}$; $f_y(x, y) = \frac{y}{\sqrt{x^2 + y^2}}$
9. $\frac{\partial z}{\partial x} = 2xe^{2y}$; $\frac{\partial z}{\partial y} = 2x^2e^{2y}$
11. $h_x(x, y) = -2xe^{-(x^2+y^2)}$; $h_y(x, y) = -2ye^{-(x^2+y^2)}$
13. $\frac{\partial z}{\partial x} = -\frac{2y}{x^2 - y^2}$; $\frac{\partial z}{\partial y} = \frac{2x}{x^2 - y^2}$
15. $f_x(x, y) = 6x + y$, 13; $f_y(x, y) = x - 2y$, 0
17. $f_x(x, y) = 3ye^{3xy}$, 12; $f_y(x, y) = 3xe^{3xy}$, 0
19. $f_x(x, y) = -\frac{y^2}{(x - y)^2}$, $-\frac{1}{4}$; $f_y(x, y) = \frac{x^2}{(x - y)^2}$, $\frac{1}{4}$
21. $f_x(x, y) = \frac{3}{3x + 5y}$, 1; $f_y(x, y) = \frac{5}{3x + 5y}$, $\frac{5}{3}$
23. (a) 2 (b) 1 **25.** (a) -2 (b) -2
27. $w_x = y^2z^4$
$w_y = 2xyz^4$
$w_z = 4xy^2z^3$
29. $w_x = -\frac{2z}{(x + y)^2}$
$w_y = -\frac{2z}{(x + y)^2}$
$w_z = \frac{2}{x + y}$

31. $w_x = 2z^2 + 3yz,\ 2$
$w_y = 3xz - 12yz,\ 30$
$w_z = 4xz + 3xy - 6y^2,\ -1$

33. $w_x = \dfrac{x}{\sqrt{x^2 + y^2 + z^2}},\ \dfrac{2}{3}$
$w_y = \dfrac{y}{\sqrt{x^2 + y^2 + z^2}},\ -\dfrac{1}{3}$
$w_z = \dfrac{z}{\sqrt{x^2 + y^2 + z^2}},\ \dfrac{2}{3}$

35. $w_x = 4xy^3z^2e^{2x^2},\ -8\sqrt{e}$
$w_y = 3y^2z^2e^{2x^2},\ 12\sqrt{e}$
$w_z = 2y^3ze^{2x^2},\ -4\sqrt{e}$

37. $w_x = \dfrac{5}{5x + 2y^3 - 3z},\ \dfrac{1}{5}$
$w_y = \dfrac{6y^2}{5x + 2y^3 - 3z},\ \dfrac{6}{25}$
$w_z = -\dfrac{3}{5x + 2y^3 - 3z},\ -\dfrac{3}{25}$

39. $(-6, 4)$ **41.** $(1, 1)$

43. $\dfrac{\partial^2 z}{\partial x^2} = 6x$
$\dfrac{\partial^2 z}{\partial y^2} = -8$
$\dfrac{\partial^2 z}{\partial y \partial x} = \dfrac{\partial^2 z}{\partial x \partial y} = 0$

45. $\dfrac{\partial^2 z}{\partial x^2} = 2$
$\dfrac{\partial^2 z}{\partial x \partial y} = \dfrac{\partial^2 z}{\partial y \partial x} = -2$
$\dfrac{\partial^2 z}{\partial y^2} = 6$

47. $\dfrac{\partial^2 z}{\partial x^2} = 108x^2(3x^4 - 2y^3)(11x^4 - 2y^3)$
$\dfrac{\partial^2 z}{\partial y^2} = 36y(2y^3 - 3x^4)(3x^4 - 8y^3)$
$\dfrac{\partial^2 z}{\partial x \partial y} = \dfrac{\partial^2 z}{\partial y \partial x} = 432x^3y^2(2y^3 - 3x^4)$

49. $\dfrac{\partial^2 z}{\partial x^2} = -\dfrac{y}{x^3}$
$\dfrac{\partial^2 z}{\partial x dy} = -\dfrac{x^2 - y^2}{2x^2y^2}$
$\dfrac{\partial^2 z}{\partial y dx} = -\dfrac{x^2 - y^2}{2x^2y^2}$
$\dfrac{\partial^2 z}{\partial y^2} = \dfrac{x}{y^3}$

51. $f_{xx}(x, y) = 12x^2 - 6y^2,\ 12$
$f_{xy}(x, y) = -12xy,\ 0$
$f_{yy}(x, y) = -6x^2 + 2,\ -4$
$f_{yx}(x, y) = -12xy,\ 0$

53. $f_{xx}(x, y) = e^{x^2}(4x^2y^3 + 2y^3),\ -6e$
$f_{xy}(x, y) = f_{yx}(x, y) = 6xy^2e^{x^2},\ 6e$
$f_{yy}(x, y) = 6ye^{x^2},\ -6e$

55. $f_{xx}(x, y, z) = 2$
$f_{xy}(x, y, z) = f_{yx}(x, y, z) = -3$
$f_{xz}(x, y, z) = f_{yy}(x, y, z) = f_{zx}(x, y, z) = 0$
$f_{yz}(x, y, z) = f_{zy}(x, y, z) = 4$
$f_{zz}(x, y, z) = 6z$

57. $f_{xx}(x, y, z) = -\dfrac{8yz}{(x + y)^3}$
$f_{xy}(x, y, z) = \dfrac{4z(x - y)}{(x + y)^3}$
$f_{xz}(x, y, z) = \dfrac{4y}{(x + y)^2}$
$f_{yy}(x, y, z) = \dfrac{8xz}{(x + y)^3}$
$f_{yx}(x, y, z) = \dfrac{4z(x - y)}{(x + y)^3}$
$f_{yz}(x, y, z) = -\dfrac{4x}{(x + y)^2}$
$f_{zz}(x, y, z) = 0$
$f_{zx}(x, y, z) = \dfrac{4y}{(x + y)^2}$
$f_{zy}(x, y, z) = -\dfrac{4x}{(x + y)^2}$

59. (a) At (120, 160), $\dfrac{\partial C}{\partial x} \approx 154.77$; At (120, 160), $\dfrac{\partial C}{\partial y} \approx 193.33$
(b) Racing bikes; Explanations will vary. Sample answer: The absolute value of dC/dy is greater than the absolute value of dC/dx at (120, 160).

61. (a) About 113.72 (b) About 97.47

63. Complementary

65. (a) $\dfrac{\partial z}{\partial x} = 0.62$; $\dfrac{\partial z}{\partial y} = -0.41$
(b) For every increase of 1 billion dollars in expenditures on amusement parks and campgrounds, the expenditures for spectator sports will increase by 0.62 billion dollars. For every increase of 1 billion dollars in expenditures on live entertainment (excluding sports), the expenditures for spectator sports will decrease by 0.41 billion dollars.

67. $IQ_M(M, C) = \dfrac{100}{C}$, $IQ_M(12, 10) = 10$; For a child who has a current mental age of 12 years and a chronological age of 10 years, the IQ is increasing at a rate of 10 IQ points for every increase of 1 year in the child's mental age.
$IQ_C(M, C) = \dfrac{-100M}{C^2}$, $IQ_C(12, 10) = -12$; For a child who has a current mental age of 12 years and a chronological age of 10 years, the IQ is decreasing at a rate of 12 IQ points for every increase of 1 year in the child's chronological age.

69. $V_I(0.03, 0.28) \approx -14{,}478.99$
$V_R(0.03, 0.28) \approx -1391.17$
The rate of inflation has the greater negative influence on the growth of the investment because $|-14{,}478.99| > |-1391.17|$.

71. (a) $U_x = -10x + y$ (b) $U_y = x - 6y$
(c) When $x = 2$ and $y = 3$, $U_x = -17$ and $U_y = -16$. The person should consume one more unit of product y, because the rate of decrease of satisfaction is less for y.
(d)

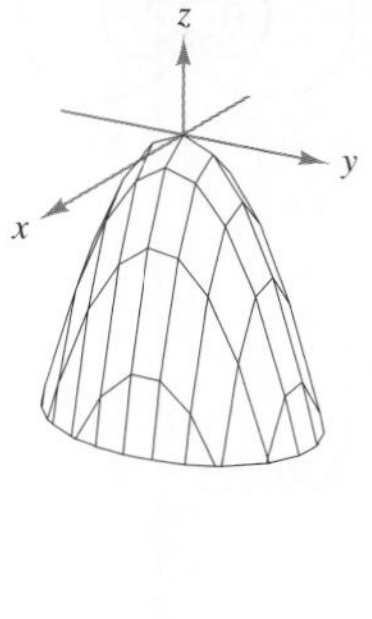

The slope of U in the x-direction is 0 when $y = 10x$ and negative when $y < 10x$. The slope of U in the y-direction is 0 when $x = 6y$ and negative when $x < 6y$.

Section 7.5 *(page 461)*

Skills Warm Up *(page 461)*

1. (3, 2) **2.** (11, 6) **3.** (1, 4) **4.** (4, 4)
5. (5, 2) **6.** (3, −2) **7.** (0, 0), (−1, 0)
8. (−2, 0), (2, −2)

9. $\frac{\partial z}{\partial x} = 12x^2$ $\quad \frac{\partial^2 z}{\partial y^2} = -6$
$\frac{\partial z}{\partial y} = -6y$ $\quad \frac{\partial^2 z}{\partial x \partial y} = 0$
$\frac{\partial^2 z}{\partial x^2} = 24x$ $\quad \frac{\partial^2 z}{\partial y \partial x} = 0$

10. $\frac{\partial z}{\partial x} = 10x^4$ $\quad \frac{\partial^2 z}{\partial y^2} = -6y$
$\frac{\partial z}{\partial y} = -3y^2$ $\quad \frac{\partial^2 z}{\partial x \partial y} = 0$
$\frac{\partial^2 z}{\partial x^2} = 40x^3$ $\quad \frac{\partial^2 z}{\partial y \partial x} = 0$

11. $\frac{\partial z}{\partial x} = 4x^3 - \frac{\sqrt{xy}}{2x}$ $\quad \frac{\partial^2 z}{\partial y^2} = \frac{\sqrt{xy}}{4y^2}$
$\frac{\partial z}{\partial y} = -\frac{\sqrt{xy}}{2y} + 2$ $\quad \frac{\partial^2 z}{\partial x \partial y} = -\frac{\sqrt{xy}}{4xy}$
$\frac{\partial^2 z}{\partial x^2} = 12x^2 + \frac{\sqrt{xy}}{4x^2}$ $\quad \frac{\partial^2 z}{\partial y \partial x} = -\frac{\sqrt{xy}}{4xy}$

12. $\frac{\partial z}{\partial x} = 4x - 3y$ $\quad \frac{\partial^2 z}{\partial y^2} = 2$
$\frac{\partial z}{\partial y} = 2y - 3x$ $\quad \frac{\partial^2 z}{\partial x \partial y} = -3$
$\frac{\partial^2 z}{\partial x^2} = 4$ $\quad \frac{\partial^2 z}{\partial y \partial x} = -3$

13. $\frac{\partial z}{\partial x} = y^3 e^{xy^2}$ $\quad \frac{\partial^2 z}{\partial y^2} = 4x^2y^3e^{xy^2} + 6xye^{xy^2}$
$\frac{\partial z}{\partial y} = 2xy^2e^{xy^2} + e^{xy^2}$ $\quad \frac{\partial^2 z}{\partial x \partial y} = 2xy^4e^{xy^2} + 3y^2e^{xy^2}$
$\frac{\partial^2 z}{\partial x^2} = y^5e^{xy^2}$ $\quad \frac{\partial^2 z}{\partial y \partial x} = 2xy^4e^{xy^2} + 3y^2e^{xy^2}$

14. $\frac{\partial z}{\partial x} = e^{xy}(xy + 1)$ $\quad \frac{\partial^2 z}{\partial y^2} = x^3e^{xy}$
$\frac{\partial z}{\partial y} = x^2 e^{xy}$ $\quad \frac{\partial^2 z}{\partial x \partial y} = xe^{xy}(xy + 2)$
$\frac{\partial^2 z}{\partial x^2} = ye^{xy}(xy + 2)$ $\quad \frac{\partial^2 z}{\partial y \partial x} = xe^{xy}(xy + 2)$

1. Critical point: (−2, −4)
No relative extrema
(−2, −4, 1) is a saddle point.
3. Critical point: (0, 0)
Relative minimum: (0, 0, 1)
5. Critical point: (1, 3)
Relative minimum: (1, 3, 0)
7. Critical point: (−1, 1)
Relative minimum: (−1, 1, −4)
9. Critical point: (8, 16)
Relative maximum: (8, 16, 74)
11. Critical point: (1, 1)
Relative minimum: (1, 1, 11)
13. Critical points: (0, 0), $\left(\frac{4}{3}, \frac{4}{3}\right)$
Saddle point: (0, 0, 1)
Relative maximum: $\left(\frac{4}{3}, \frac{4}{3}, \frac{59}{27}\right)$
15. Critical point: (0, 0)
Saddle point: (0, 0, 0)
17. Critical points: $\left(\frac{1}{2}, \frac{1}{2}\right), \left(-\frac{1}{2}, -\frac{1}{2}\right)$
Relative maximum: $\left(\frac{1}{2}, \frac{1}{2}, e^{1/2}\right)$
Relative minimum: $\left(-\frac{1}{2}, -\frac{1}{2}, -e^{1/2}\right)$
19. Insufficient information **21.** $f(x_0, y_0)$ is a saddle point.
23. $f(x_0, y_0)$ is a relative minimum.
25. Relative minima: $(a, 0, 0)$, $(0, b, 0)$
Second-Partials Test fails at $(a, 0)$ and $(0, b)$.
27. Saddle point: (0, 0, 0)
Second-Partials Test fails at (0, 0).
29. Relative minimum: (0, 0, 0)
Second-Partials Test fails at (0, 0).
31. Relative minimum: (1, −3, 0)
33. 15, 15, 15 **35.** 20, 20, 20 **37.** $x_1 = 3, x_2 = 6$
39. $p_1 = 2500, p_2 = 3000$ **41.** $x_1 \approx 94, x_2 \approx 157$
43. 32 in. × 16 in. × 16 in.
45. Base dimensions: 3 ft × 3 ft
Height: 2 ft; Minimum cost: $5.40
47. $x = 1.25, y = 2.5$; $4.625 million
49. 500 smallmouth bass; 200 largemouth bass
51. Proof **53.** True

Quiz Yourself *(page 464)*

1. (a)

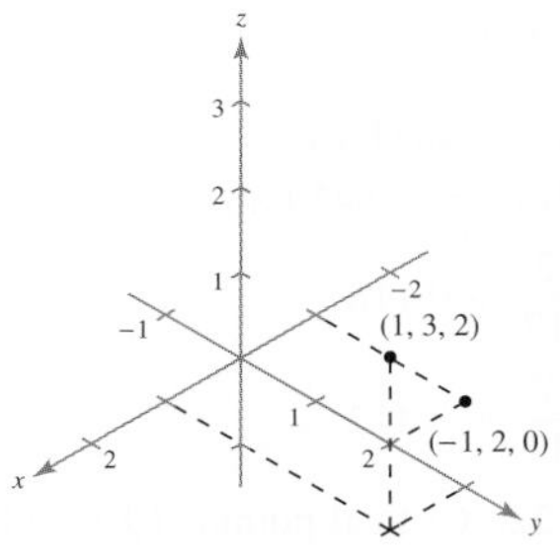

(b) 3 (c) $\left(0, \frac{5}{2}, 1\right)$

2. (a)

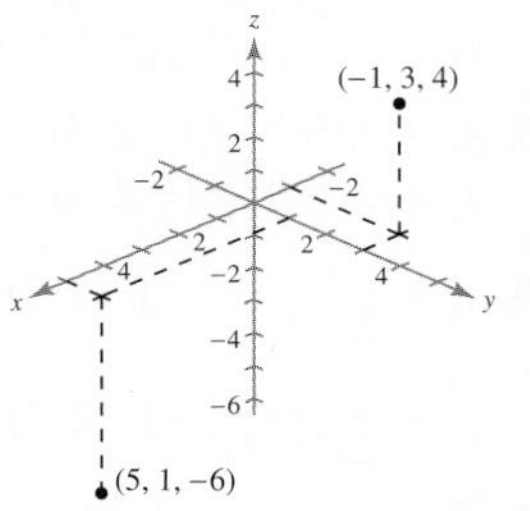

(b) $2\sqrt{35}$ (c) (2, 2, −1)

3. (a)

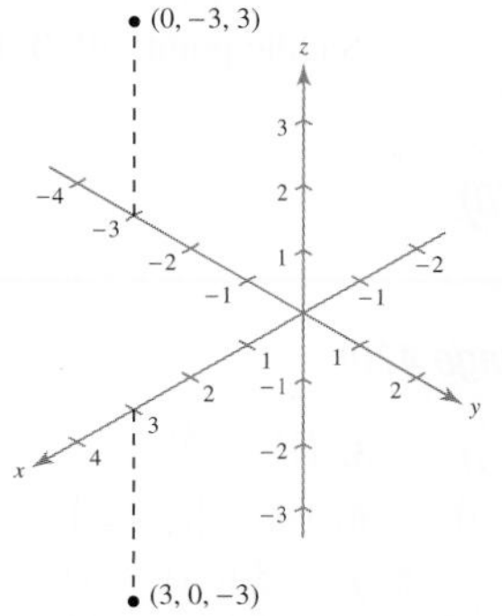

(b) $3\sqrt{6}$ (c) $\left(\frac{3}{2}, -\frac{3}{2}, 0\right)$

4. $(x - 2)^2 + (y + 1)^2 + (z - 3)^2 = 16$
5. $(x - 1)^2 + (y - 4)^2 + (z + 2)^2 = 11$
6. Center: (4, 1, 3); radius: 7

7.

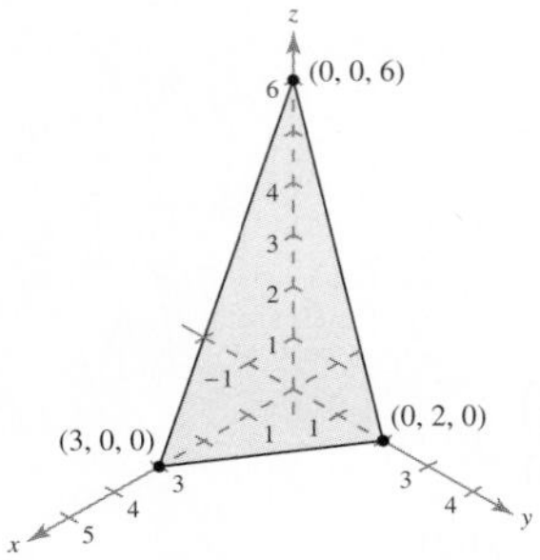

8.

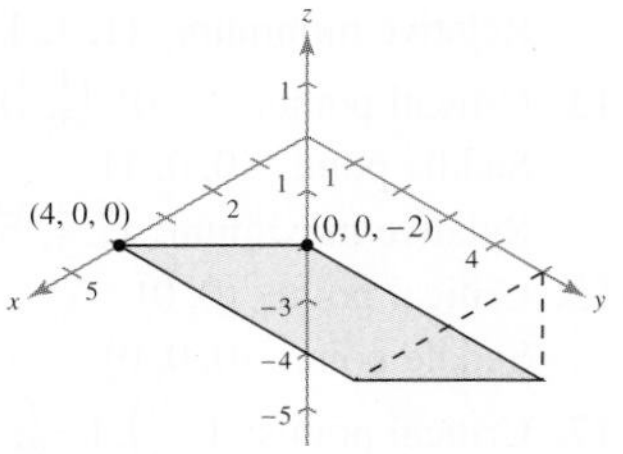

9.

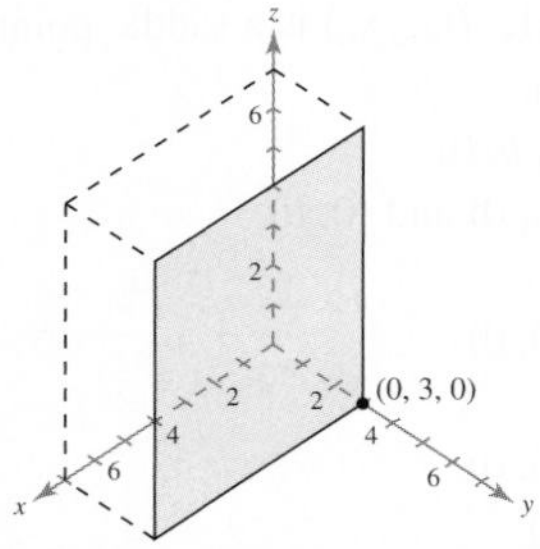

10. Ellipsoid

11. Hyperboloid of two sheets **12.** Elliptic paraboloid

13. $f(1, 0) = 1$
$f(4, -1) = -5$

14. $f(1, 0) = 2$
$f(4, -1) = 3\sqrt{7}$

15. $f(1, 0) = 0$
$f(4, -1) = \ln 6 \approx 1.79$

16. (a) Between 30° and 50° (b) Between 40° and 80°
(c) Between 70° and 90°

17. $f_x = 2x - 3$; $f_x(-2, 3) = -7$
$f_y = 4y - 1$; $f_y(-2, 3) = 11$

18. $f_x = \dfrac{y(3 + y)}{(x + y)^2}$; $f_x(-2, 3) = 18$
$f_y = \dfrac{-2xy - y^2 - 3x}{(x + y)^2}$; $f_y(-2, 3) = 9$

19. $f_x = 3x^2e^{2y}$; $f_x(-2, 3) = 12e^6 \approx 4841.15$
$f_y = 2x^3e^{2y}$; $f_y(-2, 3) = -16e^6 \approx -6454.86$

20. $f_x = \dfrac{2}{2x + 7y}$; $f_x(-2, 3) = \dfrac{2}{17} \approx 0.118$
$f_y = \dfrac{7}{2x + 7y}$; $f_y(-2, 3) = \dfrac{7}{17} \approx 0.412$

21. Critical point: $(1, 0)$
Relative minimum:
$(1, 0, -3)$

22. Critical points: $(0, 0)$, $\left(\frac{4}{3}, \frac{4}{3}\right)$
Relative maximum:
$\left(\frac{4}{3}, \frac{4}{3}, \frac{59}{27}\right)$
Saddle point: $(0, 0, 1)$

23. $x = 80$, $y = 20$; \$20,000

Section 7.6 *(page 470)*

Skills Warm Up *(page 470)*

1. $\left(\frac{7}{8}, \frac{1}{12}\right)$ **2.** $\left(-\frac{1}{24}, -\frac{7}{8}\right)$ **3.** $\left(\frac{55}{12}, -\frac{25}{12}\right)$
4. $\left(\frac{22}{23}, -\frac{3}{23}\right)$ **5.** $\left(\frac{5}{3}, \frac{1}{3}, 0\right)$ **6.** $\left(\frac{14}{19}, -\frac{10}{19}, -\frac{32}{57}\right)$

7. $f_x = 2xy + y^2$
$f_y = x^2 + 2xy$

8. $f_x = 50y^2(x + y)$
$f_y = 50y(x + y)(x + 2y)$

9. $f_x = 3x^2 - 4xy + yz$
$f_y = -2x^2 + xz$
$f_z = xy$

10. $f_x = yz + z^2$
$f_y = xz + z^2$
$f_z = xy + 2xz + 2yz$

1. $f(5, 5) = 25$ **3.** $f(4, 4) = 32$ **5.** $f(\sqrt{2}, 1) = 1$
7. $f(25, 50) = 2600$ **9.** $f(1, 1) = 2$
11. $f(2, 2) = e^4$ **13.** $f(9, 6, 9) = 432$
15. $f\left(\frac{1}{3}, \frac{1}{3}, \frac{1}{3}\right) = \frac{1}{3}$ **17.** $f\left(\frac{\sqrt{3}}{3}, \frac{\sqrt{3}}{3}, \frac{\sqrt{3}}{3}\right) = \sqrt{3}$
19. 20, 20, 20 **21.** 40, 40, 40 **23.** $3\sqrt{2}$ **25.** $\sqrt{3}$
27. 15 units × 10 units × 6 units **29.** 12 ft × 12 ft × 18 ft
31. $x_1 = 145$ units, $x_2 = 855$ units
33. (a) $f\left(\frac{3125}{6}, \frac{6250}{3}\right) \approx 147{,}314$ (b) 1.473
(c) 184,142 units (d) 515,599 units
35. (a) $x \approx 317$ units, $y \approx 68$ units (b) Answers will vary.
37. (a) 50 ft × 120 ft (b) \$2400
39. $x = \sqrt[3]{0.065} \approx 0.402$ L
$y = \frac{1}{2}\sqrt[3]{0.065} \approx 0.201$ L
$z = \frac{1}{3}\sqrt[3]{0.065} \approx 0.134$ L
41. About 190.7 g
43. (a) Cable television: \$1200
Newspaper: \$600
Radio: \$900
(b) About 3718 responses

Section 7.7 *(page 478)*

Skills Warm Up *(page 478)*

1. 5.0225 **2.** 0.0189
3. $S_a = 2a - 4 - 4b$
$S_b = 12b - 8 - 4a$
4. $S_a = 8a - 6 - 2b$
$S_b = 18b - 4 - 2a$
5. 15 **6.** 42 **7.** $\frac{25}{12}$ **8.** 14 **9.** 31 **10.** 95

1. $S = 1.6$; $S = 0.8259$ **3.** $S = 6.46$; $S = 0.125$

5. $y = x + \frac{2}{3}$

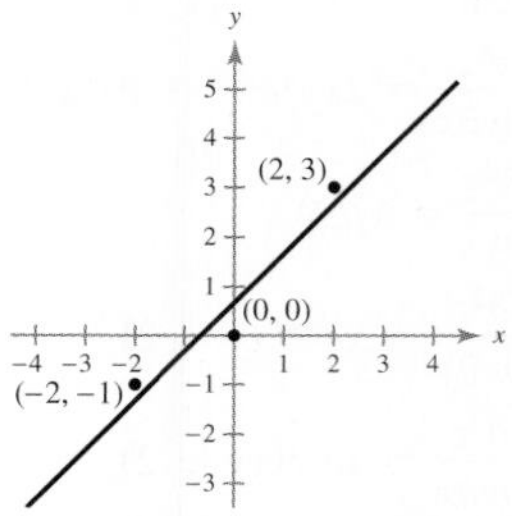

7. $y = -2.3x - 0.9$

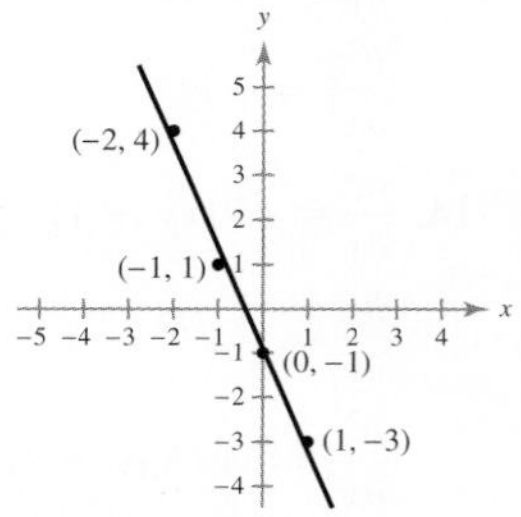

9. $y = 0.8x + 2$ **11.** $y = -1.1824x + 6.385$
13. (a) $y = 4.13t + 11.6$ (b) About \$69.4 billion (c) 2018
15. (a) $y = 0.138x + 22.1$ (b) 44.18 bushels/acre

17.

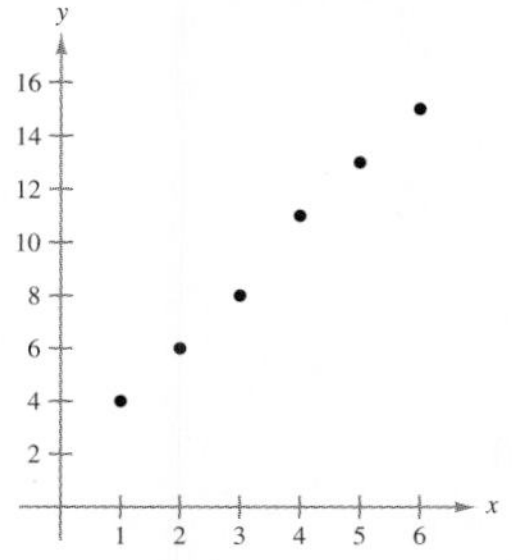

Positive correlation,
$r \approx 0.9981$

19.

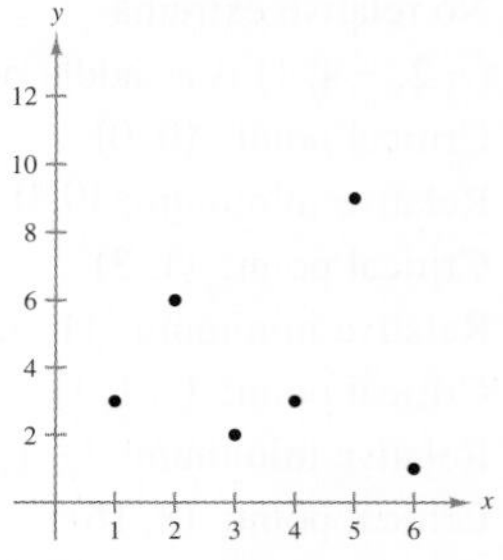

No correlation, $r = 0$

21.

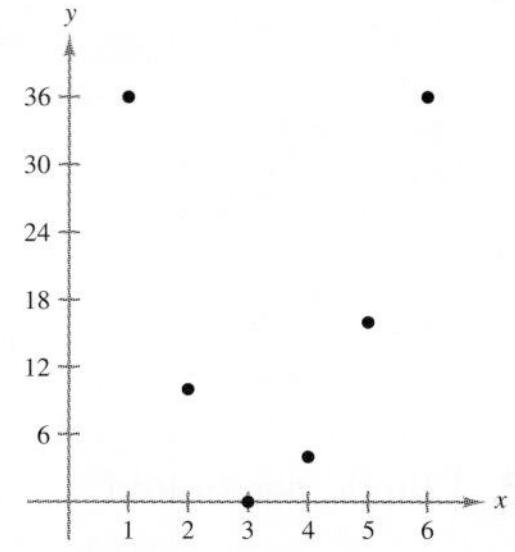

No correlation, $r \approx 0.0750$

23. False; The data modeled by $y = 3.29x - 4.17$ have a positive correlation.

25. True **27.** True **29.** Answers will vary.

Section 7.8 *(page 486)*

Skills Warm Up *(page 486)*

1. 1 **2.** 6 **3.** 42 **4.** $\frac{1}{2}$ **5.** $\frac{19}{4}$

6. $\frac{16}{3}$ **7.** $\frac{1}{7}$ **8.** 4 **9.** ln 5 **10.** $\ln(e - 1)$

11. $\frac{e}{2}(e^4 - 1)$ **12.** $\frac{1}{2}\left(1 - \frac{1}{e^2}\right)$

13.

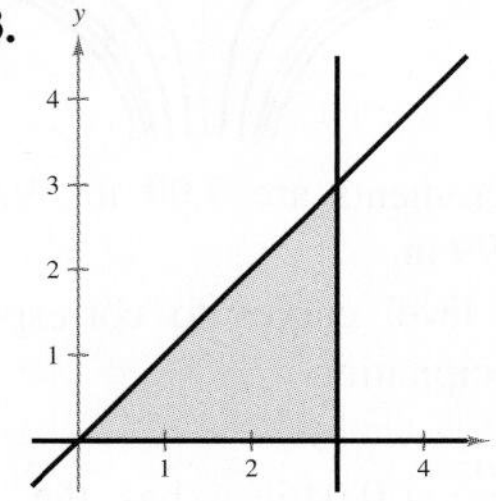

14.

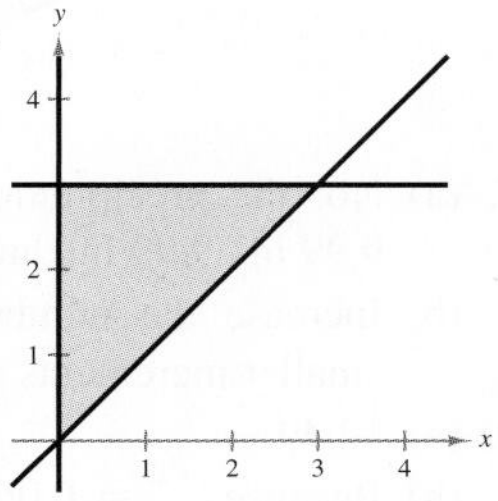

15.

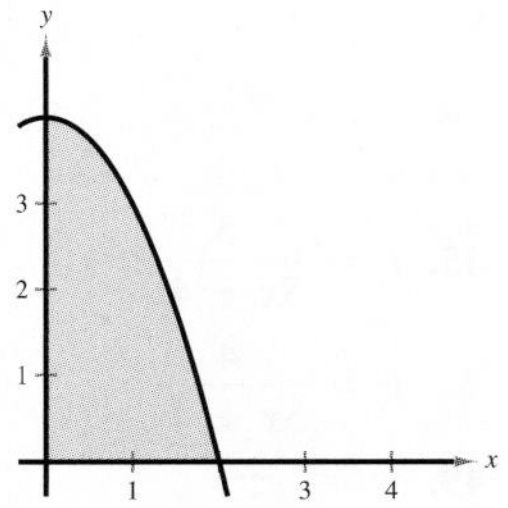

16.

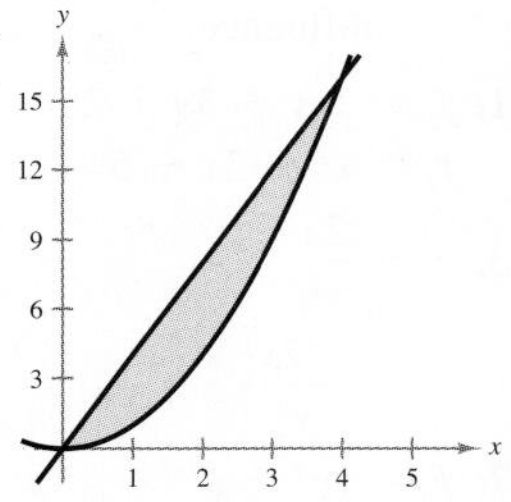

1. $\frac{3x^2}{2}$ **3.** $\frac{x}{2}(x^2 - 1)$ **5.** $2y^4 + y^3 - 2y^2 - 16y$

7. $x^2\sqrt{x} + x^{3/2} - x^5 - x^9$ **9.** $\frac{y^3}{2}$ **11.** 3 **13.** 36

15. 64 **17.** $\frac{1}{2}$ **19.** $\frac{21}{2}$ **21.** $\frac{1}{3}$ **23.** 4 **25.** 24

27. 8 **29.** $\frac{16}{3}$ **31.** 36 **33.** 5 **35.** 2

37.

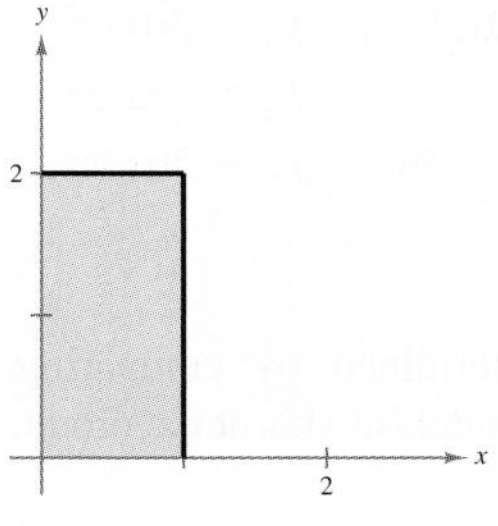

$$\int_0^1\int_0^2 dy\,dx = \int_0^2\int_0^1 dx\,dy = 2$$

39.

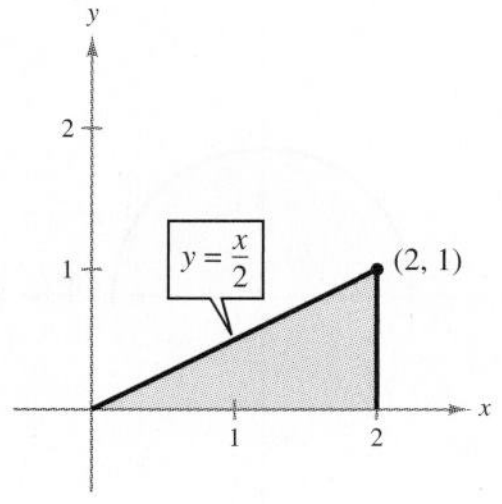

$$\int_0^1\int_{2y}^2 dx\,dy = \int_0^2\int_0^{x/2} dy\,dx = 1$$

41.

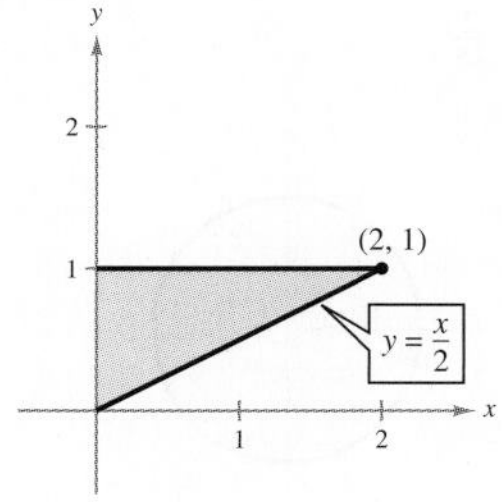

$$\int_0^2\int_{x/2}^1 dy\,dx = \int_0^1\int_0^{2y} dx\,dy = 1$$

43.

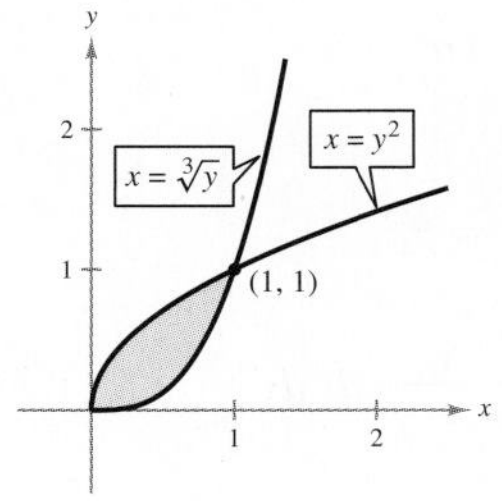

$$\int_0^1\int_{y^2}^{\sqrt[3]{y}} dx\,dy = \int_0^1\int_{x^3}^{\sqrt{x}} dy\,dx = \frac{5}{12}$$

45. (a) Answers will vary; $\frac{1}{2}(e^9 - 1) \approx 4051.042$

(b) Answers will vary; $\frac{1}{2}(1 - e^{-4}) \approx 0.491$

47. 0.6588 **49.** 8.1747 **51.** 0.4521

53. 1.1190 **55.** True

Section 7.9 *(page 494)*

Skills Warm Up *(page 494)*

1.

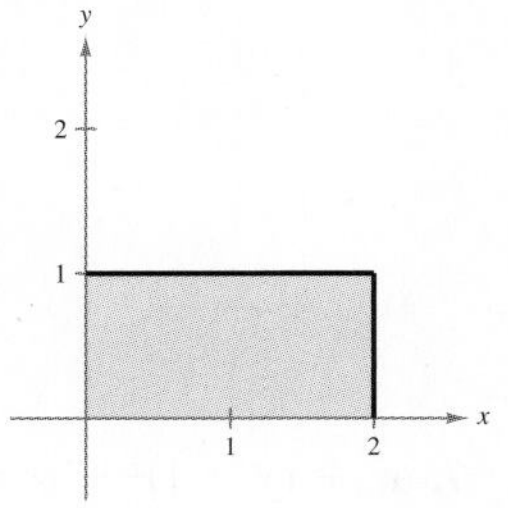

2.

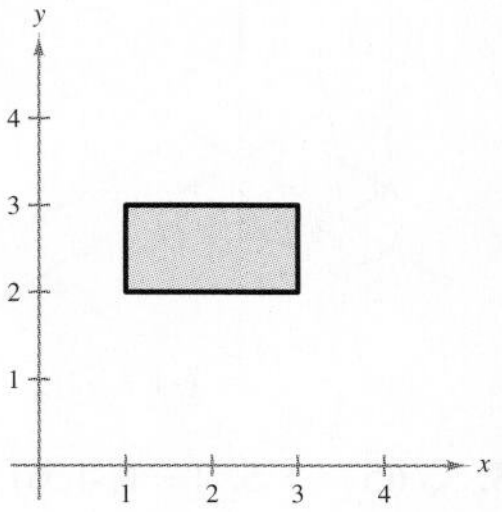

3.

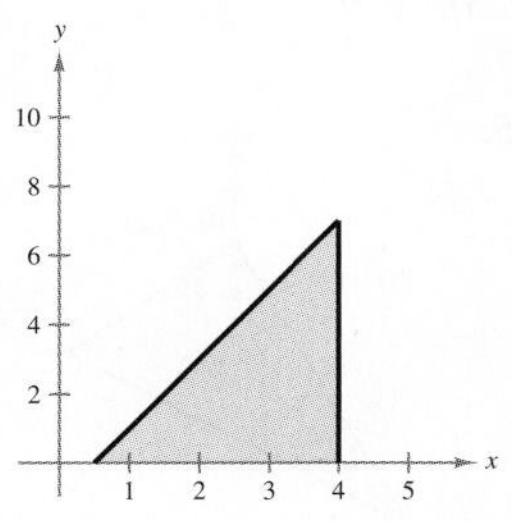

4.

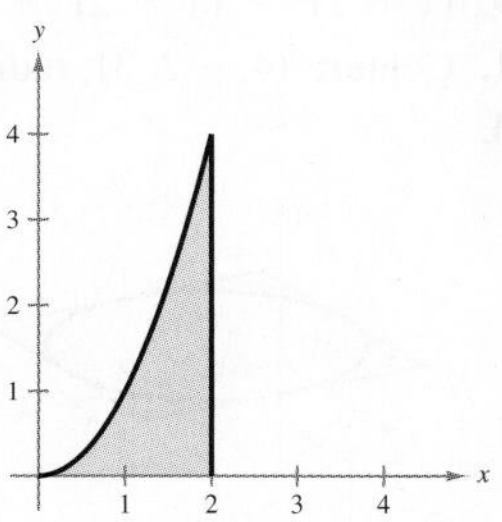

5. 1 **6.** 6 **7.** $\frac{1}{3}$ **8.** $\frac{40}{3}$ **9.** $\frac{28}{3}$ **10.** $\frac{7}{6}$

CHAPTER 7

1.

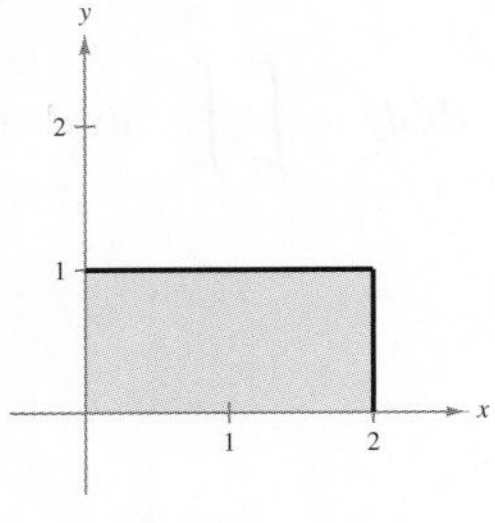

10

3.

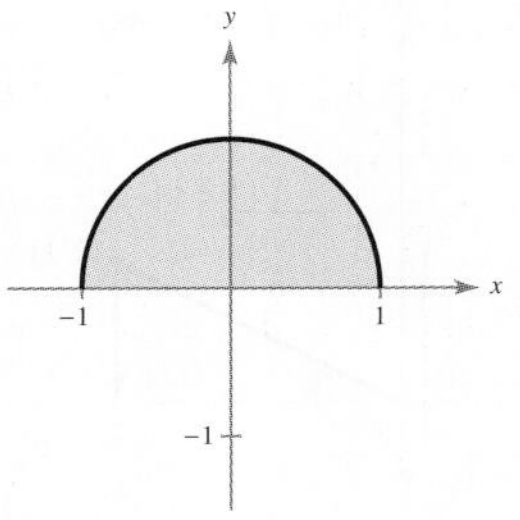

$\frac{2}{15}$

5.

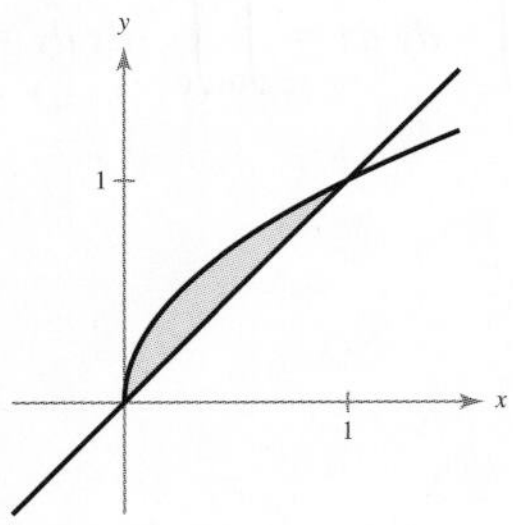

$\frac{3}{35}$

7.

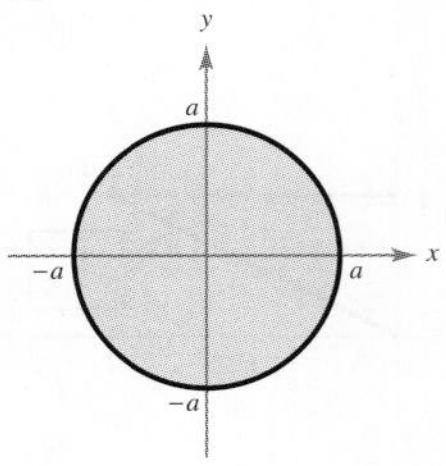

πa^2

9. $\int_0^3 \int_0^5 xy\,dy\,dx = \int_0^5 \int_0^3 xy\,dx\,dy = \frac{225}{4}$

11. $\int_0^2 \int_x^{2x} \frac{y}{x^2+y^2}\,dy\,dx = \int_0^2 \int_{y/2}^{y} \frac{y}{x^2+y^2}\,dx\,dy + \int_2^4 \int_{y/2}^{2} \frac{y}{x^2+y^2}\,dx\,dy = \ln\frac{5}{2}$

13. 4 **15.** 12 **17.** 4 **19.** $\frac{40}{3}$ **21.** $\frac{81}{2}$
23. $\frac{134}{3}$ **25.** 10,000 **27.** $\frac{3}{2}$ **29.** $\frac{8}{3}$
31. \$13,400 **33.** \$75,125 **35.** 25,645.24

Review Exercises for Chapter 7 *(page 500)*

1.

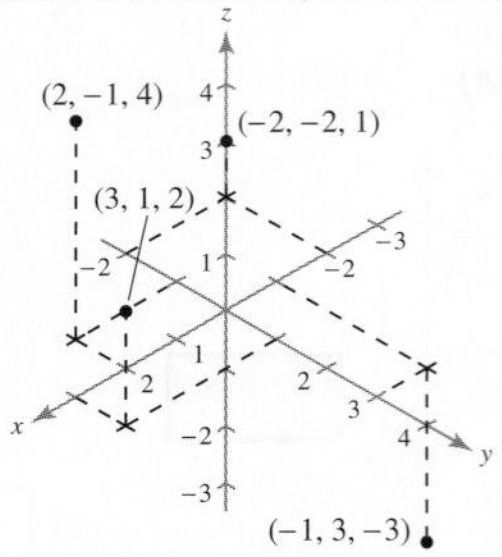

3. $\sqrt{65}$ **5.** $(-1, 4, 6)$ **7.** $x^2 + (y-1)^2 + z^2 = 25$
9. $(x-2)^2 + (y+2)^2 + (z+3)^2 = 9$
11. Center: $(4, -2, 3)$; radius: 7

13.

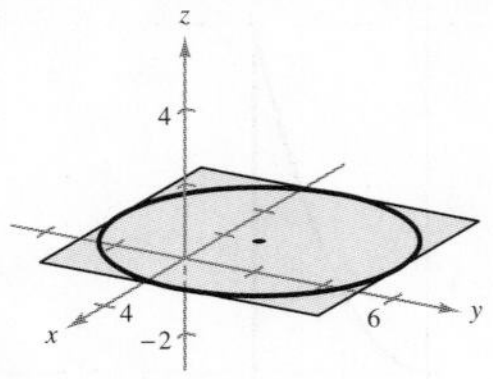

15.

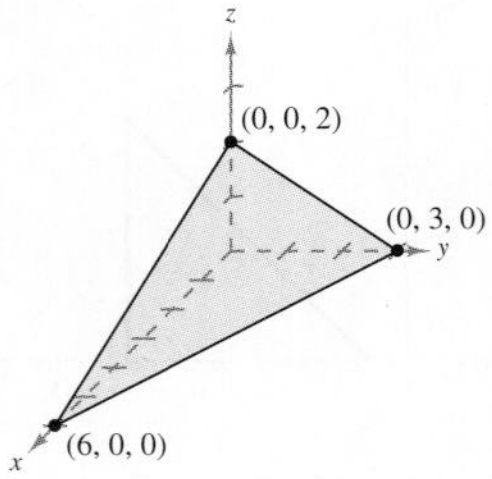

17.

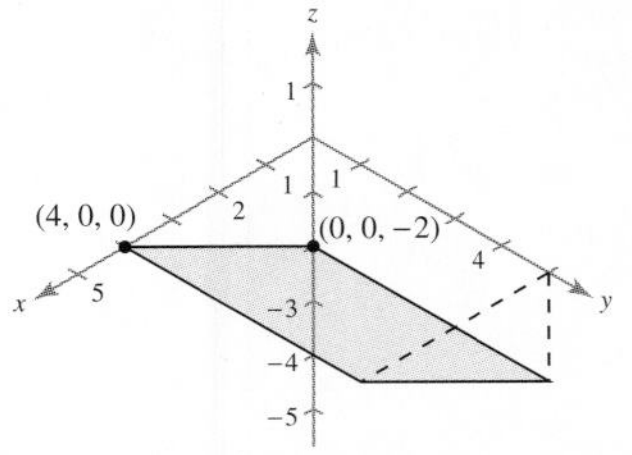

19. Sphere **21.** Ellipsoid **23.** Elliptic paraboloid
25. Top half of a circular cone
27. (a) 18 (b) 0 (c) -245 (d) -32
29. Domain: all points (x, y) inside or on the circle $x^2 + y^2 = 1$
Range: $[0, 1]$
31. Domain: all points (x, y)
Range: $(0, \infty)$
33. The level curves are lines of slope $-\frac{2}{5}$.

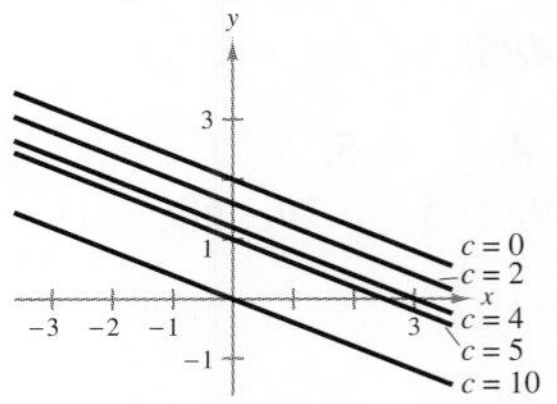

35. The level curves are hyperbolas.

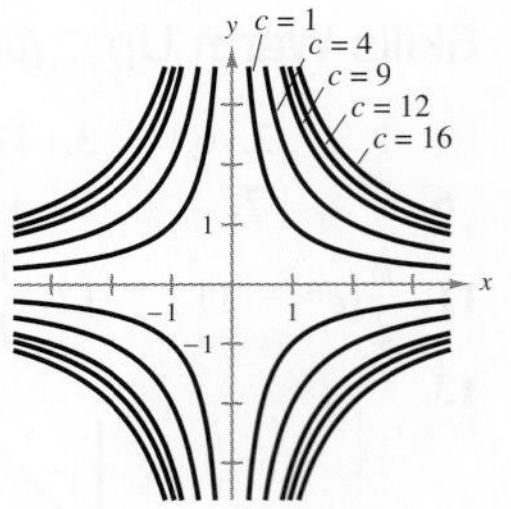

37. (a) No; the precipitation increments are 7.99 in., 9.99 in., 9.99 in., 9.99 in., and 19.99 in.
(b) Increase the number of level curves to correspond to smaller increments of precipitation.
39. (a) \$2.49
(b) Because $\frac{\partial z}{\partial y} = 0.060 > \frac{\partial z}{\partial x} = 0.046$, y has the greater influence.
41. $f_x = 2xy + 3y + 2$
$f_y = x^2 + 3x - 5$
43. $z_x = \frac{2x}{y^2}$
$z_y = \frac{-2x^2}{y^3}$
45. $f_x = \frac{5}{5x+4y}$
$f_y = \frac{4}{5x+4y}$
47. $f_x = ye^x + e^y$
$f_y = xe^y + e^x$
49. $w_x = yz^2$
$w_y = xz^2$
$w_z = 2xyz$
51. (a) -9 (b) -6 **53.** (a) -2 (b) -2
55. $f_{xx} = 6$
$f_{yy} = 12y$
$f_{xy} = f_{yx} = -1$
57. $f_{xx} = f_{yy} = f_{xy} = f_{yx} = \frac{-1}{4(1+x+y)^{3/2}}$
59. $f_{xx} = 10yz^3$ $f_{yx} = 1 + 10xz^3$ $f_{zx} = 30xyz^2$
$f_{xy} = 1 + 10xz^3$ $f_{yy} = -18yz$ $f_{zy} = 15x^2z^2 - 9y^2$
$f_{xz} = 30xyz^2$ $f_{yz} = 15x^2z^2 - 9y^2$ $f_{zz} = 30x^2yz$
61. (a) $C_x(500, 250) = 99.50$
$C_y(500, 250) = 140$
(b) Downhill skis; this is determined by comparing the marginal costs for the two models of skis at the production level (500, 250).

63. Critical point: $(0, 0)$
Relative minimum: $(0, 0, 0)$

65. Critical point: $(-2, 3)$
Saddle point: $(-2, 3, 1)$

67. Critical points: $(0, 0), \left(\frac{1}{6}, \frac{1}{12}\right)$
Relative minimum: $\left(\frac{1}{6}, \frac{1}{12}, -\frac{1}{432}\right)$
Saddle point: $(0, 0, 0)$

69. Critical points: $(1, 1), (-1, -1). (1, -1), (-1, 1)$
Relative minimum: $(1, 1, -2)$
Relative maximum: $(-1, -1, 6)$
Saddle points: $(1, -1, 2), (-1, 1, 2)$

71. $x_1 = 50, x_2 = 200$

73. At $(3, 6)$, the relative maximum is 36.

75. At $(2, 2)$, the relative minimum is 8.

77. At $\left(\frac{4}{3}, \frac{2}{3}, \frac{4}{3}\right)$, the relative maximum is $\frac{32}{27}$.

79. $x_1 = 378$ units; $x_2 = 623$ units

81. $y = \frac{60}{59}x - \frac{15}{59}$

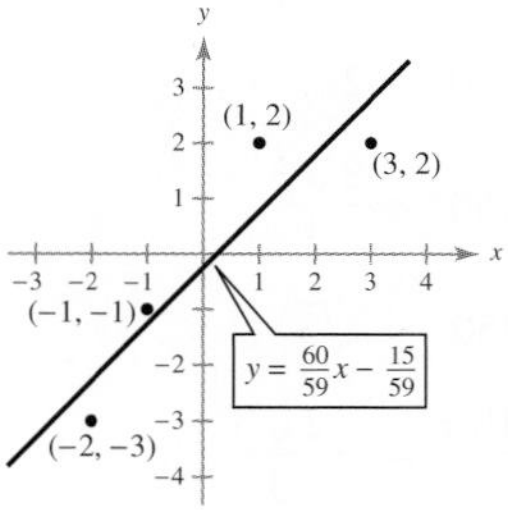

83. (a) $y = -2.6x + 347$
(b) 126 cameras
(c) About \$56.54

85. 1 **87.** $\frac{7}{4}$ **89.** $\frac{32}{3}$ **91.** $\frac{9}{2}$ **93.** 20 **95.** 8

97. $\frac{4096}{9}$ **99.** 3 **101.** \$5700 **103.** About \$155.69/ft^2

Test Yourself (page 504)

1. (a)

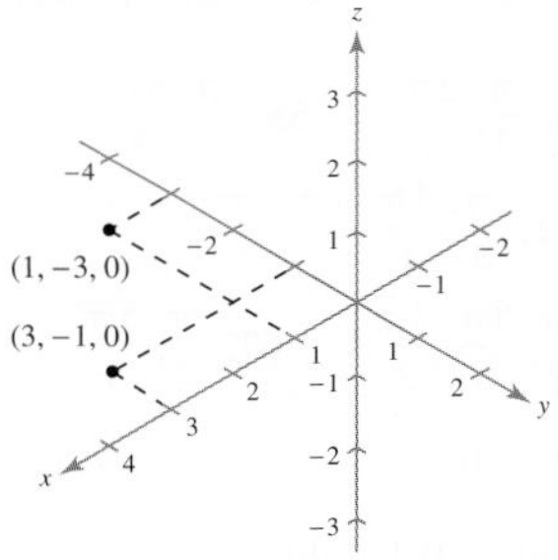

(b) $2\sqrt{2}$
(c) $(2, -2, 0)$

2. (a)

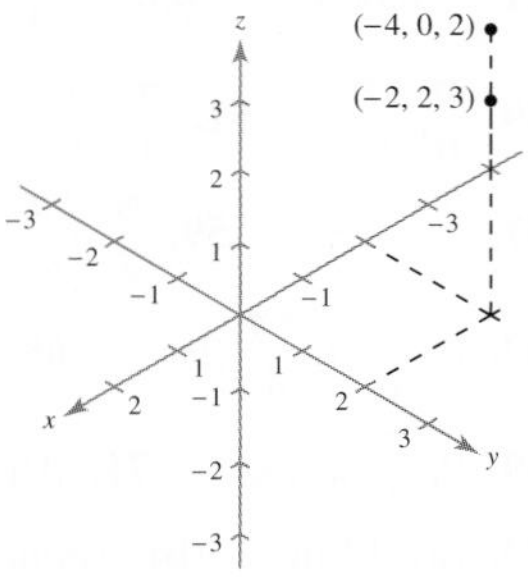

(b) 3
(c) $(-3, 1, 2.5)$

3. (a)

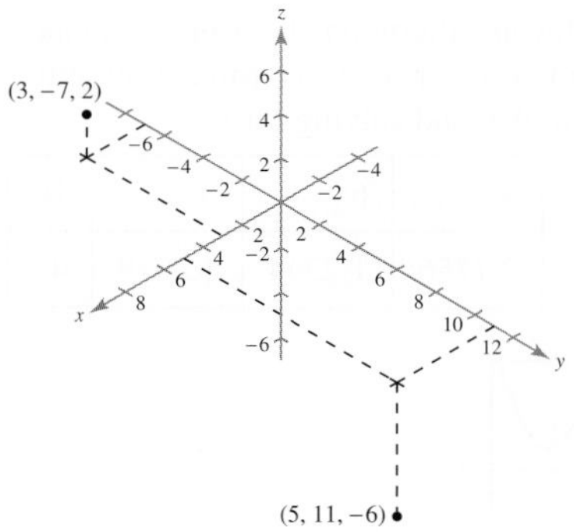

(b) $14\sqrt{2}$
(c) $(4, 2, -2)$

4. Center: $(10, -5, 5)$; radius: 5

5. Hyperboloid of one sheet

6. Elliptic cone **7.** Hyperbolic paraboloid

8. $f(3, 3) = 19$
$f(1, 4) = 6$

9. $f(3, 3) = \frac{3}{2}$
$f(1, 4) = -9$

10. $f(3, 3) = 0$
$f(1, 4) = 4 \ln \frac{1}{4} \approx -5.5$

11. $f_x = 6x + 9y^2$; $f_x(10, -1) = 69$
$f_y = 18xy$; $f_y(10, -1) = -180$

12. $f_x = (x + y)^{1/2} + \dfrac{x}{2(x + y)^{1/2}}$; $f_x(10, -1) = \dfrac{14}{3}$
$f_y = \dfrac{x}{2(x + y)^{1/2}}$; $f_y(10, -1) = \dfrac{5}{3}$

13. Critical point: $(1, -2)$; Relative minimum: $(1, -2, -23)$

14. Critical points: $(0, 0), (1, 1), (-1, -1)$
Saddle point: $(0, 0, 0)$
Relative maxima: $(1, 1, 2), (-1, -1, 2)$

15. About 128,613 units **16.** $y = 0.52x + 1.4$

17. $\frac{3}{2}$ **18.** 1 **19.** $\frac{4}{3}$ units^2 **20.** 48 **21.** $\frac{11}{6}$

Chapter 8

Section 8.1 (page 511)

Skills Warm Up (page 511)

1. 35 cm^2 **2.** 12 in.^{2} **3.** $c = 13$ **4.** $b = 4$
5. $b = 15$ **6.** $a = 6$ **7.** Equilateral triangle
8. Isosceles triangle **9.** Right triangle
10. Isosceles triangle and right triangle

1. $405°, -315°$ **3.** $240°, -480°$ **5.** $300°, -60°$

7. $\dfrac{19\pi}{9}, -\dfrac{17\pi}{9}$ **9.** $\dfrac{28\pi}{15}, -\dfrac{32\pi}{15}$ **11.** $\dfrac{\pi}{6}$ **13.** $\dfrac{3\pi}{2}$

15. $\dfrac{15\pi}{4}$ **17.** $-\dfrac{2\pi}{15}$ **19.** $-\dfrac{4\pi}{5}$ **21.** $\dfrac{11\pi}{6}$

23. 450° **25.** 420° **27.** −15° **29.** 48°

31. 570° **33.** $c = 10, \theta = 60°$ **35.** $a = 4\sqrt{3}, \theta = 30°$

37. $\theta = 40°$ **39.** $s = \sqrt{3}, \theta = 60°$ **41.** $4\sqrt{3}$ in.^{2}

43. $\dfrac{25\sqrt{3}}{4}$ ft^2 **45.** 18 ft

47.

r	8 ft	15 in.	85 cm	24 in.	$\dfrac{12,963}{\pi}$ mi
s	12 ft	24 in.	200.28 cm	96 in.	8642 mi
θ	1.5	1.6	$\dfrac{3\pi}{4}$	4	$\dfrac{2\pi}{3}$

49. $\dfrac{35\pi}{12} \approx 9.16$ inches

51. (a) About 500 revolutions per minute
(b) About 20 minutes

53. $\dfrac{4900\pi}{3}$ ft^2

55. False. An obtuse angle is between 90° and 180°.

57. True **59.** Answers will vary.

Section 8.2 *(page 522)*

Skills Warm Up *(page 522)*

1. $\frac{7\pi}{4}$ **2.** $-\frac{5\pi}{3}$ **3.** $-\frac{5\pi}{4}$ **4.** $\frac{13\pi}{6}$
5. $x = 0, 1$ **6.** $x = 0, -\frac{1}{2}$ **7.** $x = -1, 3$
8. $x = 2, 3$ **9.** $t = 10$ **10.** $t = \frac{7}{2}$
11. $t = \frac{445}{8}$ **12.** $t = 7$

1. $\sin\theta = \frac{3}{5}$, $\csc\theta = \frac{5}{3}$
$\cos\theta = \frac{4}{5}$, $\sec\theta = \frac{5}{4}$
$\tan\theta = \frac{3}{4}$, $\cot\theta = \frac{4}{3}$

3. $\sin\theta = -\frac{5}{13}$, $\csc\theta = -\frac{13}{5}$
$\cos\theta = -\frac{12}{13}$, $\sec\theta = -\frac{13}{12}$
$\tan\theta = \frac{5}{12}$, $\cot\theta = \frac{12}{5}$

5. $\sin\theta = \frac{2\sqrt{5}}{5}$, $\csc\theta = \frac{\sqrt{5}}{2}$
$\cos\theta = -\frac{\sqrt{5}}{5}$, $\sec\theta = -\sqrt{5}$
$\tan\theta = -2$, $\cot\theta = -\frac{1}{2}$

7.
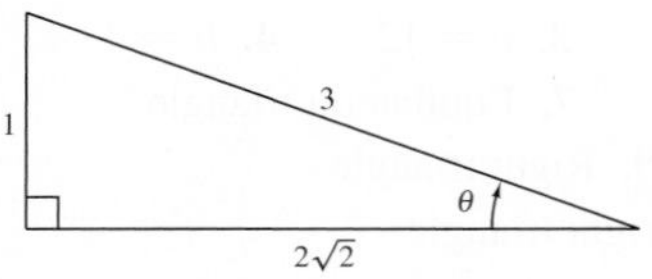

$\csc\theta = 3$
$\cos\theta = \frac{2\sqrt{2}}{3}$, $\sec\theta = \frac{3\sqrt{2}}{4}$
$\tan\theta = \frac{\sqrt{2}}{4}$, $\cot\theta = 2\sqrt{2}$

9.
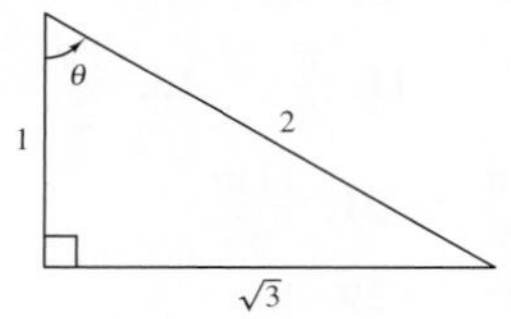

$\sin\theta = \frac{\sqrt{3}}{2}$, $\csc\theta = \frac{2\sqrt{3}}{3}$
$\cos\theta = \frac{1}{2}$
$\tan\theta = \sqrt{3}$, $\cot\theta = \frac{\sqrt{3}}{3}$

11.
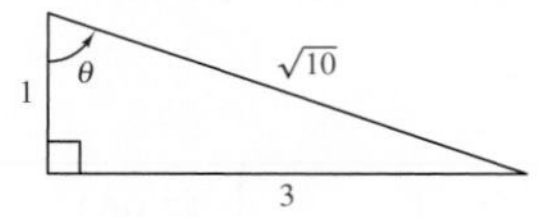

$\sin\theta = \frac{3\sqrt{10}}{10}$, $\csc\theta = \frac{\sqrt{10}}{3}$
$\cos\theta = \frac{\sqrt{10}}{10}$, $\sec\theta = \sqrt{10}$
$\cot\theta = \frac{1}{3}$

13. Quadrant IV **15.** Quadrant I **17.** Quadrant II

19. $\sin 60° = \frac{\sqrt{3}}{2}$, $\csc 60° = \frac{2\sqrt{3}}{3}$
$\cos 60° = \frac{1}{2}$, $\sec 60° = 2$
$\tan 60° = \sqrt{3}$, $\cot 60° = \frac{\sqrt{3}}{3}$

21. $\sin\frac{\pi}{4} = \frac{\sqrt{2}}{2}$, $\csc\frac{\pi}{4} = \sqrt{2}$
$\cos\frac{\pi}{4} = \frac{\sqrt{2}}{2}$, $\sec\frac{\pi}{4} = \sqrt{2}$
$\tan\frac{\pi}{4} = 1$, $\cot\frac{\pi}{4} = 1$

23. $\sin\left(-\frac{\pi}{2}\right) = -1$, $\csc\left(-\frac{\pi}{2}\right) = -1$
$\cos\left(-\frac{\pi}{2}\right) = 0$, $\sec\left(-\frac{\pi}{2}\right) =$ undefined
$\tan\left(-\frac{\pi}{2}\right) =$ undefined, $\cot\left(-\frac{\pi}{2}\right) = 0$

25. $\sin 225° = -\frac{\sqrt{2}}{2}$, $\csc 225° = -\sqrt{2}$
$\cos 225° = -\frac{\sqrt{2}}{2}$, $\sec 225° = -\sqrt{2}$
$\tan 225° = 1$, $\cot 225° = 1$

27. $\sin 300° = -\frac{\sqrt{3}}{2}$, $\csc 300° = -\frac{2\sqrt{3}}{3}$
$\cos 300° = \frac{1}{2}$, $\sec 300° = 2$
$\tan 300° = -\sqrt{3}$, $\cot 300° = -\frac{\sqrt{3}}{3}$

29. $\sin 750° = \frac{1}{2}$, $\csc 750° = 2$
$\cos 750° = \frac{\sqrt{3}}{2}$, $\sec 750° = \frac{2\sqrt{3}}{3}$
$\tan 750° = \frac{\sqrt{3}}{3}$, $\cot 750° = \sqrt{3}$

31. $\sin\frac{10\pi}{3} = -\frac{\sqrt{3}}{2}$, $\csc\frac{10\pi}{3} = -\frac{2\sqrt{3}}{3}$
$\cos\frac{10\pi}{3} = -\frac{1}{2}$, $\sec\frac{10\pi}{3} = -2$
$\tan\frac{10\pi}{3} = \sqrt{3}$, $\cot\frac{10\pi}{3} = \frac{\sqrt{3}}{3}$

33. 0.1736 **35.** 0.3640 **37.** −0.3420 **39.** 1.7321
41. −0.6052 **43.** $\frac{100\sqrt{3}}{3}$ **45.** $\frac{25\sqrt{3}}{3}$ **47.** 15.5572
49. $\frac{\pi}{6}, \frac{5\pi}{6}$ **51.** $\frac{\pi}{3}, \frac{4\pi}{3}$ **53.** $\frac{\pi}{3}, \frac{2\pi}{3}$ **55.** $\frac{\pi}{3}, \frac{5\pi}{3}$
57. $\frac{5\pi}{4}, \frac{7\pi}{4}$ **59.** $\frac{\pi}{3}, \frac{2\pi}{3}$ **61.** $\frac{\pi}{4}, \frac{3\pi}{4}, \frac{5\pi}{4}, \frac{7\pi}{4}$
63. $0, \frac{\pi}{4}, \pi, \frac{5\pi}{4}, 2\pi$ **65.** $\frac{\pi}{6}, \frac{\pi}{2}, \frac{5\pi}{6}, \frac{3\pi}{2}$ **67.** $\frac{\pi}{4}, \frac{5\pi}{4}$
69. $0, \frac{\pi}{2}, \pi, 2\pi$ **71.** About 19.3 ft
73. (a) 12 mi (b) About 6.9 mi (c) 6 mi
75. About 443.2 m; about 323.3 m
77. About 11.5370° **79.** About 1.3 mi
81. (a) 102.6°F (b) 102.1°F (c) 100.6°F
(d) At 4 P.M. the following afternoon, the patient's temperature should return to normal. This is determined by setting the function equal to 98.6 and solving for t.

83.

x	0	2	4	6	8	10
$f(x)$	0	2.7021	2.7756	1.2244	1.2979	4

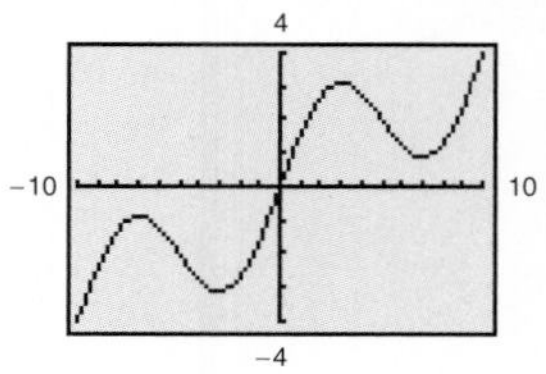

85. True
87. False. Because $\sin 45° = \cos 45°$, $\sin^2 45° - \cos^2 45° = 0$.

Section 8.3 *(page 531)*

Skills Warm Up *(page 531)*

1. 14 **2.** 10 **3.** 0 **4.** 0 **5.** 1
6. $-\frac{\sqrt{3}}{3}$ **7.** $-\frac{1}{2}$ **8.** $-\frac{\sqrt{3}}{2}$ **9.** $\frac{1}{2}$
10. $-\frac{\sqrt{3}}{2}$ **11.** 0.9659 **12.** -0.6428
13. -0.9962 **14.** 0.6428 **15.** 0.9744
16. 0.3090 **17.** -0.6494 **18.** -0.8391

1. Period: π
Amplitude: 2

3. Period: 4π
Amplitude: $\frac{3}{2}$

5. Period: 2
Amplitude: $\frac{1}{2}$

7. Period: $\frac{2\pi}{3}$
Amplitude: 1

9. Period: $\frac{\pi}{3}$
Amplitude: $\frac{3}{2}$

11. Period: 3π
Amplitude: $\frac{1}{2}$

13. Period: $\frac{1}{2}$
Amplitude: 3

15. π **17.** $\frac{2\pi}{5}$ **19.** 6 **21.** c; π **22.** e; π

23. f; 2 **24.** a; 2π **25.** b; 4π **26.** d; 2π

27.

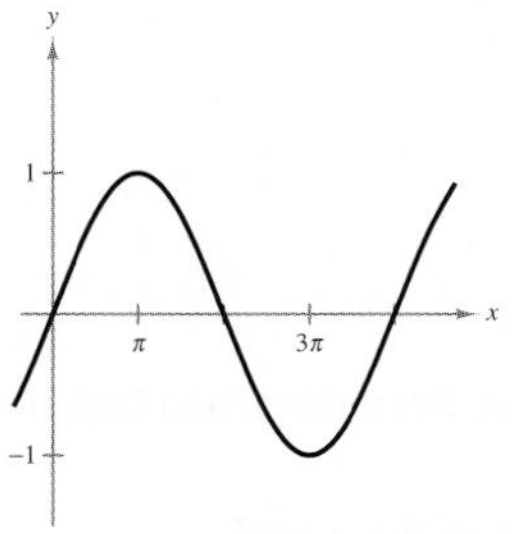

29.

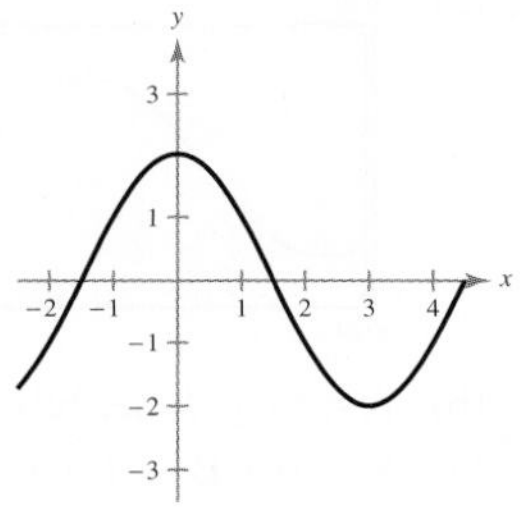

31.

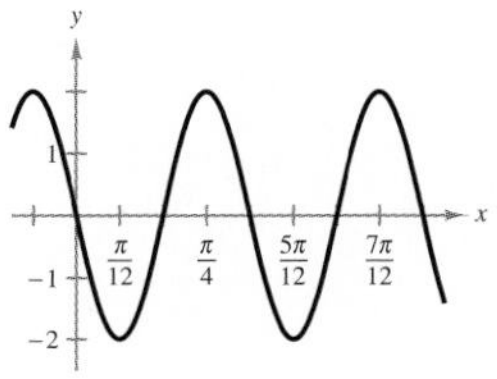

33.

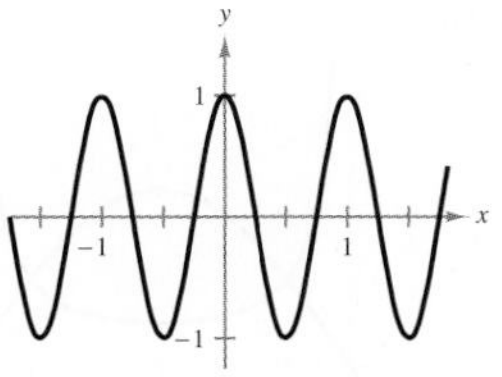

35.

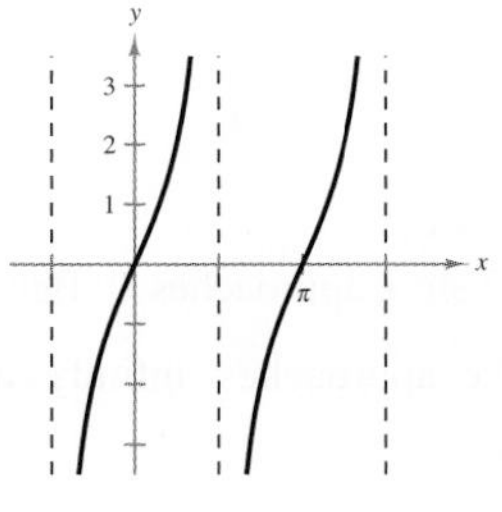

37.

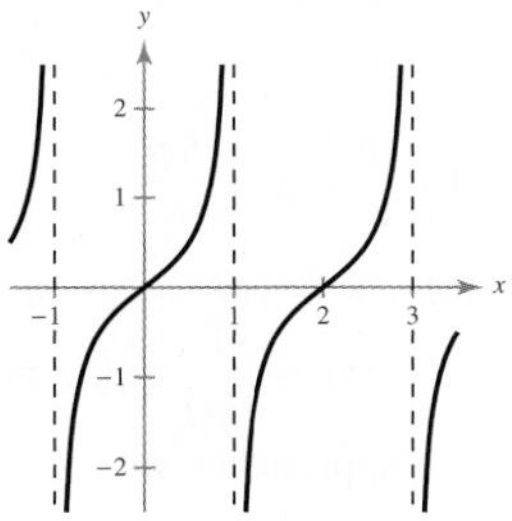

39.

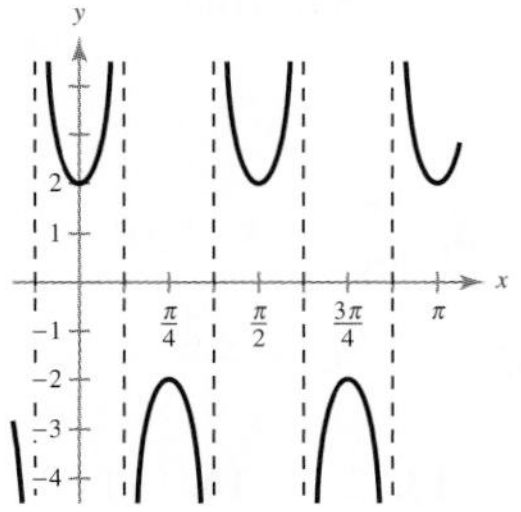

41.

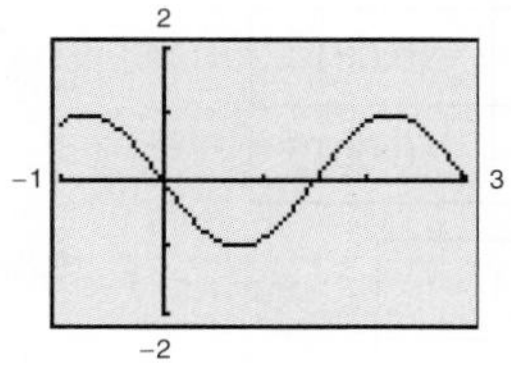

43.

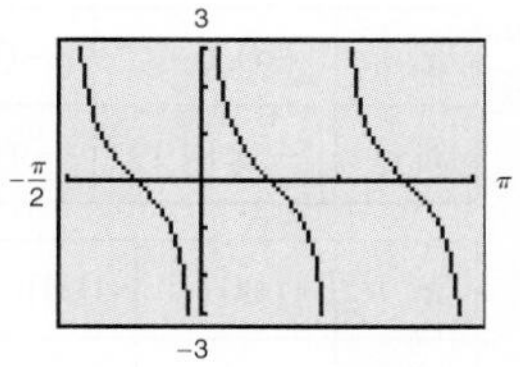

45.

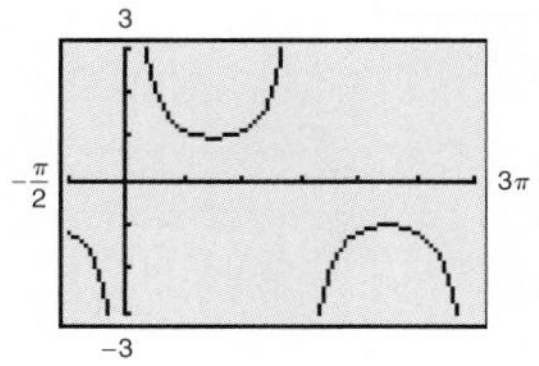

47.

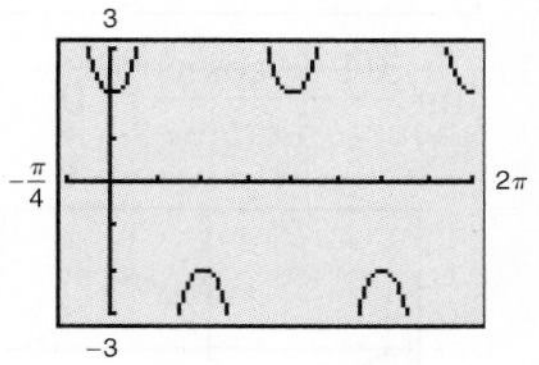

49.

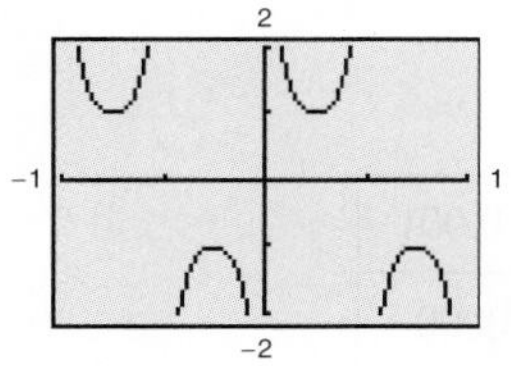

51.

x	-0.1	-0.01	-0.001
$f(x)$	1.9471	1.9995	2.0000

x	0.001	0.01	0.1
$f(x)$	2.0000	1.9995	1.9471

$$\lim_{x\to 0} \frac{\sin 4x}{2x} = 2$$

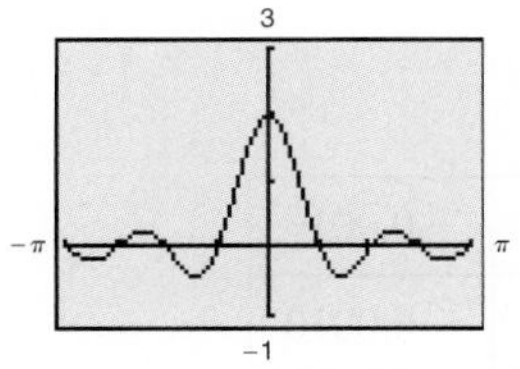

53.

x	-0.1	-0.01	-0.001
$f(x)$	0.7316	0.7498	0.75

x	0.001	0.01	0.1
$f(x)$	0.75	0.7498	0.7316

$$\lim_{x\to 0} \frac{\tan 3x}{\tan 4x} = 0.75$$

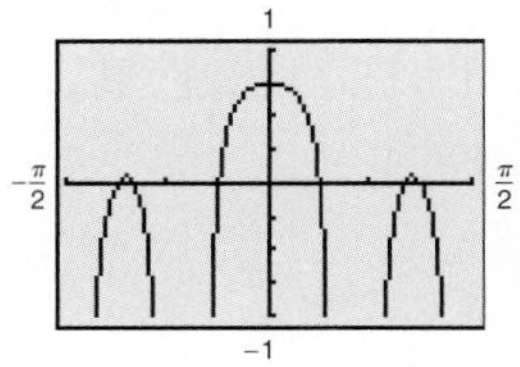

55.

x	-0.1	-0.01	-0.001
$f(x)$	-0.1499	-0.0150	-0.0015

x	0.001	0.01	0.1
$f(x)$	0.0015	0.0150	0.1499

$\lim_{x\to 0} \frac{3(1-\cos x)}{x} = 0$

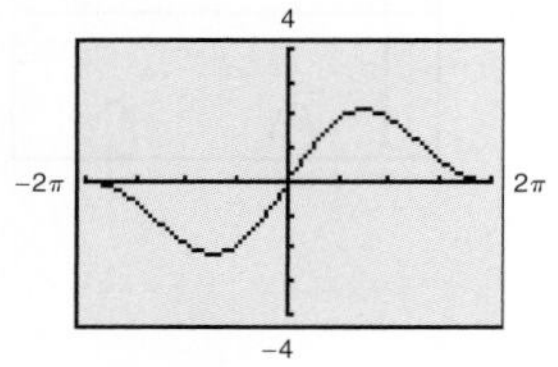

57.

x	-0.1	-0.01	-0.001
$f(x)$	2.0271	2.0003	2.0000

x	0.001	0.01	0.1
$f(x)$	2.0000	2.0003	2.0271

$\lim_{x\to 0} \frac{\tan 2x}{x} = 2$

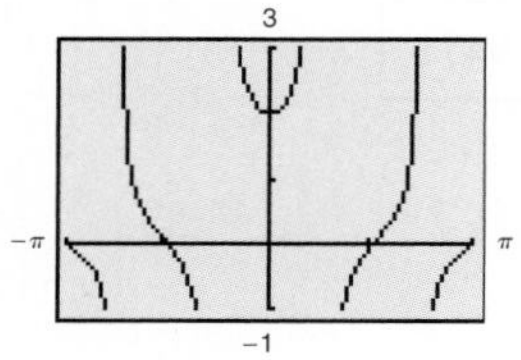

59.

x	-0.1	-0.01	-0.001
$f(x)$	-0.0997	-0.0100	-0.0010

x	0.001	0.01	0.1
$f(x)$	0.0010	0.0100	0.0997

$\lim_{x\to 0} \frac{\sin^2 x}{x} = 0$

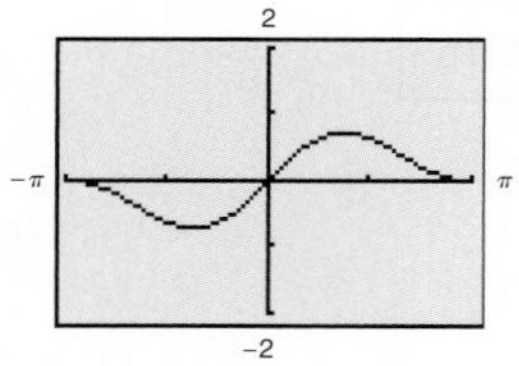

61. $a = 2, d = 1$ **63.** $a = -4, d = 4$

65. b **66.** d **67.** a **68.** c

69. (a)

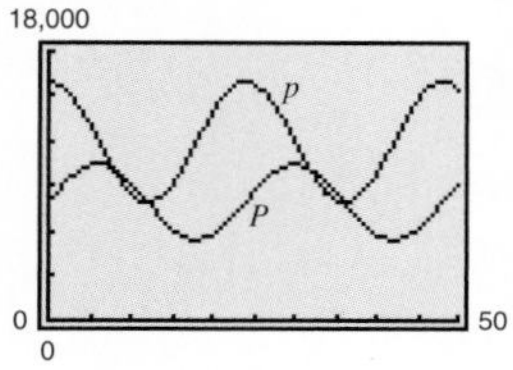

(b) As the population of the prey increases, the population of the predator increases as well. At some point, the predator eliminates the prey faster than the prey can reproduce, and the prey population decreases rapidly. As the prey becomes scarce, the predator population decreases, releasing the prey from predator pressure, and the cycle begins again.

71. $P(8930) \approx 0.9977$
$E(8930) \approx -0.4339$
$I(8930) \approx -0.6182$

73. (a) 6 sec (b) 10
(c)

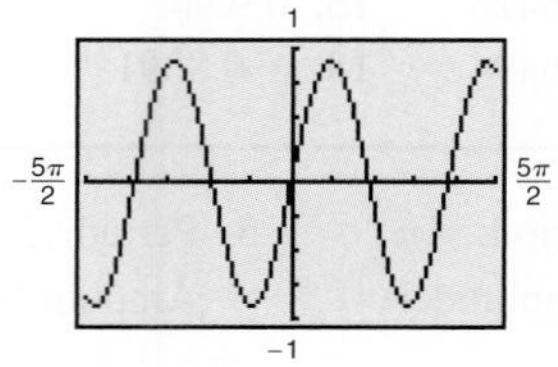

75. (a) $\frac{1}{440}$ (b) 440
(c)

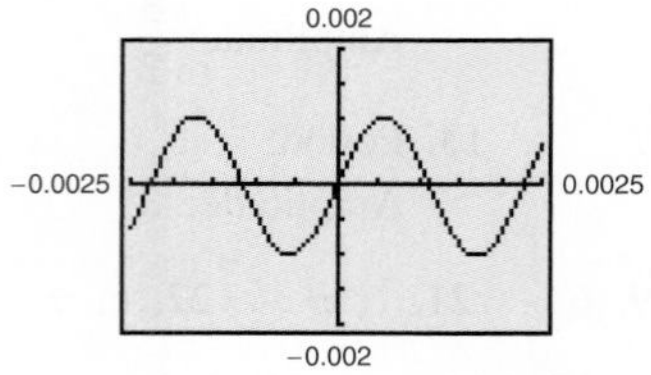

77. (a)

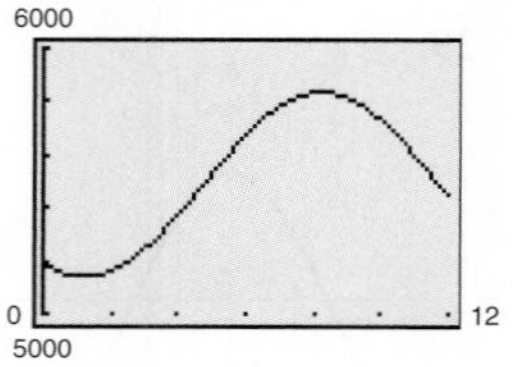

(b) Yes; May, June, July, August, September, and October

79. (a) A (b) B (c) B
(d) The frequency is the inverse of the period.

81. (a)

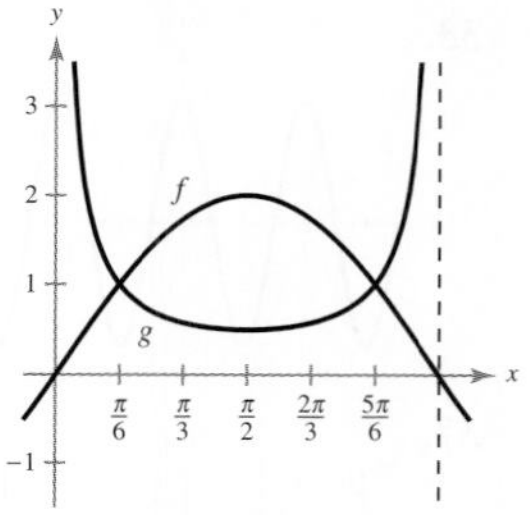

(b) $\frac{\pi}{6} < x < \frac{5\pi}{6}$

(c) As x approaches π, $f(x) = 2 \sin x$ approaches 0. Because $g(x) = \frac{1}{f(x)}$, $g(x) = 0.5 \csc x$ approaches infinity as x approaches π.

83. False. The amplitude is $|-3| = 3$. **85.** True

Quiz Yourself *(page 535)*

1. $\frac{\pi}{12}$ **2.** $\frac{7\pi}{12}$ **3.** $-\frac{4\pi}{9}$ **4.** $\frac{7\pi}{36}$ **5.** $120°$

6. $48°$ **7.** $-240°$ **8.** $165°$ **9.** $-\frac{\sqrt{2}}{2}$

10. $-\frac{1}{2}$ **11.** $-\frac{\sqrt{3}}{3}$ **12.** 1 **13.** 2 **14.** -1

15. $\frac{\pi}{4}, \frac{5\pi}{4}$ **16.** $0, 2\pi$ **17.** $\frac{\pi}{3}, \frac{2\pi}{3}, \frac{4\pi}{3}, \frac{5\pi}{3}$

18. $\theta = 30°;\ a = 5\sqrt{3}$ **19.** $\theta = 40°;\ a \approx 10.285$

20. $\theta = 65°;\ a \approx 5.5957$ **21.** $d \approx 286.8$ ft

22. (a)

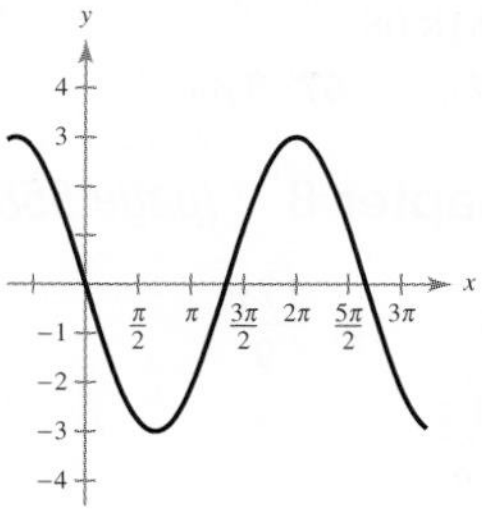

(b) $\frac{8\pi}{3}$

23. (a)

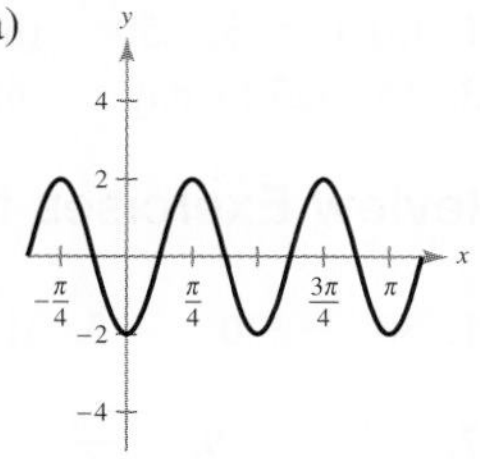

(b) $\frac{\pi}{2}$

24. (a)

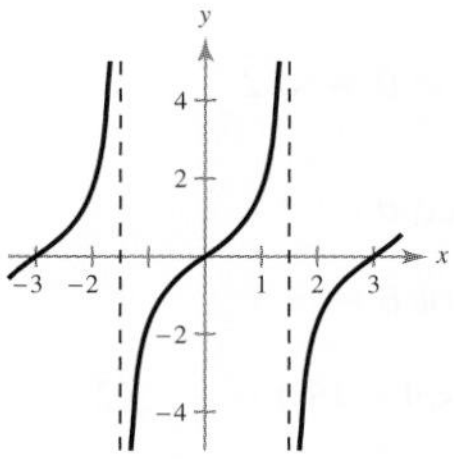

(b) 3

25. (a)

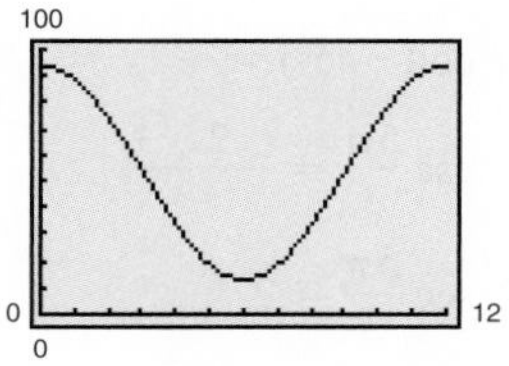

(b) Maximum: December
Minimum: June

Section 8.4 *(page 542)*

Skills Warm Up *(page 542)*

1. $f'(x) = 9x^2 - 4x + 4$ **2.** $g'(x) = 12x^2(x^3 + 4)^3$

3. $f'(x) = 3x^2 + 2x + 1$ **4.** $g'(x) = \dfrac{2(5 - x^2)}{(x^2 + 5)^2}$

5. Relative minimum: $(-2, -3)$

6. Relative maximum: $\left(-2, \frac{22}{3}\right)$
Relative minimum: $\left(2, -\frac{10}{3}\right)$

7. $x = \frac{\pi}{3}, x = \frac{2\pi}{3}$ **8.** $x = \frac{2\pi}{3}, x = \frac{4\pi}{3}$

9. $x = \pi$ **10.** No solution

1. $\frac{1}{3}\cos\frac{x}{3}$ **3.** $12x^2 \sec^2 4x^3$ **5.** $5\sec^2 5t$

7. $2\sin x\cos x$ **9.** $\pi\tan\pi x\sec\pi x$

11. $-2\csc^2(2x + 1)$ **13.** $\dfrac{\sec^2 2x}{\sqrt{\tan 2x}}$

15. $-t^2\sin t + 2t\cos t$ **17.** $-\dfrac{t\sin t + \cos t}{t^2}$

19. $e^{x^2}\sec x(\tan x + 2x)$ **21.** $-3\sin 3x + 2\sin x\cos x$

23. $\sin\frac{1}{x} - \frac{1}{x}\cos\frac{1}{x}$ **25.** $16\sec^2 4x\tan 4x$

27. $2e^{2x}(\cos 2x + \sin 2x)$ **29.** $-2\cos x\sin x = -\sin 2x$

31. $-4\cos x\sin x = -2\sin 2x$

33. $2\sin 2x + 2\sin x\cos x = 3\sin 2x$

35. $\sec^2 x - 1 = \tan^2 x$

37. $\sin^2 x\cos x - \sin^4 x\cos x = \sin^2 x\cos^3 x$

39. $\dfrac{2\cos x}{\sin x} = 2\cot x$ **41.** $y = 2x + \frac{\pi}{2} - 1$

43. $y = 4x - 4\pi$ **45.** $y = -2x + \frac{3}{2}\pi - 1$

47. $y = 0$ **49.** $\dfrac{\cos x}{2\sin 2y}; 0$ **51.** $\frac{5}{4}$; one complete cycle

53. 2; two complete cycles **55.** 1; one complete cycle

57. Relative maximum: $(4, 1)$
Relative minima: $(2, -1), (6, -1)$

59. Relative maximum: $(\pi, 1)$
Relative minima: $\left(\frac{\pi}{2}, 0\right), \left(\frac{3\pi}{2}, 0\right)$

61. Relative maximum: $\left(\frac{\pi}{3}, \frac{3\sqrt{3}}{2}\right)$
Relative minimum: $\left(\frac{5\pi}{3}, -\frac{3\sqrt{3}}{2}\right)$

63. Relative maximum: $\left(\frac{\pi}{4}, 1.5509\right)$
Relative minimum: $\left(\frac{5\pi}{4}, -35.8885\right)$

65. Relative maximum: $\left(\frac{5\pi}{3}, \frac{5\pi}{3} + \sqrt{3}\right)$
Relative minimum: $\left(\frac{\pi}{3}, \frac{\pi}{3} - \sqrt{3}\right)$

67. $-x\sin x + 2\cos x$ **69.** $-4x^2\cos x^2 - 2\sin x^2$

71. (a) $h'(t)$ is a maximum when $t = 0$, or at midnight.
(b) $h'(t)$ is a minimum when $t = \frac{1}{2}$, or at noon.

73. August; about 5836 thousand, or 5,836,000, workers

75. June; about 14 hr **77.** (a)–(d) Proofs

79. (a)

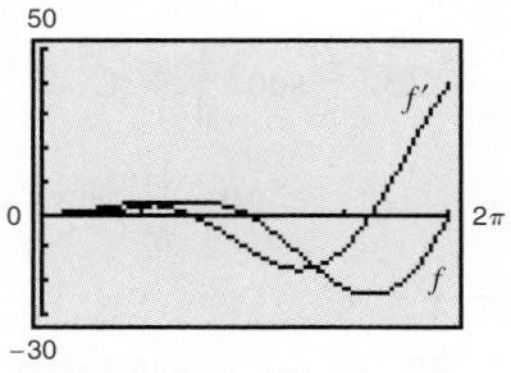

(b) 2.2889, 5.0870
(c) $f' > 0$ on $(0, 2.2889), (5.0870, 2\pi)$
$f' < 0$ on $(2.2889, 5.0870)$
(d) Relative maximum: $(2.2889, 3.9453)$
Relative minimum: $(5.0870, -24.0830)$

81. (a)

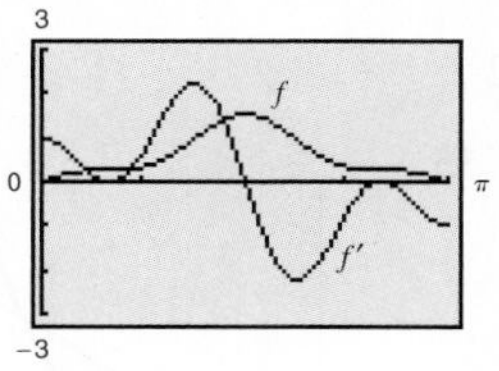

(b) 0.5236, $\pi/2$, 2.6180
(c) $f' > 0$ on $(0, 0.5236), (0.5236, \pi/2)$
$f' < 0$ on $(\pi/2, 2.6180), (2.6180, \pi)$
(d) Relative maximum: $\left(\frac{\pi}{2}, 1.5333\right)$

CHAPTER 8

83. (a)

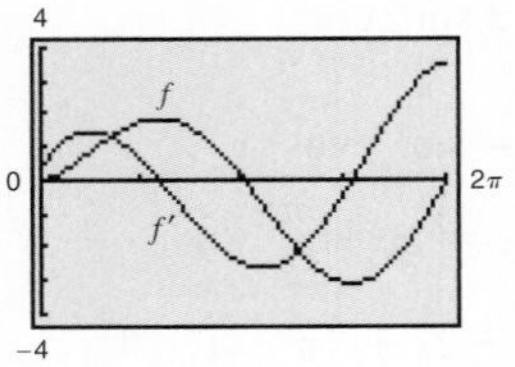

(b) 1.8366, 4.8158

(c) $f' > 0$ on $(0, 1.8366)$, $(4.8158, 2\pi)$
$f' < 0$ on $(1.8366, 4.8158)$

(d) Relative maximum: $(1.8366, 1.8493)$
Relative minimum: $(4.8158, -3.0869)$

85. False. $y' = \frac{1}{2}\sin x(1 - \cos x)^{-1/2}$

87. False. $y' = x \sin 2x + \sin^2 x$ **89.** Answers will vary.

Section 8.5 *(page 551)*

Skills Warm Up *(page 551)*

1. $-\frac{\sqrt{2}}{2}$ **2.** $-\frac{1}{2}$ **3.** $-\frac{\sqrt{3}}{2}$ **4.** $\frac{\sqrt{3}}{2}$

5. $-\frac{\sqrt{3}}{3}$ **6.** $-\frac{\sqrt{3}}{3}$ **7.** -1 **8.** 0

9. $\tan x$ **10.** $\cot x$ **11.** $\sin^2 x$ **12.** $\cos^2 x$

13. 1 **14.** $\cos x$ **15.** $\csc x$ **16.** $\cos x \sin x$

17. $\frac{88}{3}$ **18.** $\frac{4}{3}$ **19.** 4 **20.** $\frac{17}{4}$

1. $-4\cos x + C$ **3.** $-\frac{1}{2}\cos 2x + C$ **5.** $\sin x^4 + C$

7. $5\tan\frac{x}{5} + C$ **9.** $\frac{1}{2}\sec 2x + C$ **11.** $\frac{1}{4}\tan^4 x + C$

13. $\ln|\tan x| + C$ **15.** $\ln|\sec x - 1| + C$

17. $-\ln|1 + \cos x| + C$ **19.** $\frac{1}{\pi}\ln|\sin \pi x| + C$

21. $\frac{1}{2}\ln|\csc 2x - \cot 2x| + C$ **23.** $\frac{2}{3}\sec^6\frac{x}{4} + C$

25. $\frac{1}{2}\tan^2 x + C$ **27.** $-\cos e^x + C$ **29.** $\frac{1}{\pi}e^{\sin \pi x} + C$

31. $-\cos^2 x + x + C$ or $\sin^2 x + x + C$

33. $\frac{1}{2}x\sin 2x + \frac{1}{4}\cos 2x + C$ **35.** $6x\tan x + 6\ln|\cos x| + C$

37. $-\frac{t}{3}\csc 3t + \frac{1}{9}\ln|\csc 3t - \cot 3t| + C$

39. $\frac{3\sqrt{3}}{8} \approx 0.6495$ **41.** $2(\sqrt{3} - 1) \approx 1.4641$

43. $\frac{1}{2}$ **45.** $-\ln(\cos 1) \approx 0.6156$ **47.** 4

49. $\frac{\pi^2}{2} + 2 \approx 6.9348$ **51.** 2

53.

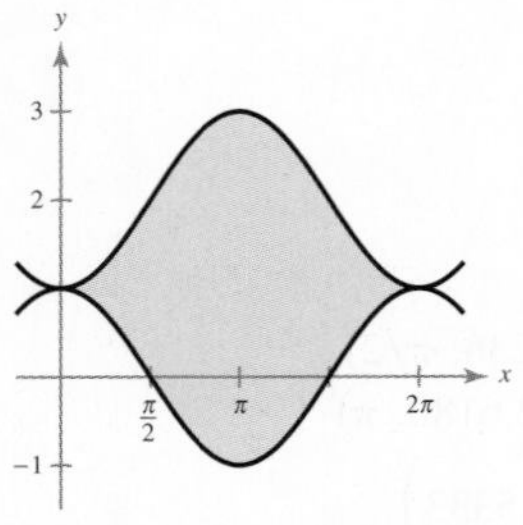

$4\pi \approx 12.566$ square units

55.

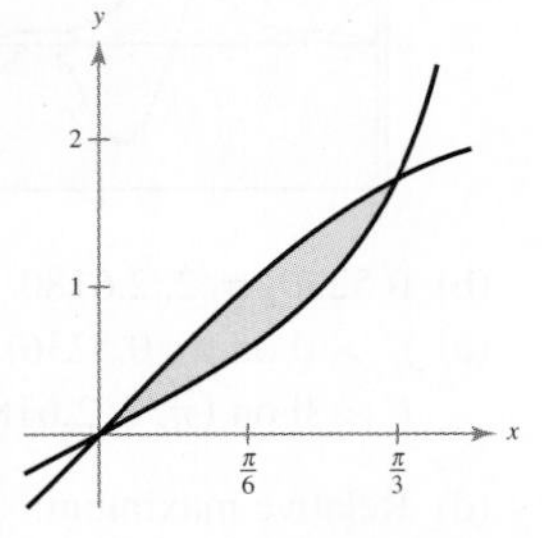

$1 - \ln 2 \approx 0.3069$ square unit

57. 17.69 in.

59. (a) About 1061.8 trillion Btu
(b) About 576.9 trillion Btu
(c) About 579.6 trillion Btu

61. (a) $C \approx \$27.50$ (b) $C \approx \$18.08$

63. 182,427 lb/day **65.** 0.9777 **67.** True

Review Exercises for Chapter 8 *(page 558)*

1. $30°, -330°$ **3.** $315°, -45°$ **5.** $\frac{17\pi}{9}$

7. $-\frac{\pi}{3}$ **9.** $-\frac{8\pi}{3}$ **11.** $\frac{11\pi}{18}$

13. $240°$ **15.** $-120°$ **17.** $b = 4\sqrt{3}, \theta = 60°$

19. $a = \frac{5\sqrt{3}}{2}, c = 5, \theta = 60°$ **21.** About 15.38 ft

23. $\sin\theta = \frac{\sqrt{2}}{2}$, $\csc\theta = \sqrt{2}$
$\cos\theta = -\frac{\sqrt{2}}{2}$, $\sec\theta = -\sqrt{2}$
$\tan\theta = -1$, $\cot\theta = -1$

25. $\sin(-45°) = -\frac{\sqrt{2}}{2}$, $\csc(-45°) = -\sqrt{2}$
$\cos(-45°) = \frac{\sqrt{2}}{2}$, $\sec(-45°) = \sqrt{2}$
$\tan(-45°) = -1$, $\cot(-45°) = -1$

27. $\sin\frac{5\pi}{3} = -\frac{\sqrt{3}}{2}$, $\csc\frac{5\pi}{3} = -\frac{2\sqrt{3}}{3}$
$\cos\frac{5\pi}{3} = \frac{1}{2}$, $\sec\frac{5\pi}{3} = 2$
$\tan\frac{5\pi}{3} = -\sqrt{3}$, $\cot\frac{5\pi}{3} = -\frac{\sqrt{3}}{3}$

29. $\sin(-225°) = \frac{\sqrt{2}}{2}$, $\csc(-225°) = \sqrt{2}$
$\cos(-225°) = -\frac{\sqrt{2}}{2}$, $\sec(-225°) = -\sqrt{2}$
$\tan(-225°) = -1$, $\cot(-225°) = -1$

31. $\sin\left(-\frac{11\pi}{6}\right) = \frac{1}{2}$, $\csc\left(-\frac{11\pi}{6}\right) = 2$
$\cos\left(-\frac{11\pi}{6}\right) = \frac{\sqrt{3}}{2}$, $\sec\left(-\frac{11\pi}{6}\right) = \frac{2\sqrt{3}}{3}$
$\tan\left(-\frac{11\pi}{6}\right) = \frac{\sqrt{3}}{3}$, $\cot\left(-\frac{11\pi}{6}\right) = \sqrt{3}$

33. 0.6494 **35.** 3.2361 **37.** -0.3420

39. -0.2588 **41.** $r \approx 146.19$ **43.** $x \approx 68.69$

45. $\frac{2\pi}{3}, \frac{4\pi}{3}$ **47.** $\frac{7\pi}{6}, \frac{3\pi}{2}, \frac{11\pi}{6}$ **49.** $\frac{\pi}{3}, \pi, \frac{5\pi}{3}$

51. About 81.18 ft

53. Period: $\frac{\pi}{2}$
Amplitude: 2

55. Period: $\frac{4\pi}{3}$
Amplitude: 2

57.

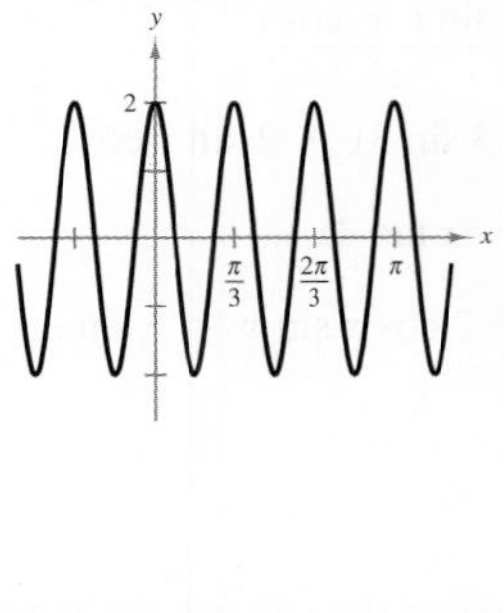

59.

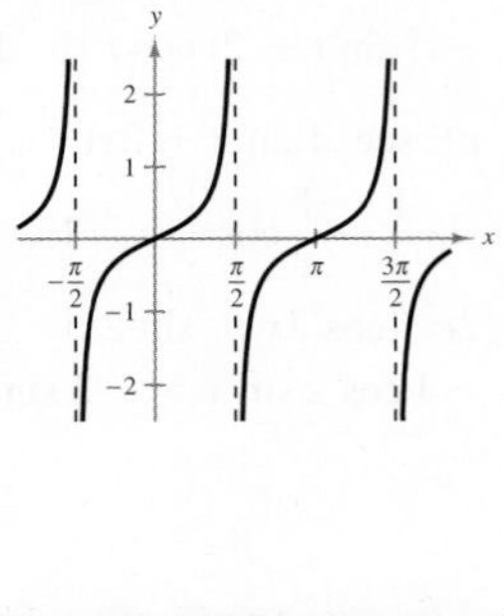

61.

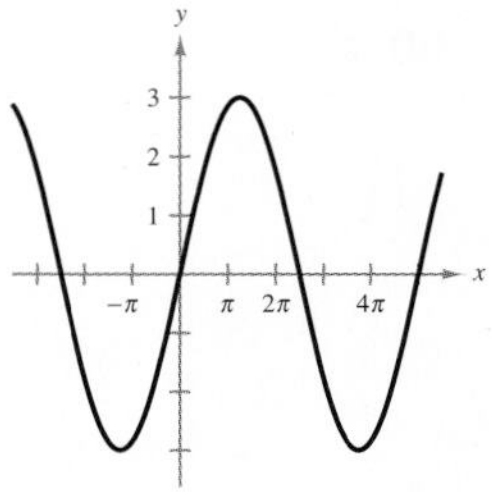

63.

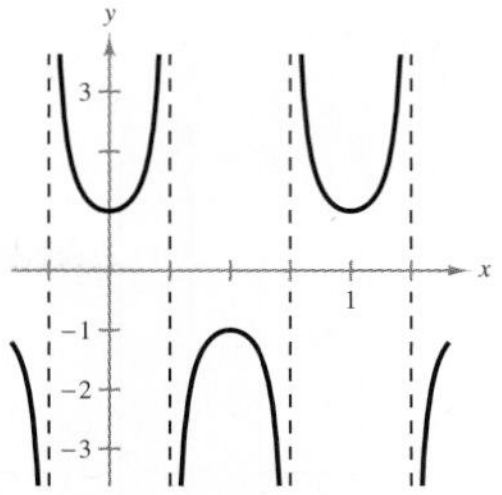

65. (a)

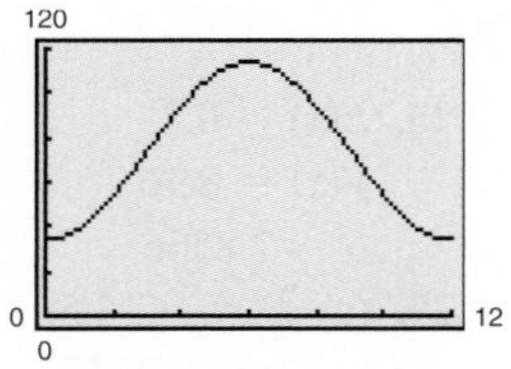

(b) Yes; May and June

67. $5\pi \cos 5\pi x$ **69.** $9x^2 \sec^2 3x^3$ **71.** $-\frac{6x}{5} \csc^2 \frac{3x^2}{5}$

73. $-\frac{\pi \sin 2\pi x}{\sqrt{\cos 2\pi x}}$ **75.** $-x \sec^2 x - \tan x$

77. $-\frac{x \sin x + 2 \cos x}{x^3}$ **79.** $2 \sin x \cos x + 1 = \sin 2x + 1$

81. $-4 \cot x \csc^4 x$ **83.** $e^x(\cot x - \csc^2 x)$

85. $y = -2x + \frac{\pi}{2}$ **87.** $y = -1$ **89.** $y = \frac{1}{2}$

91. Relative maximum: $(2, 1)$
Relative minimum: $(6, -1)$

93. Relative maximum: $\left(\frac{\pi}{6}, \frac{\pi + 6\sqrt{3}}{12}\right)$

Relative minimum: $\left(\frac{5\pi}{6}, \frac{5\pi - 6\sqrt{3}}{12}\right)$

95. Relative maxima: $\left(\frac{\pi}{2}, 2\right), \left(\frac{3\pi}{2}, 0\right)$

Relative minima: $\left(\frac{7\pi}{6}, -\frac{1}{4}\right), \left(\frac{11\pi}{6}, -\frac{1}{4}\right)$

97. June; 114 units **99.** $-\frac{1}{3} \cos 3x + C$

101. $-\frac{1}{5} \csc 5x + C$ **103.** $-\frac{1}{2} \cot x^2 + C$

105. $\frac{1}{2} \ln|\tan 2x| + C$ **107.** $-\frac{1}{3} \ln|\cos 3x| + C$

109. $\frac{1}{4} \sin^4 x + C$ **111.** $\pi + 2$ **113.** $\frac{2\sqrt{3}}{3}$

115. 0 **117.** 2 **119.** $\frac{2}{3}$ **121.** $\frac{5}{3}$

123. About 22.3 in. **125.** About \$37 billion

Test Yourself *(page 562)*

	Function	*θ (deg)*	*θ (rad)*	*Function Value*
1.	sin	67.5°	$\frac{3\pi}{8}$	0.9239
2.	cos	36°	$\frac{\pi}{5}$	0.8090
3.	tan	15°	$\frac{\pi}{12}$	0.2679
4.	cot	−30°	$-\frac{\pi}{6}$	$-\sqrt{3}$
5.	sec	−40°	$-\frac{2\pi}{9}$	1.3054
6.	csc	−225°	$-\frac{5\pi}{4}$	$\sqrt{2}$

7. About 27.37 in.

8. $\frac{\pi}{4}, \frac{3\pi}{4}$ **9.** $\frac{\pi}{4}, \frac{3\pi}{4}, \frac{5\pi}{4}, \frac{7\pi}{4}$ **10.** $\frac{\pi}{6}, \frac{7\pi}{6}$

11.

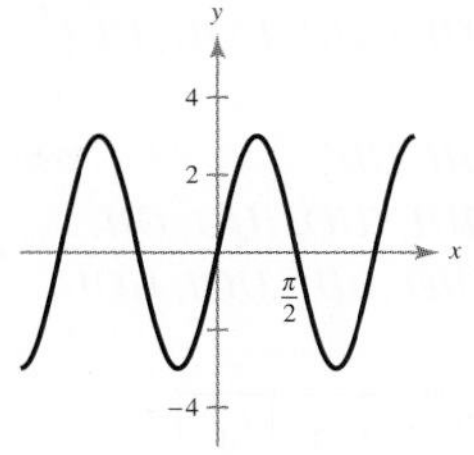

12.

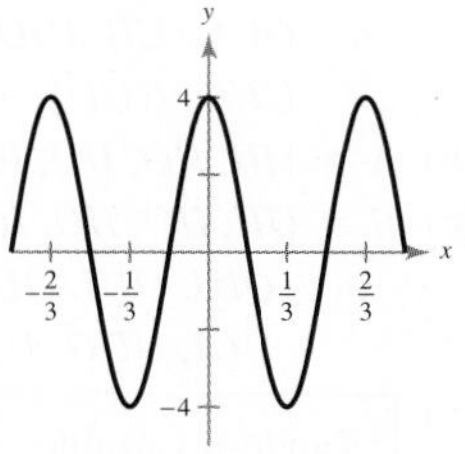

13.

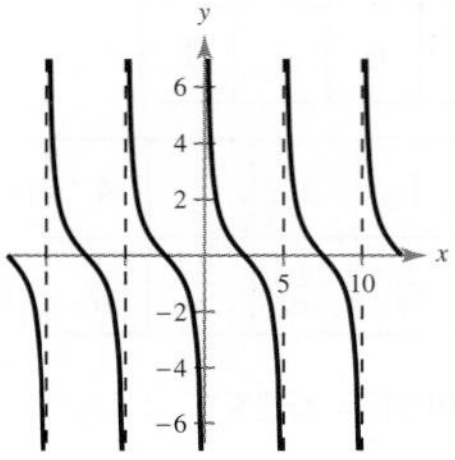

14. (a) $y' = \sin 2x - \sin x$
(b) Relative maxima: $\left(\frac{\pi}{3}, \frac{1}{4}\right), \left(\frac{5\pi}{3}, \frac{1}{4}\right)$
Relative minimum: $(\pi, -2)$

15. (a) $y' = \sec\left(x - \frac{\pi}{4}\right) \tan\left(x - \frac{\pi}{4}\right)$
(b) Relative maximum: $\left(\frac{5\pi}{4}, -1\right)$
Relative minimum: $\left(\frac{\pi}{4}, 1\right)$

16. (a) $y' = \frac{\cos(x + \pi)}{[3 - \sin(x + \pi)]^2}$
(b) Relative maximum: $\left(\frac{3\pi}{2}, \frac{1}{2}\right)$
Relative minimum: $\left(\frac{\pi}{2}, \frac{1}{4}\right)$

17. $-\frac{1}{5} \cos 5x + C$ **18.** $4 \tan \frac{x}{4} + C$

19. $-\frac{1}{2} \ln|\csc x^2 + \cot x^2| + C$ **20.** $\frac{2 - \sqrt{2}}{2\pi}$

21. $\frac{9}{2}$ **22.** $\frac{\sqrt{3}}{3} - \frac{1}{2} \approx 0.077$ **23.** $\frac{1}{8}$

24. (a) \$243.6 thousand (b) \$29.06 thousand

Chapter 9

Section 9.1 *(page 570)*

Skills Warm Up *(page 570)*

1. 1 **2.** 1 **3.** 2 **4.** 2 **5.** $\frac{1}{2}$
6. 1 **7.** 37.50% **8.** 81.82%
9. 54.17% **10.** 43.75%

1. (a) $S = \{HHH, HHT, HTH, HTT, THH, THT, TTH, TTT\}$
(b) $A = \{HHH, HHT, HTH, THH\}$
(c) $B = \{HTT, THT, TTH, TTT\}$

3. (a) $S = \{III, IIO, IIU, IOI, IOO, IOU, IUI, IUO, IUU, OII, OIO, OIU, OOI, OOO, OOU, OUI, OUO, OUU, UII, UIO, UIU, UOI, UOO, UOU, UUI, UUO, UUU\}$

(b) $A = \{III, IIO, IIU, IOI, IUI, OII, UII\}$

(c) $B = \{III, IIO, IIU, IOI, IOU, IUI, IUO, IUU, OII, OIU, OUI, OUU, UII, UIO, UIU, UOI, UOU, UUI, UUO, UUU\}$

5. (a)

Random variable	0	1	2	3	4
Frequency	1	4	6	4	1

(b)

Random variable, x	0	1	2	3	4
Probability, $P(x)$	$\frac{1}{16}$	$\frac{4}{16}$	$\frac{6}{16}$	$\frac{4}{16}$	$\frac{1}{16}$

7. (a) $\frac{7}{24}$ (b) $\frac{17}{24}$ (c) $\frac{4}{9}$ **9.** $P(3) = 0.25$

11.

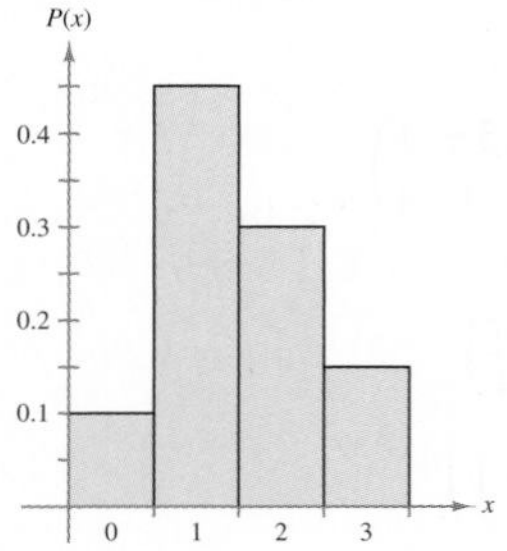

The table represents a probability distribution.

13. The table does not represent a probability distribution because the sum of the probabilities does not equal 1 and $P(4) < 0$.

15.

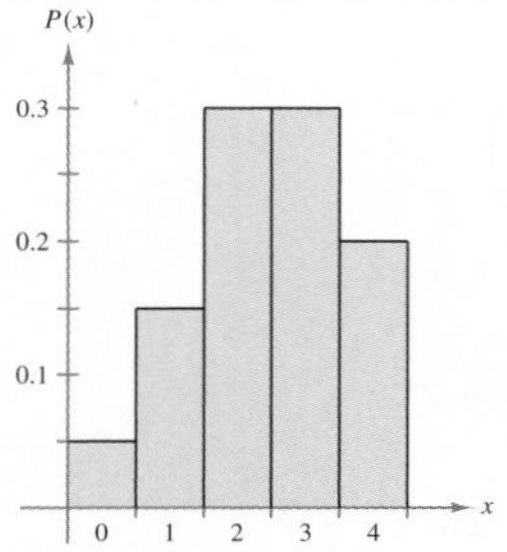

(a) $\frac{3}{4}$ (b) $\frac{4}{5}$

17.

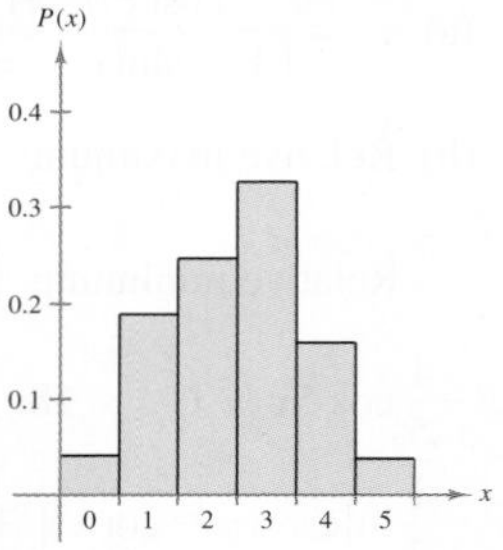

(a) 0.803 (b) 0.197

19. (a)

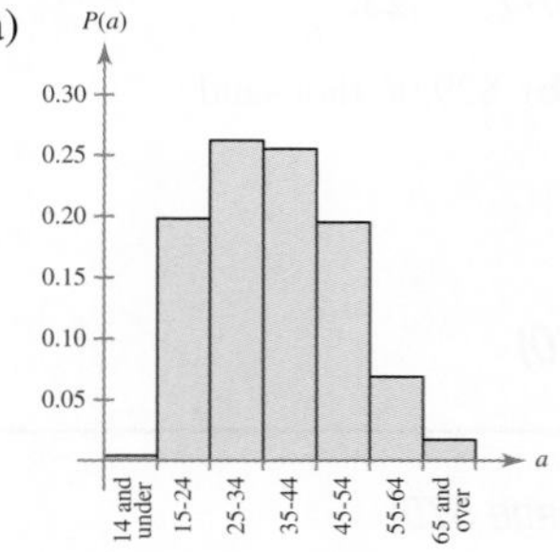

(b) 0.715 (c) 0.536 (d) 0.202

21. (a) $S = \{gggg, gggb, ggbg, gbgg, bggg, ggbb, gbgb, gbbg, bgbg, bbgg, bggb, gbbb, bgbb, bbgb, bbbg, bbbb\}$

(b)

x	0	1	2	3	4
$P(x)$	$\frac{1}{16}$	$\frac{4}{16}$	$\frac{6}{16}$	$\frac{4}{16}$	$\frac{1}{16}$

(c)

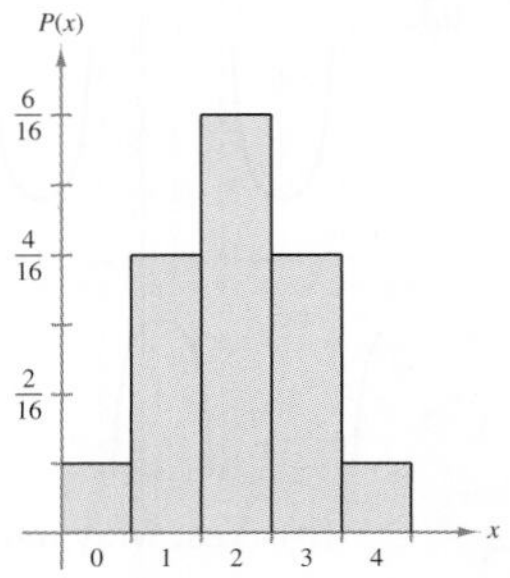

(d) $\frac{15}{16}$

23. $E(x) = 3$
$V(x) = 0.875$
$\sigma \approx 0.9354$

25. $E(x) = 0.8$
$V(x) = 8.16$
$\sigma \approx 2.8566$

27. (a) Mean: 2.5
Variance: 1.25
$\sigma \approx 1.1180$

(b) Mean: 5
Variance: 2.5
$\sigma \approx 1.5811$

29. (a) $E(x) = 18.5$
$\sigma \approx 8.0777$

(b) $91,575

31. $201

33. (a)

x	0	1	2	3	4
$P(x)$	$\frac{40}{160}$	$\frac{61}{160}$	$\frac{40}{160}$	$\frac{17}{160}$	$\frac{2}{160}$

(b)

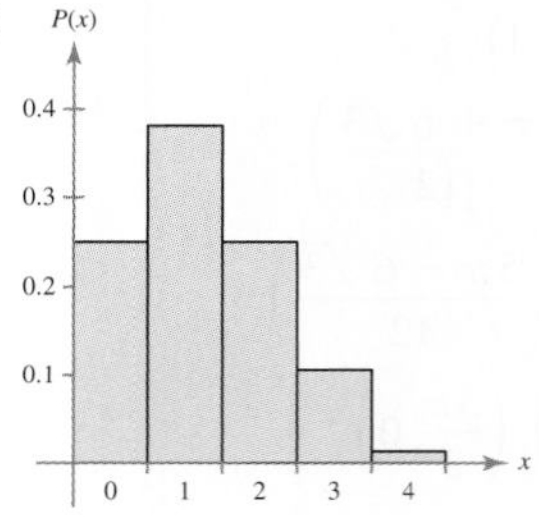

(c) $\frac{59}{80}$

(d) $E(x) = \dfrac{5}{4}$; $V(x) = \dfrac{39}{40}$; $\sigma = \dfrac{\sqrt{390}}{20}$

On average, you can expect the player to get 1.25 hits per game. The variance and standard deviation are measures of how spread out the data are.

35. $\approx -\$0.0526$ **37.** City 1 **39.** Answers will vary.

Section 9.2 *(page 578)*

Skills Warm Up *(page 578)*

1. Yes **2.** No **3.** No **4.** Yes
5. 1 **6.** $\frac{1}{4}$ **7.** 1

1.

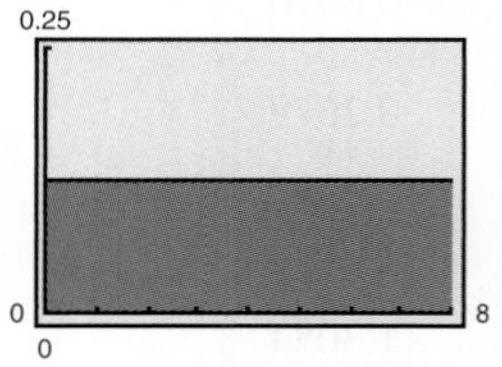

$f(x)$ is a probability density function.

$$\int_0^8 \frac{1}{8}\,dx = \left[\frac{1}{8}x\right]_0^8 = 1$$

3.

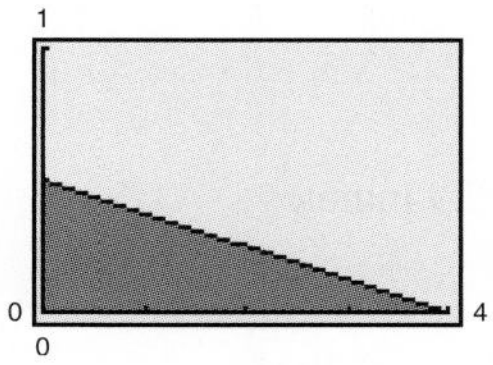

$f(x)$ is a probability density function.

$$\int_0^4 \frac{4-x}{8}\,dx = \left[\frac{1}{2}x - \frac{1}{16}x^2\right]_0^4 = 1$$

5.

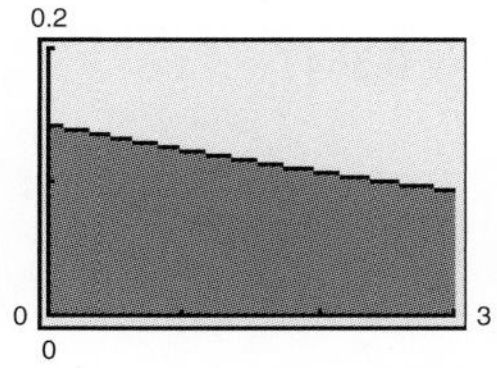

$f(x)$ is not a probability density function because

$$\int_0^3 \frac{1}{7}e^{-x/7}\,dx = \left[-e^{-x/7}\right]_0^3 \approx 0.349 \neq 1.$$

7.

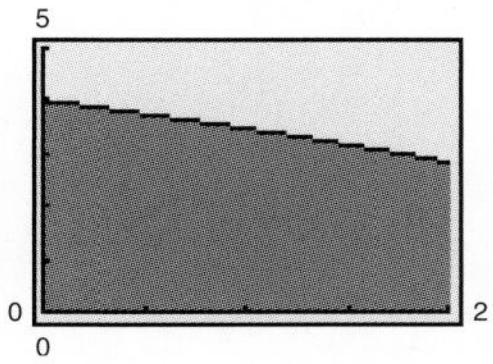

$f(x)$ is not a probability density function because

$$\int_0^2 2\sqrt{4-x}\,dx = \left[-\frac{4}{3}(4-x)^{3/2}\right]_0^2 \approx 6.90 \neq 1.$$

9.

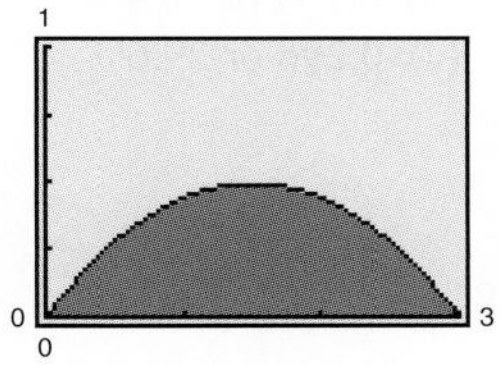

f is a probability density function.

$$\int_0^3 \left(\frac{2}{3}x - \frac{2}{9}x^2\right) dx = \left[\frac{x^2}{3} - \frac{2x^3}{27}\right]_0^3 = 1$$

11. $\frac{2}{15}$ **13.** $\frac{3}{32}$ **15.** $\frac{1}{2}$

17.

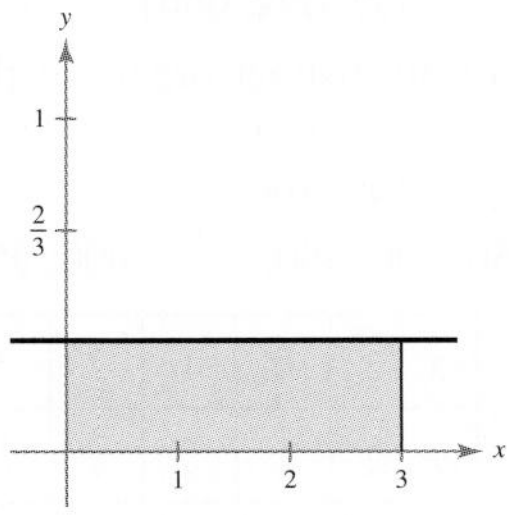

(a) $\frac{2}{3}$ (b) $\frac{1}{3}$
(c) $\frac{2}{3}$ (d) 0

19.

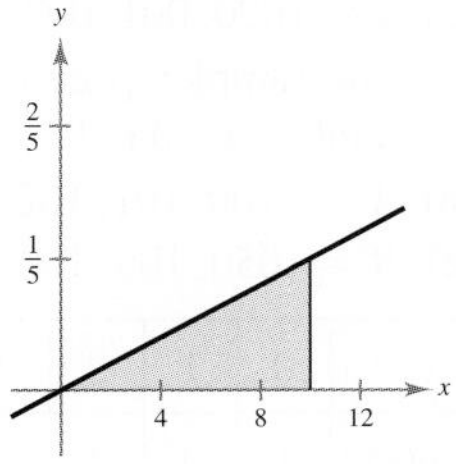

(a) $\frac{9}{25}$ (b) $\frac{1}{5}$
(c) $\frac{9}{25}$ (d) $\frac{24}{25}$

21.

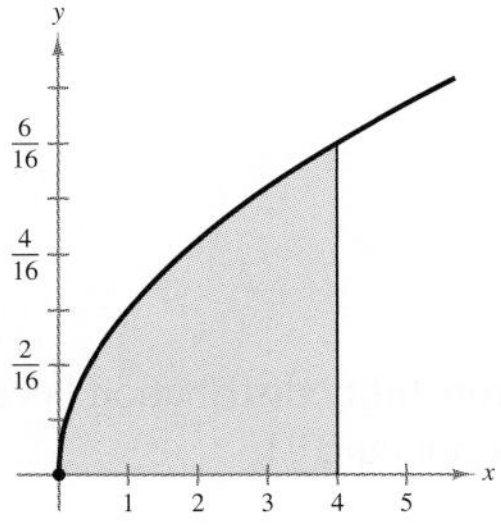

(a) $\frac{\sqrt{2}}{4} \approx 0.354$

(b) $1 - \frac{\sqrt{2}}{4} \approx 0.646$

(c) $\frac{1}{8}\left(3\sqrt{3} - 1\right) \approx 0.525$

(d) $\frac{3\sqrt{3}}{8} \approx 0.650$

23.

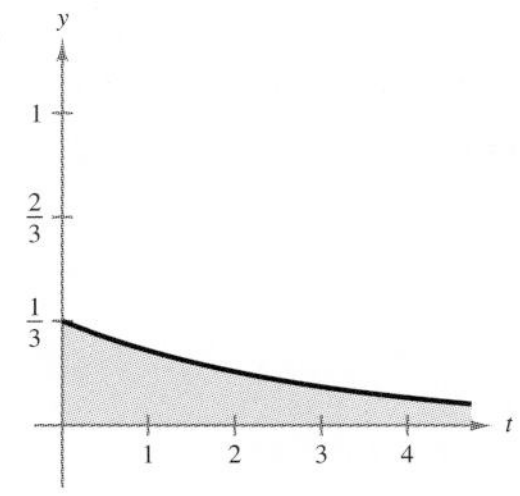

(a) $1 - e^{-2/3} \approx 0.4866$
(b) $e^{-2/3} \approx 0.5134$
(c) $e^{-1/3} - e^{-4/3} \approx 0.4529$
(d) 0

25. (a) $\frac{1}{6}$ (b) $\frac{2}{5}$

27. (a) $1 - \frac{1}{3}\sqrt{6} \approx 0.1835$ (b) $\frac{25}{81}\sqrt{5} - \frac{7}{81} \approx 0.6037$

29. (a) $1 - e^{-2/3} \approx 0.487$
(b) $e^{-2/3} - e^{-4/3} \approx 0.250$
(c) $e^{-2/3} \approx 0.513$

31. (a) $1 - e^{-6/5} \approx 0.699$
(b) $e^{-2/5} - e^{-6/5} = 0.369$
(c) $e^{-8/5} \approx 0.202$

33. (a) 0.75. There is a 75% probability that the community will receive up to 10 inches of rain.
(b) 0.25. There is a 25% probability that the community will receive 10 to 15 inches of rain.
(c) 0.25. There is a 25% probability that the community will receive up to 5 inches of rain.
(d) About 0.095. There is a probability of approximately 9.5% that the community will receive 12 to 15 inches of rain.

35. About 23.34 tons

Section 9.3 *(page 587)*

Skills Warm Up *(page 587)*

1. 8 **2.** 3 ln 2 **3.** $\frac{4}{3}$ **4.** $\frac{4}{3}$
5. (a) $\frac{1}{4}$ (b) $\frac{1}{2}$ **6.** (a) $\frac{1}{2}$ (b) $\frac{11}{16}$

1. (a) $\frac{3}{2}$ (b) $\frac{3}{4}$ (c) $\frac{\sqrt{3}}{2}$
(d)

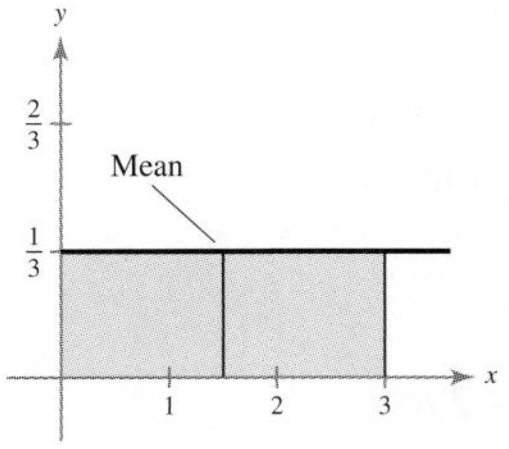

3. (a) 4 (b) 2 (c) $\sqrt{2}$
(d)

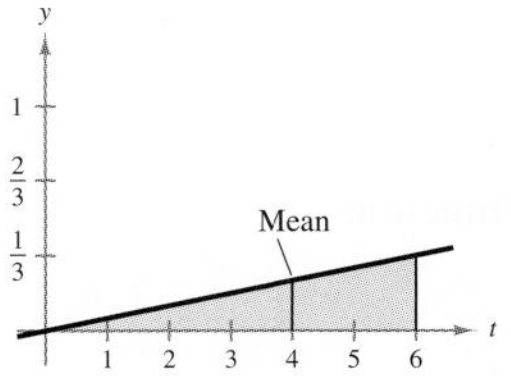

5. (a) $\frac{4}{3}\ln 4 \approx 1.848$ (b) $4 - \frac{16}{9}(\ln 4)^2 \approx 0.583$
(c) $\sqrt{0.583} \approx 0.76$
(d)

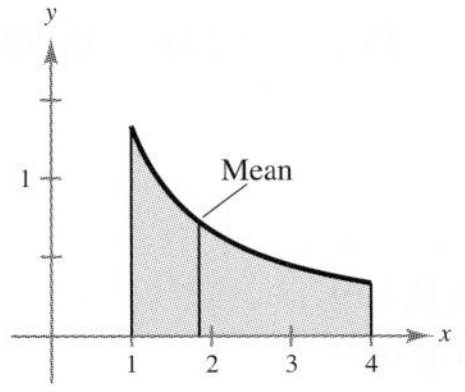

7. (a) $\frac{2}{3}$ (b) $\frac{2}{9}$ (c) $\frac{\sqrt{2}}{3}$

(d)

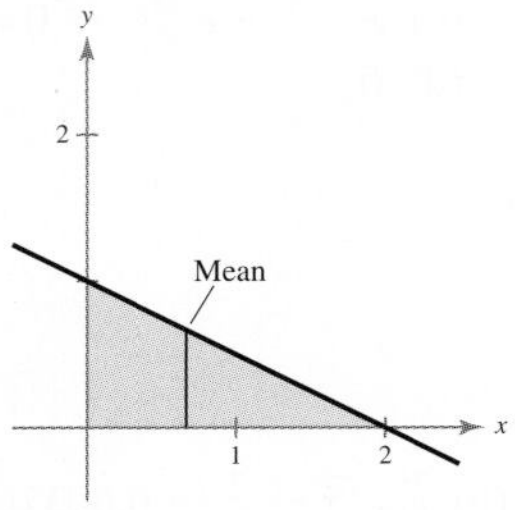

9. (a) $\frac{1}{2}$ (b) $\frac{1}{20}$ (c) $\frac{\sqrt{5}}{10}$

(d)

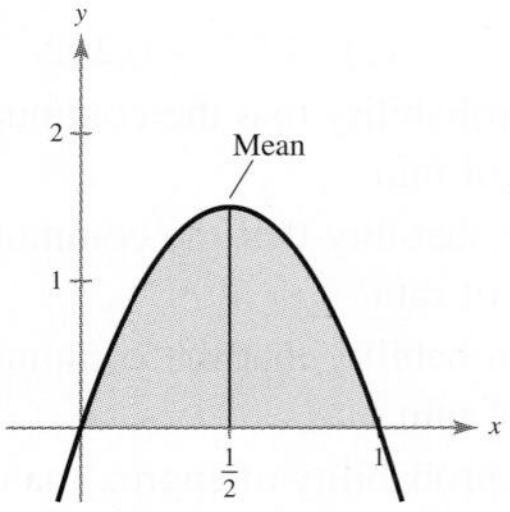

11. (a) $\frac{8}{5}$ (b) $\frac{192}{175}$ (c) $\frac{8\sqrt{21}}{35}$

(d)

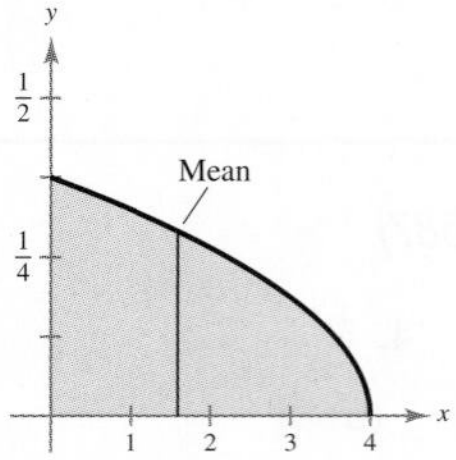

13. Mean: $\frac{11}{2}$ = median

15. Mean: $\frac{1}{6}$

Median: $\frac{2 - \sqrt{2}}{4} \approx 0.1464$

17. Mean: 9; Median: $9 \ln 2 \approx 6.238$

19. Uniform probability density function

Mean: 5

Variance: $\frac{25}{3}$

Standard deviation: $\frac{5\sqrt{3}}{3} \approx 2.887$

21. Exponential probability density function

Mean: 8

Variance: 64

Standard deviation: 8

23. Normal probability density function

Mean: 100

Variance: 121

Standard deviation: 11

25. Mean: 0

Standard deviation: 1

$P(0 \le x \le 0.85) \approx 0.3023$

27. Mean: 6

Standard deviation: 6

$P(x \ge 2.23) \approx 0.6896$

29. Mean: 8

Standard deviation: 2

$P(3 \le x \le 13) \approx 0.9876$

31. (a) About 0.309 (b) About 0.159

(c) About 0.841 (d) About 0.669

33. (a) $f(t) = \frac{1}{10}$, $[0, 10]$

(b) Mean: 10:05 A.M.

Standard deviation: $\frac{5\sqrt{3}}{3} \approx 2.9$ minutes

(c) $\frac{3}{10}$

35. (a) $f(t) = \frac{1}{2}e^{-t/2}$ (b) $1 - e^{-1/2} \approx 0.3935$

37. (a) $f(t) = \frac{1}{5}e^{-t/5}$ (b) $1 - e^{-2} \approx 0.8647$

39. (a) Mean: $\frac{7}{2}$

Standard deviation: $\frac{7\sqrt{5}}{10}$

(b) $\frac{7}{2}$ (c) About 0.6260

41. $\frac{15}{8}$

43. (a) Mean: 6

Standard deviation: $3\sqrt{2} \approx 4.243$

(b) About 0.6151

45. (a) -2.5 (b) 0.9938

47. (a) 0.6827 (b) No

49. (a)

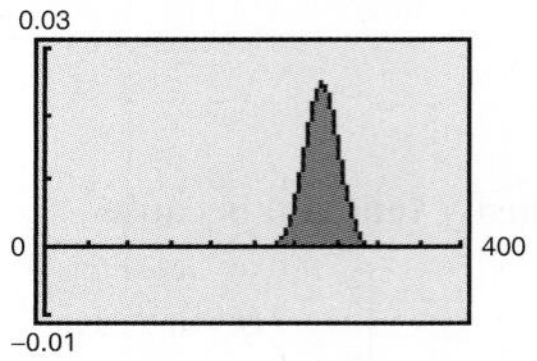

(b) 0.757 or 75.7%

(c) 0.191 or 19.1%

51. (a)

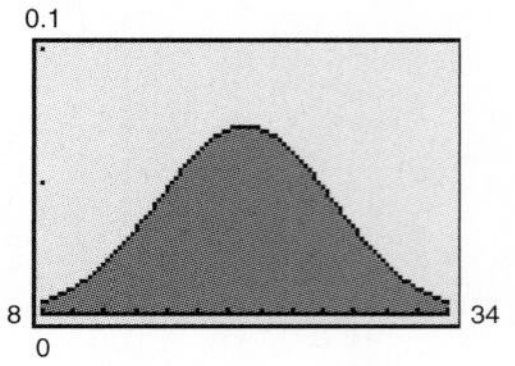

(b) 0.184 or 18.4%

(c) 0.296 or 29.6%

Review Exercises for Chapter 9 *(page 593)*

1. (a) $S = \{1, 2, 3, 4, 5, 6, 7, 8, 9, 10, 11, 12\}$

(b) $A = \{9, 10, 11, 12\}$

(c) $B = \{1, 3, 5, 7, 9, 11\}$

3. (a) $S = \{000, 001, 002, 003, . . . , 997, 998, 999\}$

The sample space consists of all distinct orders of three digits from 0 to 9.

(b) $A = \{100, 101, 102, 103, . . . , 198, 199\}$

(c) $B = \{050, 100, 150, 200, 250, 300, 350, . . . , 900, 950\}$

5.

x	0	1	2	3
$n(x)$	1	3	3	1

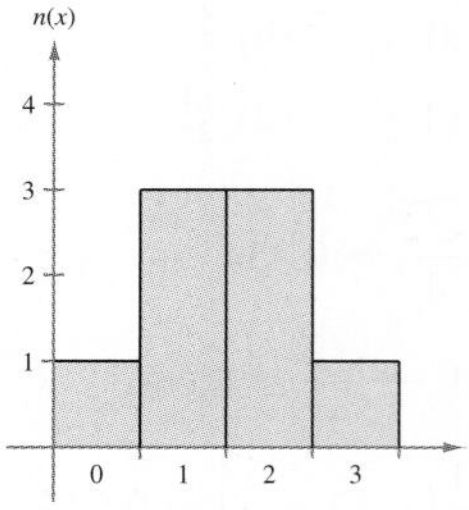

7.

x	0	1	2	3
$P(x)$	$\frac{1}{8}$	$\frac{3}{8}$	$\frac{3}{8}$	$\frac{1}{8}$

9. The table does not represent a probability distribution because the sum of the probabilities does not equal 1.

11. 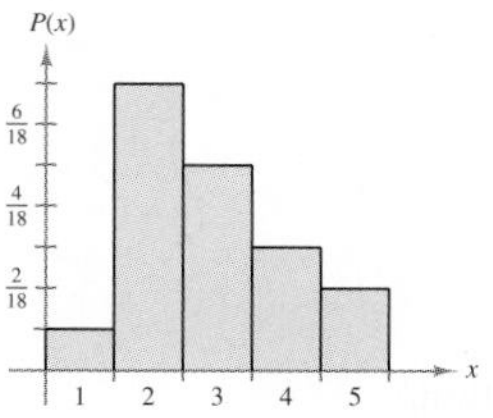

(a) $\frac{5}{6}$ (b) $\frac{5}{9}$

13. $E(x) = 2$; $V(x) = \dfrac{7}{5}$; $\sigma = \dfrac{\sqrt{35}}{5}$

15. $E(x) = 1.888$; $V(x) = 0.391456$; $\sigma \approx 0.626$

17. (a) $E(x) = 20.5$; $\sigma \approx 7.05$ (b) $80,975

19. $E(x) = 1.83$; $V(x) \approx 1.1611$; $\sigma \approx 1.0775$

21. 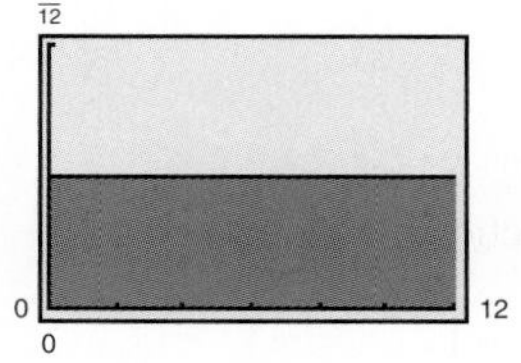

$f(x)$ is a probability density function.

$$\int_0^{12} \frac{1}{12}\,dx = \left[\frac{x}{12}\right]_0^{12} = 1$$

23. 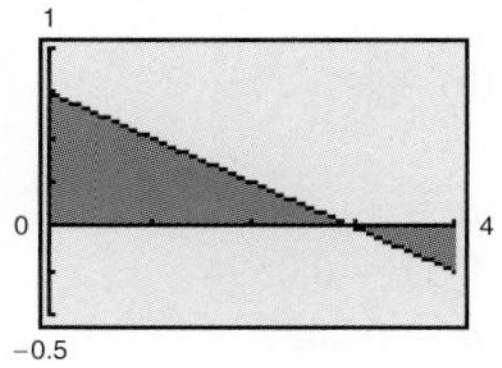

Although $\displaystyle\int_0^4 \frac{1}{4}(3 - x)\,dx = \left[\frac{3}{4}x - \frac{x^2}{8}\right]_0^4 = 1$, $f(x)$ is not a probability density function because $f(x) < 0$ over the interval $(3, 4]$.

25. 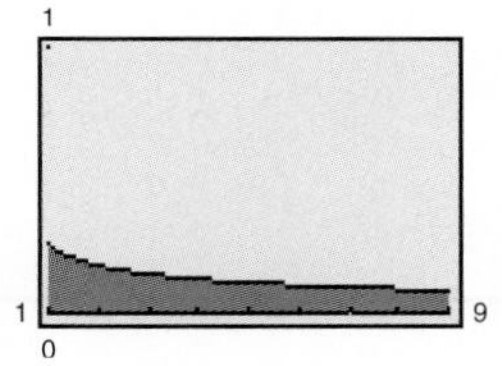

$f(x)$ is a probability density function.

$$\int_1^9 \frac{1}{4\sqrt{x}}\,dx = \frac{1}{4}\Big[2\sqrt{x}\Big]_1^9 = 1$$

27. 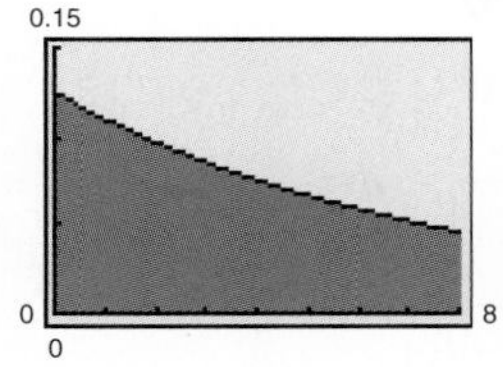

$f(x)$ is not a probability density function because

$$\int_0^8 \frac{1}{8}e^{-x/8}\,dx = \left[-e^{-x/8}\right]_0^8 \approx 0.632 \neq 1.$$

29. 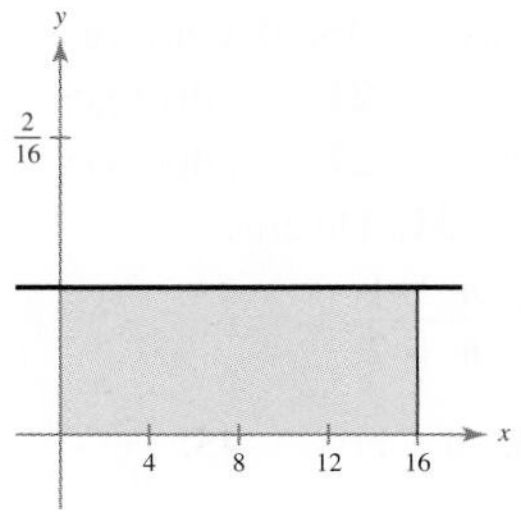

(a) $\frac{5}{16}$ (b) $\frac{1}{16}$
(c) $\frac{11}{16}$ (d) $\frac{1}{4}$

31. 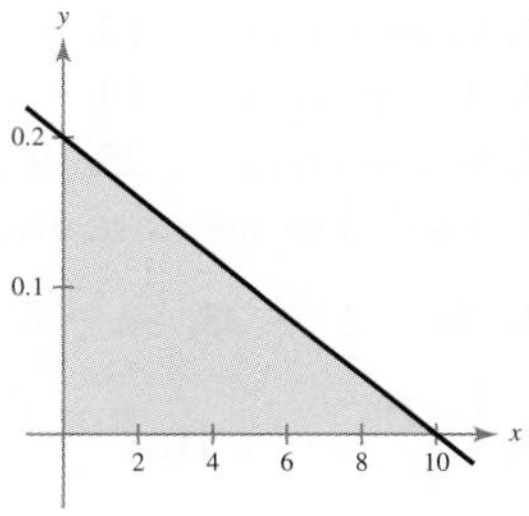

(a) $\frac{9}{25}$ (b) $\frac{9}{100}$
(c) $\frac{3}{4}$ (d) $\frac{3}{100}$

33. 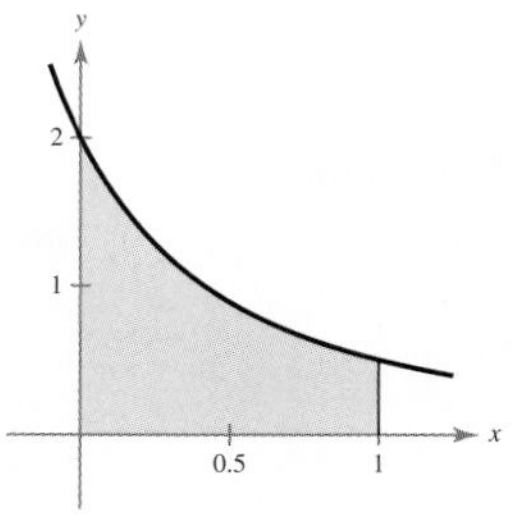

(a) $\frac{2}{3}$
(b) $\frac{16}{35}$
(c) $\frac{1}{3}$
(d) $\frac{40}{143}$

35. (a) $1 - e^{-5/12} \approx 0.341$ (b) $e^{-3/4} - e^{-1} \approx 0.104$

37. (a) $\dfrac{7}{2}$ (b) $\dfrac{49}{12}$ (c) $\dfrac{7\sqrt{3}}{6}$

(d) 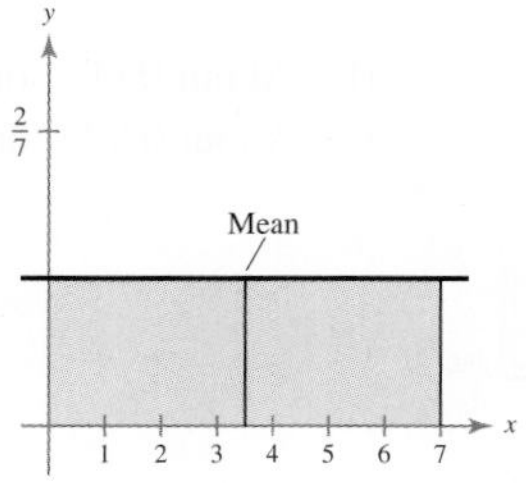

39. (a) $\frac{3}{2}\ln 3 \approx 1.648$ (b) $3 - \left(\frac{3}{2}\ln 3\right)^2 \approx 0.284$

(c) $\sqrt{3 - \left(\frac{3}{2}\ln 3\right)^2} \approx 0.533$

(d) 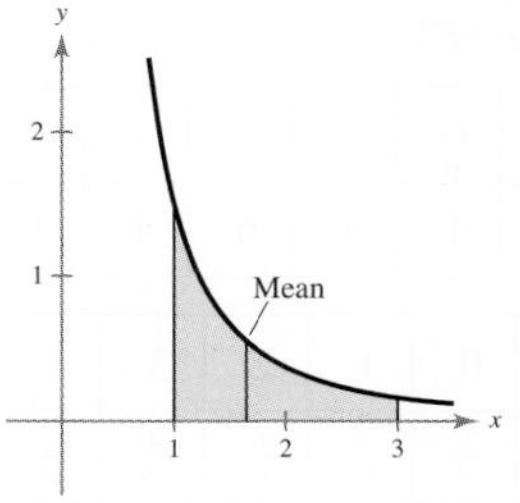

41. (a) $\dfrac{3}{2}$ (b) $\dfrac{9}{20}$ (c) $\dfrac{3\sqrt{5}}{10}$

(d) 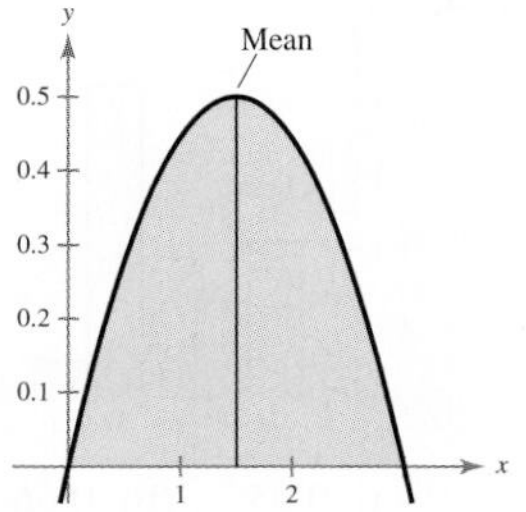

43. 7 **45.** $4 \ln 2 \approx 2.7726$

47. Uniform probability density function
Mean: 1
Variance: $\frac{1}{3}$
Standard deviation: $\dfrac{\sqrt{3}}{3}$

49. Exponential probability density function
Mean: 6
Variance: 36
Standard deviation: 6

51. Normal probability density function
Mean: 16
Variance: 9
Standard deviation: 3

53. (a) $f(t) = \frac{1}{20}$, $[0, 20]$
(b) Mean: 7:10 A.M.
Standard deviation: $\dfrac{10\sqrt{3}}{3} \approx 5.8$ min
(c) $\frac{1}{5}$

55. (a) $f(t) = \frac{1}{15}e^{-t/15}$, $0 \le t < \infty$
(b) $1 - e^{-2/3} \approx 0.4866$

57. (a) Mean: 4
Standard deviation: $\dfrac{4\sqrt{5}}{5}$
(b) 4
(c) About 0.626, or 62.6%

59. (a)

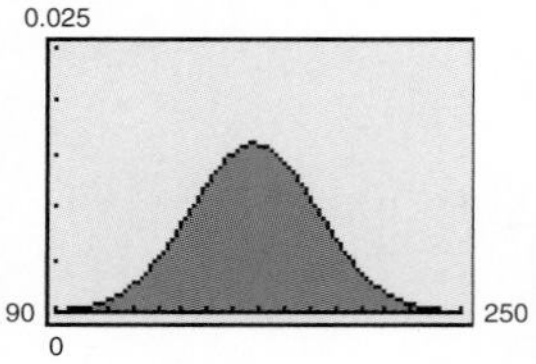

(b) About 0.006, or 0.6%
(c) About 0.973, or 97.3%

Test Yourself *(page 596)*

1. (a) $S = \{TTTT, TTTF, TTFT, TTFF, TFTT, TFTF, TFFT, TFFF, FTTT, FTTF, FTFT, FTFF, FFTT, FFTF, FFFT, FFFF\}$

(b)

Random variable, x	0	1	2	3	4
Frequency of x, $n(x)$	1	4	6	4	1

(c)

Random variable, x	0	1	2	3	4
Probability, $P(x)$	$\frac{1}{16}$	$\frac{4}{16}$	$\frac{6}{16}$	$\frac{4}{16}$	$\frac{1}{16}$

2. $P(\text{red non-face card}) = \frac{5}{13}$

3.

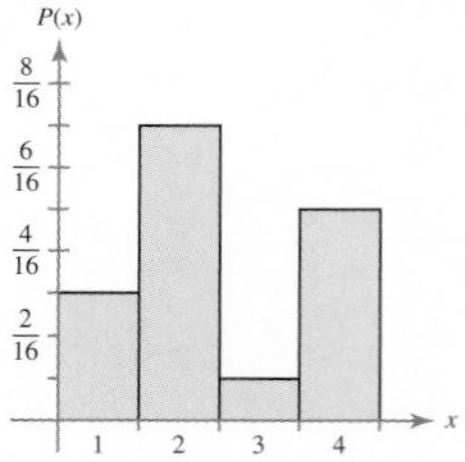

(a) $\frac{5}{8}$ (b) $\frac{3}{8}$

4.

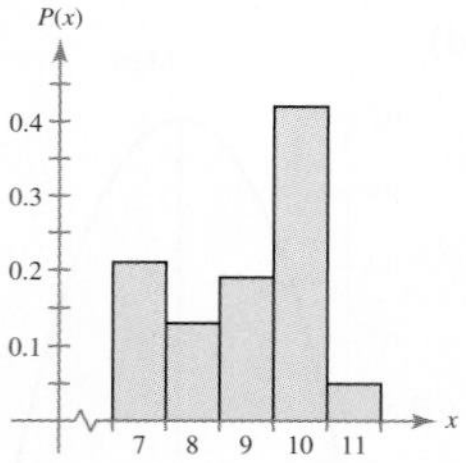

(a) 0.95 (b) 0.66

5. $E(x) = 1.8$
$V(x) = 1.16$
$\sigma \approx 1.0770$

6. $E(x) = -0.056$
$V(x) \approx 1.8649$
$\sigma \approx 1.3656$

7.

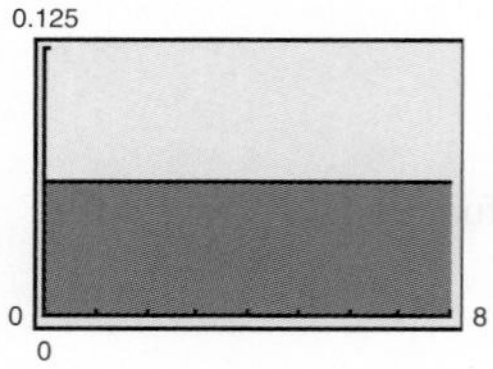

$f(x)$ is not a probability density function because
$$\int_0^8 \frac{1}{16}\,dx = \left[\frac{1}{16}x\right]_0^8 = \frac{1}{2} \ne 1.$$

8.

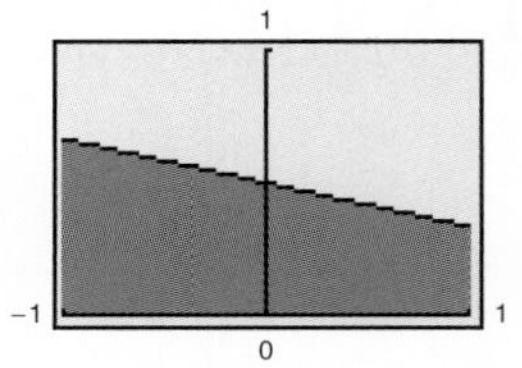

$f(x)$ is a probability density function.
$$\int_{-1}^{1} \frac{3-x}{6}\,dx = \left[-\frac{x}{12}(x-6)\right]_{-1}^{1} = 1$$

9.

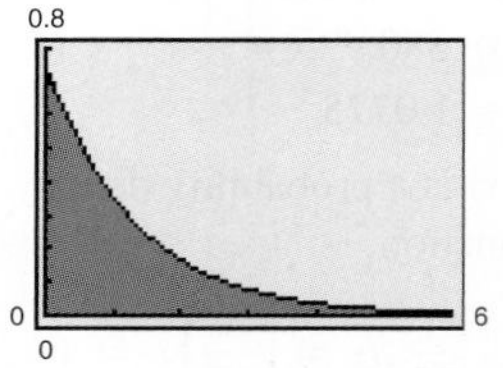

$f(x)$ is a probability density function.
$$\int_0^\infty \frac{3}{4}e^{-3x/4}\,dx = \left[-e^{-3x/4}\right]_0^\infty = 1$$

10. (a) $\frac{1}{9}$ (b) $\frac{5}{9}$ **11.** (a) $\frac{7}{16}$ (b) $\frac{225}{256}$

12. (a) 0.6321 (b) 0.3679

13. $\mu = 7$
$V(x) = \frac{49}{3}$
$\sigma \approx 4.041$

14. $\mu = \frac{5}{8}$
$V(x) \approx 0.0594$
$\sigma \approx 0.2437$

15. $\mu = 1$
$V(x) = 1$
$\sigma = 1$

16. $P(100 \le x \le 120) \approx 0.683$

Chapter 10

Section 10.1 *(page 604)*

Skills Warm Up *(page 604)*

1. 0 **2.** 2 **3.** ∞ **4.** 0

5. $\dfrac{n-2}{n}$, $n \ne -2$ **6.** $\dfrac{n-3}{n-4}$, $n \ne -4$

7. $\dfrac{3n^2+1}{n^3}$ **8.** $\dfrac{2n+1}{(n-1)(n+2)}$, $n \ge 2$

1. 1, 3, 5, 7, 9 **3.** 3, 9, 27, 81, 243 **5.** $\frac{1}{2}, \frac{2}{3}, \frac{3}{4}, \frac{4}{5}, \frac{5}{6}$

7. $3, \frac{9}{2}, \frac{27}{6}, \frac{81}{24}, \frac{243}{120}$ **9.** $-1, \frac{1}{4}, -\frac{1}{9}, \frac{1}{16}, -\frac{1}{25}$

11. 0; converges **13.** 3; converges **15.** 0; converges

17. 1; converges **19.** $\frac{1}{2}$; converges **21.** ∞; diverges

23. 0; converges **25.** ∞; diverges **27.** ∞; diverges

29. Limit does not exist; diverges **31.** Diverges

33. $3n - 2$ **35.** $5n - 6$ **37.** $\dfrac{n+1}{n+2}$

39. $\dfrac{(-1)^{n-1}}{2^{n-2}}$ **41.** $\dfrac{n+1}{n}$ **43.** $2(-1)^n$ **45.** $\dfrac{4n}{n!}$

47. 2, 5, 8, 11, 14, 17, . . . ; Pattern: each term is 1 less than 3 times n.

49. $1, \frac{5}{3}, \frac{7}{3}, 3, \frac{11}{3}, \frac{13}{3}$, . . . ; Pattern: each term is $\frac{1}{3}$ more than $\frac{2}{3}$ times n.

51. $3, -\frac{3}{2}, \frac{3}{4}, -\frac{3}{8}, \frac{3}{16}, -\frac{3}{32}, \ldots$; Pattern: each term is 3 times the quantity $\left(-\frac{1}{2}\right)$ raised to the $(n-1)$ power.

53. $2, 6, 18, 54, 162, 486, \ldots$; Pattern: each term is 3 raised to the n power times $\frac{2}{3}$.

55. Geometric, $20\left(\frac{1}{2}\right)^{n-1}$ **57.** Arithmetic, $\frac{2}{3}n + 2$

59. $\dfrac{3n+1}{4n}$

61. \$9045.00, \$9090.23, \$9135.68, \$9181.35, \$9227.26, \$9273.40, \$9319.76, \$9366.36, \$9413.20, \$9460.26

63. (a)

Year	1	2	3
Balance	\$2200	\$4620	\$7282

Year	4	5	6
Balance	\$10,210.20	\$13,431.22	\$16,974.34

(b) \$126,005.00 (c) \$973,703.62

65. (a) $a_n = 1.7n + 337$

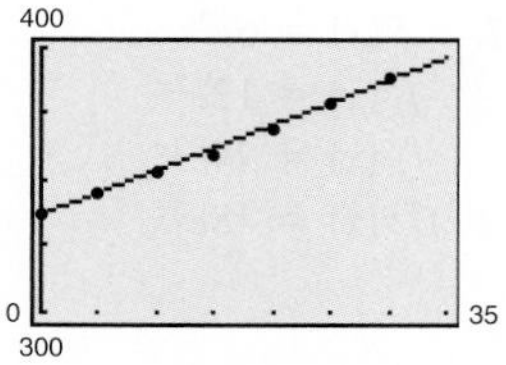

(b) 405 parts per million

67. (a) $a_n = 72.6n + 333$

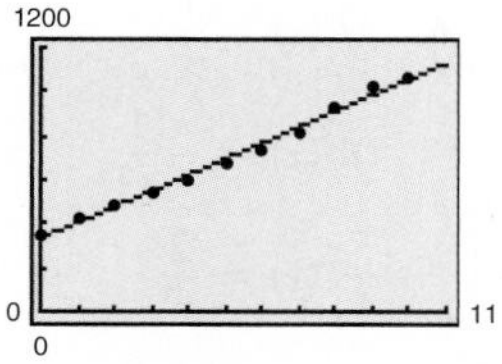

(b) About 1495 stores

69. (a) $a_n = -0.01394n^3 + 0.2648n^2 - 0.573n + 9.52$

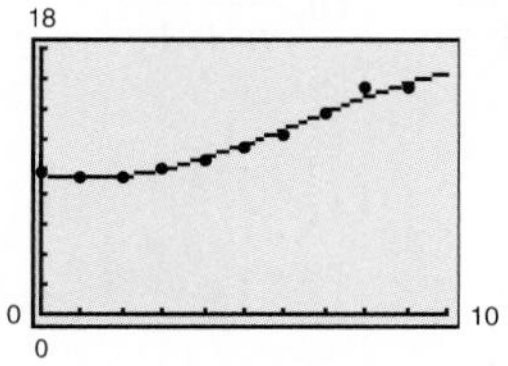

(b) About \$13.46 billion

71. (a) $h_n = 12\left(\frac{2}{3}\right)^n$

(b) 8 ft
$\frac{16}{3} \approx 5.3$ ft
$\frac{32}{9} \approx 3.6$ ft
$\frac{64}{27} \approx 2.4$ ft
$\frac{128}{81} \approx 1.6$ ft
$\frac{256}{243} \approx 1.1$ ft

(c) This sequence converges because $\lim_{n\to\infty} 12\left(\frac{2}{3}\right)^n = 0$.

73. (a) \$37,970.10
(b) The sequence diverges because $\lim_{n\to\infty} 32{,}800(1.05)^{n-1} = \infty$.

75. Answers will vary.

Section 10.2 *(page 616)*

Skills Warm Up *(page 616)*

1. $\frac{77}{60}$ **2.** $\frac{73}{24}$ **3.** $\frac{31}{16}$ **4.** $\frac{40}{9}$ **5.** $\frac{21}{8}$
6. $\frac{31}{32}$ **7.** $\frac{3}{4}$ **8.** 0 **9.** 1 **10.** $\frac{1}{2}$

1. $\sum_{n=1}^{5} 2n$ **3.** $\sum_{n=1}^{7} \dfrac{1}{2n-1}$

5. $S_1 = 1$
$S_2 = \frac{5}{4} = 1.25$
$S_3 = \frac{49}{36} \approx 1.361$
$S_4 = \frac{205}{144} \approx 1.424$
$S_5 = \frac{5269}{3600} \approx 1.464$

7. $S_1 = 3$
$S_2 = -\frac{3}{2} = -1.5$
$S_3 = \frac{21}{4} = 5.25$
$S_4 = -\frac{39}{8} = -4.875$
$S_5 = \frac{165}{16} = 10.3125$

9. Geometric series: $|r| = \frac{5}{2} > 1$

11. Geometric series: $|r| = 1.055 > 1$

13. nth-Term Test: $\lim_{n\to\infty} \dfrac{n}{n+1} = 1 \neq 0$

15. nth-Term Test: $\lim_{n\to\infty} \dfrac{n^2}{n^2+1} = 1 \neq 0$ **17.** $|r| = \frac{3}{4} < 1$

19. $|r| = 0.9 < 1$ **21.** $\frac{21}{2}$ **23.** 8 **25.** $\frac{1}{2}$ **27.** $\frac{40}{3}$

29. $S_4 = \frac{3}{7}\left[8 - \left(\frac{1}{8}\right)^4\right] \approx 3.428$
$S_6 = \frac{3}{7}\left[8 - \left(\frac{1}{8}\right)^6\right] \approx 3.429$
$S_{10} = \frac{3}{7}\left[8 - \left(\frac{1}{8}\right)^{10}\right] \approx 3.429$

31. $S_5 = \frac{5}{3}\left[2 - \left(\frac{1}{2}\right)^5\right] \approx 3.281$
$S_7 = \frac{5}{3}\left[2 - \left(\frac{1}{2}\right)^7\right] \approx 3.320$
$S_{10} = \frac{5}{3}\left[2 + \left(\frac{1}{2}\right)^{10}\right] \approx 3.335$

33. $\lim_{n\to\infty} \dfrac{n+10}{10n+1} = \dfrac{1}{10} \neq 0$; diverges

35. $\lim_{n\to\infty} \dfrac{n!+1}{n!} = 1 \neq 0$; diverges

37. $\lim_{n\to\infty} \dfrac{3n-1}{2n+1} = \dfrac{3}{2} \neq 0$; diverges

39. Geometric series: $|r| = 1.075 > 1$; diverges

41. Geometric series: $|r| = \frac{1}{5} < 1$; converges **43.** $\frac{4}{9}$ **45.** $\frac{9}{11}$

47. (a) $80{,}000(1 - 0.9^n)$ (b) 80,000 **49.** About 152.42 ft

51. (a) $\sum_{n=1}^{60} 100\left(1 + \dfrac{0.09}{12}\right)^n$ (b) \$7598.98 **53.** Proof

55. $\sum_{n=0}^{\infty} 500(0.75)^n = \$2000$ million

57. $V = 225{,}000(0.7)^n$; \$37,815.75 **59.** \$3,801,020.63

61. $\sum_{n=1}^{\infty} \left(\dfrac{1}{2}\right)^n = -1 + \sum_{n=0}^{\infty} \left(\dfrac{1}{2}\right)^n = -1 + \dfrac{1}{1 - \frac{1}{2}} = 1$

63. $P(2) = \dfrac{1}{8}$; $\sum_{n=0}^{\infty} \dfrac{1}{2}\left(\dfrac{1}{2}\right)^n = \dfrac{\frac{1}{2}}{1 - \frac{1}{2}} = 1$

65. $(0.85)^{12} 500 \approx 71.12$ ppm **67.** 6 **69.** About 0.5431

71. False. $\lim_{n\to\infty} \dfrac{1}{n} = 0$, but $\sum_{n=1}^{\infty} \dfrac{1}{n}$ diverges.

Section 10.3 *(page 624)*

Skills Warm Up *(page 624)*

1. $\dfrac{1}{n+1}$ **2.** $n + 1$ **3.** $\dfrac{3n}{n+1}$ **4.** $\dfrac{n+1}{n^2}$
5. 1 **6.** 5 **7.** 1 **8.** $\frac{1}{3}$
9. Geometric series **10.** Not a geometric series

1. p-series **3.** Not a p-series **5.** Not a p-series
7. p-series **9.** Converges **11.** Diverges
13. Converges **15.** Diverges **17.** Diverges
19. Converges **21.** Converges **23.** Converges
25. Diverges **27.** Converges **29.** Diverges

31. Converges

33. $\lim_{n\to\infty}\left|\frac{a_{n+1}}{a_n}\right| = \lim_{n\to\infty}\frac{1/[(n+1)^{3/2}]}{1/(n^{3/2})}$
$= \lim_{n\to\infty}\left(\frac{n}{n+1}\right)^{3/2} = 1$

35. $\lim_{n\to\infty}\left|\frac{a_{n+1}}{a_n}\right| = \lim_{n\to\infty}\frac{1/(n+1)^3}{1/n^3}$
$= \lim_{n\to\infty}\left(\frac{n}{n+1}\right)^{3} = 1$

37. About 1.1777; maximum error $\le \frac{1}{32}$.

39. About 1.1256; maximum error $\le \frac{\sqrt{10}}{2500} \approx 0.0013$.

41. a; diverges: $p = \frac{3}{4} < 1$

42. d; diverges: $p = 1$, harmonic series

43. e; converges: $p = \frac{5}{2} > 1$ **44.** b; diverges: $p = \frac{2}{5} < 1$

45. f; converges: $p = \frac{3}{2} > 1$ **46.** c; converges: $p = 2 > 1$

47. Diverges; nth-Term Test

49. Converges; p-Series Test; about 18.38

51. Converges; Geometric Series Test; $-\frac{3}{7} \approx -0.43$

53. Converges; Geometric Series Test; $\frac{3}{5}$

55. Converges; p-Series Test; about 0.4429

57. Diverges; Geometric Series Test

59. Diverges; Ratio Test **61.** Diverges; nth-Term Test

63. Converges; Geometric Series Test; $\frac{10}{3}$

65. No; although the terms approach zero, the series diverges because the partial sums approach infinity.

Quiz Yourself *(page 626)*

1. $-\frac{1}{4}, \frac{1}{16}, -\frac{1}{64}, \frac{1}{256}, -\frac{1}{1024}$ **2.** $\frac{1}{2}, \frac{3}{5}, \frac{2}{3}, \frac{5}{7}, \frac{3}{4}$

3. $-5, 5, -5, 5, -5$ **4.** $-1, 0, \frac{1}{6}, \frac{1}{12}, \frac{1}{40}$

5. 0; converges **6.** $\frac{1}{2}$; converges **7.** 0; converges

8. Does not exist; diverges **9.** $\frac{n-1}{n^2}$

10. $(-1)^n \cdot 3^{1/n}$ **11.** $2^{(-1)^n}$

12. $S_1 = 0$
$S_2 = \frac{1}{2} = 0.5$
$S_3 = \frac{5}{6} = 0.8\overline{3}$
$S_4 = \frac{23}{24} = 0.958\overline{3}$
$S_5 = \frac{119}{120} = 0.991\overline{6}$

13. $S_1 = 1$
$S_2 = 0$
$S_3 = \frac{3}{4} = 0.75$
$S_4 = \frac{1}{4} = 0.25$
$S_5 = \frac{9}{16} = 0.5625$

14. 12 **15.** 1 **16.** $\frac{50}{9}$

17. Diverges by the nth-Term Test because
$\lim_{n\to\infty}\frac{2n^2-1}{n^2+1} = 2 \ne 0.$

18. Converges by the Geometric Series Test because
$|r| = \left|\frac{1}{2}\right| < 1.$

19. Diverges by the Geometric Series Test because
$|r| = \left|\frac{5}{3}\right| > 1.$

20. Converges by the p-Series Test because $p = \frac{5}{3} > 1$.

21. Converges by the Ratio Test because
$\lim_{n\to\infty}\left|\frac{a_{n+1}}{a_n}\right| = \lim_{n\to\infty}\frac{n+1}{n^2} = 0 < 1.$

22. Diverges by the Ratio Test because
$\lim_{n\to\infty}\left|\frac{a_{n+1}}{a_n}\right| = \lim_{n\to\infty}\frac{2(n+1)}{3} = \infty.$

23. (a) $\sum_{n=1}^{36} 200(1.005)^n$ (b) \$7906.56

Section 10.4 *(page 635)*

Skills Warm Up *(page 635)*

1. $f(g(x)) = (x-1)^2$
$g(f(x)) = x^2 - 1$

2. $f(g(x)) = 6x + 3$
$g(f(x)) = 6x + 1$

3. $f(g(x)) = \sqrt{x^2+4}$
$g(f(x)) = x + 4,\ x \ge -4$

4. $f(g(x)) = e^{x^2}$
$g(f(x)) = e^{2x}$

5. $f'(x) = 5e^x$
$f''(x) = 5e^x$
$f'''(x) = 5e^x$
$f^{(4)}(x) = 5e^x$

6. $f'(x) = \frac{1}{x}$
$f''(x) = -\frac{1}{x^2}$
$f'''(x) = \frac{2}{x^3}$
$f^{(4)}(x) = -\frac{6}{x^4}$

7. $f'(x) = 6e^{2x}$
$f''(x) = 12e^{2x}$
$f'''(x) = 24e^{2x}$
$f^{(4)}(x) = 48e^{2x}$

8. $f'(x) = \frac{1}{x}$
$f''(x) = -\frac{1}{x^2}$
$f'''(x) = \frac{2}{x^3}$
$f^{(4)}(x) = -\frac{6}{x^4}$

9. $\frac{n+1}{3}$ **10.** $\frac{n+3}{n+1}$

1. $c = 0;\ 1, \frac{x}{4}, \left(\frac{x}{4}\right)^2, \left(\frac{x}{4}\right)^3, \left(\frac{x}{4}\right)^4$

3. $c = -1;\ -1, (x+1), -\frac{(x+1)^2}{2}, \frac{(x+1)^3}{6}, -\frac{(x+1)^4}{24}$

5. 2 **7.** 1 **9.** ∞ **11.** 0 **13.** 4

15. 5 **17.** 1 **19.** 3 **21.** $\frac{1}{2}$ **23.** ∞

25. $e\sum_{n=0}^{\infty}\frac{(x-1)^n}{n!},\quad R = \infty$ **27.** $\sum_{n=0}^{\infty}\frac{(3x)^n}{n!},\quad R = \infty$

29. $\sum_{n=0}^{\infty}(-1)^n x^n,\quad R = 1$

31. $1 + \frac{1}{2}(x-1) + \sum_{n=2}^{\infty}\frac{(-1)^{n+1}\,1\cdot 3\cdot 5\cdots(2n-3)(x-1)^n}{2^n\cdot n!},$
$R = 1$

33. $\sum_{n=0}^{\infty}(-1)^n(n+1)x^n,\quad R = 1$

35. $\sum_{n=0}^{\infty}\frac{(-1)^n(n+2)(n+1)x^n}{2},\quad R = 1$

37. (a)–(d) $R = 2$ **39.** (a)–(d) $R = 1$

41. $\sum_{n=0}^{\infty}\frac{x^{3n}}{n!}$ **43.** $3\sum_{n=0}^{\infty}\frac{x^{3n+2}}{n!}$ **45.** $\sum_{n=0}^{\infty}(-1)^n x^{4n}$

47. $\sum_{n=0}^{\infty} (-1)^n 3x^{2n+1}$ **49.** $\sum_{n=0}^{\infty} \frac{(-1)^n (x-1)^{n+1}}{n+1}$

51. $\sum_{n=1}^{\infty} (-1)^{n+1} n(x-1)^{n-1}$

53. (a) $\frac{8}{3}$ (b) $\frac{8}{13}$

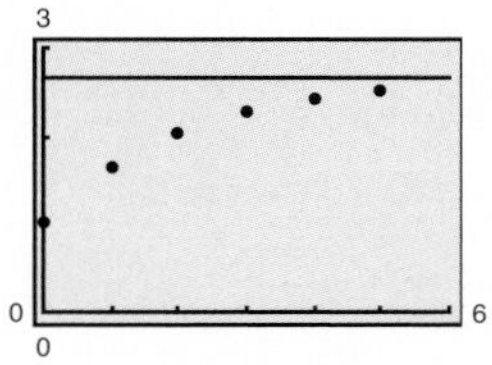

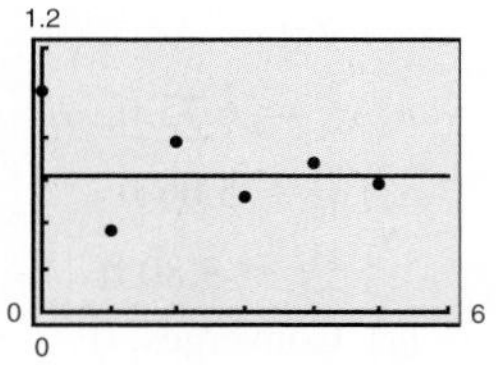

(c) In part (a), the rate of convergence is slower but all of the partial sums approach the line from below. In part (b), the partial sums approach the sum more quickly, but they oscillate above and below the line.

Section 10.5 *(page 642)*

Skills Warm Up *(page 642)*

1. $\sum_{n=0}^{\infty} \frac{3^n x^n}{n!}$ **2.** $\sum_{n=0}^{\infty} \frac{(-1)^n 3^n x^n}{n!}$

3. $4\sum_{n=0}^{\infty} (-1)^n (x-1)^n$ **4.** $\ln 5 + \sum_{n=1}^{\infty} \frac{(-1)^{n-1}(x-1)^n}{n}$

5. $1 + \frac{x}{4} - \frac{3x^2}{4^2 2!} + \frac{3 \cdot 7x^3}{4^3 3!} - \frac{3 \cdot 7 \cdot 11x^4}{4^4 4!} + \cdots$

6. $1 + \frac{x}{2} - \frac{x^2}{2^2 2!} + \frac{1 \cdot 3x^3}{2^3 3!} - \frac{1 \cdot 3 \cdot 5x^4}{2^4 4!} + \cdots$

7. $\frac{47}{60}$ **8.** $\frac{311}{576}$ **9.** $\frac{5}{12}$ **10.** $\frac{77}{192}$

1.

x	0	$\frac{1}{4}$	$\frac{1}{2}$	$\frac{3}{4}$	1
$f(x)$	1.0000	1.1331	1.2840	1.4550	1.6487
$S_1(x)$	1.0000	1.1250	1.2500	1.3750	1.5000
$S_2(x)$	1.0000	1.1328	1.2813	1.4453	1.6250
$S_3(x)$	1.0000	1.1331	1.2839	1.4541	1.6458
$S_4(x)$	1.0000	1.1331	1.2840	1.4549	1.6484

3. (a) $S_1(x) = 1 + x$ (b) $S_2(x) = 1 + x + \frac{x^2}{2}$

(c) $S_3(x) = 1 + x + \frac{x^2}{2} + \frac{x^3}{6}$

(d) $S_4(x) = 1 + x + \frac{x^2}{2} + \frac{x^3}{6} + \frac{x^4}{24}$

5. (a) $S_1(x) = 1 + 2x$ (b) $S_2(x) = 1 + 2x + 2x^2$

(c) $S_3(x) = 1 + 2x + 2x^2 + \frac{4x^3}{3}$

(d) $S_4(x) = 1 + 2x + 2x^2 + \frac{4x^3}{3} + \frac{2x^4}{3}$

7. (a) $S_1(x) = x$ (b) $S_2(x) = x - \frac{x^2}{2}$

(c) $S_3(x) = x - \frac{x^2}{2} + \frac{x^3}{3}$ (d) $S_4(x) = x - \frac{x^2}{2} + \frac{x^3}{3} - \frac{x^4}{4}$

9. (a) $S_1(x) = 1 + \frac{x}{2}$ (b) $S_2(x) = 1 + \frac{x}{2} - \frac{x^2}{8}$

(c) $S_3(x) = 1 + \frac{x}{2} - \frac{x^2}{8} + \frac{x^3}{16}$

(d) $S_4(x) = 1 + \frac{x}{2} - \frac{x^2}{8} + \frac{x^3}{16} - \frac{5x^4}{128}$

11. (a) $S_1(x) = 1 - 2x$ (b) $S_2(x) = 1 - 2x + 3x^2$

(c) $S_3(x) = 1 - 2x + 3x^2 - 4x^3$

(d) $S_4(x) = 1 - 2x + 3x^2 - 4x^3 + 5x^4$

13. (a) $S_1(x) = x$ (b) $S_2(x) = x - x^2$

(c) $S_3(x) = x - x^2 + x^3$ (d) $S_4(x) = x - x^2 + x^3 - x^4$

15. (a) $S_2(x) = 1 - x^2$ (b) $S_4(x) = 1 - x^2 + x^4$

(c) $S_6(x) = 1 - x^2 + x^4 - x^6$

(d) $S_8(x) = 1 - x^2 + x^4 - x^6 + x^8$

17. $S_4(x) = 1 - \frac{x}{3} + \frac{2x^2}{9} - \frac{14x^3}{81} + \frac{35x^4}{243}$ **19.** d

20. c **21.** a **22.** b **23.** 0.607; 0.00000155

25. 0.4055; 0.0000087 **27.** $\frac{1}{6!} \approx 0.00139$ **29.** 7

31. About 0.82143 **33.** 0.481

35. (a) $P_4(x) \approx 1 + x + \frac{x^2}{2} + \frac{x^3}{6} + \frac{x^4}{24}$

$Q_5(x) \approx x + x^2 + \frac{x^3}{2} + \frac{x^4}{6} + \frac{x^5}{24}$

$Q_5(x) = xP_4(x)$

(b) $P_6(x) \approx x^2 + x^3 + \frac{x^4}{2} + \frac{x^5}{6} + \frac{x^6}{24}$; degree 6

(c) $P_3(x) \approx \frac{1}{x} + 1 + \frac{x}{2} + \frac{x^2}{6} + \frac{x^3}{24}$; degree 3

Section 10.6 *(page 649)*

Skills Warm Up *(page 649)*

1. $f(2.4) = -0.04$, $f'(2.4) = 2.8$ **2.** $f(-0.6) = 0.064$, $f'(-0.6) = 3.48$

3. $f(0.35) = 0.01$, $f'(0.35) = 4.03$ **4.** $f(1.4) = 0.30$, $f'(1.4) = 12.88$

5. $4.9 \le x \le 5.1$ **6.** $0.798 \le x \le 0.802$

7. $5.97 \le x \le 6.03$ **8.** $-3.505 \le x \le -3.495$

9. $\left(\frac{3+\sqrt{13}}{2}, 2+\sqrt{13}\right)\left(\frac{3-\sqrt{13}}{2}, 2-\sqrt{13}\right)$

10. $\left(\frac{1-\sqrt{5}}{2}, \frac{3-\sqrt{5}}{2}\right), \left(\frac{1+\sqrt{5}}{2}, \frac{3+\sqrt{5}}{2}\right)$

1. $x_2 = 2.25$, $x_3 \approx 2.2361$

3. Newton's Method: 0.682, graphing utility: 0.682

5. Newton's Method: 1.25, graphing utility: 1.25

7. Newton's Method: 0.567, graphing utility: 0.567

9. Newton's Method: -1.597, 1.118
Graphing utility: -1.597, 1.118

11. Newton's Method: -1.380, 0.819
Graphing utility: -1.380, 0.819

13. 2.893 **15.** 0.567

17. 11.8033

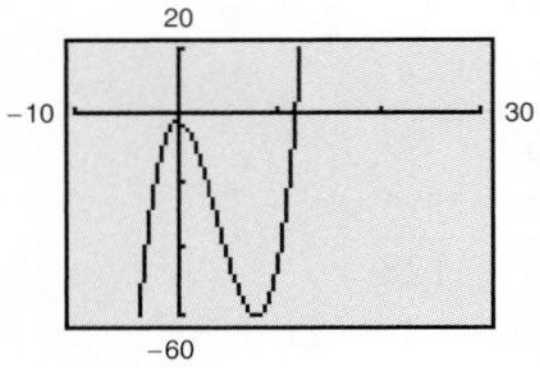

19. −0.8937, 2.0720

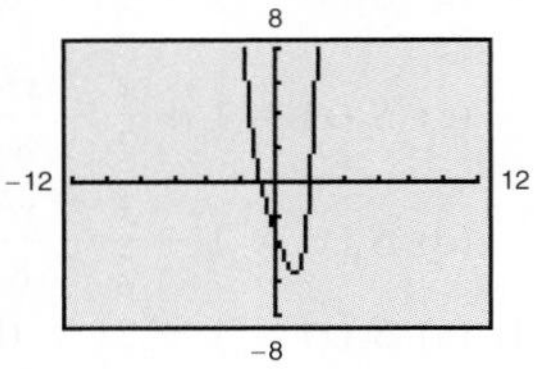

21. 0.9, 1.1, 1.9

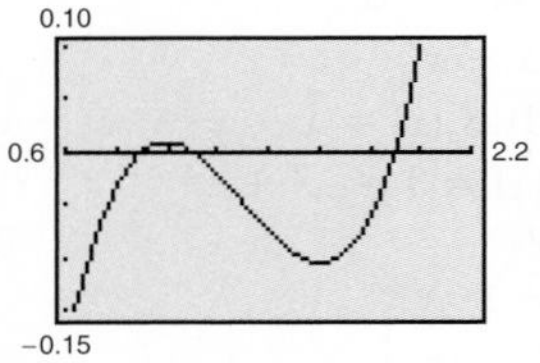

23. 1.1459, 7.8541

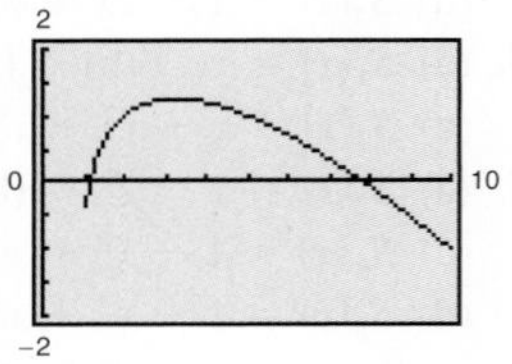

25. −1.999, 0.542

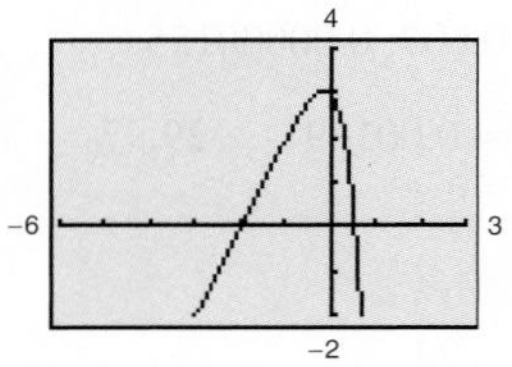

27. 2.208

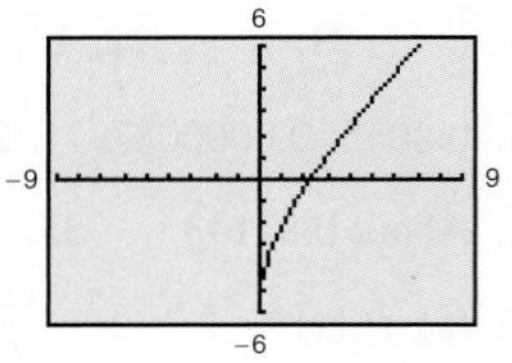

29. 0.8655

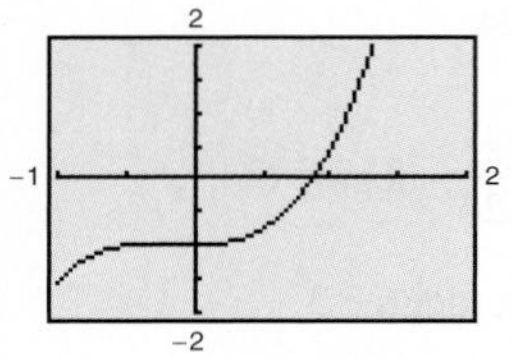

31. Newton's Method fails because $f'(x_1) = 0$.

33. Newton's Method fails because $1 = x_1 = x_3 = \ldots$; $0 = x_2 = x_4 = \ldots$. So, the limit does not exist.

35. $x_{n+1} = \dfrac{x_n^2 + a}{2x_n}$ **37.** 2.646 **39.** 1.565

41. $f(x) = \dfrac{1}{x} - a$

$f'(x) = -\dfrac{1}{x^2}$

Newton's Method: $x_{n+1} = x_n - \dfrac{f(x_n)}{f'(x_n)}$

$$x_{n+1} = x_n - \frac{\frac{1}{x_n} - a}{-\frac{1}{x_n^2}}$$

$$= x_n(2 - ax_n)$$

43. (1.939, 0.240) **45.** $x \approx 1.563$ miles down the coast

47. 131 units **49.** $x \approx 40.45$ (about 4045 units of product)

51. False. Let $f(x) = \dfrac{x^2 - 1}{x - 1}$.

Review Exercises for Chapter 10 *(page 656)*

1. 7, 10, 13, 16, 19 **3.** $\frac{3}{2}, \frac{9}{4}, \frac{27}{8}, \frac{81}{16}, \frac{243}{32}$ **5.** $4, 8, \frac{32}{3}, \frac{32}{3}, \frac{128}{15}$

7. 0; converges **9.** ∞; diverges **11.** 5; converges

13. 0; converges **15.** ∞; diverges

17. $2n + 5$, $n = 1, 2, 3, \ldots$ **19.** $\dfrac{1}{n!}$, $n = 1, 2, 3, \ldots$

21. $(-1)^n \dfrac{2^n}{3^{n+1}}$, $n = 0, 1, 2, \ldots$

23. \$1002.50, \$1005.01, \$1007.52, \$1010.04, \$1012.56, \$1015.09, \$1017.63, \$1020.18, \$1022.73, \$1025.28

25. (a) $16\left(\frac{3}{4}\right)^n$

(b) 12 ft

9 ft

$\frac{27}{4} = 6.75$ ft

$\frac{81}{16} \approx 5.06$ ft

$\frac{243}{64} \approx 3.80$ ft

(c) Converges; 0

27. $S_1 = \frac{3}{2}$

$S_2 = \frac{15}{4} = 3.75$

$S_3 = \frac{57}{8} = 7.125$

$S_4 = \frac{195}{16} \approx 12.188$

$S_5 = \frac{633}{32} \approx 19.781$

29. $S_1 = \frac{1}{2} = 0.5$

$S_2 = \frac{11}{24} \approx 0.4583$

$S_3 = \frac{331}{720} \approx 0.4597$

$S_4 = \frac{18,535}{40,320} \approx 0.4597$

$S_5 = \frac{1,668,151}{3,628,800} \approx 0.4597$

31. Diverges **33.** Converges

35. $\displaystyle\lim_{n\to\infty} \frac{2n}{n + 5} = 2 \neq 0$ **37.** $\displaystyle\lim_{n\to\infty} \frac{n^2}{n^2 + 1} = 1 \neq 0$

39. $S_3 = \frac{130}{27} \approx 4.8148$

$S_5 = \frac{1330}{243} \approx 5.4733$

$S_{10} = \frac{350,198}{59,049} \approx 5.9306$

41. Diverges **43.** Diverges **45.** Converges to $\frac{13}{4}$

47. (a) $60,000[1 - (0.85)^n]$ units (b) 60,000 units

49. \$2000 **51.** \$2181.82

53. (a) $V_n = 120,000(0.7)^n$ (b) \$20,168.40

55. Converges **57.** Converges

59. Converges **61.** Diverges **63.** Converges

65. b; diverges: $p = \frac{1}{2} < 1$ **66.** d; converges: $p = \frac{4}{3} > 1$

67. a; diverges: $p = 1$, harmonic series

68. c; converges: $p = \frac{3}{2} > 1$

69. $c = 2$; $\frac{1}{2}(x - 2), \frac{1}{4}(x - 2)^2, \frac{1}{8}(x - 2)^3, \frac{1}{16}(x - 2)^4, \frac{1}{32}(x - 2)^5$

71. $c = 0$; $-\dfrac{x}{3}, \dfrac{x^2}{9}, -\dfrac{x^3}{27}, \dfrac{x^4}{81}, -\dfrac{x^5}{243}$ **73.** $R = 10$

75. $R = \infty$ **77.** $R = \frac{1}{3}$ **79.** $R = 2$

81. $\displaystyle\sum_{n=0}^{\infty} \left(-\frac{1}{2}\right)^n \frac{x^n}{n!}$; $R = \infty$

83. $1 + \displaystyle\sum_{n=1}^{\infty} (-1)^n \left[\frac{1 \cdot 3 \cdot 5 \ldots (2n - 1)}{2^n\, n!}\right](x - 1)^n$, $R = 1$

85. $1 + \displaystyle\sum_{n=1}^{\infty} (-1)^{n-1} \left[\frac{1 \cdot 3 \cdot 7 \ldots (4n - 5)}{4^n\, n!}\right] x^n$, $R = 1$

87. $\ln 2 + \displaystyle\sum_{n=1}^{\infty} (-1)^{n+1} \frac{(x/2)^n}{n}$ **89.** $1 + 2x^2 + x^4$

91. $x^2 \displaystyle\sum_{n=0}^{\infty} \frac{x^n}{n!} = \sum_{n=0}^{\infty} \frac{x^{n+2}}{n!}$

93. $x^2 \displaystyle\sum_{n=0}^{\infty} (-1)^n x^n = \sum_{n=0}^{\infty} (-1)^n x^{n+2}$

95. (a) $S_1(x) = 1 - 4x$

(b) $S_2(x) = 1 - 4x + 8x^2$

(c) $S_3(x) = 1 - 4x + 8x^2 - \dfrac{32x^3}{3}$

(d) $S_4(x) = 1 - 4x + 8x^2 - \dfrac{32x^3}{3} + \dfrac{32x^4}{3}$

97. (a) $S_1(x) = \ln(2) + \dfrac{x}{2}$

(b) $S_2(x) = \ln(2) + \dfrac{x}{2} - \dfrac{x^2}{8}$

(c) $S_3(x) = \ln(2) + \dfrac{x}{2} - \dfrac{x^2}{8} + \dfrac{x^3}{24}$

(d) $S_4(x) = \ln(2) + \dfrac{x}{2} - \dfrac{x^2}{8} + \dfrac{x^3}{24} - \dfrac{x^4}{64}$

99. (a) $S_1(x) = \dfrac{1}{9} - \dfrac{2x}{27}$

(b) $S_2(x) = \dfrac{1}{9} - \dfrac{2x}{27} + \dfrac{x^2}{27}$

(c) $S_3(x) = \dfrac{1}{9} - \dfrac{2x}{27} + \dfrac{x^2}{27} - \dfrac{4x^3}{243}$

(d) $S_4(x) = \dfrac{1}{9} - \dfrac{2x}{27} + \dfrac{x^2}{27} - \dfrac{4x^3}{243} + \dfrac{5x^4}{729}$

101. 1.5169; 0.00000066 **103.** 0.9163; 0.0000087

105. 0.301 **107.** 0.1233

109. Newton's Method: 0.322, graphing utility: 0.322

111. Newton's Method: -1.562, graphing utility: -1.562

113. Newton's Method: -1.453, 1.164
Graphing utility: -1.453, 1.164

115. 1.341 **117.** 0.773

119. -2.926

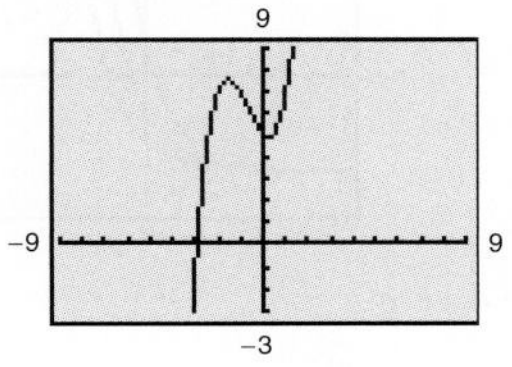

Test Yourself *(page 660)*

1. 2, 6, 10, 14, 18 **2.** 2, 4, 8, 16, 32 **3.** $4, \frac{5}{2}, 2, \frac{7}{4}, \frac{8}{5}$

4. $-1, \frac{1}{2}, -\frac{1}{6}, \frac{1}{24}, -\frac{1}{120}$ **5.** 0; converges

6. $\frac{1}{3}$; converges **7.** Limit does not exist; diverges

8. 0; converges **9.** $\dfrac{(-1)^{n+1}n}{n^2+1}$

10. Converges by the Ratio Test because
$$\lim_{n\to\infty}\left|\frac{a_{n+1}}{a_n}\right| = \lim_{n\to\infty}\frac{4}{n+1} = 0 < 1.$$

11. Diverges by the nth-Term Test because $\displaystyle\lim_{n\to\infty}\frac{n+1}{n-3} = 1 \neq 0.$

12. Converges by the Geometric Series Test because
$|r| = \left|\frac{1}{5}\right| < 1.$

13. Diverges by the p-Series Test because $p = \frac{1}{6} < 1.$

14. Diverges by the Geometric Series Test because
$|r| = \left|\frac{5}{3}\right| > 1.$

15. Converges by the p-Series Test because $p = \frac{5}{4} > 1.$

16. Converges by the Geometric Series Test because
$|r| = \left|-\frac{1}{6}\right| < 1.$

17. Diverges by the Ratio Test because
$$\lim_{n\to\infty}\left|\frac{a_{n+1}}{a_n}\right| = \lim_{n\to\infty}\frac{n+2}{5} = \infty.$$

18. $\frac{7}{3}$

19. (a) $c = 0$ (c) $R = 3$

(b) $S_0 = -1$

$S_1 = -1 + \dfrac{x}{3}$

$S_2 = -1 + \dfrac{x}{3} - \dfrac{x^2}{9}$

$S_3 = -1 + \dfrac{x}{3} - \dfrac{x^2}{9} + \dfrac{x^3}{27}$

$S_4 = -1 + \dfrac{x}{3} - \dfrac{x^2}{9} + \dfrac{x^3}{27} - \dfrac{x^4}{81}$

20. (a) $c = 0$ (c) $R = \infty$

(b) $S_0 = 1$

$S_1 = 1 + \dfrac{x}{2}$

$S_2 = 1 + \dfrac{x}{2} + \dfrac{x^2}{6}$

$S_3 = 1 + \dfrac{x}{2} + \dfrac{x^2}{6} + \dfrac{x^3}{24}$

$S_4 = 1 + \dfrac{x}{2} + \dfrac{x^2}{6} + \dfrac{x^3}{24} + \dfrac{x^4}{120}$

21. (a) $c = 3$ (c) $R = 1$

(b) $S_0 = \dfrac{1}{16}$

$S_1 = \dfrac{1}{16} - \dfrac{(x-3)}{25}$

$S_2 = \dfrac{1}{16} - \dfrac{(x-3)}{25} + \dfrac{(x-3)^2}{36}$

$S_3 = \dfrac{1}{16} - \dfrac{(x-3)}{25} + \dfrac{(x-3)^2}{36} - \dfrac{(x-3)^3}{49}$

$S_4 = \dfrac{1}{16} - \dfrac{(x-3)}{25} + \dfrac{(x-3)^2}{36} - \dfrac{(x-3)^3}{49} + \dfrac{(x-3)^4}{64}$

22. $e\displaystyle\sum_{n=0}^{\infty}\frac{3^n x^n}{n!}$ **23.** $1 + \dfrac{2x}{3} - \dfrac{x^2}{9} + \dfrac{4x^3}{81} - \dfrac{7x^4}{243} + \cdots$

24. (a) $S_1(x) = 1 + \dfrac{x}{2}$ (b) $S_2(x) = 1 + \dfrac{x}{2} + \dfrac{x^2}{8}$

(c) $S_3(x) = 1 + \dfrac{x}{2} + \dfrac{x^2}{8} + \dfrac{x^3}{48}$

(d) $S_4(x) = 1 + \dfrac{x}{2} + \dfrac{x^2}{8} + \dfrac{x^3}{48} + \dfrac{x^4}{384}$

25. (a) $S_1(x) = 1 - \dfrac{x}{2}$ (b) $S_2(x) = 1 - \dfrac{x}{2} + \dfrac{x^2}{4}$

(c) $S_3(x) = 1 - \dfrac{x}{2} + \dfrac{x^2}{4} - \dfrac{x^3}{8}$

(d) $S_4(x) = 1 - \dfrac{x}{2} + \dfrac{x^2}{4} - \dfrac{x^3}{8} + \dfrac{x^4}{16}$

26. (a) $S_1(x) = \dfrac{1}{16} - \dfrac{x}{32}$ (b) $S_2(x) = \dfrac{1}{16} - \dfrac{x}{32} + \dfrac{3x^2}{256}$

(c) $S_3(x) = \dfrac{1}{16} - \dfrac{x}{32} + \dfrac{3x^2}{256} - \dfrac{x^3}{256}$

(d) $S_4(x) = \dfrac{1}{16} - \dfrac{x}{32} + \dfrac{3x^2}{256} - \dfrac{x^3}{256} + \dfrac{5x^4}{4096}$

27. (a) $a_n = 2.7n + 283$ (b) 337 million

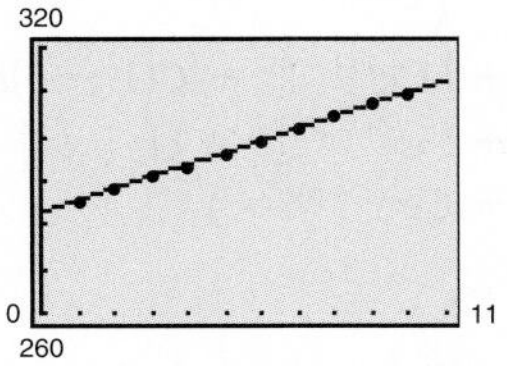

28. Newton's Method: 1.213, graphing utility: 1.213

Chapter 11

Section 11.1 (page 666)

Skills Warm Up (page 666)

1. $y' = 6x + 2$
   $y'' = 6$
2. $y' = -6x^2 - 8$
   $y'' = -12x$
3. $y' = -6e^{2x}$
   $y'' = -12e^{2x}$
4. $y' = -6xe^{x^2}$
   $y'' = -6e^{x^2}(2x^2 + 1)$
5. $k = 2 \ln 3 - \ln \frac{17}{2} \approx 0.0572$
6. $k = \ln 10 - \dfrac{\ln 41}{2} \approx 0.4458$

1. $y' = 4Ce^{4x} = 4y$

3. $y' = 6x^2$ and $y' - \dfrac{3}{x}y = 6x^2 - \dfrac{3}{x}(2x^3) = 0$

5. $y' = 2Cx - 3$ and
$xy' - 3x - 2y = x(2Cx - 3) - 3x - 2(Cx^2 - 3x) = 0$

7. $y' = \ln x + 1 + C$ and
$x(y' - 1) - (y - 4) = x(\ln x + 1 + C - 1)$
$- (x \ln x + Cx + 4 - 4) = 0$

9. $y'' = 2$ and $x^2y'' - 2y = x^2(2) - 2(x^2) = 0$

11. $y'' = -C_1 \sin x + C_2 \cos x$ and
$y'' + y = -C_1 \sin x + C_2 \cos x + C_1 \sin x - C_2 \cos x = 0$

13. Solution **15.** Not a solution **17.** Not a solution

19. Solution **21.** $y' = -2Ce^{-2x}$, so $y' + 2y = 0$; $y = 3e^{-2x}$.

23. $y' = C_2(1/x)$ and $y'' = -C_2(1/x^2)$,
so $xy'' + y' = 0$; $y = 5 + \frac{1}{2} \ln|x|$.

25.

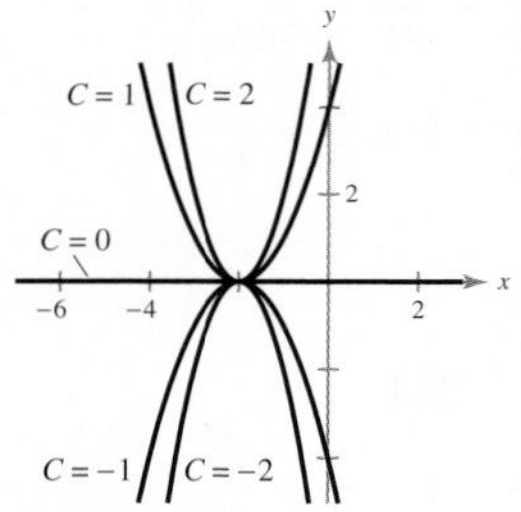

27. $y = x^3 + C$

29. $y = \ln|1 + x| + C$

31. $y = \frac{1}{3}\sqrt{(x^2 + 6)^3} + C$

33. $y = \frac{1}{4} \sin 4x + C$

35. $y^2 = \frac{1}{4}x^3$

37. $y = 3e^x$

39. (a) $N = 750 - 650e^{-0.0484t}$ (b) $N \approx 214$

41. $s = 25 - \dfrac{13}{\ln 3} \ln \dfrac{h}{2}$, $\dfrac{ds}{dh} = -\dfrac{13}{\ln 3}\left(\dfrac{1}{h}\right)$

So, the equation is a solution of $\dfrac{ds}{dh} = \dfrac{k}{h}$, where $k = -\dfrac{13}{\ln 3}$.

43.
$$y = a + Ce^{k(1-b)t}$$
$$\frac{dy}{dt} = Ck(1 - b)e^{k(1-b)t}$$
$$a + b(y - a) + \frac{1}{k}\frac{dy}{dt} = a + b[(a + Ce^{k(1-b)t}) - a] + \frac{1}{k}[Ck(1 - b)e^{k(1-b)t}]$$
$$= a + bCe^{k(1-b)t} + C(1 - b)e^{k(1-b)t}$$
$$= a + Ce^{k(1-b)t}[b + (1 - b)]$$
$$= a + Ce^{k(1-b)t} = y$$

45. True

Section 11.2 (page 672)

Skills Warm Up (page 672)

1. $\frac{2}{5}x^{5/2} + C$
2. $\frac{1}{4}t^4 - \frac{3}{4}t^{4/3} + C$
3. $2 \ln|x - 5| + C$
4. $\frac{1}{4} \ln|2y^2 + 1| + C$
5. $\frac{1}{2}e^{2y} + C$
6. $-\frac{1}{2}e^{1-x^2} + C$
7. $C = -10$
8. $C = 5$
9. $k = \dfrac{\ln 5}{2} \approx 0.8047$
10. $k = -2 \ln 3 - \ln 2 \approx -2.8904$

1. Yes; $(y + 3)\,dy = x\,dx$ **3.** Yes; $dy = \left(\dfrac{1}{x} + 1\right) dx$

5. No, the variables cannot be separated.

7. $y = x^2 + C$ **9.** $r = Ce^{0.05s}$ **11.** $y^4 = 2x^2 - 4x + C$

13. $y = \sqrt[3]{x + C}$ **15.** $y^2 = -\frac{1}{6}x^3 + C$ **17.** $y = Ce^{x^2/2}$

19. $y = \ln|t^3 + t + C|$ **21.** $y = 1 - \left(C - \dfrac{x}{2}\right)^2$

23. $y = C(2 + x)^2$ **25.** $y^2 = -2 \cos x + C$

27. (a) $y = \frac{1}{2}x^2 + C$
(b)

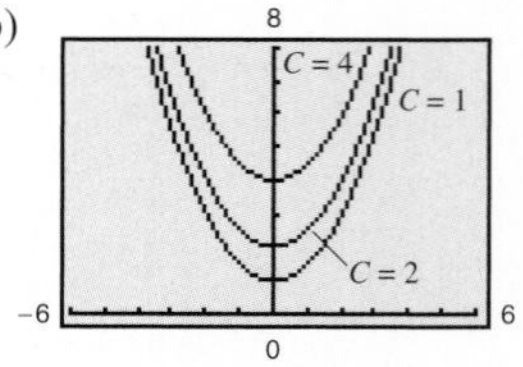

29. (a) $y = Ce^x - 3$
(b)

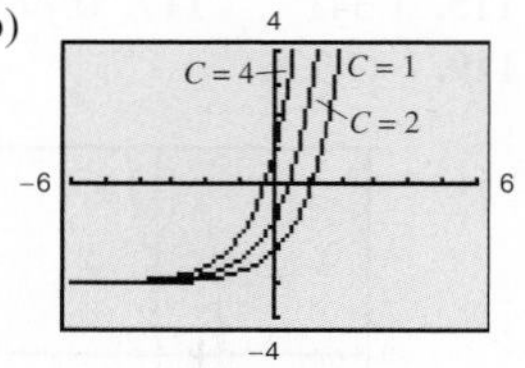

31. $y^2 = 2e^x + 14$ **33.** $y = -4 - e^{-x^2/2}$

35. $y = 5\sqrt{x^2 - 16} - 17$ **37.** $y = e^{\sin x}$

39. $16y^2 + 9x^2 = 25$

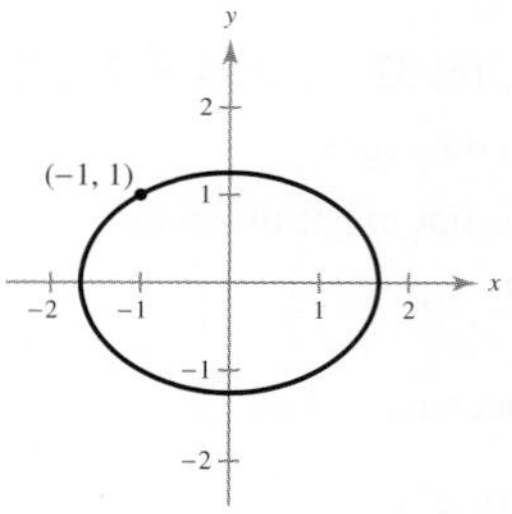

41. $v = 34.56(1 - e^{-0.1t})$

43. $V = \left(\frac{1}{3}kt + C\right)^3$

45. 98.9%

47. (a) $w = 1200 - 1140e^{-kt}$
(b) $w = 1200 - 1140e^{-0.8t}$

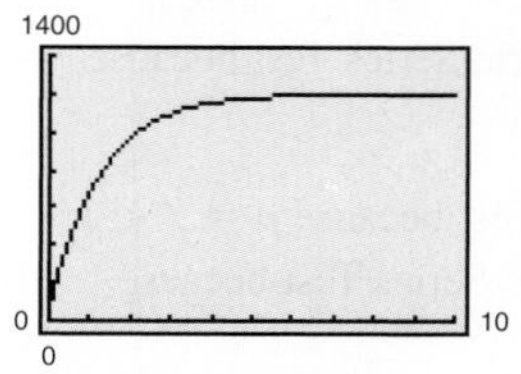

$w = 1200 - 1140e^{-0.9t}$

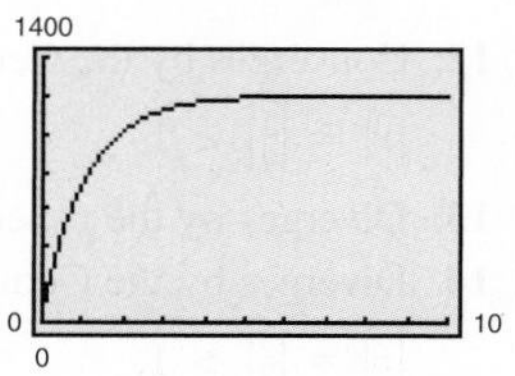

$w = 1200 - 1140e^{-t}$

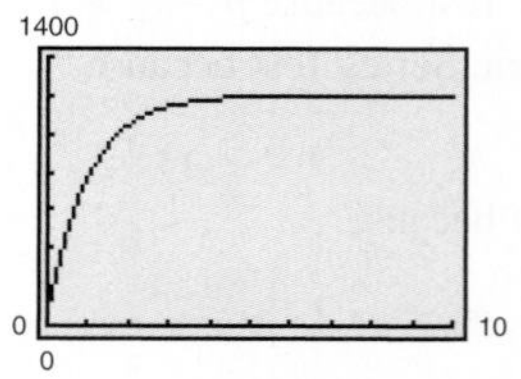

(c) ≈ 1.31 yr; ≈ 1.16 yr; ≈ 1.05 yr (d) 1200 lb

Quiz Yourself (page 674)

1. $y' = -\frac{1}{2}Ce^{-x/2}$ and $2y' + y = 2\left(-\frac{1}{2}Ce^{-x/2}\right) + Ce^{-x/2} = 0$

2. $y'' = -C_1 \cos x - C_2 \sin x$ and
$y'' + y = -C_1 \cos x - C_2 \sin x + C_1 \cos x + C_2 \sin x = 0$

3. $y' = -\dfrac{1}{x^2}$ and $y'' = \dfrac{2}{x^3}$; $xy'' + 2y' = \dfrac{2}{x^2} - \dfrac{2}{x^2} = 0$

4. $y' = \dfrac{3}{5}x^2 - 1 + \dfrac{C}{2\sqrt{x}}$ and
$2xy' - y = \dfrac{6}{5}x^3 - 2x + C\sqrt{x} - \dfrac{1}{5}x^3 + x - C\sqrt{x} = x^3 - x$

5. $y'' = -9C_1 \sin 3x - 9C_2 \cos 3x$ and
$y'' + 9y = -9C_1 \sin 3x - 9C_2 \cos 3x + 9(C_1 \sin 3x + C_2 \cos 3x) = 0$
$y = 2 \sin 3x - \frac{1}{3} \cos 3x$

6. $y' = C_1 + 3C_2x^2$, $y'' = 6C_2x$, and
$x^2y'' - 3xy' + 3y = x^2(6C_2x) - 3x(C_1 + 3C_2x^2) + 3(C_1x + C_2x^3) = 0$
$y = -2x + \frac{1}{2}x^3$

7. $y = -2x^2 + 4x + C$ **8.** $y = 1 + Ce^{x^2/2 + 2x}$

9. $y^2 = \ln|2x + 1| + C$ **10.** $y^3 + y = \dfrac{x^2}{2} + C$

11. (a) $y = \pm\sqrt{\frac{1}{3}x^3 + x + C}$
(b)
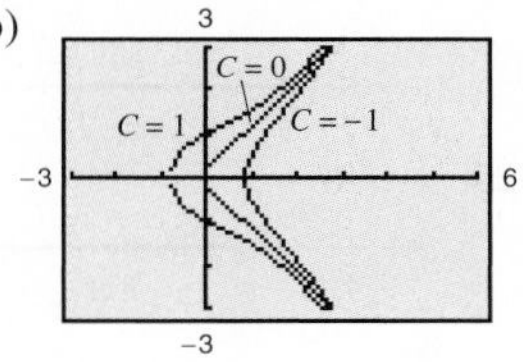

12. (a) $y = C(x - 3)$
(b)
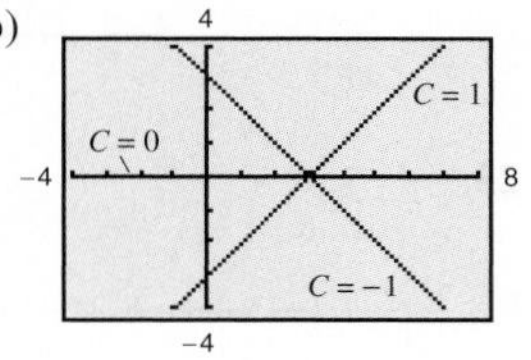

13. $y = \frac{1}{2}e^{-2x} + \frac{1}{2}$ **14.** $y = -3e^{-(\cos \pi x)/\pi}$

15. $y = 2e^{x^3}$
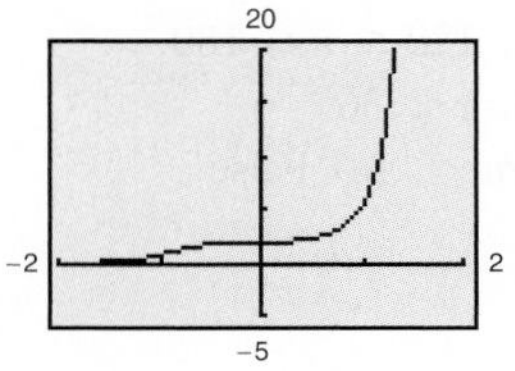

16. $v = 20 - 20e^{-1.386t}$

Section 11.3 (page 678)

Skills Warm Up (page 678)

1. $e^x + 1$ **2.** $e^{3x} + 1$ **3.** $\dfrac{1}{x^3}$ **4.** x^2e^x

5. $2e^{2x} + C$ **6.** $\frac{1}{6}e^{3x^2} + C$ **7.** $\frac{1}{2}\ln|2x + 5| + C$

8. $\frac{1}{2}\ln|x^2 + 2x + 3| + C$ **9.** $\frac{1}{12}(4x - 3)^3 + C$

10. $-\frac{1}{6}(1 - x^2)^3 + C$

1. $y' + \dfrac{-3}{2x^2}y = \dfrac{x}{2}$ **3.** $y' + \dfrac{1}{x}y = e^x$

5. $y' + \dfrac{1}{1 - x}y = \dfrac{1}{x - 1}$ **7.** $y = 2 + Ce^{-3x}$

9. $y = \frac{1}{3}e^x(e^{3x} + C)$ **11.** $y = \frac{1}{2}x^2 + 3\ln|x| + C$

13. $5e^{-x^2}(e^{x^2} + C)$ **15.** $y = \dfrac{x^3 - 3x + C}{3(x - 1)}$

17. $y = e^{1/x^2}\left(-\dfrac{1}{2x^2} + C\right)$ **19.** $y = Ce^{-x} + 4$

21. $y = Ce^{x^2} - 1$ **23.** c **24.** d **25.** a **26.** b

27. $y = 3e^x$ **29.** $xy = 4$ **31.** $y = 1 + 5e^{-x^3}$

33. $y = \dfrac{3}{4}x^2 - \dfrac{5}{3}x + \dfrac{7}{12x^2}$

35. $S = t + 95(1 - e^{-t/5})$

t	1	2	3	4	5
S	18.22	33.32	45.86	56.31	65.05

t	6	7	8	9	10
S	72.39	78.57	83.82	88.30	92.14

37. $v = \dfrac{1}{k}[9.8 + Ce^{kt}]$

39. (a) $N = 40 + Ce^{-kt}$ (b) $N = 40 - 30.57e^{-0.0188t}$

41. Answers will vary.

Section 11.4 (page 685)

Skills Warm Up (page 685)

1. $y = \frac{3}{2}x^2 + C$ **2.** $y^2 = 3x + C$ **3.** $y = Ce^{x^2}$

4. $y^4 = \frac{1}{2}(x - 4)^2 + C$ **5.** $y = 2 + Ce^{-2x}$

6. $y = xe^{-2x} + Ce^{-2x}$ **7.** $y = 1 + Ce^{-x^2/2}$

8. $y = \dfrac{1}{4}x^2 + Cx^{-2}$ **9.** $\dfrac{dy}{dx} = kx^2$ **10.** $\dfrac{dx}{dt} = k(x - t)$

1. $y = \dfrac{360}{8 + 41t}$
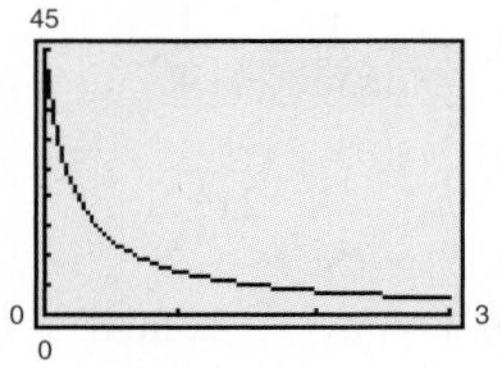

3. $y = 1 - e^{-1.386t}$ **5.** $y = 200e^{-0.6931e^{-0.4397t}}$

7. $\dfrac{y(2 - y)}{(1 - y)^2} = \dfrac{19}{81}e^{0.50634t}$ **9.** $y = e^{(x \ln 2)/3} \approx e^{0.2310x}$

11. $y = 4e^{-(x \ln 4)/4} \approx 4e^{-0.3466x}$ **13.** $y = \frac{1}{2}e^{(\ln 2)x} \approx \frac{1}{2}e^{0.6931x}$

15. $\dfrac{dy}{dt} = ky$, $y = 20e^{-0.2231t}$ **17.** 269,972 people

19. ≈ 3.15 h **21.** $S = L(1 - e^{-kt})$ **23.** 34 beavers

25. (a) $N = \dfrac{500}{1 + Ce^{-kt}}$ (b) $N = \dfrac{500}{1 + 4e^{-0.2452t}}$
(c) 121 deer (d) ≈ 9.1 yr

27. (a) $Q = 25e^{-0.05t}$ (b) ≈ 10.22 min

29. $P = Ce^{kt} - \dfrac{N}{k}$ **31.** $A = \dfrac{P}{r}(e^{rt} - 1)$

33. \$10,708,538.49 **35.** (a) $C = C_0e^{-Rt/V}$ (b) 0

37. (a) $C(t) = \dfrac{Q}{R}(1 - e^{-Rt/V})$ (b) $\dfrac{Q}{R}$

Review Exercises for Chapter 11 *(page 691)*

1. $\frac{dy}{dx} = \frac{1}{2}Ce^{x/2}$ and $2\frac{dy}{dx} = 2\left(\frac{1}{2}Ce^{x/2}\right) = Ce^{x/2} = y$

3. $y' = 6xe^{x^2}$ and $y' - 2xy = 6xe^{x^2} - 2x(3e^{x^2}) = 0$

5. $y' = 2x$ and $\frac{1}{2}y' - \frac{1}{x}y = \frac{1}{2}(2x) - \frac{1}{x}(x^2) = 0$

7. $y' = \frac{3x^2}{5} - 1 + \frac{C}{2\sqrt{x}}$ and

$2xy' - y = \frac{6x^3}{5} - 2x + C\sqrt{x} - \frac{x^3}{5} + x - C\sqrt{x} = x^3 - x$

9. $y' = C_1e^x - C_2e^{-x}$, $y'' = C_1e^x + C_2e^{-x}$, and

$y'' - y = C_1e^x + C_2e^{-x} - C_1e^x - C_2e^{-x} = 0$

11. Not a solution **13.** Solution **15.** $y = e^{-5x}$

17.

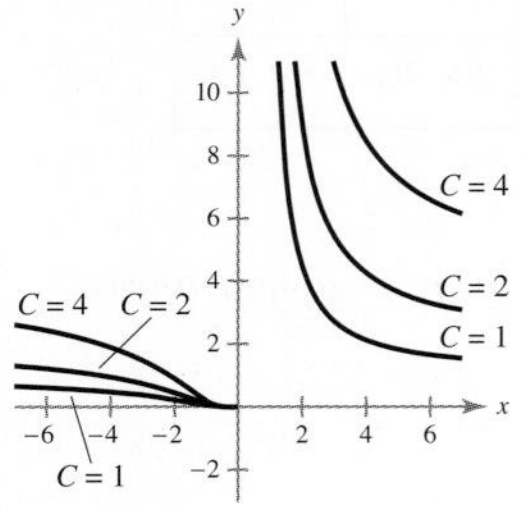

19. $y = \frac{1}{2}(x^2 + 1)$

21. $y = \frac{2}{3}x^3 + 5x + C$

23. $y = 3\sin x + C$

25. $y = x + 3\ln|x| + C$

27. Yes; $\frac{dy}{y} = \frac{dx}{x + 3}$

29. No, the variables cannot be separated.

31. $y = 2x^2 + C$

33. $y^4 = 5x + C$ **35.** $y = \frac{1}{x^2 + C}$ **37.** $y = Ce^{(x^2/2 + x)} - 1$

39. $y = Ce^{1/(4x^2)} - 2$ **41.** $y^2 = 2\sin x + C$

43. (a) $y = x^3 + C$

(b)

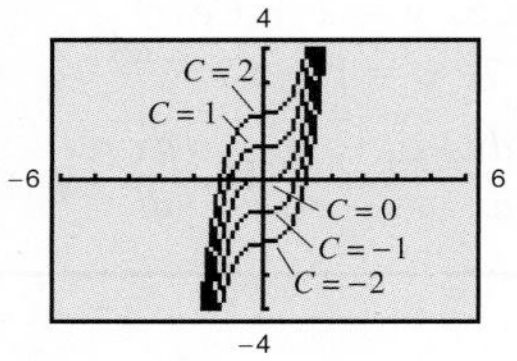

45. $y^2 = -2e^x + 6$

47. $y = e^{\sin x^2}$

49. $6x^2 - 5y^2 = 1$

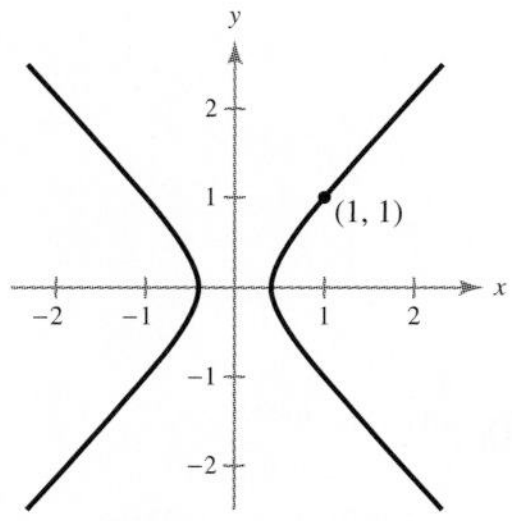

51. $\approx 88.6\%$

53. $y' - \frac{y}{x^2} = -\frac{1}{4}x^2$

55. $y' - y = \frac{1}{2x^2}$

57. $y = Ce^{-4x} + 2$

59. $y = 2x^2 - 3x\ln|x| + Cx$

61. $y = e^{-6x}\left(\frac{1}{8}e^{8x} + C\right)$

63. $y = x^2 + 2x + \frac{C}{x}$

65. $y = 6 - 3e^{-x}$ **67.** $s = 25 - \frac{13\ln(h/2)}{\ln 3}$, $2 < h < 15$

69. $y = Cx^{-k}$

71. $\frac{dy}{dt} = ky^{1/3}$, $y = \left(\frac{2}{3}kt + C\right)^{3/2}$, $y = (-5t + 9)^{3/2}$

73. About 383°F **75.** About 50 pelicans

77. (a) $y = 10 - 8e^{-t/10}$ (b) About 7.1 gal

Test Yourself *(page 694)*

1. $y' = -2e^{-2x}$ and $3y' + 2y = 3(-2e^{-2x}) + 2e^{-2x} = -4e^{-2x}$

2. $y'' = \frac{2}{(x+1)^3}$ and $y'' - \frac{2y}{(x+1)^2} = \frac{2}{(x+1)^3} - \frac{2[1/(x+1)]}{(x+1)^2} = 0$

3. $y^2 = x^2 + C$ **4.** $y = Ce^{4x^2} - 2$

5. $y^3 = \frac{1}{\pi}\sin \pi x + C$ **6.** $y = \frac{2}{5} + Ce^{-5x}$

7. $y = xe^{2x} + Ce^{2x}$ **8.** $y = \frac{1}{3}x^2 + \frac{C}{x}$

9. $y = 1 - e^{-x^3/3}$ **10.** $y = e^{e^{x^2}}$ **11.** $y = \frac{1}{14}(\ln x)^2 - 2$

12. (a) $w = 200 - 190e^{-kt}$

(b)

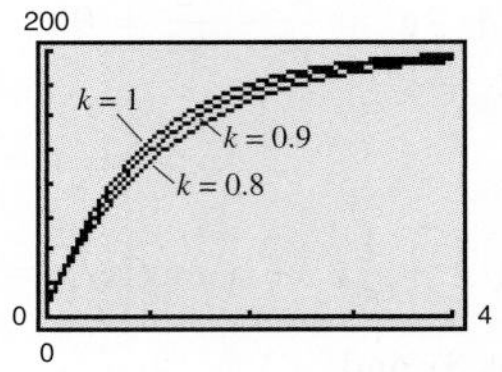

(c) ≈ 1.7 yr
≈ 1.5 yr
≈ 1.3 yr

(d) 200 lb

13. (a) $y = 1 - e^{-0.25t}$ (b) About 3 years (c) About 63%

Appendix A

Section A.1 *(page A7)*

1. Rational **3.** Irrational **5.** Rational **7.** Rational

9. Irrational **11.** (a) Yes (b) No (c) Yes

13. (a) Yes (b) No (c) No

15. $x \geq 12$

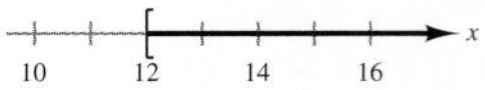

17. $x < -\frac{1}{2}$

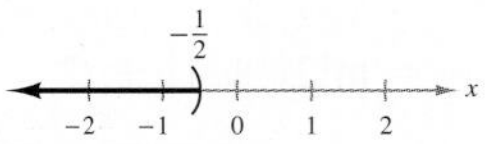

19. $x > 1$

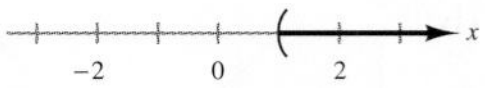

21. $-\frac{1}{2} < x < \frac{7}{2}$

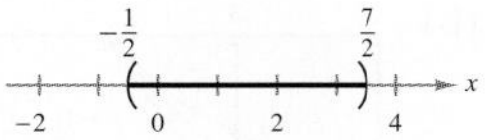

23. $-\frac{3}{4} < x < -\frac{1}{4}$

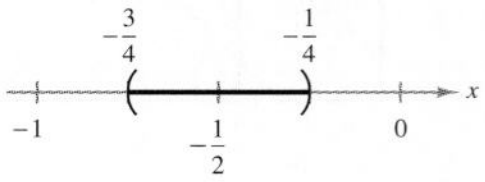

25. $x > 6$

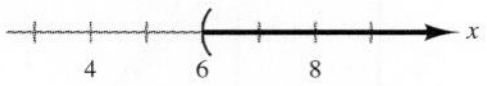

27. $-\frac{3}{2} < x < 2$

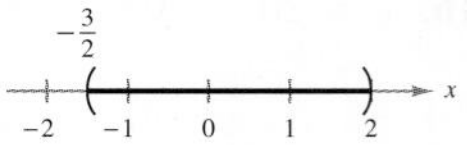

29. $4.1 \leq E \leq 4.25$

31. $p \leq 0.4$

33. $120 \leq r \leq 180$

35. $x \geq 36$

37. (a) False (b) True (c) True (d) False

Section A.2 *(page A12)*

1. (a) 51 (b) −51 (c) 51

3. (a) 14.99 (b) −14.99 (c) 14.99

5. (a) $\frac{128}{75}$ (b) $-\frac{128}{75}$ (c) $\frac{128}{75}$ **7.** $|x| \leq 2$

9. $|x| > 2$ **11.** $|x - 5| \leq 3$ **13.** $|x - 2| > 2$

15. $|x - 5| < 3$ **17.** $|y - a| \leq 2$

19. $-4 < x < 4$

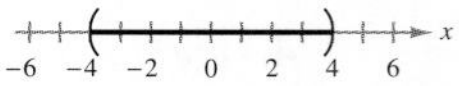

21. $x < -6$ or $x > 6$

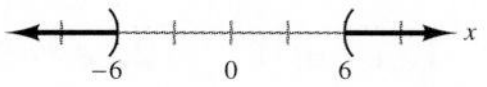

23. $3 < x < 7$

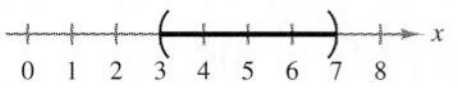

25. $x \leq -7$ or $x \geq 13$

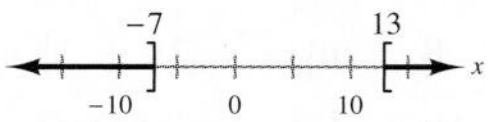

27. $x < 6$ or$x > 14$

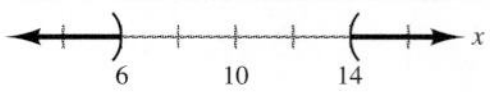

29. $4 < x < 5$

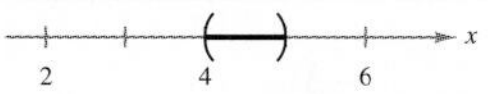

31. $a - b \leq x \leq a + b$

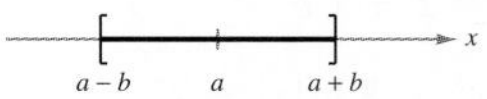

33. $\frac{a - 8b}{3} < x < \frac{a + 8b}{3}$

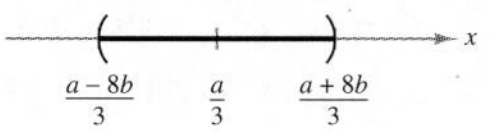

35. 16 **37.** 1.25 **39.** $\frac{1}{8}$ **41.** $|p - 33.15| \le 2$
43. 20.75 oz and 19.25 oz
45. (a) $|4750 - E| \le 500$, $|4750 - E| \le 237.50$
(b) At variance
47. (a) $|20{,}000 - E| \le 500$, $|20{,}000 - E| \le 1000$
(b) At variance
49. $\$11{,}759.40 \le C \le \$17{,}639.10$

Section A.3 *(page A18)*

1. -54 **3.** $\frac{1}{2}$ **5.** 4 **7.** 44 **9.** 5 **11.** 9
13. $\frac{1}{2}$ **15.** $\frac{1}{4}$ **17.** 908.3483 **19.** -3.7798
21. $\dfrac{3}{4y^{14}}$ **23.** $10x^4$ **25.** $7x^5$ **27.** $\frac{5}{2}(x + y)^5, x \ne -y$
29. $3x, x > 0$ **31.** $2\sqrt{2}$ **33.** $3x\sqrt[3]{2x^2}$
35. $\dfrac{2x^3z}{y}\sqrt[3]{\dfrac{18z^2}{y}}$ **37.** $3x(x + 2)(x - 2)$ **39.** $\dfrac{2x^3 + 1}{x^{1/2}}$
41. $3(x + 1)^{1/2}(x + 2)(x - 1)$ **43.** $\frac{1}{3}x(5x^3 + 1)$
45. $x \ge 4$ **47.** $(-\infty, \infty)$ **49.** $(-\infty, 4) \cup (4, \infty)$
51. $x \ne 1, x \ge -2$ **53.** \$19,121.84 **55.** \$11,345.46
57. $\dfrac{\sqrt{2}}{2}\pi$ sec or about 2.22 sec

Section A.4 *(page A24)*

1. $\dfrac{1}{6}, 1$ **3.** $\dfrac{3}{2}$ **5.** $-2 \pm \sqrt{3}$ **7.** $\dfrac{-3 \pm \sqrt{41}}{4}$
9. $(x - 2)^2$ **11.** $(2x + 1)^2$ **13.** $(3x - 1)(x - 1)$
15. $(3x - 2)(x - 1)$ **17.** $(x - 2y)^2$
19. $(3 + y)(3 - y)(9 + y^2)$ **21.** $(x - 2)(x^2 + 2x + 4)$
23. $(y + 4)(y^2 - 4y + 16)$ **25.** $(x - y)(x^2 + xy + y^2)$
27. $(x - 4)(x - 1)(x + 1)$ **29.** $(2x - 3)(x^2 + 2)$
31. $(x - 2)(2x^2 - 1)$ **33.** $(x + 4)(x - 4)(x^2 + 1)$
35. 0, 5 **37.** ± 3 **39.** $\pm\sqrt{3}$ **41.** 0, 6 **43.** $-2, 1$
45. $-1, 6$ **47.** $-1, -\frac{2}{3}$ **49.** -4 **51.** ± 2
53. $1, \pm 2$ **55.** $(-\infty, -2] \cup [2, \infty)$
57. $(-\infty, 3] \cup [4, \infty)$ **59.** $(-\infty, -1] \cup \left[-\frac{1}{5}, \infty\right)$
61. $(x + 1)(x^2 - 4x - 2)$ **63.** $(x + 1)(2x^2 - 3x + 1)$
65. $-2, -1, 4$ **67.** 1, 2, 3 **69.** $-\frac{2}{3}, -\frac{1}{2}, 3$
71. 4 **73.** $-2, -1, \frac{1}{4}$
75. Two solutions; The solutions of the equation are ± 2000, but the minimum average cost occurs at the positive value, 2000; 2000 units

Section A.5 *(page A31)*

1. $\dfrac{x - 4}{x + 6}, x \ne 3$ **3.** $\dfrac{x + 5}{2x + 3}, x \ne 2$ **5.** $\dfrac{x + 3}{x - 2}$
7. $\dfrac{x^2 - 3}{x}$ **9.** $\dfrac{5x^2 - 9x + 8}{(x - 3)(3x + 4)}$ **11.** $-\dfrac{x}{x^2 - 4}$
13. $\dfrac{1}{(x + 2)(x - 1)}$ **15.** $-\dfrac{x - 2}{x^2 + 1}$ **17.** $\dfrac{x + 2}{(x + 1)^{3/2}}$
19. $-\dfrac{3t}{2\sqrt{1 + t}}$ **21.** $\dfrac{x(x^2 + 2)}{(x^2 + 1)^{3/2}}$ **23.** $\dfrac{2}{x^2\sqrt{x^2 + 2}}$
25. $\dfrac{1}{2\sqrt{x}(x + 1)^{3/2}}$ **27.** $\dfrac{3x(x + 2)}{(2x + 3)^{3/2}}$ **29.** $\dfrac{\sqrt{10}}{5}$
31. $\dfrac{4x\sqrt{x - 1}}{x - 1}$ **33.** $\dfrac{49\sqrt{x^2 - 9}}{x + 3}$ **35.** $\dfrac{\sqrt{14} + 2}{2}$
37. $\sqrt{6} - \sqrt{5}$ **39.** $\sqrt{x} - \sqrt{x - 2}$
41. $\dfrac{1}{\sqrt{x + 2} + \sqrt{2}}$ **43.** \$200.38

Appendix B *(page A39)*

1. 17.5 square units
3. Left Riemann sum: 0.518
Right Riemann sum: 0.768
5. Left Riemann sum: 0.746
Right Riemann sum: 0.646
7. Left Riemann sum: 0.859
Right Riemann sum: 0.659
9. (a)

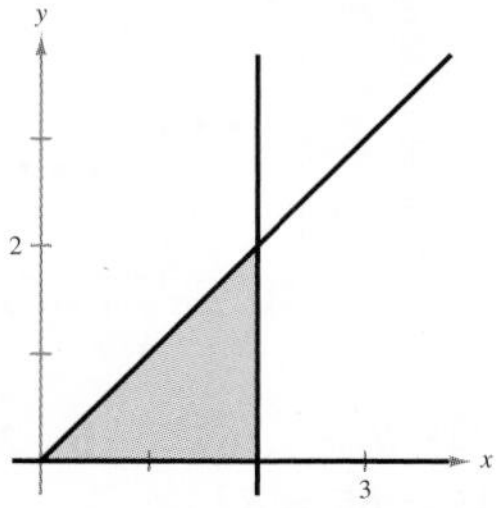

(b) Answers will vary.
(c) Answers will vary.
(d) Answers will vary.

(e)

n	5	10	50	100
Left sum, S_L	1.6	1.8	1.96	1.98
Right sum, S_R	2.4	2.2	2.04	2.02

(f) Answers will vary.

11. $\displaystyle\int_0^5 3\,dx$
13. $\displaystyle\int_{-4}^{4} (4 - |x|)\,dx = \int_{-4}^{0} (4 + x)\,dx + \int_0^4 (4 - x)\,dx$
15. $\displaystyle\int_{-2}^{2} (4 - x^2)\,dx$ **17.** $\displaystyle\int_0^2 \sqrt{x + 1}\,dx$

19.

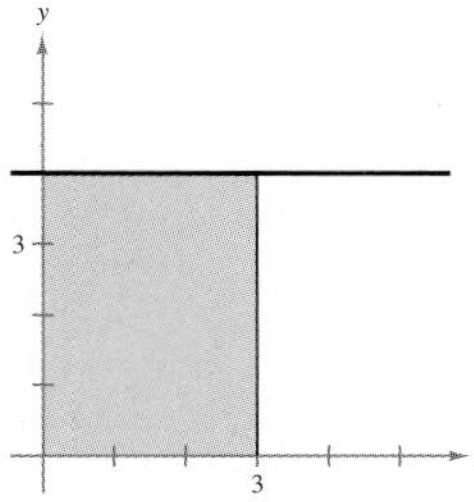

Rectangle

$A = 12$

21.

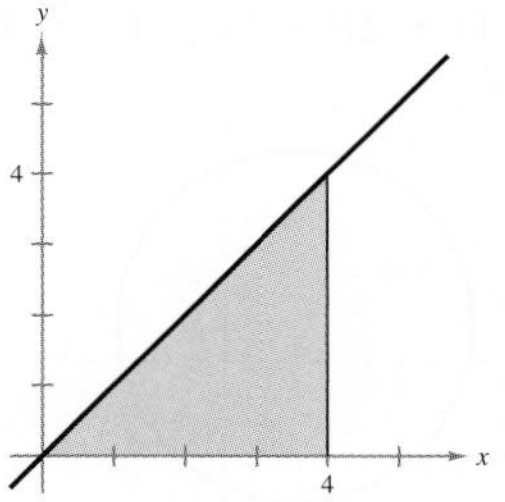

Triangle

$A = 8$

23.

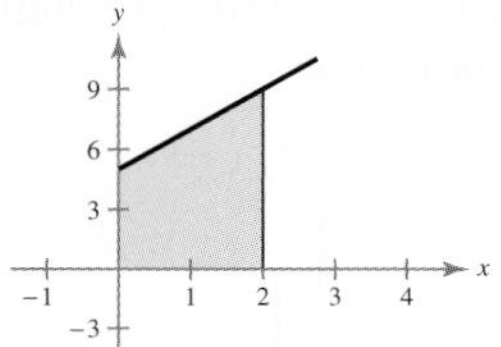

Trapezoid

$A = 14$

25.

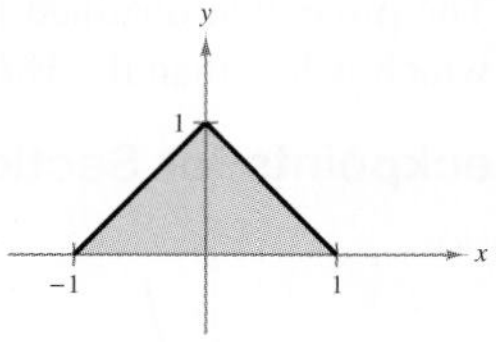

Triangle

$A = 1$

27.

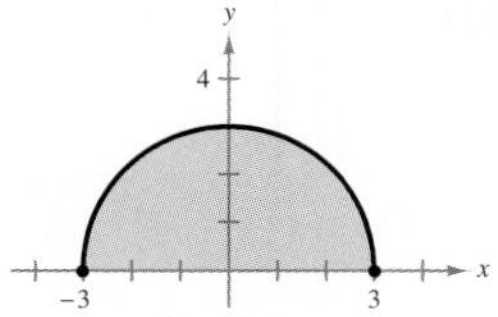

Semicircle

$A = 9\pi/2$

29. Answers will vary. **31.** >

Answers to Checkpoints

Chapter 1

Checkpoints for Section 1.1

1

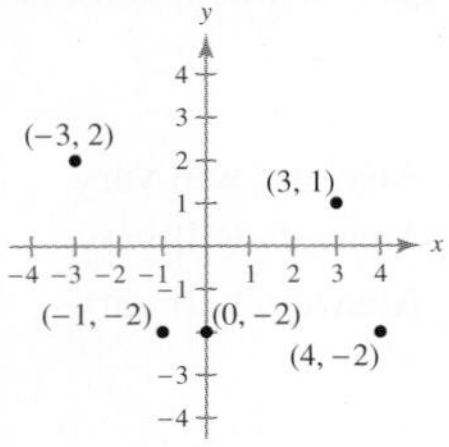

2

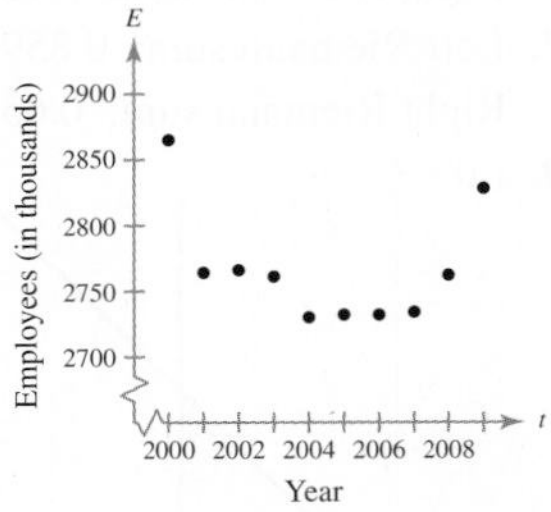

3 5 **4** $d_1 = \sqrt{20}, d_2 = \sqrt{45}, d_3 = \sqrt{65}$ **5** 25 yd

$d_1^2 + d_2^2 = 20 + 45 = 65 = d_3^2$

6 $(-2, 5)$ **7** \$87.5 billion

8 $(-1, -4), (1, -2), (1, 2), (-1, 0)$

Checkpoints for Section 1.2

1

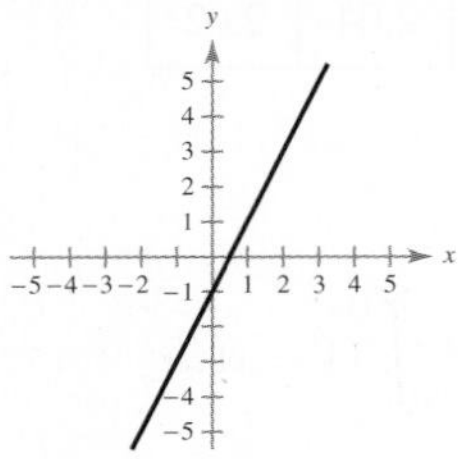

2

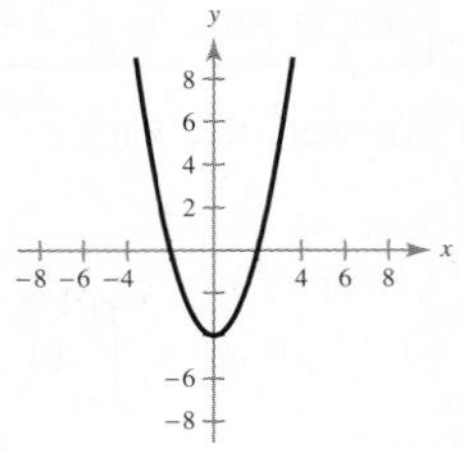

3 x-intercepts: $(3, 0), (-1, 0)$, y-intercept: $(0, -3)$

4 $(x + 2)^2 + (y - 1)^2 = 25$

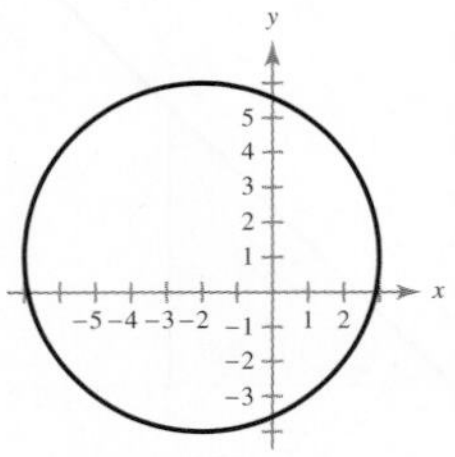

5 12,500 units **6** 4 million units at \$122/unit

7 The projection obtained from the model is \$10,814.3 million, which is less than the *Value Line* projection.

Checkpoints for Section 1.3

1 (a)

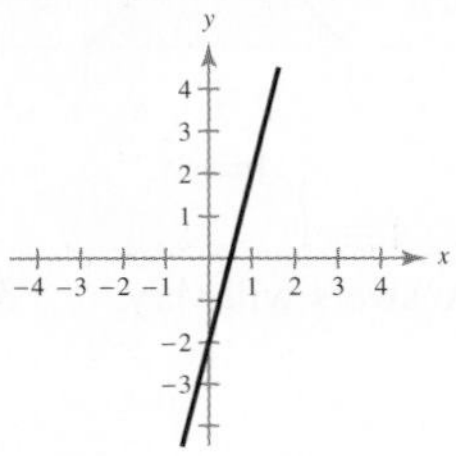

(b)

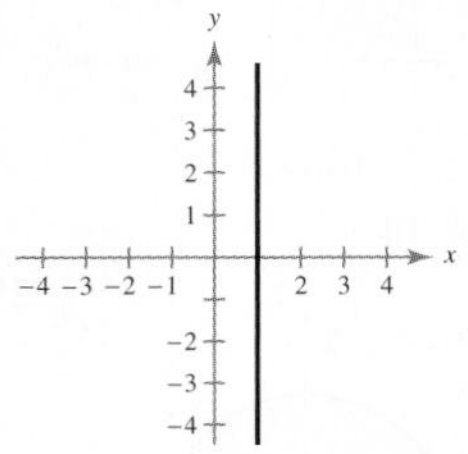

(c)

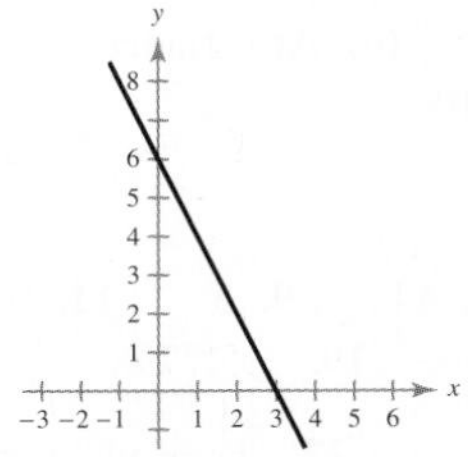

2 Yes, $\frac{27}{312} \approx 0.08654 > \frac{1}{12} = 0.08\overline{3}$.

3 The y-intercept $(0, 1500)$ tells you that the original value of the copier is \$1500. The slope of $m = -300$ tells you that the value decreases by \$300/yr.

4 (a) 2 (b) $-\frac{1}{2}$ (c) 0 **5** $y = 2x + 4$

6 $y = 0.52t - 2.67$; \$2.01

7 (a) $y = \frac{1}{2}x$ (b) $y = -2x + 5$

8 $V = -1375t + 12{,}000$

Checkpoints for Section 1.4

1 (a) Yes, $y = x - 1$. (b) No, $y = \pm\sqrt{4 - x^2}$.
(c) No, $y = \pm\sqrt{2 - x}$. (d) Yes, $y = x^2$.

2 (a) Domain: $[-1, \infty)$; Range: $[0, \infty)$
(b) Domain: $(-\infty, \infty)$; Range: $[0, \infty)$

3 $f(0) = 1, f(1) = -3, f(4) = -3$
No, f is not one-to-one.

4 (a) $x^2 + 2x\,\Delta x + (\Delta x)^2 + 3$ (b) $2x + \Delta x, \Delta x \neq 0$

5 (a) $2x^2 + 5$ (b) $4x^2 + 4x + 3$

6 (a) $f^{-1}(x) = 5x$ (b) $f^{-1}(x) = x + 6$

7 $f^{-1}(x) = \sqrt{x - 2}$

8
$$f(x) = x^2 + 4$$
$$y = x^2 + 4$$
$$x = y^2 + 4$$
$$x - 4 = y^2$$
$$\pm\sqrt{x - 4} = y$$

Checkpoints for Section 1.5

1 6 **2** (a) 4 (b) Does not exist (c) 4

3 (a) 5 (b) 6 (c) 25 (d) −2 **4** 5 **5** 12

6 7 **7** $\frac{1}{4}$ **8** (a) −1 (b) 1 **9** 1

10 $\lim_{x \to 1^-} f(x) = 18$ and $\lim_{x \to 1^+} f(x) = 20$

$\lim_{x \to 1^-} f(x) \neq \lim_{x \to 1^+} f(x)$

11 Does not exist

Checkpoints for Section 1.6

1 (a) f is continuous on the entire real line.
(b) f is continuous on the entire real line.
(c) f is continuous on the entire real line.

2 (a) f is continuous on $(-\infty, 1)$ and $(1, \infty)$.
(b) f is continuous on $(-\infty, 2)$ and $(2, \infty)$.
(c) f is continuous on the entire real line.

3 f is continuous on $[2, \infty)$. **4** f is continuous on $[-1, 5]$.

5
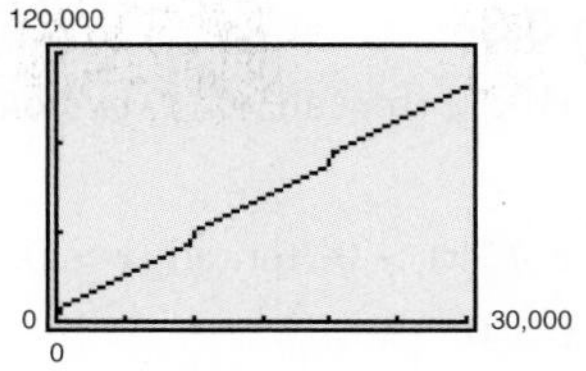

6 $A = 10{,}000(1 + 0.0075)^{[\![4t]\!]}$
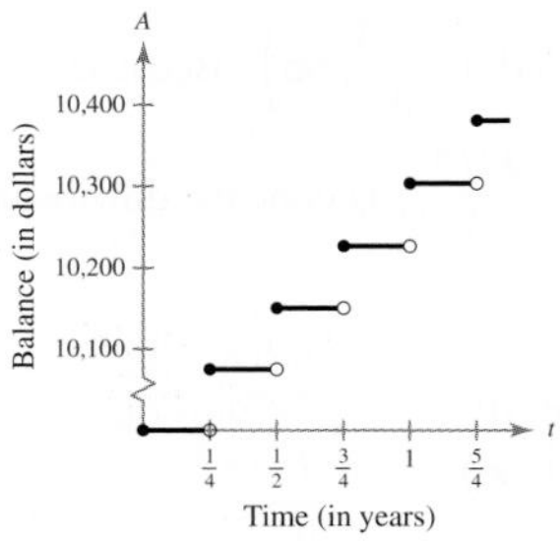

Chapter 2

Checkpoints for Section 2.1

1 3

2 For the months to the left of July on the graph, the tangent lines have positive slopes. For the months to the right of July, the tangent lines have negative slopes. The average daily temperature is increasing prior to July and decreasing after July.

3 4 **4** 2

5 $m = 8x$
At $(0, 1)$, $m = 0$.
At $(1, 5)$, $m = 8$.

6 $2x - 5$ **7** $-\dfrac{4}{t^2}$

Checkpoints for Section 2.2

1 (a) 0 (b) 0 (c) 0 (d) 0

2 (a) $4x^3$ (b) $-\dfrac{3}{x^4}$ (c) $2w$

3 $f'(x) = 3x^2$
$m = f'(-1) = 3;$
$m = f^{-1}(0) = 0;$
$m = f^{-1}(1) = 3$

4 (a) $8x$ (b) $\dfrac{8}{\sqrt{x}}$ **5** (a) $\dfrac{1}{4}$ (b) $-\dfrac{2}{5}$

6 (a) $-\dfrac{9}{2x^3}$ (b) $-\dfrac{9}{8x^3}$ **7** (a) $\dfrac{\sqrt{5}}{2\sqrt{x}}$ (b) $\dfrac{1}{4x^{3/4}}$

8 (a) $4x + 5$ (b) $4x^3 - 2$ **9** -1

10 $y = -x + 2$ **11** $0.34/yr

Checkpoints for Section 2.3

1 (a) $0.5\overline{6}$ mg/ml/min (b) 0 mg/ml/min
(c) -1.5 mg/ml/min

2 (a) -16 ft/sec (b) -48 ft/sec (c) -80 ft/sec

3 When $t = 1.75$, $h'(1.75) = -56$ ft/sec.
When $t = 2$, $h'(2) = -64$ ft/sec.

4 (a) $\frac{3}{2}$ sec (b) -32 ft/sec

5 When $x = 100$, $\dfrac{dP}{dx} = \$16/\text{unit}$.
Actual gain $= \$16.06$

6 $p = -0.057x + 23.82$, $R = -0.057x^2 + 23.82x$

7 Revenue: $R = 2000x - 4x^2$
Marginal revenue: $\dfrac{dR}{dx} = 2000 - 8x$; $0/unit

8 $\dfrac{dP}{dx} = \$1.44/\text{unit}$
Actual increase in profit $\approx \$1.44$

Checkpoints for Section 2.4

1 $-27x^2 + 12x + 24$ **2** $\dfrac{2x^2 - 1}{x^2}$

3 (a) $18x^2 + 30x$ (b) $12x + 15$ **4** $-\dfrac{22}{(5x - 2)^2}$

5 $y = \frac{8}{25}x - \frac{4}{5}$;
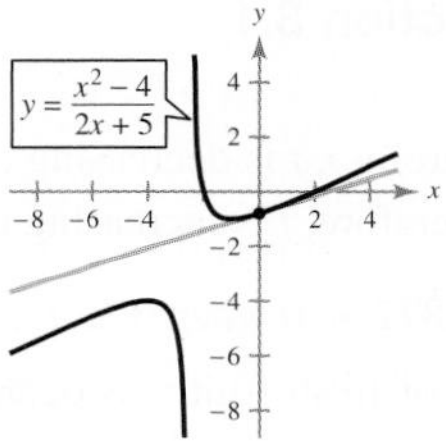

6 $\dfrac{-3x^2 + 4x + 8}{x^2(x + 4)^2}$ **7** (a) $\dfrac{2}{5}x + \dfrac{4}{5}$ (b) $3x^3$

8

t	0	1	2	3	4	5	6	7
$\dfrac{dP}{dt}$	0	-50	-16	-6	-2.77	-1.48	-0.88	-0.56

As t increases, the rate at which the blood pressure drops decreases.

Checkpoints for Section 2.5

1 (a) $u = g(x) = x + 1$, $y = f(u) = \dfrac{1}{\sqrt{u}}$
(b) $u = g(x) = x^2 + 2x + 5$, $y = f(u) = u^3$

2 $6x^2(x^3 + 1)$ **3** $4(2x + 3)(x^2 + 3x)^3$ **4** $y = \frac{1}{3}x + \frac{8}{3}$

5 $-\dfrac{8}{(2x + 1)^2}$ **6** $\dfrac{x(3x^2 + 2)}{\sqrt{x^2 + 1}}$ **7** $-\dfrac{12(x + 1)}{(x - 5)^3}$

8 About $3.48/yr

Checkpoints for Section 2.6

1 $f'(x) = 18x^2 - 4x$, $f''(x) = 36x - 4$, $f'''(x) = 36$, $f^{(4)}(x) = 0$

2 18 **3** $\dfrac{120}{x^6}$

4 Height = 144 ft
Velocity = 0 ft/sec
Acceleration = -32 ft/sec^2

5 -9.8 m/sec^2

6
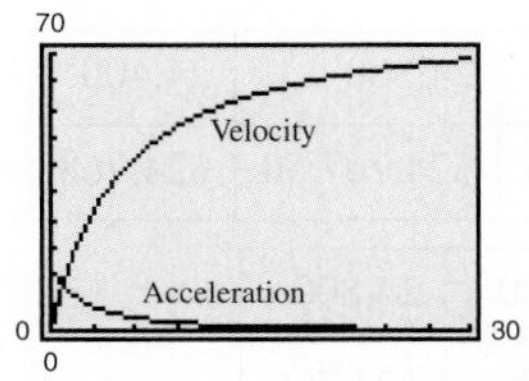

Acceleration approaches zero.

CHECKPOINTS

Checkpoints for Section 2.7

1 $-\frac{2}{x^3}$

2 (a) $12x^2$ (b) $6y\frac{dy}{dx}$ (c) $1 + 5\frac{dy}{dx}$ (d) $y^3 + 3xy^2\frac{dy}{dx}$

3 $\frac{dy}{dx} = -\frac{x-2}{y-1}$ **4** $\frac{3}{4}$ **5** $\frac{5}{9}$

6 $\frac{dx}{dp} = -\frac{2}{p^2(0.002x+1)}$

Checkpoints for Section 2.8

1 9 **2** $12\pi \approx 37.7\ \text{ft}^2/\text{sec}$

3 \$1500/day **4** \$28,400/wk

Chapter 3

Checkpoints for Section 3.1

1 $f'(x) = 4x^3$
$f'(x) < 0$ if $x < 0$; therefore, f is decreasing on $(-\infty, 0)$.
$f'(x) > 0$ if $x > 0$; therefore, f is increasing on $(0, \infty)$.

2 $\frac{dF}{dt} = -1.5348t + 2.872 < 0$ when $3 \le t \le 8$, which implies that the consumption of fresh fruit was decreasing from 2003 through 2008.

3 $x = \frac{1}{2}$

4 Increasing on $(-\infty, -2)$ and $(2, \infty)$
Decreasing on $(-2, 2)$

5 Increasing on $(0, \infty)$
Decreasing on $(-\infty, 0)$

6 Increasing on $(-\infty, -1)$ and $(1, \infty)$
Decreasing on $(-1, 0)$ and $(0, 1)$

7 Because $f'(x) = -3x^2 = 0$ when $x = 0$ and because f is decreasing on $(-\infty, 0) \cup (0, \infty)$, f is decreasing on $(-\infty, \infty)$.

8 (0, 3000)

Checkpoints for Section 3.2

1 Relative maximum at $(-1, 5)$
Relative minimum at $(1, -3)$

2 Relative minimum at $(3, -27)$

3 Relative maximum at (1, 1)
Relative minimum at (0, 0)

4 Absolute maximum at (0, 10)
Absolute minimum at $(4, -6)$

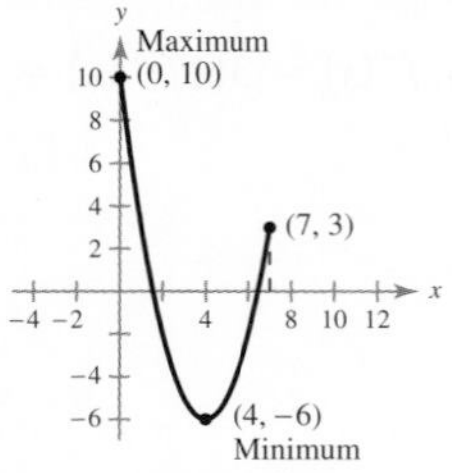

5

x (units)	24,000	24,200	24,300	24,400
P (profit)	\$24,760	\$24,766	\$24,767.50	\$24,768

x (units)	24,500	24,600	24,800
P (profit)	\$24,767.50	\$24,766	\$24,760

Checkpoints for Section 3.3

1 (a) $f'' = -4$; because $f''(x) < 0$ for all x, f is concave downward for all x.
(b) $f''(x) = \frac{1}{2x^{3/2}}$; because $f''(x) > 0$ for all $x > 0$, f is concave upward for all $x > 0$.

2 Because $f''(x) > 0$ for $x < -\frac{2\sqrt{3}}{3}$ and $x > \frac{2\sqrt{3}}{3}$, f is concave upward on $\left(-\infty, -\frac{2\sqrt{3}}{3}\right)$ and $\left(\frac{2\sqrt{3}}{3}, \infty\right)$. Because $f''(x) < 0$ for $-\frac{2\sqrt{3}}{3} < x < \frac{2\sqrt{3}}{3}$, f is concave downward on $\left(-\frac{2\sqrt{3}}{3}, \frac{2\sqrt{3}}{3}\right)$.

3 f is concave upward on $(-\infty, 0)$.
f is concave downward on $(0, \infty)$.
Point of inflection: (0, 0)

4 f is concave upward on $(-\infty, 0)$ and $(1, \infty)$.
f is concave downward on (0, 1).
Points of inflection: (0, 1), (1, 0)

5 Relative minimum: $(3, -26)$

6 Point of diminishing returns: $x = \$150$ thousand

Checkpoints for Section 3.4

1

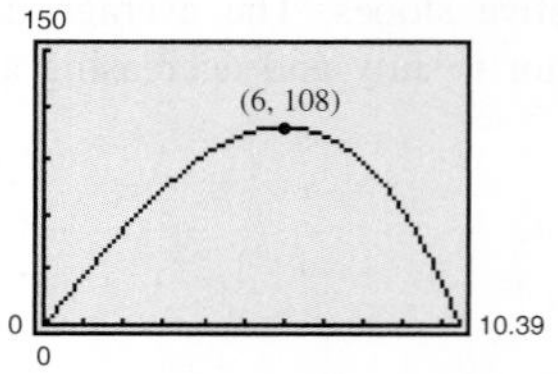

Maximum volume = 108 in.^{3}

2 $\left(\sqrt{\frac{1}{2}}, \frac{7}{2}\right)$ and $\left(-\sqrt{\frac{1}{2}}, \frac{7}{2}\right)$ **3** 8 in. by 12 in.

Checkpoints for Section 3.5

1 125 units yield a maximum revenue of \$1,562,500.

2 400 units **3** \$6.25/unit **4** \$4.00

5 Demand is elastic when $0 < x < 144$.
Demand is inelastic when $144 < x < 324$.
Demand is of unit elasticity when $x = 144$.

Checkpoints for Section 3.6

1 (a) $\lim_{x\to 2^-} \frac{1}{x-2} = -\infty$ (b) $\lim_{x\to 2^+} \frac{1}{x-2} = \infty$
(c) $\lim_{x\to -3^-} \frac{-1}{x+3} = -\infty$ (d) $\lim_{x\to -3^+} \frac{-1}{x+3} = \infty$

2 $x = 0, x = 4$ **3** $x = 3$

4 $\lim_{x\to 2^-} \frac{x^2-4x}{x-2} = \infty$; $\lim_{x\to 2^+} \frac{x^2-4x}{x-2} = -\infty$

5 2

6 (a) $y = 0$ (b) $y = \frac{1}{2}$ (c) No horizontal asymptote

7 $C = 0.75x + 25,000$
$\overline{C} = 0.75 + \frac{25,000}{x}$
$\lim_{x\to\infty} \overline{C} = \$0.75/\text{unit}$

8 No, the cost function is not defined at $p = 100$, which implies that it is not possible to remove 100% of the pollutants.

Checkpoints for Section 3.7

1

	$f(x)$	$f'(x)$	$f''(x)$	Shape of graph
x in $(-\infty, -1)$		$-$	$+$	Decreasing, concave upward
$x = -1$	-32	0	$+$	Relative minimum
x in $(-1, 1)$		$+$	$+$	Increasing, concave upward
$x = 1$	-16	$+$	0	Point of inflection
x in $(1, 3)$		$+$	$-$	Increasing, concave downward
$x = 3$	0	0	$-$	Relative maximum
x in $(3, \infty)$		$-$	$-$	Decreasing, concave downward

2

	$f(x)$	$f'(x)$	$f''(x)$	Shape of graph
x in $(-\infty, 0)$		$-$	$+$	Decreasing, concave upward
$x = 0$	5	0	0	Point of inflection
x in $(0, 2)$		$-$	$-$	Decreasing, concave downward
$x = 2$	-11	$-$	0	Point of inflection
x in $(2, 3)$		$-$	$+$	Decreasing, concave upward
$x = 3$	-22	0	$+$	Relative minimum
x in $(3, \infty)$		$+$	$+$	Increasing, concave upward

3

	$f(x)$	$f'(x)$	$f''(x)$	Shape of graph
x in $(-\infty, 0)$		$+$	$-$	Increasing, concave downward
$x = 0$	0	0	$-$	Relative maximum
x in $(0, 1)$		$-$	$-$	Decreasing, concave downward
$x = 1$	Undef.	Undef.	Undef.	Vertical asymptote
x in $(1, 2)$		$-$	$+$	Decreasing, concave upward
$x = 2$	4	0	$+$	Relative minimum
x in $(2, \infty)$		$+$	$+$	Increasing, concave upward

4

	$f(x)$	$f'(x)$	$f''(x)$	Shape of graph
x in $(-\infty, -1)$		$+$	$+$	Increasing, concave upward
$x = -1$	Undef.	Undef.	Undef.	Vertical asymptote
x in $(-1, 0)$		$+$	$-$	Increasing, concave downward
$x = 0$	-1	0	$-$	Relative maximum
x in $(0, 1)$		$-$	$-$	Decreasing, concave downward
$x = 1$	Undef.	Undef.	Undef.	Vertical asymptote
x in $(1, \infty)$		$-$	$+$	Decreasing, concave upward

5

	$f(x)$	$f'(x)$	$f''(x)$	Shape of graph
x in $(0, 1)$		$-$	$+$	Decreasing, concave upward
$x = 1$	-4	0	$+$	Relative minimum
x in $(1, \infty)$		$+$	$+$	Increasing, concave upward

Checkpoints for Section 3.8

1 $dy = 0.32$; $\Delta y = 0.32240801$ **2** $dR = \$22$; $\Delta R = \$21$

3 $dP = \$10.96$; $\Delta P = \$10.98$

4 (a) $dy = 12x^2\,dx$ (b) $dy = \frac{2}{3}\,dx$

(c) $dy = (6x - 2)\,dx$ (d) $dy = -\dfrac{2}{x^3}\,dx$

Chapter 4

Checkpoints for Section 4.1

1 (a) 243 (b) 3 (c) 64 (d) 8 (e) $\frac{1}{2}$ (f) $\sqrt{10}$

2 (a) 5.453×10^{-13} (b) 1.621×10^{-13}

(c) 2.629×10^{-14}

3

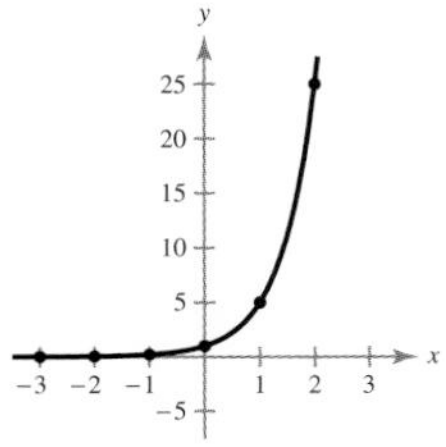

4

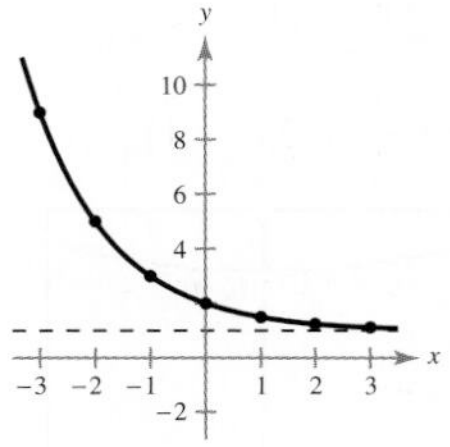

Checkpoints for Section 4.2

1

x	-2	-1	0	1	2
$g(x)$	$e^2 \approx 7.389$	$e \approx 2.718$	1	$\frac{1}{e} \approx 0.368$	$\frac{1}{e^2} \approx 0.135$

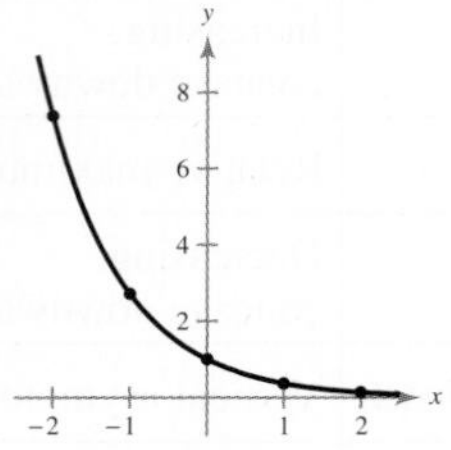

2 After 0 h, $y = 1.25$ g.
After 1 h, $y \approx 1.338$ g.
After 10 h, $y \approx 1.498$ g.

$$\lim_{t \to \infty} \frac{1.50}{1 + 0.2e^{-0.5t}} = 1.50 \text{ g}$$

3 (a) \$4870.38 (b) \$4902.71
(c) \$4918.66 (d) \$4919.21
All else being equal, the more often interest is compounded, the greater the balance.

4 (a) 7% (b) 7.12% (c) 7.19% (d) 7.23%

5 \$16,712.90

Checkpoints for Section 4.3

1 At (0, 2), the slope is 2. At $(1, 2e)$, the slope is $2e$.

2 (a) $3e^{3x}$
(b) $-\dfrac{6x^2}{e^{2x^3}}$
(c) $8xe^{x^2}$
(d) $-\dfrac{2}{e^{2x}}$

3 (a) 0
(b) $3e^{3x+1}$
(c) $xe^x(x + 2)$
(d) $\frac{1}{2}(e^x - e^{-x})$
(e) $\dfrac{e^x(x - 2)}{x^3}$
(f) $e^x(x^2 + 2x - 1)$

4

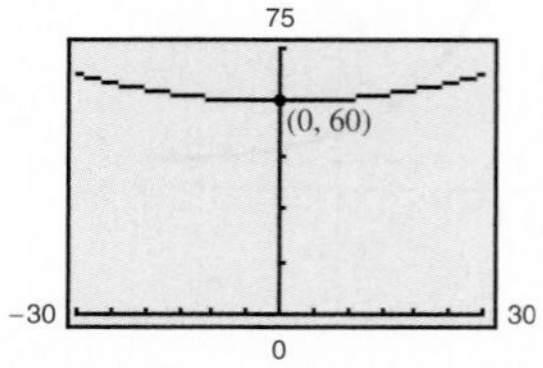

5 \$18.39/unit (80,000 units); \$1,471,517.77

6

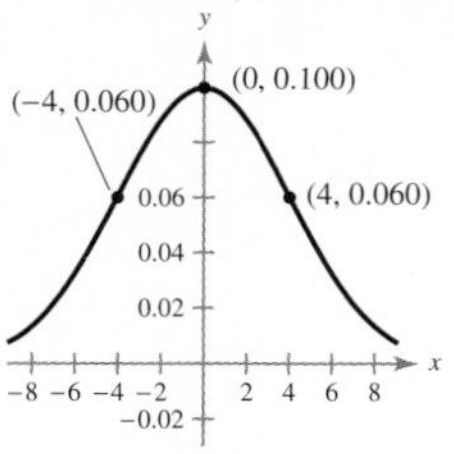

Points of inflection: $(-4, 0.060)$, $(4, 0.060)$

Checkpoints for Section 4.4

1

x	-1.5	-1	-0.5	0	0.5	1
$f(x)$	-0.693	0	0.405	0.693	0.916	1.099

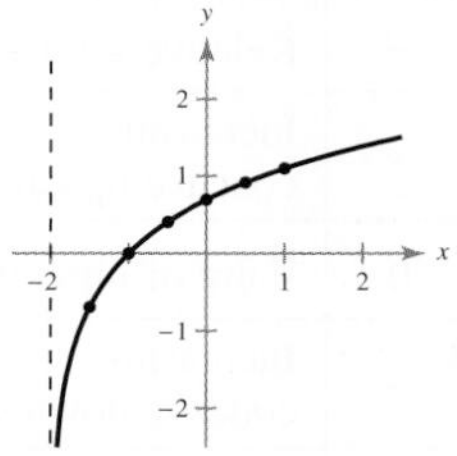

2 (a) 3 (b) $x + 1$

3 (a) $\ln 2 - \ln 5$ (b) $\frac{1}{3}\ln(x + 2)$
(c) $\ln x - \ln 5 - \ln y$ (d) $\ln x + 2\ln(x + 1)$

4 (a) $\ln x^4y^3$ (b) $\ln \dfrac{x + 1}{(x + 3)^2}$ **5** (a) $\ln 6$ (b) $5 \ln 5$

6 (a) e^4 (b) e^3 **7** 7.9 yr

Checkpoints for Section 4.5

1 $\dfrac{1}{x}$ **2** (a) $\dfrac{2x}{x^2 - 4}$ (b) $x(1 + 2\ln x)$ (c) $\dfrac{2\ln x - 1}{x^3}$

3 $\dfrac{1}{3(x + 1)}$ **4** $\dfrac{2}{x} + \dfrac{x}{x^2 + 1}$ **5** $y = 4x - 4$

6 Relative minimum: $(2, 2 - 2\ln 2) \approx (2, 0.6137)$

7 $\dfrac{dp}{dt} = -1.3\%/\text{mo}$
The average score would decrease at a greater rate than the model in Example 7.

8 (a) 4 (b) -2 (c) -5 (d) 3

9 (a) 2.322 (b) 2.631 (c) 3.161 (d) -0.5

10

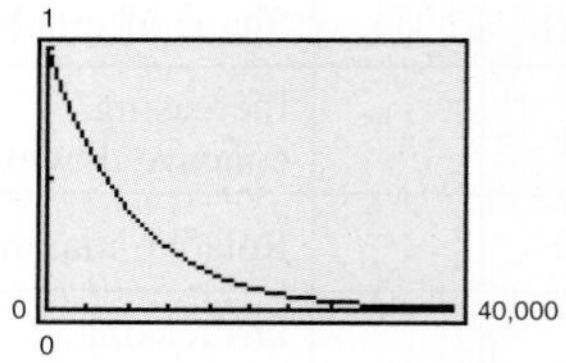

As time increases, the derivative approaches 0. The rate of change of the amount of carbon isotopes is proportional to the amount present.

Checkpoints for Section 4.6

1 About 2113.7 yr **2** $y = 25e^{0.6931t}$

3 $r = \frac{1}{8}\ln 2 \approx 0.0866$ or 8.66% **4** About 12.42 mo

Chapter 5

Checkpoints for Section 5.1

1 (a) $5x + C$ (b) $-r + C$ (c) $2t + C$ **2** $\frac{5}{2}x^2 + C$

3 (a) $-\dfrac{1}{x} + C$ (b) $\dfrac{3}{4}x^{4/3} + C$

4 (a) $\frac{1}{2}x^2 + 4x + C$ (b) $x^4 - \frac{5}{2}x^2 + 2x + C$

5 $\frac{2}{3}x^{3/2} + 4x^{1/2} + C$

6 General solution: $F(x) = 2x^2 + 2x + C$
Particular solution: $F(x) = 2x^2 + 2x + 4$

7 $s(t) = -16t^2 + 32t + 48$. The ball hits the ground 3 seconds after it is thrown, with a velocity of -64 feet per second.

8 $C = -0.01x^2 + 28x + 12.01$
$C(200) = \$5212.01$

Checkpoints for Section 5.2

1 (a) $\dfrac{(x^3+6x)^3}{3}+C$ (b) $\dfrac{2}{3}(x^2-2)^{3/2}+C$

2 $\frac{1}{36}(3x^4+1)^3+C$ **3** $\frac{1}{9}(x^3-3x)^3+C$

4 $2x^9+\frac{12}{5}x^5+2x+C$ **5** $\frac{5}{3}(x^2-1)^{3/2}+C$

6 $-\frac{1}{3}(1-2x)^{3/2}+C$ **7** $\frac{1}{3}(x^2+4)^{3/2}+C$

8 About \$34,068

Checkpoints for Section 5.3

1 (a) $3e^x+C$ (b) $e^{5x}+C$ (c) $e^x-\dfrac{x^2}{2}+C$

2 $\frac{1}{2}e^{2x+3}+C$ **3** $2e^{x^2}+C$

4 (a) $2\ln|x|+C$ (b) $\ln|x^3|+C$ (c) $\ln|2x+1|+C$

5 $\frac{1}{4}\ln|4x+1|+C$ **6** $\frac{3}{2}\ln(x^2+4)+C$

7 (a) $4x-3\ln|x|-\dfrac{2}{x}+C$ (b) $2\ln(1+e^x)+C\,dx$

(c) $\dfrac{x^2}{2}+x+3\ln|x+1|+C$

Checkpoints for Section 5.4

1 $\frac{1}{2}(3)(12)=18$

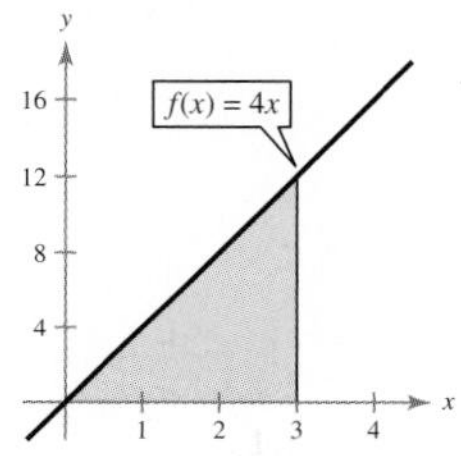

2 $\frac{22}{3}$ units^2 **3** 68

4 (a) $\frac{1}{4}(e^4-1)\approx 13.3995$ (b) $-\ln 5+\ln 2\approx -0.9163$

5 $\frac{13}{2}$ **6** (a) About \$14.18 (b) \$141.79

7 \$13.70 **8** (a) $\frac{2}{5}$ (b) 0 **9** About \$12,295.62

Checkpoints for Section 5.5

1 $\frac{8}{3}$ units^2

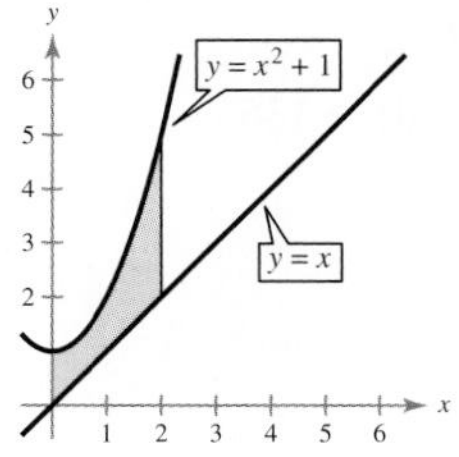

2 $\frac{32}{3}$ units^2 **3** $\frac{9}{2}$ units^2

4 $\frac{253}{12}$ units^2

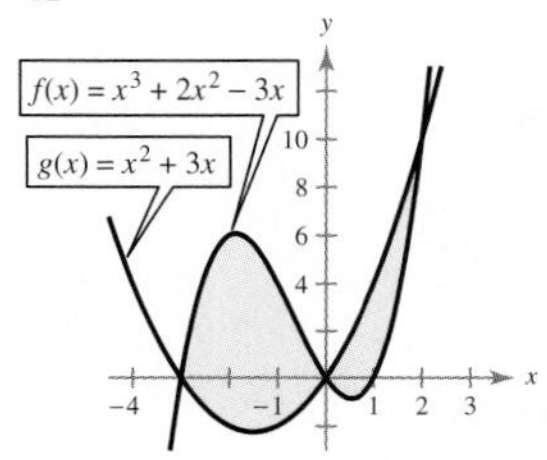

5 Consumer surplus: 40
Producer surplus: 20

6 The company can save \$47.52 million.

Checkpoints for Section 5.6

1 $\frac{37}{8}$ units^2 **2** 0.436 unit^2 **3** 5.642 units^2

4 About 1.463

Chapter 6

Checkpoints for Section 6.1

1 $\dfrac{1}{2}xe^{2x}-\dfrac{1}{4}e^{2x}+C$ **2** $\dfrac{x^2}{2}\ln x-\dfrac{1}{4}x^2+C$

3 $x\ln 2x-x+C$ **4** $e^x(x^3-3x^2+6x-6)+C$

5 $e-2$ **6** \$538,145 **7** \$721,632.08

Checkpoints for Section 6.2

1 $\frac{2}{3}(x-4)\sqrt{2+x}+C$ (Formula 19)

2 $\sqrt{x^2+16}-4\ln\left|\dfrac{4+\sqrt{x^2+16}}{x}\right|+C$ (Formula 25)

3 $\dfrac{1}{4}\ln\left|\dfrac{x-2}{x+2}\right|+C$ (Formula 21)

4 $\frac{1}{3}[1-\ln(1+e)+\ln 2]\approx 0.12663$ (Formula 39)

5 $x(\ln x)^2+2x-2x\ln x+C$ (Formula 44)

6 About 18.2%

Checkpoints for Section 6.3

1 3.2608 **2** 3.1956 **3** 1.154

Checkpoints for Section 6.4

1 (a) Converges; $\frac{1}{2}$ (b) Diverges

2 1 **3** $\frac{1}{2}$ **4** 0.0013 or 0.13%

5 No, you do not have enough money to start the scholarship fund because you need \$125,000. (\$125,000 > \$120,000)

Chapter 7

Checkpoints for Section 7.1

1

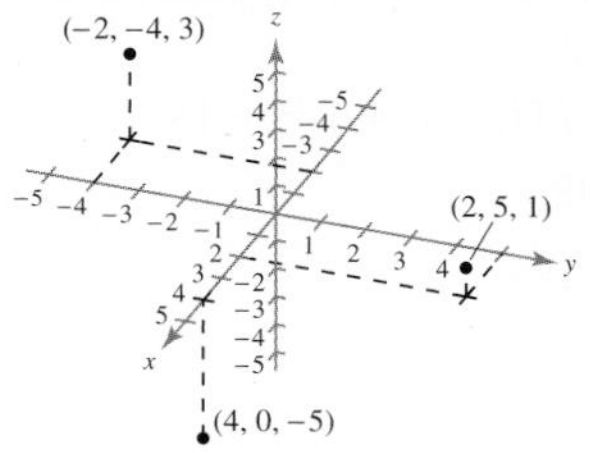

2 $2\sqrt{6}$ **3** $\left(-\frac{5}{2}, 2, -2\right)$

4 $(x-4)^2+(y-3)^2+(z-2)^2=25$

5 $(x-1)^2+(y-3)^2+(z-2)^2=38$

6 Center: $(-3, 4, -1)$; radius: 6

7 $(x+1)^2+(y-2)^2=16$

Checkpoints for Section 7.2

1 x-intercept: $(4, 0, 0)$;
y-intercept: $(0, 2, 0)$;
z-intercept: $(0, 0, 8)$

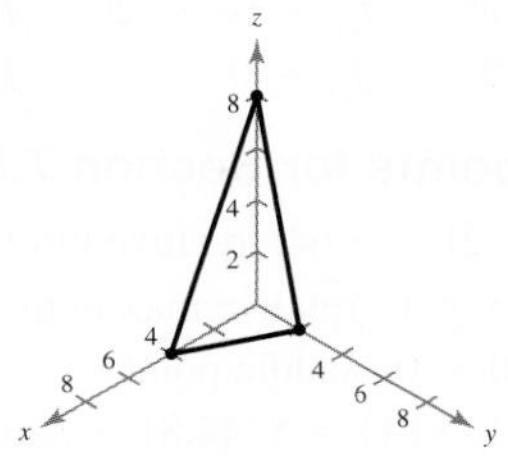

2 xy-trace: circle, $x^2 + y^2 = 1$
yz-trace: hyperbola, $y^2 - z^2 = 1$
xz-trace: hyperbola, $x^2 - z^2 = 1$
$z = 3$ trace: circle, $x^2 + y^2 = 10$
Hyperboloid of one sheet

3 (a) Elliptic paraboloid (b) Elliptic cone

Checkpoints for Section 7.3

1 (a) 0 (b) $\frac{9}{4}$

2 Domain: $x^2 + y^2 \le 9$
Range: $0 \le z \le 3$

3 For each value of c, the equation $f(x, y) = c$ is a circle (or point) in the xy-plane.

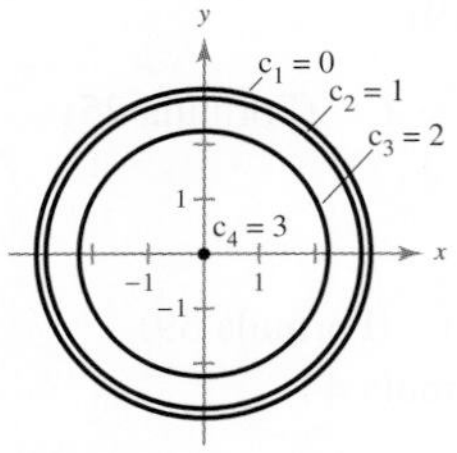

4 $f(1500, 1000) \approx 127{,}542$ units
$f(1000, 1500) \approx 117{,}608$ units
x, person-hours, has a greater effect on production.

5 (a) $M = \$421.60$/mo
(b) Total paid $= (30 \times 12) \times 421.60$
$= \$151{,}776$

Checkpoints for Section 7.4

1 $\frac{\partial z}{\partial x} = 4x - 8xy^3$

$\frac{\partial z}{\partial y} = -12x^2y^2 + 4y^3$

2 $f_x(x, y) = 2xy^3$; $f_x(1, 2) = 16$
$f_y(x, y) = 3x^2y^2$; $f_y(1, 2) = 12$

3 (a) $f_x(1, -1, 49) = 8$ (b) $f_y(1, -1, 49) = -18$

4 Substitute product relationship

5 $\frac{\partial w}{\partial x} = xy + 2xy\ln(xz)$

$\frac{\partial w}{\partial y} = x^2 \ln xz$

$\frac{\partial w}{\partial z} = \frac{x^2y}{z}$

6 $f_{xx} = 8y^2$
$f_{yy} = 8x^2 + 8$
$f_{xy} = 16xy$
$f_{yx} = 16xy$

7 $f_{xx} = 0$ $f_{xy} = e^y$ $f_{xz} = 2$
$f_{yx} = e^y$ $f_{yy} = xe^y + 2$ $f_{yz} = 0$
$f_{zx} = 2$ $f_{zy} = 0$ $f_{zz} = 0$

Checkpoints for Section 7.5

1 $f(-8, 2) = -64$: relative minimum

2 $f(0, 0) = 1$: relative maximum

3 $f(0, 0) = 0$: saddle point

4 $P(3.11, 3.81) = \$744.81$ maximum profit

5 $V\left(\frac{4}{3}, \frac{2}{3}, \frac{8}{3}\right) = \frac{64}{27}$ units^3

Checkpoints for Section 7.6

1 $V\left(\frac{4}{3}, \frac{2}{3}, \frac{8}{3}\right) = \frac{64}{27}$ units^3 **2** $f(187.5, 50) \approx 13{,}474$ units

3 About 26,740 units

4 $P(3.35, 4.26) = \$758.08$ maximum profit

Checkpoints for Section 7.7

1 For $f(x)$, $S = 10$. For $g(x)$, $S = 0.76$.
The quadratic model is a better fit.

2 $f(x) = \frac{6}{5}x + \frac{23}{10}$

3 $y = 16{,}194.4t + 132{,}405$
About 342,932,200 subscribers

Checkpoints for Section 7.8

1 (a) $\frac{1}{4}x^4 + 2x^3 - 2x - \frac{1}{4}$ (b) $\ln|y^2 + y| - \ln|2y|$

2 $\frac{25}{2}$ **3** $\int_2^4 \int_1^5 dx\,dy = 8$ **4** $\frac{4}{3}$

5 (a)

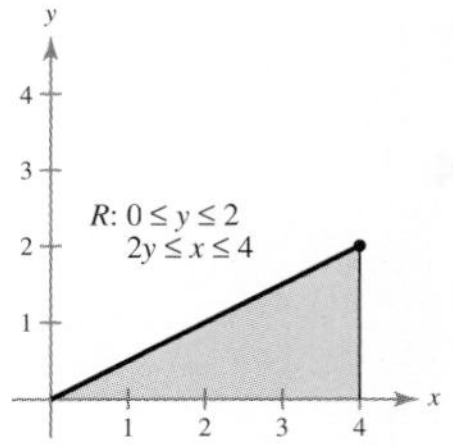

(b) $\int_0^4 \int_0^{x/2} dy\,dx$

(c) $\int_0^2 \int_{2y}^4 dx\,dy = 4 = \int_0^4 \int_0^{x/2} dy\,dx$

6 $\int_{-1}^3 \int_{x^2}^{2x+3} dy\,dx = \frac{32}{3}$

Checkpoints for Section 7.9

1 $\frac{16}{3}$ **2** $e - 1$ **3** $\frac{176}{15}$

4 Integration by parts **5** 3

Chapter 8

Checkpoints for Section 8.1

1 (a) 150° (b) 30° (c) 135° (d) 30°

2 (a) $\frac{5\pi}{4}$ (b) $-\frac{\pi}{4}$ (c) $\frac{4\pi}{3}$ (d) $\frac{5\pi}{6}$

3 (a) 300° (b) 210° (c) 270° (d) −135°

4 $\frac{1}{2}$ ft^2

Checkpoints for Section 8.2

1 $\sin\theta = \frac{1}{2}$

$\cos\theta = \frac{\sqrt{3}}{2}$

$\tan\theta = \frac{\sqrt{3}}{3}$

2 (a) $\frac{1}{2}$ (b) $-\frac{\sqrt{2}}{2}$ (c) $-\sqrt{3}$

3 (a) $-\frac{1}{2}$ (b) $\sqrt{2}$ (c) $\frac{\sqrt{6} - \sqrt{2}}{4}$
(d) 1 (e) 1 (f) 1

4 About 52.5 ft **5** About 245.76°

6 (a) $\frac{\pi}{4}, \frac{7\pi}{4}$ (b) $\frac{2\pi}{3}, \frac{5\pi}{3}$ (c) $\frac{7\pi}{6}, \frac{11\pi}{6}$

7 $\theta = 0, \frac{2\pi}{3}, \pi, \frac{4\pi}{3}, 2\pi$

Checkpoints for Section 8.3

1

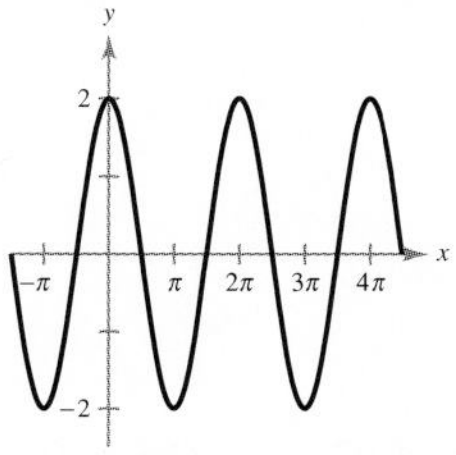

2

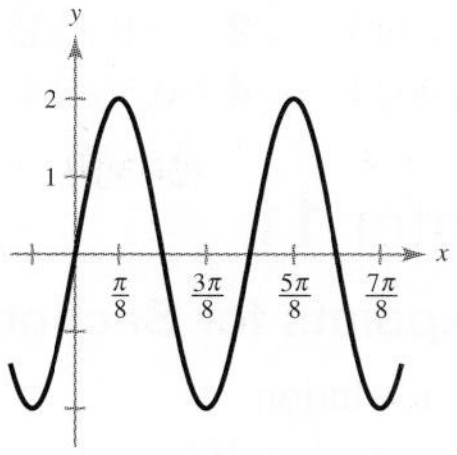

3

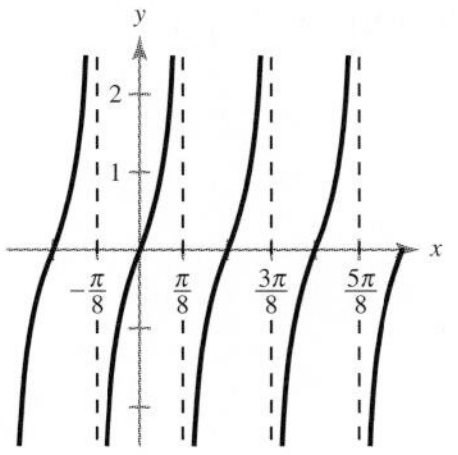

4

x	-0.10	-0.05	-0.01
$\frac{1-\cos x}{x}$	-0.0500	-0.0250	-0.0050

x	0.01	0.05	0.10
$\frac{1-\cos x}{x}$	0.0050	0.0250	0.0500

$$\lim_{x\to 0} \frac{1-\cos x}{x} = 0$$

5

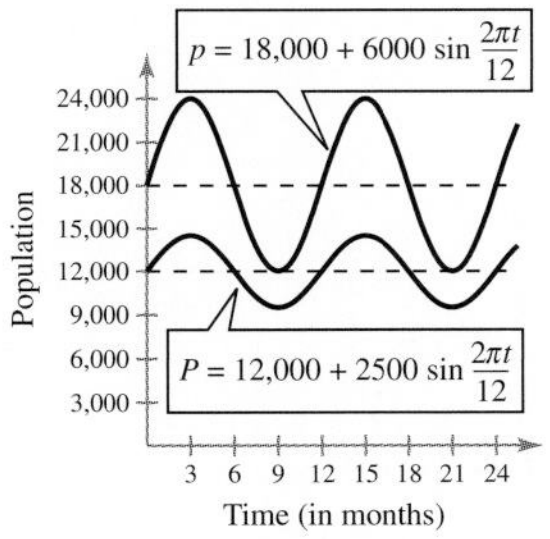

The predator population has an amplitude of 2500 and oscillates about the line $y = 12{,}000$. The prey population has an amplitude of 6000 and oscillates about the line $y = 18{,}000$.

6

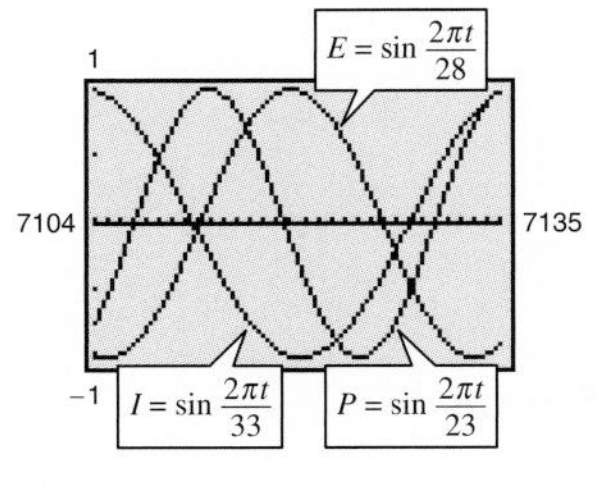

Checkpoints for Section 8.4

1–3 (a) $-4 \sin 4x$ (b) $2x \cos(x^2 - 1)$ (c) $\frac{1}{2} \sec^2 \frac{x}{2}$
(d) $-6x^2 \sin x^3$ (e) $3 \sin^2 x \cos x$

4 (a) $4 \sec(4x) \tan(4x)$ (b) $-2x \csc^2 x^2$

5 (a) $-\frac{\sin 2x}{\sqrt{\cos 2x}}$ (b) $\frac{\sec^2 3x}{\sqrt[3]{\tan^2 3x}}$

6 (a) $-x^2 \sin x + 2x \cos x$ (b) $2t \cos 2t + \sin 2t$

7 Relative maximum: $\left(\frac{7\pi}{6}, \frac{7\pi + 6\sqrt{3}}{12}\right)$

Relative minimum: $\left(\frac{11\pi}{6}, \frac{11\pi - 6\sqrt{3}}{12}\right)$

8 Relative maximum: $\left(\frac{\pi}{6}, \frac{3\sqrt{3}}{4}\right)$

Relative minimum: $\left(\frac{5\pi}{6}, -\frac{3\sqrt{3}}{4}\right)$

9 About 1721 lb/day **10** About $-3.9°$/h

Checkpoints for Section 8.5

1–2 (a) $-5 \cos x + C$ (b) $\sin x^4 + C$

3 $\frac{1}{5} \tan 5x + C$ **4** $-\csc 2x + C$ **5** $-\frac{1}{8} \cos^4 2x + C$

6 $\ln|\sin x| + C$ **7** 1 **8** 1

9 $\frac{1}{2} \ln|\sec 2x + \tan 2x| + C$ **10** 82.68°

Chapter 9

Checkpoints for Section 9.1

1 (a) $S = \{(1, 1), (1, 2), (1, 3), (1, 4), (1, 5), (1, 6), (2, 1), (2, 2), (2, 3), (2, 4), (2, 5), (2, 6), (3, 1), (3, 2), (3, 3), (3, 4), (3, 5), (3, 6), (4, 1), (4, 2), (4, 3), (4, 4), (4, 5), (4, 6), (5, 1), (5, 2), (5, 3), (5, 4), (5, 5), (5, 6), (6, 1), (6, 2), (6, 3), (6, 4), (6, 5), (6, 6)\}$

(b) $E = \{(1, 6), (2, 5), (2, 6), (3, 4), (3, 5), (3, 6), (4, 3), (4, 4), (4, 5), (4, 6), (5, 2), (5, 3), (5, 4), (5, 5), (5, 6), (6, 1), (6, 2), (6, 3), (6, 4), (6, 5), (6, 6)\}$

2

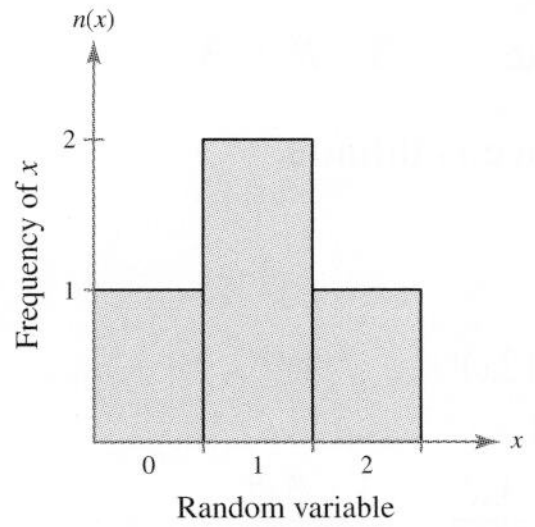

3

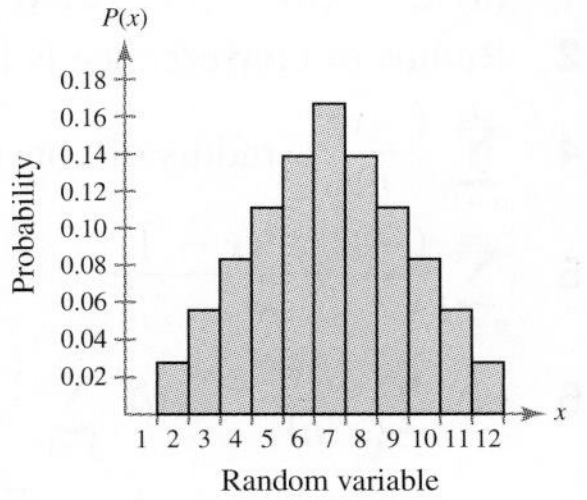

4 7 **5** 2.37 units/day

6 $V(x) = 1.2$, $\sigma \approx 1.095$

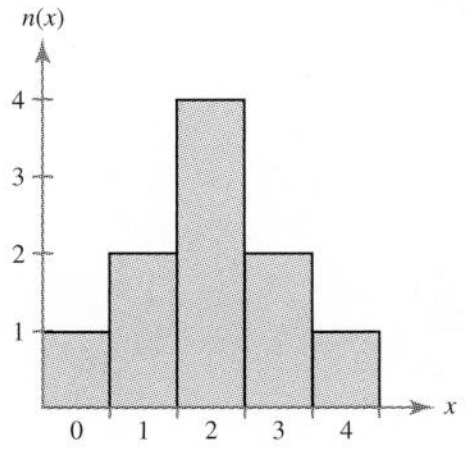

Checkpoints for Section 9.2

1 $\int_0^2 \frac{1}{2}x\,dx = 1$ **2** $\int_0^\infty 2e^{-2x}\,dx = 1$ **3** $\frac{3}{16}$

4 0.4 **5** 0.148

Checkpoints for Section 9.3

1 2 **2** $V(x) = \frac{4}{5}$, $\sigma \approx \sqrt{\frac{4}{5}} \approx 0.8944$ **3** 84.6%

4 Mean: 0.5; median: 0.35 **5** $\mu = 1$; $\sigma \approx 0.577$

6 About 0.567 or 56.7% **7** About 0.159 or 15.9%

Chapter 10

Checkpoints for Section 10.1

1 (a) $a_1 = 2, a_2 = 5, a_3 = 8, a_4 = 11$
(b) $b_1 = \frac{1}{2}, b_2 = \frac{2}{5}, b_3 = \frac{3}{10}, b_4 = \frac{4}{17}$

2 (a) 0 (b) 2 **3** 0

4 $\frac{1}{n}$; converges **5** $\frac{(-1)^{n+1}n^2}{(n+1)!}$

6 (a) $A_1 = \$1007.50$ (b) $A_{16} = \$1126.99$
$A_2 = \$1015.06$
$A_3 = \$1022.67$

Checkpoints for Section 10.2

1 $\sum_{i=1}^{4} 4\left(-\frac{1}{2}\right)^i$ **2** Converges to $\frac{1}{3}$

3 (a) 1 (b) $\frac{4}{3}$ (c) $-\frac{2}{3}$ **4** (a) Diverges (b) Diverges

5 $S_5 \approx 5.556, S_{50} = 5.\overline{5}, S_{500} = 5.\overline{5}$ **6** \$1021.17

7 (a) Converges to $\frac{5}{3}$ (b) Diverges (c) Converges to $\frac{5}{3}$

8 About 40,000 units **9** 140 ft

Checkpoints for Section 10.3

1 (a) p-series with $p = \pi$ (b) p-series with $p = \frac{3}{2}$
(c) Geometric series

2 (a) Converges (b) Converges (c) Diverges

3 Converges to approximately 20.086

4 Diverges **5** Diverges

Checkpoints for Section 10.4

1 (a) 2 (b) -3 (c) 0

2 Radius of convergence is infinite. **3** $R = 3$

4 $\sum_{n=0}^{\infty} \frac{(-x)^n}{n!}$; radius of convergence is infinite.

5 $\sum_{n=1}^{\infty} \frac{(-1)^{n+1}(x-1)^n}{n}$

6 (a) $\sum_{n=0}^{\infty} \frac{(2x)^n}{n!}$ (b) $\sum_{n=0}^{\infty} \frac{(-1)^n(2x)^n}{n!}$

7 $(1+x)^{1/2} = 1 + \frac{x}{2} - \frac{x^2}{2^2 2!} + \frac{3x^3}{2^3 3!} - \frac{3 \cdot 5x^4}{2^4 4!} + \cdots$

8 (a) $1 - \sum_{n=0}^{\infty} \frac{(-x)^n}{n!}$ (b) $e^2 \sum_{n=0}^{\infty} \frac{(-x)^n}{n!}$
(c) $\sum_{n=1}^{\infty} \frac{(-1)^{n-1}(x-1)^n}{n}$

Checkpoints for Section 10.5

1 $S_{12}(x) = 1 + 2x^3 + 2x^6 + \frac{4x^9}{3} + \frac{2x^{12}}{3}$

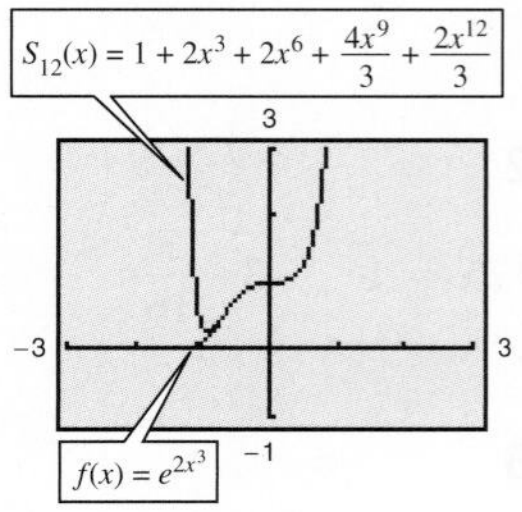

2 $e^{-0.5} \approx 0.607$ with a maximum error of 0.0003. **3** 1.970

Checkpoints for Section 10.6

1 1.732143 **2** -0.453398
3 1.319074 **4** 0.567143

Chapter 11

Checkpoints for Section 11.1

1 Not a solution

2 Because $y' = 2Cx$, and $xy' - 2y = x(2Cx) - 2Cx^2 = 0$, $y = Cx^2$ is a solution. Particular solution: $y = \frac{1}{16}x^2$

3

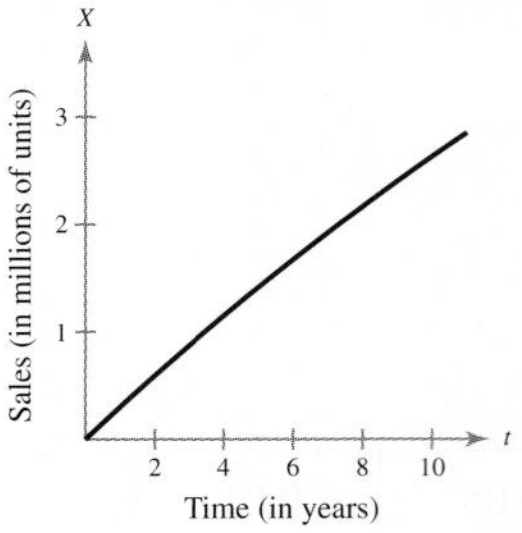

4

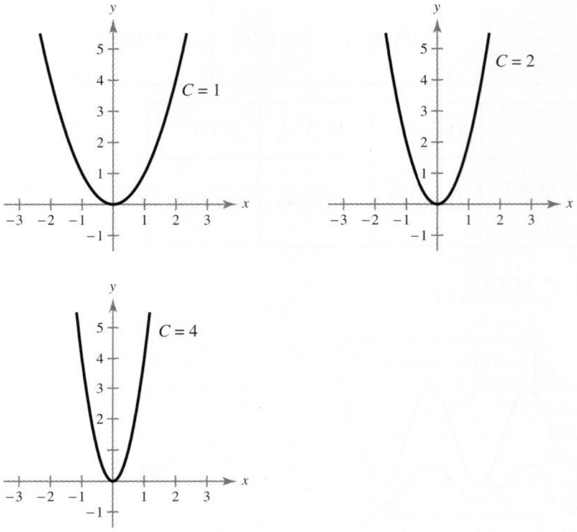

Checkpoints for Section 11.2

1 $y^2 = \frac{2x^3}{3} + C$ **2** $y^2 = (x+1)^2 + C$

3 $y = \pm\sqrt{-x^2 + C}$

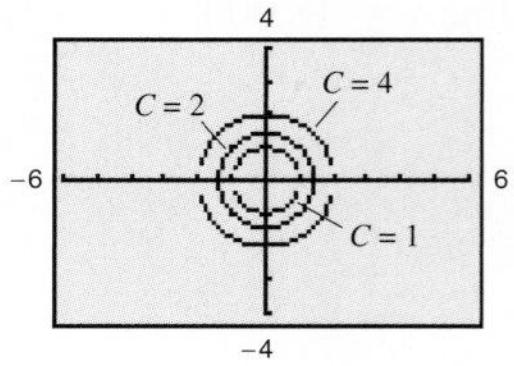

4 $y^2 = 6 - 2e^x$ **5** $y = 65 - Ce^{-kx}$

6 \$31,424,909.75

Checkpoints for Section 11.3

1 $y = -10 + Ce^x$ **2** $y = x \ln x + Cx$

3 $A = 1 - e^{-0.05t}$

Checkpoints for Section 11.4

1 $y = \dfrac{80}{7t + 2}$ **2** About 440,000 people

3 About 76 wolves **4** About 79%

5 About 17.6 gallons

Appendix A

Checkpoints for Section A.1

1 $x < 5$ or $(-\infty, 5)$

2 $x < -2$ or $x > 5$; $(-\infty, -2) \cup (5, \infty)$

3 $200 \le x \le 400$; so the daily production levels during the month varied between a low of 200 units and a high of 400 units.

Checkpoints for Section A.2

1 8; 8; -8 **2** $2 \le x \le 10$

3 $\$4027.50 \le C \le \$11{,}635$

Checkpoints for Section A.3

1 $\frac{4}{9}$ **2** 8 **3** (a) $3x^6$ (b) $8x^{7/2}$ (c) $4x^{4/3}$

4 (a) $x(x^2 - 2)$ (b) $2x^{1/2}(1 + 4x)$

5 $\dfrac{(3x - 1)^{3/2}(13x - 2)}{(x + 2)^{1/2}}$ **6** $\dfrac{x^2(5 + x^3)}{3}$

7 (a) $[2, \infty)$ (b) $(2, \infty)$ (c) $(-\infty, \infty)$

Checkpoints for Section A.4

1 (a) $\dfrac{-2 \pm \sqrt{2}}{2}$ (b) 4 (c) No real zeros

2 (a) $x = -3$ and $x = 5$ (b) $x = -1$
(c) $x = \frac{3}{2}$ and $x = 2$

3 $(-\infty, -2] \cup [1, \infty)$ **4** $-1, \frac{1}{2}, 2$

Checkpoints for Section A.5

1 $\dfrac{x - 2}{x + 1}, \; x \ne -10$ **2** (a) $\dfrac{x^2 + 2}{x}$ (b) $\dfrac{3x + 1}{(x + 1)(2x + 1)}$

3 $\dfrac{3x + 4}{(x + 2)(x - 2)}$ **4** $-\dfrac{x + 1}{3x(x + 2)}$

5 $\dfrac{2(4x^2 + 5x - 3)}{x^2(x + 3)}$ **6** (a) $\dfrac{3x + 8}{4(x + 2)^{3/2}}$ (b) $\dfrac{1}{\sqrt{x^2 + 4}}$

7 (a) $\dfrac{5\sqrt{2}}{4}$ (b) $\dfrac{x + 2}{4\sqrt{x + 2}}$ (c) $\dfrac{\sqrt{6} + \sqrt{3}}{3}$
(d) $\dfrac{\sqrt{x + 2} - \sqrt{x}}{2}$

Answers to Tech Tutors

Tech Tutor

Section 1.3 *(page 29)*

The lines appear perpendicular in the setting $-9 \leq x \leq 9$ and $-6 \leq y \leq 6$.

Section 1.6 *(page 61)*

Most calculators set in connected mode will join the two branches of the graph with a nearly vertical line near $x = 2$. This line is not part of the graph.

Section 4.5 *(page 287)*

Answers will vary.

Section 6.3 *(page 395)*

1.46265

Section 8.4 *(page 537)*

Answers will vary.

Section 8.5 *(page 546)*

Answers will vary.

Section 11.3 *(page 676)*

$y = x^2(\ln x + 1)$

Index

G

H

I

L

Q

R

S

T

Applications (continued from front endsheets)

General

Basic Differentiation Rules

1. $\frac{d}{dx}[cu] = cu'$
2. $\frac{d}{dx}[u \pm v] = u' \pm v'$
3. $\frac{d}{dx}[uv] = uv' + vu'$
4. $\frac{d}{dx}\left[\frac{u}{v}\right] = \frac{vu' - uv'}{v^2}$
5. $\frac{d}{dx}[c] = 0$
6. $\frac{d}{dx}[u^n] = nu^{n-1}u'$
7. $\frac{d}{dx}[x] = 1$
8. $\frac{d}{dx}[\ln u] = \frac{u'}{u}$
9. $\frac{d}{dx}[e^u] = e^u u'$
10. $\frac{d}{dx}[\log_a u] = \frac{u'}{(\ln a)u}$
11. $\frac{d}{dx}[a^u] = (\ln a)a^u u'$
12. $\frac{d}{dx}[\sin u] = (\cos u)u'$
13. $\frac{d}{dx}[\cos u] = -(\sin u)u'$
14. $\frac{d}{dx}[\tan u] = (\sec^2 u)u'$
15. $\frac{d}{dx}[\cot u] = -(\csc^2 u)u'$
16. $\frac{d}{dx}[\sec u] = (\sec u \tan u)u'$
17. $\frac{d}{dx}[\csc u] = -(\csc u \cot u)u'$

Basic Integration Formulas

1. $\int kf(u)\, du = k\int f(u)\, du$
2. $\int [f(u) \pm g(u)]\, du = \int f(u)\, du \pm \int g(u)\, du$
3. $\int du = u + C$
4. $\int a^u\, du = \left(\frac{1}{\ln a}\right)a^u + C$
5. $\int e^u\, du = e^u + C$
6. $\int \ln u\, du = u(-1 + \ln u) + C$
7. $\int \sin u\, du = -\cos u + C$
8. $\int \cos u\, du = \sin u + C$
9. $\int \tan u\, du = -\ln|\cos u| + C$
10. $\int \cot u\, du = \ln|\sin u| + C$
11. $\int \sec u\, du = \ln|\sec u + \tan u| + C$
12. $\int \csc u\, du = -\ln|\csc u + \cot u| + C$
13. $\int \sec^2 u\, du = \tan u + C$
14. $\int \csc^2 u\, du = -\cot u + C$

Trigonometric Identities

Pythagorean Identities

$\sin^2\theta + \cos^2\theta = 1$

$\tan^2\theta + 1 = \sec^2\theta$

$\cot^2\theta + 1 = \csc^2\theta$

Reduction Formulas

$\sin(-\theta) = -\sin\theta$

$\cos(-\theta) = \cos\theta$

$\tan(-\theta) = -\tan\theta$

$\sin\theta = -\sin(\theta - \pi)$

$\cos\theta = -\cos(\theta - \pi)$

$\tan\theta = \tan(\theta - \pi)$

Sum or Difference of Two Angles

$\sin(\theta \pm \phi) = \sin\theta\cos\phi \pm \cos\theta\sin\phi$

$\cos(\theta \pm \phi) = \cos\theta\cos\phi \mp \sin\theta\sin\phi$

$\tan(\theta \pm \phi) = \frac{\tan\theta \pm \tan\phi}{1 \mp \tan\theta\tan\phi}$

Double Angle

$\sin 2\theta = 2\sin\theta\cos\theta$

$\cos 2\theta = 2\cos^2\theta - 1 = 1 - 2\sin^2\theta$

Half Angle

$\sin^2\theta = \frac{1}{2}(1 - \cos 2\theta)$

$\cos^2\theta = \frac{1}{2}(1 + \cos 2\theta)$